LIBRARY OF FUNCTIONS

Identity Function
$f(x) = x$

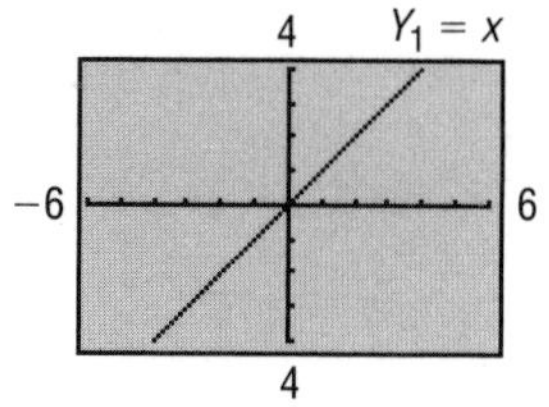

Square Function
$f(x) = x^2$

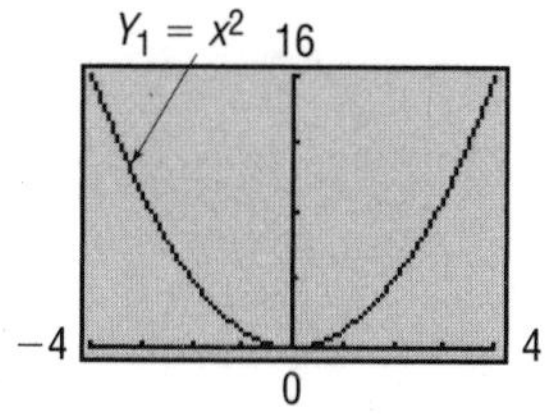

Cube Function
$f(x) = x^3$

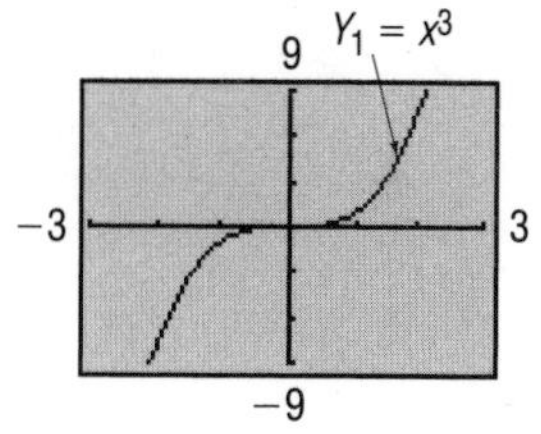

Square Root Function
$f(x) = \sqrt{x}$

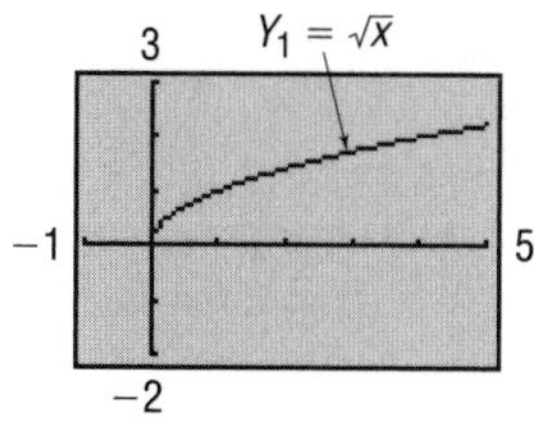

Reciprocal Function
$f(x) = 1/x$

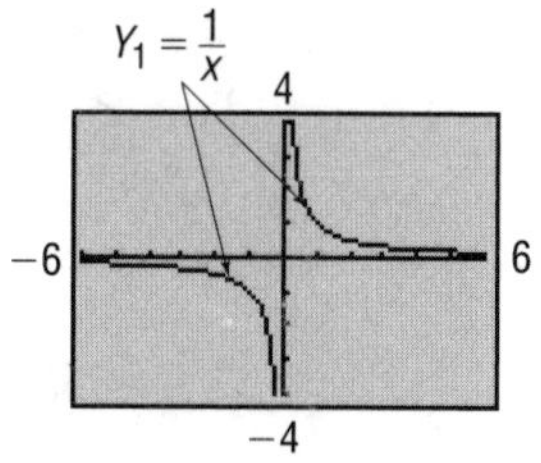

Absolute Value Function
$f(x) = |x|$

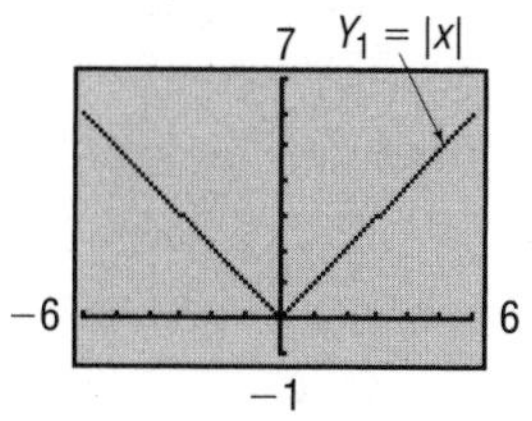

Exponential Function
$f(x) = e^x$

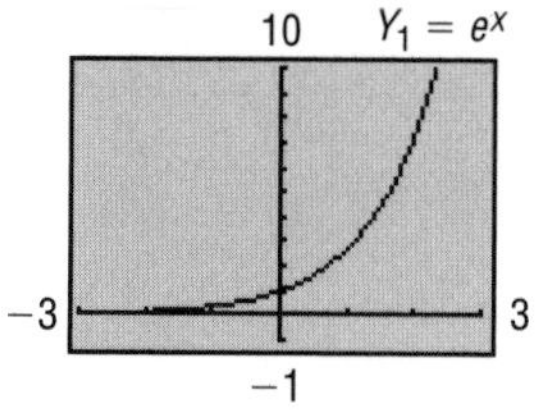

Natural Logarithm Function
$f(x) = \ln x$

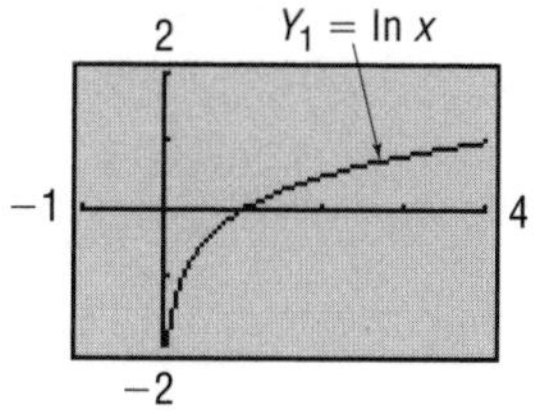

Sine Function
$f(x) = \sin x$

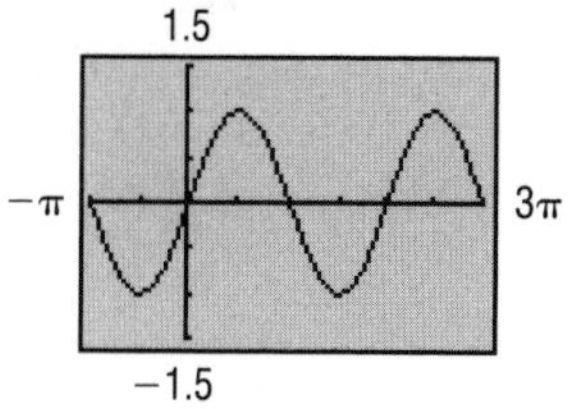

Cosine Function
$f(x) = \cos x$

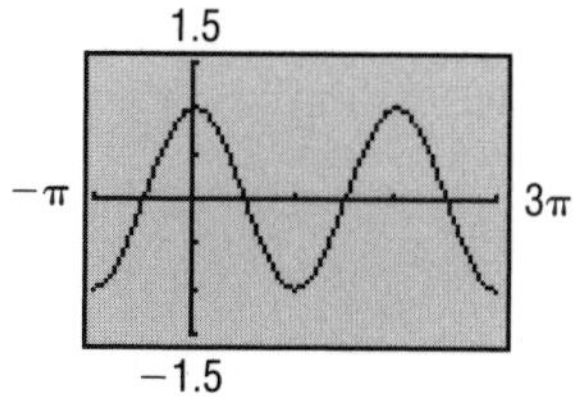

Tangent Function
$f(x) = \tan x$

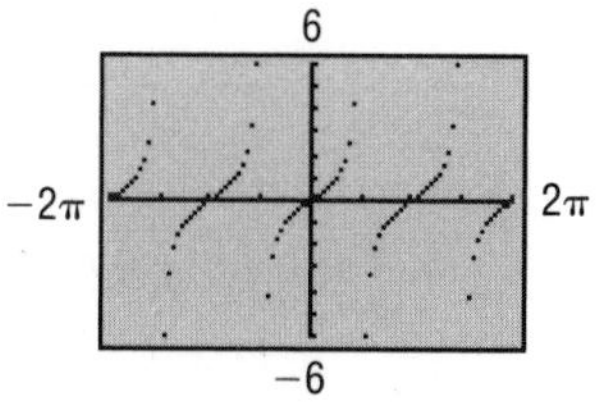

Cosecant Function
$f(x) = \csc x$

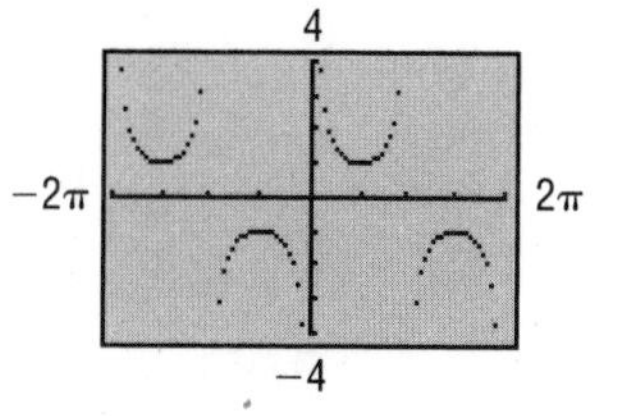

Secant Function
$f(x) = \sec x$

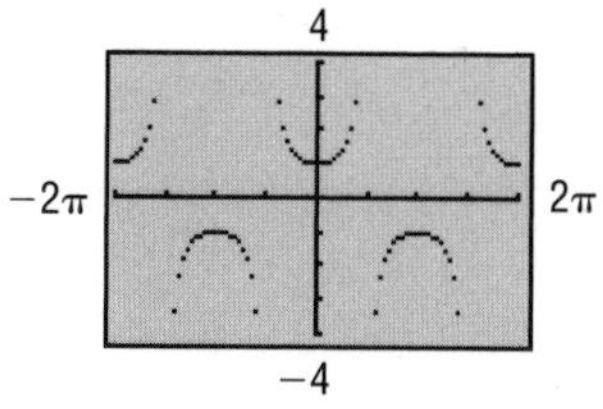

Cotangent Function
$f(x) = \cot x$

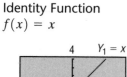

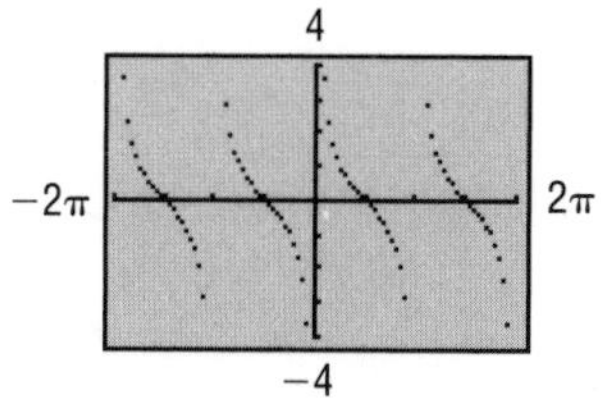

FORMULAS/EQUATIONS

Distance Formula

If $P_1 = (x_1, y_1)$ and $P_2 = (x_2, y_2)$, the distance from P_1 to P_2 is

$$d(P_1, P_2) = \sqrt{(x_2 - x_1)^2 + (y_2 - y_1)^2}$$

Equation of a Circle

The equation of a circle of radius r with center at (h, k) is

$$(x - h)^2 + (y - k)^2 = r^2$$

Slope Formula

The slope m of the line containing the points $P_1 = (x_1, y_1)$ and $P_2 = (x_2, y_2)$ is

$$m = \frac{y_2 - y_1}{x_2 - x_1} \qquad \text{if } x_1 \neq x_2$$

$$m \text{ is undefined} \qquad \text{if } x_1 = x_2$$

Point-Slope Equation of a Line

The equation of a line with slope m containing the point (x_1, y_1) is

$$y - y_1 = m(x - x_1)$$

Slope-Intercept Equation of a Line

The equation of a line with slope m and y-intercept b is

$$y = mx + b$$

Quadratic Formula

The solutions of the equation $ax^2 + bx + c = 0$, $a \neq 0$, are

$$x = \frac{-b \pm \sqrt{b^2 - 4ac}}{2a}$$

If $b^2 - 4ac > 0$, there are two real unequal solutions.
If $b^2 - 4ac = 0$, there is a repeated real solution.
If $b^2 - 4ac < 0$, there are two complex solutions that are not real.

GEOMETRY FORMULAS

Circle

r = Radius, A = Area, C = Circumference

$A = \pi r^2 \qquad C = 2\pi r$

Triangle

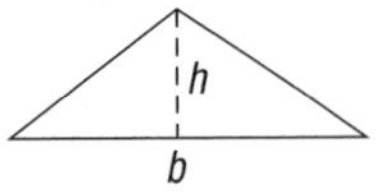

b = Base, h = Altitude (Height), A = Area

$A = \frac{1}{2}bh$

Rectangle

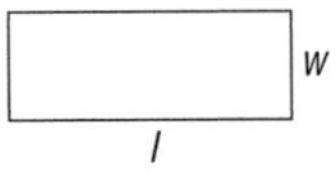

l = Length, w = Width, A = Area, P = Perimeter

$A = lw \qquad P = 2l + 2w$

Rectangular Box

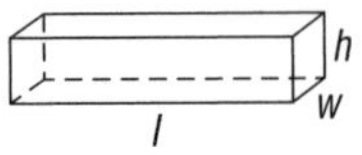

l = Length, w = Width, h = Height, V = Volume

$V = lwh$

Sphere

r = Radius, V = Volume, S = Surface area

$V = \frac{4}{3}\pi r^3 \qquad S = 4\pi r^2$

Precalculus

GRAPHING AND DATA ANALYSIS

Second Edition

Michael Sullivan
Chicago State University

Michael Sullivan, III
Joliet Junior College

Prentice Hall

Prentice Hall
Upper Saddle River, NJ 07458

Library of Congress Cataloging-in-Publication Data

Sullivan, Michael
Precalculus : graphing and data analysis / Michael Sullivan III.-- 2nd ed.
p. cm.
Includes index.
ISBN 0-13-026927-1
1. Function. I. Sullivan, Michael - II. Title.

QA331.3 .S928 2001
512'.1--dc21

00-039994

Editor in Chief: Sally Yagan
Assistant Vice President of Production and Manufacturing: David W. Riccardi
Executive Managing Editor: Kathleen Schiaparelli
Senior Managing Editor: Linda Mihatov Behrens
Production Editor: Bob Walters
Manufacturing Buyer: Alan Fischer
Manufacturing Manager: Trudy Pisciotti
Marketing Manager: Patrice Lumumba Jones
Marketing Assistant: Vince Jansen
Director of Marketing: John Tweeddale
Associate Editor, Mathematics/Statistics Media: Audra J. Walsh
Editorial Assistant/Supplements Editor: Meisha Welch
Art Director: Maureen Eide
Assistant to the Art Director: John Christiana
Interior Designer: Jill Little
Art Editor: Grace Hazeldine
Art Manager: Gus Vibal
Director of Creative Services: Paul Belfanti
Photo Editor: Beth Boyd
Cover Photo: Photomasaic™ by Robert Silvers / www.Photomasaic.com
Art Studio: Academy Artworks

Prentice Hall

Upper Saddle River, New Jersey 07458

Printed in the United States of America

10 9 8 7

ISBN: 0-13-026927-1

Prentice-Hall International (UK) Limited, *London*
Prentice-Hall of Australia Pty. Limited, *Sydney*
Prentice-Hall of Canada Inc., *Toronto*
Prentice-Hall Hispanoamericana, S.A., *Mexico*
Prentice-Hall of India Private Limited, *New Delhi*
Prentice-Hall of Japan, Inc., *Tokyo*
Pearson Education Asia Pte. Ltd.
Editora Prentice-Hall do Brasil, Ltda., *Rio de Janeiro*

For Michael S. and Kevin (Sullivan)
Shannon, Patrick, and Ryan (Murphy)
and Kaleigh (O'Hara)
The Next Generation

CONTENTS

CHAPTER 8 APPLICATIONS OF TRIGONOMETRIC FUNCTIONS 507

CHAPTER 9 POLAR COORDINATES; VECTORS 559

CHAPTER 10 ANALYTIC GEOMETRY 635

CHAPTER 11 SYSTEMS OF EQUATIONS AND INEQUALITIES 709

PREFACE TO THE INSTRUCTOR

As professors at respectively an urban public university and a community college, Michael Sullivan and Michael Sullivan III are aware of the varied needs of Precalculus students, ranging from those having little mathematical background and fear of mathematics courses to those who have had a strong mathematical education and are highly motivated. For some of your students, this will be their last course in mathematics, while others might decide to further their mathematical education. This text is written for both groups. As the author of precalculus, engineering calculus, finite math, and business calculus texts, and as a teacher, Michael understands what students must know if they are to be focused and successful in upper level math courses. However, as a father of four, including the co-author, he also understands the realities of college life. Michael Sullivan III believes passionately in the value of technology as a tool for learning that enhances understanding without sacrificing important skills. Both authors have taken great pains to insure that the text contains solid, student-friendly examples and problems, as well as a clear and seamless writing style. We encourage you to share with us your experiences teaching from this text.

In the Second Edition

The second edition builds upon a strong foundation by integrating new features and techniques that further enhance student interest and involvement. The elements of the previous edition that have proved successful remain, while many changes, some obvious, others subtle, have been made. The text has been streamlined to increase accessibility. A huge benefit of authoring a successful series is the broad-based feedback upon which improvements and additions are ultimately based. Virtually every change to this edition is the result of thoughtful comments and suggestions made by colleagues and students who have used the previous edition. This feedback has proved invaluable and has been used to make changes that improve the flow and usability of the text. For example, some topics have been moved to better reflect the way teachers approach the course. One significant change is the inclusion of the "Field Trip to Motorola" chapter projects. These projects take the incorporation of real life in mathematics to a higher level. The supplements package has been enhanced through upgrading traditional supplements and adding innovative media components such as MathPak, an Integrated Learning Environment. MathPak combines all of the text's key supplements into one easy-to-navigate software package.

Changes to the Second Edition

Chapter 1

- "Data in Ordered Pairs" has been moved to the section where linear curve fitting is introduced.
- "Circles" is now covered in a stand-alone section.
- The section on inequalties now contains a discussion of interval notation.

Chapter 2

- "More about Functions" has been divided into "Characteristics of Functions" and "Library of Function: Piecewise Defined Functions".

Chapters 3 & 4

- Chapter 3 has been divided into two parts: Polynomial and Rational Functions and the Zeros of a Polynomial Function.
- The section on Rational Functions has been split into two sections making the material more manageable.

Chapter 5

- Chapter 4 of the 1st Edition. Certain types of exponential and logarithmic equations appear earlier in the chapter.

Chapter 6

- A rewrite of the introduction to trigonometric functions makes the distinction between the functions of a real number and that of an angle clearer.
- The section on graphing of trigonometric functions is divided in two sections—one graphing the sine and cosine functions and the other graphing the four remaining trigonometric functions.
- The section on Inverse Trigonometric Functions have been moved to Chapter 7.
- The section on Right Triangle Trigonometry now appears in Chapter 8.

Chapter 7

- Now includes two sections on the section on the Inverse Trigonometric Functions and two section on Trigonometric Equations.

Chapter 8

- The first section discusses right triangle trigonometry and applications.

Chapter 9

- Static equilibrium problems were added.
- This chapter now includes a new section on the cross product.

Chapter 10

- Asymptotes are introduced earlier in the section on hyperbolas.

Chapter 11

- Now includes two introductory sections on systems of linear equations.
- The section on determinants includes a discussion on minors and cofactors.
- Systems of Inequalities and Linear Programming are combined into one section.

Chapter 12

- The first five sections of Chapter 11 from the first edition.
- The section on sequences utilizes the power of recursive functions and the graphing calculator to discuss amortization.

Chapter 13

- The second part of Chapter 11 from the 1st Edition.
- Now contains two sections on probability: classical and empirical.

Chapter 14

- A new section on the Area Problem; the Integral has been added.

Appendix

- Rewritten to provide a more thorough review of prerequisite material.
- Can be used as a "just-in-time" review by students.

Acknowledgments

Textbooks are written by authors, but evolve from an idea into final form through the efforts of many people. Special thanks to Don Dellen, who first suggested this book and the other book in this series. Dòn's extensive contributions to publishing and mathematics are well known; we all miss him dearly.

We would like to thank Motorola and its people who helped make the projects in this new edition possible. We would like to extend special thanks to Iwona Turlik, Vice President and Director of the Motorola Advanced Technology Center (MATC), for giving us the opportunity to share with students examples of their experience in applying mathematics to engineering tasks. We would also like to thank the authors of these projects:

- Tomasz Klosowiak the Automotive and Industrial Electronics Group of Integrated Electronic Systems Segment
- Nick Buris, Brian Classon, Terri Fry, and Margot Karam, Motorola Laboratories, Communications Systems and Technologies Labs (CSTL)
- Bill Oslon, Andrew Skipor, John St. Peter, Tom Tirpak, and George Valliath, Motorola Laboratories, Motorola Advanced Technology Center (MATC)
- Jocelyn Carter-Miller, Corporate Vice President and Chief Marketing Officer
- Sheila MB. Griffen, Vice President and Director, Corporate Strategic Marketing, Chief Marketing Office
- Sue Eddins, Curriculum and Assessment Leader for Mathematics and Chuck Hamberg, Mathematics Faculty of the Illinois Mathematics and Science Academy for their generous help and contributions to Chapters 1 and 2.

Special thanks also go to the following:

- Douglas Fekete, Intellectual Property Department.
- Jim Coffiing, Director Communications Future Business and Technology
- Anne Stuessy, Director Communications Future Business and Technology
- Vesna Arsic, Director Corporate Marketing Strategy, Chief Marketing Office
- Rosemarie Broda, Chief Marketing Office
- Rita Browne, CSTL
- David Broth, Vice President and Director of the Communications Systems and Technologies Labs
- Joseph Nowack and Bruce Eastmond, CSTL managers
- Tom Babin, Kevin Jelley and Bill Olson, MATC managers

Last, but not least, for his dedication to this project and the daunting task of managing it, we thank Andrew Skipor.

There are many colleagues we would like to thank for their input, encouragement, patience, and support. They have our deepest thanks and appreciation. We apologize for any omissions…

James Africh, *College of DuPage*
Steve Agronsky, *Cal Poly State University*
Grant Alexander, *Joliet Junior College*
Dave Anderson, *South Suburban College*
Joby Milo Anthony, *University of Central Florida*
James E. Arnold, *University of Wisconsin-Milwaukee*
Agnes Azzolino, *Middlesex County College*
Wilson P. Banks, *Illinois State University*
Dale R. Bedgood, *East Texas State University*
Beth Beno, *South Suburban College*
Carolyn Bernath, *Tallahassee Community College*
William H, Beyer, *University of Akron*
Anita Blackwelder, *Florida State University*
Richelle Blair, *Lakeland Community College*
Trudy Bratten, *Grossmont College*
Joanne Brunner, *Joliet Junior College*
Mary Butler, *Lincoln Public Schools*
William J. Cable, *University of Wisconsin-Stevens Point*
Lois Calamia, *Brookdale Community College*
Jim Campbell, *Lincoln Public Schools*
Roger Carlsen, *Moraine Valley Community College*
Elena Catoiu, *Joliet Junior College*
John Collado, *South Suburban College*
Nelson Collins, *Joliet Junior College*
Jim Cooper, *Joliet Junior College*
Denise Corbett, *East Carolina University*
Theodore C. Coskey, *South Seattle Community College*
John Davenport, *East Texas State University*
Faye Dang, *Joliet Junior College*
Duane E. Deal, *Ball State University*
Vivian Dennis, *Eastfield College*
Guesna Dohrman, *Tallahassee Community College*
Karen R. Dougan, *University of Florida*
Louise Dyson, *Clark College*
Paul D. East, *Lexington Community College*

Don Edmondson, *University of Texas-Austin*
Erica Egizio, *Joliet Junior College*
Christopher Ennis, *University of Minnesota*
Ralph Esparza, Jr., *Richland College*
Garret J. Etgen, *University of Houston*
W.A. Ferguson, *University of Illinois-Urbana/Champaign*
Iris B. Fetta, *Clemson University*
Mason Flake, *student at Edison Community College*
Merle Friel, *Humboldt State University*
Richard A. Fritz, *Moraine Valley Community College*
Carolyn Funk, *South Suburban College*
Dewey Furness, *Ricke College*
Dawit Getachew, *Chicago State University*
Wayne Gibson, *Rancho Santiago College*
Sudhir Kumar Goel, *Valdosta State University*
Joan Goliday, *Sante Fe Community College*
Frederic Gooding, *Goucher College*
Sue Graupner, *Lincoln Public Schools*
Ken Gurganus, *University of North Carolina*
James E. Hall, *University of Wisconsin-Madison*
Judy Hall, *West Virginia University*
Edward R. Hancock, *DeVry Institute of Technology*
Julia Hassett, *DeVry Institute-Dupage*
Michah Heibel, *Lincoln Public Schools*
Brother Herron, *Brother Rice High School*
Lee Hruby, *Naperville North High School*
Kim Hughes, *California State College-San Bernardino*
Ron Jamison, *Brigham Young University*
Richard A. Jensen, *Manatee Community College*
Sandra G. Johnson, *St. Cloud State University*
Moana H. Karsteter, *Tallahassee Community College*
Arthur Kaufman, *College of Staten Island*
Thomas Kearns, *North Kentucky University*
Shelia Kellenbarger, *Lincoln Public Schools*
Keith Kuchar, *Manatee Community College*
Tor Kwembe, *Chicago State University*
Linda J. Kyle, *Tarrant Country Jr. College*
H.E. Lacey, *Texas A & M University*
Matt Larson, *Lincoln Public Schools*
Christopher Lattin, *Oakton Community College*
Adele LeGere, *Oakton Community College*
Stanley Lukawecki, *Clemson University*
Janice C. Lyon, *Tallahassee Community College*
Virginia McCarthy, *Iowa State University*
Jean McArthur, *Joliet Junior College*
Tom McCollow, *DeVry Institute of Technology*
Laurence Maher, *North Texas State University*
Jay A. Malmstrom, *Oklahoma City Community College*
Sherry Martina, *Naperville North High School*
James Maxwell, *Oklahoma State University-Stillwater*
Judy Meckley, *Joliet Junior College*
Carolyn Meitler, *Concordia University*
Samia Metwali, *Erie Community College*
Rich Meyers, *Joliet Junior College*
Eldon Miller, *University of Mississippi*
James Miller, *West Virginia University*
Michael Miller, *Iowa State University*
Kathleen Miranda, *SUNY at Old Westbury*
Thomas Monaghan, *Naperville North High School*
Craig Morse, *Naperville North High School*
A. Muhundan, *Manatee Community College*
Jane Murphy, *Middlesex Community College*
Bill Naegele, *South Suburban College*
James Nymann, *University of Texas-El Paso*
Sharon O'Donnell, *Chicago State University*
Seth F. Oppenheimer, *Mississippi State University*
Linda Padilla, *Joliet Junior College*
E. James Peake, *Iowa State University*
Thomas Radin, *San Joaquin Delta College*
Ken A. Rager, *Metropolitan State College*
Elsi Reinhardt, *Truckee Meadows Community College*
Jane Ringwald, *Iowa State University*
Stephen Rodi, *Austin Community College*
Bill Rogge, *Lincoln Public Schools*
Howard L. Rolf, *Baylor University*
Edward Rozema, *University of Tennessee at Chattanooga*
Dennis C. Runde, *Manatee Community College*
John Sanders, *Chicago State University*
Susan Sandmeyer, *Jamestown Community College*
A.K. Shamma, *University of West Florida*
Martin Sherry, *Lower Columbia College*
Anita Sikes, *Delgado Community College*
Timothy Sipka, *Alma College*
Lori Smellegar, *Manatee Community College*
John Spellman, *Southwest Texas State University*
Becky Stamper, *Western Kentucky University*
Judy Staver, *Florida Community College-South*
Neil Stephens, *Hinsdale South High School*
Diane Tesar, *South Suburban College*
Tommy Thompson, *Brookhaven College*
Richard J. Tondra, *Iowa State University*
Marvel Townsend, *University of Florida*
Jim Trudnowski, *Carroll College*
Robert Tuskey, *Joliet Junior College*
Richard G. Vinson, *University of South Alabama*
Mary Voxman, *University of Idaho*
Donna Wandke, *Naperville North High School*
Darlene Whitkenack, *Northern Illinois University*
Christine Wilson, *West Virginia University*
Carlton Woods, *Auburn University*
George Zazi, *Chicago State University*

Recognition and thanks are due particularly to the following individuals for their valuable assistance in the preparation of this edition: Jerome Grant for his support and commitment; Sally Yagan, for her genuine interest and insightful direction; Bob Walters for his organizational skill as production supervisor; Patrice Lumumba Jones for his innovative marketing efforts; the entire Prentice-Hall sales staff for their confidence; and to Katy Murphy, Jon Weerts, and Sue Osberg for checking the answers to all the exercises.

Michael Sullivan

Michael Sullivan, III

PREFACE TO THE STUDENT

As you begin your study of Precalculus, you may feel overwhelmed by the number of theorems, definitions, procedures, and equations that confront you. You may even wonder whether or not you can learn all of this material in the time allotted. These concerns are normal. Keep in mind that many elements of Precalculus are all around us as we go through our daily routines. Many of the concepts you will learn to express mathematically, you already know intuitively. For many of you, this may be your last math course, while for others, it is just the first in a series of many. Either way, this text was written with you in mind. One of the co-authors, Michael Sullivan, has taught Precalculus courses for over thirty years. He is also the father of four college graduates, including this text's other co-author, who called home from time to time frustrated and with questions. We both know what you're going through. So we have written a text that doesn't overwhelm, or unnecessarily complicate Precalculus, but at the same time gives you the skills and practice you need to be successful.

This text is designed to help you the student, master the terminology and basic concepts of Precalculus. These aims have helped to shape every aspect of the book. Many learning aids are built into the format of the text to make your study of the material easier and more rewarding. This book is meant to be a "machine for learning," that can help you focus your efforts and get the most from the time and energy you invest.

Please do not hesitate to contact us through Prentice Hall with any suggestions or comments that would improve this text.

Best Wishes!

Michael Sullivan

Michael Sullivan, III

OVERVIEW

MOTOROLA PROJECTS

Everyone seems to have a cell phone or pager... Focusing on this type of product, we visit the Motorola Corporation. **"Field Trip to Motorola"** highlights an individual's use of mathematics on the job at Motorola. **"Interview at Motorola"** is a short biography chronicling that individual's educational and career path. The **"Project at Motorola"** concludes the chapter, leading you through an assignment like the one described at the beginning of the chapter.

CHAPTER

2 Functions

Outline

FIELD TRIP TO MOTOROLA

During the past decade the availability and usage of wireless Internet services has increased many fold. The industry has developed a number of pricing proposals for such services. Marketing data have indicated that subscribers of wireless Internet services have tended to desire flat rate fee structures as compared with rates based totally on usage.

97

Page 97

INTERVIEW AT MOTOROLA

Jocelyn Carter-Miller is Corporate Vice President and Chief Marketing Officer (CMO) for Motorola, Inc., an over $30 billion global provider of integrated communications and embedded electronics solutions. As CMO she has helped build the Motorola brand and image, and has developed high performance marketing organizations and processes. Jocelyn also heads motorola.com, the Motorola electronic commerce and information website. In this new role she and her team have developed a strategy for serving Motorola's broad and diverse constituencies offering a full range of electronic services.

In her previous roles as Vice President-Latin American and Caribbean Operations and Director of European, Middle East and African Operations, Jocelyn headed international wireless data communications operations for Motorola—creating profitable opportunities through strategic alliances, value-added applications, and new product and service launches. She also developed skills in managing complex, high risk ventures in countries like Brazil and Russia, setting standards for her company's practices in emerging markets.

Prior to her career at Motorola, Jocelyn served as Vice President, Marketing and Product Development for Mattel, where she broke new ground driving record sales of Barbie and other toys using integrated product, entertainment, promotional, and licensing programs.

Jocelyn builds strong relationships and new opportunities through her involvement on outside boards and community organizations. She serves on the board of the Principal Financial Group, and on the non-profit boards of the Association of National Advertisers, the University of Chicago Women's Business Group Advisory Board, and the Smart School Charter Middle School.

Jocelyn holds a Ma ministration degree in from the University o of Science degree in accounting Illinois-Urbana-Champaign, an Accountant. She is married to dent of Edventures, an educati ment firm, and has two daughter

Jocelyn has won numerous aw national publications, and regu ness and community groups. Sh Melissa Giavagnoli the book *Relationships and Opportuniti* will be published in June 2000 lishers. Through their website celyn and Melissa facilitate m and mutually beneficial opport expert Networlders alike.

Page 98

PROJECT AT MOTOROLA

During the past decade the availability and usage of wireless Internet services has increased. The industry has developed a number of pricing proposals for such services. Marketing data have indicated that subscribers of wireless Internet services have tended to desire flat fee rate structures as compared with rates based totally on usage. The Computer Resource Department of Indigo Media (hypothetical) has entered into a contractual agreement for wireless Internet services. As a part of the contractual agreement, employees are able to sign up for their own wireless services. Three pricing options are available:

Silver-plan: \$20/month for up to 200 K-bytes of service plus \$0.16 for each additional K-byte of service

Gold-plan: \$50/month for up to 1000 K-bytes of service plus \$0.08 for each additional K-byte of service

Platinum-plan: \$100/month for up to 3000 K-bytes of service plus \$0.04 for each additional K-byte of service

You have been requested to write a report that answers the following questions in order to aid employees in choosing the appropriate pricing plan.

(a) If C is the monthly charge for x K-bytes of service, express C as a function of x for each of the three plans.

(b) Graph each of the three functions found in (a).

(c) For how many K-bytes of service is the Silver-plan the best pricing option? When is the Gold-plan best? When is the Platinum-plan best? Explain your reasoning.

(d) Write a report that summarizes your findings.

Page 170

CLEAR WRITING STYLE

The Sullivan's **accessible writing style** is apparent throughout, often utilizing various approaches to the same concept. Authors who write clearly make potentially difficult concepts intuitive and class time more productive.

Figure 3

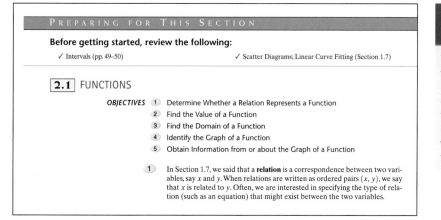

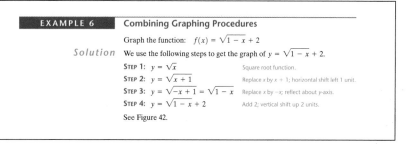

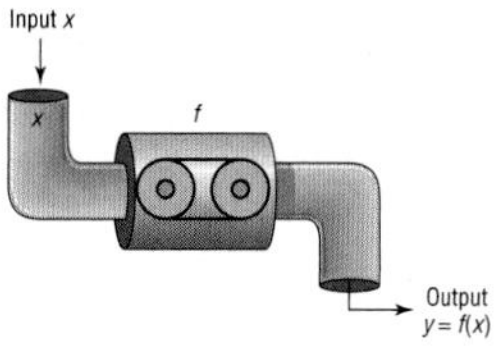

Sometimes it is helpful to think of a function f as a machine that receives as input a number from the domain, manipulates it, and outputs the value. See Figure 3.

The restrictions on this input/output machine are as follows:

1. It only accepts numbers from the domain of the function.
2. For each input, there is exactly one output (which may be repeated for different inputs).

For a function $y = f(x)$, the variable x is called the **independent variable**, because it can be assigned any of the permissible numbers from the domain. The variable y is called the **dependent variable**, because its value depends on x.

Any symbol can be used to represent the independent and dependent variables. For example, if f is the *cube function*, then f can be defined by $f(x) = x^3$ or $f(t) = t^3$ or $f(z) = z^3$. All three functions are the same: Each tells us to cube the independent variable. In practice, the symbols used for the independent and dependent variables are based on common usage, such as using C for cost in business.

2 The variable x is also called the **argument** of the function. Thinking of the independent variable as an argument can sometimes make it easier to find the value of a function. For example, if f is the function defined by $f(x) = x^3$, then f tells us to cube the argument. Thus, $f(2)$ means to cube 2, $f(a)$ means to cube the number a, and $f(x + h)$ means to cube the quantity $x + h$.

Page 102

PREPARING FOR THIS SECTION

Before getting started, review the following:

✓ Intervals (pp. 49–50) ✓ Scatter Diagrams; Linear Curve Fitting (Section 1.7)

2.1 FUNCTIONS

OBJECTIVES

1. Determine Whether a Relation Represents a Function
2. Find the Value of a Function
3. Find the Domain of a Function
4. Identify the Graph of a Function
5. Obtain Information from or about the Graph of a Function

1 In Section 1.7, we said that a **relation** is a correspondence between two variables, say x and y. When relations are written as ordered pairs (x, y), we say that x is related to y. Often, we are interested in specifying the type of relation (such as an equation) that might exist between the two variables.

Page 98

PREPARING FOR THIS SECTION

The **"Preparing for this Section"** feature provides you and your instructor with a list of skills and concepts needed to approach the section, along with page references. You can use this feature to determine what you should review before tackling each section.

STEP-BY-STEP EXAMPLES

Step-by-step examples ensure that you follow the entire solution process and give you an opportunity to check your understanding of each step.

EXAMPLE 6 **Combining Graphing Procedures**

Graph the function: $f(x) = \sqrt{1 - x} + 2$

Solution We use the following steps to get the graph of $y = \sqrt{1 - x} + 2$.

STEP 1: $y = \sqrt{x}$ Square root function.

STEP 2: $y = \sqrt{x + 1}$ Replace x by $x + 1$; horizontal shift left 1 unit.

STEP 3: $y = \sqrt{-x + 1} = \sqrt{1 - x}$ Replace x by $-x$; reflect about y-axis.

STEP 4: $y = \sqrt{1 - x} + 2$ Add 2; vertical shift up 2 units.

See Figure 42.

Page 144

OVERVIEW

REAL-WORLD DATA

Real-world data is incorporated into examples and exercise sets to emphasize that mathematics is a tool used to understand the world around us. As you use these problems and examples, you will see the relevance and utility of the skills being covered.

1 9 9 9 TAX RATE SCHEDULES

SCHEDULE X—IF YOUR FILING STATUS IS SINGLE				SCHEDULE Y-1—USE IF YOUR FILING STATUS IS MARRIED FILING JOINTLY OR QUALIFYING WIDOW(ER)			
If the amount on Form 1040, line 37, is: Over—	But not over—	Enter on Form 1040, line 38	of the amount over—	If the amount on Form 1040, line 37, is: Over—	But not over—	Enter on Form 1040, line 38	of the amount over—
\$0	\$25,750	______15%	\$0	\$0	\$43,050	______15%	\$0
25,750	62,450	\$3,862.50 + 28%	25,750	43,050	104,050	\$6,457.50 + 28%	43,050
62,450	130,250	14,138.50 + 31%	62,450	104,050	158,550	23,537.50 + 31%	104,050
130,250	283,150	35,156.50 + 36%	130,250	158,550	283,150	40,432.50 + 36%	158,550
283,150	______	90,200.50 + 39.6%	283,150	283,150	______	85,288.50 + 39.6%	283,150

Page 166

EXAMPLE 1 Finding the Value of the Six Trigonometric Functions

Figure 19
$\theta = t$ radians

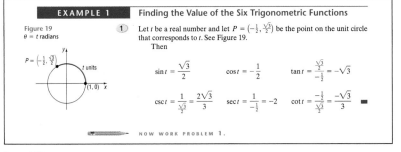

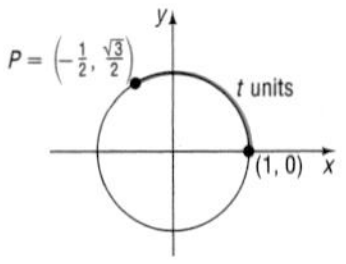

1 Let t be a real number and let $P = \left(-\frac{1}{2}, \frac{\sqrt{3}}{2}\right)$ be the point on the unit circle that corresponds to t. See Figure 19.

Then

$$\sin t = \frac{\sqrt{3}}{2} \qquad \cos t = -\frac{1}{2} \qquad \tan t = \frac{\frac{\sqrt{3}}{2}}{-\frac{1}{2}} = -\sqrt{3}$$

$$\csc t = \frac{1}{\frac{\sqrt{3}}{2}} = \frac{2\sqrt{3}}{3} \qquad \sec t = \frac{1}{-\frac{1}{2}} = -2 \qquad \cot t = \frac{-\frac{1}{2}}{\frac{\sqrt{3}}{2}} = \frac{-\sqrt{3}}{3}$$ ■

NOW WORK PROBLEM 1.

Page 382

"NOW WORK" PROBLEMS

Many examples end with the phrase **"Now Work Problem —."** Sending you to the exercise set to work a similar problem provides the opportunity to immediately check your understanding. The corresponding "Now Work" problem is easily identified in the exercise sets by the pencil icon and yellow exercise number.

In Problems 1–10, t is a real number and $P = (x, y)$ is the point on the unit circle that corresponds to t. Find the exact value of the six trigonometric functions of t.

1. $\left(\frac{1}{4}, \frac{\sqrt{15}}{4}\right)$
2. $\left(\frac{3}{8}, \frac{\sqrt{55}}{8}\right)$
3. $\left(-\frac{2}{5}, \frac{\sqrt{21}}{5}\right)$
4. $\left(-\frac{1}{5}, \frac{2\sqrt{6}}{5}\right)$
5. $\left(\frac{-\sqrt{35}}{6}, -\frac{1}{6}\right)$
6. $\left(\frac{-\sqrt{39}}{8}, \frac{5}{8}\right)$
7. $\left(\frac{2\sqrt{2}}{3}, -\frac{1}{3}\right)$
8. $\left(\frac{-\sqrt{5}}{3}, -\frac{2}{3}\right)$
9. $\left(-\frac{3\sqrt{5}}{7}, \frac{2}{7}\right)$
10. $\left(-\frac{3\sqrt{11}}{10}, -\frac{1}{10}\right)$

Page 392

SOLUTIONS

Solutions, both algebraic and graphical, are provided whenever appropriate throughout the text.

EXAMPLE 1 Graphing a Quadratic Function

Graph the function $f(x) = 2x^2 + 8x + 5$. Find the vertex and axis of symmetry.

Graphing Solution

Before graphing, notice that the leading coefficient, 2, is positive, and therefore the graph will open up and the vertex will be the lowest point. Now graph $Y_1 = f(x) = 2x^2 + 8x + 5$. See Figure 6. We observe that the graph does in fact open up. To estimate the vertex of the parabola, we use the MINIMUM command. The vertex is the point $(-2, -3)$, and the axis of symmetry is the line $x = -2$.

Figure 6

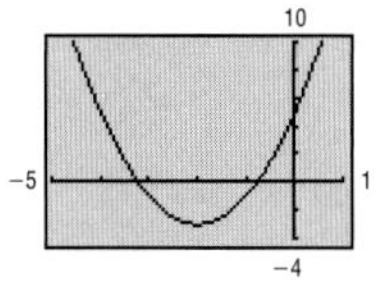

Algebraic Solution

We begin by completing the square on the right side.

$$\begin{aligned} f(x) &= 2x^2 + 8x + 5 \\ &= 2(x^2 + 4x) + 5 && \text{Factor out the 2 from } 2x^2 + 8x. \\ &= 2(x^2 + 4x + 4) + 5 - 8 && \text{Complete the square of } 2(x^2 + 4x). \\ &= 2(x + 2)^2 - 3 && \text{Notice that the factor of 2 requires that 8 be added and subtracted.} \end{aligned} \quad (2)$$

Page 174

2.3 EXERCISES

In Problems 1–8, match each graph to the function listed whose graph most resembles the one given.

A. Constant function	*B. Linear function*	*C. Square function*
D. Cube function	*E. Square root function*	*F. Reciprocal function*
G. Absolute value function	*H. Greatest-integer function*	

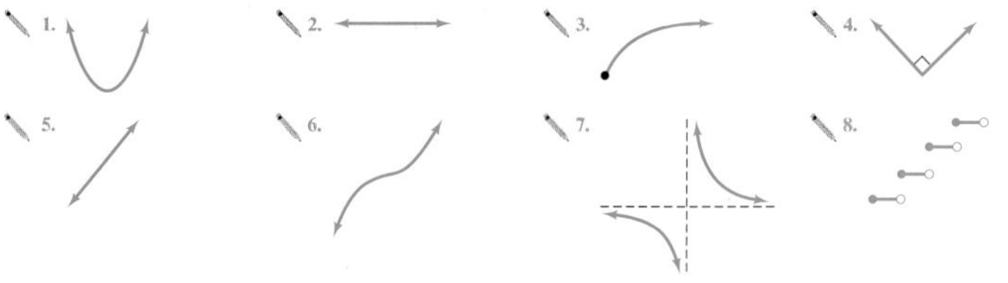

Page 134

END-OF-SECTION EXERCISES

Sullivan's exercises are unparalleled in terms of thorough coverage and accuracy. Each **end-of-section exercise** set begins with visual- and concept-based problems, starting you out with the basics of the section. Well-thought-out exercises better prepare you for exams.

Figure 50

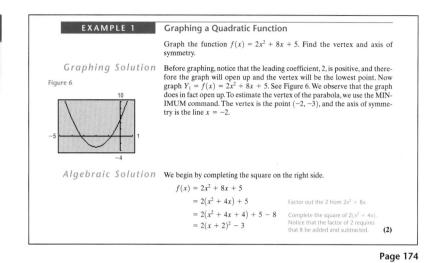

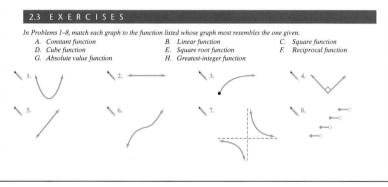

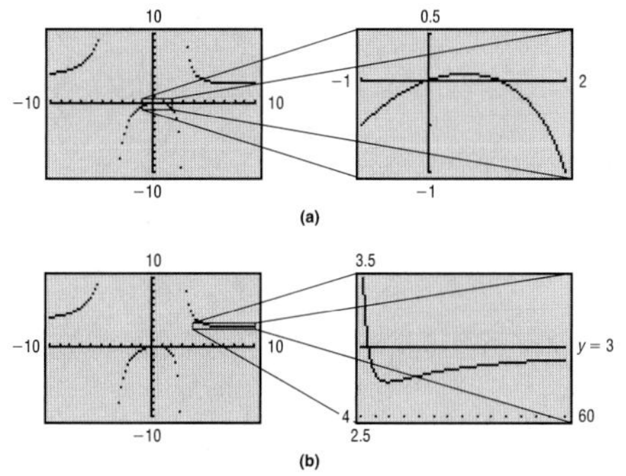

Page 224

GRAPHING UTILITIES AND TECHNIQUES

Increase your understanding, visualize, discover, explore, and solve problems using a graphing utility. Sullivan uses the graphing utility to further your understanding of concepts not to circumvent essential math skills.

OVERVIEW

MODELING

Many examples and exercises connect real-world situations to mathematical concepts. Learning to work with **models** is a skill that transfers to many disciplines.

EXAMPLE 11 **Fitting a Quadratic Function to Data**

The data in Table 2 represent the enrollment (in millions) in public high schools for the years 1986–1996.

TABLE 2

Year, x	Enrollment, E
1986	13.0
1987	12.7
1988	12.2
1989	12.1
1990	11.9
1991	12.2
1992	12.3
1993	12.6
1994	13.5
1995	13.7
1996	14.1

Source: U.S. Bureau of the Census

(a) Using a graphing utility, draw a scatter diagram of the data. Comment on the type of relation that may exist between the two variables.
(b) Find the quadratic function of best fit.
a graphing utility, draw the quadratic function of best fit on your diagram.
the function found in part (b), determine when enrollment was west.
are the result found in part (d) to the data.

Page 184

EXAMPLE 2 **Graphing $y = \frac{1}{x^2}$**

Analyze the graph $H(x) = \frac{1}{x^2}$.

Solution See Figure 39. The domain of $H(x) = 1/x^2$ consists of all real numbers x except 0. The graph has no y-intercept, because x can never equal 0. The graph has no x-intercept because the equation $H(x) = 0$ has no solution. Therefore, the graph of H will not cross either coordinate axis.

Because

$$H(-x) = \frac{1}{(-x)^2} = \frac{1}{x^2} = H(x)$$

H is an even function, so its graph is symmetric with respect to the y-axis.

Look again at Figure 39. What happens to the function as the values of x get closer and closer to 0? We use a TABLE to answer the question. See Table 8. The first four rows show that as x approaches 0 the values of $H(x)$ become larger and larger positive numbers. When this happens, we say that $H(x)$ is **unbounded in the positive direction**. We symbolize this by writing $H(x) \to \infty$ [read as "$H(x)$ **approaches infinity**"]. In calculus, **limits** are used to convey these ideas. We use the symbolism $\lim_{x\to 0} H(x) = \infty$, read as "the limit of $H(x)$ as x approaches 0 is infinity" to mean that $H(x) \to \infty$ as $x \to 0$.

Look at the last-three rows of Table 8. As $x \to \infty$, the values of $H(x)$ approach 0. This is symbolized in calculus by writing $\lim_{x\to\infty} H(x) = 0$. Figure 40 shows the graph of $y = 1/x^2$ drawn by hand.

Figure 39

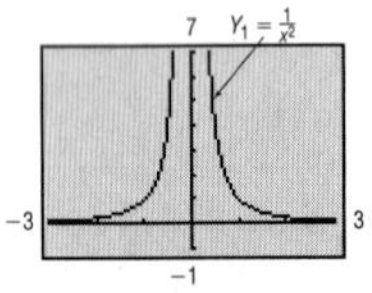

TABLE 8

X	Y1
.1	100
.01	10000
.001	1E6
1E-4	1E8
10	.01
100	1E-4
1000	1E-6

Y1=1/X²

Page 210

PREVIEW OF CALCULUS

Certain concepts and skills are very important in the study of Calculus. These ideas are highlighted by the icon throughout the text. Using this text as a review for your Calculus course will help you do well.

DISCUSSION WRITING AND READING PROBLEMS

These **problems** are designed to get you to "think outside the box," therefore fostering an intuitive understanding of key mathematical concepts. In this example, matching the graph to the functions insures that you understand functions at a fundamental level.

73. Match each of the following functions with the graphs that best describe the situation.
(a) The cost of building a house as a function of its square footage
(b) The height of an egg dropped from a 300-foot building as a function of time
(c) The height of a human as a function of time
(d) The demand for Big Macs as a function of price
(e) The height of a child on a swing as a function of time

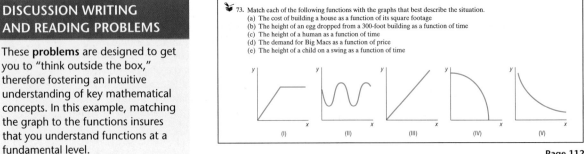

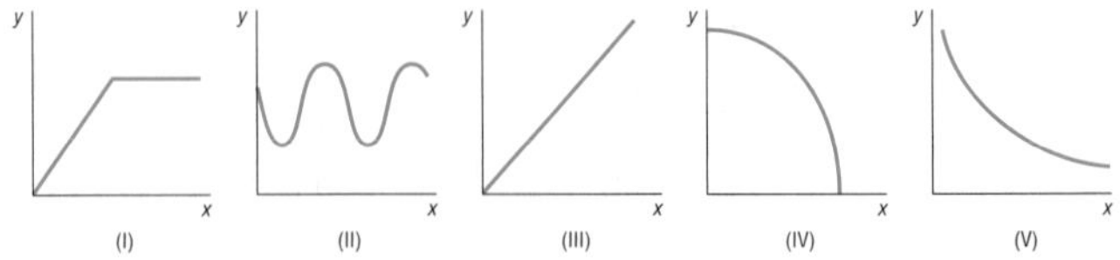

Page 112

CHAPTER REVIEW

The Chapter Review helps check your understanding of the chapter materials in several ways. **"Things to Know"** gives a general overview of review topics. The **"How To"** section provides a concept-by-concept listing of operations you are expected to perform. The **"Review Exercises"** then serve as a chance to practice the concepts presented within the chapter. Several of the Review Exercises are numbered in blue. These exercises can be combined to create a practice Chapter Test. Since these problems are odd numbered, you can check your answers in the back of the book.
The review materials are designed to make you, the student, confident in knowing the chapter material.

CHAPTER REVIEW

Library of Functions

Linear function (p. 128)
$f(x) = mx + b$ Graph is a line with slope m and y-intercept b.

Constant function (p. 128)
$f(x) = b$ Graph is a horizontal line with y-intercept b (see Figure 20).

Identity function (p. 128)
$f(x) = x$ Graph is a line with slope 1 and y-intercept 0 (see Figure 21).

Square function (p. 129)
$f(x) = x^2$ Graph is a parabola with intercept at $(0, 0)$ (see Figure 22).

Cube function (p. 129)
$f(x) = x^3$ See Figure 23.

Square root function (p. 130)
$f(x) = \sqrt{x}$ See Figure 24.

Reciprocal function (p. 130)
$f(x) = 1/x$ See Figure 25.

Absolute value function (p. 130)
$f(x) = |x|$ See Figure 26.

Greatest integer function (p. 131)
$f(x) = \text{int}(x)$ See Figure 27.

Things To Know

Function (p. 99) A relation between two sets of real numbers so that each number x in the first set, the domain, has corresponding to it exactly one number y in the second set. The range is the set of y values of the funtion for the x values in the domain.
x is the independent variable; y is the dependent variable.
A function f may be defined implicitly by an equation involving x and y or explicitly by writing $y = f(x)$.
A function can also be characterized as a set of ordered pairs (x, y) or $(x, f(x))$ in which no two distinct pairs have the same first element.

Function notation (p. 101) $y = f(x)$
f is a symbol for the function.
x is the argument, or independent variable.
y is the dependent variable.
$f(x)$ is the value of the function at x, or the image of x.

Page 166

How To

Determine whether a relation represents a function (p. 98)
Find the value of a function (p. 102)
Find the domain of a function (p. 104)
Identify the graph of a function (p. 105)
Obtain information from or about the graph of a function (p. 106)
Find the average rate of change of a function (p. 116)
Use a graphing utility to locate local maxima and minima (p. 120)
Use a graphing utility to determine where a function is increasing and decreasing (p. 122)
Determine even or odd functions from a graph (p. 122)
Identify even or odd functions fro the equation (p. 123)
Graph functions listed in the libra of functions (p. 128)
Graph piecewise-defined functio (p. 132)
Graph functions using horizontal and vertical shifts (p. 137)
Graph functions using compressions and stretches (p. 139)
Graph functions using reflections about the x-axis or y-axis (p. 141)

Fill-in-the-Blank Items

1. If f is a function defined by the equation $y = f(x)$, then x is calle ______ variable.
2. A set of points in the xy-plane is the graph of a function if and onl graph in at most one point.
3. The average rate of change of a function equals the ______ graph.
4. A(n) ______ function f is one for which $f(-x) = f(x)$ for function f is one for which $f(-x) = -f(x)$ for every x in the domai
5. Suppose that the graph of a function f is known. Then the graph of $y =$ shift of the graph of f to the ______ a distance of 2 units.
6. If $f(x) = x + 1$ and $g(x) = x^3$, then ______ $= (x + 1)^3$

Page 167

True/False Items

T F 1. Every relation is a function.
T F 2. Vertical lines intersect the graph of a function in no more than one point.
T F 3. The y-intercept of the graph of the function $y = f(x)$, whose domain is all real numbers, is $f(0)$.
T F 4. A function f is decreasing on an open interval I if, for any choice of x_1 and x_2 in I, with $x_1 < x_2$, we have $f(x_1) < f(x_2)$.
T F 5. Even functions have graphs that are symmetric with respect to the origin.
T F 6. The graph of $y = f(-x)$ is the reflection about the y-axis of the graph of $y = f(x)$.
T F 7. $f(g(x)) = f(x) \cdot g(x)$.
T F 8. The domain of the composite function $(f \circ g)(x)$ is the same as that of $g(x)$.

Review Exercises

Blue problem numbers indicate the authors' suggestions for use in a Practice Test.

1. Given that f is a linear function, $f(4) = -5$ and $f(0) = 3$, write the equation that defines f.
2. Given that g is a linear function with slope $= -4$ and $g(-2) = 2$, write the equation that defines g.
3. A function f is defined by
$$f(x) = \frac{Ax + 5}{6x - 2}$$
If $f(1) = 4$, find A.
4. A function g is defined by
$$g(x) = \frac{A}{x} + \frac{8}{x^2}$$
If $g(-1) = 0$, find A.
5. Tell which of the following graphs are graphs of functions.

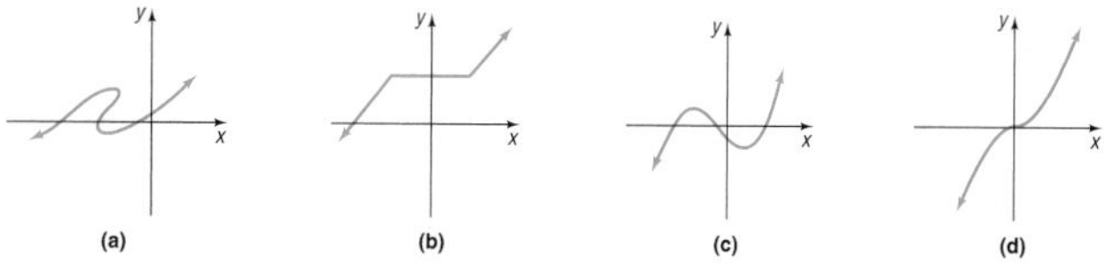

Page 168

Sullivan M@thPak

An Integrated Learning Environment

Home Next

Chapter 1:

- Preparing for This Chapter
- Chapter Internet Project
- Chapter Destinations
- Chapter Solutions Manual
- Chapter Quiz 1
- Chapter Quiz 2
- Chapter Test
- Chapter Sections
- MathPro Explorer CD
- Graphing Calculator Help
- Full Student Solutions

Other Options:

- Help
- Your Profile
- Feedback
- Site Search

→ Syllabus

GET IT TOGETHER!

M@THP@K integrates and organizes all major student supplements into an easy-to-use format at a price that can't be beat!

Here's just a sample of what you'll find in MathPak:

- **MultiMedia MathPro 4.0**
  This interactive tutorial program offers unlimited practice on College Algebra content. Watch the author work the problems via videos,view other examples,and see a fully worked out solution to the problem you are working on.

- **Graphing Calculator Manuals**
  Includes step by step procedures and screen shots for working with your TI- 82,TI-83,TI-85,TI-86, TI-89,TI-92,HP48G, CFX-9850 GaPlus, and SharpE 9600c.

- **Full Student Solutions Manual**
  Includes step by step solutions for all the odd numbered exercises in the text.

1.1 Real Numbers

- Reading Quiz
- MathPro Objectives
- PowerPoints

System Requirements:

- 32MB of random access memory (RAM); 64MB or more recommended
- 200MB free hard disk space
- CD-ROM drive
- QuickTime™ 3.0 or better
- Internet Browser 4.0 or higher
- Internet Access 28.8k or better
- LiveMath Plug-In
- Shockwave® Plug-In

MADE WITH macromedia® QuickTime™

ISBN: 0-13-088274-7

Sullivan M@thPak

Helping Students *Get it Together*

Additional Media

Sullivan Companion Website

www.prenhall.com/sullivan

This text-specific website beautifully complements the text. Here students can find chapter tests, section-specific links, and PowerPoint downloads in addition to other helpful features.

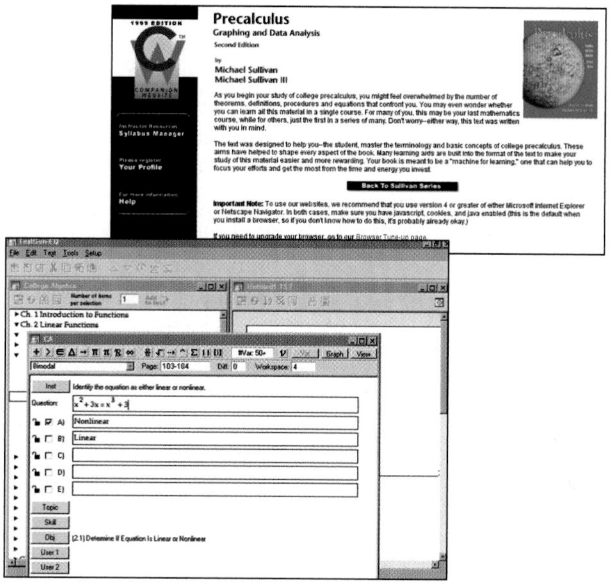

Test Gen-EQ

CD-Rom (Windows/Macintosh)

- Algorithmically driven, text-specific testing program
- Networkable for administering tests and capturing grades
- Edit existing test items or add your own questions to create a nearly unlimited number of tests and drill worksheets

ISBN: 0-13-027309-0

OTHER SUPPLEMENTS

Student Supplements

Student Solutions Manual

Worked solutions to all odd-numbered exercises from the text and complete solutions for chapter review problems and chapter tests. ISBN: 0-13-028759-8

Lecture Videos

The instructional tapes, in a lecture format, feature Michael Sullivan III presenting worked-out examples for each section of the text. ISBN: 0-13-089035-9

New York Times Themes of the Times

A *free* newspaper from Prentice Hall and *The New York Times.* Interesting and current articles on mathematics which invite discussion and writing about mathematics.

Mathematics on the Internet

Free guide providing a brief history of the Internet, discussing the use of the World Wide Web, and describing how to find your way within the Internet and how to find others on it.

Instructor Supplements

Instructor's Resource Manual

Contains complete step-by-step worked-out solutions to all even-numbered exercises in the textbook. ISBN: 0-13-028761-X

Test Item File

Hard copy of the algorithmic computerized testing materials. ISBN: 0-13-028750-4

LIST OF APPLICATIONS

PHOTO AND ILLUSTRATION CREDITS

Chapter 1	Pages 1 PhotoDisc, Inc., 2 Courtesy of Motorola, Inc.
Chapter 2	Pages 97 PhotoDisc, Inc., 98 Courtesy of Motorola, Inc.
Chapter 3	Pages 171 Buccina Studios/PhotoDisc, Inc., 172, 234 (bottom), 235 Courtesy of Motorola, Inc.
Chapter 4	Pages 237 PhotoDisc, Inc., 238 Courtesy of Motorola, Inc.; page 250 (bottom) Corbis; page 278 Courtesy of Motorola, Inc.
Chapter 5	Pages 279 PhotoDisc, Inc., 280 Courtesy of Motorola, Inc.; page 322 The Granger Collection; page 366 Courtesy of Motorola, Inc.
Chapter 6	Pages 367 PhotoDisc, Inc., 368 Courtesy of Motorola, Inc.; page 444 Figures 1 and 2 Courtesy of Motorola, Inc.
Chapter 7	Pages 445 Steve Mason/PhotoDisc, Inc., 446, 506 Courtesy of Motorola, Inc.
Chapter 8	Pages 507 PhotoDisc, Inc., 508, 556, 557 Courtesy of Motorola, Inc.
Chapter 9	Pages 559 PhotoDisc, Inc., 560 Courtesy of Motorola, Inc.; page 583 (bottom) Corbis; page 592 Corbis; page 604 Library of Congress; page 633 (bottom) Courtesy of Motorola, Inc.
Chapter 10	Pages 635 PhotoDisc, Inc., 636, 708 Courtesy of Motorola, Inc.
Chapter 11	Pages 709 PhotoDisc, Inc., 710 Courtesy of Motorola, Inc.; page 776 (bottom) Corbis; page 784 Corbis.
Chapter 12	Pages 809 Jack Hellingsworth/PhotoDisc, Inc., 810 Courtesy of Motorola, Inc.; page 837 The Granger Collection; page 850 Corbis; page 855 Courtesy of Motorola, Inc.
Chapter 13	Pages 857 PhotoDisc, Inc., 858 Courtesy of Motorola, Inc.; page 883 The Granger Collection.
Chapter 14	Pages 901 PhotoDisc, Inc., 902, 941 Courtesy of Motorola, Inc.
Appendix	Page 955 Giraudon/Art Resource, N.Y.

CHAPTER

1 Graphs

Outline

FIELD TRIP TO MOTOROLA

During the 1990's mobile phones became a significant means of communication. Industry data indicates that in 1994 there were about 37.5 million mobile phones in use globally. In 1998, that figure rose to about 70 million mobile phones and is projected to rise to 102.5 million by 2002. Industry analysts base such predictions on a *mathematical model*, that is, a relationship (usually an equation) that involves one or more variables.

INTERVIEW AT MOTOROLA

Sheila Griffin's expertise in marketing, marketing research, and the wireless communications industry has earned her a worldwide reputation in the communications industry. Currently she is Vice President and Director of Corporate Strategic Marketing in the Chief Marketing Office at Motorola.

In 1994, Sheila joined the newly formed Multimedia Group of Motorola's Messaging, Information, and Media Sector as Director of the Multimedia Group Strategy Office & Investments. Sheila was also a valued member of the Board of Directors of Virtus. During her tenure with Multimedia, she also assumed responsibility for overseeing Motorola's relationship with and sponsorship of the Massachusetts Institute of Technology Media Lab and retains that responsibility today.

Prior to her appointment as Director of Multimedia Group Strategy Office & Investments, Sheila held various positions within Motorola such as Director of Corporate Advertising Worldwide and Director of Global Market Research.

Sheila has been very active in educational endeavors. In 1985, she was appointed by Governor James Thompson to the Board of Trustees of the Illinois Mathematics and Science Academy and re-appointed to the IMSA Board in 1995.

Sheila received a Bachelor's degree in Marketing and her MBA in Marketing and in Finance from the University of Illinois at Chicago. She resides with her husband and two children in St. Charles, IL.

PREPARING FOR THIS SECTION

Before getting started, review the following:

- ✓ Algebra Review (Appendix, Section 1)
- ✓ Geometry Review (Appendix, Section 9)

1.1 RECTANGULAR COORDINATES; GRAPHING UTILITIES

OBJECTIVES
1. Use the Distance Formula
2. Use the Midpoint Formula

We locate a point on the real number line by assigning it a single real number, called the *coordinate of the point*. For work in a two-dimensional plane, we locate points by using two numbers.

We begin with two real number lines located in the same plane: one horizontal and the other vertical. We call the horizontal line the ***x*-axis**; the vertical line, the ***y*-axis**; and the point of intersection, the **origin *O***. We assign coordinates to every point on these number lines as shown in Figure 1, using a convenient scale. In mathematics, we usually use the same scale on each axis; in applications, a different scale is often used on each axis.

Figure 1

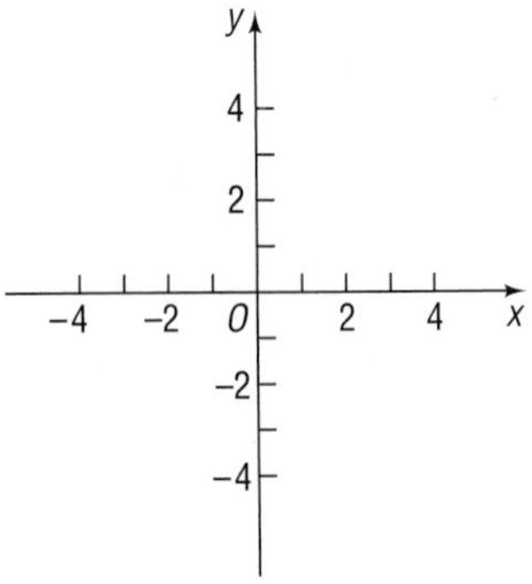

The origin O has a value of 0 on both the x-axis and the y-axis. We follow the usual convention that points on the x-axis to the right of O are associated with positive real numbers, and those to the left of O are associated with negative real numbers. Those on the y-axis above O are associated with positive real numbers, and those below O are associated with negative real numbers. In Figure 1, the x-axis and y-axis are labeled as x and y, respectively, and we have used an arrow at the end of each axis to denote the positive direction.

Figure 2

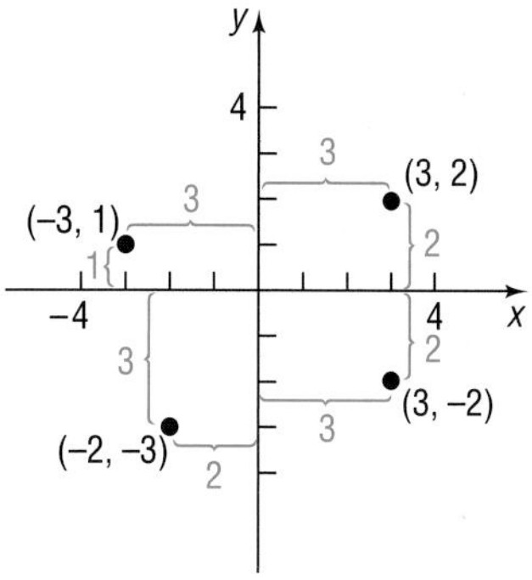

The coordinate system described here is called a **rectangular** or **Cartesian*** **coordinate system**. The plane formed by the x-axis and y-axis is sometimes called the ***xy*-plane**, and the x-axis and y-axis are referred to as the **coordinate axes**.

Any point P in the xy-plane can then be located by using an **ordered pair** (x, y) of real numbers. Let x denote the signed distance of P from the y-axis (*signed* in the sense that, if P is to the right of the y-axis, then $x > 0$, and if P is to the left of the y-axis, then $x < 0$); and let y denote the signed distance of P from the x-axis. The ordered pair (x, y), also called the **coordinates** of P, then gives us enough information to locate the point P in the plane.

For example, to locate the point whose coordinates are $(-3, 1)$, go 3 units along the x-axis to the left of O and then go straight up 1 unit. See Figure 2. We **plot** this point by placing a dot at this location. The points with coordinates $(-3, 1)$, $(-2, -3)$, $(3, -2)$, and $(3, 2)$ are plotted in Figure 2.

The origin has coordinates $(0, 0)$. Any point on the x-axis has coordinates of the form $(x, 0)$, and any point on the y-axis has coordinates of the form $(0, y)$.

Figure 3

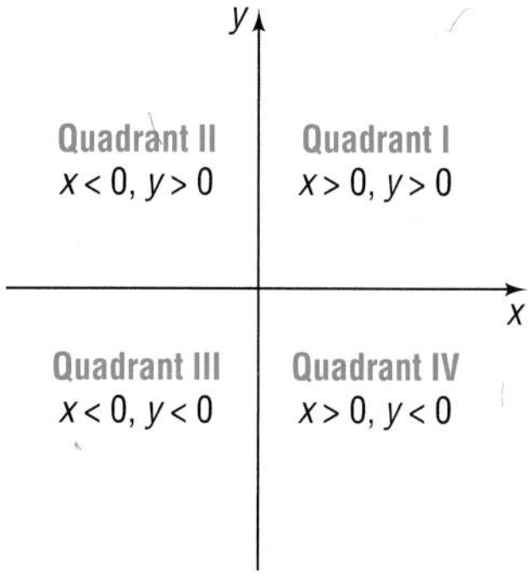

If (x, y) are the coordinates of a point P, then x is called the ***x*-coordinate**, or **abscissa**, of P and y is the ***y*-coordinate**, or **ordinate**, of P. We identify the point P by its coordinates (x, y) by writing $P = (x, y)$. Usually, we will simply say "the point (x, y)" rather than "the point whose coordinates are (x, y)."

The coordinate axes divide the xy-plane into four sections called **quadrants**, as shown in Figure 3. In quadrant I, both the x-coordinate and the y-coordinate of all points are positive; in quadrant II, x is negative and y is positive; in quadrant III, both x and y are negative; and in quadrant IV, x is positive and y is negative. Points on the coordinate axes belong to no quadrant.

NOW WORK PROBLEM 1.

GRAPHING UTILITIES

All graphing utilities, that is, all graphing calculators and all computer software graphing packages, graph equations by plotting points on a screen. The screen itself actually consists of small rectangles, called **pixels**. The more pixels the screen has, the better the resolution. Most graphing calculators have 48 pixels per square inch; most computer screens have 32 to 108 pixels per square inch. When a point to be plotted lies inside a pixel, the pixel is turned on (lights up). Thus, the graph of an equation is a collection of pixels. Figure 4 shows how the graph of $y = 2x$ looks on a TI-83 graphing calculator.

Figure 4
$y = 2x$

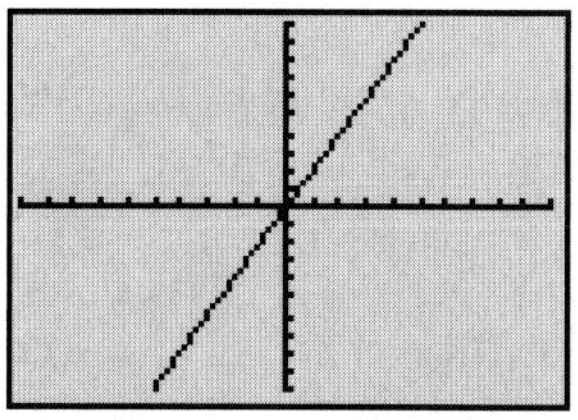

The screen of a graphing utility will display the coordinate axes of a rectangular coordinate system. However, you must set the scale on each axis. You must also include the smallest and largest values of x and y that you want included in the graph. This is called **setting the viewing rectangle** or **viewing window**. Figure 5 illustrates a typical viewing window.

Figure 5

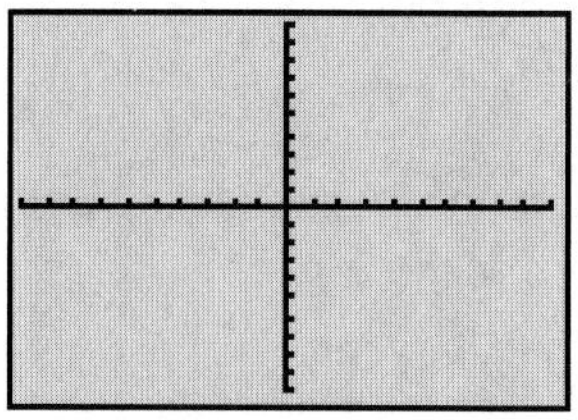

To select the viewing window, we must give values to the following expressions:

*Named after René Descartes (1596–1650), a French mathematician, philosopher, and theologian.

Xmin:	the smallest value of x
Xmax:	the largest value of x
Xscl:	the number of units per tick mark on the x-axis
Ymin:	the smallest value of y
Ymax:	the largest value of y
Yscl:	the number of units per tick mark on the y-axis

Figure 6 illustrates these settings and their relation to the Cartesian coordinate system.

Figure 6

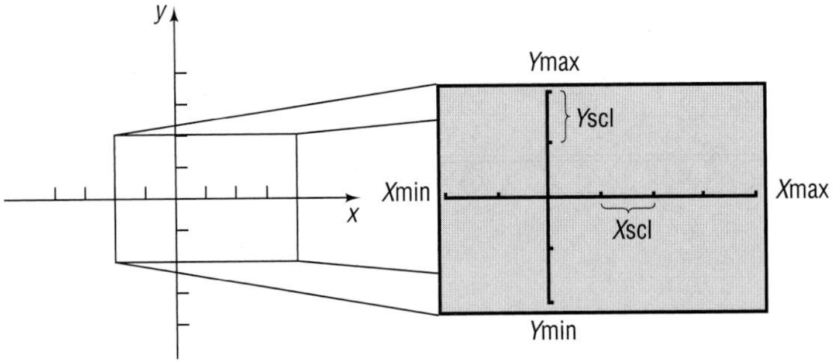

If the scale used on each axis is known, we can determine the minimum and maximum values of x and y shown on the screen by counting the tick marks. Look again at Figure 5. For a scale of 1 on each axis, the minimum and maximum values of x are -10 and 10, respectively; the minimum and maximum values of y are also -10 and 10. If the scale is 2 on each axis, then the minimum and maximum values of x are -20 and 20, respectively; and the minimum and maximum values of y are -20 and 20, respectively.

Conversely, if we know the minimum and maximum values of x and y, we can determine the scales being used by counting the tick marks displayed. We shall follow the practice of showing the minimum and maximum values of x and y in our illustrations so that you will know how the viewing window was set. See Figure 7.

Figure 7

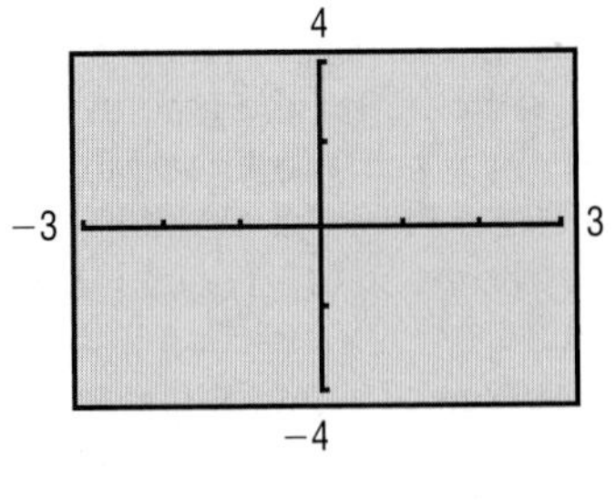

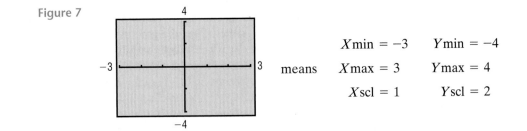

means

$X\text{min} = -3$	$Y\text{min} = -4$
$X\text{max} = 3$	$Y\text{max} = 4$
$X\text{scl} = 1$	$Y\text{scl} = 2$

EXAMPLE 1 Finding the Coordinates of a Point Shown on a Graphing Utility Screen

Find the coordinates of the point shown in Figure 8. Assume that the coordinates are integers.

Figure 8

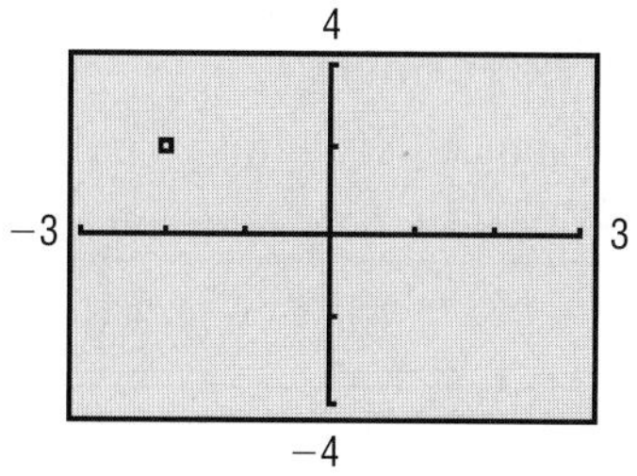

Solution First we note that the viewing window used in Figure 8 is

$$X\text{min} = -3 \qquad Y\text{min} = -4$$
$$X\text{max} = 3 \qquad Y\text{max} = 4$$
$$X\text{scl} = 1 \qquad Y\text{scl} = 2$$

The point shown is 2 tick units to the left on the horizontal axis (scale = 1) and 1 tick up on the vertical axis (scale = 2). Thus, the coordinates of the point shown are $(-2, 2)$. ■

NOW WORK PROBLEMS 5 AND 15.

DISTANCE BETWEEN POINTS

1 If the same units of measurement, such as inches, centimeters, and so on, are used for both the x-axis and the y-axis, then all distances in the xy-plane can be measured using this unit of measurement.

EXAMPLE 2 Finding the Distance between Two Points

Find the distance d between the points $(1, 3)$ and $(5, 6)$.

Solution First we plot the points $(1, 3)$ and $(5, 6)$ as shown in Figure 9(a). Then we draw a horizontal line from $(1, 3)$ to $(5, 3)$ and a vertical line from $(5, 3)$ to $(5, 6)$, forming a right triangle, as in Figure 9(b). One leg of the triangle is of length 4 and the other is of length 3. By the Pythagorean Theorem, the square of the distance d that we seek is

$$d^2 = 4^2 + 3^2 = 16 + 9 = 25$$
$$d = 5$$

Figure 9

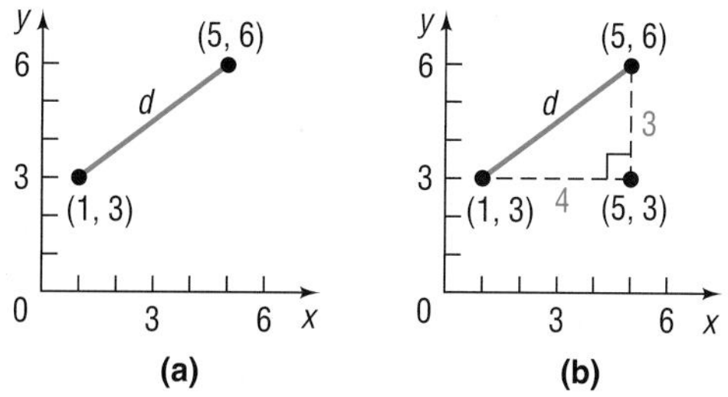

■

The **distance formula** provides a straightforward method for computing the distance between two points.

Theorem **Distance Formula**

The distance between two points $P_1 = (x_1, y_1)$ and $P_2 = (x_2, y_2)$, denoted by $d(P_1, P_2)$, is

$$d(P_1, P_2) = \sqrt{(x_2 - x_1)^2 + (y_2 - y_1)^2} \qquad \textbf{(1)}$$

Figure 10

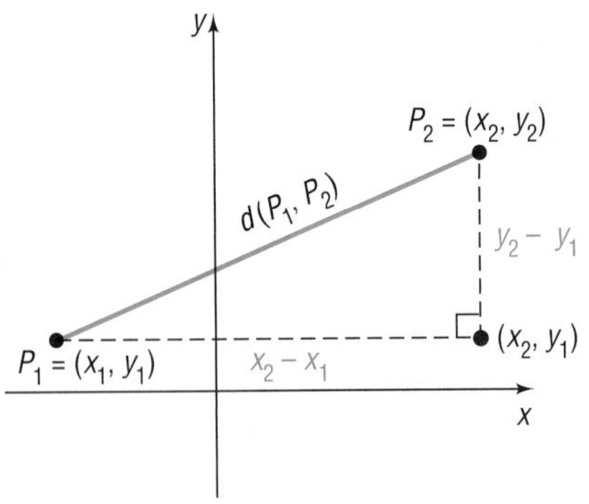

That is, to compute the distance between two points, find the difference of the x-coordinates, square it, and add this to the square of the difference of the y-coordinates. The square root of this sum is the distance. See Figure 10.

Proof of the Distance Formula Let (x_1, y_1) denote the coordinates of point P_1, and let (x_2, y_2) denote the coordinates of point P_2. Assume that the line joining P_1 and P_2 is neither horizontal nor vertical. Refer to Figure 11(a). The coordinates of P_3 are (x_2, y_1). The horizontal distance from P_1 to P_3 is the absolute value of the difference of the x-coordinates, $|x_2 - x_1|$. The vertical distance from P_3 to P_2 is the absolute value of the difference of the y-coordinates, $|y_2 - y_1|$. See Figure 11(b). The distance $d(P_1, P_2)$ that we seek is the length of the hypotenuse of the right triangle, so, by the Pythagorean Theorem, it follows that

$$\begin{aligned}[d(P_1, P_2)]^2 &= |x_2 - x_1|^2 + |y_2 - y_1|^2 \\ &= (x_2 - x_1)^2 + (y_2 - y_1)^2 \\ d(P_1, P_2) &= \sqrt{(x_2 - x_1)^2 + (y_2 - y_1)^2}\end{aligned}$$

Figure 11

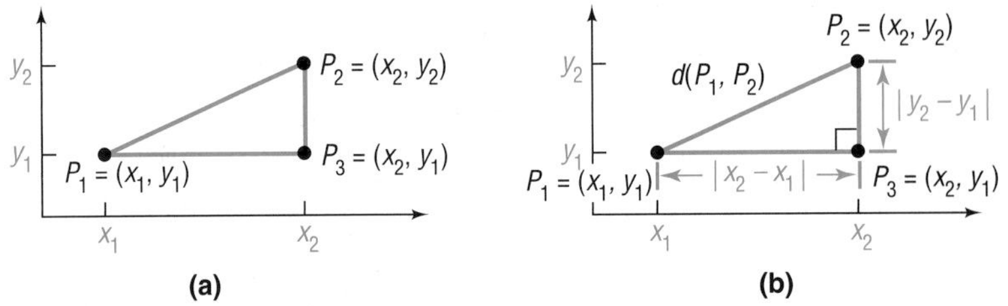

Now, if the line joining P_1 and P_2 is horizontal, then the y-coordinate of P_1 equals the y-coordinate of P_2; that is, $y_1 = y_2$. Refer to Figure 12(a). In this case, the distance formula (1) still works, because, for $y_1 = y_2$, it reduces to

$$d(P_1, P_2) = \sqrt{(x_2 - x_1)^2 + 0^2} = \sqrt{(x_2 - x_1)^2} = |x_2 - x_1|$$

A similar argument holds if the line joining P_1 and P_2 is vertical. See Figure 12(b). Thus, the distance formula is valid in all cases.

Figure 12

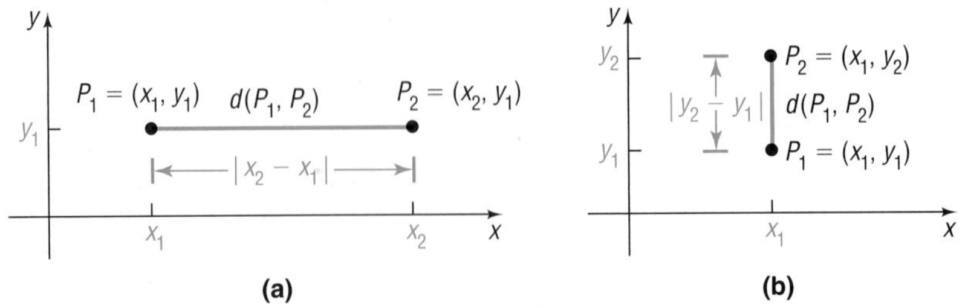

EXAMPLE 3 Finding the Length of a Line Segment

Find the length of the line segment shown in Figure 13.

Figure 13

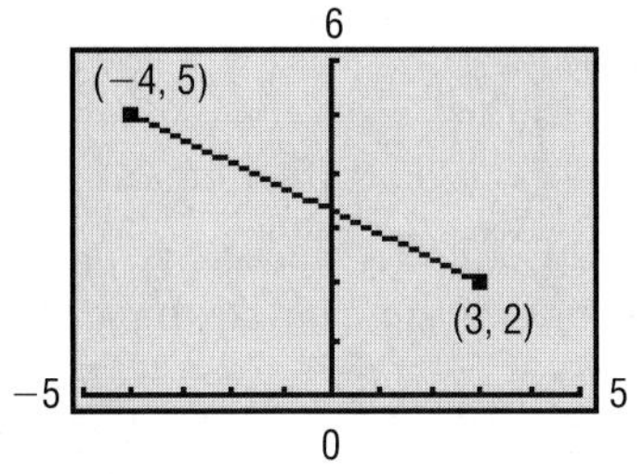

Solution The length of the line segment is the distance between the points (−4, 5) and (3, 2). Using the distance formula (1), the length d is

$$d = \sqrt{[3 - (-4)]^2 + (2 - 5)^2} = \sqrt{7^2 + (-3)^2}$$
$$= \sqrt{49 + 9} = \sqrt{58} \approx 7.62$$

NOW WORK PROBLEM 21.

The distance between two points $P_1 = (x_1, y_1)$ and $P_2 = (x_2, y_2)$ is never a negative number. Furthermore, the distance between two points is 0 only when the points are identical, that is, when $x_1 = x_2$ and $y_1 = y_2$. Also, because $(x_2 - x_1)^2 = (x_1 - x_2)^2$ and $(y_2 - y_1)^2 = (y_1 - y_2)^2$, it makes no difference whether the distance is computed from P_1 to P_2 or from P_2 to P_1; that is, $d(P_1, P_2) = d(P_2, P_1)$.

Rectangular coordinates enable us to translate geometry problems into algebra problems, and vice versa. The next example shows how algebra (the distance formula) can be used to solve geometry problems.

EXAMPLE 4 Using Algebra to Solve Geometry Problems

Consider the three points $A = (-2, 1)$, $B = (2, 3)$, and $C = (3, 1)$.

(a) Plot each point and form the triangle ABC.
(b) Find the length of each side of the triangle.
(c) Verify that the triangle is a right triangle.
(d) Find the area of the triangle.

Solution (a) Points A, B, C, and triangle ABC are plotted in Figure 14.

Figure 14

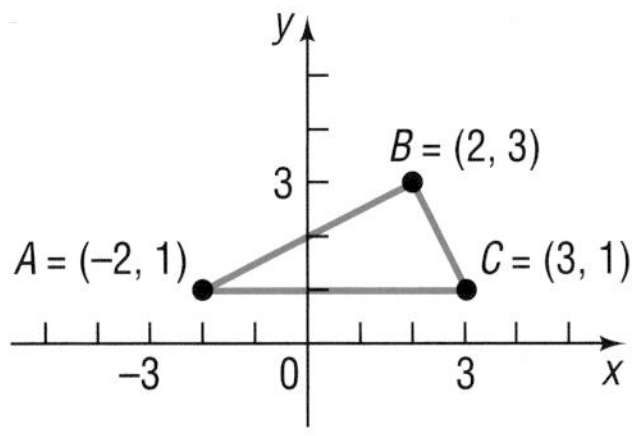

(b) $d(A, B) = \sqrt{[2 - (-2)]^2 + (3 - 1)^2} = \sqrt{16 + 4} = \sqrt{20} = 2\sqrt{5}$

$d(B, C) = \sqrt{(3 - 2)^2 + (1 - 3)^2} = \sqrt{1 + 4} = \sqrt{5}$

$d(A, C) = \sqrt{[3 - (-2)]^2 + (1 - 1)^2} = \sqrt{25 + 0} = 5$

(c) To show that the triangle is a right triangle, we need to show that the sum of the squares of the lengths of two of the sides equals the square of the length of the third side. (Why is this sufficient?) Looking at Figure 14, it seems reasonable to conjecture that the right angle is at vertex B. We shall check to see whether

$$[d(A, B)]^2 + [d(B, C)]^2 = [d(A, C)]^2$$

We find that

$$[d(A,B)]^2 + [d(B,C)]^2 = (2\sqrt{5})^2 + (\sqrt{5})^2$$
$$= 20 + 5 = 25 = [d(A,C)]^2$$

so it follows from the converse of the Pythagorean Theorem that triangle ABC is a right triangle.

(d) Because the right angle is at B, the sides AB and BC form the base and altitude of the triangle. Its area is

$$\text{Area} = \frac{1}{2}(\text{Base})(\text{Altitude}) = \frac{1}{2}(2\sqrt{5})(\sqrt{5}) = 5 \text{ square units}$$

NOW WORK PROBLEM 39.

MIDPOINT FORMULA

2 We now derive a formula for the coordinates of the **midpoint of a line segment**. Let $P_1 = (x_1, y_1)$ and $P_2 = (x_2, y_2)$ be the endpoints of a line segment, and let $M = (x, y)$ be the point on the line segment that is the same distance from P_1 as it is from P_2. See Figure 15. The triangles P_1AM and MBP_2 are congruent.* [Do you see why? Angle AP_1M = angle BMP_2,† angle P_1MA = angle MP_2B, and $d(P_1, M) = d(M, P_2)$ is given. Thus, we have angle–side–angle.] Hence, corresponding sides are equal in length. That is,

Figure 15

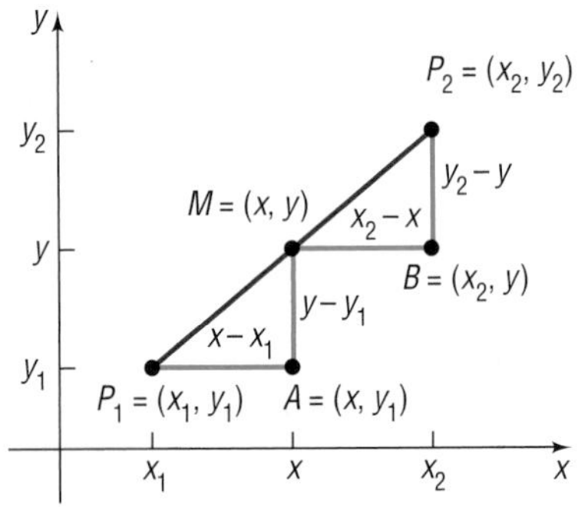

$$x - x_1 = x_2 - x \quad \text{and} \quad y - y_1 = y_2 - y$$
$$2x = x_1 + x_2 \qquad 2y = y_1 + y_2$$
$$x = \frac{x_1 + x_2}{2} \qquad y = \frac{y_1 + y_2}{2}$$

Theorem

Midpoint Formula

The midpoint (x, y) of the line segment from $P_1 = (x_1, y_1)$ to $P_2 = (x_2, y_2)$ is

$$(x, y) = \left(\frac{x_1 + x_2}{2}, \frac{y_1 + y_2}{2}\right) \qquad \textbf{(2)}$$

To find the midpoint of a line segment, we average the x-coordinates and the y-coordinates of the endpoints.

*The following statement is a postulate from geometry. Two triangles are congruent if their sides are the same length (SSS), or if two sides and the included angle are the same (SAS), or if two angles and the included sides are the same (ASA).

†Another postulate from geometry states that the transversal $\overline{P_1P_2}$ forms equal corresponding angles with the parallel lines $\overline{P_1A}$ and $\overline{MB}$.

EXAMPLE 5 Finding the Midpoint of a Line Segment

Find the midpoint of a line segment from $P_1 = (-5, 3)$ to $P_2 = (3, 1)$. Plot the points P_1 and P_2 and their midpoint. Check your answer.

Figure 16

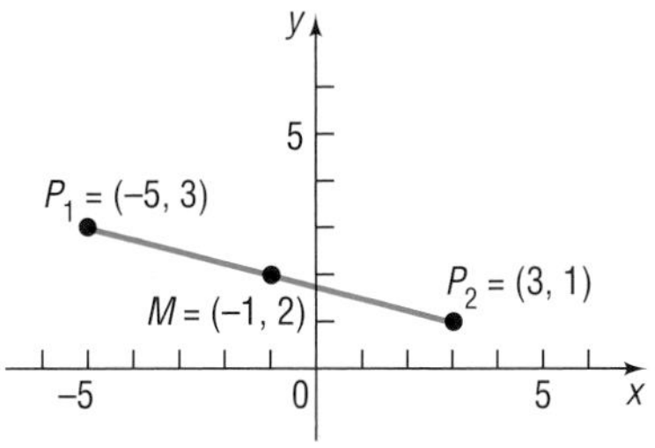

Solution We apply the midpoint formula (2) using $x_1 = -5, x_2 = 3, y_1 = 3$, and $y_2 = 1$. Then the coordinates (x, y) of the midpoint M are

$$x = \frac{x_1 + x_2}{2} = \frac{-5 + 3}{2} = -1 \quad \text{and} \quad y = \frac{y_1 + y_2}{2} = \frac{3 + 1}{2} = 2$$

That is, $M = (-1, 2)$. See Figure 16. ■

Check: Because M is the midpoint, we check the answer by verifying that $d(P_1, M) = d(M, P_2)$:

$$d(P_1, M) = \sqrt{[-1 - (-5)]^2 + (2 - 3)^2} = \sqrt{16 + 1} = \sqrt{17}$$

$$d(M, P_2) = \sqrt{[3 - (-1)]^2 + (1 - 2)^2} = \sqrt{16 + 1} = \sqrt{17}$$ ■

NOW WORK PROBLEM 49.

1.1 EXERCISES

In Problems 1 and 2, plot each point in the xy-plane. Tell in which quadrant or on what coordinate axis each point lies.

1. (a) $A = (-3, 2)$ (b) $B = (6, 0)$ (c) $C = (-2, -2)$
(d) $D = (6, 5)$ (e) $E = (0, -3)$ (f) $F = (6, -3)$

2. (a) $A = (1, 4)$ (b) $B = (-3, -4)$ (c) $C = (-3, 4)$
(d) $D = (4, 1)$ (e) $E = (0, 1)$ (f) $F = (-3, 0)$

3. Plot the points $(2, 0)$, $(2, -3)$, $(2, 4)$, $(2, 1)$, and $(2, -1)$. Describe the set of all points of the form $(2, y)$, where y is a real number.

4. Plot the points $(0, 3)$, $(1, 3)$, $(-2, 3)$, $(5, 3)$, and $(-4, 3)$. Describe the set of all points of the form $(x, 3)$, where x is a real number.

In Problems 5–8, determine the coordinates of the points shown. Tell in which quadrant each point lies. Assume that the coordinates are integers.

5.

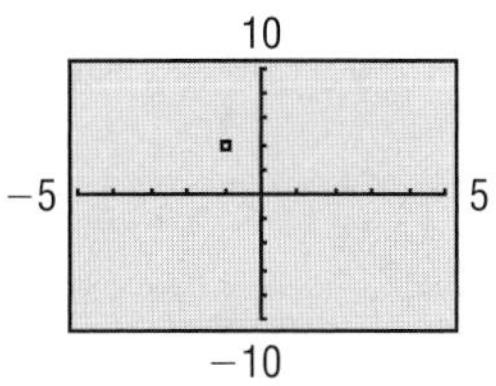

6.

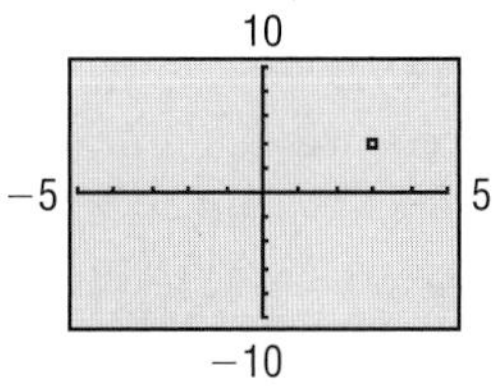

7.

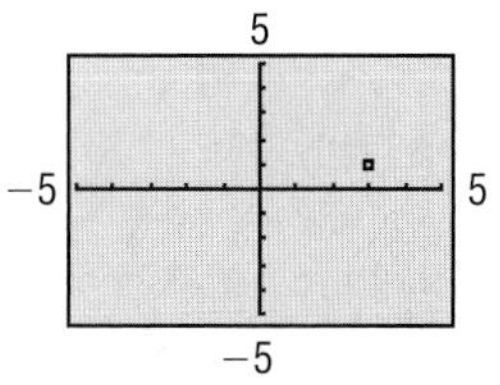

8.

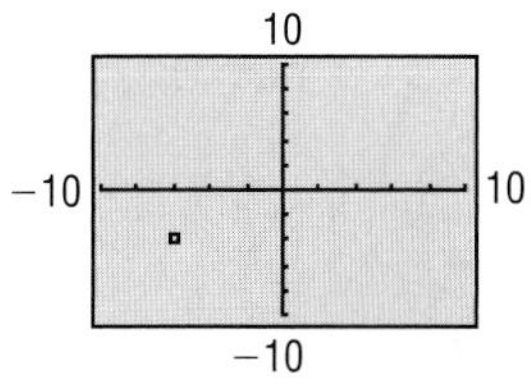

In Problems 9–14, select a setting so that each of the given points will lie within the viewing window.

9. $(-10, 5)$, $(3, -2)$, $(4, -1)$

10. $(5, 0)$, $(6, 8)$, $(-2, -3)$

11. $(40, 20)$, $(-20, -80)$, $(10, 40)$

12. $(-80, 60)$, $(20, -30)$, $(-20, -40)$

13. $(0, 0)$, $(100, 5)$, $(5, 150)$

14. $(0, -1)$, $(100, 50)$, $(-10, 30)$

In Problems 15–20, determine the viewing window used.

15.
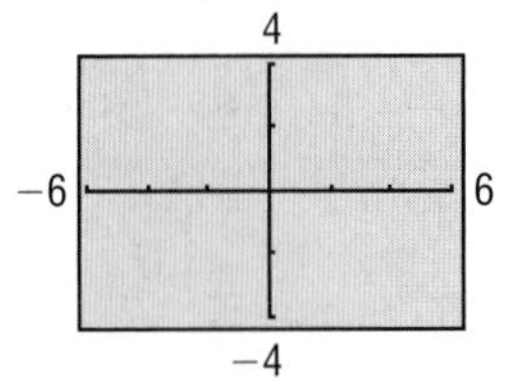

16.
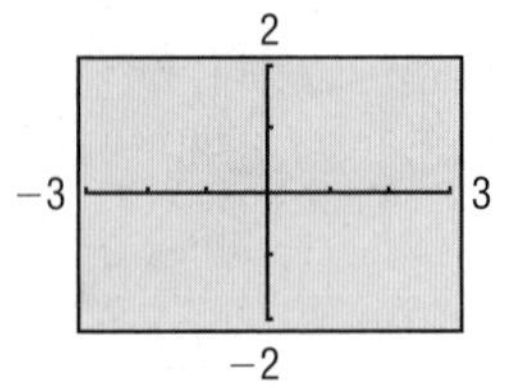

17.
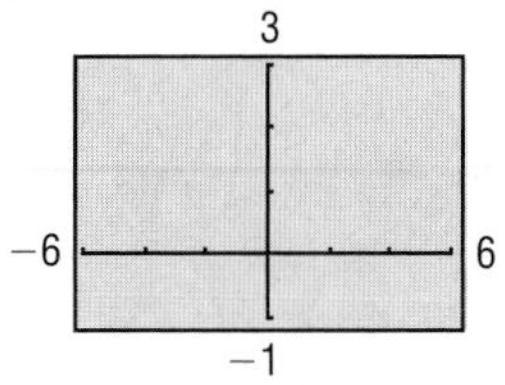

18.
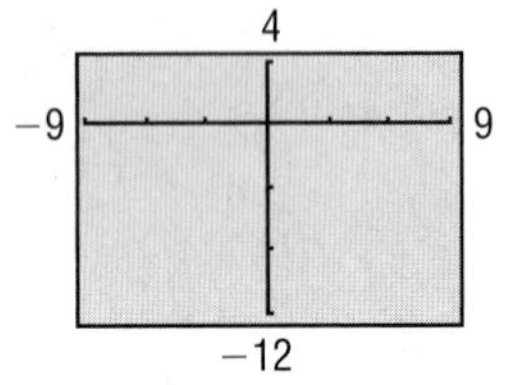

19.
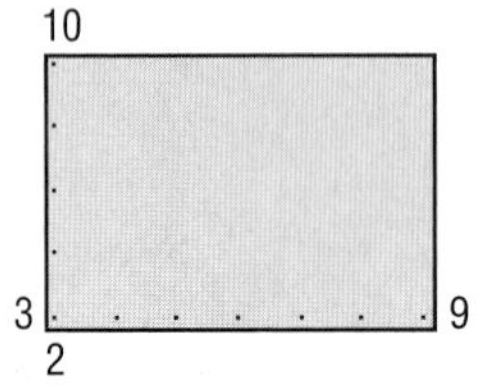

20.
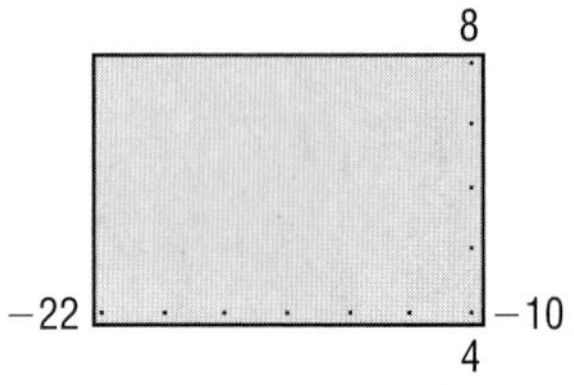

In Problems 21–34, find the distance $d(P_1, P_2)$ between the points P_1 and P_2.

21.
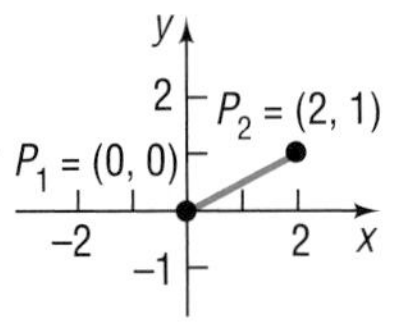

22.
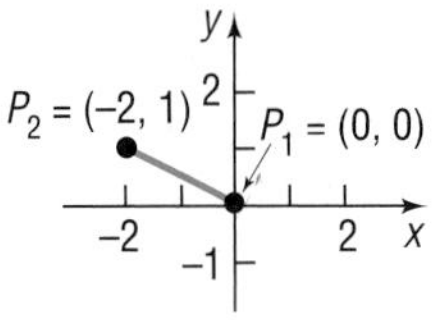

23.
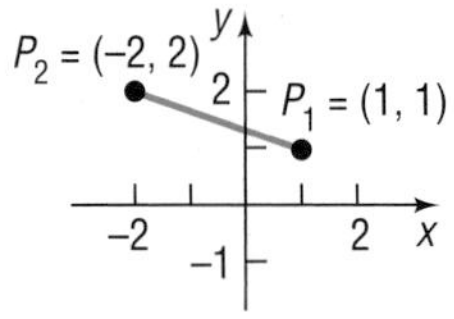

24.
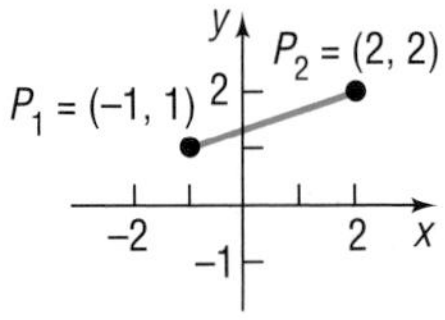

25. $P_1 = (3, -4);\quad P_2 = (5, 4)$

26. $P_1 = (-1, 0);\quad P_2 = (2, 4)$

27. $P_1 = (-3, 2);\quad P_2 = (6, 0)$

28. $P_1 = (2, -3);\quad P_2 = (4, 2)$

29. $P_1 = (4, -3);\quad P_2 = (6, 4)$

30. $P_1 = (-4, -3);\quad P_2 = (6, 2)$

31. $P_1 = (-0.2, 0.3);\quad P_2 = (2.3, 1.1)$

32. $P_1 = (1.2, 2.3);\quad P_2 = (-0.3, 1.1)$

33. $P_1 = (a, b);\quad P_2 = (0, 0)$

34. $P_1 = (a, a);\quad P_2 = (0, 0)$

In Problems 35–38, find the length of the line segment. Assume that the endpoints of each line segment have integer coordinates.

35.
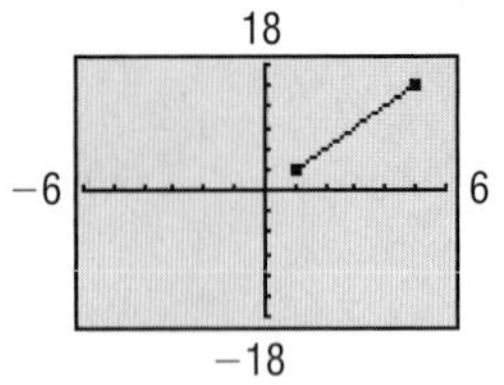

36.
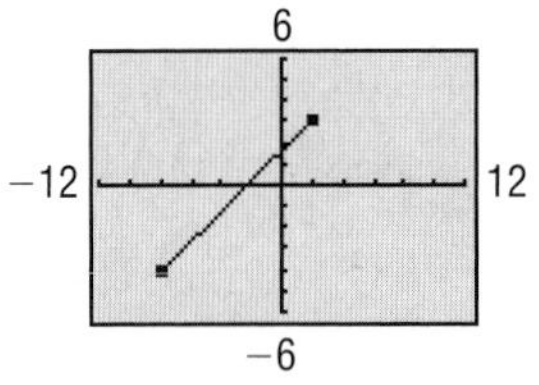

37.
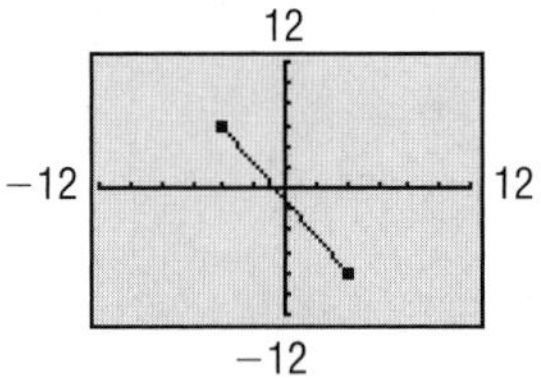

38.
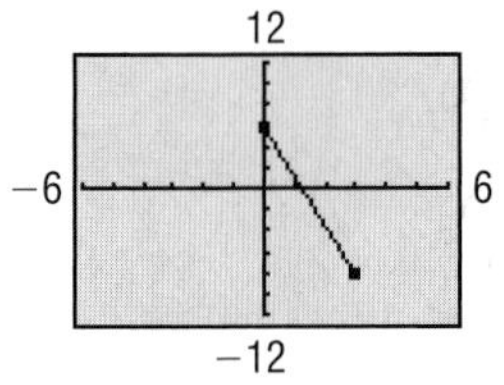

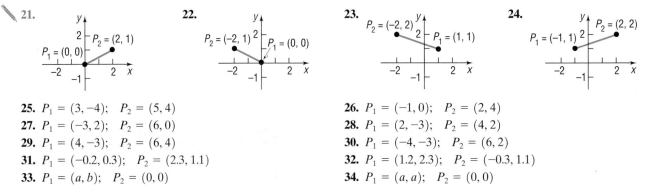

In Problems 39–44, plot each point and form the triangle ABC. Verify that the triangle is a right triangle. Find its area.

39. $A = (-2, 5);\quad B = (1, 3);\quad C = (-1, 0)$

40. $A = (-2, 5);\quad B = (12, 3);\quad C = (10, -11)$

41. $A = (-5, 3);\quad B = (6, 0);\quad C = (5, 5)$

42. $A = (-6, 3);\quad B = (3, -5);\quad C = (-1, 5)$

43. $A = (4, -3);\quad B = (0, -3);\quad C = (4, 2)$

44. $A = (4, -3);\quad B = (4, 1);\quad C = (2, 1)$

45. Find all points having an x-coordinate of 2 whose distance from the point $(-2, -1)$ is 5.

46. Find all points having a y-coordinate of -3 whose distance from the point $(1, 2)$ is 13.

47. Find all points on the x-axis that are 5 units from the point $(4, -3)$.

48. Find all points on the y-axis that are 5 units from the point $(4, 4)$.

In Problems 49–58, find the midpoint of the line segment joining the points P_1 and P_2.

49. $P_1 = (5, -4);\quad P_2 = (3, 2)$

50. $P_1 = (-1, 0);\quad P_2 = (2, 4)$

51. $P_1 = (-3, 2);\quad P_2 = (6, 0)$

52. $P_1 = (2, -3);\quad P_2 = (4, 2)$

53. $P_1 = (4, -3);\quad P_2 = (6, 1)$

54. $P_1 = (-4, -3);\quad P_2 = (2, 2)$

55. $P_1 = (-0.2, 0.3);\quad P_2 = (2.3, 1.1)$

56. $P_1 = (1.2, 2.3);\quad P_2 = (-0.3, 1.1)$

57. $P_1 = (a, b);\quad P_2 = (0, 0)$

58. $P_1 = (a, a);\quad P_2 = (0, 0)$

59. The **medians** of a triangle are the line segments from each vertex to the midpoint of the opposite side (see the figure). Find the lengths of the medians of the triangle with vertices at $A = (0, 0)$, $B = (0, 6)$, and $C = (4, 4)$.

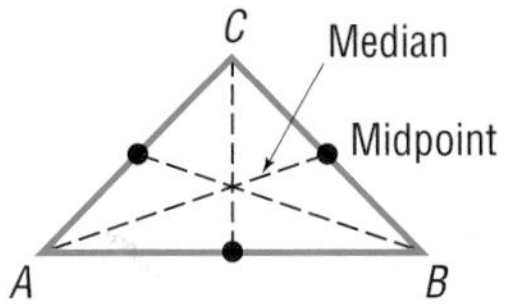

60. An **equilateral triangle** is one in which all three sides are of equal length. If two vertices of an equilateral triangle are $(0, 4)$ and $(0, 0)$, find the third vertex. How many of these triangles are possible?

In Problems 61–64, find the length of each side of the triangle determined by the three points P_1, P_2, and P_3. State whether the triangle is an isosceles triangle, a right triangle, neither of these, or both. (An **isosceles triangle** *is one in which at least two of the sides are of equal length.)*

61. $P_1 = (2, 1);\quad P_2 = (-4, 1);\quad P_3 = (-4, -3)$

62. $P_1 = (-1, 4);\quad P_2 = (6, 2);\quad P_3 = (4, -5)$

63. $P_1 = (-2, -1);\quad P_2 = (0, 7);\quad P_3 = (3, 2)$

64. $P_1 = (7, 2);\quad P_2 = (-4, 0);\quad P_3 = (4, 6)$

65. Geometry Find the midpoint of each diagonal of a square with side of length s. Draw the conclusion that the diagonals of a square intersect at their midpoints. [**Hint:** Use $(0, 0)$, $(0, s)$, $(s, 0)$, and (s, s) as the vertices of the square.]

66. Geometry Verify that the points $(0, 0)$, $(a, 0)$ and $(a/2, \sqrt{3}a/2)$ are the vertices of an equilateral triangle. Then show that the midpoints of the three sides are the vertices of a second equilateral triangle. (See Problem 60.)

67. Baseball A major league baseball "diamond" is actually a square, 90 feet on a side (see the figure). What is the distance directly from home plate to second base (the diagonal of the square)?

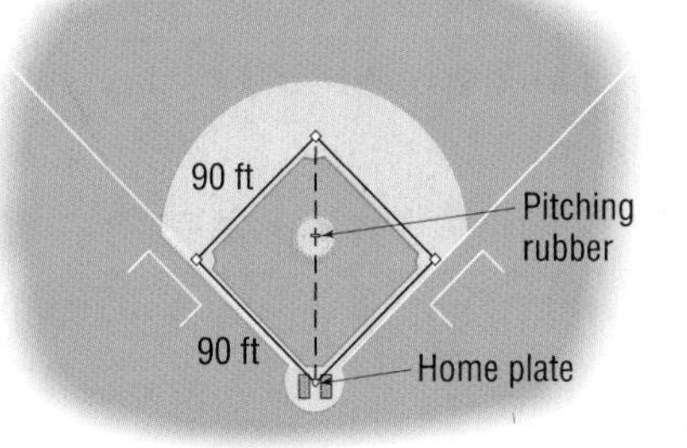

68. Little League Baseball The layout of a Little League playing field is a square, 60 feet on a side.* How far is it directly from home plate to second base (the diagonal of the square)?

69. Baseball Refer to Problem 67. Overlay a rectangular coordinate system on a major league baseball diamond so that the origin is at home plate, the positive x-axis lies in the direction from home plate to first base, and the positive y-axis lies in the direction from home plate to third base.

(a) What are the coordinates of first base, second base, and third base? Use feet as the unit of measurement.

(b) If the right fielder is located at $(310, 15)$, how far is it from there to second base?

(c) If the center fielder is located at $(300, 300)$, how far is it from there to third base?

70. Little League Baseball Refer to Problem 68. Overlay a rectangular coordinate system on a Little League baseball diamond so that the origin is at home plate, the positive x-axis lies in the direction from home plate to first base, and the positive y-axis lies in the direction from home plate to third base.

(a) What are the coordinates of first base, second base, and third base? Use feet as the unit of measurement.

(b) If the right fielder is located at $(180, 20)$, how far is it from there to second base?

(c) If the center fielder is located at $(220, 220)$, how far is it from there to third base?

71. A Dodge Intrepid and a Mack truck leave an intersection at the same time. The Intrepid heads east at an average speed of 30 miles per hour, while the truck heads south at an average speed of 40 miles per hour. Find an expression for their distance apart d (in miles) at the end of t hours.

72. A hot-air balloon, headed due east at an average speed of 15 miles per hour and at a constant altitude of 100 feet, passes over an intersection (see the figure). Find an expression for its distance d (measured in feet) from the intersection t seconds later.

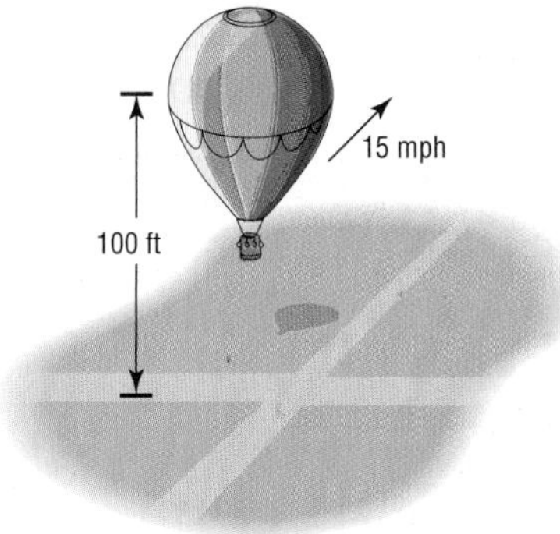

* *Source: Little League Baseball, Official Regulations and Playing Rules*, 2000.

PREPARING FOR THIS SECTION

Before getting started, review the following:

✓ Solving Equations (Appendix, Section 6)

1.2 GRAPHS OF EQUATIONS

OBJECTIVES
1. Graph Equations by Hand
2. Graph Equations Using a Graphing Utility
3. Find Intercepts
4. Test an Equation for Symmetry with Respect to the (a) x-Axis, (b) y-Axis, and (c) Origin

1 The **graph of an equation** in two variables x and y consists of the set of points in the xy-plane whose coordinates (x, y) satisfy the equation. Let's look at some examples.

EXAMPLE 1 Determining Whether a Point Is on the Graph of an Equation

Determine if the following points are on the graph of the equation $2x - y = 6$.

(a) $(2, 3)$ (b) $(2, -2)$

Solution (a) For the point $(2, 3)$, we check to see if $x = 2$, $y = 3$ satisfies the equation $2x - y = 6$.

$$2x - y = 2(2) - 3 = 4 - 3 = 1 \neq 6$$

The equation is not satisfied, so the point $(2, 3)$ is not on the graph.

(b) For the point $(2, -2)$, we have

$$2x - y = 2(2) - (-2) = 4 + 2 = 6$$

The equation is satisfied, so the point $(2, -2)$ is on the graph. ■

NOW WORK PROBLEM 1.

EXAMPLE 2 Graphing an Equation by Hand

Graph the equation: $y = 2x + 5$

Figure 17
$y = 2x + 5$

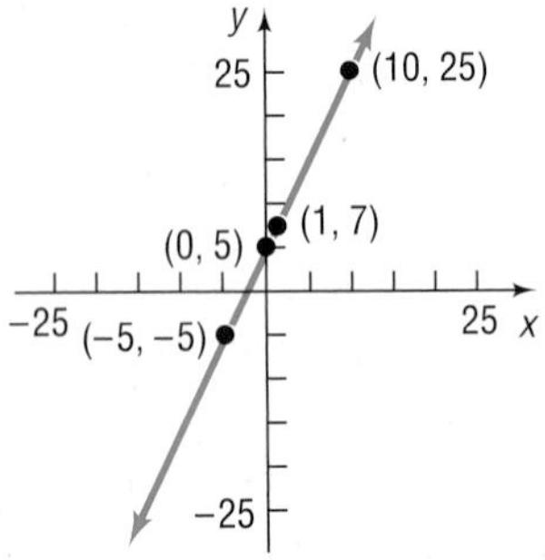

Solution We want to find all points (x, y) that satisfy the equation. To locate some of these points (and thus get an idea of the pattern of the graph), we assign some numbers to x and find corresponding values for y.

If	Then	Point on Graph
$x = 0$	$y = 2(0) + 5 = 5$	$(0, 5)$
$x = 1$	$y = 2(1) + 5 = 7$	$(1, 7)$
$x = -5$	$y = 2(-5) + 5 = -5$	$(-5, -5)$
$x = 10$	$y = 2(10) + 5 = 25$	$(10, 25)$

By plotting these points and then connecting them, we obtain the graph of the equation (a *line*), as shown in Figure 17. ■

EXAMPLE 3 Graphing an Equation by Hand

Graph the equation: $y = x^2$

Solution Table 1 provides several points on the graph. In Figure 18 we plot these points and connect them with a smooth curve to obtain the graph (a *parabola*).

TABLE 1

x	$y = x^2$	(x, y)
−4	16	(−4, 16)
−3	9	(−3, 9)
−2	4	(−2, 4)
−1	1	(−1, 1)
0	0	(0, 0)
1	1	(1, 1)
2	4	(2, 4)
3	9	(3, 9)
4	16	(4, 16)

Figure 18
$y = x^2$

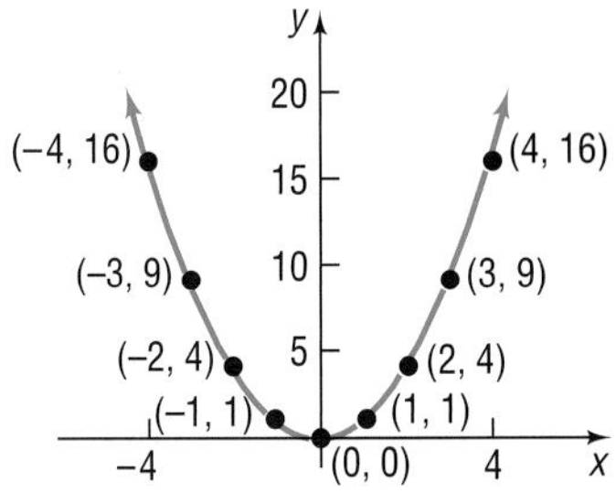

The graphs of the equations shown in Figures 17 and 18 do not show all points. For example, in Figure 17 the point (20, 45) is a part of the graph of $y = 2x + 5$, but it is not shown. Since the graph of $y = 2x + 5$ could be extended out as far as we please, we use arrows to indicate that the pattern shown continues. It is important when illustrating a graph to present enough of the graph so that any viewer of the illustration will "see" the rest of it as an obvious continuation of what is actually there. This is referred to as a **complete graph**.

So, one way to obtain a complete graph of an equation is to plot a sufficient number of points on the graph until a pattern becomes evident. Then these points are connected with a smooth curve following the suggested pattern. But how many points are sufficient? Sometimes knowledge about the equation tells us. For example, we will learn in Section 1.6 that, if an equation is of the form $y = mx + b$, then its graph is a line. In this case, two points would suffice to obtain the graph.

One purpose of this book is to investigate the properties of equations in order to decide whether a graph is complete. Sometimes we shall graph equations by plotting a sufficient number of points on the graph until a pattern becomes evident; then we connect these points with a smooth curve following the suggested pattern. (Shortly, we shall investigate various techniques that will enable us to graph an equation without plotting so many points.) Other times we shall graph equations using a graphing utility.

USING A GRAPHING UTILITY TO GRAPH EQUATIONS

2 From Examples 2 and 3, we see that a graph can be obtained by plotting points in a rectangular coordinate system and connecting them. Graphing utilities perform these same steps when graphing an equation. For example,

the TI-83 determines 95 evenly spaced input values,* uses the equation to determine the output values, plots these points on the screen, and finally (if in the connected mode) draws a line between consecutive points.

To graph an equation in two variables x and y using a graphing utility requires that the equation be written in the form $y = \{\text{expression in } x\}$. If the original equation is not in this form, replace it by equivalent equations until the form $y = \{\text{expression in } x\}$ is obtained. In general, there are four ways to obtain equivalent equations.

Procedures That Result in Equivalent Equations

1. Interchange the two sides of the equation:

   Replace $3x + 5 = y$ by $y = 3x + 5$

2. Simplify the sides of the equation by combining like terms, eliminating parentheses, and so on:

   Replace $(2y + 2) + 6 = 2x + 5(x + 1)$

   by $2y + 8 = 7x + 5$

3. Add or subtract the same expression on both sides of the equation:

   Replace $y + 3x - 5 = 4$

   by $y + 3x - 5 + 5 = 4 + 5$

4. Multiply or divide both sides of the equation by the same nonzero expression:

   Replace $3y = 6 - 2x$

   by $\frac{1}{3} \cdot 3y = \frac{1}{3}(6 - 2x)$

EXAMPLE 4 Expressing an Equation in the Form $y = \{\text{expression in } x\}$

Solve for y: $2y + 3x - 5 = 4$

Solution We replace the original equation by a succession of equivalent equations.

$$2y + 3x - 5 = 4$$

$$2y + 3x - 5 + 5 = 4 + 5 \quad \text{Add 5 to both sides.}$$

$$2y + 3x = 9 \quad \text{Simplify}$$

$$2y + 3x - 3x = 9 - 3x \quad \text{Subtract } 3x \text{ from both sides.}$$

$$2y = 9 - 3x \quad \text{Simplify}$$

$$\frac{2y}{2} = \frac{9 - 3x}{2} \quad \text{Divide both sides by 2.}$$

$$y = \frac{9 - 3x}{2} \quad \text{Simplify}$$

■

*These input values depend on the values of $X\text{min}$ and $X\text{max}$. For example, if $X\text{min} = -10$ and $X\text{max} = 10$, then the first input value will be -10 and the next input value will be $-10 + (10 - (-10))/94 = -9.7872$, and so on.

Now we are ready to graph equations using a graphing utility. Most graphing utilities require the following steps:

STEPS FOR GRAPHING AN EQUATION USING A GRAPHING UTILITY

STEP 1: Solve the equation for y in terms of x.

STEP 2: Get into the graphing mode of your graphing utility. The screen will usually display $y =$, prompting you to enter the expression involving x that you found in Step 1. (Consult your manual for the correct way to enter the expression; for example, $y = x^2$ might be entered as x^2 or as $x*x$ or as $x\ x^Y 2$).

STEP 3: Select the viewing window. Without prior knowledge about the behavior of the graph of the equation, it is common to select the **standard viewing window*** initially. The viewing window is then adjusted based on the graph that appears. In this text the standard viewing window will be

$$X\text{min} = -10 \qquad Y\text{min} = -10$$
$$X\text{max} = 10 \qquad Y\text{max} = 10$$
$$X\text{scl} = 1 \qquad Y\text{scl} = 1$$

STEP 4: Execute.

STEP 5: Adjust the viewing window until a complete graph is obtained.

EXAMPLE 5 Graphing an Equation on a Graphing Utility

Graph the equation: $6x^2 + 3y = 36$

Solution **STEP 1:** We solve for y in terms of x.

$$6x^2 + 3y = 36$$
$$3y = -6x^2 + 36 \quad \text{Subtract } 6x^2 \text{ from both sides of the equation.}$$
$$y = -2x^2 + 12 \quad \text{Divide both sides of the equation by 3 and simplify.}$$

Figure 19

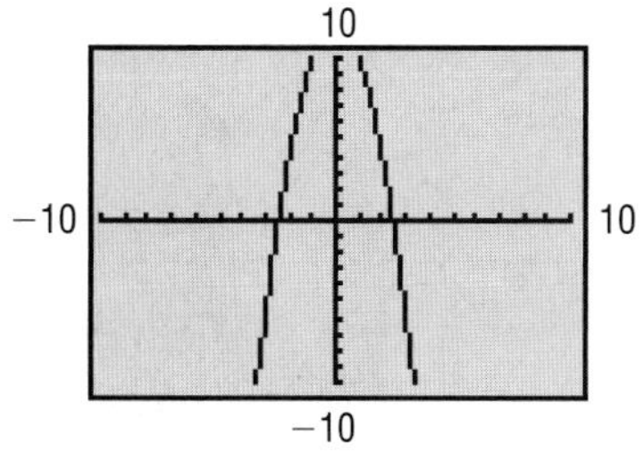

STEP 2: From the graphing mode, enter the expression $-2x^2 + 12$ after the prompt $y =$.

STEP 3: Set the viewing window to the standard viewing window.

STEP 4: Execute. The screen should look like Figure 19.

*Some graphing utilities have a ZOOM-STANDARD feature that automatically sets the viewing window to the standard viewing window and graphs the equation.

Figure 20

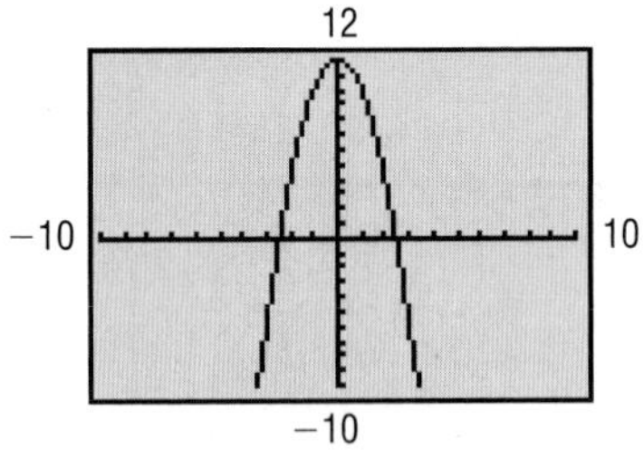

STEP 5: The graph of $y = -2x^2 + 12$ is not complete. The value of Ymax must be increased so that the top portion of the graph is visible. After increasing the value of Ymax to 12, we obtain the graph in Figure 20.† The graph is now complete. ■

Look again at Figure 20. Although a complete graph is shown, the graph might be improved by adjusting the values of Xmin and Xmax. Figure 21 shows the graph of $y = -2x^2 + 12$ using $X\text{min} = -4$ and $X\text{max} = 4$. Do you think this is a better choice for the viewing window?

Figure 21

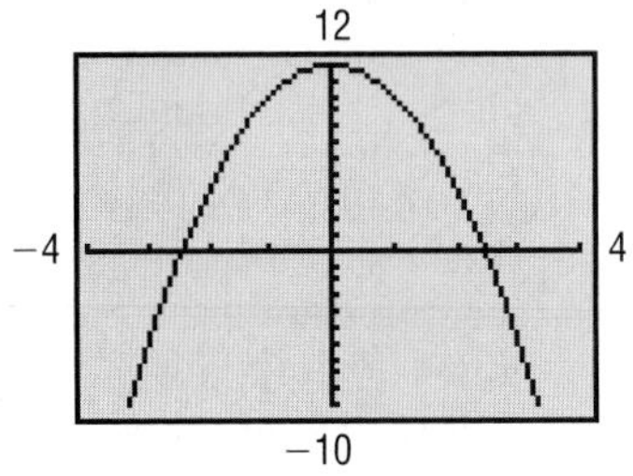

NOW WORK PROBLEMS 7(a) AND 7(b).

We said earlier that we would discuss techniques that reduce the number of points required to graph an equation by hand. Two such techniques involve finding *intercepts* and checking for *symmetry*. These techniques also may be used in setting the initial viewing window.

INTERCEPTS

3 The points, if any, at which a graph crosses or touches the coordinate axes are called the **intercepts**. See Figure 22. The x-coordinate of a point at which the graph crosses or touches the x-axis is an ***x*-intercept**, and the y-coordinate of a point at which the graph crosses or touches the y-axis is a ***y*-intercept**.

Figure 22

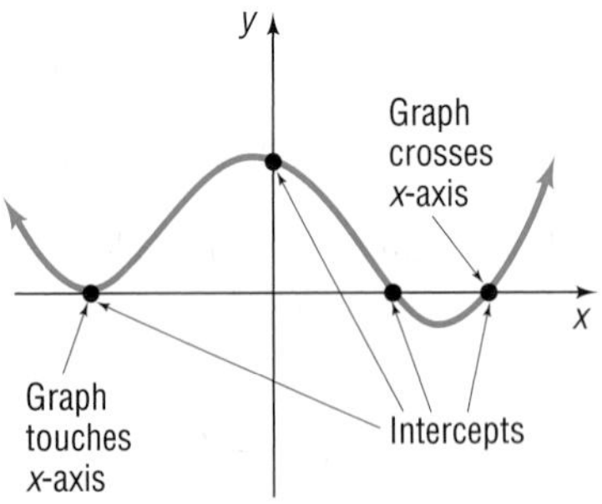

EXAMPLE 6 Finding Intercepts from a Graph

Find the intercepts of the graph in Figure 23. What are its x-intercepts? What are its y-intercepts?

†Some graphing utilities have a ZOOM-FIT feature that determines the appropriate Ymin and Ymax for a given Xmin and Xmax. Consult your owner's manual for the appropriate keystrokes.

Figure 23

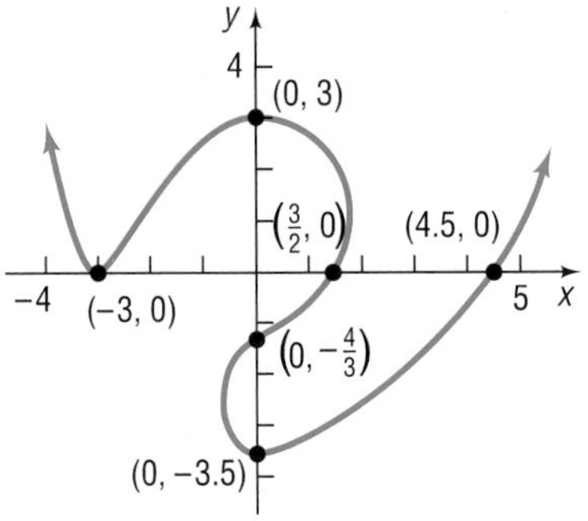

Solution The intercepts of the graph are the points

$$(-3, 0),\quad (0, 3),\quad \left(\frac{3}{2}, 0\right),\quad \left(0, -\frac{4}{3}\right),\quad (0, -3.5),\quad (4.5, 0)$$

The x-intercepts are $-3, \frac{3}{2}$, and 4.5; the y-intercepts are $-3.5, -\frac{4}{3}$, and 3. ■

NOW WORK PROBLEM 33(a).

The intercepts of the graph of an equation can be found by using the fact that points on the x-axis have y-coordinates equal to 0 and points on the y-axis have x-coordinates equal to 0.

PROCEDURE FOR FINDING INTERCEPTS

1. To find the x-intercept(s), if any, of the graph of an equation, let $y = 0$ in the equation and solve for x.
2. To find the y-intercept(s), if any, of the graph of an equation, let $x = 0$ in the equation and solve for y.

Because the x-intercepts of the graph of an equation are those x-values for which $y = 0$, they are also called the **zeros** (or **roots**) of the equation.

EXAMPLE 7 Finding Intercepts from an Equation

Find the x-intercept(s) and the y-intercept(s) of the graph of $y = x^2 - 4$.

Solution To find the x-intercept(s), we let $y = 0$ and obtain the equation

$$x^2 - 4 = 0$$
$$(x + 2)(x - 2) = 0 \quad \text{Factor}$$
$$x + 2 = 0 \quad \text{or} \quad x - 2 = 0 \quad \text{Zero-Product Property*}$$
$$x = -2 \quad \text{or} \quad x = 2$$

The equation has two solutions, −2 or 2. The x-intercepts (or zeros) are −2 and 2.

To find the y-intercept(s), we let $x = 0$ in the equation.

$$y = x^2 - 4$$
$$= 0^2 - 4 = -4$$

The y-intercept is −4. ■

* See page 978 in Section 6 of the Appendix.

TABLE 2

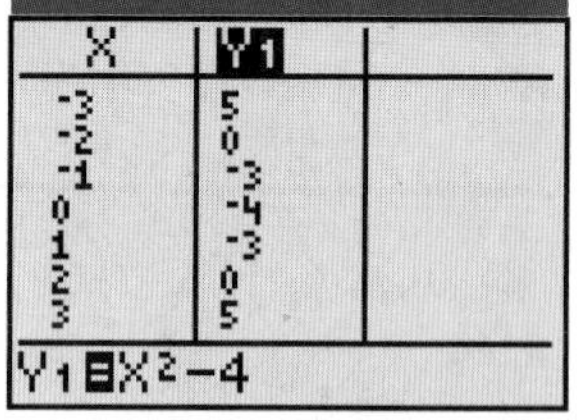

X	Y1
-3	5
-2	0
-1	-3
0	-4
1	-3
2	0
3	5

Y1 = X²−4

Sometimes the TABLE feature of a graphing utility will reveal the intercepts of an equation. (Check your manual to see if your graphing utility has a TABLE feature.) Table 2 shows a table for the equation $y = x^2 - 4$. Can you find the intercepts in the table?

NOW WORK PROBLEM 63 (LIST THE INTERCEPTS).

EXAMPLE 8 **Finding Intercepts Using a Graphing Utility**

Use a graphing utility to find the intercepts of the equation $y = x^3 - 8$.

Solution Figure 24(a) shows the graph of $y = x^3 - 8$.

The eVALUEate feature of a TI-83 graphing calculator† accepts as input a value of x and determines the value of y. If we let $x = 0$, we find that the y-intercept is -8. See Figure 24(b).

The ZERO feature of a TI-83 is used to find the x-intercept(s). See Figure 24(c). The x-intercept is 2.

Figure 24

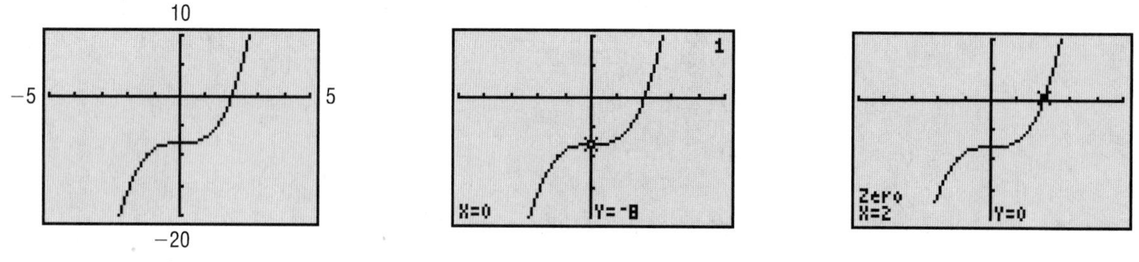

(a) (b) (c)

NOW WORK PROBLEM 63 (LIST THE INTERCEPTS USING A GRAPHING UTILITY).

SYMMETRY

4 Another helpful tool for graphing equations by hand involves *symmetry*, particularly symmetry with respect to the x-axis, the y-axis, and the origin.

> A graph is said to be **symmetric with respect to the x-axis** if, for every point (x, y) on the graph, the point $(x, -y)$ is also on the graph.
>
> A graph is said to be **symmetric with respect to the y-axis** if, for every point (x, y) on the graph, the point $(-x, y)$ is also on the graph.
>
> A graph is said to be **symmetric with respect to the origin** if, for every point (x, y) on the graph, the point $(-x, -y)$ is also on the graph.

Figure 25 illustrates the definition. Notice that, when a graph is symmetric with respect to the x-axis, the part of the graph above the x-axis is a reflection or mirror image of the part below it, and vice versa. And when a

†Consult your manual to determine the appropriate keystrokes for these features.

Figure 25

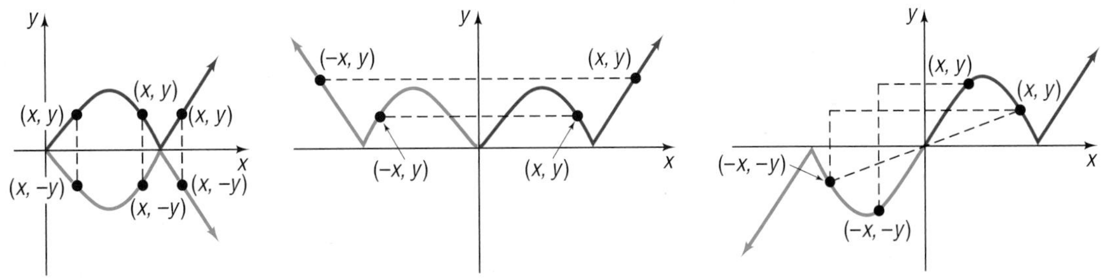

Symmetric with respect to the x-axis

Symmetric with respect to the y-axis

Symmetric with respect to the origin

graph is symmetric with respect to the y-axis, the part of the graph to the right of the y-axis is a reflection of the part to the left of it, and vice versa. Symmetry with respect to the origin may be viewed in two ways:

1. As a reflection about the y-axis, followed by a reflection about the x-axis
2. As a projection along a line through the origin so that the distances from the origin are equal

EXAMPLE 9 Symmetric Points

(a) If a graph is symmetric with respect to the x-axis and the point $(4, 2)$ is on the graph, then the point $(4, -2)$ is also on the graph.

(b) If a graph is symmetric with respect to the y-axis and the point $(4, 2)$ is on the graph, then the point $(-4, 2)$ is also on the graph.

(c) If a graph is symmetric with respect to the origin and the point $(4, 2)$ is on the graph, then the point $(-4, -2)$ is also on the graph. ■

NOW WORK PROBLEMS 23 AND 33(b).

When the graph of an equation is symmetric with respect to a coordinate axis or the origin, the number of points that you need to plot in order to see the pattern is reduced. For example, if the graph of an equation is symmetric with respect to the y-axis, then, once points to the right of the y-axis are plotted, an equal number of points on the graph can be obtained by reflecting them about the y-axis. Because of this, before we graph an equation, we first want to determine whether it has any symmetry. The following tests are used for this purpose.

TESTS FOR SYMMETRY

To test the graph of an equation for symmetry with respect to the

x-Axis Replace y by $-y$ in the equation. If an equivalent equation results, the graph of the equation is symmetric with respect to the x-axis.

y-Axis Replace x by $-x$ in the equation. If an equivalent equation results, the graph of the equation is symmetric with respect to the y-axis.

Origin Replace x by $-x$ and y by $-y$ in the equation. If an equivalent equation results, the graph of the equation is symmetric with respect to the origin.

EXAMPLE 10 Graphing an Equation by Finding Intercepts and Checking for Symmetry

Graph the equation $y = x^3$ by hand. Find any intercepts and check for symmetry first.

Solution First, we seek the intercepts. When $x = 0$, then $y = 0$; and when $y = 0$, then $x = 0$. The origin $(0, 0)$ is the only intercept. Now we test for symmetry.

x-Axis: Replace y by $-y$. Since the result, $-y = x^3$, is not equivalent to $y = x^3$, the graph is not symmetric with respect to the x-axis.

y-Axis: Replace x by $-x$. Since the result, $y = (-x)^3 = -x^3$, is not equivalent to $y = x^3$, the graph is not symmetric with respect to the y-axis.

Origin: Replace x by $-x$ and y by $-y$. Since the result, $-y = (-x)^3 = -x^3$, is equivalent to $y = x^3$ (multiply both sides by -1), the graph is symmetric with respect to the origin.

Figure 26
$y = x^3$

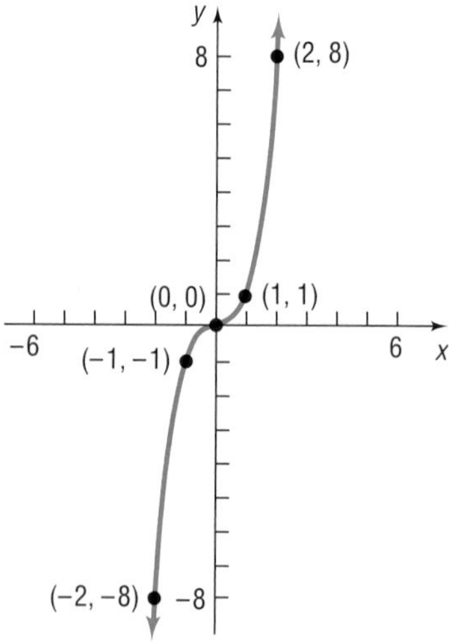

To graph by hand, we use the equation to obtain several points on the graph. Because of the symmetry, we only need to locate points on the graph for which $x \geq 0$. See Table 3. Points on the graph could also be obtained using the TABLE feature on a graphing utility. Using a TI-83, we obtain Table 4. Do you see the symmetry with respect to the origin from the table? Figure 26 shows the graph.

TABLE 3

x	$y = x^3$	(x, y)
0	0	(0, 0)
1	1	(1, 1)
2	8	(2, 8)
3	27	(3, 27)

TABLE 4

X	Y1
-3	-27
-2	-8
-1	-1
0	0
1	1
2	8
3	27

Y1=X^3

EXAMPLE 11 Graphing an Equation

Graph the equation $x = y^2$. Find any intercepts and check for symmetry first.

Solution The lone intercept is $(0, 0)$. The graph is symmetric with respect to the x-axis since $(-y)^2 = y^2 = x$. Figure 27 shows the graph drawn by hand.

Figure 27
Symmetry with respect to the x-axis

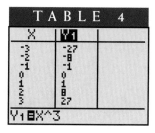

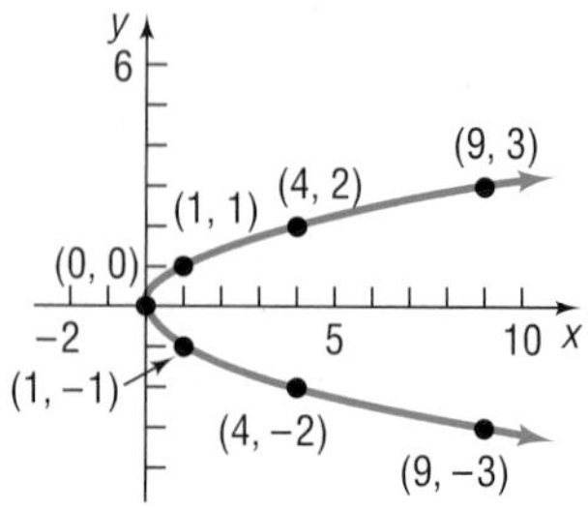

To graph this equation using a graphing utility, we must write the equation in the form $y = \{\text{expression in } x\}$. So, to graph $x = y^2$, we must solve for y.

$$x = y^2$$

$$y^2 = x$$

$$y = \pm\sqrt{x}$$ Refer to equations (1) and (2) on p. 1000.

To graph $x = y^2$, we need to graph both $Y_1 = \sqrt{x}$ and $Y_2 = -\sqrt{x}$ on the same screen. Figure 28 shows the result. Table 5 shows various values of y for a given value of x when $Y_1 = \sqrt{x}$ and $Y_2 = -\sqrt{x}$. Notice that when $x < 0$ we get an error. Can you explain why?

Figure 28

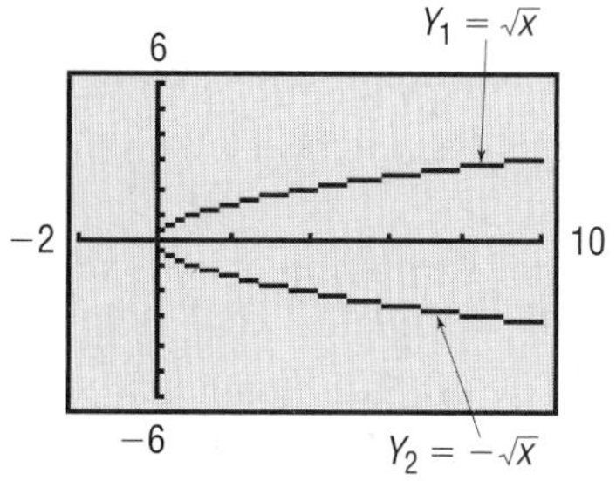

TABLE 5

X	Y1	Y2
-1	ERROR	ERROR
0	0	0
1	1	-1
2	1.4142	-1.414
3	1.7321	-1.732
4	2	-2
5	2.2361	-2.236

Y1 ■ √(X)

Look again at either Figure 27 or 28. If we restrict y so that $y \geq 0$, the equation $x = y^2$, $y \geq 0$, may be written equivalently as $y = \sqrt{x}$. The portion of the graph of $x = y^2$ in quadrant I is therefore the graph of $y = \sqrt{x}$. See Figure 29.

Figure 29
$y = \sqrt{x}$

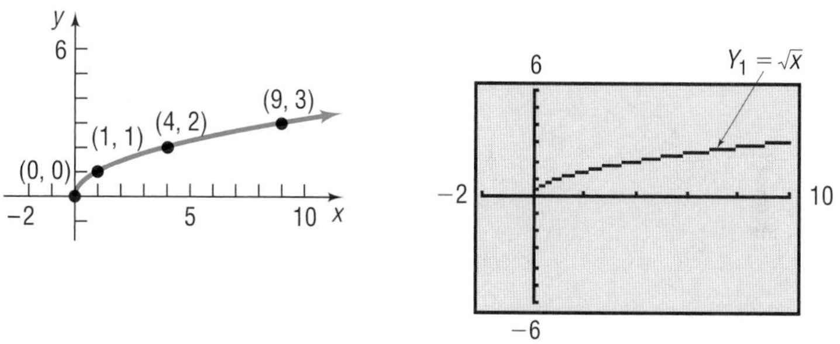

NOW WORK PROBLEM 63 (TEST FOR SYMMETRY).

EXAMPLE 12 Graphing the Equation $y = 1/x$

Consider the equation: $y = 1/x$

(a) Graph this equation using a graphing utility. Set the viewing window as

$$X\text{min} = -3 \qquad Y\text{min} = -4$$
$$X\text{max} = 3 \qquad Y\text{max} = 4$$
$$X\text{scl} = 1 \qquad Y\text{scl} = 1$$

(b) Use algebra to find any intercepts and test for symmetry.

(c) Draw the graph by hand.

Figure 30

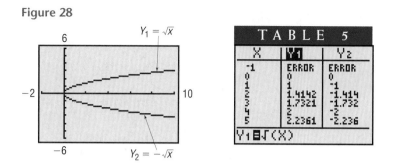

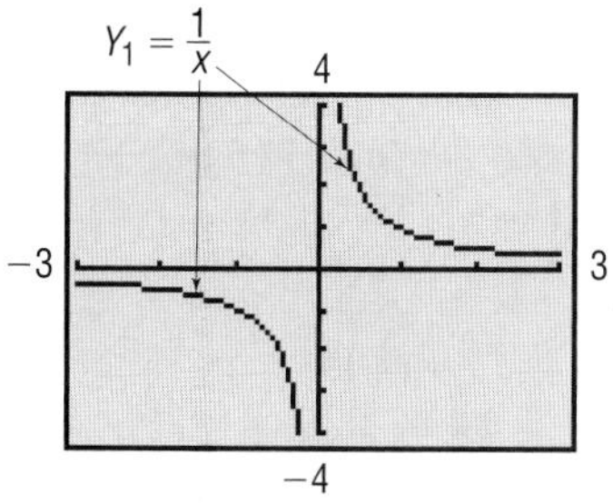

Solution (a) Figure 30 illustrates the graph. We infer from the graph that there are no intercepts; we may also infer that symmetry with respect to the origin is a possibility. The TRACE feature on a graphing utility can provide further evidence of symmetry with respect to the origin. Using TRACE, we observe that for any ordered pair (x, y) the ordered pair $(-x, -y)$ is also a point on the graph. For example, the points $(0.95744681, 1.0444444)$ and $(-0.95744681, -1.0444444)$ both lie on the graph.

(b) We check for intercepts first. If we let $x = 0$, we obtain a 0 denominator, which is not defined. We conclude that there is no y-intercept. If we let $y = 0$, we get the equation $1/x = 0$, which has no solution. We conclude that there is no x-intercept. The graph of $y = 1/x$ does not cross or touch the coordinate axes.

Next we check for symmetry.

x-Axis: Replacing y by $-y$ yields $-y = 1/x$, which is not equivalent to $y = 1/x$.

y-Axis: Replacing x by $-x$ yields $y = -1/x$, which is not equivalent to $y = 1/x$.

Origin: Replacing x by $-x$ and y by $-y$ yields $-y = -1/x$, which is equivalent to $y = 1/x$.

The graph is symmetric with respect to the origin. This confirms the inferences drawn in part (a) of the solution.

(c) We can use the equation to obtain some points on the graph. Because of symmetry, we need only find points (x, y) for which x is positive. Also, from the equation $y = 1/x$ we infer that, if x is a large and positive number, then $y = 1/x$ is a positive number close to 0. We also infer that if x is a positive number close to 0 then $y = 1/x$ is a large and positive number. Armed with this information, we can graph the equation. Table 6 and Figure 31 illustrate some of these points and the graph of $y = 1/x$. Observe how the absence of intercepts and the existence of symmetry with respect to the origin were utilized.

TABLE 6

x	$y = 1/x$	(x, y)
$\frac{1}{10}$	10	$(\frac{1}{10}, 10)$
$\frac{1}{3}$	3	$(\frac{1}{3}, 3)$
$\frac{1}{2}$	2	$(\frac{1}{2}, 2)$
1	1	$(1, 1)$
2	$\frac{1}{2}$	$(2, \frac{1}{2})$
3	$\frac{1}{3}$	$(3, \frac{1}{3})$
10	$\frac{1}{10}$	$(10, \frac{1}{10})$

Figure 31

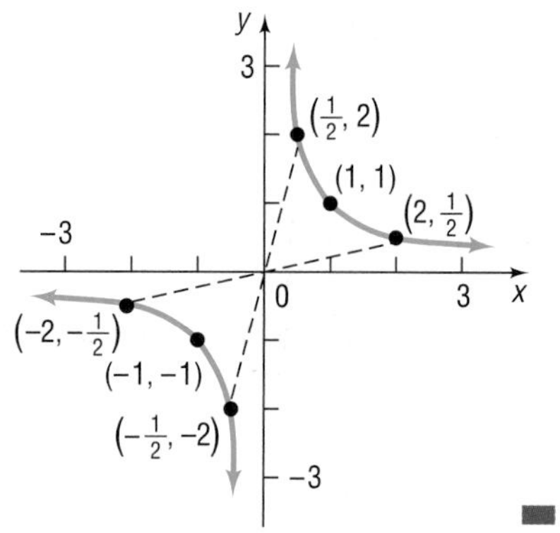

1.2 EXERCISES

In Problems 1–6, tell whether the given points are on the graph of the equation.

1. Equation: $y = x^4 - \sqrt{x}$
   Points: $(0, 0); (1, 1); (-1, 0)$
2. Equation: $y = x^3 - 2\sqrt{x}$
   Points: $(0, 0); (1, 1); (1, -1)$
3. Equation: $y^2 = x^2 + 9$
   Points: $(0, 3); (3, 0); (-3, 0)$
4. Equation: $y^3 = x + 1$
   Points: $(1, 2); (0, 1); (-1, 0)$
5. Equation: $x^2 + y^2 = 4$
   Points: $(0, 2); (-2, 2)(\sqrt{2}, \sqrt{2})$
6. Equation: $x^2 + 4y^2 = 4$
   Points: $(0, 1); (2, 0); (2, \frac{1}{2})$

In Problems 7–22, graph each equation using the following viewing windows:

(a)	(b)	(c)	(d)
Xmin = −5	Xmin = −10	Xmin = −10	Xmin = −5
Xmax = 5	Xmax = 10	Xmax = 10	Xmax = 5
Xscl = 1	Xscl = 1	Xscl = 2	Xscl = 1
Ymin = −4	Ymin = −8	Ymin = −8	Ymin = −20
Ymax = 4	Ymax = 8	Ymax = 8	Ymax = 20
Yscl = 1	Yscl = 1	Yscl = 2	Yscl = 5

7. $y = x + 2$
8. $y = x - 2$
9. $y = -x + 2$
10. $y = -x - 2$
11. $y = 2x + 2$
12. $y = 2x - 2$
13. $y = -2x + 2$
14. $y = -2x - 2$
15. $y = x^2 + 2$
16. $y = x^2 - 2$
17. $y = -x^2 + 2$
18. $y = -x^2 - 2$
19. $3x + 2y = 6$
20. $3x - 2y = 6$
21. $-3x + 2y = 6$
22. $-3x - 2y = 6$

In Problems 23–32, plot each point. Then plot the point that is symmetric to it with respect to (a) the x-axis; (b) the y-axis; (c) the origin.

23. $(3, 4)$ **24.** $(5, 3)$ **25.** $(-2, 1)$ **26.** $(4, -2)$ **27.** $(1, 1)$
28 $(-1, -1)$ **29.** $(-3, -4)$ **30.** $(4, 0)$ **31.** $(0, -3)$ **32.** $(-3, 0)$

In Problems 33–44, the graph of an equation is given.
(a) List the intercepts of the graph.
(b) Based on the graph, tell whether the graph is symmetric with respect to the x-axis, y-axis, and/or origin.

33.
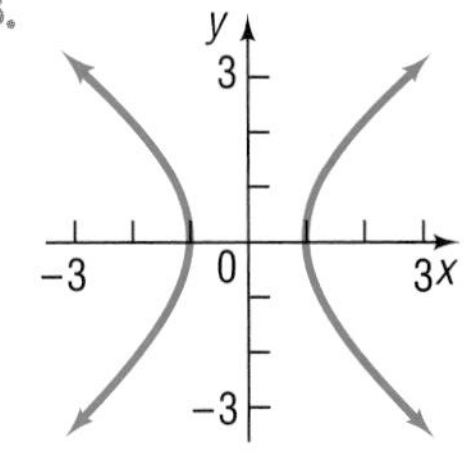

34.
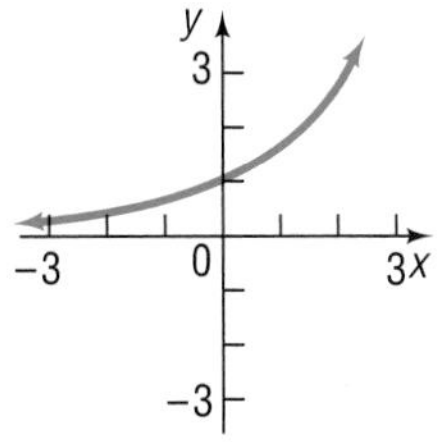

35.
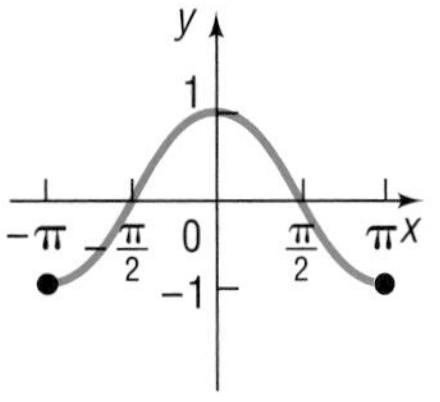

36.
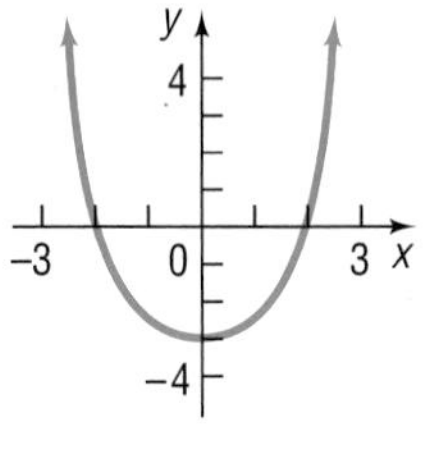

37.
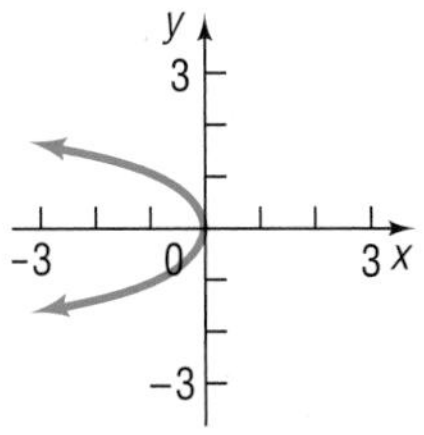

38.
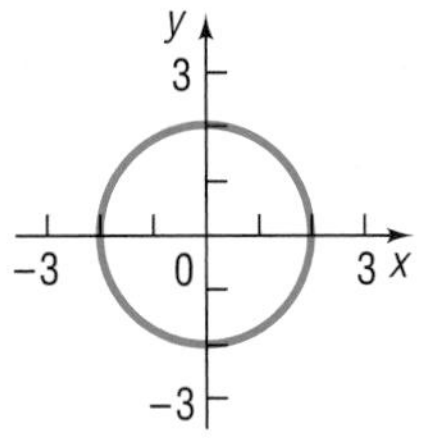

39.
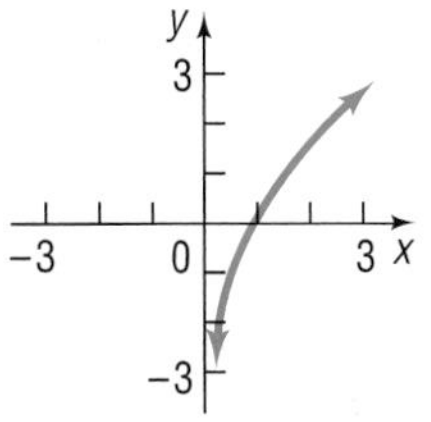

40.
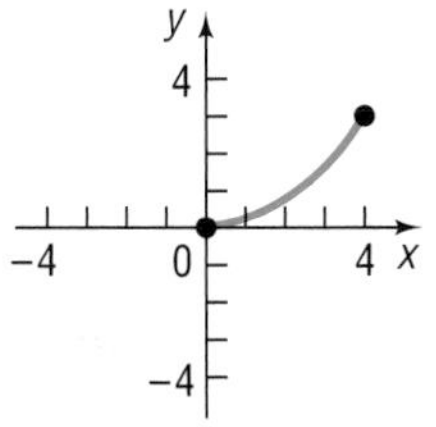

41.
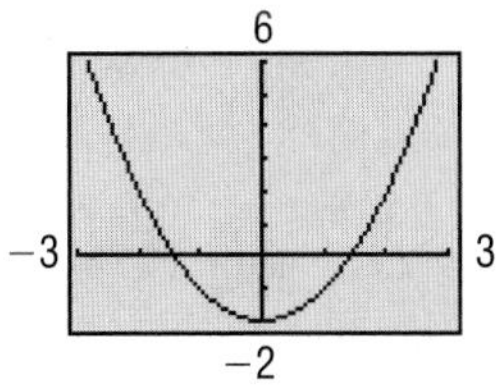

42.
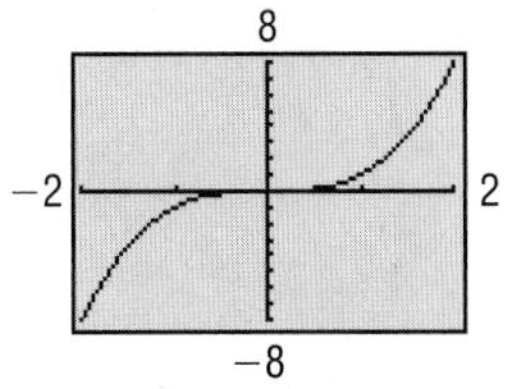

43.
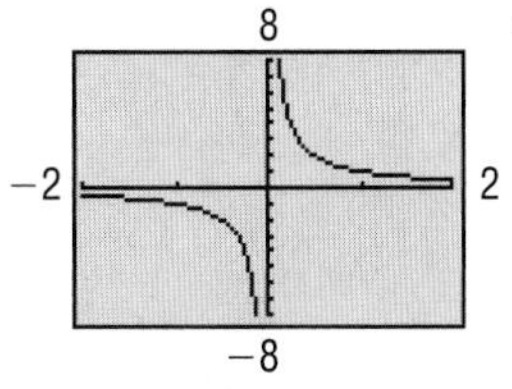

44.
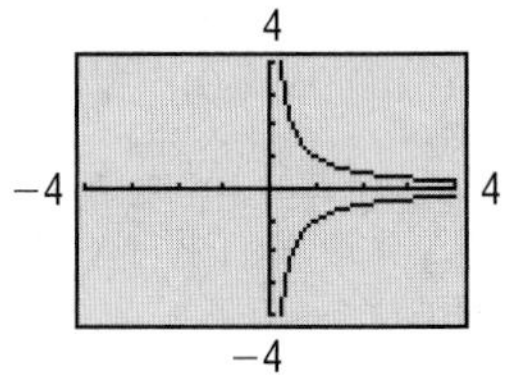

45. If $(a, 2)$ is a point on the graph of $y = 5x + 4$, what is a?
46. If $(2, b)$ is a point on the graph of $y = x^2 + 3x$, what is b?
47. If (a, b) is a point on the graph of $2x + 3y = 6$, write an equation that relates a to b.
48. If $(2, 0)$ and $(0, 5)$ are points on the graph of $y = mx + b$, what are m and b?

In Problems 49–52, use the graph on the right.

49. Draw the graph to make it symmetric with respect to the x-axis.
50. Draw the graph to make it symmetric with respect to the y-axis.
51. Draw the graph to make it symmetric with respect to the origin.
52. Draw the graph to make it symmetric with respect to the x-axis, y-axis, and origin.

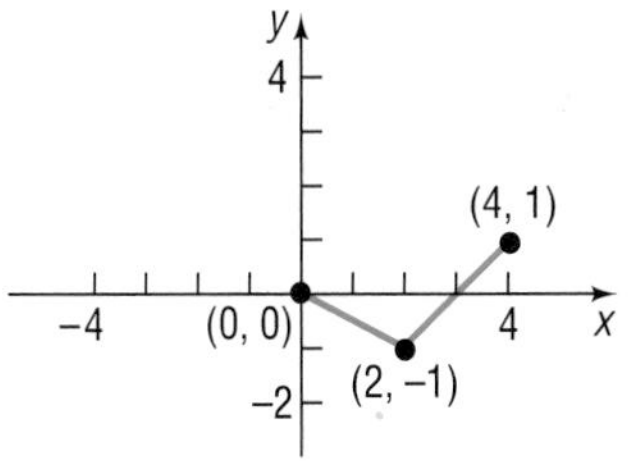

In Problems 53–56, use the graph on the right.

53. Draw the graph to make it symmetric with respect to the x-axis.
54. Draw the graph to make it symmetric with respect to the y-axis.
55. Draw the graph to make it symmetric with respect to the origin.
56. Draw the graph to make it symmetric with respect to the x-axis, y-axis, and origin.

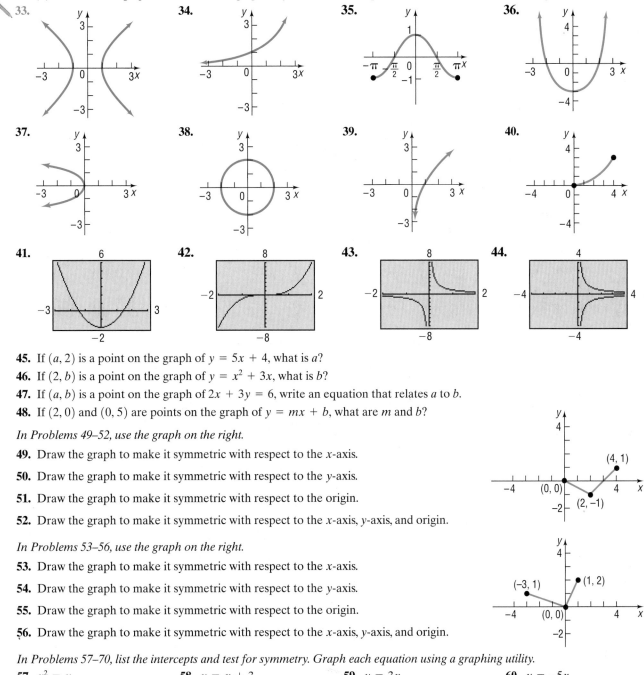

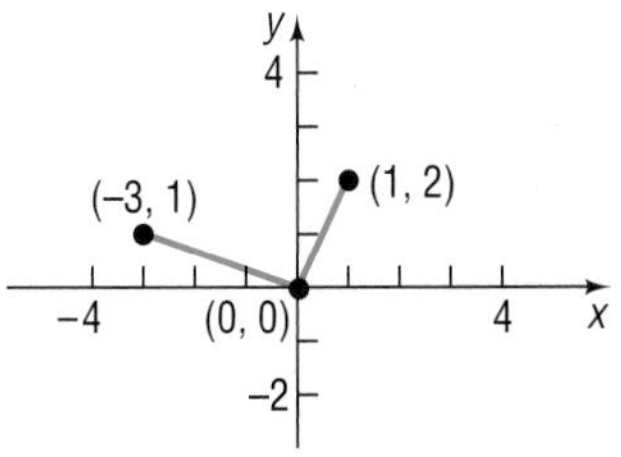

In Problems 57–70, list the intercepts and test for symmetry. Graph each equation using a graphing utility.

57. $x^2 = y$ **58.** $y = x + 3$ **59.** $y = 3x$ **60.** $y = -5x$
61. $x + y - 9 = 0$ **62.** $y - x - 4 = 0$ **63.** $9x^2 + 4y = 36$ **64.** $4x^2 + y = 4$
65. $y = x^3 - 27$ **66.** $y = x^4 - 1$ **67.** $y = x^2 - 3x - 4$ **68.** $y = x^2 + 4$
69. $y = \dfrac{x}{x^2 + 9}$ **70.** $y = \dfrac{x^2 - 4}{x}$

In Problem 71, you may use a graphing utility, but it is not required.

71. (a) Graph $y = \sqrt{x^2}$, $y = x$, $y = |x|$, and $y = (\sqrt{x})^2$, noting which graphs are the same.
(b) Explain why the graphs of $y = \sqrt{x^2}$ and $y = |x|$ are the same.
(c) Explain why the graphs of $y = x$ and $y = (\sqrt{x})^2$ are not the same.
(d) Explain why the graphs of $y = \sqrt{x^2}$ and $y = x$ are not the same.

72. Make up an equation with the intercepts $(2, 0)$, $(4, 0)$, and $(0, 1)$. Compare your equation with a friend's equation. Comment on any similarities.

73. An equation is being tested for symmetry with respect to the x-axis, the y-axis, and the origin. Explain why, if two of these symmetries are present, the remaining one must also be present.

74. Draw a graph that contains the points $(-2, -1)$, $(0, 1)$, $(1, 3)$, and $(3, 5)$. Compare your graph with those of other students. Are most of the graphs almost straight lines? How many are "curved"? Discuss the various ways that these points might be connected.

PREPARING FOR THIS SECTION

Before getting started, review the following:

- ✓ Solving Equations (Appendix, Section 6)
- ✓ Factoring Polynomials (Appendix, Section 5)
- ✓ The Quadratic Formula (p. 1001 of the Appendix, Section 10)
- ✓ Absolute Value (pp. 947–948 of the Appendix, Section 1)
- ✓ Integer Exponents (Appendix, Section 2)
- ✓ Radicals; Rational Exponents (Appendix, Section 8)

1.3 SOLVING EQUATIONS

OBJECTIVES

1. Solve Equations Using a Graphing Utility
2. Solve Linear Equations
3. Solve Quadratic Equations by Factoring
4. Solve Quadratic Equations Using the Quadratic Formula
5. Solve Equations Involving Absolute Value
6. Solve Radical Equations

SOLVING EQUATIONS USING A GRAPHING UTILITY

We shall see as we proceed through this book that some equations can be solved using algebraic techniques that result in exact solutions being obtained. Whenever algebraic techniques exist, we shall discuss them and use them to obtain exact solutions.

For many equations, though, there are no algebraic techniques that lead to a solution. For such equations, a graphing utility can often be used to investigate possible solutions. When a graphing utility is used to solve an equation, usually *approximate* solutions are obtained. Unless otherwise stated, we shall follow the practice of giving approximate solutions *rounded to two decimal places*.

1 The ZERO (or ROOT) feature of a graphing utility can be used to find the solutions of an equation when one side of the equation is 0. In using this feature to solve equations, we make use of the fact that the x-intercepts (or zeros) of the graph of an equation are found by letting $y = 0$ and solving the equation for x. Solving an equation for x when one side of the equation

is 0 is equivalent to finding where the graph of the corresponding equation crosses or touches the x-axis.

EXAMPLE 1 Using ZERO (or ROOT) to Approximate Solutions of an Equation

Find the solution(s) of the equation $x^2 - 6x + 7 = 0$. Round answers to two decimal places.

Solution The solutions of the equation $x^2 - 6x + 7 = 0$ are the same as the x-intercepts of the graph of $Y_1 = x^2 - 6x + 7$. We begin by graphing the equation. See Figure 32(a).

Figure 32

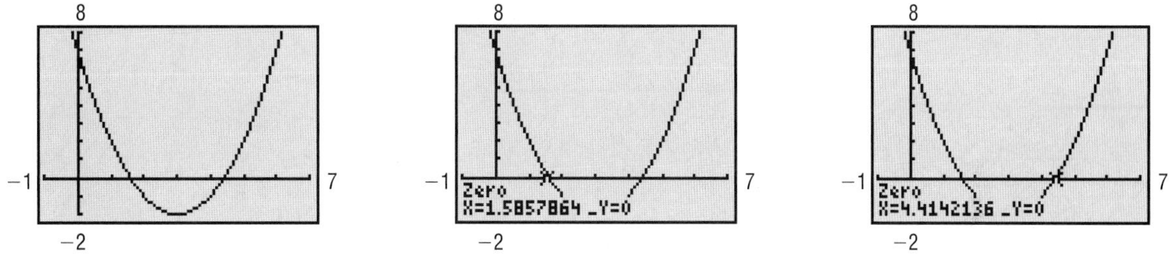

(a) (b) (c)

From the graph there appear to be two x-intercepts (solutions to the equation): one between 1 and 2, the other between 4 and 5.

Using the ZERO (or ROOT) feature of our graphing utility, we determine that the x-intercepts, and thus the solutions to the equation, are $x = 1.59$ and $x = 4.41$, rounded to two decimal places. See Figure 32(b) and (c). ■

NOW WORK PROBLEM 89.

A second method for solving equations using a graphing utility involves the INTERSECT feature of the graphing utility. This feature is used most effectively when one side of the equation is not 0.

EXAMPLE 2 Using INTERSECT to Approximate Solutions of an Equation

Find the solution(s) to the equation $3(x - 2) = 5(x - 1)$. Round answers to two decimal places.

Solution We begin by graphing each side of the equation as follows: graph $Y_1 = 3(x - 2)$ and $Y_2 = 5(x - 1)$. See Figure 33(a).

Figure 33

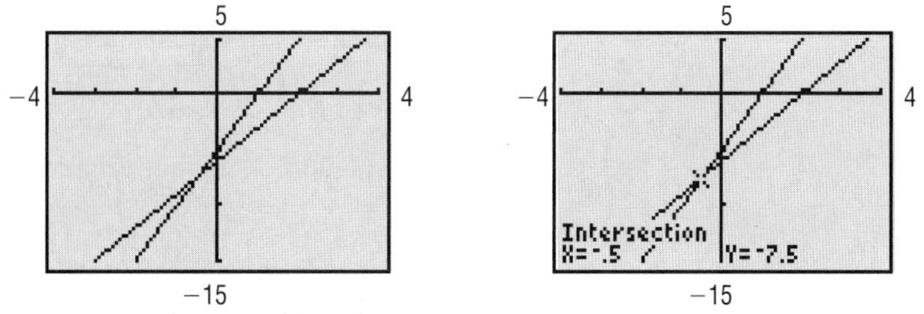

(a) (b)

At the point of intersection of the graphs, the value of the y-coordinate is the same. We conclude that the x-coordinate of the point of intersection represents the solution to the equation. Do you see why? The INTERSECT feature on a graphing utility determines the point of intersection of the graphs. Using this feature, we find that the graphs intersect at $(-0.5, -7.5)$. See Figure 33(b). The solution of the equation is therefore $x = -0.5$. ■

Check: We can verify our solution by evaluating each side of the equation with -0.5 STOred in x. See Figure 34. Since the left side of the equation equals the right side of the equation, the solution checks.

Figure 34

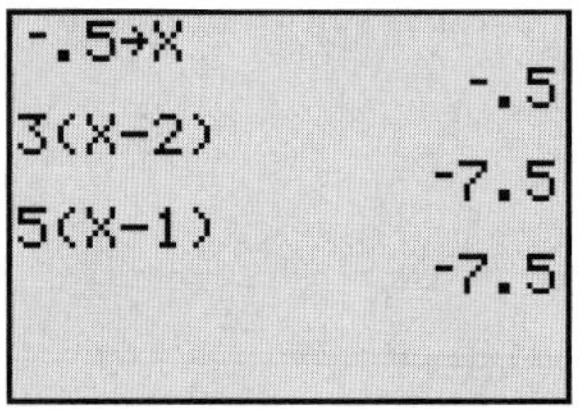

■

NOW WORK PROBLEM 93.

SUMMARY

The steps to follow for approximating solutions of equations are given next.

STEPS FOR APPROXIMATING SOLUTIONS OF EQUATIONS USING ZERO (OR ROOT)

STEP 1: Write the equation in the form {expression in x} $= 0$.
STEP 2: Graph $Y_1 =$ {expression in x}.
STEP 3: Use ZERO (or ROOT) to determine each x-intercept of the graph.

STEPS FOR APPROXIMATING SOLUTIONS OF EQUATIONS USING INTERSECT

STEP 1: Graph $Y_1 =$ {expression in x on left side of equation};
Graph $Y_2 =$ {expression in x on right side of equation}.
STEP 2: Use INTERSECT to determine each x-coordinate of the point(s) of intersection, if any.

We now discuss equations that can be solved algebraically to obtain exact solutions.

LINEAR EQUATIONS

2 *Linear equations* are equations such as

$$3x + 12 = 0, \qquad -2x + 5 = 0, \qquad 4x - 3 = 0$$

A **linear equation in one variable** is equivalent to an equation of the form

$$ax + b = 0$$

where a and b are real numbers and $a \neq 0$.

Sometimes a linear equation is called a **first-degree equation**, because the left side is a polynomial in x of degree 1.

EXAMPLE 3 Solving a Linear Equation Algebraically

Solve the equation: $3(x-2) = 5(x-1)$

Solution

$$3(x-2) = 5(x-1)$$
$$3x - 6 = 5x - 5 \quad \text{Remove parentheses}$$
$$3x - 6 - 5x = 5x - 5 - 5x \quad \text{Subtract } 5x \text{ from each side.}$$
$$-2x - 6 = -5 \quad \text{Simplify}$$
$$-2x - 6 + 6 = -5 + 6 \quad \text{Add 6 to each side.}$$
$$-2x = 1 \quad \text{Simplify}$$
$$\frac{-2x}{-2} = \frac{1}{-2} \quad \text{Divide each side by } -2.$$
$$x = -\frac{1}{2} \quad \text{Simplify}$$

Check:

$$3(x-2) = 3\left(-\frac{1}{2} - 2\right) = 3\left(-\frac{5}{2}\right) = -\frac{15}{2}$$
$$5(x-1) = 5\left(-\frac{1}{2} - 1\right) = 5\left(-\frac{3}{2}\right) = -\frac{15}{2}$$

Since the two expressions are equal, the solution $x = -\frac{1}{2}$ checks.

Notice that this is the solution obtained in Example 2 using the INTERSECT feature of a graphing utility.

NOW WORK PROBLEM 5.

The next example illustrates the solution of an equation that does not appear to be linear, but leads to a linear equation upon simplification.

EXAMPLE 4 Solving Equations

Solve the equation: $(2x+1)(x-1) = (x+5)(2x-5)$

Graphing Solution

Graph $Y_1 = (2x+1)(x-1)$ and $Y_2 = (x+5)(2x-5)$. See Figure 35. Using INTERSECT, we find the point of intersection to be (4, 27). The solution of the equation is $x = 4$.

Figure 35

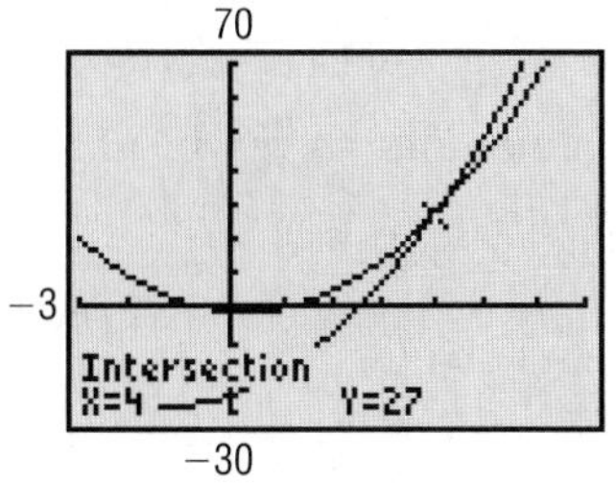

Algebraic Solution

$$
\begin{aligned}
(2x + 1)(x - 1) &= (x + 5)(2x - 5) \\
2x^2 - x - 1 &= 2x^2 + 5x - 25 && \text{Multiply and combine like terms.} \\
-x - 1 &= 5x - 25 && \text{Subtract } 2x^2 \text{ from each side.} \\
-x &= 5x - 24 && \text{Add 1 to each side.} \\
-6x &= -24 && \text{Subtract } 5x \text{ from each side.} \\
x &= 4 && \text{Divide both sides by } -6.
\end{aligned}
$$

Check: $(2x + 1)(x - 1) = (8 + 1)(4 - 1) = (9)(3) = 27$

$(x + 5)(2x - 5) = (4 + 5)(8 - 5) = (9)(3) = 27$

Since the two expressions are equal, the solution $x = 4$ checks.

NOW WORK PROBLEM 39.

QUADRATIC EQUATIONS

Quadratic equations are equations such as

$$2x^2 + x + 8 = 0, \quad 3x^2 - 5x + 6 = 0, \quad x^2 - 9 = 0$$

A general definition is given next.

A **quadratic equation** is an equation equivalent to one of the form

$$ax^2 + bx + c = 0 \tag{1}$$

where a, b, and c are real numbers and $a \neq 0$.

A quadratic equation written in the form $ax^2 + bx + c = 0$ is said to be in **standard form**.

Sometimes, a quadratic equation is called a **second-degree equation**, because the left side is a polynomial of degree 2. We shall discuss two algebraic ways of solving quadratic equations: by factoring and by using the quadratic formula.*

3 When a quadratic equation is written in standard form, $ax^2 + bx + c = 0$, it may be possible to factor the expression on the left side as the product of two first-degree polynomials. Then, by setting each factor equal to 0 and solving the resulting linear equations, we obtain the solutions of the quadratic equation.

Let's look at an example.

EXAMPLE 5 Solving a Quadratic Equation by Graphing and by Factoring

Solve the equation: $x^2 = 12 - x$

Graphing Solution Graph $Y_1 = x^2$ and $Y_2 = 12 - x$. See Figure 36(a). From the graph it appears that there are two points of intersection: one near -4, the other near 3. Using

* A third way, by completing the square, is discussed in Section 10 of the Appendix.

Figure 36

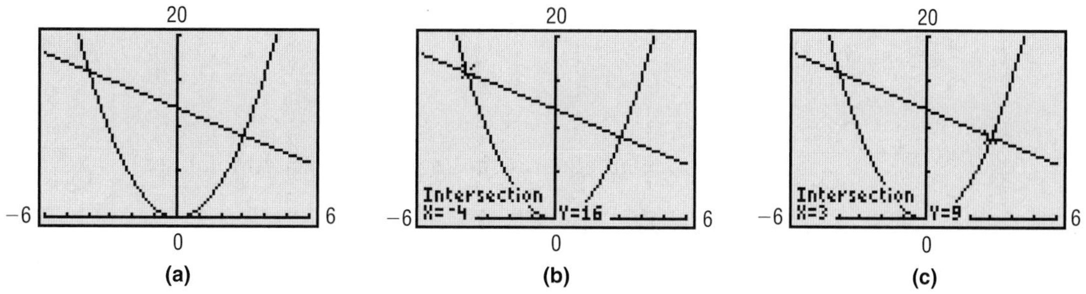

INTERSECT (twice), the points of intersection are $(-4, 16)$ and $(3, 9)$, so the solutions of the equation are $x = -4$ and $x = 3$. See Figure 36(b) and (c).

Algebraic Solution We put the equation in standard form by adding $x - 12$ to each side.

$$x^2 = 12 - x$$

$$x^2 + x - 12 = 0$$

The left side of the equation may now be factored as

$$(x + 4)(x - 3) = 0$$

Using the Zero-Product Property,

$$x + 4 = 0 \quad \text{or} \quad x - 3 = 0$$

$$x = -4 \qquad\qquad x = 3$$

The solution set is $\{-4, 3\}$. ■

When the left side factors into two linear equations with the same solution, the quadratic equation is said to have a **repeated solution**. We also call this solution a **root of multiplicity 2**, or a **double root**.

EXAMPLE 6 Solving a Quadratic Equation by Graphing and by Factoring

Solve the equation: $x^2 - 6x + 9 = 0$

Graphing Solution Figure 37 shows the graph of the equation

$$Y_1 = x^2 - 6x + 9$$

From the graph it appears that there is one x-intercept, 3. Using ZERO (or ROOT), we find that the only x-intercept is $x = 3$. The equation has only the repeated solution 3.

Figure 37

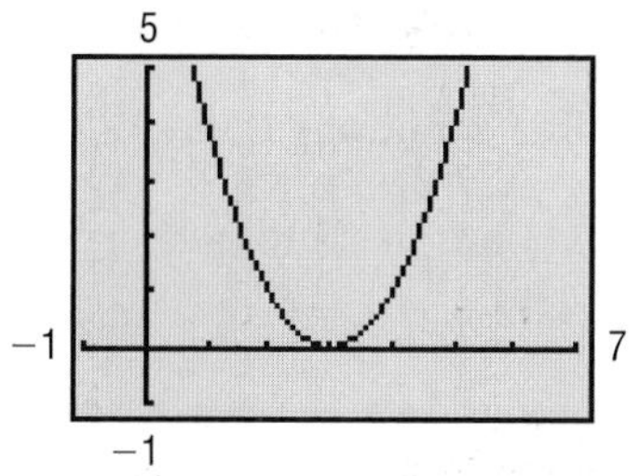

Algebraic Solution This equation is already in standard form, and the left side can be factored.

$$x^2 - 6x + 9 = 0$$
$$(x - 3)(x - 3) = 0$$
$$x = 3 \quad \text{or} \quad x = 3$$

This equation has only the repeated solution 3. ■

NOW WORK PROBLEMS 25 AND 33.

4 Quadratic equations may also be solved by using the *quadratic formula*. See Section 10 of the Appendix for the derivation.

Theorem **Quadratic Formula**

Consider the quadratic equation

$$ax^2 + bx + c = 0 \qquad a \neq 0$$

If $b^2 - 4ac < 0$, this equation has no real solution.
If $b^2 - 4ac \geq 0$, the real solution(s) of this equation is (are) given by the **quadratic formula**:

$$x = \frac{-b \pm \sqrt{b^2 - 4ac}}{2a}$$

■

The quantity $\boldsymbol{b^2 - 4ac}$ is called the **discriminant** of the quadratic equation, because its value tells us whether the equation has real solutions. In fact, it also tells us how many solutions to expect.

Discriminant of a Quadratic Equation

For a quadratic equation $ax^2 + bx + c = 0$:

1. If $b^2 - 4ac > 0$, there are two unequal real solutions.
2. If $b^2 - 4ac = 0$, there is a repeated real solution, a root of multiplicity 2.
3. If $b^2 - 4ac < 0$, there is no real solution.*

When asked to find the real solutions, if any, of a quadratic equation, always evaluate the discriminant first to see how many real solutions there are.

EXAMPLE 7 Solving a Quadratic Equation by Graphing and by Using the Quadratic Formula

Find the real solutions, if any, of the equation $3x^2 - 5x + 1 = 0$.

Solution The equation is in standard form, so we compare it to $ax^2 + bx + c = 0$ to find a, b, and c.

*We consider quadratic equations whose discriminant is negative in Section 4.2, which may be covered anytime after completing this section without any loss of continuity.

$$3x^2 - 5x + 1 = 0$$
$$ax^2 + bx + c = 0$$

With $a = 3$, $b = -5$, and $c = 1$, we evaluate the discriminant $b^2 - 4ac$.

$$b^2 - 4ac = (-5)^2 - 4(3)(1) = 25 - 12 = 13$$

Since $b^2 - 4ac > 0$, there are two unequal real solutions.

Graphing Solution Figure 38 shows the graph of the equation

$$Y_1 = 3x^2 - 5x + 1$$

Figure 38

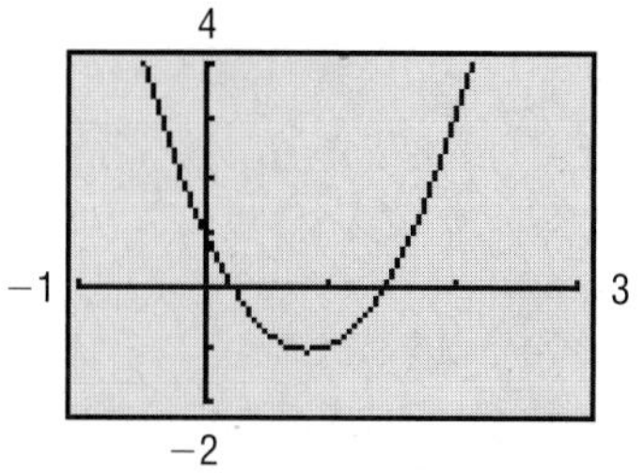

As expected, we see that there are two x-intercepts: one between 0 and 1, the other between 1 and 2. The solutions to the equation are 0.23 and 1.43, rounded to two decimal places.

Algebraic Solution We use the quadratic formula with $a = 3$, $b = -5$, $c = 1$, and $b^2 - 4ac = 13$.

$$x = \frac{-b \pm \sqrt{b^2 - 4ac}}{2a} = \frac{5 \pm \sqrt{13}}{6}$$

The solution set is $\left\{\frac{5 - \sqrt{13}}{6}, \frac{5 + \sqrt{13}}{6}\right\}$. These solutions are exact. ■

NOW WORK PROBLEM 77.

EXAMPLE 8 Solving Quadratic Equations by Graphing and by Using the Quadratic Formula

Find the real solutions, if any, of the equation

$$3x^2 + 2 = 4x$$

Solution The equation, as given, is not in standard form.

$$3x^2 + 2 = 4x$$
$$3x^2 - 4x + 2 = 0 \quad \text{Put in standard form.}$$
$$ax^2 + bx + c = 0 \quad \text{Compare to standard form.}$$

With $a = 3$, $b = -4$, and $c = 2$, we find that

$$b^2 - 4ac = 16 - 24 = -8$$

Since $b^2 - 4ac < 0$, the equation has no real solution.

Graphing Solution We use the standard form of the equation and graph

$$Y_1 = 3x^2 - 4x + 2$$

See Figure 39. We see that there are no x-intercepts, so the equation has no real solution, as expected.

Figure 39

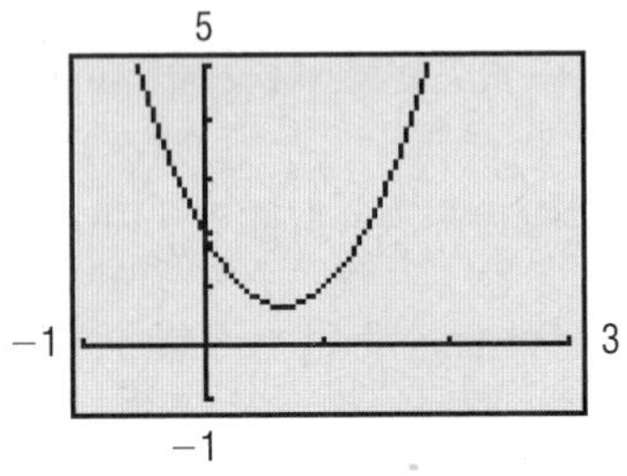

NOW WORK PROBLEM 83.

Equations Involving Absolute Value

5 Recall that, on the real number line, the absolute value of a equals the distance from the origin to the point whose coordinate is a. For example, there are two points whose distance from the origin is 5 units, −5 and 5. So the equation $|x| = 5$ will have the solution set $\{-5, 5\}$. This leads to the following result:

Equations Involving Absolute Value

If a is a positive real number and if u is any algebraic expression, then

$$|u| = a \quad \text{is equivalent to} \quad u = a \quad \text{or} \quad u = -a \qquad \textbf{(2)}$$

EXAMPLE 9 **Solving an Equation Involving Absolute Value**

Solve the equation $|x + 4| = 13$.

Graphing Solution For this equation, graph $Y_1 = |x + 4|$ and $Y_2 = 13$ on the same screen and find their point(s) of intersection, if any. See Figure 40. Using the INTERSECT command twice, we find the points of intersection to be $(-17, 13)$ and $(9, 13)$. The solution set is $\{-17, 9\}$.

Figure 40

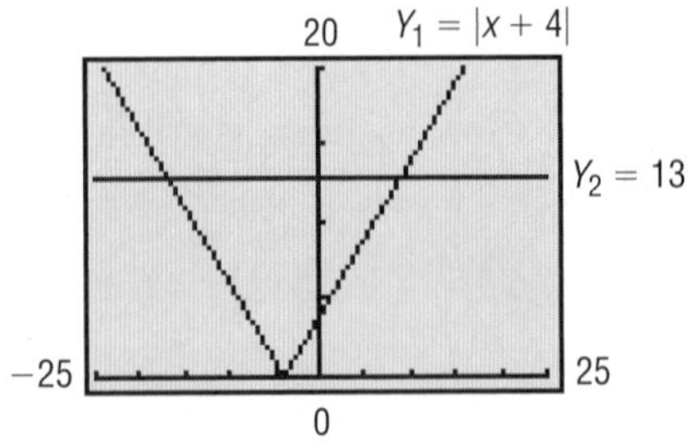

Algebraic Solution This follows the form of equation (2), where $u = x + 4$. There are two possibilities.

$$x + 4 = 13 \quad \text{or} \quad x + 4 = -13$$
$$x = 9 \qquad\qquad x = -17$$

The solution set is $\{-17, 9\}$. ■

NOW WORK PROBLEM 61.

Equations Containing Radicals

6 When the variable in an equation occurs in a square root, cube root, and so on, that is, when it occurs in a radical, the equation is called a **radical equation**. Sometimes a suitable operation will change a radical equation to one that is linear or quadratic. A commonly used procedure is to isolate the most complicated radical on one side of the equation and then eliminate it by raising each side to a power equal to the index of the radical. Care must be taken, however, because apparent solutions that are not, in fact, solutions of the original equation may result. These are called **extraneous solutions**. Therefore, we need to check all answers when working with radical equations.

EXAMPLE 10 Solving a Radical Equation

Find the real solutions of the equation $\sqrt[3]{2x - 4} - 2 = 0$.

Graphing Solution Figure 41 shows the graph of the equation $Y_1 = \sqrt[3]{2x - 4} - 2$. From the graph, we see one x-intercept near 6. Using ZERO (or ROOT), we find that the x-intercept is 6. The only solution is $x = 6$.

Figure 41

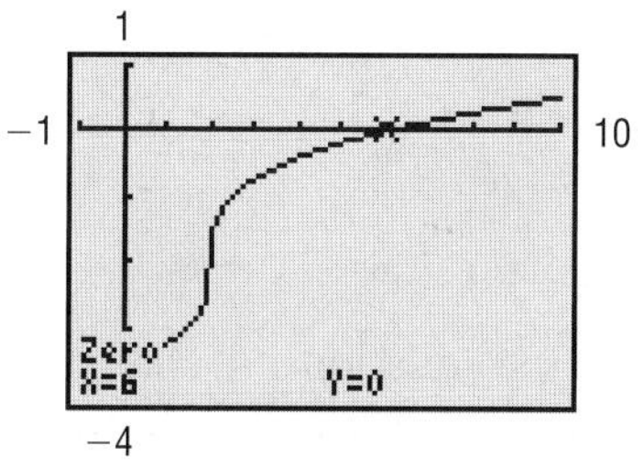

Algebraic Solution The equation contains a radical whose index is 3. We isolate it on the left side.

$$\sqrt[3]{2x - 4} - 2 = 0$$
$$\sqrt[3]{2x - 4} = 2$$

Now raise each side to the third power (the index of the radical is 3) and solve.

$$\left(\sqrt[3]{2x - 4}\right)^3 = 2^3$$
$$2x - 4 = 8 \qquad \text{Simplify}$$
$$2x = 12 \qquad \text{Add 4 to each side.}$$
$$x = 6 \qquad \text{Divide both sides by 2.}$$

Check: $\sqrt[3]{2(6) - 4} - 2 = \sqrt[3]{12 - 4} - 2 = \sqrt[3]{8} - 2 = 2 - 2 = 0$.
The solution is $x = 6$. ■

NOW WORK PROBLEM 49.

Sometimes, we need to raise each side to a power more than once in order to solve a radical equation algebraically.

EXAMPLE 11 Solving a Radical Equation

Find the real solutions of the equation $\sqrt{2x + 3} - \sqrt{x + 2} = 2$.

Graphing Solution

Graph $Y_1 = \sqrt{2x + 3} - \sqrt{x + 2}$ and $Y_2 = 2$. See Figure 42. From the graph it appears that there is one point of intersection. Using INTERSECT, the point of intersection is $(23, 2)$, so the solution is $x = 23$.

Figure 42

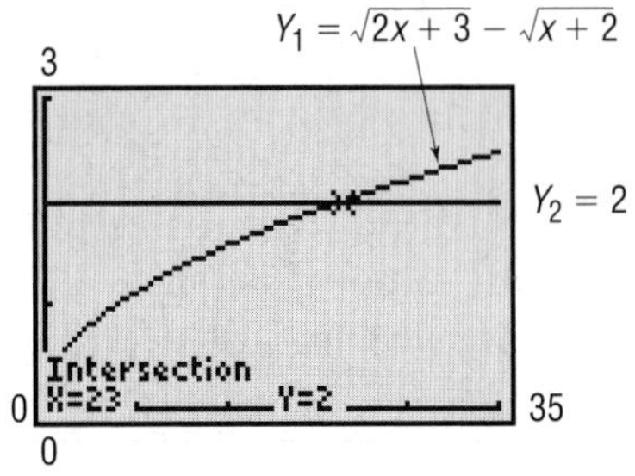

Algebraic Solution

First, we choose to isolate the more complicated radical expression (in this case, $\sqrt{2x + 3}$) on the left side.

$$\sqrt{2x + 3} = \sqrt{x + 2} + 2$$

Now square both sides (the index of the radical is 2).

$$(\sqrt{2x + 3})^2 = (\sqrt{x + 2} + 2)^2$$

$$2x + 3 = (\sqrt{x + 2})^2 + 4\sqrt{x + 2} + 4 \qquad \text{Remove parentheses.}$$

$$2x + 3 = x + 2 + 4\sqrt{x + 2} + 4 \qquad \text{Simplify}$$

$$2x + 3 = x + 6 + 4\sqrt{x + 2} \qquad \text{Combine like terms.}$$

Because the equation still contains a radical, we isolate the remaining radical on the right side and again square both sides.

$$x - 3 = 4\sqrt{x + 2} \qquad \text{Subtract } x + 6 \text{ from both sides.}$$

$$(x - 3)^2 = 16(x + 2) \qquad \text{Square both sides.}$$

$$x^2 - 6x + 9 = 16x + 32 \qquad \text{Remove parentheses.}$$

$$x^2 - 22x - 23 = 0 \qquad \text{Put in standard form.}$$

$$(x - 23)(x + 1) = 0 \qquad \text{Factor}$$

$$x = 23 \quad \text{or} \quad x = -1$$

The original equation appears to have the solution set $\{-1, 23\}$. However, we have not yet checked.

Check: $\sqrt{2(23) + 3} - \sqrt{23 + 2} = \sqrt{49} - \sqrt{25} = 7 - 5 = 2$

$\sqrt{2(-1) + 3} - \sqrt{-1 + 2} = \sqrt{1} - \sqrt{1} = 1 - 1 = 0$

The equation has only one solution, 23; the solution -1 is extraneous. ■

NOW WORK PROBLEM 55.

1.3 EXERCISES

In Problems 1–72, solve each equation algebraically. Verify your solution using a graphing utility.

1. $3x + 2 = x + 6$ **2.** $2x + 7 = 3x + 5$ **3.** $2t - 6 = 3 - t$ **4.** $5y + 6 = -18 - y$

5. $6 - x = 2x + 9$ **6.** $3 - 2x = 2 - x$ **7.** $3 + 2n = 5n + 7$ **8.** $3 - 2m = 3m + 1$

9. $2(3 + 2x) = 3(x - 4)$ **10.** $3(2 - x) = 2x - 1$ **11.** $8x - (2x + 1) = 3x - 10$ **12.** $5 - (2x - 1) = 10$

13. $\frac{2}{3}p = \frac{1}{2}p + \frac{1}{3}$ **14.** $\frac{1}{2} - \frac{1}{3}p = \frac{4}{3}$ **15.** $0.9t = 0.4 + 0.1t$ **16.** $0.9t = 1 + t$

17. $\dfrac{x+1}{3} + \dfrac{x+2}{7} = 5$ **18.** $\dfrac{2x+1}{3} + 16 = 3x$ **19.** $\dfrac{2}{y} + \dfrac{4}{y} = 3$ **20.** $\dfrac{4}{y} - 5 = \dfrac{5}{2y}$

21. $x^2 = 9x$ **22.** $x^2 = -4x$ **23.** $x^2 - 25 = 0$ **24.** $x^2 - 9 = 0$

25. $z^2 + z - 12 = 0$ **26.** $v^2 + 7v + 12 = 0$ **27.** $2x^2 - 5x - 3 = 0$ **28.** $3x^2 + 5x + 2 = 0$

29. $3t^2 - 48 = 0$ **30.** $2y^2 - 50 = 0$ **31.** $x(x - 7) + 12 = 0$ **32.** $x(x + 1) = 12$

33. $4x^2 + 9 = 12x$ **34.** $25x^2 + 16 = 40x$ **35.** $6(p^2 - 1) = 5p$ **36.** $2(2u^2 - 4u) + 3 = 0$

37. $6x - 5 = \dfrac{6}{x}$ **38.** $x + \dfrac{12}{x} = 7$

39. $(x + 7)(x - 1) = (x + 1)^2$ **40.** $(x + 2)(x - 3) = (x - 3)^2$

41. $x(2x - 3) = (2x + 1)(x - 4)$ **42.** $x(1 + 2x) = (2x - 1)(x - 2)$

43. $\sqrt{2t - 1} = 1$ **44.** $\sqrt{3t + 4} = 2$ **45.** $\sqrt{3t + 1} = -4$ **46.** $\sqrt{5t + 4} = -3$

47. $\sqrt[3]{1 - 2x} - 3 = 0$ **48.** $\sqrt[3]{1 - 2x} - 1 = 0$ **49.** $\sqrt{15 - 2x} = x$ **50.** $\sqrt{12 - x} = x$

51. $x = 2\sqrt{x - 1}$ **52.** $x = 2\sqrt{-x - 1}$ **53.** $3 + \sqrt{3x + 1} = x$ **54.** $2 + \sqrt{12 - 2x} = x$

55. $\sqrt{2x + 3} - \sqrt{x + 1} = 1$ **56.** $\sqrt{3x + 7} + \sqrt{x + 2} = 1$

57. $\sqrt{3x + 1} - \sqrt{x - 1} = 2$ **58.** $\sqrt{3x - 5} - \sqrt{x + 7} = 2$

59. $|2x| = 8$ **60.** $|3x| = 15$ **61.** $|2x + 3| = 5$ **62.** $|3x - 1| = 2$

63. $|1 - 4t| = 5$ **64.** $|1 - 2z| = 3$ **65.** $|-2x| = 8$ **66.** $|-x| = 1$

67. $|-2|x = 4$ **68.** $|3|x = 9$ **69.** $\frac{2}{3}|x| = 8$ **70.** $\frac{3}{4}|x| = 9$

71. $\left|\dfrac{x}{3} + \dfrac{2}{5}\right| = 2$ **72.** $\left|\dfrac{x}{2} - \dfrac{1}{3}\right| = 1$

In Problems 73–76, solve each equation. The letters a, b, and c are constants.

73. $ax - b = c, \quad a \neq 0$ **74.** $1 - ax = b, \quad a \neq 0$

75. $\dfrac{x}{a} + \dfrac{x}{b} = c, \quad a \neq 0, b \neq 0, a \neq -b$ **76.** $\dfrac{a}{x} + \dfrac{b}{x} = c, \quad c \neq 0$

In Problems 77–88, find the real solutions, if any, of each equation. Use the quadratic formula. Verify your results using a graphing utility.

77. $x^2 - 4x + 2 = 0$ **78.** $x^2 + 4x + 2 = 0$ **79.** $x^2 - 4x - 1 = 0$ **80.** $x^2 + 6x + 1 = 0$

81. $2x^2 - 5x + 3 = 0$ **82.** $2x^2 + 5x + 3 = 0$ **83.** $4y^2 - y + 2 = 0$ **84.** $4t^2 + t + 1 = 0$

85. $4x^2 = 1 - 2x$ **86.** $2x^2 = 1 - 2x$ **87.** $9t^2 - 6t + 1 = 0$ **88.** $4u^2 - 6u + 9 = 0$

In Problems 89–94, use a graphing utility to approximate the real solutions, if any, of each equation rounded to two decimal places.

89. $x^2 - 4x + 2 = 0$ **90.** $x^2 + 4x + 2 = 0$ **91.** $x^2 + \sqrt{3}x = 3$ **92.** $x^2 = 2 - \sqrt{2}x$

93. $\pi x^2 = x + \pi$ **94.** $\pi x^2 + \pi x = 2$

Problems 95–100 list some formulas that occur in applications. Solve each formula for the indicated variable.

95. Electricity $\dfrac{1}{R} = \dfrac{1}{R_1} + \dfrac{1}{R_2}$ for R **96. Finance** $A = P(1 + rt)$ for r

97. Mechanics $F = \dfrac{mv^2}{R}$ for R **98. Chemistry** $PV = nRT$ for T

99. Mathematics $S = \dfrac{a}{1 - r}$ for r **100. Mechanics** $v = -gt + v_0$ for t

101. Physics: Using Sound to Measure Distance The distance to the surface of the water in a well can sometimes be found by dropping an object into the well and measuring the time elapsed until a sound is heard. If t_1 is the time (measured in seconds) that it takes for the object to strike the water, then t_1 will obey the equation $s = 16t_1^2$, where s is the distance (measured in feet). It follows that $t_1 = \sqrt{s}/4$. Suppose that t_2 is the time that it takes for the sound of the impact to reach your ears. Because sound waves are known to travel at a speed of approximately 1100 feet per second, the time t_2 to travel the distance s will be $t_2 = s/1100$. Now $t_1 + t_2$ is the total time that elapses from the moment that the object is dropped to the moment that a sound is heard. Thus, we have the equation

$$\text{Total time elapsed} = \frac{\sqrt{s}}{4} + \frac{s}{1100}$$

Find the distance to the water's surface if the total time elapsed from dropping a rock to hearing it hit water is 4 seconds by graphing

$$Y_1 = \frac{\sqrt{x}}{4} + \frac{x}{1100}, \qquad Y_2 = 4$$

for $0 \le x \le 300$ and $0 \le Y_1 \le 5$ and finding the point of intersection.

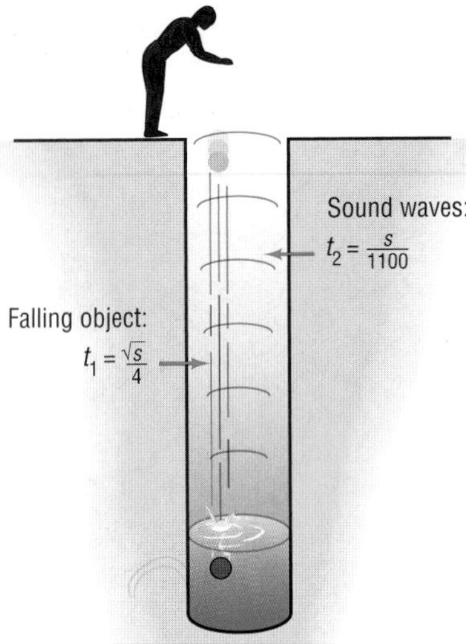

102. Make up a radical equation that has no solution.

103. Make up a radical equation that has an extraneous solution.

104. Discuss the step in the solving process for radical equations that leads to the possibility of extraneous solutions. Why is there no such possibility for linear and quadratic equations?

1.4 SETTING UP EQUATIONS; APPLICATIONS

OBJECTIVES
1. Translate Verbal Descriptions into Mathematical Expressions
2. Set Up Applied Problems
3. Solve Interest Problems
4. Solve Mixture Problems
5. Solve Uniform Motion Problems
6. Solve Constant Rate Job Problems

Applied (word) problems do not come in the form "Solve the equation...." Instead, they supply information using words, a verbal description of the real problem. So, to solve applied problems, we must be able to translate the verbal description into the language of mathematics. We do this by using variables to represent unknown quantities and then finding relationships (such as equations) that involve these variables. The process of doing all this is called **mathematical modeling**.

Any solution to the mathematical problem must be checked against the mathematical problem, the verbal description, and the real problem. See Figure 43 for an illustration of the **modeling process**.

1 Let's look at a few examples that will help you to translate certain words into mathematical symbols.

Figure 43

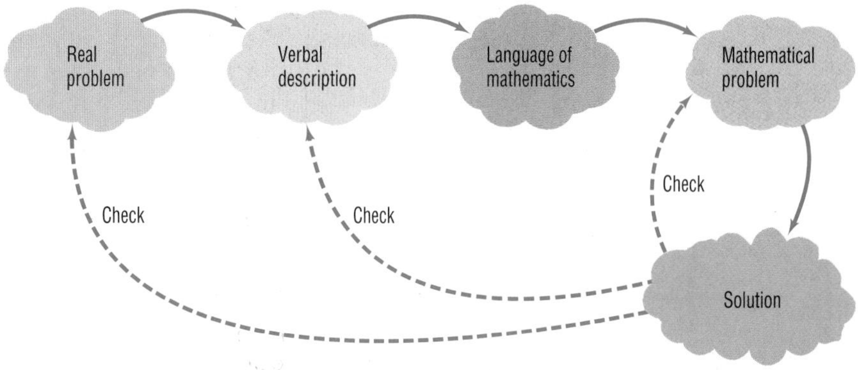

EXAMPLE 1

Translating Verbal Descriptions into Mathematical Expressions

(a) The area of a rectangle is the product of its length times its width.

Translation: If A is used to represent the area, l the length, and w the width, then $A = lw$.

(b) For uniform motion, the velocity of an object equals the distance traveled divided by the time required.

Translation: If v is the velocity, s the distance, and t the time, then $v = s/t$.

(c) A total of \$5000 is invested, some in stocks and some in bonds. If the amount invested in stocks is x, express the amount invested in bonds in terms of x.

Translation: If x is the amount invested in stocks, then the amount invested in bonds is $5000 - x$, since their sum is $x + (5000 - x) = 5000$.

(d) Let x denote a number.

The number 5 times as large as x is $5x$.

The number 3 less than x is $x - 3$.

The number that exceeds x by 4 is $x + 4$.

The number that, when added to x, gives 5 is $5 - x$. ■

NOW WORK PROBLEM 1.

2 Always check the units used to measure the variables of an applied problem. In Example 1(a), if l is measured in feet, then w also must be expressed in feet, and A will be expressed in square feet. In Example 1(b), if v is measured in miles per hour, then the distance s must be expressed in miles and the time t must be expressed in hours. It is a good practice to check units to be sure that they are consistent and make sense.

Although each situation has unique features, we can provide an outline of the steps to follow in setting up applied problems.

STEPS FOR SETTING UP APPLIED PROBLEMS

STEP 1: Read the problem carefully, perhaps two or three times. Pay particular attention to the question being asked in order to identify what you are looking for. If you can, determine realistic possibilities for the answer.

STEP 2: Assign a letter (variable) to represent what you are looking for, and, if necessary, express any remaining unknown quantities in terms of this variable.

STEP 3: Make a list of all the known facts, and translate them into mathematical expressions. These may take the form of an equation (or, later, an inequality) involving the variable. If possible, draw an appropriately labeled diagram to assist you. Sometimes a table or chart helps.

STEP 4: Solve the equation for the variable, and then answer the question.

STEP 5: Check the answer with the facts in the problem. If it agrees, congratulations! If it does not agree, try again.

Let's look at an example.

EXAMPLE 2 Investments

A total of \$18,000 is invested, some in stocks and some in bonds. If the amount invested in bonds is half that invested in stocks, how much is invested in each category?

Solution

STEP 1: We are being asked to find the amount of two investments. These amounts must total \$18,000. (Do you see why?)

STEP 2: If we let x equal the amount invested in stocks, then $18{,}000 - x$ is the amount invested in bonds. [Look back at Example 1(c) to see why.]

STEP 3: We set up a table:

Amount in Stocks	Amount in Bonds	Reason
x	$18{,}000 - x$	Total invested is \$18,000

Since the total amount invested in bonds $(18{,}000 - x)$ is half that in stocks (x), we obtain the equation $18{,}000 - x = \frac{1}{2}x$.

STEP 4:

$$\begin{aligned}
18{,}000 - x &= \tfrac{1}{2}x \\
18{,}000 &= x + \tfrac{1}{2}x \\
18{,}000 &= \tfrac{3}{2}x \\
\left(\tfrac{2}{3}\right)18{,}000 &= \left(\tfrac{2}{3}\right)\left(\tfrac{3}{2}x\right) \\
12{,}000 &= x
\end{aligned}$$

Thus, \$12,000 is invested in stocks and \$18,000 − \$12,000 = \$6000 is invested in bonds.

STEP 5: The total invested is \$12,000 + \$6000 = \$18,000, and the amount in bonds (\$6000) is half that in stocks (\$12,000). ■

NOW WORK PROBLEM 11.

INTEREST

3 The next example involves **interest**. Interest is money paid for the use of money. The total amount borrowed (whether by an individual from a bank in the form of a loan or by a bank from an individual in the form of a savings account) is called the **principal**. The **rate of interest**, expressed as a percent,

is the amount charged for the use of the principal for a given period of time, usually on a yearly (that is, per annum) basis.

Simple Interest Formula

If a principal of P dollars is borrowed for a period of t years at a per annum interest rate r, expressed as a decimal, the interest I charged is

$$I = Prt \quad \textbf{(1)}$$

Interest charged according to formula (1) is called **simple interest**.

EXAMPLE 3 Financial Planning

Candy has \$70,000 to invest and requires an overall rate of return of 9%. She can invest in a safe, government-insured certificate of deposit, but it only pays 8%. To obtain 9%, she agrees to invest some of her money in noninsured corporate bonds paying 12%. How much should be placed in each investment to achieve her goals?

Solution **STEP 1:** The question is asking for two dollar amounts: the principal to invest in the corporate bonds and the principal to invest in the certificate of deposit.

STEP 2: We let x represent the amount (in dollars) to be invested in the bonds. Then $70{,}000 - x$ is the amount that will be invested in the certificate. (Do you see why?)

STEP 3: We set up a table.

	Principal \$	Rate %	Time yr	Interest \$
Bonds	x	$12\% = 0.12$	1	$0.12x$
Certificate	$70{,}000 - x$	$8\% = 0.08$	1	$0.08(70{,}000 - x)$
Total	70,000	$9\% = 0.09$	1	$0.09(70{,}000) = 6300$

Since the total interest from the investments is equal to $0.09(70{,}000) = 6300$, we must have the equation

$$0.12x + 0.08(70{,}000 - x) = 6300$$

(Note that the units are consistent: the unit is dollars on each side.)

STEP 4:
$$\begin{aligned} 0.12x + 5600 - 0.08x &= 6300 \\ 0.04x &= 700 \\ x &= 17{,}500 \end{aligned}$$

Candy should place \$17,500 in the bonds and \$70,000 − \$17,500 = \$52,500 in the certificate.

STEP 5: The interest on the bonds after 1 year is 0.12(\$17,500) = \$2100; the interest on the certificate after 1 year is 0.08(\$52,500) = \$4200. The total annual interest is \$6300, the required amount. ■

NOW WORK PROBLEM 23.

Mixture Problems

4 Oil refineries sometimes produce gasoline that is a blend of two or more types of fuel; bakeries occasionally blend two or more types of flour for their bread. These problems are referred to as **mixture problems** because they combine two or more quantities to form a mixture.

EXAMPLE 4 Blending Coffees

The manager of a Starbucks store decides to experiment with a new blend of coffee. She will mix some B grade Columbian coffee that sells for \$5 per pound with some A grade Arabica coffee that sells for \$10 per pound to get 100 pounds of the new blend. The selling price of the new blend is to be \$7 per pound, and there is to be no difference in revenue from selling the new blend versus selling the other types. How many pounds of the B grade Columbian and A grade Arabica coffees are required?

Solution Let x represent the number of pounds of the B grade Columbian coffee. Then $100 - x$ equals the number of pounds of the A grade Arabica coffee. See Figure 44.

Figure 44

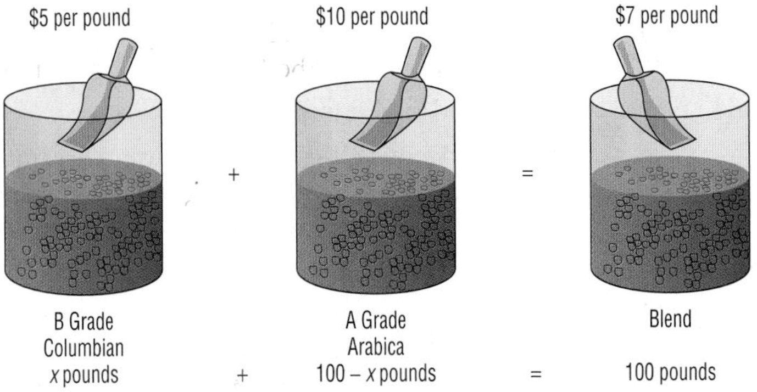

Since there is to be no difference in revenue between selling the A and B grades separately versus the blend, we have

$$\begin{pmatrix}\text{Price per pound}\\\text{of B grade}\end{pmatrix}\begin{pmatrix}\text{\# Pounds}\\\text{B grade}\end{pmatrix}+\begin{pmatrix}\text{Price per pound}\\\text{of A grade}\end{pmatrix}\begin{pmatrix}\text{\# Pounds}\\\text{A grade}\end{pmatrix}=\begin{pmatrix}\text{Price per pound}\\\text{of blend}\end{pmatrix}\begin{pmatrix}\text{\# Pounds}\\\text{blend}\end{pmatrix}$$

$$\$5 \cdot x + \$10 \cdot (100 - x) = (\$7)(100)$$

We have the equation

$$\begin{aligned} 5x + 10(100 - x) &= 700 \\ 5x + 1000 - 10x &= 700 \\ -5x &= -300 \\ x &= 60 \end{aligned}$$

The manager should blend 60 pounds of B grade Columbian coffee with $100 - 60 = 40$ pounds of A grade Arabica coffee to get the desired blend. ■

Check: The 60 pounds of B grade coffee would sell for $(\$5)(60) = \$300$, and the 40 pounds of A grade coffee would sell for $(\$10)(40) = \$400$; the total revenue, \$700, equals the revenue obtained from selling the blend, as desired. ■

NOW WORK PROBLEM 27.

Uniform Motion

5 Objects that move at a constant velocity are said to be in **uniform motion**. When the average velocity of an object is known, it can be interpreted as its constant velocity. For example, a bicyclist traveling at an average velocity of 25 miles per hour is in uniform motion.

Uniform Motion Formula

If an object moves at an average velocity v, the distance s covered in time t is given by the formula

$$s = vt \qquad (2)$$

That is, Distance = Velocity · Time.

EXAMPLE 5 **Physics: Uniform Motion**

Tanya, who is a long-distance runner, runs at an average velocity of 8 miles per hour (mph). Two hours after Tanya leaves your house, you leave in your Honda and follow the same route. If your average velocity is 40 mph, how long will it be before you catch up to Tanya? How far will each of you be from your home?

Solution Refer to Figure 45. We use t to represent the time (in hours) that it takes the Honda to catch up with Tanya. When this occurs, the total time elapsed for Tanya is $t + 2$ hours.

Figure 45

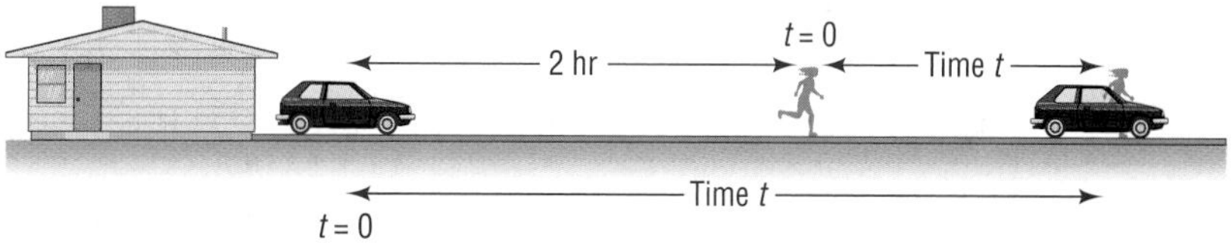

Set up the following table:

	Velocity mph	Time hr	Distance mi
Tanya	8	$t + 2$	$8(t + 2)$
Honda	40	t	$40t$

Since the distance traveled is the same, we are led to the following equation:

$$\begin{aligned} 8(t + 2) &= 40t \\ 8t + 16 &= 40t \\ 32t &= 16 \\ t &= \frac{1}{2} \text{ hour} \end{aligned}$$

It will take the Honda $\frac{1}{2}$ hour to catch up to Tanya. Each of you will have gone 20 miles. ■

Check: In 2.5 hours, Tanya travels a distance of $(2.5)(8) = 20$ miles. In $\frac{1}{2}$ hour, the Honda travels a distance of $\left(\frac{1}{2}\right)(40) = 20$ miles. ■

EXAMPLE 6 **Physics: Uniform Motion**

A motorboat heads upstream a distance of 24 miles on the Illinois River, whose current is running at 3 miles per hour. The trip up and back takes 6 hours. Assuming that the motorboat maintained a constant speed relative to the water, what was its speed relative to the water?

Solution See Figure 46. We use v to represent the constant speed of the motorboat relative to the water. Then the true speed going upstream is $v - 3$ miles per hour, and the true speed going downstream is $v + 3$ miles per hour. Since Distance = Velocity × Time, then Time = Distance/Velocity. We set up a table.

Figure 46

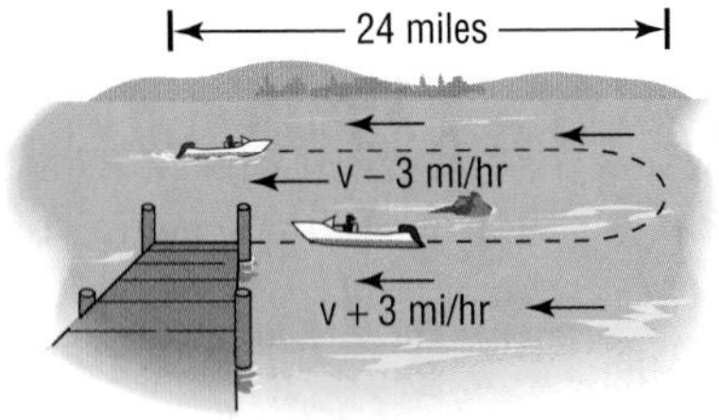

	Velocity mph	Distance mi	Time = Distance/Velocity hr
Upstream	$v - 3$	24	$\frac{24}{v-3}$
Downstream	$v + 3$	24	$\frac{24}{v+3}$

Since the total time up and back is 6 hours, we have

$$\frac{24}{v-3} + \frac{24}{v+3} = 6$$

$$\frac{24(v+3) + 24(v-3)}{(v-3)(v+3)} = 6 \quad \text{Add the expressions on the left side.}$$

$$\frac{48v}{v^2 - 9} = 6 \quad \text{Simplify}$$

$$48v = 6(v^2 - 9) \quad \text{Clear fractions.}$$

$$8v = v^2 - 9 \quad \text{Divide each side by 6.}$$

$$v^2 - 8v - 9 = 0 \quad \text{Place the quadratic equation in standard form.}$$

$$(v - 9)(v + 1) = 0 \quad \text{Factor}$$

$$v = 9 \quad \text{or} \quad v = -1$$

We discard the solution $v = -1$ mile per hour, so the speed of the motorboat relative to the water is 9 miles per hour. ■

NOW WORK PROBLEM 33.

Constant Rate Jobs

6 This section involves jobs that are performed at a **constant rate**. Our assumption is that, if a job can be done in t units of time, $1/t$ of the job is done in 1 unit of time. Let's look at an example.

EXAMPLE 7 Working Together to Do a Job

At 10 AM Danny is asked by his father to weed the garden. From past experience, Danny knows that this will take him 4 hours, working alone. His older brother, Mike, when it is his turn to do this job, requires 6 hours. Since Mike wants to go golfing with Danny and has a reservation for 1 PM, he agrees to help Danny. Assuming no gain or loss of efficiency, when will they finish if they work together? Can they make the golf date?

Solution We set up Table 7. In 1 hour, Danny does $\frac{1}{4}$ of the job, and in 1 hour, Mike does $\frac{1}{6}$ of the job. Let t be the time (in hours) that it takes them to do the job together. In 1 hour, then, $1/t$ of the job is completed. We reason as follows:

$$\left(\begin{array}{c}\text{Part done by Danny}\\ \text{in 1 hour}\end{array}\right) + \left(\begin{array}{c}\text{Part done by Mike}\\ \text{in 1 hour}\end{array}\right) = \left(\begin{array}{c}\text{Part done together}\\ \text{in 1 hour}\end{array}\right)$$

TABLE 7

	Hours to Do Job	Part of Job Done in 1 Hour
Danny	4	$\frac{1}{4}$
Mike	6	$\frac{1}{6}$
Together	t	$\frac{1}{t}$

From Table 7,

$$\frac{1}{4} + \frac{1}{6} = \frac{1}{t}$$

$$\frac{3+2}{12} = \frac{1}{t}$$

$$\frac{5}{12} = \frac{1}{t}$$

$$5t = 12$$

$$t = \frac{12}{5}$$

Working together, the job can be done in $\frac{12}{5}$ hours, or 2 hours, 24 minutes. They should make the golf date, since they will finish at 12:24 PM. ■

NOW WORK PROBLEM 37.

OTHER APPLIED PROBLEMS

The next two examples illustrate problems that you will probably see again in a slightly different form if you study calculus.

EXAMPLE 8 Preview of a Calculus Problem

From each corner of a square piece of sheet metal, remove a square of side 9 centimeters. Turn up the edges to form an open box. If the box is to hold 144 cubic centimeters, what should be the dimensions of the piece of sheet metal?

Solution We use Figure 47 as a guide. We have labeled by x the length of a side of the square piece of sheet metal. The box will be of height 9 centimeters, and its square base will have $x - 18$ as the length of a side. The volume (Length $\times$ Width $\times$ Height) of the box is therefore

$$9(x-18)(x-18) = 9(x-18)^2$$

Figure 47

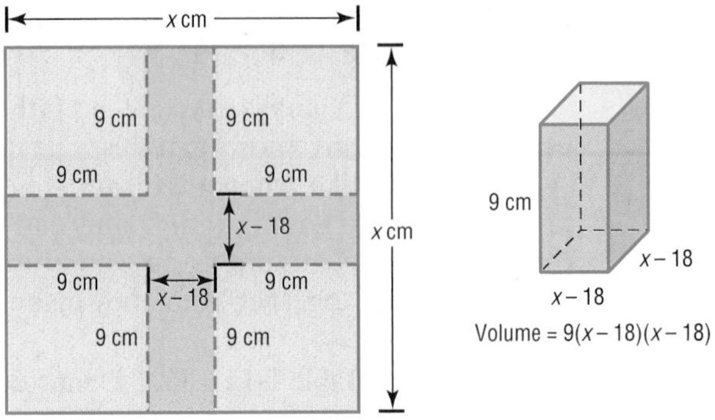

Since the volume of the box is to be 144 cubic centimeters, we have

$$9(x - 18)^2 = 144$$
$$(x - 18)^2 = 16$$
$$x - 18 = \pm 4$$
$$x = 18 \pm 4$$
$$x = 22 \quad \text{or} \quad x = 14$$

We discard the solution $x = 14$ (do you see why?) and conclude that the sheet metal should be 22 centimeters by 22 centimeters. ■

Check: If we begin with a piece of sheet metal 22 centimeters by 22 centimeters, cut out a 9 centimeter square from each corner, and fold up the edges, we get a box whose dimensions are 9 by 4 by 4. The volume of the box is $9 \times 4 \times 4 = 144$ cubic centimeters, as required. ■

NOW WORK PROBLEM 43.

EXAMPLE 9 Preview of a Calculus Problem

A piece of wire 8 feet in length is to be cut into two pieces. Each piece will then be bent into a square. Where should the cut in the wire be made if the sum of the areas of these squares is to be 2 square feet?

Solution We use Figure 48 as a guide. We label the length of one of the pieces of wire after it has been cut, x. The remaining piece will be of length $8 - x$. If each length is bent into a square, then one of the squares has a side of length $x/4$ and the other a side of length $(8 - x)/4$. Since the sum of the areas of these two squares is 2, we have the equation

Figure 48

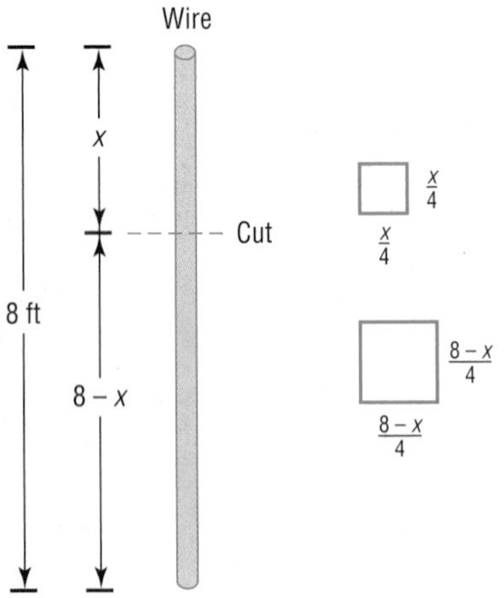

$$\left(\frac{x}{4}\right)^2 + \left(\frac{8 - x}{4}\right)^2 = 2$$
$$\frac{x^2}{16} + \frac{64 - 16x + x^2}{16} = 2$$
$$2x^2 - 16x + 64 = 32$$
$$2x^2 - 16x + 32 = 0 \quad \text{Put in standard form.}$$
$$x^2 - 8x + 16 = 0 \quad \text{Divide by 2.}$$
$$(x - 4)^2 = 0 \quad \text{Factor}$$
$$x = 4$$

Since $x = 4, 8 - x = 4$, and the original piece of wire should be cut into two pieces, each of length 4 feet. ■

Check: If the length of each piece of wire is 4 feet, then each piece can be formed into a square whose side is 1 foot. The area of each square is then 1 square foot, so the sum of the areas is 2 square feet, as required. ■

1.4 EXERCISES

In Problems 1–10, translate each sentence into a mathematical equation. Be sure to identify the meaning of all symbols.

1. **Geometry** The area of a circle is the product of the number π times the square of the radius.
2. **Geometry** The circumference of a circle is the product of the number π times twice the radius.
3. **Geometry** The area of a square is the square of the length of a side.
4. **Geometry** The perimeter of a square is four times the length of a side.
5. **Physics** Force equals the product of mass times acceleration.
6. **Physics** Pressure is force per unit area.
7. **Physics** Work equals force times distance.
8. **Physics** Kinetic energy is one-half the product of the mass times the square of the velocity.
9. **Business** The total variable cost of manufacturing x dishwashers is \$150 per dishwasher times the number of dishwashers manufactured.
10. **Business** The total revenue derived from selling x dishwashers is \$250 per dishwasher times the number of dishwashers manufactured.
11. **Finance** A total of \$20,000 is to be invested, some in bonds and some in certificates of deposit (CDs). If the amount invested in bonds is to exceed that in CDs by \$2000, how much will be invested in each type of instrument?
12. **Finance** A total of \$10,000 is to be divided up between Yani and Diane, with Diane to receive \$2000 less than Yani. How much will each receive?
13. **Finance** An inheritance of \$900,000 is to be divided among David, Paige, and Dan in the following manner: Paige is to receive $\frac{3}{4}$ of what David gets, while Dan gets $\frac{1}{2}$ of what David gets. How much does each receive?
14. **Sharing the Cost of a Pizza** Carole and Canter agree to share the cost of an \$18 pizza based on how much each ate. If Carole ate $\frac{2}{3}$ the amount that Canter ate, how much should each pay?

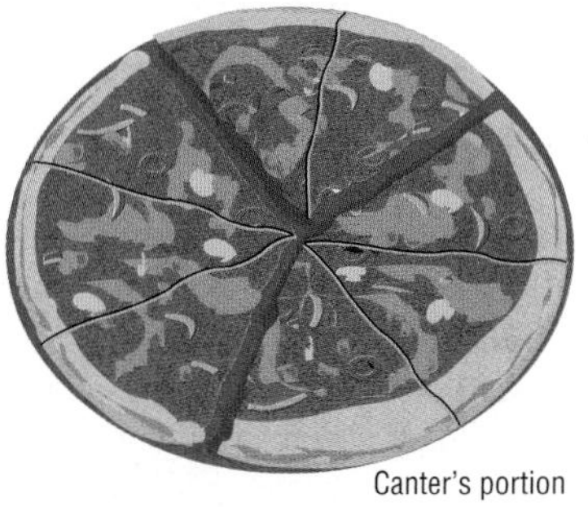

15. **Geometry** The perimeter of a rectangle is 60 feet. Find its length and width if the length is 8 feet longer than the width.
16. **Geometry** The perimeter of a rectangle is 42 meters. Find its length and width if the length is twice the width.
17. **Computing Grades** Going into the final exam, which will count as two tests, Brooke has test scores of 80, 83, 71, 61, and 95. What score does Brooke need on the final in order to have an average score of 80?
18. **Computing Grades** Going into the final exam, which will count as two-thirds of the final grade, Mike has test scores of 86, 80, 84, and 90. What score does Mike need on the final in order to earn a B, which requires an average score of 80? What does he need to earn an A, which requires an average of 90?
19. **Business: Discount Pricing** A builder of tract homes reduced the price of a model by 15%. If the new price is \$125,000, what was its original price? How much can be saved by purchasing the model?
20. **Business: Discount Pricing** A car dealer, at a year-end clearance, reduces the list price of last year's models by 15%. If a certain four-door model has a discounted price of \$8000, what was its list price? How much can be saved by purchasing last year's model?
21. **Business: Marking up the Price of Books** A college book store marks up the price that it pays the publisher for a book by 25%. If the selling price of a book is \$56.00, how much did the book store pay for this book?
22. **Personal Finance: Cost of a Car** The suggested list price of a new car is \$12,000. The dealer's cost is 85% of list. How much will you pay if the dealer is willing to accept \$100 over cost for the car?
23. **Financial Planning** Betsy, a recent retiree, requires \$6000 per year in extra income. She has \$50,000 to invest and can invest in B-rated bonds paying 15% per year or in a certificate of deposit (CD) paying 7% per year. How much money should be invested in each to realize exactly \$6000 in interest per year?
24. **Financial Planning** After 2 years, Betsy (see Problem 23) finds that she will now require \$7000 per year. Assuming that the remaining information is the same, how should the money be reinvested?
25. **Banking** A bank loaned out \$12,000, part of it at the rate of 8% per year and the rest at the rate of 18% per year. If the interest received totaled \$1000, how much was loaned at 8%?

26. **Banking** Wendy, a loan officer at a bank, has $1,000,000 to lend and is required to obtain an average return of 18% per year. If she can lend at the rate of 19% or at the rate of 16%, how much can she lend at the 16% rate and still meet her requirement?

27. **Blending Teas** The manager of a store that specializes in selling tea decides to experiment with a new blend. She will mix some Earl Gray tea that sells for $5 per pound with some Orange Pekoe tea that sells for $3 per pound to get 100 pounds of the new blend. The selling price of the new blend is to be $4.50 per pound, and there is to be no difference in revenue from selling the new blend versus selling the other types. How many pounds of the Earl Gray tea and Orange Pekoe tea are required?

28. **Business: Blending Coffee** A coffee manufacturer wants to market a new blend of coffee that will cost $3.90 per pound by mixing two coffees that sell for $2.75 and $5 per pound, respectively. What amounts of each coffee should be blended to obtain the desired mixture?

[**Hint:** Assume that the total weight of the desired blend is 100 pounds.]

29. **Business: Mixing Nuts** A nut store normally sells cashews for $4.00 per pound and peanuts for $1.50 per pound. But at the end of the month the peanuts had not sold well, so, in order to sell 60 pounds of peanuts, the manager decided to mix the 60 pounds of peanuts with some cashews and sell the mixture for $2.50 per pound. How many pounds of cashews should be mixed with the peanuts to ensure no change in the profit?

30. **Business: Mixing Candy** A candy store sells boxes of candy containing caramels and cremes. Each box sells for $12.50 and holds 30 pieces of candy (all pieces are the same size). If the caramels cost $0.25 to produce and the cremes cost $0.45 to produce, how many of each should be in a box to make a profit of $3?

31. **Chemistry: Mixing Acids** How many ounces of pure water should be added to 20 ounces of a 40% solution of muriatic acid to obtain a 30% solution of muriatic acid?

32. **Chemistry: Mixing Acids** How many cubic centimeters of pure hydrochloric acid should be added to 20 cc of a 30% solution of hydrochloric acid to obtain a 50% solution?

33. **Physics: Uniform Motion** A Metra commuter train leaves Union Station in Chicago at 12 noon. Two hours later, an Amtrak train leaves on the same track, traveling at an average speed that is 50 miles per hour faster than the Metra train. At 3 PM, the Amtrak train is 10 miles behind the commuter train. How fast is each going?

34. **Physics: Uniform Motion** Two cars enter the Florida Turnpike at Commercial Boulevard at 8:00 AM, each heading for Wildwood. One car's average speed is 10 miles per hour more than the other's. The faster car arrives at Wildwood at 11:00 AM, $\frac{1}{2}$ hour before the other car. What is the average speed of each car? How far did each travel?

35. **Physics: Uniform Motion** A motorboat can maintain a constant speed of 16 miles per hour relative to the water. The boat makes a trip upstream to a certain point in 20 minutes; the return trip takes 15 minutes. What is the speed of the current? (See the figure.)

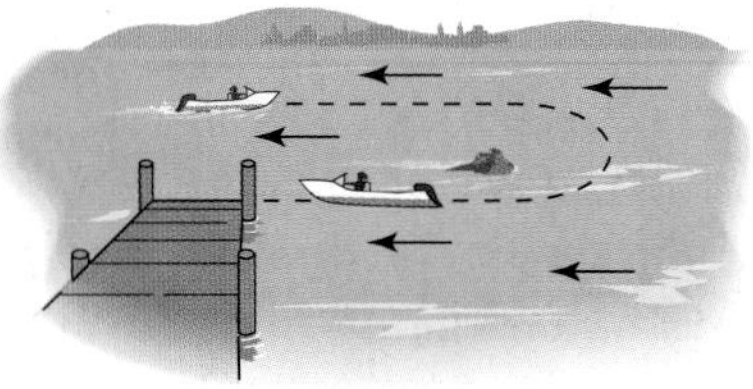

36. **Physics: Uniform Motion** A motorboat heads upstream on a river that has a current of 3 miles per hour. The trip upstream takes 5 hours, while the return trip takes 2.5 hours. What is the speed of the motorboat? (Assume that the motorboat maintains a constant speed relative to the water.)

37. **Working Together on a Job** Trent can deliver his newspapers in 30 minutes. It takes Lois 20 minutes to do the same route. How long would it take them to deliver the newspapers if they work together?

38. **Working Together on a Job** Patrick, by himself, can paint four rooms in 10 hours. If he hires April to help, they can do the same job together in 6 hours. If he lets April work alone, how long will it take her to paint four rooms?

39. **Dimensions of a Window** The area of the opening of a rectangular window is to be 143 square feet. If the length is to be 2 feet more than the width, what are the dimensions?

40. **Dimensions of a Window** The area of a rectangular window is to be 306 square centimeters. If the length exceeds the width by 1 centimeter, what are the dimensions?

41. **Geometry** Find the dimensions of a rectangle whose perimeter is 26 meters and whose area is 40 square meters.

42. **Watering a Field** An adjustable water sprinkler that sprays water in a circular pattern is placed at the center of a square field whose area is 1250 square feet (see the figure). What is the shortest radius setting that can be used if the field is to be completely enclosed within the circle?

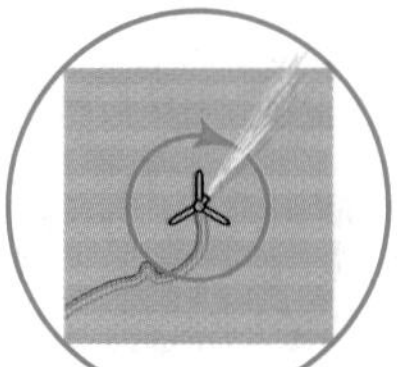

43. **Constructing a Box** An open box is to be constructed from a square piece of sheet metal by removing a square of side 1 foot from each corner and turning up the edges. If the box is to hold 4 cubic feet, what should be the dimensions of the sheet metal?

44. **Constructing a Box** Rework Problem 43 if the piece of sheet metal is a rectangle whose length is twice its width.

45. **Physics: Uniform Motion** A motorboat maintained a constant speed of 15 miles per hour relative to the water in going 10 miles upstream and then returning. The total

time for the trip was 1.5 hours. Use this information to find the speed of the current.

46. Dimensions of a Patio A contractor orders 8 cubic yards of premixed cement, all of which is to be used to pour a rectangular patio that will be 4 inches thick. If the length of the patio is specified to be twice the width, what will be the patio dimensions? (1 cubic yard = 27 cubic feet)

47. Enclosing a Garden A gardener has 46 feet of fencing to be used to enclose a rectangular garden that has a border 2 feet wide surrounding it (see the figure).
(a) If the length of the garden is to be twice its width, what will be the dimensions of the garden?
(b) What is the area of the garden?
(c) If the length and width of the garden were to be the same, what would be the dimensions of the garden?
(d) What would be the area of the square garden?

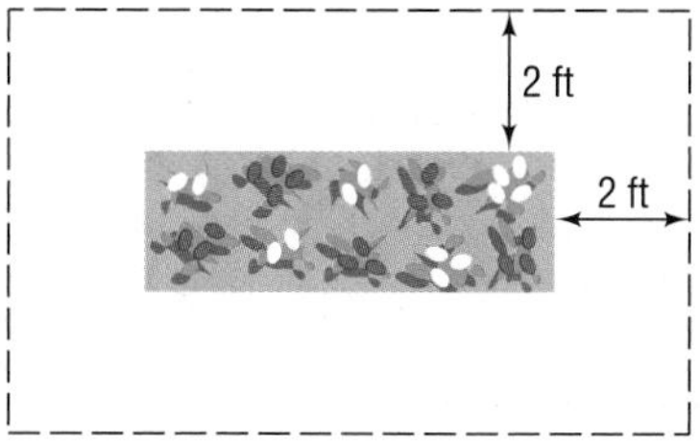

48. Construction A pond is enclosed by a wooden deck that is 3 feet wide. The fence surrounding the deck is 100 feet long.
(a) If the pond is square, what are its dimensions?
(b) If the pond is rectangular and the length of the pond is three times its width, what are the dimensions of the pond?
(c) If the pond is circular, what is the diameter of the pond?
(d) Which pond has the most area?

49. Constructing a Border around a Pool A pool in the shape of a circle measures 10 feet across. One cubic yard of concrete is to be used to create a circular border of uniform width around the pool. If the border is to have a depth of 3 inches, how wide will the border be? (1 cubic yard = 27 cubic feet)

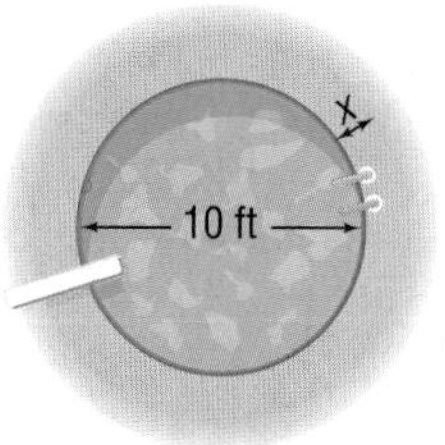

50. Constructing a Border around a Pool Rework Problem 49 if the depth of the border is 4 inches.

51. Constructing a Border around a Garden A landscaper, who just completed a rectangular flower garden measuring 6 feet by 10 feet, orders 1 cubic yard of premixed cement, all of which is to be used to create a border of uniform width around the garden. If the border is to have a depth of 3 inches, how wide will the border be? (1 cubic yard = 27 cubic feet)

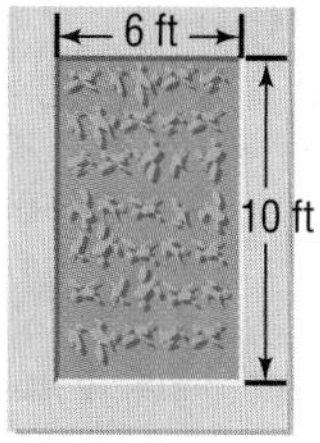

52. Constructing a Coffee Can A 39 ounce can of Hills Bros.® coffee requires 188.5 square inches of aluminum. If its height is 7 inches, what is its radius? (The surface area A of a right circular cylinder is $A = 2\pi r^2 + 2\pi rh$, where r is the radius and h is the height.)

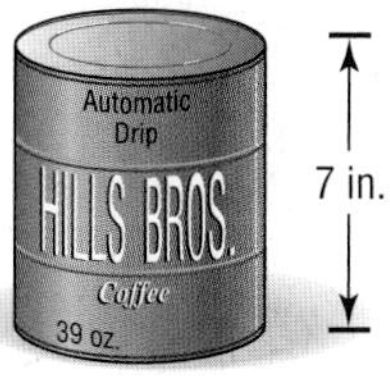

53. Mixing Water and Antifreeze How much water should be added to 1 gallon of pure antifreeze to obtain a solution that is 60% antifreeze?

54. Mixing Water and Antifreeze The cooling system of a certain foreign-made car has a capacity of 15 liters. If the system is filled with a mixture that is 40% antifreeze, how much of this mixture should be drained and replaced by pure antifreeze so that the system is filled with a solution that is 60% antifreeze?

55. Cement Mix A 20 pound bag of Economy brand cement mix contains 25% cement and 75% sand. How much pure cement must be added to produce a cement mix that is 40% cement?

56. Chemistry: Salt Solutions How much water must be evaporated from 240 gallons of a 3% salt solution to produce a 5% salt solution?

57. Reducing the Size of a Candy Bar A jumbo chocolate bar with a rectangular shape measures 12 centimeters in length, 7 centimeters in width, and 3 centimeters in thickness. Due to escalating costs of cocoa, management decides to reduce the volume of the bar by 10%. To accomplish this reduction, management decides that the new bar should have the same 3 centimeter thickness, but the length and width each should be reduced an equal number of centimeters. What should be the dimensions of the new candy bar?

58. Reducing the Size of a Candy Bar Rework Problem 57 if the reduction is to be 20%.

59. Purity of Gold The purity of gold is measured in karats, with pure gold being 24 karats. Other purities of gold are expressed as proportional parts of pure gold. Thus, 18 karat gold is $\frac{18}{24}$, or 75% pure gold; 12 karat gold is $\frac{12}{24}$, or 50% pure gold; and so on. How much 12 karat gold should be mixed with pure gold to obtain 60 grams of 16 karat gold?

60. **Chemistry: Sugar Molecules** A sugar molecule has twice as many atoms of hydrogen as it does oxygen and one more atom of carbon than oxygen. If a sugar molecule has a total of 45 atoms, how many are oxygen? How many are hydrogen?

61. **Running a Race** Mike can run the mile in 6 minutes, and Dan can run the mile in 9 minutes. If Mike gives Dan a head start of 1 minute, how far from the start will Mike pass Dan? (See the figure.) How long does it take?

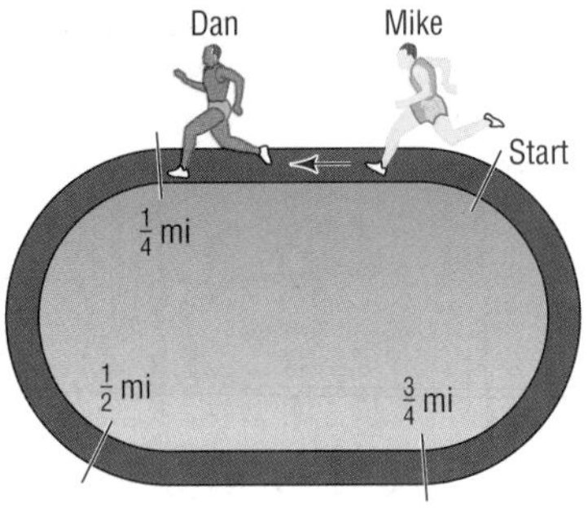

62. **Football** A tight end can run the 100 yard dash in 12 seconds. A defensive back can do it in 10 seconds. The tight end catches a pass at his own 20 yard line with the defensive back at the 15 yard line. (See the figure.) If no other players are nearby, at what yard line will the defensive back catch up to the tight end?

[**Hint:** At time $t = 0$, the defensive back is 5 yards behind the tight end.]

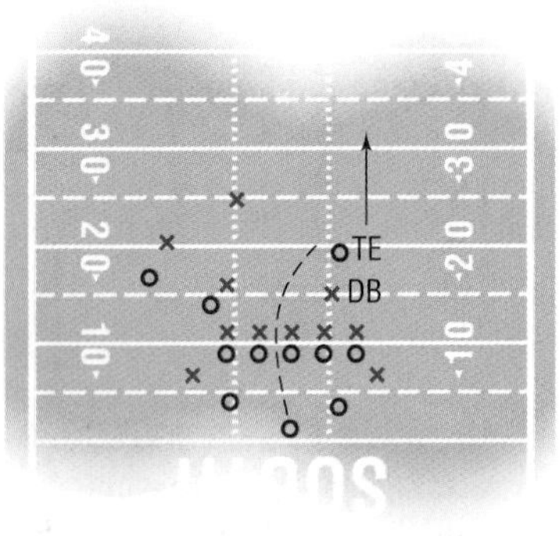

63. **Emptying Oil Tankers** An oil tanker can be emptied by the main pump in 4 hours. An auxiliary pump can empty the tanker in 9 hours. If the main pump is started at 9 AM, when should the auxiliary pump be started so that the tanker is emptied by noon?

64. **Using Two Pumps** A 5 horsepower (hp) pump can empty a pool in 5 hours. A smaller, 2 hp pump empties the same pool in 8 hours. The pumps are used together to begin emptying this pool. After two hours, the 2 hp pump breaks down. How long will it take the larger pump to empty the pool?

65. **Emptying a Tub** A bathroom tub will fill in 15 minutes with both faucets open and the stopper in place. With both faucets closed and the stopper removed, the tub will empty in 20 minutes. How long will it take for the tub to fill if both faucets are open and the stopper is removed?

66. **Range of an Airplane** An air rescue plane averages 300 miles per hour in still air. It carries enough fuel for 5 hours of flying time. If, upon takeoff, it encounters a wind of 30 miles per hour and the direction of the airplane is with the wind in one direction and against it in the other, how far can it fly and return safely? (Assume that the wind remains constant.)

67. **Home Equity Loans** Suppose that you obtain a home equity loan of $100,000 that requires only a monthly interest payment at 10% per annum, with the principal due after 5 years. You decide to invest part of the loan in a 5 year CD that pays 9% per annum compounded and paid monthly and part in a B+ rated bond due in 5 years that pays 12% per annum compounded and paid monthly. What is the most that you can invest in the CD to ensure that the monthly home equity loan payment is made?

68. **Comparing Olympic Heroes** In the 1984 Olympics, Carl Lewis of the United States won the gold medal in the 100 meter race with a time of 9.99 seconds. In the 1896 Olympics, Thomas Burke, also of the United States, won the gold medal in the 100 meter race in 12.0 seconds. If they ran in the same race repeating their respective times, by how many meters would Lewis beat Burke?

69. **Computing Average Speed** In going from Chicago to Atlanta, a car averages 45 miles per hour, and in going from Atlanta to Miami, it averages 55 miles per hour. If Atlanta is halfway between Chicago and Miami, what is the average speed from Chicago to Miami? Discuss an intuitive solution. Write a paragraph defending your intuitive solution. Then solve the problem algebraically. Is your intuitive solution the same as the algebraic one? If not, find the flaw.

70. **Speed of a Plane** On a recent flight from Phoenix to Kansas City, a distance of 919 nautical miles, the plane arrived 20 minutes early. On leaving the aircraft, I asked the captain, "What was our tail wind?" He replied, "I don't know, but our ground speed was 550 knots." How can you determine if enough information is provided to find the tail wind? If possible, find the tail wind. (1 knot = 1 nautical mile per hour)

71. **Critical Thinking** You are the manager of a clothing store and have just purchased 100 dress shirts for $20.00 each. After 1 month of selling the shirts at the regular price, you plan to have a sale giving 40% off the original selling price. However, you still want to make a profit of $4 on each shirt at the sale price. What should you price the shirts at initially to ensure this? If, instead of 40% off at the sale, you give 50% off, by how much is your profit reduced?

72. **Critical Thinking** Make up a word problem that requires solving a linear equation as part of its solution. Exchange problems with a friend. Write a critique of your friend's problem.

73. **Critical Thinking** Without solving, explain what is wrong with the following mixture problem: How many liters of 25% ethanol should be added to 20 liters of 48% ethanol to obtain a solution of 58% ethanol? Now go through an algebraic solution. What happens?

1.5 SOLVING INEQUALITIES

OBJECTIVES
1 Use Interval Notation
2 Use Properties of Inequalities
3 Solve Linear Inequalities Algebraically and Graphically
4 Solve Combined Inequalities Algebraically and Graphically
5 Solve Absolute Value Inequalities Algebraically and Graphically

Suppose that a and b are two real numbers and $a < b$. We shall use the notation $a < x < b$ to mean that x is a number *between* a and b. The expression $a < x < b$ is equivalent to the two inequalities $a < x$ and $x < b$. Similarly, the expression $a \le x \le b$ is equivalent to the two inequalities $a \le x$ and $x \le b$. The remaining two possibilities, $a \le x < b$ and $a < x \le b$, are defined similarly.

Although it is acceptable to write $3 \ge x \ge 2$, it is preferable to reverse the inequality symbols and write instead $2 \le x \le 3$ so that, as you read from left to right, the values go from smaller to larger.

A statement such as $2 \le x \le 1$ is false because there is no number x for which $2 \le x$ and $x \le 1$. Finally, we never mix inequality symbols, as in $2 \le x \ge 3$.

INTERVALS

1 Let a and b represent two real numbers with $a < b$:

A **closed interval**, denoted by **[*a*, *b*]**, consists of all real numbers x for which $a \le x \le b$.

An **open interval**, denoted by **(*a*, *b*)**, consists of all real numbers x for which $a < x < b$.

The **half-open**, or **half-closed, intervals** are **(*a*, *b*]**, consisting of all real numbers x for which $a < x \le b$, and **[*a*, *b*)**, consisting of all real numbers x for which $a \le x < b$.

In each of these definitions, a is called the **left endpoint** and b the **right endpoint** of the interval.

The symbol ∞ (read as "infinity") is not a real number, but a notational device used to indicate unboundedness in the positive direction. The symbol $-\infty$ (read as "negative infinity") also is not a real number, but a notational device used to indicate unboundedness in the negative direction. Using the symbols ∞ and $-\infty$, we can define five other kinds of intervals:

$[a, \infty)$	Consists of all real numbers x for which $x \ge a$ $(a \le x < \infty)$
(a, ∞)	Consists of all real numbers x for which $x > a$ $(a < x < \infty)$
$(-\infty, a]$	Consists of all real numbers x for which $x \le a$ $(-\infty < x \le a)$
$(-\infty, a)$	Consists of all real numbers x for which $x < a$ $(-\infty < x < a)$
$(-\infty, \infty)$	Consists of all real numbers x $(-\infty < x < \infty)$

Note that ∞ (and $-\infty$) is never included as an endpoint, since it is not a real number.

Table 8 summarizes interval notation, corresponding inequality notation, and their graphs.

TABLE 8		
Interval	**Inequality**	**Graph**
The open interval (a, b)	$a < x < b$	
The closed interval $[a, b]$	$a \le x \le b$	
The half-open interval $[a, b)$	$a \le x < b$	
The half-open interval $(a, b]$	$a < x \le b$	
The interval $[a, \infty)$	$x \ge a$	
The interval (a, ∞)	$x > a$	
The interval $(-\infty, a]$	$x \le a$	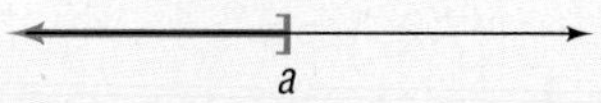
The interval $(-\infty, a)$	$x < a$	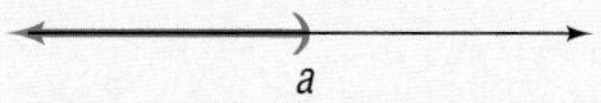
The interval $(-\infty, \infty)$	All real numbers	

EXAMPLE 1 Writing Inequalities Using Interval Notation

Write each inequality using interval notation.

(a) $1 \le x \le 3$ (b) $-4 < x < 0$ (c) $x > 5$ (d) $x \le 1$

Solution (a) $1 \le x \le 3$ describes all numbers x between 1 and 3, inclusive. In interval notation, we write $[1, 3]$.

(b) In interval notation, $-4 < x < 0$ is written $(-4, 0)$.

(c) $x > 5$ consists of all numbers x greater than 5. In interval notation, we write $(5, \infty)$.

(d) In interval notation, $x \le 1$ is written $(-\infty, 1]$. ■

EXAMPLE 2 Writing Intervals Using Inequality Notation

Write each interval as an inequality involving x.

(a) $[1, 4)$ (b) $(2, \infty)$ (c) $[2, 3]$ (d) $(-\infty, -3]$

Solution (a) $[1, 4)$ consists of all numbers x for which $1 \le x < 4$.

(b) $(2, \infty)$ consists of all numbers x for which $x > 2$ $(2 < x < \infty)$.

(c) $[2, 3]$ consists of all numbers x for which $2 \le x \le 3$.

(d) $(-\infty, -3]$ consists of all numbers x for which $x \le -3$ $(-\infty < x \le -3)$. ■

NOW WORK PROBLEMS 1 AND 9.

PROPERTIES OF INEQUALITIES

2 The product of two positive real numbers is positive, the product of two negative real numbers is positive, and the product of 0 and 0 is 0. Thus, for any

real number a, the value of a^2 is 0 or positive; that is, a^2 is nonnegative. This is called the **nonnegative property**.

For any real number a, we have the following:

Nonnegative Property

$$a^2 \geq 0 \tag{1}$$

If we add the same number to both sides of an inequality, we obtain an equivalent inequality. For example, since $3 < 5$, then $3 + 4 < 5 + 4$ or $7 < 9$. This is called the **addition property** of inequalities.

Addition Property of Inequalities

$$\text{If } a < b, \text{then } a + c < b + c \tag{2a}$$

$$\text{If } a > b, \text{then } a + c > b + c \tag{2b}$$

The addition property states that the sense, or direction, of an inequality remains unchanged if the same number is added to each side.

EXAMPLE 3 Addition Property of Inequalities

(a) If $x < -5$, then $x + 5 < -5 + 5$ or $x + 5 < 0$.

(b) If $x > 2$, then $x + (-2) > 2 + (-2)$ or $x - 2 > 0$. ■

NOW WORK PROBLEM 17.

We will use two examples to arrive at our next property.

EXAMPLE 4 Multiplying an Inequality by a Positive Number

Express as an inequality the result of multiplying each side of the inequality $3 < 7$ by 2.

Solution We begin with

$$3 < 7$$

Multiplying each side by 2 yields the numbers 6 and 14, so we have

$$6 < 14$$

■

EXAMPLE 5 Multiplying an Inequality by a Negative Number

Express as an inequality the result of multiplying each side of the inequality $9 > 2$ by -4.

Solution We begin with

$$9 > 2$$

Multiplying each side by -4 yields the numbers -36 and -8, so we have

$$-36 < -8$$

■

Note that the effect of multiplying both sides of $9 > 2$ by the negative number -4 is that the direction of the inequality symbol is reversed.

Examples 4 and 5 illustrate the following general **multiplication properties** for inequalities:

Multiplication Properties for Inequalities

If $a < b$ and if $c > 0$, then $ac < bc$. If $a < b$ and if $c < 0$, then $ac > bc$.	**(3a)**
If $a > b$ and if $c > 0$, then $ac > bc$. If $a > b$ and if $c < 0$, then $ac < bc$.	**(3b)**

The multiplication properties state that the sense, or direction, of an inequality *remains the same* if each side is multiplied by a *positive* real number, while the direction is *reversed* if each side is multiplied by a *negative* real number.

EXAMPLE 6 Multiplication Property of Inequalities

(a) If $2x < 6$, then $\frac{1}{2}(2x) < \frac{1}{2}(6)$ or $x < 3$.

(b) If $\frac{x}{-3} > 12$, then $-3\left(\frac{x}{-3}\right) < -3(12)$ or $x < -36$.

(c) If $-4x > -8$, then $\frac{-4x}{-4} < \frac{-8}{-4}$ or $x < 2$.

(d) If $-x < 8$, then $(-1)(-x) > (-1)(8)$ or $x > -8$. ■

NOW WORK PROBLEM 23.

SOLVING INEQUALITIES

3 An **inequality in one variable** is a statement involving two expressions, at least one containing the variable, separated by one of the inequality symbols $<, \leq, >$, or $\geq$. To **solve an inequality** means to find all values of the variable for which the statement is true. These values are called **solutions** of the inequality.

For example, the following are all inequalities involving one variable, x:

$$x + 5 < 8, \qquad 2x - 3 \geq 4, \qquad x^2 - 1 \leq 3, \qquad \frac{x+1}{x-2} > 0$$

Two inequalities having exactly the same solution set are called **equivalent inequalities**. As with equations, one method for solving an inequality is to replace it by a series of equivalent inequalities until an inequality with an obvious solution, such as $x < 3$, is obtained. We obtain equivalent inequalities by applying some of the same properties as those used to find equivalent equations. The addition property and the multiplication properties form the basis for the following procedures.

Procedures That Leave the Inequality Symbol Unchanged

1. Simplify both sides of the inequality by combining like terms and eliminating parentheses:

   Replace $(x + 2) + 6 > 2x + 5(x + 1)$

   by $x + 8 > 7x + 5$

2. Add or subtract the same expression on both sides of the inequality:

   Replace $3x - 5 < 4$

   by $(3x - 5) + 5 < 4 + 5$

3. Multiply or divide both sides of the inequality by the same *positive* expression:

   Replace $4x > 16$ by $\dfrac{4x}{4} > \dfrac{16}{4}$

Procedures That Reverse the Sense or Direction of the Inequality Symbol

1. Interchange the two sides of the inequality:

   Replace $3 < x$ by $x > 3$

2. Multiply or divide both sides of the inequality by the same *negative* expression:

   Replace $-2x > 6$ by $\dfrac{-2x}{-2} < \dfrac{6}{-2}$

To solve an inequality using a graphing utility, we follow these steps:

Steps for Solving Inequalities Graphically

Step 1: Write the inequality in one of the following forms:

$$Y_1 < Y_2, \quad Y_1 > Y_2, \quad Y_1 \leq Y_2, \quad Y_1 \geq Y_2$$

Step 2: Graph Y_1 and Y_2 on the same screen.

Step 3: If the inequality is of the form $Y_1 < Y_2$ (strict), determine the interval of x-coordinates for which Y_1 is below Y_2.
If the inequality is of the form $Y_1 > Y_2$ (strict), determine the interval of x-coordinates for which Y_1 is above Y_2.
If the inequality is not strict ($\leq$ or $\geq$), include the x-coordinates of the points of intersection in the solution.

EXAMPLE 7 Solving an Inequality

Solve the inequality $4x + 7 \geq 2x - 3$ and graph the solution set.

Graphing Solution

We graph $Y_1 = 4x + 7$ and $Y_2 = 2x - 3$ on the same screen. See Figure 49. Using the INTERSECT command, we find that Y_1 and Y_2 intersect at $x = -5$. The graph of Y_1 is above that of Y_2, $Y_1 > Y_2$, to the right of the point of intersection. Since the inequality is not strict, the solution set is $\{x \mid x \geq -5\}$ or, using interval notation, $[-5, \infty)$.

Figure 49

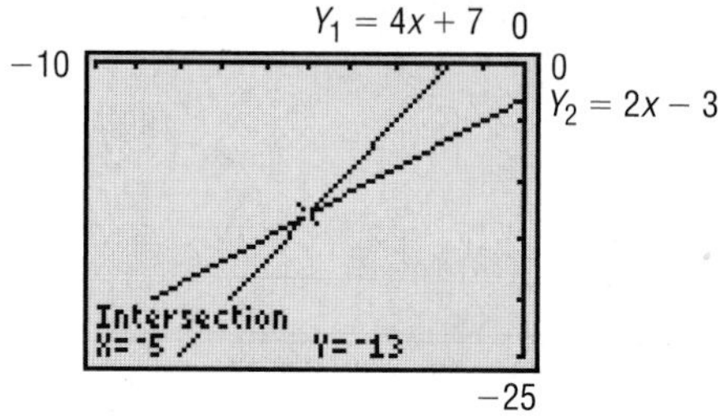

Algebraic Solution

$$4x + 7 \geq 2x - 3$$

$$4x + 7 - 7 \geq 2x - 3 - 7 \quad \text{Subtract 7 from both sides.}$$

$$4x \geq 2x - 10 \quad \text{Simplify}$$

$$4x - 2x \geq 2x - 10 - 2x \quad \text{Subtract } 2x \text{ from both sides.}$$

$$2x \geq -10 \quad \text{Simplify}$$

$$\frac{2x}{2} \geq \frac{-10}{2} \quad \text{Divide both sides by 2. (The direction of the inequality symbol is unchanged.)}$$

$$x \geq -5 \quad \text{Simplify}$$

Figure 50
$x \geq -5$ or $[-5, \infty)$

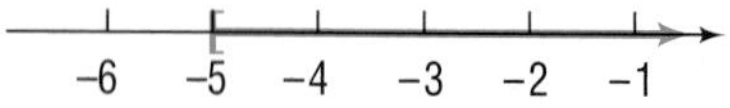

The solution set is $\{x \mid x \geq -5\}$ or, using interval notation, all numbers in the interval $[-5, \infty)$.

See Figure 50 for the graph. ■

NOW WORK PROBLEM 31.

EXAMPLE 8 Solving Combined Inequalities

4 Solve the inequality $-5 < 3x - 2 < 1$ and draw a graph to illustrate the solution.

Graphing Solution

To solve a combined inequality, we graph each part: $Y_1 = -5$, $Y_2 = 3x - 2$, and $Y_3 = 1$. We seek the values of x for which the graph of Y_2 is between the graphs of Y_1 and Y_3. See Figure 51. The point of intersection of Y_1 and Y_2 is $(-1, -5)$, and the point of intersection of Y_2 and Y_3 is $(1, 1)$. The inequality is true for all values of x between these two intersection points. Since the inequality is strict, the solution set is $\{x \mid -1 < x < 1\}$ or, using interval notation, $(-1, 1)$.

Figure 51

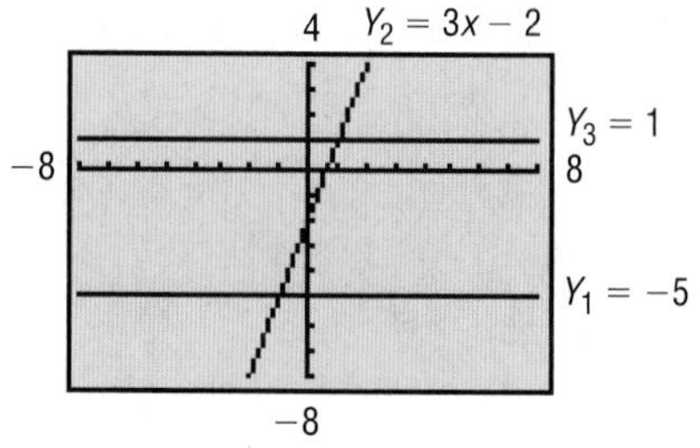

Algebraic Solution

Recall that the inequality

$$-5 < 3x - 2 < 1$$

is equivalent to the two inequalities

$$-5 < 3x - 2 \quad \text{and} \quad 3x - 2 < 1$$

We will solve each of these inequalities separately.

$-5 < 3x - 2$		$3x - 2 < 1$
$-5 + 2 < 3x - 2 + 2$	Add 2 to both sides.	$3x - 2 + 2 < 1 + 2$
$-3 < 3x$	Simplify	$3x < 3$
$\frac{-3}{3} < \frac{3x}{3}$	Divide both sides by 3.	$\frac{3x}{3} < \frac{3}{3}$
$-1 < x$	Simplify	$x < 1$

The solution set of the original pair of inequalities consists of all x for which

$$-1 < x \quad \text{and} \quad x < 1$$

Figure 52
$-1 < x < 1$ or $(-1, 1)$

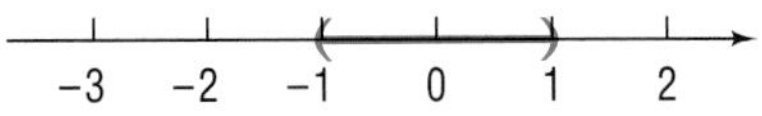

This may be written more compactly as $\{x \mid -1 < x < 1\}$. In interval notation, the solution is $(-1, 1)$. See Figure 52 for the graph. ■

We observe in the preceding process that the two inequalities that we solved required exactly the same steps. A shortcut to solving the original inequality algebraically is to deal with the two inequalities at the same time, as follows:

$-5 < 3x - 2 < 1$	
$-5 + 2 < 3x - 2 + 2 < 1 + 2$	Add 2 to each part.
$-3 < 3x < 3$	Simplify
$\frac{-3}{3} < \frac{3x}{3} < \frac{3}{3}$	Divide each part by 3.
$-1 < x < 1$	Simplify

NOW WORK PROBLEM 51.

5 Let's look at an inequality involving absolute value.

EXAMPLE 9 Solving an Inequality Involving Absolute Value

Solve the inequality: $|x| < 4$

Graphing Solution

Figure 53

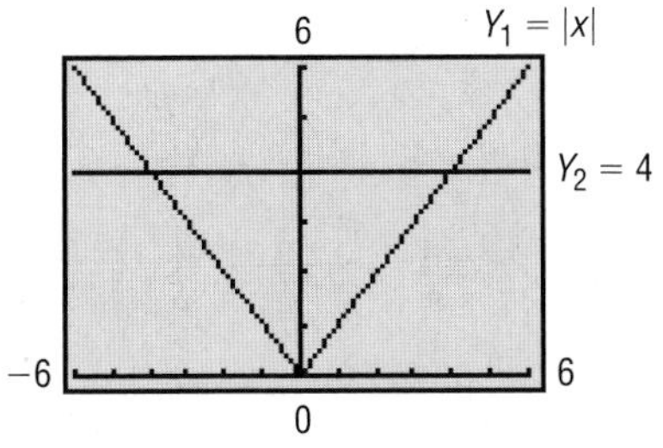

We graph $Y_1 = |x|$ and $Y_2 = 4$ on the same screen. See Figure 53. Using the INTERSECT command twice, we find that Y_1 and Y_2 intersect at $x = -4$ and at $x = 4$. The graph of Y_1 is below that of Y_2, $Y_1 < Y_2$, between the points of intersection. Since the inequality is strict, the solution set is $\{x \mid -4 < x < 4\}$ or, using interval notation, $(-4, 4)$.

Algebraic Solution

We are looking for all points whose coordinate x is a distance less than 4 units from the origin. See Figure 54 for an illustration. Because any x between -4 and 4 satisfies the condition $|x| < 4$, the solution set consists of all numbers x for which $-4 < x < 4$, that is, all x in the interval $(-4, 4)$. ■

Figure 54
$-4 < x < 4$ or $(-4, 4)$

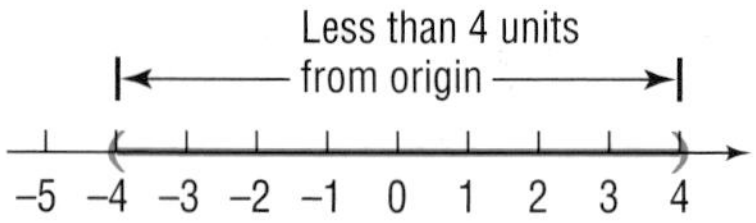

We are led to the following results:

Inequalities Involving Absolute Value

If a is any positive number and if u is any algebraic expression, then

$\lvert u\rvert < a$	is equivalent to	$-a < u < a$	**(4)**
$\lvert u\rvert \le a$	is equivalent to	$-a \le u \le a$	**(5)**

In other words, $|u| < a$ is equivalent to $-a < u$ and $u < a$.

EXAMPLE 10 Solving an Inequality Involving Absolute Value

Solve the inequality $|2x + 4| \le 3$ and graph the solution set.

Graphing Solution

We graph $Y_1 = |2x + 4|$ and $Y_2 = 3$ on the same screen. See Figure 55. Using the INTERSECT command twice, we find that Y_1 and Y_2 intersect at $x = -3.5$ and at $x = -0.5$. The graph of Y_1 is below that of Y_2, $Y_1 < Y_2$, between the points of intersection. Since the inequality is not strict, the solution set is $\{x \mid -3.5 \le x \le -0.5\}$ or, using interval notation, $[-3.5, -0.5]$.

Figure 55

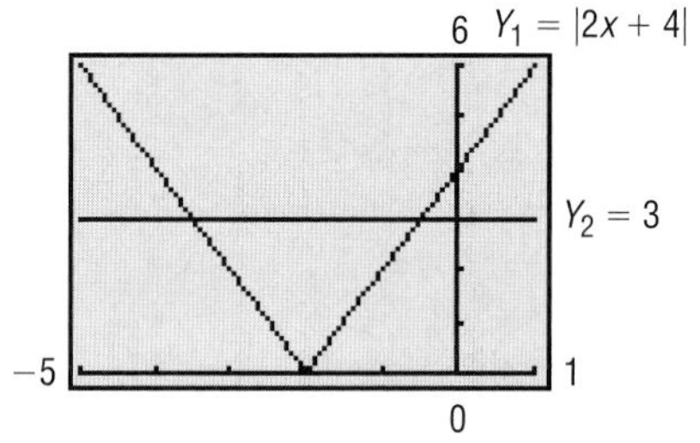

Algebraic Solution

$$|2x + 4| \le 3$$ This follows the form of statement (5); the expression $u = 2x + 4$ is inside the absolute value bars.

$$-3 \le 2x + 4 \le 3$$ Apply statement (5).

$$-3 - 4 \le 2x + 4 - 4 \le 3 - 4$$ Subtract 4 from each part.

$$-7 \le 2x \le -1$$ Simplify

$$\frac{-7}{2} \le \frac{2x}{2} \le \frac{-1}{2}$$ Divide each part by 2.

$$-\frac{7}{2} \le x \le -\frac{1}{2}$$ Simplify

The solution set is $\left\{x \mid -\frac{7}{2} \le x \le -\frac{1}{2}\right\}$, that is, all x in the interval $\left[-\frac{7}{2}, -\frac{1}{2}\right]$. See Figure 56. ■

Figure 56
$-\frac{7}{2} \le x \le -\frac{1}{2}$ or $\left[-\frac{7}{2}, -\frac{1}{2}\right]$

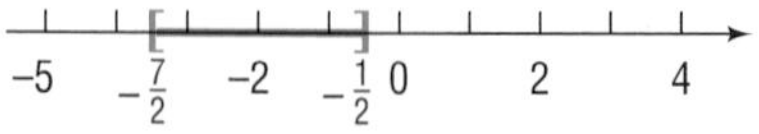

NOW WORK PROBLEM 65.

EXAMPLE 11

Solving an Inequality Involving Absolute Value

Solve the inequality $|x| > 3$ and graph the solution set.

Graphing Solution

We graph $Y_1 = |x|$ and $Y_2 = 3$ on the same screen. See Figure 57. Using the INTERSECT command twice, we find that Y_1 and Y_2 intersect at $x = -3$ and at $x = 3$. The graph of Y_1 is above that of Y_2, $Y_1 > Y_2$, to the left of $x = -3$ and to the right of $x = 3$. Since the inequality is strict, the solution set is $\{x | x < -3 \text{ or } x > 3\}$. Using interval notation, the solution is $(-\infty, -3)$ or $(3, \infty)$.

Figure 57

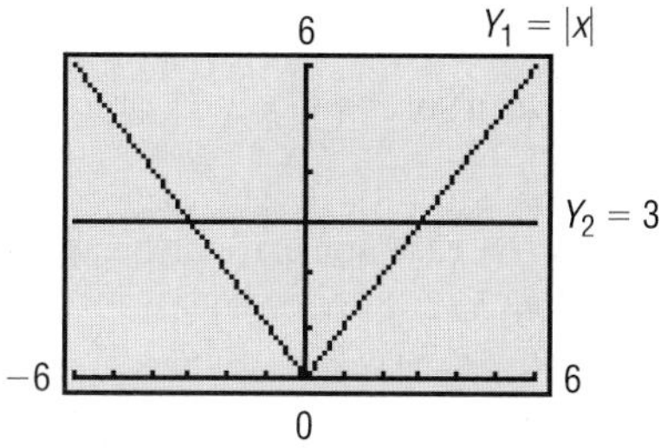

Algebraic Solution

We are looking for all points whose coordinate x is a distance greater than 3 units from the origin. Figure 58 illustrates the situation. We conclude that any x less than -3 or greater than 3 satisfies the condition $|x| > 3$. Consequently, the solution set consists of all numbers x for which $x < -3$ or $x > 3$, that is, all x in $(-\infty, -3)$ or $(3, \infty)$. ■

Figure 58

$x < -3$ or $x > 3$; $(-\infty, -3)$ or $(3, \infty)$

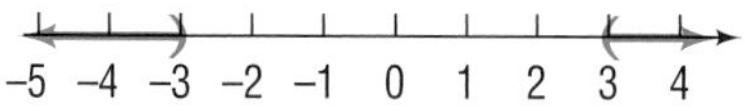

Inequalities Involving Absolute Value

If a is any positive number and u is any algebraic expression, then

$\lvert u\rvert > a$	is equivalent to	$u < -a$ or $u > a$	**(6)**
$\lvert u\rvert \geq a$	is equivalent to	$u \leq -a$ or $u \geq a$	**(7)**

EXAMPLE 12

Solving an Inequality Involving Absolute Value

Solve the inequality $|2x - 5| > 3$ and graph the solution set.

Graphing Solution

We graph $Y_1 = |2x - 5|$ and $Y_2 = 3$ on the same screen. See Figure 59. Using the INTERSECT command twice, we find that Y_1 and Y_2 intersect at $x = 1$ and at $x = 4$. The graph of Y_1 is above that of Y_2, $Y_1 > Y_2$, to the left of $x = 1$ and to the right of $x = 4$. Since the inequality is strict, the solution set is $\{x | x < 1 \text{ or } x > 4\}$. Using interval notation, the solution is $(-\infty, 1)$ or $(4, \infty)$.

Figure 59

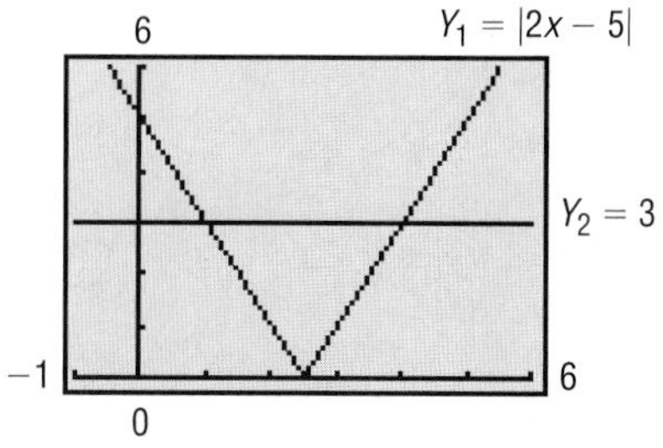

Algebraic Solution

$\lvert 2x-5\rvert > 3$			This follows the form of statement (6); the expression $u = 2x - 5$ is inside the absolute value bars.
$2x - 5 < -3$	or	$2x - 5 > 3$	Apply statement (6).
$2x - 5 + 5 < -3 + 5$	or	$2x - 5 + 5 > 3 + 5$	Add 5 to each part.
$2x < 2$	or	$2x > 8$	Simplify
$\frac{2x}{2} < \frac{2}{2}$	or	$\frac{2x}{2} > \frac{8}{2}$	Divide each part by 2.
$x < 1$	or	$x > 4$	Simplify

Figure 60
$x < 1$ or $x > 4$; $(-\infty, 1)$ or $(4, \infty)$

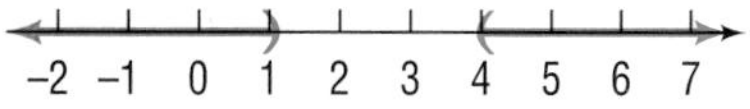

The solution set is $\{x \mid x < 1 \text{ or } x > 4\}$, that is, all x in $(-\infty, 1)$ or $(4, \infty)$. See Figure 60. ■

WARNING A common error to be avoided is to attempt to write the solution $x < 1$ or $x > 4$ as $1 > x > 4$, which is incorrect, since there are no numbers x for which $1 > x$ *and* $x > 4$.

NOW WORK PROBLEM 69.

1.5 EXERCISES

In Problems 1–8, write each inequality using interval notation, and illustrate each inequality using the real number line.

1. $0 \le x \le 4$ **2.** $-1 < x < 5$ **3.** $4 \le x < 6$ **4.** $-2 < x < 0$
5. $x \ge 4$ **6.** $x \le 5$ **7.** $x < -4$ **8.** $x > 1$

In Problems 9–16, write each interval as an inequality involving x, and illustrate each inequality using the real number line.

9. $[2, 5]$ **10.** $(1, 2)$ **11.** $(-3, -2)$ **12.** $[0, 1)$
13. $[4, \infty)$ **14.** $(-\infty, 2]$ **15.** $(-\infty, -3)$ **16.** $(-8, \infty)$

In Problems 17–30, fill in the blank with the correct inequality symbol.

17. If $x < 5$, then $x - 5$ ________ 0.
18. If $x < -4$, then $x + 4$ ________ 0.
19. If $x > -4$, then $x + 4$ ________ 0.
20. If $x > 6$, then $x - 6$ ________ 0.
21. If $x \ge -4$, then $3x$ ________ -12.
22. If $x \le 3$, then $2x$ ________ 6.
23. If $x < 6$, then $-2x$ ________ -12.
24. If $x > -2$, then $-4x$ ________ 8.
25. If $x \ge 5$, then $-4x$ ________ -20.
26. If $x \le -4$, then $-3x$ ________ 12.
27. If $2x > 6$, then x ________ 3.
28. If $3x \le 12$, then x ________ 4.
29. If $-\frac{1}{2}x \le 3$, then x ________ -6.
30. If $-\frac{1}{4}x > 1$, then x ________ -4.

In Problems 31–74, solve each inequality (a) graphically and (b) algebraically. Graph the solution set.

31. $x + 1 < 5$ **32.** $x - 6 < 1$ **33.** $1 - 2x \le 3$
34. $2 - 3x \le 5$ **35.** $3x - 7 > 2$ **36.** $2x + 5 > 1$
37. $3x - 1 \ge 3 + x$ **38.** $2x - 2 \ge 3 + x$ **39.** $-2(x + 3) < 8$
40. $-3(1 - x) < 12$ **41.** $4 - 3(1 - x) \le 3$ **42.** $8 - 4(2 - x) \le -2x$
43. $\frac{1}{2}(x - 4) > x + 8$ **44.** $3x + 4 > \frac{1}{3}(x - 2)$ **45.** $\frac{x}{2} \ge 1 - \frac{x}{4}$
46. $\frac{x}{3} \ge 2 + \frac{x}{6}$ **47.** $0 \le 2x - 6 \le 4$ **48.** $4 \le 2x + 2 \le 10$
49. $-5 \le 4 - 3x \le 2$ **50.** $-3 \le 3 - 2x \le 9$ **51.** $-3 < \frac{2x - 1}{4} < 0$
52. $0 < \frac{3x + 2}{2} < 4$ **53.** $1 < 1 - \frac{1}{2}x < 4$ **54.** $0 < 1 - \frac{1}{3}x < 1$

55. $(x + 2)(x - 3) > (x - 1)(x + 1)$

56. $(x - 1)(x + 1) > (x - 3)(x + 4)$

57. $x(4x + 3) \leq (2x + 1)^2$

58. $x(9x - 5) \leq (3x - 1)^2$

59. $\frac{1}{2} \leq \frac{x+1}{3} < \frac{3}{4}$

60. $\frac{1}{3} < \frac{x+1}{2} \leq \frac{2}{3}$

61. $|2x| < 8$

62. $|3x| < 15$

63. $|3x| > 12$

64. $|2x| > 6$

65. $|x - 2| < 1$

66. $|x + 4| < 2$

67. $|3t - 2| \leq 4$

68. $|2u + 5| \leq 7$

69. $|x - 3| \geq 2$

70. $|x + 4| \geq 2$

71. $|1 - 4x| < 5$

72. $|1 - 2x| < 3$

73. $|1 - 2x| > 3$

74. $|2 - 3x| > 1$

75. Express the fact that x differs from 2 by less than $\frac{1}{2}$ as an inequality involving an absolute value. Solve for x.

76. Express the fact that x differs from -1 by less than 1 as an inequality involving an absolute value. Solve for x.

77. Express the fact that x differs from -3 by more than 2 as an inequality involving an absolute value. Solve for x.

78. Express the fact that x differs from 2 by more than 3 as an inequality involving an absolute value. Solve for x.

79. A young adult may be defined as someone older than 21, but less than 30 years of age. Express this statement using inequalities.

80. Middle-aged may be defined as being 40 or more and less than 60. Express the statement using inequalities.

81. **Body Temperature** Normal human body temperature is 98.6°F. If a temperature x that differs from normal by at least 1.5° is considered unhealthy, write the condition for an unhealthy temperature x as an inequality involving an absolute value, and solve for x.

82. **Household Voltage** In the United States, normal household voltage is 115 volts. However, it is not uncommon for actual voltage to differ from normal voltage by at most 5 volts. Express this situation as an inequality involving an absolute value. Use x as the actual voltage and solve for x.

83. **Life Expectancy** Metropolitan Life Insurance Co. reported that an average 25-year-old male in 1996 could expect to live at least 48.4 more years and an average 25-year-old female in 1996 could expect to live at least 54.7 more years.

(a) To what age can an average 25-year-old male expect to live? Express your answer as an inequality.

(b) To what age can an average 25-year-old female expect to live? Express your answer as an inequality.

(c) Who can expect to live longer, a male or a female? By how many years?

84. **General Chemistry** For a certain ideal gas, the volume V (in cubic centimeters) equals 20 times the temperature T (in degrees Celsius). If the temperature varies from 80° to 120°C inclusive, what is the corresponding range of the volume of the gas?

85. **Real Estate** A real estate agent agrees to sell a large apartment complex according to the following commission schedule: \$45,000 plus 25% of the selling price in excess of \$900,000. Assuming that the complex will sell at some price between \$900,000 and \$1,100,000 inclusive, over what range does the agent's commission vary? How does the commission vary as a percent of selling price?

86. **Sales Commission** A used car salesperson is paid a commission of \$25 plus 40% of the selling price in excess of owner's cost. The owner claims that used cars typically sell for at least owner's cost plus \$70 and at most owner's cost plus \$300. For each sale made, over what range can the salesperson expect the commission to vary?

87. **Federal Tax Withholding** The percentage method of withholding for federal income tax (1998)* states that a single person whose weekly wages, after subtracting withholding allowances, are over \$517, but not over \$1105, shall have \$69.90 plus 28% of the excess over \$517 withheld. Over what range does the amount withheld vary if the weekly wages vary from \$525 to \$600 inclusive?

88. **Federal Tax Withholding** Rework Problem 87 if the weekly wages vary from \$600 to \$700 inclusive.

89. **Electricity Rates** Commonwealth Edison Company's summer charge for electricity is 10.494 ¢ per kilowatt-hour.† In addition, each monthly bill contains a customer charge of \$9.36. If last summer's bills ranged from a low of \$80.24 to a high of \$271.80, over what range did usage vary (in kilowatt-hours)?

90. **Water Bills** The Village of Oak Lawn charges homeowners \$21.60 per quarter-year plus \$1.70 per 1000 gallons for water usage in excess of 12,000 gallons.‡ In 2000, one homeowner's quarterly bill ranged from a high of \$65.75 to a low of \$28.40. Over what range did water usage vary?

* *Source: Employer's Tax Guide*. Department of the Treasury, Internal Revenue Service, 1998.

† *Source:* Commonwealth Edison Co., Chicago, Illinois, 1998.

‡ *Source:* Village of Oak Lawn, Illinois, 2000.

91. Markup of a New Car The markup over dealer's cost of a new car ranges from 12% to 18%. If the sticker price is $8800, over what range will the dealer's cost vary?

92. IQ Tests A standard intelligence test has an average score of 100. According to statistical theory, of the people who take the test, the 2.5% with the highest scores will have scores of more than 1.96σ above the average, where σ (sigma, a number called the *standard deviation*) depends on the nature of the test. If $\sigma = 12$ for this test and there is (in principle) no upper limit to the score possible on the test, write the interval of possible test scores of the people in the top 2.5%.

93. Computing Grades In your Economics 101 class, you have scores of 68, 82, 87, and 89 on the first four of five tests. To get a grade of B, the average of the first five test scores must be greater than or equal to 80 and less than 90. Solve an inequality to find the range of the score that you need on the last test to get a B.

94. Computing Grades Repeat Problem 93 if the fifth test counts double.

95. A car that averages 25 miles per gallon has a tank that holds 20 gallons of gasoline. After a trip that covered at least 300 miles, the car ran out of gasoline. What is the range of the amount of gasoline (in gallons) that was in the tank at the start of the trip?

96. Repeat Problem 95 if the same car runs out of gasoline after a trip of no more than 250 miles.

97. Arithmetic Mean If $a < b$, show that $a < (a + b)/2 < b$. The number $(a + b)/2$ is called the **arithmetic mean** of a and b.

98. Refer to Problem 97. Show that the arithmetic mean of a and b is equidistant from a and b.

99. Geometric Mean If $0 < a < b$, show that $a < \sqrt{ab} < b$. The number $\sqrt{ab}$ is called the **geometric mean** of a and b.

100. Refer to Problems 97 and 99. Show that the geometric mean of a and b is less than the arithmetic mean of a and b.

101. Harmonic Mean For $0 < a < b$, let h be defined by

$$\frac{1}{h} = \frac{1}{2}\left(\frac{1}{a} + \frac{1}{b}\right)$$

Show that $a < h < b$. The number h is called the **harmonic mean** of a and b.

102. Refer to Problems 97, 99, and 101. Show that the harmonic mean of a and b equals the geometric mean squared divided by the arithmetic mean.

103. Make up an inequality that has no solution. Make up one that has exactly one solution.

104. The inequality $x^2 + 1 < -5$ has no solution. Explain why.

105. Do you prefer to use inequality notation or interval notation to express the solution to an inequality? Give your reasons. Are there particular circumstances when you prefer one to the other? Cite examples.

106. How would you explain to a fellow student the underlying reason for the multiplication property for inequalities (page 52); that is, the sense or direction of an inequality remains the same if each side is multiplied by a positive real number, whereas the direction is reversed if each side is multiplied by a negative real number.

1.6 LINES

OBJECTIVES

1. Calculate and Interpret the Slope of a Line
2. Graph Lines
3. Find the Equation of Vertical Lines
4. Use the Point–Slope Form of a Line; Identify Horizontal Lines
5. Find the Equation of a Line given Two Points
6. Write the Equation of a Line in Slope–Intercept Form
7. Identify the Slope and y-intercept of a Line from Its Equation
8. Write the Equation of a Line in General Form
9. Define Parallel and Perpendicular Lines
10. Find Equations of Parallel Lines
11. Find Equations of Perpendicular Lines

In this section we study a certain type of equation that contains two variables, called a *linear equation*, and its graph, a *line*.

Slope of a Line

1 Consider the staircase illustrated in Figure 61. Each step contains exactly the same horizontal **run** and the same vertical **rise**. The ratio of the rise to the run, called the *slope*, is a numerical measure of the steepness of the staircase. For example, if the run is increased and the rise remains the same, the staircase becomes less steep. If the run is kept the same, but the rise is increased, the staircase becomes more steep. This important characteristic of a line is best defined using rectangular coordinates.

Figure 61

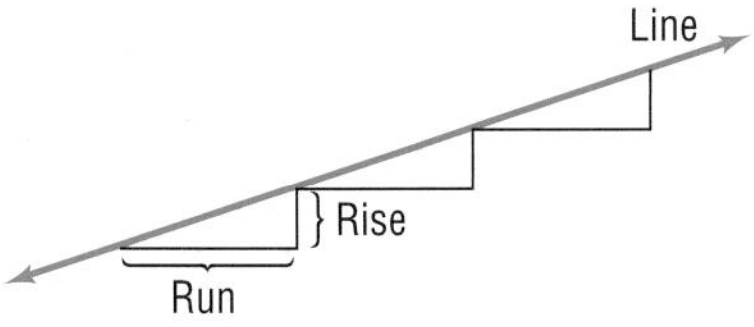

Let $P = (x_1, y_1)$ and $Q = (x_2, y_2)$ be two distinct points with $x_1 \neq x_2$. The **slope *m*** of the nonvertical line L containing P and Q is defined by the formula

$$m = \frac{y_2 - y_1}{x_2 - x_1}, \qquad x_1 \neq x_2 \qquad \textbf{(1)}$$

If $x_1 = x_2$, L is a **vertical line** and the slope m of L is **undefined** (since this results in division by 0).

Figure 62(a) provides an illustration of the slope of a nonvertical line; Figure 62(b) illustrates a vertical line.

Figure 62

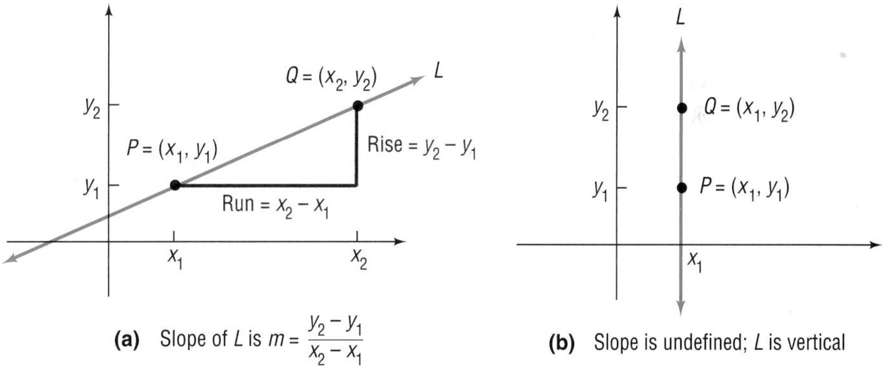

(a) Slope of L is $m = \dfrac{y_2 - y_1}{x_2 - x_1}$ **(b)** Slope is undefined; L is vertical

As Figure 62(a) illustrates, the slope m of a nonvertical line may be viewed as

$$m = \frac{y_2 - y_1}{x_2 - x_1} = \frac{\text{Rise}}{\text{Run}}$$

We can also express the slope m of a nonvertical line as

$$m = \frac{y_2 - y_1}{x_2 - x_1} = \frac{\text{Change in } y}{\text{Change in } x} = \frac{\Delta y}{\Delta x}$$

That is, the slope m of a nonvertical line L measures the amount that y changes as x changes from x_1 to x_2. This is called the **average rate of change** of y with respect to x.

Two comments about computing the slope of a nonvertical line may prove helpful:

1. Any two distinct points on the line can be used to compute the slope of the line. (See Figure 63 for justification.)
2. The slope of a line may be computed from $P = (x_1, y_1)$ to $Q = (x_2, y_2)$ or from Q to P because

$$\frac{y_2 - y_1}{x_2 - x_1} = \frac{y_1 - y_2}{x_1 - x_2}$$

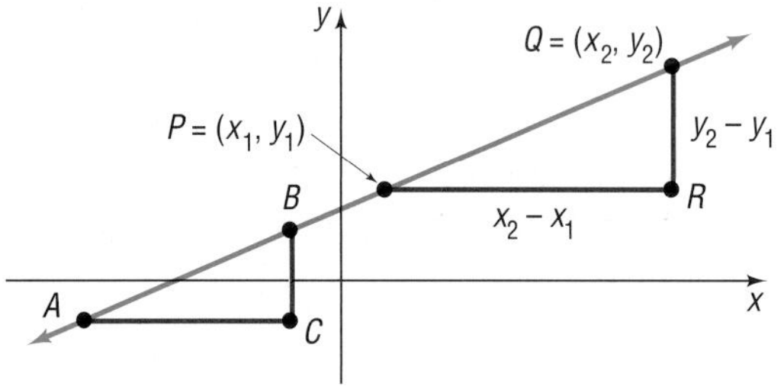

Figure 63
Triangles ABC and PQR are similar (equal angles). Hence, ratios of corresponding sides are proportional. Thus,

$$\text{Slope using } P \text{ and } Q = \frac{y_2 - y_1}{x_2 - x_1}$$
$$= \text{Slope using } A \text{ and } B = \frac{d(B, C)}{d(A, C)}$$

EXAMPLE 1

Finding and Interpreting the Slope of a Line Containing Two Points

The slope m of the line containing the points (1, 2) and (5, −3) may be computed as

$$m = \frac{-3 - 2}{5 - 1} = \frac{-5}{4} \quad \text{or as} \quad m = \frac{2 - (-3)}{1 - 5} = \frac{5}{-4} = \frac{-5}{4}$$

For every 4-unit change in x, y will change by −5 units. Thus, if x increases by 4 units, y will decrease by 5 units. ■

NOW WORK PROBLEMS 1 AND 7.

Figure 64

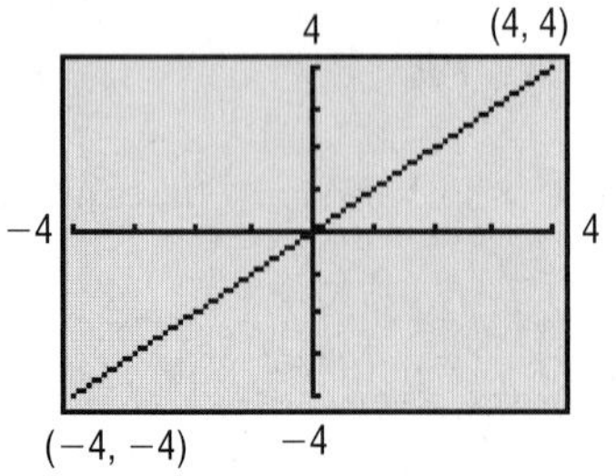

Figure 65

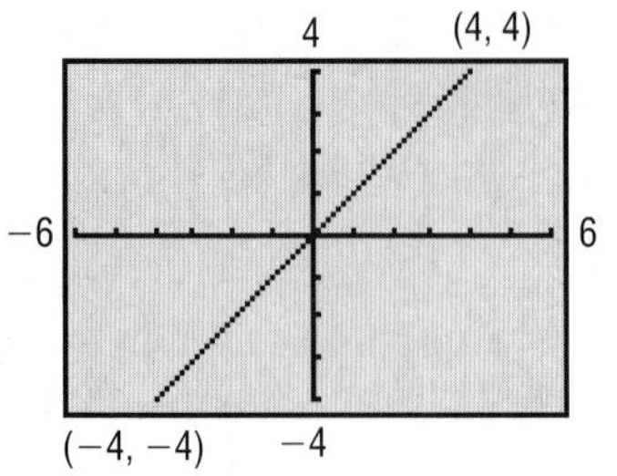

SQUARE SCREENS

To get an undistorted view of slope on a graphing utility, the same scale must be used on each axis. However, most graphing utilities have a rectangular screen. Because of this, using the same interval for both x and y will result in a distorted view. For example, Figure 64 shows the graph of the line $y = x$ connecting the points (−4, −4) and (4, 4).

We expect the line to bisect the first and third quadrants, but it doesn't. We need to adjust the selections for Xmin, Xmax, Ymin, and Ymax so that a **square screen** results. On most graphing utilities, this is accomplished by setting the ratio of x to y at 3:2.* In other words,

$$2(X\text{max} - X\text{min}) = 3(Y\text{max} - Y\text{min})$$

Figure 65 shows the graph of the line $y = x$ on a square screen using a TI-83. Notice that the line now bisects the first and third quadrants. Compare this illustration to Figure 64.

*Most graphing utilities have a feature that automatically squares the viewing window. Consult your owner's manual for the appropriate key strokes.

To get a better idea of the meaning of the slope m of a line, consider the following:

Seeing the Concept On the same square screen, graph the following equations:

$Y_1 = 0$	Slope of line is 0.
$Y_2 = \frac{1}{4}x$	Slope of line is $\frac{1}{4}$.
$Y_3 = \frac{1}{2}x$	Slope of line is $\frac{1}{2}$.
$Y_4 = x$	Slope of line is 1.
$Y_5 = 2x$	Slope of line is 2.
$Y_6 = 6x$	Slope of line is 6.

See Figure 66. ■

Figure 66

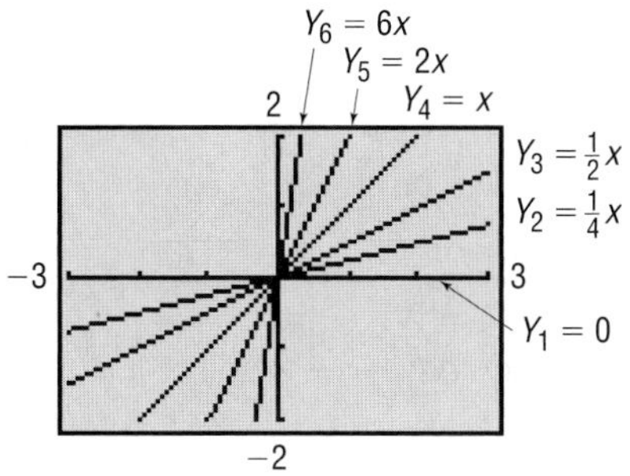

Seeing the Concept On the same square screen, graph the following equations:

$Y_1 = 0$	Slope of line is 0.
$Y_2 = -\frac{1}{4}x$	Slope of line is $-\frac{1}{4}$.
$Y_3 = -\frac{1}{2}x$	Slope of line is $-\frac{1}{2}$.
$Y_4 = -x$	Slope of line is −1.
$Y_5 = -2x$	Slope of line is −2.
$Y_6 = -6x$	Slope of line is −6.

See Figure 67. ■

Figure 67

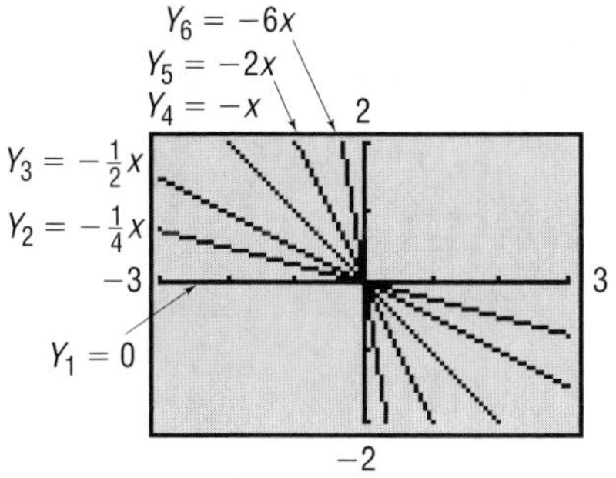

Figures 66 and 67 illustrate the following facts:

1. When the slope of a line is positive, the line slants upward from left to right.
2. When the slope of a line is negative, the line slants downward from left to right.
3. When the slope is 0, the line is horizontal.

Figures 66 and 67 also illustrate that the closer the line is to the vertical position, the greater the magnitude of the slope.

2 The next example illustrates how the slope of a line can be used to graph the line.

EXAMPLE 2 Graphing a Line Given a Point and a Slope

Draw a graph of the line that contains the point $(3, 2)$ and has a slope of:

(a) $\frac{3}{4}$ (b) $-\frac{4}{5}$

Solution (a) Slope = Rise/Run. The fact that the slope is $\frac{3}{4}$ means that for every horizontal movement (run) of 4 units to the right there will be a vertical movement (rise) of 3 units. If we start at the given point $(3, 2)$ and move 4 units to the right and 3 units up, we reach the point $(7, 5)$. By drawing the line through this point and the point $(3, 2)$, we have the graph. See Figure 68.

Figure 68
Slope $= \frac{3}{4}$

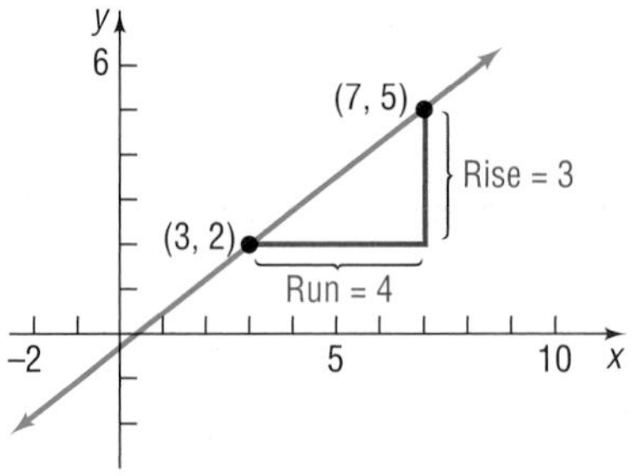

(b) The fact that the slope is

$$-\frac{4}{5} = \frac{-4}{5} = \frac{\text{Rise}}{\text{Run}}$$

means that for every horizontal movement of 5 units to the right there will be a corresponding vertical movement of −4 units (a downward movement). If we start at the given point (3, 2) and move 5 units to the right and then 4 units down, we arrive at the point (8, −2). By drawing the line through these points, we have the graph. See Figure 69.

Figure 69
Slope $= -\frac{4}{5}$

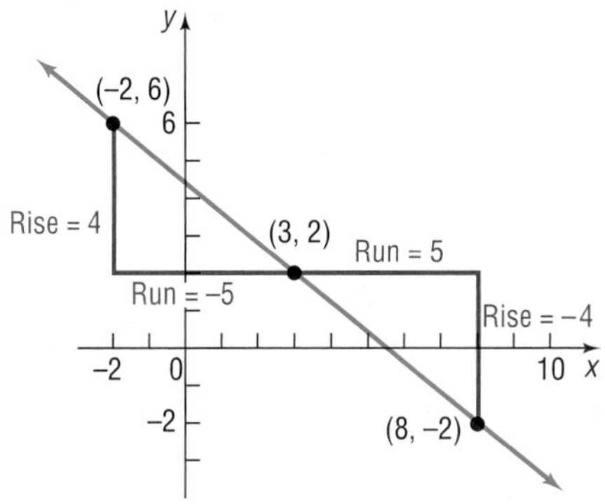

Alternatively, we can set

$$-\frac{4}{5} = \frac{4}{-5} = \frac{\text{Rise}}{\text{Run}}$$

so that for every horizontal movement of −5 units (a movement to the left) there will be a corresponding vertical movement of 4 units (upward). This approach brings us to the point (−2, 6), which is also on the graph shown in Figure 69. ■

NOW WORK PROBLEM 15.

Equations of Lines

3 Now that we have discussed the slope of a line, we are ready to derive equations of lines. As we shall see, there are several forms of the equation of a line. Let's start with an example.

EXAMPLE 3 Graphing a Line

Graph the equation: $x = 3$

Solution Using a graphing utility, we need to express the equation in the form $y = \{\text{expression in } x\}$. But $x = 3$ cannot be put into this form. Consult your manual to determine the methodology required to draw vertical lines. Figure 70(a) shows the graph that you should obtain.

To graph $x = 3$ by hand, recall that we are looking for all points (x, y) in the plane for which $x = 3$. Thus, no matter what y-coordinate is used, the corresponding x-coordinate always equals 3. Consequently, the graph of the equation $x = 3$ is a vertical line with x-intercept 3 and undefined slope. See Figure 70(b).

Figure 70
$x = 3$

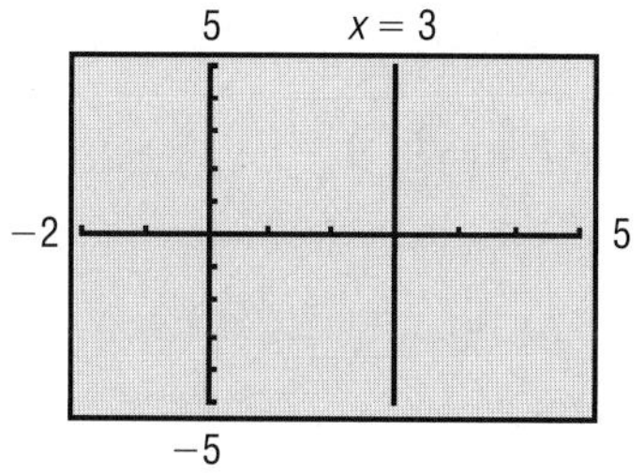

(a)

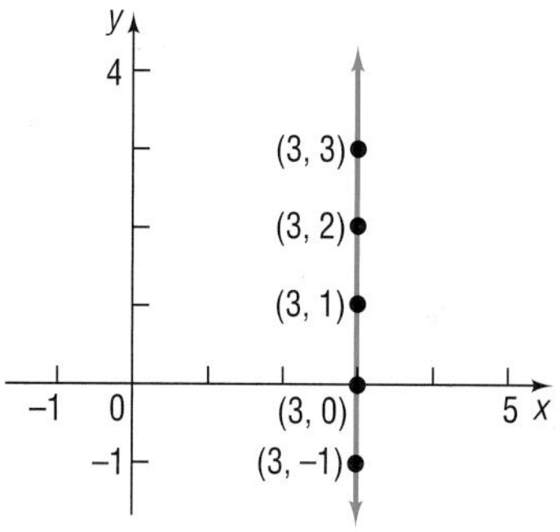

(b)

As suggested by Example 3, we have the following result:

Theorem

Equation of a Vertical Line

A vertical line is given by an equation of the form

$$x = a$$

where a is the x-intercept.

4 Now let L be a nonvertical line with slope m and containing the point (x_1, y_1). See Figure 71. For any other point (x, y) on L, we have

$$m = \frac{y - y_1}{x - x_1} \quad \text{or} \quad y - y_1 = m(x - x_1)$$

Figure 71

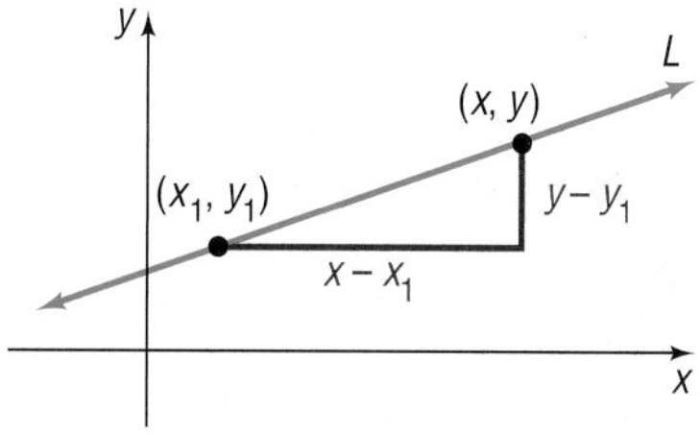

Theorem **Point-Slope Form of an Equation of a Line**

An equation of a nonvertical line of slope m that contains the point (x_1, y_1) is

$$y - y_1 = m(x - x_1) \tag{2}$$

EXAMPLE 4 Using the Point–Slope Form of a Line

An equation of the line with slope 4 and containing the point $(1, 2)$ can be found by using the point-slope form with $m = 4$, $x_1 = 1$, and $y_1 = 2$.

$$y - y_1 = m(x - x_1)$$
$$y - 2 = 4(x - 1)$$
$$y = 4x - 2$$

See Figure 72.

Figure 72
$y = 4x - 2$

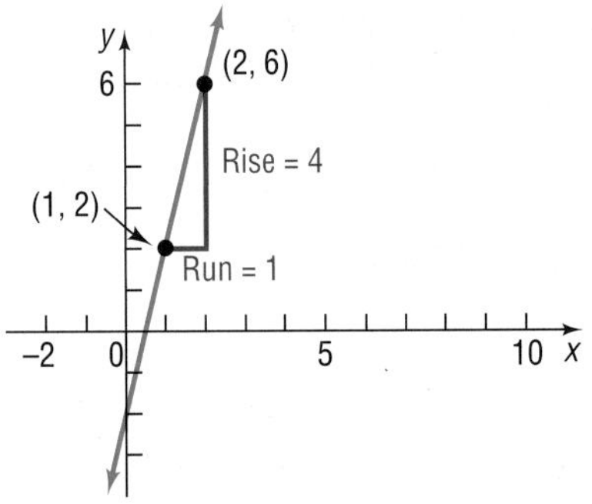

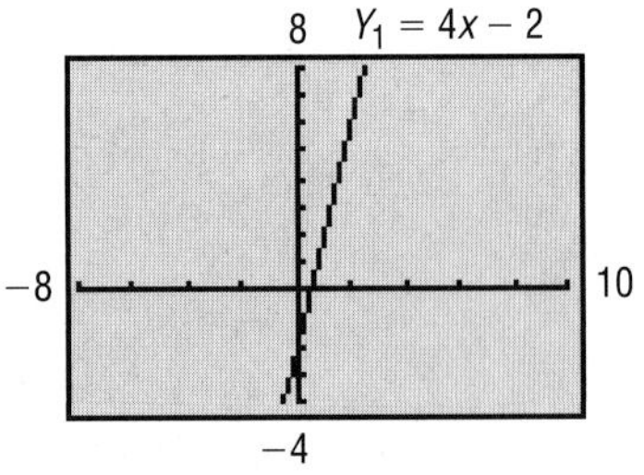

EXAMPLE 5 Finding the Equation of a Horizontal Line

Find an equation of the horizontal line containing the point $(3, 2)$.

Solution The slope of a horizontal line is 0. To get an equation, we use the point–slope form with $m = 0$, $x_1 = 3$, and $y_1 = 2$.

$$y - y_1 = m(x - x_1)$$
$$y - 2 = 0 \cdot (x - 3)$$
$$y - 2 = 0$$
$$y = 2$$

See Figure 73 for the graph.

Figure 73
$y = 2$

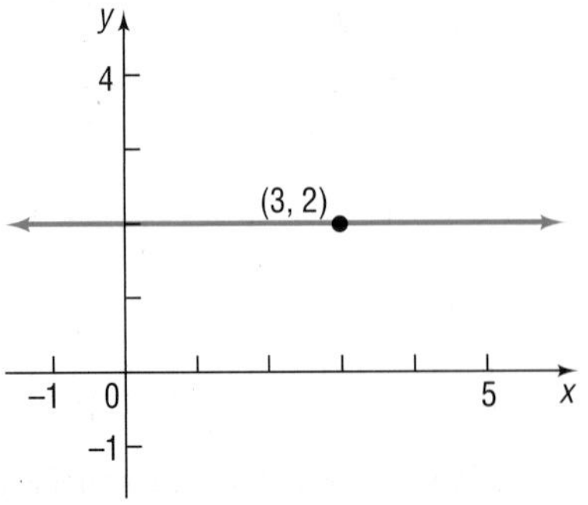

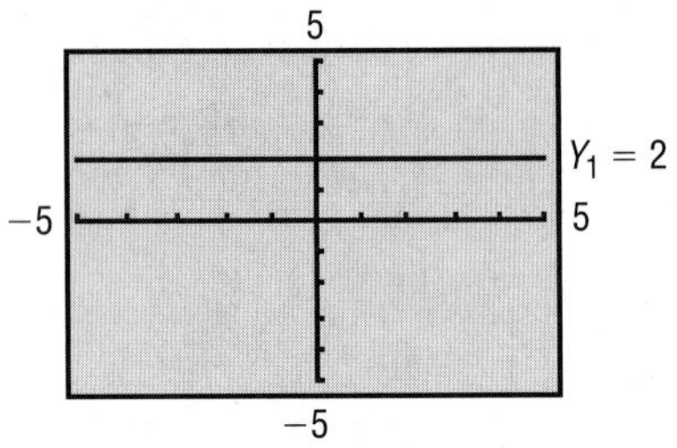

As suggested by Example 5, we have the following result:

Theorem

Equation of a Horizontal Line

A horizontal line is given by an equation of the form

$$y = b$$

where b is the y-intercept.

EXAMPLE 6 Finding an Equation of a Line Given Two Points

5 Find an equation of the line L containing the points $(2, 3)$ and $(-4, 5)$. Graph the line L.

Solution Since two points are given, we first compute the slope of the line.

$$m = \frac{5 - 3}{-4 - 2} = \frac{2}{-6} = -\frac{1}{3}$$

We use the point $(2, 3)$ and the fact that the slope $m = -\frac{1}{3}$ to get the point–slope form of the equation of the line.

$$y - 3 = -\frac{1}{3}(x - 2)$$

See Figure 74 for the graph.

Figure 74
$y - 3 = -\frac{1}{3}(x - 2)$

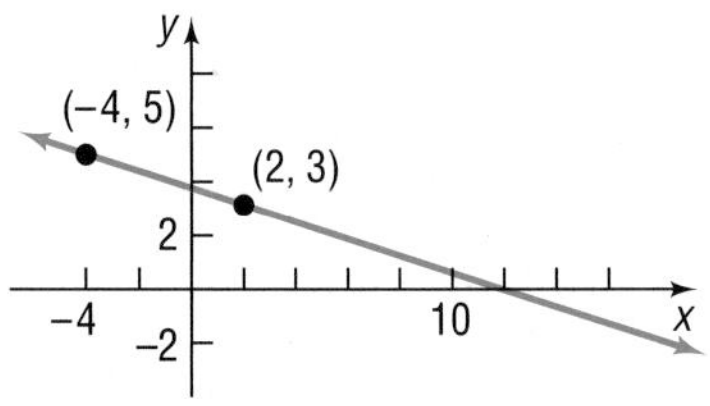

In the solution to Example 6, we could have used the other point, $(-4, 5)$, instead of the point $(2, 3)$. The equation that results, although it looks different, is equivalent to the equation that we obtained in the example. (Try it for yourself.)

6 Another useful equation of a line is obtained when the slope m and y-intercept b are known. In this event, we know both the slope m of the line and a point $(0, b)$ on the line; because of this, we may use the point–slope form, equation (2), to obtain the following equation:

$$y - b = m(x - 0) \quad \text{or} \quad y = mx + b$$

Theorem

Slope–Intercept Form of an Equation of a Line

An equation of a line L with slope m and y-intercept b is

$$y = mx + b \qquad \textbf{(3)}$$

Figure 75
$y = mx + 2$

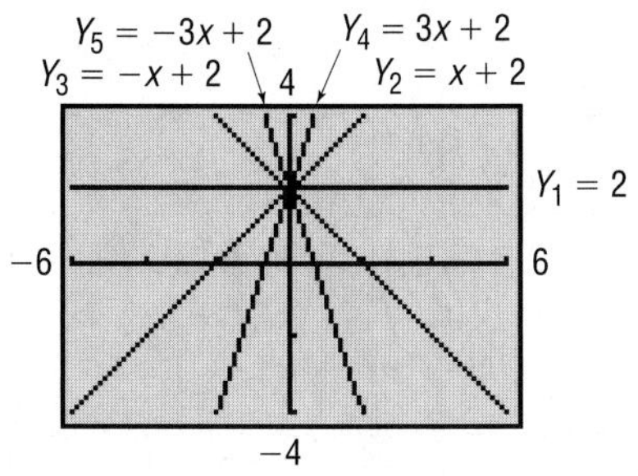

Seeing the Concept To see the role that the slope m plays, graph the following lines on the same square screen.

$$Y_1 = 2$$
$$Y_2 = x + 2$$
$$Y_3 = -x + 2$$
$$Y_4 = 3x + 2$$
$$Y_5 = -3x + 2$$

See Figure 75. What do you conclude about the lines $y = mx + 2$?

Figure 76
$y = 2x + b$

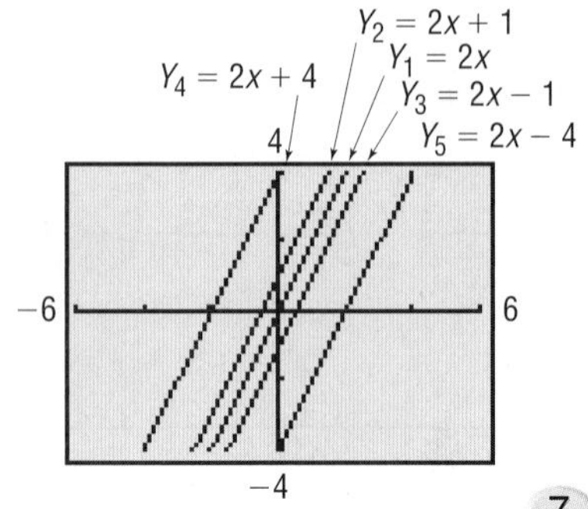

Seeing the Concept To see the role of the y-intercept b, graph the following lines on the same square screen.

$$Y_1 = 2x$$
$$Y_2 = 2x + 1$$
$$Y_3 = 2x - 1$$
$$Y_4 = 2x + 4$$
$$Y_5 = 2x - 4$$

See Figure 76. What do you conclude about the lines $y = 2x + b$? ■

7 When the equation of a line is written in slope–intercept form, it is easy to find the slope m and y-intercept b of the line. For example, suppose that the equation of a line is

$$y = -2x + 3$$

Compare it to $y = mx + b$

$$\begin{array}{ccccc} y & = & -2x & + & 3 \\ & & \uparrow & & \uparrow \\ y & = & mx & + & b \end{array}$$

The slope of this line is -2 and its y-intercept is 3.

NOW WORK PROBLEM 55.

EXAMPLE 7 Finding the Slope and y-Intercept

Find the slope m and y-intercept b of the equation $2x + 4y = 8$. Graph the equation.

Solution To obtain the slope and y-intercept, we transform the equation into its slope–intercept form. We need to solve for y.

$$2x + 4y = 8$$
$$4y = -2x + 8$$
$$y = -\tfrac{1}{2}x + 2 \qquad y = mx + b$$

Figure 77
$2x + 4y = 8$

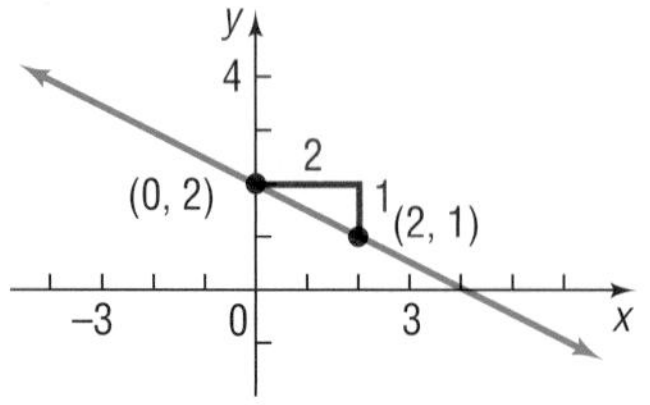

The coefficient of x, $-\frac{1}{2}$, is the slope, and the y-intercept is 2. We can graph the line in two ways:

1. Use the fact that the y-intercept is 2 and the slope is $-\frac{1}{2}$. Then, starting at the point $(0, 2)$, go to the right 2 units and then down 1 unit to the point $(2, 1)$. See Figure 77.

   Or:

2. Locate the intercepts. Because the y-intercept is 2, we know that one intercept is $(0, 2)$. To obtain the x-intercept, let $y = 0$ and solve for x. When $y = 0$, we have

$$2x + 4 \cdot 0 = 8$$
$$2x = 8$$
$$x = 4$$

Figure 78
$2x + 4y = 8$

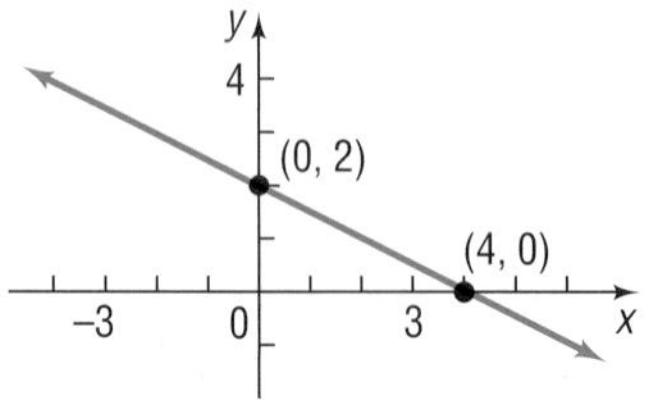

The intercepts are $(4, 0)$ and $(0, 2)$. See Figure 78. ■

NOW WORK PROBLEM 61.

8 The form of the equation of the line in Example 7, $2x + 4y = 8$, is called the *general form*.

The equation of a line L is in **general form** when it is written as

$$Ax + By = C \quad \textbf{(4)}$$

where A, B, and C are three real numbers, and A and B are not both 0.

Every line has an equation that is equivalent to an equation written in general form. For example, a vertical line whose equation is

$$x = a$$

can be written in the general form

$$1 \cdot x + 0 \cdot y = a \quad A = 1, B = 0, C = a$$

A horizontal line whose equation is

$$y = b$$

can be written in the general form

$$0 \cdot x + 1 \cdot y = b \quad A = 0, B = 1, C = b$$

Lines that are neither vertical nor horizontal have general equations of the form

$$Ax + By = C \quad A \neq 0 \text{ and } B \neq 0$$

Because the equation of every line can be written in general form, any equation equivalent to (4) is called a **linear equation**.

Parallel and Perpendicular Lines

9 When two lines (in the plane) do not intersect (that is, they have no points in common) the lines are said to be **parallel**. Look at Figure 79. There we have drawn two lines and have constructed two right triangles by drawing sides parallel to the coordinate axes. These lines are parallel if and only if the right triangles are similar. (Do you see why? Two angles are equal.) And the triangles are similar if and only if the ratios of corresponding sides are equal.

Figure 79
The lines are parallel if and only if their slopes are equal.

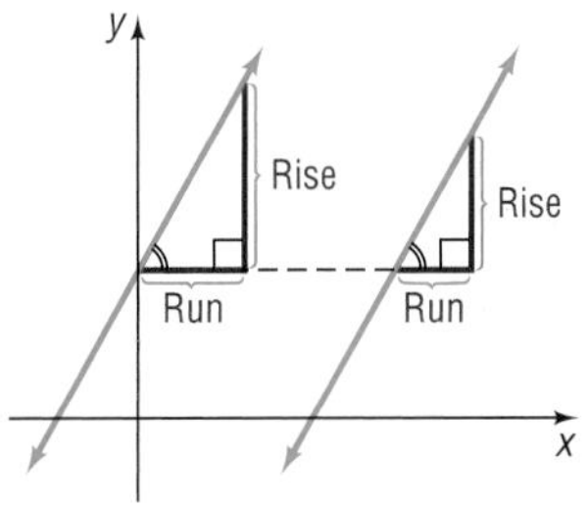

This suggests the following result:

Theorem Two nonvertical lines are parallel if and only if their slopes are equal and they have different y-intercepts.

The use of the words "if and only if" in the preceding theorem means that actually two statements are being made, one the converse of the other.

If two nonvertical lines are parallel, then their slopes are equal and they have different y-intercepts.

If two nonvertical lines have equal slopes and they have different y-intercepts, then they are parallel.

EXAMPLE 8 Showing That Two Lines Are Parallel

Show that the lines given by the following equations are parallel:

$$L:\ 2x + 3y = 6, \qquad M:\ 4x + 6y = 0$$

Solution To determine whether these lines have equal slopes and different y-intercepts, we write each equation in slope–intercept form:

L:	M:
$2x + 3y = 6$	$4x + 6y = 0$
$3y = -2x + 6$	$6y = -4x$
$y = -\frac{2}{3}x + 2$	$y = -\frac{2}{3}x$
Slope $= -\frac{2}{3}$; y-intercept $= 2$	Slope $= -\frac{2}{3}$; y-intercept $= 0$

Because these lines have the same slope, $-\frac{2}{3}$, but different y-intercepts, the lines are parallel. See Figure 80.

Figure 80
Parallel lines

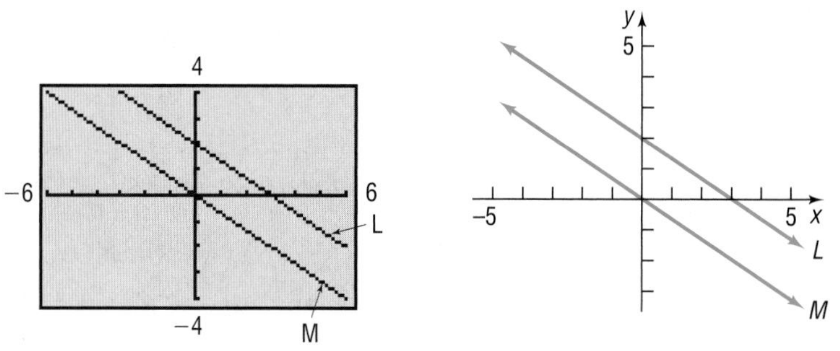

EXAMPLE 9 Finding a Line That Is Parallel to a Given Line

10 Find an equation for the line that contains the point $(2, -3)$ and is parallel to the line $2x + y = 6$.

Solution The slope of the line that we seek equals the slope of the line $2x + y = 6$, since the two lines are to be parallel. We begin by writing the equation of the line $2x + y = 6$ in slope–intercept form.

$$2x + y = 6$$
$$y = -2x + 6$$

The slope is -2. Since the line that we seek contains the point $(2, -3)$, we use the point–slope form to obtain

$$y + 3 = -2(x - 2) \quad \text{Point–slope form.}$$
$$2x + y = 1 \quad \text{General form.}$$
$$y = -2x + 1 \quad \text{Slope–intercept form.}$$

This line is parallel to the line $2x + y = 6$ and contains the point $(2, -3)$. See Figure 81.

Figure 81

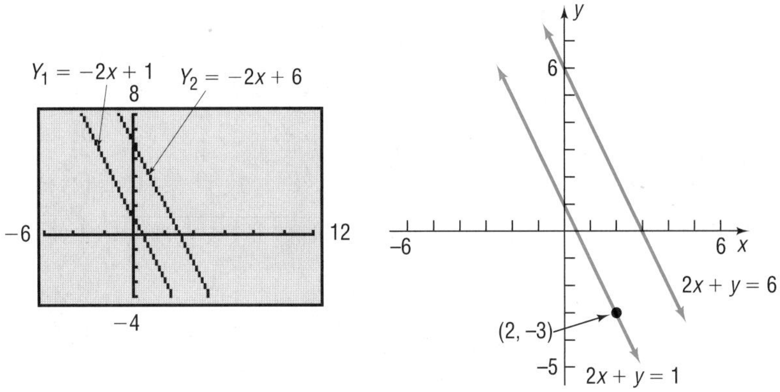

NOW WORK PROBLEM 43.

When two lines intersect at a right angle (90°), they are said to be **perpendicular**. See Figure 82.

Figure 82
Perpendicular lines

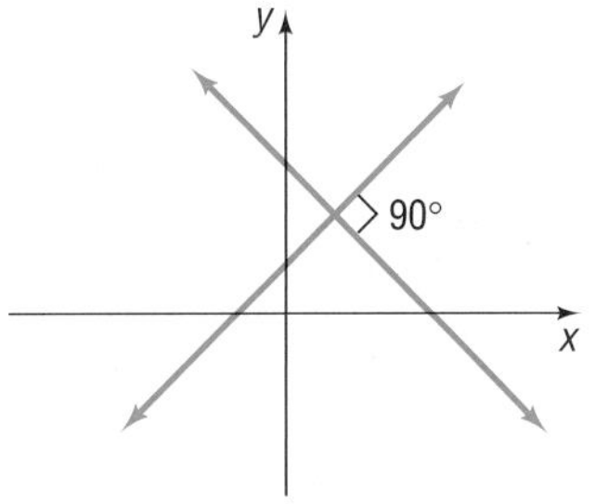

The following result gives a condition, in terms of their slopes, for two lines to be perpendicular.

Theorem Two nonvertical lines are perpendicular if and only if the product of their slopes is -1.

Here we shall prove the "only if" part of the statement:

If two nonvertical lines are perpendicular, then the product of their slopes is -1.

You are left to prove the "if" part of the theorem. That is:

If two nonvertical lines have slopes whose product is -1, then the lines are perpendicular.

Figure 83

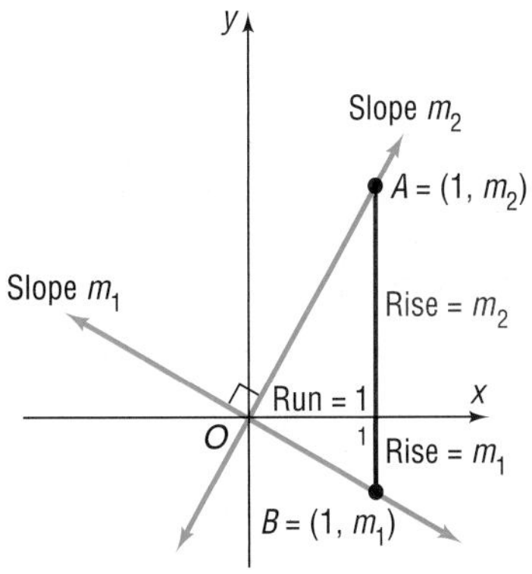

Proof Let m_1 and m_2 denote the slopes of the two lines. There is no loss in generality (that is, neither the angle nor the slopes are affected) if we situate the lines so that they meet at the origin. See Figure 83. The point $A = (1, m_2)$ is on the line having slope m_2, and the point $B = (1, m_1)$ is on the line having slope m_1. (Do you see why this must be true?)

Suppose that the lines are perpendicular. Then triangle OAB is a right triangle. As a result of the Pythagorean Theorem, it follows that

$$[d(O, A)]^2 + [d(O, B)]^2 = [d(A, B)]^2 \tag{5}$$

By the distance formula, we can write each of these distances as

$$[d(O, A)]^2 = (1 - 0)^2 + (m_2 - 0)^2 = 1 + m_2^2$$
$$[d(O, B)]^2 = (1 - 0)^2 + (m_1 - 0)^2 = 1 + m_1^2$$
$$[d(A, B)]^2 = (1 - 1)^2 + (m_2 - m_1)^2 = m_2^2 - 2m_1m_2 + m_1^2$$

Using these facts in equation (5), we get

$$(1 + m_2^2) + (1 + m_1^2) = m_2^2 - 2m_1m_2 + m_1^2$$

which, upon simplification, can be written as

$$m_1m_2 = -1$$

Thus, if the lines are perpendicular, the product of their slopes is -1. ■

You may find it easier to remember the condition for two nonvertical lines to be perpendicular by observing that the equality $m_1m_2 = -1$ means that m_1 and m_2 are negative reciprocals of each other; that is, either $m_1 = -1/m_2$ or $m_2 = -1/m_1$.

EXAMPLE 10 Finding the Slope of a Line Perpendicular to Another Line

If a line has slope $\frac{3}{2}$, any line having slope $-\frac{2}{3}$ is perpendicular to it. ■

EXAMPLE 11 Finding the Equation of a Line Perpendicular to a Given Line

11 Find an equation of the line that contains the point $(1, -2)$ and is perpendicular to the line $x + 3y = 6$. Graph the two lines.

Solution We first write the equation of the given line in slope–intercept form to find its slope.

$$x + 3y = 6$$
$$3y = -x + 6$$
$$y = -\frac{1}{3}x + 2$$

The given line has slope $-\frac{1}{3}$. Any line perpendicular to this line will have slope 3. Because we require the point $(1, -2)$ to be on this line with slope 3, we use the point–slope form of the equation of a line.

$$y - (-2) = 3(x - 1) \quad \text{Point–slope form.}$$
$$y + 2 = 3(x - 1)$$

This equation is equivalent to the forms

$$3x - y = 5 \qquad \text{General form.}$$

$$y = 3x - 5 \qquad \text{Slope–intercept form.}$$

Figure 84 shows the graphs.

Figure 84

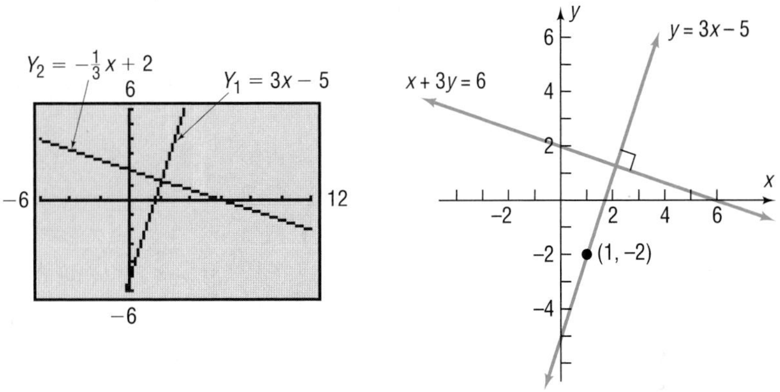

WARNING Be sure to use a square screen when you graph perpendicular lines. Otherwise, the angle between the two lines will appear distorted.

NOW WORK PROBLEM 49.

1.6 EXERCISES

In Problems 1–4, (a) find the slope of the line and (b) interpret the slope.

1.

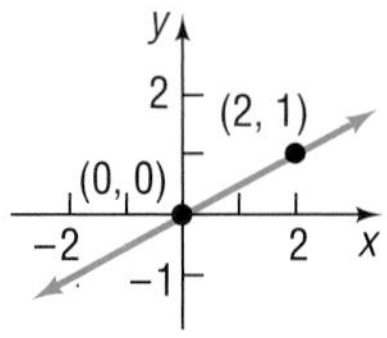

2.

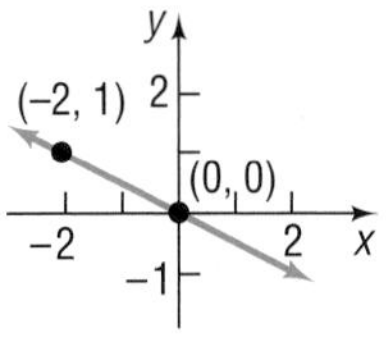

3.

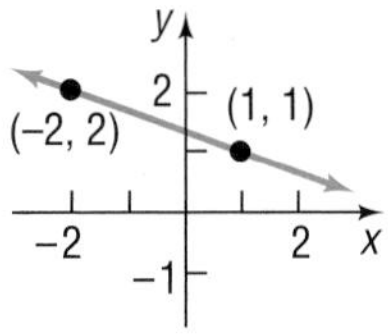

4.

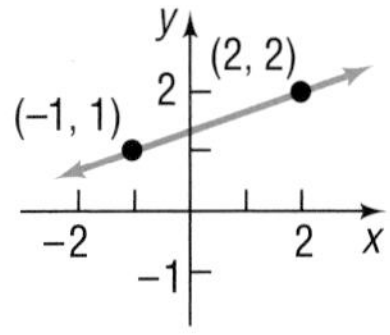

In Problems 5–14, plot each pair of points and determine the slope of the line containing them. Graph the line.

5. $(2, 3); (4, 0)$
6. $(4, 2); (3, 4)$
7. $(-2, 3); (2, 1)$
8. $(-1, 1); (2, 3)$
9. $(-3, -1); (2, -1)$
10. $(4, 2); (-5, 2)$
11. $(-1, 2); (-1, -2)$
12. $(2, 0); (2, 2)$
13. $(\sqrt{2}, 3); (1, \sqrt{3})$
14. $(-2\sqrt{2}, 0); (4, \sqrt{5})$

In Problems 15–22, graph, by hand, the line containing the point P and having slope m.

15. $P = (1, 2)$; $m = 3$
16. $P = (2, 1)$; $m = 4$
17. $P = (2, 4)$; $m = \frac{-3}{4}$
18. $P = (1, 3)$; $m = \frac{-2}{5}$
19. $P = (-1, 3)$; $m = 0$
20. $P = (2, -4)$; $m = 0$
21. $P = (0, 3)$; slope undefined
22. $P = (-2, 0)$; slope undefined

In Problems 23–30, find an equation of each line. Express your answer using either the general form or the slope–intercept form of the equation of a line, whichever you prefer.

23.

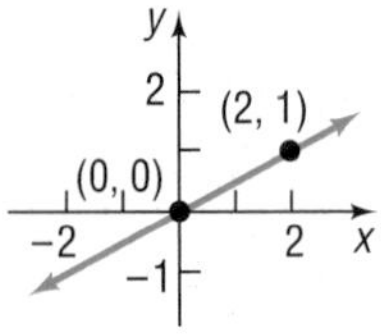

24.

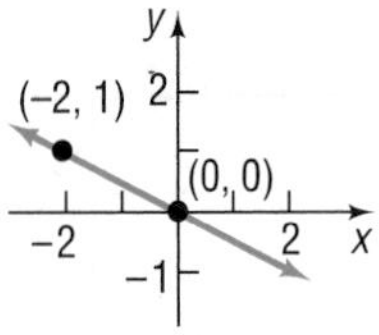

25.

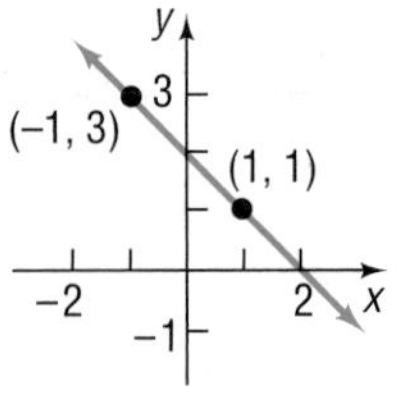

26.

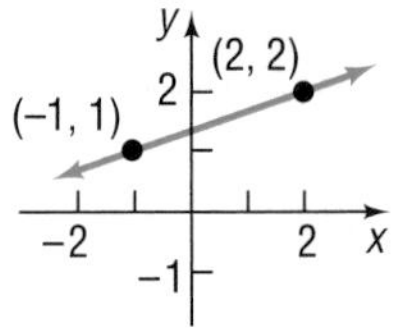

27.

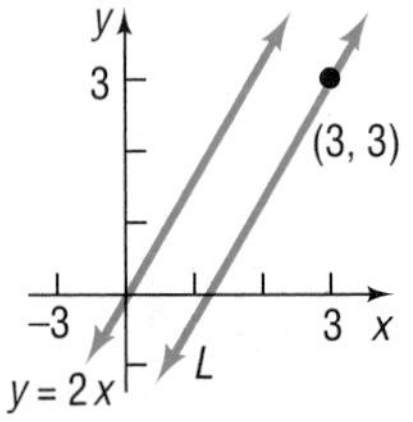

L is parallel to $y = 2x$

28.

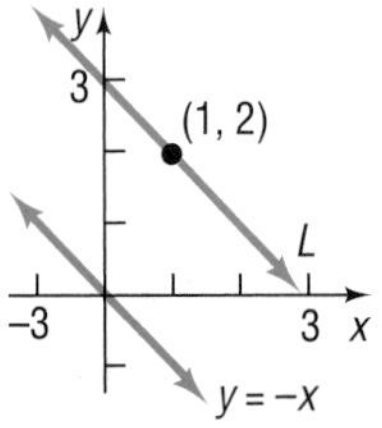

L is parallel to $y = -x$

29.

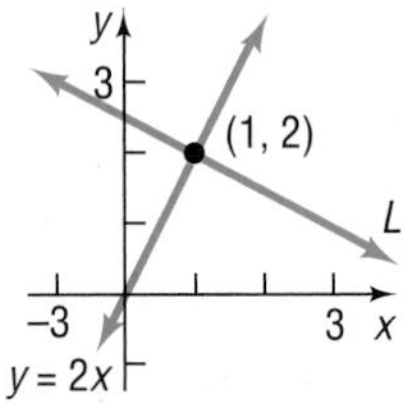

L is perpendicular to $y = 2x$

30.

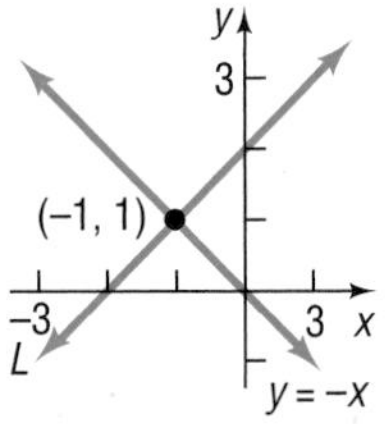

L is perpendicular to $y = -x$

In Problems 31–54, find an equation for the line with the given properties. Express your answer using either the general form or the slope–intercept form of the equation of a line, whichever you prefer.

31. Slope $= 3$; containing the point $(-2, 3)$

32. Slope $= 2$; containing the point $(4, -3)$

33. Slope $= -\frac{2}{3}$; containing the point $(1, -1)$

34. Slope $= \frac{1}{2}$; containing the point $(3, 1)$

35. Containing the points $(1, 3)$ and $(-1, 2)$

36. Containing the points $(-3, 4)$ and $(2, 5)$

37. Slope $= -3$; y-intercept $= 3$

38. Slope $= -2$; y-intercept $= -2$

39. x-intercept $= 2$; y-intercept $= -1$

40. x-intercept $= -4$; y-intercept $= 4$

41. Slope undefined; containing the point $(2, 4)$

42. Slope undefined; containing the point $(3, 8)$

43. Parallel to the line $y = 2x$; containing the point $(-1, 2)$

44. Parallel to the line $y = -3x$; containing the point $(-1, 2)$

45. Parallel to the line $2x - y = -2$; containing the point $(0, 0)$

46. Parallel to the line $x - 2y = -5$; containing the point $(0, 0)$

47. Parallel to the line $x = 5$; containing the point $(4, 2)$

48. Parallel to the line $y = 5$; containing the point $(4, 2)$

49. Perpendicular to the line $y = \frac{1}{2}x + 4$; containing the point $(1, -2)$

50. Perpendicular to the line $y = 2x - 3$; containing the point $(1, -2)$

51. Perpendicular to the line $2x + y = 2$; containing the point $(-3, 0)$

52. Perpendicular to the line $x - 2y = -5$; containing the point $(0, 4)$

53. Perpendicular to the line $x = 8$; containing the point $(3, 4)$

54. Perpendicular to the line $y = 8$; containing the point $(3, 4)$

In Problems 55–74, find the slope and y-intercept of each line. Graph the line by hand. Check your graph using a graphing utility.

55. $y = 2x + 3$

56. $y = -3x + 4$

57. $\frac{1}{2}y = x - 1$

58. $\frac{1}{3}x + y = 2$

59. $y = \frac{1}{2}x + 2$

60. $y = 2x + \frac{1}{2}$

61. $x + 2y = 4$

62. $-x + 3y = 6$

63. $2x - 3y = 6$

64. $3x + 2y = 6$

65. $x + y = 1$

66. $x - y = 2$

67. $x = -4$

68. $y = -1$

69. $y = 5$

70. $x = 2$

71. $y - x = 0$

72. $x + y = 0$

73. $2y - 3x = 0$

74. $3x + 2y = 0$

75. Find an equation of the x-axis.

76. Find an equation of the y-axis.

In Problems 77–80, match each graph with the correct equation:

(a) $y = x$ (b) $y = 2x$ (c) $y = x/2$ (d) $y = 4x$

77.

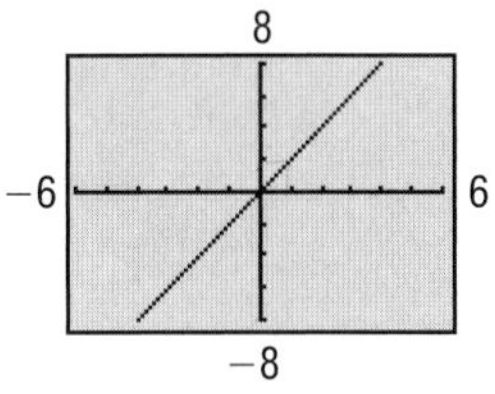

78.

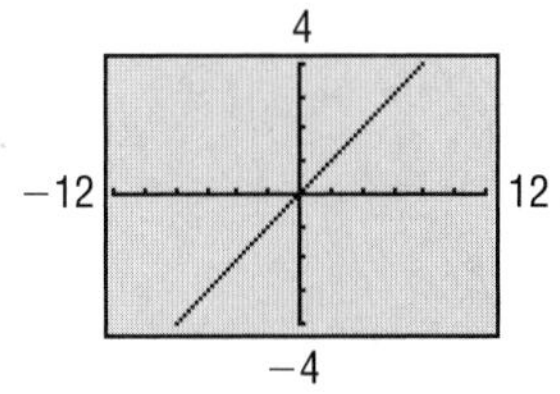

79.

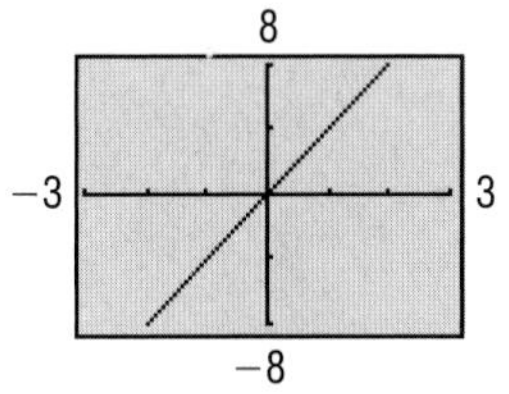

80.

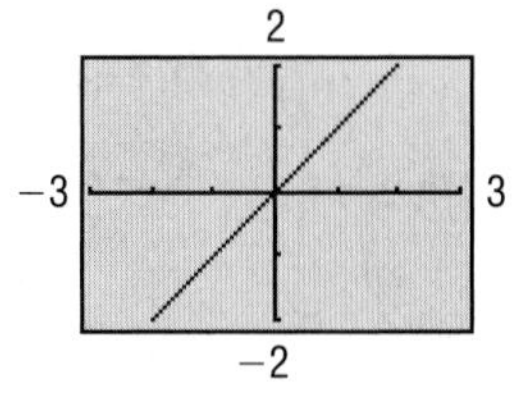

In Problems 81–84, write an equation of each line. Express your answer using either the general form or the slope–intercept form of the equation of a line, whichever you prefer.

81.

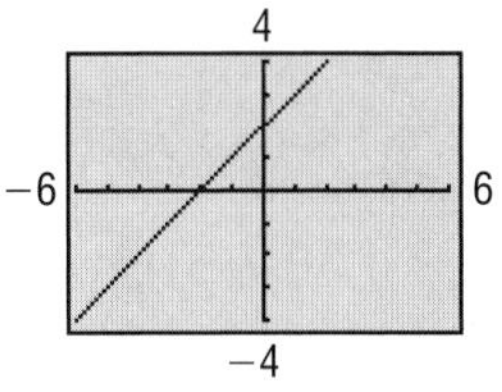

82.

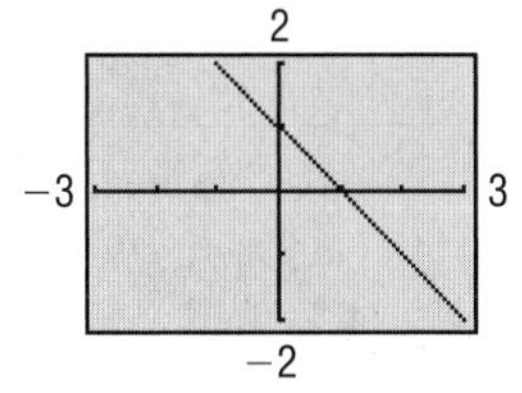

83.

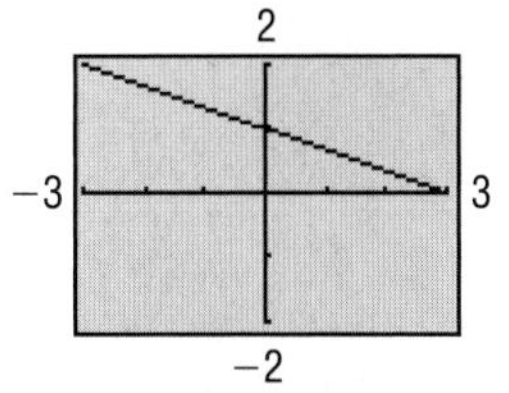

84.

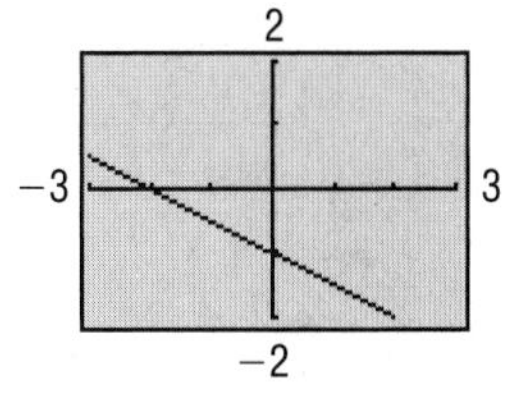

85. Measuring Temperature The relationship between Celsius (°C) and Fahrenheit (°F) degrees for measuring temperature is linear. Find an equation relating °C and °F if 0°C corresponds to 32°F and 100°C corresponds to 212°F. Use the equation to find the Celsius measure of 70°F.

86. Measuring Temperature The Kelvin (K) scale for measuring temperature is obtained by adding 273 to the Celsius temperature.
(a) Write an equation relating K and °C.
(b) Write an equation relating K and °F (see Problem 85).

87. Business: Computing Profit Each Sunday, a newspaper agency sells x copies of a certain newspaper for \$1.00 per copy. The cost to the agency of each newspaper is \$0.50. The agency pays a fixed cost for storage, delivery, and so on, of \$100 per Sunday.
(a) Write an equation that relates the profit P, in dollars, to the number x of copies sold. Graph this equation.
(b) What is the profit to the agency if 1000 copies are sold?
(c) What is the profit to the agency if 5000 copies are sold?

88. Business: Computing Profit Repeat Problem 87 if the cost to the agency is \$0.45 per copy and the fixed cost is \$125 per Sunday.

89. Cost of Electricity In 1999, Florida Power and Light Company supplied electricity in the summer months to residential customers for a monthly customer charge of \$5.65 plus 6.543 ¢ per kilowatt-hour supplied in the month for the first 750 kilowatt-hours used.* Write an equation that relates the monthly charge, C, in dollars, to the number x of kilowatt-hours used in the month. Graph this equation. What is the monthly charge for using 300 kilowatt-hours? For using 750 kilowatt-hours?

90. Show that the line containing the points (a, b) and (b, a) is perpendicular to the line $y = x$. Also show that the midpoint of (a, b) and (b, a) lies on the line $y = x$.

91. The equation $2x - y = C$ defines a **family of lines**, one line for each value of C. On one set of coordinate axes, graph the members of the family when $C = -2, C = 0$, and $C = 4$. Can you draw a conclusion from the graph about each member of the family?

92. Rework Problem 91 for the family of lines $Cx + y = -4$.

93. Which form of the equation of a line do you prefer to use? Justify your position with an example that shows that your choice is better than another. Give reasons.

94. Can every line be written in slope–intercept form? Explain.

95. Does every line have two distinct intercepts? Explain. Are there lines that have no intercepts? Explain.

96. What can you say about two lines that have equal slopes and equal y-intercepts?

97. What can you say about two lines with the same x-intercept and the same y-intercept? Assume that the x-intercept is not 0.

98. If two lines have the same slope, but different x-intercepts, can they have the same y-intercept?

99. If two lines have the same y-intercept, but different slopes, can they have the same x-intercept? What is the only way that this can happen?

100. The accepted symbol used to denote the slope of a line is the letter m. Investigate the origin of this symbolism. Begin by consulting a French dictionary and looking up the French word *monter*. Write a brief essay on your findings.

* *Source:* Florida Power and Light Co., Miami, Florida, 1999.

101. The term *grade* is used to describe the inclination of a road. How does this term relate to the notion of slope of a line? Is a 4% grade very steep? Investigate the grades of some mountainous roads and determine their slopes. Write a brief essay on your findings.

102. **Carpentry** Carpenters use the term *pitch* to describe the steepness of staircases and roofs. How does pitch relate to slope? Investigate typical pitches used for stairs and for roofs. Write a brief essay on your findings.

1.7 SCATTER DIAGRAMS; LINEAR CURVE FITTING

OBJECTIVES
1. Draw and Interpret Scatter Diagrams
2. Distinguish between Linear and Nonlinear Relations
3. Use a Graphing Utility to Find the Line of Best Fit

SCATTER DIAGRAMS

1 A **relation** is a correspondence between two variables, say x and y, and can be written as a set of ordered pairs (x, y). When relations are written as ordered pairs, we say that x is related to y. Another way of interpreting the ordered pair (x, y) is to say y depends on x. In this sense, y is referred to as the **dependent** variable and x is called the **independent** variable. Often we are interested in specifying the type of relation (such as an equation) that might exist between two variables. The first step in finding this relation is to plot the ordered pairs using rectangular coordinates. The resulting graph is called a **scatter diagram**.

EXAMPLE 1 Drawing a Scatter Diagram

The data listed in Table 9 represent the apparent temperature versus the relative humidity in a room whose actual temperature is 72° Fahrenheit.

(a) Draw a scatter diagram by hand.
(b) Use a graphing utility to draw a scatter diagram.
(c) Describe what happens to the apparent temperature as the relative humidity increases.

Solution (a) To draw a scatter diagram by hand, we plot the ordered pairs listed in Table 9 with the relative humidity as the x-coordinate (independent variable) and the apparent temperature as the y-coordinate (dependent variable). See Figure 85(a). Notice that the points in a scatter diagram are not connected.

(b) Figure 85(b) shows a scatter diagram using a graphing utility.
(c) We see from the scatter diagrams that, as the relative humidity increases, the apparent temperature increases.

TABLE 9

Relative Humidity (%), *x*	Apparent Temperature, *y*	(*x*, *y*)
0	64	(0, 64)
10	65	(10, 65)
20	67	(20, 67)
30	68	(30, 68)
40	70	(40, 70)
50	71	(50, 71)
60	72	(60, 72)
70	73	(70, 73)
80	74	(80, 74)
90	75	(90, 75)
100	76	(100, 76)

Figure 85

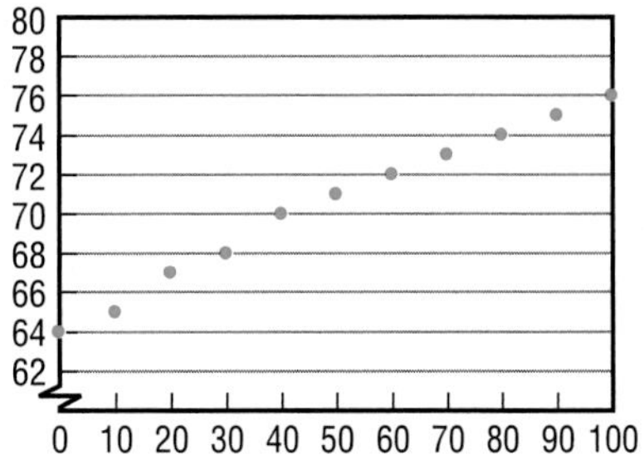

(a)

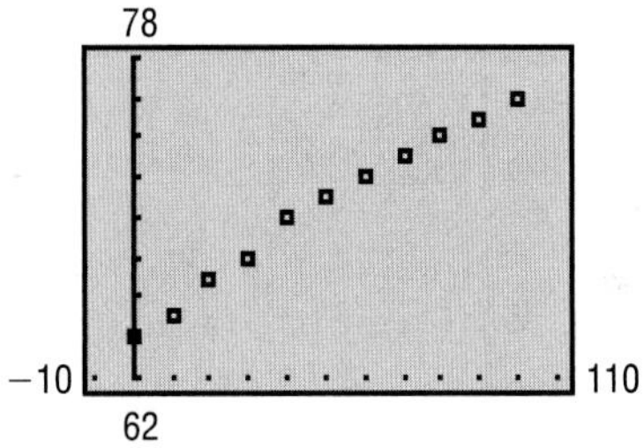

(b)

NOW WORK PROBLEM 7(a).

CURVE FITTING

2 Scatter diagrams are used to help us see the type of relation that may exist between two variables. In this text, we will discuss a variety of different relations that may exist between two variables. For now, we concentrate on distinguishing between linear and nonlinear relations. See Figure 86.

Figure 86

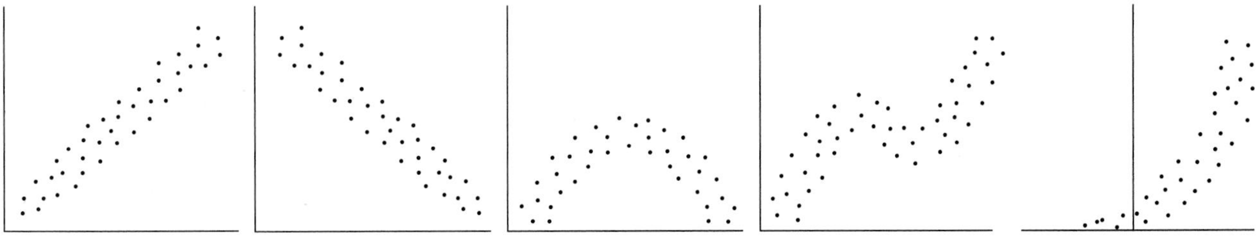

(a) Linear $y = mx + b, m > 0$ **(b)** Linear $y = mx + b, m < 0$ **(c)** Nonlinear **(d)** Nonlinear **(e)** Nonlinear

EXAMPLE 2 Distinguishing between Linear and Nonlinear Relations

Determine whether the relation between the two variables in Figure 87 is linear or nonlinear.

Figure 87

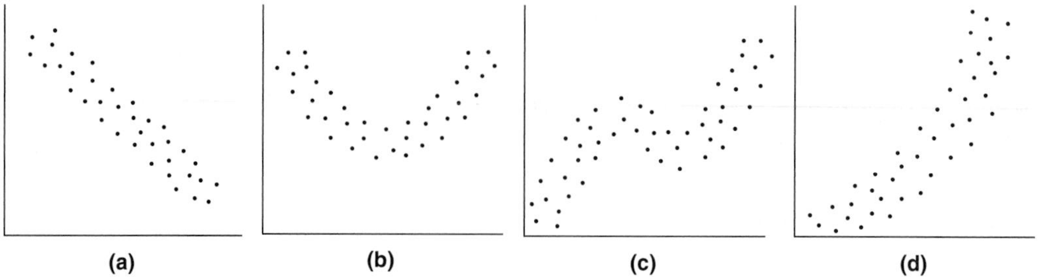

Solution (a) Linear (b) Nonlinear (c) Nonlinear (d) Nonlinear ■

NOW WORK PROBLEM 1.

In this section we will study data whose scatter diagrams imply that a linear relation exists between the two variables. Nonlinear data will be discussed in later chapters.

Suppose that the scatter diagram of a set of data appears to be linearly related as in Figure 86(a) or (b). We might wish to find an equation of a line that relates the two variables. One way to obtain an equation for such data is to draw a line through two points on the scatter diagram and estimate the equation of the line.

EXAMPLE 3 Find an Equation for Linearly Related Data

Using the data in Table 9 from Example 1:

(a) Select two points from the data and find an equation of the line containing the points.

(b) Graph the line on the scatter diagram obtained in Example 1(b).

Solution (a) Select two points, say (10, 65) and (70, 73). (You should select your own two points and complete the solution.) The slope of the line joining the points (10, 65) and (70, 73) is

$$m = \frac{73 - 65}{70 - 10} = \frac{8}{60} = \frac{2}{15}$$

The equation of the line with slope $2/15$ and passing through (10, 65) is found using the point–slope form with $m = 2/15$, $x_1 = 10$, and $y_1 = 65$.

$$y - y_1 = m(x - x_1)$$

$$y - 65 = \frac{2}{15}(x - 10)$$

$$y = \frac{2}{15}x + \frac{191}{3}$$

Figure 88

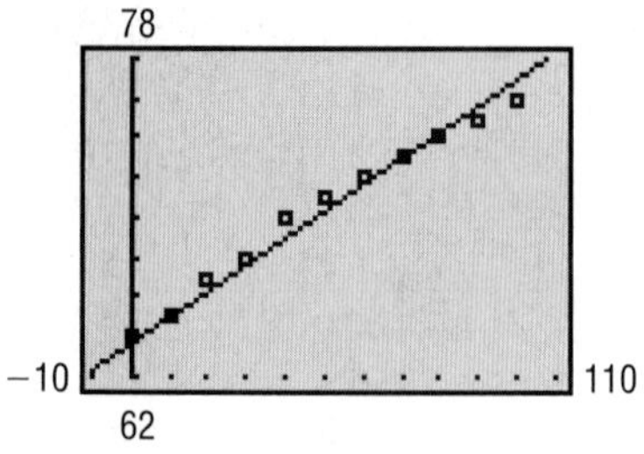

(b) Figure 88 shows the scatter diagram with the graph of the line obtained in part (a). ■

NOW WORK PROBLEM 7(b) AND (c).

Line of Best Fit

3 The line obtained in Example 3 depends on the selection of points, which will vary from person to person. So the line that we found might be different from the line that you found. Although the line that we found in Example 3 appears to "fit" the data well, there may be a line that "fits it better." Do you think your line fits the data better? Is there a line of *best fit*? As it turns out, there is a method for finding the line that best fits linearly related data (called the *line of best fit*).*

EXAMPLE 4 Finding the Line of Best Fit

Using the data in Table 9 from Example 1:

(a) Find the line of best fit using a graphing utility.

(b) Graph the line of best fit on the scatter diagram drawn in Example 1(b).

(c) Interpret the slope.

(d) Use the line of best fit to predict the apparent temperature of a room whose temperature is 72°F and relative humidity is 45%.

Figure 89

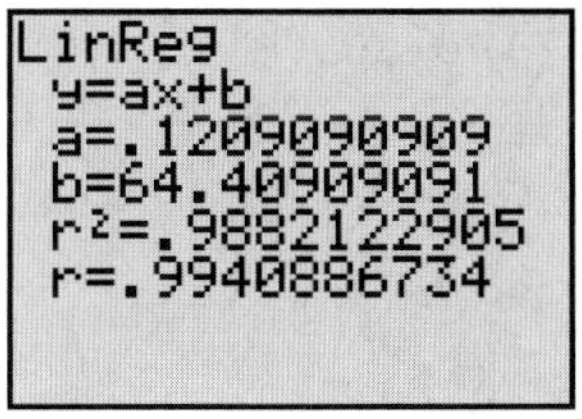

Figure 90

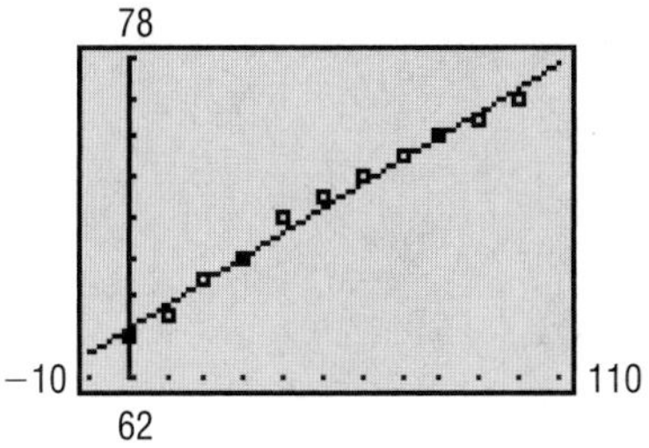

Solution (a) Graphing utilities contain built-in programs that find the line of best fit for a collection of points in a scatter diagram. (Look in your owner's manual for details on how to execute the program.) Upon executing the LINear REGression program, we obtain the results shown in Figure 89. The output that the utility provides shows us the equation $y = ax + b$, where a is the slope of the line and b is the y-intercept. The line of best fit that relates relative humidity to apparent temperature may be expressed as the line $y = 0.121x + 64.409$.

(b) Figure 90 shows the graph of the line of best fit, along with the scatter diagram obtained in Example 1(b).

(c) The slope of the line of best fit is 0.121, which means that, for every 1% increase in the relative humidity, apparent room temperature increases 0.121°F.

(d) Letting $x = 45$ in the equation of the line of best fit, we obtain $y = 0.121(45) + 64.409 \approx 70°\text{F}$. ■

NOW WORK PROBLEM 7(d) AND (e).

Does the line of best fit appear to be a good fit? In other words, does the line appear to accurately describe the relation between temperature and relative humidity?

And just how "good" is this line of best fit? The answers are given by what is called the *correlation coefficient*. Look again at Figure 89. The last line of output is $r = 0.994$. This number, called the **correlation coefficient, *r***, $-1 \le r \le 1$, is a measure of the strength of the *linear relation* that exists between two variables. The closer that $|r|$ is to 1, the more perfect the linear relationship is. If r is close to 0, there is little or no *linear* relationship between the variables. A negative value of r, $r < 0$, indicates that as x increases y decreases; a positive value of r, $r > 0$, indicates that as x increases y does also. The data given in Example 1, having a correlation coefficient of 0.994, are indicative of a strong linear relationship with positive slope.

*We shall not discuss in this book the underlying mathematics of lines of best fit. Most books in statistics and many in linear algebra discuss this topic.

1.7 EXERCISES

In Problems 1–6, examine the scatter diagram and determine whether the type of relation, if any, that may exist is linear or nonlinear.

1.

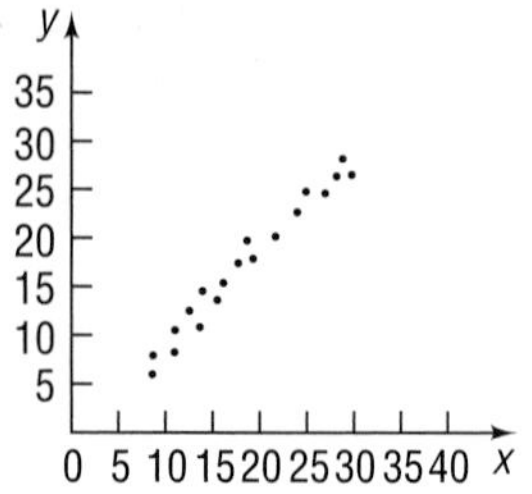

2.

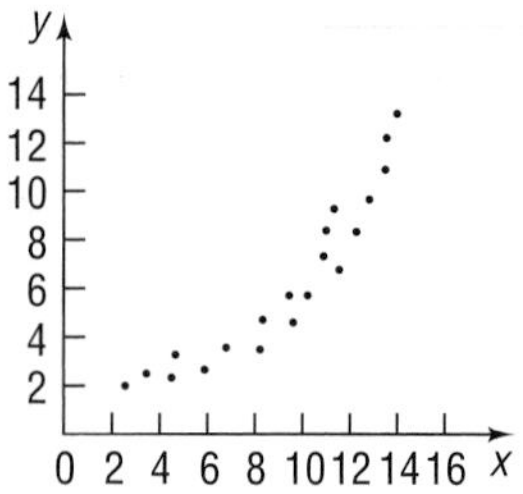

3.

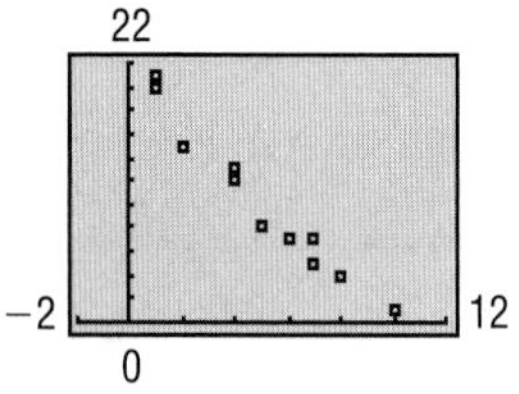

4.

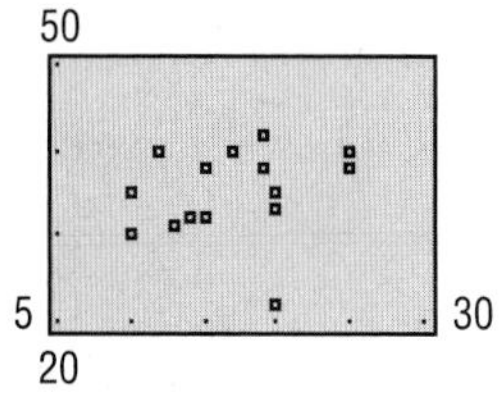

5.

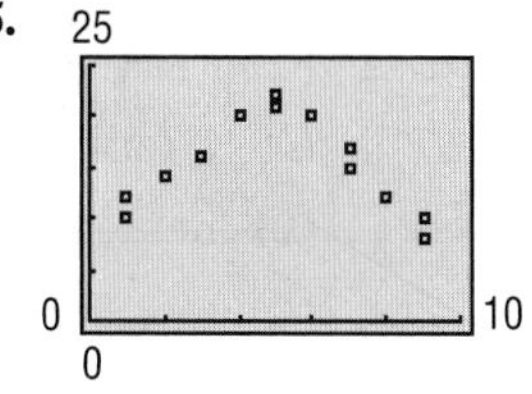

6.

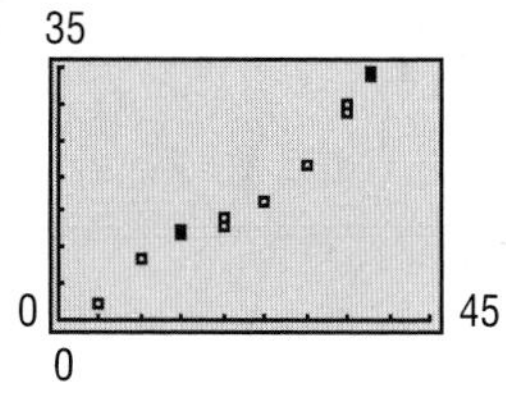

In Problems 7–14:

(a) Draw a scatter diagram.
*(b) Select two points from the scatter diagram and find the equation of the line through the points selected.**
(c) Graph the line found in part (b) on the scatter diagram.
(d) Use a graphing utility to find the line of best fit.
(e) Use a graphing utility to graph the line of best fit on the scatter diagram.

7.

x	3	4	5	6	7	8	9
y	4	6	7	10	12	14	16

8.

x	3	5	7	9	11	13
y	0	2	3	6	9	11

9.

x	−2	−1	0	1	2
y	−4	0	1	4	5

10.

x	−2	−1	0	1	2
y	7	6	3	2	0

11.

x	20	30	40	50	60
y	100	95	91	83	70

12.

x	5	10	15	20	25
y	2	4	7	11	18

13.

x	−20	−17	−15	−14	−10
y	100	120	118	130	140

14.

x	−30	−27	−25	−20	−14
y	10	12	13	13	18

15. Consumption and Disposable Income An economist wishes to estimate a linear equation that relates personal consumption expenditures (C) and disposable income (I). Both C and I are in thousands of dollars. She interviews eight heads of households for families of size 4 and obtains the data on the right.

Let I represent the independent variable and C the dependent variable.

(a) Use a graphing utility to draw a scatter diagram.
(b) Use a graphing utility to find the line of best fit to the data.
(c) Interpret the slope. The slope of this line is called the **marginal propensity to consume.**
(d) Predict the consumption of a family whose disposable income is $42 thousand.

I(000)	C(000)
20	16
20	18
18	13
27	21
36	27
37	26
45	36
50	39

* Answers will vary.

16. Marginal Propensity to Save The same economist as in Problem 15 wants to estimate a linear equation that relates savings (S) and disposable income (I). Let $S = I - C$ be the dependent variable and I the independent variable.

(a) Use a graphing utility to draw a scatter diagram.

(b) Use a graphing utility to find the line of best fit to the data.

(c) Interpret the slope. The slope of this line is called the **marginal propensity to save.**

(d) Predict the savings of a family whose income is $42 thousand.

17. Mortgage Qualification The amount of money that a lending institution will allow you to borrow mainly depends on the interest rate and your annual income. The following data represent the annual income, I, required by a bank in order to lend L dollars at an interest rate of 7.5% for 30 years.

Annual income, I (dollars)	Loan amount, L (dollars)
15,000	44,600
20,000	59,500
25,000	74,500
30,000	89,400
35,000	104,300
40,000	119,200
45,000	134,100
50,000	149,000
55,000	163,900
60,000	178,800
65,000	193,700
70,000	208,600

Source: Information Please Almanac

Let I represent the independent variable and L the dependent variable.

(a) Use a graphing utility to draw a scatter diagram of the data.

(b) Use a graphing utility to find the line of best fit to the data.

(c) Graph the line of best fit on the scatter diagram drawn in part (a).

(d) Interpret the slope.

(e) Determine the loan amount that an individual would qualify for if her income is $42,000.

18. Mortgage Qualification The amount of money that a lending institution will allow you to borrow mainly depends on the interest rate and your annual income. The following data represent the annual income, I, required by a bank in order to lend L dollars at an interest rate of 8.5% for 30 years.

Annual Income, I ($)	Loan Amount, L ($)
15,000	40,600
20,000	54,100
25,000	67,700
30,000	81,200
35,000	94,800
40,000	108,300
45,000	121,900
50,000	135,400
55,000	149,000
60,000	162,500
65,000	176,100
70,000	189,600

Source: Information Please Almanac, 1999.

Let I represent the independent variable and L the dependent variable.

(a) Use a graphing utility to draw a scatter diagram of the data.

(b) Use a graphing utility to find the line of best fit to the data.

(c) Graph the line of best fit on the scatter diagram drawn in part (a).

(d) Interpret the slope.

(e) Determine the loan amount that an individual would qualify for if her income is $42,000.

19. Apparent Room Temperature The following data represent the apparent temperature versus the relative humidity in a room whose actual temperature is 65° Fahrenheit.

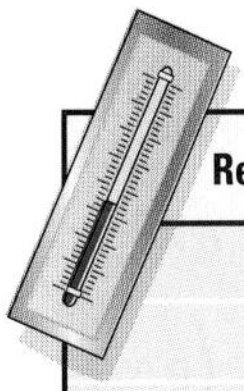

Relative Humidity, h	Apparent Temperature, T
0	59
10	60
20	61
30	61
40	62
50	63
60	64
70	65
80	65
90	66
100	67

Source: National Oceanic and Atmospheric Administration

Let h represent the independent variable and T the dependent variable.

(a) Use a graphing utility to draw a scatter diagram of the data.

(b) Use a graphing utility to find the line of best fit to the data.

(c) Graph the line of best fit on the scatter diagram drawn in part (a).

(d) Interpret the slope.

(e) Determine the apparent temperature of a room whose actual temperature is 65°F if the relative humidity is 75%.

20. Apparent Room Temperature The following data represent the apparent temperature versus the relative humidity in a room whose actual temperature is 75° Fahrenheit.

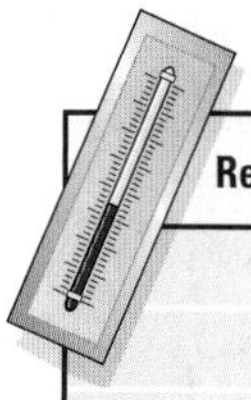

Relative Humidity, *h*	Apparent Temperature, *T*
0	68
10	69
20	71
30	72
40	74
50	75
60	76
70	76
80	77
90	78
100	79

Source: National Oceanic and Atmospheric Administration

Let h represent the independent variable and T the dependent variable.

(a) Use a graphing utility to draw a scatter diagram of the data.

(b) Use a graphing utility to find the line of best fit to the data.

(c) Graph the line of best fit on the scatter diagram drawn in part (a).

(d) Interpret the slope.

(e) Determine the apparent temperature of a room whose actual temperature is 75°F if the relative humidity is 75%.

21. Average Miles per Car The following data represent the average miles driven per car (in thousands) in the United States for the years 1985 to 1996.

Let the year, x, represent the independent variable and average miles per car, M, represent the dependent variable.

Year, *x*	Average Miles Per Car, *M*
1985	9.4
1986	9.5
1987	9.7
1988	10.0
1989	10.2
1990	10.3
1991	10.3
1992	10.6
1993	10.5
1994	10.8
1995	11.1
1996	11.3

Source: U.S. Federal Highway Administration

(a) Use a graphing utility to draw a scatter diagram of the data.

(b) Use a graphing utility to find the line of best fit to the data.

(c) Graph the line of best fit on the scatter diagram drawn in part (a).

(d) Interpret the slope.

(e) Predict the average number of miles driven per car in 1997.

22. Employment and Labor Force The following data represent the size of the civilian labor force (people aged 16 years and older, excluding those serving in the military) and the number of employed people in the United States for the years 1990 to 1997. Treat the size of the labor force as the independent variable and the number employed as the dependent variable. Both the size of the labor force and the number employed are measured in thousands of people.

Year	Civilian Labor Force	Number Employed
1990	125,840	118,793
1991	126,346	117,718
1992	128,105	118,492
1993	129,200	120,259
1994	131,056	123,060
1995	132,304	124,900
1996	133,943	126,708
1997	136,297	129,558

Source: U.S. Bureau of Labor Statistics

(a) Use a graphing utility to draw a scatter diagram.
(b) Use a graphing utility to find the line of best fit to the data.
(c) Graph the line of best fit on the scatter diagram drawn in part (a).
(d) Interpret the slope.
(e) Predict the number employed if the civilian labor force is 140,000 thousand.

23. Per Capita Personal Income The following data represent per capita personal income, x, and per capita personal consumption expenditures, y, for the years 1990 to 1997 in constant (1992) dollars. (These are dollars adjusted for inflation).

Year	Per Capita Personal Income, x	Per Capita Personal Consumption Expenditures, y
1990	17,996	16,532
1991	17,809	16,249
1992	18,113	16,520
1993	18,221	16,825
1994	18,431	17,207
1995	18,861	17,460
1996	19,116	17,750
1997	19,493	18,170

Source: U.S. Bureau of Economic Analysis

(a) Use a graphing utility to draw a scatter diagram.
(b) Use a graphing utility to find the line of best fit to the data.
(c) Graph the line of best fit on the scatter diagram drawn in part (a).
(d) Interpret the slope.
(e) Predict per capita personal consumption expenditures if per capita personal income is $19,200.

24. Government Expenditures The following data represent total federal government expenditures (in billions of dollars) of the United States for the years 1991 to 1998.

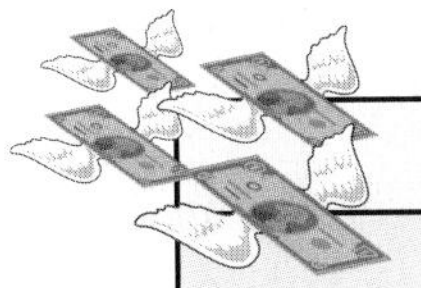

Year	Total Expenditures
1991	1366.1
1992	1381.7
1993	1374.8
1994	1392.8
1995	1407.8
1996	1412.0
1997	1414.0
1998	1441.0

Source: U.S. Office of Management and Budget

(a) Use a graphing utility to draw a scatter diagram.
(b) Use a graphing utility to find the line of best fit to the data.
(c) Graph the line of best fit on the scatter diagram drawn in part (a).
(d) Interpret the slope.
(e) Predict total federal government expenditures of the United States for 1999.

PREPARING FOR THIS SECTION

Before getting started, review the following:

✓ Completing the Square (pp. 998–999 in the Appendix, Section 10)

1.8 CIRCLES

OBJECTIVES
1. Write the Standard Form of the Equation of a Circle
2. Graph a Circle by Hand and by Using a Graphing Utility
3. Find the Center and Radius of a Circle from an Equation in General Form

1 One advantage of a coordinate system is that it enables us to translate a geometric statement into an algebraic statement, and vice versa. Consider, for example, the following geometric statement that defines a circle.

A **circle** is a set of points in the xy-plane that are a fixed distance r from a fixed point (h, k). The fixed distance r is called the **radius**, and the fixed point (h, k) is called the **center** of the circle.

Figure 91

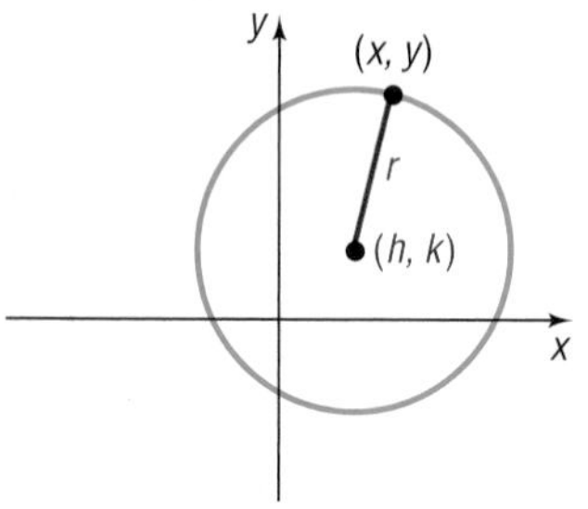

Figure 91 shows the graph of a circle. To find the equation of this graph, we let (x, y) represent the coordinates of any point on a circle with radius r and center (h, k). Then the distance between the points (x, y) and (h, k) must always equal r. That is, by the distance formula (Section 1.1),

$$\sqrt{(x-h)^2 + (y-k)^2} = r$$

or, equivalently,

$$(x-h)^2 + (y-k)^2 = r^2$$

The **standard form of an equation of a circle** with radius r and center (h, k) is

$$(x-h)^2 + (y-k)^2 = r^2 \qquad \textbf{(1)}$$

The standard form of an equation of a circle of radius r with center at the origin $(0, 0)$ is

$$x^2 + y^2 = r^2$$

If the radius $r = 1$, the circle whose center is at the origin is called the **unit circle** and has the equation

$$x^2 + y^2 = 1$$

See Figure 92.

Figure 92
Unit circle $x^2 + y^2 = 1$

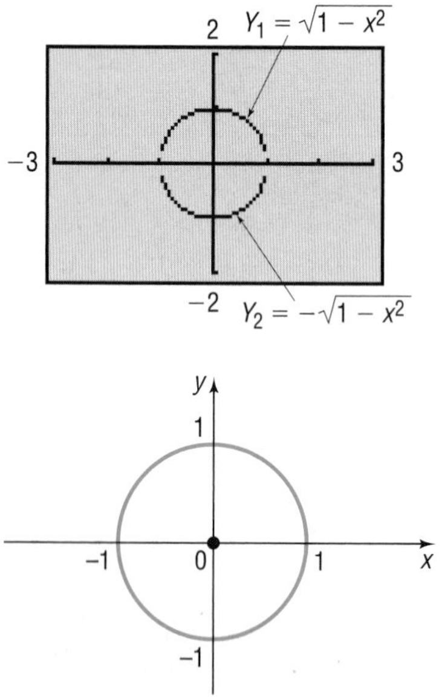

EXAMPLE 1 Writing the Standard Form of the Equation of a Circle

Write the standard form of the equation of the circle with radius 5 and center $(-3, 6)$.

Solution Using the form of equation (1) and substituting the values $r = 5$, $h = -3$, and $k = 6$, we have

$$(x-h)^2 + (y-k)^2 = r^2$$

$$(x+3)^2 + (y-6)^2 = 25$$

NOW WORK PROBLEM 1.

The graph of any equation of the form of equation (1) is that of a circle with radius r and center (h, k).

EXAMPLE 2 Graphing a Circle

2 Graph the equation: $(x + 3)^2 + (y - 2)^2 = 16$

Solution The graph of the equation is a circle. To graph a circle on a graphing utility, we must write the equation in the form $y = \{\text{expression involving } x\}$.* Thus, we must solve for y in the equation

$$(x + 3)^2 + (y - 2)^2 = 16$$

$$(y - 2)^2 = 16 - (x + 3)^2 \quad \text{Subtract } (x+3)^2 \text{ from both sides.}$$

$$y - 2 = \pm\sqrt{16 - (x + 3)^2} \quad \text{Take the square root of both sides.}$$

$$y = 2 \pm \sqrt{16 - (x + 3)^2} \quad \text{Add 2 to both sides.}$$

Figure 93

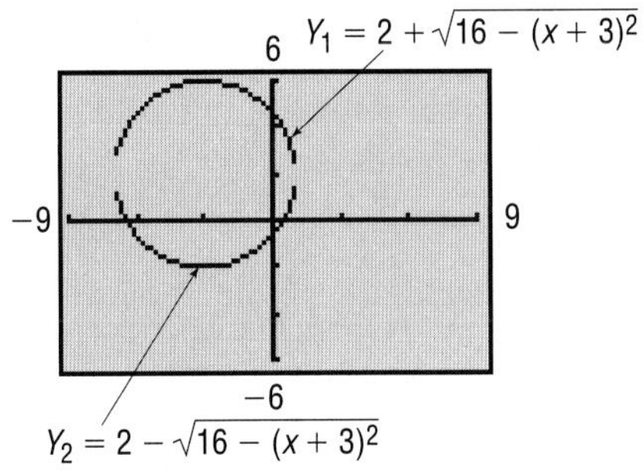

To graph the circle, we graph the top half,

$$Y_1 = 2 + \sqrt{16 - (x + 3)^2}$$

and the bottom half,

$$Y_2 = 2 - \sqrt{16 - (x + 3)^2}$$

Also, be sure to use a square screen. Otherwise, the circle will appear distorted. Figure 93 shows the graph on a TI-83.

To graph the equation by hand, we first compare the given equation to the standard form of the equation of a circle. The comparison yields information about the circle.

Figure 94

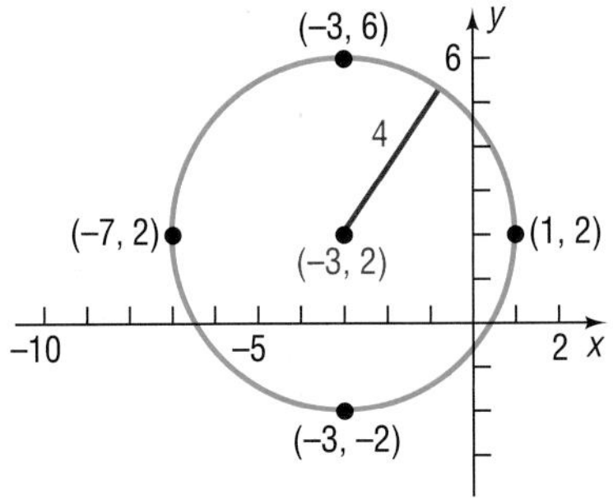

$$(x + 3)^2 + (y - 2)^2 = 16$$

$$(x - (-3))^2 + (y - 2)^2 = 4^2$$

$$\uparrow \qquad\qquad \uparrow \qquad \uparrow$$

$$(x - h)^2 + (y - k)^2 = r^2$$

We see that $h = -3$, $k = 2$, and $r = 4$. The circle has center $(-3, 2)$ and a radius of 4 units. To graph this circle, we first plot the center $(-3, 2)$. Since the radius is 4, we can locate four points on the circle by plotting 4 units to the left, to the right, up, and down from the center. These four points can then be used as guides to obtain the graph. See Figure 94. ■

NOW WORK PROBLEM 17.

If we eliminate the parentheses from the standard form of the equation of the circle given in Example 2, we get

$$(x + 3)^2 + (y - 2)^2 = 16$$

$$x^2 + 6x + 9 + y^2 - 4y + 4 = 16$$

which we find, upon simplifying, is equivalent to

$$x^2 + y^2 + 6x - 4y - 3 = 0$$

*Some graphing utilities (e.g., TI-82, TI-83, TI-85, and TI-86) have a CIRCLE function that allows the user to enter only the coordinates of the center of the circle and its radius to graph the circle.

It can be shown that any equation of the form

$$x^2 + y^2 + ax + by + c = 0$$

has a graph that is a circle or a point, or it has no graph at all. For example, the graph of the equation $x^2 + y^2 = 0$ is the single point $(0, 0)$. The equation $x^2 + y^2 + 5 = 0$, or $x^2 + y^2 = -5$, has no graph, because sums of squares of real numbers are never negative. When its graph is a circle, the equation

$$\boxed{x^2 + y^2 + ax + by + c = 0}$$

is referred to as the **general form of the equation of a circle**.

3 If an equation of a circle is in the general form, we use the method of completing the square to put the equation in standard form so that we can identify its center and radius.

EXAMPLE 3 Graphing a Circle Whose Equation Is in General Form

Graph the equation: $x^2 + y^2 + 4x - 6y + 12 = 0$

Solution We complete the square in both x and y to put the equation in standard form. Group the expression involving x, group the expression involving y, and put the constant on the right side of the equation. The result is

$$(x^2 + 4x) + (y^2 - 6y) = -12$$

Next, complete the square of each expression in parentheses. Remember that any number added on the left side of the equation must be added on the right.

$$(x^2 + 4x + 4) + (y^2 - 6y + 9) = -12 + 4 + 9$$

$$\left(\tfrac{4}{2}\right)^2 = 4 \qquad \left(\tfrac{-6}{2}\right)^2 = 9$$

$$(x + 2)^2 + (y - 3)^2 = 1 \quad \text{Factor}$$

We recognize this equation as the standard form of the equation of a circle with radius 1 and center $(-2, 3)$.

To graph the equation using a graphing utility, we need to solve for y.

$$(y - 3)^2 = 1 - (x + 2)^2$$

$$y - 3 = \pm\sqrt{1 - (x + 2)^2}$$

$$y = 3 \pm \sqrt{1 - (x + 2)^2}$$

See Figure 95(a).

To graph the equation by hand, we use the center $(-2, 3)$ and the radius 1. Figure 95(b) illustrates the graph draw by hand.

Figure 95
$x^2 + y^2 + 4x - 6y + 12 = 0$

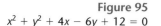

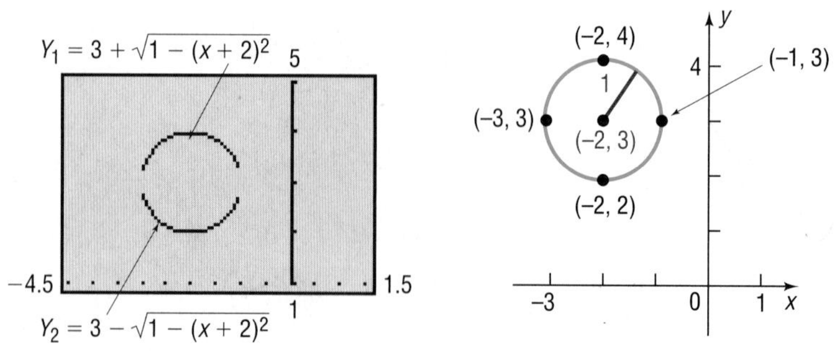

NOW WORK PROBLEM 19.

EXAMPLE 4 Finding the General Equation of a Circle

Find the general equation of the circle whose center is $(1, -2)$ and whose graph contains the point $(4, -2)$.

Solution To find the equation of a circle, we need to know its center and its radius. Here, we know that the center is $(1, -2)$. Since the point $(4, -2)$ is on the graph, the radius r will equal the distance from $(4, -2)$ to the center $(1, -2)$. See Figure 96. Thus,

$$r = \sqrt{(4-1)^2 + [-2-(-2)]^2} = \sqrt{9} = 3$$

Figure 96
$x^2 + y^2 - 2x + 4y - 4 = 0$

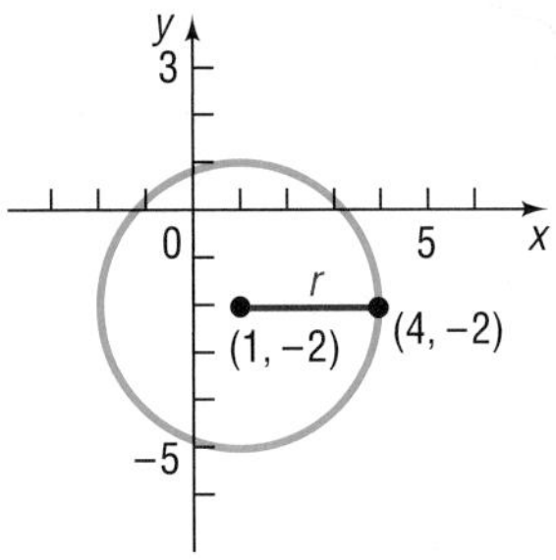

The standard form of the equation of the circle is

$$(x-1)^2 + (y+2)^2 = 9$$

Eliminating the parentheses and rearranging terms, we get the general equation

$$x^2 + y^2 - 2x + 4y - 4 = 0$$

OVERVIEW

The discussion in Sections 1.6 and 1.8 about lines and circles dealt with two main types of problems that can be generalized as follows:

1. Given an equation, classify it and graph it.
2. Given a graph, or information about a graph, find its equation.

This text deals with both types of problems. We shall study various equations, classify them, and graph them. Although the second type of problem is usually more difficult to solve than the first, in many instances a graphing utility can be used to solve such problems.

1.8 EXERCISES

In Problems 1–4, find the center and radius of each circle. Write the standard form of the equation.

1.

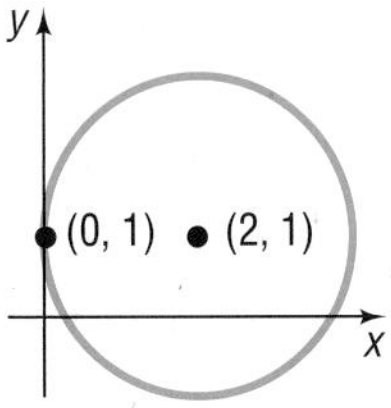

2.

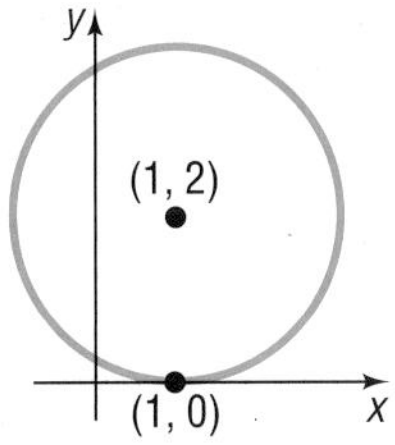

3.

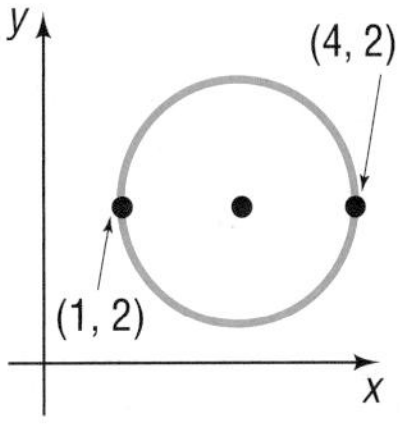

4.

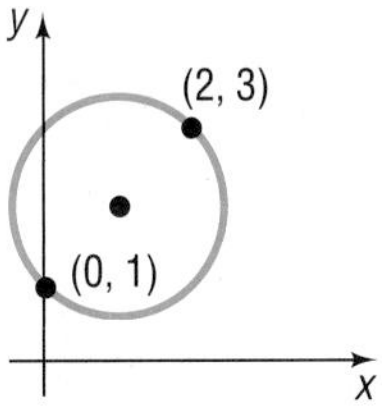

In Problems 5–14, write the standard form of the equation and the general form of the equation of each circle of radius r and center (h, k). By hand, graph each circle.

5. $r = 1;\ (h, k) = (1, -1)$

6. $r = 2;\ (h, k) = (-2, 1)$

7. $r = 2;\ (h, k) = (0, 2)$

8. $r = 3;\ (h, k) = (1, 0)$

9. $r = 5;\ (h, k) = (4, -3)$

10. $r = 4;\ (h, k) = (2, -3)$

11. $r = 2;\ (h, k) = (0, 0)$

12. $r = 3;\ (h, k) = (0, 0)$

13. $r = \frac{1}{2};\ (h, k) = \left(\frac{1}{2}, 0\right)$

14. $r = \frac{1}{2};\ (h, k) = \left(0, -\frac{1}{2}\right)$

In Problems 15–24, find the center (h, k) and radius r of each circle. By hand, graph each circle.

15. $x^2 + y^2 = 4$

16. $x^2 + (y - 1)^2 = 1$

17. $(x - 3)^2 + y^2 = 4$

18. $(x + 1)^2 + (y - 1)^2 = 2$

19. $x^2 + y^2 + 4x - 4y - 1 = 0$

20. $x^2 + y^2 - 6x + 2y + 9 = 0$

21. $x^2 + y^2 - x + 2y + 1 = 0$

22. $x^2 + y^2 + x + y - \frac{1}{2} = 0$

23. $2x^2 + 2y^2 - 12x + 8y - 24 = 0$

24. $2x^2 + 2y^2 + 8x + 7 = 0$

In Problems 25–30, find the general form of the equation of each circle.

25. Center at the origin and containing the point $(-2, 3)$

26. Center $(1, 0)$ and containing the point $(-3, 2)$

27. Center $(2, 3)$ and tangent to the x-axis

28. Center $(-3, 1)$ and tangent to the y-axis

29. With endpoints of a diameter at $(1, 4)$ and $(-3, 2)$

30. With endpoints of a diameter at $(4, 3)$ and $(0, 1)$

In Problems 31–34, match each graph with the correct equation.

(a) $(x - 3)^2 + (y + 3)^2 = 9$

(b) $(x + 1)^2 + (y - 2)^2 = 4$

(c) $(x - 1)^2 + (y + 2)^2 = 4$

(d) $(x + 3)^2 + (y - 3)^2 = 9$

31.

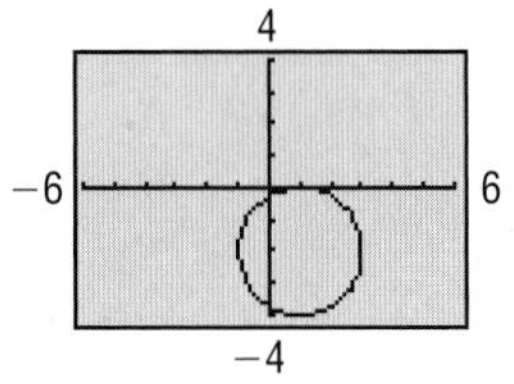

32.

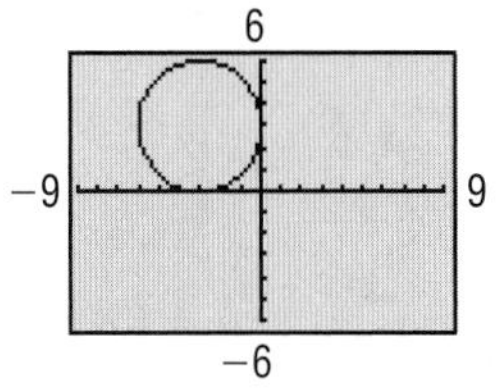

33.

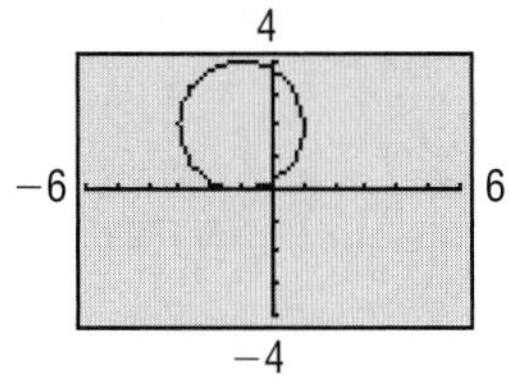

34.

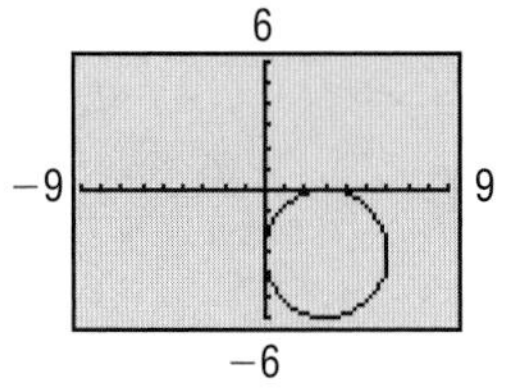

In Problems 35–38, find the standard form of the equation of each circle.

35.

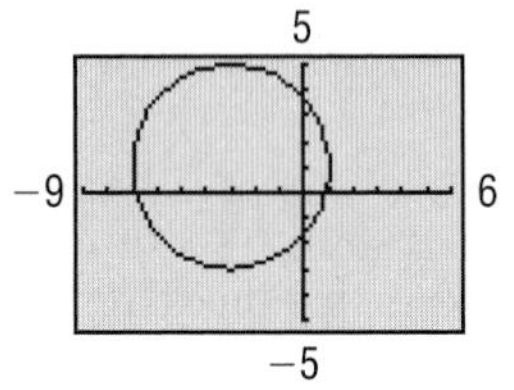

36.

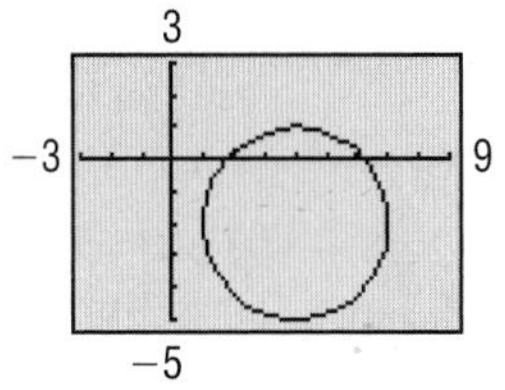

37.

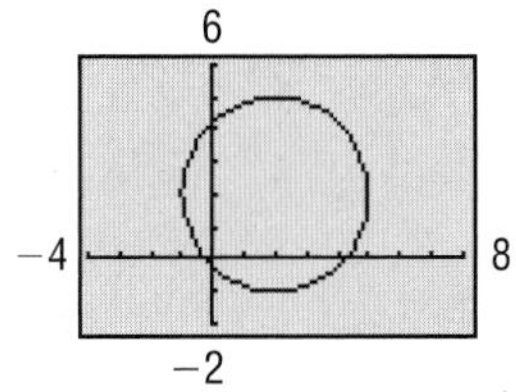

38.

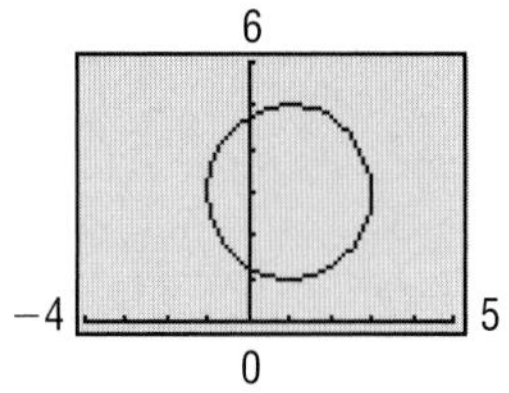

39. The **tangent line** to a circle may be defined as the line that intersects the circle in a single point, called the **point of tangency** (see the figure).

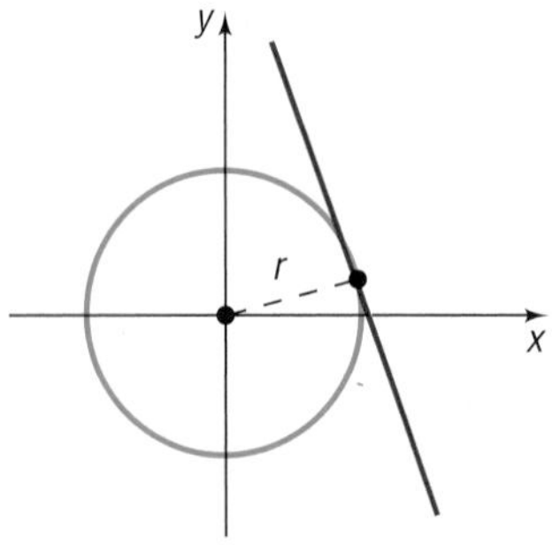

If the equation of the circle is $x^2 + y^2 = r^2$ and the equation of the tangent line is $y = mx + b$, show that:

(a) $r^2(1 + m^2) = b^2$

[**Hint:** The quadratic equation $x^2 + (mx + b)^2 = r^2$ has exactly one solution.]

(b) The point of tangency is $(-r^2 m/b, r^2/b)$.

(c) The tangent line is perpendicular to the line containing the center of the circle and the point of tangency.

40. The Greek Method for Finding Tangents The Greek method for finding the equation of the tangent line to a circle used the fact that at any point on a circle the lines containing the center and the tangent line are perpendicular (see Problem 39). Use this method to find an equation of the tangent line to the circle $x^2 + y^2 = 9$ at the point $(1, 2\sqrt{2})$.

41. Use the Greek method described in Problem 40 to find an equation of the tangent line to the circle $x^2 + y^2 - 4x + 6y + 4 = 0$ at the point $(3, 2\sqrt{2} - 3)$.

42. Refer to Problem 39. The line $x - 2y = -4$ is tangent to a circle at $(0, 2)$. The line $y = 2x - 7$ is tangent to the same circle at $(3, -1)$. Find the center of the circle.

43. Find an equation of the line containing the centers of the two circles

$$x^2 + y^2 - 4x + 6y + 4 = 0$$

and

$$x^2 + y^2 + 6x + 4y + 9 = 0$$

44. If a circle of radius 2 is made to roll along the x-axis, what is an equation for the path of the center of the circle?

CHAPTER REVIEW

Things To Know

Formulas

Distance formula (p. 6)	$d = \sqrt{(x_2 - x_1)^2 + (y_2 - y_1)^2}$
Midpoint formula (p. 8)	$(x, y) = \left(\frac{x_1 + x_2}{2}, \frac{y_1 + y_2}{2}\right)$
Slope (p. 61)	$m = \frac{y_2 - y_1}{x_2 - x_1}$, if $x_1 \neq x_2$; undefined if $x_1 = x_2$
Parallel lines (p. 69)	Equal slopes $(m_1 = m_2)$ and different y-intercepts
Perpendicular lines (p. 71)	Product of slopes is -1 $(m_1 \cdot m_2 = -1)$

Equations of Lines and Circles

Vertical line (p. 65)	$x = a$
Horizontal line (p. 67)	$y = b$
Point–slope form of the equation of a line (p. 66)	$y - y_1 = m(x - x_1)$; m is the slope of the line, (x_1, y_1) is a point on the line
Slope-intercept form of the equation of a line (p. 67)	$y = mx + b$; m is the slope of the line, b is the y-intercept
General form of the equation of a line (p. 69)	$Ax + By = C$, A, B not both 0
Standard form of the equation of a circle (p. 84)	$(x - h)^2 + (y - k)^2 = r^2$; r is the radius of the circle, (h, k) is the center of the circle
Equation of the unit circle (p. 84)	$x^2 + y^2 = 1$
General form of the equation of a circle (p. 86)	$x^2 + y^2 + ax + by + c = 0$

Quadratic equation and quadratic formula (p. 30)

If $ax^2 + bx + c = 0$, $a \neq 0$, and if $b^2 - 4ac \geq 0$, then $x = \dfrac{-b \pm \sqrt{b^2 - 4ac}}{2a}$.

Discriminant (p. 30)

If $b^2 - 4ac > 0$, there are two distinct real solutions.

If $b^2 - 4ac = 0$, there is one repeated real solution.

If $b^2 - 4ac < 0$, there are no real solutions.

Properties of Inequalities

Addition property (p. 51)	If $a < b$, then $a + c < b + c$. If $a > b$, then $a + c > b + c$.
Multiplication properties (p. 52)	(a) If $a < b$ and if $c > 0$, then $ac < bc$. If $a < b$ and if $c < 0$, then $ac > bc$. (b) If $a > b$ and if $c > 0$, then $ac > bc$. If $a > b$ and if $c < 0$, then $ac < bc$.

Absolute Value Equations and Inequalities

If $|u| = a$, $a > 0$, then $u = -a$ or $u = a$. (p. 32)

If $|u| \leq a$, $a > 0$, then $-a \leq u \leq a$. (p. 56)

If $|u| \geq a$, $a > 0$, then $u \leq -a$ or $u \geq a$. (p. 57)

How To

Use the distance formula (p. 5)

Use the midpoint formula (p. 8)

Graph equations by plotting points (p. 12)

Graph equations using a graphing utility (p. 13)

Find intercepts (p. 16)

Test an equation for symmetry (p. 18)

Solve equations using a graphing utility (p. 24)

Solve linear equations (p. 26)

Solve quadratic equations by factoring (p. 28)

Solve quadratic equations using the quadratic formula (p. 30)

Solve equations involving absolute value (p. 32)

Solve radical equations (p. 33)

Translate verbal descriptions into mathematical expressions (p. 36)

Set up applied problems (p. 37)

Solve interest problems (p. 38)

Solve mixture problems (p. 40)

Solve uniform motion problems (p. 41)

Solve constant rate job problems (p. 42)

Use interval notation (p. 49)

Use properties of inequalities (p. 50)

Solve linear inequalities algebraically and graphically (p. 52)

Solve combined inequalities algebraically and graphically (p. 54)

Solve absolute value inequalities algebraically and graphically (p. 55)

Calculate and interpret the slope of a line (p. 61)

Graph lines (p. 63)

Find the equation of vertical lines (p. 64)

Use the point–slope form of a line; identify horizontal lines (p. 65)

Find the equation of a line given two points (p. 67)

Write the equation of a line in slope–intercept form (p. 67)

Identify the slope and y-intercept of a line from its equation (p. 68)

Write the equation of a line in general form (p. 69)

Define parallel and perpendicular lines (p. 69)

Find equations of parallel lines (p. 70)

Find equations of perpendicular lines (p. 72)

Draw and interpret scatter diagrams (p. 76)

Distinguish between linear and nonlinear relations (p. 77)

Use a graphing utility to find the line of best fit (p. 79)

Write the standard form of the equation of a circle (p. 83)

Graph a circle by hand and by using a graphing utility (p. 85)

Find the center and radius of a circle from an equation in general form (p. 86)

Fill-in-the-Blank Items

1. If (x, y) are the coordinates of a point P in the xy-plane, then x is called the ______________ of P and y is the ______________ of P.

2. If three distinct points P, Q, and R all lie on a line and if $d(P, Q) = d(Q, R)$, then Q is called the ______________ of the line segment from P to R.

3. If for every point (x, y) on a graph the point $(-x, y)$ is also on the graph, then the graph is symmetric with respect to the ______________.

4. When a quadratic equation has a repeated solution, it is called a(n) ______________ root or a root of ______________ ______________.

5. When an apparent solution does not satisfy the original equation, it is called a(n) ______________ solution.

6. If each side of an inequality is multiplied by a(n) ______________ number, then the direction of the inequality symbol is reversed.

7. The equation $|x^2| = 4$ has two real solutions, ______________ and ______________.

8. The set of points in the xy-plane that are a fixed distance from a fixed point is called a(n) ______________. The fixed distance is called the ______________; the fixed point is called the ______________.

9. The slope of a vertical line is ______________; the slope of a horizontal line is ______________.

10. Two distinct nonvertical lines have slopes m_1 and m_2, respectively. The lines are parallel if ______________; the lines are perpendicular if ______________.

True/False Items

T F 1. The distance between two points is sometimes a negative number.

T F 2. The graph of the equation $y = x^4 + x^2 + 1$ is symmetric with respect to the y-axis.

T F 3. Vertical lines have undefined slope.

T F 4. The slope of the line $2y = 3x + 5$ is 3.

T F 5. Perpendicular lines have slopes that are reciprocals of one another.

T F 6. The radius of the circle $x^2 + y^2 = 9$ is 3.

T F 7. Equations can have no solutions, one solution, or more than one solution.

T F 8. Quadratic equations always have two real solutions.

T F 9. If the discriminant of a quadratic equation is positive, then the equation has two real solutions that are unequal.

T F 10. The square of any real number is always nonnegative.

T F 11. The expression $x^2 + x + 1$ is positive for any real number x.

12. If $a < b$ and $c < 0$, which of the following statements are true?

T F (a) $a \pm c < b \pm c$

T F (b) $a \cdot c < b \cdot c$

T F (c) $a/c > b/c$

Review Exercises

Blue problem numbers indicate the authors' suggestions for use in a Practice Test.

In Problems 1–28, find all real solutions, if any, of each equation (a) graphically and (b) algebraically.

1. $2 - \frac{x}{3} = 6$
2. $\frac{x}{4} - 2 = 6$
3. $-2(5 - 3x) + 8 = 4 + 5x$
4. $(6 - 3x) - 2(1 + x) = 6x$
5. $\frac{3x}{4} - \frac{x}{3} = \frac{1}{12}$
6. $\frac{4 - 2x}{3} + \frac{1}{6} = 2x$
7. $\frac{x}{x - 1} = \frac{5}{6}, \quad x \neq 1$
8. $\frac{4x - 5}{3 - 7x} = 4, \quad x \neq \frac{3}{7}$
9. $x(1 - x) = -6$
10. $x(1 + x) = 6$
11. $\frac{1}{2}\left(x - \frac{1}{3}\right) = \frac{3}{4} - \frac{x}{6}$
12. $\frac{1 - 3x}{4} = \frac{x + 6}{3} + \frac{1}{2}$
13. $(x - 1)(2x + 3) = 3$
14. $x(2 - x) = 3(x - 4)$
15. $4x + 3 = 4x^2$
16. $5x = 4x^2 + 1$
17. $\sqrt[3]{x - 1} = 2$
18. $\sqrt[4]{1 + x} = 3$
19. $x(x + 1) - 2 = 0$
20. $2x^2 - 3x + 1 = 0$
21. $\sqrt{2x - 3} + x = 3$
22. $\sqrt{2x - 1} = x - 2$
23. $\sqrt{x + 1} + \sqrt{x - 1} = \sqrt{2x + 1}$
24. $\sqrt{2x - 1} - \sqrt{x - 5} = 3$
25. $|2x + 3| = 7$
26. $|3x - 1| = 5$
27. $|2 - 3x| = 7$
28. $|1 - 2x| = 3$

In Problems 29–38, solve each inequality (a) graphically and (b) algebraically. Graph the solution set.

29. $\frac{2x - 3}{5} + 2 \leq \frac{x}{2}$
30. $\frac{5 - x}{3} \leq 6x - 4$
31. $-9 \leq \frac{2x + 3}{-4} \leq 7$
32. $-4 < \frac{2x - 2}{3} < 6$
33. $6 > \frac{3 - 3x}{12} > 2$
34. $6 > \frac{5 - 3x}{2} \geq -3$
35. $|3x + 4| < \frac{1}{2}$
36. $|1 - 2x| < \frac{1}{3}$
37. $|2x - 5| \geq 9$
38. $|3x + 1| \geq 10$

In Problems 39–48, find an equation of the line having the given characteristics. Express your answer using either the general form or the slope–intercept form of the equation of a line, whichever you prefer.

39. Slope $= -2$; containing the point $(3, -1)$
40. Slope $= 0$; containing the point $(-5, 4)$
41. Slope undefined; containing the point $(-3, 4)$
42. x-Intercept $= 2$; containing the point $(4, -5)$
43. y-Intercept $= -2$; containing the point $(5, -3)$
44. Containing the points $(3, -4)$ and $(2, 1)$
45. Parallel to the line $2x - 3y = -4$; containing the point $(-5, 3)$
46. Parallel to the line $x + y = 2$; containing the point $(1, -3)$

47. Perpendicular to the line $x + y = 2$; containing the point $(4, -3)$
48. Perpendicular to the line $3x - y = -4$; containing the point $(-2, 4)$

In Problems 49–54, graph each line by hand, labeling any intercepts.

49. $4x - 5y = -20$
50. $3x + 4y = 12$
51. $\frac{1}{2}x - \frac{1}{3}y = -\frac{1}{6}$
52. $-\frac{3}{4}x + \frac{1}{2}y = 0$
53. $\sqrt{2}x + \sqrt{3}y = \sqrt{6}$
54. $\frac{x}{3} + \frac{y}{4} = 1$

In Problems 55–62, list the x- and y-intercepts and test for symmetry.

55. $2x = 3y^2$
56. $y = 5x$
57. $4x^2 + y^2 = 1$
58. $x^2 - 9y^2 = 9$
59. $y = x^4 + 2x^2 + 1$
60. $y = x^3 - x$
61. $x^2 + x + y^2 + 2y = 0$
62. $x^2 + 4x + y^2 - 2y = 0$

In Problems 63–66, find the center and radius of each circle. Graph each circle by hand.

63. $x^2 + y^2 - 2x + 4y - 4 = 0$
64. $x^2 + y^2 + 4x - 4y - 1 = 0$
65. $3x^2 + 3y^2 - 6x + 12y = 0$
66. $2x^2 + 2y^2 - 4x = 0$

67. Find the slope of the line containing the points $(7, 4)$ and $(-3, 2)$. What is the distance between these points? What is their midpoint?
68. Find the slope of the line containing the points $(2, 5)$ and $(6, -3)$. What is the distance between these points? What is their midpoint?
69. Show that the points $A = (3, 4)$, $B = (1, 1)$, and $C = (-2, 3)$ are the vertices of an isosceles triangle.
70. Show that the points $A = (-2, 0)$, $B = (-4, 4)$, and $C = (8, 5)$ are the vertices of a right triangle in two ways:
    (a) By using the converse of the Pythagorean Theorem
    (b) By using the slopes of the lines joining the vertices
71. Show that the points $A = (2, 5)$, $B = (6, 1)$, and $C = (8, -1)$ lie on a line by using slopes.
72. Show that the points $A = (1, 5)$, $B = (2, 4)$, and $C = (-3, 5)$ lie on a circle with center $(-1, 2)$. What is the radius of this circle?
73. The endpoints of the diameter of a circle are $(-3, 2)$ and $(5, -6)$. Find the center and radius of the circle. Write the general equation of this circle.
74. Find two numbers y such that the distance from $(-3, 2)$ to $(5, y)$ is 10.

For Problems 75–78, (a) use a graphing utility to draw a scatter diagram, (b) use a graphing utility to find the line of best fit to the data, and (c) interpret the slope.

75.

x	3	4	5	6	7	8	9
y	3	5	6	8	10	11	13

76.

x	10	12	13	15	16	18	20
y	34	27	26	23	20	18	17

77.

x	100	110	125	130	140	145	150	160	170	175
y	300	340	365	380	400	410	425	430	450	460

78.

x	200	220	230	235	245	250	265	275	280	300
y	1000	990	975	960	955	940	935	920	910	895

79. **Lightning and Thunder** A flash of lightning is seen and the resulting thunderclap is heard 3 seconds later. If the speed of sound averages 1100 feet per second, how far away is the storm?
80. **Physics: Intensity of Light** The intensity I (in candlepower) of a certain light source obeys the equation $I = 900/x^2$, where x is the distance (in meters) from the light. Over what range of distances can an object be placed from this light source so that the range of intensity of light is from 1600 to 3600 candlepower, inclusive?
81. **Extent of Search and Rescue** A search plane has a cruising speed of 250 miles per hour and carries enough fuel for at most 5 hours of flying. If there is a wind that averages 30 miles per hour and the direction of search is with the wind one way and against it the other, how far can the search plane travel?
82. **Extent of Search and Rescue** If the search plane described in Problem 81 is able to add a supplementary fuel tank that allows for an additional 2 hours of flying, how much farther can the plane extend its search?
83. **Rescue at Sea** A life raft, set adrift from a sinking ship 150 miles offshore, travels directly toward a Coast Guard station at the rate of 5 miles per hour. At the time that the raft is set adrift, a rescue helicopter is dispatched from the Coast Guard station. If the helicopter's average speed is 90 miles per hour, how long will it take the helicopter to reach the life raft?

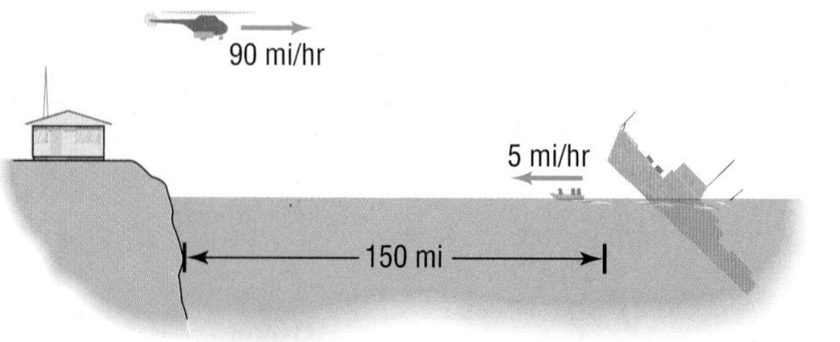

84. Physics: Uniform Motion Two bees leave two locations 150 meters apart and fly, without stopping, back and forth between these two locations at average speeds of 3 meters per second and 5 meters per second, respectively. How long is it until the bees meet for the first time? How long is it until they meet for the second time?

85. Working Together to Get a Job Done Clarissa and Shawna, working together, can paint the exterior of a house in 6 days. Clarissa by herself can complete this job in 5 days less than Shawna. How long will it take Clarissa to complete the job by herself?

86. Emptying a Tank Two pumps of different sizes, working together, can empty a fuel tank in 5 hours. The larger pump can empty this tank in 4 hours less than the smaller one. If the larger one is out of order, how long will it take the smaller one to do the job alone?

87. Chemistry: Mixing Acids For a certain experiment, a student requires 100 cubic centimeters of a solution that is 8% HCl. The storeroom has only solutions that are 15% HCl and 5% HCl. How many cubic centimeters of each available solution should be mixed to get 100 cubic centimeters of 8% HCl?

88. Chemistry: Salt Solutions How much water must be evaporated from 32 ounces of a 4% salt solution to make a 6% salt solution?

89. **Business: Theater Attendance** The manager of the Coral Theater wants to know whether the majority of its patrons is adults or children. During a week in July, 5200 tickets were sold and the receipts totaled \$20,335. The adult admission is \$4.75, and the children's admission is \$2.50. How many adult patrons were there?

90. Business: Blending Coffee A coffee manufacturer wants to market a new blend of coffee that will cost \$6 per pound by mixing two coffees that sell for \$4.50 and \$8 per pound, respectively. What amounts of each coffee should be blended to obtain the desired mixture?
[**Hint:** Assume that the total weight of the desired blend is 100 pounds.]

91. Physics: Uniform Motion Refer to the figure below. A man is walking at an average speed of 4 miles per hour alongside a railroad track. A freight train, going in the same direction at an average speed of 30 miles per hour, requires 5 seconds to pass the man. How long is the freight train? Give your answer in feet.

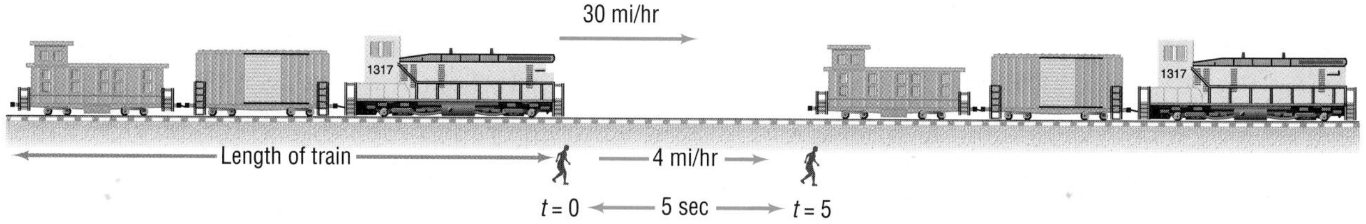

92. One formula stating the relationship between the length l and width w of a rectangle of "pleasing proportion" is $l^2 = w(l + w)$. How should a 4 foot by 8 foot sheet of plasterboard be cut so that the result is a rectangle of "pleasing proportion" with a width of 4 feet?

93. Business: Determining the Cost of a Charter A group of 20 senior citizens can charter a bus for a one-day excursion trip for \$15 per person. The charter company agrees to reduce the price of each ticket by 10¢ for each additional passenger in excess of 20 who goes on the trip, up to a maximum of 44 passengers (the capacity of the bus). If the final bill from the charter company was \$482.40, how many seniors went on the trip, and how much did each pay?

94. Time Required for Copying A new copying machine can do a certain job in 1 hour less than an older copier. Together they can do this job in 72 minutes. How long would it take the older copier by itself to do the job?

95. **Relating Algebra and Calculus Scores** The following data represent scores in an algebra achievement test and calculus achievement test for the same student. Treat the algebra test score as the independent variable and the calculus test score as the dependent variable.

Algebra Score	Calculus Score
17	73
21	66
11	64
16	61
15	70
11	71
24	90
27	68
19	84
8	52

(a) Use a graphing utility to draw a scatter diagram.
(b) Use a graphing utility to find the line of best fit to the data.

(c) Interpret the slope.
(d) Predict the score that a student would receive on the calculus achievement test if she scored a 20 on the algebra achievement test.

96. Relating Emission Levels of Hydrocarbons and Carbon Monoxide The following data represent emissions levels for different vehicles. Measurements are given in grams per meter. Treat hydrocarbons (HC) as the independent variable and carbon monoxide (CO) as the dependent variable.
(a) Use a graphing utility to draw a scatter diagram.
(b) Use a graphing utility to find the line of best fit to the data.
(c) Interpret the slope.
(d) Predict the level of CO in a vehicle's exhaust if the level of HC is 0.67.

HC	CO
0.65	14.7
0.55	12.3
0.72	14.6
0.83	15.1
0.57	5.0
0.51	4.1
0.43	3.8
0.37	4.1

97. Value of a Portfolio The following data represent the value of the Vanguard Index Trust-500 Portfolio for 1993–1997.

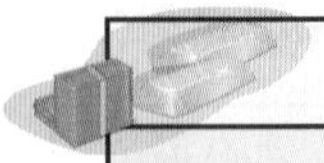

Year	Value (Dollars)
1993	40.97
1994	43.83
1995	42.97
1996	57.60
1997	69.17

Source: Vanguard Index Trust, Annual Report 1997.

(a) By hand, draw a scatter diagram of the data.
(b) What is the trend in the data? In other words, as time passes, what is happening to the value of the Vanguard Index Trust-500 Portfolio?
(c) Treating the year as the x-coordinate and the value of the Vanguard Index Trust-500 Portfolio as the y-coordinate, use a graphing utility to draw a scatter diagram of the data.
(d) What is the slope of the line joining the points $(1993, 40.97)$ and $(1995, 42.97)$?
(e) Interpret this slope.
(f) What is the slope of the line joining the points $(1995, 42.97)$ and $(1997, 69.17)$?
(g) Interpret this slope.
(h) If you were managing this trust, which of the two slopes would you use to convince someone to invest? Why?

98. Make up four problems that you might be asked to do given the two points $(-3, 4)$ and $(6, 1)$. Each problem should involve a different concept. Be sure that your directions are clearly stated.

99. Describe each of the following graphs. Give justification.
(a) $x = 0$ (b) $y = 0$
(c) $x + y = 0$ (d) $xy = 0$
(e) $x^2 + y^2 = 0$

100. In a 100-meter race, Todd crosses the finish line 5 meters ahead of Scott. To even things up, Todd suggests to Scott that they race again, this time with Todd lining up 5 meters behind the start.
(a) Assuming that Todd and Scott run at the same pace as before, does the second race end in a tie?
(b) If not, who wins?
(c) By how many meters does he win?
(d) How far back should Todd start so that the race ends in a tie?
After running the race a second time, Scott, to even things up, suggests to Todd that he (Scott) line up 5 meters in front of the start.
(e) Assuming again that they run at the same pace as in the first race, does the third race result in a tie?
(f) If not, who wins?
(g) By how many meters?
(h) How far up should Scott start so that the race ends in a tie?

PROJECT AT MOTOROLA

During the 1990's mobile phones became a significant means of communication. Industry data indicated that in 1994 there were about 37.5 million mobile phones in use globally. In 1998, that figure rose to about 70 million mobile phones and is projected to rise to 102.5 million by 2002. Industry analysts base such predictions on a *mathematical model*, that is, a relationship (usually an equation) that involves one or more variables. One method of creating a mathematical model is to use the data available and construct an equation that describes the behavior of the data. The mathematical model is then used to forecast future values of one of the variables, in this case mobile phone usage.

(a) Construct a scatter diagram of the following data treating year as the independent variable and mobile phone usage (in millions) as the dependent variable. Label each axis and create a title for the graph.

Year, t	Mobile phone usage, m (in millions)
1994	37.5
1998	70
2002	102.5

(b) Using a graphing utility, find the line of best fit to the data.

(c) Interpret the values of the slope and the y-intercept.

(d) Predict mobile phone usage for 2006.

(e) Write a report to the director of strategic marketing that discusses your predictions. The report must address the adequacy of the model. That is, you must discuss whether you feel this model can be used to make predictions about future mobile phone usage. Consider factors such as the cost of phones, population growth, etc. when writing your report.

CHAPTER

2 Functions

Outline

FIELD TRIP TO MOTOROLA

During the past decade the availability and usage of wireless Internet services has increased many fold. The industry has developed a number of pricing proposals for such services. Marketing data have indicated that subscribers of wireless Internet services have tended to desire flat rate fee structures as compared with rates based totally on usage.

INTERVIEW AT MOTOROLA

Jocelyn Carter-Miller is Corporate Vice President and Chief Marketing Officer (CMO) for Motorola, Inc., an over $30 billion global provider of integrated communications and embedded electronics solutions. As CMO she has helped build the Motorola brand and image, and has developed high performance marketing organizations and processes. Jocelyn also heads motorola.com, the Motorola electronic commerce and information website. In this new role she and her team have developed a strategy for serving Motorola's broad and diverse constituencies offering a full range of electronic services.

In her previous roles as Vice President-Latin American and Caribbean Operations and Director of European, Middle East and African Operations, Jocelyn headed international wireless data communications operations for Motorola—creating profitable opportunities through strategic alliances, value-added applications, and new product and service launches. She also developed skills in managing complex, high risk ventures, in countries like Brazil and Russia, setting standards for her company's practices in emerging markets.

Prior to her career at Motorola, Jocelyn served as Vice President, Marketing and Product Development for Mattel, where she broke new ground driving record sales of Barbie and other toys using integrated product, entertainment, promotional, and licensing programs.

Jocelyn builds strong relationships and new opportunities through her involvement on outside boards and community organizations. She serves on the board of the Principal Financial Group, and on the non-profit boards of the Association of National Advertisers, the University of Chicago Women's Business Group Advisory Board, and the Smart School Charter Middle School.

Jocelyn holds a Master of Business Administration degree in marketing and finance from the University of Chicago, a Bachelor of Science degree in accounting from the University of Illinois-Urbana-Champaign, and is a Certified Public Accountant. She is married to Edward Miller, President of Edventures, an educational reform development firm, and has two daughters Alexis and Kimberly.

Jocelyn has won numerous awards, been featured in national publications, and regularly addresses business and community groups. She has co-authored with Melissa Giavagnoli the book *Networlding: Building Relationships and Opportunities for Success*, which will be published in June 2000 by Jossey-Bass Publishers. Through their website Networlding.com, Jocelyn and Melissa facilitate meaningful connections and mutually beneficial opportunities for novice and expert Networlders alike.

PREPARING FOR THIS SECTION

Before getting started, review the following:

- ✓ Intervals (pp. 49–50)
- ✓ Scatter Diagrams; Linear Curve Fitting (Section 1.7)

2.1 FUNCTIONS

OBJECTIVES

1. Determine Whether a Relation Represents a Function
2. Find the Value of a Function
3. Find the Domain of a Function
4. Identify the Graph of a Function
5. Obtain Information from or about the Graph of a Function

1 In Section 1.7, we said that a **relation** is a correspondence between two variables, say x and y. When relations are written as ordered pairs (x, y), we say that x is related to y. Often, we are interested in specifying the type of relation (such as an equation) that might exist between the two variables.

For example, the relation between the revenue R resulting from the sale of x items selling for \$10 each may be expressed by the equation $R = 10x$. If we know how many items have been sold, then we can calculate the revenue by using the equation $R = 10x$. This equation is an example of a *function*.

As another example, suppose that an icicle falls off a building from a height of 64 feet above the ground. According to a law of physics, the distance s (in feet) of the icicle from the ground after t seconds is given (approximately) by the formula $s = 64 - 16t^2$. When $t = 0$ seconds, the icicle is $s = 64$ feet above the ground. After 1 second, the icicle is $s = 64 - 16(1)^2 = 48$ feet above the ground. After 2 seconds, the icicle strikes the ground. The formula $s = 64 - 16t^2$ provides a way of finding the distance s when the time t $(0 \le t \le 2)$ is prescribed. There is a correspondence between each time t in the interval $0 \le t \le 2$ and the distance s. We say that the distance s is a *function* of the time t because:

1. There is a correspondence between the set of times and the set of distances.
2. There is exactly one distance s obtained for any time t in the interval $0 \le t \le 2$.

Let's now look at the definition of a function.

DEFINITION OF FUNCTION

Let X and Y be two nonempty sets of real numbers.* A **function** from X into Y is a relation that associates with each element of X a unique element of Y.

Figure 1

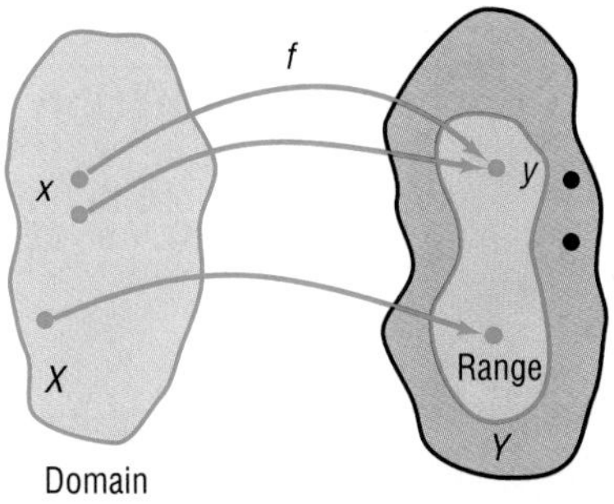

The set X is called the **domain** of the function. For each element x in X, the corresponding element y in Y is called the **value** of the function at x, or the image of x. The set of all images of the elements of the domain is called the **range** of the function. See Figure 1.

Since there may be some elements in Y that are not the image of some x in X, it follows that the range of a function may be a subset of Y, as shown in Figure 1.

Not all relations between two sets are functions.

EXAMPLE 1 Determining Whether a Relation Represents a Function

Determine whether the following relations represent functions.

(a) For this relation, the domain represents the employees of Sara's Pre-Owned Car Mart and the range represents their base salary.

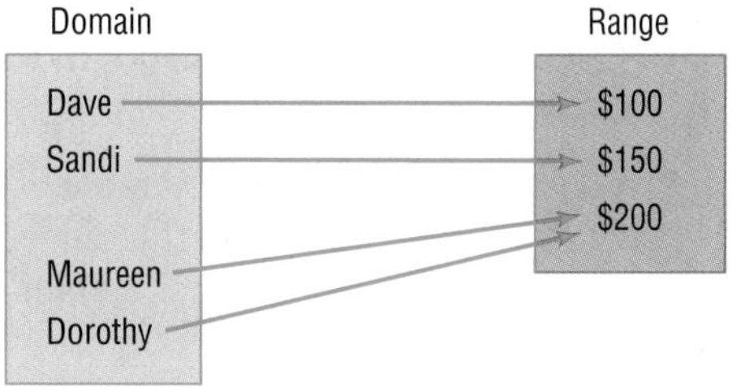

*The two sets X and Y can also be sets of complex numbers (discussed in Section 4.2), and then we have defined a complex function. In the broad definition (due to Lejeune Dirichlet), X and Y can be any two sets.

(b) For this relation, the domain represents the employees of Sara's Pre-Owned Car Mart and the range represents their phone number(s).

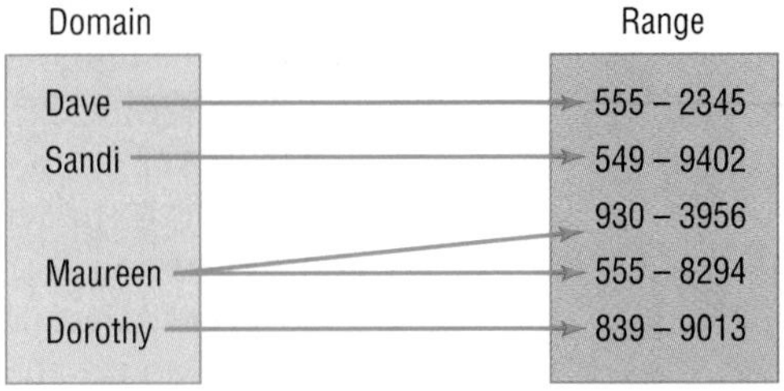

Solution (a) The relation is a function because each element in the domain corresponds to a unique element in the range. Notice that more than one element in the domain can correspond to the same element in the range.

(b) The relation is not a function because each element in the domain does not correspond to a unique element in the range. Maureen has two telephone numbers; therefore, if Maureen is chosen from the domain, a unique telephone number cannot be assigned to her. ■

We may think of a function as a set of ordered pairs (x, y) in which no two distinct pairs have the same first element. The set of all first elements x is the domain of the function, and the set of all second elements y is its range. Associated with each element x in the domain, there is a unique element y in the range.

EXAMPLE 2 Determining Whether a Relation Represents a Function

Determine whether each relation represents a function. For those that are functions, state the domain and range.

(a) $\{(1, 4), (2, 5), (3, 6), (4, 7)\}$
(b) $\{(1, 4), (2, 4), (3, 5), (6, 10)\}$
(c) $\{(-3, 9), (-2, 4), (0, 0), (1, 1), (-3, 8)\}$

Solution (a) This relation is a function because there are no distinct ordered pairs with the same first element. The domain of this function is $\{1, 2, 3, 4\}$ and its range is $\{4, 5, 6, 7\}$.

(b) This relation is a function because there are no distinct ordered pairs with the same first element. The domain of this function is $\{1, 2, 3, 6\}$ and its range is $\{4, 5, 10\}$.

(c) This relation is not a function because there is a first element, -3, that corresponds to two different second elements, 9 and 8. ■

In Example 2(b), notice that 1 and 2 in the domain each have the same image in the range. This does not violate the definition of a function; two different first elements can have the same second element. A violation of the definition occurs when two ordered pairs have the same first element and different second elements, as in Example 2(c).

NOW WORK PROBLEMS 1 AND 5.

Examples 2(a) and 2(b) demonstrate that a function may be defined by a set of ordered pairs. A function may also be defined by an equation in two variables, usually denoted x and y.

EXAMPLE 3 Example of a Function

Consider the function defined by the equation

$$y = 2x - 5, \qquad 1 \le x \le 6$$

The domain $1 \le x \le 6$ specifies that the number x is restricted to the real numbers from 1 to 6, inclusive. The equation $y = 2x - 5$ specifies that the number x is to be multiplied by 2 and then 5 is to be subtracted from the result to get y. For example, if $x = \frac{3}{2}$, then $y = 2 \cdot \frac{3}{2} - 5 = -2$. ■

FUNCTION NOTATION

Functions are often denoted by letters such as f, F, g, G, and so on. If f is a function, then for each number x in its domain the corresponding image in the range is designated by the symbol $f(x)$, read as "f of x" or as "f at x." We refer to $f(x)$ as the **value of f at the number x**; $f(x)$ is the number that results when x is given and the function f is applied; $f(x)$ does *not* mean "f times x." For example, the function given in Example 3 may be written as $y = f(x) = 2x - 5$, $1 \le x \le 6$. Then $f\left(\frac{3}{2}\right) = -2$.

Figure 2 illustrates some other functions. Note that in every function illustrated for each x in the domain there is one value in the range.

Figure 2

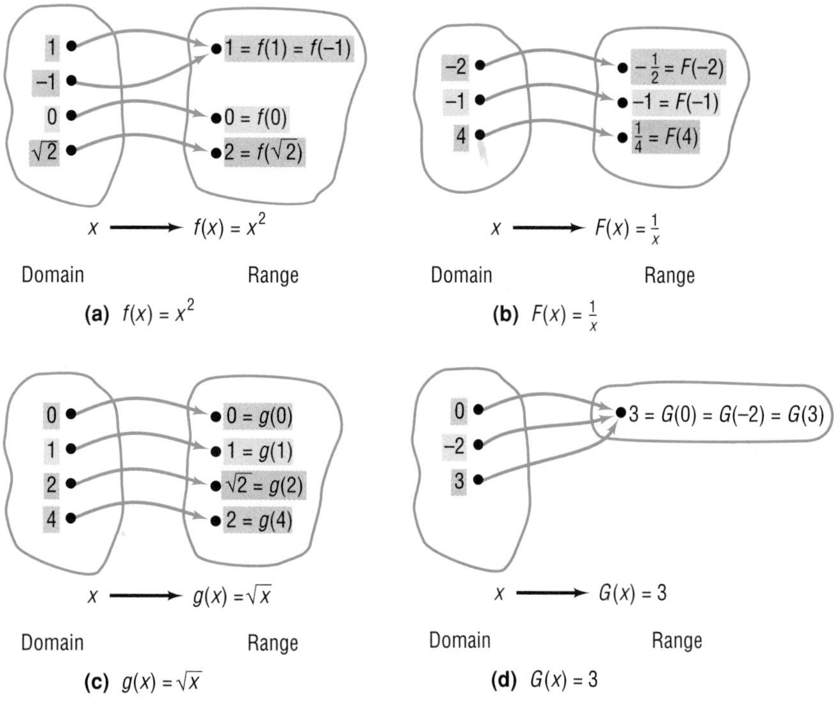

Figure 3

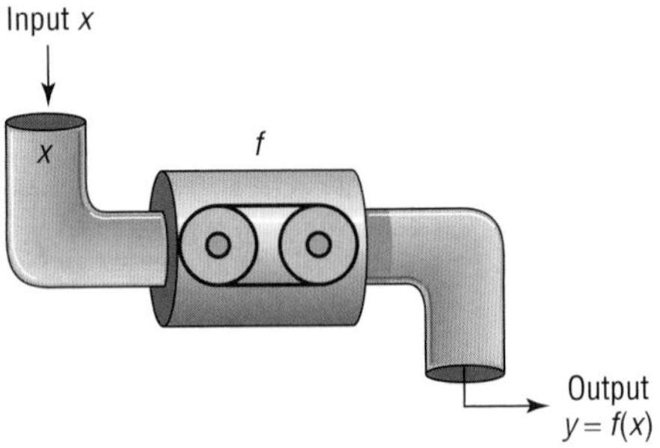

Sometimes it is helpful to think of a function f as a machine that receives as input a number from the domain, manipulates it, and outputs the value. See Figure 3.

The restrictions on this input/output machine are as follows:

1. It only accepts numbers from the domain of the function.
2. For each input, there is exactly one output (which may be repeated for different inputs).

For a function $y = f(x)$, the variable x is called the **independent variable**, because it can be assigned any of the permissible numbers from the domain. The variable y is called the **dependent variable**, because its value depends on x.

Any symbol can be used to represent the independent and dependent variables. For example, if f is the *cube function*, then f can be defined by $f(x) = x^3$ or $f(t) = t^3$ or $f(z) = z^3$. All three functions are the same: Each tells us to cube the independent variable. In practice, the symbols used for the independent and dependent variables are based on common usage, such as using C for cost in business.

2 The variable x is also called the **argument** of the function. Thinking of the independent variable as an argument can sometimes make it easier to find the value of a function. For example, if f is the function defined by $f(x) = x^3$, then f tells us to cube the argument. Thus, $f(2)$ means to cube 2, $f(a)$ means to cube the number a, and $f(x + h)$ means to cube the quantity $x + h$.

EXAMPLE 4 Finding Values of a Function

For the function G defined by $G(x) = 2x^2 - 3x$, evaluate:

(a) $G(3)$ (b) $G(x) + G(3)$ (c) $G(-x)$
(d) $-G(x)$ (e) $G(x + 3)$

Solution (a) We substitute 3 for x in the equation for G to get

$$G(3) = 2(3)^2 - 3(3) = 18 - 9 = 9$$

(b) $G(x) + G(3) = (2x^2 - 3x) + (9) = 2x^2 - 3x + 9$

(c) We substitute $-x$ for x in the equation for G:

$$G(-x) = 2(-x)^2 - 3(-x) = 2x^2 + 3x$$

(d) $-G(x) = -(2x^2 - 3x) = -2x^2 + 3x$

(e) $G(x + 3) = 2(x + 3)^2 - 3(x + 3)$ Notice the use of parentheses here.

$$= 2(x^2 + 6x + 9) - 3x - 9$$
$$= 2x^2 + 12x + 18 - 3x - 9$$
$$= 2x^2 + 9x + 9$$

Notice in this example that $G(x + 3) \neq G(x) + G(3)$ and $G(-x) \neq -G(x)$.

NOW WORK PROBLEM 13.

Graphing calculators have special keys that enable you to find the value of certain commonly used functions. For example, you should be able to find the square function $f(x) = x^2$, the square root function $f(x) = \sqrt{x}$, the reciprocal function $f(x) = 1/x = x^{-1}$, and many others that will be discussed later in this book (such as $\ln x$ and $\log x$). Verify the results of Example 5, which follows, on your calculator.

EXAMPLE 5 Finding Values of a Function on a Calculator

(a) $f(x) = x^2;\quad f(1.234) = 1.522756$
(b) $F(x) = 1/x;\quad F(1.234) = 0.8103727715$
(c) $g(x) = \sqrt{x};\quad g(1.234) = 1.110855526$ ■

Graphing calculators can also be used to evaluate any function that you wish. Figure 4 shows the result obtained in Example 4(a) on a TI-83 graphing calculator with the function to be evaluated, $G(x) = 2x^2 - 3x$, in Y_1.*

Figure 4

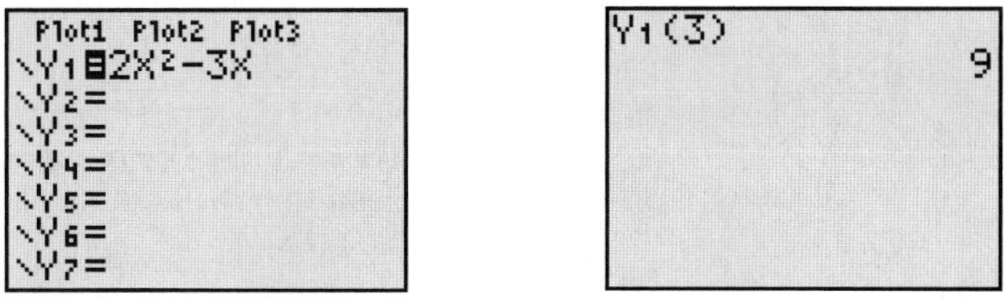

Implicit Form of a Function

In general, when a function f is defined by an equation in x and y, we say that the function f is given **implicitly**. If it is possible to solve the equation for y in terms of x, then we write $y = f(x)$ and say that the function is given **explicitly**. For example,

Implicit Form	**Explicit Form**
$3x + y = 5$	$y = f(x) = -3x + 5$
$x^2 - y = 6$	$y = f(x) = x^2 - 6$
$xy = 4$	$y = f(x) = 4/x$

Not all equations in x and y define a function $y = f(x)$. If an equation is solved for y and two or more values of y can be obtained for a given x, then the equation does not define a function. For example, consider the equation $x^2 + y^2 = 1$, which defines the unit circle. If we solve for y, we obtain $y = \pm\sqrt{1 - x^2}$, so two values of y result for numbers x between -1 and 1. Thus, $x^2 + y^2 = 1$ does not define a function.

COMMENT: The explicit form of a function is the form required by a graphing calculator. Now do you see why it is necessary to graph a circle in two "pieces"?

We list next a summary of some important facts to remember about a function f.

SUMMARY OF IMPORTANT FACTS ABOUT FUNCTIONS

1. To each x in the domain of f, there is one and only one image $f(x)$ in the range; however, an element in the range can result from more than one x in the domain.
2. f is the symbol that we use to denote the function. It is symbolic of the domain and the equation that we use to get from an x in the domain to $f(x)$ in the range.
3. If $y = f(x)$, then x is called the independent variable or argument of f, and y is called the dependent variable or the value of f at x.

*Consult your owner's manual for the required keystrokes.

Domain of a Function

3 Often, the domain of a function f is not specified; instead, only the equation defining the function is given. In such cases, we agree that the domain of f is the largest set of real numbers for which the value $f(x)$ is a real number. Thus, the domain of f is the same as the domain of the variable x in the expression $f(x)$.

EXAMPLE 6 Finding the Domain of a Function

Find the domain of each of the following functions:

(a) $f(x) = x^2 + 5x$ (b) $g(x) = \dfrac{3x}{x^2 - 4}$ (c) $h(t) = \sqrt{4 - 3t}$

Solution (a) The function tells us to square a number and then add five times the number. Since these operations can be performed on any real number, we conclude that the domain of f is all real numbers.

(b) The function g tells us to divide $3x$ by $x^2 - 4$. Since division by 0 is not defined, the denominator $x^2 - 4$ can never be 0. Thus, x can never equal -2 or 2. The domain of the function g is $\{x \mid x \neq -2, x \neq 2\}$.

(c) The function h tells us to take the square root of $4 - 3t$. But only nonnegative numbers have real square roots. Hence, we require that

$$4 - 3t \geq 0$$
$$-3t \geq -4$$
$$t \leq \tfrac{4}{3}$$

The domain of h is $\left\{t \mid t \leq \frac{4}{3}\right\}$ or the interval $\left(-\infty, \frac{4}{3}\right]$. ■

NOW WORK PROBLEM 55.

If x is in the domain of a function f, we shall say that **f is defined at x**, or **$f(x)$ exists**. If x is not in the domain of f, we say that **f is not defined at x**, or **$f(x)$ does not exist**. For example, if $f(x) = x/(x^2 - 1)$, then $f(0)$ exists, but $f(1)$ and $f(-1)$ do not exist. (Do you see why?)

We have not said much about finding the range of a function. The reason is that when a function is defined by an equation it is often difficult to find the range. Therefore, we shall usually be content to find just the domain of a function when only the rule for the function is given. We shall express the domain of a function using inequalities, interval notation, set notation, or words, whichever is most convenient.

Graph of a Function

In applications, a graph often demonstrates more clearly the relationship between two variables than, say, an equation or table would. For example, Table 1 shows the price per share of Merck stock at the end of each month during 1998. If we plot these data using the date as the x-coordinate and the price as the y-coordinate and then connect the points, we obtain Figure 5.

We can see from the graph that the price of the stock was falling during July and August and was rising from September through November. The graph also shows that the lowest price occurred at the end of August, whereas the highest occurred at the end of November. Equations and tables, on

TABLE 1

Date	Closing Price
1/31/98	115.589
2/28/98	125.621
3/31/98	126.686
4/30/98	119.089
5/31/98	115.63
6/30/98	132.702
7/31/98	122.595
8/31/98	115.029
9/30/98	129.117
10/31/98	134.598
11/30/98	154.591
12/31/98	147.5

Courtesy of A.G. Edwards & Sons, Inc.

Figure 5
Monthly closing prices of Merck stock in 1998.

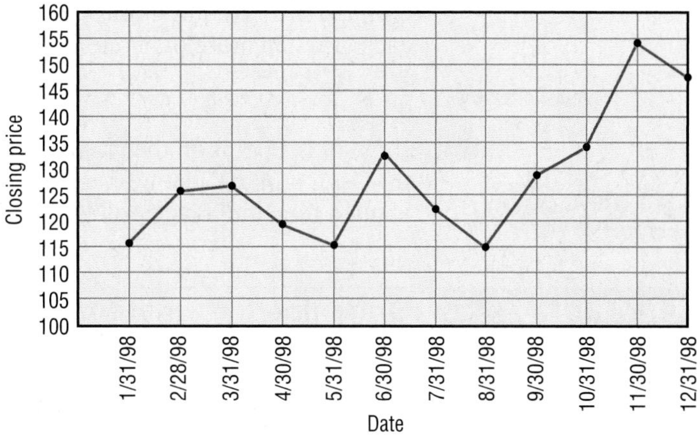

the other hand, usually require some calculations and interpretation before this kind of information can be "seen."

Look again at Figure 5. The graph shows that for each date on the horizontal axis there is only one price on the vertical axis. Thus, the graph represents a function, although the exact rule for getting from date to price is not given.

When a function is defined by an equation in x and y, the **graph** of the function is the graph of the equation, that is, the set of points (x, y) in the xy-plane that satisfies the equation.

4 Not every collection of points in the xy-plane represents the graph of a function. Remember, for a function, each number x in the domain has one and only one image y. This means that the graph of a function cannot contain two points with the same x-coordinate and different y-coordinates. Therefore, the graph of a function must satisfy the following **vertical-line test.**

Theorem **Vertical-line Test**

A set of points in the xy-plane is the graph of a function if and only if every vertical line intersects the graph in at most one point.

It follows that, if any vertical line intersects a graph at more than one point, the graph is not the graph of a function.

EXAMPLE 7 Identifying the Graph of a Function

Which of the graphs in Figure 6 are graphs of functions?

Figure 6

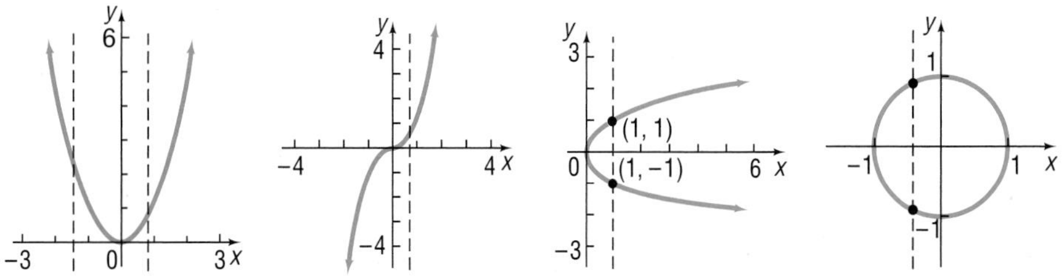

(a) $y = x^2$ **(b)** $y = x^3$ **(c)** $x = y^2$ **(d)** $x^2 + y^2 = 1$

Solution The graphs in Figures 6(a) and 6(b) are graphs of functions, because every vertical line intersects each graph in at most one point. The graphs in Figures 6(c) and 6(d) are not graphs of functions, because a vertical line intersects each graph in more than one point. ■

NOW WORK PROBLEM 43.

5 If (x, y) is a point on the graph of a function f, then y is the value of f at x; that is, $y = f(x)$. The next example illustrates how to obtain information about a function if its graph is given.

EXAMPLE 8 Obtaining Information from the Graph of a Function

Figure 7

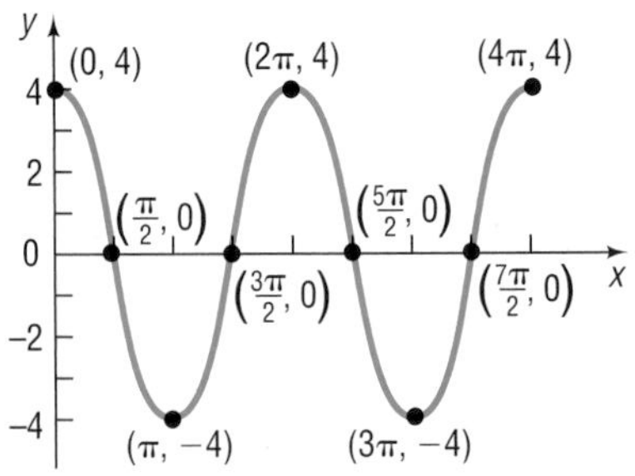

Let f be the function whose graph is given in Figure 7. (The graph of f might represent the distance that the bob of a pendulum is from its *at-rest* position. Negative values of y mean that the pendulum is to the left of the at-rest position, and positive values of y mean that the pendulum is to the right of the at-rest position.)

(a) What is $f(0)$, $f(3\pi/2)$, and $f(3\pi)$?
(b) What is the domain of f?
(c) What is the range of f?
(d) List the intercepts. (Recall that these are the points, if any, where the graph crosses or touches the coordinate axes.)
(e) How often does the line $y = 2$ intersect the graph?
(f) For what values of x does $f(x) = -4$?

Solution (a) Since $(0, 4)$ is on the graph of f, the y-coordinate 4 is the value of f at the x-coordinate 0; that is, $f(0) = 4$. In a similar way, we find that when $x = 3\pi/2$ then $y = 0$, so $f(3\pi/2) = 0$. When $x = 3\pi$, then $y = -4$, so $f(3\pi) = -4$.

(b) To determine the domain of f, we notice that the points on the graph of f will have x-coordinates between 0 and 4π, inclusive; and for each number x between 0 and 4π there is a point $(x, f(x))$ on the graph. The domain of f is $\{x | 0 \le x \le 4\pi\}$ or the interval $[0, 4\pi]$.

(c) The points on the graph all have y-coordinates between -4 and 4, inclusive; and for each such number y there is at least one number x in the domain. The range of f is $\{y | -4 \le y \le 4\}$ or the interval $[-4, 4]$.

(d) The intercepts are $(0, 4)$, $(\pi/2, 0)$, $(3\pi/2, 0)$, $(5\pi/2, 0)$, and $(7\pi/2, 0)$.

(e) Draw the horizontal line $y = 2$ on the graph in Figure 7. Then we find that it intersects the graph four times.

(f) Since $(\pi, -4)$ and $(3\pi, -4)$ are the only points on the graph for which $y = f(x) = -4$, we have $f(x) = -4$ when $x = \pi$ and $x = 3\pi$. ■

When the graph of a function is given, its domain may be viewed as the shadow created by the graph on the x-axis by vertical beams of light. Its range can be viewed as the shadow created by the graph on the y-axis by horizontal beams of light. Try this technique with the graph given in Figure 7.

NOW WORK PROBLEMS 39 AND 41.

EXAMPLE 9 Obtaining Information about the Graph of a Function

Consider the function: $f(x) = \dfrac{x}{x+2}$

(a) Is the point $(1, 1/2)$ on the graph of f?
(b) If $x = -1$, what is $f(x)$? What point is on the graph of f?
(c) If $f(x) = 2$, what is x? What point is on the graph of f?

Solution (a) When $x = 1$, then $f(x) = f(1) = 1/(1 + 2) = 1/3$. The point $(1, 1/3)$ is on the graph of f; the point $(1, 1/2)$ is not.

(b) If $x = -1$, then $f(x) = f(-1) = -1/(-1 + 2) = -1$, so the point $(-1, -1)$ is on the graph of f.

(c) If $f(x) = 2$, then

$$\frac{x}{x+2} = 2$$

$$x = 2(x + 2)$$

$$x = 2x + 4$$

$$x = -4$$

The point $(-4, 2)$ is on the graph of f. ■

NOW WORK PROBLEM 35.

APPLICATIONS

When we use functions in applications, the domain may be restricted by physical or geometric considerations. For example, the domain of the function f defined by $f(x) = x^2$ is the set of all real numbers. However, if f is used to obtain the area of a square when the length x of a side is known, then we must restrict the domain of f to the positive real numbers, since the length of a side can never be 0 or negative.

EXAMPLE 10 Area of a Circle

Express the area of a circle as a function of its radius.

Solution We know that the formula for the area A of a circle of radius r is $A = \pi r^2$. If we use r to represent the independent variable and A to represent the dependent variable, the function expressing this relationship is

$$A(r) = \pi r^2$$

In this setting, the domain is $\{r \mid r > 0\}$. (Do you see why?) ■

NOW WORK PROBLEM 91.

EXAMPLE 11 Construction Cost

TABLE 2

Square Feet, x	Cost, C
1500	165,000
1600	176,300
1700	183,000
1700	189,500
1830	201,700
1970	217,000
2050	220,000
2100	237,400

Sally, a builder of homes, is interested in finding a function that relates the cost C of building a house and x, the number of square feet of the house, so that she can easily provide estimates to clients for various-sized homes. The data that she obtained from constructing homes last year are listed in Table 2.

(a) Does the relation defined by the set of ordered pairs (x, C) represent a function?

(b) Using a graphing utility, draw a scatter diagram of the data.

(c) Using a graphing utility, find the line of best fit relating square feet and cost.

(d) Interpret the slope of the line of best fit.

(e) Express the relationship found in part (c) using function notation.

(f) What is the domain of the function?

(g) What is the cost to build a 2000 square foot house?

(h) Give some reasons to explain why the cost of building a 1700 square foot house might be \$183,000 in one case and \$189,500 in another. Does this also explain why, perhaps, $C(2000) > C(2050)$?

Solution

(a) No, the relation (x, C) does not represent a function because the first element 1700 has, corresponding to it, two second elements, 183,000 and 189,500.

(b) See Figure 8 for a scatter diagram of the data.

Figure 8

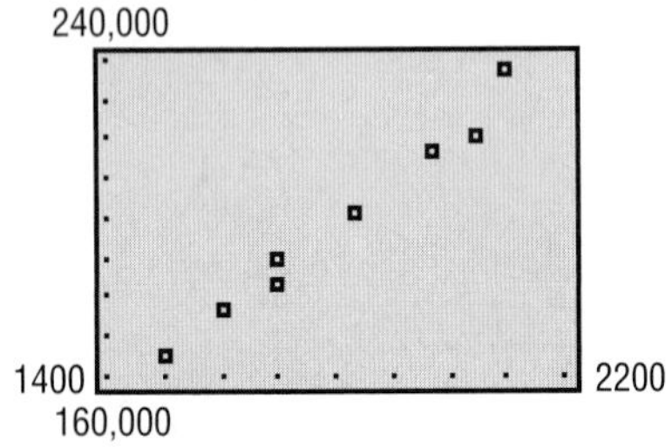

(c) The line of best fit is $C = 112.09x - 3734.02$.

(d) The slope of the line of best fit, \$112.09, represents the cost per square foot to build a house. Thus, for each additional square foot of house being built, the cost increases by \$112.09.

(e) Using function notation, $C(x) = 112.09x - 3734.02$.

(f) The domain is $\{x \mid x > 0\}$, the interval $(0, \infty)$, since a house cannot have 0 or negative square feet.

(g) The cost to build a 2000 square foot house is

$$C(2000) = 112.09(2000) - 3734.02 \approx \$220{,}446$$

(h) One explanation is that the fixtures placed in one house are more expensive than those placed in the other house. ■

Observe in the solution to Example 11 that we used the symbol C in two ways: It is used to name the function and to symbolize the dependent variable. This double use is common in applications and should not cause any difficulty.

Refer back to Example 11. Notice that the data in Table 2 do not represent a function since the same first element is paired with two different second elements $[(1700, 183{,}000), (1700, 189{,}500)]$. However, when we find the line of best fit for these data, we find a function that relates the data. Thus, **curve fitting** is a process whereby we find a functional relationship between two or more variables even though the data may not represent a function.

NOW WORK PROBLEM 85.

EXAMPLE 12 Average Cost Function

The average cost A of manufacturing x computers per day is given by the function

$$A(x) = 0.56x^2 - 34.39x + 1212.57 + \frac{20{,}000}{x}$$

Determine the average cost of manufacturing the following:

(a) 30 computers in a day
(b) 40 computers in a day
(c) 50 computers in a day
(d) Graph the function $A = A(x), 0 < x \leq 80$.
(e) Create a TABLE with TblStart $= 1$ and ΔTbl $= 1$. Which value of x minimizes the average cost?

Solution   (a) The average cost of manufacturing $x = 30$ computers is

$$A(30) = 0.56(30)^2 - 34.39(30) + 1212.57 + \frac{20{,}000}{30} = \$1351.54$$

(b) The average cost of manufacturing $x = 40$ computers is

$$A(40) = 0.56(40)^2 - 34.39(40) + 1212.57 + \frac{20{,}000}{40} = \$1232.97$$

(c) The average cost of manufacturing $x = 50$ computers is

$$A(50) = 0.56(50)^2 - 34.39(50) + 1212.57 + \frac{20{,}000}{50} = \$1293.07$$

(d) See Figure 9 for the graph of $A = A(x)$.
(e) With the function $A = A(x)$ in Y_1, we create Table 3. We scroll down until we find a value of x for which Y_1 is smallest. Table 4 shows that manufacturing $x = 41$ computers minimizes the average cost at \$1231.74 per computer.

Figure 9

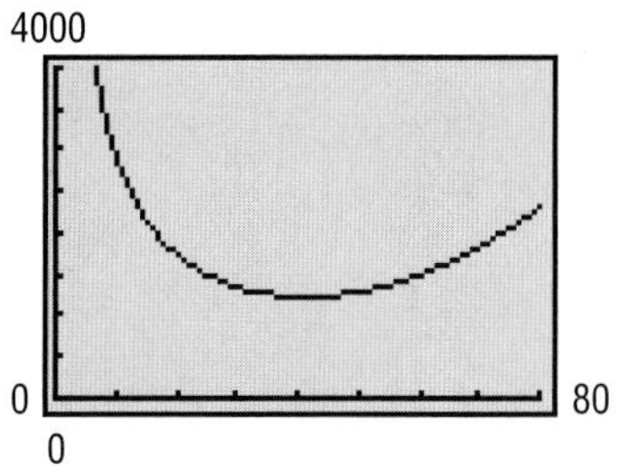

T A B L E	3
X	Y1
1	21179
2	11146
3	7781.1
4	6084
5	5054.6
6	4359.7
7	3856.4
Y1 $=.56X^2 - 34.39X...$	

T A B L E	4
X	Y1
38	1240.7
39	1235.9
40	1233
41	1231.7
42	1232.2
43	1234.4
44	1238.1
Y1 $=1231.74487805$	

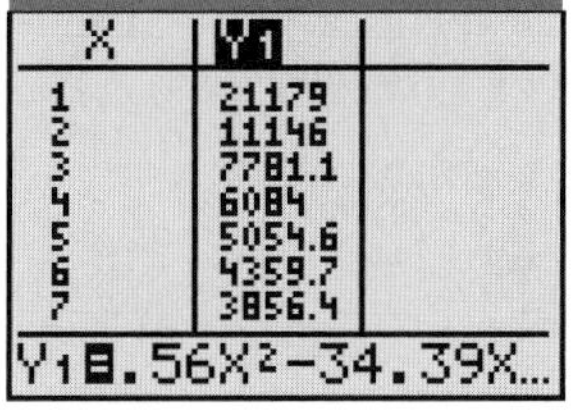

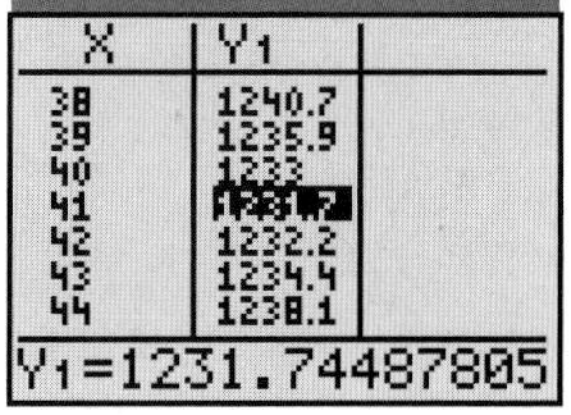

SUMMARY

We list here some of the important vocabulary introduced in this section, with a brief description of each term.

Function	A relation between two sets of real numbers so that each number x in the first set, the domain, has corresponding to it exactly one number y in the second set. A set of ordered pairs (x, y) or $(x, f(x))$ in which no two distinct pairs have the same first element. The range is the set of y values of the function for the x values in the domain. A function f may be defined implicitly by an equation involving x and y or explicitly by writing $y = f(x)$.
Unspecified domain	If a function f is defined by an equation and no domain is specified, then the domain will be taken to be the largest set of real numbers for which the equation defines a real number.
Function notation	$y = f(x)$ f is a symbol for the function. x is the independent variable or argument. y is the dependent variable. $f(x)$ is the value of the function at x, or the image of x.

Graph of a function	The collection of points (x, y) that satisfies the equation $y = f(x)$. A collection of points is the graph of a function provided that every vertical line intersects the graph in at most one point (vertical-line test).

2.1 EXERCISES

In Problems 1–12, determine whether each relation represents a function. For each function, state the domain and range.

1.

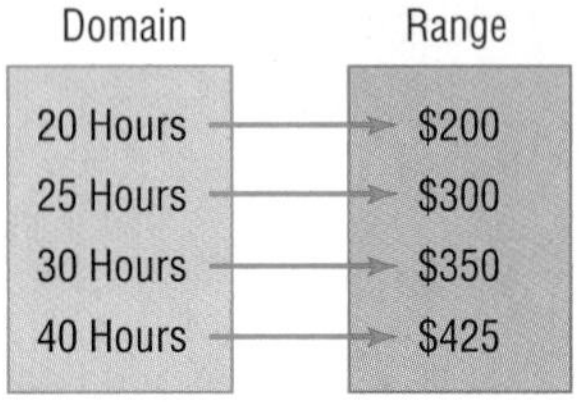

2.

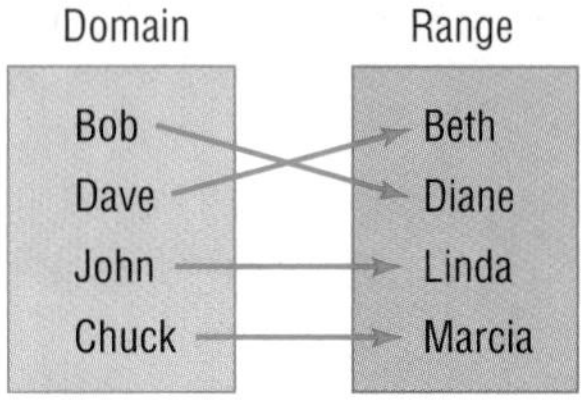

3.

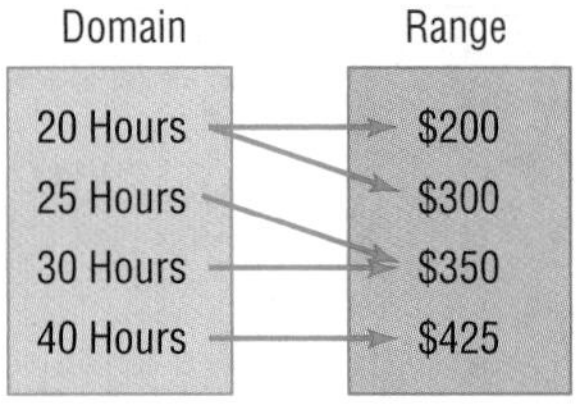

4.

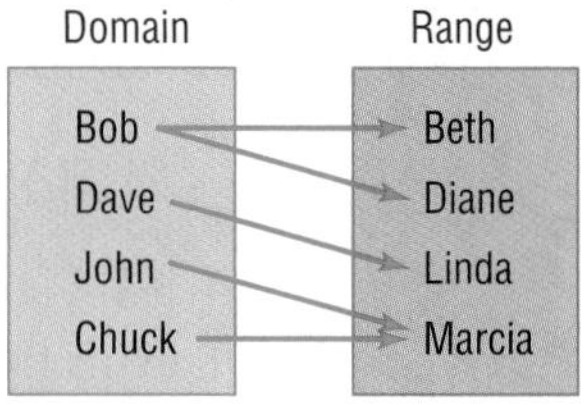

5. $\{(2, 6), (-3, 6), (4, 9), (1, 10)\}$

6. $\{(-2, 5), (-1, 3), (3, 7), (4, 12)\}$

7. $\{(1, 3), (2, 3), (3, 3), (4, 3)\}$

8. $\{(0, -2), (1, 3), (2, 3), (3, 7)\}$

9. $\{(-2, 4), (-2, 6), (0, 3), (3, 7)\}$

10. $\{(-4, 4), (-3, 3), (-2, 2), (-1, 1), (-4, 0)\}$

11. $\{(-2, 4), (-1, 1), (0, 0), (1, 1)\}$

12. $\{(-2, 16), (-1, 4), (0, 3), (1, 4)\}$

In Problems 13–20, find the following values for each function:

(a) $f(0)$; *(b)* $f(1)$; *(c)* $f(-1)$; *(d)* $f(-x)$; *(e)* $-f(x)$; *(f)* $f(x + 1)$ *(g)* $f(2x)$ *(h)* $f(x + h)$

13. $f(x) = 3x^2 + 2x - 4$

14. $f(x) = -2x^2 + x - 1$

15. $f(x) = \dfrac{x}{x^2 + 1}$

16. $f(x) = \dfrac{x^2 - 1}{x + 4}$

17. $f(x) = |x| + 4$

18. $f(x) = \sqrt{x^2 + x}$

19. $f(x) = \dfrac{2x + 1}{3x - 5}$

20. $f(x) = 1 - \dfrac{1}{(x + 2)^2}$

In Problems 21–34, use the graph of the function f given in the figure.

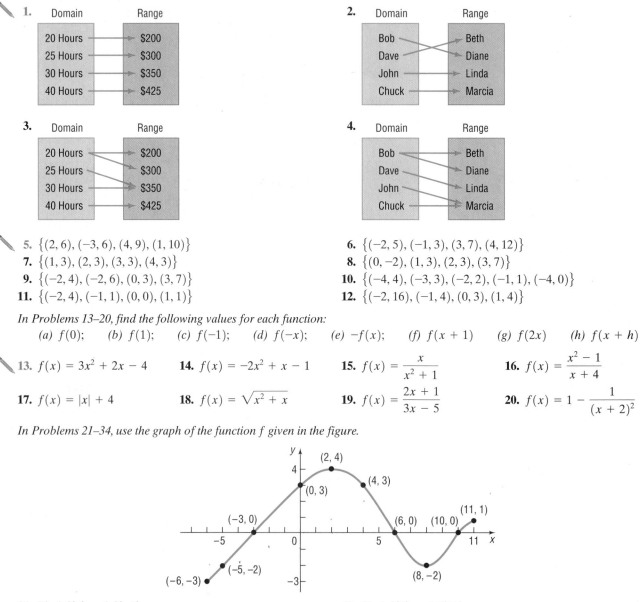

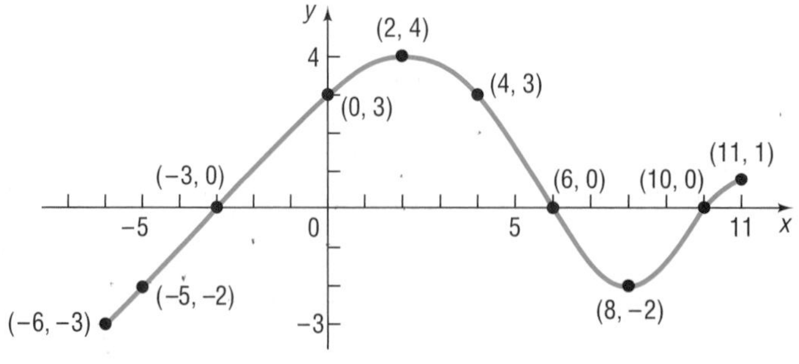

21. Find $f(0)$ and $f(-6)$.

22. Find $f(6)$ and $f(11)$.

23. Is $f(2)$ positive or negative?

24. Is $f(8)$ positive or negative?

25. For what numbers x is $f(x) = 0$?

26. For what numbers x is $f(x) > 0$?

27. What is the domain of f?

28. What is the range of f?

29. What are the x-intercepts?

30. What are the y-intercepts?

31. How often does the line $y = \frac{1}{2}$ intersect the graph?

32. How often does the line $y = 3$ intersect the graph?

33. For what values of x does $f(x) = 3$?

34. For what values of x does $f(x) = -2$?

In Problems 35–38, answer the questions about the given function.

35. $f(x) = \dfrac{x+2}{x-6}$
(a) Is the point $(3, 14)$ on the graph of f?
(b) If $x = 4$, what is $f(x)$? What point is on the graph of f?
(c) If $f(x) = 2$, what is x? What point is on the graph of f?
(d) What is the domain of f?

36. $f(x) = \dfrac{x^2+2}{x+4}$
(a) Is the point $\left(1, \frac{3}{5}\right)$ on the graph of f?
(b) If $x = 0$, what is $f(x)$? What point is on the graph of f?
(c) If $f(x) = \frac{1}{2}$, what is x? What points are on the graph of f?
(d) What is the domain of f?

37. $f(x) = \dfrac{2x^2}{x^4+1}$
(a) Is the point $(-1, 1)$ on the graph of f?
(b) If $x = 2$, what is $f(x)$? What point is on the graph of f?
(c) If $f(x) = 1$, what is x? What points are on the graph of f?
(d) What is the domain of f?

38. $f(x) = \dfrac{2x}{x-2}$
(a) Is the point $\left(\frac{1}{2}, -\frac{2}{3}\right)$ on the graph of f?
(b) If $x = 4$, what is $f(x)$? What point is on the graph of f?
(c) If $f(x) = 1$, what is x? What point is on the graph of f?
(d) What is the domain of f?

In Problems 39–50, determine whether the graph is that of a function by using the vertical-line test. If it is, use the graph to find:
(a) Its domain and range *(b) The intercepts, if any* *(c) Any symmetry with respect to the x-axis, y-axis, or origin*

39.
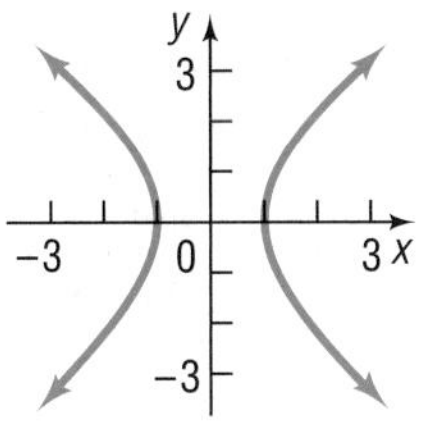

40.
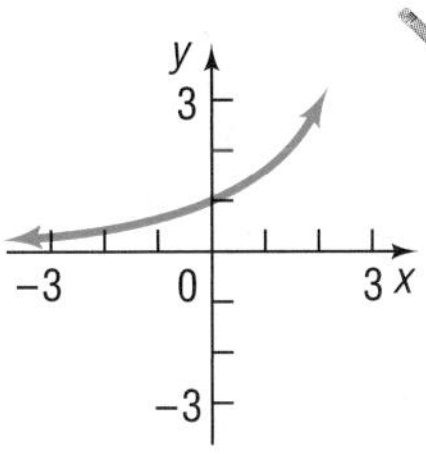

41.
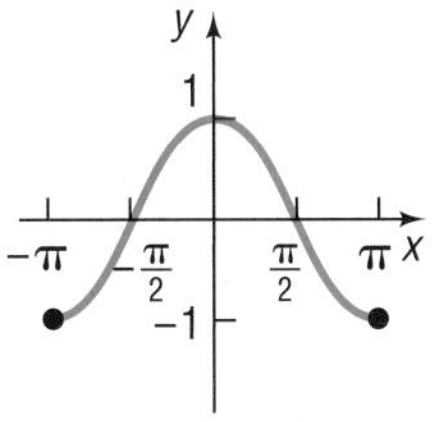

42.
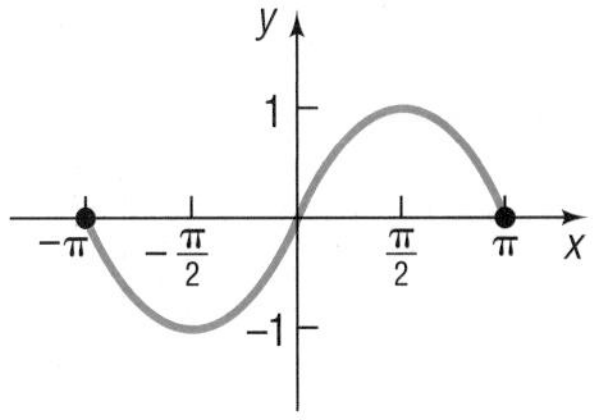

43.
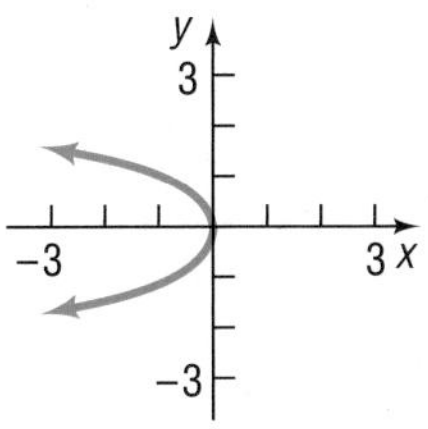

44.
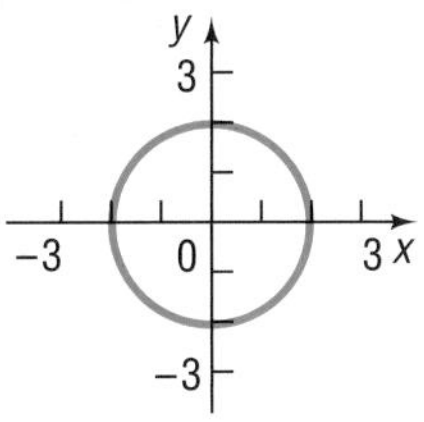

45.
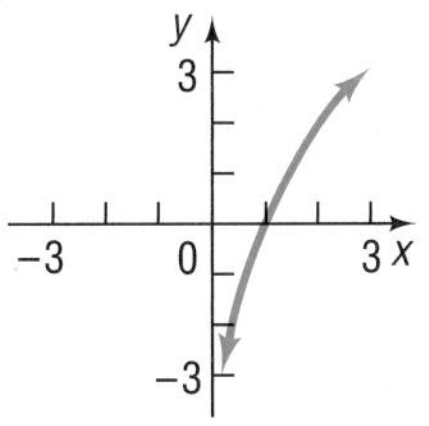

46.
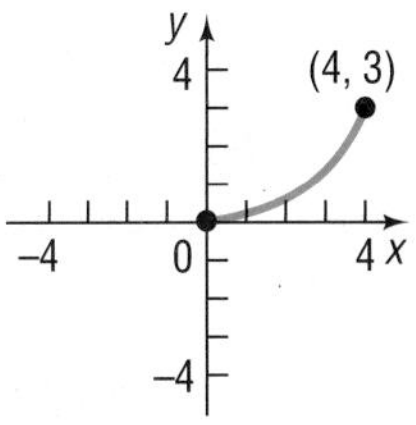

47.
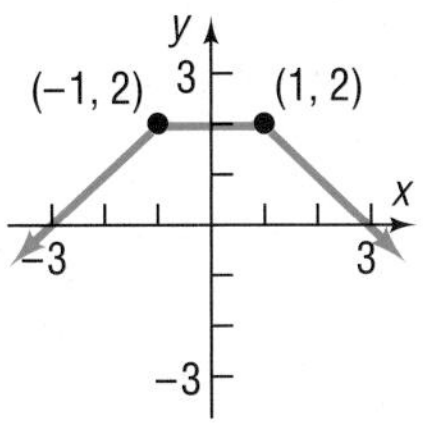

48.
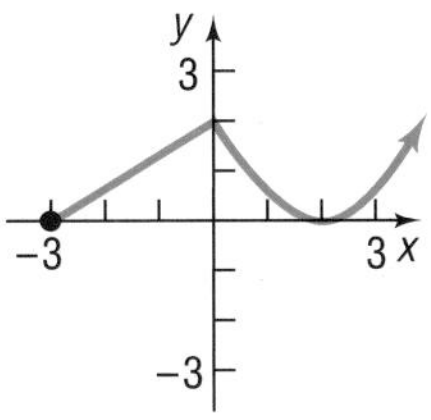

49.
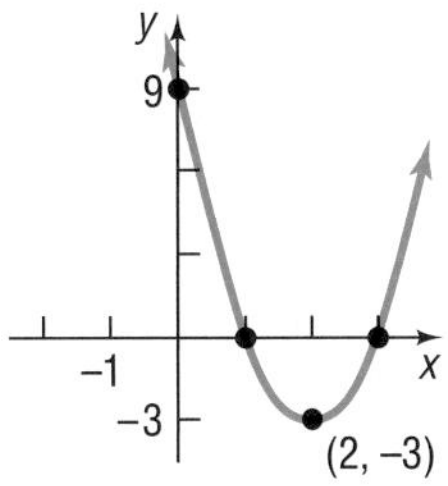

50.
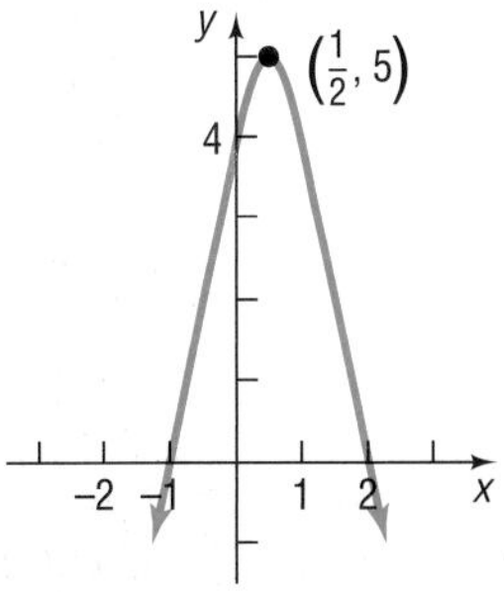

In Problems 51–64, find the domain of each function.

51. $f(x) = -5x + 4$

52. $f(x) = x^2 + 2$

53. $f(x) = \dfrac{x}{x^2+1}$

54. $f(x) = \dfrac{x^2}{x^2+1}$

55. $g(x) = \dfrac{x}{x^2-16}$

56. $h(x) = \dfrac{2x}{x^2-4}$

57. $F(x) = \dfrac{x-2}{x^3+x}$

58. $G(x) = \dfrac{x+4}{x^3-4x}$

59. $h(x) = \sqrt{3x - 12}$ **60.** $G(x) = \sqrt{1 - x}$ **61.** $f(x) = \dfrac{4}{\sqrt{x - 9}}$ **62.** $f(x) = \dfrac{x}{\sqrt{x - 4}}$

63. $p(x) = \sqrt{\dfrac{2}{x - 1}}$ **64.** $q(x) = \sqrt{-x - 2}$

In Problems 65–72, tell whether the set of ordered pairs (x, y) defined by each equation is a function.

65. $y = x^2 + 2x$ **66.** $y = x^3 - 3x$ **67.** $y = \dfrac{2}{x}$ **68.** $y = \dfrac{3}{x} - 3$

69. $y^2 = 1 - x^2$ **70.** $y = \pm\sqrt{1 - 2x}$ **71.** $x^2 + y = 1$ **72.** $x + 2y^2 = 1$

73. Match each of the following functions with the graphs that best describe the situation.
(a) The cost of building a house as a function of its square footage
(b) The height of an egg dropped from a 300-foot building as a function of time
(c) The height of a human as a function of time
(d) The demand for Big Macs as a function of price
(e) The height of a child on a swing as a function of time

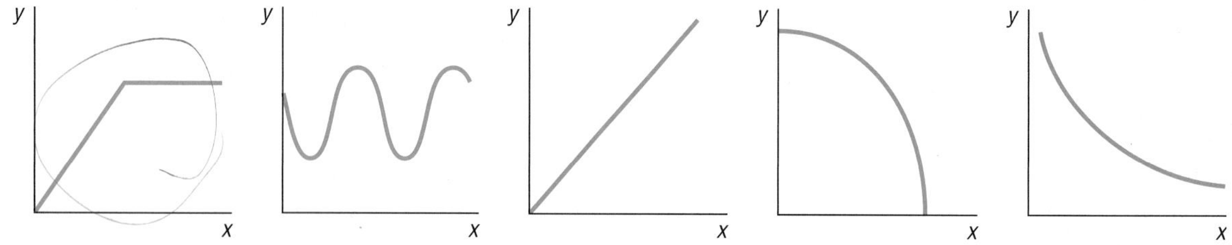

(I) (II) (III) (IV) (V)

74. Match each of the following functions with the graph that best describes the situation.
(a) The temperature of a bowl of soup as a function of time
(b) The number of hours of daylight per day over a two-year period
(c) The population of Florida as a function of time
(d) The distance of a car traveling at a constant velocity as a function of time
(e) The height of a golf ball hit with a 7-iron as a function of time

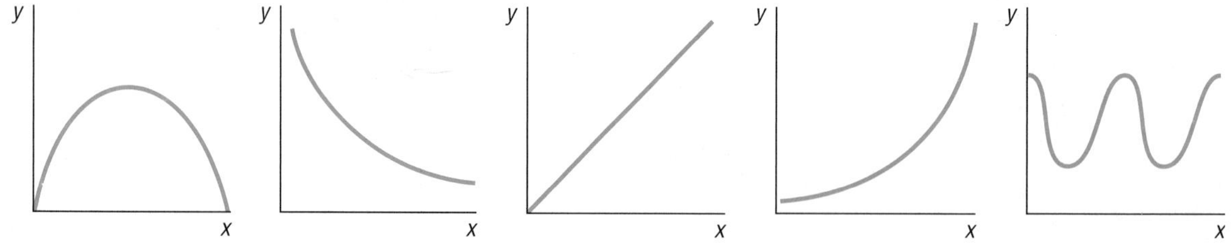

(I) (V) (III) (IV) (V)

75. Consider the following scenario: Barbara decides to take a walk. She leaves home, walks 2 blocks in 5 minutes at a constant speed, and realizes that she forgot to lock the door. So Barbara runs home in 1 minute. While at her doorstep, it takes her 1 minute to find her keys and lock the door. Barbara walks 5 blocks in 15 minutes and then decides to jog home. It takes her 7 minutes to get home. Draw a graph of Barbara's distance from home (in blocks) as a function of time.

76. Consider the following scenario: Jayne enjoys riding her bicycle through the woods. At the forest preserve, she gets on her bicycle and rides up a 2000-foot incline in 10 minutes. She then travels down the incline in 3 minutes. The next 5000 feet is level terrain and she covers the distance in 20 minutes. She rests for 15 minutes. Jayne then travels 10,000 feet in 30 minutes. Draw a graph of Jayne's distance traveled (in feet) as a function of time.

77. The following is a graph that represents the distance d (in miles) that Kevin is from home as a function of time t (in hours). Answer the questions based on the graph.

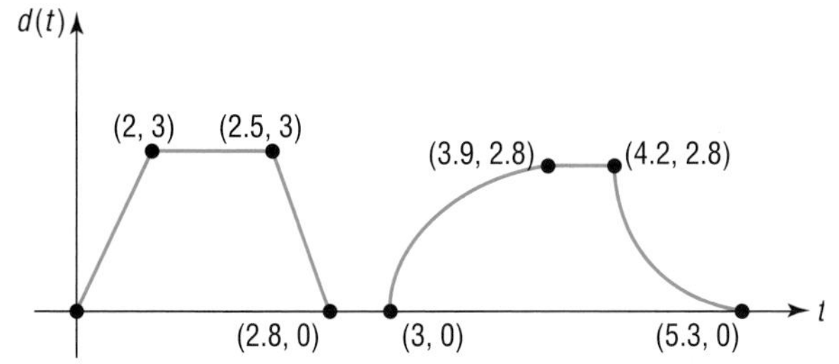

(a) How many times did Kevin return home?
Describe Kevin's distance from home:
(b) From $t = 0$ to $t = 2$
(c) From $t = 2$ to $t = 2.5$
(d) From $t = 2.5$ to $t = 2.8$
(e) From $t = 2.8$ to $t = 3$
(f) From $t = 3$ to $t = 3.9$
(g) From $t = 3.9$ to $t = 4.2$
(h) From $t = 4.2$ to $t = 5.3$
(i) What is the farthest distance that Kevin is from home?

78. The following graph represents the speed v (in miles per hour) of Michael's car as a function of time t (in minutes).

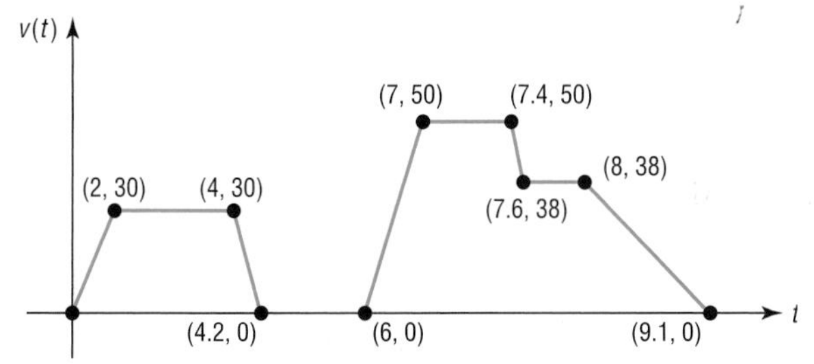

(a) Over what interval of time is Michael traveling fastest?
(b) Over what interval(s) of time is Michael's speed zero?
(c) What is happening to Michael's speed between 0 and 2 minutes?
(d) What is happening to Michael's speed between 4.2 and 6 minutes?
(e) Describe Michael's speed between 7 and 7.4 minutes. How does this result compare with that of part (c)?
(f) When is Michael's speed constant?

79. If $f(x) = 2x^3 + Ax^2 + 4x - 5$ and $f(2) = 5$, what is the value of A?

80. If $f(x) = 3x^2 - Bx + 4$ and $f(-1) = 12$, what is the value of B?

81. If $f(x) = (3x + 8)/(2x - A)$ and $f(0) = 2$, what is the value of A?

82. If $f(x) = (2x - B)/(3x + 4)$ and $f(2) = \frac{1}{2}$, what is the value of B?

83. If $f(x) = (2x - A)/(x - 3)$ and $f(4) = 0$, what is the value of A? Where is f not defined?

84. If $f(x) = (x - B)/(x - A)$, $f(2) = 0$, and $f(1)$ is undefined, what are the values of A and B?

85. Demand for Jeans The marketing manager at Levi–Strauss wishes to find a function that relates the demand D of men's jeans and p, the price of the jeans. The following data were obtained based on a price history of jeans.

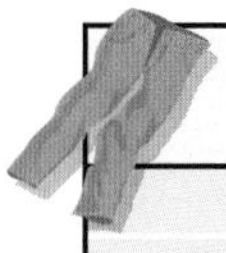

Price ($/Pair), p	Demand (Pairs of Jeans Sold per Day), D
20	60
22	57
23	56
23	53
27	52
29	49
30	44

(a) Does the relation defined by the set of ordered pairs (p, D) represent a function?
(b) Using a graphing utility, draw a scatter diagram of the data.
(c) Using a graphing utility, find the line of best fit relating price and quantity demanded.
(d) Interpret the slope of the line of best fit.
(e) Express the relationship found in part (c) using function notation.
(f) What is the domain of the function?
(g) How many jeans will be demanded if the price is $28 a pair?

86. Advertising and Sales Revenue A marketing firm wishes to find a function that relates the sales S of a product and A, the amount spent on advertising the product. The data are obtained from past experience. Advertising and sales are measured in thousands of dollars.

Advertising Expenditures, A	Sales, S
20	335
22	339
22.5	338
24	343
24	341
27	350
28.3	351

(a) Does the relation defined by the set of ordered pairs (A, S) represent a function?
(b) Using a graphing utility, draw a scatter diagram of the data.
(c) Using a graphing utility, find the line of best fit relating advertising expenditures and sales.
(d) Interpret the slope of the line of best fit.

(e) Express the relationship found in part (c) using function notation.
(f) What is the domain of the function?
(g) Predict sales if advertising expenditures are \$25,000.

87. Distance and Time Using the following data, find a function that relates the distance, s, in miles, driven by a Ford Taurus, and t, the time the Taurus has been driven.

Time (Hours), t	Distance (Miles), s
0	0
1	30
2	55
3	83
4	100
5	150
6	210
7	260
8	300

(a) Does the relation defined by the set of ordered pairs (t, s) represent a function?
(b) Using a graphing utility, draw a scatter diagram of the data.
(c) Using a graphing utility, find the line of best fit relating time and distance.
(d) Interpret the slope of the line of best fit.
(e) Express the relationship found in part (c) using function notation.
(f) What is the domain of the function?
(g) Predict the distance the car is driven after 11 hours.

88. High School versus College GPA An administrator at Southern Illinois University wants to find a function that relates a student's college grade point average G to the high school grade point average, x. She randomly selects 8 students and obtains the following data:

High School GPA, x	College GPA, G
2.73	2.43
2.92	2.97
3.45	3.63
3.78	3.81
2.56	2.83
2.98	2.81
3.67	3.45
3.10	2.93

(a) Does the relation defined by the set of ordered pairs (x, G) represent a function?
(b) Using a graphing utility, draw a scatter diagram of the data.
(c) Using a graphing utility, find the line of best fit relating high school GPA and college GPA.
(d) Interpret the slope of the line of best fit.
(e) Express the relationship found in part (c) using function notation.
(f) What is the domain of the function?
(g) Predict a student's college GPA if her high school GPA is 3.23.

89. Effect of Gravity on Earth If a rock falls from a height of 20 meters on Earth, the height H (in meters) after x seconds is approximately

$$H(x) = 20 - 4.9x^2$$

(a) Using a graphing utility, graph $H(x)$.
(b) What is the height of the rock when $x = 1$ second? $x = 1.1$ seconds? $x = 1.2$ seconds? $x = 1.3$ seconds?
(c) When is the height of the rock 15 meters? When is it 10 meters? When is it 5 meters?
(d) When does the rock strike the ground?

90. Effect of Gravity on Jupiter If a rock falls from a height of 20 meters on the planet Jupiter, its height H (in meters) after x seconds is approximately

$$H(x) = 20 - 13x^2$$

(a) Using a graphing utility, graph $H(x)$.
(b) What is the height of the rock when $x = 1$ second? $x = 1.1$ seconds? $x = 1.2$ seconds?
(c) When is the height of the rock 15 meters? When is it 10 meters? When is it 5 meters?
(d) When does the rock strike the ground?

91. Geometry Express the area A of a rectangle as a function of the length x if the length is twice the width of the rectangle.

92. Geometry Express the area A of an isosceles right triangle as a function of the length x of one of the two equal sides.

93. Express the gross salary G of a person who earns \$10 per hour as a function of the number x of hours worked.

94. Tiffany, a commissioned salesperson, earns \$100 base pay plus \$10 per item sold. Express her gross salary G as a function of the number x of items sold.

95. Motion of a Golf Ball A golf ball is hit with an initial velocity of 130 feet per second at an inclination of 30° to the horizontal. In physics, it is established that the height h of the golf ball is given by the function

$$h(x) = \frac{-32x^2}{130^2} + x$$

where x is the horizontal distance that the golf ball has traveled. See the figure.

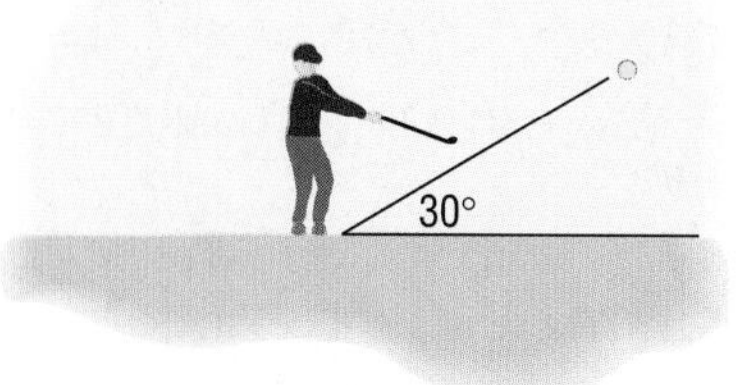

(a) Determine the height of the golf ball after it has traveled 100 feet.
(b) 300 feet
(c) 500 feet
(d) Graph the function $h = h(x)$.
(e) Algebraically, determine the distance that the ball has traveled when the height of the ball is 90 feet. Verify your results graphically.
(f) Create a TABLE with TblStart $= 0$ and ΔTbl $= 25$. To the nearest 25 feet, how far does the ball travel before it reaches a maximum height? What is the maximum height?
(g) Adjust the value of ΔTbl until you determine the distance that the ball travels before it reaches a maximum height to within 1 foot.
(h) Find the domain of h.
Hint: For what values of x is $h(x) \geq 0$?

96. Cross-sectional Area The cross-sectional area of a beam cut from a log with radius 1 foot is given by the function $A(x) = 4x\sqrt{1 - x^2}$, where x represents the length of half the base of the beam. See the figure.

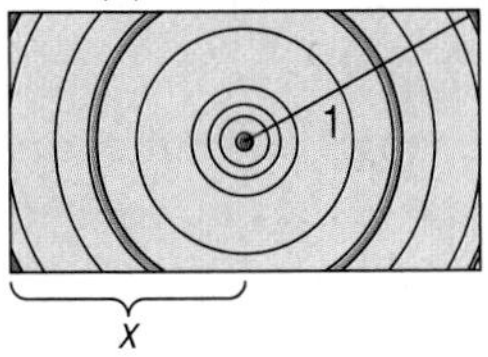

(a) Find the domain of A.
Determine the cross-sectional area of the beam if the length of half the base of the beam is as follows:
(b) One-third foot
(c) One-half of a foot
(d) Two-thirds of a foot
(e) Graph the function $A = A(x)$.
(f) Create a TABLE with TblStart $= 0$ and ΔTbl $= 0.1$ for $0 \leq x \leq 1$. Which value of x maximizes the cross-sectional area? What should be the length of base of the beam to maximize the cross-sectional area?

97. Cost of Trans-Atlantic Travel A Boeing 747 crosses the Atlantic Ocean (3000 miles) with an airspeed of 500 miles per hour. The cost C (in dollars) per passenger is given by the function

$$C(x) = 100 + \frac{x}{10} + \frac{36{,}000}{x}$$

where x is the ground speed (airspeed $\pm$ wind).
(a) What is the cost per passenger for quiescent (no wind) conditions?
(b) What is the cost per passenger with a head wind of 50 miles per hour?
(c) What is the cost per passenger with a tail wind of 100 miles per hour?
(d) What is the cost per passenger with a head wind of 100 miles per hour?
(e) Graph the function $C = C(x)$.
(f) Create a TABLE with TblStart $= 0$ and ΔTbl $= 50$. What ground speed minimizes the cost per passenger to the nearest 50 miles per hour?

98. Effect of Elevation on Weight If an object weighs m pounds at sea level, then its weight W (in pounds) at a height of h miles above sea level is given approximately by

$$W(h) = m\left(\frac{4000}{4000 + h}\right)^2$$

(a) If Amy weighs 120 pounds at sea level, how much will she weigh on Pike's Peak, which is 14,110 feet above sea level?
(b) Use a graphing utility to graph the function $W = W(h)$. Use $m = 120$ pounds.
(c) Create a Table with TblStart $= 0$ and ΔTbl $= 0.5$ to see how weight W varies as h changes from 0 to 5 miles.
(d) At what height will Amy weigh 119.95 pounds?
(e) Does your answer to part (d) seem reasonable?

99. Some functions f have the property that $f(a + b) = f(a) + f(b)$ for all real numbers a and b. Which of the following functions have this property?
(a) $h(x) = 2x$
(b) $g(x) = x^2$
(c) $F(x) = 5x - 2$
(d) $G(x) = 1/x$

100. Draw the graph of a function whose domain is $\{x \mid -3 \leq x \leq 8,\ x \neq 5\}$ and whose range is $\{y \mid -1 \leq y \leq 2, y \neq 0\}$. What point(s) in the rectangle $-3 \leq x \leq 8, -1 \leq y \leq 2$ cannot be on the graph? Compare your graph with those of other students. What differences do you see?

101. Are the functions $f(x) = x - 1$ and $g(x) = (x^2 - 1)/(x + 1)$ the same? Explain.

102. Describe how you would proceed to find the domain and range of a function if you were given its graph. How would your strategy change if, instead, you were given the equation defining the function?

103. How many x-intercepts can the graph of a function have? How many y-intercepts can it have?

104. Is a graph that consists of a single point the graph of a function? Can you write the equation of such a function?

105. Is there a function whose graph is symmetric with respect to the x-axis?

106. Investigate when, historically, the use of the function notation $y = f(x)$ first appeared.

PREPARING FOR THIS SECTION

Before getting started, review the following:

- ✓ Intervals (pp. 49–50)
- ✓ Slope of a Line (pp. 61–63)
- ✓ Point–slope Form of a Line (p. 66)
- ✓ Tests for Symmetry of an Equation (p. 18)

2.2 CHARACTERISTICS OF FUNCTIONS

OBJECTIVES

1. Find the Average Rate of Change of a Function
2. Use a Graphing Utility to Locate Local Maxima and Minima
3. Use a Graphing Utility to Determine Where a Function Is Increasing and Decreasing
4. Determine Even or Odd Functions from a Graph
5. Identify Even or Odd Functions from the Equation

AVERAGE RATE OF CHANGE

1 In Section 1.6 we said that the slope of a line could be interpreted as the average rate of change. Often we are interested in the rate at which functions change. To find the average rate of change of a function between any two points on its graph, we calculate the slope of the line containing the two points.

EXAMPLE 1 **Finding the Average Rate of Change of a Function**

TABLE 5

Year	Average Cost of Tuition and Fees ($)
1991	2159
1992	2410
1993	2604
1994	2820
1995	2977
1996	3151
1997	3321

Source: U.S. Center for National Education Statistics

The data in Table 5 represent the average cost of tuition and required fees (in dollars) at public four-year colleges.

(a) Draw a scatter diagram of the data, treating the year as the independent variable.

(b) Draw a line through the points (1991, 2159) and (1992, 2410) on the scatter diagram found in part (a).

(c) Find the average rate of change of the cost of tuition and required fees from 1991 to 1992.

(d) Draw a line through the points (1995, 2977) and (1996, 3151) on the scatter diagram found in part (a).

(e) Find the average rate of change of the cost of tuition and required fees from 1995 to 1996.

(f) What is happening to the average rate of change as time passes?

Solution

(a) We plot the ordered pairs (1991, 2159), (1992, 2410), and so on, using rectangular coordinates. See Figure 10.

(b) Draw the line through (1991, 2159) and (1992, 2410). See Figure 11(a).

(c) The average rate of change is found by computing the slope of the line containing the points (1991, 2159) and (1992, 2410).

$$\text{Average rate of change} = \frac{2410 - 2159}{1992 - 1991} = \frac{251}{1} = \$251/\text{year}$$

Figure 10

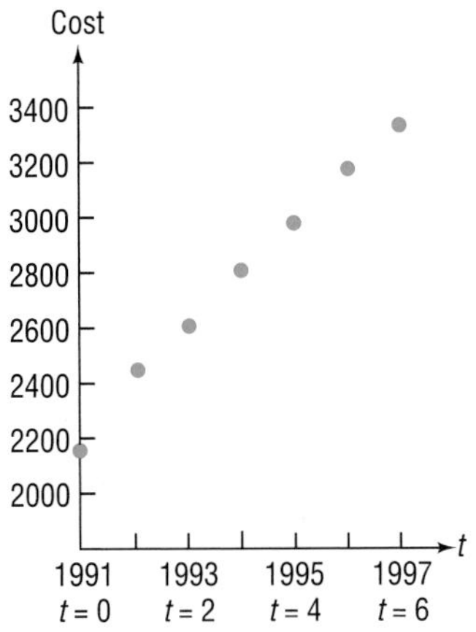

Figure 11

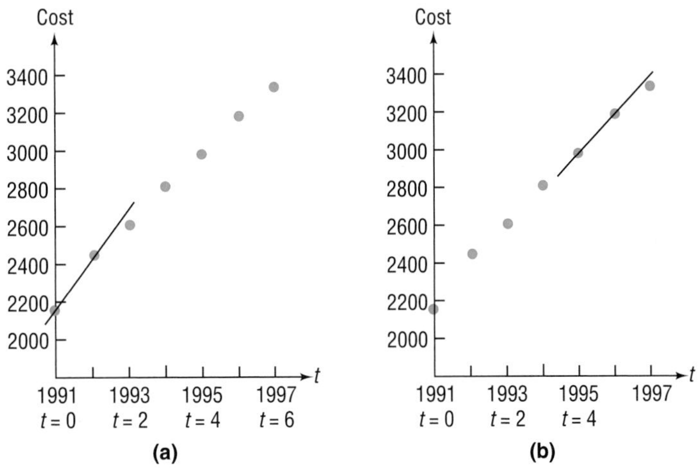

(d) Draw the line through (1995, 2977) and (1996, 3151). See Figure 11(b).

(e) The average rate of change is found by computing the slope of the line containing the points (1995, 2977) and (1996, 3151).

$$\text{Average rate of change} = \frac{3151 - 2977}{1996 - 1995} = \frac{174}{1} = \$174/\text{year}$$

(f) The cost of tuition and fees is increasing over time, but the rate of the increase is declining. ■

If we know the function C that relates the year t to the cost C of tuition and fees, then the average rate of change from 1991 to 1992 may be expressed as

$$\text{Average rate of change} = \frac{C(1992) - C(1991)}{1992 - 1991} = \frac{251}{1} = \$251/\text{year}$$

Expressions like this occur frequently in calculus.

If c is in the domain of a function $y = f(x)$, the **average rate of change of f from c to x** is defined as

$$\text{Average rate of change} = \frac{\Delta y}{\Delta x} = \frac{f(x) - f(c)}{x - c}, \qquad x \neq c \qquad \textbf{(1)}$$

This expression is also called the **difference quotient** of f at c.

The average rate of change of a function has an important geometric interpretation. Look at the graph of $y = f(x)$ in Figure 12. We have labeled two points on the graph: $(c, f(c))$ and $(x, f(x))$. The line containing these two points is called a **secant line**; its slope is

$$m_{\text{sec}} = \frac{f(x) - f(c)}{x - c}$$

Figure 12

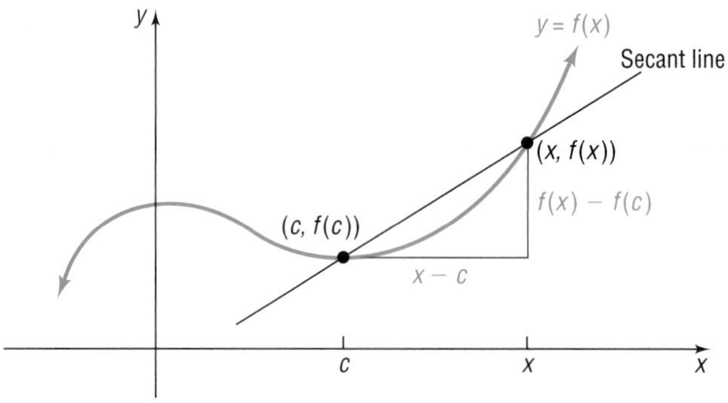

Theorem **Slope of the Secant Line**

The average rate of change of a function equals the slope of the secant line containing two points on its graph.

EXAMPLE 2 Finding the Average Rate of Change of a Function

(a) Find the average rate of change of $f(x) = 2x^2 - 3x$ from 1 to x.

(b) Use this result to find the slope of the secant line containing $(1, f(1))$ and $(2, f(2))$.

(c) Find the equation of this secant line.

(d) Graph f and the secant line on the same viewing window.

Solution (a) The average rate of change of f from 1 to x is

$$\frac{\Delta y}{\Delta x} = \frac{f(x) - f(1)}{x - 1} \qquad x \neq 1$$

$$= \frac{(2x^2 - 3x) - (2(1)^2 - 3(1))}{x - 1}$$

$$= \frac{2x^2 - 3x + 1}{x - 1}$$

$$= \frac{(2x - 1)(x - 1)}{x - 1}$$

$$= 2x - 1 \qquad x \neq 1;\ \text{cancel } x - 1$$

(b) The slope of the secant line containing $(1, f(1))$ and $(2, f(2))$ is the average rate of change of f from 1 to 2. Using $x = 2$ in part (a), we obtain $m_{sec} = 2(2) - 1 = 3$.

(c) Use the point–slope formula to find the equation of the secant line.

$$y - y_1 = m_{sec}(x - x_1)$$

$$y - 1 = 3(x - 1) \qquad x_1 = 1,\ y_1 = f(1) = 2 \cdot 1^2 - 3 \cdot 1 = 1;\ m_{sec} = 3$$

$$y - 1 = 3x - 3 \qquad \text{Simplify.}$$

$$y = 3x - 4 \qquad \text{Slope-intercept form of the secant line}$$

(d) See Figure 13 for the graph of f and the secant line.

Figure 13

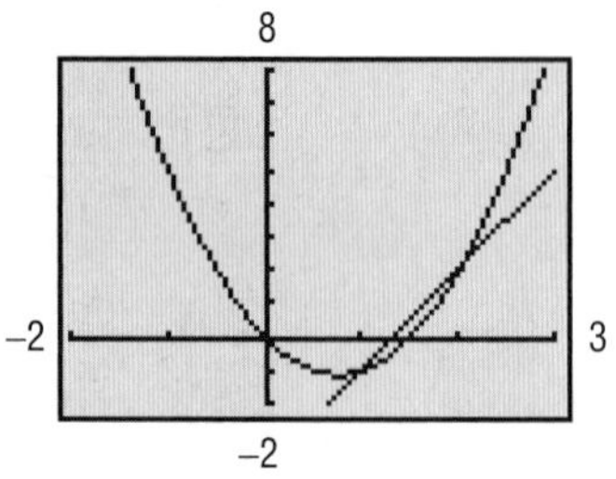

NOW WORK PROBLEM 29.

INCREASING AND DECREASING FUNCTIONS

Consider the graph given in Figure 14. If you look from left to right along the graph of the function, you will notice that parts of the graph are rising, parts are falling, and parts are horizontal. In such cases, the function is described as *increasing*, *decreasing*, or *constant*, respectively.

Figure 14

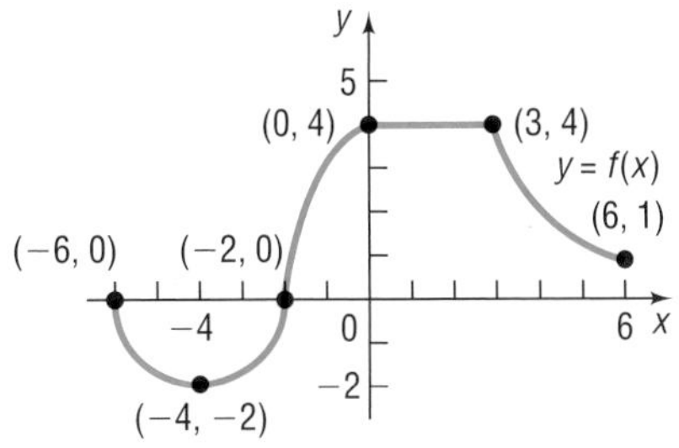

EXAMPLE 3 Determining Where a Function Is Increasing, Decreasing, or Constant from Its Graph

Where is the function in Figure 14 increasing? Where is it decreasing? Where is it constant?

Solution To answer the question of where a function is increasing, where it is decreasing, and where it is constant, we use nonstrict inequalities involving the independent variable x, or we use open intervals* of x-coordinates. The graph in Figure 14 is rising (increasing) from the point $(-4, -2)$ to the point $(0, 4)$, so we conclude that it is increasing on the open interval $(-4, 0)$ (or for $-4 < x < 0$). The graph is falling (decreasing) from the point $(-6, 0)$ to the point $(-4, -2)$ and from the point $(3, 4)$ to the point $(6, 1)$. We conclude that the graph is decreasing on the open intervals $(-6, -4)$ and $(3, 6)$ (or for $-6 < x < -4$ and $3 < x < 6$). The graph is constant on the open interval $(0, 3)$ (or $0 < x < 3$). ■

More precise definitions follow:

A function f is **increasing** on an open interval I if, for any choice of x_1 and x_2 in I, with $x_1 < x_2$, we have $f(x_1) < f(x_2)$.

A function f is **decreasing** on an open interval I if, for any choice of x_1 and x_2 in I, with $x_1 < x_2$, we have $f(x_1) > f(x_2)$.

A function f is **constant** on an open interval I if, for all choices of x in I, the values $f(x)$ are equal.

*The open interval (a, b) consists of all real numbers x for which $a < x < b$. Refer to Section 1.5, if necessary.

Figure 15 illustrates the definitions. The graph of an increasing function goes up from left to right, the graph of a decreasing function goes down from left to right, and the graph of a constant function remains at a fixed height.

Figure 15

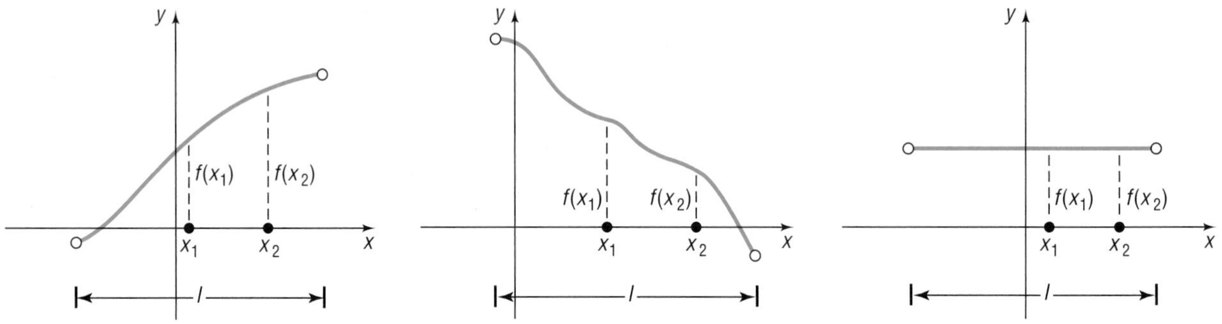

(a) For $x_1 < x_2$ in I, $f(x_1) < f(x_2)$; f is increasing on I

(b) For $x_1 < x_2$ in I, $f(x_1) > f(x_2)$; f is decreasing on I

(c) For all x in I, the values of f are equal; f is constant on I

NOW WORK PROBLEMS 1, 3, AND 5.

Local Maximum; Local Minimum

2 When the graph of a function is increasing to the left of $x = c$ and decreasing to the right of $x = c$, then at c the value of f is largest. This value is called a *local maximum* of f.

When the graph of a function is decreasing to the left of $x = c$ and is increasing to the right of $x = c$, then at c the value of f is the smallest. This value is called a *local minimum* of f.

If f has a local maximum at c, then the value of f at c is greater than the values of f near c. If f has a local minimum at c, then the value of f at c is less than the values of f near c. The word *local* is used to suggest that it is only near c that the value $f(c)$ is largest or smallest. See Figure 16.

Figure 16
f has a local maximum at x_1 and x_3; f has a local minimum at x_2

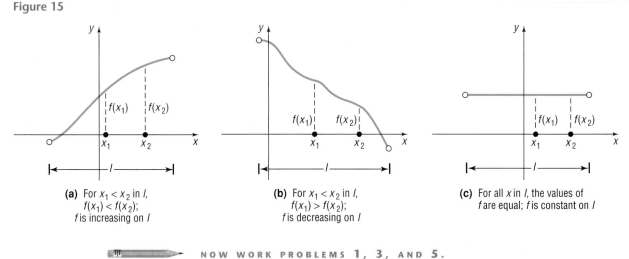

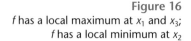

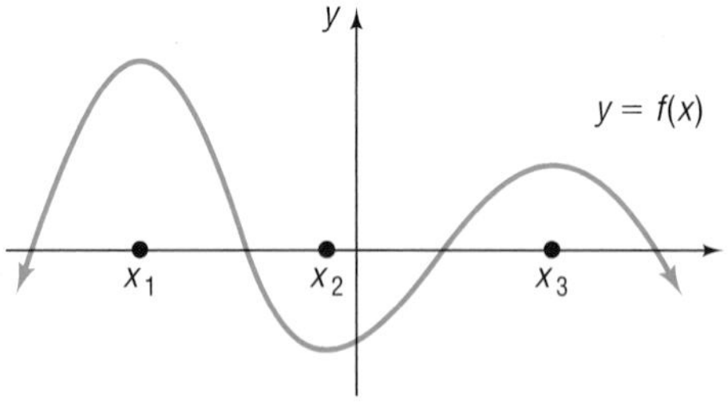

A function f has a **local maximum at c** if there is an open interval I containing c so that, for all $x \neq c$ in I, $f(x) < f(c)$. We call $f(c)$ a **local maximum of f**.

A function f has a **local minimum at c** if there is an open interval I containing c so that, for all $x \neq c$ in I, $f(x) > f(c)$. We call $f(c)$ a **local minimum of f**.

EXAMPLE 4 Finding Local Maxima and Local Minima from a Graph

Figure 17

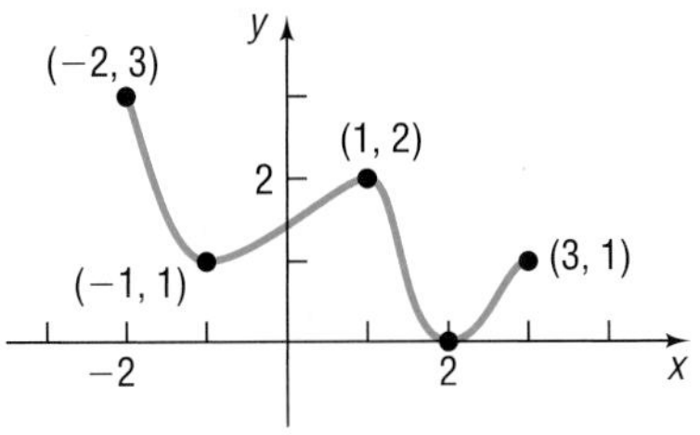

Figure 17 shows the graph of a function f.

(a) At what number(s), if any, does f have a local maximum?
(b) What are the local maxima?
(c) At what number(s), if any, does f have a local minimum?
(d) What are the local minima?

Solution The domain of f is the closed interval $[-2, 3]$, $-2 \leq x \leq 3$. Since a local maximum (or local minimum) at a number c requires that there be an open interval containing c, we exclude the endpoints -2 and 3 from consideration.

(a) f has a local maximum at 1.
(b) The local maximum is $f(1) = 2$.
(c) f has a local minimum at -1 and at 2.
(d) The local minima are $f(-1) = 1$ and $f(2) = 0$.

NOW WORK PROBLEMS 7 AND 9.

To locate the exact value at which a function f has a local maximum or a local minimum usually requires calculus. However, a graphing utility may be used to approximate these values by using the MAXIMUM and MINIMUM features.*

EXAMPLE 5 Using a Graphing Utility to Locate Local Maxima and Minima and to Determine Where a Function Is Increasing and Decreasing

(a) Use a graphing utility to graph $f(x) = 6x^3 - 12x + 5$ for $-2 < x < 2$. Determine where f has a local maximum and where f has a local minimum.
(b) Determine where f is increasing and where it is decreasing.

Solution (a) Graphing utilities have a feature that finds the maximum or minimum point of a graph within a given interval. Graph the function f for $-2 < x < 2$. Using MAXIMUM, we find that the local maximum is 11.53 and it occurs at $x = -0.82$, rounded to two decimal places. Using MINIMUM, we find that the local minimum is -1.53 and it occurs at $x = 0.82$, rounded to two decimal places. See Figures 18(a) and (b).

Figure 18

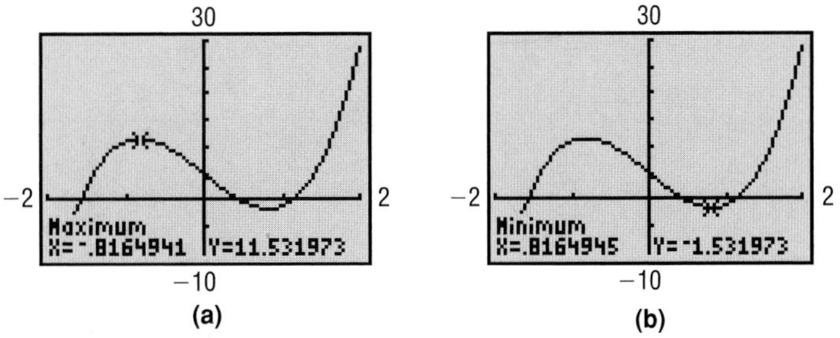

*Consult your owner's manual for the appropriate keystrokes.

3 (b) Looking at Figures 18(a) and (b), we see that the graph of f is rising (increasing) from $x = -2$ to $x = -0.82$ and from $x = 0.82$ to $x = 2$, so f is increasing on the open intervals $(-2, -0.82)$ and $(0.82, 2)$ (or for $-2 < x < -0.82$ and $0.82 < x < 2$). The graph is falling (decreasing) from $x = -0.82$ to $x = 0.82$, so f is decreasing on the open interval $(-0.82, 0.82)$ (or for $-0.82 < x < 0.82$).

NOW WORK PROBLEM 51.

EVEN AND ODD FUNCTIONS

4 A function f is even if and only if whenever the point (x, y) is on the graph of f then the point $(-x, y)$ is also on the graph. Algebraically, we define an even function as follows:

A function f is **even** if for every number x in its domain the number $-x$ is also in the domain and

$$f(-x) = f(x)$$

A function f is odd if and only if whenever the point (x, y) is on the graph of f then the point $(-x, -y)$ is also on the graph. Algebraically, we define an odd function as follows:

A function f is **odd** if for every number x in its domain the number $-x$ is also in the domain and

$$f(-x) = -f(x)$$

Refer to Section 1.2, where the tests for symmetry are listed. The following results are then evident.

Theorem A function is even if and only if its graph is symmetric with respect to the y-axis. A function is odd if and only if its graph is symmetric with respect to the origin.

EXAMPLE 6 Determining Even and Odd Functions from the Graph

Determine whether each graph given in Figure 19 is the graph of an even function, an odd function, or a function that is neither even nor odd.

Figure 19

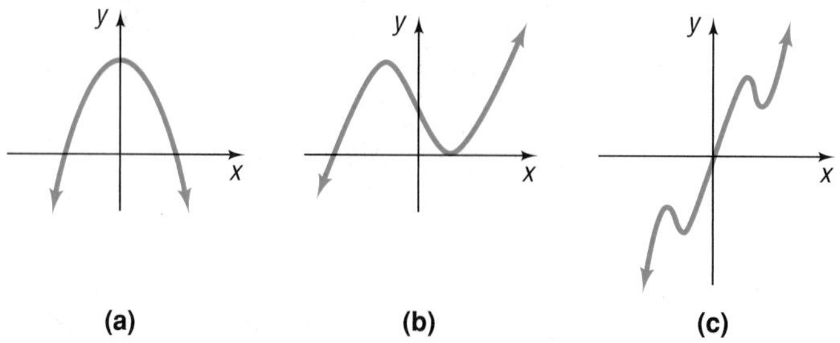

Solution The graph in Figure 19(a) is that of an even function, because the graph is symmetric with respect to the y-axis. The function whose graph is given in Figure 19(b) is neither even nor odd, because the graph is neither symmetric with respect to the y-axis nor symmetric with respect to the origin. The function whose graph is given in Figure 19(c) is odd, because its graph is symmetric with respect to the origin. ■

NOW WORK PROBLEM 11.

A graphing utility can be used to conjecture whether a function is even, odd, or neither. As stated, when the graph of an even function contains the point (x, y), it must also contain the point $(-x, y)$. Therefore, if TRACE indicates that both the point (x, y) and the point $(-x, y)$ are on the graph for every x, then we would conjecture that the function is even.*

In addition, the graph of an odd function contains the points $(-x, -y)$ and (x, y). TRACE could be used in the same way to conjecture that the function is odd.*

5 In the next example, we use algebraic techniques to verify whether a given function is even, odd, or neither.

EXAMPLE 7 Identifying Even and Odd Functions Algebraically

Determine whether each of the following functions is even, odd, or neither. Then determine whether the graph is symmetric with respect to the y-axis or with respect to the origin.

(a) $f(x) = x^2 - 5$
(b) $g(x) = x^3 - 1$
(c) $h(x) = 5x^3 - x$
(d) $F(x) = |x|$

Solution (a) To determine whether f is even, odd, or neither, we replace x by $-x$ in $f(x) = x^2 - 5$. Then

$$f(-x) = (-x)^2 - 5 = x^2 - 5 = f(x)$$

Since $f(-x) = f(x)$, we conclude that f is an even function, and the graph is symmetric with respect to the y-axis.

(b) We replace x by $-x$. Then

$$g(-x) = (-x)^3 - 1 = -x^3 - 1$$

Since $g(-x) \neq g(x)$ and $g(-x) \neq -g(x) = -(x^3 - 1) = -x^3 + 1$, we conclude that g is neither even nor odd. The graph is not symmetric with respect to the y-axis nor with respect to the origin.

(c) We replace x by $-x$ in $h(x) = 5x^3 - x$. Then

$$h(-x) = 5(-x)^3 - (-x) = -5x^3 + x = -(5x^3 - x) = -h(x)$$

Since $h(-x) = -h(x)$, h is an odd function, and the graph of h is symmetric with respect to the origin.

*$-X$-min and X-max must be equal for this to work.

(d) We replace x by $-x$ in $F(x) = |x|$. Then

$$F(-x) = |-x| = |-1| \cdot |x| = |x| = F(x)$$

Since $F(-x) = F(x)$, F is an even function, and the graph of F is symmetric with respect to the y-axis. ■

NOW WORK PROBLEM 37.

2.2 EXERCISES

In Problems 1–10, use the graph of the function f given below to answer each question.

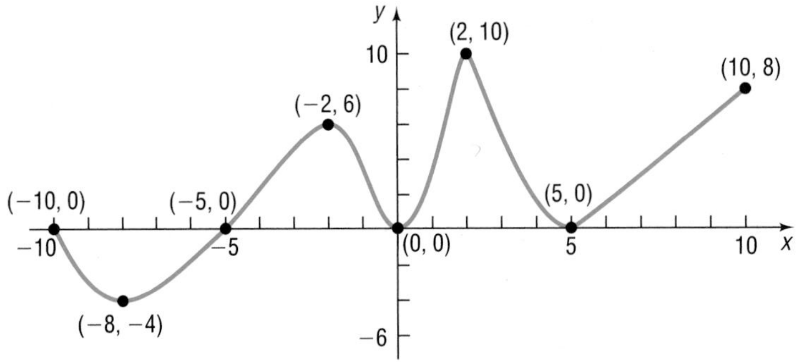

1. Is f increasing on the interval $(-8, -2)$?

2. Is f decreasing on the interval $(-8, -4)$?

3. Is f increasing on the interval $(2, 10)$?

4. Is f decreasing on the interval $(2, 5)$?

5. List the interval(s) on which f is increasing.

6. List the interval(s) on which f is decreasing.

7. Is there a local maximum at 2? If yes, what is the local maximum at 2?

8. Is there a local maximum at 5? If yes, what is the local maximum at 5?

9. List the numbers at which f has a local maximum. What are these local maxima?

10. List the numbers at which f has a local minimum. What are these local minima?

In Problems 11–20, the graph of a function is given. Use the graph to find:
(a) The intercepts, if any
(b) Its domain and range
(c) The intervals on which it is increasing, decreasing, or constant
(d) Whether it is even, odd, or neither

11.

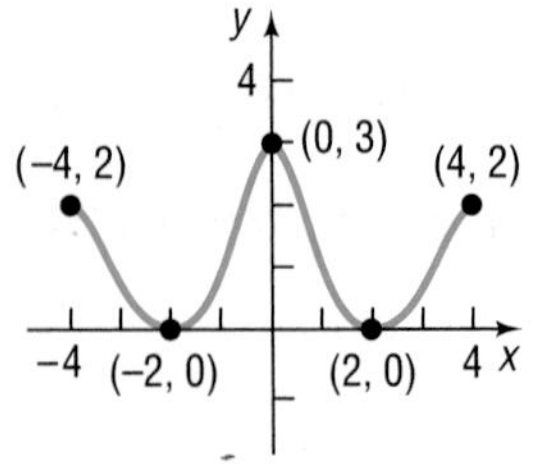

12.

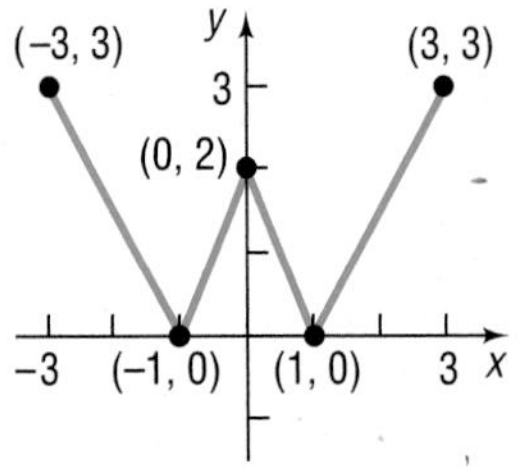

13.

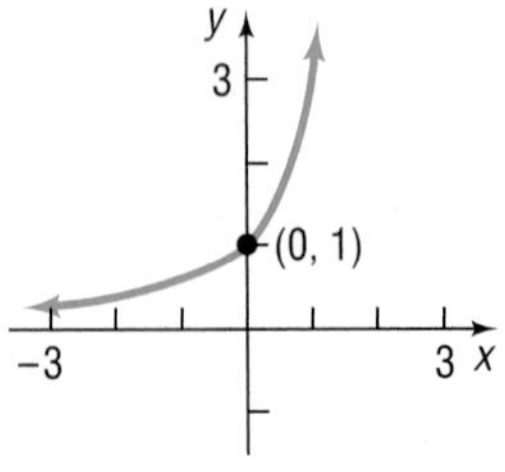

14.

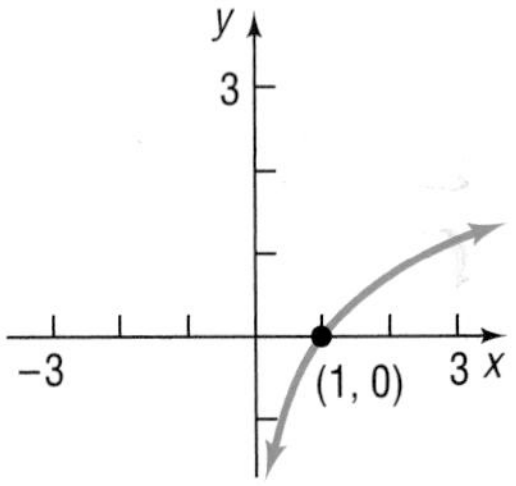

15.

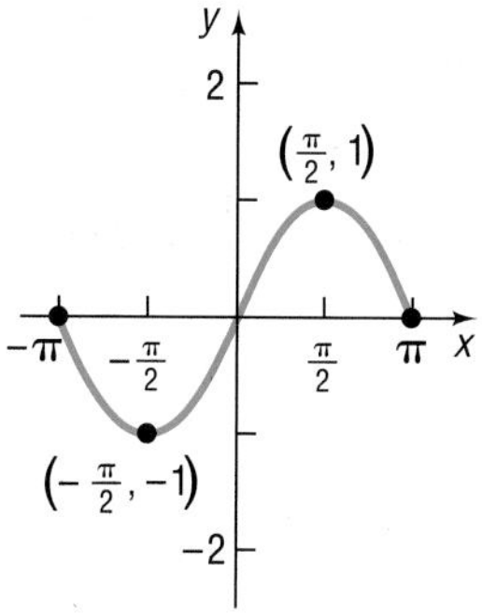

16.

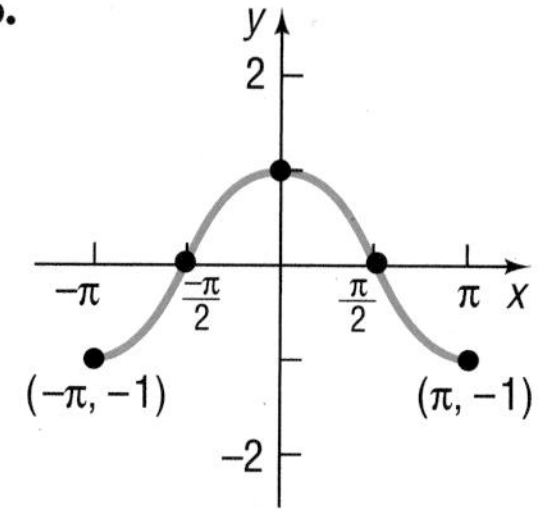

17.

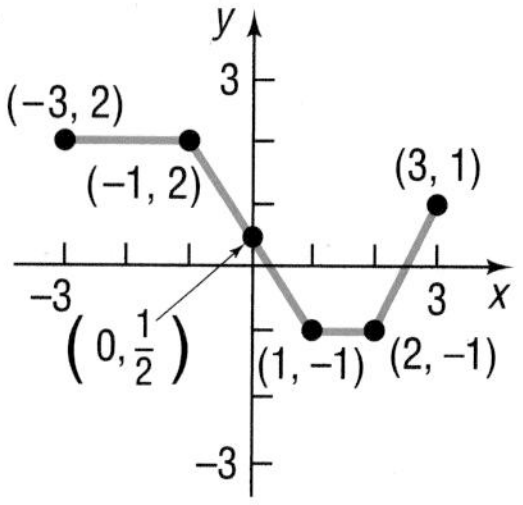

18.

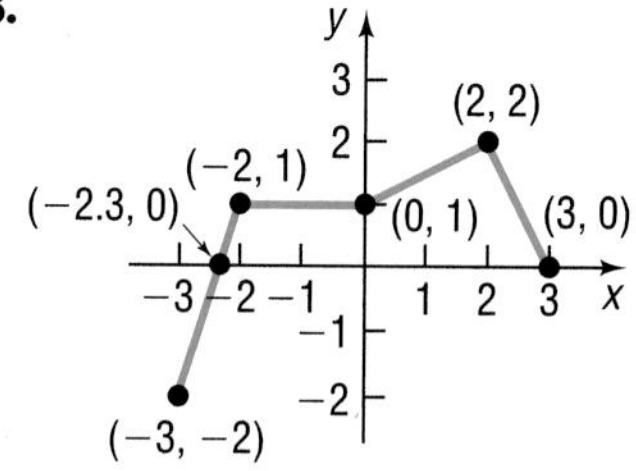

In Problems 19 and 20, assume that the entire graph is shown.

19.

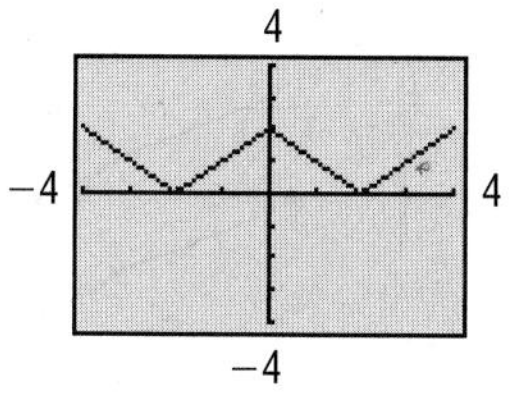

20.

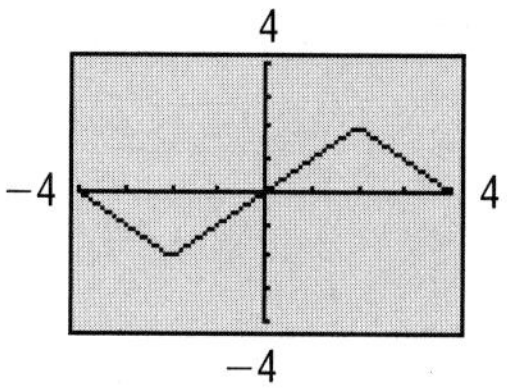

In Problems 21–24, the graph of a function f is given. Use the graph to find

(a) The numbers, if any, at which f has a local maximum. What are these local maxima?

(b) The numbers, if any, at which f has a local minimum. What are these local minima?

21.

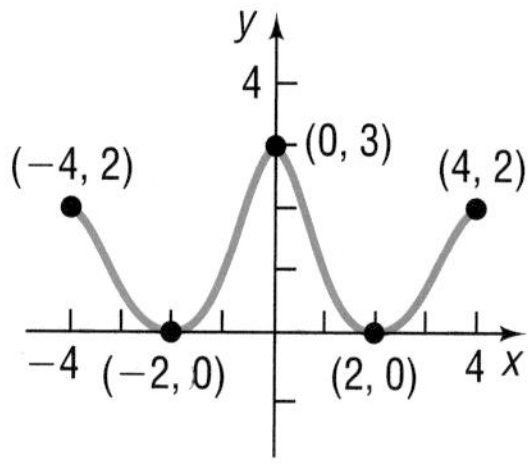

22.

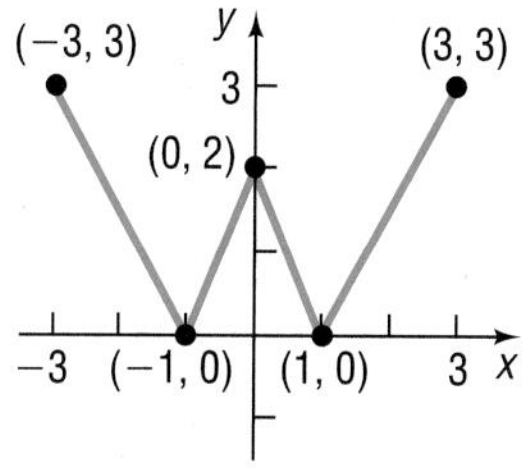

23.

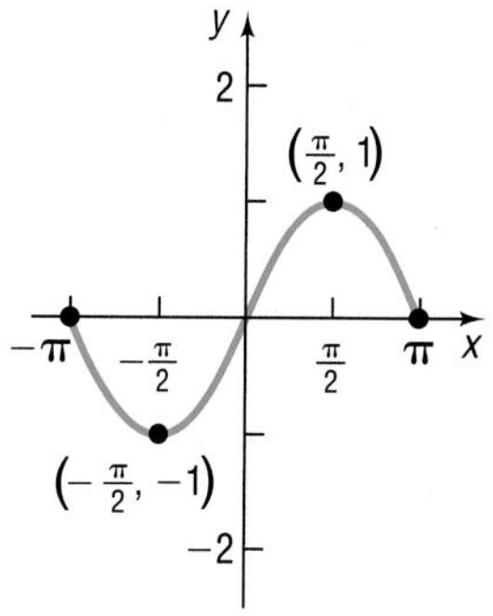

24.

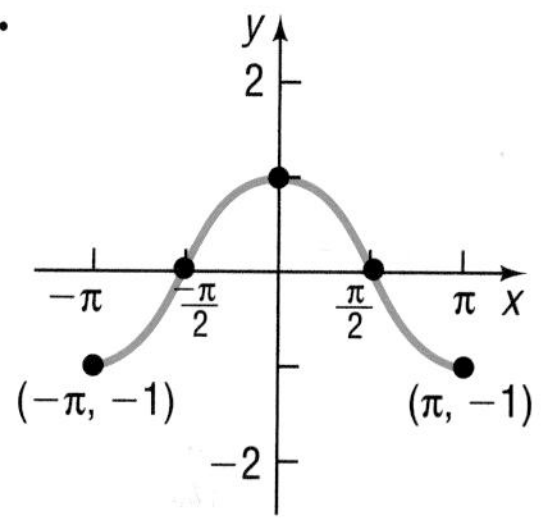

In Problems 25–36, (a) for each function find the average rate of change of f from 1 to x:

$$\frac{f(x) - f(1)}{x - 1}, \qquad x \neq 1$$

(b) Use the result from part (a) to compute the average rate of change from $x = 1$ to $x = 2$. Be sure to simplify.

(c) Find an equation of the secant line containing $(1, f(1))$ and $(2, f(2))$.

(d) Graph f and the secant line on the same viewing window.

25. $f(x) = 5x$ **26.** $f(x) = -4x$ **27.** $f(x) = 1 - 3x$ **28.** $f(x) = x^2 + 1$

29. $f(x) = x^2 - 2x$ **30.** $f(x) = x - 2x^2$ **31.** $f(x) = x^3 - x$ **32.** $f(x) = x^3 + x$

33. $f(x) = \dfrac{2}{x + 1}$ **34.** $f(x) = \dfrac{1}{x^2}$ **35.** $f(x) = \sqrt{x}$ **36.** $f(x) = \sqrt{x + 3}$

In Problems 37–48, determine algebraically whether each function is even, odd, or neither. Use a graphing utility to graph each function, then use TRACE to verify your results.

37. $f(x) = 4x^3$ **38.** $f(x) = 2x^4 - x^2$ **39.** $g(x) = -3x^2 - 5$ **40.** $h(x) = 3x^3 + 5$

41. $F(x) = \sqrt[3]{x}$ **42.** $G(x) = \sqrt{x}$ **43.** $f(x) = x + |x|$ **44.** $f(x) = \sqrt[3]{2x^2 + 1}$

45. $g(x) = \dfrac{1}{x^2}$ **46.** $h(x) = \dfrac{x}{x^2 - 1}$ **47.** $h(x) = \dfrac{-x^3}{3x^2 - 9}$ **48.** $F(x) = \dfrac{2x}{|x|}$

49. How many x-intercepts can a function defined on an interval have if it is increasing on that interval? Explain.

50. How many y-intercepts can a function have? Explain.

In Problems 51–58, use a graphing utility to graph each function over the indicated interval and approximate any local maxima and local minima. Determine where the function is increasing and where it is decreasing. Round answers to two decimal places.

51. $f(x) = x^3 - 3x + 2 \quad (-2, 2)$

52. $f(x) = x^3 - 3x^2 + 5 \quad (-1, 3)$

53. $f(x) = x^5 - x^3 \quad (-2, 2)$

54. $f(x) = x^4 - x^2 \quad (-2, 2)$

55. $f(x) = -0.2x^3 - 0.6x^2 + 4x - 6 \quad (-6, 4)$

56. $f(x) = -0.4x^3 + 0.6x^2 + 3x - 2 \quad (-4, 5)$

57. $f(x) = 0.25x^4 + 0.3x^3 - 0.9x^2 + 3 \quad (-3, 2)$

58. $f(x) = -0.4x^4 - 0.5x^3 + 0.8x^2 - 2 \quad (-3, 2)$

Problems 59–64 require the following definition of a **secant line:** *The slope of the secant line containing the two points $(x, f(x))$ and $(x + h, f(x + h))$ on the graph of a function $y = f(x)$ may be given as*

$$m_{\text{sec}} = \frac{f(x+h) - f(x)}{(x+h) - x} = \frac{f(x+h) - f(x)}{h}$$

(a) Express the slope of the secant line of each function in terms of x and h. Be sure to simplify your answer.
(b) Find m_{sec} for $h = 0.5, 0.1,$ and 0.01 at $x = 1$. What value does m_{sec} approach as h approaches 0?
(c) Find the equation for the secant line at $x = 1$ with $h = 0.01$.
(d) Graph f and the secant line found in part (c) on the same viewing window.

59. $f(x) = 2x + 5$

60. $f(x) = -3x + 2$

61. $f(x) = x^2 + 2x$

62. $f(x) = 2x^2 + x$

63. $f(x) = \dfrac{1}{x + 1}$

64. $f(x) = \dfrac{2}{x - 2}$

65. Revenue from Selling Bikes The following data represent the total revenue that would be received from selling x bicycles at Tunney's Bicycle Shop.

Number of Bicycles, x	Total Revenue, R (Dollars)
0	0
25	28,000
60	45,000
102	53,400
150	59,160
190	62,360
223	64,835
249	66,525

(a) Draw a scatter diagram of the data, treating the number of bicycles produced as the independent variable.
(b) Draw a line through the points (0, 0) and (25, 28,000) on the scatter diagram found in part (a).
(c) Find the average rate of change of revenue from 0 to 25 bicycles.
(d) Interpret the average rate of change found in part (c).
(e) Draw a line through the points (190, 62,360) and (223, 64,835) on the scatter diagram found in part (a).
(f) Find the average rate of change of revenue from 190 to 223 bicycles.
(g) Interpret the average rate of change found in part (f).

66. Cost of Manufacturing Bikes The following data represent the monthly cost of producing bicycles at Tunney's Bicycle Shop.

Number of Bicycles, x	Total Cost of Production, C (Dollars)
0	24,000
25	27,750
60	31,500
102	35,250
150	39,000
190	42,750
223	46,500
249	50,250

(a) Draw a scatter diagram of the data, treating the number of bicycles produced as the independent variable.
(b) Draw a line through the points (0, 24,000) and (25, 27,750) on the scatter diagram found in part (a).
(c) Find the average rate of change of the cost from 0 to 25 bicycles.
(d) Interpret the average rate of change found in part (c).
(e) Draw a line through the points (190, 42,750) and (223, 46,500) on the scatter diagram found in part (a).
(f) Find the average rate of change of the cost from 190 to 223 bicycles.
(g) Interpret the average rate of change found in part (f).

67. Growth of Bacteria The following data represent the population of an unknown bacteria.

Time (Days)	Population
0	50
1	153
2	234
3	357
4	547
5	839
6	1280

(a) Draw a scatter diagram of the data, treating time as the independent variable.
(b) Draw a line through the points (0, 50) and (1, 153) on the scatter diagram found in part (a).
(c) Find the average rate of change of the population from 0 to 1 days.
(d) Interpret the average rate of change found in part (c).
(e) Draw a line through the points (5, 839) and (6, 1280) on the scatter diagram found in part (a).
(f) Find the average rate of change of the population from 5 to 6 days.
(g) Interpret the average rate of change found in part (f).
(h) What is happening to the average rate of change of the population as time passes?

68. Falling Objects Suppose that you drop a ball from a cliff 1000 feet high. You measure the distance s that the ball has fallen after time t using a motion detector and obtain the following data.

Time, t (Seconds)	Distance, s (Feet)
0	0
1	16
2	64
3	144
4	256
5	400
6	576
7	784

(a) Draw a scatter diagram of the data, treating time as the independent variable.
(b) Draw a line through the points (0, 0) and (2, 64).
(c) Find the average rate of change of the ball from 0 to 2 seconds; that is, find the slope of the line in part (b).
(d) Interpret the average rate of change found in part (c).
(e) Draw a line through the points (5, 400) and (7, 784).
(f) Find the average rate of change of the ball from 5 to 7 seconds; that is, find the slope of the line in part (e).
(g) Interpret the average rate of change found in part (f).
(h) What is happening to the average rate of change of the distance of the ball has fallen as time passes?

69. Maximizing the Volume of a Box An open box with a square base is to be made from a square piece of cardboard 24 inches on a side by cutting out a square from each corner and turning up the sides. (See the illustration.) The volume V of the box as a function of the length x of the side of the square cut from each corner is

$$V(x) = x(24 - 2x)^2$$

Graph V and determine where V is largest.

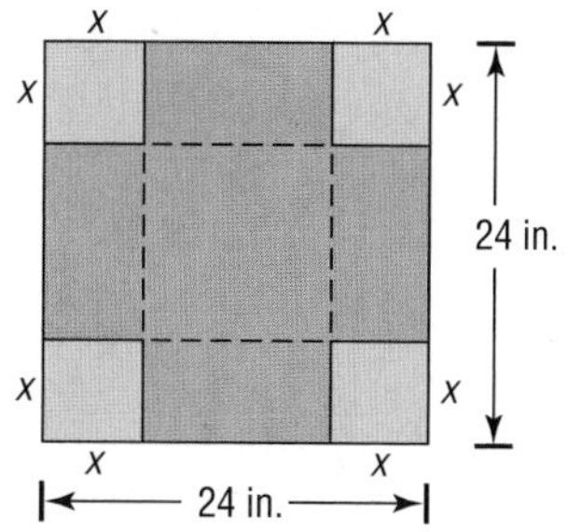

70. Minimizing the Material Needed to Make a Box An open box with a square base is required to have a volume of 10 cubic feet. The amount A of material used to make such a box as a function of the length x of a side of the square base is

$$A(x) = x^2 + \frac{40}{x}$$

Graph A and determine where A is smallest.

71. Maximum Height of a Ball The height s of a ball (in feet) thrown with an initial velocity of 80 feet per second from an initial height of 6 feet is given as a function of the time t (in seconds) by

$$s(t) = -16t^2 + 80t + 6$$

(a) Graph s.
(b) Determine the time at which height is maximum.
(c) What is the maximum height?

72. Minimum Average Cost The average cost of producing x riding lawn mowers per hour is given by

$$A(x) = 0.3x^2 + 21x - 251 + \frac{2500}{x}$$

(a) Graph A.
(b) Determine the number of riding lawn mowers to produce in order to minimize average cost.
(c) What is the minimum average cost?

73. Can you think of a function that is both even and odd?

PREPARING FOR THIS SECTION

Before getting started, review the following:

✓ Graphs of Certain Equations (Example 3, p. 13; Example 10, p. 20; Figure 29, p. 21; Example 12, p. 21)

2.3 LIBRARY OF FUNCTIONS; PIECEWISE-DEFINED FUNCTIONS

OBJECTIVES 1 Graph the Functions Listed in the Library

2 Graph Piecewise-defined Functions

LIBRARY OF FUNCTIONS

1 We now give names to some of the functions that we have encountered. In going through this list, pay special attention to the characteristics of each function, particularly to the shape of each graph. Knowing these graphs will lay the foundation for later graphing techniques.

Linear Functions

$$f(x) = mx + b \qquad m \text{ and } b \text{ are real numbers}$$

The domain of a **linear function** f consists of all real numbers. The graph of this function is a nonvertical line with slope m and y-intercept b. A linear function is increasing if $m > 0$, decreasing if $m < 0$, and constant if $m = 0$.

Constant Function

$$f(x) = b \qquad b \text{ is a real number}$$

Figure 20

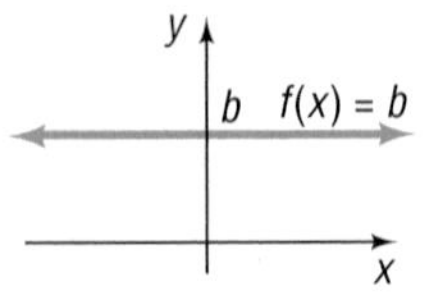

See Figure 20.

A **constant function** is a special linear function ($m = 0$). Its domain is the set of all real numbers; its range is the set consisting of a single number b. Its graph is a horizontal line whose y-intercept is b. The constant function is an even function whose graph is constant over its domain.

Identity Function

$$f(x) = x$$

See Figure 21.

Figure 21

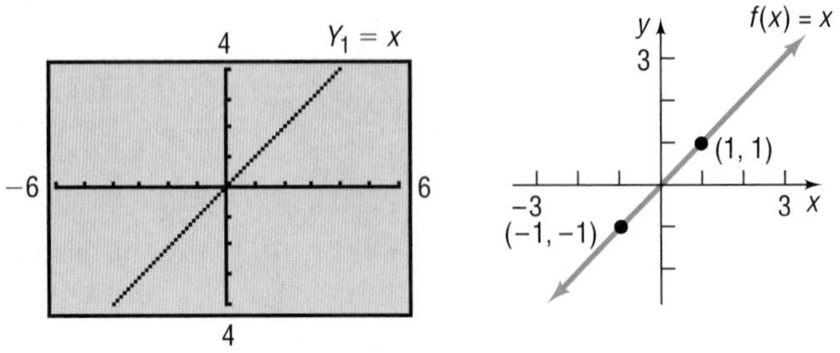

The **identity function** is also a special linear function. Its domain and range are the set of all real numbers. Its graph is a line whose slope is $m = 1$ and whose y-intercept is 0. The line consists of all points for which the x-coordinate equals the y-coordinate. The identity function is an odd function that is increasing over its domain. Note that the graph bisects quadrants I and III.

Square Function

$$f(x) = x^2$$

See Figure 22.

Figure 22

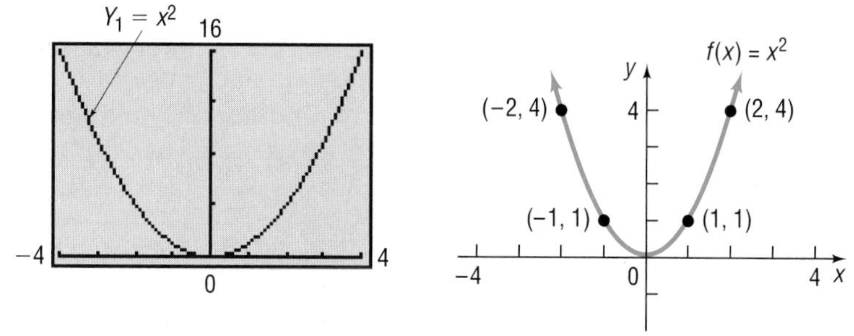

The domain of the **square function** f is the set of all real numbers; its range is the set of nonnegative real numbers. The graph of this function is a parabola whose intercept is at $(0, 0)$. The square function is an even function that is decreasing on the interval $(-\infty, 0)$ and increasing on the interval $(0, \infty)$.

Cube Function

$$f(x) = x^3$$

See Figure 23.

Figure 23

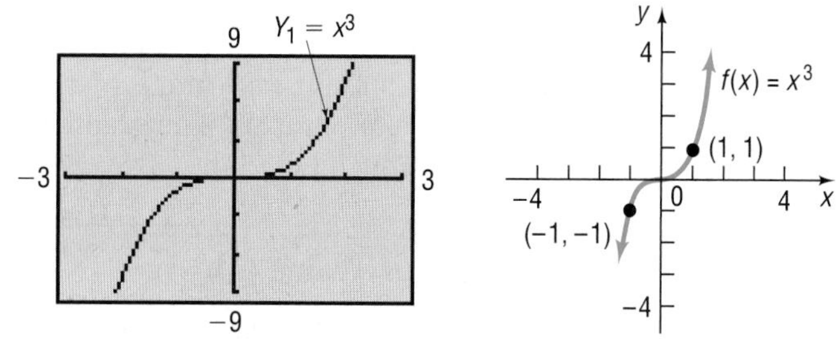

The domain and range of the **cube function** are the set of all real numbers. The intercept of the graph is at $(0, 0)$. The cube function is odd and is increasing on the interval $(-\infty, \infty)$.

Square Root Function

$$f(x) = \sqrt{x}$$

See Figure 24.

Figure 24

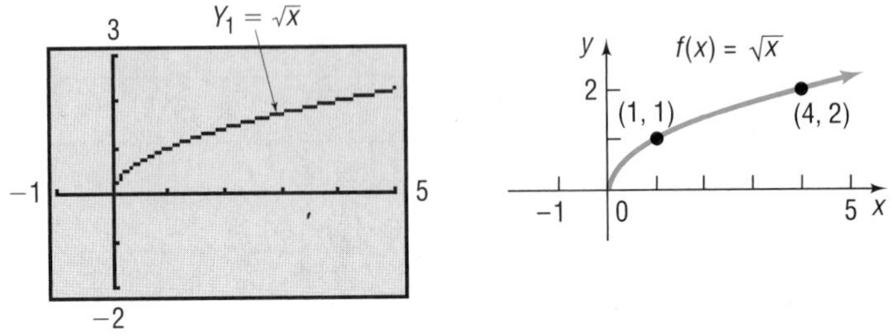

The domain and range of the **square root function** are the set of nonnegative real numbers. The intercept of the graph is at $(0, 0)$. The square root function is neither even nor odd and is increasing on the interval $(0, \infty)$.

Reciprocal Function

$$f(x) = \frac{1}{x}$$

Refer to Example 12, page 21, for a discussion of the equation $y = 1/x$. See Figure 25.

Figure 25

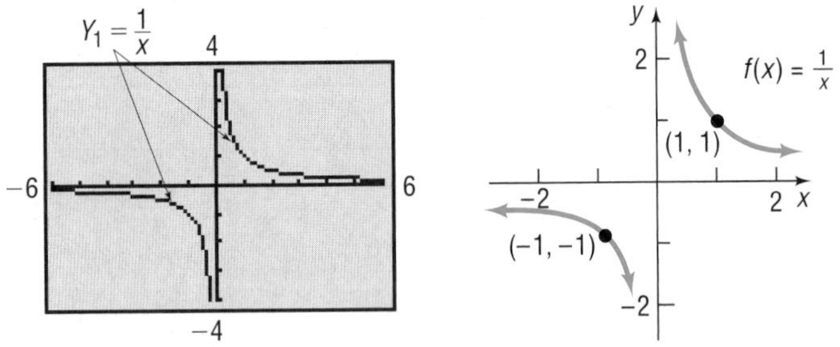

The domain and range of the **reciprocal function** are the set of all nonzero real numbers. The graph has no intercepts. The reciprocal function is decreasing on the intervals $(-\infty, 0)$ and $(0, \infty)$ and is an odd function.

Absolute Value Function

$$f(x) = |x|$$

See Figure 26.

Figure 26

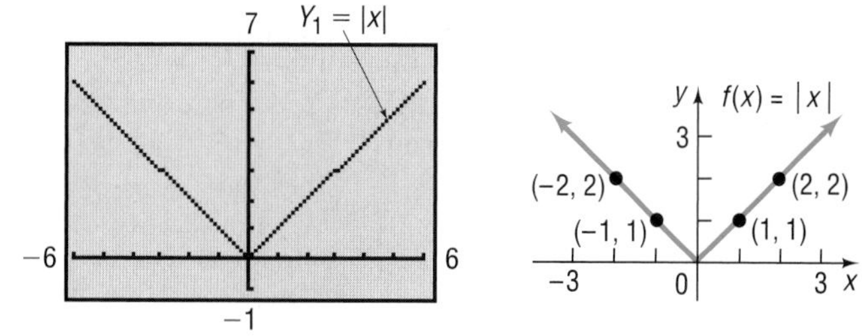

The domain of the **absolute value function** is the set of all real numbers; its range is the set of nonnegative real numbers. The intercept of the graph is at $(0, 0)$. If $x \geq 0$, then $f(x) = x$, and the graph of f is part of the line $y = x$; if $x < 0$, then $f(x) = -x$, and the graph of f is part of the line $y = -x$. The absolute value function is an even function; it is decreasing on the interval $(-\infty, 0)$ and increasing on the interval $(0, \infty)$.

COMMENT: If your utility has no built-in absolute value function, you can still graph $f(x) = |x|$ by using the fact that $|x| = \sqrt{(x^2)}$.

The notation $\text{int}(x)$ stands for the largest integer less than or equal to x. For example,

$$\text{int}(1) = 1 \quad \text{int}(2.5) = 2 \quad \text{int}\left(\tfrac{1}{2}\right) = 0 \quad \text{int}\left(\tfrac{-3}{4}\right) = -1 \quad \text{int}(\pi) = 3$$

This type of correspondence occurs frequently enough in mathematics that we give it a name.

Greatest-integer Function

$$f(x) = \text{int}(x) = \text{greatest integer less than or equal to } x$$

We obtain the graph of $f(x) = \text{int}(x)$ by plotting several points. See Table 6. For values of x, $-1 \leq x < 0$, the value of $f(x) = \text{int}(x)$ is -1; for values of x, $0 \leq x < 1$, the value of f is 0. See Figure 27 for the graph.

TABLE 6

X	Y1
-1	-1
-.75	-1
-.5	-1
-.25	-1
0	0
.25	0
.5	0

Y1=int(X)

Figure 27

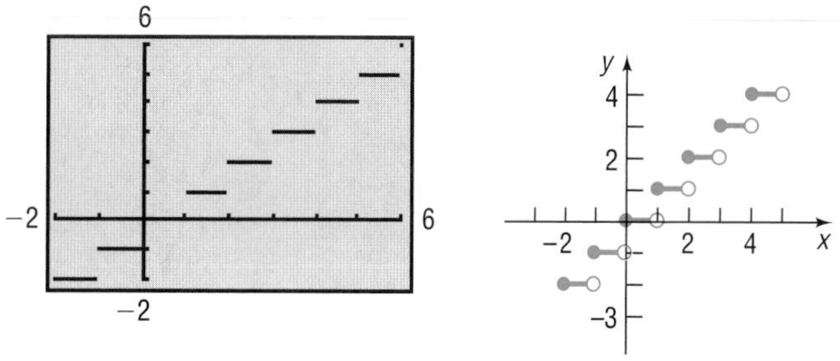

(a) (b)

The domain of the **greatest-integer function** is the set of all real numbers; its range is the set of integers. The y-intercept of the graph is at 0. The x-intercepts lie in the interval $[0, 1)$. The greatest-integer function is neither even nor odd. It is constant on every interval of the form $[k, k + 1)$, for k an integer. In Figure 27(b), we use a solid dot to indicate, for example, that at $x = 1$ the value of f is $f(1) = 1$; we use an open circle to illustrate that the function does not assume the value of 0 at $x = 1$.

From the graph of the greatest-integer function, we can see why it is also called a **step function**. At $x = 0, x = \pm 1, x = \pm 2$, and so on, this function exhibits what is called a *discontinuity*; that is, at integer values, the graph suddenly "steps" from one value to another without taking on any of the intermediate values. For example, to the immediate left of $x = 3$, the y-coordinates are 2, and to the immediate right of $x = 3$, the y-coordinates are 3.

COMMENT: When graphing a function, you can choose either the **connected mode**, in which points plotted on the screen are connected, making the graph appear without any breaks, or the **dot mode**, in which only the points plotted appear. When graphing the greatest-integer function with a graphing utility, it is necessary to be in the **dot mode**. This is to prevent the utility from "connecting the dots" when $f(x)$ changes from one integer value to the next. See Figure 28.

Figure 28

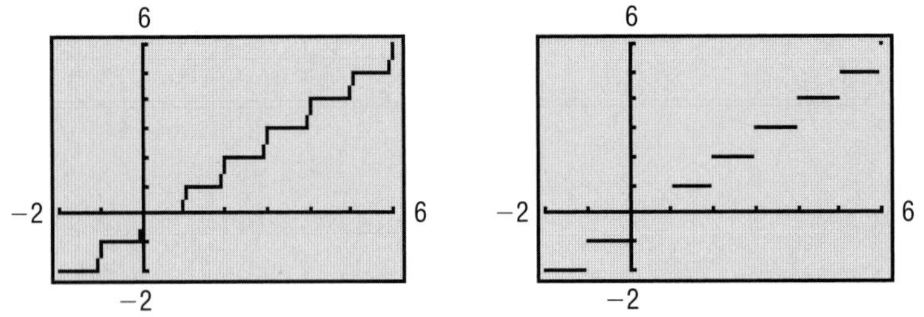

(a) Connected mode (b) Dot mode

The functions that we have discussed so far are basic. Whenever you encounter one of them, you should see a mental picture of its graph. For example, if you encounter the function $f(x) = x^2$, you should see in your mind's eye a picture like Figure 22.

NOW WORK PROBLEMS 1–8.

PIECEWISE-DEFINED FUNCTIONS

2 Sometimes a function is defined differently on different parts of its domain. For example, the absolute value function $f(x) = |x|$ is actually defined by two equations: $f(x) = x$ if $x \geq 0$ and $f(x) = -x$ if $x < 0$. For convenience, we generally combine these equations into one expression as

$$f(x) = |x| = \begin{cases} x & \text{if } x \geq 0 \\ -x & \text{if } x < 0 \end{cases}$$

When functions are defined by more than one equation, they are called **piecewise-defined** functions.

Let's look at another example of a piecewise-defined function.

EXAMPLE 1 Analyzing a Piecewise-defined Function

For the following function f,

$$f(x) = \begin{cases} -x + 1 & \text{if } -1 \leq x < 1 \\ 2 & \text{if } x = 1 \\ x^2 & \text{if } x > 1 \end{cases}$$

(a) Find $f(0)$, $f(1)$, and $f(2)$.

(b) Determine the domain of f.

(c) Graph f.

(d) Use the graph to find the range of f.

Solution (a) To find $f(0)$, we observe that when $x = 0$ the equation for f is given by $f(x) = -x + 1$. So we have

$$f(0) = -0 + 1 = 1$$

When $x = 1$, the equation for f is $f(x) = 2$. Thus,

$$f(1) = 2$$

When $x = 2$, the equation for f is $f(x) = x^2$. So

$$f(2) = 2^2 = 4$$

(b) To find the domain of f, we look at its definition. We conclude that the domain of f is $\{x \mid x \geq -1\}$, or $[-1, \infty)$.

(c) On a graphing utility, the procedure for graphing a piecewise-defined function varies depending on the particular utility. In general, you need to enter each piece as a function with a restricted domain. See Figure 29(a).

Figure 29

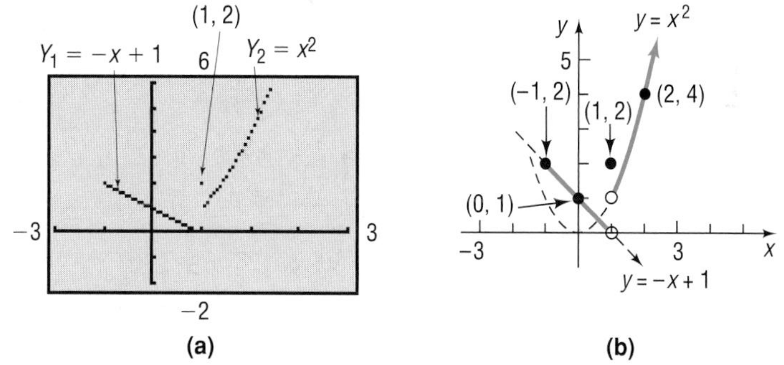

In graphing piecewise-defined functions using a graphing utility, it is usually better to be in the dot mode, since such functions may have breaks (discontinuities).

To graph f on paper, we graph "each piece." First we graph the line $y = -x + 1$ and keep only the part for which $-1 \leq x < 1$. Then we plot the point $(1, 2)$, because when $x = 1$, $f(x) = 2$. Finally, we graph the parabola $y = x^2$ and keep only the part for which $x > 1$. See Figure 29(b).

(d) From the graph, we conclude that the range of f is $\{y \mid y > 0\}$, or $(0, \infty)$. ■

NOW WORK PROBLEM 19.

EXAMPLE 2 Cost of Electricity

In June, 1999, Commonwealth Edison Company supplied electricity to residences for a monthly customer charge of \$8.02 plus 8.77¢ per kilowatt-hour (kWhr) for the first 400 kWhr supplied in the month and 6.574¢ per kWhr for all usage over 400 kWhr in the month.*

(a) What is the charge for using 300 kWhr in a month?

(b) What is the charge for using 700 kWhr in a month?

(c) If C is the monthly charge for x kWhr, express C as a function of x.

* *Source:* Commonwealth Edison Co., Chicago, Illinois, 1999.

Solution (a) For 300 kWhr, the charge is \$8.02 plus 8.77¢ = \$0.0877 per kWhr. That is,

$$\text{Charge} = \$8.02 + \$0.0877(300) = \$34.33$$

(b) For 700 kWhr, the charge is \$8.02 plus 8.77¢ per kWhr for the first 400 kWhr plus 6.574¢ per kWhr for the 300 kWhr in excess of 400. That is,

$$\text{Charge} = \$8.02 + \$0.0877(400) + \$0.06574(300) = \$62.82$$

(c) If $0 \le x \le 400$, the monthly charge C (in dollars) can be found by multiplying x times \$0.0877 and adding the monthly customer charge of \$8.02. Thus, if $0 \le x \le 400$, then $C(x) = 0.0877x + 8.02$. For $x > 400$, the charge is $0.0877(400) + 8.02 + 0.06574(x - 400)$, since $x - 400$ equals the usage in excess of 400 kWhr, which costs \$0.06574 per kWhr. That is, if $x > 400$, then

$$\begin{aligned} C(x) &= 0.0877(400) + 8.02 + 0.06574(x - 400) \\ &= 43.10 + 0.06574(x - 400) \\ &= 0.06574x + 16.80 \end{aligned}$$

The rule for computing C follows two equations:

$$C(x) = \begin{cases} 0.0877x + 8.02 & \text{if } 0 \le x \le 400 \\ 0.06574x + 16.80 & \text{if } x > 400 \end{cases}$$

See Figure 30 for the graph.

Figure 30

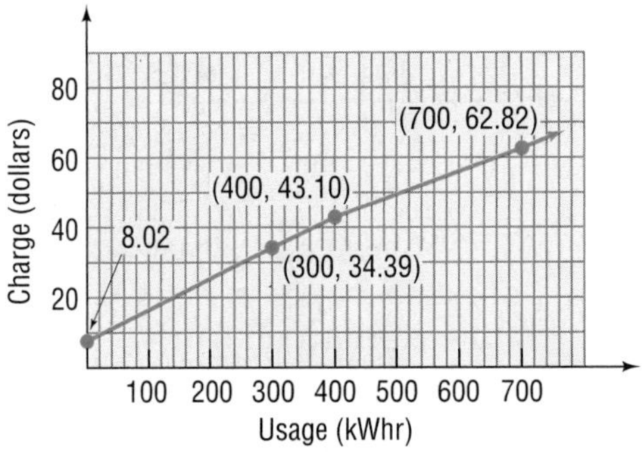

2.3 EXERCISES

In Problems 1–8, match each graph to the function listed whose graph most resembles the one given.

A. *Constant function* B. *Linear function* C. *Square function*
D. *Cube function* E. *Square root function* F. *Reciprocal function*
G. *Absolute value function* H. *Greatest-integer function*

1. 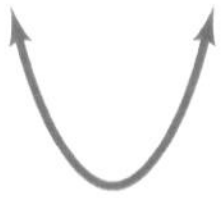

2. 

3. 

4. 

5. 

6. 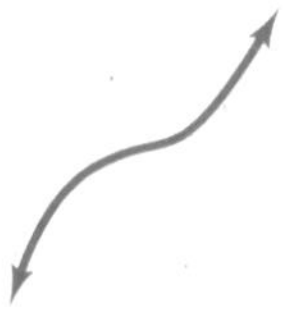

7. 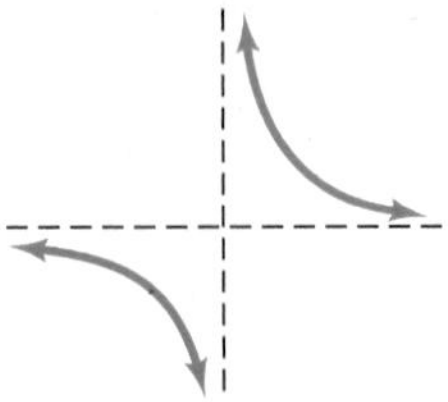

8. 

In Problems 9–14, sketch the graph of each function.

9. $f(x) = x$

10. $f(x) = x^2$

11. $f(x) = x^3$

12. $f(x) = \sqrt{x}$

13. $f(x) = 1/x$

14. $f(x) = |x|$

15. If $f(x) = \begin{cases} x^2 & \text{if } x < 0 \\ 2 & \text{if } x = 0 \\ 2x + 1 & \text{if } x > 0 \end{cases}$

find: (a) $f(-2)$ (b) $f(0)$ (c) $f(2)$

16. If $f(x) = \begin{cases} x^3 & \text{if } x < 0 \\ 3x + 2 & \text{if } x \geq 0 \end{cases}$

find: (a) $f(-1)$ (b) $f(0)$ (c) $f(1)$

17. If $f(x) = \text{int}(2x)$, find: (a) $f(1.2)$ (b) $f(1.6)$ (c) $f(-1.8)$

18. If $f(x) = \text{int}(x/2)$, find: (a) $f(1.2)$ (b) $f(1.6)$ (c) $f(-1.8)$

In Problems 19–30:

(a) Find the domain of each function.
(b) Locate any intercepts.
(c) Graph each function by hand.
(d) Based on the graph, find the range.
(e) Verify your results using a graphing utility.

19. $f(x) = \begin{cases} 2x & \text{if } x \neq 0 \\ 1 & \text{if } x = 0 \end{cases}$

20. $f(x) = \begin{cases} 3x & \text{if } x \neq 0 \\ 4 & \text{if } x = 0 \end{cases}$

21. $f(x) = \begin{cases} -2x + 3 & x < 1 \\ 3x - 2 & x \geq 1 \end{cases}$

22. $f(x) = \begin{cases} x + 3 & x < -2 \\ -2x - 3 & x \geq -2 \end{cases}$

23. $f(x) = \begin{cases} x + 3 & -2 \leq x < 1 \\ 5 & x = 1 \\ -x + 2 & x > 1 \end{cases}$

24. $f(x) = \begin{cases} 2x + 5 & -3 \leq x < 0 \\ -3 & x = 0 \\ -5x & x > 0 \end{cases}$

25. $f(x) = \begin{cases} 1 + x & \text{if } x < 0 \\ x^2 & \text{if } x \geq 0 \end{cases}$

26. $f(x) = \begin{cases} 1/x & \text{if } x < 0 \\ \sqrt{x} & \text{if } x \geq 0 \end{cases}$

27. $f(x) = \begin{cases} |x| & \text{if } -2 \leq x < 0 \\ 1 & \text{if } x = 0 \\ x^3 & \text{if } x > 0 \end{cases}$

28. $f(x) = \begin{cases} 3 + x & \text{if } -3 \leq x < 0 \\ 3 & \text{if } x = 0 \\ \sqrt{x} & \text{if } x > 0 \end{cases}$

29. $h(x) = 2\,\text{int}(x)$

30. $f(x) = \text{int}(2x)$

In Problems 31–34, the graph of a piecewise-defined function is given. Write a definition for each function.

31.

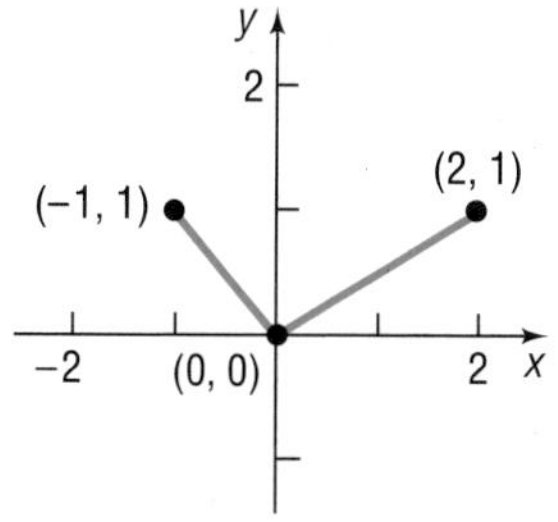

32.

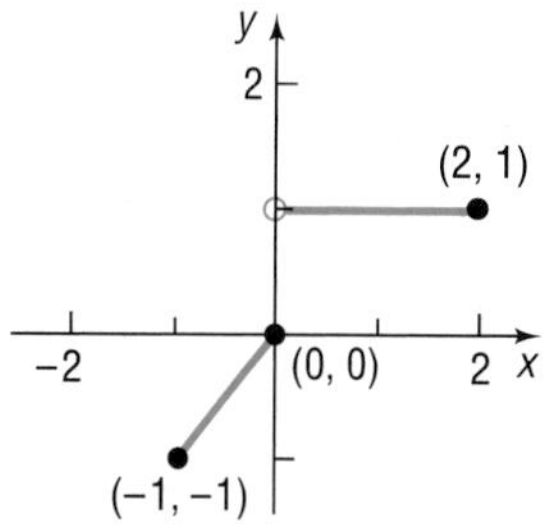

33.

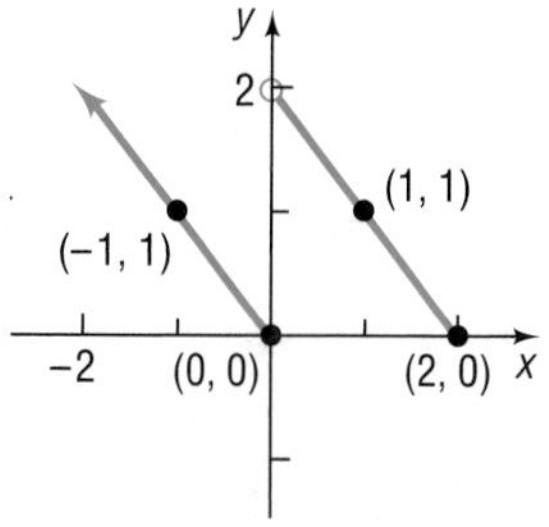

34.

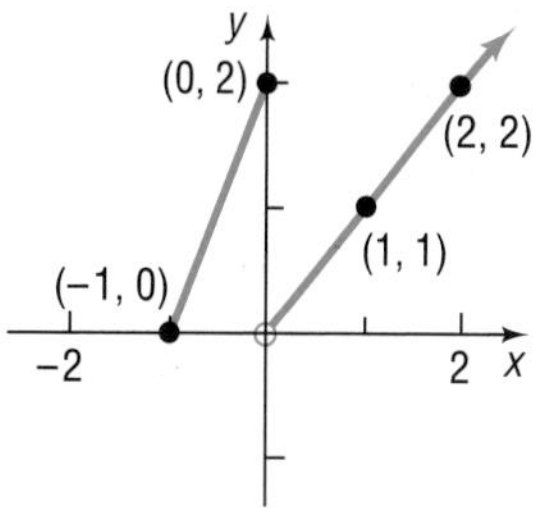

35. Cost of Natural Gas In April, 2000, the Peoples Gas Company had the following rate schedule* for natural gas usage in single-family residences:

Monthly service charge	\$9.45
Per therm service charge	
1st 50 therms	\$0.36375/therm
Over 50 therms	\$0.11445/therm
Gas charge	\$0.3128/therm

(a) What is the charge for using 50 therms in a month?
(b) What is the charge for using 500 therms in a month?
(c) Construct a function that relates the monthly charge C for x therms of gas.
(d) Graph this function.

36. Cost of Natural Gas In March, 2000, Nicor Gas had the following rate schedule† for natural gas usage in single-family residences:

Monthly customer charge	\$6.45
Distribution charge	
1st 20 therms	\$0.2012/therm
Next 30 therms	\$0.1117/therm
Over 50 therms	\$0.0374/therm
Gas supply charge	\$0.3209/therm

(a) What is the charge for using 40 therms in a month?
(b) What is the charge for using 202 therms in a month?
(c) Construct a function that gives the monthly charge C for x therms of gas.
(d) Graph this function.

37. Wind Chill The wind chill factor represents the equivalent air temperature at a standard wind speed that would produce the same heat loss as the given temperature and wind speed. One formula for computing the equivalent temperature is

$$W = \begin{cases} t & 0 \le v < 1.79 \\ 33 - \dfrac{(10.45 + 10\sqrt{v} - v)(33 - t)}{22.04} & 1.79 \le v \le 20 \\ 33 - 1.5958(33 - t) & v \ge 20 \end{cases}$$

where v represents the wind speed (in meters per second) and t represents the air temperature (°C). Compute the wind chill for the following:
(a) An air temperature of 10°C and a wind speed of 1 meter per second (m/sec)
(b) An air temperature of 10°C and a wind speed of 5 m/sec.
(c) An air temperature of 10°C and a wind speed of 15 m/sec.
(d) An air temperature of 10°C and a wind speed of 25 m/sec.
(e) Explain the physical meaning of the equation corresponding to $0 \le v < 1.79$.
(f) Explain the physical meaning of the equation corresponding to $v > 20$.

38. Wind Chill Redo Problem 37(a)–(d) for an air temperature of −10°C.

39. Exploration Graph $y = x^2$. Then on the same screen graph $y = x^2 + 2$, followed by $y = x^2 + 4$, followed by $y = x^2 - 2$. What pattern do you observe? Can you predict the graph of $y = x^2 - 4$? Of $y = x^2 + 5$.

40. Exploration Graph $y = x^2$. Then on the same screen graph $y = (x - 2)^2$, followed by $y = (x - 4)^2$, followed by $y = (x + 2)^2$. What pattern do you observe? Can you predict the graph of $y = (x + 4)^2$? Of $y = (x - 5)^2$?

41. Exploration Graph $y = |x|$. Then on the same screen graph $y = 2|x|$, followed by $y = 4|x|$, followed by $y = \frac{1}{2}|x|$. What pattern do you observe? Can you predict the graph of $y = \frac{1}{4}|x|$? Of $y = 5|x|$?

42. Exploration Graph $y = x^2$. Then on the same screen graph $y = -x^2$. What pattern do you observe? Now try $y = |x|$ and $y = -|x|$. What do you conclude?

43. Exploration Graph $y = \sqrt{x}$. Then on the same screen graph $y = \sqrt{-x}$. What pattern do you observe? Now try $y = 2x + 1$ and $y = 2(-x) + 1$. What do you conclude?

44. Exploration Graph $y = x^3$. Then on the same screen graph $y = (x - 1)^3 + 2$. Could you have predicted the result?

45. Exploration Graph $y = x^2$, $y = x^4$, and $y = x^6$ on the same screen. What do you notice is the same about each graph? What do you notice that is different?

46. Exploration Graph $y = x^3$, $y = x^5$, and $y = x^7$ on the same screen. What do you notice is the same about each graph? What do you notice that is different?

47. Consider the equation

$$y = \begin{cases} 1 & \text{if } x \text{ is rational} \\ 0 & \text{if } x \text{ is irrational} \end{cases}$$

Is this a function? What is its domain? What is its range? What is its y-intercept, if any? What are its x-intercepts, if any? Is it even, odd, or neither? How would you describe its graph?

48. Define some functions that pass through $(0, 0)$ and $(1, 1)$ and are increasing for $x \ge 0$. Begin your list with $y = \sqrt{x}$, $y = x$, and $y = x^2$. Can you propose a general result about such functions?

* *Source:* The Peoples Gas Company, Chicago, Illinois.
† *Source:* Nicor Gas, Aurora, Illinois.

2.4 GRAPHING TECHNIQUES: TRANSFORMATIONS

OBJECTIVES
1. Graph Functions Using Horizontal and Vertical Shifts
2. Graph Functions Using Compressions and Stretches
3. Graph Functions Using Reflections about the x-Axis or y-Axis

At this stage, if you were asked to graph any of the functions defined by $y = x$, $y = x^2$, $y = x^3$, $y = \sqrt{x}$, $y = |x|$, or $y = 1/x$, your response should be, "Yes, I recognize these functions and know the general shapes of their graphs." (If this is not your answer, review the previous section and Figures 21 through 26).

Sometimes we are asked to graph a function that is "almost" like one that we already know how to graph. In this section, we look at some of these functions and develop techniques for graphing them. Collectively, these techniques are referred to as **transformations**.

1 VERTICAL SHIFTS

EXPLORATION On the same screen, graph each of the following functions:

$$Y_1 = x^2$$

$$Y_2 = x^2 + 1$$

$$Y_3 = x^2 + 2$$

$$Y_4 = x^2 - 1$$

$$Y_5 = x^2 - 2$$

What do you observe?

Figure 31

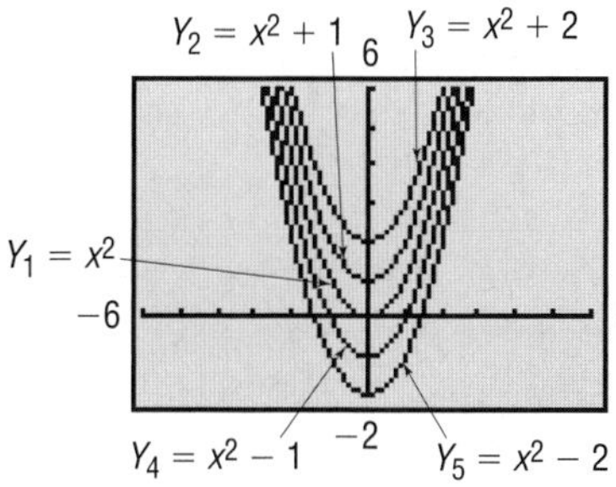

RESULT Figure 31 illustrates the graphs. You should have observed a general pattern. With $Y_1 = x^2$ on the screen, the graph of $Y_2 = x^2 + 1$ is identical to that of $Y_1 = x^2$, except that it is shifted vertically up 1 unit. Similarly, $Y_3 = x^2 + 2$ is identical to that of $Y_1 = x^2$, except that it is shifted vertically up 2 units. The graph of $Y_4 = x^2 - 1$ is identical to that of $Y_1 = x^2$, except that it is shifted vertically down 1 unit. ■

We are led to the following conclusion:

> If a real number c is added to the right side of a function $y = f(x)$, the graph of the new function $y = f(x) + c$ is the graph of f **shifted vertically** up (if $c > 0$) or down (if $c < 0$).

Let's look at an example.

EXAMPLE 1 Vertical Shift Down

Use the graph of $f(x) = x^2$ to obtain the graph of $h(x) = x^2 - 4$.

Solution Table 7 lists some points on the graphs of $f = Y_1$ and $h = Y_2$. Notice that each y-coordinate of h is 4 units less than the corresponding y-coordinate of f. The graph of h is identical to that of f, except that it is shifted down 4 units. See Figure 32.

TABLE 7

X	Y1	Y2
-3	9	5
-2	4	0
-1	1	-3
0	0	-4
1	1	-3
2	4	0
3	9	5

Y2=X²-4

Figure 32

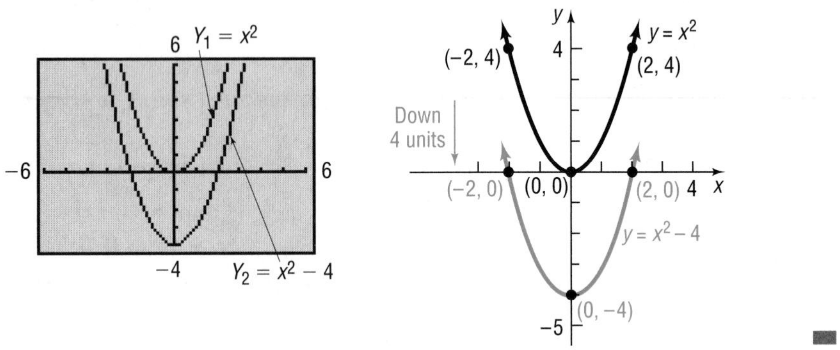

NOW WORK PROBLEM 33.

HORIZONTAL SHIFTS

EXPLORATION On the same screen, graph each of the following functions:

$$Y_1 = x^2$$
$$Y_2 = (x - 1)^2$$
$$Y_3 = (x - 3)^2$$
$$Y_4 = (x + 2)^2$$

What do you observe?

Figure 33

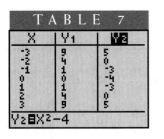

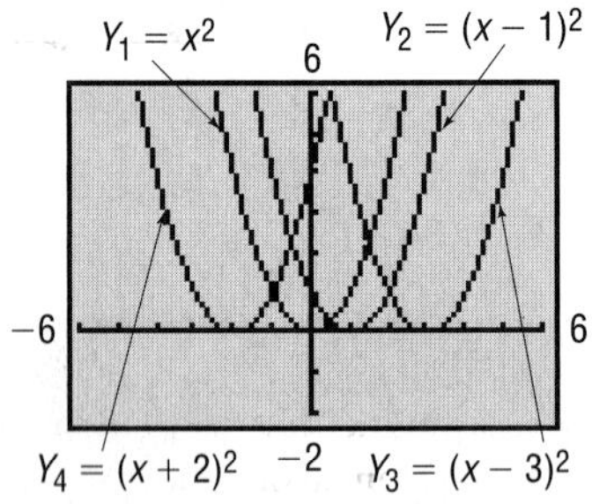

RESULT Figure 33 illustrates the graphs.

You should have observed the following pattern. With the graph of $Y_1 = x^2$ on the screen, the graph of $Y_2 = (x - 1)^2$ is identical to that of $Y_1 = x^2$, except it is shifted horizontally to the right 1 unit. Similarly, the graph of $Y_3 = (x - 3)^2$ is identical to that of $Y_1 = x^2$, except it is shifted horizontally to the right 3 units. Finally, the graph of $Y_4 = (x + 2)^2$ is identical to that of $Y_1 = x^2$, except it is shifted horizontally to the left 2 units. ■

We are led to the following conclusion.

> If the argument x of a function f is replaced by $x - c$, c a real number, the graph of the new function $g(x) = f(x - c)$ is the graph of f **shifted horizontally** left (if $c < 0$) or right (if $c > 0$).

NOW WORK PROBLEM 37.

Vertical and horizontal shifts are sometimes combined.

EXAMPLE 2 Combining Vertical and Horizontal Shifts

Graph the function: $f(x) = (x + 3)^2 - 5$

Solution We graph f in steps. First, we note that the rule for f is basically a square function, so we begin with the graph of $y = x^2$ as shown in Figure 34(a). Next, to get the graph of $y = (x + 3)^2$, we shift the graph of $y = x^2$ horizontally 3 units to the left. See Figure 34(b). Finally, to get the graph of $y = (x + 3)^2 - 5$, we shift the graph of $y = (x + 3)^2$ vertically down 5 units. See Figure 34(c). Note the points plotted on each graph. Using key points can be helpful in keeping track of the transformation that has taken place.

Figure 34

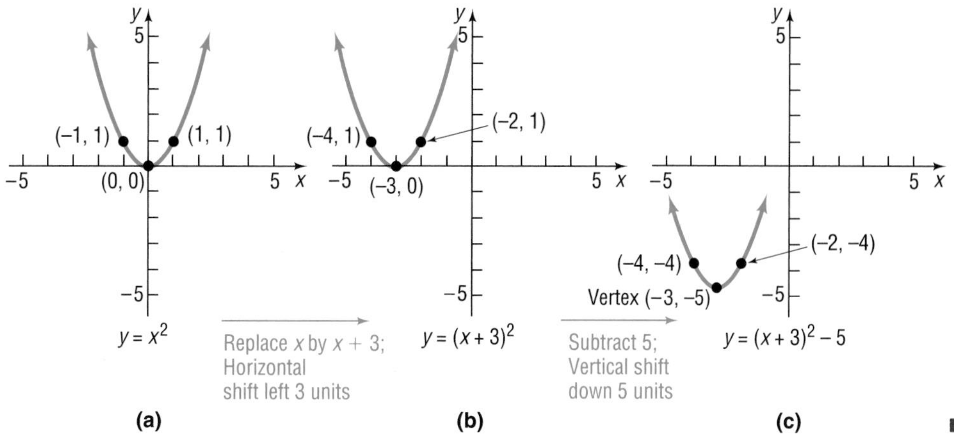

Check: Graph $Y_1 = f(x) = (x + 3)^2 - 5$ and compare the graph to Figure 34(c).

In Example 2, if the vertical shift had been done first, followed by the horizontal shift, the final graph would have been the same. (Try it for yourself.)

NOW WORK PROBLEM 39.

2 Compressions and Stretches

Exploration On the same screen, graph each of the following functions:

$$Y_1 = |x|$$
$$Y_2 = 2|x|$$
$$Y_3 = 3|x|$$
$$Y_4 = \tfrac{1}{2}|x|$$

Figure 35

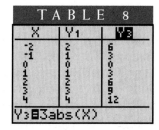

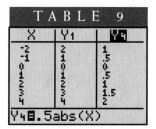

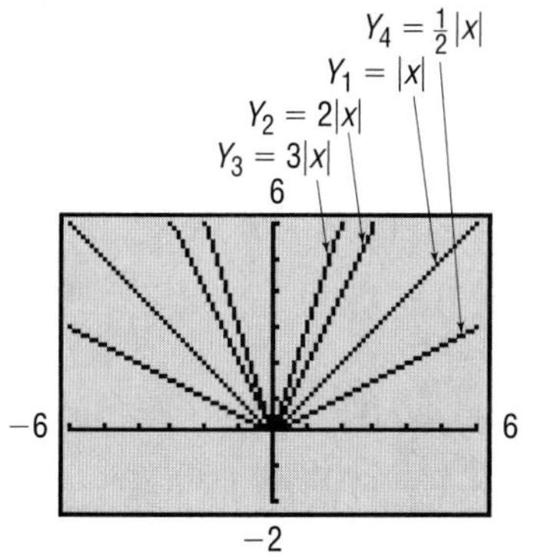

Result Figure 35 illustrates the graphs. You should have observed the following pattern. The graphs of $Y_2 = 2|x|$ and $Y_3 = 3|x|$ can be obtained from the graph of $Y_1 = |x|$ by multiplying each y-coordinate of $Y_1 = |x|$ by factors of 2 and 3, respectively. This is sometimes referred to as a vertical *stretch* using factors of 2 and 3.

The graph of $Y_4 = \frac{1}{2}|x|$ can be obtained from the graph of $Y_1 = |x|$ by multiplying each y-coordinate by $\frac{1}{2}$. This is sometimes referred to as a vertical *compression* using a factor of $\frac{1}{2}$.

Look at Tables 8 and 9, where $Y_1 = |x|$, $Y_3 = 3|x|$, and $Y_4 = \frac{1}{2}|x|$. Notice that the values for Y_3 in Table 8 are three times the values of Y_1 for each x value. Therefore, the graph of Y_3 will be vertically *stretched* by a factor of 3. Likewise, the values of Y_4 in Table 9 are half the values of Y_1 for each x value. Therefore, the graph of Y_4 will be vertically *compressed* by a factor of $\frac{1}{2}$.

TABLE 8

X	Y1	Y3
-2	2	6
-1	1	3
0	0	0
1	1	3
2	2	6
3	3	9
4	4	12

Y3=3abs(X)

TABLE 9

X	Y1	Y4
-2	2	1
-1	1	.5
0	0	0
1	1	.5
2	2	1
3	3	1.5
4	4	2

Y4=.5abs(X)

> When the right side of a function $y = f(x)$ is multiplied by a positive number k, the graph of the new function $y = kf(x)$ is a **vertically compressed** (if $0 < k < 1$) or **stretched** (if $k > 1$) version of the graph of $y = f(x)$.

NOW WORK PROBLEM 41.

What happens if the argument x of a function $y = f(x)$ is multiplied by a positive number k, creating a new function $y = f(kx)$? To find the answer, we look at the following Exploration.

EXPLORATION On the same screen, graph each of the following functions:

$$Y_1 = f(x) = x^2$$
$$Y_2 = f(2x) = (2x)^2$$
$$Y_3 = f\left(\frac{1}{2}x\right) = \left(\frac{1}{2}x\right)^2$$

RESULT You should have obtained the graphs shown in Figure 36. The graph of $Y_2 = (2x)^2$ is the graph of $Y_1 = x^2$ compressed horizontally. Look at Table 10(a).

Figure 36

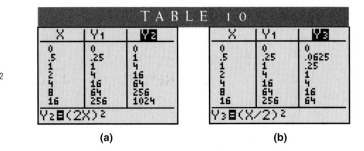

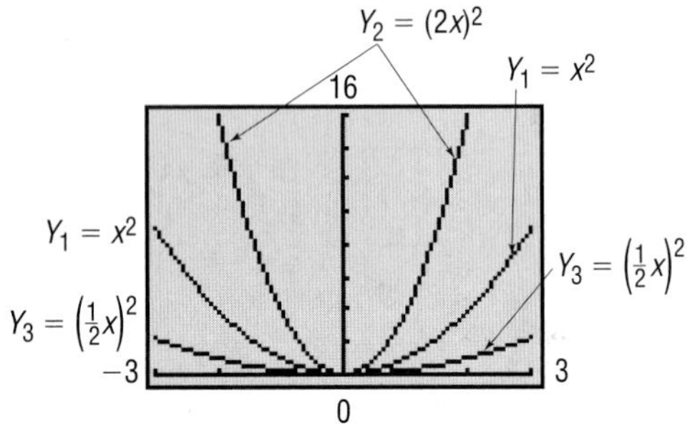

TABLE 10

(a)

X	Y1	Y2
0	0	0
.5	.25	1
1	1	4
2	4	16
4	16	64
8	64	256
16	256	1024

Y2■(2X)²

(b)

X	Y1	Y3
0	0	0
.5	.25	.0625
1	1	.25
2	4	1
4	16	4
8	64	16
16	256	64

Y3■(X/2)²

Notice that (1, 1), (2, 4), (4, 16), and (16, 256) are points on the graph of $Y_1 = x^2$. Also, (0.5, 1), (1, 4), (2, 16), and (8, 256) are points on the graph of $Y_2 = (2x)^2$. For each y-coordinate, the x-coordinate on the graph of Y_2 is $\frac{1}{2}$ the x-coordinate on Y_1. The graph of $Y_2 = (2x)^2$ is obtained by multiplying the x-coordinate of each point on the graph of $Y_1 = x^2$ by $\frac{1}{2}$. The graph of $Y_3 = \left(\frac{1}{2}x\right)^2$ is the graph of $Y_1 = x^2$ stretched horizontally. Look at Table 10(b). Notice that (0.5, 0.25), (1, 1), (2, 4), and (4, 16) are points on the graph of $Y_1 = x^2$. Also, (1, 0.25), (2, 1), (4, 4), and (8, 16) are points on the graph of $Y_3 = \left(\frac{1}{2}x\right)^2$. For each y-coordinate, the x-coordinate on the graph of Y_3 is 2 times the x-coordinate on Y_1. The graph of $Y_3 = \left(\frac{1}{2}x\right)^2$ is obtained by multiplying the x-coordinate of each point on the graph of $Y_1 = x^2$ by a factor of 2. ■

> If the argument of x of a function $y = f(x)$ is multiplied by a positive number k, the graph of the new function $y = f(kx)$ is obtained by multiplying each x-coordinate of $y = f(x)$ by $1/k$. A **horizontal compression** results if $k > 1$, and a **horizontal stretch** occurs if $0 < k < 1$.

Let's look at an example.

EXAMPLE 3 Graphing Using Stretches and Compressions

Figure 37

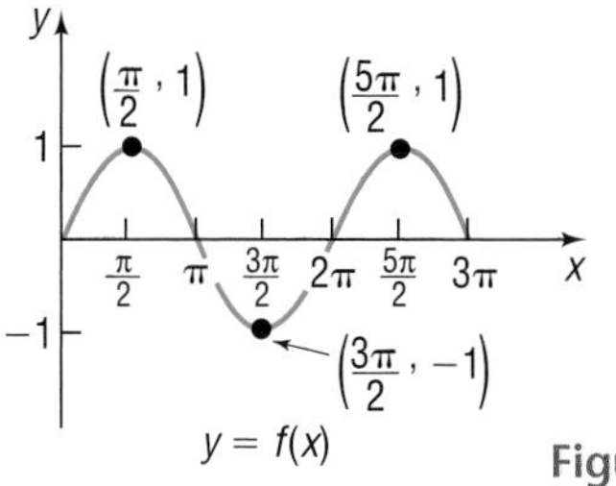

The graph of $y = f(x)$ is given in Figure 37. Use this graph to find the graphs of:

(a) $y = 3f(x)$ (b) $y = f(3x)$

Solution (a) The graph of $y = 3f(x)$ is obtained by multiplying each y-coordinate of $y = f(x)$ by a factor of 3. See Figure 38(a).

(b) The graph of $y = f(3x)$ is obtained from the graph of $y = f(x)$ by multiplying each x-coordinate of $y = f(x)$ by a factor of $\frac{1}{3}$. See Figure 38(b).

Figure 38

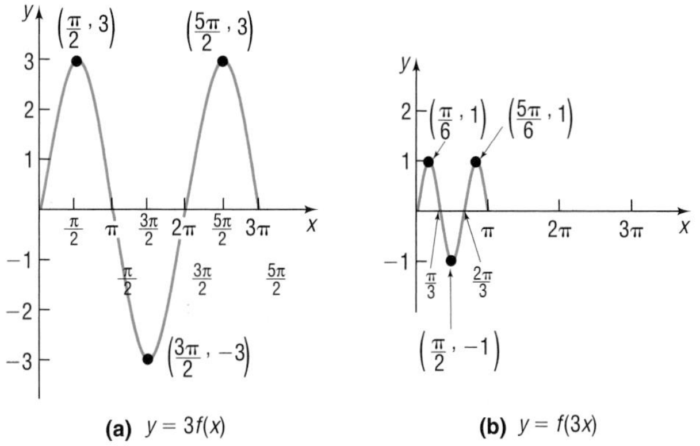

NOW WORK PROBLEMS 63(e) AND (g).

3 REFLECTIONS ABOUT THE *x*-AXIS AND THE *y*-AXIS

EXPLORATION **Reflection about the *x*-Axis**

(a) Graph $Y_1 = x^2$, followed by $Y_2 = -x^2$.

(b) Graph $Y_1 = |x|$, followed by $Y_2 = -|x|$.

(c) Graph $Y_1 = x^2 - 4$, followed by $Y_2 = -(x^2 - 4) = -x^2 + 4$.

RESULT See Tables 11(a), (b), and (c) and Figures 39(a), (b), and (c). In each instance, the second graph is the reflection about the x-axis of the first graph.

TABLE 11

X	Y1	Y2
-3	9	-9
-2	4	-4
-1	1	-1
0	0	0
1	1	-1
2	4	-4
3	9	-9

Y2 = -X²

(a)

X	Y1	Y2
-3	3	-3
-2	2	-2
-1	1	-1
0	0	0
1	1	-1
2	2	-2
3	3	-3

Y2 = -abs(X)

(b)

X	Y1	Y2
-3	5	-5
-2	0	0
-1	-3	3
0	-4	4
1	-3	3
2	0	0
3	5	-5

Y2 = -X²+4

(c)

Figure 39

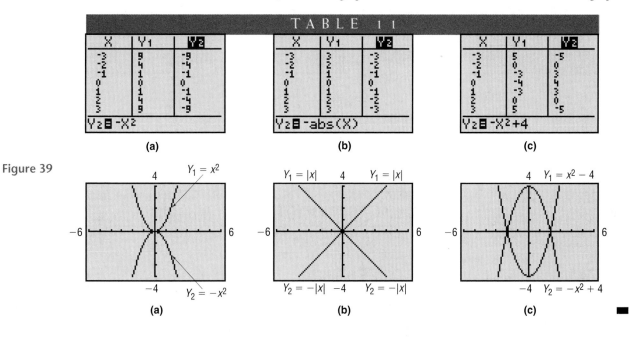

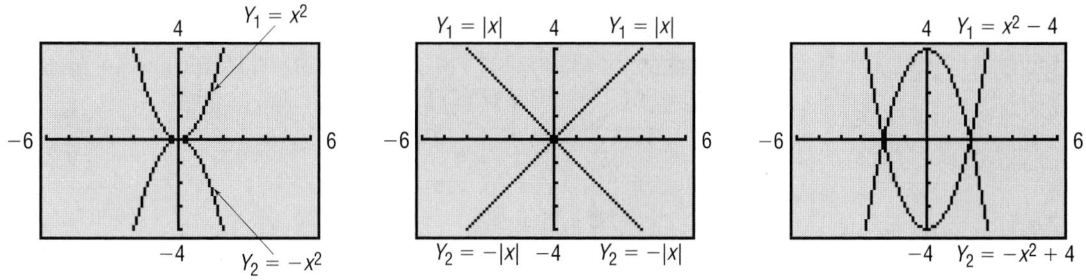

(a) (b) (c)

When the right side of the function $y = f(x)$ is multiplied by -1, the graph of the new function $y = -f(x)$ is the **reflection about the x-axis** of the graph of the function $y = f(x)$.

NOW WORK PROBLEM 45.

EXPLORATION **Reflection about the y-axis**

(a) Graph $Y_1 = \sqrt{x}$, followed by $Y_2 = \sqrt{-x}$.

(b) Graph $Y_1 = x + 1$, followed by $Y_2 = -x + 1$.

(c) Graph $Y_1 = x^4 + x$, followed by $Y_2 = (-x)^4 + (-x) = x^4 - x$.

RESULT See Tables 12(a), (b), and (c) and Figures 40(a), (b), and (c). In each instance, the second graph is the reflection about the y-axis of the first graph.

TABLE 12

X	Y1	Y2
-3	ERROR	1.7321
-2	ERROR	1.4142
-1	ERROR	1
0	0	0
1	1	ERROR
2	1.4142	ERROR
3	1.7321	ERROR

Y2=√(-X)

(a)

X	Y1	Y2
-3	-2	4
-2	-1	3
-1	0	2
0	1	1
1	2	0
2	3	-1
3	4	-2

Y2=-X+1

(b)

X	Y1	Y2
-3	78	84
-2	14	18
-1	0	2
0	0	0
1	2	0
2	18	14
3	84	78

Y2=X^4-X

(c)

Figure 40

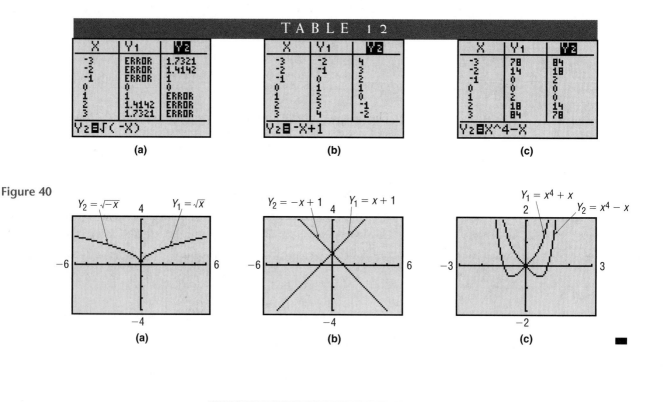

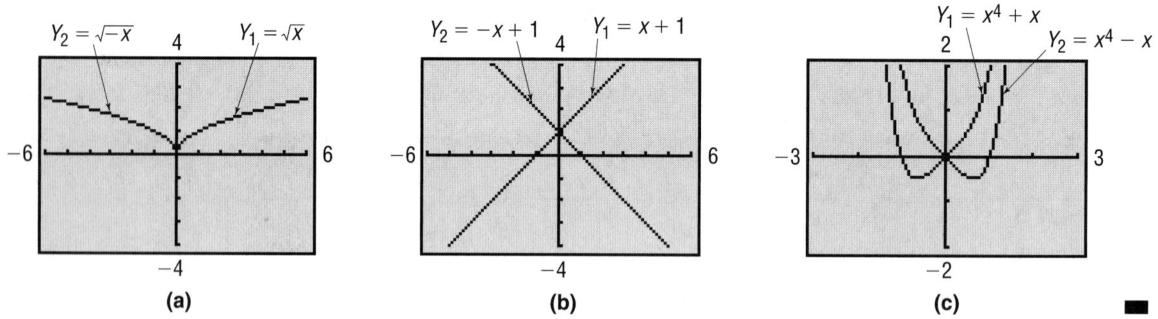

When the graph of the function $y = f(x)$ is known, the graph of the new function $y = f(-x)$ is the **reflection about the y-axis** of the graph of the function $y = f(x)$.

SUMMARY OF GRAPHING TECHNIQUES

Table 13 summarizes the graphing procedures that we have just discussed.

TABLE 13

To Graph:	Draw the Graph of *f* and:	Functional Change to $f(x)$
Vertical shifts		
$y = f(x) + c, \quad c > 0$	Raise the graph of f by c units.	Add c to $f(x)$
$y = f(x) - c, \quad c > 0$	Lower the graph of f by c units.	Subtract c from $f(x)$.
Horizontal shifts		
$y = f(x + c), \quad c > 0$	Shift the graph of f to the left c units.	Replace x by $x + c$.
$y = f(x - c), \quad c > 0$	Shift the graph of f to the right c units.	Replace x by $x - c$.
Compressing or stretching		
$y = kf(x), \quad k > 0$	Multiply each y-coordinate of $y = f(x)$ by k.	Multiply $f(x)$ by k.
$y = f(kx), \quad k > 0$	Multiply each x-coordinate of $y = f(x)$ by $\frac{1}{k}$.	Replace x by kx.
Reflection about the x-axis		
$y = -f(x)$	Reflect the graph of f about the x-axis.	Multiply $f(x)$ by -1.
Reflection about the y-axis		
$y = f(-x)$	Reflect the graph of f about the y-axis.	Replace x by $-x$.

The examples that follow combine some of the procedures outlined in this section to get the required graph.

EXAMPLE 4 Determining the Function Obtained from a Series of Transformations

Find the function that is finally graphed after the following three transformations are applied to the graph of $y = |x|$.

1. Shift left 2 units. 2. Shift up 3 units. 3. Reflect about the y-axis.

Solution

1. Shift left 2 units: Replace x by $x + 2$. $\quad y = |x + 2|$
2. Shift up 3 units: Add 3. $\quad y = |x + 2| + 3$
3. Reflect about the y-axis: Replace x by $-x$. $\quad y = |-x + 2| + 3$ ■

NOW WORK PROBLEM 25.

EXAMPLE 5 Combining Graphing Procedures

Graph the function: $f(x) = \dfrac{3}{x - 2} + 1$

Solution We use the following steps to obtain the graph of f:

Step 1: $y = \dfrac{1}{x}$ Reciprocal function.

Step 2: $y = \dfrac{3}{x}$ Multiply by 3; vertical stretch of the graph of $y = \dfrac{1}{x}$ by a factor of 3.

STEP 3: $y = \dfrac{3}{x-2}$ Replace x by $x - 2$; horizontal shift to the right 2 units.

STEP 4: $y = \dfrac{3}{x-2} + 1$ Add 1; vertical shift up 1 unit.

See Figure 41.

Figure 41

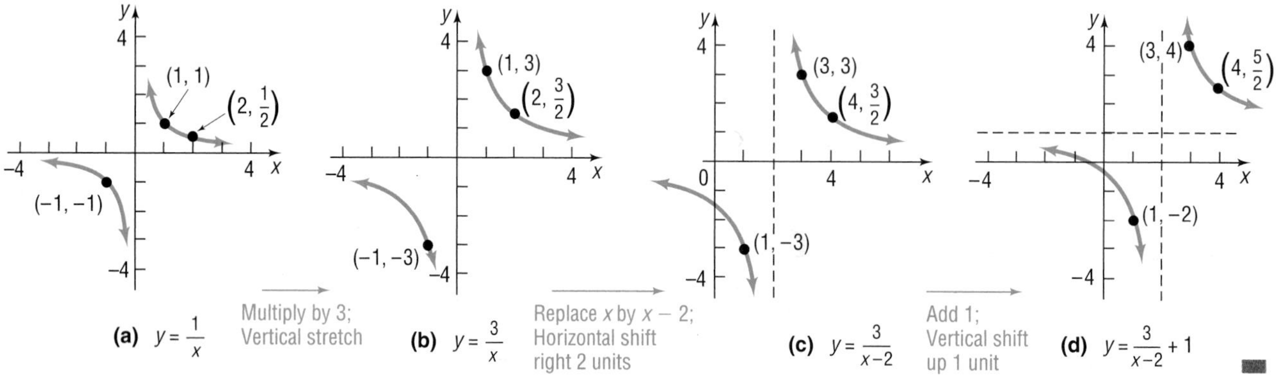

Check: Graph $Y_1 = f(x) = \dfrac{3}{x-2} + 1$ and compare the graph to Figure 41(d).

Other orderings of the steps shown in Example 5 would also result in the graph of f. For example, try this one:

STEP 1: $y = \dfrac{1}{x}$ Reciprocal function.

STEP 2: $y = \dfrac{1}{x-2}$ Replace x by $x - 2$; horizontal shift to the right 2 units.

STEP 3: $y = \dfrac{3}{x-2}$ Multiply by 3; vertical stretch of the graph of $y = \dfrac{1}{x-2}$ by factor of 3.

STEP 4: $y = \dfrac{3}{x-2} + 1$ Add 1; vertical shift up 1 unit.

NOW WORK PROBLEM 51.

EXAMPLE 6 Combining Graphing Procedures

Graph the function: $f(x) = \sqrt{1-x} + 2$

Solution We use the following steps to get the graph of $y = \sqrt{1-x} + 2$.

STEP 1: $y = \sqrt{x}$ Square root function.

STEP 2: $y = \sqrt{x+1}$ Replace x by $x + 1$; horizontal shift left 1 unit.

STEP 3: $y = \sqrt{-x+1} = \sqrt{1-x}$ Replace x by $-x$; reflect about y-axis.

STEP 4: $y = \sqrt{1-x} + 2$ Add 2; vertical shift up 2 units.

See Figure 42.

Figure 42

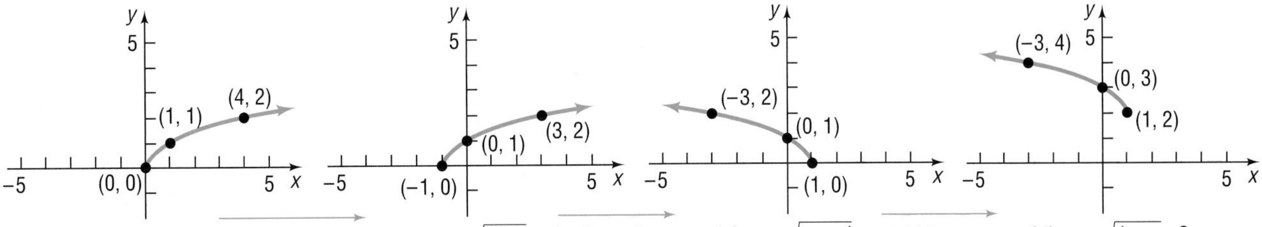

(a) $y = \sqrt{x}$ Replace x by $x + 1$; Horizontal shift left 1 unit **(b)** $y = \sqrt{x+1}$ Replace x by $-x$; Reflect about y-axis **(c)** $y = \sqrt{-x+1} = \sqrt{1-x}$ Add 2; Vertical shift up 2 units **(d)** $y = \sqrt{1-x} + 2$

Check: Graph $Y_1 = f(x) = \sqrt{1 - x} + 2$ and compare the graph to Figure 42(d).

2.4 EXERCISES

In Problems 1–12, match each graph to one of the following functions without using a graphing utility.

A. $y = x^2 + 2$	*B.* $y = -x^2 + 2$	*C.* $y = \lvert x \rvert + 2$	*D.* $y = -\lvert x \rvert + 2$
E. $y = (x - 2)^2$	*F.* $y = -(x + 2)^2$	*G.* $y = \lvert x - 2 \rvert$	*H.* $y = -\lvert x + 2 \rvert$
I. $y = 2x^2$	*J.* $y = -2x^2$	*K.* $y = 2\lvert x \rvert$	*L.* $y = -2\lvert x \rvert$

1.
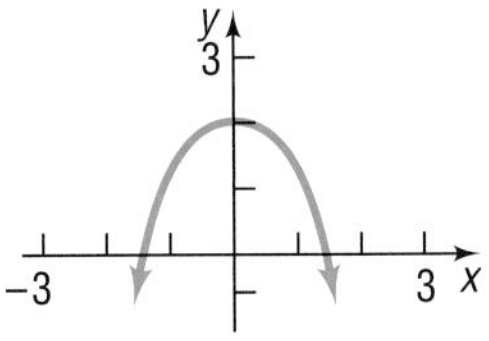

2.
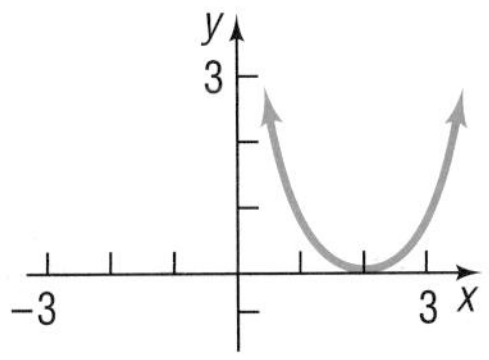

3.
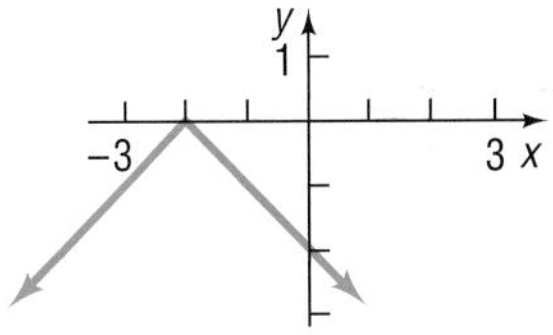

4.
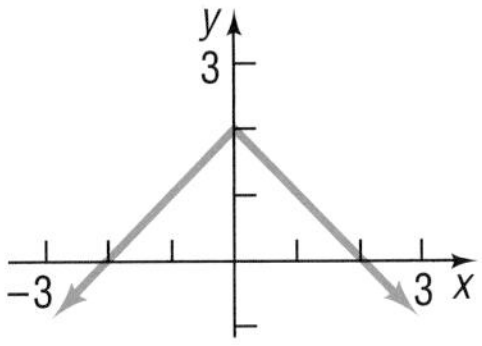

5.
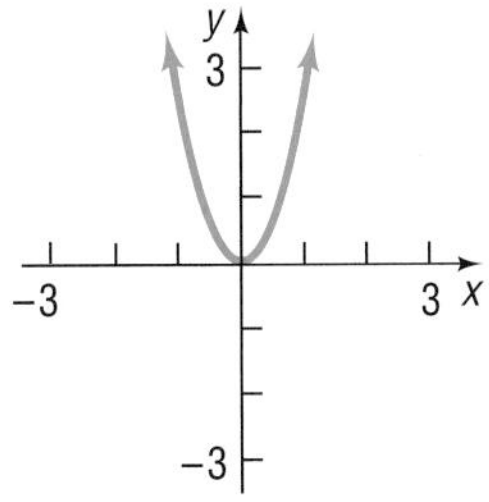

6.
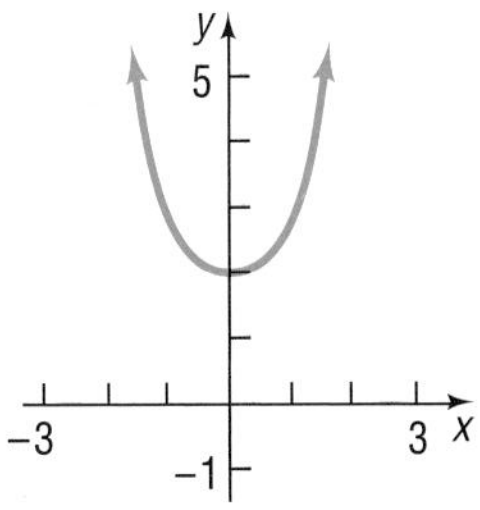

7.
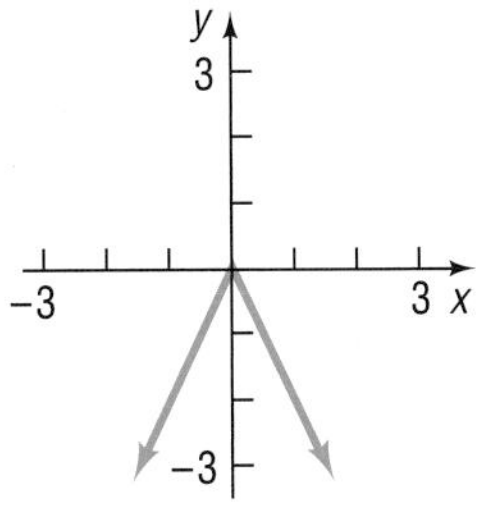

8.
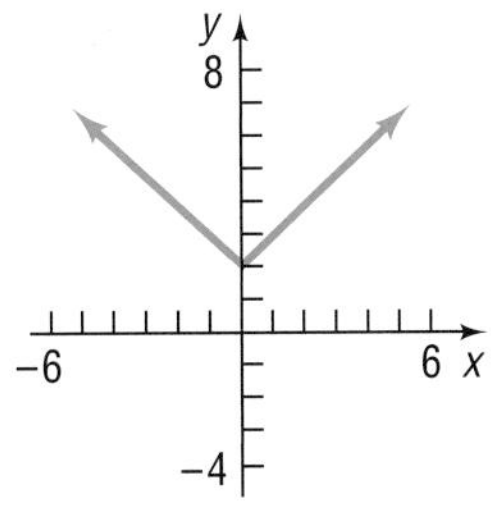

9.
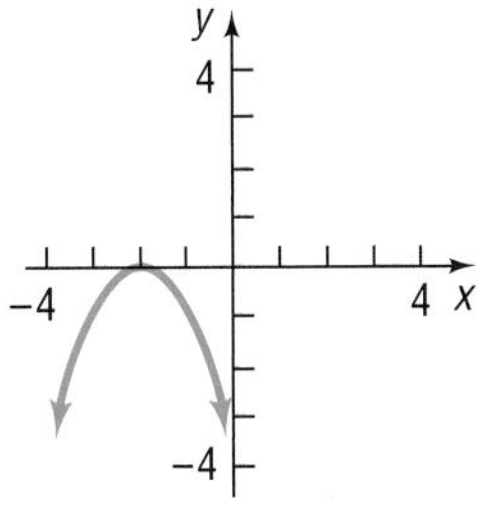

10.
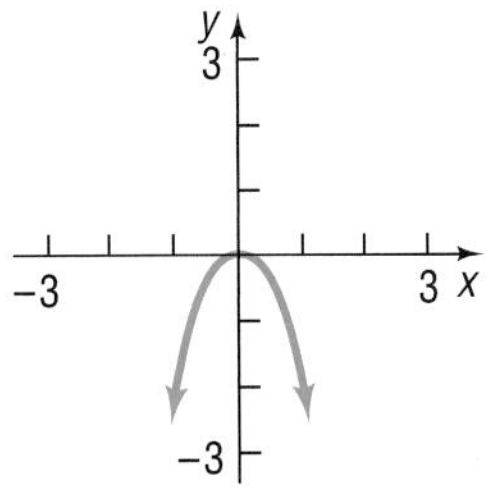

11.
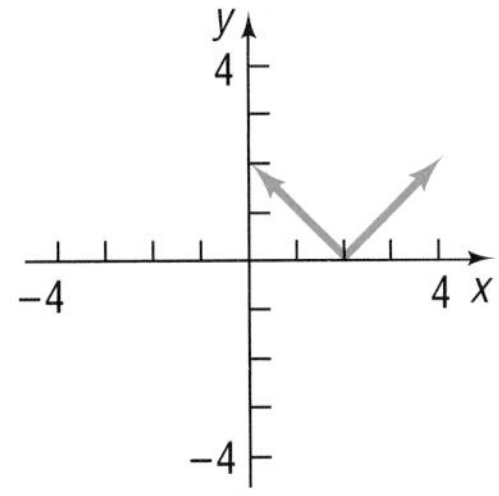

12.
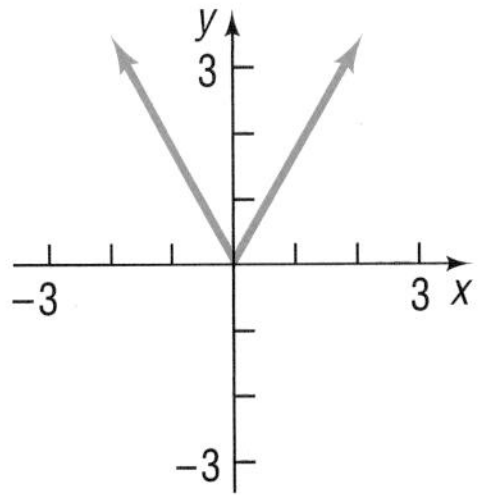

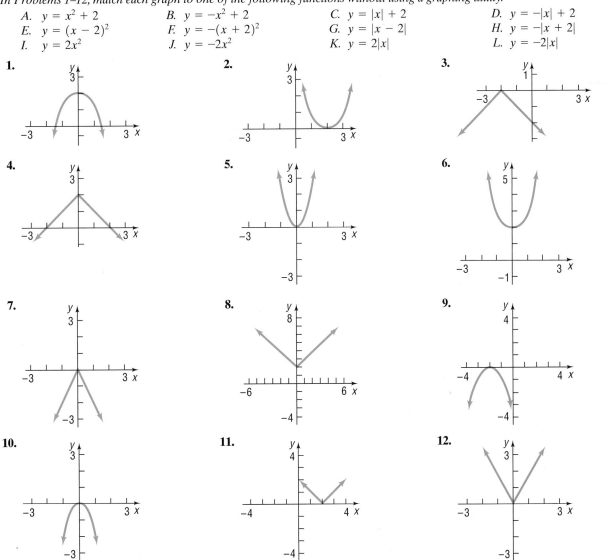

In Problems 13–16, match each graph to one of the following functions without using a graphing utility.

A. $y = x^3$ *B.* $y = (x + 2)^3$ *C.* $y = -2x^3$ *D.* $y = x^3 + 2$

13.

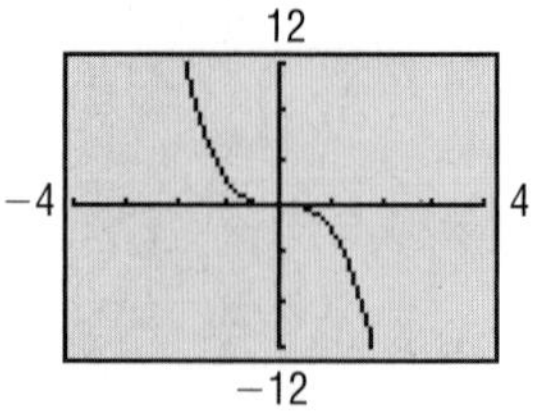

14.

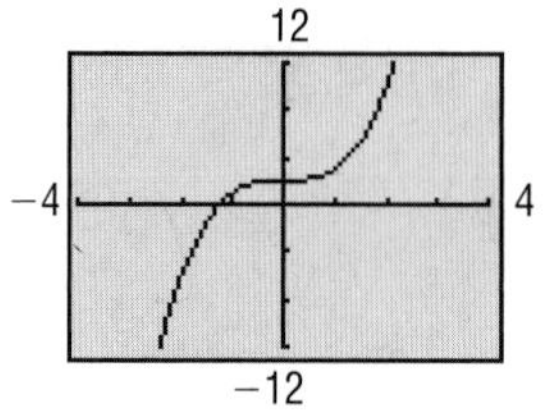

15.

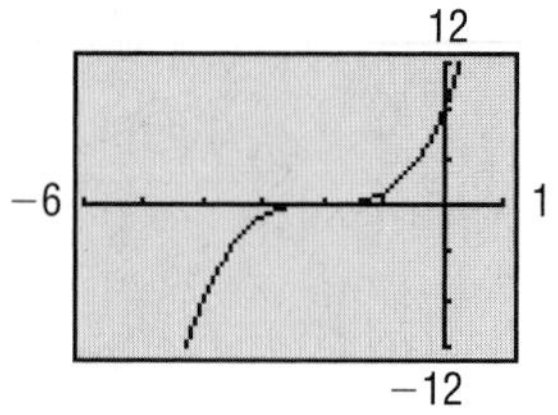

16.

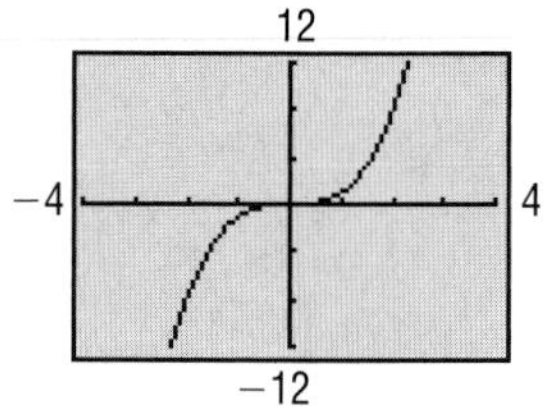

In Problems 17–24, write the function whose graph is the graph of $y = x^3$*, but is:*

17. Shifted to the right 4 units

18. Shifted to the left 4 units

19. Shifted up 4 units

20. Shifted down 4 units

21. Reflected about the y-axis

22. Reflected about the x-axis

23. Vertically stretched by a factor of 4

24. Horizontally stretched by a factor of 4

In Problems 25–28, find the function that is finally graphed after the following transformations are applied to the graph of $y = \sqrt{x}$.

25. (1) Shift up 2 units
(2) Reflect about the x-axis
(3) Reflect about the y-axis

26. (1) Reflect about the x-axis
(2) Shift right 3 units
(3) Shift down 2 units

27. (1) Reflect about the x-axis
(2) Shift up 2 units
(3) Shift left 3 units

28. (1) Shift up 2 units
(2) Reflect about the y-axis
(3) Shift left 3 units

29. If $(3, 0)$ is a point on the graph of $y = f(x)$, which of the following must be on the graph of $y = -f(x)$?
(a) $(0, 3)$ (b) $(0, -3)$
(c) $(3, 0)$ (d) $(-3, 0)$

30. If $(3, 0)$ is a point on the graph of $y = f(x)$, which of the following must be on the graph of $y = f(-x)$?
(a) $(0, 3)$ (b) $(0, -3)$
(c) $(3, 0)$ (d) $(-3, 0)$

31. If $(0, 3)$ is a point on the graph of $y = f(x)$, which of the following must be on the graph of $y = 2f(x)$?
(a) $(0, 3)$ (b) $(0, 2)$
(c) $(0, 6)$ (d) $(6, 0)$

32. If $(3, 0)$ is a point on the graph of $y = f(x)$, which of the following must be on the graph of $y = \frac{1}{2}f(x)$?
(a) $(3, 0)$ (b) $(3/2, 0)$
(c) $(0, 3/2)$ (d) $(1/2, 0)$

In Problems 33–62, graph each function by hand using the techniques of shifting, compressing, stretching, and/or reflecting. Start with the graph of the basic function (for example, $y = x^2$*) and show all stages. Verify your answer by using a graphing utility.*

33. $f(x) = x^2 - 1$

34. $f(x) = x^2 + 4$

35. $g(x) = x^3 + 1$

36. $g(x) = x^3 - 1$

37. $h(x) = \sqrt{x - 2}$

38. $h(x) = \sqrt{x + 1}$

39. $f(x) = (x - 1)^3 + 2$

40. $f(x) = (x + 2)^3 - 3$

41. $g(x) = 4\sqrt{x}$

42. $g(x) = \frac{1}{2}\sqrt{x}$

43. $h(x) = \dfrac{1}{2x}$

44. $h(x) = \dfrac{4}{x}$

45. $f(x) = -|x|$

46. $f(x) = -\sqrt{x}$

47. $g(x) = \sqrt{-x}$

48. $g(x) = -x^3$

49. $h(x) = \text{int}(-x)$

50. $h(x) = \dfrac{1}{-x}$

51. $f(x) = 2(x + 1)^2 - 3$

52. $f(x) = 3(x - 2)^2 + 1$

53. $g(x) = \sqrt{x - 2} + 1$

54. $g(x) = |x + 1| - 3$

55. $h(x) = \sqrt{-x} - 2$

56. $h(x) = \dfrac{4}{x} + 2$

57. $f(x) = -(x + 1)^3 - 1$

58. $f(x) = -4\sqrt{x - 1}$

59. $g(x) = 2|1 - x|$

60. $g(x) = 4\sqrt{2 - x}$

61. $h(x) = 2\,\text{int}(x - 1)$

62. $h(x) = -x^3 + 2$

In Problems 63–68, the graph of a function f is illustrated. Use the graph of f as the first step toward graphing each of the following functions:

(a) $F(x) = f(x) + 3$ (b) $G(x) = f(x + 2)$ (c) $P(x) = -f(x)$ (d) $H(x) = f(x + 1) - 2$
(e) $Q(x) = \frac{1}{2}f(x)$ (f) $g(x) = f(-x)$ (g) $h(x) = f(2x)$

63.

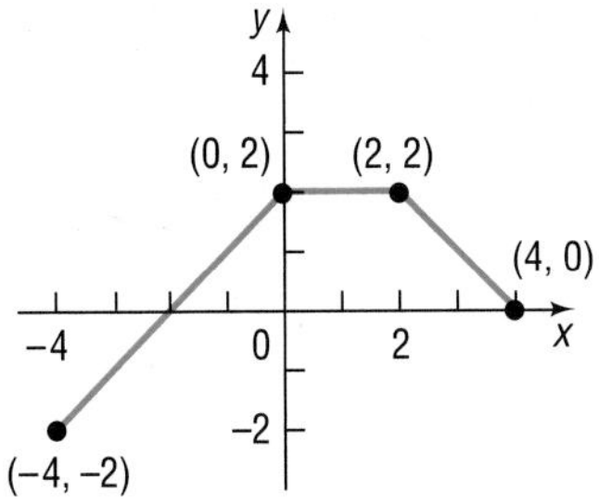

64.

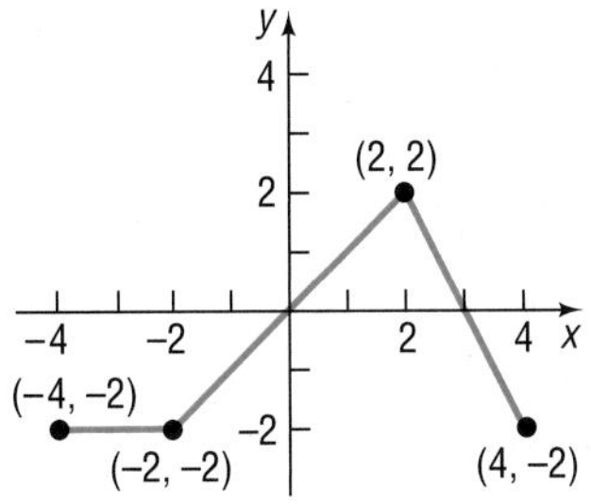

65.

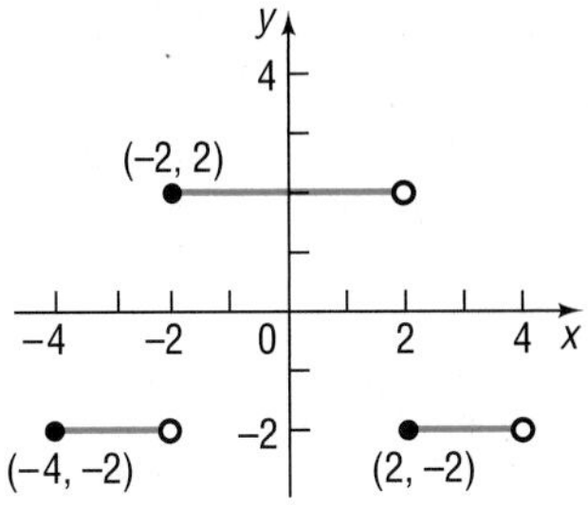

66.

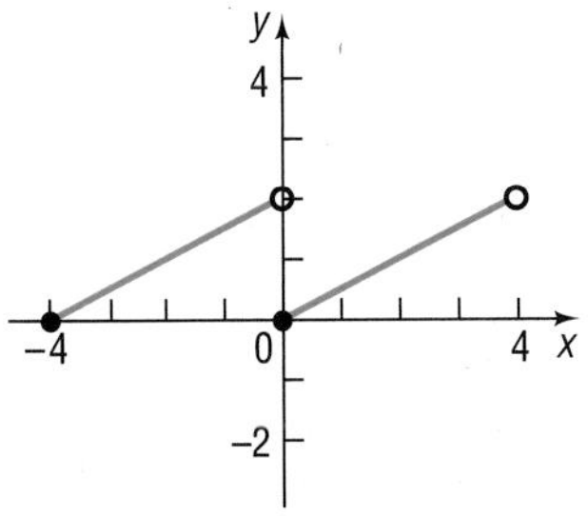

67.

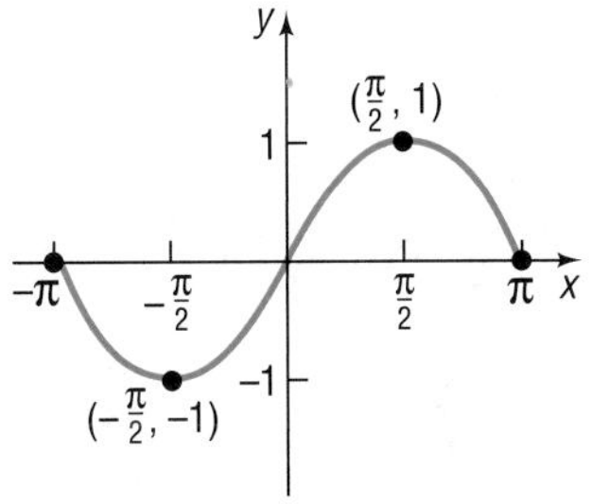

68.

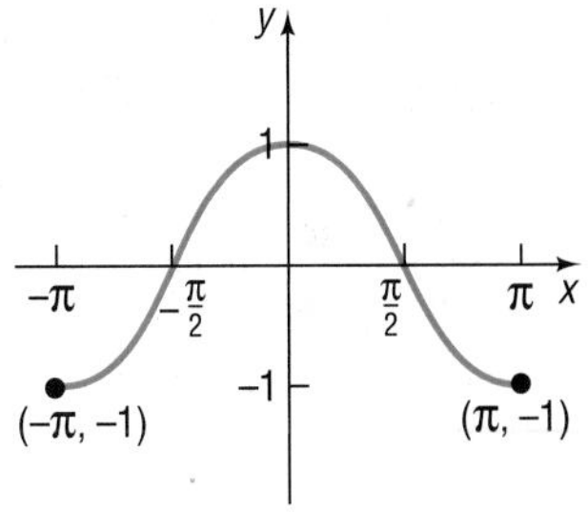

69. Exploration
(a) Use a graphing utility to graph $y = x + 1$ and $y = |x + 1|$.
(b) Graph $y = 4 - x^2$ and $y = |4 - x^2|$.
(c) Graph $y = x^3 + x$ and $y = |x^3 + x|$.
(d) What do you conclude about the relationship between the graphs of $y = f(x)$ and $y = |f(x)|$?

70. Exploration
(a) Use a graphing utility to graph $y = x + 1$ and $y = |x| + 1$.
(b) Graph $y = 4 - x^2$ and $y = 4 - |x|^2$.
(c) Graph $y = x^3 + x$ and $y = |x|^3 + |x|$.
(d) What do you conclude about the relationship between the graphs of $y = f(x)$ and $y = f(|x|)$?

71. The graph of a function f is illustrated in the figure.
(a) Draw the graph of $y = |f(x)|$.
(b) Draw the graph of $y = f(|x|)$.

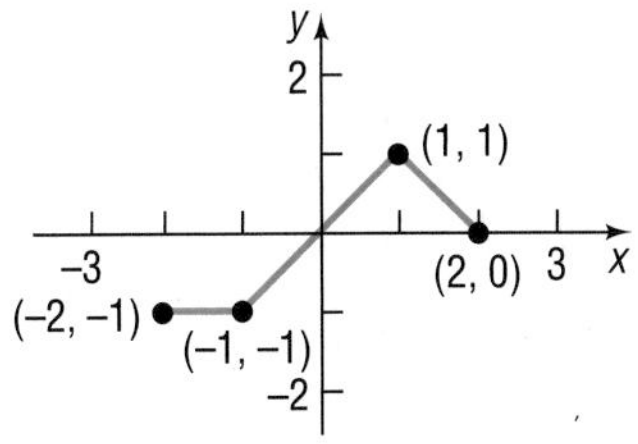

72. The graph of a function f is illustrated in the figure.
(a) Draw the graph of $y = |f(x)|$.
(b) Draw the graph of $y = f(|x|)$.

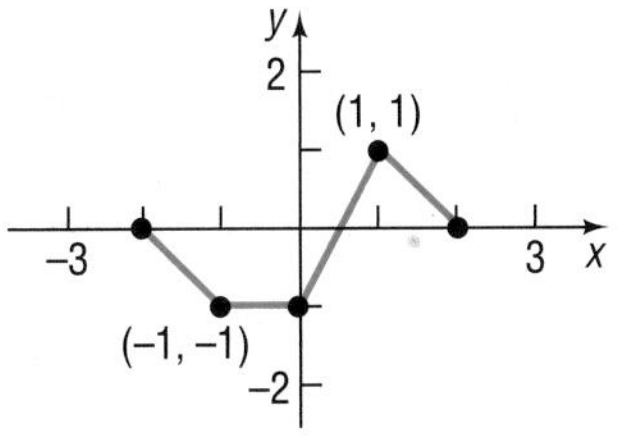

In Problems 73–78, complete the square of each quadratic expression. Then graph each function by hand using the technique of shifting. Verify your results using a graphing utility. (If necessary, refer to Section 10 of the Appendix to review completing the square.)

73. $f(x) = x^2 + 2x$

74. $f(x) = x^2 - 6x$

75. $f(x) = x^2 - 8x + 1$

76. $f(x) = x^2 + 4x + 2$

77. $f(x) = x^2 + x + 1$

78. $f(x) = x^2 - x + 1$

79. The equation $y = (x - c)^2$ defines a *family of parabolas*, one parabola for each value of c. On one set of coordinate axes, graph the members of the family for $c = 0$, $c = 3$, and $c = -2$.

80. Repeat Problem 79 for the family of parabolas $y = x^2 + c$.

81. Temperature Measurements The relationship between the Celsius (°C) and Fahrenheit (°F) scales for measuring temperature is given by the equation

$$F = \frac{9}{5}C + 32$$

The relationship between the Celsius (°C) and Kelvin (K) scales is $K = C + 273$. Graph the equation $F = \frac{9}{5}C + 32$ using degrees Fahrenheit on the y-axis and degrees Celsius on the x-axis. Use the techniques introduced in this section to obtain the graph showing the relationship between Kelvin and Fahrenheit temperatures.

82. Period of a Pendulum The period T (in seconds) of a simple pendulum is a function of its length l (in feet) defined by the equation

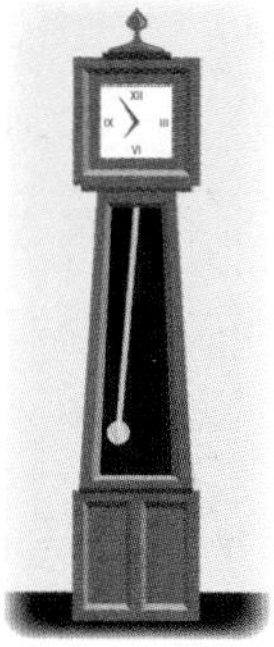

$$T = 2\pi\sqrt{\frac{l}{g}}$$

where $g \approx 32.2$ feet per second per second is the acceleration of gravity.

(a) Use a graphing utility to graph the function $T = T(l)$.

(b) Now graph the functions $T = T(l + 1)$, $T = T(l + 2)$, and $T = T(l + 3)$.

(c) Discuss how adding to the length l changes the period T.

(d) Now graph the functions $T = T(2l)$, $T = T(3l)$, and $T = T(4l)$.

(e) Discuss how multiplying the length l by factors of 2, 3, and 4 changes the period T.

83. Cigar Company Profits The daily profits of a cigar company from selling x cigars are given by

$$p(x) = -0.05x^2 + 100x - 2000$$

The government wishes to impose a tax on cigars (sometimes called a *sin tax*) that gives the company the option of either paying a flat tax of \$10,000 per day or a tax of 10% on profits. As chief financial officer (CFO) of the company, you need to decide which tax is the better option for the company.

(a) On the same screen, graph $Y_1 = p(x) - 10{,}000$ and $Y_2 = (1 - 0.10)p(x)$.

(b) Based on the graph, which option would you select? Why?

(c) Using the terminology learned in this section, describe each graph in terms of the graph of $p(x)$.

(d) Suppose that the government offered the options of a flat tax of \$4800 or a tax of 10% on profits. Which would you select? Why?

2.5 OPERATIONS ON FUNCTIONS; COMPOSITE FUNCTIONS

OBJECTIVES

1. Form the Sum, Difference, Product, and Quotient of Two Functions
2. Form the Composite Function and Find Its Domain

1 In this section, we introduce some operations on functions. We shall see that functions, like numbers, can be added, subtracted, multiplied, and divided. For example, if $f(x) = x^2 + 9$ and $g(x) = 3x + 5$, then

$$f(x) + g(x) = (x^2 + 9) + (3x + 5) = x^2 + 3x + 14$$

The new function $y = x^2 + 3x + 14$ is called the *sum function* $f + g$. Similarly,

$$f(x) \cdot g(x) = (x^2 + 9)(3x + 5) = 3x^3 + 5x^2 + 27x + 45$$

The new function $y = 3x^3 + 5x^2 + 27x + 45$ is called the *product function* $f \cdot g$.

The general definitions are given next.

If f and g are functions:

The **sum** $\boldsymbol{f + g}$ is the function defined by

$$(f + g)(x) = f(x) + g(x)$$

The domain of $f + g$ consists of the numbers x that are in the domains of both f and g.

The **difference** $\boldsymbol{f - g}$ is the function defined by

$$(f - g)(x) = f(x) - g(x)$$

The domain of $f - g$ consists of the numbers x that are in the domains of both f and g.

The **product** $\boldsymbol{f \cdot g}$ is the function defined by

$$(f \cdot g)(x) = f(x) \cdot g(x)$$

The domain of $f \cdot g$ consists of the numbers x that are in the domains of both f and g.

The **quotient** $\boldsymbol{f/g}$ is the function defined by

$$\left(\frac{f}{g}\right)(x) = \frac{f(x)}{g(x)}, \qquad g(x) \neq 0$$

The domain of f/g consists of the numbers x for which $g(x) \neq 0$ that are in the domains of both f and g.

EXAMPLE 1 Operations on Functions

Let f and g be two functions defined as

$$f(x) = \sqrt{x + 2} \quad \text{and} \quad g(x) = \sqrt{3 - x}$$

Find the following, and determine the domain in each case.

(a) $(f + g)(x)$ (b) $(f - g)(x)$ (c) $(f \cdot g)(x)$ (d) $(f/g)(x)$

Solution

(a) $(f + g)(x) = f(x) + g(x) = \sqrt{x + 2} + \sqrt{3 - x}$

(b) $(f - g)(x) = f(x) - g(x) = \sqrt{x + 2} - \sqrt{3 - x}$

(c) $(f \cdot g)(x) = f(x) \cdot g(x) = (\sqrt{x + 2})(\sqrt{3 - x}) = \sqrt{(x + 2)(3 - x)}$

(d) $\left(\frac{f}{g}\right)(x) = \frac{f(x)}{g(x)} = \frac{\sqrt{x + 2}}{\sqrt{3 - x}} = \frac{\sqrt{(x + 2)(3 - x)}}{3 - x}$

The domain of f consists of all numbers x for which $x \geq -2$; the domain of g consists of all numbers x for which $x \leq 3$. The numbers x common to both these domains are those for which $-2 \leq x \leq 3$. As a result, the numbers x

for which $-2 \le x \le 3$, the interval $[-2, 3]$, comprise the domain of the sum function $f + g$, the difference function $f - g$, and the product function $f \cdot g$. For the quotient function f/g, we must exclude from this set the number 3, because the denominator, g, has the value 0 when $x = 3$. The domain of f/g consists of all x for which $-2 \le x < 3$, the interval $[-2, 3)$. ■

NOW WORK PROBLEM 1.

In calculus, it is sometimes helpful to view a complicated function as the sum, difference, product, or quotient of simpler functions. For example,

$F(x) = x^2 + \sqrt{x}$ is the sum of $f(x) = x^2$ and $g(x) = \sqrt{x}$.
$H(x) = (x^2 - 1)/(x^2 + 1)$ is the quotient of $f(x) = x^2 - 1$ and $g(x) = x^2 + 1$.

COMPOSITE FUNCTIONS

2 Consider the function $y = (2x + 3)^2$. If we write $y = f(u) = u^2$ and $u = g(x) = 2x + 3$, then, by a substitution process, we can obtain the original function: $y = f(u) = f(g(x)) = (2x + 3)^2$. This process is called **composition**.

In general, suppose that f and g are two functions and that x is a number in the domain of g. By evaluating g at x, we get $g(x)$. If $g(x)$ is in the domain of f, then we may evaluate f at $g(x)$ and thereby obtain the expression $f(g(x))$. The correspondence from x to $f(g(x))$ is called a *composite function* $f \circ g$.

Given two functions f and g, the **composite function**, denoted by $f \circ g$ (read as "f composed with g"), is defined by

$$(f \circ g)(x) = f(g(x))$$

The domain of $f \circ g$ is the set of all numbers x in the domain of g such that $g(x)$ is in the domain of f.

Look carefully at Figure 43. Only those x's in the domain of g for which $g(x)$ is in the domain of f can be in the domain of $f \circ g$. The reason is that if $g(x)$ is not in the domain of f then $f(g(x))$ is not defined. Because of this, the domain of $f \circ g$ is a subset of the domain of g; the range of $f \circ g$ is a subset of the range of f.

Figure 43

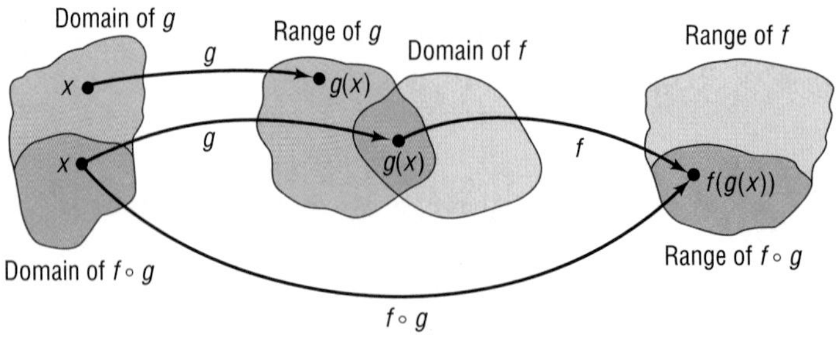

Figure 44 provides a second illustration of the definition. Notice that the "inside" function g in $f(g(x))$ is done first.

Figure 44

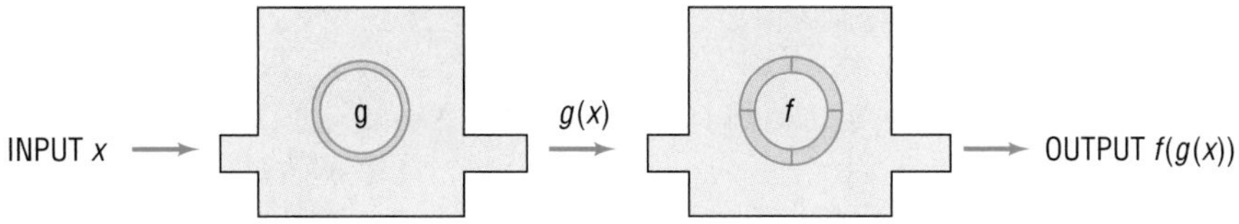

Let's look at some examples.

EXAMPLE 2 Evaluating a Composite Function

Suppose that $f(x) = 2x^2 - 3$ and $g(x) = 4x$. Find:

(a) $(f \circ g)(1)$ (b) $(g \circ f)(1)$ (c) $(f \circ f)(-2)$ (d) $(g \circ g)(-1)$

Solution (a) $(f \circ g)(1) = f(g(1)) = f(4) = 2 \cdot 16 - 3 = 29$

$g(x) = 4x$, $g(1) = 4$; $f(x) = 2x^2 - 3$

(b) $(g \circ f)(1) = g(f(1)) = g(-1) = 4 \cdot (-1) = -4$

$f(x) = 2x^2 - 3$, $f(1) = -1$; $g(x) = 4x$

(c) $(f \circ f)(-2) = f(f(-2)) = f(5) = 2 \cdot 25 - 3 = 47$

$f(-2) = 5$

(d) $(g \circ g)(-1) = g(g(-1)) = g(-4) = 4 \cdot (-4) = -16$

$g(-1) = -4$

Figure 45

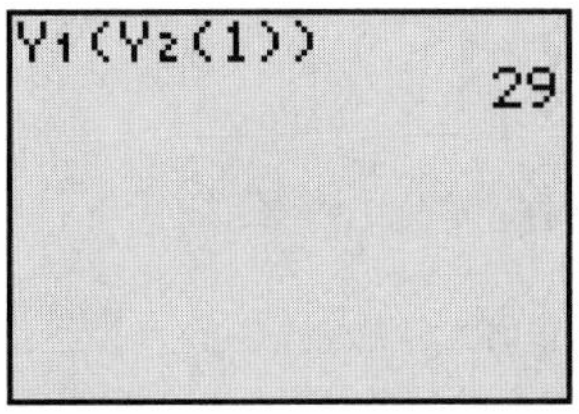

Graphing calculators can be used to evaluate composite functions.* Let $Y_1 = f(x) = 2x^2 - 3$ and $Y_2 = g(x) = 4x$. Then, using a TI-83 graphing calculator, $(f \circ g)(1)$ would be found as shown in Figure 45. Notice that this is the result obtained in Example 2(a).

NOW WORK PROBLEM 13.

Look back at Figure 43. In determining the domain of the composite function $(f \circ g)(x) = f(g(x))$, keep the following two thoughts in mind about the input x.

1. $g(x)$ must be defined so any x not in the domain of g must be excluded.
2. $f(g(x))$ must be defined so any x for which $g(x)$ is not in the domain of f must be excluded.

EXAMPLE 3 Finding the Domain of $f \circ g$

Find the domain of $(f \circ g)(x)$ if $f(x) = \dfrac{1}{x + 2}$ and $g(x) = \dfrac{4}{x - 1}$.

Solution For $(f \circ g)(x) = f(g(x))$, we first note that the domain of g is $\{x \mid x \neq 1\}$, so we exclude 1 from the domain of $f \circ g$. Next, we note that the domain of f is $\{x \mid x \neq -2\}$. Thus, $g(x)$ cannot equal -2, so we solve the equation

*Consult your owner's manual for the appropriate keystrokes.

$$\frac{4}{x-1} = -2$$
$$4 = -2(x-1)$$
$$4 = -2x + 2$$
$$2x = -2$$
$$x = -1$$

We also exclude -1 from the domain of $f \circ g$. The domain of $f \circ g$ is $\{x \mid x \neq -1, x \neq 1\}$.

Check: For $x = 1$, $g(x) = \dfrac{4}{x-1}$ is not defined, so $(f \circ g)(x) = f(g(x))$ is not defined.

For $x = -1$, $g(-1) = 4/-2 = -2$, and $(f \circ g)(-1) = f(g(-1)) = f(-2)$ is not defined.

NOW WORK PROBLEM 23.

EXAMPLE 4 Finding a Composite Function

Suppose that $f(x) = \dfrac{1}{x+2}$ and $g(x) = \dfrac{4}{x-1}$. Find the following composite functions, and then find the domain of each composite function.

(a) $f \circ g$ (b) $g \circ f$ (c) $f \circ f$ (d) $g \circ g$

Solution The domain of f is $\{x \mid x \neq -2\}$ and the domain of g is $\{x \mid x \neq 1\}$.

(a) $$(f \circ g)(x) = f(g(x)) = f\left(\frac{4}{x-1}\right) = \frac{1}{\dfrac{4}{x-1} + 2} = \frac{x-1}{4 + 2(x-1)} = \frac{x-1}{2x+2}$$

Multiply by $\dfrac{x-1}{x-1}$.

The domain of $f \circ g$ consists of those x in the domain of $g(x \neq 1)$ for which $g(x) = 4/(x-1) \neq -2$ or, equivalently, $x \neq -1$. Thus, the domain of $f \circ g$ is $\{x \mid x \neq -1, x \neq 1\}$. (See Example 3.)

(b) $$(g \circ f)(x) = g(f(x)) = g\left(\frac{1}{x+2}\right) = \frac{4}{\dfrac{1}{x+2} - 1} = \frac{4(x+2)}{1-(x+2)} = \frac{4(x+2)}{-x-1}$$

The domain of $g \circ f$ consists of those x in the domain of $f(x \neq -2)$ for which

$$f(x) = \frac{1}{x+2} \neq 1$$

$$\frac{1}{x+2} = 1$$
$$1 = x + 2$$
$$x = -1$$

or, equivalently,

$$x \neq -1$$

The domain of $g \circ f$ is $\{x \mid x \neq -1, x \neq -2\}$.

(c) $(f \circ f)(x) = f(f(x)) = f\left(\dfrac{1}{x+2}\right) = \dfrac{1}{\dfrac{1}{x+2}+2} = \dfrac{x+2}{1+2(x+2)} = \dfrac{x+2}{2x+5}$

The domain of $f \circ f$ consists of those x in the domain of $f(x \neq -2)$ for which

$$f(x) = \frac{1}{x+2} \neq -2$$

$\dfrac{1}{x+2} = -2$

$1 = -2(x+2)$

$1 = -2x - 4$

$2x = -5$

$x = -\frac{5}{2}$

or, equivalently,

$$x \neq -\tfrac{5}{2}$$

The domain of $f \circ f$ is $\{x \mid x \neq -\frac{5}{2}, x \neq -2\}$.

(d) $(g \circ g)(x) = g(g(x)) = g\left(\dfrac{4}{x-1}\right) = \dfrac{4}{\dfrac{4}{x-1}-1} = \dfrac{4(x-1)}{4-(x-1)} = \dfrac{4(x-1)}{-x+5}$

The domain of $g \circ g$ consists of those x in the domain of $g(x \neq 1)$ for which

$$g(x) = \frac{4}{x-1} \neq 1$$

$\dfrac{4}{x-1} = 1$

$4 = x - 1$

$x = 5$

or, equivalently,

$$x \neq 5$$

The domain of $g \circ g$ is $\{x \mid x \neq 1, x \neq 5\}$.

NOW WORK PROBLEMS 35 AND 37.

Examples 4(a) and 4(b) illustrate that, in general, $f \circ g \neq g \circ f$. However, sometimes $f \circ g$ does equal $g \circ f$, as shown in the next example.

EXAMPLE 5 Showing That Two Composite Functions Are Equal

If $f(x) = 3x - 4$ and $g(x) = \frac{1}{3}(x+4)$, show that

$$(f \circ g)(x) = (g \circ f)(x) = x$$

for every x.

Solution

$$\begin{aligned}(f \circ g)(x) &= f(g(x)) \\ &= f\left(\frac{x+4}{3}\right) && g(x) = \frac{1}{3}(x+4) = \frac{x+4}{3}. \\ &= 3\left(\frac{x+4}{3}\right) - 4 && \text{Substitute } g(x) \text{ into the rule for } f,\ f(x) = 3x - 4. \\ &= x + 4 - 4 = x\end{aligned}$$

$$\begin{aligned}(g \circ f)(x) &= g(f(x)) \\ &= g(3x - 4) && f(x) = 3x - 4. \\ &= \tfrac{1}{3}[(3x-4)+4] && \text{Substitute } f(x) \text{ into the rule for } g,\ g(x) = \tfrac{1}{3}(x+4). \\ &= \tfrac{1}{3}(3x) = x\end{aligned}$$

Thus, $(f \circ g)(x) = (g \circ f)(x) = x$.

In Section 5.1, we shall see that there is an important relationship between functions f and g for which $(f \circ g)(x) = (g \circ f)(x) = x$.

EXPLORATION Using a graphing calculator, let $Y_1 = f(x) = 3x - 4$, $Y_2 = g(x) = \frac{1}{3}(x + 4)$, $Y_3 = f \circ g$, and $Y_4 = g \circ f$. Using the viewing window $-3 \le x \le 3, -2 \le y \le 2$, graph only Y_3 and Y_4. What do you see? TRACE to verify that $Y_3 = Y_4$. ■

NOW WORK PROBLEM 47.

CALCULUS APPLICATION

Some techniques in calculus require that we be able to determine the components of a composite function. For example, the function $H(x) = \sqrt{x + 1}$ is the composition of the functions f and g, where $f(x) = \sqrt{x}$ and $g(x) = x + 1$, because $H(x) = (f \circ g)(x) = f(g(x)) = f(x + 1) = \sqrt{x + 1}$.

EXAMPLE 6 Finding the Components of a Composite Function

Find functions f and g such that $f \circ g = H$ if $H(x) = (x^2 + 1)^{50}$.

Solution The function H takes $x^2 + 1$ and raises it to the power 50. A natural way to decompose H is to raise the function $g(x) = x^2 + 1$ to the power 50. Thus, if we let $f(x) = x^{50}$ and $g(x) = x^2 + 1$, then

$$\begin{aligned}(f \circ g)(x) &= f(g(x)) \\ &= f(x^2 + 1) \\ &= (x^2 + 1)^{50} = H(x)\end{aligned}$$

See Figure 46. ■

Figure 46

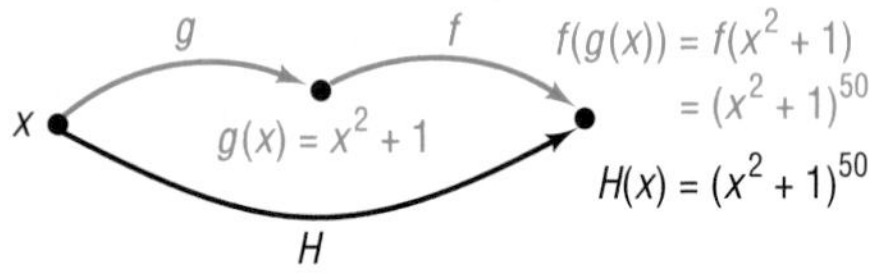

Other functions f and g may be found for which $f \circ g = H$ in Example 6. For example, if $f(x) = x^2$ and $g(x) = (x^2 + 1)^{25}$, then

$$(f \circ g)(x) = f(g(x)) = f((x^2 + 1)^{25}) = [(x^2 + 1)^{25}]^2 = (x^2 + 1)^{50}$$

Although the functions f and g found as a solution to Example 6 are not unique, there is usually a "natural" selection for f and g that comes to mind first.

EXAMPLE 7 Finding the Components of a Composite Function

Find functions f and g such that $f \circ g = H$ if $H(x) = 1/(x + 1)$.

Solution Here H is the reciprocal of $g(x) = x + 1$. If we let $f(x) = 1/x$ and $g(x) = x + 1$, we find that

$$(f \circ g)(x) = f(g(x)) = f(x + 1) = \frac{1}{x + 1} = H(x)$$ ■

2.5 EXERCISES

In Problems 1–10, for the given functions f and g, find the following functions and state the domain of each.

(a) $f + g$ (b) $f - g$ (c) $f \cdot g$ (d) f/g

1. $f(x) = 3x + 4; \quad g(x) = 2x - 3$

2. $f(x) = 2x + 1; \quad g(x) = 3x - 2$

3. $f(x) = x - 1;\ \ g(x) = 2x^2$

4. $f(x) = 2x^2 + 3;\ \ g(x) = 4x^3 + 1$

5. $f(x) = \sqrt{x};\ \ g(x) = 3x - 5$

6. $f(x) = |x|;\ \ g(x) = x$

7. $f(x) = 1 + \dfrac{1}{x};\ \ g(x) = \dfrac{1}{x}$

8. $f(x) = 2x^2 - x;\ \ g(x) = 2x^2 + x$

9. $f(x) = \dfrac{2x + 3}{3x - 2};\ \ g(x) = \dfrac{4x}{3x - 2}$

10. $f(x) = \sqrt{x + 1};\ \ g(x) = \dfrac{2}{x}$

11. Given $f(x) = 3x + 1$ and $(f + g)(x) = 6 - \frac{1}{2}x$, find the function g.

12. Given $f(x) = 1/x$ and $(f/g)(x) = (x + 1)/(x^2 - x)$, find the function g.

In Problems 13–22, for the given functions f and g, find

(a) $(f \circ g)(4)$ *(b)* $(g \circ f)(2)$ *(c)* $(f \circ f)(1)$ *(d)* $(g \circ g)(0)$

Verify your results using a graphing utility.

13. $f(x) = 2x;\ \ g(x) = 3x^2 + 1$

14. $f(x) = 3x + 2;\ \ g(x) = 2x^2 - 1$

15. $f(x) = 4x^2 - 3;\ \ g(x) = 3 - \frac{1}{2}x^2$

16. $f(x) = 2x^2;\ \ g(x) = 1 - 3x^2$

17. $f(x) = \sqrt{x};\ \ g(x) = 2x$

18. $f(x) = \sqrt{x + 1};\ \ g(x) = 3x$

19. $f(x) = |x|;\ \ g(x) = \dfrac{1}{x^2 + 1}$

20. $f(x) = |x - 2|;\ \ g(x) = \dfrac{3}{x^2 + 2}$

21. $f(x) = \dfrac{3}{x^2 + 1};\ \ g(x) = \sqrt{x}$

22. $f(x) = x^3;\ \ g(x) = \dfrac{2}{x^2 + 1}$

In Problems 23–30, find the domain of the composite function $f \circ g$.

23. $f(x) = \dfrac{3}{x - 1};\ \ g(x) = \dfrac{2}{x}$

24. $f(x) = \dfrac{1}{x + 3};\ \ g(x) = \dfrac{-2}{x}$

25. $f(x) = \dfrac{x}{x - 1};\ \ g(x) = \dfrac{-4}{x}$

26. $f(x) = \dfrac{x}{x + 3};\ \ g(x) = \dfrac{2}{x}$

27. $f(x) = \sqrt{x};\ \ g(x) = 2x + 3$

28. $f(x) = \sqrt{x - 2};\ \ g(x) = 1 - 2x$

29. $f(x) = \sqrt{x + 1};\ \ g(x) = \dfrac{2}{x - 1}$

30. $f(x) = \sqrt{3 - x};\ \ g(x) = \dfrac{2}{x - 2}$

In Problems 31–46, for the given functions f and g, find:

(a) $f \circ g$ *(b)* $g \circ f$ *(c)* $f \circ f$ *(d)* $g \circ g$

State the domain of each composite function.

31. $f(x) = 2x + 3;\ \ g(x) = 3x$

32. $f(x) = -x;\ \ g(x) = 2x - 4$

33. $f(x) = 3x + 1;\ \ g(x) = x^2$

34. $f(x) = x + 1;\ \ g(x) = x^2 + 4$

35. $f(x) = x^2;\ \ g(x) = x^2 + 4$

36. $f(x) = x^2 + 1;\ \ g(x) = 2x^2 + 3$

37. $f(x) = \dfrac{3}{x - 1};\ \ g(x) = \dfrac{2}{x}$

38. $f(x) = \dfrac{1}{x + 3};\ \ g(x) = \dfrac{-2}{x}$

39. $f(x) = \dfrac{x}{x - 1};\ \ g(x) = \dfrac{-4}{x}$

40. $f(x) = \dfrac{x}{x + 3};\ \ g(x) = \dfrac{2}{x}$

41. $f(x) = \sqrt{x};\ \ g(x) = 2x + 3$

42. $f(x) = \sqrt{x - 2};\ \ g(x) = 1 - 2x$

43. $f(x) = \sqrt{x + 1};\ \ g(x) = \dfrac{2}{x - 1}$

44. $f(x) = \sqrt{3 - x};\ \ g(x) = \dfrac{2}{x - 2}$

45. $f(x) = ax + b;\ \ g(x) = cx + d$

46. $f(x) = \dfrac{ax + b}{cx + d};\ \ g(x) = mx$

In Problems 47–54, show that $(f \circ g)(x) = (g \circ f)(x) = x$.

47. $f(x) = 2x;\ \ g(x) = \frac{1}{2}x$

48. $f(x) = 4x;\ \ g(x) = \frac{1}{4}x$

49. $f(x) = x^3;\ \ g(x) = \sqrt[3]{x}$

50. $f(x) = x + 5;\ \ g(x) = x - 5$

51. $f(x) = 2x - 6;\ \ g(x) = \frac{1}{2}(x + 6)$

52. $f(x) = 4 - 3x;\ \ g(x) = \frac{1}{3}(4 - x)$

53. $f(x) = ax + b;\ g(x) = \frac{1}{a}(x - b),\quad a \neq 0$

54. $f(x) = \frac{1}{x};\ g(x) = \frac{1}{x}$

In Problems 55–60, find functions f and g so that $f \circ g = H$.

55. $H(x) = (2x + 3)^4$

56. $H(x) = (1 + x^2)^3$

57. $H(x) = \sqrt{x^2 + 1}$

58. $H(x) = \sqrt{1 - x^2}$

59. $H(x) = |2x + 1|$

60. $H(x) = |2x^2 + 3|$

61. If $f(x) = 2x^3 - 3x^2 + 4x - 1$ and $g(x) = 2$, find $(f \circ g)(x)$ and $(g \circ f)(x)$.

62. If $f(x) = x/(x - 1)$, find $(f \circ f)(x)$.

63. If $f(x) = 2x^2 + 5$ and $g(x) = 3x + a$, find a so that the graph of $f \circ g$ crosses the y-axis at 23.

64. If $f(x) = 3x^2 - 7$ and $g(x) = 2x + a$, find a so that the graph of $f \circ g$ crosses the y-axis at 68.

65. Surface Area of a Balloon The surface area S (in square meters) of a hot-air balloon is given by

$$S(r) = 4\pi r^2$$

where r is the radius of the balloon (in meters). If the radius r is increasing with time t (in seconds) according to the formula $r(t) = \frac{2}{3}t^3$, $t \geq 0$, find the surface area S of the balloon as a function of the time t.

66. Volume of a Balloon The volume V (in cubic meters) of the hot-air balloon described in Problem 65 is given by $V(r) = \frac{4}{3}\pi r^3$. If the radius r is the same function of t as in Problem 65, find the volume V as a function of the time t.

67. Automobile Production The number N of cars produced at a certain factory in 1 day after t hours of operation is given by $N(t) = 100t - 5t^2$, $0 \leq t \leq 10$. If the cost C (in dollars) of producing N cars is $C(N) = 15{,}000 + 8000N$, find the cost C as a function of the time t of operation of the factory.

68. Environmental Concerns The spread of oil leaking from a tanker is in the shape of a circle. If the radius r (in feet) of the spread after t hours is $r(t) = 200\sqrt{t}$, find the area A of the oil slick as a function of the time t.

69. Production Cost The price p of a certain product and the quantity x sold obey the demand equation

$$p = -\tfrac{1}{4}x + 100 \qquad 0 \leq x \leq 400$$

Suppose that the cost C of producing x units is

$$C = \frac{\sqrt{x}}{25} + 600$$

Assuming that all items produced are sold, find the cost C as a function of the price p.
[**Hint:** Solve for x in the demand equation and then form the composite.]

70. Cost of a Commodity The price p of a certain commodity and the quantity x sold obey the demand equation

$$p = -\tfrac{1}{5}x + 200, \qquad 0 \leq x \leq 1000$$

Suppose that the cost C of producing x units is

$$C = \frac{\sqrt{x}}{10} + 400$$

Assuming that all items produced are sold, find the cost C as a function of the price p.

71. Volume of a Cylinder The volume V of a right circular cylinder of height h and radius r is $V = \pi r^2 h$. If the height is twice the radius, express the volume V as a function of r.

72. Volume of a Cone The volume V of a right circular cone is $V = \frac{1}{3}\pi r^2 h$. If the height is twice the radius, express the volume V as a function of r.

2.6 MATHEMATICAL MODELS: CONSTRUCTING FUNCTIONS

OBJECTIVES 1 Construct and Analyze Functions

1 Real-world problems often result in mathematical models that involve functions. These functions need to be constructed or built based on the information given. In constructing functions, we must be able to translate the verbal description into the language of mathematics. We do this by assigning symbols to represent the independent and dependent variables and then finding the function or rule that relates these variables.

EXAMPLE 1 Area of a Rectangle with Fixed Perimeter

The perimeter of a rectangle is 50 feet. Express its area A as a function of the length x of a side.

Figure 47

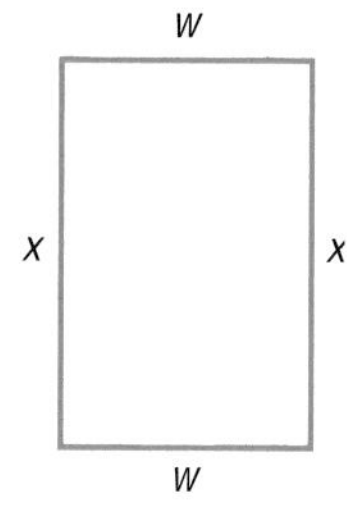

Solution Consult Figure 47. If the length of the rectangle is x and if w is its width, then the sum of the lengths of the sides is the perimeter, 50.

$$x + w + x + w = 50$$
$$2x + 2w = 50$$
$$x + w = 25$$
$$w = 25 - x$$

The area A is length times width, so

$$A = xw = x(25 - x)$$

The area A as a function of x is

$$A(x) = x(25 - x)$$

■

Note that we use the symbol A as the dependent variable and also as the name of the function that relates the length x to the area A. As we mentioned earlier, this double usage is common in applications and should cause no difficulties.

EXAMPLE 2 Economics: Demand Equations

In economics, revenue R is defined as the amount of money derived from the sale of a product and is equal to the unit selling price p of the product times the number x of units actually sold. That is,

$$R = xp$$

In economics, the Law of Demand states that p and x are related: As one increases, the other decreases. Suppose that p and x are related by the following **demand equation:**

$$p = -\tfrac{1}{10}x + 20, \qquad 0 \le x \le 200$$

Express the revenue R as a function of the number x of units sold.

Solution Since $R = xp$ and $p = -\frac{1}{10}x + 20$, it follows that

$$R(x) = xp = x\left(-\tfrac{1}{10}x + 20\right) = -\tfrac{1}{10}x^2 + 20x$$

■

NOW WORK PROBLEM 3.

EXAMPLE 3 Finding the Distance from the Origin to a Point on a Graph

Let $P = (x, y)$ be a point on the graph of $y = x^2 - 1$.

(a) Express the distance d from P to the origin O as a function of x.
(b) What is d if $x = 0$?
(c) What is d if $x = 1$?
(d) What is d if $x = \sqrt{2}/2$?
(e) Use a graphing utility to graph the function $d = d(x)$, $x \ge 0$. Rounded to two decimal places, find the value of x at which d has a local minimum. [This gives the point(s) on the graph of $y = x^2 - 1$ closest to the origin.]

Figure 48

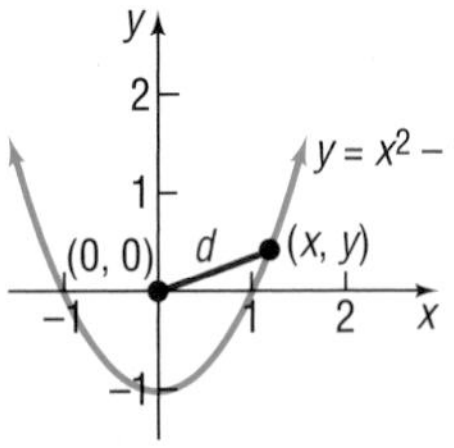

Figure 49

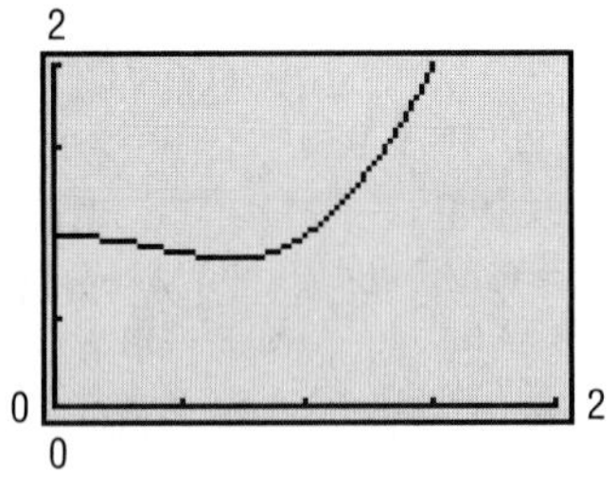

Solution (a) Figure 48 illustrates the graph. The distance d from P to O is

$$d = \sqrt{(x-0)^2 + (y-0)^2} = \sqrt{x^2 + y^2}$$

Since P is a point on the graph of $y = x^2 - 1$, we have

$$d(x) = \sqrt{x^2 + (x^2 - 1)^2} = \sqrt{x^4 - x^2 + 1}$$

We have expressed the distance d as a function of x.

(b) If $x = 0$, the distance d is

$$d(0) = \sqrt{1} = 1$$

(c) If $x = 1$, the distance d is

$$d(1) = \sqrt{1 - 1 + 1} = 1$$

(d) If $x = \sqrt{2}/2$, the distance d is

$$d\left(\frac{\sqrt{2}}{2}\right) = \sqrt{\left(\frac{\sqrt{2}}{2}\right)^4 - \left(\frac{\sqrt{2}}{2}\right)^2 + 1} = \sqrt{\frac{1}{4} - \frac{1}{2} + 1} = \frac{\sqrt{3}}{2}$$

(e) Figure 49 shows the graph of $Y_1 = \sqrt{x^4 - x^2 + 1}$. Using the MINIMUM feature on a graphing utility, we find that when $x \approx 0.70$ the value of d is smallest ($d \approx 0.86$ rounded to two decimal places). ■

NOW WORK PROBLEM 9.

EXAMPLE 4 Filling a Swimming Pool

A rectangular swimming pool 20 meters long and 10 meters wide is 4 meters deep at one end and 1 meter deep at the other. Figure 50 illustrates a cross-sectional view of the pool. Water is being pumped into the pool at the deep end.

Figure 50

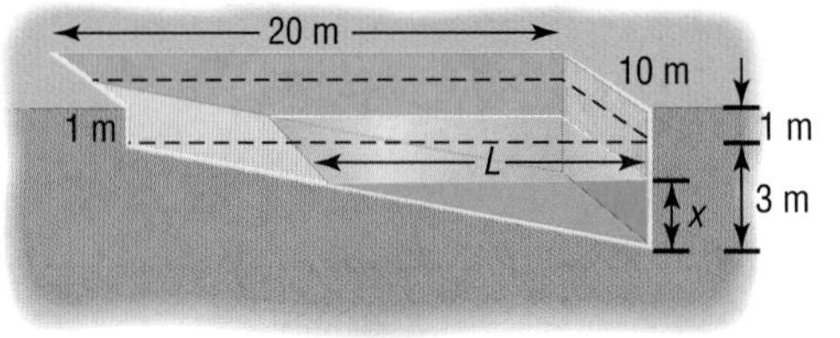

(a) Find a function that expresses the volume V of water in the pool as a function of the height x of the water at the deep end.

(b) Find the volume when the height is 1 meter.

(c) Find the volume when the height is 2 meters.

(d) Use a graphing utility to graph the function $V = V(x)$. At what height is the volume 20 cubic meters? 100 cubic meters?

Solution (a) Let L denote the distance (in meters) measured at water level from the deep end to the short end. Notice that L and x form the sides of a triangle that is similar to the triangle whose sides are 20 meters by 3 meters. Thus, L and x are related by the equation

$$\frac{L}{x} = \frac{20}{3} \quad \text{or} \quad L = \frac{20x}{3}, \qquad 0 \le x \le 3$$

The volume V of water in the pool at any time is

$$V = \begin{pmatrix}\text{cross-sectional}\\ \text{triangular area}\end{pmatrix}(\text{width}) = \left(\tfrac{1}{2}Lx\right)(10) \quad \text{cubic meters}$$

Since $L = 20x/3$, we have

$$V(x) = \left(\frac{1}{2} \cdot \frac{20x}{3} \cdot x\right)(10) = \frac{100}{3}x^2 \quad \text{cubic meters}$$

(b) When the height x of the water is 1 meter, the volume $V = V(x)$ is

$$V(1) = \frac{100}{3} \cdot 1^2 = 33.3 \text{ cubic meters}$$

(c) When the height x of the water is 2 meters, the volume $V = V(x)$ is

$$V(2) = \frac{100}{3} \cdot 2^2 = \frac{400}{3} = 133.3 \text{ cubic meters}$$

Figure 51

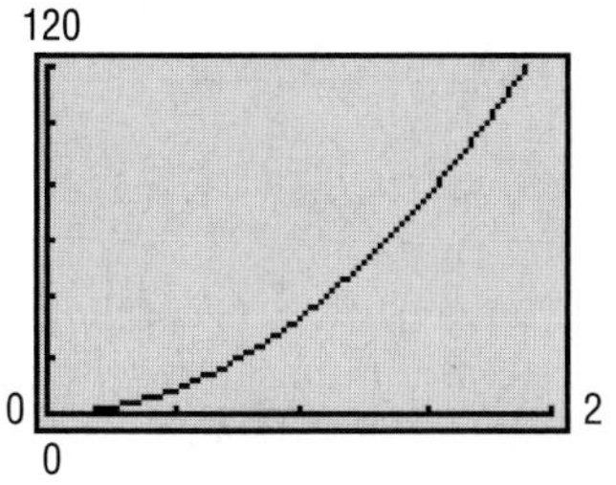

(d) See Figure 51. When $x \approx 0.77$ meter, the volume is 20 cubic meters. When $x \approx 1.73$ meters, the volume is 100 cubic meters. ■

EXAMPLE 5 Area of a Rectangle

A rectangle has one corner on the graph of $y = 25 - x^2$, another at the origin, a third on the positive y-axis, and the fourth on the positive x-axis. See Figure 52.

Figure 52

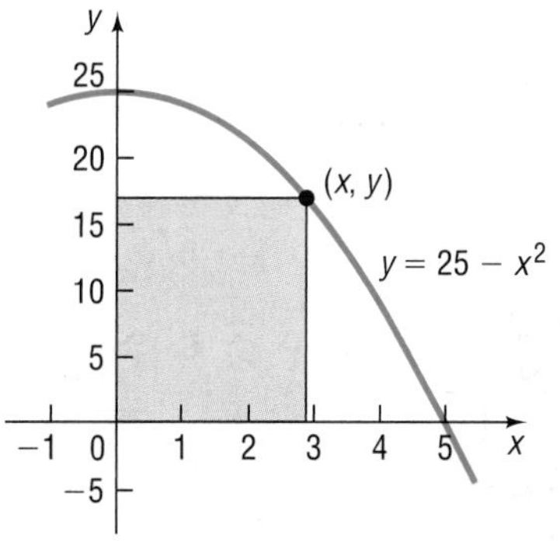

(a) Express the area A of the rectangle as a function of x.
(b) What is the domain of A?
(c) Graph $A = A(x)$.
(d) For what value of x is the area largest?

Solution

(a) The area A of the rectangle is $A = xy$, where $y = 25 - x^2$. Substituting this expression for y, we obtain $A(x) = x(25 - x^2) = 25x - x^3$.

(b) Since x represents a side of the rectangle, we have $x > 0$. In addition, the area must be positive, so $y = 25 - x^2 > 0$, which implies that $-5 < x < 5$. Combining these restrictions, we have the domain of A as $\{x \mid 0 < x < 5\}$ or $(0, 5)$ using interval notation.

(c) See Figure 53 for the graph of $A(x)$.

(d) Using MAXIMUM, we find the area is a maximum of 48.11 at $x = 2.89$, each rounded to two decimal places. See Figure 54.

Figure 53

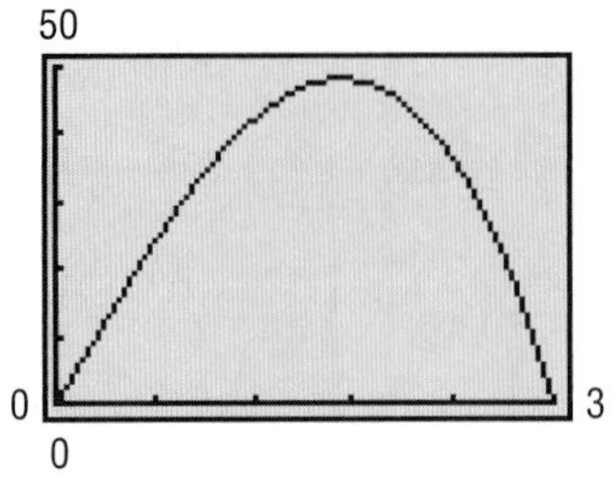

Figure 54

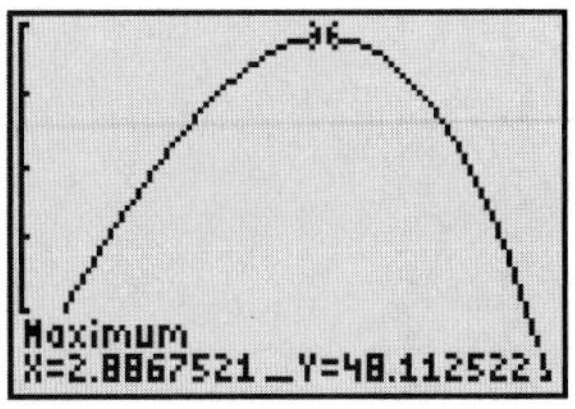

NOW WORK PROBLEM 15.

EXAMPLE 6 Minimum Payments and Interest Charged for Credit Cards

(a) Holders of credit cards issued by banks, department stores, oil companies, and so on, receive bills each month that state minimum amounts that must be paid by a certain due date. The minimum due depends on the total amount owed. One such credit card company uses the following rules: For a bill of less than \$10, the entire amount is due. For a bill of at least \$10 but less than \$500, the minimum due is \$10. There is a minimum of \$30 due on a bill of at least \$500 but less than \$1000, a minimum of \$50 due on a bill of at least \$1000 but less than \$1500, and a minimum of \$70 due on bills of \$1500 or more. Find the function f that describes the minimum payment due on a bill of x dollars. Graph f.

(b) The card holder may pay any amount between the minimum due and the total owed. The organization issuing the card charges the card holder interest of 1.5% per month for the first \$1000 owed and 1% per month on any unpaid balance over \$1000. Find the function g that gives the amount of interest charged per month on a balance of x dollars. Graph g.

Solution (a) The function f that describes the minimum payment due on a bill of x dollars is

$$f(x) = \begin{cases} x & \text{if } 0 \le x < 10 \\ 10 & \text{if } 10 \le x < 500 \\ 30 & \text{if } 500 \le x < 1000 \\ 50 & \text{if } 1000 \le x < 1500 \\ 70 & \text{if } 1500 \le x \end{cases}$$

To graph this function f, we proceed as follows: For $0 \le x < 10$, draw the graph of $y = x$; for $10 \le x < 500$, draw the graph of the constant function $y = 10$; for $500 \le x < 1000$, draw the graph of the constant function $y = 30$; and so on. The graph of f is given in Figure 55.

Figure 55

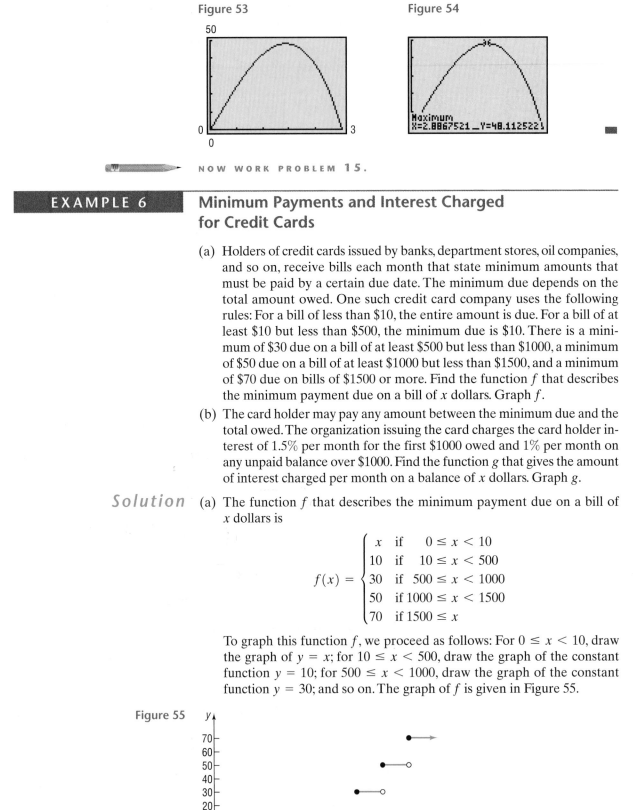

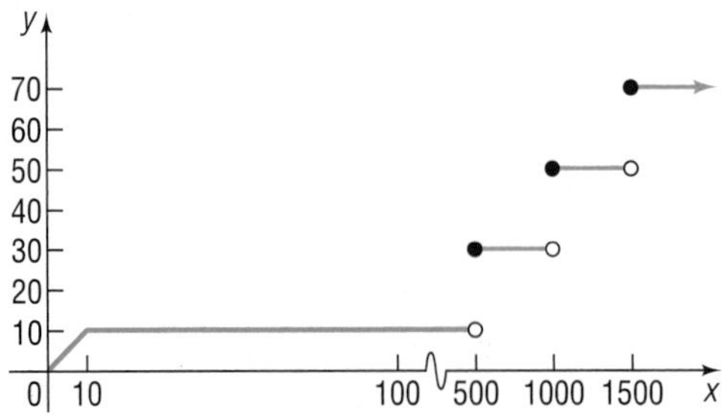

Figure 56

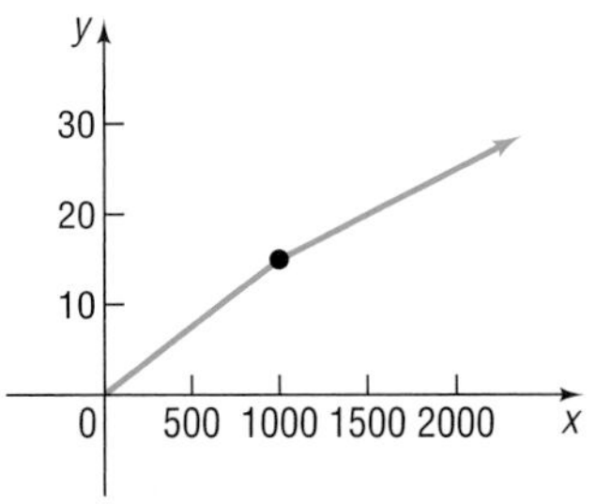

(b) If $g(x)$ is the amount of interest charged per month on a balance of x, then $g(x) = 0.015x$ for $0 \le x \le 1000$. The amount of the unpaid balance above \$1000 is $x - 1000$. If the balance due is $x > 1000$, then the interest is $0.015(1000) + 0.01(x - 1000) = 15 + 0.01x - 10 = 5 + 0.01x$; so

$$g(x) = \begin{cases} 0.015x & \text{if } 0 \le x \le 1000 \\ 5 + 0.01x & \text{if } x > 1000 \end{cases}$$

See Figure 56. ■

EXAMPLE 7

Making a Playpen*

Figure 57

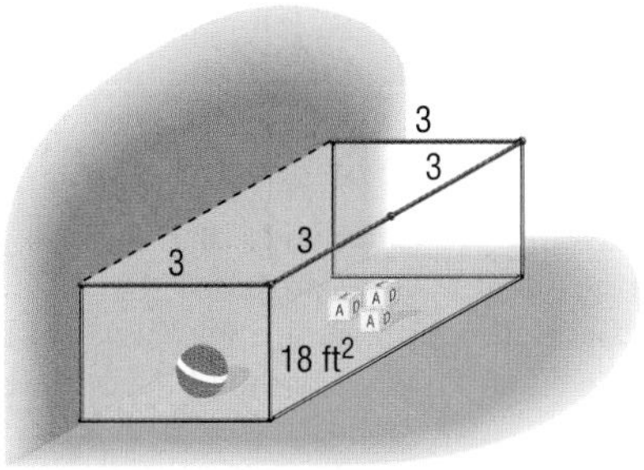

A manufacturer of children's playpens makes a square model that can be opened at one corner and attached at right angles to a wall or, perhaps, the side of a house. If each side is 3 feet in length, the open configuration doubles the available area in which the child can play from 9 square feet to 18 square feet. See Figure 57.

Now suppose that we place hinges at the outer corners to allow for a configuration like the one shown in Figure 58.

Figure 58

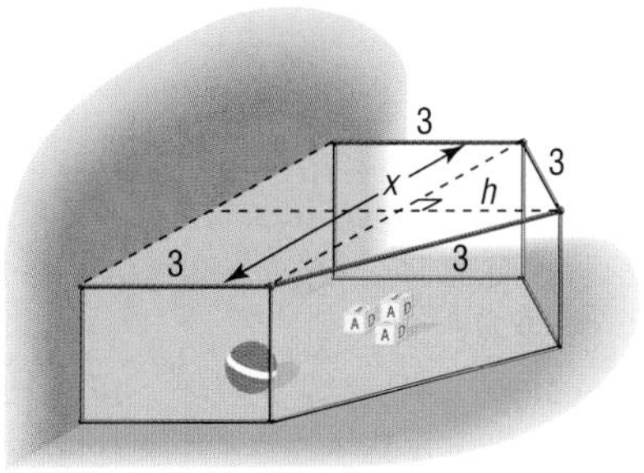

(a) Express the area A of this configuration as a function of the distance x between the two parallel sides.

(b) Find the domain of A.

(c) Find A if $x = 5$.

(d) Graph $A = A(x)$. For what value of x is the area largest? What is the maximum area?

Solution (a) Refer to Figure 58. The area A that we seek consists of the area of a rectangle (with width 3 and length x) and the area of an isosceles triangle (with base x and two equal sides of length 3). The height h of the triangle may be found using the Pythagorean Theorem.

$$h^2 = 3^2 - \left(\frac{x}{2}\right)^2 = 9 - \frac{x^2}{4} = \frac{36 - x^2}{4}$$

$$h = \tfrac{1}{2}\sqrt{36 - x^2}$$

The area A enclosed by the playpen is

$$A = \text{area of rectangle} + \text{area of triangle} = 3x + \tfrac{1}{2}x\left(\tfrac{1}{2}\sqrt{36 - x^2}\right)$$

$$A(x) = 3x + \frac{x\sqrt{36 - x^2}}{4}$$

Now the area A is expressed as a function of x.

(b) To find the domain of A, we note first that $x > 0$, since x is a length. Also, the expression under the radical must be positive, so

$$36 - x^2 > 0$$
$$x^2 < 36$$
$$-6 < x < 6$$

*Adapted from Proceedings, *Summer Conference for College Teachers on Applied Mathematics* (University of Missouri, Rolla), 1971.

Figure 59

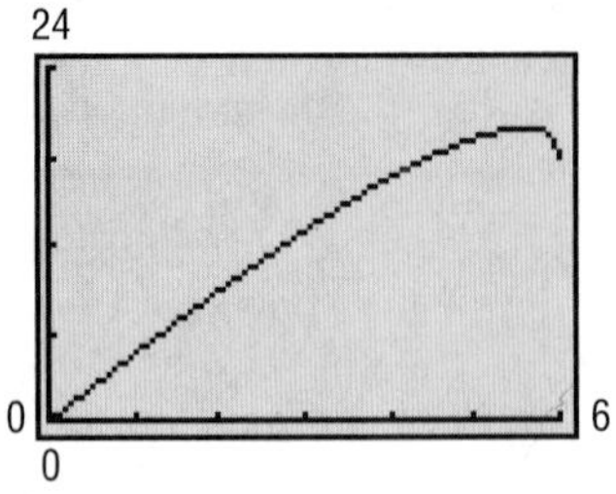

Combining these restrictions, we find that the domain of A is $0 < x < 6$, or $(0, 6)$ using interval notation.

(c) If $x = 5$, the area is

$$A(5) = 3(5) + \frac{5}{4}\sqrt{36 - (5)^2} \approx 19.15 \text{ square feet}$$

If the width of the playpen is 5 feet, its area is 19.15 square feet.

(d) See Figure 59. The maximum area is about 19.82 square feet, obtained when x is about 5.58 feet. ■

2.6 EXERCISES

1. **Volume of a Cylinder** The volume V of a right circular cylinder of height h and radius r is $V = \pi r^2 h$. If the height is twice the radius, express the volume V as a function of r.

2. **Volume of a Cone** The volume V of a right circular cone is $V = \frac{1}{3}\pi r^2 h$. If the height is twice the radius, express the volume V as a function of r.

3. **Demand Equation** The price p and the quantity x sold of a certain product obey the demand equation

$$p = -\tfrac{1}{6}x + 100, \qquad 0 \le x \le 600$$

(a) Express the revenue R as a function of x. (Remember, $R = xp$.)
(b) What is the revenue if 200 units are sold?
(c) Graph the revenue function using a graphing utility.
(d) What quantity x maximizes revenue? What is the maximum revenue?
(e) What price should the company charge to maximize revenue?

4. **Demand Equation** The price p and the quantity x sold of a certain product obey the demand equation

$$p = -\tfrac{1}{3}x + 100, \qquad 0 \le x \le 300$$

(a) Express the revenue R as a function of x.
(b) What is the revenue if 100 units are sold?
(c) Graph the revenue function using a graphing utility.
(d) What quantity x maximizes revenue? What is the maximum revenue?
(e) What price should the company charge to maximize revenue?

5. **Demand Equation** The price p and the quantity x sold of a certain product obey the demand equation

$$x = -5p + 100, \qquad 0 \le p \le 20$$

(a) Express the revenue R as a function of x.
(b) What is the revenue if 15 units are sold?
(c) Graph the revenue function using a graphing utility.
(d) What quantity x maximizes revenue? What is the maximum revenue?
(e) What price should the company charge to maximize revenue?

6. **Demand Equation** The price p and the quantity x sold of a certain product obey the demand equation

$$x = -20p + 500, \qquad 0 \le p \le 25$$

(a) Express the revenue R as a function of x.
(b) What is the revenue if 20 units are sold?
(c) Graph the revenue function using a graphing utility.
(d) What quantity x maximizes revenue? What is the maximum revenue?
(e) What price should the company charge to maximize revenue?

7. **Enclosing a Rectangular Field** David has available 400 yards of fencing and wishes to enclose a rectangular area.
(a) Express the area A of the rectangle as a function of the width x of the rectangle.
(b) What is the domain of A?
(c) Graph $A = A(x)$ using a graphing utility. For what value of x is the area largest?

8. **Enclosing a Rectangular Field along a River** Beth has 3000 feet of fencing available to enclose a rectangular field. One side of the field lies along a river, so only three sides require fencing.
(a) Express the area A of the rectangle as a function of x, where x is the length of the side parallel to the river.
(b) Graph $A = A(x)$ using a graphing utility. For what value of x is the area largest?

9. Let $P = (x, y)$ be a point on the graph of $y = x^2 - 8$.
(a) Express the distance d from P to the origin as a function of x.
(b) What is d if $x = 0$?
(c) What is d if $x = 1$?
(d) Use a graphing utility to graph $d = d(x)$.
(e) For what values of x is d smallest?

10. Let $P = (x, y)$ be a point on the graph of $y = x^2 - 8$.
(a) Express the distance d from P to the point $(0, -1)$ as a function of x.
(b) What is d if $x = 0$?
(c) What is d if $x = -1$?
(d) Use a graphing utility to graph $d = d(x)$.
(e) For what values of x is d smallest?

11. Let $P = (x, y)$ be a point on the graph of $y = \sqrt{x}$.
(a) Express the distance d from P to the point $(1, 0)$ as a function of x.
(b) Use a graphing utility to graph $d = d(x)$.
(c) For what values of x is d smallest?

12. Let $P = (x, y)$ be a point on the graph of $y = 1/x$.
(a) Express the distance d from P to the origin as a function of x.
(b) Use a graphing utility to graph $d = d(x)$.
(c) For what values of x is d smallest?

13. A right triangle has one vertex on the graph of $y = x^3$, $x > 0$, at (x, y), another at the origin, and the third on the positive y-axis at $(0, y)$, as shown in the figure. Express the area A of the triangle as a function of x.

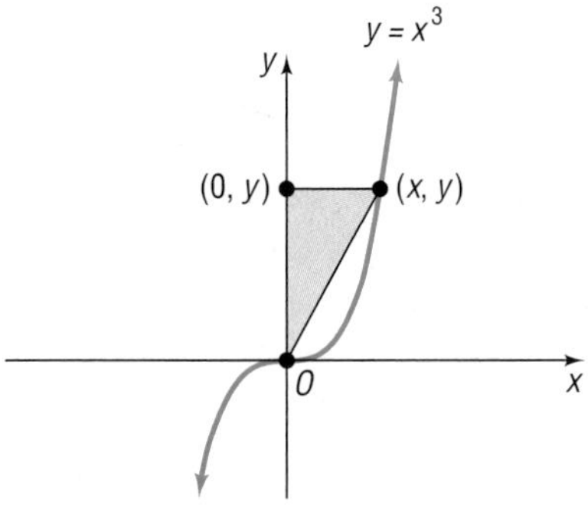

14. A right triangle has one vertex on the graph of $y = 9 - x^2$, $x > 0$, at (x, y), another at the origin, and the third on the positive x-axis at $(x, 0)$. Express the area A of the triangle as a function of x.

15. A rectangle has one corner on the graph of $y = 16 - x^2$, another at the origin, a third on the positive y-axis, and the fourth on the positive x-axis (see the figure).

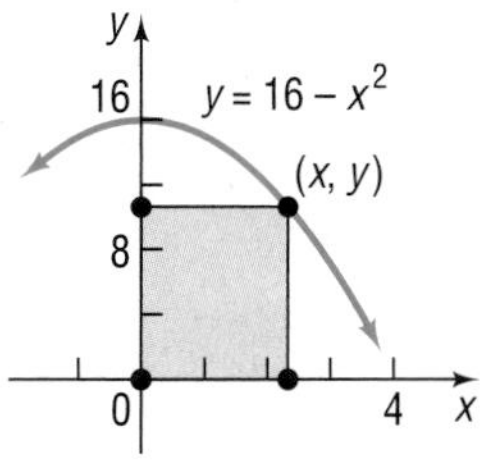

(a) Express the area A of the rectangle as a function of x.
(b) What is the domain of A?
(c) Graph $A = A(x)$. For what value of x is A largest?

16. A rectangle is inscribed in a semicircle of radius 2 (see the figure). Let $P = (x, y)$ be the point in quadrant I that is a vertex of the rectangle and is on the circle.

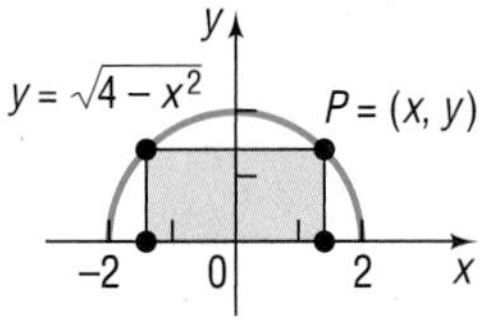

(a) Express the area A of the rectangle as a function of x.
(b) Express the perimeter p of the rectangle as a function of x.
(c) Graph $A = A(x)$. For what value of x is A largest?
(d) Graph $p = p(x)$. For what value of x is p largest?

17. A rectangle is inscribed in a circle of radius 2 (see the figure). Let $P = (x, y)$ be the point in quadrant I that is a vertex of the rectangle and is on the circle.

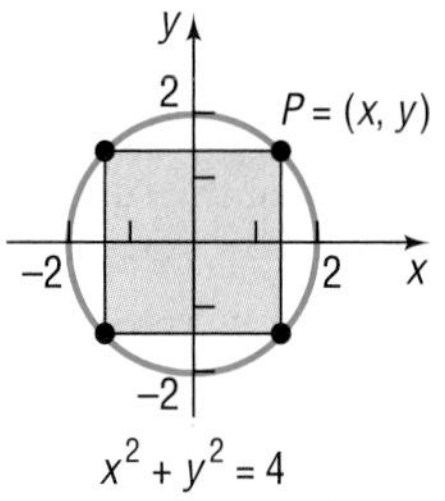

(a) Express the area A of the rectangle as a function of x.
(b) Express the perimeter p of the rectangle as a function of x.
(c) Graph $A = A(x)$. For what value of x is A largest?
(d) Graph $p = p(x)$. For what value of x is p largest?

18. A circle of radius r is inscribed in a square (see the figure).

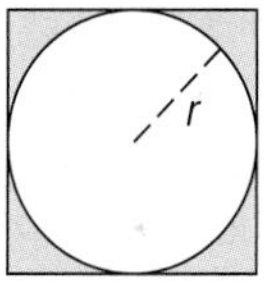

(a) Express the area A of the square as a function of the radius r of the circle.
(b) Express the perimeter p of the square as a function of r.

19. A wire 10 meters long is to be cut into two pieces. One piece will be shaped as a square, and the other piece will be shaped as a circle (see the figure).

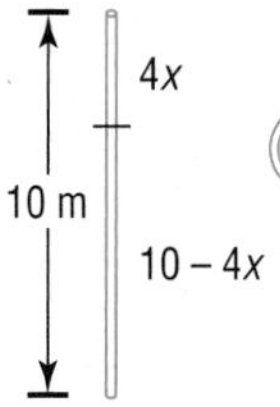

(a) Express the total area A enclosed by the pieces of wire as a function of the length x of a side of the square.
(b) What is the domain of A?
(c) Graph $A = A(x)$. For what value of x is A smallest?

20. A wire 10 meters long is to be cut into two pieces. One piece will be shaped as an equilateral triangle, and the other piece will be shaped as a circle.

(a) Express the total area A enclosed by the pieces of wire as a function of the length x of a side of the equilateral triangle.
(b) What is the domain of A?
(c) Graph $A = A(x)$. For what value of x is A smallest?

21. A wire of length x is bent into the shape of a circle.
(a) Express the circumference of the circle as a function of x.
(b) Express the area of the circle as a function of x.

22. A wire of length x is bent into the shape of a square.
(a) Express the perimeter of the square as a function of x.
(b) Express the area of the square as a function of x.

23. A semicircle of radius r is inscribed in a rectangle so that the diameter of the semicircle is the length of the rectangle (see the figure).

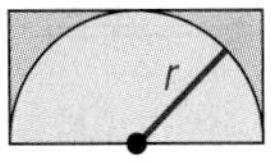

(a) Express the area A of the rectangle as a function of the radius r of the semicircle.
(b) Express the perimeter p of the rectangle as a function of r.

24. An equilateral triangle is inscribed in a circle of radius r. See the figure. Express the circumference C of the circle as a function of the length x of a side of the triangle.
[**Hint:** First show that $r^2 = x^2/3$.]

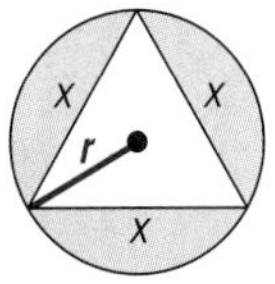

25. An equilateral triangle is inscribed in a circle of radius r. See the figure. Express the area A within the circle, but outside the triangle, as a function of the length x of a side of the triangle.

26. Cost of Transporting Goods A trucking company transports goods between Chicago and New York, a distance of 960 miles. The company's policy is to charge, for each pound, \$0.50 per mile for the first 100 miles, \$0.40 per mile for the next 300 miles, \$0.25 per mile for the next 400 miles, and no charge for the remaining 160 miles.
(a) Graph the relationship between the cost of transportation in dollars and mileage over the entire 960-mile route.
(b) Find the cost as a function of mileage for hauls between 100 and 400 miles from Chicago.
(c) Find the cost as a function of mileage for hauls between 400 and 800 miles from Chicago.

27. Car Rental Costs An economy car rented in Florida from National Car Rental® on a weekly basis costs \$95 per week. Extra days cost \$24 per day until the day rate exceeds the weekly rate, in which case the weekly rate applies. Find the cost C of renting an economy car as a piecewise-defined function of the number x of days used, where $7 \le x \le 14$. Graph this function.
[**Note:** Any part of a day counts as a full day.]

28. Rework Problem 27 for a luxury car, which costs \$219 on a weekly basis with extra days at \$45 per day.

29. Two cars leave an intersection at the same time. One is headed south at a constant speed of 30 miles per hour, and the other is headed west at a constant speed of 40 miles per hour (see the figure). Express the distance d between the cars as a function of the time t.
[**Hint:** At $t = 0$, the cars leave the intersection.]

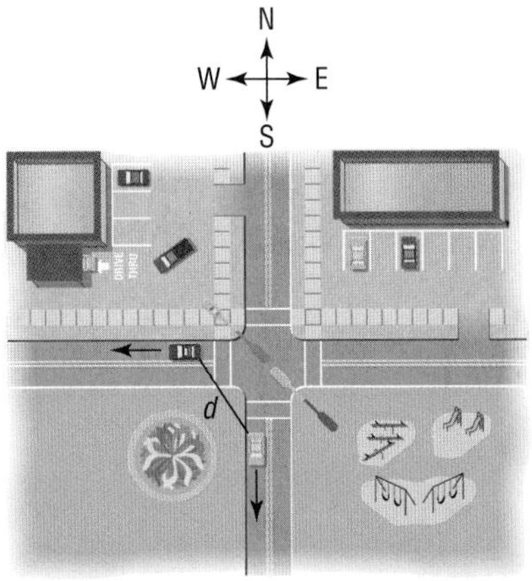

30. Two cars are approaching an intersection. One is 2 miles south of the intersection and is moving at a constant speed of 30 miles per hour. At the same time, the other car is 3 miles east of the intersection and is moving at a constant speed of 40 miles per hour.
(a) Express the distance d between the cars as a function of time t.
[**Hint:** At $t = 0$, the cars are 2 miles south and 3 miles east of the intersection, respectively.]
(b) Use a graphing utility to graph $d = d(t)$. For what value of t is d smallest?

31. Constructing an Open Box An open box with a square base is to be made from a square piece of cardboard 24 inches on a side by cutting out a square from each corner and turning up the sides (see the figure).

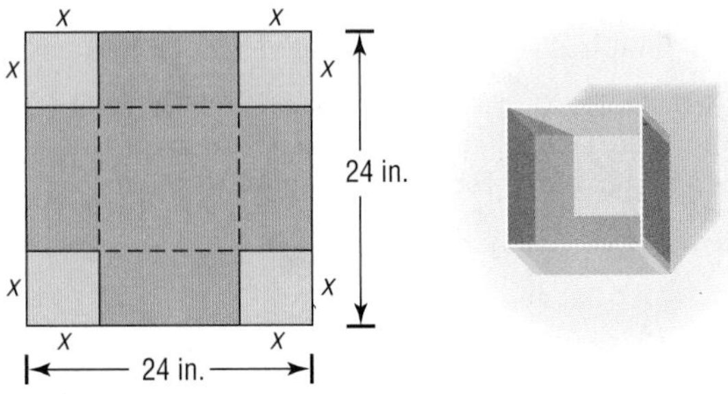

(a) Express the volume V of the box as a function of the length x of the side of the square cut from each corner.
(b) What is the volume if a 3-inch square is cut out?
(c) What is the volume if a 10-inch square is cut out?
(d) Graph $V = V(x)$. For what value of x is V largest?

32. Constructing an Open Box A open box with a square base is required to have a volume of 10 cubic feet.
(a) Express the amount A of material used to make such a box as a function of the length x of a side of the square base.
(b) How much material is required for a base 1 foot by 1 foot?
(c) How much material is required for a base 2 feet by 2 feet?
(d) Graph $A = A(x)$. For what value of x is A smallest?

33. Inscribing a Cylinder in a Sphere Inscribe a right circular cylinder of height h and radius r in a sphere of fixed radius R. See the illustration. Express the volume V of the cylinder as a function of h.
[**Hint:** $V = \pi r^2 h$. Note also the right triangle.]

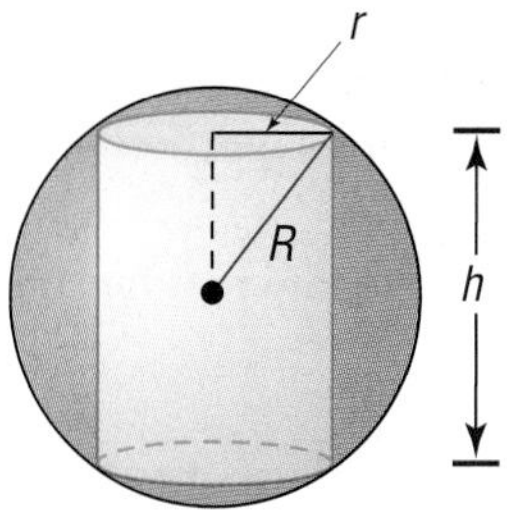

34. Inscribing a Cylinder in a Cone Inscribe a right circular cylinder of height h and radius r in a cone of fixed radius R and fixed height H. See the illustration. Express the volume V of the cylinder as a function of r.
[**Hint:** $V = \pi r^2 h$. Note also the similar triangles.]

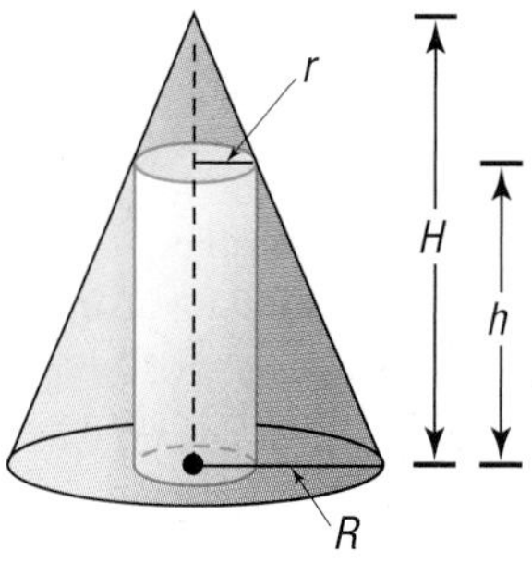

35. Installing Cable TV MetroMedia Cable is asked to provide service to a customer whose house is located 2 miles from the road along which the cable is buried. The nearest connection box for the cable is located 5 miles down the road (see the figure).

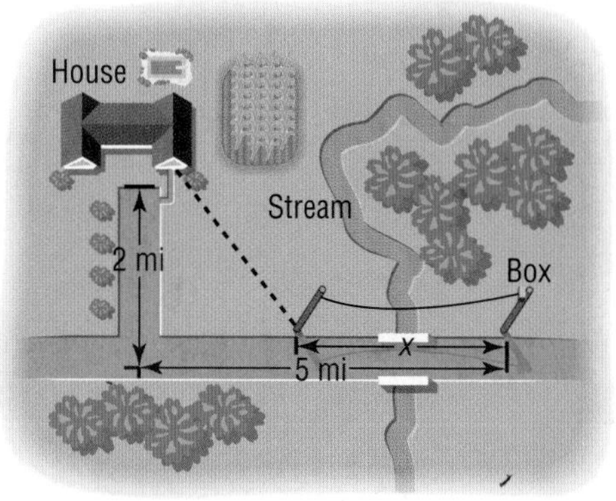

(a) If the installation cost is \$10 per mile along the road and \$14 per mile off the road, express the total cost C of installation as a function of the distance x (in miles) from the connection box to the point where the cable installation turns off the road. Give the domain.
(b) Compute the cost if $x = 1$ mile.
(c) Compute the cost if $x = 3$ miles.
(d) Graph the function $C = C(x)$.
(e) Create a TABLE with TblStart $= 0$ and ΔTbl $= 0.5$. To the nearest 1/2 mile, which value of x results in the least cost?
(f) Using MINIMUM, what value of x results in the least cost?

36. Time Required to Go from an Island to a Town An island is 3 miles from the nearest point P on a straight shoreline. A town is located 20 miles down the shore from P.
(a) If a person has a boat that averages 12 miles per hour and the same person can run 5 miles per hour, express the time T that it takes to go from the island to town as a function of x, where x is the distance from P to where the person lands the boat. Give the domain.
(b) How long will it take to travel from the island to town if you land the boat 8 miles from P?
(c) How long will it take if you land the boat 12 miles from P?
(d) Graph the function $T = T(x)$.
(e) Create a TABLE with TblStart $= 0$ and ΔTbl $= 1$. To the nearest mile, determine which value of x results in the least time.
(f) Using MINIMUM, what value of x results in the least time?
(g) The least time occurs by heading directly to town from the island. Explain why this solution makes sense.

37. Federal Income Tax Two 1999 Tax Rate Schedules are given in the accompanying table. If x equals the amount on Form 1040, line 37, and y equals the tax due, construct a function f for each schedule.

1999 TAX RATE SCHEDULES

SCHEDULE X—IF YOUR FILING STATUS IS SINGLE				SCHEDULE Y-1—USE IF YOUR FILING STATUS IS MARRIED FILING JOINTLY OR QUALIFYING WIDOW(ER)			
If the amount on Form 1040, line 37, is: Over—	But not over—	Enter on Form 1040, line 38	of the amount over—	If the amount on Form 1040, line 37, is: Over—	But not over—	Enter on Form 1040, line 38	of the amount over—
$0	$25,750	______15%	$0	$0	$43,050	______15%	$0
25,750	62,450	$3,862.50 + 28%	25,750	43,050	104,050	$6,457.50 + 28%	43,050
62,450	130,250	14,138.50 + 31%	62,450	104,050	158,550	23,537.50 + 31%	104,050
130,250	283,150	35,156.50 + 36%	130,250	158,550	283,150	40,432.50 + 36%	158,550
283,150	______	90,200.50 + 39.6%	283,150	283,150	______	85,288.50 + 39.6%	283,150

CHAPTER REVIEW

Library of Functions

Linear function (p. 128)
$f(x) = mx + b$ Graph is a line with slope m and y-intercept b.

Constant function (p. 128)
$f(x) = b$ Graph is a horizontal line with y-intercept b (see Figure 20).

Identity function (p. 128)
$f(x) = x$ Graph is a line with slope 1 and y-intercept 0 (see Figure 21).

Square function (p. 129)
$f(x) = x^2$ Graph is a parabola with intercept at $(0, 0)$ (see Figure 22).

Cube function (p. 129)
$f(x) = x^3$ See Figure 23.

Square root function (p. 130)
$f(x) = \sqrt{x}$ See Figure 24.

Reciprocal function (p. 130)
$f(x) = 1/x$ See Figure 25.

Absolute value function (p. 130)
$f(x) = |x|$ See Figure 26.

Greatest integer function (p. 131)
$f(x) = \text{int}(x)$ See Figure 27.

Things To Know

Function (p. 99)

A relation between two sets of real numbers so that each number x in the first set, the domain, has corresponding to it exactly one number y in the second set. The range is the set of y values of the funtion for the x values in the domain.

x is the independent variable; y is the dependent variable.

A function f may be defined implicitly by an equation involving x and y or explicitly by writing $y = f(x)$.

A function can also be characterized as a set of ordered pairs (x, y) or $(x, f(x))$ in which no two distinct pairs have the same first element.

Function notation (p. 101)

$y = f(x)$

f is a symbol for the function.

x is the argument, or independent variable.

y is the dependent variable.

$f(x)$ is the value of the function at x, or the image of x.

Domain (p. 104) If unspecified, the domain of a function f is the largest set of real numbers for which $f(x)$ is a real number.

Vertical-line test (p. 105) A set of points in the plane is the graph of a function if and only if every vertical line intersects the graph in at most one point.

Average rate of change of a function (p. 116) The average rate of change of f from c to x is

$$\frac{\Delta y}{\Delta x} = \frac{f(x) - f(c)}{x - c}, \qquad x \neq c$$

Increasing function (p. 119) A function f is increasing on an open interval I if, for any choice of x_1 and x_2 in I, with $x_1 < x_2$, we have $f(x_1) < f(x_2)$.

Decreasing function (p. 118) A function f is decreasing on an open interval I if, for any choice of x_1 and x_2 in I, with $x_1 < x_2$, we have $f(x_1) > f(x_2)$.

Constant function (p. 119) A function f is constant on an interval I if, for all choices of x in I, the values of $f(x)$ are equal.

Local maximum (p. 120) A function f has a local maximum at c if there is an open interval I containing c so that, for all $x \neq c$ in I, $f(x) < f(c)$.

Local minimum (p. 120) A function f has a local minimum at c if there is an open interval I containing c so that, for all $x \neq c$ in I, $f(x) > f(c)$.

Even function *f* (p. 122) $f(-x) = f(x)$ for every x in the domain ($-x$ must also be in the domain).

Odd function *f* (p. 122) $f(-x) = -f(x)$ for every x in the domain ($-x$ must also be in the domain).

How To

Determine whether a relation represents a function (p. 98)

Find the value of a function (p. 102)

Find the domain of a function (p. 104)

Identify the graph of a function (p. 105)

Obtain information from or about the graph of a function (p. 106)

Find the average rate of change of a function (p. 116)

Use a graphing utility to locate local maxima and minima (p. 120)

Use a graphing utility to determine where a function is increasing and decreasing (p. 122)

Determine even or odd functions from a graph (p. 122)

Identify even or odd functions from the equation (p. 123)

Graph functions listed in the library of functions (p. 128)

Graph piecewise-defined functions (p. 132)

Graph functions using horizontal and vertical shifts (p. 137)

Graph functions using compressions and stretches (p. 139)

Graph functions using reflections about the x-axis or y-axis (p. 141)

Form the sum, difference, product and quotient of two functions (p. 148)

Form the composite function and find its domain (p. 150)

Construct and analyze functions (p. 156)

Fill-in-the-Blank Items

1. If f is a function defined by the equation $y = f(x)$, then x is called the ______________ variable and y is the ______________ variable.
2. A set of points in the xy-plane is the graph of a function if and only if every ______________ line intersects the graph in at most one point.
3. The average rate of change of a function equals the ______________ of the secant line containing two points on its graph.
4. A(n) ______________ function f is one for which $f(-x) = f(x)$ for every x in the domain of f; a(n) ______________ function f is one for which $f(-x) = -f(x)$ for every x in the domain of f.
5. Suppose that the graph of a function f is known. Then the graph of $y = f(x - 2)$ may be obtained by a(n) ______________ shift of the graph of f to the ______________ a distance of 2 units.
6. If $f(x) = x + 1$ and $g(x) = x^3$, then ______________ $= (x + 1)^3$.

True/False Items

T F **1.** Every relation is a function.

T F **2.** Vertical lines intersect the graph of a function in no more than one point.

T F **3.** The y-intercept of the graph of the function $y = f(x)$, whose domain is all real numbers, is $f(0)$.

T F **4.** A function f is decreasing on an open interval I if, for any choice of x_1 and x_2 in I, with $x_1 < x_2$, we have $f(x_1) < f(x_2)$.

T F **5.** Even functions have graphs that are symmetric with respect to the origin.

T F **6.** The graph of $y = f(-x)$ is the reflection about the y-axis of the graph of $y = f(x)$.

T F **7.** $f(g(x)) = f(x) \cdot g(x)$.

T F **8.** The domain of the composite function $(f \circ g)(x)$ is the same as that of $g(x)$.

Review Exercises

Blue problem numbers indicate the authors' suggestions for use in a Practice Test.

1. Given that f is a linear function, $f(4) = -5$ and $f(0) = 3$, write the equation that defines f.
2. Given that g is a linear function with slope $= -4$ and $g(-2) = 2$, write the equation that defines g.
3. A function f is defined by

$$f(x) = \frac{Ax + 5}{6x - 2}$$

If $f(1) = 4$, find A.

4. A function g is defined by

$$g(x) = \frac{A}{x} + \frac{8}{x^2}$$

If $g(-1) = 0$, find A.

5. Tell which of the following graphs are graphs of functions.

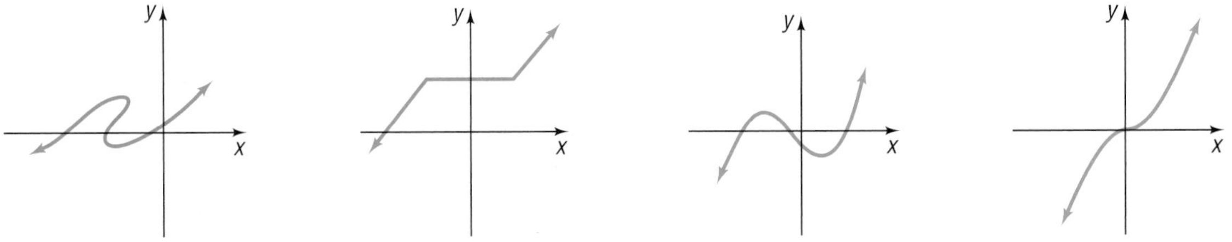

(a) (b) (c) (d)

6. Use the graph of the function f shown to find:
   (a) The domain and range of f
   (b) $f(-1)$
   (c) The intercepts of f
   (d) The intervals on which f is increasing, decreasing, or constant
   (e) Whether the function is even, odd, or neither

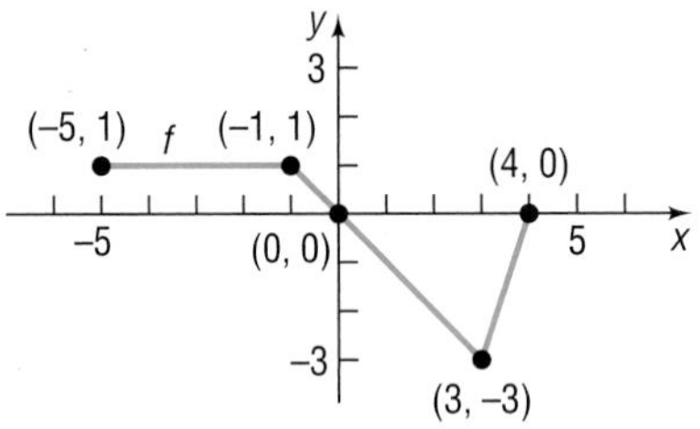

In Problems 7–12, find the following for each function:

(a) $f(-x)$ *(b)* $-f(x)$ *(c)* $f(x + 2)$ *(d)* $f(x - 2)$ *(e)* $f(2x)$

7. $f(x) = \dfrac{3x}{x^2 - 4}$
8. $f(x) = \dfrac{x^2}{x + 2}$
9. $f(x) = \sqrt{x^2 - 4}$
10. $f(x) = |x^2 - 4|$
11. $f(x) = \dfrac{x^2 - 4}{x^2}$
12. $f(x) = \dfrac{x^3}{x^2 - 4}$

In Problems 13–20, find the domain of each function.

13. $f(x) = \dfrac{x}{x^2 - 9}$
14. $f(x) = \dfrac{3x^2}{x - 2}$
15. $f(x) = \sqrt{2 - x}$
16. $f(x) = \sqrt{x + 2}$
17. $h(x) = \dfrac{\sqrt{x}}{|x|}$
18. $g(x) = \dfrac{|x|}{x}$
19. $f(x) = \dfrac{x}{x^2 + 2x - 3}$
20. $F(x) = \dfrac{1}{x^2 - 3x - 4}$

In Problems 21–24:

(a) *Find the domain of each function.*
(b) *Locate any intercepts.*
(c) *Graph each function by hand.*
(d) *Based on the graph, find the range.*
(e) *Verify your results using a graphing utility.*

21. $f(x) = \begin{cases} 3x & -2 < x \le 1 \\ x + 1 & x > 1 \end{cases}$

22. $f(x) = \begin{cases} x - 1 & -3 < x < 0 \\ 3x - 1 & x \ge 0 \end{cases}$

23. $f(x) = \begin{cases} x & -4 \le x < 0 \\ 1 & x = 0 \\ 3x & x > 0 \end{cases}$

24. $f(x) = \begin{cases} x^2 & -2 \le x \le 2 \\ 2x - 1 & x > 2 \end{cases}$

In Problems 25–28, find the average rate of change from 2 to x for each function f. Be sure to simplify.

25. $f(x) = 2 - 5x$
26. $f(x) = 2x^2 + 7$
27. $f(x) = 3x - 4x^2$
28. $f(x) = x^2 - 3x + 2$

In Problems 29–36, determine (algebraically) whether the given function is even, odd, or neither.

29. $f(x) = x^3 - 4x$
30. $g(x) = \dfrac{4 + x^2}{1 + x^4}$
31. $h(x) = \dfrac{1}{x^4} + \dfrac{1}{x^2} + 1$
32. $F(x) = \sqrt{1 - x^3}$
33. $G(x) = 1 - x + x^3$
34. $H(x) = 1 + x + x^2$
35. $f(x) = \dfrac{x}{1 + x^2}$
36. $g(x) = \dfrac{1 + x^2}{x^3}$

In Problems 37–48, graph each function using the techniques of shifting, compressing or stretching, and reflections. Identify any intercepts on the graph. State the domain and, based on the graph, find the range.

37. $F(x) = |x| - 4$
38. $f(x) = |x| + 4$
39. $g(x) = -2|x|$
40. $g(x) = \frac{1}{2}|x|$
41. $h(x) = \sqrt{x - 1}$
42. $h(x) = \sqrt{x} - 1$
43. $f(x) = \sqrt{1 - x}$
44. $f(x) = -\sqrt{x + 3}$
45. $h(x) = (x - 1)^2 + 2$
46. $h(x) = (x + 2)^2 - 3$
47. $g(x) = 3(x - 1)^3 + 1$
48. $g(x) = -2(x + 2)^3 - 8$

In Problems 49–52, use a graphing utility to graph each function over the indicated interval. Approximate any local maxima and local minima. Determine where the function is increasing and where it is decreasing.

49. $f(x) = 2x^3 - 5x + 1 \quad (-3, 3)$
50. $f(x) = -x^3 + 3x - 5 \quad (-3, 3)$
51. $f(x) = 2x^4 - 5x^3 + 2x + 1 \quad (-2, 3)$
52. $f(x) = -x^4 + 3x^3 - 4x + 3 \quad (-2, 3)$

In Problems 53–58, for the given functions f and g find:

(a) $(f \circ g)(2)$ (b) $(g \circ f)(-2)$ (c) $(f \circ f)(4)$ (d) $(g \circ g)(-1)$

Verify your results using a graphing utility.

53. $f(x) = 3x - 5; \quad g(x) = 1 - 2x^2$
54. $f(x) = 4 - x; \quad g(x) = 1 + x^2$
55. $f(x) = \sqrt{x + 2}; \quad g(x) = 2x^2 + 1$
56. $f(x) = 1 - 3x^2; \quad g(x) = \sqrt{4 - x}$
57. $f(x) = \dfrac{1}{x^2 + 4}; \quad g(x) = 3x - 2$
58. $f(x) = \dfrac{2}{1 + 2x^2}; \quad g(x) = 3x$

In Problems 59–64, find $f \circ g$, $g \circ f$, $f \circ f$, and $g \circ g$ for each pair of functions. State the domain of each.

59. $f(x) = 2 - x; \quad g(x) = 3x + 1$
60. $f(x) = 2x - 1; \quad g(x) = 2x + 1$
61. $f(x) = 3x^2 + x + 1; \quad g(x) = |3x|$
62. $f(x) = \sqrt{3x}; \quad g(x) = 1 + x + x^2$
63. $f(x) = \dfrac{x + 1}{x - 1}; \quad g(x) = \dfrac{1}{x}$
64. $f(x) = \sqrt{x - 3}; \quad g(x) = \dfrac{3}{x}$

65. For the following graph of the function f draw the graph of:
(a) $y = f(-x)$ (b) $y = -f(x)$ (c) $y = f(x + 2)$
(d) $y = f(x) + 2$ (e) $y = 2f(x)$ (f) $y = f(3x)$

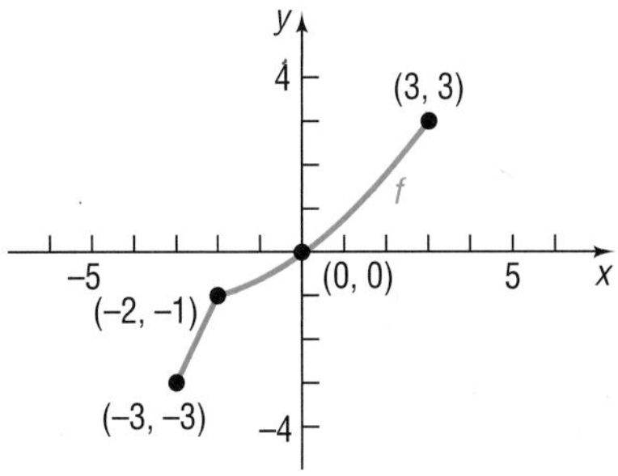

66. Repeat Problem 65 for the following graph of the function g.

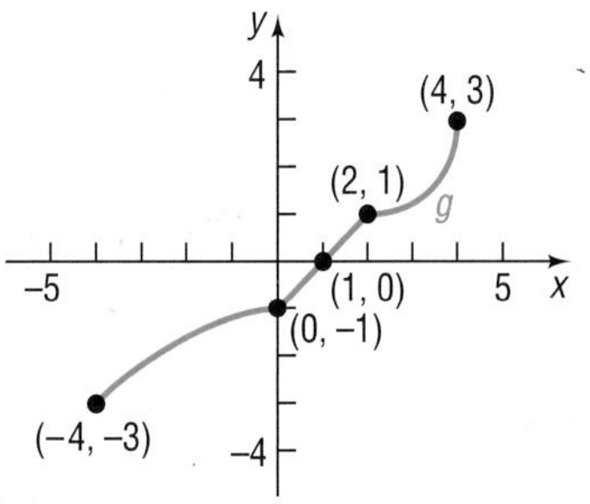

67. Find the point on the line $y = x$ that is closest to the point $(3, 1)$.
[**Hint:** Find the minimum value of the function $f(x) = d^2$, where d is the distance from $(3, 1)$ to a point on the line.]

68. Find the point on the line $y = x + 1$ that is closest to the point $(4, 1)$.

69. A rectangle has one vertex on the graph of $y = 10 - x^2$, $x > 0$, another at the origin, one on the positive x-axis, and one on the positive y-axis. Find the largest area A that can be enclosed by the rectangle.

70. Constructing a Closed Box A closed box with a square base is required to have a volume of 10 cubic feet.
(a) Express the amount A of material used to make such a box as a function of the length x of a side of the square base.
(b) How much material is required for a base 1 foot by 1 foot?
(c) How much material is required for a base 2 feet by 2 feet?
(d) Graph $A = A(x)$. For what value of x is A smallest?

71. Spheres The volume V of a sphere of radius r is $V = \frac{4}{3}\pi r^3$; the surface area S of this sphere is $S = 4\pi r^2$. Express the volume V as a function of the surface area S. If the surface area doubles, how does the volume change?

72. Productivity versus Earnings The following data represent the average hourly earnings and productivity (output per hour) of production workers for the years 1986–1995. Let productivity x be the independent variable and average hourly earnings y be the dependent variable.

Productivity	Average Hourly Earnings
94.2	8.76
94.1	8.98
94.6	9.28
95.3	9.66
96.1	10.01
96.7	10.32
100	10.57
100.2	10.83
100.7	11.12
100.8	11.44

Source: Bureau of Labor Statistics.

(a) Draw a scatter diagram of the data.
(b) Draw a line through the points (94.2, 8.76) and (96.7, 10.32) on the scatter diagram found in part (a).
(c) Find the average rate of change of hourly earnings for productivity from 94.2 to 96.7.
(d) Interpret the average rate of change found in part (c).
(e) Draw a line through the points (96.7, 10.32) and (100.8, 11.44) on the scatter diagram found in part (a).
(f) Find the average rate of change of hourly earnings for productivity from 96.7 to 100.8.
(g) Interpret the average rate of change found in part (f).
(h) What is happening to the average rate of change of hourly earnings as productivity increases?

PROJECT AT MOTOROLA

During the past decade the availability and usage of wireless Internet services has increased. The industry has developed a number of pricing proposals for such services. Marketing data have indicated that subscribers of wireless Internet services have tended to desire flat fee rate structures as compared with rates based totally on usage. The Computer Resource Department of Indigo Media (hypothetical) has entered into a contractual agreement for wireless Internet services. As a part of the contractual agreement, employees are able to sign up for their own wireless services. Three pricing options are available:

Silver-plan: \$20/month for up to 200 K-bytes of service plus \$0.16 for each additional K-byte of service

Gold-plan: \$50/month for up to 1000 K-bytes of service plus \$0.08 for each additional K-byte of service

Platinum-plan: \$100/month for up to 3000 K-bytes of service plus \$0.04 for each additional K-byte of service

You have been requested to write a report that answers the following questions in order to aid employees in choosing the appropriate pricing plan.

(a) If C is the monthly charge for x K-bytes of service, express C as a function of x for each of the three plans.
(b) Graph each of the three functions found in (a).
(c) For how many K-bytes of service is the Silver-plan the best pricing option? When is the Gold-plan best? When is the Platinum-plan best? Explain your reasoning.
(d) Write a report that summarizes your findings.

CHAPTER

3 Polynomial and Rational Functions

Outline

FIELD TRIP TO MOTOROLA

First-, second-, and third-degree polynomial functions are widely used at Motorola to model manufacturing processes and product design features, based on the measurements made in factories and testing labs. Equipment cycle time characterization is one example. By performing a set of designed experiments, engineers can develop empirical models to predict the assembly time for a given product on a given type of machine. Trade-off analyses, e.g., to optimize production costs, are frequently performed using ratios of functions that express the costs and benefits of a particular course of action.

INTERVIEW AT MOTOROLA

Tom Tirpak works in the Virtual Design and Manufacturing Group at the Motorola Advanced Technology Center in Schaumburg, Illinois. Since joining Motorola in 1991, Tom and his team have developed methods and software for improving the cycle time, quality, and cost of electronics manufacturing and product design operations. In cooperation with Motorola University, he has taught classes on "Surface Mount Technology (SMT) Manufacturing Optimization" and "Factory Physics." Tom is a member of several professional societies, including the Institute for Operations Research and Management Science (INFORMS) and Tau Beta Pi. Tom received his bachelor's and master's degrees in General Engineering, and doctorate in Electrical and Computer Engineering from the University of Illinois at Urbana-Champaign, specializing in Robotics and Control Theory. Math concepts that he uses most frequently on the job include: fitting curves for measured data, identifying input and output variables for a model, constructing functions to express the benefit of certain decisions, determining the statistical validity of a proposed model, and solving systems of inequalities for an optimum solution.

PREPARING FOR THIS SECTION

Before getting started, review the following concepts:

- ✓ Intercepts (pp. 16–18)
- ✓ Solving Quadratic Equations (pp. 28–32)
- ✓ Completing the Square (Appendix, Section 10, pp. 998–999)

3.1 QUADRATIC FUNCTIONS; CURVE FITTING

OBJECTIVES

1. Graph a Quadratic Function by Hand and by Using a Graphing Utility
2. Identify the Vertex and Axis of Symmetry of a Quadratic Function
3. Determine the Maximum or Minimum Value of a Quadratic Function
4. Use the Maximum or Minimum Value of a Quadratic Function to Solve Applied Problems
5. Find the Quadratic Function of Best Fit to Data

QUADRATIC FUNCTIONS

A *quadratic function* is a function that is defined by a second-degree polynomial in one variable.

A **quadratic function** is a function of the form

$$f(x) = ax^2 + bx + c \quad \textbf{(1)}$$

where a, b, and c are real numbers and $a \neq 0$. The domain of a quadratic function consists of all real numbers.

Many applications require a knowledge of quadratic functions. For example, suppose that Texas Instruments collects the data shown in Table 1,

TABLE 1

Price per Calculator, p (Dollars)	Number of Calculators, x
60	11,100
65	10,115
70	9,652
75	8,731
80	8,087
85	7,205
90	6,439

which relate the number of calculators sold at the price p per calculator. Since the price of a product determines the quantity that will be purchased, we treat price as the independent variable.

Using the LINear REGression option of a graphing utility, we find that a linear relationship between the number of calculators x and the price p per calculator is given by the equation

$$x = 20{,}208.43 - 152.63p$$

Then the revenue R derived from selling x calculators at the price p per calculator is

$$\begin{aligned} R &= xp \\ R(p) &= (20{,}208.43 - 152.63p)p \\ &= -152.63p^2 + 20{,}208.43p \end{aligned}$$

Figure 1

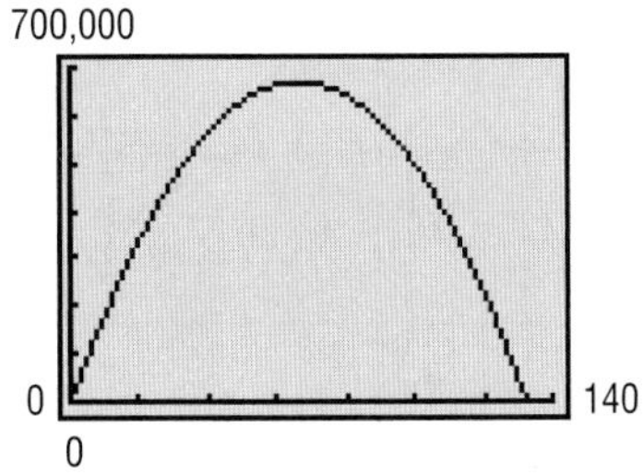

So the revenue R is a quadratic function of the price p. Figure 1 illustrates the graph of this revenue function. Since both x and p must be nonnegative, we only show the graph of R in quadrant I.

A second situation in which a quadratic function appears involves the motion of a projectile. Based on Newton's second law of motion (force equals mass times acceleration, $F = ma$), it can be shown that, ignoring air resistance, the path of a projectile propelled upward at an inclination to the horizontal is the graph of a quadratic function. See Figure 2 for an illustration.

Figure 2
Path of a cannonball

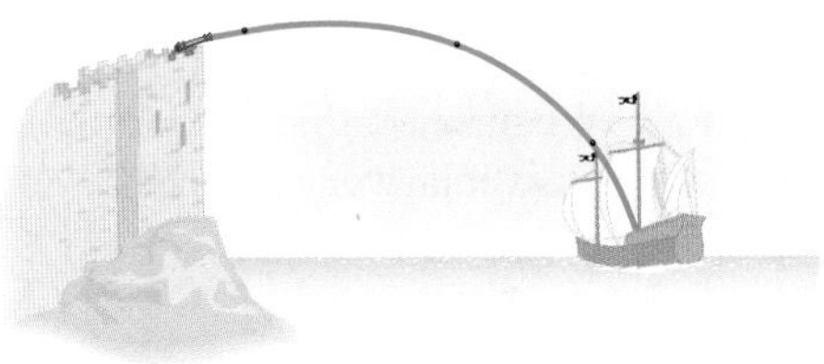

GRAPHING QUADRATIC FUNCTIONS

1 We know how to graph a quadratic function of the form $f(x) = x^2$. Figure 3 shows the graph of three functions of the form $f(x) = ax^2, a > 0$, for $a = 1$, $a = \frac{1}{2}$, and $a = 3$. Notice that the larger the value of a, the "narrower" the graph, and the smaller the value of a, the "wider" the graph.

Figure 3

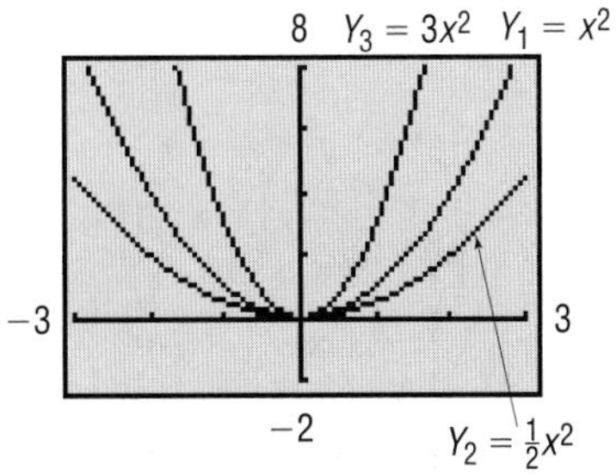

Figure 4

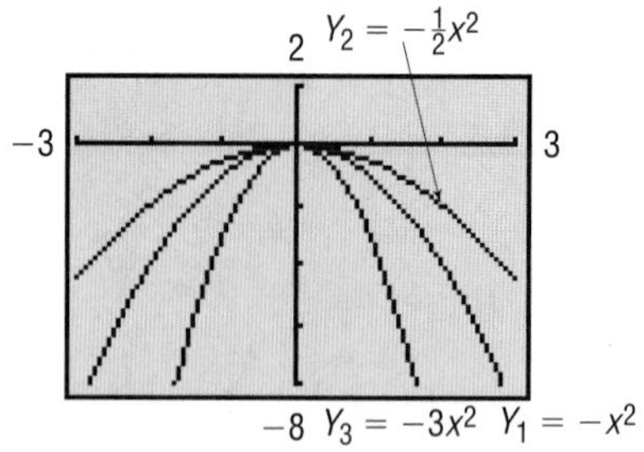

Figure 4 shows the graphs of $f(x) = ax^2$ for $a < 0$. Notice that these graphs are reflections about the x-axis of the graphs in Figure 3. Based on the results of these two figures, we can draw some general conclusions about the graph of $f(x) = ax^2$. First, as $|a|$ increases, the graph becomes *narrower* (a vertical stretch), and as $|a|$ gets closer to zero, the graph gets *wider* (a vertical compression). Second, if a is positive, then the graph opens *up*, and if a is negative, then the graph opens *down*.

The graphs in Figures 3 and 4 are typical of the graphs of all quadratic functions, which we call **parabolas**.* Refer to Figure 5, where two parabolas are pictured. The one on the left **opens up** and has a lowest point; the one on the right **opens down** and has a highest point. The lowest or highest point of a parabola is called the **vertex**. The vertical line passing through the vertex in each parabola in Figure 5 is called the **axis of symmetry** (usually abbreviated to **axis**) of the parabola. Because the parabola is symmetric about its axis, the axis of symmetry of a parabola can be used to find additional points when graphing the parabola by hand.

Figure 5
Graphs of a quadratic function, $f(x) = ax^2 + bx + c,\quad a \neq 0$

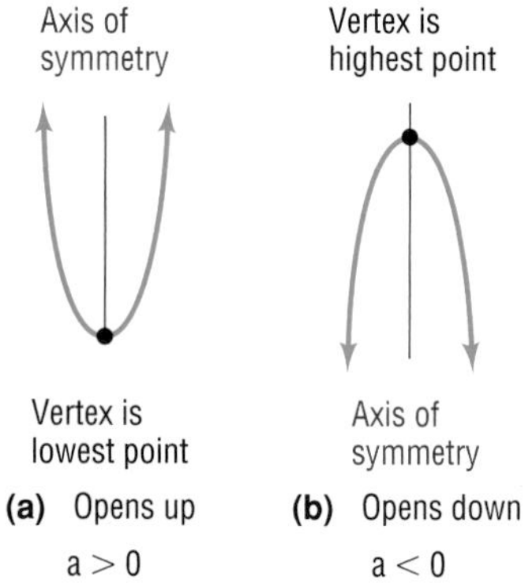

The parabolas shown in Figure 5 are the graphs of a quadratic function $f(x) = ax^2 + bx + c, a \neq 0$. Notice that the coordinate axes are not included in the figure. Depending on the values of a, b, and c, the axes could be placed anywhere. The important fact is that, except possibly for compression or stretching, the shape of the graph of a quadratic function will look like one of the parabolas in Figure 5.

EXAMPLE 1 Graphing a Quadratic Function

Graph the function $f(x) = 2x^2 + 8x + 5$. Find the vertex and axis of symmetry.

Graphing Solution

Before graphing, notice that the leading coefficient, 2, is positive, and therefore the graph will open up and the vertex will be the lowest point. Now graph $Y_1 = f(x) = 2x^2 + 8x + 5$. See Figure 6. We observe that the graph does in fact open up. To estimate the vertex of the parabola, we use the MINIMUM command. The vertex is the point $(-2, -3)$, and the axis of symmetry is the line $x = -2$.

Figure 6

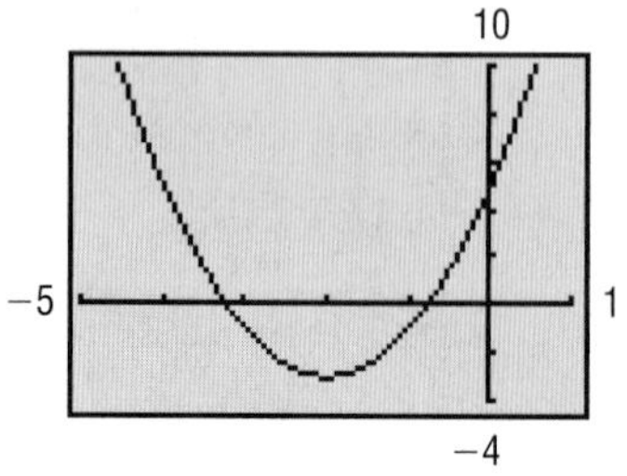

Algebraic Solution

We begin by completing the square on the right side.

$$\begin{aligned} f(x) &= 2x^2 + 8x + 5 \\ &= 2(x^2 + 4x) + 5 && \text{Factor out the 2 from } 2x^2 + 8x. \\ &= 2(x^2 + 4x + 4) + 5 - 8 && \text{Complete the square of } 2(x^2 + 4x). \\ &= 2(x + 2)^2 - 3 && \text{Notice that the factor of 2 requires that 8 be added and subtracted.} \end{aligned} \quad (2)$$

*We shall study parabolas using a geometric definition later in this book.

The graph of f can be obtained in three stages, as shown in Figure 7. Now compare this graph to the graph in Figure 5(a). The graph of $f(x) = 2x^2 + 8x + 5$ is a parabola that opens up and has its vertex (lowest point) at $(-2, -3)$. Its axis of symmetry is the line $x = -2$.

Figure 7

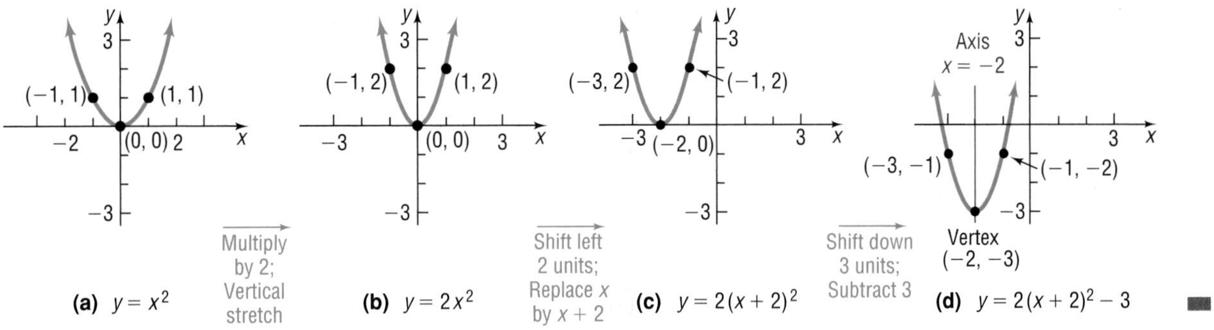

(a) $y = x^2$ (b) $y = 2x^2$ (c) $y = 2(x+2)^2$ (d) $y = 2(x+2)^2 - 3$

NOW WORK PROBLEM 21.

2 The method used in the Algebraic Solution to Example 1 can be used to graph any quadratic function $f(x) = ax^2 + bx + c$, $a \neq 0$, as follows:

$$f(x) = ax^2 + bx + c$$

$$= a\left(x^2 + \frac{b}{a}x\right) + c \qquad \text{Factor out } a \text{ from } ax^2 + bx.$$

$$= a\left(x^2 + \frac{b}{a}x + \frac{b^2}{4a^2}\right) + c - a\left(\frac{b^2}{4a^2}\right) \qquad \text{Complete the square by adding and subtracting } a(b^2/4a^2). \text{ Look closely at this step!}$$

$$= a\left(x + \frac{b}{2a}\right)^2 + c - \frac{b^2}{4a}$$

$$= a\left(x + \frac{b}{2a}\right)^2 + \frac{4ac - b^2}{4a} \qquad c - \frac{b^2}{4a} = c \cdot \frac{4a}{4a} - \frac{b^2}{4a} = \frac{4ac - b^2}{4a}$$

Based on these results, we conclude the following:

If $h = -b/2a$ and $k = (4ac - b^2)/4a$, then

$$f(x) = ax^2 + bx + c = a(x - h)^2 + k \qquad \textbf{(3)}$$

The graph of f is the parabola $y = ax^2$ shifted horizontally h units and vertically k units. As a result, the vertex is at (h, k), and the graph opens up if $a > 0$ and down if $a < 0$. The axis of symmetry is the vertical line $x = h$.

For example, compare equation (3) with equation (2) of Example 1.

$$\begin{aligned} f(x) &= 2(x+2)^2 - 3 \\ &= a(x-h)^2 + k \end{aligned}$$

We conclude that $a = 2$, so the graph opens up. Also, we find that $h = -2$ and $k = -3$, so its vertex is at $(-2, -3)$.

It is not required to complete the square to obtain the vertex. In almost every case, it is easier to obtain the vertex of a quadratic function f by remembering that its x-coordinate is $h = -b/2a$. The y-coordinate can then be found by evaluating f at $-b/2a$.

We summarize these remarks as follows:

Characteristics of the Graph of a Quadratic Function

$$f(x) = ax^2 + bx + c$$

$$\text{Vertex} = \left(\frac{-b}{2a}, f\left(\frac{-b}{2a}\right)\right) \qquad \text{Axis of symmetry: the line } x = \frac{-b}{2a} \tag{4}$$

Parabola opens up if $a > 0$; the vertex is a minimum point.
Parabola opens down if $a < 0$; the vertex is a maximum point.

EXAMPLE 2 Locating the Vertex without Graphing

Without graphing, locate the vertex and axis of symmetry of the parabola defined by $f(x) = -3x^2 + 6x + 1$. Does it open up or down?

Solution For this quadratic function, $a = -3$, $b = 6$, and $c = 1$. The x-coordinate of the vertex is

$$h = \frac{-b}{2a} = \frac{-6}{-6} = 1$$

The y-coordinate of the vertex is therefore

$$k = f\left(\frac{-b}{2a}\right) = f(1) = -3 + 6 + 1 = 4$$

The vertex is located at the point $(1, 4)$. The axis of symmetry is the line $x = 1$. Finally, because $a = -3 < 0$, the parabola opens down. ■

The information we gathered in Example 2, together with the location of the intercepts, usually provides enough information to graph $f(x) = ax^2 + bx + c$, $a \neq 0$, by hand. The y-intercept is the value of f at $x = 0$, that is, $f(0) = c$. The x-intercepts, if there are any, are found by solving the equation

$$f(x) = ax^2 + bx + c = 0$$

This equation has two, one, or no real solutions, depending on whether the discriminant $b^2 - 4ac$ is positive, 0, or negative. Depending on the value of the discriminant, the graph of f has x-intercepts, as follows:

THE X-INTERCEPTS OF A QUADRATIC FUNCTION

1. If the discriminant $b^2 - 4ac > 0$, the graph of $f(x) = ax^2 + bx + c$ has two distinct x-intercepts and so will cross the x-axis in two places.
2. If the discriminant $b^2 - 4ac = 0$, the graph of $f(x) = ax^2 + bx + c$ has one x-intercept and touches the x-axis at its vertex.
3. If the discriminant $b^2 - 4ac < 0$, the graph of $f(x) = ax^2 + bx + c$ has no x-intercept and so will not cross or touch the x-axis.

Figure 8 illustrates these possibilities for parabolas that open up.

Figure 8
$f(x) = ax^2 + bx + c, \quad a > 0$

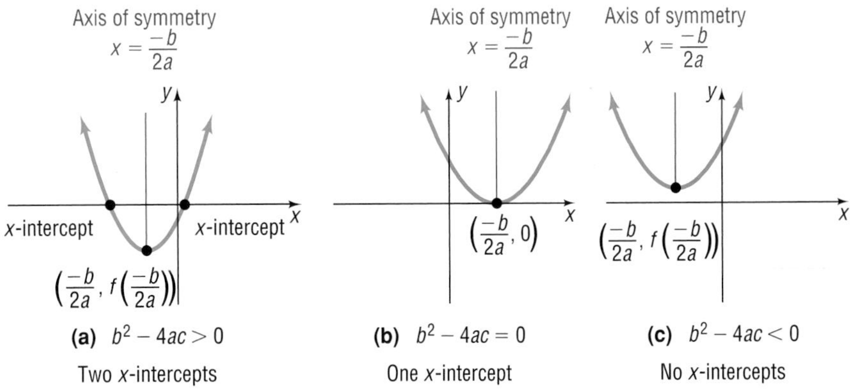

EXAMPLE 3 Graphing a Quadratic Function by Hand Using Its Vertex, Axis, and Intercepts

Use the information from Example 2 and the locations of the intercepts to graph $f(x) = -3x^2 + 6x + 1$.

Solution In Example 2, we found the vertex to be at $(1, 4)$ and the axis of symmetry to be $x = 1$. The y-intercept is found by letting $x = 0$. The y-intercept is $f(0) = 1$. The x-intercepts are found by solving the equation $f(x) = 0$. This results in the equation

$$-3x^2 + 6x + 1 = 0 \qquad a = -3, b = 6, c = 1$$

The discriminant $b^2 - 4ac = (6)^2 - 4(-3)(1) = 36 + 12 = 48 > 0$, so the equation has two real solutions and the graph has two x-intercepts. Using the quadratic formula, we find that

$$x = \frac{-b + \sqrt{b^2 - 4ac}}{2a} = \frac{-6 + \sqrt{48}}{-6} = \frac{-6 + 4\sqrt{3}}{-6} \approx -0.15$$

and

$$x = \frac{-b - \sqrt{b^2 - 4ac}}{2a} = \frac{-6 - \sqrt{48}}{-6} = \frac{-6 - 4\sqrt{3}}{-6} \approx 2.15$$

The x-intercepts are approximately -0.15 and 2.15.

The graph is illustrated in Figure 9. Notice how we used the y-intercept and the axis of symmetry, $x = 1$, to obtain the additional point $(2, 1)$ on the graph.

Figure 9
$f(x) = -3x^2 + 6x + 1$

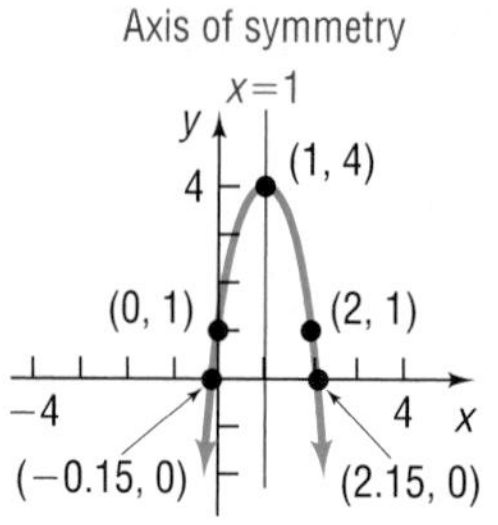

GRAPH THE FUNCTION IN EXAMPLE 3 USING THE METHOD PRESENTED IN THE ALGEBRAIC SOLUTION TO EXAMPLE 1.

Check: Use a graphing utility to verify Figure 9.

NOW WORK PROBLEM 29.

If the graph of a quadratic function has only one x-intercept or none, it is usually necessary to plot an additional point to obtain the graph by hand.

EXAMPLE 4 Graphing a Quadratic Function by Hand Using Its Vertex, Axis, and Intercepts

Graph $f(x) = x^2 - 6x + 9$ by determining whether the graph opens up or down. Find its vertex, axis of symmetry, y-intercept, and x-intercept, if any.

Figure 10

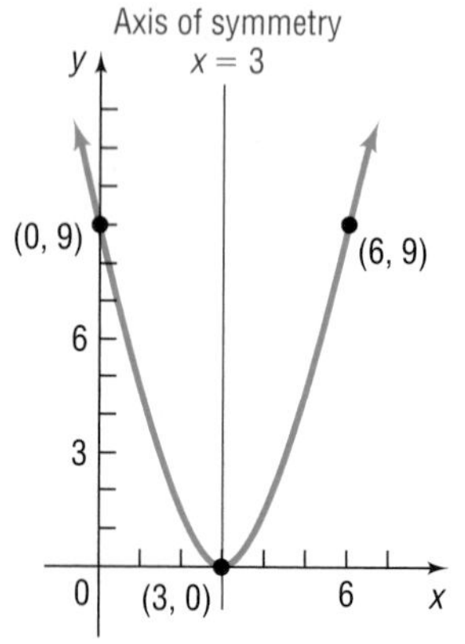

Solution For $f(x) = x^2 - 6x + 9$, we have $a = 1$, $b = -6$, and $c = 9$. Since $a = 1 > 0$, the parabola opens up. The x-coordinate of the vertex is

$$h = \frac{-b}{2a} = \frac{-(-6)}{2(1)} = 3$$

The y-coordinate of the vertex is

$$k = f(3) = (3)^2 - 6(3) + 9 = 0$$

So the vertex is at $(3, 0)$. The axis of symmetry is the line $x = 3$. The y-intercept is $f(0) = 9$. Since the vertex $(3, 0)$ lies on the x-axis, the graph touches the x-axis at the x-intercept. By using the axis of symmetry and the y-intercept at $(0, 9)$, we can locate the additional point $(6, 9)$ on the graph. See Figure 10.

GRAPH THE FUNCTION IN EXAMPLE 4 USING THE METHOD PRESENTED IN THE ALGEBRAIC SOLUTION TO EXAMPLE 1.

Check: Use a graphing utility to verify Figure 10.

NOW WORK PROBLEM 35.

EXAMPLE 5 **Graphing a Quadratic Function by Hand Using Its Vertex, Axis, and Intercepts**

Graph $f(x) = 2x^2 + x + 1$ by determining whether the graph opens up or down. Find its vertex, axis of symmetry, y-intercept, and x-intercepts, if any.

Solution For $f(x) = 2x^2 + x + 1$, we have $a = 2$, $b = 1$, and $c = 1$. Since $a = 2 > 0$, the parabola opens up. The x-coordinate of the vertex is

$$h = \frac{-b}{2a} = -\frac{1}{4}$$

The y-coordinate of the vertex is

$$k = f\left(-\frac{1}{4}\right) = 2\left(\frac{1}{16}\right) + \left(-\frac{1}{4}\right) + 1 = \frac{7}{8}$$

So the vertex is at $\left(-\frac{1}{4}, \frac{7}{8}\right)$. The axis of symmetry is the line $x = -\frac{1}{4}$. The y-intercept is $f(0) = 1$. The x-intercept(s), if any, obey the equation $2x^2 + x + 1 = 0$. Since the discriminant $b^2 - 4ac = (1)^2 - 4(2)(1) = -7 < 0$, this equation has no real solutions, and therefore the graph has no x-intercepts. We use the point $(0, 1)$ and the axis of symmetry $x = -\frac{1}{4}$ to locate the additional point $\left(-\frac{1}{2}, 1\right)$ on the graph. See Figure 11. ■

Figure 11

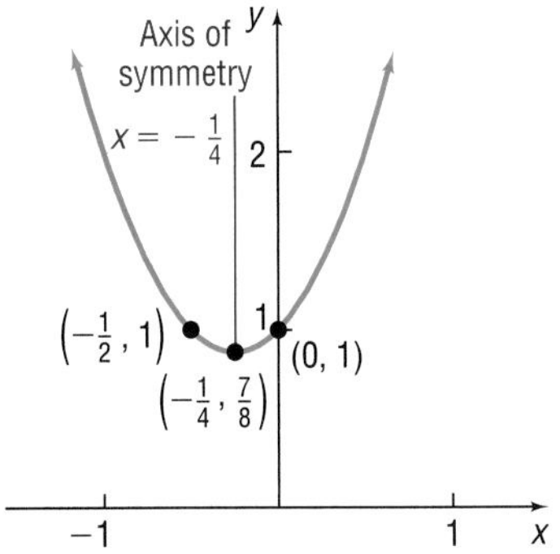

SUMMARY

Steps for Graphing a Quadratic Function $f(x) = ax^2 + bx + c, a \neq 0$ by Hand

Option 1

STEP 1: Complete the square in x to write the quadratic function in the form $f(x) = a(x - h)^2 + k$.

STEP 2: Graph the function in stages using transformations.

Option 2

STEP 1: Determine the vertex, $\left(\frac{-b}{2a}, f\left(\frac{-b}{2a}\right)\right)$.

STEP 2: Determine the axis of symmetry, $x = \frac{-b}{2a}$.

STEP 3: Determine the y-intercept, $f(0)$.

STEP 4: (a) If $b^2 - 4ac > 0$, then the graph of the quadratic function has two x-intercepts, which are found by solving the equation $ax^2 + bx + c = 0$.
(b) If $b^2 - 4ac = 0$, the vertex is the x-intercept.
(c) If $b^2 - 4ac < 0$, there are no x-intercepts.

STEP 5: Determine an additional point if $b^2 - 4ac \leq 0$ by using the y-intercept and the axis of symmetry.

STEP 6: Plot the points and draw the graph.

QUADRATIC MODELS

3 When a mathematical model leads to a quadratic function, the characteristics of that quadratic function can provide important information about the model. For example, for a quadratic revenue function, we can find the maximum revenue; for a quadratic cost function, we can find the minimum cost.

To see why, recall that the graph of a quadratic function $f(x) = ax^2 + bx + c$ is a parabola with vertex at $(-b/2a, f(-b/2a))$. This vertex is the highest point on the graph if $a < 0$ and the lowest point on the graph if $a > 0$. If the vertex is the highest point ($a < 0$), then $f(-b/2a)$ is the **maximum value** of f. If the vertex is the lowest point ($a > 0$), then $f(-b/2a)$ is the **minimum value** of f.

EXAMPLE 6 Finding the Maximum or Minimum Value of a Quadratic Function

Determine whether the quadratic function

$$f(x) = x^2 - 4x - 5$$

has a maximum or minimum value. Then find the maximum or minimum value.

Solution We compare $f(x) = x^2 - 4x - 5$ to $f(x) = ax^2 + bx + c$. We conclude that $a = 1, b = -4$, and $c = -5$. Since $a > 0$, the graph of f opens up, so the vertex is a minimum point. The minimum value occurs at

$$x = \frac{-b}{2a} \underset{\uparrow}{=} \frac{-(-4)}{2(1)} = \frac{4}{2} = 2$$

$a = 1, b = -4$

The minimum value is

$$f\left(\frac{-b}{2a}\right) = f(2) = 2^2 - 4(2) - 5 = 4 - 8 - 5 = -9$$

NOW WORK PROBLEM 47.

EXAMPLE 7 Maximizing Revenue

4 The marketing department at Texas Instruments has found that, when certain calculators are sold at a price of p dollars per unit, the revenue R (in dollars) as a function of the price p is

$$R(p) = -152.63p^2 + 20{,}208.43p$$

What unit price should be established in order to maximize revenue? If this price is charged, what is the maximum revenue?

Solution The revenue R is

$$R(p) = -152.63p^2 + 20{,}208.43p = ap^2 + bp + c$$

Figure 12
$R(p) = -152.63p^2 + 20{,}208.43p$

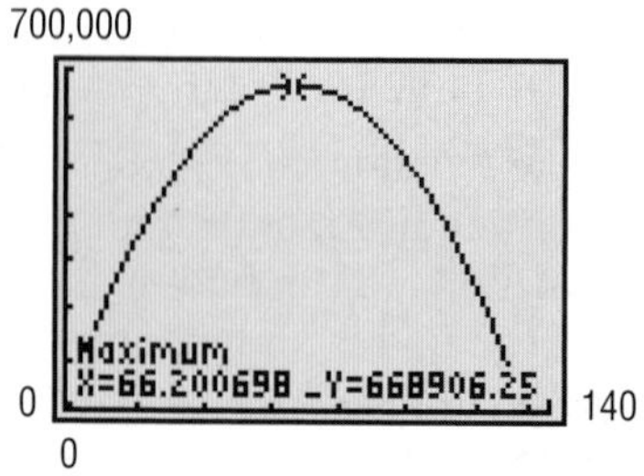

The function R is a quadratic function with $a = -152.63, b = 20{,}208.43$, and $c = 0$. Because $a < 0$, the vertex is the highest point of the parabola. The revenue R is therefore a maximum when the price p is

$$p = \frac{-b}{2a} = \frac{-20{,}208.43}{2(-152.63)} = \frac{-20{,}208.43}{-305.26} = \$66.20$$

The maximum revenue R is

$$R(66.20) = -152.63(66.20)^2 + 20{,}208.43(66.20) \approx \$668{,}906.25$$

Figure 12 verifies the result.

EXAMPLE 8 Maximizing the Area Enclosed by a Fence

A farmer has 2000 yards of fence to enclose a rectangular field. What are the dimensions of the rectangle that encloses the most area?

Figure 13

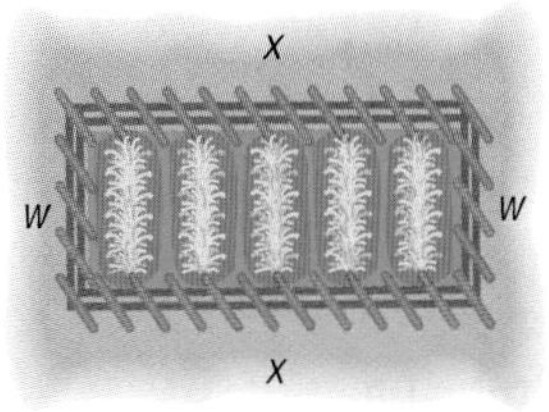

Solution Figure 13 illustrates the situation. The available fence represents the perimeter of the rectangle. If x is the length and w is the width, then

$$2x + 2w = 2000 \qquad (5)$$

The area A of the rectangle is

$$A = xw$$

To express A in terms of a single variable, we solve equation (5) for w and substitute the result in $A = xw$. Then A involves only the variable x. [You could also solve equation (5) for x and express A in terms of w alone. Try it!]

$$2x + 2w = 2000 \quad \text{Equation (5)}$$

$$2w = 2000 - 2x \quad \text{Solve for } w.$$

$$w = \frac{2000 - 2x}{2} = 1000 - x$$

Then the area A is

$$A = xw = x(1000 - x) = -x^2 + 1000x$$

Now, A is a quadratic function of x.

$$A(x) = -x^2 + 1000x \quad a = -1,\ b = 1000,\ c = 0$$

Since $a < 0$, the vertex is a maximum point on the graph of A. The maximum value occurs at

$$x = \frac{-b}{2a} = \frac{-1000}{2(-1)} = 500$$

The maximum value of A is

$$A\left(\frac{-b}{2a}\right) = A(500) = -500^2 + 1000(500) = -250{,}000 + 500{,}000 = 250{,}000$$

The most area that can be enclosed by 2000 yards of fence is 250,000 square yards. ■

NOW WORK PROBLEM 61.

EXAMPLE 9 Analyzing the Motion of a Projectile

A projectile is fired from a cliff 500 feet above the water at an inclination of 45° to the horizontal, with a muzzle velocity of 400 feet per second. In physics, it is established that the height h of the projectile above the water is given by

$$h(x) = \frac{-32x^2}{(400)^2} + x + 500$$

where x is the horizontal distance of the projectile from the base of the cliff. See Figure 14.

Figure 14

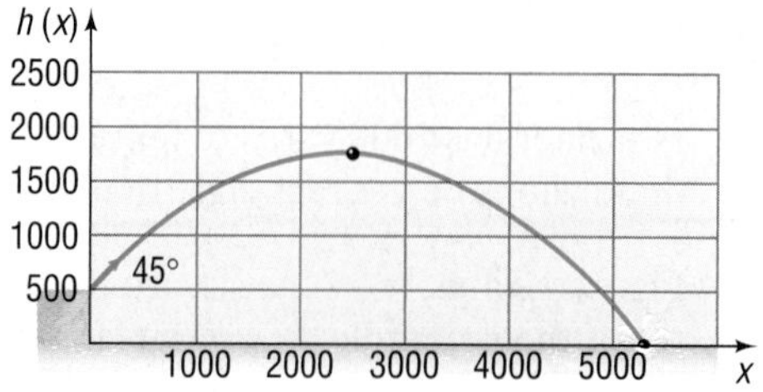

(a) Find the maximum height of the projectile.

(b) How far from the base of the cliff will the projectile strike the water?

Solution (a) The height of the projectile is given by a quadratic function.

$$h(x) = \frac{-32x^2}{(400)^2} + x + 500 = \frac{-1}{5000}x^2 + x + 500$$

We are looking for the maximum value of h. Since the maximum value is obtained at the vertex, we compute

$$x = \frac{-b}{2a} = \frac{-1}{2(-1/5000)} = \frac{5000}{2} = 2500$$

The maximum height of the projectile is

$$h(2500) = \frac{-1}{5000}(2500)^2 + 2500 + 500 = -1250 + 2500 + 500 = 1750 \text{ ft}$$

(b) The projectile will strike the water when the height is zero. To find the distance x traveled, we need to solve the equation

$$h(x) = \frac{-1}{5000}x^2 + x + 500 = 0$$

We use the quadratic formula with

$$b^2 - 4ac = 1 - 4\left(\frac{-1}{5000}\right)(500) = 1.4$$

$$x = \frac{-1 \pm \sqrt{1.4}}{2(-1/5000)} \approx \begin{cases} -458 \\ 5458 \end{cases}$$

We discard the negative solution and find that the projectile will strike the water a distance of about 5458 feet from the base of the cliff. ■

SEEING THE CONCEPT Graph

$$h(x) = \frac{-1}{5000}x^2 + x + 500, \qquad 0 \le x \le 5500$$

Use MAXIMUM to find the maximum height of the projectile, and use ROOT or ZERO to find the distance from the base of the cliff to where the projectile strikes the water. Compare your results with those obtained in the text. TRACE the path of the projectile. How far from the base of the cliff is the projectile when its height is 1000 ft? 1500 ft? ■

NOW WORK PROBLEM 65.

EXAMPLE 10 The Golden Gate Bridge

The Golden Gate Bridge, a suspension bridge, spans the entrance to San Francisco Bay. Its 746-foot-tall towers are 4200 feet apart. The bridge is suspended from two huge cables more than 3 feet in diameter; the 90-foot-wide roadway is 220 feet above the water. The cables are parabolic in shape and touch the road surface at the center of the bridge. Find the height of the cable at a distance of 1000 feet from the center.

Solution We begin by choosing the placement of the coordinate axes so that the x-axis coincides with the road surface and the origin coincides with the center of the bridge. As a result, the twin towers will be vertical (height $746 - 220 = 526$ feet above the road) and located 2100 feet from the center. Also, the cable, which has the shape of a parabola, will extend from the towers, open up, and have its vertex at $(0, 0)$. As illustrated in Figure 15, the choice of placement of the axes enables us to identify the equation of the parabola as $y = ax^2$, $a > 0$. We can also see that the points $(-2100, 526)$ and $(2100, 526)$ are on the graph.

Figure 15

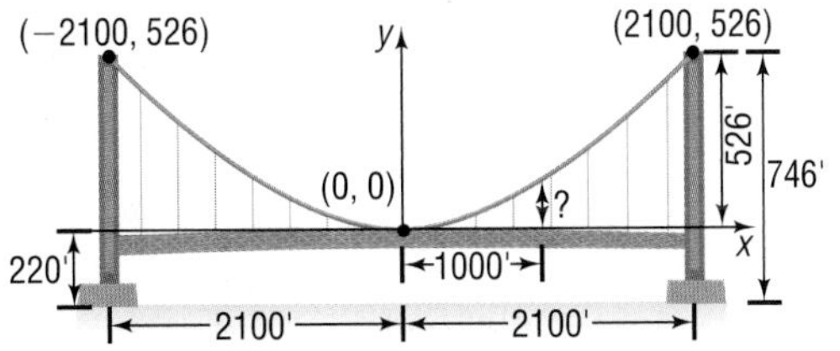

Based on these facts, we can find the value of a in $y = ax^2$.

$$y = ax^2$$
$$526 = a(2100)^2$$
$$a = \frac{526}{(2100)^2}$$

The equation of the parabola is therefore

$$y = \frac{526}{(2100)^2}x^2$$

The height of the cable when $x = 1000$ is

$$y = \frac{526}{(2100)^2}(1000)^2 \approx 119.3 \text{ feet}$$

The cable is 119.3 feet high at a distance of 1000 feet from the center of the bridge. ■

NOW WORK PROBLEM 67.

FITTING A QUADRATIC FUNCTION TO DATA

5 In Section 1.7 we found the line of best fit for data that appeared to be linearly related. It was noted that data may also follow a nonlinear relation. Figures 16(a) and (b) show scatter diagrams of data that follow a quadratic relation.

Figure 16

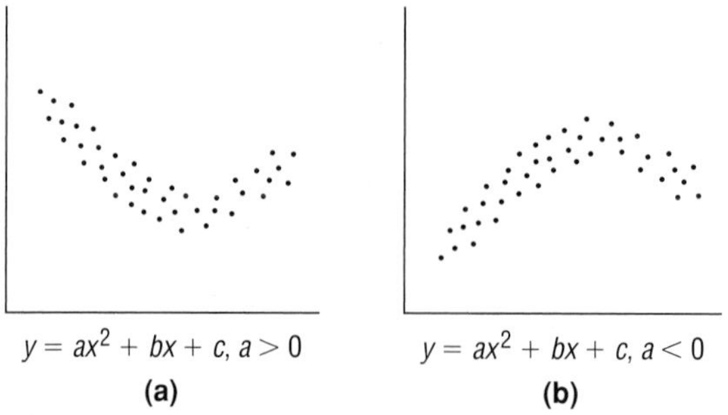

EXAMPLE 11 Fitting a Quadratic Function to Data

The data in Table 2 represent the enrollment (in millions) in public high schools for the years 1986–1996.

TABLE 2

Year, x	Enrollment, E
1986	13.0
1987	12.7
1988	12.2
1989	12.1
1990	11.9
1991	12.2
1992	12.3
1993	12.6
1994	13.5
1995	13.7
1996	14.1

Source: U.S. Bureau of the Census

(a) Using a graphing utility, draw a scatter diagram of the data. Comment on the type of relation that may exist between the two variables.
(b) Find the quadratic function of best fit.
(c) Using a graphing utility, draw the quadratic function of best fit on your scatter diagram.
(d) Using the function found in part (b), determine when enrollment was the lowest.
(e) Compare the result found in part (d) to the data.

Solution

(a) Figure 17 shows the scatter diagram, from which it appears that the data follow a quadratic relation, with $a > 0$.
(b) Upon executing the QUADratic REGression program, we obtain the results shown in Figure 18. The output that the utility provides shows us the equation $y = ax^2 + bx + c$. The quadratic function of best fit is $E(x) = 0.05967x^2 - 237.486x + 236{,}295.5$, where x represents the year and E represents the enrollment.
(c) Figure 19 shows the graph of the quadratic function found in part (b) drawn on the scatter diagram.

Figure 17

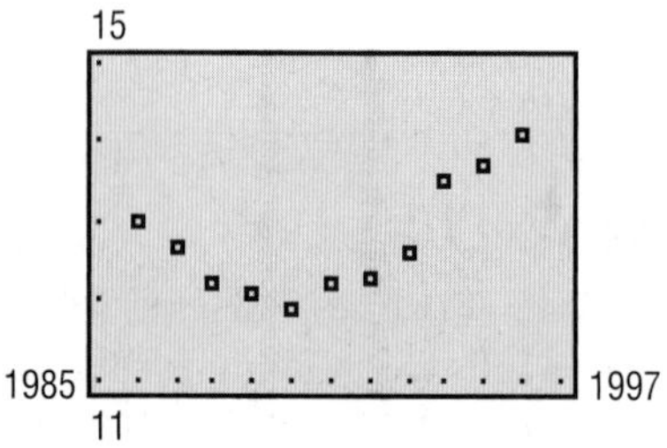

Figure 18

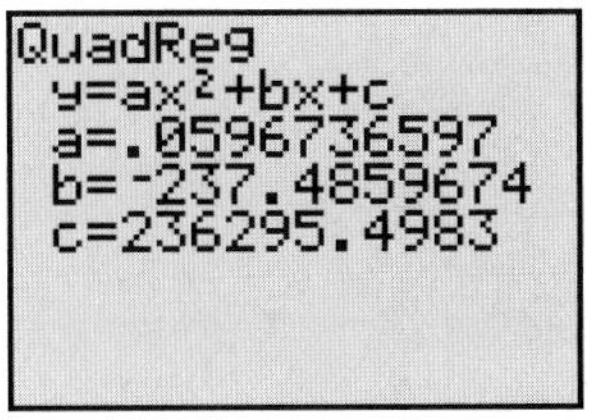

Figure 19

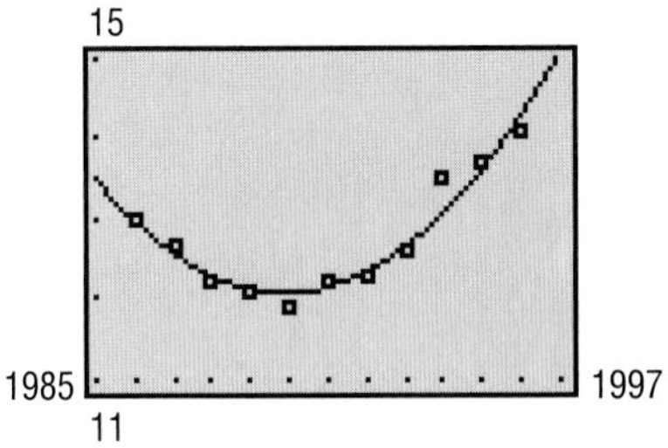

(d) Based on the quadratic function of best fit, the enrollment was lowest in the year

$$h = \frac{-b}{2a} = \frac{-(-237.486)}{2(0.05967)} \approx 1990$$

(e) From the data, enrollment is lowest in 1990, so the result obtained in part (d) agrees with the data. ■

3.1 EXERCISES

In Problems 1–8, match each graph to one the following functions without using a graphing utility.

A. $f(x) = x^2 - 1$ B. $f(x) = -x^2 - 1$ C. $f(x) = x^2 - 2x + 1$ D. $f(x) = x^2 + 2x + 1$
E. $f(x) = x^2 - 2x + 2$ F. $f(x) = x^2 + 2x$ G. $f(x) = x^2 - 2x$ H. $f(x) = x^2 + 2x + 2$

1.
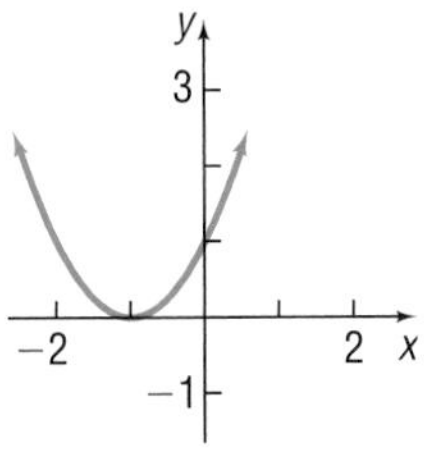

2.
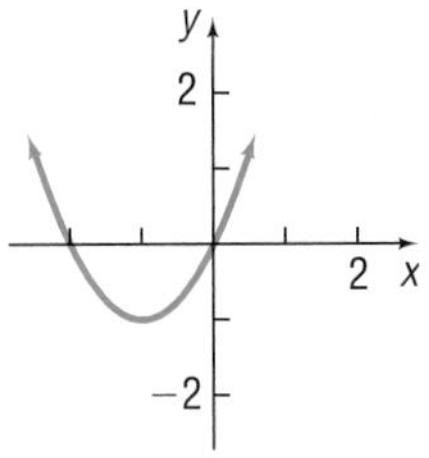

3.
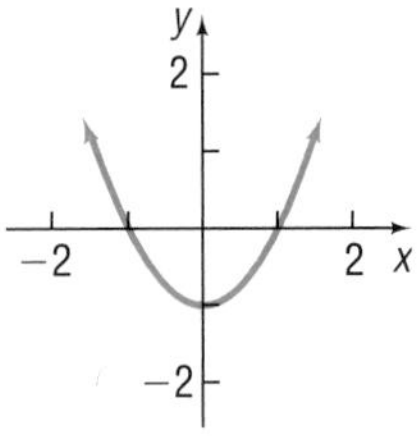

4.
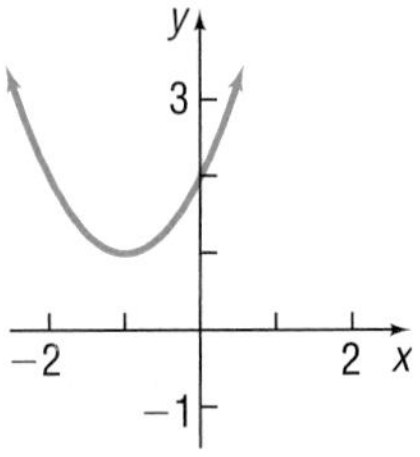

5.
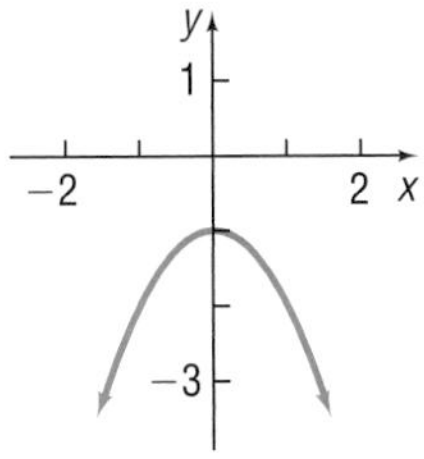

6.
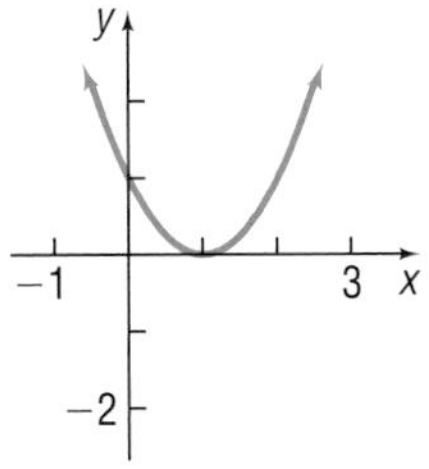

7.
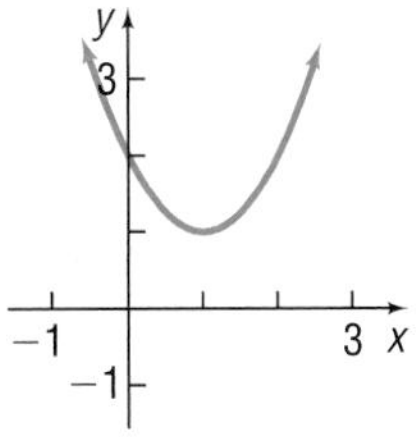

8.
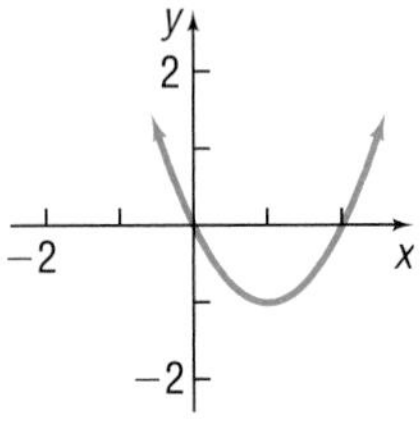

In Problems 9–12, match each graph to one of the following functions without using a graphing utility.

A. $f(x) = (1/3)x^2 + 2$ B. $f(x) = 3x^2 + 2$
C. $f(x) = x^2 + 5x + 1$ D. $f(x) = -x^2 + 5x + 1$

9.
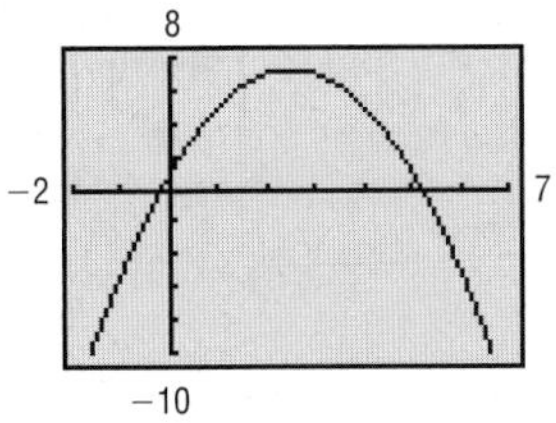

10.
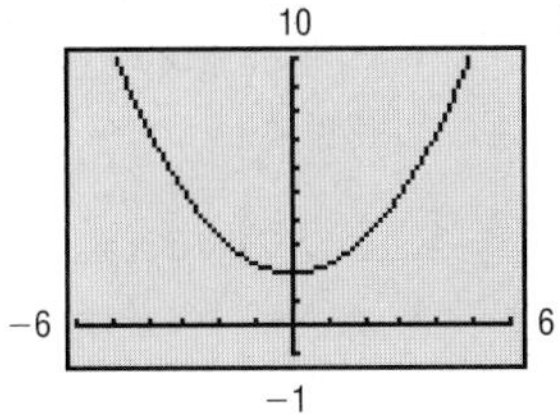

11.
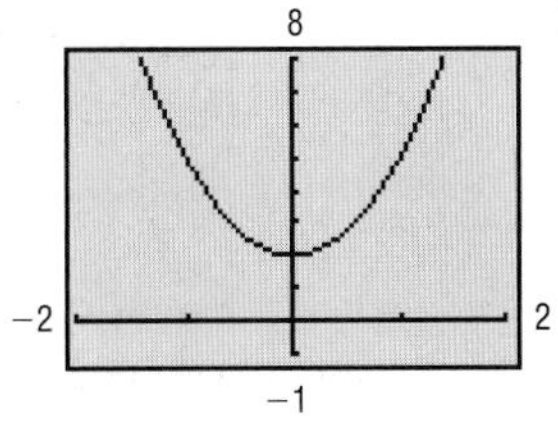

12.
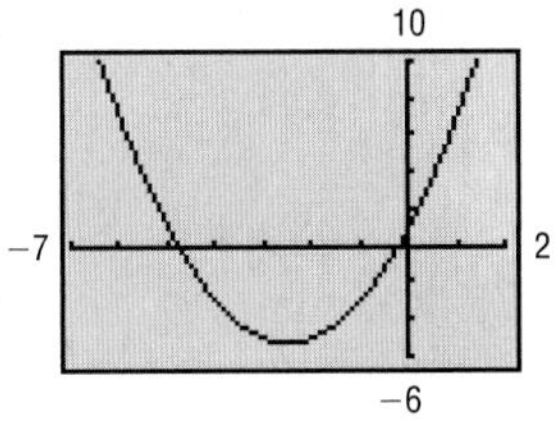

In Problems 13–28, graph the function f by starting with the graph of $y = x^2$ and using transformations (shifting, compressing, stretching, and/or reflection). Verify your results using a graphing utility.
[**Hint:** If necessary, write f in the form $f(x) = a(x - h)^2 + k$.]

13. $f(x) = \frac{1}{4}x^2$

14. $f(x) = 2x^2$

15. $f(x) = \frac{1}{4}x^2 - 2$

16. $f(x) = 2x^2 - 3$

17. $f(x) = \frac{1}{4}x^2 + 2$

18. $f(x) = 2x^2 + 4$

19. $f(x) = \frac{1}{4}x^2 + 1$

20. $f(x) = -2x^2 - 2$

21. $f(x) = x^2 + 4x + 2$

22. $f(x) = x^2 - 6x - 1$

23. $f(x) = 2x^2 - 4x + 1$

24. $f(x) = 3x^2 + 6x$

25. $f(x) = -x^2 - 2x$

26. $f(x) = -2x^2 + 6x + 2$

27. $f(x) = \frac{1}{2}x^2 + x - 1$

28. $f(x) = \frac{2}{3}x^2 + \frac{4}{3}x - 1$

In Problems 29–44, graph each quadratic function by hand by determining whether its graph opens up or down and by finding its vertex, axis of symmetry, y-intercept, and x-intercepts, if any. Verify your results using a graphing utility.

29. $f(x) = -x^2 - 6x$

30. $f(x) = -x^2 + 4x$

31. $f(x) = 2x^2 - 8x$

32. $f(x) = 3x^2 + 18x$

33. $f(x) = x^2 + 2x - 8$

34. $f(x) = x^2 - 2x - 3$

35. $f(x) = x^2 + 2x + 1$

36. $f(x) = x^2 + 6x + 9$

37. $f(x) = 2x^2 - x + 2$

38. $f(x) = 4x^2 - 2x + 1$

39. $f(x) = -2x^2 + 2x - 3$

40. $f(x) = -3x^2 + 3x - 2$

41. $f(x) = 3x^2 + 6x + 2$

42. $f(x) = 2x^2 + 5x + 3$

43. $f(x) = -4x^2 - 6x + 2$

44. $f(x) = 3x^2 - 8x + 2$

In Problems 45–52, determine, without graphing, whether the given quadratic function has a maximum value or a minimum value and then find the value.

45. $f(x) = 2x^2 + 12x$

46. $f(x) = -2x^2 + 12x$

47. $f(x) = 2x^2 + 12x - 3$

48. $f(x) = 4x^2 - 8x + 3$

49. $f(x) = -x^2 + 10x - 4$

50. $f(x) = -2x^2 + 8x + 3$

51. $f(x) = -3x^2 + 12x + 1$

52. $f(x) = 4x^2 - 4x$

Answer Problems 53 and 54 using the following: A quadratic function of the form $f(x) = ax^2 + bx + c$ with $b^2 - 4ac > 0$ may also be written in the form $f(x) = a(x - r_1)(x - r_2)$, where r_1 and r_2 are the x-intercepts of the graph of the quadratic function.

53. (a) Find a quadratic function whose x-intercepts are -3 and 1 with $a = 1$; $a = 2$; $a = -2$; $a = 5$.
(b) How does the value of a affect the intercepts?
(c) How does the value of a affect the axis of symmetry?
(d) How does the value of a affect the vertex?
(e) Compare the x-coordinate of the vertex with the midpoint of the x-intercepts. What might you conclude?

54. (a) Find a quadratic function whose x-intercepts are -5 and 3 with $a = 1$; $a = 2$; $a = -2$; $a = 5$.
(b) How does the value of a affect the intercepts?
(c) How does the value of a affect the axis of symmetry?
(d) How does the value of a affect the vertex?
(e) Compare the x-coordinate of the vertex with the midpoint of the x-intercepts. What might you conclude?

55. Maximizing Revenue Suppose that the manufacturer of a gas clothes dryer has found that, when the unit price is p dollars, the revenue R (in dollars) is

$$R(p) = -4p^2 + 4000p$$

What unit price should be established for the dryer to maximize revenue? What is the maximum revenue?

56. Maximizing Revenue The John Deere company has found that the revenue from sales of heavy-duty tractors is a function of the unit price p that it charges. If the revenue R is

$$R(p) = -\frac{1}{2}p^2 + 1900p$$

what unit price p should be charged to maximize revenue? What is the maximum revenue?

57. Demand Equation The price p and the quantity x sold of a certain product obey the demand equation

$$p = -\tfrac{1}{6}x + 100, \qquad 0 \le x \le 600$$

(a) Express the revenue R as a function of x. (Remember, $R = xp$.)
(b) What is the revenue if 200 units are sold?
(c) What quantity x maximizes revenue? What is the maximum revenue?
(d) What price should the company charge to maximize revenue?

58. Demand Equation The price p and the quantity x sold of a certain product obey the demand equation

$$p = -\tfrac{1}{3}x + 100, \qquad 0 \le x \le 300$$

(a) Express the revenue R as a function of x.
(b) What is the revenue if 100 units are sold?
(c) What quantity x maximizes revenue? What is the maximum revenue?
(d) What price should the company charge to maximize revenue?

59. Demand Equation The price p and the quantity x sold of a certain product obey the demand equation

$$x = -5p + 100, \qquad 0 \le p \le 20$$

(a) Express the revenue R as a function of x.
(b) What is the revenue if 15 units are sold?
(c) What quantity x maximizes revenue? What is the maximum revenue?
(d) What price should the company charge to maximize revenue?

60. Demand Equation The price p and the quantity x sold of a certain product obey the demand equation

$$x = -20p + 500, \qquad 0 \le p \le 25$$

(a) Express the revenue R as a function of x.
(b) What is the revenue if 20 units are sold?
(c) What quantity x maximizes revenue? What is the maximum revenue?
(d) What price should the company charge to maximize revenue?

61. Enclosing a Rectangular Field David has available 400 yards of fencing and wishes to enclose a rectangular area.
(a) Express the area A of the rectangle as a function of the width x of the rectangle.
(b) For what value of x is the area largest?
(c) What is the maximum area?

62. Enclosing a Rectangular Field Beth has 3000 feet of fencing available to enclose a rectangular field.
(a) Express the area A of the rectangle as a function of x, where x is the length of the rectangle.
(b) For what value of x is the area largest?
(c) What is the maximum area?

63. Enclosing the Most Area with a Fence A farmer with 4000 meters of fencing wants to enclose a rectangular plot that borders on a river. If the farmer does not fence the side along the river, what is the largest area that can be enclosed? (See the figure.)

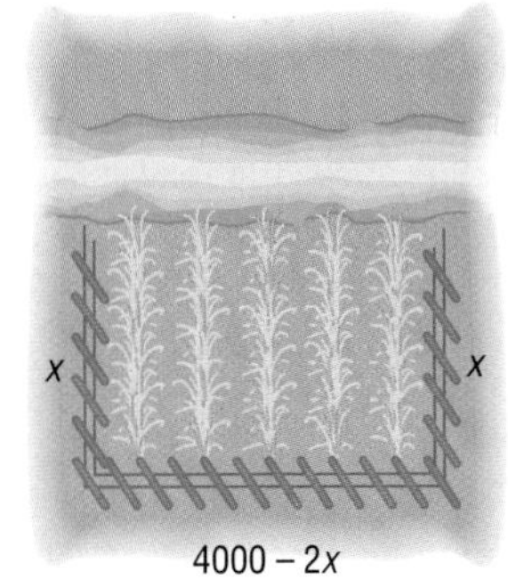

64. Enclosing the Most Area with a Fence A farmer with 2000 meters of fencing wants to enclose a rectangular plot that borders on a straight highway. If the farmer does not fence the side along the highway, what is the largest area that can be enclosed?

65. Analyzing the Motion of a Projectile A projectile is fired from a cliff 200 feet above the water at an inclination of 45° to the horizontal, with a muzzle velocity of 50 feet per second. The height h of the projectile above the water is given by

$$h(x) = \frac{-32x^2}{(50)^2} + x + 200$$

where x is the horizontal distance of the projectile from the base of the cliff.
(a) How far from the base of the cliff is the height of the projectile a maximum?
(b) Find the maximum height of the projectile.
(c) How far from the base of the cliff will the projectile strike the water?
(d) When the height of the projectile is 100 feet above the water, how far is it from the cliff?
(e) Using a graphing utility, graph the function h, $0 \le x \le 200$.
(f) Use a graphing utility to verify the solutions found in parts (a)–(d).

66. Analyzing the Motion of a Projectile A projectile is fired at an inclination of 45° to the horizontal, with a muzzle velocity of 100 feet per second. The height h of the projectile is given by

$$h(x) = \frac{-32x^2}{(100)^2} + x$$

where x is the horizontal distance of the projectile from the firing point.
(a) How far from the firing point is the height of the projectile a maximum?
(b) Find the maximum height of the projectile.
(c) How far from the firing point will the projectile strike the ground?

(d) When the height of the projectile is 50 feet above the ground, how far has it traveled horizontally?
(e) Using a graphing utility, graph the function h, $0 \le x \le 350$.
(f) Use a graphing utility to verify the results obtained in parts (a)–(d).

67. **Suspension Bridge** A suspension bridge with weight uniformly distributed along its length has twin towers that extend 75 meters above the road surface and are 400 meters apart. The cables are parabolic in shape and are suspended from the tops of the towers. The cables touch the road surface at the center of the bridge. Find the height of the cables at a point 100 meters from the center. (Assume that the road is level.)

68. **Architecture** A parabolic arch has a span of 120 feet and a maximum height of 25 feet. Choose suitable rectangular coordinate axes and find the equation of the parabola. Then calculate the height of the arch at points 10 feet, 20 feet, and 40 feet from the center.

69. **Constructing Rain Gutters** A rain gutter is to be made of aluminum sheets that are 12 inches wide by turning up the edges 90°. What depth will provide maximum cross-sectional area and hence allow the most water to flow?

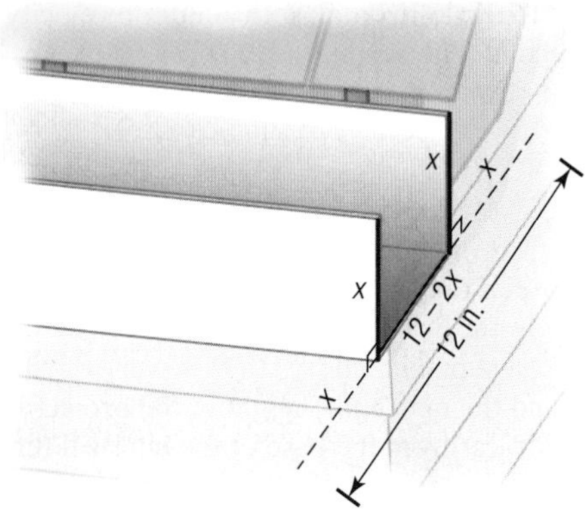

70. **Norman Windows** A Norman window has the shape of a rectangle surmounted by a semicircle of diameter equal to the width of the rectangle (see the figure). If the perimeter of the window is 20 feet, what dimensions will admit the most light (maximize the area)?
[**Hint:** Circumference of a circle $= 2\pi r$; area of a circle $= \pi r^2$, where r is the radius of the circle.]

71. **Constructing a Stadium** A track and field playing area is in the shape of a rectangle with semicircles at each end (see the figure). The inside perimeter of the track is to be 1500 meters. What should the dimensions of the rectangle be so that the area of the rectangle is a maximum?

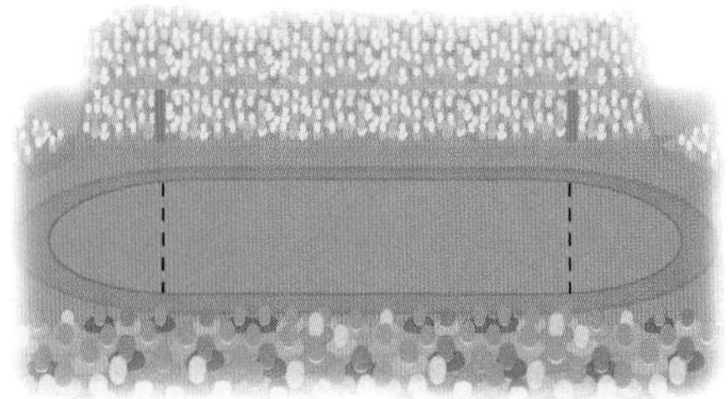

72. **Architecture** A special window has the shape of a rectangle surmounted by an equilateral triangle (see the figure). If the perimeter of the window is 16 feet, what dimensions will admit the most light?
[**Hint:** Area of an equilateral triangle $= (\sqrt{3}/4)x^2$, where x is the length of a side of the triangle.]

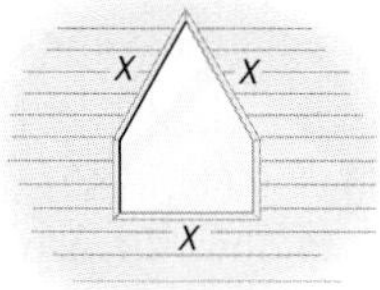

73. **Life Cycle Hypothesis** An individual's income varies with his or her age. The following table shows the median income I of individuals of different age groups within the United States for 1995. For each age group, let the class midpoint represent the independent variable, x. For the class "65 years and older," we will assume that the class midpoint is 69.5.

Age	Class Midpoint, x	Median Income, I
15–24 years	19.5	\$20,979
25–34 years	29.5	\$34,701
35–44 years	39.5	\$43,465
45–54 years	49.5	\$48,058
55–64 years	59.5	\$38,077
65 years and older	69.5	\$19,096

Source: U.S. Census Bureau

(a) Using a graphing utility, draw a scatter diagram of the data. Comment on the type of relation that may exist between the two variables.
(b) Find the quadratic function of best fit.
(c) Using a graphing utility, draw the quadratic function of best fit on your scatter diagram.
(d) Using the function found in part (b), determine the age at which an individual can expect to earn the most income.
(e) Predict the peak income earned.
(f) Compare your results in parts (d) and (e) to the data and comment.

74. Life Cycle Hypothesis An individual's income varies with his or her age. The following table shows the median income I of individuals of different age groups within the United States for 1996. For each age group, the class midpoint represents the independent variable, x. For the age group "65 years and older," we will assume that the class midpoint is 69.5.

Age	Midpoint, x	Median Income, I
15–24 years	19.5	$21,438
25–34 years	29.5	$35,888
35–44 years	39.5	$44,420
45–54 years	49.5	$50,472
55–64 years	59.5	$39,815
65 years and older	69.5	$19,448

Source: U.S. Census Bureau

(a) Using a graphing utility, draw a scatter diagram of the data. Comment on the type of relation that may exist between the two variables.
(b) Find the quadratic function of best fit.
(c) Using a graphing utility, draw the quadratic function of best fit on your scatter diagram.
(d) Using the function found in part (b), determine the age at which an individual can expect to earn the most income.
(e) Predict the peak income earned.
(f) Compare your results in parts (d) and (e) to the data and comment.
(g) Compare the results obtained in 1996 to those of 1995 found in Problem 73.

75. Price of Crude Oil The following data represent the price of crude oil (in cents per million British thermal units) for the years 1990–1996.

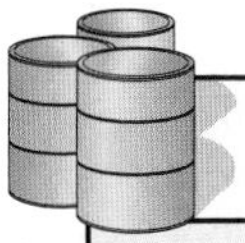

Year, x	Price, p
1990	3.69
1991	2.93
1992	2.76
1993	2.40
1994	2.17
1995	2.34
1996	2.90

Source: U.S. Energy Information Administration

(a) Using a graphing utility, draw a scatter diagram of the data. Comment on the type of relation that may exist between the two variables.
(b) Find the quadratic function of best fit.
(c) Draw the quadratic function of best fit on the scatter diagram.
(d) Use the function found in part (b) to predict the price in 1997.
(e) Using the function found in part (b), determine when price was lowest.
(f) Compare your result to the data.

76. Imports of Crude Oil The following data represent the imports of crude oil (1000 barrels per day) for the years 1980–1997.

Year, x	Imports, I	Year, x	Imports, I
1980	5263	1989	5843
1981	4396	1990	5894
1982	3488	1991	5782
1983	3329	1992	6083
1984	3426	1993	6787
1985	3201	1994	7063
1986	4178	1995	7230
1987	4674	1996	7508
1988	5107	1997	7996

Source: U.S. Energy Information Administration

(a) Using a graphing utility, draw a scatter diagram of the data. Comment on the type of relation that may exist between the two variables.
(b) Find the quadratic function of best fit.
(c) Draw the quadratic function of best fit on the scatter diagram.
(d) Use the function found in part (b) to predict the number of barrels of imported crude oil in 1998.
(e) Using the function found in part (b), determine the year in which imports of crude oil were lowest.
(f) Compare your result to the data.

77. Miles per Gallon An engineer collects data showing the speed s of a Ford Taurus and its average miles per gallon, M. See the table.

Speed, s	Miles per Gallon, M
30	18
35	20
40	23
40	25
45	25
50	28
55	30
60	29
65	26
65	25
70	25

(a) Using a graphing utility, draw a scatter diagram of the data. Comment on the type of relation that may exist between the two variables.
(b) Find the quadratic function of best fit.
(c) Using a graphing utility, draw the quadratic function of best fit on your scatter diagram.
(d) Using the function found in part (b), determine the speed that maximizes miles per gallon.
(e) Compare your result in part (d) to the data.
(f) Using the function found in part (b), predict the miles per gallon of the car if you travel an average of 63 miles per hour.

78. Height of a Ball A physicist throws a ball at an inclination of 45° to the horizontal. The following data represent the height of the ball h at the instant that it has traveled x feet horizontally.

Distance, *x*	Height, *h*
20	25
40	40
60	55
80	65
100	71
120	77
140	77
160	75
180	71
200	64

(a) Using a graphing utility, draw a scatter diagram of the data. Comment on the type of relation that may exist between the two variables.
(b) Find the quadratic function of best fit.
(c) Using a graphing utility, draw the quadratic function of best fit on your scatter diagram.
(d) Using the function found in part (b), how far will the ball travel before it reaches its maximum height?
(e) Using the function found in part (b), determine the maximum height of the ball.
(f) Compare your results in parts (d) and (e) to the data.
(g) Determine the horizontal distance the ball will travel based on the function found in part (b).

79. Chemical Reactions A self-catalytic chemical reaction results in the formation of a compound that causes the formation ratio to increase. If the reaction rate V is given by

$$V(x) = kx(a - x), \qquad 0 \le x \le a$$

where k is a positive constant, a is the initial amount of the compound, and x is the variable amount of the compound, for what value of x is the reaction rate a maximum?

80. A rectangle has one vertex on the line $y = 10 - x$, $x > 0$, another at the origin, one on the positive x-axis, and one on the positive y-axis. Find the largest area A that can be enclosed by the rectangle.

81. Calculus: Simpson's Rule The figure shows the graph of $y = ax^2 + bx + c$. Suppose that the points $(-h, y_0)$, $(0, y_1)$, and (h, y_2) are on the graph. It can be shown that the area enclosed by the parabola, the x-axis, and the lines $x = -h$ and $x = h$ is

$$\text{Area} = \frac{h}{3}\left(2ah^2 + 6c\right)$$

Show that this area may also be given by

$$\text{Area} = \frac{h}{3}\left(y_0 + 4y_1 + y_2\right)$$

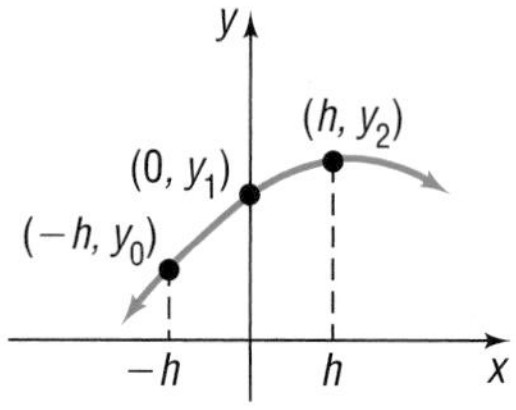

82. Use the result obtained in Problem 81 to find the area enclosed by $f(x) = -5x^2 + 8$, the x-axis, and the lines $x = -1$ and $x = 1$.

83. Use the result obtained in Problem 81 to find the area enclosed by $f(x) = 2x^2 + 8$, the x-axis, and the lines $x = -2$ and $x = 2$.

84. Use the result obtained in Problem 81 to find the area enclosed by $f(x) = x^2 + 3x + 5$, the x-axis, and the lines $x = -4$ and $x = 4$.

85. Use the result obtained in Problem 81 to find the area enclosed by $f(x) = -x^2 + x + 4$, the x-axis, and the lines $x = -1$ and $x = 1$.

86. CBL Experiment As a ball bounces up and down, the maximum height that the ball attains continually decreases from one bounce to the next. For a given bounce, plotting the height of the ball against time results in a parabola. The motion of the ball will be analyzed. (Activity 9, Real-World Math with the CBL System, 1994.)

87. CBL Experiment A cart is pushed up a ramp and allowed to return down the ramp. Plotting the distance that the cart is from a motion detector against time results in the graph of a parabola. The parabola will be analyzed to determine the characteristics of the cart's motion. (Activity 8, Real-World Math with the CBL System, 1994.)

88. Let $f(x) = ax^2 + bx + c$, where a, b, and c are odd integers. If x is an integer, show that $f(x)$ must be an odd integer.
[**Hint:** x is either an even integer or an odd integer.]

89. Make up a quadratic function that opens down and has only one x-intercept. Compare yours with others in the class. What are the similarities? What are the differences?

90. On one set of coordinate axes, graph the family of parabolas $f(x) = x^2 + 2x + c$ for $c = -3$, $c = 0$, and $c = 1$. Describe the characteristics of a member of this family.

91. On one set of coordinate axes, graph the family of parabolas $f(x) = x^2 + bx + 1$ for $b = -4$, $b = 0$, and $b = 4$. Describe the general characteristics of this family.

PREPARING FOR THIS SECTION

Before getting started, review the following concept:

✓ Monomials (Appendix, Section 3, p. 957)

3.2 POWER FUNCTIONS; CURVE FITTING

OBJECTIVES 1 Graph Transformations of Power Functions
2 Find the Power Function of Best Fit to Data

POWER FUNCTIONS

A *power function* is a function that is defined by a single monomial.

A **power function of degree *n*** is a function of the form

$$f(x) = ax^n \tag{1}$$

where a is a real number, $a \neq 0$, and $n > 0$ is an integer.

Examples of power functions are

$f(x) = 3x$	$f(x) = -5x^2$	$f(x) = 8x^3$	$f(x) = -5x^4$
degree 1	degree 2	degree 3	degree 4

In this section we discuss characteristics of power functions and graph transformations of power functions.

The graph of a power function of degree 1, $f(x) = ax$, is a line with slope a that passes through the origin. The graph of a power function of degree 2, $f(x) = ax^2$, is a parabola, with vertex at the origin, that opens up if $a > 0$ and down if $a < 0$.

If we know how to graph a power function of the form $f(x) = x^n$, then a compression or stretch and, perhaps, a reflection about the x-axis will enable us to obtain the graph of $g(x) = ax^n$ for any real number a. Consequently, we shall concentrate on graphing power functions of the form $f(x) = x^n$.

We begin with power functions of even degree of the form $f(x) = x^n$, $n \geq 2$ and *n even*.

EXPLORATION Using your graphing utility and the viewing window $-2 \leq x \leq 2$, $-4 \leq y \leq 16$, graph the function $Y_1 = f(x) = x^4$. On the same screen, graph $Y_2 = g(x) = x^8$. Now, also on the same screen, graph $Y_3 = h(x) = x^{12}$. What do you notice about the graphs as the magnitude of the exponent increases? Repeat this procedure for the viewing window $-1 \leq x \leq 1$, $0 \leq y \leq 1$. What do you notice? See Figures 20(a) and (b).

Figure 20

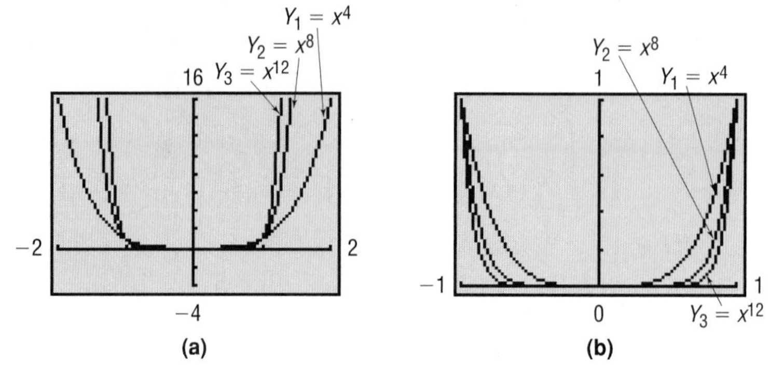

The domain of $f(x) = x^n$, $n \geq 2$ and n-even is the set of all real numbers, and the range is the set of nonnegative real numbers. Such a power function is an even function (do you see why?), so its graph is symmetric with respect to the y-axis. Its graph always contains the origin $(0, 0)$ and the points $(-1, 1)$ and $(1, 1)$.

TABLE 3

X	Y2	Y3
-1	1	1
1	1	1
.5	.00391	2.4E-4
.1	1E-8	1E-12
.01	1E-16	1E-24
.001	1E-24	1E-36
0	0	0

Y2=X^8

For large n, it appears that the graph coincides with the x-axis near the origin, but it does not; the graph actually touches the x-axis only at the origin. See Table 3, where $Y_2 = x^8$ and $Y_3 = x^{12}$. For x close to 0, the values of y are positive and close to 0. Also, for large n, it may appear that for $x < -1$ or for $x > 1$ the graph is vertical, but it is not; it is only increasing very rapidly. If you use TRACE along one of the graphs, these distinctions will be clear.

To summarize:

CHARACTERISTICS OF POWER FUNCTIONS, $y = x^n$, n IS AN EVEN INTEGER

1. The graph is symmetric with respect to the y-axis.
2. The domain is the set of all real numbers. The range is the set of nonnegative real numbers.
3. The graph always contains the points $(0, 0)$, $(1, 1)$, and $(-1, 1)$.
4. As the exponent n increases in magnitude, the graph becomes more vertical when $x < -1$ or $x > 1$; but for x near the origin, the graph tends to flatten out and lie closer to the x-axis.

Now we consider power functions of odd degree of the form $f(x) = x^n$, n *odd*.

EXPLORATION Using your graphing utility and the viewing window $-2 \leq x \leq 2, -16 \leq y \leq 16$, graph the function $Y_1 = f(x) = x^3$. On the same screen, graph $Y_2 = g(x) = x^7$ and $Y_3 = h(x) = x^{11}$. What do you notice about the graphs as the magnitude of the exponent increases? Repeat this procedure for the viewing window $-1 \leq x \leq 1, -1 \leq y \leq 1$. What do you notice? The graphs on your screen should look like Figures 21(a) and (b).

Figure 21

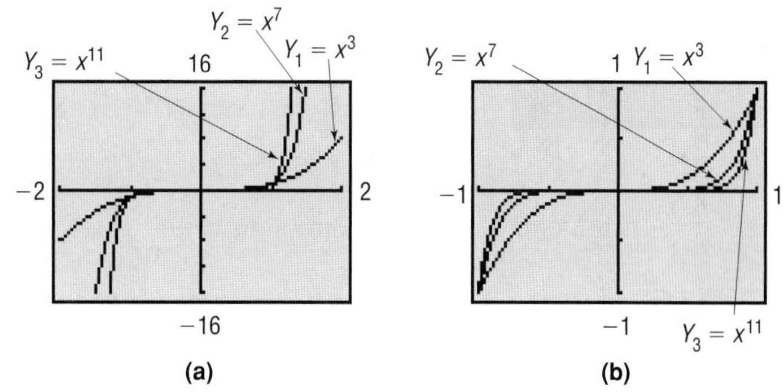

The domain and range of $f(x) = x^n, n \geq 3$ and n odd, are the set of real numbers. Such a power function is an odd function (do you see why?), so its graph is symmetric with respect to the origin. Its graph always contains the origin $(0, 0)$ and the points $(-1, -1)$ and $(1, 1)$.

It appears that the graph coincides with the x-axis near the origin, but it does not; the graph actually touches the x-axis only at the origin. Also, it appears that as x increases the graph is vertical, but it is not; it is increasing very rapidly. TRACE along the graphs to verify these distinctions.

To summarize:

CHARACTERISTICS OF POWER FUNCTIONS, $y = x^n$, n IS AN ODD INTEGER

1. The graph is symmetric with respect to the origin.
2. The domain and range are the set of all real numbers.
3. The graph always contains the points $(0, 0)$, $(1, 1)$, and $(-1, -1)$.
4. As the exponent n increases in magnitude, the graph becomes more vertical when $x < -1$ or $x > 1$; but for x near the origin, the graph tends to flatten out and lie closer to the x-axis.

1 The methods of shifting, compression, stretching, and reflection studied in Section 2.4, when used with the facts just presented, will enable us to graph and analyze a variety of functions that are transformations of power functions.

EXAMPLE 1 Graphing Transformations of Power Functions

Graph: $f(x) = 1 - x^5$

Solution Figure 22 shows the required stages.

Figure 22

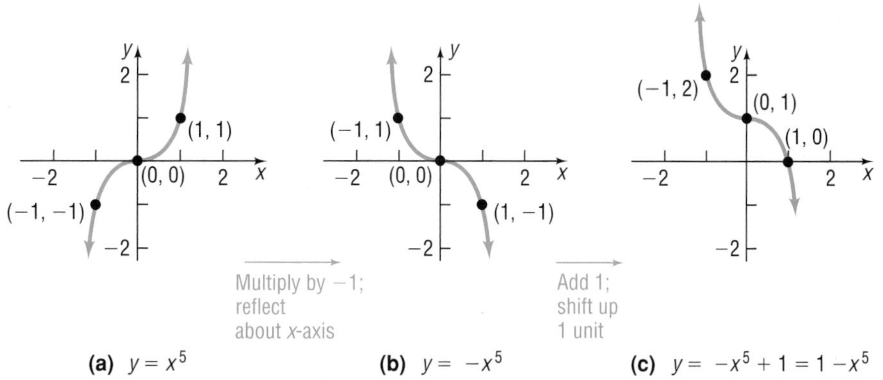

Check: Verify the result of Example 1 by graphing $Y_1 = f(x) = 1 - x^5$. ■

EXAMPLE 2 Graphing Transformations of Power Functions

Graph: $f(x) = \frac{1}{2}(x - 1)^4$

Solution Figure 23 shows the required stages.

Figure 23

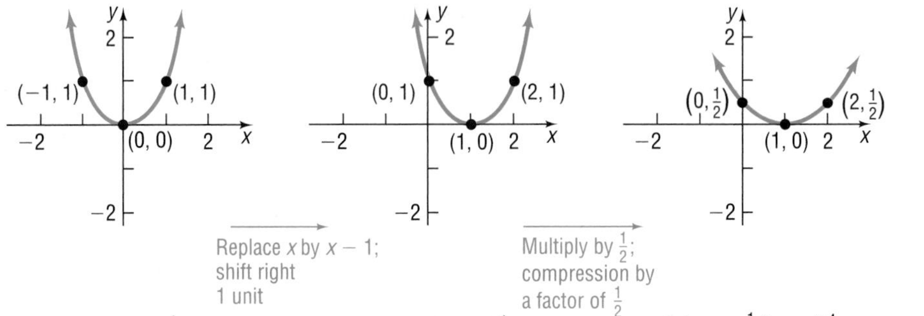

(a) $y = x^4$ (b) $y = (x - 1)^4$ (c) $y = \frac{1}{2}(x - 1)^4$ ■

Check: Verify the result of Example 2 by graphing $Y_2 = f(x) = \frac{1}{2}(x - 1)^4$. ■

NOW WORK PROBLEM 1.

MODELING POWER FUNCTIONS: CURVE FITTING

2 Figure 24 shows a scatter diagram that follows a power function.

Many situations lead to data that can be modeled using a power function. One such situation involves falling objects.

Figure 24

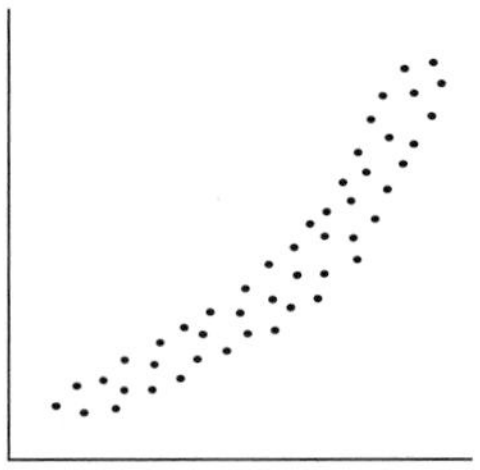

$y = ax^b, a > 0, b > 1$
$x, y > 0$

EXAMPLE 3 Fitting Data to a Power Function

TABLE 4

Time t (seconds)	Distance s (meters)
1.528	11.46
2.015	19.99
3.852	72.41
4.154	84.45
4.625	104.23

Scott drops a ball from various heights and records the time that it takes for the ball to hit the ground. Using a motion detector connected to his graphing calculator, he collects the data shown in Table 4.

(a) Using a graphing utility, draw a scatter diagram using time t as the independent variable and distance s as the dependent variable.

(b) Use a graphing utility to find the power function of best fit.

(c) Graph the power function found in part (b) on the scatter diagram.

(d) Predict how long it will take the ball to fall 100 meters.

(e) Physics theory states that the distance an object falls is directly proportional to the time squared. Rewrite the model found in part (b) so that it is of the form $s = \frac{1}{2}gt^2$, where s is distance, g is acceleration due to gravity, and t is time. What is Scott's estimate of g? (It is known that the acceleration due to gravity is approximately 9.8 meters/sec^2.)

Solution (a) After entering the data into the graphing utility, we obtain the scatter diagram shown in Figure 25.

(b) A graphing utility fits the data to a power function of the form $y = ax^b$ by using the PoWeR REGression option. See Figure 26. The distance s that the ball has fallen is related to time t by the power function

$$s(t) = 4.93t^{1.993}$$

Notice that the value of r is close to 1, indicating a good fit.*

Figure 25

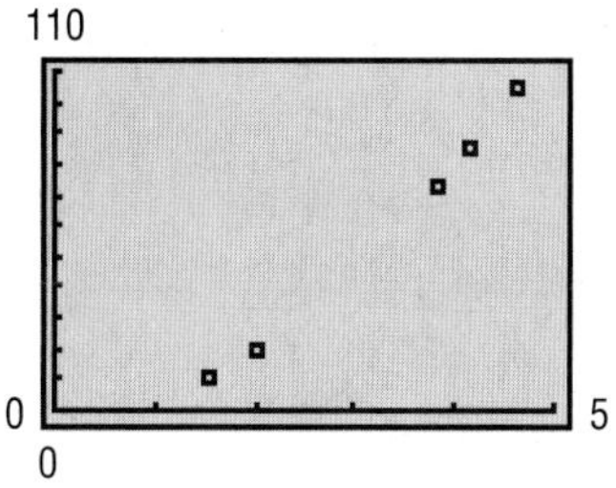

Figure 26

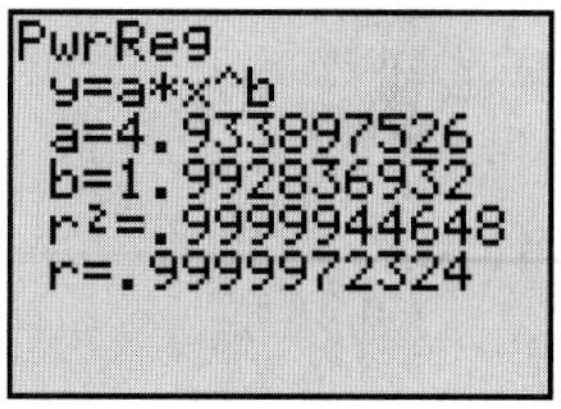

(c) See Figure 27 for the graph of $s(t)$.

(d) If $s = 100$ meters, then $Y_1 = s(t) = 4.93t^{1.993}$ INTERSECTs $Y_2 = 100$ when $t = 4.53$ seconds, rounded to two decimal places. See Figure 28.

Figure 27

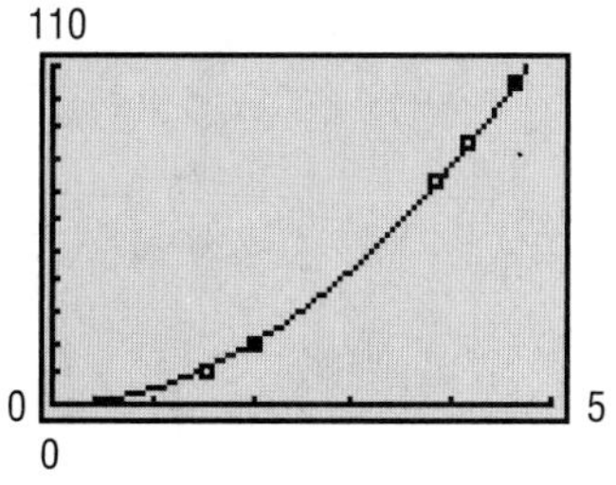

Figure 28

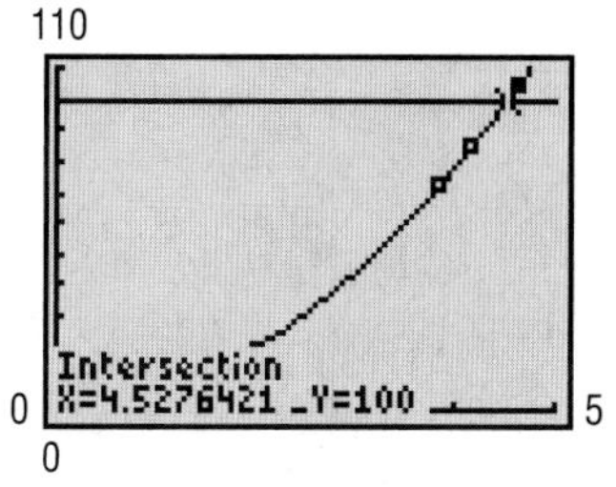

(e) The value of g obeys the equation

$$\frac{1}{2}g = 4.93$$

$$g = 9.86 \text{ meters/sec}^2$$

Scott's estimate for g is 9.86 meters/sec^2. ∎

*For power functions, a value for r is given, since such functions can be expressed as linear functions using logarithms (see Chapter 5).

3.2 EXERCISES

In Problems 1–16, use transformations of the graph of $y = x^4$ or $y = x^5$ to graph each function. Verify your results using a graphing utility.

1. $f(x) = (x + 1)^4$ **2.** $f(x) = (x - 2)^5$ **3.** $f(x) = x^5 - 3$

4. $f(x) = x^4 + 2$ **5.** $f(x) = \frac{1}{2}x^4$ **6.** $f(x) = 3x^5$

7. $f(x) = -x^5$ **8.** $f(x) = -x^4$ **9.** $f(x) = (x - 1)^5 + 2$

10. $f(x) = (x + 2)^4 - 3$ **11.** $f(x) = 2(x + 1)^4 + 1$ **12.** $f(x) = \frac{1}{2}(x - 1)^5 - 2$

13. $f(x) = 4 - (x - 2)^5$ **14.** $f(x) = 3 - (x + 2)^4$ **15.** $f(x) = -\frac{1}{2}(x - 2)^4 - 1$

16. $f(x) = 1 - 2(x + 1)^5$

17. CBL Experiment David is conducting an experiment to estimate the acceleration of an object due to gravity. David takes a ball and drops it from different heights and records the time that it takes for the ball to hit the ground. Using an optic laser connected to a stop watch in order to determine the time, he collects the following data:

Time (Seconds)	Distance (Feet)
1.003	16
1.365	30
1.769	50
2.093	70
2.238	80

(a) Draw a scatter diagram using time as the independent variable and distance as the dependent variable.
(b) Find the power function of best fit to the data.
(c) Graph the function found in part (b) of the scatter diagram.
(d) Predict how long it will take an object to fall 100 feet.
(e) Physics theory states that the distance that an object falls is directly proportional to the time squared. Rewrite the model found in part (b) so that it is of the form $s = \frac{1}{2}gt^2$, where s is distance, g is the acceleration due to gravity, and t is time. What is David's estimate of g? (It is known that the acceleration due to gravity is approximately 32 feet/sec^2.)

18. CBL Experiment Paul, David's friend, doesn't believe that David's estimate of acceleration due to gravity is correct. Paul repeats the experiment described in Problem 17 and obtains the data shown:
(a) Draw a scatter diagram using time as the dependent variable and distance as the independent variable.
(b) Find the power function of best fit to the data.
(c) Graph the function found in part (b) on the scatter diagram.
(d) Predict how long it will take an object to fall 100 feet. Why do you think this estimate differs from the one in Problem 17?

Time (Seconds)	Distance (Feet)
0.7907	10
1.1160	20
1.4760	35
1.6780	45
1.9380	60

(e) Use the function found in part (b) to estimate g. Why do you think this estimate is differs from the one found in part (e) of Problem 17?

19. Pendulums The *period* of a pendulum is the time required for one oscillation; the pendulum is usually referred to as *simple* when the angle made to the vertical is less than 5°. An experiment is conducted in which simple pendulums are constructed with different lengths, l, and the corresponding periods T are recorded. The following data are collected:

Length l (in feet)	Period T (in seconds)
1	1.10
2	1.55
3	1.89
4	2.24
5	2.51
6	2.76
7	2.91

(a) Using a graphing utility, draw a scatter diagram of the data with length as the independent variable and period as the dependent variable.
(b) Find the power function of best fit to the data.
(c) Graph the power function found in part (b) on the scatter diagram.

(d) Predict the period* of a simple pendulum whose length is known to be 2.3 feet.

20. CBL Experiment Cathy is conducting an experiment to measure the relation between a light bulb's intensity and the distance from the light source. She measures a 100-watt light bulb's intensity 1 meter from the bulb and at 0.1-meter intervals up to 2 meters from the bulb and obtains the data shown to the right.

(a) Using a graphing utility, draw a scatter diagram with distance as the independent variable and intensity as the dependent variable.

(b) Find the power function of best fit to the data.

(c) Graph this function on the scatter diagram.

(d) Predict the intensity of a 100-watt light bulb 2.3 meters away.

Distance (Meters)	Intensity
1.0	0.29645
1.1	0.25215
1.2	0.20547
1.3	0.17462
1.4	0.15342
1.5	0.13521
1.6	0.11450
1.7	0.10243
1.8	0.09231
1.9	0.08321
2.0	0.07342

PREPARING FOR THIS SECTION

Before getting started, review the following concepts:

- ✓ Polynomials (Appendix, Section 3, pp. 958–959)
- ✓ Intercepts (pp. 16–18)

3.3 POLYNOMIAL FUNCTIONS; CURVE FITTING

OBJECTIVES

1. Identify Polynomials and Their Degree
2. Identify the Zeros of a Polynomial and Their Multiplicity
3. Analyze the Graph of a Polynomial
4. Find the Cubic Function of Best Fit to Data

1 *Polynomial functions* are among the simplest expressions in algebra. They are easy to evaluate: only addition and repeated multiplication are required. Because of this, they are often used to approximate other, more complicated functions. In this section, we investigate characteristics of this important class of function.

A **polynomial function** is a function of the form

$$f(x) = a_n x^n + a_{n-1} x^{n-1} + \cdots + a_1 x + a_0 \qquad \textbf{(1)}$$

where $a_n, a_{n-1}, \ldots, a_1, a_0$ are real numbers and n is a nonnegative integer. The domain consists of all real numbers.

*In physics, it is proved that $T = \dfrac{2\pi}{\sqrt{32}}\sqrt{l}$, provided that friction is ignored. Since $\sqrt{l} = l^{1/2} = l^{0.5}$ and $\dfrac{2\pi}{\sqrt{32}} \approx 1.1107$, the power function obtained is reasonably close to what is expected.

A polynomial function is a function whose rule is given by a polynomial in one variable. The degree of a polynomial function is the degree of the polynomial in one variable.

EXAMPLE 1 Identifying Polynomial Functions

Determine which of the following are polynomial functions. For those that are, state the degree; for those that are not, tell why not.

(a) $f(x) = 2 - 3x^4$ (b) $g(x) = \sqrt{x}$ (c) $h(x) = \dfrac{x^2 - 2}{x^3 - 1}$

(d) $F(x) = 0$ (e) $G(x) = 8$

Solution

(a) f is a polynomial function of degree 4.

(b) g is not a polynomial function. The variable x is raised to the $\frac{1}{2}$ power, which is not a nonnegative integer.

(c) h is not a polynomial function. It is the ratio of two polynomials, and the polynomial in the denominator is of positive degree.

(d) F is the zero polynomial function; it is not assigned a degree.

(e) G is a nonzero constant function, a polynomial function of degree 0 since $G(x) = 8 = 8x^0$. ■

NOW WORK PROBLEMS 1 AND 5.

We have already discussed in detail polynomial functions of degrees 0, 1, and 2. See Table 5 for a summary of the characteristics of the graphs of these polynomial functions.

TABLE 5

Degree	Form	Name	Graph
No degree	$f(x) = 0$	Zero function	The x-axis
0	$f(x) = a_0, \quad a_0 \neq 0$	Constant function	Horizontal line with y-intercept a_0
1	$f(x) = a_1x + a_0, \quad a_1 \neq 0$	Linear function	Nonvertical, nonhorizontal line with slope a_1 and y-intercept a_0
2	$f(x) = a_2x^2 + a_1x + a_0, \quad a_2 \neq 0$	Quadratic function	Parabola: Graph opens up if $a_2 > 0$; graph opens down if $a_2 < 0$

GRAPHING POLYNOMIALS

If you take a course in calculus, you will learn that the graph of every polynomial function is both smooth and continuous. By **smooth**, we mean that the graph contains no sharp corners or cusps; by **continuous**, we mean that the graph has no gaps or holes and can be drawn without lifting pencil from paper. See Figures 29(a) and (b).

Figure 30 shows the graph of a polynomial function with four x-intercepts. Notice that at the x-intercepts the graph must either cross the x-axis or touch the x-axis. Consequently, between consecutive x-intercepts the graph is either above the x-axis or below the x-axis. We will make use of this characteristic of the graph of a polynomial shortly.

Figure 29

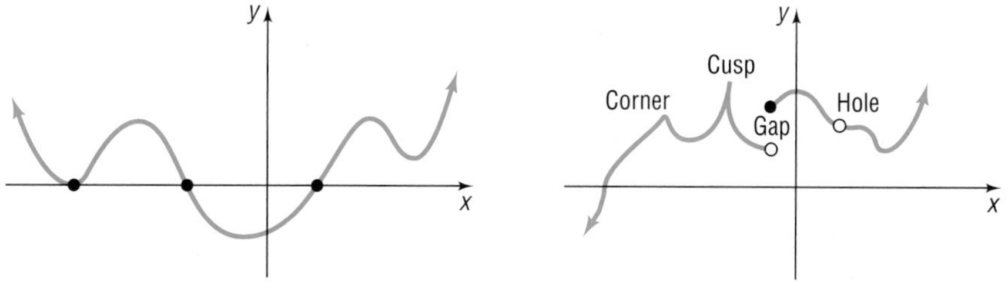

(a) Graph of a polynomial function: smooth, continuous

(b) Cannot be the graph of a polynomial function

Figure 30

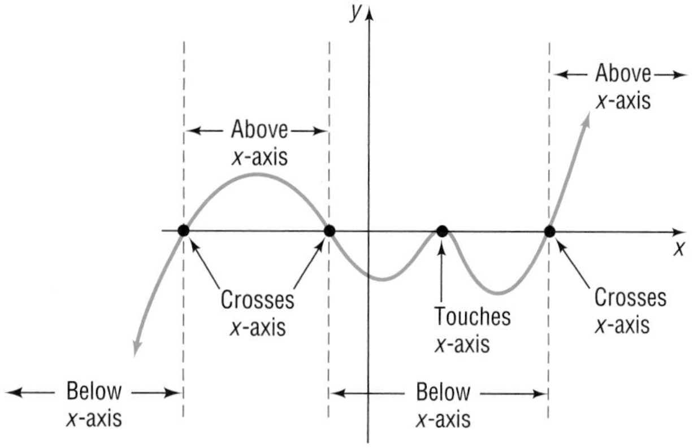

If a polynomial function f is factored completely, it is easy to solve the equation $f(x) = 0$ and locate the x-intercepts of the graph. For example, if $f(x) = (x - 1)^2(x + 3)$, then the solutions of the equation

$$f(x) = (x - 1)^2(x + 3) = 0$$

are identified as 1 and -3. Based on this result, we make the following observations:

If f is a polynomial function and r is a real number for which $f(r) = 0$, then r is called a (real) **zero of f**, or **root of f**. If r is a (real) zero of f, then

(a) r is an x-intercept of the graph of f.
(b) $(x - r)$ is a factor of f.

EXAMPLE 2 Finding a Polynomial from Its Zeros

(a) Find a polynomial of degree 3 whose zeros are -3, 2, and 5.
(b) Graph the polynomial found in part (a) to verify your result.

Solution (a) If r is a zero of a polynomial f, then $x - r$ is a factor of f. This means that $x - (-3) = x + 3$, $x - 2$, and $x - 5$ are factors of f. As a result, any polynomial of the form

$$f(x) = a(x + 3)(x - 2)(x - 5)$$

where a is any nonzero real number, qualifies.

Figure 31

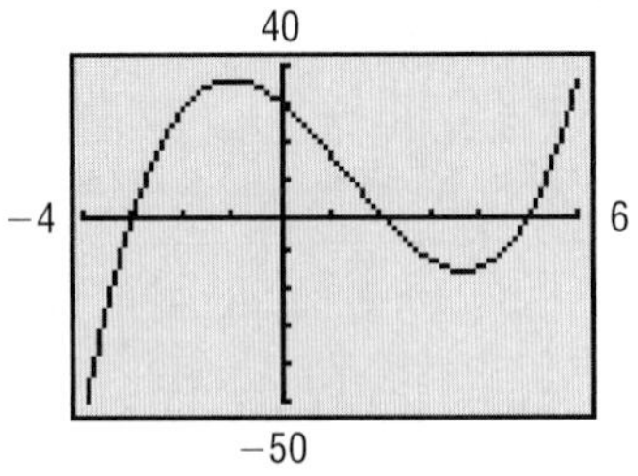

(b) The value of a causes a stretch, compression, or reflection, but does not affect the x-intercepts. We choose to graph f with $a = 1$.

$$f(x) = (x + 3)(x - 2)(x - 5) = x^3 - 4x^2 - 11x + 30$$

Figure 31 shows the graph of f. Notice that the x-intercepts are -3, 2, and 5.

SEEING THE CONCEPT Graph the function found in Example 2 for $a = 2$ and $a = -1$. Does the value of a affect the zeros of f? How does the value of a affect the graph of f?

NOW WORK PROBLEM 11.

2 If the same factor $x - r$ occurs more than once, then r is called a **repeated**, or **multiple, zero of *f***. More precisely, we have the following definition.

If $(x - r)^m$ is a factor of a polynomial f and $(x - r)^{m+1}$ is not a factor of f, then r is called a **zero of multiplicity *m* of *f***.

EXAMPLE 3 Identifying Zeros and Their Multiplicities

For the polynomial

$$f(x) = 5(x - 2)(x + 3)^2\left(x - \frac{1}{2}\right)^4$$

2 is a zero of multiplicity 1.

-3 is a zero of multiplicity 2.

$\frac{1}{2}$ is a zero of multiplicity 4.

In Example 3 notice that, if you add the multiplicities $(1 + 2 + 4 = 7)$, you obtain the degree of the polynomial.

Suppose that it is possible to factor completely a polynomial function and, as a result, locate all the x-intercepts of its graph (the real zeros of the function). The following example illustrates the role that the multiplicity of the x-intercept plays.

EXAMPLE 4 Investigating the Role of Multiplicity

For the polynomial $f(x) = x^2(x - 2)$:

(a) Find the x- and y-intercepts of the graph.
(b) Using a graphing utility, graph the polynomial.
(c) For each x-intercept, determine whether it is of odd or even multiplicity.

Solution (a) The y-intercept is $f(0) = 0^2(0 - 2) = 0$. The x-intercepts satisfy the equation

$$f(x) = x^2(x - 2) = 0$$

from which we find that

$$x^2 = 0 \quad \text{or} \quad x - 2 = 0$$

$$x = 0 \qquad\qquad x = 2$$

The x-intercepts are 0 and 2.

(b) See Figure 32 for the graph of f.

Figure 32

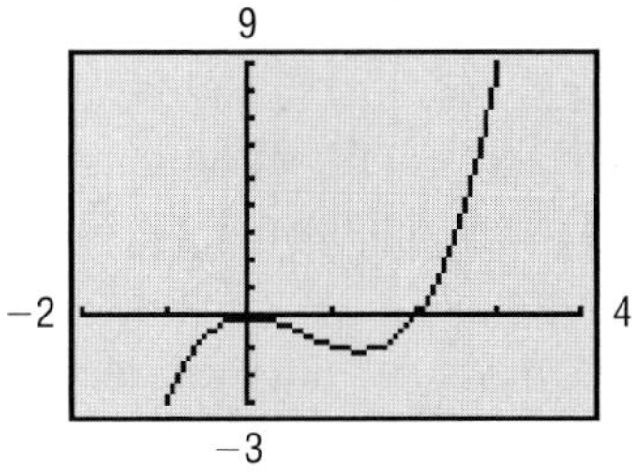

(c) We can see from the factored form of f that 0 is a zero or root of multiplicity 2, and 2 is a zero or root of multiplicity 1; so 0 is of even multiplicity and 2 is of odd multiplicity. ■

TABLE 6

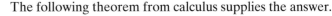

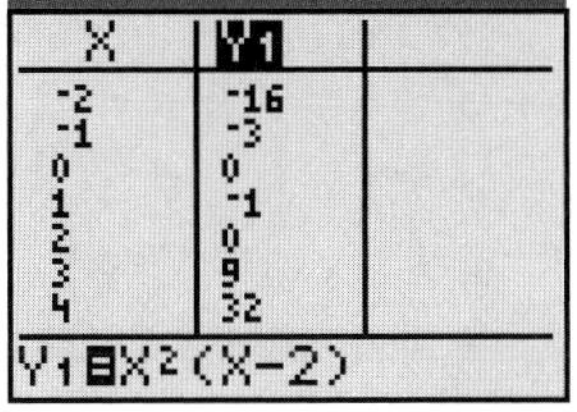

X	Y1
-2	-16
-1	-3
0	0
1	-1
2	0
3	9
4	32

Y1=X²(X−2)

We can use a TABLE to further analyze the graph. See Table 6. The sign of $f(x)$ is the same on each side of the root 0, so the graph of f just touches the x-axis at 0 (a root of even multiplicity). The sign of $f(x)$ changes from one side of the root 2 to the other, so the graph of f crosses the x-axis at 2 (a root of odd multiplicity). These observations suggest the following result:

If r Is a Zero of Even Multiplicity

Sign of $f(x)$ does not change from one side to the other side of r.	Graph **touches** x-axis at r.

If r Is a Zero of Odd Multiplicity

Sign of $f(x)$ changes from one side to the other side of r.	Graph **crosses** x-axis at r.

NOW WORK PROBLEM 17.

Look again at Figure 32. We can use a graphing utility to determine the graph's local minimum in the interval $0 < x < 2$. After utilizing MINIMUM, we find that the graph's local minimum in the interval $0 < x < 2$ is $(1.33, -1.19)$, rounded to two decimal places. Local minima and local maxima are points where the graph changes direction (that is, changes from an increasing function to a decreasing function, or vice versa). We call these points **turning points**.

Look again at Figure 32. The graph of $f(x) = x^2(x - 2) = x^3 - 2x^2$, a polynomial of degree 3, has two turning points.

Exploration Graph $Y_1 = x^3$, $Y_2 = x^3 - x$, and $Y_3 = x^3 + 3x^2 + 4$. How many turning points do you see? How does the number of turning points relate to the degree? Graph $Y_1 = x^4$, $Y_2 = x^4 - \frac{4}{3}x^3$, and $Y_3 = x^4 - 2x^2$. How many turning points do you see? How does the number of turning points compare to the degree? ■

The following theorem from calculus supplies the answer.

Theorem If f is a polynomial function of degree n, then f has at most $n - 1$ turning points.

One last remark about Figure 32. Notice that the graph of $f(x) = x^2(x - 2)$ looks somewhat like the graph of $y = x^3$. In fact, for very large values of x, either positive or negative, there is little difference.

EXPLORATION For each pair of functions Y_1 and Y_2 given below in parts (a), (b), and (c), graph Y_1 and Y_2 on the same viewing window. Create a TABLE or TRACE for large positive and large negative values of x. What do you notice about the graphs of Y_1 and Y_2 as x becomes very large and positive or very large and negative?

(a) $Y_1 = x^2(x - 2)$; $Y_2 = x^3$
(b) $Y_1 = x^4 - 3x^3 + 7x - 3$; $Y_2 = x^4$
(c) $Y_1 = -2x^3 + 4x^2 - 8x + 10$; $Y_2 = -2x^3$

The behavior of the graph of a function for large values of x, either positive or negative, is referred to as its **end behavior**.

Theorem **End Behavior**

For large values of x, either positive or negative, the graph of the polynomial

$$f(x) = a_n x^n + a_{n-1}x^{n-1} + \cdots + a_1 x + a_0$$

resembles the graph of the power function

$$y = a_n x^n$$

The following summarizes some features of the graph of a polynomial function.

SUMMARY

Graph of a Polynomial Function $f(x) = a_n x^n + a_{n-1}x^{n-1} + \cdots + a_1 x + a_0, \quad a_n \neq 0$

Degree of the polynomial f: n
Maximum number of turning points: $n - 1$
At zero of even multiplicity: The graph of f touches x-axis.
At zero of odd multiplicity: The graph of f crosses x-axis.
Between zeros, the graph of f is either above or below the x-axis.
End behavior: For large $|x|$, the graph of f behaves like the graph of $y = a_n x^n$.

EXAMPLE 5 Analyzing the Graph of a Polynomial Function

3 For the polynomial $f(x) = x^4 - 3x^3 - 4x^2$:

(a) Using a graphing utility, graph f.
(b) Find the x- and y-intercepts.
(c) Determine whether each x-intercept is of odd or even multiplicity.

(d) Find the power function that the graph of f resembles for large values of $|x|$.

(e) Determine the number of turning points on the graph of f.

(f) Determine the local maxima and local minima, if any exist, rounded to two decimal places.

Figure 33

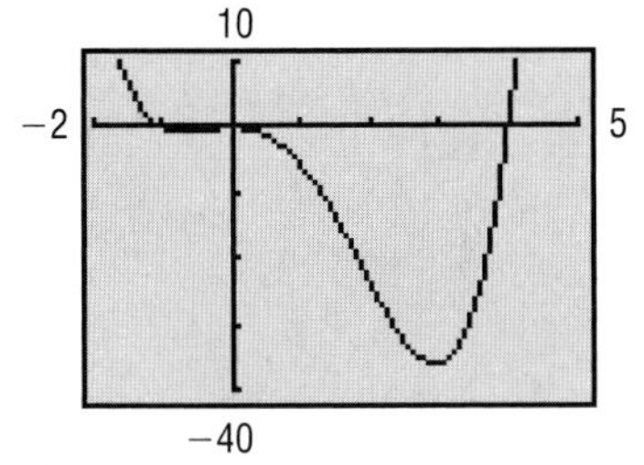

Solution

(a) See Figure 33 for the graph of f.

(b) The y-intercept is $f(0) = 0$. Factoring, we find that

$$f(x) = x^4 - 3x^3 - 4x^2 = x^2(x^2 - 3x - 4) = x^2(x - 4)(x + 1)$$

We find the x-intercepts by solving the equation

$$f(x) = x^2(x - 4)(x + 1) = 0$$

So

$$x^2 = 0 \quad \text{or} \quad x - 4 = 0 \quad \text{or} \quad x + 1 = 0$$

$$x = 0 \qquad\qquad x = 4 \qquad\qquad x = -1$$

The x-intercepts are -1, 0, and 4.

(c) The intercept 0 is a zero of even multiplicity, 2, so the graph of f will touch the x-axis at 0; 4 and -1 are zeros of odd multiplicity, 1, so the graph of f will cross the x-axis at 4 and -1. Look again at Figure 33. The graph is below the x-axis for $-1 < x < 0$ and $0 < x < 4$.

(d) The graph of f behaves like $y = x^4$ for large $|x|$.

(e) Since f is of degree 4, the graph can have at most three turning points. From the graph, we see that it has three turning points: one between -1 and 0, one at $(0, 0)$, and one between 2 and 4.

(f) Rounded to two decimal places, the local maximum is $(0, 0)$ and the local minima are $(-0.68, -0.69)$ and $(2.93, -36.10)$. ■

EXAMPLE 6 Analyzing the Graph of a Polynomial Function

Follow the instructions of Example 5 for the following polynomial:

$$f(x) = x^3 + 2.48x^2 - 4.3155x + 1.484406$$

Figure 34

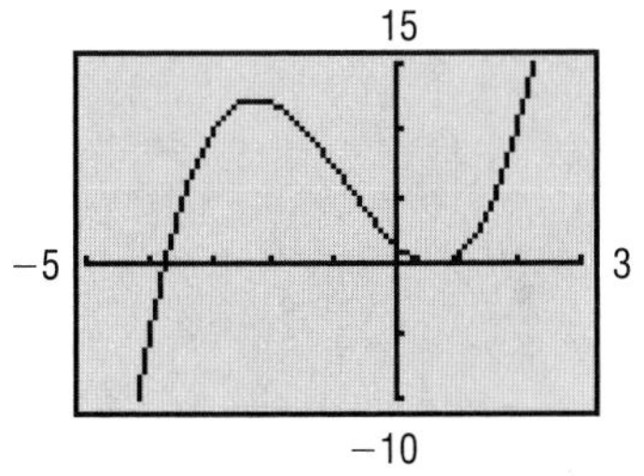

Solution

(a) See Figure 34 for the graph of f.

(b) The y-intercept is $f(0) = 1.484406$. In Example 5 we could easily factor $f(x)$ to find the x-intercepts. However, it is not readily apparent how $f(x)$ factors in this example. Therefore, we use a graphing utility and find the x-intercepts to be -3.74 and 0.63.

(c) The x-intercept -3.74 is of odd multiplicity since the graph of f crosses the x-axis at -3.74; the x-intercept 0.63 is of even multiplicity since the graph of f touches the x-axis at 0.63, rounded to two decimal places.

(d) The graph behaves like $y = x^3$ for large $|x|$.

(e) Since f is of degree 3, the graph can have at most two turning points. From the graph we see that it has two turning points: one between -3 and -2, the other at $(0.63, 0)$.

(f) Rounded to two decimal places, the local maximum is $(-2.28, 12.36)$ and the local minimum is $(0.63, 0)$. ■

NOW WORK PROBLEMS 29 AND 49.

Modeling Polynomial Functions: Curve Fitting

4 In Section 1.7 we found the line of best fit from data; in Section 3.1, we found the quadratic function of best fit; and in Section 3.2 we found the power function of best fit. It is also possible to find polynomial functions of best fit. However, most statisticians do not recommend finding polynomials of best fit of degree higher than 3.*

Data that follow a cubic relation should look like Figure 35(a) or (b).

Figure 35

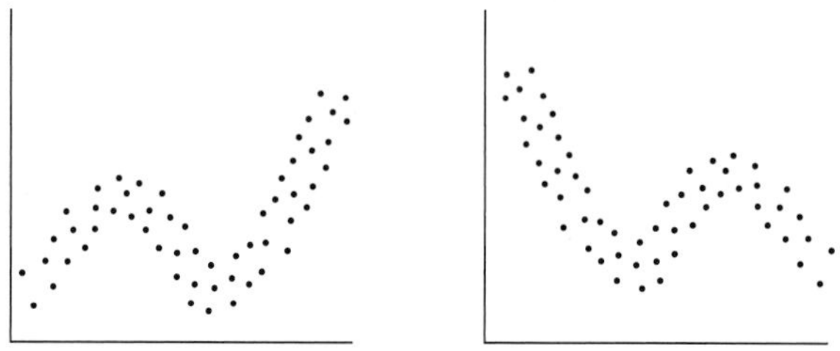

$y = ax^3 + bx^2 + cx + d, a > 0$ (a)

$y = ax^3 + bx^2 + cx + d, a < 0$ (b)

EXAMPLE 7 A Cubic Function of Best Fit

The data in Table 7 represent the average fuel consumption for cars (in gallons) in the United States for 1984–1996, where 1 represents 1984, 2 represents 1985, and so on.

TABLE 7

Year, x	Average Fuel Consumption, C
1984, 1	495
1985, 2	505
1986, 3	507
1987, 4	500
1988, 5	487
1989, 6	486
1990, 7	461
1991, 8	443
1992, 9	455
1993, 10	462
1994, 11	462
1995, 12	530
1996, 13	531

Source: U.S. Federal Highway Administration

*Two points determine a unique linear function. Three noncollinear points determine a unique quadratic function. Four points determine a unique cubic function, and n points determine a unique polynomial of degree $n - 1$. Therefore, higher-degree polynomials will always "fit" data at least as well as lower-degree polynomials. However, higher-degree polynomials yield highly erratic predictions. Since the ultimate goal of curve fitting is not necessarily to find the model that best fits the data, but instead to explain relationships between two or more variables, polynomial models of degree 3 or less are usually used.

(a) Draw a scatter diagram of the data. Comment on the type of relation that may exist between the two variables.

(b) Find the cubic function of best fit.

(c) Draw the cubic function of best fit on the scatter diagram.

(d) Use the function found in part (b) to predict the average fuel consumption for cars in 1999 ($x = 16$).

(e) Do you think the function found in part (b) will be useful in predicting the average fuel consumption for cars in 2010? Why?

Solution (a) Figure 36 shows the scatter diagram. A cubic relation with $a > 0$ may exist between the two variables.

(b) Upon executing the CUBIC REGression program, we obtain the results shown in Figure 37. The output that the utility provides shows us the equation $y = ax^3 + bx^2 + cx + d$. The cubic function of best fit is

$$C(x) = 0.398x^3 - 6.793x^2 + 26.313x + 476.692$$

where x represents the year and C represents the average fuel consumption.

(c) Figure 38 shows the graph of the cubic function of best fit on the scatter diagram. The function seems to fit the data well.

Figure 36

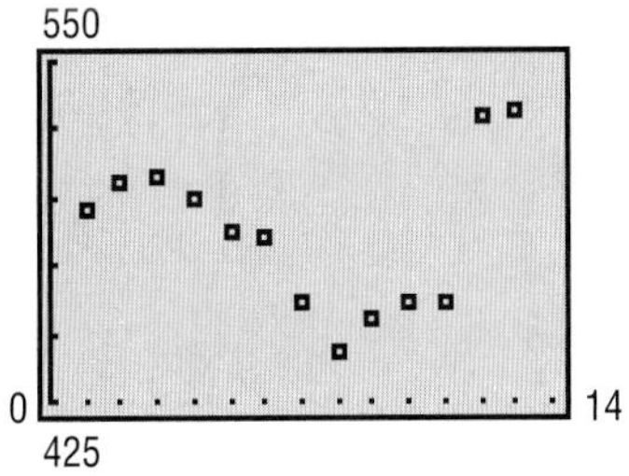

Figure 37

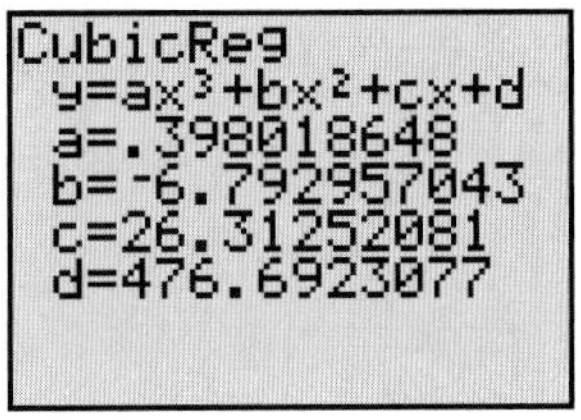

Figure 38

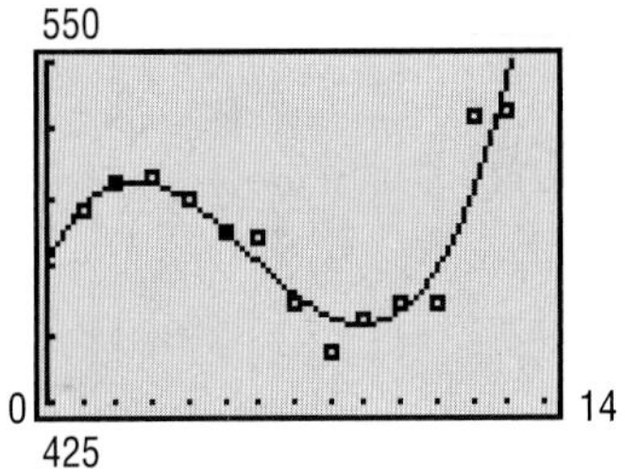

(d) We evaluate the function $C(x)$ for $x = 16$.

$$C(x) = 0.398(16)^3 - 6.793(16)^2 + 26.313(16) + 476.692 \approx 789$$

So we predict that the average annual fuel consumption for cars will be 789 gallons in 1999.

(e) The graph of the function will continually increase for years beyond 1996, indicating that the average fuel consumption for cars will continue to increase. While it is likely that people will continue to drive more in the years to come, the increases will probably start to subside, since cars are getting better gas mileage. So the function probably will not be useful in predicting the average fuel consumption per car in 2010. ■

3.3 EXERCISES

In Problems 1–10, determine which functions are polynomial functions. For those that are, state the degree. For those that are not, tell why not.

1. $f(x) = 4x + x^3$

2. $f(x) = 5x^2 + 4x^4$

3. $g(x) = \dfrac{1 - x^2}{2}$

4. $h(x) = 3 - \dfrac{1}{2}x$

5. $f(x) = 1 - \dfrac{1}{x}$

6. $f(x) = x(x - 1)$

7. $g(x) = x^{3/2} - x^2 + 2$

8. $h(x) = \sqrt{x}(\sqrt{x} - 1)$

9. $F(x) = 5x^4 - \pi x^3 + \frac{1}{2}$

10. $F(x) = \frac{x^2 - 5}{x^3}$

In Problems 11–16, form a polynomial whose zeros and degree are given.

11. Zeros: $-1, 1, 3$; degree 3

12. Zeros: $-2, 2, 3$; degree 3

13. Zeros: $-3, 0, 4$; degree 3

14. Zeros: $-4, 0, 2$; degree 3

15. Zeros: $-4, -1, 2, 3$; degree 4

16. Zeros: $-3, -1, 2, 5$; degree 4

In Problems 17–26, for each polynomial function, list each real zero and its multiplicity. Determine whether the graph crosses or touches the x-axis at each x-intercept. Do not graph f.

17. $f(x) = 3(x - 7)(x + 3)^2$

18. $f(x) = 4(x + 4)(x + 3)^3$

19. $f(x) = 4(x^2 + 1)(x - 2)^3$

20. $f(x) = 2(x - 3)(x + 4)^3$

21. $f(x) = -2\left(x + \frac{1}{2}\right)^2(x^2 + 4)^2$

22. $f(x) = \left(x - \frac{1}{3}\right)^2(x - 1)^3$

23. $f(x) = (x - 5)^3(x + 4)^2$

24. $f(x) = (x + \sqrt{3})^2(x - 2)^4$

25. $f(x) = 3(x^2 + 8)(x^2 + 9)^2$

26. $f(x) = -2(x^2 + 3)^3$

In Problems 27–62, for each polynomial function f:

(a) Using a graphing utility, graph f.
(b) Find the x- and y-intercepts.
(c) Determine whether each x-intercept is of odd or even multiplicity.
(d) Find the power function that the graph of f resembles for large values of |x|.
(e) Determine the number of turning points on the graph of f.
(f) Determine the local maxima and local minima, if any exist, rounded to two decimal places.

27. $f(x) = (x - 1)^2$

28. $f(x) = (x - 2)^3$

29. $f(x) = x^2(x - 3)$

30. $f(x) = x(x + 2)^2$

31. $f(x) = 6x^3(x + 4)$

32. $f(x) = 5x(x - 1)^3$

33. $f(x) = -4x^2(x + 2)$

34. $f(x) = -\frac{1}{2}x^3(x + 4)$

35. $f(x) = x(x - 2)(x + 4)$

36. $f(x) = x(x + 4)(x - 3)$

37. $f(x) = 4x - x^3$

38. $f(x) = x - x^3$

39. $f(x) = x^2(x - 2)(x + 2)$

40. $f(x) = x^2(x - 3)(x + 4)$

41. $f(x) = x^2(x - 2)^2$

42. $f(x) = x^3(x - 3)$

43. $f(x) = x^2(x - 3)(x + 1)$

44. $f(x) = x^2(x - 3)(x - 1)$

45. $f(x) = x(x + 2)(x - 4)(x - 6)$

46. $f(x) = x(x - 2)(x + 2)(x + 4)$

47. $f(x) = x^2(x - 2)(x^2 + 3)$

48. $f(x) = x^2(x^2 + 1)(x + 4)$

49. $f(x) = x^3 + 0.2x^2 - 1.5876x - 0.31752$

50. $f(x) = x^3 - 0.8x^2 - 4.6656x + 3.73248$

51. $f(x) = x^3 + 2.56x^2 - 3.31x + 0.89$

52. $f(x) = x^3 - 2.91x^2 - 7.668x - 3.8151$

53. $f(x) = x^4 - 2.5x^2 + 0.5625$

54. $f(x) = x^4 - 18.5x^2 + 50.2619$

55. $f(x) = x^4 + 0.65x^3 - 16.6319x^2 + 14.209335x - 3.1264785$

56. $f(x) = x^4 + 3.45x^3 - 11.6639x^2 - 5.864241x - 0.69257738$

57. $f(x) = \pi x^3 + \sqrt{2}x^2 - x - 2$

58. $f(x) = -2x^3 + \pi x^2 + \sqrt{3}x + 1$

59. $f(x) = 2x^4 - \pi x^3 + \sqrt{5}x - 4$

60. $f(x) = -1.2x^4 + 0.5x^2 - \sqrt{3}x + 2$

61. $f(x) = -2x^5 - \sqrt{2}x^2 - x - \sqrt{2}$

62. $f(x) = \pi x^5 + \pi x^4 + \sqrt{3}x + 1$

63. Consult illustrations (a)–(d). Construct a polynomial function that might have this graph. (More than one answer may be possible.)

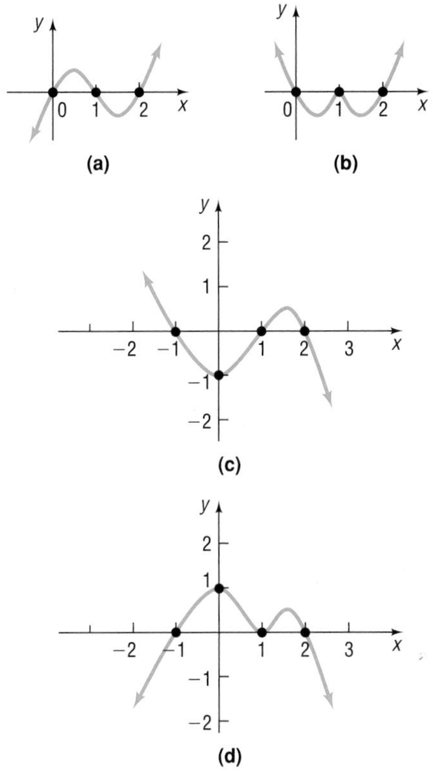

64. Cost of Printing The following data represent the weekly cost of printing textbooks C (in thousands) and the number x of texts printed (in thousands).

Number of Text Books, x	Cost, C
0	100
5	128.1
10	144
13	153.5
17	161.2
18	162.6
20	166.3
23	178.9
25	190.2
27	221.8

(a) Draw a scatter diagram of the data. Comment on the type of relation that may exist between the two variables.
(b) Find the average rate of change of cost from 10,000 to 13,000 textbooks.
(c) What is the average rate of change in the cost of producing from 18,000 to 20,000 textbooks?
(d) Find the cubic function of best fit to the data.
(e) Graph the cubic function of best fit on the scatter diagram.
(f) Use the function found in (d) to predict the cost of printing 22,000 texts per week.
(g) Interpret the y-intercept.

65. Motor Vehicle Thefts The following data represent the number of motor vehicle thefts (in thousands) in the United States for the years 1987–1997, where 1 represents 1987, 2 represents 1988, and so on.

Year, x	Motor Vehicle Thefts, M
1987, 1	1289
1988, 2	1433
1989, 3	1565
1990, 4	1636
1991, 5	1662
1992, 6	1611
1993, 7	1563
1994, 8	1539
1995, 9	1472
1996, 10	1394
1997, 11	1354

Source: U.S. Federal Bureau of Investigation

(a) Draw a scatter diagram of the data. Comment on the type of relation that may exist between the two variables.
(b) Find the cubic function of best fit.
(c) Graph the cubic function of best fit on the scatter diagram.
(d) Use the function found in (b) to predict the number of motor vehicle thefts in 1998.
(e) Check the prediction of part (d) against actual data. Do you think that the function given in part (b) will be useful in predicting the number of motor vehicle thefts in 1999?

66. Larceny Thefts The following data represent the number of larceny thefts (in thousands) in the United States for the years 1983–1993, where 1 represents 1983, 2 represents 1984, and so on.

Year, x	Larceny Thefts, L
1983, 1	6713
1984, 2	6592
1985, 3	6926
1986, 4	7257
1987, 5	7500
1988, 6	7706
1989, 7	7872
1990, 8	7946
1991, 9	8142
1992, 10	7915
1993, 11	7821

Source: U.S. Federal Bureau of Investigation

(a) Using a graphing utility, draw a scatter diagram of the data. Comment on the type of relation that may exist between the two variables.
(b) Find the cubic function of best fit.
(c) Using a graphing utility, draw the cubic function of best fit on your scatter diagram.
(d) Use the function found in part (b) to predict the number of larcency thefts in 1994.
(e) Do you think the function found in part (b) will be useful in predicting the number of larceny thefts in 1999?

67. **Cost of Manufacturing** The following data represent the cost C of manufacturing Chevy Cavaliers (in thousands of dollars) and the number x of Cavaliers produced in one hour.

Number of Cavaliers Produced, x	Cost, C
0	10
1	23
2	31
3	38
4	43
5	50
6	59
7	70
8	85
9	105
10	135

(a) Draw a scatter diagram of the data. Comment on the type of relation that may exist between the two variables.
(b) Find the average rate of change of cost from four to five Cavaliers.
(c) What is the average rate of change of cost from eight to nine Cavaliers?
(d) Find the cubic function of best fit to the data.
(e) Graph the cubic function of best fit on the scatter diagram.
(f) Use the function found in (d) to predict the cost of manufacturing 11 Cavaliers in one hour.
(g) Interpret the y-intercept.

68. Can the graph of a polynomial function have no y-intercept? Can it have no x-intercepts? Explain.

69. Write a few paragraphs that provide a general strategy for graphing a polynomial function. Be sure to mention the following: degree, intercepts, and turning points.

70. Make up a polynomial that has the following characteristics: crosses the x-axis at -1 and 4, touches the x-axis at 0 and 2, and is above the x-axis between 0 and 2. Give your polynomial to a fellow classmate and ask for a written critique of your polynomial.

71. Make up two polynomials, not of the same degree, with the following characteristics: crosses the x-axis at -2, touches the x-axis at 1, and is above the x-axis between -2 and 1. Give your polynomials to a fellow classmate and ask for a written critique of your polynomials.

72. The graph of a polynomial function is always smooth and continuous. Name a function studied earlier that is smooth and not continuous. Name one that is continuous, but not smooth.

73. Which of the following statements are true regarding the graph of the cubic polynomial $f(x) = x^3 + bx^2 + cx + d$? (Give reasons for your conclusions.)
(a) It intersects the y-axis in one and only one point.
(b) It intersects the x-axis in at most three points.
(c) It intersects the x-axis at least once.
(d) For $|x|$ very large, it behaves like the graph of $y = x^3$.
(e) It is symmetric with respect to the origin.
(f) It passes through the origin.

PREPARING FOR THIS SECTION

Before getting started, review the following concepts:

✓ Rational Expressions (Appendix, Section 7)

✓ Polynomial Division (Appendix, Section 4, pp. 963–965)

3.4 RATIONAL FUNCTIONS I

OBJECTIVES
1. Find the Domain of a Rational Function
2. Determine the Vertical Asymptotes of a Rational Function
3. Determine the Horizontal or Oblique Asymptotes of a Rational Function

Ratios of integers are called *rational numbers*. Similarly, ratios of polynomial functions are called *rational functions*.

A **rational function** is a function of the form

$$R(x) = \frac{p(x)}{q(x)}$$

where p and q are polynomial functions and q is not the zero polynomial. The domain consists of all real numbers except those for which the denominator q is 0.

EXAMPLE 1 Finding the Domain of a Rational Function

1 (a) The domain of $R(x) = \dfrac{2x^2 - 4}{x + 5}$ consists of all real numbers x except -5.

(b) The domain of $R(x) = \dfrac{1}{x^2 - 4}$ consists of all real numbers x except -2 and 2.

(c) The domain of $R(x) = \dfrac{x^3}{x^2 + 1}$ consists of all real numbers.

(d) The domain of $R(x) = \dfrac{-x^2 + 2}{3}$ consists of all real numbers.

(e) The domain of $R(x) = \dfrac{x^2 - 1}{x - 1}$ consists of all real numbers x except 1.

It is important to observe that the functions

$$R(x) = \frac{x^2 - 1}{x - 1} \quad \text{and} \quad f(x) = x + 1$$

are not equal, since the domain of R is $\{x \mid x \neq 1\}$ and the domain of f is all real numbers.

NOW WORK PROBLEM 3.

If $R(x) = p(x)/q(x)$ is a rational function and if p and q have no common factors, then the rational function R is said to be in **lowest terms**. For a rational function $R(x) = p(x)/q(x)$ in lowest terms, the zeros, if any, of the numerator are the x-intercepts of the graph of R and so will play a major role in the graph of R. The zeros of the denominator of R [that is, the numbers x, if any, for which $q(x) = 0$], although not in the domain of R, also play a major role in the graph of R. We will discuss this role shortly.

We have already discussed the characteristics of the rational function $f(x) = 1/x$. (Refer to Example 12, page 21). The next rational function that we take up is $H(x) = 1/x^2$.

EXAMPLE 2 Graphing $y = \frac{1}{x^2}$

Analyze the graph $H(x) = \frac{1}{x^2}$.

Solution See Figure 39. The domain of $H(x) = 1/x^2$ consists of all real numbers x except 0. The graph has no y-intercept, because x can never equal 0. The graph has no x-intercept because the equation $H(x) = 0$ has no solution. Therefore, the graph of H will not cross either coordinate axis.

Figure 39

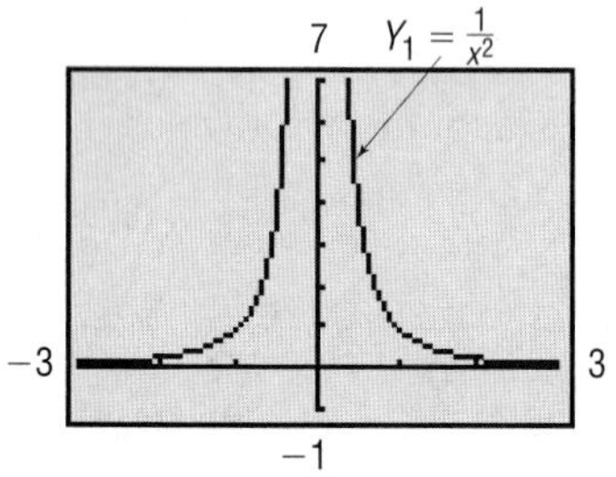

TABLE 8

X	Y1
.1	100
.01	10000
.001	1E6
1E-4	1E8
10	.01
100	1E-4
1000	1E-6

Y1=1/X²

Because

$$H(-x) = \frac{1}{(-x)^2} = \frac{1}{x^2} = H(x)$$

H is an even function, so its graph is symmetric with respect to the y-axis.

Look again at Figure 39. What happens to the function as the values of x get closer and closer to 0? We use a TABLE to answer the question. See Table 8. The first four rows show that as x approaches 0 the values of $H(x)$ become larger and larger positive numbers. When this happens, we say that $H(x)$ is **unbounded in the positive direction**. We symbolize this by writing $H(x) \to \infty$ [read as "$H(x)$ **approaches infinity**"]. In calculus, **limits** are used to convey these ideas. We use the symbolism $\lim_{x \to 0} H(x) = \infty$, read as "the limit of $H(x)$ as x approaches 0 is infinity" to mean that $H(x) \to \infty$ as $x \to 0$.

Look at the last-three rows of Table 8. As $x \to \infty$, the values of $H(x)$ approach 0. This is symbolized in calculus by writing $\lim_{x \to \infty} H(x) = 0$. Figure 40 shows the graph of $y = 1/x^2$ drawn by hand.

Figure 40 $H(x) = \frac{1}{x^2}$

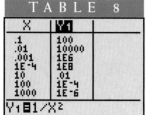

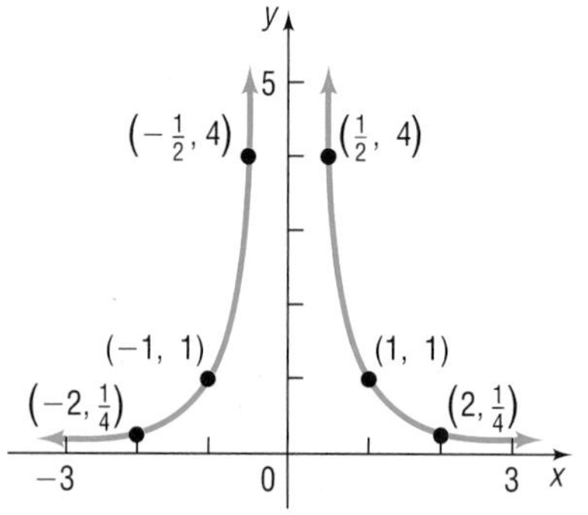

Sometimes transformations (shifting, compressing, stretching, and reflection) can be used to graph a rational function.

EXAMPLE 3 Using Transformations to Graph a Rational Function

Graph the rational function: $R(x) = \frac{1}{(x-2)^2} + 1$

Solution First, we take note of the fact that the domain of R consists of all real numbers except $x = 2$. To graph R, we start with the graph of $y = 1/x^2$. See Figure 41 for the steps.

Figure 41

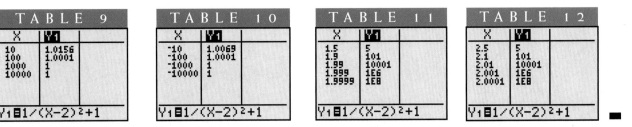

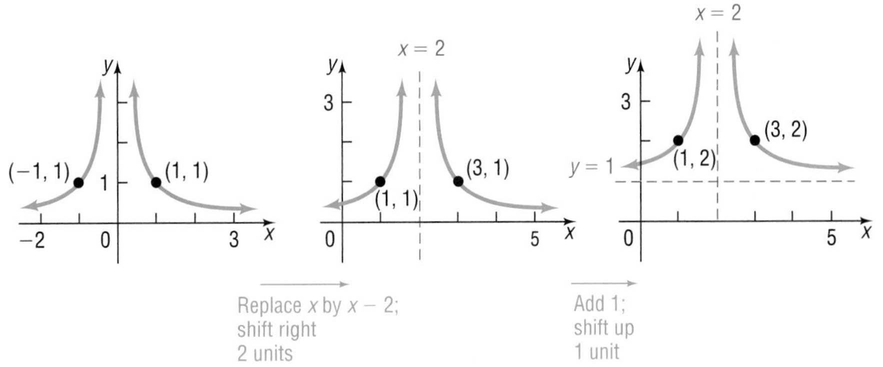

(a) $y = \dfrac{1}{x^2}$ (b) $y = \dfrac{1}{(x-2)^2}$ (c) $y = \dfrac{1}{(x-2)^2} + 1$

Check: Graph $Y_1 = 1/(x-2)^2 + 1$ using a graphing utility to verify the graph obtained in Figure 41(c).

NOW WORK PROBLEM 23.

ASYMPTOTES

Notice that the y-axis in Figure 41(a) is transformed into the vertical line $x = 2$ in Figure 41(c) and the x-axis in Figure 41(a) is transformed into the horizontal line $y = 1$ in Figure 41(c). The **Exploration** that follows will help us analyze the role of these lines.

EXPLORATION Using a graphing utility and the TABLE feature, evaluate the function $H(x) = \dfrac{1}{(x-2)^2} + 1$ at $x = 10, 100, 1000,$ and 10,000. What happens to the values of H as x becomes unbounded in the positive direction, symbolized by $\lim_{x\to\infty} H(x)$?

Evaluate H at $x = -10, -100, -1000,$ and $-10{,}000$. What happens to the values of H as x becomes unbounded in the negative direction, symbolized by $\lim_{x\to-\infty} H(x)$?

Evaluate H at $x = 1.5, 1.9, 1.99, 1.999,$ and 1.9999. What happens to the values of H as x approaches 2 for $x < 2$, symbolized by $\lim_{x\to 2^-} H(x)$?

Evaluate H at $x = 2.5, 2.1, 2.01, 2.001,$ and 2.0001. What happens to the values of H as x approaches 2 for $x > 2$, symbolized by $\lim_{x\to 2^+} H(x)$?

RESULT Table 9 shows the values of $Y_1 = H(x)$ as x approaches ∞. Notice that the values of H are approaching 1, so $\lim_{x\to\infty} H(x) = 1$. Table 10 shows the values of $Y_1 = H(x)$ as x approaches $-\infty$. Again the values of H are approaching 1, so $\lim_{x\to-\infty} H(x) = 1$. From Table 11 we see that, as x approaches 2 for $x < 2$, the values of H are increasing without bound, so $\lim_{x\to 2^-} H(x) = \infty$. Finally, Table 12 reveals that, as x approaches 2 for $x > 2$, the values of H are increasing without bound, so $\lim_{x\to 2^+} H(x) = \infty$.

TABLE 9

X	Y1
10	1.0156
100	1.0001
1000	1
10000	1

Y1=1/(X−2)²+1

TABLE 10

X	Y1
-10	1.0069
-100	1.0001
-1000	1
-10000	1

Y1=1/(X−2)²+1

TABLE 11

X	Y1
1.5	5
1.9	101
1.99	10001
1.999	1E6
1.9999	1E8

Y1=1/(X−2)²+1

TABLE 12

X	Y1
2.5	5
2.1	101
2.01	10001
2.001	1E6
2.0001	1E8

Y1=1/(X−2)²+1

The results of the **Exploration** reveal an important characteristic of rational functions. The vertical line $x = 2$ and the horizontal line $y = 1$ are called *asymptotes* of the graph of H, which we define as follows:

Let R denote a function:

If, as $x \to -\infty$ or as $x \to \infty$, the values of $R(x)$ approach some fixed number L, then the line $y = L$ is a **horizontal asymptote** of the graph of R.

If, as x approaches some number c, the values $|R(x)| \to \infty$, then the line $x = c$ is a **vertical asymptote** of the graph of R.

Even though the asymptotes of a function are not part of the graph of the function, they provide information about how the graph looks. Figure 42 illustrates some of the possibilities.

Figure 42

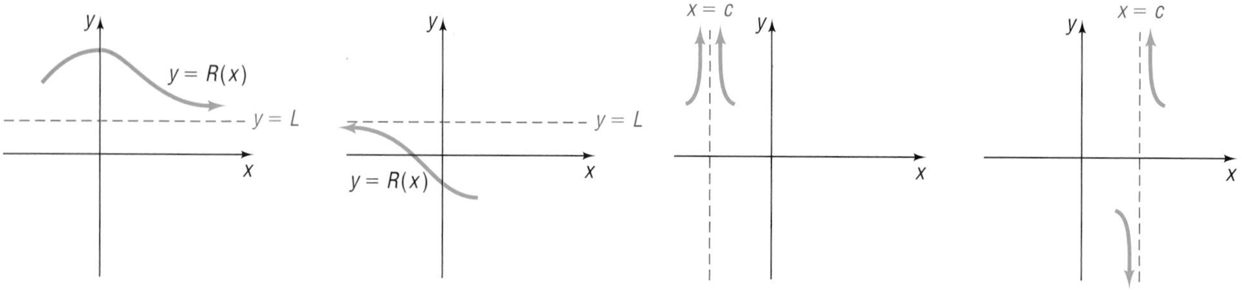

(a) End behavior: As $x \to \infty$, the values of $R(x)$ approach L. [$\lim_{x\to\infty} R(x) = L$] that is, the points on the graph of R are getting closer to the line $y = L$; $y = L$ is a horizontal asymptote.

(b) End behavior: As $x \to -\infty$, the values of $R(x)$ approach L. [$\lim_{x\to-\infty} R(x) = L$] that is, the points on the graph of R are getting closer to the line $y = L$; $y = L$ is a horizontal asymptote.

(c) As x approaches c, the values of $|R(x)| \to \infty$, [$\lim_{x\to c^-} R(x) = \infty$, $\lim_{x\to c^+} R(x) = \infty$,] that is, the points on the graph of R are getting closer to the line $x = c$; $x = c$ is a vertical asymptote.

(d) As x approaches c, the values of $|R(x)| \to \infty$, [$\lim_{x\to c^-} R(x) = -\infty$; $\lim_{x\to c^+} R(x) = \infty$,] that is, the points on the graph of R are getting closer to the line $x = c$; $x = c$ is a vertical asymptote.

Figure 43
Oblique asymptote

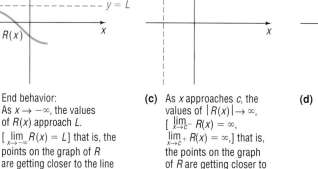

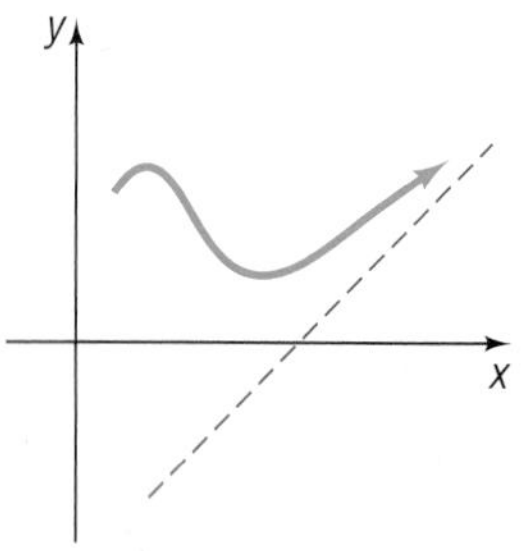

An asymptote is a line that a certain part of the graph of a function gets closer and closer to but never touches. However, at other parts of the graph, the function may intersect a nonvertical asymptote. The graph of the function will never intersect a vertical asymptote. Notice that a horizontal asymptote, when it occurs, describes a certain behavior of the graph as $x \to \infty$ or as $x \to -\infty$, that is, its end behavior. A vertical asymptote, when it occurs, describes a certain behavior of the graph when x is close to some number c.

If an asymptote is neither horizontal nor vertical, it is called **oblique**. Figure 43 shows an oblique asymptote.

Finding Asymptotes

2 The vertical asymptotes, if any, of a rational function $R(x) = p(x)/q(x)$, in lowest terms, are found by factoring the denominator of $q(x)$. Suppose that $x - r$ is a factor of the denominator. Now, as x approaches r, symbolized as $x \to r$, the values of $x - r$ approach 0, causing the ratio to become un-

bounded, that is, causing $|R(x)| \to \infty$. Based on the definition, we conclude that the line $x = r$ is a vertical asymptote.

Theorem

Locating Vertical Asymptotes

A rational function $R(x) = p(x)/q(x)$, *in lowest terms*, will have a vertical asymptote $x = r$ if $x - r$ is a factor of the denominator q. That is, if r is a real zero of the denominator of a rational function $R(x) = p(x)/q(x)$, in lowest terms, then R will have the vertical asymptote $x = r$.

WARNING: If a rational function is not in lowest terms, an application of this theorem may result in an incorrect listing of vertical asymptotes.

EXAMPLE 4 Finding Vertical Asymptotes

Find the vertical asymptotes, if any, of the graph of each rational function.

(a) $R(x) = \dfrac{x}{x^2 - 4}$ (b) $F(x) = \dfrac{x + 3}{x - 1}$

(c) $H(x) = \dfrac{x^2}{x^2 + 1}$ (d) $G(x) = \dfrac{x^2 - 9}{x^2 + 4x - 21}$

Solution

(a) The zeros of the denominator $x^2 - 4$ are -2 and 2. Hence, the lines $x = -2$ and $x = 2$ are the vertical asymptotes of the graph of R.

(b) The only zero of the denominator is 1. Hence, the line $x = 1$ is the only vertical asymptote of the graph of F.

(c) The denominator has no real zeros. Hence, the graph of H has no vertical asymptotes.

(d) Factor $G(x)$ to determine if it is in lowest terms.

$$G(x) = \frac{x^2 - 9}{x^2 + 4x - 21} = \frac{(x + 3)(x - 3)}{(x + 7)(x - 3)} = \frac{x + 3}{x + 7}, \qquad x \neq 3$$

The only zero of the denominator of $G(x)$ in lowest terms is -7. Hence, the line $x = -7$ is the only vertical asymptote of the graph of G.

EXPLORATION Graph each of the following rational functions:

$$R(x) = \frac{1}{x - 1} \quad R(x) = \frac{1}{(x - 1)^2} \quad R(x) = \frac{1}{(x - 1)^3} \quad R(x) = \frac{1}{(x - 1)^4}$$

Each has the vertical asymptote $x = 1$. What happens to the value of $R(x)$ as x approaches 1 from the right side of the vertical asymptote; that is, what is $\lim_{x \to 1^+} R(x)$? What happens to the value of $R(x)$ as x approaches 1 from the left side of the vertical asymptote; that is, what is $\lim_{x \to 1^-} R(x)$? How does the multiplicity of the zero in the denominator affect the graph of R?

3 The procedure for finding horizontal and oblique asymptotes is somewhat more involved. To find such asymptotes, we need to know how the values of a function behave as $x \to -\infty$ or as $x \to \infty$.

If a rational function $R(x)$ is **proper**, that is, if the degree of the numerator is less than the degree of the denominator, then as $x \to -\infty$ or as

$x \to \infty$, the values of $R(x)$ approach 0. Consequently, the line $y = 0$ (the x-axis) is a horizontal asymptote of the graph.

Theorem If a rational function is proper, the line $y = 0$ is a horizontal asymptote of its graph.

EXAMPLE 5 Finding Horizontal Asymptotes

Find the horizontal asymptotes, if any, of the graph of

$$R(x) = \frac{x - 12}{4x^2 + x + 1}$$

Solution The rational function R is proper, since the degree of the numerator, 1, is less than the degree of the denominator, 2. We conclude that the line $y = 0$ is a horizontal asymptote of the graph of R.

To see why $y = 0$ is a horizontal asymptote of the function R in Example 5, we need to investigate the behavior of R as $x \to -\infty$ and $x \to \infty$. When x is unbounded, the numerator of R, which is $x - 12$, can be approximated by the power function $y = x$, while the denominator of R, which is $4x^2 + x + 1$, can be approximated by the power function $y = 4x^2$. Applying these ideas to $R(x)$, we find

$$R(x) = \frac{x - 12}{4x^2 + x + 1} \underset{\substack{\uparrow \\ \text{For } x \text{ unbounded}}}{\approx} \frac{x}{4x^2} = \frac{1}{4x} \underset{\substack{\uparrow \\ \text{As } x \to -\infty \text{ or } x \to \infty}}{\to} 0$$

This shows that the line $y = 0$ is a horizontal asymptote of the graph of R. We verify these results in Tables 13(a) and (b). Notice, as $x \to -\infty$ or $x \to \infty$, that the values of $R(x)$ approach 0.

TABLE 13

X	Y1
-10	-.0563
-100	-.0028
-1000	-3E-4
-10000	-3E-5
-1E5	-3E-6
-1E6	-3E-7
-1E7	-3E-8

Y1 = (X−12)/(4X²+...

(a)

X	Y1
10	-.0049
100	.00219
1000	2.5E-4
10000	2.5E-5
100000	2.5E-6
1E6	2.5E-7
1E7	2.5E-8

Y1 = (X−12)/(4X²+...

(b)

If a rational function $R(x) = p(x)/q(x)$ is **improper**, that is, if the degree of the numerator is greater than or equal to the degree of the denominator, we must use long division to write the rational function as the sum of a polynomial $f(x)$ plus a proper rational function $r(x)/q(x)$. That is, we write

$$R(x) = \frac{p(x)}{q(x)} = f(x) + \frac{r(x)}{q(x)}$$

where $f(x)$ is a polynomial and $r(x)/q(x)$ is a proper rational function. Since $r(x)/q(x)$ is proper, then $r(x)/q(x) \to 0$ as $x \to -\infty$ or as $x \to \infty$. As a result,

$$R(x) = \frac{p(x)}{q(x)} \to f(x), \qquad \text{as } x \to -\infty \text{ or as } x \to \infty$$

The possibilities are listed next.

1. If $f(x) = b$, a constant, then the line $y = b$ is a horizontal asymptote of the graph of R.
2. If $f(x) = ax + b$, $a \neq 0$, then the line $y = ax + b$ is an oblique asymptote of the graph of R.
3. In all other cases, the graph of R approaches the graph of f, and there are no horizontal or oblique asymptotes.

The following examples demonstrate these conclusions.

EXAMPLE 6 Finding Horizontal or Oblique Asymptotes

Find the horizontal or oblique asymptotes, if any, of the graph of

$$H(x) = \frac{3x^4 - x^2}{x^3 - x^2 + 1}$$

Solution The rational function H is improper, since the degree of the numerator, 4, is larger than the degree of the denominator, 3. To find any horizontal or oblique asymptotes, we use long division.

$$\begin{array}{r} 3x + 3 \\ x^3 - x^2 + 1 \overline{\smash{)}\,3x^4 - x^2 } \\ \underline{3x^4 - 3x^3 + 3x} \\ 3x^3 - x^2 - 3x \\ \underline{3x^3 - 3x^2 + 3} \\ 2x^2 - 3x - 3 \end{array}$$

As a result,

$$H(x) = \frac{3x^4 - x^2}{x^3 - x^2 + 1} = 3x + 3 + \frac{2x^2 - 3x - 3}{x^3 - x^2 + 1}$$

Then, as $x \to -\infty$ or as $x \to \infty$, the remainder will behave as follows:

$$\frac{2x^2 - 3x - 3}{x^3 - x^2 + 1} \approx \frac{2x^2}{x^3} = \frac{2}{x} \to 0$$

As $x \to -\infty$ or as $x \to \infty$, we have $H(x) \to 3x + 3$. We conclude that the graph of the rational function H has an oblique asymptote $y = 3x + 3$. We verify these results in Tables 14(a) and (b) with $Y_1 = H(x)$ and $Y_2 = 3x + 3$. As $x \to -\infty$ or $x \to \infty$, the difference in the values between Y_1 and Y_2 is indistinguishable.

TABLE 14

X	Y1
-10	-.0563
-100	-.0028
-1000	-3E-4
-10000	-3E-5
-1E5	-3E-6
-1E6	-3E-7
-1E7	-3E-8

Y1■(X-12)/(4X²+...

(a)

X	Y1	Y2
10	33.185	33
100	303.02	303
1000	3003	3003
10000	30003	30003
100000	300003	300003
1E6	3E6	3E6
1E7	3E7	3E7

Y1■(3X^4-X²)/(X...

(b)

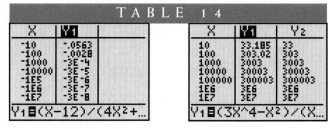

EXAMPLE 7 Finding Horizontal or Oblique Asymptotes

Find the horizontal or oblique asymptotes, if any, of the graph of

$$R(x) = \frac{8x^2 - x + 2}{4x^2 - 1}$$

Solution The rational function R is improper, since the degree of the numerator, 2, equals the degree of the denominator, 2. To find any horizontal or oblique asymptotes, we use long division.

$$\begin{array}{r} 2 \\ 4x^2 - 1\overline{)8x^2 - x + 2} \\ \underline{8x^2 \quad\quad - 2} \\ -x + 4 \end{array}$$

As a result,

$$R(x) = \frac{8x^2 - x + 2}{4x^2 - 1} = 2 + \frac{-x + 4}{4x^2 - 1}$$

Then, as $x \to -\infty$ or as $x \to \infty$, the remainder will behave as follows:

$$\frac{-x + 4}{4x^2 - 1} \approx \frac{-x}{4x^2} = \frac{-1}{4x} \to 0$$

As $x \to -\infty$ or as $x \to \infty$, we have $R(x) \to 2$. We conclude that $y = 2$ is a horizontal asymptote of the graph. ■

Check: Verify the results of Example 7 by creating a TABLE with $Y_1 = R(x)$ and $Y_2 = 2$. ■

In Example 7, we note that the quotient 2 obtained by long division is the quotient of the leading coefficients of the numerator polynomial and the denominator polynomial $\left(\frac{8}{4}\right)$. This means that we can avoid the long division process for rational functions whose numerator and denominator *are of the same degree* and conclude that the quotient of the leading coefficients will give us the horizontal asymptote.

NOW WORK PROBLEM 39.

EXAMPLE 8 Finding Horizontal or Oblique Asymptotes

Find the horizontal or oblique asymptotes, if any, of the graph of

$$G(x) = \frac{2x^5 - x^3 + 2}{x^3 - 1}$$

Solution The rational function G is improper, since the degree of the numerator, 5, is larger than the degree of the denominator, 3. To find any horizontal or oblique asymptotes, we use long division.

$$
\begin{array}{r}
2x^2 - 1 \\
x^3 - 1 \overline{)\, 2x^5 - x^3 + 2} \\
\underline{2x^5 - 2x^2 } \\
-x^3 + 2x^2 + 2 \\
\underline{-x^3 + 1} \\
2x^2 + 1
\end{array}
$$

As a result,

$$G(x) = \frac{2x^5 - x^3 + 2}{x^3 - 1} = 2x^2 - 1 + \frac{2x^2 + 1}{x^3 - 1}$$

Then, as $x \to -\infty$ or as $x \to \infty$, the remainder will behave as follows:

$$\frac{2x^2 + 1}{x^3 - 1} \approx \frac{2x^2}{x^3} = \frac{2}{x} \to 0$$

As $x \to -\infty$ or as $x \to \infty$, we have $G(x) \to 2x^2 - 1$. We conclude that, for large values of $|x|$, the graph of G approaches the graph of $y = 2x^2 - 1$. That is, the graph of G will look like the graph of $y = 2x^2 - 1$ as $x \to -\infty$ or $x \to \infty$. Since $y = 2x^2 - 1$ is not a linear function, G has no horizontal or oblique asymptotes. ■

We now summarize the procedure for finding horizontal and oblique asymptotes.

SUMMARY

FINDING HORIZONTAL AND OBLIQUE ASYMPTOTES OF A RATIONAL FUNCTION R

Consider the rational function

$$R(x) = \frac{p(x)}{q(x)} = \frac{a_n x^n + a_{n-1}x^{n-1} + \cdots + a_1 x + a_0}{b_m x^m + b_{m-1}x^{m-1} + \cdots + b_1 x + b_0}$$

in which the degree of the numerator is n and the degree of the denominator is m.

1. If $n < m$, then R is a proper rational function, and the graph of R will have the horizontal asymptote $y = 0$ (the x-axis).
2. If $n \geq m$, then R is improper. Here long division is used.
   (a) If $n = m$, the quotient obtained will be a number $L(= a_n/b_m)$, and the line $y = L(= a_n/b_m)$ is a horizontal asymptote.
   (b) If $n = m + 1$, the quotient obtained is of the form $ax + b$ (a polynomial of degree 1), and the line $y = ax + b$ is an oblique asymptote.
   (c) If $n > m + 1$, the quotient obtained is a polynomial of degree 2 or higher, and R has neither a horizontal nor an oblique asymptote. In this case, for x unbounded, the graph of R will behave like the graph of the quotient.

NOTE: The graph of a rational function either has one horizontal or one oblique asymptote or else has no horizontal and no oblique asymptote.

3.4 EXERCISES

In Problems 1–12, find the domain of each rational function.

1. $R(x) = \dfrac{4x}{x-3}$

2. $R(x) = \dfrac{5x^2}{3+x}$

3. $H(x) = \dfrac{-4x^2}{(x-2)(x+4)}$

4. $G(x) = \dfrac{6}{(x+3)(4-x)}$

5. $F(x) = \dfrac{3x(x-1)}{2x^2-5x-3}$

6. $Q(x) = \dfrac{-x(1-x)}{3x^2+5x-2}$

7. $R(x) = \dfrac{x}{x^3-8}$

8. $R(x) = \dfrac{x}{x^4-1}$

9. $H(x) = \dfrac{3x^2+x}{x^2+4}$

10. $G(x) = \dfrac{x-3}{x^4+1}$

11. $R(x) = \dfrac{3(x^2-x-6)}{4(x^2-9)}$

12. $F(x) = \dfrac{-2(x^2-4)}{3(x^2+4x+4)}$

In Problems 13–22, use the graph shown to find:

(a) The domain and range of each function
(b) The intercepts, if any
(c) Horizontal asymptotes, if any
(d) Vertical asymptotes, if any
(e) Oblique asymptotes, if any

13.

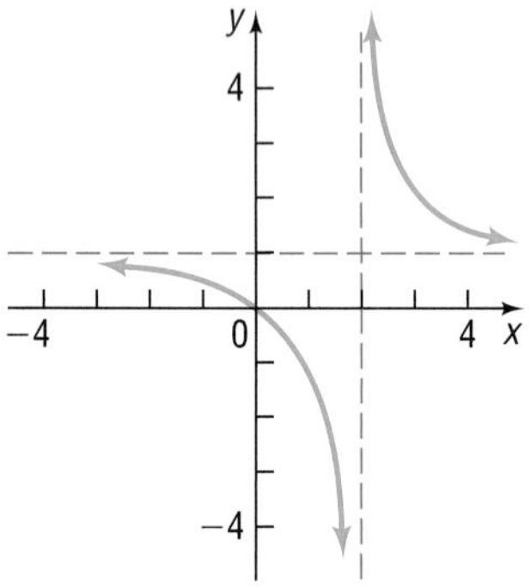

14.

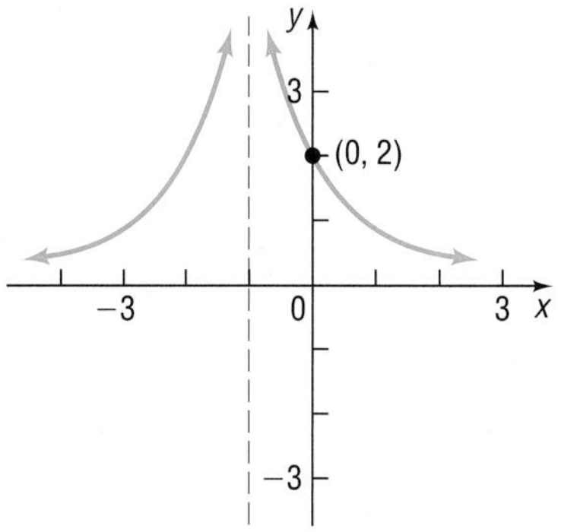

15.

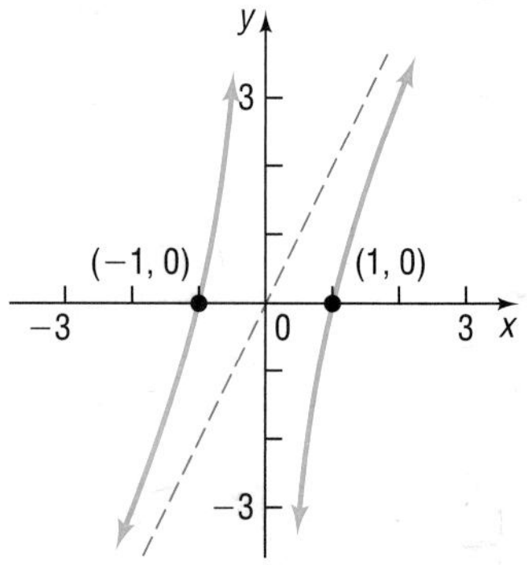

16.

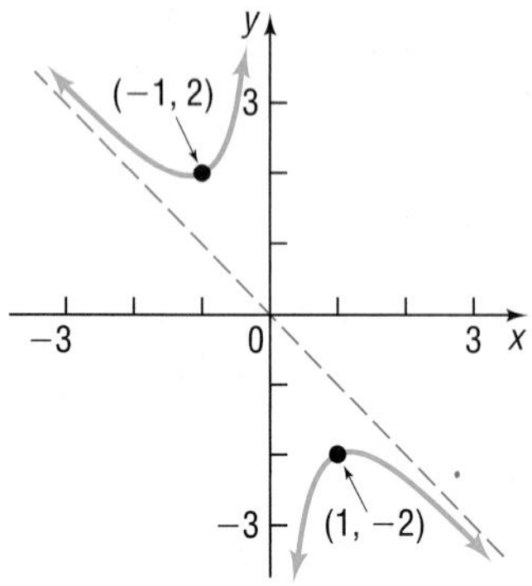

17.

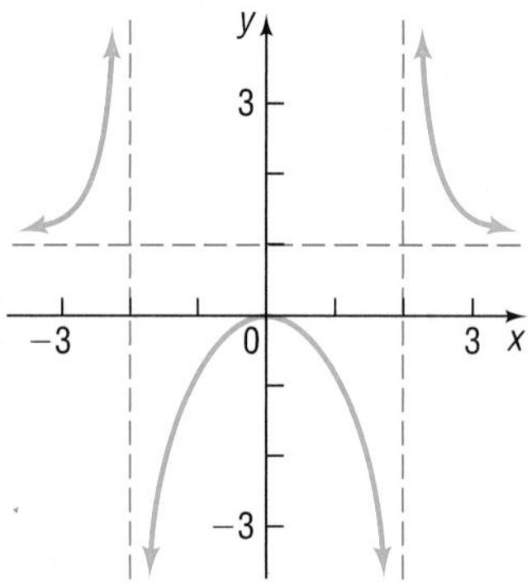

18.

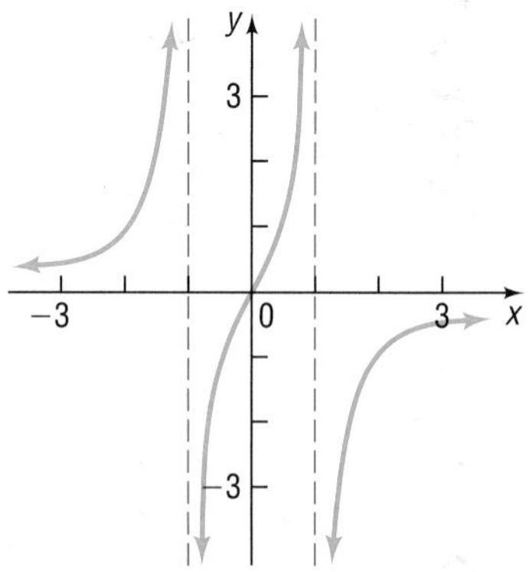

19.

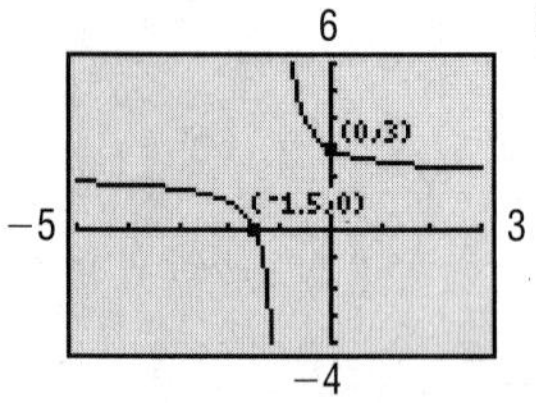

20.

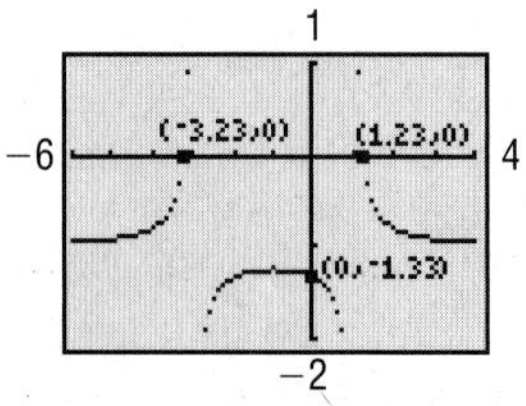

21.

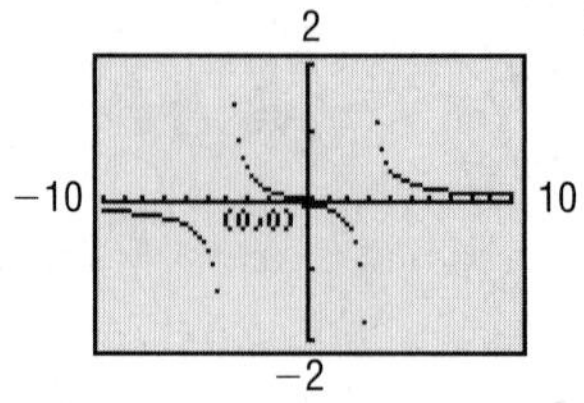

22.

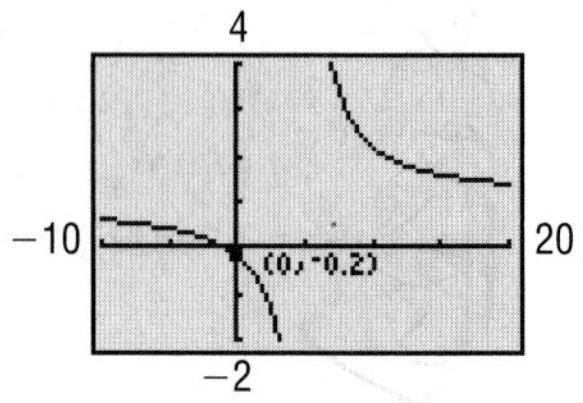

In Problems 23–34, show the steps required to graph each rational function using transformations. Verify your results using a graphing utility.

23. $R(x) = \dfrac{1}{(x-1)^2}$

24. $R(x) = \dfrac{3}{x}$

25. $H(x) = \dfrac{-2}{x+1}$

26. $G(x) = \dfrac{2}{(x+2)^2}$

27. $R(x) = \dfrac{1}{x^2+4x+4}$

28. $R(x) = \dfrac{1}{x-1} + 1$

29. $F(x) = 1 - \dfrac{1}{x}$

30. $Q(x) = 1 + \dfrac{1}{x}$

31. $R(x) = \dfrac{x^2-4}{x^2}$

32. $R(x) = \dfrac{x-4}{x}$

33. $G(x) = 1 + \dfrac{2}{(x-3)^2}$

34. $F(x) = 2 - \dfrac{1}{x+1}$

In Problems 35–46, find the vertical, horizontal, and oblique asymptotes, if any, of each rational function without graphing. Verify your results using a graphing utility.

35. $R(x) = \dfrac{3x}{x+4}$

36. $R(x) = \dfrac{3x+5}{x-6}$

37. $H(x) = \dfrac{x^4+2x^2+1}{x^2-x+1}$

38. $G(x) = \dfrac{-x^2+1}{x+5}$

39. $T(x) = \dfrac{x^3}{x^4-1}$

40. $P(x) = \dfrac{4x^5}{x^3-1}$

41. $Q(x) = \dfrac{5-x^2}{3x^4}$

42. $F(x) = \dfrac{-2x^2+1}{2x^3+4x^2}$

43. $R(x) = \dfrac{3x^4+4}{x^3+3x}$

44. $R(x) = \dfrac{6x^2+x+12}{3x^2-5x-2}$

45. $G(x) = \dfrac{x^3-1}{x-x^2}$

46. $F(x) = \dfrac{x-1}{x-x^3}$

47. If the graph of a rational function R has the vertical asymptote $x = 4$, then the factor $x - 4$ must be present in the denominator of R. Explain why.

48. If the graph of a rational function R has the horizontal asymptote $y = 2$, then the degree of the numerator of R equals the degree of the denominator of R. Explain why.

49. Can the graph of a rational function have both a horizontal and an oblique asymptote? Explain.

50. Make up a rational function that has $y = 2x + 1$ as an oblique asymptote. Explain the methodology that you used.

PREPARING FOR THIS SECTION

Before getting started, review the following concepts:

- ✓ Intercepts (pp. 16–18)
- ✓ Symmetry (pp. 18–19)
- ✓ Even and Odd Functions (pp. 122–124)

3.5 RATIONAL FUNCTIONS II: ANALYZING GRAPHS

OBJECTIVES 1 Analyze the Graph of a Rational Function
2 Solve Applied Problems Involving Rational Functions

1 Graphing utilities make the task of graphing rational functions less time consuming. However, the results of algebraic analysis must be taken into account before drawing conclusions based on the graph provided by the utility. We will use the information collected in the last section in conjunction with the graphing utility to analyze the graph of a rational function $R(x) = p(x)/q(x)$. The analysis will require the following steps:

ANALYZING THE GRAPH OF A RATIONAL FUNCTION

STEP 1: Find the domain of the rational function.

STEP 2: Locate the intercepts, if any, of the graph. The x-intercepts, if any, of $R(x) = p(x)/q(x)$ satisfy the equation $p(x) = 0$. The y-intercept, if there is one, is $R(0)$.

STEP 3: Test for symmetry. Replace x by $-x$ in $R(x)$. If $R(-x) = R(x)$, there is symmetry with respect to the y-axis; if $R(-x) = -R(x)$, there is symmetry with respect to the origin.

STEP 4: Write R in lowest terms and find the real zeros of the denominator. With R in lowest terms, each zero will give rise to a vertical asymptote.

STEP 5: Locate the horizontal or oblique asymptotes, if any, using the procedure given earlier. Determine points, if any, at which the graph of R intersects these asymptotes.

STEP 6: Graph R using a graphing utility.

STEP 7: Determine additional points on the graph of R and use the results obtained in Steps 1 through 6 to graph R by hand.

EXAMPLE 1 Analyzing the Graph of a Rational Function

Analyze the graph of the rational function: $R(x) = \dfrac{x - 1}{x^2 - 4}$

Solution First, we factor both the numerator and the denominator of R.

$$R(x) = \frac{x - 1}{(x + 2)(x - 2)}$$

R is in lowest terms.

STEP 1: The domain of R is $\{x \mid x \neq -2, x \neq 2\}$.

STEP 2: We locate the x-intercepts by finding the zeros of the numerator. By inspection, 1 is the only x-intercept. The y-intercept is $R(0) = \frac{1}{4}$.

STEP 3: Because

$$R(-x) = \frac{-x - 1}{x^2 - 4} = \frac{-(x + 1)}{x^2 - 4}$$

we conclude that R is neither even nor odd. There is no symmetry with respect to the y-axis or the origin.

STEP 4: Since R is in lowest terms, the graph of R has two vertical asymptotes: the lines $x = -2$ and $x = 2$.

STEP 5: The degree of the numerator is less than the degree of the denominator, so R is proper and the line $y = 0$ (the x-axis) is a horizontal asymptote of the graph. To determine if the graph of R intersects the horizontal asymptote, we solve the equation $R(x) = 0$:

$$\frac{x - 1}{x^2 - 4} = 0$$

$$x - 1 = 0$$

$$x = 1$$

The only solution is $x = 1$, so the graph of R intersects the horizontal asymptote at $(1, 0)$.

Figure 44

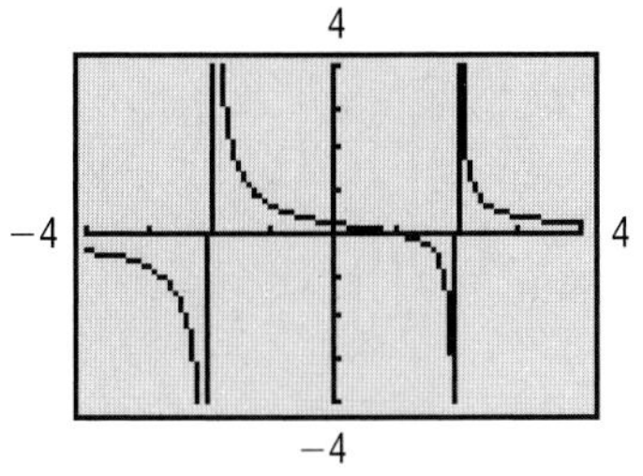

Connected mode
(a)

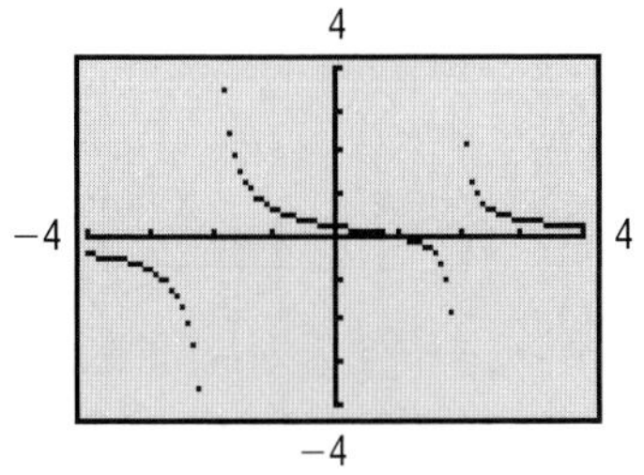

Dot mode
(b)

TABLE 15

X	Y1
-3	-.8
-2	ERROR
0	.25
1	0
1.5	-.2857
2	ERROR
3	.4

Y1■(X−1)/(X²−4)

STEP 6: The analysis just completed helps us to set the viewing window to obtain a complete graph. Figure 44(a) shows the graph of $R(x) = \dfrac{x-1}{x^2-4}$ in connected mode, and Figure 44(b) shows it in dot mode. Notice in Figure 44(a) that the graph has vertical lines at $x = -2$ and $x = 2$. This is due to the fact that, when the graphing utility is in connected mode, it will "connect the dots" between consecutive pixels. We know that the graph of R does not cross the lines $x = -2$ and $x = 2$, since R is not defined at $x = -2$ or $x = 2$. When graphing rational functions, dot mode should be used if extraneous vertical lines appear. You should confirm that all the algebraic conclusions that we arrived at in Steps 1 through 5 are part of the graph. For example, the graph has vertical asymptotes at $x = -2$ and $x = 2$, and the graph has a horizontal asymptote at $y = 0$. The y-intercept is $\frac{1}{4}$ and the x-intercept is 1.

STEP 7: We use the table feature of our graphing utility to find additional key points on the graph of R. To determine key points, use the zero of the numerator, 1, and the zeros of the denominator, −2 and 2, to divide the x-axis into four intervals:

$$-\infty < x < -2 \qquad -2 < x < 1 \qquad 1 < x < 2 \qquad 2 < x < \infty$$

Evaluate the function at any value of x within each interval and at the endpoints. See Table 15. Using the points obtained in Table 15 along with the information gathered in Steps 1 through 6, we obtain the graph of R shown in Figure 45.

Figure 45

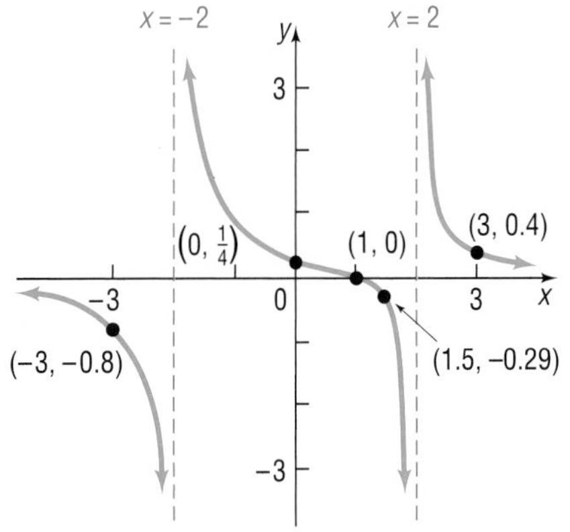

NOW WORK PROBLEM 1.

EXAMPLE 2 Analyzing the Graph of a Rational Function

Analyze the graph of the rational function: $R(x) = \dfrac{x^2 - 1}{x}$

Solution

STEP 1: The domain of R is $\{x \mid x \neq 0\}$.

STEP 2: The graph has two x-intercepts: −1 and 1. There is no y-intercept, since x cannot equal 0.

STEP 3: Since $R(-x) = -R(x)$, the function is odd and the graph is symmetric with respect to the origin.

STEP 4: Since R is in lowest terms, the graph of $R(x)$ has the line $x = 0$ (the y-axis) as a vertical asymptote.

STEP 5: The rational function R is improper since the degree of the numerator, 2, is larger than the degree of the denominator, 1. To find any horizontal or oblique asymptotes, we use long division.

$$\begin{array}{r} x \phantom{{}-1} \\ x\overline{)\,x^2 - 1} \\ \underline{x^2 \phantom{{}-1}} \\ -1 \end{array}$$

The quotient is x, so the line $y = x$ is an oblique asymptote of the graph. To determine whether the graph of R intersects the asymptote $y = x$, we solve the equation $R(x) = x$.

$$R(x) = \frac{x^2 - 1}{x} = x$$

$$x^2 - 1 = x^2$$

$$-1 = 0 \qquad \text{Impossible.}$$

We conclude that the equation $(x^2 - 1)/x = x$ has no solution, so the graph of $R(x)$ does not intersect the line $y = x$.

STEP 6: See Figure 46. We see from the graph that there is no y-intercept and two x-intercepts, -1 and 1. The symmetry with respect to the origin is also evident. We can also see that there is a vertical asymptote at $x = 0$. Finally, it is not necessary to graph this function in dot mode since no extraneous vertical lines are present.

Figure 46

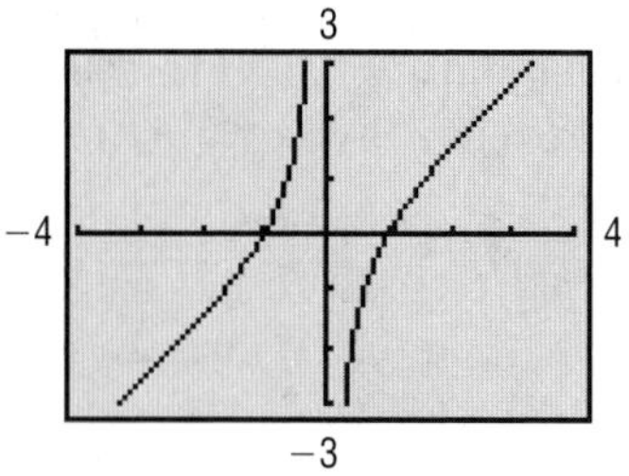

STEP 7: We use the table feature of our graphing utility to find additional key points on the graph of R. To determine key points, use the zeros of the numerator, -1 and 1, and the zero of the denominator, 0, to divide the x-axis into four intervals

$$-\infty < x < -1 \qquad -1 < x < 0 \qquad 0 < x < 1 \qquad 1 < x < \infty$$

Evaluate the function at any value of x within each interval and at the endpoints. See Table 16. Using the points obtained in Table 16 along with the information gathered in Steps 1 through 6, we obtain the graph of R shown in Figure 47. Notice how the oblique asymptote is used as a guide in graphing the rational function by hand.

TABLE 16

X	Y1	
-2	-1.5	
-1	0	
-.5	1.5	
.5	-1.5	
1	0	
2	1.5	

Y1■(X²−1)/X

Figure 47

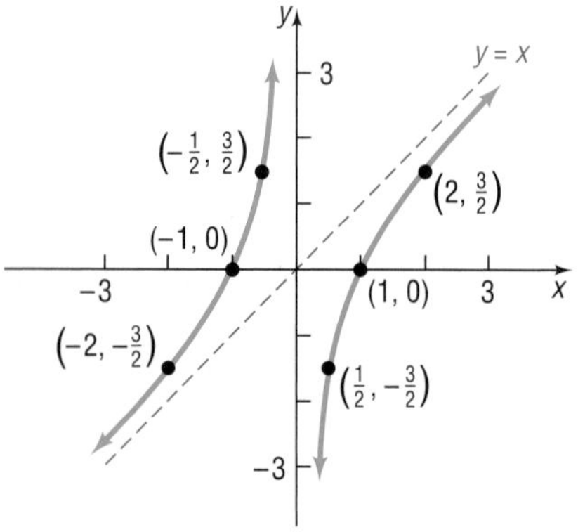

NOW WORK PROBLEM 9.

EXAMPLE 3 Analyzing the Graph of a Rational Function

Analyze the graph of the rational function: $R(x) = \dfrac{3x^2 - 3x}{x^2 + x - 12}$

Solution We factor R to get

$$R(x) = \frac{3x(x-1)}{(x+4)(x-3)}$$

R is in lowest terms.

STEP 1: The domain of R is $\{x \mid x \neq -4, x \neq 3\}$.

STEP 2: The graph has two x-intercepts: 0 and 1. The y-intercept is $R(0) = 0$.

STEP 3: Because

$$R(-x) = \frac{-3x(-x-1)}{(-x+4)(-x-3)} = \frac{3x(x+1)}{(x-4)(x+3)}$$

we conclude that R is neither even nor odd. There is no symmetry with respect to the y-axis or the origin.

STEP 4: Since R is in lowest terms, the graph of R has two vertical asymptotes: $x = -4$ and $x = 3$.

STEP 5: Since the degree of the numerator equals the degree of the denominator, the graph has a horizontal asymptote. To find it, we form the quotient of the leading coefficient of the numerator, 3, and the leading coefficient of the denominator, 1. Thus, the graph of R has the horizontal asymptote $y = 3$. To find out whether the graph of R intersects the asymptote, we solve the equation $R(x) = 3$.

$$R(x) = \frac{3x^2 - 3x}{x^2 + x - 12} = 3$$

$$3x^2 - 3x = 3x^2 + 3x - 36$$

$$-6x = -36$$

$$x = 6$$

The graph intersects the line $y = 3$ at $x = 6$, and $(6, 3)$ is a point on the graph of R.

Figure 48

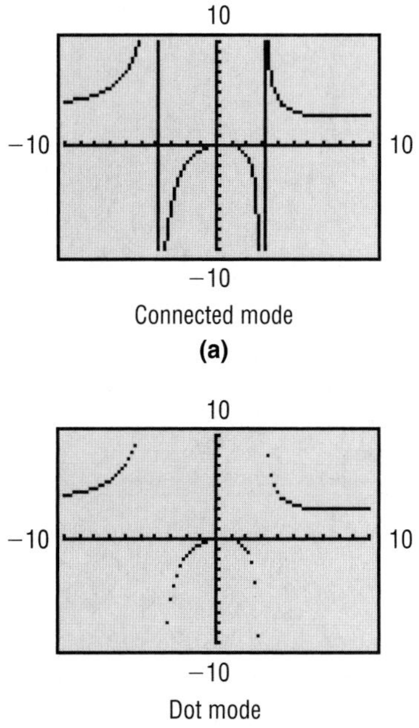

STEP 6: Figure 48(a) shows the graph of R in connected mode. Notice the extraneous vertical lines at $x = -4$ and $x = 3$ (the vertical asymptotes). As a result, we also graph R in dot mode. See Figure 48(b).

STEP 7: We use the TABLE feature of our graphing utility to find additional key points on the graph of R. To determine key points, use the zeros of the numerator, 0 and 1, and the zeros of the denominator, -4 and 3, to divide the x-axis into five intervals

$$-\infty < x < -4 \quad -4 < x < 0 \quad 0 < x < 1 \quad 1 < x < 3 \quad 3 < x < \infty$$

Evaluate the function at any value of x within each interval. We also evaluate the function on each side of $x = 6$, because we do not know whether the graph of R crosses or touches the horizontal asymptote, $y = 3$. See Table 17. From the table we conclude that the graph of R crosses the horizontal asymptote $y = 3$ at $(6, 3)$, since $R(4) = 4.5$ and $R(7) \approx 2.86$. Therefore, the graph of R approaches the horizontal asymptote $y = 3$ from below as x approaches ∞. Using the points obtained in Table 17 along with the information gathered in Steps 1 through 6, we obtain the graph of R shown in Figure 49.

TABLE 17

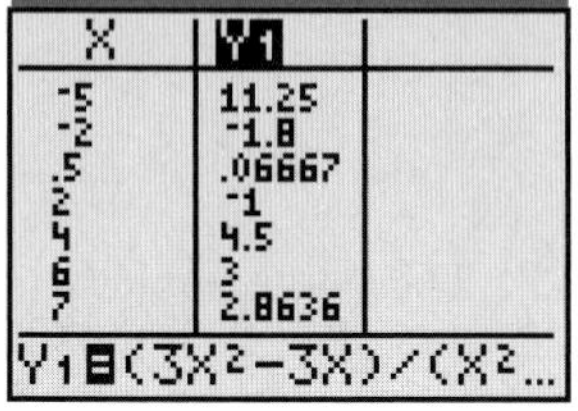

X	Y1
-5	11.25
-2	-1.8
.5	.06667
2	-1
4	4.5
6	3
7	2.8636

Y1 ■ (3X²−3X)/(X²...

Figure 49

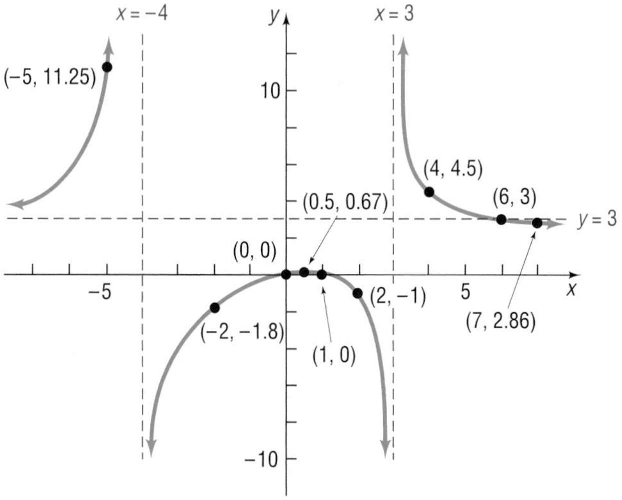

Figure 49 does not display the graph between the two x-intercepts, 0 and 1. Nor does it show the graph crossing the horizontal asymptote at $(6, 3)$. To see these parts better, we graph R for $-1 \leq x \leq 2$ [Figure 50(a)] and for $4 \leq x \leq 60$ [Figure 50(b)].

Figure 50

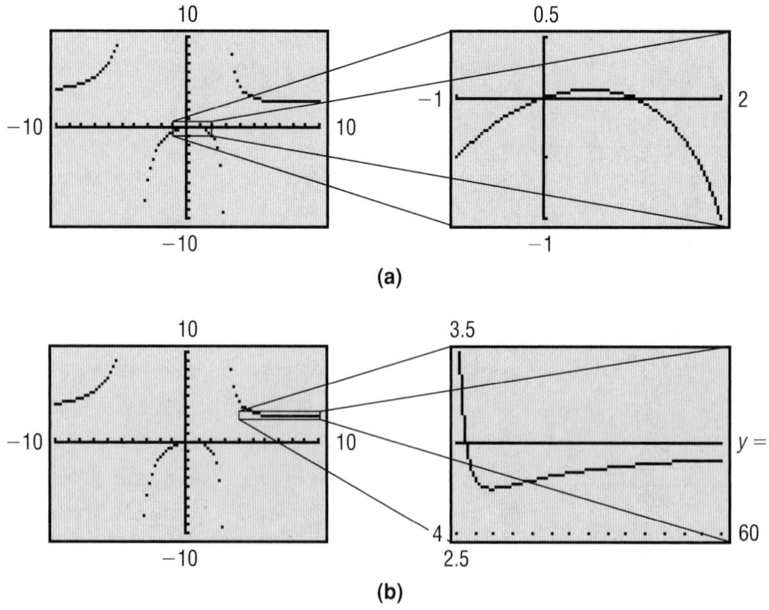

These graphs now reflect the behavior produced by the analysis in Step 7. Furthermore, we observe two turning points, one between 0 and 1 and the other to the right of 4. Rounded to two decimal places, these turning points are $(0.52, 0.07)$ and $(11.48, 2.75)$.

EXAMPLE 4 Analyzing the Graph of a Rational Function with a Hole

Analyze the graph of the rational function: $R(x) = \dfrac{2x^2 - 5x + 2}{x^2 - 4}$

Solution We factor R and obtain

$$R(x) = \frac{(2x - 1)(x - 2)}{(x + 2)(x - 2)}$$

In lowest terms,

$$R(x) = \frac{2x - 1}{x + 2}, \qquad x \neq 2$$

Step 1: The domain of R is $\{x \mid x \neq -2, x \neq 2\}$.

Step 2: The graph has one x-intercept: 0.5. The y-intercept is $R(0) = -0.5$.

Step 3: Because

$$R(-x) = \frac{2x^2 + 5x + 2}{x^2 - 4}$$

we conclude that R is neither even nor odd. There is no symmetry with respect to the y-axis or the origin.

Step 4: The graph has one vertical asymptote, $x = -2$, since $x + 2$ is the only factor of the denominator of $R(x)$ *in lowest terms*. However, the rational function is undefined at both $x = 2$ and $x = -2$.

Step 5: Since the degree of the numerator equals the degree of the denominator, the graph has a horizontal asymptote. To find it, we form the quotient of the leading coefficient of the numerator, 2, and the leading coefficient of the denominator, 1. Thus, the graph of R has the horizontal asymptote $y = 2$. To find whether the graph of R intersects the asymptote, we solve the equation $R(x) = 2$.

$$R(x) = \frac{2x - 1}{x + 2} = 2$$

$$2x - 1 = 2(x + 2)$$

$$2x - 1 = 2x + 4$$

$$-1 = 4 \qquad \text{Impossible.}$$

The graph does not intersect the line $y = 2$.

Figure 51

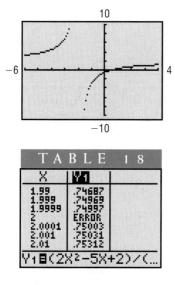

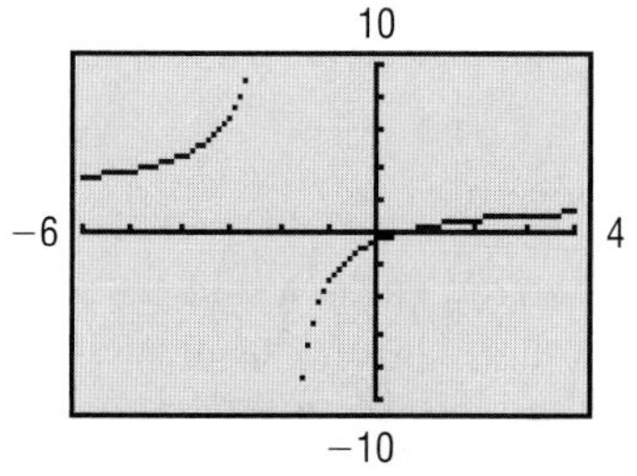

Step 6: Figure 51 shows the graph of $R(x)$. Notice that the graph has one vertical asymptote at $x = -2$. Also, the function appears to be continuous at $x = 2$.

Step 7: The analysis presented thus far does not explain the behavior of the graph at $x = 2$. We use the TABLE feature of our graphing utility to determine the behavior of the graph of R as x approaches 2. See Table 18. From the table, we conclude that the value of R approaches 0.75 as x approaches 2. This result is further verified by evaluating R in lowest terms at $x = 2$. We conclude that there is a hole in the graph at (2, 0.75). To find additional points on the graph of R, we create Table 19. Using the points obtained in Table 19 along with the information gathered in Steps 1 through 6, we obtain the graph of R shown in Figure 52.

TABLE 18

X	Y1
1.99	.74687
1.999	.74969
1.9999	.74997
2	ERROR
2.0001	.75003
2.001	.75031
2.01	.75312

Y1■(2X²-5X+2)/(...

TABLE 19

X	Y1
-3	7
-2	ERROR
-1	-3
0	-.5
1	.33333
2	ERROR
3	1

Y1■(2X²-5X+2)/(...

Figure 52

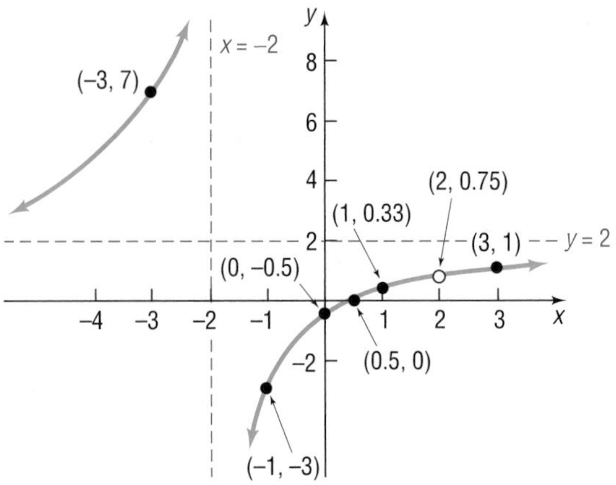

As Example 4 shows, the zeros of the denominator of a rational function give rise to either vertical asymptotes or holes on the graph.

NOW WORK PROBLEM 27.

We now discuss the problem of finding a rational function from its graph.

EXAMPLE 5 Constructing a Rational Function from Its Graph

Make up a rational function that might have the graph shown in Figure 53.

Figure 53

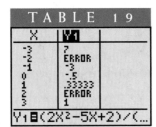

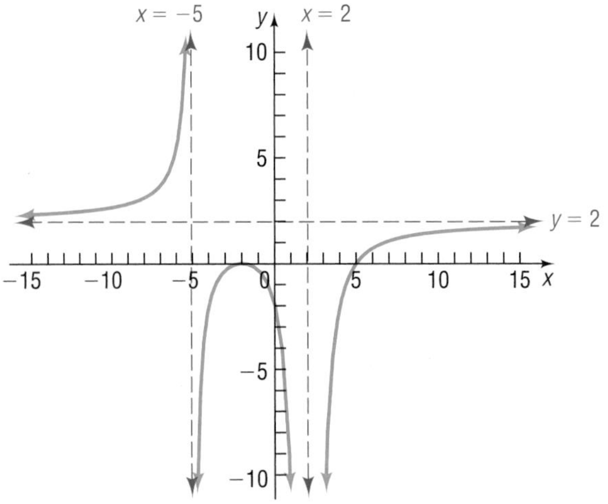

Solution The numerator of a rational function $R(x) = p(x)/q(x)$ in lowest terms determines the x-intercepts of its graph. The graph shown in Figure 53 has x-intercepts -2 (even multiplicity; graph touches the x-axis) and 5 (odd multiplicity; graph crosses the x-axis). So, one possibility for the numerator is $p(x) = (x + 2)^2(x - 5)$. The denominator of a rational function in lowest terms determines the vertical asymptotes of its graph. The vertical asymptotes of the graph are $x = -5$ and $x = 2$. Since $R(x)$ approaches ∞ from the left of $x = -5$ and $R(x)$ approaches $-\infty$ from the right of $x = -5$, we know that $(x + 5)$ is a factor of odd multiplicity in $q(x)$. Also, $R(x)$ approaches $-\infty$ from both sides of $x = 2$, so $(x - 2)$ is a factor of even multiplicity in $q(x)$. A possibility for the denominator is $q(x) = (x + 5)(x - 2)^2$. So far we have $R(x) = \dfrac{(x + 2)^2(x - 5)}{(x + 5)(x - 2)^2}$.

Figure 54

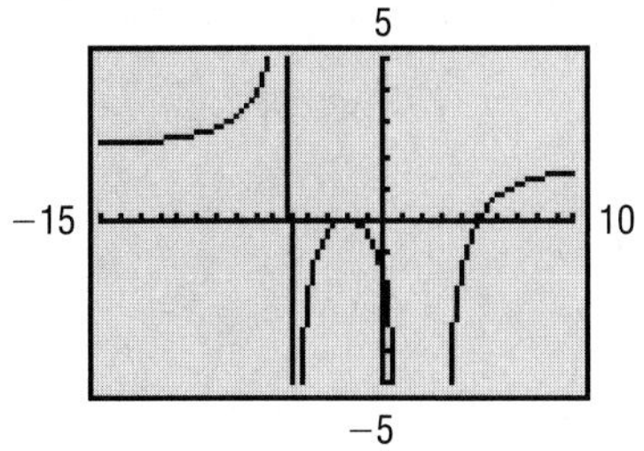

However, the horizontal asymptote of the graph given in Figure 53 is $y = 2$, so we know that the degree of the numerator must equal the degree in the denominator and the quotient of leading coefficients must be 2/1. This leads to $R(x) = \dfrac{2(x + 2)^2(x - 5)}{(x + 5)(x - 2)^2}$. Figure 54 shows the graph of R drawn on a graphing utility. Since Figure 54 looks similar to Figure 53, we have found a rational function in R for the graph in Figure 53. ■

NOW WORK PROBLEM 39.

2 Applications Involving Rational Functions

EXAMPLE 6 Finding the Least Cost of a Can

Reynolds Metal Company manufactures aluminum cans in the shape of a cylinder with a capacity of 500 cubic centimeters ($\frac{1}{2}$ liter). The top and bottom of the can are made of a special aluminum alloy that costs 0.05¢ per square centimeter. The sides of the can are made of material that costs 0.02¢ per square centimeter.

(a) Express the cost of material for the can as a function of the radius r of the can.

(b) Use a graphing utility to graph the function $C = C(r)$.

(c) What value of r will result in the least cost?

(d) What is this least cost?

Figure 55

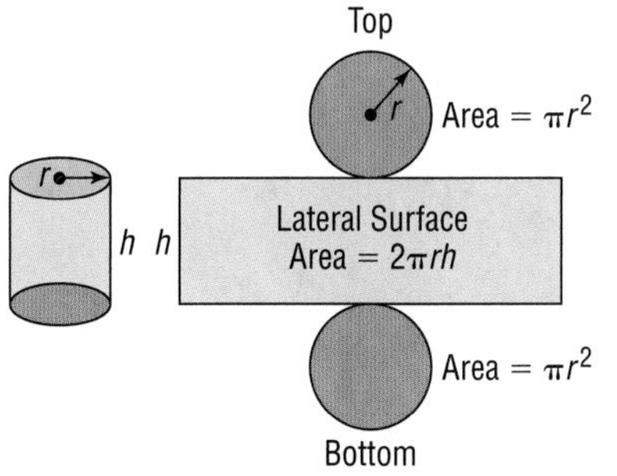

Solution (a) Figure 55 illustrates the situation. Notice that the material required to produce a cylindrical can of height h and radius r consists of a rectangle of area $2\pi rh$ and two circles, each of area πr^2. The total cost C (in cents) of manufacturing the can is therefore

$$\begin{aligned} C &= \text{Cost of top and bottom} + \text{Cost of side} \\ &= \underbrace{2(\pi r^2)}_{\substack{\text{Total area}\\\text{of top and}\\\text{bottom}}}\ \underbrace{(0.05)}_{\substack{\text{Cost/unit}\\\text{area}}} + \underbrace{(2\pi rh)}_{\substack{\text{Total}\\\text{area of}\\\text{side}}}\ \underbrace{(0.02)}_{\substack{\text{Cost/unit}\\\text{area}}} \\ &= 0.10\pi r^2 + 0.04\pi rh \end{aligned}$$

But we have the additional restriction that the height h and radius r must be chosen so that the volume V of the can is 500 cubic centimeters. Since $V = \pi r^2 h$, we have

$$500 = \pi r^2 h \quad \text{or} \quad h = \frac{500}{\pi r^2}$$

Substituting this expression for h, the cost C, in cents, as a function of the radius r is

$$C(r) = 0.10\pi r^2 + 0.04\pi r \frac{500}{\pi r^2} = 0.10\pi r^2 + \frac{20}{r} = \frac{0.10\pi r^3 + 20}{r}$$

Figure 56

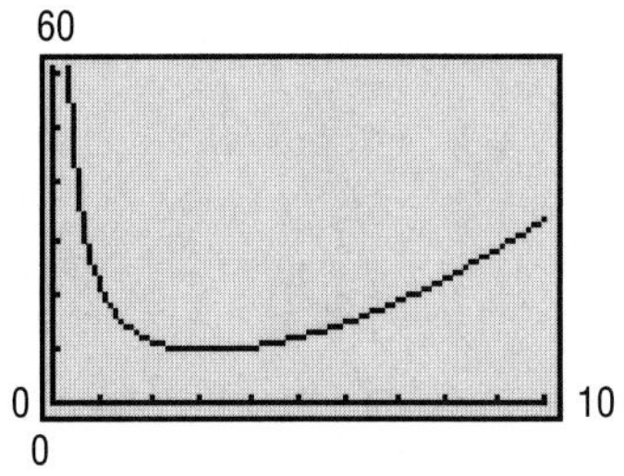

(b) See Figure 56 for the graph of $C(r)$.

(c) Using the MINIMUM command, the cost is least for a radius of about 3.17 centimeters.

(d) The least cost is $C(3.17) \approx 9.47$¢. ■

3.5 EXERCISES

In Problems 1–38, follow Steps 1 through 7 on page 220 to analyze the graph of each function.

1. $R(x) = \dfrac{x+1}{x(x+4)}$

2. $R(x) = \dfrac{x}{(x-1)(x+2)}$

3. $R(x) = \dfrac{3x+3}{2x+4}$

4. $R(x) = \dfrac{2x+4}{x-1}$

5. $R(x) = \dfrac{3}{x^2-4}$

6. $R(x) = \dfrac{6}{x^2-x-6}$

7. $P(x) = \dfrac{x^4+x^2+1}{x^2-1}$

8. $Q(x) = \dfrac{x^4-1}{x^2-4}$

9. $H(x) = \dfrac{x^3-1}{x^2-9}$

10. $G(x) = \dfrac{x^3+1}{x^2+2x}$

11. $R(x) = \dfrac{x^2}{x^2+x-6}$

12. $R(x) = \dfrac{x^2+x-12}{x^2-4}$

13. $G(x) = \dfrac{x}{x^2-4}$

14. $G(x) = \dfrac{3x}{x^2-1}$

15. $R(x) = \dfrac{3}{(x-1)(x^2-4)}$

16. $R(x) = \dfrac{-4}{(x+1)(x^2-9)}$

17. $H(x) = 4\dfrac{x^2-1}{x^4-16}$

18. $H(x) = \dfrac{x^2+4}{x^4-1}$

19. $F(x) = \dfrac{x^2-3x-4}{x+2}$

20. $F(x) = \dfrac{x^2+3x+2}{x-1}$

21. $R(x) = \dfrac{x^2+x-12}{x-4}$

22. $R(x) = \dfrac{x^2-x-12}{x+5}$

23. $F(x) = \dfrac{x^2+x-12}{x+2}$

24. $G(x) = \dfrac{x^2-x-12}{x+1}$

25. $R(x) = \dfrac{x(x-1)^2}{(x+3)^3}$

26. $R(x) = \dfrac{(x-1)(x+2)(x-3)}{x(x-4)^2}$

27. $R(x) = \dfrac{x^2+x-12}{x^2-x-6}$

28. $R(x) = \dfrac{x^2+3x-10}{x^2+8x+15}$

29. $R(x) = \dfrac{6x^2-7x-3}{2x^2-7x+6}$

30. $R(x) = \dfrac{8x^2+26x+15}{2x^2-x-15}$

31. $R(x) = \dfrac{x^2+5x+6}{x+3}$

32. $R(x) = \dfrac{x^2+x-30}{x+6}$

33. $f(x) = x + \dfrac{1}{x}$

34. $f(x) = 2x + \dfrac{9}{x}$

35. $f(x) = x^2 + \dfrac{1}{x}$

36. $f(x) = 2x^2 + \dfrac{9}{x}$

37. $f(x) = x + \dfrac{1}{x^3}$

38. $f(x) = 2x + \dfrac{9}{x^3}$

39. Consult illustration (a). Make up a rational function that might have this graph. (More than one answer might be possible.) Repeat these instructions for illustrations (b), (c), and (d).

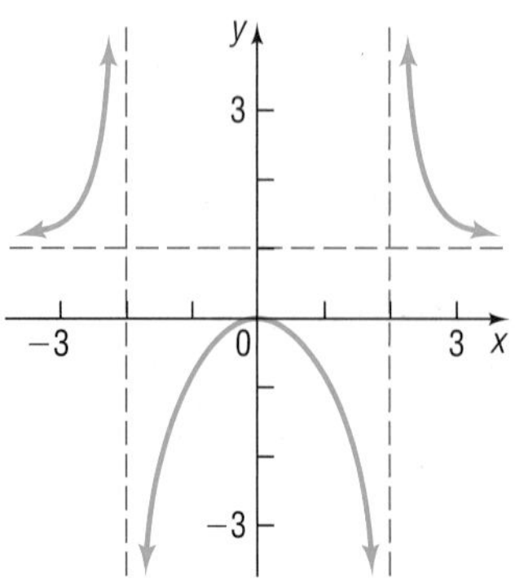

(a)

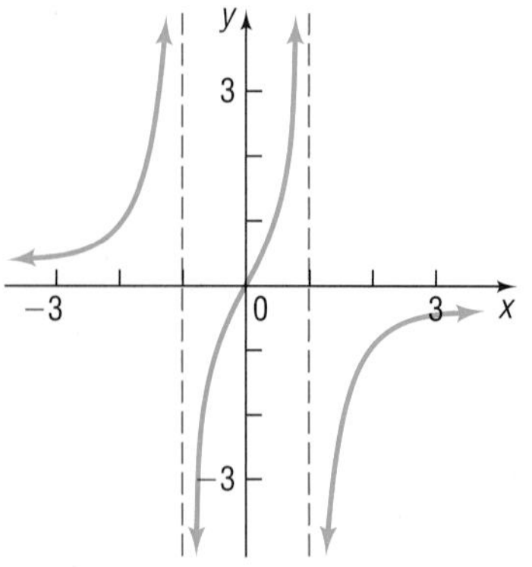

(b)

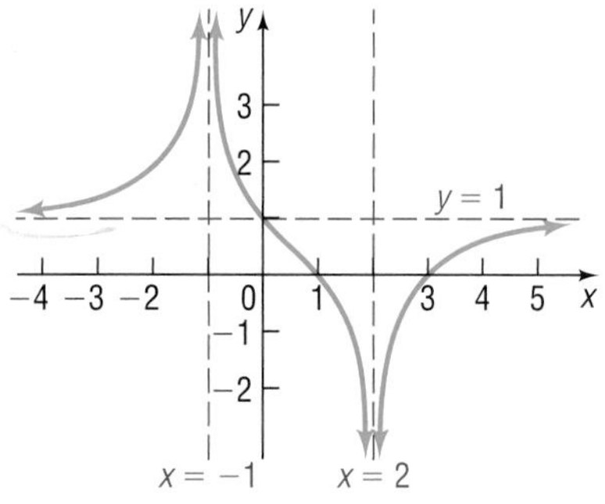

(c)

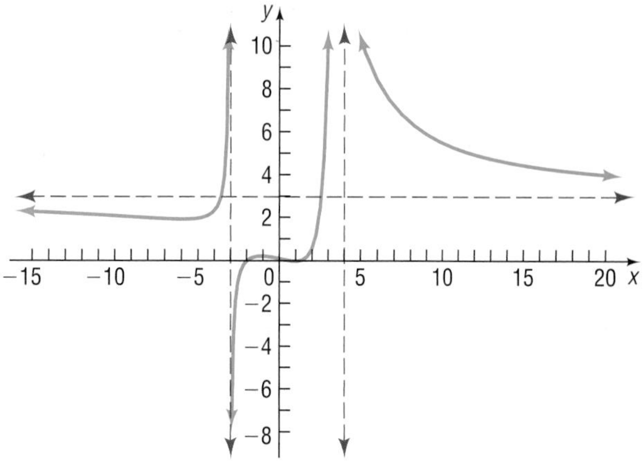

(d)

40. Write a few paragraphs that provide a general strategy for graphing a rational function. Be sure to mention the following: proper, improper, intercepts, and asymptotes.

41. Make up a rational function that has the following characteristics: crosses the x-axis at 2; touches the x-axis at -1; one vertical asymptote at -5 and another at 6; and one horizontal asymptote, $y = 3$. Compare yours to a fellow classmate's. How do they differ? What are the similarities?

42. Make up a rational function that has the following characteristics: crosses the x-axis at 3; touches the x-axis at -2; one vertical asymptote, $x = 1$; and one horizontal asymptote, $y = 2$. Give your rational function to a fellow classmate and ask for a written critique of your rational function.

43. Gravity In physics, it is established that the acceleration due to gravity, g, at a height h meters above sea level is given by

$$g(h) = \frac{3.99 \times 10^{14}}{(6.374 \times 10^6 + h)^2}$$

where 6.374×10^6 is the radius of Earth in meters.

(a) What is the acceleration due to gravity at sea level?
(b) The Sears Tower in Chicago, Illinois, is 443 meters tall. What is the acceleration due to gravity at the top of the Sears Tower?
(c) The peak of Mount Everest is 8848 meters above sea level. What is the acceleration due to gravity on the peak of Mount Everest?
(d) Find the horizontal asymptote of $g(h)$.
(e) Using your graphing utility, graph $g(h)$.
(f) Solve $g(h) = 0$. How do you interpret your answer?

44. Population Model A rare species of insect was discovered in the Amazon Rain Forest. To protect the species, environmentalists declare the insect endangered and transplant the insects into a protected area. The population of the insect t months after being transplanted is given by

$$P(t) = \frac{50(1 + 0.5t)}{(2 + 0.01t)}$$

(a) How many insects were discovered? In other words, what was the population when $t = 0$?
(b) What will the population be after 5 years?
(c) Using your graphing utility, graph $P(t)$.
(d) Determine the horizontal asymptote of $P(t)$. What is the largest population that the protected area can sustain?
(e) TRACE $P(t)$ for large values of t to verify your answer to part (d).

45. Drug Concentration The concentration C of a certain drug in a patient's bloodstream t hours after injection is given by

$$C(t) = \frac{t}{2t^2 + 1}$$

(a) Using your graphing utility, graph $C(t)$.
(b) Determine the time at which the concentration is highest.
(c) Find the horizontal asymptote of $C(t)$. What happens to the concentration of the drug as t increases?

46. Drug Concentration The concentration C of a certain drug in a patient's bloodstream t minutes after injection is given by

$$C(t) = \frac{50t}{t^2 + 25}$$

(a) Using your graphing utility, graph $C(t)$.
(b) Determine the time at which the concentration is highest.
(c) Find the horizontal asymptote of $C(t)$. What happens to the concentration of the drug as t increases?

47. Average Cost In Problem 67, Exercise 3.3, the cost function for manufacturing Chevy Cavaliers was found to be

$$C(x) = 0.2x^3 - 2.3x^2 + 14.3x + 10.2$$

Economists define the **average cost function** as

$$\bar{C}(x) = \frac{C(x)}{x}$$

(a) Find the average cost function.
(b) What is the average cost of producing six Cavaliers per hour?
(c) What is the average cost of producing nine Cavaliers per hour?
(d) Using your graphing utility, graph the average cost function.
(e) Using your graphing utility, find the number of Cavaliers that should be produced per hour to minimize average cost.
(f) What is the minimum average cost?

48. Average Cost In Problem 64, Exercise 3.3, the cost function for printing x thousand textbooks was found to be

$$C(x) = 0.015x^3 - 0.595x^2 + 9.15x + 98.43$$

(a) Find the average cost function (refer to Problem 47).
(b) What is the average cost of printing 13 thousand textbooks per week?
(c) What is the average cost of printing 25 thousand textbooks per week?
(d) Using your graphing utility, graph the average cost function.
(e) Using your graphing utility, find the number of textbooks that should be printed to minimize average cost.
(f) What is the minimum average cost?

49. Minimizing Surface Area United Parcel Service has contracted you to design a closed box with a square base that has a volume of 10,000 cubic inches. See the illustration.

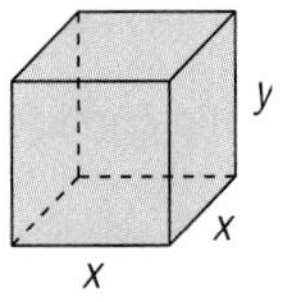

(a) Find a function for the surface area of the box.
(b) Using a graphing utility, graph the function found in part (a).
(c) What is the minimum amount of cardboard that can be used to construct the box?
(d) What are the dimensions of the box that minimize surface area?
(e) Why might UPS be interested in designing a box that minimizes surface area?

50. Minimizing Surface Area United Parcel Service has contracted you to design a closed box with a square base that has a volume of 5000 cubic inches. See the illustration.

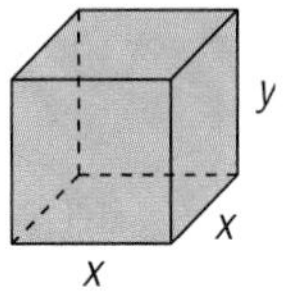

(a) Find a function for the surface area of the box.
(b) Using a graphing utility, graph the function found in part (a).
(c) What is the minimum amount of cardboard that can be used to construct the box?
(d) What are the dimensions of the box that minimize surface area?
(e) Why might UPS be interested in designing a box that minimizes surface area?

51. Cost of a Can A can in the shape of a right circular cylinder is required to have a volume of 500 cubic centimeters. The top and bottom are made of material that costs 6¢ per square centimeter, while the sides are made of material that costs 4¢ per square centimeter.
(a) Express the total cost C of the material as a function of the radius r of the cylinder. (Refer to Figure 55.)
(b) Graph $C = C(r)$. For what value of r is the cost C least?

52. Material Needed to Make a Drum A steel drum in the shape of a right circular cylinder is required to have a volume of 100 cubic feet.
(a) Express the amount A of material required to make the drum as a function of the radius r of the cylinder.
(b) How much material is required if the drum is of radius 3 feet?
(c) Of radius 2 feet?
(d) Of radius 4 feet?
(e) Graph $A = A(r)$. For what value of r is A smallest?

CHAPTER REVIEW

Things To Know

Quadratic function (p. 172)

$f(x) = ax^2 + bx + c$

Graph is a parabola that opens up if $a > 0$ and opens down if $a < 0$.

Vertex: $\left(\frac{-b}{2a}, f\left(\frac{-b}{2a}\right)\right)$

Axis of symmetry: $x = \frac{-b}{2a}$

y-intercept: $f(0)$

x-intercept(s): If any, found by solving the equation $ax^2 + bx + c = 0$.

Power function (p. 191)

$f(x) = x^n, \quad n \geq 2$ even	Even function Passes through $(-1, 1)$, $(0, 0)$, $(1, 1)$ Opens up
$f(x) = x^n, \quad n \geq 3$ odd	Odd function Passes through $(-1, -1)$, $(0, 0)$, $(1, 1)$ Increasing

Polynomial function (p. 197)

$f(x) = a_n x^n + a_{n-1}x^{n-1} + \cdots + a_1 x + a_0, \quad a_n \neq 0$	At most $n - 1$ turning points End behavior: Behaves like $y = a_n x^n$ for large $\lvert x \rvert$

Rational function (p. 209)

$R(x) = \dfrac{p(x)}{q(x)}$	p, q are polynomial functions.

How To

Graph a quadratic function by hand and by using a graphing utility (p. 173)

Identify the vertex and axis of symmetry of a quadratic function (p. 175)

Determine the maximum or minimum value of a quadratic function (p. 179)

Use the maximum or minimum value of a quadratic function to solve applied problems (p. 180)

Find the quadratic function of best fit to data (p. 183)

Graph transformations of power functions (p. 193)

Find the power function of best fit to data (p. 194)

Identify polynomials and their degree (p. 197)

Identify the zeros of a polynomial and their multiplicity (p. 200)

Analyze the graph of a polynomial (p. 202)

Find the cubic function of best fit to data (p. 204)

Find the domain of a rational function (p. 209)

Determine the vertical asymptotes of a rational function (p. 212)

Determine the horizontal or oblique asymptotes of a rational function (p. 213)

Analyze the graph of a rational function (p. 219)

Solve applied problems involving rational functions (p. 227)

Fill-in-the-Blank Items

1. The graph of a quadratic function is called a(n)________________.
2. The function $f(x) = ax^2 + bx + c$, $a \neq 0$, has a minimum value if ________________. The minimum value occurs at $x =$ ________________.
3. A number r for which $f(r) = 0$ is called a(n)________________ of the function f.
4. The line ________________ is a horizontal asymptote of $R(x) = \dfrac{x^3 - 1}{x^3 + 1}$.
5. The line ________________ is a vertical asymptote of $R(x) = \dfrac{x^3 - 1}{x^3 + 1}$.

True/False Items

T F 1. The graph of $f(x) = 2x^2 + 3x - 4$ opens up.

T F 2. The minimum value of $f(x) = -x^2 + 4x + 5$ is $f(2)$.

T F 3. The graph of $R(x) = \dfrac{x^2}{x - 1}$ has exactly one vertical asymptote.

T F 4. The graph of $f(x) = x^2(x - 3)(x + 4)$ has exactly three x-intercepts.

T F 5. The graph of a polynomial function sometimes has a hole.

T F 6. The graph of a rational function sometimes has a hole.

Review Exercises

Blue problem numbers indicate the authors' suggestions for use in a Practice Test.

In Problems 1–10, (a) graph each quadratic function by hand by determining whether its graph opens up or down and by finding its vertex, axis of symmetry, y-intercept, and x-intercepts, if any; (b) verify your results using a graphing utility.

1. $f(x) = \frac{1}{4}x^2 - 16$
2. $f(x) = -\frac{1}{2}x^2 - 2$
3. $f(x) = -4x^2 + 4x$
4. $f(x) = 9x^2 - 6x + 3$
5. $f(x) = \frac{9}{2}x^2 + 3x + 1$
6. $f(x) = -x^2 + x + \frac{1}{2}$
7. $f(x) = 3x^2 - 4x - 1$
8. $f(x) = -2x^2 - x + 4$
9. $f(x) = x^2 - 4x + 6$
10. $f(x) = x^2 + 2x - 3$

In Problems 11–16, graph each function using transformations (shifting, compressing, stretching, and reflection). Show all the stages. Verify your results using a graphing utility.

11. $f(x) = (x + 2)^3$
12. $f(x) = -x^3 + 3$
13. $f(x) = -(x - 1)^4$
14. $f(x) = (x - 1)^4 - 2$
15. $f(x) = (x - 1)^4 + 2$
16. $f(x) = (1 - x)^3$

In Problems 17–22, determine whether the given quadratic function has a maximum value or a minimum value, and then find the value.

17. $f(x) = 3x^2 - 6x + 4$
18. $f(x) = 2x^2 + 8x + 5$
19. $f(x) = -x^2 + 8x - 4$
20. $f(x) = -x^2 - 10x - 3$
21. $f(x) = -3x^2 + 12x + 4$
22. $f(x) = -2x^2 + 4$

In Problems 23–30, for each polynomial function f:
(a) Using a graphing utility, graph f.
(b) Find the x- and y-intercepts.
(c) Determine whether each x-intercept is of odd or even multiplicity.
(d) Find the power function that the graph of f resembles for large values of |x|.
(e) Determine the number of turning points on the graph of f.
(f) Determine the local maxima and local minima, if any exist, rounded to two decimal places.

23. $f(x) = x(x + 2)(x + 4)$
24. $f(x) = x(x - 2)(x - 4)$
25. $f(x) = (x - 2)^2(x + 4)$
26. $f(x) = (x - 2)(x + 4)^2$
27. $f(x) = x^3 - 4x^2$
28. $f(x) = x^3 + 4x$
29. $f(x) = (x - 1)^2(x + 3)(x + 1)$
30. $f(x) = (x - 4)(x + 2)^2(x - 2)$

In Problems 31–42, discuss each rational function following the seven steps on page 220.

31. $R(x) = \dfrac{2x - 6}{x}$
32. $R(x) = \dfrac{4 - x}{x}$
33. $H(x) = \dfrac{x + 2}{x(x - 2)}$
34. $H(x) = \dfrac{x}{x^2 - 1}$
35. $R(x) = \dfrac{x^2 + x - 6}{x^2 - x - 6}$
36. $R(x) = \dfrac{x^2 - 6x + 9}{x^2}$
37. $F(x) = \dfrac{x^3}{x^2 - 4}$
38. $F(x) = \dfrac{3x^3}{(x - 1)^2}$
39. $R(x) = \dfrac{2x^4}{(x - 1)^2}$
40. $R(x) = \dfrac{x^4}{x^2 - 9}$
41. $G(x) = \dfrac{x^2 - 4}{x^2 - x - 2}$
42. $F(x) = \dfrac{(x - 1)^2}{x^2 - 1}$

43. Landscaping A landscape engineer has 200 feet of border to enclose a rectangular pond. What dimensions will result in the largest pond?

44. Geometry Find the length and width of a rectangle whose perimeter is 20 feet and whose area is 16 square feet.

45. A rectangle has one vertex on the line $y = 10 - x$, $x > 0$, another at the origin, one on the positive x-axis, and one on the positive y-axis. Find the largest area A that can be enclosed by the rectangle.

46. Minimizing Cost Callaway Golf Company has determined that the daily per unit cost C of manufacturing x additional Big Bertha-type golf clubs may be expressed by the quadratic function

$$C(x) = 5x^2 - 620x + 20{,}000$$

(a) How many clubs should be manufactured to minimize the additional cost per club?
(b) At this level of production, what is the additional cost per club?

47. Minimizing Cost Scott–Jones Publishing Company has found that the per unit cost C for paper, printing, and binding of x additional textbooks is given by the quadratic function

$$C(x) = 0.003x^2 - 30x + 111{,}800$$

(a) How many books should be manufactured for the additional cost per text to be a minimum?
(b) At this level of production, what is the additional cost per text?

48. Parabolic Arch Bridge A horizontal bridge is in the shape of a parabolic arch. Given the information shown in the figure, what is the height h of the arch 2 feet from shore?

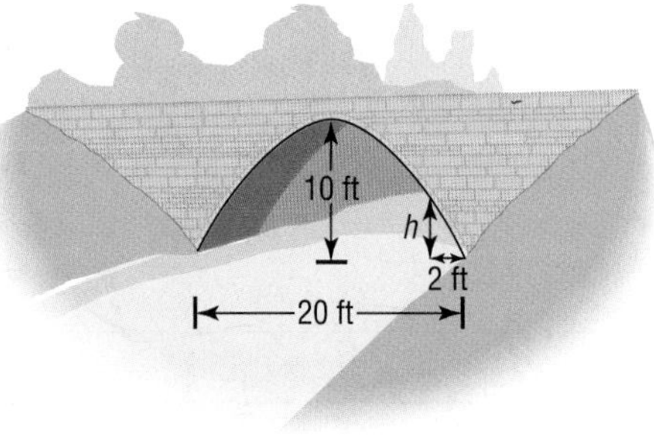

49. Life Cycle Hypothesis An individual's income varies with his or her age. The following table shows the median income I of individuals of different age groups within the United States for 1997. For each age group, the midpoint represents the independent variable, x. For the age group "65 years and older," we assume that the midpoint is 69.5.

Age	Midpoint, *x*	Median Income, *I*
15–24 years	19.5	$22,583
25–34 years	29.5	$38,174
35–44 years	39.5	$46,359
45–54 years	49.5	$51,875
55–64 years	59.5	$41,356
65 years and older	69.5	$20,761

Source: U.S. Census Bureau

(a) Draw a scatter diagram of the data. Comment on the type of relation that may exist between the two variables.
(b) Find the quadratic function of best fit.
(c) Graph the quadratic function of best fit on the scatter diagram.
(d) Using the function found in part (b), determine the age at which an individual can expect to earn the most income.
(e) Predict the peak income earned.
(f) Compare your results in parts (d) and (e) to the data and comment.
(g) Compare the results obtained in 1997 to those of 1995 and 1996 found in Problems 73 and 74 in Section 3.1.

50. Relating the Length and Period of a Pendulum Tom constructs simple pendulums with different lengths, l, and uses a light probe to record the corresponding period, T. The following data are collected:

Length *l* (in feet)	Period *T* (in seconds)
0.5	0.79
1.2	1.20
1.8	1.51
2.3	1.65
3.7	2.14
4.9	2.43

(a) Using a graphing utility, draw a scatter diagram of the data with length as the independent variable and period as the dependent variable.
(b) Find the power function of best fit.
(c) Graph this power function on the scatter diagram.
(d) Use the function $T(l)$ to predict the period of a pendulum whose length is 4 feet.

51. AIDS Cases in the United States The following data represent the cumulative number of reported AIDS cases in the United States for 1990–1997.

Year, *t*	Number of AIDS Cases, *A*
1990, 1	193,878
1991, 2	251,638
1992, 3	326,648
1993, 4	399,613
1994, 5	457,280
1995, 6	528,215
1996, 7	594,760
1997, 8	653,253

Source: U.S. Center for Disease Control and Prevention

(a) Using a graphing utility, draw a scatter diagram of the data, treating year as the independent variable.
(b) Find the cubic function of best fit.
(c) Graph the cubic function of best fit on the scatter diagram given in part (b).
(d) Use the function $A(t)$ to predict the cumulative number of AIDS cases reported in the United States in 2000.
(e) Use the function $A(t)$ given in part (b) to predict the year in which the cumulative number of AIDS cases reported in the United States reaches 850,000.
(f) Do you think that the function given in part (b) will be useful in predicting the number of AIDS cases in 2010?

52. Construct a polynomial function with the following characteristics: degree 6; four real zeros, one of multiplicity 3; y-intercept 3; behaves like $y = -5x^6$ for large values of x. Is this polynomial unique? Compare your polynomial with those of other students. What terms will be the same as everyone else's? Add some more characteristics, such as symmetry or naming the real zeros. How does this modify the polynomial?

53. Construct a rational function with the following characteristics: three real zeros, one of multiplicity 2; y-intercept 1; vertical asymptotes $x = -2$ and $x = 3$; oblique asymptote $y = 2x + 1$. Is this rational function unique? Compare yours with those of other students. What will be the same as everyone else's? Add some more characteristics, such as symmetry or naming the real zeros. How does this modify the rational function?

54. The illustration shows the graph of a polynomial function.
(a) Is the degree of the polynomial even or odd?
(b) Is the leading coefficient positive or negative?
(c) Is the function even, odd, or neither?
(d) Why is x^2 necessarily a factor of the polynomial?
(e) What is the minimum degree of the polynomial?
(f) Formulate five different polynomials whose graphs could look like the one shown. Compare yours to those of other students. What similarities do you see? What differences?

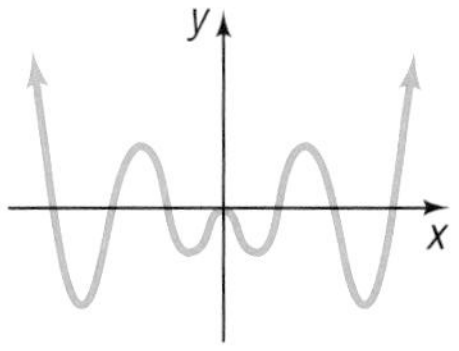

PROJECT AT MOTOROLA

How many cellular phones can I make?

Motorola uses several types of specialized robots to assemble electronic products accurately and quickly. Resisitors, capacitors, and other parts are picked from the feeder carriage by one of the vacuum nozzles on the rotating turret. The X-Y table, which holds a panel containing one or more boards, moves to the placement location for a given part, while the turret advances one position. The average time to pick and place one part is called the "tact time", e.g., 0.1 sec. per part. The actual throughput, measured in boards per hour or in placements per hour (PPH), depends on a number of factors, including the number and variety of parts on the panel, the time to load-unload a panel in the machine, and the time to visually inspect "fiducial marks" and calibrate the position of the panel with respect to the coordinate system of the machine.

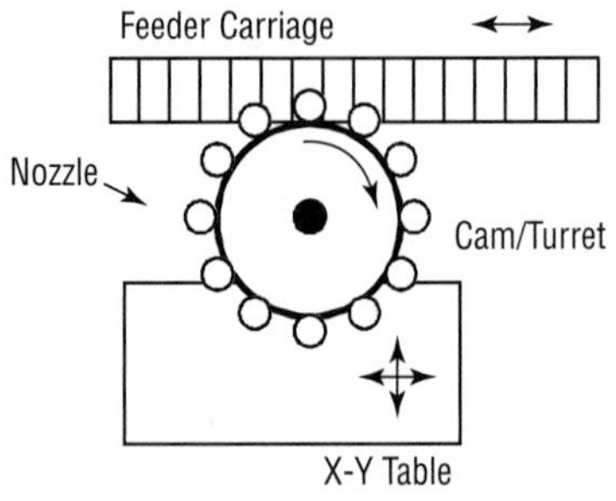

Throughput can be estimated with the following equation, where n is the number of boards per panel, p is the number of parts per board, C is the "tact time", L is the load-unload time, and M is the mark reading time.

$$TH(n) = \frac{n}{Cnp + L + M}$$

$$f(n) = \frac{L + M}{Cpn^2 + (L + M)n}$$

As shown for three different machines in the graph on page 235, throughput is highly dependent on the number of placements per board. Manufacturing engineers work with product designers to select the best number of boards per panel, i.e., the one that results in the highest possible throughput.

1. Write a rational function for TH in terms of the variable n, where the "tact time" is 0.2 sec. per part, the load-unload time is 5 sec., the mark reading time is 1 sec., and there are 50 parts per board.
2. Graph the function $Y_1 = TH(n)$, and determine the minimum number of boards per panel, such that the throughput is at least 300 boards per hour.
3. If it were possible to fit 10 boards per panel, what would the throughput be? What would be the highest possible throughput, if there were no limit on the number of boards per panel?
4. The function $f(n, p)$ represents the percent increase in throughput for a one unit increase in the number of boards per panel. Graph the function

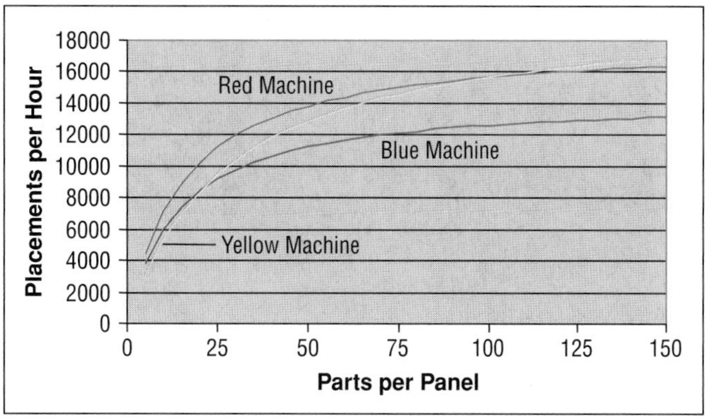

$Y_1 = f(n)$ for the values of "tact time", etc., given in question 1.,

5. Is $f(n)$ a proper rational function?

6. At what value of n is it is not possible to increase throughput by more than 2%, when one more board is added to the panel?

7. Assuming that all machines (blue, red, and yellow) cost the same to operate, which one is the best choice for a panel with one board? With three boards? Why?

CHAPTER

4 The Zeros of a Polynomial Function

Outline

FIELD TRIP TO MOTOROLA

Complex number theory may appear on the surface to be an abstract concept with little practical value. However, in the field of electrical engineering it is a powerful tool that aids in the design of new products. Some examples in the area of communications include cellular phones, pagers, two-way radios, cordless telephones, police and fire dispatch systems, and satellites. All of these products contain electrical circuits that must operate according to detailed design specifications. Engineers use circuit analysis techniques based on complex number theory to gain understanding into circuit operation in order to optimize their designs. Software tools (circuit simulators) are also available that automate this analytical process and reduce the design time. Without complex number theory the design process would be a much more difficult task.

INTERVIEW AT MOTOROLA

John St. Peter is currently an electrical engineer at Motorola's Advanced Technology Center (MATC). John started his career at Motorola with the Cellular Infrastructure Group (CIG) in 1987 after receiving his BSEE from the University of Illinois at Chicago. At CIG John worked as an RF development engineer where he designed feed-forward linear RF power amplifiers for cellular base stations. While working at CIG John continued his college education and in 1993 he received his MSEE from the University of Illinois at Chicago. His thesis work involved the use of numerical methods (FDTD) to model multiconductor transmission lines. John's current project at MATC involves research into RF applications of high density interconnect substrates in order to produce small wireless products. John has always enjoyed mathematics as a subject in school and decided to pursue a career in electrical engineering based on this. He especially enjoys solving engineering problems through applied mathematics.

PREPARING FOR THIS SECTION

Before getting started, review the following concepts:

- ✓ Polynomial Division; Synthetic Division (Appendix, Section 4)
- ✓ Polynomial Functions (Section 3.3, pp. 197–201)
- ✓ Quadratic Formula (Section 1.3, pp. 30–32)

4.1 THE REAL ZEROS OF A POLYNOMIAL FUNCTION

OBJECTIVES

1. Use the Remainder and Factor Theorems
2. Use Descartes' Rule of Signs to Determine the Number of Positive and the Number of Negative Real Zeros of a Polynomial Function
3. Use the Rational Zeros Theorem to List the Potential Rational Zeros of a Polynomial Function
4. Find the Real Zeros of a Polynomial Function
5. Solve Polynomial Equations
6. Use the Theorem for Bounds on Zeros
7. Use the Intermediate Value Theorem

In this section, we discuss techniques that can be used to find the real zeros of a polynomial function. Recall that if r is a real zero of a polynomial function f then $f(r) = 0$; r is an x-intercept of the graph of f; and r is a solution of the equation $f(x) = 0$. For polynomial and rational functions, we have seen the importance of the zeros for graphing. In most cases, however, the zeros of a polynomial function are difficult to find using algebraic methods. No nice formulas like the quadratic formula are available to help us find zeros for polynomials of degree 3 or higher. Formulas do exist for solving any third- or fourth-degree polynomial equation, but they are somewhat complicated. No general formulas exist for polynomial equations of degree 5 or higher. Refer to the Historical Feature at the end of the section for more information.

REMAINDER AND FACTOR THEOREMS

1 When we divide one polynomial (the dividend) by another (the divisor), we obtain a quotient polynomial and a remainder, the remainder being either the zero polynomial or a polynomial whose degree is less than the degree of the divisor. To check our work, we verify that

$$\text{(Quotient)(Divisor)} + \text{Remainder} = \text{Dividend}$$

This checking routine is the basis for a famous theorem called the **division algorithm* for polynomials**, which we now state without proof.

Theorem

Division Algorithm for Polynomials

If $f(x)$ and $g(x)$ denote polynomial functions and if $g(x)$ is not the zero polynomial, then there are unique polynomial functions $q(x)$ and $r(x)$ such that

$$\frac{f(x)}{g(x)} = q(x) + \frac{r(x)}{g(x)} \quad \text{or} \quad \underset{\text{dividend}}{f(x)} = \underset{\text{quotient}}{q(x)}\,\underset{\text{divisor}}{g(x)} + \underset{\text{remainder}}{r(x)} \tag{1}$$

where $r(x)$ is either the zero polynomial or a polynomial of degree less than that of $g(x)$.

In equation (1), $f(x)$ is the **dividend**, $g(x)$ is the **divisor**, $q(x)$ is the **quotient**, and $r(x)$ is the **remainder**.

If the divisor $g(x)$ is a first-degree polynomial of the form

$$g(x) = x - c, \qquad c \text{ a real number}$$

then the remainder $r(x)$ is either the zero polynomial or a polynomial of degree 0. As a result, for such divisors, the remainder is some number, say R, and we may write

$$f(x) = (x - c)q(x) + R \tag{2}$$

This equation is an identity in x and is true for all real numbers x. Suppose that $x = c$. Then equation (2) becomes

$$f(c) = (c - c)q(c) + R$$

$$f(c) = R$$

Substitute $f(c)$ for R in equation (2) to obtain

$$f(x) = (x - c)q(x) + f(c) \tag{3}$$

We have now proved the **Remainder Theorem.**

* A systematic process in which certain steps are repeated a finite number of times is called an **algorithm**. For example, long division is an algorithm.

Remainder Theorem Let f be a polynomial function. If $f(x)$ is divided by $x - c$, then the remainder is $f(c)$.

EXAMPLE 1 Using the Remainder Theorem

Find the remainder if $f(x) = x^3 - 4x^2 + 2x - 5$ is divided by

(a) $x - 3$ (b) $x + 2$

Solution (a) We could use long division or synthetic division, but it is easier to use the Remainder Theorem, which says that the remainder is $f(3)$.

$$f(3) = (3)^3 - 4(3)^2 + 2(3) - 5 = 27 - 36 + 6 - 5 = -8$$

The remainder is -8.

(b) To find the remainder when $f(x)$ is divided by $x + 2 = x - (-2)$, we evaluate $f(-2)$.

$$f(-2) = (-2)^3 - 4(-2)^2 + 2(-2) - 5 = -8 - 16 - 4 - 5 = -33$$

The remainder is -33.

An important and useful consequence of the Remainder Theorem is the **Factor Theorem**.

Factor Theorem Let f be a polynomial function. Then $x - c$ is a factor of $f(x)$ if and only if $f(c) = 0$.

The Factor Theorem actually consists of two separate statements:

1. If $f(c) = 0$, then $x - c$ is a factor of $f(x)$.
2. If $x - c$ is a factor of $f(x)$, then $f(c) = 0$.

The proof requires two parts.

Proof

1. Suppose that $f(c) = 0$. Then, by equation (3), we have

$$f(x) = (x - c)q(x)$$

for some polynomial $q(x)$. That is, $x - c$ is a factor of $f(x)$.

2. Suppose that $x - c$ is a factor of $f(x)$. Then there is a polynomial function q such that

$$f(x) = (x - c)q(x)$$

Replacing x by c, we find that

$$f(c) = (c - c)q(c) = 0 \cdot q(c) = 0$$

This completes the proof.

One use of the Factor Theorem is to determine whether a polynomial has a particular factor.

EXAMPLE 2 Using the Factor Theorem

Use the Factor Theorem to determine whether the function $f(x) = 2x^3 - x^2 + 2x - 3$ has the factor

(a) $x - 1$ (b) $x + 3$

Solution The Factor Theorem states that if $f(c) = 0$ then $x - c$ is a factor.

(a) Because $x - 1$ is of the form $x - c$ with $c = 1$, we find the value of $f(1)$.

$$f(1) = 2(1)^3 - (1)^2 + 2(1) - 3 = 2 - 1 + 2 - 3 = 0$$

See also Figure 1. By the Factor Theorem, $x - 1$ is a factor of $f(x)$.

(b) We first need to write $x + 3$ in the form $x - c$. Since $x + 3 = x - (-3)$, we find the value of $f(-3)$. See Figure 1. Because $f(-3) = -72 \neq 0$, we conclude from the Factor Theorem that $x - (-3) = x + 3$ is not a factor of $f(x)$. ■

Figure 1

$Y_1 = 2x^3 - x^2 + 2x - 3$

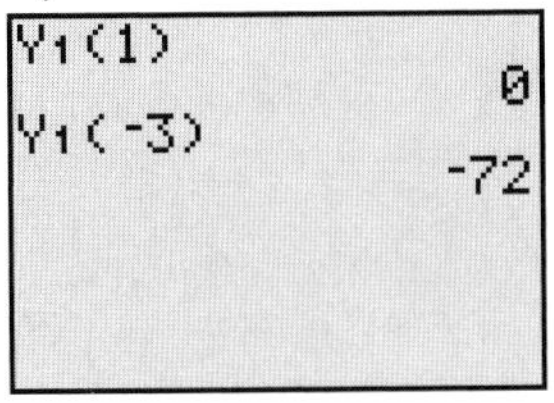

In Example 2, we found that $x - 1$ was a factor of f. To write f in factored form, we can use long division or synthetic division. Using synthetic division, we find that

$$\begin{array}{r|rrrr} 1 & 2 & -1 & 2 & -3 \\ & & 2 & 1 & 3 \\ \hline & 2 & 1 & 3 & 0 \end{array}$$

The quotient is $q(x) = 2x^2 + x + 3$ with a remainder of 0, as expected. We can write f in factored form as

$$f(x) = 2x^3 - x^2 + 2x - 3 = (x - 1)(2x^2 + x + 3).$$

NOW WORK PROBLEM 1.

THE NUMBER AND LOCATION OF REAL ZEROS

The next theorem concerns the number of real zeros that a polynomial function may have. In counting the zeros of a polynomial, we count each zero as many times as its multiplicity.

Theorem **Number of Real Zeros**

A polynomial function cannot have more real zeros than its degree. ■

Proof The proof is based on the Factor Theorem. If r is a zero of a polynomial function f, then $f(r) = 0$ and, hence, $x - r$ is a factor of $f(x)$. Each zero corresponds to a factor of degree 1. Because f cannot have more first-degree factors than its degree, the result follows. ■

2 **Descartes' Rule of Signs** provides information about the number and location of the real zeros of a polynomial function written in standard form (descending powers of x). It requires that we count the number of variations in sign of the coefficients of $f(x)$ and $f(-x)$.

For example, the following polynomial function has two variations in the signs of coefficients.

$$\begin{aligned} f(x) &= -3x^7 + 4x^4 + 3x^2 - 2x - 1 \\ &= -3x^7 + 0x^6 + 0x^5 + 4x^4 + 0x^3 + 3x^2 - 2x - 1 \end{aligned}$$

– to +   + to –

Notice that we ignored the zero coefficients in $0x^6$, $0x^5$ and $0x^3$ in counting the number of variations in the sign of $f(x)$. Replacing x by $-x$, we get

$$\begin{aligned} f(-x) &= -3(-x)^7 + 4(-x)^4 + 3(-x)^2 - 2(-x) - 1 \\ &= 3x^7 + 4x^4 + 3x^2 + 2x - 1 \end{aligned}$$

+ to –

which has one variation in sign.

Theorem **Descartes' Rule of Signs**

Let f denote a polynomial function.

The number of positive real zeros of f either equals the number of variations in sign of the nonzero coefficients of $f(x)$ or else equals that number less an even integer.

The number of negative real zeros of f either equals the number of variations in sign of the nonzero coefficients of $f(-x)$ or else equals that number less an even integer.

We shall not prove Descartes' Rule of Signs. Let's see how it is used.

EXAMPLE 3 Using the Number of Real Zeros Theorem and Descartes' Rule of Signs

Discuss the real zeros of $f(x) = 3x^6 - 4x^4 + 3x^3 + 2x^2 - x - 3$.

Solution Because the polynomial is of degree 6 by the Number of Real Zeros Theorem, there are at most six real zeros. Since there are three variations in sign of the nonzero coefficients of $f(x)$, by Descartes' Rule of Signs we expect either three (or one) positive real zeros. To continue, we look at $f(-x)$:

$$f(-x) = 3x^6 - 4x^4 - 3x^3 + 2x^2 + x - 3$$

There are three variations in sign, so we expect either three (or one) negative real zeros. Equivalently, we now know that the graph of f has either three (or one) positive x-intercepts and three (or one) negative x-intercepts.

Figure 2

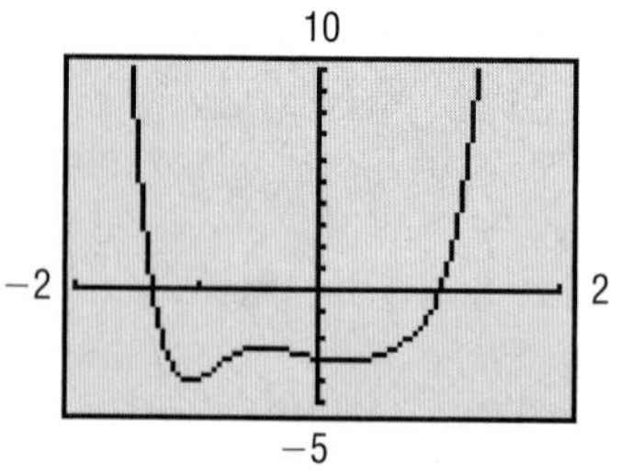

We see in Figure 2 that f has one negative zero near −1.3 and one positive zero near 1. The conclusions of Descartes' Rule of Signs are confirmed by the graph.

NOW WORK PROBLEM 11.

Rational Zeros Theorem

3 The next result, called the **Rational Zeros Theorem**, provides information about the rational zeros of a polynomial *with integer coefficients.*

Theorem

Rational Zeros Theorem

Let f be a polynomial function of degree 1 or higher of the form

$$f(x) = a_n x^n + a_{n-1}x^{n-1} + \cdots + a_1 x + a_0, \qquad a_n \neq 0, a_0 \neq 0$$

where each coefficient is an integer. If p/q, in lowest terms, is a rational zero of f, then p must be a factor of a_0, and q must be a factor of a_n.

EXAMPLE 4 Listing Potential Rational Zeros

List the potential rational zeros of

$$f(x) = 2x^3 + 11x^2 - 7x - 6$$

Solution Because f has integer coefficients, we may use the Rational Zeros Theorem. First, we list all the integers p that are factors of $a_0 = -6$ and all the integers q that are factors of $a_3 = 2$.

$$p: \quad \pm 1, \pm 2, \pm 3, \pm 6$$

$$q: \quad \pm 1, \pm 2$$

Now we form all possible ratios p/q.

$$\frac{p}{q}: \quad \pm 1, \pm 2, \pm 3, \pm 6, \pm\frac{1}{2}, \pm\frac{3}{2}$$

If f has a rational zero, it will be found in this list, which contains 12 possibilities.

NOW WORK PROBLEM 23.

Be sure that you understand what the Rational Zeros Theorem says: For a polynomial with integer coefficients, *if* there is a rational zero, it is one of those listed. It may be the case that the function does not have any rational zeros.

The Rational Zeros Theorem provides a list of potential rational zeros of a function f. If we graph f, we can get a better sense of the location of the x-intercepts and test to see if they are rational. We can also use the potential rational zeros to select our initial viewing window to graph f and then adjust the window based on the results. The graphs shown throughout the text will be those obtained after setting the final viewing window.

EXAMPLE 5 Finding the Rational Zeros of a Polynomial Function

Continue working with Example 4 to find the rational zeros of

$$f(x) = 2x^3 + 11x^2 - 7x - 6$$

Solution We gather all the information that we can about the zeros.

1. There are at most three real zeros.
2. By Descartes' Rule of Signs, there is one positive real zero. Also, because
$$f(-x) = -2x^3 + 11x^2 + 7x - 6$$
there are two (or no) negative real zeros.
3. We list the potential rational zeros obtained in Example 4:
$$\pm 1, \pm 2, \pm 3, \pm 6, \pm \tfrac{1}{2}, \pm \tfrac{3}{2}.$$

We could, of course, test each potential rational zero to see if the value of f there is zero. This is not very efficient. The graph of f will tell us approximately where the real zeros are. So we only need to test those rational zeros that are nearby. Figure 3 shows the graph of f. We see that f has three zeros: one near -6, one between -1 and 0, and one near 1. From our original list of potential rational zeros, we will test the following:

Figure 3

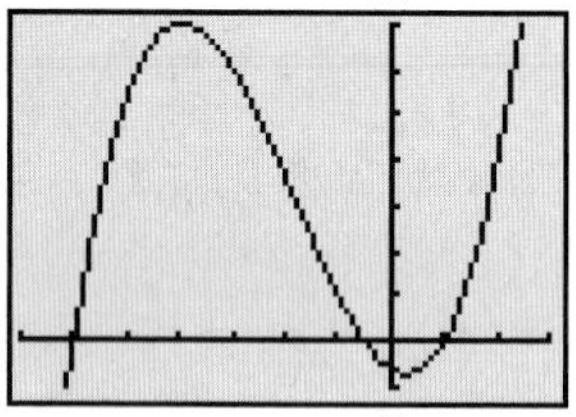

Near -6: test -6; Between -1 and 0: test $-\dfrac{1}{2}$; Near 1: test 1

$$f(-6) = 0 \qquad f\left(-\frac{1}{2}\right) = 0 \qquad f(1) = 0$$

The three zeros of f are $-6, -\frac{1}{2}$, and 1; each is a rational zero. ■

We use the Factor Theorem to factor the function f given in Example 5. Each zero gives rise to a factor, so $x - (-6) = x + 6$, $x - (-1/2) = x + 1/2$, and $x - 1$ are factors of f. Since the leading coefficient of f is 2 and f is of degree 3, we have
$$f(x) = 2x^3 + 11x^2 - 7x - 6 = 2(x + 6)(x + 1/2)(x - 1)$$

EXAMPLE 6 Finding the Real Zeros of a Polynomial Function

4 Find the real zeros of $f(x) = 3x^4 - 2x^3 + 4x^2 - 8x - 32$. Factor f over the real numbers.

Solution We gather all the information that we can about the zeros.

1. There are at most four real zeros.
2. By Descartes' Rule of Signs, there are three (or one) positive real zeros. Also, because
$$f(-x) = 3x^4 + 2x^3 + 4x^2 + 8x - 32$$
there is one negative real zero.
3. To obtain the list of potential rational zeros, we write the factors p of $a_0 = -32$ and the factors q of $a_5 = 3$.
$$p\colon \ \pm 1, \pm 2, \pm 4, \pm 8, \pm 16, \pm 32$$
$$q\colon \ \pm 1, \pm 3$$
The potential rational zeros consist of all possible quotients p/q:
$$\frac{p}{q}\colon \ \pm 1, \pm 2, \pm 4, \pm 8, \pm 16, \pm 32, \pm \frac{1}{3}, \pm \frac{2}{3}, \pm \frac{4}{3}, \pm \frac{8}{3}, \pm \frac{16}{3}, \pm \frac{32}{3}$$

Figure 4

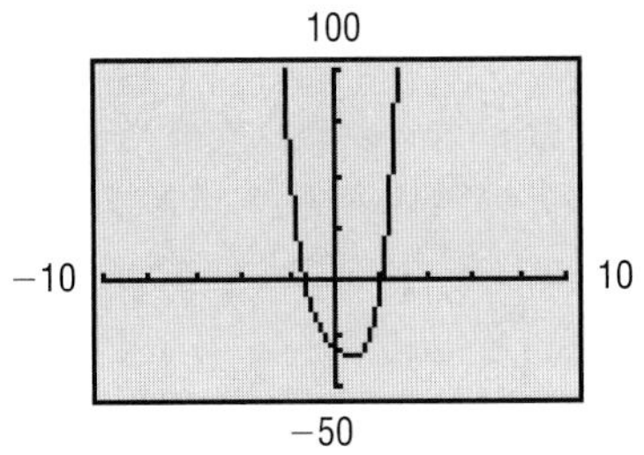

Figure 4 shows the graph of f. The graph has the characteristics that we expect of the given polynomial of degree 4: no more than three turning points, y-intercept -32, and behaves like $y = 3x^4$ for large $|x|$. The information found in parts 1 and 2 is also confirmed by the graph. The two real zeros consist of one negative real zero, located between -2 and -1, and one positive real zero, near 2.

Since 2 appears to be a zero and 2 is a potential rational zero, we evaluate f at 2 and find that $f(2) = 0$. We use synthetic division to factor f.

2)	3	−2	4	−8	−32
		6	8	24	32
	3	4	12	16	0

We now have

$$f(x) = 3x^4 - 2x^3 + 4x^2 - 8x - 32 = (x - 2)(3x^3 + 4x^2 + 12x + 16)$$

Any solution of the equation $3x^3 + 4x^2 + 12x + 16 = 0$ is a zero of f. Because of this, we call the equation $3x^3 + 4x^2 + 12x + 16 = 0$ a **depressed equation** of f. Since the degree of the depressed equation is less than that of the original function, we work with the depressed equation to find the remaining zeros of f.

We check the potential rational zero $-4/3$ and find that $f(-4/3) = 0$. Using synthetic division,

−4/3)	3	4	12	16
		−4	0	−16
	3	0	12	0

We now have

$$f(x) = (x - 2)\left(x + \frac{4}{3}\right)(3x^2 + 12) = 3(x - 2)\left(x + \frac{4}{3}\right)(x^2 + 4)$$

The depressed equation $x^2 + 4 = 0$ is a quadratic equation that has no real solutions. The real zeros of f are $-4/3$ and 2. The factored form of f is

$$f(x) = 3(x - 2)\left(x + \frac{4}{3}\right)(x^2 + 4)$$

■

NOW WORK PROBLEM 41.

EXAMPLE 7 Solving a Polynomial Equation

5 Solve the equation $3x^4 - 2x^3 + 4x^2 - 8x - 32 = 0$.

Solution The solutions of this equation are the zeros of the polynomial function

$$f(x) = 3x^4 - 2x^3 + 4x^2 - 8x - 32$$

Using the result of Example 6, the real zeros of f are $-4/3$ and 2. These are the real solutions of the equation $3x^4 - 2x^3 + 4x^2 - 8x - 32 = 0$. ■

NOW WORK PROBLEM 65.

In Example 6, the quadratic factor $x^2 + 4$ that appears in the factored form of $f(x)$ is called *irreducible*, because the polynomial $x^2 + 4$ cannot be factored over the real numbers. In general, we say that a quadratic factor

$ax^2 + bx + c$ is **irreducible** if it cannot be factored over the real numbers, that is, if it is prime over the real numbers.

Refer back to Examples 5 and 6. The polynomial function of Example 5 has three real zeros, and its factored form contains three linear factors. The polynomial function of Example 6 has two real zeros, and its factored form contains two linear factors and one irreducible quadratic factor.

Theorem Every polynomial function (with real coefficients) can be uniquely factored into a product of linear factors and/or irreducible quadratic factors.

We shall prove this result in Section 4.3, and, in fact, we shall draw several additional conclusions about the zeros of a polynomial function. One conclusion is worth noting now. If a polynomial (with real coefficients) is of odd degree, then it must contain at least one linear factor. (Do you see why?) This means that it must have at least one real zero.

Corollary A polynomial function (with real coefficients) of odd degree has at least one real zero.

One challenge in using a graphing utility is to set the viewing window so that a complete graph is obtained. The next theorem is a tool that can be used to find bounds on the zeros. This will assure that the function does not have any zeros outside these bounds. Then using these bounds to set Xmin and Xmax assures that all the x-intercepts appear in the viewing window.

BOUNDS ON ZEROS

6 The search for the real zeros of a polynomial function can be reduced somewhat if *bounds* on the zeros are found. A number M is a **bound** on the zeros of a polynomial if every zero r lies between $-M$ and M, inclusive. That is, M is a bound to the zeros of a polynomial f if

$$-M \leq \text{any zero of } f \leq M$$

Theorem **Bounds on Zeros**

Let f denote a polynomial function whose leading coefficient is 1.

$$f(x) = x^n + a_{n-1}x^{n-1} + \cdots + a_1x + a_0$$

A bound M on the zeros of f is the smaller of the two numbers

$$\text{Max}\{1, |a_0| + |a_1| + \cdots + |a_{n-1}|\}, \; 1 + \text{Max}\{|a_0|, |a_1|, \ldots, |a_{n-1}|\} \quad \textbf{(4)}$$

where Max{ } means "choose the largest entry in { }."

An example will help to make the theorem clear.

EXAMPLE 8 Using the Theorem for Finding Bounds on Zeros

Find a bound to the zeros of each polynomial.

(a) $f(x) = x^5 + 3x^3 - 9x^2 + 5$ (b) $g(x) = 4x^5 - 2x^3 + 2x^2 + 1$

Solution (a) The leading coefficient of f is 1.

$$f(x) = x^5 + 3x^3 - 9x^2 + 5 \quad a_4 = 0, a_3 = 3, a_2 = -9, a_1 = 0, a_0 = 5$$

We evaluate the expressions in equation (4).

$$\text{Max}\{1, |a_0| + |a_1| + \cdots + |a_{n-1}|\} = \text{Max}\{1, |5| + |0| + |-9| + |3| + |0|\}$$
$$= \text{Max}\{1, 17\} = 17$$
$$1 + \text{Max}\{|a_0|, |a_1|, \ldots, |a_{n-1}|\} = 1 + \text{Max}\{|5|, |0|, |-9|, |3|, |0|\}$$
$$= 1 + 9 = 10$$

The smaller of the two numbers, 10, is the bound. Every zero of f lies between -10 and 10.

(b) First we write g so that its leading coefficient is 1.

$$g(x) = 4x^5 - 2x^3 + 2x^2 + 1 = 4\left(x^5 - \tfrac{1}{2}x^3 + \tfrac{1}{2}x^2 + \tfrac{1}{4}\right)$$

Next we evaluate the two expressions in equation (4) with $a_4 = 0$, $a_3 = -\frac{1}{2}$, $a_2 = \frac{1}{2}$, $a_1 = 0$, and $a_0 = \frac{1}{4}$.

$$\text{Max}\{1, |a_0| + |a_1| + \cdots + |a_{n-1}|\} = \text{Max}\left\{1, \left|\tfrac{1}{4}\right| + |0| + \left|\tfrac{1}{2}\right| + \left|-\tfrac{1}{2}\right| + |0|\right\}$$
$$= \text{Max}\left\{1, \tfrac{5}{4}\right\} = \tfrac{5}{4}$$
$$1 + \text{Max}\{|a_0|, |a_1|, \ldots, |a_{n-1}|\} = 1 + \text{Max}\left\{\left|\tfrac{1}{4}\right|, |0|, \left|\tfrac{1}{2}\right|, \left|-\tfrac{1}{2}\right|, |0|\right\}$$
$$= 1 + \tfrac{1}{2} = \tfrac{3}{2}$$

The smaller of the two numbers, $\frac{5}{4}$, is the bound. Every zero of g lies between $-\frac{5}{4}$ and $\frac{5}{4}$. ■

EXAMPLE 9 Obtaining Graphs Using Bounds on Zeros

Obtain a graph for each polynomial.

(a) $f(x) = x^5 + 3x^3 - 9x^2 + 5$ (b) $g(x) = 4x^5 - 2x^3 + 2x^2 + 1$

Solution (a) Based on Example 8, every zero lies between -10 and 10. Using Xmin $= -10$ and Xmax $= 10$, we graph $Y_1 = f(x) = x^5 + 3x^3 - 9x^2 + 5$. Figure 5(a) shows the graph obtained using ZOOM-FIT. Figure 5(b) shows the graph after adjusting the viewing window to improve the graph.

Figure 5

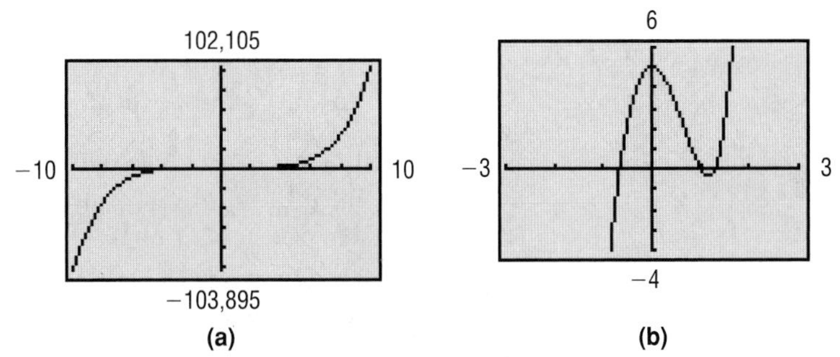

(b) Based on Example 8, every zero lies between $-5/4$ and $5/4$. Using $X\text{min} = -5/4$ and $X\text{max} = 5/4$, we graph $Y_1 = g(x) = 4x^5 - 2x^3 + 2x^2 + 1$. Figure 6 shows the graph after using ZOOM-FIT. No adjustment of the viewing window is needed.

Figure 6

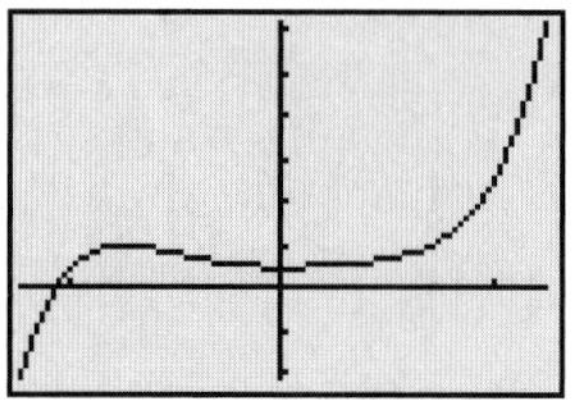

NOW WORK PROBLEM 35.

SUMMARY

Steps for Finding the Zeros of a Polynomial

Step 1: Use the degree of the polynomial to determine the maximum number of real zeros.

Step 2: Use Descartes' Rule of Signs to determine the possible number of positive real zeros and negative real zeros.

Step 3: If the polynomial has integer coefficients, use the Rational Zeros Theorem to identify those rational numbers that are potential zeros.

Step 4: Use the Upper and Lower Bounds Test to obtain a graph of the polynomial.

Step 5: (a) Conjecture possible zeros based on the graph and the list of potential rational zeros.
(b) Test each potential rational zero.
(c) Each time that a zero (and thus a factor) is found, repeat Step 5(b) on the depressed equation. Remember to use factoring techniques and the quadratic formula (if possible).

Let's work one more example that shows the STEPS listed above.

EXAMPLE 10 Finding the Zeros of a Polynomial

Find all the real zeros of the polynomial function

$$f(x) = x^5 - 1.8x^4 - 17.79x^3 + 31.672x^2 + 37.95x - 8.7121$$

Solution **Step 1:** There are at most five real zeros.

Step 2: By Descartes' Rule of Signs, there are three (or one) positive real zeros. Also, because

$$f(-x) = -x^5 - 1.8x^4 + 17.79x^3 + 31.672x^2 - 37.95x - 8.7121$$

there are two (or no) negative real zeros.

Step 3: Since there are noninteger coefficients, the Rational Zeros Theorem does not apply.

Step 4: We determine the bounds of f. The leading coefficient of f is 1 with $a_4 = -1.8$, $a_3 = -17.79$, $a_2 = 31.672$, $a_1 = 37.95$, and $a_0 = -8.7121$.

We evaluate the expressions using equation (4).

$$\text{Max}\{1, |-8.7121| + |37.95| + |31.672| + |-17.79| + |-1.8|\} = \text{Max}\{1, 97.9241\} = 97.9241$$

$$1 + \text{Max}\{|-8.7121|, |37.95|, |31.672|, |-17.79|, |-1.8|\} = 1 + 37.95 = 38.95$$

The smaller of the two numbers, 38.95, is the bound. Every real zero of f lies between -38.95 and 38.95. Figure 7(a) shows the graph of f with $X\text{min} = -38.95$ and $X\text{max} = 38.95$. Figure 7(b) shows a graph of f after adjusting the viewing window to improve the graph.

Figure 7

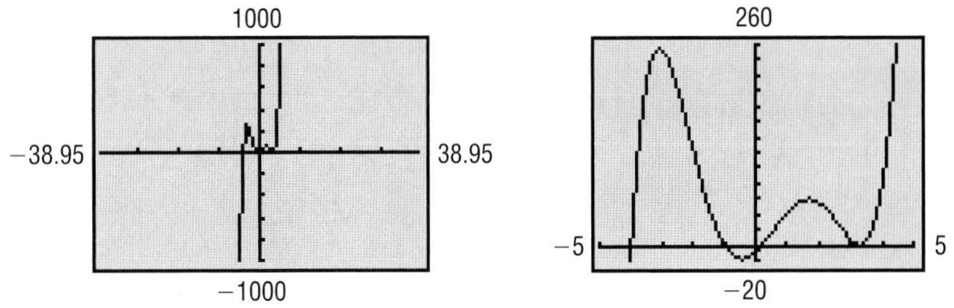

(a) (b)

STEP 5: From Figure 7(b), we see that f appears to have four x-intercepts: one near -4, one near -1, one between 0 and 1, and one near 3. The x-intercept near 3 might be a zero of even multiplicity since the graph seems to touch the x-axis at that point. From Step 2, we learned that the function would have two (or no) negative real zeros and three (or one) positive real zeros.

We use the Factor Theorem to determine if -4 and -1 are zeros. Since $f(-4) = f(-1) = 0$, we know that -4 and -1 are the two negative zeros. Using ZERO (or ROOT), we find that the remaining zeros are 0.20 and 3.30, rounded to two decimal places.

There are no real zeros on the graph that have not already been identified. So either 3.30 is a zero of multiplicity 2 or there are two distinct zeros, each of which is 3.30, rounded to two decimal places. (Example 11 provides the answer.)

Intermediate Value Theorem

7 The Intermediate Value Theorem requires that the function be *continuous*. Although it requires calculus to explain the meaning precisely, the *idea* of a continuous function is easy to understand. Very basically, a function f is continuous when its graph can be drawn without lifting pencil from paper, that is, when the graph contains no "holes" or "jumps" or "gaps." For example, polynomial functions are continuous.

Intermediate Value Theorem

Let f denote a continuous function. If $a < b$ and if $f(a)$ and $f(b)$ are of opposite sign, then f has at least one zero between a and b.

Although the proof of this result requires advanced methods in calculus, it is easy to "see" why the result is true. Look at Figure 8.

Figure 8
If $f(a) < 0$ and $f(b) > 0$ and if f is continuous, there is at least one x-intercept between a and b.

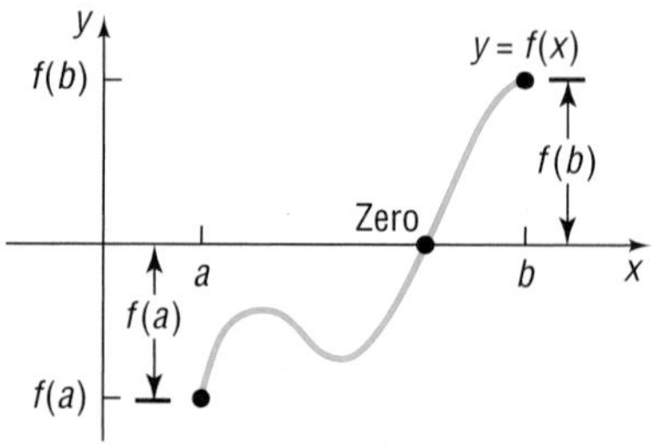

The Intermediate Value Theorem together with the TABLE feature of a graphing utility provides a basis for finding zeros.

EXAMPLE 11 Using the Intermediate Value Theorem and a Graphing Utility to Locate Zeros

Continue working with Example 10 to determine whether there is a repeated zero or two distinct zeros near 3.30.

Solution We use the TABLE feature of a graphing utility. See Table 1. Since $f(3.29) = 0.00956 > 0$ and $f(3.30) = -0.0001 < 0$, by the Intermediate Value Theorem there is a zero between 3.29 and 3.30. Similarly, the table indicates another zero between 3.30 and 3.31. Now we know that the five zeros of f are distinct. ■

TABLE 1

X	Y1
3.27	.08567
3.28	.03829
3.29	.00956
3.3	-1E-4
3.31	.0097
3.32	.03936
3.33	.08931

Y1=X^5-1.8X^4-1...

HISTORICAL FEATURE

Tartaglia (1500–1557)

Formulas for the solution of third- and fourth-degree polynomial equations exist, and, although not very practical, they do have an interesting history.

In the 1500s in Italy, mathematical contests were a popular pastime, and persons possessing methods for solving problems kept them secret. (Solutions that were published were already common knowledge.) Niccolo of Brescia (1500–1557), commonly referred to as Tartaglia ("the stammerer"), had the secret for solving cubic (third-degree) equations, which gave him a decided advantage in the contests. Girolamo Cardano (1501–1576) found out that Tartaglia had the secret and, being interested in cubics, he requested it from Tartaglia. The reluctant Tartaglia hesitated for some time, but finally, swearing Cardano to secrecy with midnight oaths by candlelight, told him the secret. Cardano then published the solution in his book *Ars Magna* (1545), giving Tartaglia the credit, but rather compromising the secrecy. Tartaglia exploded into bitter recriminations, and each wrote pamphlets that reflected on the other's mathematics, moral character, and ancestry. See Problems 93–97.

The quartic (fourth-degree) equation was solved by Cardano's student Lodovico Ferrari, and this solution also was included, with credit and this time with permission, in the *Ars Magna*.

Attempts were made to solve the fifth-degree equation in similar ways, all of which failed. In the early 1800s, P. Ruffini, Niels Abel, and Evariste Galois all found ways to show that it is not possible to solve fifth-degree equations by formula, but the proofs required the introduction of new methods. Galois' methods eventually developed into a large part of modern algebra.

4.1 EXERCISES

In Problems 1–10, use the Factor Theorem to determine whether x–c is a factor of f. If it is, write f in factored form.

1. $f(x) = 4x^3 - 3x^2 - 8x + 4; \quad c = 3$
2. $f(x) = -4x^3 + 5x^2 + 8; \quad c = -2$
3. $f(x) = 3x^4 - 6x^3 - 5x + 10; \quad c = 1$
4. $f(x) = 4x^4 - 15x^2 - 4; \quad c = 2$
5. $f(x) = 3x^6 + 2x^3 - 176; \quad c = -2$
6. $f(x) = 2x^6 - 18x^4 + x^2 - 9; \quad c = -3$
7. $f(x) = 4x^6 - 64x^4 + x^2 - 16; \quad c = 4$
8. $f(x) = x^6 - 16x^4 + x^2 - 16; \quad c = -4$
9. $f(x) = 2x^4 - x^3 + 2x - 1; \quad c = -\frac{1}{2}$
10. $f(x) = 3x^4 + x^3 - 3x + 1; \quad c = -\frac{1}{3}$

In Problems 11–22, tell the maximum number of real zeros that each polynomial function may have. Then use Descartes' Rule of Signs to determine how many positive and how many negative zeros each polynomial function may have. Do not attempt to find the zeros.

11. $f(x) = -4x^7 + x^3 - x^2 + 2$
12. $f(x) = 5x^4 + 2x^2 - 6x - 5$
13. $f(x) = 2x^6 - 3x^2 - x + 1$
14. $f(x) = -3x^5 + 4x^4 + 2$
15. $f(x) = 3x^3 - 2x^2 + x + 2$
16. $f(x) = -x^3 - x^2 + x + 1$
17. $f(x) = -x^4 + x^2 - 1$
18. $f(x) = x^4 + 5x^3 - 2$
19. $f(x) = x^5 + x^4 + x^2 + x + 1$
20. $f(x) = x^5 - x^4 + x^3 - x^2 + x - 1$
21. $f(x) = x^6 - 1$
22. $f(x) = x^6 + 1$

In Problems 23–34, list the potential rational zeros of each polynomial function. Do not attempt to find the zeros.

23. $f(x) = 3x^4 - 3x^3 + x^2 - x + 1$
24. $f(x) = x^5 - x^4 + 2x^2 + 3$
25. $f(x) = x^5 - 6x^2 + 9x - 3$
26. $f(x) = 2x^5 - x^4 - x^2 + 1$
27. $f(x) = -4x^3 - x^2 + x + 2$
28. $f(x) = 6x^4 - x^2 + 2$
29. $f(x) = 3x^4 - x^2 + 2$
30. $f(x) = -4x^3 + x^2 + x + 2$
31. $f(x) = 2x^5 - x^3 + 2x^2 + 4$
32. $f(x) = 3x^5 - x^2 + 2x + 3$
33. $f(x) = 6x^4 + 2x^3 - x^2 + 2$
34. $f(x) = -6x^3 - x^2 + x + 3$

In Problems 35–40, find the bounds to the zeros of each polynomial function. Obtain a complete graph of f.

35. $f(x) = 2x^3 + x^2 - 1$
36. $f(x) = 3x^3 - 2x^2 + x + 4$
37. $f(x) = x^3 - 5x^2 - 11x + 11$
38. $f(x) = 2x^3 - x^2 - 11x - 6$
39. $f(x) = x^4 + 3x^3 - 5x^2 + 9$
40. $f(x) = 4x^4 - 12x^3 + 27x^2 - 54x + 81$

In Problems 41–58, find the real zeros of f. Use the real zeros to factor f.

41. $f(x) = x^3 + 2x^2 - 5x - 6$
42. $f(x) = x^3 + 8x^2 + 11x - 20$
43. $f(x) = 2x^3 - 13x^2 + 24x - 9$
44. $f(x) = 2x^3 - 5x^2 - 4x + 12$
45. $f(x) = 3x^3 + 4x^2 + 4x + 1$
46. $f(x) = 3x^3 - 7x^2 + 12x - 28$
47. $f(x) = x^3 - 8x^2 + 17x - 6$
48. $f(x) = x^3 + 6x^2 + 6x - 4$
49. $f(x) = x^4 + x^3 - 3x^2 - x + 2$
50. $f(x) = x^4 - x^3 - 6x^2 + 4x + 8$
51. $f(x) = 2x^4 + 17x^3 + 35x^2 - 9x - 45$
52. $f(x) = 4x^4 - 15x^3 - 8x^2 + 15x + 4$
53. $f(x) = 2x^4 - 3x^3 - 21x^2 - 2x + 24$
54. $f(x) = 2x^4 + 11x^3 - 5x^2 - 43x + 35$
55. $f(x) = 4x^4 + 7x^2 - 2$
56. $f(x) = 4x^4 + 15x^2 - 4$
57. $f(x) = 4x^5 - 8x^4 - x + 2$
58. $f(x) = 4x^5 + 12x^4 - x - 3$

In Problems 59–64, find the real zeros of f rounded to two decimal places.

59. $f(x) = x^3 + 3.2x^2 - 16.83x - 5.31$
60. $f(x) = x^3 + 3.2x^2 - 7.25x - 6.3$
61. $f(x) = x^4 - 1.4x^3 - 33.71x^2 + 23.94x + 292.41$
62. $f(x) = x^4 + 1.2x^3 - 7.46x^2 - 4.692x + 15.2881$
63. $f(x) = x^3 + 19.5x^2 - 1021x + 1000.5$
64. $f(x) = x^3 + 42.2x^2 - 664.8x + 1490.4$

In Problems 65–74, find the real solutions of each equation.

65. $x^4 - x^3 + 2x^2 - 4x - 8 = 0$
66. $2x^3 + 3x^2 + 2x + 3 = 0$
67. $3x^3 + 4x^2 - 7x + 2 = 0$
68. $2x^3 - 3x^2 - 3x - 5 = 0$
69. $3x^3 - x^2 - 15x + 5 = 0$
70. $2x^3 - 11x^2 + 10x + 8 = 0$
71. $x^4 + 4x^3 + 2x^2 - x + 6 = 0$
72. $x^4 - 2x^3 + 10x^2 - 18x + 9 = 0$
73. $x^3 - \frac{2}{3}x^2 + \frac{8}{3}x + 1 = 0$
74. $x^3 - \frac{2}{3}x^2 + 3x - 2 = 0$

In Problems 75–80, use the Intermediate Value Theorem to show that each function has a zero in the given interval. Approximate the zero rounded to two decimal places.

75. $f(x) = 8x^4 - 2x^2 + 5x - 1; \quad [0, 1]$
76. $f(x) = x^4 + 8x^3 - x^2 + 2; \quad [-1, 0]$
77. $f(x) = 2x^3 + 6x^2 - 8x + 2; \quad [-5, -4]$
78. $f(x) = 3x^3 - 10x + 9; \quad [-3, -2]$
79. $f(x) = x^5 - x^4 + 7x^3 - 7x^2 - 18x + 18; \quad [1.4, 1.5]$
80. $f(x) = x^5 - 3x^4 - 2x^3 + 6x^2 + x + 2; \quad [1.7, 1.8]$

81. Cost of Manufacturing In Problem 67 of Section 3.3, you found the cost function for manufacturing Chevy Cavaliers. Use the methods learned in this section to determine how many Cavaliers can be manufactured at a total cost of \$3,000,000. In other words, solve the equation $C(x) = 3000$.

82. Cost of Printing In Problem 64 of Section 3.3, you found the cost function for printing textbooks. Use the methods learned in this section to determine how many textbooks can be printed at a total cost of \$200,000. In other words, solve the equation $C(x) = 200$.

83. Find k such that $f(x) = x^3 - kx^2 + kx + 2$ has the factor $x - 2$.

84. Find k such that $f(x) = x^4 - kx^3 + kx^2 + 1$ has the factor $x + 2$.

85. What is the remainder when

$$f(x) = 2x^{20} - 8x^{10} + x - 2$$

is divided by $x - 1$?

86. What is the remainder when

$$f(x) = -3x^{17} + x^9 - x^5 + 2x$$

is divided by $x + 1$?

87. Is $\frac{1}{3}$ a zero of $f(x) = 2x^3 + 3x^2 - 6x + 7$? Explain.

88. Is $\frac{1}{3}$ a zero of $f(x) = 4x^3 - 5x^2 - 3x + 1$? Explain.

89. Is $\frac{3}{5}$ a zero of $f(x) = 2x^6 - 5x^4 + x^3 - x + 1$? Explain.

90. Is $\frac{2}{3}$ a zero of $f(x) = x^7 + 6x^5 - x^4 + x + 2$? Explain.

91. What is the length of the edge of a cube if, after a slice 1 inch thick is cut from one side, the volume remaining is 294 cubic inches?

92. What is the length of the edge of a cube if its volume could be doubled by an increase of 6 centimeters in one edge, an increase of 12 centimeters in a second edge, and a decrease of 4 centimeters in the third edge?

In Problems 93–97, develop the Tartaglia–Cardano solution of the cubic equation and show why it is not altogether practical.

93. Show that the general cubic equation $y^3 + by^2 + cy + d = 0$ can be transformed into an equation of the form $x^3 + px + q = 0$ by using the substitution $y = x - b/3$.

94. In the equation $x^3 + px + q = 0$, replace x by $H + K$. Let $3HK = -p$, and show that $H^3 + K^3 = -q$. [**Hint:** $3H^2K + 3HK^2 = 3HKx$.]

95. Based on Problem 94, we have the two equations

$$3HK = -p \quad \text{and} \quad H^3 + K^3 = -q$$

Solve for K in $3HK = -p$ and substitute into $H^3 + K^3 = -q$. Then show that

$$H = \sqrt[3]{\frac{-q}{2} + \sqrt{\frac{q^2}{4} + \frac{p^3}{27}}}$$

[**Hint:** Look for an equation that is quadratic in form.]

96. Use the solution for H from Problem 95 and the equation $H^3 + K^3 = -q$ to show that

$$K = \sqrt[3]{\frac{-q}{2} - \sqrt{\frac{q^2}{4} + \frac{p^3}{27}}}$$

97. Use the results from Problems 94–96 to show that the solution of $x^3 + px + q = 0$ is

$$x = \sqrt[3]{\frac{-q}{2} + \sqrt{\frac{q^2}{4} + \frac{p^3}{27}}} + \sqrt[3]{\frac{-q}{2} - \sqrt{\frac{q^2}{4} + \frac{p^3}{27}}}$$

98. Use the result of Problem 97 to solve the equation $x^3 - 6x - 9 = 0$.

99. Use the result of Problem 97 to solve the equation $x^3 + 3x - 14 = 0$.

100. Use the methods of this chapter to solve the equation $x^3 + 3x - 14 = 0$.

101. A function f has the property that $f(2 + x) = f(2 - x)$ for all x. If f has exactly four real zeros, find their sum.

PREPARING FOR THIS SECTION

Before getting started, review the following:

✓ Quadratic Equations (pp. 28–32)

4.2 COMPLEX NUMBERS; QUADRATIC EQUATIONS WITH A NEGATIVE DISCRIMINANT

OBJECTIVES
1. Add, Subtract, Multiply, and Divide Complex Numbers
2. Solve Quadratic Equations with a Negative Discriminant

One property of a real number is that its square is nonnegative. For example, there is no real number x for which

$$x^2 = -1$$

To remedy this situation, we introduce a number called the **imaginary unit**, which we denote by $\boldsymbol{i}$ and whose square is -1. Thus,

$$i^2 = -1$$

This should not surprise you. If our universe were to consist only of integers, there would be no number x for which $2x = 1$. This unfortunate circumstance was remedied by introducing numbers such as $\frac{1}{2}$ and $\frac{2}{3}$, the *rational numbers*. If our universe were to consist only of rational numbers, there would be no x whose square equals 2. That is, there would be no number x for which $x^2 = 2$. To remedy this, we introduced numbers such as $\sqrt{2}$ and $\sqrt[3]{5}$, the *irrational numbers*. The *real numbers*, you will recall, consist of the rational numbers and the irrational numbers. Now, if our universe were to consist only of real numbers, then there would be no number x whose square is -1. To remedy this, we introduce a number i, whose square is -1.

In the progression outlined, each time that we encountered a situation that was unsuitable, we introduced a new number system to remedy this situation. And each new number system contained the earlier number system as a subset. The number system that results from introducing the number i is called the **complex number system**.

Complex numbers are numbers of the form $\boldsymbol{a + bi}$, where a and b are real numbers. The real number a is called the **real part** of the number $a + bi$; the real number b is called the **imaginary part** of $a + bi$.

For example, the complex number $-5 + 6i$ has the real part -5 and the imaginary part 6.

When a complex number is written in the form $a + bi$, where a and b are real numbers, we say it is in **standard form**. However, if the imaginary part of a complex number is negative, such as in the complex number $3 + (-2)i$, we agree to write it instead in the form $3 - 2i$.

Also, the complex number $a + 0i$ is usually written merely as a. This serves to remind us that the real numbers are a subset of the complex numbers. The complex number $0 + bi$ is usually written as bi. Sometimes the complex number bi is called a **pure imaginary number**.

1 Equality, addition, subtraction, and multiplication of complex numbers are defined so as to preserve the familiar rules of algebra for real numbers. Thus, two complex numbers are equal if and only if their real parts are equal and their imaginary parts are equal. That is,

Equality of Complex Numbers

$$a + bi = c + di \quad \text{if and only if } a = c \text{ and } b = d \qquad \textbf{(1)}$$

Two complex numbers are added by forming the complex number whose real part is the sum of the real parts and whose imaginary part is the sum of the imaginary parts. That is,

Sum of Complex Numbers

$$(a + bi) + (c + di) = (a + c) + (b + d)i \quad \textbf{(2)}$$

To subtract two complex numbers, we use this rule:

Difference of Complex Numbers

$$(a + bi) - (c + di) = (a - c) + (b - d)i \quad \textbf{(3)}$$

EXAMPLE 1 Adding and Subtracting Complex Numbers

(a) $(3 + 5i) + (-2 + 3i) = [3 + (-2)] + (5 + 3)i = 1 + 8i$
(b) $(6 + 4i) - (3 + 6i) = (6 - 3) + (4 - 6)i = 3 + (-2)i = 3 - 2i$ ■

Some graphing calculators have the capability of handling complex numbers.* For example, Figure 9 shows the results of Example 1 using a TI-83 graphing calculator.

Figure 9

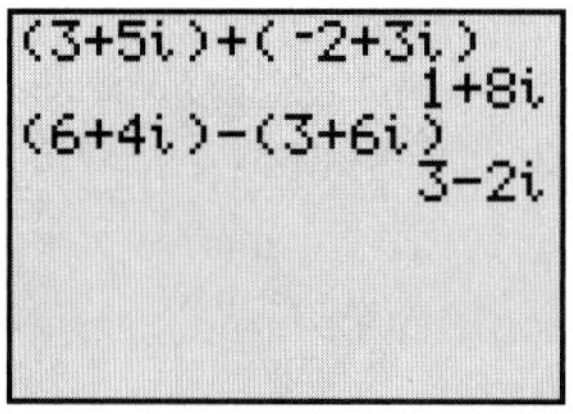

NOW WORK PROBLEM 5.

Products of complex numbers are calculated as illustrated in Example 2.

EXAMPLE 2 Multiplying Complex Numbers

$$(5 + 3i)\cdot(2 + 7i) = \underset{\uparrow \text{ Distributive property}}{5\cdot(2 + 7i) + 3i(2 + 7i)} = \underset{\uparrow \text{ Distributive property}}{10 + 35i + 6i + 21i^2}$$

$$= 10 + 41i + 21(-1) \quad (\uparrow\ i^2 = -1)$$

$$= -11 + 41i$$ ■

*Consult your user's manual for the appropriate keystrokes.

Based on the procedure of Example 2, we define the **product** of two complex numbers by the following formula:

Product of Complex Numbers

$$(a + bi) \cdot (c + di) = (ac - bd) + (ad + bc)i \qquad \textbf{(4)}$$

Do not bother to memorize formula (4). Instead, whenever it is necessary to multiply two complex numbers, follow the usual rules for multiplying two binomials, as in Example 2, remembering that $i^2 = -1$. For example,

$$(2i)(2i) = 4i^2 = -4$$

$$(2 + i)(1 - i) = 2 - 2i + i - i^2 = 3 - i$$

Figure 10

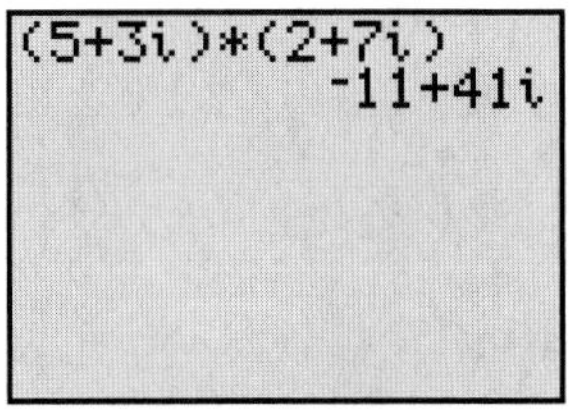

Graphing calculators may also be used to multiply complex numbers. Figure 10 shows the result obtained in Example 2 using a TI-83 graphing calculator.

NOW WORK PROBLEM 11.

Algebraic properties for addition and multiplication, such as the commutative, associative, and distributive properties, hold for complex numbers. Of these, the property that every nonzero complex number has a multiplicative inverse, or reciprocal, requires a closer look.

CONJUGATES

If $z = a + bi$ is a complex number, then its **conjugate**, denoted by $\bar{z}$, is defined as

$$\bar{z} = \overline{a + bi} = a - bi$$

For example, $\overline{2 + 3i} = 2 - 3i$ and $\overline{-6 - 2i} = -6 + 2i$.

EXAMPLE 3 Multiplying a Complex Number by Its Conjugate

Find the product of the complex number $z = 3 + 4i$ and its conjugate $\bar{z}$.

Solution Since $\bar{z} = 3 - 4i$, we have

$$z\bar{z} = (3 + 4i)(3 - 4i) = 9 + 12i - 12i - 16i^2 = 9 + 16 = 25$$

The result obtained in Example 3 has an important generalization.

Theorem The product of a complex number and its conjugate is a nonnegative real number. That is, if $z = a + bi$, then

$$z\bar{z} = a^2 + b^2 \qquad \textbf{(5)}$$

Proof If $z = a + bi$, then

$$z\bar{z} = (a + bi)(a - bi) = a^2 - (bi)^2 = a^2 - b^2i^2 = a^2 + b^2 \quad \blacksquare$$

To express the reciprocal of a nonzero complex number z in standard form, multiply the numerator and denominator by its conjugate $\bar{z}$. That is, if $z = a + bi$ is a nonzero complex number, then

$$\frac{1}{a + bi} = \frac{1}{z} = \frac{1}{z} \cdot \frac{\bar{z}}{\bar{z}} = \frac{\bar{z}}{z\bar{z}} = \frac{a - bi}{(a + bi)(a - bi)}$$

$$= \frac{a - bi}{a^2 + b^2}$$

↑ Use (5).

$$= \frac{a}{a^2 + b^2} - \frac{b}{a^2 + b^2}i$$

EXAMPLE 4 Writing the Reciprocal of a Complex Number in Standard Form

Write $\dfrac{1}{3 + 4i}$ in standard form $a + bi$; that is, find the reciprocal of $3 + 4i$.

Solution The idea is to multiply the numerator and denominator by the conjugate of $3 + 4i$, that is, the complex number $3 - 4i$. The result is

$$\frac{1}{3 + 4i} = \frac{1}{3 + 4i} \cdot \frac{3 - 4i}{3 - 4i} = \frac{3 - 4i}{9 + 16} = \frac{3}{25} - \frac{4}{25}i \quad \blacksquare$$

Figure 11

A graphing calculator can be used to verify the result of Example 4. See Figure 11.

To express the quotient of two complex numbers in standard form, we multiply the numerator and denominator of the quotient by the conjugate of the denominator.

EXAMPLE 5 Writing the Quotient of Complex Numbers in Standard Form

Write each of the following in standard form.

(a) $\dfrac{1 + 4i}{5 - 12i}$ (b) $\dfrac{2 - 3i}{4 - 3i}$

Solution (a)

$$\frac{1 + 4i}{5 - 12i} = \frac{1 + 4i}{5 - 12i} \cdot \frac{5 + 12i}{5 + 12i} = \frac{5 + 20i + 12i + 48i^2}{25 + 144}$$

$$= \frac{-43 + 32i}{169} = \frac{-43}{169} + \frac{32}{169}i$$

(b)

$$\frac{2 - 3i}{4 - 3i} = \frac{2 - 3i}{4 - 3i} \cdot \frac{4 + 3i}{4 + 3i} = \frac{8 - 12i + 6i - 9i^2}{16 + 9}$$

$$= \frac{17 - 6i}{25} = \frac{17}{25} - \frac{6}{25}i \quad \blacksquare$$

NOW WORK PROBLEM 19.

EXAMPLE 6 Writing Other Expressions in Standard Form

If $z = 2 - 3i$ and $w = 5 + 2i$, write each of the following expressions in standard form.

(a) $\dfrac{z}{w}$ (b) $\overline{z + w}$ (c) $z + \bar{z}$

Solution (a) $\dfrac{z}{w} = \dfrac{z \cdot \bar{w}}{w \cdot \bar{w}} = \dfrac{(2 - 3i)(5 - 2i)}{(5 + 2i)(5 - 2i)} = \dfrac{10 - 15i - 4i + 6i^2}{25 + 4}$

$= \dfrac{4 - 19i}{29} = \dfrac{4}{29} - \dfrac{19}{29}i$

(b) $\overline{z + w} = \overline{(2 - 3i) + (5 + 2i)} = \overline{7 - i} = 7 + i$

(c) $z + \bar{z} = (2 - 3i) + (2 + 3i) = 4$

The conjugate of a complex number has certain general properties that we shall find useful later.

For a real number $a = a + 0i$, the conjugate is $\bar{a} = \overline{a + 0i} = a - 0i = a$. That is,

Theorem The conjugate of a real number is the real number itself.

Other properties of the conjugate that are direct consequences of the definition are given next. In each statement, z and w represent complex numbers.

Theorem The conjugate of the conjugate of a complex number is the complex number itself.

$$\overline{(\bar{z})} = z \tag{6}$$

The conjugate of the sum of two complex numbers equals the sum of their conjugates.

$$\overline{z + w} = \bar{z} + \bar{w} \tag{7}$$

The conjugate of the product of two complex numbers equals the product of their conjugates.

$$\overline{z \cdot w} = \bar{z} \cdot \bar{w} \tag{8}$$

We leave the proofs of equations (6), (7), and (8) as exercises.

POWERS OF i

The **powers of *i*** follow a pattern that is useful to know.

$$i^1 = i \qquad\qquad i^5 = i^4 \cdot i = 1 \cdot i = i$$
$$i^2 = -1 \qquad\qquad i^6 = i^4 \cdot i^2 = -1$$
$$i^3 = i^2 \cdot i = -i \qquad\qquad i^7 = i^4 \cdot i^3 = -i$$
$$i^4 = i^2 \cdot i^2 = (-1)(-1) = 1 \qquad\qquad i^8 = i^4 \cdot i^4 = 1$$

And so on. The powers of i repeat with every fourth power.

EXAMPLE 7 Evaluating Powers of i

(a) $i^{27} = i^{24} \cdot i^3 = \left(i^4\right)^6 \cdot i^3 = 1^6 \cdot i^3 = -i$
(b) $i^{101} = i^{100} \cdot i^1 = \left(i^4\right)^{25} \cdot i = 1^{25} \cdot i = i$

EXAMPLE 8 Writing the Power of a Complex Number in Standard Form

Write $(2 + i)^3$ in standard form.

Solution We use the special product formula for $(x + a)^3$.

$$(x + a)^3 = x^3 + 3ax^2 + 3a^2x + a^3$$

Using this special product formula,

$$\begin{aligned}(2 + i)^3 &= 2^3 + 3 \cdot i \cdot 2^2 + 3 \cdot i^2 \cdot 2 + i^3\\ &= 8 + 12i + 6(-1) + (-i)\\ &= 2 + 11i\end{aligned}$$

NOW WORK PROBLEM 33.

QUADRATIC EQUATIONS WITH A NEGATIVE DISCRIMINANT

2 Quadratic equations with a negative discriminant have no real number solution. However, if we extend our number system to allow complex numbers, quadratic equations will always have a solution. Since the solution to a quadratic equation involves the square root of the discriminant, we begin with a discussion of square roots of negative numbers.

If N is a positive real number, we define the **principal square root of $-N$**, denoted by $\sqrt{-N}$, as

$$\sqrt{-N} = \sqrt{N}\,i$$

where i is the imaginary unit and $i^2 = -1$.

EXAMPLE 9 Evaluating the Square Root of a Negative Number

(a) $\sqrt{-1} = \sqrt{1}i = i$ (b) $\sqrt{-4} = \sqrt{4}i = 2i$
(c) $\sqrt{-8} = \sqrt{8}i = 2\sqrt{2}i$

EXAMPLE 10 Solving Equations

Solve each equation in the complex number system.

(a) $x^2 = 4$ (b) $x^2 = -9$

Solution (a) $x^2 = 4$

$x = \pm\sqrt{4} = \pm 2$

The equation has two solutions, -2 and 2.

(b) $x^2 = -9$

$x = \pm\sqrt{-9} = \pm\sqrt{9}i = \pm 3i$

The equation has two solutions, $-3i$ and $3i$. ■

NOW WORK PROBLEM 45.

WARNING: When working with square roots of negative numbers, do not set the square root of a product equal to the product of the square roots (which can be done with positive numbers). To see why, look at this calculation: We know that $\sqrt{100} = 10$. However, it is also true that $100 = (-25)(-4)$, so

$$10 = \sqrt{100} = \sqrt{(-25)(-4)} \neq \sqrt{-25}\sqrt{-4} = (\sqrt{25}i)(\sqrt{4}i) = (5i)(2i) = 10i^2 = -10$$

↑ Here is the error. ■

Because we have defined the square root of a negative number, we can now restate the quadratic formula without restriction.

Theorem In the complex number system, the solutions of the quadratic equation $ax^2 + bx + c = 0$, where a, b, and c are real numbers and $a \neq 0$, are given by the formula

$$x = \frac{-b \pm \sqrt{b^2 - 4ac}}{2a} \quad \textbf{(9)}$$

■

EXAMPLE 11 Solving Quadratic Equations in the Complex Number System

Solve the equation $x^2 - 4x + 8 = 0$ in the complex number system.

Solution Here $a = 1$, $b = -4$, $c = 8$, and $b^2 - 4ac = 16 - 4(1)(8) = -16$. Using equation (9), we find that

$$x = \frac{-(-4) \pm \sqrt{-16}}{2(1)} = \frac{4 \pm \sqrt{16}i}{2} = \frac{4 \pm 4i}{2} = 2 \pm 2i$$

The equation has the solution set $\{2 - 2i, 2 + 2i\}$. ■

Check:

$2 + 2i$: $(2 + 2i)^2 - 4(2 + 2i) + 8 = 4 + 8i + 4i^2 - 8 - 8i + 8$

$= 4 - 4 = 0$

$2 - 2i$: $(2 - 2i)^2 - 4(2 - 2i) + 8 = 4 - 8i + 4i^2 - 8 + 8i + 8$

$= 4 - 4 = 0$ ■

Figure 12

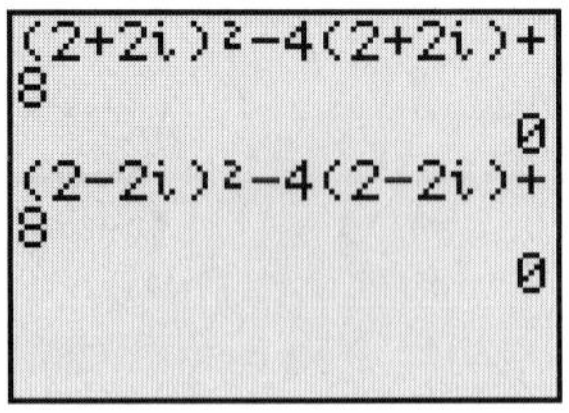

Figure 12 shows the check of the solution using a TI-83 graphing calculator.

NOW WORK PROBLEM 51.

The discriminant $b^2 - 4ac$ of a quadratic equation still serves as a way to determine the character of the solutions.

DISCRIMINANT OF A QUADRATIC EQUATION

In the complex number system, consider a quadratic equation $ax^2 + bx + c = 0$ with real coefficients.

1. If $b^2 - 4ac > 0$, the equation has two unequal real solutions.
2. If $b^2 - 4ac = 0$, the equation has a repeated real solution, a double root.
3. If $b^2 - 4ac < 0$, the equation has two complex solutions that are not real. The solutions are conjugates of each other.

The third conclusion in the display is a consequence of the fact that if $b^2 - 4ac = -N < 0$ then, by the quadratic formula, the solutions are

$$x = \frac{-b + \sqrt{b^2 - 4ac}}{2a} = \frac{-b + \sqrt{-N}}{2a} = \frac{-b + \sqrt{N}i}{2a} = \frac{-b}{2a} + \frac{\sqrt{N}}{2a}i$$

and

$$x = \frac{-b - \sqrt{b^2 - 4ac}}{2a} = \frac{-b - \sqrt{-N}}{2a} = \frac{-b - \sqrt{N}i}{2a} = \frac{-b}{2a} - \frac{\sqrt{N}}{2a}i$$

which are conjugates of each other.

EXAMPLE 12 Determining the Character of the Solution of a Quadratic Equation

Without solving, determine the character of the solution of each equation.

(a) $3x^2 + 4x + 5 = 0$ (b) $2x^2 + 4x + 1 = 0$
(c) $9x^2 - 6x + 1 = 0$

Solution

(a) Here $a = 3, b = 4$, and $c = 5$, so $b^2 - 4ac = 16 - 4(3)(5) = -44$. The solutions are complex numbers that are not real and are conjugates of each other.

(b) Here $a = 2, b = 4$, and $c = 1$, so $b^2 - 4ac = 16 - 8 = 8$. The solutions are two unequal real numbers.

(c) Here $a = 9, b = -6$, and $c = 1$, so $b^2 - 4ac = 36 - 4(9)(1) = 0$. The solution is a repeated real number, that is, a double root. ■

4.2 EXERCISES

In Problems 1–38, write each expression in the standard form $a + bi$. Verify your results using a graphing utility.

1. $(2 - 3i) + (6 + 8i)$ **2.** $(4 + 5i) + (-8 + 2i)$ **3.** $(-3 + 2i) - (4 - 4i)$
4. $(3 - 4i) - (-3 - 4i)$ **5.** $(2 - 5i) - (8 + 6i)$ **6.** $(-8 + 4i) - (2 - 2i)$

7. $3(2 - 6i)$
8. $-4(2 + 8i)$
9. $2i(2 - 3i)$
10. $3i(-3 + 4i)$
11. $(3 - 4i)(2 + i)$
12. $(5 + 3i)(2 - i)$
13. $(-6 + i)(-6 - i)$
14. $(-3 + i)(3 + i)$
15. $\dfrac{10}{3 - 4i}$
16. $\dfrac{13}{5 - 12i}$
17. $\dfrac{2 + i}{i}$
18. $\dfrac{2 - i}{-2i}$
19. $\dfrac{6 - i}{1 + i}$
20. $\dfrac{2 + 3i}{1 - i}$
21. $\left(\dfrac{1}{2} + \dfrac{\sqrt{3}}{2}i\right)^2$
22. $\left(\dfrac{\sqrt{3}}{2} - \dfrac{1}{2}i\right)^2$
23. $(1 + i)^2$
24. $(1 - i)^2$
25. i^{23}
26. i^{14}
27. i^{-15}
28. i^{-23}
29. $i^6 - 5$
30. $4 + i^3$
31. $6i^3 - 4i^5$
32. $4i^3 - 2i^2 + 1$
33. $(1 + i)^3$
34. $(3i)^4 + 1$
35. $i^7(1 + i^2)$
36. $2i^4(1 + i^2)$
37. $i^6 + i^4 + i^2 + 1$
38. $i^7 + i^5 + i^3 + i$

In Problems 39–44, perform the indicated operations and express your answer in the form $a + bi$.

39. $\sqrt{-4}$
40. $\sqrt{-9}$
41. $\sqrt{-25}$
42. $\sqrt{-64}$
43. $\sqrt{(3 + 4i)(4i - 3)}$
44. $\sqrt{(4 + 3i)(3i - 4)}$

In Problems 45–64, solve each equation in the complex number system. Check your results using a graphing utility.

45. $x^2 + 4 = 0$
46. $x^2 - 4 = 0$
47. $x^2 - 16 = 0$
48. $x^2 + 25 = 0$
49. $x^2 - 6x + 13 = 0$
50. $x^2 + 4x + 8 = 0$
51. $x^2 - 6x + 10 = 0$
52. $x^2 - 2x + 5 = 0$
53. $8x^2 - 4x + 1 = 0$
54. $10x^2 + 6x + 1 = 0$
55. $5x^2 + 2x + 1 = 0$
56. $13x^2 + 6x + 1 = 0$
57. $x^2 + x + 1 = 0$
58. $x^2 - x + 1 = 0$
59. $x^3 - 8 = 0$
60. $x^3 + 27 = 0$
61. $x^4 - 16 = 0$
62. $x^4 - 1 = 0$
63. $x^4 + 13x^2 + 36 = 0$
64. $x^4 + 3x^2 - 4 = 0$

In Problems 65–70, without solving, determine the character of the solutions of each equation. Verify your answer using a graphing utility.

65. $3x^2 - 3x + 4 = 0$
66. $2x^2 - 4x + 1 = 0$
67. $2x^2 + 3x - 4 = 0$
68. $x^2 + 2x + 6 = 0$
69. $9x^2 - 12x + 4 = 0$
70. $4x^2 + 12x + 9 = 0$
71. $2 + 3i$ is a solution of a quadratic equation with real coefficients. Find the other solution.
72. $4 - i$ is a solution of a quadratic equation with real coefficients. Find the other solution.

In Problems 73–76, $z = 3 - 4i$ and $w = 8 + 3i$. Write each expression in the standard form $a + bi$.

73. $z + \bar{z}$
74. $w - \bar{w}$
75. $z\bar{z}$
76. $\overline{z - w}$
77. Use $z = a + bi$ to show that $z + \bar{z} = 2a$ and that $z - \bar{z} = 2bi$.
78. Use $z = a + bi$ to show that $(\bar{\bar{z}}) = z$.
79. Use $z = a + bi$ and $w = c + di$ to show that $\overline{z + w} = \bar{z} + \bar{w}$.
80. Use $z = a + bi$ and $w = c + di$ to show that $\overline{z \cdot w} = \bar{z} \cdot \bar{w}$.
81. Explain to a friend how you would add two complex numbers and how you would multiply two complex numbers. Explain any differences in the two explanations.
82. Write a brief paragraph that compares the method used to rationalize denominators and the method used to write the quotient of two complex numbers in standard form.

4.3 COMPLEX ZEROS; FUNDAMENTAL THEOREM OF ALGEBRA

OBJECTIVES
1. Utilize the Conjugate Pairs Theorem to Find the Complex Zeros of a Polynomial
2. Find a Polynomial Function with Specified Zeros
3. Find the Complex Zeros of a Polynomial

In Section 4.1 we found the **real** zeros of a polynomial function. In this section we will find the **complex** zeros of a polynomial function. Since the set of real numbers is a subset of the set of complex numbers, finding the complex zeros of a function requires finding all zeros of the form $a + bi$. These zeros will be real if $b = 0$.

A variable in the complex number system is referred to as a **complex variable**. A **complex polynomial function** f of degree n is a function of the form

$$f(x) = a_n x^n + a_{n-1} x^{n-1} + \cdots + a_1 x + a_0 \qquad (1)$$

where $a_n, a_{n-1}, \ldots, a_1, a_0$ are complex numbers, $a_n \neq 0$, n is a nonnegative integer, and x is a complex variable. As before, a_n is called the **leading coefficient** of f. A complex number r is called a (complex) **zero** of f if $f(r) = 0$.

We have learned that some quadratic equations have no real solutions, but that in the complex number system every quadratic equation has a solution, either real or complex. The next result, proved by Karl Friedrich Gauss (1777–1855) when he was 22 years old,* gives an extension to complex polynomials. In fact, this result is so important and useful that it has become known as the **Fundamental Theorem of Algebra**.

Fundamental Theorem of Algebra

Every complex polynomial function $f(x)$ of degree $n \geq 1$ has at least one complex zero.

■

We shall not prove this result, as the proof is beyond the scope of this book. However, using the Fundamental Theorem of Algebra and the Factor Theorem, we can prove the following result:

Theorem

Every complex polynomial function $f(x)$ of degree $n \geq 1$ can be factored into n linear factors (not necessarily distinct) of the form

$$f(x) = a_n(x - r_1)(x - r_2)\cdot \ldots \cdot(x - r_n) \qquad (2)$$

where $a_n, r_1, r_2, \ldots, r_n$ are complex numbers.

■

*In all, Gauss gave four different proofs of this theorem, the first one in 1799 being the subject of his doctoral dissertation.

Proof Let

$$f(x) = a_n x^n + a_{n-1}x^{n-1} + \cdots + a_1 x + a_0$$

By the Fundamental Theorem of Algebra, f has at least one zero, say r_1. Then, by the Factor Theorem, $x - r_1$ is a factor, and

$$f(x) = (x - r_1)q_1(x)$$

where $q_1(x)$ is a complex polynomial of degree $n - 1$ whose leading coefficient is a_n. Again by the Fundamental Theorem of Algebra, the complex polynomial $q_1(x)$ has at least one zero, say r_2. By the Factor Theorem, $q_1(x)$ has the factor $x - r_2$, so

$$q_1(x) = (x - r_2)q_2(x)$$

where $q_2(x)$ is a complex polynomial of degree $n - 2$ whose leading coefficient is a_n. Consequently,

$$f(x) = (x - r_1)(x - r_2)q_2(x)$$

Repeating this argument n times, we finally arrive at

$$f(x) = (x - r_1)(x - r_2)\cdot\ldots\cdot(x - r_n)q_n(x)$$

where $q_n(x)$ is a complex polynomial of degree $n - n = 0$ whose leading coefficient is a_n. Thus, $q_n(x) = a_n x^0 = a_n$, and so

$$f(x) = a_n(x - r_1)(x - r_2)\cdot\ldots\cdot(x - r_n)$$

We conclude every complex polynomial function $f(x)$ of degree $n \geq 1$ has exactly n (not necessarily distinct) zeros. ■

Complex Zeros of Polynomials with Real Coefficients

1 We can use the Fundamental Theorem of Algebra to obtain valuable information about the complex zeros of polynomials whose coefficients are real numbers.

Conjugate Pairs Theorem

Let $f(x)$ be a polynomial whose coefficients are real numbers. If $r = a + bi$ is a zero of f, then the complex conjugate $\bar{r} = a - bi$ is also a zero of f.

In other words, for polynomials whose coefficients are real numbers, the zeros occur in conjugate pairs.

Proof Let

$$f(x) = a_n x^n + a_{n-1}x^{n-1} + \cdots + a_1 x + a_0$$

where $a_n, a_{n-1}, \ldots, a_1, a_0$ are real numbers and $a_n \neq 0$. If $r = a + bi$ is a zero of f, then $f(r) = f(a + bi) = 0$, so

$$a_n r^n + a_{n-1}r^{n-1} + \cdots + a_1 r + a_0 = 0$$

We take the conjugate of both sides to get

$$\overline{a_n r^n + a_{n-1}r^{n-1} + \cdots + a_1 r + a_0} = \overline{0}$$

$$\overline{a_n r^n} + \overline{a_{n-1}r^{n-1}} + \cdots + \overline{a_1 r} + \overline{a_0} = \overline{0}$$ The conjugate of a sum equals the sum of the conjugates (see Section 4.2).

$$\overline{a_n}(\overline{r})^n + \overline{a_{n-1}}(\overline{r})^{n-1} + \cdots + \overline{a}_1\overline{r} + \overline{a}_0 = \overline{0}$$ The conjugate of a product equals the product of the conjugates.

$$a_n(\overline{r})^n + a_{n-1}(\overline{r})^{n-1} + \cdots + a_1\overline{r} + a_0 = 0$$ The conjugate of a real number equals the real number.

This last equation states that $f(\overline{r}) = 0$; that is, $\overline{r} = a - bi$ is a zero of f. ■

The value of this result should be clear. Once we know that, say, $3 + 4i$ is a zero of a polynomial with real coefficients, then we know that $3 - 4i$ is also a zero. This result has an important corollary.

Corollary A polynomial f of odd degree with real coefficients has at least one real zero.

■

Proof Because complex zeros occur as conjugate pairs in a polynomial with real coefficients, there will always be an even number of zeros that are not real numbers. Consequently, since f is of odd degree, one of its zeros has to be a real number. ■

For example, the polynomial $f(x) = x^5 - 3x^4 + 4x^3 - 5$ has at least one zero that is a real number, since f is of degree 5 (odd) and has real coefficients.

EXAMPLE 1 Using the Conjugate Pairs Theorem

A polynomial f of degree 5 whose coefficients are real numbers has the zeros $1, 5i$, and $1 + i$. Find the remaining two zeros.

Solution Since complex zeros appear as conjugate pairs, it follows that $-5i$, the conjugate of $5i$, and $1 - i$, the conjugate of $1 + i$, are the two remaining zeros.

■

NOW WORK PROBLEM 1.

EXAMPLE 2 Finding a Polynomial Function Whose Zeros Are Given

2 (a) Find a polynomial f of degree 4 whose coefficients are real numbers and that has the zeros $1, 1$, and $-4 + i$.

(b) Graph the polynomial found in part (a) to verify your result.

Solution (a) Since $-4 + i$ is a zero, by the Conjugate Pairs Theorem, $-4 - i$ must also be a zero of f. Because of the Factor Theorem, if $f(c) = 0$, then $x - c$ is a factor of $f(x)$. So we can now write f as

$$f(x) = a(x - 1)(x - 1)[x - (-4 + i)][x - (-4 - i)]$$

where a is any real number. If we let $a = 1$, we obtain

$$\begin{aligned} f(x) &= (x-1)(x-1)[x-(-4+i)][x-(-4-i)] \\ &= (x^2-2x+1)[x^2-(-4+i)x-(-4-i)x+(-4+i)(-4-i)] \\ &= (x^2-2x+1)(x^2+4x-ix+4x+ix+16+4i-4i-i^2) \\ &= (x^2-2x+1)(x^2+8x+17) \\ &= (x^4+8x^3+17x^2-2x^3-16x^2-34x+x^2+8x+17) \\ &= x^4+6x^3+2x^2-26x+17 \end{aligned}$$

Figure 13

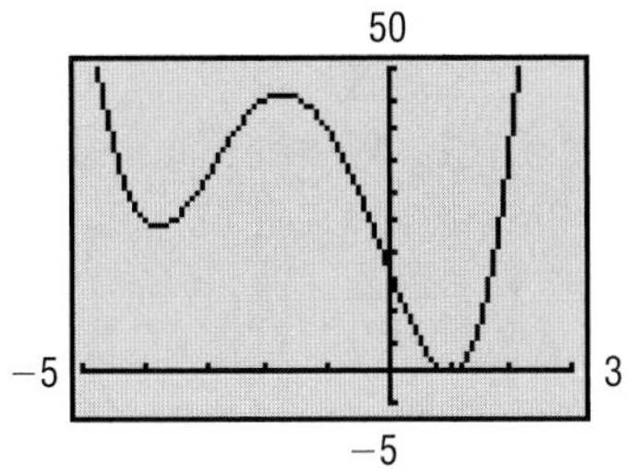

(b) A quick analysis of the polynomial f tells us what to expect:

At most three turning points.
For large $|x|$, the graph will behave like $y = x^4$.
A repeated real zero at 1 so that the graph will touch the x-axis at 1.
The only x-intercept is at 1.

Figure 13 shows the complete graph. (Do you see why? The graph has exactly three turning points and the degree of the polynomial is 4.) ■

EXPLORATION Graph the function found in Example 2 for $a = 2$ and $a = -1$. Does the value of a affect the zeros of f? How does the value of a affect the graph of f? ■

Now we can prove the theorem we conjectured earlier in Section 4.2.

Theorem Every polynomial function with real coefficients can be uniquely factored over the real numbers into a product of linear factors and/or irreducible quadratic factors. ■

Proof Every complex polynomial f of degree n has exactly n zeros and can be factored into a product of n linear factors. If its coefficients are real, then those zeros that are complex numbers will always occur as conjugate pairs. As a result, if $r = a + bi$ is a complex zero, then so is $\bar{r} = a - bi$. Consequently, when the linear factors $x - r$ and $x - \bar{r}$ of f are multiplied, we have

$$(x - r)(x - \bar{r}) = x^2 - (r + \bar{r})x + r\bar{r} = x^2 - 2ax + a^2 + b^2$$

This second-degree polynomial has real coefficients and is irreducible (over the real numbers). Thus, the factors of f are either linear or irreducible quadratic factors. ■

EXAMPLE 3 Finding the Complex Zeros of a Polynomial

3 Find the complex zeros of the polynomial function

$$f(x) = 3x^4 + 5x^3 + 25x^2 + 45x - 18$$

Solution **STEP 1:** The degree of f is 4. So f will have four complex zeros.

STEP 2: Descartes' Rule of Signs provides information about the real zeros. For this polynomial, there is one positive real zero. Because

$$f(-x) = 3x^4 - 5x^3 + 25x^2 - 45x - 18$$

there are three (or one) negative real zeros.

STEP 3: The Rational Zeros Theorem provides information about the potential rational zeros of polynomials with integer coefficients. For this polynomial (which has integer coefficients), the potential rational zeros are

$$\pm\frac{1}{3}, \pm\frac{2}{3}, \pm 1, \pm 2, \pm 3, \pm 6, \pm 9, \pm 18$$

Figure 14

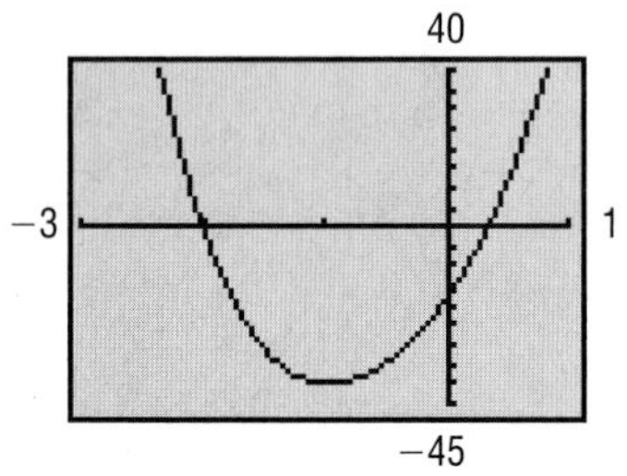

STEP 4: Figure 14 shows the graph of f. The graph has the characteristics that we expect of this polynomial of degree 4: It behaves like $y = 3x^4$ for large $|x|$ and has y-intercept -18. There are x-intercepts near -2 and between 0 and 1.

STEP 5: Since $f(-2) = 0$, we know that -2 is a zero of f. We use synthetic division to factor f.

$-2)$	3	5	25	45	−18
		−6	2	−54	18
	3	−1	27	−9	0

So $f(x) = (x + 2)(3x^3 - x^2 + 27x - 9)$. From the graph of f and the list of potential rational zeros, it appears that 1/3 may also be a zero of f. Since $f\left(\frac{1}{3}\right) = 0$, we know that $\frac{1}{3}$ is a zero of f. We use synthetic division on the depressed equation of f to factor.

$1/3)$	3	−1	27	−9
		1	0	9
	3	0	27	0

Using the bottom row of the synthetic division, we find

$$f(x) = (x + 2)\left(x - \frac{1}{3}\right)(3x^2 + 27) = 3(x + 2)\left(x - \frac{1}{3}\right)(x^2 + 9)$$

The factor $x^2 + 9$ does not have any real zeros; its complex zeros are $\pm 3i$. The complex zeros of $f(x) = 3x^4 + 5x^3 + 25x^2 + 45x - 18$ are $-2, 1/3, 3i, -3i$. ■

NOW WORK PROBLEM 27.

4.3 EXERCISES

In Problems 1–10, information is given about a polynomial $f(x)$ whose coefficients are real numbers. Find the remaining zeros of f.

1. Degree 3; zeros: $3, 4 - i$
2. Degree 3; zeros: $4, 3 + i$
3. Degree 4; zeros: $i, 1 + i$
4. Degree 4; zeros: $1, 2, 2 + i$
5. Degree 5; zeros: $1, i, 2i$
6. Degree 5; zeros: $0, 1, 2, i$
7. Degree 4; zeros: $i, 2, -2$
8. Degree 4; zeros: $2 - i, -i$
9. Degree 6; zeros: $2, 2 + i, -3 - i, 0$
10. Degree 6; zeros: $i, 3 - 2i, -2 + i$

In Problems 11–16, form a polynomial $f(x)$ with real coefficients having the given degree and zeros.

11. Degree 4; zeros: $3 + 2i$; 4, multiplicity 2
12. Degree 4; zeros: $i, 1 + 2i$
13. Degree 5; zeros: 2, multiplicity 1; $-i$; $1 + i$
14. Degree 6; zeros: $i, 4 - i$; $2 + i$
15. Degree 4; zeros: 3, multiplicity 2; $-i$
16. Degree 5; zeros: 1, multiplicity 3; $1 + i$

In Problems 17–24, use the given zero to find the remaining zeros of each function. Graph the function to verify your results.

17. $f(x) = x^3 - 4x^2 + 4x - 16$; zero: $2i$

18. $g(x) = x^3 + 3x^2 + 25x + 75$; zero: $-5i$

19. $f(x) = 2x^4 + 5x^3 + 5x^2 + 20x - 12$; zero: $-2i$

20. $h(x) = 3x^4 + 5x^3 + 25x^2 + 45x - 18$; zero: $3i$

21. $h(x) = x^4 - 9x^3 + 21x^2 + 21x - 130$; zero: $3 - 2i$

22. $f(x) = x^4 - 7x^3 + 14x^2 - 38x - 60$; zero: $1 + 3i$

23. $h(x) = 3x^5 + 2x^4 + 15x^3 + 10x^2 - 528x - 352$; zero: $-4i$

24. $g(x) = 2x^5 - 3x^4 - 5x^3 - 15x^2 - 207x + 108$; zero: $3i$

In Problems 25–34, find the complex zeros of each polynomial function. Check your results by evaluating the function at each of the zeros.

25. $f(x) = x^3 - 1$

26. $f(x) = x^4 - 1$

27. $f(x) = x^3 - 8x^2 + 25x - 26$

28. $f(x) = x^3 + 13x^2 + 57x + 85$

29. $f(x) = x^4 + 5x^2 + 4$

30. $f(x) = x^4 + 13x^2 + 36$

31. $f(x) = x^4 + 2x^3 + 22x^2 + 50x - 75$

32. $f(x) = x^4 + 3x^3 - 19x^2 + 27x - 252$

33. $f(x) = 3x^4 - x^3 - 9x^2 + 159x - 52$

34. $f(x) = 2x^4 + x^3 - 35x^2 - 113x + 65$

In Problems 35 and 36, tell why the facts given are contradictory.

35. $f(x)$ is a polynomial of degree 3 whose coefficients are real numbers; its zeros are $4 + i$, $4 - i$, and $2 + i$.

36. $f(x)$ is a polynomial of degree 3 whose coefficients are real numbers; its zeros are 2, i, and $3 + i$.

37. $f(x)$ is a polynomial of degree 4 whose coefficients are real numbers; three of its zeros are 2, $1 + 2i$, and $1 - 2i$. Explain why the remaining zero must be a real number.

38. $f(x)$ is a polynomial of degree 4 whose coefficients are real numbers; two of its zeros are -3 and $4 - i$. Explain why one of the remaining zeros must be a real number. Write down one of the missing zeros.

PREPARING FOR THIS SECTION

Before getting started, review the following:

✓ Solving Inequalities (Section 1.5)

4.4 POLYNOMIAL AND RATIONAL INEQUALITIES

OBJECTIVES

1. Solve Polynomial Inequalities Graphically and Algebraically
2. Solve Rational Inequalities Graphically and Algebraically

1 In this section we consider inequalities that involve polynomials of degree 2 and higher, as well as some that involve rational expressions. We will solve these inequalities using both a graphing approach and an algebraic approach.

To solve polynomial and rational inequalities algebraically, we follow these steps:

STEPS FOR SOLVING POLYNOMIAL AND RATIONAL INEQUALITIES ALGEBRAICALLY

STEP 1: Write the inequality so that a polynomial or rational expression f is on the left side and zero is on the right side in one of the following forms:

$$f(x) > 0 \qquad f(x) \geq 0 \qquad f(x) < 0 \qquad f(x) \leq 0$$

For rational expressions, be sure that the left side is written as a single quotient.

STEP 2: Determine the numbers at which the expression f on the left side equals zero and, if the expression is rational, the numbers at which the expression f on the left side is undefined.

STEP 3: Use the numbers found in Step 2 to separate the real number line into intervals.

STEP 4: Select a **test number** in each interval and evaluate f at the test number.

(a) If the value of f is positive, then $f(x) > 0$ for all numbers x in the interval.

(b) If the value of f is negative, then $f(x) < 0$ for all numbers x in the interval.

If the inequality is not strict, include the solutions of $f(x) = 0$ in the solution set.

EXAMPLE 1 Solving Quadratic Inequalities

Solve the inequality $x^2 \leq 4x + 12$, and graph the solution set.

Graphing Solution

We graph $Y_1 = x^2$ and $Y_2 = 4x + 12$ on the same screen. See Figure 15. Using the INTERSECT command, we find that Y_1 and Y_2 intersect at $x = -2$ and at $x = 6$. The graph of Y_1 is below that of Y_2, $Y_1 < Y_2$, between the points of intersection. Since the inequality is not strict, the solution set is $\{x \mid -2 \leq x \leq 6\}$ or, using interval notation, $[-2, 6]$.

Figure 15

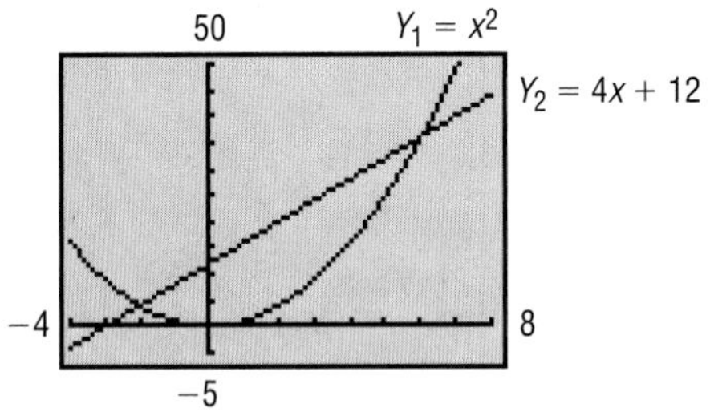

Algebraic Solution

STEP 1: Rearrange the inequality so that 0 is on the right side.

$$x^2 \leq 4x + 12$$

$$x^2 - 4x - 12 \leq 0 \quad \text{Subtract } 4x + 12 \text{ from both sides of the inequality.}$$

STEP 2: Find the zeros of $f(x) = x^2 - 4x - 12$ by solving the equation $x^2 - 4x - 12 = 0$.

$$x^2 - 4x - 12 = 0$$

$$(x + 2)(x - 6) = 0 \quad \text{Factor.}$$

$$x = -2 \quad \text{or} \quad x = 6$$

STEP 3: We use the zeros of f to separate the real number line into three intervals.

$$-\infty < x < -2 \qquad -2 < x < 6 \qquad 6 < x < \infty$$

STEP 4: We select a test number in each interval found in Step 3 and evaluate $f(x) = x^2 - 4x - 12$ at each test number to determine if $f(x)$ is positive or negative. See Table 2.

TABLE 2

Interval	Test Number	$f(x)$	Positive/Negative
$-\infty < x < -2$	-3	$f(-3) = (-3)^2 - 4(-3) - 12 = 9$	Positive
$-2 < x < 6$	0	$f(0) = -12$	Negative
$6 < x < \infty$	7	$f(7) = 9$	Positive

Since we want to know where $f(x)$ is negative, we conclude that the solutions are all x such that $-2 < x < 6$. However, because the original inequality is not strict, numbers x that satisfy the equation $x^2 = 4x + 12$ are also solutions of the inequality $x^2 \leq 4x + 12$. Thus, we include -2 and 6. The solution set of the given inequality is $\{x \mid -2 \leq x \leq 6\}$ or, using interval notation, $[-2, 6]$.

Figure 16
$-2 \leq x \leq 6$

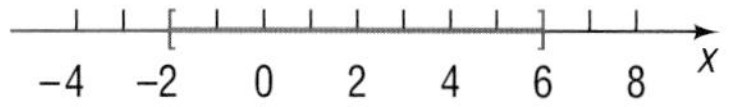

Figure 16 shows the graph of the solution set.

NOW WORK PROBLEMS 3 AND 7.

EXAMPLE 2 Solving a Polynomial Inequality

Solve the inequality $x^4 > x$, and graph the solution set.

Graphing Solution

We graph $Y_1 = x^4$ and $Y_2 = x$ on the same screen. See Figure 17. Using the INTERSECT command, we find that Y_1 and Y_2 intersect at $x = 0$ and at $x = 1$. The graph of Y_1 is above that of Y_2, $Y_1 > Y_2$, to the left of $x = 0$ and to the right of $x = 1$. Since the inequality is strict, the solution set is $\{x \mid x < 0$ or $x > 1\}$ or, using interval notation, $(-\infty, 0)$ or $(1, \infty)$.

Figure 17

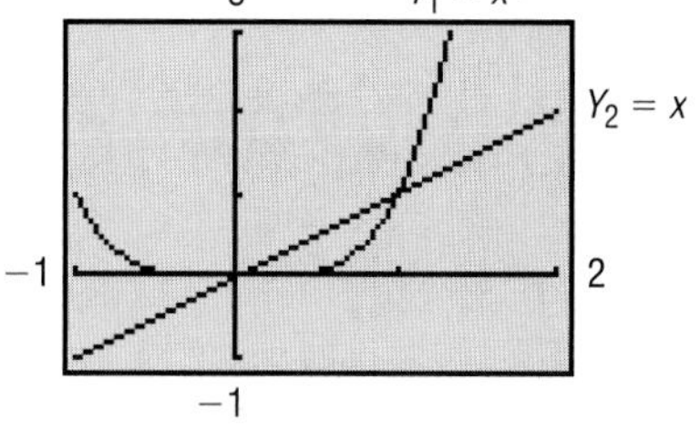

Algebraic Solution

STEP 1: Rearrange the inequality so that 0 is on the right side.

$$x^4 > x$$

$$x^4 - x > 0 \quad \text{Subtract } x \text{ from both sides of the inequality.}$$

STEP 2: Find the zeros of $f(x) = x^4 - x$ by solving $x^4 - x = 0$.

$$x^4 - x = 0$$

$$x(x^3 - 1) = 0 \quad \text{Factor out } x.$$

$$x(x-1)(x^2 + x + 1) = 0 \quad \text{Factor the difference of two cubes.}$$

$$x = 0 \quad \text{or} \quad x - 1 = 0 \quad \text{or} \quad x^2 + x + 1 = 0 \quad \text{Set each factor equal to zero and solve.}$$

$$x = 0 \quad \text{or} \quad x = 1$$

The equation $x^2 + x + 1 = 0$ has no real solutions. (Do you see why?)

STEP 3: We use the zeros to separate the real number line into three intervals.

$$-\infty < x < 0 \qquad 0 < x < 1 \qquad 1 < x < \infty$$

STEP 4: We select a test number in each interval found in Step 3 and evaluate $f(x) = x^4 - x$ at each test number to determine if $f(x)$ is positive or negative. See Table 3.

TABLE 3

Interval	Test Number	f(x)	Positive/Negative
$-\infty < x < 0$	-1	$f(-1) = (-1)^4 - (-1) = 2$	Positive
$0 < x < 1$	$1/2$	$f(1/2) = -7/16$	Negative
$1 < x < \infty$	2	$f(2) = 14$	Positive

Since we want to know where $f(x)$ is positive, we conclude that the solutions are numbers x for which $-\infty < x < 0$ or $1 < x < \infty$. Because the original inequality is strict, numbers x that satisfy the equation $x^4 = x$ are not solutions. Thus, the solution set of the given inequality is $\{x \mid -\infty < x < 0 \text{ or } 1 < x < \infty\}$ or, using interval notation, $(-\infty, 0)$ or $(1, \infty)$.

Figure 18
$x < 0$ or $x > 1$

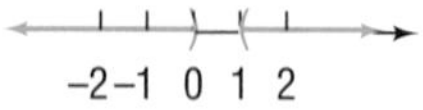

Figure 18 shows the graph of the solution set. ■

NOW WORK PROBLEM 19.

2 Let's solve a rational inequality.

EXAMPLE 3 Solving a Rational Inequality

Solve the inequality $\dfrac{4x + 5}{x + 2} \geq 3$, and graph the solution set.

Graphing Solution We first note that the domain of the variable consists of all real numbers except -2. We graph $Y_1 = (4x + 5)/(x + 2)$ and $Y_2 = 3$ on the same screen. See Figure 19. Using the INTERSECT command, we find that Y_1 and Y_2 intersect at $x = 1$. The graph of Y_1 is above that of Y_2, $Y_1 > Y_2$, to the left of $x = -2$ and to the right of $x = 1$. Since the inequality is not strict, the solution set is $\{x \mid -\infty < x < -2 \text{ or } 1 \leq x < \infty\}$. Using interval notation, the solution set is $(-\infty, -2)$ or $[1, \infty)$.

Figure 19

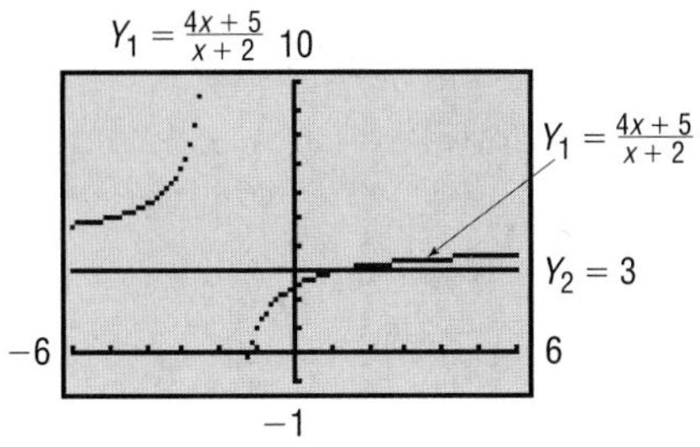

Algebraic Solution

STEP 1: We first note that the domain of the variable consists of all real numbers except −2. We rearrange terms so that 0 is on the right side.

$$\frac{4x+5}{x+2} - 3 \geq 0 \quad \text{Subtract 3 from both sides of the inequality.}$$

STEP 2: Find the zeros and values of x at which $f(x) = \frac{4x+5}{x+2} - 3$ is undefined. To find these numbers, we must express f as a quotient.

$$\begin{aligned} f(x) &= \frac{4x+5}{x+2} - 3 && \text{Least Common Denominator: } x + 2. \\ &= \frac{4x+5}{x+2} - 3\frac{x+2}{x+2} && \text{Multiply } -3 \text{ by } \frac{x+2}{x+2}. \\ &= \frac{4x+5-3x-6}{x+2} && \text{Write as a single quotient.} \\ &= \frac{x-1}{x+2} && \text{Combine like terms.} \end{aligned}$$

The zero of f is 1. Also, f is undefined for $x = -2$.

STEP 3: We use the numbers found in Step 2 to separate the real number line into three intervals.

$$-\infty < x < -2 \qquad -2 < x < 1 \qquad 1 < x < \infty$$

STEP 4: We select a test number in each interval found in Step 3 and evaluate $f(x) = \frac{4x+5}{x+2} - 3$ at each test number to determine if $f(x)$ is positive or negative. See Table 4.

TABLE 4

Interval	Test Number	$f(x)$	Positive/Negative
$-\infty < x < -2$	−3	$f(-3) = 4$	Positive
$-2 < x < 1$	0	$f(0) = -1/2$	Negative
$1 < x < \infty$	2	$f(2) = 1/4$	Positive

We want to know where $f(x)$ is positive. We conclude that the solutions are all x such that $-\infty < x < -2$ or $1 < x < \infty$. Because the original inequality is not strict, numbers x that satisfy the equation

Figure 20
$-\infty < x < -2$ or $1 \le x < \infty$
$(-\infty, -2)$ or $[1, \infty)$

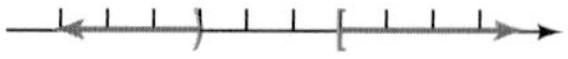

$\dfrac{x-1}{x+2} = 0$ are also solutions of the inequality. Since $\dfrac{x-1}{x+2} = 0$ only if $x = 1$, we conclude that the solution set is $\{x \mid -\infty < x < -2$ or $1 \le x < \infty\}$ or, using interval notation, $(-\infty, -2)$ or $[1, \infty)$.

Figure 20 shows the graph of the solution set. ■

NOW WORK PROBLEM 37.

EXAMPLE 4 Minimum Sales Requirements

TABLE 5

Pounds of Cookies (in Hundreds), x	Profit, P
0	−20,000
50	−5990
75	412
120	10,932
200	26,583
270	36,948
340	44,381
420	49,638
525	49,225
610	44,381
700	34,220

Tami is considering leaving her $30,000 a year job and buying a cookie company. According to the financial records of the firm, the relationship between pounds of cookies sold and profit is as exhibited by Table 5.

(a) Draw a scatter diagram of the data in Table 5 with the pounds of cookies sold as the independent variable.

(b) Find the quadratic function of best fit using a graphing utility.

(c) Using the function found in part (b), determine the number of pounds of cookies that Tami must sell in order for the profits to exceed $30,000 a year and therefore to make it worthwhile for her to quit her job.

(d) Using the function found in part (b), determine the number of pounds of cookies that Tami should sell in order to maximize profits.

(e) Using the function found in part (b), determine the maximum profit that Tami can expect to earn.

Solution (a) Figure 21 shows the scatter diagram.

Figure 21

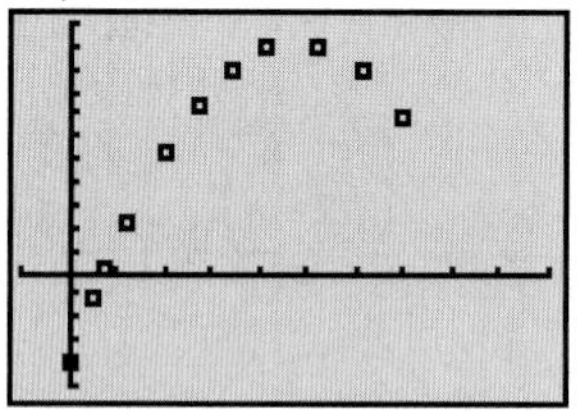

(b) Using a graphing utility, the quadratic function of best fit is

$$p(x) = -0.31x^2 + 295.86x - 20042.52$$

See Figure 22.

Figure 22

(c) We want profit to exceed \$30,000; therefore, we want to solve the inequality

$$-0.31x^2 + 295.86x - 20042.52 > 30{,}000$$

We graph

$$Y_1 = p(x) = -0.31x^2 + 295.86x - 20042.52 \quad \text{and} \quad Y_2 = 30{,}000$$

on the same screen. See Figure 23.

Figure 23

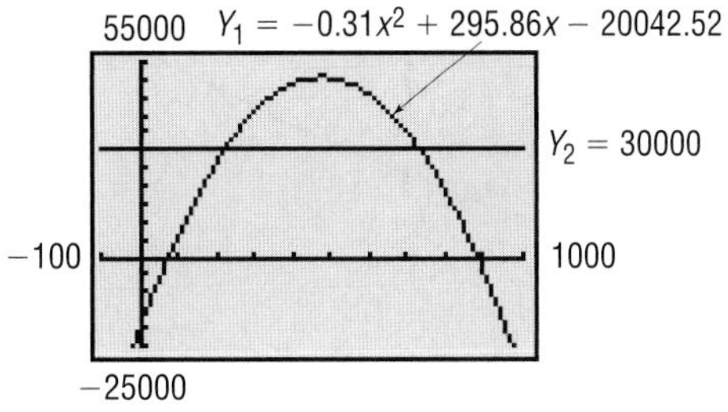

Using the INTERSECT command, we find that Y_1 and Y_2 intersect at $x = 220$ and at $x = 735$. The graph of Y_1 is above that of Y_2, $Y_1 > Y_2$, between the points of intersection. Since the inequality is strict, the solution set is $\{x \mid 220 < x < 735\}$.

(d) The function $p(x)$ found in part (b) is a quadratic function whose graph opens down. The vertex is therefore the highest point. The x-coordinate of the vertex is

$$x = \frac{-b}{2a} = \frac{-295.86}{2(-0.31)} = 477$$

To maximize profit, 47,700 pounds of cookies must be sold.

(e) The maximum profit is

$$p(477) = -0.31(477)^2 + 295.86(477) - 20042.52 = \$50{,}549$$

4.4 EXERCISES

In Problems 1–58, solve each inequality (a) graphically and (b) algebraically.

1. $(x-5)(x+2) < 0$
2. $(x-5)(x+2) > 0$
3. $x^2 - 4x > 0$
4. $x^2 + 8x > 0$
5. $x^2 - 9 < 0$
6. $x^2 - 1 < 0$
7. $x^2 + x > 12$
8. $x^2 + 7x < -12$
9. $2x^2 < 5x + 3$
10. $6x^2 < 6 + 5x$
11. $x(x-7) > 8$
12. $x(x+1) > 20$
13. $4x^2 + 9 < 6x$
14. $25x^2 + 16 < 40x$
15. $6(x^2 - 1) > 5x$
16. $2(2x^2 - 3x) > -9$
17. $(x-1)(x^2+x+1) > 0$
18. $(x+2)(x^2-x+1) > 0$
19. $(x-1)(x-2)(x-3) < 0$
20. $(x+1)(x+2)(x+3) < 0$
21. $x^3 - 2x^2 - 3x > 0$
22. $x^3 + 2x^2 - 3x > 0$
23. $x^4 > x^2$
24. $x^4 < 4x^2$
25. $x^3 > x^2$
26. $x^3 < 3x^2$
27. $x^4 > 1$
28. $x^3 > 1$
29. $x^2 - 7x - 8 < 0$
30. $x^2 + 12x + 32 \geq 0$
31. $x^3 + x - 12 \geq 0$
32. $x^3 - 3x + 1 \leq 0$
33. $x^4 - 3x^2 - 4 > 0$
34. $x^4 - 5x^2 + 6 < 0$
35. $x^3 - 4 \geq 3x^2 + 5x - 3$
36. $x^4 - 4x \leq -x^2 + 2x + 1$
37. $\dfrac{x+1}{x-1} > 0$
38. $\dfrac{x-3}{x+1} > 0$
39. $\dfrac{(x-1)(x+1)}{x} < 0$

40. $\dfrac{(x-3)(x+2)}{x-1} < 0$

41. $\dfrac{(x-2)^2}{x^2-1} \geq 0$

42. $\dfrac{(x+5)^2}{x^2-4} \geq 0$

43. $6x - 5 < \dfrac{6}{x}$

44. $x + \dfrac{12}{x} < 7$

45. $\dfrac{x+4}{x-2} \leq 1$

46. $\dfrac{x+2}{x-4} \geq 1$

47. $\dfrac{3x-5}{x+2} \leq 2$

48. $\dfrac{x-4}{2x+4} \geq 1$

49. $\dfrac{1}{x-2} < \dfrac{2}{3x-9}$

50. $\dfrac{5}{x-3} > \dfrac{3}{x+1}$

51. $\dfrac{2x+5}{x+1} > \dfrac{x+1}{x-1}$

52. $\dfrac{1}{x+2} > \dfrac{3}{x+1}$

53. $\dfrac{x^2(3+x)(x+4)}{(x+5)(x-1)} > 0$

54. $\dfrac{x(x^2+1)(x-2)}{(x-1)(x+1)} > 0$

55. $\dfrac{2x^2-x-1}{x-4} \leq 0$

56. $\dfrac{3x^2+2x-1}{x+2} > 0$

57. $\dfrac{x^2+3x-1}{x+3} > 0$

58. $\dfrac{x^2-5x+3}{x-5} < 0$

59. For what positive numbers will the cube of a number exceed four times its square?

60. For what positive numbers will the square of a number exceed twice the number?

61. What is the domain of the variable in the expression $\sqrt{x^2-16}$?

62. What is the domain of the variable in the expression $\sqrt{x^3-3x^2}$?

63. What is the domain of the variable in the expression $\sqrt{\dfrac{x-2}{x+4}}$?

64. What is the domain of the variable in the expression $\sqrt{\dfrac{x-1}{x+4}}$?

65. **Physics** A ball is thrown vertically upward with an initial velocity of 80 feet per second. The distance s (in feet) of the ball from the ground after t seconds is $s = 80t - 16t^2$.

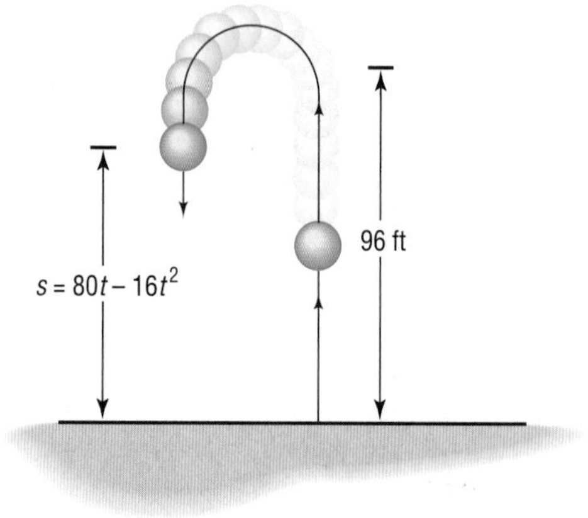

(a) For what time interval is the ball more than 96 feet above the ground? (See the figure.)
(b) Using a graphing utility, graph the relation between s and t.
(c) What is the maximum height of the ball?
(d) After how many seconds does the ball reach the maximum height?

66. **Physics** A ball is thrown vertically upward with an initial velocity of 96 feet per second. The distance s (in feet) of the ball from the ground after t seconds is $s = 96t - 16t^2$.
(a) For what time interval is the ball more than 112 feet above the ground?
(b) Using a graphing utility, graph the relation between s and t.
(c) What is the maximum height of the ball?
(d) After how many seconds does the ball reach the maximum height?

67. **Business** The monthly revenue achieved by selling x wristwatches is figured to be $x(40 - 0.2x)$ dollars. The wholesale cost of each watch is \$32.
(a) How many watches must be sold each month to achieve a profit (revenue − cost) of at least \$50?
(b) Using a graphing utility, graph the revenue function.
(c) What is the maximum revenue that this firm could earn?
(d) How many wristwatches should the firm sell to maximize revenue?
(e) Using a graphing utility, graph the profit function.
(f) What is the maximum profit that this firm can earn?
(g) How many watches should the firm sell to maximize profit?
(h) Provide a reasonable explanation as to why the answers found in parts (d) and (g) differ. Is the shape of the revenue function reasonable in your opinion? Why?

68. Business The monthly revenue achieved by selling x boxes of candy is figured to be $x(5 - 0.05x)$ dollars. The wholesale cost of each box of candy is \$1.50.
(a) How many boxes must be sold each month to achieve a profit of at least \$60?
(b) Using a graphing utility, graph the revenue function.
(c) What is the maximum revenue that this firm could earn?
(d) How many boxes of candy should the firm sell to maximize revenue?
(e) Using a graphing utility, graph the profit function.
(f) What is the maximum profit that this firm can earn?
(g) How many boxes of candy should the firm sell to maximize profit?
(h) Provide a reasonable explanation as to why the answers found in parts (d) and (g) differ. Is the shape of the revenue function reasonable in your opinion? Why?

69. Cost of Manufacturing In Problem 67 of Section 3.3, a cubic function of best fit relating the cost C of manufacturing x Chevy Cavaliers in a day was found. Budget constraints will not allow Chevy to spend more than \$97,000 per day. Determine the number of Cavaliers that could be produced in a day.

70. Cost of Printing In Problem 64 of Section 3.3, a cubic function of best fit relating the cost C of printing x textbooks in a week was found. Budget constraints will not allow the printer to spend more than \$170,000 per week. Determine the number of textbooks that could be printed in a week.

71. Minimum Sales Requirements Marissa is thinking of leaving her \$1000 a week job and buying a computer resale shop. According to the financial records of the firm, the profits (in dollars) of the company for different amounts of computers sold and the corresponding profits are as follows:

Number of Computers Sold, x	Profit, p
0	−1500
4	−522
7	54
12	775
18	1184
23	1132
29	653

(a) Using a graphing utility, draw a scatter diagram of the data with the number of computers sold as the independent variable.
(b) Find the quadratic function of best fit using a graphing utility.
(c) Using the function found in part (b), determine the number of computers that Marissa must sell in order for the profits to exceed \$1000 per week and therefore to make it worthwhile for her to quit her job.
(d) Using the function found in part (b), determine the number of computers that Marissa should sell in order to maximize profits.
(e) Using the function found in part (b), determine the maximum profit that Marissa can expect to earn.

72. Minimum Sales Requirements Barry is considering the purchase of a gas station. According to the financial records of the gas station, its monthly sales (in thousands of gallons of gasoline) and the corresponding profits are as follows:

Thousands of Gallons of Gasoline, x	Profit, p
50	3947
54	4214
74	4942
92	4838
82	5003
75	4965
100	4521
88	4933
63	4665

(a) Using a graphing utility, draw a scatter diagram of the data with the number of gallons of gasoline sold as the independent variable.
(b) Find the quadratic function of best fit using a graphing utility.
(c) Using the function found in part (b), determine the number of gallons of gasoline that Barry must sell in order for the profits to exceed \$4000 a month and therefore to make it worthwhile for him to quit his job.
(d) Using the function found in part (b), determine the number of gallons of gasoline that Barry should sell in order to maximize profits.
(e) Using the function found in part (b), determine the maximum profit Barry can expect to earn.

73. Prove that if a, b are real numbers and $a \geq 0, b \geq 0$, then

$$a \leq b \quad \text{is equivalent to} \quad \sqrt{a} \leq \sqrt{b}$$

[**Hint:** $b - a = (\sqrt{b} - \sqrt{a})(\sqrt{b} + \sqrt{a})$.]

74. Make up an inequality that has no real solution. Make up one that has exactly one real solution.

75. The inequality $x^2 + 1 < -5$ has no real solution. Explain why.

CHAPTER REVIEW

Things To Know

Zeros of a polynomial f (p. 238) — Numbers for which $f(x) = 0$; these are the x-intercepts of the graph of f.

Remainder Theorem (p. 240) — If a polynomial $f(x)$ is divided by $x - c$, then the remainder is $f(c)$.

Factor Theorem (p. 240) — $x - c$ is a factor of a polynomial $f(x)$ if and only if $f(c) = 0$.

Descartes' Rule of Signs (p. 242) — Let f denote a polynomial function. The number of positive zeros of f either equals the number of variations in sign of the nonzero coefficients of $f(x)$ or else equals that number less some even integer. The number of negative zeros of f either equals the number of variations in sign of the nonzero coefficients of $f(-x)$ or else equals that number less some even integer.

Rational Zeros Theorem (p. 243) — Let f be a polynomial function of degree 1 or higher of the form

$$f(x) = a_n x^n + a_{n-1}x^{n-1} + \cdots + a_1 x + a_0, \quad a_n \neq 0, a_0 \neq 0$$

where each coefficient is an integer. If p/q, in lowest terms, is a rational zero of f, then p must be a factor of a_0 and q must be a factor of a_n.

Quadratic equation and quadratic formula (p. 259) — If $ax^2 + bx + c = 0$, $a \neq 0$, then $x = \dfrac{-b \pm \sqrt{b^2 - 4ac}}{2a}$.

Discriminant (p. 260) — If $b^2 - 4ac > 0$, there are two distinct real solutions.
If $b^2 - 4ac = 0$, there is one repeated real solution.
If $b^2 - 4ac < 0$, there are two distinct complex solutions that are not real; the solutions are conjugates of each other.

Fundamental Theorem of Algebra (p. 262) — Every complex polynomial function $f(x)$ of degree $n \geq 1$ has at least one complex zero.

Conjugate Pairs Theorem (p. 263) — Let $f(x)$ be a polynomial whose coefficients are real numbers. If $r = a + bi$ is a zero of f, then its complex conjugate $\bar{r} = a - bi$ is also a zero of f.

How To

Use the Remainder and Factor Theorems (p. 239)

Use Descartes' Rule of Signs (p. 241)

Use the Rational Zeros Theorem (p. 243)

Find the real zeros of a polynomial function (p. 244)

Solve polynomial equations (p. 245)

Use the Theorem for Bounds on Zeros (p. 246)

Use the Intermediate Value Theorem (p. 249)

Add, subtract, multiply and divide complex numbers (p. 253)

Solve quadratic equations with a negative discriminant (p. 258)

Utilize the conjugate pairs theorem (p. 263)

Find a polynomial function with specified zeros (p. 264)

Find the complex zeros of a polynomial (p. 265)

Solve polynomial inequalities graphically and algebraically (p. 267)

Solve rational inequalities graphically and algebraically (p. 270)

Fill-in-the-Blank Items

1. In the process of polynomial division, (Divisor)(Quotient) + ________ = ________.
2. When a polynomial function f is divided by $x - c$, the remainder is ________.
3. A polynomial function f has the factor $x - c$ if and only if ________.
4. The polynomial function $f(x) = x^5 - 2x^3 + x^2 - x + 1$ has at most ________ real zeros.
5. The possible rational zeros of $f(x) = 2x^5 - x^3 + x^2 - x + 1$ are ________.
6. In the complex number $5 + 2i$, the number 5 is called the ________ part; the number 2 is called the ________ part; the number i is called the ________ ________.
7. If $3 + 4i$ is a zero of a polynomial of degree 5 with real coefficients, then so is ________.
8. The equation $|x^2| = 4$ has four solutions: ________, ________, ________, and ________.
9. If a function f whose domain is all real numbers is even and 4 is a zero of f, then ________ is also a zero.

True/False Items

T F 1. Every polynomial of degree 3 with real coefficients has exactly three real zeros.

T F 2. If $2 - 3i$ is a zero of a polynomial with real coefficients, then so is $-2 + 3i$.

T F 3. If f is a polynomial function of degree 4 and if $f(2) = 5$, then

$$\frac{f(x)}{x-2} = p(x) + \frac{5}{x-2}$$

where $p(x)$ is a polynomial of degree 3.

T F 4. The conjugate of $2 + 5i$ is $-2 - 5i$.

T F 5. A polynomial of degree n with real coefficients has exactly n complex zeros. At most n of them are real numbers.

Review Exercises

Blue problem numbers indicate the authors' suggestions for use in a Practice Test.

In Problems 1 and 2, use Descartes' Rule of Signs to determine how many positive and negative zeros each polynomial function may have. Do not attempt to find the zeros.

1. $f(x) = 12x^8 - x^7 + 8x^4 - 2x^3 + x + 3$

2. $f(x) = -6x^5 + x^4 + 5x^3 + x + 1$

3. List all the potential rational zeros at $f(x) = 12x^8 - x^7 + 6x^4 - x^3 + x - 3$.

4. List all the potential rational zeros of $f(x) = -6x^5 + x^4 + 2x^3 - x + 1$.

In Problems 5–10, follow the steps on page 248 to find all the real zeros of each polynomial function.

5. $f(x) = x^3 - 3x^2 - 6x + 8$

6. $f(x) = x^3 - x^2 - 10x - 8$

7. $f(x) = 4x^3 + 4x^2 - 7x + 2$

8. $f(x) = 4x^3 - 4x^2 - 7x - 2$

9. $f(x) = x^4 - 4x^3 + 9x^2 - 20x + 20$

10. $f(x) = x^4 + 6x^3 + 11x^2 + 12x + 18$

In Problems 11–16, determine the real zeros of the polynomial function. Approximate all irrational zeros rounded to two decimal places.

11. $f(x) = 2x^3 - 11.84x^2 - 9.116x + 82.46$

12. $f(x) = 12x^3 + 39.8x^2 - 4.4x - 3.4$

13. $g(x) = 15x^4 - 21.5x^3 - 1718.3x^2 + 5308x + 3796.8$

14. $g(x) = 3x^4 + 67.93x^3 + 486.265x^2 + \cdot 1121.32x + 412.195$

15. $f(x) = 3x^3 + 18.02x^2 + 11.0467x - 53.8756$

16. $f(x) = x^3 - 3.16x^2 - 39.4611x + 151.638$

In Problems 17–20, find the real solutions of each equation.

17. $2x^4 + 2x^3 - 11x^2 + x - 6 = 0$

18. $3x^4 + 3x^3 - 17x^2 + x - 6 = 0$

19. $2x^4 + 7x^3 + x^2 - 7x - 3 = 0$

20. $2x^4 + 7x^3 - 5x^2 - 28x - 12 = 0$

In Problems 21–24, find bounds to the zeros of each polynomial function. Obtain a complete graph of f.

21. $f(x) = x^3 - x^2 - 4x + 2$

22. $f(x) = x^3 + x^2 - 10x - 5$

23. $f(x) = 2x^3 - 7x^2 - 10x + 35$

24. $f(x) = 3x^3 - 7x^2 - 6x + 14$

In Problems 25–28, use the Intermediate Value Theorem to show that each polynomial has a zero in the given interval. Approximate the zero rounded to two decimal places.

25. $f(x) = 3x^3 - x - 1$; $[0, 1]$

26. $f(x) = 2x^3 - x^2 - 3$; $[1, 2]$

27. $f(x) = 8x^4 - 4x^3 - 2x - 1$; $[0, 1]$

28. $f(x) = 3x^4 + 4x^3 - 8x - 2$; $[1, 2]$

In Problems 29–38, write each expression in the standard form $a + bi$. Verify your results using a graphing utility.

29. $(6 + 3i) - (2 - 4i)$

30. $(8 - 3i) + (-6 + 2i)$

31. $4(3 - i) + 3(-5 + 2i)$

32. $2(1 + i) - 3(2 - 3i)$

33. $\dfrac{3}{3 + i}$

34. $\dfrac{4}{2 - i}$

35. i^{50}

36. i^{29}

37. $(2 + 3i)^3$

38. $(3 - 2i)^3$

In Problems 39–42, information is given about a complex polynomial $f(x)$ whose coefficients are real numbers. Find the remaining zeros of f.

39. Degree 3; zeros: $4 + i, 6$

40. Degree 3; zeros: $3 + 4i, 5$

41. Degree 4; zeros: $i, 1 + i$

42. Degree 4; zeros: $1, 2, 1 + i$

In Problems 43–56, solve each equation in the complex number system.

43. $x^2 + x + 1 = 0$
44. $x^2 - x + 1 = 0$
45. $2x^2 + x - 2 = 0$
46. $3x^2 - 2x - 1 = 0$
47. $x^2 + 3 = x$
48. $2x^2 + 1 = 2x$
49. $x(1 - x) = 6$
50. $x(1 + x) = 2$
51. $x^4 + 2x^2 - 8 = 0$
52. $x^4 + 8x^2 - 9 = 0$
53. $x^3 - x^2 - 8x + 12 = 0$
54. $x^3 - 3x^2 - 4x + 12 = 0$
55. $3x^4 - 4x^3 + 4x^2 - 4x + 1 = 0$
56. $x^4 + 4x^3 + 2x^2 - 8x - 8 = 0$

In Problems 57–66, solve each inequality (a) graphically and (b) algebraically.

57. $2x^2 + 5x - 12 < 0$
58. $3x^2 - 2x - 1 \geq 0$
59. $\dfrac{6}{x + 3} \geq 1$
60. $\dfrac{-2}{1 - 3x} < 1$
61. $\dfrac{2x - 6}{1 - x} < 2$
62. $\dfrac{3 - 2x}{2x + 5} \geq 2$
63. $\dfrac{(x - 2)(x - 1)}{x - 3} > 0$
64. $\dfrac{x + 1}{x(x - 5)} \leq 0$
65. $\dfrac{x^2 - 8x + 12}{x^2 - 16} > 0$
66. $\dfrac{x(x^2 + x - 2)}{x^2 + 9x + 20} \leq 0$

PROJECT AT MOTOROLA

Alternating Current (AC) Circuit Analysis

AC circuit analysis techniques are used on electrical circuits that are energized with sinusoidal sources. A common example of an AC source is your electric power at home. Its effective amplitude is 120 volts and its frequency (f) is 60 cycles/second (60 Hz). The figure below is an electric circuit with sinusoidal voltage source V_s with amplitude 10 volts and frequency $5000/\pi$ Hz. It is connected to two complex load "impedances" $Z_1 = R_1 - X_1 i$ and $Z_2 = R_2 + X_2 i$ that are connected in "series" with each other. Each load contains a resistor (R) connected to a capacitor (C) or an inductor (L).

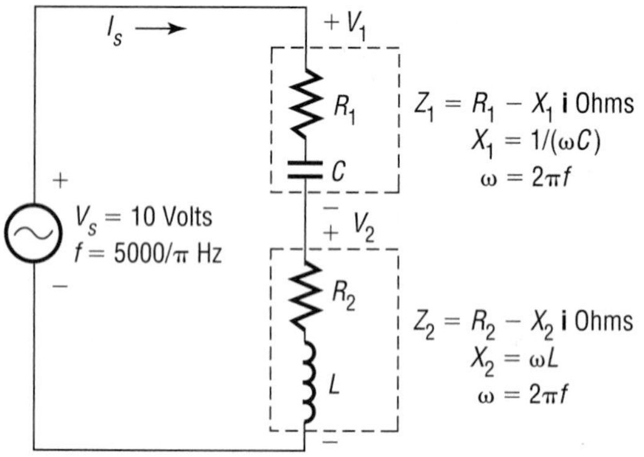

Resistors are electronic components designed to resist or limit electric current flow. The greater the resistance the more effectively it limits current. Electric power is lost or dissipated as heat in resistors. Capacitors and inductors limit current also and their ability to do this is measured as "reactance". Electric power is not lost but is stored in ideal capacitors and inductors. The capacitive and inductive reactance values X_1 and X_2 are given by $X_1 = 1/(\omega C)$ and $X_2 = \omega L$, where $\omega = 2\pi f$, C is capacitance in Farads (F), and L is inductance in Henrys (H). Reactance is unique to resistance in that it is dependent upon the frequency of operation.

1. Given resistors R_1 and R_2 are each 5 Ohms, capacitor C is F, and inductor L is 0. 0015 H calculate X_1 and X_2 for the given frequency of operation. Write out Z_1 and Z_2 in standard complex form. Impedance, Z_t, is formed by adding Z_1 and Z_2 as shown in the figure below. What is Z_t? What two circuit elements (resistors, capacitors, inductors) do you think Z_t contains?

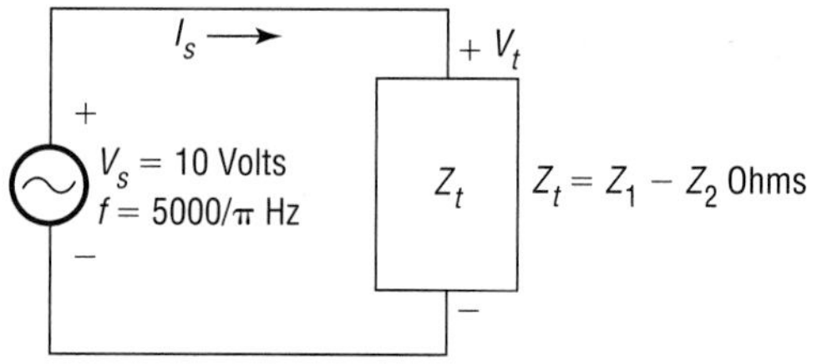

2. The current flow (I_s) is sinusoidal and its frequency is the same as the voltage source. Calculate I_s using the formula $I_s = V_s/Z_t$ (Amps).

3. As I_s flows through Z_1 a voltage drop V_1 is created as given by $V_1 = I_s Z_1$. Calculate the voltage across Z_1.

4. Find the voltage drop V_2 across $Z_2(V_2 = I_s Z_2)$.

5. Does $V_s = V_1 + V_2$? Do you think this is reasonable?

6. Find the voltage drop V_t across $Z_t(V_t = I_s Z_t)$. Does $V_t = V_1 + V_2$? Is this a reasonable result?

7. The power dissipated in load Z_1 in Watts is given by $P_1 = (1/2)R_e\{V_1 I_{s*}\}$ where I_{s*} is the conjugate of I_s and $R_e\{V_1 I_{s*}\}$ is the real part of $V_1 I_{s*}$. Calculate P_1.

8. The power dissipated in load Z_2 in Watts is given by $P_2 = (1/2)R_e\{V_2 I_{s*}\}$. Calculate P_2.

9. Calculate P_t the power dissipated in load Z_t.

10. Does $P_t = P_1 + P_2$? Do you think this is reasonable?

11. What do you think will happen to Z_t, I_s, and P_t as frequency goes to zero? To infinity?

CHAPTER

5 Exponential and Logarithmic Functions

Outline

FIELD TRIP TO MOTOROLA

Many physical and chemical processes exhibit exponential or logarithmic behavior. These functions are the cornerstones of science and engineering. For example, chemical reactions often can be modelled by an exponential function. Metal fatigue, very complex when one introduces temperature and the rate of cycling, can be described with logarithmic functions.

INTERVIEW AT MOTOROLA

Andrew Skipor is a Principal Staff Engineer with the Design Reliability Group in Motorola Lab's, Motorola Advanced Technology Center. He has been with Motorola over 16 years. His interests include development of methods to evaluate package reliability, numerical simulation, experimental methods such as Moiré interferometry, nonlinear solder mechanics, fatigue, PCB/HDI mechanics. He has five patents, an engineering award, and numerous internal and external publications and is the Motorola Chairman of the Motorola-IEEE/CPMT Graduate Student Fellowship for Research in Electronic Packaging. He received his Ph.D. in engineering mechanics from the University of Illinois at Chicago, supported in part by the Motorola Distinguished Scholar Program. Throughout his career at Motorola, he has used advanced topics in mathematics that directly depend on a sound foundation in algebra, trigonometry, and analytic geometry.

PREPARING FOR THIS SECTION

Before getting started, review the following:

- ✓ Functions (Section 2.1)
- ✓ Increasing/Decreasing Functions (pp. 119–120)

5.1 ONE-TO-ONE FUNCTIONS; INVERSE FUNCTIONS

OBJECTIVES

1. Determine Whether a Function Is One-to-One
2. Obtain the Graph of the Inverse Function from the Graph of the Function
3. Find an Inverse Function

In Section 2.1, we said that a function f can be thought of as a machine that receives as input a number, say x, from the domain, manipulates it, and outputs the value $f(x)$. The *inverse* of f receives as input a number $f(x)$, manipulates it, and outputs the value x. If the function f is the set of ordered pairs (x, y), then the inverse of f is the set of ordered pairs (y, x).

EXAMPLE 1 Finding the Inverse of a Function

Find the inverse of the following functions.

(a) Let the domain of the function represent the employees of Yolanda's Preowned Car Mart and let the range represent their base salaries.

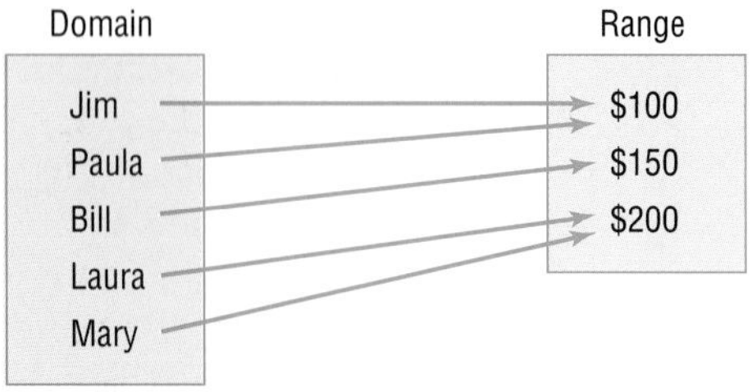

(b) Let the domain of the function represent the employees of Yolanda's Preowned Car Mart and let the range represent their spouse's names.

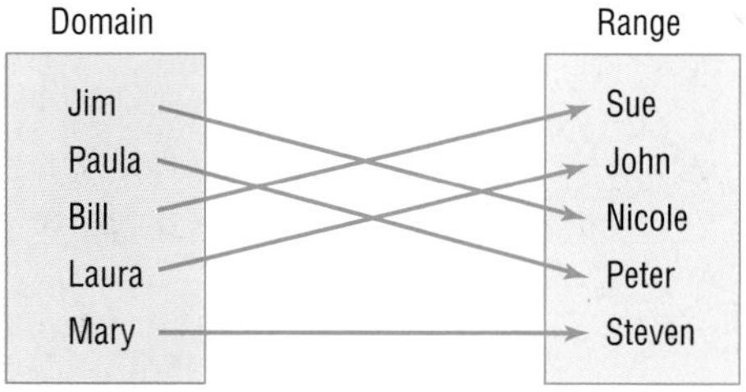

(c) $\{(-3, -27), (-2, -8), (-1, -1), (0, 0), (1, 1), (2, 8), (3, 27)\}$
(d) $\{(-3, 9), (-2, 4), (-1, 1), (0, 0), (1, 1), (2, 4), (3, 9)\}$

Solution (a) The elements in the domain represent inputs to the function, and the elements in the range represent the outputs. To find the inverse, interchange the elements in the domain with the elements in the range. For example, the function receives as input Bill and outputs \$150. So the inverse receives an input \$150 and outputs Bill. The inverse of the given function takes the form

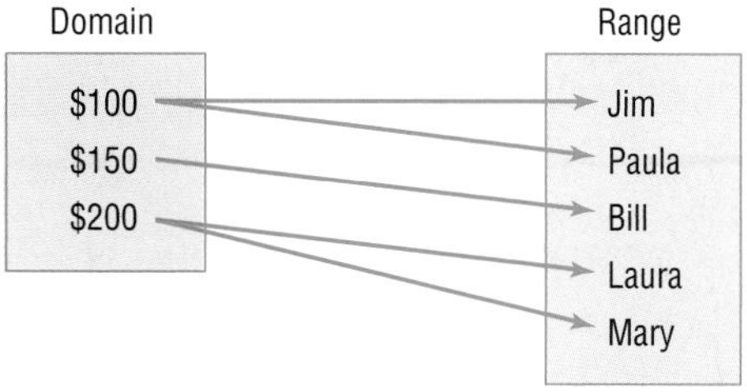

(b) The inverse of the given function is

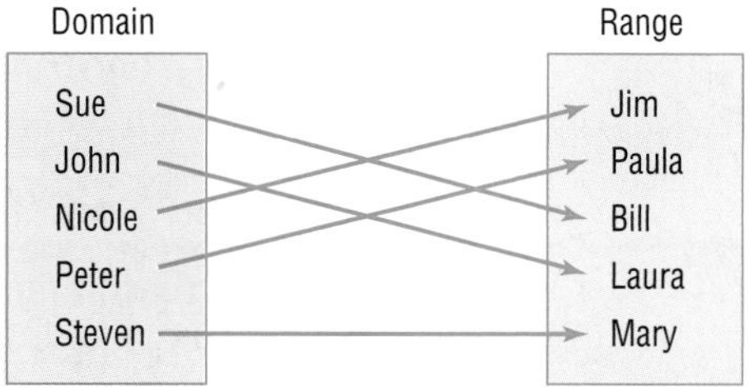

(c) The inverse of the given function is found by interchanging the entries in each ordered pair and so is given by

$$\{(-27, -3), (-8, -2), (-1, -1), (0, 0), (1, 1), (8, 2), (27, 3)\}$$

(d) The inverse of the given function is

$$\{(9, -3), (4, -2), (1, -1), (0, 0), (1, 1), (4, 2), (9, 3)\}$$

1 We notice that the inverses found in Examples 1(b) and (c) represent functions, since each element in the domain corresponds to a unique element in the range. The inverses found in Examples 1(a) and (d) do not represent functions, since each element in the domain does not correspond to a unique element in the range. Compare the function in Example 1(c) with the function in Example 1(d). For the function in Example 1(c), to every unique x-coordinate there corresponds a unique y-coordinate; for the function in Example 1(d), every unique x-coordinate does not correspond to a unique

y-coordinate (both -3 and 3 correspond to 9). Functions where unique x-coordinates correspond to unique y-coordinates are called *one-to-one* functions. So, for the inverse of a function f to be a function itself, f must be one-to-one.

A function f is said to be **one-to-one** if, for any choice of numbers x_1 and x_2 in the domain of f, if $f(x_1) = f(x_2)$, then $x_1 = x_2$.

In other words, if a function f is one-to-one, then for each x in the domain of f there is exactly one y in the range, and no y in the range is the image of more than one x in the domain. See Figure 1.

Figure 1

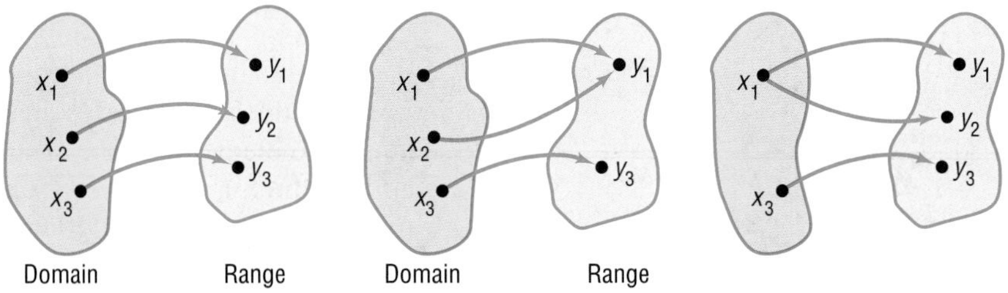

(a) One-to-one function: Each x in the domain has one and only one image in the range

(b) Not a one-to-one function: y_1 is the image of both x_1 and x_2

(c) Not a function: x_1 has two images, y_1 and y_2

NOW WORK PROBLEM 1.

If the graph of a function f is known, there is a simple test, called the **horizontal-line test**, to determine whether f is one-to-one.

Theorem

Horizontal-line Test

If every horizontal line intersects the graph of a function f in at most one point, then f is one-to-one.

The reason that this test works can be seen in Figure 2, where the horizontal line $y = h$ intersects the graph at two distinct points, (x_1, h) and (x_2, h). Since h is the image of both x_1 and x_2, $x_1 \neq x_2$, f is not one-to-one.

Figure 2
$f(x_1) = f(x_2) = h$, but $x_1 \neq x_2$; f is not a one-to-one function.

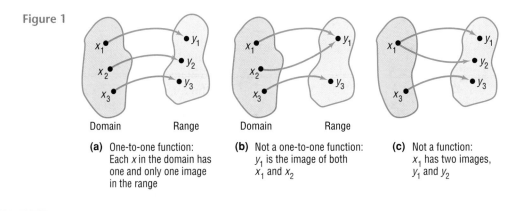

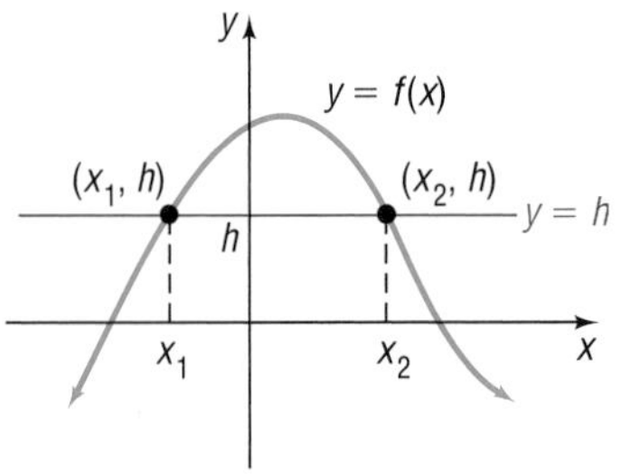

EXAMPLE 2 Using the Horizontal-line Test

For each function, use the graph to determine whether the function is one-to-one.

(a) $f(x) = x^2$ (b) $g(x) = x^3$

Solution (a) Figure 3(a) illustrates the horizontal-line test for $f(x) = x^2$. The horizontal line $y = 1$ intersects the graph of f twice, at $(1, 1)$ and at $(-1, 1)$, so f is not one-to-one.

(b) Figure 3(b) illustrates the horizontal-line test for $g(x) = x^3$. Because every horizontal line will intersect the graph of g exactly once, it follows that g is one-to-one.

Figure 3

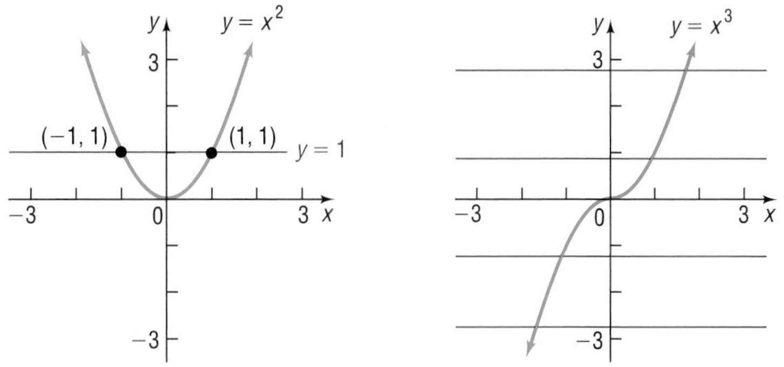

(a) A horizontal line intersects the graph twice; thus, f is not one-to-one

(b) Every horizontal line intersects the graph exactly once; thus, g is one-to-one

NOW WORK PROBLEM 9.

Let's look more closely at the one-to-one function $g(x) = x^3$. This function is an increasing function. Because an increasing (or decreasing) function will always have different y values for unequal x values, it follows that a function that is increasing (decreasing) on its domain is also a one-to-one function.

Theorem An increasing (decreasing) function is a one-to-one function.

Inverse of a Function $y = f(x)$

For a function $y = f(x)$ to have an inverse *function*, f must be one-to-one. Then for each x in its domain there is exactly one y in its range; furthermore, to each y in the range, there corresponds exactly one x in the domain. The correspondence from the range of f onto the domain of f is, therefore, also a function. It is this function that is the *inverse of f*.

We mentioned in Chapter 2 that a function $y = f(x)$ can be thought of as a rule that tells us to do something to the argument x. For example, the function $f(x) = 2x$ multiplies the argument by 2. An *inverse function* of f undoes whatever f does. For example, the function $g(x) = \frac{1}{2}x$, which divides the argument by 2, is an inverse of $f(x) = 2x$. See Figure 4.

Figure 4

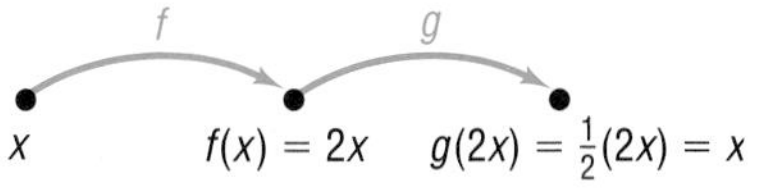

To put it another way, if we think of f as an input/output machine that processes an input x into $f(x)$, then the inverse function reverses this process, taking $f(x)$ back to x. A definition is given next.

Let f denote a one-to-one function $y = f(x)$. The **inverse of f**, denoted by f^{-1}, is a function such that $f^{-1}(f(x)) = x$ for every x in the domain of f and $f(f^{-1}(x)) = x$ for every x in the domain of f^{-1}.

WARNING: Be careful! The -1 used in f^{-1} is not an exponent. That is, f^{-1} does *not* mean the reciprocal of f; it means the inverse of f.

Figure 5

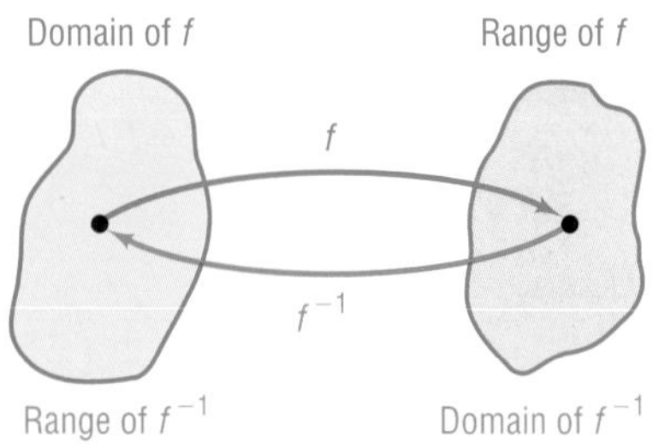

Figure 5 illustrates the definition.

Two facts are now apparent about a function f and its inverse f^{-1}.

Domain of f = Range of f^{-1}	Range of f = Domain of f^{-1}

Look again at Figure 5 to visualize the relationship. If we start with x, apply f, and then apply f^{-1}, we get x back again. If we start with x, apply f^{-1}, and then apply f, we get the number x back again. To put it simply, what f does, f^{-1} undoes, and vice versa:

Input x —Apply f→ $f(x)$ —Apply f^{-1}→ $f^{-1}(f(x)) = x$

Input x —Apply f^{-1}→ $f^{-1}(x)$ —Apply f→ $f(f^{-1}(x)) = x$

In other words,

$$f^{-1}(f(x)) = x \quad \text{and} \quad f(f^{-1}(x)) = x$$

The preceding conditions can be used to verify that a function is, in fact, the inverse of f, as Example 3 demonstrates.

EXAMPLE 3 Verifying Inverse Functions

(a) We verify that the inverse of $g(x) = x^3$ is $g^{-1}(x) = \sqrt[3]{x}$ by showing that

$$g^{-1}(g(x)) = g^{-1}(x^3) = \sqrt[3]{x^3} = x$$

and

$$g(g^{-1}(x)) = g(\sqrt[3]{x}) = (\sqrt[3]{x})^3 = x$$

(b) We verify that the inverse of $h(x) = 3x$ is $h^{-1}(x) = \frac{1}{3}x$ by showing that

$$h^{-1}(h(x)) = h^{-1}(3x) = \tfrac{1}{3}(3x) = x$$

and

$$h(h^{-1}(x)) = h(\tfrac{1}{3}x) = 3(\tfrac{1}{3}x) = x$$

(c) We verify that the inverse of $f(x) = 2x + 3$ is $f^{-1}(x) = \frac{1}{2}(x - 3)$ by showing that

$$f^{-1}(f(x)) = f^{-1}(2x + 3) = \tfrac{1}{2}[(2x + 3) - 3] = \tfrac{1}{2}(2x) = x$$

and

$$f(f^{-1}(x)) = f(\tfrac{1}{2}(x - 3)) = 2[\tfrac{1}{2}(x - 3)] + 3 = (x - 3) + 3 = x$$

EXPLORATION Simultaneously graph $Y_1 = x$, $Y_2 = x^3$, and $Y_3 = \sqrt[3]{x}$ on a square screen, using the viewing rectangle $-3 \le x \le 3, -2 \le y \le 2$. What do you observe about the graphs of $Y_2 = x^3$, its inverse $Y_3 = \sqrt[3]{x}$, and the line $Y_1 = x$?

Do you see the symmetry of the graph of Y_2 and its inverse Y_3 with respect to the line $Y_1 = x$?

NOW WORK PROBLEM 21.

2 Geometric Interpretation

For the functions in Example 3(c), we list points on the graph of $f = Y_1$ and on the graph of $f^{-1} = Y_2$ in Table 1.

We notice that whenever (a, b) is on the graph of f then (b, a) is on the graph of f^{-1}. Figure 6 shows these points plotted. Also shown is the graph of $y = x$, which you should observe is a line of symmetry of the points.

TABLE 1

X	Y1
-5	-7
-4	-5
-3	-3
-2	-1
-1	1
0	3
1	5

Y1=2X+3

X	Y2
-7	-5
-5	-4
-3	-3
-1	-2
1	-1
3	0
5	1

Y2=(1/2)(X-3)

Figure 6

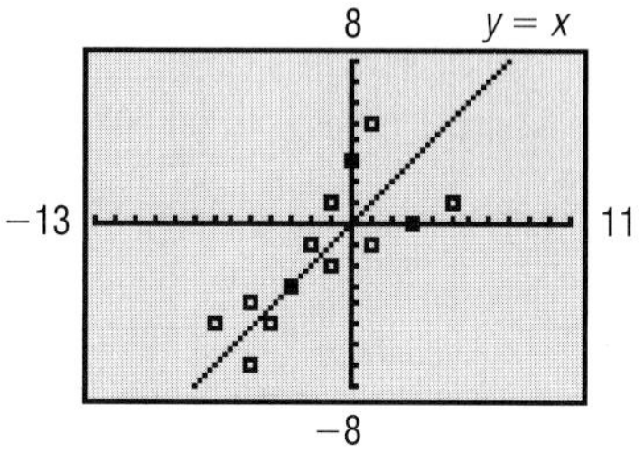

Figure 7

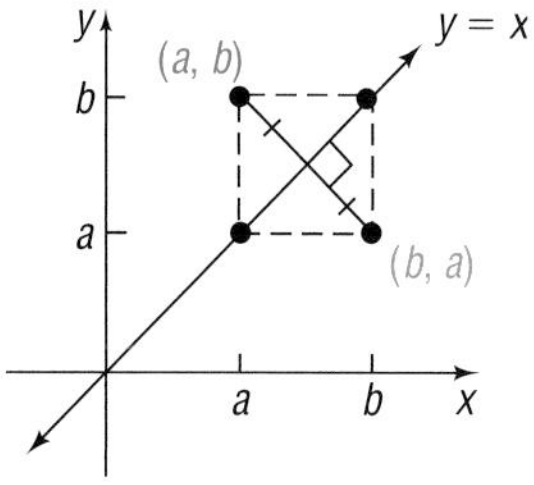

Suppose that (a, b) is a point on the graph of the one-to-one function f defined by $y = f(x)$. Then $b = f(a)$. This means that $a = f^{-1}(b)$, so (b, a) is a point on the graph of the inverse function f^{-1}. The relationship between the point (a, b) on f and the point (b, a) on f^{-1} is shown in Figure 7. The line segment containing (a, b) and (b, a) is perpendicular to the line $y = x$ and is bisected by the line $y = x$. (Do you see why?) It follows that the point (b, a) on f^{-1} is the reflection about the line $y = x$ of the point (a, b) on f.

Theorem The graph of a function f and the graph of its inverse f^{-1} are symmetric with respect to the line $y = x$.

Figure 8 illustrates this result. Notice that, once the graph of f is known, the graph of f^{-1} may be obtained by reflecting the graph of f about the line $y = x$.

Figure 8

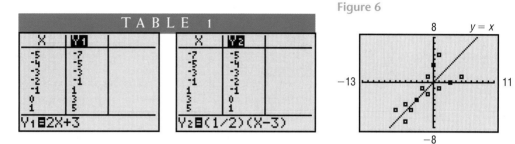

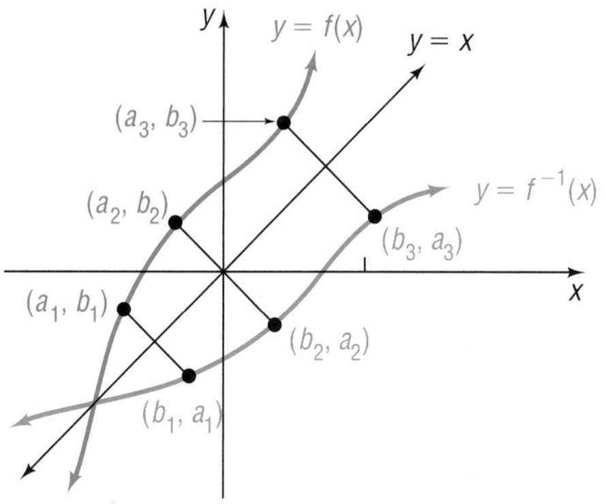

EXAMPLE 4 Graphing the Inverse Function

The graph in Figure 9(a) (p. 286) is that of a one-to-one function $y = f(x)$. Draw the graph of its inverse.

Solution We begin by adding the graph of $y = x$ to Figure 9(a). Since the points $(-2, -1)$, $(-1, 0)$, and $(2, 1)$ are on the graph of f, we know that the points

Figure 9

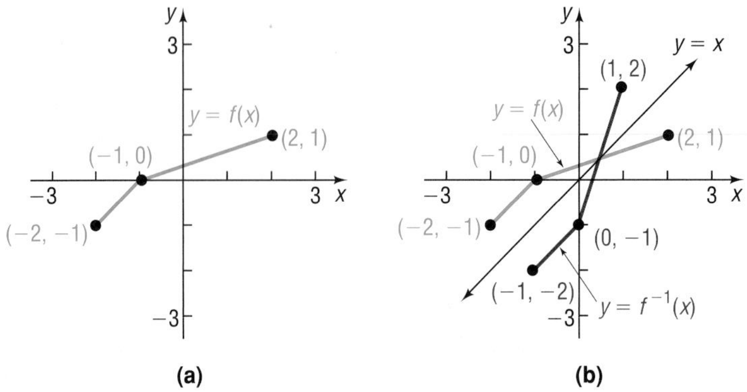

$(-1, -2)$, $(0, -1)$, and $(1, 2)$ must be on the graph of f^{-1}. Keeping in mind that the graph of f^{-1} is the reflection about the line $y = x$ of the graph of f, we can draw f^{-1}. See Figure 9(b). ■

NOW WORK PROBLEM 15.

3 Finding the Inverse Function

The fact that the graph of a one-to-one function f and its inverse are symmetric with respect to the line $y = x$ tells us more. It says that we can obtain f^{-1} by interchanging the roles of x and y in f. Look again at Figure 8. If f is defined by the equation

$$y = f(x)$$

then f^{-1} is defined by the equation

$$x = f(y)$$

The equation $x = f(y)$ defines f^{-1} *implicitly*. If we can solve this equation for y, we will have the *explicit* form of f^{-1}, that is,

$$y = f^{-1}(x)$$

Let's use this procedure to find the inverse of $f(x) = 2x + 3$. (Since f is a linear function and is increasing, we know that f is one-to-one and so has an inverse function.)

EXAMPLE 5 Finding the Inverse Function

Find the inverse of $f(x) = 2x + 3$. Also find the domain and range of f and f^{-1}. Graph f and f^{-1} on the same coordinate axes.

Solution In the equation $y = 2x + 3$, interchange the variables x and y. The result,

$$x = 2y + 3$$

is an equation that defines the inverse f^{-1} implicitly. To find the explicit form, we solve for y.

$$2y + 3 = x$$
$$2y = x - 3$$
$$y = \tfrac{1}{2}(x - 3)$$

The explicit form of the inverse f^{-1} is therefore

$$f^{-1}(x) = \tfrac{1}{2}(x - 3)$$

which we verified in Example 3(c).

Next we find

$$\text{Domain } f = \text{Range } f^{-1} = (-\infty, \infty)$$
$$\text{Range } f = \text{Domain } f^{-1} = (-\infty, \infty)$$

The graphs of $Y_1 = f(x) = 2x + 3$ and its inverse $Y_2 = f^{-1}(x) = \frac{1}{2}(x - 3)$ are shown in Figure 10. Note the symmetry of the graphs with respect to the line $Y_3 = x$.

Figure 10

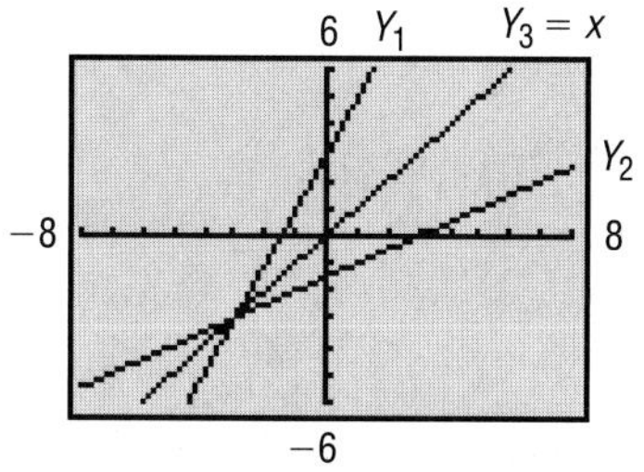

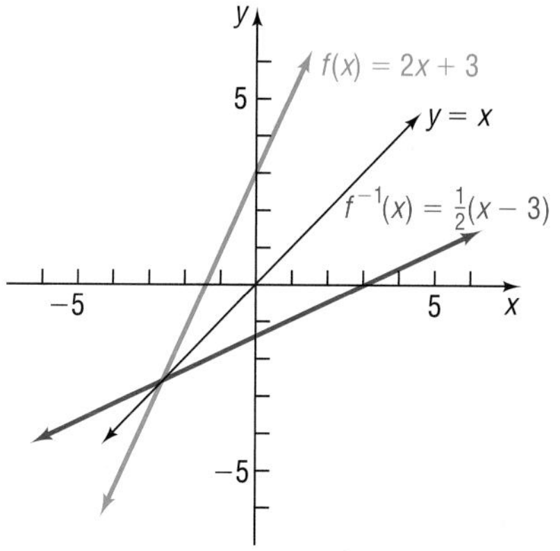

We outline next the steps to follow for finding the inverse of a one-to-one function.

PROCEDURE FOR FINDING THE INVERSE OF A ONE-TO-ONE FUNCTION

STEP 1: In $y = f(x)$, interchange the variables x and y to obtain

$$x = f(y)$$

This equation defines the inverse function f^{-1} implicitly.

STEP 2: If possible, solve the implicit equation for y in terms of x to obtain the explicit form of f^{-1}.

$$y = f^{-1}(x)$$

STEP 3: Check the result by showing that

$$f^{-1}(f(x)) = x \quad \text{and} \quad f(f^{-1}(x)) = x$$

EXAMPLE 6

Finding the Inverse Function

The function

$$f(x) = \frac{2x + 1}{x - 1}, \qquad x \neq 1$$

is one-to-one. Find its inverse and check the result.

Solution **STEP 1:** Interchange the variables x and y in

$$y = \frac{2x + 1}{x - 1}$$

to obtain

$$x = \frac{2y + 1}{y - 1}$$

STEP 2: Solve for y.

$$x = \frac{2y+1}{y-1}$$

$$x(y-1) = 2y+1 \quad \text{Multiply both sides by } y-1.$$

$$xy - x = 2y+1 \quad \text{Apply the distributive property.}$$

$$xy - 2y = x+1 \quad \text{Subtract } 2y \text{ from both sides; add } x \text{ to both sides.}$$

$$(x-2)y = x+1 \quad \text{Factor.}$$

$$y = \frac{x+1}{x-2} \quad \text{Divide by } x-2.$$

The inverse is

$$f^{-1}(x) = \frac{x+1}{x-2}, \qquad x \neq 2 \quad \text{Replace } y \text{ by } f^{-1}(x).$$

STEP 3: Check:

$$f^{-1}(f(x)) = f^{-1}\left(\frac{2x+1}{x-1}\right) = \frac{\frac{2x+1}{x-1}+1}{\frac{2x+1}{x-1}-2} = \frac{2x+1+x-1}{2x+1-2(x-1)} = \frac{3x}{3} = x$$

$$f(f^{-1}(x)) = f\left(\frac{x+1}{x-2}\right) = \frac{2\left(\frac{x+1}{x-2}\right)+1}{\frac{x+1}{x-2}-1} = \frac{2(x+1)+x-2}{x+1-(x-2)} = \frac{3x}{3} = x$$

EXPLORATION In Example 6, we found that, if $f(x) = (2x+1)/(x-1)$, then $f^{-1}(x) = (x+1)/(x-2)$. Compare the vertical and horizontal asymptotes of f and f^{-1}. What did you find? Are you surprised?

NOW WORK PROBLEM 33.

If a function is not one-to-one, then its inverse is not a function. Sometimes, though, an appropriate restriction on the domain of such a function will yield a new function that is one-to-one. Let's look at an example of this common practice.

EXAMPLE 7 Finding the Inverse Function

Find the inverse of $y = f(x) = x^2$ if $x \geq 0$.

Solution The function $y = x^2$ is not one-to-one. [Refer to Example 2(a).] However, if we restrict this function to only that part of its domain for which $x \geq 0$, as indicated, we have a new function that is increasing and therefore is one-to-one. As a result, the function defined by $y = f(x) = x^2$, $x \geq 0$, has an inverse function, f^{-1}.

We follow the steps given previously to find f^{-1}.

STEP 1: In the equation $y = x^2$, $x \geq 0$, interchange the variables x and y. The result is

$$x = y^2, \qquad y \geq 0$$

This equation defines (implicitly) the inverse function.

STEP 2: We solve for y to get the explicit form of the inverse. Since $y \geq 0$, only one solution for y is obtained.

$$y = \sqrt{x}$$

So $f^{-1}(x) = \sqrt{x}$.

STEP 3: Check: $f^{-1}(f(x)) = f^{-1}(x^2) = \sqrt{x^2} = |x| = x$, since $x \geq 0$

$$f(f^{-1}(x)) = f(\sqrt{x}) = (\sqrt{x})^2 = x.$$

Figure 11 illustrates the graphs of $Y_1 = f(x) = x^2$, $x \geq 0$, and $Y_2 = f^{-1}(x) = \sqrt{x}$.

Figure 11

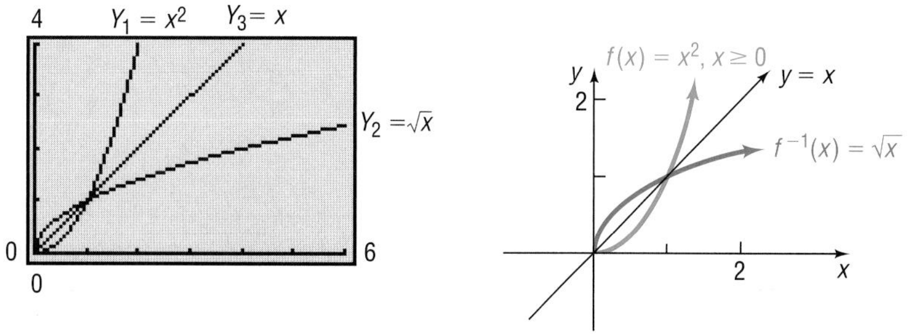

SUMMARY

1. If a function f is one-to-one, then it has an inverse function f^{-1}.
2. Domain f = Range f^{-1}; Range f = Domain f^{-1}.
3. To verify that f^{-1} is the inverse of f, show that $f^{-1}(f(x)) = x$ and $f(f^{-1}(x)) = x$.
4. The graphs of f and f^{-1} are symmetric with respect to the line $y = x$.

5.1 EXERCISES

In Problems 1–8, (a) find the inverse and (b) determine whether the inverse represents a function.

1.

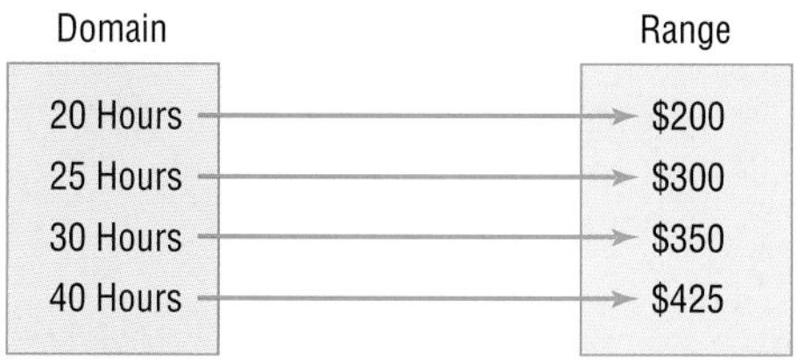

2.

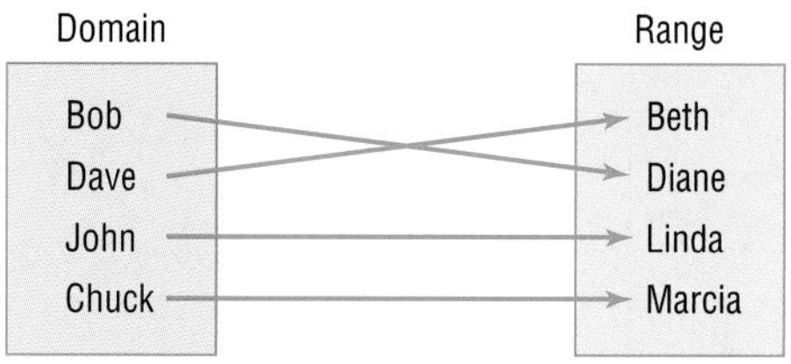

3.

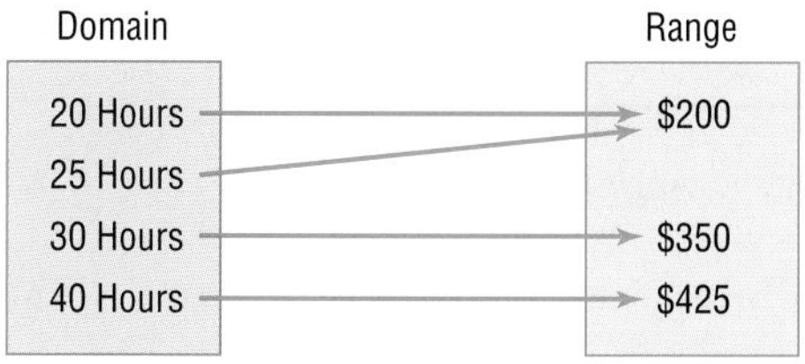

4.

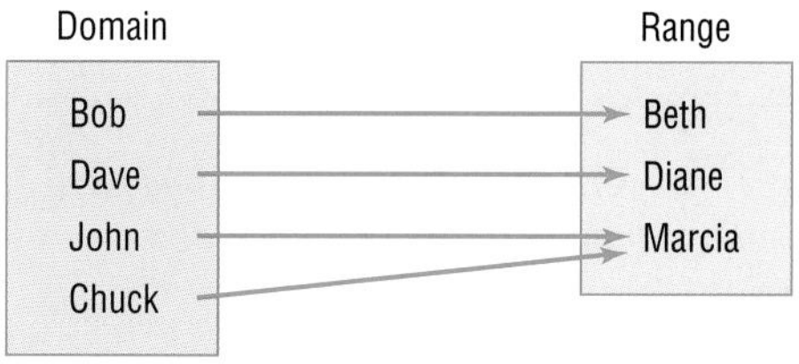

5. $\{(2, 6), (-3, 6), (4, 9), (1, 10)\}$

6. $\{(-2, 5), (-1, 3), (3, 7), (4, 12)\}$

7. $\{(0, 0), (1, 1), (2, 16), (3, 81)\}$

8. $\{(1, 2), (2, 8), (3, 18), (4, 32)\}$

In Problems 9–14, the graph of a function f is given. Use the horizontal line test to determine whether f is one-to-one.

9.

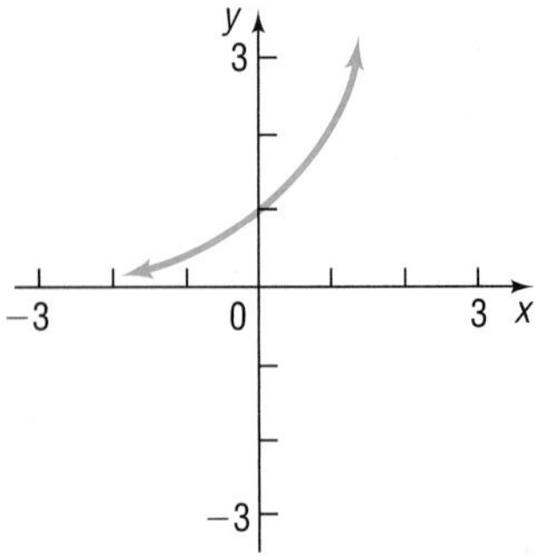

10.

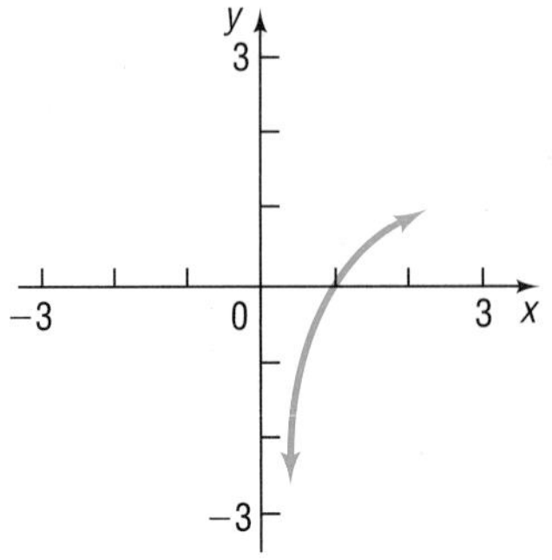

11.

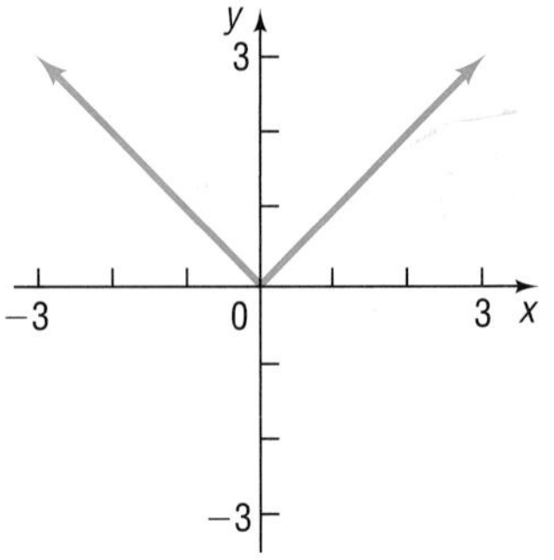

12.

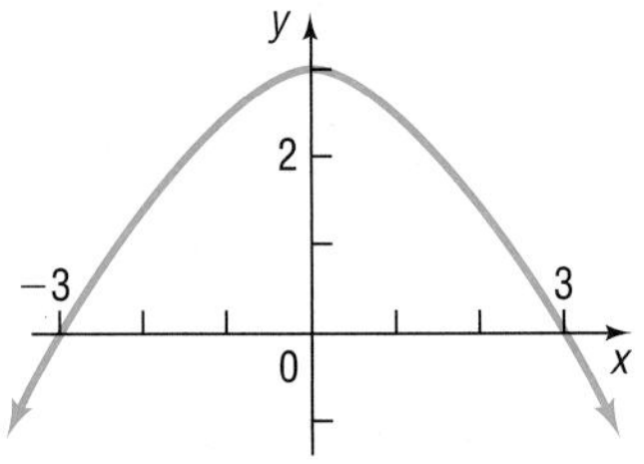

13.

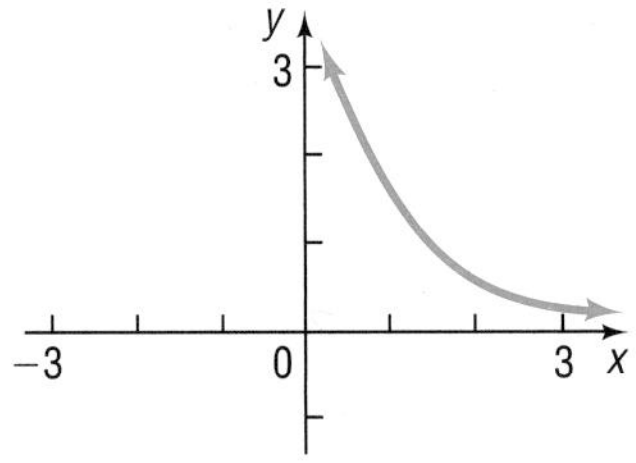

14.

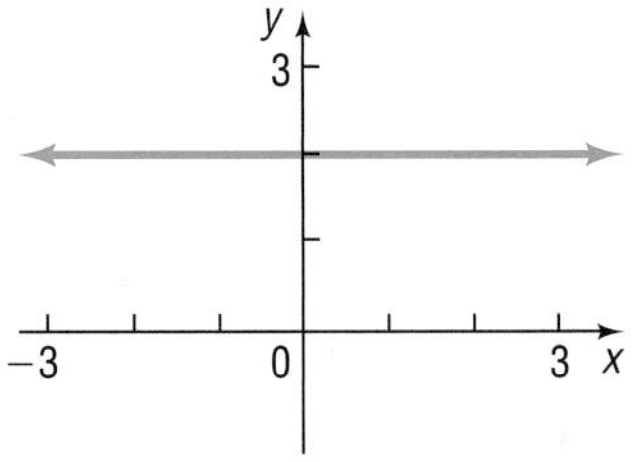

In Problems 15–20, the graph of a one-to-one function f is given. Draw the graph of the inverse function f^{-1}. For convenience (and as a hint), the graph of $y = x$ is also given.

15.

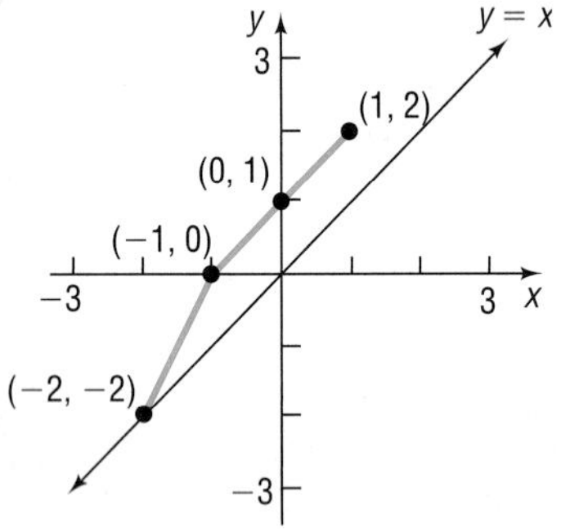

16.

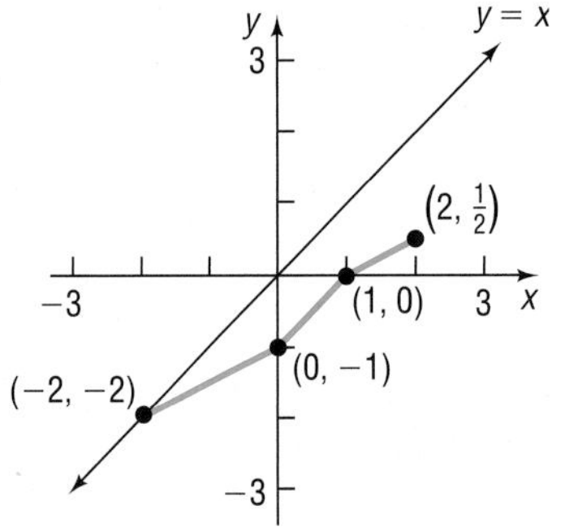

17.

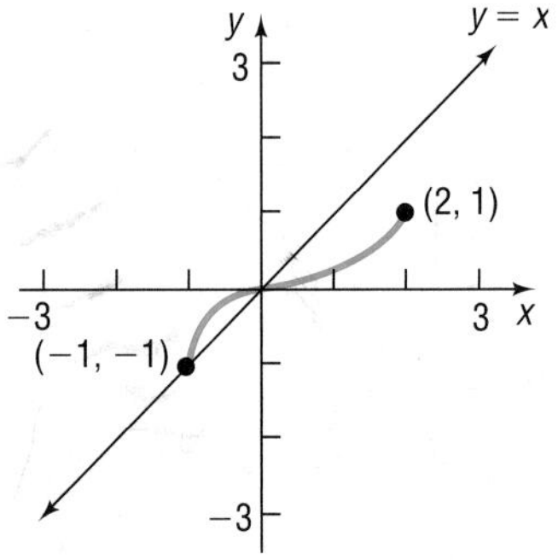

18.

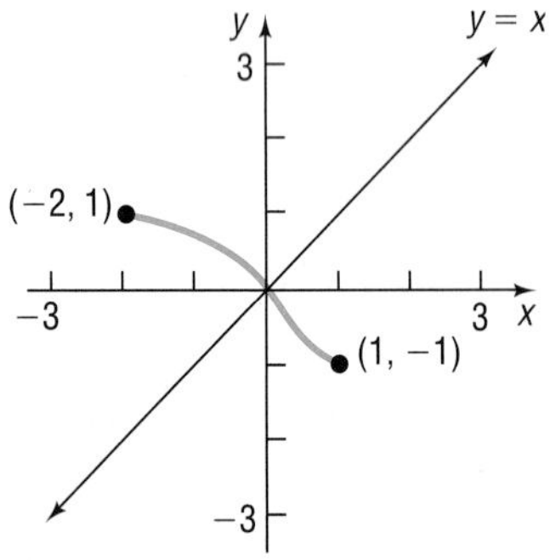

19.

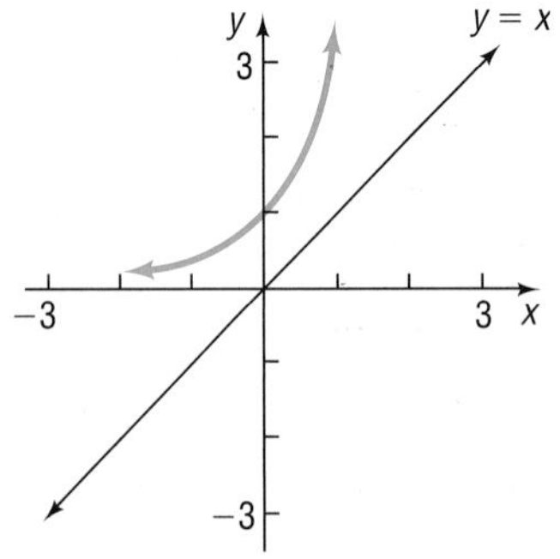

20.

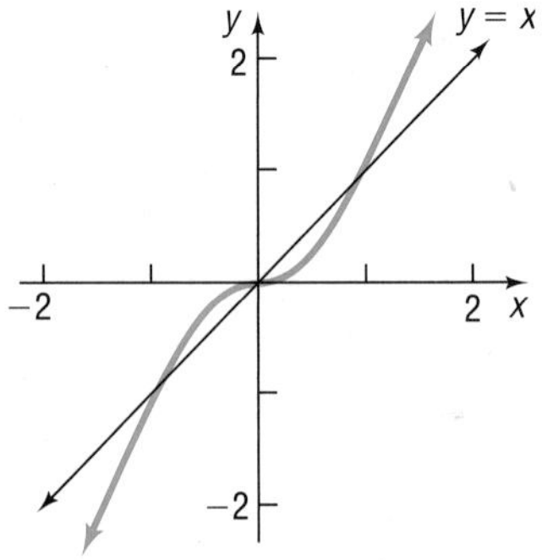

In Problems 21–30, verify that the functions f and g are inverses of each other by showing that $f(g(x)) = x$ and $g(f(x)) = x$. Using a graphing utility, simultaneously graph f, g, and $y = x$ on the same square screen.

21. $f(x) = 3x + 4;\quad g(x) = \frac{1}{3}(x - 4)$

22. $f(x) = 3 - 2x;\quad g(x) = -\frac{1}{2}(x - 3)$

23. $f(x) = 4x - 8;\quad g(x) = \dfrac{x}{4} + 2$

24. $f(x) = 2x + 6;\quad g(x) = \frac{1}{2}x - 3$

25. $f(x) = x^3 - 8;\quad g(x) = \sqrt[3]{x + 8}$

26. $f(x) = (x - 2)^2,\ \ x \geq 2;\quad g(x) = \sqrt{x} + 2,\ x \geq 0$

27. $f(x) = \frac{1}{x}$; $g(x) = \frac{1}{x}$

28. $f(x) = x$; $g(x) = x$

29. $f(x) = \frac{2x + 3}{x + 4}$; $g(x) = \frac{4x - 3}{2 - x}$

30. $f(x) = \frac{x - 5}{2x + 3}$; $g(x) = \frac{3x + 5}{1 - 2x}$

In Problems 31–42, the function f is one-to-one. Find its inverse and check your answer. State the domain and range of f and f^{-1}. By hand, graph f, f^{-1}, and $y = x$ on the same coordinate axes. Check your results using a graphing utility.

31. $f(x) = 3x$

32. $f(x) = -4x$

33. $f(x) = 4x + 2$

34. $f(x) = 1 - 3x$

35. $f(x) = x^3 - 1$

36. $f(x) = x^3 + 1$

37. $f(x) = x^2 + 4, \quad x \geq 0$

38. $f(x) = x^2 + 9, \quad x \geq 0$

39. $f(x) = \frac{4}{x}$

40. $f(x) = -\frac{3}{x}$

41. $f(x) = \frac{1}{x - 2}$

42. $f(x) = \frac{4}{x + 2}$

In Problems 43–54, the function f is one-to-one. Find its inverse and check your answer. State the domain and range of f and f^{-1}. Using a graphing utility, simultaneously graph f, f^{-1}, and $y = x$ on the same square screen.

43. $f(x) = \frac{2}{3 + x}$

44. $f(x) = \frac{4}{2 - x}$

45. $f(x) = (x + 2)^2, \quad x \geq -2$

46. $f(x) = (x - 1)^2, \quad x \geq 1$

47. $f(x) = \frac{2x}{x - 1}$

48. $f(x) = \frac{3x + 1}{x}$

49. $f(x) = \frac{3x + 4}{2x - 3}$

50. $f(x) = \frac{2x - 3}{x + 4}$

51. $f(x) = \frac{2x + 3}{x + 2}$

52. $f(x) = \frac{-3x - 4}{x - 2}$

53. $f(x) = 2\sqrt[3]{x}$

54. $f(x) = \frac{4}{\sqrt{x}}$

55. Find the inverse of the linear function
$f(x) = mx + b, \qquad m \neq 0$

56. Find the inverse of the function
$f(x) = \sqrt{r^2 - x^2}, \qquad 0 \leq x \leq r$

57. Can an even function be one-to-one? Explain.

58. Is every odd function one-to-one? Explain.

59. A function f has an inverse function. If the graph of f lies in quadrant I, in which quadrant does the graph of f^{-1} lie?

60. A function f has an inverse function. If the graph of f lies in quadrant II, in which quadrant does the graph of f^{-1} lie?

61. The function $f(x) = |x|$ is not one-to-one. Find a suitable restriction on the domain of f so that the new function that results is one-to-one. Then find the inverse of f.

62. The function $f(x) = x^4$ is not one-to-one. Find a suitable restriction on the domain of f so that the new function that results is one-to-one. Then find the inverse of f.

63. Temperature Conversion To convert from x degrees Celsius to y degrees Fahrenheit, we use the formula $y = f(x) = \frac{9}{5}x + 32$. To convert from x degrees Fahrenheit to y degrees Celsius, we use the formula $y = g(x) = \frac{5}{9}(x - 32)$. Show that f and g are inverse functions.

64. Demand for Corn The demand for corn obeys the equation $p(x) = 300 - 50x$, where p is the price per bushel (in dollars) and x is the number of bushels produced, in millions. Express the production amount x as a function of the price p.

65. Period of a Pendulum The period T (in seconds) of a simple pendulum is a function of its length l (in feet), given by $T(l) = 2\pi\sqrt{l/g}$, where $g \approx 32.2$ feet per second per second is the acceleration of gravity. Express the length l as a function of the period T.

66. Give an example of a function whose domain is the set of real numbers and that is neither increasing nor decreasing on its domain, but is one-to-one.
[**Hint:** Use a piecewise-defined function.]

67. Given

$$f(x) = \frac{ax + b}{cx + d}$$

find $f^{-1}(x)$. If $c \neq 0$, under what conditions on a, b, c, and d is $f = f^{-1}$?

68. We said earlier that finding the range of a function f is not easy. However, if f is one-to-one, we can find its range by finding the domain of the inverse function f^{-1}. Use this technique to find the range of each of the following one-to-one functions:

(a) $f(x) = \frac{2x + 5}{x - 3}$ (b) $g(x) = 4 - \frac{2}{x}$

(c) $F(x) = \frac{3}{4 - x}$

69. If the graph of a function and its inverse intersect, where must this necessarily occur? Can they intersect anywhere else? Must they intersect?

70. Can a one-to-one function and its inverse be equal? What must be true about the graph of f for this to happen? Give some examples to support your conclusion.

71. Draw the graph of a one-to-one function that contains the points $(-2, -3)$, $(0, 0)$, and $(1, 5)$. Now draw the graph of its inverse. Compare your graph to those of other students. Discuss any similarities. What differences do you see?

PREPARING FOR THIS SECTION

Before getting started, review the following:

✓ Exponents (pp. 951–953 and 991–993)
✓ Solving Equations (Section 1.3)
✓ Graphing Techniques: Transformations (Section 2.4)

5.2 EXPONENTIAL FUNCTIONS

OBJECTIVES
1 Evaluate Exponential Functions
2 Graph Exponential Functions
3 Define the Number *e*
4 Solve Exponential Equations

1 In the Appendix, Section 8, we give a definition for raising a real number a to a rational power. Based on that discussion, we gave meaning to expressions of the form

$$a^r$$

where the base a is a positive real number and the exponent r is a rational number.

But what is the meaning of a^x, where the base a is a positive real number and the exponent x is an irrational number? Although a rigorous definition requires methods discussed in calculus, the basis for the definition is easy to follow: Select a rational number r that is formed by truncating (removing) all but a finite number of digits from the irrational number x. Then it is reasonable to expect that

$$a^x \approx a^r$$

For example, take the irrational number $\pi = 3.14159\ldots$. Then, an approximation to a^π is

$$a^\pi \approx a^{3.14}$$

where the digits after the hundredths position have been removed from the value for π. A better approximation would be

$$a^\pi \approx a^{3.14159}$$

where the digits after the hundred-thousandths position have been removed. Continuing in this way, we can obtain approximations to a^π to any desired degree of accuracy.

Graphing calculators can easily evaluate expressions of the form a^x as follows. Enter the base a, press the caret key (^), enter the exponent x, and press enter.

EXAMPLE 1 **Using a Calculator to Evaluate Powers of 2**

Using a calculator, evaluate:

(a) $2^{1.4}$ (b) $2^{1.41}$ (c) $2^{1.414}$ (d) $2^{1.4142}$ (e) $2^{\sqrt{2}}$

Solution Figure 12 shows the solution to part (a) using a TI-83 graphing calculator.

Figure 12

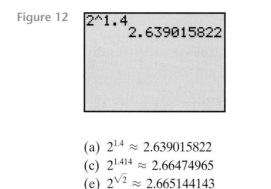

(a) $2^{1.4} \approx 2.639015822$ (b) $2^{1.41} \approx 2.657371628$
(c) $2^{1.414} \approx 2.66474965$ (d) $2^{1.4142} \approx 2.665119089$
(e) $2^{\sqrt{2}} \approx 2.665144143$

NOW WORK PROBLEM 1.

It can be shown that the familiar laws for rational exponents hold for real exponents.

Theorem

Laws of Exponents

If s, t, a, and b are real numbers with $a > 0$ and $b > 0$, then

$$a^s \cdot a^t = a^{s+t} \qquad (a^s)^t = a^{st} \qquad (ab)^s = a^s \cdot b^s$$
$$1^s = 1 \qquad a^{-s} = \frac{1}{a^s} = \left(\frac{1}{a}\right)^s \qquad a^0 = 1 \tag{1}$$

We are now ready for the following definition:

An **exponential function** is a function of the form

$$f(x) = a^x$$

where a is a positive real number $(a > 0)$ and $a \neq 1$. The domain of f is the set of all real numbers.

We exclude the base $a = 1$, because this function is simply the constant function $f(x) = 1^x = 1$. We also need to exclude bases that are negative, because, otherwise, we would have to exclude many values of x from the domain, such as $x = \frac{1}{2}$ and $x = \frac{3}{4}$. [Recall that $(-2)^{1/2} = \sqrt{-2}$, $(-3)^{3/4} = \sqrt[4]{(-3)^3} = \sqrt[4]{-27}$, and so on, are not defined in the system of real numbers.]

Graphs of Exponential Functions

2 First, we graph the exponential function $f(x) = 2^x$.

EXAMPLE 2 **Graphing an Exponential Function**

Graph the exponential function: $f(x) = 2^x$

Solution The domain of $f(x) = 2^x$ consists of all real numbers. We begin by locating some points on the graph of $f(x) = 2^x$, as listed in Table 2.

TABLE 2

X	Y1
-10	9.8E-4
-3	.125
-1	.5
0	1
1	2
3	8
10	1024

Y1 = 2^X

Since $2^x > 0$ for all x, the range of f is $(0, \infty)$. From this, we conclude that the graph has no x-intercepts, and, in fact, the graph will lie above the x-axis. As Table 2 indicates, the y-intercept is 1. Table 2 also indicates that as $x \to -\infty$ the value of $f(x) = 2^x$ gets closer and closer to 0. We conclude that the x-axis $(y = 0)$ is a horizontal asymptote to the graph as $x \to -\infty$.

To determine the end behavior for x large and positive, look again at Table 2. As $x \to \infty$, $f(x) = 2^x$ grows very quickly, causing the graph of $f(x) = 2^x$ to rise very rapidly. It is apparent that f is an increasing function and hence is one-to-one.

Figure 13 shows the graph of $f(x) = 2^x$. Notice that all the conclusions given earlier are confirmed by the graph. ■

Figure 13

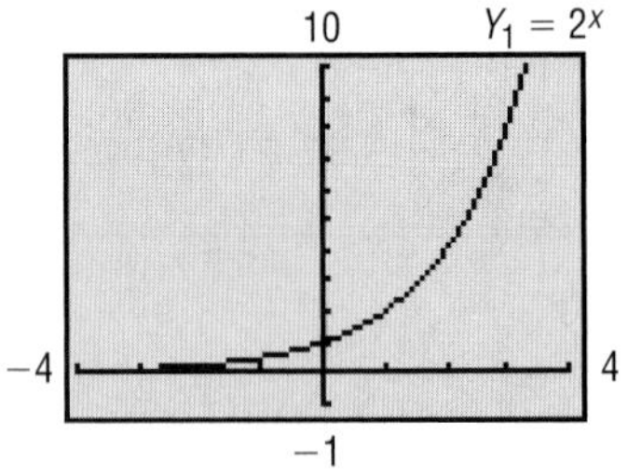

As we shall see, graphs that look like the one in Figure 13 occur very frequently in a variety of situations. For example, look at the graph in Figure 14, which illustrates the closing price of a share of Dell Computer stock. Investors might conclude from this graph that the price of Dell Computer is *behaving exponentially*; that is, the graph exhibits rapid, or exponential, growth. We shall have more to say about situations that lead to exponential growth later in this chapter. For now, we continue to seek properties of the exponential functions.

The graph of $f(x) = 2^x$ in Figure 13 is typical of all exponential functions that have a base larger than 1. Such functions are increasing functions and hence are one-to-one. Their graphs lie above the x-axis, pass through the point $(0, 1)$, and thereafter rise rapidly as $x \to \infty$. As $x \to -\infty$, the x-axis $(y = 0)$ is a horizontal asymptote. There are no vertical asymptotes. Finally, the graphs are smooth and continuous, with no corners or gaps.

Figure 15 illustrates the graphs of two more exponential functions whose bases are larger than 1. Notice that for the larger base the graph is steeper when $x > 0$. Figure 16 shows that when $x < 0$ the graph of the equation with the larger base is closer to the x-axis.

Figure 14

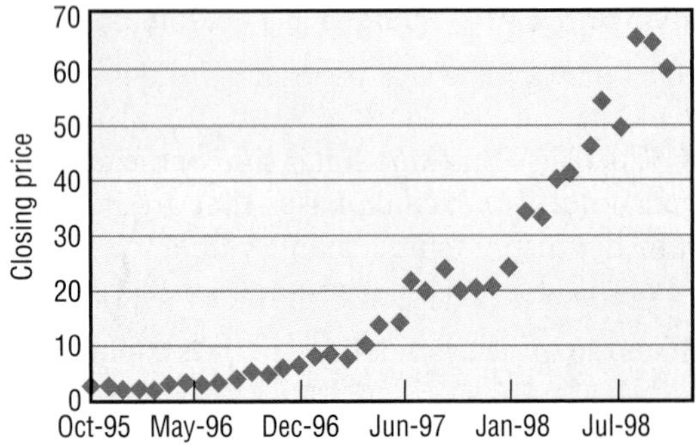

Figure 15

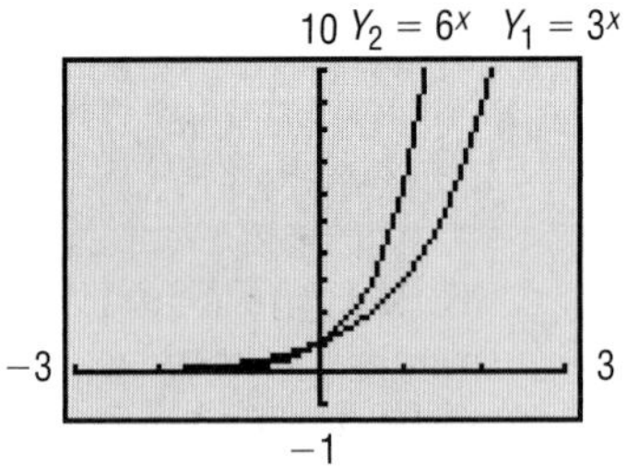

Figure 16

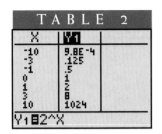

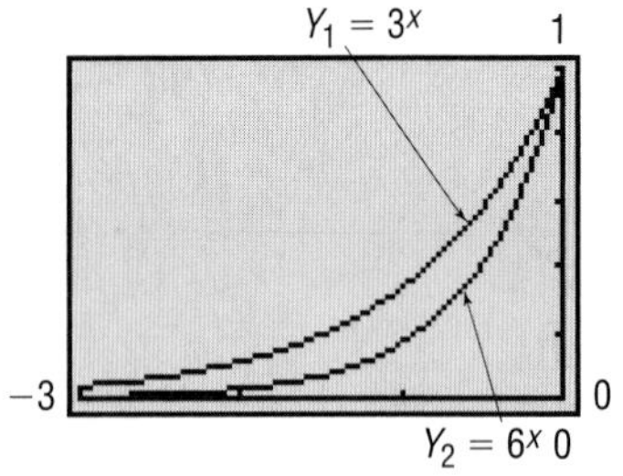

The following display summarizes the information that we have about $f(x) = a^x$, $a > 1$.

Facts about the Graph of an Exponential Function
$f(x) = a^x$, $a > 1$

1. The domain is all real numbers; the range is the set of positive real numbers.
2. There are no x-intercepts; the y-intercept is 1.

3. The x-axis ($y = 0$) is a horizontal asymptote as $x \to -\infty$.
4. $f(x) = a^x$, $a > 1$, is an increasing function and is one-to-one.
5. The graph of f contains the points $(0, 1)$ and $(1, a)$.
6. The graph of f is smooth and continuous, with no corners or gaps.

Now we consider $f(x) = a^x$ when $0 < a < 1$.

EXAMPLE 3 Graphing an Exponential Function

Graph the exponential function: $f(x) = \left(\frac{1}{2}\right)^x$

Solution The domain of $f(x) = \left(\frac{1}{2}\right)^x$ consists of all real numbers. As before, we locate some points on the graph, as listed in Table 3. Since $\left(\frac{1}{2}\right)^x > 0$ for all x, the range of f is $(0, \infty)$. The graph lies above the x-axis and so has no x-intercepts. The y-intercept is 1. As $x \to -\infty$, $f(x) = \left(\frac{1}{2}\right)^x$ grows very quickly. As $x \to \infty$, the values of $f(x)$ approach 0. The x-axis ($y = 0$) is a horizontal asymptote as $x \to \infty$. It is apparent that f is a decreasing function and hence is one-to-one. Figure 17 illustrates the graph. ■

TABLE 3

X	Y1
-10	1024
-3	8
-1	2
0	1
1	.5
3	.125
10	9.8E-4

Y1=(1/2)^X

Figure 17

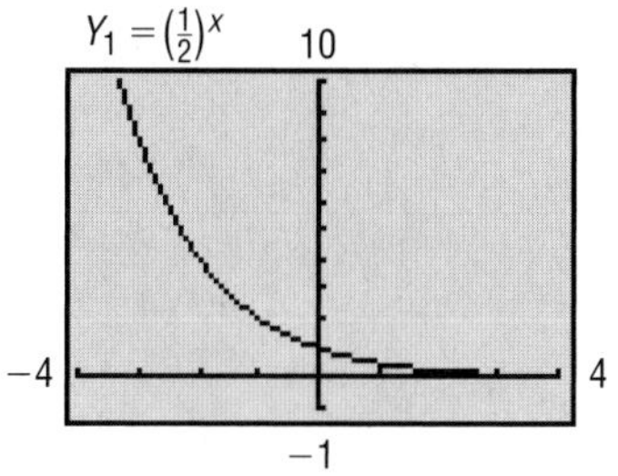

We could have obtained the graph of $y = \left(\frac{1}{2}\right)^x$ from the graph of $y = 2^x$ using transformations. If $f(x) = 2^x$, then $f(-x) = 2^{-x} = 1/2^x = \left(\frac{1}{2}\right)^x$. The graph of $y = \left(\frac{1}{2}\right)^x = 2^{-x}$ is a reflection about the y-axis of the graph of $y = 2^x$. Compare Figures 13 and 17.

Seeing the Concept Using a graphing utility, simultaneously graph:

(a) $Y_1 = 3^x, Y_2 = \left(\dfrac{1}{3}\right)^x$ (b) $Y_1 = 6^x, Y_2 = \left(\dfrac{1}{6}\right)^x$

Conclude that the graph of $Y_2 = \left(\dfrac{1}{a}\right)^x$, for $a > 0$, is the reflection about the y-axis of the graph of $Y_1 = a^x$. ■

The graph of $f(x) = \left(\frac{1}{2}\right)^x$ in Figure 17 is typical of all exponential functions that have a base between 0 and 1. Such functions are decreasing and one-to-one. Their graphs lie above the x-axis and pass through the point $(0, 1)$. The graphs rise rapidly as $x \to -\infty$. As $x \to \infty$, the x-axis is a horizontal asymptote. There are no vertical asymptotes. Finally, the graphs are smooth and continuous, with no corners or gaps.

Figure 18 illustrates the graphs of two more exponential functions whose bases are between 0 and 1. Notice that the smaller base results in a graph that is steeper when $x < 0$. Figure 19 shows that when $x > 0$ the graph of the equation with the smaller base is closer to the x-axis.

Figure 18

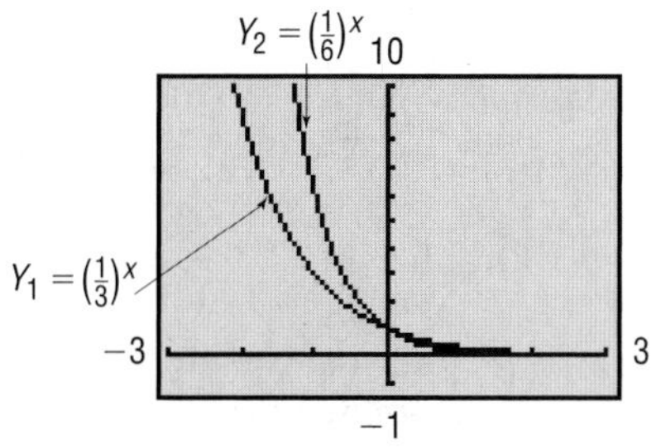

Figure 19

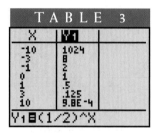

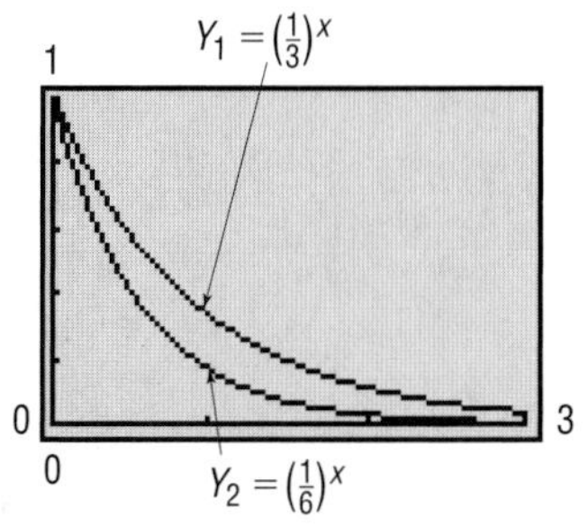

The following display summarizes the information that we have about the function $f(x) = a^x, 0 < a < 1$.

Facts about the Graph of an Exponential Function
$f(x) = a^x, 0 < a < 1$

1. The domain is all real numbers; the range is the set of positive real numbers.
2. There are no x-intercepts; the y-intercept is 1.
3. The x-axis ($y = 0$) is a horizontal asymptote as $x \to \infty$.
4. $f(x) = a^x, 0 < a < 1$, is a decreasing function and is one-to-one.
5. The graph of f contains the points $(0, 1)$ and $(1, a)$.
6. The graph of f is smooth and continuous, with no corners or gaps.

EXAMPLE 4 Graphing Exponential Functions Using Transformations

Graph $f(x) = 2^{-x} - 3$ and determine the domain, range, and horizontal asymptote of f.

Solution We begin with the graph of $y = 2^x$. Figure 20 shows the various steps.

Figure 20

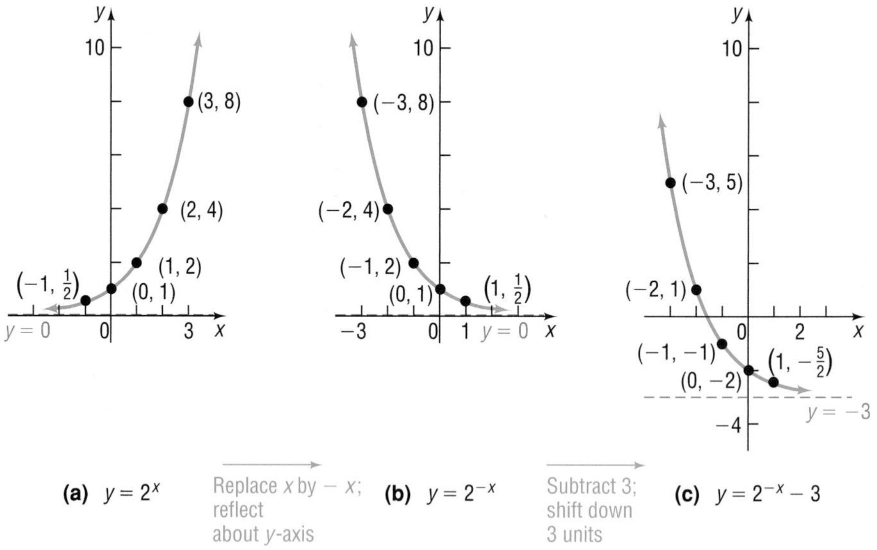

As Figure 20(c) illustrates, the domain of $f(x) = 2^{-x} - 3$ is $(-\infty, \infty)$ and the range is $(-3, \infty)$. The horizontal asymptote of f is the line $y = -3$. ■

Check: Graph $Y_1 = 2^{-x} - 3$ to verify the graph obtained in Figure 20(c).

NOW WORK PROBLEM 27.

The Base *e*

3 As we shall see shortly, many problems that occur in nature require the use of an exponential function whose base is a certain irrational number, symbolized by the letter e.

Let's look now at one way of arriving at this important number e.

The **number *e*** is defined as the number that the expression

$$\left(1 + \frac{1}{n}\right)^n \qquad (2)$$

approaches as $n \to \infty$. In calculus, this is expressed using limit notation as

$$e = \lim_{n \to \infty} \left(1 + \frac{1}{n}\right)^n$$

Table 4 illustrates what happens to the defining expression (2) as n takes on increasingly large values. The last number in the last column in the table is correct to nine decimal places and is the same as the entry given for e on your calculator (if expressed correctly to nine decimal places).

TABLE 4

n	$\frac{1}{n}$	$1 + \frac{1}{n}$	$\left(1 + \frac{1}{n}\right)^n$
1	1	2	2
2	0.5	1.5	2.25
5	0.2	1.2	2.48832
10	0.1	1.1	2.59374246
100	0.01	1.01	2.704813829
1,000	0.001	1.001	2.716923932
10,000	0.0001	1.0001	2.718145927
100,000	0.00001	1.00001	2.718268237
1,000,000	0.000001	1.000001	2.718280469
1,000,000,000	10^{-9}	$1 + 10^{-9}$	2.718281827

TABLE 5

X	Y1
-2	.13534
-1	.36788
0	1
1	2.7183
2	7.3891

Y1 = e^(X)

The exponential function $f(x) = e^x$, whose base is the number e, occurs with such frequency in applications that it is usually referred to as *the* exponential function. Indeed, graphing calculators have the key $\boxed{e^x}$ or $\boxed{\exp(x)}$, which may be used to evaluate the exponential function for a given value of x. Now use your calculator to find e^x for $x = -2$, $x = -1$, $x = 0$, $x = 1$, and $x = 2$, as we have done to create Table 5. The graph of the exponential function $f(x) = e^x$ is given in Figure 21. Since $2 < e < 3$, the graph of $y = e^x$ lies between the graphs of $y = 2^x$ and $y = 3^x$. [See Figure 21(c)].

Figure 21

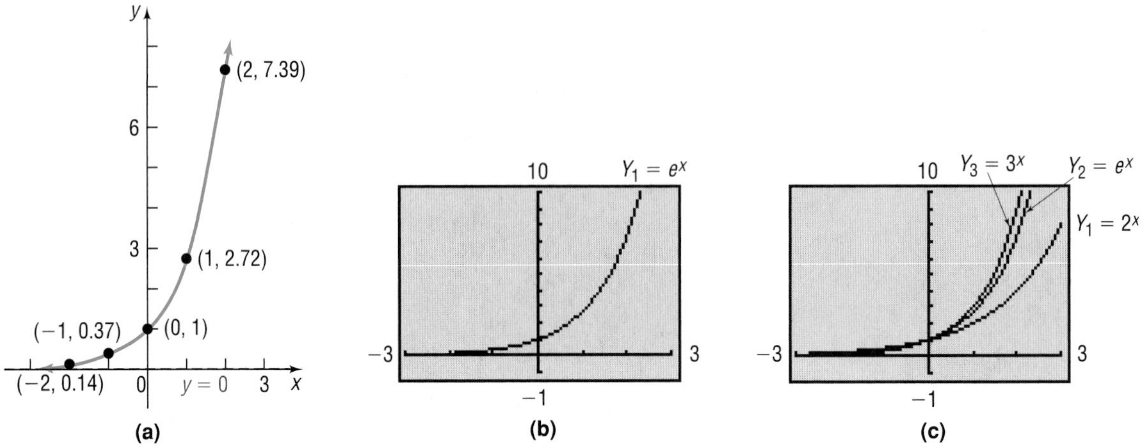

EXAMPLE 5 Graphing Exponential Functions Using Transformations

Graph $f(x) = -2e^{x-3}$ and determine the domain, range, and horizontal asymptote of f.

Solution We begin with the graph of $y = e^x$. Figure 22 shows the various steps.

Figure 22

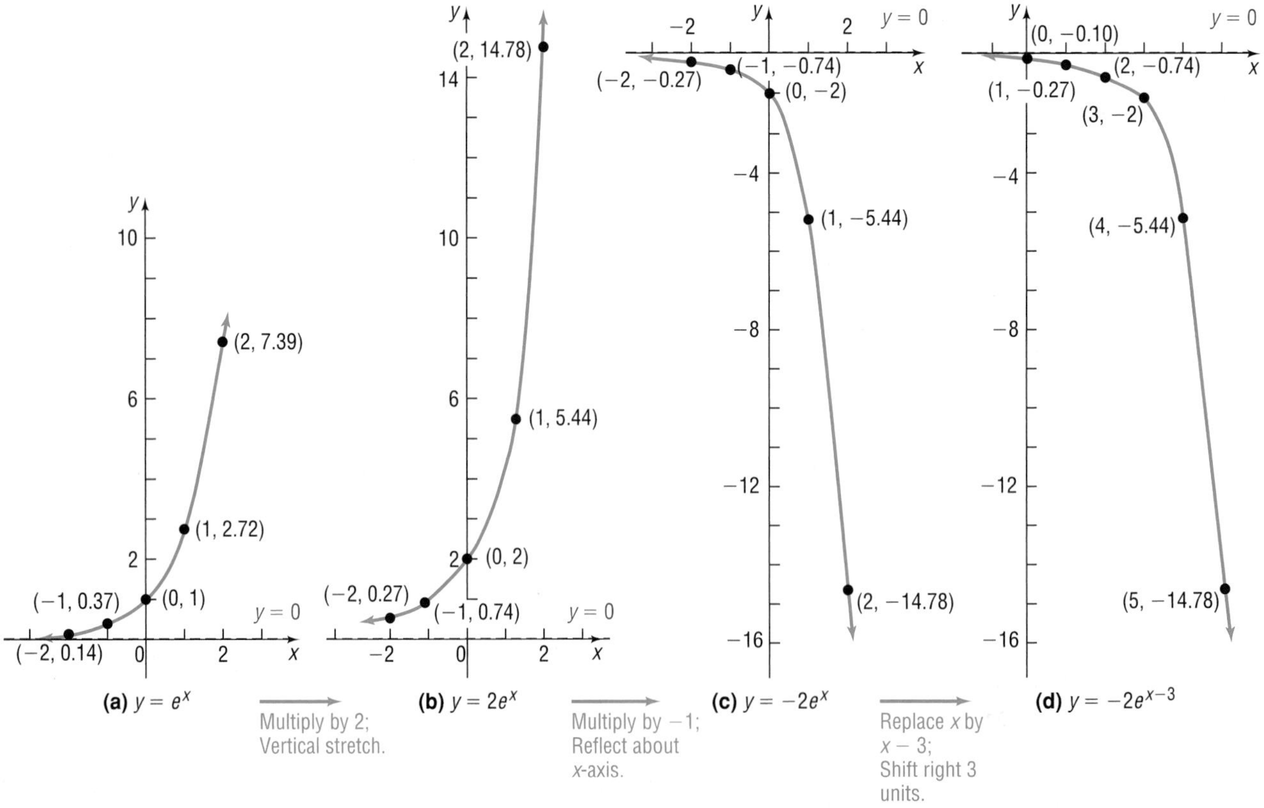

As Figure 22(d) illustrates, the domain of $f(x) = -2e^{x-3}$ is $(-\infty, \infty)$ and the range is $(-\infty, 0)$. The horizontal asymptote is the line $y = 0$. ■

Check: Graph $Y_1 = -2e^{x-3}$ to verify the graph obtained in Figure 22(d).

NOW WORK PROBLEM 35.

Exponential Equations

4 Equations that involve terms of the form a^x, $a > 0$, $a \neq 1$, are often referred to as **exponential equations.** Such equations can sometimes be solved by appropriately applying the Laws of Exponents and equation (3).

$$\text{If } a^u = a^v, \quad \text{then} \quad u = v \tag{3}$$

Equation (3) is a consequence of the fact that exponential functions are one-to-one. To use equation (3), each side of the equality must be written with the same base.

EXAMPLE 6 Solving an Exponential Equation

Solve: $3^{x+1} = 81$

Solution Since $81 = 3^4$, we can write the equation as

$$3^{x+1} = 81 = 3^4$$

Now we have the same base, 3, on each side, so we can apply (3) to obtain

$$x + 1 = 4$$

$$x = 3$$

Figure 23

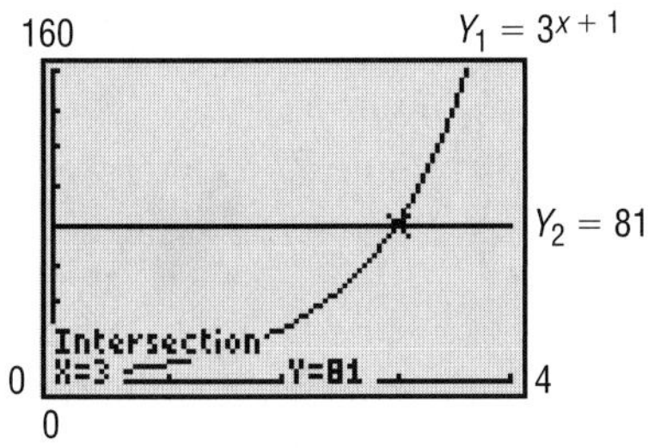

Check: We verify the solution by graphing $Y_1 = 3^{x+1}$ and $Y_2 = 81$ to determine where the graphs intersect. See Figure 23. The graphs intersect at $x = 3$.

NOW WORK PROBLEM 43.

EXAMPLE 7 Solving an Exponential Equation

Solve: $e^{-x^2} = (e^x)^2 \cdot \dfrac{1}{e^3}$

Solution We use Laws of Exponents first to get the base e on the right side.

$$(e^x)^2 \cdot \frac{1}{e^3} = e^{2x} \cdot e^{-3} = e^{2x-3}$$

As a result,

$$e^{-x^2} = e^{2x-3}$$

$$-x^2 = 2x - 3 \qquad \text{Apply (3).}$$

$$x^2 + 2x - 3 = 0 \qquad \text{Place the quadratic equation in standard form.}$$

$$(x + 3)(x - 1) = 0 \qquad \text{Factor.}$$

$$x = -3 \quad \text{or} \quad x = 1 \qquad \text{Use the Zero-Product Property.}$$

You should verify these solutions using a graphing utility.

Many applications involve the exponential functions. Let's look at one.

EXAMPLE 8 Exponential Probability

Between 9:00 PM and 10:00 PM cars arrive at Burger King's drive-thru at the rate of 12 cars per hour (0.2 car per minute). The following formula from statistics can be used to determine the probability that a car will arrive within t minutes of 9:00 PM.

$$F(t) = 1 - e^{-0.2t}$$

(a) Determine the probability that a car will arrive within 5 minutes of 9 PM (that is, before 9:05 PM).

(b) Determine the probability that a car will arrive within 30 minutes of 9 PM (before 9:30 PM).

(c) Graph F using your graphing utility.

(d) What value does F approach as t becomes unbounded in the positive direction?

Solution

Figure 24

(a) The probability that a car will arrive within 5 minutes is found by evaluating $F(t)$ at $t = 5$.

$$F(5) = 1 - e^{-0.2(5)}$$

We evaluate this expression in Figure 24. We conclude that there is a 63% probability that a car will arrive within 5 minutes.

(b) The probability that a car will arrive within 30 minutes is found by evaluating $F(t)$ at $t = 30$.

$$F(30) = 1 - e^{-0.2(30)} \approx 0.9975$$

There is a 99.75% probability that a car will arrive within 30 minutes.

Figure 25

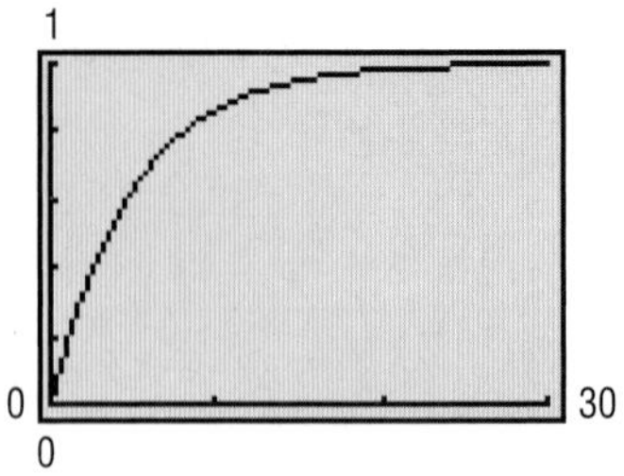

(c) See Figure 25 for the graph of F.

(d) As time passes, the probability that a car will arrive increases. The value that F approaches can be found by letting $t \to \infty$. Since $e^{-0.2t} = 1/e^{0.2t}$, it follows that $e^{-0.2t} \to 0$ as $t \to \infty$. Thus, F approaches 1 as t gets large.

NOW WORK PROBLEM 67.

SUMMARY

Properties of the Exponential Function

$f(x) = a^x, \quad a > 1$ Domain: $(-\infty, \infty)$; Range: $(0, \infty)$; x-intercepts: none; y-intercept: 1; horizontal asymptote: x-axis as $x \to -\infty$; increasing; one-to-one; smooth; continuous

See Figure 13 for a typical graph.

$f(x) = a^x, \quad 0 < a < 1$ Domain: $(-\infty, \infty)$; Range: $(0, \infty)$; x-intercepts: none; y-intercept: 1; horizontal asymptote: x-axis as $x \to \infty$; decreasing; one-to-one; smooth; continuous

See Figure 17 for a typical graph.

5.2 EXERCISES

In Problems 1–10, approximate each number using a calculator. Express your answer rounded to three decimal places.

1. (a) $3^{2.2}$ (b) $3^{2.23}$ (c) $3^{2.236}$ (d) $3^{\sqrt{5}}$
2. (a) $5^{1.7}$ (b) $5^{1.73}$ (c) $5^{1.732}$ (d) $5^{\sqrt{3}}$
3. (a) $2^{3.14}$ (b) $2^{3.141}$ (c) $2^{3.1415}$ (d) 2^{π}
4. (a) $2^{2.7}$ (b) $2^{2.71}$ (c) $2^{2.718}$ (d) 2^{e}
5. (a) $3.1^{2.7}$ (b) $3.14^{2.71}$ (c) $3.141^{2.718}$ (d) π^{e}
6. (a) $2.7^{3.1}$ (b) $2.71^{3.14}$ (c) $2.718^{3.141}$ (d) e^{π}

7. $e^{1.2}$ **8.** $e^{-1.3}$ **9.** $e^{-0.85}$ **10.** $e^{2.1}$

In Problems 11–18, the graph of an exponential function is given. Match each graph to one of the following functions without the aid of a graphing utility.

A. $y = 3^x$ *B.* $y = 3^{-x}$ *C.* $y = -3^x$ *D.* $y = -3^{-x}$
E. $y = 3^x - 1$ *F.* $y = 3^{x-1}$ *G.* $y = 3^{1-x}$ *H.* $y = 1 - 3^x$

11.
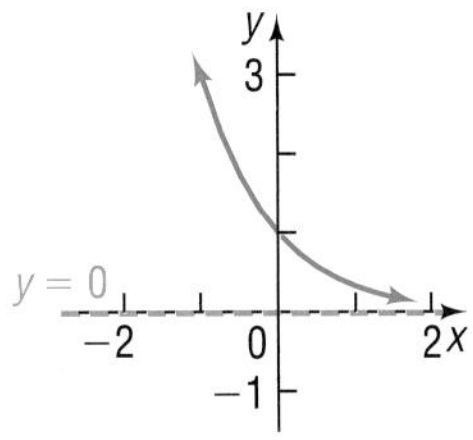

12.
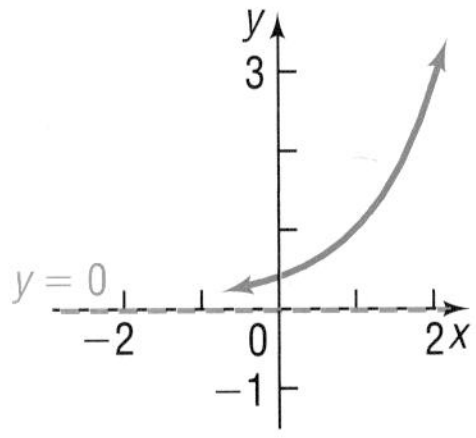

13.
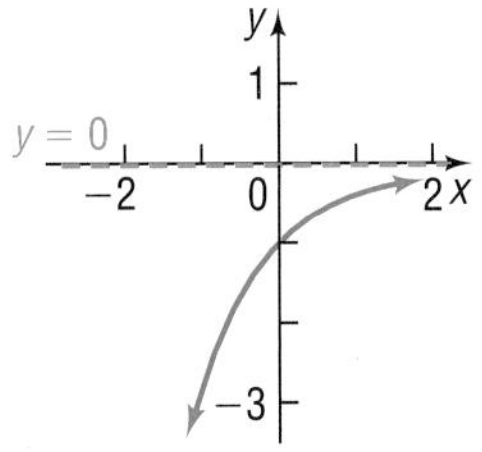

14.
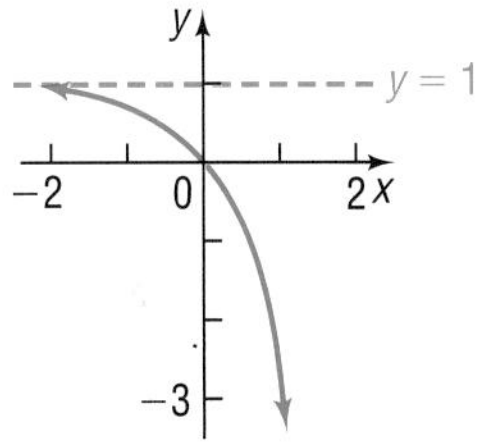

15.
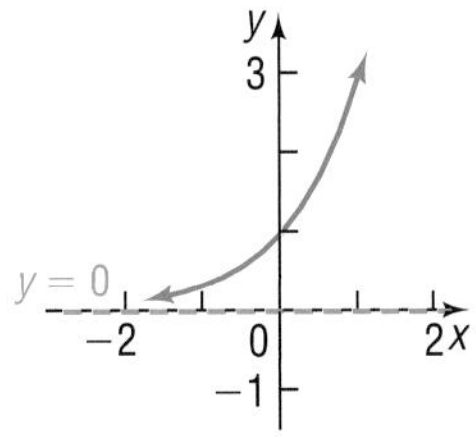

16.
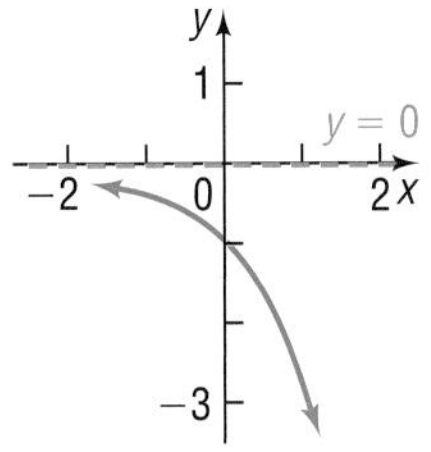

17.
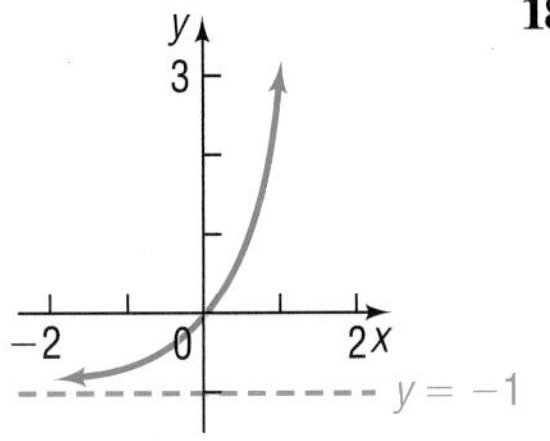

18.
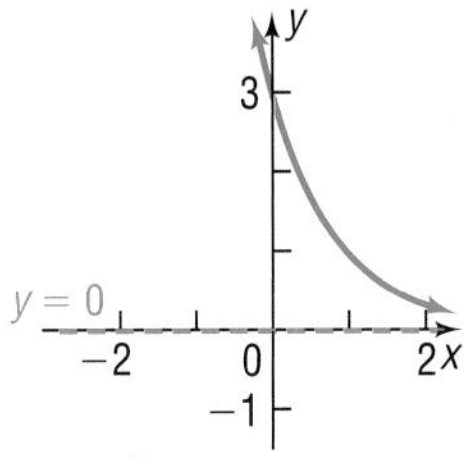

In Problems 19–24, the graph of an exponential function is given. Match each graph to one of the following functions:

A. $y = 4^x$ *B.* $y = 4^{-x}$ *C.* $y = 4^{x-1}$ *D.* $y = 4^x - 1$
E. $y = -4^x$ *F.* $y = 1 - 4^x$

19.
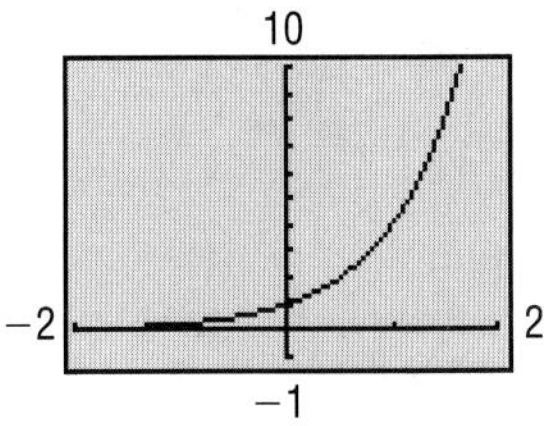

20.
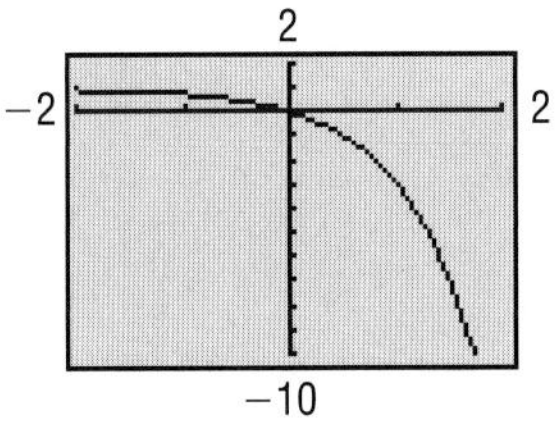

21.
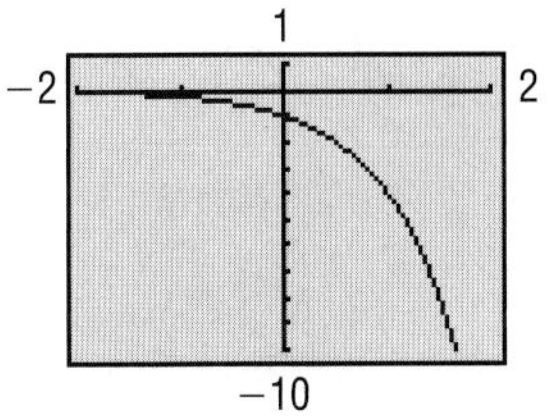

22.
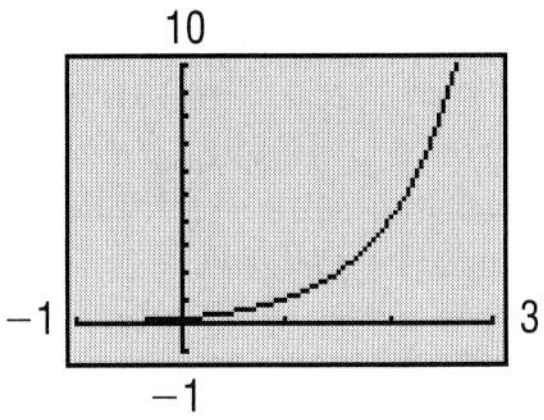

23.
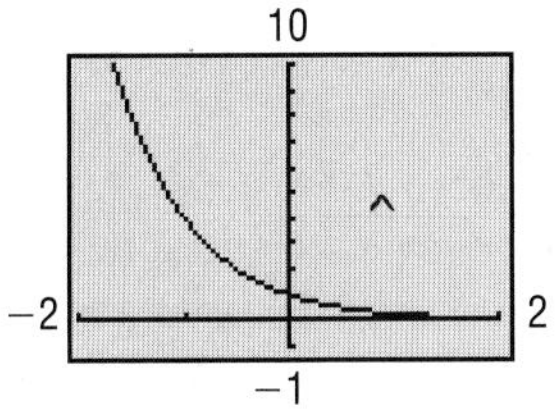

24.
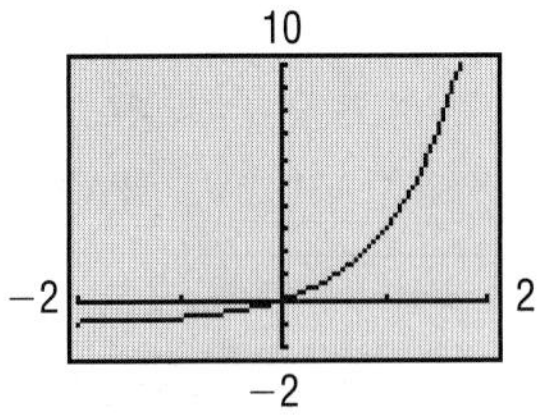

In Problems 25–42, use transformations to graph each function. Determine the domain, range, and horizontal asymptote of each function. Verify your results using a graphing utility.

25. $f(x) = 2^x + 1$ **26.** $f(x) = 2^{x+2}$ 27. $f(x) = 3^{-x} - 2$ **28.** $f(x) = -3^x + 1$
29. $f(x) = 2 + 3(4^x)$ **30.** $f(x) = 1 - 3(2^x)$ **31.** $f(x) = 2 + 3^{x/2}$ **32.** $f(x) = 1 - 2^{-x/3}$

33. $f(x) = 5 - 2(3^{-(x+1)})$ **34.** $f(x) = 1 + 3^{-(x-4)}$ **35.** $f(x) = e^{-x}$ **36.** $f(x) = -e^{x}$
37. $f(x) = e^{x+2}$ **38.** $f(x) = e^{x} - 1$ **39.** $f(x) = 5 - e^{-x}$ **40.** $f(x) = 9 - 3e^{-x}$
41. $f(x) = 2 - e^{-x/2}$ **42.** $f(x) = 7 - 3e^{-2x}$

In Problems 43–56, solve each equation. Verify your solution using a graphing utility.

43. $2^{2x+1} = 4$ **44.** $5^{1-2x} = \frac{1}{5}$ **45.** $3^{x^3} = 9^{x}$ **46.** $4^{x^2} = 2^{x}$
47. $8^{x^2-2x} = \frac{1}{2}$ **48.** $9^{-x} = \frac{1}{3}$ **49.** $2^{x} \cdot 8^{-x} = 4^{x}$ **50.** $\left(\frac{1}{2}\right)^{1-x} = 4$
51. $\left(\frac{1}{5}\right)^{2-x} = 25$ **52.** $4^{x} - 2^{x} = 0$ **53.** $4^{x} = 8$ **54.** $9^{2x} = 27$
55. $e^{x^2} = \left(e^{3x}\right) \cdot \dfrac{1}{e^2}$ **56.** $\left(e^{4}\right)^{x} \cdot e^{x^2} = e^{12}$

57. If $4^{x} = 7$, what does 4^{-2x} equal?
58. If $2^{x} = 3$, what does 4^{-x} equal?
59. If $3^{-x} = 2$, what does 3^{2x} equal?
60. If $5^{-x} = 3$, what does 5^{3x} equal?

61. Optics If a single pane of glass obliterates 3% of the light passing through it, then the percent p of light that passes through n successive panes is given approximately by the function

$$p(n) = 100e^{-0.03n}$$

(a) What percent of light will pass through 10 panes?
(b) What percent of light will pass through 25 panes?

62. Atmospheric Pressure The atmospheric pressure p on a balloon or plane decreases with increasing height. This pressure, measured in millimeters of mercury, is related to the number of kilometers h above sea level by the function

$$p(h) = 760e^{-0.145h}$$

(a) Find the atmospheric pressure at a height of 2 kilometers (over a mile).
(b) What is it at a height of 10 kilometers (over 30,000 feet)?

63. Space Satellites The number of watts w provided by a space satellite's power supply over a period of d days is given by the function

$$w(d) = 50e^{-0.004d}$$

(a) How much power will be available after 30 days?
(b) How much power will be available after 1 year (365 days)?

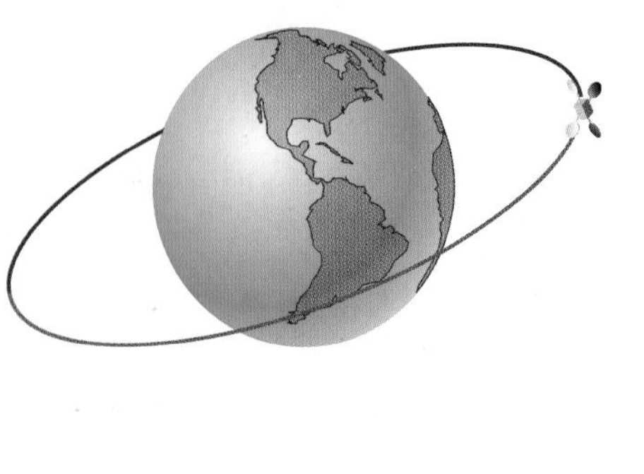

64. Healing of Wounds The normal healing of wounds can be modeled by an exponential function. If A_0 represents the original area of the wound and if A equals the area of the wound after n days, then the function

$$A(n) = A_0e^{-0.35n}$$

describes the area of a wound on the nth day following an injury when no infection is present to retard the healing. Suppose that a wound initially had an area of 100 square millimeters.
(a) If healing is taking place, how large should the area of the wound be after 3 days?
(b) How large should it be after 10 days?

65. Drug Medication The function

$$D(h) = 5e^{-0.4h}$$

can be used to find the number of milligrams D of a certain drug that is in a patient's bloodstream h hours after the drug has been administered. How many milligrams will be present after 1 hour? After 6 hours?

66. Spreading of Rumors A model for the number of people N in a college community who have heard a certain rumor is

$$N = P\left(1 - e^{-0.15d}\right)$$

where P is the total population of the community and d is the number of days that have elapsed since the rumor began. In a community of 1000 students, how many students will have heard the rumor after 3 days?

67. Exponential Probability Between 12:00 PM and 1:00 PM, cars arrive at Citibank's drive-thru at the rate of 6 cars per hour (0.1 car per minute). The following formula from statistics can be used to determine the probability that a car will arrive within t minutes of 12:00 PM.

$$F(t) = 1 - e^{-0.1t}$$

(a) Determine the probability that a car will arrive within 10 minutes of 12:00 PM (that is, before 12:10 PM).

(b) Determine the probability that a car will arrive within 40 minutes of 12:00 PM (before 12:40 PM).
(c) Graph F using your graphing utility.
(d) What value does F approach as t becomes unbounded in the positive direction?

68. Exponential Probability Between 5:00 PM and 6:00 PM, cars arrive at Jiffy Lube at the rate of 9 cars per hour (0.15 car per minute). The following formula from statistics can be used to determine the probability that a car will arrive within t minutes of 5:00 PM.

$$F(t) = 1 - e^{-0.15t}$$

(a) Determine the probability that a car will arrive within 15 minutes of 5:00 PM (that is, before 5:15 PM).
(b) Determine the probability that a car will arrive within 30 minutes of 5:00 PM (before 5:30 PM).
(c) Graph F using your graphing utility.
(d) What value does F approach as t becomes unbounded in the positive direction?

69. Poisson Probability Between 5:00 PM and 6:00 PM cars, arrive at McDonald's drive-thru at the rate of 20 cars per hour. The following formula from statistics can be used to determine the probability that x cars will arrive between 5:00 PM and 6:00 PM.

$$P(x) = \frac{20^x e^{-20}}{x!}$$

where

$$x! = x \cdot (x - 1) \cdot (x - 2) \cdot \cdots \cdot 3 \cdot 2 \cdot 1$$

(a) Determine the probability that $x = 15$ cars will arrive between 5:00 PM and 6:00 PM.
(b) Determine the probability that $x = 20$ cars will arrive between 5:00 PM and 6:00 PM.

70. Poisson Probability People enter a line for the *Demon Roller Coaster* at the rate of 4 per minute. The following formula from statistics can be used to determine the probability that x people will arrive within the next minute.

$$P(x) = \frac{4^x e^{-4}}{x!}$$

where

$$x! = x \cdot (x - 1) \cdot (x - 2) \cdot \cdots \cdot 3 \cdot 2 \cdot 1$$

(a) Determine the probability that $x = 5$ people will arrive within the next minute.
(b) Determine the probability that $x = 8$ people will arrive within the next minute.

71. Relative Humidity The relative humidity is the ratio (expressed as a percent) of the amount of water vapor in the air to the maximum amount that it can hold at a specific temperature. The relative humidity, R, is found using the following formula:

$$R = 10^{\left(\frac{2345}{T} - \frac{2345}{D} + 2\right)}$$

where T is the air temperature (in Kelvins) and D is the dew point temperature (in Kelvins).
[**Note:** Kelvins are found by adding 273 to Celsius degrees.]
(a) Determine the relative humidity if the air temperature is 10° Celsius and the dew point temperature is 5° Celsius.
(b) Determine the relative humidity if the air temperature is 20° Celsius and the dew point temperature is 15° Celsius.
(c) What is the relative humidity if the air temperature and the dew point temperature are the same?

72. Learning Curve Suppose that a student has 500 vocabulary words to learn. If the student learns 15 words after 5 minutes, the function

$$L(t) = 500(1 - e^{-0.0061t})$$

approximates the number of words L that the student will learn after t minutes.
(a) How many words will the student learn after 30 minutes?
(b) How many words will the student learn after 60 minutes?

73. Alternating Current in a *RL* Circuit The equation governing the amount of current I (in amperes) after time t (in seconds) in a single RL circuit consisting of a resistance R (in ohms), an inductance L (in henrys), and an electromotive force E (in volts) is

$$I = \frac{E}{R}\left[1 - e^{-(R/L)t}\right]$$

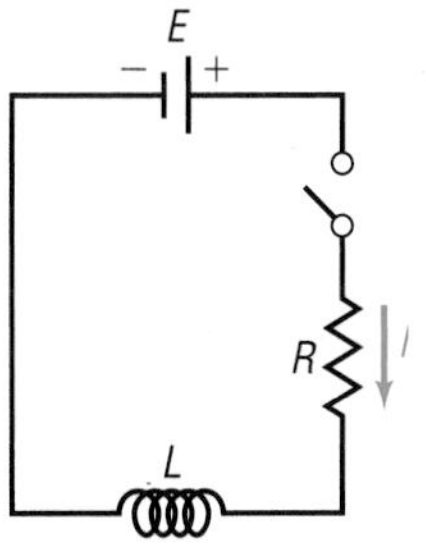

(a) If $E = 120$ volts, $R = 10$ ohms, and $L = 5$ henrys, how much current I_1 is flowing after 0.3 second? After 0.5 second? After 1 second?
(b) What is the maximum current?
(c) Graph this function $I = I_1(t)$, measuring I along the y-axis and t along the x-axis.

(continued on page 304)

(d) If $E = 120$ volts, $R = 5$ ohms, and $L = 10$ henrys, how much current I_2 is flowing after 0.3 second? After 0.5 second? After 1 second?
(e) What is the maximum current?
(f) Graph this function $I = I_2(t)$ on the same viewing window as $I_1(t)$.

74. Alternating Current in a *RC* Circuit The equation governing the amount of current I (in amperes) after time t (in microseconds) in a single RC circuit consisting of a resistance R (in ohms), a capacitance C (in microfarads), and an electromotive force E (in volts) is

$$I = \frac{E}{R}e^{-t/(RC)}$$

(a) If $E = 120$ volts, $R = 2000$ ohms, and $C = 1.0$ microfarad, how much current I_1 is flowing initially ($t = 0$)? After 1000 microseconds? After 3000 microseconds?
(b) What is the maximum current?
(c) Graph this function $I = I_1(t)$, measuring I along the y-axis and t along the x-axis.
(d) If $E = 120$ volts, $R = 1000$ ohms, and $C = 2.0$ microfarads, how much current I_2 is flowing initially? After 1000 microseconds? After 3000 microseconds?
(e) What is the maximum current?
(f) Graph this function $I = I_2(t)$ on the same viewing window as $I_1(t)$.

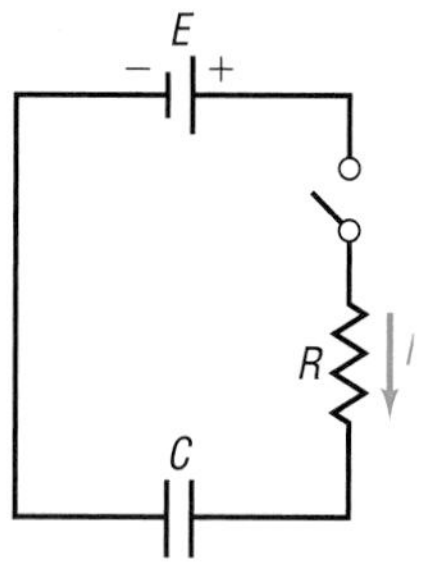

75. Another Formula for *e* Use a calculator to compute the values of

$$2 + \frac{1}{2!} + \frac{1}{3!} + \cdots + \frac{1}{n!}$$

for $n = 4, 6, 8$, and 10. Compare each result with e.
[**Hint:** $1! = 1, 2! = 2 \cdot 1, 3! = 3 \cdot 2 \cdot 1$,
$n! = n(n-1) \cdot \cdots \cdot (3)(2)(1)$]

76. Another Formula for *e* Use a calculator to compute the various values of the expression. Compare the values to e.

$$2 + \cfrac{1}{1 + \cfrac{1}{2 + \cfrac{2}{3 + \cfrac{3}{4 + \cfrac{4}{\text{etc.}}}}}}$$

77. Difference Quotient If $f(x) = a^x$, show that

$$\frac{f(x+h) - f(x)}{h} = a^x\left(\frac{a^h - 1}{h}\right)$$

78. If $f(x) = a^x$, show that $f(A + B) = f(A) \cdot f(B)$.

79. If $f(x) = a^x$, show that $f(-x) = \dfrac{1}{f(x)}$.

80. If $f(x) = a^x$, show that $f(\alpha x) = [f(x)]^\alpha$.

Problems 81 and 82 provide definitions for two other transcendental functions.

81. The **hyperbolic sine function**, designated by $\sinh x$, is defined as

$$\sinh x = \frac{1}{2}\left(e^x - e^{-x}\right)$$

(a) Show that $f(x) = \sinh x$ is an odd function.
(b) Graph $f(x) = \sinh x$ using a graphing utility.

82. The **hyperbolic cosine function**, designated by $\cosh x$, is defined as

$$\cosh x = \frac{1}{2}\left(e^x + e^{-x}\right)$$

(a) Show that $f(x) = \cosh x$ is an even function.
(b) Graph $f(x) = \cosh x$ using a graphing utility.
(c) Refer to Problem 81. Show that, for every x, $(\cosh x)^2 - (\sinh x)^2 = 1$.

83. Historical Problem Pierre de Fermat (1601–1665) conjectured that the function

$$f(x) = 2^{(2^x)} + 1$$

for $x = 1, 2, 3, \ldots$, would always have a value equal to a prime number. But Leonhard Euler (1707–1783) showed that this formula fails for $x = 5$. Use a calculator to determine the prime numbers produced by f for $x = 1, 2, 3, 4$. Then show that $f(5) = 641 \times 6{,}700{,}417$, which is not prime.

84. The bacteria in a 4-liter container double every minute. After 60 minutes the container is full. How long did it take to fill half the container?

85. Explain in your own words what the number e is. Provide at least two applications that require the use of this number.

86. Do you think that there is a power function that increases more rapidly than an exponential function whose base is greater than 1? Explain.

PREPARING FOR THIS SECTION

Before getting started, review the following:

✓ Solving Inequalities (Section 1.5)

✓ Polynomial and Rational Inequalities (Section 4.4)

5.3 LOGARITHMIC FUNCTIONS

OBJECTIVES
1. Change Exponential Expressions to Logarithmic Expressions
2. Change Logarithmic Expressions to Exponential Expressions
3. Evaluate Logarithmic Functions
4. Determine the Domain of a Logarithmic Function
5. Graph Logarithmic Functions
6. Solve Logarithmic Equations

Recall that a one-to-one function $y = f(x)$ has an inverse function that is defined (implicitly) by the equation $x = f(y)$. In particular, the exponential function $y = f(x) = a^x$, $a > 0$, $a \neq 1$, is one-to-one and hence has an inverse function that is defined implicitly by the equation

$$x = a^y, \qquad a > 0, \quad a \neq 1$$

This inverse function is so important that it is given a name, the *logarithmic function.*

The **logarithmic function to the base *a***, where $a > 0$ and $a \neq 1$, is denoted by $y = \log_a x$ (read as "y is the logarithm to the base a of x") and is defined by

$$y = \log_a x \quad \text{if and only if} \quad x = a^y$$

The domain of the logarithmic function $y = \log_a x$ is $x > 0$.

EXAMPLE 1 Relating Logarithms to Exponents

(a) If $y = \log_3 x$, then $x = 3^y$. For example, $2 = \log_3 9$ is equivalent to $9 = 3^2$.

(b) If $y = \log_5 x$, then $x = 5^y$. For example, $-1 = \log_5\left(\frac{1}{5}\right)$ is equivalent to $\frac{1}{5} = 5^{-1}$. ■

EXAMPLE 2 Changing Exponential Expressions to Logarithmic Expressions

1 Change each exponential expression to an equivalent expression involving a logarithm.

(a) $1.2^3 = m$ (b) $e^b = 9$ (c) $a^4 = 24$

Solution We use the fact that $y = \log_a x$ and $x = a^y$, $a > 0$, $a \neq 1$, are equivalent.

(a) If $1.2^3 = m$, then $3 = \log_{1.2} m$. (b) If $e^b = 9$, then $b = \log_e 9$.
(c) If $a^4 = 24$, then $4 = \log_a 24$.

NOW WORK PROBLEM 1.

EXAMPLE 3 Changing Logarithmic Expressions to Exponential Expressions

2 Change each logarithmic expression to an equivalent expression involving an exponent.

(a) $\log_a 4 = 5$ (b) $\log_e b = -3$ (c) $\log_3 5 = c$

Solution (a) If $\log_a 4 = 5$, then $a^5 = 4$. (b) If $\log_e b = -3$, then $e^{-3} = b$.
(c) If $\log_3 5 = c$, then $3^c = 5$.

NOW WORK PROBLEM 13.

3 To find the exact value of a logarithm, we write the logarithm in exponential notation and use the fact that if $a^u = a^v$ then $u = v$.

EXAMPLE 4 Finding the Exact Value of a Logarithmic Expression

Find the exact value of:

(a) $\log_2 16$ (b) $\log_3 \frac{1}{27}$

Solution (a)

$$y = \log_2 16$$
$$2^y = 16 \quad \text{Change to exponential form.}$$
$$2^y = 2^4 \quad 16 = 2^4$$
$$y = 4 \quad \text{Equate exponents.}$$

Therefore, $\log_2 16 = 4$.

(b)

$$y = \log_3 \frac{1}{27}$$
$$3^y = \frac{1}{27} \quad \text{Change to exponential form.}$$
$$3^y = 3^{-3} \quad \frac{1}{27} = \frac{1}{3^3} = 3^{-3}$$
$$y = -3 \quad \text{Equate exponents.}$$

Therefore, $\log_3 \frac{1}{27} = -3$.

NOW WORK PROBLEM 25.

Domain of a Logarithmic Function

4 The logarithmic function $y = \log_a x$ has been defined as the inverse of the exponential function $y = a^x$. That is, if $f(x) = a^x$, then $f^{-1}(x) = \log_a x$. Based on the discussion given in Section 5.1 on inverse functions, we know that, for a function f and its inverse f^{-1},

$$\text{Domain } f^{-1} = \text{Range } f \quad \text{and} \quad \text{Range } f^{-1} = \text{Domain } f$$

Consequently, it follows that

Domain of logarithmic function = Range of exponential function = $(0, \infty)$
Range of logarithmic function = Domain of exponential function = $(-\infty, \infty)$

In the next box, we summarize some properties of the logarithmic function:

$$y = \log_a x \quad (\text{defining equation: } x = a^y)$$
$$\text{Domain: } 0 < x < \infty \qquad \text{Range: } -\infty < y < \infty$$

Notice that the domain of a logarithmic function consists of the *positive* real numbers. The argument of a logarithmic function must be greater than zero.

EXAMPLE 5 Finding the Domain of a Logarithmic Function

Find the domain of each logarithmic function.

(a) $F(x) = \log_2(1 - x)$ (b) $g(x) = \log_5\left(\frac{1 + x}{1 - x}\right)$

(c) $h(x) = \log_{1/2}|x|$

Solution (a) The domain of F consists of all x for which $(1 - x) > 0$, that is, all $x < 1$, or using interval notation, $(-\infty, 1)$.

(b) The domain of g is restricted to

$$\frac{1 + x}{1 - x} > 0$$

Solving this inequality, we find that the domain of g consists of all x between -1 and 1, that is, $-1 < x < 1$, or using interval notation, $(-1, 1)$.

(c) Since $|x| > 0$, provided that $x \neq 0$, the domain of h consists of all nonzero real numbers, or using interval notation, $(-\infty, 0)$ or $(0, \infty)$. ■

NOW WORK PROBLEMS 39 AND 45.

Graphs of Logarithmic Functions

5 Since exponential functions and logarithmic functions are inverses of each other, the graph of a logarithmic function $y = \log_a x$ is the reflection about the line $y = x$ of the graph of the exponential function $y = a^x$, as shown in Figure 26.

Figure 26

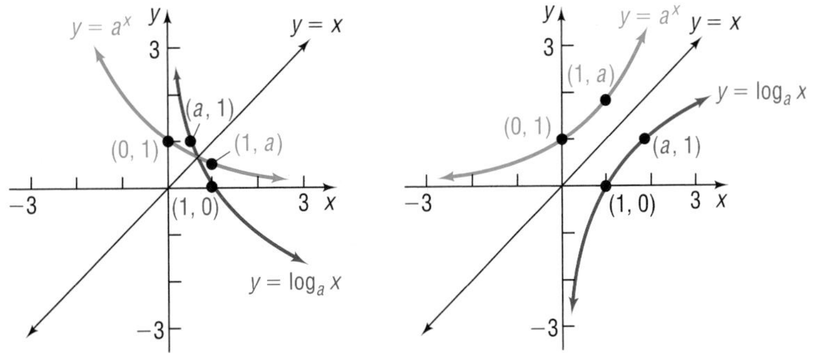

(a) $0 < a < 1$ (b) $a > 1$

Facts about the Graph of a Logarithmic Function $f(x) = \log_a x$

1. The domain is the set of positive real numbers; the range is all real numbers.
2. The x-intercept of the graph is 1. There is no y-intercept.
3. The y-axis ($x = 0$) is a vertical asymptote of the graph.
4. A logarithmic function is decreasing if $0 < a < 1$ and increasing if $a > 1$.
5. The graph of f contains the points $(1, 0)$ and $(a, 1)$.
6. The graph is smooth and continuous, with no corners or gaps.

If the base of a logarithmic function is the number e, then we have the **natural logarithm function**. This function occurs so frequently in applications that it is given a special symbol, **ln** (from the Latin, *logarithmus naturalis*). Thus,

$$y = \ln x \quad \text{if and only if} \quad x = e^y \tag{1}$$

Since $y = \ln x$ and the exponential function $y = e^x$ are inverse functions, we can obtain the graph of $y = \ln x$ by reflecting the graph of $y = e^x$ about the line $y = x$. See Figure 27.

Table 6 displays other points on the graph of $f(x) = \ln x$. Notice for $x < 0$ that we obtain an error message. Do you recall why?

Figure 27

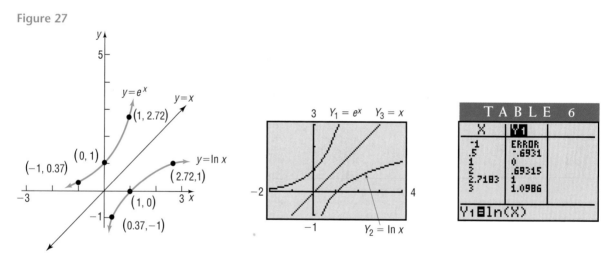

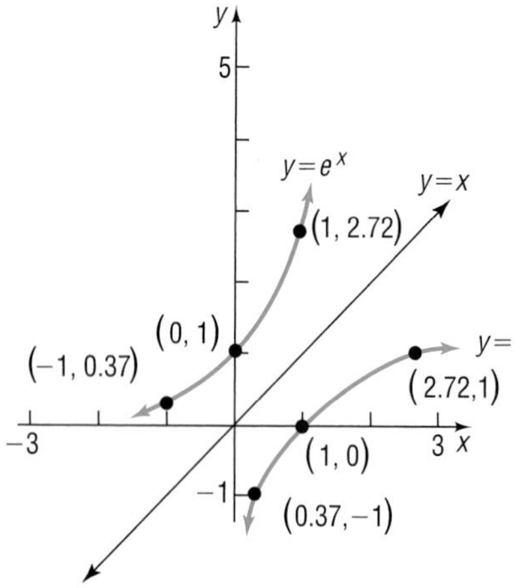

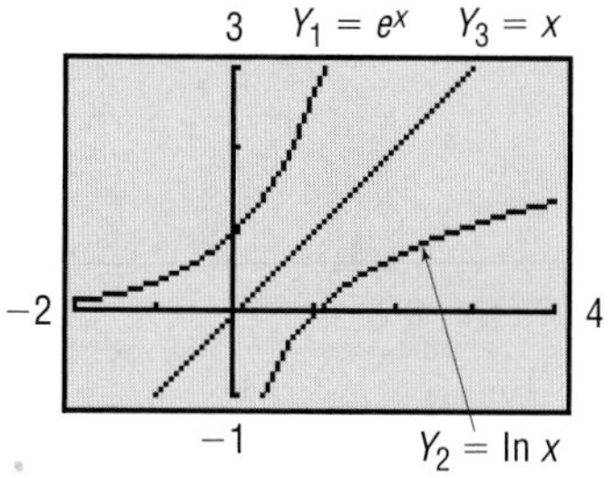

TABLE 6

X	Y1
-1	ERROR
.5	-.6931
1	0
2	.69315
2.7183	1
3	1.0986

Y1=ln(X)

EXAMPLE 6 Graphing Logarithmic Functions Using Transformations

Graph $f(x) = -\ln(x + 2)$ by starting with the graph of $y = \ln x$. Determine the domain, range, and vertical asymptote.

Solution The domain consists of all x for which

$$x + 2 > 0 \quad \text{or} \quad x > -2$$

To obtain the graph of $y = -\ln(x + 2)$, we use the steps illustrated in Figure 28.

Figure 28

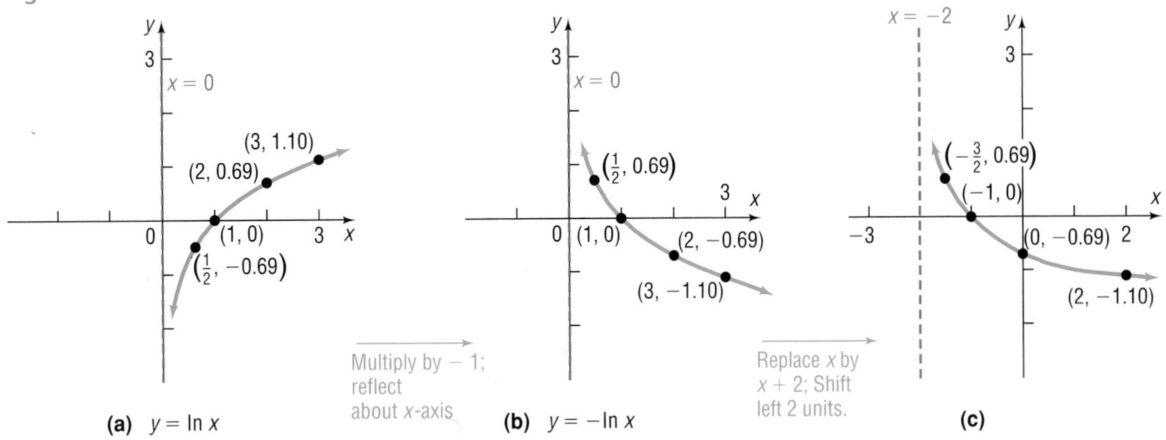

The range of $f(x) = -\ln(x + 2)$ is $(-\infty, \infty)$, and the vertical asymptote is $x = -2$. [Do you see why? The original asymptote ($x = 0$) is shifted to the left 2 units.]

Check: Graph $Y_1 = -\ln(x + 2)$ using a graphing utility to verify Figure 28(c).

EXAMPLE 7 Graphing Logarithmic Functions Using Transformations

Graph $f(x) = \ln(1 - x)$. Determine the domain, range, and vertical asymptote of f.

Solution The domain consists of all x for which

$$1 - x > 0 \quad \text{or} \quad x < 1$$

To obtain the graph of $y = \ln(1 - x)$, we use the steps illustrated in Figure 29.

Figure 29

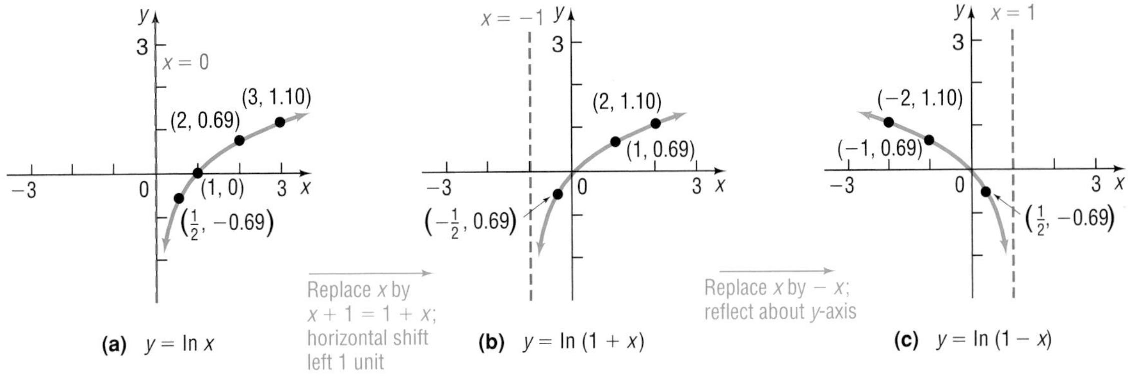

The range of $f(x) = \ln(1 - x)$ is $(-\infty, \infty)$, and the vertical asymptote is $x = 1$.

Check: Graph $Y_1 = \ln(1 - x)$ using a graphing utility to verify Figure 29(c).

NOW WORK PROBLEM 61.

Logarithmic Equations

6 Equations that contain logarithms are called **logarithmic equations.** Care must be taken when solving logarithmic equations algebraically. Be sure to check each apparent solution in the original equation and discard any that are extraneous. In the expression $\log_a M$, remember that a and M are positive and $a \neq 1$.

Some logarithmic equations can be solved by changing from a logarithmic expression to an exponential expression.

EXAMPLE 8 Solving a Logarithmic Equation

Solve: (a) $\log_3(4x - 7) = 2$ (b) $\log_x 64 = 2$

Solution (a) We can obtain an exact solution by changing the logarithm to exponential form.

$$\log_3(4x - 7) = 2$$

$$4x - 7 = 3^2 \quad \text{Change to exponential form.}$$

$$4x - 7 = 9$$

$$4x = 16$$

$$x = 4$$

(b) We can obtain an exact solution by changing the logarithm to exponential form.

$$\log_x 64 = 2$$

$$x^2 = 64 \quad \text{Change to exponential form.}$$

$$x = \sqrt{64} = 8 \quad \text{In } \log_a M,\ a > 0, \text{ so } x > 0.$$

■

EXAMPLE 9 Using Logarithms to Solve Exponential Equations

Solve: $e^{2x} = 5$

Solution We can obtain an exact solution by changing the exponential equation to logarithmic form.

$$e^{2x} = 5$$

$$\ln 5 = 2x \quad \text{Change to a logarithmic expression using (1).}$$

$$x = \frac{\ln 5}{2} \approx 0.805$$

■

NOW WORK PROBLEMS 73 AND 85.

EXAMPLE 10 Alcohol and Driving

The concentration of alcohol in a person's blood is measurable. Recent medical research suggests that the risk R (given as a percent) of having an accident while driving a car can be modeled by the equation

$$R = 6e^{kx}$$

where x is the variable concentration of alcohol in the blood and k is a constant.

(a) Suppose that a concentration of alcohol in the blood of 0.04 results in a 10% risk ($R = 10$) of an accident. Find the constant k in the equation. Graph $R = 6e^{kx}$ using this value of k.

(b) Using this value of k, what is the risk if the concentration is 0.17?

(c) Using the same value of k, what concentration of alcohol corresponds to a risk of 100%?

(d) If the law asserts that anyone with a risk of having an accident of 20% or more should not have driving privileges, at what concentration of alcohol in the blood should a driver be arrested and charged with a DUI (Driving Under the Influence)?

Solution (a) For a concentration of alcohol in the blood of 0.04 and a risk of 10%, we let $x = 0.04$ and $R = 10$ in the equation and solve for k.

$$R = 6e^{kx}$$
$$10 = 6e^{k(0.04)}$$
$$\frac{10}{6} = e^{0.04k} \quad \text{Divide both sides by 6.}$$
$$0.04k = \ln\frac{10}{6} = 0.5108256 \quad \text{Change to a logarithmic expression.}$$
$$k = 12.77 \quad \text{Solve for } k.$$

See Figure 30 for the graph of $R = 6e^{12.77x}$.

Figure 30

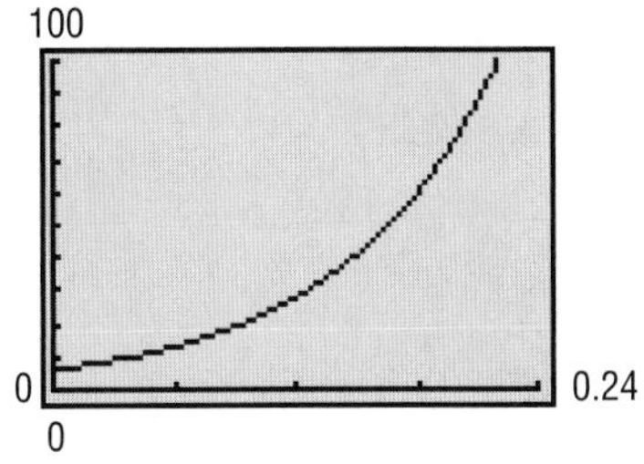

(b) Using $k = 12.77$ and $x = 0.17$ in the equation, we find the risk R to be

$$R = 6e^{kx} = 6e^{(12.77)(0.17)} = 52.6$$

For a concentration of alcohol in the blood of 0.17, the risk of an accident is about 52.6%. Verify this answer using eVALUEate.

(c) Using $k = 12.77$ and $R = 100$ in the equation, we find the concentration x of alcohol in the blood to be

$$R = 6e^{kx}$$
$$100 = 6e^{12.77x}$$
$$\frac{100}{6} = e^{12.77x} \quad \text{Divide both sides by 6.}$$
$$12.77x = \ln\frac{100}{6} = 2.8134 \quad \text{Change to a logarithmic expression.}$$
$$x = 0.22 \quad \text{Solve for } x.$$

For a concentration of alcohol in the blood of 0.22, the risk of an accident is 100%. Verify this answer using a graphing utility.

(d) Using $k = 12.77$ and $R = 20$ in the equation, we find the concentration x of alcohol in the blood to be

$$R = 6e^{kx}$$
$$20 = 6e^{12.77x}$$
$$\frac{20}{6} = e^{12.77x}$$
$$12.77x = \ln\frac{20}{6} = 1.204$$
$$x = 0.094$$

A driver with a concentration of alcohol in the blood of 0.094 or more should be arrested and charged with DUI. Verify using a graphing utility. ■

NOTE: Most states use 0.08 or 0.10 as the blood alcohol content at which a DUI citation is given.

SUMMARY

Properties of the Logarithmic Function

$f(x) = \log_a x, \quad a > 1$ ($y = \log_a x$ means $x = a^y$) — Domain: $(0, \infty)$; Range: $(-\infty, \infty)$; x-intercept: 1; y-intercept: none; vertical asymptote: $x = 0$ (y-axis); increasing; one-to-one

See Figure 26(b) for a typical graph.

$f(x) = \log_a x, \quad 0 < a < 1$ ($y = \log_a x$ means $x = a^y$) — Domain: $(0, \infty)$; Range: $(-\infty, \infty)$; x-intercept: 1; y-intercept: none; vertical asymptote: $x = 0$ (y-axis); decreasing; one-to-one

See Figure 26(a) for a typical graph.

5.3 EXERCISES

In Problems 1–12, change each exponential expression to an equivalent expression involving a logarithm.

1. $9 = 3^2$
2. $16 = 4^2$
3. $a^2 = 1.6$
4. $a^3 = 2.1$
5. $1.1^2 = M$
6. $2.2^3 = N$
7. $2^x = 7.2$
8. $3^x = 4.6$
9. $x^{\sqrt{2}} = \pi$
10. $x^\pi = e$
11. $e^x = 8$
12. $e^{2.2} = M$

In Problems 13–24, change each logarithmic expression to an equivalent expression involving an exponent.

13. $\log_2 8 = 3$
14. $\log_3\left(\frac{1}{9}\right) = -2$
15. $\log_a 3 = 6$
16. $\log_b 4 = 2$
17. $\log_3 2 = x$
18. $\log_2 6 = x$
19. $\log_2 M = 1.3$
20. $\log_3 N = 2.1$
21. $\log_{\sqrt{2}} \pi = x$
22. $\log_\pi x = \frac{1}{2}$
23. $\ln 4 = x$
24. $\ln x = 4$

In Problems 25–36, find the exact value of each logarithm without using a calculator.

25. $\log_2 1$
26. $\log_8 8$
27. $\log_5 25$
28. $\log_3\left(\frac{1}{9}\right)$
29. $\log_{1/2} 16$
30. $\log_{1/3} 9$
31. $\log_{10} \sqrt{10}$
32. $\log_5 \sqrt[3]{25}$
33. $\log_{\sqrt{2}} 4$
34. $\log_{\sqrt{3}} 9$
35. $\ln \sqrt{e}$
36. $\ln e^3$

In Problems 37–46, find the domain of each function.

37. $f(x) = \ln(x - 3)$
38. $g(x) = \ln(x - 1)$
39. $F(x) = \log_2 x^2$
40. $H(x) = \log_5 x^3$
41. $h(x) = \log_{1/2}(x^2 - 2x + 1)$
42. $G(x) = \log_{1/2}(x^2 - 1)$
43. $f(x) = \ln\left(\dfrac{1}{x + 1}\right)$
44. $g(x) = \ln\left(\dfrac{1}{x - 5}\right)$
45. $g(x) = \log_5\left(\dfrac{x + 1}{x}\right)$
46. $h(x) = \log_3\left(\dfrac{x}{x - 1}\right)$

In Problems 47–50, use a calculator to evaluate each expression. Round your answer to three decimal places.

47. $\ln \dfrac{5}{3}$
48. $\dfrac{\ln 5}{3}$
49. $\dfrac{\ln(10/3)}{0.04}$
50. $\dfrac{\ln(2/3)}{-0.1}$
51. Find a so that the graph of $f(x) = \log_a x$ contains the point $(2, 2)$.
52. Find a so that the graph of $f(x) = \log_a x$ contains the point $\left(\frac{1}{2}, -4\right)$.

In Problems 53–60, the graph of a logarithmic function is given. Match each graph to one of the following functions:

A. $y = \log_3 x$	*B.* $y = \log_3(-x)$	*C.* $y = -\log_3 x$	*D.* $y = -\log_3(-x)$
E. $y = \log_3 x - 1$	*F.* $y = \log_3(x - 1)$	*G.* $y = \log_3(1 - x)$	*H.* $y = 1 - \log_3 x$

53.

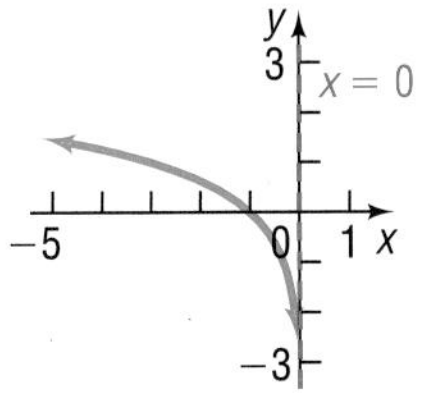

54.

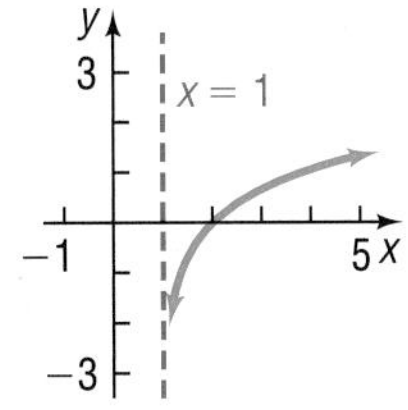

55.

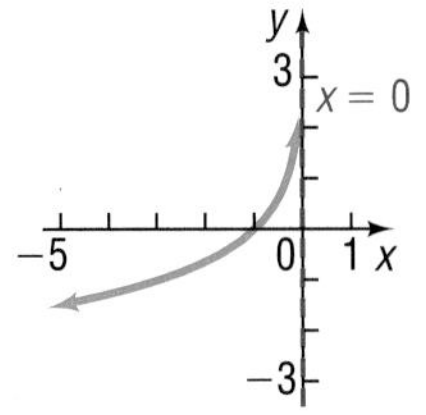

56.

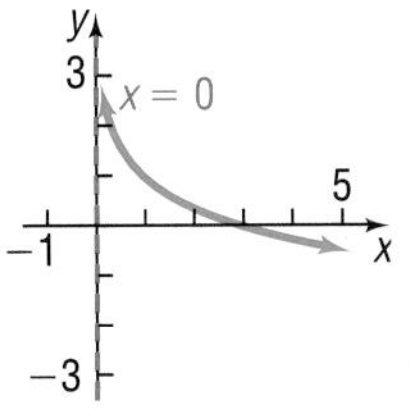

57.

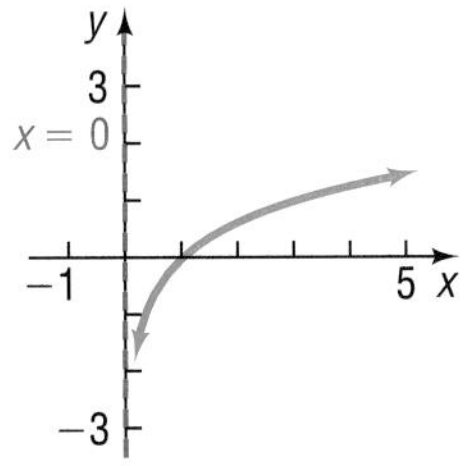

58.

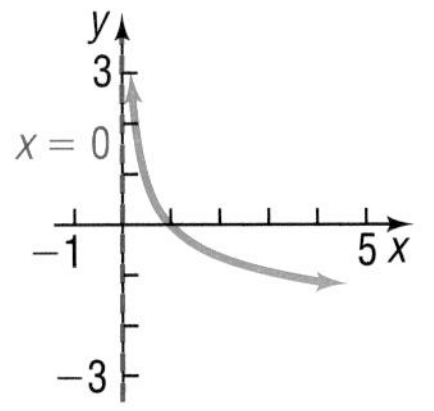

59.

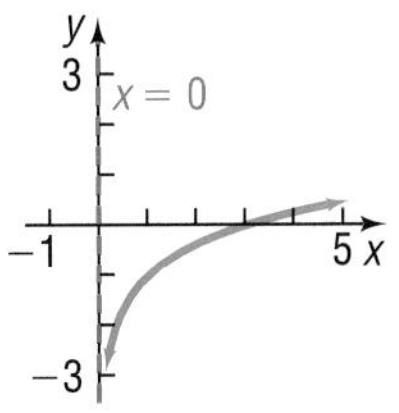

60.

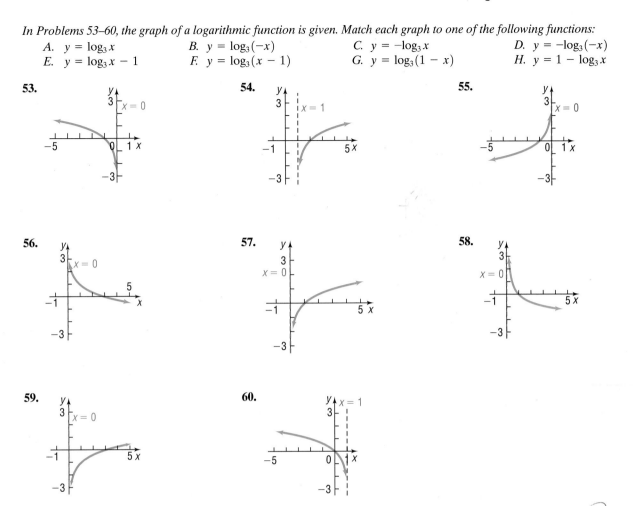

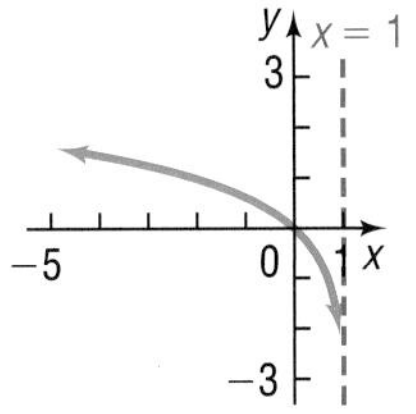

In Problems 61–72, use transformations to graph each function. Determine the domain, range, and vertical asymptote of each function. Verify your results using a graphing utility.

61. $f(x) = \ln(x + 4)$	**62.** $f(x) = \ln(x - 3)$	**63.** $f(x) = \ln(-x)$	**64.** $f(x) = -\ln(-x)$
65. $g(x) = \ln(2x)$	**66.** $h(x) = \ln(\frac{1}{2}x)$	**67.** $f(x) = 3\ln x$	**68.** $f(x) = -2\ln x$
69. $g(x) = \ln(3 - x)$	**70.** $h(x) = \ln(4 - x)$	**71.** $f(x) = -\ln(x - 1)$	**72.** $f(x) = 2 - \ln x$

In Problems 73–88, solve each equation.

73. $\log_3 x = 2$	**74.** $\log_5 x = 3$	**75.** $\log_2(2x + 1) = 3$	**76.** $\log_3(3x - 2) = 2$
77. $\log_x 4 = 2$	**78.** $\log_x(\frac{1}{8}) = 3$	**79.** $\ln e^x = 5$	**80.** $\ln e^{-2x} = 8$
81. $\log_4 64 = x$	**82.** $\log_5 625 = x$	**83.** $\log_3 243 = 2x + 1$	**84.** $\log_6 36 = 5x + 3$
85. $e^{3x} = 10$	**86.** $e^{-2x} = \frac{1}{3}$	**87.** $e^{2x+5} = 8$	**88.** $e^{-2x+1} = 13$
89. $\log_3(x^2 + 1) = 2$	**90.** $\log_5(x^2 + x + 4) = 2$	**91.** $\log_2 8^x = -3$	**92.** $\log_3 3^x = -1$

In Problems 93–98, (a) graph each function and state its domain, range, and asymptote; (b) determine the inverse function; (c) use the graph obtained in (a) to graph the inverse and state its domain, range, and asymptote.

93. $f(x) = 2^x$	**94.** $f(x) = 5^x$	**95.** $f(x) = 2^{x+3}$	**96.** $f(x) = 5^x - 2$
97. $f(x) = 1 + 2^{x-3}$	**98.** $f(x) = 5^{x+1} - 2$		

99. Optics If a single pane of glass obliterates 10% of the light passing through it, then the percent P of light that passes through n successive panes is given approximately by the equation

$$P = 100e^{-0.1n}$$

(a) How many panes are necessary to block at least 50% of the light?
(b) How many panes are necessary to block at least 75% of the light?

100. Chemistry The pH of a chemical solution is given by the formula

$$\text{pH} = -\log_{10}\left[\text{H}^+\right]$$

where $\left[\text{H}^+\right]$ is the concentration of hydrogen ions in moles per liter. Values of pH range from 0 (acidic) to 14 (alkaline).
(a) Find the pH of a 1-liter container of water with 0.0000001 mole of hydrogen ion.
(b) Find the hydrogen ion concentration of a mildly acidic solution with a pH of 4.2.

101. Space Satellites The number of watts w provided by a space satellite's power supply after d days is given by the formula

$$w = 50e^{-0.004d}$$

(a) How long will it take for the available power to drop to 30 watts?
(b) How long will it take for the available power to drop to only 5 watts?

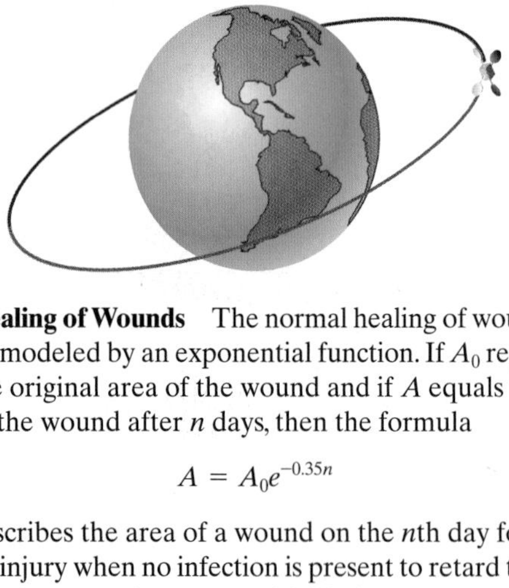

102. Healing of Wounds The normal healing of wounds can be modeled by an exponential function. If A_0 represents the original area of the wound and if A equals the area of the wound after n days, then the formula

$$A = A_0e^{-0.35n}$$

describes the area of a wound on the nth day following an injury when no infection is present to retard the healing. Suppose that a wound initially had an area of 100 square millimeters.
(a) If healing is taking place, how many days should pass before the wound is one-half its original size?
(b) How long before the wound is 10% of its original size?

103. Exponential Probability Between 12:00 PM and 1:00 PM, cars arrive at Citibank's drive-thru at the rate of 6 cars per hour (0.1 car per minute). The following formula from statistics can be used to determine the probability that a car will arrive within t minutes of 12:00 PM.

$$F(t) = 1 - e^{-0.1t}$$

(a) Determine how many minutes are needed for the probability to reach 50%.
(b) Determine how many minutes are needed for the probability to reach 80%.
(c) Is it possible for the probability to equal 100%? Explain.

104. Exponential Probability Between 5:00 PM and 6:00 PM, cars arrive at Jiffy Lube at the rate of 9 cars per hour (0.15 car per minute). The following formula from statistics can be used to determine the probability that a car will arrive within t minutes of 5:00 PM.

$$F(t) = 1 - e^{-0.15t}$$

(a) Determine how many minutes are needed for the probability to reach 50%.
(b) Determine how many minutes are needed for the probability to reach 80%.

105. Drug Medication The formula

$$D = 5e^{-0.4h}$$

can be used to find the number of milligrams D of a certain drug that is in a patient's bloodstream h hours after the drug has been administered. When the number of milligrams reaches 2, the drug is to be administered again. What is the time between injections?

106. Spreading of Rumors A model for the number of people N in a college community who have heard a certain rumor is

$$N = P\left(1 - e^{-0.15d}\right)$$

where P is the total population of the community and d is the number of days that have elapsed since the rumor began. In a community of 1000 students, how many days will elapse before 450 students have heard the rumor?

107. Current in a *RL* Circuit The equation governing the amount of current I (in amperes) after time t (in seconds) in a simple RL circuit consisting of a resistance R (in ohms), an inductance L (in henrys), and an electromotive force E (in volts) is

$$I = \frac{E}{R}\left[1 - e^{-(R/L)t}\right]$$

If $E = 12$ volts, $R = 10$ ohms, and $L = 5$ henrys, how long does it take to obtain a current of 0.5 ampere? Of 1.0 ampere? Graph the equation.

108. Learning Curve Psychologists sometimes use the function

$$L(t) = A(1 - e^{-kt})$$

to measure the amount L learned at time t. The number A represents the amount to be learned, and the number k measures the rate of learning. Suppose that a student has an amount A of 200 vocabulary words to learn. A psychologist determines that the student learned 20 vocabulary words after 5 minutes.
(a) Determine the rate of learning k.
(b) Approximately how many words will the student have learned after 10 minutes?
(c) After 15 minutes?
(d) How long does it take for the student to learn 180 words?

Problems 109–112 use the following discussion: The ***loudness*** $L(x)$*, measured in decibels, of a sound of intensity x, measured in watts per square meter, is defined as* $L(x) = 10 \log \frac{x}{I_0}$*, where* $I_0 = 10^{-12}$ *watt per square meter is the least intense sound that a human ear can detect. Determine the loudness, in decibels, of each of the following sounds.*

109. Loudness of Sound Normal conversation: intensity of $x = 10^{-7}$ watt per square meter.

110. Loudness of Sound Heavy city traffic: intensity of $x = 10^{-3}$ watt per square meter.

111. Loudness of Sound Amplified rock music: intensity of 10^{-1} watt per square meter.

112. Loudness of Sound Diesel truck traveling 40 miles per hour 50 feet away: intensity 10 times that of a passenger car traveling 50 miles per hour 50 feet away whose loudness is 70 decibels.

Problems 113 and 114 use the following discussion: The ***Richter scale*** *is one way of converting seismographic readings into numbers that provide an easy reference for measuring the magnitude M of an earthquake. All earthquakes are compared to a* ***zero-level earthquake*** *whose seismographic reading measures 0.001 millimeter at a distance of 100 kilometers from the epicenter. An earthquake whose seismographic reading measures x millimeters has* ***magnitude*** $M(x)$ *given by*

$$M(x) = \log\left(\frac{x}{x_0}\right)$$

where $x_0 = 10^{-3}$ *is the reading of a zero-level earthquake the same distance from its epicenter. Determine the magnitude of the following earthquakes.*

113. Magnitude of an Earthquake Mexico City in 1985: seismographic reading of 125,892 millimeters 100 kilometers from the center.

114. Magnitude of an Earthquake San Francisco in 1906: seismographic reading of 7943 millimeters 100 kilometers from the center.

115. Alcohol and Driving The concentration of alcohol in a person's blood is measurable. Suppose that the risk R (given as a percent) of having an accident while driving a car can be modeled by the equation

$$R = 3e^{kx}$$

where x is the variable concentration of alcohol in the blood and k is a constant.
(a) Suppose that a concentration of alcohol in the blood of 0.06 results in a 10% risk ($R = 10$) of an accident. Find the constant k in the equation.
(b) Using this value of k, what is the risk if the concentration is 0.17?
(c) Using the same value of k, what concentration of alcohol corresponds to a risk of 100%?
(d) If the law asserts that anyone with a risk of having an accident of 15% or more should not have driving privileges, at what concentration of alcohol in the blood should a driver be arrested and charged with a DUI?
(e) Compare this situation with that of Example 10. If you were a lawmaker, which situation would you support? Give your reasons.

116. Is there any function of the form $y = x^{\alpha}, 0 < \alpha < 1$, that increases more slowly than a logarithmic function whose base is greater than 1? Explain.

5.4 PROPERTIES OF LOGARITHMS

OBJECTIVES
1. Write a Logarithmic Expression as a Sum or Difference of Logarithms
2. Write a Logarithmic Expression as a Single Logarithm
3. Evaluate Logarithms Whose Base Is Neither 10 nor e
4. Graph Logarithmic Functions Whose Base Is Neither 10 nor e

Logarithms have some very useful properties that can be derived directly from the definition and the laws of exponents.

EXAMPLE 1 Establishing Properties of Logarithms

(a) Show that $\log_a 1 = 0$. (b) Show that $\log_a a = 1$.

Solution (a) This fact was established when we graphed $y = \log_a x$ (see Figure 26). Algebraically, for $y = \log_a 1$, we have $a^y = 1 = a^0$, so $y = 0$.

(b) For $y = \log_a a$, we have $a^y = a = a^1$, so $y = 1$. ■

To summarize:

$$\log_a 1 = 0 \qquad \log_a a = 1$$

Theorem

Properties of Logarithms

In the properties given next, M and a are positive real numbers, with $a \neq 1$, and r is any real number.

The number $\log_a M$ is the exponent to which a must be raised to obtain M. That is,

$$a^{\log_a M} = M \qquad \textbf{(1)}$$

The logarithm to the base a of a raised to a power equals that power. That is,

$$\log_a a^r = r \qquad \textbf{(2)}$$

■

Proof of Property (1) Let $x = \log_a M$. Change this logarithmic expression to the equivalent exponential expression.

$$a^x = M$$

But $x = \log_a M$, so

$$a^{\log_a M} = M$$ ■

Proof of Property (2) Let $x = a^r$. Change this exponential expression to the equivalent logarithmic expression

$$\log_a x = r$$

But $x = a^r$, so

$$\log_a a^r = r$$

■

EXAMPLE 2 **Using Properties (1) and (2)**

(a) $2^{\log_2 \pi} = \pi$ (b) $\log_{0.2} 0.2^{-\sqrt{2}} = -\sqrt{2}$ (c) $\ln e^{kt} = kt$ ■

Other useful properties of logarithms are given next.

Theorem **Properties of Logarithms**

In the following properties, M, N, and a are positive real numbers, with $a \neq 1$, and r is any real number.

The Log of a Product Equals the Sum of the Logs

$$\log_a(MN) = \log_a M + \log_a N \quad \textbf{(3)}$$

The Log of a Quotient Equals the Difference of the Logs

$$\log_a\left(\frac{M}{N}\right) = \log_a M - \log_a N \quad \textbf{(4)}$$

$$\log_a\left(\frac{1}{N}\right) = -\log_a N \quad \textbf{(5)}$$

$$\log_a M^r = r\log_a M \quad \textbf{(6)}$$

■

We shall derive properties (3) and (6) and leave the derivations of properties (4) and (5) as exercises (see Problems 81 and 82).

Proof of Property (3) Let $A = \log_a M$ and let $B = \log_a N$. These expressions are equivalent to the exponential expressions

$$a^A = M \quad \text{and} \quad a^B = N$$

Now

$$\begin{aligned} \log_a(MN) = \log_a(a^A a^B) &= \log_a a^{A+B} && \text{Law of exponents} \\ &= A + B && \text{Property (2) of logarithms} \\ &= \log_a M + \log_a N \end{aligned}$$

■

Proof of Property (6) Let $A = \log_a M$. This expression is equivalent to

$$a^A = M$$

Now

$$\begin{aligned} \log_a M^r = \log_a(a^A)^r &= \log_a a^{rA} && \text{Law of exponents} \\ &= rA && \text{Property (2) of logarithms} \\ &= r\log_a M \end{aligned}$$

■

1 Logarithms can be used to transform products into sums, quotients into differences, and powers into factors. Such transformations prove useful in certain types of calculus problems.

EXAMPLE 3 Writing a Logarithmic Expression as a Sum of Logarithms

Write $\log_a(x\sqrt{x^2+1})$ as a sum of logarithms. Express all powers as factors.

Solution

$$\begin{aligned}\log_a(x\sqrt{x^2+1}) &= \log_a x + \log_a \sqrt{x^2+1} && \text{Property (3)}\\ &= \log_a x + \log_a(x^2+1)^{1/2}\\ &= \log_a x + \tfrac{1}{2}\log_a(x^2+1) && \text{Property (6)}\end{aligned}$$

EXAMPLE 4 Writing a Logarithmic Expression as a Difference of Logarithms

Write

$$\ln \frac{x^2}{(x-1)^3}$$

as a difference of logarithms. Express all powers as factors.

Solution

$$\ln \frac{x^2}{(x-1)^3} \underset{\text{Property (4)}}{=} \ln x^2 - \ln(x-1)^3 \underset{\text{Property (6)}}{=} 2\ln x - 3\ln(x-1)$$

NOW WORK PROBLEM 19.

EXAMPLE 5 Writing a Logarithmic Expression as a Sum and Difference of Logarithms

Write

$$\log_a \frac{x^3\sqrt{x^2+1}}{(x+1)^4}$$

as a sum and difference of logarithms. Express all powers as factors.

Solution

$$\begin{aligned}\log_a \frac{x^3\sqrt{x^2+1}}{(x+1)^4} &= \log_a(x^3\sqrt{x^2+1}) - \log_a(x+1)^4 && \text{Property (4)}\\ &= \log_a x^3 + \log_a\sqrt{x^2+1} - \log_a(x+1)^4 && \text{Property (3)}\\ &= \log_a x^3 + \log_a(x^2+1)^{1/2} - \log_a(x+1)^4\\ &= 3\log_a x + \tfrac{1}{2}\log_a(x^2+1) - 4\log_a(x+1) && \text{Property (6)}\end{aligned}$$

2 Another use of properties (3) through (6) is to write sums and/or differences of logarithms with the same base as a single logarithm. This skill will be needed to solve certain logarithmic equations discussed in the next section.

EXAMPLE 6 Writing Expressions as a Single Logarithm

Write each of the following as a single logarithm.

(a) $\log_a 7 + 4\log_a 3$ (b) $\frac{2}{3}\ln 8 - \ln(3^4 - 8)$
(c) $\log_a x + \log_a 9 + \log_a(x^2 + 1) - \log_a 5$

Solution

(a)
$$\begin{aligned}\log_a 7 + 4\log_a 3 &= \log_a 7 + \log_a 3^4 && \text{Property (6)}\\ &= \log_a 7 + \log_a 81\\ &= \log_a(7 \cdot 81) && \text{Property (3)}\\ &= \log_a 567\end{aligned}$$

(b)
$$\begin{aligned}\tfrac{2}{3}\ln 8 - \ln(3^4 - 8) &= \ln 8^{2/3} - \ln(81 - 8) && \text{Property (6)}\\ &= \ln 4 - \ln 73\\ &= \ln\left(\tfrac{4}{73}\right) && \text{Property (4)}\end{aligned}$$

(c)
$$\begin{aligned}\log_a x + \log_a 9 + \log_a(x^2 + 1) - \log_a 5 &= \log_a(9x) + \log_a(x^2 + 1) - \log_a 5\\ &= \log_a[9x(x^2 + 1)] - \log_a 5\\ &= \log_a\left[\frac{9x(x^2 + 1)}{5}\right]\end{aligned}$$

■

WARNING: A common error made by some students is to express the logarithm of a sum as the sum of logarithms.

$\log_a(M + N)$ is not equal to $\log_a M + \log_a N$

Correct statement $\log_a(MN) = \log_a M + \log_a N$ Property (3)

Another common error is to express the difference of logarithms as the quotient of logarithms.

$\log_a M - \log_a N$ is not equal to $\dfrac{\log_a M}{\log_a N}$

Correct statement $\log_a M - \log_a N = \log_a\left(\dfrac{M}{N}\right)$ Property (4)

A third common error is to express a logarithm raised to a power as the product of the power times the logarithm.

$(\log_a M)^r$ is not equal to $r\log_a M$

Correct statement $\log_a M^r = r\log_a M$ Property (6) ■

NOW WORK PROBLEM 29.

There remain two other properties of logarithms that we need to know. They are consequences of the fact that the logarithmic function $y = \log_a x$ is one-to-one.

Theorem In the following properties, M, N, and a are positive real numbers, with $a \neq 1$.

If $M = N$, then $\log_a M = \log_a N$.	**(7)**
If $\log_a M = \log_a N$, then $M = N$.	**(8)**

■

Properties (7) and (8) are useful for solving *logarithmic equations*, a topic discussed in the next section.

Using a Calculator to Evaluate and Graph Logarithms with Bases Other Than *e* or 10

3 Logarithms to the base 10, called **common logarithms**, were used to facilitate arithmetic computations before the widespread use of calculators. (See the Historical Feature at the end of this section.) Natural logarithms, that is, logarithms whose base is the number e, remain very important because they arise frequently in the study of natural phenomena.

Common logarithms are usually abbreviated by writing **log**, with the base understood to be 10, just as natural logarithms are abbreviated by **ln**, with the base understood to be e.

Graphing calculators have both [log] and [ln] keys to calculate the common logarithm and natural logarithm of a number. Let's look at an example to see how to calculate logarithms having a base other than 10 or e.

EXAMPLE 7 Evaluating Logarithms Whose Base Is Neither 10 nor *e*

Evaluate $\log_2 7$.

Solution Let $y = \log_2 7$. Then $2^y = 7$, so

$$2^y = 7$$

$$\ln 2^y = \ln 7 \quad \text{Property (7)}$$

$$y \ln 2 = \ln 7 \quad \text{Property (6)}$$

$$y = \frac{\ln 7}{\ln 2} \quad \text{Solve for } y.$$

$$\approx 2.8074 \quad \text{Use calculator ([ln] key).}$$

Example 7 shows how to evaluate a logarithm whose base is 2 by changing to logarithms involving the base e. In general, we use the **Change-of-Base Formula**.

Theorem **Change-of-Base Formula**

If $a \neq 1, b \neq 1$, and M are positive real numbers, then

$$\log_a M = \frac{\log_b M}{\log_b a} \quad \textbf{(9)}$$

Proof We derive this formula as follows: Let $y = \log_a M$. Then $a^y = M$, so

$$\log_b a^y = \log_b M \quad \text{Property (7)}$$
$$y \log_b a = \log_b M \quad \text{Property (6)}$$
$$y = \frac{\log_b M}{\log_b a} \quad \text{Solve for } y.$$
$$\log_a M = \frac{\log_b M}{\log_b a} \quad y = \log_a M$$

Since calculators have only keys for [log] and [ln], in practice, the Change-of-Base Formula uses either $b = 10$ or $b = e$. Thus,

$$\log_a M = \frac{\log M}{\log a} \quad \text{and} \quad \log_a M = \frac{\ln M}{\ln a} \qquad \textbf{(10)}$$

EXAMPLE 8 Using the Change-of-Base Formula

Calculate:

(a) $\log_5 89$ (b) $\log_{\sqrt{2}} \sqrt{5}$

Solution (a) $\log_5 89 = \dfrac{\log 89}{\log 5} \approx \dfrac{1.94939}{0.69897} = 2.7889$

or

$$\log_5 89 = \frac{\ln 89}{\ln 5} \approx \frac{4.4886}{1.6094} = 2.7889$$

(b) $\log_{\sqrt{2}} \sqrt{5} = \dfrac{\log \sqrt{5}}{\log \sqrt{2}} = \dfrac{\frac{1}{2}\log 5}{\frac{1}{2}\log 2} \approx 2.3219$

or

$$\log_{\sqrt{2}} \sqrt{5} = \frac{\ln \sqrt{5}}{\ln \sqrt{2}} = \frac{\frac{1}{2}\ln 5}{\frac{1}{2}\ln 2} \approx 2.3219$$

NOW WORK PROBLEM 45.

4 We also use the Change-of-Base Formula to graph logarithmic functions whose base is neither 10 nor e.

EXAMPLE 9 Graphing a Logarithmic Function Whose Base Is Neither 10 nor *e*

Use a graphing utility to graph $y = \log_2 x$.

Figure 31

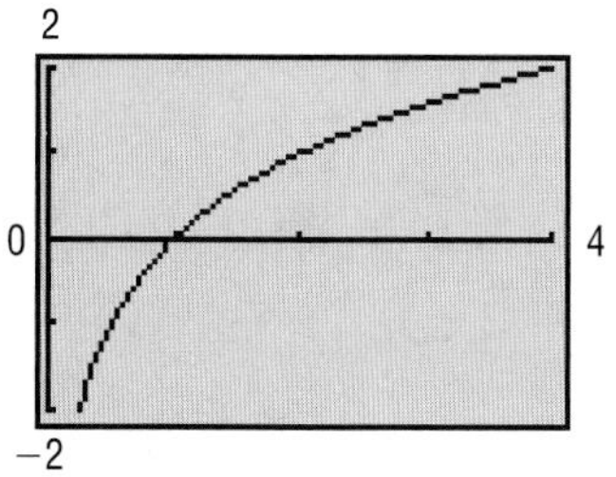

Solution Since graphing utilities only have logarithms with the base 10 or the base e, we need to use the Change-of-Base Formula to express $y = \log_2 x$ in terms of logarithms with base 10 or base e. We can graph either $y = \ln x/\ln 2$ or $y = \log x/\log 2$ to obtain the graph of $y = \log_2 x$. See Figure 31.

Check: Verify that $y = \ln x/\ln 2$ and $y = \log x/\log 2$ result in the same graph by graphing each on the same screen.

NOW WORK PROBLEM 53.

SUMMARY

Properties of Logarithms

In the list that follows, $a > 0$, $a \neq 1$, and $b > 0$, $b \neq 1$; also, $M > 0$ and $N > 0$.

Definition	$y = \log_a x$ means $x = a^y$
Properties of logarithms	$\log_a 1 = 0; \quad \log_a a = 1$
	$a^{\log_a M} = M; \quad \log_a a^r = r$
	$\log_a (MN) = \log_a M + \log_a N$
	$\log_a \left(\dfrac{M}{N}\right) = \log_a M - \log_a N$
	$\log_a \left(\dfrac{1}{N}\right) = -\log_a N$
	$\log_a M^r = r \log_a M$
Change-of-Base Formula	$\log_a M = \dfrac{\log_b M}{\log_b a}$

HISTORICAL FEATURE

John Napier (1550–1617)

Logarithms were invented about 1590 by John Napier (1550–1617) and Joost Bürgi (1552–1632), working independently. Napier, whose work had the greater influence, was a Scottish lord, a secretive man whose neighbors were inclined to believe him to be in league with the devil. His approach to logarithms was very different from ours; it was based on the relationship between arithmetic and geometric sequences (see Chapter 12 on induction and sequences) and not on the inverse function relationship of logarithms to exponential functions (described in Section 5.3). Napier's tables, published in 1614, listed what would now be called *natural logarithms* of sines and were rather difficult to use. A London professor, Henry Briggs, became interested in the tables and visited Napier. In their conversations, they developed the idea of common logarithms, which were published in 1617. Their importance for calculation was immediately recognized, and by 1650 they were being printed as far away as China. They remained an important calculation tool until the advent of the inexpensive handheld calculator about 1972, which has decreased their calculational, but not their theoretical, importance.

A side effect of the invention of logarithms was the popularization of the decimal system of notation for real numbers.

5.4 EXERCISES

In Problems 1–12, suppose that $\ln 2 = a$ and $\ln 3 = b$. Use properties of logarithms to write each logarithm in terms of a and b.

1. $\ln 6$ **2.** $\ln \frac{2}{3}$ **3.** $\ln 1.5$ **4.** $\ln 0.5$

5. $\ln(2e)$ **6.** $\ln\left(\dfrac{3}{e}\right)$ **7.** $\ln 12$ **8.** $\ln 24$

9. $\ln \sqrt[5]{18}$ **10.** $\ln \sqrt[4]{48}$ **11.** $\log_2 3$ **12.** $\log_3 2$

In Problems 13–28, write each expression as a sum and/or difference of logarithms. Express powers as factors.

13. $\log_a(u^2v^3)$ **14.** $\log_2\left(\dfrac{a}{b^2}\right)$ **15.** $\log \dfrac{1}{M^3}$ **16.** $\log(10u^2)$

17. $\log_5 \sqrt{\dfrac{a^3}{b}}$ **18.** $\log_6\left(\dfrac{ab^4}{\sqrt[3]{c^2}}\right)$ **19.** $\ln(x^2\sqrt{1-x})$ **20.** $\ln(x\sqrt{1+x^2})$

21. $\log_2\left(\dfrac{x^3}{x-3}\right)$ **22.** $\log_5\left(\dfrac{\sqrt[3]{x^2+1}}{x^2-1}\right)$ **23.** $\log\left[\dfrac{x(x+2)}{(x+3)^2}\right]$ **24.** $\log\left[\dfrac{x^3\sqrt{x+1}}{(x-2)^2}\right]$

25. $\ln\left[\dfrac{x^2-x-2}{(x+4)^2}\right]^{1/3}$ **26.** $\ln\left[\dfrac{(x-4)^2}{x^2-1}\right]^{2/3}$ **27.** $\ln\dfrac{5x\sqrt{1-3x}}{(x-4)^3}$ **28.** $\ln\left[\dfrac{5x^2\sqrt[3]{1-x}}{4(x+1)^2}\right]$

In Problems 29–38, write each expression as a single logarithm.

29. $3\log_5 u + 4\log_5 v$

30. $\log_3 u^2 - \log_3 v$

31. $\log_{1/2}\sqrt{x} - \log_{1/2} x^3$

32. $\log_2\left(\dfrac{1}{x}\right) + \log_2\left(\dfrac{1}{x^2}\right)$

33. $\ln\left(\dfrac{x}{x-1}\right) + \ln\left(\dfrac{x+1}{x}\right) - \ln(x^2-1)$

34. $\log\left(\dfrac{x^2+2x-3}{x^2-4}\right) - \log\left(\dfrac{x^2+7x+6}{x+2}\right)$

35. $8\log_2\sqrt{3x-2} - \log_2\left(\dfrac{4}{x}\right) + \log_2 4$

36. $21\log_3\sqrt[3]{x} + \log_3(9x^2) - \log_3 25$

37. $2\log_a(5x^3) - \frac{1}{2}\log_a(2x+3)$

38. $\frac{1}{3}\log(x^3+1) + \frac{1}{2}\log(x^2+1)$

39. Write the exponential model $y = ab^x$ as a linear model. [**Hint:** Take the logarithm of both sides.]

40. Write the power model $y = ax^b$ as a linear model.

In Problems 41–44, find the exact value of each expression.

41. $3^{\log_3 5 - \log_3 4}$ **42.** $5^{\log_5 6 + \log_5 7}$ **43.** $e^{\log_{e^2} 16}$ **44.** $e^{\log_{e^2} 9}$

In Problems 45–52, use the Change-of-Base Formula and a calculator to evaluate each logarithm. Round your answer to three decimal places.

45. $\log_3 21$ **46.** $\log_5 18$ **47.** $\log_{1/3} 71$ **48.** $\log_{1/2} 15$

49. $\log_{\sqrt{2}} 7$ **50.** $\log_{\sqrt{5}} 8$ **51.** $\log_\pi e$ **52.** $\log_\pi \sqrt{2}$

In Problems 53–58, graph each function using a graphing utility and the Change-of-Base Formula.

53. $y = \log_4 x$ **54.** $y = \log_5 x$ **55.** $y = \log_2(x+2)$ **56.** $y = \log_4(x-3)$

57. $y = \log_{x-1}(x+1)$ **58.** $y = \log_{x+2}(x-2)$

In Problems 59–68, express y as a function of x. The constant C is a positive number.

59. $\ln y = \ln x + \ln C$

60. $\ln y = \ln(x + C)$

61. $\ln y = \ln x + \ln(x+1) + \ln C$

62. $\ln y = 2\ln x - \ln(x+1) + \ln C$

63. $\ln y = 3x + \ln C$

64. $\ln y = -2x + \ln C$

65. $\ln(y-3) = -4x + \ln C$

66. $\ln(y+4) = 5x + \ln C$

67. $3\ln y = \frac{1}{2}\ln(2x+1) - \frac{1}{3}\ln(x+4) + \ln C$

68. $2\ln y = -\frac{1}{2}\ln x + \frac{1}{3}\ln(x^2+1) + \ln C$

69. Find the value of $\log_2 3 \cdot \log_3 4 \cdot \log_4 5 \cdot \log_5 6 \cdot \log_6 7 \cdot \log_7 8$.

70. Find the value of $\log_2 4 \cdot \log_4 6 \cdot \log_6 8$.

71. Find the value of $\log_2 3 \cdot \log_3 4 \cdot \ldots \cdot \log_n(n+1) \cdot \log_{n+1} 2$.

72. Find the value of $\log_2 2 \cdot \log_2 4 \cdot \ldots \cdot \log_2 2^n$.

73. Show that $\log_a(x + \sqrt{x^2-1}) + \log_a(x - \sqrt{x^2-1}) = 0$.

74. Show that $\log_a(\sqrt{x} + \sqrt{x-1}) + \log_a(\sqrt{x} - \sqrt{x-1}) = 0$.

75. Show that $\ln(1 + e^{2x}) = 2x + \ln(1 + e^{-2x})$.

76. If $f(x) = \log_a x$, show that $\dfrac{f(x+h) - f(x)}{h} = \log_a\left(1 + \dfrac{h}{x}\right)^{1/h}, h \neq 0$.

77. If $f(x) = \log_a x$, show that $-f(x) = \log_{1/a} x$.

78. If $f(x) = \log_a x$, show that $f(AB) = f(A) + f(B)$.

79. If $f(x) = \log_a x$, show that $f(1/x) = -f(x)$

80. If $f(x) = \log_a x$, show that $f(x^\alpha) = \alpha f(x)$.

81. Show that $\log_a(M/N) = \log_a M - \log_a N$, where a, M, and N are positive real numbers, with $a \neq 1$.

82. Show that $\log_a(1/N) = -\log_a N$, where a and N are positive real numbers, with $a \neq 1$.

83. Find the domain of $f(x) = \log_a x^2$ and the domain of $g(x) = 2\log_a x$. Since $\log_a x^2 = 2\log_a x$, how do you reconcile the fact that the domains are not equal? Write a brief explanation.

5.5 LOGARITHMIC AND EXPONENTIAL EQUATIONS

OBJECTIVES
1. Solve Logarithmic Equations Using the Properties of Logarithms
2. Solve Exponential Equations
3. Solve Logarithmic and Exponential Equations Using a Graphing Utility

Logarithmic Equations

1 In Section 5.3 we solved logarithmic equations by changing a logarithm to exponential form. Often, however, some manipulation of the equation (usually using the properties of logarithms) is required before we can change to exponential form.

Our practice will be to solve equations, whenever possible, by finding exact solutions using algebraic methods. In such cases, we will also verify the solution obtained by using a graphing utility. When algebraic methods cannot be used, approximate solutions will be obtained using a graphing utility. The reader is encouraged to pay particular attention to the form of equations for which exact solutions are possible.

EXAMPLE 1 Solving a Logarithmic Equation

Solve: $2\log_5 x = \log_5 9$

Solution Because each logarithm is to the same base, 5, we can obtain an exact solution as follows:

$$2\log_5 x = \log_5 9$$

$$\log_5 x^2 = \log_5 9 \quad \text{Property (6), Section 5.4}$$

$$x^2 = 9 \quad \text{Property (8), Section 5.4}$$

$$x = 3 \quad \text{or} \quad \cancel{x = -3}$$

Recall that logarithms of negative numbers are not defined, so, in the expression $2\log_5 x$, x must be positive. Therefore, -3 is extraneous and we discard it.

Figure 32

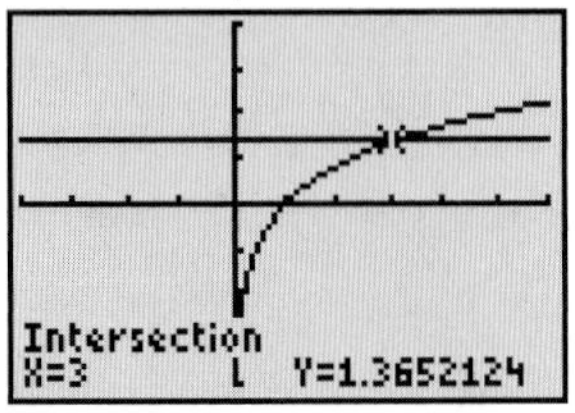

The equation has only one solution, 3.

Check: Verify that 3 is the only solution using a graphing utility. Graph $Y_1 = 2\log_5 x = \dfrac{2\log x}{\log 5}$, and $Y_2 = \log_5 9 = \dfrac{\log 9}{\log 5}$, and determine the point of intersection. See Figure 32.

NOW WORK PROBLEM 5.

EXAMPLE 2 Solving a Logarithmic Equation

Solve: $\log_4(x + 3) + \log_4(2 - x) = 1$

Solution To obtain an exact solution, we need to express the left side as a single logarithm. Then we will change the expression to exponential form.

$$\log_4(x+3) + \log_4(2-x) = 1$$

$$\log_4[(x+3)(2-x)] = 1 \quad \text{Property (3), Section 5.4}$$

$$(x+3)(2-x) = 4^1 = 4 \quad \text{Change to an exponential expression.}$$

$$-x^2 - x + 6 = 4 \quad \text{Simplify.}$$

$$x^2 + x - 2 = 0 \quad \text{Place the quadratic equation in standard form.}$$

$$(x+2)(x-1) = 0 \quad \text{Factor.}$$

$$x = -2 \quad \text{or} \quad x = 1 \quad \text{Zero-Product Property}$$

Since the arguments of each logarithmic expression in the equation are positive for both $x = -2$ and $x = 1$, neither is extraneous. You should verify that both of these are solutions using a graphing utility. ■

NOW WORK PROBLEM 9.

Exponential Equations

2 In Sections 5.2 and 5.3 we solved certain exponential equations by expressing each side of the equation with the same base. However, many exponential equations cannot be rewritten so each side has the same base. In such cases, sometimes properties of logarithms along with algebraic techniques can be used to obtain a solution.

EXAMPLE 3 Solving an Exponential Equation

Solve: $4^x - 2^x - 12 = 0$

Solution We note that $4^x = (2^2)^x = 2^{2x} = (2^x)^2$, so the equation is actually quadratic in form, and we can rewrite it as

$$(2^x)^2 - 2^x - 12 = 0 \quad \text{Let } u = 2^x\text{; then } u^2 - u - 12 = 0.$$

Now we can factor as usual.

$$(2^x - 4)(2^x + 3) = 0 \quad (u-4)(u+3) = 0$$

$$2^x - 4 = 0 \quad \text{or} \quad 2^x + 3 = 0 \quad u - 4 = 0 \text{ or } u + 3 = 0$$

$$2^x = 4 \qquad 2^x = -3 \quad u = 2^x = 4 \qquad u = 2^x = -3$$

The equation on the left has the solution $x = 2$, since $2^x = 4 = 2^2$; the equation on the right has no solution, since $2^x > 0$ for all x. ■

In the preceding example, we were able to write the exponential expression using the same base after utilizing some algebra, obtaining an exact solution to the equation. When this is not possible, logarithms can sometimes obtain the solution.

EXAMPLE 4 Solving an Exponential Equation

Solve: $2^x = 5$

Solution We write the exponential equation as the equivalent logarithmic equation.

$$2^x = 5$$

$$x = \log_2 5 = \frac{\ln 5}{\ln 2}$$

↑ Change-of-Base Formula (10), Section 5.4

Alternatively, we can solve the equation $2^x = 5$ by taking the natural logarithm (or common logarithm) of each side. Taking the natural logarithm,

$$2^x = 5$$

$$\ln 2^x = \ln 5$$

$$x \ln 2 = \ln 5 \qquad \text{Property (6), Section 5.4}$$

$$x = \frac{\ln 5}{\ln 2}$$

Using a calculator, the solution, rounded to three decimal places, is

$$x = \frac{\ln 5}{\ln 2} \approx 2.322$$

■

NOW WORK PROBLEM 17.

EXAMPLE 5 Solving an Exponential Equation

Solve: $8 \cdot 3^x = 5$

Solution

$$8 \cdot 3^x = 5$$

$$3^x = \tfrac{5}{8} \qquad \text{Solve for } 3^x.$$

$$x = \log_3\left(\tfrac{5}{8}\right) = \frac{\ln \frac{5}{8}}{\ln 3} \qquad \text{Solve for } x.$$

The solution, rounded to three decimal places, is

$$x = \frac{\ln\left(\frac{5}{8}\right)}{\ln 3} \approx -0.428$$

■

EXAMPLE 6 Solving an Exponential Equation

Solve: $5^{x-2} = 3^{3x+2}$

Solution Because the bases are different, we take the natural logarithm of each side and apply appropriate properties of logarithms. The result is an equation in x that we can solve.

$$5^{x-2} = 3^{3x+2}$$

$$\ln 5^{x-2} = \ln 3^{3x+2} \quad \text{Property (7)}$$

$$(x-2)\ln 5 = (3x+2)\ln 3 \quad \text{Property (6)}$$

$$(\ln 5)x - 2\ln 5 = (3\ln 3)x + 2\ln 3 \quad \text{Distribute.}$$

$$(\ln 5)x - (3\ln 3)x = 2\ln 3 + 2\ln 5 \quad \text{Place terms involving } x \text{ on the left.}$$

$$(\ln 5 - 3\ln 3)x = 2\ln 3 + 2\ln 5 \quad \text{Factor.}$$

$$x = \frac{2(\ln 3 + \ln 5)}{\ln 5 - 3\ln 3} \approx -3.212$$

NOW WORK PROBLEM 25.

Graphing Utility Solutions

3 The techniques introduced in this section apply only to certain types of logarithmic and exponential equations. Solutions for other types are usually studied in calculus, using numerical methods. However, we can use a graphing utility to approximate the solution.

EXAMPLE 7 Solving Equations Using a Graphing Utility

Solve: $\log_3 x + \log_4 x = 4$

Express the solution(s) rounded to two decimal places.

Figure 33

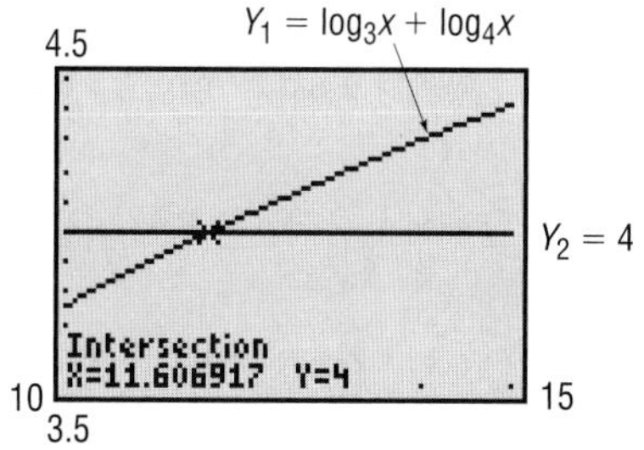

Solution The solution is found by graphing

$$Y_1 = \log_3 x + \log_4 x = \frac{\log x}{\log 3} + \frac{\log x}{\log 4} \quad \text{and} \quad Y_2 = 4$$

(Remember that you must use the Change-of-Base Formula to graph Y_1.) Y_1 is an increasing function (do you know why?), and so there is only one point of intersection for Y_1 and Y_2. Figure 33 shows the graphs of Y_1 and Y_2. Using the INTERSECT command, the solution is 11.61, rounded to two decimal places.

Can you discover an algebraic solution to Example 7?
[**Hint:** Factor $\log x$ from Y_1.]

EXAMPLE 8 Solving Equations Using a Graphing Utility

Solve: $x + e^x = 2$

Express the solution(s) rounded to two decimal places.

Figure 34

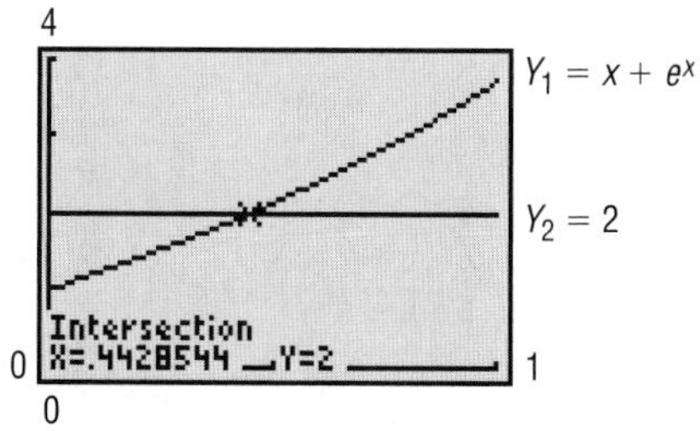

Solution The solution is found by graphing $Y_1 = x + e^x$ and $Y_2 = 2$. Y_1 is an increasing function (do you know why?), and so there is only one point of intersection for Y_1 and Y_2. Figure 34 shows the graphs of Y_1 and Y_2. Using the INTERSECT command, the solution is 0.44 rounded to two decimal places.

5.5 EXERCISES

In Problems 1–44, solve each equation. Verify your solution using a graphing utility.

1. $\log_4(x+2) = \log_4 8$

2. $\log_5(2x+3) = \log_5 3$

3. $\frac{1}{2}\log_3 x = 2\log_3 2$

4. $-2\log_4 x = \log_4 9$

5. $2\log_5 x = 3\log_5 4$

6. $3\log_2 x = -\log_2 27$

7. $3\log_2(x-1) + \log_2 4 = 5$

8. $2\log_3(x+4) - \log_3 9 = 2$

9. $\log x + \log(x+15) = 2$

10. $\log_4 x + \log_4(x-3) = 1$

11. $\ln x + \ln(x+2) = 4$

12. $\ln(x+1) - \ln x = 2$

13. $2^{2x} + 2^x - 12 = 0$

14. $3^{2x} + 3^x - 2 = 0$

15. $3^{2x} + 3^{x+1} - 4 = 0$

16. $2^{2x} + 2^{x+2} - 12 = 0$

17. $2^x = 10$

18. $3^x = 14$

19. $8^{-x} = 1.2$

20. $2^{-x} = 1.5$

21. $3^{1-2x} = 4^x$

22. $2^{x+1} = 5^{1-2x}$

23. $\left(\frac{3}{5}\right)^x = 7^{1-x}$

24. $\left(\frac{4}{3}\right)^{1-x} = 5^x$

25. $1.2^x = (0.5)^{-x}$

26. $(0.3)^{1+x} = 1.7^{2x-1}$

27. $\pi^{1-x} = e^x$

28. $e^{x+3} = \pi^x$

29. $5(2^{3x}) = 8$

30. $0.3(4^{0.2x}) = 0.2$

31. $\log_a(x-1) - \log_a(x+6) = \log_a(x-2) - \log_a(x+3)$

32. $\log_a x + \log_a(x-2) = \log_a(x+4)$

33. $\log_{1/3}(x^2+x) - \log_{1/3}(x^2-x) = -1$

34. $\log_4(x^2-9) - \log_4(x+3) = 3$

35. $\log_2(x+1) - \log_4 x = 1$
[**Hint:** Change $\log_4 x$ to base 2.]

36. $\log_2(3x+2) - \log_4 x = 3$

37. $\log_{16} x + \log_4 x + \log_2 x = 7$

38. $\log_9 x + 3\log_3 x = 14$

39. $(\sqrt[3]{2})^{2-x} = 2^{x^2}$

40. $\log_2 x^{\log_2 x} = 4$

41. $\dfrac{e^x + e^{-x}}{2} = 1$
[**Hint:** Multiply each side by e^x.]

42. $\dfrac{e^x + e^{-x}}{2} = 3$

43. $\dfrac{e^x - e^{-x}}{2} = 2$

44. $\dfrac{e^x - e^{-x}}{2} = -2$

In Problems 45–60, use a graphing utility to solve each equation. Express your answer rounded to two decimal places.

45. $\log_5 x + \log_3 x = 1$

46. $\log_2 x + \log_6 x = 3$

47. $\log_5(x+1) - \log_4(x-2) = 1$

48. $\log_2(x-1) - \log_6(x+2) = 2$

49. $e^x = -x$

50. $e^{2x} = x + 2$

51. $e^x = x^2$

52. $e^x = x^3$

53. $\ln x = -x$

54. $\ln(2x) = -x + 2$

55. $\ln x = x^3 - 1$

56. $\ln x = -x^2$

57. $e^x + \ln x = 4$

58. $e^x - \ln x = 4$

59. $e^{-x} = \ln x$

60. $e^{-x} = -\ln x$

PREPARING FOR THIS SECTION

Before getting started, review the following:

✓ Simple Interest (pp. 38–39)

5.6 COMPOUND INTEREST

OBJECTIVES
1. Determine the Future Value of a Lump Sum of Money
2. Calculate Effective Rates of Return
3. Determine the Present Value of a Lump Sum of Money
4. Determine the Time Required to Double or Triple a Lump Sum of Money

1 Interest is money paid for the use of money. The total amount borrowed (whether by an individual from a bank in the form of a loan or by a bank from

an individual in the form of a savings account) is called the **principal**. The **rate of interest**, expressed as a percent, is the amount charged for the use of the principal for a given period of time, usually on a yearly (that is, per annum) basis.

Simple Interest Formula

If a principal of P dollars is borrowed for a period of t years at a per annum interest rate r, expressed as a decimal, the interest I charged is

$$I = Prt \tag{1}$$

Interest charged according to formula (1) is called **simple interest**.

In working with problems involving interest, we use the term **payment period** as follows:

Annually	Once per year	Monthly	12 times per year
Semiannually	Twice per year	Daily	365 times per year*
Quarterly	Four times per year		

When the interest due at the end of a payment period is added to the principal so that the interest computed at the end of the next payment period is based on this new principal amount (old principal + interest), the interest is said to have been **compounded**. **Compound interest** is interest paid on previously earned interest.

EXAMPLE 1 Computing Compound Interest

A credit union pays interest of 8% per annum compounded quarterly on a certain savings plan. If \$1000 is deposited in such a plan and the interest is left to accumulate, how much is in the account after 1 year?

Solution We use the simple interest formula, $I = Prt$. The principal P is \$1000 and the rate of interest is $8\% = 0.08$. After the first quarter of a year, the time t is $\frac{1}{4}$ year, so the interest earned is

$$I = Prt = (\$1000)(0.08)(\tfrac{1}{4}) = \$20$$

The new principal is $P + I = \$1000 + \$20 = \$1020$. At the end of the second quarter, the interest on this principal is

$$I = (\$1020)(0.08)(\tfrac{1}{4}) = \$20.40$$

At the end of the third quarter, the interest on the new principal of $\$1020 + \$20.40 = \$1040.40$ is

$$I = (\$1040.40)(0.08)(\tfrac{1}{4}) = \$20.81$$

Finally, after the fourth quarter, the interest is

$$I = (\$1061.21)(0.08)(\tfrac{1}{4}) = \$21.22$$

Thus, after 1 year the account contains \$1082.43. ■

* Most banks use a 360-day "year." Why do you think they do?

The pattern of the calculations performed in Example 1 leads to a general formula for compound interest. To fix our ideas, let P represent the principal to be invested at a per annum interest rate r that is compounded n times per year. (For computing purposes, r is expressed as a decimal.) The interest earned after each compounding period is the principal times r/n. The amount A after one compounding period is

$$A = P + P\left(\frac{r}{n}\right) = P\left(1 + \frac{r}{n}\right)$$

After two compounding periods, the amount A, based on the new principal $P(1 + r/n)$, is

$$A = \underbrace{P\left(1 + \frac{r}{n}\right)}_{\text{New principal.}} + \underbrace{P\left(1 + \frac{r}{n}\right)\left(\frac{r}{n}\right)}_{\text{Interest on new principal.}} = P\left(1 + \frac{r}{n}\right)\left(1 + \frac{r}{n}\right) = P\left(1 + \frac{r}{n}\right)^2$$

After three compounding periods,

$$A = P\left(1 + \frac{r}{n}\right)^2 + P\left(1 + \frac{r}{n}\right)^2\left(\frac{r}{n}\right) = P\left(1 + \frac{r}{n}\right)^2\left(1 + \frac{r}{n}\right) = P\left(1 + \frac{r}{n}\right)^3$$

Continuing this way, after n compounding periods (1 year),

$$A = P\left(1 + \frac{r}{n}\right)^n$$

Because t years will contain $n \cdot t$ compounding periods, after t years we have

$$A = P\left(1 + \frac{r}{n}\right)^{nt}$$

Theorem

Compound Interest Formula

The amount A after t years due to a principal P invested at an annual interest rate r compounded n times per year is

$$A = P\left(1 + \frac{r}{n}\right)^{nt} \qquad \textbf{(2)}$$

The amount A is typically referred to as the **future value** of the account, while P is called the **present value**.

EXPLORATION To see the effects of compounding interest monthly on an initial deposit of \$1, graph $Y_1 = \left(1 + \frac{r}{12}\right)^{12x}$ with $r = 0.06$ and $r = 0.12$ for $0 \le x \le 30$. What is the future value of \$1 in 30 years when the interest rate per annum is $r = 0.06$ (6%)? What is the future value of \$1 in 30 years when the interest rate per annum is $r = 0.12$ (12%)? Does doubling the interest rate double the future value?

NOW WORK PROBLEM 1.

EXAMPLE 2 Comparing Investments Using Different Compounding Periods

Investing \$1000 at an annual rate of 10% compounded annually, quarterly, monthly, and daily will yield the following amounts after 1 year:

Annual compounding: $$A = P(1 + r) = (\$1000)(1 + 0.10) = \$1100.00$$

Quarterly compounding: $$A = P\left(1 + \frac{r}{4}\right)^4 = (\$1000)(1 + 0.025)^4 = \$1103.81$$

Monthly compounding: $$A = P\left(1 + \frac{r}{12}\right)^{12} = (\$1000)(1 + 0.00833)^{12} = \$1104.71$$

Daily compounding: $$A = P\left(1 + \frac{r}{365}\right)^{365} = (\$1000)(1 + 0.000274)^{365} = \$1105.16$$ ■

From Example 2 we can see that the effect of compounding more frequently is that the amount after 1 year is higher: \$1000 compounded 4 times a year at 10% results in \$1103.81; \$1000 compounded 12 times a year at 10% results in \$1104.71; and \$1000 compounded 365 times a year at 10% results in \$1105.16. This leads to the following question: What would happen to the amount after 1 year if the number of times that the interest is compounded were increased without bound?

Let's find the answer. Suppose that P is the principal, r is the per annum interest rate, and n is the number of times that the interest is compounded each year. The amount after 1 year is

$$A = P\left(1 + \frac{r}{n}\right)^n$$

Now suppose that the number n of times that the interest is compounded per year gets larger and larger; that is, suppose that $n \to \infty$. Then

$$A = P\left(1 + \frac{r}{n}\right)^n = P\left[1 + \frac{1}{n/r}\right]^n = P\left(\left[1 + \frac{1}{n/r}\right]^{n/r}\right)^r \underset{h = \frac{n}{r}}{=} P\left[\left(1 + \frac{1}{h}\right)^h\right]^r \quad \textbf{(3)}$$

In (3), as $n \to \infty$, then $h = n/r \to \infty$, and the expression in brackets equals e. [Refer to (2) on p. 297]. Thus, $A \to Pe^r$. Table 7 compares $(1 + r/n)^n$, for large values of n, to e^r for $r = 0.05$, $r = 0.10$, $r = 0.15$, and $r = 1$. The larger that n gets, the closer $(1 + r/n)^n$ gets to e^r. Thus, no matter how frequent the compounding, the amount after 1 year has the definite ceiling Pe^r.

When interest is compounded so that the amount after 1 year is Pe^r, we say the interest is **compounded continuously**.

TABLE 7

	$\left(1+\frac{r}{n}\right)^n$			
	$n = 100$	$n = 1000$	$n = 10{,}000$	e^r
$r = 0.05$	1.0512579	1.05127	1.051271	1.0512711
$r = 0.10$	1.1051157	1.1051654	1.1051703	1.1051709
$r = 0.15$	1.1617037	1.1618212	1.1618329	1.1618342
$r = 1$	2.7048138	2.7169239	2.7181459	2.7182818

Theorem **Continuous Compounding**

The amount A after t years due to a principal P invested at an annual interest rate r compounded continuously is

$$A = Pe^{rt} \tag{4}$$

EXAMPLE 3 Using Continuous Compounding

The amount A that results from investing a principal P of \$1000 at an annual rate r of 10% compounded continuously for a time t of 1 year is

$$A = \$1000e^{0.10} = (\$1000)(1.10517) = \$1105.17$$

NOW WORK PROBLEM 9.

2 The **effective rate of interest** is the equivalent annual simple rate of interest that would yield the same amount as compounding after 1 year. For example, based on Example 3, a principal of \$1000 will result in \$1105.17 at a rate of 10% compounded continuously. To get this same amount using a simple rate of interest would require that interest of \$1105.17 − \$1000.00 = \$105.17 be earned on the principal. Since \$105.17 is 10.517% of \$1000, a simple rate of interest of 10.517% is needed to equal 10% compounded continuously. The effective rate of interest of 10% compounded continuously is 10.517%.

Based on the results of Examples 2 and 3, we find the following comparisons:

	Annual Rate	Effective Rate
Annual compounding	10%	10%
Quarterly compounding	10%	10.381%
Monthly compounding	10%	10.471%
Daily compounding	10%	10.516%
Continuous compounding	10%	10.517%

NOW WORK PROBLEM 21.

EXAMPLE 4 Computing the Value of an IRA

On January 2, 2000, \$2000 is placed in an Individual Retirement Account (IRA) that will pay interest of 10% per annum compounded continuously. What will the IRA be worth on January 1, 2020?

Solution The amount A after 20 years is

$$A = Pe^{rt} = \$2000e^{(0.10)(20)} = \$14{,}778.11$$

Exploration How long will it be until $A = \$4000$? \$6000? [**Hint:** Graph $Y_1 = 2000e^{0.1x}$ and $Y_2 = 4000$. Use INTERSECT to find x.]

Time is money

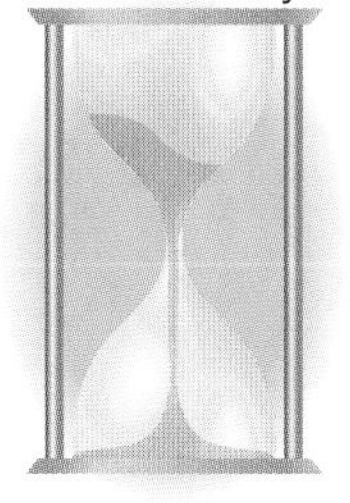

3 When people engaged in finance speak of the "time value of money," they are usually referring to the *present value* of money. The **present value** of A dollars to be received at a future date is the principal that you would need to invest now so that it would grow to A dollars in the specified time period. The present value of money to be received at a future date is always less than the amount to be received, since the amount to be received will equal the present value (money invested now) *plus* the interest accrued over the time period.

We use the compound interest formula (2) to get a formula for present value. If P is the present value of A dollars to be received after t years at a per annum interest rate r compounded n times per year, then, by formula (2),

$$A = P\left(1 + \frac{r}{n}\right)^{nt}$$

To solve for P, we divide both sides by $(1 + r/n)^{nt}$, and the result is

$$\frac{A}{(1 + r/n)^{nt}} = P \quad \text{or} \quad P = A\left(1 + \frac{r}{n}\right)^{-nt}$$

Theorem **Present Value Formulas**

The present value P of A dollars to be received after t years, assuming a per annum interest rate r compounded n times per year, is

$$P = A\left(1 + \frac{r}{n}\right)^{-nt} \tag{5}$$

If the interest is compounded continuously, then

$$P = Ae^{-rt} \tag{6}$$

To prove (6), solve formula (4) for P.

EXAMPLE 5 Computing the Value of a Zero-Coupon Bond

A zero-coupon (noninterest-bearing) bond can be redeemed in 10 years for \$1000. How much should you be willing to pay for it now if you want a return of:

(a) 8% compounded monthly? (b) 7% compounded continuously?

Solution (a) We are seeking the present value of \$1000. We use formula (5) with $A = \$1000$, $n = 12$, $r = 0.08$, and $t = 10$.

$$P = A\left(1 + \frac{r}{n}\right)^{-nt}$$

$$= \$1000\left(1 + \frac{0.08}{12}\right)^{-12(10)}$$

$$= \$450.52$$

For a return of 8% compounded monthly, you should pay \$450.52 for the bond.

(b) Here we use formula (6) with $A = \$1000$, $r = 0.07$, and $t = 10$.

$$P = Ae^{-rt}$$

$$= \$1000e^{-(0.07)(10)}$$

$$= \$496.59$$

For a return of 7% compounded continuously, you should pay \$496.59 for the bond. ■

NOW WORK PROBLEM 11.

EXAMPLE 6 Rate of Interest Required to Double an Investment

4 What annual rate of interest compounded annually should you seek if you want to double your investment in 5 years?

Algebraic Solution If P is the principal and we want P to double, the amount A will be $2P$. We use the compound interest formula with $n = 1$ and $t = 5$ to find r.

$$2P = P(1 + r)^5$$

$$2 = (1 + r)^5$$

$$1 + r = \sqrt[5]{2}$$

$$r = \sqrt[5]{2} - 1 = 1.148698 - 1 = 0.148698$$

The annual rate of interest needed to double the principal in 5 years is 14.87%.

Graphing Solution We solve the equation

$$2 = (1 + r)^5$$

for r by graphing the two functions $Y_1 = 2$ and $Y_2 = (1 + x)^5$. The x-coordinate of their point of intersection is the rate r that we seek. See Figure 35.

Figure 35

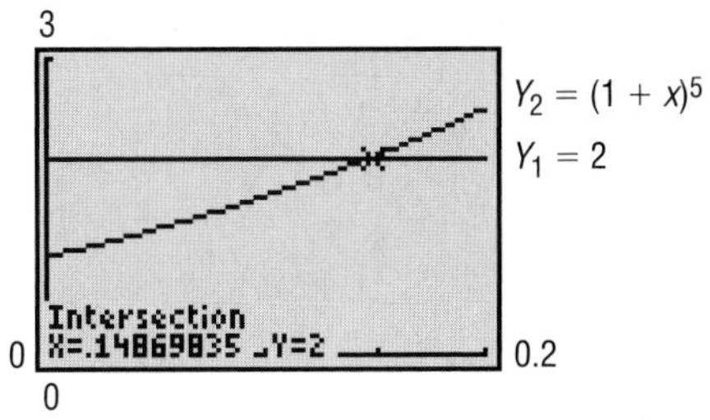

Using the INTERSECT command, we find that the point of intersection of Y_1 and Y_2 is $(0.14869835, 2)$, so $r = 0.148698$ as before. ■

NOW WORK PROBLEM 23.

EXAMPLE 7 Doubling and Tripling Time for an Investment

(a) How long will it take for an investment to double in value if it earns 5% compounded continuously?

(b) How long will it take to triple at this rate?

Algebraic Solution (a) If P is the initial investment and we want P to double, the amount A will be $2P$. We use formula (4) for continuously compounded interest with $r = 0.05$. Then

$$A = Pe^{rt}$$

$$2P = Pe^{0.05t}$$

$$2 = e^{0.05t}$$

$$0.05t = \ln 2$$

$$t = \frac{\ln 2}{0.05} = 13.86$$

It will take about 14 years to double the investment.

Graphing Solution We solve the equation

$$2 = e^{0.05t}$$

for t by graphing the two functions $Y_1 = 2$ and $Y_2 = e^{0.05x}$. Their point of intersection is $(13.86, 2)$. See Figure 36.

Figure 36

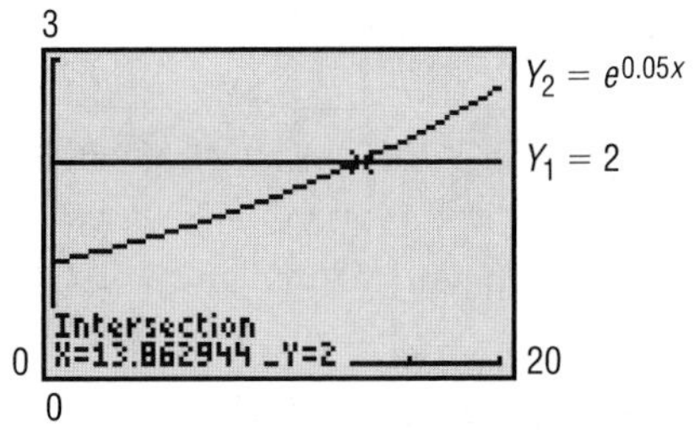

Algebraic Solution (b) To triple the investment, we set $A = 3P$ in formula (4).

$$A = Pe^{rt}$$

$$3P = Pe^{0.05t}$$

$$3 = e^{0.05t}$$

$$0.05t = \ln 3$$

$$t = \frac{\ln 3}{0.05} = 21.97$$

It will take about 22 years to triple the investment.

Graphing Solution We solve the equation

$$3 = e^{0.05t}$$

for t by graphing the two functions $Y_1 = 3$ and $Y_2 = e^{0.05x}$. Their point of intersection is (21.97, 3). See Figure 37.

Figure 37

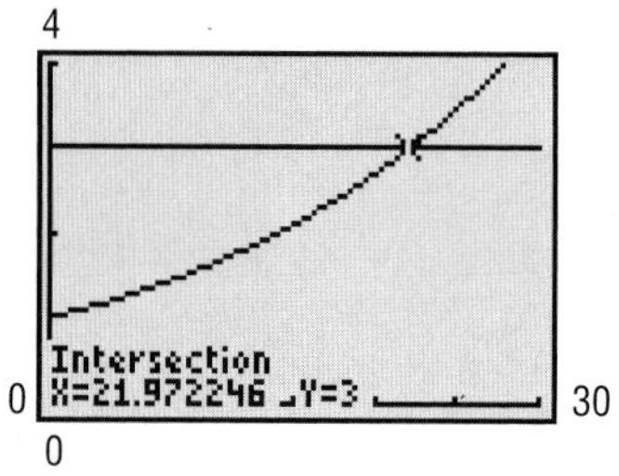

NOW WORK PROBLEM 29.

5.6 EXERCISES

In Problems 1–10, find the amount that results from each investment.

1. \$100 invested at 4% compounded quarterly after a period of 2 years
2. \$50 invested at 6% compounded monthly after a period of 3 years
3. \$500 invested at 8% compounded quarterly after a period of $2\frac{1}{2}$ years
4. \$300 invested at 12% compounded monthly after a period of $1\frac{1}{2}$ years
5. \$600 invested at 5% compounded daily after a period of 3 years
6. \$700 invested at 6% compounded daily after a period of 2 years
7. \$10 invested at 11% compounded continuously after a period of 2 years
8. \$40 invested at 7% compounded continuously after a period of 3 years
9. \$100 invested at 10% compounded continuously after a period of $2\frac{1}{4}$ years
10. \$100 invested at 12% compounded continuously after a period of $3\frac{3}{4}$ years

In Problems 11–20, find the principal needed now to get each amount; that is, find the present value.

11. To get \$100 after 2 years at 6% compounded monthly
12. To get \$75 after 3 years at 8% compounded quarterly
13. To get \$1000 after $2\frac{1}{2}$ years at 6% compounded daily
14. To get \$800 after $3\frac{1}{2}$ years at 7% compounded monthly
15. To get \$600 after 2 years at 4% compounded quarterly
16. To get \$300 after 4 years at 3% compounded daily
17. To get \$80 after $3\frac{1}{4}$ years at 9% compounded continuously
18. To get \$800 after $2\frac{1}{2}$ years at 8% compounded continuously
19. To get \$400 after 1 year at 10% compounded continuously
20. To get \$1000 after 1 year at 12% compounded continuously
21. Find the effective rate of interest for $5\frac{1}{4}$% compounded quarterly.
22. What interest rate compounded quarterly will give an effective interest rate of 7%?
23. What annual rate of interest is required to double an investment in 3 years? Verify your answer using a graphing utility.
24. What annual rate of interest is required to double an investment in 10 years? Verify your answer using a graphing utility.

In Problems 25–28, which of the two rates would yield the larger amount in 1 year?
[**Hint:** Start with a principal of \$10,000 in each instance.]

25. 6% compounded quarterly or $6\frac{1}{4}$% compounded annually
26. 9% compounded quarterly or $9\frac{1}{4}$% compounded annually
27. 9% compounded monthly or 8.8% compounded daily
28. 8% compounded semiannually or 7.9% compounded daily
29. How long does it take for an investment to double in value if it is invested at 8% per annum compounded monthly? Compounded continuously? Verify your answer using a graphing utility.
30. How long does it take for an investment to double in value if it is invested at 10% per annum compounded monthly? Compounded continuously? Verify your answer using a graphing utility.
31. If Tanisha has \$100 to invest at 8% per annum compounded monthly, how long will it be before she has \$150? If the compounding is continuous, how long will it be?
32. If Angela has \$100 to invest at 10% per annum compounded monthly, how long will it be before she has \$175? If the compounding is continuous, how long will it be?
33. How many years will it take for an initial investment of \$10,000 to grow to \$25,000? Assume a rate of interest of 6% compounded continuously.
34. How many years will it take for an initial investment of \$25,000 to grow to \$80,000? Assume a rate of interest of 7% compounded continuously.
35. What will a \$90,000 house cost 5 years from now if the inflation rate over that period averages 3% compounded annually?
36. Sears charges 1.25% per month on the unpaid balance for customers with charge accounts (interest is compounded monthly). A customer charges \$200 and does not pay her bill for 6 months. What is the bill at that time?
37. Jerome will be buying a used car for \$15,000 in 3 years. How much money should he ask his parents for now so that, if he invests it at 5% compounded continuously, he will have enough to buy the car?
38. John will require \$3000 in 6 months to pay off a loan that has no prepayment privileges. If he has the \$3000 now, how much of it should he save in an account paying 3% compounded monthly so that in 6 months he will have exactly \$3000?
39. George is contemplating the purchase of 100 shares of a stock selling for \$15 per share. The stock pays no dividends. The history of the stock indicates that it should grow at an annual rate of 15% per year. How much will the 100 shares of stock be worth in 5 years?
40. Tracy is contemplating the purchase of 100 shares of a stock selling for \$15 per share. The stock pays no dividends. Her broker says that the stock will be worth \$20 per share in 2 years. What is the annual rate of return on this investment?
41. A business purchased for \$650,000 in 1994 is sold in 1997 for \$850,000. What is the annual rate of return for this investment?
42. Tanya has just inherited a diamond ring appraised at \$5000. If diamonds have appreciated in value at an annual rate of 8%, what was the value of the ring 10 years ago when the ring was purchased?
43. Jim places \$1000 in a bank account that pays 5.6% compounded continuously. After 1 year, will he have enough money to buy a computer system that costs \$1060? If another bank will pay Jim 5.9% compounded monthly, is this a better deal?
44. On January 1, Kim places \$1000 in a certificate of deposit that pays 6.8% compounded continuously and matures in 3 months. Then Kim places the \$1000 and the interest in a passbook account that pays 5.25% compounded monthly. How much does Kim have in the passbook account on May 1?
45. Will invests \$2000 in a bond trust that pays 9% interest compounded semiannually. His friend Henry invests \$2000 in a certificate of deposit (CD) that pays $8\frac{1}{2}$% compounded continuously. Who has more money after 20 years, Will or Henry?

46. Suppose that April has access to an investment that will pay 10% interest compounded continuously. Which is better: To be given \$1000 now so that she can take advantage of this investment opportunity or to be given \$1325 after 3 years?

47. Colleen and Bill have just purchased a house for \$150,000, with the seller holding a second mortgage of \$50,000. They promise to pay the seller \$50,000 plus all accrued interest 5 years from now. The seller offers them three interest options on the second mortgage:
(a) Simple interest at 12% per annum
(b) $11\frac{1}{2}\%$ interest compounded monthly
(c) $11\frac{1}{4}\%$ interest compounded continuously
Which option is best; that is, which results in the least interest on the loan?

48. The First National Bank advertises that it pays interest on savings accounts at the rate of 4.25% compounded daily. Find the effective rate if the bank uses (a) 360 days or (b) 365 days in determining the daily rate.

Problems 49–52 involve zero-coupon bonds. A zero-coupon bond is a bond that is sold now at a discount and will pay its face value at some time when it matures; no interest payments are made.

49. A zero-coupon bond can be redeemed in 20 years for \$10,000. How much should you be willing to pay for it now if you want a return of:
(a) 10% compounded monthly?
(b) 10% compounded continuously?

50. A child's grandparents are considering buying a \$40,000 face value zero-coupon bond at birth so that she will have enough money for her college education 17 years later. If they want a rate of return of 8% compounded annually, what should they pay for the bond?

51. How much should a \$10,000 face value zero-coupon bond, maturing in 10 years, be sold for now if its rate of return is to be 8% compounded annually?

52. If Pat pays \$12,485.52 for a \$25,000 face value zero-coupon bond that matures in 8 years, what is his annual rate of return?

53. Explain in your own words what the term *compound interest* means. What does *continuous compounding* mean?

54. Explain in your own words the meaning of *present value*.

55. Time to Double or Triple an Investment The formula

$$y = \frac{\ln m}{n \ln\left(1 + \frac{r}{n}\right)}$$

can be used to find the number of years y required to multiply an investment m times when r is the per annum interest rate compounded n times a year.
(a) How many years will it take to double the value of an IRA that compounds annually at the rate of 12%?
(b) How many years will it take to triple the value of a savings account that compounds quarterly at an annual rate of 6%?
(c) Give a derivation of this formula.

56. Time to Reach an Investment Goal The formula

$$y = \frac{\ln A - \ln P}{r}$$

can be used to find the number of years y required for an investment P to grow to a value A when compounded continuously at an annual rate r.
(a) How long will it take to increase an initial investment of \$1000 to \$8000 at an annual rate of 10%?
(b) What annual rate is required to increase the value of a \$2000 IRA to \$30,000 in 35 years?
(c) Give a derivation of this formula.

57. Critical Thinking You have just contracted to buy a house and will seek financing in the amount of \$100,000. You go to several banks. Bank 1 will lend you \$100,000 at the rate of 8.75% amortized over 30 years with a loan origination fee of 1.75%. Bank 2 will lend you \$100,000 at the rate of 8.375% amortized over 15 years with a loan origination fee of 1.5%. Bank 3 will lend you \$100,000 at the rate of 9.125% amortized over 30 years with no loan origination fee. Bank 4 will lend you \$100,000 at the rate of 8.625% amortized over 15 years with no loan origination fee. Which loan would you take? Why? Be sure to have sound reasons for your choice. Use the information in the table to assist you. If the amount of the monthly payment does not matter to you, which loan would you take? Again, have sound reasons for your choice. Compare your final decision with others in the class. Discuss.

	Monthly Payment	Loan Origination Fee
Bank 1	\$786.70	\$1,750.00
Bank 2	\$977.42	\$1,500.00
Bank 3	\$813.63	\$0.00
Bank 4	\$990.68	\$0.00

5.7 GROWTH AND DECAY

OBJECTIVES
1. Find Equations of Populations That Obey the Law of Uninhibited Growth
2. Find Equations of Populations That Obey the Law of Decay
3. Use Newton's Law of Cooling
4. Use Logistic Growth Models

1 Many natural phenomena have been found to follow the law that an amount A varies with time t according to

$$A = A_0 e^{kt} \tag{1}$$

Here A_0 is the original amount ($t = 0$) and $k \neq 0$ is a constant.

If $k > 0$, then equation (1) states that the amount A is increasing over time; if $k < 0$, the amount A is decreasing over time. In either case, when an amount A varies over time according to equation (1), it is said to follow the **exponential law** or the **law of uninhibited growth** ($k > 0$) **or decay** ($k < 0$). See Figure 38.

Figure 38

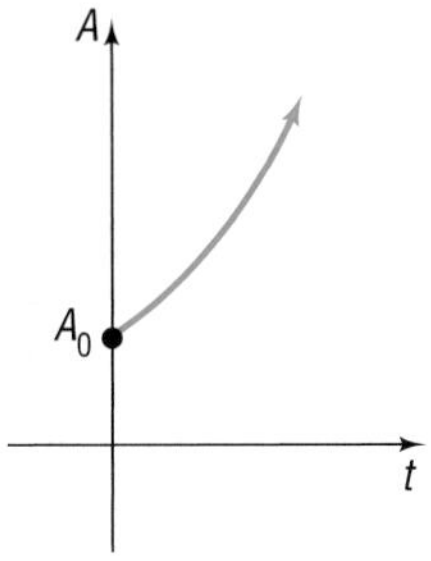

(a) $A(t) = A_0 e^{kt}, k > 0$

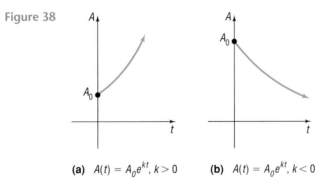

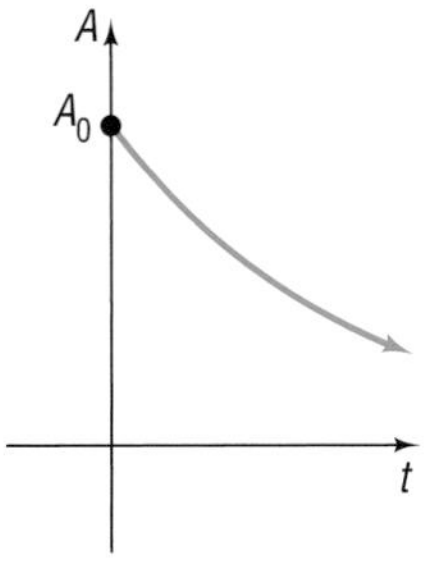

(b) $A(t) = A_0 e^{kt}, k < 0$

For example, we saw in Section 5.6 that continuously compounded interest follows the law of uninhibited growth. In this section we shall look at three additional phenomena that follow the exponential law.

UNINHIBITED GROWTH

Cell division is a universal process in the growth of many living organisms, such as amoebas, plants, and human skin cells. Based on an ideal situation in which no cells die and no by-products are produced, the number of cells present at a given time follows the law of uninhibited growth. Actually, however, after enough time has passed, growth at an exponential rate will cease due to the influence of factors such as lack of living space and dwindling food supply. The law of uninhibited growth accurately reflects only the early stages of the cell division process.

The cell division process begins with a culture containing N_0 cells. Each cell in the culture grows for a certain period of time and then divides into two identical cells. We assume that the time needed for each cell to divide in two is constant and does not change as the number of cells increases. These new cells then grow, and eventually each divides in two, and so on.

A model that gives the number N of cells in the culture after a time t has passed (in the early stages of growth) is given next.

Uninhibited Growth of Cells

$$N(t) = N_0 e^{kt}, \qquad k > 0 \tag{2}$$

Here N_0 is the initial number of cells and k is a positive constant that represents the growth rate of the cells.

In using formula (2) to model the growth of cells, we are using a function that yields positive real numbers, even though we are counting the number of cells, which must be an integer. This is a common practice in many applications.

EXAMPLE 1 Bacterial Growth

A colony of bacteria increases according to the law of uninhibited growth.

(a) If the number of bacteria doubles in 3 hours, find the function that gives the number of cells in the culture.

(b) How long will it take for the size of the colony to triple?

(c) Using a graphing utility, verify that the population doubles in 3 hours.

(d) Using a graphing utility, approximate the time that it takes for the population to double a second time (that is, increase four times). Does this answer seem reasonable?

Solution (a) Using formula (2), the number N of cells at a time t is

$$N(t) = N_0 e^{kt}$$

where N_0 is the initial number of bacteria present and k is a positive number. We first seek the number k. The number of cells doubles in 3 hours, so we have

$$N(3) = 2N_0$$

But $N(3) = N_0 e^{k(3)}$, so

$$N_0 e^{k(3)} = 2N_0$$

$$e^{3k} = 2$$

$$3k = \ln 2 \quad \text{Write the exponential equation as a logarithm.}$$

$$k = \tfrac{1}{3}\ln 2 \approx \tfrac{1}{3}(0.6931) = 0.2310$$

Formula (2) for this growth process is therefore

$$N(t) = N_0 e^{0.2310t}$$

(b) The time t needed for the size of the colony to triple requires that $N = 3N_0$. We substitute $3N_0$ for N to get

$$3N_0 = N_0 e^{0.2310t}$$

$$3 = e^{0.2310t}$$

$$0.2310t = \ln 3$$

$$t = \frac{1}{0.2310}\ln 3 \approx \frac{1.0986}{0.2310} = 4.756 \text{ hours}$$

It will take about 4.756 hours for the size of the colony to triple.

(c) Figure 39 shows the graph of $Y_1 = 2$ and $Y_2 = e^{0.2310x}$, where x represents the time in hours. Using INTERSECT, it takes 3 hours for the population to double.

Figure 39

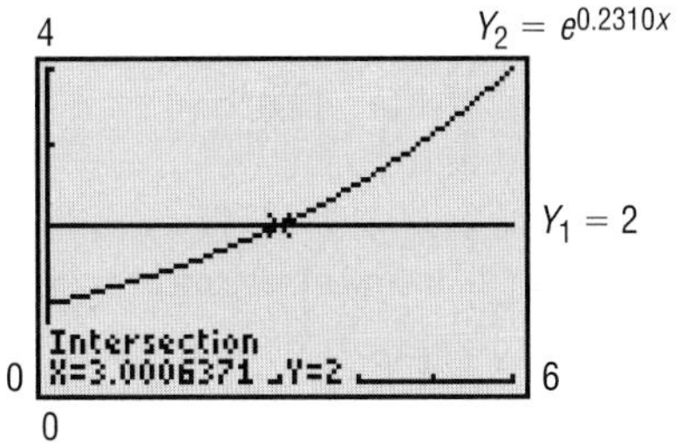

(d) Figure 40 shows the graph of $Y_1 = 4$ and $Y_2 = e^{0.2310x}$, where x represents the time in hours. Using INTERSECT, it takes 6 hours for the population to double a second time. This answer makes sense: if a population doubles in 3 hours, it will double a second time in 3 more hours.

Figure 40

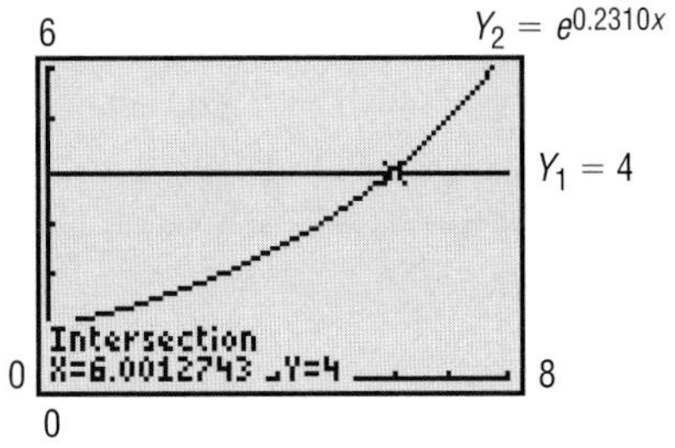

NOW WORK PROBLEM 1.

Radioactive Decay

2 Radioactive materials follow the law of uninhibited decay. The amount A of a radioactive material present at time t is given by the following model:

Uninhibited Radioactive Decay

$$A(t) = A_0 e^{kt}, \qquad k < 0 \tag{3}$$

Here A_0 is the original amount of radioactive material and k is a negative number that represents the rate of decay.

All radioactive substances have a specific **half-life**, which is the time required for half of the radioactive substance to decay. In **carbon dating**, we use the fact that all living organisms contain two kinds of carbon, carbon 12 (a stable carbon) and carbon 14 (a radioactive carbon, with a half-life of 5600 years). While an organism is living, the ratio of carbon 12 to carbon 14 is constant. But when an organism dies, the original amount of carbon 12 present remains unchanged, whereas the amount of carbon 14 begins to decrease. This change in the amount of carbon 14 present relative to the amount of carbon 12 present makes it possible to calculate when an organism died.

EXAMPLE 2 Estimating the Age of Ancient Tools

Traces of burned wood along with ancient stone tools in an archaeological dig in Chile were found to contain approximately 1.67% of the original amount of carbon 14.

(a) If the half-life of carbon 14 is 5600 years, approximately when was the tree cut and burned?

(b) Using a graphing utility, graph the relation between the percentage of carbon 14 remaining and time.

(c) Determine the time that elapses until half of the carbon 14 remains. This answer should equal the half-life of carbon 14.

(d) Verify the answer found in part (a).

Solution (a) Using formula (3), the amount A of carbon 14 present at time t is

$$A(t) = A_0 e^{kt}$$

where A_0 is the original amount of carbon 14 present and k is a negative number. We first seek the number k. To find it, we use the fact that after 5600 years half of the original amount of carbon 14 remains, so $A(5600) = \frac{1}{2}A_0$. Thus,

$$\frac{1}{2}A_0 = A_0 e^{k(5600)}$$

$$\frac{1}{2} = e^{5600k}$$

$$5600k = \ln\frac{1}{2}$$

$$k = \frac{1}{5600}\ln\frac{1}{2} \approx -0.000124$$

Formula (3) therefore becomes

$$A(t) = A_0 e^{-0.000124t}$$

If the amount A of carbon 14 now present is 1.67% of the original amount, it follows that

$$0.0167A_0 = A_0 e^{-0.000124t}$$

$$0.0167 = e^{-0.000124t}$$

$$-0.000124t = \ln 0.0167$$

$$t = \frac{1}{-0.000124}\ln 0.0167 \approx 33{,}000 \text{ years}$$

The tree was cut and burned about 33,000 years ago. Some archaeologists use this conclusion to argue that humans lived in the Americas 33,000 years ago, much earlier than is generally accepted.

(b) Figure 41 shows the graph of $y = e^{-0.000124x}$, where y is the fraction of carbon 14 present and x is the time.

Figure 41

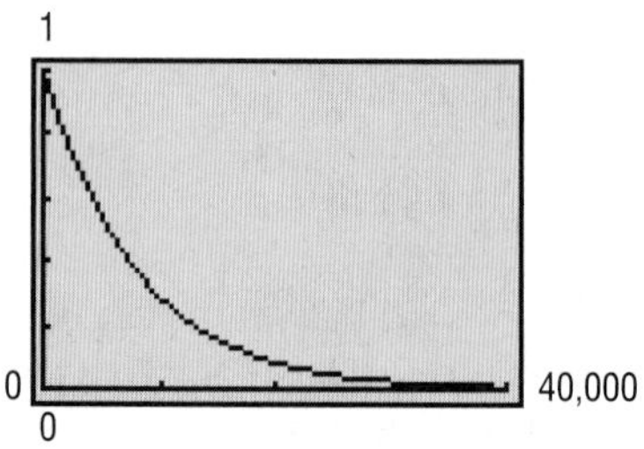

(c) By graphing $Y_1 = 0.5$ and $Y_2 = e^{-0.000124x}$, where x is time, and using INTERSECT, we find that it takes 5590 years until half the carbon 14 remains.

(d) By graphing $Y_1 = 0.0167$ and $Y_2 = e^{-0.000124x}$, where x is time, and using INTERSECT, we find that it takes 33,003 years until 1.67% of the carbon 14 remains. ■

NOW WORK PROBLEM 3.

NEWTON'S LAW OF COOLING

3 **Newton's Law of Cooling*** states that the temperature of a heated object decreases exponentially over time toward the temperature of the surrounding medium. That is, the temperature u of a heated object at a given time t can be modeled by the following function:

Newton's Law of Cooling

$$u(t) = T + (u_0 - T)e^{kt} \qquad k < 0 \qquad \textbf{(4)}$$

Here T is the constant temperature of the surrounding medium, u_0 is the initial temperature of the heated object, and k is a negative constant.

EXAMPLE 3 Using Newton's Law of Cooling

An object is heated to 100°C (degrees Celsius) and is then allowed to cool in a room whose air temperature is 30°C.

(a) If the temperature of the object is 80°C after 5 minutes, when will its temperature be 50°C?

(b) Using a graphing utility, graph the relation found between the temperature and time.

(c) Using a graphing utility, verify that after 18.6 minutes the temperature is 50°C.

(d) Using a graphing utility, determine the elapsed time before the object is 35°C.

(e) What do you notice about the temperature as time passes?

Solution (a) Using formula (4) with $T = 30$ and $u_0 = 100$, the temperature (in degrees Celsius) of the object at time t (in minutes) is

$$u(t) = 30 + (100 - 30)e^{kt} = 30 + 70e^{kt} \qquad \textbf{(5)}$$

where k is a negative constant. To find k, we use the fact that $u = 80$ when $t = 5$. Then

$$80 = 30 + 70e^{k(5)}$$
$$50 = 70e^{5k}$$
$$e^{5k} = \frac{50}{70}$$
$$5k = \ln\frac{5}{7}$$
$$k = \frac{1}{5}\ln\frac{5}{7} \approx -0.0673$$

*Named after Sir Isaac Newton (1642–1727), one of the cofounders of calculus.

Formula (4) therefore becomes

$$u(t) = 30 + 70e^{-0.0673t}$$

We want to find t when $u = 50°C$, so

$$50 = 30 + 70e^{-0.0673t}$$

$$20 = 70e^{-0.0673t}$$

$$e^{-0.0673t} = \frac{20}{70}$$

$$-0.0673t = \ln\frac{2}{7}$$

$$t = \frac{1}{-0.0673}\ln\frac{2}{7} \approx 18.6 \text{ minutes}$$

The temperature of the object will be 50°C after about 18.6 minutes.

Figure 42

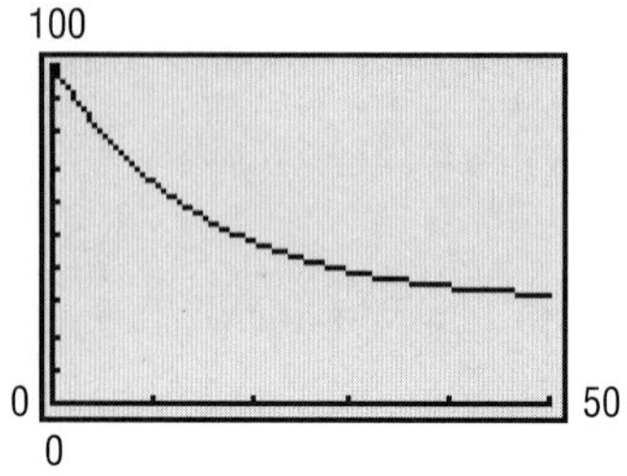

(b) Figure 42 shows the graph of $y = 30 + 70e^{-0.0673x}$, where y is the temperature and x is the time.

(c) By graphing $Y_1 = 50$ and $Y_2 = 30 + 70e^{-0.0673x}$, where x is time, and using INTERSECT, we find that it takes $x = 18.6$ minutes for the temperature to cool to 50°C.

(d) By graphing $Y_1 = 35$ and $Y_2 = 30 + 70e^{-0.0673x}$, where x is time, and using INTERSECT, we find that it takes $x = 39.21$ minutes for the temperature to cool to 35°C.

(e) As x increases, the value of $e^{-0.0673x}$ approaches zero, so the value of y approaches 30°C. The temperature of the object approaches 30°C. ■

Logistic Models

4 The exponential growth model $A(t) = A_0e^{kt}$, $k > 0$, assumes uninhibited growth, meaning that the value of the function grows without limit. Recall that we stated that cell division could be modeled using this function, assuming that no cells die and no by-products are produced. However, cell division would eventually be limited by factors such as living space and food supply. The **logistic growth model** is an exponential function that can model situations where the growth of the dependent variable is limited.

Other situations that lead to a logistic growth model include population growth and the sales of a product due to advertising. See Problems 23 through 27. The logistic growth model is given next.

Logistic Growth Model

$$P(t) = \frac{c}{1 + ae^{-bt}}$$

Here a, b, and c are constants with $c > 0$ and $b > 0$.

The number c is called the **carrying capacity** because the value $P(t)$ approaches c as t approaches infinity; that is, $\lim_{t\to\infty} P(t) = c$.

EXAMPLE 4 Fruit Fly Population

Fruit flies are placed in a half-pint milk bottle with a banana (for food) and yeast plants (for food and to provide a stimulus to lay eggs). Suppose that the fruit fly population after t days is given by

$$P(t) = \frac{230}{1 + 56.5e^{-0.37t}}$$

(a) Using a graphing utility, graph $P(t)$.

(b) What is the carrying capacity of the half-pint bottle? That is, what is $P(t)$ as $t \to \infty$?

(c) How many fruit flies were initially placed in the half-pint bottle?

(d) When will the population of fruit flies be 180?

Figure 43

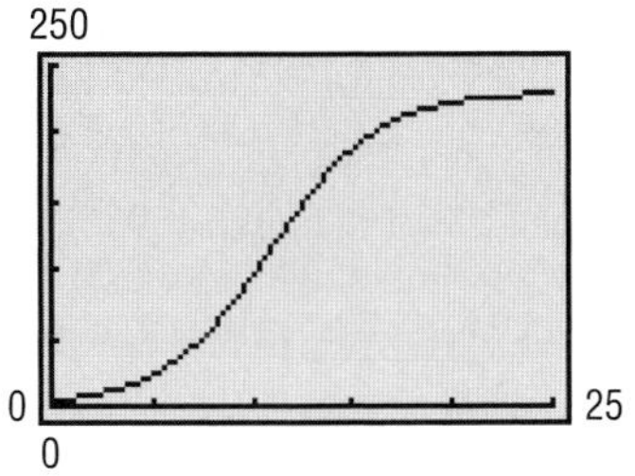

Solution

(a) See Figure 43 for the graph of $P(t)$.

(b) As $t \to \infty$, $e^{-0.37t} \to 0$ and $P(t) \to 230/1$. The carrying capacity of the half-pint bottle is 230 fruit flies.

(c) To find the initial number of fruit flies in the half-pint bottle, we evaluate $P(0)$.

$$P(0) = \frac{230}{1 + 56.5e^{-0.37(0)}} = \frac{230}{1 + 56.5} = 4$$

So initially there were four fruit flies in the half-pint bottle.

(d) To determine when the population of fruit flies will be 180, we solve the equation

$$\frac{230}{1 + 56.5e^{-0.37t}} = 180$$

$$230 = 180\left(1 + 56.5e^{-0.37t}\right)$$

$$1.2778 = 1 + 56.5e^{-0.37t} \qquad \text{Divide both sides by 180.}$$

$$0.2778 = 56.5e^{-0.37t} \qquad \text{Subtract 1 from both sides.}$$

$$0.0049 = e^{-0.37t} \qquad \text{Divide both sides by 56.5.}$$

$$\ln(0.0049) = -0.37t \qquad \text{Rewrite as a logarithmic expression.}$$

$$t \approx 14.4 \text{ days} \qquad \text{Divide both sides by } -0.37.$$

It will take approximately 14.4 days for the population to reach 180 fruit flies.

Figure 44

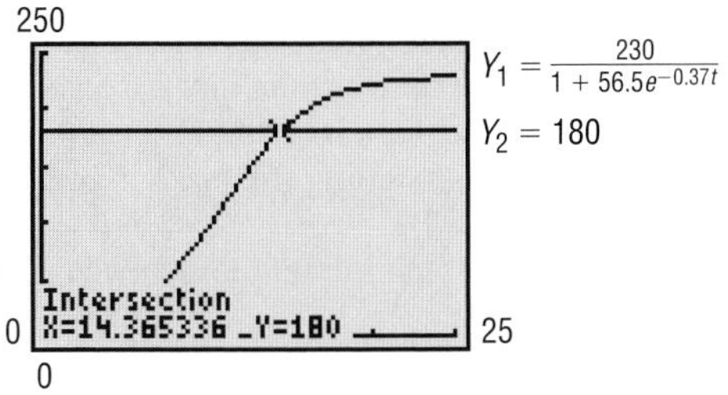

We could also solve this problem by graphing $Y_1 = \dfrac{230}{1 + 56.5e^{-0.37t}}$ and $Y_2 = 180$, using INTERSECT to find the solution shown in Figure 44. ■

EXPLORATION On the same viewing rectangle, graph $Y_1 = \dfrac{500}{1 + 24e^{-0.03t}}$ and $Y_2 = \dfrac{500}{1 + 24e^{-0.08t}}$. What effect does b have on the logistic growth function?

5.7 EXERCISES

1. **Growth of an Insect Population** The size P of a certain insect population at time t (in days) obeys the function $P(t) = 500e^{0.02t}$.
(a) After how many days will the population reach 1000? 2000?
(b) Using a graphing utility, graph the relation between P and t. Verify your answers in part (a) using INTERSECT.

2. **Growth of Bacteria** The number N of bacteria present in a culture at time t (in hours) obeys the function $N(t) = 1000e^{0.01t}$.
(a) After how many hours will the population equal 1500? 2000?
(b) Using a graphing utility, graph the relation between N and t. Verify your answers in part (a) using INTERSECT.

3. **Radioactive Decay** Strontium 90 is a radioactive material that decays according to the function $A(t) = A_0e^{-0.0244t}$, where A_0 is the initial amount present and A is the amount present at time t (in years). What is the half-life of strontium 90?

4. **Radioactive Decay** Iodine 131 is a radioactive material that decays according to the function $A(t) = A_0e^{-0.087t}$, where A_0 is the initial amount present and A is the amount present at time t (in days). What is the half-life of iodine 131?

5. (a) Use the information in Problem 3 to determine how long it takes for 100 grams of strontium 90 to decay to 10 grams.
(b) Graph the function $A(t) = 100e^{-0.0244t}$ and verify your answer found in part (a).

6. (a) Use the information in Problem 4 to determine how long it takes for 100 grams of iodine 131 to decay to 10 grams.
(b) Graph the function $A(t) = 100e^{-0.087t}$ and verify your answer found in part (a).

7. **Growth of a Colony of Mosquitoes** The population of a colony of mosquitoes obeys the law of uninhibited growth. If there are 1000 mosquitoes initially and there are 1800 after 1 day, what is the size of the colony after 3 days? How long is it until there are 10,000 mosquitoes?

8. **Bacterial Growth** A culture of bacteria obeys the law of uninhibited growth. If 500 bacteria are present initially and there are 800 after 1 hour, how many will be present in the culture after 5 hours? How long is it until there are 20,000 bacteria?

9. **Population Growth** The population of a southern city follows the exponential law. If the population doubled in size over an 18-month period and the current population is 10,000, what will the population be 2 years from now?

10. **Population Growth** The population of a midwestern city follows the exponential law. If the population decreased from 900,000 to 800,000 from 1993 to 1995, what will the population be in 1997?

11. **Radioactive Decay** The half-life of radium is 1690 years. If 10 grams is present now, how much will be present in 50 years?

12. **Radioactive Decay** The half-life of radioactive potassium is 1.3 billion years. If 10 grams is present now, how much will be present in 100 years? In 1000 years?

13. **Estimating the Age of a Tree** A piece of charcoal is found to contain 30% of the carbon 14 that it originally had.
(a) When did the tree from which the charcoal came die? Use 5600 years as the half-life of carbon 14.
(b) Using a graphing utility, graph the relation between the percentage of carbon 14 remaining and time.
(c) Using INTERSECT, determine the time that elapses until half of the carbon 14 remains.
(d) Verify the answer found in part (a).

14. **Estimating the Age of a Fossil** A fossilized leaf contains 70% of its normal amount of carbon 14.
(a) How old is the fossil?
(b) Using a graphing utility, graph the relation between the percentage of carbon 14 remaining and time.
(c) Using INTERSECT, determine the time that elapses until half of the carbon 14 remains.
(d) Verify the answer found in part (a).

15. **Cooling Time of a Pizza** A pizza baked at 450°F is removed from the oven at 5:00 PM into a room that is a constant 70°F. After 5 minutes, the pizza is at 300°F.
(a) At what time can you begin eating the pizza if you want its temperature to be 135°F?
(b) Using a graphing utility, graph the relation between temperature and time.
(c) Using INTERSECT, determine the time that needs to elapse before the pizza is 160°F.

(d) TRACE the function for large values of time. What do you notice about y, the temperature?

16. Newton's Law of Cooling A thermometer reading 72°F is placed in a refrigerator where the temperature is a constant 38°F.
(a) If the thermometer reads 60°F after 2 minutes, what will it read after 7 minutes?
(b) How long will it take before the thermometer reads 39°F?
(c) Using a graphing utility, graph the relation between temperature and time.
(d) Using INTERSECT, determine the time needed to elapse before the thermometer reads 45°F.
(e) TRACE the function for large values of time. What do you notice about y, the temperature?

17. Newton's Law of Cooling A thermometer reading 8°C is brought into a room with a constant temperature of 35°C.
(a) If the thermometer reads 15°C after 3 minutes, what will it read after being in the room for 5 minutes? For 10 minutes?
(b) Graph the relation between temperature and time. TRACE to verify that your answers are correct.
[**Hint:** You need to construct a formula similar to equation (4).]

18. Thawing Time of a Steak A frozen steak has a temperature of 28°F. It is placed in a room with a constant temperature of 70°F. After 10 minutes, the temperature of the steak has risen to 35°F. What will the temperature of the steak be after 30 minutes? How long will it take the steak to thaw to a temperature of 45°F? [See the hint given for Problem 17.] Graph the relation between temperature and time. TRACE to verify that your answer is correct.

19. Decomposition of Salt in Water Salt (NaCl) decomposes in water into sodium (NA^+) and chloride (Cl^-) ions according to the law of uninhibited decay. If the initial amount of salt is 25 kilograms and, after 10 hours, 15 kilograms of salt is left, how much salt is left after 1 day? How long does it take until $\frac{1}{2}$ kilogram of salt is left?

20. Voltage of a Conductor The voltage of a certain conductor decreases over time according to the law of uninhibited decay. If the initial voltage is 40 volts, and 2 seconds later it is 10 volts, what is the voltage after 5 seconds?

21. Radioactivity from Chernobyl After the release of radioactive material into the atmosphere from a nuclear power plant at Chernobyl (Ukraine) in 1986, the hay in Austria was contaminated by iodine 131 (half-life 8 days). If it is all right to feed the hay to cows when 10% of the iodine 131 remains, how long do the farmers need to wait to use this hay?

22. Pig Roasts The hotel Bora-Bora is having a pig roast. At noon, the chef put the pig in a large earthen oven. The pig's original temperature was 75°F. At 2:00 PM the chef checked the pig's temperature and was upset because it had reached only 100°F. If the oven's temperature remains a constant 325°F, at what time may the hotel serve its guests, assuming that pork is done when it reaches 175°F?

23. Proportion of the Population That Owns a VCR The logistic growth model

$$P(t) = \frac{0.9}{1 + 6e^{-0.32t}}$$

relates the proportion of U.S. households that own a VCR to the year. Let $t = 0$ represent 1984, $t = 1$ represent 1985, and so on.
(a) What proportion of the U.S. households owned a VCR in 1984?
(b) Determine the maximum proportion of households that will own a VCR.
(c) Using a graphing utility, graph $P(t)$.
(d) When will 0.8 (80%) of U.S. households own a VCR?

24. Market Penetration of Intel's Coprocessor The logistic growth model

$$P(t) = \frac{0.90}{1 + 3.5e^{-0.339t}}$$

relates the proportion of new personal computers sold at Best Buy that have Intel's latest coprocessor t months after it has been introduced.
(a) What proportion of new personal computers sold at Best Buy will have Intel's latest coprocessor when it is first introduced (that is, at $t = 0$)?
(b) Determine the maximum proportion of new personal computers sold at Best Buy that will have Intel's latest coprocessor.
(c) Using a graphing utility, graph $P(t)$.
(d) When will 0.75 (75%) of new personal computers sold at Best Buy have Intel's latest coprocessor?

25. Population of a Bacteria Culture The logistic growth model

$$P(t) = \frac{1000}{1 + 32.33e^{-0.439t}}$$

represents the population of a bacterium after t hours.
(a) Using a graphing utility, graph $P(t)$.

(b) What is the carrying capacity of the environment?
(c) What was the initial amount of bacteria in the population?
(d) When will the amount of bacteria be 800?

26. **Population of a Endangered Species** Often environmentalists will capture an endangered species and transport the species to a controlled environment where the species can produce offspring and regenerate its population. Suppose that six American bald eagles are captured, transported to Montana, and set free. Based on experience, the environmentalists expect the population to grow according to the model

$$P(t) = \frac{500}{1 + 83.33e^{-0.162t}}$$

(a) Using a graphing utility, graph $P(t)$.
(b) What is the carrying capacity of the environment?
(c) What is the predicted population of the American bald eagle in 20 years?
(d) When will the population be 300?

27. **The *Challenger* Disaster*** After the *Challenger* disaster in 1986, a study of the 23 launches that preceded the fatal flight was made. A mathematical model was developed involving the relationship between the Fahrenheit temperature x around the O-rings and the number y of eroded or leaky primary O-rings. The model stated that

$$y = \frac{6}{1 + e^{-(5.085-0.1156x)}}$$

where the number 6 indicates the 6 primary O-rings on the spacecraft.
(a) What is the predicted number of eroded or leaky primary O-rings at a temperature of 100°F?
(b) What is the predicted number of eroded or leaky primary O-rings at a temperature of 60°F?
(c) What is the predicted number of eroded or leaky primary O-rings at a temperature of 30°F?
(d) Graph the equation. At what temperature is the predicted number of eroded or leaky O-rings 1? 3? 5?

* Linda Tappin, "Analyzing Data Relating to the *Challenger* Disaster," *Mathematics Teacher*, Vol. 87, No. 6, September 1994, pp. 423–426.

PREPARING FOR THIS SECTION

Before getting started, review the following:

✓ Linear Curve Fitting (Section 1.7)

5.8 EXPONENTIAL, LOGARITHMIC, AND LOGISTIC CURVE FITTING

OBJECTIVES
1 Use a Graphing Utility to Obtain the Exponential Function of Best Fit
2 Use a Graphing Utility to Obtain the Logarithmic Function of Best Fit
3 Use a Graphing Utility to Obtain the Logistic Function of Best Fit
4 Determine the Model of Best Fit

In Section 1.7 we discussed how to find the linear equation of best fit $(y = ax + b)$, and in Chapter 3 we discussed how to find the quadratic function of best fit $(y = ax^2 + bx + c)$, the power function of best fit $(y = ax^b)$, and the cubic function of best fit $(y = ax^3 + bx^2 + cx + d)$.

In this section we will discuss how to use a graphing utility to find equations of best fit that describe the relation between two variables when the relation is thought to be exponential $(y = ab^x)$, logarithmic $(y = a + b \ln x)$, or logistic $\left(y = \dfrac{c}{1 + ae^{-bx}}\right)$. As before, we draw a scatter diagram of the data to help to determine the appropriate model to use.

Figure 45 shows scatter diagrams that will typically be observed for the three models. Below each scatter diagram are any restrictions on the values of the parameters.

Figure 45

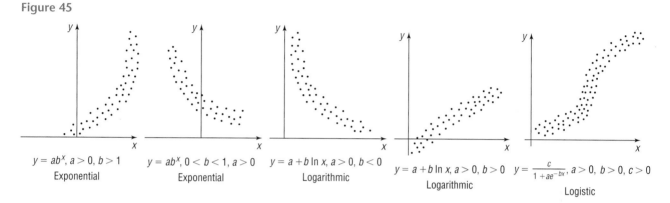

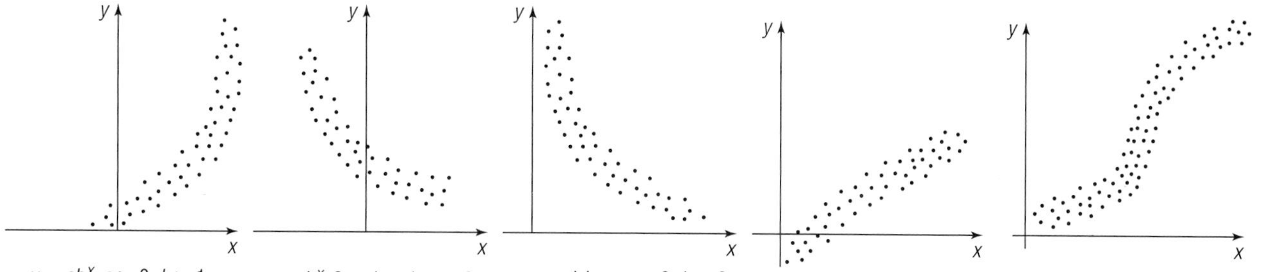

Most graphing utilities have REGression options that fit data to a specific type of curve. Once the data have been entered and a scatter diagram obtained, the type of curve that you want to fit to the data is selected. Then the REGression option is used to obtain the curve of "best fit" of the type selected.

The correlation coefficient r will appear only if the model can be written as a linear expression. Thus, r will appear for the linear, power, exponential, and logarithmic models, since these models can be written as a linear expression (see Problems 39 and 40 in Section 5.4). Remember, the closer $|r|$ is to 1, the better the fit.

Let's look at some examples.

Exponential Curve Fitting

1 We saw in Section 5.6 that the future value of money behaves exponentially, and we saw in Section 5.7 that growth and decay models also behave exponentially. The next example shows how data can lead to an exponential model.

EXAMPLE 1 Fitting a Curve to an Exponential Model

Beth is interested in finding a function that explains the closing price (adjusted for stock splits) of Dell Computer stock at the end of each year. She obtains the following data:

Year	Closing Price	Year	Closing Price
1989	0.0573	1995	1.082
1990	0.1927	1996	3.3203
1991	0.2669	1997	10.5
1992	0.75	1998	36.5938
1993	0.3535	1999	51
1994	0.6406		

Source: NASDAQ

(a) Using a graphing utility, draw a scatter diagram.

(b) Using a graphing utility, fit an exponential model to the data.

(c) Express the function in the form $A = A_0 e^{kt}$.

(d) Graph the exponential function found in part (b) or (c) on the scatter diagram.

(e) Using the solution to part (b) or (c), predict the closing price of Dell Computer stock at the end of the year 2000.

(f) Interpret the value of k found in part (c).

Figure 46

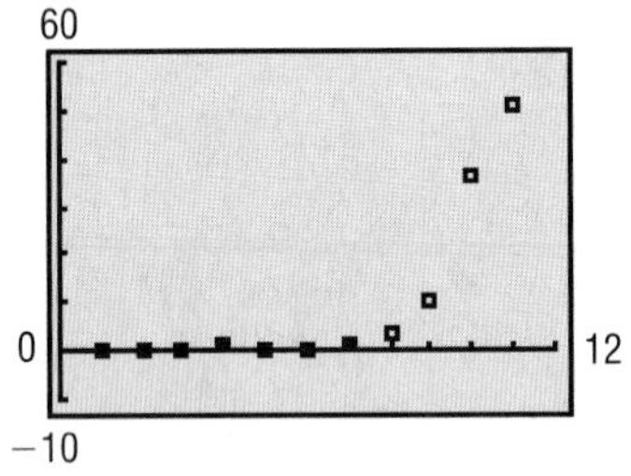

Figure 47

Figure 48

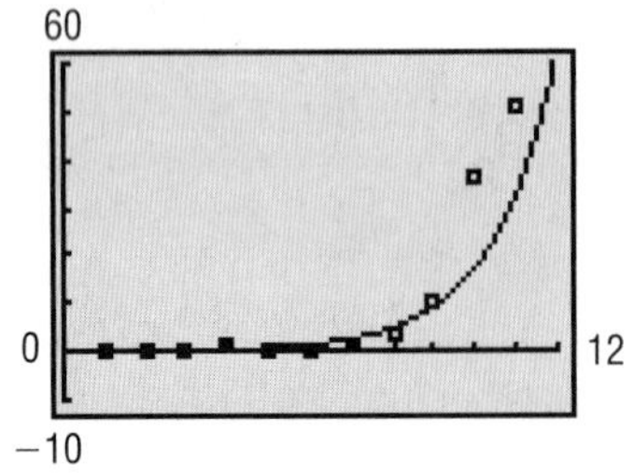

Solution

(a) Enter the data into the graphing utility, letting 1 represent 1989, 2 represent 1990, and so on. We obtain the scatter diagram shown in Figure 46.

(b) A graphing utility fits the data to an exponential model of the form $y = ab^x$ by using the EXPonential REGression option. See Figure 47. Thus, $y = ab^x = 0.03028(1.8905)^x$.

(c) To express $y = ab^x$ in the form $A = A_0 e^{kt}$, we proceed as follows:

$$ab^x = A_0 e^{kt}$$

We set the coefficients equal to each other and the exponential expressions equal to each other.

$$a = A_0 \qquad b^x = e^{kt} = \left(e^k\right)^t$$

$$b = e^k$$

Since $y = ab^x = 0.03028(1.8905)^x$, we find that $a = 0.03028$ and $b = 1.8905$. Thus,

$$a = A_0 = 0.03028 \quad \text{and} \quad b = 1.8905 = e^k$$

$$k = \ln(1.8905) = 0.6368$$

As a result, $A = A_0 e^{kt} = 0.03028e^{0.6368t}$.

(d) See Figure 48 for the graph of $A(t) = 0.0328e^{0.6368t}$.

(e) Let $t = 12$ (the year 2000) in the function found in part (c). The predicted closing price of Dell Computer stock at the end of the year 2000 is

$$A = 0.03028e^{0.6368(12)} = \$63.08$$

(f) The value of k represents the annual interest rate compounded continuously.

$$A = A_0 e^{kt} = 0.03028e^{0.6368t}$$

$$Pe^{rt} = 0.03028e^{0.6368t} \qquad \text{Equation (4), Section 5.6}$$

$$r = 0.6368$$

The price of Dell Computer stock has grown at an annual rate of 63.68% (compounded continuously) between 1989 and 1999. ■

NOW WORK PROBLEM 7.

Logarithmic Curve Fitting

2 Many relations between variables do not follow an exponential model, but, instead, the independent variable is related to the dependent variable using a logarithmic model.

EXAMPLE 2 Fitting a Curve to a Logarithmic Model

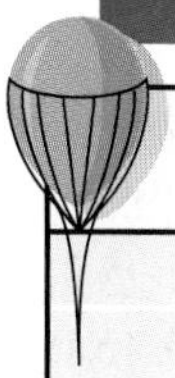

Atmospheric Pressure, p	Height, h
760	0
740	0.184
725	0.328
700	0.565
650	1.079
630	1.291
600	1.634
580	1.862
550	2.235

Jodi, a meteorologist, is interested in finding a function that explains the relation between the height of a weather balloon (in kilometers) and the atmospheric pressure (measured in millimeters of mercury) on the balloon. She collects the data shown in the table.

(a) Using a graphing utility, draw a scatter diagram of the data with atmospheric pressure as the independent variable.

(b) Using a graphing utility, fit a logarithmic model to the data.

(c) Draw the logarithmic function found in part (b) on the scatter diagram.

(d) Use the function found in part (b) to predict the height of the weather balloon if the atmospheric pressure is 560 millimeters of mercury.

Solution

(a) After entering the data into the graphing utility, we obtain the scatter diagram shown in Figure 49.

Figure 49

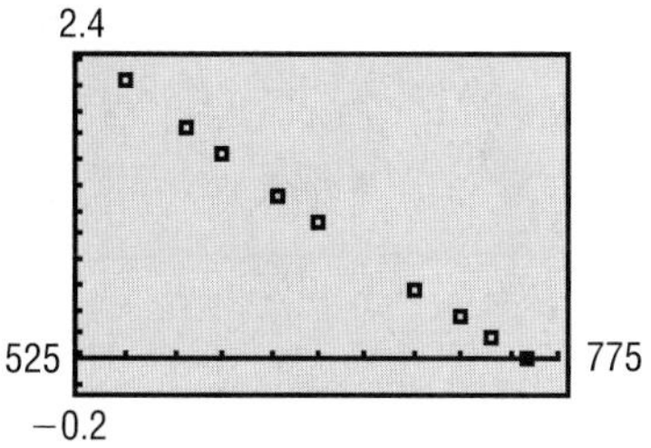

(b) A graphing utility fits the data to a logarithmic model of the form $y = a + b\ln x$ by using the Logarithm REGression option. See Figure 50. The logarithmic function of best fit to the data is

$$h(p) = 45.7863 - 6.9025 \ln p$$

where h is the height of the weather balloon and p is the atmospheric pressure. Notice that $|r|$ is close to 1, indicating a good fit.

(c) Figure 51 shows the graph of $h(p) = 45.7863 - 6.9025 \ln p$ on the scatter diagram.

Figure 50

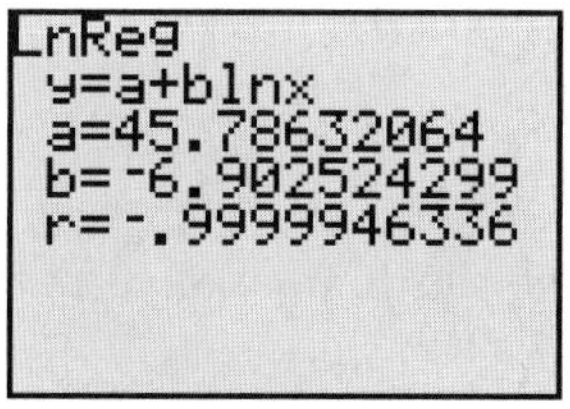

Figure 51

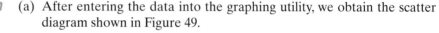

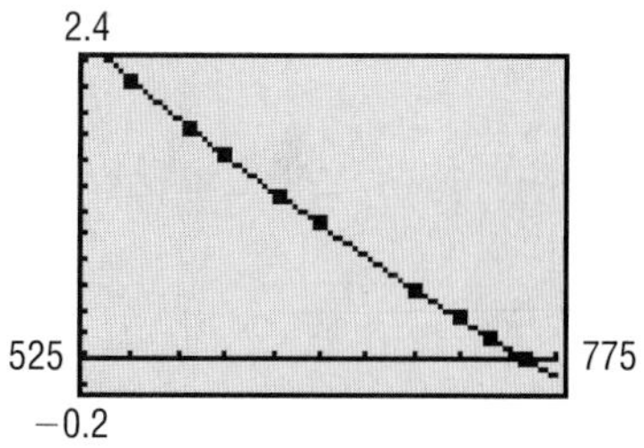

(d) Using the function found in part (b), Jodi predicts the height of the weather balloon when the atmospheric pressure is 560 to be

$$h(560) = 45.7863 - 6.9025 \ln 560$$
$$\approx 2.108 \text{ kilometers}$$

NOW WORK PROBLEM 9.

LOGISTIC CURVE FITTING

3 Logistic growth models can be used to model situations where the value of the dependent variable is limited. Many real-world situations conform to this scenario. For example, the population of the human race is limited by the availability of natural resources, such as food and shelter. When the value of the dependent variable is limited, a logistic growth model is often appropriate.

EXAMPLE 3 Fitting a Curve to a Logistic Growth Model

Time (in hours)	Yeast Biomass
0	9.6
1	18.3
2	29.0
3	47.2
4	71.1
5	119.1
6	174.6
7	257.3
8	350.7
9	441.0
10	513.3
11	559.7
12	594.8
13	629.4
14	640.8
15	651.1
16	655.9
17	659.6
18	661.8

The data in the table, obtained from R. Pearl ["The Growth of Population," *Quarterly Review of Biology* 2 (1927): 532–548], represent the amount of yeast biomass after t hours in a culture.

(a) Using a graphing utility, draw a scatter diagram of the data with time as the independent variable.

(b) Using a graphing utility, fit a logistic growth model to the data.

(c) Using a graphing utility, graph the function found in part (b) on the scatter diagram.

(d) What is the predicted carrying capacity of the culture?

(e) Use the function found in part (b) to predict the population of the culture at $t = 19$ hours.

Solution (a) See Figure 52 for a scatter diagram of the data.

Figure 52

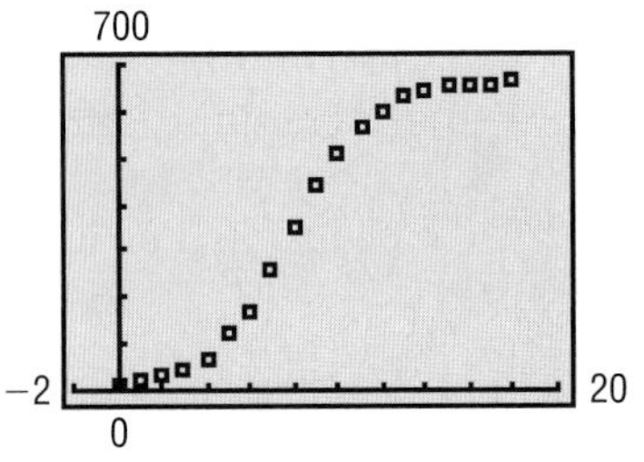

(b) A graphing utility fits a logistic growth model of the form $y = \dfrac{c}{1 + ae^{-bx}}$ by using the LOGISTIC regression option. See Figure 53. The logistic growth function of best fit to the data is

$$y = \frac{663.0}{1 + 71.6e^{-0.5470x}}$$

where y is the amount of yeast biomass in the culture and x is the time.

(c) See Figure 54 for a graph of the function on the scatter diagram.

Figure 53

Figure 54

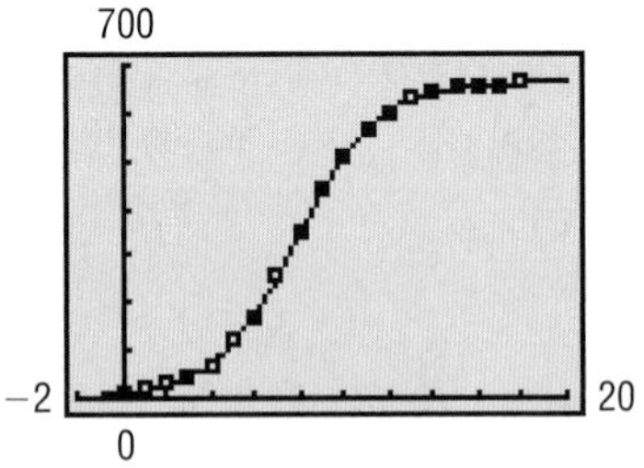

(d) Based on the logistic growth function found in part (b), the carrying capacity of the culture is 663.

(e) Using the logistic growth function found in part (b), the predicted amount of yeast biomass at $t = 19$ hours is

$$y = \frac{663.0}{1 + 71.6e^{-0.5470(19)}} = 661.5$$

NOW WORK PROBLEM 13.

Choosing the Best Model

4 We have discussed seven different models thus far that can be used to explain the relation between two variables, x and y (see Table 8).

TABLE 8

Model	Functional Form	Section
1. Linear	$y = ax + b$	1.7
2. Quadratic	$y = ax^2 + bx + c$	3.1
3. Power	$y = ax^b$	3.2
4. Cubic	$y = ax^3 + bx^2 + cx + d$	3.3
5. Exponential	$y = ab^x$	5.8
6. Logarithmic	$y = a + b \ln x$	5.8
7. Logistic	$y = \dfrac{c}{1 + ae^{-bx}}$	5.8

How can we be certain the *model* we used is the "best" *model* to explain this relation? For example, why did we use an exponential model in Example 1 of this section?

Unfortunately, there is no such thing as *the* correct model. Modeling is not only a science but also an art form. Selecting an appropriate model requires experience and skill in the field in which you are modeling. For example, knowledge of economics is imperative when trying to determine a model to predict unemployment. The main reason for this is that there are theories in the field that can help the modeler to select appropriate relations and variables.

Examining the methodologies used to determine the most appropriate model are beyond the scope of this text. However, we can use scatter diagrams and the correlation coefficient r to decide the "best" model among linear, power, exponential, and logarithmic models. For our purposes, we will say that the "best" model is the one that yields the value of r for which $|r|$ is closest to 1.

When there is no theoretical basis for choosing a particular model, the best approach to follow is to first draw a scatter diagram and obtain a general idea of the shape of the data. Then fit several models to the data whose graph exhibits the same behavior as the scatter diagram and determine which seems to explain the relation between the two variables best. Again, we will define "best" among linear, power, exponential, and logarithmic models as the one that yields the value of $|r|$ closest to 1.

EXAMPLE 4 Selecting the Best Model

Suppose that an astronomer claims to have discovered a tenth planet that is 4.2 billion miles from the sun. How long would it take for this planet to completely orbit the sun?

Solution To solve this problem, we need to develop a model that relates the distance a planet is from the sun and the time it takes to make a complete orbit, called the *sidereal year*. The data in the following table show the distances that the planets are from the sun and their sidereal years.

Planet	Distance from Sun (millions of miles)	Sidereal Year
Mercury	36	0.24
Venus	67	0.62
Earth	93	1.00
Mars	142	1.88
Jupiter	483	11.9
Saturn	887	29.5
Uranus	1,785	84.0
Neptune	2,797	165.0
Pluto	3,675	248.0

Figure 55

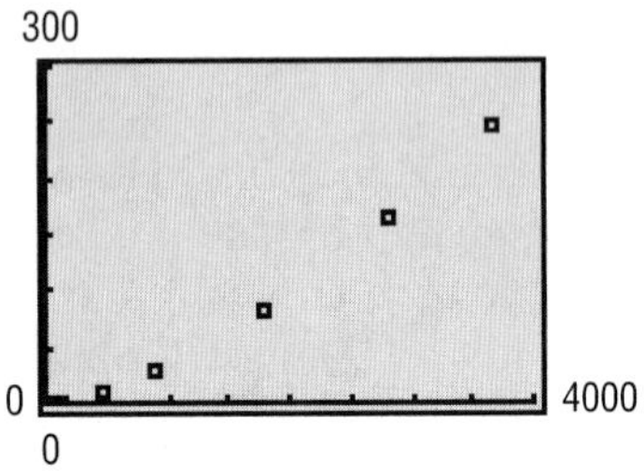

The first step in determining the best model is to draw a scatter diagram. Figure 55 shows a scatter diagram for the planet data using distance from the sun as the independent variable and sidereal year as the dependent variable. From the scatter diagram in Figure 55 it appears that the data follow either a linear model, an exponential model $(k > 0)$, or a power model $(b > 1)$. Therefore, we shall find functions for all three models and use the value of r to determine which is best. Figure 56(a) shows the results obtained from the graphing utility for an exponential model, Figure 56(b) shows the results for a power model, and Figure 56(c) shows the results for a linear model.

Figure 56

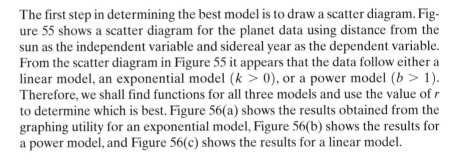

(a) (b) (c)

By looking at the value of r, we conclude that the power model is superior to both the exponential model and the linear model. We conclude that the relation between the distance that a planet is from the sun (x) and its sidereal year (y) is $y = 0.00112x^{1.49938}$.*

* If we square both sides of this equation we obtain

$$y^2 \approx 0.0000013x^3$$

which implies the square of the sidereal year is proportional to the cube of the distance from the sun to the planet. This is Kepler's Third Law of Planetary Motion, named after Johann Kepler. Kepler discovered this relation in 1618 through experimentation without the aid of data analysis.

We can predict the sidereal year of the newly discovered planet by substituting $x = 4200$ (4.2 billion = 4200 million) into the function just found.

$$y = 0.00112(4200)^{1.49938}$$
$$\approx 303.3 \text{ sidereal years}$$

NOW WORK PROBLEM 15.

5.8 EXERCISES

1. Biology A certain bacterium initially increases according to the law of uninhibited growth. A biologist collects the following data for these bacteria:

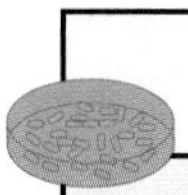

Time (Hours)	Population
0	1000
1	1415
2	2000
3	2828
4	4000
5	5656
6	8000

(a) Using a graphing utility, draw a scatter diagram.
(b) Using a graphing utility, fit an exponential curve to the data.
(c) Express the curve in the form $N(t) = N_0e^{kt}$.
(d) Graph the exponential function found in part (b) or (c) on the scatter diagram.
(e) Using the function found in part (b), predict the population at $t = 7$ hours.

2. Biology A colony of bacteria initially increases according to the law of uninhibited growth. A biologist collects the following data for these bacteria:

Time (Days)	Population
0	50
1	153
2	234
3	357
4	547
5	839
6	1280

(a) Using a graphing utility, draw a scatter diagram.
(b) Using a graphing utility, fit an exponential curve to the data.
(c) Express the curve in the form $N(t) = N_0e^{kt}$.
(d) Graph the exponential function found in part (b) or (c) on the scatter diagram.
(e) Using the function found in part (b), predict the population at $t = 7$ days.

3. Chemistry A chemist has a 100-gram sample of a radioactive material. He records the amount of radioactive material every week for 6 weeks and obtains the following data:

Week	Weight (in Grams)
0	100.0
1	88.3
2	75.9
3	69.4
4	59.1
5	51.8
6	45.5

(a) Using a graphing utility, draw a scatter diagram with week as the independent variable.
(b) Using a graphing utility, fit an exponential curve to the data.
(c) Express the curve in the form $A(t) = A_0e^{kt}$.
(d) Graph the exponential function found in part (b) or (c) on the scatter diagram.
(e) From the result found in part (b), determine the half-life of the radioactive material.
(f) How much radioactive material will be left after 50 weeks?

4. Chemistry A chemist has a 1000-gram sample of a radioactive material. She records the amount of radioactive material remaining in the sample every day for a week and obtains the data on the following page:

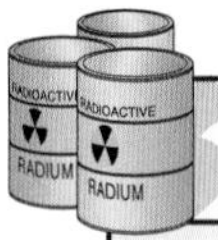

Day	Weight (in Grams)
0	1000.0
1	897.1
2	802.5
3	719.8
4	651.1
5	583.4
6	521.7
7	468.3

(a) Using a graphing utility, draw a scatter diagram with day as the independent variable.
(b) Using a graphing utility, fit an exponential curve to the data.
(c) Express the curve in the form $A(t) = A_0e^{kt}$.
(d) Graph the exponential function found in part (b) or (c) on the scatter diagram.
(e) From the result found in part (b), find the half-life of the radioactive material.
(f) How much radioactive material will be left after 20 days?

5. **Finance** The following data represent the amount of money an investor has in an investment account each year for 10 years. She wishes to determine the effective rate of return on her investment.

Year	Value of Account
1990	\$10,000
1991	\$10,573
1992	\$11,260
1993	\$11,733
1994	\$12,424
1995	\$13,269
1996	\$13,968
1997	\$14,823
1998	\$15,297
1999	\$16,539

(a) Using a graphing utility, draw a scatter diagram with time as the independent variable and the value of the account as the dependent variable.
(b) Using a graphing utility, fit an exponential curve to the data.
(c) Based on the answer in part (b), what was the effective rate of return from this account over the past 10 years?
(d) If the investor plans on retiring in 2020, what will be the predicted value of this account?

6. **Finance** The following data show the amount of money an investor has in an investment account each year for 7 years. He wishes to determine the effective rate of return on his investment.

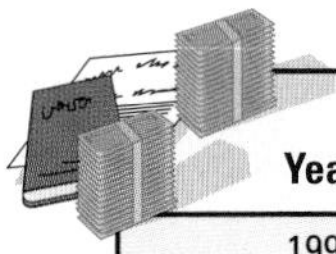

Year	Value of Account
1993	\$20,000
1994	\$21,516
1995	\$23,355
1996	\$24,885
1997	\$27,434
1998	\$30,053
1999	\$32,622

(a) Using a graphing utility, draw a scatter diagram with time as the independent variable and the value of the account as the dependent variable.
(b) Using a graphing utility, fit an exponential curve to the data.
(c) Based on the answer to part (b), what was the effective rate of return from this account over the past 7 years?
(d) If the investor plans on retiring in 2020, what will be the predicted value of this account?

7. **Microsoft Stock** The following data represent the closing price (adjusted for stock splits) of Microsoft Corporation stock at the end of each year.

Year	Closing Price	Year	Closing Price
1989	1.2083	1995	10.9688
1990	2.0903	1996	20.6562
1991	4.354	1997	32.3125
1992	5.3359	1998	69.3438
1993	5.0391	1999	119.125
1994	7.6406		

Source: NASDAQ

(a) Using a graphing utility, draw a scatter diagram.
(b) Using a graphing utility, fit an exponential model to the data.
(c) Express the function in the form $A(t) = A_0e^{kt}$.
(d) Graph the exponential function found in part (b) or (c) on the scatter diagram.
(e) Using the solution to part (b) or (c), predict the closing price of Microsoft stock at the end of 2000 ($t = 12$).

8. Intel The following data represent the closing price (adjusted for stock splits) of Intel Corporation stock at the end of each year.

Year	Closing Price	Year	Closing Price
1989	2.1516	1995	14.1567
1990	2.401	1996	32.6633
1991	3.0559	1997	35.0487
1992	5.4257	1998	59.1787
1993	7.7332	1999	85.0625
1994	7.967		

Source: NASDAQ

(a) Using a graphing utility, draw a scatter diagram.
(b) Using a graphing utility, fit an exponential model to the data.
(c) Express the function in the form $A(t) = A_0e^{kt}$.
(d) Graph the exponential function found in part (b) or (c) on the scatter diagram.
(e) Using the solution to part (b) or (c), predict the closing price of Intel stock at the end of 2000.

9. Wind Chill Factor The following data represent the wind speed (mph) and wind chill factor at an air temperature of 15°F.

Wind Speed (mph)	Wind Chill Factor
5	12
10	−3
15	−11
20	−17
25	−22
30	−25
35	−27

Source: Information Please Almanac

(a) Using a graphing utility, draw a scatter diagram with wind speed as the independent variable.
(b) Using a graphing utility, fit a logarithmic model to the data.
(c) Using a graphing utility, draw the logarithmic function found in part (b) on the scatter diagram.
(d) Use the function found in part (b) to predict the wind chill factor if the air temperature is 15°F and the wind speed is 23 mph.

10. Wind Chill Factor The following data represent the wind speed (mph) and wind chill factor at an air temperature of 5°F.

Wind Speed (mph)	Wind Chill Factor
5	0
10	−15
15	−25
20	−31
25	−36
30	−41
35	−43

Source: Information Please Almanac

(a) Using a graphing utility, draw a scatter diagram with wind speed as the independent variable.
(b) Using a graphing utility, fit a logarithmic model to the data.
(c) Using a graphing utility, draw the logarithmic function found in part (b) on the scatter diagram.
(d) Use the function found in part (b) to predict the wind chill factor if the air temperature is 5°F and the wind speed is 23 mph.

11. Population Model The following data obtained from the U.S. Census Bureau represent the population of the United States. An ecologist is interested in finding a function that describes the population of the United States.

Year	Population
1900	76,212,168
1910	92,228,496
1920	106,021,537
1930	123,202,624
1940	132,164,569
1950	151,325,798
1960	179,323,175
1970	203,302,031
1980	226,542,203
1990	248,709,873

(a) Using a graphing utility, draw a scatter diagram of the data using the year as the independent variable and population as the dependent variable.
(b) Using a graphing utility, fit a logistic model to the data.
(c) Using a graphing utility, draw the function found in part (b) on the scatter diagram.
(d) Based on the function found in part (b), what is the carrying capacity of the United States?
(e) Use the function found in part (b) to predict the population of the United States in 2000.

12. Population Model The following data obtained from the U.S. Census Bureau represent the world population. An ecologist is interested in finding a function that describes the world population.

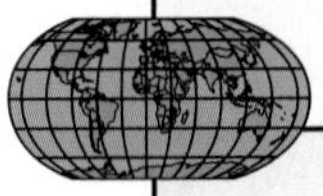

Year	Population (in Billions)
1981	4.533
1982	4.614
1983	4.695
1984	4.775
1985	4.856
1986	4.941
1987	5.029
1988	5.117
1989	5.205
1990	5.295
1991	5.381
1992	5.469
1993	5.556
1994	5.644
1995	5.732

(a) Using a graphing utility, draw a scatter diagram of the data using year as the independent variable and population as the dependent variable.
(b) Using a graphing utility, fit a logistic model to the data.
(c) Using a graphing utility, draw the function found in part (b) on the scatter diagram.
(d) Based on the function found in part (b), what is the carrying capacity of the world?
(e) Use the function found in part (b) to predict the population of the world in 2000.

13. Population Model The following data obtained from the U.S. Census Bureau represent the population of Illinois. An urban economist is interested in finding a model that describes the population of Illinois.
(a) Using a graphing utility, draw a scatter diagram of the data using year as the independent variable and population as the dependent variable.
(b) Using a graphing utility, fit a logistic model to the data.
(c) Using a graphing utility, draw the function found in part (b) on the scatter diagram.
(d) Based on the function found in part (b), what is the carrying capacity of Illinois?
(e) Use the function found in part (b) to predict the population of Illinois in 2000.

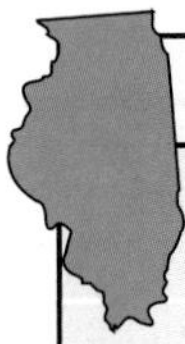

Year	Population
1900	4,821,550
1910	5,638,591
1920	6,485,280
1930	7,630,654
1940	7,897,241
1950	8,712,176
1960	10,081,158
1970	11,110,285
1980	11,427,409
1990	11,430,602

14. Population Model The following data obtained from the U.S. Census Bureau represent the population of Pennsylvania. An urban economist is interested in finding a model that describes the population of Pennsylvania.

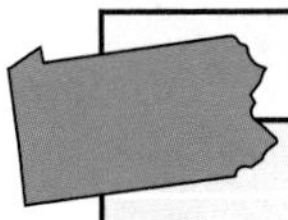

Year	Population
1900	6,302,115
1910	7,665,111
1920	8,720,017
1930	9,631,350
1940	9,900,180
1950	10,498,012
1960	11,319,366
1970	11,800,766
1980	11,864,720
1990	11,881,643

(a) Using a graphing utility, draw a scatter diagram of the data using year as the independent variable and population as the dependent variable.
(b) Using a graphing utility, fit a logistic model to the data.
(c) Using a graphing utility, draw the function found in part (b) on the scatter diagram.
(d) Based on the function found in part (b), what is the carrying capacity of Pennsylvania?
(e) Use the function found in part (b) to predict the population of Pennsylvania in 2000.

For Problems 15–20, the independent variable is x and the dependent variable is y. (a) For each set of data, draw a scatter diagram of the data. (b) Use an exponential model, logarithmic model, power model, and linear model to find the "best" model that explains the relationship between the variables.

15.

x	1	2	3	4	5
y	164	269	450	740	1220

16.

x	1	2	3	4	5
y	110	245	553	1228	2728

17.

x	10	20	30	40	50
y	6.94	13.78	20.45	21.23	22.95

18.

x	10	20	30	40	50
y	4.95	6.46	7.23	7.78	8.23

19.

x	1	1.3	1.7	2.1	2.4	2.7
y	9.92	4.56	1.42	0.37	0.12	0.032

20.

x	1	1.3	1.7	2.1	2.4	2.7
y	1.72	1.94	2.48	3.28	4.01	5.07

21. Economics The data that follow give the levels of the Consumer Price Index (CPI) in December of the year indicated for food items.

(a) Using a graphing utility, draw a scatter diagram of the data using the year as the independent variable and the CPI as the dependent variable.

(b) Use an exponential model, logarithmic model, power model, and linear model to find the "best" model to describe the relation between the year and the CPI.

(c) Use this model to predict the CPI for food items for 1998.

Year	CPI
1988	121.2
1989	128.0
1990	134.8
1991	137.2
1992	139.0
1993	143.0
1994	147.1
1995	150.2
1996	156.5
1997	158.9

Source: Bureau of Labor Statistics.

22. Economics The following data give the level of the Consumer Price Index in December of the indicated year for housing.

Year	CPI
1988	120.5
1989	125.3
1990	130.9
1991	135.5
1992	139.1
1993	142.9
1994	145.9
1995	150.3
1996	154.6
1997	158.3

Source: Bureau of Labor Statistics.

(a) Using a graphing utility, draw a scatter diagram of the data using the year as the independent variable and the CPI as the dependent variable.

(b) Use an exponential model, logarithmic model, power model, and linear model to find the "best" model to describe the relation between the year and the CPI.

(c) Use this model to predict the CPI for housing for 1998.

23. Economics The following data represent the Gross Domestic Product (GDP) in billions of dollars.

Year	Gross Domestic Product
1992	6244.4
1993	6558.1
1994	6947.0
1995	7269.9
1996	7661.6
1997	8110.9

Source: U.S. Bureau of Economic Analysis.

(a) Using a graphing utility, draw a scatter diagram of the data using the year as the independent variable and the GDP as the dependent variable.

(b) Use an exponential model, logarithmic model, power model, and linear model to find the "best" model to describe the relation between the year and the GDP.

(c) Use this model to predict the GDP for 1998.

24. Economics The following data represent the Gross National Product (GNP) in billions of dollars.

Year	Gross National Product
1992	6255.5
1993	6576.8
1994	6955.2
1995	7287.1
1996	7674.0
1997	8102.9

Source: U.S. Bureau of Economic Analysis.

(a) Using a graphing utility, draw a scatter diagram of the data using the year as the independent variable and the GNP as the dependent variable.

(b) Use an exponential model, logarithmic model, power model, and linear model to find the "best" model to describe the relation between the year and the GNP.

(c) Use this model to predict the GNP for 1998.

CHAPTER REVIEW

Things To Know

One-to-one function f (p. 282)
If $f(x_1) = f(x_2)$, then $x_1 = x_2$ for any choice of x_1 and x_2 in the domain.

Horizontal-line test (p. 282)
If every horizontal line intersects the graph of a function f in at most one point, then f is one-to-one.

Inverse function f^{-1} of f (p. 283)
Domain of f = Range of f^{-1}; Range of f = Domain of f^{-1}.
$f^{-1}(f(x)) = x$ and $f(f^{-1}(x)) = x$.
Graphs of f and f^{-1} are symmetric with respect to the line $y = x$.

Properties of the exponential function (p. 294–295 and 296)

$f(x) = a^x, \quad a > 1$ — Domain: $(-\infty, \infty)$; Range: $(0, \infty)$; x-intercepts: none; y-intercept: 1; horizontal asymptote: x-axis as $x \to -\infty$; increasing; one-to-one; smooth; continuous
See Figure 13 for a typical graph.

$f(x) = a^x, \quad 0 < a < 1$ — Domain: $(-\infty, \infty)$; Range: $(0, \infty)$; x-intercepts: none; y-intercept: 1; horizontal asymptote: x-axis as $x \to \infty$; decreasing; one-to-one; smooth; continuous
See Figure 17 for a typical graph.

Number e (p. 297)
Value approached by the expression $\left(1 + \frac{1}{n}\right)^n$ as $n \to \infty$; that is, $\lim_{n\to\infty}\left(1 + \frac{1}{n}\right)^n = e$

Properties of the logarithmic function (p. 308)

$f(x) = \log_a x, \quad a > 1$ $(y = \log_a x$ means $x = a^y)$ — Domain: $(0, \infty)$; Range: $(-\infty, \infty)$; x-intercept: 1; y-intercept: none; vertical asymptote: $x = 0$ (y-axis); increasing; one-to-one; smooth; continuous
See Figure 26(b) for a typical graph.

$f(x) = \log_a x, 0 < a < 1$ $(y = \log_a x$ means $x = a^y)$ — Domain: $(0, \infty)$; Range: $(-\infty, \infty)$; x-intercept: 1; y-intercept: none; vertical asymptote: $x = 0$ (y-axis); decreasing; one-to-one; smooth; continuous
See Figure 26(a) for a typical graph.

Natural logarithm (p. 308)
$y = \ln x$ means $x = e^y$.

Properties of logarithms (p. 316–317)
$\log_a 1 = 0 \qquad \log_a a = 1 \qquad a^{\log_a M} = M \qquad \log_a a^r = r$

$\log_a(MN) = \log_a M + \log_a N \qquad \log_a\left(\frac{M}{N}\right) = \log_a M - \log_a N$

$\log_a\left(\frac{1}{N}\right) = -\log_a N \qquad \log_a M^r = r \log_a M$

Formulas

Change-of-Base Formula (p. 320)	$\log_a M = \dfrac{\log_b M}{\log_b a}$
Compound interest (p. 330)	$A = P\left(1 + \dfrac{r}{n}\right)^{nt}$
Continuous compounding (p. 332)	$A = Pe^{rt}$
Present value (p. 333)	$P = A\left(1 + \dfrac{r}{n}\right)^{-nt}$ or $P = Ae^{-rt}$
Growth and decay (p. 339)	$A(t) = A_0 e^{kt}$
Logistic growth (p. 344)	$P(t) = \dfrac{c}{1 + ae^{-bt}}$

How To

Determine whether a function is one-to-one (p. 281)

Obtain the graph of the inverse function from the graph of a function (p. 285)

Find an inverse function (p. 286)

Evaluate exponential functions (p. 292)

Graph exponential functions (p. 293)

Define the number e (p. 297)

Solve exponential equations (p. 299)

Change exponential expressions to logarithmic expressions (p. 305)

Change logarithmic expressions to exponential expressions (p. 306)

Evaluate logarithmic functions (p. 306)

Determine the domain of a logarithmic function (p. 306)

Graph logarithmic functions (p. 307)

Solve logarithmic equations (p. 310)

Write a logarithmic expression as a sum or difference of logarithms (p. 318)

Write a logarithmic expression as a single logarithm (p. 318)

Evaluate logarithms whose base is neither 10 nor e (p. 320)

Graph logarithmic functions whose base is neither 10 nor e (p. 321)

Solve logarithmic equations using the properties of logarithms (p. 324)

Solve exponential equations (p. 325)

Solve logarithmic and exponential equations using a graphing utility (p. 327)

Determine the future value of a lump sum of money (p. 328)

Calculate effective rates of return (p. 332)

Determine the present value of a lump sum of money (p. 333)

Determine the time required to double or triple a lump sum of money (p. 334)

Find equations of populations that obey the law of uninhibited growth (p. 339)

Find equations of populations that obey the law of decay (p. 341)

Use Newton's Law of Cooling (p. 343)

Use logistic growth models (p. 344)

Use a graphing utility to obtain the exponential function of best fit (p. 349)

Use a graphing utility to obtain the logarithmic function of best fit (p. 350)

Use a graphing utility to obtain the logistic function of best fit (p. 351)

Determine the model of best fit (p. 353)

Fill-in-the-Blank Items

1. If every horizontal line intersects the graph of a function f at no more than one point, then f is a(n) ____________ function.
2. If f^{-1} denotes the inverse of a function f, then the graphs of f and f^{-1} are symmetric with respect to the line ____________.
3. The graph of every exponential function $f(x) = a^x, a > 0, a \neq 1$, passes through the two points ____________.
4. If the graph of an exponential function $f(x) = a^x, a > 0, a \neq 1$, is decreasing, then its base must be less than ____________.
5. If $3^x = 3^4$, then $x =$ ____________.
6. The logarithm of a product equals the ____________ of the logarithms.
7. For every base, the logarithm of ____________ equals 0.
8. If $\log_8 M = \log_5 7/\log_5 8$, then $M =$ ____________.
9. The domain of the logarithmic function $f(x) = \log_a x$ consists of ____________.
10. The graph of every logarithmic function $f(x) = \log_a x, a > 0, a \neq 1$, passes through the two points ____________.
11. If the graph of a logarithmic function $f(x) = \log_a x, a > 0, a \neq 1$, is increasing, then its base must be larger than ____________.
12. If $\log_3 x = \log_3 7$, then $x =$ ____________.

True/False Items

T F 1. If f and g are inverse functions, then the domain of f is the same as the domain of g.

T F 2. If f and g are inverse functions, then their graphs are symmetric with respect to the line $y = x$.

T F 3. The graph of every exponential function $f(x) = a^x, a > 0, a \neq 1$, will contain the points $(0, 1)$ and $(1, a)$.

T F 4. The graphs of $y = 3^{-x}$ and $y = \left(\frac{1}{3}\right)^x$ are identical.

T F **5.** The present value of \$1000 to be received after 2 years at 10% per annum compounded continuously is approximately \$1205.

T F **6.** If $y = \log_a x$, then $y = a^x$.

T F **7.** The graph of every logarithmic function $f(x) = \log_a x$, $a > 0$, $a \neq 1$, will contain the points $(1, 0)$ and $(a, 1)$.

T F **8.** $a^{\log_M a} = M$, where $a > 0$, $a \neq 1$, $M > 0$.

T F **9.** $\log_a(M + N) = \log_a M + \log_a N$, where $a > 0$, $a \neq 1$, $M > 0$, $N > 0$.

T F **10.** $\log_a M - \log_a N = \log_a(M/N)$, where $a > 0$, $a \neq 1$, $M > 0$, $N > 0$.

Review Exercises

Blue problem numbers indicate the authors' suggestions for use in a Practice Test.

In Problems 1–6, the function f is one-to-one. Find the inverse of each function and check your answer. Find the domain and range of f and f^{-1}. Use a graphing utility to simultaneously graph f, f^{-1}, and $y = x$ on the same square screen.

1. $f(x) = \dfrac{2x + 3}{5x - 2}$
2. $f(x) = \dfrac{2 - x}{3 + x}$
3. $f(x) = \dfrac{1}{x - 1}$
4. $f(x) = \sqrt{x - 2}$
5. $f(x) = \dfrac{3}{x^{1/3}}$
6. $f(x) = x^{1/3} + 1$

In Problems 7–12, evaluate each expression. Do not use a graphing utility.

7. $\log_2\left(\frac{1}{8}\right)$
8. $\log_3 81$
9. $\ln e^{\sqrt{2}}$
10. $e^{\ln 0.1}$
11. $2^{\log_2 0.4}$
12. $\log_2 2^{\sqrt{3}}$

In Problems 13–18, write each expression as the sum and/or difference of logarithms. Express powers as factors.

13. $\log_3\left(\dfrac{uv^2}{w}\right)$
14. $\log_2(a^2\sqrt{b})^4$
15. $\log(x^2\sqrt{x^3 + 1})$
16. $\log_5\left(\dfrac{x^2 + 2x + 1}{x^2}\right)$
17. $\ln\left(\dfrac{x\sqrt[3]{x^2 + 1}}{x - 3}\right)$
18. $\ln\left(\dfrac{2x + 3}{x^2 - 3x + 2}\right)^2$

In Problems 19–24, write each expression as a single logarithm.

19. $3\log_4 x^2 + \dfrac{1}{2}\log_4 \sqrt{x}$
20. $-2\log_3\left(\dfrac{1}{x}\right) + \dfrac{1}{3}\log_3 \sqrt{x}$
21. $\ln\left(\dfrac{x - 1}{x}\right) + \ln\left(\dfrac{x}{x + 1}\right) - \ln(x^2 - 1)$
22. $\log(x^2 - 9) - \log(x^2 + 7x + 12)$
23. $2\log 2 + 3\log x - \dfrac{1}{2}[\log(x + 3) + \log(x - 2)]$
24. $\dfrac{1}{2}\ln(x^2 + 1) - 4\ln\dfrac{1}{2} - \dfrac{1}{2}[\ln(x - 4) + \ln x]$

In Problems 25 and 26, use the Change-of-Base Formula and a calculator to evaluate each logarithm. Round your answer to three decimal places.

25. $\log_4 19$
26. $\log_2 21$

In Problems 27–32, find y as a function of x. The constant C is a positive number.

27. $\ln y = 2x^2 + \ln C$
28. $\ln(y - 3) = \ln 2x^2 + \ln C$
29. $\ln(y - 3) + \ln(y + 3) = x + C$
30. $\ln(y - 1) + \ln(y + 1) = -x + C$
31. $e^{y+C} = x^2 + 4$
32. $e^{3y-C} = (x + 4)^2$

In Problems 33–42, use transformations to graph each function. Determine the domain, range, and any asymptotes. Verify your results using a graphing utility.

33. $f(x) = 2^{x-3}$
34. $f(x) = -2^x + 3$
35. $f(x) = \dfrac{1}{2}(3^{-x})$
36. $f(x) = 1 + 3^{2x}$

37. $f(x) = 1 - e^x$
38. $f(x) = 3 + \ln x$
39. $f(x) = 3e^x$
40. $f(x) = \frac{1}{2}\ln x$
41. $f(x) = 3 - e^{-x}$
42. $f(x) = 4 - \ln(-x)$

In Problems 43–62, solve each equation. Verify your result using a graphing utility.

43. $4^{1-2x} = 2$
44. $8^{6+3x} = 4$
45. $3^{x^2+x} = \sqrt{3}$
46. $4^{x-x^2} = \frac{1}{2}$
47. $\log_x 64 = -3$
48. $\log_{\sqrt{2}} x = -6$
49. $5^x = 3^{x+2}$
50. $5^{x+2} = 7^{x-2}$
51. $9^{2x} = 27^{3x-4}$
52. $25^{2x} = 5^{x^2-12}$
53. $\log_3 \sqrt{x-2} = 2$
54. $2^{x+1} \cdot 8^{-x} = 4$
55. $8 = 4^{x^2} \cdot 2^{5x}$
56. $2^x \cdot 5 = 10^x$
57. $\log_6(x+3) + \log_6(x+4) = 1$
58. $\log_{10}(7x - 12) = 2\log_{10} x$
59. $e^{1-x} = 5$
60. $e^{1-2x} = 4$
61. $2^{3x} = 3^{2x+1}$
62. $2^{x^3} = 3^{x^2}$

In Problems 63 and 64, use the following result: If x is the atmospheric pressure (measured in millimeters of mercury), then the formula for the altitude h(x) (measured in meters above sea level) is

$$h(x) = (30T + 8000)\log\left(\frac{P_0}{x}\right)$$

where T is the temperature (in degrees Celsius) and P_0 is the atmospheric pressure at sea level, which is approximately 760 millimeters of mercury.

63. Finding the Altitude of an Airplane At what height is a Piper Cub whose instruments record an outside temperature of 0°C and a barometric pressure of 300 millimeters of mercury?

64. Finding the Height of a Mountain How high is a mountain if instruments placed on its peak record a temperature of 5°C and a barometric pressure of 500 millimeters of mercury?

65. Amplifying Sound An amplifier's power output P (in watts) is related to its decibel voltage gain d by the formula $P = 25e^{0.1d}$.

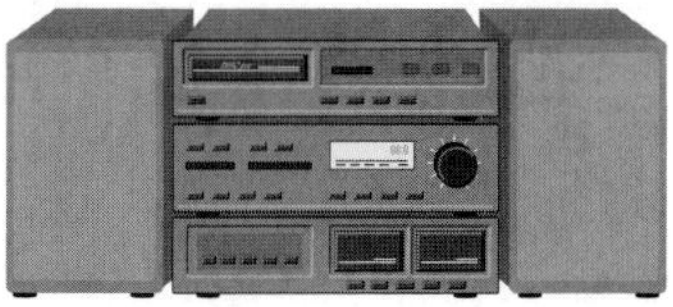

(a) Find the power output for a decibel voltage gain of 4 decibels.
(b) For a power output of 50 watts, what is the decibel voltage gain?

66. Limiting Magnitude of a Telescope A telescope is limited in its usefulness by the brightness of the star it is aimed at and by the diameter of its lens. One measure of a star's brightness is its *magnitude*: the dimmer the star, the larger its magnitude. A formula for the limiting magnitude L of a telescope, that is, the magnitude of the dimmest star that it can be used to view, is given by

$$L = 9 + 5.1\log d$$

where d is the diameter (in inches) of the lens.

(a) What is the limiting magnitude of a 3.5-inch telescope?
(b) What diameter is required to view a star of magnitude 14?

67. Salvage Value The number of years n for a piece of machinery to depreciate to a known salvage value can be found using the formula

$$n = \frac{\log s - \log i}{\log(1 - d)}$$

where s is the salvage value of the machinery, i is its initial value, and d is the annual rate of depreciation.
(a) How many years will it take for a piece of machinery to decline in value from \$90,000 to \$10,000 if the annual rate of depreciation is 0.20 (20%)?
(b) How many years will it take for a piece of machinery to lose half of its value if the annual rate of depreciation is 15%?

68. Funding a College Education A child's grandparents purchase a \$10,000 bond fund that matures in 18 years to be used for her college education. The bond fund pays 4% interest compounded semiannually. How much will the bond fund be worth at maturity?

69. Funding a College Education A child's grandparents wish to purchase a bond fund that matures in 18 years to be used for her college education. The bond fund pays 4% interest compounded semiannually. How much should they purchase so that the bond fund will be worth \$85,000 at maturity?

70. Funding an IRA First Colonial Bankshares Corporation advertised the following IRA investment plans.

Target IRA Plans

For each $5000 Maturity Value Desired	
Deposit:	**At a Term of:**
$620.17	20 Years
$1045.02	15 Years
$1760.92	10 Years
$2967.26	5 Years

(a) Assuming continuous compounding, what was the annual rate of interest that they offered?
(b) First Colonial Bankshares claims that $4000 invested today will have a value of over $32,000 in 20 years. Use the answer found in part (a) to find the actual value of $4000 in 20 years. Assume continuous compounding.

71. Estimating the Date that a Prehistoric Man Died The bones of a prehistoric man found in the desert of New Mexico contain approximately 5% of the original amount of carbon 14. If the half-life of carbon 14 is 5600 years, approximately how long ago did the man die?

72. Temperature of a Skillet A skillet is removed from an oven whose temperature is 450°F and placed in a room whose temperature is 70°F. After 5 minutes, the temperature of the skillet is 400°F. How long will it be until its temperature is 150°F?

73. World Population According to the U.S. Census Bureau, the growth rate of the world's population in 1997 was $k = 1.33\% = 0.0133$. The population of the world in 1997 was 5,840,445,216. Letting $t = 0$ represent 1997, use the uninhibited growth model to predict the world's population in the year 2000.

74. Radioactive Decay The half-life of radioactive cobalt is 5.27 years. If 100 grams of radioactive cobalt is present now, how much will be present in 20 years? In 40 years?

75. Logistic Growth The logistic growth model

$$P(t) = \frac{0.8}{1 + 1.67e^{-0.16t}}$$

represents the proportion of new computers sold that utilize the Microsoft Windows 98 operating system. Let $t = 0$ represent 1998, $t = 1$ represent 1999, and so on.
(a) What proportion of new computers sold in 1998 utilized Windows 98?
(b) Determine the maximum proportion of new computers sold that will utilize Windows 98.
(c) Using a graphing utility, graph $P(t)$.
(d) When will 75% of new computers sold utilize Windows 98?

76. CBL Experiment The following data were collected by placing a temperature probe in a portable heater, removing the probe, and then recording temperature over time.

According to Newton's Law of Cooling, these data should follow an exponential model.

(a) Using a graphing utility, draw a scatter diagram for the data.
(b) Using a graphing utility, fit an exponential model to the data.
(c) Graph the exponential function found in part (b) on the scatter diagram.
(d) Predict how long it will take for the probe to reach a temperature of 110°F.

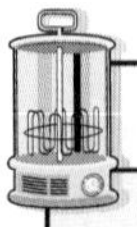

Time	Temperature (F°)
0	165.07
1	164.77
2	163.99
3	163.22
4	162.82
5	161.96
6	161.20
7	160.45
8	159.35
9	158.61
10	157.89
11	156.83
12	156.11
13	155.08
14	154.40
15	153.72

77. The following data represent the per capita usage of carrots.

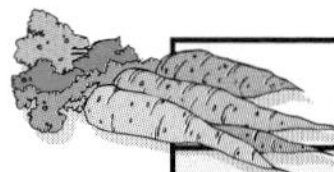

Year	Per Capita Usage
1985 ($t = 1$)	6.5
1990 ($t = 6$)	8.3
1991 ($t = 7$)	7.7
1992 ($t = 8$)	8.3
1993 ($t = 9$)	8.2
1994 ($t = 10$)	8.7
1995 ($t = 11$)	9.0
1996 ($t = 12$)	10.2

Source: U.S. Department of Agriculture.

(a) Using a graphing utility, draw a scatter diagram of the data using time as the independent variable and per capita consumption as the dependent variable.
(b) Use an exponential model, logarithmic model, power model, and linear model to find the "best" model to describe the relation between time and per capita consumption. Record each model's correlation coefficient.
(c) Use this model to predict the per capita consumption of carrots in 1997.

78. AIDS Infections The following data represent the cumulative number of HIV/AIDS cases reported worldwide for the years 1980–1995. Let $x = 1$ represent 1980, $x = 2$ represent 1981, and so on.
(a) Using a graphing utility, draw a scatter diagram of the data using the year as the independent variable and the number of HIV infections as the dependent variable.
(b) Using a graphing utility, determine the logistic function of best fit.
(c) Graph the function found in part (b) on the scatter diagram.
(d) Based on the function found in part (b), what is the maximum number of HIV infections predicted to be?
(e) Using a graphing utility, determine the exponential function of best fit.
(f) Graph the function found in part (e) on the scatter diagram.
(g) Based on the graphs drawn in parts (c) and (f), which model appears to best describe the relation between the year and the number of HIV infections reported?

Year	HIV Infections (Millions)
1980, 1	0.2
1981, 2	0.6
1982, 3	1.1
1983, 4	1.8
1984, 5	2.7
1985, 6	3.9
1986, 7	5.3
1987, 8	6.9
1988, 9	8.7
1989, 10	10.7
1990, 11	13.0
1991, 12	15.5
1992, 13	18.5
1993, 14	21.9
1994, 15	25.9
1995, 16	30.6

Source: Global AIDS Policy Coalition, Harvard School of Public Health, Cambridge, Mass., Private Communication, January, 18, 1996.

PROJECT AT MOTOROLA

Thermal Fatigue of Solder Interconnects

What happens to an electronic package when it is subjected to repeated temperature changes? That question is important if you want your electronic product to be reliable. Every time you use your cell phone, pager, computer, or start your car, the electronics inside begin to warm up. When materials warm up they expand. However, different materials expand at different rates. For example, the glass-epoxy laminate, called a printed circuit board, has a coefficient of thermal expansion (CTE) around 12 to 15 $\times 10^{-6}$/deg C, while the silicon IC located inside an electronic package has a CTE equal to 2.5 $\times 10^{-6}$/deg C. These differences will induce inelastic deformation to the solder interconnect. The solder interconnect makes the electrical connection from the PCB to the electronic package and is usually made up of a low melting point (183 deg C) alloy comprised of tin and lead. After using your portable product, computer, or turning the car engine off, the electronics will cool off. These temperature cycles result in repeated expansion and contraction of the material used to make the electronic assemblies. The greater the temperature change or the greater the difference in CTE between materials, the greater the inelastic strain imparted to the solder joint. A decrease in strain will increase fatigue life. Let's look at the result of an experiment.

EXPERIMENTAL FATIGUE DATA

Solder Joint Strain, εp	Fatigue Cycles, Nf
.01	10,000
.035	1000
.1	100
.4	10
1.5	1

Using the experimental fatigue life data given above, answer the following:

1. Draw a scatter diagram of the data with solder joint strain as the independent variable.

2. Let $X = \ln(\varepsilon p)$. Draw a scatter diagram of the transformed data with X as the independent variable and fatigue cycles as the dependent variable. What happens to the shape of the scatter diagram?
3. Let $Y = \ln(Nf)$. Draw a scatter diagram of the transformed data with X as the independent variable and Y as the dependent variable. What type of relation would best describe the data?
4. Find the line of best fit to the transformed data using a graphing utility. Graph the line of best fit on the scatter diagram drawn in (3).
5. Write the equation from (4) in the form $Nf = e^b(\varepsilon p)^n_m$. [**Hint**: The line of best fit is $\ln(Nf) = m\ln(\varepsilon p) + b$. Think of a property of logarithms that will eliminate the natural logarithm].
6. If the solder joint strain is 0.02 what is the expected fatigue life? What is the inelastic solder joint strain if the fatigue life is 3000 cycles?
7. Rewrite the function from part (5) so that the inelastic solder joint strain is a function of the fatigue life. Compute the solder joint strain when the fatigue life equals 3000 cycles.

Example of an electronic package (top view) soldered to a PCB used to evaluate fatigue life.

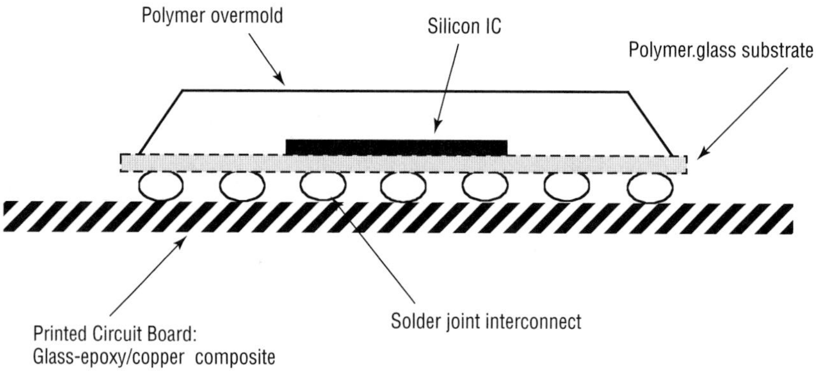

Schematic diagram of the inside of an electronic assembly.

CHAPTER

6 Trigonometric Functions

Outline

6.1 Angles and Their Measure

6.2 Trigonometric Functions: Unit Circle Approach

6.3 Properties of the Trigonometric Functions

6.4 Graphs of the Sine and Cosine Functions

6.5 Graphs of the Tangent, Cotangent, Cosecant, and Secant Functions

6.6 Sinusoidal Graphs; Sinusoidal Curve Fitting

Chapter Review

Project at Motorola

FIELD TRIP TO MOTOROLA

For many centuries, trigonometry has been the basis for science, astronomy, and navigation. Today trigonometry is employed in various disciplines of common life such as civil engineering, acoustics, analog and digital wireline and wireless communications, and many other applications related to wave theory.

INTERVIEW AT MOTOROLA

Margot Karam is a senior staff engineer working as a researcher for Motorola Labs in Schaumburg, Illinois. She is part of a dynamic team designing a digital cellular phone able to operate in every country of the world. Her work consists in providing a solution to process fast data at low cost.

Before joining Motorola Labs in the USA, she worked in a product group of Motorola in France as a designer with a focus on delivering high-quality, testable, integrated digital circuits. She is proud to work for such an international company known as #1 provider for the best quality products.

6.1 ANGLES AND THEIR MEASURE

OBJECTIVES

1. Convert between Degrees, Minutes, Seconds, and Decimal Forms for Angles
2. Find the Arc Length of a Circle
3. Convert from Degrees to Radians
4. Convert from Radians to Degrees
5. Find the Linear Speed of an Object Traveling in Circular Motion

Figure 1

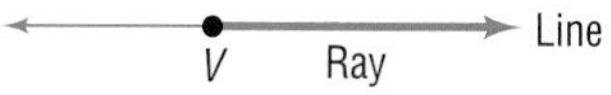

A **ray**, or **half-line**, is that portion of a line that starts at a point V on the line and extends indefinitely in one direction. The starting point V of a ray is called its **vertex**. See Figure 1.

If two rays are drawn with a common vertex, they form an **angle**. We call one of the rays of an angle the **initial side** and the other the **terminal side**. The angle that is formed is identified by showing the direction and amount of rotation from the initial side to the terminal side. If the rotation is in the counterclockwise direction, the angle is **positive**; if the rotation is clockwise, the angle is **negative**. See Figure 2. Lowercase Greek letters, such as α (alpha), β (beta), γ (gamma), and θ (theta), will be used to denote angles. Notice in Figure 2(a) that the angle α is positive because the direction of the rotation from the initial side to the terminal side is counterclockwise. The angle β in Figure 2(b) is negative because the rotation is clockwise. The angle γ in Figure 2(c) is positive. Notice that the angle α in Figure 2(a) and the angle γ have the same initial side and the same terminal side and both are positive. However, α and γ are unequal, because the amount of rotation required to go from the initial side to the terminal side is greater for angle γ than for angle α.

Figure 2

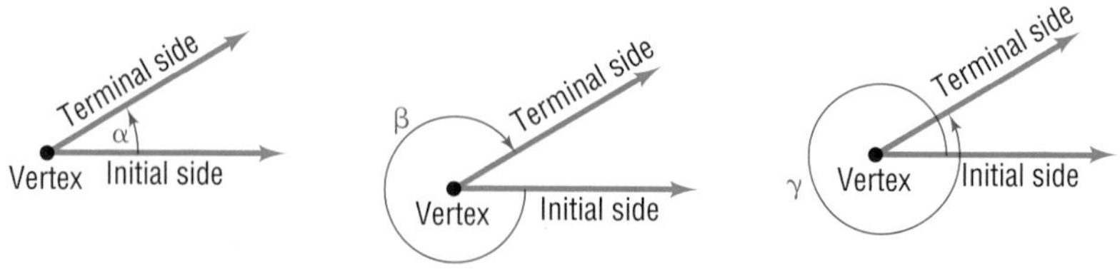

(a) Positive angle Counterclockwise rotation

(b) Negative angle Clockwise rotation

(c) Positive angle Counterclockwise rotation

An angle θ is said to be in **standard position** if its vertex is at the origin of a rectangular coordinate system and its initial side coincides with the positive x-axis. See Figure 3.

Figure 3

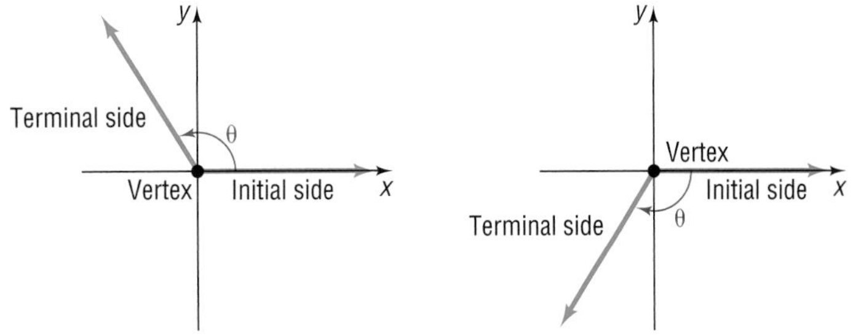

(a) θ in standard position; θ positive

(b) θ in standard position; θ negative

When an angle θ is in standard position, the terminal side will lie either in a quadrant, in which case we say that θ **lies in that quadrant**, or on the x-axis or the y-axis, in which case we say that θ is a **quadrantal angle**. For example, the angle θ in Figure 4(a) lies in quadrant II, the angle θ in Figure 4(b) lies in quadrant IV, and the angle θ in Figure 4(c) is a quadrantal angle.

Figure 4

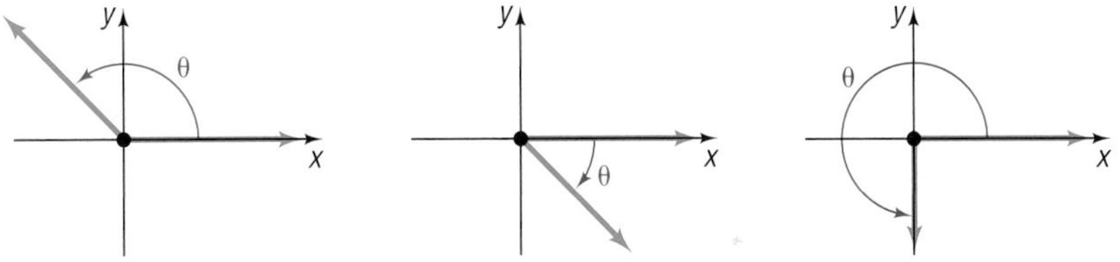

(a) θ lies in quadrant II **(b)** θ lies in quadrant IV **(c)** θ is a quadrantal angle

We measure angles by determining the amount of rotation needed for the initial side to become coincident with the terminal side. There are two commonly used measures for angles: *degrees* and *radians*.

Degrees

The angle formed by rotating the initial side exactly once in the counterclockwise direction until it coincides with itself (1 revolution) is said to measure 360 degrees, abbreviated 360°. Thus **one degree, 1°**, is $\frac{1}{360}$ revolution. A **right angle** is an angle that measures 90°, or $\frac{1}{4}$ revolution; a **straight angle** is an angle that measures 180°, or $\frac{1}{2}$ revolution. See Figure 5. As Figure 5(b) shows, it is customary to indicate a right angle by using the symbol ⊾.

Figure 5

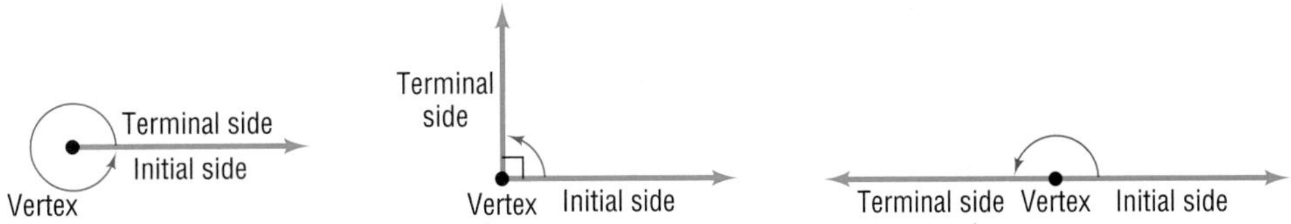

(a) 1 revolution counterclockwise, 360°

(b) $\frac{1}{4}$ revolution counterclockwise, 90°

(c) $\frac{1}{2}$ revolution counterclockwise, 180°

It is also customary to refer to an angle that measures θ degrees as an angle of θ degrees.

EXAMPLE 1 Drawing an Angle

Draw each angle.

(a) 45° (b) -90° (c) 225° (d) 405°

Solution (a) An angle of 45° is $\frac{1}{2}$ of a right angle. See Figure 6.

(b) An angle of -90° is $\frac{1}{4}$ revolution in the clockwise direction. See Figure 7.

Figure 6

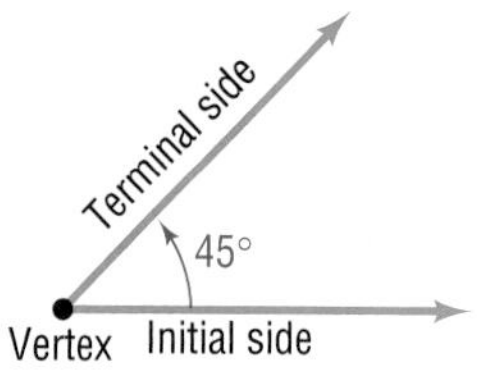

Figure 7

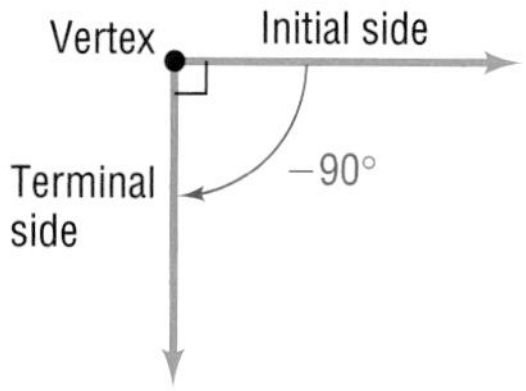

(c) An angle of 225° consists of a rotation through 180° followed by a rotation through 45°. See Figure 8.

(d) An angle of 405° consists of 1 revolution (360°) followed by a rotation through 45°. See Figure 9.

Figure 8

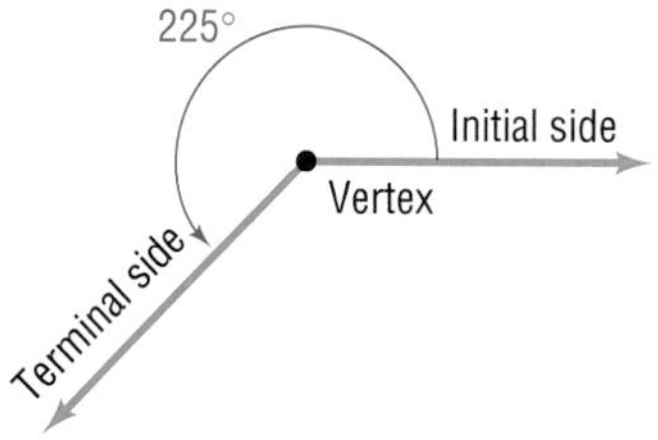

Figure 9

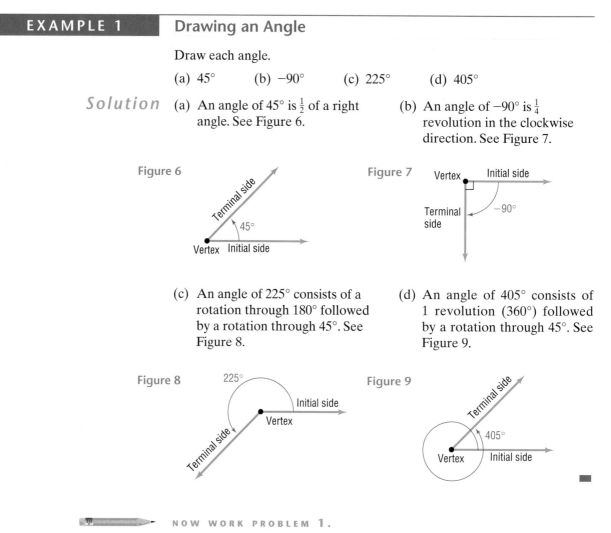

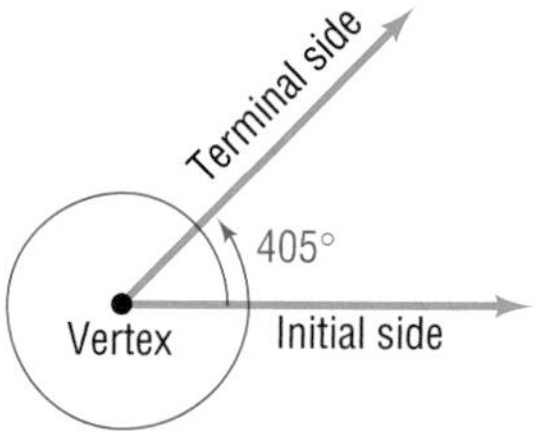

NOW WORK PROBLEM 1.

1 Although subdivisions of a degree may be obtained by using decimals, we also may use the notion of *minutes* and *seconds*. **One minute**, denoted by **1′**, is defined as $\frac{1}{60}$ degree. **One second**, denoted by **1″**, is defined as $\frac{1}{60}$ minute, or equivalently, $\frac{1}{3600}$ degree. An angle of, say, 30 degrees, 40 minutes, 10 seconds is written compactly as $30^\circ 40' 10''$. To summarize:

$$1 \text{ counterclockwise revolution} = 360^\circ$$
$$60' = 1^\circ \qquad 60'' = 1' \qquad \textbf{(1)}$$

It is sometimes necessary to convert from the degree, minute, second notation (D°M′S″) to a decimal form, and vice versa. Check your calculator; it should be capable of doing the conversion for you.

Before getting started, though, you must set the mode to degrees, because there are two common ways to measure angles: degree mode and radian mode. (We will define radians shortly.) Usually, a menu is used to change from one mode to another. Check your owner's manual to find out how your particular calculator works.

EXAMPLE 2 Converting between Degrees, Minutes, Seconds, and Decimal Forms

(a) Convert $50°6'21''$ to a decimal in degrees.
(b) Convert $21.256°$ to the $D°M'S''$ form.

Algebraic Solution

(a) Because $1' = \frac{1}{60}°$ and $1'' = \frac{1}{60}' = \left(\frac{1}{60} \cdot \frac{1}{60}\right)°$, we convert as follows:

$$\begin{aligned} 50°6'21'' &= \left(50 + 6 \cdot \tfrac{1}{60} + 21 \cdot \tfrac{1}{60} \cdot \tfrac{1}{60}\right)° \\ &\approx (50 + 0.1 + 0.005833)° \\ &= 50.105833° \end{aligned}$$

(b) We start with the decimal part of $21.256°$, that is, $0.256°$.

$$0.256° = (0.256)(1°) \underset{\uparrow}{=} (0.256)(60') = 15.36'$$

$1° = 60'$

Now we work with the decimal part of $15.36'$, that is, $0.36'$.

$$0.36' = (0.36)(1') = (0.36)(60'') = 21.6'' \approx 22''$$

Thus,

$$\begin{aligned} 21.256° &= 21° + 0.256° = 21° + 15.36' = 21° + 15' + 0.36' \\ &= 21° + 15' + 21.6'' \approx 21°15'22'' \end{aligned}$$

Graphing Solution

(a) Figure 10 shows the solution using a TI-83 graphing calculator.

(b) Figure 11 shows the solution using a TI-83 graphing calculator.

Figure 10

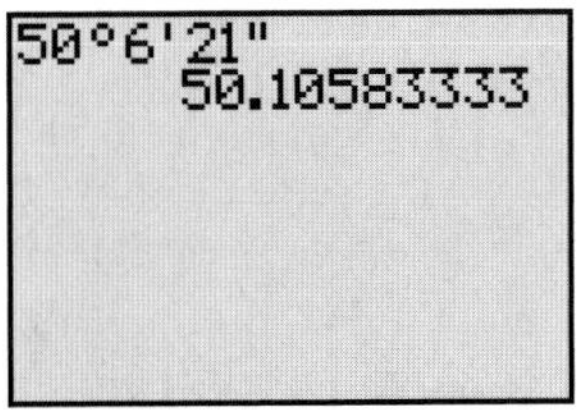

Figure 11

NOW WORK PROBLEMS 61 AND 67.

In certain applications, such as describing the exact location of a star or the precise position of a boat at sea, angles measured in degrees, minutes, and even seconds are used. For calculation purposes, these are transformed to decimal form. In other applications, especially those in calculus, angles are measured using *radians*.

RADIANS

Consider a circle of radius r. Construct an angle whose vertex is at the center of this circle, called a **central angle**, and whose rays subtend (intersect) an arc on the circle whose length equals r. See Figure 12(a). The measure of such an angle is **1 radian**. For a circle of radius 1, the rays of a central angle with measure 1 radian would subtend an arc of length 1. For a circle of radius 3, the rays of a central angle with measure 1 radian would subtend an arc of length 3. See Figure 12(b).

Figure 12

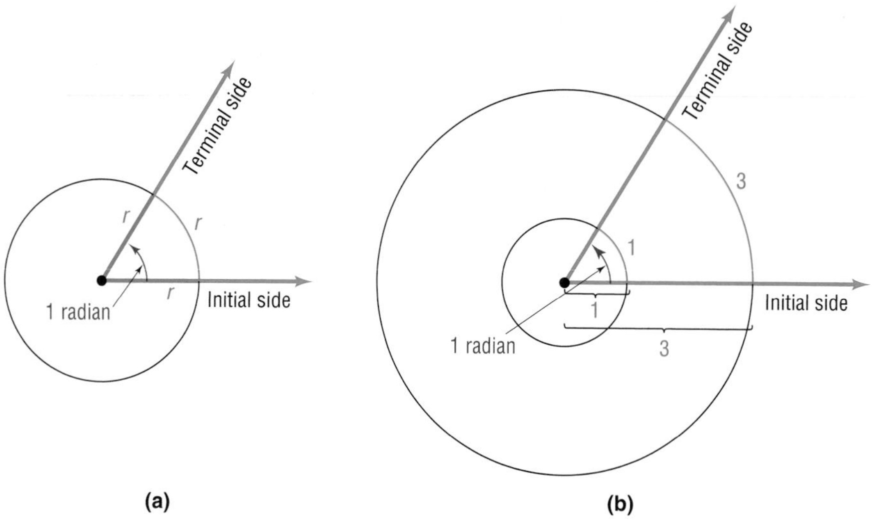

Figure 13 $\dfrac{\theta}{\theta_1} = \dfrac{s}{s_1}$

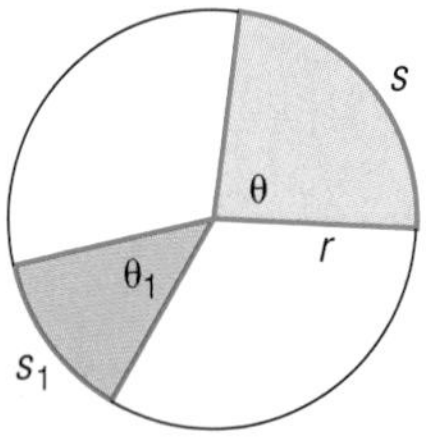

2 Now consider a circle and two central angles, θ and θ_1, measured in radians. Suppose that these central angles subtend arcs of lengths s and s_1, respectively, as shown in Figure 13. From geometry, we know that the ratio of the measures of the angles equals the ratio of the corresponding lengths of the arcs subtended by these angles; that is,

$$\frac{\theta}{\theta_1} = \frac{s}{s_1} \tag{2}$$

Suppose that $\theta_1 = 1$ radian. Refer again to Figure 13. The amount of arc s_1 subtended by the central angle $\theta_1 = 1$ radian equals the radius r of the circle. Then $s_1 = r$, so formula (2) reduces to

$$\frac{\theta}{1} = \frac{s}{r} \quad \text{or} \quad s = r\theta \tag{3}$$

Theorem **Arc Length**

For a circle of radius r, a central angle of θ radians subtends an arc whose length s is

$$s = r\theta \tag{4}$$

NOTE: Formulas must be consistent with regard to the units used. In equation (4), we write

$$s = r\theta$$

To see the units, however, we must go back to equation (3) and write

$$\frac{\theta \text{ radians}}{1 \text{ radian}} = \frac{s \text{ length units}}{r \text{ length units}}$$

$$s \text{ length units} = r \text{ length units} \, \frac{\theta \text{ radians}}{1 \text{ radian}}$$

Since the radians cancel, we are left with

$$s \text{ length units} = (r \text{ length units})\theta \quad s = r\theta$$

where θ appears to be "dimensionless" but, in fact, is measured in radians. So, in using the formula $s = r\theta$, the dimension for θ is radians, and any convenient unit of length (such as inches or meters) may be used for s and r.

EXAMPLE 3 Finding the Length of Arc of a Circle

Find the length of the arc of a circle of radius 2 meters subtended by a central angle of 0.25 radian.

Solution We use equation (4) with $r = 2$ meters and $\theta = 0.25$. The length s of the arc is

$$s = r\theta = 2(0.25) = 0.5 \text{ meter}$$

NOW WORK PROBLEM 37.

RELATIONSHIP BETWEEN DEGREES AND RADIANS

Figure 14
1 revolution $= 2\pi$ radians

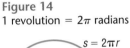

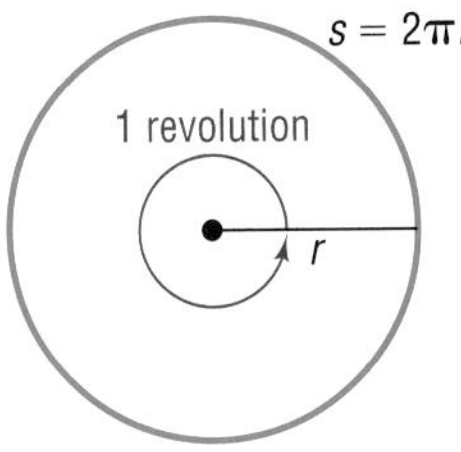

Consider a circle of radius r. A central angle of 1 revolution will subtend an arc equal to the circumference of the circle (Figure 14). Because the circumference of a circle equals $2\pi r$, we use $s = 2\pi r$ in equation (4) to find that, for an angle θ of 1 revolution,

$$s = r\theta$$
$$2\pi r = r\theta$$
$$\theta = 2\pi \text{ radians}$$

From this we have,

$$1 \text{ revolution} = 2\pi \text{ radians} \tag{5}$$

so

$$360^\circ = 2\pi \text{ radians}$$

or

$$180^\circ = \pi \text{ radians} \tag{6}$$

Divide both sides of equation (6) by 180. Then

$$1 \text{ degree} = \frac{\pi}{180} \text{ radian}$$

Divide both sides of (6) by π. Then

$$\frac{180}{\pi} \text{ degrees} = 1 \text{ radian}$$

We have the following two conversion formulas:

$$1 \text{ degree} = \frac{\pi}{180} \text{ radian} \qquad 1 \text{ radian} = \frac{180}{\pi} \text{ degrees} \tag{7}$$

EXAMPLE 4 **Converting from Degrees to Radians**

3 Convert each angle in degrees to radians.

(a) $60°$ (b) $150°$ (c) $-45°$ (d) $90°$

Solution (a) $60° = 60 \cdot 1 \text{ degree} = 60 \cdot \frac{\pi}{180} \text{ radian} = \frac{\pi}{3} \text{ radians}$

(b) $150° = 150 \cdot \frac{\pi}{180} \text{ radian} = \frac{5\pi}{6} \text{ radians}$

(c) $-45° = -45 \cdot \frac{\pi}{180} \text{ radian} = -\frac{\pi}{4} \text{ radian}$

(d) $90° = 90 \cdot \frac{\pi}{180} \text{ radian} = \frac{\pi}{2} \text{ radians}$ ■

Example 4 illustrates that angles that are fractions of a revolution are expressed in radian measure as fractional multiples of π, rather than as decimals. For example, a right angle, as in Example 4(d), is left in the form $\pi/2$ radians, which is exact, rather than using the approximation $\pi/2 \approx 3.1416/2 = 1.5708$ radians. A graphing calculator may be used to convert an angle measured in degrees to an angle in radians. Consult your manual for the appropriate keystrokes.

NOW WORK PROBLEM 13.

EXAMPLE 5 **Converting Radians to Degrees**

4 Convert each angle in radians to degrees.

(a) $\frac{\pi}{6}$ radian (b) $\frac{3\pi}{2}$ radians (c) $-\frac{3\pi}{4}$ radians (d) $\frac{7\pi}{3}$ radians

Solution (a) $\frac{\pi}{6} \text{ radian} = \frac{\pi}{6} \cdot 1 \text{ radian} = \frac{\pi}{6} \cdot \frac{180}{\pi} \text{ degrees} = 30°$

(b) $\frac{3\pi}{2} \text{ radians} = \frac{3\pi}{2} \cdot \frac{180}{\pi} \text{ degrees} = 270°$

(c) $-\frac{3\pi}{4} \text{ radians} = -\frac{3\pi}{4} \cdot \frac{180}{\pi} \text{ degrees} = -135°$

(d) $\frac{7\pi}{3} \text{ radians} = \frac{7\pi}{3} \cdot \frac{180}{\pi} \text{ degrees} = 420°$ ■

A graphing calculator may also be used to convert an angle measured in radians to an angle in degrees. Consult your owner's manual for the appropriate keystrokes.

NOW WORK PROBLEM 25.

Table 1 lists the degree and radian measures of some commonly encountered angles. You should learn to feel equally comfortable using degree or radian measure for these angles.

TABLE 1

Degrees	0°	30°	45°	60°	90°	120°	135°	150°	180°
Radians	0	$\frac{\pi}{6}$	$\frac{\pi}{4}$	$\frac{\pi}{3}$	$\frac{\pi}{2}$	$\frac{2\pi}{3}$	$\frac{3\pi}{4}$	$\frac{5\pi}{6}$	π
Degrees	210°	225°	240°	270°	300°	315°	330°	360°	
Radians	$\frac{7\pi}{6}$	$\frac{5\pi}{4}$	$\frac{4\pi}{3}$	$\frac{3\pi}{2}$	$\frac{5\pi}{3}$	$\frac{7\pi}{4}$	$\frac{11\pi}{6}$	2π	

EXAMPLE 6 Finding the Distance Between Two Cities

See Figure 15(a). The latitude of a location L is the angle formed by a ray drawn from the center of Earth to the Equator and a ray drawn from the center of Earth to L. See Figure 15(b). Glasgow, Montana, is due north of Albuquerque, New Mexico. Find the distance between Glasgow (48°, 9′ north latitude) and Albuquerque (35°, 5′ north latitude). Assume that the radius of Earth is 3960 miles.

Figure 15

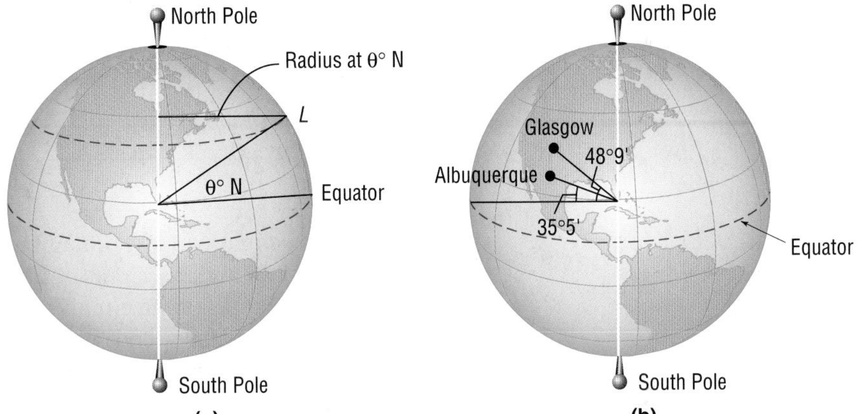

Solution The measure of the central angle between the two cities is $48°9' - 35°5' = 13°4'$. We use equation (4), $s = r\theta$, but first we must convert the angle of $13°4'$ to radians.

$$\theta = 13°4' = 13.0667° = 13.0667 \cdot \frac{\pi}{180} \text{ radian} = 0.228 \text{ radian}$$

We use $\theta = 0.228$ radian and $r = 3960$ miles in equation (4). The distance between the two cities is

$$s = r\theta = 3960 \cdot 0.228 \approx 903 \text{ miles}$$

When an angle is measured in degrees, the degree symbol will always be shown. However, when an angle is measured in radians, we will follow the usual practice and omit the word *radians*. So, if the measure of an angle is given as $\pi/6$, it is understood to mean $\pi/6$ radian.

CIRCULAR MOTION

5 We have already defined the average speed of an object as the distance traveled divided by the elapsed time. Suppose that an object moves around a circle of radius r at a constant speed. If s is the distance traveled in time t around this circle, then the **linear speed** v of the object is defined as

$$v = \frac{s}{t} \tag{8}$$

Figure 16

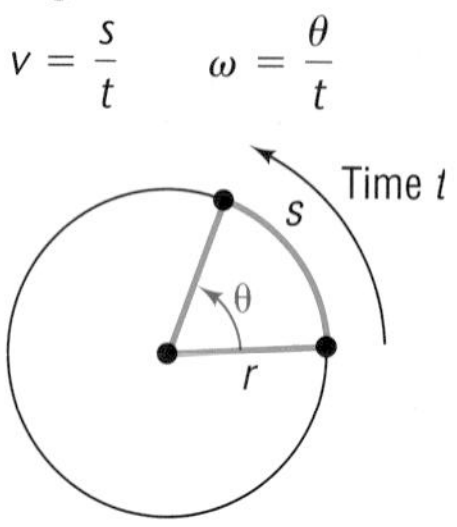

As this object travels around the circle, suppose that θ (measured in radians) is the central angle swept out in time t (see Figure 16). Then the **angular speed** ω (the Greek letter omega) of this object is the angle (measured in radians) swept out divided by the elapsed time; that is,

$$\omega = \frac{\theta}{t} \tag{9}$$

Angular speed is the way the turning rate of an engine is described. For example, an engine idling at 900 rpm (revolutions per minute) is one that rotates at an angular speed of

$$900\,\frac{\text{revolutions}}{\text{minute}} = 900\,\frac{\cancel{\text{revolutions}}}{\text{minute}} \cdot 2\pi\,\frac{\text{radians}}{\cancel{\text{revolution}}} = 1800\pi\,\frac{\text{radians}}{\text{minute}}$$

There is an important relationship between linear speed and angular speed. In the formula $s = r\theta$, divide each side by t.

$$\frac{s}{t} = r\,\frac{\theta}{t}$$

Then, using equations (8) and (9), we obtain

$$v = r\omega \tag{10}$$

When using equation (10), remember that $v = s/t$ (the linear speed) has the dimensions of length per unit of time (such as feet per second or miles per hour), r (the radius of the circular motion) has the same length dimension as s, and ω (the angular speed) has the dimensions of radians per unit of time. As noted earlier, we leave the radian dimension off the numerical value of the angular speed ω so that both sides of the equation will be dimensionally consistent (with "length per unit of time"). If the angular speed is given in terms of *revolutions* per unit of time (as is often the case), be sure to convert it to *radians* per unit of time before attempting to use equation (10).

EXAMPLE 7 Finding Linear Speed

A child is spinning a rock at the end of a 2-foot rope at the rate of 180 revolutions per minute (rpm). Find the linear speed of the rock when it is released.

Figure 17

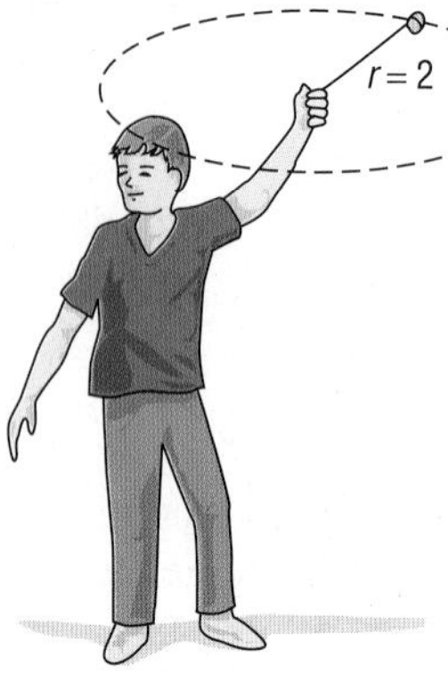

Solution Look at Figure 17. The rock is moving around a circle of radius $r = 2$ feet. The angular speed ω of the rock is

$$\omega = 180\,\frac{\text{revolutions}}{\text{minute}} = 180\,\frac{\cancel{\text{revolutions}}}{\text{minute}} \cdot 2\pi\,\frac{\text{radians}}{\cancel{\text{revolution}}} = 360\pi\,\frac{\text{radians}}{\text{minute}}$$

From equation (10), the linear speed v of the rock is

$$v = r\omega = 2\text{ feet} \cdot 360\pi\,\frac{\text{radians}}{\text{minute}} = 720\pi\,\frac{\text{feet}}{\text{minute}} \approx 2262\,\frac{\text{feet}}{\text{minute}}$$

The linear speed of the rock when it is released is 2262 ft/min $\approx$ 25.7 mi/hr. ■

NOW WORK PROBLEM 75.

6.1 EXERCISES

In Problems 1–12, draw each angle.

1. $30°$ **2.** $60°$ **3.** $135°$ **4.** $-120°$
5. $450°$ **6.** $540°$ **7.** $3\pi/4$ **8.** $4\pi/3$
9. $-\pi/6$ **10.** $-2\pi/3$ **11.** $16\pi/3$ **12.** $21\pi/4$

In Problems 13–24, convert each angle in degrees to radians. Express your answer as a multiple of π.

13. $30°$ **14.** $120°$ **15.** $240°$ **16.** $330°$
17. $-60°$ **18.** $-30°$ **19.** $180°$ **20.** $270°$
21. $-135°$ **22.** $-225°$ **23.** $-90°$ **24.** $-180°$

In Problems 25–36, convert each angle in radians to degrees.

25. $\pi/3$ **26.** $5\pi/6$ **27.** $-5\pi/4$ **28.** $-2\pi/3$
29. $\pi/2$ **30.** 4π **31.** $\pi/12$ **32.** $5\pi/12$
33. $-\pi/2$ **34.** $-\pi$ **35.** $-\pi/6$ **36.** $-3\pi/4$

In Problems 37–44, s denotes the length of the arc of a circle of radius r subtended by the central angle θ. Find the missing quantity.

37. $r = 10$ meters, $\theta = \frac{1}{2}$ radian, $s = ?$
38. $r = 6$ feet, $\theta = 2$ radians, $s = ?$
39. $\theta = \frac{1}{3}$ radian, $s = 2$ feet, $r = ?$
40. $\theta = \frac{1}{4}$ radian, $s = 6$ centimeters, $r = ?$
41. $r = 5$ miles, $s = 3$ miles, $\theta = ?$
42. $r = 6$ meters, $s = 8$ meters, $\theta = ?$
43. $r = 2$ inches, $\theta = 30°$, $s = ?$
44. $r = 3$ meters, $\theta = 120°$, $s = ?$

In Problems 45–52, convert each angle in degrees to radians. Express your answer in decimal form, rounded to two decimal places.

45. $17°$ **46.** $73°$ **47.** $-40°$ **48.** $-51°$
49. $125°$ **50.** $200°$ **51.** $340°$ **52.** $350°$

In Problems 53–60, convert each angle in radians to degrees. Express your answer in decimal form, rounded to two decimal places.

53. 3.14 **54.** π **55.** 10.25 **56.** 0.75
57. 2 **58.** 3 **59.** 6.32 **60.** $\sqrt{2}$

In Problems 61–66, convert each angle to a decimal in degrees. Round your answer to two decimal places.

61. $40°10'25''$ **62.** $61°42'21''$ **63.** $1°2'3''$ **64.** $73°40'40''$
65. $9°9'9''$ **66.** $98°22'45''$

In Problems 67–72, convert each angle to D°M'S" form. Round your answer to the nearest second.

67. $40.32°$ **68.** $61.24°$ **69.** $18.255°$ **70.** $29.411°$
71. $19.99°$ **72.** $44.01°$

73. Minute Hand of a Clock The minute hand of a clock is 6 inches long. How far does the tip of the minute hand move in 15 minutes? How far does it move in 25 minutes?

74. Movement of a Pendulum A pendulum swings through an angle of 20° each second. If the pendulum is 40 inches long, how far does its tip move each second?

75. An object is traveling around a circle with a radius of 5 centimeters. If in 20 seconds a central angle of $\frac{1}{3}$ radian is swept out, what is the angular speed of the object? What is its linear speed?

76. An object is traveling around a circle with a radius of 2 meters. If in 20 seconds the object travels 5 meters, what is its angular speed? What is its linear speed?

77. Bicycle Wheels The diameter of each wheel of a bicycle is 26 inches. If you are traveling at a speed of 35 miles per hour on this bicycle, through how many revolutions per minute are the wheels turning?

78. Car Wheels The radius of each wheel of a car is 15″. If the wheels are turning at the rate of 3 revolutions per second, how fast is the car moving? Express your answer in inches per second and in miles per hour.

In Problems 79–82, the latitude of a location, L, is the angle formed by a ray drawn from the center of Earth to the Equator and from the center of Earth to L. See the figure.

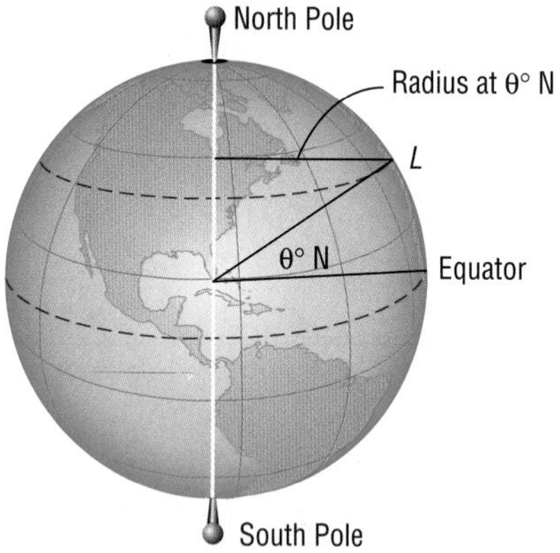

79. Distance between Cities Memphis, Tennessee, is due north of New Orleans, Louisiana. Find the distance between Memphis (35°9′ north latitude) and New Orleans (29°57′ north latitude). Assume that the radius of Earth is 3960 miles.

80. Distance between Cities Charleston, West Virginia, is due north of Jacksonville, Florida. Find the distance between Charleston (38°21′ north latitude) and Jacksonville (30°20′ north latitude). Assume that the radius of Earth is 3960 miles.

81. Linear Speed on Earth Earth rotates on an axis through its poles. The distance from the axis to a location on Earth 30° north latitude is about 3429.5 miles. Therefore, a location on Earth at 30° north latitude is spinning on a circle of radius 3429.5 miles. Compute the linear speed on the surface of Earth at 30° north latitude.

82. Linear Speed on Earth Earth rotates on an axis through its poles. The distance from the axis to a location on Earth 40° north latitude is about 3033.5 miles. Therefore, a location on Earth at 40° north latitude is spinning on a circle of radius 3033.5 miles. Compute the linear speed on the surface of Earth at 40° north latitude.

83. Speed of the Moon The mean distance of the Moon from Earth is 2.39×10^5 miles. Assuming that the orbit of the Moon around Earth is circular and that 1 revolution takes 27.3 days, find the linear speed of the Moon. Express your answer in miles per hour.

84. Speed of Earth The mean distance of Earth from the Sun is 9.29×10^7 miles. Assuming that the orbit of Earth around the Sun is circular and that 1 revolution takes 365 days, find the linear speed of Earth. Express your answer in miles per hour.

85. Pulleys Two pulleys, one with radius 2 inches and the other with radius 8 inches, are connected by a belt. (See the figure.) If the 2-inch pulley is caused to rotate at 3 revolutions per minute, determine the revolutions per minute of the 8-inch pulley.
[**Hint:** The linear speeds of the pulleys, that is, the speed of the belt, are the same.]

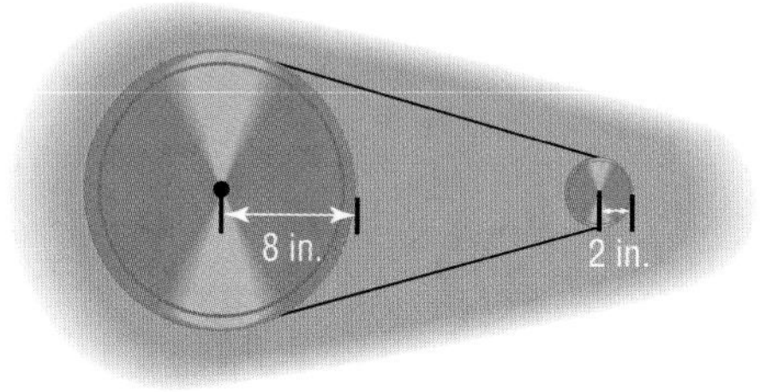

86. Ferris Wheels A neighborhood carnival has a Ferris wheel whose radius is 30 feet. You measure the time it takes for one revolution to be 70 seconds. What is the linear speed of this Ferris wheel? What is the angular speed?

87. Computing the Speed of a River Current To approximate the speed of the current of a river, a circular paddle wheel with radius 4 feet is lowered into the water. If the current causes the wheel to rotate at a speed of 10 revolutions per minute, what is the speed of the current? Express your answer in miles per hour.

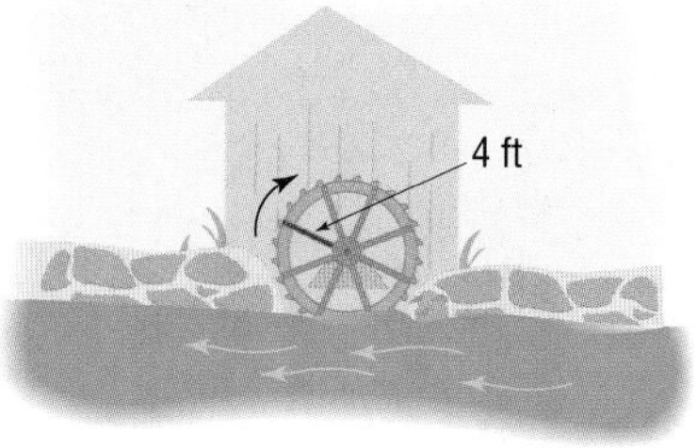

88. Spin Balancing Tires A spin balancer rotates the wheel of a car at 480 revolutions per minute. If the diameter of the wheel is 26 inches, what road speed is being tested? Express your answer in miles per hour. At how many revolutions per minute should the balancer be set to test a road speed of 80 miles per hour?

89. The Cable Cars of San Francisco At the Cable Car Museum you can see the four cable lines that are used to pull cable cars up and down the hills of San Francisco. Each cable travels at a speed of 9.55 miles per hour, caused by a rotating wheel whose diameter is 8.5 feet. How fast is the wheel rotating? Express your answer in revolutions per minute.

90. Difference in Time of Sunrise Naples, Florida, is approximately 90 miles due west of Ft. Lauderdale. How

much sooner would a person in Ft. Lauderdale first see the rising Sun than a person in Naples?
[**Hint:** Consult the figure. When a person at Q sees the first rays of the Sun, a person at P is still in the dark. The person at P sees the first rays after Earth has rotated so that P is at the location Q. Now use the fact that at the latitude of Ft. Lauderdale in 24 hours a length of arc of $2\pi(3150)$ miles is subtended.]

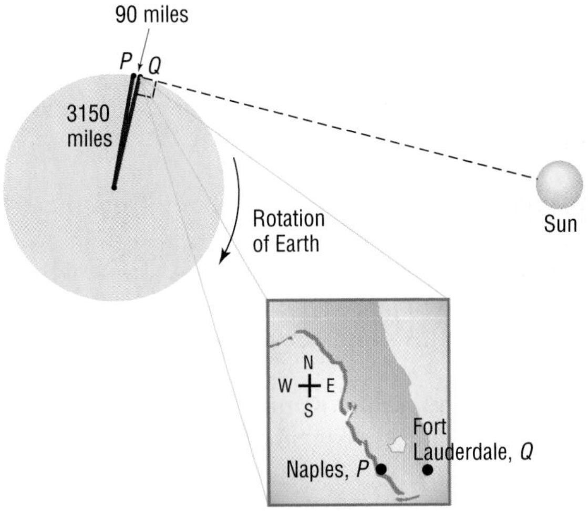

91. **Keeping up with the Sun** How fast would you have to travel on the surface of Earth at the equator to keep up with the Sun (that is, so that the Sun would appear to remain in the same position in the sky)?

92. **Nautical Miles** A **nautical mile** equals the length of arc subtended by a central angle of 1 minute on a great circle* on the surface of Earth. (See the figure.) If the radius of Earth is taken as 3960 miles, express 1 nautical mile in terms of ordinary, or **statute**, miles (5280 feet).

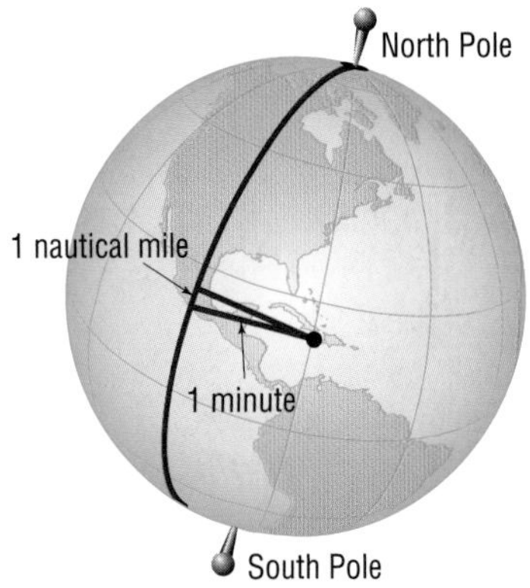

93. **Pulleys** Two pulleys, one with radius r_1 and the other with radius r_2, are connected by a belt. The pulley with radius r_1 rotates at ω_1 revolutions per minute, whereas the pulley with radius r_2 rotates at ω_2 revolutions per minute. Show that $r_1/r_2 = \omega_2/\omega_1$.

94. Do you prefer to measure angles using degrees or radians? Provide justification and a rationale for your choice.

95. Discuss why ships and airplanes use nautical miles to measure distance. Explain the difference between a nautical mile and a statute mile.

96. Investigate the way that speed bicycles work. In particular, explain the differences and similarities between 5-speed and 9-speed derailleurs. Be sure to include a discussion of linear speed and angular speed.

* Any circle drawn on the surface of Earth that divides Earth into two equal hemispheres.

PREPARING FOR THIS SECTION

Before getting started, review the following:

- ✓ Unit Circle (p. 84)
- ✓ Symmetry (pp. 18–19)
- ✓ Functions (Section 2.1)
- ✓ Pythagorean Theorem (Appendix, Section 9)

6.2 TRIGONOMETRIC FUNCTIONS: UNIT CIRCLE APPROACH

OBJECTIVES

1. Find the Exact Value of the Trigonometric Functions Using a Point on the Unit Circle
2. Find the Exact Value of the Trigonometric Functions of Quadrantal Angles
3. Find the Exact Value of the Trigonometric Functions of 45°
4. Find the Exact Value of the Trigonometric Functions of 60°
5. Find the Exact Value of the Trigonometric Functions of 30°
6. Use a Calculator to Approximate the Value of the Trigonometric Functions

We are now ready to introduce trigonometric functions. The approach that we take uses the unit circle.

THE UNIT CIRCLE

Recall that the unit circle is a circle whose radius is 1 and whose center is at the origin of a rectangular coordinate system. Also recall that any circle of radius r has circumference of length $2\pi r$. Therefore, the unit circle (radius $= 1$) has a circumference of length 2π. In other words, for 1 revolution around the unit circle the length of arc is 2π units.

The following discussion sets the stage for defining the trigonometric functions.

Let $t \geq 0$ be any real number and let s be the distance from the origin to t on the real number line. See the red portion of Figure 18(a). Now look at the unit circle in Figure 18(a). Beginning at the the point $(1, 0)$ on the unit circle, travel $s = t$ units in the counterclockwise direction along the circle, to arrive at the point $P = (x, y)$. In this sense, the length t is being **wrapped** around the unit circle.

If $t < 0$, we begin at the point $(1, 0)$ on the unit circle and travel $s = |t|$ units in the clockwise direction to arrive at the point $P = (x, y)$. See Figure 18(b).

Figure 18

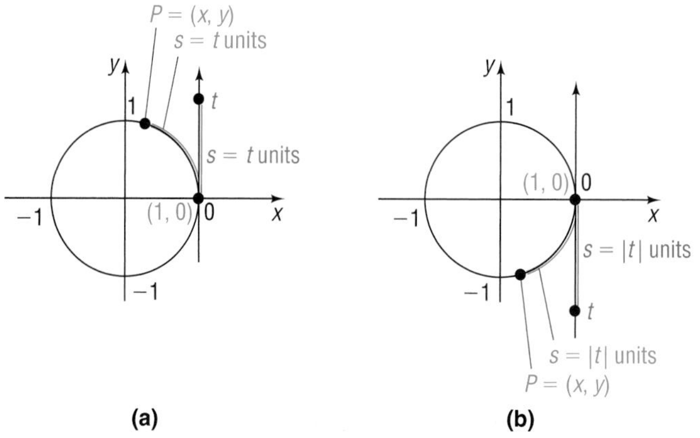

If $t > 2\pi$ or if $t < -2\pi$, it will be necessary to travel around the unit circle more than once before arriving at point P. Do you see why?

Let's describe this process another way. Picture a string of length $s = |t|$ units being wrapped around a circle of radius 1 unit. We start wrapping the string around the circle at the point $(1, 0)$. If $t \geq 0$, we wrap the string in the counterclockwise direction; if $t < 0$, we wrap the string in the clockwise direction. The point $P = (x, y)$ is the point where the string ends.

This discussion tells us that, for any real number t, we can locate a unique point $P = (x, y)$ on the unit circle. We call this point P **the point *P* on the unit circle that corresponds to *t*.** This is the important idea here. No matter what real number t is chosen, there is a unique point P on the unit circle corresponding to it. We use the coordinates of the point $P = (x, y)$ on the unit circle corresponding to the real number t to define the **six trigonometric functions of *t*.**

Let t be a real number and let $P = (x, y)$ be the point on the unit circle that corresponds to t.

The **sine function** associates with t the y-coordinate of P and is denoted by

$$\sin t = y$$

The **cosine function** associates with t the x-coordinate of P and is denoted by

$$\cos t = x$$

If $x \neq 0$, the **tangent function** is defined as

$$\tan t = \frac{y}{x}$$

If $y \neq 0$, the **cosecant function** is defined as

$$\csc t = \frac{1}{y}$$

If $x \neq 0$, the **secant function** is defined as

$$\sec t = \frac{1}{x}$$

If $y \neq 0$, the **cotangent function** is defined as

$$\cot t = \frac{x}{y}$$

Notice in these definitions that if $x = 0$, that is, if the point $P = (0, y)$ is on the y-axis, then the tangent function and the secant function are undefined. Also, if $y = 0$, that is, if the point $P = (x, 0)$ is on the x-axis, then the cosecant function and the cotangent function are undefined.

Because we use the unit circle in these definitions of the trigonometric functions, they are also sometimes referred to as **circular functions**.

EXAMPLE 1 **Finding the Value of the Six Trigonometric Functions**

1 Let t be a real number and let $P = \left(-\frac{1}{2}, \frac{\sqrt{3}}{2}\right)$ be the point on the unit circle that corresponds to t. See Figure 19.

Figure 19
$\theta = t$ radians

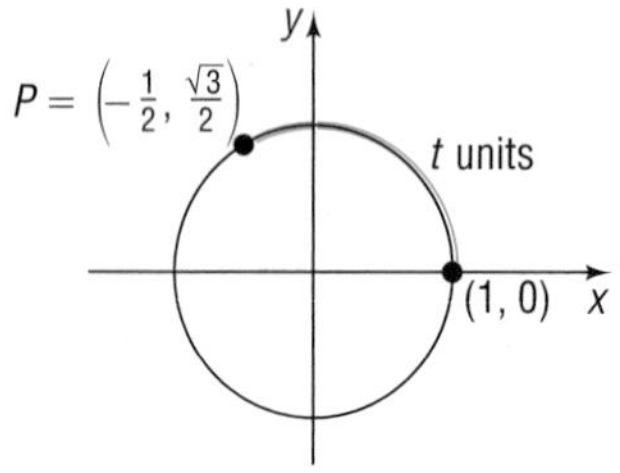

Then

$$\sin t = \frac{\sqrt{3}}{2} \qquad \cos t = -\frac{1}{2} \qquad \tan t = \frac{\frac{\sqrt{3}}{2}}{-\frac{1}{2}} = -\sqrt{3}$$

$$\csc t = \frac{1}{\frac{\sqrt{3}}{2}} = \frac{2\sqrt{3}}{3} \qquad \sec t = \frac{1}{-\frac{1}{2}} = -2 \qquad \cot t = \frac{-\frac{1}{2}}{\frac{\sqrt{3}}{2}} = \frac{-\sqrt{3}}{3}$$

NOW WORK PROBLEM 1.

TRIGONOMETRIC FUNCTIONS OF ANGLES

Let $P = (x, y)$ be the point on the unit circle corresponding to the real number t. See Figure 20(a). Let θ be the angle in standard position, measured in radians, whose terminal side is the ray from the origin through P. See Figure 20(b). Since the unit circle has radius 1 unit, from the formula for arc length, $s = r\theta$, we find that

$$s = r\theta = \theta \quad (r = 1)$$

So, if $s = |t|$ units, then $\theta = t$ radians. See Figures 20(c) and (d).

Figure 20

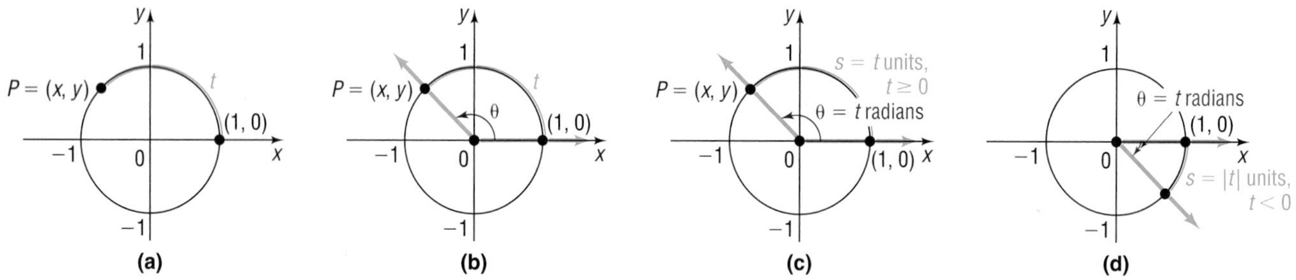

On the unit circle, the point $P = (x, y)$ that corresponds to the real number t is the point P on the terminal side of the angle $\theta = t$ radians. As a result, we can say that

$$\underset{\text{Real number}}{\sin t} = \underset{\theta = t \text{ radians}}{\sin \theta}$$

and so on. We can now define the trigonometric functions of the angle θ.

If $\theta = t$ radians, the **six trigonometric functions of the angle θ** are defined as

$\sin\theta = \sin t$	$\cos\theta = \cos t$	$\tan\theta = \tan t$
$\csc\theta = \csc t$	$\sec\theta = \sec t$	$\cot\theta = \cot t$

Even though the distinction between trigonometric functions of real numbers and trigonometric functions of angles is important, it is customary to refer to trigonometric functions of real numbers and trigonometric functions of angles collectively as *the trigonometric functions*. We shall follow this practice from now on.

If an angle θ is measured in degrees, we shall use the degree symbol when writing a trigonometric function of θ, as, for example, in $\sin 30°$ and $\tan 45°$. If an angle θ is measured in radians, then no symbol is used when writing a trigonometric function of θ, as, for example, in $\cos\pi$ and $\sec(\pi/3)$.

Finally, since the values of the trigonometric functions of an angle θ are determined by the coordinates of the point $P = (x, y)$ on the unit circle corresponding to θ, the units used to measure the angle θ are irrelevant. For example, it does not matter whether we write $\theta = \pi/2$ radians or $\theta = 90°$. The point on the unit circle corresponding to this angle is $P = (0, 1)$. Hence,

$$\sin\frac{\pi}{2} = \sin 90° = 1 \quad \text{and} \quad \cos\frac{\pi}{2} = \cos 90° = 0$$

EVALUATING THE TRIGONOMETRIC FUNCTIONS

2 To find the exact value of a trigonometric function of an angle θ or a real number t requires that we locate the point $P = (x, y)$ on the unit circle that corresponds to t. This is not always easy to do. In the examples that follow, we will evaluate the trigonometric functions of certain angles or real numbers for which this process is relatively easy. A calculator will be used to evaluate the trigonometric functions of most angles.

EXAMPLE 2 Finding the Exact Value of the Six Trigonometric Functions of Quadrantal Angles

Find the exact value of the six trigonometric functions at:

(a) $\theta = 0 = 0°$ (b) $\theta = \pi/2 = 90°$

(c) $\theta = \pi = 180°$ (d) $\theta = 3\pi/2 = 270°$

Figure 21(a)

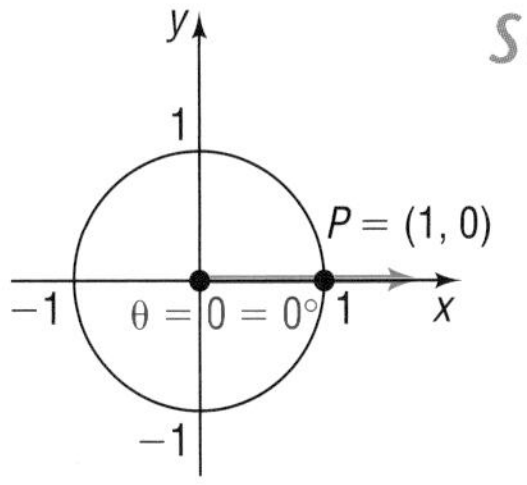

Solution (a) The point on the unit circle that corresponds to $\theta = 0 = 0°$ is $P = (1, 0)$. See Figure 21(a). Then

$$\sin 0 = \sin 0° = y = 0 \qquad \cos 0 = \cos 0° = x = 1$$

$$\tan 0 = \tan 0° = y/x = 0 \qquad \sec 0 = \sec 0° = 1/x = 1$$

Since the y-coordinate of P is 0, $\csc 0$ and $\cot 0$ are not defined.

Figure 21(b)

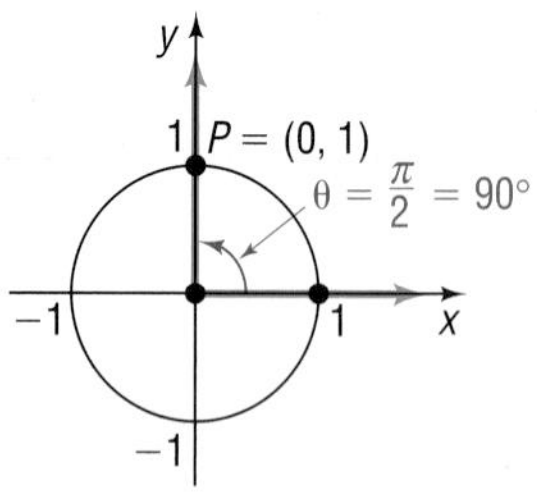

(b) The point on the unit circle that corresponds to $\theta = \pi/2 = 90°$ is $P = (0, 1)$. See Figure 21(b). Then

$$\sin\frac{\pi}{2} = \sin 90° = y = 1 \qquad \cos\frac{\pi}{2} = \cos 90° = x = 0$$

$$\csc\frac{\pi}{2} = \csc 90° = 1/y = 1 \qquad \cot\frac{\pi}{2} = \cot 90° = x/y = 0$$

Since the x-coordinate of P is 0, $\tan(\pi/2)$ and $\sec(\pi/2)$ are not defined.

Figure 21(c)

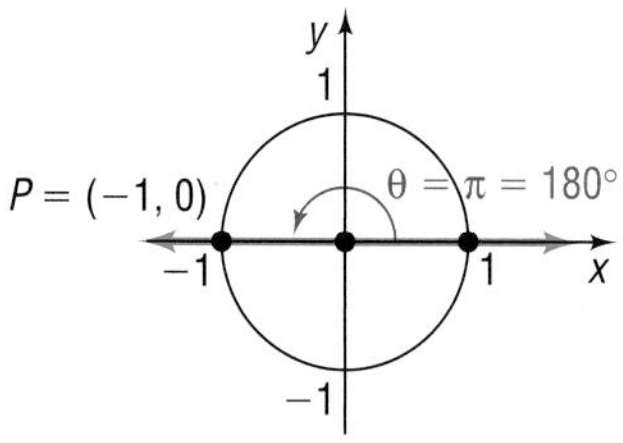

(c) The point on the unit circle that corresponds to $\theta = \pi = 180°$ is $P = (-1, 0)$. See Figure 21(c). Then

$$\sin\pi = \sin 180° = y = 0 \qquad \cos\pi = \cos 180° = x = -1$$

$$\tan\pi = \tan 180° = y/x = 0 \qquad \sec\pi = \sec 180° = 1/x = -1$$

Since the y-coordinate of P is 0, $\csc\pi$ and $\cot\pi$ are not defined.

Figure 21(d)

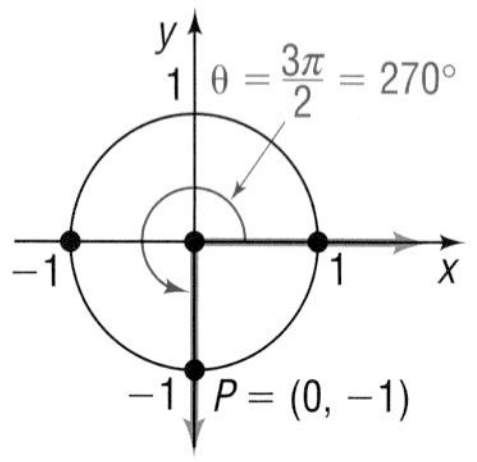

(d) The point on the unit circle that corresponds to $\theta = 3\pi/2 = 270°$ is $P = (0, -1)$. See Figure 21(d). Then

$$\sin\frac{3\pi}{2} = \sin 270° = y = -1 \qquad \cos\frac{3\pi}{2} = \cos 270° = x = 0$$

$$\csc\frac{3\pi}{2} = \csc 270° = 1/y = -1 \qquad \cot\frac{3\pi}{2} = \cot 270° = x/y = 0$$

Since the x-coordinate of P is 0, $\tan(3\pi/2)$ and $\sec(3\pi/2)$ are not defined. ■

Table 2 summarizes the values of the trigonometric functions found in Example 2.

TABLE 2

QUADRANTAL ANGLES

θ (radians)	θ (degrees)	$\sin\theta$	$\cos\theta$	$\tan\theta$	$\csc\theta$	$\sec\theta$	$\cot\theta$
0	0°	0	1	0	Not defined	1	Not defined
$\pi/2$	90°	1	0	Not defined	1	Not defined	0
π	180°	0	−1	0	Not defined	−1	Not defined
$3\pi/2$	270°	−1	0	Not defined	−1	Not defined	0

NOW WORK PROBLEM 27.

EXAMPLE 3 Finding Exact Values of Trigonometric Functions ($\theta = \pi/4 = 45°$)

3 Find the exact value of the six trigonometric functions of $\frac{\pi}{4} = 45°$.

Figure 22
$\theta = \pi/4 = 45°$

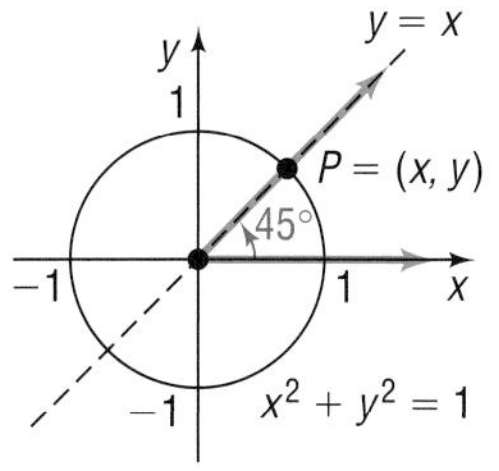

Solution We seek the coordinates of the point $P = (x, y)$ on the unit circle that corresponds to $\theta = \pi/4 = 45°$. See Figure 22. First, we observe that P lies on the line $y = x$. (Do you see why? Since $\theta = 45° = \frac{1}{2} \cdot 90°$, P must lie on the line that bisects quadrant I.) Since $P = (x, y)$ lies on the unit circle, $x^2 + y^2 = 1$, and it follows that

$$x^2 + y^2 = 1 \quad y = x,\ x > 0,\ y > 0$$
$$x^2 + x^2 = 1$$
$$2x^2 = 1$$
$$x = \frac{1}{\sqrt{2}} = \frac{\sqrt{2}}{2}, \qquad y = \frac{\sqrt{2}}{2}$$

Thus,

$$\sin\frac{\pi}{4} = \sin 45° = \frac{\sqrt{2}}{2} \qquad \cos\frac{\pi}{4} = \cos 45° = \frac{\sqrt{2}}{2} \qquad \tan\frac{\pi}{4} = \tan 45° = \frac{\sqrt{2}/2}{\sqrt{2}/2} = 1$$

$$\csc\frac{\pi}{4} = \csc 45° = \frac{1}{\sqrt{2}/2} = \sqrt{2} \qquad \sec\frac{\pi}{4} = \sec 45° = \frac{1}{\sqrt{2}/2} = \sqrt{2} \qquad \cot\frac{\pi}{4} = \cot 45° = \frac{\sqrt{2}/2}{\sqrt{2}/2} = 1$$

■

EXAMPLE 4 Finding the Exact Value of a Trigonometric Expression

Find the exact value of each expression.

(a) $\sin 45° \cos 180°$ (b) $\tan\frac{\pi}{4} - \sin\frac{3\pi}{2}$

Solution (a) $\sin 45° \cos 180° = \frac{\sqrt{2}}{2} \cdot (-1) = \frac{-\sqrt{2}}{2}$

↑ From Example 4 ↑ From Table 2

(b) $\tan\frac{\pi}{4} - \sin\frac{3\pi}{2} = 1 - (-1) = 2$

↑ From Example 4 ↑ From Table 2

■

NOW WORK PROBLEM 15.

TRIGONOMETRIC FUNCTIONS OF 30° AND 60°

Consider a right triangle in which one of the angles is 30°. It then follows that the other angle is 60°. Figure 23(a) illustrates such a triangle with hypotenuse of length 1. Our problem is to determine a and b.

We begin by placing next to this triangle another triangle congruent to the first, as shown in Figure 23(b). Notice that we now have a triangle whose angles are each 60°. This triangle is therefore equilateral, so each side is of length 1. In particular, the base is $2a = 1$, and so $a = \frac{1}{2}$. By the Pythagorean Theorem, b satisfies the equation $a^2 + b^2 = c^2$, so we have

Figure 23

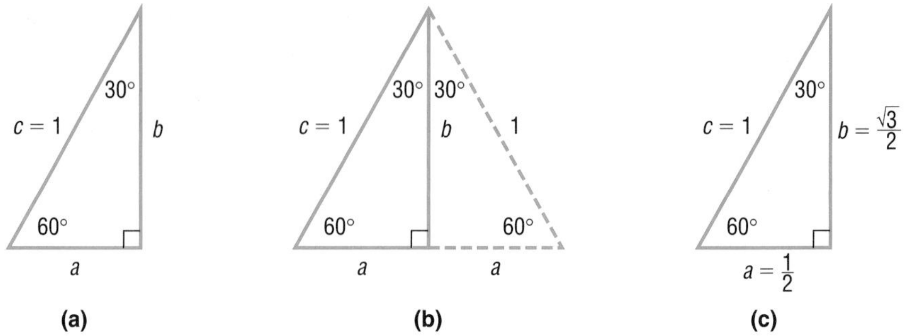

$$
\begin{aligned}
a^2 + b^2 &= c^2 \qquad a = \tfrac{1}{2},\ c = 1 \\
\tfrac{1}{4} + b^2 &= 1 \\
b^2 &= 1 - \tfrac{1}{4} = \tfrac{3}{4} \\
b &= \tfrac{\sqrt{3}}{2}
\end{aligned}
$$

This results in Figure 23(c).

EXAMPLE 5 Finding Exact Values of the Trigonometric Functions ($\theta = \pi/3 = 60°$)

4 Find the exact value of the six trigonometric functions of $\pi/3 = 60°$.

Solution Position the triangle in Figure 23(c) so that the 60° angle is in the standard position. See Figure 24.

Figure 24

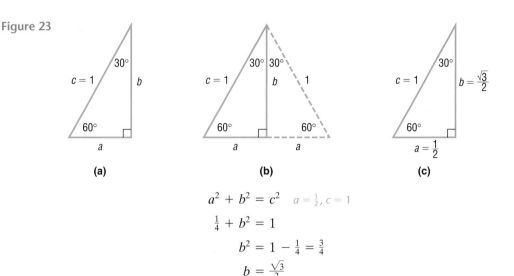

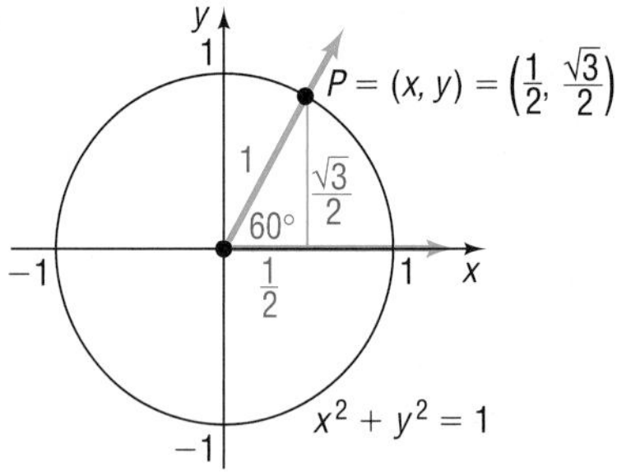

The point on the unit circle that corresponds to $\theta = \pi/3 = 60°$ is $P = \left(1/2, \sqrt{3}/2\right)$. Then

$$\sin\frac{\pi}{3} = \sin 60° = \frac{\sqrt{3}}{2} \qquad \cos\frac{\pi}{3} = \cos 60° = \frac{1}{2}$$

$$\csc\frac{\pi}{3} = \csc 60° = \frac{1}{\sqrt{3}/2} = \frac{2}{\sqrt{3}} = \frac{2\sqrt{3}}{3} \qquad \sec\frac{\pi}{3} = \sec 60° = \frac{1}{1/2} = 2$$

$$\tan\frac{\pi}{3} = \tan 60° = \frac{\sqrt{3}/2}{1/2} = \sqrt{3} \qquad \cot\frac{\pi}{3} = \cot 60° = \frac{1/2}{\sqrt{3}/2} = \frac{1}{\sqrt{3}} = \frac{\sqrt{3}}{3}$$

■

EXAMPLE 6 Finding Exact Values of the Trigonometric Functions ($\theta = \pi/6 = 30°$)

5 Find the exact value of the trigonometric functions of $\pi/6 = 30°$.

Solution Position the triangle in Figure 23(c) so that the 30° angle is in the standard position. See Figure 25.

Figure 25

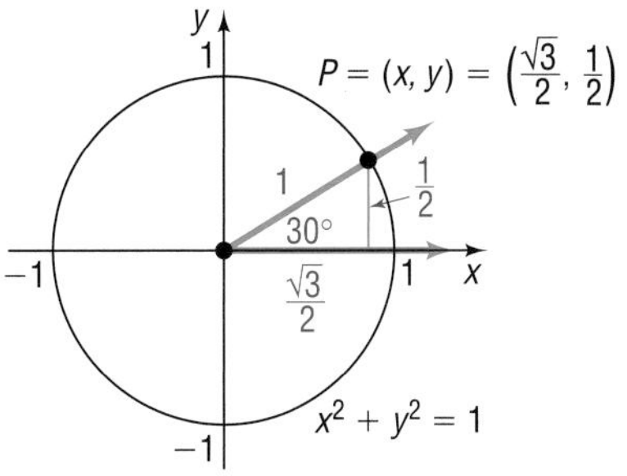

The point on the unit circle that corresponds to $\theta = \pi/6 = 30°$ is $P = \left(\sqrt{3}/2, 1/2\right)$. Then

$$\sin\frac{\pi}{6} = \sin 30° = \frac{1}{2} \qquad \cos\frac{\pi}{6} = \cos 30° = \frac{\sqrt{3}}{2}$$

$$\csc\frac{\pi}{6} = \csc 30° = \frac{1}{1/2} = 2 \qquad \sec\frac{\pi}{6} = \sec 30° = \frac{1}{\sqrt{3}/2} = \frac{2}{\sqrt{3}} = \frac{2\sqrt{3}}{3}$$

$$\tan\frac{\pi}{6} = \tan 30° = \frac{1/2}{\sqrt{3}/2} = \frac{1}{\sqrt{3}} = \frac{\sqrt{3}}{3} \qquad \cot\frac{\pi}{6} = \cot 30° = \frac{\sqrt{3}/2}{1/2} = \sqrt{3}$$

Table 3 summarizes the information just derived for $\pi/6\,(30°)$, $\pi/4\,(45°)$, and $\pi/3\,(60°)$. Until you memorize the entries in Table 3, you should draw an appropriate diagram to determine the values given in the table.

TABLE 3

θ (radians)	θ (degrees)	$\sin\theta$	$\cos\theta$	$\tan\theta$	$\csc\theta$	$\sec\theta$	$\cot\theta$
$\pi/6$	30°	$1/2$	$\sqrt{3}/2$	$\sqrt{3}/3$	2	$2\sqrt{3}/3$	$\sqrt{3}$
$\pi/4$	45°	$\sqrt{2}/2$	$\sqrt{2}/2$	1	$\sqrt{2}$	$\sqrt{2}$	1
$\pi/3$	60°	$\sqrt{3}/2$	$1/2$	$\sqrt{3}$	$2\sqrt{3}/3$	2	$\sqrt{3}/3$

NOW WORK PROBLEM 21.

EXAMPLE 7 Constructing a Rain Gutter

Figure 26

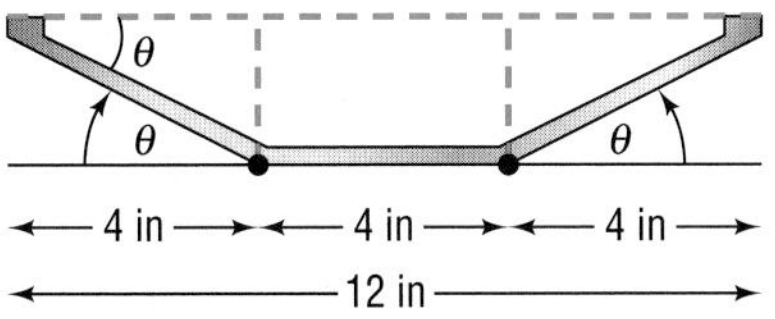

A rain gutter is to be constructed of aluminum sheets 12 inches wide. After marking off a length of 4 inches from each edge, this length is bent up at an angle θ. See Figure 26. The area A of the opening may be expressed as a function of θ as

$$A(\theta) = 16\sin\theta(\cos\theta + 1)$$

Find the area A of the opening for $\theta = 30°$, $\theta = 45°$, and $\theta = 60°$.

Solution

$$\text{For } \theta = 30°\text{:} \quad A(30°) = 16\sin 30°\,(\cos 30° + 1)$$
$$= 16\left(\frac{1}{2}\right)\left(\frac{\sqrt{3}}{2} + 1\right) = 4\sqrt{3} + 8$$

The area of the opening for $\theta = 30°$ is about 14.9 square inches.

$$\text{For } \theta = 45°: \quad A(45°) = 16 \sin 45° (\cos 45° + 1) = 16\left(\frac{\sqrt{2}}{2}\right)\left(\frac{\sqrt{2}}{2} + 1\right) = 8 + 8\sqrt{2}$$

The area of the opening for $\theta = 45°$ is about 19.3 square inches.

$$\text{For } \theta = 60°: \quad A(60°) = 16 \sin 60° (\cos 60° + 1) = 16\left(\frac{\sqrt{3}}{2}\right)\left(\frac{1}{2} + 1\right) = 12\sqrt{3}$$

The area of the opening for $\theta = 60°$ is about 20.8 square inches. ■

Exact Values for Integral Multiples of Quadrantal Angles and 30°, 45°, and 60°

Any integral multiple of a quadrantal angle is also a quadrantal angle. For example, $540° = 6(90°)$ and the point that corresponds to $\theta = 540°$ is $P = (-1, 0)$, so the trigonometric functions of 540° are the same as the trigonometric functions of 180°.

The even integral multiples of 45° are quandrantal angles (90°, 180°, and so on). Consider the odd integral multiples of 45° ($\pi/4$): 135°($3\pi/4$), 225°($5\pi/4$), and 315°($7\pi/4$). Using the symmetry of the unit circle with respect to the x-axis, y-axis, and origin, we can find the points corresponding to these angles. See Figure 27.

Figure 27

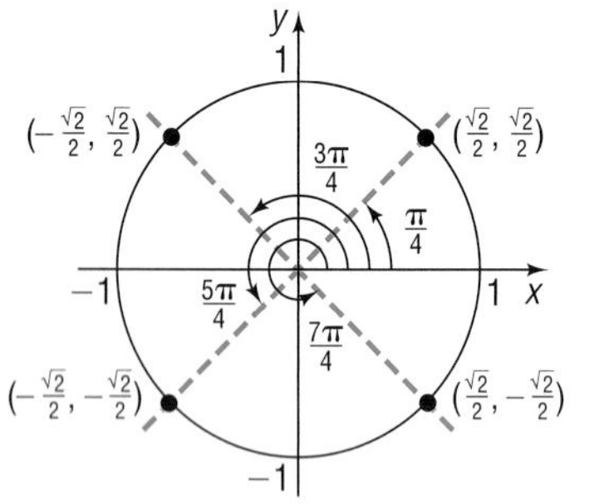

Based on Figure 27, we conclude that the point $P = (x, y)$ on the unit circle that corresponds to $\theta = \frac{3\pi}{4}$ is $\left(-\frac{\sqrt{2}}{2}, \frac{\sqrt{2}}{2}\right)$, the point $P = (x, y)$ on the unit circle that corresponds to $\theta = \frac{5\pi}{4}$ is $\left(-\frac{\sqrt{2}}{2}, -\frac{\sqrt{2}}{2}\right)$, and the point $P = (x, y)$ on the unit circle that corresponds to $\theta = \frac{7\pi}{4}$ is $\left(\frac{\sqrt{2}}{2}, -\frac{\sqrt{2}}{2}\right)$. Then, for example, $\sin\frac{3\pi}{4} = \frac{\sqrt{2}}{2}$, $\sin\frac{5\pi}{4} = -\frac{\sqrt{2}}{2}$, $\sin\frac{7\pi}{4} = -\frac{\sqrt{2}}{2}$, and so on.

Consider the following integral multiples of 30° ($\pi/6$): 150°($5\pi/6$), 210°($7\pi/6$), and 330°($11\pi/6$). Using the symmetry of the unit circle with respect to the x-axis, y-axis, and origin, we can find the points corresponding to these angles. See Figure 28.

Figure 28

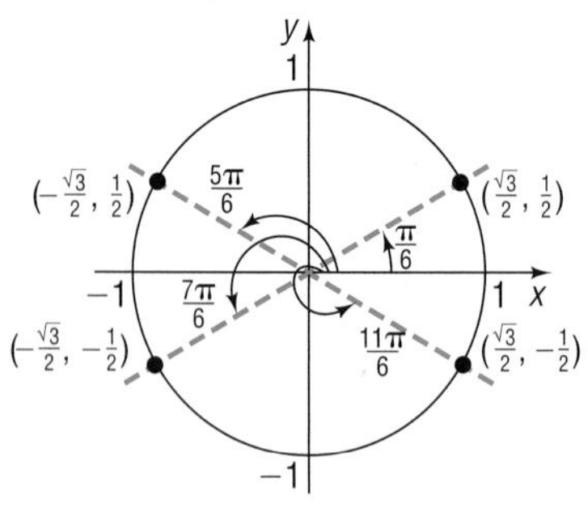

Based on Figure 28, we conclude that the point $P = (x, y)$ on the unit circle that corresponds to $\theta = \frac{5\pi}{6}$ is $\left(-\frac{\sqrt{3}}{2}, \frac{1}{2}\right)$, the point $P = (x, y)$ on the unit circle that corresponds to $\theta = \frac{7\pi}{6}$ is $\left(-\frac{\sqrt{3}}{2}, -\frac{1}{2}\right)$, and the point $P = (x, y)$ on the unit circle that corresponds to $\theta = \frac{11\pi}{6}$ is $\left(\frac{\sqrt{3}}{2}, -\frac{1}{2}\right)$. Then, for example, $\sin\frac{5\pi}{6} = \frac{1}{2}$, $\sin\frac{7\pi}{6} = -\frac{1}{2}$, $\sin\frac{11\pi}{6} = -\frac{1}{2}$, and so on.

Figure 29

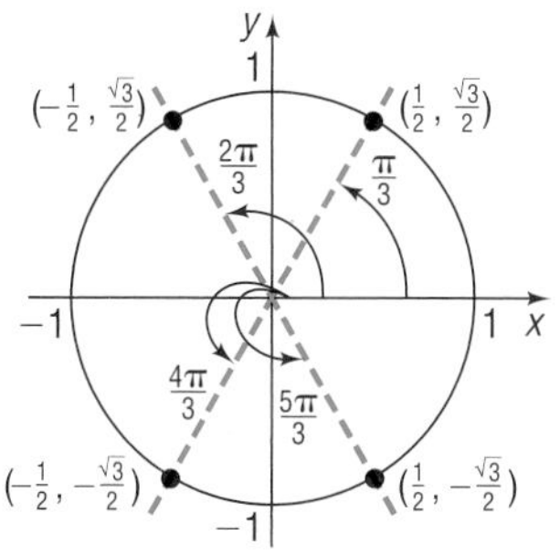

Consider the following integral multiples of $60°(\pi/3)$: $120°(2\pi/3)$, $240°(4\pi/3)$, and $300°(5\pi/3)$. Using the symmetry of the unit circle with respect to the x-axis, y-axis, and origin, we can find the points corresponding to these angles. See Figure 29.

Based on Figure 29, we conclude that the point $P = (x, y)$ on the unit circle that corresponds to $\theta = \frac{2\pi}{3}$ is $\left(-\frac{1}{2}, \frac{\sqrt{3}}{2}\right)$, the point $P = (x, y)$ on the unit circle that corresponds to $\theta = \frac{4\pi}{3}$ is $\left(-\frac{1}{2}, -\frac{\sqrt{3}}{2}\right)$, and the point $P = (x, y)$ on the unit circle that corresponds to $\theta = \frac{5\pi}{3}$ is $\left(\frac{1}{2}, -\frac{\sqrt{3}}{2}\right)$. Then, for example, $\sin\frac{2\pi}{3} = \frac{\sqrt{3}}{2}$, $\sin\frac{4\pi}{3} = -\frac{\sqrt{3}}{2}$, $\sin\frac{5\pi}{3} = -\frac{\sqrt{3}}{2}$, and so on.

EXAMPLE 8 Finding the Exact Value of Trigonometric Functions

Find the exact value of the following:

(a) $\cos\left(\frac{7\pi}{4}\right)$ (b) $\tan\left(\frac{2\pi}{3}\right)$ (c) $\csc\left(\frac{7\pi}{6}\right)$

Solution (a) The point $P = (x, y)$ on the unit circle that corresponds to $\theta = \frac{7\pi}{4}$ is

$$P = \left(\frac{\sqrt{2}}{2}, -\frac{\sqrt{2}}{2}\right), \text{ so } \cos\left(\frac{7\pi}{4}\right) = \frac{\sqrt{2}}{2}.$$

(b) The point $P = (x, y)$ on the unit circle that corresponds to $\theta = \frac{2\pi}{3}$ is

$$P = \left(-\frac{1}{2}, \frac{\sqrt{3}}{2}\right), \text{ so } \tan\left(\frac{2\pi}{3}\right) = \frac{\frac{\sqrt{3}}{2}}{-\frac{1}{2}} = -\sqrt{3}.$$

(c) The point $P = (x, y)$ on the unit circle that corresponds to $\theta = \frac{7\pi}{6}$ is

$$P = \left(-\frac{\sqrt{3}}{2}, -\frac{1}{2}\right), \text{ so } \csc\left(\frac{7\pi}{6}\right) = \frac{1}{-\frac{1}{2}} = -2.$$

■

NOW WORK PROBLEM 31.

Can you see how to use Figure 27 to find the value of $\sin\left(-\frac{\pi}{4}\right)$? The point on the unit circle that corresponds to $\theta = -\frac{\pi}{4}$ is $\left(\frac{\sqrt{2}}{2}, -\frac{\sqrt{2}}{2}\right)$, so $\sin\left(-\frac{\pi}{4}\right) = -\frac{\sqrt{2}}{2}$. In a similar manner, we can find $\tan\left(\frac{9\pi}{4}\right)$ by observing that the point $\left(\frac{\sqrt{2}}{2}, \frac{\sqrt{2}}{2}\right)$ is the point on the unit circle corresponding to $\theta = \frac{9\pi}{4}$. Thus, $\tan\left(\frac{9\pi}{4}\right) = \frac{\sqrt{2}/2}{\sqrt{2}/2} = 1.$

NOW WORK PROBLEMS 37 AND 45.

USING A CALCULATOR TO FIND VALUES OF TRIGONOMETRIC FUNCTIONS

6 Before getting started, you must first decide whether to enter the angle in the calculator using radians or degrees and then set the calculator to the correct MODE. Your calculator has the keys marked [sin], [cos], and [tan]. To find the values of the remaining three trigonomtric functions, secant, cosecant, and cotangent, we use the fact that if $P = (x, y)$ is a point on the unit circle on the terminal side of θ, then

$$\sec\theta = \frac{1}{x} = \frac{1}{\cos\theta} \qquad \csc\theta = \frac{1}{y} = \frac{1}{\sin\theta} \qquad \cot\theta = \frac{x}{y} = \frac{1}{y/x} = \frac{1}{\tan\theta}$$

EXAMPLE 9 Using a Calculator to Approximate the Value of Trigonometric Functions

Use a calculator to find the approximate value of:

(a) $\cos 48°$ (b) $\csc 21°$ (c) $\tan\frac{\pi}{12}$

Express your answer rounded to two decimal places.

Solution (a) First, we set the MODE to receive degrees. See Figure 30(a). Figure 30(b) shows the solution using a TI-83 graphing calculator.

Figure 30

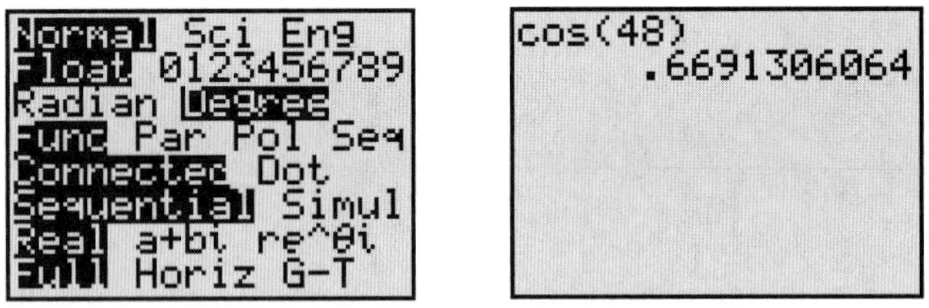

(a) (b)

Then

$$\cos 48° = 0.67$$

rounded to two decimal places.

(b) Most calculators do not have a csc key. The manufacturers assume that the user knows some trigonometry. To find the value of $\csc 21°$, we use the fact that $\csc 21° = 1/(\sin 21°)$. Figure 31 shows the solution using a TI-83 graphing calculator. Then

Figure 31

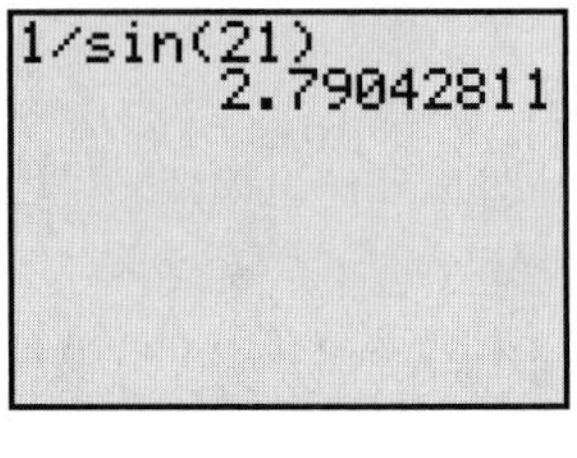

$$\csc 21° = 2.79$$

rounded to two decimal places.

Figure 32

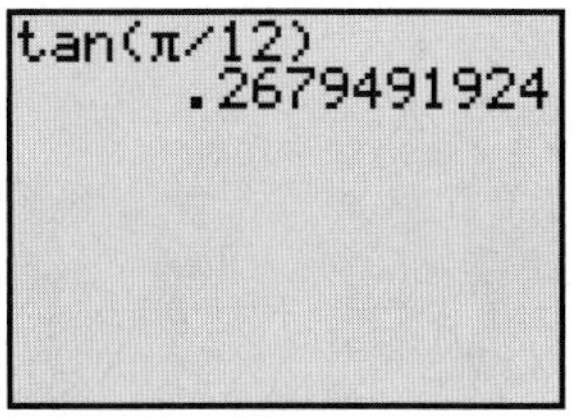

(c) Set the MODE to receive radians. Figure 32 shows the solution using a TI-83 graphing calculator. Then

$$\tan\frac{\pi}{12} = 0.27$$

rounded to two decimal places. ■

NOTE: In Figure 32, the parentheses are necessary. Without them, the calculator will evaluate $\tan \pi$ first, then divide by 12.

NOW WORK PROBLEM 53.

Figure 33

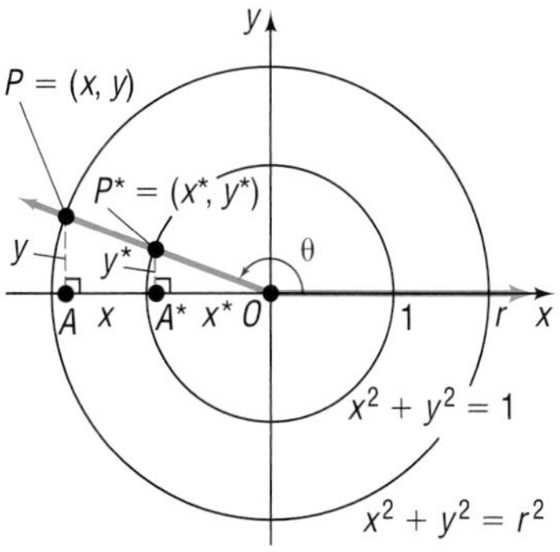

USING A CIRCLE OF RADIUS r TO EVALUATE THE TRIGONOMETRIC FUNCTIONS

Until now, to find the exact value of a trigonometric function of an angle θ required that we locate the corresponding point P on the unit circle. In fact, though, any circle whose center is at the origin can be used.

Let θ be any nonquadrantal angle placed in standard position. Let $P^* = (x^*, y^*)$ be the point where the terminal side of θ intersects the unit circle. See Figure 33.

Let $P = (x, y)$ be the point on the terminal side of θ that is also on the circle $x^2 + y^2 = r^2$. Refer again to Figure 33. Notice that the triangles OA^*P^* and OAP are similar; as a result, the ratios of corresponding sides are equal.

$$\frac{y^*}{1} = \frac{y}{r} \qquad \frac{x^*}{1} = \frac{x}{r} \qquad \frac{y^*}{x^*} = \frac{y}{x}$$

$$\frac{1}{y^*} = \frac{r}{y} \qquad \frac{1}{x^*} = \frac{r}{x} \qquad \frac{x^*}{y^*} = \frac{x}{y}$$

These results lead us to formulate the following theorem:

Theorem

For an angle θ in standard position, let $P = (x, y)$ be the point on the terminal side of θ that is also on the circle $x^2 + y^2 = r^2$. Then

$$\sin\theta = \frac{y}{r} \qquad \cos\theta = \frac{x}{r} \qquad \tan\theta = \frac{y}{x}, \quad x \neq 0$$

$$\csc\theta = \frac{r}{y}, \quad y \neq 0 \qquad \sec\theta = \frac{r}{x}, \quad x \neq 0 \qquad \cot\theta = \frac{x}{y}, \quad y \neq 0$$

■

EXAMPLE 10 Finding the Exact Value of the Six Trigonometric Functions

Find the exact value of each of the six trigonometric functions of an angle θ if $(4, -3)$ is a point on its terminal side.

Figure 34

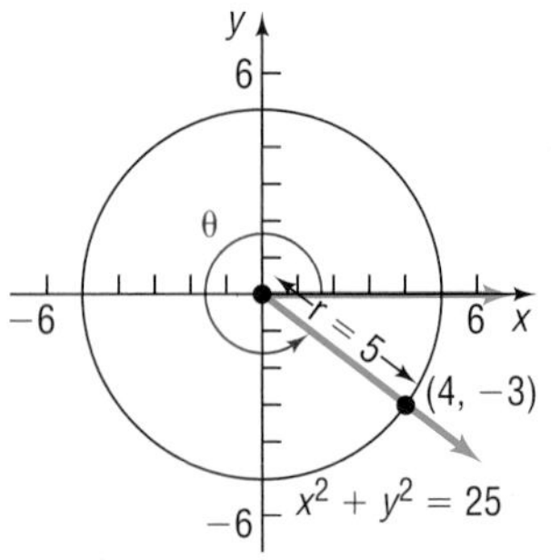

Solution Figure 34 illustrates the situation for θ a positive angle. For the point $(x, y) = (4, -3)$, we have $x = 4$ and $y = -3$. Since $r = \sqrt{x^2 + y^2} = \sqrt{16 + 9} = 5$, the point $(4, -3)$ is also on the circle $x^2 + y^2 = 25$. Then

$$\sin\theta = \frac{y}{r} = -\frac{3}{5} \qquad \cos\theta = \frac{x}{r} = \frac{4}{5} \qquad \tan\theta = \frac{y}{x} = -\frac{3}{4}$$

$$\csc\theta = \frac{r}{y} = -\frac{5}{3} \qquad \sec\theta = \frac{r}{x} = \frac{5}{4} \qquad \cot\theta = \frac{x}{y} = -\frac{4}{3}$$

NOW WORK PROBLEM 89.

HISTORICAL FEATURE

The name *sine* for the sine function is due to a medieval confusion. The name comes from the Sanskrit word *jiva* (meaning chord), first used in India by Aryabhata the Elder (AD 510). He really meant half-chord, but abbreviated it. This was brought into Arabic as *ji ba*, which was meaningless. Because the proper Arabic word *jaib* would be written the same way (short vowels are not written out in Arabic), *ji ba* was pronounced as *jaib*, which meant bosom or hollow, and *jaib* remains as the Arabic word for sine to this day. Scholars translating the Arabic works into Latin found that the word *sinus* also meant bosom or hollow, and from *sinus* we get the word *sine*.

The name *tangent*, due to Thomas Finck (1583), can be understood by looking at Figure 35. The line segment $\overline{DC}$ is tangent to the circle at C. If $d(O, B) = d(O, C) = 1$, then the length of the line segment $\overline{DC}$ is

$$d(D, C) = \frac{d(D, C)}{1} = \frac{d(D, C)}{d(O, C)} = \tan\alpha$$

The old name for the tangent is *umbra versa* (meaning turned shadow), referring to the use of the tangent in solving height problems with shadows.

The names of the remaining functions came about as follows. If α and β are complementary angles, then $\cos\alpha = \sin\beta$. Because β is the complement of α, it was natural to write the cosine of α as *sin co* α. Probably for reasons involving ease of pronunciation, the *co* migrated to the front, and then cosine received a three-letter abbreviation to match sin, sec, and tan. The two other cofunctions were similarly treated, except that the long forms *cotan* and *cosec* survive to this day in some countries.

Figure 35

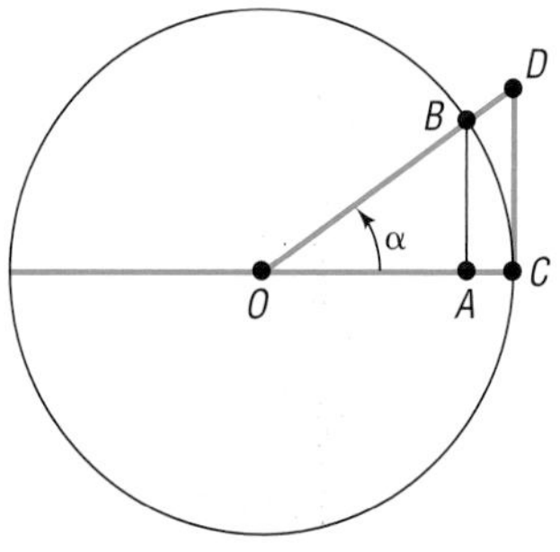

6.2 EXERCISES

In Problems 1–10, t is a real number and P = (x, y) is the point on the unit circle that corresponds to t. Find the exact value of the six trigonometric functions of t.

1. $\left(\frac{1}{4}, \frac{\sqrt{15}}{4}\right)$ **2.** $\left(\frac{3}{8}, \frac{\sqrt{55}}{8}\right)$ **3.** $\left(-\frac{2}{5}, \frac{\sqrt{21}}{5}\right)$ **4.** $\left(-\frac{1}{5}, \frac{2\sqrt{6}}{5}\right)$

5. $\left(\frac{-\sqrt{35}}{6}, -\frac{1}{6}\right)$ **6.** $\left(\frac{-\sqrt{39}}{8}, \frac{5}{8}\right)$ **7.** $\left(\frac{2\sqrt{2}}{3}, -\frac{1}{3}\right)$ **8.** $\left(\frac{-\sqrt{5}}{3}, -\frac{2}{3}\right)$

9. $\left(-\frac{3\sqrt{5}}{7}, \frac{2}{7}\right)$ **10.** $\left(-\frac{3\sqrt{11}}{10}, -\frac{1}{10}\right)$

In Problems 11–30, find the exact value of each expression. Do not use a calculator.

11. $\sin 45° + \cos 60°$ **12.** $\sin 30° - \cos 45°$ **13.** $\sin 90° + \tan 45°$

14. $\cos 180° - \sin 180°$ **15.** $\sin 45° \cos 45°$ **16.** $\tan 45° \cos 30°$

17. $\csc 45° \tan 60°$

18. $\sec 30° \cot 45°$

19. $4 \sin 90° - 3 \tan 180°$

20. $5 \cos 90° - 8 \sin 270°$

21. $2 \sin \frac{\pi}{3} - 3 \tan \frac{\pi}{6}$

22. $2 \sin \frac{\pi}{4} + 3 \tan \frac{\pi}{4}$

23. $\sin \frac{\pi}{4} - \cos \frac{\pi}{4}$

24. $\tan \frac{\pi}{3} + \cos \frac{\pi}{3}$

25. $2 \sec \frac{\pi}{4} + 4 \cot \frac{\pi}{3}$

26. $3 \csc \frac{\pi}{3} + \cot \frac{\pi}{4}$

27. $\tan \pi - \cos 0$

28. $\sin \frac{3\pi}{2} + \tan \pi$

29. $\csc \frac{\pi}{2} + \cot \frac{\pi}{2}$

30. $\sec \pi - \csc \frac{\pi}{2}$

In Problems 31–52, find the exact value of the six trigonometric functions of the given angle. If any are not defined, say "not defined." Do not use a calculator.

31. $\frac{2\pi}{3}$

32. $\frac{5\pi}{6}$

33. $210°$

34. $240°$

35. $\frac{5\pi}{3}$

36. $\frac{11\pi}{6}$

37. $7\pi/3$

38. $13\pi/6$

39. $405°$

40. $390°$

41. $-\pi/6$

42. $-\pi/3$

43. $-45°$

44. $-60°$

45. $5\pi/2$

46. 3π

47. $-180°$

48. $-270°$

49. $-\pi/2$

50. -5π

51. $480°$

52. $-150°$

In Problems 53–70, use a calculator to find the approximate value of each expression rounded to two decimal places.

53. $\sin 28°$

54. $\cos 14°$

55. $\tan 21°$

56. $\sin 15°$

57. $\sec 41°$

58. $\csc 55°$

59. $\cot 70°$

60. $\tan 80°$

61. $\sin \frac{\pi}{10}$

62. $\cos \frac{\pi}{8}$

63. $\tan \frac{5\pi}{12}$

64. $\sin \frac{3\pi}{10}$

65. $\sec \frac{\pi}{12}$

66. $\csc \frac{5\pi}{13}$

67. $\sin 1$

68. $\tan 1$

69. $\sin 1°$

70. $\tan 1°$

In Problems 71–82, $f(\theta) = \sin\theta$ and $g(\theta) = \cos\theta$. Find the exact value of each function below if $\theta = 60°$. Do not use a calculator.

71. $f(\theta)$

72. $g(\theta)$

73. $f\left(\frac{\theta}{2}\right)$

74. $g\left(\frac{\theta}{2}\right)$

75. $[f(\theta)]^2$

76. $[g(\theta)]^2$

77. $f(2\theta)$

78. $g(2\theta)$

79. $2f(\theta)$

80. $2g(\theta)$

81. $f(-\theta)$

82. $g(-\theta)$

83. Use a calculator in radian mode to complete the following table.
What can you conclude about the ratio $(\sin\theta)/\theta$ as θ approaches 0?

θ	0.5	0.4	0.2	0.1	0.01	0.001	0.0001	0.00001
$\sin\theta$								
$\frac{\sin\theta}{\theta}$								

84. Use a calculator in radian mode to complete the following table.
What can you conclude about the ratio $(\cos\theta - 1)/\theta$ as θ approaches 0?

θ	0.5	0.4	0.2	0.1	0.01	0.001	0.0001	0.00001
$\cos\theta - 1$								
$\frac{\cos\theta - 1}{\theta}$								

In Problems 85–88, use the figure to approximate the value of the six trigonometric functions at t to the nearest tenth. Then use a calculator to approximate each of the six trigonometric functions at t.

85. (a) $t = 1$ (b) $t = 5.1$ (c) $t = 2.4$

86. (a) $t = 2$ (b) $t = 4$ (c) $t = 5.9$

87. (a) $t = 1.5$ (b) $t = 4.3$ (c) $t = 5.3$

88. (a) $t = 2.7$ (b) $t = 3.9$ (c) $t = 6.1$

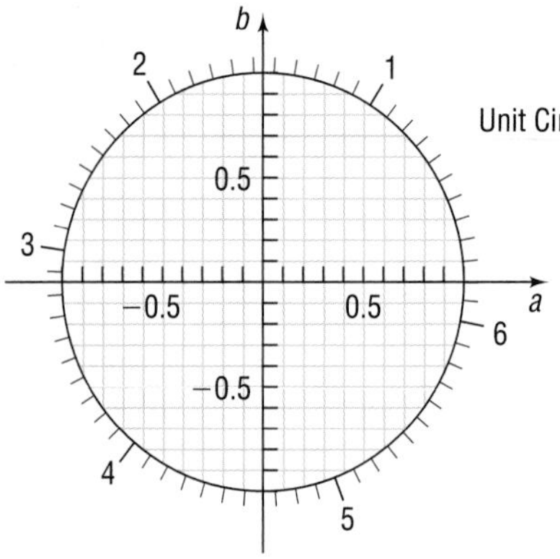

In Problems 89–98, a point on the terminal side of an angle θ is given. Find the exact value of the six trigonometric functions of θ.

89. $(-3, 4)$ **90.** $(5, -12)$ **91.** $(2, -3)$ **92.** $(-1, -2)$

93. $(-2, -2)$ **94.** $(1, -1)$ **95.** $(-3, -2)$ **96.** $(2, 2)$

97. $\left(\frac{1}{3}, -\frac{1}{4}\right)$ **98.** $(-0.3, -0.4)$

99. Find the exact value of $\sin 45° + \sin 135° + \sin 225° + \sin 315°$.

100. Find the exact value of $\tan 60° + \tan 150°$.

101. If $\sin\theta = 0.1$, find $\sin(\theta + \pi)$.

102. If $\cos\theta = 0.3$, find $\cos(\theta + \pi)$.

103. If $\tan\theta = 3$, find $\tan(\theta + \pi)$.

104. If $\cot\theta = -2$, find $\cot(\theta + \pi)$.

105. If $\sin\theta = \frac{1}{5}$, find $\csc\theta$.

106. If $\cos\theta = \frac{2}{3}$, find $\sec\theta$.

Projectile Motion *The path of a projectile fired at an inclination θ to the horizontal with initial speed v_0 is a parabola (see the figure).*

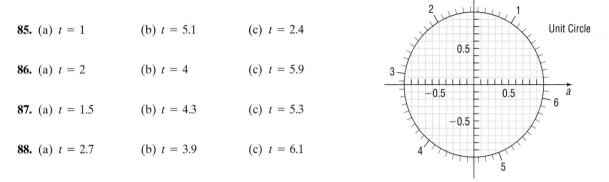

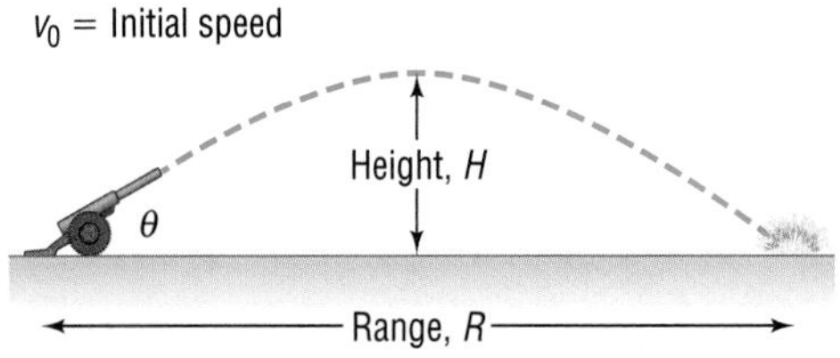

The range R of the projectile, that is, the horizontal distance that the projectile travels, is found by using the formula

$$R = \frac{v_0^2 \sin(2\theta)}{g}$$

where $g \approx 32.2$ feet per second per second ≈ 9.8 meters per second per second is the acceleration due to gravity. The maximum height H of the projectile is

$$H = \frac{v_0^2 \sin^2\theta}{2g}$$

In Problems 107–110, find the range R and maximum height H.

107. The projectile is fired at an angle of 45° to the horizontal with an initial speed of 100 feet per second.

108. The projectile is fired at an angle of 30° to the horizontal with an initial speed of 150 meters per second.

109. The projectile is fired at an angle of 25° to the horizontal with an initial speed of 500 meters per second.

110. The projectile is fired at an angle of 50° to the horizontal with an initial speed of 200 feet per second.

111. Inclined Plane If friction is ignored, the time t (in seconds) required for a block to slide down an inclined plane (see the figure) is given by the formula

$$t = \sqrt{\frac{2a}{g \sin\theta \cos\theta}}$$

where a is the length (in feet) of the base and $g \approx 32$ feet per second per second is the acceleration of gravi-

ty. How long does it take a block to slide down an inclined plane with base $a = 10$ feet when:
(a) $\theta = 30°$? (b) $\theta = 45°$? (c) $\theta = 60°$?

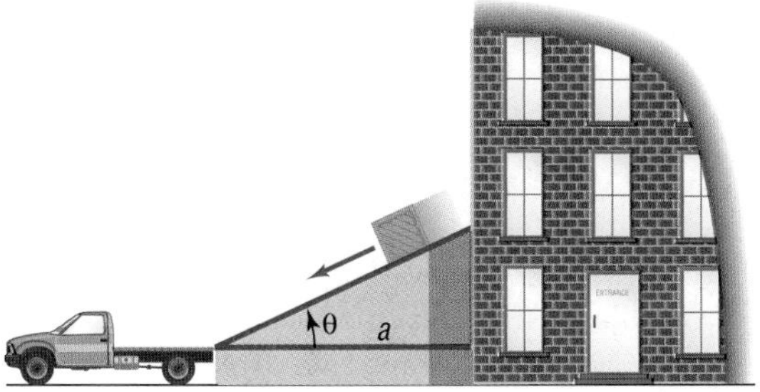

112. Piston Engines In a certain piston engine, the distance x (in centimeters) from the center of the drive shaft to the head of the piston is given by

$$x = \cos\theta + \sqrt{16 + 0.5\cos(2\theta)}$$

where θ is the angle between the crank and the path of the piston head (see the figure). Find x when $\theta = 30°$ and when $\theta = 45°$.

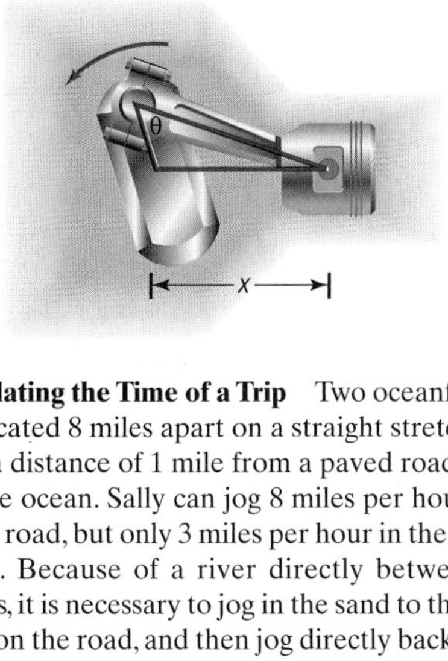

113. Calculating the Time of a Trip Two oceanfront homes are located 8 miles apart on a straight stretch of beach, each a distance of 1 mile from a paved road that parallels the ocean. Sally can jog 8 miles per hour along the paved road, but only 3 miles per hour in the sand on the beach. Because of a river directly between the two houses, it is necessary to jog in the sand to the road, continue on the road, and then jog directly back in the sand to get from one house to the other. See the illustration. The time T to get from one house to the other as a function of the angle θ shown in the illustration is

$$T(\theta) = 1 + \frac{2}{3\sin\theta} - \frac{1}{4\tan\theta}, \qquad 0° < \theta < 90°$$

(a) Calculate the time T for $\theta = 30°$. How long is Sally on the paved road?
(b) Calculate the time T for $\theta = 45°$. How long is Sally on the paved road?
(c) Calculate the time T for $\theta = 60°$. How long is Sally on the paved road?
(d) Calculate the time T for $\theta = 90°$. Describe the path taken. Why can't the formula for T be used?

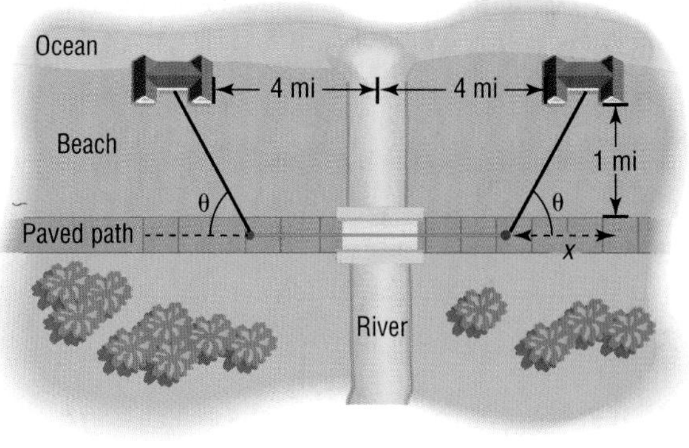

114. Designing Fine Decorative Pieces A designer of decorative art plans to market solid gold spheres encased in clear crystal cones. Each sphere is of fixed radius R and will be enclosed in a cone of height h and radius r. See the illustration. Many cones can be used to enclose the sphere, each having a different slant angle θ. The volume V of the cone can be expressed as a function of the slant angle θ of the cone as

$$V(\theta) = \frac{1}{3}\pi R^3 \frac{(1 + \sec\theta)^3}{\tan^2\theta}, \qquad 0° < \theta < 90°$$

What volume V is required to enclose a sphere of radius 2 centimeters in a cone whose slant angle θ is 30°? 45°? 60°?

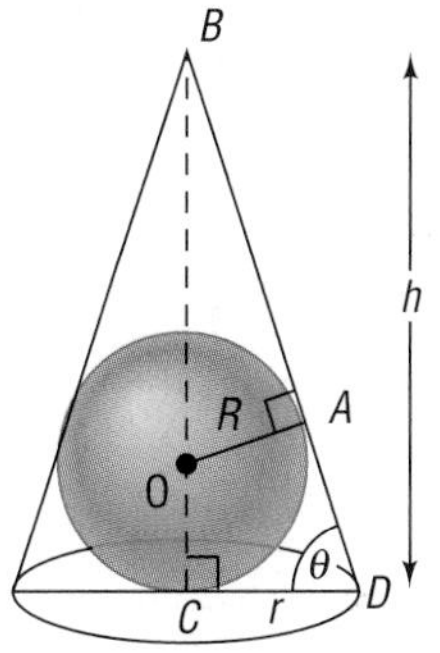

115. Projectile Motion An object is propelled upward at an angle θ, $45° < \theta < 90°$, to the horizontal with an initial velocity of v_0 feet per second from the base of a plane that makes an angle of 45° with the horizontal. See the illustration. If air resistance is ignored, the distance R that it travels up the inclined plane is given by

$$R = \frac{v_0^2\sqrt{2}}{32}\left[\sin(2\theta) - \cos(2\theta) - 1\right]$$

(a) Find the distance R that the object travels along the inclined plane if the initial velocity is 32 feet per second and $\theta = 60°$.

(b) Graph $R = R(\theta)$ if the initial velocity is 32 feet per second.
(c) What value of θ makes R largest?

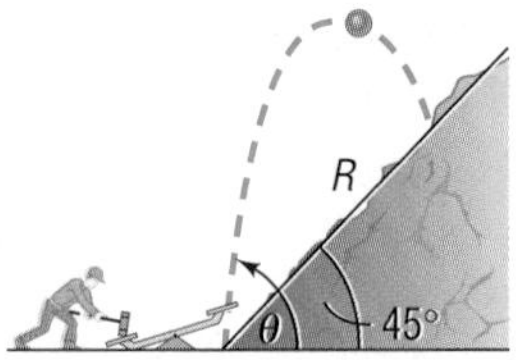

116. If $\theta\,(0 < \theta < \pi)$ is the angle between a horizontal ray directed to the right (say, the positive x-axis) and a nonhorizontal, nonvertical line L, show that the slope m of L equals $\tan\theta$. The angle θ is called the **inclination** of L. [**Hint:** See the illustration, where we have drawn the line L^* parallel to L and passing through the origin. Use the fact that L^* intersects the unit circle at the point $(\cos\theta, \sin\theta)$.]

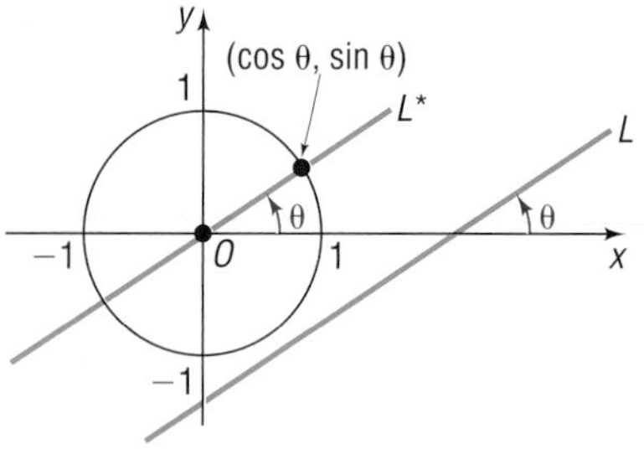

117. Write a brief paragraph that explains how to quickly compute the trigonometric functions of 30°, 45°, and 60°.

118. Write a brief paragraph that explains how to quickly compute the trigonometric functions of 0°, 90°, 180°, and 270°.

119. How would you explain the meaning of the sine function to a fellow student who has just completed college algebra?

PREPARING FOR THIS SECTION

Before getting started, review the following:

- ✓ Domain and Range of a Function (pp. 99 and 104)
- ✓ Even and Odd Functions (pp. 122–124)
- ✓ Identity (p. 976)

6.3 PROPERTIES OF THE TRIGONOMETRIC FUNCTIONS

OBJECTIVES

1. Determine the Domain and Range of the Trigonometric Functions
2. Determine the Period of the Trigonometric Functions
3. Determine the Signs of the Trigonometric Functions
4. Find the Value of the Trigonometric Functions Utilizing Fundamental Identities
5. Use Even–Odd Properties to Find the Exact Value of the Trigonometric Functions

DOMAIN AND RANGE OF THE TRIGONOMETRIC FUNCTIONS

Figure 36

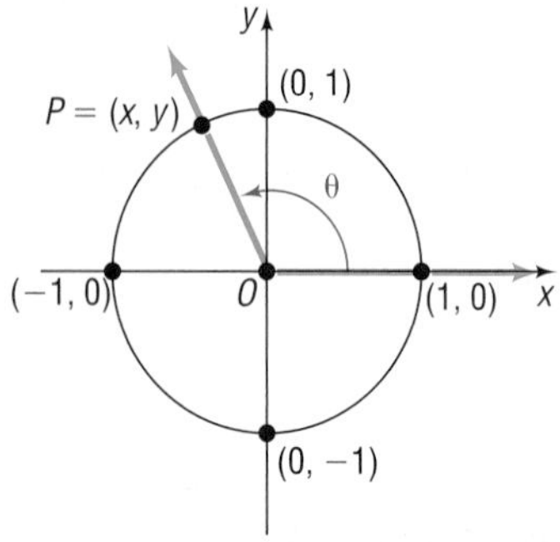

1 Let θ be an angle in standard position, and let $P = (x, y)$ be the point on the unit circle that corresponds to θ. See Figure 36. Then, by definition,

$$\sin\theta = y \qquad \cos\theta = x \qquad \tan\theta = \frac{y}{x}, \quad x \neq 0$$

$$\csc\theta = \frac{1}{y}, \quad y \neq 0 \qquad \sec\theta = \frac{1}{x}, \quad x \neq 0 \qquad \cot\theta = \frac{x}{y}, \quad y \neq 0$$

For $\sin\theta$ and $\cos\theta$, θ can be any angle, so it follows that the domain of the sine function and cosine function is the set of all real numbers.

> The domain of the sine function is the set of all real numbers.
> The domain of the cosine function is the set of all real numbers.

If $x = 0$, then the tangent function and the secant function are not defined. That is, for the tangent function and secant function, the x-coordinate of $P = (x, y)$ cannot be 0. On the unit circle, there are two such points $(0, 1)$ and $(0, -1)$. These two points correspond to the angles $\pi/2\,(90°)$ and $3\pi/2\,(270°)$ or, more generally, to any angle that is an odd multiple of $\pi/2\,(90°)$, such as $\pi/2\,(90°)$, $3\pi/2\,(270°)$, $5\pi/2\,(450°)$, $-\pi/2\,(-90°)$, $-3\pi/2\,(-270°)$, and so on. Such angles must therefore be excluded from the domain of the tangent function and secant function.

> The domain of the tangent function is the set of all real numbers, except odd multiples of $\pi/2\,(90°)$.
> The domain of the secant function is the set of all real numbers, except odd multiples of $\pi/2\,(90°)$.

If $y = 0$, then the cotangent function and the cosecant function are not defined. For the cotangent function and cosecant function, the y-coordinate of $P = (x, y)$ cannot be 0. On the unit circle, there are two such points, $(1, 0)$ and $(-1, 0)$. These two points correspond to the angles $0\,(0°)$ and $\pi\,(180°)$ or, more generally, to any angle that is an integral multiple of $\pi\,(180°)$, such as $0\,(0°)$, $\pi\,(180°)$, $2\pi\,(360°)$, $3\pi\,(540°)$, $-\pi\,(-180°)$, and so on. Such angles must therefore be excluded from the domain of the cotangent function and cosecant function.

> The domain of the cotangent function is the set of all real numbers, except integral multiples of $\pi\,(180°)$.
> The domain of the cosecant function is the set of all real numbers, except integral multiples of $\pi\,(180°)$.

Next, we determine the range of each of the six trigonometric functions. Refer again to Figure 36. Let $P = (x, y)$ be the point on the unit circle that corresponds to the angle θ. It follows that $-1 \le x \le 1$ and $-1 \le y \le 1$. Consequently, since $\sin\theta = y$ and $\cos\theta = x$, we have

$$-1 \le \sin\theta \le 1 \qquad -1 \le \cos\theta \le 1$$

The range of both the sine function and the cosine function consists of all real numbers between -1 and 1, inclusive. Using absolute value notation, we have $|\sin\theta| \le 1$ and $|\cos\theta| \le 1$.

Similarly, if θ is not a multiple of $\pi\,(180°)$, then $\csc\theta = 1/y$. Since $y = \sin\theta$ and $|y| = |\sin\theta| \le 1$, it follows that $|\csc\theta| = 1/|\sin\theta| = 1/|y| \ge 1$. The range of the cosecant function consists of all real numbers less than or equal to -1 or greater than or equal to 1. That is,

$$\csc\theta \le -1 \quad \text{or} \quad \csc\theta \ge 1$$

If θ is not an odd multiple of $\pi/2\,(90°)$, then, by definition, $\sec\theta = 1/x$. Since $x = \cos\theta$ and $|x| = |\cos\theta| \le 1$, it follows that $|\sec\theta| = 1/|\cos\theta| =$

$1/|x| \geq 1$. The range of the secant function consists of all real numbers less than or equal to -1 or greater than or equal to 1.

$$\sec\theta \leq -1 \quad \text{or} \quad \sec\theta \geq 1$$

The range of both the tangent function and the cotangent function consists of all real numbers. You are asked to prove this in Problems 111 and 112.

$$-\infty < \tan\theta < \infty \qquad -\infty < \cot\theta < \infty$$

Table 4 summarizes these results.

TABLE 4

Function	Symbol	Domain	Range
sine	$f(\theta) = \sin\theta$	All real numbers	All real numbers from -1 to 1, inclusive
cosine	$f(\theta) = \cos\theta$	All real numbers	All real numbers from -1 to 1, inclusive
tangent	$f(\theta) = \tan\theta$	All real numbers, except odd multiples of $\pi/2(90°)$	All real numbers
cosecant	$f(\theta) = \csc\theta$	All real numbers, except integral multiples of $\pi(180°)$	All real numbers greater than or equal to 1 or less than or equal to -1
secant	$f(\theta) = \sec\theta$	All real numbers, except odd multiples of $\pi/2(90°)$	All real numbers greater than or equal to 1 or less than or equal to -1
cotangent	$f(\theta) = \cot\theta$	All real numbers, except integral multiples of $\pi(180°)$	All real numbers

NOW WORK PROBLEM 87.

Figure 37

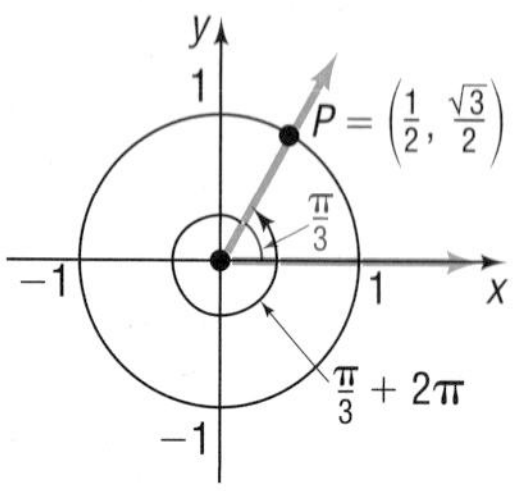

Figure 38

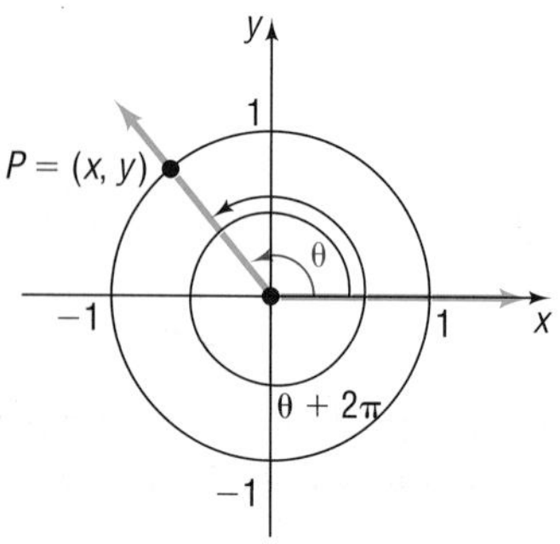

PERIOD OF THE TRIGONOMETRIC FUNCTIONS

2 Look at Figure 37. This figure shows that for an angle of $\pi/3$ radians the corresponding point P on the unit circle is $(1/2, \sqrt{3}/2)$. Notice that, for an angle of $\pi/3 + 2\pi$ radians, the corresponding point P on the unit circle is also $(1/2, \sqrt{3}/2)$. Then

$$\sin\frac{\pi}{3} = \frac{\sqrt{3}}{2} \quad \text{and} \quad \sin\left(\frac{\pi}{3} + 2\pi\right) = \frac{\sqrt{3}}{2}$$

$$\cos\frac{\pi}{3} = \frac{1}{2} \quad \text{and} \quad \cos\left(\frac{\pi}{3} + 2\pi\right) = \frac{1}{2}$$

This example illustrates a more general situation. For a given angle θ, measured in radians, suppose that we know the corresponding point $P = (x, y)$ on the unit circle. Now add 2π to θ. The point on the unit circle corresponding to $\theta + 2\pi$ is identical to the point P on the unit circle corresponding to θ. See Figure 38. The values of the trigonometric functions of $\theta + 2\pi$ are equal to the values of the corresponding trigonometric functions of θ.

If we add (or subtract) integral multiples of 2π to θ, the trigonometric values remain unchanged. That is, for all θ,

$$\sin(\theta + 2\pi k) = \sin\theta \qquad \cos(\theta + 2\pi k) = \cos\theta \tag{1}$$

where k is any integer.

Functions that exhibit this kind of behavior are called *periodic functions*.

Seeing the Concept To see the periodic behavior of the sine function, create a TABLE with $Y_1 = \sin x$, $Y_2 = \sin(x + 2\pi)$, $Y_3 = \sin(x - 2\pi)$, and $Y_4 = \sin(x + 4\pi)$ and TblStart $= 0$ and ΔTbl $= \pi/6$. ■

A function f is called **periodic** if there is a positive number p such that, whenever θ is in the domain of f, so is $\theta + p$, and

$$f(\theta + p) = f(\theta)$$

If there is a smallest such number p, this smallest value is called the **(fundamental) period** of f.

Based on equation (1), the sine and cosine functions are periodic. In fact, the sine and cosine functions have period 2π. You are asked to prove this fact in Problems 113 and 114. The secant and cosecant functions are also periodic with period 2π, and the tangent and cotangent functions are periodic with period π. You are asked to prove these statements in Problems 115 through 118.

Periodic Properties

$$\sin(\theta + 2\pi k) = \sin\theta \qquad \cos(\theta + 2\pi k) = \cos\theta \qquad \tan(\theta + \pi k) = \tan\theta$$

$$\csc(\theta + 2\pi k) = \csc\theta \qquad \sec(\theta + 2\pi k) = \sec\theta \qquad \cot(\theta + \pi k) = \cot\theta$$

where k is any integer.

Because the sine, cosine, secant, and cosecant functions have period 2π, once we know their values for $0 \le \theta < 2\pi$, we know all their values; similarly, since the tangent and cotangent functions have period π, once we know their values for $0 \le \theta < \pi$, we know all their values.

EXAMPLE 1 Finding Exact Values Using Periodic Properties

Find the exact value of:

(a) $\sin\dfrac{17\pi}{4}$ (b) $\cos(5\pi)$ (c) $\tan\dfrac{5\pi}{4}$

Solution (a) It is best to sketch the angle first, as shown in Figure 39(a). Since the period of the sine function is 2π, each full revolution can be ignored. This leaves the angle $\pi/4$. Thus,

$$\sin \frac{17\pi}{4} = \sin\left(\frac{\pi}{4} + 4\pi\right) = \sin \frac{\pi}{4} = \frac{\sqrt{2}}{2}$$

(b) See Figure 39(b). Since the period of the cosine function is 2π, each full revolution can be ignored. This leaves the angle π. Thus,

$$\cos(5\pi) = \cos(\pi + 4\pi) = \cos \pi = -1$$

(c) See Figure 39(c). Since the period of the tangent function is π, each half-revolution can be ignored. This leaves the angle $\pi/4$. Thus,

$$\tan \frac{5\pi}{4} = \tan\left(\frac{\pi}{4} + \pi\right) = \tan \frac{\pi}{4} = 1$$

Figure 39

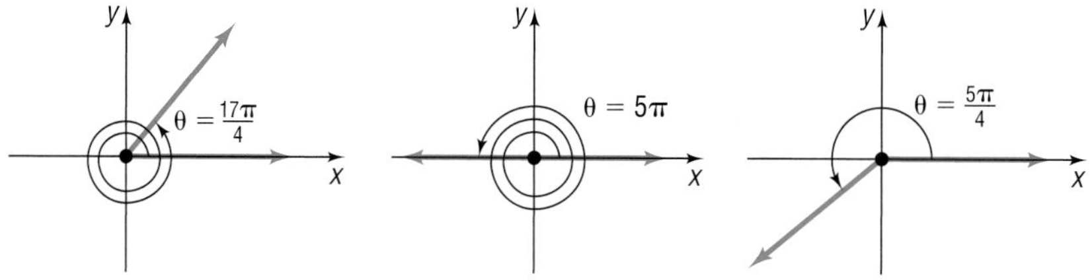

(a) (b) (c)

The periodic properties of the trigonometric functions will be very helpful to us when we study their graphs later in the chapter.

NOW WORK PROBLEM 1.

THE SIGNS OF THE TRIGONOMETRIC FUNCTIONS

3 Let $P = (x, y)$ be the point on the unit circle that corresponds to the angle θ. If we know in which quadrant the point P lies, then we can determine the signs of the trigonometric functions of θ. For example, if $P = (x, y)$ lies in quadrant IV, as shown in Figure 40, then we know that $x > 0$ and $y < 0$. Consequently,

Figure 40

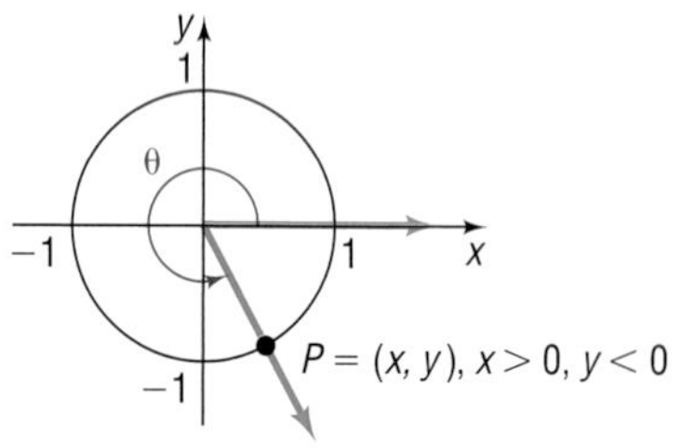

$$\sin\theta = y < 0 \qquad \cos\theta = x > 0 \qquad \tan\theta = \frac{y}{x} < 0$$

$$\csc\theta = \frac{1}{y} < 0 \qquad \sec\theta = \frac{1}{x} > 0 \qquad \cot\theta = \frac{x}{y} < 0$$

Table 5 lists the signs of the six trigonometric functions for each quadrant. See also Figure 41.

TABLE 5

Quadrant of P	$\sin\theta$, $\csc\theta$	$\cos\theta$, $\sec\theta$	$\tan\theta$, $\cot\theta$
I	Positive	Positive	Positive
II	Positive	Negative	Negative
III	Negative	Negative	Positive
IV	Negative	Positive	Negative

Figure 41

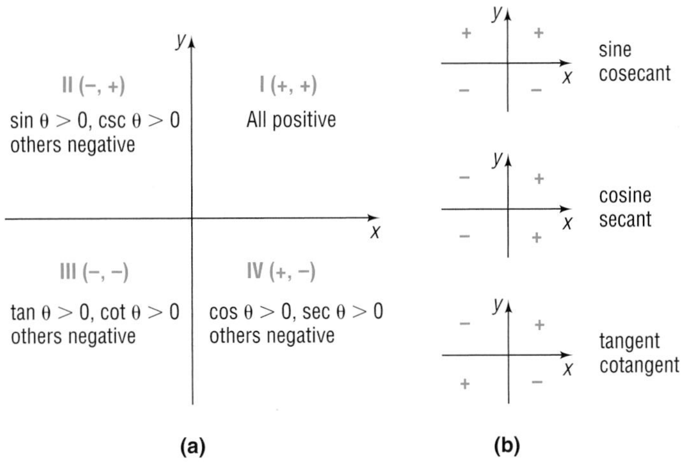

EXAMPLE 2 Finding the Quadrant in Which an Angle θ Lies

If $\sin\theta < 0$ and $\cos\theta < 0$, name the quadrant in which the angle θ lies.

Solution Let $P = (x, y)$ be the point on the unit circle corresponding to θ. Then $\sin\theta = y < 0$ and $\cos\theta = x < 0$. Thus, $P = (x, y)$ must be in quadrant III, so θ lies in quadrant III. ■

NOW WORK PROBLEM 17.

FUNDAMENTAL IDENTITIES

4 If $P = (x, y)$ is the point on the unit circle corresponding to θ, then

$$\sin\theta = y \qquad \cos\theta = x \qquad \tan\theta = \frac{y}{x}, \quad \text{if } x \neq 0$$

$$\csc\theta = \frac{1}{y}, \quad \text{if } y \neq 0 \qquad \sec\theta = \frac{1}{x}, \quad \text{if } x \neq 0 \qquad \cot\theta = \frac{x}{y}, \quad \text{if } y \neq 0$$

Based on these definitions, we have the **reciprocal identities:**

Reciprocal Identities

$$\csc\theta = \frac{1}{\sin\theta} \qquad \sec\theta = \frac{1}{\cos\theta} \qquad \cot\theta = \frac{1}{\tan\theta} \qquad \textbf{(2)}$$

Two other fundamental identities are the **quotient identities.**

Quotient Identities

$$\tan\theta = \frac{\sin\theta}{\cos\theta} \qquad \cot\theta = \frac{\cos\theta}{\sin\theta} \qquad \textbf{(3)}$$

The proofs of formulas (2) and (3) follow from the definitions of the trigonometric functions. (See Problems 119 and 120.)

SEEING THE CONCEPT To see the identity $\tan\theta = (\sin\theta)/(\cos\theta)$, create a TABLE with $Y_1 = \tan x$, $Y_2 = (\sin x)/(\cos x)$, TblStart $= 0$ and ΔTbl $= \pi/6$. ■

If $\sin\theta$ and $\cos\theta$ are known, formulas (2) and (3) make it easy to find the values of the remaining trigonometric functions.

EXAMPLE 3 Finding Exact Values Using Identities When Sine and Cosine Are Given

Given $\sin\theta = \sqrt{5}/5$ and $\cos\theta = 2\sqrt{5}/5$, find the exact values of the four remaining trigonometric functions of θ using identities.

Solution Based on a quotient identity from formula (3), we have

$$\tan\theta = \frac{\sin\theta}{\cos\theta} = \frac{\sqrt{5}/5}{2\sqrt{5}/5} = \frac{1}{2}$$

Then we use the reciprocal identities from formula (2) to get

$$\csc\theta = \frac{1}{\sin\theta} = \frac{1}{\sqrt{5}/5} = \frac{5}{\sqrt{5}} = \sqrt{5} \qquad \sec\theta = \frac{1}{\cos\theta} = \frac{1}{2\sqrt{5}/5} = \frac{5}{2\sqrt{5}} = \frac{\sqrt{5}}{2}$$

$$\cot\theta = \frac{1}{\tan\theta} = \frac{1}{\frac{1}{2}} = 2$$

■

NOW WORK PROBLEM 25.

The equation of the unit circle is $x^2 + y^2 = 1$. If $P = (x, y)$ is the point on the unit circle that corresponds to the angle θ, then

$$y^2 + x^2 = 1$$

But $y = \sin\theta$ and $x = \cos\theta$, so

$$(\sin\theta)^2 + (\cos\theta)^2 = 1 \qquad \textbf{(4)}$$

It is customary to write $\sin^2\theta$ instead of $(\sin\theta)^2$, $\cos^2\theta$ instead of $(\cos\theta)^2$, and so on. With this notation, we can rewrite equation (4) as

$$\sin^2\theta + \cos^2\theta = 1 \qquad \textbf{(5)}$$

If $\cos\theta \neq 0$, we can divide each side of equation (5) by $\cos^2\theta$.

$$\frac{\sin^2\theta}{\cos^2\theta} + 1 = \frac{1}{\cos^2\theta}$$

$$\left(\frac{\sin\theta}{\cos\theta}\right)^2 + 1 = \left(\frac{1}{\cos\theta}\right)^2$$

Now use formulas (2) and (3) to get

$$\tan^2\theta + 1 = \sec^2\theta \qquad \textbf{(6)}$$

Similarly, if $\sin\theta \neq 0$, we can divide equation (5) by $\sin^2\theta$ and use formulas (2) and (3) to get the result:

$$1 + \cot^2\theta = \csc^2\theta \qquad \textbf{(7)}$$

Collectively, the identities in equations (5), (6), and (7) are referred to as the **Pythagorean identities.**

Let's pause here to summarize the fundamental identities.

Fundamental Identities

$$\tan\theta = \frac{\sin\theta}{\cos\theta} \qquad \cot\theta = \frac{\cos\theta}{\sin\theta}$$

$$\cot\theta = \frac{1}{\tan\theta} \qquad \sec\theta = \frac{1}{\cos\theta} \qquad \csc\theta = \frac{1}{\sin\theta}$$

$$\sin^2\theta + \cos^2\theta = 1 \qquad \tan^2\theta + 1 = \sec^2\theta \qquad 1 + \cot^2\theta = \csc^2\theta$$

The Pythagorean identity

$$\sin^2\theta + \cos^2\theta = 1$$

can be solved for $\sin\theta$ in terms of $\cos\theta$ (or vice versa) as follows:

$$\sin^2\theta = 1 - \cos^2\theta$$

$$\sin\theta = \pm\sqrt{1 - \cos^2\theta}$$

where the $+$ sign is used if $\sin\theta > 0$ and the $-$ sign is used if $\sin\theta < 0$.

EXAMPLE 4 Finding Exact Values Given One Value and the Sign of Another

Given that $\sin\theta = \frac{1}{3}$ and $\cos\theta < 0$, find the exact value of each of the remaining five trigonometric functions.

Solution We solve this problem in two ways: the first way uses the definition of the trigonometric functions; the second method uses the fundamental identities.

Solution 1 *Using the Definition*

Suppose that $P = (x, y)$ is the point on the unit circle that corresponds to θ. See Figure 42. Since $\sin\theta = \frac{1}{3} = y$ and $\cos\theta = x < 0$, we have

Figure 42

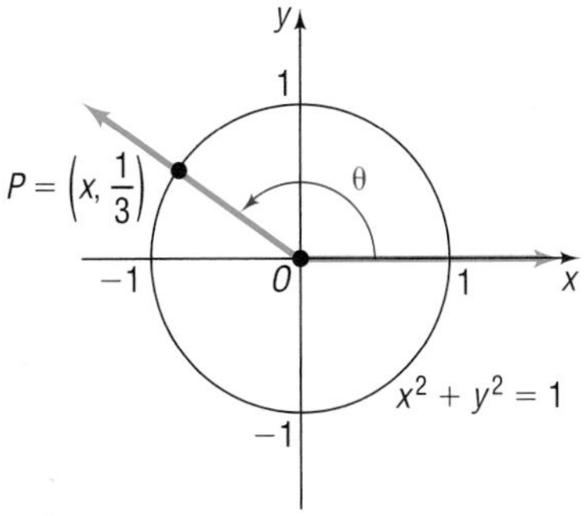

$$x^2 + y^2 = 1 \qquad y = 1/3,\ x < 0$$

$$x^2 + \left(\frac{1}{3}\right)^2 = 1$$

$$x^2 = \frac{8}{9}$$

$$x = -\frac{2\sqrt{2}}{3}$$

Since $x = -\frac{2\sqrt{2}}{3}$ and $y = \frac{1}{3}$, we find that

$$\cos\theta = x = -\frac{2\sqrt{2}}{3} \qquad \tan\theta = \frac{y}{x} = \frac{\frac{1}{3}}{\frac{-2\sqrt{2}}{3}} = -\frac{1}{2\sqrt{2}} = -\frac{\sqrt{2}}{4}$$

$$\csc\theta = \frac{1}{y} = \frac{1}{\frac{1}{3}} = 3 \qquad \sec\theta = \frac{1}{x} = \frac{1}{\frac{-2\sqrt{2}}{3}} = -\frac{3}{2\sqrt{2}} = -\frac{3\sqrt{2}}{4} \qquad \cot\theta = \frac{x}{y} = \frac{\frac{-2\sqrt{2}}{3}}{\frac{1}{3}} = -2\sqrt{2}$$

Solution 2 Using Identities

First, we solve equation (5) for $\cos\theta$.

$$\sin^2\theta + \cos^2\theta = 1$$
$$\cos^2\theta = 1 - \sin^2\theta$$
$$\cos\theta = \pm\sqrt{1 - \sin^2\theta}$$

Because $\cos\theta < 0$, we choose the minus sign.

$$\cos\theta = -\sqrt{1 - \sin^2\theta} \underset{\uparrow}{=} -\sqrt{1 - \frac{1}{9}} = -\sqrt{\frac{8}{9}} = -\frac{2\sqrt{2}}{3}$$

$\sin\theta = \frac{1}{3}$

Now we know the values of $\sin\theta$ and $\cos\theta$, so we can use formulas (2) and (3) to get

$$\tan\theta = \frac{\sin\theta}{\cos\theta} = \frac{1/3}{-2\sqrt{2}/3} = \frac{1}{-2\sqrt{2}} = \frac{-\sqrt{2}}{4} \qquad \cot\theta = \frac{1}{\tan\theta} = -2\sqrt{2}$$

$$\sec\theta = \frac{1}{\cos\theta} = \frac{1}{-2\sqrt{2}/3} = \frac{-3}{2\sqrt{2}} = \frac{-3\sqrt{2}}{4} \qquad \csc\theta = \frac{1}{\sin\theta} = \frac{1}{1/3} = 3$$

NOW WORK PROBLEM 33.

EVEN–ODD PROPERTIES

5 Recall that a function f is even if $f(-\theta) = f(\theta)$ for all θ in the domain of f; a function f is odd if $f(-\theta) = -f(\theta)$ for all θ in the domain of f. We will now show that the trigonometric functions sine, tangent, cotangent, and cosecant are odd functions, whereas the functions cosine and secant are even functions.

Theorem **Even–Odd Properties**

$$\sin(-\theta) = -\sin\theta \qquad \cos(-\theta) = \cos\theta \qquad \tan(-\theta) = -\tan\theta$$
$$\csc(-\theta) = -\csc\theta \qquad \sec(-\theta) = \sec\theta \qquad \cot(-\theta) = -\cot\theta$$

Proof Let $P = (x, y)$ be the point on the unit circle that corresponds to the angle θ. (See Figure 43.) The point Q on the unit circle that corresponds

Figure 43

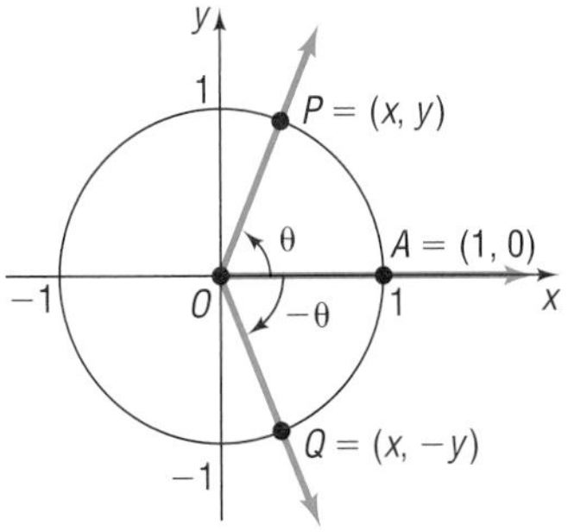

to the angle $-\theta$ will have coordinates $(x, -y)$. Using the definition for the trigonometric functions, we have

$$\sin\theta = y \qquad \cos\theta = x \qquad \sin(-\theta) = -y \qquad \cos(-\theta) = x$$

so

$$\sin(-\theta) = -y = -\sin\theta \qquad \cos(-\theta) = x = \cos\theta$$

Now, using these results and some of the fundamental identities, we have

$$\tan(-\theta) = \frac{\sin(-\theta)}{\cos(-\theta)} = \frac{-\sin\theta}{\cos\theta} = -\tan\theta \qquad \cot(-\theta) = \frac{1}{\tan(-\theta)} = \frac{1}{-\tan\theta} = -\cot\theta$$

$$\sec(-\theta) = \frac{1}{\cos(-\theta)} = \frac{1}{\cos\theta} = \sec\theta \qquad \csc(-\theta) = \frac{1}{\sin(-\theta)} = \frac{1}{-\sin\theta} = -\csc\theta$$

Seeing the Concept To see that the cosine function is even, create a TABLE with $Y_1 = \cos x$, $Y_2 = \cos(-x)$, and TblStart $= 0$, and ΔTbl $= \pi/6$. To see that the sine function is odd, repeat this procedure with $Y_1 = -\sin x$ and $Y_2 = \sin(-x)$.

EXAMPLE 5 Finding Exact Values Using Even–Odd Properties

Find the exact value of:

(a) $\sin(-45°)$ (b) $\cos(-\pi)$ (c) $\cot(-3\pi/2)$ (d) $\tan(-37\pi/4)$

Solution (a) $\sin(-45°) = -\sin 45° = -\frac{\sqrt{2}}{2}$ (Odd function)

(b) $\cos(-\pi) = \cos\pi = -1$ (Even function)

(c) $\cot\left(-\frac{3\pi}{2}\right) = -\cot\frac{3\pi}{2} = 0$ (Odd function)

(d) $\tan\left(-\frac{37\pi}{4}\right) = -\tan\frac{37\pi}{4} = -\tan\left(\frac{\pi}{4} + 9\pi\right) = -\tan\frac{\pi}{4} = -1$ (Odd function; Period is π)

NOW WORK PROBLEM 49.

6.3 EXERCISES

In Problems 1–16, use the fact that the trigonometric functions are periodic to find the exact value of each expression. Do not use a calculator.

1. $\sin 405°$ **2.** $\cos 420°$ **3.** $\tan 405°$ **4.** $\sin 390°$

5. $\csc 450°$ **6.** $\sec 540°$ **7.** $\cot 390°$ **8.** $\sec 420°$

9. $\cos\frac{33\pi}{4}$ **10.** $\sin\frac{9\pi}{4}$ **11.** $\tan(21\pi)$ **12.** $\csc\frac{9\pi}{2}$

13. $\sec\frac{17\pi}{4}$ **14.** $\cot\frac{17\pi}{4}$ **15.** $\tan\frac{19\pi}{6}$ **16.** $\sec\frac{25\pi}{6}$

In Problems 17–24, name the quadrant in which the angle θ lies.

17. $\sin\theta > 0$, $\cos\theta < 0$ **18.** $\sin\theta < 0$, $\cos\theta > 0$ **19.** $\sin\theta < 0$, $\tan\theta < 0$

20. $\cos\theta > 0$, $\tan\theta > 0$ **21.** $\cos\theta > 0$, $\tan\theta < 0$ **22.** $\cos\theta < 0$, $\tan\theta > 0$

23. $\sec\theta < 0$, $\sin\theta > 0$ **24.** $\csc\theta > 0$, $\cos\theta < 0$

In Problems 25–32, $\sin\theta$ and $\cos\theta$ are given. Find the exact value of each of the four remaining trigonometric functions. In Problems 31 and 32, round your answer to four decimal places.

25. $\sin\theta = 2\sqrt{5}/5,\quad \cos\theta = \sqrt{5}/5$

26. $\sin\theta = -\sqrt{5}/5,\quad \cos\theta = -2\sqrt{5}/5$

27. $\sin\theta = \frac{1}{2},\quad \cos\theta = \sqrt{3}/2$

28. $\sin\theta = \sqrt{3}/2,\quad \cos\theta = \frac{1}{2}$

29. $\sin\theta = -\frac{1}{3},\quad \cos\theta = 2\sqrt{2}/3$

30. $\sin\theta = 2\sqrt{2}/3,\quad \cos\theta = -\frac{1}{3}$

31. $\sin\theta = 0.2588,\quad \cos\theta = 0.9659$

32. $\sin\theta = 0.6428,\quad \cos\theta = 0.7660$

In Problems 33–48, find the exact value of each of the remaining trigonometric functions of θ.

33. $\sin\theta = \frac{12}{13},\quad \theta$ in quadrant II

34. $\cos\theta = \frac{3}{5},\quad \theta$ in quadrant IV

35. $\cos\theta = -\frac{4}{5},\quad \theta$ in quadrant III

36. $\sin\theta = -\frac{5}{13},\quad \theta$ in quadrant III

37. $\sin\theta = \frac{5}{13},\quad 90^\circ < \theta < 180^\circ$

38. $\cos\theta = \frac{4}{5},\quad 270^\circ < \theta < 360^\circ$

39. $\cos\theta = -\frac{1}{3},\quad \frac{\pi}{2} < \theta < \pi$

40. $\sin\theta = -\frac{2}{3},\quad \pi < \theta < 3\pi/2$

41. $\sin\theta = \frac{2}{3},\quad \tan\theta < 0$

42. $\cos\theta = -\frac{1}{4},\quad \tan\theta > 0$

43. $\sec\theta = 2,\quad \sin\theta < 0$

44. $\csc\theta = 3,\quad \cot\theta < 0$

45. $\tan\theta = \frac{3}{4},\quad \sin\theta < 0$

46. $\cot\theta = \frac{4}{3},\quad \cos\theta < 0$

47. $\tan\theta = -\frac{1}{3},\quad \sin\theta > 0$

48. $\sec\theta = -2,\quad \tan\theta > 0$

In Problems 49–66, use the even–odd properties to find the exact value of each expression. Do not use a calculator.

49. $\sin(-60^\circ)$

50. $\cos(-30^\circ)$

51. $\tan(-30^\circ)$

52. $\sin(-135^\circ)$

53. $\sec(-60^\circ)$

54. $\csc(-30^\circ)$

55. $\sin(-90^\circ)$

56. $\cos(-270^\circ)$

57. $\tan\left(-\frac{\pi}{4}\right)$

58. $\sin(-\pi)$

59. $\cos\left(-\frac{\pi}{4}\right)$

60. $\sin\left(-\frac{\pi}{3}\right)$

61. $\tan(-\pi)$

62. $\sin\left(-\frac{3\pi}{2}\right)$

63. $\csc\left(-\frac{\pi}{4}\right)$

64. $\sec(-\pi)$

65. $\sec\left(-\frac{\pi}{6}\right)$

66. $\csc\left(-\frac{\pi}{3}\right)$

In Problems 67–78, find the exact value of each expression. Do not use a calculator.

67. $\sin(-\pi) + \cos(5\pi)$

68. $\tan\left(-\frac{5\pi}{6}\right) - \cot\frac{7\pi}{2}$

69. $\sec(-\pi) + \csc\left(-\frac{\pi}{2}\right)$

70. $\tan(-6\pi) + \cos\frac{9\pi}{4}$

71. $\sin\left(-\frac{9\pi}{4}\right) - \tan\left(-\frac{9\pi}{4}\right)$

72. $\cos\left(-\frac{17\pi}{4}\right) - \sin\left(-\frac{3\pi}{2}\right)$

73. $\sin^2 40^\circ + \cos^2 40^\circ$

74. $\sec^2 18^\circ - \tan^2 18^\circ$

75. $\sin 80^\circ \csc 80^\circ$

76. $\tan 10^\circ \cot 10^\circ$

77. $\tan 40^\circ - \dfrac{\sin 40^\circ}{\cos 40^\circ}$

78. $\cot 20^\circ - \dfrac{\cos 20^\circ}{\sin 20^\circ}$

79. If $\sin\theta = 0.3$, find the value of: $\sin\theta + \sin(\theta + 2\pi) + \sin(\theta + 4\pi)$.

80. If $\cos\theta = 0.2$, find the value of: $\cos\theta + \cos(\theta + 2\pi) + \cos(\theta + 4\pi)$.

81. If $\tan\theta = 3$, find the value of: $\tan\theta + \tan(\theta + \pi) + \tan(\theta + 2\pi)$.

82. If $\cot\theta = -2$, find the value of: $\cot\theta + \cot(\theta - \pi) + \cot(\theta - 2\pi)$.

83. Find the exact value of $\sin 1^\circ + \sin 2^\circ + \sin 3^\circ + \cdots + \sin 358^\circ + \sin 359^\circ$.

84. Find the exact value of $\cos 1^\circ + \cos 2^\circ + \cos 3^\circ + \cdots + \cos 358^\circ + \cos 359^\circ$.

85. What is the domain of the sine function?

86. What is the domain of the cosine function?

87. For what numbers θ is $f(\theta) = \tan\theta$ not defined?

88. For what numbers θ is $f(\theta) = \cot\theta$ not defined?

89. For what numbers θ is $f(\theta) = \sec\theta$ not defined?

90. For what numbers θ is $f(\theta) = \csc\theta$ not defined?

91. What is the range of the sine function?

92. What is the range of the cosine function?

93. What is the range of the tangent function?

94. What is the range of the cotangent function?

95. What is the range of the secant function?

96. What is the range of the cosecant function?

97. Is the sine function even, odd, or neither? Is its graph symmetric? With respect to what?

98. Is the cosine function even, odd, or neither? Is its graph symmetric? With respect to what?

99. Is the tangent function even, odd, or neither? Is its graph symmetric? With respect to what?

100. Is the cotangent function even, odd, or neither? Is its graph symmetric? With respect to what?

101. Is the secant function even, odd, or neither? Is its graph symmetric? With respect to what?

102. Is the cosecant function even, odd, or neither? Is its graph symmetric? With respect to what?

In Problems 103–108, use the periodic and even–odd properties.

103. If $f(\theta) = \sin\theta$ and $f(a) = 1/3$, find the exact value of:
(a) $f(-a)$ (b) $f(a) + f(a + 2\pi) + f(a + 4\pi)$

104. If $f(\theta) = \cos\theta$ and $f(a) = 1/4$, find the exact value of:
(a) $f(-a)$ (b) $f(a) + f(a + 2\pi) + f(a - 2\pi)$

105. If $f(\theta) = \tan\theta$ and $f(a) = 2$, find the exact value of:
(a) $f(-a)$ (b) $f(a) + f(a + \pi) + f(a + 2\pi)$

106. If $f(\theta) = \cot\theta$ and $f(a) = -3$, find the exact value of:
(a) $f(-a)$ (b) $f(a) + f(a + \pi) + f(a + 4\pi)$

107. If $f(\theta) = \sec\theta$ and $f(a) = -4$, find the exact value of:
(a) $f(-a)$ (b) $f(a) + f(a + 2\pi) + f(a + 4\pi)$

108. If $f(\theta) = \csc\theta$ and $f(a) = 2$, find the exact value of:
(a) $f(-a)$ (b) $f(a) + f(a + 2\pi) + f(a + 4\pi)$

109. Calculating the Time of a Trip From a parking lot, you want to walk to a house on the ocean. The house is located 1500 feet down a paved path that parallels the ocean, which is 500 feet away. See the illustration. Along the path you can walk 300 feet per minute, but in the sand on the beach you can only walk 100 feet per minute.

The time T to get from the parking lot to the beachhouse can be expressed as a function of the angle θ shown in the illustration and is

$$T(\theta) = 5 - \frac{5}{3\tan\theta} + \frac{5}{\sin\theta}, \qquad 0 < \theta < \frac{\pi}{2}$$

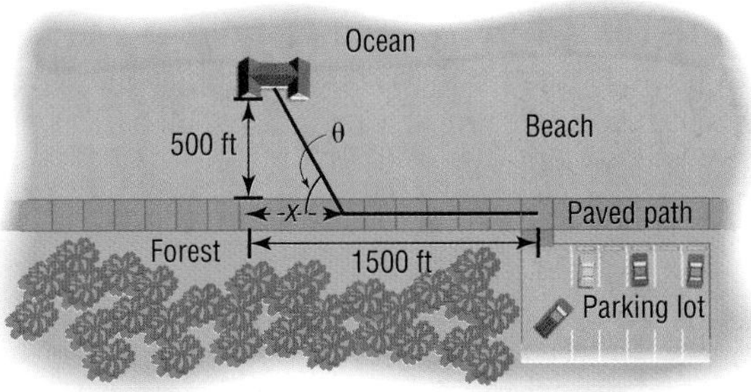

Calculate the time T if you walk directly from the parking lot to the house.
[**Hint:** $\tan\theta = 500/1500$.]

110. Calculating the Time of a Trip Two oceanfront homes are located 8 miles apart on a straight stretch of beach, each a distance of 1 mile from a paved road that parallels the ocean. Sally can jog 8 miles per hour along the paved road, but only 3 miles per hour in the sand on the beach. Because of a river directly between the two houses, it is necessary to jog in the sand to the road, continue on the road, and then jog directly back in the sand to get from one house to the other. See the illustration. The time T to get from one house to the other as a function of the angle θ shown in the illustration is

$$T(\theta) = 1 + \frac{2}{3\sin\theta} - \frac{1}{4\tan\theta} \qquad 0 < \theta < \frac{\pi}{2}$$

(a) Calculate the time T for $\tan\theta = 1/4$.
(b) Describe the path taken.
(c) Explain why θ must be larger than 14°.

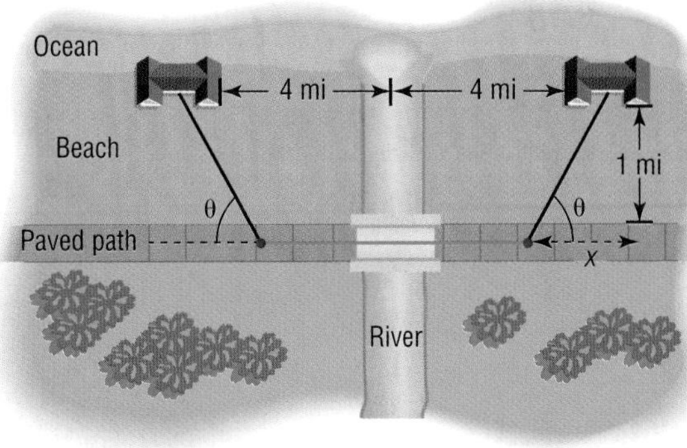

111. Show that the range of the tangent function is the set of all real numbers.

112. Show that the range of the cotangent function is the set of all real numbers.

113. Show that the period of $f(\theta) = \sin\theta$ is 2π.
[**Hint:** Assume that $0 < p < 2\pi$ exists so that $\sin(\theta + p) = \sin\theta$ for all θ. Let $\theta = 0$ to find p. Then let $\theta = \pi/2$ to obtain a contradiction.]

114. Show that the period of $f(\theta) = \cos\theta$ is 2π.

115. Show that the period of $f(\theta) = \sec\theta$ is 2π.

116. Show that the period of $f(\theta) = \csc\theta$ is 2π.

117. Show that the period of $f(\theta) = \tan\theta$ is π.

118. Show that the period of $f(\theta) = \cot\theta$ is π.

119. Prove the reciprocal identities given in formula (2).

120. Prove the quotient identities given in formula (3).

121. Establish the identity:

$$(\sin\theta\cos\phi)^2 + (\sin\theta\sin\phi)^2 + \cos^2\theta = 1$$

122. Write down five characteristics of the tangent function. Explain the meaning of each.

123. Describe your understanding of the meaning of a periodic function.

PREPARING FOR THIS SECTION

Before getting started, review the following:

✓ Graphing Techniques: Transformations (Section 2.4)

6.4 GRAPHS OF THE SINE AND COSINE FUNCTIONS

OBJECTIVES
1. Graph Transformations of the Sine Function
2. Graph Transformations of the Cosine Function

Since we want to graph the trigonometric functions in the xy-plane, we shall use the traditional symbols x for the independent variable (or argument) and y for the dependent variable (or value at x) for each function. So we write the six trigonometric functions as

$y = f(x) = \sin x$	$y = f(x) = \cos x$	$y = f(x) = \tan x$
$y = f(x) = \csc x$	$y = f(x) = \sec x$	$y = f(x) = \cot x$

Here the independent variable x represents an angle, measured in radians. In calculus, x will usually be treated as a real number. As we said earlier, these are equivalent ways of viewing x.

TABLE 6

x	$y = \sin x$	(x, y)
0	0	$(0, 0)$
$\pi/6$	$\frac{1}{2}$	$(\pi/6, \frac{1}{2})$
$\pi/2$	1	$(\pi/2, 1)$
$5\pi/6$	$\frac{1}{2}$	$(5\pi/6, \frac{1}{2})$
π	0	$(\pi, 0)$
$7\pi/6$	$-\frac{1}{2}$	$(7\pi/6, -\frac{1}{2})$
$3\pi/2$	-1	$(3\pi/2, -1)$
$11\pi/6$	$-\frac{1}{2}$	$(11\pi/6, -\frac{1}{2})$
2π	0	$(2\pi, 0)$

THE GRAPH OF $y = \sin x$

Since the sine function has period 2π, we need to graph $y = \sin x$ only on the interval $[0, 2\pi]$. The remainder of the graph will consist of repetitions of this portion of the graph.

We begin by constructing Table 6, which lists some points on the graph of $y = \sin x$, $0 \le x \le 2\pi$. As the table shows, the graph of $y = \sin x$, $0 \le x \le 2\pi$, begins at the origin. As x increases from 0 to $\pi/2$, the value of $y = \sin x$ increases from 0 to 1; as x increases from $\pi/2$ to π to $3\pi/2$, the value of y decreases from 1 to 0 to -1; as x increases from $3\pi/2$ to 2π, the value of y increases from -1 to 0. Figure 44(a) shows the graph drawn by hand. With the viewing window set as shown in Figure 44(b), we graph $y = \sin x, 0 \le x \le 2\pi$. See Figure 44(c).

Figure 44

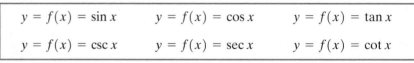

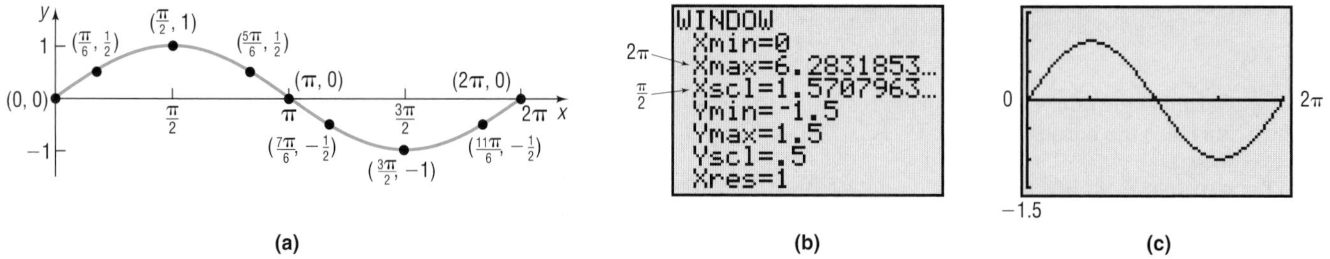

The graph in Figure 44 is one period, or **cycle**, of the graph of $y = \sin x$. To obtain a more complete graph of $y = \sin x$, we repeat this period in each direction, as shown in Figure 45.

Figure 45
$y = \sin x, -\infty < x < \infty$

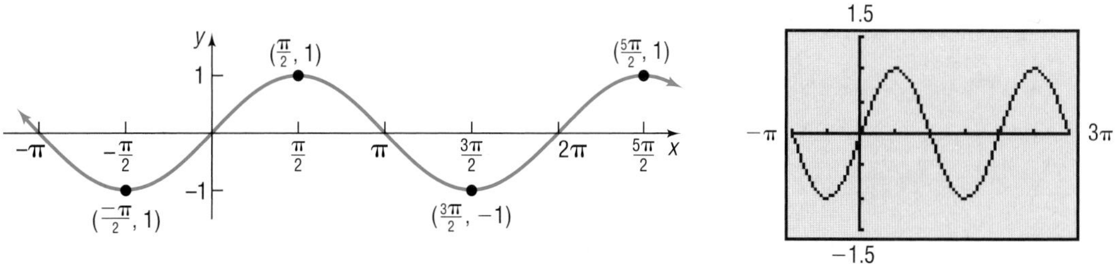

The graph of $y = \sin x$ illustrates some of the facts that we already know about the sine function.

CHARACTERISTICS OF THE SINE FUNCTION

1. The domain is the set of all real numbers.
2. The range consists of all real numbers from −1 to 1, inclusive.
3. The sine function is an odd function, as the symmetry of the graph with respect to the origin indicates.
4. The sine function is periodic, with period 2π.
5. The x-intercepts are $\ldots, -2\pi, -\pi, 0, \pi, 2\pi, 3\pi, \ldots$; the y-intercept is 0.
6. The maximum value is 1 and occurs at $x = \ldots, -3\pi/2, \pi/2, 5\pi/2, 9\pi/2, \ldots$; the minimum value is −1 and occurs at $x = \ldots, -\pi/2, 3\pi/2, 7\pi/2, 11\pi/2, \ldots$.

NOW WORK PROBLEMS 1, 3, AND 5.

1 The graphing techniques introduced in Chapter 2 may be used to graph functions that are transformations of the sine function (refer to Section 2.4).

EXAMPLE 1 Graphing Variations of $y = \sin x$ Using Transformations

Use the graph of $y = \sin x$ to graph $y = \sin\left(x - \frac{\pi}{4}\right)$.

Solution Figure 46 illustrates the steps.

Figure 46

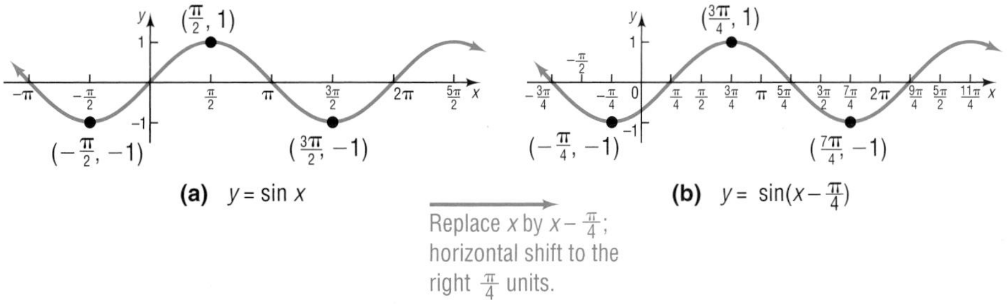

(a) $y = \sin x$

(b) $y = \sin(x - \frac{\pi}{4})$

EXAMPLE 2 Graphing Variations of $y = \sin x$ Using Transformations

Use the graph of $y = \sin x$ to graph $y = -\sin x + 2$.

Solution Figure 47 illustrates the steps.

Figure 47

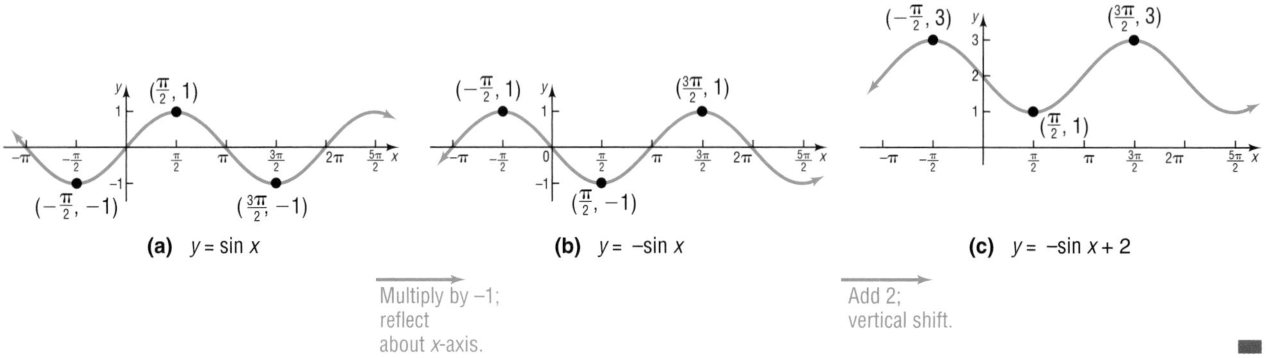

NOW WORK PROBLEM 21.

TABLE 7

x	$y = \sin x$	(x, y)
0	1	$(0, 1)$
$\pi/3$	$\frac{1}{2}$	$(\pi/3, \frac{1}{2})$
$\pi/2$	0	$(\pi/2, 0)$
$2\pi/3$	$-\frac{1}{2}$	$(2\pi/3, -\frac{1}{2})$
π	-1	$(\pi, -1)$
$4\pi/3$	$-\frac{1}{2}$	$(4\pi/3, -\frac{1}{2})$
$3\pi/2$	0	$(3\pi/2, 0)$
$5\pi/3$	$\frac{1}{2}$	$(5\pi/3, \frac{1}{2})$
2π	1	$(2\pi, 1)$

THE GRAPH OF $y = \cos x$

The cosine function also has period 2π. We proceed as we did with the sine function by constructing Table 7, which lists some points on the graph of $y = \cos x$, $0 \le x \le 2\pi$. As the table shows, the graph of $y = \cos x$, $0 \le x \le 2\pi$, begins at the point $(0, 1)$. As x increases from 0 to $\pi/2$ to π, the value of y decreases from 1 to 0 to -1; as x increases from π to $3\pi/2$ to 2π, the value of y increases from -1 to 0 to 1. Figure 48(a) shows the graph drawn by hand. We set the viewing window as shown in Figure 48(b) and graph $y = \cos x, 0 \le x \le 2\pi$. See Figure 48(c).

Figure 48
$y = \cos x,\ 0 \le x \le 2\pi$

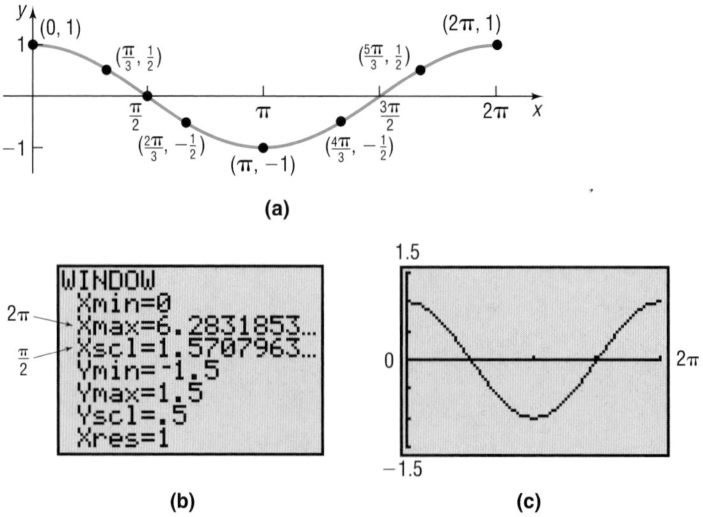

A more complete graph of $y = \cos x$ is obtained by repeating this period in each direction, as shown in Figure 49.

Figure 49
$y = \cos x, -\infty < x < \infty$

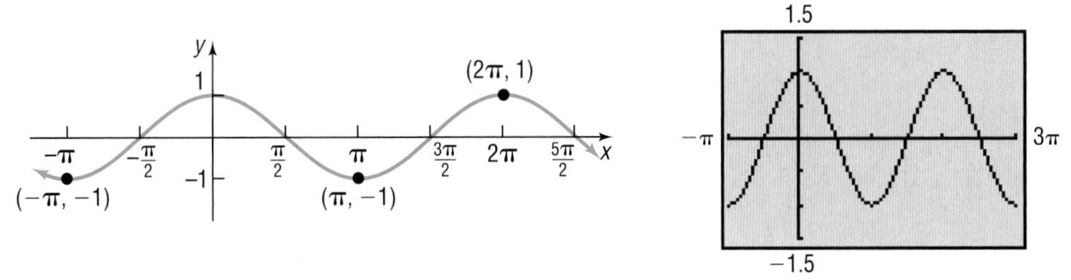

The graph of $y = \cos x$ illustrates some of the facts that we already know about the cosine function.

CHARACTERISTICS OF THE COSINE FUNCTION

1. The domain is the set of all real numbers.
2. The range consists of all real numbers from -1 to 1, inclusive.
3. The cosine function is an even function, as the symmetry of the graph with respect to the y-axis indicates.
4. The cosine function is periodic, with period 2π.
5. The x-intercepts are $\ldots, -3\pi/2, -\pi/2, \pi/2, 3\pi/2, 5\pi/2, \ldots$; the y-intercept is 1.
6. The maximum value is 1 and occurs at $x = \ldots, -2\pi, 0, 2\pi, 4\pi, 6\pi, \ldots$; the minimum value is -1 and occurs at $x = \ldots, -\pi, \pi, 3\pi, 5\pi, \ldots$.

2 Again, the graphing techniques from Chapter 2 may be used to graph transformations of the cosine function.

EXAMPLE 3 Graphing Variations of $y = \cos x$ Using Transformations

Use the graph of $y = \cos x$ to graph $y = 2\cos x$.

Solution Figure 50 illustrates the steps.

Figure 50

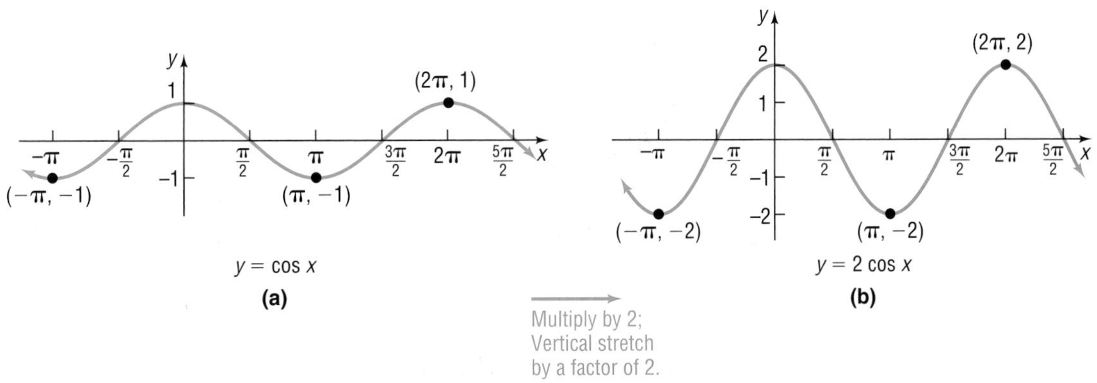

EXAMPLE 4 Graphing Variations of $y = \cos x$ Using Transformations

Use the graph of $y = \cos x$ to graph $y = \cos(3x)$.

Solution Figure 51 illustrates the graph, which is a horizontal compression of the graph of $y = \cos x$. (Multiply each x-coordinate by $\frac{1}{3}$.) Notice that, due to this compression, the period of $y = \cos(3x)$ is $2\pi/3$, whereas the period of $y = \cos x$ is 2π. We will comment more on this in Section 6.6.

Figure 51

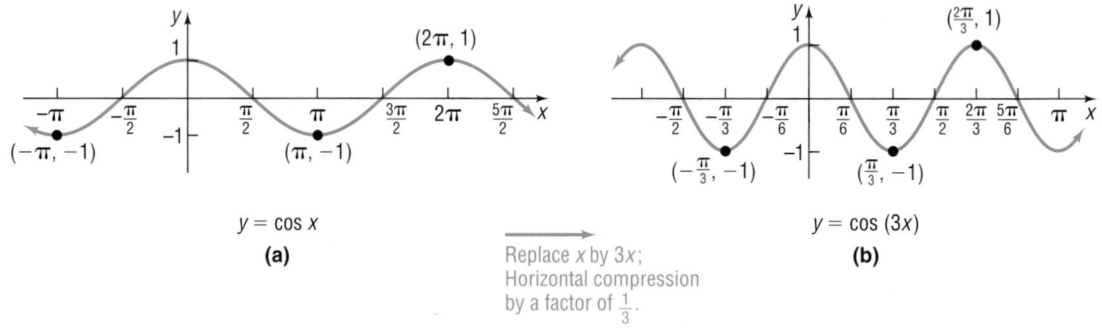

NOW WORK PROBLEM 29.

6.4 EXERCISES

In Problems 1–10, if necessary, refer to the graphs to answer each question.

1. What is the y-intercept of $y = \sin x$?
2. What is the y-intercept of $y = \cos x$?
3. For what numbers x, $-\pi \le x \le \pi$, is the graph of $y = \sin x$ increasing?
4. For what numbers x, $-\pi \le x \le \pi$, is the graph of $y = \cos x$ decreasing?
5. What is the largest value of $y = \sin x$?
6. What is the smallest value of $y = \cos x$?
7. For what numbers x, $0 \le x \le 2\pi$, does $\sin x = 0$?
8. For what numbers x, $0 \le x \le 2\pi$, does $\cos x = 0$?
9. For what numbers x, $-2\pi \le x \le 2\pi$, does $\sin x = 1$? What about $\sin x = -1$?
10. For what numbers x, $-2\pi \le x \le 2\pi$, does $\cos x = 1$? What about $\cos x = -1$?

In Problems 11 and 12, match the graph to a function. Three answers are possible.

A. $y = -\sin x$ *B.* $y = -\cos x$ *C.* $y = \sin\left(x - \frac{\pi}{2}\right)$

D. $y = -\cos\left(x - \frac{\pi}{2}\right)$ *E.* $y = \sin(x + \pi)$ *F.* $y = \cos(x + \pi)$

11.

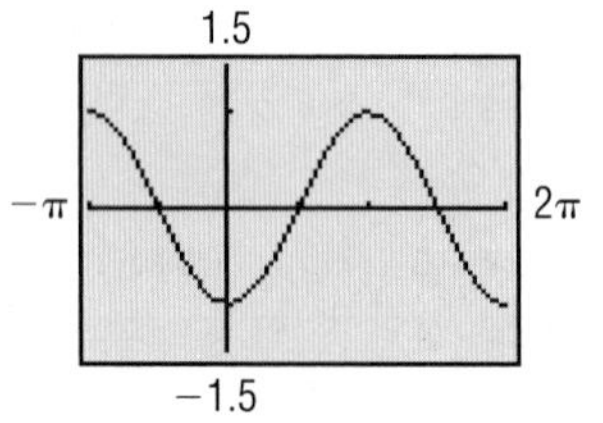

12.

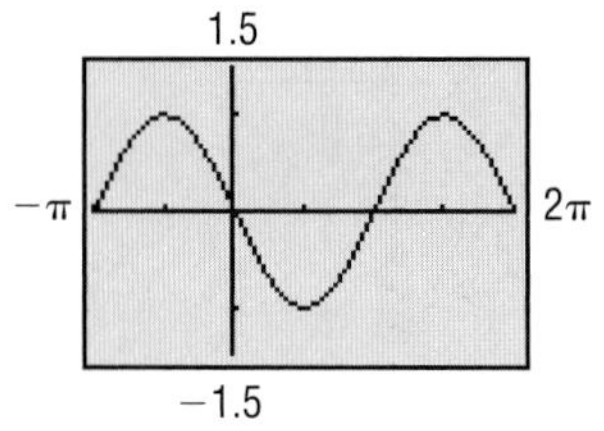

In Problems 13–16, match each function to its graph.

A. $y = \sin(2x)$ *B.* $y = \sin(4x)$ *C.* $y = \cos(2x)$ *D.* $y = \cos(4x)$

13.

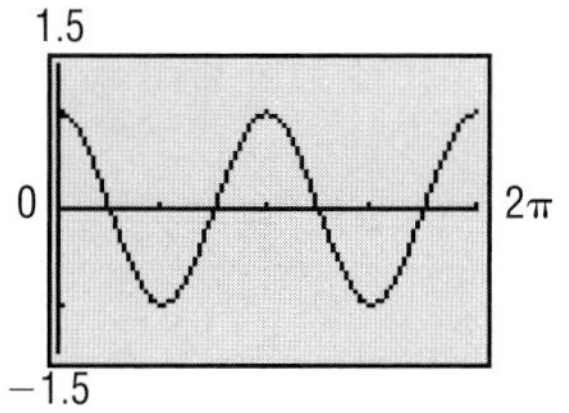

14.

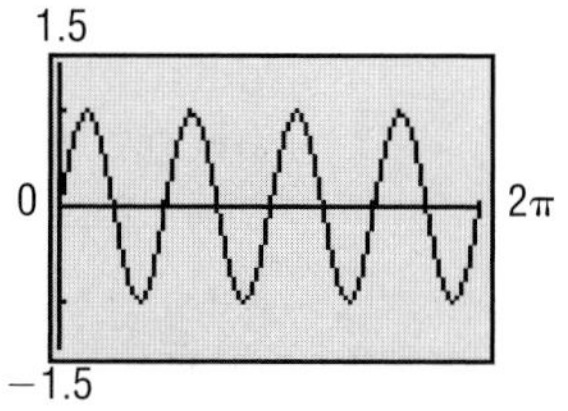

15.

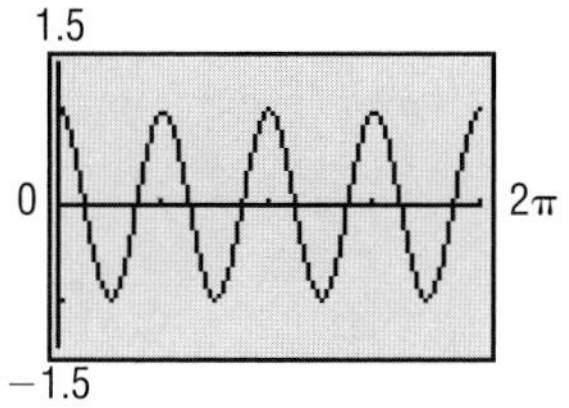

16.

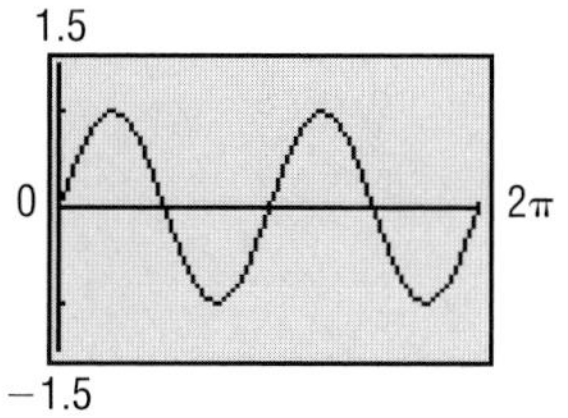

In Problems 17–32, use transformations to graph each function. Verify your results using a graphing utility.

17. $y = 3\sin x$

18. $y = 4\cos x$

19. $y = \cos\left(x + \dfrac{\pi}{4}\right)$

20. $y = \sin(x - \pi)$

21. $y = \sin x - 1$

22. $y = \cos x + 1$

23. $y = -2\sin x$

24. $y = -3\cos x$

25. $y = \sin(\pi x)$

26. $y = \cos\left(\dfrac{\pi}{2}x\right)$

27. $y = 2\sin x + 2$

28. $y = 3\cos x + 3$

29. $y = -2\cos\left(x - \dfrac{\pi}{2}\right)$

30. $y = -3\sin\left(x + \dfrac{\pi}{2}\right)$

31. $y = 3\sin(\pi - x)$

32. $y = 2\cos(\pi - x)$

33. Exploration Graph $y = \sin x$, $y = 2\sin x$, $y = \frac{1}{2}\sin x$, and $y = 8\sin x$. What do you conclude about the graph of $y = A\sin x$, $A > 0$?

34. Exploration Graph $y = \sin x$, $y = \sin(2x)$, $y = \sin(4x)$, and $y = \sin(\frac{1}{2}x)$. What do you conclude about the graph of $y = \sin(\omega x)$?

35. Exploration Graph $y = \sin x$, $y = \sin[x - (\pi/3)]$, $y = \sin[x - (\pi/4)]$, and $y = \sin[x - (\pi/6)]$. What do you conclude about the graph of $y = \sin(x - \phi)$, $\phi > 0$?

36. Graph $y = |\sin x|$, $-2\pi \le x \le 2\pi$.

37. Graph $y = |\cos x|$, $-2\pi \le x \le 2\pi$.

38. Graph

$$y = \sin x \quad \text{and} \quad y = \cos\left(x - \frac{\pi}{2}\right)$$

Do you think that $\sin x = \cos\left(x - \dfrac{\pi}{2}\right)$?

6.5 GRAPHS OF THE TANGENT, COTANGENT, COSECANT, AND SECANT FUNCTIONS

OBJECTIVES

1 Graph Transformations of the Tangent Function and Cotangent Function

2 Graph Transformations of the Secant Function and Cosecant Function

THE GRAPHS OF $y = \tan x$ AND $y = \cot x$

1 Because the tangent function has period π, we only need to determine the graph over some interval of length π. The rest of the graph will consist of repetitions of that graph. Because the tangent function is not defined at ...,

$-3\pi/2, -\pi/2, \pi/2, 3\pi/2, \ldots$, we will concentrate on the interval $(-\pi/2, \pi/2)$, of length π, and construct Table 8, which lists some points on the graph of $y = \tan x$, $-\pi/2 < x < \pi/2$. We plot the points in the table and connect them with a smooth curve. See Figure 52 for a partial graph of $y = \tan x$, where $-\pi/3 \le x \le \pi/3$.

TABLE 8

x	$y = \tan x$	(x, y)
$-\pi/3$	$-\sqrt{3} \approx -1.73$	$(-\pi/3, -\sqrt{3})$
$-\pi/4$	-1	$(-\pi/4, -1)$
$-\pi/6$	$-\sqrt{3}/3 \approx -0.58$	$(-\pi/6, -\sqrt{3}/3)$
0	0	$(0, 0)$
$\pi/6$	$\sqrt{3}/3 \approx 0.58$	$(\pi/6, \sqrt{3}/3)$
$\pi/4$	1	$(\pi/4, 1)$
$\pi/3$	$\sqrt{3} \approx 1.73$	$(\pi/3, \sqrt{3})$

Figure 52
$y = \tan x, -\pi/3 \le x \le \pi/3$

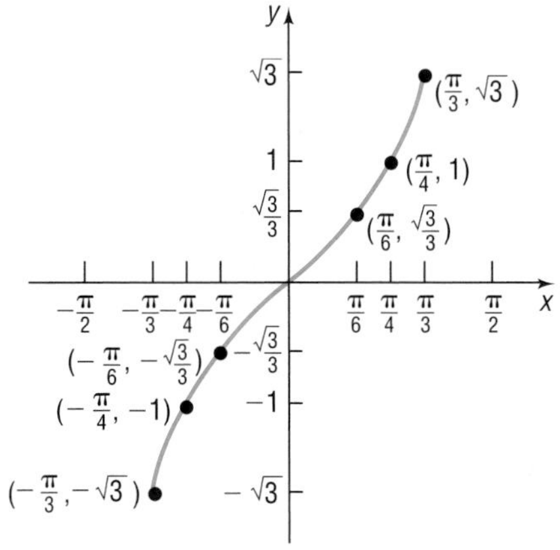

To complete one period of the graph of $y = \tan x$, we need to investigate the behavior of the function as x approaches $-\pi/2$ and $\pi/2$. We must be careful, though, because $y = \tan x$ is not defined at these numbers. To determine this behavior, we use the identity

$$\tan x = \frac{\sin x}{\cos x}$$

If x is close to $\pi/2 \approx 1.5708$, but remains less than $\pi/2$, then $\sin x$ will be close to 1 and $\cos x$ will be positive and close to 0. (Refer back to the graphs of the sine function and the cosine function.) Hence, the ratio $(\sin x)/(\cos x)$ will be positive and large. In fact, as x approaches $\pi/2$, with $x < \pi/2$, $\sin x$ approaches 1 and $\cos x$ approaches 0, so $\tan x$ approaches ∞ $\left(\lim_{x \to \frac{\pi}{2}^-} \tan x = \infty\right)$. In other words, the vertical line $x = \pi/2$ is a vertical asymptote to the graph of $y = \tan x$.

If x is close to $-\pi/2$, but remains greater than $-\pi/2$, then $\sin x$ will be close to -1 and $\cos x$ will be positive and close to 0. Hence, the ratio $(\sin x)/(\cos x)$ approaches $-\infty$ $\left(\lim_{x \to -\frac{\pi}{2}^+} \tan x = -\infty\right)$. In other words, the vertical line $x = -\pi/2$ is also a vertical asymptote to the graph.

TABLE 9

X	Y1
1.5	14.101
1.57	1255.8
1.5707	10381
1.5708 ← $\frac{\pi}{2}$	ERROR
-1.5	-14.1
-1.57	-1256
-1.571 ← $-\frac{\pi}{2}$	ERROR

Y1≡tan(X)

Table 9 confirms these conclusions.

Figure 53 shows the graph of $y = \tan x$, $-\infty < x < \infty$. Notice that we used dot mode when graphing $y = \tan x$ using a graphing utility. Do you know why?

The graph of $y = \tan x$ illustrates some of the facts that we already know about the tangent function.

Figure 53
$y = \tan x$, $-\infty < x < \infty$, x not equal to odd multiples of $\pi/2$

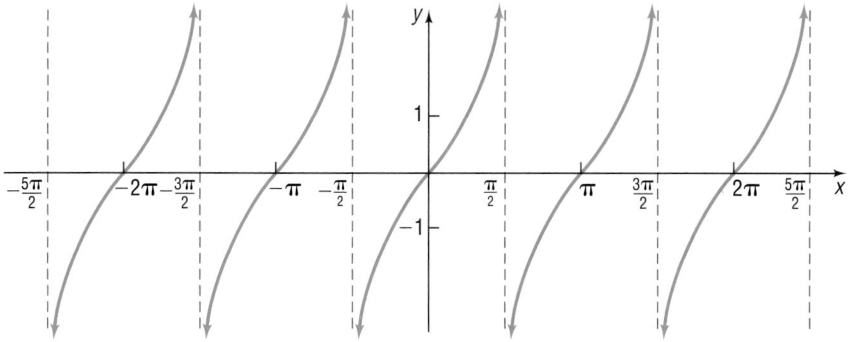

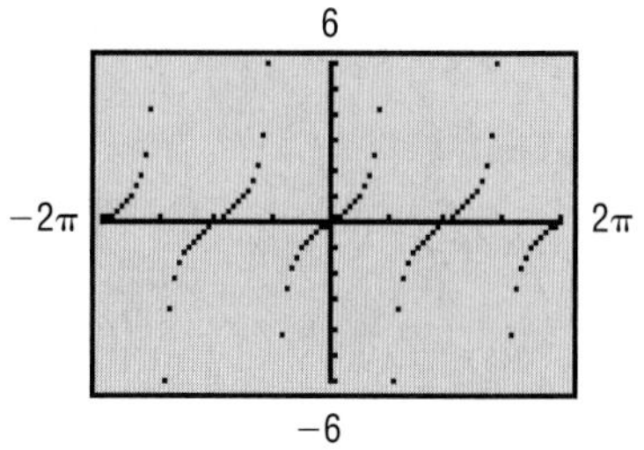

CHARACTERISTICS OF THE TANGENT FUNCTION

1. The domain is the set of all real numbers, except odd multiples of $\pi/2$.
2. The range consists of all real numbers.
3. The tangent function is an odd function, as the symmetry of the graph with respect to the origin indicates.
4. The tangent function is periodic, with period π.
5. The x-intercepts are $\dots, -2\pi, -\pi, 0, \pi, 2\pi, 3\pi, \dots$; the y-intercept is 0.
6. Vertical asymptotes occur at $x = \dots, -3\pi/2, -\pi/2, \pi/2, 3\pi/2, \dots$.

NOW WORK PROBLEMS 1 AND 9.

EXAMPLE 1 Graphing Variations of $y = \tan x$ Using Transformations

Graph $y = -\tan\left(x + \dfrac{\pi}{4}\right)$.

Solution We start with the graph of $y = \tan x$. See Figure 54.

Figure 54

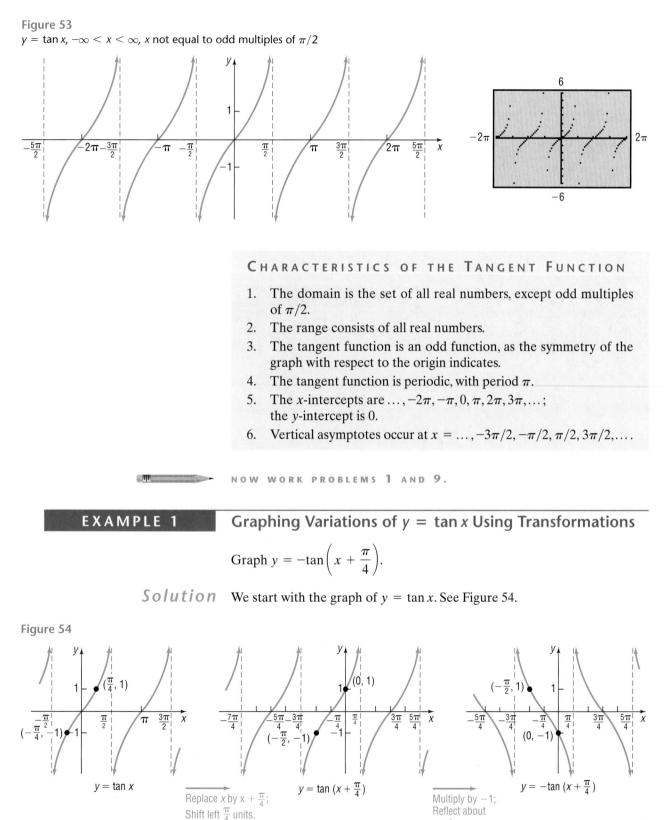

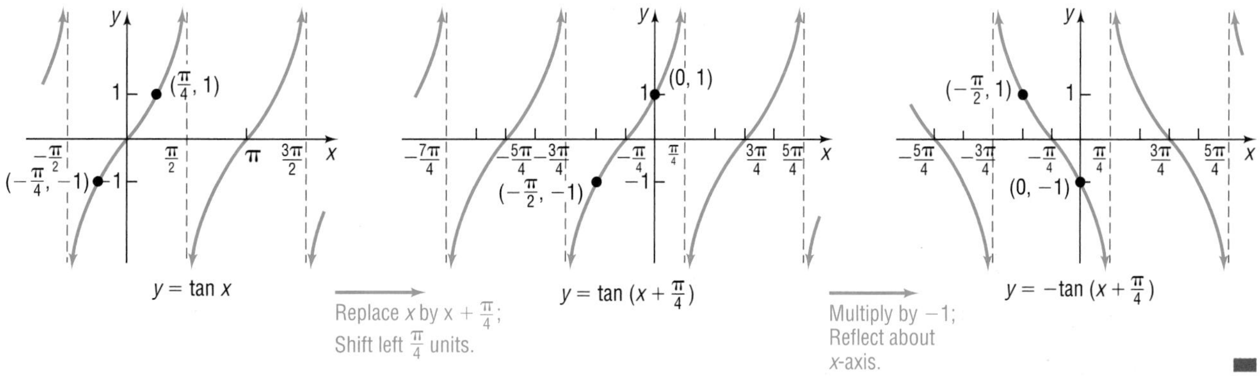

Check: Graph $Y_1 = -\tan(x + \pi/4)$ using a graphing utility.

NOW WORK PROBLEM 19.

We obtain the graph of $y = \cot x$ as we did the graph of $y = \tan x$. The period of $y = \cot x$ is π. Because the cotangent function is not defined for integral multiples of π, we will concentrate on the interval $(0, \pi)$. Table 10 lists some points on the graph of $y = \cot x$, $0 < x < \pi$. As x approaches 0, but remains greater than 0, the value of $\cos x$ will be close to 1 and the value of $\sin x$ will be positive and close to 0. Hence, the ratio $(\cos x)/(\sin x) = \cot x$ will be positive and large; so as x approaches 0, with $x > 0$, $\cot x$ approaches ∞ $\left(\lim_{x\to 0^+} \cot x = \infty\right)$. Similarly, as x approaches π, but remains less than π, the value of $\cos x$ will be close to -1, and the value of $\sin x$ will be positive and close to 0. Hence, the ratio $(\cos x)/(\sin x) = \cot x$ will be negative and approach $-\infty$ as x approaches π $\left(\lim_{x\to \pi^-} \cot x = -\infty\right)$. Figure 55 shows the graph.

TABLE 10

x	$y = \cot x$	(x, y)
$\pi/6$	$\sqrt{3}$	$(\pi/6, \sqrt{3})$
$\pi/4$	1	$(\pi/4, 1)$
$\pi/3$	$\sqrt{3}/3$	$(\pi/3, \sqrt{3}/3)$
$\pi/2$	0	$(\pi/2, 0)$
$2\pi/3$	$-\sqrt{3}/3$	$(2\pi/3, -\sqrt{3}/3)$
$3\pi/4$	-1	$(3\pi/4, -1)$
$5\pi/6$	$-\sqrt{3}$	$(5\pi/6, -\sqrt{3})$

Figure 55
$y = \cot x$, $-\infty < x < \infty$, x not equal to integral multiples of π, $-\infty < y < \infty$

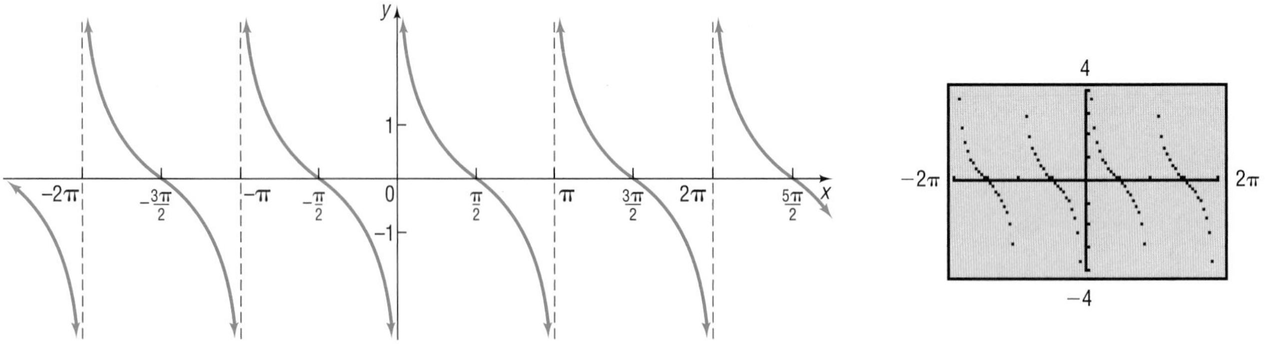

NOW WORK PROBLEM 25.

THE GRAPHS OF $y = \csc x$ AND $y = \sec x$

2 The cosecant and secant functions, sometimes referred to as **reciprocal functions**, are graphed by making use of the reciprocal identities

$$\csc x = \frac{1}{\sin x} \quad \text{and} \quad \sec x = \frac{1}{\cos x}$$

For example, the value of the cosecant function $y = \csc x$ at a given number x equals the reciprocal of the corresponding value of the sine function, provided that the value of the sine function is not 0. If the value of $\sin x$ is 0, then, at such numbers x, the cosecant function is not defined. In fact, the graph of the cosecant function has vertical asymptotes at integral multiples of π. Figure 56(a) shows the graph drawn by hand, and Figure 56(b) shows the graph using a graphing utility.

Figure 56
$y = \csc x$, $-\infty < x < \infty$, x not equal to integral multiples of π, $|y| \geq 1$

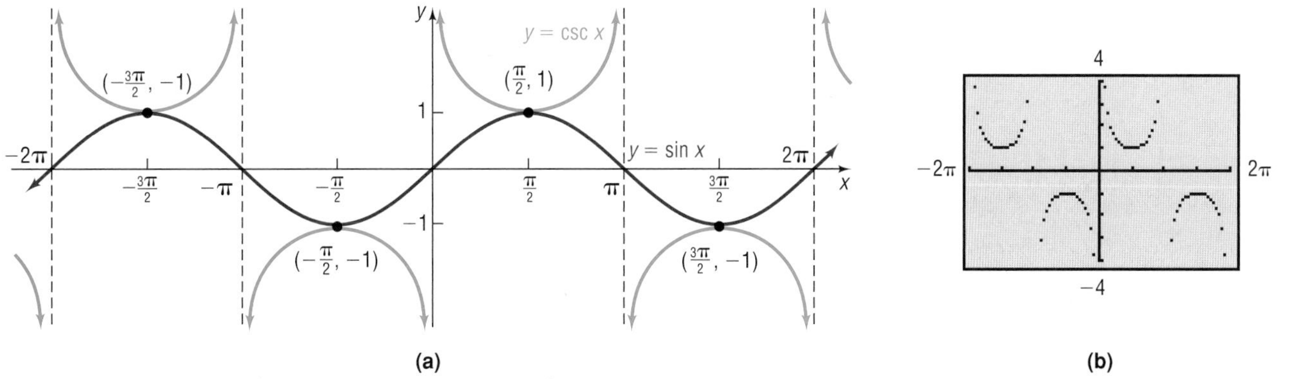

EXAMPLE 2 Graphing Variations of $y = \csc x$ Using Transformations

Graph $y = 2\csc(x - \pi/2)$.

Solution Figure 57 shows the required steps.

Figure 57

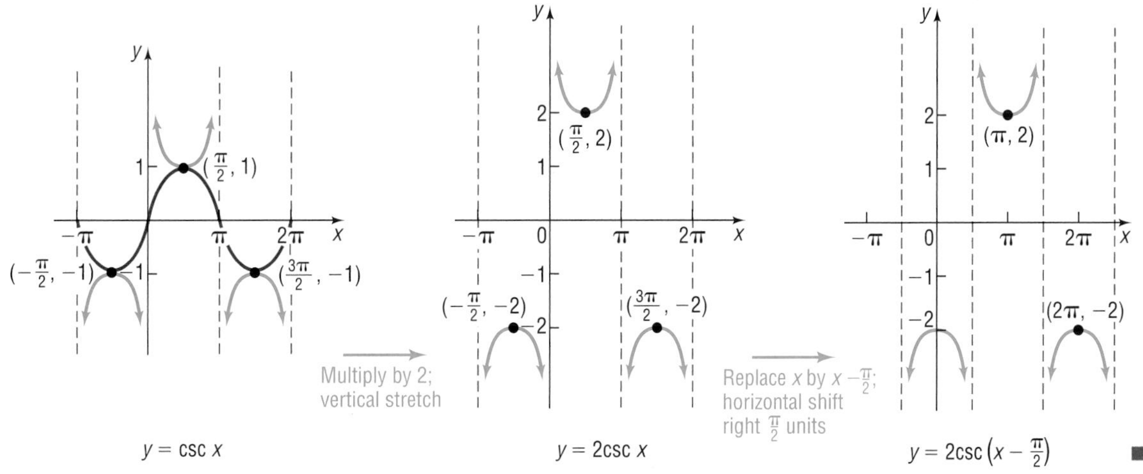

Check: Graph $Y_1 = 2\csc(x - \pi/2)$ using a graphing utility.

Using the idea of reciprocals, we can similarly obtain the graph of $y = \sec x$. See Figure 58.

Figure 58
$y = \sec x$, $-\infty < x < \infty$, x not equal to odd multiples of $\pi/2$, $|y| \geq 1$

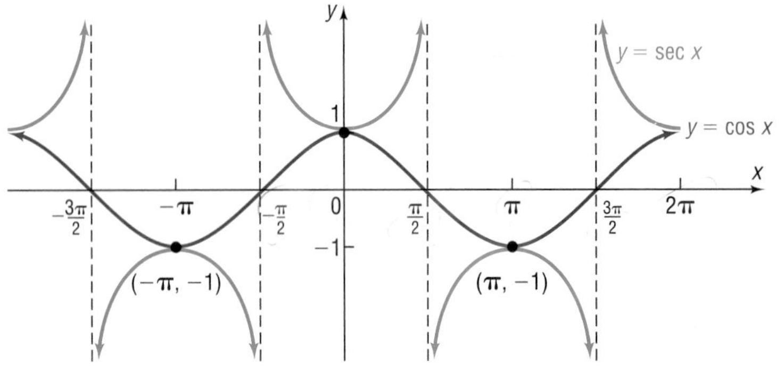

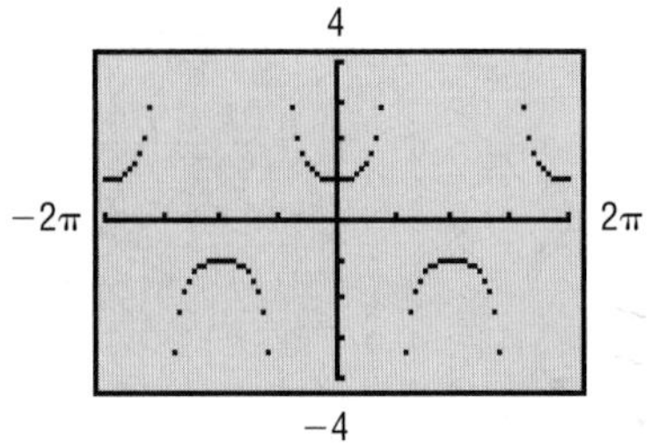

6.5 EXERCISES

In Problems 1–10, if necessary, refer to the graphs to answer each question.

1. What is the y-intercept of $y = \tan x$?
2. What is the y-intercept of $y = \cot x$?
3. What is the y-intercept of $y = \sec x$?
4. What is the y-intercept of $y = \csc x$?
5. For what numbers x, $-2\pi \leq x \leq 2\pi$, does $\sec x = 1$? What about $\sec x = -1$?
6. For what numbers x, $-2\pi \leq x \leq 2\pi$, does $\csc x = 1$? What about $\csc x = -1$?
7. For what numbers x, $-2\pi \leq x \leq 2\pi$, does the graph of $y = \sec x$ have vertical asymptotes?
8. For what numbers x, $-2\pi \leq x \leq 2\pi$, does the graph of $y = \csc x$ have vertical asymptotes?
9. For what numbers x, $-2\pi \leq x \leq 2\pi$, does the graph of $y = \tan x$ have vertical asymptotes?
10. For what numbers x, $-2\pi \leq x \leq 2\pi$, does the graph of $y = \cot x$ have vertical asymptotes?

In Problems 11–14, match each function to its graph.

A. $y = -\tan x$ *B.* $y = \tan\left(x + \dfrac{\pi}{2}\right)$ *C.* $y = \tan(x + \pi)$ *D.* $y = -\tan\left(x - \dfrac{\pi}{2}\right)$

11.

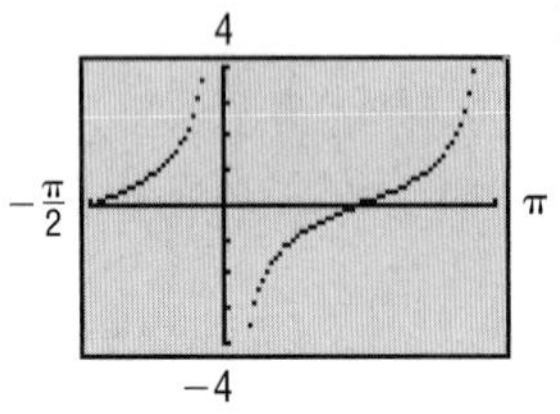

12.

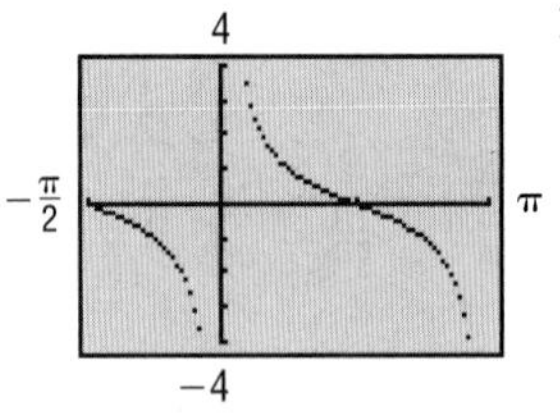

13.

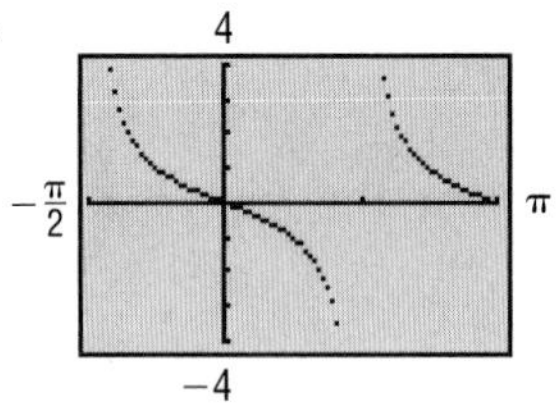

14.

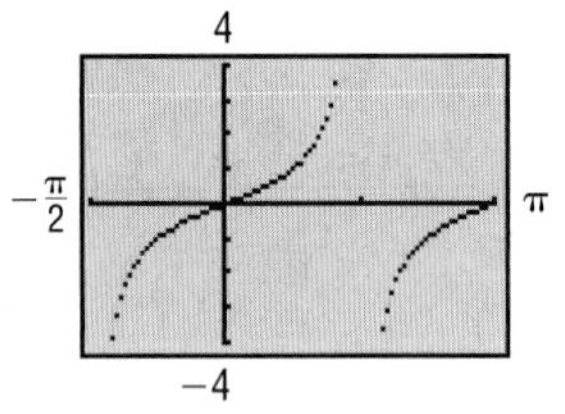

In Problems 15–29, use tranformations to graph each function. Verify your results using a graphing utility.

15. $y = -\sec x$
16. $y = -\cot x$
17. $y = \sec\left(x - \dfrac{\pi}{2}\right)$
18. $y = \csc(x - \pi)$
19. $y = \tan(x - \pi)$
20. $y = \cot(x - \pi)$
21. $y = 3\tan(2x)$
22. $y = 4\tan(\frac{1}{2}x)$
23. $y = \sec(2x)$
24. $y = \csc(\frac{1}{2}x)$
25. $y = \cot(\pi x)$
26. $y = \cot(2x)$
27. $y = -3\tan(4x)$
28. $y = -3\tan(2x)$
29. $y = 2\sec(\frac{1}{2}x)$
30. $y = 2\sec(3x)$
31. $y = -3\csc\left(x + \dfrac{\pi}{4}\right)$
32. $y = -2\tan\left(x + \dfrac{\pi}{4}\right)$
33. $y = \dfrac{1}{2}\cot\left(x - \dfrac{\pi}{4}\right)$
34. $y = 3\sec\left(x + \dfrac{\pi}{2}\right)$

35. Carrying a Ladder around a Corner A ladder of length L is carried horizontally around a corner from a hall 3 feet wide into a hall 4 feet wide. See the illustration.

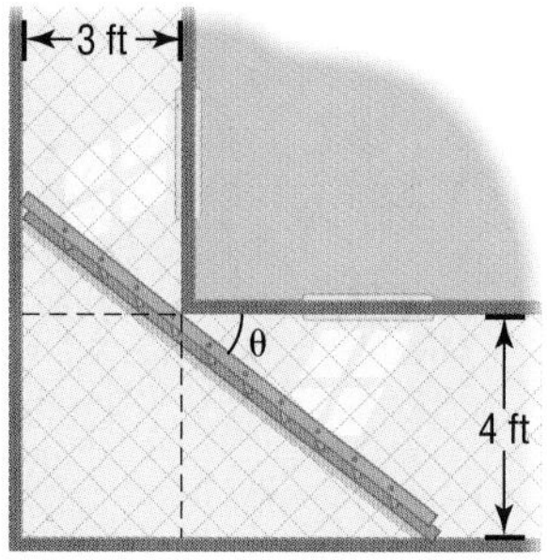

(a) Show that the length L of the ladder as a function of the angle θ is

$$L(\theta) = 3\sec\theta + 4\csc\theta$$

(b) Graph $L, 0 < \theta < \frac{\pi}{2}$.

(c) Where is L the least?

(d) What is the length of the longest ladder that can be carried around the corner? Why is this also the least value of L?

36. Graph

$$y = \tan x \quad \text{and} \quad y = -\cot\left(x + \frac{\pi}{2}\right)$$

Do you think that $\tan x = -\cot\left(x + \frac{\pi}{2}\right)$?

6.6 SINUSOIDAL GRAPHS; SINUSOIDAL CURVE FITTING

OBJECTIVES

1. Determine the Amplitude and Period of Sinusoidal Functions
2. Find an Equation for a Sinusoidal Graph
3. Determine the Phase Shift of a Sinusoidal Function
4. Graph Sinusoidal Functions
5. Find a Sinusoidal Function from Data

The graph of $y = \cos x$, when compared to the graph of $y = \sin x$, suggests that the graph of $y = \sin x$ is the same as the graph of $y = \cos x$ after a horizontal shift of $\pi/2$ units to the right. See Figure 59.

Figure 59

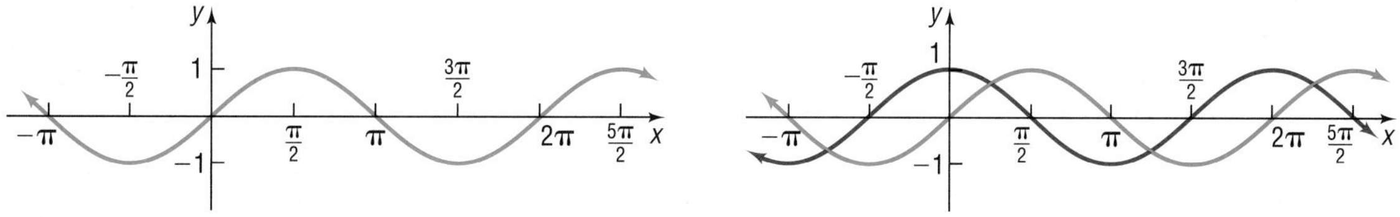

(a) $y = \sin x$ **(b)** $y = \cos x$ $y = \cos\left(x - \frac{\pi}{2}\right)$

Based on Figure 59, we conjecture that

$$\sin x = \cos\left(x - \frac{\pi}{2}\right)$$

(We shall prove this fact in Chapter 7.) Because of this similarity, the graphs of sine functions and cosine functions are referred to as **sinusoidal graphs**.

Let's look at some general characteristics of sinusoidal graphs.

1 In Example 3 of Section 6.4, we obtained the graph of $y = 2\cos x$, which we reproduce in Figure 60. Notice that the values of $y = 2\cos x$ lie between -2 and 2, inclusive.

Figure 60
$y = 2\cos x$

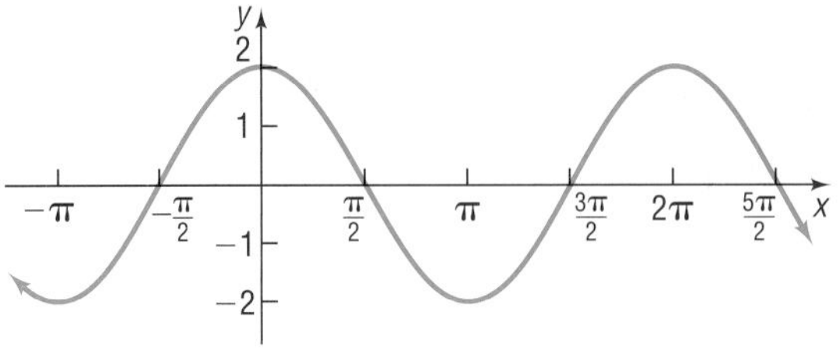

In general, the values of the functions $y = A\sin x$ and $y = A\cos x$, where $A \neq 0$, will always satisfy the inequalities.

$$-|A| \leq A\sin x \leq |A| \quad \text{and} \quad -|A| \leq A\cos x \leq |A|$$

respectively. The number $|A|$ is called the **amplitude** of $y = A\sin x$ or $y = A\cos x$. See Figure 61.

Figure 61

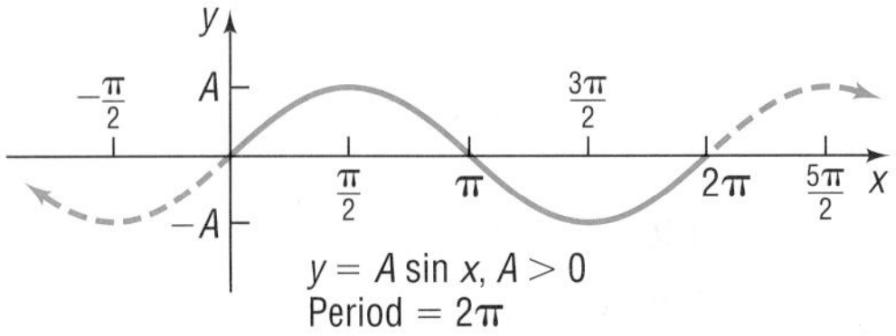

In Example 4 of Section 6.4, we obtained the graph of $y = \cos(3x)$, which we reproduce in Figure 62. Notice that the period of this function is $2\pi/3$.

Figure 62
$y = \cos(3x)$

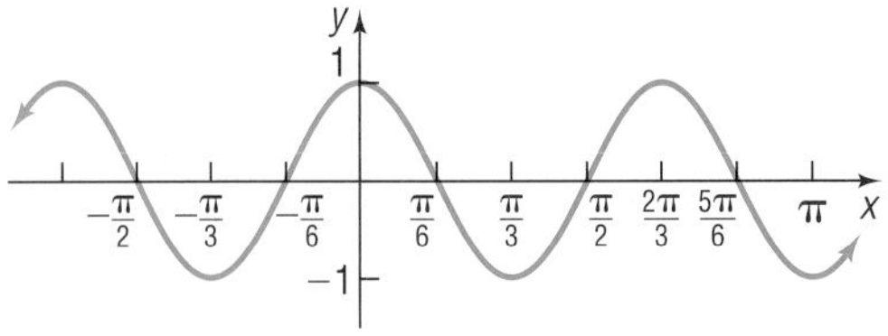

In general, if $\omega > 0$, the functions $y = \sin(\omega x)$ and $y = \cos(\omega x)$ will have period $T = 2\pi/\omega$. To see why, recall that the graph of $y = \sin(\omega x)$ is obtained from the graph of $y = \sin x$ by performing a horizontal compression or stretch by a factor $1/\omega$. This horizontal compression replaces the interval $[0, 2\pi]$, which contains one period of the graph of $y = \sin x$, by the interval $[0, 2\pi/\omega]$, which contains one period of the graph of $y = \sin(\omega x)$. Thus, the period of the functions $y = \sin(\omega x)$ and $y = \cos(\omega x)$, $\omega > 0$, is $2\pi/\omega$.

For example, for the function $y = \cos(3x)$, graphed in Figure 62, $\omega = 3$, so the period is $2\pi/3$.

One period of the graph of $y = \sin(\omega x)$ or $y = \cos(\omega x)$ is called a **cycle.** Figure 63 illustrates the general situation. The solid portion of the graph is one cycle.

Figure 63

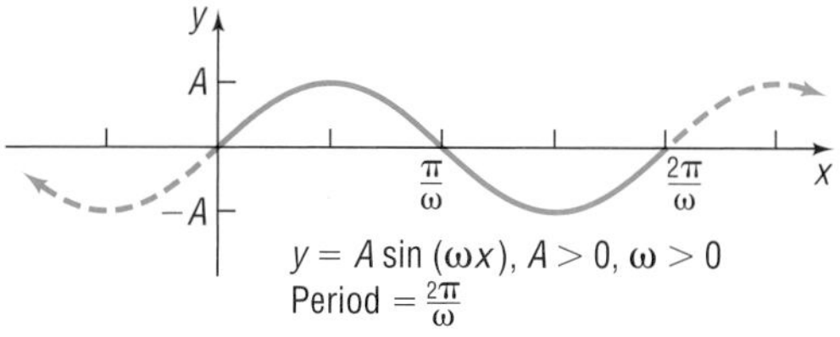

If $\omega < 0$ in $y = \sin(\omega x)$ or $y = \cos(\omega x)$, we use the even–odd properties of the sine and cosine functions as follows:

$$-\sin(\omega x) = \sin(-\omega x) \quad \text{and} \quad \cos(\omega x) = \cos(-\omega x)$$

This gives us an equivalent form in which the coefficient of x is positive. For example,

$$\sin(-2x) = -\sin(2x) \quad \text{and} \quad \cos(-\pi x) = \cos(\pi x)$$

Theorem If $\omega > 0$, the amplitude and period of $y = A\sin(\omega x)$ and $y = A\cos(\omega x)$ are given by

$$\text{Amplitude} = |A| \qquad \text{Period} = T = \frac{2\pi}{\omega} \tag{1}$$

EXAMPLE 1 Finding the Amplitude and Period of a Sinusoidal Function

Determine the amplitude and period of $y = 3\sin(4x)$.

Solution Comparing $y = 3\sin(4x)$ to $y = A\sin(\omega x)$, we find that $A = 3$ and $\omega = 4$. From equation (1),

$$\text{Amplitude} = |A| = 3 \qquad \text{Period} = T = 2\pi/\omega = 2\pi/4 = \pi/2$$

NOW WORK PROBLEM 5.

In Section 6.4, we graphed sine and cosine functions using tranformations. We now introduce another method that can be used to graph these functions.

Figure 64 shows one cycle of the graphs of $y = \sin x$ and $y = \cos x$ on the interval $[0, 2\pi]$. Notice that each graph consists of four parts corresponding to the four subintervals: $\left[0, \frac{\pi}{2}\right], \left[\frac{\pi}{2}, \pi\right], \left[\pi, \frac{3\pi}{2}\right]$, and $\left[\frac{3\pi}{2}, 2\pi\right]$. Each of these subintervals is of length $\pi/2$ (the period 2π divided by 4) and the endpoints of these intervals give rise to five key points, as shown in Figure 64.

Figure 64

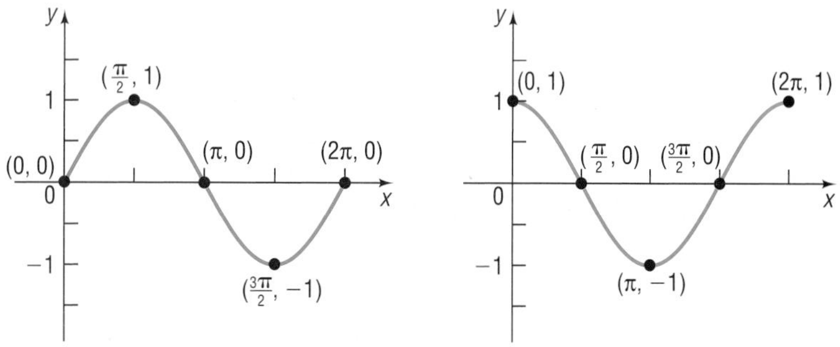

When graphing a sinusoidal function of the form $y = A\sin(\omega x)$ or $y = A\cos(\omega x)$, we use the amplitude to determine the maximum and minimum values of the function. The period is used to divide the x-axis into four subintervals. The endpoints of the subintervals give rise to five key points on the graph, which are used to sketch one cycle. Finally, extend the graph in either direction to make it complete. Let's look at an example.

EXAMPLE 2 Graphing a Sinusoidal Function

Graph $y = 3\sin(4x)$.

Solution From Example 1, the amplitude is 3 and the period is $\frac{\pi}{2}$. The graph of $y = 3\sin(4x)$ will lie between -3 and 3 on the y-axis. One cycle will begin at $x = 0$ and end at $x = \pi/2$.

We divide the interval $\left[0, \frac{\pi}{2}\right]$ into four subintervals, each of length $\pi/2 \div 4 = \pi/8$: $\left[0, \frac{\pi}{8}\right], \left[\frac{\pi}{8}, \frac{\pi}{4}\right], \left[\frac{\pi}{4}, \frac{3\pi}{8}\right]$, and $\left[\frac{3\pi}{8}, \frac{\pi}{2}\right]$. The endpoints of these intervals give rise to five key points on the graph: $(0, 0)$, $\left(\frac{\pi}{8}, 3\right)$, $\left(\frac{\pi}{4}, 0\right)$, $\left(\frac{3\pi}{8}, -3\right)$, and $\left(\frac{\pi}{2}, 0\right)$. We plot these five points and fill in the graph of the sine curve as shown in Figure 65(a). If we extend the graph in either direction, we obtain the complete graph shown in Figure 65(b).

Figure 65

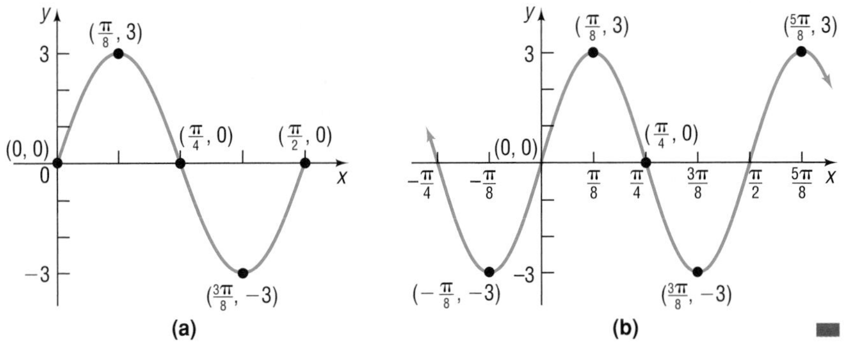

Check: Graph $y = 3\sin(4x)$ using transformations. Which graphing method do you prefer?

Check: Graph $Y_1 = 3\sin(4x)$ using a graphing utility. *Hint:* Use the amplitude to set Ymin, Ymax. Use the period to set Xmin, Xmax.

NOW WORK PROBLEM 11.

EXAMPLE 3 Finding the Amplitude and Period of a Sinusoidal Function

Determine the amplitude and period of $y = -4\cos(\pi x)$, and graph the function.

Solution Comparing $y = -4\cos(\pi x)$ with $y = A\cos(\omega x)$, we find that $A = -4$ and $\omega = \pi$. The amplitude is $|A| = |-4| = 4$, and the period is $T = 2\pi/\omega = 2\pi/\pi = 2$.

The graph of $y = 4\cos(\pi x)$ will lie between -4 and 4 on the y-axis. One cycle will begin at $x = 0$ and end at $x = 2$. We divide the interval $[0, 2]$ into four subintervals, each of length $2 \div 4 = 1/2$: $[0, 1/2], [1/2, 1], [1, 3/2]$, and $[3/2, 2]$. The five key points on the graph are $(0, -4), (1/2, 0), (1, 4), (3/2, 0)$, and $(2, -4)$. We plot these five points and fill in the graph of the cosine function as shown in Figure 66(a). Extending the graph in either direction, we obtain Figure 66(b).

Figure 66
$y = -4\cos(\pi x)$

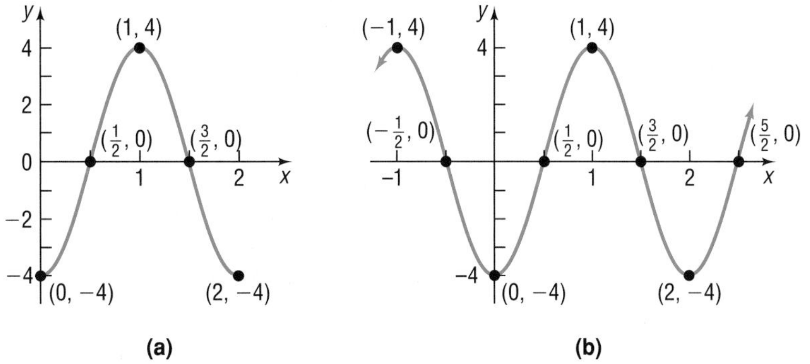

Check: Graph $y = -4\cos(\pi x)$ using transformations. Which graphing method do you prefer?

Check: Graph $Y_1 = -4\cos(\pi x)$ using a graphing utility.

EXAMPLE 4 Finding the Amplitude and Period of a Sinusoidal Function

Determine the amplitude and period of $y = 2\sin(-\frac{\pi}{2}x)$, and graph the function.

Solution Since the sine function is odd, we use the equivalent form:

$$y = -2\sin\left(\frac{\pi}{2}x\right)$$

Comparing $y = -2\sin(\frac{\pi}{2}x)$ to $y = A\sin(\omega x)$, we find that $A = -2$ and $\omega = \pi/2$. The amplitude is $|A| = 2$, and the period is $T = 2\pi/\omega = 2\pi/\frac{\pi}{2} = 4$. The graph of $y = -2\sin(\frac{\pi}{2}x)$ will lie between -2 and 2 on the y-axis. One cycle will begin at $x = 0$ and end at $x = 4$. We divide the interval $[0, 4]$ into four subintervals, each of length $4 \div 4 = 1$: $[0, 1], [1, 2], [2, 3]$, and $[3, 4]$. The five key points on the graph are $(0, 0), (1, -2), (2, 0), (3, 2)$, and $(4, 0)$. We plot these five points and fill in the graph of the sine function as shown in Figure 67(a). Extending the graph in either direction, we obtain Figure 67(b).

Figure 67

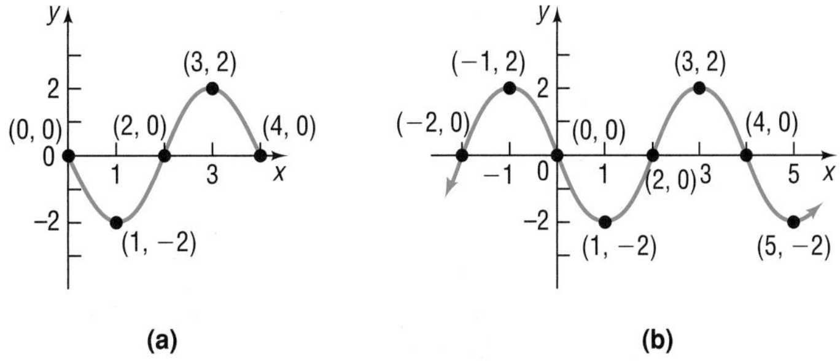

Check: Graph $y = 2\sin(-\frac{\pi}{2}x)$ using transformations. Which graphing method do you prefer?

Check: Graph $Y_1 = 2\sin(-\pi x/2)$ using a graphing utility.

NOW WORK PROBLEM 27.

2 We can also use the ideas of amplitude and period to identify a sinusoidal function when its graph is given.

EXAMPLE 5 Finding an Equation for a Sinusoidal Graph

Find an equation for the graph shown in Figure 68.

Figure 68

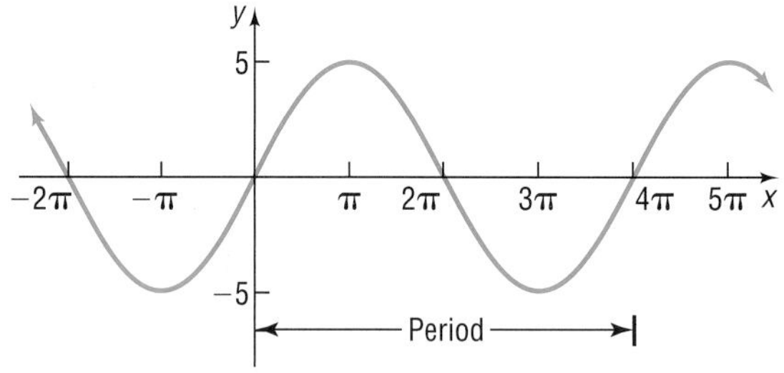

Solution This graph can be viewed as the graph of a sine function* with amplitude $A = 5$. The period T is observed to be 4π. By equation (1),

$$T = \frac{2\pi}{\omega}$$

$$4\pi = \frac{2\pi}{\omega}$$

$$\omega = \frac{2\pi}{4\pi} = \frac{1}{2}$$

The sine function whose graph is given in Figure 68 is

$$y = A\sin(\omega x) = 5\sin\frac{x}{2}$$

Check: Graph $Y_1 = 5\sin\frac{x}{2}$ and compare the result with Figure 68.

EXAMPLE 6 Finding an Equation for a Sinusoidal Graph

Find an equation for the graph shown in Figure 69.

Figure 69

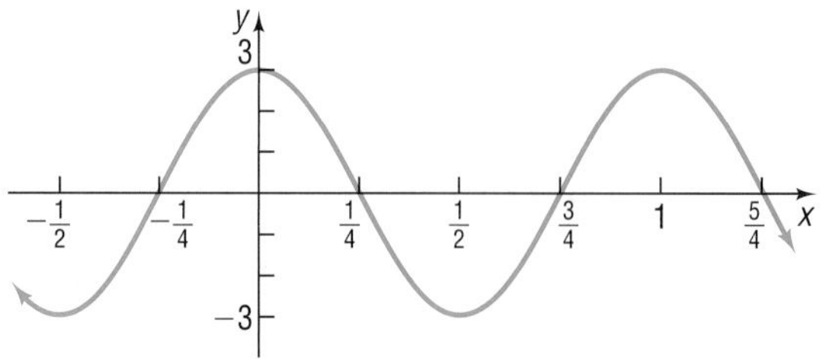

*The equation could also be viewed as a cosine function with a horizontal shift, but viewing it as a sine function is easier.

Solution From the graph we conclude that it is easiest to view the equation as a cosine function with amplitude $A = 3$ and period $T = 1$. Then $2\pi/\omega = 1$, so $\omega = 2\pi$. The cosine function whose graph is given in Figure 69 is

$$y = A\cos(\omega x) = 3\cos(2\pi x)$$

Check: Graph $Y_1 = 3\cos(2\pi x)$ and compare the result with Figure 69.

NOW WORK PROBLEM 37.

Phase Shift

3 We have seen that the graph of $y = A\sin(\omega x)$, $\omega > 0$, has amplitude $|A|$ and period $T = 2\pi/\omega$. One cycle can be drawn as x varies from 0 to $2\pi/\omega$ or, equivalently, as ωx varies from 0 to 2π. See Figure 70.

Figure 70
One cycle $y = A\sin(\omega x)$, $A > 0$, $\omega > 0$

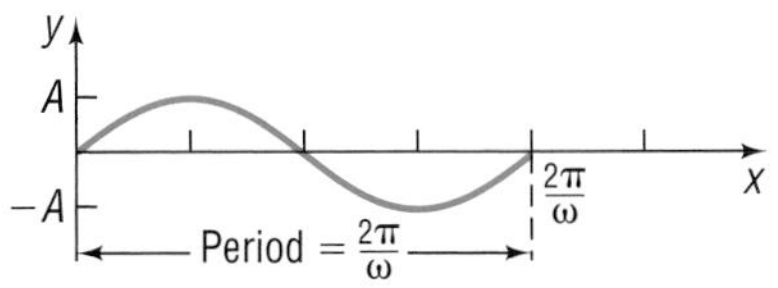

We now want to discuss the graph of

$$y = A\sin(\omega x - \phi) = A\sin\left[\omega\left(x - \frac{\phi}{\omega}\right)\right]$$

where $\omega > 0$ and ϕ (the Greek letter phi) are real numbers. The graph will be a sine curve of amplitude $|A|$. As $\omega x - \phi$ varies from 0 to 2π, one period will be traced out. This period will begin when

$$\omega x - \phi = 0 \quad \text{or} \quad x = \frac{\phi}{\omega}$$

and will end when

$$\omega x - \phi = 2\pi \quad \text{or} \quad x = \frac{2\pi}{\omega} + \frac{\phi}{\omega}$$

Figure 71
One cycle $y = A\sin(\omega x - \phi)$, $A > 0$, $\omega > 0$, $\phi > 0$

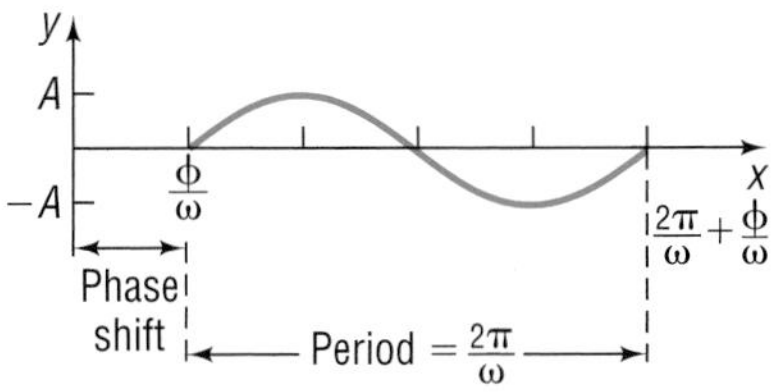

See Figure 71.

We see that the graph of $y = A\sin(\omega x - \phi) = A\sin\left[\omega\left(x - \frac{\phi}{\omega}\right)\right]$ is the same as the graph of $y = A\sin(\omega x)$, except that it has been shifted ϕ/ω units (to the right if $\phi > 0$ and to the left if $\phi < 0$). This number ϕ/ω is called the **phase shift** of the graph of $y = A\sin(\omega x - \phi)$.

For the graphs of $y = A\sin(\omega x - \phi)$ or $y = A\cos(\omega x - \phi)$, $\omega > 0$,

Amplitude $= |A|$	Period $= T = \dfrac{2\pi}{\omega}$	Phase shift $= \dfrac{\phi}{\omega}$

The phase shift is to the left if $\phi < 0$ and to the right if $\phi > 0$.

EXAMPLE 7 Finding the Amplitude, Period, and Phase Shift of a Sinusoidal Function

4 Find the amplitude, period, and phase shift of $y = 3\sin(2x - \pi)$, and graph the function.

Solution Comparing

$$y = 3\sin(2x - \pi) = 3\sin[2(x - \pi/2)]$$

to

$$y = A\sin(\omega x - \phi) = A\sin[\omega(x - \phi/\omega)]$$

we find that $A = 3$, $\omega = 2$, and $\phi = \pi$. The graph is a sine curve with amplitude $A = 3$, period $T = 2\pi/\omega = 2\pi/2 = \pi$, and phase shift $= \dfrac{\phi}{\omega} = \dfrac{\pi}{2}$.

The graph of $y = 3\sin(2x - \pi)$ will lie between -3 and 3 on the y-axis. One cycle will begin at $x = \dfrac{\phi}{\omega} = \dfrac{\pi}{2}$ and end at $x = \dfrac{2\pi}{\omega} + \dfrac{\phi}{\omega} = \pi + \dfrac{\pi}{2} = \dfrac{3\pi}{2}$. We divide the interval $\left[\dfrac{\pi}{2}, \dfrac{3\pi}{2}\right]$ into four subintervals, each of length $\pi \div 4 = \pi/4$: $\left[\dfrac{\pi}{2}, \dfrac{3\pi}{4}\right]$, $\left[\dfrac{3\pi}{4}, \pi\right]$, $\left[\pi, \dfrac{5\pi}{4}\right]$, and $\left[\dfrac{5\pi}{4}, \dfrac{3\pi}{2}\right]$. The five key points on the graph are $\left(\dfrac{\pi}{2}, 0\right)$, $\left(\dfrac{3\pi}{4}, 3\right)$, $(\pi, 0)$, $\left(\dfrac{5\pi}{4}, -3\right)$, and $\left(\dfrac{3\pi}{2}, 0\right)$. We plot these five points and fill in the graph of the sine function as shown in Figure 72(a). Extending the graph in either direction, we obtain Figure 72(b).

Figure 72

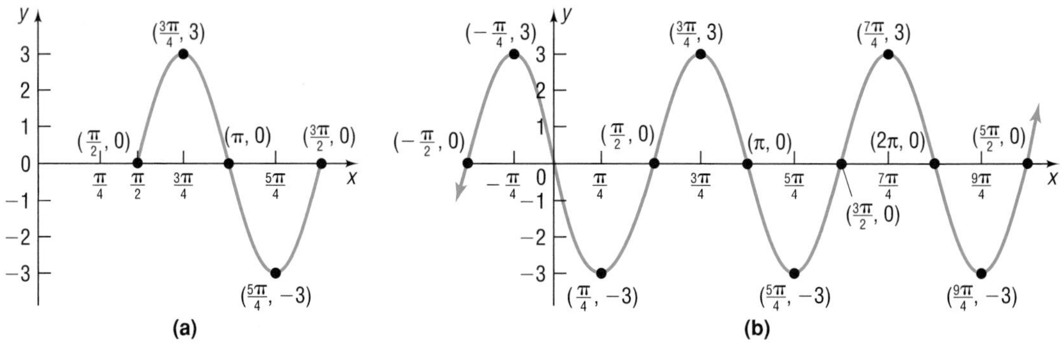

The graph of $y = 3\sin(2x - \pi) = 3\sin[2(x - \pi/2)]$ may also be obtained using transformations. See Figure 73.

Figure 73

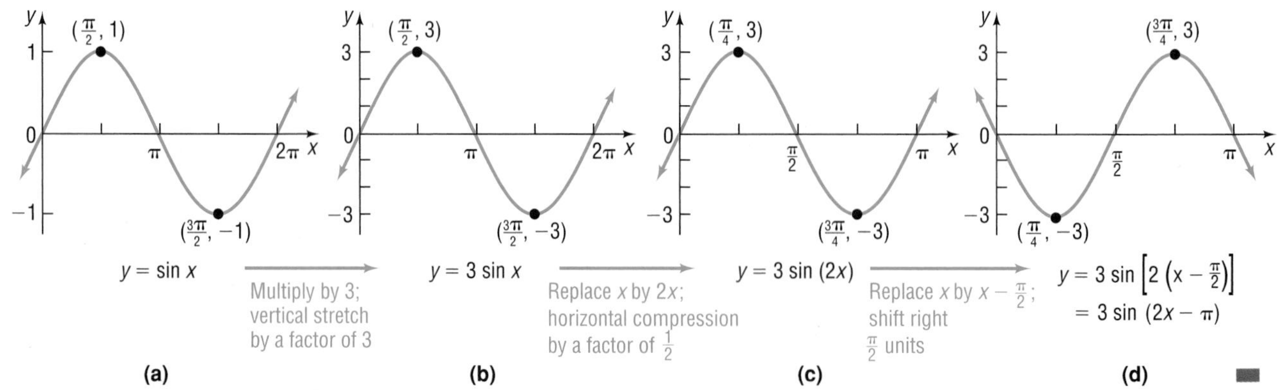

Check: Graph $Y_1 = 3\sin(2x - \pi)$ using a graphing utility.

EXAMPLE 8 Finding the Amplitude, Period, and Phase Shift of a Sinusoidal Function

Find the amplitude, period, and phase shift of $y = 2\cos(4x + 3\pi)$, and graph the function.

Solution Comparing

$$y = 2\cos(4x + 3\pi) = 2\cos\left[4\left(x + \frac{3\pi}{4}\right)\right]$$

to

$$y = A\cos(\omega x - \phi) = A\cos\left[\omega\left(x - \frac{\phi}{\omega}\right)\right]$$

we see that $A = 2$, $\omega = 4$, and $\phi = -3\pi$. The graph is a cosine curve with amplitude $A = 2$, period $T = 2\pi/\omega = 2\pi/4 = \pi/2$, and phase shift $= \dfrac{\phi}{\omega} = \dfrac{-3\pi}{4}$.

The graph of $y = 2\cos(4x + 3\pi)$ will lie between -2 and 2 on the y-axis. One cycle will begin at $x = \dfrac{\phi}{\omega} = \dfrac{-3\pi}{4}$ and end at $x = \dfrac{2\pi}{\omega} + \dfrac{\phi}{\omega} = \dfrac{\pi}{2} + \left(\dfrac{-3\pi}{4}\right) = \dfrac{-\pi}{4}$. We divide the interval $\left[\dfrac{-3\pi}{4}, \dfrac{-\pi}{4}\right]$ into four subintervals, each of the length $\pi/2 \div 4 = \pi/8$: $\left[\dfrac{-3\pi}{4}, \dfrac{-5\pi}{8}\right]$, $\left[\dfrac{-5\pi}{8}, \dfrac{-\pi}{2}\right]$, $\left[\dfrac{-\pi}{2}, \dfrac{-3\pi}{8}\right]$, and $\left[\dfrac{-3\pi}{8}, \dfrac{-\pi}{4}\right]$. The five key points on the graph are $\left(\dfrac{-3\pi}{4}, 2\right)$, $\left(\dfrac{-5\pi}{8}, 0\right)$, $\left(\dfrac{-\pi}{2}, -2\right)$, $\left(\dfrac{-3\pi}{8}, 0\right)$, and $\left(\dfrac{-\pi}{4}, 2\right)$. We plot these five points and fill in the graph of the cosine function as shown in Figure 74(a). Extending the graph in either direction, we obtain Figure 74(b).

Figure 74

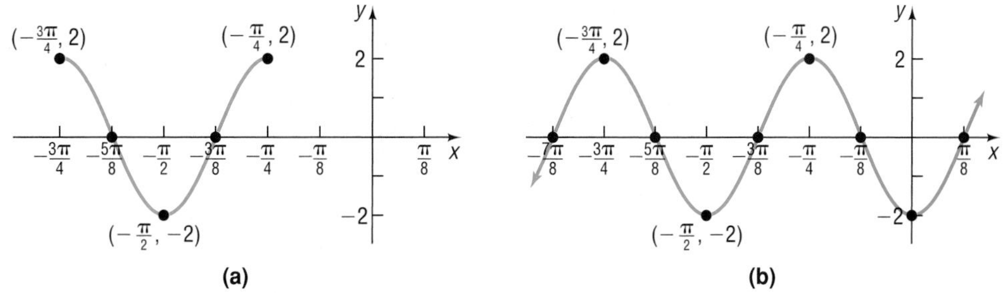

The graph of $y = 2\cos(4x + 3\pi) = 2\cos\left[4\left(x + \dfrac{3\pi}{4}\right)\right]$ may also be obtained using transformations. See Figure 75.

Figure 75

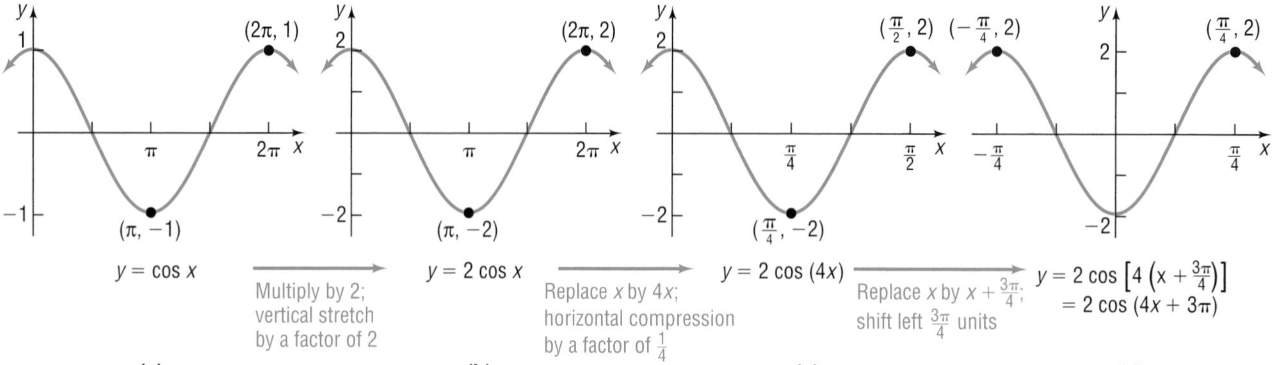

(a) (b) (c) (d)

Check: Graph $Y_1 = 2\cos(4x + 3\pi)$ using a graphing utility.

NOW WORK PROBLEM 51.

SUMMARY

Steps for Graphing Sinusoidal Functions

To graph sinusoidal functions of the form $y = A\sin(\omega x - \phi)$ or $y = A\cos(\omega x - \phi)$:

STEP 1: Determine the amplitude $|A|$ and period $T = \dfrac{2\pi}{\omega}$.

STEP 2: Determine the starting point of one cycle of the graph, $\dfrac{\phi}{\omega}$.

STEP 3: Determine the ending point of one cycle of the graph, $\dfrac{2\pi}{\omega} + \dfrac{\phi}{\omega}$.

STEP 4: Divide the interval $\left[\dfrac{\phi}{\omega}, \dfrac{2\pi}{\omega} + \dfrac{\phi}{\omega}\right]$ into four subintervals, each of length $\dfrac{2\pi}{\omega} \div 4$.

STEP 5: Use the endpoints of the subintervals to find the five key points on the graph.

STEP 6: Fill in one cycle of the graph.

STEP 7: Extend the graph in each direction to make it complete.

FINDING SINUSOIDAL FUNCTIONS FROM DATA

5 Scatter diagrams of data sometimes take the form of a sinusoidal function. Let's look at an example.

The data given in Table 11 represent the average monthly temperatures in Denver, Colorado. Since the data represent *average* monthly temperatures collected over so many years, the data will not vary much from year to year and so will essentially repeat each year. In other words, the data are periodic. Figure 76 shows the scatter diagram of these data repeated over two years, where $x = 1$ represents January, $x = 2$ represents February, and so on.

Notice that the scatter diagram looks like the graph of a sinusoidal function. We choose to fit the data to a sine function of the form

$$y = A\sin(\omega x - \phi) + B$$

where A, B, ω, and ϕ are constants.

TABLE 11

Month, *x*	Average Monthly Temperature, °F
January, 1	29.7
February, 2	33.4
March, 3	39.0
April, 4	48.2
May, 5	57.2
June, 6	66.9
July, 7	73.5
August, 8	71.4
September, 9	62.3
October, 10	51.4
November, 11	39.0
December, 12	31.0

Source: U.S. National Oceanic and Atmospheric Administration

Figure 76

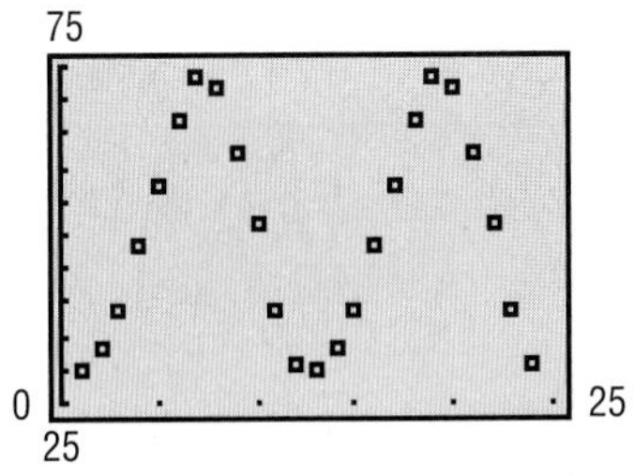

EXAMPLE 9 Finding a Sinusoidal Function from Temperature Data

Fit a sine function to the data in Table 11.

Solution We begin with a scatter diagram of the data for one year. See Figure 77. The data will be fitted to a sine function of the form

$$y = A\sin(\omega x - \phi) + B$$

Figure 77

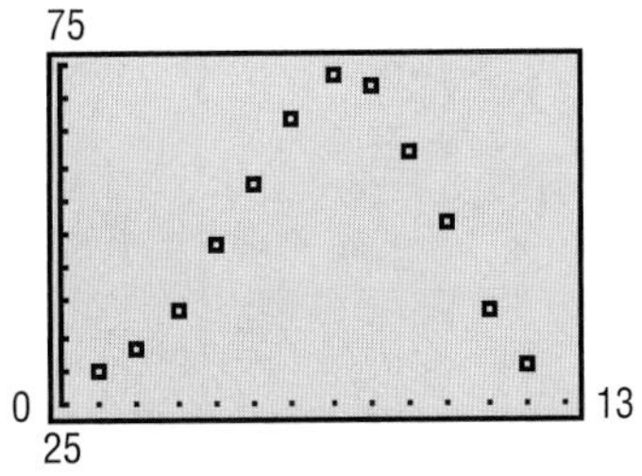

STEP 1: To find the amplitude A, we compute

$$\begin{aligned}\text{Amplitude} &= (\text{largest data value} - \text{smallest data value})/2 \\ &= (73.5 - 29.7)/2 = 21.9\end{aligned}$$

To see the remaining steps in this process, we superimpose the graph of the function $Y_1 = 21.9\sin x$, where x represents months, on the scatter diagram. To see both, the viewing window has been adjusted. See Figure 78.

Figure 78

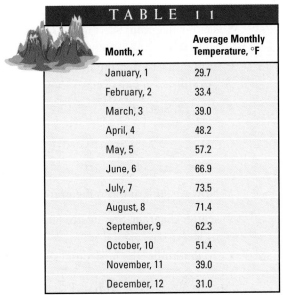

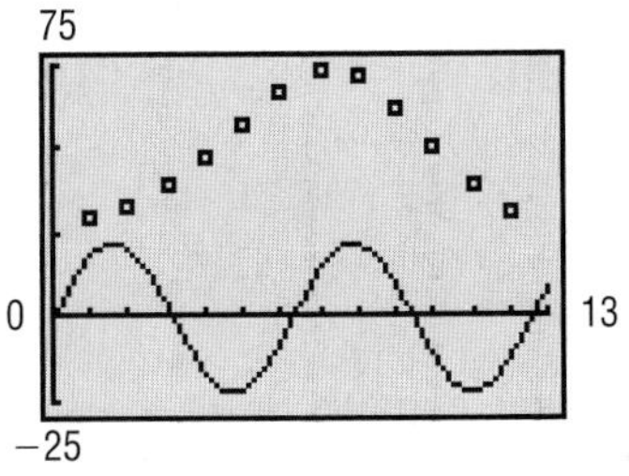

To fit the data, the graph needs to be shifted vertically, shifted horizontally, and stretched horizontally.

STEP 2: We determine the vertical shift by finding the average of the highest and lowest data value.

$$\text{Vertical shift} = (73.5 + 29.7)/2 = 51.6$$

Figure 79

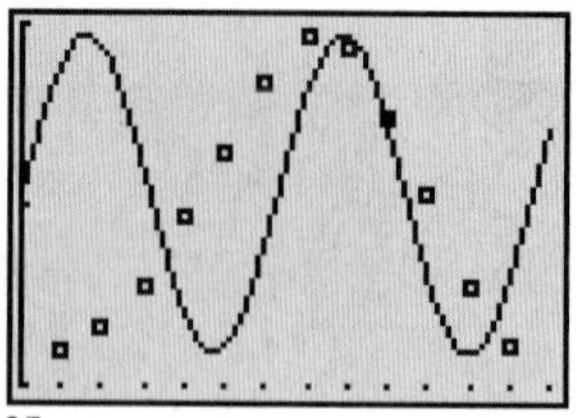

Now we superimpose the graph of $Y_1 = 21.9 \sin x + 51.6$ on the scatter diagram. See Figure 79. We see that the graph needs to be shifted horizontally and stretched horizontally.

STEP 3: It is easier to find the horizontal stretch factor first. Since the temperatures repeat every 12 months, the period of the function is $T = 12$. Since $T = \frac{2\pi}{\omega} = 12$,

$$\omega = \frac{2\pi}{12} = \frac{\pi}{6}$$

Now we superimpose the graph of $Y_1 = 21.9 \sin \frac{\pi x}{6} + 51.6$ on the scatter diagram. See Figure 80. We see that the graph still needs to be shifted horizontally.

Figure 80

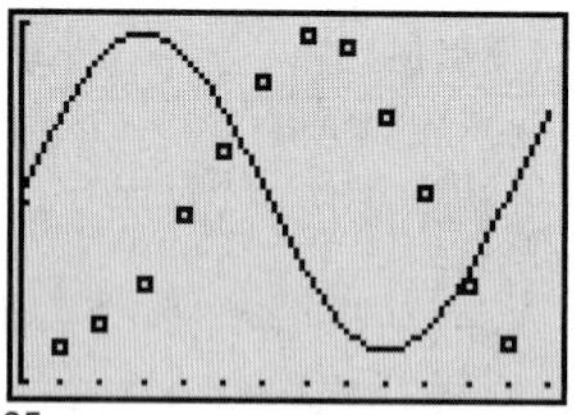

STEP 4: To determine the horizontal shift, we solve the equation

$$y = 21.9 \sin\left(\frac{\pi}{6}x - \phi\right) + 51.6$$

for ϕ by letting $y = 29.7$ and $x = 1$ (the average temperature in Denver in January).*

$$29.7 = 21.9 \sin\left(\frac{\pi}{6} \cdot 1 - \phi\right) + 51.6$$

$$-21.9 = 21.9 \sin\left(\frac{\pi}{6} - \phi\right) \quad \text{Subtract 51.6 from both sides of the equation.}$$

$$-1 = \sin\left(\frac{\pi}{6} - \phi\right) \quad \text{Divide both sides of the equation by 21.9.}$$

$$-\frac{\pi}{2} = \frac{\pi}{6} - \phi \quad \sin\theta = -1 \text{ when } \theta = -\frac{\pi}{2}.$$

$$\phi = \frac{2\pi}{3} \quad \text{Solve for } \phi.$$

Figure 81

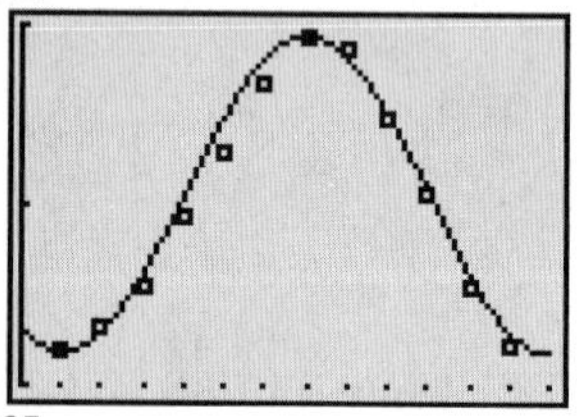

The sine function that fits the data is

$$y = 21.9 \sin\left(\frac{\pi}{6}x - \frac{2\pi}{3}\right) + 51.6$$

The graph of $Y_1 = 21.9 \sin(\pi x/6 - 2\pi/3) + 51.6$ and the scatter diagram of the data are shown in Figure 81. ■

The steps to fit a sine function

$$y = A \sin(\omega x - \phi) + B$$

to sinusoidal data follow:

*The data point selected here to find ϕ is arbitrary. Selecting a different data point will usually result in a different value for ϕ. To maintain consistency, we will always choose the data point for which y is smallest (in this case, January gives the lowest temperature).

STEPS FOR FITTING DATA TO A SINE FUNCTION $y = A\sin(\omega x - \phi) + B$

STEP 1: Determine A, the amplitude of the function.

$$\text{Amplitude} = \frac{\text{largest data value} - \text{smallest data value}}{2}$$

STEP 2: Determine B, the vertical shift of the function.

$$\text{Vertical shift} = \frac{\text{largest data value} + \text{smallest data value}}{2}$$

STEP 3: Determine ω. Since the period T, the time it takes for the data to repeat, is $T = \dfrac{2\pi}{\omega}$, we have

$$\omega = \frac{2\pi}{T}$$

STEP 4: Determine the horizontal shift of the function by solving the equation

$$y = A\sin(\omega x - \phi) + B$$

for ϕ by choosing an ordered pair (x, y) from the data. Since answers will vary depending on the ordered pair selected, we will always choose the ordered pair for which y is smallest in order to maintain consistency.

Certain graphing utilities (such as a TI-83 and TI-86) have the capability of finding the sine function of best fit for sinusoidal data. At least four data points are required for this process.

EXAMPLE 10 Finding the Sine Function of Best Fit

Use a graphing utility to find the sine function of best fit for the data in Table 11. Graph this function with the scatter diagram of the data.

Solution Enter the data from Table 11 and execute the SINe REGression program. The result is shown in Figure 82.

Figure 82

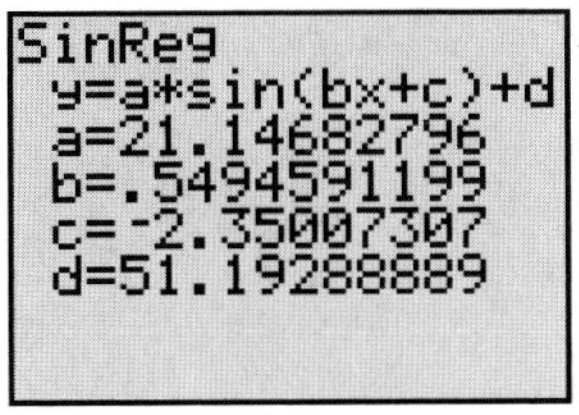

The output that the utility provides shows us the equation

$$y = a\sin(bx + c) + d$$

The sinusoidal function of best fit is

$$y = 21.15\sin(0.55x - 2.35) + 51.19$$

where x represents the month and y represents the average temperature.

Figure 83 shows the graph of the sinusoidal function of best fit on the scatter diagram.

Figure 83

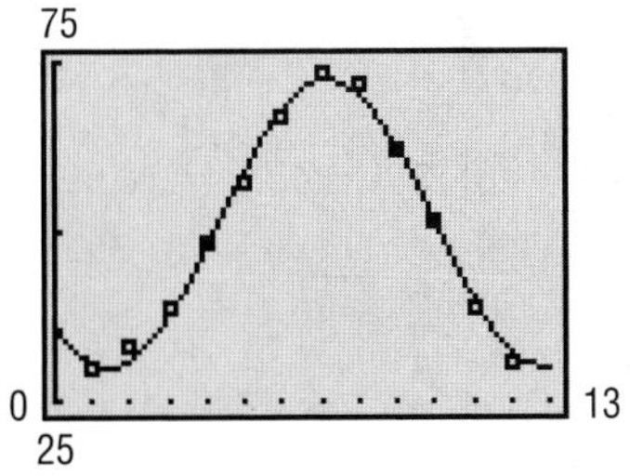

6.6 EXERCISES

In Problems 1–10, determine the amplitude and period of each function without graphing.

1. $y = 2\sin x$ **2.** $y = 3\cos x$ **3.** $y = -4\cos(2x)$ **4.** $y = -\sin\left(\frac{1}{2}x\right)$

5. $y = 6\sin(\pi x)$ **6.** $y = -3\cos(3x)$ **7.** $y = -\frac{1}{2}\cos\left(\frac{3}{2}x\right)$ **8.** $y = \frac{4}{3}\sin\left(\frac{2}{3}x\right)$

9. $y = \frac{5}{3}\sin\left(-\frac{2\pi}{3}x\right)$ **10.** $y = \frac{9}{5}\cos\left(-\frac{3\pi}{2}x\right)$

In Problems 11–20, match the given function to one of the graphs (A)–(J). Do not use a graphing utility.

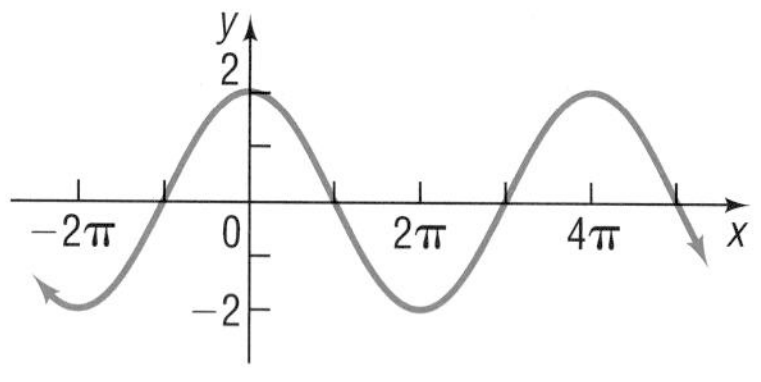

(A)

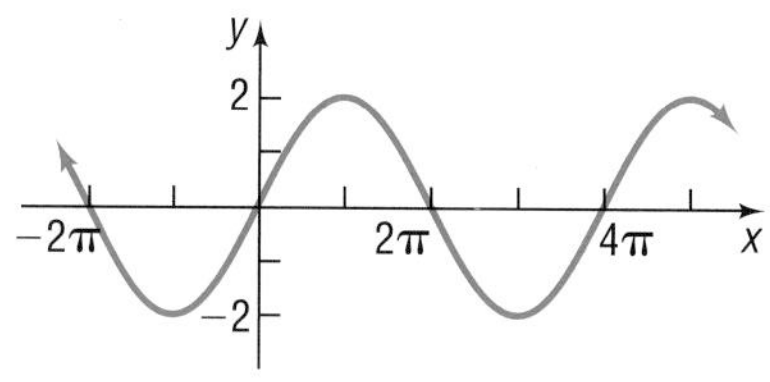

(B)

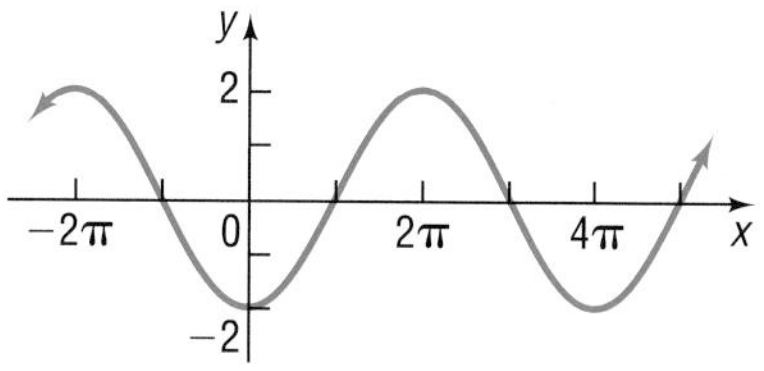

(C)

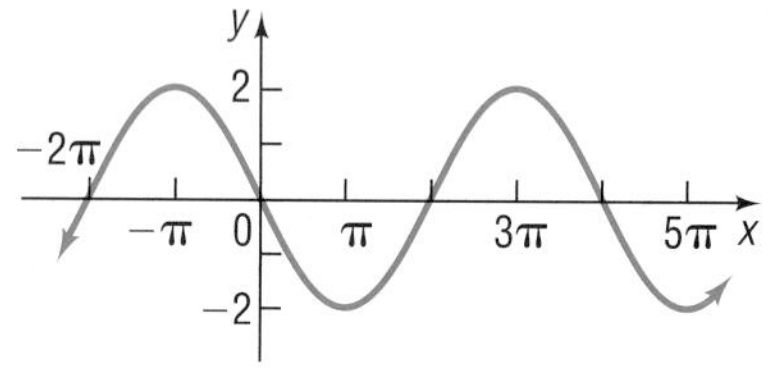

(D)

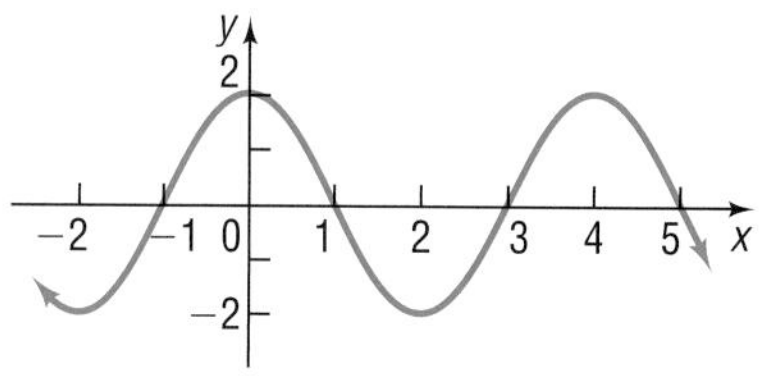

(E)

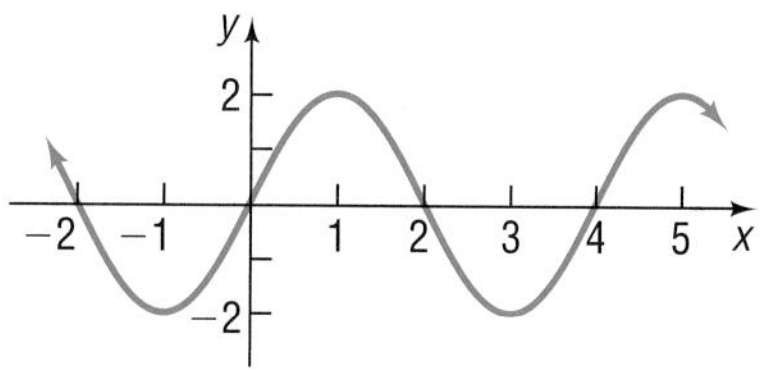

(F)

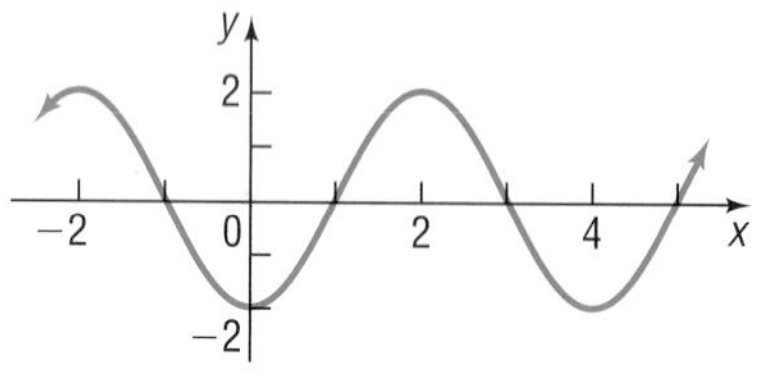

(G)

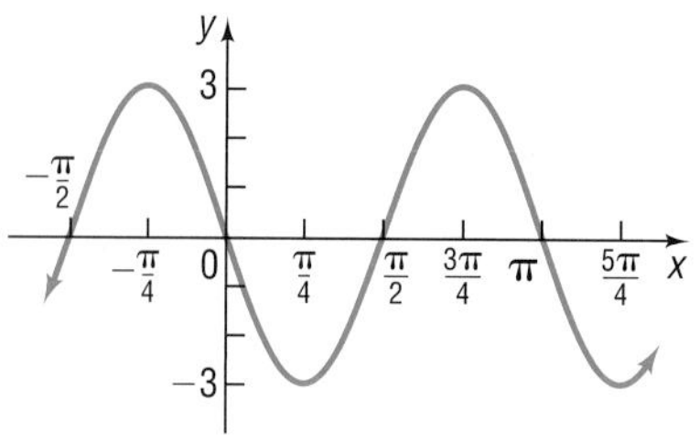

(H)

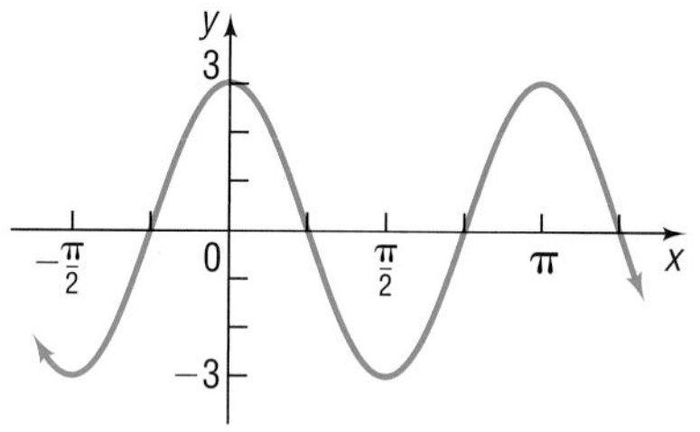

(I)

11. $y = 2\sin\left(\dfrac{\pi}{2}x\right)$ **12.** $y = 2\cos\left(\dfrac{\pi}{2}x\right)$ **13.** $y = 2\cos\left(\frac{1}{2}x\right)$

14. $y = 3\cos(2x)$ **15.** $y = -3\sin(2x)$ **16.** $y = 2\sin\left(\frac{1}{2}x\right)$

17. $y = -2\cos\left(\frac{1}{2}x\right)$ **18.** $y = -2\cos\left(\dfrac{\pi}{2}x\right)$ **19.** $y = 3\sin(2x)$

20. $y = -2\sin\left(\frac{1}{2}x\right)$

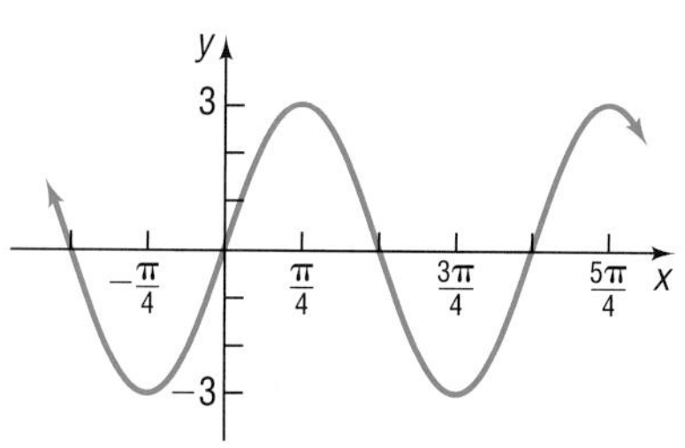

(J)

In Problems 21–26, match the given function to one of the graphs (A)–(F).

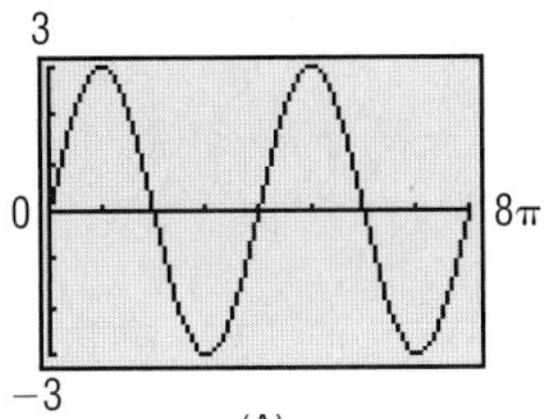

(A)

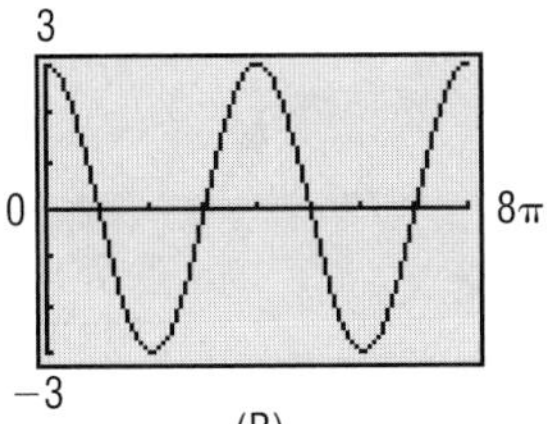

(B)

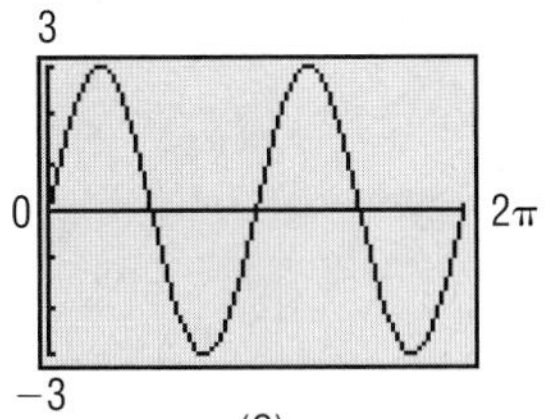

(C)

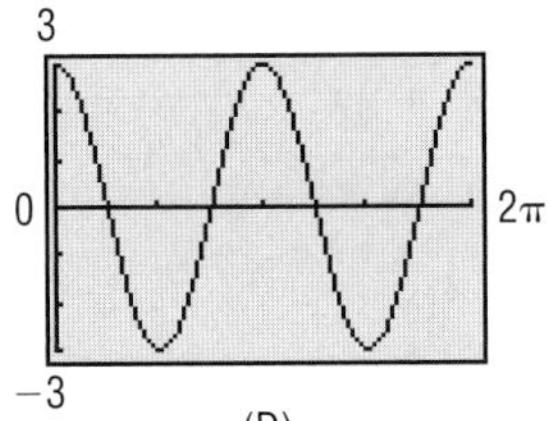

(D)

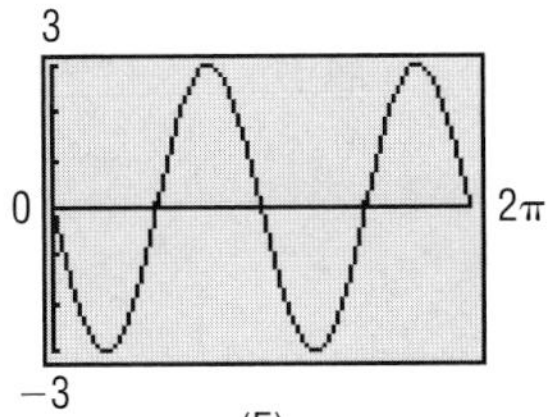

(E)

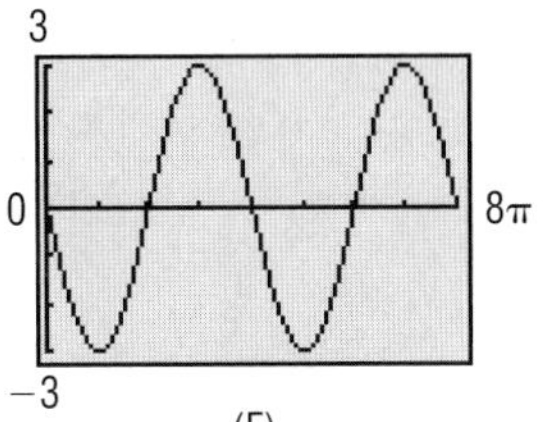

(F)

21. $y = 3\sin\left(\frac{1}{2}x\right)$ **22.** $y = -3\sin(2x)$ **23.** $y = 3\cos(2x)$

24. $y = 3\sin(2x)$ **25.** $y = 3\cos\left(\frac{1}{2}x\right)$ **26.** $y = -3\sin\left(\frac{1}{2}x\right)$

In Problems 27–36, graph each function by hand. Verify the result using a graphing utility.

27. $y = 5\sin(4x)$ **28.** $y = 4\cos(6x)$ **29.** $y = 5\cos(\pi x)$ **30.** $y = 2\sin(\pi x)$

31. $y = -2\cos(2\pi x)$ **32.** $y = -5\cos(2\pi x)$ **33.** $y = -4\sin\left(\frac{1}{2}x\right)$ **34.** $y = -2\cos\left(\frac{1}{2}x\right)$

35. $y = \frac{3}{2}\sin\left(-\frac{2}{3}x\right)$ **36.** $y = \frac{4}{3}\cos\left(-\frac{1}{3}x\right)$

In Problems 37–50, find an equation for each graph.

37.

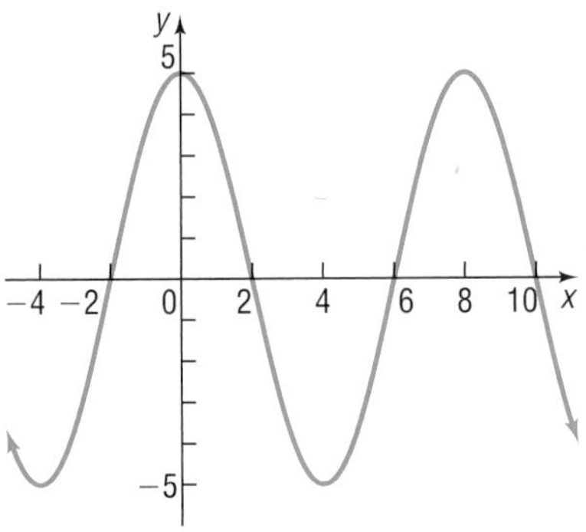

38.

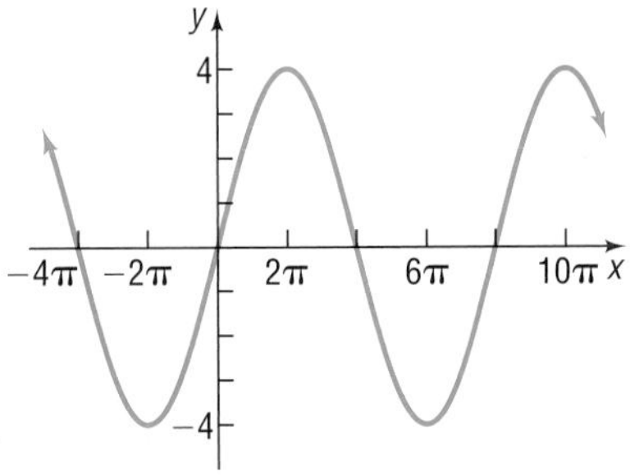

39.

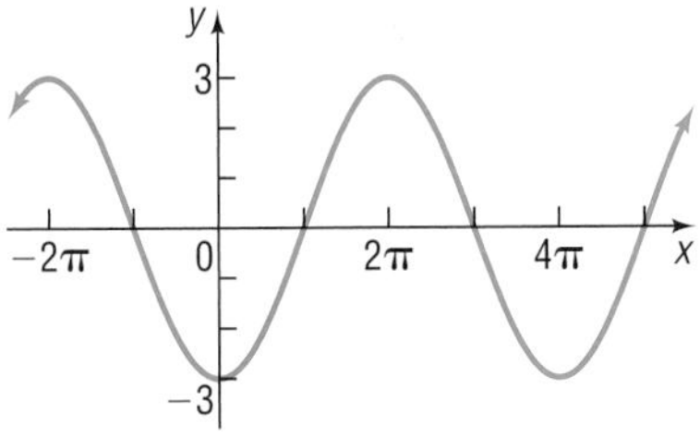

40.

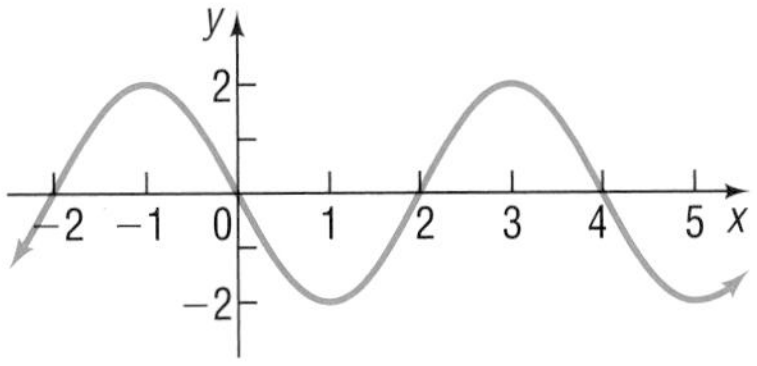

41.

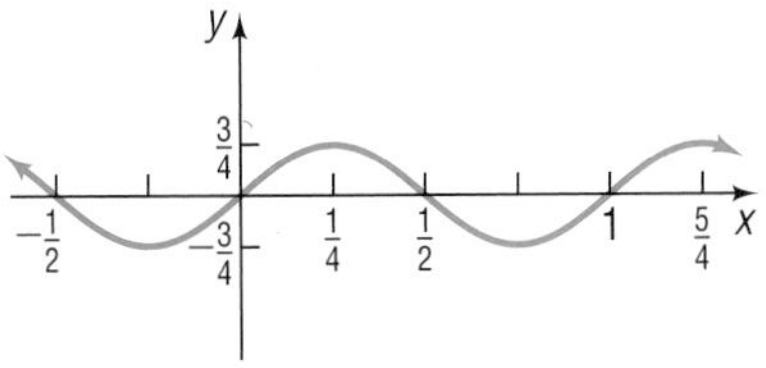

42.

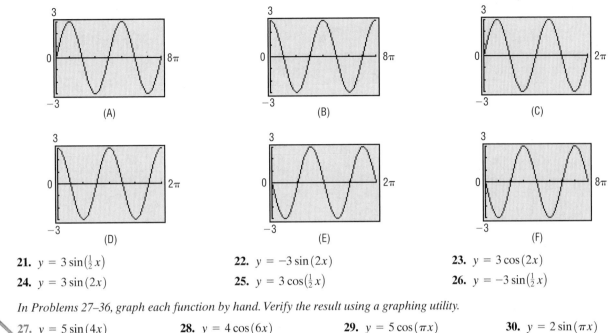

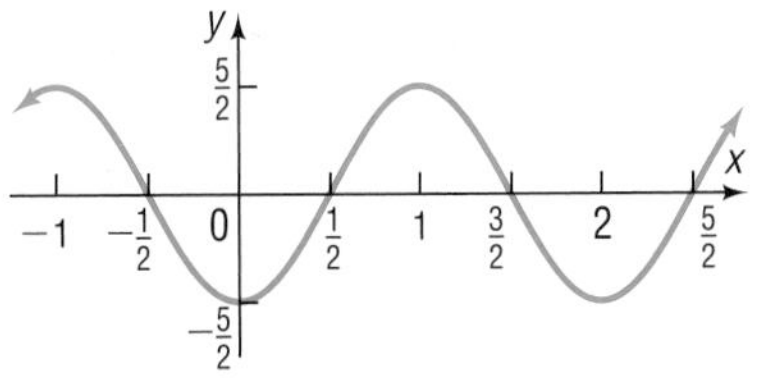

43.

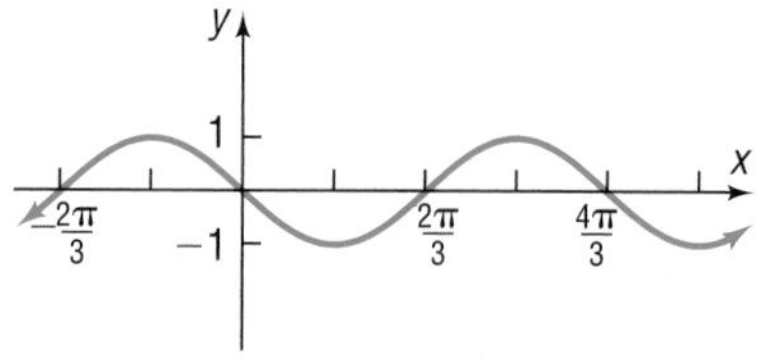

44.

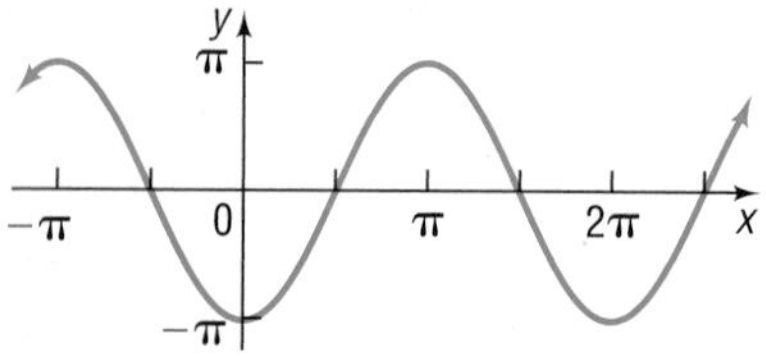

45.

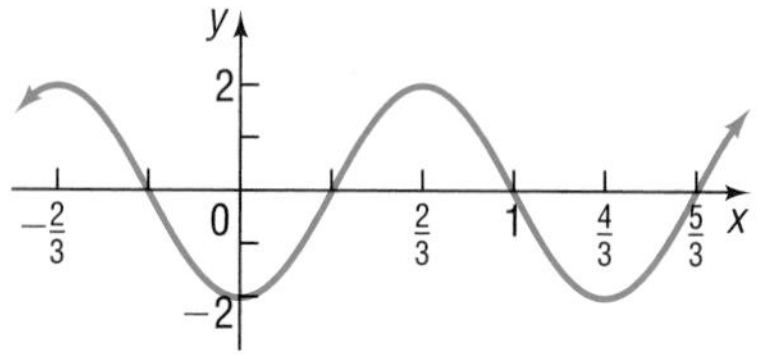

46.

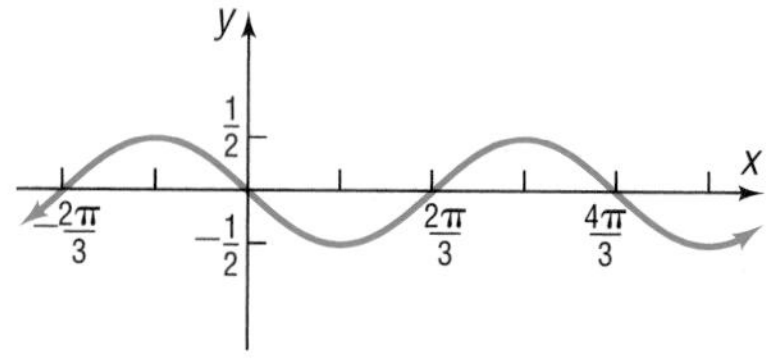

47.

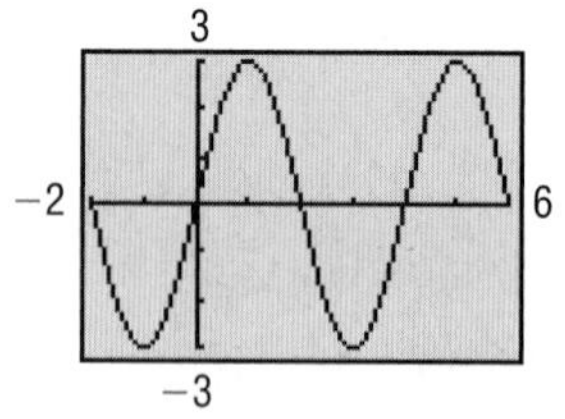

48.

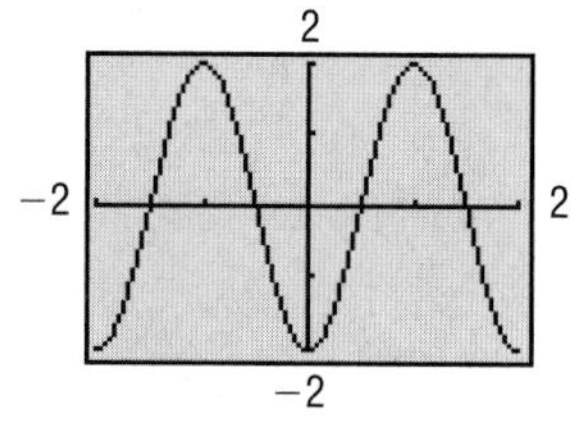

49.

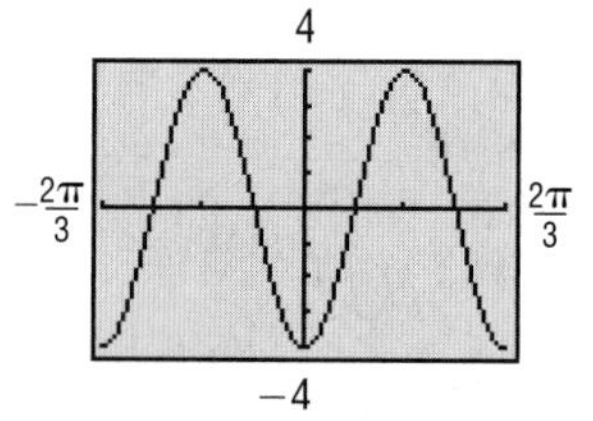

50.

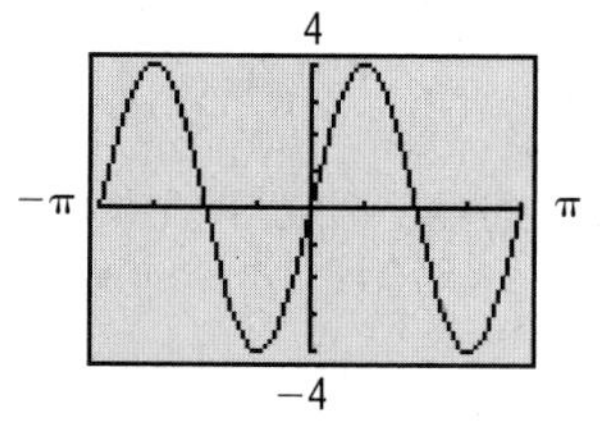

In Problems 51–62, find the amplitude, period, and phase shift of each function. Graph the function by hand and show at least one period. Verify your results using a graphing utility.

51. $y = 4\sin(2x - \pi)$

52. $y = 3\sin(3x - \pi)$

53. $y = 2\cos\left(3x + \frac{\pi}{2}\right)$

54. $y = 3\cos(2x + \pi)$

55. $y = -3\sin\left(2x + \frac{\pi}{2}\right)$

56. $y = -2\cos\left(2x - \frac{\pi}{2}\right)$

57. $y = 4\sin(\pi x + 2)$

58. $y = 2\cos(2\pi x + 4)$

59. $y = 3\cos(\pi x - 2) + 1$

60. $y = 2\cos(2\pi x - 4) - 3$

61. $y = 3\sin\left(-2x + \frac{\pi}{2}\right) - 2$

62. $y = 3\cos\left(-2x + \frac{\pi}{2}\right) + 5$

In Problems 63–70, write the equation of a sine function that has the given characteristics.

63. Amplitude: 3
Period: π

64. Amplitude: 2
Period: 4π

65. Amplitude: 3
Period: 2

66. Amplitude: 4
Period: 1

67. Amplitude: 2
Period: π
Phase shift: $\frac{1}{2}$

68. Amplitude: 3
Period: $\pi/2$
Phase shift: 2

69. Amplitude: 3
Period: 3π
Phase shift: $-\frac{1}{3}$

70. Amplitude: 2
Period: π
Phase shift: −2

71. Alternating Current (ac) Circuits The current I, in amperes, flowing through an ac (alternating current) circuit at time t is

$$I = 220\sin(60\pi t), \qquad t \geq 0$$

What is the period? What is the amplitude? Graph this function over two periods.

72. Alternating Current (ac) Circuits The current I, in amperes, flowing through an ac (alternating current) circuit at time t is

$$I = 120\sin(30\pi t), \qquad t \geq 0$$

What is the period? What is the amplitude? Graph this function over two periods.

73. Alternating Current (ac) Circuits The current I, in amperes, flowing through an ac (alternating current) circuit at time t is

$$I = 120\sin\left(30\pi t - \frac{\pi}{3}\right), \qquad t \geq 0$$

What is the period? What is the amplitude? What is the phase shift? Graph this function over two periods.

74. Alternating Current (ac) Circuits The current I, in amperes, flowing through an ac (alternating current) circuit at time t is

$$I = 220\sin\left(60\pi t - \frac{\pi}{6}\right), \qquad t \geq 0$$

What is the period? What is the amplitude? What is the phase shift? Graph this function over two periods.

75. Alternating Current (ac) Generators The voltage V produced by an ac generator is

$$V = 220\sin(120\pi t)$$

(a) What is the amplitude? What is the period?
(b) Graph V over two periods, beginning at $t = 0$.
(c) If a resistance of $R = 10$ ohms is present, what is the current I?
[**Hint:** Use Ohm's Law, $V = IR$.]
(d) What is the amplitude and period of the current I?
(e) Graph I over two periods, beginning at $t = 0$.

76. Alternating Current (ac) Generators The voltage V produced by an ac generator is

$$V = 120\sin(120\pi t)$$

(a) What is the amplitude? What is the period?
(b) Graph V over two periods, beginning at $t = 0$.
(c) If a resistance of $R = 20$ ohms is present, what is the current I?
[**Hint:** Use Ohm's Law, $V = IR$.]
(d) What is the amplitude and period of the current I?
(e) Graph I over two periods, beginning at $t = 0$.

77. Alternating Current (ac) Generators The voltage V produced by an ac generator is sinusoidal. As a function of time, the voltage V is

$$V = V_0\sin(2\pi f t)$$

where f is the **frequency**, the number of complete oscillations (cycles) per second. [In the United States and Canada, f is 60 hertz (Hz).] The **power** P delivered to a resistance R at any time t is defined as

$$P = \frac{V^2}{R}$$

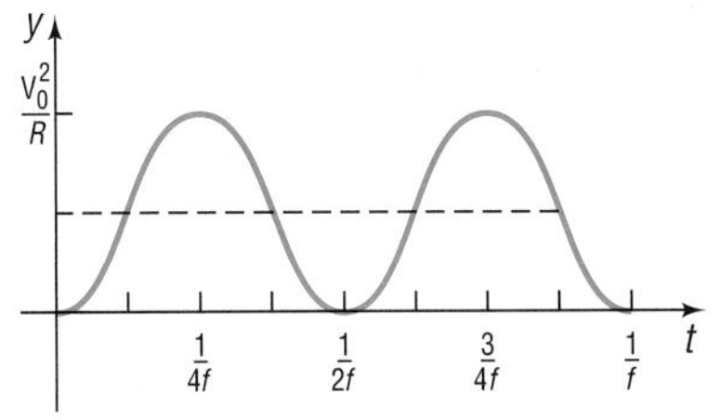

Power in an ac generator

(a) Show that $P = \dfrac{V_0^2}{R}\sin^2(2\pi f t)$.
(b) The graph of P is shown in the figure. Express P as a sinusoidal function.
(c) Deduce that

$$\sin^2(2\pi f t) = \frac{1}{2}[1 - \cos(4\pi f t)]$$

78. Biorhythms In the theory of biorhythms, a sine function of the form

$$P = 100\sin(\omega t)$$

is used to measure the percent P of a person's potential at time t, where t is measured in days and $t = 0$ is the person's birthday. Three characteristics are commonly measured:

Physical potential: period of 23 days
Emotional potential: period of 28 days
Intellectual potential: period of 33 days

(a) Find ω for each characteristic.
(b) Graph all three functions.
(c) Is there a time t when all three characteristics have 100% potential? When is it?
(d) Suppose that you are 20 years old today ($t = 7305$ days). Describe your physical, emotional, and intellectual potential for the next 30 days.

79. Monthly Temperature The following data represent the average monthly temperatures for Juneau, Alaska.

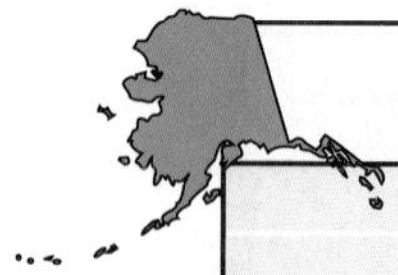

Month, x	Average Monthly Temperature, °F
January, 1	24.2
February, 2	28.4
March, 3	32.7
April, 4	39.7
May, 5	47.0
June, 6	53.0
July, 7	56.0
August, 8	55.0
September, 9	49.4
October, 10	42.2
November, 11	32.0
December, 12	27.1

Source: U.S. National Oceanic and Atmospheric Administration.

(a) Using a graphing utility, draw a scatter diagram of the data for one period.
(b) By hand, find a sinusoidal function of the form $y = A\sin(\omega x - \phi) + B$ that fits the data.
(c) Draw the sinusoidal function found in part (b) on the scatter diagram.
(d) Use a graphing utility to find the sinusoidal function of best fit.
(e) Draw the sinusoidal function of best fit on the scatter diagram.

80. Monthly Temperature The following data represent the average monthly temperatures for Washington, D.C.

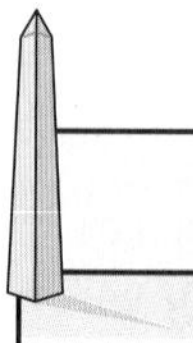

Month, x	Average Monthly Temperature, °F
January, 1	34.6
February, 2	37.5
March, 3	47.2
April, 4	56.5
May, 5	66.4
June, 6	75.6
July, 7	80.0
August, 8	78.5
September, 9	71.3
October, 10	59.7
November, 11	49.8
December, 12	39.4

Source: U.S. National Oceanic and Atmospheric Administration.

(a) Using a graphing utility, draw a scatter diagram of the data for one period.
(b) By hand, find a sinusoidal function of the form $y = A\sin(\omega x - \phi) + B$ that fits the data.
(c) Draw the sinusoidal function found in part (b) on the scatter diagram.
(d) Use a graphing utility to find the sinusoidal function of best fit.
(e) Draw the sinusoidal function of best fit on the scatter diagram.

81. Monthly Temperature The following data represent the average monthly temperatures for Indianapolis, Indiana.

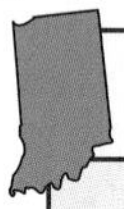

Month, x	Average Monthly Temperature, °F
January, 1	25.5
February, 2	29.6
March, 3	41.4
April, 4	52.4
May, 5	62.8
June, 6	71.9
July, 7	75.4
August, 8	73.2
September, 9	66.6
October, 10	54.7
November, 11	43.0
December, 12	30.9

Source: U.S. National Oceanic and Atmospheric Administration.

(a) Using a graphing utility, draw a scatter diagram of the data for one period.
(b) By hand, find a sinusoidal function of the form $y = A\sin(\omega x - \phi) + B$ that fits the data.
(c) Draw the sinusoidal function found in part (b) on the scatter diagram.
(d) Use a graphing utility to find the sinusoidal function of best fit.
(e) Draw the sinusoidal function of best fit on the scatter diagram.

82. Monthly Temperature The following data represent the average monthly temperatures for Baltimore, Maryland.

Month, x	Average Monthly Temperature, °F
January, 1	31.8
February, 2	34.8
March, 3	44.1
April, 4	53.4
May, 5	63.4
June, 6	72.5
July, 7	77.0
August, 8	75.6
September, 9	68.5
October, 10	56.6
November, 11	46.8
December, 12	36.7

Source: U.S. National Oceanic and Atmospheric Administration.

(a) Using a graphing utility, draw a scatter diagram of the data for one period.
(b) By hand, find a sinusoidal function of the form $y = A\sin(\omega x - \phi) + B$ that fits the data.
(c) Draw the sinusoidal function found in part (b) on the scatter diagram.
(d) Use a graphing utility to find the sinusoidal function of best fit.
(e) Draw the sinusoidal function of best fit on the scatter diagram.

83. Tides Suppose that the length of time between consecutive high tides is approximately 12.5 hours. According to the National Oceanic and Atmospheric Administration, on Saturday, June 28, 1997, in Savannah, Georgia, high tide occurred at 3:38 AM (03.6333 hours) and low tide occurred at 10:08 AM (10.1333 hours). Water heights are measured as the amounts above or below the mean lower low water. The height, h, of the water at high tide was 8.2 feet and the height of the water at low tide was −0.6 feet.
(a) Approximately, when will the next high tide occur?
(b) By hand, find a sinusoidal function of the form $y = A\sin(\omega x - \phi) + B$ that fits the data, where x is the time of the day.
(c) Draw a graph of the function found in part (b).
(d) Use the function found in part (b) to predict the height of the water at the next high tide.

84. Tides Suppose that the length of time between consecutive high tides is approximately 12.5 hours. According to the National Oceanic and Atmospheric Administration, on Saturday, June 28, 1997, in Juneau, Alaska, high tide occurred at 8:11 AM (08.1833 hours) and low tide occurred at 2:14 PM (14.2333 hours). Water heights are measured as the amounts above or below the mean lower low water. The height, h, of the water at high tide was 13.2 feet and the height of the water at low tide was 2.2 feet.
(a) Approximately when will the next high tide occur?
(b) By hand, find a sinusoidal function of the form $y = A\sin(\omega x - \phi) + B$ that fits the data, where x is the time of day.
(c) Draw a graph of the function found in part (b).
(d) Use the function found in part (b) to predict the height of the water at the next high tide.

Problems 85–88 require the following discussion: Since the number of hours of sunlight in a day cycles annually, the number of hours of sunlight in a day for a given location can be modeled by a sinusoidal function. The longest day of the year (in terms of hours of sunlight) occurs on the day of the summer solstice. The summer solstice is the time when the sun is farthest north (for locations in the northern hemisphere). In 2000, the summer solstice is June 20 (the 172nd day of the year) at 9:48 PM (EDT). The shortest day of the year occurs on the day of the winter solstice. The winter solstice is the time when the sun is farthest south (again for locations in the northern hemisphere). In 2000, the winter solstice is December 21 (the 356th day of the year) at 8:37 AM (EST).

The vernal (spring) and autumnal (fall) equinoxes are the times when all locations on Earth receive 12 hours of daylight. In 2000, the vernal equinox is March 20 (80th day of the year) at 2:35 AM (EST) and the autumnal equinox is September 22 (266th day of the year) at 1:27 PM (EDT).

85. Hours of Daylight According to the *Old Farmer's Almanac*, in Miami, Florida, the number of hours of sunlight on the day of the summer solstice is 12.75, and the number of hours of sunlight on the day of the winter solstice is 10.583.
(a) By hand, find a sinusoidal function of the form $y = A\sin(\omega x - \phi) + B$ that fits the data, where x is the day of the year.
(b) Draw a graph of the function found in part (a).
(c) Use the function found in part (a) to predict the number of hours of sunlight on April 1.
(d) Look up the number of hours of sunlight on April 1, 2000, in the *Old Farmer's Almanac*, and compare the actual hours of daylight to the results found in part (c).

86. Hours of Daylight According to the *Old Farmer's Almanac*, in Detroit, Michigan, the number of hours of sunlight on the day of the summer solstice is 13.65, and the number of hours of sunlight on the day of the winter solstice is 9.067.
(a) By hand, find a sinusoidal function of the form $y = A\sin(\omega x - \phi) + B$ that fits the data.
(b) Draw a graph of the function found in part (a).
(c) Use the function found in part (a) to predict the number of hours of sunlight on April 1.
(d) Look up the number of hours of sunlight on April 1, 2000, in the *Old Farmer's Almanac*, and compare the actual hours of daylight to the results found in part (c).

87. Hours of Daylight According to the *Old Farmer's Almanac*, in Anchorage, Alaska, the number of hours of sunlight on the day of the summer solstice is 16.233, and the number of hours of sunlight on the day of the winter solstice is 5.45.
(a) By hand, find a sinusoidal function of the form $y = A\sin(\omega x - \phi) + B$ that fits the data.
(b) Draw a graph of the function found in part (a).
(c) Use the function found in part (a) to predict the number of hours of sunlight on April 1.
(d) Look up the number of hours of sunlight on April 1, 2000, in the *Old Farmer's Almanac*, and compare the actual hours of daylight to the results found in part (c).

88. Hours of Daylight According to the *Old Farmer's Almanac*, in Honolulu, Hawaii, the number of hours of sunlight on the day of the summer solstice is 12.767, and the

number of hours of sunlight on the day of the winter solstice is 10.783.
(a) By hand, find a sinusoidal function of the form $y = A\sin(\omega x - \phi) + B$ that fits the data.
(b) Draw a graph of the function found in part (a).
(c) Use the function found in part (a) to predict the number of hours of sunlight on April 1.
(d) Look up the number of hours of sunlight on April 1, 2000, in the *Old Farmer's Almanac* and compare the actual hours of daylight to the results found in part (c).

89. Explain how the amplitude and period of a sinusoidal graph are used to establish the scale on each coordinate axis.

90. Find an application in your major field that leads to a sinusoidal graph. Write a paper about your findings.

91. CBL Experiment Part 1: A student puts his or her thumb over a light probe and points it at a light source. The student then begins lifting his or her thumb from the sensor and replacing it repeatedly. Light intensity is plotted over time, and the period and frequency are determined. Part 2: A light probe is pointed toward a fluorescent light bulb, and its light intensity is recorded. The frequency and period of the light bulb are recorded. (Activity 15, Real-World Math with the CBL System, 1994.)

CHAPTER REVIEW

Things To Know

Definitions

Angle in standard position (p. 369) — Vertex is at the origin; initial side is along the positive x-axis

Degree (1°) (p. 369) — $1° = \frac{1}{360}$ revolution

Radian (1) (p. 371) — The measure of a central angle whose rays subtend an arc whose length is the radius of the circle

Trigonometric functions (p. 381) — $P = (x, y)$ is the point on the unit circle corresponding to $\theta = t$ radians:

$$\sin t = \sin\theta = y \qquad \cos t = \cos\theta = x$$

$$\tan t = \tan\theta = \frac{y}{x},\quad x \neq 0 \qquad \cot t = \cot\theta = \frac{x}{y},\quad y \neq 0$$

$$\csc t = \csc\theta = \frac{1}{y},\quad y \neq 0 \qquad \sec t = \sec\theta = \frac{1}{x},\quad x \neq 0$$

Periodic function (p. 399) — $f(\theta + p) = f(\theta)$, for all θ, $p > 0$, where the smallest such p is the fundamental period

Formulas

1 revolution $= 360°$

$= 2\pi$ radians (p. 373)

$s = r\theta$ (p. 372) — θ is measured in radians; s is the length of arc subtended by the central angle θ of the circle of radius r.

$v = r\omega$ (p. 376) — v is the linear speed around the circle of radius r; ω is the angular speed (measured in radians per unit time).

TABLE OF VALUES

θ (Radians)	θ (Degrees)	$\sin\theta$	$\cos\theta$	$\tan\theta$	$\csc\theta$	$\sec\theta$	$\cot\theta$
0	0°	0	1	0	Not defined	1	Not defined
$\pi/6$	30°	1/2	$\sqrt{3}/2$	$\sqrt{3}/3$	2	$2\sqrt{3}/3$	$\sqrt{3}$
$\pi/4$	45°	$\sqrt{2}/2$	$\sqrt{2}/2$	1	$\sqrt{2}$	$\sqrt{2}$	1
$\pi/3$	60°	$\sqrt{3}/2$	1/2	$\sqrt{3}$	$2\sqrt{3}/3$	2	$\sqrt{3}/3$
$\pi/2$	90°	1	0	Not defined	1	Not defined	0
π	180°	0	−1	0	Not defined	−1	Not defined
$3\pi/2$	270°	−1	0	Not defined	−1	Not defined	0

Identities (p. 403)

$$\tan\theta = \frac{\sin\theta}{\cos\theta}, \quad \cot\theta = \frac{\cos\theta}{\sin\theta}$$

$$\cot\theta = \frac{1}{\tan\theta}, \quad \sec\theta = \frac{1}{\cos\theta}, \quad \csc\theta = \frac{1}{\sin\theta}$$

$$\sin^2\theta + \cos^2\theta = 1, \quad \tan^2\theta + 1 = \sec^2\theta, \quad 1 + \cot^2\theta = \csc^2\theta$$

Properties of the Trigonometric Functions

$y = \sin x$ (p. 409)
Domain: $-\infty < x < \infty$
Range: $-1 \le y \le 1$
Periodic: period $= 2\pi(360°)$
Odd function

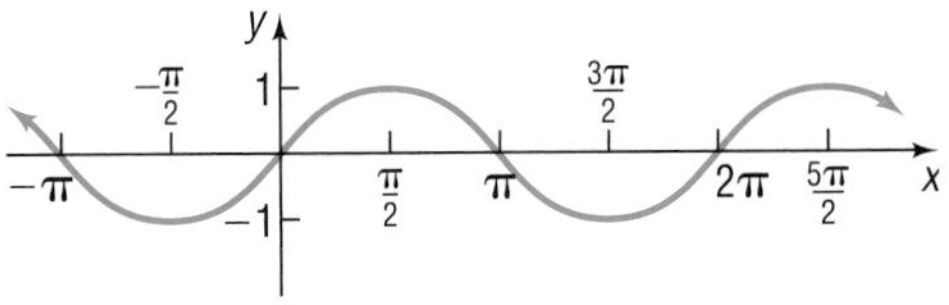

$y = \cos x$ (p. 411)
Domain: $-\infty < x < \infty$
Range: $-1 \le y \le 1$
Periodic: period $= 2\pi(360°)$
Even function

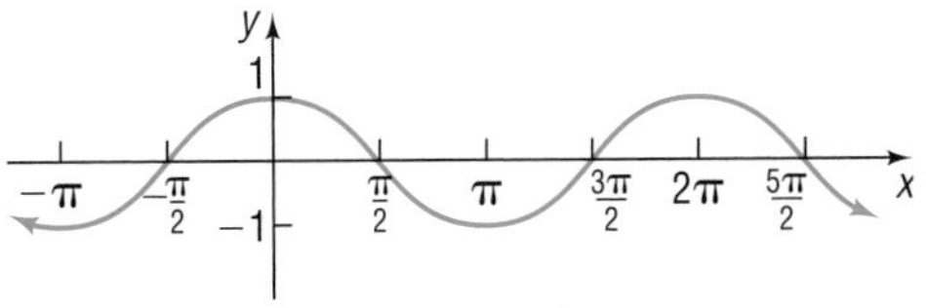

$y = \tan x$ (p. 415)
Domain: $-\infty < x < \infty$, except odd multiples of $\pi/2(90°)$
Range: $-\infty < y < \infty$
Periodic: period $= \pi(180°)$
Odd function

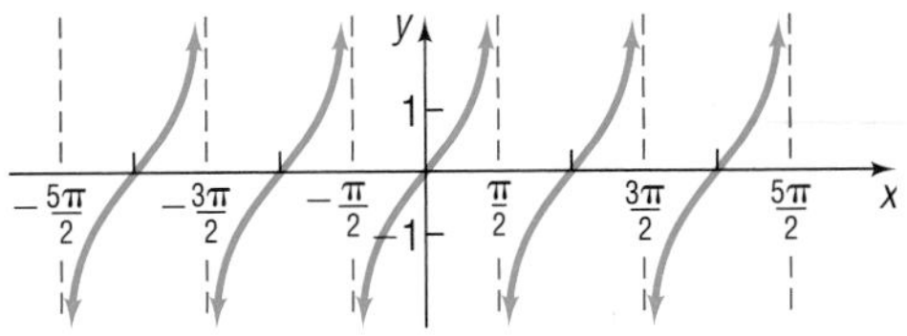

$y = \cot x$ (p. 416)
Domain: $-\infty < x < \infty$, except integral multiples of $\pi(180°)$
Range: $-\infty < y < \infty$
Periodic: period $= \pi(180°)$
Odd function

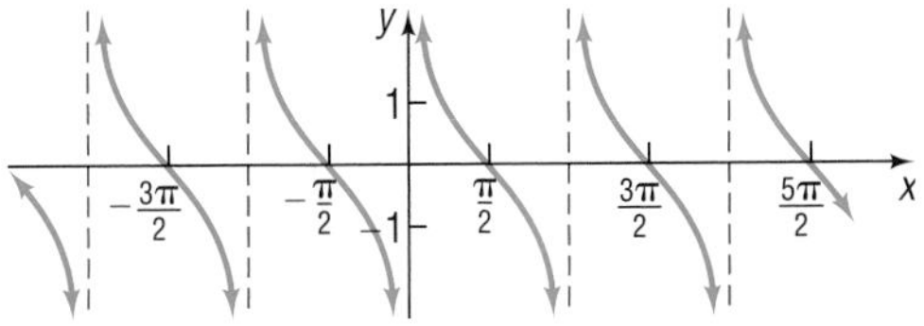

$y = \csc x$ (p. 417)
Domain: $-\infty < x < \infty$, except integral multiples of $\pi(180°)$
Range: $|y| \ge 1$
Periodic: period $= 2\pi(360°)$
Odd function

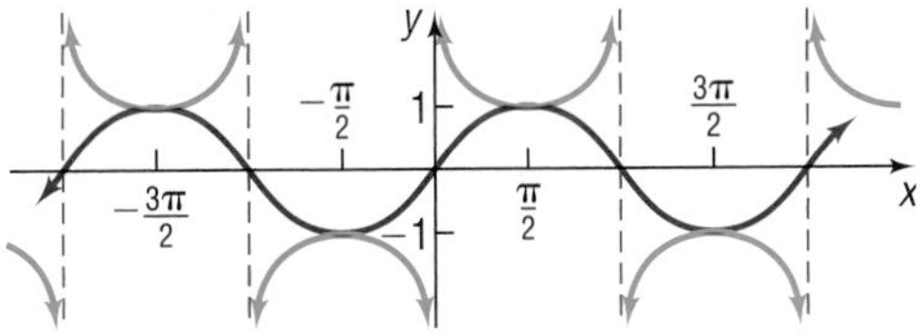

$y = \sec x$ (p. 418)
Domain: $-\infty < x < \infty$, except odd multiples of $\pi/2(90°)$
Range: $|y| \ge 1$
Periodic: period $= 2\pi(360°)$
Even function

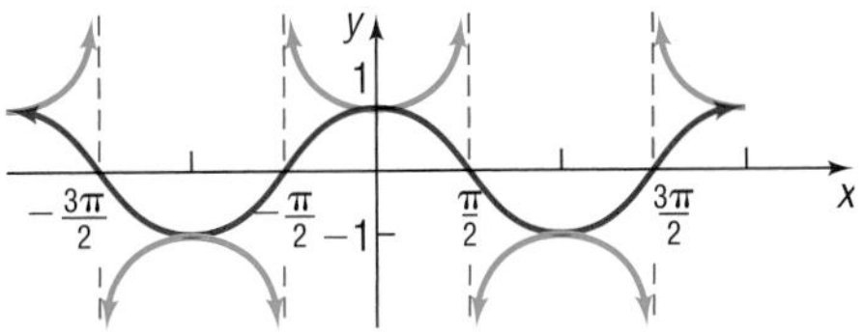

Sinusoidal graphs

$y = A\sin(\omega x), \quad \omega > 0$ — Period $= 2\pi/\omega$ (p. 421)

$y = A\cos(\omega x), \quad \omega > 0$ — Amplitude $= |A|$ (p. 421)

$y = A\sin(\omega x - \phi) = A\sin[\omega(x - \phi/\omega)]$ — Phase shift $= \phi/\omega$ (p. 425)

$y = A\cos(\omega x - \phi) = A\cos[\omega(x - \phi/\omega)]$

How To

Convert between degrees, minutes, seconds, and decimal forms for angles (p. 370)

Find the arc length of a circle (p. 372)

Convert from degrees to radians (p. 374)

Convert from radians to degrees (p. 374)

Find the linear speed of an object traveling in circular motion (p. 375)

Find the exact value of the trigonometric functions using a point on the unit circle (p. 382)

Find the exact value of the trigonometric functions of quadrantal angles (p. 383)

Find the exact value of the trigonometric functions of 45° (p. 384)

Find the exact value of the trigonometric functions of 30° (p. 386)

Find the exact value of the trigonometric functions of 60° (p. 386)

Use a calculator to approximate the value of the trigonometric functions (p. 390)

Determine the domain and range of the trigonometric functions (p. 396)

Determine the period of the trigonometric functions (p. 398)

Determine the signs of the trigonometric functions (p. 400)

Find the value of the trigonometric functions utilizing fundamental identities (p. 401)

Use even–odd properties to find the exact value of the trigonometric functions (p. 404)

Graph transformations of the sine function (p. 409)

Graph transformations of the cosine function (p. 411)

Graph transformations of the tangent and cotangent function (p. 413)

Graph transformations of the secant and cosecant function (p. 416)

Determine the amplitude and period of sinusoidal functions (p. 420)

Find an equation for a sinusoidal graph (p. 424)

Determine the phase shift of a sinusoidal function (p. 425)

Graph sinusoidal functions (p. 425)

Find a sinusoidal function from data (p. 428)

Fill-in-the-Blank Items

1. Two rays drawn with a common vertex form a(n) ________. One of the rays is called the ________ ________; the other is called the ________ ________.
2. In the formula $s = r\theta$ for measuring the length s of arc along a circle of radius r, the angle θ must be measured in ________.
3. 180 degrees = ________ radians.
4. An angle is in ________ ________ if its vertex is at the origin and its initial side coincides with the positive x-axis.
5. The sine, cosine, cosecant, and secant functions have period ________; the tangent and cotangent functions have period ________.
6. Which of the trigonometric functions have graphs that are symmetric with respect to the y-axis?
7. Which of the trigonometric functions have graphs that are symmetric with respect to the origin?
8. The following function has amplitude 3 and period 2: $y =$ ________ sin ________ x.
9. The function $y = 3\sin(6x)$ has amplitude ________ and period ________.

True/False Items

T F 1. In the formula $s = r\theta$, r is the radius of a circle and s is the arc subtended by a central angle θ, where θ is measured in degrees.

T F 2. $|\sin\theta| \leq 1$

T F 3. $1 + \tan^2\theta = \csc^2\theta$

T F 4. The only even trigonometric functions are the cosine and secant functions.

T F 5. The graphs of $y = \tan x$, $y = \cot x$, $y = \sec x$, and $y = \csc x$ have infinitely many vertical asymptotes.

T F 6. The graphs of $y = \sin x$ and $y = \cos x$ are identical except for a horizontal shift.

T F 7. For $y = 2\sin(\pi x)$, the amplitude is 2 and the period is $\pi/2$.

Review Exercises

Blue problem numbers indicate the authors' suggestions for use in a Practice Test.

In Problems 1–4, convert each angle in degrees to radians. Express your answer as a multiple of π.

1. $135°$
2. $210°$
3. $18°$
4. $15°$

In Problems 5–8, convert each angle in radians to degrees.

5. $3\pi/4$
6. $2\pi/3$
7. $-5\pi/2$
8. $-3\pi/2$

In Problems 9–26, find the exact value of each expression. Do not use a calculator.

9. $\tan\frac{\pi}{4} - \sin\frac{\pi}{6}$
10. $\cos\frac{\pi}{3} + \sin\frac{\pi}{2}$
11. $3\sin 45° - 4\tan\frac{\pi}{6}$
12. $4\cos 60° + 3\tan\frac{\pi}{3}$
13. $6\cos\frac{3\pi}{4} + 2\tan\left(-\frac{\pi}{3}\right)$
14. $3\sin\frac{2\pi}{3} - 4\cos\frac{5\pi}{2}$
15. $\sec\left(-\frac{\pi}{3}\right) - \cot\left(-\frac{5\pi}{4}\right)$
16. $4\csc\frac{3\pi}{4} - \cot\left(-\frac{\pi}{4}\right)$
17. $\tan\pi + \sin\pi$
18. $\cos\frac{\pi}{2} - \csc\left(-\frac{\pi}{2}\right)$
19. $\cos 180° - \tan(-45°)$
20. $\sin 270° + \cos(-180°)$
21. $\sin^2 20° + \frac{1}{\sec^2 20°}$
22. $\frac{1}{\cos^2 40°} - \frac{1}{\cot^2 40°}$
23. $\sec 50° \cos 50°$
24. $\tan 10° \cot 10°$
25. $\frac{\cos 400°}{\cos(-40°)}$
26. $\frac{\tan(-20°)}{\tan 200°}$

In Problems 27–42, find the exact value of each of the remaining trigonometric functions.

27. $\sin\theta = -\frac{4}{5}, \quad \cos\theta > 0$
28. $\cos\theta = -\frac{3}{5}, \quad \sin\theta < 0$
29. $\tan\theta = \frac{12}{5}, \quad \sin\theta < 0$
30. $\cot\theta = \frac{12}{5}, \quad \cos\theta < 0$
31. $\sec\theta = -\frac{5}{4}, \quad \tan\theta < 0$
32. $\csc\theta = -\frac{5}{3}, \quad \cot\theta < 0$
33. $\sin\theta = \frac{12}{13}, \quad \theta$ in quadrant II
34. $\cos\theta = -\frac{3}{5}, \quad \theta$ in quadrant III
35. $\sin\theta = -\frac{5}{13}, \quad 3\pi/2 < \theta < 2\pi$
36. $\cos\theta = \frac{12}{13}, \quad 3\pi/2 < \theta < 2\pi$
37. $\tan\theta = \frac{1}{3}, \quad 180° < \theta < 270°$
38. $\tan\theta = -\frac{2}{3}, \quad 90° < \theta < 180°$
39. $\sec\theta = 3, \quad 3\pi/2 < \theta < 2\pi$
40. $\csc\theta = -4, \quad \pi < \theta < 3\pi/2$
41. $\cot\theta = -2, \quad \pi/2 < \theta < \pi$
42. $\tan\theta = -2, \quad 3\pi/2 < \theta < 2\pi$

In Problems 43–54, graph each function by hand. Each graph should contain at least one period. Verify your results using a graphing utility.

43. $y = 2\sin(4x)$
44. $y = -3\cos(2x)$
45. $y = -2\cos\left(x + \frac{\pi}{2}\right)$
46. $y = 3\sin(x - \pi)$
47. $y = \tan(x + \pi)$
48. $y = -\tan\left(x - \frac{\pi}{2}\right)$
49. $y = -2\tan(3x)$
50. $y = 4\tan(2x)$
51. $y = \cot\left(x + \frac{\pi}{8}\right)$
52. $y = -4\cot(2x)$
53. $y = \sec\left(x - \frac{\pi}{4}\right)$
54. $y = \csc\left(x + \frac{\pi}{4}\right)$

In Problems 55–58, determine the amplitude and period of each function without graphing.

55. $y = 4\cos x$
56. $y = \sin(2x)$
57. $y = -8\sin\left(\frac{\pi}{2}x\right)$
58. $y = -2\cos(3\pi x)$

In Problems 59–66, find the amplitude, period, and phase shift of each function. Graph each function by hand. Show at least one period.

59. $y = 4\sin(3x)$

60. $y = 2\cos\left(\frac{1}{3}x\right)$

61. $y = -2\sin\left(\frac{\pi}{2}x + \frac{1}{2}\right)$

62. $y = -6\sin(2\pi x - 2)$

63. $y = \frac{1}{2}\sin\left(\frac{3}{2}x - \pi\right)$

64. $y = \frac{3}{2}\cos(6x + 3\pi)$

65. $y = -\frac{2}{3}\cos(\pi x - 6) + \frac{2}{3}$

66. $y = -7\sin\left(\frac{\pi}{3}x + \frac{4}{3}\right) - 2$

In Problems 67–70, find a function for the given graph.

67.

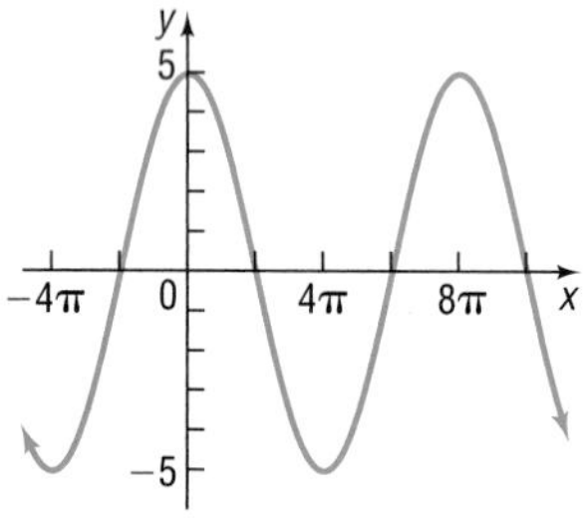

68.

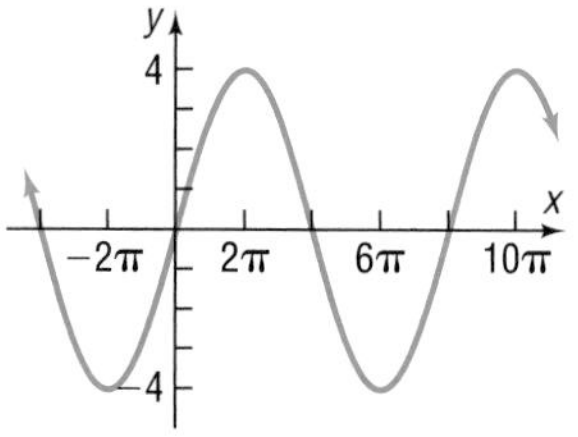

69.

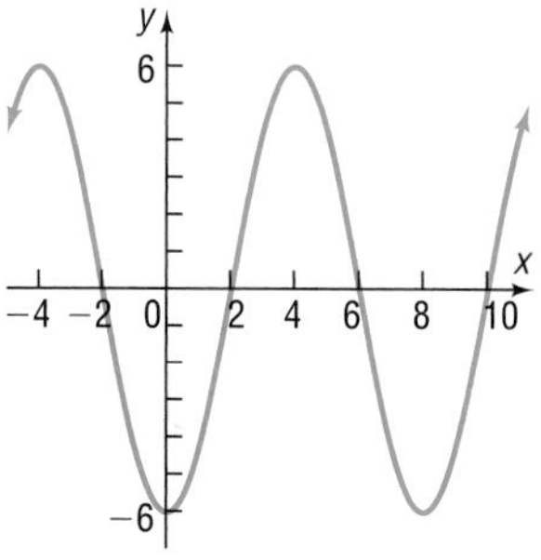

70.

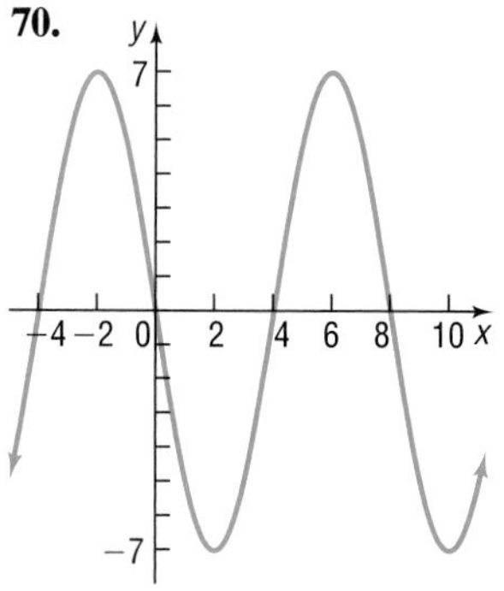

71. Find the length of arc subtended by a central angle of 30° on a circle of radius 2 feet.

72. The minute hand of a clock is 8 inches long. How far does the tip of the minute hand move in 30 minutes? How far does it move in 20 minutes?

73. Angular Speed of a Race Car A race car is driven around a circular track at a constant speed of 180 miles per hour. If the diameter of the track is $\frac{1}{2}$ mile, what is the angular speed of the car? Express your answer in revolutions per hour (which is equivalent to laps per hour).

74. Merry-Go-Rounds A neighborhood carnival has a merry-go-round whose radius is 25 feet. If the time for one revolution is 30 seconds, how fast is the merry-go-round going?

75. **Lighthouse Beacons** The Montauk Point Lighthouse on Long Island has dual beams (two light sources opposite each other). Ships at sea observe a blinking light every 5 seconds. What angular speed is required to do this?

76. Spin Balancing Tires The radius of each wheel of a car is 16 inches. At how many revolutions per minute should a spin balancer be set to balance the tires at a speed of 90 miles per hour? Is the setting different for a wheel of radius 14 inches? What is this setting?

77. Alternating Voltage The electromotive force E, in volts, in a certain ac circuit obeys the equation

$$E(t) = 120\sin(120\pi t), \qquad t \geq 0$$

where t is measured in seconds.
(a) What is the maximum value of E?
(b) What is the period?
(c) Graph the function over two periods.

78. Alternating Current The current I, in amperes, flowing through an ac (alternating current) circuit at time t is

$$I(t) = 220\sin\left(30\pi t + \frac{\pi}{6}\right), \qquad t \geq 0$$

(a) What is the period?
(b) What is the amplitude?
(c) What is the phase shift?
(d) Graph this function over two periods.

79. **Monthly Temperature** The following data represent the average monthly temperatures for Phoenix, Arizona.
(a) Using a graphing utility, draw a scatter diagram of the data for one period.
(b) Find a sinusoidal function of the form $y = A\sin(\omega x - \phi) + B$ that fits the data.
(c) Draw the sinusoidal function found in part (b) on the scatter diagram.
(d) Use a graphing utility to find the sinusoidal function of best fit.
(e) Draw the sinusoidal function of best fit on the scatter diagram.

Month, m	Average Monthly Temperature, T
January, 1	51
February, 2	55
March, 3	63
April, 4	67
May, 5	77
June, 6	86
July, 7	90
August, 8	90
September, 9	84
October, 10	71
November, 11	59
December, 12	52

Source: U.S. National Oceanic and Atmospheric Administration.

80. Monthly Temperature The following data represent the average monthly temperatures for Chicago, Illinois.

Month, *m*	Average Monthly Temperature, T
January, 1	25
February, 2	28
March, 3	36
April, 4	48
May, 5	61
June, 6	72
July, 7	74
August, 8	75
September, 9	66
October, 10	55
November, 11	39
December, 12	28

Source: U.S. National Oceanic and Atmospheric Administration.

(a) Using a graphing utility, draw a scatter diagram of the data for one period.
(b) Find a sinusoidal function of the form $y = A\sin(\omega x - \phi) + B$ that fits the data.
(c) Draw the sinusoidal function found in part (b) on the scatter diagram.
(d) Use a graphing utility to find the sinusoidal function of best fit.
(e) Draw the sinusoidal function of best fit on the scatter diagram.

81. Hours of Daylight According to the *Old Farmer's Almanac*, in Las Vegas, Nevada, the number of hours of sunlight on the day of the summer solstice is 13.367, and the number of hours of sunlight on the day of the winter solstice is 9.667.
(a) Find a sinusoidal function of the form $y = A\sin(\omega x - \phi) + B$ that fits the data.
(b) Draw a graph of the function found in part (a).
(c) Use the function found in part (a) to predict the number of hours of sunlight on April 1.
(d) Look up the number of hours of sunlight on April 1, 2000, in the *Old Farmer's Almanac* and compare the actual hours of daylight to the results found in part (c).

82. Hours of Daylight According to the *Old Farmer's Almanac*, in Seattle, Washington, the number of hours of sunlight on the day of the summer solstice is 13.967, and the number of hours of sunlight on the day of the winter solstice is 8.417.
(a) Find a sinusoidal function of the form $y = A\sin(\omega x - \phi) + B$ that fits the data.
(b) Draw a graph of the function found in part (a).
(c) Use the function found in part (a) to predict the number of hours of sunlight on April 1.
(d) Look up the number of hours of sunlight on April 1, 2000, in the *Old Farmer's Almanac* and compare the actual hours of daylight to the results found in part (c).

PROJECT AT MOTOROLA

Digital Transmission Over the Air

Digital communications is a revolutionary technology of the century. For many years, Motorola has been one of the leading companies to employ digital communication in wireless devices, such as cell phones.

Figure 1 shows a simplified overview of a digital communication transmission over the air. The information source to transmit can be audio, video, or data. The information source may be formatted into a digital sequence of symbols from a finite set $\{\alpha_n\} = \{0, 1\}$. So, 0110100 is an example of a digital sequence. The period of the symbols is denoted by T.

The principal of digital communication systems is that during the finite interval of time T, the information symbol is represented by one digital waveform from a

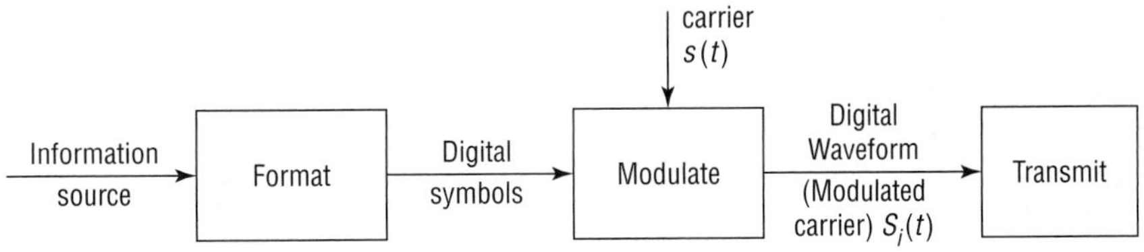

Figure 1 Simplified Overview of a Digital Communication Transmission

finite set of digital waveforms before it is sent. This technique is called **modulation.**

Modulation techniques use a carrier which is modulated by the information to be transmitted. The modulated carrier is transformed into an electromagnetic field and propagated in the air through the antenna. The unmodulated carrier can be represented in its general form by a sinusoidal function $s(t) = A\sin(\omega_0 t + \phi)$, where A is the amplitude, ω_0 is the radian frequency, and ϕ is the phase.

Let's assume $A = 1$, $\phi = 0$ and $\omega_0 = 2\pi f_0$ radian where f_0 is the frequency of the unmodulated carrier. $T_0 = 1/f_0$ is the period of the unmodulated carrier.

1. Write $s(t)$ using these assumptions.
2. What is the period, T_0, of the unmodulated carrier?
3. Evaluate $s(t)$ for $t = 0, 1/(4f_0), 1/(2f_0), 3/(4f_0)$ and $1/f_0$.
4. Graph $s(t)$ for $0 \le t \le 12T_0$. That is, graph 12 cycles of the function.
5. For what values of t does the function reach its maximum value? [HINT: Express t in terms of f_0].

Three modulation techniques are used for transmission over the air: amplitude modulation, frequency modulation and phase modulation. In this study we are interested in phase modulation. Figure 2 illustrates this process. An information symbol is mapped onto a phase which modulates the carrier. The modulated carrier is expressed by $S_i(t) = \sin(2\pi f_0 t + \psi_i)$.

Let's assume the following mapping scheme:

$\{\alpha_n\}$	$\rightarrow$	$\{\psi_n\}$
0		$\psi_0 = 0$
1		$\psi_1 = \pi$

6. Map the binary sequence $M = 010$ into a phase sequence P.
7. What is the expression of the modulated carrier $S_0(t)$ for $\psi_i = \psi_0$ and $S_1(t)$ for $\psi_i = \psi_1$?
8. Let's assume that in the sequence M, the period of each symbol is $T = 4T_0$. For each of the three intervals $[0, 4T_0]$, $[4T_0, 8T_0]$, $[8T_0, 12T_0]$, indicate which of $S_0(t)$ or $S_1(t)$ is the modulated carrier. On the same graph illustrate M, P and the modulated carrier for $0 \le t \le 12T_0$.

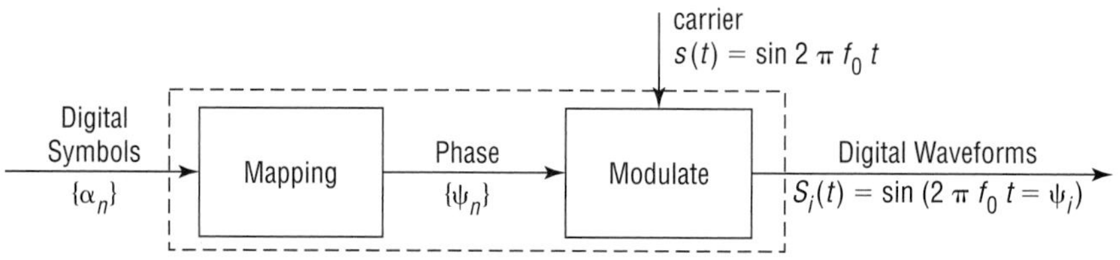

Figure 2 Principle of Phase Modulation

CHAPTER

7 Analytic Trigonometry

Outline

FIELD TRIP AT MOTOROLA

Trigonometry has a surprisingly wide application. Though it begins as the study of the relation between angles and lines, many functions can be represented as a combination of simple trigonometric functions. This immediately finds application to electric fields, radio waves, light, heat, sound, vibrations in mechanical structures, and so on. A good understanding of this subject will lay a foundation for a wide range of practical subjects.

INTERVIEW AT MOTOROLA

George Valliath is a manager in the Motorola Advanced Technology Center, in Schaumburg Illinois. He works in the area of advanced product concepts in Visual Communications, and holds a Ph.D. in Electrical Engineering from the University of Rhode Island, Kingston, Rhode Island. George has four publications and eight issued patents. He holds mathematics as an extremely important subject, both as a tool for analyzing real world problems and as method for training the mind to think logically and clearly.

PREPARING FOR THIS SECTION

Before getting started, review the following:

✓ Fundamental Identities (pp. 403 and 404)

7.1 TRIGONOMETRIC IDENTITIES

OBJECTIVES 1 Establish Identities

We saw in the previous chapter that the trigonometric functions lend themselves to a wide variety of identities. Before establishing some additional identities, let's review the definition of an *identity.*

Two functions f and g are said to be **identically equal** if

$$f(x) = g(x)$$

for every value of x for which both functions are defined. Such an equation is referred to as an **identity.** An equation that is not an identity is called a **conditional equation.**

For example, the following are identities:

$$(x - 1)^2 = x^2 - 2x + 1 \qquad \sin^2 x + \cos^2 x = 1 \qquad \csc x = \frac{1}{\sin x}$$

The following are conditional equations:

$2x + 5 = 0$ True only if $x = -\frac{5}{2}$.

$\sin x = 0$ True only if $x = k\pi$, k an integer.

$\sin x = \cos x$ True only if $x = \frac{\pi}{4} + 2k\pi$ or $x = \frac{5\pi}{4} + 2k\pi$, k an integer.

The following boxes summarize the trigonometric identities that we have established thus far.

Quotient Identities

$$\tan\theta = \frac{\sin\theta}{\cos\theta} \qquad \cot\theta = \frac{\cos\theta}{\sin\theta}$$

Reciprocal Identities

$$\csc\theta = \frac{1}{\sin\theta} \qquad \sec\theta = \frac{1}{\cos\theta} \qquad \cot\theta = \frac{1}{\tan\theta}$$

Pythagorean Identities

$$\sin^2\theta + \cos^2\theta = 1 \qquad \tan^2\theta + 1 = \sec^2\theta$$
$$1 + \cot^2\theta = \csc^2\theta$$

Even–Odd Identities

$$\sin(-\theta) = -\sin\theta \qquad \cos(-\theta) = \cos\theta \qquad \tan(-\theta) = -\tan\theta$$
$$\csc(-\theta) = -\csc\theta \qquad \sec(-\theta) = \sec\theta \qquad \cot(-\theta) = -\cot\theta$$

This list of identities comprises what we shall refer to as the **basic trigonometric identities.** These identities should not merely be memorized, but should be *known* (just as you know your name, rather than have it memorized). In fact, minor variations of a basic identity are often used. For example, we might want to use $\sin^2\theta = 1 - \cos^2\theta$ or $\cos^2\theta = 1 - \sin^2\theta$ instead of $\sin^2\theta + \cos^2\theta = 1$. For this reason, among others, you need to know these relationships and be quite comfortable with variations of them.

1 In the examples that follow, the directions will read "Establish the identity. . . ." As you will see, this is accomplished by starting with one side of the given equation (usually the one containing the more complicated expression) and, using appropriate basic identities and algebraic manipulations, arriving at the other side. The selection of an appropriate basic identity to obtain the desired result is learned only through experience and lots of practice.

EXAMPLE 1 Establishing an Identity

Establish the identity: $\csc\theta \cdot \tan\theta = \sec\theta$

Solution We start with the left side, because it contains the more complicated expression, and apply a reciprocal identity and a quotient identity.

$$\csc\theta \cdot \tan\theta = \frac{1}{\cancel{\sin\theta}} \cdot \frac{\cancel{\sin\theta}}{\cos\theta} = \frac{1}{\cos\theta} = \sec\theta$$

Having arrived at the right side, the identity is established. ■

COMMENT: A graphing utility can be used to provide evidence of an identity. For example, if we graph $Y_1 = \csc\theta \cdot \tan\theta$ and $Y_2 = \sec\theta$, the graphs appear to be the same. This provides evidence that $Y_1 = Y_2$. However, it does not prove their equality. A graphing utility *cannot be used to establish an identity*—identities must be established algebraically.

NOW WORK PROBLEM 1.

EXAMPLE 2 Establishing an Identity

Establish the identity: $\sin^2(-\theta) + \cos^2(-\theta) = 1$

Solution We begin with the left side and apply even–odd identities.

$$\begin{aligned}\sin^2(-\theta) + \cos^2(-\theta) &= [\sin(-\theta)]^2 + [\cos(-\theta)]^2 \\ &= (-\sin\theta)^2 + (\cos\theta)^2 && \text{Even–odd identities.} \\ &= (\sin\theta)^2 + (\cos\theta)^2 \\ &= 1 && \text{Pythagorean identity.}\end{aligned}$$

■

EXAMPLE 3 Establishing an Identity

Establish the identity: $\dfrac{\sin^2(-\theta) - \cos^2(-\theta)}{\sin(-\theta) - \cos(-\theta)} = \cos\theta - \sin\theta$

Solution We begin with two observations: The left side appears to contain the more complicated expression. Also, the left side contains expressions with the argument $-\theta$, whereas the right side contains expressions with the argument θ. We decide, therefore, to start with the left side and apply even–odd identities.

$$\begin{aligned}\frac{\sin^2(-\theta) - \cos^2(-\theta)}{\sin(-\theta) - \cos(-\theta)} &= \frac{[\sin(-\theta)]^2 - [\cos(-\theta)]^2}{\sin(-\theta) - \cos(-\theta)} \\ &= \frac{(-\sin\theta)^2 - (\cos\theta)^2}{-\sin\theta - \cos\theta} && \text{Even–odd identities.} \\ &= \frac{(\sin\theta)^2 - (\cos\theta)^2}{-\sin\theta - \cos\theta} && \text{Simplify.} \\ &= \frac{(\sin\theta - \cos\theta)\cancel{(\sin\theta + \cos\theta)}}{-\cancel{(\sin\theta + \cos\theta)}} && \text{Factor.} \\ &= \cos\theta - \sin\theta && \text{Cancel and simplify.}\end{aligned}$$

■

EXAMPLE 4 Establishing an Identity

Establish the identity: $\dfrac{1 + \tan\theta}{1 + \cot\theta} = \tan\theta$

Solution

$$\frac{1 + \tan\theta}{1 + \cot\theta} = \frac{1 + \tan\theta}{1 + \dfrac{1}{\tan\theta}} = \frac{1 + \tan\theta}{\dfrac{\tan\theta + 1}{\tan\theta}} = \frac{\tan\theta\cancel{(1 + \tan\theta)}}{\cancel{\tan\theta + 1}} = \tan\theta$$

■

NOW WORK PROBLEM 9.

When sums or differences of quotients appear, it is usually best to rewrite them as a single quotient, especially if the other side of the identity consists of only one term.

EXAMPLE 5 Establishing an Identity

Establish the identity: $\dfrac{\sin\theta}{1+\cos\theta} + \dfrac{1+\cos\theta}{\sin\theta} = 2\csc\theta$

Solution The left side is more complicated, so we start with it and proceed to add.

$$\frac{\sin\theta}{1+\cos\theta} + \frac{1+\cos\theta}{\sin\theta} = \frac{\sin^2\theta + (1+\cos\theta)^2}{(1+\cos\theta)(\sin\theta)} \quad \text{Add the quotients.}$$

$$= \frac{\sin^2\theta + 1 + 2\cos\theta + \cos^2\theta}{(1+\cos\theta)(\sin\theta)} \quad \text{Remove parentheses in numerator.}$$

$$= \frac{(\sin^2\theta + \cos^2\theta) + 1 + 2\cos\theta}{(1+\cos\theta)(\sin\theta)} \quad \text{Regroup.}$$

$$= \frac{2 + 2\cos\theta}{(1+\cos\theta)(\sin\theta)} \quad \text{Pythagorean Identity.}$$

$$= \frac{2\cancel{(1+\cos\theta)}}{\cancel{(1+\cos\theta)}(\sin\theta)} \quad \text{Factor and cancel.}$$

$$= \frac{2}{\sin\theta}$$

$$= 2\csc\theta \quad \text{Reciprocal identity.}$$

■

Sometimes it helps to write one side in terms of sines and cosines only.

EXAMPLE 6 Establishing an Identity

Establish the identity: $\dfrac{\tan\theta + \cot\theta}{\sec\theta\csc\theta} = 1$

Solution

$$\frac{\tan\theta + \cot\theta}{\sec\theta\csc\theta} = \frac{\dfrac{\sin\theta}{\cos\theta} + \dfrac{\cos\theta}{\sin\theta}}{\dfrac{1}{\cos\theta}\,\dfrac{1}{\sin\theta}} = \frac{\dfrac{\sin^2\theta + \cos^2\theta}{\cos\theta\sin\theta}}{\dfrac{1}{\cos\theta\sin\theta}}$$

Change to sines and cosines. Add the quotients in the numerator.

$$= \frac{1}{\cos\theta\sin\theta} \cdot \frac{\cos\theta\sin\theta}{1} = 1$$

Divide quotient; $\sin^2\theta + \cos^2\theta = 1$

■

NOW WORK PROBLEM 51.

Sometimes, multiplying the numerator and denominator by an appropriate factor will result in a simplification.

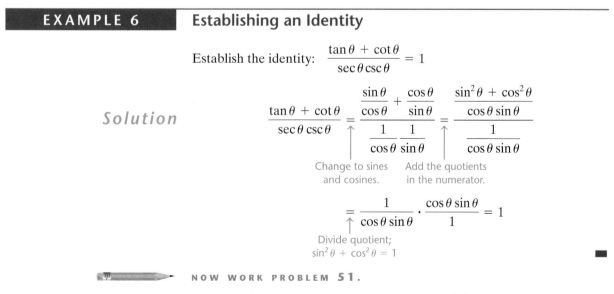

EXAMPLE 7 **Establishing an Identity**

Establish the identity: $\frac{1 - \sin\theta}{\cos\theta} = \frac{\cos\theta}{1 + \sin\theta}$

Solution We start with the left side and multiply the numerator and the denominator by $1 + \sin\theta$. (Alternatively, we could multiply the numerator and denominator of the right side by $1 - \sin\theta$).

$$\frac{1 - \sin\theta}{\cos\theta} = \frac{1 - \sin\theta}{\cos\theta} \cdot \frac{1 + \sin\theta}{1 + \sin\theta} \quad \text{Multiply numerator and denominator by } 1 + \sin\theta.$$

$$= \frac{1 - \sin^2\theta}{\cos\theta(1 + \sin\theta)}$$

$$= \frac{\cos^2\theta}{\cos\theta(1 + \sin\theta)} \quad 1 - \sin^2\theta = \cos^2\theta.$$

$$= \frac{\cos\theta}{1 + \sin\theta} \quad \text{Cancel.}$$

NOW WORK PROBLEM 35.

Although a lot of practice is the only real way to learn how to establish identities, the following guidelines should prove helpful.

GUIDELINES FOR ESTABLISHING IDENTITIES

1. It is almost always preferable to start with the side containing the more complicated expression.
2. Rewrite sums or differences of quotients as a single quotient.
3. Sometimes rewriting one side in terms of sines and cosines only will help.
4. Always keep your goal in mind. As you manipulate one side of the expression, you must keep in mind the form of the expression on the other side.

WARNING: Be careful not to handle identities to be established as if they were conditional equations. You *cannot* establish an identity by such methods as adding the same expression to each side and obtaining a true statement. This practice is not allowed, because the original statement is precisely the one that you are trying to establish. You do not know until it has been established that it is, in fact, true.

7.1 EXERCISES

In Problems 1–80, establish each identity.

1. $\csc\theta \cdot \cos\theta = \cot\theta$

2. $\sec\theta \cdot \sin\theta = \tan\theta$

3. $1 + \tan^2(-\theta) = \sec^2\theta$

4. $1 + \cot^2(-\theta) = \csc^2\theta$

5. $\cos\theta(\tan\theta + \cot\theta) = \csc\theta$

6. $\sin\theta(\cot\theta + \tan\theta) = \sec\theta$

7. $\tan\theta\cot\theta - \cos^2\theta = \sin^2\theta$

8. $\sin\theta\csc\theta - \cos^2\theta = \sin^2\theta$

9. $(\sec\theta - 1)(\sec\theta + 1) = \tan^2\theta$

10. $(\csc\theta - 1)(\csc\theta + 1) = \cot^2\theta$

11. $(\sec\theta + \tan\theta)(\sec\theta - \tan\theta) = 1$

12. $(\csc\theta + \cot\theta)(\csc\theta - \cot\theta) = 1$

13. $\cos^2\theta(1 + \tan^2\theta) = 1$

14. $(1 - \cos^2\theta)(1 + \cot^2\theta) = 1$

15. $(\sin\theta + \cos\theta)^2 + (\sin\theta - \cos\theta)^2 = 2$

16. $\tan^2\theta\cos^2\theta + \cot^2\theta\sin^2\theta = 1$

17. $\sec^4\theta - \sec^2\theta = \tan^4\theta + \tan^2\theta$

18. $\csc^4\theta - \csc^2\theta = \cot^4\theta + \cot^2\theta$

19. $\sec\theta - \tan\theta = \dfrac{\cos\theta}{1 + \sin\theta}$

20. $\csc\theta - \cot\theta = \dfrac{\sin\theta}{1 + \cos\theta}$

21. $3\sin^2\theta + 4\cos^2\theta = 3 + \cos^2\theta$

22. $9\sec^2\theta - 5\tan^2\theta = 5 + 4\sec^2\theta$

23. $1 - \dfrac{\cos^2\theta}{1 + \sin\theta} = \sin\theta$

24. $1 - \dfrac{\sin^2\theta}{1 - \cos\theta} = -\cos\theta$

25. $\dfrac{1 + \tan\theta}{1 - \tan\theta} = \dfrac{\cot\theta + 1}{\cot\theta - 1}$

26. $\dfrac{\csc\theta - 1}{\csc\theta + 1} = \dfrac{1 - \sin\theta}{1 + \sin\theta}$

27. $\dfrac{\sec\theta}{\csc\theta} + \dfrac{\sin\theta}{\cos\theta} = 2\tan\theta$

28. $\dfrac{\csc\theta - 1}{\cot\theta} = \dfrac{\cot\theta}{\csc\theta + 1}$

29. $\dfrac{1 + \sin\theta}{1 - \sin\theta} = \dfrac{\csc\theta + 1}{\csc\theta - 1}$

30. $\dfrac{\cos\theta + 1}{\cos\theta - 1} = \dfrac{1 + \sec\theta}{1 - \sec\theta}$

31. $\dfrac{1 - \sin\theta}{\cos\theta} + \dfrac{\cos\theta}{1 - \sin\theta} = 2\sec\theta$

32. $\dfrac{\cos\theta}{1 + \sin\theta} + \dfrac{1 + \sin\theta}{\cos\theta} = 2\sec\theta$

33. $\dfrac{\sin\theta}{\sin\theta - \cos\theta} = \dfrac{1}{1 - \cot\theta}$

34. $1 - \dfrac{\sin^2\theta}{1 + \cos\theta} = \cos\theta$

35. $\dfrac{1 - \sin\theta}{1 + \sin\theta} = (\sec\theta - \tan\theta)^2$

36. $\dfrac{1 - \cos\theta}{1 + \cos\theta} = (\csc\theta - \cot\theta)^2$

37. $\dfrac{\cos\theta}{1 - \tan\theta} + \dfrac{\sin\theta}{1 - \cot\theta} = \sin\theta + \cos\theta$

38. $\dfrac{\cot\theta}{1 - \tan\theta} + \dfrac{\tan\theta}{1 - \cot\theta} = 1 + \tan\theta + \cot\theta$

39. $\tan\theta + \dfrac{\cos\theta}{1 + \sin\theta} = \sec\theta$

40. $\dfrac{\sin\theta\cos\theta}{\cos^2\theta - \sin^2\theta} = \dfrac{\tan\theta}{1 - \tan^2\theta}$

41. $\dfrac{\tan\theta + \sec\theta - 1}{\tan\theta - \sec\theta + 1} = \tan\theta + \sec\theta$

42. $\dfrac{\sin\theta - \cos\theta + 1}{\sin\theta + \cos\theta - 1} = \dfrac{\sin\theta + 1}{\cos\theta}$

43. $\dfrac{\tan\theta - \cot\theta}{\tan\theta + \cot\theta} = \sin^2\theta - \cos^2\theta$

44. $\dfrac{\sec\theta - \cos\theta}{\sec\theta + \cos\theta} = \dfrac{\sin^2\theta}{1 + \cos^2\theta}$

45. $\dfrac{\tan\theta - \cot\theta}{\tan\theta + \cot\theta} + 1 = 2\sin^2\theta$

46. $\dfrac{\tan\theta - \cot\theta}{\tan\theta + \cot\theta} + 2\cos^2\theta = 1$

47. $\dfrac{\sec\theta + \tan\theta}{\cot\theta + \cos\theta} = \tan\theta\sec\theta$

48. $\dfrac{\sec\theta}{1 + \sec\theta} = \dfrac{1 - \cos\theta}{\sin^2\theta}$

49. $\dfrac{1 - \tan^2\theta}{1 + \tan^2\theta} + 1 = 2\cos^2\theta$

50. $\dfrac{1 - \cot^2\theta}{1 + \cot^2\theta} + 2\cos^2\theta = 1$

51. $\dfrac{\sec\theta - \csc\theta}{\sec\theta\csc\theta} = \sin\theta - \cos\theta$

52. $\dfrac{\sin^2\theta - \tan\theta}{\cos^2\theta - \cot\theta} = \tan^2\theta$

53. $\sec\theta - \cos\theta - \sin\theta\tan\theta = 0$

54. $\tan\theta + \cot\theta - \sec\theta\csc\theta = 0$

55. $\dfrac{1}{1 - \sin\theta} + \dfrac{1}{1 + \sin\theta} = 2\sec^2\theta$

56. $\dfrac{1 + \sin\theta}{1 - \sin\theta} - \dfrac{1 - \sin\theta}{1 + \sin\theta} = 4\tan\theta\sec\theta$

57. $\dfrac{\sec\theta}{1 - \sin\theta} = \dfrac{1 + \sin\theta}{\cos^3\theta}$

58. $\dfrac{1 - \sin\theta}{1 + \sin\theta} = (\sec\theta - \tan\theta)^2$

59. $\dfrac{(\sec\theta - \tan\theta)^2 + 1}{\csc\theta(\sec\theta - \tan\theta)} = 2\tan\theta$

60. $\dfrac{\sec^2\theta - \tan^2\theta + \tan\theta}{\sec\theta} = \sin\theta + \cos\theta$

61. $\dfrac{\sin\theta + \cos\theta}{\cos\theta} - \dfrac{\sin\theta - \cos\theta}{\sin\theta} = \sec\theta\csc\theta$

62. $\dfrac{\sin\theta + \cos\theta}{\sin\theta} - \dfrac{\cos\theta - \sin\theta}{\cos\theta} = \sec\theta\csc\theta$

63. $\dfrac{\sin^3\theta + \cos^3\theta}{\sin\theta + \cos\theta} = 1 - \sin\theta\cos\theta$

64. $\dfrac{\sin^3\theta + \cos^3\theta}{1 - 2\cos^2\theta} = \dfrac{\sec\theta - \sin\theta}{\tan\theta - 1}$

65. $\dfrac{\cos^2\theta - \sin^2\theta}{1 - \tan^2\theta} = \cos^2\theta$

66. $\dfrac{\cos\theta + \sin\theta - \sin^3\theta}{\sin\theta} = \cot\theta + \cos^2\theta$

67. $\dfrac{(2\cos^2\theta - 1)^2}{\cos^4\theta - \sin^4\theta} = 1 - 2\sin^2\theta$

68. $\dfrac{1 - 2\cos^2\theta}{\sin\theta\cos\theta} = \tan\theta - \cot\theta$

69. $\dfrac{1 + \sin\theta + \cos\theta}{1 + \sin\theta - \cos\theta} = \dfrac{1 + \cos\theta}{\sin\theta}$

70. $\dfrac{1 + \cos\theta + \sin\theta}{1 + \cos\theta - \sin\theta} = \sec\theta + \tan\theta$

71. $(a\sin\theta + b\cos\theta)^2 + (a\cos\theta - b\sin\theta)^2 = a^2 + b^2$

72. $(2a\sin\theta\cos\theta)^2 + a^2(\cos^2\theta - \sin^2\theta)^2 = a^2$

73. $\dfrac{\tan\alpha + \tan\beta}{\cot\alpha + \cot\beta} = \tan\alpha\tan\beta$

74. $(\tan\alpha + \tan\beta)(1 - \cot\alpha\cot\beta) + (\cot\alpha + \cot\beta)(1 - \tan\alpha\tan\beta) = 0$

75. $(\sin\alpha + \cos\beta)^2 + (\cos\beta + \sin\alpha)(\cos\beta - \sin\alpha) = 2\cos\beta(\sin\alpha + \cos\beta)$

76. $(\sin\alpha - \cos\beta)^2 + (\cos\beta + \sin\alpha)(\cos\beta - \sin\alpha) = -2\cos\beta(\sin\alpha - \cos\beta)$

77. $\ln|\sec\theta| = -\ln|\cos\theta|$

78. $\ln|\tan\theta| = \ln|\sin\theta| - \ln|\cos\theta|$

79. $\ln|1 + \cos\theta| + \ln|1 - \cos\theta| = 2\ln|\sin\theta|$

80. $\ln|\sec\theta + \tan\theta| + \ln|\sec\theta - \tan\theta| = 0$

81. Write a few paragraphs outlining your strategy for establishing identities.

PREPARING FOR THIS SECTION

Before getting started, review the following:

✓ Distance Formula (p. 6)

✓ Values of the Trigonometric Functions of Certain Angles (pp. 384 and 387)

7.2 SUM AND DIFFERENCE FORMULAS

OBJECTIVES 1 Use Sum and Difference Formulas to Find Exact Values

2 Use Sum and Difference Formulas to Establish Identities

In this section, we continue our derivation of trigonometric identities by obtaining formulas that involve the sum or difference of two angles, such as $\cos(\alpha + \beta)$, $\cos(\alpha - \beta)$, or $\sin(\alpha + \beta)$. These formulas are referred to as the **sum and difference formulas.** We begin with the formulas for $\cos(\alpha + \beta)$ and $\cos(\alpha - \beta)$.

Theorem **Sum and Difference Formulas for Cosines**

$$\cos(\alpha + \beta) = \cos\alpha\cos\beta - \sin\alpha\sin\beta \qquad \textbf{(1)}$$

$$\cos(\alpha - \beta) = \cos\alpha\cos\beta + \sin\alpha\sin\beta \qquad \textbf{(2)}$$

In words, formula (1) states that the cosine of the sum of two angles equals the cosine of the first times the cosine of the second minus the sine of the first times the sine of the second.

Proof We will prove formula (2) first. Although this formula is true for all numbers α and β, we shall assume in our proof that $0 < \beta < \alpha < 2\pi$. We begin with a circle with center at the origin $(0, 0)$ and radius of 1 unit (the unit circle), and we place the angles α and β in standard position, as shown in Figure 1(a). The point $P_1 = (x_1, y_1)$ lies on the terminal side of β, and the point $P_2 = (x_2, y_2)$ lies on the terminal side of α.

Figure 1

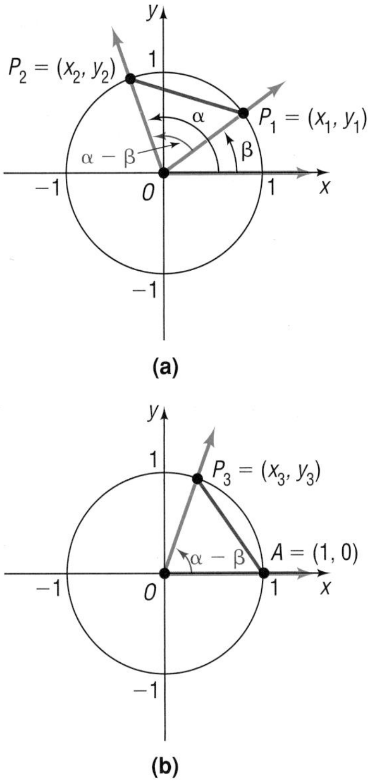

Now, place the angle $\alpha - \beta$ in standard position, as shown in Figure 1(b), where the point A has coordinates $(1, 0)$ and the point $P_3 = (x_3, y_3)$ is on the terminal side of the angle $\alpha - \beta$.

Looking at triangle OP_1P_2 in Figure 1(a) and triangle OAP_3 in Figure 1(b), we see that these triangles are congruent. (Do you see why? Two sides and the included angle, $\alpha - \beta$, are equal.) Hence, the unknown side of each triangle must be equal; that is,

$$d(A, P_3) = d(P_1, P_2)$$

Using the distance formula, we find that

$$\sqrt{(x_3 - 1)^2 + y_3^2} = \sqrt{(x_2 - x_1)^2 + (y_2 - y_1)^2}$$

$$(x_3 - 1)^2 + y_3^2 = (x_2 - x_1)^2 + (y_2 - y_1)^2 \quad \text{Square each side.}$$

$$x_3^2 - 2x_3 + 1 + y_3^2 = x_2^2 - 2x_1x_2 + x_1^2 + y_2^2 - 2y_1y_2 + y_1^2 \quad \textbf{(3)}$$

Since $P_1 = (x_1, y_1)$, $P_2 = (x_2, y_2)$, and $P_3 = (x_3, y_3)$ are points on the unit circle $x^2 + y^2 = 1$, it follows that

$$x_1^2 + y_1^2 = 1 \qquad x_2^2 + y_2^2 = 1 \qquad x_3^2 + y_3^2 = 1$$

Consequently, equation (3) simplifies to

$$x_3^2 + y_3^2 - 2x_3 + 1 = (x_2^2 + y_2^2) + (x_1^2 + y_1^2) - 2x_1x_2 - 2y_1y_2$$

$$1 - 2x_3 + 1 = 1 + 1 - 2x_1x_2 - 2y_1y_2$$

$$2 - 2x_3 = 2 - 2x_1x_2 - 2y_1y_2$$

$$x_3 = x_1x_2 + y_1y_2 \quad \textbf{(4)}$$

Since $P_1 = (x_1, y_1)$ is a point on the unit circle that is on the terminal side of angle β, it follows by definition that

$$\sin\beta = y_1 \qquad \cos\beta = x_1 \quad \textbf{(5)}$$

Similarly,

$$\sin\alpha = y_2 \qquad \cos\alpha = x_2 \qquad \cos(\alpha - \beta) = x_3 \quad \textbf{(6)}$$

Using equations (5) and (6) in equation (4), we get

$$\cos(\alpha - \beta) = \cos\alpha\cos\beta + \sin\alpha\sin\beta$$

which is formula (2).

The proof of formula (1) follows from formula (2). We use the fact that $\alpha + \beta = \alpha - (-\beta)$. Then

$$\cos(\alpha + \beta) = \cos[\alpha - (-\beta)]$$

$$= \cos\alpha\cos(-\beta) + \sin\alpha\sin(-\beta) \quad \text{Use formula (2)}$$

$$= \cos\alpha\cos\beta - \sin\alpha\sin\beta \quad \text{Even–odd identities.} \ \blacksquare$$

1 One use of formulas (1) and (2) is to obtain the exact value of the cosine of an angle that can be expressed as the sum or difference of angles whose sine and cosine are known exactly.

EXAMPLE 1 Using the Sum Formula to Find Exact Values

Find the exact value of $\cos 75°$.

Solution Since $75° = 45° + 30°$, we use formula (1) to obtain

$$\cos 75° = \cos(45° + 30°) = \cos 45° \cos 30° - \sin 45° \sin 30°$$

↑ Formula (1)

$$= \frac{\sqrt{2}}{2} \cdot \frac{\sqrt{3}}{2} - \frac{\sqrt{2}}{2} \cdot \frac{1}{2} = \frac{1}{4}(\sqrt{6} - \sqrt{2})$$

EXAMPLE 2 Using the Difference Formula to Find Exact Values

Find the exact value of $\cos(\pi/12)$.

Solution

$$\cos\frac{\pi}{12} = \cos\left(\frac{3\pi}{12} - \frac{2\pi}{12}\right) = \cos\left(\frac{\pi}{4} - \frac{\pi}{6}\right)$$

$$= \cos\frac{\pi}{4}\cos\frac{\pi}{6} + \sin\frac{\pi}{4}\sin\frac{\pi}{6} \quad \text{Use formula (2).}$$

$$= \frac{\sqrt{2}}{2} \cdot \frac{\sqrt{3}}{2} + \frac{\sqrt{2}}{2} \cdot \frac{1}{2} = \frac{1}{4}(\sqrt{6} + \sqrt{2})$$

NOW WORK PROBLEM 3.

Another use of formulas (1) and (2) is to establish other identities. One important pair of identities is given next.

$$\cos\left(\frac{\pi}{2} - \theta\right) = \sin\theta \quad \textbf{(7a)}$$

$$\sin\left(\frac{\pi}{2} - \theta\right) = \cos\theta \quad \textbf{(7b)}$$

Seeing the Concept Graph $Y_1 = \cos(\pi/2 - \theta)$ and $Y_2 = \sin\theta$ in radian mode on the same screen. Does this support the result 7(a)? How would you support the result 7(b)?

Proof To prove formula (7a), we use the formula for $\cos(\alpha - \beta)$ with $\alpha = \pi/2$ and $\beta = \theta$.

$$\cos\left(\frac{\pi}{2} - \theta\right) = \cos\frac{\pi}{2}\cos\theta + \sin\frac{\pi}{2}\sin\theta$$

$$= 0 \cdot \cos\theta + 1 \cdot \sin\theta$$

$$= \sin\theta$$

To prove formula (7b), we make use of the identity (7a) just established.

$$\sin\left(\frac{\pi}{2} - \theta\right) = \cos\left[\frac{\pi}{2} - \left(\frac{\pi}{2} - \theta\right)\right] = \cos\theta$$

↑ Use (7a).

Also, since

$$\cos\left(\frac{\pi}{2} - \theta\right) = \cos\left[-\left(\theta - \frac{\pi}{2}\right)\right] \underset{\substack{\uparrow \\ \text{Even Property} \\ \text{of Cosine}}}{=} \cos\left(\theta - \frac{\pi}{2}\right)$$

and

$$\cos\left(\frac{\pi}{2} - \theta\right) \underset{\substack{\uparrow \\ 7(a)}}{=} \sin\theta$$

it follows that $\cos(\theta - \pi/2) = \sin\theta$. Thus, the graphs of $y = \cos(\theta - \pi/2)$ and $y = \sin\theta$ are identical, a fact that we conjectured earlier in Section 6.4.

NOW WORK PROBLEM 31.

FORMULAS FOR $\sin(\alpha + \beta)$ AND $\sin(\alpha - \beta)$

Having established the identities in formulas (7a) and (7b), we now can derive the sum and difference formulas for $\sin(\alpha + \beta)$ and $\sin(\alpha - \beta)$.

Proof

$$\sin(\alpha + \beta) = \cos\left[\frac{\pi}{2} - (\alpha + \beta)\right] \quad \text{Formula (7a).}$$

$$= \cos\left[\left(\frac{\pi}{2} - \alpha\right) - \beta\right]$$

$$= \cos\left(\frac{\pi}{2} - \alpha\right)\cos\beta + \sin\left(\frac{\pi}{2} - \alpha\right)\sin\beta \quad \text{Formula (2).}$$

$$= \sin\alpha\cos\beta + \cos\alpha\sin\beta \quad \text{Formulas (7a) and (7b).}$$

$$\sin(\alpha - \beta) = \sin[\alpha + (-\beta)]$$

$$= \sin\alpha\cos(-\beta) + \cos\alpha\sin(-\beta)$$

$$= \sin\alpha\cos\beta + \cos\alpha(-\sin\beta) \quad \text{Even–odd identities.}$$

$$= \sin\alpha\cos\beta - \cos\alpha\sin\beta$$

■

The following results have been proved:

Theorem **Sum and Difference Formulas for Sines**

$$\sin(\alpha + \beta) = \sin\alpha\cos\beta + \cos\alpha\sin\beta \quad \textbf{(8)}$$

$$\sin(\alpha - \beta) = \sin\alpha\cos\beta - \cos\alpha\sin\beta \quad \textbf{(9)}$$

■

In words, formula (8) states that the sine of the sum of two angles equals the sine of the first times the cosine of the second plus the cosine of the first times the sine of the second.

EXAMPLE 3 Using the Sum Formula to Find Exact Values

Find the exact value of $\sin(7\pi/12)$.

Solution

$$\sin\frac{7\pi}{12} = \sin\left(\frac{3\pi}{12} + \frac{4\pi}{12}\right) = \sin\left(\frac{\pi}{4} + \frac{\pi}{3}\right)$$

$$= \sin\frac{\pi}{4}\cos\frac{\pi}{3} + \cos\frac{\pi}{4}\sin\frac{\pi}{3} \quad \text{Formula (8).}$$

$$= \frac{\sqrt{2}}{2}\cdot\frac{1}{2} + \frac{\sqrt{2}}{2}\cdot\frac{\sqrt{3}}{2} = \frac{1}{4}(\sqrt{2} + \sqrt{6})$$

NOW WORK PROBLEM 9.

EXAMPLE 4 Using the Difference Formula to Find Exact Values

Find the exact value of $\sin 80^\circ \cos 20^\circ - \cos 80^\circ \sin 20^\circ$.

Solution The form of the expression $\sin 80^\circ \cos 20^\circ - \cos 80^\circ \sin 20^\circ$ is that of the right side of the formula for $\sin(\alpha - \beta)$ with $\alpha = 80^\circ$ and $\beta = 20^\circ$. Thus,

$$\sin 80^\circ \cos 20^\circ - \cos 80^\circ \sin 20^\circ = \sin(80^\circ - 20^\circ) = \sin 60^\circ = \frac{\sqrt{3}}{2}$$

NOW WORK PROBLEM 19.

EXAMPLE 5 Finding Exact Values

If it is known that $\sin\alpha = \frac{4}{5}$, $\pi/2 < \alpha < \pi$, and that $\sin\beta = -2/\sqrt{5} = -2\sqrt{5}/5$, $\pi < \beta < 3\pi/2$, find the exact value of

(a) $\cos\alpha$ (b) $\cos\beta$ (c) $\cos(\alpha + \beta)$ (d) $\sin(\alpha + \beta)$

Solution (a) Since $\sin\alpha = \frac{4}{5} = \frac{y}{r}$ and $\frac{\pi}{2} < \alpha < \pi$, we let $y = 4$ and $r = 5$. See Figure 2. Since $(x, 4)$ is on the circle $x^2 + y^2 = 25$ and is in Quadrant II, we have

Figure 2
Given $\sin\alpha = \frac{4}{5}$, $\pi/2 < \alpha < \pi$

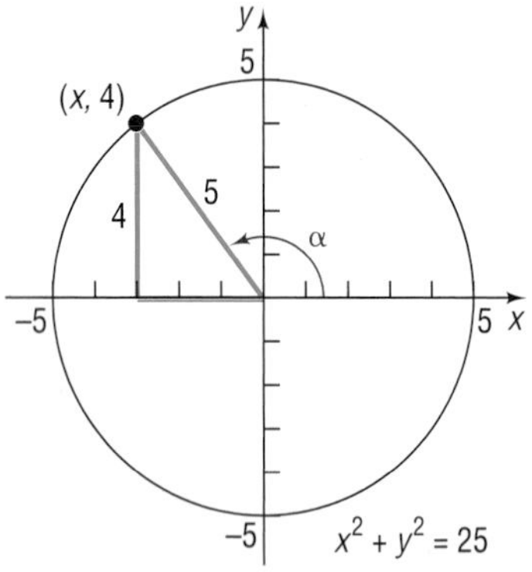

$$x^2 + 4^2 = 25, \qquad x < 0$$

$$x^2 = 25 - 16 = 9$$

$$x = -3$$

We now determine the value of $\cos\alpha$.

$$\cos\alpha = \frac{x}{r} = -\frac{3}{5}$$

Alternatively, we can find $\cos\alpha$ using identities, as follows:

$$\cos\alpha = -\sqrt{1 - \sin^2\alpha} = -\sqrt{1 - \frac{16}{25}} = -\sqrt{\frac{9}{25}} = -\frac{3}{5}$$

↑ α in quadrant II, $\cos\alpha < 0$

Figure 3
Given $\sin\beta = -2/\sqrt{5}$, $\pi < \beta < 3\pi/2$

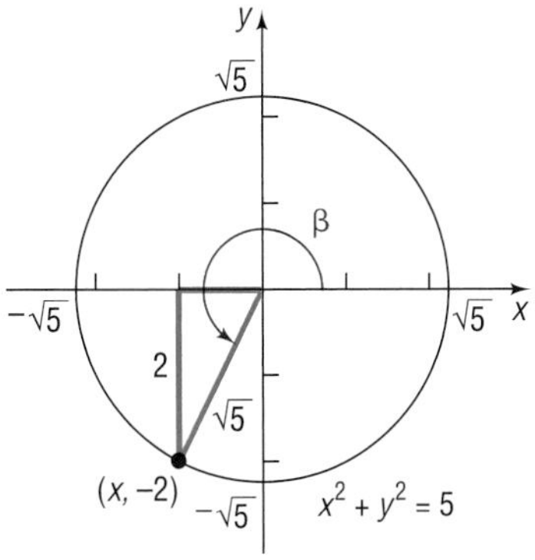

(b) Since $\sin\beta = -2/\sqrt{5} = y/r$ and $\pi < \beta < 3\pi/2$, we let $y = -2$ and $r = \sqrt{5}$. See Figure 3. Since $(x, -2)$ is on the circle $x^2 + y^2 = 5$ and is in Quadrant III, we have

$$x^2 + (-2)^2 = 5, \qquad x < 0$$
$$x^2 = 5 - 4 = 1$$
$$x = -1$$

We now determine the value of $\cos\beta$.

$$\cos\beta = \frac{x}{r} = -\frac{1}{\sqrt{5}} = -\frac{\sqrt{5}}{5}$$

Alternatively, we can find $\cos\beta$ using identities, as follows:

$$\cos\beta = -\sqrt{1 - \sin^2\beta} = -\sqrt{1 - \frac{4}{5}} = -\sqrt{\frac{1}{5}} = -\frac{\sqrt{5}}{5}$$

(c) Using the results found in parts (a) and (b) and formula (1), we have

$$\cos(\alpha + \beta) = \cos\alpha\cos\beta - \sin\alpha\sin\beta$$
$$= -\frac{3}{5}\left(-\frac{\sqrt{5}}{5}\right) - \frac{4}{5}\left(-\frac{2\sqrt{5}}{5}\right) = \frac{11\sqrt{5}}{25}$$

(d) $\sin(\alpha + \beta) = \sin\alpha\cos\beta + \cos\alpha\sin\beta$

$$= \frac{4}{5}\left(-\frac{\sqrt{5}}{5}\right) + \left(-\frac{3}{5}\right)\left(-\frac{2\sqrt{5}}{5}\right) = \frac{2\sqrt{5}}{25}$$ ■

NOW WORK PROBLEMS 23 (a), (b), AND (c).

EXAMPLE 6 Establishing an Identity

2 Establish the identity: $\dfrac{\cos(\alpha - \beta)}{\sin\alpha\sin\beta} = \cot\alpha\cot\beta + 1$

Solution

$$\frac{\cos(\alpha - \beta)}{\sin\alpha\sin\beta} = \frac{\cos\alpha\cos\beta + \sin\alpha\sin\beta}{\sin\alpha\sin\beta}$$
$$= \frac{\cos\alpha\cos\beta}{\sin\alpha\sin\beta} + \frac{\sin\alpha\sin\beta}{\sin\alpha\sin\beta}$$
$$= \frac{\cos\alpha}{\sin\alpha}\,\frac{\cos\beta}{\sin\beta} + 1$$
$$= \cot\alpha\cot\beta + 1$$ ■

NOW WORK PROBLEM 43.

FORMULAS FOR $\tan(\alpha + \beta)$ AND $\tan(\alpha - \beta)$

We use the identity $\tan\theta = (\sin\theta)/(\cos\theta)$ and the sum formulas for $\sin(\alpha + \beta)$ and $\cos(\alpha + \beta)$ to derive a formula for $\tan(\alpha + \beta)$.

Proof $\quad \tan(\alpha + \beta) = \dfrac{\sin(\alpha + \beta)}{\cos(\alpha + \beta)} = \dfrac{\sin\alpha\cos\beta + \cos\alpha\sin\beta}{\cos\alpha\cos\beta - \sin\alpha\sin\beta}$

Now we divide the numerator and denominator by $\cos\alpha\cos\beta$.

$$\tan(\alpha+\beta) = \frac{\dfrac{\sin\alpha\cos\beta + \cos\alpha\sin\beta}{\cos\alpha\cos\beta}}{\dfrac{\cos\alpha\cos\beta - \sin\alpha\sin\beta}{\cos\alpha\cos\beta}} = \frac{\dfrac{\sin\alpha\cancel{\cos\beta}}{\cos\alpha\cancel{\cos\beta}} + \dfrac{\cancel{\cos\alpha}\sin\beta}{\cancel{\cos\alpha}\cos\beta}}{\dfrac{\cancel{\cos\alpha}\cancel{\cos\beta}}{\cancel{\cos\alpha}\cancel{\cos\beta}} - \dfrac{\sin\alpha\sin\beta}{\cos\alpha\cos\beta}}$$

$$= \frac{\dfrac{\sin\alpha}{\cos\alpha} + \dfrac{\sin\beta}{\cos\beta}}{1 - \dfrac{\sin\alpha\sin\beta}{\cos\alpha\cos\beta}} = \frac{\tan\alpha + \tan\beta}{1 - \tan\alpha\tan\beta}$$

We use the sum formula for $\tan(\alpha+\beta)$ and even–odd properties to get the difference formula.

$$\tan(\alpha-\beta) = \tan[\alpha + (-\beta)] = \frac{\tan\alpha + \tan(-\beta)}{1 - \tan\alpha\tan(-\beta)} = \frac{\tan\alpha - \tan\beta}{1 + \tan\alpha\tan\beta}$$ ■

We have proved the following results:

Theorem **Sum and Difference Formulas for Tangents**

$$\tan(\alpha+\beta) = \frac{\tan\alpha + \tan\beta}{1 - \tan\alpha\tan\beta} \quad \textbf{(10)}$$

$$\tan(\alpha-\beta) = \frac{\tan\alpha - \tan\beta}{1 + \tan\alpha\tan\beta} \quad \textbf{(11)}$$

■

In words, formula (10) states that the tangent of the sum of two angles equals the tangent of the first plus the tangent of the second divided by 1 minus their product.

EXAMPLE 7 Establishing an Identity

Prove the identity: $\tan(\theta+\pi) = \tan\theta$

Solution

$$\tan(\theta+\pi) = \frac{\tan\theta + \tan\pi}{1 - \tan\theta\tan\pi} = \frac{\tan\theta + 0}{1 - \tan\theta\cdot 0} = \tan\theta$$ ■

The result obtained in Example 7 verifies that the tangent function is periodic with period π, a fact that we mentioned earlier.

WARNING: Be careful when using formulas (10) and (11). These formulas can be used only for angles α and β for which $\tan\alpha$ and $\tan\beta$ are defined, that is, all angles except odd multiples of $\pi/2$. ■

EXAMPLE 8 Establishing an Identity

Prove the identity: $\tan\left(\theta + \frac{\pi}{2}\right) = -\cot\theta$

Solution We cannot use formula (10), since $\tan(\pi/2)$ is not defined. Instead, we proceed as follows:

$$\tan\left(\theta + \frac{\pi}{2}\right) = \frac{\sin\left(\theta + \frac{\pi}{2}\right)}{\cos\left(\theta + \frac{\pi}{2}\right)} = \frac{\sin\theta\cos\frac{\pi}{2} + \cos\theta\sin\frac{\pi}{2}}{\cos\theta\cos\frac{\pi}{2} - \sin\theta\sin\frac{\pi}{2}}$$

$$= \frac{(\sin\theta)(0) + (\cos\theta)(1)}{(\cos\theta)(0) - (\sin\theta)(1)} = \frac{\cos\theta}{-\sin\theta} = -\cot\theta$$

SUMMARY

The following box summarizes the sum and difference formulas.

Sum and Difference Formulas

$$\cos(\alpha + \beta) = \cos\alpha\cos\beta - \sin\alpha\sin\beta \qquad \cos(\alpha - \beta) = \cos\alpha\cos\beta + \sin\alpha\sin\beta$$

$$\sin(\alpha + \beta) = \sin\alpha\cos\beta + \cos\alpha\sin\beta \qquad \sin(\alpha - \beta) = \sin\alpha\cos\beta - \cos\alpha\sin\beta$$

$$\tan(\alpha + \beta) = \frac{\tan\alpha + \tan\beta}{1 - \tan\alpha\tan\beta} \qquad \tan(\alpha - \beta) = \frac{\tan\alpha - \tan\beta}{1 + \tan\alpha\tan\beta}$$

7.2 EXERCISES

In Problems 1–12, find the exact value of each trigonometric function.

1. $\sin\frac{5\pi}{12}$ **2.** $\sin\frac{\pi}{12}$ 3. $\cos\frac{7\pi}{12}$ **4.** $\tan\frac{7\pi}{12}$

5. $\cos 165°$ **6.** $\sin 105°$ **7.** $\tan 15°$ **8.** $\tan 195°$

9. $\sin\frac{17\pi}{12}$ **10.** $\tan\frac{19\pi}{12}$ **11.** $\sec\left(-\frac{\pi}{12}\right)$ **12.** $\cot\left(-\frac{5\pi}{12}\right)$

In Problems 13–22, find the exact value of each expression.

13. $\sin 20° \cos 10° + \cos 20° \sin 10°$ **14.** $\sin 20° \cos 80° - \cos 20° \sin 80°$

15. $\cos 70° \cos 20° - \sin 70° \sin 20°$ **16.** $\cos 40° \cos 10° + \sin 40° \sin 10°$

17. $\dfrac{\tan 20° + \tan 25°}{1 - \tan 20° \tan 25°}$

18. $\dfrac{\tan 40° - \tan 10°}{1 + \tan 40° \tan 10°}$

19. $\sin \dfrac{\pi}{12} \cos \dfrac{7\pi}{12} - \cos \dfrac{\pi}{12} \sin \dfrac{7\pi}{12}$

20. $\cos \dfrac{5\pi}{12} \cos \dfrac{7\pi}{12} - \sin \dfrac{5\pi}{12} \sin \dfrac{7\pi}{12}$

21. $\cos \dfrac{\pi}{12} \cos \dfrac{5\pi}{12} + \sin \dfrac{5\pi}{12} \sin \dfrac{\pi}{12}$

22. $\sin \dfrac{\pi}{18} \cos \dfrac{5\pi}{18} + \cos \dfrac{\pi}{18} \sin \dfrac{5\pi}{18}$

In Problems 23–28, find the exact value of each of the following under the given conditions:

(a) $\sin(\alpha + \beta)$ *(b)* $\cos(\alpha + \beta)$ *(c)* $\sin(\alpha - \beta)$ *(d)* $\tan(\alpha - \beta)$

23. $\sin \alpha = \frac{3}{5}, \quad 0 < \alpha < \pi/2; \quad \cos \beta = 2\sqrt{5}/5, \quad -\pi/2 < \beta < 0$

24. $\cos \alpha = \sqrt{5}/5, \quad 0 < \alpha < \pi/2; \quad \sin \beta = -\frac{4}{5}, \quad -\pi/2 < \beta < 0$

25. $\tan \alpha = -\frac{4}{3}, \quad \pi/2 < \alpha < \pi; \quad \cos \beta = \frac{1}{2}, \quad 0 < \beta < \pi/2$

26. $\tan \alpha = \frac{5}{12}, \quad \pi < \alpha < 3\pi/2; \quad \sin \beta = -\frac{1}{2}, \quad \pi < \beta < 3\pi/2$

27. $\sin \alpha = \frac{5}{13}, \quad -3\pi/2 < \alpha < -\pi; \quad \tan \beta = -\sqrt{3}, \quad \pi/2 < \beta < \pi$

28. $\cos \alpha = \frac{1}{2}, \quad -\pi/2 < \alpha < 0; \quad \sin \beta = \frac{1}{3}, \quad 0 < \beta < \pi/2$

29. If $\sin \theta = \frac{1}{3}$, θ in quadrant II, find the exact value of:

(a) $\cos \theta$ (b) $\sin\left(\theta + \frac{\pi}{6}\right)$ (c) $\cos\left(\theta - \frac{\pi}{3}\right)$ (d) $\tan\left(\theta + \frac{\pi}{4}\right)$

30 If $\cos \theta = \frac{1}{4}$, θ in quadrant IV, find the exact value of:

(a) $\sin \theta$ (b) $\sin\left(\theta - \frac{\pi}{6}\right)$ (c) $\cos\left(\theta + \frac{\pi}{3}\right)$ (d) $\tan\left(\theta - \frac{\pi}{4}\right)$

In Problems 31–56, establish each identity.

31. $\sin\left(\dfrac{\pi}{2} + \theta\right) = \cos \theta$

32. $\cos\left(\dfrac{\pi}{2} + \theta\right) = -\sin \theta$

33. $\sin(\pi - \theta) = \sin \theta$

34. $\cos(\pi - \theta) = -\cos \theta$

35. $\sin(\pi + \theta) = -\sin \theta$

36. $\cos(\pi + \theta) = -\cos \theta$

37. $\tan(\pi - \theta) = -\tan \theta$

38. $\tan(2\pi - \theta) = -\tan \theta$

39. $\sin\left(\dfrac{3\pi}{2} + \theta\right) = -\cos \theta$

40. $\cos\left(\dfrac{3\pi}{2} + \theta\right) = \sin \theta$

41. $\sin(\alpha + \beta) + \sin(\alpha - \beta) = 2 \sin \alpha \cos \beta$

42. $\cos(\alpha + \beta) + \cos(\alpha - \beta) = 2 \cos \alpha \cos \beta$

43. $\dfrac{\sin(\alpha + \beta)}{\sin \alpha \cos \beta} = 1 + \cot \alpha \tan \beta$

44. $\dfrac{\sin(\alpha + \beta)}{\cos \alpha \cos \beta} = \tan \alpha + \tan \beta$

45. $\dfrac{\cos(\alpha + \beta)}{\cos \alpha \cos \beta} = 1 - \tan \alpha \tan \beta$

46. $\dfrac{\cos(\alpha - \beta)}{\sin \alpha \cos \beta} = \cot \alpha + \tan \beta$

47. $\dfrac{\sin(\alpha + \beta)}{\sin(\alpha - \beta)} = \dfrac{\tan \alpha + \tan \beta}{\tan \alpha - \tan \beta}$

48. $\dfrac{\cos(\alpha + \beta)}{\cos(\alpha - \beta)} = \dfrac{1 - \tan \alpha \tan \beta}{1 + \tan \alpha \tan \beta}$

49. $\cot(\alpha + \beta) = \dfrac{\cot \alpha \cot \beta - 1}{\cot \beta + \cot \alpha}$

50. $\cot(\alpha - \beta) = \dfrac{\cot \alpha \cot \beta + 1}{\cot \beta - \cot \alpha}$

51. $\sec(\alpha + \beta) = \dfrac{\csc \alpha \csc \beta}{\cot \alpha \cot \beta - 1}$

52. $\sec(\alpha - \beta) = \dfrac{\sec \alpha \sec \beta}{1 + \tan \alpha \tan \beta}$

53. $\sin(\alpha - \beta) \sin(\alpha + \beta) = \sin^2 \alpha - \sin^2 \beta$

54. $\cos(\alpha - \beta) \cos(\alpha + \beta) = \cos^2 \alpha - \sin^2 \beta$

55. $\sin(\theta + k\pi) = (-1)^k \sin \theta$, k any integer

56. $\cos(\theta + k\pi) = (-1)^k \cos \theta$, k any integer

57. Calculus Show that the difference quotient for $f(x) = \sin x$ is given by

$$\frac{f(x+h)-f(x)}{h} = \frac{\sin(x+h)-\sin x}{h} = \cos x \cdot \frac{\sin h}{h} - \sin x \cdot \frac{1-\cos h}{h}$$

58. Calculus Show that the difference quotient for $f(x) = \cos x$ is given by

$$\frac{f(x+h)-f(x)}{h} = \frac{\cos(x+h)-\cos x}{h} = -\sin x \cdot \frac{\sin h}{h} - \cos x \cdot \frac{1-\cos h}{h}$$

59. Explain why formula (11) cannot be used to show that

$$\tan\left(\frac{\pi}{2}-\theta\right) = \cot\theta$$

Establish this identity by using formulas (7a) and (7b).

60. If $\tan\alpha = x+1$ and $\tan\beta = x-1$, show that $2\cot(\alpha-\beta) = x^2$.

61. Geometry: Angle between Two Lines Let L_1 and L_2 denote two nonvertical intersecting lines, and let θ denote the acute angle between L_1 and L_2 (see the figure). Show that

$$\tan\theta = \frac{m_2 - m_1}{1 + m_1 m_2}$$

where m_1 and m_2 are the slopes of L_1 and L_2, respectively. [**Hint:** Use the facts that $\tan\theta_1 = m_1$ and $\tan\theta_2 = m_2$.]

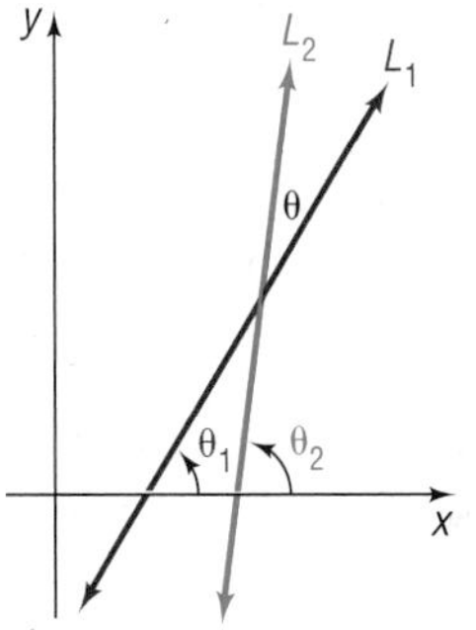

62. If $\alpha + \beta + \gamma = 180°$ and

$$\cot\theta = \cot\alpha + \cot\beta + \cot\gamma, \; 0 < \theta < 90°,$$

show that

$$\sin^3\theta = \sin(\alpha-\theta)\sin(\beta-\theta)\sin(\gamma-\theta)$$

7.3 DOUBLE-ANGLE AND HALF-ANGLE FORMULAS

OBJECTIVES
1. Use Double-Angle Formulas to Find Exact Values
2. Use Double-Angle and Half-Angle Formulas to Establish Identities
3. Use Half-Angle Formulas to Find Exact Values

In this section we derive formulas for $\sin(2\theta)$, $\cos(2\theta)$, $\sin(\frac{1}{2}\theta)$, and $\cos(\frac{1}{2}\theta)$ in terms of $\sin\theta$ and $\cos\theta$. They are easily derived using the sum formulas.

Double-Angle Formulas

In the sum formulas for $\sin(\alpha+\beta)$ and $\cos(\alpha+\beta)$, let $\alpha = \beta = \theta$. Then

$$\sin(\alpha+\beta) = \sin\alpha\cos\beta + \cos\alpha\sin\beta$$
$$\sin(\theta+\theta) = \sin\theta\cos\theta + \cos\theta\sin\theta$$
$$\sin(2\theta) = 2\sin\theta\cos\theta \qquad \textbf{(1)}$$

and

$$\cos(\alpha+\beta) = \cos\alpha\cos\beta - \sin\alpha\sin\beta$$
$$\cos(\theta+\theta) = \cos\theta\cos\theta - \sin\theta\sin\theta$$
$$\cos(2\theta) = \cos^2\theta - \sin^2\theta \qquad \textbf{(2)}$$

An application of the Pythagorean identity $\sin^2\theta + \cos^2\theta = 1$ results in two other ways to write formula (2) for $\cos(2\theta)$.

$$\cos(2\theta) = \cos^2\theta - \sin^2\theta = (1-\sin^2\theta) - \sin^2\theta = 1 - 2\sin^2\theta$$

and

$$\cos(2\theta) = \cos^2\theta - \sin^2\theta = \cos^2\theta - (1 - \cos^2\theta) = 2\cos^2\theta - 1$$

We have established the following **double-angle formulas:**

Theorem **Double-Angle Formulas**

$$\sin(2\theta) = 2\sin\theta\cos\theta \quad \textbf{(3)}$$
$$\cos(2\theta) = \cos^2\theta - \sin^2\theta \quad \textbf{(4a)}$$
$$\cos(2\theta) = 1 - 2\sin^2\theta \quad \textbf{(4b)}$$
$$\cos(2\theta) = 2\cos^2\theta - 1 \quad \textbf{(4c)}$$

EXAMPLE 1 Finding Exact Values Using the Double-Angle Formula

1 If $\sin\theta = \frac{3}{5}$, $\pi/2 < \theta < \pi$, find the exact value of:
(a) $\sin(2\theta)$ (b) $\cos(2\theta)$

Solution (a) Because $\sin(2\theta) = 2\sin\theta\cos\theta$ and we already know that $\sin\theta = \frac{3}{5}$, we only need to find $\cos\theta$. Since $\sin\theta = \frac{3}{5} = \frac{y}{r}$, we let $y = 3$ and $r = 5$. Since $\pi/2 < \theta < \pi$, the point $(x, 3)$ is on the circle $x^2 + y^2 = 25$ and is in Quadrant II. See Figure 4. Thus,

Figure 4

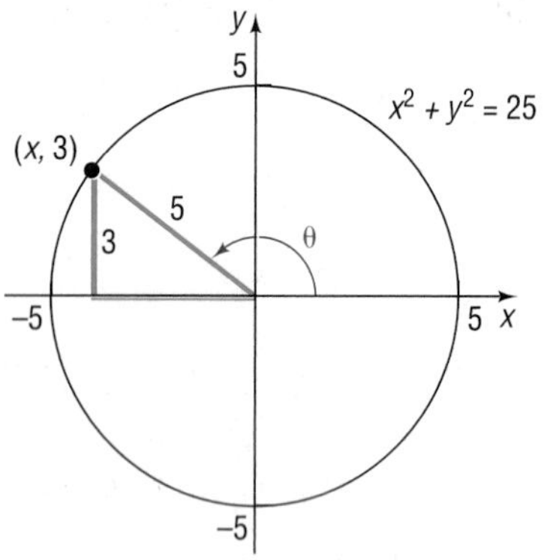

$$x^2 + 3^2 = 25, \qquad x < 0$$
$$x^2 = 25 - 9 = 16$$
$$x = -4$$

We now know that $\cos\theta = -\frac{4}{5}$. Now we use formula (3) to obtain

$$\sin(2\theta) = 2\sin\theta\cos\theta = 2\left(\tfrac{3}{5}\right)\left(-\tfrac{4}{5}\right) = -\tfrac{24}{25}$$

(b) Because we are given $\sin\theta = \frac{3}{5}$, it is easiest to use formula (4b) to get $\cos(2\theta)$.

$$\cos(2\theta) = 1 - 2\sin^2\theta = 1 - 2\left(\tfrac{9}{25}\right) = 1 - \tfrac{18}{25} = \tfrac{7}{25}$$

WARNING: In finding $\cos(2\theta)$ in Example 1(b), we chose to use a version of the double-angle formula, formula (4b). Note that we are unable to use the Pythagorean identity $\cos(2\theta) = \pm\sqrt{1 - \sin^2(2\theta)}$, with $\sin(2\theta) = -\frac{24}{25}$, because we have no way of knowing which sign to choose.

NOW WORK PROBLEMS 1(a) AND (b).

EXAMPLE 2 Establishing Identities

2 (a) Develop a formula for $\tan(2\theta)$ in terms of $\tan\theta$.
(b) Develop a formula for $\sin(3\theta)$ in terms of $\sin\theta$ and $\cos\theta$.

Solution (a) In the sum formula for $\tan(\alpha + \beta)$, let $\alpha = \beta = \theta$. Then

$$\tan(\alpha + \beta) = \frac{\tan\alpha + \tan\beta}{1 - \tan\alpha\tan\beta}$$

$$\tan(\theta + \theta) = \frac{\tan\theta + \tan\theta}{1 - \tan\theta\tan\theta}$$

$$\tan(2\theta) = \frac{2\tan\theta}{1 - \tan^2\theta} \tag{5}$$

(b) To get a formula for $\sin 3\theta$, we use the sum formula and write 3θ as $2\theta + \theta$.

$$\sin(3\theta) = \sin(2\theta + \theta) = \sin(2\theta)\cos\theta + \cos(2\theta)\sin\theta$$

Now use the double-angle formulas to get

$$\begin{aligned}\sin(3\theta) &= (2\sin\theta\cos\theta)(\cos\theta) + (\cos^2\theta - \sin^2\theta)(\sin\theta)\\ &= 2\sin\theta\cos^2\theta + \sin\theta\cos^2\theta - \sin^3\theta\\ &= 3\sin\theta\cos^2\theta - \sin^3\theta\end{aligned}$$

The formula obtained in Example 2(b) can also be written as

$$\begin{aligned}\sin(3\theta) = 3\sin\theta\cos^2\theta - \sin^3\theta &= 3\sin\theta(1 - \sin^2\theta) - \sin^3\theta\\ &= 3\sin\theta - 4\sin^3\theta\end{aligned}$$

That is, $\sin(3\theta)$ is a third-degree polynomial in the variable $\sin\theta$. In fact, $\sin(n\theta)$, n a positive odd integer, can always be written as a polynomial of degree n in the variable $\sin\theta$.*

NOW WORK PROBLEM 47.

Other Variations of the Double-Angle Formulas

By rearranging the double-angle formulas (4b) and (4c), we obtain other formulas that we will use a little later in this section.

We begin with formula (4b) and proceed to solve for $\sin^2\theta$.

$$\cos(2\theta) = 1 - 2\sin^2\theta$$

$$2\sin^2\theta = 1 - \cos(2\theta)$$

$$\sin^2\theta = \frac{1 - \cos(2\theta)}{2} \tag{6}$$

Similarly, using formula (4c), we proceed to solve for $\cos^2\theta$:

$$\cos(2\theta) = 2\cos^2\theta - 1$$

$$2\cos^2\theta = 1 + \cos(2\theta)$$

$$\cos^2\theta = \frac{1 + \cos(2\theta)}{2} \tag{7}$$

Formulas (6) and (7) can be used to develop a formula for $\tan^2\theta$.

$$\tan^2\theta = \frac{\sin^2\theta}{\cos^2\theta} = \frac{\dfrac{1 - \cos 2\theta}{2}}{\dfrac{1 + \cos 2\theta}{2}}$$

* Due to the work done by P.L. Chebyshëv, these polynomials are sometimes called *Chebyshëv polynomials.*

$$\tan^2 \theta = \frac{1 - \cos(2\theta)}{1 + \cos(2\theta)} \quad \textbf{(8)}$$

Formulas (6) through (8) do not have to be memorized since their derivations are so straightforward.

Formulas (6) and (7) are important in calculus. The next example illustrates a problem that arises in calculus requiring the use of formula (7).

EXAMPLE 3 Establishing an Identity

Write an equivalent expression for $\cos^4 \theta$ that does not involve any powers of sine or cosine greater than 1.

Solution The idea here is to apply formula (7) twice.

$$\begin{aligned}
\cos^4 \theta = (\cos^2 \theta)^2 &= \left(\frac{1 + \cos(2\theta)}{2}\right)^2 && \text{Formula (7)} \\
&= \frac{1}{4}[1 + 2\cos(2\theta) + \cos^2(2\theta)] \\
&= \frac{1}{4} + \frac{1}{2}\cos(2\theta) + \frac{1}{4}\cos^2(2\theta) \\
&= \frac{1}{4} + \frac{1}{2}\cos(2\theta) + \frac{1}{4}\left\{\frac{1 + \cos[2(2\theta)]}{2}\right\} && \text{Formula (7)} \\
&= \frac{1}{4} + \frac{1}{2}\cos(2\theta) + \frac{1}{8}[1 + \cos(4\theta)] \\
&= \frac{3}{8} + \frac{1}{2}\cos(2\theta) + \frac{1}{8}\cos(4\theta)
\end{aligned}$$

NOW WORK PROBLEM 23.

Identities, such as the double-angle formulas, can sometimes be used to rewrite expressions in a more suitable form. Let's look at an example.

EXAMPLE 4 Projectile Motion

Figure 5

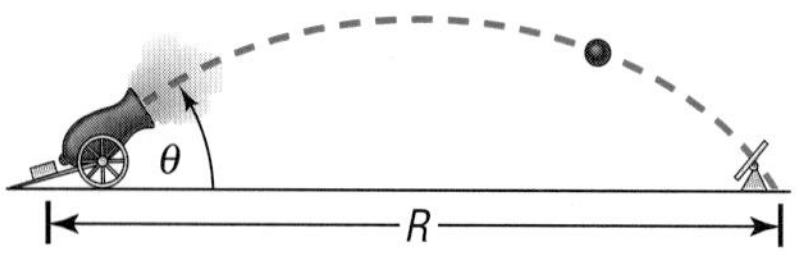

An object is propelled upward at an angle θ to the horizontal with an initial velocity of v_0 feet per second. See Figure 5. If air resistance is ignored, the **range** R, the horizontal distance R that the object travels, is given by

$$R = \frac{1}{16} v_0^2 \sin\theta \cos\theta$$

(a) Show that $R = \frac{1}{32} v_0^2 \sin(2\theta)$.

(b) Find the angle θ for which R is a maximum.

Solution (a) We rewrite the given expression for the range using the double-angle formula $\sin(2\theta) = 2\sin\theta\cos\theta$. Then

$$R = \frac{1}{16} v_0^2 \sin\theta\cos\theta = \frac{1}{16} v_0^2 \frac{2\sin\theta\cos\theta}{2} = \frac{1}{32} v_0^2 \sin(2\theta)$$

(b) For a fixed initial speed v_0, the angle θ of inclination to the horizontal determines the value of R. Since the largest value of a sine function is 1, occurring when the argument 2θ is 90°, it follows that for maximum R we must have

$$2\theta = 90^\circ$$
$$\theta = 45^\circ$$

An inclination to the horizontal of 45° results in maximum range. ■

Check: Graph $Y_1 = \frac{1}{32}v_0^2\sin(2\theta)$ with $v_0 = 1$, in degree mode. Use MAXIMUM to determine the angle θ that maximizes the range R. ■

HALF-ANGLE FORMULAS

Another important use of formulas (6) through (8) is to prove the **half-angle formulas.** In formulas (6) through (8), let $\theta = \alpha/2$. Then

$$\sin^2\frac{\alpha}{2} = \frac{1 - \cos\alpha}{2} \qquad \cos^2\frac{\alpha}{2} = \frac{1 + \cos\alpha}{2} \qquad \textbf{(9)}$$

$$\tan^2\frac{\alpha}{2} = \frac{1 - \cos\alpha}{1 + \cos\alpha}$$

If we solve for the trigonometric functions on the left sides of equations (9), we obtain the half-angle formulas.

Theorem **Half-Angle Formulas**

$$\sin\frac{\alpha}{2} = \pm\sqrt{\frac{1 - \cos\alpha}{2}} \qquad \textbf{(10a)}$$

$$\cos\frac{\alpha}{2} = \pm\sqrt{\frac{1 + \cos\alpha}{2}} \qquad \textbf{(10b)}$$

$$\tan\frac{\alpha}{2} = \pm\sqrt{\frac{1 - \cos\alpha}{1 + \cos\alpha}} \qquad \textbf{(10c)}$$

where the + or − sign is determined by the quadrant of the angle $\alpha/2$. ■

We use the half-angle formulas in the next example.

EXAMPLE 5 Finding Exact Values Using Half-Angle Formulas

3 Find the exact value of:

(a) $\cos 15^\circ$ (b) $\sin(-15^\circ)$

Solution (a) Because $15^\circ = 30^\circ/2$, we can use the half-angle formula for $\cos(\alpha/2)$ with $\alpha = 30^\circ$. Also, because 15° is in quadrant I, $\cos 15^\circ > 0$, and we choose the + sign in using formula (10b).

$$\cos 15° = \cos\frac{30°}{2} = \sqrt{\frac{1+\cos 30°}{2}}$$

$$= \sqrt{\frac{1+\sqrt{3}/2}{2}} = \sqrt{\frac{2+\sqrt{3}}{4}} = \frac{\sqrt{2+\sqrt{3}}}{2}$$

(b) We use the fact that $\sin(-15°) = -\sin 15°$ and then apply formula (10a).

$$\sin(-15°) = -\sin\frac{30°}{2} = -\sqrt{\frac{1-\cos 30°}{2}}$$

$$= -\sqrt{\frac{1-\sqrt{3}/2}{2}} = -\sqrt{\frac{2-\sqrt{3}}{4}} = -\frac{\sqrt{2-\sqrt{3}}}{2}$$ ■

It is interesting to compare the answer found in Example 5(a) with the answer to Example 2 of Section 7.2. There we calculated

$$\cos\frac{\pi}{12} = \cos 15° = \frac{1}{4}(\sqrt{6}+\sqrt{2})$$

Based on these results, we conclude that

$$\frac{1}{4}(\sqrt{6}+\sqrt{2}) \quad \text{and} \quad \frac{\sqrt{2+\sqrt{3}}}{2}$$

are equal. (Since each expression is positive, you can verify this equality by squaring each expression.) Thus, two very different looking, yet correct, answers can be obtained, depending on the approach taken to solve a problem.

NOW WORK PROBLEM 13.

EXAMPLE 6 Finding Exact Values Using Half-Angle Formulas

If $\cos\alpha = -\frac{3}{5}$, $\pi < \alpha < 3\pi/2$, find the exact value of:

(a) $\sin\frac{\alpha}{2}$ (b) $\cos\frac{\alpha}{2}$ (c) $\tan\frac{\alpha}{2}$

Solution First, we observe that if $\pi < \alpha < 3\pi/2$, then $\pi/2 < \alpha/2 < 3\pi/4$. As a result, $\alpha/2$ lies in quadrant II.

(a) Because $\alpha/2$ lies in quadrant II, $\sin(\alpha/2) > 0$, so we use the + sign in formula (10a) to get

$$\sin\frac{\alpha}{2} = \sqrt{\frac{1-\cos\alpha}{2}} = \sqrt{\frac{1-\left(-\frac{3}{5}\right)}{2}}$$

$$= \sqrt{\frac{\frac{8}{5}}{2}} = \sqrt{\frac{4}{5}} = \frac{2}{\sqrt{5}} = \frac{2\sqrt{5}}{5}$$

(b) Because $\alpha/2$ lies in quadrant II, $\cos(\alpha/2) < 0$, so we use the − sign in formula (10b) to get

$$\cos\frac{\alpha}{2} = -\sqrt{\frac{1+\cos\alpha}{2}} = -\sqrt{\frac{1+\left(-\frac{3}{5}\right)}{2}}$$

$$= -\sqrt{\frac{\frac{2}{5}}{2}} = -\frac{1}{\sqrt{5}} = -\frac{\sqrt{5}}{5}$$

(c) Because $\alpha/2$ lies in quadrant II, $\tan(\alpha/2) < 0$, so we use the $-$ sign in formula (10c) to get

$$\tan\frac{\alpha}{2} = -\sqrt{\frac{1-\cos\alpha}{1+\cos\alpha}} = -\sqrt{\frac{1-\left(-\frac{3}{5}\right)}{1+\left(-\frac{3}{5}\right)}} = -\sqrt{\frac{\frac{8}{5}}{\frac{2}{5}}} = -2$$ ■

Another way to solve Example 6(c) is to use the solutions found in parts (a) and (b).

$$\tan\frac{\alpha}{2} = \frac{\sin(\alpha/2)}{\cos(\alpha/2)} = \frac{2\sqrt{5}/5}{-\sqrt{5}/5} = -2$$

NOW WORK PROBLEMS 1(c) AND (d).

There is a formula for $\tan(\alpha/2)$ that does not contain $+$ and $-$ signs, making it more useful than Formula 10(c). Because

$$1 - \cos\alpha = 2\sin^2\left(\frac{\alpha}{2}\right) \quad \text{Formula (9)}$$

and

$$\sin\alpha = \sin\left[2\left(\frac{\alpha}{2}\right)\right] = 2\sin\left(\frac{\alpha}{2}\right)\cos\left(\frac{\alpha}{2}\right) \quad \text{Double-angle formula}$$

we have

$$\frac{1-\cos\alpha}{\sin\alpha} = \frac{2\sin^2\left(\frac{\alpha}{2}\right)}{2\sin\left(\frac{\alpha}{2}\right)\cos\left(\frac{\alpha}{2}\right)} = \frac{\sin\left(\frac{\alpha}{2}\right)}{\cos\left(\frac{\alpha}{2}\right)} = \tan\left(\frac{\alpha}{2}\right)$$

Since it also can be shown that

$$\frac{1-\cos\alpha}{\sin\alpha} = \frac{\sin\alpha}{1+\cos\alpha}$$

we have the following two half-angle formulas:

Half-Angle Formulas for $\tan(\alpha/2)$

$$\tan\left(\frac{\alpha}{2}\right) = \frac{1-\cos\alpha}{\sin\alpha} = \frac{\sin\alpha}{1+\cos\alpha} \quad \textbf{(11)}$$

With this formula, the solution to Example 6(c) can be given as

$$\cos\alpha = -\tfrac{3}{5}$$

$$\sin\alpha = -\sqrt{1-\cos^2\alpha} = -\sqrt{1-\tfrac{9}{25}} = -\sqrt{\tfrac{16}{25}} = -\tfrac{4}{5}$$

Then, by equation (11),

$$\tan\frac{\alpha}{2} = \frac{1-\cos\alpha}{\sin\alpha} = \frac{1-\left(-\frac{3}{5}\right)}{-\frac{4}{5}} = \frac{\frac{8}{5}}{-\frac{4}{5}} = -2$$

7.3 EXERCISES

In Problems 1–12, use the information given about the angle θ, $0 \le \theta \le 2\pi$, to find the exact value of:

(a) $\sin(2\theta)$ (b) $\cos(2\theta)$ (c) $\sin\frac{\theta}{2}$ (d) $\cos\frac{\theta}{2}$

1. $\sin\theta = \frac{3}{5}$, $0 < \theta < \pi/2$
2. $\cos\theta = \frac{3}{5}$, $0 < \theta < \pi/2$
3. $\tan\theta = \frac{4}{3}$, $\pi < \theta < 3\pi/2$
4. $\tan\theta = \frac{1}{2}$, $\pi < \theta < 3\pi/2$
5. $\cos\theta = -\sqrt{6}/3$, $\pi/2 < \theta < \pi$
6. $\sin\theta = -\sqrt{3}/3$, $3\pi/2 < \theta < 2\pi$
7. $\sec\theta = 3$, $\sin\theta > 0$
8. $\csc\theta = -\sqrt{5}$, $\cos\theta < 0$
9. $\cot\theta = -2$, $\sec\theta < 0$
10. $\sec\theta = 2$, $\csc\theta < 0$
11. $\tan\theta = -3$, $\sin\theta < 0$
12. $\cot\theta = 3$, $\cos\theta < 0$

In Problems 13–22, use the half-angle formulas to find the exact value of each trigonometric function.

13. $\sin 22.5°$
14. $\cos 22.5°$
15. $\tan\frac{7\pi}{8}$
16. $\tan\frac{9\pi}{8}$
17. $\cos 165°$
18. $\sin 195°$
19. $\sec\frac{15\pi}{8}$
20. $\csc\frac{7\pi}{8}$
21. $\sin\left(-\frac{\pi}{8}\right)$
22. $\cos\left(-\frac{3\pi}{8}\right)$
23. Show that $\sin^4\theta = \frac{3}{8} - \frac{1}{2}\cos(2\theta) + \frac{1}{8}\cos(4\theta)$.
24. Develop a formula for $\cos(3\theta)$ as a third-degree polynomial in the variable $\cos\theta$.
25. Show that $\sin(4\theta) = (\cos\theta)(4\sin\theta - 8\sin^3\theta)$.
26. Develop a formula for $\cos(4\theta)$ as a fourth-degree polynomial in the variable $\cos\theta$.
27. Find an expression for $\sin(5\theta)$ as a fifth-degree polynomial in the variable $\sin\theta$.
28. Find an expression for $\cos(5\theta)$ as a fifth-degree polynomial in the variable $\cos\theta$.

In Problems 29–48, establish each identity.

29. $\cos^4\theta - \sin^4\theta = \cos(2\theta)$
30. $\dfrac{\cot\theta - \tan\theta}{\cot\theta + \tan\theta} = \cos(2\theta)$
31. $\cot(2\theta) = \dfrac{\cot^2\theta - 1}{2\cot\theta}$
32. $\cot(2\theta) = \frac{1}{2}(\cot\theta - \tan\theta)$
33. $\sec(2\theta) = \dfrac{\sec^2\theta}{2 - \sec^2\theta}$
34. $\csc(2\theta) = \frac{1}{2}\sec\theta\csc\theta$
35. $\cos^2(2\theta) - \sin^2(2\theta) = \cos(4\theta)$
36. $(4\sin\theta\cos\theta)(1 - 2\sin^2\theta) = \sin(4\theta)$
37. $\dfrac{\cos(2\theta)}{1 + \sin(2\theta)} = \dfrac{\cot\theta - 1}{\cot\theta + 1}$
38. $\sin^2\theta\cos^2\theta = \frac{1}{8}[1 - \cos(4\theta)]$
39. $\sec^2\dfrac{\theta}{2} = \dfrac{2}{1 + \cos\theta}$
40. $\csc^2\dfrac{\theta}{2} = \dfrac{2}{1 - \cos\theta}$
41. $\cot^2\dfrac{\theta}{2} = \dfrac{\sec\theta + 1}{\sec\theta - 1}$
42. $\tan\dfrac{\theta}{2} = \csc\theta - \cot\theta$
43. $\cos\theta = \dfrac{1 - \tan^2(\theta/2)}{1 + \tan^2(\theta/2)}$
44. $1 - \frac{1}{2}\sin(2\theta) = \dfrac{\sin^3\theta + \cos^3\theta}{\sin\theta + \cos\theta}$
45. $\dfrac{\sin(3\theta)}{\sin\theta} - \dfrac{\cos(3\theta)}{\cos\theta} = 2$
46. $\dfrac{\cos\theta + \sin\theta}{\cos\theta - \sin\theta} - \dfrac{\cos\theta - \sin\theta}{\cos\theta + \sin\theta} = 2\tan(2\theta)$
47. $\tan(3\theta) = \dfrac{3\tan\theta - \tan^3\theta}{1 - 3\tan^2\theta}$
48. $\tan\theta + \tan(\theta + 120°) + \tan(\theta + 240°) = 3\tan(3\theta)$

49. If $x = 2\tan\theta$, express $\sin(2\theta)$ as a function of x.

50. If $x = 2\tan\theta$, express $\cos(2\theta)$ as a function of x.

51. Find the value of the number C:

$$\tfrac{1}{2}\sin^2\theta + C = -\tfrac{1}{4}\cos(2\theta)$$

52. Find the value of the number C:

$$\tfrac{1}{2}\cos^2\theta + C = \tfrac{1}{4}\cos(2\theta)$$

53. Graph $f(x) = \sin^2 x = [1 - \cos(2x)]/2$ for $0 \le x \le 2\pi$ by using transformations.

54. Repeat Problem 53 for $g(x) = \cos^2 x$.

55. Use the fact that

$$\cos\frac{\pi}{12} = \frac{1}{4}(\sqrt{6} + \sqrt{2})$$

to find $\sin(\pi/24)$ and $\cos(\pi/24)$.

56. Show that

$$\cos\frac{\pi}{8} = \frac{\sqrt{2 - \sqrt{2}}}{2}$$

and use it to find $\sin(\pi/16)$ and $\cos(\pi/16)$.

57. Show that

$$\sin^3\theta + \sin^3(\theta + 120°) + \sin^3(\theta + 240°) = -\tfrac{3}{4}\sin(3\theta)$$

58. If $\tan\theta = a\tan(\theta/3)$, express $\tan(\theta/3)$ in terms of a.

In Problems 59 and 60, establish each identity.

59. $\ln|\sin\theta| = \tfrac{1}{2}(\ln|1 - \cos(2\theta)| - \ln 2)$

60. $\ln|\cos\theta| = \tfrac{1}{2}(\ln|1 + \cos(2\theta)| - \ln 2)$

61. Projectile Motion An object is propelled upward at an angle θ, $45° < \theta < 90°$, to the horizontal with an initial velocity of v_0 feet per second from the base of a plane that makes an angle of 45° with the horizontal. See the illustration. If air resistance is ignored, the distance R that it travels up the inclined plane is given by

$$R(\theta) = \frac{v_0^2\sqrt{2}}{16}\cos\theta(\sin\theta - \cos\theta)$$

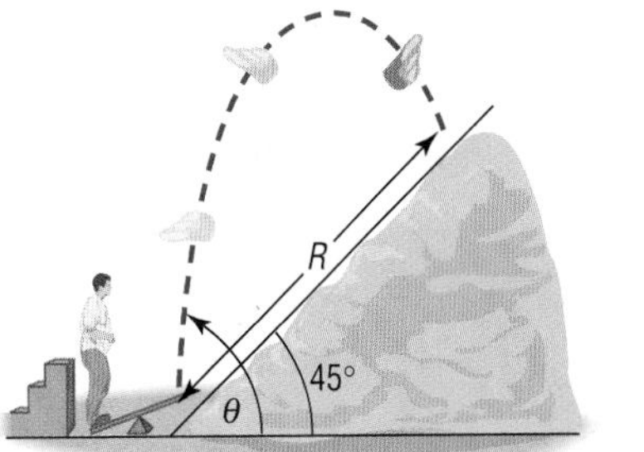

(a) Show that

$$R(\theta) = \frac{v_0^2\sqrt{2}}{32}[\sin(2\theta) - \cos(2\theta) - 1]$$

(b) Use a graphing utility to graph $R = R(\theta)$. (Use $v_0 = 32$ feet per second.)
(c) What value of θ makes R the largest? (Use $v_0 = 32$ feet per second.)

62. Sawtooth Curve An oscilloscope often displays a sawtooth curve. This curve can be approximated by sinusoidal curves of varying periods and amplitudes. A first approximation to the sawtooth curve is given by

$$f(x) = \frac{1}{2}\sin(2\pi x) + \frac{1}{4}\sin(4\pi x)$$

Show that $f(x) = \sin(2\pi x)\cos^2(\pi x)$.

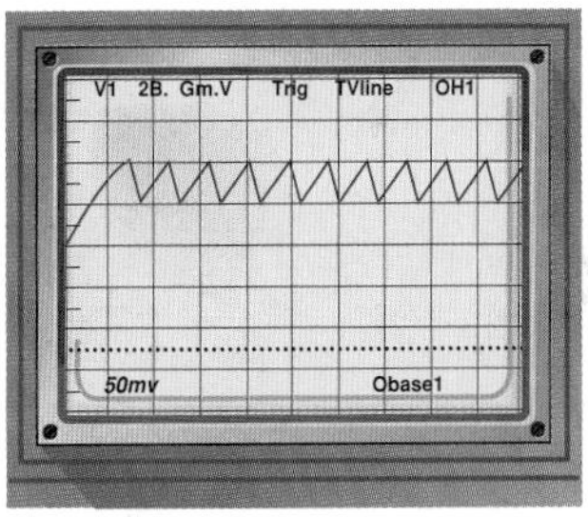

63. If $z = \tan(\alpha/2)$, show that

$$\sin\alpha = \frac{2z}{1 + z^2}$$

64. If $z = \tan(\alpha/2)$, show that

$$\cos\alpha = \frac{1 - z^2}{1 + z^2}$$

65. Go to the library and research Chebyshëv polynomials. Write a report on your findings.

7.4 PRODUCT-TO-SUM AND SUM-TO-PRODUCT FORMULAS

OBJECTIVES 1 Express Products as Sums
2 Express Sums as Products

1 The sum and difference formulas can be used to derive formulas for writing the products of sines and/or cosines as sums or differences. These identities are usually called the **Product-to-Sum Formulas.**

Theorem **Product-to-Sum Formulas**

$$\sin\alpha\sin\beta = \tfrac{1}{2}[\cos(\alpha-\beta) - \cos(\alpha+\beta)] \quad \textbf{(1)}$$

$$\cos\alpha\cos\beta = \tfrac{1}{2}[\cos(\alpha-\beta) + \cos(\alpha+\beta)] \quad \textbf{(2)}$$

$$\sin\alpha\cos\beta = \tfrac{1}{2}[\sin(\alpha+\beta) + \sin(\alpha-\beta)] \quad \textbf{(3)}$$

These formulas do not have to be memorized. Instead, you should remember how they are derived. Then, when you want to use them, either look them up or derive them, as needed.

To derive formulas (1) and (2), write down the sum and difference formulas for the cosine:

$$\cos(\alpha-\beta) = \cos\alpha\cos\beta + \sin\alpha\sin\beta \quad \textbf{(4)}$$

$$\cos(\alpha+\beta) = \cos\alpha\cos\beta - \sin\alpha\sin\beta \quad \textbf{(5)}$$

Subtract equation (5) from equation (4) to get

$$\cos(\alpha-\beta) - \cos(\alpha+\beta) = 2\sin\alpha\sin\beta$$

from which

$$\sin\alpha\sin\beta = \tfrac{1}{2}[\cos(\alpha-\beta) - \cos(\alpha+\beta)]$$

Now, add equations (4) and (5) to get

$$\cos(\alpha-\beta) + \cos(\alpha+\beta) = 2\cos\alpha\cos\beta$$

from which

$$\cos\alpha\cos\beta = \tfrac{1}{2}[\cos(\alpha-\beta) + \cos(\alpha+\beta)]$$

To derive Product-to-Sum Formula (3), use the Sum and Difference Formulas for sine in a similar way. (You are asked to do this in Problem 41.)

EXAMPLE 1 Expressing Products as Sums

Express each of the following products as a sum containing only sines or cosines.

(a) $\sin(6\theta)\sin(4\theta)$ (b) $\cos(3\theta)\cos\theta$ (c) $\sin(3\theta)\cos(5\theta)$

Solution (a) We use formula (1) to get

$$\sin(6\theta)\sin(4\theta) = \tfrac{1}{2}[\cos(6\theta-4\theta) - \cos(6\theta+4\theta)]$$
$$= \tfrac{1}{2}[\cos(2\theta) - \cos(10\theta)]$$

(b) We use formula (2) to get

$$\cos(3\theta)\cos\theta = \tfrac{1}{2}[\cos(3\theta - \theta) + \cos(3\theta + \theta)]$$
$$= \tfrac{1}{2}[\cos(2\theta) + \cos(4\theta)]$$

(c) We use formula (3) to get

$$\sin(3\theta)\cos(5\theta) = \tfrac{1}{2}[\sin(3\theta + 5\theta) + \sin(3\theta - 5\theta)]$$
$$= \tfrac{1}{2}[\sin(8\theta) + \sin(-2\theta)] = \tfrac{1}{2}[\sin(8\theta) - \sin(2\theta)]$$ ■

NOW WORK PROBLEM 1.

2 The **Sum-to-Product Formulas** are given next.

Theorem **Sum-to-Product Formulas**

$$\sin\alpha + \sin\beta = 2\sin\frac{\alpha+\beta}{2}\cos\frac{\alpha-\beta}{2} \quad \textbf{(6)}$$

$$\sin\alpha - \sin\beta = 2\sin\frac{\alpha-\beta}{2}\cos\frac{\alpha+\beta}{2} \quad \textbf{(7)}$$

$$\cos\alpha + \cos\beta = 2\cos\frac{\alpha+\beta}{2}\cos\frac{\alpha-\beta}{2} \quad \textbf{(8)}$$

$$\cos\alpha - \cos\beta = -2\sin\frac{\alpha+\beta}{2}\sin\frac{\alpha-\beta}{2} \quad \textbf{(9)}$$

■

We will derive formula (6) and leave the derivations of formulas (7) through (9) as exercises (see Problems 42 through 44).

Proof

$$2\sin\frac{\alpha+\beta}{2}\cos\frac{\alpha-\beta}{2} = 2\cdot\frac{1}{2}\left[\sin\left(\frac{\alpha+\beta}{2}+\frac{\alpha-\beta}{2}\right) + \sin\left(\frac{\alpha+\beta}{2}-\frac{\alpha-\beta}{2}\right)\right]$$

↑ Product-to-Sum Formula (3)

$$= \sin\frac{2\alpha}{2} + \sin\frac{2\beta}{2} = \sin\alpha + \sin\beta$$ ■

EXAMPLE 2 Expressing Sums (or Differences) as a Product

Express each sum or difference as a product of sines and/or cosines.

(a) $\sin(5\theta) - \sin(3\theta)$ (b) $\cos(3\theta) + \cos(2\theta)$

Solution (a) We use formula (7) to get

$$\sin(5\theta) - \sin(3\theta) = 2\sin\frac{5\theta - 3\theta}{2}\cos\frac{5\theta + 3\theta}{2}$$
$$= 2\sin\theta\cos(4\theta)$$

(b)
$$\cos(3\theta) + \cos(2\theta) = 2\cos\frac{3\theta + 2\theta}{2}\cos\frac{3\theta - 2\theta}{2} \quad \text{Formula (8)}$$
$$= 2\cos\frac{5\theta}{2}\cos\frac{\theta}{2}$$

NOW WORK PROBLEM 11.

7.4 EXERCISES

In Problems 1–10, express each product as a sum containing only sines or cosines.

1. $\sin(4\theta)\sin(2\theta)$
2. $\cos(4\theta)\cos(2\theta)$
3. $\sin(4\theta)\cos(2\theta)$
4. $\sin(3\theta)\sin(5\theta)$
5. $\cos(3\theta)\cos(5\theta)$
6. $\sin(4\theta)\cos(6\theta)$
7. $\sin\theta\sin(2\theta)$
8. $\cos(3\theta)\cos(4\theta)$
9. $\sin\frac{3\theta}{2}\cos\frac{\theta}{2}$
10. $\sin\frac{\theta}{2}\cos\frac{5\theta}{2}$

In Problems 11–18, express each sum or difference as a product of sines and/or cosines.

11. $\sin(4\theta) - \sin(2\theta)$
12. $\sin(4\theta) + \sin(2\theta)$
13. $\cos(2\theta) + \cos(4\theta)$
14. $\cos(5\theta) - \cos(3\theta)$
15. $\sin\theta + \sin(3\theta)$
16. $\cos\theta + \cos(3\theta)$
17. $\cos\frac{\theta}{2} - \cos\frac{3\theta}{2}$
18. $\sin\frac{\theta}{2} - \sin\frac{3\theta}{2}$

In Problems 19–36, establish each identity.

19. $\dfrac{\sin\theta + \sin(3\theta)}{2\sin(2\theta)} = \cos\theta$
20. $\dfrac{\cos\theta + \cos(3\theta)}{2\cos(2\theta)} = \cos\theta$
21. $\dfrac{\sin(4\theta) + \sin(2\theta)}{\cos(4\theta) + \cos(2\theta)} = \tan(3\theta)$
22. $\dfrac{\cos\theta - \cos(3\theta)}{\sin(3\theta) - \sin\theta} = \tan(2\theta)$
23. $\dfrac{\cos\theta - \cos(3\theta)}{\sin\theta + \sin(3\theta)} = \tan\theta$
24. $\dfrac{\cos\theta - \cos(5\theta)}{\sin\theta + \sin(5\theta)} = \tan(2\theta)$
25. $\sin\theta[\sin\theta + \sin(3\theta)] = \cos\theta[\cos\theta - \cos(3\theta)]$
26. $\sin\theta[\sin(3\theta) + \sin(5\theta)] = \cos\theta[\cos(3\theta) - \cos(5\theta)]$
27. $\dfrac{\sin(4\theta) + \sin(8\theta)}{\cos(4\theta) + \cos(8\theta)} = \tan(6\theta)$
28. $\dfrac{\sin(4\theta) - \sin(8\theta)}{\cos(4\theta) - \cos(8\theta)} = -\cot(6\theta)$
29. $\dfrac{\sin(4\theta) + \sin(8\theta)}{\sin(4\theta) - \sin(8\theta)} = -\dfrac{\tan(6\theta)}{\tan(2\theta)}$
30. $\dfrac{\cos(4\theta) - \cos(8\theta)}{\cos(4\theta) + \cos(8\theta)} = \tan(2\theta)\tan(6\theta)$
31. $\dfrac{\sin\alpha + \sin\beta}{\sin\alpha - \sin\beta} = \tan\dfrac{\alpha + \beta}{2}\cot\dfrac{\alpha - \beta}{2}$
32. $\dfrac{\cos\alpha + \cos\beta}{\cos\alpha - \cos\beta} = -\cot\dfrac{\alpha + \beta}{2}\cot\dfrac{\alpha - \beta}{2}$
33. $\dfrac{\sin\alpha + \sin\beta}{\cos\alpha + \cos\beta} = \tan\dfrac{\alpha + \beta}{2}$
34. $\dfrac{\sin\alpha - \sin\beta}{\cos\alpha - \cos\beta} = -\cot\dfrac{\alpha + \beta}{2}$
35. $1 + \cos(2\theta) + \cos(4\theta) + \cos(6\theta) = 4\cos\theta\cos(2\theta)\cos(3\theta)$
36. $1 - \cos(2\theta) + \cos(4\theta) - \cos(6\theta) = 4\sin\theta\cos(2\theta)\sin(3\theta)$

37. Touch-Tone Phones On a Touch-Tone phone, each button produces a unique sound. The sound produced is the sum of two tones, given by

$$y = \sin[2\pi l t] \quad \text{and} \quad y = \sin[2\pi h t]$$

where l and h are the low and high frequencies (cycles per second) shown on the illustration. For example, if you touch 7, the low frequency is $l = 852$ cycles per second and the high frequency is $h = 1209$ cycles per second. The sound emitted by touching 7 is

$$y = \sin[2\pi(852)t] + \sin[2\pi(1209)t]$$

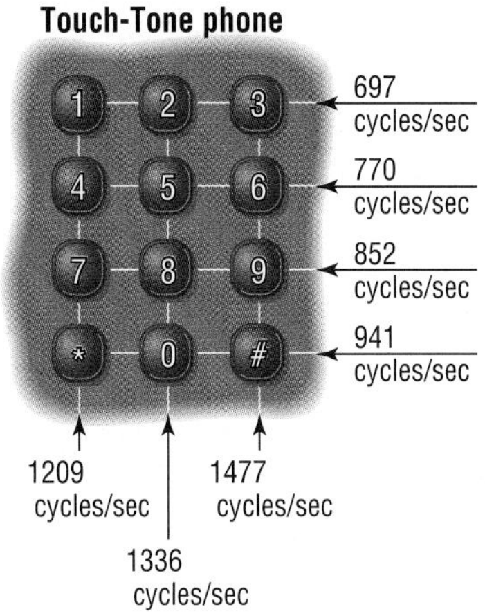

(a) Write this sound as a product of sines and/or cosines.
(b) Determine the maximum value of y.
(c) Use a graphing utility to graph the sound emitted by touching 7.

38. Touch-Tone Phones
(a) Write the sound emitted by touching the # key as a product of sines and/or cosines.
(b) Determine the maximum value of y.
(c) Use a graphing utility to graph the sound emitted by touching the # key.

39. If $\alpha + \beta + \gamma = \pi$, show that

$$\sin(2\alpha) + \sin(2\beta) + \sin(2\gamma) = 4\sin\alpha\sin\beta\sin\gamma$$

40. If $\alpha + \beta + \gamma = \pi$, show that

$$\tan\alpha + \tan\beta + \tan\gamma = \tan\alpha\tan\beta\tan\gamma$$

41. Derive formula (3).

42. Derive formula (7).

43. Derive formula (8).

44. Derive formula (9).

PREPARING FOR THIS SECTION

Before getting started, review the following:

- ✓ One-to-One Functions, Inverse Functions (Section 5.1)
- ✓ Definition of the Trigonometric Functions (p. 381)
- ✓ Values of the Trigonometric Functions of Certain Angles (pp. 384 and 387)
- ✓ Domain and Range of the Trigonometric Functions (pp. 397–398)
- ✓ Theorem on Trigonometric Functions (p. 391)

7.5 THE INVERSE TRIGONOMETRIC FUNCTIONS (I)

OBJECTIVES 1 Find the Exact Value of an Inverse Trigonometric Function
2 Find the Approximate Value of an Inverse Trigonometric Function

In Section 5.1 we discussed inverse functions, and we noted that if a function is one-to-one it will have an inverse function. We also observed that if a function is not one-to-one it may be possible to restrict its domain in some suitable manner so that the restricted function is one-to-one.

Next, we review some characteristics of a function f and its inverse function f^{-1}.

1. $f^{-1}(f(x)) = x$ for every x in the domain of f and $f(f^{-1}(x)) = x$ for every x in the domain of f^{-1}.
2. Domain of f = range of f^{-1} and range of f = domain of f^{-1}.
3. The graph of f and the graph of f^{-1} are symmetric with respect to the line $y = x$.
4. If a function $y = f(x)$ has an inverse, the equation of the inverse is $x = f(y)$. The solution of this equation is $y = f^{-1}(x)$.

The Inverse Sine Function

In Figure 6, we reproduce the graph of $y = \sin x$. Because every horizontal line $y = b$, where b is between -1 and 1, intersects the graph of $y = \sin x$ infinitely many times, it follows from the horizontal-line test that the function $y = \sin x$ is not one-to-one.

Figure 6
$y = \sin x, -\infty < x < \infty, -1 \leq y \leq 1$

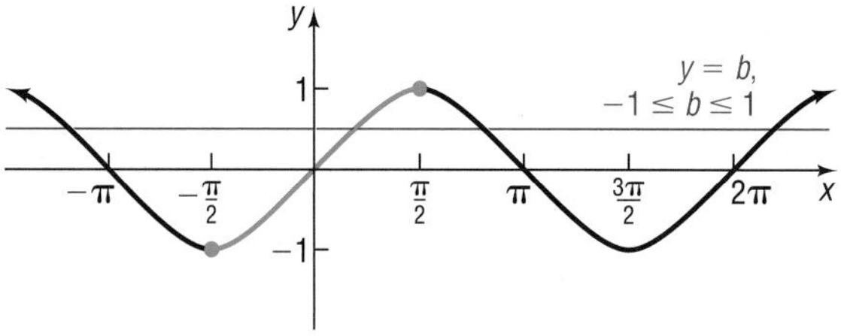

However, if we restrict the domain of $y = \sin x$ to the interval $[-\pi/2, \pi/2]$, the restricted function

$$y = \sin x, \quad -\frac{\pi}{2} \leq x \leq \frac{\pi}{2}$$

is one-to-one and, hence, will have an inverse function.* See Figure 7.

Figure 7
$y = \sin x, -\pi/2 \leq x \leq \pi/2, -1 \leq y \leq 1$

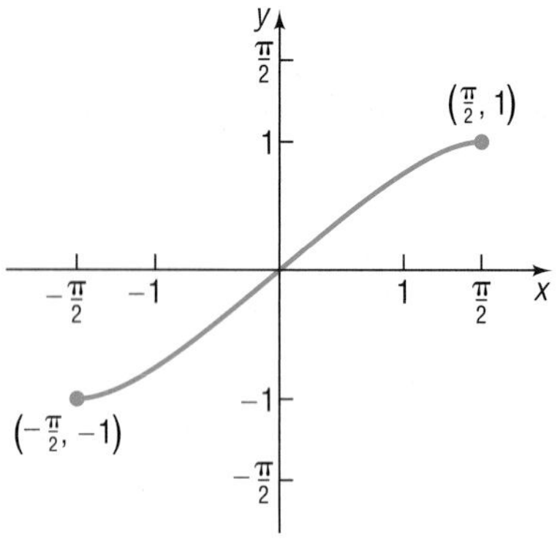

An equation for the inverse is obtained by interchanging x and y. The implicit form of the inverse function is $x = f(y) = \sin y, -\frac{\pi}{2} \leq y \leq \frac{\pi}{2}$. The explicit form is called the **inverse sine** of x and is symbolized by $y = f^{-1}(x) = \sin^{-1} x$.

$$y = \sin^{-1} x \quad \text{means} \quad x = \sin y \qquad \textbf{(1)}$$
$$\text{where} \quad -1 \leq x \leq 1 \quad \text{and} \quad -\frac{\pi}{2} \leq y \leq \frac{\pi}{2}$$

Because $y = \sin^{-1} x$ means $x = \sin y$, we read $y = \sin^{-1} x$ as "y is the angle or real number whose sine equals x." Alternatively, we can say that "y is the inverse sine of x." Be careful about the notation used. The superscript -1 that appears in $y = \sin^{-1} x$ is not an exponent, but is reminiscent of the symbolism f^{-1} used to denote the inverse of a function f. [To avoid this notation, some books use the notation $y = \arcsin x$ instead of $y = \sin^{-1} x$.]

The inverse of a function f receives as input an element from the range of f and returns as output an element in the domain of f. The restricted sine

*Although there are many other ways to restrict the domain and obtain a one-to-one function, mathematicians have agreed on a consistent use of the interval $[-\pi/2, \pi/2]$ in order to define the inverse of $y = \sin x$.

function, $y = f(x) = \sin x$, receives as input an angle or real number x in the interval $\left[-\frac{\pi}{2}, \frac{\pi}{2}\right]$ and outputs a real number in the interval $[-1, 1]$. Therefore, the inverse sine function receives as input a real number in the interval $[-1, 1]$ and outputs an angle or real number in the interval $\left[-\frac{\pi}{2}, \frac{\pi}{2}\right]$. Since the domain of f = range of f^{-1} and the range of f = domain of f^{-1}, the domain of the inverse sine function, $y = f^{-1}(x) = \sin^{-1} x$ is $[-1, 1]$ or $-1 \leq x \leq 1$, and the range of the inverse sine function is $\left[-\frac{\pi}{2}, \frac{\pi}{2}\right]$ or $-\frac{\pi}{2} \leq y \leq \frac{\pi}{2}$. The graph of the inverse sine function can be obtained by reflecting the restricted portion of the graph of $y = f(x) = \sin x$ about the line $y = x$ as shown in Figure 8(a). Figure 8(b) shows the graph obtained using a graphing utility.

Figure 8
$y = \sin^{-1} x, -1 \leq x \leq 1, -\pi/2 \leq y \leq \pi/2$

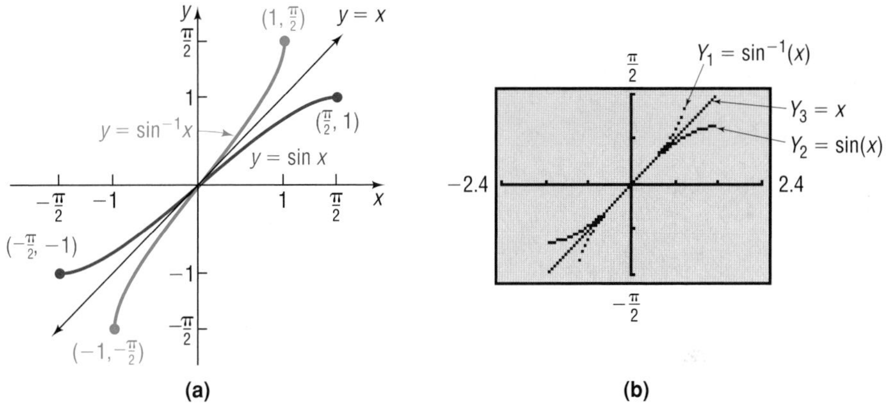

When we discussed functions and their inverses in Section 5.1, we found that $f^{-1}(f(x)) = x$ and $f(f^{-1}(x)) = x$. In terms of the sine function and its inverse, these properties are of the form

$$f^{-1}(f(x)) = \sin^{-1}(\sin x) = x, \quad \text{where } -\frac{\pi}{2} \leq x \leq \frac{\pi}{2} \tag{2a}$$

$$f(f^{-1}(x)) = \sin(\sin^{-1} x) = x, \quad \text{where } -1 \leq x \leq 1 \tag{2b}$$

Figure 9

For example, $\sin^{-1}\left[\sin\left(\frac{\pi}{8}\right)\right] = \frac{\pi}{8}$ and $\sin[\sin^{-1}(0.8)] = 0.8$. See Figure 9.

However, $\sin^{-1}\left[\sin\left(\frac{5\pi}{8}\right)\right] \neq \frac{5\pi}{8}$, because $\frac{5\pi}{8}$ is not in the restricted domain of $f(x) = \sin x$. See Figure 10. Also, $\sin\left[\sin^{-1}(1.8)\right] \neq 1.8$, because 1.8 is not in the domain of the inverse sine function. See Figure 11.

Figure 10

Figure 11

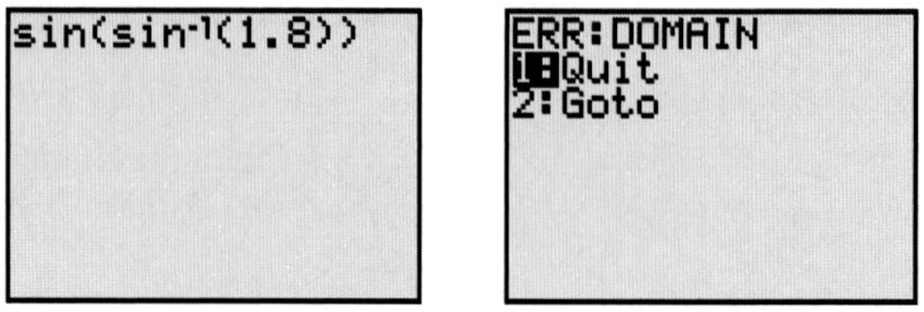

1 For some numbers x it is possible to find the exact value of $y = \sin^{-1} x$.

EXAMPLE 1 Finding the Exact Value of an Inverse Sine Function

Find the exact value of: $\sin^{-1} 1$

Solution Let $\theta = \sin^{-1} 1$. We seek the angle $\theta, -\pi/2 \leq \theta \leq \pi/2$, whose sine equals 1.

$$\theta = \sin^{-1} 1, \qquad -\frac{\pi}{2} \leq \theta \leq \frac{\pi}{2}$$

$$\sin\theta = 1, \qquad -\frac{\pi}{2} \leq \theta \leq \frac{\pi}{2} \qquad \text{By definition of } y = \sin^{-1} x$$

Now look at Table 1 and Figure 12.

TABLE 1

θ	$\sin\theta$
$-\pi/2$	-1
$-\pi/3$	$-\sqrt{3}/2$
$-\pi/4$	$-\sqrt{2}/2$
$-\pi/6$	$-1/2$
0	0
$\pi/6$	$1/2$
$\pi/4$	$\sqrt{2}/2$
$\pi/3$	$\sqrt{3}/2$
$\pi/2$	1

Figure 12

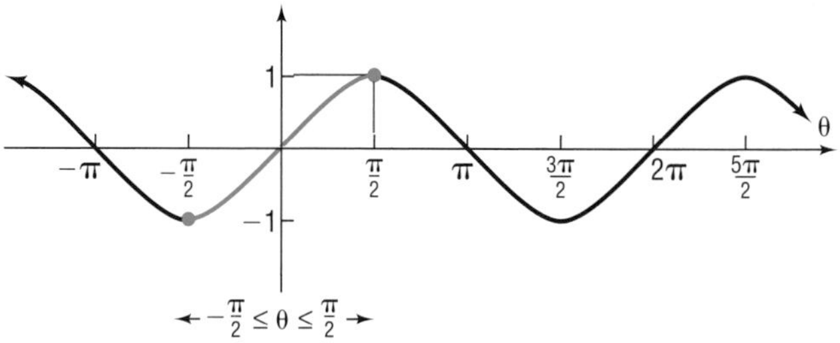

We see that the only angle θ within the interval $[-\pi/2, \pi/2]$ whose sine is 1 is $\pi/2$. [Note that $\sin(5\pi/2)$ also equals 1, but $5\pi/2$ lies outside the interval $[-\pi/2, \pi/2]$ and hence is not admissible.] So, since $\sin(\pi/2) = 1$ and $\pi/2$ is in $[-\pi/2, \pi/2]$, we conclude that

$$\sin^{-1} 1 = \frac{\pi}{2}$$

NOW WORK PROBLEM 1.

EXAMPLE 2 Finding the Exact Value of an Inverse Sine Function

Find the exact value of: $\sin^{-1}\left(-\frac{1}{2}\right)$

Solution Let $\theta = \sin^{-1}\left(-\frac{1}{2}\right)$. We seek the angle θ, $-\pi/2 \leq \theta \leq \pi/2$, whose sine equals $-\frac{1}{2}$.

$$\theta = \sin^{-1}\left(-\frac{1}{2}\right), \qquad -\frac{\pi}{2} \leq \theta \leq \frac{\pi}{2}$$

$$\sin\theta = -\frac{1}{2}, \qquad -\frac{\pi}{2} \leq \theta \leq \frac{\pi}{2}$$

(Refer to Table 1 and Figure 12 if necessary.) The only angle within the interval $[-\pi/2, \pi/2]$, whose sine is $-\frac{1}{2}$, is $-\pi/6$. So, since $\sin(-\pi/6) = -1/2$ and $-\pi/6$ is in $[-\pi/2, \pi/2]$, we conclude that

$$\sin^{-1}\left(-\frac{1}{2}\right) = -\frac{\pi}{6}$$

NOW WORK PROBLEM 7.

2 For most numbers x, the value $y = \sin^{-1}(x)$ must be approximated.

EXAMPLE 3 Finding an Approximate Value of an Inverse Sine Function

Find the approximate value of:

(a) $\sin^{-1}\frac{1}{3}$ (b) $\sin^{-1}\left(-\frac{1}{4}\right)$

Express the answer in radians rounded to two decimal places.

Solution Because we want the angle measured in radians, we first set the mode to radians.

(a) Figure 13 shows the solution using a TI-83 graphing calculator.

Figure 13

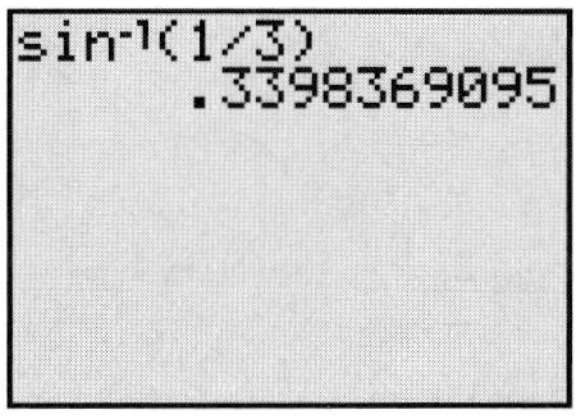

We have $\sin^{-1}\frac{1}{3} = 0.34$, rounded to two decimal places.

(b) Figure 14 shows the solution using a TI-83 graphing calculator.

Figure 14

We have $\sin^{-1}\left(-\frac{1}{4}\right) = -0.25$, rounded to two decimal places.

NOW WORK PROBLEM 21.

THE INVERSE COSINE FUNCTION

In Figure 15 we reproduce the graph of $y = \cos x$. Because every horizontal line $y = b$, where b is between -1 and 1, intersects the graph of $y = \cos x$ infinitely many times, it follows that the cosine function is not one-to-one.

Figure 15
$y = \cos x, -\infty < x < \infty, -1 \le y \le 1$

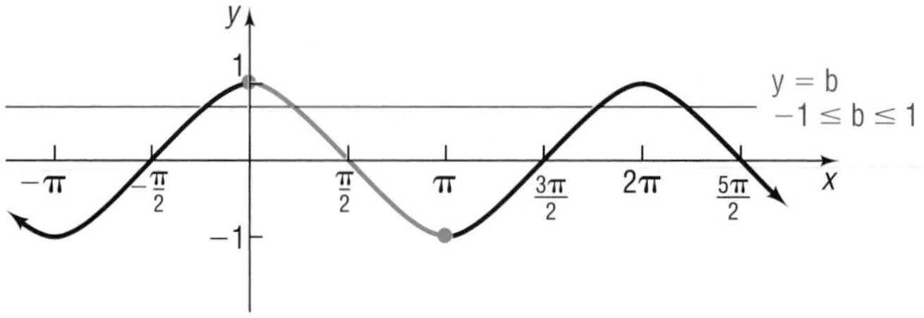

Figure 16
$y = \cos x, 0 \le x \le \pi, -1 \le y \le 1$

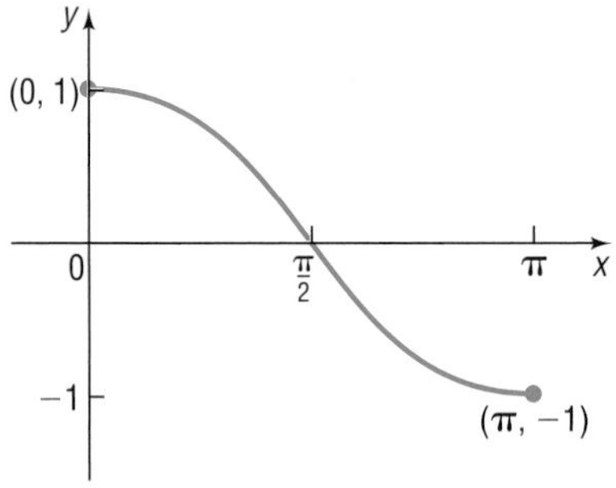

However, if we restrict the domain of $y = \cos x$ to the interval $[0, \pi]$, the restricted function

$$y = \cos x \qquad 0 \le x \le \pi$$

is one-to-one and hence will have an inverse function.* See Figure 16.

An equation for the inverse is obtained by interchanging x and y. The implicit form of the inverse function is $x = f(y) = \cos y$, $0 \le y \le \pi$. The explicit form is called the **inverse cosine** of x and is symbolized by $y = f^{-1}(x) = \cos^{-1} x$ (or by $y = \arccos x$).

$$y = \cos^{-1} x \quad \text{means} \quad x = \cos y \qquad \textbf{(3)}$$
$$\text{where} \quad -1 \le x \le 1 \quad \text{and} \quad 0 \le y \le \pi$$

Here, y is the angle whose cosine is x. The domain of the function $y = \cos^{-1} x$ is $-1 \le x \le 1$, and its range is $0 \le y \le \pi$. (Do you know why?) The graph of $y = \cos^{-1} x$ can be obtained by reflecting the restricted portion of the graph of $y = \cos x$ about the line $y = x$, as shown in Figure 17(a). Figure 17(b) shows the graph using a graphing utility.

Figure 17
$y = \cos^{-1} x, -1 \le x \le 1, 0 \le y \le \pi$

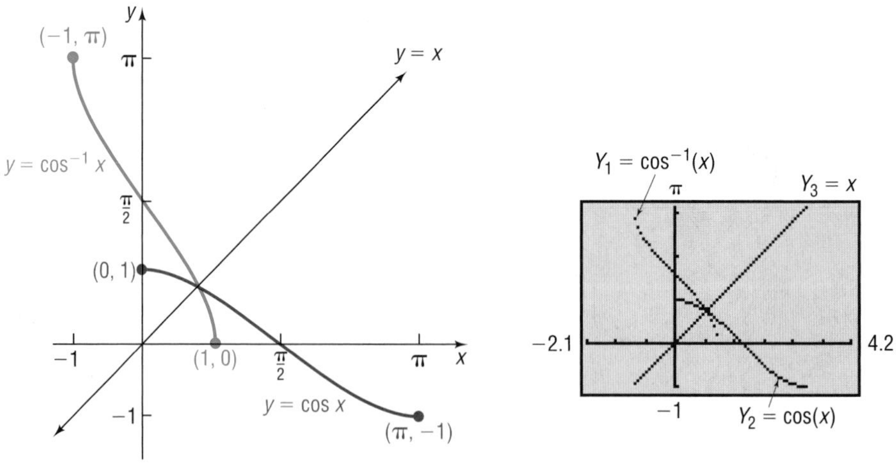

(a) (b)

For the cosine function and its inverse, the following properties hold:

$$f^{-1}(f(x)) = \cos^{-1}(\cos x) = x, \quad \text{where } 0 \le x \le \pi \qquad \textbf{(4a)}$$
$$f(f^{-1}(x)) = \cos(\cos^{-1} x) = x, \quad \text{where } -1 \le x \le 1 \qquad \textbf{(4b)}$$

*This is the generally accepted restriction to define the inverse.

EXAMPLE 4 Finding the Exact Value of an Inverse Cosine Function

Find the exact value of: $\cos^{-1} 0$

Solution Let $\theta = \cos^{-1} 0$. We seek the angle θ, $0 \leq \theta \leq \pi$, whose cosine equals 0.

$$\theta = \cos^{-1} 0, \qquad 0 \leq \theta \leq \pi$$
$$\cos\theta = 0, \qquad 0 \leq \theta \leq \pi$$

Look at Table 2 and Figure 18.

TABLE 2

θ	$\cos\theta$
0	1
$\pi/6$	$\sqrt{3}/2$
$\pi/4$	$\sqrt{2}/2$
$\pi/3$	$1/2$
$\pi/2$	0
$2\pi/3$	$-1/2$
$3\pi/4$	$-\sqrt{2}/2$
$5\pi/6$	$-\sqrt{3}/2$
π	-1

Figure 18

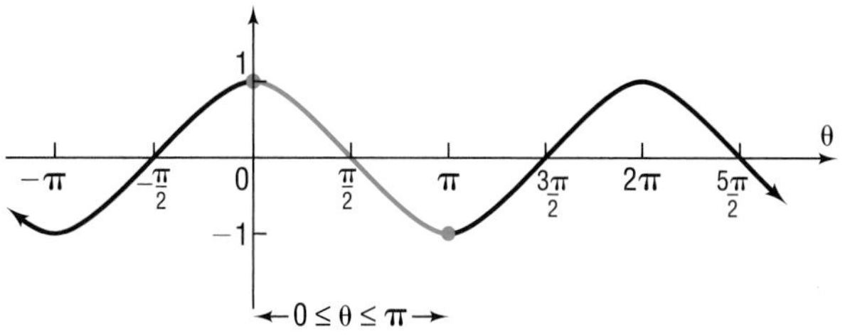

We see that the only angle θ within the interval $[0, \pi]$, whose cosine is 0, is $\pi/2$. [Note that $\cos(3\pi/2)$ also equals 0, but $3\pi/2$ lies outside the interval $[0, \pi]$ and hence is not admissible.] So, since $\cos(\pi/2) = 0$ and $\pi/2$ is in $[0, \pi]$, we conclude that

$$\cos^{-1} 0 = \frac{\pi}{2}$$

EXAMPLE 5 Finding the Exact Value of an Inverse Cosine Function

Find the exact value of: $\cos^{-1}(\sqrt{2}/2)$

Solution Let $\theta = \cos^{-1}(\sqrt{2}/2)$. We seek the angle θ, $0 \leq \theta \leq \pi$, whose cosine equals $\sqrt{2}/2$.

$$\theta = \cos^{-1}\frac{\sqrt{2}}{2}, \qquad 0 \leq \theta \leq \pi$$
$$\cos\theta = \frac{\sqrt{2}}{2}, \qquad 0 \leq \theta \leq \pi$$

Look at Table 2 and Figure 19.

Figure 19

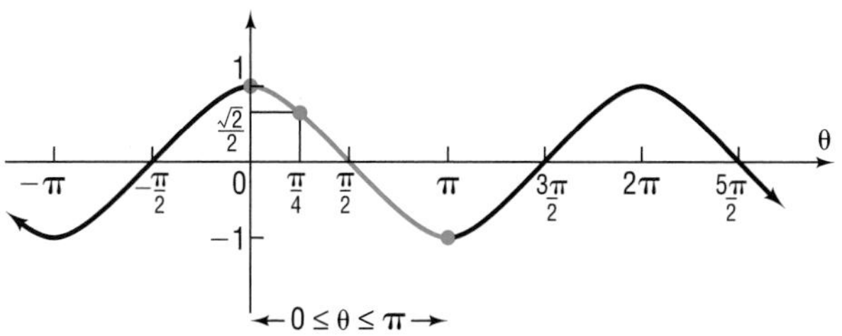

We see that the only angle θ within the interval $[0, \pi]$, whose cosine is $\sqrt{2}/2$, is $\pi/4$. So, since $\cos(\pi/4) = \sqrt{2}/2$ and $\pi/4$ is in $[0, \pi]$, we conclude that

$$\cos^{-1}\frac{\sqrt{2}}{2} = \frac{\pi}{4}$$

NOW WORK PROBLEM 11.

EXAMPLE 6 Finding the Exact Value of a Composite Function

Find the exact value of: (a) $\cos^{-1}\left[\cos\left(\frac{\pi}{12}\right)\right]$ (b) $\cos\left[\cos^{-1}(-0.4)\right]$

Solution (a) $\cos^{-1}\left[\cos\left(\frac{\pi}{12}\right)\right] = \frac{\pi}{12}$ By Property (4a)

(b) $\cos\left[\cos^{-1}(-0.4)\right] = -0.4$ By Property (4b)

NOW WORK PROBLEM 47.

THE INVERSE TANGENT FUNCTION

In Figure 20 we reproduce the graph of $y = \tan x$. Because every horizontal line intersects the graph infinitely many times, it follows that the tangent function is not one-to-one.

Figure 20
$y = \tan x$, $-\infty < x < \infty$, x not equal to odd multiples of $\pi/2$, $-\infty < y < \infty$

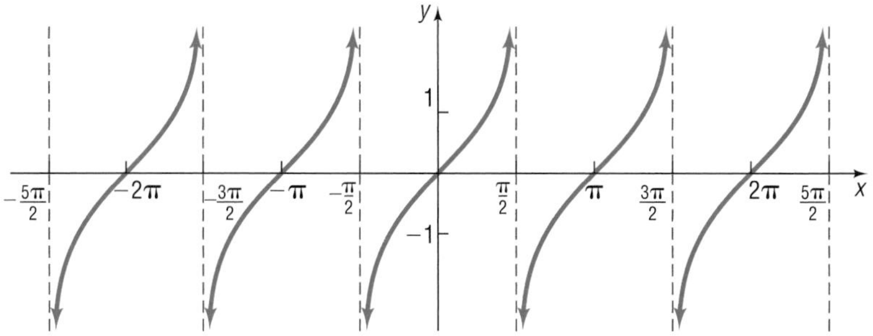

Figure 21
$y = \tan x$, $-\pi/2 < x < \pi/2$, $-\infty < y < \infty$

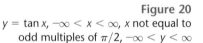

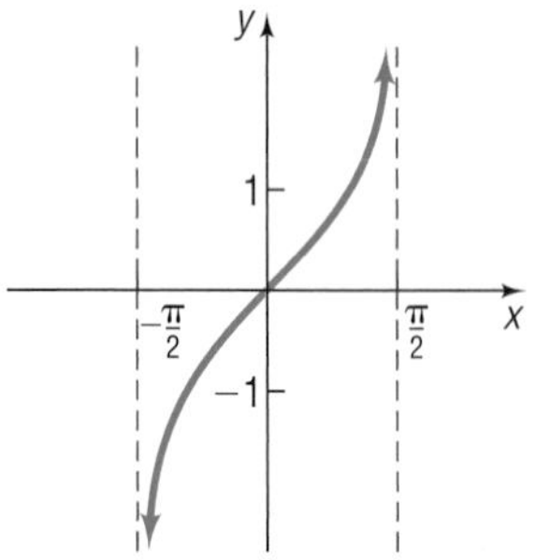

However, if we restrict the domain of $y = \tan x$ to the interval $(-\pi/2, \pi/2)$,* the restricted function

$$y = \tan x, \qquad -\frac{\pi}{2} < x < \frac{\pi}{2}$$

is one-to-one and hence has an inverse function. See Figure 21.

An equation for the inverse is obtained by interchanging x and y. The implicit form of the inverse function is $x = f(y) = \tan y$, $-\frac{\pi}{2} < y < \frac{\pi}{2}$. The explicit form is called the **inverse tangent** of x and is symbolized by $y = f^{-1}(x) = \tan^{-1} x$ (or by $y = \arctan x$).

$$y = \tan^{-1} x \quad \text{means} \quad x = \tan y \qquad \textbf{(5)}$$
$$\text{where} \quad -\infty < x < \infty \quad \text{and} \quad -\frac{\pi}{2} < y < \frac{\pi}{2}$$

Here, y is the angle whose tangent is x. The domain of the function $y = \tan^{-1} x$ is $-\infty < x < \infty$, and its range is $-\pi/2 < y < \pi/2$. The graph of $y = \tan^{-1} x$ can be obtained by reflecting the restricted portion of the graph of $y = \tan x$ about the line $y = x$, as shown in Figure 22(a). Figure 22(b) shows the graph using a graphing utility.

*This is the generally accepted restriction.

Figure 22
$y = \tan^{-1} x, -\infty < x < \infty, -\pi/2 < y < \pi/2$

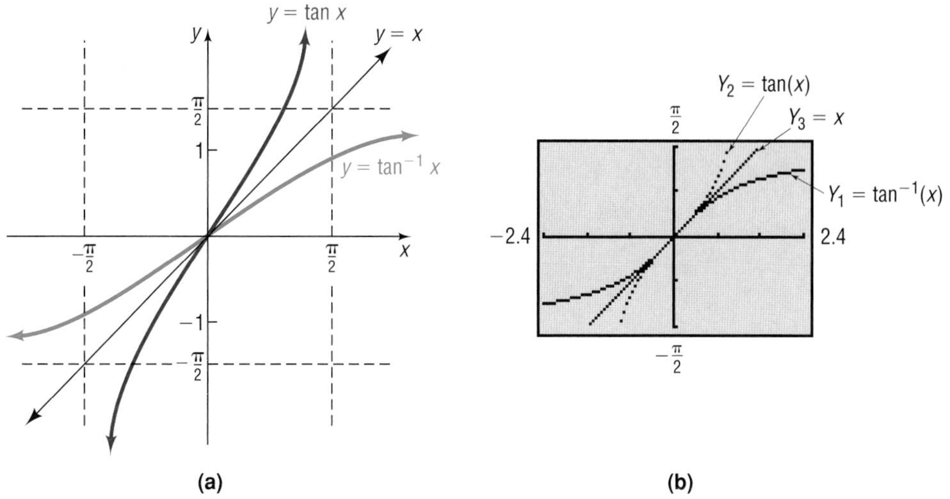

For the tangent function and its inverse, the following properties hold:

$$f^{-1}(f(x)) = \tan^{-1}(\tan x) = x, \quad \text{where } -\frac{\pi}{2} < x < \frac{\pi}{2}$$
$$f(f^{-1}(x)) = \tan(\tan^{-1} x) = x, \quad \text{where } -\infty < x < \infty$$

EXAMPLE 7 Finding the Exact Value of an Inverse Tangent Function

Find the exact value of: $\tan^{-1} 1$

Solution Let $\theta = \tan^{-1} 1$. We seek the angle θ, $-\pi/2 < \theta < \pi/2$, whose tangent equals 1.

$$\theta = \tan^{-1} 1, \quad -\frac{\pi}{2} < \theta < \frac{\pi}{2}$$

$$\tan\theta = 1, \quad -\frac{\pi}{2} < \theta < \frac{\pi}{2}$$

TABLE 3

θ	$\tan\theta$
$-\pi/2$	Undefined
$-\pi/3$	$-\sqrt{3}$
$-\pi/4$	-1
$-\pi/6$	$-\sqrt{3}/3$
0	0
$\pi/6$	$\sqrt{3}/3$
$\pi/4$	1
$\pi/3$	$\sqrt{3}$
$\pi/2$	Undefined

Look at Table 3 or Figure 21. The only angle θ within the interval $(-\pi/2, \pi/2)$ whose tangent is 1 is $\pi/4$. So, since $\tan(\pi/4) = 1$ and $\pi/4$ is in $(-\pi/2, \pi/2)$, we conclude that

$$\tan^{-1} 1 = \frac{\pi}{4}$$

EXAMPLE 8 **Finding the Exact Value of an Inverse Tangent Function**

Find the exact value of: $\tan^{-1}(-\sqrt{3})$

Solution Let $\theta = \tan^{-1}(-\sqrt{3})$. We seek the angle $\theta, -\pi/2 < \theta < \pi/2$, whose tangent equals $-\sqrt{3}$.

$$\theta = \tan^{-1}(-\sqrt{3}), \qquad -\frac{\pi}{2} < \theta < \frac{\pi}{2}$$

$$\tan\theta = -\sqrt{3}, \qquad -\frac{\pi}{2} < \theta < \frac{\pi}{2}$$

Look at Table 3 or Figure 21 if necessary. The only angle θ within the interval $(-\pi/2, \pi/2)$, whose tangent is $-\sqrt{3}$, is $-\pi/3$. So, since $\tan(-\pi/3) = -\sqrt{3}$ and $-\pi/3$ is in $(-\pi/2, \pi/2)$, we conclude that

$$\tan^{-1}(-\sqrt{3}) = -\frac{\pi}{3}$$

NOW WORK PROBLEM 5.

THE REMAINING INVERSE TRIGONOMETRIC FUNCTIONS

The inverse cotangent, inverse secant, and inverse cosecant functions are defined as follows:

$$y = \cot^{-1} x \quad \text{means} \quad x = \cot y \tag{6}$$
$$\text{where} \quad -\infty < x < \infty \quad \text{and} \quad 0 < y < \pi$$

$$y = \sec^{-1} x \quad \text{means} \quad x = \sec y \tag{7}$$
$$\text{where} \quad |x| \ge 1 \quad \text{and} \quad 0 \le y \le \pi, \quad y \ne \frac{\pi}{2}^{*}$$

$$y = \csc^{-1} x \quad \text{means} \quad x = \csc y \tag{8}$$
$$\text{where} \quad |x| \ge 1 \quad \text{and} \quad -\frac{\pi}{2} \le y \le \frac{\pi}{2}, \quad y \ne 0^{\dagger}$$

You are encouraged to review the graphs of the cotangent, cosecant, and secant functions in Figure 55 (p. 416), Figure 56 (p. 417), and Figure 58 (p. 418) in Section 6.5 to help you see the basis for these definitions.

EXAMPLE 9 **Finding the Exact Value of an Inverse Cosecant Function**

Find the exact value of: $\csc^{-1} 2$

Solution Let $\theta = \csc^{-1} 2$. We seek the angle θ, $-\frac{\pi}{2} \le \theta \le \frac{\pi}{2}, \theta \ne 0$, whose cosecant equals 2.

$$\theta = \csc^{-1} 2, \qquad -\frac{\pi}{2} \le \theta \le \frac{\pi}{2}, \quad \theta \ne 0$$

$$\csc\theta = 2, \qquad -\frac{\pi}{2} \le \theta \le \frac{\pi}{2}, \quad \theta \ne 0$$

*Most books use this definition. A few use the restriction $0 \le y < \frac{\pi}{2}, \pi \le y < \frac{3\pi}{2}$.

†Most books use this definition. A few use the restriction $-\pi < y \le -\frac{\pi}{2}, 0 < y \le \frac{\pi}{2}$.

The only angle θ in the interval $-\frac{\pi}{2} \le \theta \le \frac{\pi}{2}\pi, \theta \ne 0$, whose cosecant is 2 is $\frac{\pi}{6}$, so $\csc^{-1}2 = \pi/6$. ■

NOW WORK PROBLEM 15.

Most calculators do not have keys for evaluating the inverse cotangent, cosecant, and secant functions. The easiest way to evaluate them is to convert to an inverse trigonometric function whose range is the same as the one to be evaluated. In this regard, notice that $y = \cot^{-1} x$ and $y = \sec^{-1} x$ (except where undefined) each has the same range as $y = \cos^{-1} x$; $y = \csc^{-1} x$, except where undefined, has the same range as $y = \sin^{-1} x$.

EXAMPLE 10 Approximating the Value of Inverse Trigonometric Functions

Use a calculator to approximate each expression in radians rounded to two decimal places.

(a) $\sec^{-1} 3$ (b) $\csc^{-1}(-4)$ (c) $\cot^{-1}\frac{1}{2}$ (d) $\cot^{-1}(-2)$

Solution First, set your calculator to radian mode.

(a) Let $\theta = \sec^{-1} 3$. Then $\sec\theta = 3$ and $0 \le \theta \le \pi, \theta \ne \pi/2$. Since $\cos\theta = \frac{1}{3}$ and $\theta = \cos^{-1}\frac{1}{3}$, we have

$$\sec^{-1} 3 = \theta = \cos^{-1}\tfrac{1}{3} \approx 1.23$$

↑ Use a calculator.

Figure 23
$\cot\theta = \frac{1}{2}, 0 < \theta < \pi$

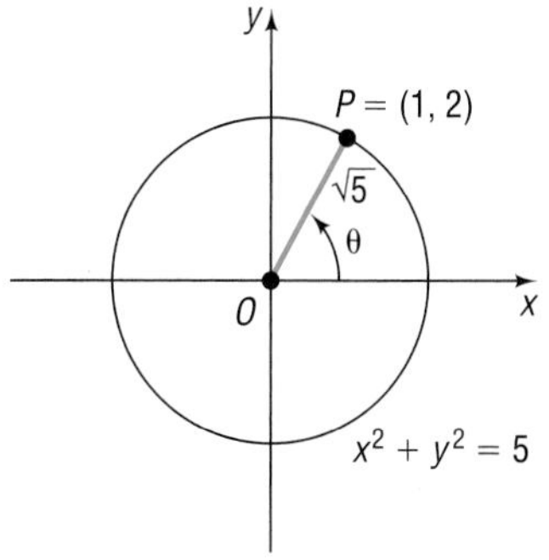

(b) Let $\theta = \csc^{-1}(-4)$. Then $\csc\theta = -4, -\pi/2 \le \theta \le \pi/2, \theta \ne 0$. Since $\sin\theta = -\frac{1}{4}$ and $\theta = \sin^{-1}(-\frac{1}{4})$, we have

$$\csc^{-1}(-4) = \theta = \sin^{-1}(-\tfrac{1}{4}) \approx -0.25$$

(c) Let $\theta = \cot^{-1}\frac{1}{2}$. Then $\cot\theta = \frac{1}{2}, 0 < \theta < \pi$. From these facts we know that θ is in quadrant I. We seek $\cos\theta$ given that $\cot\theta = 1/2$, $0 < \theta < \pi/2$. Since $\cot\theta = 1/2 = x/y, 0 < \theta < \pi/2$, we let $x = 1$ and $y = 2$. The point $P = (x, y) = (1, 2)$ is on the circle $x^2 + y^2 = 5$, since $r = d(O, P) = \sqrt{1^2 + 2^2} = \sqrt{5}$. See Figure 23. Then $\cos\theta = x/r = 1/\sqrt{5}$, so $\theta = \cos^{-1}(1/\sqrt{5})$. As a result,

$$\cot^{-1}\frac{1}{2} = \theta = \cos^{-1}\left(\frac{1}{\sqrt{5}}\right) \approx 1.11$$

Figure 24
$\cot\theta = -2, 0 < \theta < \pi$

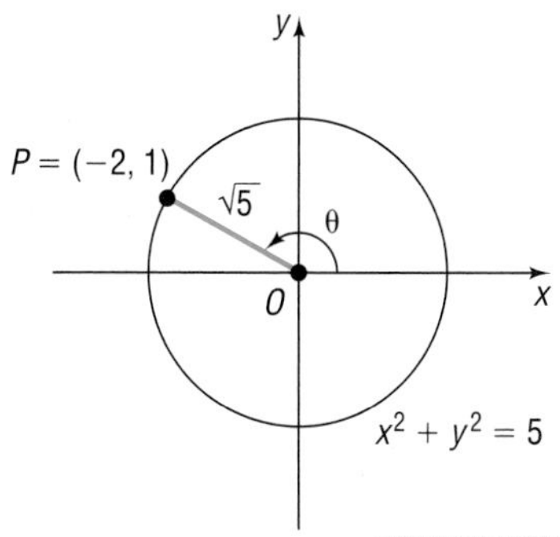

(d) Let $\theta = \cot^{-1}(-2)$. Then $\cot\theta = -2, 0 < \theta < \pi$. From these facts we know that θ lies in quadrant II. We seek $\cos\theta$ given that $\cot\theta = -2$, $\pi/2 < \theta < \pi$. Since $\cot\theta = -2 = x/y, \pi/2 < \theta < \pi$, we let $x = -2$ and $y = 1$. The point $P = (x, y) = (-2, 1)$ is on the circle $x^2 + y^2 = 5$, since $r = d(O, P) = \sqrt{(-2)^2 + 1^2} = \sqrt{5}$. See Figure 24. Then $\cos\theta = x/r = -2/\sqrt{5}$, so $\theta = \cos^{-1}(-2/\sqrt{5})$. As a result,

$$\cot^{-1}(-2) = \theta = \cos^{-1}\left(-\frac{2}{\sqrt{5}}\right) \approx 2.68$$ ■

NOW WORK PROBLEM 33.

7.5 EXERCISES

In Problems 1–20, find the exact value of each expression.

1. $\sin^{-1} 0$
2. $\cos^{-1} 1$
3. $\sin^{-1}(-1)$
4. $\cos^{-1}(-1)$
5. $\tan^{-1} 0$
6. $\tan^{-1}(-1)$
7. $\sin^{-1} \frac{\sqrt{2}}{2}$
8. $\tan^{-1} \frac{\sqrt{3}}{3}$
9. $\tan^{-1} \sqrt{3}$
10. $\sin^{-1}\left(-\frac{\sqrt{3}}{2}\right)$
11. $\cos^{-1}\left(-\frac{\sqrt{3}}{2}\right)$
12. $\sin^{-1}\left(-\frac{\sqrt{2}}{2}\right)$
13. $\cot^{-1} \sqrt{3}$
14. $\cot^{-1} 1$
15. $\csc^{-1}(-1)$
16. $\csc^{-1} \sqrt{2}$
17. $\sec^{-1} \frac{2\sqrt{3}}{3}$
18. $\sec^{-1}(-2)$
19. $\cot^{-1}\left(-\frac{\sqrt{3}}{3}\right)$
20. $\csc^{-1}\left(-\frac{2\sqrt{3}}{3}\right)$

In Problems 21–44, use a calculator to find the value of each expression rounded to two decimal places.

21. $\sin^{-1} 0.1$
22. $\cos^{-1} 0.6$
23. $\tan^{-1} 5$
24. $\tan^{-1} 0.2$
25. $\cos^{-1} \frac{7}{8}$
26. $\sin^{-1} \frac{1}{8}$
27. $\tan^{-1}(-0.4)$
28. $\tan^{-1}(-3)$
29. $\sin^{-1}(-0.12)$
30. $\cos^{-1}(-0.44)$
31. $\cos^{-1} \frac{\sqrt{2}}{3}$
32. $\sin^{-1} \frac{\sqrt{3}}{5}$
33. $\sec^{-1} 4$
34. $\csc^{-1} 5$
35. $\cot^{-1} 2$
36. $\sec^{-1}(-3)$
37. $\csc^{-1}(-3)$
38. $\cot^{-1}\left(-\frac{1}{2}\right)$
39. $\cot^{-1}(-\sqrt{5})$
40. $\cot^{-1}(-8.1)$
41. $\csc^{-1}\left(-\frac{3}{2}\right)$
42. $\sec^{-1}\left(-\frac{4}{3}\right)$
43. $\cot^{-1}\left(-\frac{3}{2}\right)$
44. $\cot^{-1}(-\sqrt{10})$

In Problems 45–52, find the exact value of the expression.

45. $\sin\left[\sin^{-1}(0.54)\right]$
46. $\tan\left[\tan^{-1}(7.4)\right]$
47. $\cos^{-1}\left[\cos\left(\frac{4\pi}{5}\right)\right]$
48. $\sin^{-1}\left[\sin\left(-\frac{\pi}{10}\right)\right]$
49. $\tan\left[\tan^{-1}(-3.5)\right]$
50. $\cos\left[\cos^{-1}(-0.05)\right]$
51. $\sin^{-1}\left[\sin\left(-\frac{3\pi}{7}\right)\right]$
52. $\tan^{-1}\left[\tan\left(\frac{2\pi}{5}\right)\right]$

53. **Being the First to See the Rising Sun** Cadillac Mountain, elevation 1530 feet, is located in Acadia National Park, Maine, and is the highest peak on the east coast of the United States. It is said that a person standing on the summit will be the first person in the United States to see the rays of the rising Sun. How much sooner would a person atop Cadillac Mountain see the first rays than a person standing below, at sea level?
[**Hint:** Consult the figure. When the person at D sees the first rays of the Sun, the person at P does not. The person at P sees the first rays of the Sun only after Earth has rotated so that P is at location Q. Compute the length of the arc subtended by the central angle θ. Then use the fact that, at the latitude of Cadillac Mountain, in 24 hours a length of 2π (2710) miles is subtended, and find the time it takes to subtend this length.]

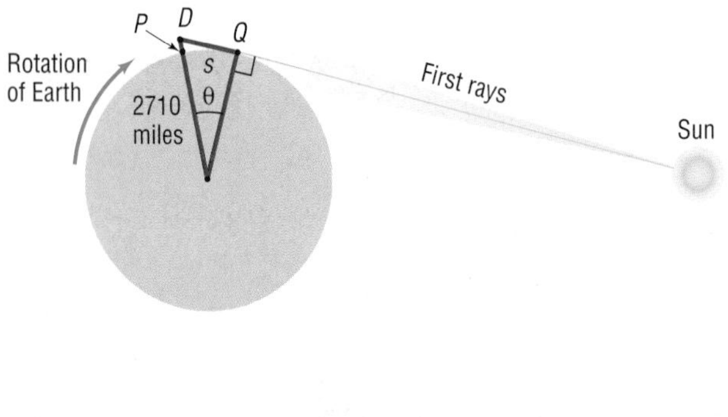

54. For what numbers x does $\sin(\sin^{-1} x) = x$?
55. For what numbers x does $\cos(\cos^{-1} x) = x$?
56. For what numbers x does $\sin^{-1}(\sin x) = x$?
57. For what numbers x does $\cos^{-1}(\cos x) = x$?
58. Using a graphing utility, graph $y = \cot^{-1} x$.
59. Using a graphing utility, graph $y = \sec^{-1} x$.
60. Using a graphing utility, graph $y = \csc^{-1} x$.
61. Explain in your own words how you would use your calculator to find the value of $\cot^{-1} 10$.
62. Consult three books on calculus and write down the definition in each of $y = \sec^{-1} x$ and $y = \csc^{-1} x$. Compare these with the definition given in this book.

PREPARING FOR THIS SECTION

Before getting started, review the following concepts:

✓ Sum and Difference Formulas (pp. 452, 455)

7.6 THE INVERSE TRIGONOMETRIC FUNCTIONS (II)

OBJECTIVES
1. Find the Exact Value of Expressions Involving Inverse Trigonometric Functions
2. Write a Trigonometric Expression as an Algebraic Expression
3. Establish Identities Involving Inverse Trigonometric Expressions

1 In this section we continue our discussion of the inverse trigonometric functions.

EXAMPLE 1 Finding the Exact Value of Expressions Involving Inverse Trigonometric Functions

Find the exact value of: $\sin^{-1}(\sin 5\pi/4)$

Solution

$$\sin^{-1}\left(\sin\frac{5\pi}{4}\right) = \sin^{-1}\left(-\frac{\sqrt{2}}{2}\right) = -\frac{\pi}{4}$$

Notice in the solution to Example 1 that we did not use Property (2a), page 475. This is because the argument of the sine function is not in the interval $\left[-\frac{\pi}{2}, \frac{\pi}{2}\right]$, as required. To use Property (2a), we use the fact that $\sin\frac{5\pi}{4} = -\sin\frac{\pi}{4}$. Then

$$\sin^{-1}\left(\sin\frac{5\pi}{4}\right) = \sin^{-1}\left(-\sin\frac{\pi}{4}\right) \underset{\substack{\uparrow \\ y = \sin x \text{ is odd}}}{=} \sin^{-1}[\sin(-\pi/4)] \underset{\substack{\uparrow \\ \text{Property (2a)}}}{=} -\frac{\pi}{4}$$

NOW WORK PROBLEM 13.

EXAMPLE 2 Finding the Exact Value of Expressions Involving Inverse Trigonometric Functions

Find the exact value of: $\cos\left[\tan^{-1}(-1)\right]$

Solution Let $\theta = \tan^{-1}(-1)$. We seek the angle $\theta, -\pi/2 < \theta < \pi/2$, whose tangent equals -1.

$$\tan\theta = -1, \qquad -\frac{\pi}{2} < \theta < \frac{\pi}{2}$$

$$\theta = -\frac{\pi}{4}$$

Now

$$\cos\left[\tan^{-1}(-1)\right] = \cos\theta = \cos\left(-\frac{\pi}{4}\right) = \frac{\sqrt{2}}{2}$$

NOW WORK PROBLEM 1.

It is not necessary to be able to find the angle in order to solve a problem like Example 2.

EXAMPLE 3 Finding the Exact Value of Expressions Involving Inverse Trigonometric Functions

Find the exact value of: $\sin\left(\tan^{-1}\frac{1}{2}\right)$

Figure 25
$\tan\theta = \frac{1}{2}$

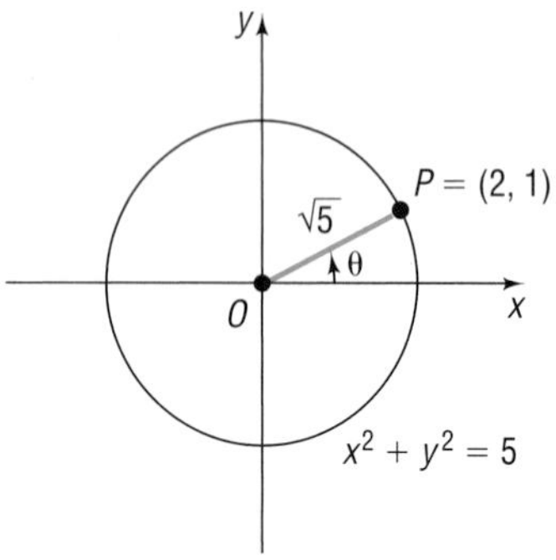

Solution Let $\theta = \tan^{-1}\frac{1}{2}$. Then $\tan\theta = \frac{1}{2}$, where $-\pi/2 < \theta < \pi/2$. Because $\tan\theta > 0$, it follows that $0 < \theta < \pi/2$. We seek $\sin\theta$ given that $\tan\theta = 1/2$, $0 < \theta < \pi/2$. Since $\tan\theta = 1/2 = y/x$, $0 < \theta < \pi/2$, we let $x = 2$ and $y = 1$. The point $P = (x, y) = (2, 1)$ is on the circle $x^2 + y^2 = 5$, since $r = d(O, P) = \sqrt{2^2 + 1^2} = \sqrt{5}$. See Figure 25. Then $\sin\theta = y/r = 1/\sqrt{5}$, and

$$\sin\left(\tan^{-1}\frac{1}{2}\right) = \sin\theta = \frac{1}{\sqrt{5}} = \frac{\sqrt{5}}{5}$$

EXAMPLE 4 Finding the Exact Value of Expressions Involving Inverse Trigonometric Functions

Find the exact value of: $\cos\left[\sin^{-1}\left(-\frac{1}{3}\right)\right]$

Solution Let $\theta = \sin^{-1}\left(-\frac{1}{3}\right)$. Then $\sin\theta = -\frac{1}{3}$ and $-\pi/2 \le \theta \le \pi/2$. Because $\sin\theta < 0$, it follows that $-\pi/2 \le \theta < 0$. We seek $\cos\theta$ given that $\sin\theta = -1/3, -\pi/2 \le \theta < 0$. Since $\sin\theta = -1/3 = y/r, -\pi/2 < \theta < 0$, we let $y = -1$ and $r = 3$. The point $P = (x, y) = (x, -1)$ lies in Quadrant IV and is on a circle of radius 3, $x^2 + y^2 = 9$. See Figure 26. Then

Figure 26
$\sin\theta = -\frac{1}{3}$

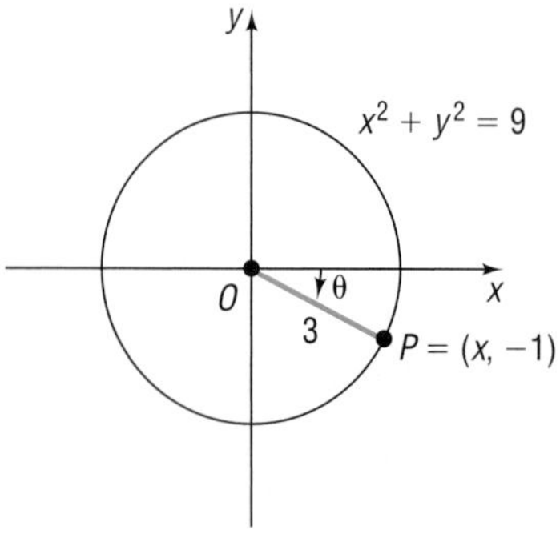

$$x^2 + y^2 = 9 \qquad x > 0,\ y = -1$$
$$x^2 + (-1)^2 = 9$$
$$x^2 = 8$$
$$x = 2\sqrt{2}$$

Then, $\cos\theta = x/r = 2\sqrt{2}/3$ and

$$\cos\left[\sin^{-1}\left(-\frac{1}{3}\right)\right] = \cos\theta = \frac{2\sqrt{2}}{3}$$

EXAMPLE 5 Finding the Exact Value of Expressions Involving Inverse Trigonometric Functions

Find the exact value of: $\tan\left[\cos^{-1}\left(-\frac{1}{3}\right)\right]$

Solution Let $\theta = \cos^{-1}\left(-\frac{1}{3}\right)$. Then $\cos\theta = -\frac{1}{3}$ and $0 \le \theta \le \pi$. Because $\cos\theta < 0$, it follows that $\pi/2 < \theta \le \pi$. We seek $\tan\theta$ given that $\cos\theta = -1/3$, $\pi/2 < \theta \le \pi$. Since $\cos\theta = -1/3 = x/r$, $\pi/2 < \theta \le \pi$, we let $x = -1$ and $r = 3$. The point $P = (x, y) = (-1, y)$ lies in Quadrant II and is on a circle of radius 3, $x^2 + y^2 = 9$. See Figure 27. Then

Figure 27
$\cos\theta = -\frac{1}{3}$

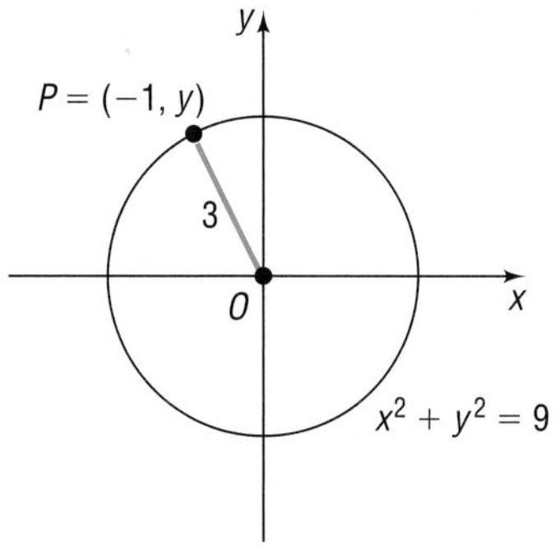

$$x^2 + y^2 = 9 \qquad x = -1,\ y = 0$$

$$(-1)^2 + y^2 = 9$$

$$y^2 = 8$$

$$y = 2\sqrt{2}$$

Now, $\tan\theta = y/x = 2\sqrt{2}/-1 = -2\sqrt{2}$ so that

$$\tan\left[\cos^{-1}\left(-\frac{1}{3}\right)\right] = \tan\theta = -2\sqrt{2}$$

NOW WORK PROBLEM 19.

EXAMPLE 6 Finding the Exact Value of Expressions Involving Inverse Trigonometric Functions

Find the exact value of: $\sin\left(\cos^{-1}\frac{1}{2} + \sin^{-1}\frac{3}{5}\right)$

Solution Let $\alpha = \cos^{-1}\frac{1}{2}$ and $\beta = \sin^{-1}\frac{3}{5}$. Then

$$\cos\alpha = \tfrac{1}{2}, \quad 0 \le \alpha \le \pi, \quad \text{and} \quad \sin\beta = \tfrac{3}{5}, \quad -\frac{\pi}{2} \le \beta \le \frac{\pi}{2}$$

We use Pythagorean identities to obtain $\sin\alpha$ and $\cos\beta$. Since $\sin\alpha > 0$ and $\cos\beta > 0$, (Do you know why?), we find

$$\sin\alpha = \sqrt{1 - \cos^2\alpha} = \sqrt{1 - \frac{1}{4}} = \sqrt{\frac{3}{4}} = \frac{\sqrt{3}}{2}$$

$$\cos\beta = \sqrt{1 - \sin^2\beta} = \sqrt{1 - \frac{9}{25}} = \sqrt{\frac{16}{25}} = \frac{4}{5}$$

As a result,

$$\sin\left(\cos^{-1}\frac{1}{2} + \sin^{-1}\frac{3}{5}\right) = \sin(\alpha + \beta) = \sin\alpha\cos\beta + \cos\alpha\sin\beta$$

$$= \frac{\sqrt{3}}{2}\cdot\frac{4}{5} + \frac{1}{2}\cdot\frac{3}{5} = \frac{4\sqrt{3} + 3}{10}$$

NOW WORK PROBLEM 31.

EXAMPLE 7 Writing a Trigonometric Expression as an Algebraic Expression

2 Write $\sin(\sin^{-1} u + \cos^{-1} v)$ as an algebraic expression containing u and v (that is, without any trigonometric functions).

Solution Let $\alpha = \sin^{-1} u$ and $\beta = \cos^{-1} v$. Then

$$\sin\alpha = u, \qquad -\frac{\pi}{2} \le \alpha \le \frac{\pi}{2} \quad \text{and} \quad \cos\beta = v, \qquad 0 \le \beta \le \pi$$

Since $-\pi/2 \le \alpha \le \pi/2$, we know that $\cos\alpha \ge 0$. As a result,

$$\cos\alpha = \sqrt{1 - \sin^2\alpha} = \sqrt{1 - u^2}$$

Similarly, since $0 \le \beta \le \pi$, we know that $\sin\beta \ge 0$. Thus,

$$\sin\beta = \sqrt{1 - \cos^2\beta} = \sqrt{1 - v^2}$$

Now

$$\begin{aligned}\sin(\sin^{-1} u + \cos^{-1} v) = \sin(\alpha + \beta) &= \sin\alpha\cos\beta + \cos\alpha\sin\beta \\ &= uv + \sqrt{1 - u^2}\sqrt{1 - v^2}\end{aligned}$$

NOW WORK PROBLEM 53.

EXAMPLE 8 Establishing an Identity Involving Inverse Trigonometric Functions

3 Show that $\sin(\tan^{-1} v) = \dfrac{v}{\sqrt{1 + v^2}}$.

Solution Let $\theta = \tan^{-1} v$, so that $\tan\theta = v$, $-\pi/2 < \theta < \pi/2$. As a result, we know that $\sec\theta > 0$.

$$\sin(\tan^{-1} v) = \sin\theta = \sin\theta \cdot \frac{\cos\theta}{\cos\theta} \underset{\frac{\sin\theta}{\cos\theta} = \tan\theta}{=} \tan\theta\cos\theta = \frac{\tan\theta}{\sec\theta} \underset{\substack{\sec^2\theta = 1 + \tan^2\theta \\ \sec\theta > 0}}{=} \frac{\tan\theta}{\sqrt{1 + \tan^2\theta}} = \frac{v}{\sqrt{1 + v^2}}$$

NOW WORK PROBLEM 59.

7.6 EXERCISES

In Problems 1–28, find the exact value of each expression.

1. $\cos\left(\sin^{-1}\frac{\sqrt{2}}{2}\right)$
2. $\sin\left(\cos^{-1}\frac{1}{2}\right)$
3. $\tan\left[\cos^{-1}\left(-\frac{\sqrt{3}}{2}\right)\right]$
4. $\tan\left[\sin^{-1}\left(-\frac{1}{2}\right)\right]$
5. $\sec\left(\cos^{-1}\frac{1}{2}\right)$
6. $\cot\left[\sin^{-1}\left(-\frac{1}{2}\right)\right]$
7. $\csc(\tan^{-1} 1)$
8. $\sec(\tan^{-1}\sqrt{3})$
9. $\sin[\tan^{-1}(-1)]$
10. $\cos\left[\sin^{-1}\left(-\frac{\sqrt{3}}{2}\right)\right]$
11. $\sec\left[\sin^{-1}\left(-\frac{1}{2}\right)\right]$
12. $\csc\left[\cos^{-1}\left(-\frac{\sqrt{3}}{2}\right)\right]$
13. $\cos^{-1}\left(\cos\frac{5\pi}{4}\right)$
14. $\tan^{-1}\left(\tan\frac{2\pi}{3}\right)$
15. $\sin^{-1}\left[\sin\left(-\frac{7\pi}{6}\right)\right]$
16. $\cos^{-1}\left[\cos\left(-\frac{\pi}{3}\right)\right]$
17. $\tan\left(\sin^{-1}\frac{1}{3}\right)$
18. $\tan\left(\cos^{-1}\frac{1}{3}\right)$
19. $\sec\left(\tan^{-1}\frac{1}{2}\right)$
20. $\cos\left(\sin^{-1}\frac{\sqrt{2}}{3}\right)$
21. $\cot\left[\sin^{-1}\left(-\frac{\sqrt{2}}{3}\right)\right]$
22. $\csc[\tan^{-1}(-2)]$
23. $\sin[\tan^{-1}(-3)]$
24. $\cot\left[\cos^{-1}\left(-\frac{\sqrt{3}}{3}\right)\right]$
25. $\sec\left(\sin^{-1}\frac{2\sqrt{5}}{5}\right)$
26. $\csc\left(\tan^{-1}\frac{1}{2}\right)$
27. $\sin^{-1}\left(\cos\frac{3\pi}{4}\right)$
28. $\cos^{-1}\left(\sin\frac{7\pi}{6}\right)$

In Problems 29–52, find the exact value of each expression.

29. $\sin\left(\sin^{-1}\frac{1}{2} + \cos^{-1}0\right)$

30. $\sin\left(\sin^{-1}\frac{\sqrt{3}}{2} + \cos^{-1}1\right)$

31. $\sin\left[\sin^{-1}\frac{3}{5} - \cos^{-1}\left(-\frac{4}{5}\right)\right]$

32. $\sin\left[\sin^{-1}\left(-\frac{4}{5}\right) - \tan^{-1}\frac{3}{4}\right]$

33. $\cos\left(\tan^{-1}\frac{4}{3} + \cos^{-1}\frac{5}{13}\right)$

34. $\sin\left[\tan^{-1}\frac{5}{12} - \sin^{-1}\left(-\frac{3}{5}\right)\right]$

35. $\sec\left(\sin^{-1}\frac{5}{13} - \tan^{-1}\frac{3}{4}\right)$

36. $\sec\left(\tan^{-1}\frac{4}{3} + \cot^{-1}\frac{5}{12}\right)$

37. $\cot\left(\sec^{-1}\frac{5}{3} + \frac{\pi}{6}\right)$

38. $\cos\left(\frac{\pi}{4} - \csc^{-1}\frac{5}{3}\right)$

39. $\sin\left(2\sin^{-1}\frac{1}{2}\right)$

40. $\sin\left[2\sin^{-1}\frac{\sqrt{3}}{2}\right]$

41. $\cos\left(2\sin^{-1}\frac{3}{5}\right)$

42. $\cos\left(2\cos^{-1}\frac{4}{5}\right)$

43. $\tan\left[2\cos^{-1}\left(-\frac{3}{5}\right)\right]$

44. $\tan\left(2\tan^{-1}\frac{3}{4}\right)$

45. $\sin\left(2\cos^{-1}\frac{4}{5}\right)$

46. $\cos\left[2\tan^{-1}\left(-\frac{4}{3}\right)\right]$

47. $\sin^2\left(\frac{1}{2}\cos^{-1}\frac{3}{5}\right)$

48. $\cos^2\left(\frac{1}{2}\sin^{-1}\frac{3}{5}\right)$

49. $\sec\left(2\tan^{-1}\frac{3}{4}\right)$

50. $\csc\left[2\sin^{-1}\left(-\frac{3}{5}\right)\right]$

51. $\cot^2\left(\frac{1}{2}\tan^{-1}\frac{4}{3}\right)$

52. $\cot^2\left(\frac{1}{2}\cos^{-1}\frac{5}{13}\right)$

In Problems 53–58, write each trigonometric expression as an algebraic expression containing u and v.

53. $\cos(\cos^{-1}u + \sin^{-1}v)$

54. $\sin(\sin^{-1}u - \cos^{-1}v)$

55. $\sin(\tan^{-1}u - \sin^{-1}v)$

56. $\cos(\tan^{-1}u + \tan^{-1}v)$

57. $\tan(\sin^{-1}u - \cos^{-1}v)$

58. $\sec(\tan^{-1}u + \cos^{-1}v)$

In Problems 59–70, establish each identity.

59. Show that $\sec(\tan^{-1}v) = \sqrt{1 + v^2}$.

60. Show that $\tan(\sin^{-1}v) = v/\sqrt{1 - v^2}$.

61. Show that $\tan(\cos^{-1}v) = \sqrt{1 - v^2}/v$.

62. Show that $\sin(\cos^{-1}v) = \sqrt{1 - v^2}$.

63. Show that $\cos(\sin^{-1}v) = \sqrt{1 - v^2}$.

64. Show that $\cos(\tan^{-1}v) = 1/\sqrt{1 + v^2}$.

65. Show that $\sin^{-1}v + \cos^{-1}v = \pi/2$.

66. Show that $\tan^{-1}v + \cot^{-1}v = \pi/2$.

67. Show that $\tan^{-1}(1/v) = \pi/2 - \tan^{-1}v$, if $v > 0$.

68. Show that $\cot^{-1}e^v = \tan^{-1}e^{-v}$.

69. Show that $\sin(\sin^{-1}v + \cos^{-1}v) = 1$.

70. Show that $\cos(\sin^{-1}v + \cos^{-1}v) = 0$.

In Problems 71–76, use the following: The formula

$$D = 24\left[1 - \frac{\cos^{-1}(\tan i \tan\theta)}{\pi}\right]$$

can be used to approximate the number of hours of daylight when the declination of the sun is $i°$ at a location $\theta°$ north latitude for any date between the vernal equinox and autumnal equinox. The declination of the sun is defined as the angle i between the equatorial plane and any ray of light from the sun. The latitude of a location is the angle θ between the Equator and the location on the surface of Earth, with the vertex of the angle located at the center of Earth. See the figure. To use the formula, $\cos^{-1}(\tan i \tan\theta)$ must be expressed in radians.

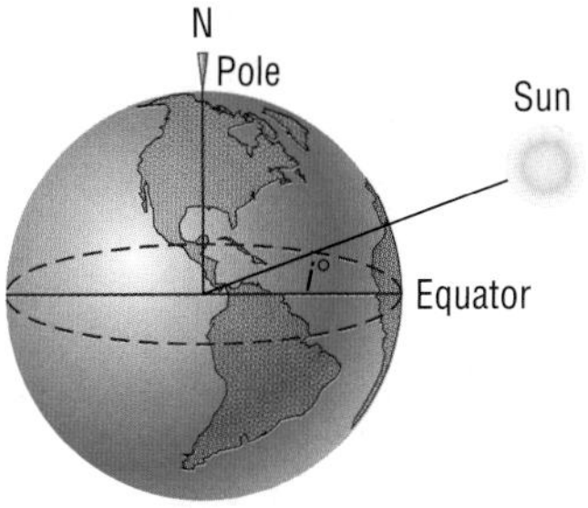

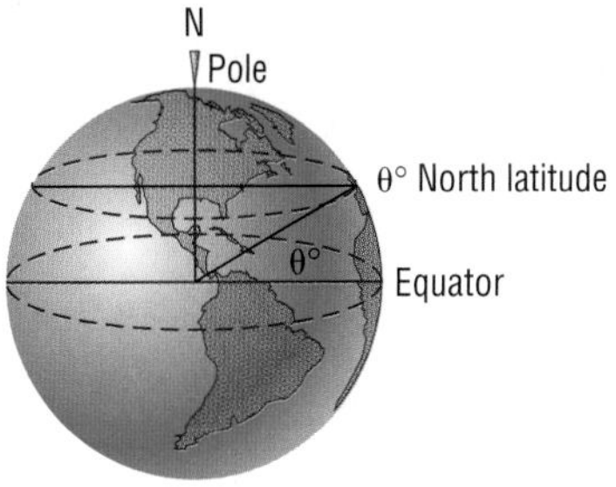

71. Approximate the number of hours of daylight in Houston, Texas, (29°45′ north latitude) for the following dates:
(a) Summer solstice ($i = 23.5°$)
(b) Vernal equinox ($i = 0°$)
(c) July 4 ($i = 22°48'$)

72. Approximate the number of hours of daylight in New York, New York, (40°45′ north latitude) for the following dates:
(a) Summer solstice ($i = 23.5°$)
(b) Vernal equinox ($i = 0°$)
(c) July 4 ($i = 22°48'$)

73. Approximate the number of hours of daylight in Honolulu, Hawaii, (21°18′ north latitude) for the following dates:
(a) Summer solstice ($i = 23.5°$)
(b) Vernal equinox ($i = 0°$)
(c) July 4 ($i = 22°48'$)

74. Approximate the number of hours of daylight in Anchorage, Alaska, (61°10′ north latitude) for the following dates:
(a) Summer solstice ($i = 23.5°$)
(b) Vernal equinox ($i = 0°$)
(c) July 4 ($i = 22°48'$)

75. Approximate the number of hours of daylight at the Equator (0° north latitude) for the following dates:
(a) Summer solstice ($i = 23.5°$)
(b) Vernal equinox ($i = 0°$)
(c) July 4 ($i = 22°48'$)
(d) What do you conclude about the number of hours of daylight throughout the year for a location at the Equator?

76. Approximate the number of hours of daylight for any location that is 66°30′ north latitude for the following dates:
(a) Summer solstice ($i = 23.5°$)
(b) Vernal equinox ($i = 0°$)
(c) July 4 ($i = 22°48'$)
(d) The number of hours of daylight on the winter solstice may be found by computing the number of hours of daylight on the summer solstice and subtracting this result from 24 hours, due to the symmetry of the orbital path of Earth around the Sun. Compute the number of hours of daylight for this location on the winter solstice. What do you conclude about daylight for a location at 66°30′ north latitude?

PREPARING FOR THIS SECTION

Before getting started, review the following:

✓ Solving Equations (Section 1.3)

✓ Values of the Trigonometric Functions of Certain Angles (pp. 384 and 387)

7.7 TRIGONOMETRIC EQUATIONS (I)

OBJECTIVES 1 Solve Equations Involving a Single Trigonometric Function

1 The first four sections of this chapter were devoted to trigonometric identities—that is, equations involving trigonometric functions that are satisfied by every value in the domain of the variable. In the remaining two sections, we discuss **trigonometric equations**—that is, equations involving trigonometric functions that are satisfied only by some values of the variable (or, possibly, are not satisfied by any values of the variable). The values that satisfy the equation are called **solutions** of the equation.

EXAMPLE 1 Checking Whether a Given Number Is a Solution of a Trigonometric Equation

Determine whether $\theta = \pi/4$ is a solution of the equation $\sin\theta = \frac{1}{2}$. Is $\theta = \pi/6$ a solution?

Solution Replace θ by $\pi/4$ in the given equation. The result is

$$\sin\frac{\pi}{4} = \frac{\sqrt{2}}{2} \neq \frac{1}{2}$$

We conclude that $\pi/4$ is not a solution.

Next, replace θ by $\pi/6$ in the equation. The result is

$$\sin\frac{\pi}{6} = \frac{1}{2}$$

We conclude that $\pi/6$ is a solution of the given equation. ■

The equation given in Example 1 has other solutions besides $\theta = \pi/6$. For example, $\theta = 5\pi/6$ is also a solution, as is $\theta = 13\pi/6$. (You should check

Figure 28

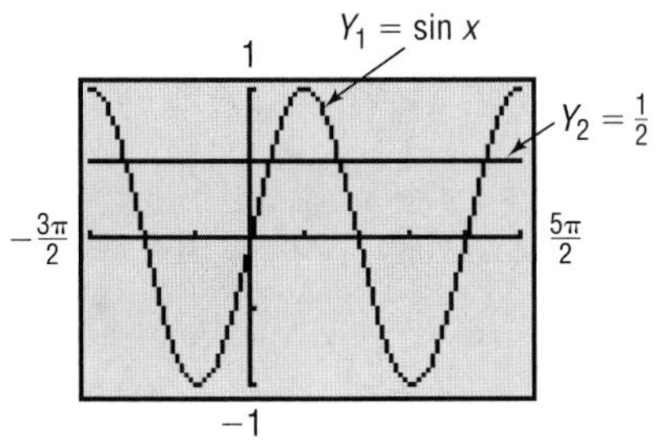

this for yourself.) In fact, the equation has an infinite number of solutions due to the periodicity of the sine function, as can be seen in Figure 28.

As before, our practice will be to solve equations, whenever possible, by finding exact solutions. In such cases, we will also verify the solution obtained by using a graphing utility. When traditional methods cannot be used, approximate solutions will be obtained using a graphing utility. The reader is encouraged to pay particular attention to the form of equations for which exact solutions are possible.

Unless the domain of the variable is restricted, we need to find *all* the solutions of a trigonometric equation. As the next example illustrates, finding all the solutions can be accomplished by first finding solutions over an interval whose length equals the period of the function and then adding multiples of that period to the solutions found. Let's look at some examples.

EXAMPLE 2 Finding All the Solutions of a Trigonometric Equation

Solve the equation: $\cos\theta = \frac{1}{2}$

Solution The period of the cosine function is 2π. In the interval $[0, 2\pi]$, there are two angles θ for which $\cos\theta = \frac{1}{2}$: $\theta = \pi/3$ and $\theta = 5\pi/3$. See Figure 29.

Figure 29
$\cos\theta = \frac{1}{2}$

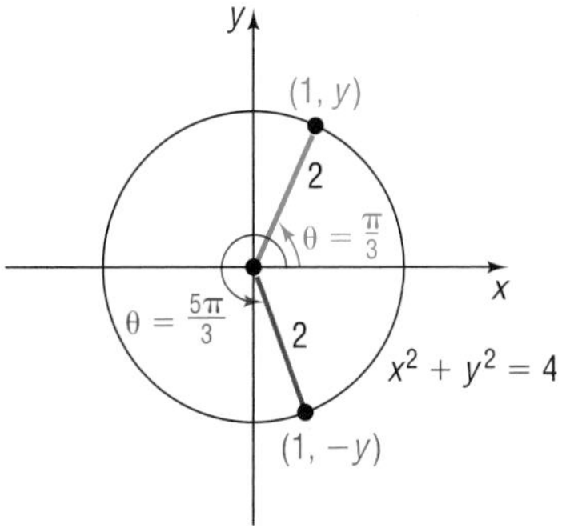

Because the cosine function has period 2π, all the solutions of $\cos\theta = \frac{1}{2}$ may be given by

$$\theta = \frac{\pi}{3} + 2k\pi \quad \text{or} \quad \theta = \frac{5\pi}{3} + 2k\pi \quad k \text{ any integer.}$$

Some of the solutions are

$$\underbrace{\frac{\pi}{3}, \frac{5\pi}{3}}_{k=0}, \underbrace{\frac{7\pi}{3}, \frac{11\pi}{3}}_{k=1}, \underbrace{\frac{13\pi}{3}, \frac{17\pi}{3}}_{k=2}, \text{ and so on}$$

Figure 30

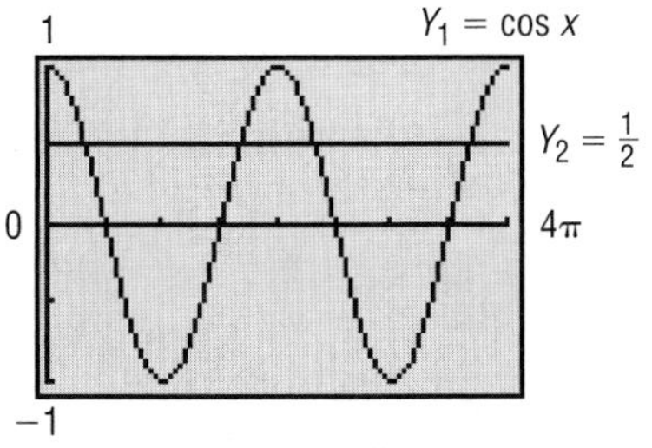

Check: We can verify the solutions by graphing $Y_1 = \cos x$ and $Y_2 = \frac{1}{2}$ to determine where the graphs intersect. (Be sure to graph in radian mode.) See Figure 30. The graph of Y_1 intersects the graph of Y_2 at $x = 1.05$ ($\approx\pi/3$), 5.24 ($\approx 5\pi/3$), 7.33 ($\approx 7\pi/3$), and 11.52 ($\approx 11\pi/3$), rounded to two decimal places.

NOW WORK PROBLEM 1.

In most of our work, we shall be interested only in finding solutions of trigonometric equations for $0 \le \theta < 2\pi$.

EXAMPLE 3 Solving a Linear Trigonometric Equation

Solve the equation: $2\sin\theta + \sqrt{3} = 0, \quad 0 \le \theta < 2\pi$

Solution We solve the equation for $\sin\theta$:

$$2\sin\theta + \sqrt{3} = 0$$

$$2\sin\theta = -\sqrt{3} \quad \text{Subtract } \sqrt{3} \text{ from both sides.}$$

$$\sin\theta = -\frac{\sqrt{3}}{2} \quad \text{Divide both sides by 2.}$$

The period of the sine function is 2π. In the interval $[0, 2\pi)$, there are two angles θ for which $\sin\theta = -\frac{\sqrt{3}}{2}$: $\theta = \frac{4\pi}{3}$ and $\theta = \frac{5\pi}{3}$. ■

EXAMPLE 4 Solving a Trigonometric Equation

Solve the equation: $\sin(2\theta) = \frac{1}{2}, 0 \le \theta < 2\pi$

Solution The period of the sine function is 2π. In the interval $[0, 2\pi)$, the sine function has the value $\frac{1}{2}$ when the argument is $\pi/6$ and $5\pi/6$. See Figure 31. Consequently, because the argument is 2θ in the equation $\sin(2\theta) = \frac{1}{2}$, we have

Figure 31
$\sin(2\theta) = \frac{1}{2}, 0 \le \theta < 2\pi$

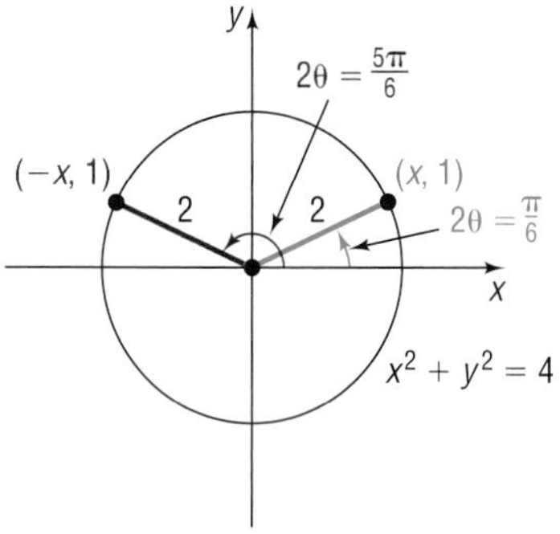

$$2\theta = \frac{\pi}{6} + 2k\pi \quad \text{or} \quad 2\theta = \frac{5\pi}{6} + 2k\pi \quad k \text{ any integer.}$$

$$\theta = \frac{\pi}{12} + k\pi \qquad \theta = \frac{5\pi}{12} + k\pi$$

In the interval $[0, 2\pi)$, the solutions of $\sin(2\theta) = \frac{1}{2}$ are

$$\pi/12, \quad \pi/12 + \pi = 13\pi/12, \quad 5\pi/12, \quad \text{and} \quad 5\pi/12 + \pi = 17\pi/12$$

You should verify the solutions using a graphing utility. ■

WARNING: In solving a trigonometric equation for $\theta, 0 \le \theta < 2\pi$, in which the argument is not θ (as in Example 4), you must write down all the solutions first and then list those that are in the interval $[0, 2\pi)$. Otherwise, solutions may be lost. For example, in solving $\sin(2\theta) = \frac{1}{2}$, if you merely write the solutions $2\theta = \frac{\pi}{6}$, and $2\theta = \frac{5\pi}{6}$, you will find only $\theta = \pi/12$ and $\theta = \frac{5\pi}{12}$ and miss the other solutions.

NOW WORK PROBLEM 7.

EXAMPLE 5 Solving a Trigonometric Equation

Solve the equation: $\tan\left(\theta - \frac{\pi}{2}\right) = 1, \quad 0 \le \theta < 2\pi$

Solution The period of the tangent function is π. In the interval $[0, \pi)$, the tangent function has the value 1 when the argument is $\pi/4$. Because the argument is $\theta - \frac{\pi}{2}$ in the given equation, we have

$$\theta - \frac{\pi}{2} = \frac{\pi}{4} + k\pi \qquad k \text{ any integer.}$$

$$\theta = \frac{3\pi}{4} + k\pi$$

In the interval $[0, 2\pi)$, $\theta = \frac{3\pi}{4}$ and $\theta = \frac{3\pi}{4} + \pi = \frac{7\pi}{4}$ are the only solutions. You should verify the solutions using a graphing utility. ■

The next example illustrates how to solve trigonometric equations using a calculator. Remember that the function keys on a calculator will only give values consistent with the definition of the function.

EXAMPLE 6 Solving a Trigonometric Equation with a Calculator

Use a calculator to solve the equation $\sin\theta = 0.3, 0 \le \theta < 2\pi$. Round any solution(s) to two decimal places.

Solution First we choose the radian mode. Then we find $\theta = \sin^{-1}(0.3)$. See Figure 32.

Figure 32

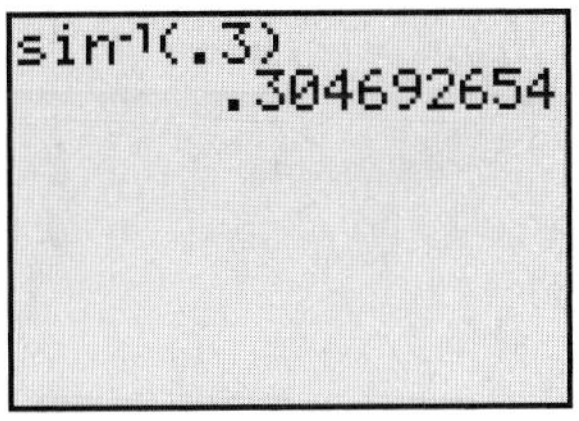

Figure 33
$\sin\theta = 0.3$

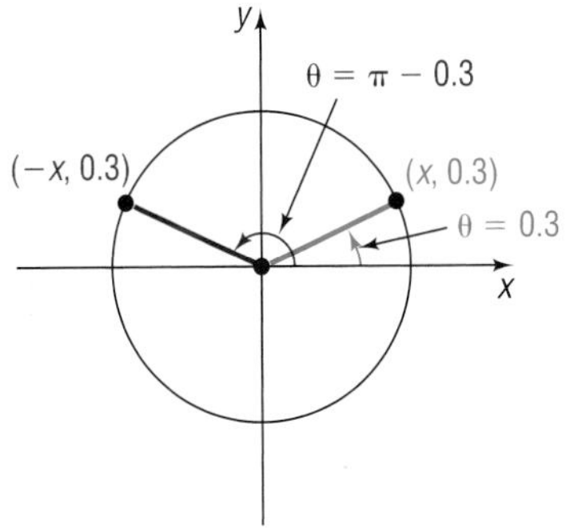

Because of the definition of $y = \sin^{-1} x$, the angle we obtain, 0.30 radian, rounded to two decimal places, is the angle $-\pi/2 \le \theta \le \pi/2$ for which $\sin\theta = 0.3$. Another angle for which $\sin\theta = 0.3$ is $\pi - 0.30$. See Figure 33. The angle $\pi - 0.30$ is the angle in quadrant II, where $\sin\theta = 0.3$. The solutions $\sin\theta = 0.3, 0 \le \theta < 2\pi$, are

$$\theta = 0.30 \text{ radian} \quad \text{or} \quad \theta = \pi - 0.30 = 2.84 \text{ radians}$$

each rounded to two decimal places. ■

A second method for solving $\sin\theta = 0.3, 0 \le \theta < 2\pi$, would be to graph $Y_1 = \sin x$ and $Y_2 = 0.3$ for $0 \le x < 2\pi$ and find their point(s) of intersection. Try this method for yourself to verify the results obtained in Example 6.

WARNING: Example 6 illustrates that caution must be exercised when solving trigonometric equations on a calculator. Remember that the calculator supplies an angle only within the restrictions of the definition of the inverse trigonometric function. To find the remaining solutions, you must identify other quadrants, if any, in which the angle may be located.

NOW WORK PROBLEM 25.

7.7 EXERCISES

In Problems 1–24, solve each equation on the interval $0 \leq \theta < 2\pi$.

1. $\sin\theta = \frac{1}{2}$
2. $\tan\theta = 1$
3. $\tan\theta = -\frac{\sqrt{3}}{3}$
4. $\cos\theta = -\frac{\sqrt{3}}{2}$
5. $\cos\theta = 0$
6. $\sin\theta = \frac{\sqrt{2}}{2}$
7. $\sin(3\theta) = -1$
8. $\tan\frac{\theta}{2} = \sqrt{3}$
9. $\cos(2\theta) = -\frac{1}{2}$
10. $\tan(2\theta) = -1$
11. $\sec\frac{3\theta}{2} = -2$
12. $\cot\frac{2\theta}{3} = -\sqrt{3}$
13. $\cos\left(2\theta - \frac{\pi}{2}\right) = -1$
14. $\sin\left(3\theta + \frac{\pi}{18}\right) = 1$
15. $\tan\left(\frac{\theta}{2} + \frac{\pi}{3}\right) = 1$
16. $\cos\left(\frac{\theta}{3} - \frac{\pi}{4}\right) = \frac{1}{2}$
17. $2\sin\theta + 1 = 0$
18. $\cos\theta + 1 = 0$
19. $\tan\theta + 1 = 0$
20. $\sqrt{3}\cot\theta + 1 = 0$
21. $4\sec\theta + 6 = -2$
22. $5\csc\theta - 3 = 2$
23. $3\sqrt{2}\cos\theta + 2 = -1$
24. $4\sin\theta + 3\sqrt{3} = \sqrt{3}$

In Problems 25–32, use a calculator to solve each equation on the interval $0 \leq \theta < 2\pi$. Round answers to two decimal places.

25. $\sin\theta = 0.4$
26. $\cos\theta = 0.6$
27. $\tan\theta = 5$
28. $\cot\theta = 2$
29. $\cos\theta = -0.9$
30. $\sin\theta = -0.2$
31. $\sec\theta = -4$
32. $\csc\theta = -3$

*The following discussion of **Snell's Law of Refraction** (named after Willebrord Snell, 1580–1626) is needed for Problems 33–39. Light, sound, and other waves travel at different speeds, depending on the media (air, water, wood, and so on) through which they pass. Suppose that light travels from a point A in one medium, where its speed is v_1, to a point B in another medium, where its speed is v_2. Refer to the figure, where the angle θ_1 is called the **angle of incidence** and the angle θ_2 is the **angle of refraction.** Snell's Law,* which can be proved using calculus, states that*

$$\frac{\sin\theta_1}{\sin\theta_2} = \frac{v_1}{v_2}$$

*The ratio v_1/v_2 is called the **index of refraction.** Some values are given in the following table.*

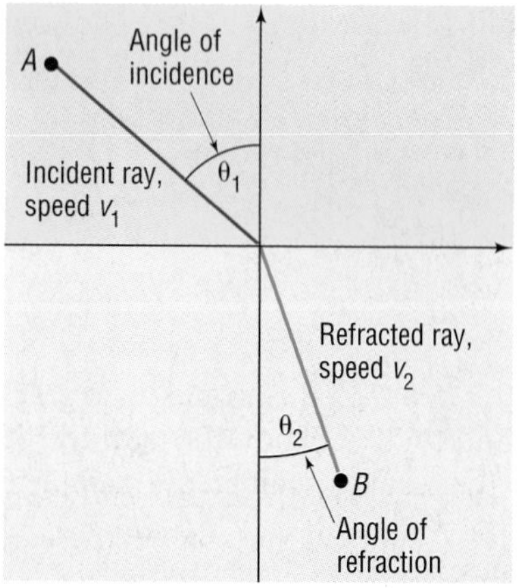

SOME INDEXES OF REFRACTION

Medium	Index of Refraction
Water	1.33
Ethyl alcohol	1.36
Carbon bisulfide	1.63
Air (1 atm and 20°C)	1.0003
Methylene iodide	1.74
Fused quartz	1.46
Glass, crown	1.52
Glass, dense flint	1.66
Sodium chloride	1.53

For light of wavelength 589 nanometers, measured with respect to a vacuum. The index with respect to air is negligibly different in most cases.

*Because this law was also deduced by René Descartes in France, it is also known as Descartes' Law.

33. The index of refraction of light in passing from a vacuum into water is 1.33. If the angle of incidence is 40°, determine the angle of refraction.

34. The index of refraction of light in passing from a vacuum into dense glass is 1.66. If the angle of incidence is 50°, determine the angle of refraction.

35. Ptolemy, who lived in the city of Alexandria in Egypt during the second century AD, gave the measured values in the table below for the angle of incidence θ_1 and the angle of refraction θ_2 for a light beam passing from air into water. Do these values agree with Snell's Law? If so, what index of refraction results? (These data are interesting as the oldest recorded physical measurements.)†

θ_1	θ_2	θ_1	θ_2
10°	7°45′	50°	35°0′
20°	15°30′	60°	40°30′
30°	22°30′	70°	45°30′
40°	29°0′	80°	50°0′

36. The speed of yellow sodium light (wavelength of 589 nanometers) in a certain liquid is measured to be 1.92×10^8 meters per second. What is the index of refraction of this liquid, with respect to air, for sodium light?*
[**Hint:** The speed of light in air is approximately 2.99×10^8 meters per second.]

37. A beam of light with a wavelength of 589 nanometers traveling in air makes an angle of incidence of 40° upon a slab of transparent material, and the refracted beam makes an angle of refraction of 26°. Find the index of refraction of the material.*

38. A light ray with a wavelength of 589 nanometers (produced by a sodium lamp) traveling through air makes an angle of incidence of 30° on a smooth, flat slab of crown glass. Find the angle of refraction.*

39. A light beam passes through a thick slab of material whose index of refraction is n_2. Show that the emerging beam is parallel to the incident beam.*

40. Explain in your own words how you would use your calculator to solve the equation $\sin x = 0.3$, $0 \le x < 2\pi$. How would you modify your approach in order to solve the equation $\cot x = 5$, $0 < x < 2\pi$?

† Adapted from Halliday and Resnick, *Physics, Parts 1 & 2,* 3rd ed. New York: Wiley, 1978, p. 953.
* Adapted from Serway, *Physics,* 3rd ed. Philadelphia: W. B. Saunders, p. 805.

PREPARING FOR THIS SECTION

Before getting started, review the following:

- ✓ Solving Equations Using a Graphing Utility (pp. 24–26)
- ✓ The Quadratic Formula (p. 30)
- ✓ Solving Quadratic Equations by Graphing Utility Factoring (pp. 28–30)

7.8 TRIGONOMETRIC EQUATIONS (II)

OBJECTIVES

1. Solve Trigonometric Equations That Are Quadratic in Form
2. Solve Trigonometric Equations Using Identities
3. Solve Trigonometric Equations Linear in Sine and Cosine
4. Solve Trigonometric Equations Using a Graphing Utility

1 In this section we continue our study of trigonometric equations. Many trigonometric equations can be solved by applying techniques that we already know, such as applying the quadratic formula (if the equation is a second-degree polynomial) or factoring.

EXAMPLE 1 **Solving a Trigonometric Equation Quadratic in Form**

Solve the equation: $2\sin^2\theta - 3\sin\theta + 1 = 0, \quad 0 \le \theta < 2\pi$

Solution The equation that we wish to solve is a quadratic equation (in $\sin\theta$) that can be factored:

$$2\sin^2\theta - 3\sin\theta + 1 = 0 \quad 2x^2 - 3x + 1 = 0, \quad x = \sin\theta$$

$$(2\sin\theta - 1)(\sin\theta - 1) = 0 \quad (2x - 1)(x - 1) = 0$$

$$2\sin\theta - 1 = 0 \quad \text{or} \quad \sin\theta - 1 = 0$$

$$\sin\theta = \tfrac{1}{2} \qquad \sin\theta = 1$$

Solving each equation in the interval $[0, 2\pi)$, we obtain

$$\theta = \frac{\pi}{6} \qquad \theta = \frac{5\pi}{6} \qquad \theta = \frac{\pi}{2}$$

NOW WORK PROBLEM 3.

2 When a trigonometric equation contains more than one trigonometric function, identities sometimes can be used to obtain an equivalent equation that contains only one trigonometric function.

EXAMPLE 2 Solving a Trigonometric Equation Using Identities

Solve the equation: $3\cos\theta + 3 = 2\sin^2\theta, \quad 0 \le \theta < 2\pi$

Solution The equation in its present form contains sines and cosines. However, a form of the Pythagorean Identity can be used to transform the equation into an equivalent expression containing only cosines:

$$3\cos\theta + 3 = 2\sin^2\theta$$

$$3\cos\theta + 3 = 2(1 - \cos^2\theta) \quad \sin^2\theta = 1 - \cos^2\theta$$

$$3\cos\theta + 3 = 2 - 2\cos^2\theta$$

$$2\cos^2\theta + 3\cos\theta + 1 = 0 \quad \text{Quadratic in } \cos\theta$$

$$(2\cos\theta + 1)(\cos\theta + 1) = 0 \quad \text{Factor}$$

$$2\cos\theta + 1 = 0 \quad \text{or} \quad \cos\theta + 1 = 0$$

$$\cos\theta = -\tfrac{1}{2} \qquad \cos\theta = -1$$

Solving each equation in the interval $[0, 2\pi)$, we obtain

$$\theta = \frac{2\pi}{3} \qquad \theta = \frac{4\pi}{3} \qquad \theta = \pi$$

Check: Graph $Y_1 = 3\cos x + 3$ and $Y_2 = 2\sin^2 x$, $0 \le x \le 2\pi$ and find the points of intersection. How close are your approximate solutions to the exact ones found in this example?

EXAMPLE 3 Solving a Trigonometric Equation Using Identities

Solve the equation: $\cos(2\theta) + 3 = 5\cos\theta, 0 \le \theta < 2\pi$

Solution First, we observe that the given equation contains two cosine functions, but with different arguments, θ and 2θ. We use the Double-Angle Formula $\cos(2\theta) = 2\cos^2\theta - 1$ to obtain an equivalent equation containing only $\cos\theta$.

$$\cos(2\theta) + 3 = 5\cos\theta$$
$$(2\cos^2\theta - 1) + 3 = 5\cos\theta$$
$$2\cos^2\theta - 5\cos\theta + 2 = 0$$
$$(\cos\theta - 2)(2\cos\theta - 1) = 0$$
$$\cos\theta = 2 \quad \text{or} \quad \cos\theta = \tfrac{1}{2}$$

For any angle θ, $-1 \le \cos\theta \le 1$; therefore, the equation $\cos\theta = 2$ has no solution. The solutions of $\cos\theta = \frac{1}{2}$, $0 \le \theta < 2\pi$, are

$$\theta = \frac{\pi}{3} \qquad \theta = \frac{5\pi}{3}$$

Check: Graph $Y_1 = \cos(2x) + 3$ and $Y_2 = 5\cos x$, $0 \le x \le 2\pi$, and find the points of intersection. Compare your results with those of Example 3.

NOW WORK PROBLEM 13.

EXAMPLE 4 Solving a Trigonometric Equation Using Identities

Solve the equation: $\cos^2\theta + \sin\theta = 2$, $0 \le \theta < 2\pi$

Solution We use a form of the Pythagorean Identity.

$$\cos^2\theta + \sin\theta = 2$$
$$(1 - \sin^2\theta) + \sin\theta = 2 \qquad \cos^2\theta = 1 - \sin^2\theta$$
$$\sin^2\theta - \sin\theta + 1 = 0$$

This is a quadratic equation in $\sin\theta$. The discriminant is $b^2 - 4ac = 1 - 4 = -3 < 0$. Therefore, the equation has no real solution.

Figure 34 shows the graphs of $Y_1 = \cos^2 x + \sin x$ and $Y_2 = 2$. The graphs do not intersect, so the equation $Y_1 = Y_2$ has no real solution.

Figure 34

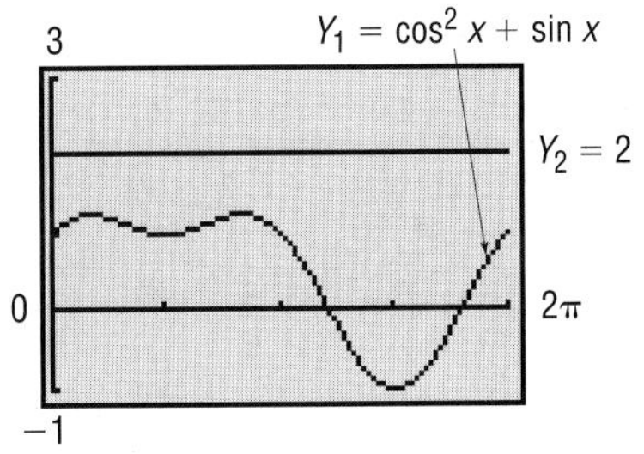

EXAMPLE 5 Solving a Trigonometric Equation Using Identities

Solve the equation: $\sin\theta\cos\theta = -\frac{1}{2}$, $0 \le \theta < 2\pi$

Solution The left side of the given equation is in the form of the Double-Angle Formula $2\sin\theta\cos\theta = \sin(2\theta)$, except for a factor of 2. We multiply each side by 2.

$$\sin\theta\cos\theta = -\tfrac{1}{2}$$
$$2\sin\theta\cos\theta = -1$$
$$\sin(2\theta) = -1$$

The argument here is 2θ. So we need to write all the solutions of this equation and then list those that are in the interval $[0, 2\pi)$.

$$2\theta = \frac{3\pi}{2} + 2k\pi \quad \text{k any integer.}$$

$$\theta = \frac{3\pi}{4} + k\pi$$

The solutions in the interval $[0, 2\pi)$ are

$$\theta = \frac{3\pi}{4} \qquad \theta = \frac{7\pi}{4}$$

3 Sometimes it is necessary to square both sides of an equation in order to obtain expressions that allow the use of identities. Remember, however, that when squaring both sides extraneous solutions may be introduced. As a result, apparent solutions must be checked.

EXAMPLE 6 Other Methods for Solving a Trigonometric Equation

Solve the equation: $\sin\theta + \cos\theta = 1, \quad 0 \le \theta < 2\pi$

Solution A Attempts to use available identities do not lead to equations that are easy to solve. (Try it yourself.) Given the form of this equation, we decide to square each side.

$$\sin\theta + \cos\theta = 1$$

$$(\sin\theta + \cos\theta)^2 = 1$$

$$\sin^2\theta + 2\sin\theta\cos\theta + \cos^2\theta = 1$$

$$2\sin\theta\cos\theta = 0 \quad \sin^2\theta + \cos^2\theta = 1$$

$$\sin\theta\cos\theta = 0$$

Setting each factor equal to zero, we obtain

$$\sin\theta = 0 \quad \text{or} \quad \cos\theta = 0$$

The apparent solutions are

$$\theta = 0 \qquad \theta = \pi \qquad \theta = \frac{\pi}{2} \qquad \theta = \frac{3\pi}{2}$$

Because we squared both sides of the original equation, we must check these apparent solutions to see if any are extraneous.

$\theta = 0$: $\sin 0 + \cos 0 = 0 + 1 = 1$ A solution

$\theta = \pi$: $\sin\pi + \cos\pi = 0 + (-1) = -1$ Not a solution

$\theta = \frac{\pi}{2}$: $\sin\frac{\pi}{2} + \cos\frac{\pi}{2} = 1 + 0 = 1$ A solution

$\theta = \frac{3\pi}{2}$: $\sin\frac{3\pi}{2} + \cos\frac{3\pi}{2} = -1 + 0 = -1$ Not a solution

Therefore, $\theta = 3\pi/2$ and $\theta = \pi$ are extraneous. The only solutions are $\theta = 0$ and $\theta = \pi/2$.

We can solve the equation given in Example 6 in another way.

Solution B We start with the equation

$$\sin\theta + \cos\theta = 1$$

and divide each side by $\sqrt{2}$. (The reason for this choice will become apparent shortly.) Then

$$\frac{1}{\sqrt{2}}\sin\theta + \frac{1}{\sqrt{2}}\cos\theta = \frac{1}{\sqrt{2}}$$

The left side now resembles the formula for the sine of the sum of two angles, one of which is θ. The other angle is unknown (call it ϕ.) Then

$$\sin(\theta + \phi) = \sin\theta\cos\phi + \cos\theta\sin\phi = \frac{1}{\sqrt{2}} \quad \textbf{(1)}$$

where

$$\cos\phi = \frac{1}{\sqrt{2}}, \qquad \sin\phi = \frac{1}{\sqrt{2}}, \qquad 0 \le \phi < 2\pi$$

The angle ϕ is therefore $\pi/4$. As a result, equation (1) becomes

$$\sin\left(\theta + \frac{\pi}{4}\right) = \frac{1}{\sqrt{2}}$$

We solve this equation to get

$$\theta + \frac{\pi}{4} = \frac{\pi}{4} \quad \text{or} \quad \theta + \frac{\pi}{4} = \frac{3\pi}{4}$$

$$\theta = 0 \qquad\qquad \theta = \frac{\pi}{2}$$

These solutions agree with the solutions found earlier. ■

This second method of solution can be used to solve any linear equation in the variables $\sin\theta$ and $\cos\theta$.

EXAMPLE 7 Solving a Trigonometric Equation Linear in $\sin\theta$ and $\cos\theta$

Solve:

$$a\sin\theta + b\cos\theta = c, \qquad 0 \le \theta < 2\pi \quad \textbf{(2)}$$

where a, b, and c are constants and either $a \ne 0$ or $b \ne 0$.

Solution We divide each side of equation (2) by $\sqrt{a^2 + b^2}$. Then

$$\frac{a}{\sqrt{a^2 + b^2}}\sin\theta + \frac{b}{\sqrt{a^2 + b^2}}\cos\theta = \frac{c}{\sqrt{a^2 + b^2}} \quad \textbf{(3)}$$

Figure 35

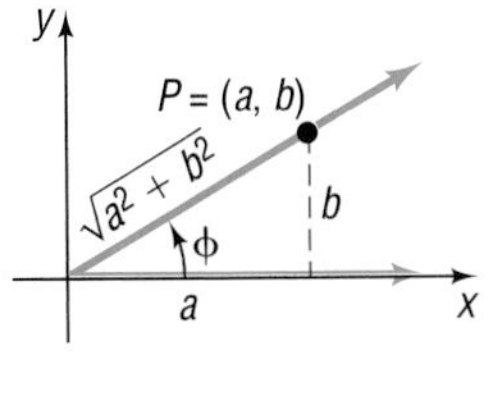

There is a unique angle ϕ, $0 \le \phi < 2\pi$, for which

$$\cos\phi = \frac{a}{\sqrt{a^2 + b^2}} \quad \text{and} \quad \sin\phi = \frac{b}{\sqrt{a^2 + b^2}} \quad \textbf{(4)}$$

(see Figure 35). Equation (3) may be written as

$$\sin\theta\cos\phi + \cos\theta\sin\phi = \frac{c}{\sqrt{a^2 + b^2}}$$

or, equivalently,

$$\sin(\theta + \phi) = \frac{c}{\sqrt{a^2 + b^2}} \quad \textbf{(5)}$$

where ϕ satisfies equation (4).

If $|c| > \sqrt{a^2 + b^2}$, then $\sin(\theta + \phi) > 1$ or $\sin(\theta + \phi) < -1$, and equation (5) has no solution.

If $|c| \le \sqrt{a^2 + b^2}$, then the solutions of equation (5) are

$$\theta + \phi = \sin^{-1}\frac{c}{\sqrt{a^2 + b^2}} \quad \text{or} \quad \theta + \phi = \pi - \sin^{-1}\frac{c}{\sqrt{a^2 + b^2}}$$

Because the angle ϕ is determined by equations (4), these are the solutions to equation (2). ■

NOW WORK PROBLEM 27.

4 GRAPHING UTILITY SOLUTIONS

The techniques introduced in this section apply only to certain types of trigonometric equations. Solutions for other types are usually studied in calculus, using numerical methods. In the next example, we show how a graphing utility may be used to obtain solutions.

EXAMPLE 8 Solving Trigonometric Equations Using a Graphing Utility

Solve: $5\sin x + x = 3$

Express the solution(s) correct to two decimal places.

Solution This type of trigonometric equation cannot be solved by previous methods. A graphing utility, though, can be used here. The solution(s) of this equation is the same as the points of intersection of the graphs of $Y_1 = 5\sin x + x$ and $Y_2 = 3$. See Figure 36.

Figure 36

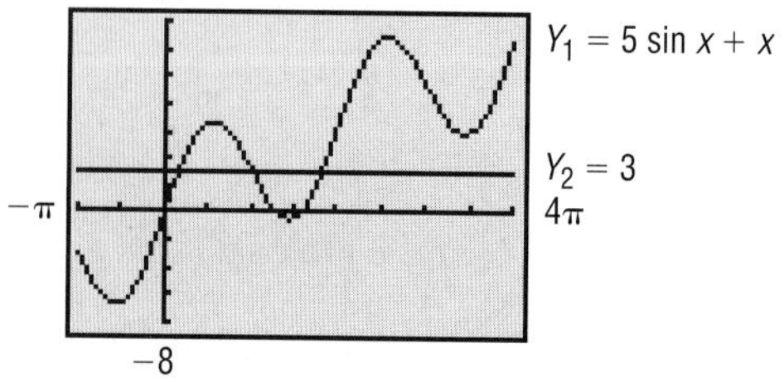

There are three points of intersection; the x-coordinates are the solutions that we seek. Using INTERSECT, we find

$$x = 0.52 \qquad x = 3.18 \qquad x = 5.71$$

rounded to two decimal places. ■

NOW WORK PROBLEM 39.

7.8 EXERCISES

In Problems 1–32, solve each equation on the interval $0 \le \theta < 2\pi$.

1. $2\cos^2\theta + \cos\theta = 0$
2. $\sin^2\theta - 1 = 0$
3. $2\sin^2\theta - \sin\theta - 1 = 0$
4. $2\cos^2\theta + \cos\theta - 1 = 0$
5. $(\tan\theta - 1)(\sec\theta - 1) = 0$
6. $(\cot\theta + 1)(\csc\theta - \frac{1}{2}) = 0$
7. $\cos\theta = \sin\theta$
8. $\cos\theta + \sin\theta = 0$
9. $\tan\theta = 2\sin\theta$
10. $\sin(2\theta) = \cos\theta$
11. $\sin\theta = \csc\theta$
12. $\tan\theta = \cot\theta$
13. $\cos(2\theta) = \cos\theta$
14. $\sin(2\theta)\sin\theta = \cos\theta$
15. $\sin(2\theta) + \sin(4\theta) = 0$
16. $\cos(2\theta) + \cos(4\theta) = 0$
17. $\cos(4\theta) - \cos(6\theta) = 0$
18. $\sin(4\theta) - \sin(6\theta) = 0$
19. $1 + \sin\theta = 2\cos^2\theta$
20. $\sin^2\theta = 2\cos\theta + 2$
21. $\tan^2\theta = \frac{3}{2}\sec\theta$
22. $\csc^2\theta = \cot\theta + 1$
23. $3 - \sin\theta = \cos(2\theta)$
24. $\cos(2\theta) + 5\cos\theta + 3 = 0$
25. $\sec^2\theta + \tan\theta = 0$
26. $\sec\theta = \tan\theta + \cot\theta$
27. $\sin\theta - \sqrt{3}\cos\theta = 1$
28. $\sqrt{3}\sin\theta + \cos\theta = 1$
29. $\tan(2\theta) + 2\sin\theta = 0$
30. $\tan(2\theta) + 2\cos\theta = 0$
31. $\sin\theta + \cos\theta = \sqrt{2}$
32. $\sin\theta + \cos\theta = -\sqrt{2}$

In Problems 33–38, solve each equation for x, $-\pi \le x \le \pi$. Express the solution(s) rounded to two decimal places.

33. Solve the equation $\cos x = e^x$ by graphing $Y_1 = \cos x$ and $Y_2 = e^x$ and finding their point(s) of intersection.

34. Solve the equation $\cos x = e^x$ by graphing $Y_1 = \cos x - e^x$ and finding the x-intercept(s).

35. Solve the equation $2\sin x = 0.7x$ by graphing $Y_1 = 2\sin x$ and $Y_2 = 0.7x$ and finding their point(s) of intersection.

36. Solve the equation $2\sin x = 0.7x$ by graphing $Y_1 = 2\sin x - 0.7x$ and finding the x-intercept(s).

37. Solve the equation $\cos x = x^2$ by graphing $Y_1 = \cos x$ and $Y_2 = x^2$ and finding their point(s) of intersection.

38. Solve the equation $\cos x = x^2$ by graphing $Y_1 = \cos x - x^2$ and finding the x-intercept(s).

In Problems 39–50, use a graphing utility to solve each equation. Express the solution(s) rounded to two decimal places.

39. $x + 5\cos x = 0$
40. $x - 4\sin x = 0$
41. $22x - 17\sin x = 3$
42. $19x + 8\cos x = 2$
43. $\sin x + \cos x = x$
44. $\sin x - \cos x = x$
45. $x^2 - 2\cos x = 0$
46. $x^2 + 3\sin x = 0$
47. $x^2 - 2\sin(2x) = 3x$
48. $x^2 = x + 3\cos(2x)$
49. $6\sin x - e^x = 2, \quad x > 0$
50. $4\cos(3x) - e^x = 1, \quad x > 0$

51. Constructing a Rain Gutter A rain gutter is to be constructed of aluminum sheets 12 inches wide. After marking off a length of 4 inches from each edge, this length is bent up at an angle θ. See the illustration. The area A of the opening as a function of θ is given by

$$A(\theta) = 16\sin\theta(\cos\theta + 1), \quad 0° < \theta < 90°$$

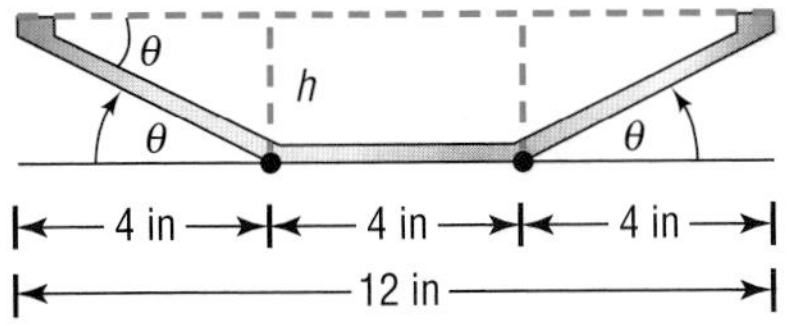

(a) In calculus, you will be asked to find the angle θ that maximizes A by solving the equation

$$\cos(2\theta) + \cos\theta = 0, \quad 0° < \theta < 90°$$

Solve this equation for θ by using the Double-Angle Formula.
(b) Solve the equation for θ by writing the sum of the two cosines as a product.
(c) What is the maximum area A of the opening?
(d) Graph A, $0° \le \theta \le 90°$, and find the angle θ that maximizes the area A. Also find the maximum area. Compare the results to the answers found earlier.

52. Projectile Motion An object is propelled upward at an angle θ, $45° < \theta < 90°$, to the horizontal with an initial velocity of v_0 feet per second from the base of a plane that makes an angle of 45° with the horizontal. See the illustration. If air resistance is ignored, the distance R that it travels up the inclined plane is given by

$$R = \frac{v_0^2\sqrt{2}}{32}[\sin(2\theta) - \cos(2\theta) - 1]$$

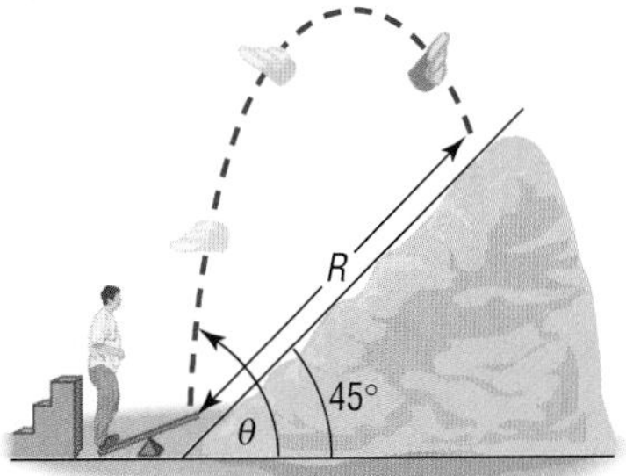

(a) In calculus, you will be asked to find the angle θ that maximizes R by solving the equation

$$\sin(2\theta) + \cos(2\theta) = 0$$

Solve this equation for θ using the method of Example 7.
(b) Solve this equation for θ by dividing each side by $\cos(2\theta)$.
(c) What is the maximum distance R if $v_0 = 32$ feet per second?
(d) Graph R, $45° \le \theta \le 90°$, and find the angle θ that maximizes the distance R. Also find the maximum distance. Use $v_0 = 32$ feet per second. Compare the results with the answers found earlier.

53. **Heat Transfer** In the study of heat transfer, the equation $x + \tan x = 0$ occurs. Graph $Y_1 = -x$ and $Y_2 = \tan x$ for $x \ge 0$. Conclude that there are an infinite number of points of intersection of these two graphs. Now find the first two positive solutions of $x + \tan x = 0$ rounded to two decimal places.

54. **Carrying a Ladder around a Corner** A ladder of length L is carried horizontally around a corner from a hall 3 feet wide into a hall 4 feet wide. See the illustration.

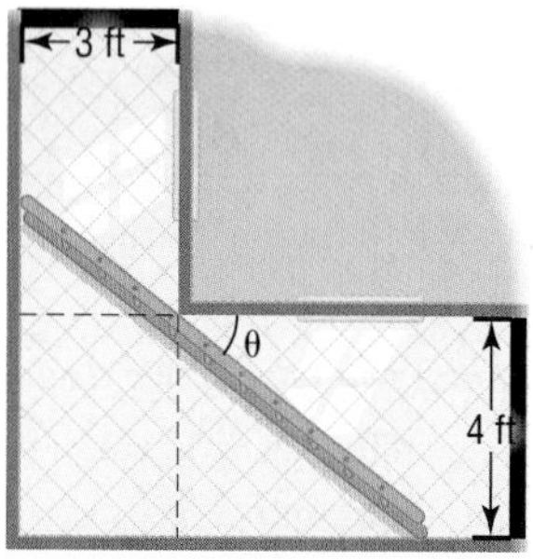

(a) Express L as a function of θ.
(b) In calculus, you will be asked to find the length of the longest ladder that can turn the corner by solving the equation

$$3 \sec\theta \tan\theta - 4 \csc\theta \cot\theta = 0, \quad 0° < \theta < 90°$$

Solve this equation for θ.
(c) What is the length of the longest ladder that can be carried around the corner?
(d) Graph L, $0° \le \theta \le 90°$, and find the angle θ that maximizes the length L. Also find the maximum length. Compare the results with the ones found in parts (b) and (c).

55. **Projectile Motion** The horizontal distance that a projectile will travel in the air is given by the equation

$$R = \frac{v_0^2 \sin(2\theta)}{g}$$

where v_0 is the initial velocity of the projectile, θ is the angle of elevation, and g is acceleration due to gravity (9.8 meters per second squared).
(a) If you can throw a baseball with an initial speed of 34.8 meters per second, at what angle of elevation θ should you direct the throw so that the ball travels a distance of 107 meters before striking the ground?
(b) Determine the maximum distance that you can throw the ball.
(c) Graph R, with $v_0 = 34.8$ meters per second.
(d) Verify the results obtained in parts (a) and (b) using ZERO or ROOT.

56. **Projectile Motion** Refer to Problem 55.
(a) If you can throw a baseball with an initial speed of 40 meters per second, at what angle of elevation θ should you direct the throw so that the ball travels a distance of 110 meters before striking the ground?
(b) Determine the maximum distance that you can throw the ball.
(c) Graph R, with $v_0 = 40$ meters per second.
(d) Verify the results obtained in parts (a) and (b) using ZERO or ROOT.

CHAPTER REVIEW

Things To Know

Sum and Difference Formulas (pp. 452, 455, and 458)

$\cos(\alpha + \beta) = \cos\alpha\cos\beta - \sin\alpha\sin\beta$

$\cos(\alpha - \beta) = \cos\alpha\cos\beta + \sin\alpha\sin\beta$

$\sin(\alpha + \beta) = \sin\alpha\cos\beta + \cos\alpha\sin\beta$

$\sin(\alpha - \beta) = \sin\alpha\cos\beta - \cos\alpha\sin\beta$

$$\tan(\alpha + \beta) = \frac{\tan\alpha + \tan\beta}{1 - \tan\alpha\tan\beta}$$

$$\tan(\alpha - \beta) = \frac{\tan\alpha - \tan\beta}{1 + \tan\alpha\tan\beta}$$

Double-Angle Formulas (pp. 462 and 463)

$\sin(2\theta) = 2\sin\theta\cos\theta$

$\cos(2\theta) = \cos^2\theta - \sin^2\theta$

$\cos(2\theta) = 1 - 2\sin^2\theta$

$\cos(2\theta) = 2\cos^2\theta - 1$

$$\tan(2\theta) = \frac{2\tan\theta}{1 - \tan^2\theta}$$

Half-Angle Formulas (pp. 465 and 467)

$$\sin^2\frac{\alpha}{2} = \frac{1-\cos\alpha}{2} \qquad \cos^2\frac{\alpha}{2} = \frac{1+\cos\alpha}{2} \qquad \tan^2\frac{\alpha}{2} = \frac{1-\cos\alpha}{1+\cos\alpha}$$

$$\sin\frac{\alpha}{2} = \pm\sqrt{\frac{1-\cos\alpha}{2}} \qquad \cos\frac{\alpha}{2} = \pm\sqrt{\frac{1+\cos\alpha}{2}} \qquad \tan\frac{\alpha}{2} = \pm\sqrt{\frac{1-\cos\alpha}{1+\cos\alpha}} = \frac{1-\cos\alpha}{\sin\alpha} = \frac{\sin\alpha}{1+\cos\alpha}$$

Product-to-Sum Formulas (p. 470)

$\sin\alpha\sin\beta = \frac{1}{2}[\cos(\alpha-\beta) - \cos(\alpha+\beta)]$

$\cos\alpha\cos\beta = \frac{1}{2}[\cos(\alpha-\beta) + \cos(\alpha+\beta)]$

$\sin\alpha\cos\beta = \frac{1}{2}[\sin(\alpha+\beta) + \sin(\alpha-\beta)]$

Sum-to-Product Formulas (p. 471)

$$\sin\alpha + \sin\beta = 2\sin\frac{\alpha+\beta}{2}\cos\frac{\alpha-\beta}{2} \qquad \sin\alpha - \sin\beta = 2\sin\frac{\alpha-\beta}{2}\cos\frac{\alpha+\beta}{2}$$

$$\cos\alpha + \cos\beta = 2\cos\frac{\alpha+\beta}{2}\cos\frac{\alpha-\beta}{2} \qquad \cos\alpha - \cos\beta = -2\sin\frac{\alpha+\beta}{2}\sin\frac{\alpha-\beta}{2}$$

Definitions of the six inverse trigonometric functions

$y = \sin^{-1}x$ means $x = \sin y$ where $-1 \le x \le 1$, $-\pi/2 \le y \le \pi/2$ (p. 474)

$y = \cos^{-1}x$ means $x = \cos y$ where $-1 \le x \le 1$, $0 \le y \le \pi$ (p. 478)

$y = \tan^{-1}x$ means $x = \tan y$ where $-\infty < x < \infty$, $-\pi/2 < y < \pi/2$ (p. 480)

$y = \cot^{-1}x$ means $x = \cot y$ where $-\infty < x < \infty$, $0 < y < \pi$ (p. 482)

$y = \sec^{-1}x$ means $x = \sec y$ where $|x| \ge 1$, $0 \le y \le \pi$, $y \ne \pi/2$ (p. 482)

$y = \csc^{-1}x$ means $x = \csc y$ where $|x| \ge 1$, $-\pi/2 \le y \le \pi/2$, $y \ne 0$ (p. 482)

How To

Establish identities (p. 447)

Use sum and difference formulas to find exact values (p. 453)

Use sum and difference formulas to establish identities (p. 457)

Use double-angle formulas to find exact values (p. 462)

Use double-angle and half-angle formulas to establish identities (p. 462)

Use half-angle formulas to find exact values (p. 465)

Express products as sums (p. 470)

Express sums as products (p. 471)

Find the exact value of an inverse trigonometric function (p. 476)

Find the approximate value of an inverse trigonometric function (p. 477)

Find the exact value of expressions involving inverse trigonometric functions (p. 485)

Write a trigonometric expression as an algebraic expression (p. 488)

Establish identities involving inverse trigonometric expressions (p. 488)

Solve equations involving a single trigonometric function (p. 490)

Solve trigonometric equations that are quadratic in form (p. 495)

Solve trigonometric equations using identities (p. 496)

Solve trigonometric equations linear in sine and cosine (p. 498)

Solve trigonometric equations using a graphing utility (p. 500)

Fill-in-the-Blank Items

1. Suppose that f and g are two functions with the same domain. If $f(x) = g(x)$ for every x in the domain, the equation is called a(n) ______________. Otherwise, it is called a(n) ______________ equation.
2. $\cos(\alpha+\beta) = \cos\alpha\cos\beta$ ______________ $\sin\alpha\sin\beta$.
3. $\sin(\alpha+\beta) = \sin\alpha\cos\beta$ ______________ $\cos\alpha\sin\beta$.
4. $\cos(2\theta) = \cos^2\theta -$ ______________ $=$ ______________ $-1 = 1 -$ ______________.
5. $\sin^2\frac{\alpha}{2} = \frac{\text{\_\_\_\_\_\_}}{2}$.
6. The function $y = \sin^{-1}x$ has domain ______________ and range ______________.
7. The value of $\sin^{-1}[\cos(\pi/2)]$ is ______________.

True/False Items

T F 1. $\sin(-\theta) + \sin\theta = 0$ for all θ.

T F 2. $\sin(\alpha + \beta) = \sin\alpha + \sin\beta + 2\sin\alpha\sin\beta$.

T F 3. $\cos(2\theta)$ has three equivalent forms: $\cos^2\theta - \sin^2\theta$, $1 - 2\sin^2\theta$, and $2\cos^2\theta - 1$.

T F 4. $\cos\dfrac{\alpha}{2} = \pm\dfrac{\sqrt{1 + \cos\alpha}}{2}$, where the + or − sign depends on the angle $\alpha/2$.

T F 5. The domain of $y = \sin^{-1} x$ is $-\pi/2 \le x \le \pi/2$.

T F 6. $\cos(\sin^{-1} 0) = 1$ and $\sin(\cos^{-1} 0) = 1$.

T F 7. Most trigonometric equations have unique solutions.

T F 8. The equation $\tan\theta = \pi/2$ has no solution.

Review Exercises

Blue problem numbers indicate the authors' suggestions for use in a Practice Test.

In Problems 1–32, establish each identity.

1. $\tan\theta\cot\theta - \sin^2\theta = \cos^2\theta$
2. $\sin\theta\csc\theta - \sin^2\theta = \cos^2\theta$
3. $\cos^2\theta(1 + \tan^2\theta) = 1$
4. $(1 - \cos^2\theta)(1 + \cot^2\theta) = 1$
5. $4\cos^2\theta + 3\sin^2\theta = 3 + \cos^2\theta$
6. $4\sin^2\theta + 2\cos^2\theta = 4 - 2\cos^2\theta$
7. $\dfrac{1 - \cos\theta}{\sin\theta} + \dfrac{\sin\theta}{1 - \cos\theta} = 2\csc\theta$
8. $\dfrac{\sin\theta}{1 + \cos\theta} + \dfrac{1 + \cos\theta}{\sin\theta} = 2\csc\theta$
9. $\dfrac{\cos\theta}{\cos\theta - \sin\theta} = \dfrac{1}{1 - \tan\theta}$
10. $1 - \dfrac{\cos^2\theta}{1 + \sin\theta} = \sin\theta$
11. $\dfrac{\csc\theta}{1 + \csc\theta} = \dfrac{1 - \sin\theta}{\cos^2\theta}$
12. $\dfrac{1 + \sec\theta}{\sec\theta} = \dfrac{\sin^2\theta}{1 - \cos\theta}$
13. $\csc\theta - \sin\theta = \cos\theta\cot\theta$
14. $\dfrac{\csc\theta}{1 - \cos\theta} = \dfrac{1 + \cos\theta}{\sin^3\theta}$
15. $\dfrac{1 - \sin\theta}{\sec\theta} = \dfrac{\cos^3\theta}{1 + \sin\theta}$
16. $\dfrac{1 - \cos\theta}{1 + \cos\theta} = (\csc\theta - \cot\theta)^2$
17. $\dfrac{1 - 2\sin^2\theta}{\sin\theta\cos\theta} = \cot\theta - \tan\theta$
18. $\dfrac{(2\sin^2\theta - 1)^2}{\sin^4\theta - \cos^4\theta} = 1 - 2\cos^2\theta$
19. $\dfrac{\cos(\alpha + \beta)}{\cos\alpha\sin\beta} = \cot\beta - \tan\alpha$
20. $\dfrac{\sin(\alpha - \beta)}{\sin\alpha\cos\beta} = 1 - \cot\alpha\tan\beta$
21. $\dfrac{\cos(\alpha - \beta)}{\cos\alpha\cos\beta} = 1 + \tan\alpha\tan\beta$
22. $\dfrac{\cos(\alpha + \beta)}{\sin\alpha\cos\beta} = \cot\alpha - \tan\beta$
23. $(1 + \cos\theta)\left(\tan\dfrac{\theta}{2}\right) = \sin\theta$
24. $\sin\theta\tan\dfrac{\theta}{2} = 1 - \cos\theta$
25. $2\cot\theta\cot(2\theta) = \cot^2\theta - 1$
26. $2\sin(2\theta)(1 - 2\sin^2\theta) = \sin(4\theta)$
27. $1 - 8\sin^2\theta\cos^2\theta = \cos(4\theta)$
28. $\dfrac{\sin(3\theta)\cos\theta - \sin\theta\cos(3\theta)}{\sin(2\theta)} = 1$
29. $\dfrac{\sin(2\theta) + \sin(4\theta)}{\cos(2\theta) + \cos(4\theta)} = \tan(3\theta)$
30. $\dfrac{\sin(2\theta) + \sin(4\theta)}{\sin(2\theta) - \sin(4\theta)} + \dfrac{\tan(3\theta)}{\tan\theta} = 0$
31. $\dfrac{\cos(2\theta) - \cos(4\theta)}{\cos(2\theta) + \cos(4\theta)} - \tan\theta\tan(3\theta) = 0$
32. $\cos(2\theta) - \cos(10\theta) = [\tan(4\theta)(\sin(2\theta) + \sin(10\theta))]$

In Problems 33–40, find the exact value of each expression.

33. $\sin 165°$
34. $\tan 105°$
35. $\cos\dfrac{5\pi}{12}$
36. $\sin\left(-\dfrac{\pi}{12}\right)$
37. $\cos 80°\cos 20° + \sin 80°\sin 20°$
38. $\sin 70°\cos 40° - \cos 70°\sin 40°$
39. $\tan\dfrac{\pi}{8}$
40. $\sin\dfrac{5\pi}{8}$

In Problems 41–50, use the information given about the angles α and β to find the exact value of:

(a) $\sin(\alpha + \beta)$ (b) $\cos(\alpha + \beta)$ (c) $\sin(\alpha - \beta)$ (d) $\tan(\alpha + \beta)$

(e) $\sin(2\alpha)$ (f) $\cos(2\beta)$ (g) $\sin\dfrac{\beta}{2}$ (h) $\cos\dfrac{\alpha}{2}$

41. $\sin\alpha = \frac{4}{5}$, $0 < \alpha < \pi/2$; $\sin\beta = \frac{5}{13}$, $\pi/2 < \beta < \pi$

42. $\cos\alpha = \frac{4}{5}$, $0 < \alpha < \pi/2$; $\cos\beta = \frac{5}{13}$, $-\pi/2 < \beta < 0$

43. $\sin\alpha = -\frac{3}{5}$, $\pi < \alpha < 3\pi/2$; $\cos\beta = \frac{12}{13}$, $3\pi/2 < \beta < 2\pi$

44. $\sin\alpha = -\frac{4}{5}$, $-\pi/2 < \alpha < 0$; $\cos\beta = -\frac{5}{13}$, $\pi/2 < \beta < \pi$

45. $\tan\alpha = \frac{3}{4}$, $\pi < \alpha < 3\pi/2$; $\tan\beta = \frac{12}{5}$, $0 < \beta < \pi/2$

46. $\tan\alpha = -\frac{4}{3}$, $\pi/2 < \alpha < \pi$, $\cot\beta = \frac{12}{5}$, $\pi < \beta < 3\pi/2$

47. $\sec\alpha = 2$, $-\pi/2 < \alpha < 0$; $\sec\beta = 3$, $3\pi/2 < \beta < 2\pi$

48. $\csc\alpha = 2$, $\pi/2 < \alpha < \pi$, $\sec\beta = -3$, $\pi/2 < \beta < \pi$

49. $\sin\alpha = -\frac{2}{3}$, $\pi < \alpha < 3\pi/2$; $\cos\beta = -\frac{2}{3}$, $\pi < \beta < 3\pi/2$

50. $\tan\alpha = -2$, $\pi/2 < \alpha < \pi$; $\cot\beta = -2$, $\pi/2 < \beta < \pi$

In Problems 51–70, find the exact value of each expression. Do not use a calculator.

51. $\sin^{-1} 1$ **52.** $\cos^{-1} 0$ **53.** $\tan^{-1} 1$ **54.** $\sin^{-1}\left(-\dfrac{1}{2}\right)$

55. $\cos^{-1}\left(-\dfrac{\sqrt{3}}{2}\right)$ **56.** $\tan^{-1}(-\sqrt{3})$ **57.** $\sin\left(\cos^{-1}\dfrac{\sqrt{2}}{2}\right)$ **58.** $\cos(\sin^{-1} 0)$

59. $\tan\left[\sin^{-1}\left(-\dfrac{\sqrt{3}}{2}\right)\right]$ **60.** $\tan\left[\cos^{-1}\left(-\dfrac{1}{2}\right)\right]$ **61.** $\sec\left(\tan^{-1}\dfrac{\sqrt{3}}{3}\right)$ **62.** $\csc\left(\sin^{-1}\dfrac{\sqrt{3}}{2}\right)$

63. $\sin\left(\tan^{-1}\dfrac{3}{4}\right)$ **64.** $\cos\left(\sin^{-1}\dfrac{3}{5}\right)$ **65.** $\tan\left[\sin^{-1}\left(-\dfrac{4}{5}\right)\right]$ **66.** $\tan\left[\cos^{-1}\left(-\dfrac{3}{5}\right)\right]$

67. $\sin^{-1}\left(\cos\dfrac{2\pi}{3}\right)$ **68.** $\cos^{-1}\left(\tan\dfrac{3\pi}{4}\right)$ **69.** $\tan^{-1}\left(\tan\dfrac{7\pi}{4}\right)$ **70.** $\cos^{-1}\left(\cos\dfrac{7\pi}{6}\right)$

In Problems 71–76, find the exact value of each expression.

71. $\cos(\sin^{-1}\frac{3}{5} - \cos^{-1}\frac{1}{2})$ **72.** $\sin(\cos^{-1}\frac{5}{13} - \cos^{-1}\frac{4}{5})$ **73.** $\tan[\sin^{-1}(-\frac{1}{2}) - \tan^{-1}\frac{3}{4}]$

74. $\cos[\tan^{-1}(-1) + \cos^{-1}(-\frac{4}{5})]$ **75.** $\sin[2\cos^{-1}(-\frac{3}{5})]$ **76.** $\cos(2\tan^{-1}\frac{4}{3})$

In Problems 77–96, solve each equation on the interval $0 \le \theta < 2\pi$. Use a graphing utility to verify your solution.

77. $\cos\theta = \frac{1}{2}$ **78.** $\sin\theta = -\sqrt{3}/2$ **79.** $2\cos\theta + \sqrt{2} = 0$

80. $\tan\theta + \sqrt{3} = 0$ **81.** $\sin(2\theta) + 1 = 0$ **82.** $\cos(2\theta) = 0$

83. $\tan(2\theta) = 0$ **84.** $\sin(3\theta) = 1$ **85.** $\sin\theta = 0.9$

86. $\tan\theta = 25$ **87.** $\sin\theta = \tan\theta$ **88.** $\cos\theta = \sec\theta$

89. $\sin\theta + \sin(2\theta) = 0$ **90.** $\cos(2\theta) = \sin\theta$ **91.** $\sin(2\theta) - \cos\theta - 2\sin\theta + 1 = 0$

92. $\sin(2\theta) - \sin\theta - 2\cos\theta + 1 = 0$ **93.** $2\sin^2\theta - 3\sin\theta + 1 = 0$ **94.** $2\cos^2\theta + \cos\theta - 1 = 0$

95. $\sin\theta - \cos\theta = 1$ **96.** $\sin\theta + 2\cos\theta = 1$

In Problems 97–102, use a graphing utility to solve each equation on the interval $0 \le x \le 2\pi$. Approximate any solutions rounded to two decimal places.

97. $2x = 5\cos x$ **98.** $2x = 5\sin x$ **99.** $2\sin x + 3\cos x = 4x$

100. $3\cos x + x = \sin x$ **101.** $\sin x = \ln x$ **102.** $\sin x = e^{-x}$

PROJECT AT MOTOROLA

Sending pictures wirelessly

The picture below on the left shows a special digital camera being developed at Motorola to allow people to take pictures and then send them anywhere in the world right from where they are. This is called an ImagePhone Camera. It attaches to a cellular phone as shown below in the picture on the right. Once attached the images can be sent wirelessly by email.

The images contain a lot of data. A digital image is made up of picture elements, or pixels for short. A common type of image is called a VGA image which is short for Video Graphic Array. The VGA array of pixels has 640 columns and 480 rows for a total of $640 \times 480 = 307{,}200$ pixels. Each pixel is composed of a red subpixel, a green subpixel and a blue subpixel which makes $307200 \times 3 = 921{,}600$ subpixels. Now each subpixel is represented by 8 bits of data which makes 7,372,800 bits of data for a single VGA picture!

If we want to send this image wirelessly it will take a long time. Present cellular systems can only send 9,600 bits per second (bps) over the air. At this rate the above VGA image will take $7{,}372{,}800/9600 = 768$ seconds or approximately 13 minutes.

The ImagePhone camera takes only 1 minute to send a VGA image instead of the expected 13 minutes. This reduction comes with a process called image compression to reduce the size of the image data. One common compression method is called JPEG compression where JPEG stands for Joint Photographic Experts Group. This method can reduce the image data by a factor ranging from 10–50. For example a compression factor of 25 (or 25:1 compression) will reduce the image transmission time from 13 minutes to 30 seconds!

How does compression work? To understand this, consider one row of the VGA image. Each row has 640 pixels corresponding to the 640 columns. If we were to draw an x-axis through these data points and plot the intensity (brightness) as a function of x across the image, we can expect an arbitrary function $f(x)$. Using a technique called **Fourier series expansion**, this function can be represented as a series of sines and cosines as follows:

$$f(x) = \frac{a_0}{2} + \sum_{n=1}^{\infty} a_n \cos(nx) + \sum_{n=1}^{\infty} b_n \sin(nx)$$

where a_n and b_n are the **Fourier coefficients.** If the function $f(x)$ were such that an infinite number of terms in the series were not required, but only a few terms (say 10), then all we would have to do is send 10 coefficients and achieve a significant reduction in the amount of data needed to represent the function $f(x)$. Note that JPEG uses a similar type of expression as the Fourier series.

As an example, consider a simple square wave that represents the information we have to send. Let us call this function $g(x)$ defined as follows:

$$g(x) = 0, \quad -\pi < x < 0, \pi < x < 2\pi, \ldots$$

$$g(x) = h, \quad 0 < x < \pi, 2\pi < x < 3\pi, \ldots$$

where $h > 0$.

Perhaps this represents intensity variation as a row in an image that depicts a solid block. Using a Fourier series expansion we can represent the square wave as shown below:

$$g(x) = \frac{h}{2} + \frac{2h}{\pi}\left[\frac{\sin x}{1} + \frac{\sin(3x)}{3} + \frac{\sin(5x)}{5} + \ldots\right]$$

or

$$g(x) = c_0 + c_1 \sin x + c_3 \sin(3x) + c_5 \sin(5x) + \cdots$$

To completely represent this square wave we do not need to send all the points on this waveform, but only the coefficients $(c_0, c_1, c_3, c_5, \ldots)$. If we have to send an infinite set of coefficients then we do not gain much. However, if we find that we only need to send the first N coefficients, where N = 15, then we have reduced the data needed to represent the square wave significantly.

The following exercises will help you understand this process better.

1. Plot the square wave $g(x)$ which is defined above by hand.
2. Using a graphing utility, plot the Fourier Series for $g(x)$ using 5 coefficients with $h = 2$.
3. Plot the Fourier Series using 10 coefficients.
4. Plot the Fourier Series using 20 coefficients.
5. Plot them all on the same chart and compare the differences in the curves plotted in 2, 3 and 4. Which would you use to represent the data?

CHAPTER

8 Applications of Trigonometric Functions

Outline

FIELD TRIP TO MOTOROLA

In Motorola Automotive we have to test most of our products (automotive electronics, especially engine controllers) in fairly severe vibration conditions in order to ensure a long and healthy life in the field. One of the tools for analyzing and monitoring any profile of vibrations (whether it is the vibration of a car engine or, for that matter, any acoustic wave) is Fourier series and Fourier transformations. The use of these transformations is based on the fundamental theorem stating that any vibration profile (or periodic function) can be expressed as a superposition of elementary sine and cosine profiles. This type of presentation is very conducive to an analysis of a specific frequency content in a complex wave, as well as to a study of the frequency response function of a system.

INTERVIEW AT MOTOROLA

Tomas Klosoniak's education has its roots in Poland, where I completed my schooling, obtaining a Masters degree and a Ph.D. in Mechanical Engineering. I have always, since school times, loved the elegance and logic of geometry, and found it to be very nicely complemented by trigonometry. When I came to America I was pleasantly

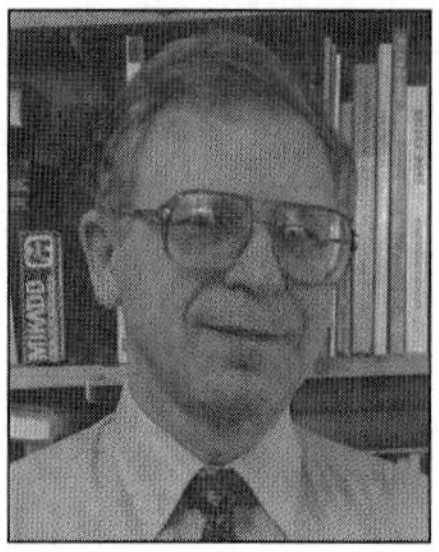

surprised that all I learned in the area of science and engineering was fully international and was perfectly applicable here in a different environment. At Motorola I work on various innovative methods to "package" and make modern electronics for an automobile so the electronics perform its function perfectly and reliably—to make our life easier.

PREPARING FOR THIS SECTION

Before getting started, review the following:

- ✓ Pythagorean Theorem (pp. 995–996)
- ✓ Definition of the Trigonometric Functions (p. 381)
- ✓ Theorem on Trigonometric Functions (p. 391)
- ✓ Trigonometric Equations (I) (pp. 490–491)

8.1 RIGHT TRIANGLE TRIGONOMETRY

OBJECTIVES
1. Find the Value of Trigonometric Functions of Acute Angles
2. Use the Complementary Angle Theorem
3. Solve Right Triangles
4. Solve Applied Problems

Figure 1

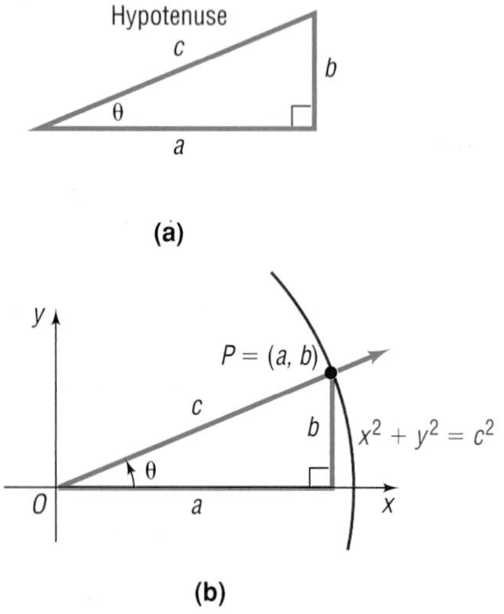

1 A triangle in which one angle is a right angle (90°) is called a **right triangle**. Recall that the side opposite the right angle is called the **hypotenuse**, and the remaining two sides are called the **legs** of the triangle. In Figure 1(a), we have labeled the hypotenuse as c to indicate that its length is c units, and, in a like manner, we have labeled the legs as a and b. Because the triangle is a right triangle, the Pythagorean Theorem tells us that

$$a^2 + b^2 = c^2$$

In Figure 1(a), we also show the angle θ. The angle θ is an **acute angle**: that is, $0° < \theta < 90°$, if θ is measured in degrees, or $0 < \theta < \pi/2$, if θ is measured in radians. Place θ in standard position, as shown in Figure 1(b). Then the coordinates of the point P are (a, b). Also, P is a point on the terminal side of θ that is also on the circle $x^2 + y^2 = c^2$. (Do you see why?).

Now use the theorem on page 391. By referring to the lengths of the sides of the triangle by the names hypotenuse (c), opposite (b), and adjacent (a), as indicated in Figure 2, we can express the trigonometric functions of θ as ratios of the sides of a right triangle.

Figure 2

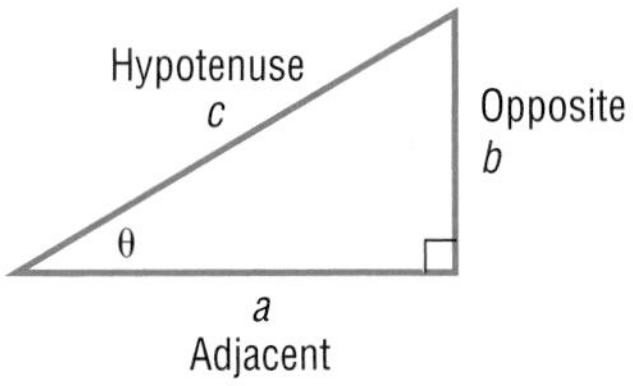

$$\sin\theta = \frac{\text{Opposite}}{\text{Hypotenuse}} = \frac{b}{c} \qquad \cos\theta = \frac{\text{Adjacent}}{\text{Hypotenuse}} = \frac{a}{c}$$
$$\tan\theta = \frac{\text{Opposite}}{\text{Adjacent}} = \frac{b}{a} \qquad \csc\theta = \frac{\text{Hypotenuse}}{\text{Opposite}} = \frac{c}{b} \qquad \textbf{(1)}$$
$$\sec\theta = \frac{\text{Hypotenuse}}{\text{Adjacent}} = \frac{c}{a} \qquad \cot\theta = \frac{\text{Adjacent}}{\text{Opposite}} = \frac{a}{b}$$

Notice that each of the trigonometric functions of the acute angle θ is positive.

EXAMPLE 1 Finding the Value of Trigonometric Functions from a Right Triangle

Figure 3

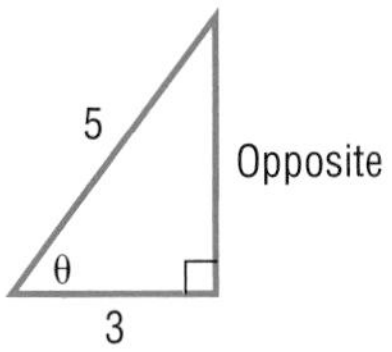

Find the exact value of the six trigonometric functions of the angle θ in Figure 3.

Solution We see in Figure 3 that the two given sides of the triangle are

$$c = \text{Hypotenuse} = 5 \qquad a = \text{Adjacent} = 3$$

To find the length of the opposite side, we use the Pythagorean Theorem.

$$(\text{Adjacent})^2 + (\text{Opposite})^2 = (\text{Hypotenuse})^2$$
$$3^2 + (\text{Opposite})^2 = 5^2$$
$$(\text{Opposite})^2 = 25 - 9 = 16$$
$$\text{Opposite} = 4$$

Now that we know the lengths of the three sides, we use the ratios in equations (1) to find the value of each of the six trigonometric functions.

$$\sin\theta = \frac{\text{Opposite}}{\text{Hypotenuse}} = \frac{4}{5} \qquad \cos\theta = \frac{\text{Adjacent}}{\text{Hypotenuse}} = \frac{3}{5} \qquad \tan\theta = \frac{\text{Opposite}}{\text{Adjacent}} = \frac{4}{3}$$
$$\csc\theta = \frac{\text{Hypotenuse}}{\text{Opposite}} = \frac{5}{4} \qquad \sec\theta = \frac{\text{Hypotenuse}}{\text{Adjacent}} = \frac{5}{3} \qquad \cot\theta = \frac{\text{Adjacent}}{\text{Opposite}} = \frac{3}{4}$$

NOW WORK PROBLEM 1.

The values of the trigonometric functions of an acute angle are ratios of the lengths of the sides of a right triangle. This way of viewing the trigonometric functions leads to many applications and, in fact, was the point of view used by early mathematicians (before calculus) in studying the subject of trigonometry.

We look at one such application next.

EXAMPLE 2 Constructing a Rain Gutter

Figure 4

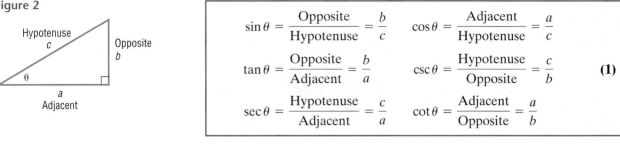

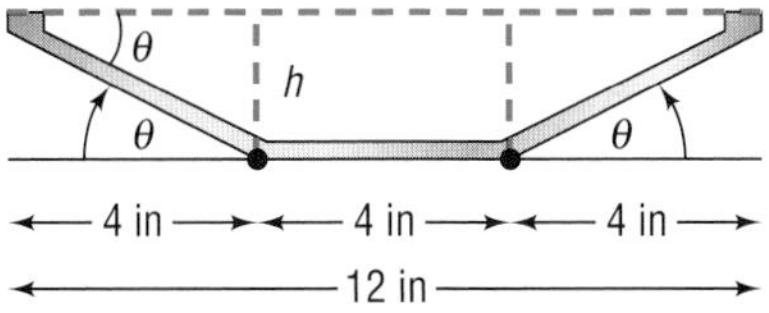

A rain gutter is to be constructed of aluminum sheets 12 inches wide. After marking off a length of 4 inches from each edge, this length is bent up at an angle θ. See Figure 4.

(a) Express the area A of the opening as a function of θ.

(b) Graph $A = A(\theta)$. Find the angle θ that makes A largest. (This bend will allow the most water to flow through the gutter.)

Solution (a) The area A of the opening is the sum of the areas of two right triangles and one rectangle. Each triangle has legs h and $\sqrt{16 - h^2}$ and hypotenuse 4. The rectangle has length 4 and height h. Then

$$A = 2 \cdot \tfrac{1}{2} \cdot h\sqrt{16 - h^2} + 4h = (4\sin\theta)(4\cos\theta) + 4(4\sin\theta)$$

$\uparrow$ $\sin\theta = \frac{h}{4}$, $\cos\theta = \frac{\sqrt{16 - h^2}}{4}$

$$A(\theta) = 16\sin\theta(\cos\theta + 1)$$

(b) See Figure 5. Using MAXIMUM, the angle θ that makes A largest is 60°.

Figure 5

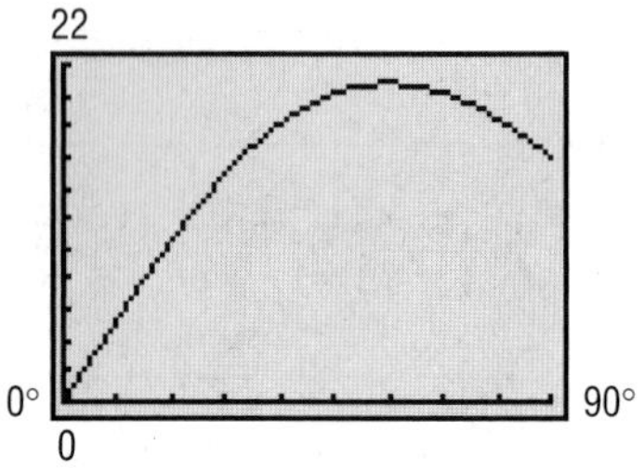

Complementary Angles: Cofunctions

2 Two acute angles are called **complementary** if their sum is a right angle. Because the sum of the angles of any triangle is 180°, it follows that, for a right triangle, the two acute angles are complementary.

Refer now to Figure 6; we have labeled the angle opposite side b as β and the angle opposite side a as α. Notice that side b is adjacent to angle α and side a is adjacent to angle β. As a result,

Figure 6

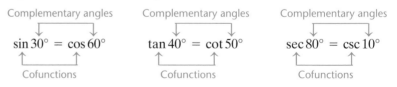

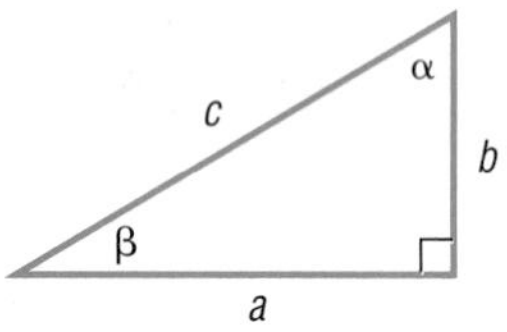

$$\sin\beta = \frac{b}{c} = \cos\alpha \qquad \cos\beta = \frac{a}{c} = \sin\alpha \qquad \tan\beta = \frac{b}{a} = \cot\alpha \qquad \textbf{(2)}$$

$$\csc\beta = \frac{c}{b} = \sec\alpha \qquad \sec\beta = \frac{c}{a} = \csc\alpha \qquad \cot\beta = \frac{a}{b} = \tan\alpha$$

Because of these relationships, the functions sine and cosine, tangent and cotangent, and secant and cosecant are called **cofunctions** of each other. The identities (2) may be expressed in words as follows:

Complementary Angle Theorem Cofunctions of complementary angles are equal.

Examples of this theorem are given next:

Complementary angles	Complementary angles	Complementary angles
$\sin 30° = \cos 60°$	$\tan 40° = \cot 50°$	$\sec 80° = \csc 10°$
Cofunctions	Cofunctions	Cofunctions

EXAMPLE 3 Using the Complementary Angle Theorem

(a) $\sin 62° = \cos(90° - 62°) = \cos 28°$

(b) $\tan\frac{\pi}{12} = \cot\left(\frac{\pi}{2} - \frac{\pi}{12}\right) = \cot\frac{5\pi}{12}$

(c) $\sin^2 40° + \sin^2 50° = \sin^2 40° + \cos^2 40° = 1$

$\uparrow$ $\sin 50° = \cos 40°$

Check: Verify the results obtained in Example 3 by evaluating each expression. ■

NOW WORK PROBLEM 11.

SOLVING RIGHT TRIANGLES; APPLICATIONS

3 In the discussion that follows, we will always label a right triangle so that side a is opposite angle α, side b is opposite angle β, and side c is the hypotenuse, as shown in Figure 7. **To solve a right triangle** means to find the missing lengths of its sides and the measurements of its angles. We shall follow the practice of expressing the lengths of the sides rounded to two decimal places and of expressing angles in degrees rounded to one decimal place.

Figure 7

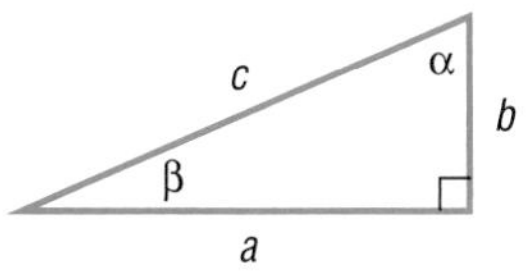

To solve a right triangle, we need to know either an angle (besides the 90° one) and a side or else two sides. Then we make use of the Pythagorean Theorem and the fact that the sum of the angles of a triangle is 180°. The sum of the unknown angles in a right triangle is therefore 90°.

For the right triangle shown in Figure 7, we have

$$c^2 = a^2 + b^2, \qquad \alpha + \beta = 90°$$

EXAMPLE 4 Solving a Right Triangle

Figure 8

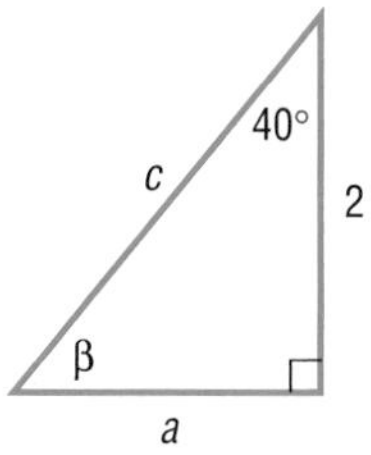

Use Figure 8. If $b = 2$ and $\alpha = 40°$, find a, c, and β.

Solution Since $\alpha = 40°$ and $\alpha + \beta = 90°$, we find that $\beta = 50°$. To find the sides a and c, we use the facts that

$$\tan 40° = \frac{a}{2} \quad \text{and} \quad \cos 40° = \frac{2}{c}$$

Now solve for a and c.

$$a = 2\tan 40° \approx 1.68 \quad \text{and} \quad c = \frac{2}{\cos 40°} \approx 2.61$$ ■

NOW WORK PROBLEM 29.

EXAMPLE 5 Solving a Right Triangle

Figure 9

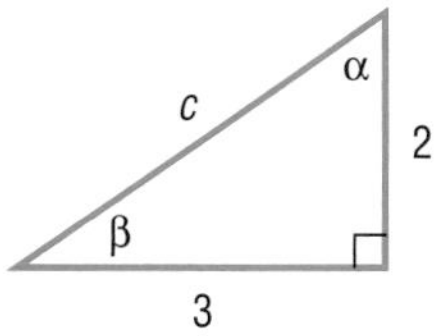

Use Figure 9. If $a = 3$ and $b = 2$, find c, α, and β.

Solution Since $a = 3$ and $b = 2$, then, by the Pythagorean Theorem, we have

$$c^2 = a^2 + b^2 = 3^2 + 2^2 = 9 + 4 = 13$$
$$c = \sqrt{13} \approx 3.61$$

To find angle α, we use the fact that

$$\tan \alpha = \frac{3}{2} \quad \text{so} \quad \alpha = \tan^{-1}\frac{3}{2}$$

Set the mode on your calculator to degrees. Then, rounded to one decimal place, we find that $\alpha = 56.3°$. Since $\alpha + \beta = 90°$, we find that $\beta = 33.7°$. ■

NOTE: In subsequent examples and in the exercises that follow, unless otherwise indicated, we will measure angles in degrees and round to one decimal place; we will round all sides to two decimal places. To avoid round-off errors when using a calculator, we will store unrounded values in memory for use in subsequent calculations.

NOW WORK PROBLEM 35.

4 One common use for trigonometry is to measure heights and distances that are either awkward or impossible to measure by ordinary means.

EXAMPLE 6 Finding the Width of a River

A surveyor can measure the width of a river by setting up a transit* at a point C on one side of the river and taking a sighting of a point A on the other side. Refer to Figure 10. After turning through an angle of 90° at C, the surveyor walks a distance of 200 meters to point B. Using the transit at B, the angle β is measured and found to be 20°. What is the width of the river?

Figure 10

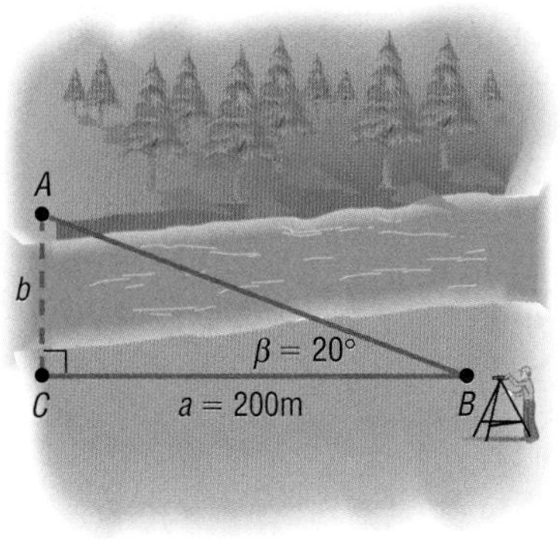

Solution We seek the length of side b. We know a and β, so we use the fact that

$$\tan\beta = \frac{b}{a}$$

to get

$$\tan 20^\circ = \frac{b}{200}$$

$$b = 200 \tan 20^\circ \approx 72.79 \text{ meters}$$

The width of the river is approximately 73 meters, rounded to the nearest meter. ■

NOW WORK PROBLEM 45.

EXAMPLE 7 Finding the Inclination of a Mountain Trail

A straight trail with a uniform inclination leads from the Alpine Hotel, elevation 8000 feet, to a scenic overlook, elevation 11,100 feet. The length of the trail is 14,100 feet. What is the inclination (grade) of the trail? That is, what is the angle β in Figure 11?

* An instrument used in surveying to measure angles.

Figure 11

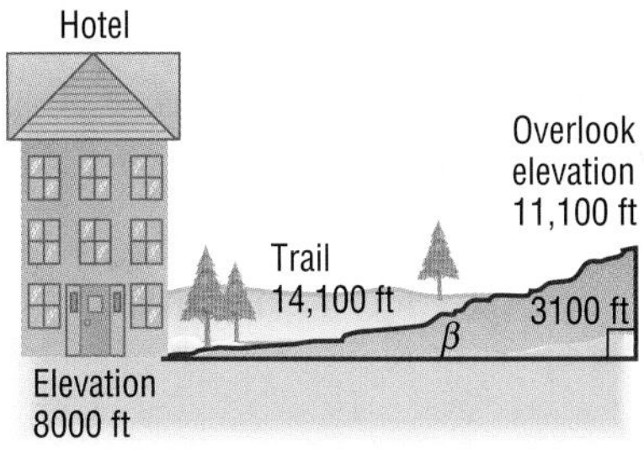

Solution As Figure 11 illustrates, the angle β obeys the equation

$$\sin \beta = \frac{3100}{14{,}100}$$

Using a calculator,

$$\beta = \sin^{-1} \frac{3100}{14{,}100} \approx 12.7^\circ$$

The inclination (grade) of the trail is approximately 12.7°. ■

Vertical heights can sometimes be measured using either the angle of elevation or the angle of depression. If a person is looking up at an object, the acute angle measured from the horizontal to a line-of-sight observation of the object is called the **angle of elevation.** See Figure 12(a).

If a person is standing on a cliff looking down at an object, the acute angle made by the line-of-sight observation of the object and the horizontal is called the **angle of depression.** See Figure 12(b).

Figure 12

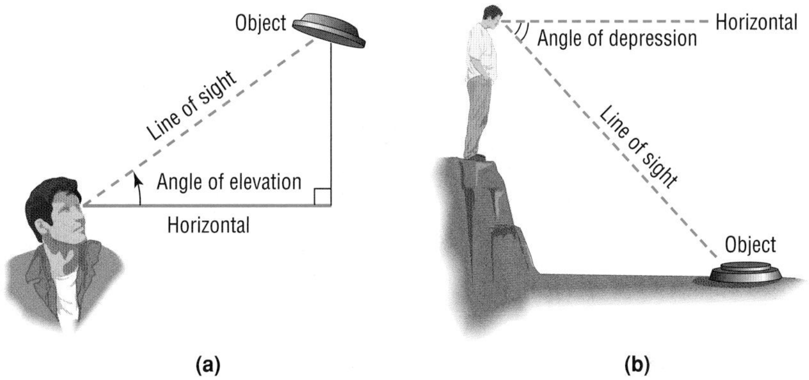

EXAMPLE 8 Finding Heights Using the Angle of Elevation

To determine the height of a radio transmission tower, a surveyor walks off a distance of 300 meters from the base of the tower. The angle of elevation is then measured and found to be 40°. If the transit is 2 meters off the ground when the sighting is taken, how high is the radio tower? See Figure 13(a).

Figure 13

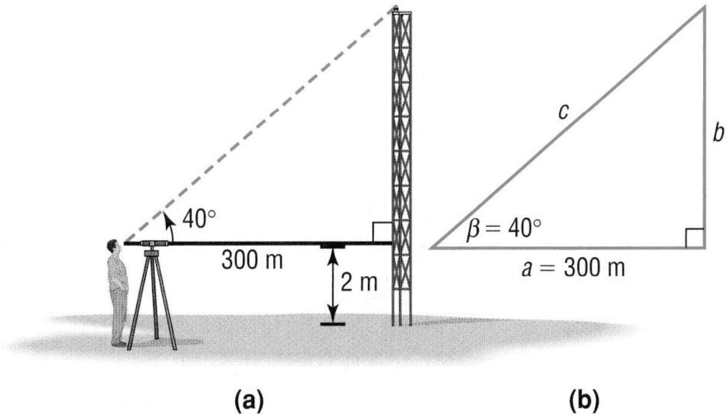

Solution Figure 13(b) shows a triangle that replicates the illustration in Figure 13(a). To find the length b, we use the fact that $\tan\beta = b/a$. Then

$$b = a\tan\beta = 300\tan 40° \approx 251.73 \text{ meters}$$

Because the transit is 2 meters high, the actual height of the tower is approximately 254 meters, rounded to the nearest meter. ■

NOW WORK PROBLEM 47.

The idea behind Example 8 can also be used to find the height of an object with a base that is not accessible to the horizontal.

EXAMPLE 9 Finding the Height of a Statue on a Building

Adorning the top of the Board of Trade building in Chicago is a statue of Ceres, the Greek goddess of wheat. From street level, two observations are taken 400 feet from the center of the building. The angle of elevation to the base of the statue is found to be 45°; the angle of elevation to the top of the statue is 47.2°. See Figure 14(a). What is the height of the statue?

Figure 14

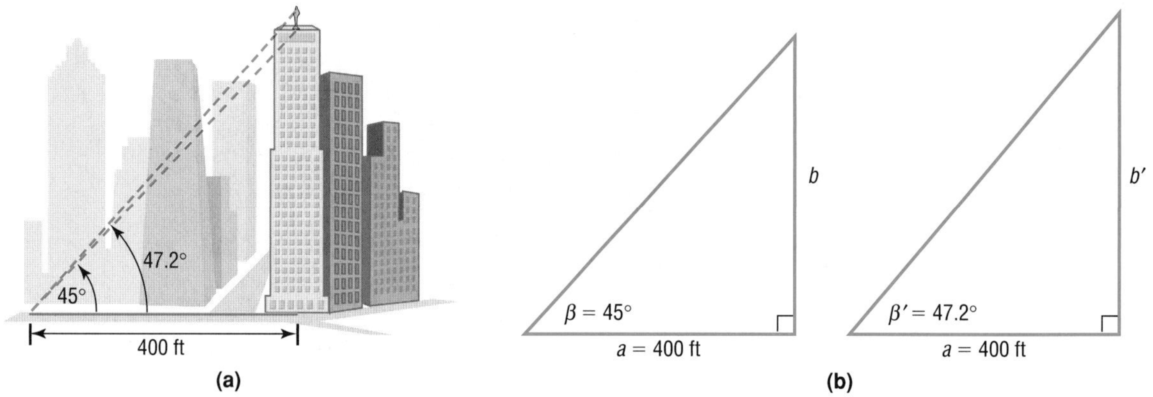

Solution Figure 14(b) shows two triangles that replicate Figure 14(a). The height of the statue of Ceres will be $b' - b$. To find b and b', we refer to Figure 14(b).

$$\tan 45° = \frac{b}{400} \qquad \tan 47.2° = \frac{b'}{400}$$

$$b = 400\tan 45° = 400 \qquad b' = 400\tan 47.2° \approx 432$$

The height of the statue is approximately $432 - 400 = 32$ feet. ■

EXAMPLE 10 The Gibb's Hill Lighthouse, Southampton, Bermuda

In operation since 1846, the Gibb's Hill Lighthouse stands 117 feet high on a hill 245 feet high, so its beam of light is 362 feet above sea level. A brochure states that the light can be seen on the horizon about 26 miles distant. Verify the accuracy of this statement.

Figure 15

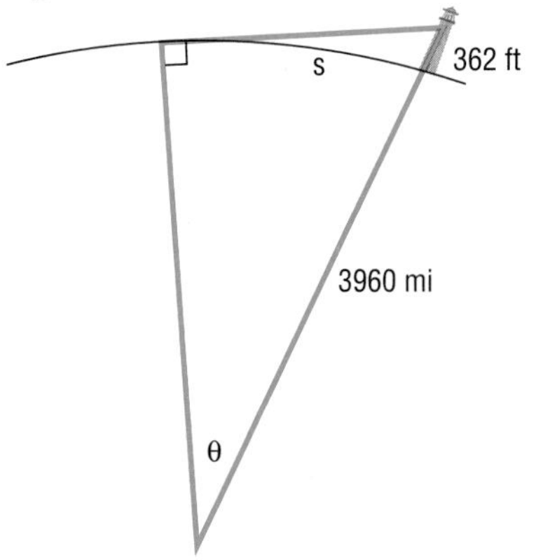

Solution Figure 15 illustrates the situation. The central angle θ obeys the equation

$$\cos\theta = \frac{3960}{3960 + \frac{362}{5280}} \approx 0.999982687 \quad \text{1 mile = 5280 feet}$$

so

$$\theta \approx 0.00588 \text{ radian} \approx 0.33715^\circ \approx 20.23'$$

The brochure does not indicate whether the distance is measured in nautical miles or statute miles. The distance s in nautical miles is 20.23, the measurement of θ in minutes. (Refer to Problem 92, Section 6.1). The distance s in statute miles is

$$s = r\theta = 3960(0.00588) = 23.3 \text{ miles}$$

In either case, it would seem that the brochure overstated the distance somewhat. ■

Figure 16

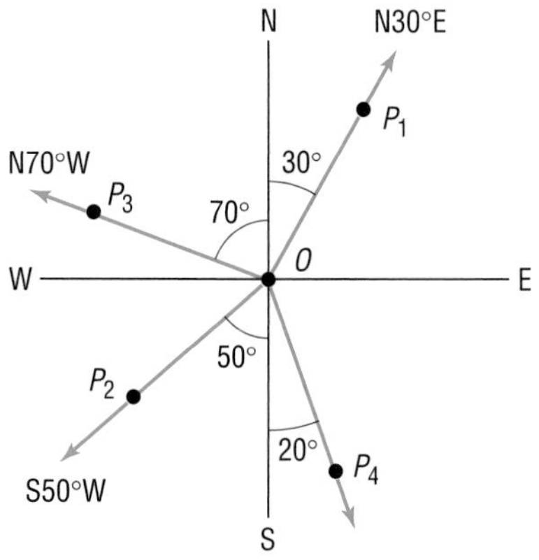

In navigation and surveying, the **direction** or **bearing** from a point O to a point P equals the acute angle θ between the ray OP and the vertical line through O, the north–south line.

Figure 16 illustrates some bearings. Notice that the bearing from O to P_1 is denoted by the symbolism N30°E, indicating that the bearing is 30° east of north. In writing the bearing from O to P, the direction north or south always appears first, followed by an acute angle, followed by east or west. In Figure 16, the bearing from O to P_2 is S50°W, and from O to P_3 it is N70°W.

EXAMPLE 11 Finding the Bearing of an Object

In Figure 16, what is the bearing from O to an object at P_4?

Solution The acute angle between the ray OP_4 and the north–south line through O is given as 20°. The bearing from O to P is S20°E. ■

EXAMPLE 12 Finding the Bearing of an Airplane

Figure 17

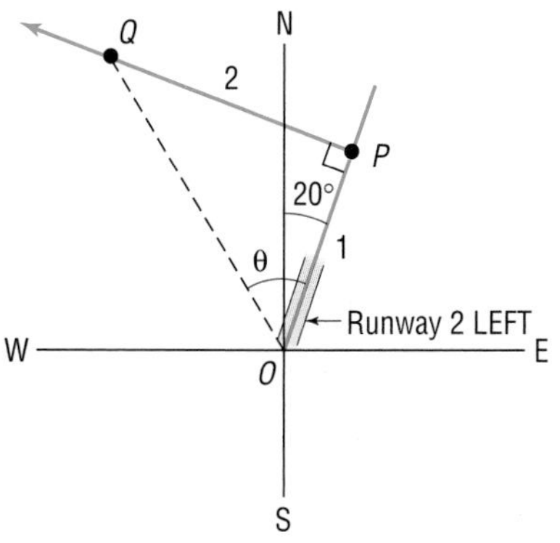

A Boeing 777 aircraft takes off from O'Hare Airport on runway 2 LEFT, which has a bearing of N20°E.* After flying for 1 mile, the pilot of the aircraft requests permission to turn 90° and head toward the northwest. The request is granted. After the plane goes 2 miles in this direction, what bearing should the control tower use to locate the aircraft?

Solution Figure 17 illustrates the situation. After flying 1 mile from the airport O (the control tower), the aircraft is at P. After turning 90° toward

*In air navigation, the term **azimuth** is employed to denote the positive angle measured clockwise from the north (N) to a ray OP. In Figure 16, the azimuth from O to P_1 is 30°; the azimuth from O to P_2 is 230°; the azimuth from O to P_3 is 290°. Runways are designated using azimuth. Thus, runway 2 LEFT means the left runway with a direction of azimuth 20° (bearing N20°E). Runway 23 is the runway with azimuth 230° and bearing S50°W.

the northwest and flying 2 miles, the aircraft is at the point Q. In triangle OPQ, the angle θ obeys the equation

$$\tan\theta = \frac{2}{1} = 2 \quad \text{so} \quad \theta = \tan^{-1}2 \approx 63.4°$$

The acute angle between north and the ray OQ is $63.4° - 20° = 43.4°$. The bearing of the aircraft from O to Q is N43.4°W. ■

8.1 EXERCISES

In Problems 1–10, find the exact value of the six trigonometric functions of the angle θ in each figure.

1.
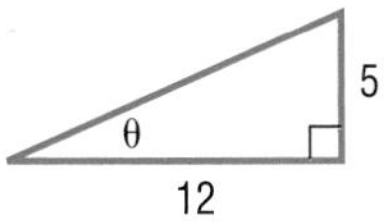

2.
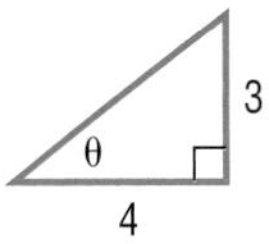

3.
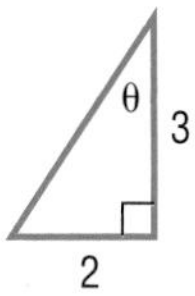

4.
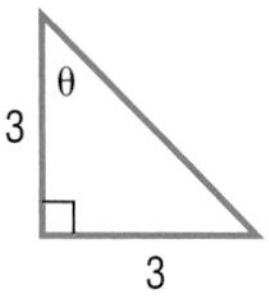

5.
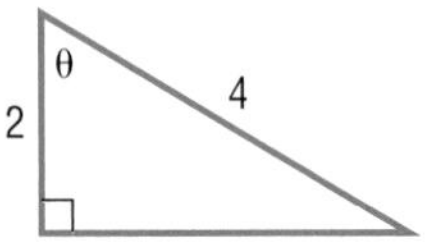

6.
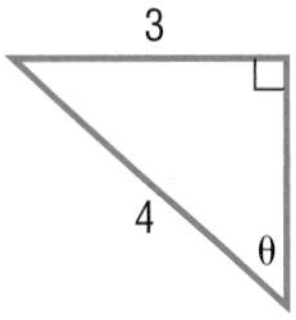

7.
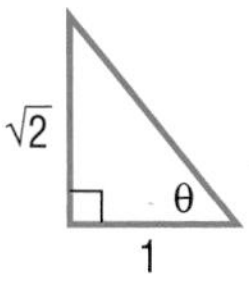

8.
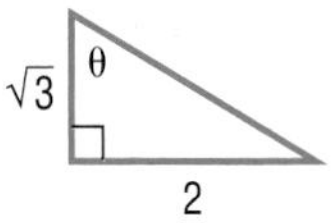

9.
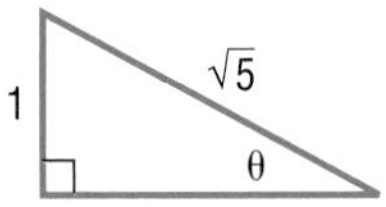

10.
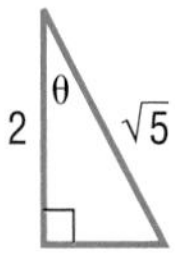

In Problems 11–20, find the exact value of each expression. Do not use a calculator.

11. $\sin 38° - \cos 52°$ **12.** $\tan 12° - \cot 78°$ **13.** $\dfrac{\cos 10°}{\sin 80°}$ **14.** $\dfrac{\cos 40°}{\sin 50°}$

15. $1 - \cos^2 20° - \cos^2 70°$ **16.** $1 + \tan^2 5° - \csc^2 85°$ **17.** $\tan 20° - \dfrac{\cos 70°}{\cos 20°}$

18. $\cot 40° - \dfrac{\sin 50°}{\sin 40°}$ **19.** $\cos 35° \sin 55° + \sin 35° \cos 55°$ **20.** $\sec 35° \csc 55° - \tan 35° \cot 55°$

21. If $\sin\theta = \frac{1}{3}$, find the exact value of: (a) $\cos(90° - \theta)$ (b) $\cos^2\theta$ (c) $\csc\theta$ (d) $\sec\left(\dfrac{\pi}{2} - \theta\right)$

22. If $\sin\theta = 0.2$, find the exact value of: (a) $\cos\left(\dfrac{\pi}{2} - \theta\right)$ (b) $\cos^2\theta$ (c) $\sec(90° - \theta)$ (d) $\csc\theta$

23. If $\tan\theta = 4$, find the exact value of: (a) $\sec^2\theta$ (b) $\cot\theta$ (c) $\cot\left(\dfrac{\pi}{2} - \theta\right)$ (d) $\csc^2\theta$

24. If $\sec\theta = 3$, find the exact value of: (a) $\cos\theta$ (b) $\tan^2\theta$ (c) $\csc(90° - \theta)$ (d) $\sin^2\theta$

25. If $\csc\theta = 4$, find the exact value of: (a) $\sin\theta$ (b) $\cot^2\theta$ (c) $\sec(90° - \theta)$ (d) $\sec^2\theta$

26. If $\cot\theta = 2$, find the exact value of: (a) $\tan\theta$ (b) $\csc^2\theta$ (c) $\tan\left(\dfrac{\pi}{2} - \theta\right)$ (d) $\sec^2\theta$

27. If $\sin\theta = 0.3$, find the exact value of $\sin\theta + \cos\left(\dfrac{\pi}{2} - \theta\right)$.

28. If $\tan\theta = 4$, find the exact value of $\tan\theta + \tan\left(\dfrac{\pi}{2} - \theta\right)$.

In Problems 29–38, use the right triangle shown. Then, using the given information, solve the triangle.

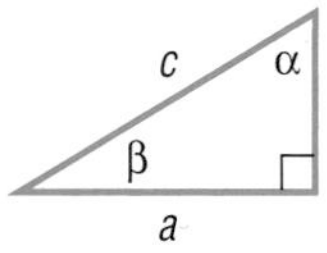

29. $b = 5$, $\beta = 20°$; find a, c, and α

30. $b = 4$, $\beta = 10°$; find a, c, and α

31. $a = 6$, $\beta = 40°$; find b, c, and α

32. $a = 7$, $\beta = 50°$; find b, c, and α

33. $b = 4$, $\alpha = 10°$; find a, c, and β

34. $b = 6$, $\alpha = 20°$; find a, c, and β

35. $a = 5$, $b = 3$; find c, α, and β

36. $a = 2$, $b = 8$; find c, α, and β

37. $a = 2$, $c = 5$; find b, α, and β

38. $b = 4$, $c = 6$; find a, α, and β

39. A right triangle has a hypotenuse of length 8 inches. If one angle is 35°, find the length of each leg.

40. A right triangle has a hypotenuse of length 10 centimeters. If one angle is 40°, find the length of each leg.

41. A right triangle contains a 25° angle. If one leg is of length 5 inches, what is the length of the hypotenuse?
[**Hint:** Two answers are possible.]

42. A right triangle contains an angle of $\pi/8$ radian. If one leg is of length 3 meters, what is the length of the hypotenuse?
[**Hint:** Two answers are possible.]

43. The hypotenuse of a right triangle is 5 inches. If one leg is 2 inches, find the degree measure of each angle.

44. The hypotenuse of a right triangle is 3 feet. If one leg is 1 foot, find the degree measure of each angle.

45. Finding the Width of a Gorge Find the distance from A to C across the gorge illustrated in the figure.

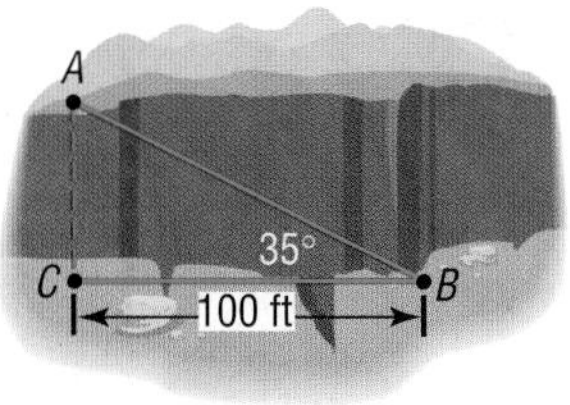

46. Finding the Distance across a Pond Find the distance from A to C across the pond illustrated in the figure.

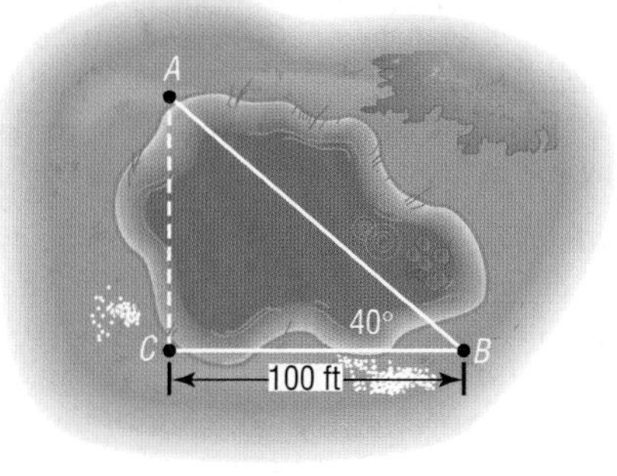

47. The Eiffel Tower The tallest tower built before the era of television masts, the Eiffel Tower was completed on March 31, 1889. Find the height of the Eiffel Tower (before a television mast was added to the top) using the information given in the illustration.

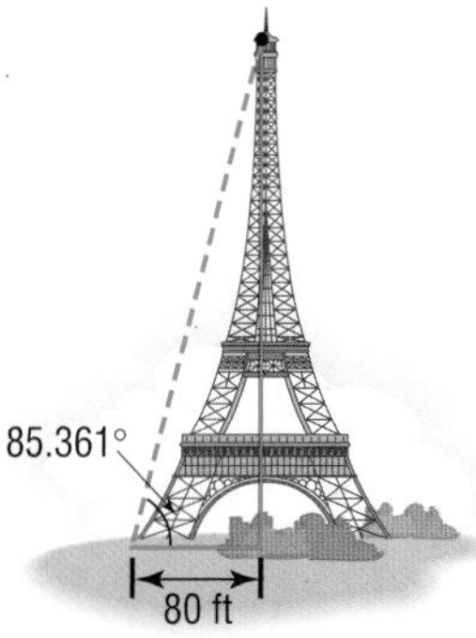

48. Finding the Distance of a Ship from Shore A ship, offshore from a vertical cliff known to be 100 feet in height, takes a sighting of the top of the cliff. If the angle of elevation is found to be 25°, how far offshore is the ship?

49. Finding the Distance to a Plateau Suppose that you are headed toward a plateau 50 meters high. If the angle of elevation to the top of the plateau is 20°, how far are you from the base of the plateau?

50. Statue of Liberty A ship is just offshore of New York City. A sighting is taken of the Statue of Liberty, which is about 305 feet tall. If the angle of elevation to the top of the statue is 20°, how far is the ship from the base of the statue?

51. Finding the Reach of a Ladder A 22-foot extension ladder leaning against a building makes a 70° angle with the ground. How far up the building does the ladder touch?

52. Finding the Height of a Building To measure the height of a building, two sightings are taken a distance of 50 feet apart. If the first angle of elevation is 40° and the second is 32°, what is the height of the building?

53. A laser beam is to be directed through a small hole in the center of a circle of radius 10 feet. The origin of the beam is 35 feet from the circle (see the figure). At what angle of elevation should the beam be aimed to ensure that it goes through the hole?

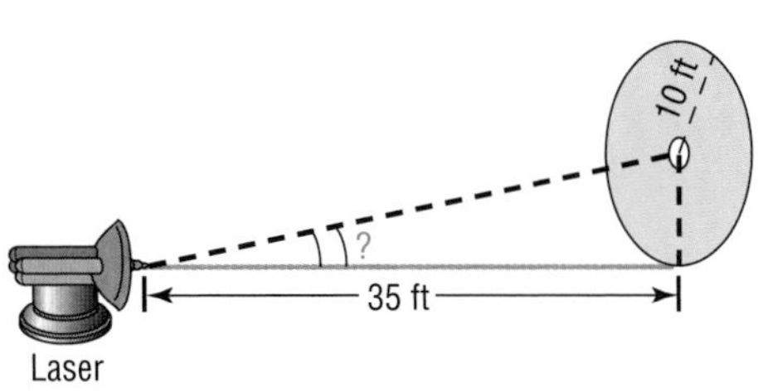

54. Finding the Angle of Elevation of the Sun At 10 AM on April 26, 2000, a building 300 feet high casts a shadow 50 feet long. What is the angle of elevation of the Sun?

55. Mt. Rushmore To measure the height of Lincoln's caricature on Mt. Rushmore, two sightings 800 feet from the base of the mountain are taken. If the angle of elevation to the bottom of Lincoln's face is 32° and the angle of elevation to the top is 35°, what is the height of Lincoln's face?

56. Finding the Distance between Two Objects A blimp, suspended in the air at a height of 500 feet, lies directly over a line from Soldier Field to the Adler Planetarium on Lake Michigan (see the figure). If the angle of depression from the blimp to the stadium is 32° and from the blimp to the planetarium is 23°, find the distance between Soldier Field and the Adler Planetarium.

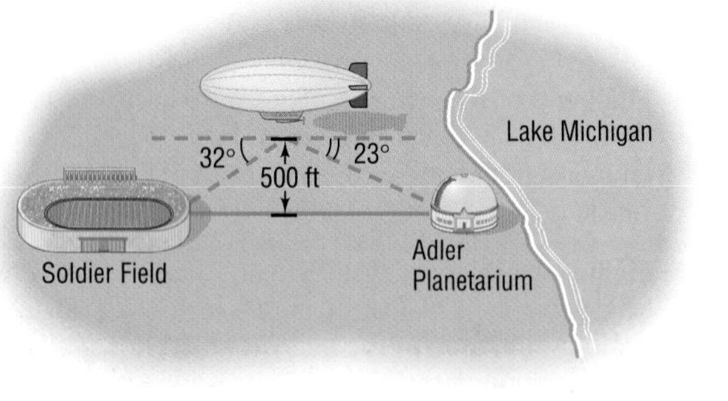

57. Finding the Length of a Guy Wire A radio transmission tower is 200 feet high. How long should a guy wire be if it is to be attached to the tower 10 feet from the top and is to make an angle of 21° with the ground?

58. Finding the Height of a Tower A guy wire 80 feet long is attached to the top of a radio transmission tower, making an angle of 25° with the ground. How high is the tower?

59. Washington Monument The angle of elevation of the Washington Monument is 35.1° at the instant it casts a shadow 789 feet long. Use this information to calculate the height of the monument.

60. Finding the Length of a Mountain Trail A straight trail with a uniform inclination of 17° leads from a hotel at an elevation of 9000 feet to a mountain lake at an elevation of 11,200 feet. What is the length of the trail?

61. Finding the Speed of a Truck A state trooper is hidden 30 feet from a highway. One second after a truck passes, the angle θ between the highway and the line of observation from the patrol car to the truck is measured. See the illustration.

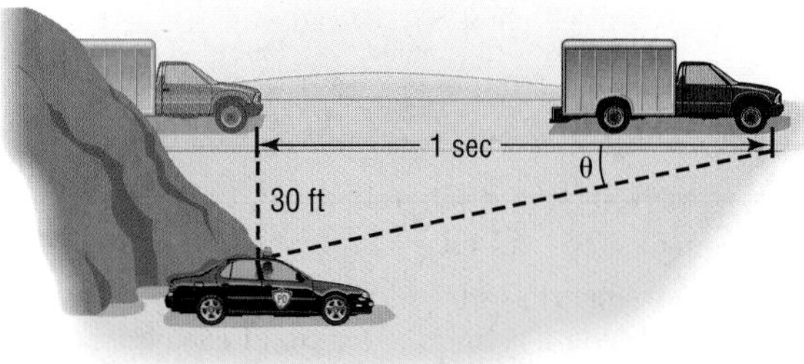

(a) If the angle measures 15°, how fast is the truck traveling? Express the answer in feet per second and in miles per hour.
(b) If the angle measures 20°, how fast is the truck traveling? Express the answer in feet per second and in miles per hour.
(c) If the speed limit is 55 miles per hour and a speeding ticket is issued for speeds of 5 miles per hour or more over the limit, for what angles should the trooper issue a ticket?

62. Security A security camera in a neighborhood bank is mounted on a wall 9 feet above the floor. What angle of depression should be used if the camera is to be directed to a spot 6 feet above the floor and 12 feet from the wall?

63. Finding the Bearing of an Aircraft A DC-9 aircraft leaves Midway Airport from runway 4 RIGHT, whose bearing is N40°E. After flying for 1/2 mile, the pilot requests permission to turn 90° and head toward the southeast. The permission is granted. After the airplane goes 1 mile in this direction, what bearing should the control tower use to locate the aircraft?

64. Finding the Bearing of a Ship A ship leaves the port of Miami with a bearing of S80°E and a speed of 15 knots. After 1 hour, the ship turns 90° toward the south. After 2 hours, maintaining the same speed, what is the bearing to the ship from port?

65. Shooting Free Throws in Basketball The eyes of a basketball player are 6 feet above the floor. The player is at the free-throw line, which is 15 feet from the center of the basket rim (see the figure). What is the angle of elevation from the player's eyes to the center of the rim?
[**Hint:** The rim is 10 feet above the floor.]

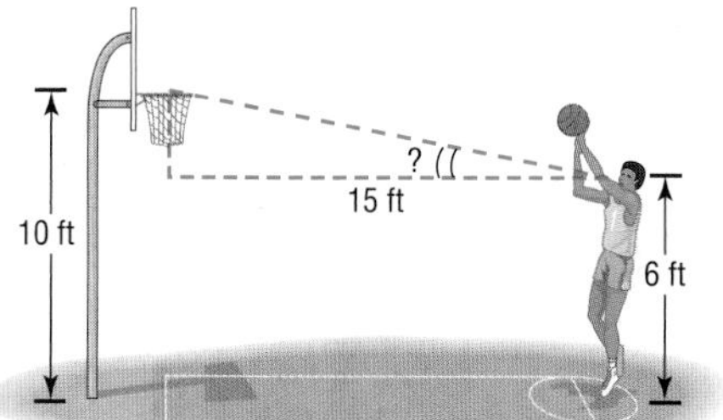

66. Finding the Pitch of a Roof A carpenter is preparing to put a roof on a garage that is 20 feet by 40 feet by 20 feet. A steel support beam 46 feet in length is positioned in the center of the garage. To support the roof, another beam will be attached to the top of the center beam (see the figure). At what angle of elevation is the new beam? In other words, what is the pitch of the roof?

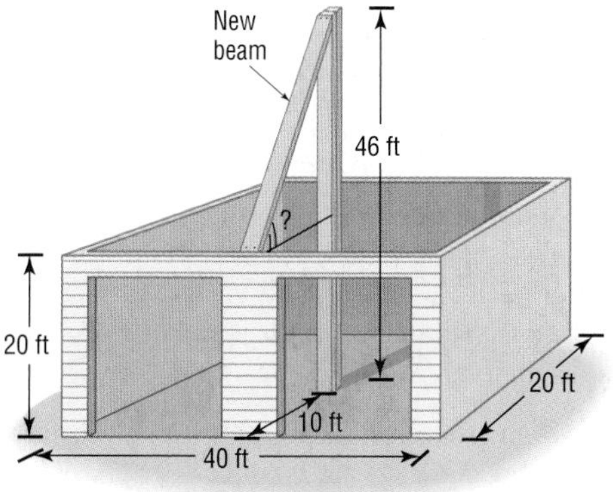

67. Calculating the Time of a Trip Two oceanfront homes are located 8 miles apart on a straight stretch of beach, each a distance of 1 mile from a paved road that parallels the ocean. Sally can jog 8 miles per hour along the paved road, but only 3 miles per hour in the sand on the beach. Because of a river directly between the two houses, it is necessary to jog in the sand to the road, continue on the road, and then jog directly back in the sand to get from one house to the other. See the illustration.
(a) Express the time T to get from one house to the other as a function of the angle θ shown in the illustration.
(b) Graph $T = T(\theta)$. What angle θ results in the least time? What is the least time? How long is Sally on the paved road?

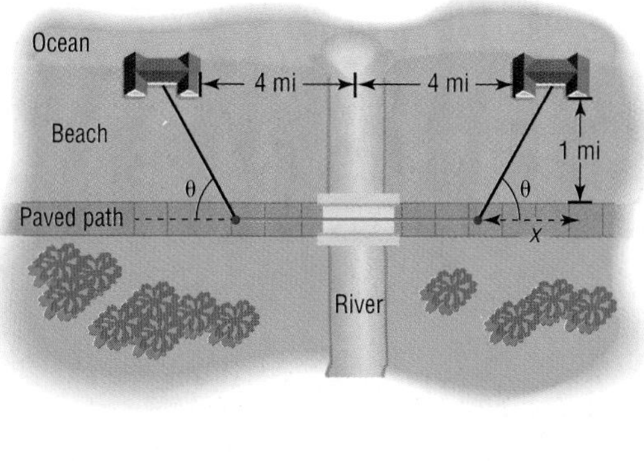

68. Designing Fine Decorative Pieces A designer of decorative art plans to market solid gold spheres enclosed in clear crystal cones. Each sphere is of fixed radius R and will be enclosed in a cone of height h and radius r. See the illustration. Many cones can be used to enclose the sphere, each having a different slant angle θ.
(a) Express the volume V of the cone as a function of the slant angle θ of the cone.
[**Hint:** The volume V of a cone of height h and radius r is $V = \frac{1}{3}\pi r^2 h$.]
(b) What slant angle θ should be used for the volume V of the cone to be a minimum? (This choice minimizes the amount of crystal required and gives maximum emphasis to the gold sphere).

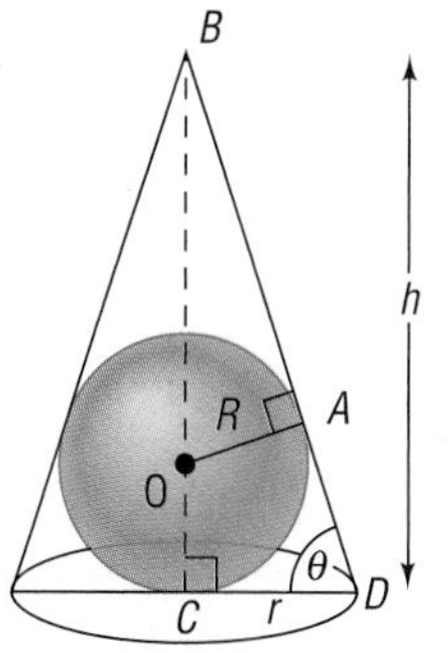

69. Calculating the Time of a Trip From a parking lot, you want to walk to a house on the ocean. The house is located 1500 feet down a paved path that parallels the ocean, which is 500 feet away. See the illustration. Along the path you can walk 300 feet per minute, but in the sand on the beach you can walk only 100 feet per minute.
(a) Calculate the time T if you walk 1500 feet along the paved path and then 500 feet in the sand to the house.
(b) Calculate the time T if you walk in the sand first for 500 feet and then walk along the sand for 1500 feet to the house.
(c) Express the time T to get from the parking lot to the beachhouse as a function of the angle θ shown in the illustration.
(d) Calculate the time T if you walk 1000 feet along the paved path and then walk directly to the house.
(e) Graph $T = T(\theta)$. For what angle θ is T least? What is the least time? What is x for this angle?

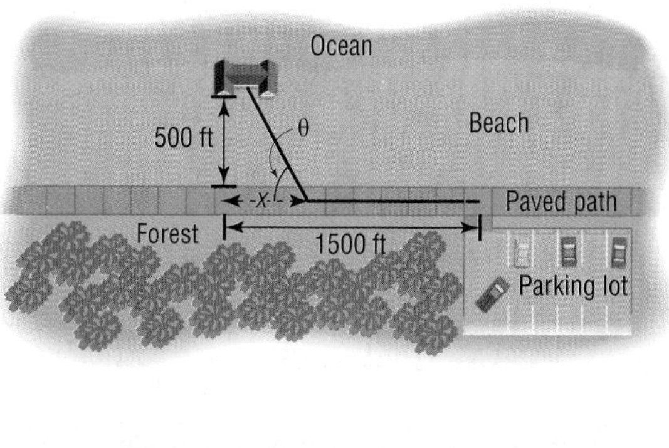

70. Carrying a Ladder around a Corner A ladder of length L is carried horizontally around a corner from a hall 3 feet wide into a hall 4 feet wide. See the illustration. Find the length L of the ladder as a function of the angle θ shown in the illustration.

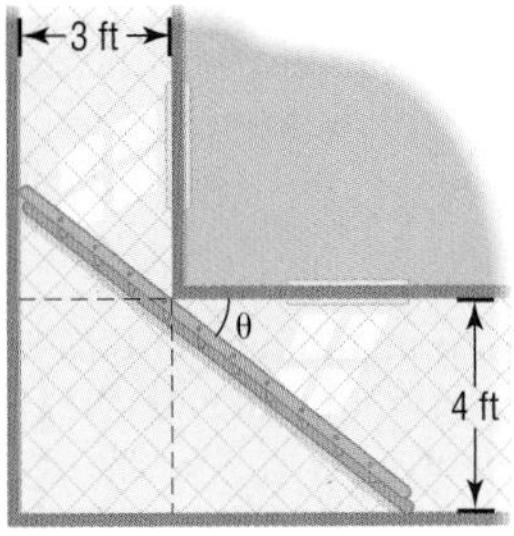

71. Constructing a Highway A highway whose primary directions are north–south is being constructed along the west coast of Florida. Near Naples, a bay obstructs the straight path of the road. Since the cost of a bridge is prohibitive, engineers decide to go around the bay. The illustration shows the path that they decide on and the measurements taken. What is the length of highway needed to go around the bay?

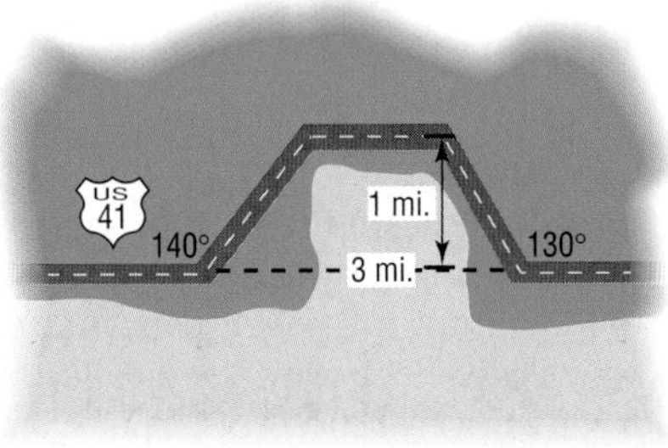

72. Surveillance Satellites A surveillance satellite circles Earth at a height of h miles above the surface. Suppose that d is the distance, in miles, on the surface of Earth that can be observed from the satellite. See the illustration.

(a) Find an equation that relates the central angle θ to the height h.

(b) Find an equation that relates the observable distance d and θ.

(c) Find an equation that relates d and h.

(d) If d is to be 2500 miles, how high must the satellite orbit above Earth?

(e) If the satellite orbits at a height of 300 miles, what distance d on the surface can be observed?

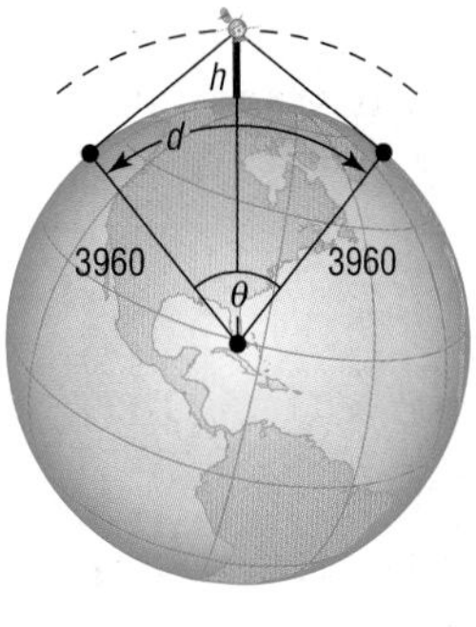

73. Photography A camera is mounted on a tripod 4 feet high at a distance of 10 feet from George, who is 6 feet tall. See the illustration. If the camera lens has angles of depression and elevation of 20°, will George's feet and head be seen by the lens? If not, how far back will the camera need to be moved to include George's feet and head?

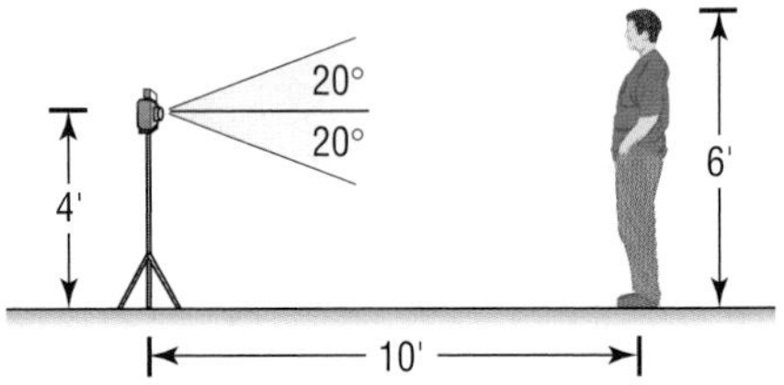

74. Construction A ramp for wheel chair accessibility is to be constructed with an angle of elevation of 15° and a final height of 5 feet. How long is the ramp?

75. The Gibb's Hill Lighthouse, Southampton, Bermuda In operation since 1846, the Gibb's Hill Lighthouse stands 117 feet high on a hill 245 feet high, so its beam of light is 362 feet above sea level. A brochure states that ships 40 miles away can see the light and planes flying at 10,000 feet can see it 120 miles away. Verify the accuracy of these statements. What assumption did the brochure make about the height of the ship?

76. Area of an Isosceles Triangle Show that the area A of an isosceles triangle, whose equal sides are of length s and the angle between them is θ, is

$$A = \tfrac{1}{2}s^2 \sin\theta$$

[**Hint:** See the illustration. The height h bisects the angle θ and is the perpendicular bisector of the base.]

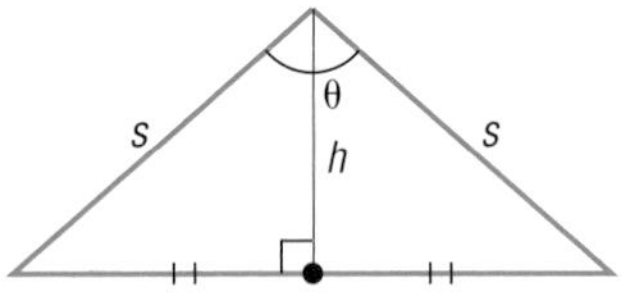

77. Suppose that the angle θ is a central angle of a circle of radius 1 (see the figure). Show that:

(a) Angle $OAC = \dfrac{\theta}{2}$

(b) $|CD| = \sin\theta$ and $|OD| = \cos\theta$

(c) $\tan\dfrac{\theta}{2} = \dfrac{\sin\theta}{1 + \cos\theta}$

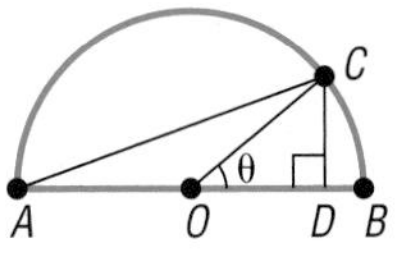

78. Show that the area of an isosceles triangle is $A = a^2 \sin\theta \cos\theta$, where a is the length of one of the two equal sides and θ is the measure of one of the two equal angles (see the figure).

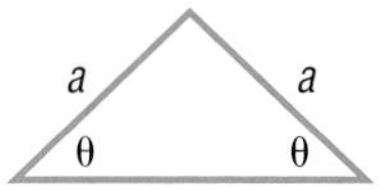

79. Let $n > 0$ be any real number, and let θ be any angle for which $0 < \theta < \pi/(1 + n)$. Then we can construct a triangle with the angles θ and $n\theta$ and included side of length 1 (do you see why?) and place it on the unit circle as illustrated. Now, drop the perpendicular from C to $D = (x, 0)$ and show that

$$x = \frac{\tan(n\theta)}{\tan\theta + \tan(n\theta)}$$

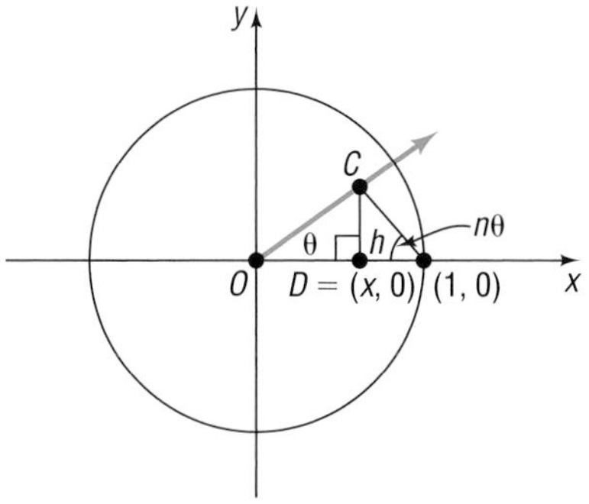

80. Refer to the given figure. The smaller circle, whose radius is a, is tangent to the larger circle, whose radius is b. The ray OA contains a diameter of each circle, and the ray OB is tangent to each circle. Show that

$$\cos\theta = \frac{\sqrt{ab}}{\frac{a + b}{2}}$$

(That is, $\cos\theta$ equals the ratio of the geometric mean of a and b to the arithmetic mean of a and b.)
[**Hint:** First show that $\sin\theta = (b - a)/(b + a)$.]

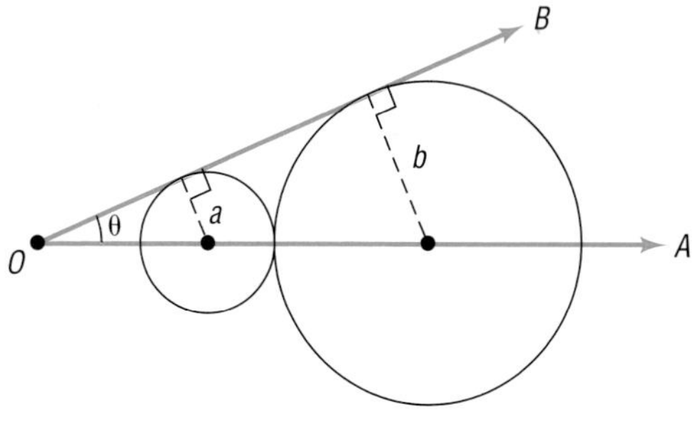

81. Geometry A rectangle is inscribed in a semicircle of radius 1. See the illustration.

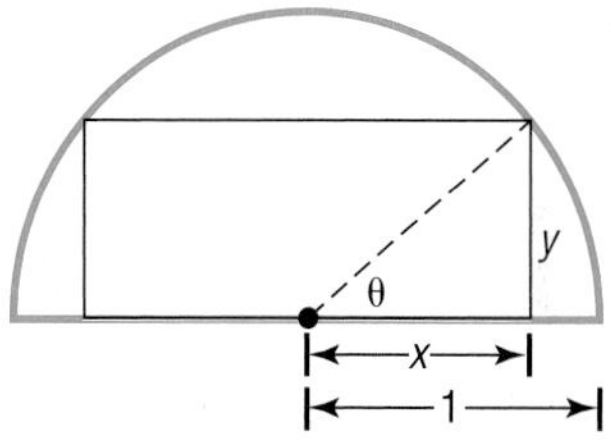

(a) Express the area A of the rectangle as a function of the angle θ shown in the illustration.
(b) Show that $A = \sin(2\theta)$.
(c) Find the angle θ that results in the largest area A.
(d) Find the dimensions of this largest rectangle.

82. If $\cos\alpha = \tan\beta$ and $\cos\beta = \tan\alpha$, where α and β are acute angles, show that

$$\sin\alpha = \sin\beta = \sqrt{\frac{3 - \sqrt{5}}{2}}$$

8.2 THE LAW OF SINES

OBJECTIVES
1 Solve SAA or ASA Triangles
2 Solve SSA Triangles
3 Solve Applied Problems

If none of the angles of a triangle is a right angle, the triangle is called **oblique.** An oblique triangle will have either three acute angles or two acute angles and one obtuse angle (an angle between 90° and 180°). See Figure 18.

Figure 18

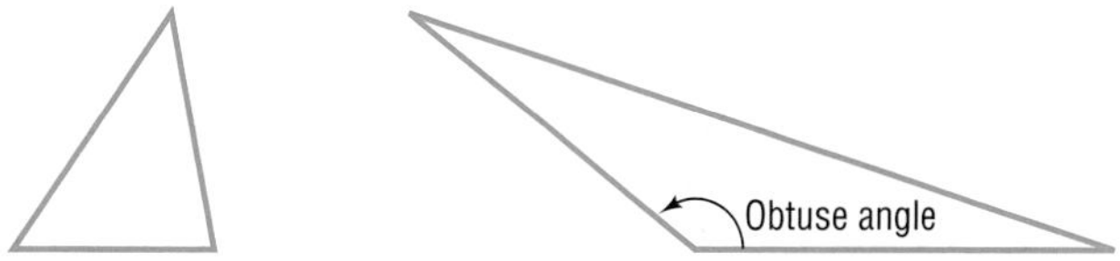

(a) All angles are acute **(b)** Two acute angles and one obtuse angle

Figure 19

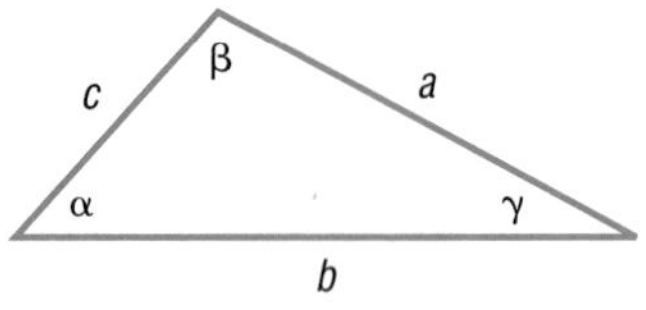

In the discussion that follows, we will always label an oblique triangle so that side a is opposite angle α, side b is opposite angle β, and side c is opposite angle γ, as shown in Figure 19.

To **solve an oblique triangle** means to find the lengths of its sides and the measurements of its angles. To do this, we shall need to know the length of one side along with two other facts: either two angles, or the other two sides, or one angle and one other side.* There are four possibilities to consider:

CASE 1: One side and two angles are known (ASA or SAA).
CASE 2: Two sides and the angle opposite one of them are known (SSA).
CASE 3: Two sides and the included angle are known (SAS).
CASE 4: Three sides are known (SSS).

Figure 20 illustrates the four cases.

Figure 20

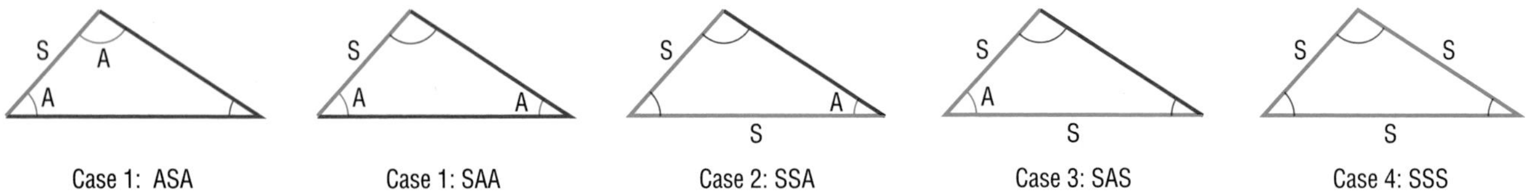

The **Law of Sines** is used to solve triangles for which Case 1 or 2 holds.

Theorem **Law of Sines**

For a triangle with sides a, b, c and opposite angles α, β, γ, respectively,

$$\frac{\sin\alpha}{a} = \frac{\sin\beta}{b} = \frac{\sin\gamma}{c} \quad (1)$$

Figure 21

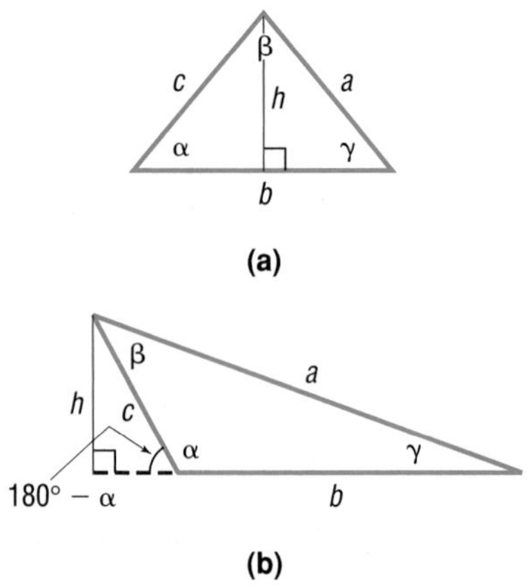

Proof To prove the Law of Sines, we construct an altitude of length h from one of the vertices of such a triangle. Figure 21(a) shows h for a triangle with three acute angles, and Figure 21(b) shows h for a triangle with an obtuse angle. In each case, the altitude is drawn from the vertex at β. Using either illustration, we have

$$\sin\gamma = \frac{h}{a}$$

from which

$$h = a\sin\gamma \quad (2)$$

*Recall from plane geometry the fact that knowing three angles of a triangle determines a family of *similar triangles*, that is, triangles that have the same shape but different sizes.

From Figure 21(a), it also follows that

$$\sin\alpha = \frac{h}{c}$$

from which

$$h = c\sin\alpha \qquad \textbf{(3)}$$

From Figure 21(b), it follows that

$$\sin(180° - \alpha) = \sin\alpha = \frac{h}{c}$$

↑ Difference formula

which again gives

$$h = c\sin\alpha$$

So, whether the triangle has three acute angles or has two acute angles and one obtuse angle, equations (2) and (3) hold. As a result, we may equate the expressions for h in equations (2) and (3) to get

$$a\sin\gamma = c\sin\alpha$$

from which

$$\frac{\sin\alpha}{a} = \frac{\sin\gamma}{c} \qquad \textbf{(4)}$$

Figure 22

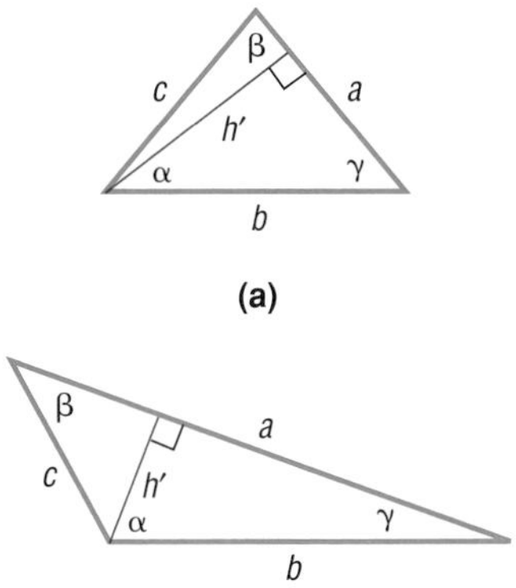

In a similar manner, by constructing the altitude h' from the vertex of angle α as shown in Figure 22, we can show that

$$\sin\beta = \frac{h'}{c} \quad \text{and} \quad \sin\gamma = \frac{h'}{b}$$

Equating the expressions for h', we find

$$h' = c\sin\beta = b\sin\gamma$$

from which

$$\frac{\sin\beta}{b} = \frac{\sin\gamma}{c} \qquad \textbf{(5)}$$

When equations (4) and (5) are combined, we have equation (1), the Law of Sines. ■

In applying the Law of Sines to solve triangles, we use the fact that the sum of the angles of any triangle equals 180°; that is,

$$\alpha + \beta + \gamma = 180° \qquad \textbf{(6)}$$

1 Our first two examples show how to solve a triangle when one side and two angles are known (Case 1: SAA or ASA).

EXAMPLE 1 Using the Law of Sines to Solve a SAA Triangle

Solve the triangle: $\alpha = 40°, \beta = 60°, a = 4$

Solution Figure 23 shows the triangle that we want to solve. The third angle γ is found using equation (6).

Figure 23

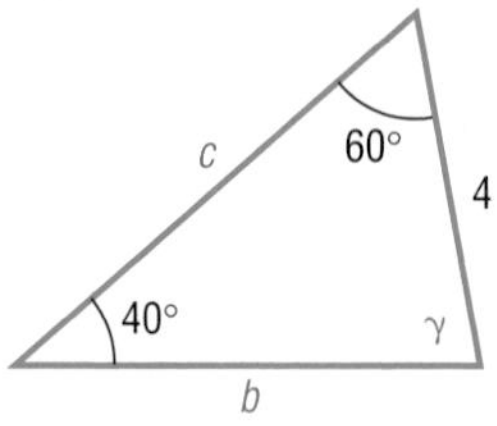

$$\alpha + \beta + \gamma = 180^\circ$$
$$40^\circ + 60^\circ + \gamma = 180^\circ$$
$$\gamma = 80^\circ$$

Now we use the Law of Sines (twice) to find the unknown sides b and c.

$$\frac{\sin\alpha}{a} = \frac{\sin\beta}{b} \qquad \frac{\sin\alpha}{a} = \frac{\sin\gamma}{c}$$

Because $a = 4$, $\alpha = 40^\circ$, $\beta = 60^\circ$, and $\gamma = 80^\circ$, we have

$$\frac{\sin 40^\circ}{4} = \frac{\sin 60^\circ}{b} \qquad \frac{\sin 40^\circ}{4} = \frac{\sin 80^\circ}{c}$$

Solving for b and c, we find that

$$b = \frac{4\sin 60^\circ}{\sin 40^\circ} \approx 5.39 \qquad c = \frac{4\sin 80^\circ}{\sin 40^\circ} \approx 6.13$$

Notice in Example 1 that we found b and c by working with the given side a. This is better than finding b first and working with a rounded value of b to find c.

NOW WORK PROBLEM 1.

EXAMPLE 2 Using the Law of Sines to Solve an ASA Triangle

Solve the triangle: $\alpha = 35^\circ, \beta = 15^\circ, c = 5$

Solution Figure 24 illustrates the triangle that we want to solve. Because we know two angles ($\alpha = 35^\circ$ and $\beta = 15^\circ$), we find the third angle using equation (6).

Figure 24

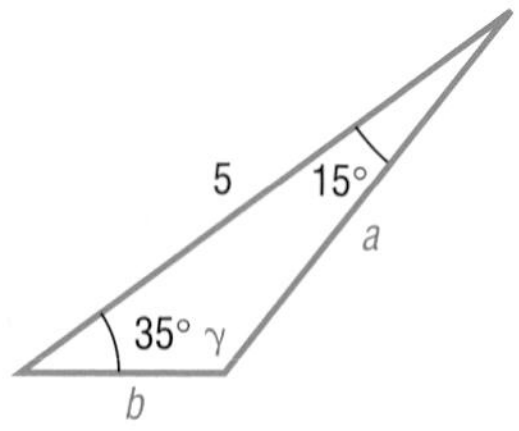

$$\alpha + \beta + \gamma = 180^\circ$$
$$35^\circ + 15^\circ + \gamma = 180^\circ$$
$$\gamma = 130^\circ$$

Now we know the three angles and one side ($c = 5$) of the triangle. To find the remaining two sides a and b, we use the Law of Sines (twice).

$$\frac{\sin\alpha}{a} = \frac{\sin\gamma}{c} \qquad \frac{\sin\beta}{b} = \frac{\sin\gamma}{c}$$

$$\frac{\sin 35^\circ}{a} = \frac{\sin 130^\circ}{5} \qquad \frac{\sin 15^\circ}{b} = \frac{\sin 130^\circ}{5}$$

$$a = \frac{5\sin 35^\circ}{\sin 130^\circ} \approx 3.74 \qquad b = \frac{5\sin 15^\circ}{\sin 130^\circ} \approx 1.69$$

NOW WORK PROBLEM 15.

THE AMBIGUOUS CASE

Figure 25

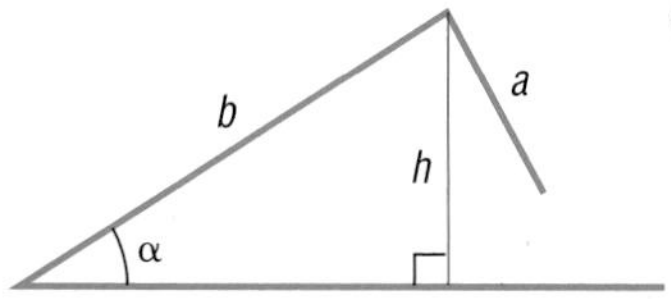

2 Case 2 (SSA), which applies to triangles for which two sides and the angle opposite one of them are known, is referred to as the **ambiguous case,** because the known information may result in one triangle, two triangles, or no triangle at all. Suppose that we are given sides a and b and angle α, as illustrated in Figure 25. The key to determining the possible triangles, if any, that may

be formed from the given information, lies primarily with the height h and the fact that $h = b \sin \alpha$.

No Triangle If $a < h = b \sin \alpha$, then side a is not sufficiently long to form a triangle. See Figure 26.

Figure 26
$a < h = b \sin \alpha$

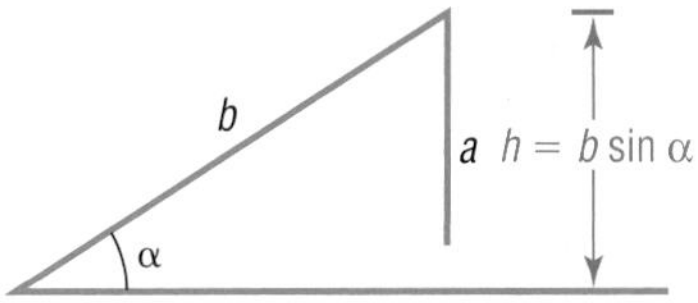

One Right Triangle If $a = h = b \sin \alpha$, then side a is just long enough to form a right triangle. See Figure 27.

Figure 27
$a = b \sin \alpha$

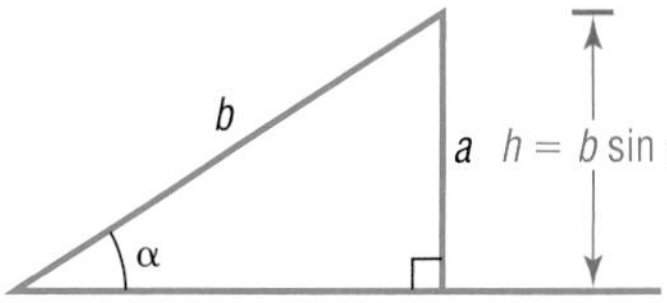

Two Triangles If $a < b$ and $h = b \sin \alpha < a$, then two distinct triangles can be formed from the given information. See Figure 28.

Figure 28
$b \sin \alpha < a$ and $a < b$

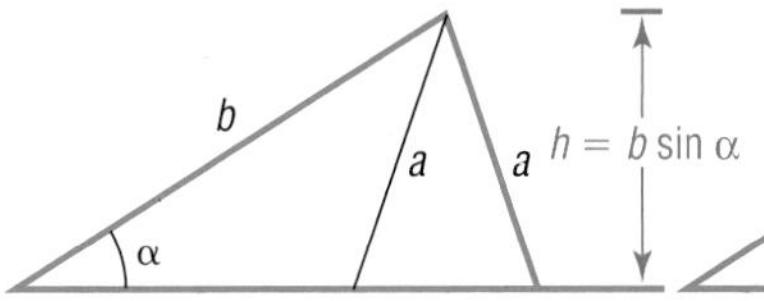

One Triangle If $a \geq b$, then only one triangle can be formed. See Figure 29.

Figure 29
$a \geq b$

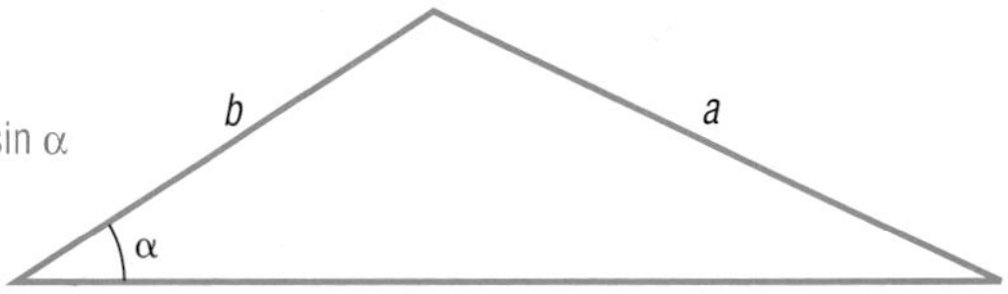

Fortunately, we do not have to rely on an illustration to draw the correct conclusion in the ambiguous case. The Law of Sines will lead us to the correct determination. Let's see how.

EXAMPLE 3 Using the Law of Sines to Solve a SSA Triangle (One Solution)

Solve the triangle: $a = 3, b = 2, \alpha = 40^\circ$

Solution See Figure 30(a). Because $a = 3, b = 2$, and $\alpha = 40^\circ$ are known, we use the Law of Sines to find the angle β.

Figure 30(a)

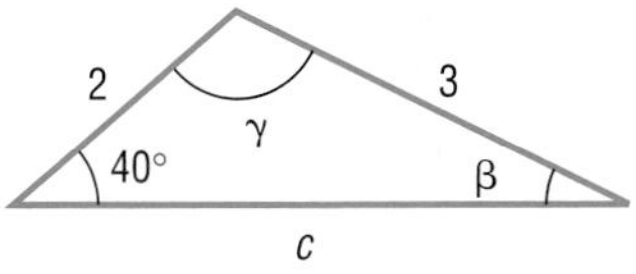

$$\frac{\sin \alpha}{a} = \frac{\sin \beta}{b}$$

Then

$$\frac{\sin 40^\circ}{3} = \frac{\sin \beta}{2}$$

$$\sin \beta = \frac{2 \sin 40^\circ}{3} \approx 0.43$$

There are two angles β, $0° < \beta < 180°$, for which $\sin\beta \approx 0.43$.

$$\beta_1 \approx 25.4° \quad \text{and} \quad \beta_2 \approx 154.6°$$

NOTE: Here we computed β using the stored value of $\sin\beta$. If you use the rounded value, $\sin\beta \approx 0.43$, you will obtain slightly different results.

The second possibility, $\beta_2 \approx 154.6°$, is ruled out, because $\alpha = 40°$, making $\alpha + \beta_2 \approx 194.6° > 180°$. Now, using $\beta_1 \approx 25.4°$, we find that

$$\gamma = 180° - \alpha - \beta_1 \approx 180° - 40° - 25.4° = 114.6°$$

The third side c may now be determined using the Law of Sines.

$$\frac{\sin\alpha}{a} = \frac{\sin\gamma}{c}$$

$$\frac{\sin 40°}{3} = \frac{\sin 114.6°}{c}$$

$$c = \frac{3\sin 114.6°}{\sin 40°} \approx 4.24$$

Figure 30(b)

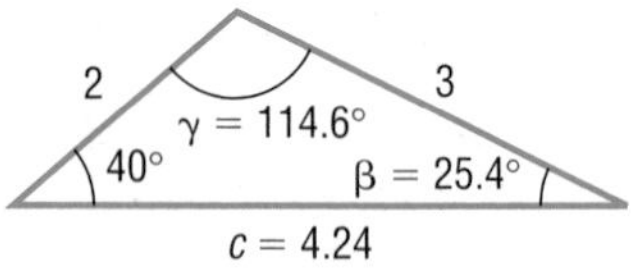

Figure 30(b) illustrates the solved triangle. ■

EXAMPLE 4 Using the Law of Sines to Solve a SSA Triangle (Two Solutions)

Solve the triangle: $a = 6, b = 8, \alpha = 35°$

Solution See Figure 31(a). Because $a = 6$, $b = 8$, and $\alpha = 35°$ are known, we use the Law of Sines to find the angle β.

Figure 31(a)

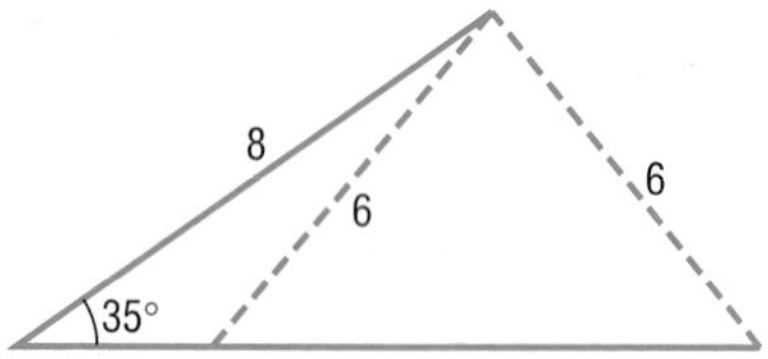

$$\frac{\sin\alpha}{a} = \frac{\sin\beta}{b}$$

Then

$$\frac{\sin 35°}{6} = \frac{\sin\beta}{8}$$

$$\sin\beta = \frac{8\sin 35°}{6} \approx 0.76$$

$$\beta_1 \approx 49.9° \quad \text{or} \quad \beta_2 \approx 130.1°$$

For both choices of β, we have $\alpha + \beta < 180°$. Hence, there are two triangles, one containing the angle $\beta_1 \approx 49.9°$ and the other containing the angle $\beta_2 \approx 130.1°$. The third angle γ is either

$$\gamma_1 = 180° - \alpha - \underset{\substack{\alpha = 35° \\ \beta_1 = 49.9°}}{\beta_1} \approx 95.1° \quad \text{or} \quad \gamma_2 = 180° - \alpha - \underset{\substack{\alpha = 35° \\ \beta_2 = 130.1°}}{\beta_2} \approx 14.9°$$

Figure 31(b)

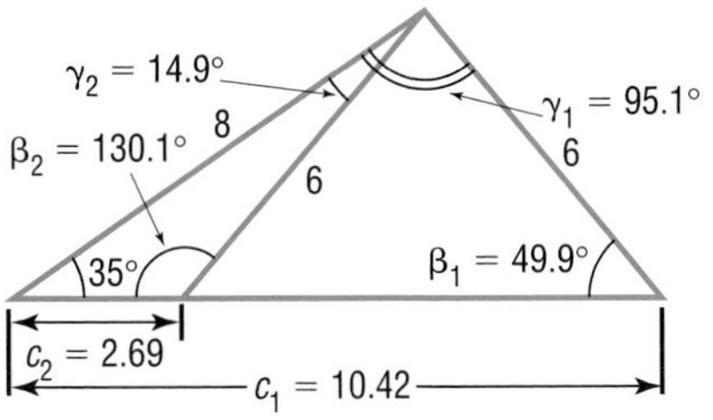

The third side c obeys the Law of Sines, so we have

$$\frac{\sin\alpha}{a} = \frac{\sin\gamma_1}{c_1} \qquad \frac{\sin\alpha}{a} = \frac{\sin\gamma_2}{c_2}$$

$$\frac{\sin 35^\circ}{6} = \frac{\sin 95.1^\circ}{c_1} \quad \text{or} \quad \frac{\sin 35^\circ}{6} = \frac{\sin 14.9^\circ}{c_2}$$

$$c_1 = \frac{6\sin 95.1^\circ}{\sin 35^\circ} \approx 10.42 \qquad c_2 = \frac{6\sin 14.9^\circ}{\sin 35^\circ} \approx 2.69$$

The two solved triangles are illustrated in Figure 31(b). ■

EXAMPLE 5 Using the Law of Sines to Solve a SSA Triangle (No Solution)

Solve the triangle: $a = 2, c = 1, \gamma = 50^\circ$

Solution Because $a = 2$, $c = 1$, and $\gamma = 50^\circ$ are known, we use the Law of Sines to find the angle α.

$$\frac{\sin\alpha}{a} = \frac{\sin\gamma}{c}$$

$$\frac{\sin\alpha}{2} = \frac{\sin 50^\circ}{1}$$

$$\sin\alpha = 2\sin 50^\circ \approx 1.53$$

Figure 32

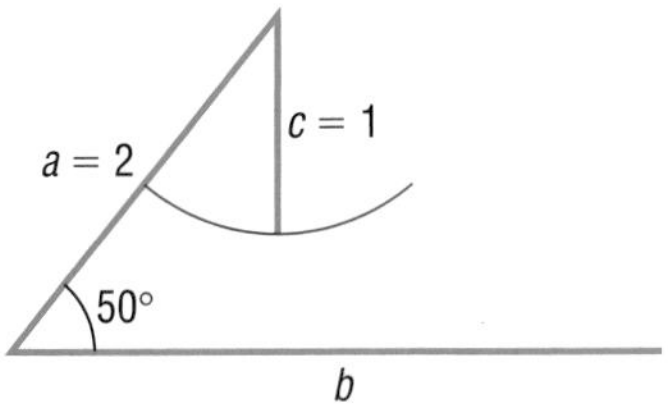

There is no angle α for which $\sin\alpha > 1$. Hence, there can be no triangle with the given measurements. Figure 32 illustrates the measurements given. Notice that, no matter how we attempt to position side c, it will never touch side b to form a triangle. ■

NOW WORK PROBLEMS 17 AND 23.

APPLICATION

3 The Law of Sines is particularly useful for solving certain applied problems.

EXAMPLE 6 Finding the Height of a Mountain

To measure the height of a mountain, a surveyor takes two sightings of the peak at a distance 900 meters apart on a direct line to the mountain.* See Figure 33(a). The first observation results in an angle of elevation of 47°, whereas the second results in an angle of elevation of 35°. If the transit is 2 meters high, what is the height h of the mountain?

*For simplicity, we assume that these sightings are at the same level.

Figure 33

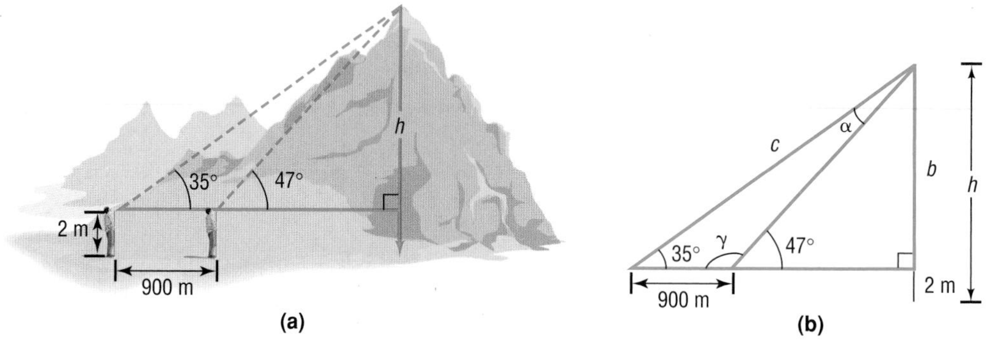

Solution Figure 33(b) shows the triangles that replicate the illustration in Figure 33(a). Since $\gamma + 47° = 180°$, we find that $\gamma = 133°$. Also, since $\alpha + \gamma + 35° = 180°$, we find that $\alpha = 145° - \gamma = 145° - 133° = 12°$. We use the Law of Sines to find c.

$$\frac{\sin \alpha}{a} = \frac{\sin \gamma}{c} \qquad \alpha = 12°, \gamma = 133°, a = 900$$

$$c = \frac{900 \sin 133°}{\sin 12°} = 3165.86$$

Using the larger right triangle, we have

$$\sin 35° = \frac{b}{c} \qquad c = 3165.86$$

$$b = 3165.86 \sin 35° = 1815.86 \approx 1816 \text{ meters}$$

The height of the peak from ground level is approximately $1816 + 2 = 1818$ meters. ■

NOW WORK PROBLEM 31.

8.2 EXERCISES

In Problems 1–8, solve each triangle.

1.

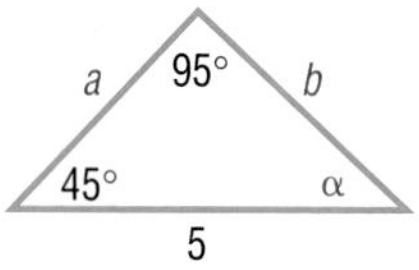

2.

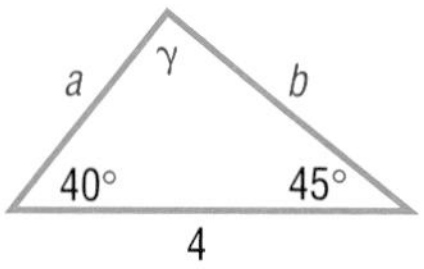

3.

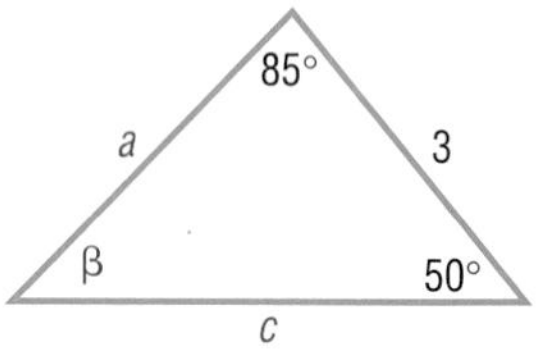

4.

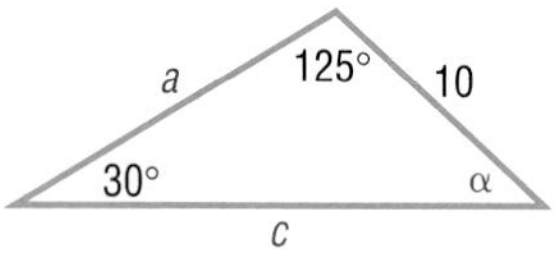

5.

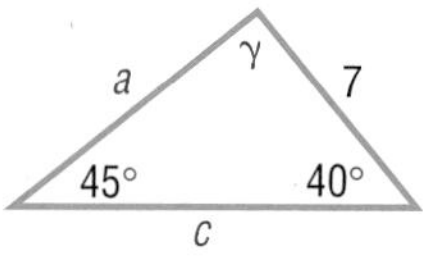

6.

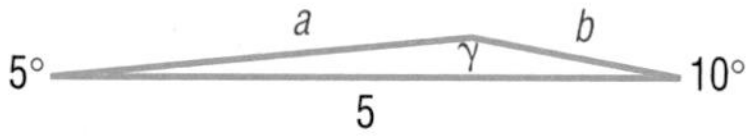

7.

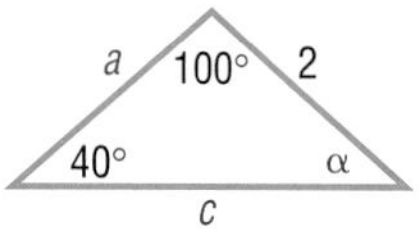

8.

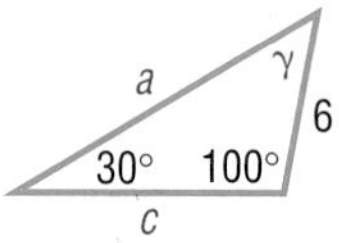

In Problems 9–16, solve each triangle.

9. $\alpha = 40°$, $\beta = 20°$, $a = 2$

10. $\alpha = 50°$, $\gamma = 20°$, $a = 3$

11. $\beta = 70°$, $\gamma = 10°$, $b = 5$

12. $\alpha = 70°$, $\beta = 60°$, $c = 4$

13. $\alpha = 110°$, $\gamma = 30°$, $c = 3$

14. $\beta = 10°$, $\gamma = 100°$, $b = 2$

15. $\alpha = 40°$, $\beta = 40°$, $c = 2$

16. $\beta = 20°$, $\gamma = 70°$, $a = 1$

In Problems 17–28, two sides and an angle are given. Determine whether the given information results in one triangle, two triangles, or no triangle at all. Solve any triangle(s) that results.

17. $a = 3$, $b = 2$, $\alpha = 50°$

18. $b = 4$, $c = 3$, $\beta = 40°$

19. $b = 5$, $c = 3$, $\beta = 100°$

20. $a = 2$, $c = 1$, $\alpha = 120°$

21. $a = 4$, $b = 5$, $\alpha = 60°$

22. $b = 2$, $c = 3$, $\beta = 40°$

23. $b = 4$, $c = 6$, $\beta = 20°$

24. $a = 3$, $b = 7$, $\alpha = 70°$

25. $a = 2$, $c = 1$, $\gamma = 100°$

26. $b = 4$, $c = 5$, $\beta = 95°$

27. $a = 2$, $c = 1$, $\gamma = 25°$

28. $b = 4$, $c = 5$, $\beta = 40°$

29. Rescue at Sea Coast Guard Station Able is located 150 miles due south of Station Baker. A ship at sea sends an SOS call that is received by each station. The call to Station Able indicates that the ship is located N55°E; the call to Station Baker indicates that the ship is located S60°E.
(a) How far is each station from the ship?
(b) If a helicopter capable of flying 200 miles per hour is dispatched from the nearest station to the ship, how long will it take to reach the ship?

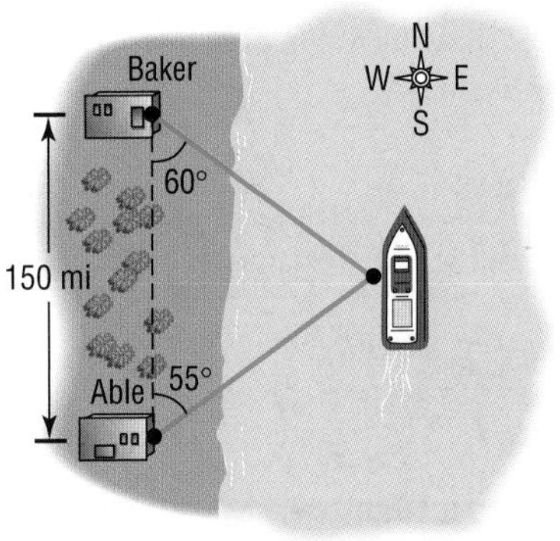

30. Surveying Consult the figure. To find the distance from the house at A to the house at B, a surveyor measures the angle BAC to be 40° and then walks off a distance of 100 feet to C and measures the angle ACB to be 50°. What is the distance from A to B?

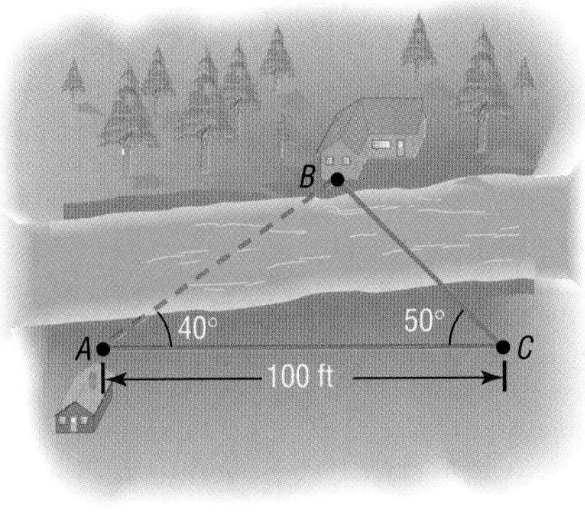

31. Finding the Length of a Ski Lift Consult the figure. To find the length of the span of a proposed ski lift from A to B, a surveyor measures the angle DAB to be 25° and then walks off a distance of 1000 feet to C and measures the angle ACB to be 15°. What is the distance from A to B?

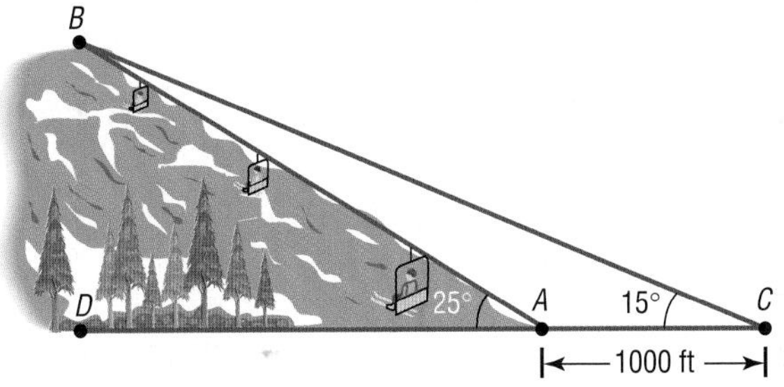

32. Finding the Height of a Mountain Use the illustration in Problem 31 to find the height BD of the mountain at B.

33. Finding the Height of an Airplane An aircraft is spotted by two observers who are 1000 feet apart. As the airplane passes over the line joining them, each observer takes a sighting of the angle of elevation to the plane, as indicated in the figure. How high is the airplane?

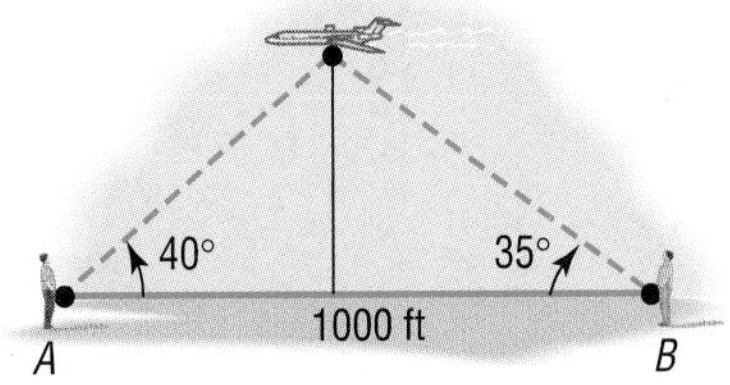

34. Finding the Height of the Bridge over the Royal Gorge The highest bridge in the world is the bridge over the Royal Gorge of the Arkansas River in Colorado.*

* *Source: Guinness Book of World Records.*

Sightings to the same point at water level directly under the bridge are taken from each side of the 880-foot-long bridge, as indicated in the figure. How high is the bridge?

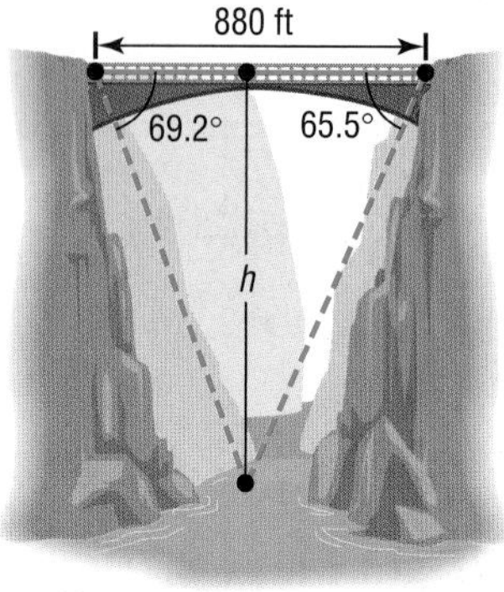

35. Navigation An airplane flies from city A to city B, a distance of 150 miles, and then turns through an angle of 40° and heads toward city C, as shown in the figure.
(a) If the distance between cities A and C is 300 miles, how far is it from city B to city C?
(b) Through what angle should the pilot turn at city C to return to city A?

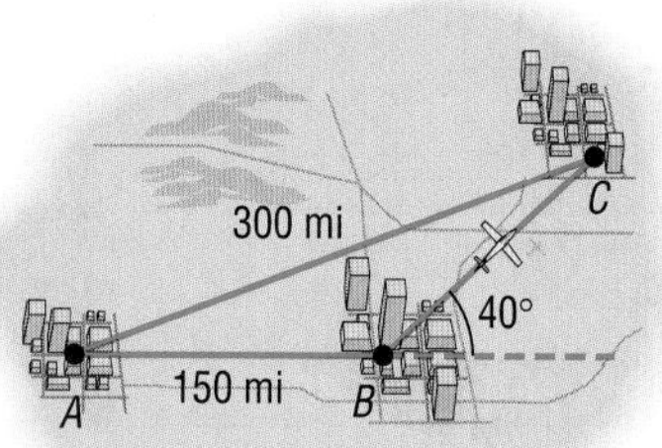

36. Time Lost due to a Navigation Error In attempting to fly from city A to city B, an aircraft followed a course that was 10° in error, as indicated in the figure. After flying a distance of 50 miles, the pilot corrected the course by turning at point C and flying 70 miles further. If the constant speed of the aircraft was 250 miles per hour, how much time was lost due to the error?

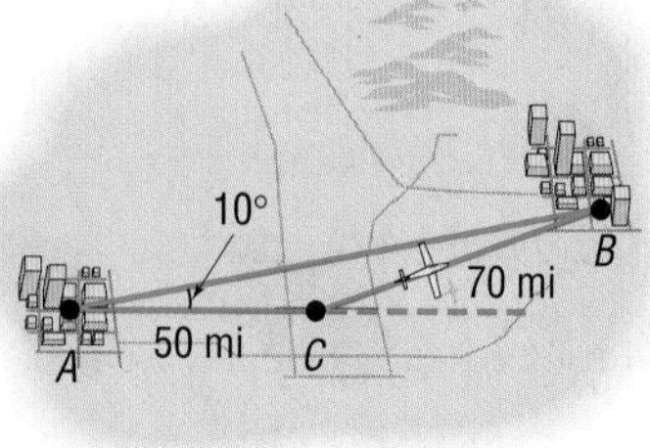

37. Finding the Lean of the Leaning Tower of Pisa The famous Leaning Tower of Pisa was originally 184.5 feet high.* At a distance of 123 feet from the base of the tower, the angle of elevation to the top of the tower is found to be 60°. Find the angle CAB indicated in the figure. Also, find the perpendicular distance from C to AB.

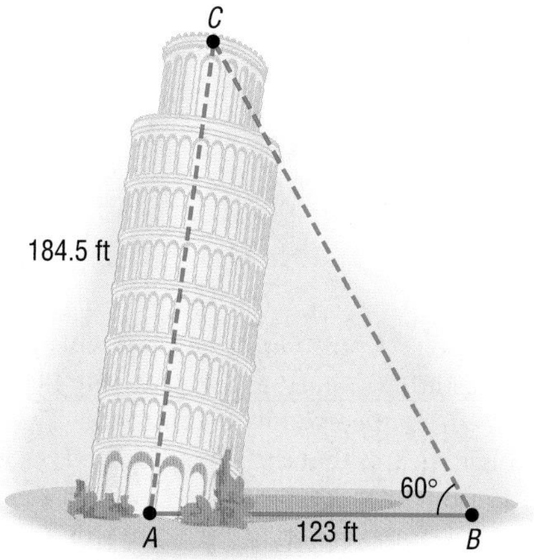

38. Crankshafts on Cars On a certain automobile, the crankshaft is 3 inches long and the connecting rod is 9 inches long (see the figure). At the time when the angle OPA is 15°, how far is the piston (P) from the center (O) of the crankshaft?

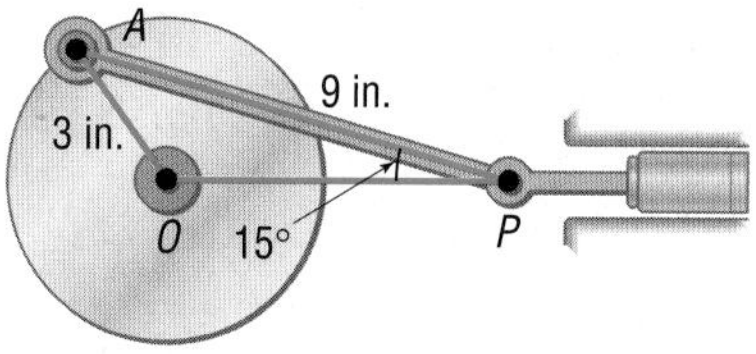

39. Constructing a Highway U.S. 41, a highway whose primary directions are north–south, is being constructed along the west coast of Florida. Near Naples, a bay obstructs the straight path of the road. Since the cost of a bridge is prohibitive, engineers decide to go around the bay. The illustration shows the path that they decide on and the measurements taken. What is the length of highway needed to go around the bay?

*In their 1986 report on the fragile seven-century-old bell tower, scientists in Pisa, Italy, said that the Leaning Tower of Pisa had increased its famous lean by 1 millimeter, or 0.04 inch. This is about the annual average, although the tilting had slowed to about half that much in the previous 2 years. (*Source:* United Press International, June 29, 1986.)

PISA, ITALY. September 1995. The Leaning Tower of Pisa has suddenly shifted, jeopardizing years of preservation work to stabilize it, Italian newspapers said Sunday. The tower, built on shifting subsoil between 1174 and 1350 as a belfry for the nearby cathedral, recently moved 0.07 inch in one night. The tower has been closed to tourists since 1990.

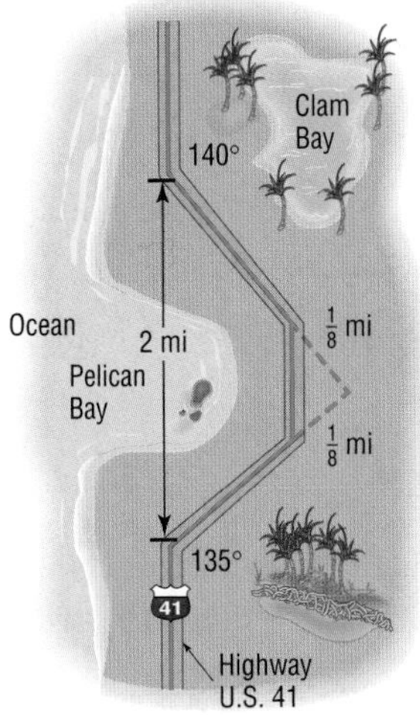

40. Determining Distances at Sea The navigator of a ship at sea spots two lighthouses that she knows to be 3 miles apart along a straight seashore. She determines that the angles formed between two line-of-sight observations of the lighthouses and the line from the ship directly to shore are 15° and 35°. See the illustration.
(a) How far is the ship from lighthouse A?
(b) How far is the ship from lighthouse B?
(c) How far is the ship from shore?

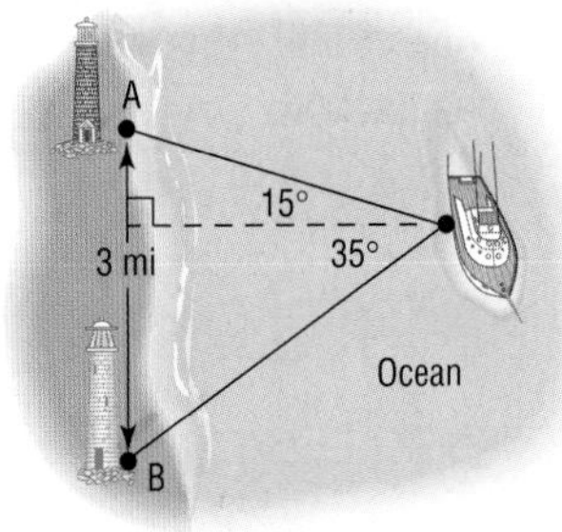

41. Calculating Distance at Sea The navigator of a ship at sea has the harbor in sight at which the ship is to dock. She spots a lighthouse that she knows is 1 mile up the coast from the mouth of the harbor, and she measures the angle between the line-of-sight observations of the harbor and lighthouse to be 20°. With the ship heading directly toward the harbor, she repeats this measurement after 5 minutes of traveling at 12 miles per hour. If the new angle is 30°, how far is the ship from the harbor?

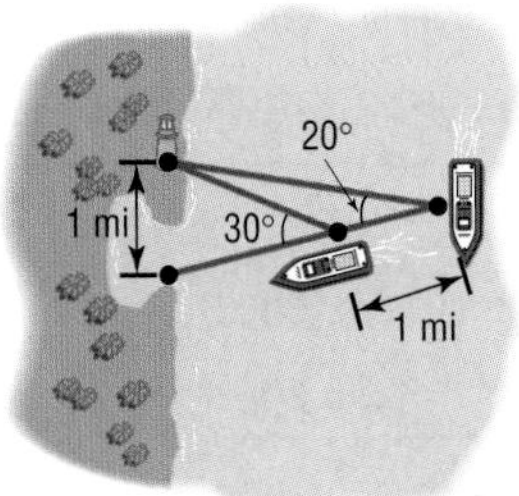

42. Finding Distances A forest ranger is walking on a path inclined at 5° to the horizontal directly toward a 100-foot-tall fire observation tower. The angle of elevation of the top of the tower is 40°. How far is the ranger from the tower at this time?

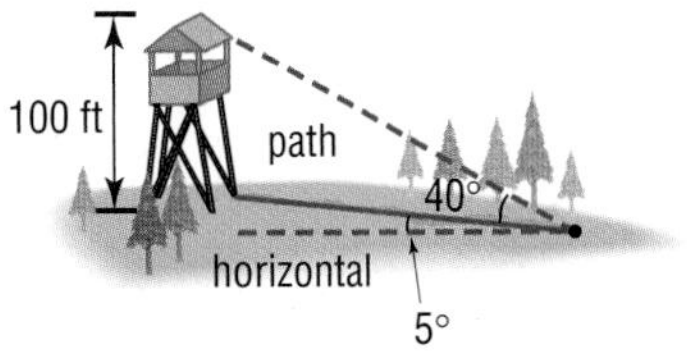

43. Great Pyramid of Cheops One of the original Seven Wonders of the World, the Great Pyramid of Cheops was built about 2580 BC. Its original height was 480 feet 11 inches, but due to the loss of its topmost stones, it is now shorter.* Find the current height of the Great Pyramid, using the information given in the illustration.

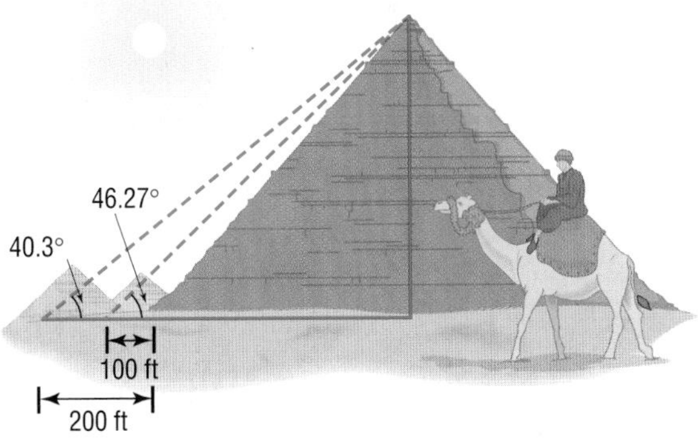

44. Determining the Height of an Aircraft Two sensors are spaced 700 feet apart along the approach to a small airport. When an aircraft is 2 miles from the airport, the angle of elevation from the first sensor to the aircraft is 20°, and from the second sensor to the aircraft it is 15°. Determine how high the aircraft is at this time.

45. Landscaping Pat needs to determine the height of a tree before cutting it down to be sure that it will not fall on a nearby fence. The angle of elevation of the tree from one position on a flat path from the tree is 30°, and from a second position 40 feet farther along this path it is 20°. What is the height of the tree?

46. Construction A loading ramp 10 feet long that makes an angle of 18° with the horizontal is to be replaced by one that makes an angle of 12° with the horizontal. How long is the new ramp?

47. Finding the Height of a Helicopter Two observers simultaneously measure the angle of elevation of a helicopter. One angle is measured as 25°, the other as 40° (see

* *Source: Guinness Book of World Records.*

the figure). If the observers are 100 feet apart and the helicopter lies over the line joining them, how high is the helicopter?

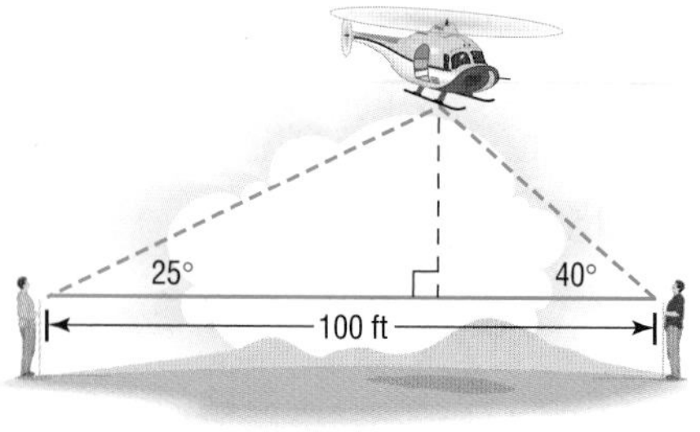

48. Mollweide's Formula For any triangle, Mollweide's Formula (named after Karl Mollweide, 1774–1825) states that

$$\frac{a+b}{c} = \frac{\cos\left[\frac{1}{2}(\alpha-\beta)\right]}{\sin\left(\frac{1}{2}\gamma\right)}$$

Derive it.

[**Hint:** Use the Law of Sines and then a sum-to-product formula. Notice that this formula involves all six parts of a triangle. As a result, it is sometimes used to check the solution of a triangle.]

49. Mollweide's Formula Another form of Mollweide's Formula is

$$\frac{a-b}{c} = \frac{\sin\left[\frac{1}{2}(\alpha-\beta)\right]}{\cos\left(\frac{1}{2}\gamma\right)}$$

Derive it.

50. For any triangle, derive the formula

$$a = b\cos\gamma + c\cos\beta$$

[**Hint:** Use the fact that $\sin\alpha = \sin(180° - \beta - \gamma)$.]

51. Law of Tangents For any triangle, derive the Law of Tangents:

$$\frac{a-b}{a+b} = \frac{\tan\left[\frac{1}{2}(\alpha-\beta)\right]}{\tan\left[\frac{1}{2}(\alpha+\beta)\right]}$$

[**Hint:** Use Mollweide's Formula.]

52. Circumscribing a Triangle Show that

$$\frac{\sin\alpha}{a} = \frac{\sin\beta}{b} = \frac{\sin\gamma}{c} = \frac{1}{2r}$$

where r is the radius of the circle circumscribing the triangle ABC whose sides are a, b, and c, as shown in the figure.

[**Hint:** Draw the diameter AB'. Then $\beta =$ angle $ABC =$ angle $AB'C$, and angle $ACB' = 90°$.]

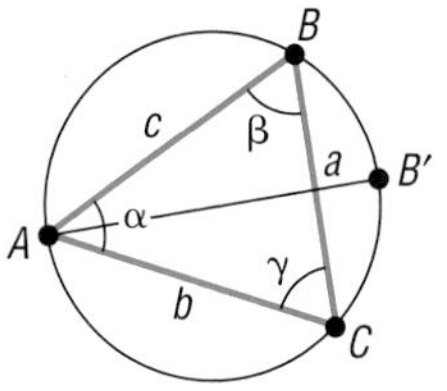

53. Make up three problems involving oblique triangles. One should result in one triangle, the second in two triangles, and the third in no triangle.

8.3 THE LAW OF COSINES

OBJECTIVES
1. Solve SAS Triangles
2. Solve SSS Triangles
3. Solve Applied Problems

In Section 8.2, we use the Law of Sines to solve Case 1 (SAA or ASA) and Case 2 (SSA) of an oblique triangle. In this section, we derive the Law of Cosines and use it to solve the remaining cases, 3 and 4.

CASE 3: Two sides and the included angle are known (SAS).
CASE 4: Three sides are known (SSS).

Theorem **Law of Cosines**

For a triangle with sides a, b, c and opposite angles α, β, γ, respectively,

$$c^2 = a^2 + b^2 - 2ab\cos\gamma \qquad \textbf{(1)}$$

$$b^2 = a^2 + c^2 - 2ac\cos\beta \qquad \textbf{(2)}$$

$$a^2 = b^2 + c^2 - 2bc\cos\alpha \qquad \textbf{(3)}$$

Figure 34

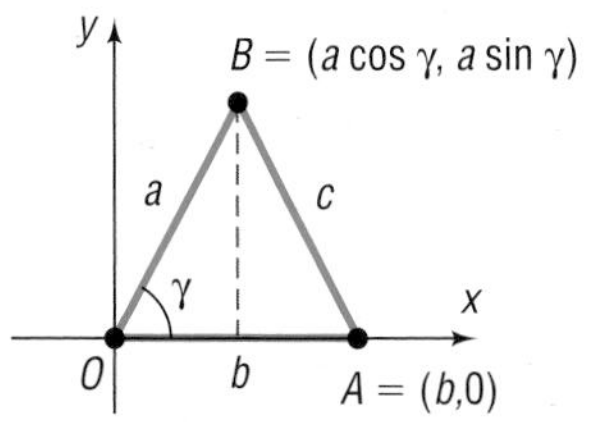

(a) Angle γ is acute

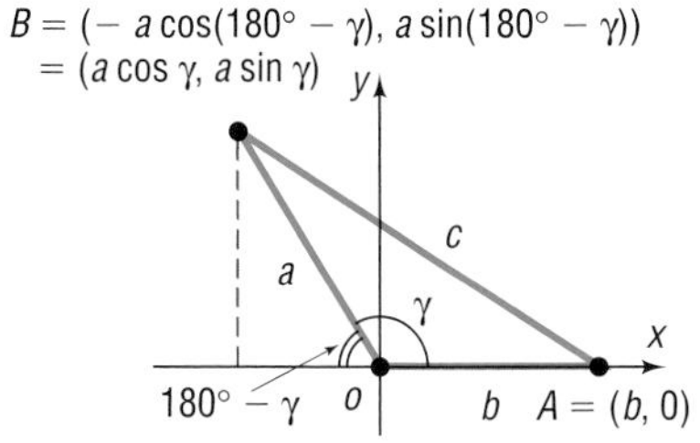

(b) Angle γ is obtuse

Proof We will prove only formula (1) here. Formulas (2) and (3) may be proved using the same argument.

We begin by strategically placing a triangle on a rectangular coordinate system so that the vertex of angle γ is at the origin and side b lies along the positive x-axis. Regardless of whether γ is acute, as in Figure 34(a), or obtuse, as in Figure 34(b), the vertex B has coordinates $(a\cos\gamma, a\sin\gamma)$. Vertex A has coordinates $(b, 0)$.

We can now use the distance formula to compute c^2.

$$\begin{aligned} c^2 &= (b - a\cos\gamma)^2 + (0 - a\sin\gamma)^2 \\ &= b^2 - 2ab\cos\gamma + a^2\cos^2\gamma + a^2\sin^2\gamma \\ &= b^2 - 2ab\cos\gamma + a^2(\cos^2\gamma + \sin^2\gamma) \\ &= a^2 + b^2 - 2ab\cos\gamma \end{aligned}$$

Each of formulas (1), (2), and (3) may be stated in words as follows:

Theorem **Law of Cosines**

The square of one side of a triangle equals the sum of the squares of the other two sides minus twice their product times the cosine of their included angle.

Observe that if the triangle is a right triangle (so that, say, $\gamma = 90°$) then formula (1) becomes the familiar Pythagorean Theorem: $c^2 = a^2 + b^2$. Thus, the Pythagorean Theorem is a special case of the Law of Cosines.

1 Let's see how to use the Law of Cosines to solve Case 3 (SAS), which applies to triangles for which two sides and the included angle are known.

EXAMPLE 1 Using the Law of Cosines to Solve a SAS Triangle

Solve the triangle: $a = 2, b = 3, \gamma = 60°$

Solution See Figure 35. The Law of Cosines makes it easy to find the third side, c.

Figure 35

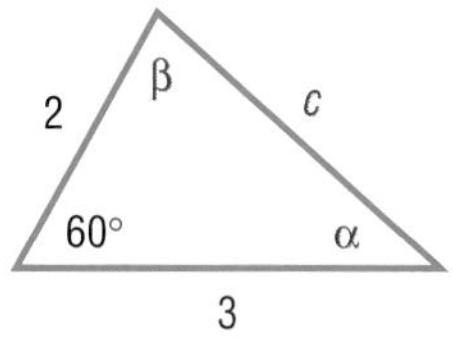

$$\begin{aligned} c^2 &= a^2 + b^2 - 2ab\cos\gamma \\ &= 4 + 9 - 2\cdot 2\cdot 3\cdot\cos 60° \\ &= 13 - \left(12\cdot\tfrac{1}{2}\right) = 7 \\ c &= \sqrt{7} \end{aligned}$$

Side c is of length $\sqrt{7}$. To find the angles α and β, we may use either the Law of Sines or the Law of Cosines. It is preferable to use the Law of Cosines, since it will lead to an equation with one solution. Using the Law of Sines would lead to an equation with two solutions that would need to be checked to determine which solution fits the given data. Thus, we choose to use formulas (2) and (3) of the Law of Cosines to find α and β.

For α:

$$a^2 = b^2 + c^2 - 2bc\cos\alpha$$

$$2bc\cos\alpha = b^2 + c^2 - a^2$$

$$\cos\alpha = \frac{b^2 + c^2 - a^2}{2bc} = \frac{9 + 7 - 4}{2 \cdot 3\sqrt{7}} = \frac{12}{6\sqrt{7}} = \frac{2\sqrt{7}}{7}$$

$$\alpha \approx 40.9°$$

For β:

$$b^2 = a^2 + c^2 - 2ac\cos\beta$$

$$\cos\beta = \frac{a^2 + c^2 - b^2}{2ac} = \frac{4 + 7 - 9}{4\sqrt{7}} = \frac{1}{2\sqrt{7}} = \frac{\sqrt{7}}{14}$$

$$\beta \approx 79.1°$$

Notice that $\alpha + \beta + \gamma = 40.9° + 79.1° + 60° = 180°$, as required. ■

NOW WORK PROBLEM 1.

2 The next example illustrates how the Law of Cosines is used when three sides of a triangle are known, Case 4 (SSS).

EXAMPLE 2 Using the Law of Cosines to Solve a SSS Triangle

Solve the triangle: $a = 4, b = 3, c = 6$

Solution See Figure 36. To find the angles α, β, and γ, we proceed as we did in the latter part of the solution to Example 1.

Figure 36

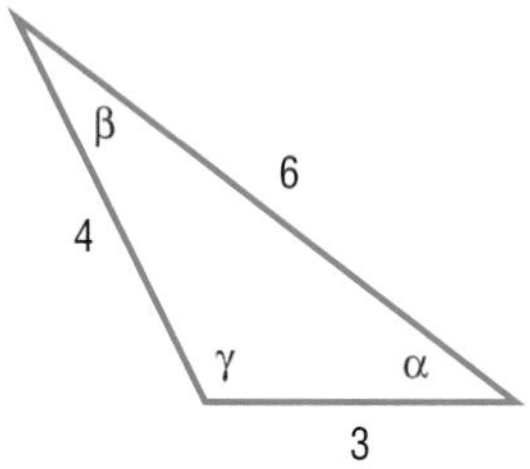

For α:

$$\cos\alpha = \frac{b^2 + c^2 - a^2}{2bc} = \frac{9 + 36 - 16}{2 \cdot 3 \cdot 6} = \frac{29}{36}$$

$$\alpha \approx 36.3°$$

For β:

$$\cos\beta = \frac{a^2 + c^2 - b^2}{2ac} = \frac{16 + 36 - 9}{2 \cdot 4 \cdot 6} = \frac{43}{48}$$

$$\beta \approx 26.4°$$

Since we know α and β,

$$\gamma = 180° - \alpha - \beta \approx 180° - 36.3° - 26.4° = 117.3°$$ ■

NOW WORK PROBLEM 7.

EXAMPLE 3 Correcting a Navigational Error

3 A motorized sailboat leaves Naples, Florida, bound for Key West, 150 miles away. Maintaining a constant speed of 15 miles per hour, but encountering heavy crosswinds and strong currents, the crew finds, after 4 hours, that the sailboat is off course by 20°.

(a) How far is the sailboat from Key West at this time?

(b) Through what angle should the sailboat turn to correct its course?

(c) How much time has been added to the trip because of this? (Assume that the speed remains at 15 miles per hour.)

Figure 37

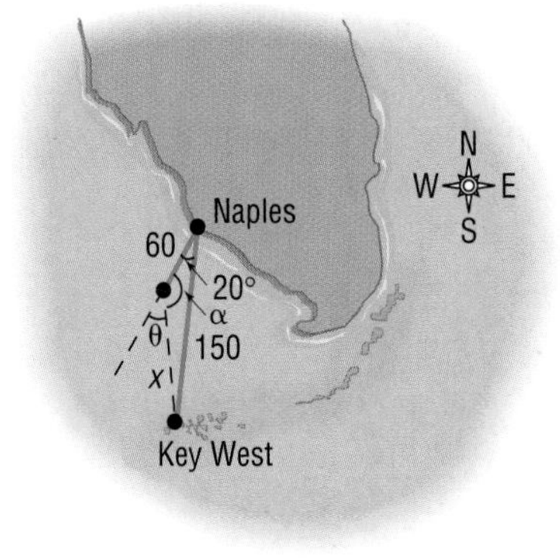

Solution See Figure 37. With a speed of 15 miles per hour, the sailboat has gone 60 miles after 4 hours. We seek the distance x of the sailboat from Key West. We also seek the angle θ that the sailboat should turn through to correct its course.

(a) To find x, we use the Law of Cosines, since we know two sides and the included angle.

$$x^2 = 150^2 + 60^2 - 2(150)(60)\cos 20° = 9186$$

$$x \approx 95.8$$

The sailboat is about 96 miles from Key West.

(b) We now know three sides of the triangle, so we can use the Law of Cosines again to find the angle α opposite the side of length 150 miles.

$$150^2 = 96^2 + 60^2 - 2(96)(60)\cos\alpha$$

$$9684 = 11{,}520\cos\alpha$$

$$\cos\alpha \approx -0.8406$$

$$\alpha \approx 147.2°$$

The sailboat should turn through an angle of

$$\theta = 180° - \alpha \approx 180° - 147.2° = 32.8°$$

The sailboat should turn through an angle of about 33° to correct its course.

(c) The total length of the trip is now $60 + 96 = 156$ miles. The extra 6 miles will only require about 0.4 hour or 24 minutes more if the speed of 15 miles per hour is maintained. ■

NOW WORK PROBLEM 27.

HISTORICAL FEATURE

The Law of Sines was known vaguely long before it was explicitly stated by Nasîr Eddîn (about AD 1250). Ptolemy (about AD 150) was aware of it in a form using a chord function instead of the sine function. But it was first clearly stated in Europe by Regiomontanus, writing in 1464.

The Law of Cosines appears first in Euclid's *Elements* (Book II), but in a well-disguised form in which squares built on the sides of triangles are added and a rectangle representing the cosine term is subtracted. It was thus known to all mathematicians because of their familiarity with Euclid's work. An early modern form of the Law of Cosines, that for finding the angle when the sides are known, was stated by François Viète (in 1593).

The Law of Tangents (see Problem 51 of Exercise 8.2) has become obsolete. In the past it was used in place of the Law of Cosines, because the Law of Cosines was very inconvenient for calculation with logarithms or slide rules. Mixing of addition and multiplication is now very easy on a calculator, however, and the Law of Tangents has been shelved along with the slide rule.

8.3 EXERCISES

In Problems 1–8, solve each triangle.

1.
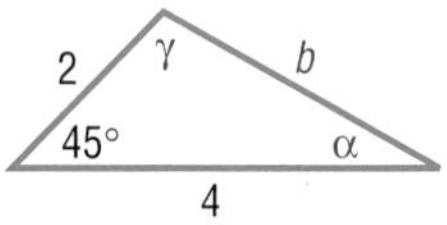

2.
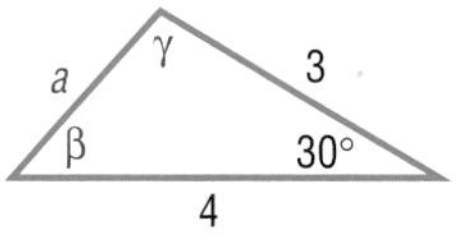

3.
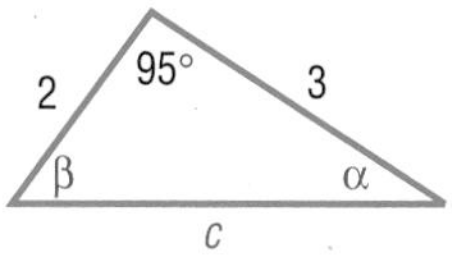

4.
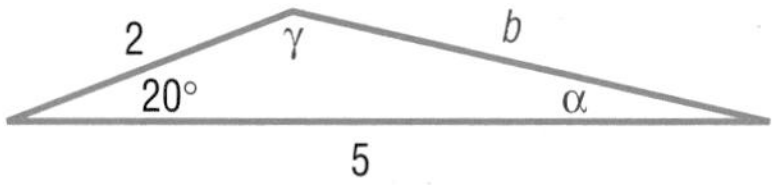

5.
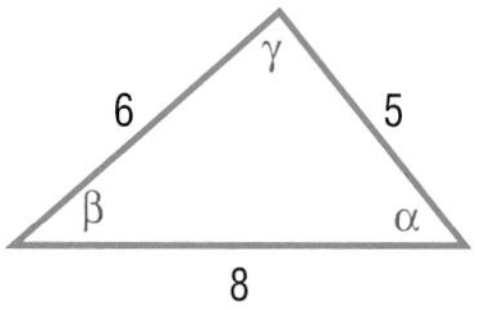

6.
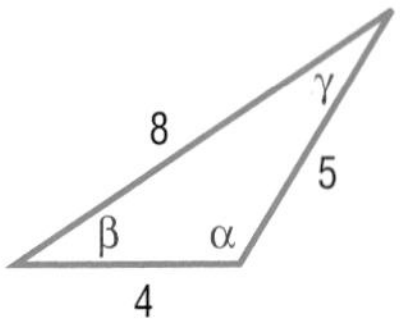

7.
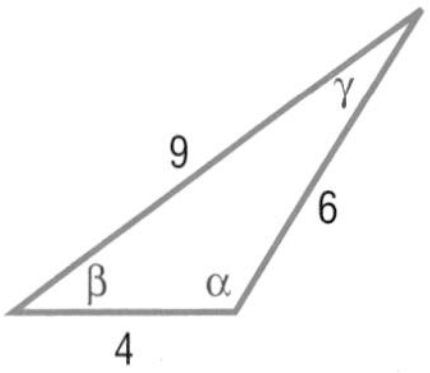

8.
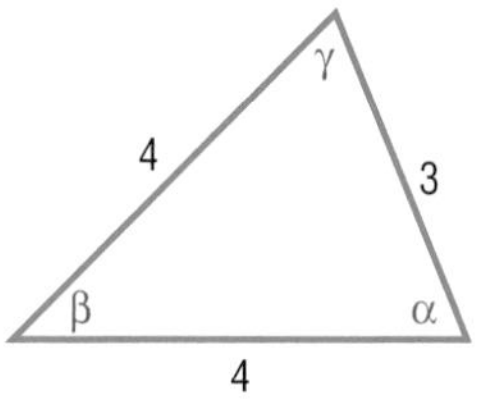

In Problems 9–24, solve each triangle.

9. $a = 3$, $b = 4$, $\gamma = 40°$

10. $a = 2$, $c = 1$, $\beta = 10°$

11. $b = 1$, $c = 3$, $\alpha = 80°$

12. $a = 6$, $b = 4$, $\gamma = 60°$

13. $a = 3$, $c = 2$, $\beta = 110°$

14. $b = 4$, $c = 1$, $\alpha = 120°$

15. $a = 2$, $b = 2$, $\gamma = 50°$

16. $a = 3$, $c = 2$, $\beta = 90°$

17. $a = 12$, $b = 13$, $c = 5$

18. $a = 4$, $b = 5$, $c = 3$

19. $a = 2$, $b = 2$, $c = 2$

20. $a = 3$, $b = 3$, $c = 2$

21. $a = 5$, $b = 8$, $c = 9$

22. $a = 4$, $b = 3$, $c = 6$

23. $a = 10$, $b = 8$, $c = 5$

24. $a = 9$, $b = 7$, $c = 10$

25. Surveying Consult the figure. To find the distance from the house at A to the house at B, a surveyor measures the angle ACB, which is found to be 70°, and then walks off the distance to each house, 50 feet and 70 feet, respectively. How far apart are the houses?

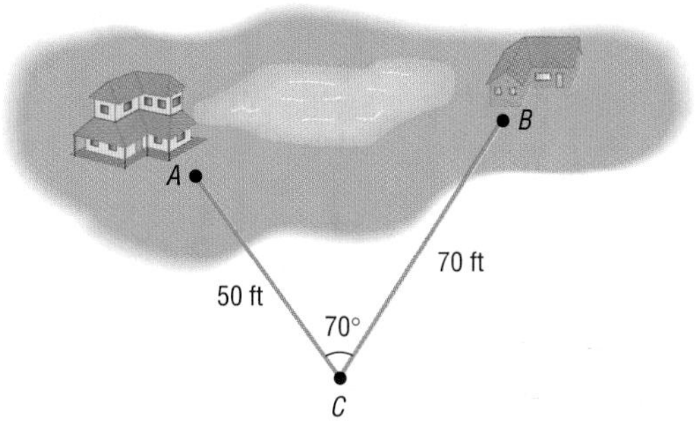

26. Navigation An airplane flies from Ft. Myers to Sarasota, a distance of 150 miles, and then turns through an angle of 50° and flies to Orlando, a distance of 100 miles (see the figure).

(a) How far is it from Ft. Myers to Orlando?

(b) Through what angle should the pilot turn at Orlando to return to Ft. Myers?

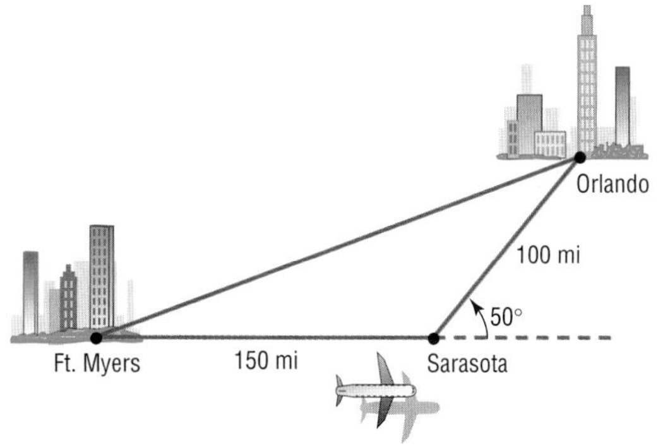

27. Revising a Flight Plan In attempting to fly from Chicago to Louisville, a distance of 330 miles, a pilot inadvertently took a course that was 10° in error, as indicated in the figure.

(a) If the aircraft maintains an average speed of 220 miles per hour and if the error in direction is discovered after 15 minutes, through what angle should the pilot turn to head toward Louisville?

(b) What new average speed should the pilot maintain so that the total time of the trip is 90 minutes?

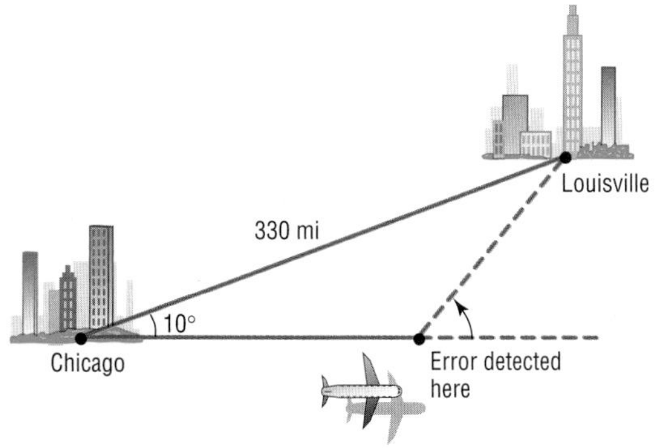

28. Avoiding a Tropical Storm A cruise ship maintains an average speed of 15 knots in going from San Juan, Puerto Rico, to Barbados, West Indies, a distance of 600 nautical miles. To avoid a tropical storm, the captain heads out of San Juan in a direction of 20° off a direct heading to Barbados. The captain maintains the 15-knot speed for 10 hours, after which time the path to Barbados becomes clear of storms.

(a) Through what angle should the captain turn to head directly to Barbados?

(b) How long will it be before the ship reaches Barbados if the same 15-knot speed is maintained?

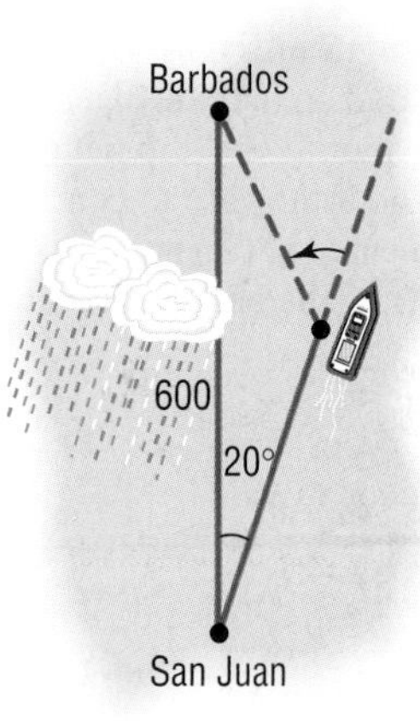

29. Major League Baseball Field A Major League baseball diamond is actually a square 90 feet on a side. The pitching rubber is located 60.5 feet from home plate on a line joining home plate and second base.

(a) How far is it from the pitching rubber to first base?

(b) How far is it from the pitching rubber to second base?

(c) If a pitcher faces home plate, through what angle does he need to turn to face first base?

30. Little League Baseball Field According to Little League baseball official regulations, the diamond is a square 60 feet on a side. The pitching rubber is located 46 feet from home plate on a line joining home plate and second base.

(a) How far is it from the pitching rubber to first base?

(b) How far is it from the pitching rubber to second base?

(c) If a pitcher faces home plate, through what angle does he need to turn to face first base?

31. Finding the Length of a Guy Wire The height of a radio tower is 500 feet, and the ground on one side of the tower slopes upward at an angle of 10° (see the figure on p. 538).

(a) How long should a guy wire be if it is to connect to the top of the tower and be secured at a point on the sloped side 100 feet from the base of the tower?

(b) How long should a second guy wire be if it is to connect to the middle of the tower and be secured at a point 100 feet from the base on the flat side?

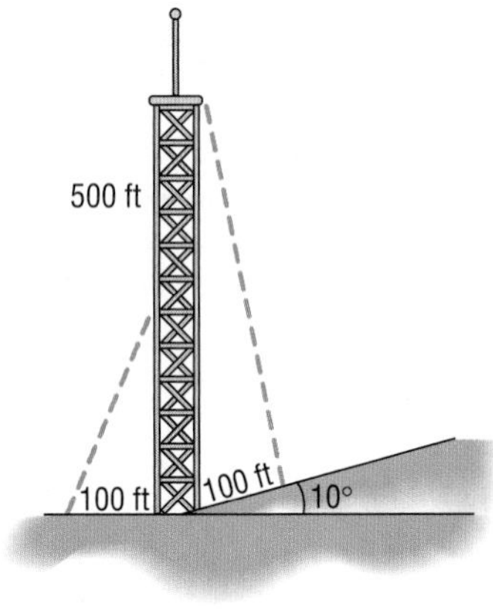

32. Finding the Length of a Guy Wire A radio tower 500 feet high is located on the side of a hill with an inclination to the horizontal of 5° (see the figure). How long should two guy wires be if they are to connect to the top of the tower and be secured at two points 100 feet directly above and directly below the base of the tower?

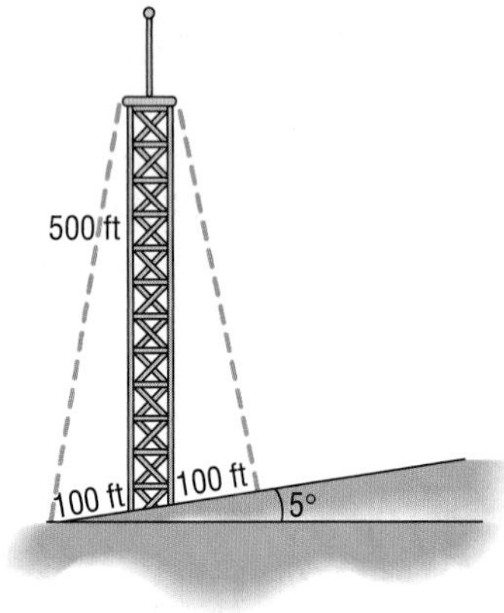

33. Wrigley Field, Home of the Chicago Cubs The distance from home plate to dead center in Wrigley Field is 400 feet (see the figure). How far is it from dead center to third base?

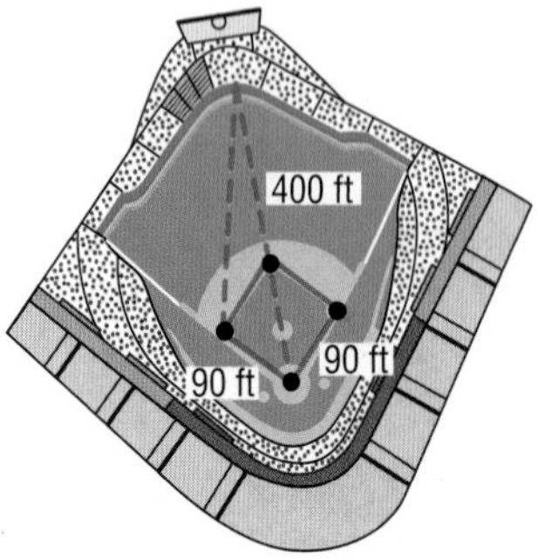

34. Little League Baseball The distance from home plate to dead center at the Oak Lawn Little League field is 280 feet. How far is it from dead center to third base? [**Hint:** The distance between the bases in Little League is 60 feet.]

35. Rods and Pistons Rod OA (see the figure) rotates about the fixed point O so that point A travels on a circle of radius r. Connected to point A is another rod AB of length $L > r$, and point B is connected to a piston. Show that the distance x between point O and point B is given by

$$x = r\cos\theta + \sqrt{r^2\cos\theta + L^2 - r^2}$$

where θ is the angle of rotation of rod OA.

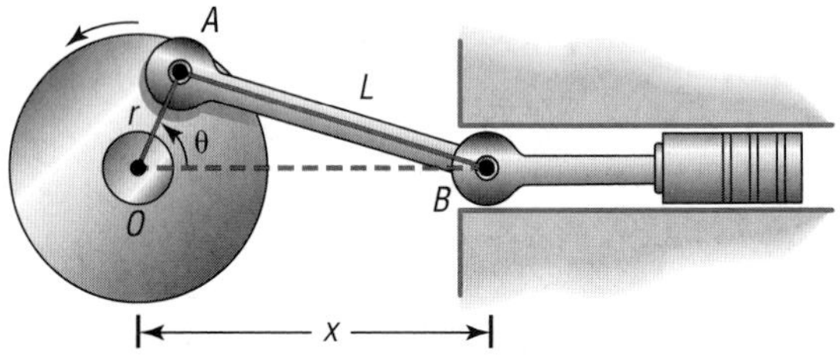

36. Geometry Show that the length d of a chord of a circle of radius r is given by the formula

$$d = 2r\sin\frac{\theta}{2}$$

where θ is the central angle formed by the radii to the ends of the chord (see the figure). Use this result to derive the fact that $\sin\theta < \theta$, where $\theta > 0$ is measured in radians.

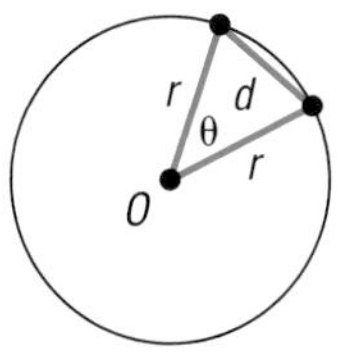

37. For any triangle, show that

$$\cos\frac{\gamma}{2} = \sqrt{\frac{s(s-c)}{ab}}$$

where $s = \frac{1}{2}(a + b + c)$.
[**Hint:** Use a half-angle formula and the Law of Cosines.]

38. For any triangle show that

$$\sin\frac{\gamma}{2} = \sqrt{\frac{(s-a)(s-b)}{ab}}$$

where $s = \frac{1}{2}(a + b + c)$.

39. Use the Law of Cosines to prove the identity

$$\frac{\cos\alpha}{a} + \frac{\cos\beta}{b} + \frac{\cos\gamma}{c} = \frac{a^2 + b^2 + c^2}{2abc}$$

40. Write down your strategy for solving an oblique triangle.

PREPARING FOR THIS SECTION

Before getting started, review the following:

✓ Geometry Review (Appendix, Section 9)

8.4 THE AREA OF A TRIANGLE

OBJECTIVES 1 Find the Area of SAS Triangles

2 Find the Area of SSS Triangles

In this section, we will derive several formulas for calculating the area A of a triangle. The most familiar of these is the following:

Theorem The area A of a triangle is

$$A = \tfrac{1}{2}bh \qquad (1)$$

where b is the base and h is an altitude drawn to that base.

Figure 38
$A = \frac{1}{2}bh$

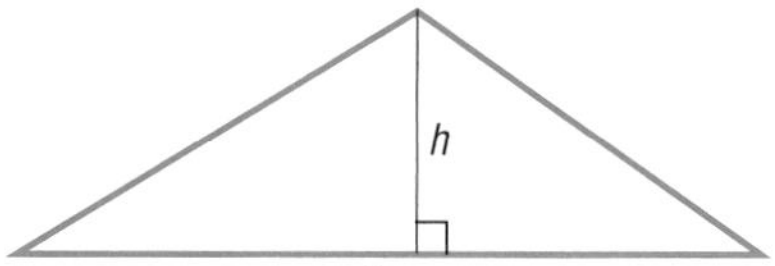

Figure 39

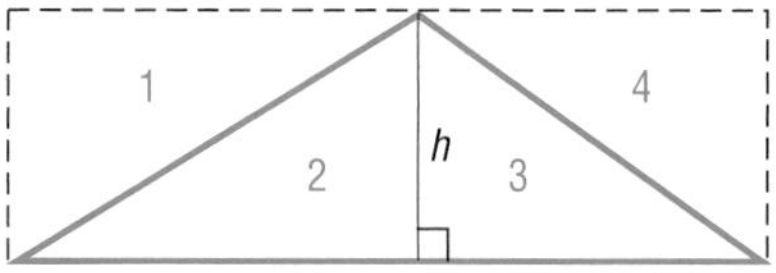

Figure 40
$h = a\sin\gamma$

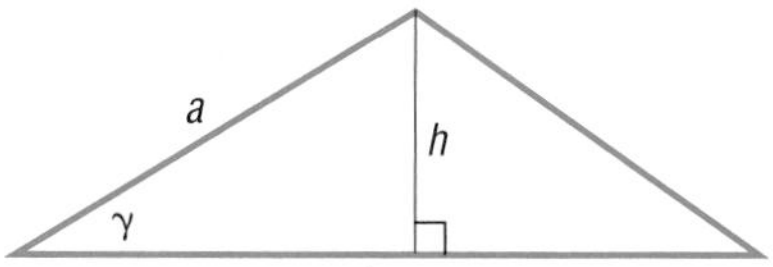

Proof The derivation of this formula is rather easy once a rectangle of base b and height h is constructed around the triangle. See Figures 38 and 39.

Triangles 1 and 2 in Figure 39 are equal in area, as are triangles 3 and 4. Consequently, the area of the triangle with base b and altitude h is exactly half the area of the rectangle, which is bh. ∎

If the base b and altitude h to that base are known, then we can find the area of such a triangle using formula (1). Usually, though, the information required to use formula (1) is not given. Suppose, for example, that we know two sides a and b and the included angle γ (see Figure 40). Then the altitude h can be found by noting that

$$\frac{h}{a} = \sin\gamma$$

so that

$$h = a\sin\gamma$$

Using this fact in formula (1) produces

$$A = \tfrac{1}{2}bh = \tfrac{1}{2}b(a\sin\gamma) = \tfrac{1}{2}ab\sin\gamma$$

We now have the formula

$$A = \tfrac{1}{2}ab\sin\gamma \qquad (2)$$

By dropping altitudes from the other two vertices of the triangle, we obtain the following corresponding formulas:

$$A = \tfrac{1}{2}bc\sin\alpha \qquad (3)$$

$$A = \tfrac{1}{2}ac\sin\beta \qquad (4)$$

It is easiest to remember these formulas using the following wording:

Theorem The area A of a triangle equals one-half the product of two of its sides times the sine of their included angle.

EXAMPLE 1 Finding the Area of a SAS Triangle

1 Find the area A of the triangle for which $a = 8$, $b = 6$, and $\gamma = 30°$.

Solution We use formula (2) to get

$$A = \frac{1}{2}ab\sin\gamma = \frac{1}{2}\cdot 8\cdot 6\sin 30° = 12$$

NOW WORK PROBLEM 1.

2 If the three sides of a triangle are known, another formula, called **Heron's Formula** (named after Heron of Alexandria), can be used to find the area of a triangle.

Theorem **Heron's Formula**

The area A of a triangle with sides a, b, and c is

$$A = \sqrt{s(s-a)(s-b)(s-c)} \quad \textbf{(5)}$$

where $s = \frac{1}{2}(a + b + c)$.

Proof The proof that we shall give uses the Law of Cosines and is quite different from the proof given by Heron.

From the Law of Cosines,

$$c^2 = a^2 + b^2 - 2ab\cos\gamma$$

and the two half-angle formulas,

$$\cos^2\frac{\gamma}{2} = \frac{1 + \cos\gamma}{2} \qquad \sin^2\frac{\gamma}{2} = \frac{1 - \cos\gamma}{2}$$

and, using $2s = a + b + c$, we find that

$$\cos^2\frac{\gamma}{2} = \frac{1 + \cos\gamma}{2} = \frac{1 + \dfrac{a^2 + b^2 - c^2}{2ab}}{2}$$

$$= \frac{a^2 + 2ab + b^2 - c^2}{4ab} = \frac{(a+b)^2 - c^2}{4ab}$$

$$= \frac{(a+b-c)(a+b+c)}{4ab} = \frac{2(s-c)\cdot 2s}{4ab} = \frac{s(s-c)}{ab} \quad \textbf{(6)}$$

$a + b - c = a + b + c - 2c$
$= 2s - 2c$

Similarly,

$$\sin^2\frac{\gamma}{2} = \frac{(s-a)(s-b)}{ab} \quad \textbf{(7)}$$

Now we use formula (2) for the area.

$$\begin{aligned} A &= \frac{1}{2}ab\sin\gamma \\ &= \frac{1}{2}ab \cdot 2\sin\frac{\gamma}{2}\cos\frac{\gamma}{2} && \sin\gamma = \sin\left[2\left(\frac{\gamma}{2}\right)\right] = 2\sin\frac{\gamma}{2}\cos\frac{\gamma}{2} \\ &= ab\sqrt{\frac{(s-a)(s-b)}{ab}}\sqrt{\frac{s(s-c)}{ab}} && \text{Use equations (6) and (7).} \\ &= \sqrt{s(s-a)(s-b)(s-c)} \end{aligned}$$

■

EXAMPLE 2 Finding the Area of a SSS Triangle

Find the area of a triangle whose sides are 4, 5, and 7.

Solution We let $a = 4$, $b = 5$, and $c = 7$. Then

$$s = \tfrac{1}{2}(a + b + c) = \tfrac{1}{2}(4 + 5 + 7) = 8$$

Heron's Formula then gives the area A as

$$A = \sqrt{s(s-a)(s-b)(s-c)} = \sqrt{8 \cdot 4 \cdot 3 \cdot 1} = \sqrt{96} = 4\sqrt{6}$$

■

NOW WORK PROBLEM 7.

HISTORICAL FEATURE

Heron's formula (also known as *Hero's Formula*) is due to Heron of Alexandria (first century AD), who had, besides his mathematical talents, a good deal of engineering skills. In various temples his mechanical devices produced effects that seemed supernatural, and visitors presumably were thus influenced to generosity. Heron's book *Metrica*, on making such devices, has survived and was discovered in 1896 in the city of Constantinople.

Heron's Formulas for the area of a triangle caused some mild discomfort in Greek mathematics, because a product with two factors was an area, while one with three factors was a volume, but four factors seemed contradictory in Heron's time.

8.4 EXERCISES

In Problems 1–8, find the area of each triangle. Round answers to two decimal places.

1.

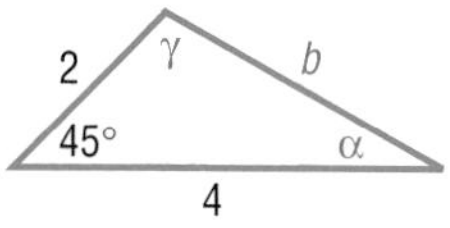

2.

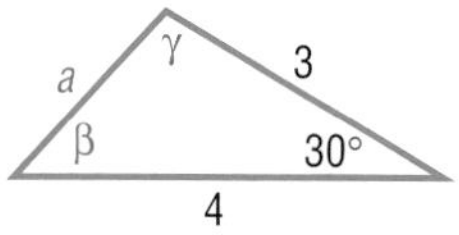

3.

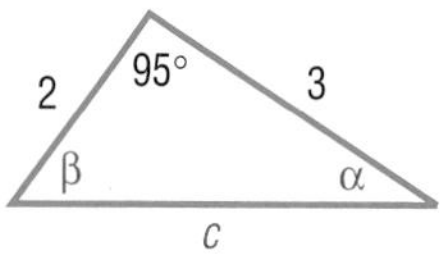

4.

5.

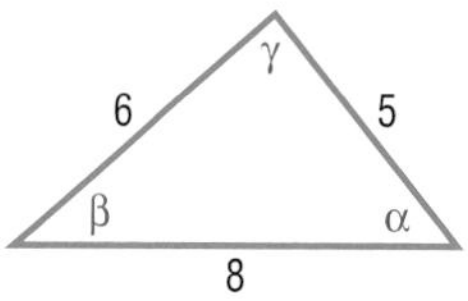

6.

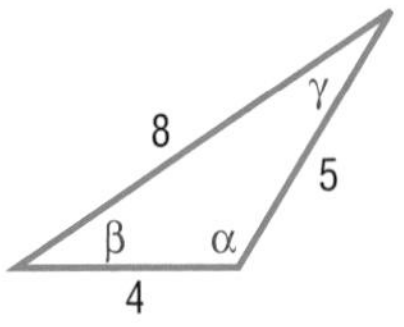

7.

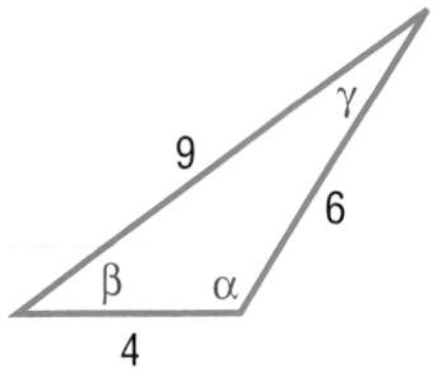

8.

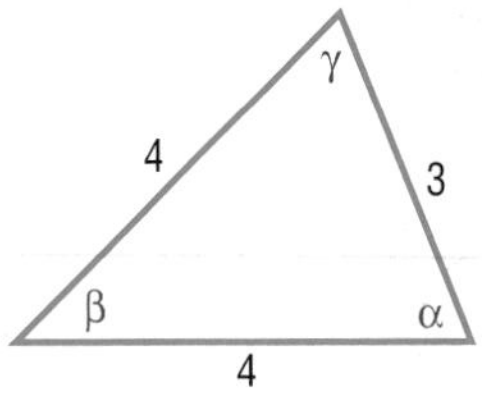

In Problems 9–24, find the area of each triangle. Round answers to two decimal places.

9. $a = 3, \quad b = 4, \quad \gamma = 40°$

10. $a = 2, \quad c = 1, \quad \beta = 10°$

11. $b = 1, \quad c = 3, \quad \alpha = 80°$

12. $a = 6, \quad b = 4, \quad \gamma = 60°$

13. $a = 3, \quad c = 2, \quad \beta = 110°$

14. $b = 4, \quad c = 1, \quad \alpha = 120°$

15. $a = 2, \quad b = 2, \quad \gamma = 50°$

16. $a = 3, \quad c = 2, \quad \beta = 90°$

17. $a = 12, \quad b = 13, \quad c = 5$

18. $a = 4, \quad b = 5, \quad c = 3$

19. $a = 2, \quad b = 2, \quad c = 2$

20. $a = 3, \quad b = 3, \quad c = 2$

21. $a = 5, \quad b = 8, \quad c = 9$

22. $a = 4, \quad b = 3, \quad c = 6$

23. $a = 10, \quad b = 8, \quad c = 5$

24. $a = 9, \quad b = 7, \quad c = 10$

25. Area of a Segment Find the area of the segment (shaded in blue in the figure) of a circle whose radius is 8 feet, formed by a central angle of 70°.
[**Hint:** The area of the sector (enclosed in red) of a circle of radius r formed by a central angle of θ radians is $1/2\, r^2\theta$. Now subtract the area of the triangle from the area of the sector to obtain the area of the segment.]

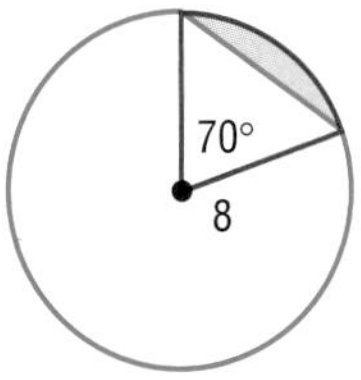

26. Area of a Segment Find the area of the segment of a circle whose radius is 5 inches, formed by a central angle of 40°.

27. Cost of a Triangular Lot The dimensions of a triangular lot are 100 feet by 50 feet by 75 feet. If the price of such land is \$3 per square foot, how much does the lot cost?

28. Amount of Materials to Make a Tent A cone-shaped tent is made from a circular piece of canvas 24 feet in diameter by removing a sector with central angle 100° and connecting the ends. What is the surface area of the tent?

29. Computing Areas Find the area of the shaded region enclosed in a semicircle of diameter 8 centimeters. The length of the chord AB is 6 centimeters.
[**Hint:** Triangle ABC is a right triangle.]

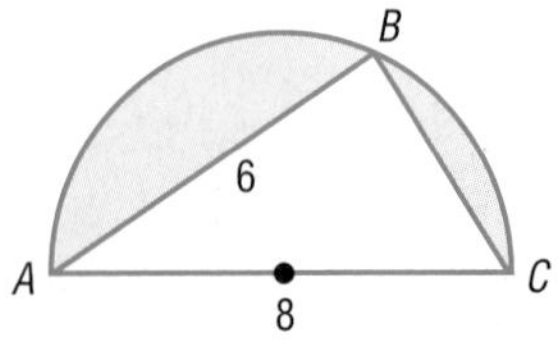

30. Computing Areas Find the area of the shaded region enclosed in a semicircle of diameter 10 inches. The length of the chord AB is 8 inches.
[**Hint:** Triangle ABC is a right triangle.]

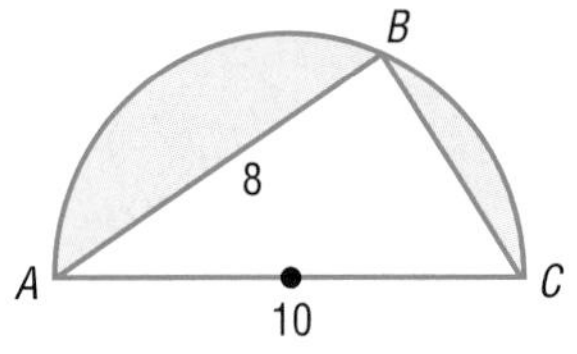

31. Area of a Triangle Prove that the area A of a triangle is given by the formula

$$A = \frac{a^2 \sin\beta \sin\gamma}{2 \sin\alpha}$$

32. Area of a Triangle Prove the two other forms of the formula given in Problem 31.

$$A = \frac{b^2 \sin\alpha \sin\gamma}{2 \sin\beta} \quad \text{and} \quad A = \frac{c^2 \sin\alpha \sin\beta}{2 \sin\gamma}$$

In Problems 33–38, use the results of Problem 31 or 32 to find the area of each triangle. Round answers to two decimal places.

33. $\alpha = 40^\circ, \quad \beta = 20^\circ, \quad a = 2$
34. $\alpha = 50^\circ, \quad \gamma = 20^\circ, \quad a = 3$
35. $\beta = 70^\circ, \quad \gamma = 10^\circ, \quad b = 5$
36. $\alpha = 70^\circ, \quad \beta = 60^\circ, \quad c = 4$
37. $\alpha = 110^\circ, \quad \gamma = 30^\circ, \quad c = 3$
38. $\beta = 10^\circ, \quad \gamma = 100^\circ, \quad b = 2$

39. Geometry Consult the figure, which shows a circle of radius r with center at O. Find the area A of the shaded region as a function of the central angle θ.

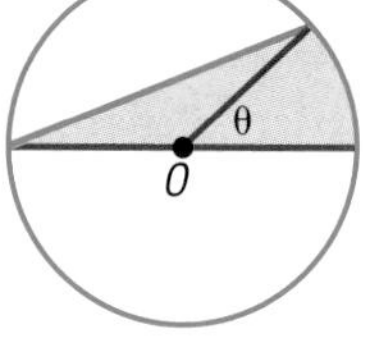

40. Approximating the Area of a Lake To approximate the area of a lake, a surveyor walks around the perimeter of the lake, taking the measurements shown in the illustration. Using this technique, what is the approximate area of the lake?
[**Hint:** Use the Law of Cosines on the three triangles shown and then find the sum of their areas.]

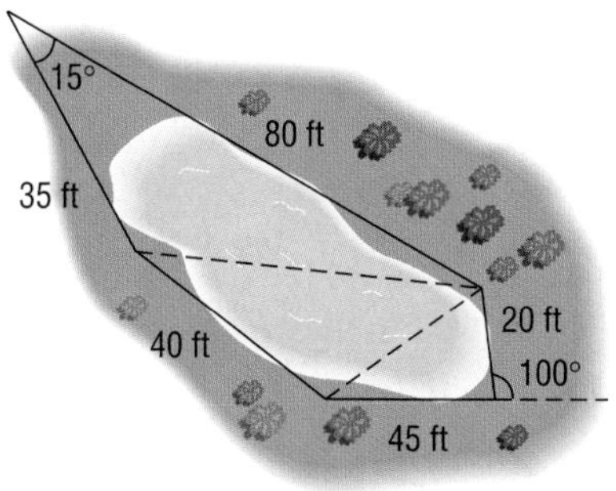

41. Refer to the figure. If $|OA| = 1$, show that:
(a) Area $\triangle OAC = \frac{1}{2}\sin\alpha\cos\alpha$
(b) Area $\triangle OCB = \frac{1}{2}|OB|^2\sin\beta\cos\beta$
(c) Area $\triangle OAB = \frac{1}{2}|OB|\sin(\alpha + \beta)$
(d) $|OB| = \dfrac{\cos\alpha}{\cos\beta}$
(e) $\sin(\alpha + \beta) = \sin\alpha\cos\beta + \cos\alpha\sin\beta$
[**Hint:** Area $\triangle OAB$ = Area $\triangle OAC$ + Area $\triangle OCB$.]

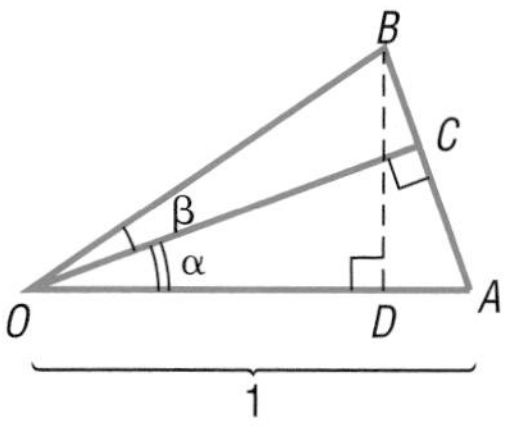

42. Refer to the figure, in which a unit circle is drawn. The line DB is tangent to the circle.
(a) Express the area of $\triangle OBC$ in terms of $\sin\theta$ and $\cos\theta$.
(b) Express the area of $\triangle OBD$ in terms of $\sin\theta$ and $\cos\theta$.
(c) The area of the sector $\widehat{OBC}$ of the circle is $\frac{1}{2}\theta$, where θ is measured in radians. Use the results of parts (a) and (b) and the fact that

$$\text{Area } \triangle OBC < \text{Area } \widehat{OBC} < \text{Area } \triangle OBD$$

to show that

$$1 < \frac{\theta}{\sin\theta} < \frac{1}{\cos\theta}$$

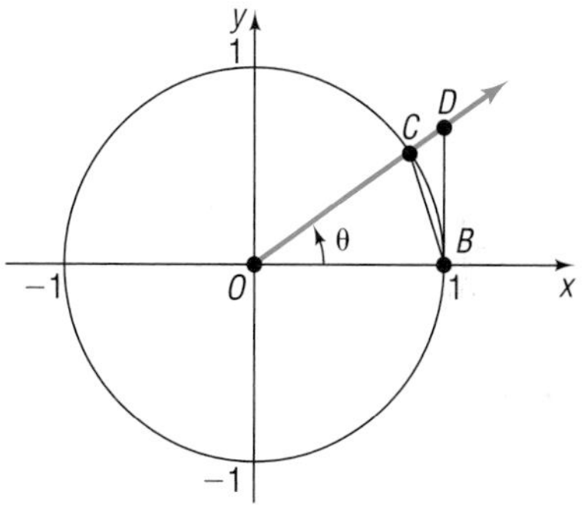

43. The Cow Problem* A cow is tethered to one corner of a square barn, 10 feet by 10 feet, with a rope 100 feet long. What is the maximum grazing area for the cow?
[**Hint:** See the illustration.]

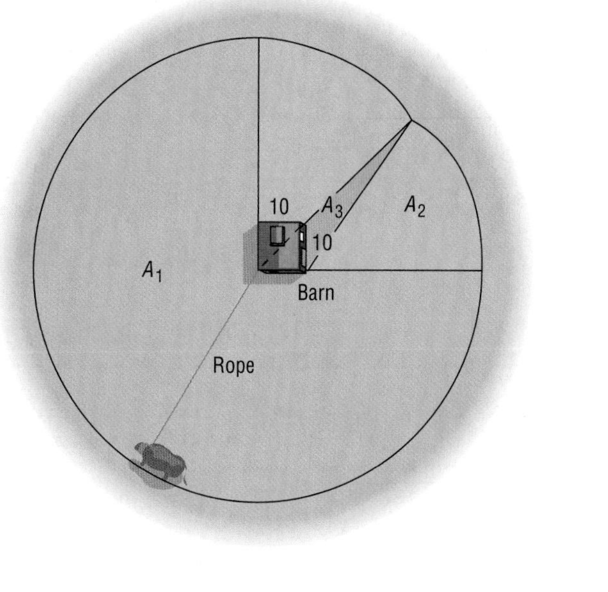

*Suggested by Professor Teddy Koukounas of SUNY at Old Westbury, who learned of it from an old farmer in Virginia. Solution provided by Professor Kathleen Miranda of SUNY at Old Westbury.

44. Another Cow Problem If the barn in Problem 43 is rectangular, 10 feet by 20 feet, what is the maximum grazing area for the cow?

45. If h_1, h_2, and h_3 are the altitudes dropped from A, B, and C, respectively, in a triangle (see the figure), show that

$$\frac{1}{h_1} + \frac{1}{h_2} + \frac{1}{h_2} = \frac{s}{K}$$

where K is the area of the triangle and $s = \frac{1}{2}(a + b + c)$. [**Hint:** $h_1 = 2K/a$.]

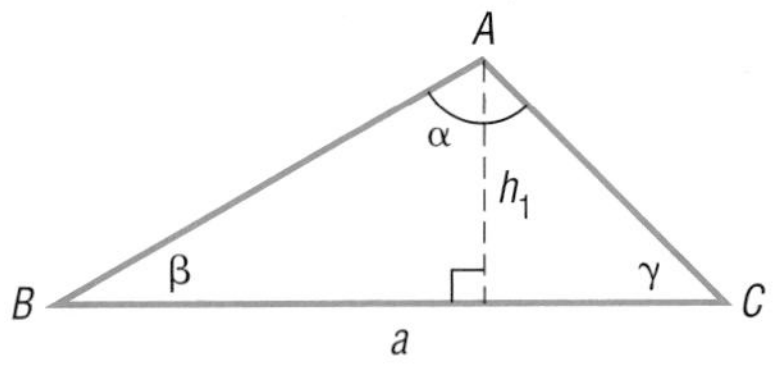

46. Show that a formula for the altitude h from a vertex to the opposite side a of a triangle is

$$h = \frac{a \sin\beta \sin\gamma}{\sin\alpha}$$

Inscribed Circle *For Problems 47–50, the lines that bisect each angle of a triangle meet in a single point O, and the perpendicular distance r from O to each side of the triangle is the same. The circle with center at O and radius r is called the* ***inscribed circle*** *of the triangle (see the figure).*

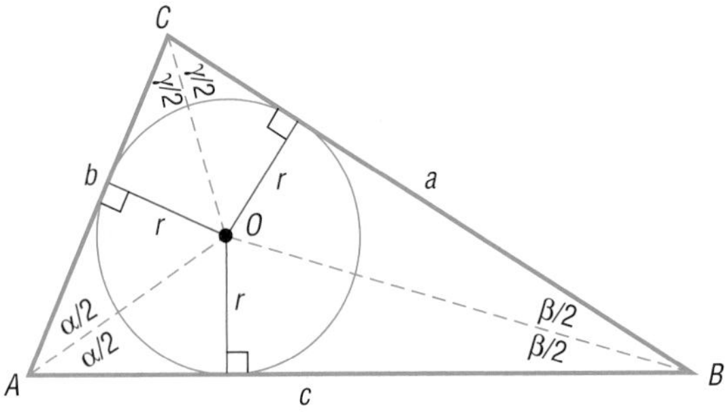

47. Apply Problem 46 to triangle OAB to show that

$$r = \frac{c \sin(\alpha/2) \sin(\beta/2)}{\cos(\gamma/2)}$$

48. Use the results of Problem 47 and Problem 38 in Section 8.3 to show that

$$\cot\frac{\gamma}{2} = \frac{s - c}{r}$$

49. Show that

$$\cot\frac{\alpha}{2} + \cot\frac{\beta}{2} + \cot\frac{\gamma}{2} = \frac{s}{r}$$

50. Show that the area K of triangle ABC is $K = rs$. Then show that

$$r = \sqrt{\frac{(s - a)(s - b)(s - c)}{s}}$$

where $s = \frac{1}{2}(a + b + c)$.

PREPARING FOR THIS SECTION

Before getting started, review the following:

✓ Sinusoidal Graphs (pp. 419–425)

8.5 SIMPLE HARMONIC MOTION; DAMPED MOTION

OBJECTIVES

1 Find an Equation for an Object in Simple Harmonic Motion
2 Analyze Simple Harmonic Motion
3 Analyze an Object in Damped Motion

SIMPLE HARMONIC MOTION

Many physical phenomena can be described as simple harmonic motion. Radio and television waves, light waves, sound waves, and water waves exhibit motion that is simple harmonic.

The swinging of a pendulum, the vibrations of a tuning fork, and the bobbing of a weight attached to a coiled spring are examples of vibrational motion. In this type of motion, an object swings back and forth over the same path. In each illustration in Figure 41, the point B is the **equilibrium (rest) position** of the vibrating object. The **amplitude** of vibration is the distance from the object's rest position to its point of greatest displacement (either point A or point C in Figure 41). The **period** of a vibrating object is the time required to complete one vibration, that is, the time it takes to go from, say, point A through B to C and back to A.

Figure 41

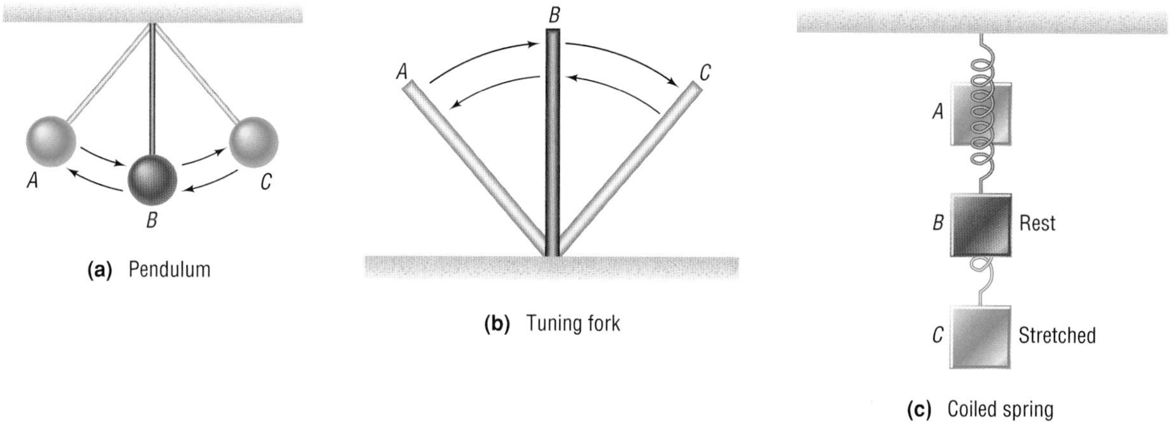

Simple harmonic motion is a special kind of vibrational motion in which the acceleration a of the object is directly proportional to the negative of its displacement d from its rest position. That is, $a = -kd$, $k > 0$.

For example, when the mass hanging from the spring in Figure 41(c) is pulled down from its rest position B to the point C, the force of the spring tries to restore the mass to its rest position. Assuming that there is no frictional force* to retard the motion, the amplitude will remain constant. The force increases in direct proportion to the distance that the mass is pulled from its rest position. Since the force increases directly, the acceleration of the mass of the object must do likewise, because (by Newton's Second Law of Motion) force is directly proportional to acceleration. Thus, the acceleration of the object varies directly with its displacement, and the motion is an example of simple harmonic motion.

Simple harmonic motion is related to circular motion. To see this relationship, consider a circle of radius a, with center at $(0, 0)$. See Figure 42. Suppose that an object initially placed at $(a, 0)$ moves counterclockwise around the circle at constant angular speed ω. Suppose further that after

*If friction is present, the amplitude will decrease with time to 0. This type of motion is an example of **damped motion**, which is discussed later in this section.

Figure 42

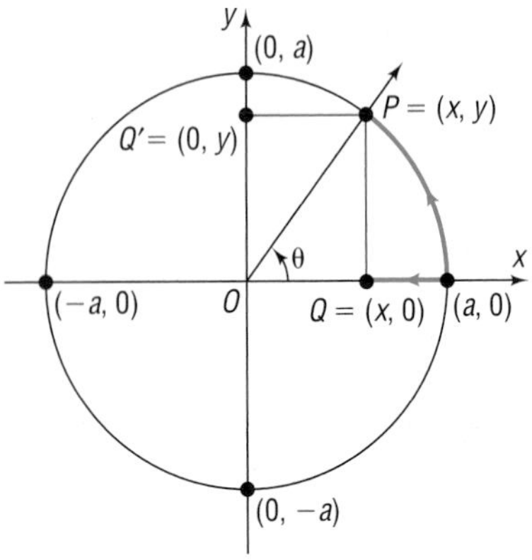

time t has elapsed the object is at the point $P = (x, y)$ on the circle. The angle θ, in radians, swept out by the ray $\overrightarrow{OP}$ in this time t is

$$\theta = \omega t$$

The coordinates of the point P at time t are

$$x = a\cos\theta = a\cos(\omega t)$$

$$y = a\sin\theta = a\sin(\omega t)$$

Corresponding to each position $P = (x, y)$ of the object moving about the circle, there is the point $Q = (x, 0)$, called the **projection of P on the x-axis.** As P moves around the circle at a constant rate, the point Q moves back and forth between the points $(a, 0)$ and $(-a, 0)$ along the x-axis with a motion that is simple harmonic. Similarly, for each point P there is a point $Q' = (0, y)$, called the **projection of P on the y-axis.** As P moves around the circle, the point Q' moves back and forth between the points $(0, a)$ and $(0, -a)$ on the y-axis with a motion that is simple harmonic. Thus, simple harmonic motion can be described as the projection of constant circular motion on a coordinate axis.

To put it another way, again consider a mass hanging from a spring where the mass is pulled down from its rest position to the point C and then released. See Figure 43(a). The graph shown in Figure 43(b) describes the distance that the object is from its rest position, d, as a function of time, t, assuming that no frictional force is present.

Figure 43

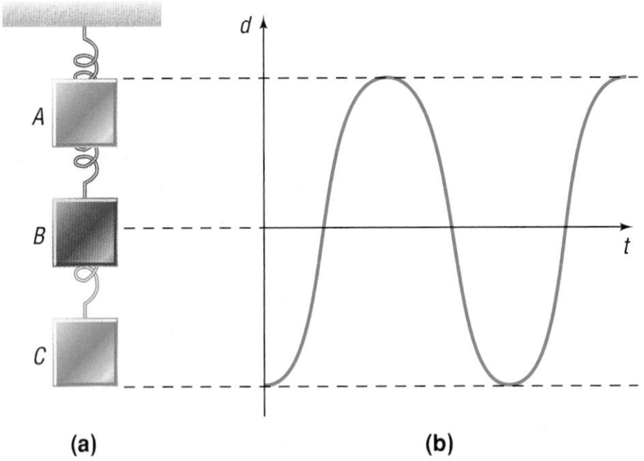

Theorem

Simple Harmonic Motion

An object that moves on a coordinate axis so that its distance d from the origin at time t is given by either

$$d = a\cos(\omega t) \quad \text{or} \quad d = a\sin(\omega t)$$

where a and $\omega > 0$ are constants, moves with simple harmonic motion. The motion has amplitude $|a|$ and period $2\pi/\omega$.

The **frequency** f of an object in simple harmonic motion is the number of oscillations per unit time. Since the period is the time required for one oscillation, it follows that frequency is the reciprocal of the period; that is,

$$f = \frac{\omega}{2\pi}, \qquad \omega > 0$$

EXAMPLE 1 Finding an Equation for an Object in Harmonic Motion

1 Suppose that the object attached to the coiled spring in Figure 41(c) is pulled down a distance of 5 inches from its rest position and then released. If the time for one oscillation is 3 seconds, write an equation that relates the distance d of the object from its rest position after time t (in seconds). Assume no friction.

Solution The motion of the object is simple harmonic. When the object is released ($t = 0$), the distance of the object from the rest position is -5 units (since the object was pulled down). Because $d = -5$ when $t = 0$, it is easier* to use the cosine function

$$d = a\cos(\omega t)$$

to describe the motion. Now the amplitude is $|-5| = 5$ and the period is 3. Thus,

$$a = -5 \quad \text{and} \quad \frac{2\pi}{\omega} = \text{period} = 3 \quad \text{or} \quad \omega = \frac{2\pi}{3}$$

An equation of the motion of the object is

$$d = -5\cos\left[\frac{2\pi}{3}t\right]$$

■

NOTE: In the solution to Example 1, we let $a = -5$, since the initial motion is down. If the initial direction were up, we would let $a = 5$.

NOW WORK PROBLEM 1.

EXAMPLE 2 Analyzing the Motion of an Object

2 Suppose that the distance d (in meters) that an object travels in time t (in seconds) satisfies the equation

$$d = 10\sin(5t)$$

(a) Describe the motion of the object.
(b) What is the maximum displacement from its resting position?
(c) What is the time required for one oscillation?
(d) What is the frequency?

Solution We observe that the given equation is of the form

$$d = a\sin(\omega t) \qquad d = 10\sin(5t)$$

where $a = 10$ and $\omega = 5$.

*No phase shift is required if a cosine function is used.

(a) The motion is simple harmonic.

(b) The maximum displacement of the object from its resting position is the amplitude: $|a| = 10$ meters.

(c) The time required for one oscillation is the period:

$$\text{Period} = \frac{2\pi}{\omega} = \frac{2\pi}{5} \text{ seconds}$$

(d) The frequency is the reciprocal of the period. Thus,

$$\text{Frequency} = f = \frac{5}{2\pi} \text{ oscillations per second}$$

NOW WORK PROBLEM 9.

Damped Motion

3 Most physical phenomena are affected by friction or other resistive forces. These forces remove energy from a moving system and thereby damp its motion. For example, when a mass hanging from a spring is pulled down a distance a and released, the friction in the spring causes the distance that the mass moves from its at-rest position to decrease over time. Thus, the amplitude of any real oscillating spring or swinging pendulum decreases with time due to air resistance, friction, and so forth. See Figure 44.

Figure 44

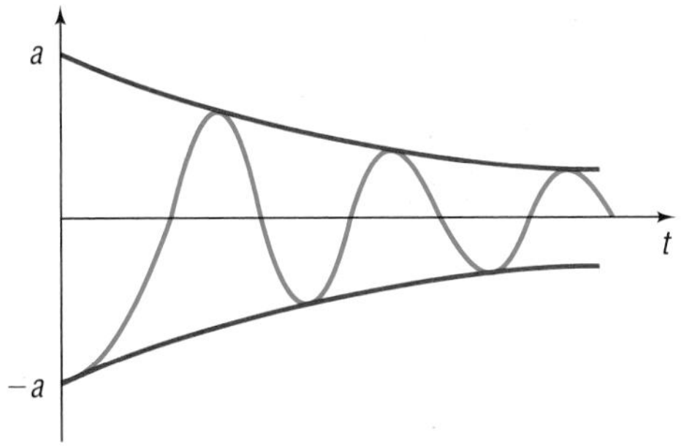

A function that describes this phenomenon maintains a sinusoidal component, but the amplitude of this component will decrease with time in order to account for the damping effect. In addition, the period of the oscillating component will be affected by the damping. The next result, from physics, describes damped motion.

Theorem **Damped Motion**

The displacement d of an oscillating object from its at rest position at time t is given by

$$d = ae^{-bt/2m} \cos\left(\sqrt{\omega^2 - \frac{b^2}{4m^2}}\, t\right)$$

where b is a **damping factor** (most physics texts call this a **damping coefficient**) and m is the mass of the oscillating object.

Notice for $b = 0$ (zero damping) that we have the formula for simple harmonic motion with amplitude $|a|$ and period $2\pi/\omega$.

EXAMPLE 3 Analyzing the Motion of a Pendulum with Damped Motion

Figure 45

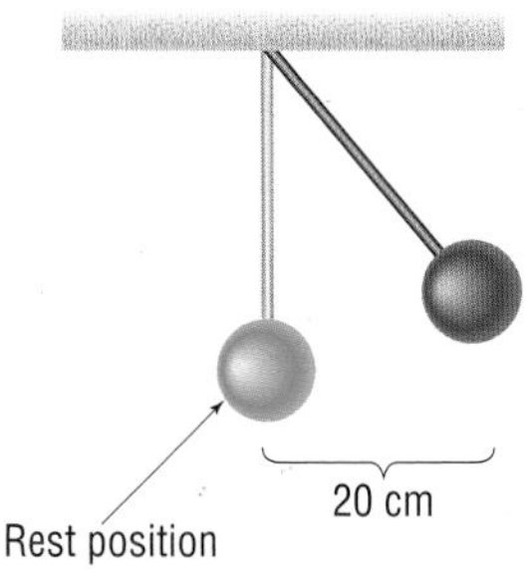

Suppose that a simple pendulum with a bob of mass 10 grams and a damping factor of 0.8 gram/second is pulled 20 centimeters from its at-rest position and released. See Figure 45. The period of the pendulum without the damping effect is 4 seconds.

(a) Find an equation that describes the position of the pendulum bob.

(b) Using a graphing utility, graph the function found in part (a).

(c) Determine the maximum displacement of the bob after the first oscillation.

(d) What happens to the displacement of the bob as time increases without bound?

Solution (a) We have $m = 10$, $a = 20$, and $b = 0.8$. Since the period of the pendulum under simple harmonic motion is 4 seconds, we have

$$4 = \frac{2\pi}{\omega}$$

$$\omega = \frac{2\pi}{4} = \frac{\pi}{2}$$

Figure 46

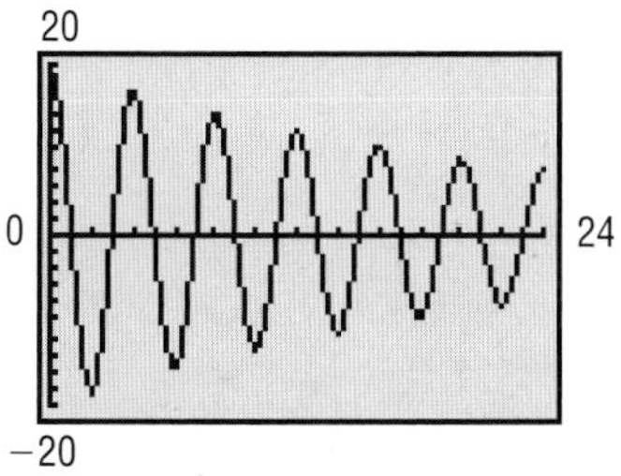

Substituting these values into the equation for damped motion, we obtain

$$d = 20e^{-0.8t/2(10)} \cos\left(\sqrt{\left(\frac{\pi}{2}\right)^2 - \frac{0.8^2}{4(10)^2}}\, t\right)$$

$$= 20e^{-0.8t/20} \cos\left(\sqrt{\frac{\pi^2}{4} - \frac{0.64}{400}}\, t\right)$$

Figure 47

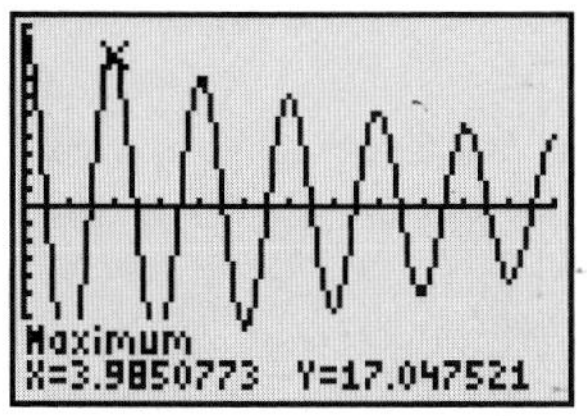

(b) See Figure 46 for the graph of d.

(c) See Figure 47. After the first oscillation, the maximum displacement is approximately 17.05 centimeters.

(d) As t increases without bound, $e^{-0.8t/20} \to 0$, so the displacement of the bob approaches zero. As a result, the pendulum will eventually come to rest. ■

8.5 EXERCISES

In Problems 1–4, an object attached to a coiled spring is pulled down a distance a from its rest position and then released. Assuming that the motion is simple harmonic with period T, write an equation that relates the distance d of the object from its rest position after t seconds. Also assume that the positive direction of the motion is up.

1. $a = 5$; $T = 2$ seconds

2. $a = 10$; $T = 3$ seconds

3. $a = 6$; $T = \pi$ seconds

4. $a = 4$; $T = \pi/2$ seconds

5. Rework Problem 1 under the same conditions except that, at time $t = 0$, the object is at its resting position and moving down.

6. Rework Problem 2 under the same conditions except that, at time $t = 0$, the object is at its resting position and moving down.

7. Rework Problem 3 under the same conditions except that, at time $t = 0$, the object is at its resting position and moving down.

8. Rework Problem 4 under the same conditions except that, at time $t = 0$, the object is at its resting position and moving down.

In Problems 9–16, the distance d (in meters) that an object travels in time t (in seconds) is given.

(a) Describe the motion of the object.
(b) What is the maximum displacement from its resting position?
(c) What is the time required for one oscillation?
(d) What is the frequency?

9. $d = 5\sin(3t)$

10. $d = 4\sin(2t)$

11. $d = 6\cos(\pi t)$

12. $d = 5\cos\left[\frac{\pi}{2}t\right]$

13. $d = -3\sin\left(\frac{1}{2}t\right)$

14. $d = -2\cos(2t)$

15. $d = 6 + 2\cos(2\pi t)$

16. $d = 4 + 3\sin(\pi t)$

In Problems 17–22, an object of mass m attached to a coiled spring with damping factor b is pulled down a distance a from its rest position and then released. Assume that the positive direction of the motion is up and the period of the first oscillation is T.

(a) Write an equation that relates the distance d of the object from its rest position after t seconds.
(b) Graph the equation found in part (a) for 5 oscillations using a graphing utility.

17. $m = 25$ grams; $a = 10$ centimeters; $b = 0.7$ gram/second; $T = 5$ seconds

18. $m = 20$ grams; $a = 15$ centimeters; $b = 0.75$ gram/second; $T = 6$ seconds

19. $m = 30$ grams; $a = 18$ centimeters; $b = 0.6$ gram/second; $T = 4$ seconds

20. $m = 15$ grams; $a = 16$ centimeters; $b = 0.65$ gram/second; $T = 5$ seconds

21. $m = 10$ grams; $a = 5$ centimeters; $b = 0.8$ gram/second; $T = 3$ seconds

22. $m = 10$ grams; $a = 5$ centimeters; $b = 0.7$ gram/second; $T = 3$ seconds

In Problems 23–28, the distance d (in meters) of the bob of a pendulum of mass m (in kilograms) from its rest position at time t (in seconds) is given.

(a) Describe the motion of the object. Be sure to give the mass and damping factor.
(b) What is the initial displacement of the bob? That is, what is the displacement at t = 0?
(c) Graph the motion using a graphing utility.
(d) What is the maximum displacement after the first oscillation?
(e) What happens to the displacement of the bob as time increases without bound?

23. $d = -20e^{-0.7t/40}\cos\left(\sqrt{\left(\frac{2\pi}{5}\right)^2 - \frac{0.49}{1600}}\,t\right)$

24. $d = -20e^{-0.8t/40}\cos\left(\sqrt{\left(\frac{2\pi}{5}\right)^2 - \frac{0.64}{1600}}\,t\right)$

25. $d = -30e^{-0.6t/80}\cos\left(\sqrt{\left(\frac{2\pi}{7}\right)^2 - \frac{0.36}{6400}}\,t\right)$

26. $d = -30e^{-0.5t/70}\cos\left(\sqrt{\left(\frac{\pi}{2}\right)^2 - \frac{0.25}{4900}}\,t\right)$

27. $d = -15e^{-0.9t/30}\cos\left(\sqrt{\left(\frac{\pi}{3}\right)^2 - \frac{0.81}{900}}\,t\right)$

28. $d = -10e^{-0.8t/50}\cos\left(\sqrt{\left(\frac{2\pi}{3}\right)^2 - \frac{0.64}{2500}}\,t\right)$

29. Simple Pendulum The following data represent the distance that the bob of a simple pendulum is from a motion detector.

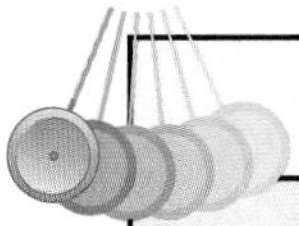

Time (Seconds)	Distance (Centimeters)
0.0	22.59
0.2	20.51
0.4	19.06
0.6	19.52
0.8	21.81
1.0	25.52
1.2	30.01
1.4	34.49
1.6	38.19
1.8	40.47
2.0	40.93
2.2	39.50
2.4	36.41
2.6	32.20
2.8	27.62
3.0	23.45
3.2	20.41

(a) Using a graphing utility, draw a scatter diagram of the data.
(b) By hand, find a sinusoidal function of the form $y = A\sin(\omega x - \phi) + B$ that fits the data.
(c) Graph the sinusoidal function found in part (b) on the scatter diagram.
(d) Use a graphing utility to find the sinusoidal function of best fit.
(e) Graph the sinusoidal function of best fit on the scatter diagram.
(f) Based on your answer to part (d), what is the period of the pendulum?

30. Simple Pendulum The following data represent the distance that the bob of a simple pendulum is from a motion detector.
(a) Using a graphing utility, draw a scatter diagram of the data.
(b) By hand, find a sinusoidal function of the form $y = A\sin(\omega x - \phi) + B$ that fits the data.
(c) Graph the sinusoidal function found in part (b) on the scatter diagram.
(d) Use a graphing utility to find the sinusoidal function of best fit.
(e) Graph the sinusoidal function of best fit on the scatter diagram.
(f) Based on your answer to part (d), what is the period of the pendulum?

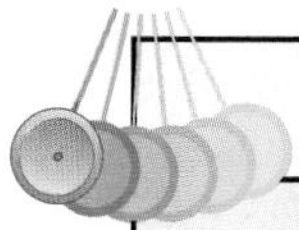

Time (Seconds)	Distance (Centimeters)
0.0	21.79
0.2	26.32
0.4	32.66
0.6	39.48
0.8	45.37
1.0	49.10
1.2	49.91
1.4	47.61
1.6	42.69
1.8	36.17
2.0	29.41
2.2	23.81
2.4	20.54
2.6	20.72

31. Charging a Capacitor If a charged capacitor is connected to a coil by closing a switch (see the figure), energy is transferred to the coil and then back to the capacitor in an oscillatory motion. The voltage V (in volts) across the capacitor will gradually diminish to 0 with time t.
(a) Graph the equation relating V and t:

$$V(t) = e^{-1.9t}\cos(\pi t), \qquad 0 \le t \le 3$$

(b) At what times t will the graph of V touch the graph of $y = e^{-1.9t}$? When does V touch the graph of $y = -e^{-1.9t}$?
(c) When will the voltage V be between -0.1 and 0.1 volt?

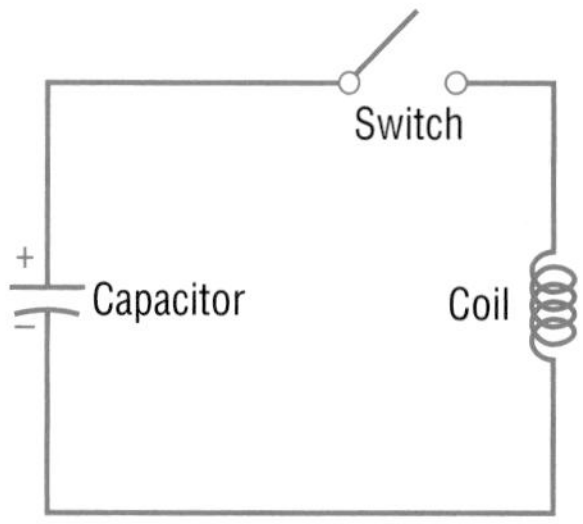

32. Graph the function $f(x) = (\sin x)/x$, $x > 0$. Based on the graph, what do you conjecture about the value of $(\sin x)/x$ for x close to 0?

33. Graph $y = x\sin x$, $y = x^2\sin x$, and $y = x^3\sin x$ for $x > 0$. What patterns do you observe?

34. Graph $y = \frac{1}{x}\sin x$, $y = \frac{1}{x^2}\sin x$, and $y = \frac{1}{x^3}\sin x$ for $x > 0$. What patterns do you observe?

35. How would you explain to a friend what simple harmonic motion is? Damped motion?

36. CBL Experiment Pendulum motion is analyzed to estimate simple harmonic motion. A plot is generated with the position of the pendulum over time. The graph is used to find a sinusoidal curve of the form $y = A\cos[B(x - C)] + D$. Determine the amplitude, period and frequency. (Activity 16, Real-World Math with the CBL System.)

37. CBL Experiment The sound from a tuning fork is collected over time. The amplitude, frequency, and period of the graph are determined. A model of the form $y = A\cos[B(x - C)]$ is fitted to the data. (Activity 23, Real-World Math with the CBL System.)

CHAPTER REVIEW

Things To Know

Acute angle (p. 508)	An angle θ whose measure is $0° < \theta < 90°$ (or $0 < \theta < \pi/2$)
Complementary angles (p. 510)	Two acute angles whose sum is $90°$ ($\pi/2$)
Cofunction (p. 510)	The following pairs of functions are cofunctions of each other: sine and cosine; tangent and cotangent; secant and cosecant

Formulas

Law of Sines (p. 522)

$$\frac{\sin\alpha}{a} = \frac{\sin\beta}{b} = \frac{\sin\gamma}{c}$$

Law of Cosines (p. 532)

$$c^2 = a^2 + b^2 - 2ab\cos\gamma$$
$$b^2 = a^2 + c^2 - 2ac\cos\beta$$
$$a^2 = b^2 + c^2 - 2bc\cos\alpha$$

Area of a triangle (pp. 539–540)

$$A = \tfrac{1}{2}bh$$
$$A = \tfrac{1}{2}ab\sin\gamma$$
$$A = \tfrac{1}{2}bc\sin\alpha$$
$$A = \tfrac{1}{2}ac\sin\beta$$
$$A = \sqrt{s(s-a)(s-b)(s-c)}, \quad \text{where} \quad s = \tfrac{1}{2}(a + b + c)$$

How To

Find the value of trigonometric functions of acute angles (p. 508)

Use the complementary angle theorem (p. 510)

Solve right triangles (p. 511)

Solve applied problems using right triangle trigonometry (p. 512)

Solve SAA or ASA triangles (p. 523)

Solve SSA triangles (p. 524)

Solve applied problems using the Law of Sines (p. 527)

Solve SAS triangles (p. 533)

Solve SSS triangles (p. 534)

Solve applied triangles using the Law of Cosines (p. 535)

Find the area of SAS triangles (p. 540)

Find the area of SSS triangles (p. 540)

Find an equation for an object in simple harmonic motion (p. 547)

Analyze simple harmonic motion (p. 547)

Analyze an object in damped motion (p. 548)

Fill-in-the-Blank Items

1. Two acute angles whose sum is a right angle are called ________________.
2. The sine and ________________ functions are cofunctions.
3. If two sides and the angle opposite one of them are known, the Law of ________________ is used to determine whether the known information results in no triangle, one triangle, or two triangles.
4. If three sides of a triangle are given, the Law of ________________ is used to solve the triangle.
5. If three sides of a triangle are given, ________________ Formula is used to find the area of the triangle.
6. The motion of an object obeys the equation $d = 4\cos(6t)$. Such motion is described as ________________ ________________.
7. The mass and damping factor of $d = 5e^{-0.5t/10}\cos\left(\sqrt{\pi^2 - \frac{(0.5)^2}{100}}\,t\right)$ are ________________ and ________________.

True/False Items

T F **1.** $\tan 62° = \cot 38°$

T F **2.** $\sin 182° = \cos 2°$

T F **3.** An oblique triangle in which two sides and an angle are given always results in at least one triangle.

T F **4.** Given three sides of a triangle, there is a formula for finding its area.

T F **5.** In a right triangle, if two sides are known, we can solve the triangle.

T F **6.** The ambiguous case refers to the fact that, when two sides and the angle opposite one of them is known, sometimes the Law of Sines cannot be used.

Review Exercises

Blue problem numbers indicate the authors' suggestions for use in a Practice Test.

In Problems 1–4, solve each triangle.

1. 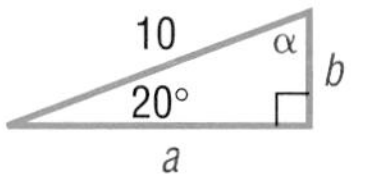

2. 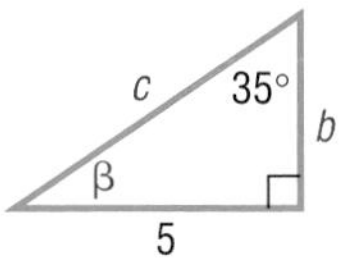

3. 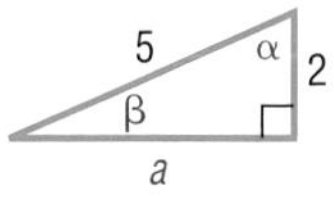

4. 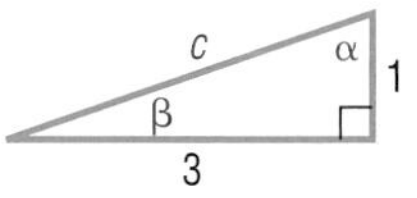

In Problems 5–24, find the remaining angle(s) and side(s) of each triangle, if it (they) exists. If no triangle exists, say "No triangle."

5. $\alpha = 50°,\ \beta = 30°,\ a = 1$
6. $\alpha = 10°,\ \gamma = 40°,\ c = 2$
7. $\alpha = 100°,\ a = 5,\ c = 2$
8. $a = 2,\ c = 5,\ \alpha = 60°$
9. $a = 3,\ c = 1,\ \gamma = 110°$
10. $a = 3,\ c = 1,\ \gamma = 20°$
11. $a = 3,\ c = 1,\ \beta = 100°$
12. $a = 3,\ b = 5,\ \beta = 80°$
13. $a = 2,\ b = 3,\ c = 1$
14. $a = 10,\ b = 7,\ c = 8$
15. $a = 1,\ b = 3,\ \gamma = 40°$
16. $a = 4,\ b = 1,\ \gamma = 100°$
17. $a = 5,\ b = 3,\ \alpha = 80°$
18. $a = 2,\ b = 3,\ \alpha = 20°$
19. $a = 1,\ b = \frac{1}{2},\ c = \frac{4}{3}$
20. $a = 3,\ b = 2,\ c = 2$
21. $a = 3,\ \alpha = 10°,\ b = 4$
22. $a = 4,\ \alpha = 20°,\ \beta = 100°$
23. $c = 5,\ b = 4,\ \alpha = 70°$
24. $a = 1,\ b = 2,\ \gamma = 60°$

In Problems 25–34, find the area of each triangle.

25. $a = 2,\ b = 3,\ \gamma = 40°$
26. $b = 5,\ c = 5,\ \alpha = 20°$
27. $b = 4,\ c = 10,\ \alpha = 70°$
28. $a = 2,\ b = 1,\ \gamma = 100°$
29. $a = 4,\ b = 3,\ c = 5$
30. $a = 10,\ b = 7,\ c = 8$
31. $a = 4,\ b = 2,\ c = 5$
32. $a = 3,\ b = 2,\ c = 2$
33. $\alpha = 50°,\ \beta = 30°,\ a = 1$
34. $\alpha = 10°,\ \gamma = 40°,\ c = 3$

35. Measuring the Length of a Lake From a stationary hot-air balloon 500 feet above the ground, two sightings of a lake are made (see the figure). How long is the lake?

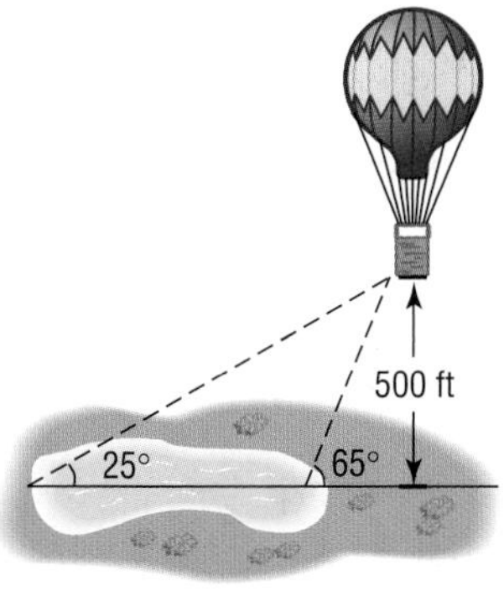

36. Finding the Speed of a Glider From a glider 200 feet above the ground, two sightings of a stationary object directly in front are taken 1 minute apart (see the figure). What is the speed of the glider?

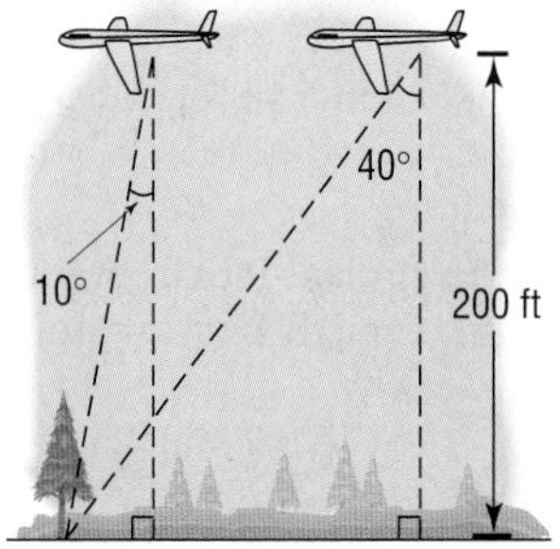

37. Finding the Width of a River Find the distance from A to C across the river illustrated in the figure.

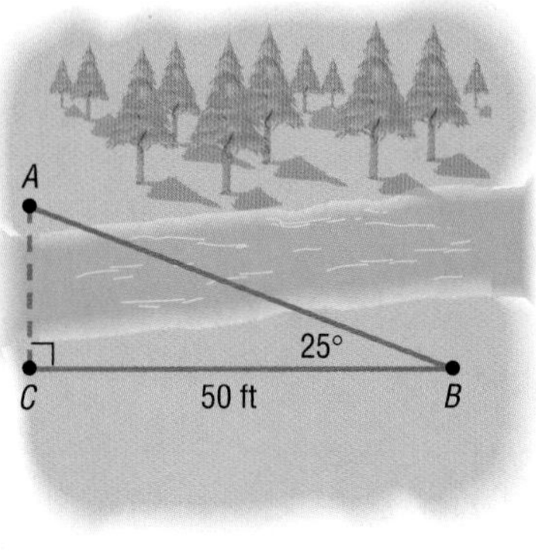

38. Finding the Height of a Building Find the height of the building shown in the figure.

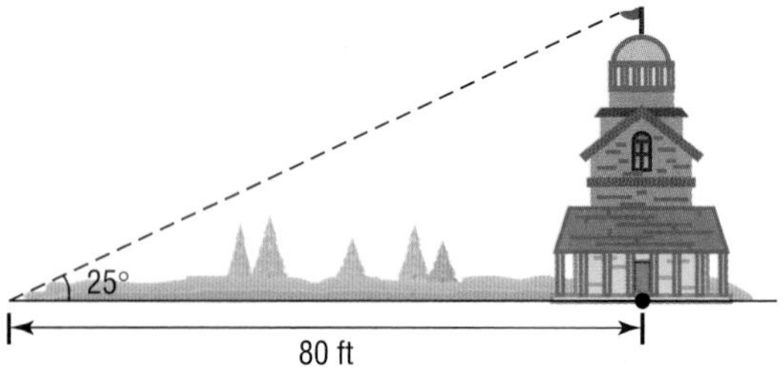

39. Finding the Distance to Shore The Sears Tower in Chicago is 1454 feet tall and is situated about 1 mile inland from the shore of Lake Michigan, as indicated in the figure. An observer in a pleasure boat on the lake directly in front of the Sears Tower looks at the top of the tower and measures the angle of elevation as 5°. How far offshore is the boat?

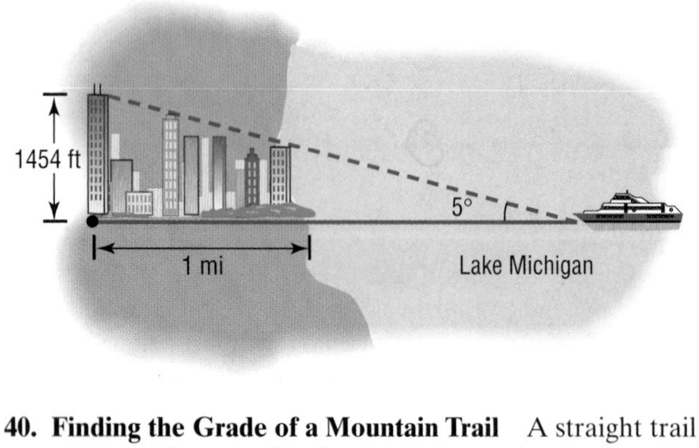

40. Finding the Grade of a Mountain Trail A straight trail with a uniform inclination leads from a hotel, elevation 5000 feet, to a lake in a valley, elevation 4100 feet. The length of the trail is 4100 feet. What is the inclination (grade) of the trail?

41. Navigation An airplane flies from city A to city B, a distance of 100 miles, and then turns through an angle of 20° and heads toward city C, as indicated in the figure. If the distance from A to C is 300 miles, how far is it from city B to city C?

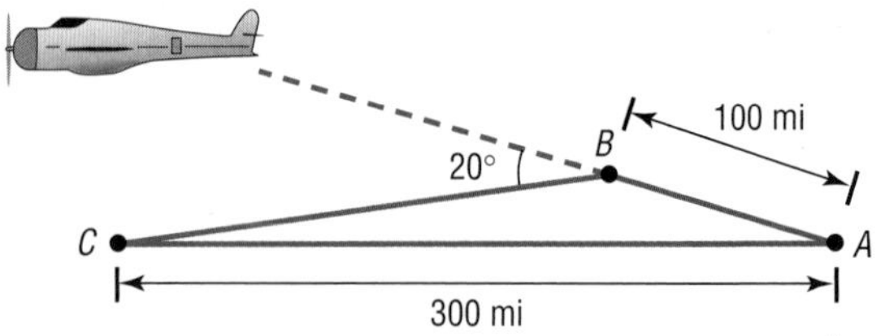

42. Correcting a Navigation Error Two cities A and B are 300 miles apart. In flying from city A to city B, a pilot inadvertently took a course that was 5° in error.

(a) If the error was discovered after flying 10 minutes at a constant speed of 420 miles per hour, through what angle should the pilot turn to correct the course? (Consult the figure.)

(b) What new constant speed should be maintained so that no time is lost due to the error? (Assume that the speed would have been a constant 420 miles per hour if no error had occurred.)

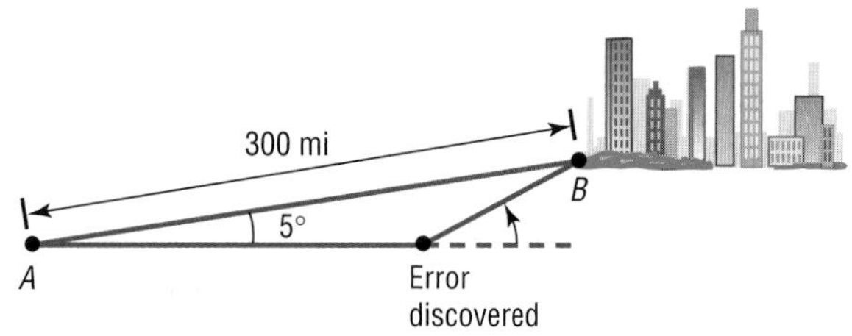

43. Determining Distances at Sea Rebecca, the navigator of a ship at sea, spots two lighthouses that she knows to be 2 miles apart along a straight shoreline. She determines that the angles formed between two line-of-sight observations of the lighthouses and the line from the ship directly to shore are 12° and 30°. See the illustration.

(a) How far is the ship from lighthouse A?

(b) How far is the ship from lighthouse B?

(c) How far is the ship from shore?

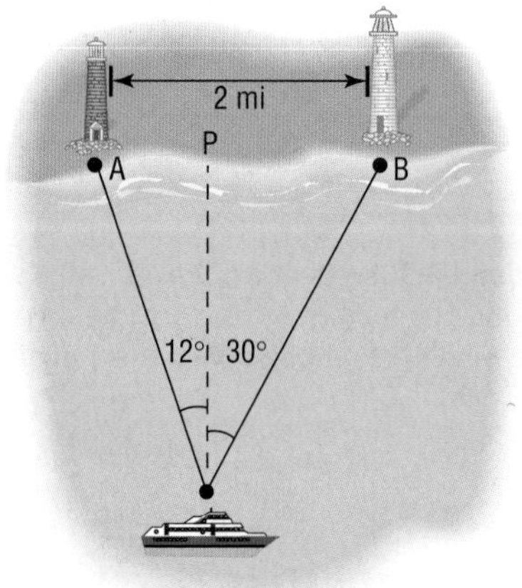

44. Constructing a Highway A highway whose primary directions are north–south is being constructed along the west coast of Florida. Near Naples, a bay obstructs the straight path of the road. Since the cost of a bridge is prohibitive, engineers decide to go around the bay. The illustration shows the path that they decide on and the measurements taken. What is the length of highway needed to go around the bay?

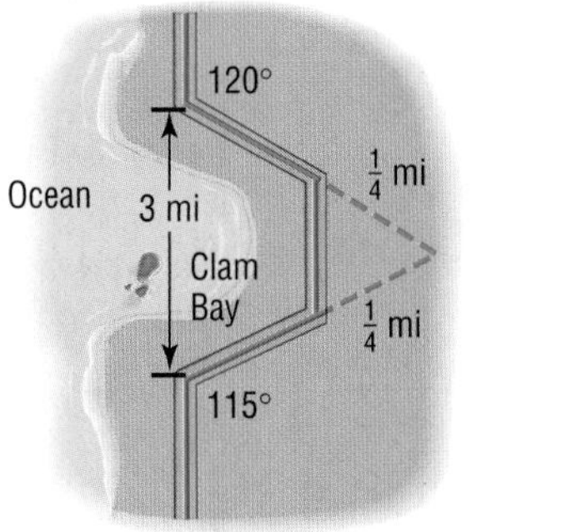

45. Correcting a Navigational Error A yacht leaves St. Thomas bound for an island in the British West Indies, 200 miles away. Maintaining a constant speed of 18 miles per hour, but encountering heavy crosswinds and strong currents, the crew finds after 4 hours that the sailboat is off course by 15°.

(a) How far is the sailboat from the island at this time?

(b) Through what angle should the sailboat turn to correct its course?

(c) How much time has been added to the trip because of this? (Assume that the speed remains at 18 miles per hour.)

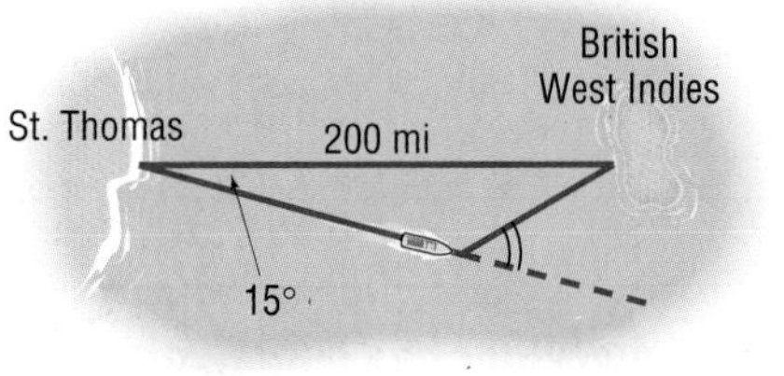

46. Surveying Two homes are located on opposite sides of a small hill. See the illustration. To measure the distance between them, a surveyor walks a distance of 50 feet from house A to point C, uses a transit to measure the angle ACB, which is found to be 80°, and then walks to house B, a distance of 60 feet. How far apart are the houses?

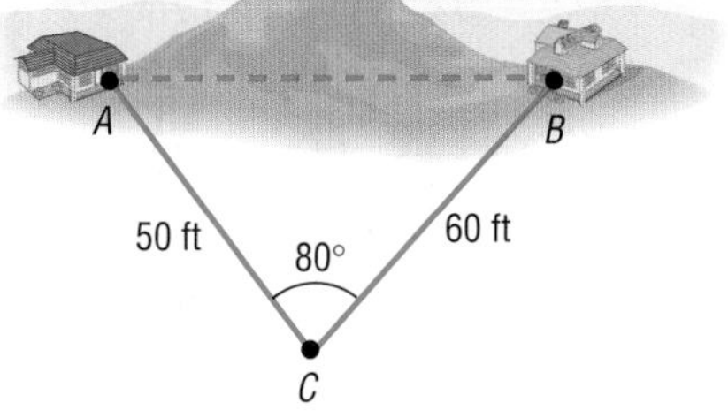

47. Approximating the Area of a Lake To approximate the area of a lake, Cindy walks around the perimeter of the lake, taking the measurements shown in the illustration. Using this technique, what is the approximate area of the lake?

[**Hint:** Use the Law of Cosines on the three triangles shown and then find the sum of their areas.]

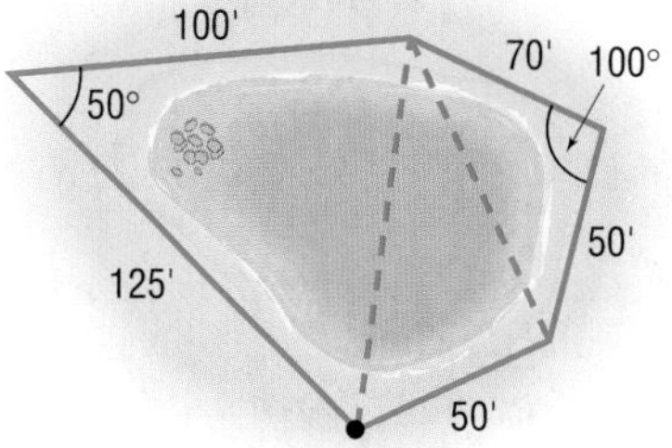

48. Calculating the Cost of Land The irregular parcel of land shown in the figure is being sold for \$100 per square foot. What is the cost of this parcel?

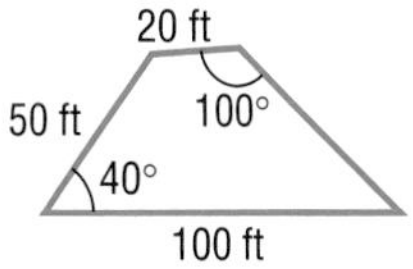

49. Area of a Segment Find the area of the segment of a circle whose radius is 6 inches formed by a central angle of 50°.

50. Finding the Bearing of a Ship The *Majesty* leaves the Port at Boston for Bermuda with a bearing of S80°E at an average speed of 10 knots. After 1 hour, the ship turns 90° toward the southwest. After 2 hours at an average speed of 20 knots, what is the bearing of the ship from Boston?

51. The drive wheel of an engine is 13 inches in diameter, and the pulley on the rotary pump is 5 inches in diameter. If the shafts of the drive wheel and the pulley are 2 feet apart, what length of belt is required to join them as shown in the figure?

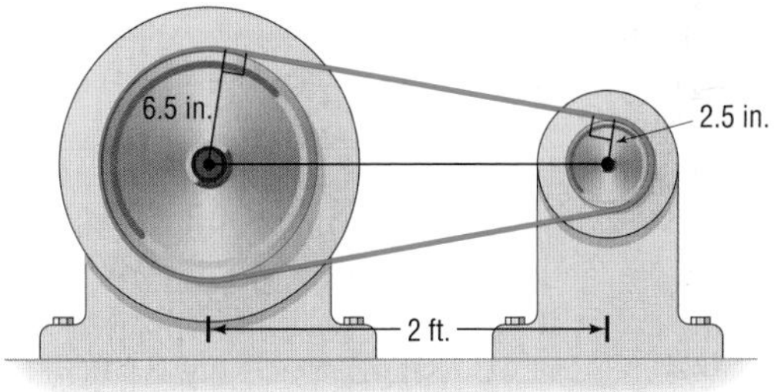

52. Rework Problem 51 if the belt is crossed, as shown in the figure.

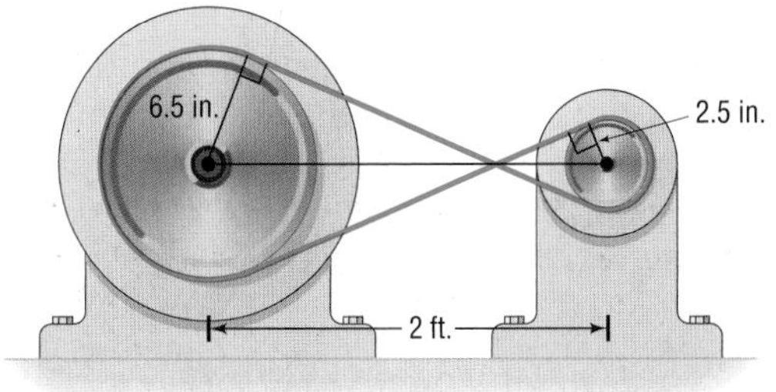

In Problems 53–56, the distance d (in feet) that an object travels in time t (in seconds) is given.

(a) *Describe the motion of the object.*
(b) *What is the maximum displacement from its resting position?*
(c) *What is the time required for one oscillation?*
(d) *What is the frequency?*

53. $d = 6\sin(2t)$ **54.** $d = 2\cos(4t)$ **55.** $d = -2\cos(\pi t)$ **56.** $d = -3\sin\left[\frac{\pi}{2}t\right]$

In Problems 57 and 58, an object of mass m attached to a coiled spring with damping factor b is pulled down a distance a from its rest position and then released. Assume that the positive direction of the motion is up and the period of the first oscillation is T.

(a) *Write an equation that relates the distance d of the object from its rest position after t seconds.*
(b) *Using a graphing utility, graph the equation found in part (a) for 5 oscillations.*

57. $m = 40$ grams; $a = 15$ centimeters; $b = 0.75$ gram/second; $T = 5$ seconds

58. $m = 25$ grams; $a = 13$ centimeters; $b = 0.65$ gram/second; $T = 4$ seconds

In Problems 59 and 60, the distance d (in meters) of the bob of a pendulum of mass m (in kilograms) from its rest position at time t (in seconds) is given.

(a) *Describe the motion of the object.*
(b) *What is the initial displacement of the bob? That is, what is the displacement at t = 0?*
(c) *Using a graphing utility, graph the motion.*
(d) *What is the maximum displacement after the first oscillation?*
(e) *What happens to the displacement of the bob as time increases without bound?*

59. $d = -15e^{-0.6t/40}\cos\left(\sqrt{\left(\frac{2\pi}{5}\right)^2 - \frac{0.36}{1600}}\,t\right)$ **60.** $d = -20e^{-0.5t/60}\cos\left(\sqrt{\left(\frac{2\pi}{3}\right)^2 - \frac{0.25}{3600}}\,t\right)$

PROJECT AT MOTOROLA

How can you build or analyze vibration profile.

Automotive electronics has to be built to one of the most stringent mechanical specifications in the industry. One of the important elements of the requirements is vibration specification. We have to design our products for a harsh vibration environment and validate our designs by testing. The picture to the right depicts some of our engine controllers on a horizontal vibration slip table. In order to properly design our products we have to understand specific vibration profiles, be able to analyze and modify them, and precisely determine their interaction with our engine controller (identify modes of vibration, resonances, transfer functions).

One of the fundamental tools for vibration analysis and characterization is the Fourier Theorem, Fourier Series and Fourier Transformations. The Fourier Theorem states that ***any physical function which varies periodically with time can be expressed by superposition of elementary sine and cosine components of various frequencies*** (see Section 8.5 for Simple Harmonic Motion). The basic expression for Fourier series is as follows:

$$f(t) = a_0 + \sum_{n=1}^{\infty}\left[a_n\cos\left(\frac{2\pi n}{T}t\right) + b_n\sin\left(\frac{2\pi n}{T}t\right)\right]$$

Where: $f(t)$-original source function; a_n and b_n coefficients (precise determination of these coefficients is beyond the scope of this text but by working in the synthetic direction it is possible to develop the feel for their influences); n the index; T-period of the original source function; t-time. As an example of the power of superposition, a summation of two sine functions $f(x) = \sin(x) + 1.2^* \sin(1.8^*x)$ is given in the picture below. It is visible that the resultant (red color) does not resemble precisely any of the components but creates a new quality.

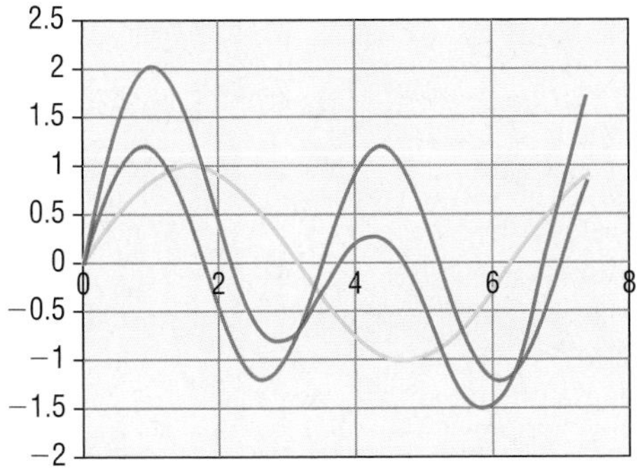

Problem: Build an approximation of a square function.

Given is an expression:

$$f_{\text{square}}(t) = \sin \pi t + \frac{\sin 3\pi t}{3} + \frac{\sin 5\pi t}{5} + \frac{\sin 7\pi t}{7} + \cdots$$

1. Build each component of a series given as a separate function. For example, $f_1 = \sin(\pi^*t)$; $f_n = (1/n)^* \sin(n\pi^*t)$. Investigate variability of each of the functions up to $n = 9$.
2. Start adding them together starting from f_1, then $f_1 + f_3$, then $f_1 + f_3 + f_5$, and so on.
3. Observe and analyze what each consecutive component does to the resultant function.
4. Try to squeeze a cosine component for example between f_1 and f_3: $f_{13} = (1/2)^* \cos(2^*\pi^*t)$. What does it do to the resultant function?
5. An example of a sum of the first four functions is given in the picture below.
6. Go to Yahoo and search for "Fourier series" and have some adventure with very interesting interactive animations.

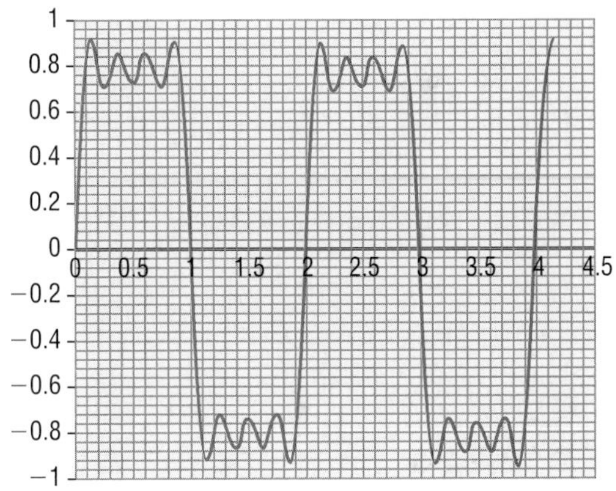

C H A P T E R

9 Polar Coordinates; Vectors

Outline

FIELD TRIP TO MOTOROLA

Cell phones and pagers receive signals from transmission towers that are often far away. Many of these signals are received directly by the cell phone or pager. Some of the signals received are the result of a reflection of the signal from flat surfaces, in most cases the ground.

It is also often the case that the cell phone or pager are not stationary; in fact, the user of such devices is typically in motion while receiving calls. Such motion may be in a car or bus or in an elevator that is going up the side of a building. In the latter case, it is important to be able to calculate and analyze the interference caused by the ground on the signal received by the phone as the phone moves along a vertical path. The result of such an analysis is used to establish certain minimum sensitivity requirements for the phone.

INTERVIEW AT MOTOROLA

Nick Buris was born in Greece and came to the United States in 1982. He studied at the National Technical University of Athens and at North Carolina State University. After his degrees, he joined the Electrical and Computer Engineering faculty at the University of Massachusetts at Amherst. During this appointment he taught and conducted research projects for the government and the defense industry. Since 1992 he has been with Motorola pursuing applied research and software tool development for electromagnetic design of wireless products. He enjoys all kinds of sports and still dreams about taking a year off so that he can sail and travel around the world. He lives in Illinois with his wife and two children.

PREPARING FOR THIS SECTION

Before getting started, review the following:

- ✓ Rectangular Coordinates (pp. 2–3)
- ✓ Definitions of the Sine and Cosine Functions (p. 381)
- ✓ Inverse Tangent Function (pp. 480–482)
- ✓ Completing the Square (Appendix, Section 10, pp. 998–999)

9.1 POLAR COORDINATES

OBJECTIVES
1 Plot Points Using Polar Coordinates
2 Convert from Polar Coordinates to Rectangular Coordinates
2 Convert from Rectangular Coordinates to Polar Coordinates

So far, we have always used a system of rectangular coordinates to plot points in the plane. Now we are ready to describe another system called *polar coordinates.* As we shall soon see, in many instances polar coordinates offer certain advantages over rectangular coordinates.

In a rectangular coordinate system, you will recall, a point in the plane is represented by an ordered pair of numbers (x, y), where x and y equal the signed distance of the point from the y-axis and x-axis, respectively. In a polar coordinate system, we select a point, called the **pole,** and then a ray with vertex at the pole, called the **polar axis.** Comparing the rectangular and polar coordinate systems, we see (in Figure 1) that the origin in rectangular coordinates coincides with the pole in polar coordinates, and the positive x-axis in rectangular coordinates coincides with the polar axis in polar coordinates.

Figure 1

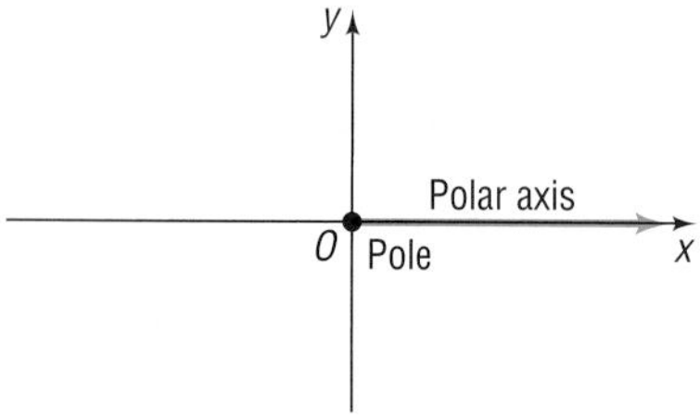

1 A point P in a polar coordinate system is represented by an ordered pair of numbers (r, θ). The number r is the distance of the point from the pole, and θ is an angle (in degrees or radians) formed by the polar axis and a ray from the pole through the point. We call the ordered pair (r, θ) the **polar coordinates** of the point. See Figure 2.

Figure 2

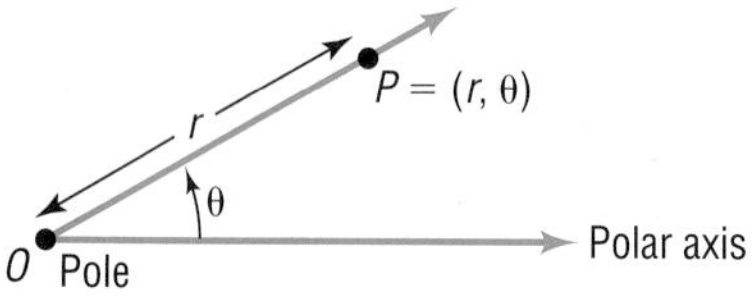

As an example, suppose that the polar coordinates of a point P are $(2, \pi/4)$. We locate P by first drawing an angle of $\pi/4$ radian, placing its ver-

Figure 3

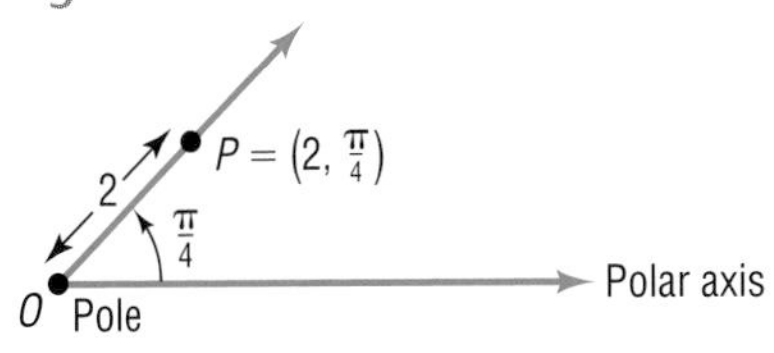

tex at the pole and its initial side along the polar axis. Then we go out a distance of 2 units along the terminal side of the angle to reach the point P. See Figure 3.

NOW WORK PROBLEM 9.

Recall that an angle measured counterclockwise is positive, whereas one measured clockwise is negative. This convention has some interesting consequences relating to polar coordinates. Let's see what these consequences are.

EXAMPLE 1 Finding Several Polar Coordinates of a Single Point

Consider again the point P with polar coordinates $(2, \pi/4)$, as shown in Figure 4(a). Because $\pi/4$, $9\pi/4$, and $-7\pi/4$ all have the same terminal side, we also could have located this point P by using the polar coordinates $(2, 9\pi/4)$ or $(2, -7\pi/4)$, as shown in Figures 4(b) and (c).

Figure 4

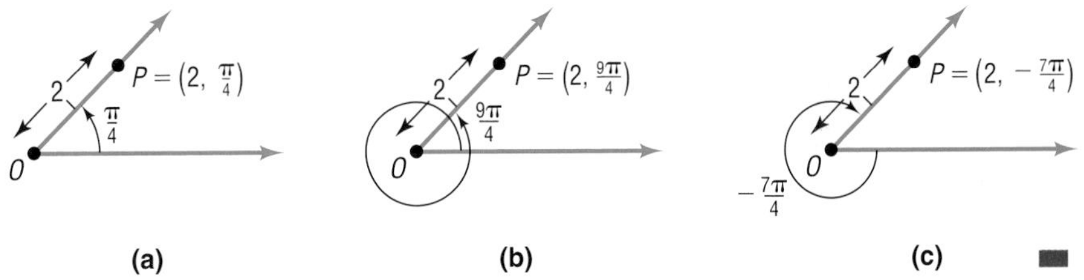

Figure 5

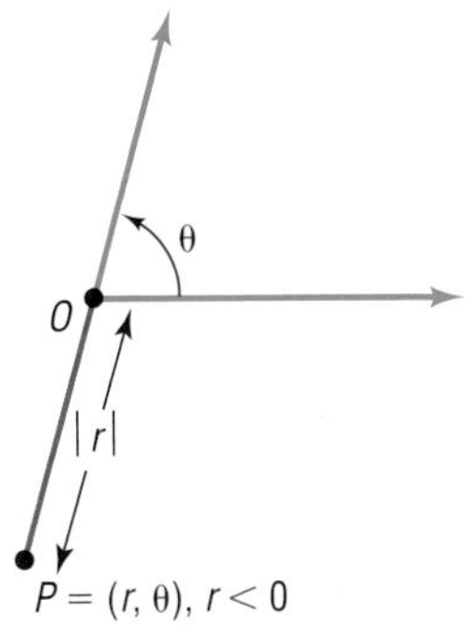

In using polar coordinates (r, θ), it is possible for the first entry r to be negative. When this happens, we follow the convention that the location of the point, instead of being on the terminal side of θ, is on the ray from the pole extending in the direction *opposite* the terminal side of θ at a distance $|r|$ from the pole. See Figure 5 for an illustration.

EXAMPLE 2 Polar Coordinates (r, θ), $r < 0$

Consider again the point P with polar coordinates $(2, \pi/4)$, as shown in Figure 6(a). This same point P can be assigned the polar coordinates $(-2, 5\pi/4)$, as indicated in Figure 6(b). To locate the point $(-2, 5\pi/4)$, we use the ray in the opposite direction of $5\pi/4$ and go out 2 units along that ray to find the point P.

Figure 6

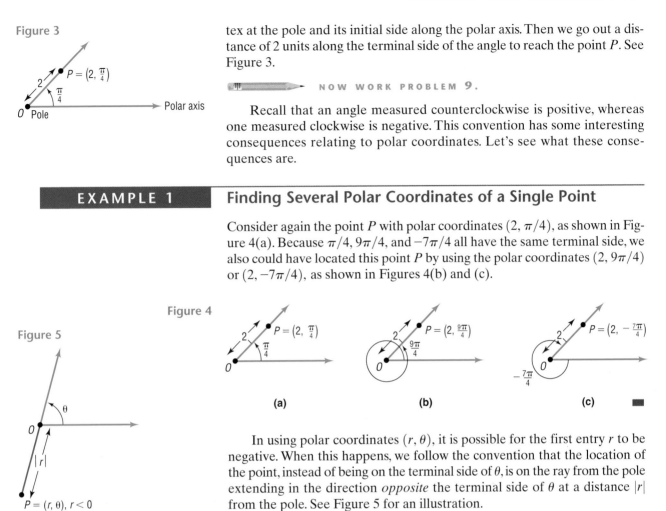

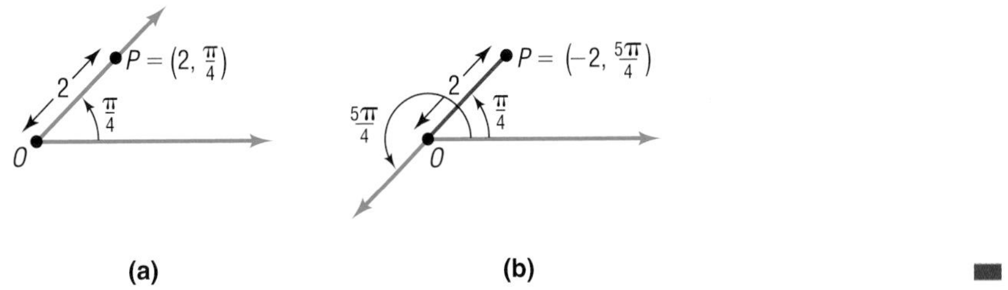

These examples show a major difference between rectangular coordinates and polar coordinates. In the former, each point has exactly one pair of rectangular coordinates; in the latter, a point can have infinitely many pairs of polar coordinates.

SUMMARY

A point with polar coordinates (r, θ) also can be represented by any of the following:

$$(r, \theta + 2k\pi) \quad \text{or} \quad (-r, \theta + \pi + 2k\pi), \qquad k \text{ any integer}$$

The polar coordinates of the pole are $(0, \theta)$, where θ can be any angle.

EXAMPLE 3 Plotting Points Using Polar Coordinates

Plot the points with the following polar coordinates:

(a) $(3, 5\pi/3)$ (b) $(2, -\pi/4)$ (c) $(3, 0)$ (d) $(-2, \pi/4)$

Solution Figure 7 shows the points.

Figure 7

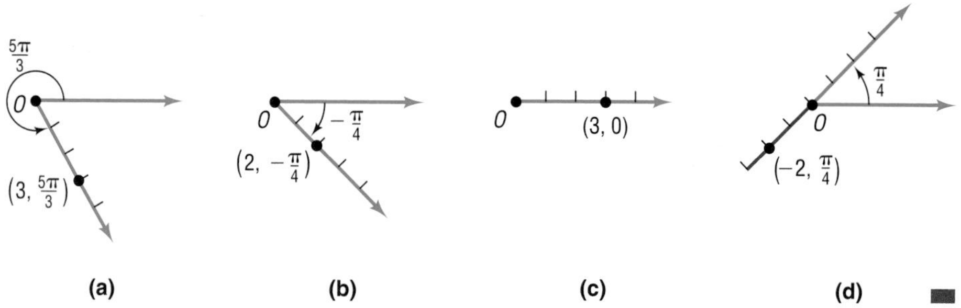

NOW WORK PROBLEMS 1 AND 17.

EXAMPLE 4 Finding Other Polar Coordinates of a Given Point

Plot the point P with polar coordinates $(3, \pi/6)$, and find other polar coordinates (r, θ) of this same point for which:

(a) $r > 0, \quad 2\pi \le \theta < 4\pi$ (b) $r < 0, \quad 0 \le \theta < 2\pi$
(c) $r > 0, \quad -2\pi \le \theta < 0$

Solution The point $(3, \pi/6)$ is plotted in Figure 8.

Figure 8

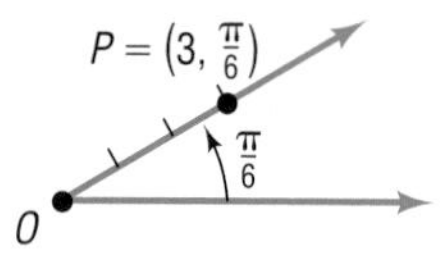

(a) We add 1 revolution (2π radians) to the angle $\pi/6$ to get $P = (3, \pi/6 + 2\pi) = (3, 13\pi/6)$. See Figure 9.

(b) We add $\frac{1}{2}$ revolution (π radians) to the angle $\pi/6$ and replace 3 by -3 to get $P = (-3, \pi/6 + \pi) = (-3, 7\pi/6)$. See Figure 10.

(c) We subtract 2π from the angle $\pi/6$ to get $P = (3, \pi/6 - 2\pi) = (3, -11\pi/6)$. See Figure 11.

Figure 9 Figure 10 Figure 11

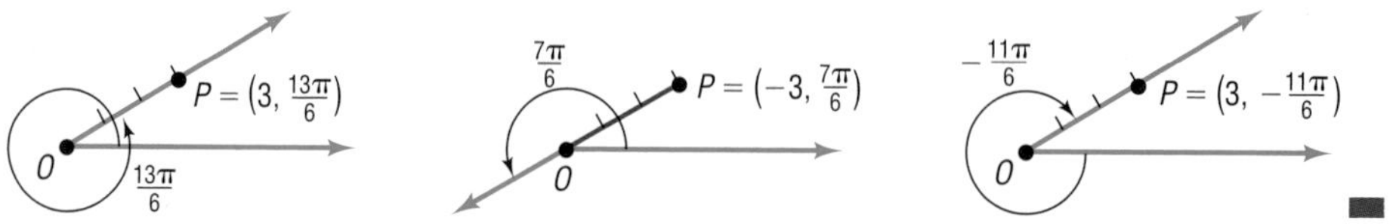

NOW WORK PROBLEM 21.

Conversion from Polar Coordinates to Rectangular Coordinates, and Vice Versa

2 It is sometimes convenient and, indeed, necessary to be able to convert coordinates or equations in rectangular form to polar form, and vice versa. To do this, we recall that the origin in rectangular coordinates is the pole in polar coordinates and that the positive x-axis in rectangular coordinates is the polar axis in polar coordinates.

Theorem **Conversion from Polar Coordinates to Rectangular Coordinates**

If P is a point with polar coordinates (r, θ), the rectangular coordinates (x, y) of P are given by

$$x = r\cos\theta \qquad y = r\sin\theta \tag{1}$$

■

Figure 12

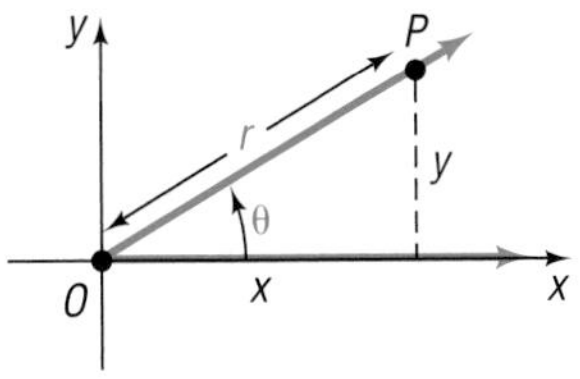

Proof Suppose that P has the polar coordinates (r, θ). We seek the rectangular coordinates (x, y) of P. Refer to Figure 12.

If $r = 0$, then, regardless of θ, the point P is the pole, for which the rectangular coordinates are $(0, 0)$. Formula (1) is valid for $r = 0$.

If $r > 0$, the point P is on the terminal side of θ and $r = d(O, P) = \sqrt{x^2 + y^2}$. Since

$$\cos\theta = \frac{x}{r} \qquad \sin\theta = \frac{y}{r}$$

we have

$$x = r\cos\theta \qquad y = r\sin\theta$$

If $r < 0$, then the point $P = (r, \theta)$ can be represented as $(-r, \pi + \theta)$, where $-r > 0$. Since

$$\cos(\pi + \theta) = -\cos\theta = \frac{x}{-r} \qquad \sin(\pi + \theta) = -\sin\theta = \frac{y}{-r}$$

we have

$$x = r\cos\theta \qquad y = r\sin\theta$$

■

EXAMPLE 5 **Converting from Polar Coordinates to Rectangular Coordinates**

Find the rectangular coordinates of the points with the following polar coordinates:

(a) $(6, \pi/6)$ (b) $(-4, -\pi/4)$

Figure 13

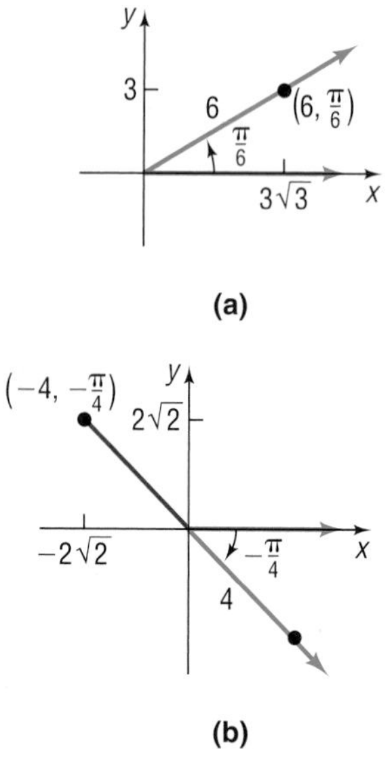

Solution We use formula (1): $x = r\cos\theta$ and $y = r\sin\theta$.

(a) Figure 13(a) shows $(6, \pi/6)$ plotted. With $r = 6$ and $\theta = \pi/6$, we have

$$x = r\cos\theta = 6\cos\frac{\pi}{6} = 6 \cdot \frac{\sqrt{3}}{2} = 3\sqrt{3}$$

$$y = r\sin\theta = 6\sin\frac{\pi}{6} = 6 \cdot \frac{1}{2} = 3$$

The rectangular coordinates of the point $(6, \pi/6)$ are $(3\sqrt{3}, 3)$.

(b) Figure 13(b) shows $(-4, -\pi/4)$ plotted. With $r = -4$ and $\theta = -\pi/4$, we have

$$x = r\cos\theta = -4\cos\left(-\frac{\pi}{4}\right) = -4 \cdot \frac{\sqrt{2}}{2} = -2\sqrt{2}$$

$$y = r\sin\theta = -4\sin\left(-\frac{\pi}{4}\right) = -4\left(-\frac{\sqrt{2}}{2}\right) = 2\sqrt{2}$$

The rectangular coordinates of the point $(-4, -\pi/4)$ are $(-2\sqrt{2}, 2\sqrt{2})$. ■

Figure 14

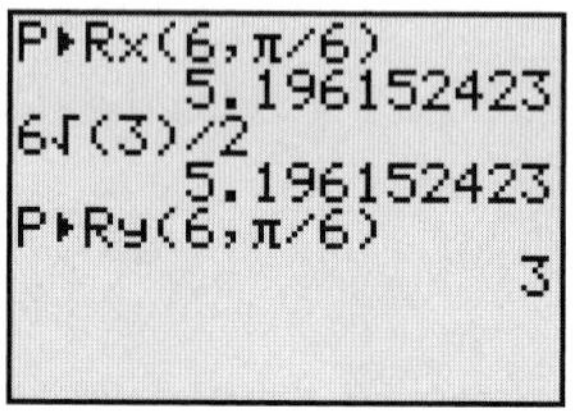

Most graphing calculators have the capability of converting from polar coordinates to rectangular coordinates. Consult you owner's manual for the proper keystrokes. Figure 14 verifies the results obtained in Example 5(a) using a TI-83. Note that the calculator is in radian mode.

NOW WORK PROBLEMS 29 AND 41.

3 To convert from rectangular coordinates (x, y) to polar coordinates (r, θ) is a little more complicated. Notice that we begin each example by plotting the given rectangular coordinates.

EXAMPLE 6 Converting from Rectangular Coordinates to Polar Coordinates

Figure 15

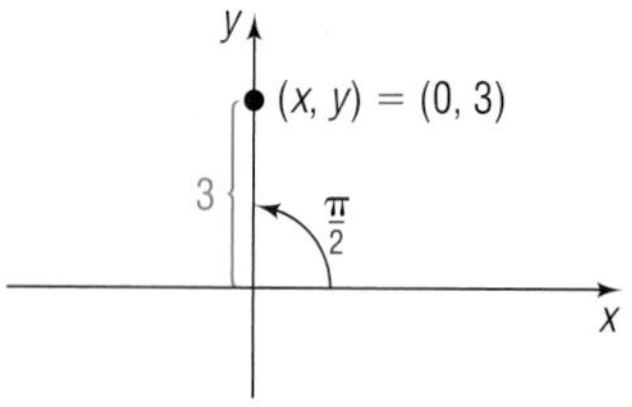

Find polar coordinates of a point whose rectangular coordinates are $(0, 3)$.

Solution See Figure 15. The point $(0, 3)$ lies on the y-axis a distance of 3 units from the origin (pole), so $r = 3$. A ray with vertex at the pole through $(0, 3)$ forms an angle $\theta = \frac{\pi}{2}$ with the polar axis. Polar coordinates for this point can be given by $(3, \pi/2)$. ■

Figure 16

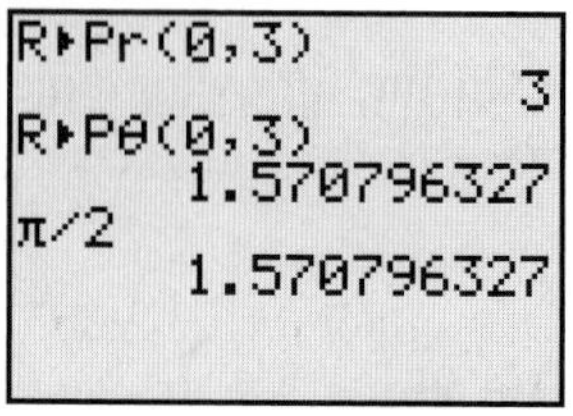

Most graphing calculators have the capability of converting from rectangular coordinates to polar coordinates. Consult you owner's manual for the proper keystrokes. Figure 16 verifies the results obtained in Example 6 using a TI-83. Note that the calculator is in radian mode.

Figure 17 shows polar coordinates of points that lie on either the x-axis or the y-axis. In each illustration, $a > 0$.

Figure 17

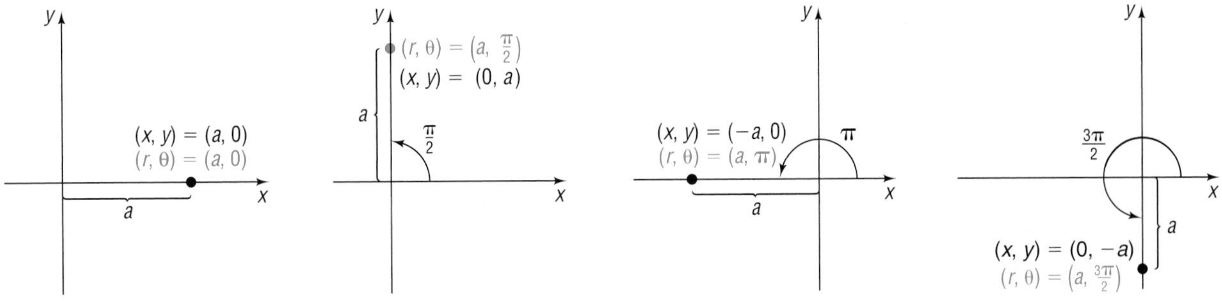

(a) $(x, y) = (a, 0), a > 0$ (b) $(x, y) = (0, a), a > 0$ (c) $(x, y) = (-a, 0), a > 0$ (d) $(x, y) = (0, -a), a > 0$

NOW WORK PROBLEM 45.

EXAMPLE 7 Converting from Rectangular Coordinates to Polar Coordinates

Find polar coordinates of a point whose rectangular coordinates are:
(a) $(2, -2)$ (b) $(-1, -\sqrt{3})$

Solution (a) See Figure 18(a). The distance r from the origin to the point $(2, -2)$ is

$$r = \sqrt{x^2 + y^2} = \sqrt{(2)^2 + (-2)^2} = \sqrt{8} = 2\sqrt{2}$$

We find θ by recalling that $\tan\theta = \frac{y}{x}$, so $\theta = \tan^{-1}\frac{y}{x}, -\frac{\pi}{2} < \theta < \frac{\pi}{2}$. Since $(2, -2)$ lies in quadrant IV, we know that $-\frac{\pi}{2} < \theta < 0$. As a result,

$$\theta = \tan^{-1}\frac{y}{x} = \tan^{-1}\frac{-2}{2} = \tan^{-1}(-1) = -\frac{\pi}{4}$$

A set of polar coordinates for this point is $(2\sqrt{2}, -\pi/4)$. Other possible representations include $(2\sqrt{2}, 7\pi/4)$ and $(-2\sqrt{2}, 3\pi/4)$.

Figure 18

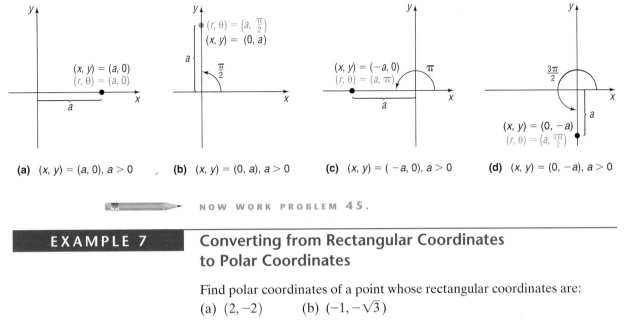

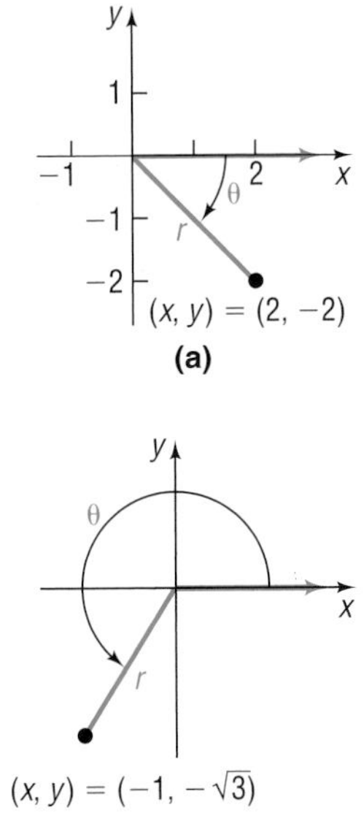

(b) See Figure 18(b). The distance r from the origin to the point $(-1, -\sqrt{3})$ is

$$r = \sqrt{(-1)^2 + (-\sqrt{3})^2} = \sqrt{4} = 2$$

To find θ, we use $\theta = \tan^{-1}\frac{y}{x}, -\frac{\pi}{2} < \theta < \frac{\pi}{2}$. Since the point $(-1, -\sqrt{3})$ lies in quadrant III and the inverse tangent function gives an angle in quadrant I, we add π to the result to obtain an angle in quadrant III.

$$\theta = \pi + \tan^{-1}\left(\frac{-\sqrt{3}}{-1}\right) = \pi + \tan^{-1}\sqrt{3} = \pi + \frac{\pi}{3} = \frac{4\pi}{3}$$

A set of polar coordinates is $(2, 4\pi/3)$. Other possible representations include $(-2, \pi/3)$ and $(2, -2\pi/3)$. ■

Check: Verify the results obtained in Example 7 using a graphing utility. Which polar representation of the rectangular coordinate did the utility provide? Can you explain why?

Figure 19 shows how to find polar coordinates of a point that lies in a quadrant when its rectangular coordinates (x, y) are given.

Figure 19

(a) $r = \sqrt{x^2 + y^2}$
$\theta = \tan^{-1}\frac{y}{x}$

(b) $r = \sqrt{x^2 + y^2}$
$\theta = \pi + \tan^{-1}\frac{y}{x}$

(c) $r = \sqrt{x^2 + y^2}$
$\theta = \pi + \tan^{-1}\frac{y}{x}$

(d) $r = \sqrt{x^2 + y^2}$
$\theta = \tan^{-1}\frac{y}{x}$

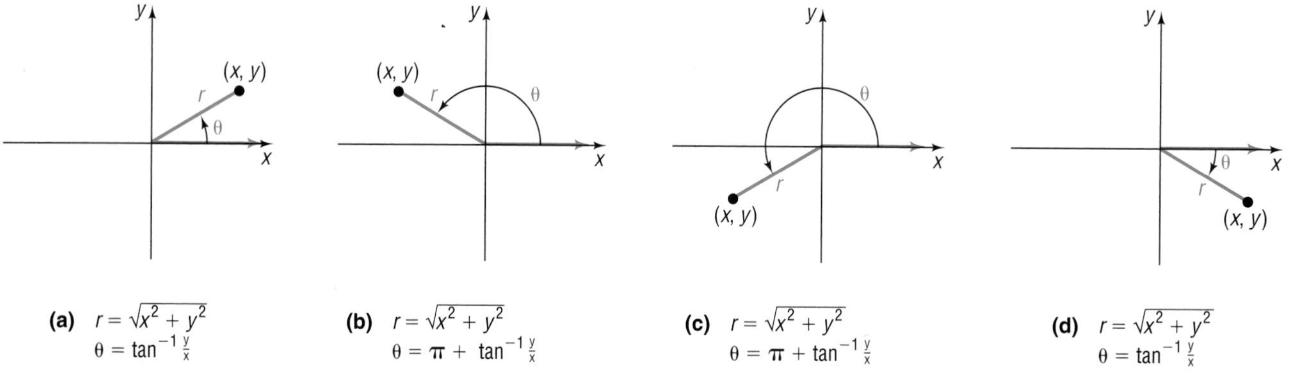

Based on the preceding discussion, we have the formulas

$$r^2 = x^2 + y^2 \qquad \tan\theta = \frac{y}{x} \qquad \text{if } x \neq 0 \qquad \textbf{(2)}$$

To use formula (2) effectively, follow these steps:

STEPS FOR CONVERTING FROM RECTANGULAR TO POLAR COORDINATES

STEP 1: Always plot the point (x, y) first, as we did in Examples 6 and 7.

STEP 2: To find r, compute the distance from the origin to (x, y).

STEP 3: To find θ, first determine the quadrant that the point lies in.

Quadrant I: $\theta = \tan^{-1}\frac{y}{x}$ Quadrant II: $\theta = \pi + \tan^{-1}\frac{y}{x}$

Quadrant III: $\theta = \pi + \tan^{-1}\frac{y}{x}$ Quadrant IV: $\theta = \tan^{-1}\frac{y}{x}$

NOW WORK PROBLEM **49**.

Formulas (1) and (2) may also be used to transform equations.

EXAMPLE 8

Transforming an Equation from Polar to Rectangular Form

Transform the equation $r = 4\sin\theta$ from polar coordinates to rectangular coordinates, and identify the graph.

Solution   If we multiply each side by r, it will be easier to apply formulas (1) and (2).

$$r = 4\sin\theta$$

$$r^2 = 4r\sin\theta \qquad \text{Multiply each side by } r.$$

$$x^2 + y^2 = 4y \qquad \text{Apply formulas (1) and (2).}$$

This is the equation of a circle; we proceed to complete the square to obtain the standard form of the equation.

$$x^2 + (y^2 - 4y) = 0 \quad \text{General form.}$$
$$x^2 + (y^2 - 4y + 4) = 4 \quad \text{Complete the square in } y.$$
$$x^2 + (y - 2)^2 = 4 \quad \text{Standard form.}$$

The center of the circle is at $(0, 2)$, and its radius is 2. ■

NOW WORK PROBLEM 65.

EXAMPLE 9 Transforming an Equation from Rectangular to Polar Form

Transform the equation $4xy = 9$ from rectangular coordinates to polar coordinates.

Solution We use formula (1).

$$4xy = 9$$
$$4(r\cos\theta)(r\sin\theta) = 9 \quad \text{Formula (1)}$$
$$4r^2\cos\theta\sin\theta = 9$$
$$2r^2\sin(2\theta) = 9 \quad \text{Double-angle formula}$$
■

9.1 EXERCISES

In Problems 1–8, match each point in polar coordinates with either A, B, C, or D on the graph.

1. $\left(2, \frac{-11\pi}{6}\right)$ **2.** $\left(-2, \frac{-\pi}{6}\right)$ **3.** $\left(-2, \frac{\pi}{6}\right)$ **4.** $\left(2, \frac{7\pi}{6}\right)$

5. $\left(2, \frac{5\pi}{6}\right)$ **6.** $\left(-2, \frac{5\pi}{6}\right)$ **7.** $\left(-2, \frac{7\pi}{6}\right)$ **8.** $\left(2, \frac{11\pi}{6}\right)$

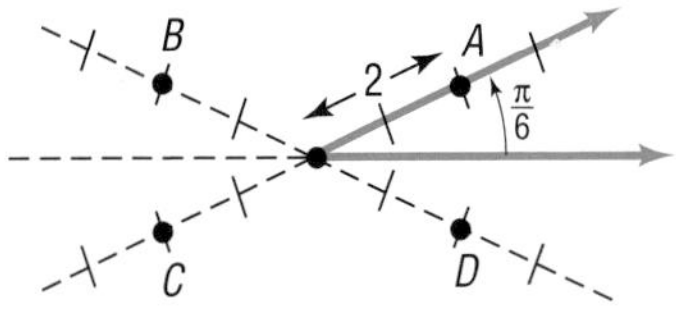

In Problems 9–20, plot each point given in polar coordinates.

9. $(3, 90°)$ **10.** $(4, 270°)$ **11.** $(-2, 0)$ **12.** $(-3, \pi)$
13. $(6, \pi/6)$ **14.** $(5, 5\pi/3)$ **15.** $(-2, 135°)$ **16.** $(-3, 120°)$
17. $(-1, -\pi/3)$ **18.** $(-3, -3\pi/4)$ **19.** $(-2, -\pi)$ **20.** $(-3, -\pi/2)$

In Problems 21–28, plot each point given in polar coordinates, and find other polar coordinates (r, θ) of the point for which:
(a) $r > 0, \ -2\pi \le \theta < 0$ *(b)* $r < 0, \ 0 \le \theta < 2\pi$ *(c)* $r > 0, \ 2\pi \le \theta < 4\pi$

21. $(5, 2\pi/3)$ **22.** $(4, 3\pi/4)$ **23.** $(-2, 3\pi)$ **24.** $(-3, 4\pi)$
25. $(1, \pi/2)$ **26.** $(2, \pi)$ **27.** $(-3, -\pi/4)$ **28.** $(-2, -2\pi/3)$

In Problems 29–44, polar coordinates of a point are given. Find the rectangular coordinates of each point.

29. $(3, \pi/2)$ **30.** $(4, 3\pi/2)$ **31.** $(-2, 0)$ **32.** $(-3, \pi)$
33. $(6, 150°)$ **34.** $(5, 300°)$ **35.** $(-2, 3\pi/4)$ **36.** $(-3, 2\pi/3)$
37. $(-1, -\pi/3)$ **38.** $(-3, -3\pi/4)$ **39.** $(-2, -180°)$ **40.** $(-3, -90°)$
41. $(7.5, 110°)$ **42.** $(-3.1, 182°)$ **43.** $(6.3, 3.8)$ **44.** $(8.1, 5.2)$

In Problems 45–56, the rectangular coordinates of a point are given. Find polar coordinates for each point.

45. $(3, 0)$ **46.** $(0, 2)$ **47.** $(-1, 0)$ **48.** $(0, -2)$
49. $(1, -1)$ **50.** $(-3, 3)$ **51.** $(\sqrt{3}, 1)$ **52.** $(-2, -2\sqrt{3})$
53. $(1.3, -2.1)$ **54.** $(-0.8, -2.1)$ **55.** $(8.3, 4.2)$ **56.** $(-2.3, 0.2)$

In Problems 57–64, the letters x and y represent rectangular coordinates. Write each equation using polar coordinates (r, θ).

57. $2x^2 + 2y^2 = 3$ **58.** $x^2 + y^2 = x$ **59.** $x^2 = 4y$ **60.** $y^2 = 2x$

61. $2xy = 1$ **62.** $4x^2y = 1$ **63.** $x = 4$ **64.** $y = -3$

In Problems 65–72, the letters r and θ represent polar coordinates. Write each equation using rectangular coordinates (x, y).

65. $r = \cos\theta$ **66.** $r = \sin\theta + 1$ **67.** $r^2 = \cos\theta$ **68.** $r = \sin\theta - \cos\theta$

69. $r = 2$ **70.** $r = 4$ **71.** $r = \dfrac{4}{1 - \cos\theta}$ **72.** $r = \dfrac{3}{3 - \cos\theta}$

73. Show that the formula for the distance d between two points $P_1 = (r_1, \theta_1)$ and $P_2 = (r_2, \theta_2)$ is

$$d = \sqrt{r_1^2 + r_2^2 - 2r_1r_2\cos(\theta_2 - \theta_1)}$$

PREPARING FOR THIS SECTION

Before getting started, review the following:

- ✓ Graphs of Equations (Section 1.2)
- ✓ Even/Odd Properties of Trigonometric Functions (pp. 404–405)
- ✓ Circles (Section 1.8)
- ✓ Difference Formulas (pp. 452 and 455)

9.2 POLAR EQUATIONS AND GRAPHS

OBJECTIVES
1. Graph and Identify Polar Equations by Converting to Rectangular Equations
2. Graph Polar Equations Using a Graphing Utility
3. Test Polar Equations for Symmetry
4. Graph Polar Equations by Plotting Points

Just as a rectangular grid may be used to plot points given by rectangular coordinates, as in Figure 20(a), we can use a grid consisting of concentric circles (with centers at the pole) and rays (with vertices at the pole) to plot points given by polar coordinates, as shown in Figure 20(b). We shall use such **polar grids** to graph *polar equations.*

Figure 20

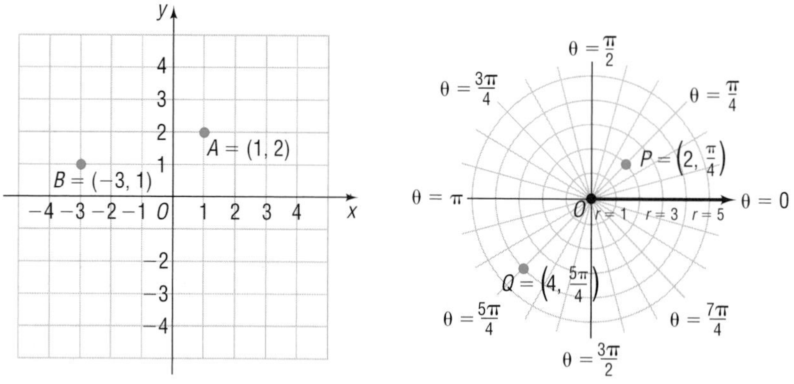

(a) Rectangular grid **(b)** Polar grid

An equation whose variables are polar coordinates is called a **polar equation.** The **graph of a polar equation** consists of all points whose polar coordinates satisfy the equation.

1 One method we can use to graph a polar equation is to convert the equation to rectangular coordinates. In the discussion that follows, (x, y) represent the rectangular coordinates of a point P, and (r, θ) represent polar coordinates of the point P.

EXAMPLE 1 Identifying and Graphing a Polar Equation by Hand (Circle)

Identify and graph the equation: $r = 3$

Solution We convert the polar equation to a rectangular equation.

$$r = 3$$

$$r^2 = 9 \quad \text{Square both sides}$$

$$x^2 + y^2 = 9 \quad \text{Formula (2), Section 9.1, p. 566}$$

The graph of $r = 3$ is a circle, with center at the pole and radius 3. See Figure 21.

Figure 21
$r = 3$ or $x^2 + y^2 = 9$

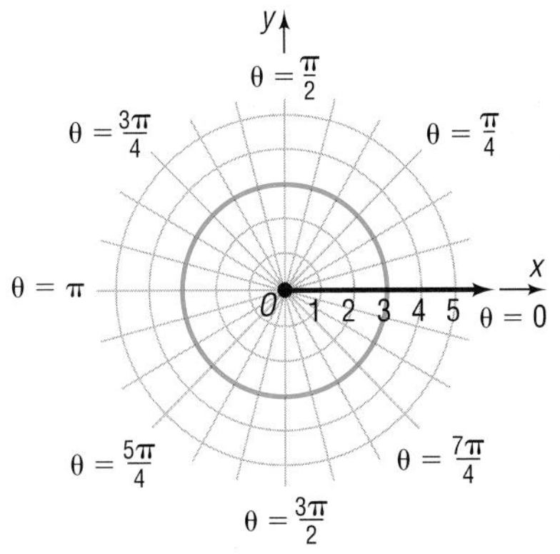

NOW WORK PROBLEM 1.

EXAMPLE 2 Identifying and Graphing a Polar Equation by Hand (Line)

Identify and graph the equation: $\theta = \pi/4$

Solution We convert the polar equation to a rectangular equation.

$$\theta = \frac{\pi}{4}$$

$$\tan\theta = \tan\frac{\pi}{4} = 1$$

$$\frac{y}{x} = 1 \quad \text{Formula (2), Section 9.1, p. 566.}$$

$$y = x$$

The graph of $\theta = \pi/4$ is a line passing through the pole making an angle of $\pi/4$ with the polar axis. See Figure 22.

Figure 22
$\theta = \frac{\pi}{4}$ or $y = x$

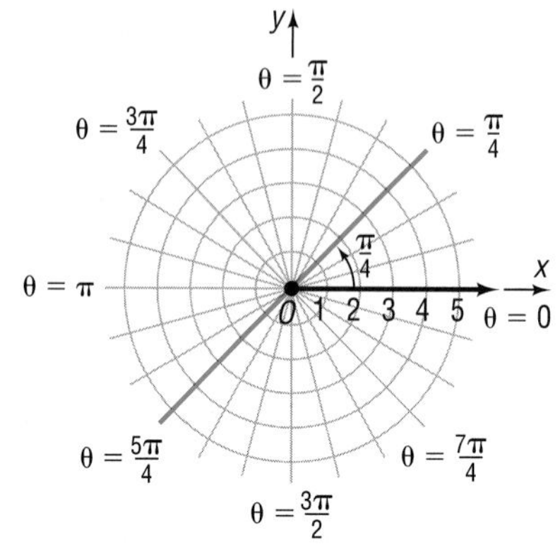

NOW WORK PROBLEM 3.

EXAMPLE 3 Identifying and Graphing a Polar Equation by Hand (Horizontal Line)

Identify and graph the equation: $r \sin\theta = 2$

Solution Since $y = r\sin\theta$ [Formula (1), p. 563], we can write the equation as

$$y = 2$$

We conclude that the graph of $r\sin\theta = 2$ is a horizontal line 2 units above the pole. See Figure 23.

Figure 23
$r\sin\theta = 2$ or $y = 2$

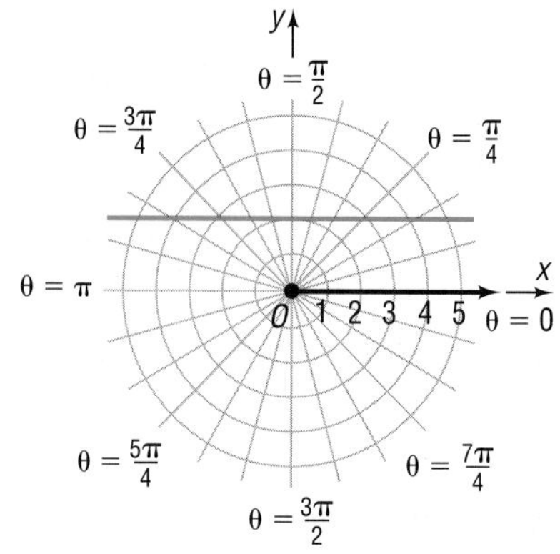

2 A second method that we can use to graph a polar equation is to graph the equation using a graphing utility.

Most graphing utilities require the following steps in order to obtain the graph of an equation:

GRAPHING A POLAR EQUATION USING A GRAPHING UTILITY

STEP 1: Solve the equation for r in terms of θ.

STEP 2: Select the viewing window in POLar mode. In addition to setting Xmin, Xmax, Xscl, and so forth, the viewing window in polar mode requires setting minimum and maximum values for θ and an increment setting for θ (θstep). Finally, a square screen should be used.

STEP 3: Enter the expression involving θ that you found in Step 1. (Consult your manual for the correct way to enter the expression.)

STEP 4: Execute.

EXAMPLE 4 Graphing a Polar Equation Using a Graphing Utility

Use a graphing utility to graph the polar equation $r \sin\theta = 2$.

Solution **STEP 1:** We solve the equation for r in terms of θ.

$$r \sin\theta = 2$$

$$r = \frac{2}{\sin\theta}$$

STEP 2: From the polar and radian mode, select a square viewing window. We will use the one given next.

$$\theta\text{min} = 0 \qquad X\text{min} = -9 \qquad Y\text{min} = -6$$

$$\theta\text{max} = 2\pi \qquad X\text{max} = 9 \qquad Y\text{max} = 6$$

$$\theta\text{step} = \pi/24 \qquad X\text{scl} = 1 \qquad Y\text{scl} = 1$$

θstep determines the number of points that the graphing utility will plot. For example, if θstep is $\pi/24$, then the graphing utility will evaluate r at $\theta = 0$ (θmin), $\pi/24$, $2\pi/24$, $3\pi/24$, and so forth, up to 2π (θmax). The smaller θstep is, the more points the graphing utility will plot. The student is encouraged to experiment with different values for θmin, θmax, and θstep to see how the graph is affected.

Figure 24

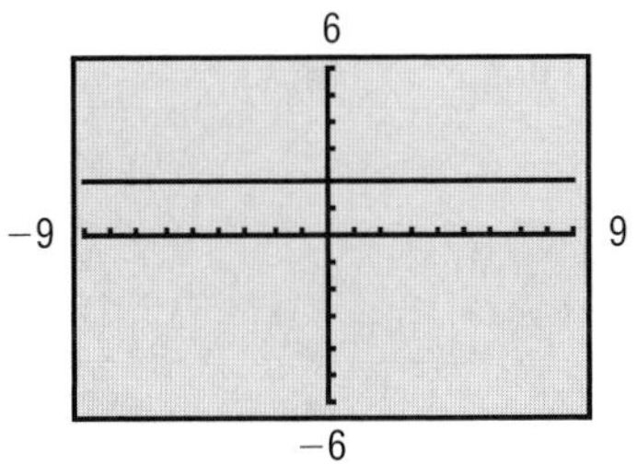

STEP 3: Enter the expression $\frac{2}{\sin\theta}$ after the prompt $r =$.

STEP 4: Execute.

The graph is shown in Figure 24. ■

EXAMPLE 5 Identifying and Graphing a Polar Equation (Vertical Line)

Identify and graph the equation: $r \cos\theta = -3$

Solution Since $x = r\cos\theta$ [Formula (1), p. 563], we can write the equation as

$$x = -3$$

We conclude that the graph of $r \cos\theta = -3$ is a vertical line 3 units to the left of the pole. Figure 25(a) shows the graph drawn by hand. Figure 25(b) shows the graph using a graphing utility with $\theta\text{min} = 0$, $\theta\text{max} = 2\pi$, and $\theta\text{step} = \pi/24$.

Figure 25
$r\cos\theta = -3$ or $x = -3$

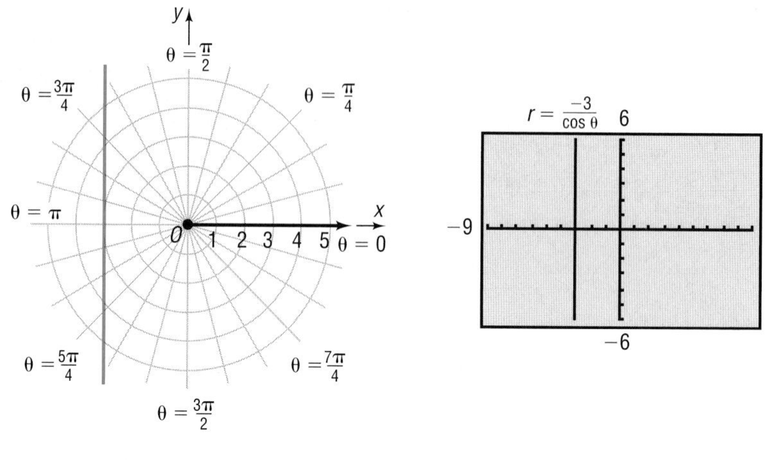

(a) (b)

Based on Examples 3, 4, and 5, we are led to the following results. (The proofs are left as exercises.)

Theorem Let a be a nonzero real number. Then the graph of the equation

$$r\sin\theta = a$$

is a horizontal line a units above the pole if $a > 0$ and $|a|$ units below the pole if $a < 0$.

The graph of the equation

$$r\cos\theta = a$$

is a vertical line a units to the right of the pole if $a > 0$ and $|a|$ units to the left of the pole if $a < 0$.

NOW WORK PROBLEM 7.

EXAMPLE 6 Identifying and Graphing a Polar Equation (Circle)

Identify and graph the equation: $r = 4\sin\theta$

Solution To transform the equation to rectangular coordinates, we multiply each side by r.

$$r^2 = 4r\sin\theta$$

Now we use the facts that $r^2 = x^2 + y^2$ and $y = r\sin\theta$. Then

$$x^2 + y^2 = 4y$$

$$x^2 + (y^2 - 4y) = 0$$

$$x^2 + (y^2 - 4y + 4) = 4 \quad \text{Complete the square in } y.$$

$$x^2 + (y - 2)^2 = 4 \quad \text{Standard equation of a circle.}$$

This is the equation of a circle with center at $(0, 2)$ in rectangular coordinates and radius 2. Figure 26(a) shows the graph drawn by hand. Figure 26(b) shows the graph using a graphing utility with $\theta\text{min} = 0$, $\theta\text{max} = 2\pi$, and $\theta\text{step} = \frac{\pi}{24}$.

Figure 26
$r = 4 \sin\theta$ or $x^2 + (y - 2)^2 = 4$

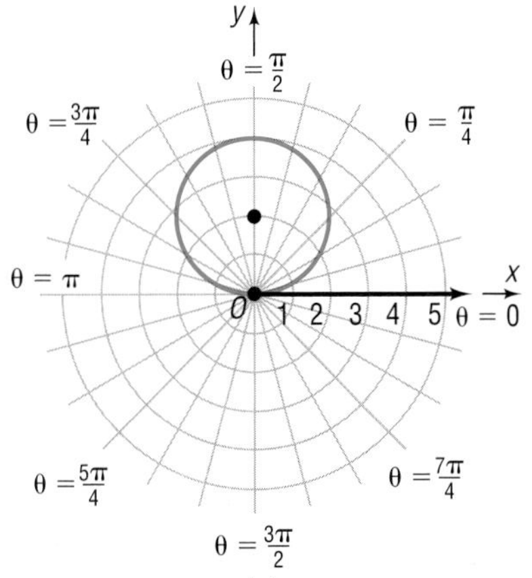

(a)

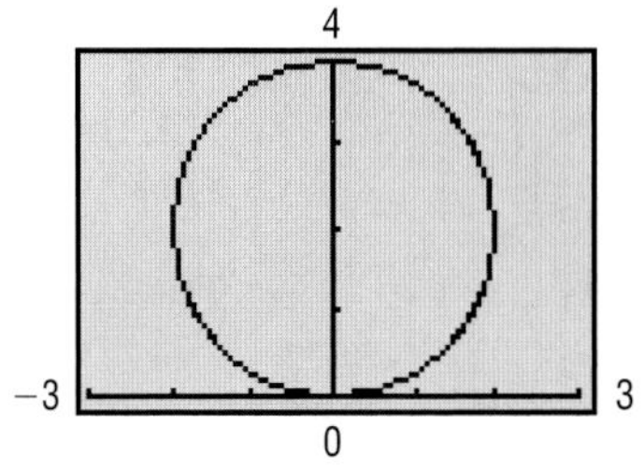

(b)

EXAMPLE 7 Identifying and Graphing a Polar Equation (Circle)

Identify and graph the equation: $r = -2\cos\theta$

Solution We proceed as in Example 6.

$$r^2 = -2r\cos\theta \quad \text{Multiply both sides by } r.$$
$$x^2 + y^2 = -2x$$
$$x^2 + 2x + y^2 = 0$$
$$(x^2 + 2x + 1) + y^2 = 1 \quad \text{Complete the square in } x.$$
$$(x + 1)^2 + y^2 = 1 \quad \text{Standard equation of a circle.}$$

This is the equation of a circle with center at $(-1, 0)$ in rectangular coordinates and radius 1. Figure 27(a) shows the graph drawn by hand. Figure 27(b) shows the graph using a graphing utility with $\theta\text{min} = 0$, $\theta\text{max} = 2\pi$, and $\theta\text{step} = \frac{\pi}{24}$.

Figure 27
$r = -2\cos\theta$ or $(x + 1)^2 + y^2 = 1$

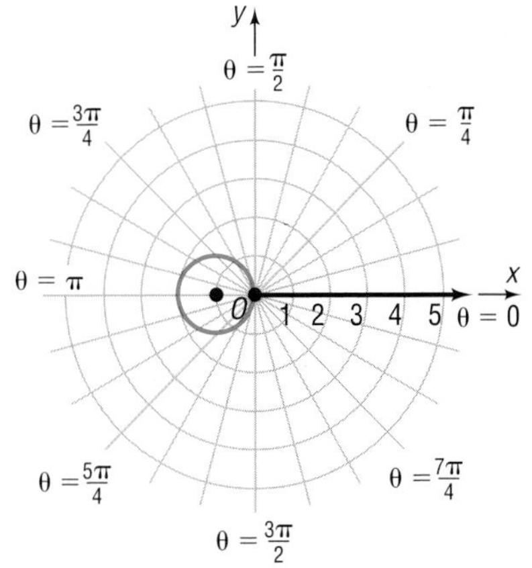

(a)

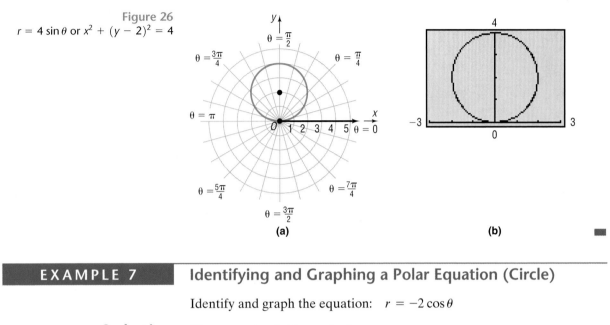

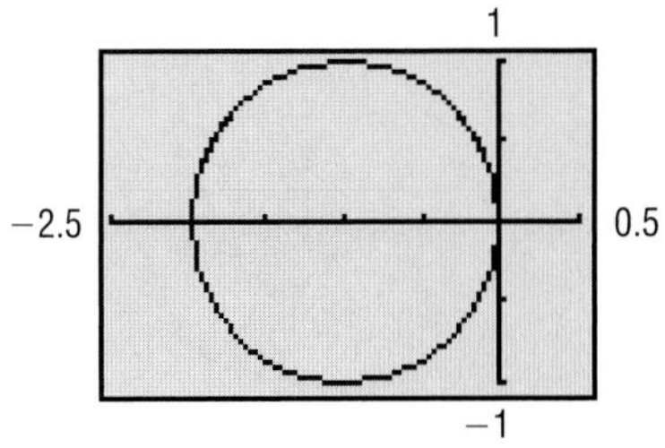

(b)

Exploration Using a square screen, graph $r_1 = \sin\theta$, $r_2 = 2\sin\theta$, and $r_3 = 3\sin\theta$. Do you see the pattern? Clear the screen and graph $r_1 = -\sin\theta$, $r_2 = -2\sin\theta$, and $r_3 = -3\sin\theta$. Do you see the pattern? Clear the screen and graph $r_1 = \cos\theta$, $r_2 = 2\cos\theta$, and $r_3 = 3\cos\theta$. Do you see the pattern? Clear the screen and graph $r_1 = -\cos\theta$, $r_2 = -2\cos\theta$, and $r_3 = -3\cos\theta$. Do you see the pattern?

Based on Examples 6 and 7 and the preceding Exploration, we are led to the following results. (The proofs are left as exercises).

Theorem Let a be a positive real number. Then,

Equation	**Description**
(a) $r = 2a\sin\theta$	Circle: radius a; center at $(0, a)$ in rectangular coordinates
(b) $r = -2a\sin\theta$	Circle: radius a; center at $(0, -a)$ in rectangular coordinates
(c) $r = 2a\cos\theta$	Circle: radius a; center at $(a, 0)$ in rectangular coordinates
(d) $r = -2a\cos\theta$	Circle: radius a; center at $(-a, 0)$ in rectangular coordinates

Each circle passes through the pole.

NOW WORK PROBLEM 9.

The method of converting a polar equation to an identifiable rectangular equation in order to graph it is not always helpful, nor is it always necessary. Usually, we set up a table that lists several points on the graph. By checking for symmetry, it may be possible to reduce the number of points needed to draw the graph.

Symmetry

3 In polar coordinates, the points (r, θ) and $(r, -\theta)$ are symmetric with respect to the polar axis (and to the x-axis). See Figure 28(a). The points (r, θ) and $(r, \pi - \theta)$ are symmetric with respect to the line $\theta = \pi/2$ (the y-axis). See Figure 28(b). The points (r, θ) and $(-r, \theta)$ are symmetric with respect to the pole (the origin). See Figure 28(c).

Figure 28

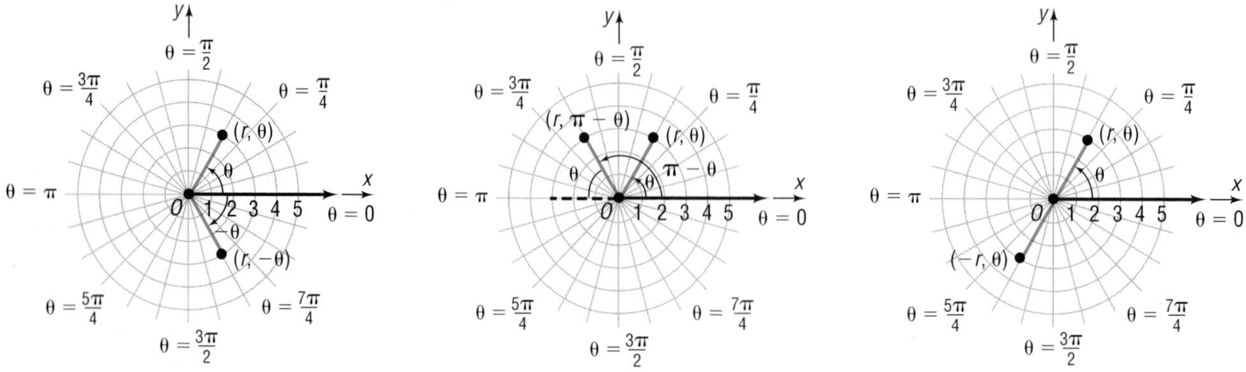

(a) Points symmetric with respect to the polar axis

(b) Points symmetric with respect to the line $\theta = \frac{\pi}{2}$

(c) Points symmetric with respect to the pole

The following tests are a consequence of these observations.

Theorem **Tests for Symmetry**

Symmetry with Respect to the Polar Axis (*x*-Axis):
In a polar equation, replace θ by $-\theta$. If an equivalent equation results, the graph is symmetric with respect to the polar axis.

Symmetry with Respect to the Line $\theta = \pi/2$ (*y*-Axis):
In a polar equation, replace θ by $\pi - \theta$. If an equivalent equation results, the graph is symmetric with respect to the line $\theta = \pi/2$.

Symmetry with Respect to the Pole (Origin):
In a polar equation, replace r by $-r$. If an equivalent equation results, the graph is symmetric with respect to the pole.

The three tests for symmetry given here are *sufficient* conditions for symmetry, but they are not *necessary* conditions. That is, an equation may fail these tests and still have a graph that is symmetric with respect to the polar axis, the line $\theta = \pi/2$, or the pole. For example, the graph of $r = \sin(2\theta)$ turns out to be symmetric with respect to the polar axis, the line $\theta = \pi/2$, and the pole, but all three tests given here fail. See also Problems 71, 72, and 73.

EXAMPLE 8 Graphing a Polar Equation (Cardioid)

4 Graph the equation: $r = 1 - \sin\theta$

Solution We check for symmetry first.

Polar Axis: Replace θ by $-\theta$. The result is

$$r = 1 - \sin(-\theta) = 1 + \sin\theta$$

The test fails, so the graph may or may not be symmetric with respect to the polar axis.

The Line $\theta = \pi/2$: Replace θ by $\pi - \theta$. The result is

$$r = 1 - \sin(\pi - \theta) = 1 - (\sin\pi\cos\theta - \cos\pi\sin\theta)$$
$$= 1 - [0 \cdot \cos\theta - (-1)\sin\theta] = 1 - \sin\theta$$

The test is satisfied, so the graph is symmetric with respect to the line $\theta = \pi/2$.

The Pole: Replace r by $-r$. Then the result is $-r = 1 - \sin\theta$, so $r = -1 + \sin\theta$. The test fails, so the graph may or may not be symmetric with respect to the pole.

TABLE 1

θ	$r = 1 - \sin\theta$
$-\pi/2$	$1 + 1 = 2$
$-\pi/3$	$1 + \sqrt{3}/2 \approx 1.87$
$-\pi/6$	$1 + \frac{1}{2} = \frac{3}{2}$
0	1
$\pi/6$	$1 - \frac{1}{2} = \frac{1}{2}$
$\pi/3$	$1 - \sqrt{3}/2 \approx 0.13$
$\pi/2$	0

Next, we identify points on the graph by assigning values to the angle θ and calculating the corresponding values of r. Due to the symmetry with respect to the line $\theta = \pi/2$, we only need to assign values to θ from $-\pi/2$ to $\pi/2$, as given in Table 1.

Now we plot the points (r, θ) from Table 1 and trace out the graph, beginning at the point $(2, -\pi/2)$ and ending at the point $(0, \pi/2)$. Then we reflect this portion of the graph about the line $\theta = \pi/2$ (the y-axis) to obtain the complete graph. Figure 29(a) shows the graph drawn by hand.

Figure 29(b) shows the graph using a graphing utility with $\theta\text{min} = 0$, $\theta\text{max} = 2\pi$, and $\theta\text{step} = \frac{\pi}{24}$.

Figure 29
$r = 1 - \sin\theta$

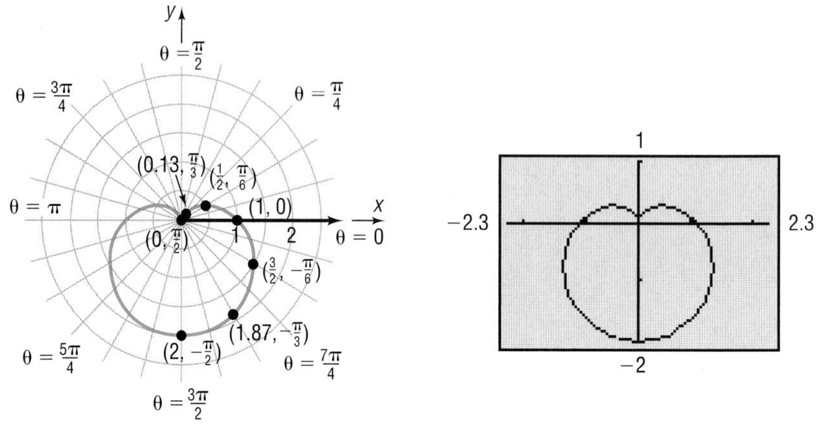

(a) (b)

EXPLORATION Graph $r_1 = 1 + \sin\theta$. Clear the screen and graph $r_1 = 1 - \cos\theta$. Clear the screen and graph $r_1 = 1 + \cos\theta$. Do you see a pattern?

The curve in Figure 29 is an example of a *cardioid* (a heart-shaped curve).

Cardioids are characterized by equations of the form

$$r = a(1 + \cos\theta) \qquad r = a(1 + \sin\theta)$$
$$r = a(1 - \cos\theta) \qquad r = a(1 - \sin\theta)$$

where $a > 0$. The graph of a cardioid passes through the pole.

NOW WORK PROBLEM 31.

EXAMPLE 9 Graphing a Polar Equation (Limaçon without Inner Loop)

Graph the equation: $r = 3 + 2\cos\theta$

Solution We check for symmetry first.

Polar Axis: Replace θ by $-\theta$. The result is

$$r = 3 + 2\cos(-\theta) = 3 + 2\cos\theta$$

The test is satisfied, so the graph is symmetric with respect to the polar axis.

The Line $\theta = \pi/2$: Replace θ by $\pi - \theta$. The result is

$$r = 3 + 2\cos(\pi - \theta) = 3 + 2(\cos\pi\cos\theta + \sin\pi\sin\theta)$$
$$= 3 - 2\cos\theta$$

The test fails, so the graph may or may not be symmetric with respect to the line $\theta = \pi/2$.

The Pole: Replace r by $-r$. The test fails, so the graph may or may not be symmetric with respect to the pole.

TABLE 2

θ	$r = 3 + 2\cos\theta$
0	5
$\pi/6$	$3 + \sqrt{3} \approx 4.73$
$\pi/3$	4
$\pi/2$	3
$2\pi/3$	2
$5\pi/6$	$3 - \sqrt{3} \approx 1.27$
π	1

Next, we identify points on the graph by assigning values to the angle θ and calculating the corresponding values of r. Due to the symmetry with respect to the polar axis, we only need to assign values to θ from 0 to π, as given in Table 2.

Now we plot the points (r, θ) from Table 2 and trace out the graph, beginning at the point $(5, 0)$ and ending at the point $(1, \pi)$. Then we reflect this portion of the graph about the polar axis (the x-axis) to obtain the complete graph. Figure 30(a) shows the graph drawn by hand. Figure 30(b) shows the graph using a graphing utility with $\theta\text{min} = 0$, $\theta\text{max} = 2\pi$, and $\theta\text{step} = \frac{\pi}{24}$.

Figure 30
$r = 3 + 2\cos\theta$

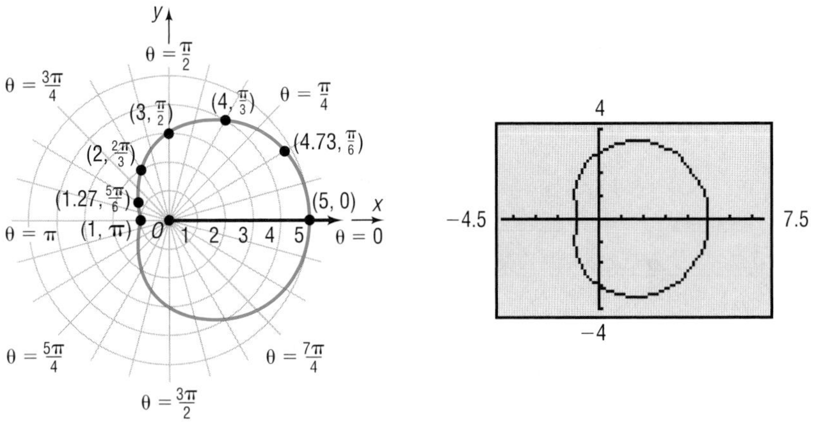

(a) (b)

EXPLORATION Graph $r_1 = 3 - 2\cos\theta$. Clear the screen and graph $r_1 = 3 + 2\sin\theta$. Clear the screen and graph $r_1 = 3 - 2\sin\theta$. Do you see a pattern?

The curve in Figure 30 is an example of a *limaçon* (the French word for *snail*) without an inner loop.

Limaçons without an inner loop are characterized by equations of the form

$$r = a + b\cos\theta \qquad r = a + b\sin\theta$$
$$r = a - b\cos\theta \qquad r = a - b\sin\theta$$

where $a > 0$, $b > 0$, and $a > b$. The graph of a limaçon without an inner loop does not pass through the pole.

NOW WORK PROBLEM 37.

EXAMPLE 10 Graphing a Polar Equation (Limaçon with Inner Loop)

Graph the equation: $r = 1 + 2\cos\theta$

Solution First, we check for symmetry.

Polar Axis: Replace θ by $-\theta$. The result is

$$r = 1 + 2\cos(-\theta) = 1 + 2\cos\theta$$

The test is satisfied, so the graph is symmetric with respect to the polar axis.

The Line $\theta = \pi/2$: Replace θ by $\pi - \theta$. The result is

$$r = 1 + 2\cos(\pi - \theta) = 1 + 2(\cos\pi\cos\theta + \sin\pi\sin\theta)$$
$$= 1 - 2\cos\theta$$

The test fails, so the graph may or may not be symmetric with respect to the line $\theta = \pi/2$.

The Pole: Replace r by $-r$. The test fails, so the graph may or may not be symmetric with respect to the pole.

TABLE 3

θ	$r = 1 + 2\cos\theta$
0	3
$\pi/6$	$1 + \sqrt{3} \approx 2.73$
$\pi/3$	2
$\pi/2$	1
$2\pi/3$	0
$5\pi/6$	$1 - \sqrt{3} \approx -0.73$
π	-1

Next, we identify points on the graph of $r = 1 + 2\cos\theta$ by assigning values to the angle θ and calculating the corresponding values of r. Due to the symmetry with respect to the polar axis, we only need to assign values to θ from 0 to π, as given in Table 3.

Now we plot the points (r, θ) from Table 3, beginning at $(3, 0)$ and ending at $(-1, \pi)$. See Figure 31(a). Finally, we reflect this portion of the graph about the polar axis (the x-axis) to obtain the complete graph. See Figure 31(b). Figure 31(c) shows the graph using a graphing utility with $\theta\text{min} = 0$, $\theta\text{max} = 2\pi$, and $\theta\text{step} = \frac{\pi}{24}$.

Figure 31

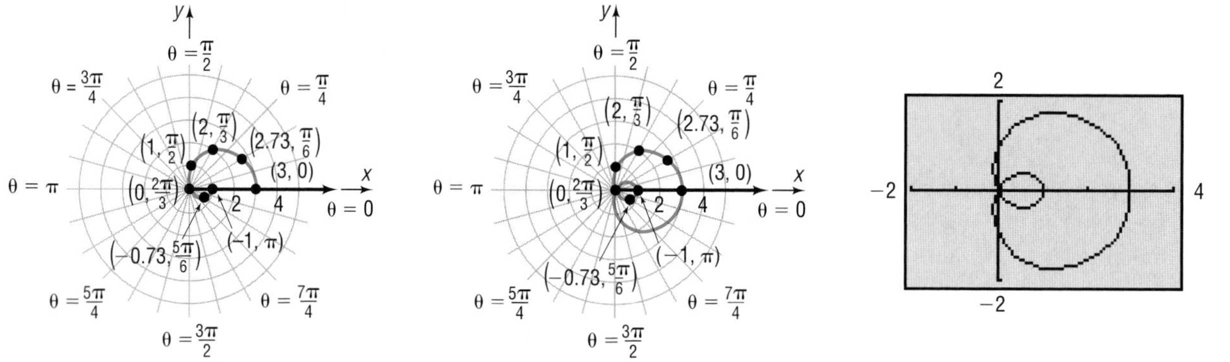

(a) (b) $r = 1 + 2\cos\theta$ (c)

EXPLORATION Graph $r_1 = 1 - 2\cos\theta$. Clear the screen and graph $r_1 = 1 + 2\sin\theta$. Clear the screen and graph $r_1 = 1 - 2\sin\theta$. Do you see a pattern?

The curve in Figure 31(b) is an example of a limaçon with an inner loop.

Limaçons with an inner loop are characterized by equations of the form

$$r = a + b\cos\theta \qquad r = a + b\sin\theta$$
$$r = a - b\cos\theta \qquad r = a - b\sin\theta$$

where $a > 0$, $b > 0$, and $a < b$. The graph of a limaçon with an inner loop will pass through the pole twice.

NOW WORK PROBLEM 39.

EXAMPLE 11 Graphing a Polar Equation (Rose)

Graph the equation: $r = 2\cos(2\theta)$

Solution We check for symmetry.

Polar Axis: If we replace θ by $-\theta$, the result is

$$r = 2\cos[2(-\theta)] = 2\cos(2\theta)$$

The test is satisfied, so the graph is symmetric with respect to the polar axis.

The Line $\theta = \pi/2$: If we replace θ by $\pi - \theta$, we obtain

$$r = 2\cos[2(\pi - \theta)] = 2\cos(2\pi - 2\theta) = 2\cos(2\theta)$$

The test is satisfied, so the graph is symmetric with respect to the line $\theta = \pi/2$.

The Pole: Since the graph is symmetric with respect to both the polar axis and the line $\theta = \pi/2$, it must be symmetric with respect to the pole.

TABLE 4

θ	$r = 2\cos(2\theta)$
0	2
$\pi/6$	1
$\pi/4$	0
$\pi/3$	-1
$\pi/2$	-2

Next, we construct Table 4. Due to the symmetry with respect to the polar axis, the line $\theta = \pi/2$, and the pole, we consider only values of θ from 0 to $\pi/2$.

We plot and connect these points in Figure 32(a). Finally, because of symmetry, we reflect this portion of the graph first about the polar axis (the x-axis) and then about the line $\theta = \pi/2$ (the y-axis) to obtain the complete graph. See Figure 32(b) for the graph drawn by hand. Figure 32(c) shows the graph using a graphing utility with θmin $= 0$, θmax $= 2\pi$, and θstep $= \frac{\pi}{24}$.

Figure 32

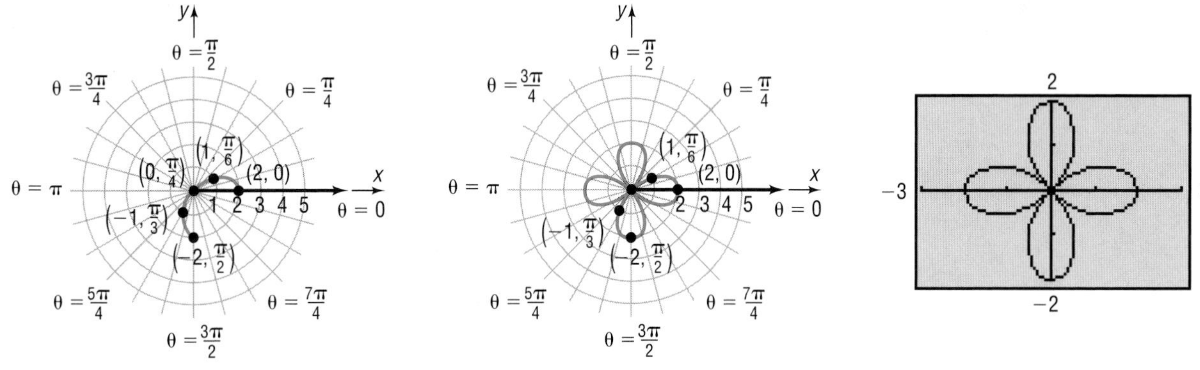

(a) (b) $r = 2\cos 2\theta$ (c)

EXPLORATION Graph $r = 2\cos(4\theta)$; clear the screen and graph $r = 2\cos(6\theta)$. How many petals did each of these graphs have?

Clear the screen and graph, in order, each on a clear screen, $r = 2\cos(3\theta)$, $r = 2\cos(5\theta)$, and $r = 2\cos(7\theta)$. What do you notice about the number of petals?

The curve in Figure 32(b) is called a *rose* with four petals.

Rose curves are characterized by equations of the form

$$r = a\cos(n\theta), \qquad r = a\sin(n\theta), \qquad a \neq 0$$

and have graphs that are rose shaped. If $n \neq 0$ is even, the rose has $2n$ petals; if $n \neq \pm 1$ is odd, the rose has n petals.

NOW WORK PROBLEM 43.

EXAMPLE 12 Graphing a Polar Equation (Lemniscate)

Graph the equation: $r^2 = 4\sin(2\theta)$

TABLE 5

θ	$r^2 = 4\sin(2\theta)$	r
0	0	0
$\pi/6$	$2\sqrt{3}$	± 1.9
$\pi/4$	4	± 2
$\pi/3$	$2\sqrt{3}$	± 1.9
$\pi/2$	0	0

Solution We leave it to you to verify that the graph is symmetric with respect to the pole. Table 5 lists points on the graph for values of $\theta = 0$ through $\theta = \pi/2$. Note that there are no points on the graph for $\pi/2 < \theta < \pi$ (quadrant II), since $\sin(2\theta) < 0$ for such values. The points from Table 5 where $r \geq 0$ are plotted in Figure 33(a). The remaining points on the graph may be obtained by using symmetry. Figure 33(b) shows the final graph drawn by hand. Figure 33(c) shows the graph using a graphing utility with $\theta\text{min} = 0$, $\theta\text{max} = 2\pi$, and $\theta\text{step} = \frac{\pi}{24}$.

Figure 33

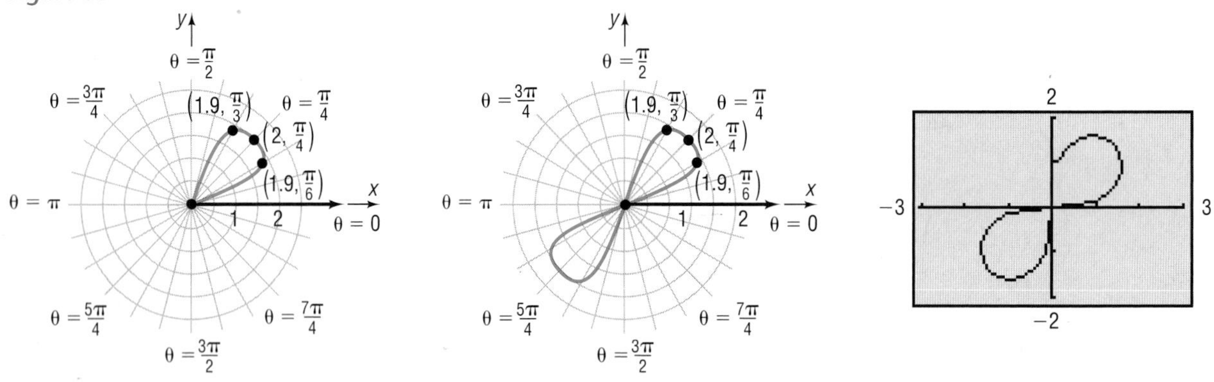

(a) (b) $r^2 = 4\sin 2\theta$ (c)

The curve in Figure 33(b) is an example of a *lemniscate.*

Lemniscates are characterized by equations of the form

$$r^2 = a^2\sin(2\theta) \qquad r^2 = a^2\cos(2\theta)$$

where $a \neq 0$, and have graphs that are propeller shaped.

NOW WORK PROBLEM 47.

EXAMPLE 13 Graphing a Polar Equation (Spiral)

Graph the equation: $r = e^{\theta/5}$

Solution The tests for symmetry with respect to the pole, the polar axis, and the line $\theta = \pi/2$ fail. Furthermore, there is no number θ for which $r = 0$. Hence, the graph does not pass through the pole. We observe that r is positive for all θ, r increases as θ increases, $r \to 0$ as $\theta \to -\infty$, and $r \to \infty$ as $\theta \to \infty$. With the help of a calculator, we obtain the values in Table 6. See Figure 34(a) for the graph drawn by hand. Figure 34(b) shows the graph using a graphing utility with θmin $= -5\pi$, θmax $= 3\pi$, and θstep $= \frac{\pi}{24}$.

TABLE 6

θ	$r = e^{\theta/5}$
$-3\pi/2$	0.39
$-\pi$	0.53
$-\pi/2$	0.73
$-\pi/4$	0.85
0	1
$\pi/4$	1.17
$\pi/2$	1.37
π	1.87
$3\pi/2$	2.57
2π	3.51

Figure 34
$r = e^{\theta/5}$

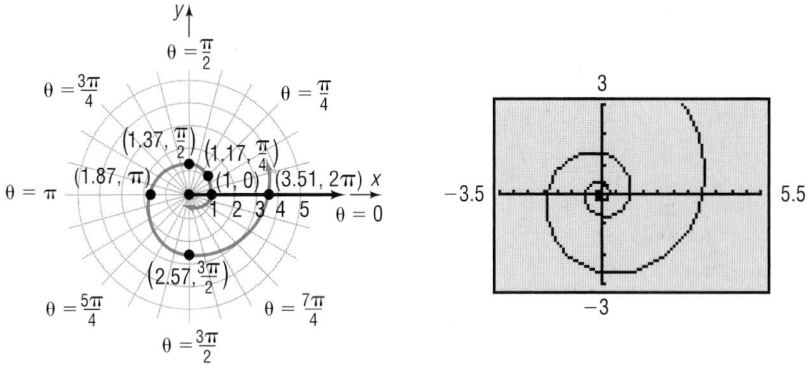

(a) (b)

The curve in Figure 34 is called a **logarithmic spiral,** since its equation may be written as $\theta = 5 \ln r$ and it spirals infinitely both toward the pole and away from it.

CLASSIFICATION OF POLAR EQUATIONS

The equations of some lines and circles in polar coordinates and their corresponding equations in rectangular coordinates are given in Table 7 on page 582. Also included are the names and the graphs of a few of the more frequently encountered polar equations.

SKETCHING QUICKLY

If a polar equation only involves a sine (or cosine) function, you can quickly obtain a sketch of its graph by making use of Table 7, periodicity, and a short table.

EXAMPLE 14 Sketching the Graph of a Polar Equation Quickly by Hand

Graph the equation: $r = 2 + 2 \sin \theta$

Solution We recognize the polar equation: Its graph is a cardioid. The period of $\sin \theta$ is 2π, so we form a table using $0 \le \theta \le 2\pi$, compute r, plot the points (r, θ), and sketch the graph of a cardioid as θ varies from 0 to 2π. See Table 8 and Figure 35 on page 583.

TABLE 7

LINES

Description	Line passing through the pole making an angle α with the polar axis	Vertical line	Horizontal line
Rectangular equation	$y = (\tan\alpha)x$	$x = a$	$y = b$
Polar equation	$\theta = \alpha$	$r\cos\theta = a$	$r\sin\theta = b$
Typical graph	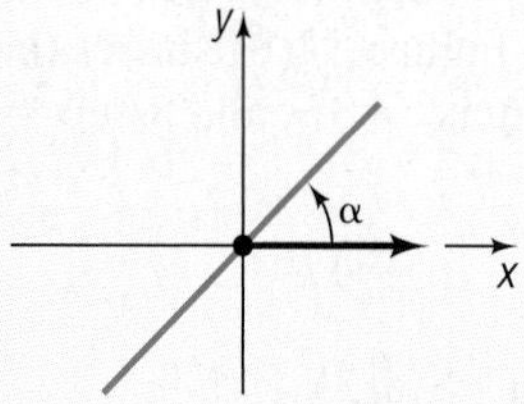	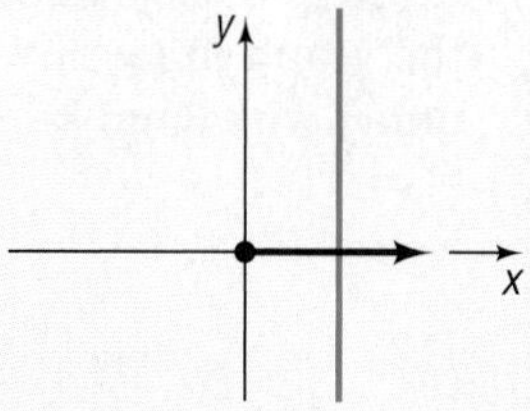	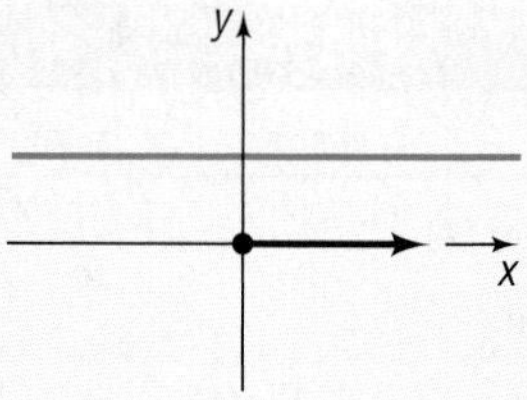

CIRCLES

Description	Center at the pole, radius a	Passing through the pole, tangent to the line $\theta = \pi/2$, center on the polar axis, radius a	Passing through the pole, tangent to the polar axis, center on the line $\theta = \pi/2$, radius a
Rectangular equation	$x^2 + y^2 = a^2, \quad a > 0$	$x^2 + y^2 = \pm 2ax, \quad a > 0$	$x^2 + y^2 = \pm 2ay, \quad a > 0$
Polar equation	$r = a, \quad a > 0$	$r = \pm 2a\cos\theta, \quad a > 0$	$r = \pm 2a\sin\theta, \quad a > 0$
Typical graph	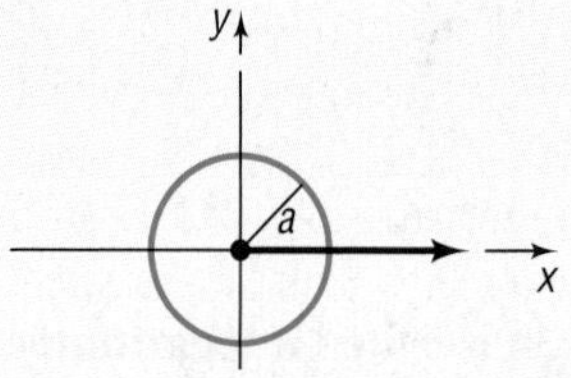	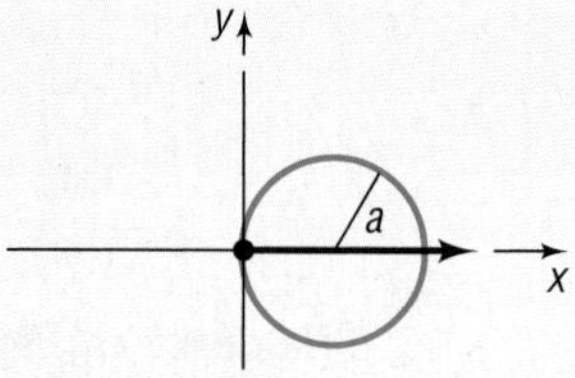	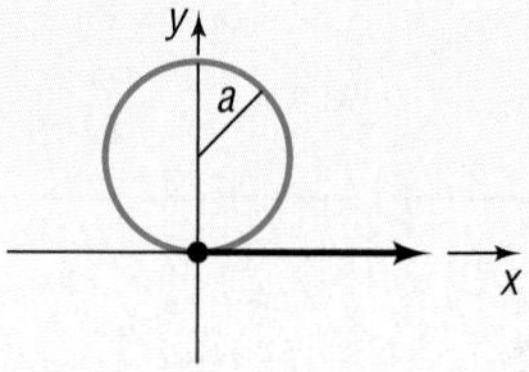

OTHER EQUATIONS

Name	Cardioid	Limaçon without inner loop	Limaçon with inner loop
Polar equations	$r = a \pm a\cos\theta, \quad a > 0$ $r = a \pm a\sin\theta, \quad a > 0$	$r = a \pm b\cos\theta, \quad 0 < b < a$ $r = a \pm b\sin\theta, \quad 0 < b < a$	$r = a \pm b\cos\theta, \quad 0 < a < b$ $r = a \pm b\sin\theta, \quad 0 < a < b$
Typical graph	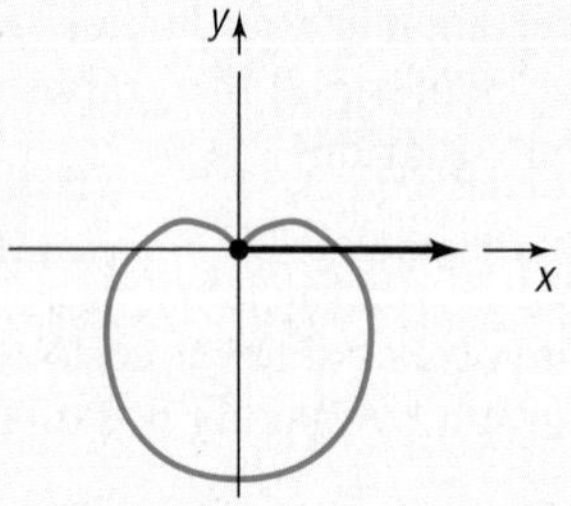	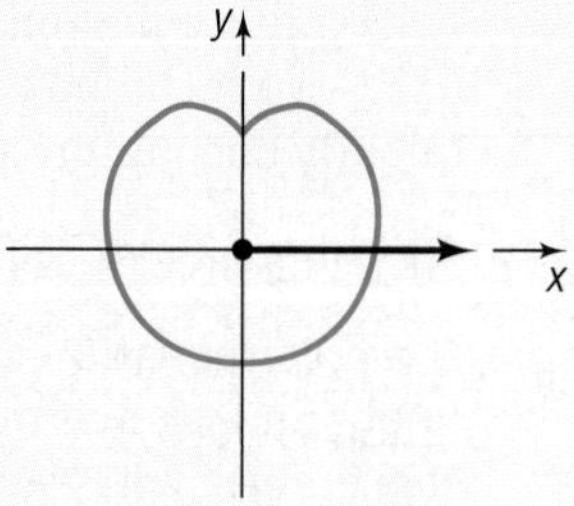	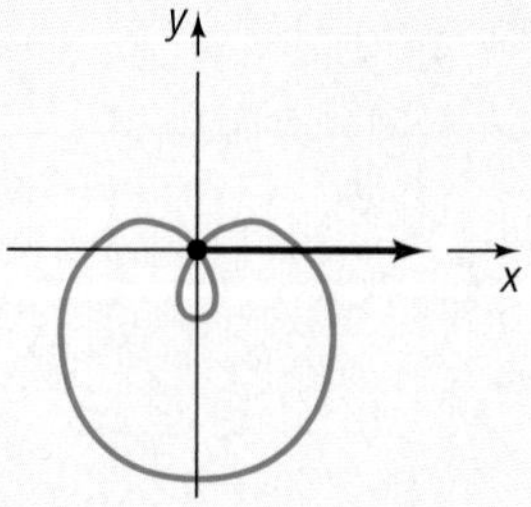
Name	Lemniscate	Rose with three petals	Rose with four petals
Polar equations	$r^2 = a^2\cos(2\theta), \quad a > 0$ $r^2 = a^2\sin(2\theta), \quad a > 0$	$r = a\sin(3\theta), \quad a > 0$ $r = a\cos(3\theta), \quad a > 0$	$r = a\sin(2\theta), \quad a > 0$ $r = a\cos(2\theta), \quad a > 0$
Typical graph	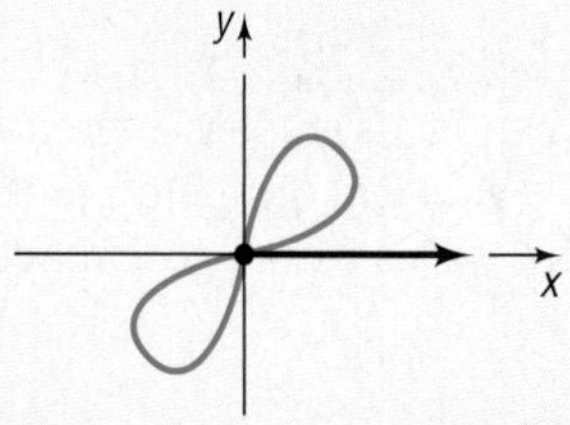	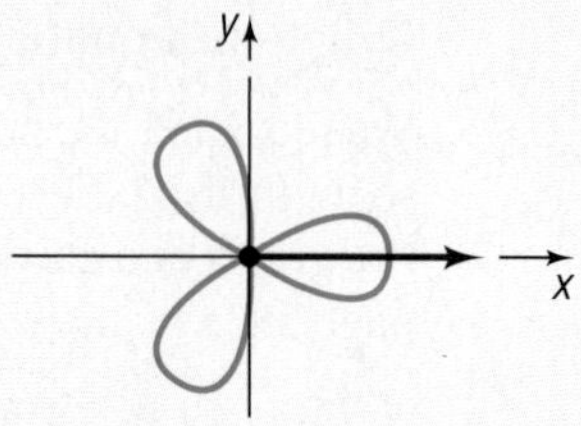	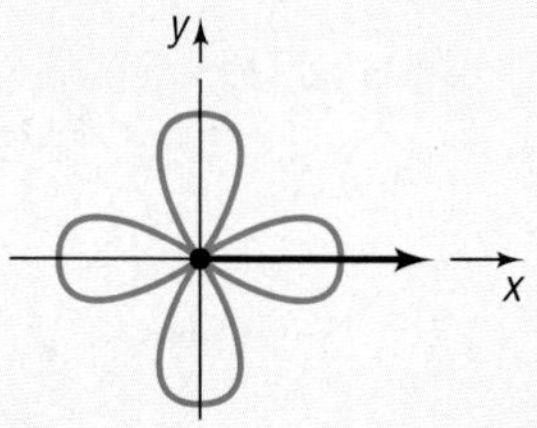

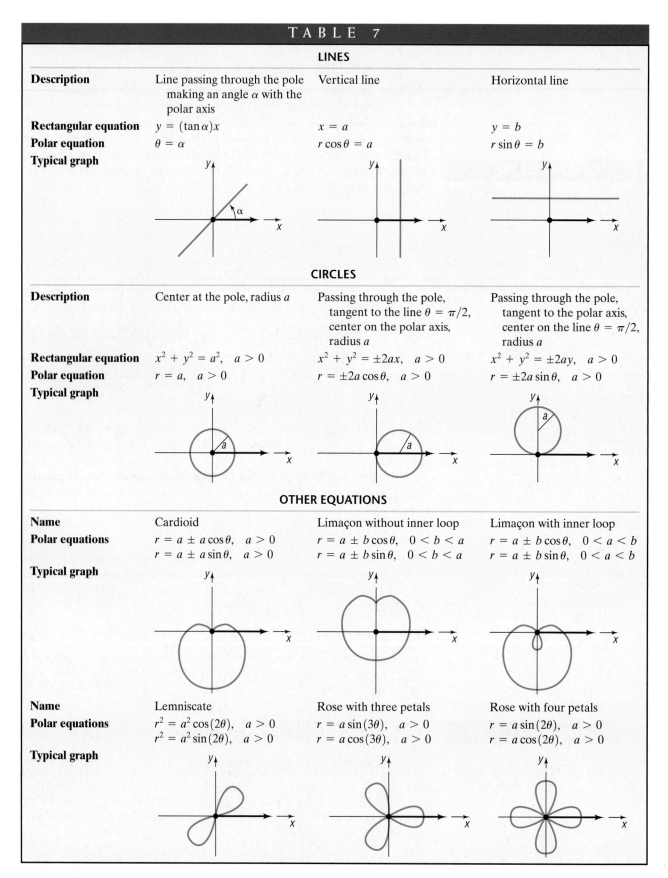

TABLE 8

θ	$r = 2 + 2\sin\theta$
0	2
$\pi/2$	4
π	2
$3\pi/2$	0
2π	2

Figure 35

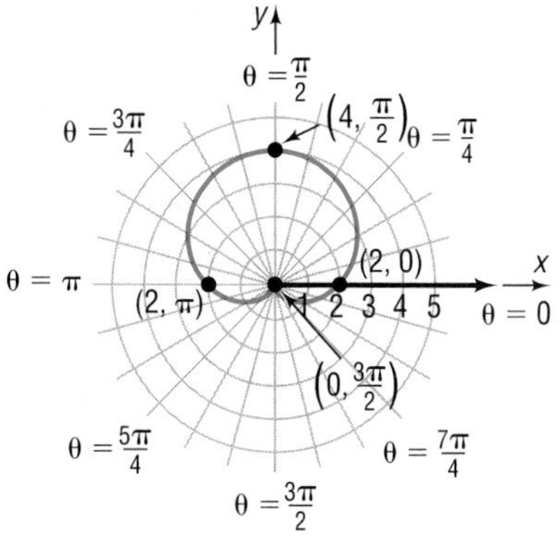

Calculus Comment

For those of you who are planning to study calculus, a comment about one important role of polar equations is in order.

In rectangular coordinates, the equation $x^2 + y^2 = 1$, whose graph is the unit circle, is not the graph of a function. In fact, it requires two functions to obtain the graph of the unit circle:

$$y_1 = \sqrt{1 - x^2} \quad \text{Upper semicircle} \qquad y_2 = -\sqrt{1 - x^2} \quad \text{Lower semicircle}$$

In polar coordinates, the equation $r = 1$, whose graph is also the unit circle, does define a function. That is, for each choice of θ there is only one corresponding value of r, that is, $r = 1$. Since many uses of calculus require that functions be used, the opportunity to express nonfunctions in rectangular coordinates as functions in polar coordinates becomes extremely useful.

Note also that the vertical line test for functions is valid only for equations in rectangular coordinates.

HISTORICAL FEATURE

Jacob Bernoulli 1654–1705

Polar coordinates seem to have been invented by Jacob Bernoulli (1654–1705) about 1691, although, as with most such ideas, earlier traces of the notion exist. Early users of calculus remained committed to rectangular coordinates, and polar coordinates did not become widely used until the early 1800s. Even then, it was mostly geometers who used them for describing odd curves. Finally, about the mid-1800s, applied mathematicians realized the tremendous simplification that polar coordinates make possible in the description of objects with circular or cylindrical symmetry. From then on their use became widespread.

9.2 EXERCISES

In Problems 1–16, identify each polar equation by transforming the equation to rectangular coordinates. Graph each polar equation by hand. Verify your results using a graphing utility.

1. $r = 4$ **2.** $r = 2$ **3.** $\theta = \pi/3$ **4.** $\theta = -\pi/4$

5. $r\sin\theta = 4$ **6.** $r\cos\theta = 4$ **7.** $r\cos\theta = -2$ **8.** $r\sin\theta = -2$

9. $r = 2\cos\theta$ **10.** $r = 2\sin\theta$ **11.** $r = -4\sin\theta$ **12.** $r = -4\cos\theta$

13. $r\sec\theta = 4$ **14.** $r\csc\theta = 8$ **15.** $r\csc\theta = -2$ **16.** $r\sec\theta = -4$

In Problems 17–24, match each of the graphs (A) through (H) to one of the following polar equations.

17. $r = 2$ **18.** $\theta = \pi/4$ **19.** $r = 2\cos\theta$ **20.** $r\cos\theta = 2$

21. $r = 1 + \cos\theta$ **22.** $r = 2\sin\theta$ **23.** $\theta = 3\pi/4$ **24.** $r\sin\theta = 2$

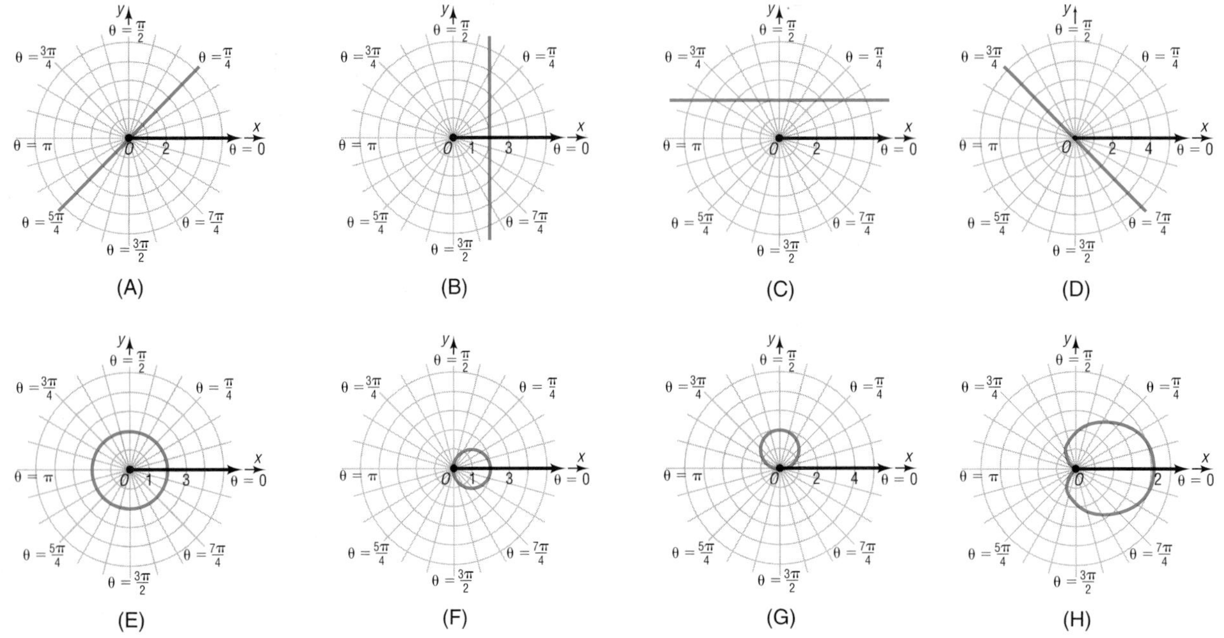

In Problems 25–30, match each of the graphs (A) through (F) to one of the following polar equations.

25. $r = 4$ **26.** $r = 3\cos\theta$ **27.** $r = 3\sin\theta$

28. $r\sin\theta = 3$ **29.** $r\cos\theta = 3$ **30.** $r = 2 + \sin\theta$

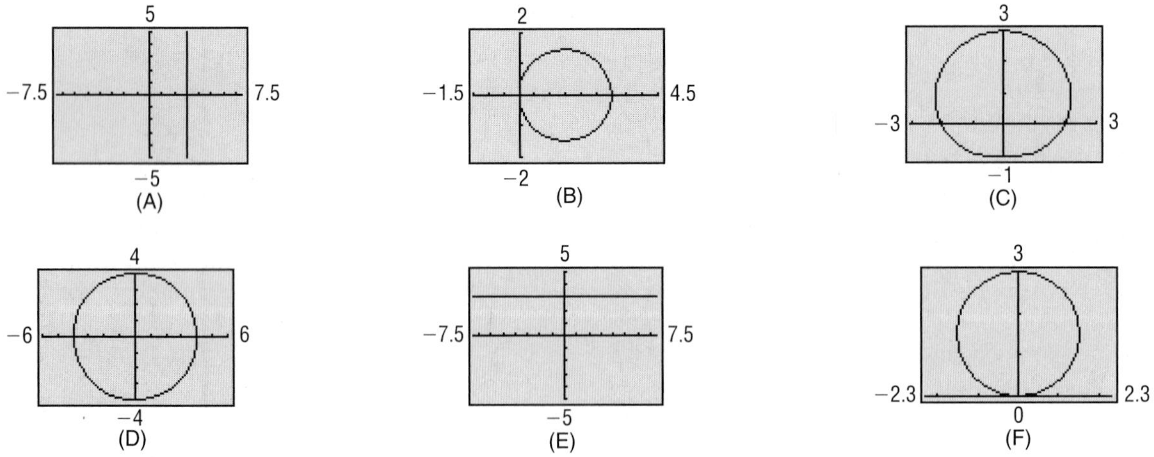

In Problems 31–54, identify and graph each polar equation by hand. Verify your results using a graphing utility.

31. $r = 2 + 2\cos\theta$ **32.** $r = 1 + \sin\theta$ **33.** $r = 3 - 3\sin\theta$ **34.** $r = 2 - 2\cos\theta$

35. $r = 2 + \sin\theta$ **36.** $r = 2 - \cos\theta$ **37.** $r = 4 - 2\cos\theta$ **38.** $r = 4 + 2\sin\theta$

39. $r = 1 + 2\sin\theta$ **40.** $r = 1 - 2\sin\theta$ **41.** $r = 2 - 3\cos\theta$ **42.** $r = 2 + 4\cos\theta$

43. $r = 3\cos(2\theta)$ **44.** $r = 2\sin(2\theta)$ **45.** $r = 4\sin(3\theta)$ **46.** $r = 3\cos(4\theta)$

47. $r^2 = 9\cos(2\theta)$ **48.** $r^2 = \sin(2\theta)$ **49.** $r = 2^\theta$ **50.** $r = 3^\theta$

51. $r = 1 - \cos\theta$ **52.** $r = 3 + \cos\theta$ **53.** $r = 1 - 3\cos\theta$ **54.** $r = 4\cos(3\theta)$

In Problems 55–64, graph each polar equation by hand. Verify your results using a graphing utility.

55. $r = \dfrac{2}{1 - \cos\theta}$ (parabola)

56. $r = \dfrac{2}{1 - 2\cos\theta}$ (hyperbola)

57. $r = \dfrac{1}{3 - 2\cos\theta}$ (ellipse)

58. $r = \dfrac{1}{1 - \cos\theta}$ (parabola)

59. $r = \theta, \quad \theta \geq 0$ (spiral of Archimedes)

60. $r = \dfrac{3}{\theta}$ (reciprocal spiral)

61. $r = \csc\theta - 2, \quad 0 < \theta < \pi$ (conchoid)

62. $r = \sin\theta\tan\theta$ (cissoid)

63. $r = \tan\theta$ (kappa curve), $-\dfrac{\pi}{2} < \theta < \dfrac{\pi}{2}$

64. $r = \cos\dfrac{\theta}{2}$

65. Show that the graph of the equation $r\sin\theta = a$ is a horizontal line a units above the pole if $a > 0$ and $|a|$ units below the pole if $a < 0$.

66. Show that the graph of the equation $r\cos\theta = a$ is a vertical line a units to the right of the pole if $a > 0$ and $|a|$ units to the left of the pole if $a < 0$.

67. Show that the graph of the equation $r = 2a\sin\theta$, $a > 0$, is a circle of radius a with center at $(0, a)$ in rectangular coordinates.

68. Show that the graph of the equation $r = -2a\sin\theta$, $a > 0$, is a circle of radius a with center at $(0, -a)$ in rectangular coordinates.

69. Show that the graph of the equation $r = 2a\cos\theta$, $a > 0$, is a circle of radius a with center at $(a, 0)$ in rectangular coordinates.

70. Show that the graph of the equation $r = -2a\cos\theta$, $a > 0$, is a circle of radius a with center at $(-a, 0)$ in rectangular coordinates.

71. Explain why the following test for symmetry is valid: Replace r by $-r$ and θ by $-\theta$ in a polar equation. If an equivalent equation results, the graph is symmetric with respect to the line $\theta = \pi/2$ (y-axis).
(a) Show that the test on page 575 fails for $r^2 = \cos\theta$, but this new test works.
(b) Show that the test on page 575 works for $r^2 = \sin\theta$, yet this new test fails.

72. Develop a new test for symmetry with respect to the pole.
(a) Find a polar equation for which this new test fails, yet the test on page 575 works.
(b) Find a polar equation for which the test on page 575 fails, yet the new test works.

73. Write down two different tests for symmetry with respect to the polar axis. Find examples in which one works and the other fails. Which test do you prefer to use? Justify your answer.

PREPARING FOR THIS SECTION

Before getting started, review the following:

- ✓ Complex Numbers (pp. 252–258)
- ✓ Definition of the Sine and Cosine Function (p. 381)
- ✓ Value of the Sine and Cosine Functions at Certain Angles (pp. 384 and 387)

9.3 THE COMPLEX PLANE; DEMOIVRE'S THEOREM

OBJECTIVES

1. Convert a Complex Number from Rectangular Form to Polar Form
2. Plot Points in the Complex Plane
3. Find Products and Quotients of Complex Numbers in Polar Form
4. Use DeMoivre's Theorem
5. Find Complex Roots

When we first introduced complex numbers, we were not prepared to give a geometric interpretation of a complex number. Now we are ready. Although we could give several interpretations, the one that follows is the easiest to understand.

Figure 36
Complex plane

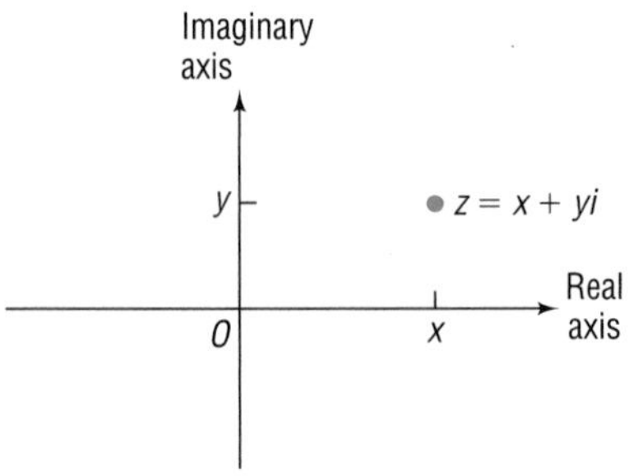

A complex number $z = x + yi$ can be interpreted geometrically as the point (x, y) in the xy-plane. Each point in the plane corresponds to a complex number and, conversely, each complex number corresponds to a point in the plane. We shall refer to the collection of such points as the **complex plane.** The x-axis will be referred to as the **real axis,** because any point that lies on the real axis is of the form $z = x + 0i = x$, a real number. The y-axis is called the **imaginary axis,** because any point that lies on it is of the form $z = 0 + yi = yi$, a pure imaginary number. See Figure 36.

Let $z = x + yi$ be a complex number. The **magnitude** or **modulus** of z, denoted by $|z|$, is defined as the distance from the origin to the point (x, y). That is,

$$|z| = \sqrt{x^2 + y^2} \qquad \textbf{(1)}$$

Figure 37

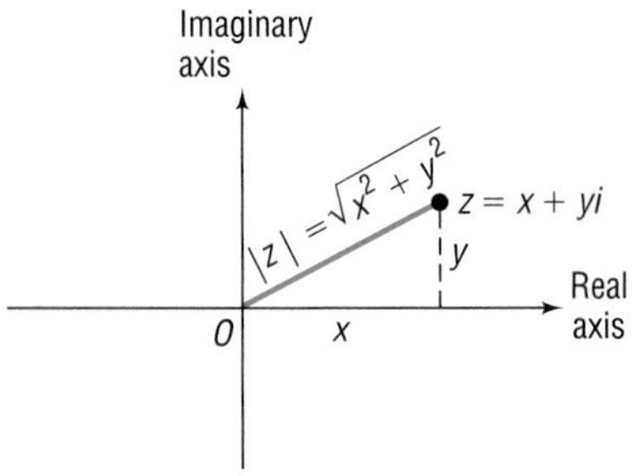

See Figure 37 for an illustration.

This definition for $|z|$ is consistent with the definition for the absolute value of a real number: If $z = x + yi$ is real, then $z = x + 0i$ and

$$|z| = \sqrt{x^2 + 0^2} = \sqrt{x^2} = |x|$$

For this reason, the magnitude of z is sometimes called the absolute value of z.

Recall (Section 4.2) that if $z = x + yi$, then its **conjugate,** denoted by $\bar{z}$, is $\bar{z} = x - yi$. Because $z\bar{z} = x^2 + y^2$, it follows from equation (1) that the magnitude of z can be written as

$$|z| = \sqrt{z\bar{z}} \qquad \textbf{(2)}$$

Polar Form of a Complex Number

1 When a complex number is written in the standard form $z = x + yi$, we say that it is in **rectangular,** or **Cartesian, form,** because (x, y) are the rectangular coordinates of the corresponding point in the complex plane. Suppose that (r, θ) are the polar coordinates of this point. Then

$$x = r\cos\theta \qquad y = r\sin\theta \qquad \textbf{(3)}$$

If $r \geq 0$ and $0 \leq \theta < 2\pi$, the complex number $z = x + yi$ may be written in **polar form** as

$$z = x + yi = (r\cos\theta) + (r\sin\theta)i = r(\cos\theta + i\sin\theta) \qquad \textbf{(4)}$$

Figure 38

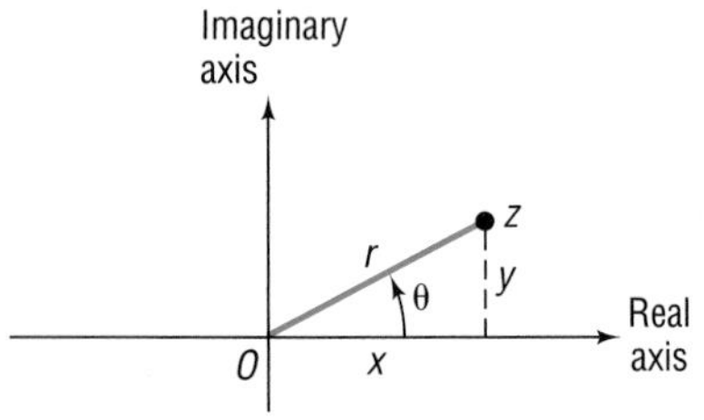

$z = x + yi = r(\cos\theta + i\sin\theta)$, $r \geq 0$, $0 \leq \theta < 2\pi$

See Figure 38.

If $z = r(\cos\theta + i\sin\theta)$ is the polar form of a complex number, the angle θ, $0 \leq \theta < 2\pi$, is called the **argument of z.**

Also, because $r \geq 0$, we have $r = \sqrt{x^2 + y^2}$. From equation (1) it follows that the magnitude of $z = r(\cos\theta + i\sin\theta)$ is

$$|z| = r$$

EXAMPLE 1 Plotting a Point in the Complex Plane and Writing a Complex Number in Polar Form

2 Plot the point corresponding to $z = \sqrt{3} - i$ in the complex plane, and write an expression for z in polar form.

Figure 39

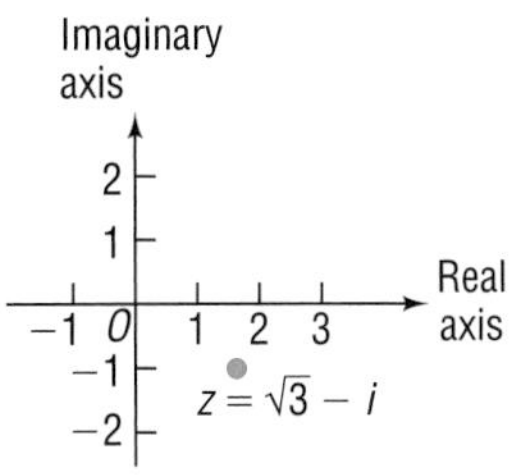

Solution The point corresponding to $z = \sqrt{3} - i$ has the rectangular coordinates $(\sqrt{3}, -1)$. The point, located in quadrant IV, is plotted in Figure 39. Because $x = \sqrt{3}$ and $y = -1$, it follows that

$$r = \sqrt{x^2 + y^2} = \sqrt{(\sqrt{3})^2 + (-1)^2} = \sqrt{4} = 2$$

and

$$\sin\theta = \frac{y}{r} = \frac{-1}{2}, \qquad \cos\theta = \frac{x}{r} = \frac{\sqrt{3}}{2}, \qquad 0 \le \theta < 2\pi$$

Then $\theta = 11\pi/6$ and $r = 2$, so the polar form of $z = \sqrt{3} - i$ is

$$z = r(\cos\theta + i\sin\theta) = 2\left(\cos\frac{11\pi}{6} + i\sin\frac{11\pi}{6}\right)$$

NOW WORK PROBLEM 1.

EXAMPLE 2 Plotting a Point in the Complex Plane and Converting from Polar to Rectangular Form

Figure 40

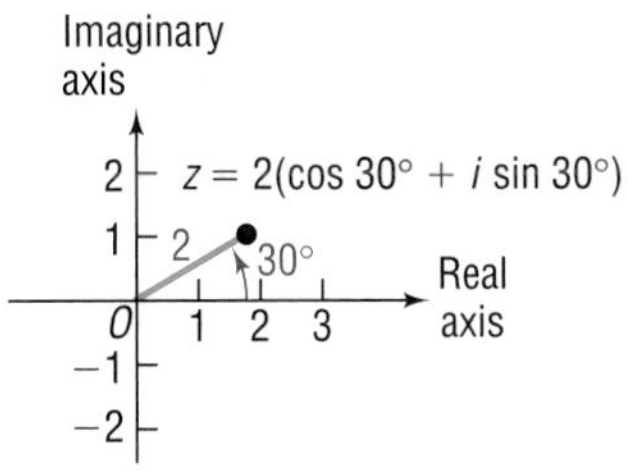

Plot the point corresponding to $z = 2(\cos 30° + i\sin 30°)$ in the complex plane, and write an expression for z in rectangular form.

Solution To plot the complex number $z = 2(\cos 30° + i\sin 30°)$, we plot the point whose polar coordinates are $(r, \theta) = (2, 30°)$, as shown in Figure 40. In rectangular form,

$$z = 2(\cos 30° + i\sin 30°) = 2\left(\frac{\sqrt{3}}{2} + \frac{1}{2}i\right) = \sqrt{3} + i$$

NOW WORK PROBLEM 13.

3 The polar form of a complex number provides an alternative for finding products and quotients of complex numbers.

Theorem Let $z_1 = r_1(\cos\theta_1 + i\sin\theta_1)$ and $z_2 = r_2(\cos\theta_2 + i\sin\theta_2)$ be two complex numbers. Then

$$z_1 z_2 = r_1 r_2[\cos(\theta_1 + \theta_2) + i\sin(\theta_1 + \theta_2)] \quad (5)$$

If $z_2 \ne 0$, then

$$\frac{z_1}{z_2} = \frac{r_1}{r_2}[\cos(\theta_1 - \theta_2) + i\sin(\theta_1 - \theta_2)] \quad (6)$$

Proof We will prove formula (5). The proof of formula (6) is left as an exercise (see Problem 56).

$$\begin{aligned} z_1 z_2 &= [r_1(\cos\theta_1 + i\sin\theta_1)][r_2(\cos\theta_2 + i\sin\theta_2)] \\ &= r_1 r_2[(\cos\theta_1 + i\sin\theta_1)(\cos\theta_2 + i\sin\theta_2)] \\ &= r_1 r_2[(\cos\theta_1\cos\theta_2 - \sin\theta_1\sin\theta_2) + i(\sin\theta_1\cos\theta_2 + \cos\theta_1\sin\theta_2)] \\ &= r_1 r_2[\cos(\theta_1 + \theta_2) + i\sin(\theta_1 + \theta_2)] \end{aligned}$$

Because the magnitude of a complex number z is r and its argument is θ, when $z = r(\cos\theta + i\sin\theta)$, we can restate this theorem as follows:

Theorem The magnitude of the product (quotient) of two complex numbers equals the product (quotient) of their magnitudes; the argument of the product (quotient) of two complex numbers is determined by the sum (difference) of their arguments.

■

Let's look at an example of how this theorem can be used.

EXAMPLE 3 Finding Products and Quotients of Complex Numbers in Polar Form

If $z = 3(\cos 20° + i\sin 20°)$ and $w = 5(\cos 100° + \sin 100°)$, find the following (leave your answers in polar form):

(a) zw (b) $\dfrac{z}{w}$

Solution (a)
$$\begin{aligned} zw &= [3(\cos 20° + i\sin 20°)][5(\cos 100° + i\sin 100°)] \\ &= (3 \cdot 5)[\cos(20° + 100°) + i\sin(20° + 100°)] \\ &= 15(\cos 120° + i\sin 120°) \end{aligned}$$

(b)
$$\begin{aligned} \frac{z}{w} &= \frac{3(\cos 20° + i\sin 20°)}{5(\cos 100° + i\sin 100°)} \\ &= \tfrac{3}{5}[\cos(20° - 100°) + i\sin(20° - 100°)] \\ &= \tfrac{3}{5}[\cos(-80°) + i\sin(-80°)] \\ &= \tfrac{3}{5}(\cos 280° + i\sin 280°) \quad \text{Argument must lie between 0° and 360°.} \end{aligned}$$

■

NOW WORK PROBLEM 23.

DEMOIVRE'S THEOREM

4 DeMoivre's Theorem, stated by Abraham DeMoivre (1667–1754) in 1730, but already known to many people by 1710, is important for the following reason: The fundamental processes of algebra are the four operations of addition, subtraction, multiplication, and division, together with powers and the extraction of roots. DeMoivre's Theorem allows these latter fundamental algebraic operations to be applied to complex numbers.

DeMoivre's Theorem, in its most basic form, is a formula for raising a complex number z to the power n, where $n \geq 1$ is a positive integer. Let's see if we can guess the form of the result.

Let $z = r(\cos\theta + i\sin\theta)$ be a complex number. Then, based on equation (5), we have

$$n = 2: \quad z^2 = r^2[\cos(2\theta) + i\sin(2\theta)]$$

$$\begin{aligned} n = 3: \quad z^3 &= z^2 \cdot z \\ &= \{r^2[\cos(2\theta) + i\sin(2\theta)]\}[r(\cos\theta + i\sin\theta)] \\ &= r^3[\cos(3\theta) + i\sin(3\theta)] \quad \text{Equation (5).} \end{aligned}$$

$$\begin{aligned} n = 4: \quad z^4 &= z^3 \cdot z \\ &= \{r^3[\cos(3\theta) + i\sin(3\theta)]\}[r(\cos\theta + i\sin\theta)] \\ &= r^4[\cos(4\theta) + i\sin(4\theta)] \quad \text{Equation (5).} \end{aligned}$$

The pattern should now be clear.

Theorem

DeMoivre's Theorem

If $z = r(\cos\theta + i\sin\theta)$ is a complex number, then

$$z^n = r^n[\cos(n\theta) + i\sin(n\theta)] \tag{7}$$

where $n \geq 1$ is a positive integer.

We will not prove DeMoivre's Theorem because the proof requires mathematical induction (which is not discussed until Section 12.4).

Let's look at some examples.

EXAMPLE 4 Using DeMoivre's Theorem

Write $[2(\cos 20° + i\sin 20°)]^3$ in the standard form $a + bi$.

Solution

$$\begin{aligned} [2(\cos 20° + i\sin 20°)]^3 &= 2^3[\cos(3 \cdot 20°) + i\sin(3 \cdot 20°)] \\ &= 8(\cos 60° + i\sin 60°) \\ &= 8\left(\frac{1}{2} + \frac{\sqrt{3}}{2}i\right) = 4 + 4\sqrt{3}i \end{aligned}$$

NOW WORK PROBLEM 31.

EXAMPLE 5 Using DeMoivre's Theorem

Write $(1 + i)^5$ in the standard form $a + bi$.

Solution To apply DeMoivre's Theorem, we must first write the complex number in polar form. Since the magnitude of $1 + i$ is $\sqrt{1^2 + 1^2} = \sqrt{2}$, we begin by writing

$$1 + i = \sqrt{2}\left(\frac{1}{\sqrt{2}} + \frac{1}{\sqrt{2}}i\right) = \sqrt{2}\left(\cos\frac{\pi}{4} + i\sin\frac{\pi}{4}\right)$$

Now

$$\begin{aligned} (1 + i)^5 &= \left[\sqrt{2}\left(\cos\frac{\pi}{4} + i\sin\frac{\pi}{4}\right)\right]^5 \\ &= (\sqrt{2})^5\left[\cos\left(5 \cdot \frac{\pi}{4}\right) + i\sin\left(5 \cdot \frac{\pi}{4}\right)\right] \\ &= 4\sqrt{2}\left(\cos\frac{5\pi}{4} + i\sin\frac{5\pi}{4}\right) \\ &= 4\sqrt{2}\left[-\frac{1}{\sqrt{2}} + \left(-\frac{1}{\sqrt{2}}\right)i\right] = -4 - 4i \end{aligned}$$

Check: Use a graphing utility to verify that $(1 + i)^5 = -4 - 4i$.

Complex Roots

5 Let w be a given complex number, and let $n \geq 2$ denote a positive integer. Any complex number z that satisfies the equation

$$z^n = w$$

is called a **complex nth root** of w. In keeping with previous usage, if $n = 2$, the solutions of the equation $z^2 = w$ are called **complex square roots** of w, and if $n = 3$, the solutions of the equation $z^3 = w$ are called **complex cube roots** of w.

Theorem

Finding Complex Roots

Let $w = r(\cos\theta_0 + i\sin\theta_0)$ be a complex number. If $w \neq 0$, there are n distinct complex roots of w, given by the formula

$$z_k = \sqrt[n]{r}\left[\cos\left(\frac{\theta_0}{n} + \frac{2k\pi}{n}\right) + i\sin\left(\frac{\theta_0}{n} + \frac{2k\pi}{n}\right)\right] \quad \textbf{(8)}$$

where $k = 0, 1, 2, \ldots, n - 1$.

Proof (Outline) We will not prove this result in its entirety. Instead, we shall show only that each z_k in equation (8) obeys the equation $z_k^n = w$, proving that each z_k is a complex nth root of w.

$$\begin{aligned} z_k^n &= \left\{\sqrt[n]{r}\left[\cos\left(\frac{\theta_0}{n} + \frac{2k\pi}{n}\right) + i\sin\left(\frac{\theta_0}{n} + \frac{2k\pi}{n}\right)\right]\right\}^n \\ &= \left(\sqrt[n]{r}\right)^n\left\{\cos\left[n\left(\frac{\theta_0}{n} + \frac{2k\pi}{n}\right)\right] + i\sin\left[n\left(\frac{\theta_0}{n} + \frac{2k\pi}{n}\right)\right]\right\} && \text{DeMoivre's Theorem} \\ &= r[\cos(\theta_0 + 2k\pi) + i\sin(\theta_0 + 2k\pi)] \\ &= r(\cos\theta_0 + i\sin\theta_0) = w && \text{Periodic Property} \end{aligned}$$

So, each z_k, $k = 0, 1, \ldots, n - 1$, is a complex nth root of w. To complete the proof, we would need to show that each z_k, $k = 0, 1, \ldots, n - 1$, is, in fact, distinct and that there are no complex nth roots of w other than those given by equation (8).

EXAMPLE 6

Finding Complex Cube Roots

Find the complex cube roots of $-1 + \sqrt{3}i$. Leave your answers in polar form, with θ in degrees.

Solution First, we express $-1 + \sqrt{3}i$ in polar form using degrees.

$$-1 + \sqrt{3}i = 2\left(-\frac{1}{2} + \frac{\sqrt{3}}{2}i\right) = 2(\cos 120° + i\sin 120°)$$

So, $r = 2$ and $\theta_0 = 120°$. The three complex cube roots of $-1 + \sqrt{3}i = 2(\cos 120° + i\sin 120°)$ are

$$\begin{aligned} z_k &= \sqrt[3]{2}\left[\cos\left(\frac{120°}{3} + \frac{360°k}{3}\right) + i\sin\left(\frac{120°}{3} + \frac{360°k}{3}\right)\right], && k = 0, 1, 2 \\ &= \sqrt[3]{2}[\cos(40° + 120°k) + i\sin(40° + 120°k)], && k = 0, 1, 2 \end{aligned}$$

So,

$$z_0 = \sqrt[3]{2}\left[\cos(40^\circ + 120^\circ \cdot 0) + i\sin(40^\circ + 120^\circ \cdot 0)\right] = \sqrt[3]{2}(\cos 40^\circ + i\sin 40^\circ)$$

$$z_1 = \sqrt[3]{2}\left[\cos(40^\circ + 120^\circ \cdot 1) + i\sin(40^\circ + 120^\circ \cdot 1)\right] = \sqrt[3]{2}(\cos 160^\circ + i\sin 160^\circ)$$

$$z_2 = \sqrt[3]{2}\left[\cos(40^\circ + 120^\circ \cdot 2) + i\sin(40^\circ + 120^\circ \cdot 2)\right] = \sqrt[3]{2}(\cos 280^\circ + i\sin 280^\circ)$$

WARNING: Most graphing utilities will only provide the answer z_0 to the calculation $(-1 + \sqrt{3}i) \wedge (\frac{1}{3})$. The following paragraph explains how to obtain z_1 and z_2 from z_0.

Notice that each of the three complex roots of $-1 + \sqrt{3}i$ has the same magnitude, $\sqrt[3]{2}$. This means that the points corresponding to each cube root lie the same distance from the origin; that is, the three points lie on a circle with center at the origin and radius $\sqrt[3]{2}$. Furthermore, the arguments of these cube roots are 40°, 160°, and 280°, the difference of consecutive pairs being $120^\circ = 360^\circ/3$. This means that the three points are equally spaced on the circle, as shown in Figure 41. These results are not coincidental. In fact, you are asked to show that these results hold for complex nth roots in Problems 53 through 55.

Figure 41

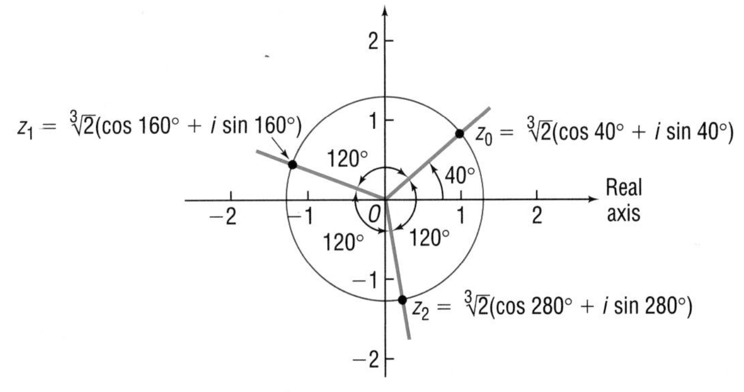

NOW WORK PROBLEM 43.

To graph the complex cube roots of $-1 + \sqrt{3}i$ on a graphing utility, we set the graphing calculator to polar, degree (since the argument is to be in degrees), and dot mode. We let $r_1 = \sqrt[3]{2}$ and use the following viewing window:

$\theta\text{min} = 40$	$X\text{min} = -3$	$Y\text{min} = -2$
$\theta\text{max} = 40 + 360 = 400$	$X\text{max} = 3$	$Y\text{max} = 2$
$\theta\text{step} = 360/3 = 120$	$X\text{scl} = 1$	$Y\text{scl} = 1$

Figure 42(a) shows the graph of the three complex cube roots. Figure 42(b) shows the circle $r_1 = \sqrt[3]{2}$ on which z_0, z_1, z_2 lie. Figure 42(c) shows the complex cube root corresponding to $k = 0$ $(0 = 40^\circ)$, obtained using TRACE.

Figure 42

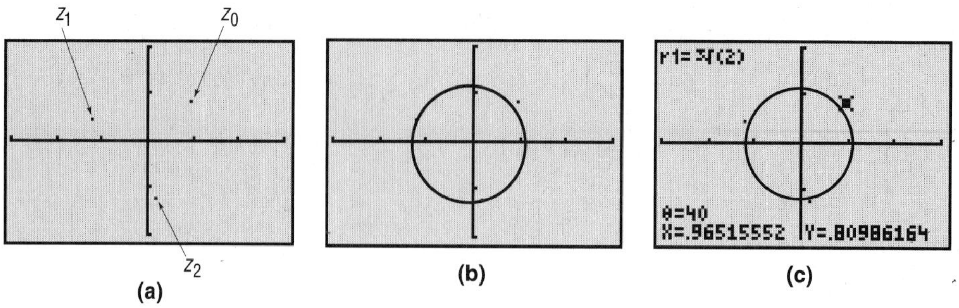

The following steps can be used to graph the complex nth roots of a complex number.

STEPS FOR GRAPHING THE COMPLEX NTH ROOTS OF A COMPLEX NUMBER $z = r(\cos\theta_0 + i\sin\theta_0)$ ON A GRAPHING UTILITY

STEP 1: Use polar and dot mode. If the argument is to be in radians, set the utility to radian mode. If the argument is to be in degrees, set the utility to degree mode.

STEP 2: Let $r_1 = \sqrt[n]{r}$, where r is the magnitude of the complex number in polar form.

STEP 3: Set the viewing window so that $\theta\text{min} = \dfrac{\theta_0}{n}$, where θ_0 is the argument of the complex number in polar form, $\theta\text{max} = \dfrac{\theta_0}{n} + 2\pi$ or $\dfrac{\theta_0}{n} + 360°$, and $\theta\text{step} = \dfrac{2\pi}{n}$ or $\dfrac{360°}{n}$.

STEP 4: Graph. To see the complex roots, use TRACE.

HISTORICAL FEATURE

John Wallis

The Babylonians, Greeks, and Arabs considered square roots of negative quantities to be impossible and equations with complex solutions to be unsolvable. The first hint that there was some connection between real solutions of equations and complex numbers came when Girolamo Cardano (1501–1576) and Tartaglia (1499–1557) found *real* roots of cubic equations by taking cube roots of *complex* quantities. For centuries thereafter, mathematicians worked with complex numbers without much belief in their actual existence. In 1673, John Wallis appears to have been the first to suggest the graphical representation of complex numbers, a truly significant idea that was not pursued further until about 1800. Several people, including Karl Friedrich Gauss (1777–1855), then rediscovered the idea, and the graphical representation helped to establish complex numbers as equal members of the number family. In practical applications, complex numbers have found their greatest uses in the area of alternating current, where they are a commonplace tool, and in the area of subatomic physics.

HISTORICAL PROBLEMS

1. The quadratic formula will work perfectly well if the coefficients are complex numbers. Solve the following using DeMoivre's Theorem where necessary.
[**Hint:** The answers are "nice."]

(a) $z^2 - (2 + 5i)z - 3 + 5i = 0$ (b) $z^2 - (1 + i)z - 2 - i = 0$

9.3 EXERCISES

In Problems 1–12, plot each complex number in the complex plane and write it in polar form. Express the argument in degrees.

1. $1 + i$ **2.** $-1 + i$ **3.** $\sqrt{3} - i$ **4.** $1 - \sqrt{3}i$

5. $-3i$ **6.** -2 **7.** $4 - 4i$ **8.** $9\sqrt{3} + 9i$

9. $3 - 4i$ **10.** $2 + \sqrt{3}i$ **11.** $-2 + 3i$ **12.** $\sqrt{5} - i$

In Problems 13–22, write each complex number in rectangular form.

13. $2(\cos 120° + i \sin 120°)$ **14.** $3(\cos 210° + i \sin 210°)$ **15.** $4\left(\cos \frac{7\pi}{4} + i \sin \frac{7\pi}{4}\right)$

16. $2\left(\cos \frac{5\pi}{6} + i \sin \frac{5\pi}{6}\right)$ **17.** $3\left(\cos \frac{3\pi}{2} + i \sin \frac{3\pi}{2}\right)$ **18.** $4\left(\cos \frac{\pi}{2} + i \sin \frac{\pi}{2}\right)$

19. $0.2(\cos 100° + i \sin 100°)$ **20.** $0.4(\cos 200° + i \sin 200°)$ **21.** $2\left(\cos \frac{\pi}{18} + i \sin \frac{\pi}{18}\right)$

22. $3\left(\cos \frac{\pi}{10} + i \sin \frac{\pi}{10}\right)$

In Problems 23–30, find zw and z/w. Leave your answers in polar form.

23. $z = 2(\cos 40° + i \sin 40°)$, $w = 4(\cos 20° + i \sin 20°)$

24. $z = \cos 120° + i \sin 120°$, $w = \cos 100° + i \sin 100°$

25. $z = 3(\cos 130° + i \sin 130°)$, $w = 4(\cos 270° + i \sin 270°)$

26. $z = 2(\cos 80° + i \sin 80°)$, $w = 6(\cos 200° + i \sin 200°)$

27. $z = 2\left(\cos \frac{\pi}{8} + i \sin \frac{\pi}{8}\right)$, $w = 2\left(\cos \frac{\pi}{10} + i \sin \frac{\pi}{10}\right)$

28. $z = 4\left(\cos \frac{3\pi}{8} + i \sin \frac{3\pi}{8}\right)$, $w = 2\left(\cos \frac{9\pi}{16} + i \sin \frac{9\pi}{16}\right)$

29. $z = 2 + 2i$, $w = \sqrt{3} - i$

30. $z = 1 - i$, $w = 1 - \sqrt{3}i$

In Problems 31–42, write each expression in the standard form a + bi. Verify your answer using a graphing utility.

31. $[4(\cos 40° + i \sin 40°)]^3$ **32.** $[3(\cos 80° + i \sin 80°)]^3$ **33.** $\left[2\left(\cos \frac{\pi}{10} + i \sin \frac{\pi}{10}\right)\right]^5$

34. $\left[\sqrt{2}\left(\cos \frac{5\pi}{16} + i \sin \frac{5\pi}{16}\right)\right]^4$ **35.** $[\sqrt{3}(\cos 10° + i \sin 10°)]^6$ **36.** $\left[\frac{1}{2}(\cos 72° + i \sin 72°)\right]^5$

37. $\left[\sqrt{5}\left(\cos \frac{3\pi}{16} + i \sin \frac{3\pi}{16}\right)\right]^4$ **38.** $\left[\sqrt{3}\left(\cos \frac{5\pi}{18} + i \sin \frac{5\pi}{18}\right)\right]^6$ **39.** $(1 - i)^5$

40. $(\sqrt{3} - i)^6$ **41.** $(\sqrt{2} - i)^6$ **42.** $(1 - \sqrt{5}i)^8$

In Problems 43–50, find all the complex roots. Leave your answers in polar form with the argument in degrees.

43. The complex cube roots of $1 + i$

44. The complex fourth roots of $\sqrt{3} - i$

45. The complex fourth roots of $4 - 4\sqrt{3}i$

46. The complex cube roots of $-8 - 8i$

47. The complex fourth roots of $-16i$

48. The complex cube roots of -8

49. The complex fifth roots of i

50. The complex fifth roots of $-i$

51. Find the four complex fourth roots of unity (1). Plot each one.

52. Find the six complex sixth roots of unity (1). Plot each one.

53. Show that each complex nth root of a nonzero complex number w has the same magnitude.

54. Use the result of Problem 53 to draw the conclusion that each complex nth root lies on a circle with center at the origin. What is the radius of this circle?

55. Refer to Problem 54. Show that the complex nth roots of a nonzero complex number w are equally spaced on the circle.

56. Prove formula (6).

PREPARING FOR THIS SECTION

Before getting started, review the following:

✓ Rectangular Coordinates (pp. 2–3) ✓ Pythagorean Theorem (p. 994)

9.4 VECTORS

OBJECTIVES
1. Graph Vectors
2. Find a Position Vector
3. Add and Subtract Vectors
4. Find a Scalar Product and the Magnitude of a Vector
5. Find a Unit Vector
6. Find a Vector from Its Direction and Magnitude
7. Work with Objects in Static Equilibrium

In simple terms, a **vector** (derived from the Latin *vehere,* meaning "to carry") is a quantity that has both magnitude and direction. It is convenient to represent a vector by using an arrow. The length of the arrow represents the **magnitude** of the vector, and the arrowhead indicates the **direction** of the vector.

Figure 43

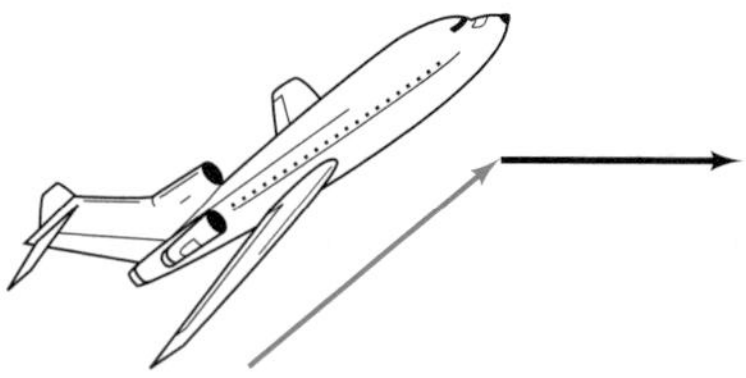

Many quantities in physics can be represented by vectors. For example, the velocity of an aircraft can be represented by an arrow that points in the direction of movement; the length of the arrow represents speed. If the aircraft speeds up, we lengthen the arrow; if the aircraft changes direction, we introduce an arrow in the new direction. See Figure 43. Based on this representation, it is not surprising that vectors and directed line segments are somehow related.

GEOMETRIC VECTORS

If P and Q are two distinct points in the xy-plane, there is exactly one line containing both P and Q [Figure 44(a)]. The points on that part of the line that joins P to Q, including P and Q, form what is called the **line segment** $\overline{PQ}$ [Figure 44(b)]. If we order the points so that they proceed from P to Q, we have a **directed line segment** from P to Q, or a **geometric vector**, which we denote by $\overrightarrow{PQ}$. In a directed line segment $\overrightarrow{PQ}$, we call P the **initial point** and Q the **terminal point,** as indicated in Figure 44(c).

Figure 44

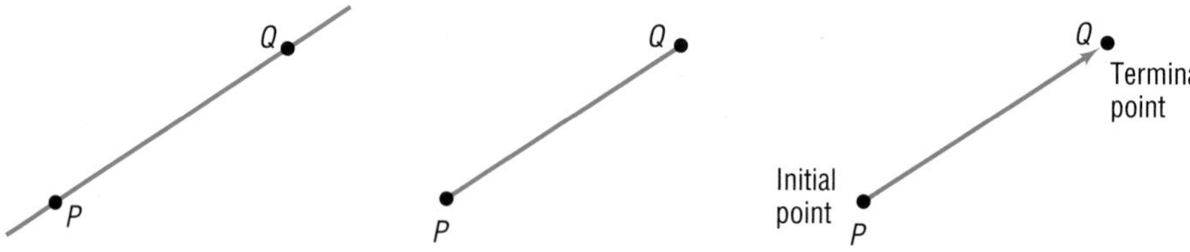

(a) Line containing P and Q **(b)** Line segment $\overline{PQ}$ **(c)** Directed line segment $\overrightarrow{PQ}$

The magnitude of the directed line segment $\overrightarrow{PQ}$ is the distance from the point P to the point Q; that is, it is the length of the line segment. The direction of $\overrightarrow{PQ}$ is from P to Q. If a vector **v*** has the same magnitude and the same direction as the directed line segment $\overrightarrow{PQ}$, then we write

$$\mathbf{v} = \overrightarrow{PQ}$$

Figure 45

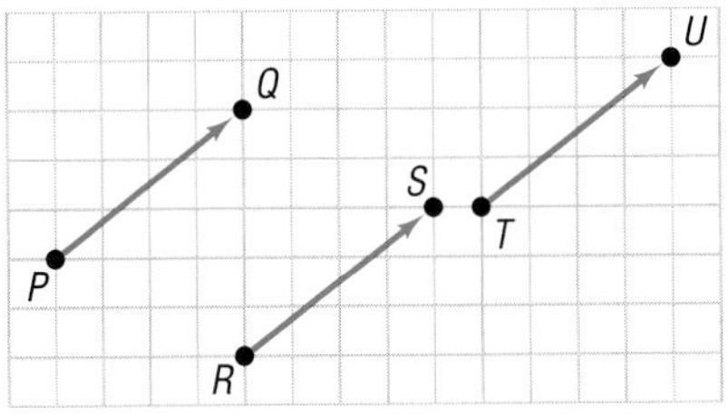

The vector **v** whose magnitude is 0 is called the **zero vector, 0.** The zero vector is assigned no direction.

Two vectors **v** and **w** are **equal,** written

$$\mathbf{v} = \mathbf{w}$$

if they have the same magnitude and the same direction.

For example, the vectors shown in Figure 45 have the same magnitude and the same direction, so they are equal, even though they have different initial points and different terminal points. As a result, we find it useful to think of a vector simply as an arrow, keeping in mind that two arrows (vectors) are equal if they have the same direction and the same magnitude (length).

Figure 46

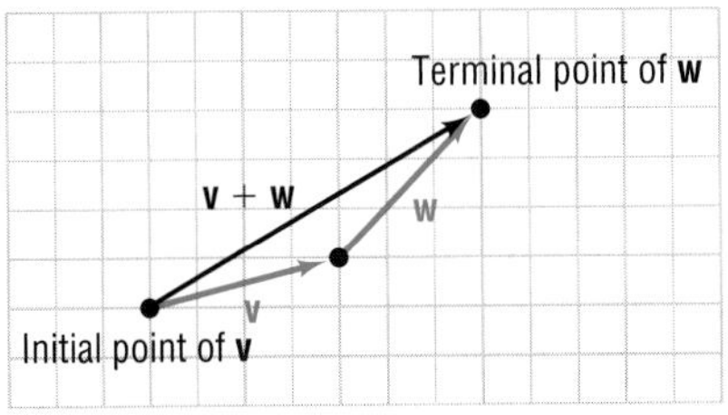

Adding Vectors

The **sum v + w** of two vectors is defined as follows: We position the vectors **v** and **w** so that the terminal point of **v** coincides with the initial point of **w,** as shown in Figure 46. The vector **v + w** is then the unique vector whose initial point coincides with the initial point of **v** and whose terminal point coincides with the terminal point of **w.**

Vector addition is **commutative.** That is, if **v** and **w** are any two vectors, then

$$\mathbf{v} + \mathbf{w} = \mathbf{w} + \mathbf{v}$$

Figure 47

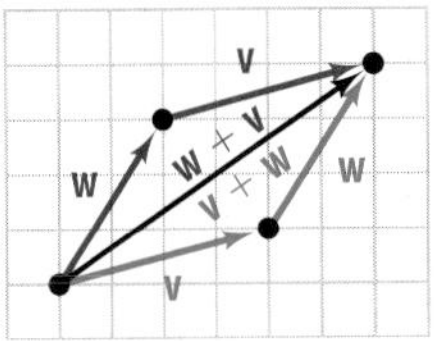

Figure 47 illustrates this fact. (Observe that the commutative property is another way of saying that opposite sides of a parallelogram are equal and parallel.)

Vector addition is also **associative.** That is, if **u, v,** and **w** are vectors, then

$$\mathbf{u} + (\mathbf{v} + \mathbf{w}) = (\mathbf{u} + \mathbf{v}) + \mathbf{w}$$

Figure 48
$(\mathbf{u} + \mathbf{v}) + \mathbf{w} = \mathbf{u} + (\mathbf{v} + \mathbf{w})$

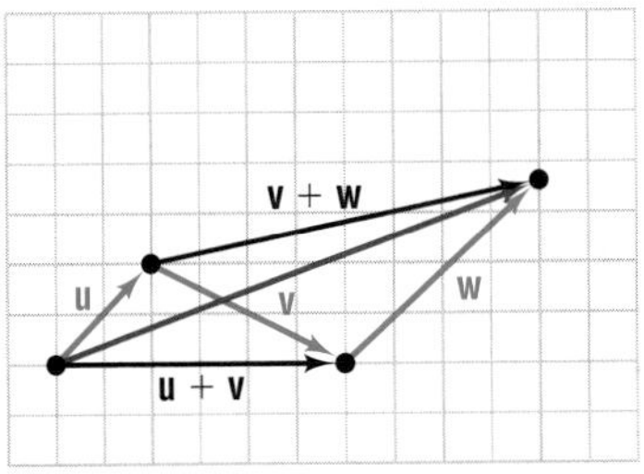

Figure 48 illustrates the associative property for vectors.

The zero vector has the property that

$$\mathbf{v} + \mathbf{0} = \mathbf{0} + \mathbf{v} = \mathbf{v}$$

for any vector **v.**

Figure 49

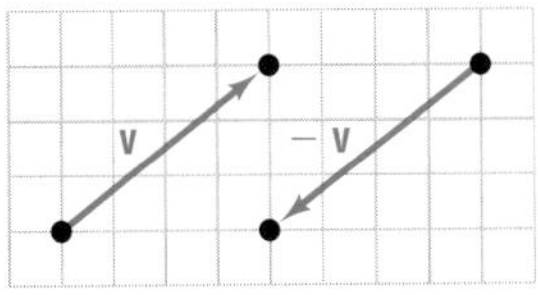

If **v** is a vector, then −**v** is the vector having the same magnitude as **v,** but whose direction is opposite to **v,** as shown in Figure 49.

Furthermore,

$$\mathbf{v} + (-\mathbf{v}) = \mathbf{0}$$

If **v** and **w** are two vectors, we define the **difference v − w** as

$$\mathbf{v} - \mathbf{w} = \mathbf{v} + (-\mathbf{w})$$

*Boldface letters will be used to denote vectors, in order to distinguish them from numbers. For handwritten work, an arrow is placed over the letter to signify a vector.

Figure 50

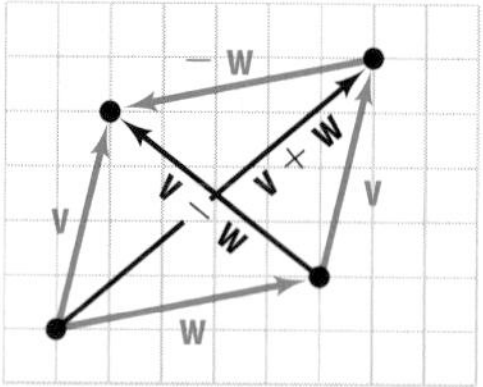

Figure 50 illustrates the relationships among **v**, **w**, **v** + **w**, and **v** − **w**.

Multiplying Vectors by Numbers

When dealing with vectors, we refer to real numbers as **scalars.** Scalars are quantities that have only magnitude. Examples from physics of scalar quantities are temperature, speed, and time. We now define how to multiply a vector by a scalar.

If α is a scalar and **v** is a vector, the **scalar product** $\alpha\mathbf{v}$ is defined as

1. If $\alpha > 0$, the product $\alpha\mathbf{v}$ is the vector whose magnitude is α times the magnitude of **v** and whose direction is the same as **v**.
2. If $\alpha < 0$, the product $\alpha\mathbf{v}$ is the vector whose magnitude is $|\alpha|$ times the magnitude of **v** and whose direction is opposite that of **v**.
3. If $\alpha = 0$ or if $\mathbf{v} = \mathbf{0}$, then $\alpha\mathbf{v} = \mathbf{0}$.

Figure 51

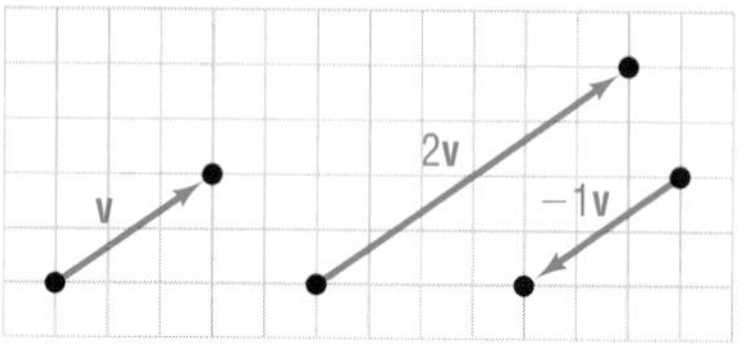

See Figure 51 for some illustrations.

For example, if **a** is the acceleration of an object of mass m due to a force **F** being exerted on it, then, by Newton's second law of motion, $\mathbf{F} = m\mathbf{a}$. Here, $m\mathbf{a}$ is the product of the scalar m and the vector **a**.

Scalar products have the following properties:

$$0\mathbf{v} = \mathbf{0} \qquad 1\mathbf{v} = \mathbf{v} \qquad -1\mathbf{v} = -\mathbf{v}$$

$$(\alpha + \beta)\mathbf{v} = \alpha\mathbf{v} + \beta\mathbf{v} \qquad \alpha(\mathbf{v} + \mathbf{w}) = \alpha\mathbf{v} + \alpha\mathbf{w}$$

$$\alpha(\beta\mathbf{v}) = (\alpha\beta)\mathbf{v}$$

EXAMPLE 1 Graphing Vectors

Figure 52

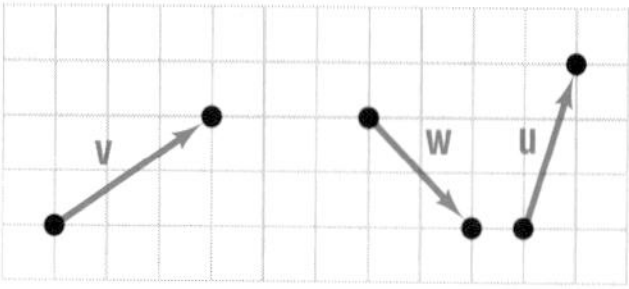

1 Use the vectors illustrated in Figure 52 to graph each of the following vectors:

(a) **v** − **w** (b) 2**v** + 3**w** (c) 2**v** − **w** + **u**

Solution Figure 53 illustrates each graph.

Figure 53

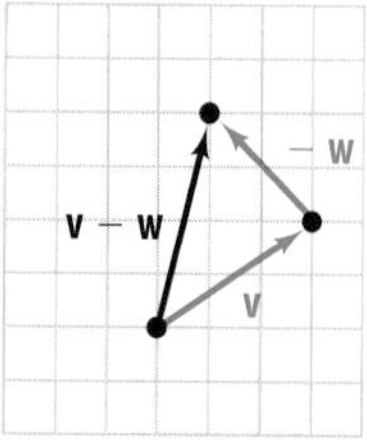

(a) v − w

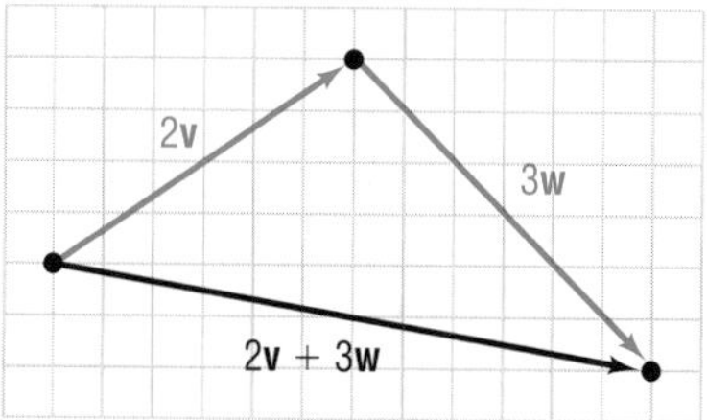

(b) 2v + 3w

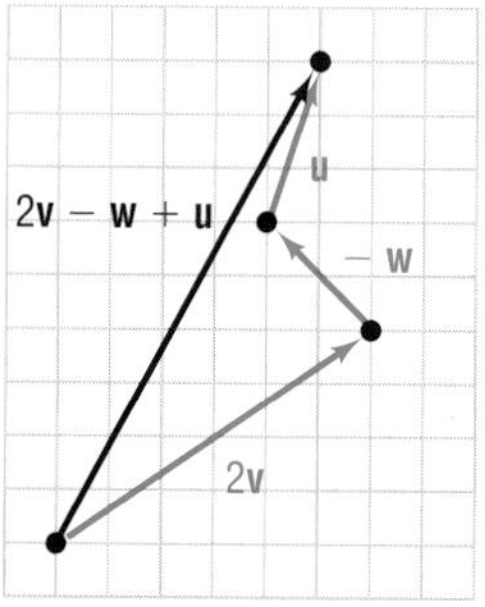

(c) 2v − w + u ■

NOW WORK PROBLEMS 1 AND 3.

Magnitudes of Vectors

If **v** is a vector, we use the symbol $\|\mathbf{v}\|$ to represent the **magnitude** of **v.** Since $\|\mathbf{v}\|$ equals the length of a directed line segment, it follows that $\|\mathbf{v}\|$ has the following properties:

Theorem **Properties of $\|\mathbf{v}\|$**

If **v** is a vector and if α is a scalar, then

(a) $\|\mathbf{v}\| \geq 0$ (b) $\|\mathbf{v}\| = 0$ if and only if $\mathbf{v} = \mathbf{0}$

(c) $\|-\mathbf{v}\| = \|\mathbf{v}\|$ (d) $\|\alpha\mathbf{v}\| = |\alpha|\,\|\mathbf{v}\|$

Property (a) is a consequence of the fact that distance is a nonnegative number. Property (b) follows, because the length of the directed line segment $\overrightarrow{PQ}$ is positive unless P and Q are the same point, in which case the length is 0. Property (c) follows because the length of the line segment $\overline{PQ}$ equals the length of the line segment $\overline{QP}$. Property (d) is a direct consequence of the definition of a scalar product.

A vector **u** for which $\|\mathbf{u}\| = 1$ is called a **unit vector.**

To compute the magnitude and direction of a vector, we need an algebraic way of representing vectors.

Algebraic Vectors

2 An **algebraic vector v** is represented as

$$\mathbf{v} = \langle a, b\rangle$$

Figure 54

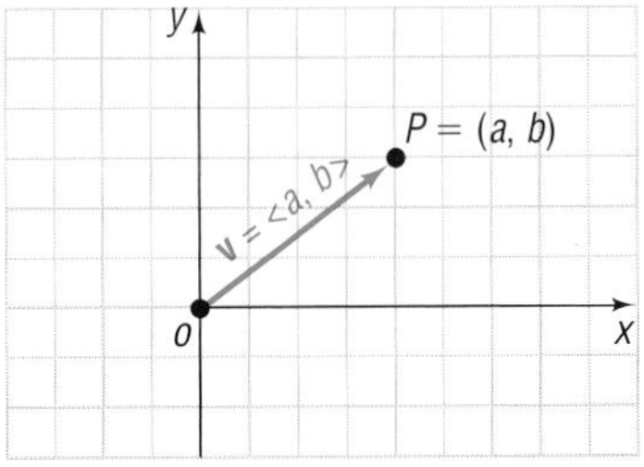

where a and b are real numbers (scalars) called the **components** of the vector **v**.

We use a rectangular coordinate system to represent algebraic vectors in the plane. If $\mathbf{v} = \langle a, b\rangle$ is an algebraic vector whose initial point is at the origin, then **v** is called a **position vector**. See Figure 54. Notice that the terminal point of the position vector $\mathbf{v} = \langle a, b\rangle$ is $P = (a, b)$.

The next result states that any vector whose initial point is not at the origin is equal to a unique position vector.

Theorem Suppose that **v** is a vector with initial point $P_1 = (x_1, y_1)$, not necessarily the origin, and terminal point $P_2 = (x_2, y_2)$. If $\mathbf{v} = \overrightarrow{P_1P_2}$, then **v** is equal to the position vector

$$\mathbf{v} = \langle x_2 - x_1, y_2 - y_1\rangle \qquad (1)$$

To see why this is true, look at Figure 55. Triangle OPA and triangle P_1P_2Q are congruent. (Do you see why? The line segments have the same magnitude, so $d(O, P) = d(P_1, P_2)$; and they have the same direction, so $\angle POA = \angle P_2P_1Q$. Since the triangles are right triangles, we have

Figure 55
$\langle a, b \rangle = \langle x_2 - x_1, y_2 - y_1 \rangle$

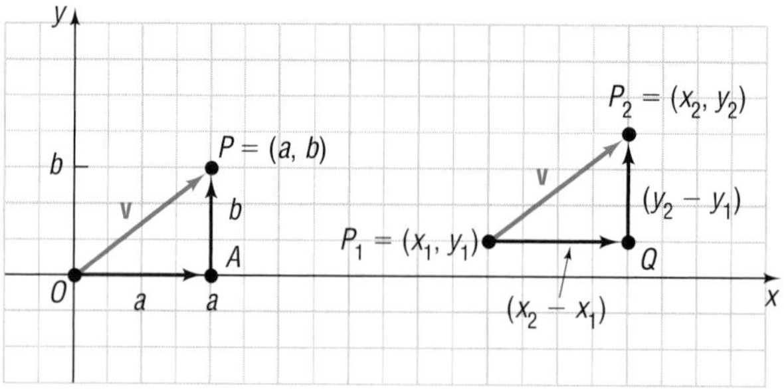

angle–side–angle.) It follows that corresponding sides are equal. As a result, $x_2 - x_1 = a$ and $y_2 - y_1 = b$, so **v** may be written as

$$\mathbf{v} = \langle a, b \rangle = \langle x_2 - x_1, y_2 - y_1 \rangle$$

Because of this result, we can replace any algebraic vector by a unique position vector, and vice versa. This flexibility is one of the main reasons for the wide use of vectors. Unless otherwise specified, from now on the term *vector* will mean the unique position vector equal to it.

EXAMPLE 2 Finding a Position Vector

Find the position vector of the vector $\mathbf{v} = \overrightarrow{P_1P_2}$ if $P_1 = (-1, 2)$ and $P_2 = (4, 6)$.

Solution By equation (1), the position vector equal to **v** is

$$\mathbf{v} = \langle 4 - (-1), 6 - 2 \rangle = \langle 5, 4 \rangle$$

See Figure 56.

Figure 56

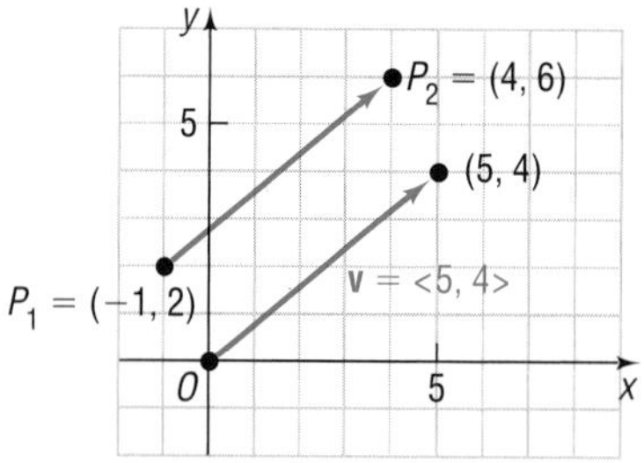

Two position vectors **v** and **w** are equal if and only if the terminal point of **v** is the same as the terminal point of **w**. This leads to the following result:

Theorem **Equality of Vectors**

Two vectors **v** and **w** are equal if and only if their corresponding components are equal. That is,

$$\text{If } \mathbf{v} = \langle a_1, b_1 \rangle \text{ and } \mathbf{w} = \langle a_2, b_2 \rangle$$
$$\text{then } \mathbf{v} = \mathbf{w} \text{ if and only if } a_1 = a_2 \text{ and } b_1 = b_2.$$

Figure 57

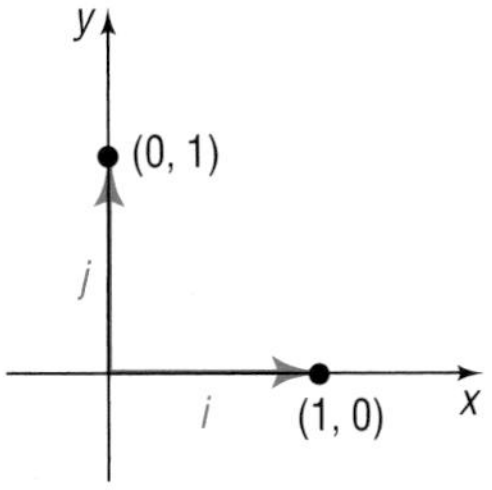

We now present an alternative representation of a vector in the plane that is common in the physical sciences. Let **i** denote the unit vector whose direction is along the positive x-axis; let **j** denote the unit vector whose direction is along the positive y-axis. Then $\mathbf{i} = \langle 1, 0 \rangle$ and $\mathbf{j} = \langle 0, 1 \rangle$, as shown in Figure 57. Any vector $\mathbf{v} = \langle a, b \rangle$ can be written using the unit vectors **i** and **j** as follows:

$$\mathbf{v} = \langle a, b \rangle = a\langle 1, 0 \rangle + b\langle 0, 1 \rangle = a\mathbf{i} + b\mathbf{j}$$

We call a and b the **horizontal** and **vertical components** of **v**, respectively.

NOW WORK PROBLEM 21.

We define addition, subtraction, scalar product, and magnitude in terms of the components of a vector.

Let $\mathbf{v} = a_1\mathbf{i} + b_1\mathbf{j} = \langle a_1, b_1 \rangle$ and $\mathbf{w} = a_2\mathbf{i} + b_2\mathbf{j} = \langle a_2, b_2 \rangle$ be two vectors, and let α be a scalar. Then

$$\mathbf{v} + \mathbf{w} = (a_1 + a_2)\mathbf{i} + (b_1 + b_2)\mathbf{j} = \langle a_1 + a_2, b_1 + b_2 \rangle \quad \textbf{(2)}$$

$$\mathbf{v} - \mathbf{w} = (a_1 - a_2)\mathbf{i} + (b_1 - b_2)\mathbf{j} = \langle a_1 - a_2, b_1 - b_2 \rangle \quad \textbf{(3)}$$

$$\alpha\mathbf{v} = (\alpha a_1)\mathbf{i} + (\alpha b_1)\mathbf{j} = \langle \alpha a_1, \alpha b_1 \rangle \quad \textbf{(4)}$$

$$\|\mathbf{v}\| = \sqrt{a_1^2 + b_1^2} \quad \textbf{(5)}$$

These definitions are compatible with the geometric ones given earlier in this section. See Figure 58.

Figure 58

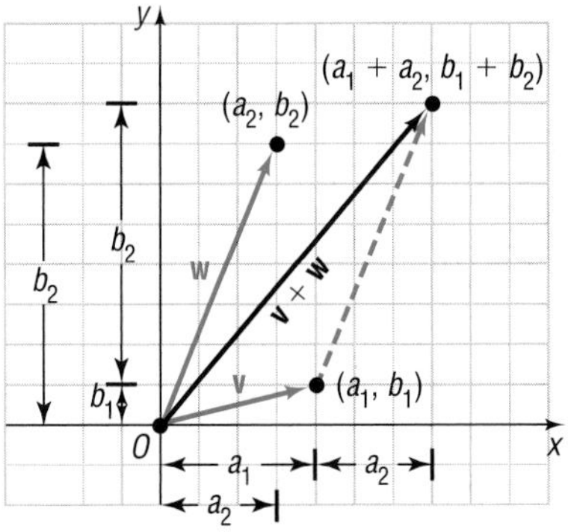

(a) Illustration of property (2)

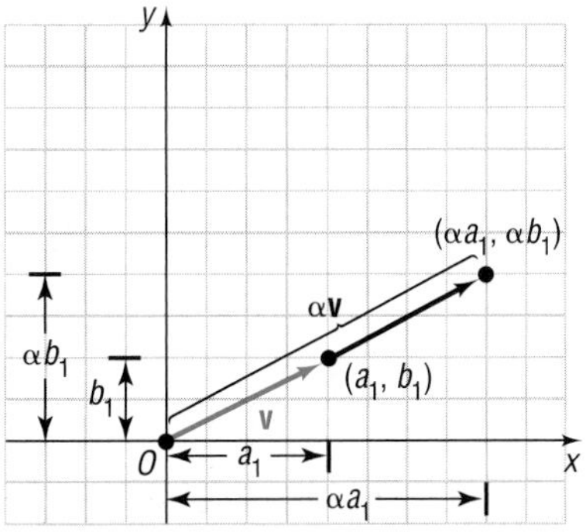

(b) Illustration of property (4), $\alpha > 0$

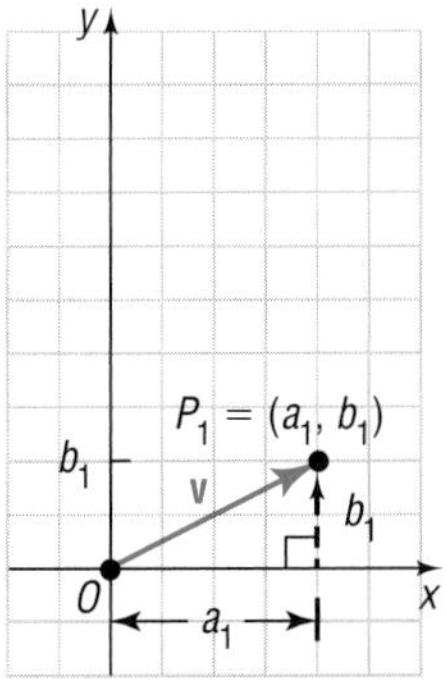

(c) Illustration of property (5):
$\|\mathbf{v}\|$ = Distance from O to P_1
$\|\mathbf{v}\| = \sqrt{a_1^2 + b_1^2}$

To add two vectors, add corresponding components. To subtract two vectors, subtract corresponding components.

EXAMPLE 3 **Adding and Subtracting Vectors**

3 If $\mathbf{v} = 2\mathbf{i} + 3\mathbf{j} = \langle 2, 3\rangle$ and $\mathbf{w} = 3\mathbf{i} - 4\mathbf{j} = \langle 3, -4\rangle$, find:

(a) $\mathbf{v} + \mathbf{w}$ (b) $\mathbf{v} - \mathbf{w}$

Solution (a) $\mathbf{v} + \mathbf{w} = (2\mathbf{i} + 3\mathbf{j}) + (3\mathbf{i} - 4\mathbf{j}) = (2 + 3)\mathbf{i} + (3 - 4)\mathbf{j} = 5\mathbf{i} - \mathbf{j}$

or

$\mathbf{v} + \mathbf{w} = \langle 2, 3\rangle + \langle 3, -4\rangle = \langle 2 + 3, 3 + (-4)\rangle = \langle 5, -1\rangle$

(b) $\mathbf{v} - \mathbf{w} = (2\mathbf{i} + 3\mathbf{j}) - (3\mathbf{i} - 4\mathbf{j}) = (2 - 3)\mathbf{i} + [3 - (-4)]\mathbf{j} = -\mathbf{i} + 7\mathbf{j}$

or

$\mathbf{v} - \mathbf{w} = \langle 2, 3\rangle - \langle 3, -4\rangle = \langle 2 - 3, 3 - (-4)\rangle = \langle -1, 7\rangle$

EXAMPLE 4 **Finding Scalar Products and Magnitudes**

4 If $\mathbf{v} = 2\mathbf{i} + 3\mathbf{j} = \langle 2, 3\rangle$ and $\mathbf{w} = 3\mathbf{i} - 4\mathbf{j} = \langle 3, -4\rangle$, find:

(a) $3\mathbf{v}$ (b) $2\mathbf{v} - 3\mathbf{w}$ (c) $\|\mathbf{v}\|$

Solution (a) $3\mathbf{v} = 3(2\mathbf{i} + 3\mathbf{j}) = 6\mathbf{i} + 9\mathbf{j}$

or

$3\mathbf{v} = 3\langle 2, 3\rangle = \langle 6, 9\rangle$

(b) $2\mathbf{v} - 3\mathbf{w} = 2(2\mathbf{i} + 3\mathbf{j}) - 3(3\mathbf{i} - 4\mathbf{j}) = 4\mathbf{i} + 6\mathbf{j} - 9\mathbf{i} + 12\mathbf{j}$

$= -5\mathbf{i} + 18\mathbf{j}$

or

$2\mathbf{v} - 3\mathbf{w} = 2\langle 2, 3\rangle - 3\langle 3, -4\rangle = \langle 4, 6\rangle - \langle 9, -12\rangle$

$= \langle 4 - 9, 6 - (-12)\rangle = \langle -5, 18\rangle$

(c) $\|\mathbf{v}\| = \|2\mathbf{i} + 3\mathbf{j}\| = \sqrt{2^2 + 3^2} = \sqrt{13}$

NOW WORK PROBLEMS 27 AND 33.

For the remainder of the section, we will express a vector $\mathbf{v}$ in the form $a\mathbf{i} + b\mathbf{j}$.

5 Recall that a unit vector $\mathbf{u}$ is one for which $\|\mathbf{u}\| = 1$. In many applications, it is useful to be able to find a unit vector $\mathbf{u}$ that has the same direction as a given vector $\mathbf{v}$.

Theorem **Unit Vector in Direction of v**

For any nonzero vector $\mathbf{v}$, the vector

$$\mathbf{u} = \frac{\mathbf{v}}{\|\mathbf{v}\|}$$

is a unit vector that has the same direction as $\mathbf{v}$.

Proof Let $\mathbf{v} = a\mathbf{i} + b\mathbf{j}$. Then $\|\mathbf{v}\| = \sqrt{a^2 + b^2}$ and

$$\mathbf{u} = \frac{\mathbf{v}}{\|\mathbf{v}\|} = \frac{a\mathbf{i} + b\mathbf{j}}{\sqrt{a^2 + b^2}} = \frac{a}{\sqrt{a^2 + b^2}}\mathbf{i} + \frac{b}{\sqrt{a^2 + b^2}}\mathbf{j}$$

The vector $\mathbf{u}$ is in the same direction as $\mathbf{v}$, since $\|\mathbf{v}\| > 0$. Furthermore,

$$\|\mathbf{u}\| = \sqrt{\frac{a^2}{a^2 + b^2} + \frac{b^2}{a^2 + b^2}} = \sqrt{\frac{a^2 + b^2}{a^2 + b^2}} = 1$$

Thus, $\mathbf{u}$ is a unit vector in the direction of $\mathbf{v}$. ■

As a consequence of this theorem, if $\mathbf{u}$ is a unit vector in the same direction as a vector $\mathbf{v}$, then $\mathbf{v}$ may be expressed as

$$\mathbf{v} = \|\mathbf{v}\|\,\mathbf{u} \qquad (6)$$

This way of expressing a vector is useful in many applications.

EXAMPLE 5 Finding a Unit Vector

Find a unit vector in the same direction as $\mathbf{v} = 4\mathbf{i} - 3\mathbf{j}$.

Solution We find $\|\mathbf{v}\|$ first.

$$\|\mathbf{v}\| = \|4\mathbf{i} - 3\mathbf{j}\| = \sqrt{16 + 9} = 5$$

Now we multiply $\mathbf{v}$ by the scalar $1/\|\mathbf{v}\| = \frac{1}{5}$. A unit vector in the same direction as $\mathbf{v}$ is

$$\frac{\mathbf{v}}{\|\mathbf{v}\|} = \frac{4\mathbf{i} - 3\mathbf{j}}{5} = \frac{4}{5}\mathbf{i} - \frac{3}{5}\mathbf{j}$$ ■

Check: This vector is, in fact, a unit vector because

$$\left(\tfrac{4}{5}\right)^2 + \left(-\tfrac{3}{5}\right)^2 = \tfrac{16}{25} + \tfrac{9}{25} = \tfrac{25}{25} = 1$$ ■

NOW WORK PROBLEM 43.

WRITING A VECTOR IN TERMS OF ITS MAGNITUDE AND DIRECTION

6 If a vector represents the speed and direction of an object, it is called a **velocity vector**. If a vector represents the direction and amount of a force acting on an object, it is called a **force vector**. In many applications, a vector is described in terms of its magnitude and direction, rather than in terms of its components. For example, a ball thrown with an initial speed of 25 miles per hour at an angle 30° to the horizontal is a velocity vector.

Suppose that we are given the magnitude $\|\mathbf{v}\|$ of a nonzero vector $\mathbf{v}$ and the angle α, $0° \le \alpha < 360°$ between $\mathbf{v}$ and $\mathbf{i}$. To express $\mathbf{v}$ in terms of $\|\mathbf{v}\|$ and α, we first find the unit vector $\mathbf{u}$ having the same direction as $\mathbf{v}$.

Figure 59

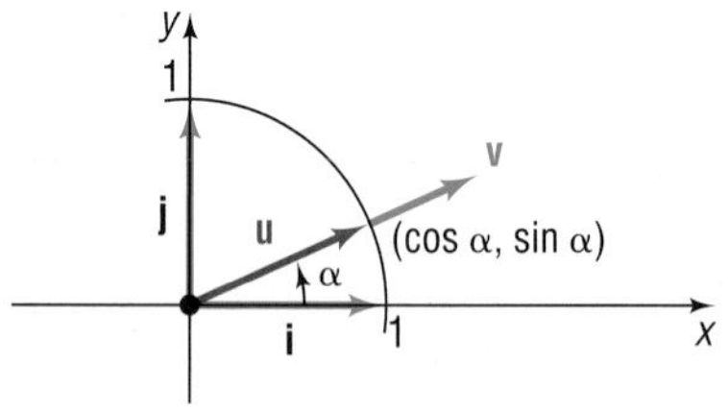

$$\mathbf{u} = \frac{\mathbf{v}}{\|\mathbf{v}\|} \quad \text{or} \quad \mathbf{v} = \|\mathbf{v}\|\mathbf{u} \tag{7}$$

Look at Figure 59. The coordinates of the terminal point of **u** are $(\cos\alpha, \sin\alpha)$. Then $\mathbf{u} = \cos\alpha\mathbf{i} + \sin\alpha\mathbf{j}$ and, from (7),

$$\mathbf{v} = \|\mathbf{v}\|(\cos\alpha\mathbf{i} + \sin\alpha\mathbf{j}) \tag{8}$$

EXAMPLE 6 Writing a Vector When Its Magnitude and Direction Are Given

A ball is thrown with an initial speed of 25 miles per hour in a direction that makes an angle of 30° with the positive x-axis. Express the velocity vector **v** in terms of **i** and **j**. What is the initial speed in the horizontal direction? What is the initial speed in the vertical direction?

Solution The magnitude of **v** is $\|\mathbf{v}\| = 25$ miles per hour, and the angle between the direction of **v** and **i**, the positive x-axis, is $\alpha = 30°$. By equation (8),

$$\mathbf{v} = \|\mathbf{v}\|(\cos\alpha\mathbf{i} + \sin\alpha\mathbf{j}) = 25(\cos 30°\mathbf{i} + \sin 30°\mathbf{j}) = 25\left(\frac{\sqrt{3}}{2}\mathbf{i} + \frac{1}{2}\mathbf{j}\right) = \frac{25}{2}(\sqrt{3}\mathbf{i} + \mathbf{j})$$

The initial speed of the ball in the horizontal direction is the horizontal component of **v**, $25\sqrt{3}/2 \approx 21.65$ miles per hour. The initial speed in the vertical direction is the vertical component of **v**, $25/2 = 12.5$ miles per hour. ■

Application: Static Equilibrium

Figure 60

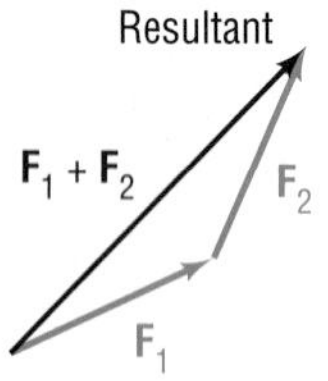

7 Because forces can be represented by vectors, two forces "combine" the way that vectors "add," If $\mathbf{F}_1$ and $\mathbf{F}_2$ are two forces simultaneously acting on an object, the vector sum $\mathbf{F}_1 + \mathbf{F}_2$ is the **resultant force**. The resultant force, produces the same effect on the object as that obtained when the two forces $\mathbf{F}_1$ and $\mathbf{F}_2$ act on the object. See Figure 60. An application of this concept is *static equilibrium*. An object is said to be in **static equilibrium** if (1) the object is at rest and (2) the sum of all forces acting on the object is zero, that is, if the resultant force is 0.

EXAMPLE 7 An Object in Static Equilibrium

A box of supplies that weighs 1200 pounds is suspended by two cables attached to the ceiling, as shown in Figure 61. What is the tension in the two cables?

Figure 61

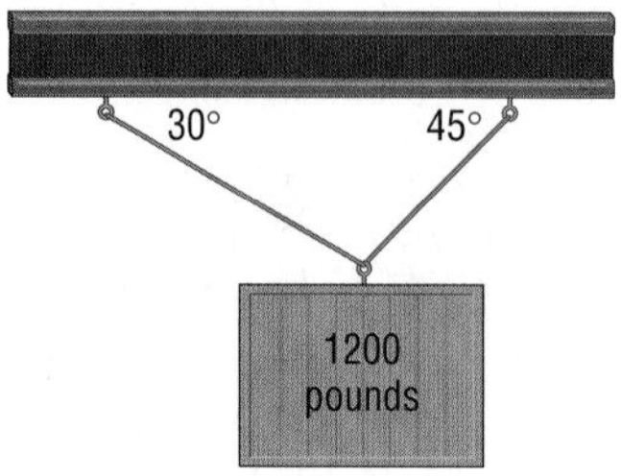

Solution We draw a force diagram with the vectors drawn as shown in Figure 62. The tensions in the cables are the magnitudes $\|\mathbf{F}_1\|$ and $\|\mathbf{F}_2\|$ of the force vectors

Figure 62

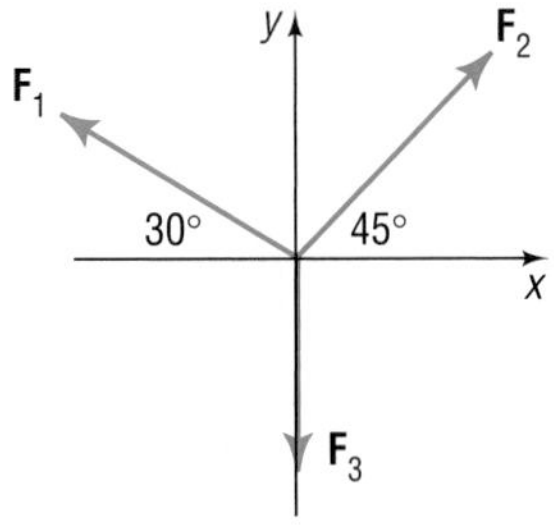

$\mathbf{F}_1$ and $\mathbf{F}_2$. The magnitude of the force vector $\mathbf{F}_3$ equals 1200 pounds, the weight of the box. Now write each force vector in terms of the unit vectors $\mathbf{i}$ and $\mathbf{j}$.

$$\mathbf{F}_1 = \|\mathbf{F}_1\|(\cos 150°\mathbf{i} + \sin 150°\mathbf{j}) = \|\mathbf{F}_1\|\left(-\frac{\sqrt{3}}{2}\mathbf{i} + \frac{1}{2}\mathbf{j}\right) = -\frac{\sqrt{3}}{2}\|\mathbf{F}_1\|\mathbf{i} + \frac{1}{2}\|\mathbf{F}_1\|\mathbf{j}$$

$$\mathbf{F}_2 = \|\mathbf{F}_2\|(\cos 45°\mathbf{i} + \sin 45°\mathbf{j}) = \|\mathbf{F}_2\|\left(\frac{\sqrt{2}}{2}\mathbf{i} + \frac{\sqrt{2}}{2}\mathbf{j}\right) = \frac{\sqrt{2}}{2}\|\mathbf{F}_2\|\mathbf{i} + \frac{\sqrt{2}}{2}\|\mathbf{F}_2\|\mathbf{j}$$

$$\mathbf{F}_3 = -1200\mathbf{j}$$

For static equilibrium the sum of the force vectors must equal zero.

$$\mathbf{F}_1 + \mathbf{F}_2 + \mathbf{F}_3 = -\frac{\sqrt{3}}{2}\|\mathbf{F}_1\|\mathbf{i} + \frac{1}{2}\|\mathbf{F}_1\|\mathbf{j} + \frac{\sqrt{2}}{2}\|\mathbf{F}_2\|\mathbf{i} + \frac{\sqrt{2}}{2}\|\mathbf{F}_2\|\mathbf{j} - 1200\mathbf{j} = \mathbf{0}$$

The $\mathbf{i}$ component and $\mathbf{j}$ component will each equal zero. This results in the two equations

$$-\frac{\sqrt{3}}{2}\|\mathbf{F}_1\| + \frac{\sqrt{2}}{2}\|\mathbf{F}_2\| = 0 \qquad \textbf{(9)}$$

$$\frac{1}{2}\|\mathbf{F}_1\| + \frac{\sqrt{2}}{2}\|\mathbf{F}_2\| - 1200 = 0 \qquad \textbf{(10)}$$

We solve equation (9) for $\|\mathbf{F}_2\|$ and obtain

$$\|\mathbf{F}_2\| = \frac{\sqrt{3}}{\sqrt{2}}\|\mathbf{F}_1\| \qquad \textbf{(11)}$$

Substituting into equation (10) and solving for $\|\mathbf{F}_1\|$, we obtain

$$\frac{1}{2}\|\mathbf{F}_1\| + \frac{\sqrt{2}}{2}\left(\frac{\sqrt{3}}{\sqrt{2}}\|\mathbf{F}_1\|\right) - 1200 = 0$$

$$\frac{1}{2}\|\mathbf{F}_1\| + \frac{\sqrt{3}}{2}\|\mathbf{F}_1\| - 1200 = 0$$

$$\frac{1 + \sqrt{3}}{2}\|\mathbf{F}_1\| = 1200$$

$$\|\mathbf{F}_1\| = \frac{2400}{1 + \sqrt{3}} \approx 878.5 \text{ pounds}$$

Substituting this value into equation (11) yields $\|\mathbf{F}_2\|$.

$$\|\mathbf{F}_2\| = \frac{\sqrt{3}}{\sqrt{2}}\|\mathbf{F}_1\| = \frac{\sqrt{3}}{\sqrt{2}}\frac{2400}{1 + \sqrt{3}} \approx 1075.9 \text{ pounds}$$

The left cable has tension equal to 878.5 pounds and the right cable has tension equal to 1075.9 pounds. ■

HISTORICAL FEATURE

Josiah Gibbs 1839–1903

The history of vectors is surprisingly complicated for such a natural concept. In the *xy*-plane, complex numbers do a good job of imitating vectors. About 1840, mathematicians became interested in finding a system that would do for three dimensions what the complex numbers do for two dimensions. Hermann Grassmann (1809–1877), in Germany, and William Rowan Hamilton (1805–1865), in Ireland, both attempted to find solutions.

Hamilton's system was the *quaternions,* which are best thought of as a real number plus a vector, and do for four dimensions what complex numbers do for two dimensions. In this system the order of multiplication matters; that is, **ab** ≠ **ba.** Also, two products of vectors emerged, the scalar (or dot) product and the vector (or cross) product.

Grassmann's abstract style, although easily read today, was almost impenetrable during the previous century, and only a few of his ideas were appreciated. Among those few were the same scalar and vector products that Hamilton had found.

About 1880, the American physicist Josiah Willard Gibbs (1839–1903) worked out an algebra involving only the simplest concepts: the vectors and the two products. He then added some calculus, and the resulting system was simple, flexible, and well adapted to expressing a large number of physical laws. This system remains in use essentially unchanged. Hamilton's and Grassmann's more extensive systems each gave birth to much interesting mathematics, but little of this mathematics is seen at elementary levels.

9.4 EXERCISES

In Problems 1–8, use the vectors in the figure at the right to graph each of the following vectors.

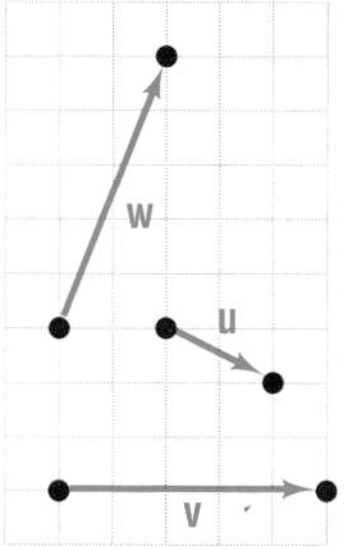

1. $\mathbf{v} + \mathbf{w}$ **2.** $\mathbf{u} + \mathbf{v}$

3. $3\mathbf{v}$ **4.** $4\mathbf{w}$

5. $\mathbf{v} - \mathbf{w}$ **6.** $\mathbf{u} - \mathbf{v}$

7. $3\mathbf{v} + \mathbf{u} - 2\mathbf{w}$ **8.** $2\mathbf{u} - 3\mathbf{v} + \mathbf{w}$

In Problems 9–16, use the figure at the right. Determine whether the given statement is true or false.

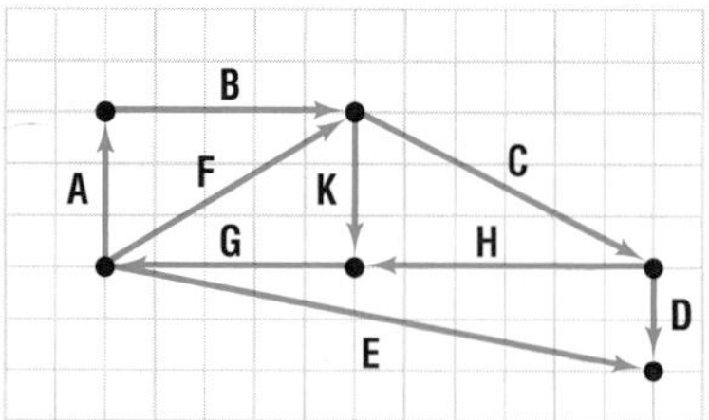

9. $\mathbf{A} + \mathbf{B} = \mathbf{F}$ **10.** $\mathbf{K} + \mathbf{G} = \mathbf{F}$

11. $\mathbf{C} = \mathbf{D} - \mathbf{E} + \mathbf{F}$ **12.** $\mathbf{G} + \mathbf{H} + \mathbf{E} = \mathbf{D}$

13. $\mathbf{E} + \mathbf{D} = \mathbf{G} + \mathbf{H}$ **14.** $\mathbf{H} - \mathbf{C} = \mathbf{G} - \mathbf{F}$

15. $\mathbf{A} + \mathbf{B} + \mathbf{K} + \mathbf{G} = \mathbf{0}$ **16.** $\mathbf{A} + \mathbf{B} + \mathbf{C} + \mathbf{H} + \mathbf{G} = \mathbf{0}$

17. If $\|\mathbf{v}\| = 4$, what is $\|3\mathbf{v}\|$? **18.** If $\|\mathbf{v}\| = 2$, what is $\|-4\mathbf{v}\|$?

In Problems 19–26, the vector **v** *has initial point P and terminal point Q. Write* **v** *in the form a***i** + *b***j**, *that is, find its position vector.*

19. $P = (0,0)$; $Q = (3,4)$ **20.** $P = (0,0)$; $Q = (-3,-5)$

21. $P = (3,2)$; $Q = (5,6)$ **22.** $P = (-3,2)$; $Q = (6,5)$

23. $P = (-2,-1)$; $Q = (6,-2)$ **24.** $P = (-1,4)$; $Q = (6,2)$

25. $P = (1,0)$; $Q = (0,1)$ **26.** $P = (1,1)$; $Q = (2,2)$

In Problems 27–32, find $\|\mathbf{v}\|$.

27. $\mathbf{v} = 3\mathbf{i} - 4\mathbf{j}$ **28.** $\mathbf{v} = -5\mathbf{i} + 12\mathbf{j}$ **29.** $\mathbf{v} = \mathbf{i} - \mathbf{j}$

30. $\mathbf{v} = -\mathbf{i} - \mathbf{j}$ **31.** $\mathbf{v} = -2\mathbf{i} + 3\mathbf{j}$ **32.** $\mathbf{v} = 6\mathbf{i} + 2\mathbf{j}$

In Problems 33–38, find each quantity if $\mathbf{v} = 3\mathbf{i} - 5\mathbf{j}$ *and* $\mathbf{w} = -2\mathbf{i} + 3\mathbf{j}$.

33. $2\mathbf{v} + 3\mathbf{w}$
34. $3\mathbf{v} - 2\mathbf{w}$
35. $\|\mathbf{v} - \mathbf{w}\|$
36. $\|\mathbf{v} + \mathbf{w}\|$
37. $\|\mathbf{v}\| - \|\mathbf{w}\|$
38. $\|\mathbf{v}\| + \|\mathbf{w}\|$

In Problems 39–44, find the unit vector having the same direction as $\mathbf{v}$.

39. $\mathbf{v} = 5\mathbf{i}$
40. $\mathbf{v} = -3\mathbf{j}$
41. $\mathbf{v} = 3\mathbf{i} - 4\mathbf{j}$
42. $\mathbf{v} = -5\mathbf{i} + 12\mathbf{j}$
43. $\mathbf{v} = \mathbf{i} - \mathbf{j}$
44. $\mathbf{v} = 2\mathbf{i} - \mathbf{j}$

45. Find a vector $\mathbf{v}$ whose magnitude is 4 and whose component in the $\mathbf{i}$ direction is twice the component in the $\mathbf{j}$ direction.

46. Find a vector $\mathbf{v}$ whose magnitude is 3 and whose component in the $\mathbf{i}$ direction is equal to the component in the $\mathbf{j}$ direction.

47. If $\mathbf{v} = 2\mathbf{i} - \mathbf{j}$ and $\mathbf{w} = x\mathbf{i} + 3\mathbf{j}$, find all numbers x for which $\|\mathbf{v} + \mathbf{w}\| = 5$.

48. If $P = (-3, 1)$ and $Q = (x, 4)$, find all numbers x such that the vector represented by $\overrightarrow{PQ}$ has length 5.

In Problems 49–54, write the vector $\mathbf{v}$ *in the form* $a\mathbf{i} + b\mathbf{j}$, *given its magnitude* $\|\mathbf{v}\|$ *and the angle* α *it makes with the positive x-axis.*

49. $\|\mathbf{v}\| = 5, \quad \alpha = 60°$
50. $\|\mathbf{v}\| = 8, \quad \alpha = 45°$
51. $\|\mathbf{v}\| = 14, \quad \alpha = 120°$
52. $\|\mathbf{v}\| = 3, \quad \alpha = 240°$
53. $\|\mathbf{v}\| = 25, \quad \alpha = 330°$
54. $\|\mathbf{v}\| = 15, \quad \alpha = 315°$

55. A child pulls a wagon with a force of 40 pounds. The handle of the wagon makes an angle of 30° with the ground. Express the force vector $\mathbf{F}$ in terms of $\mathbf{i}$ and $\mathbf{j}$.

56. A man pushes a wheelbarrow up an incline of 20° with a force of 100 pounds. Express the force vector $\mathbf{F}$ in terms of $\mathbf{i}$ and $\mathbf{j}$.

57. Resultant Force Two forces of magnitude 40 newtons (N) and 60 newtons act on an object at angles of 30° and −45° with the positive x-axis as shown in the figure. Find the direction and magnitude of the resultant force; that is, find $\mathbf{F}_1 + \mathbf{F}_2$.

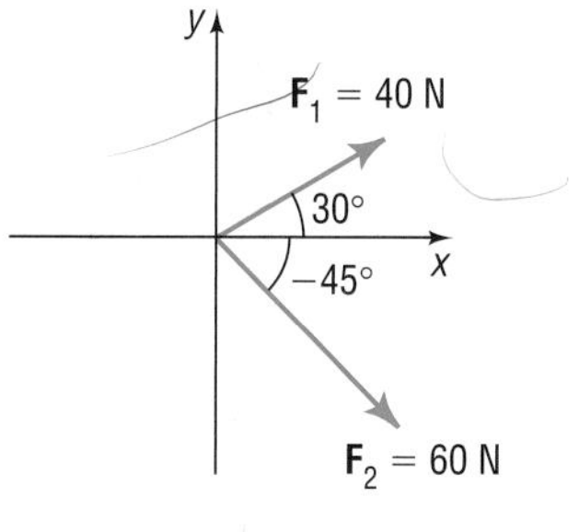

58. Resultant Force Two forces of magnitude 30 newtons (N) and 70 newtons act on an object at angles of 45° and 120° with the positive x-axis as shown in the figure. Find the direction and magnitude of the resultant force; that is, find $\mathbf{F}_1 + \mathbf{F}_2$.

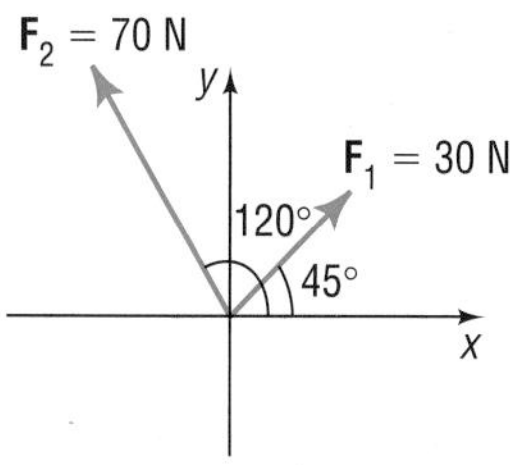

59. Static Equilibrium A weight of 1000 pounds is suspended from two cables as shown in the figure. What is the tension of the two cables?

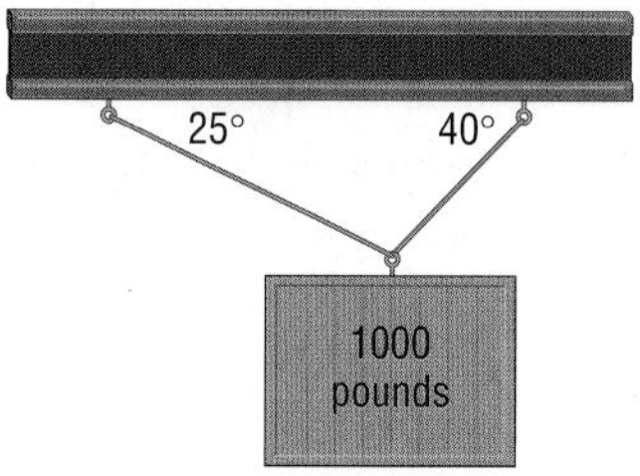

60. Static Equilibrium A weight of 800 pounds is suspended from two cables as shown in the figure. What is the tension of the two cables?

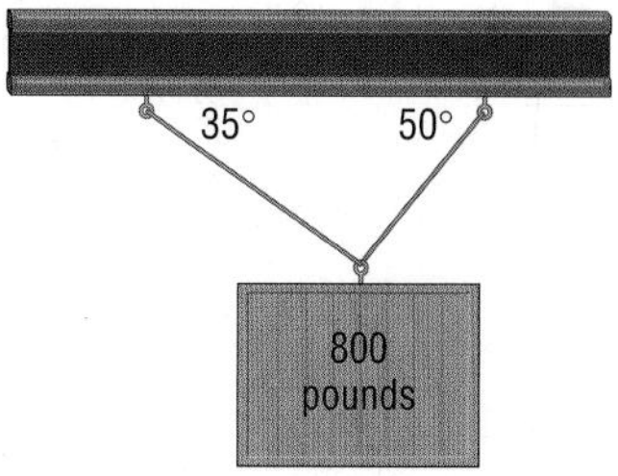

61. Static Equilibrium A tightrope walker located at a certain point deflects the rope as indicated in the figure. If the weight of the tightrope walker is 150 pounds, how much tension is in each part of the slope?

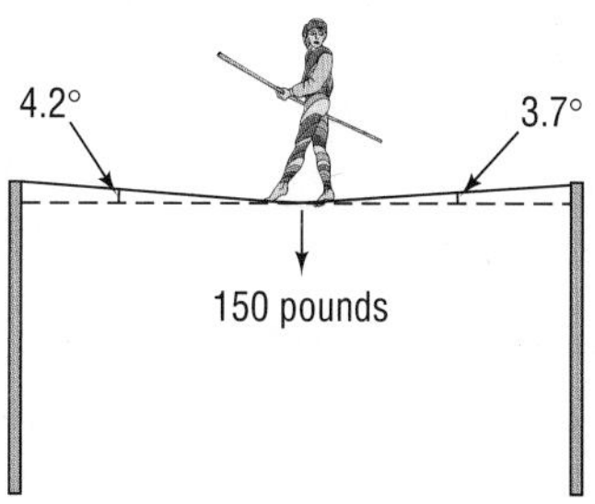

62. Static Equilibrium Repeat Problem 57 if the left angle is 3.8°, the right angle is 2.6°, and the weight of the tightrope walker is 135 pounds.

63. Show on the following graph the force needed for the object at P to be in static equilibrium.

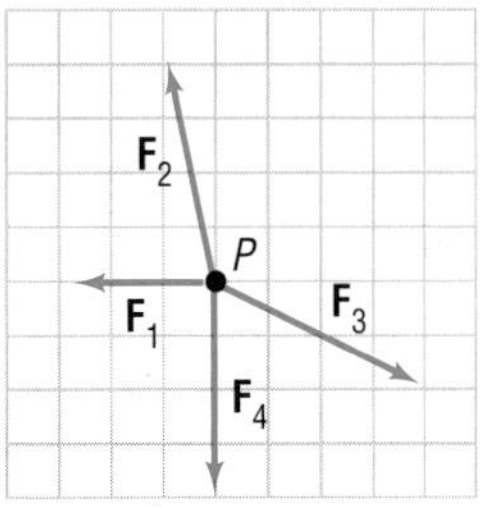

64. Explain in your own words what a vector is. Give an example of a vector.

65. Write a brief paragraph comparing the algebra of complex numbers and the algebra of vectors.

PREPARING FOR THIS SECTION

Before getting started, review the following:

✓ Law of Cosines (Section 8.3)

9.5 THE DOT PRODUCT

OBJECTIVES
1. Find the Dot Product of Two Vectors
2. Find the Angle between Two Vectors
3. Determine Whether Two Vectors Are Parallel
4. Determine Whether Two Vectors Are Orthogonal
5. Decompose a Vector into Two Orthogonal Vectors
6. Compute Work

1 The definition for a product of two vectors is somewhat unexpected. However, such a product has meaning in many geometric and physical applications.

If $\mathbf{v} = a_1\mathbf{i} + b_1\mathbf{j}$ and $\mathbf{w} = a_2\mathbf{i} + b_2\mathbf{j}$ are two vectors, the **dot product** $\mathbf{v} \cdot \mathbf{w}$ is defined as

$$\mathbf{v} \cdot \mathbf{w} = a_1 a_2 + b_1 b_2 \qquad (1)$$

EXAMPLE 1 Finding Dot Products

If $\mathbf{v} = 2\mathbf{i} - 3\mathbf{j}$ and $\mathbf{w} = 5\mathbf{i} + 3\mathbf{j}$, find:

(a) $\mathbf{v} \cdot \mathbf{w}$ (b) $\mathbf{w} \cdot \mathbf{v}$ (c) $\mathbf{v} \cdot \mathbf{v}$
(d) $\mathbf{w} \cdot \mathbf{w}$ (e) $\|\mathbf{v}\|$ (f) $\|\mathbf{w}\|$

Solution
(a) $\mathbf{v} \cdot \mathbf{w} = 2(5) + (-3)3 = 1$ (b) $\mathbf{w} \cdot \mathbf{v} = 5(2) + 3(-3) = 1$
(c) $\mathbf{v} \cdot \mathbf{v} = 2(2) + (-3)(-3) = 13$ (d) $\mathbf{w} \cdot \mathbf{w} = 5(5) + 3(3) = 34$
(e) $\|\mathbf{v}\| = \sqrt{2^2 + (-3)^2} = \sqrt{13}$ (f) $\|\mathbf{w}\| = \sqrt{5^2 + 3^2} = \sqrt{34}$ ■

Since the dot product $\mathbf{v} \cdot \mathbf{w}$ of two vectors $\mathbf{v}$ and $\mathbf{w}$ is a real number (scalar), we sometimes refer to it as the **scalar product.**

PROPERTIES

The results obtained in Example 1 suggest some general properties.

Theorem **Properties of the Dot Product**

If **u, v,** and **w** are vectors, then

Commutative Property

$$\mathbf{u} \cdot \mathbf{v} = \mathbf{v} \cdot \mathbf{u} \quad (2)$$

Distributive Property

$$\mathbf{u} \cdot (\mathbf{v} + \mathbf{w}) = \mathbf{u} \cdot \mathbf{v} + \mathbf{u} \cdot \mathbf{w} \quad (3)$$

$$\mathbf{v} \cdot \mathbf{v} = \|\mathbf{v}\|^2 \quad (4)$$

$$\mathbf{0} \cdot \mathbf{v} = 0 \quad (5)$$

■

Proof We will prove properties (2) and (4) here and leave properties (3) and (5) as exercises (see Problems 33 and 34 at the end of this section).

To prove property (2), we let $\mathbf{u} = a_1\mathbf{i} + b_1\mathbf{j}$ and $\mathbf{v} = a_2\mathbf{i} + b_2\mathbf{j}$. Then

$$\mathbf{u} \cdot \mathbf{v} = a_1a_2 + b_1b_2 = a_2a_1 + b_2b_1 = \mathbf{v} \cdot \mathbf{u}$$

To prove property (4), we let $\mathbf{v} = a\mathbf{i} + b\mathbf{j}$. Then

$$\mathbf{v} \cdot \mathbf{v} = a^2 + b^2 = \|\mathbf{v}\|^2$$ ■

One use of the dot product is to calculate the angle between two vectors.

ANGLE BETWEEN VECTORS

Figure 63

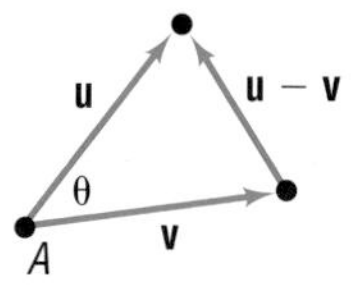

2 Let **u** and **v** be two vectors with the same initial point A. Then the vectors **u, v,** and $\mathbf{u} - \mathbf{v}$ form a triangle. The angle θ at vertex A of the triangle is the angle between the vectors **u** and **v**. See Figure 63. We wish to find a formula for calculating the angle θ.

The sides of the triangle have lengths $\|\mathbf{v}\|$, $\|\mathbf{u}\|$, and $\|\mathbf{u} - \mathbf{v}\|$, and θ is the included angle between the sides of length $\|\mathbf{v}\|$, and $\|\mathbf{u}\|$. The Law of Cosines (Section 8.3) can be used to find the cosine of the included angle:

$$\|\mathbf{u} - \mathbf{v}\|^2 = \|\mathbf{u}\|^2 + \|\mathbf{v}\|^2 - 2\|\mathbf{u}\|\,\|\mathbf{v}\|\cos\theta$$

Now we use property (4) to rewrite this equation in terms of dot products.

$$(\mathbf{u} - \mathbf{v}) \cdot (\mathbf{u} - \mathbf{v}) = \mathbf{u} \cdot \mathbf{u} + \mathbf{v} \cdot \mathbf{v} - 2\,\|\mathbf{u}\|\,\|\mathbf{v}\|\cos\theta \quad (6)$$

Then we apply the distributive property (3) twice on the left side of (6) to obtain

$$
\begin{aligned}
(\mathbf{u}-\mathbf{v}) \cdot (\mathbf{u}-\mathbf{v}) &= \mathbf{u} \cdot (\mathbf{u}-\mathbf{v}) - \mathbf{v} \cdot (\mathbf{u}-\mathbf{v}) \\
&= \mathbf{u} \cdot \mathbf{u} - \mathbf{u} \cdot \mathbf{v} - \mathbf{v} \cdot \mathbf{u} + \mathbf{v} \cdot \mathbf{v} \\
&\underset{\uparrow}{=} \mathbf{u} \cdot \mathbf{u} + \mathbf{v} \cdot \mathbf{v} - 2\,\mathbf{u} \cdot \mathbf{v} \qquad (7)
\end{aligned}
$$

Property (2)

Combining equations (6) and (7), we have

$$\mathbf{u} \cdot \mathbf{u} + \mathbf{v} \cdot \mathbf{v} - 2\,\mathbf{u} \cdot \mathbf{v} = \mathbf{u} \cdot \mathbf{u} + \mathbf{v} \cdot \mathbf{v} - 2\|\mathbf{u}\|\,\|\mathbf{v}\| \cos\theta$$

$$\mathbf{u} \cdot \mathbf{v} = \|\mathbf{u}\|\,\|\mathbf{v}\| \cos\theta$$

We have proved the following result:

Theorem

Angle between Vectors

If **u** and **v** are two nonzero vectors, the angle θ, $0 \le \theta \le \pi$, between **u** and **v** is determined by the formula

$$\cos\theta = \frac{\mathbf{u} \cdot \mathbf{v}}{\|\mathbf{u}\|\,\|\mathbf{v}\|} \qquad (8)$$

■

EXAMPLE 2 Finding the Angle θ between Two Vectors

Find the angle θ between $\mathbf{u} = 4\mathbf{i} - 3\mathbf{j}$ and $\mathbf{v} = 2\mathbf{i} + 5\mathbf{j}$.

Solution We compute the quantities $\mathbf{u} \cdot \mathbf{v}$, $\|\mathbf{u}\|$, and $\|\mathbf{v}\|$.

Figure 64

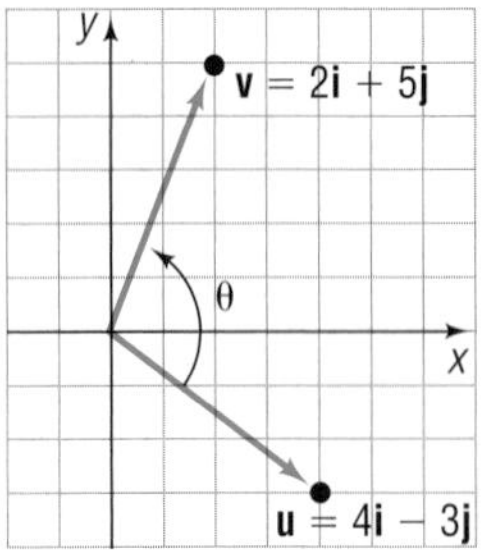

$$\mathbf{u} \cdot \mathbf{v} = 4(2) + (-3)(5) = -7$$

$$\|\mathbf{u}\| = \sqrt{4^2 + (-3)^2} = 5$$

$$\|\mathbf{v}\| = \sqrt{2^2 + 5^2} = \sqrt{29}$$

By formula (8), if θ is the angle between **u** and **v**, then

$$\cos\theta = \frac{\mathbf{u} \cdot \mathbf{v}}{\|\mathbf{u}\|\,\|\mathbf{v}\|} = \frac{-7}{5\sqrt{29}} \approx -0.26$$

We find that $\theta \approx 105°$. See Figure 64. ■

NOW WORK PROBLEM 1(a) AND (b).

EXAMPLE 3 Finding the Actual Speed and Direction of an Aircraft

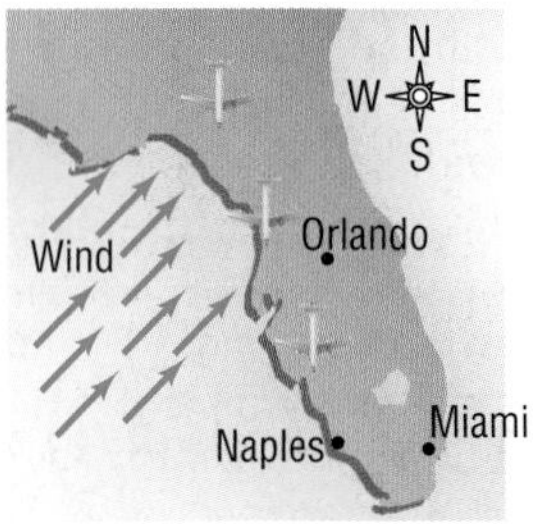

A Boeing 737 aircraft maintains a constant airspeed of 500 miles per hour in the direction due south. The velocity of the jet stream is 80 miles per hour in a northeasterly direction. Find the actual speed and direction of the aircraft relative to the ground.

Solution We set up a coordinate system in which North (N) is along the positive y-axis. See Figure 65. Let

Figure 65

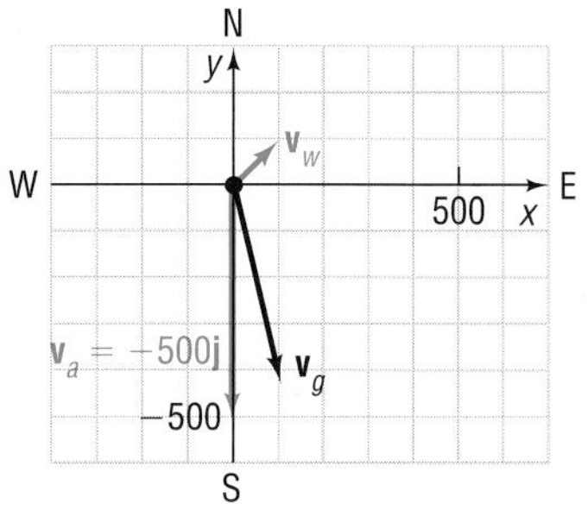

$$\mathbf{v}_a = \text{velocity of aircraft relative to the air} = -500\mathbf{j}$$
$$\mathbf{v}_g = \text{velocity of aircraft relative to ground}$$
$$\mathbf{v}_w = \text{velocity of jet stream.}$$

The velocity of the jet stream $\mathbf{v}_w$ has magnitude 80 and direction $\alpha = 45°$. We express $\mathbf{v}_w$ in terms of $\mathbf{i}$ and $\mathbf{j}$ as

$$\mathbf{v}_w = 80(\cos 45°\mathbf{i} + \sin 45°\mathbf{j}) = 80\left(\frac{\sqrt{2}}{2}\mathbf{i} + \frac{\sqrt{2}}{2}\mathbf{j}\right) = 40\sqrt{2}(\mathbf{i} + \mathbf{j})$$

The velocity of the aircraft relative to the ground is

$$\mathbf{v}_g = \mathbf{v}_a + \mathbf{v}_w = -500\mathbf{j} + 40\sqrt{2}(\mathbf{i} + \mathbf{j}) = 40\sqrt{2}\mathbf{i} + \left(40\sqrt{2} - 500\right)\mathbf{j}$$

The actual speed of the aircraft is

$$\|\mathbf{v}_g\| = \sqrt{(40\sqrt{2})^2 + \left(40\sqrt{2} - 500\right)^2} \approx 447 \text{ miles per hour.}$$

The angle θ between $\mathbf{v}_g$ and the vector $\mathbf{v}_a = -500\mathbf{j}$ (the velocity of the aircraft relative to the air) is determined by the equation

$$\cos\theta = \frac{\mathbf{v}_g \cdot \mathbf{v}_a}{\|\mathbf{v}_g\|\,\|\mathbf{v}_a\|} = \frac{(40\sqrt{2} - 500)(-500)}{(447)(500)} \approx 0.9920$$
$$\theta \approx 7.3°$$

The direction of the aircraft relative to the ground is approximately S7.3°E (about 7.3° east of south). ■

NOW WORK PROBLEM 19.

Parallel and Orthogonal Vectors

3 Two vectors $\mathbf{v}$ and $\mathbf{w}$ are said to be **parallel** if there is a nonzero scalar α so that $\mathbf{v} = \alpha\mathbf{w}$. In this case, the angle θ between $\mathbf{v}$ and $\mathbf{w}$ is 0 or π.

EXAMPLE 4 Determining Whether Vectors Are Parallel

The vectors $\mathbf{v} = 3\mathbf{i} - \mathbf{j}$ and $\mathbf{w} = 6\mathbf{i} - 2\mathbf{j}$ are parallel, since $\mathbf{v} = \frac{1}{2}\mathbf{w}$. Furthermore, since

$$\cos\theta = \frac{\mathbf{v} \cdot \mathbf{w}}{\|\mathbf{v}\|\,\|\mathbf{w}\|} = \frac{18 + 2}{\sqrt{10}\sqrt{40}} = \frac{20}{\sqrt{400}} = 1$$

the angle θ between $\mathbf{v}$ and $\mathbf{w}$ is 0. ■

4 If the angle θ between two nonzero vectors $\mathbf{v}$ and $\mathbf{w}$ is $\pi/2$, the vectors $\mathbf{v}$ and $\mathbf{w}$ are called **orthogonal.***

It follows from formula (8) that if $\mathbf{v}$ and $\mathbf{w}$ are orthogonal then $\mathbf{v} \cdot \mathbf{w} = 0$, since $\cos(\pi/2) = 0$.

On the other hand, if $\mathbf{v} \cdot \mathbf{w} = 0$, then either $\mathbf{v} = \mathbf{0}$ or $\mathbf{w} = \mathbf{0}$ or $\cos\theta = 0$. In the latter case, $\theta = \pi/2$, and $\mathbf{v}$ and $\mathbf{w}$ are orthogonal. See Figure 66. If $\mathbf{v}$ or $\mathbf{w}$ is the zero vector, then, since the zero vector has no specific direction, we adopt the convention that the zero vector is orthogonal to every vector.

Figure 66
$\mathbf{v} \cdot \mathbf{w} = 0$; $\mathbf{v}$ is orthogonal to $\mathbf{w}$

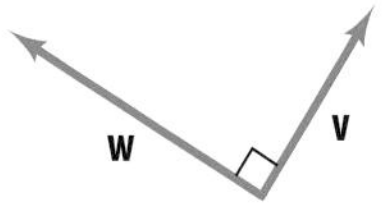

* *Orthogonal, perpendicular,* and *normal* are all terms that mean "meet at a right angle." It is customary to refer to two vectors as being *orthogonal,* two lines as being *perpendicular,* and a line and a plane or a vector and a plane as being *normal.*

Theorem Two vectors **v** and **w** are orthogonal if and only if

$$\mathbf{v} \cdot \mathbf{w} = 0$$

EXAMPLE 5 Determining Whether Two Vectors Are Orthogonal

Figure 67

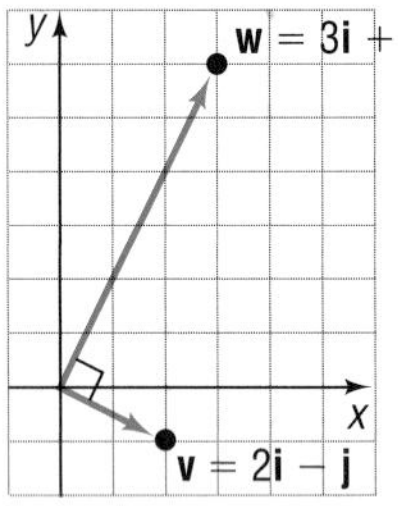

The vectors

$$\mathbf{v} = 2\mathbf{i} - \mathbf{j} \quad \text{and} \quad \mathbf{w} = 3\mathbf{i} + 6\mathbf{j}$$

are orthogonal, since

$$\mathbf{v} \cdot \mathbf{w} = 6 - 6 = 0$$

See Figure 67.

NOW WORK PROBLEM 1(c).

PROJECTION OF A VECTOR ONTO ANOTHER VECTOR

5 In many physical applications, it is necessary to find "how much" of a vector is applied in a given direction. Look at Figure 68. The force **F** due to gravity is pulling straight down (toward the center of Earth) on the block. To study the effect of gravity on the block, it is necessary to determine how much of **F** is actually pushing the block down the incline $(\mathbf{F}_1)$ and how much is pressing the block against the incline $(\mathbf{F}_2)$, at a right angle to the incline. Knowing the **decomposition** of **F** often will allow us to determine when friction is overcome and the block will slide down the incline.

Figure 68

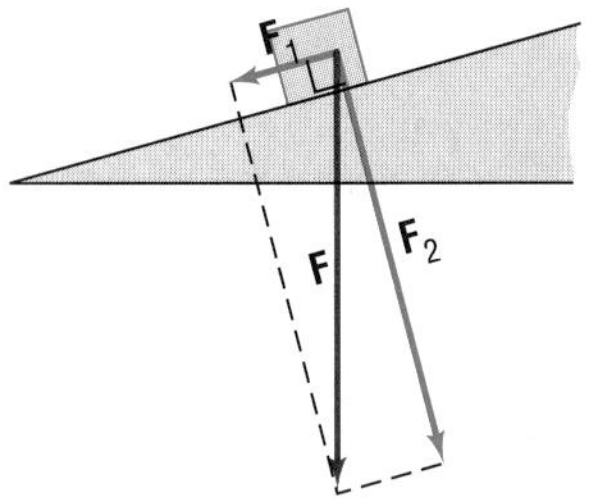

Suppose that **v** and **w** are two nonzero vectors with the same initial point P. We seek to decompose **v** into two vectors: $\mathbf{v}_1$, which is parallel to **w**, and $\mathbf{v}_2$, which is orthogonal to **w**. See Figure 69(a) and (b). The vector $\mathbf{v}_1$ is called the **vector projection of v onto w** and is denoted by $\text{proj}_{\mathbf{w}}\,\mathbf{v}$.

Figure 69

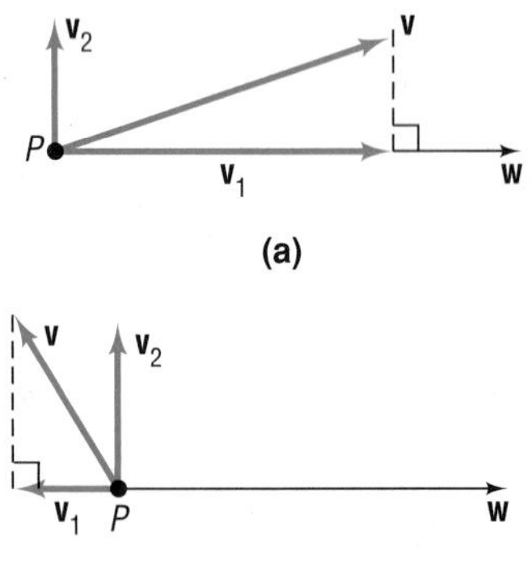

The vector $\mathbf{v}_1$ is obtained as follows: From the terminal point of **v**, drop a perpendicular to the line containing **w**. The vector $\mathbf{v}_1$ is the vector from P to the foot of this perpendicular. The vector $\mathbf{v}_2$ is given by $\mathbf{v}_2 = \mathbf{v} - \mathbf{v}_1$. Note that $\mathbf{v} = \mathbf{v}_1 + \mathbf{v}_2$, $\mathbf{v}_1$ is parallel to **w**, and $\mathbf{v}_2$ is orthogonal to **w**. This is the decomposition of **v** that we wanted.

Now we seek a formula for $\mathbf{v}_1$ that is based on a knowledge of the vectors **v** and **w**. Since $\mathbf{v} = \mathbf{v}_1 + \mathbf{v}_2$, we have

$$\mathbf{v} \cdot \mathbf{w} = (\mathbf{v}_1 + \mathbf{v}_2) \cdot \mathbf{w} = \mathbf{v}_1 \cdot \mathbf{w} + \mathbf{v}_2 \cdot \mathbf{w} \qquad (9)$$

Since $\mathbf{v}_2$ is orthogonal to **w**, we have $\mathbf{v}_2 \cdot \mathbf{w} = 0$. Since $\mathbf{v}_1$ is parallel to **w**, we have $\mathbf{v}_1 = \alpha\mathbf{w}$ for some scalar α. Equation (9) can be written as

$$\mathbf{v} \cdot \mathbf{w} = \alpha\mathbf{w} \cdot \mathbf{w} = \alpha\|\mathbf{w}\|^2$$

$$\alpha = \frac{\mathbf{v} \cdot \mathbf{w}}{\|\mathbf{w}\|^2}$$

Then

$$\mathbf{v}_1 = \alpha\mathbf{w} = \frac{\mathbf{v} \cdot \mathbf{w}}{\|\mathbf{w}\|^2}\mathbf{w}$$

Theorem If $\mathbf{v}$ and $\mathbf{w}$ are two nonzero vectors, the vector projection of $\mathbf{v}$ onto $\mathbf{w}$ is

$$\text{proj}_{\mathbf{w}}\,\mathbf{v} = \frac{\mathbf{v}\cdot\mathbf{w}}{\|\mathbf{w}\|^2}\mathbf{w}$$

The decomposition of $\mathbf{v}$ into $\mathbf{v}_1$ and $\mathbf{v}_2$, where $\mathbf{v}_1$ is parallel to $\mathbf{w}$ and $\mathbf{v}_2$ is perpendicular to $\mathbf{w}$, is

$$\mathbf{v}_1 = \text{proj}_{\mathbf{w}}\,\mathbf{v} = \frac{\mathbf{v}\cdot\mathbf{w}}{\|\mathbf{w}\|^2}\mathbf{w} \qquad \mathbf{v}_2 = \mathbf{v} - \mathbf{v}_1 \tag{10}$$

EXAMPLE 6 Decomposing a Vector into Two Orthogonal Vectors

Figure 70

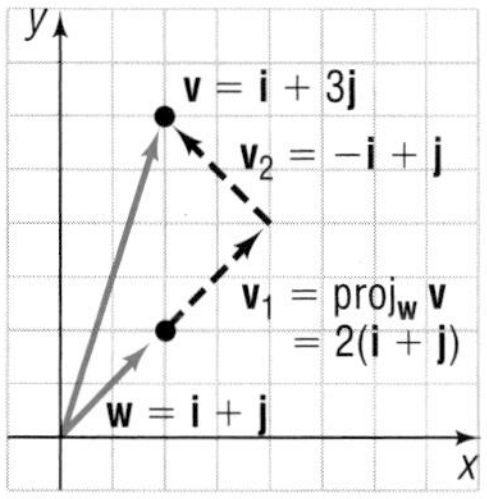

Find the vector projection of $\mathbf{v} = \mathbf{i} + 3\mathbf{j}$ onto $\mathbf{w} = \mathbf{i} + \mathbf{j}$. Decompose $\mathbf{v}$ into two vectors $\mathbf{v}_1$ and $\mathbf{v}_2$, where $\mathbf{v}_1$ is parallel to $\mathbf{w}$ and $\mathbf{v}_2$ is orthogonal to $\mathbf{w}$.

Solution We use formulas (10).

$$\mathbf{v}_1 = \text{proj}_{\mathbf{w}}\,\mathbf{v} = \frac{\mathbf{v}\cdot\mathbf{w}}{\|\mathbf{w}\|^2}\mathbf{w} = \frac{1+3}{(\sqrt{2})^2}\mathbf{w} = 2\mathbf{w} = 2(\mathbf{i}+\mathbf{j})$$

$$\mathbf{v}_2 = \mathbf{v} - \mathbf{v}_1 = (\mathbf{i} + 3\mathbf{j}) - 2(\mathbf{i} + \mathbf{j}) = -\mathbf{i} + \mathbf{j}$$

See Figure 70.

NOW WORK PROBLEM 13.

WORK DONE BY A CONSTANT FORCE

6 In elementary physics, the **work** W done by a constant force $\mathbf{F}$ in moving an object from a point A to a point B is defined as

$$W = (\text{magnitude of force})(\text{distance}) = \|\mathbf{F}\|\,\|\overrightarrow{AB}\|$$

Figure 71

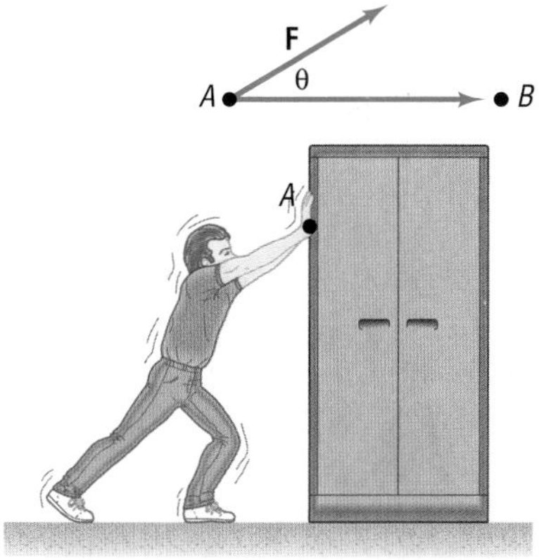

Work is commonly measured in foot-pounds or in newton-meters (joules).

In this definition, it is assumed that the force $\mathbf{F}$ is applied along the line of motion. If the constant force $\mathbf{F}$ is not along the line of motion, but, instead, is at an angle θ to the direction of motion, as illustrated in Figure 71, then the **work** W **done by F** in moving an object from A to B is defined as

$$W = \mathbf{F}\cdot\overrightarrow{AB} \tag{11}$$

This definition is compatible with the force times distance definition given above, since

$$W = (\text{amount of force in the direction of } \overrightarrow{AB})(\text{distance})$$

$$= \|\text{proj}_{\overrightarrow{AB}}\mathbf{F}\|\,\|\overrightarrow{AB}\| = \frac{\mathbf{F}\cdot\overrightarrow{AB}}{\|\overrightarrow{AB}\|^2}\|\overrightarrow{AB}\|\,\|\overrightarrow{AB}\| = \mathbf{F}\cdot\overrightarrow{AB}$$

EXAMPLE 7 Computing Work

Figure 72(a) shows a girl pulling a wagon with a force of 50 pounds. How much work is done in moving the wagon 100 feet if the handle makes an angle of 30° with the ground?

Figure 72

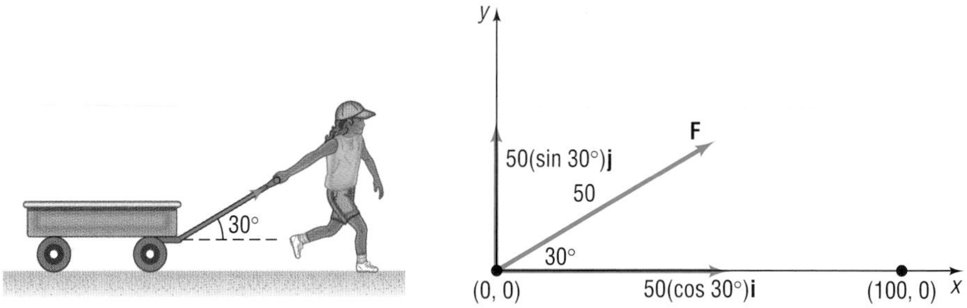

(a) (b)

Solution We position the vectors in a coordinate system in such a way that the wagon is moved from (0, 0) to (100, 0). The motion is from $A = (0, 0)$ to $B = (100, 0)$, so $\overrightarrow{AB} = 100\mathbf{i}$. The force vector **F**, as shown in Figure 72(b), is

$$\mathbf{F} = 50(\cos 30°\mathbf{i} + \sin 30°\mathbf{j}) = 50\left(\frac{\sqrt{3}}{2}\mathbf{i} + \frac{1}{2}\mathbf{j}\right) = 25(\sqrt{3}\mathbf{i} + \mathbf{j})$$

By formula (11), the work W done is

$$W = \mathbf{F} \cdot \overrightarrow{AB} = 25(\sqrt{3}\mathbf{i} + \mathbf{j}) \cdot 100\mathbf{i} = 2500\sqrt{3} \text{ foot-pounds}$$ ■

NOW WORK PROBLEM 29.

HISTORICAL PROBLEM

1. We stated in an earlier Historical Feature that complex numbers were used as vectors in the plane before the general notion of a vector was clarified. Suppose that we make the correspondence

$$\text{Vector} \leftrightarrow \text{Complex number}$$
$$a\mathbf{i} + b\mathbf{j} \leftrightarrow a + bi$$
$$c\mathbf{i} + d\mathbf{j} \leftrightarrow c + di$$

Show that

$$(a\mathbf{i} + b\mathbf{j}) \cdot (c\mathbf{i} + d\mathbf{j}) = \text{real part}\left[\overline{(a + bi)}(c + di)\right]$$

This is how the dot product was found originally. The imaginary part is also interesting. It is a determinant (see Section 11.4) and represents the area of the parallelogram whose edges are the vectors. This is close to some of Hermann Grassmann's ideas and is also connected with the scalar triple product of three-dimensional vectors.

9.5 EXERCISES

In Problems 1–10, (a) find the dot product $\mathbf{v} \cdot \mathbf{w}$; *(b) find the angle between* **v** *and* **w**; *(c) state whether the vectors are parallel, orthogonal, or neither.*

1. $\mathbf{v} = \mathbf{i} - \mathbf{j}, \quad \mathbf{w} = \mathbf{i} + \mathbf{j}$
2. $\mathbf{v} = \mathbf{i} + \mathbf{j}, \quad \mathbf{w} = -\mathbf{i} + \mathbf{j}$
3. $\mathbf{v} = 2\mathbf{i} + \mathbf{j}, \quad \mathbf{w} = \mathbf{i} + 2\mathbf{j}$
4. $\mathbf{v} = 2\mathbf{i} + 2\mathbf{j}, \quad \mathbf{w} = \mathbf{i} + 2\mathbf{j}$
5. $\mathbf{v} = \sqrt{3}\mathbf{i} - \mathbf{j}, \quad \mathbf{w} = \mathbf{i} + \mathbf{j}$
6. $\mathbf{v} = \mathbf{i} + \sqrt{3}\mathbf{j}, \quad \mathbf{w} = \mathbf{i} - \mathbf{j}$

7. $\mathbf{v} = 3\mathbf{i} + 4\mathbf{j}, \quad \mathbf{w} = 4\mathbf{i} + 3\mathbf{j}$

8. $\mathbf{v} = 3\mathbf{i} - 4\mathbf{j}, \quad \mathbf{w} = 4\mathbf{i} - 3\mathbf{j}$

9. $\mathbf{v} = 4\mathbf{i}, \quad \mathbf{w} = \mathbf{j}$

10. $\mathbf{v} = \mathbf{i}, \quad \mathbf{w} = -3\mathbf{j}$

11. Find a so that vectors $\mathbf{v} = \mathbf{i} - a\mathbf{j}$ and $\mathbf{w} = 2\mathbf{i} + 3\mathbf{j}$ are orthogonal.

12. Find b so that vectors $\mathbf{v} = \mathbf{i} + \mathbf{j}$ and $\mathbf{w} = \mathbf{i} + b\mathbf{j}$ are orthogonal.

In Problems 13–18, decompose $\mathbf{v}$ *into two vectors* $\mathbf{v}_1$ *and* $\mathbf{v}_2$, *where* $\mathbf{v}_1$ *is parallel to* $\mathbf{w}$ *and* $\mathbf{v}_2$ *is orthogonal to* $\mathbf{w}$.

13. $\mathbf{v} = 2\mathbf{i} - 3\mathbf{j}, \quad \mathbf{w} = \mathbf{i} - \mathbf{j}$

14. $\mathbf{v} = -3\mathbf{i} + 2\mathbf{j}, \quad \mathbf{w} = 2\mathbf{i} + \mathbf{j}$

15. $\mathbf{v} = \mathbf{i} - \mathbf{j}, \quad \mathbf{w} = \mathbf{i} + 2\mathbf{j}$

16. $\mathbf{v} = 2\mathbf{i} - \mathbf{j}, \quad \mathbf{w} = \mathbf{i} - 2\mathbf{j}$

17. $\mathbf{v} = 3\mathbf{i} + \mathbf{j}, \quad \mathbf{w} = -2\mathbf{i} - \mathbf{j}$

18. $\mathbf{v} = \mathbf{i} - 3\mathbf{j}, \quad \mathbf{w} = 4\mathbf{i} - \mathbf{j}$

19. Finding the Actual Speed and Direction of an Aircraft A DC-10 jumbo jet maintains an airspeed of 550 miles per hour in a southwesterly direction. The velocity of the jet stream is a constant 80 miles per hour from the west. Find the actual speed and direction of the aircraft.

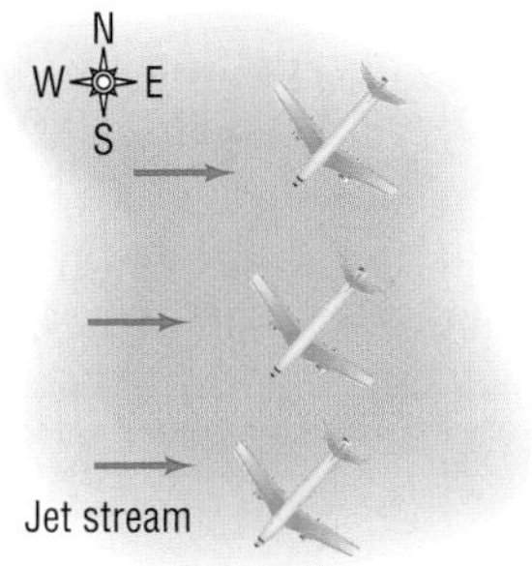

20. Finding the Correct Compass Heading The pilot of an aircraft wishes to head directly east, but is faced with a wind speed of 40 miles per hour from the northwest. If the pilot maintains an airspeed of 250 miles per hour, what compass heading should be maintained? What is the actual speed of the aircraft?

21. Correct Direction for Crossing a River A river has a constant current of 3 kilometers per hour. At what angle to a boat dock should a motorboat, capable of maintaining a constant speed of 20 kilometers per hour, be headed in order to reach a point directly opposite the dock? If the river is $\frac{1}{2}$ kilometer wide, how long will it take to cross?

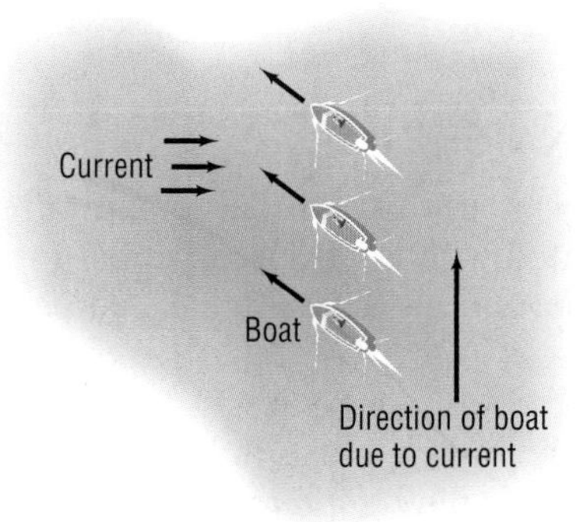

22. Correct Direction for Crossing a River Repeat Problem 21 if the current is 5 kilometers per hour.

23. Braking Load A Toyota Sienna with a gross weight of 5300 pounds is parked on a street with a slope of 8°. See the figure. Find the force required to keep the Sienna from rolling down the hill. What is the force perpendicular to the hill?

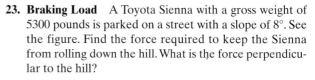

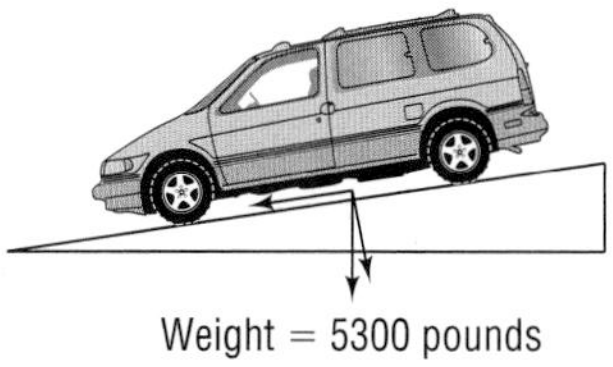

24. Braking Load A Pontiac Bonneville with a gross weight of 4500 pounds is parked on a street with a slope of 10°. Find the force required to keep the Bonneville from rolling down the hill. What is the force perpendicular to the hill?

25. Ground Speed and Direction of an Airplane An airplane has an airspeed of 500 kilometers per hour bearing N45°E. The wind velocity is 60 kilometers per hour in the direction N30°W. Find the resultant vector representing the path of the plane relative to the ground. What is the ground speed of the plane? What is its direction?

26. Ground Speed and Direction of an Airplane An airplane has an airspeed of 600 kilometers per hour bearing S30°E. The wind velocity is 40 kilometers per hour in the direction S45°E. Find the resultant vector representing the path of the plane relative to the ground. What is the ground speed of the plane? What is its direction?

27. Crossing a River A small motorboat in still water maintains a speed of 20 miles per hour. In heading directly across a river (that is, perpendicular to the current) whose current is 3 miles per hour, find a vector representing the speed and direction of the motorboat. What is the true speed of the motorboat? What is its direction?

28. Crossing a River A small motorboat in still water maintains a speed of 10 miles per hour. In heading directly across a river (that is, perpendicular to the current) whose current is 4 miles per hour, find a vector representing the speed and direction of the motorboat. What is the true speed of the motorboat? What is its direction?

29. Computing Work Find the work done by a force of 3 pounds acting in the direction 60° to the horizontal in moving an object 2 feet from $(0, 0)$ to $(2, 0)$.

30. Computing Work Find the work done by a force of 1 pound acting in the direction 45° to the horizontal in moving an object 5 feet from $(0, 0)$ to $(5, 0)$.

31. Computing Work A wagon is pulled horizontally by exerting a force of 20 pounds on the handle at an angle of 30° with the horizontal. How much work is done in moving the wagon 100 feet?

32. Find the acute angle that a constant unit force vector makes with the positive x-axis if the work done by the force in moving a particle from $(0, 0)$ to $(4, 0)$ equals 2.

33. Prove the distributive property:

$$\mathbf{u} \cdot (\mathbf{v} + \mathbf{w}) = \mathbf{u} \cdot \mathbf{v} + \mathbf{u} \cdot \mathbf{w}$$

34. Prove property (5), $\mathbf{0} \cdot \mathbf{v} = 0$.

35. If $\mathbf{v}$ is a unit vector and the angle between $\mathbf{v}$ and $\mathbf{i}$ is α, show that $\mathbf{v} = \cos\alpha\mathbf{i} + \sin\alpha\mathbf{j}$.

36. Suppose that $\mathbf{v}$ and $\mathbf{w}$ are unit vectors. If the angle between $\mathbf{v}$ and $\mathbf{i}$ is α and if the angle between $\mathbf{w}$ and $\mathbf{i}$ is β, use the idea of the dot product $\mathbf{v} \cdot \mathbf{w}$ to prove that

$$\cos(\alpha - \beta) = \cos\alpha\cos\beta + \sin\alpha\sin\beta$$

37. Show that the projection of $\mathbf{v}$ onto $\mathbf{i}$ is $(\mathbf{v} \cdot \mathbf{i})\mathbf{i}$. In fact, show that we can always write a vector $\mathbf{v}$ as

$$\mathbf{v} = (\mathbf{v} \cdot \mathbf{i})\mathbf{i} + (\mathbf{v} \cdot \mathbf{j})\mathbf{j}$$

38. (a) If $\mathbf{u}$ and $\mathbf{v}$ have the same magnitude, then show that $\mathbf{u} + \mathbf{v}$ and $\mathbf{u} - \mathbf{v}$ are orthogonal.
(b) Use this to prove that an angle inscribed in a semicircle is a right angle (see the figure).

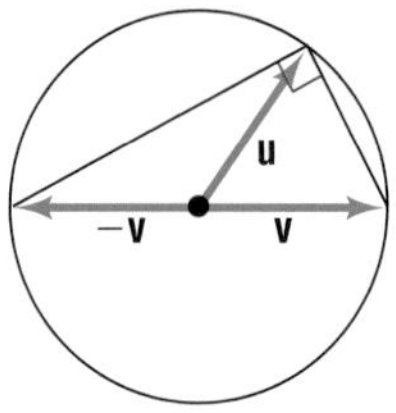

39. Let $\mathbf{v}$ and $\mathbf{w}$ denote two nonzero vectors. Show that the vector $\mathbf{v} - \alpha\mathbf{w}$ is orthogonal to $\mathbf{w}$ if $\alpha = (\mathbf{v} \cdot \mathbf{w}) / \|\mathbf{w}\|^2$.

40. Let $\mathbf{v}$ and $\mathbf{w}$ denote two nonzero vectors. Show that the vectors $\|\mathbf{w}\|\mathbf{v} + \|\mathbf{v}\|\mathbf{w}$ and $\|\mathbf{w}\|\mathbf{v} - \|\mathbf{v}\|\mathbf{w}$ are orthogonal.

41. In the definition of work given in this section, what is the work done if $\mathbf{F}$ is orthogonal to $\overrightarrow{AB}$?

42. Prove the **polarization identity**,

$$\|\mathbf{u} + \mathbf{v}\|^2 - \|\mathbf{u} - \mathbf{v}\|^2 = 4(\mathbf{u} \cdot \mathbf{v}).$$

43. Make up an application different from any found in the text that requires the dot product.

PREPARING FOR THIS SECTION

Before getting started, review the following:

✓ Distance Formula (p. 6)

9.6 VECTORS IN SPACE

OBJECTIVES
1. Find the Distance between Two Points
2. Find Position Vectors
3. Perform Operations on Vectors
4. Find the Dot Product
5. Find the Angle between Two Vectors
6. Find the Direction Angles of a Vector

Figure 73

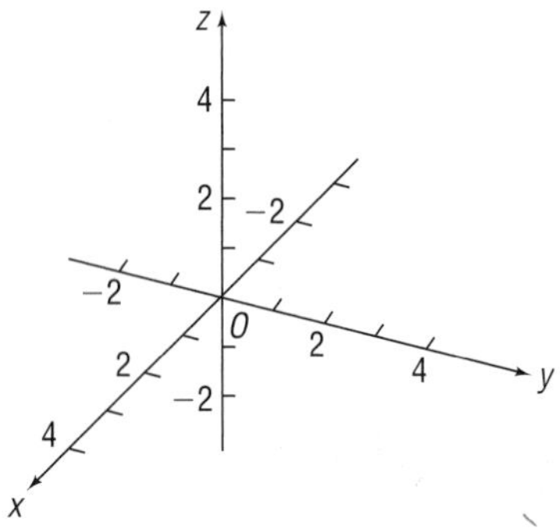

RECTANGULAR COORDINATES IN SPACE

In the plane, each point is associated with an ordered pair of real numbers. In space, each point is associated with an ordered triple of real numbers. Through a fixed point, the **origin**, O, draw three mutually perpendicular lines, the *x-axis*, the *y-axis*, and the *z-axis*. On each of these axes, select an appropriate scale and the positive direction. See Figure 73.

The direction chosen for the positive z-axis in Figure 73 makes the system *right-handed*. This conforms to the *right-hand rule*, which states that, if the index finger of the right hand points in the direction of the positive x-axis and the middle finger points in the direction of the positive y-axis, then the thumb will point in the direction of the positive z-axis. See Figure 74.

Figure 74

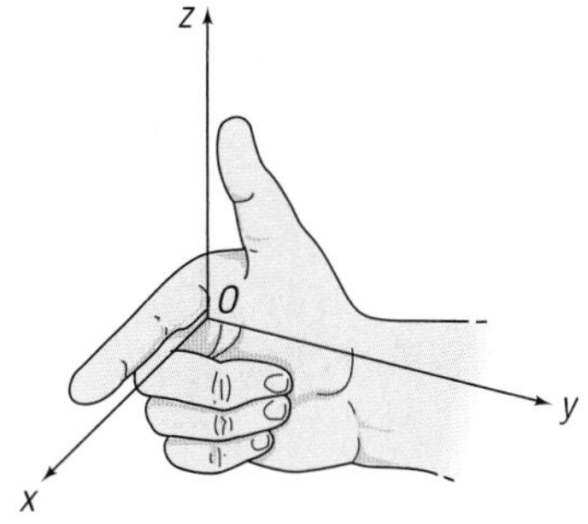

Figure 75

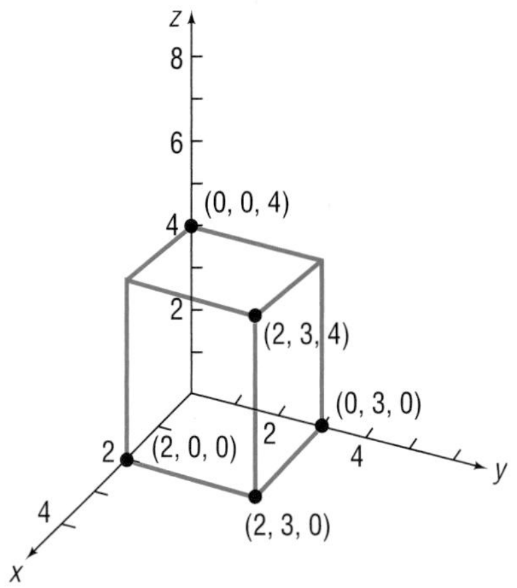

We associate with each point P an ordered triple (x, y, z) of real numbers, the *coordinates of* P. For example, the point $(2, 3, 4)$ is located by starting at the origin and moving 2 units along the positive x-axis, 3 units in the direction of the positive y-axis, and 4 units in the direction of the positive z-axis. See Figure 75.

Figure 75 also shows the location of the points $(2, 0, 0)$, $(0, 3, 0)$, $(0, 0, 4)$, and $(2, 3, 0)$. Points of the form $(x, 0, 0)$ lie on the x-axis, while points of the form $(0, y, 0)$ and $(0, 0, z)$ lie on the y-axis and z-axis, respectively. Points of the form $(x, y, 0)$ lie in a plane, called the ***xy*-plane**. Its equation is $z = 0$. Similarly, points of the form $(x, 0, z)$ lie in the ***xz*-plane** (equation $y = 0$) and points of the form $(0, y, z)$ lie in the ***yz*-plane** (equation $x = 0$). See Figure 76(a). By extension of these ideas, all points obeying the equation $z = 3$ will lie in a plane parallel to and 3 units above the xy-plane. The equation $y = 4$ represents a plane parallel to the xz-plane and 4 units to the right of the plane $y = 0$. See Figure 76(b).

Figure 76

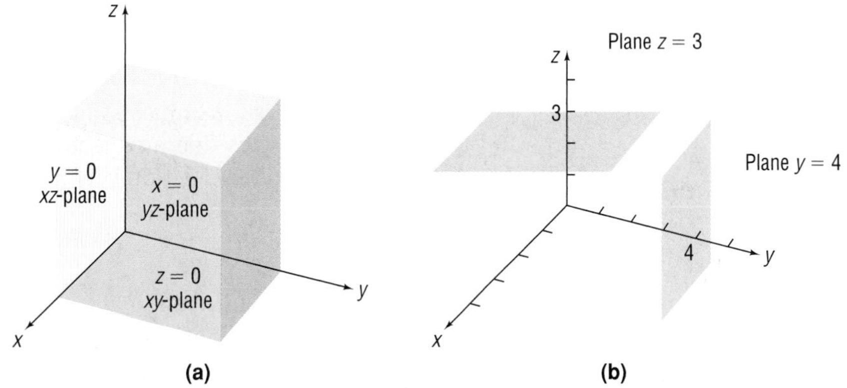

NOW WORK PROBLEM 3.

1 The formula for the distance between two points in space is an extension of the Distance Formula for points in the plane given in Chapter 1.

Theorem **Distance Formula in Space**

If $P_1 = (x_1, y_1, z_1)$ and $P_2 = (x_2, y_2, z_2)$ are two points in space, the distance d from P_1 to P_2 is

$$d = \sqrt{(x_2 - x_1)^2 + (y_2 - y_1)^2 + (z_2 - z_1)^2} \qquad (1)$$

The proof, which we omit, utilizes a double application of the Pythagorean Theorem.

EXAMPLE 1 Using the Distance Formula

Find the distance from $P_1 = (-1, 3, 2)$ to $P_2 = (4, -2, 5)$.

Solution $d = \sqrt{[4 - (-1)]^2 + [-2 - 3]^2 + [5 - 2]^2} = \sqrt{25 + 25 + 9} = \sqrt{59}$

NOW WORK PROBLEM 9.

Representing Vectors in Space

2 To represent vectors in space, we introduce the unit vectors **i**, **j**, and **k** whose directions are along the positive x-axis, positive y-axis, and positive z-axis, respectively. If **v** is a vector with initial point at the origin O and terminal point at $P = (a, b, c)$, then we can represent **v** in terms of the vectors **i**, **j**, and **k** as

Figure 77

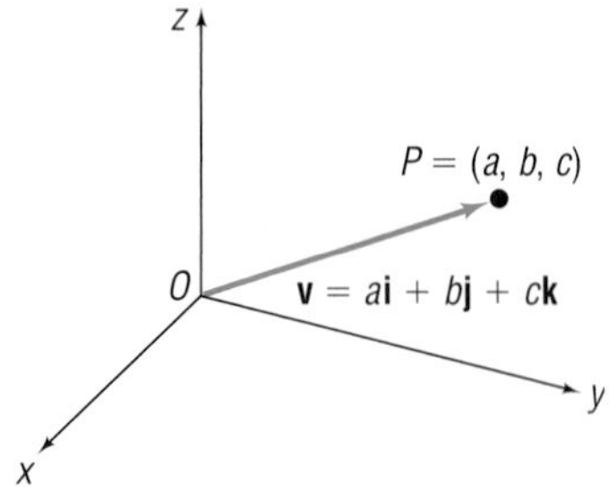

$$\mathbf{v} = a\mathbf{i} + b\mathbf{j} + c\mathbf{k}$$

The scalars a, b, and c are called the **components** of the vector $\mathbf{v} = a\mathbf{i} + b\mathbf{j} + c\mathbf{k}$, with a being the component in the direction **i**, b the component in the direction **j**, and c the component in the direction **k**. See Figure 77.

A vector whose initial point is at the origin is called a **position vector**. The next result states that any vector whose initial point is not at the origin is equal to a unique position vector.

Theorem Suppose that **v** is a vector with initial point $P_1 = (x_1, y_1, z_1)$, not necessarily the origin, and terminal point $P_2 = (x_2, y_2, z_2)$. If $\mathbf{v} = \overrightarrow{P_1P_2}$, then **v** is equal to the position vector

$$\mathbf{v} = (x_2 - x_1)\mathbf{i} + (y_2 - y_1)\mathbf{j} + (z_2 - z_1)\mathbf{k} \qquad (2)$$

Figure 78 illustrates this result.

Figure 78

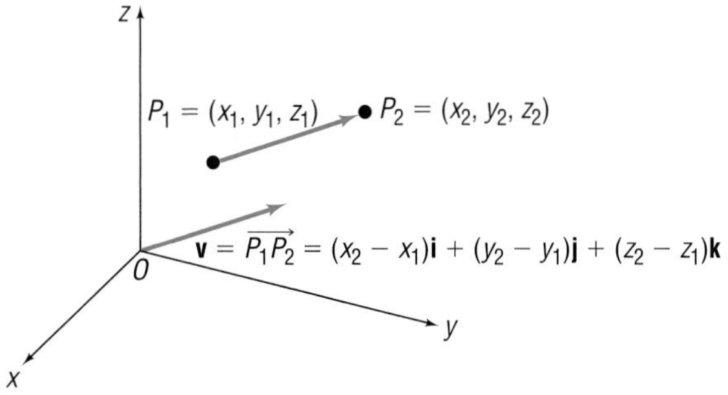

EXAMPLE 2 Finding a Position Vector

Find the position vector of the vector $\mathbf{v} = \overrightarrow{P_1P_2}$ if $P_1 = (-1, 2, 3)$ and $P_2 = (4, 6, 2)$.

Solution By equation (2), the position vector equal to **v** is

$$\mathbf{v} = [4 - (-1)]\mathbf{i} + (6 - 2)\mathbf{j} + (2 - 3)\mathbf{k} = 5\mathbf{i} + 4\mathbf{j} - \mathbf{k}$$ ■

NOW WORK PROBLEM 23.

3 Next, we define equality, addition, subtraction, scalar product, and magnitude in terms of the components of a vector.

Let $\mathbf{v} = a_1\mathbf{i} + b_1\mathbf{j} + c_1\mathbf{k}$ and $\mathbf{w} = a_2\mathbf{i} + b_2\mathbf{j} + c_2\mathbf{k}$ be two vectors, and let α be a scalar. Then

$$\mathbf{v} = \mathbf{w} \quad \text{if and only if } a_1 = a_2,\ b_1 = b_2, \text{ and } c_1 = c_2$$
$$\mathbf{v} + \mathbf{w} = (a_1 + a_2)\mathbf{i} + (b_1 + b_2)\mathbf{j} + (c_1 + c_2)\mathbf{k}$$
$$\mathbf{v} - \mathbf{w} = (a_1 - a_2)\mathbf{i} + (b_1 - b_2)\mathbf{j} + (c_1 - c_2)\mathbf{k}$$
$$\alpha\mathbf{v} = (\alpha a_1)\mathbf{i} + (\alpha b_1)\mathbf{j} + (\alpha c_1)\mathbf{k}$$
$$\|\mathbf{v}\| = \sqrt{a_1^2 + b_1^2 + c_1^2}$$

These definitions are compatible with the geometric ones given earlier in Section 9.4.

EXAMPLE 3 Adding and Subtracting Vectors

If $\mathbf{v} = 2\mathbf{i} + 3\mathbf{j} - 2\mathbf{k}$ and $\mathbf{w} = 3\mathbf{i} - 4\mathbf{j} + 5\mathbf{k}$, find:

(a) $\mathbf{v} + \mathbf{w}$ (b) $\mathbf{v} - \mathbf{w}$

Solution (a)
$$\begin{aligned}\mathbf{v} + \mathbf{w} &= (2\mathbf{i} + 3\mathbf{j} - 2\mathbf{k}) + (3\mathbf{i} - 4\mathbf{j} + 5\mathbf{k})\\ &= (2 + 3)\mathbf{i} + (3 - 4)\mathbf{j} + (-2 + 5)\mathbf{k}\\ &= 5\mathbf{i} - \mathbf{j} + 3\mathbf{k}\end{aligned}$$

(b)
$$\begin{aligned}\mathbf{v} - \mathbf{w} &= (2\mathbf{i} + 3\mathbf{j} - 2\mathbf{k}) - (3\mathbf{i} - 4\mathbf{j} + 5\mathbf{k})\\ &= (2 - 3)\mathbf{i} + [3 - (-4)]\mathbf{j} + [-2 - 5]\mathbf{k}\\ &= -\mathbf{i} + 7\mathbf{j} - 7\mathbf{k}\end{aligned}$$ ■

EXAMPLE 4 **Finding Scalar Products and Magnitudes**

If $\mathbf{v} = 2\mathbf{i} + 3\mathbf{j} - 2\mathbf{k}$ and $\mathbf{w} = 3\mathbf{i} - 4\mathbf{j} + 5\mathbf{k}$, find:

(a) $3\mathbf{v}$ (b) $2\mathbf{v} - 3\mathbf{w}$ (c) $\|\mathbf{v}\|$

Solution (a) $3\mathbf{v} = 3(2\mathbf{i} + 3\mathbf{j} - 2\mathbf{k}) = 6\mathbf{i} + 9\mathbf{j} - 6\mathbf{k}$

(b) $2\mathbf{v} - 3\mathbf{w} = 2(2\mathbf{i} + 3\mathbf{j} - 2\mathbf{k}) - 3(3\mathbf{i} - 4\mathbf{j} + 5\mathbf{k})$
$= 4\mathbf{i} + 6\mathbf{j} - 4\mathbf{k} - 9\mathbf{i} + 12\mathbf{j} - 15\mathbf{k} = -5\mathbf{i} + 18\mathbf{j} - 19\mathbf{k}$

(c) $\|\mathbf{v}\| = \|2\mathbf{i} + 3\mathbf{j} - 2\mathbf{k}\| = \sqrt{2^2 + 3^2 + (-2)^2} = \sqrt{17}$ ■

NOW WORK PROBLEMS 27 AND 33.

Recall that a unit vector $\mathbf{u}$ is one for which $\|\mathbf{u}\| = 1$. In many applications, it is useful to be able to find a unit vector $\mathbf{u}$ that has the same direction as a given vector $\mathbf{v}$.

Theorem **Unit Vector in the Direction of *v***

For any nonzero vector $\mathbf{v}$, the vector

$$\mathbf{u} = \frac{\mathbf{v}}{\|\mathbf{v}\|}$$

is a unit vector that has the same direction as $\mathbf{v}$.

■

As a consequence of this theorem, if $\mathbf{u}$ is a unit vector in the same direction as a vector $\mathbf{v}$, then $\mathbf{v}$ may be expressed as

$$\mathbf{v} = \|\mathbf{v}\|\,\mathbf{u}$$

This way of expressing a vector is useful in many applications.

EXAMPLE 5 **Finding a Unit Vector**

Find a unit vector in the same direction as $\mathbf{v} = 2\mathbf{i} - 3\mathbf{j} - 6\mathbf{k}$.

Solution We find $\|\mathbf{v}\|$ first.

$$\|\mathbf{v}\| = \|2\mathbf{i} - 3\mathbf{j} - 6\mathbf{k}\| = \sqrt{4 + 9 + 36} = \sqrt{49} = 7$$

Now we multiply $\mathbf{v}$ by the scalar $1/\|\mathbf{v}\| = \frac{1}{7}$. The result is the unit vector

$$\mathbf{u} = \frac{\mathbf{v}}{\|\mathbf{v}\|} = \frac{2\mathbf{i} - 3\mathbf{j} - 6\mathbf{k}}{7} = \frac{2}{7}\mathbf{i} - \frac{3}{7}\mathbf{j} - \frac{6}{7}\mathbf{k}$$ ■

NOW WORK PROBLEM 41.

Dot Product

4 The definition of *dot product* is an extension of the definition given for vectors in the plane.

If $\mathbf{v} = a_1\mathbf{i} + b_1\mathbf{j} + c_1\mathbf{k}$ and $\mathbf{w} = a_2\mathbf{i} + b_2\mathbf{j} + c_2\mathbf{k}$ are two vectors, the **dot product** $\mathbf{v} \cdot \mathbf{w}$ is defined as

$$\mathbf{v} \cdot \mathbf{w} = a_1a_2 + b_1b_2 + c_1c_2 \quad \textbf{(3)}$$

EXAMPLE 6 Finding Dot Products

If $\mathbf{v} = 2\mathbf{i} - 3\mathbf{j} + 6\mathbf{k}$ and $\mathbf{w} = 5\mathbf{i} + 3\mathbf{j} - \mathbf{k}$, find:

(a) $\mathbf{v} \cdot \mathbf{w}$ (b) $\mathbf{w} \cdot \mathbf{v}$ (c) $\mathbf{v} \cdot \mathbf{v}$

(d) $\mathbf{w} \cdot \mathbf{w}$ (e) $\|\mathbf{v}\|$ (f) $\|\mathbf{w}\|$

Solution

(a) $\mathbf{v} \cdot \mathbf{w} = 2(5) + (-3)3 + 6(-1) = -5$

(b) $\mathbf{w} \cdot \mathbf{v} = 5(2) + 3(-3) + (-1)(6) = -5$

(c) $\mathbf{v} \cdot \mathbf{v} = 2(2) + (-3)(-3) + 6(6) = 49$

(d) $\mathbf{w} \cdot \mathbf{w} = 5(5) + 3(3) + (-1)(-1) = 35$

(e) $\|\mathbf{v}\| = \sqrt{2^2 + (-3)^2 + 6^2} = \sqrt{49} = 7$

(f) $\|\mathbf{w}\| = \sqrt{5^2 + 3^2 + (-1)^2} = \sqrt{35}$ ■

The dot product in space has the same properties as the dot product in the plane.

Theorem

Properties of the Dot Product

If $\mathbf{u}$, $\mathbf{v}$, and $\mathbf{w}$ are vectors, then

Commutative Property

$$\mathbf{u} \cdot \mathbf{v} = \mathbf{v} \cdot \mathbf{u}$$

Distributive Property

$$\mathbf{u} \cdot (\mathbf{v} + \mathbf{w}) = \mathbf{u} \cdot \mathbf{v} + \mathbf{u} \cdot \mathbf{w}$$

$$\mathbf{v} \cdot \mathbf{v} = \|\mathbf{v}\|^2$$

$$\mathbf{0} \cdot \mathbf{v} = 0$$

■

5 The angle θ between two vectors in space follows the same formula as for two vectors in the plane.

Theorem

Angle between Vectors

If $\mathbf{u}$ and $\mathbf{v}$ are two nonzero vectors, the angle θ, $0 \le \theta \le \pi$, between $\mathbf{u}$ and $\mathbf{v}$ is determined by the formula

$$\cos\theta = \frac{\mathbf{u} \cdot \mathbf{v}}{\|\mathbf{u}\| \, \|\mathbf{v}\|} \qquad (4)$$

■

EXAMPLE 7 **Finding the Angle θ between Two Vectors**

Find the angle θ between $\mathbf{u} = 2\mathbf{i} - 3\mathbf{j} + 6\mathbf{k}$ and $\mathbf{v} = 2\mathbf{i} + 5\mathbf{j} - \mathbf{k}$.

Solution We compute the quantities $\mathbf{u} \cdot \mathbf{v}$, $\|\mathbf{u}\|$, and $\|\mathbf{v}\|$.

$$\mathbf{u} \cdot \mathbf{v} = 2(2) + (-3)(5) + 6(-1) = -17$$
$$\|\mathbf{u}\| = \sqrt{2^2 + (-3)^2 + 6^2} = \sqrt{49} = 7$$
$$\|\mathbf{v}\| = \sqrt{2^2 + 5^2 + (-1)^2} = \sqrt{30}$$

By formula (4), if θ is the angle between $\mathbf{u}$ and $\mathbf{v}$, then

$$\cos\theta = \frac{\mathbf{u} \cdot \mathbf{v}}{\|\mathbf{u}\|\,\|\mathbf{v}\|} = \frac{-17}{7\sqrt{30}} \approx -0.443$$

We find that $\theta \approx 116°$. ■

NOW WORK PROBLEM 45.

Direction Angles of Vectors in Space

6 A nonzero vector $\mathbf{v}$ in space can be described by specifying its magnitude and its three **direction angles** α, β, and γ. These direction angles are defined as

α = angle between $\mathbf{v}$ and $\mathbf{i}$, the positive x-axis, $0° \le \alpha \le 180°$

β = angle between $\mathbf{v}$ and $\mathbf{j}$, the positive y-axis, $0° \le \beta \le 180°$

γ = angle between $\mathbf{v}$ and $\mathbf{k}$, the positive z-axis, $0° \le \gamma \le 180°$

See Figure 79.

Figure 79

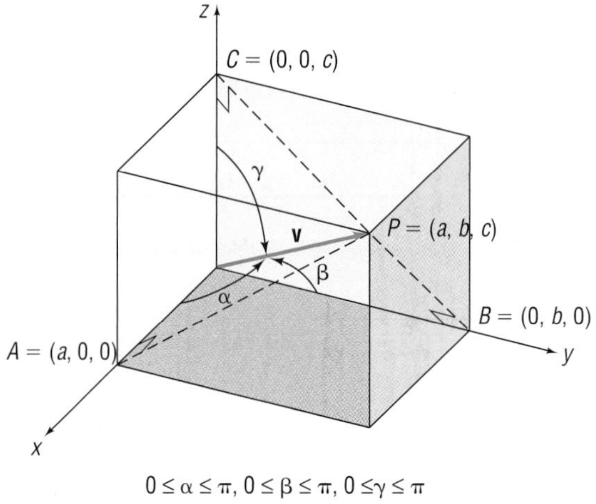

Our first goal is to find an expression for α, β, and γ in terms of the components of a vector. Let $\mathbf{v} = a\mathbf{i} + b\mathbf{j} + c\mathbf{k}$ denote a nonzero vector. The angle α between $\mathbf{v}$ and $\mathbf{i}$, the positive x-axis, obeys

$$\cos\alpha = \frac{\mathbf{v} \cdot \mathbf{i}}{\|\mathbf{v}\|\,\|\mathbf{i}\|} = \frac{a}{\|\mathbf{v}\|}$$

Similarly,

$$\cos\beta = \frac{b}{\|\mathbf{v}\|} \qquad \cos\gamma = \frac{c}{\|\mathbf{v}\|}$$

Since $\|\mathbf{v}\| = \sqrt{a^2 + b^2 + c^2}$, we have the following result:

Theorem **Direction Angles**

If $\mathbf{v} = a\mathbf{i} + b\mathbf{j} + c\mathbf{k}$ is a nonzero vector in space, the direction angles α, β, and γ obey

$$\cos\alpha = \frac{a}{\sqrt{a^2 + b^2 + c^2}} = \frac{a}{\|\mathbf{v}\|} \qquad \cos\beta = \frac{b}{\sqrt{a^2 + b^2 + c^2}} = \frac{b}{\|\mathbf{v}\|}$$

$$\cos\gamma = \frac{c}{\sqrt{a^2 + b^2 + c^2}} = \frac{c}{\|\mathbf{v}\|} \qquad \textbf{(5)}$$

■

The numbers $\cos\alpha$, $\cos\beta$, and $\cos\gamma$ are called the **direction cosines** of the vector **v**. They play the same role in space as slope does in the plane.

EXAMPLE 8 Finding the Direction Angles of a Vector

Find the direction angles of $\mathbf{v} = -3\mathbf{i} + 2\mathbf{j} - 6\mathbf{k}$.

Solution

$$\|\mathbf{v}\| = \sqrt{(-3)^2 + 2^2 + (-6)^2} = \sqrt{49} = 7$$

Using the theorem on direction angles, we get

$$\cos\alpha = \frac{-3}{7} \qquad \cos\beta = \frac{2}{7} \qquad \cos\gamma = \frac{-6}{7}$$

$$\alpha = 115^\circ \qquad \beta = 73^\circ \qquad \gamma = 149^\circ$$

■

Theorem **Property of Direction Cosines**

If α, β, and γ are the direction angles of a nonzero vector **v** in space, then

$$\cos^2\alpha + \cos^2\beta + \cos^2\gamma = 1 \qquad \textbf{(6)}$$

■

The proof is a direct consequence of equations (5).

Based on equation (6), when two direction cosines are known, the third is determined up to its sign. Knowing two direction cosines is not sufficient to uniquely determine the direction of a vector in space.

EXAMPLE 9 Finding the Direction Angle of a Vector

The vector **v** makes an angle of $\alpha = \pi/3$ with the positive x-axis, an angle of $\beta = \pi/3$ with the positive y-axis, and an acute angle γ with the positive z-axis. Find γ.

Solution By equation (6), we have

$$\cos^2\left(\frac{\pi}{3}\right) + \cos^2\left(\frac{\pi}{3}\right) + \cos^2\gamma = 1 \qquad 0 \le \gamma \le \pi$$

$$\left(\frac{1}{2}\right)^2 + \left(\frac{1}{2}\right)^2 + \cos^2\gamma = 1$$

$$\cos^2\gamma = \frac{1}{2}$$

$$\cos\gamma = \frac{\sqrt{2}}{2} \quad \text{or} \quad \cos\gamma = -\frac{\sqrt{2}}{2}$$

$$\gamma = \frac{\pi}{4} \quad \text{or} \quad \gamma = \frac{3\pi}{4}$$

Since we are requiring that γ be acute, the answer is $\gamma = \pi/4$. ■

The direction cosines of a vector give information about only the direction of the vector; they provide no information about its magnitude. For example, *any* vector parallel to the xy-plane and making an angle of $\pi/4$ radian with the positive x- and y-axes has direction cosines

$$\cos\alpha = \frac{\sqrt{2}}{2} \qquad \cos\beta = \frac{\sqrt{2}}{2} \qquad \cos\gamma = 0$$

However, if the direction angles *and* the magnitude of a vector are known, then the vector is uniquely determined.

EXAMPLE 10 Writing a Vector in Terms of Its Magnitude and Direction Cosines

Show that any nonzero vector $\mathbf{v}$ in space can be written in terms of its magnitude and direction cosines as

$$\mathbf{v} = \|\mathbf{v}\|\left[(\cos\alpha)\mathbf{i} + (\cos\beta)\mathbf{j} + (\cos\gamma)\mathbf{k}\right] \qquad \textbf{(7)}$$

Solution Let $\mathbf{v} = a\mathbf{i} + b\mathbf{j} + c\mathbf{k}$. From equation (5), we see that

$$a = \|\mathbf{v}\|\cos\alpha \qquad b = \|\mathbf{v}\|\cos\beta \qquad c = \|\mathbf{v}\|\cos\gamma$$

Substituting, we find that

$$\mathbf{v} = a\mathbf{i} + b\mathbf{j} + c\mathbf{k} = \|\mathbf{v}\|(\cos\alpha)\mathbf{i} + \|\mathbf{v}\|(\cos\beta)\mathbf{j} + \|\mathbf{v}\|(\cos\gamma)\mathbf{k}$$

$$= \|\mathbf{v}\|\left[(\cos\alpha)\mathbf{i} + (\cos\beta)\mathbf{j} + (\cos\gamma)\mathbf{k}\right]$$ ■

Example 10 shows that the direction cosines of a vector $\mathbf{v}$ are also the components of the unit vector in the direction of $\mathbf{v}$.

9.6 EXERCISES

In Problems 1–8, describe the set of points (x, y, z) defined by the equation.

1. $y = 0$
2. $x = 0$
3. $z = 2$
4. $y = 3$
5. $x = -4$
6. $z = -3$
7. $x = 1$ and $y = 2$
8. $x = 3$ and $z = 1$

In Problems 9–14, find the distance from P_1 to P_2.

9. $P_1 = (0, 0, 0)$ and $P_2 = (4, 1, 2)$
10. $P_1 = (0, 0, 0)$ and $P_2 = (1, -2, 3)$
11. $P_1 = (-1, 2, -3)$ and $P_2 = (0, -2, 1)$
12. $P_1 = (-2, 2, 3)$ and $P_2 = (4, 0, -3)$
13. $P_1 = (4, -2, -2)$ and $P_2 = (3, 2, 1)$
14. $P_1 = (2, -3, -3)$ and $P_2 = (4, 1, -1)$

In Problems 15–20, opposite vertices of a rectangular box whose edges are parallel to the coordinate axes are given. List the coordinates of the other six vertices of the box.

15. $(0, 0, 0)$; $(2, 1, 3)$
16. $(0, 0, 0)$; $(4, 2, 2)$
17. $(1, 2, 3)$; $(3, 4, 5)$
18. $(5, 6, 1)$; $(3, 8, 2)$
19. $(-1, 0, 2)$; $(4, 2, 5)$
20. $(-2, -3, 0)$; $(-6, 7, 1)$

*In Problems 21–26, the vector **v** has initial point P and terminal point Q. Write **v** in the form $a\mathbf{i} + b\mathbf{j} + c\mathbf{k}$; that is, find its position vector.*

21. $P = (0, 0, 0)$; $Q = (3, 4, -1)$
22. $P = (0, 0, 0)$; $Q = (-3, -5, 4)$
23. $P = (3, 2, -1)$; $Q = (5, 6, 0)$
24. $P = (-3, 2, 0)$; $Q = (6, 5, -1)$
25. $P = (-2, -1, 4)$; $Q = (6, -2, 4)$
26. $P = (-1, 4, -2)$; $Q = (6, 2, 2)$

In Problems 27–32, find $\|\mathbf{v}\|$.

27. $\mathbf{v} = 3\mathbf{i} - 6\mathbf{j} - 2\mathbf{k}$
28. $\mathbf{v} = -6\mathbf{i} + 12\mathbf{j} + 4\mathbf{k}$
29. $\mathbf{v} = \mathbf{i} - \mathbf{j} + \mathbf{k}$
30. $\mathbf{v} = -\mathbf{i} - \mathbf{j} + \mathbf{k}$
31. $\mathbf{v} = -2\mathbf{i} + 3\mathbf{j} - 3\mathbf{k}$
32. $\mathbf{v} = 6\mathbf{i} + 2\mathbf{j} - 2\mathbf{k}$

In Problems 33–38, find each quantity if $\mathbf{v} = 3\mathbf{i} - 5\mathbf{j} + 2\mathbf{k}$ and $\mathbf{w} = -2\mathbf{i} + 3\mathbf{j} - 2\mathbf{k}$.

33. $2\mathbf{v} + 3\mathbf{w}$
34. $3\mathbf{v} - 2\mathbf{w}$
35. $\|\mathbf{v} - \mathbf{w}\|$
36. $\|\mathbf{v} + \mathbf{w}\|$
37. $\|\mathbf{v}\| - \|\mathbf{w}\|$
38. $\|\mathbf{v}\| + \|\mathbf{w}\|$

*In Problems 39–44, find the unit vector having the same direction as **v**.*

39. $\mathbf{v} = 5\mathbf{i}$
40. $\mathbf{v} = -3\mathbf{j}$
41. $\mathbf{v} = 3\mathbf{i} - 6\mathbf{j} - 2\mathbf{k}$
42. $\mathbf{v} = -6\mathbf{i} + 12\mathbf{j} + 4\mathbf{k}$
43. $\mathbf{v} = \mathbf{i} + \mathbf{j} + \mathbf{k}$
44. $\mathbf{v} = 2\mathbf{i} - \mathbf{j} + \mathbf{k}$

*In Problems 45–52, find the dot product $\mathbf{v} \cdot \mathbf{w}$ and the angle between **v** and **w**.*

45. $\mathbf{v} = \mathbf{i} - \mathbf{j}$, $\mathbf{w} = \mathbf{i} + \mathbf{j} + \mathbf{k}$
46. $\mathbf{v} = \mathbf{i} + \mathbf{j}$, $\mathbf{w} = -\mathbf{i} + \mathbf{j} - \mathbf{k}$
47. $\mathbf{v} = 2\mathbf{i} + \mathbf{j} - 3\mathbf{k}$, $\mathbf{w} = \mathbf{i} + 2\mathbf{j} + 2\mathbf{k}$
48. $\mathbf{v} = 2\mathbf{i} + 2\mathbf{j} - \mathbf{k}$, $\mathbf{w} = \mathbf{i} + 2\mathbf{j} + 3\mathbf{k}$
49. $\mathbf{v} = 3\mathbf{i} - \mathbf{j} + 2\mathbf{k}$, $\mathbf{w} = \mathbf{i} + \mathbf{j} - \mathbf{k}$
50. $\mathbf{v} = \mathbf{i} + 3\mathbf{j} + 2\mathbf{k}$, $\mathbf{w} = \mathbf{i} - \mathbf{j} + \mathbf{k}$
51. $\mathbf{v} = 3\mathbf{i} + 4\mathbf{j} + \mathbf{k}$, $\mathbf{w} = 6\mathbf{i} + 8\mathbf{j} + 2\mathbf{k}$
52. $\mathbf{v} = 3\mathbf{i} - 4\mathbf{j} + \mathbf{k}$, $\mathbf{w} = 6\mathbf{i} - 8\mathbf{j} + 2\mathbf{k}$

In Problems 53–60, find the direction angles of each vector. Write each vector in the form of equation (7).

53. $\mathbf{v} = 3\mathbf{i} - 6\mathbf{j} - 2\mathbf{k}$
54. $\mathbf{v} = -6\mathbf{i} + 12\mathbf{j} + 4\mathbf{k}$
55. $\mathbf{v} = \mathbf{i} + \mathbf{j} + \mathbf{k}$
56. $\mathbf{v} = \mathbf{i} - \mathbf{j} - \mathbf{k}$
57. $\mathbf{v} = \mathbf{i} + \mathbf{j}$
58. $\mathbf{v} = \mathbf{j} + \mathbf{k}$
59. $\mathbf{v} = 3\mathbf{i} - 5\mathbf{j} + 2\mathbf{k}$
60. $\mathbf{v} = 2\mathbf{i} + 3\mathbf{j} - 4\mathbf{k}$

61. The Sphere In space, the collection of all points that are the same distance from some fixed point is called a **sphere**. See the illustration. The constant distance is called the **radius**, and the fixed point is the **center** of the sphere. Show that the equation of a sphere with center at (x_0, y_0, z_0) and radius r is

$$(x - x_0)^2 + (y - y_0)^2 + (z - z_0)^2 = r^2$$

[**Hint:** Use the Distance Formula (1).]

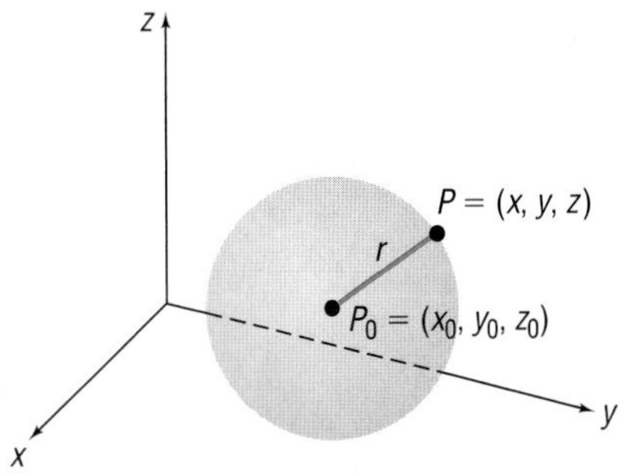

In Problems 62–64, find the equation of a sphere with radius r and center P_0.

62. $r = 1;\quad P_0 = (3, 1, 1)$

63. $r = 2;\quad P_0 = (1, 2, 2)$

64. $r = 3;\quad P_0 = (-1, 1, 2)$

In Problems 65–70, find the radius and center of each sphere. [**Hint:** Complete the square in each variable.]

65. $x^2 + y^2 + z^2 + 2x - 2y = 2$

66. $x^2 + y^2 + z^2 + 2x - 2z = -1$

67. $x^2 + y^2 + z^2 - 4x + 4y + 2z = 0$

68. $x^2 + y^2 + z^2 - 4x = 0$

69. $2x^2 + 2y^2 + 2z^2 - 8x + 4z = -1$

70. $3x^2 + 3y^2 + 3z^2 + 6x - 6y = 3$

The **work** *W done by a constant force* **F** *in moving an object from a point A in space to a point B in space is defined as* $W = \mathbf{F} \cdot \overrightarrow{AB}$. *Use this definition in Problems 71–73.*

71. Work Find the work done by a force of 3 newtons acting in the direction $2\mathbf{i} + \mathbf{j} + 2\mathbf{k}$ in moving an object 2 meters from $(0, 0, 0)$ to $(0, 2, 0)$.

72. Work Find the work done by a force of 1 newton acting in the direction $2\mathbf{i} + 2\mathbf{j} + \mathbf{k}$ in moving an object 3 meters from $(0, 0, 0)$ to $(1, 2, 2)$.

73. Work Find the work done in moving an object along a vector $\mathbf{u} = 3\mathbf{i} + 2\mathbf{j} - 5\mathbf{k}$ if the applied force is $\mathbf{F} = 2\mathbf{i} - \mathbf{j} - \mathbf{k}$.

9.7 THE CROSS PRODUCT

OBJECTIVES

1. Find the Cross Product of Two Vectors
2. Know Algebraic Properties of the Cross Product
3. Know Geometric Properties of the Cross Product
4. Find a Vector Orthogonal to Two Given Vectors
5. Find the Area of a Parallelogram

1 For vectors in space, and only for vectors in space, a second product of two vectors is defined, called the *cross product*. The cross product of two vectors in space is, in fact, also a vector that has applications in both geometry and physics.

If $\mathbf{v} = a_1\mathbf{i} + b_1\mathbf{j} + c_1\mathbf{k}$ and $\mathbf{w} = a_2\mathbf{i} + b_2\mathbf{j} + c_2\mathbf{k}$ are two vectors in space, the **cross product** $\mathbf{v} \times \mathbf{w}$ is defined as the vector

$$\mathbf{v} \times \mathbf{w} = (b_1c_2 - b_2c_1)\mathbf{i} - (a_1c_2 - a_2c_1)\mathbf{j} + (a_1b_2 - a_2b_1)\mathbf{k} \quad \textbf{(1)}$$

Notice that the cross product $\mathbf{v} \times \mathbf{w}$ of two vectors is a vector. Because of this, it is sometimes referred to as the **vector product**.

EXAMPLE 1 Finding Cross Products Using Equation (1)

If $\mathbf{v} = 2\mathbf{i} + 3\mathbf{j} + 5\mathbf{k}$ and $\mathbf{w} = \mathbf{i} + 2\mathbf{j} + 3\mathbf{k}$, then an application of equation (1) gives

$$\begin{aligned}\mathbf{v} \times \mathbf{w} &= (3 \cdot 3 - 2 \cdot 5)\mathbf{i} - (2 \cdot 3 - 1 \cdot 5)\mathbf{j} + (2 \cdot 2 - 3 \cdot 1)\mathbf{k} \\ &= (9 - 10)\mathbf{i} - (6 - 5)\mathbf{j} + (4 - 3)\mathbf{k} \\ &= -\mathbf{i} - \mathbf{j} + \mathbf{k}\end{aligned}$$

■

Determinants* may be used as an aid in computing cross products. A **2 by 2 determinant**, symbolized by

$$\begin{vmatrix} a_1 & b_1 \\ a_2 & b_2 \end{vmatrix}$$

has the value $a_1b_2 - a_2b_1$; that is,

$$\begin{vmatrix} a_1 & b_1 \\ a_2 & b_2 \end{vmatrix} = a_1b_2 - a_2b_1$$

A **3 by 3 determinant** has the value

$$\begin{vmatrix} A & B & C \\ a_1 & b_1 & c_1 \\ a_2 & b_2 & c_2 \end{vmatrix} = \begin{vmatrix} b_1 & c_1 \\ b_2 & c_2 \end{vmatrix} A - \begin{vmatrix} a_1 & c_1 \\ a_2 & c_2 \end{vmatrix} B + \begin{vmatrix} a_1 & b_1 \\ a_2 & b_2 \end{vmatrix} C$$

EXAMPLE 2 Evaluating Determinants

(a) $\begin{vmatrix} 2 & 3 \\ 1 & 2 \end{vmatrix} = 2 \cdot 2 - 1 \cdot 3 = 4 - 3 = 1$

(b) $$\begin{vmatrix} A & B & C \\ 2 & 3 & 5 \\ 1 & 2 & 3 \end{vmatrix} = \begin{vmatrix} 3 & 5 \\ 2 & 3 \end{vmatrix} A - \begin{vmatrix} 2 & 5 \\ 1 & 3 \end{vmatrix} B + \begin{vmatrix} 2 & 3 \\ 1 & 2 \end{vmatrix} C$$
$$= (9 - 10)A - (6 - 5)B + (4 - 3)C$$
$$= -A - B + C$$

NOW WORK PROBLEM 1.

The cross product of the vectors $\mathbf{v} = a_1\mathbf{i} + b_1\mathbf{j} + c_1\mathbf{k}$ and $\mathbf{w} = a_2\mathbf{i} + b_2\mathbf{j} + c_2\mathbf{k}$, that is,

$$\mathbf{v} \times \mathbf{w} = (b_1c_2 - b_2c_1)\mathbf{i} - (a_1c_2 - a_2c_1)\mathbf{j} + (a_1b_2 - a_2b_1)\mathbf{k},$$

may be written symbolically using determinants as

$$\mathbf{v} \times \mathbf{w} = \begin{vmatrix} \mathbf{i} & \mathbf{j} & \mathbf{k} \\ a_1 & b_1 & c_1 \\ a_2 & b_2 & c_2 \end{vmatrix} = \begin{vmatrix} b_1 & c_1 \\ b_2 & c_2 \end{vmatrix} \mathbf{i} - \begin{vmatrix} a_1 & c_1 \\ a_2 & c_2 \end{vmatrix} \mathbf{j} + \begin{vmatrix} a_1 & b_1 \\ a_2 & b_2 \end{vmatrix} \mathbf{k}$$

EXAMPLE 3 Using Determinants to Find Cross Products

If $\mathbf{v} = 2\mathbf{i} + 3\mathbf{j} + 5\mathbf{k}$ and $\mathbf{w} = \mathbf{i} + 2\mathbf{j} + 3\mathbf{k}$, find:

(a) $\mathbf{v} \times \mathbf{w}$ (b) $\mathbf{w} \times \mathbf{v}$ (c) $\mathbf{v} \times \mathbf{v}$ (d) $\mathbf{w} \times \mathbf{w}$

Solution

(a) $$\mathbf{v} \times \mathbf{w} = \begin{vmatrix} \mathbf{i} & \mathbf{j} & \mathbf{k} \\ 2 & 3 & 5 \\ 1 & 2 & 3 \end{vmatrix} = \begin{vmatrix} 3 & 5 \\ 2 & 3 \end{vmatrix} \mathbf{i} - \begin{vmatrix} 2 & 5 \\ 1 & 3 \end{vmatrix} \mathbf{j} + \begin{vmatrix} 2 & 3 \\ 1 & 2 \end{vmatrix} \mathbf{k} = -\mathbf{i} - \mathbf{j} + \mathbf{k}$$

(b) $$\mathbf{w} \times \mathbf{v} = \begin{vmatrix} \mathbf{i} & \mathbf{j} & \mathbf{k} \\ 1 & 2 & 3 \\ 2 & 3 & 5 \end{vmatrix} = \begin{vmatrix} 2 & 3 \\ 3 & 5 \end{vmatrix} \mathbf{i} - \begin{vmatrix} 1 & 3 \\ 2 & 5 \end{vmatrix} \mathbf{j} + \begin{vmatrix} 1 & 2 \\ 2 & 3 \end{vmatrix} \mathbf{k} = \mathbf{i} + \mathbf{j} - \mathbf{k}$$

*Determinants are discussed in detail in Section 11.4.

(c) $\mathbf{v} \times \mathbf{v} = \begin{vmatrix} \mathbf{i} & \mathbf{j} & \mathbf{k} \\ 2 & 3 & 5 \\ 2 & 3 & 5 \end{vmatrix}$

$$= \begin{vmatrix} 3 & 5 \\ 3 & 5 \end{vmatrix}\mathbf{i} - \begin{vmatrix} 2 & 5 \\ 2 & 5 \end{vmatrix}\mathbf{j} + \begin{vmatrix} 2 & 3 \\ 2 & 3 \end{vmatrix}\mathbf{k} = 0\mathbf{i} - 0\mathbf{j} + 0\mathbf{k} = \mathbf{0}$$

(d) $\mathbf{w} \times \mathbf{w} = \begin{vmatrix} \mathbf{i} & \mathbf{j} & \mathbf{k} \\ 1 & 2 & 3 \\ 1 & 2 & 3 \end{vmatrix}$

$$= \begin{vmatrix} 2 & 3 \\ 2 & 3 \end{vmatrix}\mathbf{i} - \begin{vmatrix} 1 & 3 \\ 1 & 3 \end{vmatrix}\mathbf{j} + \begin{vmatrix} 1 & 2 \\ 1 & 2 \end{vmatrix}\mathbf{k} = 0\mathbf{i} - 0\mathbf{j} + 0\mathbf{k} = \mathbf{0}$$

NOW WORK PROBLEM 9.

Algebraic Properties of the Cross Product

2 Notice in Example 3(a) and 3(b) that $\mathbf{v} \times \mathbf{w}$ and $\mathbf{w} \times \mathbf{v}$ are negatives of one another. From Examples 3(c) and 3(d), we might conjecture that the cross product of a vector with itself is the zero vector. These and other algebraic properties of cross product are given next.

Theorem **Algebraic Properties of the Cross Product**

If $\mathbf{u}$, $\mathbf{v}$, and $\mathbf{w}$ are vectors in space and if α is a scalar, then

$$\mathbf{u} \times \mathbf{u} = \mathbf{0} \tag{2}$$

$$\mathbf{u} \times \mathbf{v} = -(\mathbf{v} \times \mathbf{u}) \tag{3}$$

$$\alpha(\mathbf{u} \times \mathbf{v}) = (\alpha\mathbf{u}) \times \mathbf{v} = \mathbf{u} \times (\alpha\mathbf{v}) \tag{4}$$

$$\mathbf{u} \times (\mathbf{v} + \mathbf{w}) = (\mathbf{u} \times \mathbf{v}) + (\mathbf{u} \times \mathbf{w}) \tag{5}$$

Proof We will prove properties (2) and (4) here and leave properties (3) and (5) as exercises (see Problems 49 and 50 at the end of this section).

To prove property (2), we let $\mathbf{u} = a_1\mathbf{i} + b_1\mathbf{j} + c_1\mathbf{k}$. Then

$$\mathbf{u} \times \mathbf{u} = \begin{vmatrix} \mathbf{i} & \mathbf{j} & \mathbf{k} \\ a_1 & b_1 & c_1 \\ a_1 & b_1 & c_1 \end{vmatrix} = \begin{vmatrix} b_1 & c_1 \\ b_1 & c_1 \end{vmatrix}\mathbf{i} - \begin{vmatrix} a_1 & c_1 \\ a_1 & c_1 \end{vmatrix}\mathbf{j} + \begin{vmatrix} a_1 & b_1 \\ a_1 & b_1 \end{vmatrix}\mathbf{k}$$

$$= 0\mathbf{i} - 0\mathbf{j} + 0\mathbf{k} = \mathbf{0}$$

To prove property (4), we let $\mathbf{u} = a_1\mathbf{i} + b_1\mathbf{j} + c_1\mathbf{k}$ and $\mathbf{v} = a_2\mathbf{i} + b_2\mathbf{j} + c_2\mathbf{k}$. Then

$$\alpha(\mathbf{u} \times \mathbf{v}) \underset{\substack{\uparrow \\ \text{Apply (1)}}}{=} \alpha\left[(b_1c_2 - b_2c_1)\mathbf{i} - (a_1c_2 - a_2c_1)\mathbf{j} + (a_1b_2 - a_2b_1)\mathbf{k}\right]$$

$$= \alpha(b_1c_2 - b_2c_1)\mathbf{i} - \alpha(a_1c_2 - a_2c_1)\mathbf{j} + \alpha(a_1b_2 - a_2b_1)\mathbf{k} \tag{6}$$

Since $\alpha\mathbf{u} = \alpha a_1\mathbf{i} + \alpha b_1\mathbf{j} + \alpha c_1\mathbf{k}$, we have

$$(\alpha\mathbf{u}) \times \mathbf{v} = (\alpha b_1 c_2 - b_2\alpha c_1)\mathbf{i} - (\alpha a_1 c_2 - a_2\alpha c_1)\mathbf{j} + (\alpha a_1 b_2 - a_2\alpha b_1)\mathbf{k}$$
$$= \alpha(b_1 c_2 - b_2 c_1)\mathbf{i} - \alpha(a_1 c_2 - a_2 c_1)\mathbf{j} + \alpha(a_1 b_2 - a_2 b_1)\mathbf{k} \qquad (7)$$

Based on equations (6) and (7), the first part of property (4) follows. The second part can be proved in like fashion. ■

NOW WORK PROBLEM 11.

3 GEOMETRIC PROPERTIES OF THE CROSS PRODUCT

The cross product has several interesting geometric properties.

Theorem **Geometric Properties of the Cross Product**

Let **u** and **v** be vectors in space.

$\mathbf{u} \times \mathbf{v}$ is orthogonal to both **u** and **v**. **(8)**

$\|\mathbf{u} \times \mathbf{v}\| = \|\mathbf{u}\|\,\|\mathbf{v}\| \sin\theta$, where θ is the angle between **u** and **v**. **(9)**

$\|\mathbf{u} \times \mathbf{v}\|$ is the area of the parallelogram having $\mathbf{u} \neq \mathbf{0}$ and $\mathbf{v} \neq \mathbf{0}$ as adjacent sides. **(10)**

$\mathbf{v} \times \mathbf{w} = \mathbf{0}$ if and only if **u** and **v** are parallel. **(11)**

■

Proof of Property (8) Let $\mathbf{u} = a_1\mathbf{i} + b_1\mathbf{j} + c_1\mathbf{k}$ and $\mathbf{v} = a_2\mathbf{i} + b_2\mathbf{j} + c_2\mathbf{k}$. Then

$$\mathbf{u} \times \mathbf{v} = (b_1 c_2 - b_2 c_1)\mathbf{i} - (a_1 c_2 - a_2 c_1)\mathbf{j} + (a_1 b_2 - a_2 b_1)\mathbf{k}$$

Now we compute the dot product $\mathbf{u} \cdot (\mathbf{u} \times \mathbf{v})$.

$$\mathbf{u} \cdot (\mathbf{u} \times \mathbf{v}) = (a_1\mathbf{i} + b_1\mathbf{j} + c_1\mathbf{k}) \cdot \left[(b_1 c_2 - b_2 c_1)\mathbf{i} - (a_1 c_2 - a_2 c_1)\mathbf{j} + (a_1 b_2 - a_2 b_1)\mathbf{k}\right]$$
$$= a_1(b_1 c_2 - b_2 c_1) - b_1(a_1 c_2 - a_2 c_1) + c_1(a_1 b_2 - a_2 b_1) = 0$$

Since two vectors are orthogonal if their dot product is zero, it follows that **u** and $\mathbf{u} \times \mathbf{v}$ are orthogonal. Similarly, $\mathbf{v} \cdot (\mathbf{u} \times \mathbf{v}) = 0$, so **v** and $\mathbf{u} \times \mathbf{v}$ are orthogonal. ■

4 As long as the vectors **u** and **v** are not parallel, they will form a plane in space. See Figure 80. Based on property (8), the vector $\mathbf{u} \times \mathbf{v}$ is normal to this plane. As Figure 80 illustrates, there are two vectors normal to the plane containing **u** and **v**. It can be shown that the vector $\mathbf{u} \times \mathbf{v}$ is the one determined by the thumb of the right hand when the other fingers of the right hand are cupped so that they point in a direction from **u** to **v**. See Figure 81.*

Figure 80

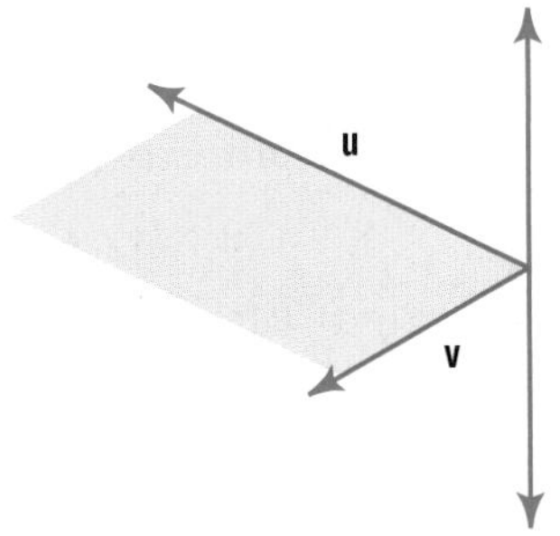

Figure 81

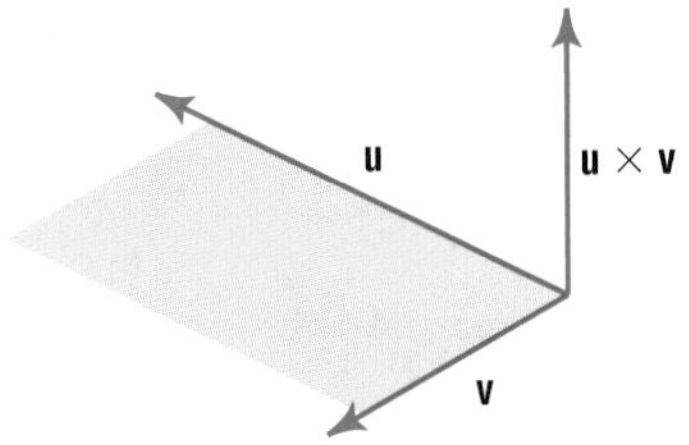

*This is a consequence of using a right-handed coordinate system.

EXAMPLE 4 Finding a Vector Orthogonal to Two Given Vectors

Find a vector that is orthogonal to $\mathbf{u} = 3\mathbf{i} - 2\mathbf{j} + \mathbf{k}$ and $\mathbf{v} = -\mathbf{i} + 3\mathbf{j} - \mathbf{k}$.

Solution Based on property (8), such a vector is $\mathbf{u} \times \mathbf{v}$.

$$\mathbf{u} \times \mathbf{v} = \begin{vmatrix} \mathbf{i} & \mathbf{j} & \mathbf{k} \\ 3 & -2 & 1 \\ -1 & 3 & -1 \end{vmatrix} = (2-3)\mathbf{i} - [-3-(-1)]\mathbf{j} + (9-2)\mathbf{k} = -\mathbf{i} + 2\mathbf{j} + 7\mathbf{k}$$

The vector $-\mathbf{i} + 2\mathbf{j} + 7\mathbf{k}$ is orthogonal to both $\mathbf{u}$ and $\mathbf{v}$. ■

Check: Two vectors are orthogonal if their dot product is zero.

$$\mathbf{u} \cdot (-\mathbf{i} + 2\mathbf{j} + 7\mathbf{k}) = (3\mathbf{i} - 2\mathbf{j} + \mathbf{k}) \cdot (-\mathbf{i} + 2\mathbf{j} + 7\mathbf{k}) = -3 - 4 + 7 = 0$$

$$\mathbf{v} \cdot (-\mathbf{i} + 2\mathbf{j} + 7\mathbf{k}) = (-\mathbf{i} + 3\mathbf{j} - \mathbf{k}) \cdot (-\mathbf{i} + 2\mathbf{j} + 7\mathbf{k}) = 1 + 6 - 7 = 0$$ ■

NOW WORK PROBLEM 35.

The proof of property (9) is left as an exercise. See Problem 52.

Figure 82

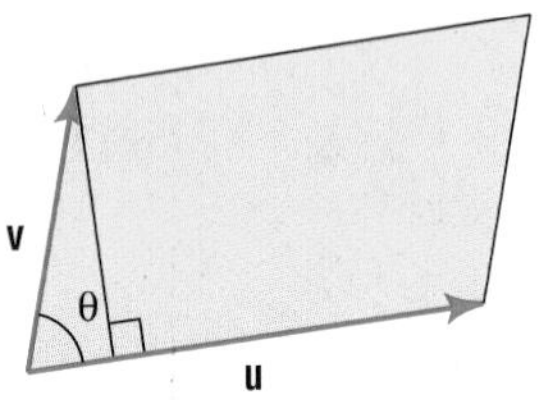

Proof of Property (10) Suppose that $\mathbf{u}$ and $\mathbf{v}$ are adjacent sides of a parallelogram. See Figure 82. Then the lengths of these sides are $\|\mathbf{u}\|$ and $\|\mathbf{v}\|$. If θ is the angle between $\mathbf{u}$ and $\mathbf{v}$, then the height of the parallelogram is $\|\mathbf{v}\| \sin\theta$ and its area is

$$\text{Area of parallelogram} = \text{Base} \times \text{Height} = \|\mathbf{u}\|\left[\|\mathbf{v}\| \sin\theta\right] \underset{\substack{\uparrow \\ \text{Property (9)}}}{=} \|\mathbf{u} \times \mathbf{v}\|$$ ■

EXAMPLE 5 Finding the Area of a Parallelogram

5 Find the area of the parallelogram whose vertices are $P_1 = (0, 0, 0)$, $P_2 = (3, -2, 1)$, $P_3 = (-1, 3, -1)$, and $P_4 = (2, 1, 0)$.

Solution Two adjacent sides* of this parallelogram are

$$\mathbf{u} = \overrightarrow{P_1P_2} = 3\mathbf{i} - 2\mathbf{j} + \mathbf{k} \quad \text{and} \quad \mathbf{v} = \overrightarrow{P_1P_3} = -\mathbf{i} + 3\mathbf{j} - \mathbf{k}$$

Since $\mathbf{u} \times \mathbf{v} = -\mathbf{i} + 2\mathbf{j} + 7\mathbf{k}$ (Example 4), the area of the parallelogram is

$$\text{Area of parallelogram} = \|\mathbf{u} \times \mathbf{v}\| = \sqrt{1 + 4 + 49} = \sqrt{54} = 3\sqrt{6}$$ ■

NOW WORK PROBLEM 43.

Proof of Property (11) The proof requires two parts. If $\mathbf{u}$ and $\mathbf{v}$ are parallel, then there is a scalar α such that $\mathbf{u} = \alpha\mathbf{v}$. Then

$$\mathbf{u} \times \mathbf{v} = (\alpha\mathbf{v}) \times \mathbf{v} \underset{\substack{\uparrow \\ \text{Property (4)}}}{=} \alpha(\mathbf{v} \times \mathbf{v}) \underset{\substack{\uparrow \\ \text{Property (2)}}}{=} \mathbf{0}$$

If $\mathbf{u} \times \mathbf{v} = \mathbf{0}$, then, by property (9), we have

$$\|\mathbf{u} \times \mathbf{v}\| = \|\mathbf{u}\| \, \|\mathbf{v}\| \sin\theta = 0$$

Since $\mathbf{u} \neq \mathbf{0}$ and $\mathbf{v} \neq \mathbf{0}$, then we must have $\sin\theta = 0$, so $\theta = 0$ or $\theta = \pi$. In either case, $\mathbf{u}$ and $\mathbf{v}$ are parallel. ■

* **Be careful!** Not all pairs of vertices give rise to a side. For example, $\overrightarrow{P_1P_4}$ is a diagonal of the parallelogram since $\overrightarrow{P_1P_3} + \overrightarrow{P_3P_4} = \overrightarrow{P_1P_4}$. Also, $\overrightarrow{P_1P_3}$ and $\overrightarrow{P_2P_4}$ are not adjacent sides; they are parallel sides.

9.7 EXERCISES

In Problems 1–8, find the value of each determinant.

1. $\begin{vmatrix} 3 & 4 \\ 1 & 2 \end{vmatrix}$

2. $\begin{vmatrix} -2 & 5 \\ 2 & -3 \end{vmatrix}$

3. $\begin{vmatrix} 6 & 5 \\ -2 & -1 \end{vmatrix}$

4. $\begin{vmatrix} -4 & 0 \\ 5 & 3 \end{vmatrix}$

5. $\begin{vmatrix} A & B & C \\ 2 & 1 & 4 \\ 1 & 3 & 1 \end{vmatrix}$

6. $\begin{vmatrix} A & B & C \\ 0 & 2 & 4 \\ 3 & 1 & 3 \end{vmatrix}$

7. $\begin{vmatrix} A & B & C \\ -1 & 3 & 5 \\ 5 & 0 & -2 \end{vmatrix}$

8. $\begin{vmatrix} A & B & C \\ 1 & -2 & -3 \\ 0 & 2 & -2 \end{vmatrix}$

In Problems 9–16, find (a) $\mathbf{v} \times \mathbf{w}$, *(b)* $\mathbf{w} \times \mathbf{v}$, *(c)* $\mathbf{w} \times \mathbf{w}$, *and (d)* $\mathbf{v} \times \mathbf{v}$.

9. $\mathbf{v} = 2\mathbf{i} - 3\mathbf{j} + \mathbf{k}$, $\mathbf{w} = 3\mathbf{i} - 2\mathbf{j} - \mathbf{k}$

10. $\mathbf{v} = -\mathbf{i} + 3\mathbf{j} + 2\mathbf{k}$, $\mathbf{w} = 3\mathbf{i} - 2\mathbf{j} - \mathbf{k}$

11. $\mathbf{v} = \mathbf{i} + \mathbf{j}$, $\mathbf{w} = 2\mathbf{i} + \mathbf{j} + \mathbf{k}$

12. $\mathbf{v} = \mathbf{i} - 4\mathbf{j} + 2\mathbf{k}$, $\mathbf{w} = 3\mathbf{i} + 2\mathbf{j} + \mathbf{k}$

13. $\mathbf{v} = 2\mathbf{i} - \mathbf{j} + 2\mathbf{k}$, $\mathbf{w} = \mathbf{j} - \mathbf{k}$

14. $\mathbf{v} = 3\mathbf{i} + \mathbf{j} + 3\mathbf{k}$, $\mathbf{w} = \mathbf{i} - \mathbf{k}$

15. $\mathbf{v} = \mathbf{i} - \mathbf{j} - \mathbf{k}$, $\mathbf{w} = 4\mathbf{i} - 3\mathbf{k}$

16. $\mathbf{v} = 2\mathbf{i} - 3\mathbf{j}$, $\mathbf{w} = 3\mathbf{j} - 2\mathbf{k}$

In Problems 17–38, use the vectors **u**, **v**, *and* **w** *given next to find each expression.*

$$\mathbf{u} = 2\mathbf{i} - 3\mathbf{j} + \mathbf{k} \qquad \mathbf{v} = -3\mathbf{i} + 3\mathbf{j} + 2\mathbf{k} \qquad \mathbf{w} = \mathbf{i} + \mathbf{j} + 3\mathbf{k}$$

17. $\mathbf{u} \times \mathbf{v}$
18. $\mathbf{v} \times \mathbf{w}$
19. $\mathbf{v} \times \mathbf{u}$
20. $\mathbf{w} \times \mathbf{v}$
21. $\mathbf{v} \times \mathbf{v}$
22. $\mathbf{w} \times \mathbf{w}$
23. $(3\mathbf{u}) \times \mathbf{v}$
24. $\mathbf{v} \times (4\mathbf{w})$
25. $\mathbf{u} \times (2\mathbf{v})$
26. $(-3\mathbf{v}) \times \mathbf{w}$
27. $\mathbf{u} \cdot (\mathbf{u} \times \mathbf{v})$
28. $\mathbf{v} \cdot (\mathbf{v} \times \mathbf{w})$
29. $\mathbf{u} \cdot (\mathbf{v} \times \mathbf{w})$
30. $(\mathbf{u} \times \mathbf{v}) \cdot \mathbf{w}$
31. $\mathbf{v} \cdot (\mathbf{u} \times \mathbf{w})$
32. $(\mathbf{v} \times \mathbf{u}) \cdot \mathbf{w}$
33. $\mathbf{u} \times (\mathbf{v} \times \mathbf{v})$
34. $(\mathbf{w} \times \mathbf{w}) \times \mathbf{v}$
35. Find a vector orthogonal to both **u** and **v**.
36. Find a vector orthogonal to both **u** and **w**.
37. Find a vector orthogonal to both **u** and $\mathbf{i} + \mathbf{j}$.
38. Find a vector orthogonal to both **u** and $\mathbf{j} + \mathbf{k}$.

In Problems 39–42, find the area of the parallelogram with one corner at P_1 *and adjacent sides* $\overrightarrow{P_1P_2}$ *and* $\overrightarrow{P_1P_3}$.

39. $P_1 = (0, 0, 0)$, $P_2 = (1, 2, 3)$, $P_3 = (-2, 3, 0)$
40. $P_1 = (0, 0, 0)$, $P_2 = (2, 3, 1)$, $P_3 = (-2, 4, 1)$
41. $P_1 = (1, 2, 0)$, $P_2 = (-2, 3, 4)$, $P_3 = (0, -2, 3)$
42. $P_1 = (-2, 0, 2)$, $P_2 = (2, 1, -1)$, $P_3 = (2, -1, 2)$

In Problems 43–46, find the area of the parallelogram with vertices P_1, P_2, P_3, *and* P_4.

43. $P_1 = (1, 1, 2)$, $P_2 = (1, 2, 3)$, $P_3 = (-2, 3, 0)$, $P_4 = (-2, 4, 1)$
44. $P_1 = (2, 1, 1)$, $P_2 = (2, 3, 1)$, $P_3 = (-2, 4, 1)$, $P_4 = (-2, 6, 1)$
45. $P_1 = (1, 2, -1)$, $P_2 = (4, 2, -3)$, $P_3 = (6, -5, 2)$, $P_4 = (9, -5, 0)$
46. $P_1 = (-1, 1, 1)$, $P_2 = (-1, 2, 2)$, $P_3 = (-3, 4, -5)$, $P_4 = (-3, 5, -4)$
47. Find a unit vector normal to the plane containing $\mathbf{v} = \mathbf{i} + 3\mathbf{j} - 2\mathbf{k}$ and $\mathbf{w} = -2\mathbf{i} + \mathbf{j} + 3\mathbf{k}$.
48. Find a unit vector normal to the plane containing $\mathbf{v} = 2\mathbf{i} + 3\mathbf{j} - \mathbf{k}$ and $\mathbf{w} = -2\mathbf{i} - 4\mathbf{j} - 3\mathbf{k}$.
49. Prove property (3).
50. Prove property (5).
51. Prove for vectors **u** and **v** that
$$\|\mathbf{u} \times \mathbf{v}\|^2 = \|\mathbf{u}\|^2\,\|\mathbf{v}\|^2 - (\mathbf{u} \cdot \mathbf{v})^2.$$
[**Hint:** Proceed as in the proof of property (4), computing first the left side and then the right side.]
52. Prove property (9).
[**Hint:** Use the result of Problem 51 and the fact that if θ is the angle between **u** and **v** then $\mathbf{u} \cdot \mathbf{v} = \|\mathbf{u}\|\,\|\mathbf{v}\| \cos\theta$.]
53. Show that if **u** and **v** are orthogonal then
$$\|\mathbf{u} \times \mathbf{v}\| = \|\mathbf{u}\|\,\|\mathbf{v}\|.$$
54. Show that if **u** and **v** are unit vectors then so is $\mathbf{u} \times \mathbf{v}$.
55. If $\mathbf{u} \cdot \mathbf{v} = \mathbf{0}$ and $\mathbf{u} \times \mathbf{v} = \mathbf{0}$, what can you conclude about **u** and **v**?

CHAPTER REVIEW

Things To Know

Relationship between polar coordinates (r, θ) and rectangular coordinates (x, y)	$x = r\cos\theta$, $y = r\sin\theta$ (p. 563) $x^2 + y^2 = r^2$, $\tan\theta = \frac{y}{x}$, $x \neq 0$ (p. 566)
Polar form of a complex number (p. 586)	If $z = x + yi$, then $z = r(\cos\theta + i\sin\theta)$, where $r = \lvert z\rvert = \sqrt{x^2 + y^2}$, $\sin\theta = \frac{y}{r}$, $\cos\theta = \frac{x}{r}$, $0 \le \theta < 2\pi$
DeMoivre's Theorem (p. 589)	If $z = r(\cos\theta + i\sin\theta)$, then $z^n = r^n[\cos(n\theta) + i\sin(n\theta)]$, where $n \ge 1$ is a positive integer
nth root of a complex number $z = r(\cos\theta_0 + i\sin\theta_0)$ (p. 590)	$\sqrt[n]{z} = \sqrt[n]{r}\left[\cos\left(\frac{\theta_0}{n} + \frac{2k\pi}{n}\right) + i\sin\left(\frac{\theta_0}{n} + \frac{2k\pi}{n}\right)\right]$, $k = 0, \dots, n-1$
Vector (p. 594)	Quantity having magnitude and direction; equivalent to a directed line segment $\overrightarrow{PQ}$
Position vector (p. 597 and 616)	Vector whose initial point is at the origin
Unit vector (p. 597)	Vector whose magnitude is 1
Dot product (p. 606 and 618)	If $\mathbf{v} = a_1\mathbf{i} + b_1\mathbf{j}$ and $\mathbf{w} = a_2\mathbf{i} + b_2\mathbf{j}$, then $\mathbf{v}\cdot\mathbf{w} = a_1a_2 + b_1b_2$. If $\mathbf{v} = a_1\mathbf{i} + b_1\mathbf{j} + c_1\mathbf{k}$ and $\mathbf{w} = a_2\mathbf{i} + b_2\mathbf{j} + c_2\mathbf{k}$, then $\mathbf{v}\cdot\mathbf{w} = a_1a_2 + b_1b_2 + c_1c_2$.
Angle θ between two nonzero vectors $\mathbf{u}$ and $\mathbf{v}$ (p. 608 and 619)	$\cos\theta = \frac{\mathbf{u}\cdot\mathbf{v}}{\lVert\mathbf{u}\rVert\,\lVert\mathbf{v}\rVert}$
Vectors in space (p. 621)	If $\mathbf{v} = a\mathbf{i} + b\mathbf{j} + c\mathbf{k}$, then $\mathbf{v} = \lVert\mathbf{v}\rVert[(\cos\alpha)\mathbf{i} + (\cos\beta)\mathbf{j} + (\cos\gamma)\mathbf{k}]$, where $\cos\alpha = \frac{a}{\lVert\mathbf{v}\rVert}$, $\cos\beta = \frac{b}{\lVert\mathbf{v}\rVert}$, $\cos\gamma = \frac{c}{\lVert\mathbf{v}\rVert}$.
Cross Product (p. 624)	If $\mathbf{v} = a_1\mathbf{i} + b_1\mathbf{j} + c_1\mathbf{k}$ and $\mathbf{w} = a_2\mathbf{i} + b_2\mathbf{j} + c_2\mathbf{k}$, then $\mathbf{v}\times\mathbf{w} = [b_1c_2 - b_2c_1]\mathbf{i} - (a_1c_2 - a_2c_1)\mathbf{j} + (a_1b_2 - a_2b_1)\mathbf{k}$
(p. 627)	$\lVert\mathbf{u}\times\mathbf{v}\rVert = \lVert\mathbf{u}\rVert\,\lVert\mathbf{v}\rVert\sin\theta$, where θ is the angle between $\mathbf{u}$ and $\mathbf{v}$.

How To

Plot points using polar coordinates (p. 560)

Convert from polar coordinates to rectangular coordinates (p. 563)

Convert from rectangular coordinates to polar coordinates (p. 564)

Graph and identify polar equations by converting to rectangular equations (p. 569)

Graph polar equations using a graphing utility (p. 570)

Test polar equations for symmetry (p. 574)

Graph polar equations by plotting points (p. 575)

Convert a complex number from rectangular form to polar form (p. 586)

Plot points in the complex plane (p. 586)

Find products and quotients of complex numbers in polar form (p. 587)

Use DeMoivre's Theorem (p. 588)

Find complex roots (p. 590)

Graph vectors (p. 596)

Find a position vector (p. 597 and 616)

Add and subtract vectors (p. 600 and 617)

Find a scalar product and the magnitude of a vector (p. 600 and 618)

Find a unit vector (p. 600 and 618)

Find a vector from its direction and magnitude (p. 601)

Work with objects in static equilibrium (p. 602)

Find the dot product of two vectors (p. 606 and 618)

Find the angle between two vectors (p. 607 and 619)

Determine whether two vectors are parallel (p. 609)

Determine whether two vectors are orthogonal (p. 609)

Decompose a vector into two orthogonal vectors (p. 610)

Compute work (p. 611)

Find the distance between two points in space (p. 615)

Find the direction angles of a vector in space (p. 620)

Find the cross product of two vectors in space (p. 624)

Know algebraic properties of the cross product (p. 626)

Know the geometric properties of the cross product (p. 627)

Find a vector orthogonal to two given vectors (p. 627)

Find the area of a parallelogram (p. 628)

Fill-in-the-Blank Items

1. In polar coordinates, the origin is called the ______________, and the positive x-axis is referred to as the ______________ ______________.
2. Another representation in polar coordinates for the point $(2, \pi/3)$ is (______________, $4\pi/3$).
3. Using polar coordinates (r, θ), the circle $x^2 + y^2 = 2x$ takes the form ______________.
4. In a polar equation, replace θ by $-\theta$. If an equivalent equation results, the graph is symmetric with respect to ______________ ______________.
5. When a complex number z is written in the polar form $z = r(\cos\theta + i\sin\theta)$, the nonnegative number r is the ______________ or ______________ of z, and the angle θ, $0 \leq \theta < 2\pi$, is the ______________ of z.
6. A vector whose magnitude is 1 is called a(n) ______________ vector.
7. If the angle between two vectors **v** and **w** is $\pi/2$, then the dot product $\mathbf{v} \cdot \mathbf{w}$ equals ______________.

True/False Items

T F **1.** The polar coordinates of a point are unique.

T F **2.** The rectangular coordinates of a point are unique.

T F **3.** The tests for symmetry in polar coordinates are necessary, but not sufficient.

T F **4.** DeMoivre's Theorem is useful for raising a complex number to a positive integer power.

T F **5.** Vectors are quantities that have magnitude and direction.

T F **6.** Force is a physical example of a vector.

T F **7.** If **u** and **v** are orthogonal vectors, then $\mathbf{u} \cdot \mathbf{v} = 0$.

T F **8.** The sum of the squares of the direction cosines of a vector in space equals 1.

Review Exercises

Blue problem numbers indicate the authors' suggestions for use in a Practice Test.

In Problems 1–6, plot each point given in polar coordinates, and find its rectangular coordinates.

1. $(3, \pi/6)$
2. $(4, 2\pi/3)$
3. $(-2, 4\pi/3)$
4. $(-1, 5\pi/4)$
5. $(-3, -\pi/2)$
6. $(-4, -\pi/4)$

In Problems 7–12, the rectangular coordinates of a point are given. Find two pairs of polar coordinates (r, θ) for each point, one with $r > 0$ and the other with $r < 0$. Express θ in radians.

7. $(-3, 3)$
8. $(1, -1)$
9. $(0, -2)$
10. $(2, 0)$
11. $(3, 4)$
12. $(-5, 12)$

In Problems 13–18, the letters x and y represent rectangular coordinates. Write each equation using polar coordinates (r, θ).

13. $3x^2 + 3y^2 = 6y$
14. $2x^2 - 2y^2 = 5y$
15. $2x^2 - y^2 = \dfrac{y}{x}$
16. $x^2 + 2y^2 = \dfrac{y}{x}$
17. $x(x^2 + y^2) = 4$
18. $y(x^2 - y^2) = 3$

In Problems 19–24, the letters r and θ represent polar coordinates. Write each polar equation as an equation in rectangular coordinates (x, y).

19. $r = 2\sin\theta$
20. $3r = \sin\theta$
21. $r = 5$
22. $\theta = \pi/4$
23. $r\cos\theta + 3r\sin\theta = 6$
24. $r^2\tan\theta = 1$

In Problems 25–30, sketch the graph of each polar equation. Be sure to test for symmetry. Verify your results using a graphing utility.

25. $r = 4\cos\theta$
26. $r = 3\sin\theta$
27. $r = 3 - 3\sin\theta$
28. $r = 2 + \cos\theta$
29. $r = 4 - \cos\theta$
30. $r = 1 - 2\sin\theta$

In Problems 31–34, write each complex number in polar form. Express each argument in degrees.

31. $-1 - i$ **32.** $-\sqrt{3} + i$ 33. $4 - 3i$ **34.** $3 - 2i$

In Problems 35–40, write each complex number in the standard form $a + bi$.

35. $2(\cos 150^\circ + i \sin 150^\circ)$ **36.** $3(\cos 60^\circ + i \sin 60^\circ)$ **37.** $3\left(\cos\frac{2\pi}{3} + i \sin\frac{2\pi}{3}\right)$

38. $4\left(\cos\frac{3\pi}{4} + i \sin\frac{3\pi}{4}\right)$ **39.** $0.1(\cos 350^\circ + i \sin 350^\circ)$ **40.** $0.5(\cos 160^\circ + i \sin 160^\circ)$

In Problems 41–46, find zw and z/w. Leave your answers in polar form.

41. $z = \cos 80^\circ + i \sin 80^\circ$, $w = \cos 50^\circ + i \sin 50^\circ$

42. $z = \cos 205^\circ + i \sin 205^\circ$, $w = \cos 85^\circ + i \sin 85^\circ$

43. $z = 3\left(\cos\frac{9\pi}{5} + i \sin\frac{9\pi}{5}\right)$, $w = 2\left(\cos\frac{\pi}{5} + i \sin\frac{\pi}{5}\right)$

44. $z = 2\left(\cos\frac{5\pi}{3} + i \sin\frac{5\pi}{3}\right)$, $w = 3\left(\cos\frac{\pi}{3} + i \sin\frac{\pi}{3}\right)$

45. $z = 5(\cos 10^\circ + i \sin 10^\circ)$, $w = \cos 355^\circ + i \sin 355^\circ$

46. $z = 4(\cos 50^\circ + i \sin 50^\circ)$, $w = \cos 340^\circ + i \sin 340^\circ$

In Problems 47–54, write each expression in the standard form $a + bi$. Verify your results using a graphing utility.

47. $[3(\cos 20^\circ + i \sin 20^\circ)]^3$ **48.** $[2(\cos 50^\circ + i \sin 50^\circ)]^3$

49. $\left[\sqrt{2}\left(\cos\frac{5\pi}{8} + i \sin\frac{5\pi}{8}\right)\right]^4$ **50.** $\left[2\left(\cos\frac{5\pi}{16} + i \sin\frac{5\pi}{16}\right)\right]^4$

51. $(1 - \sqrt{3}i)^6$ **52.** $(2 - 2i)^8$ **53.** $(3 + 4i)^4$ **54.** $(1 - 2i)^4$

55. Find all the complex cube roots of 27. **56.** Find all the complex fourth roots of -16.

*In Problems 57–64, the vector **v** is represented by the directed line segment $\overrightarrow{PQ}$. Write **v** in the form $a\mathbf{i} + b\mathbf{j}$ or in the form $a\mathbf{i} + b\mathbf{j} + c\mathbf{k}$ and find $\|\mathbf{v}\|$.*

57. $P = (1, -2)$; $Q = (3, -6)$ **58.** $P = (-3, 1)$; $Q = (4, -2)$

59. $P = (0, -2)$; $Q = (-1, 1)$ **60.** $P = (3, -4)$; $Q = (-2, 0)$

61. $P = (6, 2, 1)$; $Q = (3, 0, 2)$ **62.** $P = (4, 7, 0)$; $Q = (0, 5, 6)$

63. $P = (-1, 0, 1)$; $Q = (2, 0, 0)$ **64.** $P = (6, 2, 2)$; $Q = (2, 6, 2)$

In Problems 65–72, use the vectors $\mathbf{v} = -2\mathbf{i} + \mathbf{j}$ and $\mathbf{w} = 4\mathbf{i} - 3\mathbf{j}$.

65. Find $4\mathbf{v} - 3\mathbf{w}$. **66.** Find $-\mathbf{v} + 2\mathbf{w}$. 67. Find $\|\mathbf{v}\|$.

68. Find $\|\mathbf{v} + \mathbf{w}\|$. **69.** Find $\|\mathbf{v}\| + \|\mathbf{w}\|$. **70.** Find $\|2\mathbf{v}\| - 3\|\mathbf{w}\|$.

71. Find a unit vector in the same direction as **v**. **72.** Find a unit vector in the opposite direction of **w**.

In Problems 73–80, use the vectors $\mathbf{v} = 3\mathbf{i} + \mathbf{j} - 2\mathbf{k}$ and $\mathbf{w} = -3\mathbf{i} + 2\mathbf{j} - \mathbf{k}$ to find each expression.

73. $4\mathbf{v} - 3\mathbf{w}$ **74.** $-\mathbf{v} + 2\mathbf{w}$ **75.** $\|\mathbf{v} - \mathbf{w}\|$ **76.** $\|\mathbf{v} + \mathbf{w}\|$

77. $\|\mathbf{v}\| - \|\mathbf{w}\|$ **78.** $\|\mathbf{v}\| + \|\mathbf{w}\|$ 79. $\mathbf{v} \times \mathbf{w}$ **80.** $\mathbf{v} \cdot (\mathbf{v} \times \mathbf{w})$

81. Find a unit vector in the same direction as **v** and then in the opposite direction of **v**.

82. Find a unit vector orthogonal to both **v** and **w**.

*In Problems 83–90, find the dot product $\mathbf{v} \cdot \mathbf{w}$ and the angle between **v** and **w**.*

83. $\mathbf{v} = -2\mathbf{i} + \mathbf{j}$, $\mathbf{w} = 4\mathbf{i} - 3\mathbf{j}$ **84.** $\mathbf{v} = 3\mathbf{i} - \mathbf{j}$, $\mathbf{w} = \mathbf{i} + \mathbf{j}$

85. $\mathbf{v} = \mathbf{i} - 3\mathbf{j}$, $\mathbf{w} = -\mathbf{i} + \mathbf{j}$ **86.** $\mathbf{v} = \mathbf{i} + 4\mathbf{j}$, $\mathbf{w} = 3\mathbf{i} - 2\mathbf{j}$

87. $\mathbf{v} = \mathbf{i} + \mathbf{j} + \mathbf{k}$, $\mathbf{w} = \mathbf{i} - \mathbf{j} + \mathbf{k}$ **88.** $\mathbf{v} = \mathbf{i} - \mathbf{j} + \mathbf{k}$, $\mathbf{w} = 2\mathbf{i} + \mathbf{j} + \mathbf{k}$

89. $\mathbf{v} = 4\mathbf{i} - \mathbf{j} + 2\mathbf{k}$, $\mathbf{w} = \mathbf{i} - 2\mathbf{j} - 3\mathbf{k}$ **90.** $\mathbf{v} = -\mathbf{i} - 2\mathbf{i} + 3\mathbf{k}$, $\mathbf{w} = 5\mathbf{i} + \mathbf{j} + \mathbf{k}$

91. Find the vector projection of $\mathbf{v} = 2\mathbf{i} + 3\mathbf{j}$ onto $\mathbf{w} = 3\mathbf{i} + \mathbf{j}$.

92. Find the vector projection of $\mathbf{v} = -\mathbf{i} + 2\mathbf{j}$ onto $\mathbf{w} = 3\mathbf{i} - \mathbf{j}$.

93. Find the direction angles of the vector $\mathbf{v} = 3\mathbf{i} - 4\mathbf{j} + 2\mathbf{k}$.

94. Find the direction angles of the vector $\mathbf{v} = \mathbf{i} - \mathbf{j} + 2\mathbf{k}$.

95. Find the area of the parallelogram with vertices $P_1 = (1, 1, 1)$, $P_2 = (2, 3, 4)$, $P_3 = (6, 5, 2)$, and $P_4 = (7, 7, 5)$.

96. Find the area of the parallelogram with vertices $P_1 = (2, -1, 1)$, $P_2 = (5, 1, 4)$, $P_3 = (0, 1, 1)$, and $P_4 = (3, 3, 4)$.

97. Actual Speed and Direction of a Swimmer A swimmer can maintain a constant speed of 5 miles per hour. If the swimmer heads directly across a river that has a current moving at the rate of 2 miles per hour, what is the actual speed of the swimmer? (See the figure.) If the river is 1 mile wide, how far downstream will the swimmer end up from the point directly across the river from the starting point?

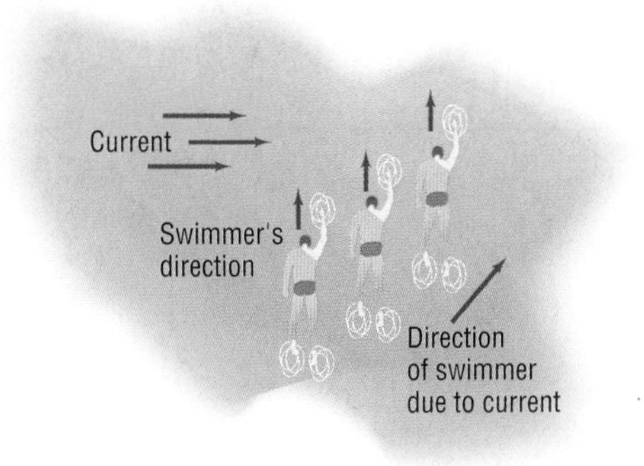

98. Actual Speed and Direction of an Airplane An airplane has an airspeed of 500 kilometers per hour in a northerly direction. The wind velocity is 60 kilometers per hour in a southeasterly direction. Find the actual speed and direction of the plane relative to the ground.

99. **Static Equilibrium** A weight of 2000 pounds is suspended from two cables as shown in the figure. What are the tensions of each cable?

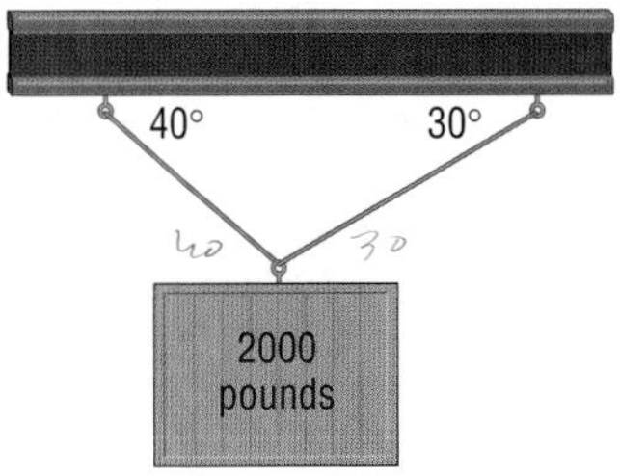

PROJECT AT MOTOROLA

Signal Fades due to interference

Consider an individual wearing a pager in an exterior elevator on a high-rise building. The pager receives direct signals from a transmitting tower far away as well as their reflections from the flat ground (see the figure).

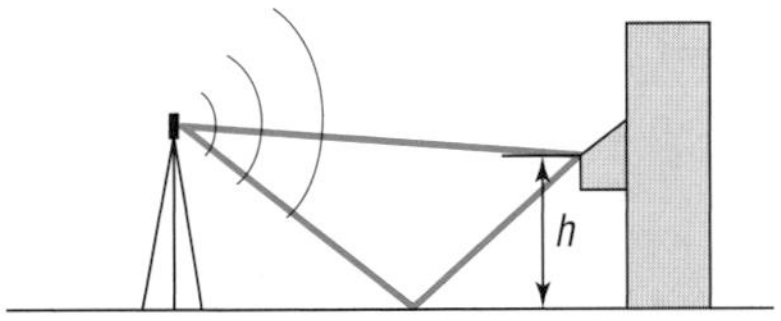

Oversimplifying things, it can be shown that the normalized signal received by the pager is given by

$$E(h)_{\text{norm}} \approx 1 + 0.25[\cos(2\pi h) + i\sin(2\pi h)]$$

where h is the height of the pager (elevator) from the ground expressed in meters.

1. Show that the magnitude of the normalized signal received by the pager oscillates between a minimum and a maximum value as the elevator moves up the building. This pattern is called a **standing wave pattern**. Calculate the ratio of the maximum to the minimum signal strength that the pager sees. This is called the **Standing Wave Ratio** (SWR).

2. Show that the distance between two consecutive minima is 1 meter.

3. Show that the distance between two consecutive maxima is 1 meter.

4. Show that the distance between a minimum and its nearby maximum is 0.5 meter.

5. What should the sensitivity of the pager be to guarantee reception of the signal regardless of the distance of the elevator from the ground?

6. Draw the graph E_{norm} in the complex plane as h varies. Explain all your answers above using this figure.

CHAPTER

10 Analytic Geometry

Outline

FIELD TRIP TO MOTOROLA

One of the many activites that NASA (National Aeronautic and Space Administration) is involved with has to with placing antennas in orbit. These antenna are usually in the shape of a paraboloid of revolution. However, in transporting objects of any kind into orbit, cost is a major factor. One factor that reduces the cost is to keep the weight as small as possible. A second factor is to keep the size of the object as small as possible.

As a result, a parabolic reflector is manufactured by NASA for use on the space shuttle by making the antenna out of material that can be folded up. Once in orbit the folded antenna is deployed and opened up to look like a parabolic reflector. One of the problems associated with this method of deployment is to be sure that when the antenna is opened up, it aligns properly with the axis of the parabola.

INTERVIEW AT MOTOROLA

Nick Buris was born in Greece and came to the United States in 1982. He studied at the National Technical University of Athens and at North Carolina State University. After earning his degrees, he joined the Electrical and Computer Engineering faculty at the University of Massachusetts at Amherst. During this appointment he taught and conducted research projects for the Government and the defense industry. Since 1992 he has been with Motorola pursuing applied research and software tool development for electromagnetic design of wireless products. He enjoys all kinds of sports and still dreams about taking a year off so that he can sail and travel around the world. He lives in Illinois with his wife and two children.

10.1 CONICS

OBJECTIVE 1 Know the Names of the Conics

1 The word *conic* derives from the word *cone*, which is a geometric figure that can be constructed in the following way: Let a and g be two distinct lines that intersect at a point V. Keep the line a fixed. Now rotate the line g about a while maintaining the same angle between a and g. The collection of points swept out (generated) by the line g is called a **(right circular) cone.** See Figure 1. The fixed line a is called the **axis** of the cone; the point V is called its **vertex;** the lines that pass through V and make the same angle with a as g are called **generators** of the cone. Thus, each generator is a line that lies entirely on the cone. The cone consists of two parts, called **nappes,** that intersect at the vertex.

Figure 1

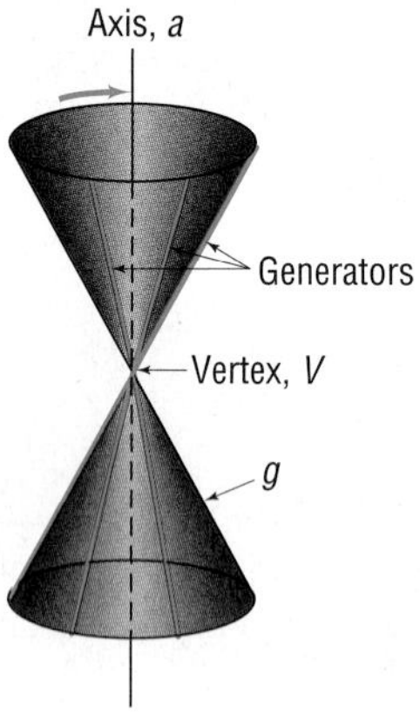

Conics, an abbreviation for **conic sections,** are curves that result from the intersection of a (right circular) cone and a plane. The conics we shall study arise when the plane does not contain the vertex, as shown in Figure 2. These conics are **circles** when the plane is perpendicular to the axis of the cone and intersects each generator; **ellipses** when the plane is tilted slightly so that it intersects each generator, but intersects only one nappe of the cone; **parabolas** when the plane is tilted further so that it is parallel to one (and only one) generator and intersects only one nappe of the cone; and **hyperbolas** when the plane intersects both nappes.

If the plane does contain the vertex, the intersection of the plane and the cone is a point, a line, or a pair of intersecting lines. These are usually called **degenerate conics.**

Figure 2

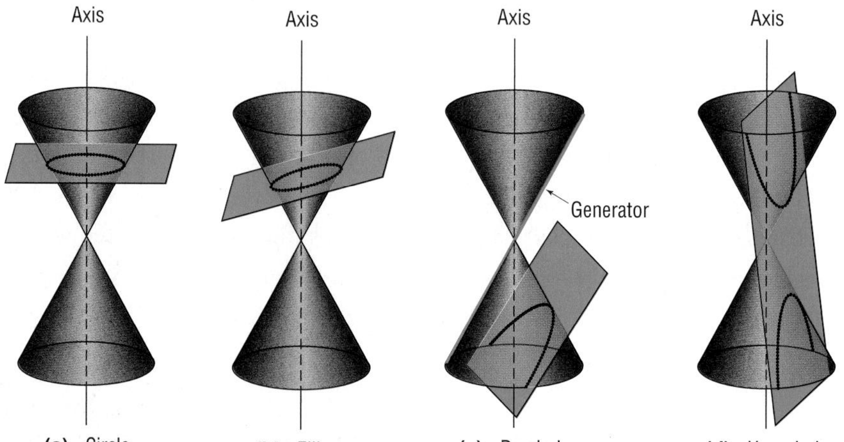

(a) Circle (b) Ellipse (c) Parabola (d) Hyperbola

PREPARING FOR THIS SECTION

Before getting started, review the following:

- ✓ Distance Formula (p. 6)
- ✓ Completing the Square (Appendix, Section 10)
- ✓ Symmetry (pp. 18–19)
- ✓ Graphing Techniques: Transformation (Section 2.4)

10.2 THE PARABOLA

OBJECTIVES
1 Find the Equation of a Parabola
2 Graph Parabolas
3 Discuss the Equation of a Parabola
4 Work with Parabolas with Vertex at (h, k)
5 Solve Applied Problems Involving Parabolas

We stated earlier (Section 3.1) that the graph of a quadratic function is a parabola. In this section, we begin with a geometric definition of a parabola and use it to obtain an equation.

A **parabola** is defined as the collection of all points P in the plane that are the same distance from a fixed point F as they are from a fixed line D. The point F is called the **focus** of the parabola, and the line D is its **directrix.** As a result, a parabola is the set of points P for which

$$d(F, P) = d(P, D) \qquad \textbf{(1)}$$

1 Figure 3 shows a parabola. The line through the focus F and perpendicular to the directrix D is called the **axis of symmetry** of the parabola. The point of intersection of the parabola with its axis of symmetry is called the **vertex** V.

Figure 3

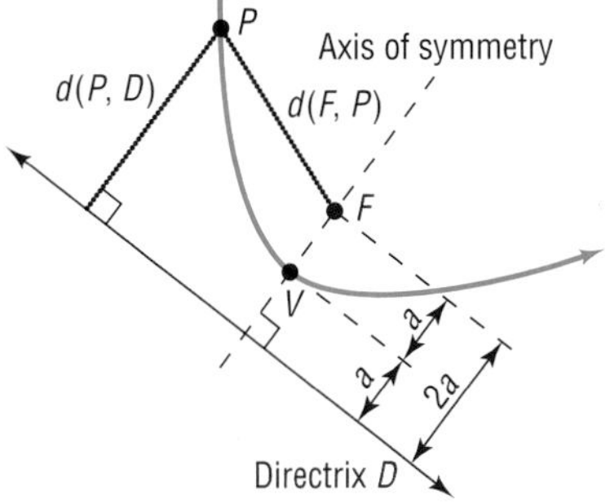

Because the vertex V lies on the parabola, it must satisfy equation (1): $d(F, V) = d(V, D)$. Thus, the vertex is midway between the focus and the directrix. We shall let a equal the distance $d(F, V)$ from F to V. Now we are ready to derive an equation for a parabola. To do this, we use a rectangular system of coordinates positioned so that the vertex V, focus F, and directrix

D of the parabola are conveniently located. If we choose to locate the vertex V at the origin $(0, 0)$, then we can conveniently position the focus F on either the x-axis or the y-axis.

First, we consider the case where the focus F is on the positive x-axis, as shown in Figure 4. Because the distance from F to V is a, the coordinates of F will be $(a, 0)$ with $a > 0$. Similarly, because the distance from V to the directrix D is also a and because D must be perpendicular to the x-axis (since the x-axis is the axis of symmetry), the equation of the directrix D must be $x = -a$. Now, if $P = (x, y)$ is any point on the parabola, then P must obey equation (1):

Figure 4
$y^2 = 4ax$

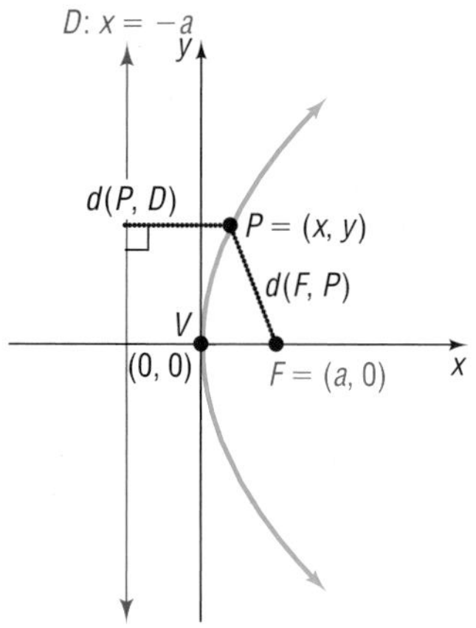

$$d(F, P) = d(P, D)$$

So we have

$$\sqrt{(x - a)^2 + (y - 0)^2} = |x + a| \qquad \text{Use the distance formula.}$$
$$(x - a)^2 + y^2 = (x + a)^2 \qquad \text{Square both sides.}$$
$$x^2 - 2ax + a^2 + y^2 = x^2 + 2ax + a^2 \qquad \text{Remove parentheses.}$$
$$y^2 = 4ax \qquad \text{Simplify.}$$

Theorem **Equation of a Parabola; Vertex at (0, 0), Focus at (a, 0), $a > 0$**

The equation of a parabola with vertex at $(0, 0)$, focus at $(a, 0)$, and directrix $x = -a$, $a > 0$, is

$$y^2 = 4ax \tag{2}$$

EXAMPLE 1 Finding the Equation of a Parabola and Graphing It by Hand

2 Find an equation of the parabola with vertex at $(0, 0)$ and focus at $(3, 0)$. Graph the equation by hand.

Solution The distance from the vertex $(0, 0)$ to the focus $(3, 0)$ is $a = 3$. Based on equation (2), the equation of this parabola is

$$y^2 = 4ax$$
$$y^2 = 12x \qquad a = 3$$

Figure 5
$y^2 = 12x$

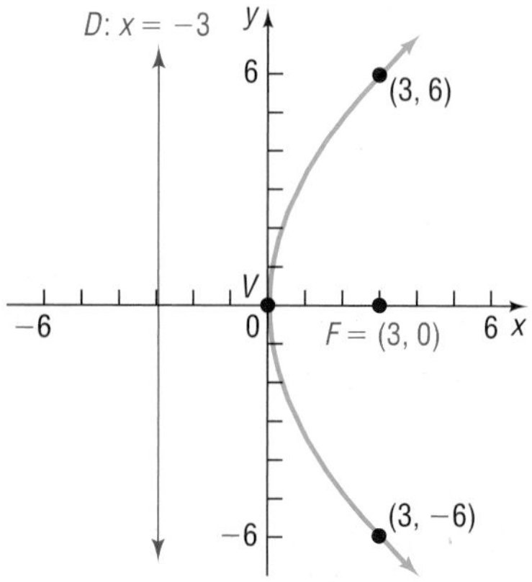

To graph this parabola by hand, it is helpful to plot the two points on the graph above and below the focus. To locate them, we let $x = 3$. Then

$$y^2 = 12x = 12(3) = 36$$
$$y = \pm 6 \qquad \text{Solve for } y.$$

The points on the parabola above and below the focus are $(3, -6)$ and $(3, 6)$. These points help in graphing the parabola because they determine the "opening." See Figure 5.

In general, the points on a parabola $y^2 = 4ax$ that lie above and below the focus $(a, 0)$ are each at a distance $2a$ from the focus. This follows from the fact that if $x = a$ then $y^2 = 4ax = 4a^2$, or $y = \pm 2a$. The line segment joining these two points is called the **latus rectum**; its length is $4a$.

NOW WORK PROBLEM 17.

EXAMPLE 2 Graphing a Parabola Using a Graphing Utility

Graph the parabola: $y^2 = 12x$

Figure 6
$y^2 = 12x$

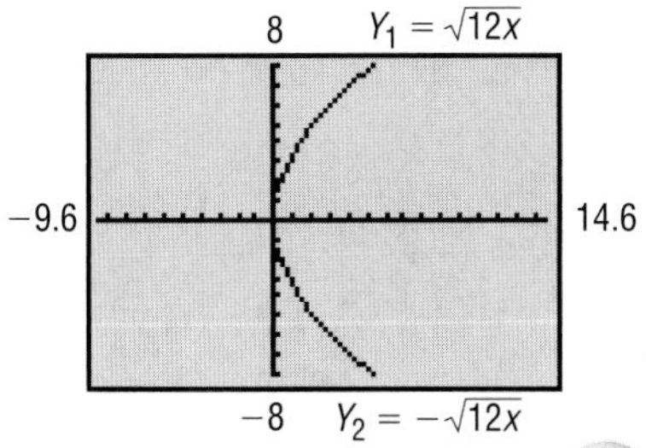

Solution To graph the parabola $y^2 = 12x$, we need to graph the two functions $Y_1 = \sqrt{12x}$ and $Y_2 = -\sqrt{12x}$ on a square screen. Figure 6 shows the graph of $y^2 = 12x$. Notice that the graph fails the vertical line test, so $y^2 = 12x$ is not a function. ■

By reversing the steps we used to obtain equation (2), it follows that the graph of an equation of the form of equation (2), $y^2 = 4ax$, is a parabola; its vertex is at $(0, 0)$, its focus is at $(a, 0)$, its directrix is the line $x = -a$, and its axis of symmetry is the x-axis.

3 For the remainder of this section, the direction "Discuss the equation" will mean to find the vertex, focus, and directrix of the parabola and graph it.

EXAMPLE 3 Discussing the Equation of a Parabola

Discuss the equation: $y^2 = 8x$

Solution Figure 7(a) shows the graph of $y^2 = 8x$ using a graphing utility. We now proceed to analyze the equation.

Figure 7
$y^2 = 8x$

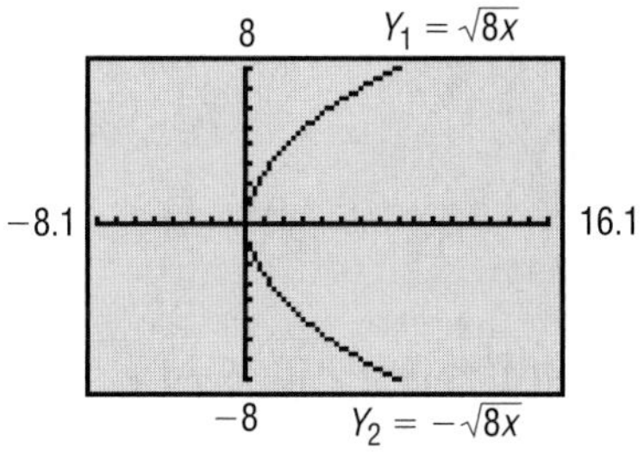

(a)

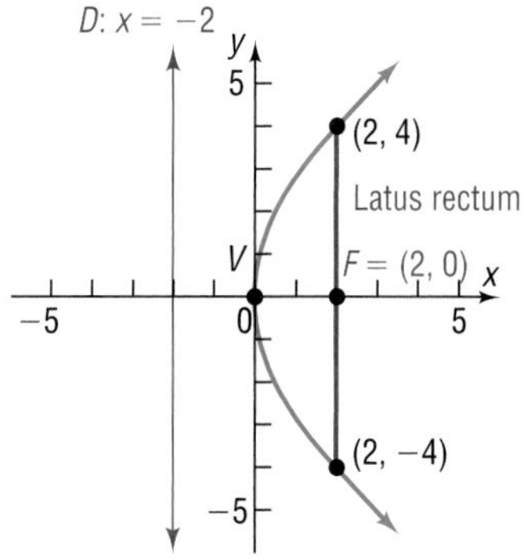

(b)

The equation $y^2 = 8x$ is of the form $y^2 = 4ax$, where $4a = 8$. Thus, $a = 2$. Consequently, the graph of the equation is a parabola with vertex at $(0, 0)$ and focus on the positive x-axis at $(2, 0)$. The directrix is the vertical line $x = -2$. The two points defining the latus rectum are obtained by letting $x = 2$. Then $y^2 = 16$, or $y = \pm 4$. Figure 7(b) shows the graph of the parabola drawn by hand. ■

Recall that we arrived at equation (2) after placing the focus on the positive x-axis. If the focus is placed on the negative x-axis, positive y-axis, or negative y-axis, a different form of the equation for the parabola results. The four forms of the equation of a parabola with vertex at $(0, 0)$ and focus on a coordinate axis a distance a from $(0, 0)$ are given in Table 1, and their graphs are given in Figure 8. Notice that each graph is symmetric with respect to its axis of symmetry.

TABLE 1

EQUATIONS OF A PARABOLA: VERTEX AT (0, 0); FOCUS ON AXIS; $a > 0$

Vertex	Focus	Directrix	Equation	Description
$(0, 0)$	$(a, 0)$	$x = -a$	$y^2 = 4ax$	Parabola, axis of symmetry is the x-axis, opens to right
$(0, 0)$	$(-a, 0)$	$x = a$	$y^2 = -4ax$	Parabola, axis of symmetry is the x-axis, opens to left
$(0, 0)$	$(0, a)$	$y = -a$	$x^2 = 4ay$	Parabola, axis of symmetry is the y-axis, opens up
$(0, 0)$	$(0, -a)$	$y = a$	$x^2 = -4ay$	Parabola, axis of symmetry is the y-axis, opens down

Figure 8

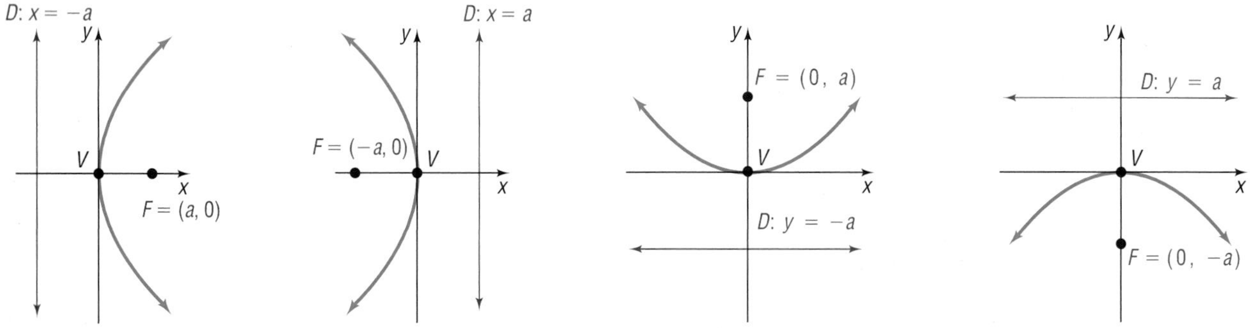

(a) $y^2 = 4ax$ **(b)** $y^2 = -4ax$ **(c)** $x^2 = 4ay$ **(d)** $x^2 = -4ay$

EXAMPLE 4 Discussing the Equation of a Parabola

Discuss the equation: $x^2 = -12y$

Solution Figure 9(a) shows the graph of $x^2 = -12y$ using a graphing utility. We now proceed to analyze the equation.

Figure 9
$x^2 = -12y$

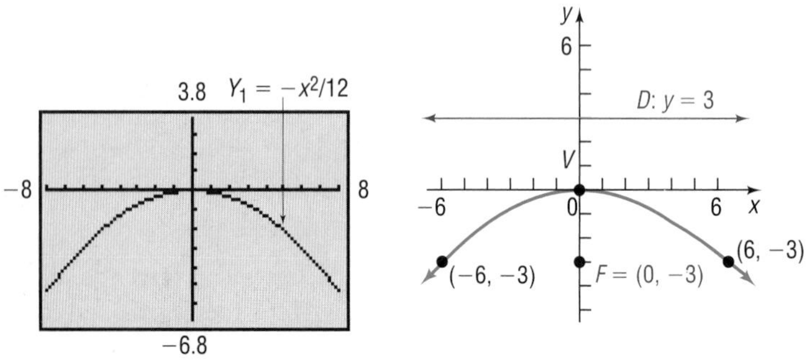

(a) **(b)**

The equation $x^2 = -12y$ is of the form $x^2 = -4ay$, with $a = 3$. Consequently, the graph of the equation is a parabola with vertex at $(0, 0)$, focus at $(0, -3)$, and directrix the line $y = 3$. The parabola opens down, and its axis of symmetry is the y-axis. To obtain the points defining the latus rectum, let $y = -3$. Then $x^2 = 36$, or $x = \pm 6$. See Figure 9(b). ■

NOW WORK PROBLEM 35.

EXAMPLE 5 Finding the Equation of a Parabola

Find the equation of the parabola with focus at $(0, 4)$ and directrix the line $y = -4$. Graph the equation by hand.

Solution A parabola whose focus is at $(0, 4)$ and whose directrix is the horizontal line $y = -4$ will have its vertex at $(0, 0)$. (Do you see why? The vertex is midway between the focus and the directrix.) Since the focus is on the positive y-axis, the equation of this parabola is of the form $x^2 = 4ay$, with $a = 4$; that is,

$$x^2 = \underset{\substack{\uparrow \\ a = 4}}{4ay} = 4(4)y = 16y$$

Figure 10 shows the graph. ■

Figure 10
$x^2 = 16y$

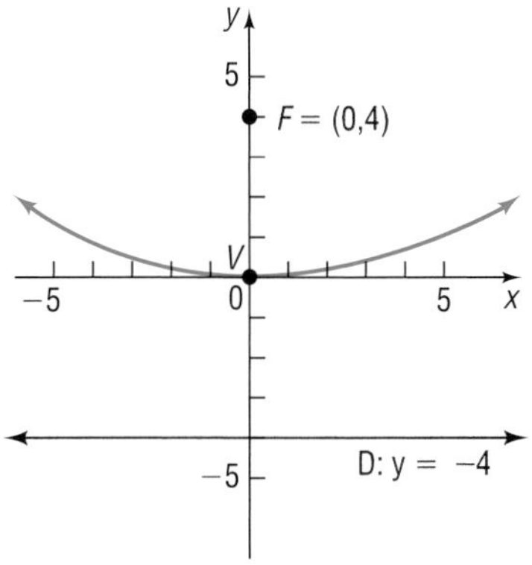

EXAMPLE 6 Finding the Equation of a Parabola

Find the equation of a parabola with vertex at $(0, 0)$ if its axis of symmetry is the x-axis and its graph contains the point $\left(-\frac{1}{2}, 2\right)$. Find its focus and directrix, and graph the equation by hand.

Solution The vertex is at the origin, the axis of symmetry is the x-axis, and the graph contains a point in the second quadrant, so the parabola opens to the left. We see from Table 1 that the form of the equation is

$$y^2 = -4ax$$

Because the point $\left(-\frac{1}{2}, 2\right)$ is on the parabola, the coordinates $x = -\frac{1}{2}$, $y = 2$ must satisfy the equation. Substituting $x = -\frac{1}{2}$ and $y = 2$ into the equation, we find that

$$4 = -4a\left(-\tfrac{1}{2}\right)$$
$$a = 2$$

The equation of the parabola is

$$y^2 = -4(2)x = -8x$$

The focus is at $(-a, 0) = (-2, 0)$ and the directrix is the line $x = 2$. Letting $x = -2$, we find $y^2 = 16$, or $y = \pm 4$. The points $(-2, 4)$ and $(-2, -4)$ define the latus rectum. See Figure 11. ■

Figure 11
$y^2 = -8x$

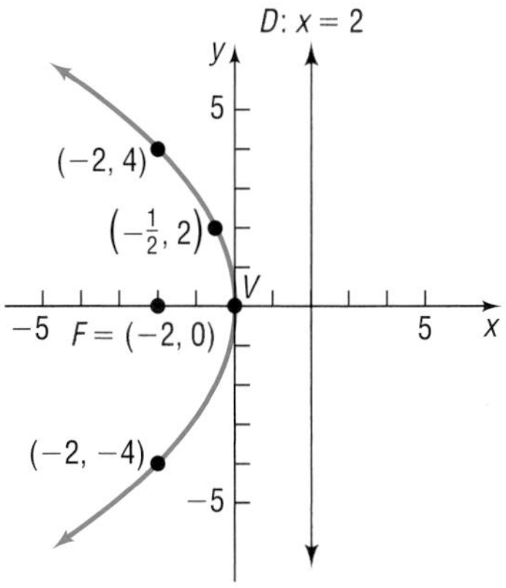

NOW WORK PROBLEM 27.

Vertex at (h, k)

4 If a parabola with vertex at the origin and axis of symmetry along a coordinate axis is shifted horizontally h units and then vertically k units, the result is a parabola with vertex at (h, k) and axis of symmetry parallel to a coordinate axis. The equations of such parabolas have the same forms as those in Table 1, but with x replaced by $x - h$ (the horizontal shift) and y replaced by $y - k$ (the vertical shift). Table 2 gives the forms of the equations of such parabolas. Figure 12(a)–(d) illustrates the graphs for $h > 0, k > 0$.

TABLE 2

PARABOLAS WITH VERTEX AT (h, k); AXIS OF SYMMETRY PARALLEL TO A COORDINATE AXIS, $a > 0$

Vertex	Focus	Directrix	Equation	Description
(h, k)	$(h + a, k)$	$x = h - a$	$(y - k)^2 = 4a(x - h)$	Parabola, axis of symmetry parallel to x-axis, opens to the right
(h, k)	$(h - a, k)$	$x = h + a$	$(y - k)^2 = -4a(x - h)$	Parabola, axis of symmetry parallel to x-axis, opens to the left
(h, k)	$(h, k + a)$	$y = k - a$	$(x - h)^2 = 4a(y - k)$	Parabola, axis of symmetry parallel to y-axis, opens up
(h, k)	$(h, k - a)$	$y = k + a$	$(x - h)^2 = -4a(y - k)$	Parabola, axis of symmetry parallel to y-axis, opens down

Figure 12

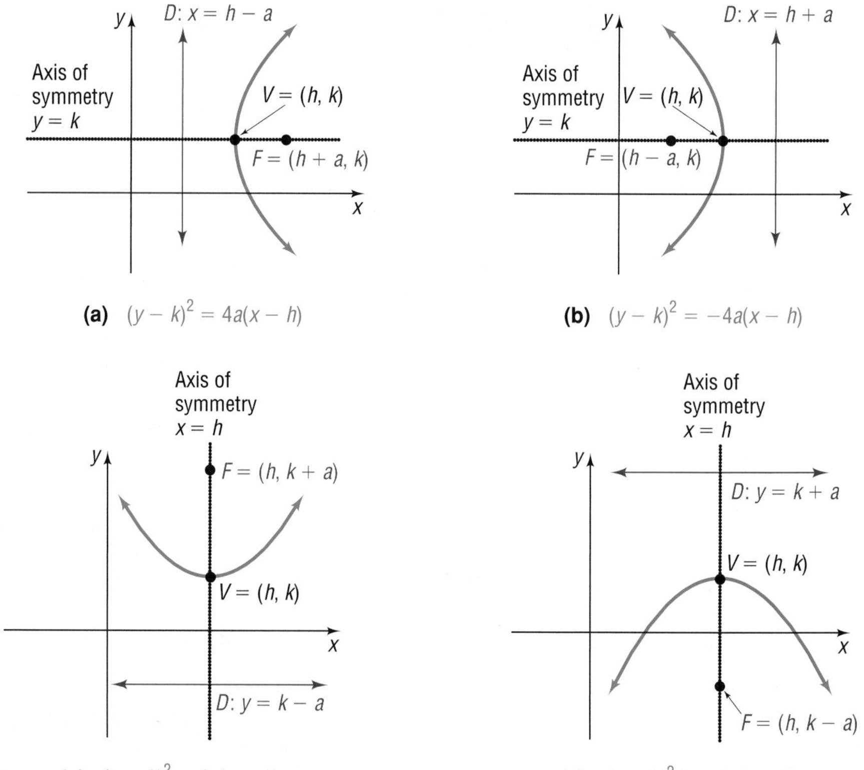

(a) $(y - k)^2 = 4a(x - h)$

(b) $(y - k)^2 = -4a(x - h)$

(c) $(x - h)^2 = 4a(y - k)$

(d) $(x - h)^2 = -4a(y - k)$

EXAMPLE 7 Finding the Equation of a Parabola, Vertex Not at Origin

Find an equation of the parabola with vertex at $(-2, 3)$ and focus at $(0, 3)$. Graph the equation by hand.

Figure 13
$(y - 3)^2 = 8(x + 2)$

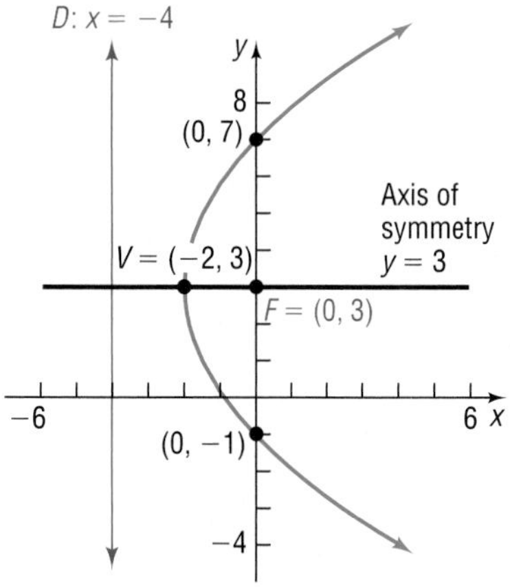

Solution The vertex $(-2, 3)$ and focus $(0, 3)$ both lie on the horizontal line $y = 3$ (the axis of symmetry). The distance a from the vertex $(-2, 3)$ to the focus $(0, 3)$ is $a = 2$. Also, because the focus lies to the right of the vertex, we know that the parabola opens to the right. Consequently, the form of the equation is

$$(y - k)^2 = 4a(x - h)$$

where $(h, k) = (-2, 3)$ and $a = 2$. Therefore, the equation is

$$(y - 3)^2 = 4 \cdot 2[x - (-2)]$$
$$(y - 3)^2 = 8(x + 2)$$

If $x = 0$, then $(y - 3)^2 = 16$. Thus, $y - 3 = \pm 4$, and $y = -1$ or $y = 7$. The points $(0, -1)$ and $(0, 7)$ define the latus rectum; the line $x = -4$ is the directrix. See Figure 13. ■

NOW WORK PROBLEM 25.

EXAMPLE 8 Using a Graphing Utility to Graph a Parabola, Vertex Not at Origin

Using a graphing utility, graph the equation: $(y - 3)^2 = 8(x + 2)$

Figure 14

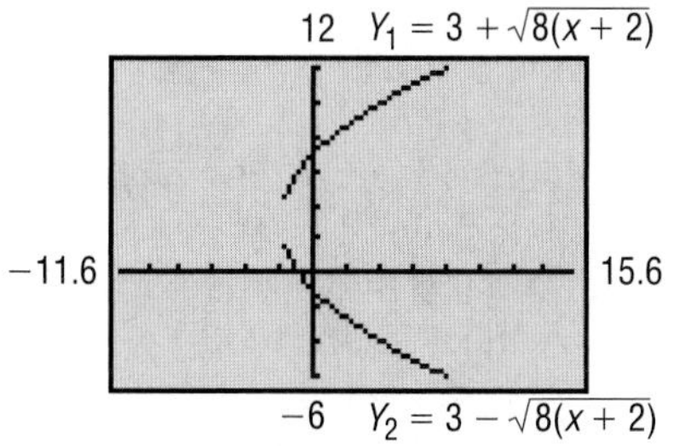

Solution First, we must solve the equation for y.

$$(y - 3)^2 = 8(x + 2)$$
$$y - 3 = \pm\sqrt{8(x + 2)} \quad \text{Take the square root of each side.}$$
$$y = 3 \pm \sqrt{8(x + 2)} \quad \text{Add 3 to both sides.}$$

Figure 14 shows the graphs of the equations $Y_1 = 3 + \sqrt{8(x + 2)}$ and $Y_2 = 3 - \sqrt{8(x + 2)}$. ■

NOW WORK PROBLEM 39.

Polynomial equations define parabolas whenever they involve two variables that are quadratic in one variable and linear in the other. To discuss this type of equation, we first complete the square of the variable that is quadratic.

EXAMPLE 9 Discussing the Equation of a Parabola

Discuss the equation: $x^2 + 4x - 4y = 0$

Solution Figure 15(a) shows the graph of $x^2 + 4x - 4y = 0$ using a graphing utility. We now proceed to analyze the equation.

To discuss the equation $x^2 + 4x - 4y = 0$, we complete the square involving the variable x.

Figure 15
$x^2 + 4x - 4y = 0$

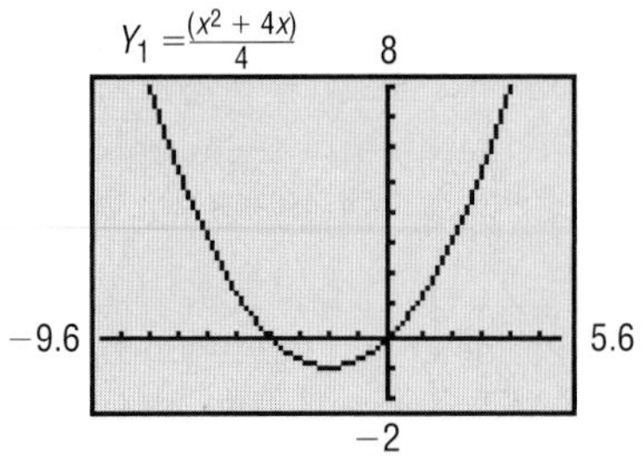

(a)

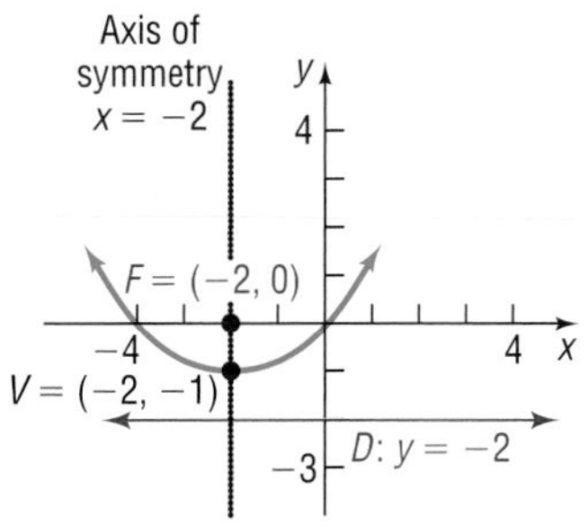

(b)

$$x^2 + 4x - 4y = 0$$

$$x^2 + 4x = 4y \quad \text{Isolate the terms involving } x \text{ on the left side.}$$

$$x^2 + 4x + 4 = 4y + 4 \quad \text{Complete the square on the left side.}$$

$$(x + 2)^2 = 4(y + 1) \quad \text{Factor.}$$

This equation is of the form $(x - h)^2 = 4a(y - k)$, with $h = -2$, $k = -1$, and $a = 1$. The graph is a parabola with vertex at $(h, k) = (-2, -1)$ that opens up. The focus is at $(-2, 0)$, and the directrix is the line $y = -2$. See Figure 15(b). ■

5 Parabolas find their way into many applications. For example, as we discussed in Section 3.1, suspension bridges have cables in the shape of a parabola. Another property of parabolas that is used in applications is their reflecting property.

Reflecting Property

Suppose that a mirror is shaped like a **paraboloid of revolution,** a surface formed by rotating a parabola about its axis of symmetry. If a light (or any other emitting source) is placed at the focus of the parabola, all the rays emanating from the light will reflect off the mirror in lines parallel to the axis of symmetry. This principle is used in the design of searchlights, flashlights, certain automobile headlights, and other such devices. See Figure 16.

Figure 16
Searchlight

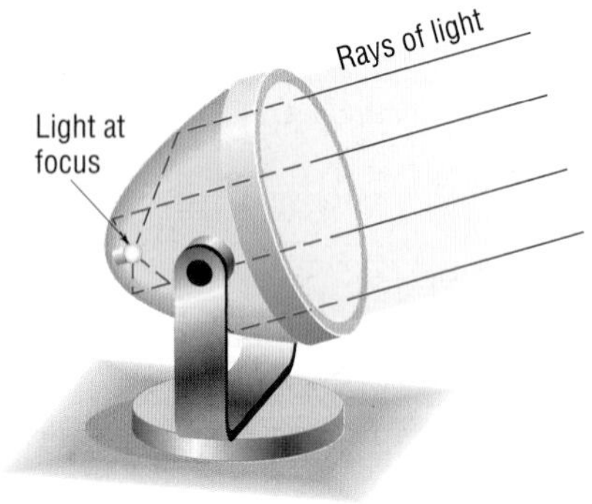

Conversely, suppose that rays of light (or other signals) emanate from a distant source so that they are essentially parallel. When these rays strike the surface of a parabolic mirror whose axis of symmetry is parallel to these rays, they are reflected to a single point at the focus. This principle is used in the design of some solar energy devices, satellite dishes, and the mirrors used in some types of telescopes. See Figure 17.

Figure 17

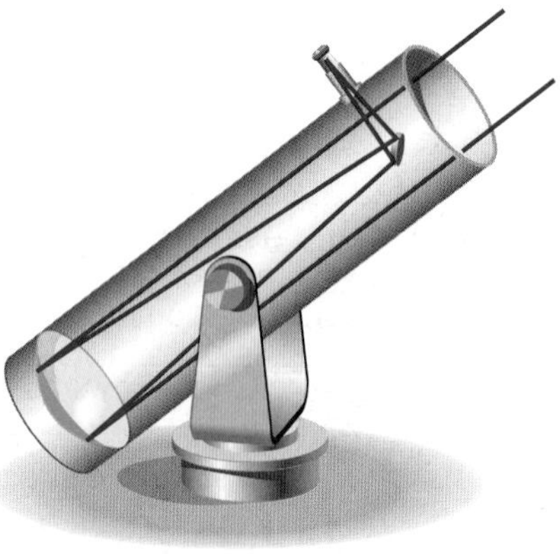

EXAMPLE 10 Satellite Dish

A satellite dish is shaped like a paraboloid of revolution. The signals that emanate from a satellite strike the surface of the dish and are reflected to a single point, where the receiver is located. If the dish is 8 feet across at its opening and is 3 feet deep at its center, at what position should the receiver be placed?

Solution Figure 18(a) shows the satellite dish. We draw the parabola used to form the dish on a rectangular coordinate system so that the vertex of the parabola is at the origin and its focus is on the positive y-axis. See Figure 18(b).

Figure 18

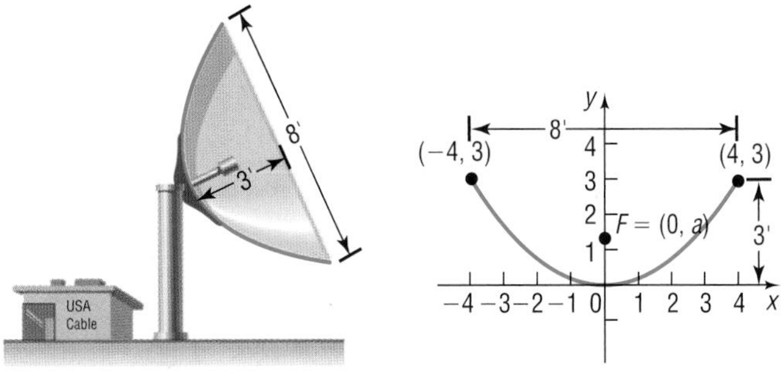

(a) (b)

The form of the equation of the parabola is

$$x^2 = 4ay$$

and its focus is at $(0, a)$. Since $(4, 3)$ is a point on the graph, we have

$$4^2 = 4a(3)$$

$$a = \frac{4}{3}$$

The receiver should be located $1\frac{1}{3}$ feet from the base of the dish, along its axis of symmetry. ■

10.2 EXERCISES

In Problems 1–8, the graph of a parabola is given. Match each graph to its equation.

A. $y^2 = 4x$	B. $x^2 = 4y$	C. $y^2 = -4x$
D. $x^2 = -4y$	E. $(y - 1)^2 = 4(x - 1)$	F. $(x + 1)^2 = 4(y + 1)$
G. $(y - 1)^2 = -4(x - 1)$	H. $(x + 1)^2 = -4(y + 1)$	

1.

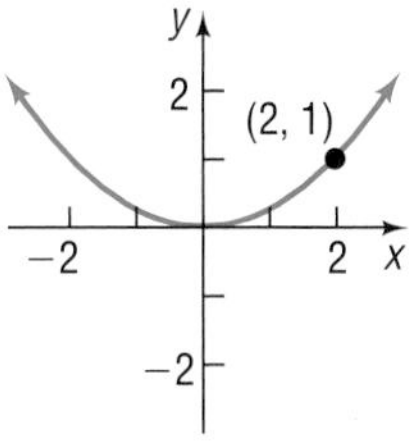

2.

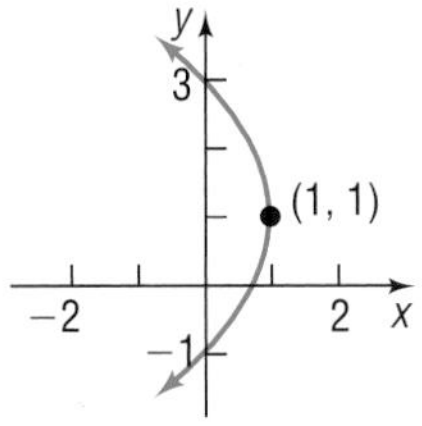

3.

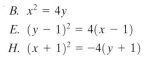

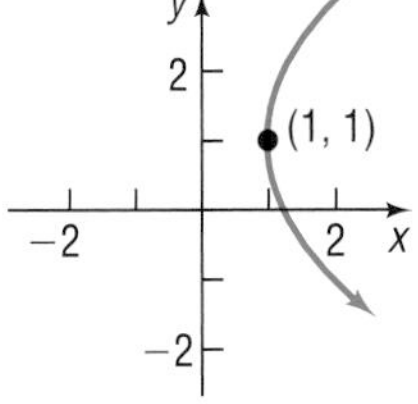

4.

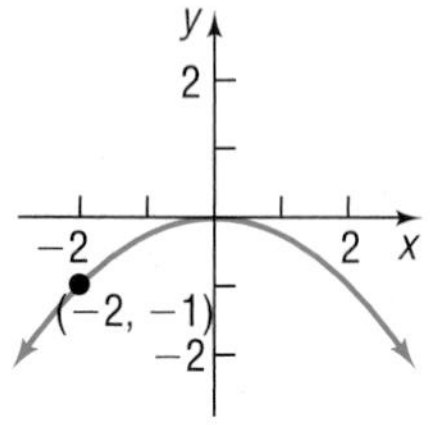

5.

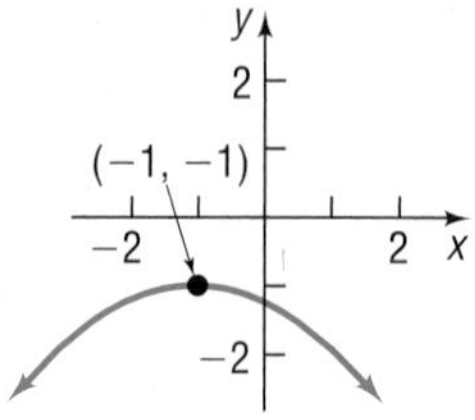

6.

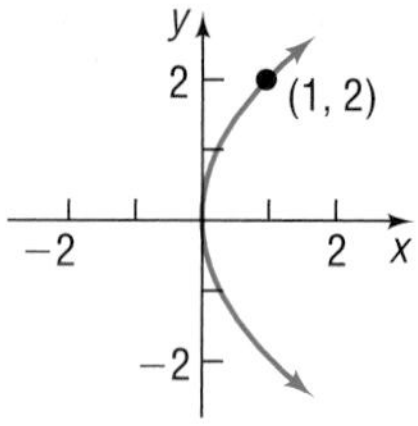

7.

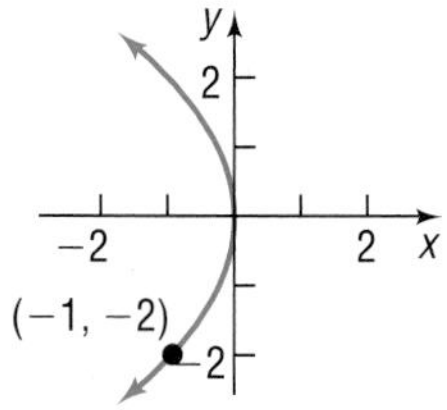

8.

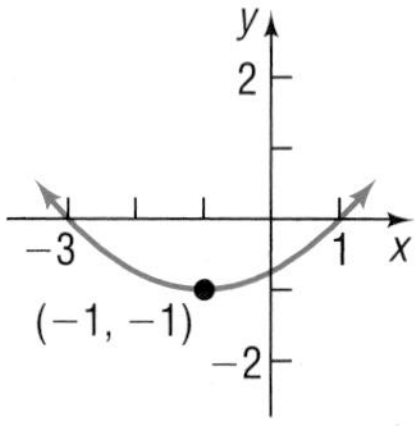

In Problems 9–16, the graph of a parabola is given. Match each graph to its equation.

A. $x^2 = 6y$ *B.* $x^2 = -6y$ *C.* $y^2 = 6x$

D. $y^2 = -6x$ *E.* $(y-2)^2 = -6(x+2)$ *F.* $(y-2)^2 = 6(x+2)$

G. $(x+2)^2 = -6(y-2)$ *H.* $(x+2)^2 = 6(y-2)$

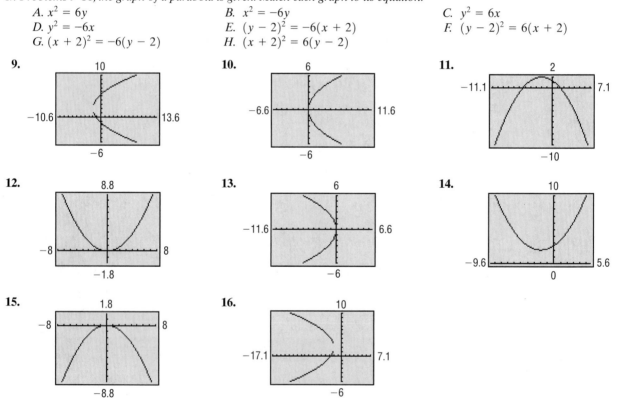

9.

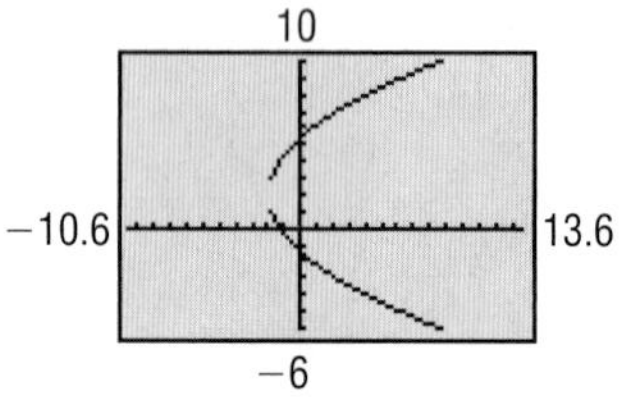

10.

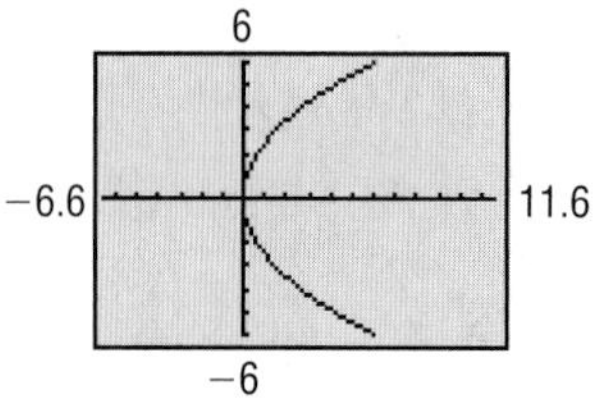

11.

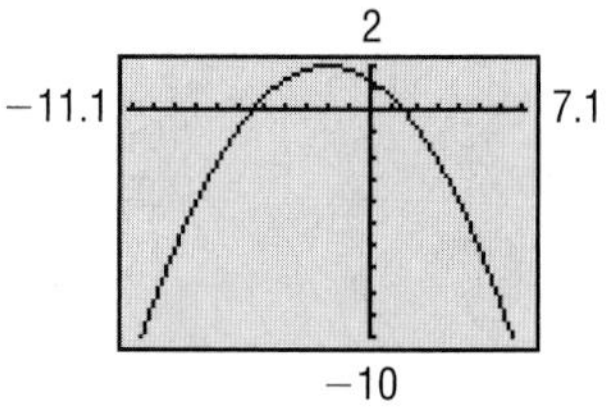

12.

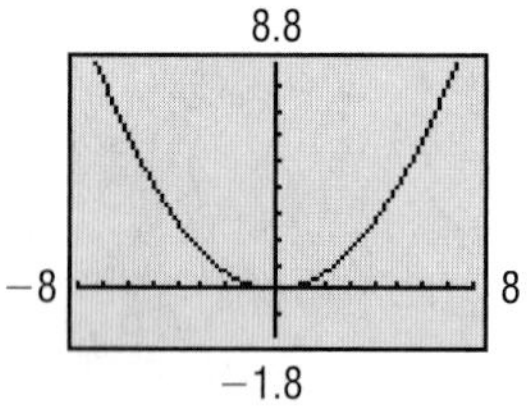

13.

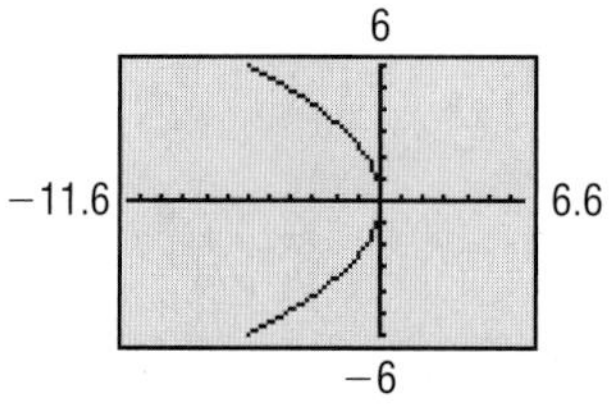

14.

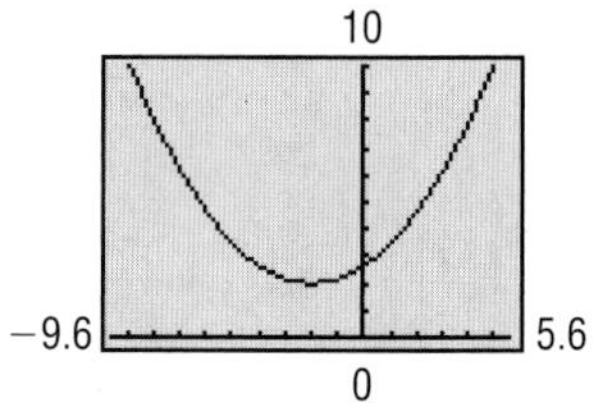

15.

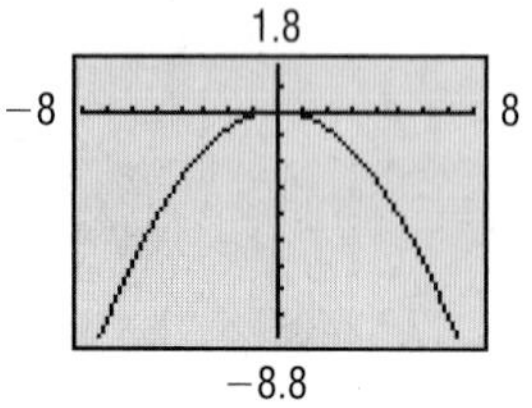

16.

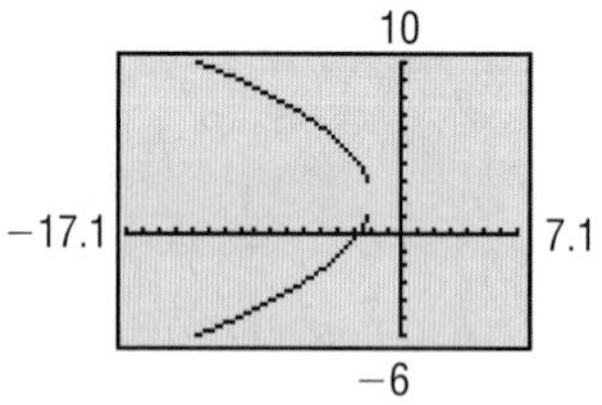

In Problems 17–32, find the equation of the parabola described. Find the two points that define the latus rectum, and graph the equation by hand.

17. Focus at $(4, 0)$; vertex at $(0, 0)$

18. Focus at $(0, 2)$; vertex at $(0, 0)$

19. Focus at $(0, -3)$; vertex at $(0, 0)$

20. Focus at $(-4, 0)$; vertex at $(0, 0)$

21. Focus at $(-2, 0)$; directrix the line $x = 2$

22. Focus at $(0, -1)$; directrix the line $y = 1$

23. Directrix the line $y = -\frac{1}{2}$; vertex at $(0, 0)$

24. Directrix the line $x = -\frac{1}{2}$; vertex at $(0, 0)$

25. Vertex at $(2, -3)$; focus at $(2, -5)$

26. Vertex at $(4, -2)$; focus at $(6, -2)$

27. Vertex at $(0, 0)$; axis of symmetry the y-axis; containing the point $(2, 3)$

28. Vertex at $(0, 0)$; axis of symmetry the x-axis; containing the point $(2, 3)$

29. Focus at $(-3, 4)$; directrix the line $y = 2$

30. Focus at $(2, 4)$; directrix the line $x = -4$

31. Focus at $(-3, -2)$; directrix the line $x = 1$

32. Focus at $(-4, 4)$; directrix the line $y = -2$

In Problems 33–50, find the vertex, focus, and directrix of each parabola. Graph the equation using a graphing utility.

33. $x^2 = 4y$

34. $y^2 = 8x$

35. $y^2 = -16x$

36. $x^2 = -4y$

37. $(y - 2)^2 = 8(x + 1)$

38. $(x + 4)^2 = 16(y + 2)$

39. $(x - 3)^2 = -(y + 1)$

40. $(y + 1)^2 = -4(x - 2)$

41. $(y + 3)^2 = 8(x - 2)$

42. $(x - 2)^2 = 4(y - 3)$

43. $y^2 - 4y + 4x + 4 = 0$

44. $x^2 + 6x - 4y + 1 = 0$

45. $x^2 + 8x = 4y - 8$

46. $y^2 - 2y = 8x - 1$

47. $y^2 + 2y - x = 0$

48. $x^2 - 4x = 2y$

49. $x^2 - 4x = y + 4$

50. $y^2 + 12y = -x + 1$

In Problems 51–58, write an equation for each parabola.

51.

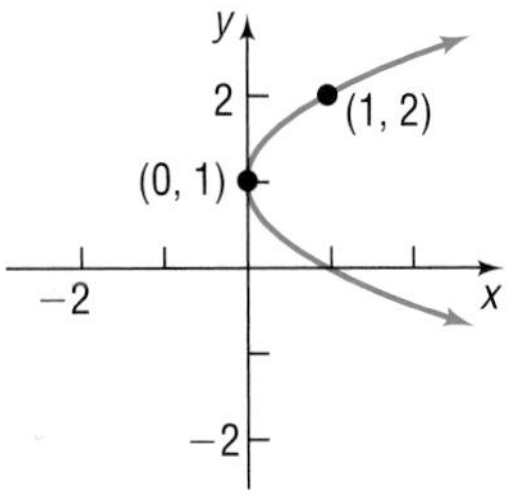

52.

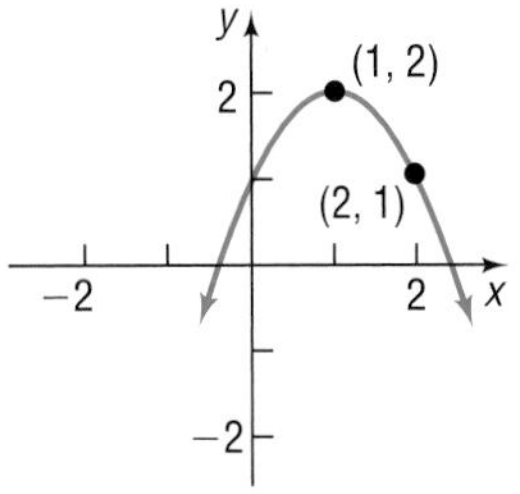

53.

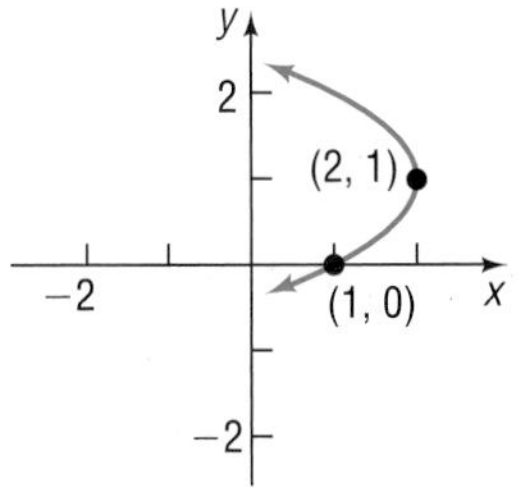

54.

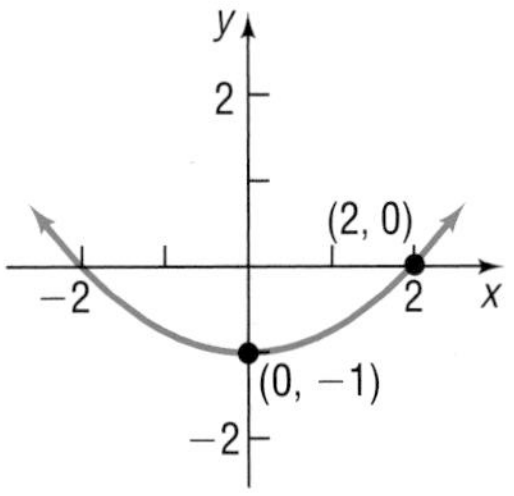

55.

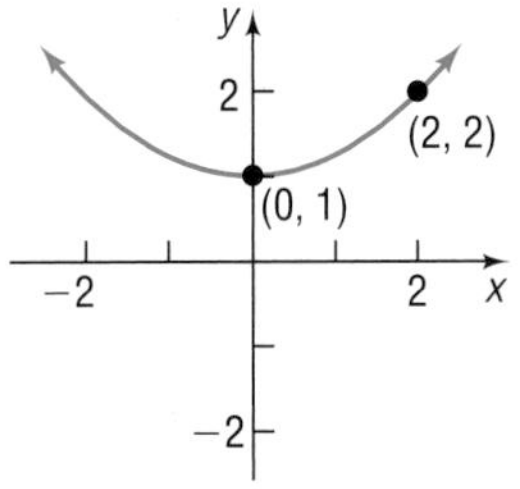

56.

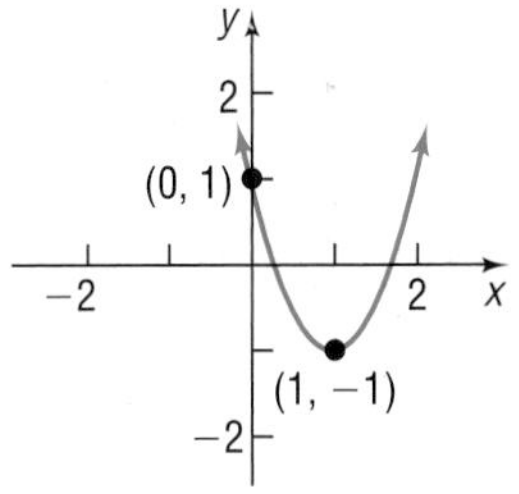

57.

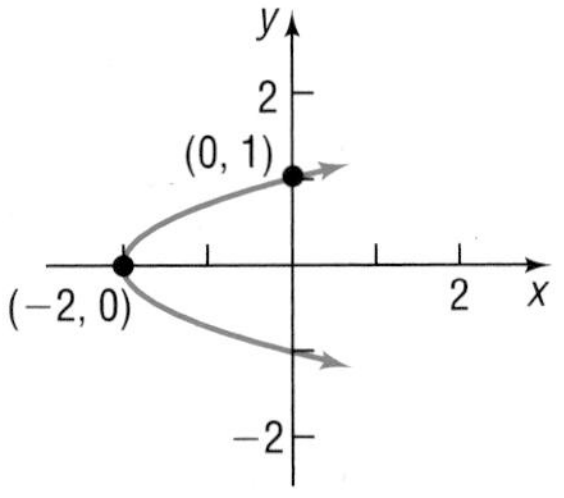

58.

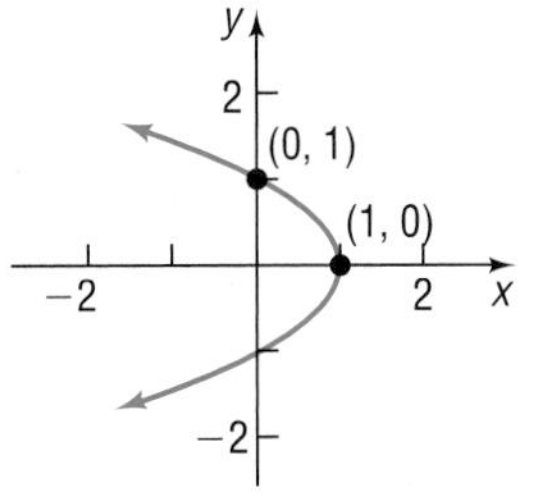

59. **Satellite Dish** A satellite dish is shaped like a paraboloid of revolution. The signals that emanate from a satellite strike the surface of the dish and are reflected to a single point, where the receiver is located. If the dish is 10 feet across at its opening and is 4 feet deep at its center, at what position should the receiver be placed?

60. **Constructing a TV Dish** A cable TV receiving dish is in the shape of a paraboloid of revolution. Find the location of the receiver, which is placed at the focus, if the dish is 6 feet across at its opening and 2 feet deep.

61. Constructing a Flashlight The reflector of a flashlight is in the shape of a paraboloid of revolution. Its diameter is 4 inches and its depth is 1 inch. How far from the vertex should the light bulb be placed so that the rays will be reflected parallel to the axis?

62. Constructing a Headlight A sealed-beam headlight is in the shape of a paraboloid of revolution. The bulb, which is placed at the focus, is 1 inch from the vertex. If the depth is to be 2 inches, what is the diameter of the headlight at its opening?

63. Suspension Bridge The cables of a suspension bridge are in the shape of a parabola, as shown in the figure. The towers supporting the cable are 600 feet apart and 80 feet high. If the cables touch the road surface midway between the towers, what is the height of the cable at a point 150 feet from the center of the bridge?

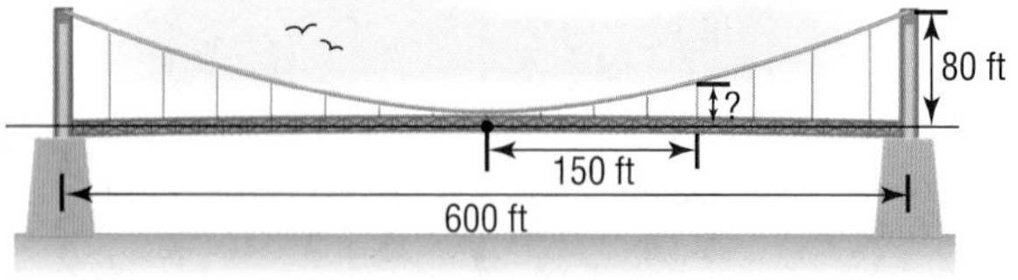

64. Suspension Bridge The cables of a suspension bridge are in the shape of a parabola. The towers supporting the cable are 400 feet apart and 100 feet high. If the cables are at a height of 10 feet midway between the towers, what is the height of the cable at a point 50 feet from the center of the bridge?

65. Searchlight A searchlight is shaped like a paraboloid of revolution. If the light source is located 2 feet from the base along the axis of symmetry and the opening is 5 feet across, how deep should the searchlight be?

66. Searchlight A searchlight is shaped like a paraboloid of revolution. If the light source is located 2 feet from the base along the axis of symmetry and the depth of the searchlight is 4 feet, what should the width of the opening be?

67. Solar Heat A mirror is shaped like a paraboloid of revolution and will be used to concentrate the rays of the sun at its focus, creating a heat source. If the mirror is 20 feet across at its opening and is 6 feet deep, where will the heat source be concentrated?

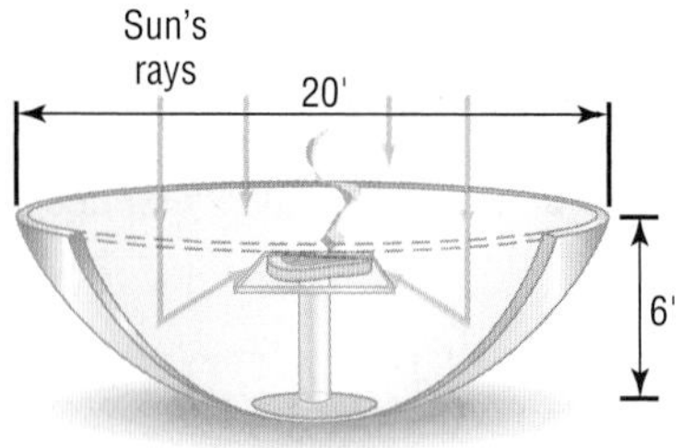

68. Reflecting Telescope A reflecting telescope contains a mirror shaped like a paraboloid of revolution. If the mirror is 4 inches across at its opening and is 3 feet deep, where will the collected light be concentrated?

69. Parabolic Arch Bridge A bridge is built in the shape of a parabolic arch. The bridge has a span of 120 feet and a maximum height of 25 feet. See the illustration. Choose a suitable rectangular coordinate system and find the height of the arch at distances of 10, 30, and 50 feet from the center.

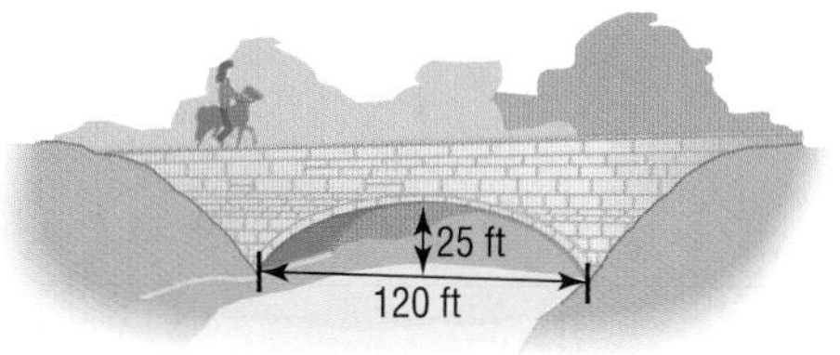

70. Parabolic Arch Bridge A bridge is to be built in the shape of a parabolic arch and is to have a span of 100 feet. The height of the arch a distance of 40 feet from the center is to be 10 feet. Find the height of the arch at its center.

71. Show that an equation of the form

$$Ax^2 + Ey = 0, \qquad A \neq 0, E \neq 0$$

is the equation of a parabola with vertex at $(0, 0)$ and axis of symmetry the y-axis. Find its focus and directrix.

72. Show that an equation of the form

$$Cy^2 + Dx = 0, \qquad C \neq 0, D \neq 0$$

is the equation of a parabola with vertex at $(0, 0)$ and axis of symmetry the x-axis. Find its focus and directrix.

73. Show that the graph of an equation of the form

$$Ax^2 + Dx + Ey + F = 0, \qquad A \neq 0$$

(a) Is a parabola if $E \neq 0$.
(b) Is a vertical line if $E = 0$ and $D^2 - 4AF = 0$.
(c) Is two vertical lines if $E = 0$ and $D^2 - 4AF > 0$.
(d) Contains no points if $E = 0$ and $D^2 - 4AF < 0$.

74. Show that the graph of an equation of the form

$$Cy^2 + Dx + Ey + F = 0, \qquad C \neq 0$$

(a) Is a parabola if $D \neq 0$.
(b) Is a horizontal line if $D = 0$ and $E^2 - 4CF = 0$.
(c) Is two horizontal lines if $D = 0$ and $E^2 - 4CF > 0$.
(d) Contains no points if $D = 0$ and $E^2 - 4CF < 0$.

PREPARING FOR THIS SECTION

Before getting started, review the following:

- ✓ Distance Formula (p. 6)
- ✓ Completing the Square (Appendix, Section 10)
- ✓ Intercepts (pp. 16–18)
- ✓ Symmetry (pp. 18–19)
- ✓ Circles (Section 1.8)
- ✓ Graphing Techniques: Transformation (Section 2.4)

10.3 THE ELLIPSE

OBJECTIVES
1. Find the Equation of an Ellipse
2. Graph Ellipses
3. Discuss the Equation of an Ellipse
4. Work with Ellipses with Center at (h, k)
5. Solve Applied Problems Involving Ellipses

An **ellipse** is the collection of all points in the plane the sum of whose distances from two fixed points, called the **foci,** is a constant.

Figure 19

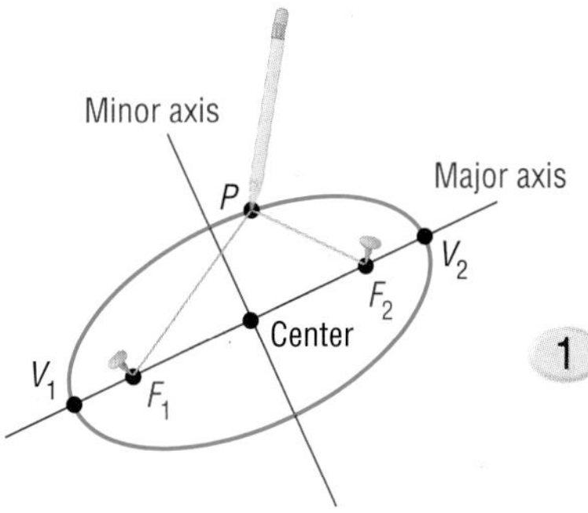

The definition actually contains within it a physical means for drawing an ellipse. Find a piece of string (the length of this string is the constant referred to in the definition). Then take two thumbtacks (the foci) and stick them on a piece of cardboard so that the distance between them is less than the length of the string. Now attach the ends of the string to the thumbtacks and, using the point of a pencil, pull the string taut. Keeping the string taut, rotate the pencil around the two thumbtacks. The pencil traces out an ellipse, as shown in Figure 19.

1 In Figure 19, the foci are labeled F_1 and F_2. The line containing the foci is called the **major axis.** The midpoint of the line segment joining the foci is called the **center** of the ellipse. The line through the center and perpendicular to the major axis is called the **minor axis.**

The two points of intersection of the ellipse and the major axis are the **vertices,** V_1 and V_2, of the ellipse. The distance from one vertex to the other is called the **length of the major axis.** The ellipse is symmetric with respect to its major axis and with respect to its minor axis.

Figure 20
$d(F_1, P) + d(F_2, P) = 2a$

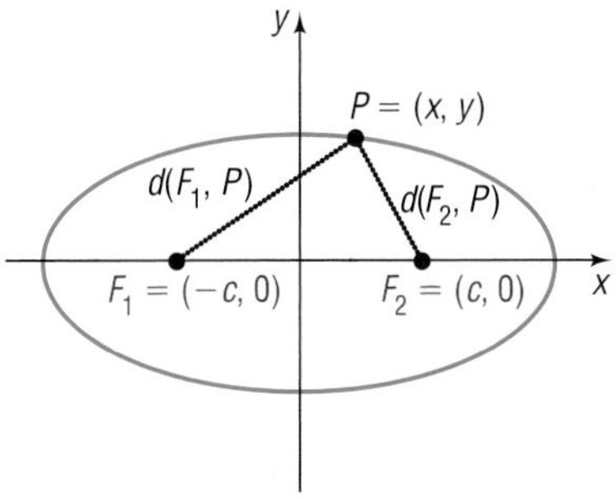

With these ideas in mind, we are now ready to find the equation of an ellipse in a rectangular coordinate system. First, we place the center of the ellipse at the origin. Second, we position the ellipse so that its major axis coincides with a coordinate axis. Suppose that the major axis coincides with the x-axis, as shown in Figure 20. If c is the distance from the center to a focus, then one focus will be at $F_1 = (-c, 0)$ and the other at $F_2 = (c, 0)$. As we shall see, it is convenient to let $2a$ denote the constant distance referred to in the definition. Thus, if $P = (x, y)$ is any point on the ellipse, we have

$$d(F_1, P) + d(F_2, P) = 2a$$ Sum of the distances from P to the foci equals a constant, $2a$.

$$\sqrt{(x+c)^2 + y^2} + \sqrt{(x-c)^2 + y^2} = 2a$$ Use the distance formula.

$$\sqrt{(x+c)^2 + y^2} = 2a - \sqrt{(x-c)^2 + y^2}$$ Isolate one radical.

$$(x + c)^2 + y^2 = 4a^2 - 4a\sqrt{(x - c)^2 + y^2} + (x - c)^2 + y^2$$ Square both sides.

$$x^2 + 2cx + c^2 + y^2 = 4a^2 - 4a\sqrt{(x - c)^2 + y^2} + x^2 - 2cx + c^2 + y^2$$ Remove parentheses

$$4cx - 4a^2 = -4a\sqrt{(x - c)^2 + y^2}$$ Simplify; Isolate the radical.

$$cx - a^2 = -a\sqrt{(x - c)^2 + y^2}$$ Divide each side by 4.

$$c^2x^2 - 2a^2cx + a^4 = a^2[(x - c)^2 + y^2]$$ Square both sides again.

$$c^2x^2 - 2a^2cx + a^4 = a^2(x^2 - 2cx + c^2 + y^2)$$ Simplify.

$$(c^2 - a^2)x^2 - a^2y^2 = a^2c^2 - a^4$$ Rearrange the terms.

$$(a^2 - c^2)x^2 + a^2y^2 = a^2(a^2 - c^2) \quad (1)$$ Multiply each side by -1; factor a^2 on the right side.

To obtain points on the ellipse off the x-axis, it must be that $a > c$. To see why, look again at Figure 20.

$$d(F_1, P) + d(F_2, P) > d(F_1, F_2)$$ The sum of the lengths of two sides of a triangle is greater than the length of the third side.

$$2a > 2c$$ $d(F_1, P) + d(F_2, P) = 2a$; $d(F_1, F_2) = 2c$.

$$a > c$$

Since $a > c$, we also have $a^2 > c^2$, so $a^2 - c^2 > 0$. Let $b^2 = a^2 - c^2$, $b > 0$. Then $a > b$ and equation (1) can be written as

$$b^2x^2 + a^2y^2 = a^2b^2$$

$$\frac{x^2}{a^2} + \frac{y^2}{b^2} = 1$$ Divide each side by a^2b^2.

Theorem

Equation of an Ellipse; Center at (0, 0); Foci at ($\pm c$, 0); Major Axis along the x-Axis

An equation of the ellipse with center at $(0, 0)$ and foci at $(-c, 0)$ and $(c, 0)$ is

$$\frac{x^2}{a^2} + \frac{y^2}{b^2} = 1, \quad \text{where } a > b > 0 \text{ and } b^2 = a^2 - c^2 \quad (2)$$

The major axis is the x-axis.

2 As you can verify, the ellipse defined by equation (2) is symmetric with respect to the x-axis, y-axis, and origin.

To find the vertices of the ellipse defined by equation (2), let $y = 0$. The vertices satisfy the equation $x^2/a^2 = 1$, the solutions of which are $x = \pm a$. Consequently, the vertices of the ellipse given by equation (2) are $V_1 = (-a, 0)$ and $V_2 = (a, 0)$. The y-intercepts of the ellipse, found by letting $x = 0$, have coordinates $(0, -b)$ and $(0, b)$. These four intercepts, $(a, 0)$, $(-a, 0)$, $(0, b)$, and $(0, -b)$, are used to graph the ellipse by hand. See Figure 21.

Figure 21

$\frac{x^2}{a^2} + \frac{y^2}{b^2} = 1$

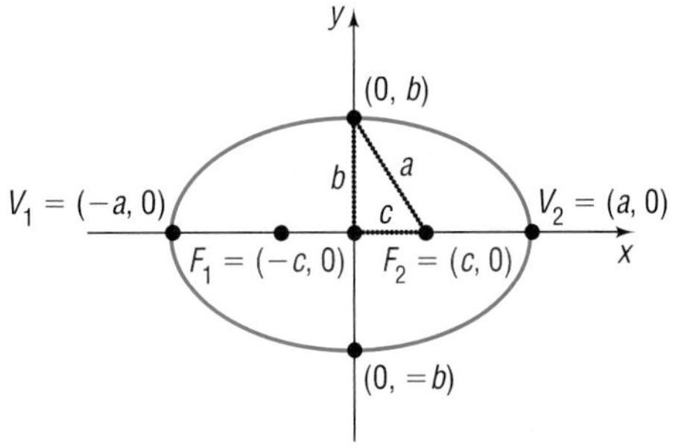

Notice in Figure 21 the right triangle formed with the points $(0, 0)$, $(c, 0)$, and $(0, b)$. Because $b^2 = a^2 - c^2$ (or $b^2 + c^2 = a^2$), the distance from the focus at $(c, 0)$ to the point $(0, b)$ is a.

EXAMPLE 1 Finding an Equation of an Ellipse

Find an equation of the ellipse with center at the origin, one focus at $(3, 0)$, and a vertex at $(-4, 0)$. Graph the equation by hand.

Solution The ellipse has its center at the origin and, since the given focus and vertex lie on the x-axis, the major axis is the x-axis. The distance from the center, $(0, 0)$, to one of the foci, $(3, 0)$, is $c = 3$. The distance from the center, $(0, 0)$, to one of the vertices, $(-4, 0)$, is $a = 4$. From equation (2), it follows that

$$b^2 = a^2 - c^2 = 16 - 9 = 7$$

so an equation of the ellipse is

$$\frac{x^2}{16} + \frac{y^2}{7} = 1$$

Figure 22 shows the graph drawn by hand. ■

Figure 22

$\frac{x^2}{16} + \frac{y^2}{7} = 1$

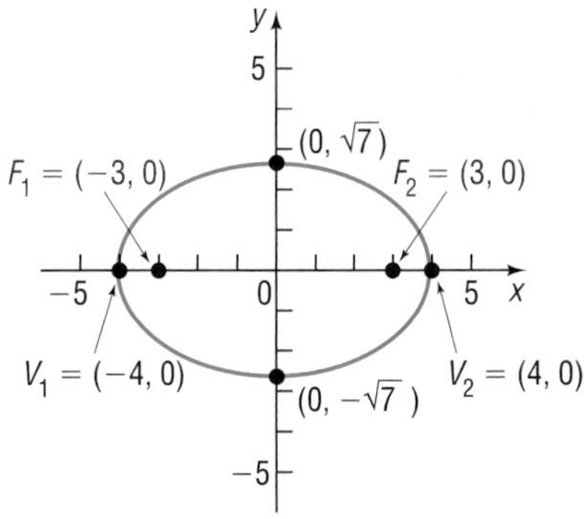

Notice in Figure 22 how we used the intercepts of the equation to graph the ellipse. Following this practice will make it easier for you to obtain an accurate graph of an ellipse when graphing by hand. It also tells you how to set the initial viewing window when using a graphing utility.

EXAMPLE 2 Graphing an Ellipse Using a Graphing Utility

Use a graphing utility to graph the ellipse: $\frac{x^2}{16} + \frac{y^2}{7} = 1$

Solution First, we must solve $\frac{x^2}{16} + \frac{y^2}{7} = 1$ for y.

$$\frac{y^2}{7} = 1 - \frac{x^2}{16} \qquad \text{Subtract } \frac{x^2}{16} \text{ from each side.}$$

$$y^2 = 7\left(1 - \frac{x^2}{16}\right) \qquad \text{Multiply both sides by 7.}$$

$$y = \pm\sqrt{7\left(1 - \frac{x^2}{16}\right)} \qquad \text{Take the square root of each side.}$$

Figure 23

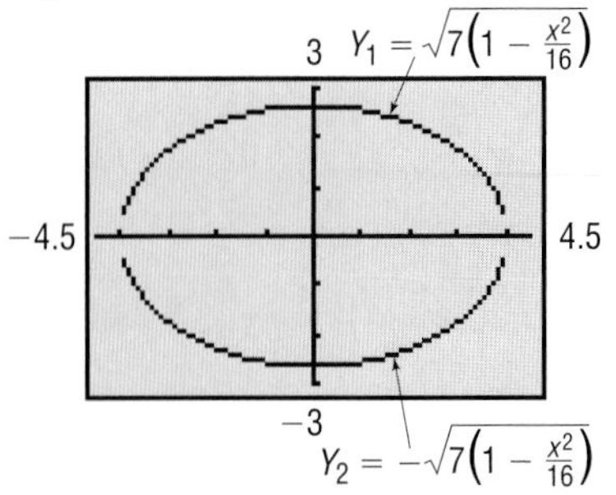

Figure 23* shows graphs of $Y_1 = \sqrt{7\left(1 - \frac{x^2}{16}\right)}$ and $Y_2 = -\sqrt{7\left(1 - \frac{x^2}{16}\right)}$.

Notice in Figure 23 that we used a square screen. As with circles and parabolas, this is done to avoid a distorted view of the graph.

An equation of the form of equation (2), with $a > b$, is the equation of an ellipse with center at the origin, foci on the x-axis at $(-c, 0)$ and $(c, 0)$, where $c^2 = a^2 - b^2$, and major axis along the x-axis.

3 For the remainder of this section, the direction "Discuss the equation" will mean to find the center, major axis, foci, and vertices of the ellipse and graph it.

EXAMPLE 3 Discussing the Equation of an Ellipse

Discuss the equation: $\frac{x^2}{25} + \frac{y^2}{9} = 1$

Solution Figure 24(a) shows the graph of $\frac{x^2}{25} + \frac{y^2}{9} = 1$ using a graphing utility. We now proceed to analyze the equation. The given equation is of the form of equation (2), with $a^2 = 25$ and $b^2 = 9$. The equation is that of an ellipse with center $(0, 0)$ and major axis along the x-axis. The vertices are at $(\pm a, 0) = (\pm 5, 0)$. Because $b^2 = a^2 - c^2$, we find that

$$c^2 = a^2 - b^2 = 25 - 9 = 16$$

The foci are at $(\pm c, 0) = (\pm 4, 0)$. Figure 24(b) shows the graph drawn by hand.

Figure 24

$\frac{x^2}{25} + \frac{y^2}{9} = 1$

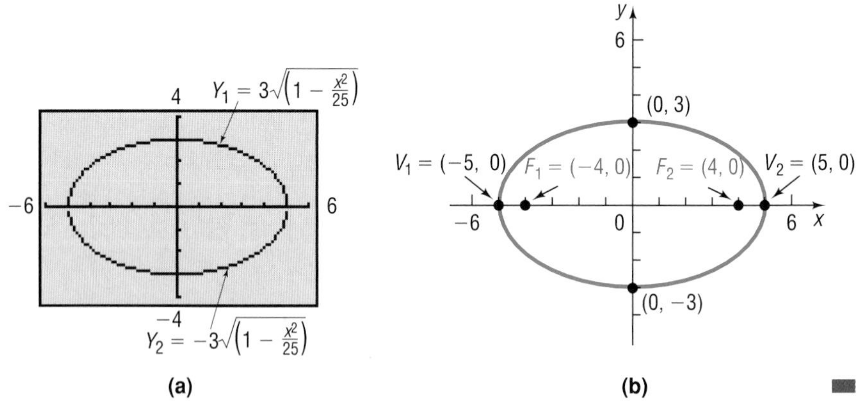

NOW WORK PROBLEMS 9 AND 19.

If the major axis of an ellipse with center at $(0, 0)$ lies on the y-axis, then the foci are at $(0, -c)$ and $(0, c)$. Using the same steps as before, the definition of an ellipse leads to the following result:

*The initial viewing window selected was $X\min = -4$, $X\max = 4$, $Y\min = -3$, $Y\max = 3$. Then we used the ZOOM-SQUARE option to obtain the window shown.

Theorem

Equation of an Ellipse; Center at (0, 0); Foci at (0, ± c); Major Axis along the *y*-Axis

An equation of the ellipse with center at $(0, 0)$ and foci at $(0, -c)$ and $(0, c)$ is

$$\frac{x^2}{b^2} + \frac{y^2}{a^2} = 1, \quad \text{where } a > b > 0 \text{ and } b^2 = a^2 - c^2 \qquad \textbf{(3)}$$

The major axis is the y-axis; the vertices are at $(0, -a)$ and $(0, a)$.

Figure 25
$\frac{x^2}{b^2} + \frac{y^2}{a^2} = 1$

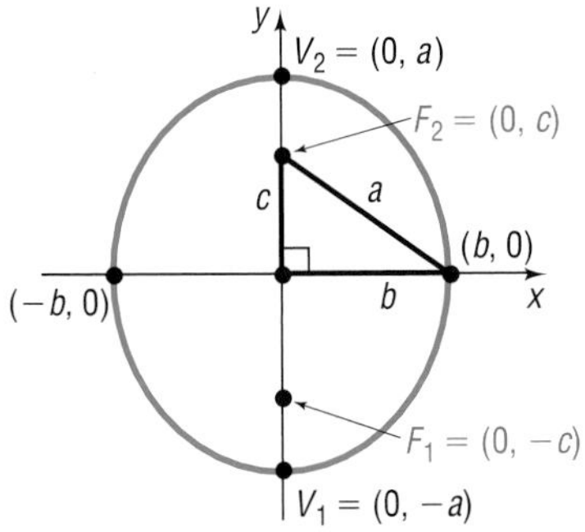

Figure 25 illustrates the graph of such an ellipse. Again, notice the right triangle with the points at $(0, 0)$, $(b, 0)$, and $(0, c)$.

Look closely at equations (2) and (3). Although they may look alike, there is a difference! In equation (2), the larger number, a^2, is in the denominator of the x^2-term, so the major axis of the ellipse is along the x-axis. In equation (3), the larger number, a^2, is in the denominator of the y^2-term, so the major axis is along the y-axis.

EXAMPLE 4 Discussing the Equation of an Ellipse

Discuss the equation: $9x^2 + y^2 = 9$

Solution Figure 26(a) shows the graph of $9x^2 + y^2 = 9$ using a graphing utility. We now proceed to analyze the equation. To put the equation in proper form, we divide each side by 9.

$$x^2 + \frac{y^2}{9} = 1$$

The larger number, 9, is in the denominator of the y^2-term so, based on equation (3), this is the equation of an ellipse with center at the origin and major axis along the y-axis. Also, we conclude that $a^2 = 9$, $b^2 = 1$, and $c^2 = a^2 - b^2 = 9 - 1 = 8$. The vertices are at $(0, \pm a) = (0, \pm 3)$, and the foci are at $(0, \pm c) = (0, \pm 2\sqrt{2})$. The graph, drawn by hand, is given in Figure 26(b).

Figure 26
$x^2 + \frac{y^2}{9} = 1$

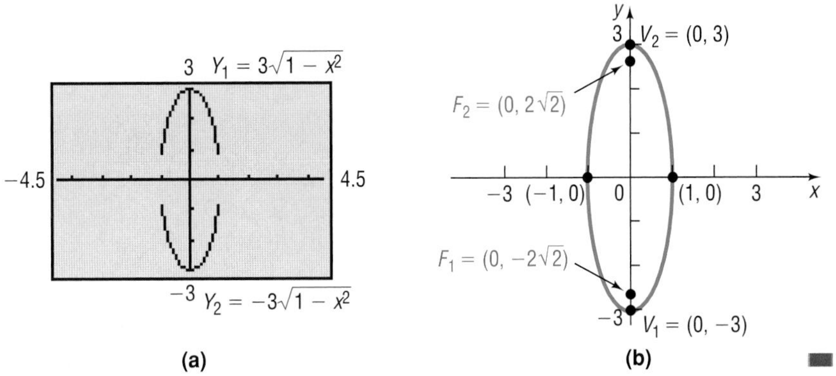

EXAMPLE 5 Finding an Equation of an Ellipse

Find an equation of the ellipse having one focus at $(0, 2)$ and vertices at $(0, -3)$ and $(0, 3)$. Graph the equation by hand.

Figure 27
$\frac{x^2}{5} + \frac{y^2}{9} = 1$

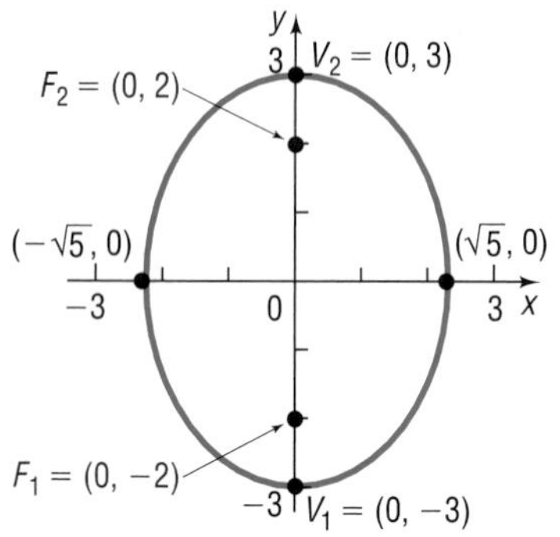

Solution Because the vertices are at $(0, -3)$ and $(0, 3)$, the center of this ellipse is at their midpoint, the origin. Also, its major axis lies on the y-axis. The distance from the center, $(0, 0)$, to one of the foci, $(0, 2)$, is $c = 2$. The distance from the center, $(0, 0)$, to one of the vertices $(0, 3)$ is $a = 3$. So $b^2 = a^2 - c^2 = 9 - 4 = 5$. The form of the equation of this ellipse is given by equation (3).

$$\frac{x^2}{b^2} + \frac{y^2}{a^2} = 1$$

$$\frac{x^2}{5} + \frac{y^2}{9} = 1$$

Figure 27 shows the graph. ■

NOW WORK PROBLEMS 13 AND 21.

The circle may be considered a special kind of ellipse. To see why, let $a = b$ in equation (2) or in equation (3). Then

$$\frac{x^2}{a^2} + \frac{y^2}{a^2} = 1$$

$$x^2 + y^2 = a^2$$

This is the equation of a circle with center at the origin and radius a. The value of c is

$$c^2 = a^2 - b^2 = 0$$

We conclude that the closer the two foci of an ellipse are to the center, the more the ellipse will look like a circle.

CENTER AT (h, k)

4 If an ellipse with center at the origin and major axis coinciding with a coordinate axis is shifted horizontally h units and then vertically k units, the result is an ellipse with center at (h, k) and major axis parallel to a coordinate axis. The equations of such ellipses have the same forms as those given in equations (2) and (3), except that x is replaced by $x - h$ (the horizontal shift) and y is replaced by $y - k$ (the vertical shift). Table 3 gives the forms of the equations of such ellipses, and Figure 28 shows their graphs.

TABLE 3

ELLIPSES WITH CENTER AT (h, k) AND MAJOR AXIS PARALLEL TO A COORDINATE AXIS

Center	Major Axis	Foci	Vertices	Equation
(h, k)	Parallel to x-axis	$(h + c, k)$	$(h + a, k)$	$\frac{(x-h)^2}{a^2} + \frac{(y-k)^2}{b^2} = 1,$
		$(h - c, k)$	$(h - a, k)$	$a > b$ and $b^2 = a^2 - c^2$
(h, k)	Parallel to y-axis	$(h, k + c)$	$(h, k + a)$	$\frac{(x-h)^2}{b^2} + \frac{(y-k)^2}{a^2} = 1,$
		$(h, k - c)$	$(h, k - a)$	$a > b$ and $b^2 = a^2 - c^2$

Figure 28

(a) $\dfrac{(x-h)^2}{a^2}+\dfrac{(y-k)^2}{b^2}=1$

(b) $\dfrac{(x-h)^2}{b^2}+\dfrac{(y-k)^2}{a^2}=1$

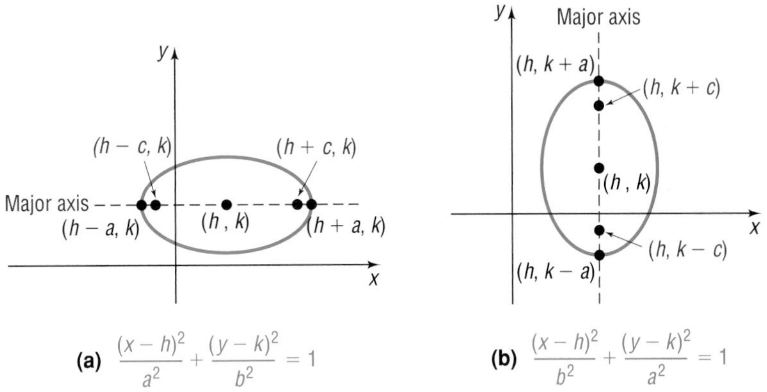

EXAMPLE 6

Finding an Equation of an Ellipse, Center Not at the Origin

Find an equation for the ellipse with center at $(2,-3)$, one focus at $(3,-3)$, and one vertex at $(5,-3)$. Graph the equation by hand.

Solution   The center is at $(h,k)=(2,-3)$, so $h=2$ and $k=-3$. Since the center, focus, and vertex all lie on the line $y=-3$, the major axis is parallel to the x-axis. The distance from the center $(2,-3)$ to a focus $(3,-3)$ is $c=1$; the distance from the center $(2,-3)$ to a vertex $(5,-3)$ is $a=3$. Thus, $b^2=a^2-c^2=9-1=8$. The form of the equation is

$$\frac{(x-h)^2}{a^2}+\frac{(y-k)^2}{b^2}=1, \quad \text{where } h=2, k=-3, a=3, b=2\sqrt{2}$$

$$\frac{(x-2)^2}{9}+\frac{(y+3)^2}{8}=1$$

To graph the equation by hand, we use the center $(h,k)=(2,-3)$ to locate the vertices. The major axis is parallel to the x-axis, so the vertices are $a=3$ units left and right of the center $(2,-3)$. Therefore, the vertices are

$$V_1=(2-3,-3)=(-1,-3) \quad \text{and} \quad V_2=(2+3,-3)=(5,-3)$$

Since $c=1$ and the major axis is parallel to the x-axis, the foci are 1 unit left and right of the center. Therefore, the foci are

$$F_1=(2-1,-3)=(1,-3) \quad \text{and} \quad F_2=(2+1,-3)=(3,-3)$$

Finally, we use the value of $b=2\sqrt{2}$ to find the two points on the ellipse above and below the center, namely,

$$\left(2,-3-2\sqrt{2}\right) \quad \text{and} \quad \left(2,-3+2\sqrt{2}\right)$$

Figure 29 shows the graph.   ■

Figure 29

$$\frac{(x-2)^2}{9}+\frac{(y+3)^2}{8}=1$$

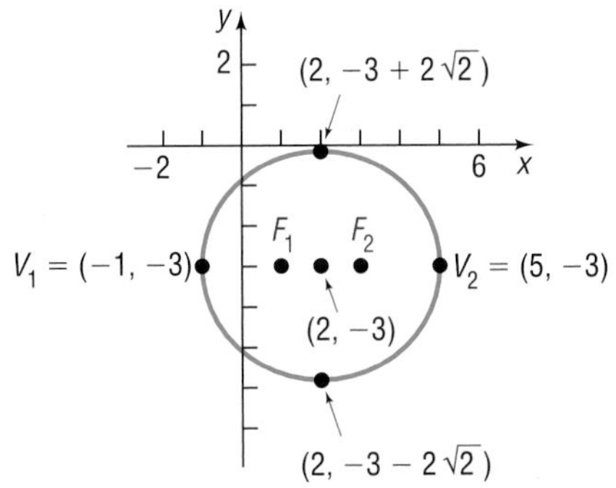

EXAMPLE 7 Using a Graphing Utility to Graph an Ellipse, Center Not at the Origin

Using a graphing utility, graph the ellipse: $\dfrac{(x-2)^2}{9} + \dfrac{(y+3)^2}{8} = 1$

Solution First, we must solve the equation $\dfrac{(x-2)^2}{9} + \dfrac{(y+3)^2}{8} = 1$ for y.

$$\frac{(y+3)^2}{8} = 1 - \frac{(x-2)^2}{9} \qquad \text{Subtract } \frac{(x-2)^2}{9} \text{ from each side.}$$

$$(y+3)^2 = 8\left[1 - \frac{(x-2)^2}{9}\right] \qquad \text{Multiply each side by 8.}$$

$$y + 3 = \pm\sqrt{8\left[1 - \frac{(x-2)^2}{9}\right]} \qquad \text{Take the square root of each side.}$$

$$y = -3 \pm \sqrt{8\left[1 - \frac{(x-2)^2}{9}\right]} \qquad \text{Subtract 3 from each side.}$$

Figure 30

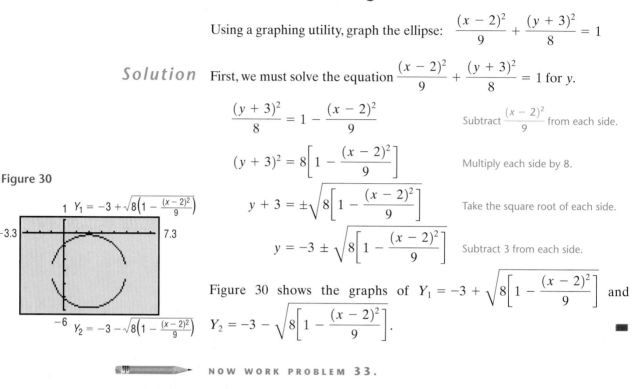

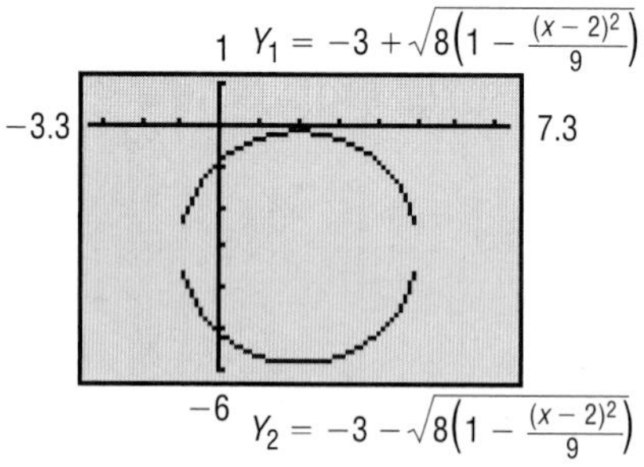

Figure 30 shows the graphs of $Y_1 = -3 + \sqrt{8\left[1 - \frac{(x-2)^2}{9}\right]}$ and $Y_2 = -3 - \sqrt{8\left[1 - \frac{(x-2)^2}{9}\right]}$.

NOW WORK PROBLEM 33.

EXAMPLE 8 Discussing the Equation of an Ellipse

Discuss the equation: $4x^2 + y^2 - 8x + 4y + 4 = 0$.

Solution We proceed to complete the squares in x and in y.

$$4x^2 + y^2 - 8x + 4y + 4 = 0$$

$$4x^2 - 8x + y^2 + 4y = -4$$

$$4(x^2 - 2x) + (y^2 + 4y) = -4$$

$$4(x^2 - 2x + 1) + (y^2 + 4y + 4) = -4 + 4 + 4 \qquad \text{Complete each square.}$$

$$4(x-1)^2 + (y+2)^2 = 4$$

$$(x-1)^2 + \frac{(y+2)^2}{4} = 1 \qquad \text{Divide each side by 4.}$$

Figure 31(a) shows the graph of $(x-1)^2 + \dfrac{(y+2)^2}{4} = 1$ using a graphing utility. We now proceed to analyze the equation.

This is the equation of an ellipse with center at $(1, -2)$ and major axis parallel to the y-axis. Since $a^2 = 4$ and $b^2 = 1$, we have $c^2 = a^2 - b^2 = 4 - 1 = 3$. The vertices are at $(h, k \pm a) = (1, -2 \pm 2)$ or $(1, 0)$ and $(1, -4)$. The foci are at $(h, k \pm c) = (1, -2 \pm \sqrt{3})$ or $(1, -2 - \sqrt{3})$ and $(1, -2 + \sqrt{3})$. Figure 31(b) shows the graph drawn by hand.

Figure 31

$$(x-1)^2 + \frac{(y+2)^2}{4} = 1$$

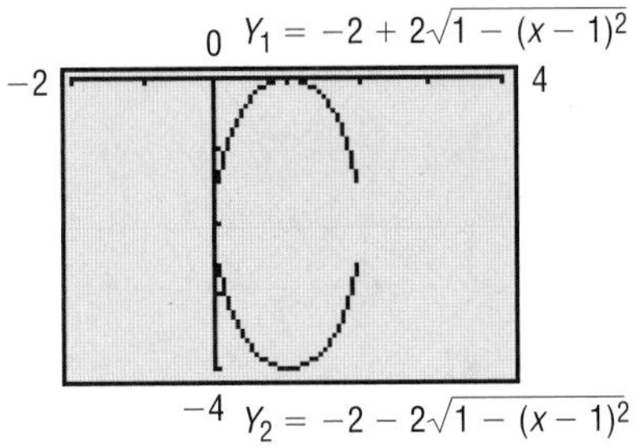

(a)

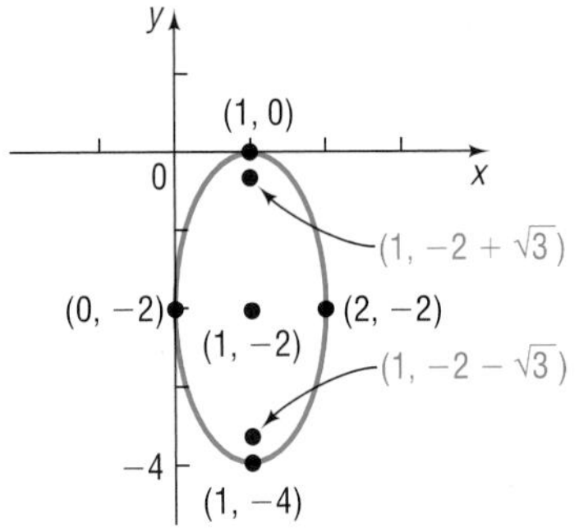

(b)

APPLICATIONS

5 Ellipses are found in many applications in science and engineering. For example, the orbits of the planets around the Sun are elliptical, with the Sun's position at a focus. See Figure 32.

Figure 32

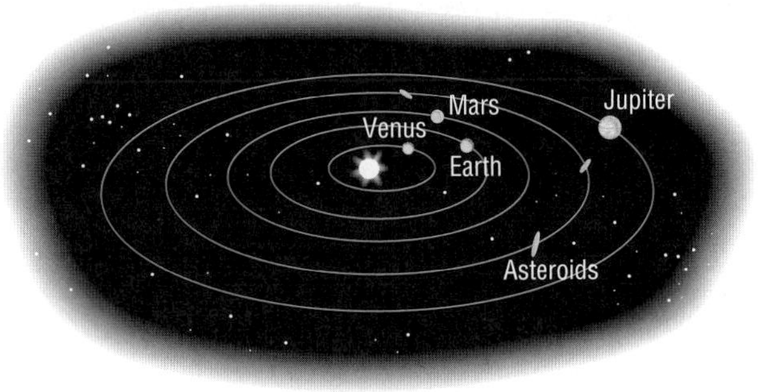

Stone and concrete bridges are often shaped as semielliptical arches. Elliptical gears are used in machinery when a variable rate of motion is required.

Ellipses also have an interesting reflection property. If a source of light (or sound) is placed at one focus, the waves transmitted by the source will reflect off the ellipse and concentrate at the other focus. This is the principle behind *whispering galleries*, which are rooms designed with elliptical ceilings. A person standing at one focus of the ellipse can whisper and be heard by a person standing at the other focus, because all the sound waves that reach the ceiling are reflected to the other person.

EXAMPLE 9 Whispering Galleries

Figure 33 shows the specifications for an elliptical ceiling in a hall designed to be a whispering gallery. In a whispering gallery, a person standing at one focus of the ellipse can whisper and be heard by another person standing at the other focus, because all the sound waves that reach the ceiling from one focus are reflected to the other focus. Where are the foci located in the hall?

Figure 33

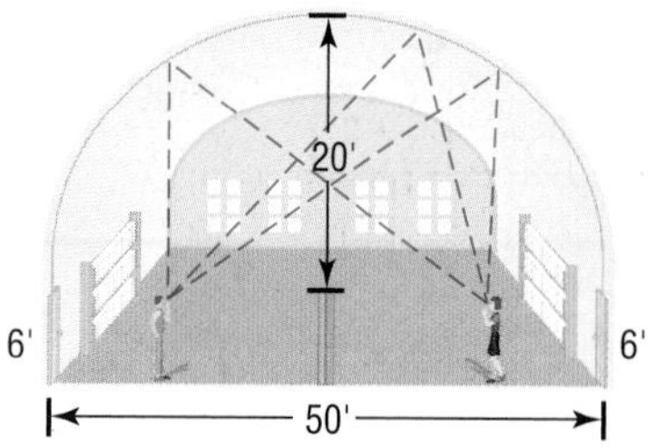

Figure 34

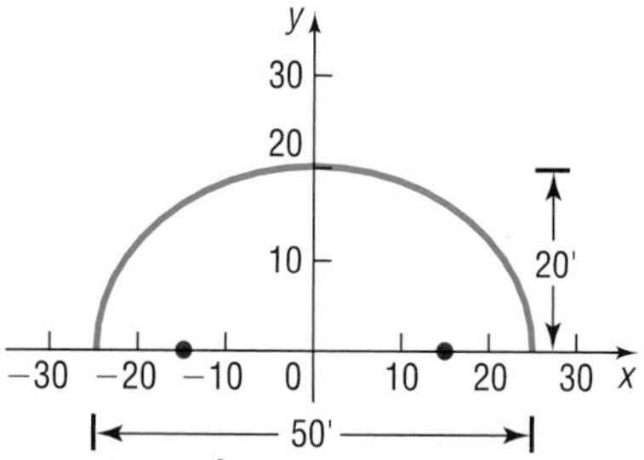

$\frac{x^2}{a^2} + \frac{y^2}{b^2} = 1 \quad a = 25, b = 20$

Solution We set up a rectangular coordinate system so that the center of the ellipse is at the origin and the major axis is along the x-axis. See Figure 34. The equation of the ellipse is

$$\frac{x^2}{a^2} + \frac{y^2}{b^2} = 1$$

where $a = 25$ and $b = 20$. Since

$$c^2 = a^2 - b^2 = 25^2 - 20^2 = 625 - 400 = 225$$

we have $c = 15$. Thus, the foci are located 15 feet from the center of the ellipse along the major axis. ■

10.3 EXERCISES

In Problems 1–4, the graph of an ellipse is given. Match each graph to its equation.

A. $\frac{x^2}{4} + y^2 = 1$ B. $x^2 + \frac{y^2}{4} = 1$ C. $\frac{x^2}{16} + \frac{y^2}{4} = 1$ D. $\frac{x^2}{4} + \frac{y^2}{16} = 1$

1.

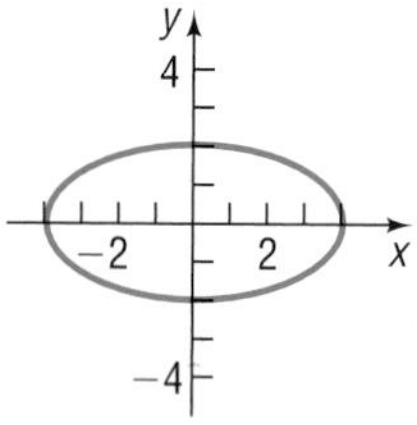

2.

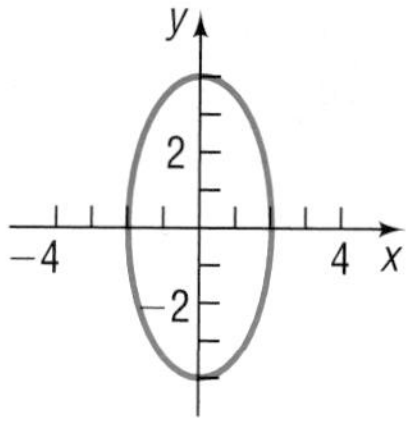

3.

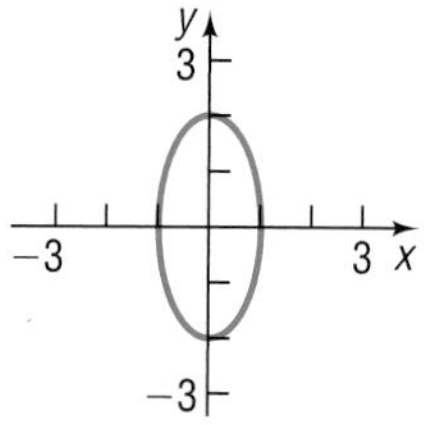

4.

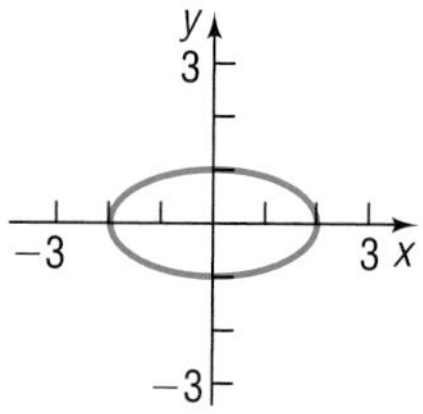

In Problems 5–8, the graph of an ellipse is given. Match each graph to its equation.

A. $\frac{(x+1)^2}{4} + \frac{(y-1)^2}{9} = 1$ B. $\frac{(x-1)^2}{4} + \frac{(y+1)^2}{9} = 1$

C. $\frac{(x-1)^2}{9} + \frac{(y+1)^2}{4} = 1$ D. $\frac{(x+1)^2}{9} + \frac{(y-1)^2}{4} = 1$

5.

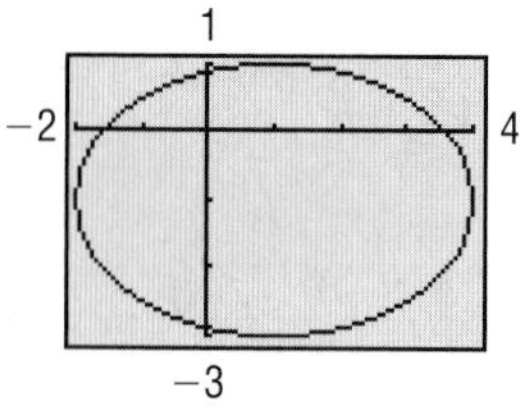

6.

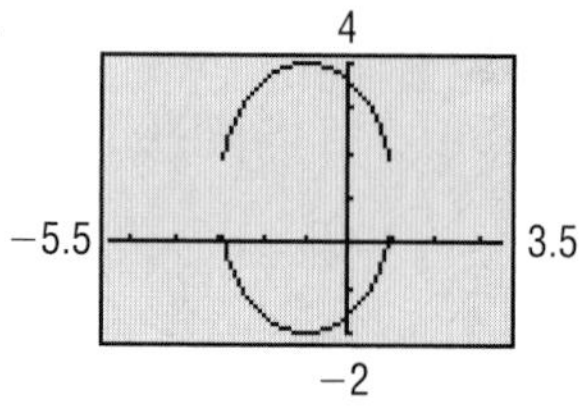

7.

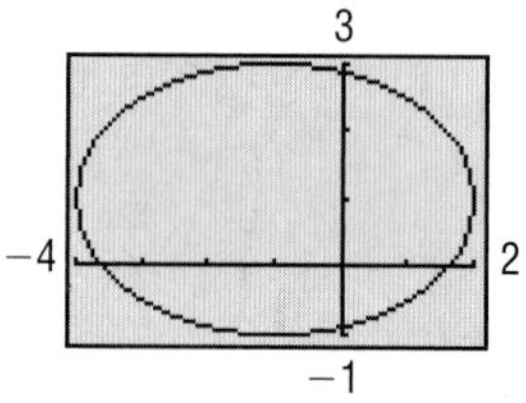

8.

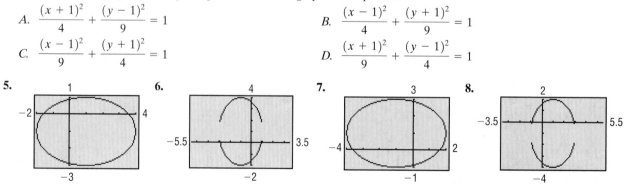

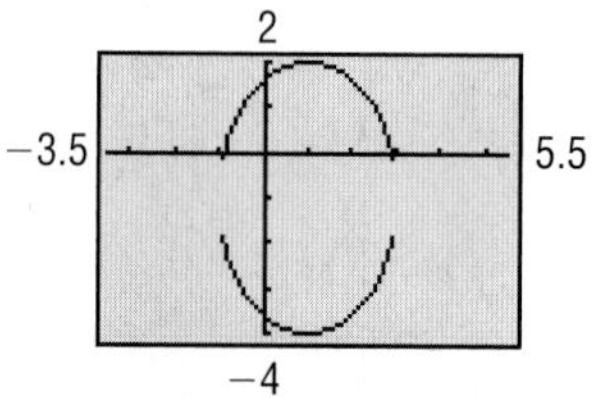

In Problems 9–18, find the vertices and foci of each ellipse. Graph each equation using a graphing utility.

9. $\frac{x^2}{25} + \frac{y^2}{4} = 1$ **10.** $\frac{x^2}{9} + \frac{y^2}{4} = 1$ **11.** $\frac{x^2}{9} + \frac{y^2}{25} = 1$ **12.** $x^2 + \frac{y^2}{16} = 1$

13. $4x^2 + y^2 = 16$ **14.** $x^2 + 9y^2 = 18$ **15.** $4y^2 + x^2 = 8$ **16.** $4y^2 + 9x^2 = 36$

17. $x^2 + y^2 = 16$ **18.** $x^2 + y^2 = 4$

In Problems 19–28, find an equation for each ellipse. Graph the equation by hand.

19. Center at $(0, 0)$; focus at $(3, 0)$; vertex at $(5, 0)$

20. Center at $(0, 0)$; focus at $(-1, 0)$; vertex at $(3, 0)$

21. Center at $(0, 0)$; focus at $(0, -4)$; vertex at $(0, 5)$

22. Center at $(0, 0)$; focus at $(0, 1)$; vertex at $(0, -2)$

23. Foci at $(\pm 2, 0)$; length of the major axis is 6

24. Focus at $(0, -4)$; vertices at $(0, \pm 8)$

25. Foci at $(0, \pm 3)$; x-intercepts are ± 2

26. Foci at $(0, \pm 2)$; length of the major axis is 8

27. Center at $(0, 0)$; vertex at $(0, 4)$; $b = 1$

28. Vertices at $(\pm 5, 0)$; $c = 2$

In Problems 29–32, write an equation for each ellipse.

29.

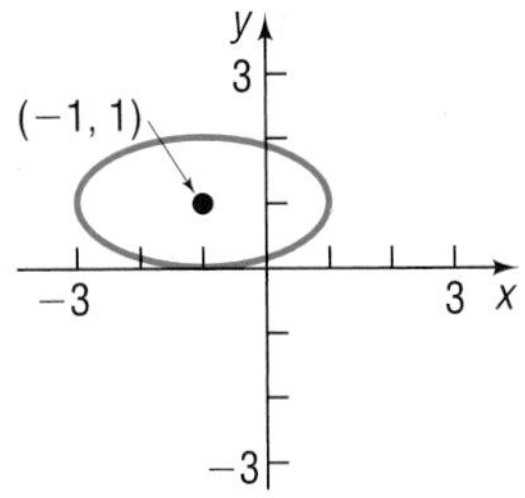

30.

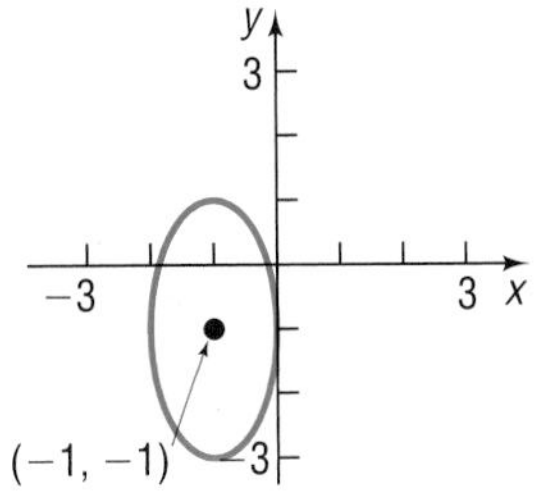

31.

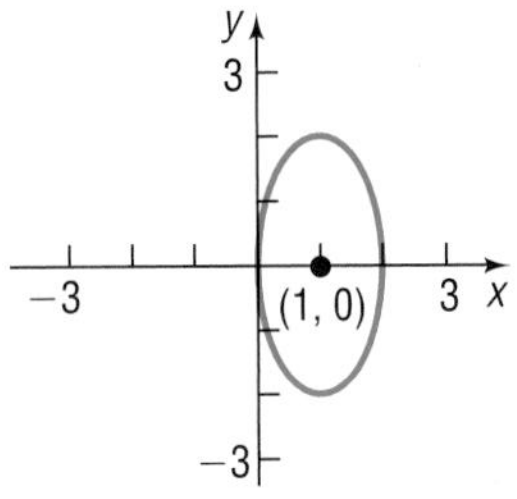

32.

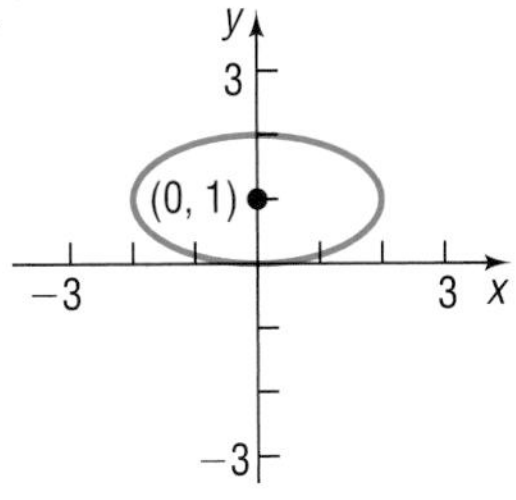

In Problems 33–44, find the center, foci, and vertices of each ellipse. Graph each equation using a graphing utility.

33. $\dfrac{(x-3)^2}{4} + \dfrac{(y+1)^2}{9} = 1$

34. $\dfrac{(x+4)^2}{9} + \dfrac{(y+2)^2}{4} = 1$

35. $(x+5)^2 + 4(y-4)^2 = 16$

36. $9(x-3)^2 + (y+2)^2 = 18$

37. $x^2 + 4x + 4y^2 - 8y + 4 = 0$

38. $x^2 + 3y^2 - 12y + 9 = 0$

39. $2x^2 + 3y^2 - 8x + 6y + 5 = 0$

40. $4x^2 + 3y^2 + 8x - 6y = 5$

41. $9x^2 + 4y^2 - 18x + 16y - 11 = 0$

42. $x^2 + 9y^2 + 6x - 18y + 9 = 0$

43. $4x^2 + y^2 + 4y = 0$

44. $9x^2 + y^2 - 18x = 0$

In Problems 45–54, find an equation for each ellipse. Graph the equation by hand.

45. Center at $(2, -2)$; vertex at $(7, -2)$; focus at $(4, -2)$

46. Center at $(-3, 1)$; vertex at $(-3, 3)$; focus at $(-3, 0)$

47. Vertices at $(4, 3)$ and $(4, 9)$; focus at $(4, 8)$

48. Foci at $(1, 2)$ and $(-3, 2)$; vertex at $(-4, 2)$

49. Foci at $(5, 1)$ and $(-1, 1)$; length of the major axis is 8

50. Vertices at $(2, 5)$ and $(2, -1)$; $c = 2$

51. Center at $(1, 2)$; focus at $(4, 2)$; contains the point $(1, 3)$

52. Center at $(1, 2)$; focus at $(1, 4)$; contains the point $(2, 2)$

53. Center at $(1, 2)$; vertex at $(4, 2)$; contains the point $(1, 3)$

54. Center at $(1, 2)$; vertex at $(1, 4)$; contains the point $(2, 2)$

In Problems 55–58, graph each function by hand. Use a graphing utility to verify your results.
[Hint: Notice that each function is half an ellipse.]

55. $f(x) = \sqrt{16 - 4x^2}$

56. $f(x) = \sqrt{9 - 9x^2}$

57. $f(x) = -\sqrt{64 - 16x^2}$

58. $f(x) = -\sqrt{4 - 4x^2}$

59. Semielliptical Arch Bridge An arch in the shape of the upper half of an ellipse is used to support a bridge that is to span a river 20 meters wide. The center of the arch is 6 meters above the center of the river (see the figure). Write an equation for the ellipse in which the x-axis coincides with the water level and the y-axis passes through the center of the arch.

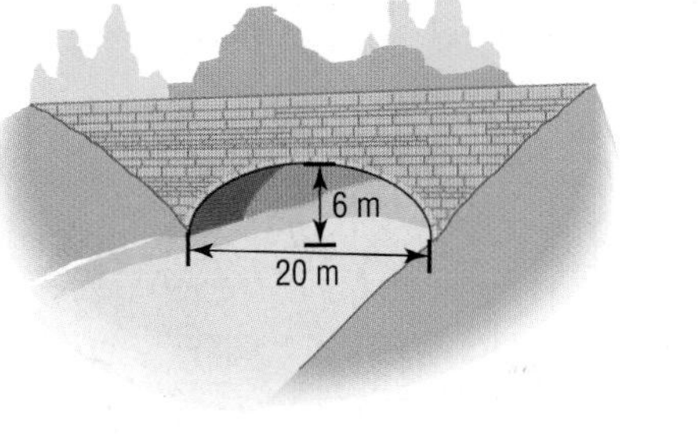

60. **Semielliptical Arch Bridge** The arch of a bridge is a semiellipse with a horizontal major axis. The span is 30 feet, and the top of the arch is 10 feet above the major axis. The roadway is horizontal and is 2 feet above the top of the arch. Find the vertical distance from the roadway to the arch at 5-foot intervals along the roadway.

61. **Whispering Gallery** A hall 100 feet in length is to be designed as a whispering gallery. If the foci are located 25 feet from the center, how high will the ceiling be at the center?

62. **Whispering Gallery** Jim, standing at one focus of a whispering gallery, is 6 feet from the nearest wall. His friend is standing at the other focus, 100 feet away. What is the length of this whispering gallery? How high is its elliptical ceiling at the center?

63. **Semielliptical Arch Bridge** A bridge is built in the shape of a semielliptical arch. The bridge has a span of 120 feet and a maximum height of 25 feet. Choose a suitable rectangular coordinate system and find the height of the arch at distances of 10, 30, and 50 feet from the center.

64. **Semielliptical Arch Bridge** A bridge is built in the shape of a semielliptical arch and is to have a span of 100 feet. The height of the arch, at a distance of 40 feet from the center, is to be 10 feet. Find the height of the arch at its center.

65. **Semielliptical Arch** An arch in the form of half an ellipse is 40 feet wide and 15 feet high at the center. Find the height of the arch at intervals of 10 feet along its width.

66. **Semielliptical Arch Bridge** An arch for a bridge over a highway is in the form of half an ellipse. The top of the arch is 20 feet above the ground level (the major axis). The highway has four lanes, each 12 feet wide; a center safety strip 8 feet wide; and two side strips, each 4 feet wide. What should the span of the bridge be (the length of its major axis) if the height 28 feet from the center is to be 13 feet?

In Problems 67–70, use the fact that the orbit of a planet about the Sun is an ellipse, with the Sun at one focus. The ***aphelion*** *of a planet is its greatest distance from the Sun and the* ***perihelion*** *is its shortest distance. The* ***mean distance*** *of a planet from the Sun is the length of the semimajor axis of the elliptical orbit. See the illustration.*

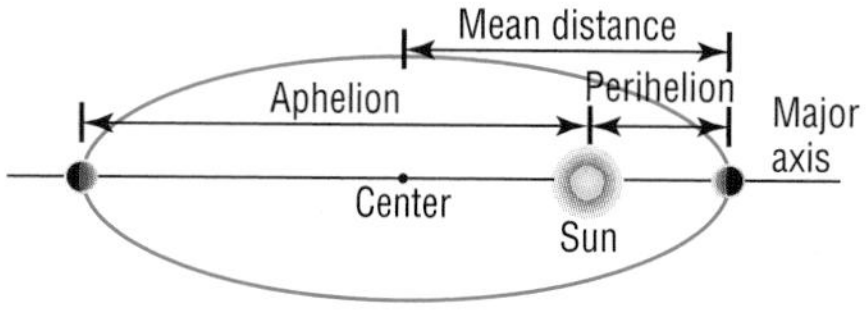

67. **Earth** The mean distance of Earth from the Sun is 93 million miles. If the aphelion of Earth is 94.5 million miles, what is the perihelion? Write an equation for the orbit of Earth around the Sun.

68. **Mars** The mean distance of Mars from the Sun is 142 million miles. If the perihelion of Mars is 128.5 million miles, what is the aphelion? Write an equation for the orbit of Mars about the Sun.

69. **Jupiter** The aphelion of Jupiter is 507 million miles. If the distance from the Sun to the center of its elliptical orbit is 23.2 million miles, what is the perihelion? What is the mean distance? Write an equation for the orbit of Jupiter around the Sun.

70. **Pluto** The perihelion of Pluto is 4551 million miles, and the distance of the Sun from the center of its elliptical orbit is 897.5 million miles. Find the aphelion of Pluto. What is the mean distance of Pluto from the Sun? Write an equation for the orbit of Pluto about the Sun.

71. Consult the figure. A racetrack is in the shape of an ellipse, 100 feet long and 50 feet wide. What is the width 10 feet from the side?

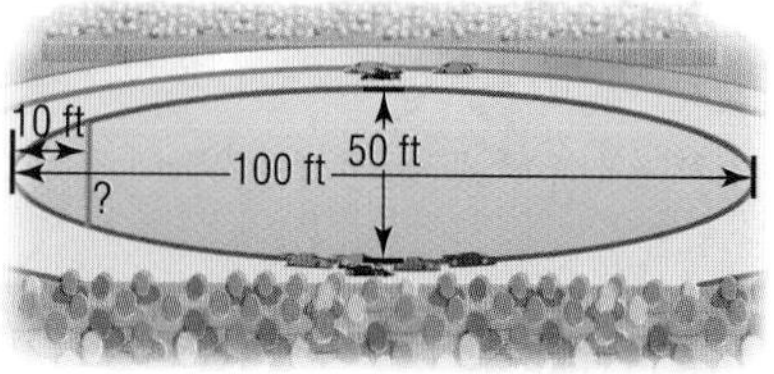

72. A racetrack is in the shape of an ellipse 80 feet long and 40 feet wide. What is the width 10 feet from the side?

73. Show that an equation of the form

$$Ax^2 + Cy^2 + F = 0, \quad A \neq 0, C \neq 0, F \neq 0$$

where A and C are of the same sign and F is of opposite sign,

(a) Is the equation of an ellipse with center at $(0, 0)$ if $A \neq C$.

(b) Is the equation of a circle with center $(0, 0)$ if $A = C$.

74. Show that the graph of an equation of the form

$$Ax^2 + Cy^2 + Dx + Ey + F = 0, \quad A \neq 0, C \neq 0$$

where A and C are of the same sign,

(a) Is an ellipse if $(D^2/4A) + (E^2/4C) - F$ is the same sign as A.

(b) Is a point if $(D^2/4A) + (E^2/4C) - F = 0$.

(c) Contains no points if $(D^2/4A) + (E^2/4C) - F$ is of opposite sign to A.

75. The **eccentricity** e of an ellipse is defined as the number c/a, where a and c are the numbers given in equation (2). Because $a > c$, it follows that $e < 1$. Write a brief paragraph about the general shape of each of the following ellipses. Be sure to justify your conclusions.

(a) Eccentricity close to 0

(b) Eccentricity $= 0.5$

(c) Eccentricity close to 1

PREPARING FOR THIS SECTION

Before getting started, review the following:

- ✓ Distance Formula (p. 6)
- ✓ Completing the Square (Appendix, Section 10)
- ✓ Symmetry (pp. 18–19)
- ✓ Asymptotes (pp. 211–212)
- ✓ Graphing Techniques: Transformations (Section 2.4)

10.4 THE HYPERBOLA

OBJECTIVES
1. Find the Equation of a Hyperbola with Center at Origin
2. Find the Asymptotes of a Hyperbola
3. Discuss the Equation of a Hyperbola
4. Work with Hyperbolas with Center at (h, k)
5. Solve Applied Problems Involving Hyperbolas

A **hyperbola** is the collection of all points in the plane the difference of whose distances from two fixed points, called the **foci,** is a constant.

Figure 35

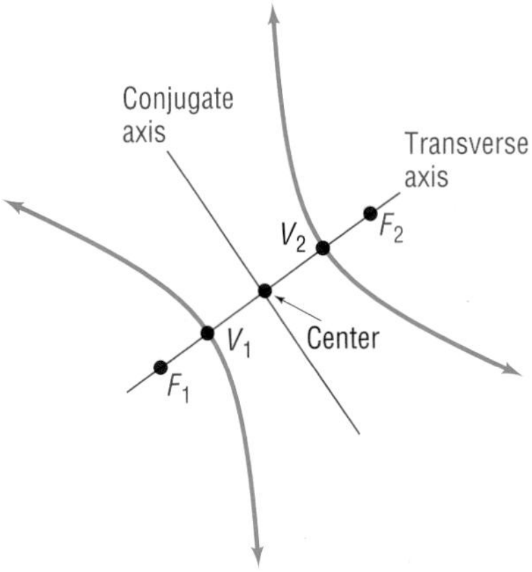

1 Figure 35 illustrates a hyperbola with foci F_1 and F_2. The line containing the foci is called the **transverse axis.** The midpoint of the line segment joining the foci is called the **center** of the hyperbola. The line through the center and perpendicular to the transverse axis is called the **conjugate axis.** The hyperbola consists of two separate curves, called **branches,** that are symmetric with respect to the transverse axis, conjugate axis, and center. The two points of intersection of the hyperbola and the transverse axis are the **vertices,** V_1 and V_2, of the hyperbola.

With these ideas in mind, we are now ready to find the equation of a hyperbola in a rectangular coordinate system. First, we place the center at the origin. Next, we position the hyperbola so that its transverse axis coincides with a coordinate axis. Suppose that the transverse axis coincides with the x-axis, as shown in Figure 36.

Figure 36
$d(F_1, P) - d(F_2, P) = \pm 2a$

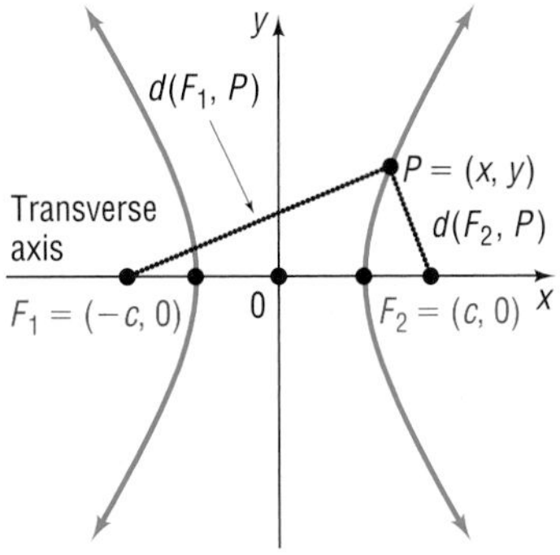

If c is the distance from the center to a focus, then one focus will be at $F_1 = (-c, 0)$ and the other at $F_2 = (c, 0)$. Now we let the constant difference of the distances from any point $P = (x, y)$ on the hyperbola to the foci F_1 and F_2 be denoted by $\pm 2a$. (If P is on the right branch, the + sign is used; if P is on the left branch, the − sign is used.) The coordinates of P must satisfy the equation

$$d(F_1, P) - d(F_2, P) = \pm 2a$$ Difference of the distances from P to the foci equals $\pm 2a$.

$$\sqrt{(x + c)^2 + y^2} - \sqrt{(x - c)^2 + y^2} = \pm 2a$$ Use the distance formula.

$$\sqrt{(x + c)^2 + y^2} = \pm 2a + \sqrt{(x - c)^2 + y^2}$$ Isolate one radical.

$$(x + c)^2 + y^2 = 4a^2 \pm 4a\sqrt{(x - c)^2 + y^2} + (x - c)^2 + y^2$$ Square both sides.

Next, we remove the parentheses.

$$x^2 + 2cx + c^2 + y^2 = 4a^2 \pm 4a\sqrt{(x - c)^2 + y^2} + x^2 - 2cx + c^2 + y^2$$

$$4cx - 4a^2 = \pm 4a\sqrt{(x - c)^2 + y^2}$$ Isolate the radical.

$$cx - a^2 = \pm a\sqrt{(x - c)^2 + y^2}$$ Divide each side by 4.

$$(cx - a^2)^2 = a^2\left[(x - c)^2 + y^2\right]$$ Square both sides.

$$c^2x^2 - 2ca^2x + a^4 = a^2(x^2 - 2cx + c^2 + y^2)$$ Simplify.

$$c^2x^2 + a^4 = a^2x^2 + a^2c^2 + a^2y^2$$ Remove parentheses.

$$(c^2 - a^2)x^2 - a^2y^2 = a^2c^2 - a^4$$ Rearrange terms.

$$(c^2 - a^2)x^2 - a^2y^2 = a^2(c^2 - a^2) \qquad \textbf{(1)}$$

To obtain points on the hyperbola off the x-axis, it must be that $a < c$. To see why, look again at Figure 36.

$$d(F_1, P) < d(F_2, P) + d(F_1, F_2)$$ Use triangle F_1PF_2.

$$d(F_1, P) - d(F_2, P) < d(F_1, F_2)$$ P is on the right branch, so $d(F_1, P) - d(F_2, P) = 2a$.

$$2a < 2c$$

$$a < c$$

Since $a < c$, we also have $a^2 < c^2$, so $c^2 - a^2 > 0$. Let $b^2 = c^2 - a^2$, $b > 0$. Then equation (1) can be written as

$$b^2x^2 - a^2y^2 = a^2b^2$$

$$\frac{x^2}{a^2} - \frac{y^2}{b^2} = 1$$

To find the vertices of the hyperbola defined by this equation, let $y = 0$. The vertices satisfy the equation $x^2/a^2 = 1$, the solutions of which are $x = \pm a$. Consequently, the vertices of the hyperbola are $V_1 = (-a, 0)$ and $V_2 = (a, 0)$. Notice that the distance from the center $(0, 0)$ to either vertex is a.

Theorem **Equation of a Hyperbola; Center at (0, 0); Foci at ($\pm c$, 0); Vertices at ($\pm a$, 0); Transverse Axis along the x-Axis**

An equation of the hyperbola with center at $(0, 0)$, foci at $(-c, 0)$ and $(c, 0)$, and vertices at $(-a, 0)$ and $(a, 0)$ is

$$\frac{x^2}{a^2} - \frac{y^2}{b^2} = 1, \qquad \text{where } b^2 = c^2 - a^2 \tag{2}$$

The transverse axis is the x-axis.

Figure 37
$\frac{x^2}{a^2} - \frac{y^2}{b^2} = 1, \quad b^2 = c^2 - a^2$

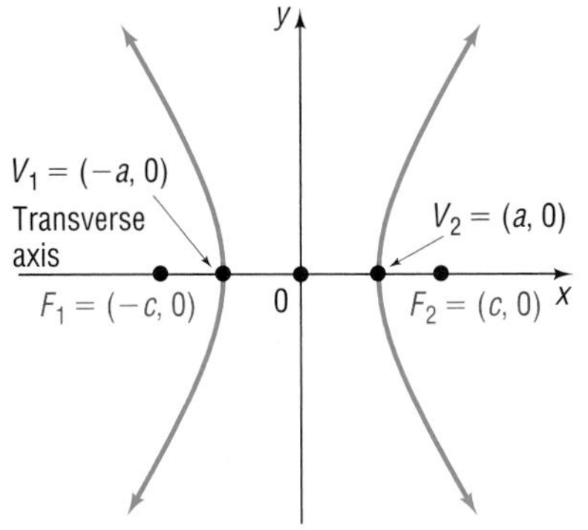

As you can verify, the hyperbola defined by equation (2) is symmetric with respect to the x-axis, y-axis, and origin. To find the y-intercepts, if any, let $x = 0$ in equation (2). This results in the equation $y^2/b^2 = -1$, which has no real solution. We conclude that the hyperbola defined by equation (2) has no y-intercepts. In fact, since $x^2/a^2 - 1 = y^2/b^2 \geq 0$, it follows that $x^2/a^2 \geq 1$. Thus, there are no points on the graph for $-a < x < a$. See Figure 37.

EXAMPLE 1 Finding an Equation of a Hyperbola

Find an equation of the hyperbola with center at the origin, one focus at $(4, 0)$, and one vertex at $(-3, 0)$.

Solution The hyperbola has its center at the origin, and the transverse axis lies on the x-axis. The distance from the center $(0, 0)$ to one focus $(4, 0)$ is $c = 4$. The distance from the center $(0, 0)$ to one vertex $(-3, 0)$ is $a = 3$. From equation (2), it follows that $b^2 = c^2 - a^2 = 16 - 9 = 7$, so an equation of the hyperbola is

$$\frac{x^2}{9} - \frac{y^2}{7} = 1$$

EXAMPLE 2 Using a Graphing Utility to Graph a Hyperbola

Using a graphing utility, graph the hyperbola: $\frac{x^2}{9} - \frac{y^2}{7} = 1$

Solution To graph the hyperbola $\frac{x^2}{9} - \frac{y^2}{7} = 1$, we need to graph the two functions $Y_1 = \sqrt{7}\sqrt{\frac{x^2}{9} - 1}$ and $Y_2 = -\sqrt{7}\sqrt{\frac{x^2}{9} - 1}$. As with graphing circles, parabolas, and ellipses on a graphing utility, we use a square screen setting so that the graph is not distorted. Figure 38 shows the graph of the hyperbola.

Figure 38

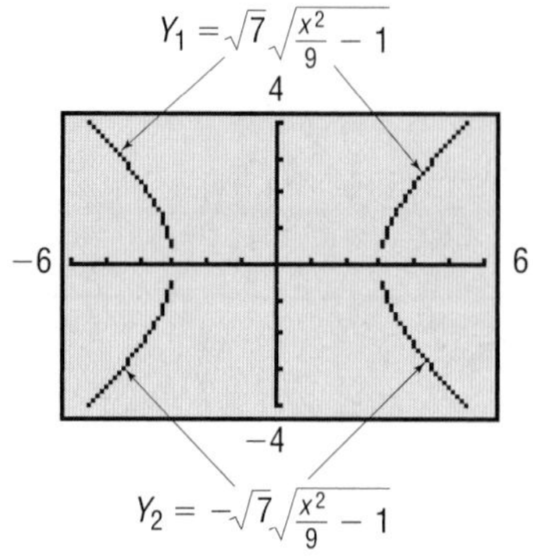

NOW WORK PROBLEM 9.

The next result gives the form of the equation of a hyperbola with center at the origin and transverse axis along the y-axis.

Theorem

Equation of a Hyperbola; Center at (0, 0); Foci at (0, ± *c*); Vertices at (0, ± *a*); Transverse Axis along the *y*-Axis

An equation of the hyperbola with center at $(0, 0)$, foci at $(0, -c)$ and $(0, c)$, and vertices at $(0, -a)$ and $(0, a)$ is

$$\frac{y^2}{a^2} - \frac{x^2}{b^2} = 1, \qquad \text{where } b^2 = c^2 - a^2 \tag{3}$$

The transverse axis is the y-axis.

Figure 39 $\frac{y^2}{a^2} - \frac{x^2}{b^2} = 1, \quad b^2 = c^2 - a^2$

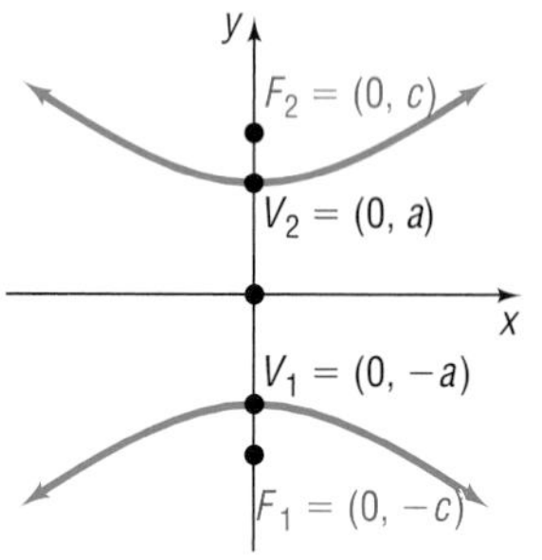

Figure 39 shows the graph of a typical hyperbola defined by equation (3).

An equation of the form of equation (2), $\frac{x^2}{a^2} - \frac{y^2}{b^2} = 1$, is the equation of a hyperbola with center at the origin, foci on the x-axis at $(-c, 0)$ and $(c, 0)$, where $c^2 = a^2 + b^2$, and transverse axis along the x-axis.

An equation of the form of equation (3), $\frac{y^2}{a^2} - \frac{x^2}{b^2} = 1$, is the equation of a hyperbola with center at the origin, foci on the y-axis at $(0, -c)$ and $(0, c)$, where $c^2 = a^2 + b^2$, and transverse axis along the y-axis.

Notice the difference in the forms of equations (2) and (3). When the y^2-term is subtracted from the x^2-term, the transverse axis is the x-axis. When the x^2-term is subtracted from the y^2-term, the transverse axis is the y-axis.

EXAMPLE 3 Finding an Equation of a Hyperbola

Find an equation of the hyperbola having one vertex at $(0, 2)$ and foci at $(0, -3)$ and $(0, 3)$.

Solution Since the foci are at $(0, -3)$ and $(0, 3)$, the center of the hyperbola is at their midpoint, the origin. Also, the transverse axis is along the y-axis. The given information also reveals that $c = 3$, $a = 2$, and $b^2 = c^2 - a^2 = 9 - 4 = 5$. The form of the equation of the hyperbola is given by equation (3):

$$\frac{y^2}{a^2} - \frac{x^2}{b^2} = 1$$

$$\frac{y^2}{4} - \frac{x^2}{5} = 1$$

NOW WORK PROBLEM 13.

Look at the equations of the hyperbolas in Examples 1 and 3. For the hyperbola in Example 1, $a^2 = 9$ and $b^2 = 7$, so $a > b$; for the hyperbola in Example 3, $a^2 = 4$ and $b^2 = 5$, so $a < b$. We conclude that, for hyperbolas, there are no requirements involving the relative sizes of a and b. Contrast this situation to the case of an ellipse, in which the relative sizes of a and b dictate which axis is the major axis. Hyperbolas have another feature to distinguish them from ellipses and parabolas: hyperbolas have asymptotes.

ASYMPTOTES

2 Recall from Section 3.4 that a horizontal or oblique asymptote of a graph is a line with the property that the distance from the line to points on the graph approaches 0 as $x \to -\infty$ or as $x \to \infty$.

Theorem **Asymptotes of a Hyperbola**

The hyperbola $\frac{x^2}{a^2} - \frac{y^2}{b^2} = 1$ has the two oblique asymptotes

$$y = \frac{b}{a}x \quad \text{and} \quad y = -\frac{b}{a}x \qquad \textbf{(4)}$$

Proof We begin by solving for y in the equation of the hyperbola.

$$\frac{x^2}{a^2} - \frac{y^2}{b^2} = 1$$

$$\frac{y^2}{b^2} = \frac{x^2}{a^2} - 1$$

$$y^2 = b^2\left(\frac{x^2}{a^2} - 1\right)$$

Since $x \neq 0$, we can rearrange the right side in the form

$$y^2 = \frac{b^2x^2}{a^2}\left(1 - \frac{a^2}{x^2}\right)$$

$$y = \pm\frac{bx}{a}\sqrt{1 - \frac{a^2}{x^2}}$$

Now, as $x \to -\infty$ or as $x \to \infty$, the term a^2/x^2 approaches 0, so the expression under the radical approaches 1. Thus, as $x \to -\infty$ or as $x \to \infty$, the value of y approaches $\pm bx/a$; that is, the graph of the hyperbola approaches the lines

$$y = -\frac{b}{a}x \quad \text{and} \quad y = \frac{b}{a}x$$

These lines are oblique asymptotes of the hyperbola. ■

Figure 40 $\frac{x^2}{a^2} - \frac{y^2}{b^2} = 1$

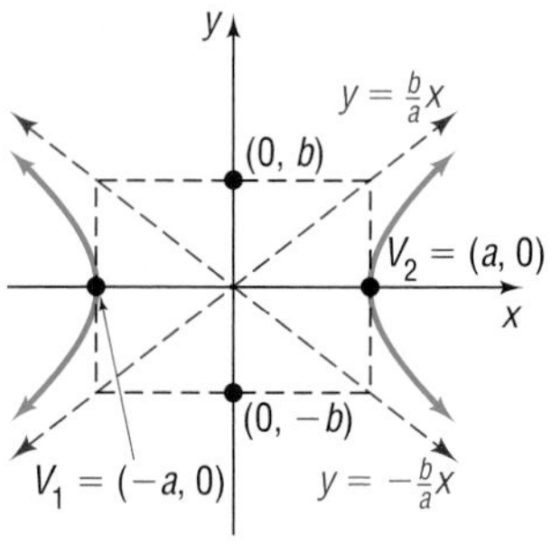

The asymptotes of a hyperbola are not part of the hyperbola, but they do serve as a guide for graphing a hyperbola by hand. For example, suppose that we want to graph the equation

$$\frac{x^2}{a^2} - \frac{y^2}{b^2} = 1$$

We begin by plotting the vertices $(-a, 0)$ and $(a, 0)$. Then we plot the points $(0, -b)$ and $(0, b)$ and use these four points to construct a rectangle, as shown in Figure 40. The diagonals of this rectangle have slopes b/a and $-b/a$, and their extensions are the asymptotes $y = (b/a)x$ and $y = -(b/a)x$ of the hyperbola.

Theorem **Asymptotes of a Hyperbola**

The hyperbola $\frac{y^2}{a^2} - \frac{x^2}{b^2} = 1$ has the two oblique asymptotes

$$y = \frac{a}{b}x \quad \text{and} \quad y = -\frac{a}{b}x \tag{5}$$

■

You are asked to prove this result in Problem 64.

3 For the remainder of this section, the direction "Discuss the equation" will mean to find the center, transverse axis, vertices, foci, and asymptotes of the hyperbola and graph it.

EXAMPLE 4 Discussing the Equation of a Hyperbola

Discuss the equation: $\frac{y^2}{4} - x^2 = 1$

Figure 41 $\frac{y^2}{4} - x^2 = 1$

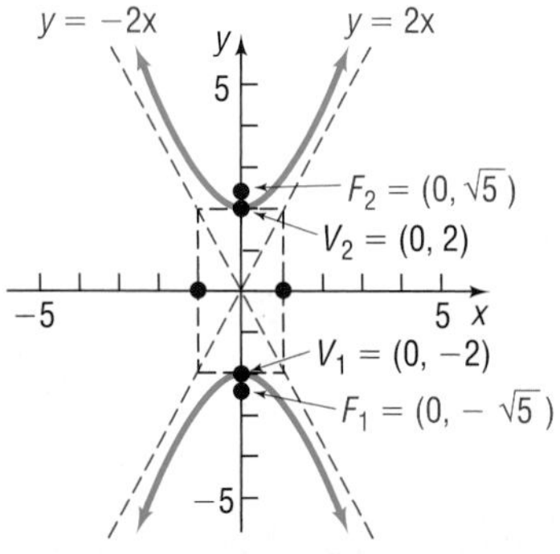

Solution Since the x^2 term is subtracted from the y^2 term, the equation is of the form of equation (3) and is a hyperbola with center at the origin and transverse axis along the y-axis. Also, comparing this equation to equation (3), we find that $a^2 = 4$, $b^2 = 1$, and $c^2 = a^2 + b^2 = 5$. The vertices are at $(0, \pm a) = (0, \pm 2)$, and the foci are at $(0, \pm c) = (0, \pm\sqrt{5})$. Using (5), the asymptotes are the lines $y = \frac{a}{b}x = 2x$ and $y = -\frac{a}{b}x = -2x$. Form the rectangle containing the points $(0, \pm a) = (0, \pm 2)$ and $(\pm b, 0) = (\pm 1, 0)$. The extensions of the diagonals of this rectangle are the asymptotes. Now graph the rectangle, the asymptotes, and the hyperbola. See Figure 41. ■

EXAMPLE 5 Discussing the Equation of a Hyperbola

Discuss the equation: $9x^2 - 4y^2 = 36$

Solution Divide each side of the equation by 36 to put the equation in proper form.

$$\frac{x^2}{4} - \frac{y^2}{9} = 1$$

We now proceed to analyze the equation. The center of the hyperbola is the origin. Since the x^2-term is first in the equation, we know that the transverse axis is along the x-axis and the vertices and foci will lie on the x-axis. Using equation (2), we find $a^2 = 4$, $b^2 = 9$, and $c^2 = a^2 + b^2 = 13$. The vertices are $a = 2$ units left and right of the center at $(\pm a, 0) = (\pm 2, 0)$; the foci are $c = \sqrt{13}$ units left and right of the center at $(\pm c, 0) = (\pm\sqrt{13}, 0)$; and the asymptotes have the equations

$$y = \frac{b}{a}x = \frac{3}{2}x \quad \text{and} \quad y = -\frac{b}{a}x = -\frac{3}{2}x$$

Figure 42(a) shows the graph of $\frac{x^2}{4} - \frac{y^2}{9} = 1$ and its asymptotes, $y = \frac{3}{2}x$ and $y = -\frac{3}{2}x$, using a graphing utility.

To graph the hyperbola by hand, form the rectangle containing the points $(\pm a, 0)$ and $(0, \pm b)$, that is, $(-2, 0)$, $(2, 0)$, $(0, -3)$, and $(0, 3)$. The extensions of the diagonals of this rectangle are the asymptotes. See Figure 42(b) for the graph.

Figure 42
$\frac{x^2}{4} - \frac{y^2}{9} = 1$

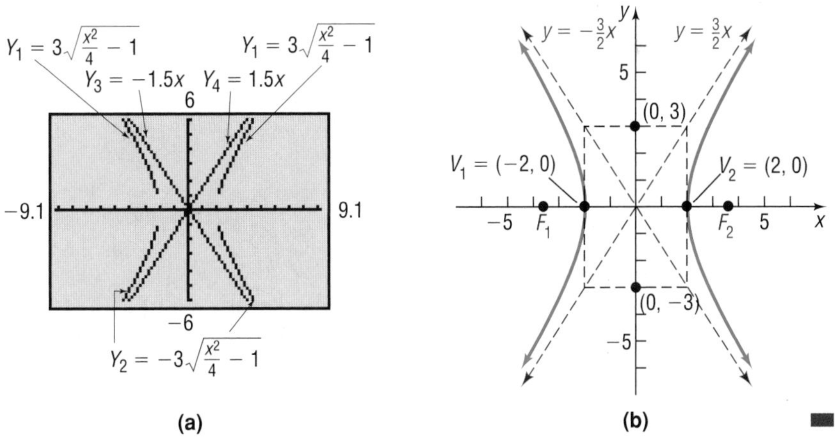

NOW WORK PROBLEM 21.

SEEING THE CONCEPT Refer to Figure 42(a). Create a TABLE using Y_1 and Y_4 with $x = 10$, 100, 1000, and 10,000. Compare the values of Y_1 and Y_4. Repeat for Y_1 and Y_3, Y_2 and Y_3, and Y_2 and Y_4.

CENTER AT (h, k)

4 If a hyperbola with center at the origin and transverse axis coinciding with a coordinate axis is shifted horizontally h units and then vertically k units, the result is a hyperbola with center at (h, k) and transverse axis parallel to a coordinate axis. Table 4 gives the forms of the equations of such hyperbolas. See Figure 43 for the graphs.

TABLE 4

HYPERBOLAS WITH CENTER AT (h, k); AND TRANSVERSE AXIS PARALLEL TO A COORDINATE AXIS

Center	Transverse Axis	Foci	Vertices	Equation	Asymptotes
(h, k)	Parallel to x-axis	$(h \pm c, k)$	$(h \pm a, k)$	$\dfrac{(x-h)^2}{a^2} - \dfrac{(y-k)^2}{b^2} = 1,\ b^2 = c^2 - a^2$	$y - k = \pm\dfrac{b}{a}(x - h)$
(h, k)	Parallel to y-axis	$(h, k \pm c)$	$(h, k \pm a)$	$\dfrac{(y-k)^2}{a^2} - \dfrac{(x-h)^2}{b^2} = 1,\ b^2 = c^2 - a^2$	$y - k = \pm\dfrac{a}{b}(x - h)$

Figure 43

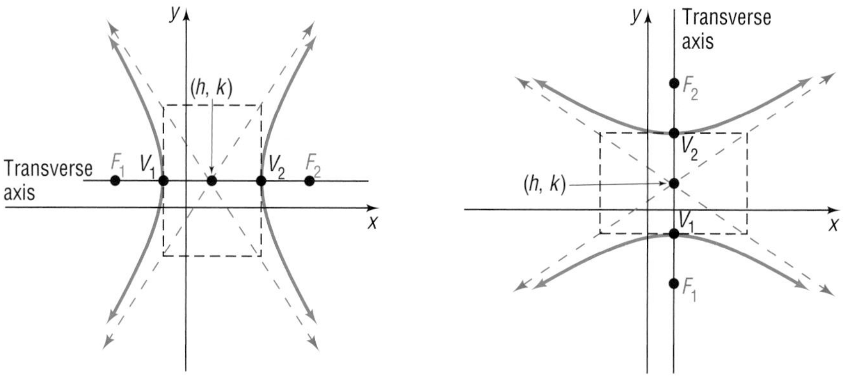

(a) $\dfrac{(x-h)^2}{a^2} - \dfrac{(y-k)^2}{b^2} = 1$ (b) $\dfrac{(y-k)^2}{a^2} - \dfrac{(x-h)^2}{b^2} = 1$

EXAMPLE 6 Finding an Equation of a Hyperbola, Center Not at the Origin

Find an equation for the hyperbola with center at $(1, -2)$, one focus at $(4, -2)$, and one vertex at $(3, -2)$. Graph the equation by hand.

Solution The center is at $(h, k) = (1, -2)$, so $h = 1$ and $k = -2$. Since the center, focus, and vertex all lie on the line $y = -2$, the transverse axis is parallel to the x-axis. The distance from the center $(1, -2)$ to the focus $(4, -2)$ is $c = 3$; the distance from the center $(1, -2)$ to the vertex $(3, -2)$ is $a = 2$. Thus, $b^2 = c^2 - a^2 = 9 - 4 = 5$. The equation is

$$\frac{(x-h)^2}{a^2} - \frac{(y-k)^2}{b^2} = 1$$

$$\frac{(x-1)^2}{4} - \frac{(y+2)^2}{5} = 1$$

See Figure 44.

Figure 44
$$\frac{(x-1)^2}{4} - \frac{(y+2)^2}{5} = 1$$

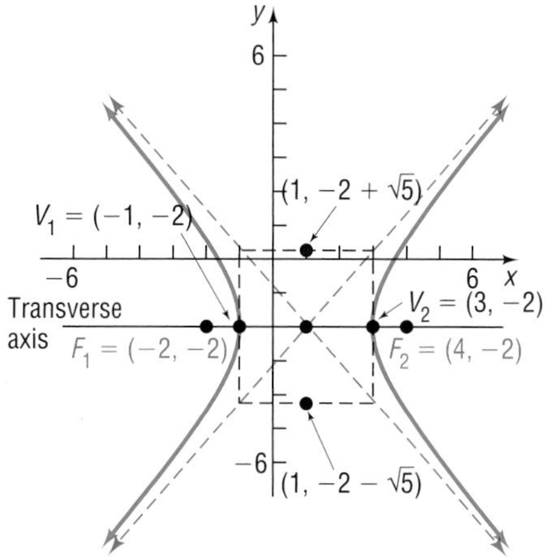

NOW WORK PROBLEM 31.

EXAMPLE 7 Using a Graphing Utility to Graph a Hyperbola, Center Not at the Origin

Using a graphing utility, graph the hyperbola: $\frac{(x-1)^2}{4} - \frac{(y+2)^2}{5} = 1$

Solution First, we must solve the equation for y.

$$\frac{(x-1)^2}{4} - \frac{(y+2)^2}{5} = 1$$
$$\frac{(y+2)^2}{5} = \frac{(x-1)^2}{4} - 1$$
$$(y+2)^2 = 5\left[\frac{(x-1)^2}{4} - 1\right]$$
$$y + 2 = \pm\sqrt{5\left[\frac{(x-1)^2}{4} - 1\right]}$$
$$y = -2 \pm \sqrt{5\left[\frac{(x-1)^2}{4} - 1\right]}$$

Figure 45

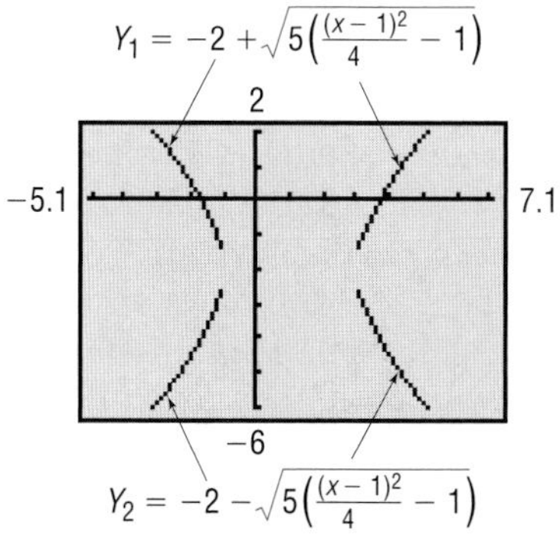

Figure 45 shows the graph of $Y_1 = -2 + \sqrt{5\left[\frac{(x-1)^2}{4} - 1\right]}$ and $Y_2 = -2 - \sqrt{5\left[\frac{(x-1)^2}{4} - 1\right]}$.

EXAMPLE 8 Discussing the Equation of a Hyperbola

Discuss the equation: $-x^2 + 4y^2 - 2x - 16y + 11 = 0$

Solution We complete the squares in x and in y.

$$-x^2 + 4y^2 - 2x - 16y + 11 = 0$$
$$-(x^2 + 2x) + 4(y^2 - 4y) = -11 \quad \text{Group terms.}$$
$$-(x^2 + 2x + 1) + 4(y^2 - 4y + 4) = -1 + 16 - 11 \quad \text{Complete each square.}$$

$$-(x + 1)^2 + 4(y - 2)^2 = 4$$

$$(y - 2)^2 - \frac{(x + 1)^2}{4} = 1 \qquad \text{Divide by 4.}$$

Figure 46(a) shows the graph of $(y - 2)^2 - \frac{(x + 1)^2}{4} = 1$ using a graphing utility. We now proceed to analyze the equation.

It is the equation of a hyperbola with center at $(-1, 2)$ and transverse axis parallel to the y-axis. Also, $a^2 = 1$ and $b^2 = 4$, so $c^2 = a^2 + b^2 = 5$. Since the transverse axis is parallel to the y-axis, the vertices and foci are located a and c units above and below the center, respectively. The vertices are at $(h, k \pm a) = (-1, 2 \pm 1)$, or $(-1, 1)$ and $(-1, 3)$. The foci are at $(h, k \pm c) = (-1, 2 \pm \sqrt{5})$. The asymptotes are $y - 2 = \frac{1}{2}(x + 1)$ and $y - 2 = -\frac{1}{2}(x + 1)$. Figure 46(b) shows the graph drawn by hand.

Figure 46

$(y - 2)^2 - \frac{(x + 1)^2}{4} = 1$

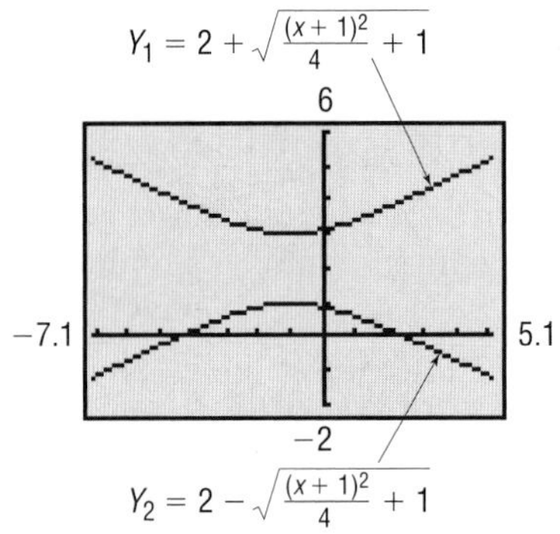

(a)

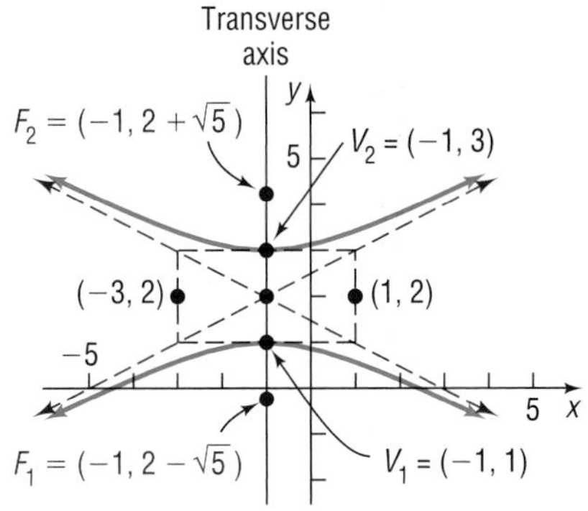

(b)

NOW WORK PROBLEM 45.

APPLICATIONS

Figure 47

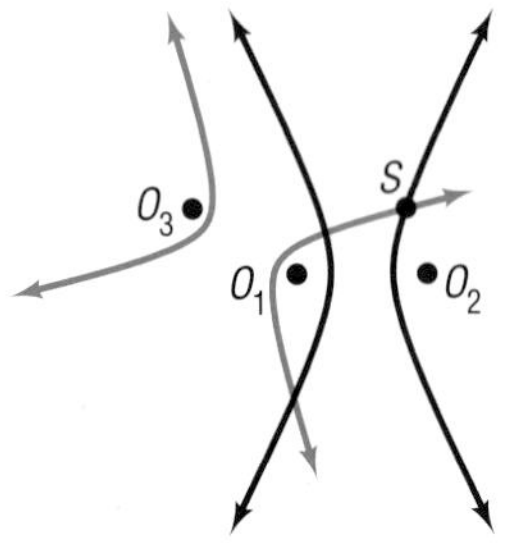

5 Suppose that a gun is fired from an unknown source S. An observer at O_1 hears the report (sound of gun shot) 1 second after another observer at O_2. Because sound travels at about 1100 feet per second, it follows that the point S must be 1100 feet closer to O_2 than to O_1. Thus, S lies on one branch of a hyperbola with foci at O_1 and O_2. (Do you see why? The difference of the distances from S to O_1 and from S to O_2 is the constant 1100.) If a third observer at O_3 hears the same report 2 seconds after O_1 hears it, then S will lie on a branch of a second hyperbola with foci at O_1 and O_3. The intersection of the two hyperbolas will pinpoint the location of S. See Figure 47 for an illustration.

Figure 48

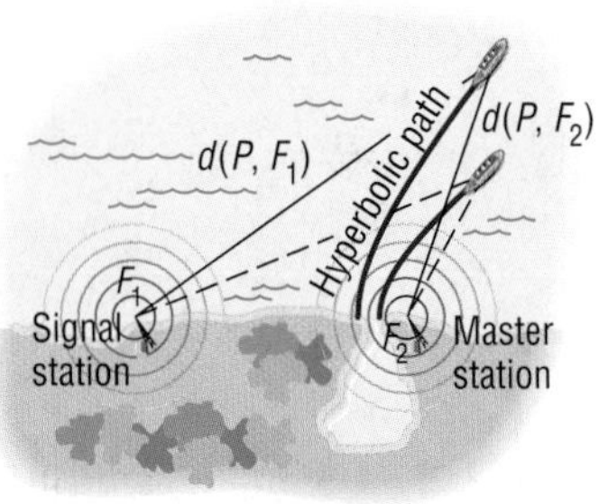

$d(P, F_1) - d(P, F_2) = \text{constant}$

LORAN

In the LOng RAnge Navigation system (LORAN), a master radio sending station and a secondary sending station emit signals that can be received by a ship at sea. See Figure 48. Because a ship monitoring the two signals will usually be nearer to one of the two stations, there will be a difference in the distance that the two signals travel, which will register as a slight time difference between the signals. As long as the time difference remains constant, the difference of the two distances will also be constant. If the ship follows a path corresponding to the fixed time difference, it will follow the path of a hyperbola whose foci are located at the positions of the two sending stations.

So for each time difference a different hyperbolic path results, each bringing the ship to a different shore location. Navigation charts show the various hyperbolic paths corresponding to different time differences.

EXAMPLE 9 LORAN

Two LORAN stations are positioned 250 miles apart along a straight shore.

(a) A ship records a time difference of 0.00054 second between the LORAN signals. Set up an appropriate rectangular coordinate system to determine where the ship would reach shore if it were to follow the hyperbola corresponding to this time difference.

(b) If the ship wants to enter a harbor located between the two stations 25 miles from the master station, what time difference should it be looking for?

(c) If the ship is 80 miles offshore when the desired time difference is obtained, what is the approximate location of the ship?

[**NOTE:** The speed of each radio signal is 186,000 miles per second.]

Figure 49

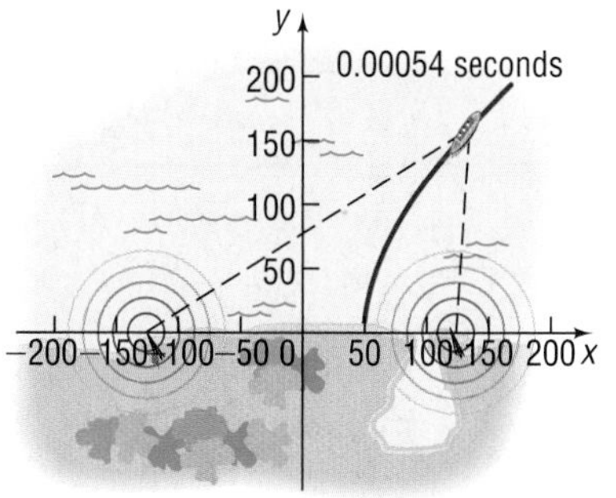

Solution (a) We set up a rectangular coordinate system so that the two stations lie on the x-axis and the origin is midway between them. See Figure 49. The ship lies on a hyperbola whose foci are the locations of the two stations. The reason for this is that the constant time difference of the signals from each station results in a constant difference in the distance of the ship from each station. Since the time difference is 0.00054 second and the speed of the signal is 186,000 miles per second, the difference of the distances from the ship to each station (foci) is

$$\text{Distance} = \text{Speed} \times \text{Time} = 186{,}000 \times 0.00054 \approx 100 \text{ miles}$$

The difference of the distances from the ship to each station, 100, equals $2a$, so $a = 50$ and the vertex of the corresponding hyperbola is at $(50, 0)$. Since the focus is at $(125, 0)$, following this hyperbola the ship would reach shore 75 miles from the master station.

(b) To reach shore 25 miles from the master station, the ship would follow a hyperbola with vertex at $(100, 0)$. For this hyperbola, $a = 100$, so the constant difference of the distances from the ship to each station is 200. The time difference that the ship should look for is

$$\text{Time} = \frac{\text{Distance}}{\text{Speed}} = \frac{200}{186{,}000} = 0.001075 \text{ second}$$

(c) To find the approximate location of the ship, we need to find the equation of the hyperbola with vertex at $(100, 0)$ and a focus at $(125, 0)$. The form of the equation of this hyperbola is

$$\frac{x^2}{a^2} - \frac{y^2}{b^2} = 1$$

where $a = 100$. Since $c = 125$, we have

$$b^2 = c^2 - a^2 = 125^2 - 100^2 = 5625$$

The equation of the hyperbola is

$$\frac{x^2}{100^2} - \frac{y^2}{5625} = 1$$

Since the ship is 80 miles from shore, we use $y = 80$ in the equation and solve for x.

Figure 50

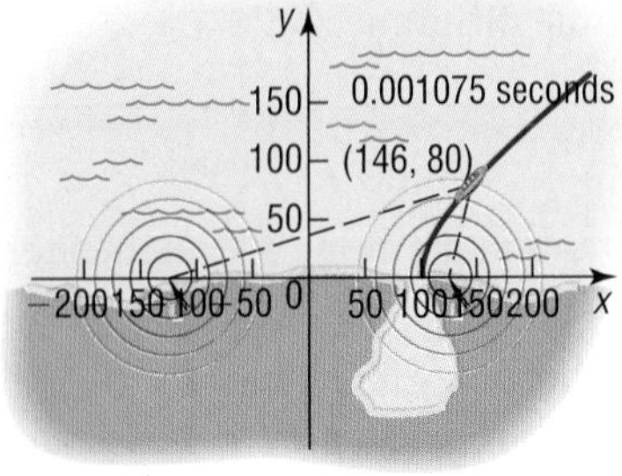

$$\frac{x^2}{100^2} - \frac{80^2}{5625} = 1$$

$$\frac{x^2}{100^2} = 1 + \frac{80^2}{5625} = 2.14$$

$$x^2 = 100^2(2.14)$$

$$x = 146$$

The ship is at the position (146, 80). See Figure 50. ■

10.4 EXERCISES

In Problems 1–4, the graph of a hyperbola is given. Match each graph to its equation.

A. $\frac{x^2}{4} - y^2 = 1$ *B.* $x^2 - \frac{y^2}{4} = 1$ *C.* $\frac{y^2}{4} - x^2 = 1$ *D.* $y^2 - \frac{x^2}{4} = 1$

1.

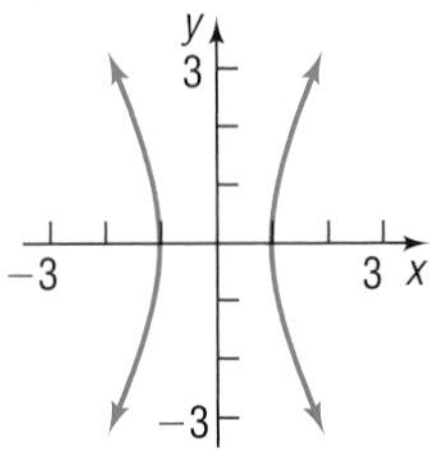

2.

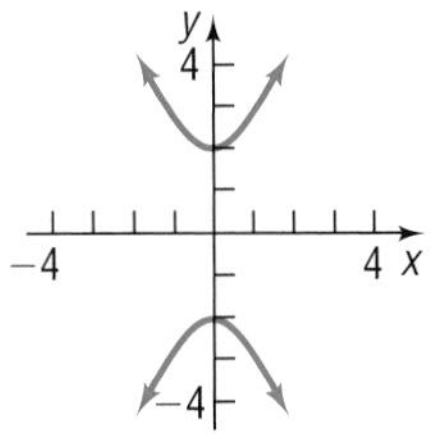

3.

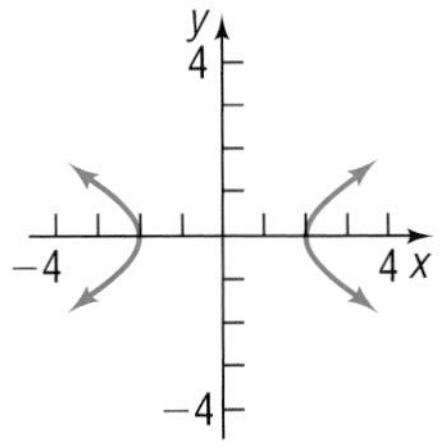

4.

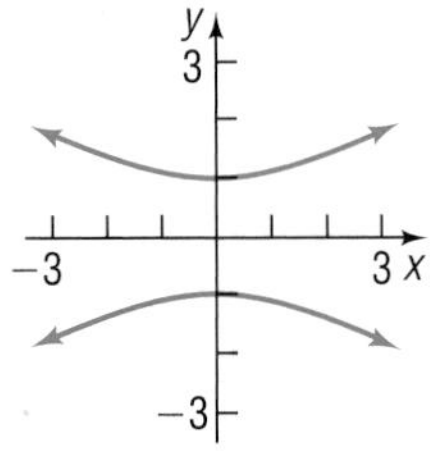

In Problems 5–8, the graph of a hyperbola is given. Match each graph to its equation.

A. $\frac{x^2}{16} - \frac{y^2}{9} = 1$ *B.* $\frac{x^2}{9} - \frac{y^2}{16} = 1$ *C.* $\frac{y^2}{16} - \frac{x^2}{9} = 1$ *D.* $\frac{y^2}{9} - \frac{x^2}{16} = 1$

5.

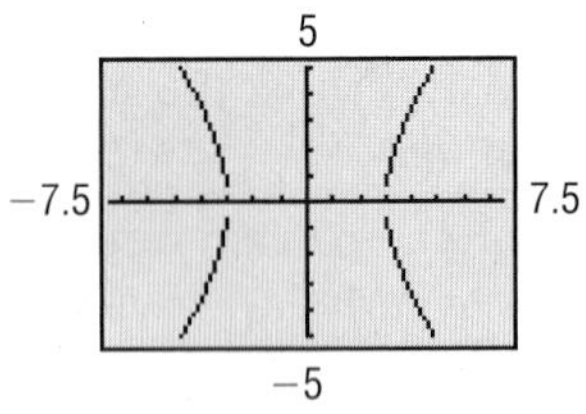

6.

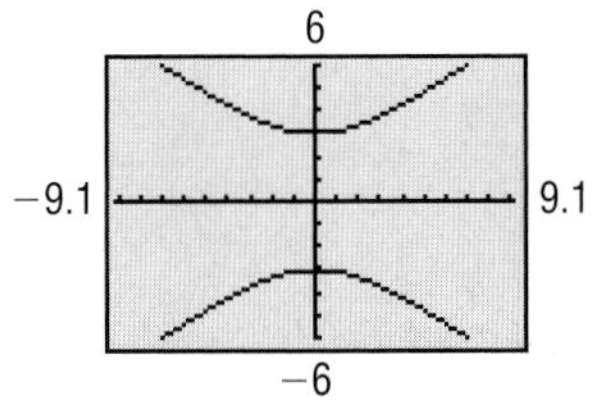

7.

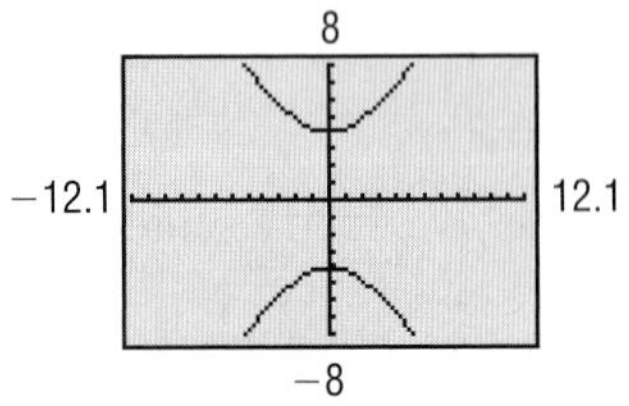

8.

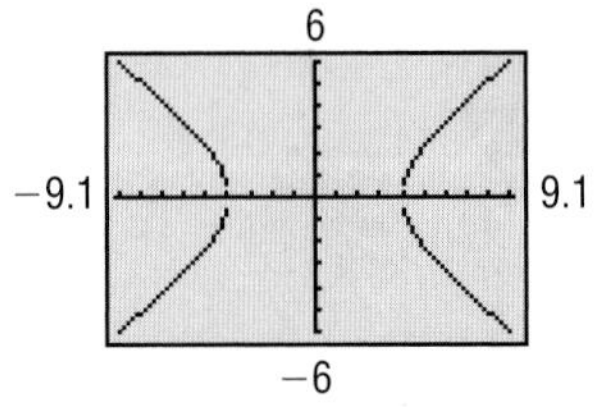

In Problems 9–18, find an equation for the hyperbola described. Graph the equation using a graphing utility.

9. Center at (0, 0); focus at (3, 0); vertex at (1, 0)

10. Center at (0, 0); focus at (0, 5); vertex at (0, 3)

11. Center at (0, 0); focus at (0, −6); vertex at (0, 4)

12. Center at (0, 0); focus at (−3, 0); vertex at (2, 0)

13. Foci at (−5, 0) and (5, 0); vertex at (3, 0)

14. Focus at (0, 6); vertices at (0, −2) and (0, 2)

15. Vertices at (0, −6) and (0, 6); asymptote the line $y = 2x$

16. Vertices at (−4, 0) and (4, 0); asymptote the line $y = 2x$

17. Foci at (−4, 0) and (4, 0); asymptote the line $y = -x$

18. Foci at (0, −2) and (0, 2); asymptote the line $y = -x$

In Problems 19–26, find the center, transverse axis, vertices, foci, and asymptotes. Graph each equation (a) by hand and (b) using a graphing utility.

19. $\frac{x^2}{25} - \frac{y^2}{9} = 1$ **20.** $\frac{y^2}{16} - \frac{x^2}{4} = 1$ **21.** $4x^2 - y^2 = 16$ **22.** $y^2 - 4x^2 = 16$

23. $y^2 - 9x^2 = 9$ **24.** $x^2 - y^2 = 4$ **25.** $y^2 - x^2 = 25$ **26.** $2x^2 - y^2 = 4$

In Problems 27–30, write an equation for each hyperbola.

27.

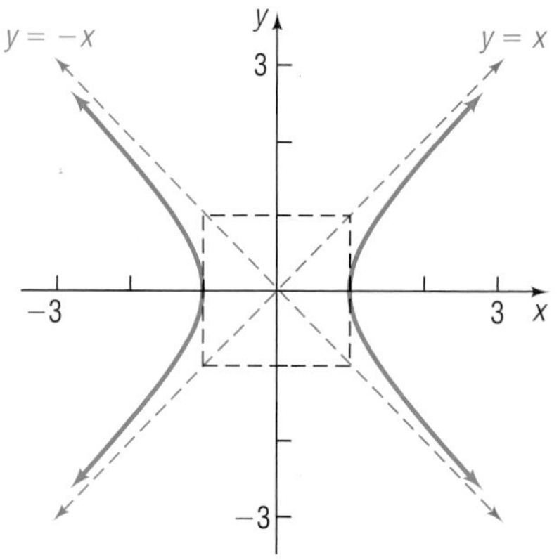

28.

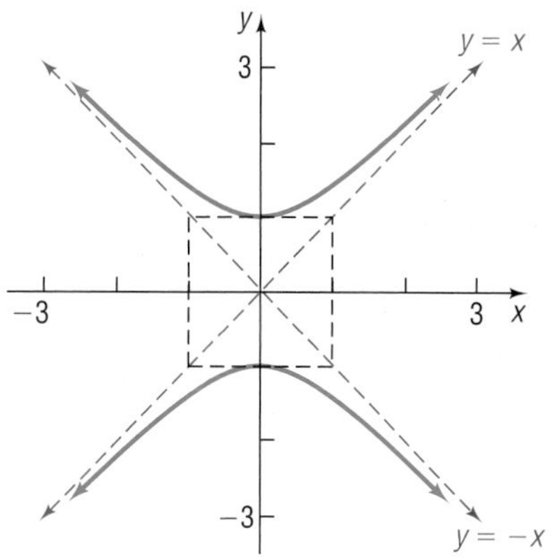

29.

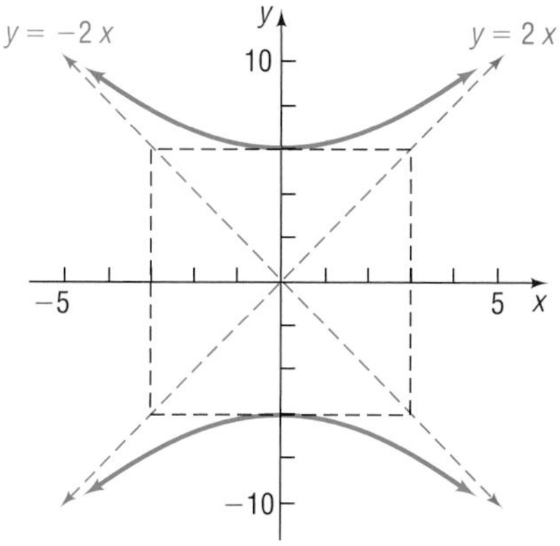

30.

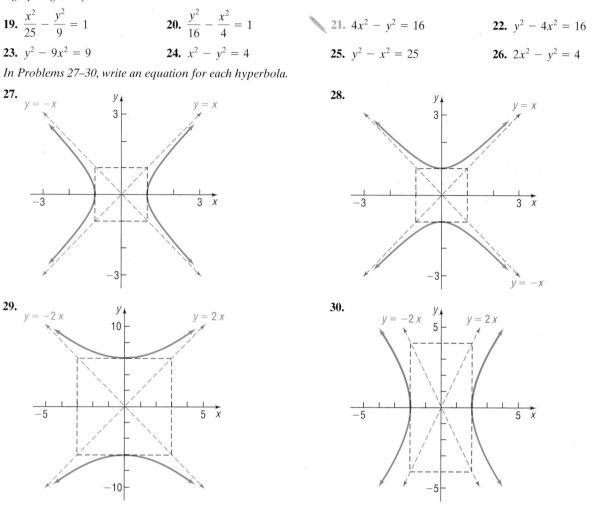

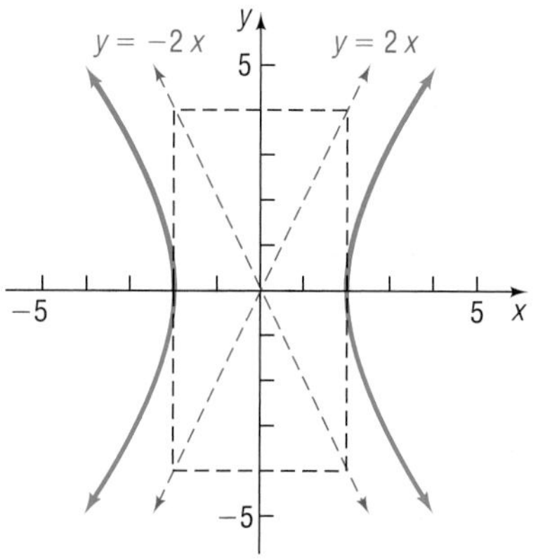

In Problems 31–38, find an equation for the hyperbola described. Graph the equation by hand.

31. Center at $(4, -1)$; focus at $(7, -1)$; vertex at $(6, -1)$

32. Center at $(-3, 1)$; focus at $(-3, 6)$; vertex at $(-3, 4)$

33. Center at $(-3, -4)$; focus at $(-3, -8)$; vertex at $(-3, -2)$

34. Center at $(1, 4)$; focus at $(-2, 4)$; vertex at $(0, 4)$

35. Foci at $(3, 7)$ and $(7, 7)$; vertex at $(6, 7)$

36. Focus at $(-4, 0)$; vertices at $(-4, 4)$ and $(-4, 2)$

37. Vertices at $(-1, -1)$ and $(3, -1)$; asymptote the line $(x - 1)/2 = (y + 1)/3$

38. Vertices at $(1, -3)$ and $(1, 1)$; asymptote the line $(x - 1)/2 = (y + 1)/3$

In Problems 39–52, find the center, transverse axis, vertices, foci, and asymptotes. Graph each equation using a graphing utility.

39. $\frac{(x-2)^2}{4} - \frac{(y+3)^2}{9} = 1$ **40.** $\frac{(y+3)^2}{4} - \frac{(x-2)^2}{9} = 1$

41. $(y - 2)^2 - 4(x + 2)^2 = 4$ **42.** $(x + 4)^2 - 9(y - 3)^2 = 9$ **43.** $(x + 1)^2 - (y + 2)^2 = 4$

44. $(y - 3)^2 - (x + 2)^2 = 4$ **45.** $x^2 - y^2 - 2x - 2y - 1 = 0$ **46.** $y^2 - x^2 - 4y + 4x - 1 = 0$

47. $y^2 - 4x^2 - 4y - 8x - 4 = 0$ **48.** $2x^2 - y^2 + 4x + 4y - 4 = 0$ **49.** $4x^2 - y^2 - 24x - 4y + 16 = 0$

50. $2y^2 - x^2 + 2x + 8y + 3 = 0$ **51.** $y^2 - 4x^2 - 16x - 2y - 19 = 0$ **52.** $x^2 - 3y^2 + 8x - 6y + 4 = 0$

In Problems 53–56, graph each function by hand. Verify your answer using a graphing utility.
[**Hint:** Notice that each function is half a hyperbola.]

53. $f(x) = \sqrt{16 + 4x^2}$ **54.** $f(x) = -\sqrt{9 + 9x^2}$ **55.** $f(x) = -\sqrt{-25 + x^2}$ **56.** $f(x) = \sqrt{-1 + x^2}$

57. LORAN Two LORAN stations are positioned 200 miles apart along a straight shore.
(a) A ship records a time difference of 0.00038 second between the LORAN signals. Set up an appropriate rectangular coordinate system to determine where the ship would reach shore if it were to follow the hyperbola corresponding to this time difference.
(b) If the ship wants to enter a harbor located between the two stations 20 miles from the master station, what time difference should it be looking for?
(c) If the ship is 50 miles offshore when the desired time difference is obtained, what is the approximate location of the ship?
[**Note:** The speed of each radio signal is 186,000 miles per second.]

58. LORAN Two LORAN stations are positioned 100 miles apart along a straight shore.
(a) A ship records a time difference of 0.00032 second between the LORAN signals. Set up an appropriate rectangular coordinate system to determine where the ship would reach shore if it were to follow the hyperbola corresponding to this time difference.
(b) If the ship wants to enter a harbor located between the two stations 10 miles from the master station, what time difference should it be looking for?
(c) If the ship is 20 miles offshore when the desired time difference is obtained, what is the approximate location of the ship?
[**Note:** The speed of each radio signal is 186,000 miles per second.]

59. Calibrating Instruments In a test of their recording devices, a team of seismologists positioned two of the devices 2000 feet apart, with the device at point A to the west of the device at point B. At a point between the devices and 200 feet from point B, a small amount of explosive was detonated and a note made of the time at which the sound reached each device. A second explosion is to be carried out at a point directly north of point B.
(a) How far north should the site of the second explosion be chosen so that the measured time difference recorded by the devices for the second detonation is the same as that recorded for the first detonation?
(b) Explain why this experiment can be used to calibrate the instruments.

60. Explain in your own words the LORAN system of navigation.

61. The **eccentricity** e of a hyperbola is defined as the number c/a, where a and c are the numbers given in equation (2). Because $c > a$, it follows that $e > 1$. Describe the general shape of a hyperbola whose eccentricity is close to 1. What is the shape if e is very large?

62. A hyperbola for which $a = b$ is called an **equilateral hyperbola.** Find the eccentricity e of an equilateral hyperbola.
[**Note:** The eccentricity of a hyperbola is defined in Problem 61.]

63. Two hyperbolas that have the same set of asymptotes are called **conjugate.** Show that the hyperbolas

$$\frac{x^2}{4} - y^2 = 1 \quad \text{and} \quad y^2 - \frac{x^2}{4} = 1$$

are conjugate. Graph each hyperbola on the same set of coordinate axes.

64. Prove that the hyperbola

$$\frac{y^2}{a^2} - \frac{x^2}{b^2} = 1$$

has the two oblique asymptotes

$$y = \frac{a}{b}x \quad \text{and} \quad y = -\frac{a}{b}x$$

65. Show that the graph of an equation of the form

$$Ax^2 + Cy^2 + F = 0, \quad A \neq 0, C \neq 0, F \neq 0$$

where A and C are of opposite sign, is a hyperbola with center at $(0, 0)$.

66. Show that the graph of an equation of the form

$$Ax^2 + Cy^2 + Dx + Ey + F = 0, \quad A \neq 0, C \neq 0$$

where A and C are of opposite sign,
(a) Is a hyperbola if $(D^2/4A) + (E^2/4C) - F \neq 0$.
(b) Is two intersecting lines if

$$(D^2/4A) + (E^2/4C) - F = 0$$

PREPARING FOR THIS SECTION

Before getting started, review the following:

- ✓ Sum Formulas for Sine and Cosine (pp. 452 and 455)
- ✓ Half-angle Formulas for Sine and Cosine (p. 465)
- ✓ Double-angle Formulas for Sine and Cosine (p. 462)

10.5 ROTATION OF AXES; GENERAL FORM OF A CONIC

OBJECTIVES
1. Identify a Conic
2. Use a Rotation of Axes to Transform Equations
3. Discuss an Equation Using a Rotation of Axes
4. Identify Conics without a Rotation of Axes

In this section, we show that the graph of a general second-degree polynomial containing two variables x and y, that is, an equation of the form

$$Ax^2 + Bxy + Cy^2 + Dx + Ey + F = 0 \qquad \textbf{(1)}$$

where A, B, and C are not simultaneously 0, is a conic. We shall not concern ourselves here with the degenerate cases of equation (1), such as $x^2 + y^2 = 0$, whose graph is a single point $(0, 0)$; or $x^2 + 3y^2 + 3 = 0$, whose graph contains no points; or $x^2 - 4y^2 = 0$, whose graph is two lines, $x - 2y = 0$ and $x + 2y = 0$.

We begin with the case where $B = 0$. In this case, the term containing xy is not present, so equation (1) has the form

$$Ax^2 + Cy^2 + Dx + Ey + F = 0$$

where either $A \neq 0$ or $C \neq 0$.

1 We have already discussed the procedure for identifying the graph of this kind of equation; we complete the squares of the quadratic expressions in x or y, or both. Once this has been done, the conic can be identified by comparing it to one of the forms studied in Sections 10.2 through 10.4.

In fact, though, we can identify the conic directly from the equation without completing the squares.

Theorem **Identifying Conics without Completing the Squares**

Excluding degenerate cases, the equation

$$Ax^2 + Cy^2 + Dx + Ey + F = 0 \qquad \textbf{(2)}$$

where A and C cannot both equal zero:

(a) Defines a parabola if $AC = 0$.

(b) Defines an ellipse (or a circle) if $AC > 0$.

(c) Defines a hyperbola if $AC < 0$.

Proof

(a) If $AC = 0$, then either $A = 0$ or $C = 0$, but not both, so the form of equation (2) is either

$$Ax^2 + Dx + Ey + F = 0, \quad A \neq 0$$

or

$$Cy^2 + Dx + Ey + F = 0, \quad C \neq 0$$

Using the results of Problems 73 and 74 in Exercise 10.2, it follows that, except for the degenerate cases, the equation is a parabola.

(b) If $AC > 0$, then A and C are of the same sign. Using the results of Problems 73 and 74 in Exercise 10.3, except for the degenerate cases, the equation is an ellipse if $A \neq C$ or a circle if $A = C$.

(c) If $AC < 0$, then A and C are of opposite sign. Using the results of Problems 65 and 66 in Exercise 10.4, except for the degenerate cases, the equation is a hyperbola. ■

We will not be concerned with the degenerate cases of equation (2). However, in practice, you should be alert to the possibility of degeneracy.

EXAMPLE 1 Identifying a Conic without Completing the Squares

Identify each equation without completing the squares.

(a) $3x^2 + 6y^2 + 6x - 12y = 0$ (b) $2x^2 - 3y^2 + 6y + 4 = 0$

(c) $y^2 - 2x + 4 = 0$

Solution (a) We compare the given equation to equation (2) and conclude that $A = 3$ and $C = 6$. Since $AC = 18 > 0$, the equation is an ellipse.

(b) Here, $A = 2$ and $C = -3$, so $AC = -6 < 0$. The equation is a hyperbola.

(c) Here, $A = 0$ and $C = 1$, so $AC = 0$. The equation is a parabola. ■

NOW WORK PROBLEM 1.

Although we can now identify the type of conic represented by any equation of the form of equation (2) without completing the squares, we will still need to complete the squares if we desire additional information about a conic.

Now we turn our attention to equations of the form of equation (1), where $B \neq 0$. To discuss this case, we first need to investigate a new procedure: *rotation of axes.*

Figure 51

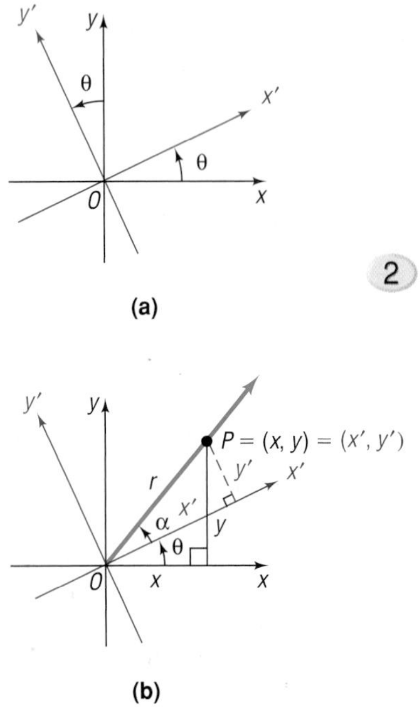

ROTATION OF AXES

2 In a **rotation of axes,** the origin remains fixed while the x-axis and y-axis are rotated through an angle θ to a new position; the new positions of the x- and y-axes are denoted by x' and y', respectively, as shown in Figure 51(a).

Now look at Figure 51(b). There the point P has the coordinates (x, y) relative to the xy-plane, while the same point P has coordinates (x', y') relative to the $x'y'$-plane. We seek relationships that will enable us to express x and y in terms of x', y', and θ.

As Figure 51(b) shows, r denotes the distance from the origin O to the point P, and α denotes the angle between the positive x'-axis and the ray from O through P. Then, using the definitions of sine and cosine, we have

$$x' = r\cos\alpha \qquad y' = r\sin\alpha \tag{3}$$

$$x = r\cos(\theta + \alpha) \qquad y = r\sin(\theta + \alpha) \tag{4}$$

Now

$$\begin{aligned} x &= r\cos(\theta + \alpha) \\ &= r(\cos\theta\cos\alpha - \sin\theta\sin\alpha) && \text{Sum formula} \\ &= (r\cos\alpha)(\cos\theta) - (r\sin\alpha)(\sin\theta) \\ &= x'\cos\theta - y'\sin\theta && \text{By equation (3)} \end{aligned}$$

Similarly,

$$\begin{aligned} y &= r\sin(\theta + \alpha) \\ &= r(\sin\theta\cos\alpha + \cos\theta\sin\alpha) \\ &= x'\sin\theta + y'\cos\theta \end{aligned}$$

Theorem

Rotation Formulas

If the x- and y-axes are rotated through an angle θ, the coordinates (x, y) of a point P relative to the xy-plane and the coordinates (x', y') of the same point relative to the new x'- and y'-axes are related by the formulas

$$x = x'\cos\theta - y'\sin\theta \qquad y = x'\sin\theta + y'\cos\theta \qquad \textbf{(5)}$$

EXAMPLE 2 Rotating Axes

Express the equation $xy = 1$ in terms of new $x'y'$-coordinates by rotating the axes through a $45°$ angle. Discuss the new equation.

Figure 52

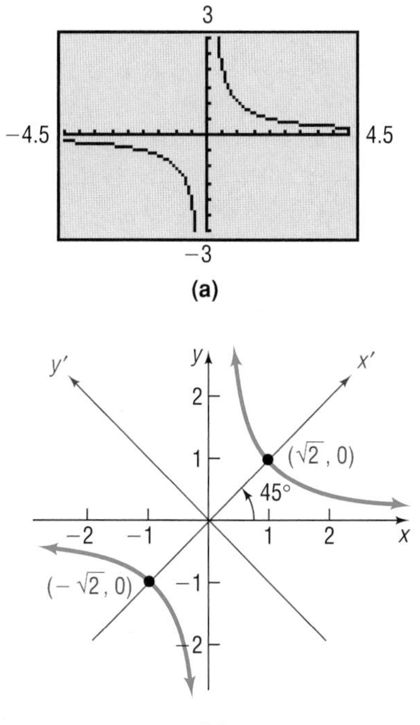

Solution Figure 52(a) shows the graph of $xy = 1$ using a graphing utility. We now proceed to express the equation in terms of new $x'y'$-coordinates and discuss the new equation.

Let $\theta = 45°$ in equation (5). Then

$$x = x'\cos 45° - y'\sin 45° = x'\frac{\sqrt{2}}{2} - y'\frac{\sqrt{2}}{2} = \frac{\sqrt{2}}{2}(x' - y')$$

$$y = x'\sin 45° + y'\cos 45° = x'\frac{\sqrt{2}}{2} + y'\frac{\sqrt{2}}{2} = \frac{\sqrt{2}}{2}(x' + y')$$

Substituting these expressions for x and y in $xy = 1$ gives

$$\left[\frac{\sqrt{2}}{2}(x' - y')\right]\left[\frac{\sqrt{2}}{2}(x' + y')\right] = 1$$

$$\frac{1}{2}(x'^2 - y'^2) = 1$$

$$\frac{x'^2}{2} - \frac{y'^2}{2} = 1$$

This is the equation of a hyperbola with center at $(0, 0)$ and transverse axis along the x'-axis. The vertices are at $(\pm\sqrt{2}, 0)$ on the x'-axis; the asymptotes are $y' = x'$ and $y' = -x'$ (which correspond to the original x- and y-axes). See Figure 52(b) for the graph drawn by hand.

As Example 2 illustrates, a rotation of axes through an appropriate angle can transform a second-degree equation in x and y containing an xy-term into one in x' and y' in which no $x'y'$-term appears. In fact, we will show that a rotation of axes through an appropriate angle will transform any equation of the form of equation (1) into an equation in x' and y' without an $x'y'$-term.

To find the formula for choosing an appropriate angle θ through which to rotate the axes, we begin with equation (1),

$$Ax^2 + Bxy + Cy^2 + Dx + Ey + F = 0, \quad B \neq 0$$

Next we rotate through an angle θ using rotation formulas (5).

$$\begin{aligned} A(x'\cos\theta - y'\sin\theta)^2 &+ B(x'\cos\theta - y'\sin\theta)(x'\sin\theta + y'\cos\theta) \\ &+ C(x'\sin\theta + y'\cos\theta)^2 + D(x'\cos\theta - y'\sin\theta) \\ &+ E(x'\sin\theta + y'\cos\theta) + F = 0 \end{aligned}$$

By expanding and collecting like terms, we obtain

$$\begin{aligned} (A\cos^2\theta + B\sin\theta\cos\theta + C\sin^2\theta)x'^2 &+ \left[B(\cos^2\theta - \sin^2\theta) + 2(C - A)(\sin\theta\cos\theta)\right]x'y' \\ &+ (A\sin^2\theta - B\sin\theta\cos\theta + C\cos^2\theta)y'^2 \\ &+ (D\cos\theta + E\sin\theta)x' \\ &+ (-D\sin\theta + E\cos\theta)y' + F = 0 \end{aligned} \qquad \textbf{(6)}$$

In equation (6), the coefficient of $x'y'$ is

$$B' = 2(C - A)(\sin\theta\cos\theta) + B(\cos^2\theta - \sin^2\theta)$$

Since we want to eliminate the $x'y'$-term, we select an angle θ so that $B' = 0$. Then

$$\begin{aligned} 2(C - A)(\sin\theta\cos\theta) + B(\cos^2\theta - \sin^2\theta) &= 0 \\ (C - A)(\sin(2\theta)) + B\cos(2\theta) &= 0 \quad \text{Double-angle formulas} \\ B\cos(2\theta) &= (A - C)(\sin(2\theta)) \\ \cot(2\theta) &= \frac{A - C}{B}, \qquad B \neq 0 \end{aligned}$$

Theorem To transform the equation

$$Ax^2 + Bxy + Cy^2 + Dx + Ey + F = 0, \qquad B \neq 0$$

into an equation in x' and y' without an $x'y'$-term, rotate the axes through an angle θ that satisfies the equation

$$\cot(2\theta) = \frac{A - C}{B} \qquad \textbf{(7)}$$

■

Equation (7) has an infinite number of solutions for θ. We shall adopt the convention of choosing the acute angle θ that satisfies (7). Then we have the following two possibilities:

If $\cot(2\theta) \geq 0$, then $0° < 2\theta \leq 90°$ so that $0° < \theta \leq 45°$.
If $\cot(2\theta) < 0$, then $90° < 2\theta < 180°$ so that $45° < \theta < 90°$.

Each of these results in a counterclockwise rotation of the axes through an acute angle θ.*

WARNING: Be careful if you use a calculator to solve equation (7).

1. If $\cot(2\theta) = 0$, then $2\theta = 90°$ and $\theta = 45°$.
2. If $\cot(2\theta) \neq 0$, first find $\cos(2\theta)$. Then use the inverse cosine function key(s) to obtain 2θ, $0° < 2\theta < 180°$. Finally, divide by 2 to obtain the correct acute angle θ.

EXAMPLE 3 Discussing an Equation Using a Rotation of Axes

3 Discuss the equation: $x^2 + \sqrt{3}xy + 2y^2 - 10 = 0$

Solution We need to solve the equation for y in order to graph the equation using a graphing utility. Rearranging the terms, we observe that the equation is quadratic in the variable y: $2y^2 + \sqrt{3}xy + (x^2 - 10) = 0$. We can solve the equation for y using the quadratic formula with $a = 2$, $b = \sqrt{3}x$, and $c = x^2 - 10$.

$$Y_1 = \frac{-\sqrt{3}x + \sqrt{(\sqrt{3}x)^2 - 4(2)(x^2 - 10)}}{2(2)} = \frac{-\sqrt{3}x + \sqrt{-5x^2 + 80}}{4}$$

and

$$Y_2 = \frac{-\sqrt{3}x - \sqrt{(\sqrt{3}x)^2 - 4(2)(x^2 - 10)}}{2(2)} = \frac{-\sqrt{3}x - \sqrt{-5x^2 + 80}}{4}$$

Figure 53(a)

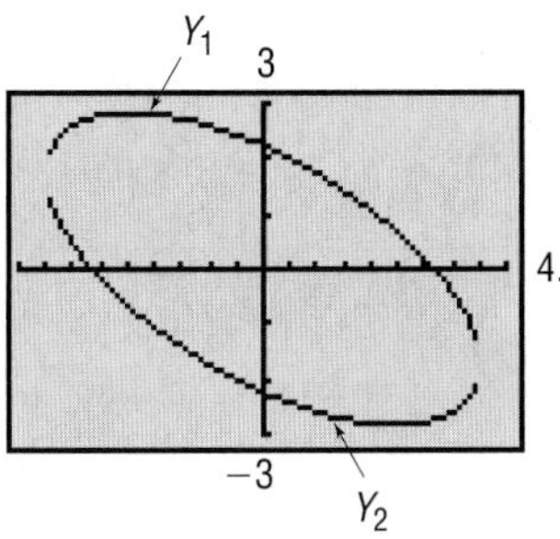

Figure 53(a) shows the graph of Y_1 and Y_2.

We now proceed to analyze the equation. Since an xy-term is present, we must rotate the axes. Using $A = 1$, $B = \sqrt{3}$, and $C = 2$ in equation (7), the appropriate acute angle θ through which to rotate the axes satisfies the equation

$$\cot(2\theta) = \frac{A - C}{B} = \frac{-1}{\sqrt{3}} = \frac{-\sqrt{3}}{3}, \qquad 0° < 2\theta < 180°$$

Since $\cot(2\theta) = -\sqrt{3}/3$, we find $2\theta = 120°$, so $\theta = 60°$. Using $\theta = 60°$ in rotation formulas (5), we find

$$x = \frac{1}{2}x' - \frac{\sqrt{3}}{2}y' = \frac{1}{2}(x' - \sqrt{3}y')$$

$$y = \frac{\sqrt{3}}{2}x' + \frac{1}{2}y' = \frac{1}{2}(\sqrt{3}x' + y')$$

Substituting these values into the original equation and simplifying, we have

$$x^2 + \sqrt{3}xy + 2y^2 - 10 = 0$$

$$\frac{1}{4}(x' - \sqrt{3}y')^2 + \sqrt{3}\left[\frac{1}{2}\left(x' - \sqrt{3}y'\right)\right]\left[\frac{1}{2}(\sqrt{3}x' + y')\right] + 2\left[\frac{1}{4}(\sqrt{3}x' + y')^2\right] = 10$$

* Any rotation (clockwise or counterclockwise) through an angle θ that satisfies $\cot(2\theta) = (A - C)/B$ will eliminate the $x'y'$-term. However, the final form of the transformed equation may be different (but equivalent), depending on the angle chosen.

Multiply both sides by 4 and expand to obtain

$$x'^2 - 2\sqrt{3}x'y' + 3y'^2 + \sqrt{3}(\sqrt{3}x'^2 - 2x'y' - \sqrt{3}y'^2) + 2(3x'^2 + 2\sqrt{3}x'y' + y'^2) = 40$$

$$10x'^2 + 2y'^2 = 40$$

$$\frac{x'^2}{4} + \frac{y'^2}{20} = 1$$

This is the equation of an ellipse with center at $(0, 0)$ and major axis along the y'-axis. The vertices are at $(0, \pm 2\sqrt{5})$ on the y'-axis. See Figure 53(b) for the graph drawn by hand.

Figure 53(b)

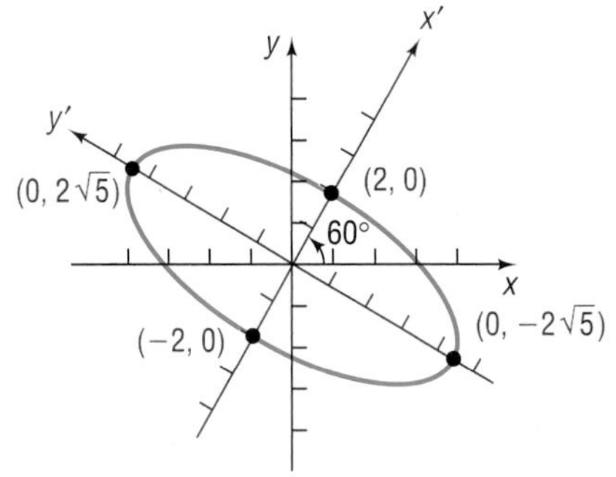

NOW WORK PROBLEM 21.

In Example 3, the acute angle θ through which to rotate the axes was easy to find because of the numbers that we used in the given equation. In general, the equation $\cot(2\theta) = (A - C)/B$ will not have such a "nice" solution. As the next example shows, we can still find the appropriate rotation formulas without using a calculator approximation by applying half-angle formulas.

EXAMPLE 4 Discussing an Equation Using a Rotation of Axes

Discuss the equation: $4x^2 - 4xy + y^2 + 5\sqrt{5}x + 5 = 0$

Solution We need to solve the equation for y in order to graph the equation using a graphing utility. Rearranging the terms, we observe that the equation is quadratic in the variable y: $y^2 - 4xy + (4x^2 + 5\sqrt{5}x + 5) = 0$. We can solve the equation for y using the quadratic formula with $a = 1$, $b = -4x$, and $c = 4x^2 + 5\sqrt{5}x + 5$.

$$Y_1 = \frac{-(-4x) + \sqrt{(-4x)^2 - 4(1)(4x^2 + 5\sqrt{5}x + 5)}}{2(1)} = 2x + \sqrt{-5(\sqrt{5}x + 1)}$$

$$Y_2 = \frac{-(-4x) - \sqrt{(-4x)^2 - 4(1)(4x^2 + 5\sqrt{5}x + 5)}}{2(1)} = 2x - \sqrt{-5(\sqrt{5}x + 1)}$$

Figure 54(a)

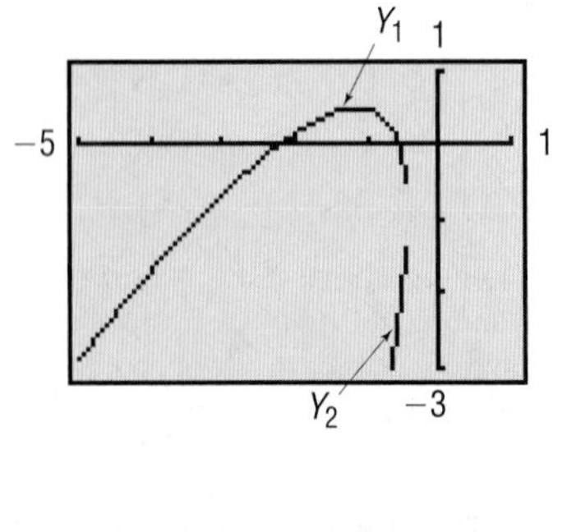

Figure 54(a) shows the graph of Y_1 and Y_2.

We now proceed to analyze the equation. Letting $A = 4$, $B = -4$, and $C = 1$ in equation (7), the appropriate angle θ through which to rotate the axes satisfies

$$\cot(2\theta) = \frac{A - C}{B} = \frac{3}{-4}$$

In order to use rotation formulas (5), we need to know the values of $\sin\theta$ and $\cos\theta$. Since we seek an acute angle θ, we know that $\sin\theta > 0$ and $\cos\theta > 0$. We use the half-angle formulas in the form

$$\sin\theta = \sqrt{\frac{1-\cos(2\theta)}{2}} \qquad \cos\theta = \sqrt{\frac{1+\cos(2\theta)}{2}}$$

Now we need to find the value of $\cos(2\theta)$. Since $\cot(2\theta) = -\frac{3}{4}$ and $90° < 2\theta < 180°$. (Do you know why?), it follows that $\cos(2\theta) = -\frac{3}{5}$. Then

$$\sin\theta = \sqrt{\frac{1-\cos(2\theta)}{2}} = \sqrt{\frac{1-\left(-\frac{3}{5}\right)}{2}} = \sqrt{\frac{4}{5}} = \frac{2}{\sqrt{5}} = \frac{2\sqrt{5}}{5}$$

$$\cos\theta = \sqrt{\frac{1+\cos(2\theta)}{2}} = \sqrt{\frac{1+\left(-\frac{3}{5}\right)}{2}} = \sqrt{\frac{1}{5}} = \frac{1}{\sqrt{5}} = \frac{\sqrt{5}}{5}$$

With these values, rotation formulas (5) give us

$$x = \frac{\sqrt{5}}{5}x' - \frac{2\sqrt{5}}{5}y' = \frac{\sqrt{5}}{5}(x' - 2y')$$

$$y = \frac{2\sqrt{5}}{5}x' + \frac{\sqrt{5}}{5}y' = \frac{\sqrt{5}}{5}(2x' + y')$$

Substituting these values in the original equation and simplifying, we obtain

$$4x^2 - 4xy + y^2 + 5\sqrt{5}x + 5 = 0$$

$$4\left[\frac{\sqrt{5}}{5}(x' - 2y')\right]^2 - 4\left[\frac{\sqrt{5}}{5}(x' - 2y')\right]\left[\frac{\sqrt{5}}{5}(2x' + y')\right] + \left[\frac{\sqrt{5}}{5}(2x' + y')\right]^2 + 5\sqrt{5}\left[\frac{\sqrt{5}}{5}(x' - 2y')\right] = -5$$

Multiply both sides by 5 and expand to obtain

$$4(x'^2 - 4x'y' + 4y'^2) - 4(2x'^2 - 3x'y' - 2y'^2) + 4x'^2 + 4x'y' + y'^2 + 25(x' - 2y') = -25$$

$$25y'^2 - 50y' + 25x' = -25$$

$$y'^2 - 2y' + x' = -1$$

$$y'^2 - 2y' + 1 = -x' \qquad \text{Complete the square in } y'.$$

$$(y' - 1)^2 = -x'$$

Figure 54(b)

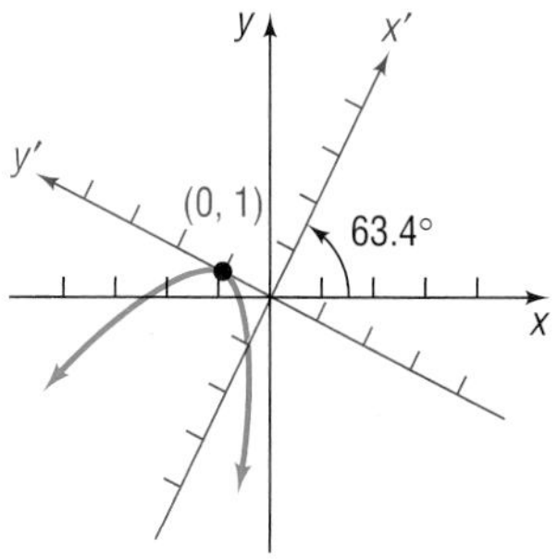

This is the equation of a parabola with vertex at (0, 1) in the $x'y'$-plane. The axis of symmetry is parallel to the x'-axis. Using a calculator to solve $\sin\theta = 2\sqrt{5}/5$, we find that $\theta \approx 63.4°$. See Figure 54(b) for the graph drawn by hand. ■

NOW WORK PROBLEM 27.

Identifying Conics without a Rotation of Axes

4 Suppose that we are required only to identify (rather than discuss) an equation of the form

$$Ax^2 + Bxy + Cy^2 + Dx + Ey + F = 0, \qquad B \neq 0 \qquad \textbf{(8)}$$

If we apply rotation formulas (5) to this equation, we obtain an equation of the form

$$A'x'^2 + B'x'y' + C'y'^2 + D'x' + E'y' + F' = 0 \tag{9}$$

where A', B', C', D', E', and F' can be expressed in terms of A, B, C, D, E, F, and the angle θ of rotation (see Problem 43 at the end of this section). It can be shown that the value of $B^2 - 4AC$ in equation (8) and the value of $B'^2 - 4A'C'$ in equation (9) are equal no matter what angle θ of rotation is chosen (see Problem 45). In particular, if the angle θ of rotation satisfies equation (7), then $B' = 0$ in equation (9), and $B^2 - 4AC = -4A'C'$. Since equation (9) then has the form of equation (2),

$$A'x'^2 + C'y'^2 + D'x' + E'y' + F' = 0$$

we can identify it without completing the squares, as we did in the beginning of this section. In fact, now we can identify the conic described by any equation of the form of equation (8) without a rotation of axes.

Theorem **Identifying Conics without a Rotation of Axes**

Except for degenerate cases, the equation

$$\boxed{Ax^2 + Bxy + Cy^2 + Dx + Ey + F = 0}$$

(a) Defines a parabola if $B^2 - 4AC = 0$.
(b) Defines an ellipse (or a circle) if $B^2 - 4AC < 0$.
(c) Defines a hyperbola if $B^2 - 4AC > 0$.

You are asked to prove this theorem in Problem 46.

EXAMPLE 5 Identifying a Conic without a Rotation of Axes

Identify the equation: $8x^2 - 12xy + 17y^2 - 4\sqrt{5}x - 2\sqrt{5}y - 15 = 0$

Solution Here $A = 8$, $B = -12$, and $C = 17$, so $B^2 - 4AC = -400$. Since $B^2 - 4AC < 0$, the equation defines an ellipse.

NOW WORK PROBLEM 33.

10.5 EXERCISES

In Problems 1–10, identify each equation without completing the squares.

1. $x^2 + 4x + y + 3 = 0$

2. $2y^2 - 3y + 3x = 0$

3. $6x^2 + 3y^2 - 12x + 6y = 0$

4. $2x^2 + y^2 - 8x + 4y + 2 = 0$

5. $3x^2 - 2y^2 + 6x + 4 = 0$

6. $4x^2 - 3y^2 - 8x + 6y + 1 = 0$

7. $2y^2 - x^2 - y + x = 0$

8. $y^2 - 8x^2 - 2x - y = 0$

9. $x^2 + y^2 - 8x + 4y = 0$

10. $2x^2 + 2y^2 - 8x + 8y = 0$

In Problems 11–20, determine the appropriate rotation formulas to use so that the new equation contains no xy-term.

11. $x^2 + 4xy + y^2 - 3 = 0$

12. $x^2 - 4xy + y^2 - 3 = 0$

13. $5x^2 + 6xy + 5y^2 - 8 = 0$

14. $3x^2 - 10xy + 3y^2 - 32 = 0$

15. $13x^2 - 6\sqrt{3}xy + 7y^2 - 16 = 0$

16. $11x^2 + 10\sqrt{3}xy + y^2 - 4 = 0$

17. $4x^2 - 4xy + y^2 - 8\sqrt{5}x - 16\sqrt{5}y = 0$

18. $x^2 + 4xy + 4y^2 + 5\sqrt{5}y + 5 = 0$

19. $25x^2 - 36xy + 40y^2 - 12\sqrt{13}x - 8\sqrt{13}y = 0$

20. $34x^2 - 24xy + 41y^2 - 25 = 0$

In Problems 21–32, graph the equation using a graphing utility. Rotate the axes so that the new equation contains no xy-term. Discuss and, by hand, graph the new equation. Refer to Problems 11–20 for Problems 21–30.

21. $x^2 + 4xy + y^2 - 3 = 0$

22. $x^2 - 4xy + y^2 - 3 = 0$

23. $5x^2 + 6xy + 5y^2 - 8 = 0$

24. $3x^2 - 10xy + 3y^2 - 32 = 0$

25. $13x^2 - 6\sqrt{3}xy + 7y^2 - 16 = 0$

26. $11x^2 + 10\sqrt{3}xy + y^2 - 4 = 0$

27. $4x^2 - 4xy + y^2 - 8\sqrt{5}x - 16\sqrt{5}y = 0$

28. $x^2 + 4xy + 4y^2 + 5\sqrt{5}y + 5 = 0$

29. $25x^2 - 36xy + 40y^2 - 12\sqrt{13}x - 8\sqrt{13}y = 0$

30. $34x^2 - 24xy + 41y^2 - 25 = 0$

31. $16x^2 + 24xy + 9y^2 - 130x + 90y = 0$

32. $16x^2 + 24xy + 9y^2 - 60x + 80y = 0$

In Problems 33–42, identify each equation without applying a rotation of axes.

33. $x^2 + 3xy - 2y^2 + 3x + 2y + 5 = 0$

34. $2x^2 - 3xy + 4y^2 + 2x + 3y - 5 = 0$

35. $x^2 - 7xy + 3y^2 - y - 10 = 0$

36. $2x^2 - 3xy + 2y^2 - 4x - 2 = 0$

37. $9x^2 + 12xy + 4y^2 - x - y = 0$

38. $10x^2 + 12xy + 4y^2 - x - y + 10 = 0$

39. $10x^2 - 12xy + 4y^2 - x - y - 10 = 0$

40. $4x^2 + 12xy + 9y^2 - x - y = 0$

41. $3x^2 - 2xy + y^2 + 4x + 2y - 1 = 0$

42. $3x^2 + 2xy + y^2 + 4x - 2y + 10 = 0$

In Problems 43–46, apply rotation formulas (5) to

$$Ax^2 + Bxy + Cy^2 + Dx + Ey + F = 0$$

to obtain the equation

$$A'x'^2 + B'x'y' + C'y'^2 + D'x' + E'y' + F' = 0$$

43. Express A', B', C', D', E', and F' in terms of A, B, C, D, E, F, and the angle θ of rotation.

44. Show that $A + C = A' + C'$, and thus show that $A + C$ is **invariant;** that is, its value does not change under a rotation of axes.

45. Refer to Problem 44. Show that $B^2 - 4AC$ is invariant.

46. Prove that, except for degenerate cases, the equation

$$Ax^2 + Bxy + Cy^2 + Dx + Ey + F = 0$$

(a) Defines a parabola if $B^2 - 4AC = 0$.
(b) Defines an ellipse (or a circle) if $B^2 - 4AC < 0$.
(c) Defines a hyperbola if $B^2 - 4AC > 0$.

47. Use rotation formulas (5) to show that distance is invariant under a rotation of axes. That is, show that the distance from $P_1 = (x_1, y_1)$ to $P_2 = (x_2, y_2)$ in the xy-plane equals the distance from $P_1 = (x'_1, y'_1)$ to $P_2 = (x'_2, y'_2)$ in the $x'y'$-plane.

48. Show that the graph of the equation $x^{1/2} + y^{1/2} = a^{1/2}$ is part of the graph of a parabola.

49. Formulate a strategy for discussing and graphing an equation of the form

$$Ax^2 + Cy^2 + Dx + Ey + F = 0$$

How does your strategy change if the equation is of the form

$$Ax^2 + Bxy + Cy^2 + Dx + Ey + F = 0$$

PREPARING FOR THIS SECTION

Before getting started, review the following:

✓ Polar Coordinates (Section 9.1)

10.6 POLAR EQUATIONS OF CONICS

OBJECTIVES
1. Discuss and Graph Polar Equations of Conics
2. Convert a Polar Equation of a Conic to a Rectangular Equation

1 In Sections 10.2, 10.3, and 10.4, we gave separate definitions for the parabola, ellipse, and hyperbola based on geometric properties and the distance formula. In this section, we present an alternative definition that simultaneously defines all these conics. As we shall see, this approach is well suited to polar coordinate representation. (Refer to Section 9.1.)

Let D denote a fixed line called the **directrix;** let F denote a fixed point called the **focus,** which is not on D; and let e be a fixed positive number called the **eccentricity.** A **conic** is the set of points P in the plane such that the ratio of the distance from F to P to the distance from D to P equals e. Thus, a conic is the collection of points P for which

$$\frac{d(F, P)}{d(D, P)} = e \qquad \textbf{(1)}$$

If $e = 1$, the conic is a **parabola.**
If $e < 1$, the conic is an **ellipse.**
If $e > 1$, the conic is a **hyperbola.**

Observe that if $e = 1$ the definition of a parabola in equation (1) is exactly the same as the definition used earlier in Section 10.2.

In the case of an ellipse, the **major axis** is a line through the focus perpendicular to the directrix. In the case of a hyperbola, the **transverse axis** is a line through the focus perpendicular to the directrix. For both an ellipse and a hyperbola, the eccentricity e satisfies

$$e = \frac{c}{a} \qquad \textbf{(2)}$$

where c is the distance from the center to the focus and a is the distance from the center to a vertex.

Just as we did earlier using rectangular coordinates, we derive equations for the conics in polar coordinates by choosing a convenient position for the focus F and the directrix D. The focus F is positioned at the pole, and the directrix D is either parallel to the polar axis or perpendicular to it.

Suppose that we start with the directrix D perpendicular to the polar axis at a distance p units to the left of the pole (the focus F). See Figure 55.

Figure 55

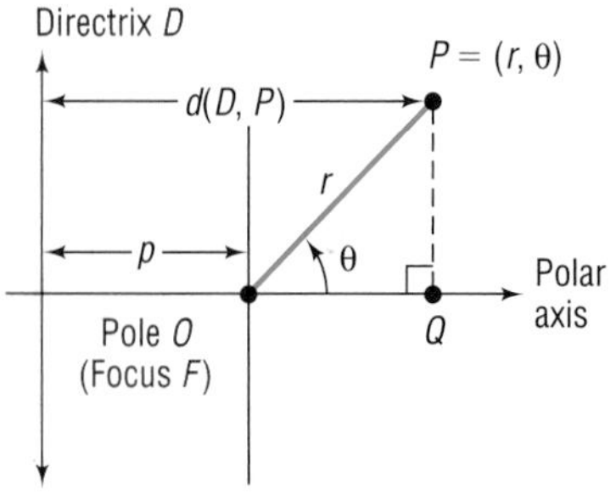

If $P = (r, \theta)$ is any point on the conic, then, by equation (1),

$$\frac{d(F, P)}{d(D, P)} = e \quad \text{or} \quad d(F, P) = e \cdot d(D, P) \tag{3}$$

Now we use the point Q obtained by dropping the perpendicular from P to the polar axis to calculate $d(D, P)$:

$$d(D, P) = p + d(O, Q) = p + r\cos\theta$$

Using this expression and the fact that $d(F, P) = d(O, P) = r$ in equation (3), we get

$$\begin{aligned} d(F, P) &= e \cdot d(D, P) \\ r &= e(p + r\cos\theta) \\ r &= ep + er\cos\theta \\ r - er\cos\theta &= ep \\ r(1 - e\cos\theta) &= ep \\ r &= \frac{ep}{1 - e\cos\theta} \end{aligned}$$

Theorem

Polar Equation of a Conic; Focus at Pole; Directrix Perpendicular to Polar Axis a Distance p to the Left of the Pole

The polar equation of a conic with focus at the pole and directrix perpendicular to the polar axis at a distance p to the left of the pole is

$$r = \frac{ep}{1 - e\cos\theta} \tag{4}$$

where e is the eccentricity of the conic.

■

EXAMPLE 1 Discussing and Graphing the Polar Equation of a Conic

Discuss and graph the equation: $r = \dfrac{4}{2 - \cos\theta}$

Solution Figure 56(a) shows the graph of the equation using a graphing utility in POLar mode with $\theta\text{min} = 0$, $\theta\text{max} = 2\pi$, and $\theta\text{step} = \pi/24$. We now proceed to discuss the equation.

Figure 56(a)

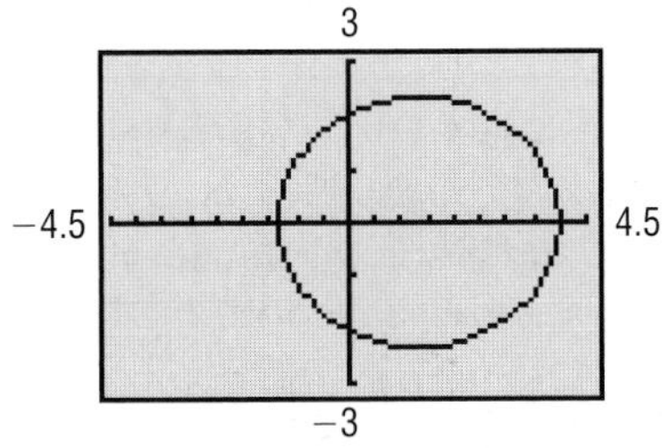

The given equation is not quite in the form of equation (4), since the first term in the denominator is 2 instead of 1. We divide the numerator and denominator by 2 to obtain

$$r = \frac{2}{1 - \frac{1}{2}\cos\theta}$$

This equation is in the form of equation (4), with

$$e = \tfrac{1}{2} \quad \text{and} \quad ep = \tfrac{1}{2}p = 2$$

Thus, $e = \frac{1}{2}$ and $p = 4$. We conclude that the conic is an ellipse, since $e = \frac{1}{2} < 1$. One focus is at the pole, and the directrix is perpendicular to the polar axis, a distance of 4 units to the left of the pole. It follows that the major axis is along the polar axis. To find the vertices, we let $\theta = 0$ and $\theta = \pi$. The vertices of the ellipse are $(4, 0)$ and $\left(\frac{4}{3}, \pi\right)$. The midpoint of the vertices, $\left(\frac{4}{3}, 0\right)$ in polar coordinates, is the center of the ellipse. [Do you see why? The vertices $(4, 0)$ and $\left(\frac{4}{3}, \pi\right)$ in polar coordinates are $(4, 0)$ and $\left(-\frac{4}{3}, 0\right)$ in rectangular coordinates. The midpoint in rectangular coordinates is $\left(\frac{4}{3}, 0\right)$, which is also $\left(\frac{4}{3}, 0\right)$ in polar coordinates.] Then a = distance from the center to a vertex $= \frac{8}{3}$. Using $a = \frac{8}{3}$ and $e = \frac{1}{2}$ in equation (2), $e = c/a$, we find $c = \frac{4}{3}$. Finally, using $a = \frac{8}{3}$ and $c = \frac{4}{3}$ in $b^2 = a^2 - c^2$, we have

Figure 56(b)

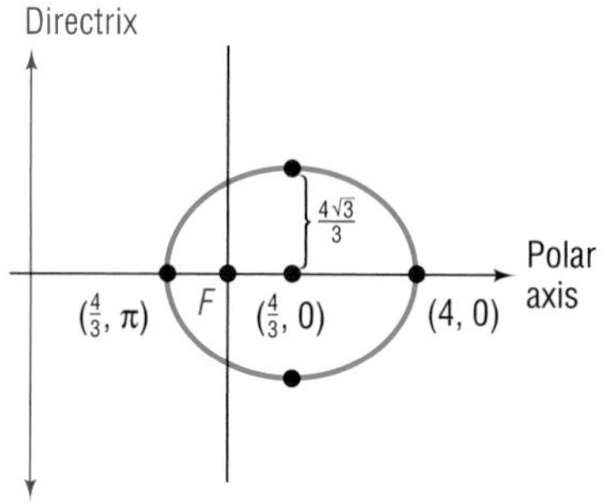

$$b^2 = a^2 - c^2 = \frac{64}{9} - \frac{16}{9} = \frac{48}{9}$$

$$b = \frac{4\sqrt{3}}{3}$$

Figure 56(b) shows the graph drawn by hand. ■

EXPLORATION Graph $r_1 = 4/(2 + \cos\theta)$ and compare the result with Figure 56(a). What do you conclude? Clear the screen and graph $r_1 = 4/(2 - \sin\theta)$ and then $r_1 = 4/(2 + \sin\theta)$. Compare each of these graphs with Figure 56(a). What do you conclude? ■

NOW WORK PROBLEM 5.

Equation (4) was obtained under the assumption that the directrix was perpendicular to the polar axis at a distance p units to the left of the pole. A similar derivation (see Problem 37), in which the directrix is perpendicular to the polar axis at a distance p units to the right of the pole, results in the equation

$$r = \frac{ep}{1 + e\cos\theta}$$

In Problems 38 and 39 you are asked to derive the polar equations of conics with focus at the pole and directrix parallel to the polar axis. Table 5 on page 687, summarizes the polar equations of conics.

EXAMPLE 2 Discussing and Graphing the Polar Equation of a Conic

Discuss and graph the equation: $r = \dfrac{6}{3 + 3\sin\theta}$

Solution Figure 57(a) shows the graph of the equation using a graphing utility in POLar mode with $\theta\text{min} = 0$, $\theta\text{max} = 2\pi$, and $\theta\text{step} = \pi/24$. We now proceed to discuss the equation.

Figure 57

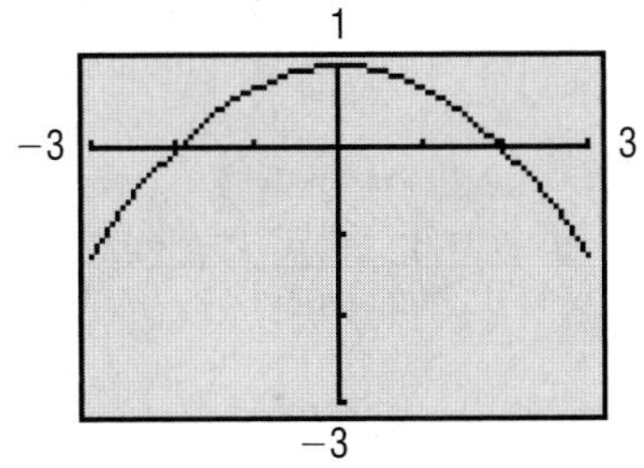

(a)

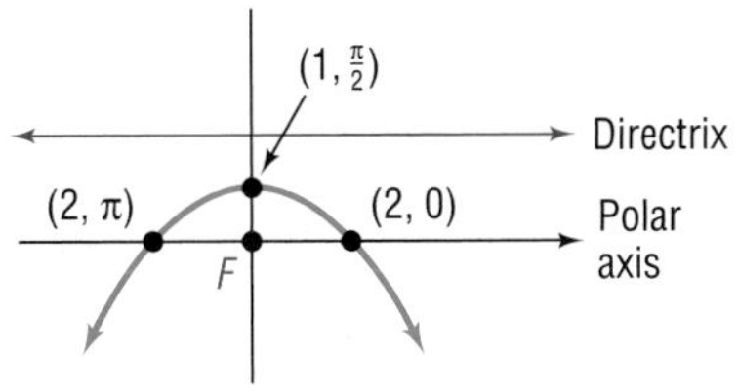

(b)

TABLE 5

POLAR EQUATIONS OF CONICS (FOCUS AT THE POLE, ECCENTRICITY e)

Equation	Description
(a) $r = \dfrac{ep}{1 - e\cos\theta}$	Directrix is perpendicular to the polar axis at a distance p units to the left of the pole.
(b) $r = \dfrac{ep}{1 + e\cos\theta}$	Directrix is perpendicular to the polar axis at a distance p units to the right of the pole.
(c) $r = \dfrac{ep}{1 + e\sin\theta}$	Directrix is parallel to the polar axis at a distance p units above the pole.
(d) $r = \dfrac{ep}{1 - e\sin\theta}$	Directrix is parallel to the polar axis at a distance p units below the pole.

Eccentricity

If $e = 1$, the conic is a parabola; the axis of symmetry is perpendicular to the directrix.

If $e < 1$, the conic is an ellipse; the axis is perpendicular to the directrix.

If $e > 1$, the conic is a hyperbola; the transverse axis is perpendicular to the directrix.

To place the equation in proper form, we divide the numerator and denominator by 3 to get

$$r = \frac{2}{1 + \sin\theta}$$

Referring to Table 5, we conclude that this equation is in the form of equation (c) with

$$e = 1 \quad \text{and} \quad ep = 2$$

Thus, $e = 1$ and $p = 2$. The conic is a parabola with focus at the pole. The directrix is parallel to the polar axis at a distance 2 units above the pole; the axis of symmetry is perpendicular to the polar axis. The vertex of the parabola is at $(1, \pi/2)$. (Do you see why?) See Figure 57(b) for the graph drawn by hand. Notice that we plotted two additional points, $(2, 0)$ and $(2, \pi)$, to assist in graphing. ■

NOW WORK PROBLEM 7.

EXAMPLE 3 Discussing and Graphing the Polar Equation of a Conic

Discuss and graph the equation: $r = \dfrac{3}{1 + 3\cos\theta}$.

Solution Figure 58(a) and (b) show the graph of the equation using a graphing utility in POLar mode with $\theta\text{min} = 0$, $\theta\text{max} = 2\pi$, and $\theta\text{step} = \pi/24$, using both dot mode and connected mode. Notice the extraneous asymptotes in the connected mode. We now proceed to discuss the equation.

This equation is in the form of equation (b) in Table 5. We conclude that

$$e = 3 \quad \text{and} \quad ep = 3p = 3$$

Thus, $e = 3$ and $p = 1$. This is the equation of a hyperbola with a focus at the pole. The directrix is perpendicular to the polar axis, 1 unit to the right of the pole. The transverse axis is along the polar axis. To find the vertices, we let $\theta = 0$ and $\theta = \pi$. The vertices are $\left(\frac{3}{4}, 0\right)$ and $\left(-\frac{3}{2}, \pi\right)$. The center, which is at the midpoint of $\left(\frac{3}{4}, 0\right)$ and $\left(-\frac{3}{2}, \pi\right)$, is $\left(\frac{9}{8}, 0\right)$. Then c = distance from the center to a focus $= \frac{9}{8}$. Since $e = 3$, it follows from equation (2), $e = c/a$, that $a = \frac{3}{8}$. Finally, using $a = \frac{3}{8}$ and $c = \frac{9}{8}$ in $b^2 = c^2 - a^2$, we find

$$b^2 = c^2 - a^2 = \frac{81}{64} - \frac{9}{64} = \frac{72}{64} = \frac{9}{8}$$

$$b = \frac{3}{2\sqrt{2}} = \frac{3\sqrt{2}}{4}$$

Figure 58(c) shows the graph drawn by hand. Notice that we plotted two additional points, $(3, \pi/2)$ and $(3, 3\pi/2)$, on the left branch and used symmetry to obtain the right branch. The asymptotes of this hyperbola were found in the usual way by constructing the rectangle shown.

Figure 58

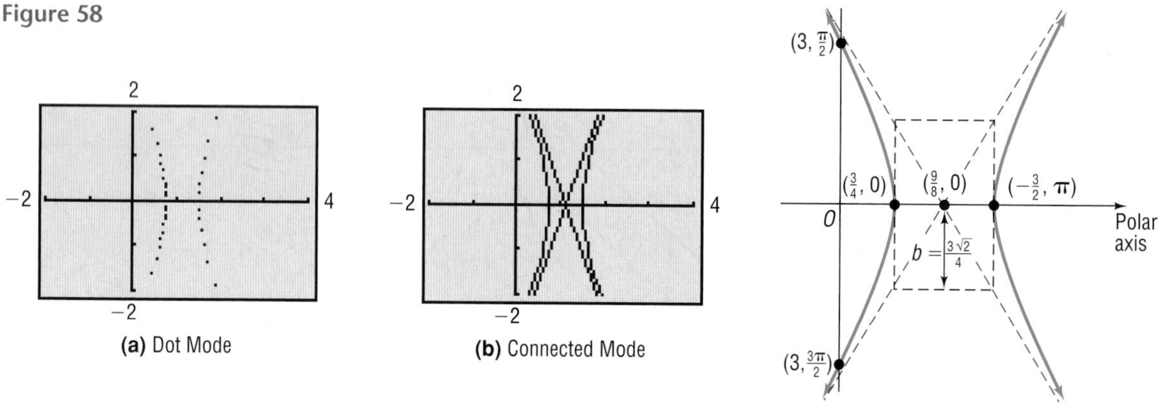

(a) Dot Mode (b) Connected Mode (c)

NOW WORK PROBLEM 11.

EXAMPLE 4 Converting a Polar Equation to a Rectangular Equation

2 Convert the polar equation

$$r = \frac{1}{3 - 3\cos\theta}$$

to a rectangular equation.

Solution The strategy here is first to rearrange the equation and square each side before using the transformation equations.

$$r = \frac{1}{3 - 3\cos\theta}$$

$$3r - 3r\cos\theta = 1$$

$$3r = 1 + 3r\cos\theta \quad \text{Rearrange the equation.}$$

$$9r^2 = (1 + 3r\cos\theta)^2 \quad \text{Square each side.}$$

$$9(x^2 + y^2) = (1 + 3x)^2 \quad \text{Use the transformation equations.}$$

$$9x^2 + 9y^2 = 9x^2 + 6x + 1$$

$$9y^2 = 6x + 1$$

This is the equation of a parabola in rectangular coordinates. ■

NOW WORK PROBLEM 19.

10.6 EXERCISES

In Problems 1–6, identify the conic that each polar equation represents. Also, give the position of the directrix.

1. $r = \dfrac{1}{1 + \cos\theta}$

2. $r = \dfrac{3}{1 - \sin\theta}$

3. $r = \dfrac{4}{2 - 3\sin\theta}$

4. $r = \dfrac{2}{1 + 2\cos\theta}$

5. $r = \dfrac{3}{4 - 2\cos\theta}$

6. $r = \dfrac{6}{8 + 2\sin\theta}$

In Problems 7–18, graph each equation using a graphing utility. Discuss each equation and graph it by hand.

7. $r = \dfrac{1}{1 + \cos\theta}$

8. $r = \dfrac{3}{1 - \sin\theta}$

9. $r = \dfrac{8}{4 + 3\sin\theta}$

10. $r = \dfrac{10}{5 + 4\cos\theta}$

11. $r = \dfrac{9}{3 - 6\cos\theta}$

12. $r = \dfrac{12}{4 + 8\sin\theta}$

13. $r = \dfrac{8}{2 - \sin\theta}$

14. $r = \dfrac{8}{2 + 4\cos\theta}$

15. $r(3 - 2\sin\theta) = 6$

16. $r(2 - \cos\theta) = 2$

17. $r = \dfrac{6\sec\theta}{2\sec\theta - 1}$

18. $r = \dfrac{3\csc\theta}{\csc\theta - 1}$

In Problems 19–30, convert each polar equation to a rectangular equation.

19. $r = \dfrac{1}{1 + \cos\theta}$

20. $r = \dfrac{3}{1 - \sin\theta}$

21. $r = \dfrac{8}{4 + 3\sin\theta}$

22. $r = \dfrac{10}{5 + 4\cos\theta}$

23. $r = \dfrac{9}{3 - 6\cos\theta}$

24. $r = \dfrac{12}{4 + 8\sin\theta}$

25. $r = \dfrac{8}{2 - \sin\theta}$

26. $r = \dfrac{8}{2 + 4\cos\theta}$

27. $r(3 - 2\sin\theta) = 6$

28. $r(2 - \cos\theta) = 2$

29. $r = \dfrac{6\sec\theta}{2\sec\theta - 1}$

30. $r = \dfrac{3\csc\theta}{\csc\theta - 1}$

In Problems 31–36, find a polar equation for each conic. For each, a focus is at the pole.

31. $e = 1$; directrix is parallel to the polar axis 1 unit above the pole

32. $e = 1$; directrix is parallel to the polar axis 2 units below the pole

33. $e = \frac{4}{5}$; directrix is perpendicular to the polar axis 3 units to the left of the pole

34. $e = \frac{2}{3}$; directrix is parallel to the polar axis 3 units above the pole

35. $e = 6$; directrix is parallel to the polar axis 2 units below the pole

36. $e = 5$; directrix is perpendicular to the polar axis 5 units to the right of the pole

37. Derive equation (b) in Table 5:

$$r = \frac{ep}{1 + e\cos\theta}$$

38. Derive equation (c) in Table 5:

$$r = \frac{ep}{1 + e\sin\theta}$$

39. Derive equation (d) in Table 5:

$$r = \frac{ep}{1 - e\sin\theta}$$

40. Orbit of Mercury The planet Mercury travels around the Sun in an elliptical orbit given approximately by

$$r = \frac{(3.442)10^7}{1 - 0.206\cos\theta}$$

where r is measured in miles and the Sun is at the pole. Find the distance from Mercury to the Sun at *aphelion* (greatest distance from the Sun) and at *perihelion* (shortest distance from the Sun). See the figure. Use the aphelion and perihelion to graph the orbit of Mercury using a graphing utility.

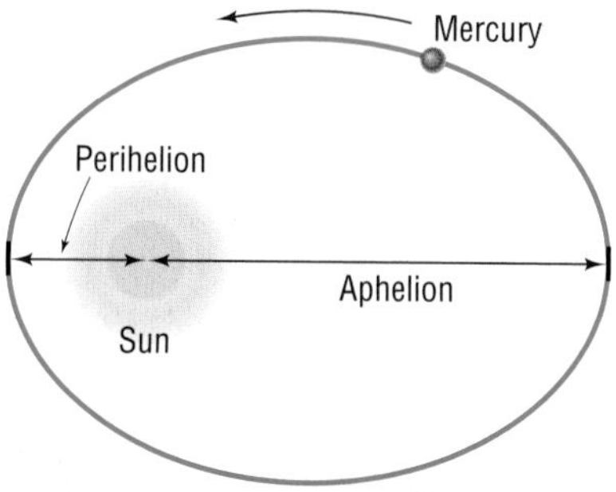

PREPARING FOR THIS SECTION

Before getting started, review the following:

✓ Amplitude and Period of Sinusoidal Graphs (p. 421)

10.7 PLANE CURVES AND PARAMETRIC EQUATIONS

OBJECTIVES
1. Graph Parametric Equations by Hand
2. Graph Parametric Equations Using a Graphing Utility
3. Find a Rectangular Equation for a Curve Defined Parametrically
4. Use Time as a Parameter in Parametric Equations
5. Find Parametric Equations for Curves Defined by Rectangular Equations

Equations of the form $y = f(x)$, where f is a function, have graphs that are intersected no more than once by any vertical line. The graphs of many of the conics and certain other, more complicated graphs do not have this characteristic. Yet each graph, like the graph of a function, is a collection of points (x, y) in the xy-plane; that is, each is a *plane curve.* In this section, we discuss another way of representing such graphs.

Let $x = f(t)$ and $y = g(t)$, where f and g are two functions whose common domain is some interval I. The collection of points defined by

$$(x, y) = (f(t), g(t))$$

is called a **plane curve.** The equations

$$x = f(t) \qquad y = g(t)$$

where t is in I, are called **parametric equations** of the curve. The variable t is called a **parameter.**

1 Parametric equations are particularly useful in describing movement along a curve. Suppose that a curve is defined by the parametric equations

Figure 59

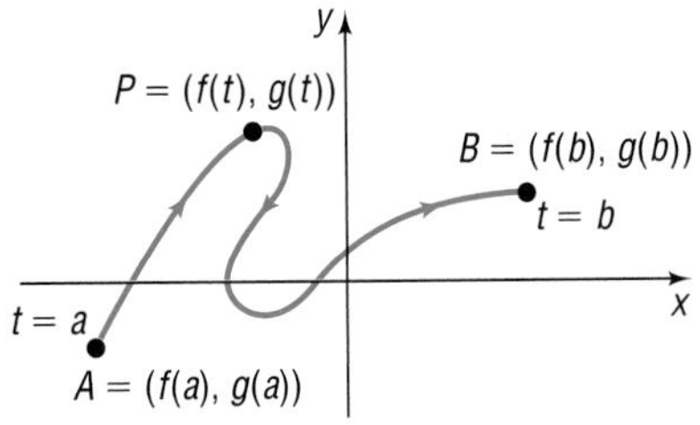

$$x = f(t), \quad y = g(t), \quad a \le t \le b$$

where f and g are each defined over the interval $a \le t \le b$. For a given value of t, we can find the value of $x = f(t)$ and $y = g(t)$, obtaining a point (x, y) on the curve. In fact, as t varies over the interval from $t = a$ to $t = b$, successive values of t give rise to a directed movement along the curve; that is, the curve is traced out in a certain direction by the corresponding succession of points (x, y). See Figure 59. The arrows show the direction, or **orientation,** along the curve as t varies from a to b.

EXAMPLE 1 Discussing a Curve Defined by Parametric Equations

Discuss the curve defined by the parametric equations

$$x = 3t^2, \quad y = 2t, \quad -2 \le t \le 2 \tag{1}$$

Solution For each number t, $-2 \le t \le 2$, there corresponds a number x and a number y. For example, when $t = -2$, then $x = 12$ and $y = -4$. When $t = 0$, then $x = 0$ and $y = 0$. Indeed, we can set up a table listing various choices of the parameter t and the corresponding values for x and y, as shown in Table 6. Plotting these points and connecting them with a smooth curve leads to Figure 60. The arrows in Figure 60 are used to indicate the orientation.

TABLE 6

t	x	y	(x, y)
−2	12	−4	(12, −4)
−1	3	−2	(3, −2)
0	0	0	(0, 0)
1	3	2	(3, 2)
2	12	4	(12, 4)

Figure 60

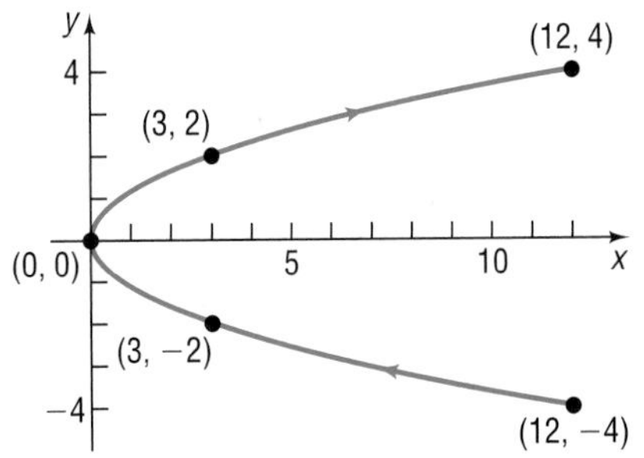

2 Most graphing utilities have the capability of graphing parametric equations. The following steps are usually required in order to obtain the graph of parametric equations. Check your owner's manual to see how yours works.

GRAPHING PARAMETRIC EQUATIONS USING A GRAPHING UTILITY

STEP 1: Set the mode to PARametric. Enter $x(t)$ and $y(t)$.

STEP 2: Select the viewing window. In addition to setting Xmin, Xmax, Xscl, and so on, the viewing window in parametric mode requires setting minimum and maximum values for the parameter t and an increment setting for t (Tstep).

STEP 3: Execute.

EXAMPLE 2 Graphing a Curve Defined by Parametric Equations Using a Graphing Utility

Graph the curve defined by the parametric equations

$$x = 3t^2, \qquad y = 2t, \qquad -2 \le t \le 2$$

Solution **STEP 1:** Enter the equations $x(t) = 3t^2$, $y(t) = 2t$ with the graphing utility in PARametric mode.

STEP 2: Select the viewing window. The interval I is $-2 \le t \le 2$, so we select the following square viewing window:

$$T\text{min} = -2 \qquad X\text{min} = 0 \qquad Y\text{min} = -5$$

$$T\text{max} = 2 \qquad X\text{max} = 15 \qquad Y\text{max} = 5$$

$$T\text{step} = 0.1 \qquad X\text{scl} = 1 \qquad Y\text{scl} = 1$$

We choose $T\text{min} = -2$ and $T\text{max} = 2$ because $-2 \le t \le 2$. Finally, the choice for Tstep will determine the number of points that the graphing utility will plot. For example, with Tstep at 0.1, the graphing utility will evaluate x and y at $t = -2, -1.9, -1.8$, and so on. The smaller the Tstep, the more points the graphing utility will plot. The reader is encouraged to experiment with different values of Tstep to see how the graph is affected.

Figure 61

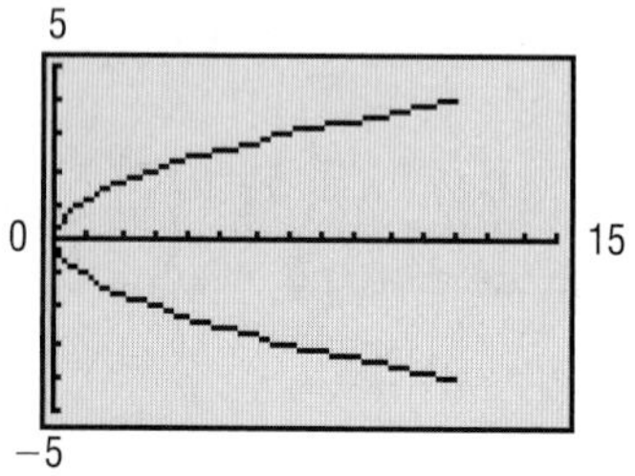

STEP 3: Execute. Notice the direction that the graph is drawn. This direction shows the orientation of the curve.

The graph shown in Figure 61 is complete. ■

EXPLORATION Graph the following parametric equations using a graphing utility with $X\text{min} = 0$, $X\text{max} = 15$, $Y\text{min} = -5$, $Y\text{max} = 5$, and $T\text{step} = 0.1$.

1. $x = \dfrac{3t^2}{4}, \quad y = t, \quad -4 \le t \le 4$
2. $x = 3t^2 + 12t + 12, \quad y = 2t + 4, \quad -4 \le t \le 0$
3. $x = 3t^{2/3}, \quad y = 2\sqrt[3]{t}, \quad -8 \le t \le 8$

Compare these graphs to the graph in Figure 61. Conclude that parametric equations defining a curve are not unique; that is, different parametric equations can represent the same graph. ■

3 The curve given in Examples 1 and 2 should be familiar. To identify it accurately, we find the corresponding rectangular equation by eliminating the parameter t from the parametric equations (1) given in Example 1:

$$x = 3t^2, \qquad y = 2t, \qquad -2 \le t \le 2$$

Noting that we can readily solve for t in $y = 2t$, obtaining $t = y/2$, we substitute this expression in the other equation.

$$x = 3t^2 = 3\left(\frac{y}{2}\right)^2 = \frac{3y^2}{4}$$
$$\uparrow\; t = \frac{y}{2}$$

This equation, $x = 3y^2/4$, is the equation of a parabola with vertex at (0, 0) and axis of symmetry along the x-axis.

EXPLORATION In FUNCtion mode graph $x = \frac{3y^2}{4}\left(Y_1 = \sqrt{\frac{4x}{3}}\right.$ and $\left.Y_2 = -\sqrt{\frac{4x}{3}}\right)$ with $X\text{min} = 0$, $X\text{max} = 15$, $Y\text{min} = -5$, $Y\text{max} = 5$. Compare this graph with Figure 61. Why do the graphs differ? ■

Note that the parameterized curve defined by equation (1) and shown in Figure 60 (or 61) is only a part of the parabola $x = 3y^2/4$. The graph of the rectangular equation obtained by eliminating the parameter will, in general, contain more points than the original parameterized curve. Care must therefore be taken when a parameterized curve is sketched by hand after eliminating the parameter. Even so, the process of eliminating the parameter t of a parameterized curve in order to identify it accurately is sometimes a better approach than merely plotting points. However, the elimination process sometimes requires a little ingenuity.

EXAMPLE 3 Finding the Rectangular Equation of a Curve Defined Parametrically

Find the rectangular equation of the curve whose parametric equations are

$$x = a\cos t \qquad y = a\sin t$$

where $a > 0$ is a constant. Graph this curve, indicating its orientation.

Solution The presence of sines and cosines in the parametric equations suggests that we use a Pythagorean identity. In fact, since

$$\cos t = \frac{x}{a} \qquad \sin t = \frac{y}{a}$$

we find that

$$\cos^2 t + \sin^2 t = 1$$

$$\left(\frac{x}{a}\right)^2 + \left(\frac{y}{a}\right)^2 = 1$$

$$x^2 + y^2 = a^2$$

Figure 62

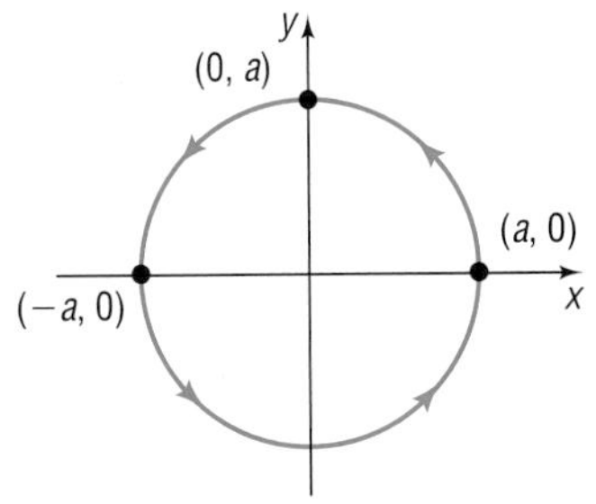

The curve is a circle with center at $(0, 0)$ and radius a. As the parameter t increases, say from $t = 0$ [the point $(a, 0)$] to $t = \pi/2$ [the point $(0, a)$] to $t = \pi$ [the point $(-a, 0)$], we see that the corresponding points are traced in a counterclockwise direction around the circle. Hence, the orientation is as indicated in Figure 62. ■

NOW WORK PROBLEMS 1 AND 13.

Let's discuss the curve in Example 3 further. The domain of each parametric equation is $-\infty < t < \infty$. Thus, the graph in Figure 62 is actually being repeated each time that t increases by 2π.

If we wanted the curve to consist of exactly 1 revolution in the counterclockwise direction, we could write

$$x = a\cos t, \quad y = a\sin t, \qquad 0 \le t \le 2\pi$$

This curve starts at $t = 0$ [the point $(a, 0)$] and, proceeding counterclockwise around the circle, ends at $t = 2\pi$ [also the point $(a, 0)$].

If we wanted the curve to consist of exactly three revolutions in the counterclockwise direction, we could write

$$x = a\cos t, \quad y = a\sin t, \quad -2\pi \le t \le 4\pi$$

or

$$x = a\cos t, \quad y = a\sin t, \quad 0 \le t \le 6\pi$$

or

$$x = a\cos t, \quad y = a\sin t, \quad 2\pi \le t \le 8\pi$$

EXAMPLE 4 Describing Parametric Equations

Find rectangular equations for and graph the following curves defined by parametric equations.

(a) $x = a\cos t, \quad y = a\sin t, \quad 0 \le t \le \pi, \quad a > 0$

(b) $x = -a\sin t, \quad y = -a\cos t, \quad 0 \le t \le \pi, \quad a > 0$

Solution (a) We eliminate the parameter t using a Pythagorean identity.

$$\left(\frac{x}{a}\right)^2 + \left(\frac{y}{a}\right)^2 = \cos^2 t + \sin^2 t = 1$$

Figure 63

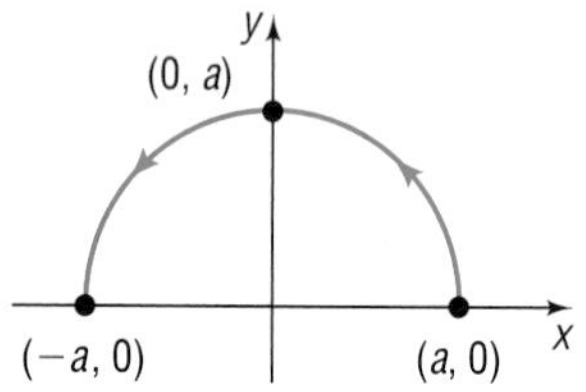

The curve defined by these parametric equations is a circle, with radius a and center at $(0, 0)$. The circle begins at the point $(a, 0)$, $t = 0$; passes through the point $(0, a)$, $t = \pi/2$; and ends at the point $(-a, 0)$, $t = \pi$. The parametric equations define an upper semicircle of radius a with a counterclockwise orientation. See Figure 63. The rectangular equation is

$$y = a\sqrt{1 - (x/a)^2}, \qquad -a \le x \le a$$

(b) We eliminate the parameter t using a Pythagorean identity.

$$\left(\frac{x}{-a}\right)^2 + \left(\frac{y}{-a}\right)^2 = \sin^2 t + \cos^2 t = 1$$

Figure 64

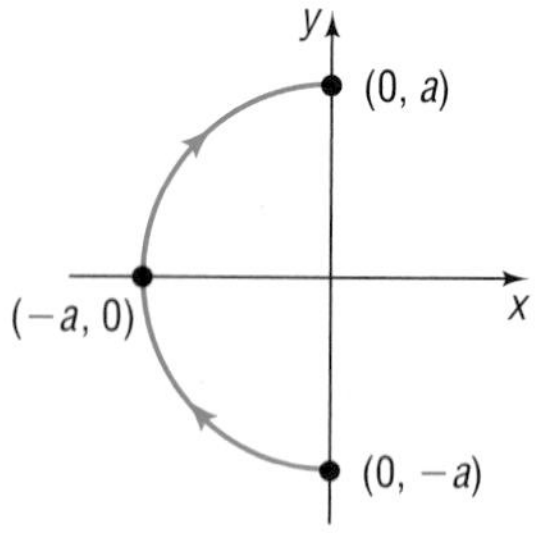

The curve defined by these parametric equations is a circle, with radius a and center at $(0, 0)$. The circle begins at the point $(0, -a)$, $t = 0$; passes through the point $(-a, 0)$, $t = \pi/2$; and ends at the point $(0, a)$, $t = \pi$. The parametric equations define a left semicircle of radius a with a clockwise orientation. See Figure 64. The rectangular equation is

$$x = -a\sqrt{1 - (y/a)^2}, \qquad -a \le y \le a$$

Example 3 illustrates the versatility of parametric equations for replacing complicated rectangular equations, while providing additional information about orientation. These characteristics make parametric equations very useful in applications, such as projectile motion.

SEEING THE CONCEPT (a) Graph $x = \cos t$, $y = \sin t$ for $0 \le t \le 2\pi$. Compare to Figure 62. Graph $x = \cos t$, $y = \sin t$ for $0 \le t \le \pi$. Compare to Figure 63. Graph $x = -\sin t$, $y = -\cos t$ for $0 \le t \le \pi$. Compare to Figure 64.

TIME AS A PARAMETER: PROJECTILE MOTION; SIMULATED MOTION

4 If we think of the parameter t as time, then the parametric equations $x = f(t)$ and $y = g(t)$ of a curve C specify how the x- and y-coordinates of a moving point vary with time.

For example, we can use parametric equations to describe the motion of an object, sometimes referred to as **curvilinear motion.** Using parametric equations, we can specify not only where the object travels, that is, its location (x, y) but also when it gets there, that is, the time t.

When an object is propelled upward at an inclination θ to the horizontal with initial speed v_0, the resulting motion is called **projectile motion.** See Figure 65(a).

In calculus it is shown that the parametric equations of the path of a projectile fired at an inclination θ to the horizontal, with an initial speed v_0, from a height h above the horizontal are

$$x = (v_0 \cos\theta)t \qquad y = -\frac{1}{2}gt^2 + (v_0 \sin\theta)t + h \qquad \textbf{(2)}$$

where t is the time and g is the constant acceleration due to gravity (approximately 32 ft/sec/sec or 9.8 m/sec/sec). See Figure 65(b).

Figure 65

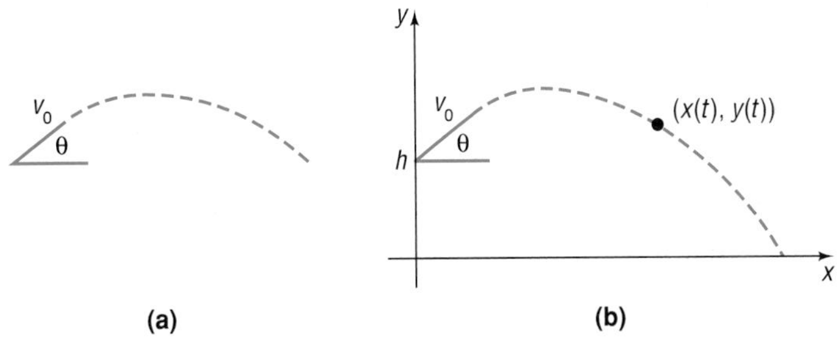

EXAMPLE 5 Projectile Motion

Figure 66

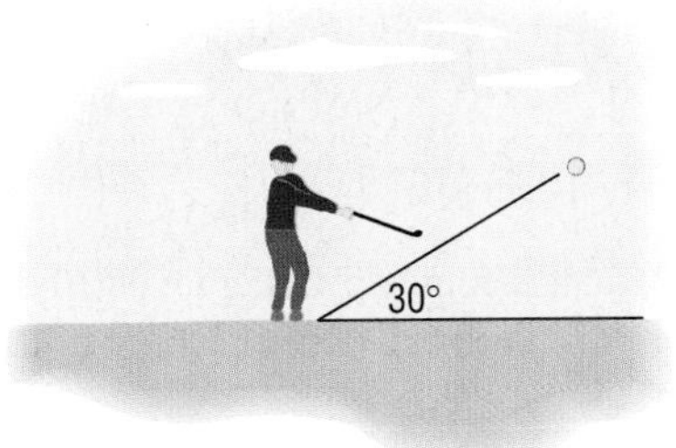

Suppose that Jim hit a golf ball with an initial velocity of 150 feet per second at an angle of 30° to the horizontal. See Figure 66.

(a) Find parametric equations that describe the position of the ball as a function of time.

(b) How long is the golf ball in the air?

(c) When is the ball at its maximum height? Determine the maximum height of the ball.

(d) Determine the distance that the ball traveled.

(e) Using a graphing utility, simulate the motion of the golf ball by simultaneously graphing the equations found in part (a).

Solution (a) We have $v_0 = 150$, $\theta = 30°$, $h = 0$ (the ball is on the ground), and $g = 32$ (since units are in feet and seconds). Substituting these values into equations (2), we find that

$$x = (v_0 \cos\theta)t = (150 \cos 30°)t = 75\sqrt{3}t$$

$$y = -\frac{1}{2}gt^2 + (v_0 \sin\theta)t + h = -\frac{1}{2}(32)t^2 + (150 \sin 30°)t + 0$$

$$= -16t^2 + 75t$$

(b) To determine the length of time that the ball is in the air, we solve the equation $y = 0$.

$$-16t^2 + 75t = 0$$
$$t(-16t + 75) = 0$$
$$t = 0 \text{ sec} \quad \text{or} \quad t = \frac{75}{16} = 4.6875 \text{ sec}$$

The ball will strike the ground after 4.6875 seconds.

(c) Notice that the height y of the ball is a quadratic function of t, so the maximum height of the ball can be found by determining the vertex of $y = -16t^2 + 75t$. The value of t at the vertex is

$$t = \frac{-b}{2a} = \frac{-75}{-32} = 2.34375 \text{ sec}$$

The ball is at its maximum height after 2.34375 seconds. The maximum height of the ball is found by evaluating the function y at $t = 2.34375$ seconds.

$$\text{Maximum height} = -16(2.34375)^2 + (75)2.34375 \approx 87.89 \text{ feet}$$

(d) Since the ball is in the air for 4.6875 seconds, the horizontal distance that the ball travels is

$$x = (75\sqrt{3})4.6875 \approx 608.92 \text{ feet}$$

(e) We enter the equations from part (a) into a graphing utility with $T\text{min} = 0$, $T\text{max} = 4.7$, and $T\text{step} = 0.1$. We use ZOOM-SQUARE to avoid any distortion to the angle of elevation. See Figure 67. ■

Figure 67

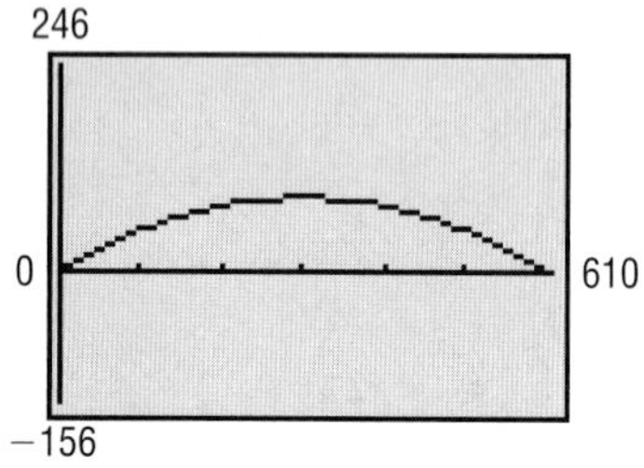

Exploration Simulate the motion of a ball thrown straight up with an initial speed of 100 feet per second from a height of 5 feet above the ground. Use PARametric mode with $T\text{min} = 0$, $T\text{max} = 6.5$, $T\text{step} = 0.1$, $X\text{min} = 0$, $X\text{max} = 5$, $Y\text{min} = 0$, and $Y\text{max} = 180$. What happens to the speed with which the graph is drawn as the ball goes up and then comes back down? How do you interpret this physically? Repeat the experiment using other values for Tstep. How does this affect the experiment?
[**Hint:** In the projectile motion equations, let $\theta = 90°$, $v_0 = 100$, $h = 5$, and $g = 32$. We use $x = 3$ instead of $x = 0$ to see the vertical motion better.]

Result See Figure 68. In Figure 68(a) the ball is going up. In Figure 68(b) the ball is near its highest point. Finally, in Figure 68(c) the ball is coming back down.

Figure 68

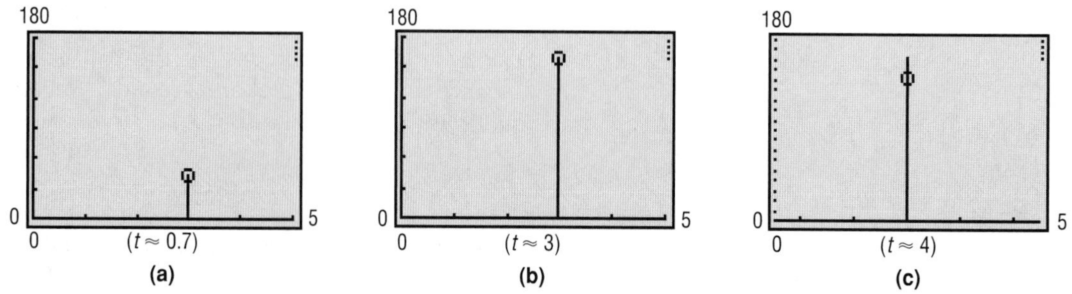

Notice that, as the ball goes up, its speed decreases, until at the highest point it is zero. Then the speed increases as the ball comes back down. ∎

NOW WORK PROBLEM 27.

A graphing utility can be used to simulate other kinds of motion as well. Let's work again Example 5 from Section 1.4.

EXAMPLE 6 Simulating Motion

Tanya, who is a long distance runner, runs at an average velocity of 8 miles per hour. Two hours after Tanya leaves your house, you leave in your Honda and follow the same route. If your average velocity is 40 miles per hour, how long will it be before you catch up to Tanya? See Figure 69. Use a simulation of the two motions to verify the answer.

Figure 69

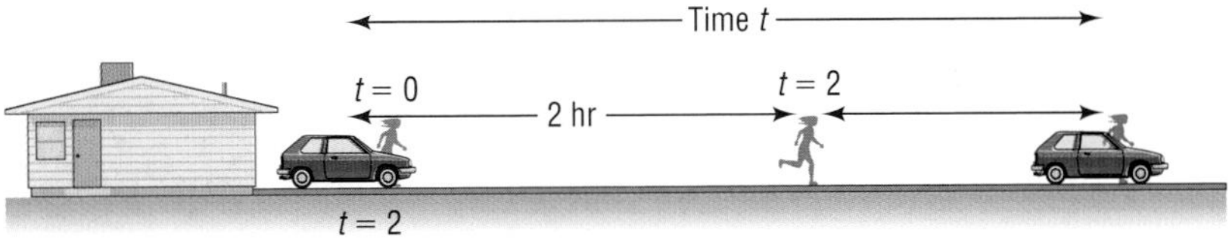

Solution We begin with two sets of parametric equations: one to describe Tanya's motion, the other to describe the motion of the Honda. We choose time $t = 0$ to be when Tanya leaves the house. If we choose $y_1 = 2$ as Tanya's path, then we can use $y_2 = 4$ as the parallel path of the Honda. The horizontal distances traversed in time t (Distance = Velocity × Time) are

$$\text{Tanya: } x_1 = 8t \qquad \text{Honda: } x_2 = 40(t - 2)$$

The Honda catches up to Tanya when $x_1 = x_2$.

$$8t = 40(t - 2)$$

$$8t = 40t - 80$$

$$-32t = -80$$

$$t = \frac{-80}{-32} = 2.5$$

The Honda catches up to Tanya 2.5 hours after Tanya leaves the house.

In PARametric mode with Tstep $= 0.01$, we simultaneously graph

$$\text{Tanya: } x_1 = 8t \qquad \text{Honda: } x_2 = 40(t - 2)$$

$$y_1 = 2 \qquad y_2 = 4$$

for $0 \le t \le 3$.

Figure 70 shows the relative position of Tanya and the Honda for $t = 0$, $t = 2$, $t = 2.25$, $t = 2.5$, and $t = 2.75$.

Figure 70

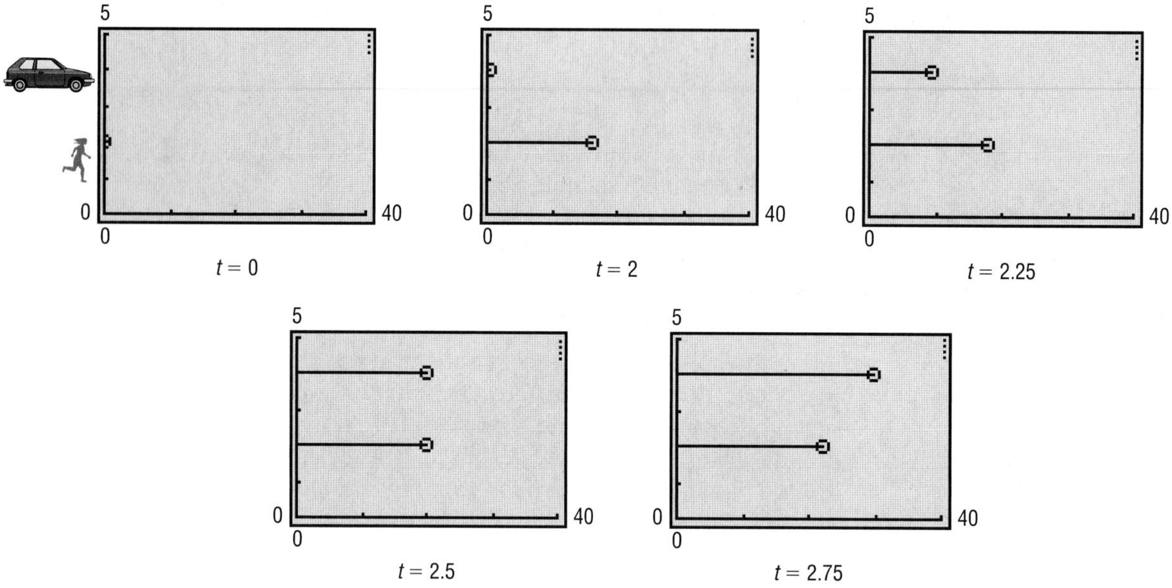

Finding Parametric Equations

We now take up the question of how to find parametric equations of a given curve.

5 If a curve is defined by the equation $y = f(x)$, where f is a function, one way of finding parametric equations is to let $x = t$. Then $y = f(t)$ and

$$x = t, \quad y = f(t), \qquad t \text{ in the domain of } f$$

are parametric equations of the curve.

EXAMPLE 7 Finding Parametric Equations for a Curve Defined by a Rectangular Equation

Find parametric equations for the equation $y = x^2 - 4$.

Solution Let $x = t$. Then the parametric equations are

$$x = t, \quad y = t^2 - 4, \qquad -\infty < t < \infty$$

Another less obvious approach to Example 7 is to let $x = t^3$. Then the parametric equations become

$$x = t^3, \quad y = t^6 - 4, \qquad -\infty < t < \infty$$

Care must be taken when using this approach, since the substitution for x must be a function that allows x to take on all the values stipulated by the domain of f. For example, letting $x = t^2$ so that $y = t^4 - 4$ does not result in equivalent parametric equations for $y = x^2 - 4$, since only points for which $x \geq 0$ are obtained.

EXAMPLE 8 Finding Parametric Equations for an Object in Motion

Find parametric equations for the ellipse

$$x^2 + \frac{y^2}{9} = 1$$

where the parameter t is time (in seconds) and

(a) The motion around the ellipse is clockwise, begins at the point $(0, 3)$, and requires 1 second for a complete revolution.

(b) The motion around the ellipse is counterclockwise, begins at the point $(1, 0)$, and requires 2 seconds for a complete revolution.

Figure 71

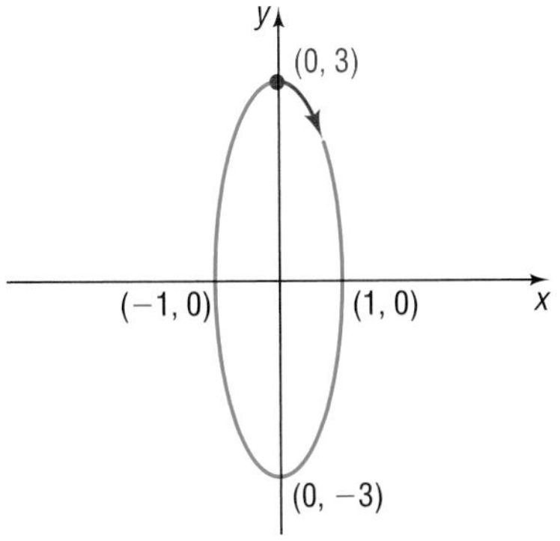

Solution (a) See Figure 71. Since the motion begins at the point $(0, 3)$, we want $x = 0$ and $y = 3$ when $t = 0$. Furthermore, since the given equation is an ellipse, we begin by letting

$$x = \sin(\omega t) \qquad \frac{y}{3} = \cos(\omega t)$$

for some constant ω. These parametric equations satisfy the equation of the ellipse. Furthermore, with this choice, when $t = 0$, we have $x = 0$ and $y = 3$.

For the motion to be clockwise, the motion will have to begin with the value of x increasing and y decreasing as t increases. This requires that $\omega > 0$. [Do you know why? If $\omega > 0$, then $x = \sin(\omega t)$ is increasing when $t > 0$ is near zero and $y = 3\cos(\omega t)$ is decreasing when $t > 0$ is near zero]. See the red part of the graph in Figure 71.

Finally, since 1 revolution requires 1 second, the period $2\pi/\omega = 1$, so $\omega = 2\pi$. Parametric equations that satisfy the conditions stipulated are

$$x = \sin(2\pi t), \quad y = 3\cos(2\pi t), \qquad 0 \le t \le 1 \tag{3}$$

Figure 72

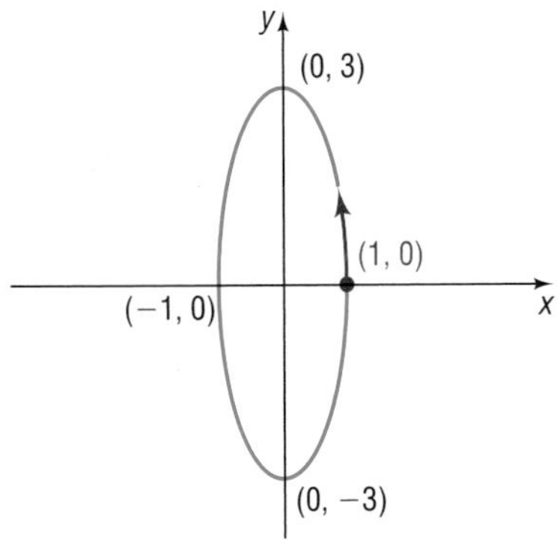

(b) See Figure 72. Since the motion begins at the point $(1, 0)$, we want $x = 1$ and $y = 0$ when $t = 0$. Furthermore, since the given equation is an ellipse, we begin by letting

$$x = \cos(\omega t) \qquad \frac{y}{3} = \sin(\omega t)$$

for some constant ω. These parametric equations satisfy the equation of the ellipse. Furthermore, with this choice, when $t = 0$, we have $x = 1$ and $y = 0$.

For the motion to be counterclockwise, the motion will have to begin with the value of x decreasing and y increasing as t increases. This requires that $\omega > 0$. [Do you know why?] Finally, since 1 revolution requires 2 seconds, the period is $2\pi/\omega = 2$, so $\omega = \pi$. The parametric equations that satisfy the conditions stipulated are

$$x = \cos(\pi t), \quad y = 3\sin(\pi t), \qquad 0 \le t \le 2 \tag{4}$$

■

Either of equations (3) or (4) can serve as parametric equations for the ellipse $x^2 + y^2/9 = 1$ given in Example 8. The direction of the motion, the beginning point, and the time for 1 revolution merely serve to help us arrive at a particular parametric representation.

NOW WORK PROBLEM 43.

The Cycloid

Suppose that a circle of radius a rolls along a horizontal line without slipping. As the circle rolls along the line, a point P on the circle will trace out a curve called a **cycloid** (see Figure 73). We now seek parametric equations* for a cycloid.

Figure 73

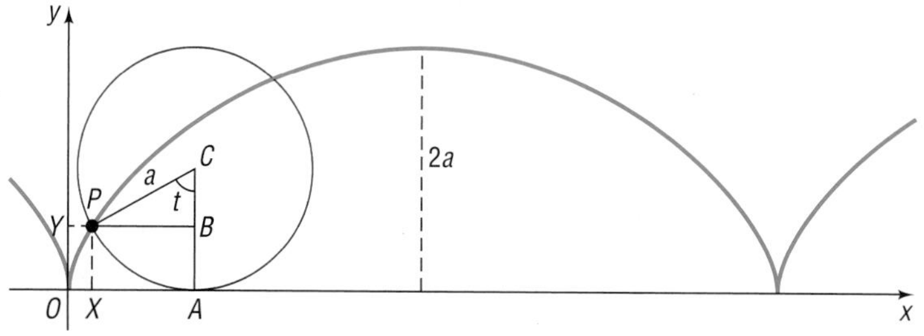

We begin with a circle of radius a and take the fixed line on which the circle rolls as the x-axis. Let the origin be one of the points at which the point P comes in contact with the x-axis. Figure 73 illustrates the position of this point P after the circle has rolled somewhat. The angle t (in radians) measures the angle through which the circle has rolled.

Since we require no slippage, it follows that

$$\text{Arc } AP = d(O, A)$$

Therefore,

$$at = d(O, A) \qquad s = r\theta$$

The x-coordinate of the point P is

$$d(O, X) = d(O, A) - d(X, A) = at - a\sin t = a(t - \sin t)$$

The y-coordinate of the point P is equal to

$$d(O, Y) = d(A, C) - d(B, C) = a - a\cos t = a(1 - \cos t)$$

The parametric equations of the cycloid are

$$x = a(t - \sin t) \qquad y = a(1 - \cos t) \qquad \textbf{(5)}$$

Exploration Graph $x = t - \sin t$, $y = 1 - \cos t$, $0 \le t \le 3\pi$, using your graphing utility with Tstep $= \pi/36$ and a square screen. Compare your results with Figure 73.

Applications to Mechanics

If a is negative in equation (5), we obtain an inverted cycloid, as shown in Figure 74(a). The inverted cycloid occurs as a result of some remarkable applications in the field of mechanics. We shall mention two of them: the *brachistochrone* and the *tautochrone*.†

* Any attempt to derive the rectangular equation of a cycloid would soon demonstrate how complicated the task is.

† In Greek, *brachistochrone* means "the shortest time" and *tautochrone* means "equal time."

Figure 74

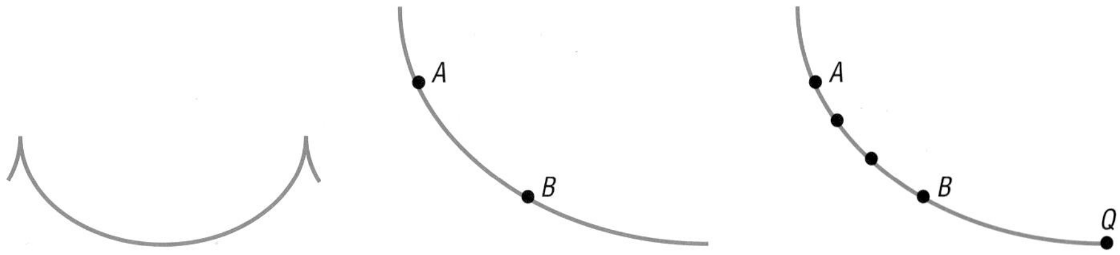

(a) Inverted cycloid (b) Curve of quickest descent (c) All reach Q at the same time

The **brachistochrone** is the curve of quickest descent. If a particle is constrained to follow some path from one point A to a lower point B (not on the same vertical line) and is acted on only by gravity, the time needed to make the descent is least if the path is an inverted cycloid. See Figure 74(b). This remarkable discovery, which is attributed to many famous mathematicians (including Johann Bernoulli and Blaise Pascal), was a significant step in creating the branch of mathematics known as the *calculus of variations.*

Figure 75

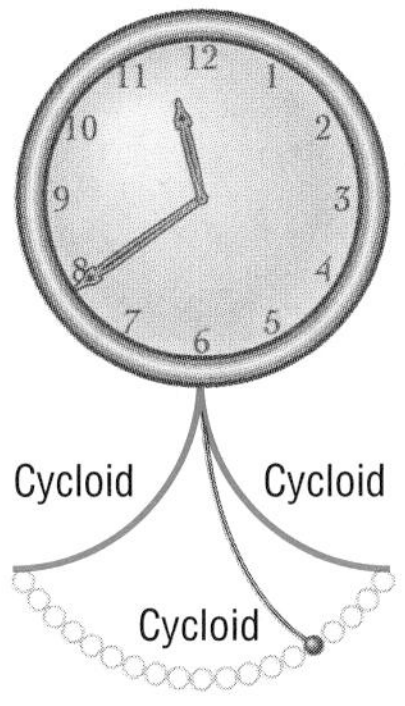

To define the **tautochrone,** let Q be the lowest point on an inverted cycloid. If several particles placed at various positions on an inverted cycloid simultaneously begin to slide down the cycloid, they will reach the point Q at the same time, as indicated in Figure 74(c). The tautochrone property of the cycloid was used by Christiaan Huygens (1629–1695), the Dutch mathematician, physicist, and astronomer, to construct a pendulum clock with a bob that swings along a cycloid (see Figure 75). In Huygen's clock, the bob was made to swing along a cycloid by suspending the bob on a thin wire constrained by two plates shaped like cycloids. In a clock of this design, the period of the pendulum is independent of its amplitude.

10.7 EXERCISES

In Problems 1–20, graph the curve whose parametric equations are given using a graphing utility. Find the rectangular equation of each curve. Graph the curve by hand and show its orientation.

1. $x = 3t + 2, \quad y = t + 1; \quad 0 \le t \le 4$

2. $x = t - 3, \quad y = 2t + 4; \quad 0 \le t \le 2$

3. $x = t + 2, \quad y = \sqrt{t}; \quad t \ge 0$

4. $x = \sqrt{2t}, \quad y = 4t; \quad t \ge 0$

5. $x = t^2 + 4, \quad y = t^2 - 4; \quad -\infty < t < \infty$

6. $x = \sqrt{t} + 4, \quad y = \sqrt{t} - 4; \quad t \ge 0$

7. $x = 3t^2, \quad y = t + 1; \quad -\infty < t < \infty$

8. $x = 2t - 4, \quad y = 4t^2; \quad -\infty < t < \infty$

9. $x = 2e^t, \quad y = 1 + e^t; \quad t \ge 0$

10. $x = e^t, \quad y = e^{-t}; \quad t \ge 0$

11. $x = \sqrt{t}, \quad y = t^{3/2}; \quad t \ge 0$

12. $x = t^{3/2} + 1, \quad y = \sqrt{t}; \quad t \ge 0$

13. $x = 2\cos t, \quad y = 3\sin t; \quad 0 \le t \le 2\pi$

14. $x = 2\cos t, \quad y = 3\sin t; \quad 0 \le t \le \pi$

15. $x = 2\cos t, \quad y = 3\sin t; \quad -\pi \le t \le 0$

16. $x = 2\cos t, \quad y = \sin t; \quad 0 \le t \le \pi/2$

17. $x = \sec t, \quad y = \tan t; \quad 0 \le t \le \pi/4$

18. $x = \csc t, \quad y = \cot t; \quad \pi/4 \le t \le \pi/2$

19. $x = \sin^2 t, \quad y = \cos^2 t; \quad 0 \le t \le 2\pi$

20. $x = t^2, \quad y = \ln t; \quad t > 0$

21. Projectile Motion Bob throws a ball straight up with an initial speed of 50 feet per second from a height of 6 feet.
(a) Find parametric equations that describe the motion of the ball as a function of time.
(b) How long is the ball in the air?
(c) When is the ball at its maximum height? Determine the maximum height of the ball.
(d) Simulate the motion of the ball by graphing the equations found in part (a).

22. Projectile Motion Alice throws a ball straight up with an initial speed of 40 feet per second from a height of 5 feet.
(a) Find parametric equations that describe the motion of the ball as a function of time.
(b) How long is the ball in the air?
(c) When is the ball at its maximum height? Determine the maximum height of the ball.
(d) Simulate the motion of the ball by graphing the equations found in part (a).

23. Catching a Train Bill's train leaves at 8:06 AM and accelerates at the rate of 2 meters per second per second. Bill, who can run 5 meters per second, arrives at the train station 5 seconds after the train has left.
(a) Find parametric equations that describe the motion of the train and Bill as a function of time.
(b) Determine algebraically whether Bill will catch the train. If so, when?
(c) Simulate the motion of the train and Bill by simultaneously graphing the equations found in part (a).

24. Catching a Bus Jodi's bus leaves at 5:30 PM and accelerates at the rate of 3 meters per second per second. Jodi, who can run 5 meters per second, arrives at the bus station 2 seconds after the bus has left.
(a) Find parametric equations that describe the motion of the bus and Jodi as a function of time.
(b) Determine algebraically whether Jodi will catch the bus. If so, when?
(c) Simulate the motion of the bus and Jodi by simultaneously graphing the equations found in part (a).

25. Projectile Motion Nolan Ryan throws a baseball with an initial speed of 145 feet per second at an angle of 20° to the horizontal. The ball leaves Nolan Ryan's hand at a height of 5 feet.
(a) Find parametric equations that describe the position of the ball as a function of time.
(b) How long is the ball in the air?
(c) When is the ball at its maximum height? Determine the maximum height of the ball.
(d) Determine the distance that the ball traveled.
(e) Using a graphing utility, simultaneously graph the equations found in part (a).

26. Projectile Motion Mark McGuire hit a baseball with an initial speed of 180 feet per second at an angle of 40° to the horizontal. The ball was hit at a height of 3 feet off the ground.
(a) Find parametric equations that describe the position of the ball as a function of time.
(b) How long is the ball in the air?
(c) When is the ball at its maximum height? Determine the maximum height of the ball.
(d) Determine the distance that the ball traveled.
(e) Using a graphing utility, simultaneously graph the equations found in part (a).

27. Projectile Motion Suppose that Adam throws a tennis ball off a cliff 300 meters high with an initial speed of 40 meters per second at an angle of 45° to the horizontal.
(a) Find parametric equations that describe the position of the ball as a function of time.
(b) How long is the ball in the air?
(c) When is the ball at its maximum height? Determine the maximum height of the ball.
(d) Determine the distance that the ball traveled.
(e) Using a graphing utility, simultaneously graph the equations found in part (a).

28. Projectile Motion Suppose that Adam throws a tennis ball off a cliff 300 meters high with an initial speed of 40 meters per second at an angle of 45° to the horizontal on the Moon (gravity on the Moon is one-sixth of that on Earth).
(a) Find parametric equations that describe the position of the ball as a function of time.
(b) How long is the ball in the air?
(c) When is the ball at its maximum height? Determine the maximum height of the ball.
(d) Determine the distance that the ball traveled.
(e) Using a graphing utility, simultaneously graph the equations found in (a).

29. Uniform Motion A Toyota Paseo (traveling east at 40 mph) and Pontiac Bonneville (traveling north at 30 mph) are heading toward the same intersection. The Paseo is 5 miles from the intersection when the Bonneville is 4 miles from the intersection. See the figure.

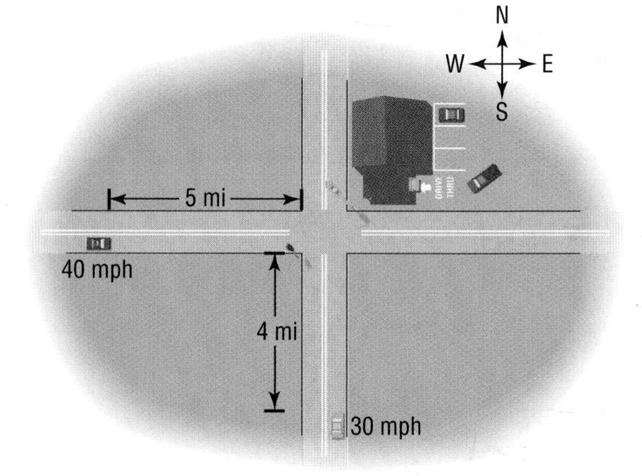

(a) Find a formula for the distance between the cars as a function of time.
(b) Graph the function in part (a) using a graphing utility.
(c) What is the minimum distance between the cars? When are the cars closest?
(d) Find parametric equations that describe the motion of the Paseo and Bonneville.
(e) Simulate the motion of the cars by simultaneously graphing the equations found in part (d).

30. Uniform Motion A Cessna (heading south at 120 mph) and a Boeing 747 (heading west at 600 mph) are flying toward each other at the same altitude. The Cessna is 100 miles from the point where the flight patterns intersect and the 747 is 550 miles from this intersection point. See the figure.

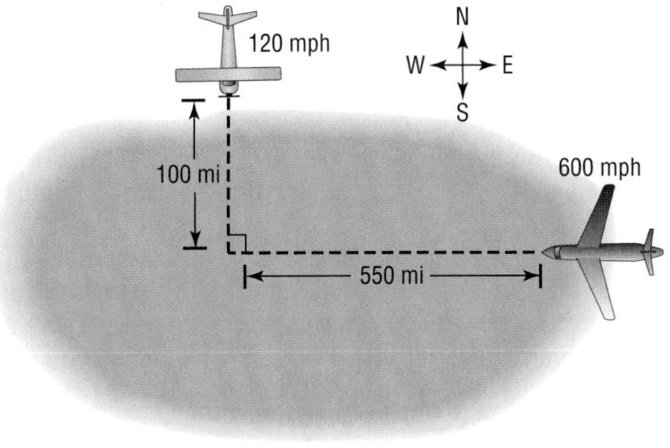

(a) Find a formula for the distance between the planes as a function of time.
(b) Graph the function in part (a) using a graphing utility.
(c) What is the minimum distance between the planes? When are the planes closest?
(d) Find parametric equations that describe the motion of the Cessna and 747.
(e) Simulate the motion of the planes by simultaneously graphing the equations found in part (d).

In Problems 31–38, find two different parametric equations for each rectangular equation.

31. $y = 4x - 1$ **32.** $y = -8x + 3$ **33.** $y = x^2 + 1$ **34.** $y = -2x^2 + 1$

35. $y = x^3$ **36.** $y = x^4 + 1$ **37.** $x = y^{3/2}$ **38.** $x = \sqrt{y}$

In Problems 39–42, find parametric equations that define the curve shown.

39.

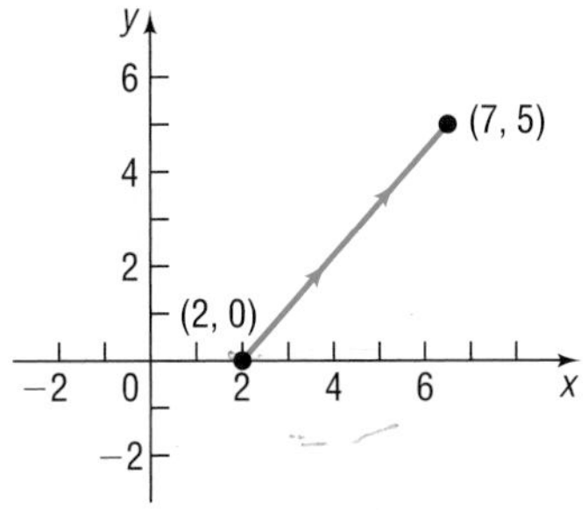

40.

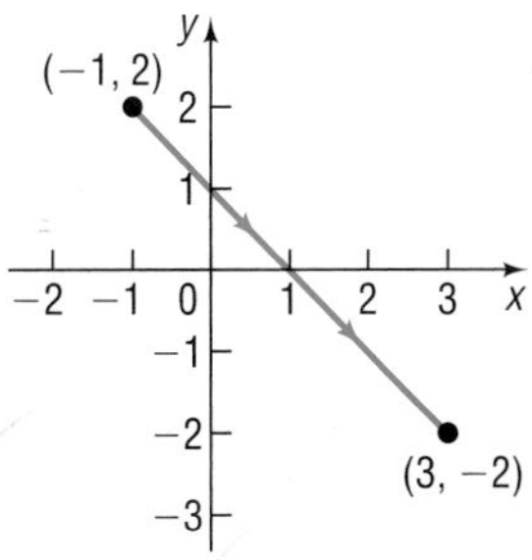

41.

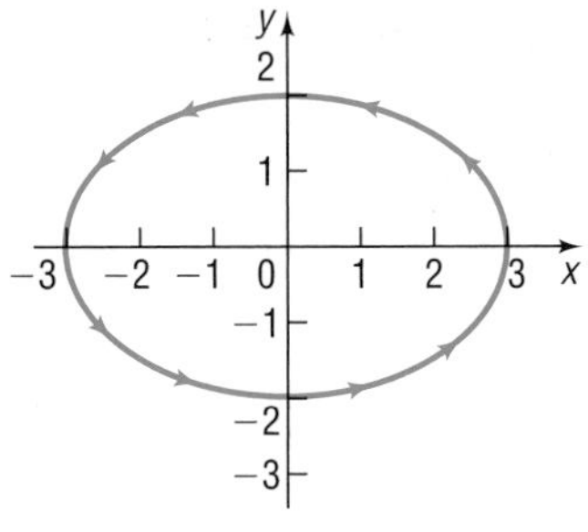

42.

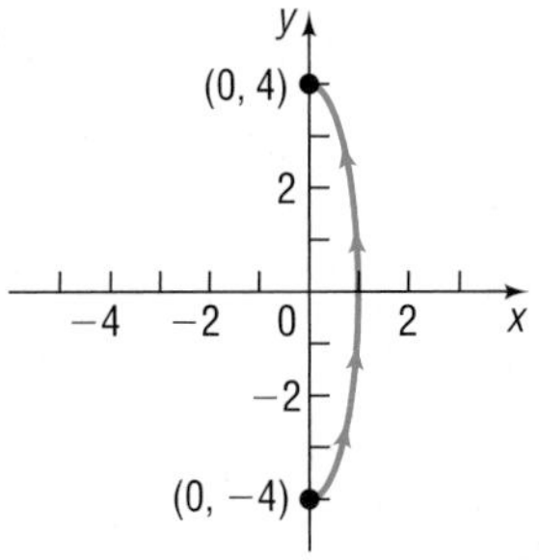

In Problems 43–46, find parametric equations for an object that moves along the ellipse $x^2/4 + y^2/9 = 1$ with the motion described.

43. The motion begins at (2, 0), is clockwise, and requires 2 seconds for a complete revolution.

44. The motion begins at (0, 3), is counterclockwise, and requires 1 second for a complete revolution.

45. The motion begins at (0, 3), is counterclockwise, and requires 1 second for a complete revolution.

46. The motion begins at (2, 0), is counterclockwise, and requires 3 seconds for a complete revolution.

In Problems 47 and 48, the parametric equations of four curves are given. Graph each of them by hand, indicating the orientation.

47. C_1: $x = t, \quad y = t^2; \quad -4 \le t \le 4$
C_2: $x = \cos t, \quad y = 1 - \sin^2 t; \quad 0 \le t \le \pi$
C_3: $x = e^t, \quad y = e^{2t}; \quad 0 \le t \le \ln 4$
C_4: $x = \sqrt{t}, \quad y = t; \quad 0 \le t \le 16$

48. C_1: $x = t, \quad y = \sqrt{1 - t^2}; \quad -1 \le t \le 1$
C_2: $x = \sin t, \quad y = \cos t; \quad 0 \le t \le 2\pi$
C_3: $x = \cos t, \quad y = \sin t; \quad 0 \le t \le 2\pi$
C_4: $x = \sqrt{1 - t^2}, \quad y = t; \quad -1 \le t \le 1$

49. Show that the parametric equations for a line passing through the points (x_1, y_1) and (x_2, y_2) are

$$x = (x_2 - x_1)t + x_1$$

$$y = (y_2 - y_1)t + y_1, \quad -\infty < t < \infty$$

What is the orientation of this line?

50. Projectile Motion The position of a projectile fired with an initial velocity v_0 feet per second and at an angle θ to the horizontal at the end of t seconds is given by the parametric equations

$$x = (v_0 \cos\theta)t \qquad y = (v_0 \sin\theta)t - 16t^2$$

See the following illustration.

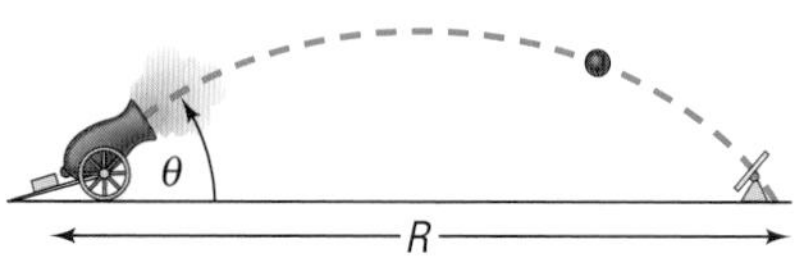

(a) Obtain the rectangular equation of the trajectory and identify the curve.
(b) Show that the projectile hits the ground $(y = 0)$ when $t = \frac{1}{16} v_0 \sin\theta$.
(c) How far has the projectile traveled (horizontally) when it strikes the ground? In other words, find the range R.
(d) Find the time t when $x = y$. Then find the horizontal distance x and the vertical distance y traveled by the projectile in this time. Then compute $\sqrt{x^2 + y^2}$. This is the distance R, the range, that the projectile travels up a plane inclined at 45° to the horizontal $(x = y)$. See the following illustration. (See also Problem 75 in Exercise 7.3.)

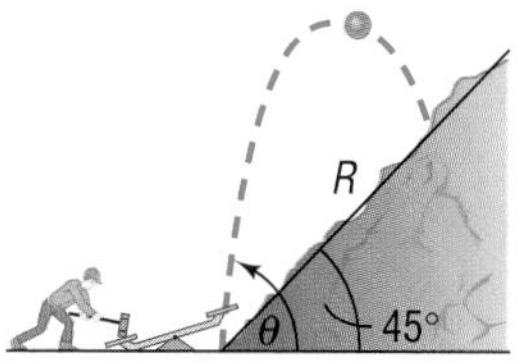

In Problems 51–54, use a graphing utility to graph the curve defined by the given parametric equations.

51. $x = t\sin t, \quad y = t\cos t$

52. $x = \sin t + \cos t, \quad y = \sin t - \cos t$

53. $x = 4\sin t - 2\sin(2t)$
$y = 4\cos t - 2\cos(2t)$

54. $x = 4\sin t + 2\sin(2t)$
$y = 4\cos t + 2\cos(2t)$

55. Hypocycloid The hypocycloid is a curve defined by the parametric equations

$$x(t) = \cos^3 t, \quad y(t) = \sin^3 t, \qquad 0 \le t \le 2\pi$$

(a) Graph the hypocycloid using a graphing utility.
(b) Find rectangular equations of the hypocycloid.

56. In Problem 55, we graphed the hypocycloid. Now graph the rectangular equations of the hypocycloid. Did you obtain a complete graph? If not, experiment until you do.

57. Look up the curves called *hypocycloid* and *epicycloid.* Write a report on what you find. Be sure to draw comparisons with the cycloid.

CHAPTER REVIEW

Things To Know

Equations

Parabola	See Tables 1 and 2 (pages 640 and 642).	
Ellipse	See Table 3 (page 654).	
Hyperbola	See Table 4 (page 668).	
General equation of a conic (page 682)	$Ax^2 + Bxy + Cy^2 + Dx + Ey + F = 0$	Parabola if $B^2 - 4AC = 0$ Ellipse (or circle) if $B^2 - 4AC < 0$ Hyperbola if $B^2 - 4AC > 0$
Conic in polar coordinates (page 684)	$\dfrac{d(F, P)}{d(P, D)} = e$	Parabola if $e = 1$ Ellipse if $e < 1$ Hyperbola if $e > 1$
Polar equations of a conic	See Table 5 (page 687).	

Parametric equations of a curve (page 690)	$x = f(t)$, $y = g(t)$, t a parameter

Definitions

Parabola (p. 637)	Set of points P in the plane for which $d(F, P) = d(P, D)$, where F is the focus and D is the directrix
Ellipse (p. 649)	Set of points P in the plane, the sum of whose distances from two fixed points (the foci) is a constant
Hyperbola (p. 661)	Set of points P in the plane, the difference of whose distances from two fixed points (the foci) is a constant

Formulas

Rotation formulas (p. 677)	$x = x' \cos\theta - y' \sin\theta$ $y = x' \sin\theta + y' \cos\theta$
Angle θ of rotation that eliminates the $x'y'$-term (p. 678)	$\cot(2\theta) = \dfrac{A - C}{B}, \quad 0 < \theta < \pi/2$

How To

Know the name of the conics (p. 636)

Find the equation of a parabola (p. 637)

Graph parabolas (p. 638)

Discuss the equation of a parabola (p. 639)

Work with parabolas with vertex at (h, k) (p. 642)

Solve applied problems involving parabolas (p. 644)

Find the equation of an ellipse (p. 649)

Graph ellipses (p. 650)

Discuss the equation of an ellipse (p. 652)

Work with ellipses with center at (h, k) (p. 654)

Solve applied problems involving ellipses (p. 657)

Find the equation of a hyperbola with center at the origin (p. 661)

Find the asymptotes of a hyperbola (p. 665)

Discuss the equation of a hyperbola (p. 666)

Work with hyperbolas with center at (h, k) (p. 668)

Solve applied problems involving hyperbolas (p. 670)

Identify a conic (p. 675)

Use a rotation of axes to transform equations (p. 676)

Discuss an equation using a rotation of axes (p. 679)

Identify conics without a rotation of axes (p. 681)

Discuss and graph polar equations of conics (p. 684)

Convert a polar equation of a conic to a rectangular equation (p. 688)

Graph parametric equations by hand (p. 690)

Graph parametric equations using a graphing utility (p. 691)

Find a rectangular equation for a curve defined parametrically (p. 692)

Use time as a parameter in parametric equations (p. 694)

Find parametric equations for curves defined by rectangular equations (p. 698)

Fill-in-the-Blank Items

1. A(n) ______________ is the collection of all points in the plane such that the distance from each point to a fixed point equals its distance to a fixed line.

2. A(n) ______________ is the collection of all points in the plane the sum of whose distances from two fixed points is a constant.

3. A(n) ______________ is the collection of all points in the plane the difference of whose distances from two fixed points is a constant.

4. For an ellipse, the foci lie on the ______________ axis; for a hyperbola, the foci lie on the ______________ axis.

5. For the ellipse $(x^2/9) + (y^2/16) = 1$, the major axis is along the ______________.

6. The equations of the asymptotes of the hyperbola $(y^2/9) - (x^2/4) = 1$ are ______________ and ______________.

7. To transform the equation

$$Ax^2 + Bxy + Cy^2 + Dx + Ey + F = 0, \qquad B \neq 0$$

into one in x' and y' without an $x'y'$-term, rotate the axes through an acute angle θ that satisfies the equation ______________.

8. The polar equation $r = \dfrac{8}{4 - 2\sin\theta}$ is a conic whose eccentricity is ______________. It is a(n) ______________ whose directrix is ______________ to the polar axis at a distance ______________ units ______________ the pole.

9. The parametric equations $x = 2\sin t$ and $y = 3\cos t$ represent a(n) ______________.

True/False Items

T F 1. On a parabola, the distance from any point to the focus equals the distance from that point to the directrix.
T F 2. The foci of an ellipse lie on its minor axis.
T F 3. The foci of a hyperbola lie on its transverse axis.
T F 4. Hyperbolas always have asymptotes, and ellipses never have asymptotes.
T F 5. A hyperbola never intersects its conjugate axis.
T F 6. A hyperbola always intersects its transverse axis.
T F 7. The equation $ax^2 + 6y^2 - 12y = 0$ defines an ellipse if $a > 0$.
T F 8. The equation $3x^2 + bxy + 12y^2 = 10$ defines a parabola if $b = -12$.
T F 9. If (r, θ) are polar coordinates, the equation $r = 2/(2 + 3\sin\theta)$ defines a hyperbola.
T F 10. Parametric equations defining a curve are unique.

Review Exercises

Blue problem numbers indicate the authors' suggestions for use in a Practice Test.

In Problems 1–20, identify each equation. If it is a parabola, gives its vertex, focus, and directrix; if it is an ellipse, give its center, vertices, and foci; if it is a hyperbola, give its center, vertices, foci, and asymptotes.

1. $y^2 = -16x$
2. $16x^2 = y$
3. $\dfrac{x^2}{25} - y^2 = 1$
4. $\dfrac{y^2}{25} - x^2 = 1$
5. $\dfrac{y^2}{25} + \dfrac{x^2}{16} = 1$
6. $\dfrac{x^2}{9} + \dfrac{y^2}{16} = 1$
7. $x^2 + 4y = 4$
8. $3y^2 - x^2 = 9$
9. $4x^2 - y^2 = 8$
10. $9x^2 + 4y^2 = 36$
11. $x^2 - 4x = 2y$
12. $2y^2 - 4y = x - 2$
13. $y^2 - 4y - 4x^2 + 8x = 4$
14. $4x^2 + y^2 + 8x - 4y + 4 = 0$
15. $4x^2 + 9y^2 - 16x - 18y = 11$
16. $4x^2 + 9y^2 - 16x + 18y = 11$
17. $4x^2 - 16x + 16y + 32 = 0$
18. $4y^2 + 3x - 16y + 19 = 0$
19. $9x^2 + 4y^2 - 18x + 8y = 32$
20. $x^2 - y^2 - 2x - 2y = 1$

In Problems 21–36, obtain an equation of the conic described. Graph the equation by hand.

21. Parabola; focus at $(-2, 0)$; directrix the line $x = 2$
22. Ellipse; center at $(0, 0)$; focus at $(0, 3)$; vertex at $(0, 5)$
23. Hyperbola; center at $(0, 0)$; focus at $(0, 4)$; vertex at $(0, -2)$
24. Parabola; vertex at $(0, 0)$; directrix the line $y = -3$
25. Ellipse; foci at $(-3, 0)$ and $(3, 0)$; vertex at $(4, 0)$
26. Hyperbola; vertices at $(-2, 0)$ and $(2, 0)$; focus at $(4, 0)$
27. Parabola; vertex at $(2, -3)$; focus at $(2, -4)$
28. Ellipse; center at $(-1, 2)$; focus at $(0, 2)$; vertex at $(2, 2)$
29. Hyperbola; center at $(-2, -3)$; focus at $(-4, -3)$; vertex at $(-3, -3)$
30. Parabola; focus at $(3, 6)$; directrix the line $y = 8$
31. Ellipse; foci at $(-4, 2)$ and $(-4, 8)$; vertex at $(-4, 10)$
32. Hyperbola; vertices at $(-3, 3)$ and $(5, 3)$; focus at $(7, 3)$
33. Center at $(-1, 2)$; $a = 3$; $c = 4$; transverse axis parallel to the x-axis
34. Center at $(4, -2)$; $a = 1$; $c = 4$; transverse axis parallel to the y-axis
35. Vertices at $(0, 1)$ and $(6, 1)$; asymptote the line $3y + 2x = 9$
36. Vertices at $(4, 0)$ and $(4, 4)$; asymptote the line $y + 2x = 10$

In Problems 37–46, identify each conic without completing the squares and without applying a rotation of axes.

37. $y^2 + 4x + 3y - 8 = 0$
38. $2x^2 - y + 8x = 0$
39. $x^2 + 2y^2 + 4x - 8y + 2 = 0$
40. $x^2 - 8y^2 - x - 2y = 0$

41. $9x^2 - 12xy + 4y^2 + 8x + 12y = 0$

42. $4x^2 + 4xy + y^2 - 8\sqrt{5}x + 16\sqrt{5}y = 0$

43. $4x^2 + 10xy + 4y^2 - 9 = 0$

44. $4x^2 - 10xy + 4y^2 - 9 = 0$

45. $x^2 - 2xy + 3y^2 + 2x + 4y - 1 = 0$

46. $4x^2 + 12xy - 10y^2 + x + y - 10 = 0$

In Problems 47–52, rotate the axes so that the new equation contains no xy-term. Discuss and graph the new equation by hand.

47. $2x^2 + 5xy + 2y^2 - \frac{9}{2} = 0$

48. $2x^2 - 5xy + 2y^2 - \frac{9}{2} = 0$

49. $6x^2 + 4xy + 9y^2 - 20 = 0$

50. $x^2 + 4xy + 4y^2 + 16\sqrt{5}x - 8\sqrt{5}y = 0$

51. $4x^2 - 12xy + 9y^2 + 12x + 8y = 0$

52. $9x^2 - 24xy + 16y^2 + 80x + 60y = 0$

In Problems 53–58, identify the conic that each polar equation represents and graph it by hand.

53. $r = \dfrac{4}{1 - \cos\theta}$

54. $r = \dfrac{6}{1 + \sin\theta}$

55. $r = \dfrac{6}{2 - \sin\theta}$

56. $r = \dfrac{2}{3 + 2\cos\theta}$

57. $r = \dfrac{8}{4 + 8\cos\theta}$

58. $r = \dfrac{10}{5 + 20\sin\theta}$

In Problems 59–62, convert each polar equation to a rectangular equation.

59. $r = \dfrac{4}{1 - \cos\theta}$

60. $r = \dfrac{6}{2 - \sin\theta}$

61. $r = \dfrac{8}{4 + 8\cos\theta}$

62. $r = \dfrac{2}{3 + 2\cos\theta}$

In Problems 63–68, graph the curve whose parametric equations are given and show its orientation. Find the rectangular equation of each curve.

63. $x = 4t - 2, \quad y = 1 - t; \quad -\infty < t < \infty$

64. $x = 2t^2 + 6, \quad y = 5 - t; \quad -\infty < t < \infty$

65. $x = 3\sin t, \quad y = 4\cos t + 2; \quad 0 \le t \le 2\pi$

66. $x = \ln t, \quad y = t^3; \quad t > 0$

67. $x = \sec^2 t, \quad y = \tan^2 t; \quad 0 \le t \le \pi/4$

68. $x = t^{3/2}, \quad y = 2t + 4; \quad t \ge 0$

69. Find an equation of the hyperbola whose foci are the vertices of the ellipse $4x^2 + 9y^2 = 36$ and whose vertices are the foci of this ellipse.

70. Find an equation of the ellipse whose foci are the vertices of the hyperbola $x^2 - 4y^2 = 16$ and whose vertices are the foci of this hyperbola.

71. Describe the collection of points in a plane so that the distance from each point to the point $(3, 0)$ is three-fourths of its distance from the line $x = \frac{16}{3}$.

72. Describe the collection of points in a plane so that the distance from each point to the point $(5, 0)$ is five-fourths of its distance from the line $x = \frac{16}{5}$.

73. Mirrors A mirror is shaped like a paraboloid of revolution. If a light source is located 1 foot from the base along the axis of symmetry and the opening is 2 feet across, how deep should the mirror be?

74. Parabolic Arch Bridge A bridge is built in the shape of a parabolic arch. The bridge has a span of 60 feet and a maximum height of 20 feet. Find the height of the arch at distances of 5, 10, and 20 feet from the center.

75. Semielliptical Arch Bridge A bridge is built in the shape of a semielliptical arch. The bridge has a span of 60 feet and a maximum height of 20 feet. Find the height of the arch at distances of 5, 10, and 20 feet from the center.

76. Whispering Galleries The figure shows the specifications for an elliptical ceiling in a hall designed to be a whispering gallery. Where are the foci located in the hall?

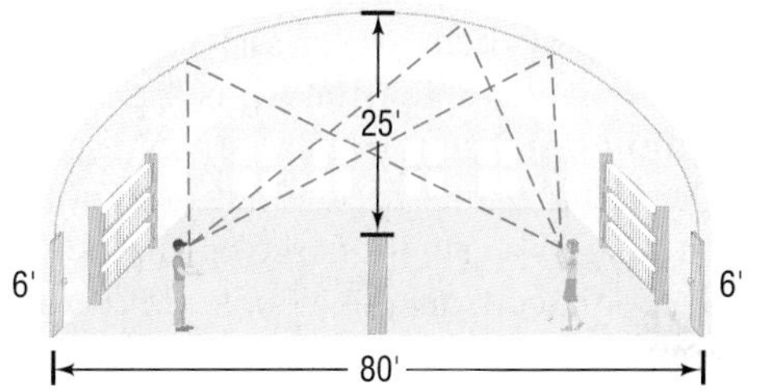

77. LORAN Two LORAN stations are positioned 150 miles apart along a straight shore.
(a) A ship records a time difference of 0.00032 second between the LORAN signals. Set up an appropriate rectangular coordinate system to determine where the ship would reach shore if it were to follow the hyperbola corresponding to this time difference.
(b) If the ship wants to enter a harbor located between the two stations 15 miles from the master station, what time difference should it be looking for?
(c) If the ship is 20 miles offshore when the desired time difference is obtained, what is the approximate location of the ship?
[**Note:** The speed of each radio signal is 186,000 miles per second.]

78. Uniform Motion Mary's train leaves at 7:15 AM and accelerates at the rate of 3 meters per second per second. Mary, who can run 6 meters per second, arrives at the train station 2 seconds after the train has left.
(a) Find parametric equations that describe the motion of the train and Mary as a function of time.

(b) Determine algebraically whether Mary will catch the train. If so, when?
(c) Simulate the motion of the train and Mary by simultaneously graphing the equations found in part (a).

79. Projectile Motion Drew Bledsoe throws a football with an initial speed of 100 feet per second at an angle of 35¡ to the horizontal. The ball leaves Drew Bledsoe's hand at a height of 6 feet.
(a) Find parametric equations that describe the position of the ball as a function of time.
(b) How long is the ball in the air?
(c) When is the ball at its maximum height? Determine the maximum height of the ball.
(d) Determine the distance that the ball travels.
(e) Using a graphing utility, simultaneously graph the equations found in part (a).

80. Formulate a strategy for discussing and graphing an equation of the form

$$Ax^2 + Bxy + Cy^2 + Dx + Ey + F = 0$$

PROJECT AT MOTOROLA

Distorted Deployable Space Reflector Antennas

A certain parabolic reflector antenna is to be used in a space application. To avoid excessive costs transporting the antenna into orbit, NASA makes these antennas from a very thin metallic cloth material attached on a frame that is deployed while in space to open up and look like a parabolic umbrella. In the correct configuration, the axis of the antenna aligns with the z-axis. To ensure that the deployed antenna is of the right shape, the following scheme is devised. Eight small, light reflecting targets are placed on the cloth with the coordinates below (when the antenna is correctly deployed). Four of them are on the x-z plane and four are on the y-z plane. Tension wires are used to adjust the position of these targets when distorted, that is, when the actual values of x, y, and z do not equal the correct values. When in the cargo bay of the space shuttle, a laser placed on the z-axis and at a distance $L = 10$ m from the vertex performs measurements and records its distance from the targets and their angle θ (see the figure). Table 1 contains the correct values of x, y, and z of the parabolic reflector.

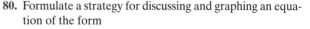

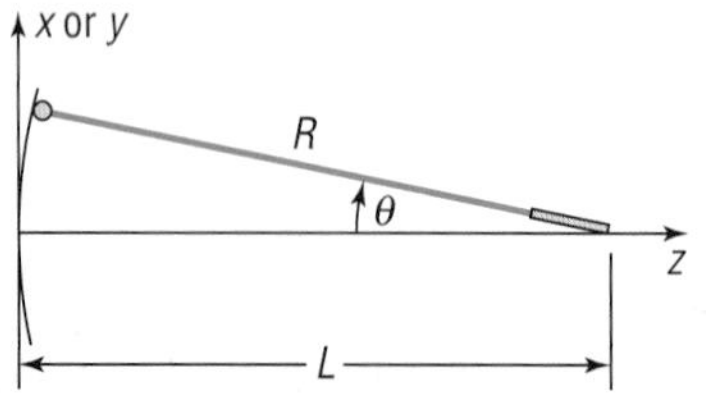

TABLE 1

Target	x	y(m)	z(m)
T_1	0	−2	0.5
T_2	0	−1	0.125
T_3	0	1	0.125
T_4	0	2	0.5
T_5	−2	0	0.5
T_6	−1	0	0.125
T_7	1	0	0.125
T_8	2	0	0.5

The laser measurements of the targets on the antenna immediately after deployment are tabulated in Table 2.

TABLE 2

Target	x(m)	R(m)	θ(deg)
T_1	0	9.551	−11.78
T_2	0	9.948	−5.65
T_3	0	9.928	5.90
T_4	0	9.708	11.89
Target	**y(m)**	**R(m)**	**θ(deg)**
T_5	0	9.722	−11.99
T_6	0	9.917	−5.85
T_7	0	9.925	5.78
T_8	0	9.551	11.78

In Problems 1–3, use any of the target points from Table 1 and the fact the equation of the parabolic reflector (assuming vertex at the origin) is $z = 4ax^2 + 4ay^2$.

1. Find the focal distance of the undistorted parabolic reflector.
2. Use the tabulated measurements of the target points (Table 2) to convert the R and θ coordinates of the targets into cartesian coordinates (z and x, or y, as appropriate). That is, find the actual values of x, y, and z.
3. Find how much the targets T_1 through T_4 have to be moved in the y and z directions and the T_5 through T_8 in the x and z directions in order to correct the distortions of the antenna.

Note: In actuality, instead of adjusting the tension wires, the laser measurements are used to determine how much the feed antenna at the focus of the parabola needs to change its electronic distribution so as to correct for the distortions (phase conjugation).

CHAPTER

11 Systems of Equations and Inequalities

Outline

FIELD TRIP TO MOTOROLA

Digital cell phones communicate using the language of computers, with voice and data sent as groups of information bits (zeros and ones) over the air. Wireless communication is very difficult—the signals are reflected off buildings, absorbed by trees and rain, and scattered by street signs before being received by the antenna on the phone. Because of these impairments, it is difficult to receive all of the information bits correctly—some will have invariably flipped from 0 to 1 or vice versa. On a voice call, the effect is digital distortion, which may make the call sound like it is coming from underwater, for example. On a data call, the effect changes the data present in the transferred files. One of the primary reasons digital phones are so powerful is that Motorola phone designers can use *error control coding* to ensure that phone calls are clear and error free.

INTERVIEW AT MOTOROLA

Brian Classon is a twenty-something-year-old research engineer for Motorola Labs in Schaumburg, IL. His job is to research and design wireless communication systems, which includes finding a way to surf the web from your phone at eye-popping speeds as well as making sure that popping popcorn in your microwave doesn't make your phone malfunction. His specialty is error control, which prevents your little sister's voice from sounding like it is coming from underwater over the phone. As a licensed Professional Engineer, his conduct is regulated by the same organization that monitors doctors and nurses. He greatly enjoys the combination of challenging problems and intellectual freedom at Motorola Labs—he currently holds two US patents, with 16 more pending.

PREPARING FOR THIS SECTION

Before getting started, review the following:

- ✓ Lines (Section 1.6)
- ✓ Linear Equations (p. 26–28)

11.1 SYSTEMS OF LINEAR EQUATIONS: TWO EQUATIONS CONTAINING TWO VARIABLES

OBJECTIVES

1. Solve Systems of Equations by Substitution
2. Solve Systems of Equations by Elimination
3. Inconsistent Systems and Dependent Equations

We begin with an example.

EXAMPLE 1 Movie Theater Ticket Sales

A movie theater sells tickets for \$8.00 each, with seniors receiving a discount of \$2.00. One evening the theater took in \$3580 in revenue. If x represents the number of tickets sold at \$8.00 and y the number of tickets sold at the discounted price of \$6.00, write an equation that relates these variables.

Solution Each nondiscounted ticket brings in \$8.00, so x tickets will bring in $8x$ dollars. Similarly, y discounted tickets bring in $6y$ dollars. Since the total brought in is \$3580, we must have

$$8x + 6y = 3580$$

■

In Example 1, suppose that we also know that 525 tickets were sold that evening. Then we have another equation relating the variables x and y.

$$x + y = 525$$

The two equations

$$8x + 6y = 3580$$
$$x + y = 525$$

form a *system* of equations.

In general, a **system of equations** is a collection of two or more equations, each containing one or more variables. Example 2 gives some samples of systems of equations.

EXAMPLE 2 Examples of Systems of Equations

(a) $\begin{cases} 2x + y = 5 & (1) \\ -4x + 6y = -2 & (2) \end{cases}$ Two equations containing two variables, x and y

(b) $\begin{cases} x + y^2 = 5 & (1) \\ 2x + y = 4 & (2) \end{cases}$ Two equations containing two variables, x and y

(c) $\begin{cases} x + y + z = 6 & (1) \\ 3x - 2y + 4z = 9 & (2) \\ x - y - z = 0 & (3) \end{cases}$ Three equations containing three variables, x, y, and z

(d) $\begin{cases} x + y + z = 5 & (1) \\ x - y = 2 & (2) \end{cases}$ Two equations containing three variables, x, y, and z

■

We use a brace, as shown, to remind us that we are dealing with a system of equations. We also will find it convenient to number each equation in the system.

A **solution** of a system of equations consists of values for the variables that are solutions of each equation of the system. To **solve** a system of equations means to find all solutions of the system.

For example, $x = 2$, $y = 1$ is a solution of the system in Example 2(a), because

$$2(2) + 1 = 5 \quad \text{and} \quad -4(2) + 6(1) = -2$$

A solution of the system in Example 2(b) is $x = 1$, $y = 2$, because

$$1 + 2^2 = 5 \quad \text{and} \quad 2(1) + 2 = 4$$

Another solution of the system in Example 2(b) is $x = \frac{11}{4}$, $y = -\frac{3}{2}$, which you can check for yourself. A solution of the system in Example 2(c) is $x = 3$, $y = 2$, $z = 1$, because

$$\begin{cases} 3 + 2 + 1 = 6 & (1) \quad x = 3, y = 2, z = 1 \\ 3(3) - 2(2) + 4(1) = 9 & (2) \\ 3 - 2 - 1 = 0 & (3) \end{cases}$$

Note that $x = 3$, $y = 3$, $z = 0$ is not a solution of the system in Example 2(c).

$$\begin{cases} 3 + 3 + 0 = 6 & (1) \quad x = 3, y = 3, z = 0 \\ 3(3) - 2(3) + 4(0) = 3 \neq 9 & (2) \\ 3 - 3 - 0 = 0 & (3) \end{cases}$$

Although these values satisfy equations (1) and (3), they do not satisfy equation (2). Any solution of the system must satisfy *each* equation of the system.

NOW WORK PROBLEM 3.

When a system of equations has at least one solution, it is said to be **consistent;** otherwise, it is called **inconsistent.**

An equation in n variables is said to be **linear** if it is equivalent to an equation of the form

$$a_1x_1 + a_2x_2 + \cdots + a_nx_n = b$$

where $x_1, x_2, \ldots, x_n$ are n distinct variables, $a_1, a_2, \ldots, a_n, b$ are constants, and at least one of the a's is not 0.

Some examples of linear equations are

$$2x + 3y = 2 \qquad 5x - 2y + 3z = 10 \qquad 8x + 8y - 2z + 5w = 0$$

If each equation in a system of equations is linear, then we have a **system of linear equations.** Thus, the systems in Examples 2(a), (c), and (d) are linear, whereas the system in Example 2(b) is nonlinear. In this chapter we shall solve linear systems in Sections 11.1–11.4. We discuss nonlinear systems in Section 11.7.

Two Linear Equations Containing Two Variables

We can view the problem of solving a system of two linear equations containing two variables as a geometry problem. The graph of each equation in such a system is a straight line. Thus, a system of two equations containing two variables represents a pair of lines. The lines either (1) intersect or (2) are parallel or (3) are **coincident** (that is, identical).

1. If the lines intersect, then the system of equations has one solution, given by the point of intersection. The system is **consistent** and the equations are **independent.**
2. If the lines are parallel, then the system of equations has no solution, because the lines never intersect. The system is **inconsistent.**
3. If the lines are coincident, then the system of equations has infinitely many solutions, represented by the totality of points on the line. The system is **consistent** and the equations are **dependent.**

Figure 1 illustrates these conclusions.

Figure 1

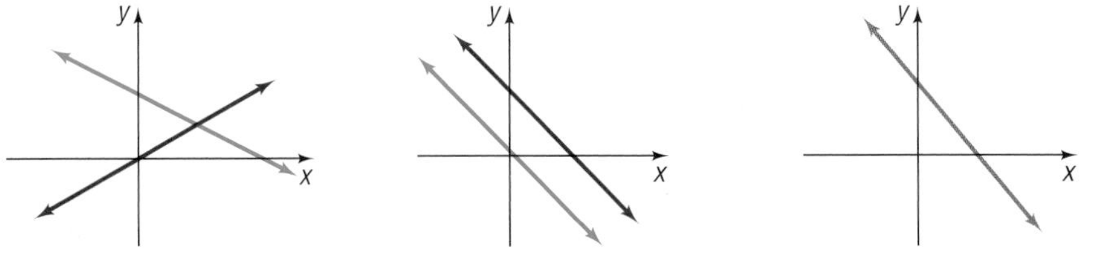

(a) Intersecting lines; system has one solution

(b) Parallel lines; system has no solution

(c) Coincident lines; system has infinitely many solutions

EXAMPLE 3

Solving a System of Linear Equations Using a Graphing Utility

Solve: $\begin{cases} 2x + y = 5 & (1) \\ -4x + 6y = 12 & (2) \end{cases}$

Solution First, we solve each equation for y. This is equivalent to writing each equation in slope–intercept form. Equation (1) in slope–intercept form is $Y_1 = -2x + 5$. Equation (2) in slope–intercept form is $Y_2 = \frac{2}{3}x + 2$. Figure 2 shows the graphs using a graphing utility. From the graph in Figure 2, we see that the lines intersect, so the system is consistent and the equations are independent. Using INTERSECT, we obtain the solution (1.125, 2.75). ■

Figure 2

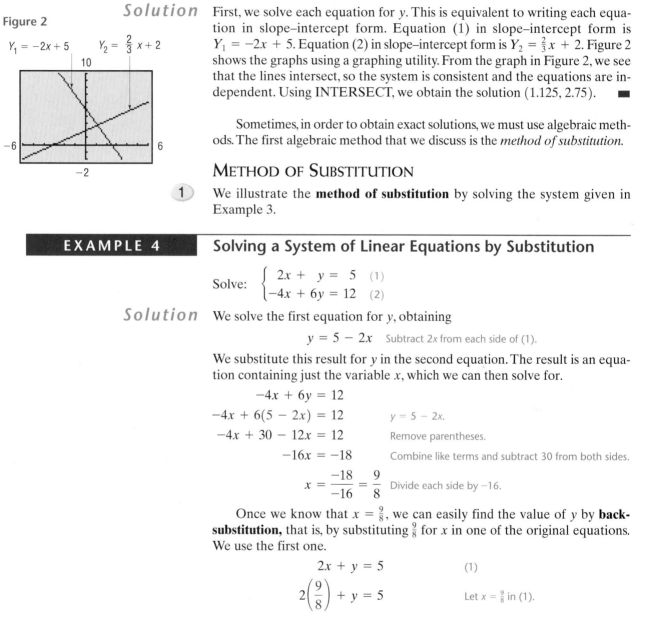

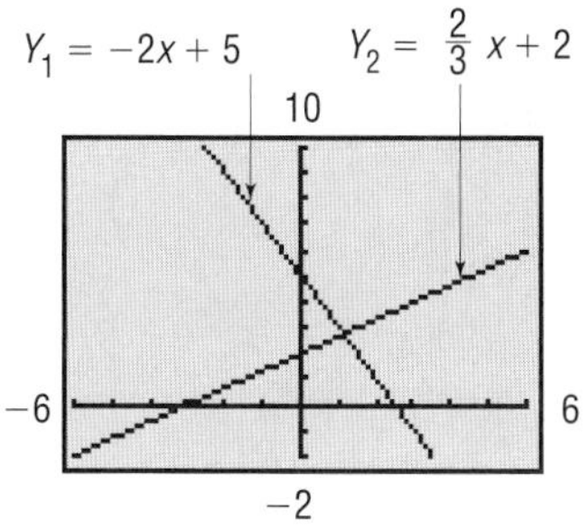

Sometimes, in order to obtain exact solutions, we must use algebraic methods. The first algebraic method that we discuss is the *method of substitution.*

Method of Substitution

1 We illustrate the **method of substitution** by solving the system given in Example 3.

EXAMPLE 4

Solving a System of Linear Equations by Substitution

Solve: $\begin{cases} 2x + y = 5 & (1) \\ -4x + 6y = 12 & (2) \end{cases}$

Solution We solve the first equation for y, obtaining

$$y = 5 - 2x \quad \text{Subtract } 2x \text{ from each side of (1).}$$

We substitute this result for y in the second equation. The result is an equation containing just the variable x, which we can then solve for.

$$\begin{aligned} -4x + 6y &= 12 & \\ -4x + 6(5 - 2x) &= 12 & y = 5 - 2x. \\ -4x + 30 - 12x &= 12 & \text{Remove parentheses.} \\ -16x &= -18 & \text{Combine like terms and subtract 30 from both sides.} \\ x &= \frac{-18}{-16} = \frac{9}{8} & \text{Divide each side by } -16. \end{aligned}$$

Once we know that $x = \frac{9}{8}$, we can easily find the value of y by **back-substitution,** that is, by substituting $\frac{9}{8}$ for x in one of the original equations. We use the first one.

$$\begin{aligned} 2x + y &= 5 & (1) \\ 2\left(\frac{9}{8}\right) + y &= 5 & \text{Let } x = \tfrac{9}{8} \text{ in (1).} \\ \frac{9}{4} + y &= 5 & \text{Simplify.} \\ y &= 5 - \frac{9}{4} & \text{Subtract } \tfrac{9}{4} \text{ from both sides.} \\ &= \frac{20}{4} - \frac{9}{4} = \frac{11}{4} & \end{aligned}$$

The solution of the system is $x = \frac{9}{8} = 1.125$, $y = \frac{11}{4} = 2.75$. ■

The method used to solve the system in Example 4 is called **substitution.** The steps to be used are outlined next.

STEPS FOR SOLVING BY SUBSTITUTION

STEP 1: Pick one of the equations and solve for one of the variables in terms of the remaining variables.

STEP 2: Substitute the result in the remaining equations.

STEP 3: If one equation in one variable results, solve this equation. Otherwise, repeat Step 1 until a single equation with one variable remains.

STEP 4: Find the values of the remaining variables by back-substitution.

STEP 5: Check the solution found.

EXAMPLE 5 Solving a System of Linear Equations by Substitution

Solve: $\begin{cases} 3x - 2y = 5 & (1) \\ 5x - y = 6 & (2) \end{cases}$

Solution **STEP 1:** After looking at the two equations, we conclude that it is easiest to solve for the variable y in equation (2).

$$\begin{aligned} 5x - y &= 6 \\ 5x - y - 6 &= 0 && \text{Subtract 6 from both sides.} \\ 5x - 6 &= y && \text{Add } y \text{ to both sides.} \end{aligned}$$

STEP 2: We substitute this result into equation (1) and simplify.

$$\begin{aligned} 3x - 2y &= 5 && (1) \\ 3x - 2(5x - 6) &= 5 && \text{Substitute } y = 5x - 6 \text{ in (1).} \\ -7x + 12 &= 5 && \text{Remove parentheses and collect like terms.} \\ -7x &= -7 && \text{Subtract 12 from both sides.} \\ x &= 1 && \text{Divide both sides by } -7. \end{aligned}$$

STEP 3: Because we now have one solution, $x = 1$, we proceed to Step 4.

STEP 4: Knowing $x = 1$, we can find y from the equation

$$y = 5x - 6 \underset{\substack{\uparrow \\ x = 1}}{=} 5(1) - 6 = -1$$

STEP 5: Check: $\begin{cases} 3(1) - 2(-1) = 3 + 2 = 5 \\ 5(1) - (-1) = 5 + 1 = 6 \end{cases}$

The solution of the system is $x = 1$, $y = -1$. ■

NOW USE SUBSTITUTION TO WORK PROBLEM 11.

Method of Elimination

2 A second method for solving a system of linear equations is the *method of elimination.* This method is usually preferred over substitution if substitution leads to fractions or if the system contains more than two variables. Elimination also provides the necessary motivation for solving systems using matrices (the subject of Section 11.3).

The idea behind the method of elimination is to keep replacing the original equations in the system with equivalent equations until a system of equations with an obvious solution is reached. When we proceed in this way, we obtain **equivalent systems of equations.** The rules for obtaining equivalent equations are the same as those studied earlier. However, we may also interchange any two equations of the system and/or replace any equation in the system by the sum (or difference) of that equation and any other equation in the system.

Rules for Obtaining an Equivalent System of Equations

1. Interchange any two equations of the system.
2. Multiply (or divide) each side of an equation by the same nonzero constant.
3. Replace any equation in the system by the sum (or difference) of that equation and a nonzero multiple of any other equation in the system.

An example will give you the idea. As you work through the example, pay particular attention to the pattern being followed.

EXAMPLE 6 Solving a System of Linear Equations by Elimination

Solve: $\begin{cases} 2x + 3y = 1 & (1) \\ -x + y = -3 & (2) \end{cases}$

We multiply each side of equation (2) by 2 so that the coefficients of x in the two equations are negatives of one another. The result is the equivalent system

$$\begin{cases} 2x + 3y = 1 & (1) \\ -2x + 2y = -6 & (2) \end{cases}$$

If we now replace equation (2) of this system by the sum of the two equations, we obtain an equation containing just the variable y, which we can solve for.

$$\begin{aligned} &\begin{cases} 2x + 3y = 1 \\ -2x + 2y = -6 \end{cases} && \begin{matrix}(1)\\(2)\end{matrix} \\ &\qquad 5y = -5 && \text{Add (1) and (2).} \\ &\qquad y = -1 \end{aligned}$$

We back-substitute by using this value for y in equation (1) and simplify to get

$$2x + 3(-1) = 1$$
$$2x = 4$$
$$x = 2$$

Thus, the solution of the original system is $x = 2$, $y = -1$. We leave it to you to check the solution. ■

The procedure used in Example 6 is called the **method of elimination.** Notice the pattern of the solution. First, we eliminated the variable x from the second equation. Then we back-substituted; that is, we substituted the value found for y back into the first equation to find x.

Let's return to the movie theater example (Example 1).

EXAMPLE 7 Movie Theater Ticket Sales

A movie theater sells tickets for \$8.00 each, with seniors receiving a discount of \$2.00. One evening the theater sold 525 tickets and took in \$3580 in revenue. How many of each type of ticket was sold?

Solution If x represents the number of tickets sold at \$8.00 and y the number of tickets sold at the discounted price of \$6.00, then the given information results in the system of equations

$$\begin{cases} 8x + 6y = 3580 & (1) \\ x + y = 525 & (2) \end{cases}$$

We use elimination and multiply the second equation by -6 and then add the equations

$$\begin{cases} 8x + 6y = 3580 & (1) \\ -6x - 6y = -3150 & (2) \end{cases}$$
$$2x = 430 \quad \text{Add the equations.}$$
$$x = 215$$

Since $x + y = 525$, then $y = 525 - x = 525 - 215 = 310$. Thus, 215 nondiscounted tickets and 310 senior discount tickets were sold. ■

NOW USE ELIMINATION TO WORK PROBLEM 11.

3 The previous examples dealt with consistent systems of equations that had a unique solution. The next two examples deal with two other possibilities that may occur, the first being a system that has no solution.

EXAMPLE 8 An Inconsistent System of Linear Equations

Solve: $\begin{cases} 2x + y = 5 & (1) \\ 4x + 2y = 8 & (2) \end{cases}$

Solution We choose to use the method of substitution and solve equation (1) for y.

$$2x + y = 5$$
$$y = 5 - 2x \quad \text{Subtract } 2x \text{ from each side.}$$

Substituting in equation (2), we get

$$4x + 2y = 8 \quad (2)$$

$$4x + 2(5 - 2x) = 8 \quad y = 5 - 2x$$

$$4x + 10 - 4x = 8 \quad \text{Remove parentheses.}$$

$$0 \cdot x = -2 \quad \text{Combine like terms.}$$

This equation has no solution. Thus, we conclude that the system itself has no solution and is therefore inconsistent. ■

Figure 3 illustrates the pair of lines whose equations form the system in Example 8. Notice that the graphs of the two equations are lines, each with slope -2; one has a y-intercept of 5, the other a y-intercept of 4. Thus, the lines are parallel and have no point of intersection. This geometric statement is equivalent to the algebraic statement that the system has no solution.

Figure 3

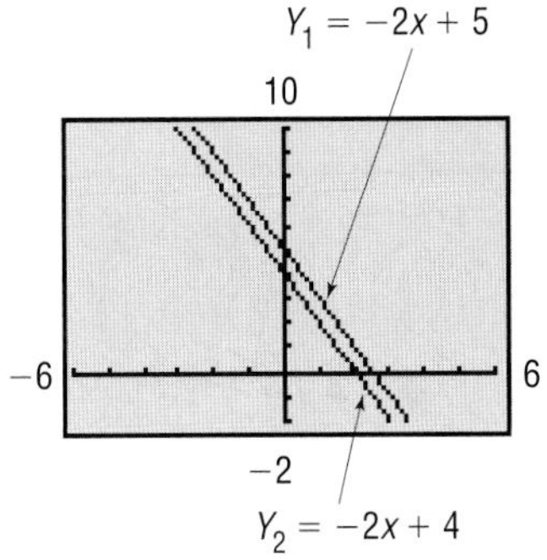

The next example is an illustration of a system with infinitely many solutions.

EXAMPLE 9 Solving a System of Linear Equations with Infinitely Many Solutions

Solve: $\begin{cases} 2x + y = 4 & (1) \\ -6x - 3y = -12 & (2) \end{cases}$

Solution We choose to use the method of elimination.

$$\begin{cases} 2x + y = 4 & (1) \\ -6x - 3y = -12 & (2) \end{cases}$$

$$\begin{cases} 6x + 3y = 12 & (1) \\ -6x - 3y = -12 & (2) \end{cases} \quad \text{Multiply each side of equation (1) by 3.}$$

$$\begin{cases} 6x + 3y = 12 & (1) \\ 0 = 0 & (2) \end{cases} \quad \text{Replace equation (2) by the sum of equations (1) and (2).}$$

The original system is equivalent to a system containing one equation, so the equations are dependent. This means that any values of x and y for which $6x + 3y = 12$ or, equivalently, $2x + y = 4$ are solutions. For example, $x = 2$, $y = 0$; $x = 0$, $y = 4$; $x = -2$, $y = 8$; $x = 4$, $y = -4$; and so on, are solutions. There are, in fact, infinitely many values of x and y for which $2x + y = 4$, so the original system has infinitely many solutions. We will write the solutions of the original systems either as

$$y = 4 - 2x$$

where x can be any real number, or as

$$x = 2 - \frac{1}{2}y$$

where y can be any real number. ■

Figure 4
$Y_1 = Y_2 = -2x + 4$

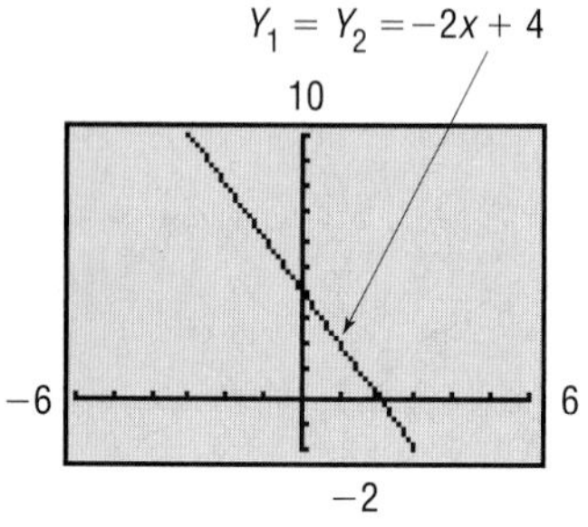

Figure 4 illustrates the situation presented in Example 9. Notice that the graphs of the two equations are lines, each with slope -2 and each with y-intercept 4. Thus, the lines are coincident. Notice also that equation (2) in the original system is just -3 times equation (1), indicating that the two equations are dependent.

For the system in Example 9, we can write down some of the infinite number of solutions by assigning values to x and then finding $y = 4 - 2x$. Thus,

If $x = -2$, then $y = 8$.

If $x = 0$, then $y = 4$.

If $x = 2$, then $y = 0$.

The pairs (x, y) are points on the line in Figure 4.

NOW WORK PROBLEMS 17 AND 21.

11.1 EXERCISES

In Problems 1–8, verify that the values of the variables listed are solutions of the system of equations.

1. $\begin{cases} 2x - y = 5 \\ 5x + 2y = 8 \end{cases}$ $x = 2, y = -1$

2. $\begin{cases} 3x - 2y = 2 \\ x - 7y = -30 \end{cases}$ $x = -2, y = 4$

3. $\begin{cases} 3x + 4y = 4 \\ \frac{1}{2}x - 3y = -\frac{1}{2} \end{cases}$ $x = 2, y = \frac{1}{2}$

4. $\begin{cases} 2x + \frac{1}{2}y = 0 \\ 3x - 4y = -\frac{19}{2} \end{cases}$ $x = -\frac{1}{2}, y = 2$

5. $\begin{cases} x - y = 3 \\ \frac{1}{2}x + y = 3 \end{cases}$ $x = 4, y = 1$

6. $\begin{cases} x - y = 3 \\ -3x + y = 1 \end{cases}$ $x = -2, y = -5$

7. $\begin{cases} 3x + 3y + 2z = 4 \\ x - y - z = 0 \\ 2y - 3z = -8 \end{cases}$ $x = 1, y = -1, z = 2$

8. $\begin{cases} 4x - z = 7 \\ 8x + 5y - z = 0 \\ -x - y + 5z = 6 \end{cases}$ $x = 2, y = -3, z = 1$

In Problems 9–32, solve each system of equations. If the system has no solution, say that it is inconsistent. Use either substitution or elimination. Verify your solution using a graphing utility.

9. $\begin{cases} x + y = 8 \\ x - y = 4 \end{cases}$

10. $\begin{cases} x + 2y = 5 \\ x + y = 3 \end{cases}$

11. $\begin{cases} 5x - y = 13 \\ 2x + 3y = 12 \end{cases}$

12. $\begin{cases} x + 3y = 5 \\ 2x - 3y = -8 \end{cases}$

13. $\begin{cases} 3x = 24 \\ x + 2y = 0 \end{cases}$

14. $\begin{cases} 4x + 5y = -3 \\ -2y = -4 \end{cases}$

15. $\begin{cases} 3x - 6y = 2 \\ 5x + 4y = 1 \end{cases}$

16. $\begin{cases} 2x + 4y = \frac{2}{3} \\ 3x - 5y = -10 \end{cases}$

17. $\begin{cases} 2x + y = 1 \\ 4x + 2y = 3 \end{cases}$

18. $\begin{cases} x - y = 5 \\ -3x + 3y = 2 \end{cases}$

19. $\begin{cases} 2x - y = 0 \\ 3x + 2y = 7 \end{cases}$

20. $\begin{cases} 3x + 3y = -1 \\ 4x + y = \frac{8}{3} \end{cases}$

21. $\begin{cases} x + 2y = 4 \\ 2x + 4y = 8 \end{cases}$

22. $\begin{cases} 3x - y = 7 \\ 9x - 3y = 21 \end{cases}$

23. $\begin{cases} 2x - 3y = -1 \\ 10x + y = 11 \end{cases}$

24. $\begin{cases} 3x - 2y = 0 \\ 5x + 10y = 4 \end{cases}$

25. $\begin{cases} 2x + 3y = 6 \\ x - y = \frac{1}{2} \end{cases}$

26. $\begin{cases} \frac{1}{2}x + y = -2 \\ x - 2y = 8 \end{cases}$

27. $\begin{cases} \frac{1}{2}x + \frac{1}{3}y = 3 \\ \frac{1}{4}x - \frac{2}{3}y = -1 \end{cases}$

28. $\begin{cases} \frac{1}{3}x - \frac{3}{2}y = -5 \\ \frac{3}{4}x + \frac{1}{3}y = 11 \end{cases}$

29. $\begin{cases} 3x - 5y = 3 \\ 15x + 5y = 21 \end{cases}$

30. $\begin{cases} 2x - y = -1 \\ x + \frac{1}{2}y = \frac{3}{2} \end{cases}$

31. $\begin{cases} \dfrac{1}{x} + \dfrac{1}{y} = 8 \\ \dfrac{3}{x} - \dfrac{5}{y} = 0 \end{cases}$

[**Hint:** Let $u = 1/x$ and $v = 1/y$, and solve for u and v. Then $x = 1/u$ and $y = 1/v$.]

32. $\begin{cases} \dfrac{4}{x} - \dfrac{3}{y} = 0 \\ \dfrac{6}{x} + \dfrac{3}{2y} = 2 \end{cases}$

In Problems 33–38, use a graphing utility to solve each system of equations. Express the solution rounded to two decimal places.

33. $\begin{cases} y = \sqrt{2}x - 20\sqrt{7} \\ y = -0.1x + 20 \end{cases}$

34. $\begin{cases} y = -\sqrt{3}x + 100 \\ y = 0.2x + \sqrt{19} \end{cases}$

35. $\begin{cases} \sqrt{2}x + \sqrt{3}y + \sqrt{6} = 0 \\ \sqrt{3}x - \sqrt{2}y + 60 = 0 \end{cases}$

36. $\begin{cases} \sqrt{5}x - \sqrt{6}y + 60 = 0 \\ 0.2x + 0.3y + \sqrt{5} = 0 \end{cases}$

37. $\begin{cases} \sqrt{3}x + \sqrt{2}y = \sqrt{0.3} \\ 100x - 95y = 20 \end{cases}$

38. $\begin{cases} \sqrt{6}x - \sqrt{5}y + \sqrt{1.1} = 0 \\ y = -0.2x + 0.1 \end{cases}$

39. Supply and Demand Suppose that the quantity supplied Q_s and quantity demanded Q_d of T-shirts at a concert is given by the following equations:

$$Q_s = -200 + 50p$$

$$Q_d = 1000 - 25p$$

where p is the price. The equilibrium price of a market is defined as the price at which quantity supplied equals quantity demanded ($Q_s = Q_d$). Find the equilibrium price for T-shirts at this concert. What is the equilibrium quantity?

40. Supply and Demand Suppose that the quantity supplied Q_s and quantity demanded Q_d of hot dogs at a baseball game is given by the following equations:

$$Q_s = -2000 + 3000p$$

$$Q_d = 10{,}000 - 1000p$$

where p is the price. The equilibrium price of a market is defined as the price at which quantity supplied equals quantity demanded ($Q_s = Q_d$). Find the equilibrium price for hot dogs at the baseball game. What is the equilibrium quantity?

41. The perimeter of a rectangular floor is 90 feet. Find the dimensions of the floor if the length is twice the width.

42. The length of fence required to enclose a rectangular field is 3000 meters. What are the dimensions of the field if it is known that the difference between its length and width is 50 meters?

43. Cost of Fast Food Four large cheeseburgers and two chocolate shakes cost a total of \$7.90. Two shakes cost 15¢ more than one cheeseburger. What is the cost of a cheeseburger? A shake?

44. Movie Theater Tickets A movie theater charges \$9.00 for adults and \$7.00 for senior citizens. On a day when 325 people paid an admission, the total receipts were \$2495. How many who paid were adults? How many were seniors?

45. Mixing Nuts A store sells cashews for \$5.00 per pound and peanuts for \$1.50 per pound. The manager decides to mix 30 pounds of peanuts with some cashews and sell the mixture for \$3.00 per pound. How many pounds of cashews should be mixed with the peanuts so that the mixture will produce the same revenue as would selling the nuts separately?

46. Financial Planning A recently retired couple need \$12,000 per year to supplement their Social Security. They have \$150,000 to invest to obtain this income. They have decided on two investment options: AA bonds yielding 10% per annum and a Bank Certificate yielding 5%.

(a) How much should be invested in each to realize exactly \$12,000?

(b) If, after two years, the couple requires \$14,000 per year in income, how should they reallocate their investment to achieve the new amount?

47. Computing Wind Speed With a tail wind, a small Piper aircraft can fly 600 miles in 3 hours. Against this same wind, the Piper can fly the same distance in 4 hours. Find the average wind speed and the average airspeed of the Piper.

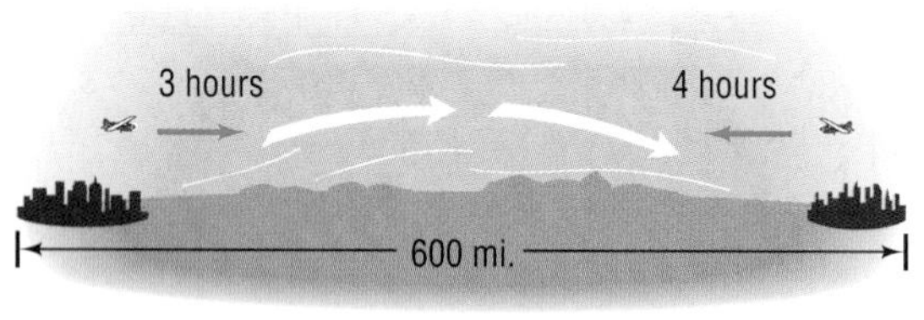
3 hours
4 hours
600 mi.

48. Computing Wind Speed The average airspeed of a single-engine aircraft is 150 miles per hour. If the aircraft flew the same distance in 2 hours with the wind as it flew in 3 hours against the wind, what was the wind speed?

49. Restaurant Management A restaurant manager wants to purchase 200 sets of dishes. One design costs \$25 per set, while another costs \$45 per set. If she only has \$7400 to spend, how many of each design should be ordered?

50. Cost of Fast Food One group of people purchased 10 hot dogs and 5 soft drinks at a cost of \$12.50. A second bought 7 hot dogs and 4 soft drinks at a cost of \$9.00. What is the cost of a single hot dog? A single soft drink?

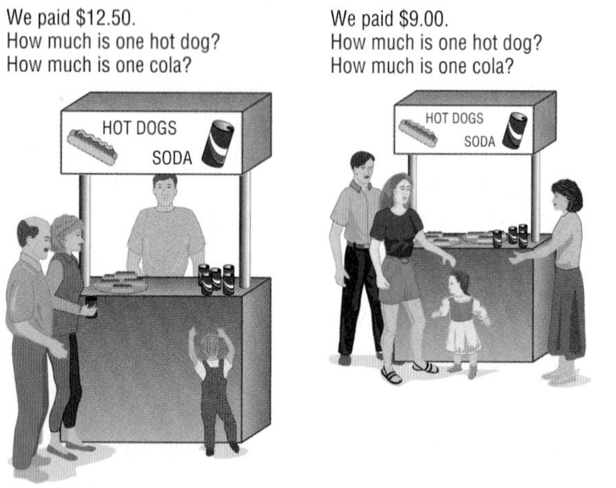
We paid \$12.50.
How much is one hot dog?
How much is one cola?
HOT DOGS
SODA
We paid \$9.00.
How much is one hot dog?
How much is one cola?
HOT DOGS
SODA

51. Computing a Refund The grocery store we use does not mark prices on its goods. My wife went to this store, bought three 1-pound packages of bacon and two cartons of eggs, and paid a total of \$7.45. Not knowing that she went to the store, I also went to the same store, purchased two 1-pound packages of bacon and three cartons of eggs, and paid a total of \$6.45. Now we want to return two 1-pound packages of bacon and two cartons of eggs. How much will be refunded?

52. Finding the Current of a Stream Pamela requires 3 hours to swim 15 miles downstream on the Illinois River. The return trip upstream takes 5 hours. Find Pamela's average speed in still water. How fast is the current? (Assume that Pamela's speed is the same in each direction.)

The point at which a company's profits equal zero is called the company's ***break-even point.*** *For Problems 53 and 54, let R represent a company's revenue, let C represent the company's costs, and let x represent the number of units produced and sold each day. Find the firm's break-even point; that is, find x so that $R = C$.*

53. $R = 8x$
$C = 4.5x + 17{,}500$

54. $R = 12x$
$C = 10x + 15{,}000$

55. Curve Fitting Find real numbers b and c such that the parabola $y = x^2 + bx + c$ passes through the points $(1, 2)$ and $(-1, 3)$.

56. Curve Fitting Find real numbers b and c such that the parabola $y = x^2 + bx + c$ passes through the points $(1, 3)$ and $(3, 5)$.

57. Make up a system of two linear equations containing two variables that has:
(a) No solution
(b) Exactly one solution
(c) Infinitely many solutions

Give the three systems to a friend to solve and critique.

58. Write a brief paragraph outlining your strategy for solving a system of two linear equations containing two variables.

59. Do you prefer the method of substitution or the method of elimination for solving a system of two linear equations containing two variables? Give reasons.

11.2 SYSTEMS OF LINEAR EQUATIONS: THREE EQUATIONS CONTAINING THREE VARIABLES

OBJECTIVES

1. Solve Systems of Three Equations Containing Three Variables
2. Inconsistent Systems
3. Dependent Equations

Just as with a system of two linear equations containing two variables, a system of three linear equations containing three variables also has either (1) exactly one solution (a consistent system with independent equations), or (2) no solution (an inconsistent system), or (3) infinitely many solutions (a consistent system with dependent equations).

We can view the problem of solving a system of three linear equations containing three variables as a geometry problem. The graph of each equation in such a system is a plane in space. Thus, a system of three linear equations containing three variables represents three planes in space. Figure 5 illustrates some of the possibilities.

Figure 5

(a) Consistent system; one solution **(b)** Consistent system; infinite number of solutions **(c)** Inconsistent system; no solution

Recall that a **solution** to a system of equations consists of values for the variables that are solutions of each equation of the system. For example, $x = 3, y = -1, z = -5$ is a solution to the system of equations

$$\begin{cases} x + y + z = -3 & (1) \quad 3 + (-1) + (-5) = -3 \\ 2x - 3y + 6z = -21 & (2) \quad 2(3) - 3(-1) + 6(-5) = 6 + 3 - 30 = -21 \\ -3x + 5y = -14 & (3) \quad -3(3) + 5(-1) = -9 - 5 = -14 \end{cases}$$

because these values of the variables are solutions of each equation.

Typically, when solving a system of three linear equations containing three variables, we use the method of elimination. Recall that the idea behind the method of elimination is to form equivalent equations until a solution is reached. We list the rules for obtaining an equivalent system of equations as a reminder.

RULES FOR OBTAINING AN EQUIVALENT SYSTEM OF EQUATIONS

1. Interchange any two equations of the system.
2. Multiply (or divide) each side of an equation by the same nonzero constant.
3. Replace any equation in the system by the sum (or difference) of that equation and a nonzero multiple of any other equation in the system.

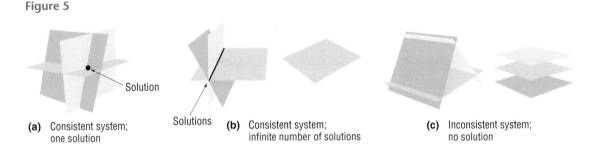

1 Let's see how elimination works on a system of three equations containing three variables.

EXAMPLE 1 Solving a System of Three Linear Equations with Three Variables

Use the method of elimination to solve the system of equations.

$$\begin{cases} x + y - z = -1 & (1) \\ 4x - 3y + 2z = 16 & (2) \\ 2x - 2y - 3z = 5 & (3) \end{cases}$$

Solution For a system of three equations, we attempt to eliminate one variable at a time, using pairs of equations. Our plan of attack on this system will be to first eliminate the variable x from equations (2) and (3). Next, we will eliminate the variable y from equation (3), leaving only the variable z. Back-substitution can then be used to obtain the values of y and then x.

We begin by multiplying each side of equation (1) by -4 and adding the result to equation (2). (Do you see why? The coefficients on x are now opposites of each other.) We also multiply equation (1) by -2 and add the result to equation (3). Notice that these two procedures result in the removal of the x-variable from equations (2) and (3).

$$\begin{array}{rl} -4x - 4y + 4z = 4 & (1) \text{ Multiply by } -4 \\ 4x - 3y + 2z = 16 & (2) \text{ and Add} \\ \hline -7y + 6z = 20 & \end{array}$$

$$\begin{array}{rl} -2x - 2y + 2z = 2 & (1) \text{ Multiply by } -2 \\ 2x - 2y - 3z = 5 & (3) \text{ and Add} \\ \hline -4y - z = 7 & \end{array}$$

$$\begin{cases} x + y - z = -1 & (1) \\ -7y + 6z = 20 & (2) \\ -4y - z = 7 & (3) \end{cases}$$

We now concentrate on equations (2) and (3), treating them as a system of two equations containing two variables. We multiply each side of equation (2) by 4 and each side of equation (3) by -7 and add these equations. The result is the new equation (3).

$$\begin{array}{rlr} -7y + 6z = 20 & \text{Multiply by } 4 & -28y + 24z = 80 \\ -4y - z = 7 & \text{Multiply by } -7 & 28y + 7z = -49 \quad \text{Add} \\ \hline & & 31z = 31 \end{array}$$

$$\begin{cases} x + y - z = -1 & (1) \\ -7y + 6z = 20 & (2) \\ 31z = 31 & (3) \end{cases}$$

We now solve equation (3) for z by dividing both sides of the equation by 31.

$$\begin{cases} x + y - z = -1 & (1) \\ -7y + 6z = 20 & (2) \\ z = 1 & (3) \end{cases}$$

Back-substitute $z = 1$ in equation (2) and solve for y.

$$\begin{aligned} -7y + 6z &= 20 && (2) \\ -7y + 6(1) &= 20 && z = 1. \\ -7y &= 14 && \text{Subtract 6 from both sides of the equation.} \\ y &= -2 && \text{Divide both sides of the equation by } -7. \end{aligned}$$

Finally, we back-substitute $y = -2$ and $z = 1$ in equation (1) and solve for x.

$$\begin{aligned} x + y - z &= -1 && (1) \\ x + (-2) - 1 &= -1 && y = -2 \text{ and } z = 1. \\ x - 3 &= -1 && \text{Simplify.} \\ x &= 2 && \text{Add 3 to both sides.} \end{aligned}$$

The solution of the original system is $x = 2$, $y = -2$, $z = 1$. You should verify this solution. ■

Look back over the solution given in Example 1. Note the pattern of removing one of the variables from two of the equations, followed by solving this system of two equations and two unknowns. Although which variables to remove is your choice, the methodology remains the same for all systems.

NOW WORK PROBLEM 1.

2 The previous example was a consistent system that had a unique solution. The next two examples deal with the two other possibilities that may occur.

EXAMPLE 2 An Inconsistent System of Linear Equations

Solve: $\begin{cases} 2x + y - z = -2 & (1) \\ x + 2y - z = -9 & (2) \\ x - 4y + z = 1 & (3) \end{cases}$

Solution Our plan of attack is the same as in Example 1. We eliminate the variable x from equations (2) and (3); then we eliminate y from equation (3), leaving only the variable z.

We begin by interchanging equations (1) and (2). This is done so that the coefficient of x in (1) is 1. Now it is easier to eliminate the x-variable from equations (2) and (3).

$$\begin{cases} x + 2y - z = -9 & (1) \\ 2x + y - z = -2 & (2) \\ x - 4y + z = 1 & (3) \end{cases}$$

Multiply each side of equation (1) by -2 and add the result to equation (2). Multiply each side of equation (1) by -1 and add the result to equation (3).

$$\begin{aligned} -2x - 4y + 2z &= 18 && (1) \text{ Multiply by } -2 \text{ and Add} \\ 2x + y - z &= -2 && (2) \\ \hline -3y + z &= 16 \end{aligned}$$

$$\begin{aligned} -x - 2y + z &= 9 && (1) \text{ Multiply by } -1 \text{ and Add} \\ x - 4y + z &= 1 && (3) \\ \hline -6y + 2z &= 10 \end{aligned}$$

$$\begin{cases} x + 2y - z = -9 & (1) \\ -3y + z = 16 & (2) \\ -6y + 2z = 10 & (3) \end{cases}$$

We now concentrate on equations (2) and (3), treating them as a system of two equations containing two variables. Multiply each side of equation (2) by -2 and add the result to equation (3).

$$\begin{array}{l} -3y + z = 16 \\ -6y + 2z = 10 \end{array} \quad \text{Multiply by } -2 \quad \begin{array}{r} 6y - 2z = -32 \\ -6y + 2z = 10 \\ \hline 0 = -22 \end{array} \quad \text{Add} \quad \begin{cases} x + 2y - z = -9 & (1) \\ -3y + z = 16 & (2) \\ 0 = -22 & (3) \end{cases}$$

Equation (3) has no solution and thus the system is inconsistent. ■

3 Now let's look at a system of equations with infinitely many solutions.

EXAMPLE 3 Solving a System of Linear Equations with Infinitely Many Solutions

$$\text{Solve: } \begin{cases} x - 2y - z = 8 & (1) \\ 2x - 3y + z = 23 & (2) \\ 4x - 5y + 5z = 53 & (3) \end{cases}$$

Solution Multiply each side of equation (1) by −2 and add the result to equation (2). Also, multiply each side of equation (1) by −4 and add the result to equation (3).

$$\begin{array}{rl} -2x + 4y + 2z = -16 & (1) \text{ Multiply by } -2 \\ 2x - 3y + z = 23 & (2) \text{ and Add} \\ \hline y + 3z = 7 & \\ \\ -4x + 8y + 4z = -32 & (1) \text{ Multiply by } -4 \\ 4x - 5y + 5z = 53 & (2) \text{ and Add} \\ \hline 3y + 9z = 21 & \end{array} \qquad \begin{cases} x - 2y - z = 8 & (1) \\ y + 3z = 7 & (2) \\ 3y + 9z = 21 & (3) \end{cases}$$

Treat equations (2) and (3) as a system of two equations containing two variables, and eliminate the y variable by multiplying both sides of equation (2) by −3 and adding the result to equation (3).

$$\begin{array}{l} y + 3z = 7 \\ 3y + 9z = 21 \end{array} \quad \text{Multiply by} -3 \quad \begin{array}{r} -3y - 9z = -21 \\ 3y + 9z = 21 \\ \hline 0 = 0 \end{array} \quad \text{Add} \quad \begin{cases} x - 2y - z = 8 & (1) \\ y + 3z = 7 & (2) \\ 0 = 0 & (3) \end{cases}$$

The original system is equivalent to a system containing two equations, so the equations are dependent and the system has infinitely many solutions. If we let z represent any real number, then, solving equation (2) for y, we determine that $y = -3z + 7$. Substitute this expression into equation (1) to determine x in terms of z.

$$\begin{array}{rl} x - 2y - z = 8 & (1) \\ x - 2(-3z + 7) - z = 8 & y = -3z + 7 \\ x + 6z - 14 - z = 8 & \text{Remove parentheses.} \\ x + 5z = 22 & \text{Combine like terms.} \\ x = -5z + 22 & \text{Solve for } x. \end{array}$$

We will write the solution to the system as

$$\begin{cases} x = -5z + 22 \\ y = -3z + 7 \end{cases}$$

where z can be any real number.

To find specific solutions to the system, choose any value of z and use the equations $x = -5z + 22$ and $y = -3z + 7$ to determine x and y. For example, if $z = 0$, then $x = 22$ and $y = 7$, and if $z = 1$, then $x = 17$ and $y = 4$. ∎

NOW WORK PROBLEM 5

Two points in the Cartesian plane determine a unique line. Given three noncollinear points, we can find the (unique) quadratic function whose graph contains these three points.

EXAMPLE 4 Curve Fitting

Find real numbers a, b, and c so that the graph of the quadratic function $y = ax^2 + bx + c$ contains the points $(-1, -4)$, $(1, 6)$, and $(3, 0)$.

Solution For the point $(-1, -4)$ we have: $-4 = a(-1)^2 + b(-1) + c$ $\quad -4 = a - b + c$

For the point $(1, 6)$ we have: $6 = a(1)^2 + b(1) + c$ $\quad 6 = a + b + c$

For the point $(3, 0)$ we have: $0 = a(3)^2 + b(3) + c$ $\quad 0 = 9a + 3b + c$

We wish to determine a, b, and c such that each equation is satisfied. That is, we want to solve the following system of three equations containing three variables:

$$\begin{cases} a - b + c = -4 & (1) \\ a + b + c = 6 & (2) \\ 9a + 3b + c = 0 & (3) \end{cases}$$

Figure 6

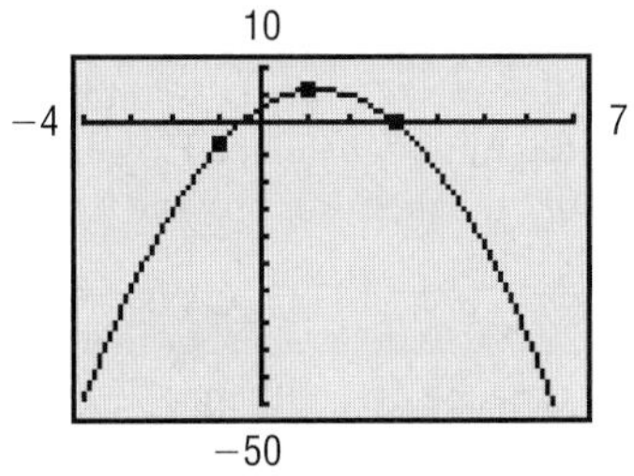

Solving this system of equations, we obtain $a = -2$, $b = 5$, and $c = 3$. So the quadratic function whose graph contains the points $(-1, -4)$, $(1, 6)$, and $(3, 0)$ is

$$y = -2x^2 + 5x + 3 \qquad y = ax^2 + bx + c, \quad a = -2, b = 5, c = 3$$

Figure 6 shows the graph of the function along with the three points. ∎

11.2 EXERCISES

In Problems 1–14, solve each system of equations. If the system has no solution, say that it is inconsistent.

1. $\begin{cases} x - y = 6 \\ 2x - 3z = 16 \\ 2y + z = 4 \end{cases}$

2. $\begin{cases} 2x + y = -4 \\ -2y + 4z = 0 \\ 3x - 2z = -11 \end{cases}$

3. $\begin{cases} x - 2y + 3z = 7 \\ 2x + y + z = 4 \\ -3x + 2y - 2z = -10 \end{cases}$

4. $\begin{cases} 2x + y - 3z = 0 \\ -2x + 2y + z = -7 \\ 3x - 4y - 3z = 7 \end{cases}$

5. $\begin{cases} x - y - z = 1 \\ 2x + 3y + z = 2 \\ 3x + 2y = 0 \end{cases}$

6. $\begin{cases} 2x - 3y - z = 0 \\ -x + 2y + z = 5 \\ 3x - 4y - z = 1 \end{cases}$

7. $\begin{cases} x - y - z = 1 \\ -x + 2y - 3z = -4 \\ 3x - 2y - 7z = 0 \end{cases}$

8. $\begin{cases} 2x - 3y - z = 0 \\ 3x + 2y + 2z = 2 \\ x + 5y + 3z = 2 \end{cases}$

9. $\begin{cases} 2x - 2y + 3z = 6 \\ 4x - 3y + 2z = 0 \\ -2x + 3y - 7z = 1 \end{cases}$

10. $\begin{cases} 3x - 2y + 2z = 6 \\ 7x - 3y + 2z = -1 \\ 2x - 3y + 4z = 0 \end{cases}$

11. $\begin{cases} x + y - z = 6 \\ 3x - 2y + z = -5 \\ x + 3y - 2z = 14 \end{cases}$

12. $\begin{cases} x - y + z = -4 \\ 2x - 3y + 4z = -15 \\ 5x + y - 2z = 12 \end{cases}$

13. $\begin{cases} x + 2y - z = -3 \\ 2x - 4y + z = -7 \\ -2x + 2y - 3z = 4 \end{cases}$

14. $\begin{cases} x + 4y - 3z = -8 \\ 3x - y + 3z = 12 \\ x + y + 6z = 1 \end{cases}$

15. Curve Fitting Find real numbers a, b, and c such that the graph of the function $y = ax^2 + bx + c$ contains the points $(-1, 4)$, $(2, 3)$, and $(0, 1)$.

16. Curve Fitting Find real numbers a, b, and c such that the graph of the function $y = ax^2 + bx + c$ contains the points $(-1, -2)$, $(1, -4)$, and $(2, 4)$.

17. Electricity: Kirchhoff's Rules An application of Kirchhoff's Rules to the circuit shown results in the following system of equations:

$$\begin{cases} I_2 = I_1 + I_3 \\ 5 - 3I_1 - 5I_2 = 0 \\ 10 - 5I_2 - 7I_3 = 0 \end{cases}$$

Find the currents I_1, I_2, and I_3.*

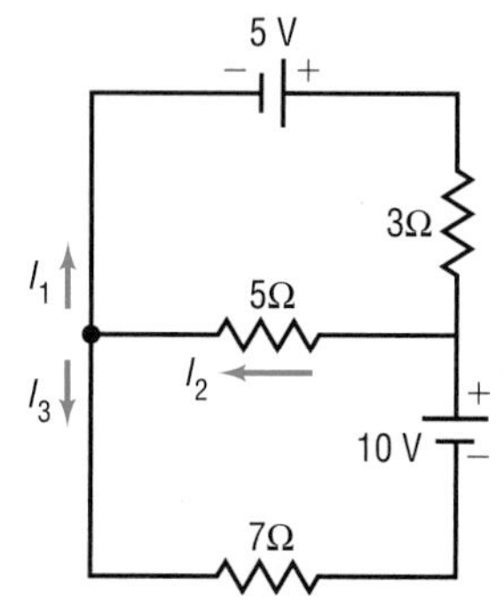

18. Electricity: Kirchhoff's Rules An application of Kirchhoff's Rules to the circuit shown results in the following system of equations:

$$\begin{cases} I_3 = I_1 + I_2 \\ 8 = 4I_3 + 6I_2 \\ 8I_1 = 4 + 6I_2 \end{cases}$$

Find the currents I_1, I_2, and I_3.†

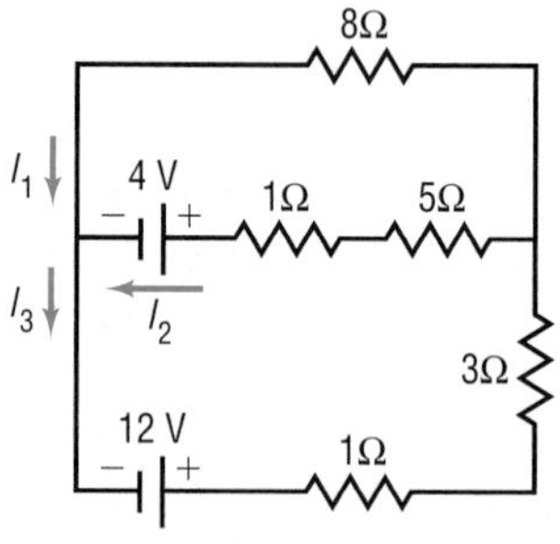

19. Theater Revenues A Broadway theater has 500 seats, divided into orchestra, main, and balcony seating. Orchestra seats sell for \$50, main seats for \$35, and balcony seats for \$25. If all the seats are sold, the gross revenue to the theater is \$17,100. If all the main and balcony seats are sold, but only half the orchestra seats are sold, the gross revenue is \$14,600. How many are there of each kind of seat?

20. Theater Revenues A movie theater charges \$8.00 for adults, \$4.50 for children, and \$6.00 for senior citizens. One day the theater sold 405 tickets and collected \$2320 in receipts. There were twice as many children's tickets sold as adult tickets. How many adults, children, and senior citizens went to the theater that day?

21. Nutrition A dietician wishes a patient to have a meal that has 66 grams of protein, 94.5 grams of carbohydrates, and 910 milligrams of calcium. The hospital food service tells the dietician that the dinner for today is chicken, corn, and 2% milk. Each serving of chicken has 30 grams of protein, 35 grams of carbohydrates, and 200 milligrams of calcium. Each serving of corn has 3 grams of protein, 16 grams of carbohydrates, and 10 milligrams of calcium. Each glass of 2% milk has 9 grams of protein, 13 grams of carbohydrates, and 300 milligrams of calcium. How many servings of each food should the dietitian provide for the patient?

22. Investments Kelly has \$20,000 to invest. As her financial planner, you recommend that she diversify into three investments: Treasury bills that yield 5% simple interest, Treasury bonds that yield 7% simple interest, and corporate bonds that yield 10% simple interest. Kelly wishes to earn \$1390 per year in income. Also, Kelly wants her investment in Treasury bills to be \$3000 more than her investment in corporate bonds. How much money should Kelly place in each investment?

* *Source:* Based on Raymond Serway, *Physics*, 3rd ed. (Philadelphia: Saunders, 1990), Prob. 26, p. 790.

† *Source:* Ibid., Prob. 27, p. 790.

11.3 SYSTEMS OF LINEAR EQUATIONS: MATRICES

OBJECTIVES
1. Write the Augmented Matrix of a System of Linear Equations
2. Write the System from the Augmented Matrix
3. Perform Row Operations on a Matrix
4. Solve Systems of Linear Equations Using Matrices
5. Use a Graphing Utility to Solve a System of Linear Equations

The systematic approach of the method of elimination for solving a system of linear equations provides another method of solution that involves a simplified notation.

Consider the following system of linear equations:

$$\begin{cases} x + 4y = 14 \\ 3x - 2y = 0 \end{cases}$$

If we choose not to write the symbols used for the variables, we can represent this system as

$$\left[\begin{array}{cc|c} 1 & 4 & 14 \\ 3 & -2 & 0 \end{array}\right]$$

where it is understood that the first column represents the coefficients of the variable x, the second column the coefficients of y, and the third column the constants on the right side of the equal signs. The vertical line serves as a reminder of the equal signs. The large square brackets are the traditional symbols used to denote a *matrix* in algebra.

A **matrix** is defined as a rectangular array of numbers,

$$\begin{array}{l} \\ \text{Row } 1 \\ \text{Row } 2 \\ \vdots \\ \text{Row } i \\ \vdots \\ \text{Row } m \end{array} \begin{array}{c} \begin{array}{ccccc} \text{Column } 1 & \text{Column } 2 & & \text{Column } j & & \text{Column } n \end{array} \\ \begin{bmatrix} a_{11} & a_{12} & \cdots & a_{1j} & \cdots & a_{1n} \\ a_{21} & a_{22} & \cdots & a_{2j} & \cdots & a_{2n} \\ \vdots & \vdots & & \vdots & & \vdots \\ a_{i1} & a_{i2} & \cdots & a_{ij} & \cdots & a_{in} \\ \vdots & \vdots & & \vdots & & \vdots \\ a_{m1} & a_{m2} & \cdots & a_{mj} & \cdots & a_{mn} \end{bmatrix} \end{array} \qquad \textbf{(1)}$$

Each number a_{ij} of the matrix has two indexes: the **row index** i and the **column index** j. The matrix shown in display (1) has m rows and n columns. The numbers a_{ij} are usually referred to as the **entries** of the matrix. For example, a_{23} refers to the entry in the second row, third column.

1 Now we will use matrix notation to represent a system of linear equations. The matrices used to represent systems of linear equations are called **augmented matrices.** In writing the augmented matrix of a system, the variables of each equation must be on the left side of the equal sign and the constants on the right side. A variable that does not appear in an equation has a coefficient of 0.

EXAMPLE 1 Writing the Augmented Matrix of a System of Linear Equations

Write the augmented matrix of each system of equations.

(a) $\begin{cases} 3x - 4y = -6 & (1) \\ 2x - 3y = -5 & (2) \end{cases}$ (b) $\begin{cases} 2x - y + z = 0 & (1) \\ x + z - 1 = 0 & (2) \\ x + 2y - 8 = 0 & (3) \end{cases}$

Solution (a) The augmented matrix is

$$\left[\begin{array}{cc|c} 3 & -4 & -6 \\ 2 & -3 & -5 \end{array}\right]$$

(b) Care must be taken that the system be written so that the coefficients of all variables are present (if any variable is missing, its coefficient is 0). Also, all constants must be to the right of the equal sign. Thus, we need to rearrange the given system as follows:

$$\begin{cases} 2x - y + z = 0 & (1) \\ x + z - 1 = 0 & (2) \\ x + 2y - 8 = 0 & (3) \end{cases}$$

$$\begin{cases} 2x - y + z = 0 & (1) \\ x + 0 \cdot y + z = 1 & (2) \\ x + 2y + 0 \cdot z = 8 & (3) \end{cases}$$

The augmented matrix is

$$\left[\begin{array}{ccc|c} 2 & -1 & 1 & 0 \\ 1 & 0 & 1 & 1 \\ 1 & 2 & 0 & 8 \end{array}\right]$$

If we do not include the constants to the right of the equal sign, that is, to the right of the vertical bar in the augmented matrix of a system of equations, the resulting matrix is called the **coefficient matrix** of the system. For the systems discussed in Example 1, the coefficient matrices are

$$\begin{bmatrix} 3 & -4 \\ 2 & -3 \end{bmatrix} \quad \text{and} \quad \begin{bmatrix} 2 & -1 & 1 \\ 1 & 0 & 1 \\ 1 & 2 & 0 \end{bmatrix}$$

NOW WORK PROBLEM 3.

EXAMPLE 2 Writing the System of Linear Equations from the Augmented Matrix

2 Write the system of linear equations corresponding to each augmented matrix.

(a) $\left[\begin{array}{cc|c} 5 & 2 & 13 \\ -3 & 1 & -10 \end{array}\right]$ (b) $\left[\begin{array}{ccc|c} 3 & -1 & -1 & 7 \\ 2 & 0 & 2 & 8 \\ 0 & 1 & 1 & 0 \end{array}\right]$

Solution (a) The matrix has two rows and so represents a system of two equations. The two columns to the left of the vertical bar indicate that the system has two variables. If x and y are used to denote these variables, the system of equations is

$$\begin{cases} 5x + 2y = 13 & (1) \\ -3x + y = -10 & (2) \end{cases}$$

(b) This matrix represents a system of three equations containing three variables. If x, y, and z are the three variables, this system is

$$\begin{cases} 3x - y - z = 7 & (1) \\ 2x + 2z = 8 & (2) \\ y + z = 0 & (3) \end{cases}$$

Row Operations on a Matrix

3 **Row operations** on a matrix are used to solve systems of equations when the system is written as an augmented matrix. There are three basic row operations.

Row Operations

1. Interchange any two rows.
2. Replace a row by a nonzero multiple of that row.
3. Replace a row by the sum of that row and a constant nonzero multiple of some other row.

These three row operations correspond to the three rules given earlier for obtaining an equivalent system of equations. Thus, when a row operation is performed on a matrix, the resulting matrix represents a system of equations equivalent to the system represented by the original matrix.

For example, consider the augmented matrix

$$\left[\begin{array}{cc|c} 1 & 2 & 3 \\ 4 & -1 & 2 \end{array}\right]$$

Suppose that we want to apply a row operation to this matrix that results in a matrix whose entry in row 2, column 1 is a 0. The row operation to use is

Multiply each entry in row 1 by -4 and add the result to the corresponding entries in row 2. **(2)**

If we use R_2 to represent the new entries in row 2 and we use r_1 and r_2 to represent the original entries in rows 1 and 2, respectively, then we can represent the row operation in statement (2) by

$$R_2 = -4r_1 + r_2$$

Then

$$\left[\begin{array}{cc|c} 1 & 2 & 3 \\ 4 & -1 & 2 \end{array}\right] \underset{\uparrow \atop R_2 = -4r_1 + r_2}{\rightarrow} \left[\begin{array}{cc|c} 1 & 2 & 3 \\ -4(1) + 4 & -4(2) + (-1) & -4(3) + 2 \end{array}\right] = \left[\begin{array}{cc|c} 1 & 2 & 3 \\ 0 & -9 & -10 \end{array}\right]$$

As desired, we now have the entry 0 in row 2, column 1.

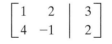

EXAMPLE 3 Applying a Row Operation to an Augmented Matrix

Apply the row operation $R_2 = -3r_1 + r_2$ to the augmented matrix

$$\left[\begin{array}{cc|c} 1 & -2 & 2 \\ 3 & -5 & 9 \end{array}\right]$$

Solution The row operation $R_2 = -3r_1 + r_2$ tells us that the entries in row 2 are to be replaced by the entries obtained after multiplying each entry in row 1 by -3 and adding the result to the corresponding entries in row 2. Thus,

$$\left[\begin{array}{cc|c} 1 & -2 & 2 \\ 3 & -5 & 9 \end{array}\right] \underset{R_2 = -3r_1 + r_2}{\rightarrow} \left[\begin{array}{cc|c} 1 & -2 & 2 \\ -3(1) + 3 & (-3)(-2) + (-5) & -3(2) + 9 \end{array}\right] = \left[\begin{array}{cc|c} 1 & -2 & 2 \\ 0 & 1 & 3 \end{array}\right]$$

NOW WORK PROBLEM 11.

EXAMPLE 4 Finding a Particular Row Operation

Using the matrix

$$\left[\begin{array}{cc|c} 1 & -2 & 2 \\ 0 & 1 & 3 \end{array}\right]$$

find a row operation that will result in a matrix with a 0 in row 1, column 2.

Solution We want a 0 in row 1, column 2. This result can be accomplished by multiplying row 2 by 2 and adding the result to row 1. That is, we apply the row operation $R_1 = 2r_2 + r_1$.

$$\left[\begin{array}{cc|c} 1 & -2 & 2 \\ 0 & 1 & 3 \end{array}\right] \underset{R_1 = 2r_2 + r_1}{\rightarrow} \left[\begin{array}{cc|c} 2(0) + 1 & 2(1) + (-2) & 2(3) + 2 \\ 0 & 1 & 3 \end{array}\right] = \left[\begin{array}{cc|c} 1 & 0 & 8 \\ 0 & 1 & 3 \end{array}\right]$$

A word about the notation that we have introduced. A row operation such as $R_1 = 2r_2 + r_1$ changes the entries in row 1. Note also that for this type of row operation we change the entries in a given row by multiplying the entries in some other row by an appropriate nonzero number and adding the results to the original entries of the row to be changed.

4 To use row operations to solve a system of linear equations, we transform the augmented matrix of the system and obtain a matrix that is in *echelon form.*

A matrix is in **echelon form** when

1. The entry in row 1, column 1 is a 1, and 0's appear below it.
2. The first nonzero entry in each row after the first row is a 1, 0's appear below it, and it appears to the right of the first nonzero entry in any row above.
3. Any rows that contain all 0's to the left of the vertical bar appear at the bottom.

For example, for a system of three equations containing three variables with a unique solution, the augmented matrix is in echelon form if it is of the form

$$\left[\begin{array}{ccc|c} 1 & a & b & d \\ 0 & 1 & c & e \\ 0 & 0 & 1 & f \end{array}\right]$$

where a, b, c, d, e, and f are real numbers. The last row of the augmented matrix states that $z = f$. We can then determine the value of y using back-substitution with $z = f$, since row 2 represents the equation $y + cz = e$. Finally, x is determined using back-substitution again.

Two advantages of solving a system of equations by writing the augmented matrix in echelon form are the following:

1. The process is algorithmic; that is, it consists of repetitive steps that can be programmed on a computer.
2. The process works on any system of linear equations, no matter how many equations or variables are present.

The next example shows how to write a matrix in echelon form.

EXAMPLE 5 Solving a System of Linear Equations Using Matrices (Echelon Form)

Solve: $\begin{cases} 4x + 3y = 11 & (1) \\ x - 3y = -1 & (2) \end{cases}$

Solution First, we write the augmented matrix that represents this system.

$$\left[\begin{array}{cc|c} 4 & 3 & 11 \\ 1 & -3 & -1 \end{array}\right]$$

The first step requires getting the entry 1 in row 1, column 1. An interchange of rows 1 and 2 is the easiest way to do this.

$$\left[\begin{array}{cc|c} 1 & -3 & -1 \\ 4 & 3 & 11 \end{array}\right]$$

Next, we want a 0 under the entry 1 in column 1. We use the row operation $R_2 = -4r_1 + r_2$.

$$\left[\begin{array}{cc|c} 1 & -3 & -1 \\ 4 & 3 & 11 \end{array}\right] \underset{\uparrow \atop R_2 = -4r_1 + r_2}{\rightarrow} \left[\begin{array}{cc|c} 1 & -3 & -1 \\ 0 & 15 & 15 \end{array}\right]$$

Now we want the entry 1 in row 2, column 2. We use $R_2 = \frac{1}{15}r_2$.

$$\left[\begin{array}{cc|c} 1 & -3 & -1 \\ 0 & 15 & 15 \end{array}\right] \underset{\uparrow \atop R_2 = \frac{1}{15}r_2}{\rightarrow} \left[\begin{array}{cc|c} 1 & -3 & -1 \\ 0 & 1 & 1 \end{array}\right]$$

The matrix on the right is the echelon form of the augmented matrix. The second row of this matrix represents the equation $y = 1$. Using $y = 1$, we back-substitute into the equation $x - 3y = -1$ (from the first row) to get

$$x - 3(1) = -1 \qquad y = 1$$
$$x = 2$$

The solution of the system is $x = 2$, $y = 1$. ■

The steps that we used to solve the system of linear equations in Example 5 can be summarized as follows:

MATRIX METHOD FOR SOLVING A SYSTEM OF LINEAR EQUATIONS (ECHELON FORM)

STEP 1: Write the augmented matrix that represents the system.

STEP 2: Perform row operations that place the entry 1 in row 1, column 1.

STEP 3: Perform row operations that leave the entry 1 in row 1, column 1 unchanged, while causing 0's to appear below it in column 1.

STEP 4: Perform row operations that place the entry 1 in row 2, column 2 and leave the entries in columns to the left unchanged. If it is impossible to place a 1 in row 2, column 2, then proceed to place a 1 in row 2, column 3. Once a 1 is in place, perform row operations to place 0's under it.

STEP 5: Now repeat Step 4, placing a 1 in the next row, but one column to the right. Continue until the bottom row or the vertical bar is reached.

STEP 6: If any rows are obtained that contain only 0's on the left side of the vertical bar, place such rows at the bottom of the matrix.

STEP 7: The matrix that results is the echelon form of the augmented matrix. Analyze the system of equations corresponding to it to solve the original system.

NOW WORK PROBLEM 27.

5 Some graphing utilities have the ability to put an augmented matrix in echelon form. The next example demonstrates this feature using a TI-83 graphing calculator.

EXAMPLE 6 Solving a System of Linear Equations Using Matrices (Echelon Form)

Solve:
$$\begin{cases} x - y + z = 8 & (1) \\ 2x + 3y - z = -2 & (2) \\ 3x - 2y - 9z = 9 & (3) \end{cases}$$

Graphing Solution The augmented matrix of the system is

$$\left[\begin{array}{rrr|r} 1 & -1 & 1 & 8 \\ 2 & 3 & -1 & -2 \\ 3 & -2 & -9 & 9 \end{array}\right]$$

We enter this matrix into our graphing utility and name it A. See Figure 7(a). Using the REF command on matrix A, we obtain the results shown in Figure 7(b). Since the entire matrix does not fit on the screen, we need to scroll right to see the rest of it. See Figure 7(c).

Figure 7

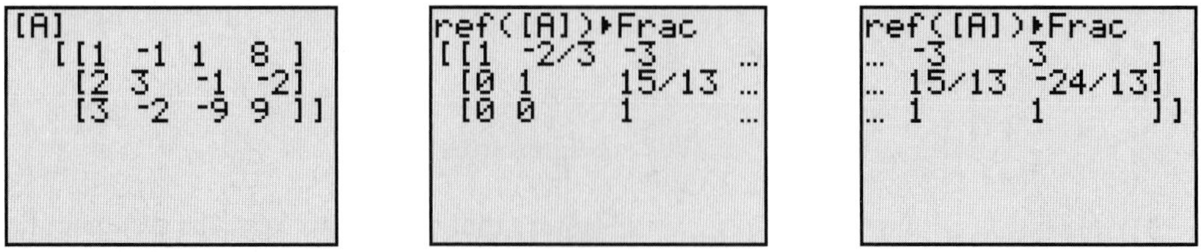

(a) (b) (c)

The system of equations represented by the matrix in echelon form is

$$\begin{cases} x - \frac{2}{3}y - 3z = 3 & (1) \\ y + \frac{15}{13}z = -\frac{24}{13} & (2) \\ z = 1 & (3) \end{cases}$$

Using $z = 1$, we back-substitute to get

$$\begin{cases} x - \frac{2}{3}y - 3(1) = 3 & (1) \\ y + \frac{15}{13}(1) = -\frac{24}{13} & (2) \end{cases} \underset{\text{Simplify.}}{\longrightarrow} \begin{cases} x - \frac{2}{3}y = 6 & (1) \\ y = \frac{-39}{13} = -3 & (2) \end{cases}$$

Solving the second equation for y, we find that $y = -3$. Back-substituting $y = -3$ into $x - \frac{2}{3}y = 6$, we find that $x = 4$. The solution of the system is $x = 4$, $y = -3$, $z = 1$.

Algebraic Solution

STEP 1: The augmented matrix of the system is

$$\left[\begin{array}{rrr|r} 1 & -1 & 1 & 8 \\ 2 & 3 & -1 & -2 \\ 3 & -2 & -9 & 9 \end{array}\right]$$

STEP 2: Because the entry 1 is already present in row 1, column 1, we can go to Step 3.

STEP 3: Perform the row operations $R_2 = -2r_1 + r_2$ and $R_3 = -3r_1 + r_3$. Each of these leaves the entry 1 in row 1, column 1 unchanged, while causing 0's to appear under it.

$$\left[\begin{array}{rrr|r} 1 & -1 & 1 & 8 \\ 2 & 3 & -1 & -2 \\ 3 & -2 & -9 & 9 \end{array}\right] \underset{\substack{\uparrow \\ R_2 = -2r_1 + r_2 \\ R_3 = -3r_1 + r_3}}{\rightarrow} \left[\begin{array}{rrr|r} 1 & -1 & 1 & 8 \\ 0 & 5 & -3 & -18 \\ 0 & 1 & -12 & -15 \end{array}\right]$$

STEP 4: The easiest way to obtain the entry 1 in row 2, column 2 without altering column 1 is to interchange rows 2 and 3 (another way would be to multiply row 2 by $\frac{1}{5}$, but this introduces fractions).

$$\left[\begin{array}{rrr|r} 1 & -1 & 1 & 8 \\ 0 & 1 & -12 & -15 \\ 0 & 5 & -3 & -18 \end{array}\right]$$

To get 0's under the 1 in row 2, column 2, perform the row operation $R_3 = -5r_2 + r_3$.

$$\left[\begin{array}{rrr|r} 1 & -1 & 1 & 8 \\ 0 & 1 & -12 & -15 \\ 0 & 5 & -3 & -18 \end{array}\right] \underset{\substack{\uparrow \\ R_3 = -5r_2 + r_3}}{\rightarrow} \left[\begin{array}{rrr|r} 1 & -1 & 1 & 8 \\ 0 & 1 & -12 & -15 \\ 0 & 0 & 57 & 57 \end{array}\right]$$

STEP 5: Continuing, we obtain a 1 in row 3, column 3 by using $R_3 = \frac{1}{57}r_3$.

$$\left[\begin{array}{rrr|r} 1 & -1 & 1 & 8 \\ 0 & 1 & -12 & -15 \\ 0 & 0 & 57 & 57 \end{array}\right] \underset{\substack{\uparrow \\ R_3 = \frac{1}{57}r_3}}{\rightarrow} \left[\begin{array}{rrr|r} 1 & -1 & 1 & 8 \\ 0 & 1 & -12 & -15 \\ 0 & 0 & 1 & 1 \end{array}\right]$$

STEP 6: The matrix on the right is the echelon form of the augmented matrix. The system of equations represented by the matrix in echelon form is

$$\begin{cases} x - y + z = 8 & (1) \\ \phantom{x - {}} y - 12z = -15 & (2) \\ z = 1 & (3) \end{cases}$$

Using $z = 1$, we back-substitute to get

$$\begin{cases} x - y + 1 = 8 & (1) \\ \phantom{x - {}} y - 12(1) = -15 & (2) \end{cases} \underset{\text{Simplify.}}{\longrightarrow} \begin{cases} x - y = 7 & (1) \\ \phantom{x - {}} y = -3 & (2) \end{cases}$$

We get $y = -3$, and back-substituting into $x - y = 7$, we find that $x = 4$. The solution of the system is $x = 4$, $y = -3$, $z = 1$. ■

Notice that the echelon form of the augmented matrix using the graphing utility differs from the echelon form in our algebraic solution, yet both matrices provide the same solution! This is because the two solutions used different row operations to obtain the echelon form. In all likelihood, the two solutions parted ways in Step 4 of the algebraic solution, where we avoided introducing fractions by interchanging rows 2 and 3.

Sometimes it is advantageous to write a matrix in **reduced row echelon form.** In this form, row operations are used to obtain entries that are 0 above (as well as below) the leading 1 in a row. For example, the echelon form obtained in the algebraic solution to Example 6 is

$$\left[\begin{array}{rrr|r} 1 & -1 & 1 & 8 \\ 0 & 1 & -12 & -15 \\ 0 & 0 & 1 & 1 \end{array}\right]$$

To write this matrix in reduced row echelon form, we proceed as follows:

$$\left[\begin{array}{rrr|r} 1 & -1 & 1 & 8 \\ 0 & 1 & -12 & -15 \\ 0 & 0 & 1 & 1 \end{array}\right] \underset{\substack{\uparrow \\ R_1 = r_2 + r_1}}{\rightarrow} \left[\begin{array}{rrr|r} 1 & 0 & -11 & -7 \\ 0 & 1 & -12 & -15 \\ 0 & 0 & 1 & 1 \end{array}\right] \underset{\substack{\uparrow \\ R_1 = 11r_3 + r_1 \\ R_2 = 12r_3 + r_2}}{\rightarrow} \left[\begin{array}{rrr|r} 1 & 0 & 0 & 4 \\ 0 & 1 & 0 & -3 \\ 0 & 0 & 1 & 1 \end{array}\right]$$

The matrix is now written in reduced row echelon form. The advantage of writing the matrix in this form is that the solution to the system, $x = 4$, $y = -3$, $z = 1$, is readily found, without the need to back-substitute. Another advantage will be seen in Section 11.5, where the inverse of a matrix is discussed.

Figure 8

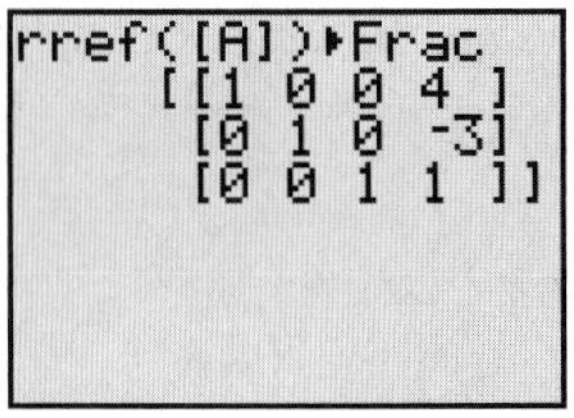

Most graphing utilities also have the ability to put a matrix in reduced row echelon form. Figure 8 shows the reduced row echelon form of the augmented matrix from Example 6 using the RREF command on a TI-83 graphing calculator.

For the remaining examples in this section, we will only provide algebraic solutions to the systems. The reader is encouraged to verify the results using a graphing utility.

NOW WORK PROBLEM 45.

The matrix method for solving a system of linear equations also identifies systems that have infinitely many solutions and systems that are inconsistent. Let's see how.

EXAMPLE 7 Solving a System of Linear Equations Using Matrices

$$\text{Solve: } \begin{cases} 6x - y - z = 4 & (1) \\ -12x + 2y + 2z = -8 & (2) \\ 5x + y - z = 3 & (3) \end{cases}$$

Solution We start with the augmented matrix of the system.

$$\left[\begin{array}{rrr|r} 6 & -1 & -1 & 4 \\ -12 & 2 & 2 & -8 \\ 5 & 1 & -1 & 3 \end{array}\right] \underset{R_1 = -1r_3 + r_1}{\rightarrow} \left[\begin{array}{rrr|r} 1 & -2 & 0 & 1 \\ -12 & 2 & 2 & -8 \\ 5 & 1 & -1 & 3 \end{array}\right] \underset{\substack{R_2 = 12r_1 + r_2 \\ R_3 = -5r_1 + r_3}}{\rightarrow} \left[\begin{array}{rrr|r} 1 & -2 & 0 & 1 \\ 0 & -22 & 2 & 4 \\ 0 & 11 & -1 & -2 \end{array}\right]$$

Obtaining a 1 in row 2, column 2 without altering column 1 can be accomplished only by $R_2 = -\frac{1}{22}r_2$ or by $R_3 = -\frac{1}{11}r_3$ and interchanging rows or by $R_2 = \frac{23}{11}r_3 + r_2$. We shall use the first of these.

$$\left[\begin{array}{rrr|r} 1 & -2 & 0 & 1 \\ 0 & -22 & 2 & 4 \\ 0 & 11 & -1 & -2 \end{array}\right] \underset{R_2 = -\frac{1}{22}r_2}{\rightarrow} \left[\begin{array}{rrr|r} 1 & -2 & 0 & 1 \\ 0 & 1 & -\frac{1}{11} & -\frac{2}{11} \\ 0 & 11 & -1 & -2 \end{array}\right] \underset{R_3 = -11r_2 + r_3}{\rightarrow} \left[\begin{array}{rrr|r} 1 & -2 & 0 & 1 \\ 0 & 1 & -\frac{1}{11} & -\frac{2}{11} \\ 0 & 0 & 0 & 0 \end{array}\right]$$

This matrix is in echelon form. Because the bottom row consists entirely of 0's, the system actually consists of only two equations.

$$\begin{cases} x - 2y = 1 & (1) \\ y - \frac{1}{11}z = -\frac{2}{11} & (2) \end{cases}$$

We shall back-substitute the solution for y from the second equation, $y = \frac{1}{11}z - \frac{2}{11}$, into the first equation to get

$$x = 2y + 1 = 2\left(\tfrac{1}{11}z - \tfrac{2}{11}\right) + 1 = \tfrac{2}{11}z + \tfrac{7}{11}$$

Thus, the original system is equivalent to the system

$$\begin{cases} x = \frac{2}{11}z + \frac{7}{11} & (1) \\ y = \frac{1}{11}z - \frac{2}{11} & (2) \end{cases}$$

where z can be any real number.

Let's look at the situation. The original system of three equations is equivalent to a system containing two equations. This means that any values of x, y, z that satisfy both

$$x = \tfrac{2}{11}z + \tfrac{7}{11} \quad \text{and} \quad y = \tfrac{1}{11}z - \tfrac{2}{11}$$

will be solutions. For example, $z = 0$, $x = \frac{7}{11}$, $y = -\frac{2}{11}$; $z = 1$, $x = \frac{9}{11}$, $y = -\frac{1}{11}$; and $z = -1$, $x = \frac{5}{11}$, $y = -\frac{3}{11}$ are some of the solutions of the original system. There are, in fact, infinitely many values of x, y, and z for which the two equations are satisfied. That is, the original system has infinitely many solutions. We will write the solution of the original system as

$$\begin{cases} x = \frac{2}{11}z + \frac{7}{11} \\ y = \frac{1}{11}z - \frac{2}{11} \end{cases}$$

where z can be any real number. ■

We can also find the solution by writing the augmented matrix in reduced row echelon form. Starting with the echelon form, we have

$$\left[\begin{array}{ccc|c} 1 & -2 & 0 & 1 \\ 0 & 1 & -\frac{1}{11} & -\frac{2}{11} \\ 0 & 0 & 0 & 0 \end{array}\right] \underset{R_1 = 2r_2 + r_1}{\rightarrow} \left[\begin{array}{ccc|c} 1 & 0 & -\frac{2}{11} & \frac{7}{11} \\ 0 & 1 & -\frac{1}{11} & -\frac{2}{11} \\ 0 & 0 & 0 & 0 \end{array}\right]$$

The matrix on the right is in reduced row echelon form. The corresponding system of equations is

$$\begin{cases} x - \frac{2}{11}z = \frac{7}{11} & (1) \\ y - \frac{1}{11}z = -\frac{2}{11} & (2) \end{cases}$$

or, equivalently,

$$\begin{cases} x = \frac{2}{11}z + \frac{7}{11} & (1) \\ y = \frac{1}{11}z - \frac{2}{11} & (2) \end{cases}$$

where z can be any real number.

NOW WORK PROBLEM 49.

EXAMPLE 8 Solving a System of Linear Equations Using Matrices

Solve: $\begin{cases} x + y + z = 6 \\ 2x - y - z = 3 \\ x + 2y + 2z = 0 \end{cases}$

Solution We proceed as follows, beginning with the augmented matrix.

$$\left[\begin{array}{ccc|c} 1 & 1 & 1 & 6 \\ 2 & -1 & -1 & 3 \\ 1 & 2 & 2 & 0 \end{array}\right] \underset{\substack{R_2 = -2r_1 + r_2 \\ R_3 = -1r_1 + r_3}}{\rightarrow} \left[\begin{array}{ccc|c} 1 & 1 & 1 & 6 \\ 0 & -3 & -3 & -9 \\ 0 & 1 & 1 & -6 \end{array}\right] \underset{\text{Interchange rows 2 and 3.}}{\rightarrow} \left[\begin{array}{ccc|c} 1 & 1 & 1 & 6 \\ 0 & 1 & 1 & -6 \\ 0 & -3 & -3 & -9 \end{array}\right] \underset{R_3 = 3r_2 + r_3}{\rightarrow} \left[\begin{array}{ccc|c} 1 & 1 & 1 & 6 \\ 0 & 1 & 1 & -6 \\ 0 & 0 & 0 & -27 \end{array}\right]$$

This matrix is in echelon form. The bottom row is equivalent to the equation

$$0x + 0y + 0z = -27$$

which has no solution. Hence, the original system is inconsistent. ■

NOW WORK PROBLEM 23.

EXAMPLE 9 Nutrition

A dietitian at Cook County Hospital wishes a patient to have a meal that has 65 grams of protein, 95 grams of carbohydrates, and 905 milligrams of calcium. The hospital food service tells the dietitian that the dinner for today is chicken *a la king*, baked potatoes, and 2% milk. Each serving of chicken *a la king* has 30 grams of protein, 35 grams of carbohydrates, and 200 milligrams of calcium. Each serving of baked potatoes contains 4 grams of protein, 33 grams of carbohydrates, and 10 milligrams of calcium. Each glass of 2% milk contains 9 grams of protein, 13 grams of carbohydrates, and 300 milligrams of calcium. How many servings of each food should the dietitian provide for the patient?

Solution Let c, p, and m represent the number of servings of chicken *a la king*, baked potatoes, and milk, respectively. The dietitian wants the patient to have 65 grams of protein. Each serving of chicken *a la king* has 30 grams of protein, so c servings will have $30c$ grams of protein. Each serving of baked potatoes contains 4 grams of protein, so p potatoes will have $4p$ grams of protein. Finally, each glass of milk has 9 grams of protein, so m glasses of milk will have $9m$ grams of protein. The same logic will result in equations for carbohydrates and calcium, and we have the following system of equations:

$$\begin{cases} 30c + 4p + 9m = 65 & \text{protein equation} \\ 35c + 33p + 13m = 95 & \text{carbohydrate equation} \\ 200c + 10p + 300m = 905 & \text{calcium equation} \end{cases}$$

We begin with the augmented matrix and proceed as follows:

$$\left[\begin{array}{ccc|c} 30 & 4 & 9 & 65 \\ 35 & 33 & 13 & 95 \\ 200 & 10 & 300 & 905 \end{array}\right] \underset{R_1 = \left(\frac{1}{30}\right)r_1}{\rightarrow} \left[\begin{array}{ccc|c} 1 & \frac{2}{15} & \frac{3}{10} & \frac{13}{6} \\ 35 & 33 & 13 & 95 \\ 200 & 10 & 300 & 905 \end{array}\right] \underset{\substack{R_2 = -35r_1 + r_2 \\ R_3 = -200r_1 + r_3}}{\rightarrow} \left[\begin{array}{ccc|c} 1 & \frac{2}{15} & \frac{3}{10} & \frac{13}{6} \\ 0 & \frac{85}{3} & \frac{5}{2} & \frac{115}{6} \\ 0 & -\frac{50}{3} & 240 & \frac{1415}{3} \end{array}\right]$$

$$\underset{R_2 = \left(\frac{3}{85}\right)r_2}{\rightarrow} \left[\begin{array}{ccc|c} 1 & \frac{2}{15} & \frac{3}{10} & \frac{13}{6} \\ 0 & 1 & \frac{3}{34} & \frac{23}{34} \\ 0 & -\frac{50}{3} & 240 & \frac{1415}{3} \end{array}\right] \underset{R_3 = \left(\frac{50}{3}\right)r_2 + r_3}{\rightarrow} \left[\begin{array}{ccc|c} 1 & \frac{2}{15} & \frac{3}{10} & \frac{13}{6} \\ 0 & 1 & \frac{3}{34} & \frac{23}{34} \\ 0 & 0 & \frac{4105}{17} & \frac{8210}{17} \end{array}\right] \underset{R_3 = \left(\frac{17}{4105}\right)r_3}{\rightarrow} \left[\begin{array}{ccc|c} 1 & \frac{2}{15} & \frac{3}{10} & \frac{13}{6} \\ 0 & 1 & \frac{3}{34} & \frac{23}{34} \\ 0 & 0 & 1 & 2 \end{array}\right]$$

The matrix is now in echelon form. The final matrix represents the system

$$\begin{cases} c + \frac{2}{15}p + \frac{3}{10}m = \frac{13}{6} & (1) \\ p + \frac{3}{34}m = \frac{23}{34} & (2) \\ m = 2 & (3) \end{cases}$$

From (3), we determine that 2 glasses of milk should be served. Back-substitute $m = 2$ into equation (2) to find that $p = \frac{1}{2}$, so $\frac{1}{2}$ of a baked potato should be served. Back-substitute these values into equation (1) and find that $c = 1.5$, so 1.5 servings of chicken *a la king* should be given to the patient to meet the dietary requirements. ■

11.3 EXERCISES

In Problems 1–10, write the augmented matrix of the given system of equations.

1. $\begin{cases} x - 5y = 5 \\ 4x + 3y = 6 \end{cases}$

2. $\begin{cases} 3x + 4y = 7 \\ 4x - 2y = 5 \end{cases}$

3. $\begin{cases} 2x + 3y - 6 = 0 \\ 4x - 6y + 2 = 0 \end{cases}$

4. $\begin{cases} 9x - y = 0 \\ 3x - y - 4 = 0 \end{cases}$

5. $\begin{cases} 0.01x - 0.03y = 0.06 \\ 0.13x + 0.10y = 0.20 \end{cases}$

6. $\begin{cases} \frac{4}{3}x - \frac{3}{2}y = \frac{3}{4} \\ -\frac{1}{4}x + \frac{1}{3}y = \frac{2}{3} \end{cases}$

7. $\begin{cases} x - y + z = 10 \\ 3x + 3y = 5 \\ x + y + 2z = 2 \end{cases}$

8. $\begin{cases} 5x - y - z = 0 \\ x + y = 5 \\ 2x - 3z = 2 \end{cases}$

9. $\begin{cases} x + y - z = 2 \\ 3x - 2y = 2 \\ 5x + 3y - z = 1 \end{cases}$

10. $\begin{cases} 2x + 3y - 4z = 0 \\ x - 5z + 2 = 0 \\ x + 2y - 3z = -2 \end{cases}$

In Problems 11–20, perform in order (a), followed by (b), followed by (c) on the given augmented matrix.

11. $\left[\begin{array}{rrr|r} 1 & -3 & -5 & -2 \\ 2 & -5 & -4 & 5 \\ -3 & 5 & 4 & 6 \end{array}\right]$ (a) $R_2 = -2r_1 + r_2$ (b) $R_3 = 3r_1 + r_3$ (c) $R_3 = 4r_2 + r_3$

12. $\left[\begin{array}{rrr|r} 1 & -3 & -3 & -3 \\ 2 & -5 & 2 & -4 \\ -3 & 2 & 4 & 6 \end{array}\right]$ (a) $R_2 = -2r_1 + r_2$ (b) $R_3 = 3r_1 + r_3$ (c) $R_3 = 7r_2 + r_3$

13. $\left[\begin{array}{rrr|r} 1 & -3 & 4 & 3 \\ 2 & -5 & 6 & 6 \\ -3 & 3 & 4 & 6 \end{array}\right]$ (a) $R_2 = -2r_1 + r_2$ (b) $R_3 = 3r_1 + r_3$ (c) $R_3 = 6r_2 + r_3$

14. $\left[\begin{array}{rrr|r} 1 & -3 & 3 & -5 \\ 2 & -5 & -3 & -5 \\ -3 & -2 & 4 & 6 \end{array}\right]$ (a) $R_2 = -2r_1 + r_2$ (b) $R_3 = 3r_1 + r_3$ (c) $R_3 = 11r_2 + r_3$

15. $\left[\begin{array}{rrr|r} 1 & -3 & 2 & -6 \\ 2 & -5 & 3 & -4 \\ -3 & -6 & 4 & 6 \end{array}\right]$ (a) $R_2 = -2r_1 + r_2$ (b) $R_3 = 3r_1 + r_3$ (c) $R_3 = 15r_2 + r_3$

16. $\left[\begin{array}{rrr|r} 1 & -3 & -4 & -6 \\ 2 & -5 & 6 & -6 \\ -3 & 1 & 4 & 6 \end{array}\right]$ (a) $R_2 = -2r_1 + r_2$ (b) $R_3 = 3r_1 + r_3$ (c) $R_3 = 8r_2 + r_3$

17. $\left[\begin{array}{rrr|r} 1 & -3 & 1 & -2 \\ 2 & -5 & 6 & -2 \\ -3 & 1 & 4 & 6 \end{array}\right]$ (a) $R_2 = -2r_1 + r_2$ (b) $R_3 = 3r_1 + r_3$ (c) $R_3 = 8r_2 + r_3$

18. $\left[\begin{array}{rrr|r} 1 & -3 & -1 & 2 \\ 2 & -5 & 2 & 6 \\ -3 & -6 & 4 & 6 \end{array}\right]$ (a) $R_2 = -2r_1 + r_2$ (b) $R_3 = 3r_1 + r_3$ (c) $R_3 = 15r_2 + r_3$

19. $\left[\begin{array}{rrr|r} 1 & -3 & -2 & 3 \\ 2 & -5 & 2 & -1 \\ -3 & -2 & 4 & 6 \end{array}\right]$ (a) $R_2 = -2r_1 + r_2$ (b) $R_3 = 3r_1 + r_3$ (c) $R_3 = 11r_2 + r_3$

20. $\left[\begin{array}{rrr|r} 1 & -3 & 5 & -3 \\ 2 & -5 & 1 & -4 \\ -3 & 3 & 4 & 6 \end{array}\right]$ (a) $R_2 = -2r_1 + r_2$ (b) $R_3 = 3r_1 + r_3$ (c) $R_3 = 6r_2 + r_3$

In Problems 21–26, the reduced row echelon form of a system of linear equations is given. Write the system of equations corresponding to the given matrix. Use x, y or x, y, z as variables. Determine whether the system is consistent or inconsistent. If it is consistent, give the solution.

21. $\left[\begin{array}{rr|r} 1 & 0 & 5 \\ 0 & 1 & -1 \end{array}\right]$

22. $\left[\begin{array}{rr|r} 1 & 0 & -4 \\ 0 & 1 & 0 \end{array}\right]$

23. $\left[\begin{array}{rrr|r} 1 & 0 & 0 & 1 \\ 0 & 1 & 0 & 2 \\ 0 & 0 & 0 & 3 \end{array}\right]$

24. $\left[\begin{array}{rrr|r} 1 & 0 & 0 & 0 \\ 0 & 1 & 0 & 0 \\ 0 & 0 & 0 & 2 \end{array}\right]$

25. $\left[\begin{array}{rrr|r} 1 & 0 & 2 & -1 \\ 0 & 1 & -4 & -2 \\ 0 & 0 & 0 & 0 \end{array}\right]$

26. $\left[\begin{array}{rrr|r} 1 & 0 & 4 & 4 \\ 0 & 1 & 3 & 2 \\ 0 & 0 & 0 & 0 \end{array}\right]$

In Problems 27–62, solve each system of equations using matrices (row operations). If the system has no solution, say that it is inconsistent. Verify your results using a graphing utility.

27. $\begin{cases} x + y = 8 \\ x - y = 4 \end{cases}$

28. $\begin{cases} x + 2y = 5 \\ x + y = 3 \end{cases}$

29. $\begin{cases} x - 5y = -13 \\ 3x + 2y = 12 \end{cases}$

30. $\begin{cases} x + 3y = 5 \\ 2x - 3y = -8 \end{cases}$

31. $\begin{cases} 3x - 6y = 24 \\ 5x + 4y = 12 \end{cases}$

32. $\begin{cases} 2x + 4y = 16 \\ 3x - 5y = 9 \end{cases}$

33. $\begin{cases} 2x + y = 1 \\ 4x + 2y = 6 \end{cases}$

34. $\begin{cases} x - y = 5 \\ -3x + 3y = 2 \end{cases}$

35. $\begin{cases} 2x - 4y = -2 \\ 3x + 2y = 3 \end{cases}$

36. $\begin{cases} 3x + 3y = 3 \\ 4x + 2y = \frac{8}{3} \end{cases}$

37. $\begin{cases} x + 2y = 4 \\ 2x + 4y = 8 \end{cases}$

38. $\begin{cases} 3x - y = 7 \\ 9x - 3y = 21 \end{cases}$

39. $\begin{cases} 2x + 3y = 6 \\ x - y = \frac{1}{2} \end{cases}$

40. $\begin{cases} \frac{1}{2}x + y = -2 \\ x - 2y = 8 \end{cases}$

41. $\begin{cases} 3x - 5y = 3 \\ 15x + 5y = 21 \end{cases}$

42. $\begin{cases} 2x - y = -1 \\ x + \frac{1}{2}y = \frac{3}{2} \end{cases}$

43. $\begin{cases} x - y = 6 \\ 2x - 3z = 16 \\ 2y + z = 4 \end{cases}$

44. $\begin{cases} 2x + y = -4 \\ -2y + 4z = 0 \\ 3x - 2z = -11 \end{cases}$

45. $\begin{cases} x - 2y + 3z = 7 \\ 2x + y + z = 4 \\ -3x + 2y - 2z = -10 \end{cases}$

46. $\begin{cases} 2x + y - 3z = 0 \\ -2x + 2y + z = -7 \\ 3x - 4y - 3z = 7 \end{cases}$

47. $\begin{cases} 2x - 2y - 2z = 2 \\ 2x + 3y + z = 2 \\ 3x + 2y = 0 \end{cases}$

48. $\begin{cases} 2x - 3y - z = 0 \\ -x + 2y + z = 5 \\ 3x - 4y - z = 1 \end{cases}$

49. $\begin{cases} -x + y + z = -1 \\ -x + 2y - 3z = -4 \\ 3x - 2y - 7z = 0 \end{cases}$

50. $\begin{cases} 2x - 3y - z = 0 \\ 3x + 2y + 2z = 2 \\ x + 5y + 3z = 2 \end{cases}$

51. $\begin{cases} 2x - 2y + 3z = 6 \\ 4x - 3y + 2z = 0 \\ -2x + 3y - 7z = 1 \end{cases}$

52. $\begin{cases} 3x - 2y + 2z = 6 \\ 7x - 3y + 2z = -1 \\ 2x - 3y + 4z = 0 \end{cases}$

53. $\begin{cases} x + y - z = 6 \\ 3x - 2y + z = -5 \\ x + 3y - 2z = 14 \end{cases}$

54. $\begin{cases} x - y + z = -4 \\ 2x - 3y + 4z = -15 \\ 5x + y - 2z = 12 \end{cases}$

55. $\begin{cases} x + 2y - z = -3 \\ 2x - 4y + z = -7 \\ -2x + 2y - 3z = 4 \end{cases}$

56. $\begin{cases} x + 4y - 3z = -8 \\ 3x - y + 3z = 12 \\ x + y + 6z = 1 \end{cases}$

57. $\begin{cases} 3x + y - z = \frac{2}{3} \\ 2x - y + z = 1 \\ 4x + 2y = \frac{8}{3} \end{cases}$

58. $\begin{cases} x + y = 1 \\ 2x - y + z = 1 \\ x + 2y + z = \frac{8}{3} \end{cases}$

59. $\begin{cases} x + 2y + z = 1 \\ 2x - y + 2z = 2 \\ 3x + y + 3z = 3 \end{cases}$

60. $\begin{cases} x + 2y - z = 3 \\ 2x - y + 2z = 6 \\ x - 3y + 3z = 4 \end{cases}$

61. $\begin{cases} x + y + z + w = 4 \\ 2x - y + z = 0 \\ 3x + 2y + z - w = 6 \\ x - 2y - 2z + 2w = -1 \end{cases}$

62. $\begin{cases} x + y + z + w = 4 \\ -x + 2y + z = 0 \\ 2x + 3y + z - w = 6 \\ -2x + y - 2z + 2w = -1 \end{cases}$

63. **Curve Fitting** Find the graph of the function $y = ax^2 + bx + c$ that contains the points $(1, 2)$, $(-2, -7)$, and $(2, -3)$.

64. **Curve Fitting** Find the graph of the function $y = ax^2 + bx + c$ that contains the points $(1, -1)$, $(3, -1)$, and $(-2, 14)$.

65. **Curve Fitting** Find the function $f(x) = ax^3 + bx^2 + cx + d$ for which $f(-3) = -112$, $f(-1) = -2$, $f(1) = 4$, and $f(2) = 13$.

66. **Curve Fitting** Find the function $f(x) = ax^3 + bx^2 + cx + d$ for which $f(-2) = -10$, $f(-1) = 3$, $f(1) = 5$, and $f(3) = 15$.

67. **Nutrition** A dietitian at Palos Community Hospital wishes a patient to have a meal that has 78 grams of protein, 59 grams of carbohydrates, and 75 milligrams of vitamin A. The hospital food service tells the dietitian that the dinner for today is salmon steak, baked eggs, and acorn squash. Each serving of salmon steak has 30 grams of protein, 20 grams of carbohydrates, and 2 milligrams of vitamin A. Each serving of baked eggs contains 15 grams of protein, 2 grams of carbohydrates, and 20 milligrams of vitamin A. Each serving of acorn squash contains 3 grams of protein, 25 grams of carbohydrates, and 32 milligrams of vitamin A. How many servings of each food should the dietitian provide for the patient?

68. **Nutrition** A dietitian at General Hospital wishes a patient to have a meal that has 47 grams of protein, 58 grams of carbohydrates, and 630 milligrams of calcium. The hospital food service tells the dietitian that the dinner for today is pork chops, corn on the cob, and 2% milk. Each serving of pork chops has 23 grams of protein, 0 grams of carbohydrates, and 10 milligrams of calcium. Each serving of corn on the cob contains 3 grams of protein, 16 grams of carbohydrates, and 10 milligrams of calcium. Each glass of 2% milk contains 9 grams of protein, 13 grams of carbohydrates, and 300 milligrams of calcium. How many servings of each food should the dietitian provide for the patient?

69. **Financial Planning** Carletta has \$10,000 to invest. As her financial consultant, you recommend that she invest in Treasury bills that yield 6%, Treasury bonds that yield 7%, and corporate bonds that yield 8%. Carletta wants to have an annual income of \$680, and the amount invested in corporate bonds must be half that invested in Treasury bills. Find the amount in each investment.

70. **Financial Planning** John has \$20,000 to invest. As his financial consultant, you recommended that he invest in Treasury bills that yield 5%, Treasury bonds that yield 7%, and corporate bonds that yield 9%. John wants to have an annual income of \$1280, and the amount invested in Treasury bills must be two times the amount invested in corporate bonds. Find the amount in each investment.

71. **Production** To manufacture an automobile requires painting, drying, and polishing. Epsilon Motor Company produces three types of cars: the Delta, the Beta, and the Sigma. Each Delta requires 10 hours for painting, 3 hours for drying, and 2 hours for polishing. A Beta requires 16 hours of painting, 5 hours of drying, and 3 hours of polishing, while a Sigma requires 8 hours for painting, 2 hours for drying, and 1 hour for polishing. If the company has 240 hours for painting, 69 hours for drying, and 41 hours for polishing per month, how many of each type of car are produced?

72. **Production** A Florida juice company completes the preparation of its products by sterilizing, filling, and labeling bottles. Each case of orange juice requires 9 minutes for sterilizing, 6 minutes for filling, and 1 minute for

labeling Each case of grapefruit juice requires 10 minutes for sterilizing, 4 minutes for filling, and 2 minutes for labeling. Each case of tomato juice requires 12 minutes for sterilizing, 4 minutes for filling, and 1 minute for labeling. If the company runs the cleaning machine for 398 minutes, the filling machine for 164 minutes, and the labeling machine for 58 minutes, how many cases of each type of juice are prepared?

73. Electricity: Kirchhoff's Rules An application of Kirchhoff's Rules to the circuit shown results in the following system of equations:

$$\begin{cases} -4 + 8 - 2I_2 = 0 \\ 8 = 5I_4 + I_1 \\ 4 = 3I_3 + I_1 \\ I_3 + I_4 = I_1 \end{cases}$$

Find the currents I_1, I_2, I_3, and I_4.*

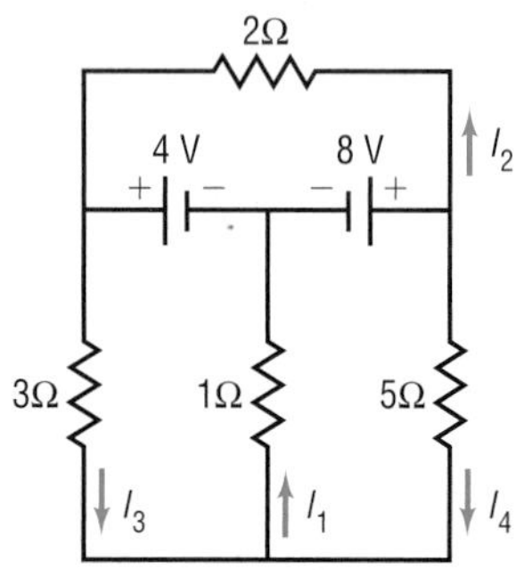

74. Electricity: Kirchhoff's Rules An application of Kirchhoff's Rules to the circuit shown results in the following system of equations:

$$\begin{cases} I_1 = I_3 + I_2 \\ 24 - 6I_1 - 3I_3 = 0 \\ 12 + 24 - 6I_1 - 6I_2 = 0 \end{cases}$$

Find the currents I_1, I_2, and I_3.†

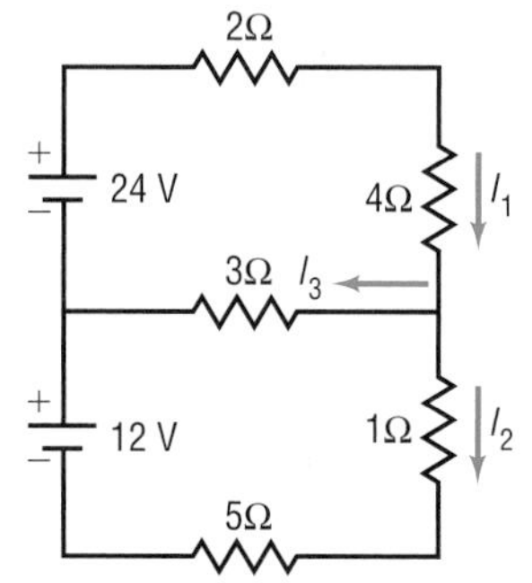

75. Write a brief paragraph or two that outlines your strategy for solving a system of linear equations using matrices.

76. When solving a system of linear equations using matrices, do you prefer to place the augmented matrix in echelon form or in reduced row echelon form? Give reasons for your choice.

77. Make up a system of three linear equations containing three variables that has:
(a) No solution
(b) Exactly one solution
(c) Infinitely many solutions
Give the three systems to a friend to solve and critique.

* *Source:* Based on Raymond Serway, *Physics,* 3rd ed. (Philadelphia: Saunders, 1990), Prob. 34, p. 79.

† *Source:* Ibid., Prob. 38, p. 791.

11.4 SYSTEMS OF LINEAR EQUATIONS: DETERMINANTS

OBJECTIVES
1. Evaluate 2 by 2 Determinants
2. Use Cramer's Rule to Solve a System of Two Equations, Two Variables
3. Evaluate 3 by 3 Determinants
4. Use Cramer's Rule to Solve a System of Three Equations, Three Variables
5. Know Some Properties of Determinants

1 In the preceding section, we described a method of using matrices to solve a system of linear equations. This section deals with yet another method for solving systems of linear equations; however, it can be used only when the number of equations equals the number of variables. Although the method will work for any system (provided that the number of equations equals the number of variables), it is most often used for systems of two equations containing two variables or three equations containing three variables. This method, called *Cramer's Rule,* is based on the concept of a *determinant.*

2 BY 2 DETERMINANTS

If a, b, c, and d are four real numbers, the symbol

$$D = \begin{vmatrix} a & b \\ c & d \end{vmatrix}$$

is called a **2 by 2 determinant.** Its value is the number $ad - bc$; that is,

$$D = \begin{vmatrix} a & b \\ c & d \end{vmatrix} = ad - bc \qquad \textbf{(1)}$$

A device that may be helpful for remembering the value of a 2 by 2 determinant is the following:

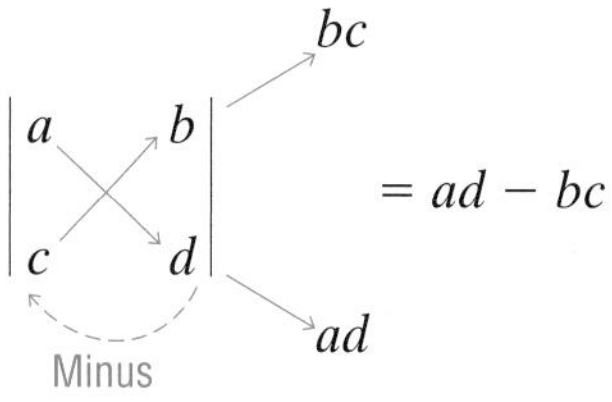

EXAMPLE 1 **Evaluating a 2 × 2 Determinant**

$$\begin{vmatrix} 3 & -2 \\ 6 & 1 \end{vmatrix} = (3)(1) - (6)(-2) = 3 - (-12) = 15$$

Graphing utilities can also be used to evaluate determinants.

EXAMPLE 2 **Using a Graphing Utility to Evaluate a 2 × 2 Determinant**

Use a graphing utility to evaluate the determinant from Example 1: $\begin{vmatrix} 3 & -2 \\ 6 & 1 \end{vmatrix}$.

Solution First, we enter the matrix whose entries are those of the determinant into the graphing utility and name it A. Using the determinant command, we obtain the result shown in Figure 9.

Figure 9

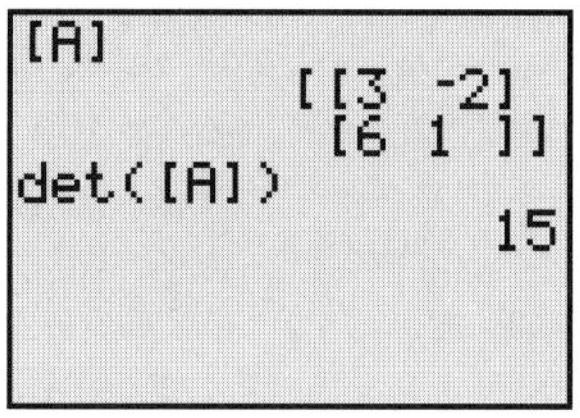

NOW WORK PROBLEM 3.

2 Let's now see the role that a 2 by 2 determinant plays in the solution of a system of two equations containing two variables. Consider the system

$$\begin{cases} ax + by = s & (1) \\ cx + dy = t & (2) \end{cases} \tag{2}$$

We shall use the method of elimination to solve this system.

Provided $d \neq 0$ and $b \neq 0$, this system is equivalent to the system

$$\begin{cases} adx + bdy = sd & \text{(1) Multiply by d.} \\ bcx + bdy = tb & \text{(2) Multiply by b.} \end{cases}$$

On subtracting the second equation from the first equation, we get

$$\begin{cases} (ad - bc)x + 0 \cdot y = sd - tb & (1) \\ bcx \quad + bdy = tb & (2) \end{cases}$$

Now, the first equation can be rewritten using determinant notation.

$$\begin{vmatrix} a & b \\ c & d \end{vmatrix} x = \begin{vmatrix} s & b \\ t & d \end{vmatrix}$$

If $D = \begin{vmatrix} a & b \\ c & d \end{vmatrix} = ad - bc \neq 0$, we can solve for x to get

$$x = \frac{\begin{vmatrix} s & b \\ t & d \end{vmatrix}}{\begin{vmatrix} a & b \\ c & d \end{vmatrix}} = \frac{\begin{vmatrix} s & b \\ t & d \end{vmatrix}}{D} \tag{3}$$

Return now to the original system (2). Provided that $a \neq 0$ and $c \neq 0$, the system is equivalent to

$$\begin{cases} acx + bcy = cs & \text{(1) Multiply by c.} \\ acx + ady = at & \text{(2) Multiply by a.} \end{cases}$$

On subtracting the first equation from the second equation, we get

$$\begin{cases} acx + \quad bcy = cs & (1) \\ 0 \cdot x + (ad - bc)y = at - cs & (2) \end{cases}$$

The second equation can now be rewritten using determinant notation.

$$\begin{vmatrix} a & b \\ c & d \end{vmatrix} y = \begin{vmatrix} a & s \\ c & t \end{vmatrix}$$

If $D = \begin{vmatrix} a & b \\ c & d \end{vmatrix} = ad - bc \neq 0$, we can solve for y to get

$$y = \frac{\begin{vmatrix} a & s \\ c & t \end{vmatrix}}{\begin{vmatrix} a & b \\ c & d \end{vmatrix}} = \frac{\begin{vmatrix} a & s \\ c & t \end{vmatrix}}{D} \tag{4}$$

Equations (3) and (4) lead us to the following result, called **Cramer's Rule.**

Theorem **Cramer's Rule for Two Equations Containing Two Variables**

The solution to the system of equations

$$\begin{cases} ax + by = s & (1) \\ cx + dy = t & (2) \end{cases} \tag{5}$$

is given by

$$x = \frac{\begin{vmatrix} s & b \\ t & d \end{vmatrix}}{\begin{vmatrix} a & b \\ c & d \end{vmatrix}}, \qquad y = \frac{\begin{vmatrix} a & s \\ c & t \end{vmatrix}}{\begin{vmatrix} a & b \\ c & d \end{vmatrix}} \tag{6}$$

provided that

$$D = \begin{vmatrix} a & b \\ c & d \end{vmatrix} = ad - bc \neq 0$$

■

In the derivation given for Cramer's Rule above, we assumed that none of the numbers a, b, c and d was 0. In Problem 58 you will be asked to complete the proof under the less stringent conditions that $D = ad - bc \neq 0$.

Now look carefully at the pattern in Cramer's Rule. The denominator in the solution (6) is the determinant of the coefficients of the variables.

$$\begin{cases} ax + by = s \\ cx + dy = t \end{cases} \qquad D = \begin{vmatrix} a & b \\ c & d \end{vmatrix}$$

In the solution for x, the numerator is the determinant, denoted by D_x, formed by replacing the entries in the first column (the coefficients of x) in D by the constants on the right side of the equal sign.

$$D_x = \begin{vmatrix} s & b \\ t & d \end{vmatrix}$$

In the solution for y, the numerator is the determinant, denoted by D_y, formed by replacing the entries in the second column (the coefficients of y) in D by the constants on the right side of the equal sign.

$$D_y = \begin{vmatrix} a & s \\ c & t \end{vmatrix}$$

Cramer's Rule then states that, if $D \neq 0$,

$$x = \frac{D_x}{D}, \qquad y = \frac{D_y}{D} \tag{7}$$

EXAMPLE 3 Solving a System of Linear Equations Using Determinants

Use Cramer's Rule, if applicable, to solve the system

$$\begin{cases} 3x - 2y = 4 & (1) \\ 6x + y = 13 & (2) \end{cases}$$

Algebraic Solution

The determinant D of the coefficients of the variables is

$$D = \begin{vmatrix} 3 & -2 \\ 6 & 1 \end{vmatrix} = (3)(1) - (6)(-2) = 15$$

Because $D \neq 0$, Cramer's Rule (7) can be used.

$$x = \frac{D_x}{D} = \frac{\begin{vmatrix} 4 & -2 \\ 13 & 1 \end{vmatrix}}{15} = \frac{30}{15} = 2 \qquad y = \frac{D_y}{D} = \frac{\begin{vmatrix} 3 & 4 \\ 6 & 13 \end{vmatrix}}{15} = \frac{15}{15} = 1$$

The solution is $x = 2$, $y = 1$.

Graphing Solution

Figure 10

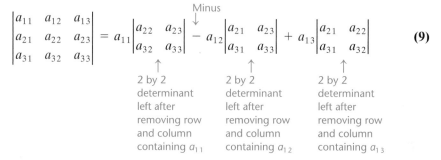

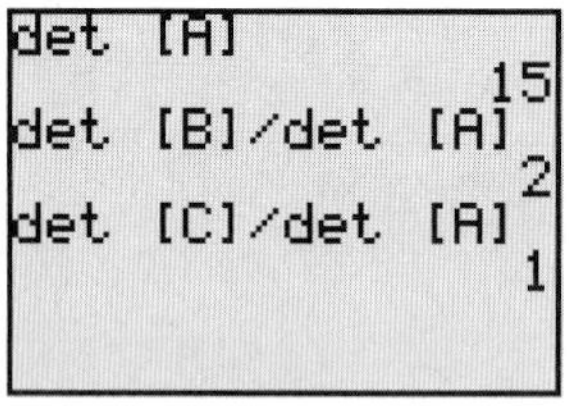

We enter the coefficient matrix into our graphing utility. Call it A and evaluate det$[A]$. Since det$[A] \neq 0$, we can use Cramer's Rule. We enter the matrices D_x and D_y into our graphing utility and call them B and C, respectively. Finally, we find x by calculating det$[B]$/det$[A]$ and y by calculating det$[C]$/det$[A]$. The results are shown in Figure 10.

The solution is $x = 2$, $y = 1$. ■

In attempting to use Cramer's Rule, if the determinant D of the coefficients of the variables is found to equal 0 (so that Cramer's Rule is not applicable), then the system either is inconsistent or has infinitely many solutions.

NOW WORK PROBLEM 11.

3 BY 3 DETERMINANTS

3 In order to use Cramer's Rule to solve a system of three equations containing three variables, we need to define a 3 by 3 determinant.

A **3 by 3 determinant** is symbolized by

$$\begin{vmatrix} a_{11} & a_{12} & a_{13} \\ a_{21} & a_{22} & a_{23} \\ a_{31} & a_{32} & a_{33} \end{vmatrix} \tag{8}$$

in which $a_{11}, a_{12}, \ldots$, are real numbers.

As with matrices, we use a double subscript to identify an entry by indicating its row and column numbers. For example, the entry a_{23} is in row 2, column 3.

The value of a 3 by 3 determinant may be defined in terms of 2 by 2 determinants by the following formula:

$$\begin{vmatrix} a_{11} & a_{12} & a_{13} \\ a_{21} & a_{22} & a_{23} \\ a_{31} & a_{32} & a_{33} \end{vmatrix} = a_{11}\begin{vmatrix} a_{22} & a_{23} \\ a_{32} & a_{33} \end{vmatrix} - a_{12}\begin{vmatrix} a_{21} & a_{23} \\ a_{31} & a_{33} \end{vmatrix} + a_{13}\begin{vmatrix} a_{21} & a_{22} \\ a_{31} & a_{32} \end{vmatrix} \tag{9}$$

Minus (↓ pointing to the − sign)

↑ 2 by 2 determinant left after removing row and column containing a_{11}

↑ 2 by 2 determinant left after removing row and column containing a_{12}

↑ 2 by 2 determinant left after removing row and column containing a_{13}

The 2 by 2 determinants shown in formula (9) are called **minors** of the 3 by 3 determinant. For an n by n determinant, the **minor** M_{ij} of element a_{ij} is the determinant resulting from removing the ith row and jth column.

EXAMPLE 4 Finding Minors of a 3 by 3 Determinant

For the determinant $A = \begin{vmatrix} 2 & -1 & 3 \\ -2 & 5 & 1 \\ 0 & 6 & -9 \end{vmatrix}$, find: (a) M_{12} (b) M_{23}

Solution (a) M_{12} is the determinant that results from removing the first row and second column from A.

$$A = \begin{vmatrix} 2 & -1 & 3 \\ -2 & 5 & 1 \\ 0 & 6 & -9 \end{vmatrix} \qquad M_{12} = \begin{vmatrix} -2 & 1 \\ 0 & -9 \end{vmatrix} = (-2)(-9) - (0)(1) = 18$$

(b) M_{23} is the determinant that results from removing the second row and third column from A.

$$A = \begin{vmatrix} 2 & -1 & 3 \\ -2 & 5 & 1 \\ 0 & 6 & -9 \end{vmatrix} \qquad M_{23} = \begin{vmatrix} 2 & -1 \\ 0 & 6 \end{vmatrix} = (2)(6) - (0)(-1) = 12$$ ■

Referring back to formula (9), we see that each element a_{ij} is multiplied by its minor, but sometimes this term is added and other times subtracted. To determine whether to add or subtract a term, we must consider the *cofactor*.

For an n by n determinant A, the **cofactor** of element a_{ij}, denoted by A_{ij}, is given by

$$A_{ij} = (-1)^{i+j} M_{ij}$$

where M_{ij} is the minor of element a_{ij}.

The exponent of $(-1)^{i+j}$ is the sum of the row and column of the element a_{ij} so if $i + j$ is even, $(-1)^{i+j}$ will equal 1, and if $i + j$ is odd, $(-1)^{i+j}$ will equal (-1).

To find the value of a determinant, multiply each element in any row or column by its cofactor and sum the results, This process is referred to as *expanding across a row or column*. For example, the value of the 3 by 3 determinant in formula (9) was found by expanding across row 1.

If we choose to expand down column 2, we obtain

$$\begin{vmatrix} a_{11} & a_{12} & a_{13} \\ a_{21} & a_{22} & a_{23} \\ a_{31} & a_{32} & a_{33} \end{vmatrix} = (-1)^{1+2} a_{12} \begin{vmatrix} a_{21} & a_{23} \\ a_{31} & a_{33} \end{vmatrix} + (-1)^{2+2} a_{22} \begin{vmatrix} a_{11} & a_{13} \\ a_{31} & a_{33} \end{vmatrix} + (-1)^{3+2} a_{32} \begin{vmatrix} a_{11} & a_{13} \\ a_{21} & a_{23} \end{vmatrix}$$

↑ Expand down column 2.

If we choose to expand across row 3, we obtain

$$\begin{vmatrix} a_{11} & a_{12} & a_{13} \\ a_{21} & a_{22} & a_{23} \\ a_{31} & a_{32} & a_{33} \end{vmatrix} = (-1)^{3+1} a_{31} \begin{vmatrix} a_{12} & a_{13} \\ a_{22} & a_{23} \end{vmatrix} + (-1)^{3+2} a_{32} \begin{vmatrix} a_{11} & a_{13} \\ a_{21} & a_{23} \end{vmatrix} + (-1)^{3+3} a_{33} \begin{vmatrix} a_{11} & a_{12} \\ a_{21} & a_{22} \end{vmatrix}$$

↑ Expand across row 3.

It can be shown that the value of a determinant does not depend on the choice of the row or column used in the expansion. However, expanding across a row or column that has element equal to 0 reduces the amount of work needed to compute the value of the determinant.

EXAMPLE 5 Evaluating a 3 × 3 Determinant

Find the value of the 3 by 3 determinant: $\begin{vmatrix} 3 & 4 & -1 \\ 4 & 6 & 2 \\ 8 & -2 & 3 \end{vmatrix}$

Solution We choose to expand across row 1.

$$\begin{vmatrix} 3 & 4 & -1 \\ 4 & 6 & 2 \\ 8 & -2 & 3 \end{vmatrix} = (-1)^{1+1}3\begin{vmatrix} 6 & 2 \\ -2 & 3 \end{vmatrix} + (-1)^{1+2}4\begin{vmatrix} 4 & 2 \\ 8 & 3 \end{vmatrix} + (-1)^{1+3}(-1)\begin{vmatrix} 4 & 6 \\ 8 & -2 \end{vmatrix}$$

$$= 3(18 + 4) - 4(12 - 16) + (-1)(-8 - 48)$$

$$= 3(22) - 4(-4) + (-1)(-56)$$

$$= 66 + 16 + 56 = 138$$

We could also find the value of the 3 by 3 determinant in Example 5 by expanding down column 3.

$$\begin{vmatrix} 3 & 4 & -1 \\ 4 & 6 & 2 \\ 8 & -2 & 3 \end{vmatrix} = (-1)^{1+3}(-1)\begin{vmatrix} 4 & 6 \\ 8 & -2 \end{vmatrix} + (-1)^{2+3}2\begin{vmatrix} 3 & 4 \\ 8 & -2 \end{vmatrix} + (-1)^{3+3}3\begin{vmatrix} 3 & 4 \\ 4 & 6 \end{vmatrix}$$

$$= -1(-8 - 48) - 2(-6 - 32) + 3(18 - 16)$$

$$= 56 + 76 + 6 = 138$$

Evaluating 3 × 3 determinants on a graphing utility follows the same procedure as evaluating 2 × 2 determinants.

NOW WORK PROBLEM 7.

SYSTEMS OF THREE EQUATIONS CONTAINING THREE VARIABLES

4 Consider the following system of three equations containing three variables.

$$\begin{cases} a_{11}x + a_{12}y + a_{13}z = c_1 \\ a_{21}x + a_{22}y + a_{23}z = c_2 \\ a_{31}x + a_{32}y + a_{33}z = c_3 \end{cases} \qquad \textbf{(10)}$$

It can be shown that if the determinant D of the coefficients of the variables is not 0, that is, if

$$D = \begin{vmatrix} a_{11} & a_{12} & a_{13} \\ a_{21} & a_{22} & a_{23} \\ a_{31} & a_{32} & a_{33} \end{vmatrix} \neq 0$$

then the unique solution of system (10) is given by

Cramer's Rule for Three Equations Containing Three Variables

$$x = \frac{D_x}{D} \qquad y = \frac{D_y}{D} \qquad z = \frac{D_z}{D}$$

where

$$D_x = \begin{vmatrix} c_1 & a_{12} & a_{13} \\ c_2 & a_{22} & a_{23} \\ c_3 & a_{32} & a_{33} \end{vmatrix} \qquad D_y = \begin{vmatrix} a_{11} & c_1 & a_{13} \\ a_{21} & c_2 & a_{23} \\ a_{31} & c_3 & a_{33} \end{vmatrix} \qquad D_z = \begin{vmatrix} a_{11} & a_{12} & c_1 \\ a_{21} & a_{22} & c_2 \\ a_{31} & a_{32} & c_3 \end{vmatrix}$$

The similarity of this pattern and the pattern observed earlier for a system of two equations containing two variables should be apparent.

EXAMPLE 6 Using Cramer's Rule

Use Cramer's Rule, if applicable, to solve the following system:

$$\begin{cases} 2x + y - z = 3 & (1) \\ -x + 2y + 4z = -3 & (2) \\ x - 2y - 3z = 4 & (3) \end{cases}$$

Solution The value of the determinant D of the coefficients of the variables is

$$D = \begin{vmatrix} 2 & 1 & -1 \\ -1 & 2 & 4 \\ 1 & -2 & -3 \end{vmatrix} = (-1)^{1+1}2\begin{vmatrix} 2 & 4 \\ -2 & -3 \end{vmatrix} + (-1)^{1+2}1\begin{vmatrix} -1 & 4 \\ 1 & -3 \end{vmatrix} + (-1)^{1+3}(-1)\begin{vmatrix} -1 & 2 \\ 1 & -2 \end{vmatrix}$$

$$= 2(2) - 1(-1) + (-1)(0)$$

$$= 4 + 1 = 5$$

Because $D \neq 0$, we proceed to find the values of D_x, D_y, and D_z.

$$D_x = \begin{vmatrix} 3 & 1 & -1 \\ -3 & 2 & 4 \\ 4 & -2 & -3 \end{vmatrix} = (-1)^{1+1}3\begin{vmatrix} 2 & 4 \\ -2 & -3 \end{vmatrix} + (-1)^{1+2}1\begin{vmatrix} -3 & 4 \\ 4 & -3 \end{vmatrix} + (-1)^{1+3}(-1)\begin{vmatrix} -3 & 2 \\ 4 & -2 \end{vmatrix}$$

$$= 3(2) - 1(-7) + (-1)(-2) = 15$$

$$D_y = \begin{vmatrix} 2 & 3 & -1 \\ -1 & -3 & 4 \\ 1 & 4 & -3 \end{vmatrix} = (-1)^{1+1}2\begin{vmatrix} -3 & 4 \\ 4 & -3 \end{vmatrix} + (-1)^{1+2}3\begin{vmatrix} -1 & 4 \\ 1 & -3 \end{vmatrix} + (-1)^{1+3}(-1)\begin{vmatrix} -1 & -3 \\ 1 & 4 \end{vmatrix}$$

$$= 2(-7) - 3(-1) + (-1)(-1)$$

$$= -14 + 3 + 1 = -10$$

$$D_z = \begin{vmatrix} 2 & 1 & 3 \\ -1 & 2 & -3 \\ 1 & -2 & 4 \end{vmatrix} = (-1)^{1+1}2\begin{vmatrix} 2 & -3 \\ -2 & 4 \end{vmatrix} + (-1)^{1+2}1\begin{vmatrix} -1 & -3 \\ 1 & 4 \end{vmatrix} + (-1)^{1+3}3\begin{vmatrix} -1 & 2 \\ 1 & -2 \end{vmatrix}$$

$$= 2(2) - 1(-1) + 3(0) = 5$$

As a result,

$$x = \frac{D_x}{D} = \frac{15}{5} = 3 \qquad y = \frac{D_y}{D} = \frac{-10}{5} = -2 \qquad z = \frac{D_z}{D} = \frac{5}{5} = 1$$

The solution is $x = 3$, $y = -2$, $z = 1$. ■

If the determinant of the coefficients of the variables of a system of three linear equations containing three variables is 0, then Cramer's Rule is not applicable. In such a case, the system either is inconsistent or has infinitely many solutions.

Solving systems of three equations containing three variables using Cramer's Rule on a graphing utility follows the same procedure as that for solving systems of two equations containing two variables.

NOW WORK PROBLEM 29.

MORE ABOUT DETERMINANTS

5 Determinants have several properties that are sometimes helpful for obtaining their value. We list some of them here.

Theorem The value of a determinant changes sign if any two rows (or any two columns) are interchanged. **(11)**

Proof for 2 by 2 Determinants

$$\begin{vmatrix} a & b \\ c & d \end{vmatrix} = ad - bc \quad \text{and} \quad \begin{vmatrix} c & d \\ a & b \end{vmatrix} = bc - ad = -(ad - bc)$$

EXAMPLE 7 Demonstrating Theorem (11)

$$\begin{vmatrix} 3 & 4 \\ 1 & 2 \end{vmatrix} = 6 - 4 = 2 \qquad \begin{vmatrix} 1 & 2 \\ 3 & 4 \end{vmatrix} = 4 - 6 = -2$$

Theorem If all the entries in any row (or any column) equal 0, the value of the determinant is 0. **(12)**

Proof Merely expand across the row (or down the column) containing the 0's.

Theorem If any two rows (or any two columns) of a determinant have corresponding entries that are equal, the value of the determinant is 0. **(13)**

You are asked to prove this result for a 3 by 3 determinant in which the entries in column 1 equal the entries in column 3 in Problem 61.

EXAMPLE 8 Demonstrating Theorem (13)

$$\begin{vmatrix} 1 & 2 & 3 \\ 1 & 2 & 3 \\ 4 & 5 & 6 \end{vmatrix} = (-1)^{1+1}1\begin{vmatrix} 2 & 3 \\ 5 & 6 \end{vmatrix} + (-1)^{1+2}2\begin{vmatrix} 1 & 3 \\ 4 & 6 \end{vmatrix} + (-1)^{1+3}3\begin{vmatrix} 1 & 2 \\ 4 & 5 \end{vmatrix}$$

$$= 1(-3) - 2(-6) + 3(-3)$$

$$= -3 + 12 - 9 = 0$$

Theorem If any row (or any column) of a determinant is multiplied by a nonzero number k, the value of the determinant is also changed by a factor of k. **(14)**

■

You are asked to prove this result for a 3 by 3 determinant using row 2 in Problem 60.

EXAMPLE 9 Demonstrating Theorem (14)

$$\begin{vmatrix} 1 & 2 \\ 4 & 6 \end{vmatrix} = 6 - 8 = -2$$

$$\begin{vmatrix} k & 2k \\ 4 & 6 \end{vmatrix} = 6k - 8k = -2k = k(-2) = k\begin{vmatrix} 1 & 2 \\ 4 & 6 \end{vmatrix}$$ ■

Theorem If the entries of any row (or any column) of a determinant are multiplied by a nonzero number k and the result is added to the corresponding entries of another row (or column), the value of the determinant remains unchanged. **(15)**

■

In Problem 62, you are asked to prove this result for a 3 by 3 determinant using rows 1 and 2.

EXAMPLE 10 Demonstrating Theorem (15)

$$\begin{vmatrix} 3 & 4 \\ 5 & 2 \end{vmatrix} \underset{\uparrow}{=} \begin{vmatrix} -7 & 0 \\ 5 & 2 \end{vmatrix} = -14$$

Multiply row 2 by -2 and add to row 1. ■

11.4 EXERCISES

In Problems 1–10, find the value of each determinant (a) by hand and (b) by using a graphing utility.

1. $\begin{vmatrix} 3 & 1 \\ 4 & 2 \end{vmatrix}$ **2.** $\begin{vmatrix} 6 & 1 \\ 5 & 2 \end{vmatrix}$ **3.** $\begin{vmatrix} 6 & 4 \\ -1 & 3 \end{vmatrix}$ **4.** $\begin{vmatrix} 8 & -3 \\ 4 & 2 \end{vmatrix}$

5. $\begin{vmatrix} -3 & -1 \\ 4 & 2 \end{vmatrix}$ **6.** $\begin{vmatrix} -4 & 2 \\ -5 & 3 \end{vmatrix}$ **7.** $\begin{vmatrix} 3 & 4 & 2 \\ 1 & -1 & 5 \\ 1 & 2 & -2 \end{vmatrix}$ **8.** $\begin{vmatrix} 1 & 3 & -2 \\ 6 & 1 & -5 \\ 8 & 2 & 3 \end{vmatrix}$

9. $\begin{vmatrix} 4 & -1 & 2 \\ 6 & -1 & 0 \\ 1 & -3 & 4 \end{vmatrix}$ **10.** $\begin{vmatrix} 3 & -9 & 4 \\ 1 & 4 & 0 \\ 8 & -3 & 1 \end{vmatrix}$

In Problems 11–40, solve each system of equations using Cramer's Rule if it is applicable (a) by hand and (b) by using a graphing utility. If Cramer's Rule is not applicable, say so.

11. $\begin{cases} x + y = 8 \\ x - y = 4 \end{cases}$ **12.** $\begin{cases} x + 2y = 5 \\ x - y = 3 \end{cases}$ **13.** $\begin{cases} 5x - y = 13 \\ 2x + 3y = 12 \end{cases}$ **14.** $\begin{cases} x + 3y = 5 \\ 2x - 3y = -8 \end{cases}$

15. $\begin{cases} 3x = 24 \\ x + 2y = 0 \end{cases}$ **16.** $\begin{cases} 4x + 5y = -3 \\ -2y = -4 \end{cases}$ **17.** $\begin{cases} 3x - 6y = 24 \\ 5x + 4y = 12 \end{cases}$ **18.** $\begin{cases} 2x + 4y = 16 \\ 3x - 5y = -9 \end{cases}$

19. $\begin{cases} 3x - 2y = 4 \\ 6x - 4y = 0 \end{cases}$ **20.** $\begin{cases} -x + 2y = 5 \\ 4x - 8y = 6 \end{cases}$ **21.** $\begin{cases} 2x - 4y = -2 \\ 3x + 2y = 3 \end{cases}$ **22.** $\begin{cases} 3x + 3y = 3 \\ 4x + 2y = \frac{8}{3} \end{cases}$

23. $\begin{cases} 2x - 3y = -1 \\ 10x + 10y = 5 \end{cases}$ **24.** $\begin{cases} 3x - 2y = 0 \\ 5x + 10y = 4 \end{cases}$ **25.** $\begin{cases} 2x + 3y = 6 \\ x - y = \frac{1}{2} \end{cases}$ **26.** $\begin{cases} \frac{1}{2}x + y = -2 \\ x - 2y = 8 \end{cases}$

27. $\begin{cases} 3x - 5y = 3 \\ 15x + 5y = 21 \end{cases}$ **28.** $\begin{cases} 2x - y = -1 \\ x + \frac{1}{2}y = \frac{3}{2} \end{cases}$ **29.** $\begin{cases} x + y - z = 6 \\ 3x - 2y + z = -5 \\ x + 3y - 2z = 14 \end{cases}$ **30.** $\begin{cases} x - y + z = -4 \\ 2x - 3y + 4z = -15 \\ 5x + y - 2z = 12 \end{cases}$

31. $\begin{cases} x + 2y - z = -3 \\ 2x - 4y + z = -7 \\ -2x + 2y - 3z = 4 \end{cases}$ **32.** $\begin{cases} x + 4y - 3z = -8 \\ 3x - y + 3z = 12 \\ x + y + 6z = 1 \end{cases}$ **33.** $\begin{cases} x - 2y + 3z = 1 \\ 3x + y - 2z = 0 \\ 2x - 4y + 6z = 2 \end{cases}$ **34.** $\begin{cases} x - y + 2z = 5 \\ 3x + 2y = 4 \\ -2x + 2y - 4z = -10 \end{cases}$

35. $\begin{cases} x + 2y - z = 0 \\ 2x - 4y + z = 0 \\ -2x + 2y - 3z = 0 \end{cases}$ **36.** $\begin{cases} x + 4y - 3z = 0 \\ 3x - y + 3z = 0 \\ x + y + 6z = 0 \end{cases}$ **37.** $\begin{cases} x - 2y + 3z = 0 \\ 3x + y - 2z = 0 \\ 2x - 4y + 6z = 0 \end{cases}$ **38.** $\begin{cases} x - y + 2z = 0 \\ 3x + 2y = 0 \\ -2x + 2y - 4z = 0 \end{cases}$

39. $\begin{cases} \dfrac{1}{x} + \dfrac{1}{y} = 8 \\ \dfrac{3}{x} - \dfrac{5}{y} = 0 \end{cases}$ **40.** $\begin{cases} \dfrac{4}{x} - \dfrac{3}{y} = 0 \\ \dfrac{6}{x} + \dfrac{3}{2y} = 2 \end{cases}$

[**Hint:** Let $u = 1/x$ and $v = 1/y$ and solve for u and v.]

In Problems 41–46, solve for x.

41. $\begin{vmatrix} x & x \\ 4 & 3 \end{vmatrix} = 5$ **42.** $\begin{vmatrix} x & 1 \\ 3 & x \end{vmatrix} = -2$ **43.** $\begin{vmatrix} x & 1 & 1 \\ 4 & 3 & 2 \\ -1 & 2 & 5 \end{vmatrix} = 2$

44. $\begin{vmatrix} 3 & 2 & 4 \\ 1 & x & 5 \\ 0 & 1 & -2 \end{vmatrix} = 0$ **45.** $\begin{vmatrix} x & 2 & 3 \\ 1 & x & 0 \\ 6 & 1 & -2 \end{vmatrix} = 7$ **46.** $\begin{vmatrix} x & 1 & 2 \\ 1 & x & 3 \\ 0 & 1 & 2 \end{vmatrix} = -4x$

In Problems 47–54, use properties of determinants to find the value of each determinant if it is known that

$$\begin{vmatrix} x & y & z \\ u & v & w \\ 1 & 2 & 3 \end{vmatrix} = 4$$

47. $\begin{vmatrix} 1 & 2 & 3 \\ u & v & w \\ x & y & z \end{vmatrix}$ **48.** $\begin{vmatrix} x & y & z \\ u & v & w \\ 2 & 4 & 6 \end{vmatrix}$ **49.** $\begin{vmatrix} x & y & z \\ -3 & -6 & -9 \\ u & v & w \end{vmatrix}$

50. $\begin{vmatrix} 1 & 2 & 3 \\ x - u & y - v & z - w \\ u & v & w \end{vmatrix}$ **51.** $\begin{vmatrix} 1 & 2 & 3 \\ x - 3 & y - 6 & z - 9 \\ 2u & 2v & 2w \end{vmatrix}$ **52.** $\begin{vmatrix} x & y & z - x \\ u & v & w - u \\ 1 & 2 & 2 \end{vmatrix}$

53. $\begin{vmatrix} 1 & 2 & 3 \\ 2x & 2y & 2z \\ u - 1 & v - 2 & w - 3 \end{vmatrix}$ **54.** $\begin{vmatrix} x + 3 & y + 6 & z + 9 \\ 3u - 1 & 3v - 2 & 3w - 3 \\ 1 & 2 & 3 \end{vmatrix}$

55. Geometry: Equation of a Line An equation of the line containing the two points (x_1, y_1) and (x_2, y_2) may be expressed as the determinant

$$\begin{vmatrix} x & y & 1 \\ x_1 & y_1 & 1 \\ x_2 & y_2 & 1 \end{vmatrix} = 0$$

Prove this result by expanding the determinant and comparing the result to the two-point form of the equation of a line.

56. Geometry: Collinear Points Using the result obtained in Problem 55, show that three distinct points (x_1, y_1), (x_2, y_2), and (x_3, y_3) are collinear (lie on the same line) if and only if

$$\begin{vmatrix} x_1 & y_1 & 1 \\ x_2 & y_2 & 1 \\ x_3 & y_3 & 1 \end{vmatrix} = 0$$

57. Show that $\begin{vmatrix} x^2 & x & 1 \\ y^2 & y & 1 \\ z^2 & z & 1 \end{vmatrix} = (y - z)(x - y)(x - z)$.

58. Complete the proof of Cramer's Rule for two equations containing two variables.
[**Hint:** In system (5), page 743, if $a = 0$, then $b \neq 0$ and $c \neq 0$, since $D = -bc \neq 0$. Now show that equation (6) provides a solution of the system when $a = 0$. There are then three remaining cases: $b = 0$, $c = 0$, and $d = 0$.

59. Interchange columns 1 and 3 of a 3 by 3 determinant. Show that the value of the new determinant is -1 times the value of the original determinant.

60. Multiply each entry in row 2 of a 3 by 3 determinant by the number k, $k \neq 0$. Show that the value of the new determinant is k times the value of the original determinant.

61. Prove that a 3 by 3 determinant in which the entries in column 1 equal those in column 3 has the value 0.

62. Prove that if row 2 of a 3 by 3 determinant is multiplied by k, $k \neq 0$, and the result is added to the entries in row 1, then there is no change in the value of the determinant.

11.5 MATRIX ALGEBRA

OBJECTIVES

1. Work with Equality and Addition of Matrices
2. Know Properties of Matrices
3. Know How to Multiply Matrices
4. Find the Inverse of a Matrix
5. Solve Systems of Equations Using Inverse Matrices

In Section 11.3, we defined a matrix as an array of real numbers and used an augmented matrix to represent a system of linear equations. There is, however, a branch of mathematics, called **linear algebra,** that deals with matrices in such a way that an algebra of matrices is permitted. In this section, we provide a survey of how this **matrix algebra** is developed.

Before getting started, we restate the definition of a matrix.

A **matrix** is defined as a rectangular array of numbers:

$$\begin{array}{l} \\ \text{Row } 1 \\ \text{Row } 2 \\ \vdots \\ \text{Row } i \\ \vdots \\ \text{Row } m \end{array}
\begin{array}{c} \begin{array}{ccccccc} \text{Column } 1 & \text{Column } 2 & & \text{Column } j & & \text{Column } n \end{array} \\
\begin{bmatrix} a_{11} & a_{12} & \cdots & a_{1j} & \cdots & a_{1n} \\ a_{21} & a_{22} & \cdots & a_{2j} & \cdots & a_{2n} \\ \vdots & \vdots & & \vdots & & \vdots \\ a_{i1} & a_{i2} & \cdots & a_{ij} & \cdots & a_{in} \\ \vdots & \vdots & & \vdots & & \vdots \\ a_{m1} & a_{m2} & \cdots & a_{mj} & \cdots & a_{mn} \end{bmatrix} \end{array}$$

Each number a_{ij} of the matrix has two indexes: the **row index** i and the **column index** j. The matrix shown above has m rows and n columns. The

$m \cdot n$ numbers a_{ij} are usually referred to as the **entries** of the matrix. For example, a_{23} refers to the entry in the second row, third column.

Let's begin with an example that illustrates how matrices can be used to conveniently represent an array of information.

EXAMPLE 1

Arranging Data in a Matrix

In a survey of 900 people, the following information was obtained:

200 males	Thought federal defense spending was too high
150 males	Thought federal defense spending was too low
45 males	Had no opinion
315 females	Thought federal defense spending was too high
125 females	Thought federal defense spending was too low
65 females	Had no opinion

We can arrange the above data in a rectangular array as follows:

	Too High	Too Low	No Opinion
Male	200	150	45
Female	315	125	65

or as the matrix

$$\begin{bmatrix} 200 & 150 & 45 \\ 315 & 125 & 65 \end{bmatrix}$$

This matrix has two rows (representing males and females) and three columns (representing "too high," "too low," and "no opinion"). ■

The matrix we developed in Example 1 has 2 rows and 3 columns. In general, a matrix with m rows and n columns is called an ***m* by *n* matrix.** The matrix we developed in Example 1 is a 2 by 3 matrix and contains $2 \cdot 3 = 6$ entries. An m by n matrix will contain $m \cdot n$ entries.

If an m by n matrix has the same number of rows as columns, that is, if $m = n$, then the matrix is referred to as a **square matrix.**

EXAMPLE 2

Examples of Matrices

(a) $\begin{bmatrix} 5 & 0 \\ -6 & 1 \end{bmatrix}$ A 2 by 2 square matrix (b) $\begin{bmatrix} 1 & 0 & 3 \end{bmatrix}$ A 1 by 3 matrix

(c) $\begin{bmatrix} 6 & -2 & 4 \\ 4 & 3 & 5 \\ 8 & 0 & 1 \end{bmatrix}$ A 3 by 3 square matrix ■

Equality and Addition of Matrices

1 We begin our discussion of matrix algebra by first defining what is meant by two matrices being equal and then defining the operations of addition and subtraction. It is important to note that these definitions require each matrix

to have the same number of rows *and* the same number of columns as a prerequisite for equality and for addition and subtraction.

We usually represent matrices by capital letters, such as A, B, C, and so on.

Two m by n matrices A and B are said to be **equal,** written as

$$A = B$$

provided that each entry a_{ij} in A is equal to the corresponding entry b_{ij} in B.

For example,

$$\begin{bmatrix} 2 & 1 \\ 0.5 & -1 \end{bmatrix} = \begin{bmatrix} \sqrt{4} & 1 \\ \frac{1}{2} & -1 \end{bmatrix} \text{ and } \begin{bmatrix} 3 & 2 & 1 \\ 0 & 1 & -2 \end{bmatrix} = \begin{bmatrix} \sqrt{9} & \sqrt{4} & 1 \\ 0 & 1 & \sqrt[3]{-8} \end{bmatrix}$$

$$\begin{bmatrix} 4 & 1 \\ 6 & 1 \end{bmatrix} \neq \begin{bmatrix} 4 & 0 \\ 6 & 1 \end{bmatrix}$$ Because the entries in row 1, column 2 are not equal

$$\begin{bmatrix} 4 & 1 & 2 \\ 6 & 1 & 2 \end{bmatrix} \neq \begin{bmatrix} 4 & 1 & 2 & 3 \\ 6 & 1 & 2 & 4 \end{bmatrix}$$ Because the matrix on the left is 2 by 3 and the matrix on the right is 2 by 4

Suppose that A and B represent two m by n matrices. We define their **sum** $\boldsymbol{A + B}$ to be the m by n matrix formed by adding the corresponding entries a_{ij} of A and b_{ij} of B. The **difference** $\boldsymbol{A - B}$ is defined as the m by n matrix formed by subtracting the entries b_{ij} in B from the corresponding entries a_{ij} in A. Addition and subtraction of matrices are allowed only for matrices having the same number m of rows and the same number n of columns. For example, a 2 by 3 matrix and a 2 by 4 matrix cannot be added or subtracted.

Graphing utilities make the sometimes tedious process of matrix algebra easy. Let's compare how a graphing utility adds and subtracts matrices with doing it by hand.

EXAMPLE 3 Adding and Subtracting Matrices

Suppose that

$$A = \begin{bmatrix} 2 & 4 & 8 & -3 \\ 0 & 1 & 2 & 3 \end{bmatrix} \quad \text{and} \quad B = \begin{bmatrix} -3 & 4 & 0 & 1 \\ 6 & 8 & 2 & 0 \end{bmatrix}$$

Find: (a) $A + B$ (b) $A - B$

Graphing Solution Enter the matrices into a graphing utility. Name them $[A]$ and $[B]$. Figure 11 shows the results of adding and subtracting $[A]$ and $[B]$.

Figure 11

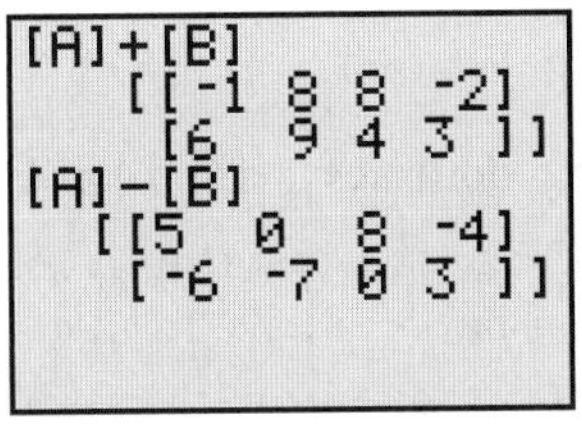

Algebraic Solution

(a) $A + B = \begin{bmatrix} 2 & 4 & 8 & -3 \\ 0 & 1 & 2 & 3 \end{bmatrix} + \begin{bmatrix} -3 & 4 & 0 & 1 \\ 6 & 8 & 2 & 0 \end{bmatrix}$

$= \begin{bmatrix} 2+(-3) & 4+4 & 8+0 & -3+1 \\ 0+6 & 1+8 & 2+2 & 3+0 \end{bmatrix}$ Add corresponding entries.

$= \begin{bmatrix} -1 & 8 & 8 & -2 \\ 6 & 9 & 4 & 3 \end{bmatrix}$

(b) $A - B = \begin{bmatrix} 2 & 4 & 8 & -3 \\ 0 & 1 & 2 & 3 \end{bmatrix} - \begin{bmatrix} -3 & 4 & 0 & 1 \\ 6 & 8 & 2 & 0 \end{bmatrix}$

$= \begin{bmatrix} 2-(-3) & 4-4 & 8-0 & -3-1 \\ 0-6 & 1-8 & 2-2 & 3-0 \end{bmatrix}$ Subtract corresponding entries.

$= \begin{bmatrix} 5 & 0 & 8 & -4 \\ -6 & -7 & 0 & 3 \end{bmatrix}$ ■

NOW WORK PROBLEM 1.

2 Many of the algebraic properties of sums of real numbers are also true for sums of matrices. Suppose that A, B, and C are m by n matrices. Then matrix addition is **commutative.** That is,

Commutative Property

$$A + B = B + A$$

Matrix addition is also **associative.** That is,

Associative Property

$$(A + B) + C = A + (B + C)$$

Although we shall not prove these results, the proofs, as the following example illustrates, are based on the commutative and associative properties for real numbers.

EXAMPLE 4 Demonstrating the Commutative Property

$$\begin{bmatrix} 2 & 3 & -1 \\ 4 & 0 & 7 \end{bmatrix} + \begin{bmatrix} -1 & 2 & 1 \\ 5 & -3 & 4 \end{bmatrix} = \begin{bmatrix} 2+(-1) & 3+2 & -1+1 \\ 4+5 & 0+(-3) & 7+4 \end{bmatrix}$$

$$= \begin{bmatrix} -1+2 & 2+3 & 1+(-1) \\ 5+4 & -3+0 & 4+7 \end{bmatrix}$$

$$= \begin{bmatrix} -1 & 2 & 1 \\ 5 & -3 & 4 \end{bmatrix} + \begin{bmatrix} 2 & 3 & -1 \\ 4 & 0 & 7 \end{bmatrix}$$ ■

A matrix whose entries are all equal to 0 is called a **zero matrix.** Each of the following matrices is a zero matrix.

$\begin{bmatrix} 0 & 0 \\ 0 & 0 \end{bmatrix}$ 2 by 2 square zero matrix $\quad \begin{bmatrix} 0 & 0 & 0 \\ 0 & 0 & 0 \end{bmatrix}$ 2 by 3 zero matrix $\quad \begin{bmatrix} 0 & 0 & 0 \end{bmatrix}$ 1 by 3 zero matrix

Zero matrices have properties similar to the real number 0. If A is an m by n matrix and 0 is an m by n zero matrix, then

$$A + 0 = A$$

In other words, the zero matrix is the additive identity in matrix algebra.

We can also multiply a matrix by a real number. If k is a real number and A is an m by n matrix, the matrix kA is the m by n matrix formed by multiplying each entry a_{ij} in A by k. The number k is sometimes referred to as a **scalar,** and the matrix kA is called a **scalar multiple** of A.

EXAMPLE 5 Operations Using Matrices

Suppose that

$$A = \begin{bmatrix} 3 & 1 & 5 \\ -2 & 0 & 6 \end{bmatrix} \qquad B = \begin{bmatrix} 4 & 1 & 0 \\ 8 & 1 & -3 \end{bmatrix} \qquad C = \begin{bmatrix} 9 & 0 \\ -3 & 6 \end{bmatrix}$$

Find: (a) $4A$ (b) $\frac{1}{3}C$ (c) $3A - 2B$

Graphing Solution Enter the matrices $[A]$, $[B]$, and $[C]$ into a graphing utility. Figure 12 shows the required computations.

Figure 12

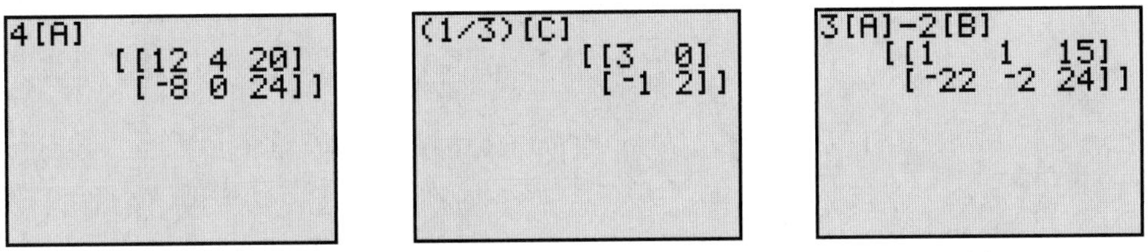

(a) (b) (c)

Algebraic Solution (a) $4A = 4\begin{bmatrix} 3 & 1 & 5 \\ -2 & 0 & 6 \end{bmatrix} = \begin{bmatrix} 4 \cdot 3 & 4 \cdot 1 & 4 \cdot 5 \\ 4(-2) & 4 \cdot 0 & 4 \cdot 6 \end{bmatrix} = \begin{bmatrix} 12 & 4 & 20 \\ -8 & 0 & 24 \end{bmatrix}$

(b) $\frac{1}{3}C = \frac{1}{3}\begin{bmatrix} 9 & 0 \\ -3 & 6 \end{bmatrix} = \begin{bmatrix} \frac{1}{3} \cdot 9 & \frac{1}{3} \cdot 0 \\ \frac{1}{3}(-3) & \frac{1}{3} \cdot 6 \end{bmatrix} = \begin{bmatrix} 3 & 0 \\ -1 & 2 \end{bmatrix}$

(c) $3A - 2B = 3\begin{bmatrix} 3 & 1 & 5 \\ -2 & 0 & 6 \end{bmatrix} - 2\begin{bmatrix} 4 & 1 & 0 \\ 8 & 1 & -3 \end{bmatrix}$

$$= \begin{bmatrix} 3 \cdot 3 & 3 \cdot 1 & 3 \cdot 5 \\ 3(-2) & 3 \cdot 0 & 3 \cdot 6 \end{bmatrix} - \begin{bmatrix} 2 \cdot 4 & 2 \cdot 1 & 2 \cdot 0 \\ 2 \cdot 8 & 2 \cdot 1 & 2(-3) \end{bmatrix}$$

$$= \begin{bmatrix} 9 & 3 & 15 \\ -6 & 0 & 18 \end{bmatrix} - \begin{bmatrix} 8 & 2 & 0 \\ 16 & 2 & -6 \end{bmatrix}$$

$$= \begin{bmatrix} 9 - 8 & 3 - 2 & 15 - 0 \\ -6 - 16 & 0 - 2 & 18 - (-6) \end{bmatrix}$$

$$= \begin{bmatrix} 1 & 1 & 15 \\ -22 & -2 & 24 \end{bmatrix}$$

■

NOW WORK PROBLEM 5.

We list next some of the algebraic properties of scalar multiplication. Let h and k be real numbers, and let A and B be m by n matrices. Then

Properties of Scalar Multiplication

$$k(hA) = (kh)A$$
$$(k + h)A = kA + hA$$
$$k(A + B) = kA + kB$$

MULTIPLICATION OF MATRICES

3 Unlike the straightforward definition for adding two matrices, the definition for multiplying two matrices is not what we might expect. In preparation for this definition, we need the following definitions:

A **row vector *R*** is a 1 by n matrix

$$R = \begin{bmatrix} r_1 & r_2 & \cdots & r_n \end{bmatrix}$$

A **column vector *C*** is an n by 1 matrix

$$C = \begin{bmatrix} c_1 \\ c_2 \\ \vdots \\ c_n \end{bmatrix}$$

The **product *RC*** of R times C is defined as the number

$$RC = \begin{bmatrix} r_1 & r_2 & \cdots & r_n \end{bmatrix} \begin{bmatrix} c_1 \\ c_2 \\ \vdots \\ c_n \end{bmatrix} = r_1 c_1 + r_2 c_2 + \cdots + r_n c_n$$

Notice that a row vector and a column vector can be multiplied only if they contain the same number of entries.

EXAMPLE 6 The Product of a Row Vector by a Column Vector

If $R = [3 \quad -5 \quad 2]$ and $C = \begin{bmatrix} 3 \\ 4 \\ -5 \end{bmatrix}$, then

$$RC = [3 \quad -5 \quad 2] \begin{bmatrix} 3 \\ 4 \\ -5 \end{bmatrix} = 3 \cdot 3 + (-5)4 + 2(-5)$$
$$= 9 - 20 - 10 = -21$$

Let's look at an application of the product of a row vector by a column vector.

EXAMPLE 7 Using Matrices to Compute Revenue

A clothing store sells men's shirts for \$25, silk ties for \$8, and wool suits for \$300. Last month, the store had sales consisting of 100 shirts, 200 ties, and 50 suits. What was the total revenue due to these sales?

Solution We set up a row vector R to represent the prices of each item and a column vector C to represent the corresponding number of items sold.

Then

$$\overset{\substack{\text{Prices}\\ \text{Shirts Ties Suits}}}{R = [25 \quad 8 \quad 300]} \qquad \overset{\text{Number sold}}{C = \begin{bmatrix} 100 \\ 200 \\ 50 \end{bmatrix}} \begin{matrix} \text{Shirts} \\ \text{Ties} \\ \text{Suits} \end{matrix}$$

The total revenue obtained is the product RC. That is,

$$RC = [25 \;\; 8 \;\; 300] \begin{bmatrix} 100 \\ 200 \\ 50 \end{bmatrix}$$

$$= \underbrace{25 \cdot 100}_{\text{Shirt revenue}} + \underbrace{8 \cdot 200}_{\text{Tie revenue}} + \underbrace{300 \cdot 50}_{\text{Suit revenue}} = \underbrace{\$19{,}100}_{\text{Total revenue}}$$

■

The definition for multiplying two matrices is based on the definition of a row vector times a column vector.

Let A denote an m by r matrix, and let B denote an r by n matrix. The **product** AB is defined as the m by n matrix whose entry in row i, column j is the product of the ith row of A and the jth column of B.

The definition of the product AB of two matrices A and B, in this order, requires that the number of columns of A equal the number of rows of B; otherwise, no product is defined.

$$\begin{matrix} A & & B \\ m \text{ by } r & & r \text{ by } n \end{matrix}$$

Must be same for AB to be defined (the r's)

AB is m by n. (the m and n)

An example will help to clarify the definition.

EXAMPLE 8 Multiplying Two Matrices

Find the product AB if

$$A = \begin{bmatrix} 2 & 4 & -1 \\ 5 & 8 & 0 \end{bmatrix} \quad \text{and} \quad B = \begin{bmatrix} 2 & 5 & 1 & 4 \\ 4 & 8 & 0 & 6 \\ -3 & 1 & -2 & -1 \end{bmatrix}$$

Solution First, we note that A is 2 by 3 and B is 3 by 4, so the product AB is defined and will be a 2 by 4 matrix.

Graphing Solution Enter the matrices A and B into a graphing utility. Figure 13 shows the product AB.

Figure 13

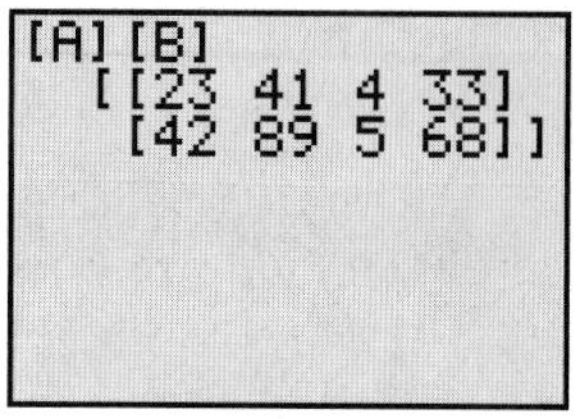

Algebraic Solution Suppose that we want the entry in row 2, column 3 of AB. To find it, we find the product of the row vector from row 2 of A and the column vector from column 3 of B.

Column 3 of B

Row 2 of A

$$\begin{bmatrix} 5 & 8 & 0 \end{bmatrix}\begin{bmatrix} 1 \\ 0 \\ -2 \end{bmatrix} = 5 \cdot 1 + 8 \cdot 0 + 0(-2) = 5$$

So far, we have

Column 3 ↓

$$AB = \begin{bmatrix} \_\_ & \_\_ & \_\_ & \_\_ \\ \_\_ & \_\_ & 5 & \_\_ \end{bmatrix} \leftarrow \text{Row 2}$$

Now, to find the entry in row 1, column 4 of AB, we find the product of row 1 of A and column 4 of B.

Column 4 of B

Row 1 of A

$$\begin{bmatrix} 2 & 4 & -1 \end{bmatrix}\begin{bmatrix} 4 \\ 6 \\ -1 \end{bmatrix} = 2 \cdot 4 + 4 \cdot 6 + (-1)(-1) = 33$$

Continuing in this fashion, we find AB.

$$AB = \begin{bmatrix} 2 & 4 & -1 \\ 5 & 8 & 0 \end{bmatrix}\begin{bmatrix} 2 & 5 & 1 & 4 \\ 4 & 8 & 0 & 6 \\ -3 & 1 & -2 & -1 \end{bmatrix}$$

$$= \begin{bmatrix} \text{Row 1 of } A \text{ times column 1 of } B & \text{Row 1 of } A \text{ times column 2 of } B & \text{Row 1 of } A \text{ times column 3 of } B & \text{Row 1 of } A \text{ times column 4 of } B \\ \text{Row 2 of } A \text{ times column 1 of } B & \text{Row 2 of } A \text{ times column 2 of } B & \text{Row 2 of } A \text{ times column 3 of } B & \text{Row 2 of } A \text{ times column 4 of } B \end{bmatrix}$$

$$= \begin{bmatrix} 2 \cdot 2 + 4 \cdot 4 + (-1)(-3) & 2 \cdot 5 + 4 \cdot 8 + (-1)1 & 2 \cdot 1 + 4 \cdot 0 + (-1)(-2) & 33 \text{ (from earlier)} \\ 5 \cdot 2 + 8 \cdot 4 + 0(-3) & 5 \cdot 5 + 8 \cdot 8 + 0 \cdot 1 & 5 \text{ (from earlier)} & 5 \cdot 4 + 8 \cdot 6 + 0(-1) \end{bmatrix}$$

$$= \begin{bmatrix} 23 & 41 & 4 & 33 \\ 42 & 89 & 5 & 68 \end{bmatrix}$$

NOW WORK PROBLEM 17.

Notice that for the matrices given in Example 8 the product BA is not defined, because B is 3 by 4 and A is 2 by 3. Try calculating BA on a graphing utility. What do you notice?

Another result that can occur when multiplying two matrices is illustrated in the next example.*

EXAMPLE 9 Multiplying Two Matrices

If

$$A = \begin{bmatrix} 2 & 1 & 3 \\ 1 & -1 & 0 \end{bmatrix} \quad \text{and} \quad B = \begin{bmatrix} 1 & 0 \\ 2 & 1 \\ 3 & 2 \end{bmatrix}$$

find: (a) AB (b) BA

Solution (a)

$$AB = \underset{\text{2 by 3}}{\begin{bmatrix} 2 & 1 & 3 \\ 1 & -1 & 0 \end{bmatrix}} \underset{\text{3 by 2}}{\begin{bmatrix} 1 & 0 \\ 2 & 1 \\ 3 & 2 \end{bmatrix}} = \underset{\text{2 by 2}}{\begin{bmatrix} 13 & 7 \\ -1 & -1 \end{bmatrix}}$$

(b)

$$BA = \underset{\text{3 by 2}}{\begin{bmatrix} 1 & 0 \\ 2 & 1 \\ 3 & 2 \end{bmatrix}} \underset{\text{2 by 3}}{\begin{bmatrix} 2 & 1 & 3 \\ 1 & -1 & 0 \end{bmatrix}} = \underset{\text{3 by 3}}{\begin{bmatrix} 2 & 1 & 3 \\ 5 & 1 & 6 \\ 8 & 1 & 9 \end{bmatrix}}$$

■

Notice in Example 9 that AB is 2 by 2 and BA is 3 by 3. Thus, it is possible for both AB and BA to be defined, yet be unequal. In fact, even if A and B are both n by n matrices, so that AB and BA are each defined and n by n, AB and BA will usually be unequal.

EXAMPLE 10 Multiplying Two Square Matrices

If

$$A = \begin{bmatrix} 2 & 1 \\ 0 & 4 \end{bmatrix} \quad \text{and} \quad B = \begin{bmatrix} -3 & 1 \\ 1 & 2 \end{bmatrix}$$

find: (a) AB (b) BA

Solution (a)

$$AB = \begin{bmatrix} 2 & 1 \\ 0 & 4 \end{bmatrix} \begin{bmatrix} -3 & 1 \\ 1 & 2 \end{bmatrix}$$

$$= \begin{bmatrix} 2(-3) + 1 \cdot 1 & 2 \cdot 1 + 1 \cdot 2 \\ 0(-3) + 4 \cdot 1 & 0 \cdot 1 + 4 \cdot 2 \end{bmatrix} = \begin{bmatrix} -5 & 4 \\ 4 & 8 \end{bmatrix}$$

(b)

$$BA = \begin{bmatrix} -3 & 1 \\ 1 & 2 \end{bmatrix} \begin{bmatrix} 2 & 1 \\ 0 & 4 \end{bmatrix}$$

$$= \begin{bmatrix} (-3)2 + 1 \cdot 0 & (-3)1 + 1 \cdot 4 \\ 1 \cdot 2 + 2 \cdot 0 & 1 \cdot 1 + 2 \cdot 4 \end{bmatrix} = \begin{bmatrix} -6 & 1 \\ 2 & 9 \end{bmatrix}$$

■

*For most of the examples that follow, we will multiply matrices by hand. You should verify each result using a graphing utility.

The preceding examples demonstrate that an important property of real numbers, the commutative property of multiplication, is not shared by matrices. Thus, in general:

Theorem Matrix multiplication is not commutative.

NOW WORK PROBLEMS 7 AND 9.

Next we give two of the properties of real numbers that are shared by matrices. Assuming that each product and sum is defined, we have the following:

Associative Property

$$A(BC) = (AB)C$$

Distributive Property

$$A(B + C) = AB + AC$$

The Identity Matrix

For an n by n square matrix, the entries located in row i, column i, $1 \leq i \leq n$, are called the **diagonal entries.** An n by n square matrix whose diagonal entries are 1's, while all other entries are 0's, is called the **identity matrix I_n.** For example,

$$I_2 = \begin{bmatrix} 1 & 0 \\ 0 & 1 \end{bmatrix} \qquad I_3 = \begin{bmatrix} 1 & 0 & 0 \\ 0 & 1 & 0 \\ 0 & 0 & 1 \end{bmatrix}$$

and so on.

EXAMPLE 11 Multiplication with an Identity Matrix

Let

$$A = \begin{bmatrix} -1 & 2 & 0 \\ 0 & 1 & 3 \end{bmatrix} \quad \text{and} \quad B = \begin{bmatrix} 3 & 2 \\ 4 & 6 \\ 5 & 2 \end{bmatrix}$$

Find: (a) AI_3 (b) I_2A (c) BI_2

Solution (a) $$AI_3 = \begin{bmatrix} -1 & 2 & 0 \\ 0 & 1 & 3 \end{bmatrix}\begin{bmatrix} 1 & 0 & 0 \\ 0 & 1 & 0 \\ 0 & 0 & 1 \end{bmatrix} = \begin{bmatrix} -1 & 2 & 0 \\ 0 & 1 & 3 \end{bmatrix} = A$$

(b) $$I_2A = \begin{bmatrix} 1 & 0 \\ 0 & 1 \end{bmatrix}\begin{bmatrix} -1 & 2 & 0 \\ 0 & 1 & 3 \end{bmatrix} = \begin{bmatrix} -1 & 2 & 0 \\ 0 & 1 & 3 \end{bmatrix} = A$$

(c) $$BI_2 = \begin{bmatrix} 3 & 2 \\ 4 & 6 \\ 5 & 2 \end{bmatrix}\begin{bmatrix} 1 & 0 \\ 0 & 1 \end{bmatrix} = \begin{bmatrix} 3 & 2 \\ 4 & 6 \\ 5 & 2 \end{bmatrix} = B$$

Example 11 demonstrates the following property:

Identity Property

If A is an m by n matrix, then

$$I_m A = A \quad \text{and} \quad AI_n = A$$

If A is an n by n square matrix, then $AI_n = I_n A = A$.

An identity matrix has properties analogous to those of the real number 1. In other words, the identity matrix is a multiplicative identity in matrix algebra.

The Inverse of a Matrix

4 Let A be a square n by n matrix. If there exists an n by n matrix A^{-1}, read "A inverse," for which

$$AA^{-1} = A^{-1}A = I_n$$

then A^{-1} is called the **inverse** of the matrix A.

As we shall soon see, not every square matrix has an inverse. When a matrix A does have an inverse A^{-1}, then A is said to be **nonsingular.** If a matrix A has no inverse, it is called **singular.***

EXAMPLE 12 Multiplying a Matrix by Its Inverse

Show that the inverse of

$$A = \begin{bmatrix} 3 & 1 \\ 2 & 1 \end{bmatrix} \quad \text{is} \quad A^{-1} = \begin{bmatrix} 1 & -1 \\ -2 & 3 \end{bmatrix}$$

Solution We need to show that $AA^{-1} = A^{-1}A = I_2$.

$$AA^{-1} = \begin{bmatrix} 3 & 1 \\ 2 & 1 \end{bmatrix}\begin{bmatrix} 1 & -1 \\ -2 & 3 \end{bmatrix} = \begin{bmatrix} 3 \cdot 1 + 1(-2) & 3(-1) + 1 \cdot 3 \\ 2 \cdot 1 + 1(-2) & 2(-1) + 1 \cdot 3 \end{bmatrix}$$

$$= \begin{bmatrix} 1 & 0 \\ 0 & 1 \end{bmatrix} = I_2$$

$$A^{-1}A = \begin{bmatrix} 1 & -1 \\ -2 & 3 \end{bmatrix}\begin{bmatrix} 3 & 1 \\ 2 & 1 \end{bmatrix} = \begin{bmatrix} 3 - 2 & 1 - 1 \\ -6 + 6 & -2 + 3 \end{bmatrix} = \begin{bmatrix} 1 & 0 \\ 0 & 1 \end{bmatrix} = I_2$$ ■

We now show one way to find the inverse of

$$A = \begin{bmatrix} 3 & 1 \\ 2 & 1 \end{bmatrix}$$

Suppose that A^{-1} is given by

$$A^{-1} = \begin{bmatrix} x & y \\ z & w \end{bmatrix} \tag{1}$$

* If the determinant of A is zero, then A is singular. (Refer to Section 11.4.)

where x, y, z, and w are four variables. Based on the definition of an inverse, if, indeed, A has an inverse, we have

$$AA^{-1} = I_2$$

$$\begin{bmatrix} 3 & 1 \\ 2 & 1 \end{bmatrix}\begin{bmatrix} x & y \\ z & w \end{bmatrix} = \begin{bmatrix} 1 & 0 \\ 0 & 1 \end{bmatrix}$$

$$\begin{bmatrix} 3x + z & 3y + w \\ 2x + z & 2y + w \end{bmatrix} = \begin{bmatrix} 1 & 0 \\ 0 & 1 \end{bmatrix}$$

Because corresponding entries must be equal, it follows that this matrix equation is equivalent to four ordinary equations.

$$\begin{cases} 3x + z = 1 \\ 2x + z = 0 \end{cases} \qquad \begin{cases} 3y + w = 0 \\ 2y + w = 1 \end{cases}$$

The augmented matrix of each system is

$$\left[\begin{array}{cc|c} 3 & 1 & 1 \\ 2 & 1 & 0 \end{array}\right] \quad \left[\begin{array}{cc|c} 3 & 1 & 0 \\ 2 & 1 & 1 \end{array}\right] \qquad \textbf{(2)}$$

The usual procedure would be to transform each augmented matrix into reduced row echelon form. Notice, though, that the left sides of the augmented matrices are equal, so the same row operations (see Section 11.3) can be used to reduce each one. Thus, we find it more efficient to combine the two augmented matrices (2) into a single matrix, as shown next, and then transform it into reduced row echelon form.

$$\left[\begin{array}{cc|cc} 3 & 1 & 1 & 0 \\ 2 & 1 & 0 & 1 \end{array}\right]$$

Now we attempt to transform the left side into an identity matrix.

$$\left[\begin{array}{cc|cc} 3 & 1 & 1 & 0 \\ 2 & 1 & 0 & 1 \end{array}\right] \underset{R_1 = -1r_2 + r_1}{\rightarrow} \left[\begin{array}{cc|cc} 1 & 0 & 1 & -1 \\ 2 & 1 & 0 & 1 \end{array}\right]$$

$$\underset{R_2 = -2r_1 + r_2}{\rightarrow} \left[\begin{array}{cc|cc} 1 & 0 & 1 & -1 \\ 0 & 1 & -2 & 3 \end{array}\right] \qquad \textbf{(3)}$$

Matrix (3) is in reduced row echelon form. Now we reverse the earlier step of combining the two augmented matrices in (2) and write the single matrix (3) as two augmented matrices.

$$\left[\begin{array}{cc|c} 1 & 0 & 1 \\ 0 & 1 & -2 \end{array}\right] \text{ and } \left[\begin{array}{cc|c} 1 & 0 & -1 \\ 0 & 1 & 3 \end{array}\right]$$

We conclude from these matrices that $x = 1$, $z = -2$, and $y = -1$, $w = 3$. Substituting these values into matrix (1), we find that

$$A^{-1} = \begin{bmatrix} 1 & -1 \\ -2 & 3 \end{bmatrix}$$

Notice in display (3) that the 2 by 2 matrix to the right of the vertical bar is, in fact, the inverse of A. Also notice that the identity matrix I_2 is the matrix that appears to the left of the vertical bar. These observations and the procedures followed above will work in general.

PROCEDURE FOR FINDING THE INVERSE OF A NONSINGULAR MATRIX

To find the inverse of an n by n nonsingular matrix A, proceed as follows:

STEP 1: Form the matrix $\left[A \mid I_n\right]$.

STEP 2: Transform the matrix $\left[A \mid I_n\right]$ into reduced row echelon form.

STEP 3: The reduced row echelon form of $\left[A \mid I_n\right]$ will contain the identity matrix I_n on the left of the vertical bar; the n by n matrix on the right of the vertical bar is the inverse of A.

In other words, if A is nonsingular, we begin with the matrix $\left[A \mid I_n\right]$ and, after transforming it into reduced row echelon form, we end up with the matrix $\left[I_n \mid A^{-1}\right]$.

Let's look at another example.

EXAMPLE 13 Finding the Inverse of a Matrix

The matrix

$$A = \begin{bmatrix} 1 & 1 & 0 \\ -1 & 3 & 4 \\ 0 & 4 & 3 \end{bmatrix}$$

is nonsingular. Find its inverse.

Graphing Solution Enter the matrix A into a graphing utility. Figure 14 shows A^{-1}.

Figure 14

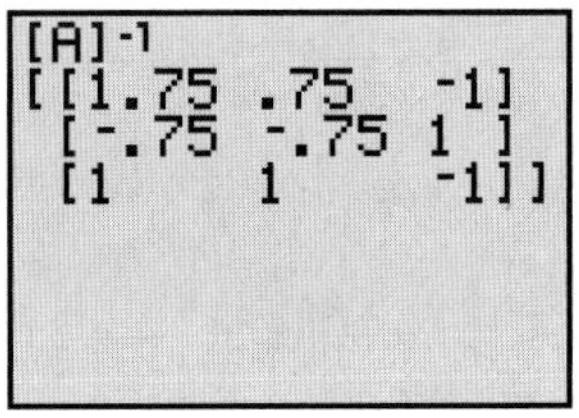

Algebraic Solution First, we form the matrix

$$\left[A \mid I_3\right] = \left[\begin{array}{rrr|rrr} 1 & 1 & 0 & 1 & 0 & 0 \\ -1 & 3 & 4 & 0 & 1 & 0 \\ 0 & 4 & 3 & 0 & 0 & 1 \end{array}\right]$$

Next, we use row operations to transform $\left[A \mid I_3\right]$ into reduced row echelon form.

$$\left[\begin{array}{ccc|ccc} 1 & 1 & 0 & 1 & 0 & 0 \\ -1 & 3 & 4 & 0 & 1 & 0 \\ 0 & 4 & 3 & 0 & 0 & 1 \end{array}\right] \underset{R_2 = r_1 + r_2}{\rightarrow} \left[\begin{array}{ccc|ccc} 1 & 1 & 0 & 1 & 0 & 0 \\ 0 & 4 & 4 & 1 & 1 & 0 \\ 0 & 4 & 3 & 0 & 0 & 1 \end{array}\right] \underset{R_2 = \frac{1}{4}r_2}{\rightarrow} \left[\begin{array}{ccc|ccc} 1 & 1 & 0 & 1 & 0 & 0 \\ 0 & 1 & 1 & \frac{1}{4} & \frac{1}{4} & 0 \\ 0 & 4 & 3 & 0 & 0 & 1 \end{array}\right]$$

$$\underset{\substack{R_1 = -1r_2 + r_1 \\ R_3 = -4r_2 + r_3}}{\rightarrow} \left[\begin{array}{ccc|ccc} 1 & 0 & -1 & \frac{3}{4} & -\frac{1}{4} & 0 \\ 0 & 1 & 1 & \frac{1}{4} & \frac{1}{4} & 0 \\ 0 & 0 & -1 & -1 & -1 & 1 \end{array}\right] \underset{R_3 = -1r_3}{\rightarrow} \left[\begin{array}{ccc|ccc} 1 & 0 & -1 & \frac{3}{4} & -\frac{1}{4} & 0 \\ 0 & 1 & 1 & \frac{1}{4} & \frac{1}{4} & 0 \\ 0 & 0 & 1 & 1 & 1 & -1 \end{array}\right]$$

$$\underset{\substack{R_1 = r_3 + r_1 \\ R_2 = -1r_3 + r_2}}{\rightarrow} \left[\begin{array}{ccc|ccc} 1 & 0 & 0 & \frac{7}{4} & \frac{3}{4} & -1 \\ 0 & 1 & 0 & -\frac{3}{4} & -\frac{3}{4} & 1 \\ 0 & 0 & 1 & 1 & 1 & -1 \end{array}\right]$$

The matrix $[A \mid I_3]$ is now in reduced row echelon form, and the identity matrix I_3 is on the left of the vertical bar. Hence, the inverse of A is

$$A^{-1} = \begin{bmatrix} \frac{7}{4} & \frac{3}{4} & -1 \\ -\frac{3}{4} & -\frac{3}{4} & 1 \\ 1 & 1 & -1 \end{bmatrix}$$

■

You can (and should) verify that this is the correct inverse by showing that $AA^{-1} = A^{-1}A = I_3$.

NOW WORK PROBLEM 23.

If transforming the matrix $[A \mid I_n]$ into reduced row echelon form does not result in the identity matrix I_n to the left of the vertical bar, then A has no inverse. The next example demonstrates such a matrix.

EXAMPLE 14 Showing That a Matrix Has No Inverse

Show that the following matrix has no inverse.

$$A = \begin{bmatrix} 4 & 6 \\ 2 & 3 \end{bmatrix}$$

Graphing Solution

Enter the matrix A. Figure 15 shows the result when we try to find its inverse. The ERRor comes about because A is singular.

Figure 15

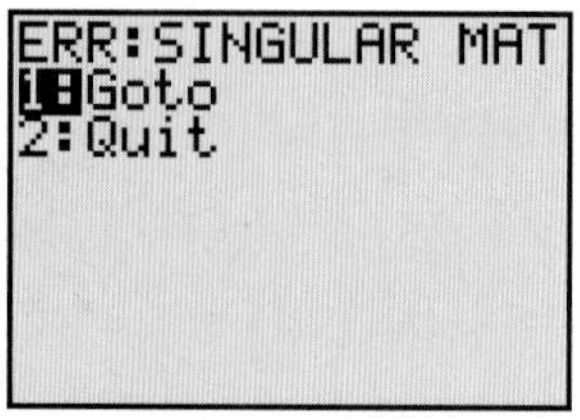

Algebraic Solution

Proceeding as in Example 13, we form the matrix

$$[A \mid I_2] = \left[\begin{array}{cc|cc} 4 & 6 & 1 & 0 \\ 2 & 3 & 0 & 1 \end{array}\right]$$

Then we use row operations to transform $[A \mid I_2]$ into reduced row echelon form.

$$[A \mid I_2] = \left[\begin{array}{cc|cc} 4 & 6 & 1 & 0 \\ 2 & 3 & 0 & 1 \end{array}\right] \underset{R_1 = \frac{1}{4}r_1}{\rightarrow} \left[\begin{array}{cc|cc} 1 & \frac{3}{2} & \frac{1}{4} & 0 \\ 2 & 3 & 0 & 1 \end{array}\right] \underset{R_2 = -2r_1 + r_2}{\rightarrow} \left[\begin{array}{cc|cc} 1 & \frac{3}{2} & \frac{1}{4} & 0 \\ 0 & 0 & -\frac{1}{2} & 1 \end{array}\right]$$

The matrix $[A \mid I_2]$ is sufficiently reduced for us to see that the identity matrix cannot appear to the left of the vertical bar. We conclude that A has no inverse. ■

Seeing the Concept Compute the determinant of A in Example 14 using a graphing utility. What is the result? Are you surprised? ■

NOW WORK PROBLEM 51.

Solving Systems of Linear Equations

5 Inverse matrices can be used to solve systems of equations in which the number of equations is the same as the number of variables.

EXAMPLE 15 Using the Inverse Matrix to Solve a System of Linear Equations

Solve the system of equations: $\begin{cases} x + y = 3 \\ -x + 3y + 4z = -3 \\ 4y + 3z = 2 \end{cases}$

Solution If we let

$$A = \begin{bmatrix} 1 & 1 & 0 \\ -1 & 3 & 4 \\ 0 & 4 & 3 \end{bmatrix} \quad X = \begin{bmatrix} x \\ y \\ z \end{bmatrix} \quad B = \begin{bmatrix} 3 \\ -3 \\ 2 \end{bmatrix}$$

then the original system of equations can be written compactly as the matrix equation

$$AX = B \tag{4}$$

We know from Example 13 that the matrix A has the inverse A^{-1}, so we multiply each side of equation (4) by A^{-1}.

$$\begin{aligned} AX &= B \\ A^{-1}(AX) &= A^{-1}B \\ (A^{-1}A)X &= A^{-1}B && \text{Associative property of multiplication} \\ I_3X &= A^{-1}B && \text{Definition of inverse matrix} \\ X &= A^{-1}B && \text{Property of identity matrix} \end{aligned} \tag{5}$$

Now we use (5) to find $X = \begin{bmatrix} x \\ y \\ z \end{bmatrix}$.

Graphing Solution Enter the matrices A and B into a graphing utility. Figure 16 shows the solution to the system of equations.

Figure 16

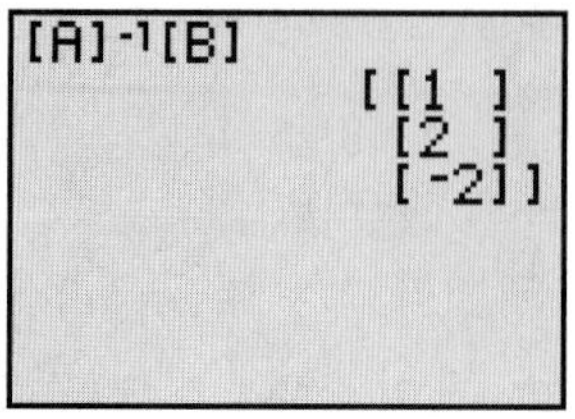

Algebraic Solution

$$X = \begin{bmatrix} x \\ y \\ z \end{bmatrix} = A^{-1}B = \underset{\uparrow \text{ Example 13}}{\begin{bmatrix} \frac{7}{4} & \frac{3}{4} & -1 \\ -\frac{3}{4} & -\frac{3}{4} & 1 \\ 1 & 1 & -1 \end{bmatrix}} \begin{bmatrix} 3 \\ -3 \\ 2 \end{bmatrix} = \begin{bmatrix} 1 \\ 2 \\ -2 \end{bmatrix}$$

Thus, $x = 1, y = 2, z = -2$. ■

The method used in Example 15 to solve a system of equations is particularly useful when it is necessary to solve several systems of equations in which the constants appearing to the right of the equal signs change, while the coefficients of the variables on the left side remain the same. See Problems 31–50 for some illustrations. Be careful; this method can only be used if the inverse exists. If it does not exist, row reduction must be used since the system is either inconsistent or dependent.

HISTORICAL FEATURE

Arthur Cayley (1821–1895)

Matrices were invented in 1857 by Arthur Cayley (1821–1895) as a way of efficiently computing the result of substituting one linear system into another (see Historical Problem 2). The resulting system had incredible richness, in the sense that a very wide variety of mathematical systems could be mimicked by the matrices. Cayley and his friend James J. Sylvester (1814–1897) spent much of the rest of their lives elaborating the theory. The torch was then passed to Georg Frobenius (1849–1917), whose deep investigations established a central place for matrices in modern mathematics. In 1924, rather to the surprise of physicists, it was found that matrices (with complex numbers in them) were exactly the right tool for describing the behavior of atomic systems. Today, matrices are used in a wide variety of applications.

HISTORICAL PROBLEMS

1. *Matrices and Complex Numbers* Frobenius emphasized in his research how matrices could be used to mimic other mathematical systems. Here, we mimic the behavior of complex numbers using matrices. Mathematicians call such a relationship an *isomorphism.*

Complex number $\longleftrightarrow$ Matrix

$$a + bi \longleftrightarrow \begin{bmatrix} a & b \\ -b & a \end{bmatrix}$$

Note that the complex number can be read off the top line of the matrix. Thus,

$$2 + 3i \longleftrightarrow \begin{bmatrix} 2 & 3 \\ -3 & 2 \end{bmatrix} \quad \text{and} \quad \begin{bmatrix} 4 & -2 \\ 2 & 4 \end{bmatrix} \longleftrightarrow 4 - 2i$$

EXAMPLE 1 Nonrepeated Linear Factors

Write the partial fraction decomposition of $\dfrac{x}{x^2 - 5x + 6}$.

Solution First, we factor the denominator,

$$x^2 - 5x + 6 = (x - 2)(x - 3)$$

and conclude that the denominator contains only nonrepeated linear factors. Then we decompose the rational expression according to equation (1):

$$\frac{x}{x^2 - 5x + 6} = \frac{A}{x - 2} + \frac{B}{x - 3} \qquad \textbf{(2)}$$

where A and B are to be determined. To find A and B, we clear the fractions by multiplying each side by $(x - 2)(x - 3) = x^2 - 5x + 6$. The result is

$$x = A(x - 3) + B(x - 2) \qquad \textbf{(3)}$$

or

$$x = (A + B)x + (-3A - 2B)$$

This equation is an identity in x. Thus, we may equate the coefficients of like powers of x to get

$$\begin{cases} 1 = A + B & \text{Equate coefficients of } x\text{: } 1x = (A + B)x. \\ 0 = -3A - 2B & \text{Equate coefficients of } x^0\text{, the constants: } 0x^0 = (-3A - 2B)x^0. \end{cases}$$

This system of two equations containing two variables, A and B, can be solved using whatever method you wish. Solving it, we get

$$A = -2 \qquad B = 3$$

From equation (2), the partial fraction decomposition is

$$\frac{x}{x^2 - 5x + 6} = \frac{-2}{x - 2} + \frac{3}{x - 3}$$

Check: The decomposition can be checked by adding the fractions.

$$\frac{-2}{x - 2} + \frac{3}{x - 3} = \frac{-2(x - 3) + 3(x - 2)}{(x - 2)(x - 3)} = \frac{x}{(x - 2)(x - 3)}$$

$$= \frac{x}{x^2 - 5x + 6}$$

NOW WORK PROBLEM 9.

The numbers to be found in the partial fraction decomposition can sometimes be found more readily by using suitable choices for x (which may include complex numbers) in the identity obtained after fractions have been cleared. In Example 1, the identity after clearing fractions, equation (3), is

$$x = A(x - 3) + B(x - 2)$$

If we let $x = 2$ in this expression, the term containing B drops out, leaving $2 = A(-1)$, or $A = -2$. Similarly, if we let $x = 3$, the term containing A drops out, leaving $3 = B$. As before, $A = -2$ and $B = 3$.

We use this method in the next example.

2 **CASE 2: Q has repeated linear factors.**

If the polynomial Q has a repeated factor, say $(x - a)^n$, $n \geq 2$ an integer, then, in the partial fraction decomposition of P/Q, we allow for the terms

$$\frac{A_1}{x - a} + \frac{A_2}{(x - a)^2} + \cdots + \frac{A_n}{(x - a)^n}$$

where the numbers $A_1, A_2, \ldots, A_n$ are to be determined.

EXAMPLE 2 Repeated Linear Factors

Write the partial fraction decomposition of $\dfrac{x + 2}{x^3 - 2x^2 + x}$.

Solution First, we factor the denominator,

$$x^3 - 2x^2 + x = x(x^2 - 2x + 1) = x(x - 1)^2$$

and find that the denominator has the nonrepeated linear factor x and the repeated linear factor $(x - 1)^2$. By Case 1, we must allow for the term A/x in the decomposition; and, by Case 2, we must allow for the terms $B/(x - 1) + C/(x - 1)^2$ in the decomposition.

Thus, we write

$$\frac{x + 2}{x^3 - 2x^2 + x} = \frac{A}{x} + \frac{B}{x - 1} + \frac{C}{(x - 1)^2} \quad \textbf{(4)}$$

Again, we clear fractions by multiplying each side by $x^3 - 2x^2 + x = x(x - 1)^2$. The result is the identity

$$x + 2 = A(x - 1)^2 + Bx(x - 1) + Cx \quad \textbf{(5)}$$

If we let $x = 0$ in this expression, the terms containing B and C drop out, leaving $2 = A(-1)^2$, or $A = 2$. Similarly, if we let $x = 1$, the terms containing A and B drop out, leaving $3 = C$. Thus, equation (5) becomes

$$x + 2 = 2(x - 1)^2 + Bx(x - 1) + 3x$$

Now let $x = 2$ (any choice other than 0 or 1 will work as well). The result is

$$4 = 2(1)^2 + B(2)(1) + 3(2)$$

$$2B = 4 - 2 - 6 = -4$$

$$B = -2$$

Thus, we have $A = 2$, $B = -2$, and $C = 3$.

From equation (4), the partial fraction decomposition is

$$\frac{x + 2}{x^3 - 2x^2 + x} = \frac{2}{x} + \frac{-2}{x - 1} + \frac{3}{(x - 1)^2}$$

■

EXAMPLE 3 Repeated Linear Factors

Write the partial fraction decomposition of $\dfrac{x^3 - 8}{x^2(x-1)^3}$.

Solution The denominator contains the repeated linear factor x^2 and the repeated linear factor $(x-1)^3$. Thus, the partial fraction decomposition takes the form

$$\frac{x^3 - 8}{x^2(x-1)^3} = \frac{A}{x} + \frac{B}{x^2} + \frac{C}{x-1} + \frac{D}{(x-1)^2} + \frac{E}{(x-1)^3} \qquad \textbf{(6)}$$

As before, we clear fractions and obtain the identity

$$x^3 - 8 = Ax(x-1)^3 + B(x-1)^3 + Cx^2(x-1)^2 + Dx^2(x-1) + Ex^2 \qquad \textbf{(7)}$$

Let $x = 0$. (Do you see why this choice was made?) Then

$$-8 = B(-1)$$
$$B = 8$$

Now let $x = 1$ in equation (7). Then

$$-7 = E$$

Use $B = 8$ and $E = -7$ in equation (7) and collect like terms.

$$\begin{aligned}
x^3 - 8 &= Ax(x-1)^3 + 8(x-1)^3 \\
&\quad + Cx^2(x-1)^2 + Dx^2(x-1) - 7x^2 \\
x^3 - 8 - 8(x^3 - 3x^2 + 3x - 1) + 7x^2 &= Ax(x-1)^3 + Cx^2(x-1)^2 + Dx^2(x-1) \\
-7x^3 + 31x^2 - 24x &= x(x-1)[A(x-1)^2 + Cx(x-1) + Dx] \\
x(x-1)(-7x+24) &= x(x-1)[A(x-1)^2 + Cx(x-1) + Dx] \\
-7x + 24 &= A(x-1)^2 + Cx(x-1) + Dx \qquad \textbf{(8)}
\end{aligned}$$

We now work with equation (8). Let $x = 0$. Then

$$24 = A$$

Now let $x = 1$ in equation (8). Then

$$17 = D$$

Use $A = 24$ and $D = 17$ in equation (8) and collect like terms.

$$\begin{aligned}
-7x + 24 &= 24(x-1)^2 + Cx(x-1) + 17x \\
-24x^2 + 48x - 24 - 17x - 7x + 24 &= Cx(x-1) \\
-24x^2 + 24x &= Cx(x-1) \\
-24x(x-1) &= Cx(x-1) \\
-24 &= C
\end{aligned}$$

We now know all the numbers A, B, C, D, and E, so, from equation (6), we have the decomposition

$$\frac{x^3 - 8}{x^2(x-1)^3} = \frac{24}{x} + \frac{8}{x^2} + \frac{-24}{x-1} + \frac{17}{(x-1)^2} + \frac{-7}{(x-1)^3}$$

■

The method employed in Example 3, although somewhat tedious, is still preferable to solving the system of five equations containing five variables that the expansion of equation (6) leads to.

NOW WORK PROBLEM 15.

3 The final two cases involve irreducible quadratic factors. As mentioned in Section 4.1, a quadratic factor is irreducible if it cannot be factored into linear factors with real coefficients. A quadratic expression $ax^2 + bx + c$ is irreducible whenever $b^2 - 4ac < 0$. For example, $x^2 + x + 1$ and $x^2 + 4$ are irreducible.

CASE 3: Q contains a nonrepeated irreducible quadratic factor.

If Q contains a nonrepeated irreducible quadratic factor of the form $ax^2 + bx + c$, then, in the partial fraction decomposition of P/Q, allow for the term

$$\frac{Ax + B}{ax^2 + bx + c}$$

where the numbers A and B are to be determined.

EXAMPLE 4 Nonrepeated Irreducible Quadratic Factor

Write the partial factor decomposition of $\dfrac{3x - 5}{x^3 - 1}$.

Solution We factor the denominator,

$$x^3 - 1 = (x - 1)(x^2 + x + 1)$$

and find that it has a nonrepeated linear factor $x - 1$ and a nonrepeated irreducible quadratic factor $x^2 + x + 1$. Thus, we allow for the term $A/(x - 1)$ by Case 1, and we allow for the term $(Bx + C)/(x^2 + x + 1)$ by Case 3. Hence, we write

$$\frac{3x - 5}{x^3 - 1} = \frac{A}{x - 1} + \frac{Bx + C}{x^2 + x + 1} \tag{9}$$

We clear fractions by multiplying each side of equation (9) by $x^3 - 1 = (x - 1)(x^2 + x + 1)$ to obtain the identity

$$3x - 5 = A(x^2 + x + 1) + (Bx + C)(x - 1) \tag{10}$$

Now let $x = 1$. Then equation (10) gives $-2 = A(3)$, or $A = -\frac{2}{3}$. We use this value of A in equation (10) and simplify.

$$3x - 5 = -\frac{2}{3}(x^2 + x + 1) + (Bx + C)(x - 1)$$

$$3(3x - 5) = -2(x^2 + x + 1) + 3(Bx + C)(x - 1) \quad \text{Multiply each side by 3.}$$

$$9x - 15 = -2x^2 - 2x - 2 + 3(Bx + C)(x - 1)$$

$$2x^2 + 11x - 13 = 3(Bx + C)(x - 1) \quad \text{Collect terms.}$$

$$(2x + 13)(x - 1) = 3(Bx + C)(x - 1) \quad \text{Factor the left side.}$$
$$2x + 13 = 3Bx + 3C$$
$$2 = 3B \quad \text{and} \quad 13 = 3C \quad \text{Equate coefficients.}$$
$$B = \frac{2}{3} \qquad C = \frac{13}{3}$$

Thus, from equation (9), we see that

$$\frac{3x - 5}{x^3 - 1} = \frac{-\frac{2}{3}}{x - 1} + \frac{\frac{2}{3}x + \frac{13}{3}}{x^2 + x + 1}$$

NOW WORK PROBLEM 17.

4 **CASE 4: Q contains repeated irreducible quadratic factors.**

If the polynomial Q contains a repeated irreducible quadratic factor $(ax^2 + bx + c)^n$, $n \geq 2$, n an integer, then, in the partial fraction decomposition of P/Q, allow for the terms

$$\frac{A_1x + B_1}{ax^2 + bx + c} + \frac{A_2x + B_2}{(ax^2 + bx + c)^2} + \cdots + \frac{A_nx + B_n}{(ax^2 + bx + c)^n}$$

where the numbers $A_1, B_1, A_2, B_2, \ldots, A_n, B_n$ are to be determined.

EXAMPLE 5 Repeated Irreducible Quadratic Factor

Write the partial fraction decomposition of $\dfrac{x^3 + x^2}{(x^2 + 4)^2}$.

Solution The denominator contains the repeated irreducible quadratic factor $(x^2 + 4)^2$, so we write

$$\frac{x^3 + x^2}{(x^2 + 4)^2} = \frac{Ax + B}{x^2 + 4} + \frac{Cx + D}{(x^2 + 4)^2} \qquad \textbf{(11)}$$

We clear fractions to obtain

$$x^3 + x^2 = (Ax + B)(x^2 + 4) + Cx + D$$

Collecting like terms yields

$$x^3 + x^2 = Ax^3 + Bx^2 + (4A + C)x + D + 4B$$

Equating coefficients, we arrive at the system

$$\begin{cases} A = 1 \\ B = 1 \\ 4A + C = 0 \\ D + 4B = 0 \end{cases}$$

The solution is $A = 1, B = 1, C = -4, D = -4$. Hence, from equation (11),

$$\frac{x^3 + x^2}{(x^2 + 4)^2} = \frac{x + 1}{x^2 + 4} + \frac{-4x - 4}{(x^2 + 4)^2}$$

NOW WORK PROBLEM 31.

11.6 EXERCISES

In Problems 1–8, tell whether the given rational expression is proper or improper. If improper, rewrite it as the sum of a polynomial and a proper rational expression.

1. $\dfrac{x}{x^2 - 1}$ **2.** $\dfrac{5x + 2}{x^3 - 1}$ **3.** $\dfrac{x^2 + 5}{x^2 - 4}$ **4.** $\dfrac{3x^2 - 2}{x^2 - 1}$

5. $\dfrac{5x^3 + 2x - 1}{x^2 - 4}$ **6.** $\dfrac{3x^4 + x^2 - 2}{x^3 + 8}$ **7.** $\dfrac{x(x - 1)}{(x + 4)(x - 3)}$ **8.** $\dfrac{2x(x^2 + 4)}{x^2 + 1}$

In Problems 9–42, write the partial fraction decomposition of each rational expression.

9. $\dfrac{4}{x(x - 1)}$ **10.** $\dfrac{3x}{(x + 2)(x - 1)}$ **11.** $\dfrac{1}{x(x^2 + 1)}$ **12.** $\dfrac{1}{(x + 1)(x^2 + 4)}$

13. $\dfrac{x}{(x - 1)(x - 2)}$ **14.** $\dfrac{3x}{(x + 2)(x - 4)}$ **15.** $\dfrac{x^2}{(x - 1)^2(x + 1)}$ **16.** $\dfrac{x + 1}{x^2(x - 2)}$

17. $\dfrac{1}{x^3 - 8}$ **18.** $\dfrac{2x + 4}{x^3 - 1}$ **19.** $\dfrac{x^2}{(x - 1)^2(x + 1)^2}$ **20.** $\dfrac{x + 1}{x^2(x - 2)^2}$

21. $\dfrac{x - 3}{(x + 2)(x + 1)^2}$ **22.** $\dfrac{x^2 + x}{(x + 2)(x - 1)^2}$ **23.** $\dfrac{x + 4}{x^2(x^2 + 4)}$ **24.** $\dfrac{10x^2 + 2x}{(x - 1)^2(x^2 + 2)}$

25. $\dfrac{x^2 + 2x + 3}{(x + 1)(x^2 + 2x + 4)}$ **26.** $\dfrac{x^2 - 11x - 18}{x(x^2 + 3x + 3)}$ **27.** $\dfrac{x}{(3x - 2)(2x + 1)}$ **28.** $\dfrac{1}{(2x + 3)(4x - 1)}$

29. $\dfrac{x}{x^2 + 2x - 3}$ **30.** $\dfrac{x^2 - x - 8}{(x + 1)(x^2 + 5x + 6)}$ **31.** $\dfrac{x^2 + 2x + 3}{(x^2 + 4)^2}$ **32.** $\dfrac{x^3 + 1}{(x^2 + 16)^2}$

33. $\dfrac{7x + 3}{x^3 - 2x^2 - 3x}$ **34.** $\dfrac{x^5 + 1}{x^6 - x^4}$ **35.** $\dfrac{x^2}{x^3 - 4x^2 + 5x - 2}$ **36.** $\dfrac{x^2 + 1}{x^3 + x^2 - 5x + 3}$

37. $\dfrac{x^3}{(x^2 + 16)^3}$ **38.** $\dfrac{x^2}{(x^2 + 4)^3}$ **39.** $\dfrac{4}{2x^2 - 5x - 3}$ **40.** $\dfrac{4x}{2x^2 + 3x - 2}$

41. $\dfrac{2x + 3}{x^4 - 9x^2}$ **42.** $\dfrac{x^2 + 9}{x^4 - 2x^2 - 8}$

11.7 SYSTEMS OF NONLINEAR EQUATIONS

OBJECTIVES
1. Solve a System of Nonlinear Equations Using Substitution
2. Solve a System of Nonlinear Equations Using Elimination

1 In Section 11.1 we observed that the solution to a system of linear equations could be found geometrically by determining the point of intersection of the equations in the system. Similarly, when solving systems of nonlinear equations, the solution(s) also represent the point(s) of intersection of the equations.

There is no general methodology for solving a system of nonlinear equations by hand. There are times when substitution is best; other times, elimination is best; and there are times when neither of these methods works. Experience and a certain degree of imagination are your allies here.

Before we begin, two comments are in order.

1. If the system contains two variables, then graph them. By graphing each equation in the system, we can get an idea of how many solutions a system has and approximately where they are located.
2. Extraneous solutions can creep in when solving nonlinear systems algebraically, so it is imperative that all apparent solutions be checked.

EXAMPLE 1 Solving a System of Nonlinear Equations

Solve the following system of equations:

$$\begin{cases} 3x - y = -2 & \text{(1) A line} \\ 2x^2 - y = 0 & \text{(2) A parabola} \end{cases}$$

Graphing Solution

We use a graphing utility to graph $Y_1 = 3x + 2$ and $Y_2 = 2x^2$. From Figure 17, we see that the system apparently has two solutions. Using INTERSECT, the solutions to the system of equations are $(-0.5, 0.5)$ and $(2, 8)$.

Algebraic Solution Using Substitution

First, we notice that the system contains two variables and that we know how to graph each equation by hand. In Figure 18, we see that the system apparently has two solutions.

Figure 17

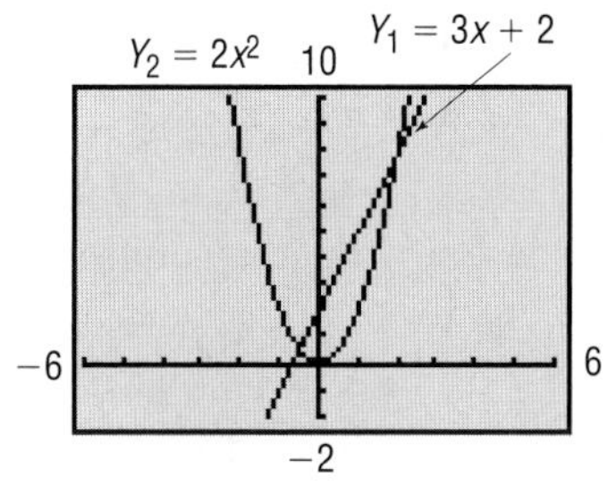

Figure 18

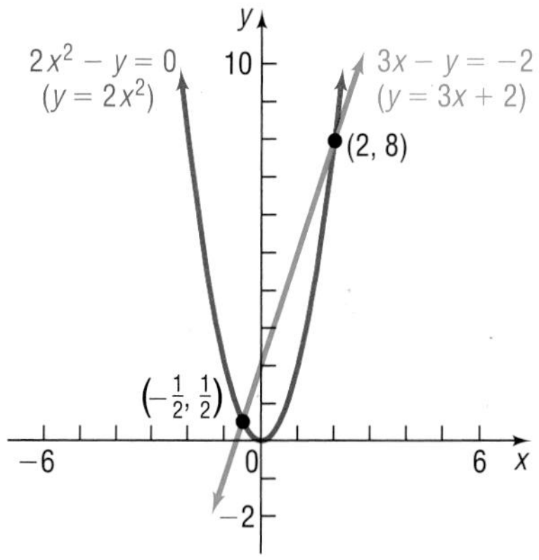

We will use substitution to solve the system. Equation (1) is easily solved for y.

$$3x - y = -2$$
$$y = 3x + 2$$

We substitute this expression for y in equation (2). The result is an equation containing just the variable x, which we can then solve.

$$2x^2 - y = 0$$
$$2x^2 - (3x + 2) = 0 \quad y = 3x + 2$$
$$2x^2 - 3x - 2 = 0 \quad \text{Simplify.}$$
$$(2x + 1)(x - 2) = 0 \quad \text{Factor.}$$

$$2x + 1 = 0 \quad \text{or} \quad x - 2 = 0 \quad \text{Zero-Product Property}$$
$$x = -\frac{1}{2} \quad \text{or} \quad x = 2$$

Using these values for x in $y = 3x + 2$, we find that

$$y = 3\left(-\frac{1}{2}\right) + 2 = \frac{1}{2} \quad \text{or} \quad y = 3(2) + 2 = 8$$

The apparent solutions are $x = -\frac{1}{2}$, $y = \frac{1}{2}$ and $x = 2$, $y = 8$.

Check For $x = -\frac{1}{2}$, $y = \frac{1}{2}$:

$$\begin{cases} 3\left(-\frac{1}{2}\right) - \frac{1}{2} = -\frac{3}{2} - \frac{1}{2} = -2 & (1) \\ 2\left(-\frac{1}{2}\right)^2 - \frac{1}{2} = 2\left(\frac{1}{4}\right) - \frac{1}{2} = 0 & (2) \end{cases}$$

For $x = 2$, $y = 8$:

$$\begin{cases} 3(2) - 8 = 6 - 8 = -2 & (1) \\ 2(2)^2 - 8 = 2(4) - 8 = 0 & (2) \end{cases}$$

Each solution checks. Now we know that the graphs in Figure 18 intersect at $\left(-\frac{1}{2}, \frac{1}{2}\right)$ and at $(2, 8)$.

NOW WORK PROBLEM 11.

2 Our next example illustrates how the method of elimination works for nonlinear systems.

EXAMPLE 2 Solving a System of Nonlinear Equations

Solve: $\begin{cases} x^2 + y^2 = 13 & \text{(1) A circle} \\ x^2 - y = 7 & \text{(2) A parabola} \end{cases}$

Graphing Solution We use a graphing utility to graph $x^2 + y^2 = 13$ and $x^2 - y = 7$. (Remember that to graph $x^2 + y^2 = 13$ requires two functions, $Y_1 = \sqrt{13 - x^2}$ and $Y_2 = -\sqrt{13 - x^2}$, and a square screen.) From Figure 19 we see that the system apparently has four solutions. Using INTERSECT, the solutions to the system of equations are $(-3, 2)$, $(3, 2)$, $(-2, -3)$, and $(2, -3)$.

Algebraic Solution Using Elimination First, we graph each equation, as shown in Figure 20. Based on the graph, we expect four solutions. By subtracting equation (2) from equation (1), the variable x can be eliminated.

Figure 19

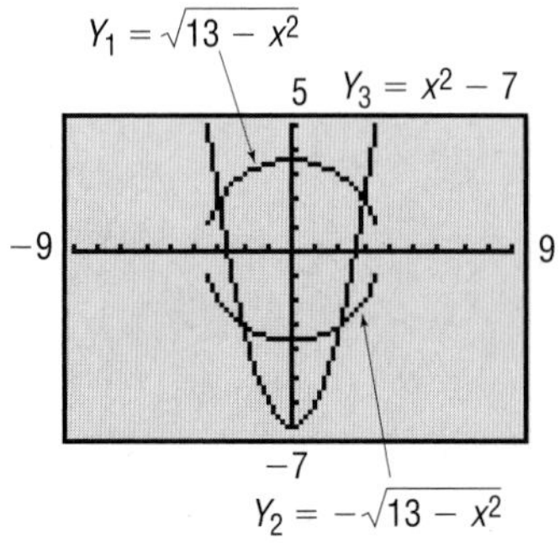

Figure 20

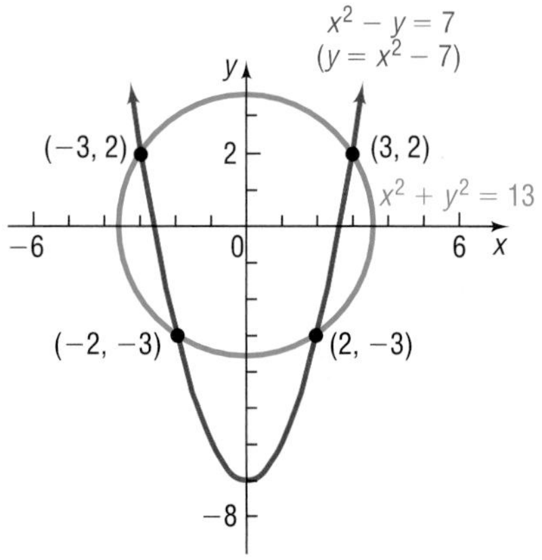

$$\begin{cases} x^2 + y^2 = 13 \\ x^2 - y \;\; = 7 \end{cases} \quad \text{Subtract}$$
$$y^2 + y = 6$$

This quadratic equation in y is easily solved by factoring.

$$y^2 + y - 6 = 0$$
$$(y + 3)(y - 2) = 0$$
$$y = -3 \quad \text{or} \quad y = 2$$

We use these values for y in equation (2) to find x. If $y = 2$, then $x^2 = y + 7 = 9$ and $x = 3$ or -3. If $y = -3$, then $x^2 = y + 7 = 4$ and $x = 2$ or -2. Thus, we have four solutions: $x = 3$, $y = 2$; $x = -3$, $y = 2$; $x = 2$, $y = -3$; and $x = -2$, $y = -3$. You should verify that, in fact, these four solutions also satisfy equation (1), so all four are solutions of the system. The four points $(3, 2)$, $(-3, 2)$, $(2, -3)$, and $(-2, -3)$, are the points of intersection of the graphs. Look again at Figure 20. ■

NOW WORK PROBLEM 9.

EXAMPLE 3 Solving a System of Nonlinear Equations

Solve:
$$\begin{cases} x^2 + x + y^2 - 3y + 2 = 0 & (1) \\ x + 1 + \dfrac{y^2 - y}{x} = 0 & (2) \end{cases}$$

Graphing Solution First, we multiply equation (2) by x to eliminate the fraction. The result is an equivalent system because x cannot be 0 [look at equation (2) to see why].

$$\begin{cases} x^2 + x + y^2 - 3y + 2 = 0 & (1) \\ x^2 + x + y^2 - y = 0 & (2) \end{cases}$$

We need to solve each equation for y. First, we solve equation (1) for y.

$$x^2 + x + y^2 - 3y + 2 = 0$$

$$y^2 - 3y = -x^2 - x - 2 \qquad \text{Rearrange so that terms involving } y \text{ are on left side.}$$

$$y^2 - 3y + \frac{9}{4} = -x^2 - x - 2 + \frac{9}{4} \qquad \text{Complete the square involving } y.$$

$$\left(y - \frac{3}{2}\right)^2 = -x^2 - x + \frac{1}{4}$$

$$y - \frac{3}{2} = \pm\sqrt{-x^2 - x + \frac{1}{4}} \qquad \text{Solve for the squared term.}$$

$$y = \frac{3}{2} \pm \sqrt{-x^2 - x + \frac{1}{4}} \qquad \text{Solve for } y.$$

Now we solve equation (2) for y.

$$x^2 + x + y^2 - y = 0$$

$$y^2 - y = -x^2 - x \qquad \text{Rearrange so that terms involving } y \text{ are on left side.}$$

$$y^2 - y + \frac{1}{4} = -x^2 - x + \frac{1}{4} \qquad \text{Complete the square involving } y.$$

$$\left(y - \frac{1}{2}\right)^2 = -x^2 - x + \frac{1}{4}$$

$$y - \frac{1}{2} = \pm\sqrt{-x^2 - x + \frac{1}{4}} \qquad \text{Solve for the squared term.}$$

$$y = \frac{1}{2} \pm \sqrt{-x^2 - x + \frac{1}{4}} \qquad \text{Solve for } y.$$

Now graph each equation using a graphing utility. See Figure 21.

Figure 21

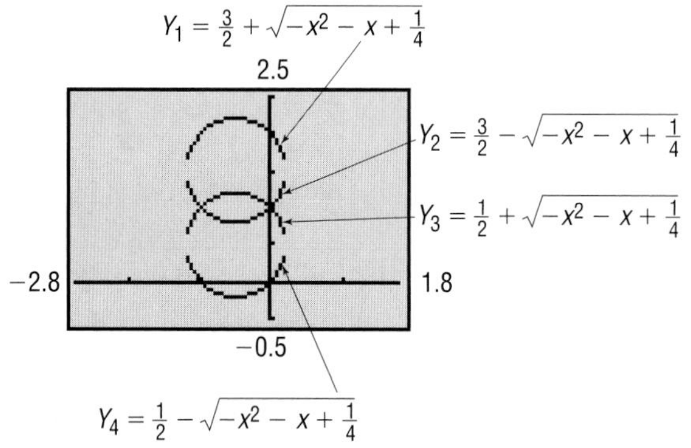

Using INTERSECT, the points of intersection are $(-1, 1)$ and $(0, 1)$. Since $x \neq 0$ [look back at the original equation (2)], the value $x = 0$ is extraneous, and we discard it. The only solution is $x = -1$ and $y = 1$.

Algebraic Solution Using Elimination

First, we multiply equation (2) by x to eliminate the fraction. The result is an equivalent system because x cannot be 0 [look at equation (2) to see why].

$$\begin{cases} x^2 + x + y^2 - 3y + 2 = 0 & (1) \\ x^2 + x + y^2 - y = 0 & (2) \end{cases}$$

Now subtract equation (2) from equation (1) to eliminate x. The result is

$$-2y + 2 = 0$$
$$y = 1$$

To find x, we back-substitute $y = 1$ in equation (1).

$$x^2 + x + 1 - 3 + 2 = 0$$
$$x^2 + x = 0$$
$$x(x + 1) = 0$$
$$x = 0 \quad \text{or} \quad x = -1$$

Because x cannot be 0, the value $x = 0$ is extraneous, and we discard it. The solution is $x = -1$, $y = 1$. ■

Check: We now check $x = -1$, $y = 1$.

$$\begin{cases} (-1)^2 + (-1) + 1^2 - 3(1) + 2 = 1 - 1 + 1 - 3 + 2 = 0 & (1) \\ -1 + 1 + \dfrac{1^2 - 1}{-1} = 0 + \dfrac{0}{-1} = 0 & (2) \end{cases}$$ ■

NOW WORK PROBLEMS 25 AND 49.

EXAMPLE 4 Solving a System of Nonlinear Equations

Solve: $\begin{cases} x^2 - y^2 = 4 & \text{(1) A hyperbola} \\ y = x^2 & \text{(2) A parabola} \end{cases}$

Graphing Solution We graph $Y_1 = x^2$ and $x^2 - y^2 = 4$ in Figure 22. You will need to graph $x^2 - y^2 = 4$ as two functions:

$$Y_2 = \sqrt{x^2 - 4} \quad \text{and} \quad Y_3 = -\sqrt{x^2 - 4}$$

From Figure 22 we see that the graphs of these two equations do not intersect. The system is inconsistent.

Figure 22

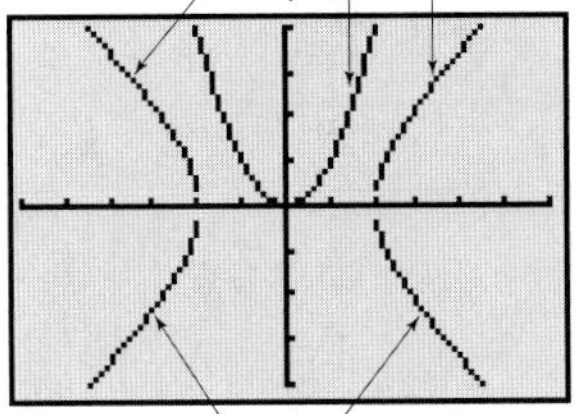

Algebraic Solution Either substitution or elimination can be used here. We use substitution and replace x^2 by y in equation (1). The result is

$$y - y^2 = 4$$
$$y^2 - y + 4 = 0$$

This is a quadratic equation whose discriminant is $(-1)^2 - 4 \cdot 1 \cdot 4 = 1 - 4 \cdot 4 = -15 < 0$. The equation has no real solutions, so the system is inconsistent. ■

EXAMPLE 5 Solving a System of Nonlinear Equations

Solve: $\begin{cases} 3xy - 2y^2 = -2 & (1) \\ 9x^2 + 4y^2 = 10 & (2) \end{cases}$

Graphing Solution To graph $3xy - 2y^2 = -2$, we need to solve for y. In this instance, it is easier to view the equation as a quadratic equation in the variable y.

$$3xy - 2y^2 = -2$$

$$2y^2 - 3xy - 2 = 0 \qquad \text{Place in standard form.}$$

$$y = \frac{-(-3x) \pm \sqrt{(-3x)^2 - 4(2)(-2)}}{2(2)} \qquad \text{Use the quadratic formula } a = 2,\ b = -3x,\ c = -2.$$

$$y = \frac{3x \pm \sqrt{9x^2 + 16}}{4} \qquad \text{Simplify.}$$

Using a graphing utility, we graph $Y_1 = \dfrac{3x + \sqrt{9x^2 + 16}}{4}$ and $Y_2 = \dfrac{3x - \sqrt{9x^2 + 16}}{4}$. From equation (2), We graph $Y_3 = \dfrac{\sqrt{10 - 9x^2}}{2}$ and $Y_4 = \dfrac{-\sqrt{10 - 9x^2}}{2}$. See Figure 23.

Figure 23

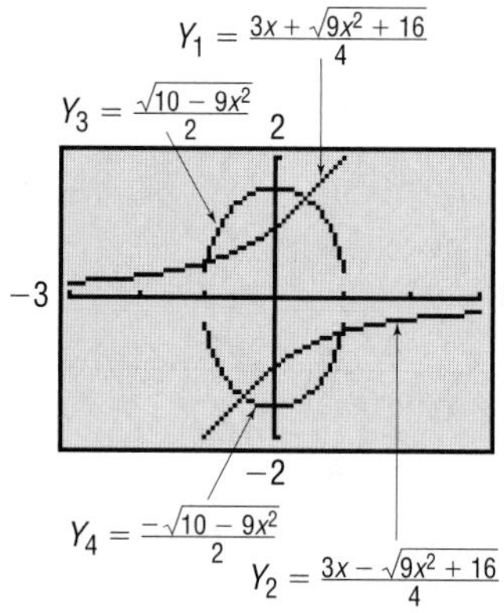

Using INTERSECT, the solutions to the system of equations are $(-1, 0.5)$, $(0.47, 1.41)$, $(1, -0.5)$, and $(-0.47, -1.41)$, each rounded to two decimal places.

Algebraic Solution

We multiply equation (1) by 2 and add the result to equation (2) to eliminate the y^2-terms.

$$\begin{cases} 6xy - 4y^2 = -4 & (1) \\ 9x^2 + 4y^2 = 10 & (2)\ \text{Add} \end{cases}$$

$$9x^2 + 6xy = 6$$

$$3x^2 + 2xy = 2 \qquad \text{Divide each side by 3.}$$

Since $x \neq 0$ (do you see why?), we can solve for y in this equation to get

$$y = \frac{2 - 3x^2}{2x}, \qquad x \neq 0 \tag{1}$$

Now substitute for y in equation (2) of the system.

$$9x^2 + 4y^2 = 10$$

$$9x^2 + 4\left(\frac{2 - 3x^2}{2x}\right)^2 = 10 \qquad y = \frac{2 - 3x^2}{2x}$$

$$9x^2 + \frac{4 - 12x^2 + 9x^4}{x^2} = 10 \qquad \text{Remove parentheses; cancel the 4's.}$$

$$9x^4 + 4 - 12x^2 + 9x^4 = 10x^2 \qquad \text{Multiply both sides by } x^2.$$

$$18x^4 - 22x^2 + 4 = 0 \qquad \text{Combine like terms.}$$

$$9x^4 - 11x^2 + 2 = 0 \qquad \text{Divide both sides by 2.}$$

This quadratic equation (in x^2) can be factored.

$$(9x^2 - 2)(x^2 - 1) = 0$$

$$9x^2 - 2 = 0 \qquad \text{or} \qquad x^2 - 1 = 0$$

$$x^2 = \frac{2}{9} \qquad\qquad x^2 = 1$$

$$x = \pm\sqrt{\frac{2}{9}} = \pm\frac{\sqrt{2}}{3} \qquad\qquad x = \pm 1$$

To find y, we use equation (1):

$$\text{If } x = \frac{\sqrt{2}}{3}: \quad y = \frac{2 - 3x^2}{2x} = \frac{2 - \frac{2}{3}}{2(\sqrt{2}/3)} = \frac{4}{2\sqrt{2}} = \sqrt{2}$$

$$\text{If } x = -\frac{\sqrt{2}}{3}: \quad y = \frac{2 - 3x^2}{2x} = \frac{2 - \frac{2}{3}}{2(-\sqrt{2}/3)} = \frac{4}{-2\sqrt{2}} = -\sqrt{2}$$

$$\text{If } x = 1: \quad y = \frac{2 - 3x^2}{2x} = \frac{2 - 3}{2} = -\frac{1}{2}$$

$$\text{If } x = -1: \quad y = \frac{2 - 3x^2}{2x} = \frac{2 - 3}{-2} = \frac{1}{2}$$

The system has four solutions. Check them for yourself. ■

NOW WORK PROBLEM 45.

EXAMPLE 6 Running a Race

In a 50-mile race, the winner crosses the finish line 1 mile ahead of the second-place runner and 4 miles ahead of the third-place runner. Assuming that each runner maintains a constant speed throughout the race, by how many miles does the second-place runner beat the third-place runner?

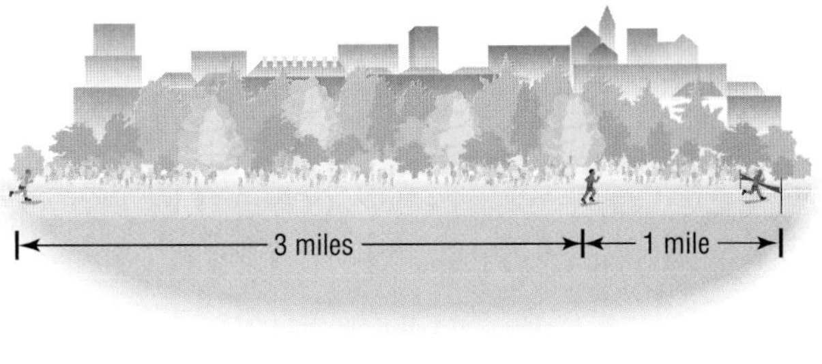

Solution Let v_1, v_2, v_3 denote the speeds of the first-, second-, and third-place runners, respectively. Let t_1 and t_2 denote the times (in hours) required for the first-place runner and second-place runner to finish the race. Then we have the system of equations

$$\begin{cases} 50 = v_1 t_1 & \text{(1) First-place runner goes 50 miles in } t_1 \text{ hours.} \\ 49 = v_2 t_1 & \text{(2) Second-place runner goes 49 miles in } t_1 \text{ hours.} \\ 46 = v_3 t_1 & \text{(3) Third-place runner goes 46 miles in } t_1 \text{ hours.} \\ 50 = v_2 t_2 & \text{(4) Second-place runner goes 50 miles in } t_2 \text{ hours.} \end{cases}$$

We seek the distance of the third-place runner from the finish at time t_2. That is, we seek

$$\begin{aligned}
50 - v_3 t_2 &= 50 - v_3\left(t_1 \cdot \frac{t_2}{t_1}\right) \\
&= 50 - (v_3 t_1) \cdot \frac{t_2}{t_1} \\
&= 50 - 46 \cdot \frac{50/v_2}{50/v_1} && \begin{cases} \text{From (3), } v_3 t_1 = 46; \\ \text{From (4), } t_2 = 50/v_2; \\ \text{From (1), } t_1 = 50/v_1. \end{cases} \\
&= 50 - 46 \cdot \frac{v_1}{v_2} \\
&= 50 - 46 \cdot \frac{50}{49} && \text{Form the quotient of (1) and (2).} \\
&\approx 3.061 \text{ miles}
\end{aligned}$$

The second-place runner will cross the finish line about 3.061 miles ahead of the third-place runner.

HISTORICAL FEATURE

Etienne Bezout (1730–1783)

Recall that, in the beginning of this section, we said imagination and experience are important in solving simultaneous nonlinear equations. Indeed, these kinds of problems lead into some of the deepest and most difficult parts of modern mathematics. Look again at the graphs in Examples 1 and 2 of this section (Figures 17 or 18 and 19 or 20). We see that Example 1 has two solutions, and Example 2 has four solutions. We might conjecture that the number of solutions is equal to the product of the degrees of the equations involved. This conjecture was indeed made by Etienne Bezout (1730–1783), but working out the details took about 150 years. To arrive at the correct number of intersections, it turns out that, we must count not only the complex intersections, but also the intersections that, in a certain sense, lie at infinity. For example, a parabola and a line lying on the axis of the parabola intersect at the vertex and at infinity. This topic is part of the study of algebraic geometry.

HISTORICAL PROBLEM

1. A papyrus dating back to 1950 BC contains the following problem: A given surface area of 100 units of area shall be represented as the sum of two squares whose sides are to each other as $1:\frac{3}{4}$. Solve for the sides by solving the system of equations

$$\begin{cases} x^2 + y^2 = 100 \\ x = \frac{3}{4}y \end{cases}$$

11.7 EXERCISES

In Problems 1–20, use a graphing utility to graph each equation of the system. Then solve the system by finding the intersection points. Express your answer correct to two decimal places. Also solve each system algebraically.

1. $\begin{cases} y = x^2 + 1 \\ y = x + 1 \end{cases}$

2. $\begin{cases} y = x^2 + 1 \\ y = 4x + 1 \end{cases}$

3. $\begin{cases} y = \sqrt{36 - x^2} \\ y = 8 - x \end{cases}$

4. $\begin{cases} y = \sqrt{4 - x^2} \\ y = 2x + 4 \end{cases}$

5. $\begin{cases} y = \sqrt{x} \\ y = 2 - x \end{cases}$

6. $\begin{cases} y = \sqrt{x} \\ y = 6 - x \end{cases}$

7. $\begin{cases} x = 2y \\ x = y^2 - 2y \end{cases}$

8. $\begin{cases} y = x - 1 \\ y = x^2 - 6x + 9 \end{cases}$

9. $\begin{cases} x^2 + y^2 = 4 \\ x^2 + 2x + y^2 = 0 \end{cases}$

10. $\begin{cases} x^2 + y^2 = 8 \\ x^2 + y^2 + 4y = 0 \end{cases}$

11. $\begin{cases} y = 3x - 5 \\ x^2 + y^2 = 5 \end{cases}$

12. $\begin{cases} x^2 + y^2 = 10 \\ y = x + 2 \end{cases}$

13. $\begin{cases} x^2 + y^2 = 4 \\ y^2 - x = 4 \end{cases}$

14. $\begin{cases} x^2 + y^2 = 16 \\ x^2 - 2y = 8 \end{cases}$

15. $\begin{cases} xy = 4 \\ x^2 + y^2 = 8 \end{cases}$

16. $\begin{cases} x^2 = y \\ xy = 1 \end{cases}$

17. $\begin{cases} x^2 + y^2 = 4 \\ y = x^2 - 9 \end{cases}$

18. $\begin{cases} xy = 1 \\ y = 2x + 1 \end{cases}$

19. $\begin{cases} y = x^2 - 4 \\ y = 6x - 13 \end{cases}$

20. $\begin{cases} x^2 + y^2 = 10 \\ xy = 3 \end{cases}$

In Problems 21–52, solve each system algebraically. Use any method you wish.

21. $\begin{cases} 2x^2 + y^2 = 18 \\ xy = 4 \end{cases}$

22. $\begin{cases} x^2 - y^2 = 21 \\ x + y = 7 \end{cases}$

23. $\begin{cases} y = 2x + 1 \\ 2x^2 + y^2 = 1 \end{cases}$

24. $\begin{cases} x^2 - 4y^2 = 16 \\ 2y - x = 2 \end{cases}$

25. $\begin{cases} x + y + 1 = 0 \\ x^2 + y^2 + 6y - x = -5 \end{cases}$

26. $\begin{cases} 2x^2 - xy + y^2 = 8 \\ xy = 4 \end{cases}$

27. $\begin{cases} 4x^2 - 3xy + 9y^2 = 15 \\ 2x + 3y = 5 \end{cases}$

28. $\begin{cases} 2y^2 - 3xy + 6y + 2x + 4 = 0 \\ 2x - 3y + 4 = 0 \end{cases}$

29. $\begin{cases} x^2 - 4y^2 + 7 = 0 \\ 3x^2 + y^2 = 31 \end{cases}$

30. $\begin{cases} 3x^2 - 2y^2 + 5 = 0 \\ 2x^2 - y^2 + 2 = 0 \end{cases}$

31. $\begin{cases} 7x^2 - 3y^2 + 5 = 0 \\ 3x^2 + 5y^2 = 12 \end{cases}$

32. $\begin{cases} x^2 - 3y^2 + 1 = 0 \\ 2x^2 - 7y^2 + 5 = 0 \end{cases}$

33. $\begin{cases} x^2 + 2xy = 10 \\ 3x^2 - xy = 2 \end{cases}$

34. $\begin{cases} 5xy + 13y^2 + 36 = 0 \\ xy + 7y^2 = 6 \end{cases}$

35. $\begin{cases} 2x^2 + y^2 = 2 \\ x^2 - 2y^2 + 8 = 0 \end{cases}$

36. $\begin{cases} y^2 - x^2 + 4 = 0 \\ 2x^2 + 3y^2 = 6 \end{cases}$

37. $\begin{cases} x^2 + 2y^2 = 16 \\ 4x^2 - y^2 = 24 \end{cases}$

38. $\begin{cases} 4x^2 + 3y^2 = 4 \\ 2x^2 - 6y^2 = -3 \end{cases}$

39. $\begin{cases} \dfrac{5}{x^2} - \dfrac{2}{y^2} + 3 = 0 \\ \dfrac{3}{x^2} + \dfrac{1}{y^2} = 7 \end{cases}$

40. $\begin{cases} \dfrac{2}{x^2} - \dfrac{3}{y^2} + 1 = 0 \\ \dfrac{6}{x^2} - \dfrac{7}{y^2} + 2 = 0 \end{cases}$

41. $\begin{cases} \dfrac{1}{x^4} + \dfrac{6}{y^4} = 6 \\ \dfrac{2}{x^4} - \dfrac{2}{y^4} = 19 \end{cases}$

42. $\begin{cases} \dfrac{1}{x^4} - \dfrac{1}{y^4} = 1 \\ \dfrac{1}{x^4} + \dfrac{1}{y^4} = 4 \end{cases}$

43. $\begin{cases} x^2 - 3xy + 2y^2 = 0 \\ x^2 + xy = 6 \end{cases}$

44. $\begin{cases} x^2 - xy - 2y^2 = 0 \\ xy + x + 6 = 0 \end{cases}$

45. $\begin{cases} xy - x^2 + 3 = 0 \\ 3xy - 4y^2 = 2 \end{cases}$

46. $\begin{cases} 5x^2 + 4xy + 3y^2 = 36 \\ x^2 + xy + y^2 = 9 \end{cases}$

47. $\begin{cases} x^3 - y^3 = 26 \\ x - y = 2 \end{cases}$

48. $\begin{cases} x^3 + y^3 = 26 \\ x + y = 2 \end{cases}$

49. $\begin{cases} y^2 + y + x^2 - x - 2 = 0 \\ y + 1 + \dfrac{x - 2}{y} = 0 \end{cases}$

50. $\begin{cases} x^3 - 2x^2 + y^2 + 3y - 4 = 0 \\ x - 2 + \dfrac{y^2 - y}{x^2} = 0 \end{cases}$

51. $\begin{cases} \log_x y = 3 \\ \log_x(4y) = 5 \end{cases}$

52. $\begin{cases} \log_x(2y) = 3 \\ \log_x(4y) = 2 \end{cases}$

In Problems 53–60, use a graphing utility to solve each system of equations. Express the solution(s) rounded to two decimal places.

53. $\begin{cases} y = x^{2/3} \\ y = e^{-x} \end{cases}$

54. $\begin{cases} y = x^{3/2} \\ y = e^{-x} \end{cases}$

55. $\begin{cases} x^2 + y^3 = 2 \\ x^3 y = 4 \end{cases}$

56. $\begin{cases} x^3 + y^2 = 2 \\ x^2 y = 4 \end{cases}$

57. $\begin{cases} x^4 + y^4 = 12 \\ xy^2 = 2 \end{cases}$

58. $\begin{cases} x^4 + y^4 = 6 \\ xy = 1 \end{cases}$

59. $\begin{cases} xy = 2 \\ y = \ln x \end{cases}$

60. $\begin{cases} x^2 + y^2 = 4 \\ y = \ln x \end{cases}$

61. The difference of two numbers is 2 and the sum of their squares is 10. Find the numbers.

62. The sum of two numbers is 7 and the difference of their squares is 21. Find the numbers.

63. The product of two numbers is 4 and the sum of their squares is 8. Find the numbers.

64. The product of two numbers is 10 and the difference of their squares is 21. Find the numbers.

65. The difference of two numbers is the same as their product, and the sum of their reciprocals is 5. Find the numbers.

66. The sum of two numbers is the same as their product, and the difference of their reciprocals is 3. Find the numbers.

67. The ratio of a to b is $\frac{2}{3}$. The sum of a and b is 10. What is the ratio of $a + b$ to $b - a$?

68. The ratio of a to b is $\frac{4}{3}$. The sum of a and b is 14. What is the ratio of $a - b$ to $a + b$?

In Problems 69–74, graph each equation by hand and find the point(s) of intersection, if any, using algebra.

69. The line $x + 2y = 0$ and the circle $(x - 1)^2 + (y - 1)^2 = 5$

70. The line $x + 2y + 6 = 0$ and the circle $(x + 1)^2 + (y + 1)^2 = 5$

71. The circle $(x - 1)^2 + (y + 2)^2 = 4$ and the parabola $y^2 + 4y - x + 1 = 0$

72. The circle $(x + 2)^2 + (y - 1)^2 = 4$ and the parabola $y^2 - 2y - x - 5 = 0$

73. The graph of $y = \dfrac{4}{x - 3}$ and the circle $x^2 - 6x + y^2 + 1 = 0$

74. The graph of $y = \dfrac{4}{x + 2}$ and the circle $x^2 + 4x + y^2 - 4 = 0$

75. Geometry The perimeter of a rectangle is 16 inches and its area is 15 square inches. What are its dimensions?

76. Geometry An area of 52 square feet is to be enclosed by two squares whose sides are in the ratio of 2:3. Find the sides of the squares.

77. Geometry Two circles have perimeters that add up to 12π centimeters and areas that add up to 20π square centimeters. Find the radius of each circle.

78. Geometry The altitude of an isosceles triangle drawn to its base is 3 centimeters, and its perimeter is 18 centimeters. Find the length of its base.

79. The Tortoise and the Hare In a 21-meter race between a tortoise and a hare, the tortoise leaves 9 minutes before the hare. The hare, by running at an average speed of 0.5 meter per hour faster than the tortoise, crosses the finish line 3 minutes before the tortoise. What are the average speeds of the tortoise and the hare?

80. Running a Race In a 1-mile race, the winner crosses the finish line 10 feet ahead of the second-place runner and 20 feet ahead of the third-place runner. Assuming that each runner maintains a constant speed throughout the race, by how many feet does the second-place runner beat the third-place runner?

81. Constructing a Box A rectangular piece of cardboard, whose area is 216 square centimeters, is made into an open box by cutting a 2-centimeter square from each corner and turning up the sides. See the figure. If the box is to have a volume of 224 cubic centimeters, what size cardboard should you start with?

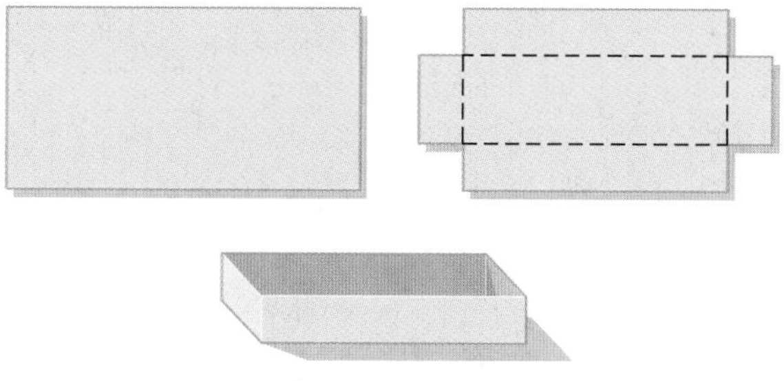

82. Constructing a Cylindrical Tube A rectangular piece of cardboard, whose area is 216 square centimeters, is made into a cylindrical tube by joining together two sides of the rectangle. (See the figure.) If the tube is to have a volume of 224 cubic centimeters, what size cardboard should you start with?

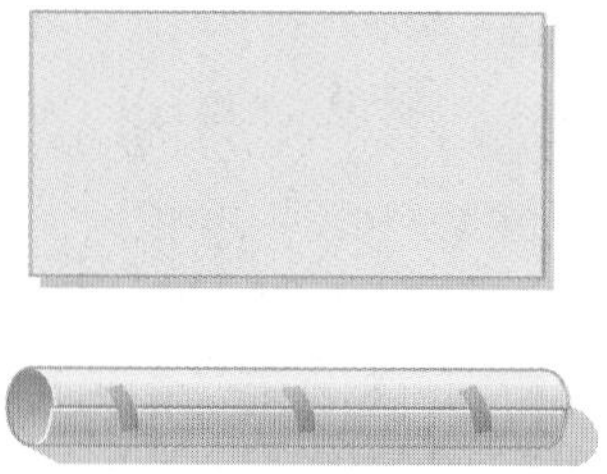

83. Fencing A farmer has 300 feet of fence available to enclose 4500 square feet in the shape of adjoining squares, with sides of length x and y. See the figure. Find x and y.

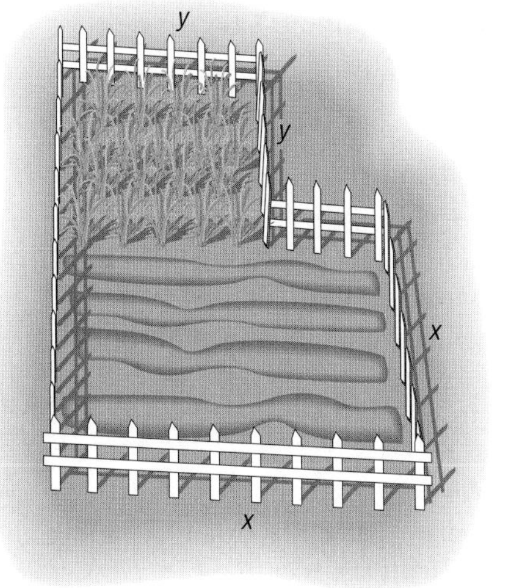

84. Bending Wire A wire 60 feet long is cut into two pieces. Is it possible to bend one piece into the shape of a square and the other into the shape of a circle so that the total area enclosed by the two pieces is 100 square feet? If this is possible, find the length of the side of the square and the radius of the circle.

85. Geometry Find formulas for the length l and width w of a rectangle in terms of its area A and perimeter P.

86. Geometry Find formulas for the base b and one of the equal sides l of an isosceles triangle in terms of its altitude h and perimeter P.

87. Descartes' Method of Equal Roots Descartes' method for finding tangents depends on the idea that, for many graphs, the tangent line at a given point is the *unique* line that intersects the graph at that point only. We will apply his method to find an equation of the tangent line to the parabola $y = x^2$ at the point $(2, 4)$; see the figure. First, we know that the equation of the tangent line must be in the form $y = mx + b$. Using the fact that the point $(2, 4)$ is on the line, we can solve for b in terms of m and get the equation $y = mx + (4 - 2m)$. We want $(2, 4)$ to be the *unique* solution to the system

$$\begin{cases} y = x^2 \\ y = mx + 4 - 2m \end{cases}$$

From this system, we get $x^2 = mx + 4 - 2m$ or $x^2 - mx + (2m - 4) = 0$. By using the quadratic formula, we get

$$x = \frac{m \pm \sqrt{m^2 - 4(2m - 4)}}{2}$$

To obtain a unique solution for x, the two roots must be equal; in other words, the discriminant $m^2 - 4(2m - 4)$ must be 0. Complete the work to get m, and write an equation of the tangent line.

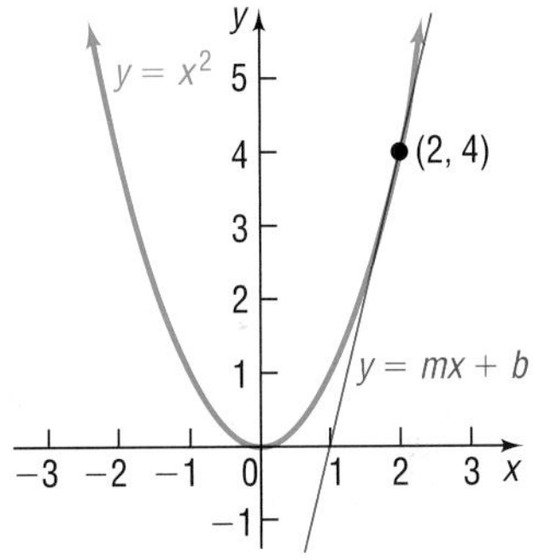

In Problems 88–94, use Descartes' method from Problem 87 to find the equation of the line tangent to each graph at the given point.

88. $x^2 + y^2 = 10$; at $(1, 3)$

89. $y = x^2 + 2$; at $(1, 3)$

90. $x^2 + y = 5$; at $(-2, 1)$

91. $2x^2 + 3y^2 = 14$; at $(1, 2)$

92. $3x^2 + y^2 = 7$; at $(-1, 2)$

93. $x^2 - y^2 = 3$; at $(2, 1)$

94. $2y^2 - x^2 = 14$; at $(2, 3)$

95. If r_1 and r_2 are two solutions of a quadratic equation $ax^2 + bx + c = 0$, then it can be shown that

$$r_1 + r_2 = -\frac{b}{c} \quad \text{and} \quad r_1 r_2 = \frac{c}{a}$$

Solve this system of equations for r_1 and r_2.

96. A circle and a line intersect at most twice. A circle and a parabola intersect at most four times. Deduce that a circle and the graph of a polynomial of degree 3 intersect at most six times. What do you conjecture about a polynomial of degree 4? What about a polynomial of degree n? Can you explain your conclusions using an algebraic argument?

97. Suppose that you are the manager of a sheet metal shop. A customer asks you to manufacture 10,000 boxes, each box being open on top. The boxes are required to have a square base and a 9-cubic-foot capacity. You construct the boxes by cutting out a square from each corner of a square piece of sheet metal and folding along the edges.
(a) What are the dimensions of the square to be cut if the area of the square piece of sheet metal is 100 square feet?
(b) Could you make the box using a smaller piece of sheet metal? Make a list of the dimensions of the box for various pieces of sheet metal.

11.8 SYSTEMS OF LINEAR INEQUALITIES; LINEAR PROGRAMMING

OBJECTIVES

1. Graph a Linear Inequality by Hand
2. Graph a Linear Inequality Using a Graphing Utility
3. Graph a System of Linear Inequalities
4. Set up a Linear Programming Problem
5. Solve a Linear Programming Problem

In Chapter 1, we discussed inequalities in one variable. In this section, we discuss linear inequalities in two variables.

Linear inequalities are inequalities in one of the forms

$$Ax + By < C \qquad Ax + By > C \qquad Ax + By \leq C \qquad Ax + By \geq C$$

where A and B are not both zero.

Figure 24

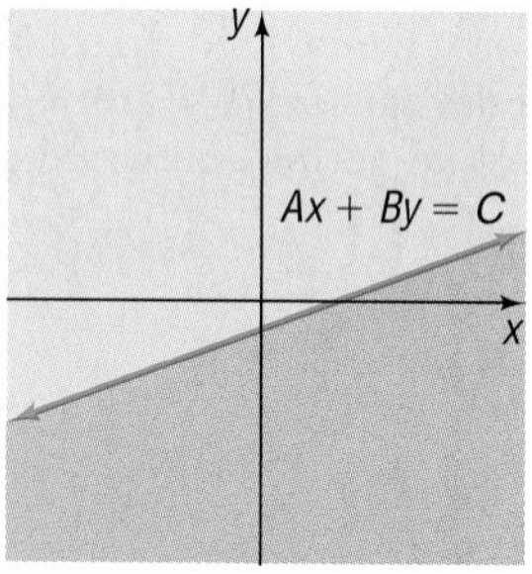

1 A linear inequality in two variables x and y is **satisfied** by an ordered pair (a, b) if, when x is replaced by a and y by b, a true statement results. A **graph of a linear inequality in two variables** x and y consists of all points (x, y) whose coordinates satisfy the inequality.

The graph of the corresponding equation of a linear inequality is a line, which separates the xy-plane into two regions, called **half-planes.** See Figure 24.

As shown, $Ax + By = C$ is the equation of the boundary line and it divides the plane into two half-planes, one for which $Ax + By < C$ and the other for which $Ax + By > C$.

Let's look at an example.

EXAMPLE 1 Graphing a Linear Inequality

Graph the linear inequality: $3x + y - 6 \leq 0$

Solution We begin with the associated problem of the graph of the linear equation

$$3x + y - 6 = 0$$

formed by replacing (for now) the $\leq$ symbol with an $=$ sign. The graph of the linear equation is a line. See Figure 25(a). This line is part of the graph of the inequality that we seek because the inequality is nonstrict. (Do you see why? We are seeking points for which $3x + y - 6$ is less than *or equal to* 0.)

Now let's test a few randomly selected points to see whether they belong to the graph of the inequality.

	$3x + y - 6$	Conclusion
$(4, -1)$	$3(4) + (-1) - 6 = 5 > 0$	Does not belong to graph
$(5, 5)$	$3(5) + 5 - 6 = 14 > 0$	Does not belong to graph
$(-1, 2)$	$3(-1) + 2 - 6 = -7 < 0$	Belongs to graph
$(-2, -2)$	$3(-2) + (-2) - 6 = -14 < 0$	Belongs to graph

Look again at Figure 25(a). Notice that the two points that belong to the graph both lie on the same side of the line, and the two points that do not belong to the graph lie on the opposite side. As it turns out, this is always the case. Thus, the graph we seek consists of all points that lie on the same

Figure 25

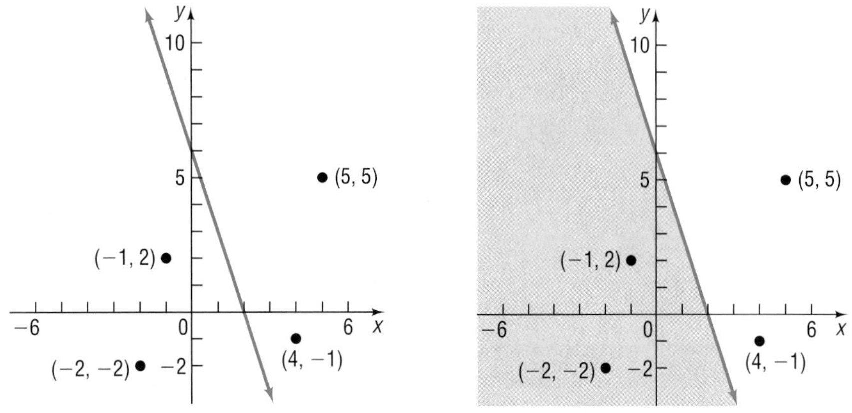

(a) $3x + y - 6 = 0$

(b) Graph of $3x + y - 6 \leq 0$

side of the line as do $(-1, 2)$ and $(-2, -2)$, that is, the shaded region in Figure 25(b). ■

The graph of any linear inequality in two variables may be obtained by graphing the equation corresponding to the inequality, using dashes if the inequality is strict ($<$ or $>$) and solid lines if it is nonstrict ($\leq$ or $\geq$). This graph will separate the xy-plane into two half-planes. In each half-plane either all points satisfy the inequality or no points satisfy the inequality. Thus, the use of a single test point is all that is required to determine whether the points of that half-plane are part of the graph or not. The steps to follow are given next.

STEPS FOR GRAPHING AN INEQUALITY BY HAND

STEP 1: Replace the inequality symbol by an equal sign and graph the resulting equation. If the inequality is strict, use dashes; if it is nonstrict, use a solid line. This graph separates the xy-plane into two half-planes.

STEP 2: Select a test point P in one of the half-planes.

(a) If the coordinates of P satisfy the inequality, then so do all the points in that half-plane. Indicate this by shading the half-plane.

(b) If the coordinates of P do not satisfy the inequality, then none of the points in that half-plane do, so shade the opposite half-plane.

2 Graphing utilities can also be used to graph linear inequalities. The steps to follow to graph an inequality using a graphing utility are given next.

STEPS FOR GRAPHING AN INEQUALITY USING A GRAPHING UTILITY

STEP 1: Replace the inequality symbol by an equal sign and graph the resulting equation.

STEP 2: Select a test point P in one of the half-planes.

(a) Use a graphing utility to determine if the test point P satisfies the inequality. If the test point satisfies the inequality, then so do all the points in this half-plane. Indicate this by using the graphing utility to shade the half-plane.

(b) If the coordinates of P do not satisfy the inequality, then none of the points in that half-plane do, so shade the opposite half-plane.

EXAMPLE 2 Graphing a Linear Inequality Using a Graphing Utility

Use a graphing utility to graph $3x + y - 6 \leq 0$.

Solution **STEP 1:** We begin by graphing the equation $3x + y - 6 = 0 \left(Y_1 = -3x + 6\right)$. See Figure 26.

STEP 2: Select a test point in one of the half-planes and determine whether it satisfies the inequality. To test the point $(-1, 2)$, for example, enter $3(-1) + 2 - 6 \leq 0$. See Figure 27(a). The 1 that appears indicates that the statement entered (the inequality) is true. When the point $(5, 5)$ is tested, a 0 appears, indicating that the statement entered is false. Thus, $(-1, 2)$ is a part of the graph of the inequality and $(5, 5)$ is not, so we shade the half-plane below Y_1. Figure 27(b) shows the graph of the inequality on a TI-83.*

Figure 26

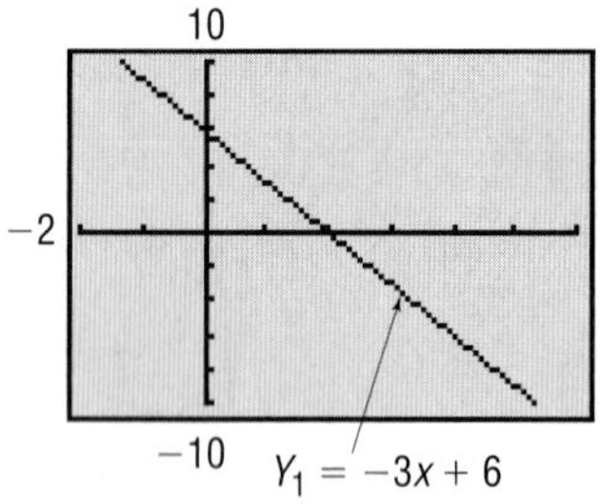

Figure 27

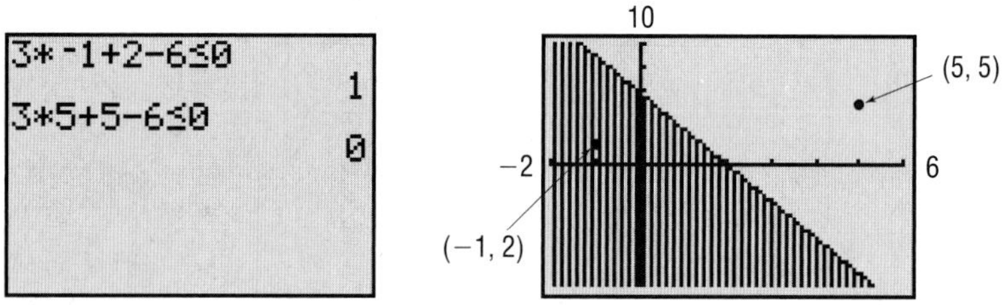

(a)

(b)

NOW WORK PROBLEM 5.

EXAMPLE 3 Graphing Linear Inequalities

Graph: (a) $y < 2$ (b) $y \geq 2x$

Solution (a) The graph of the equation $y = 2$ is a horizontal line and is not part of the graph of the inequality, so we use a dashed line. Since $(0, 0)$ satisfies the inequality, the graph consists of the half-plane below the line $y = 2$. See Figure 28.

(b) The graph of the equation $y = 2x$ is a line and is part of the graph of the inequality, so we use a solid line. Using $(3, 0)$ as a test point, we find that it does not satisfy the inequality $[0 \not\geq 2 \cdot 3]$. Thus, points in the half-plane on the opposite side of $y = 2x$ satisfy the inequality. See Figure 29.

*Consult your owner's manual for shading techniques.

Figure 28

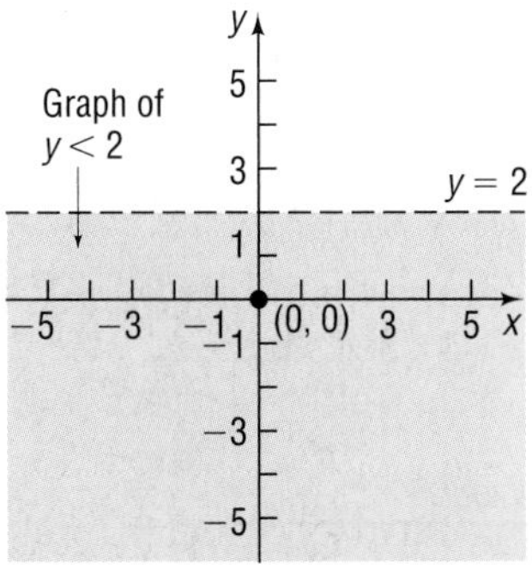

Figure 29

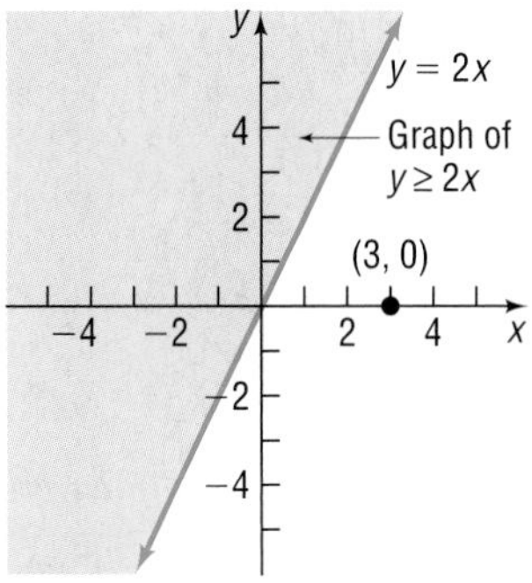

NOW WORK PROBLEM 3.

Systems of Linear Inequalities in Two Variables

3 The **graph of a system of linear inequalities** in two variables x and y is the set of all points (x, y) that simultaneously satisfies *each* of the inequalities in the system. Thus, the graph of a system of linear inequalities can be obtained by graphing each inequality individually and then determining where, if at all, they intersect.

EXAMPLE 4 Graphing a System of Linear Inequalities

Graph the system: $\begin{cases} x + y \geq 2 \\ 2x - y \leq 4 \end{cases}$

Solution First, we graph the inequality $x + y \geq 2$ as the shaded region in Figure 30(a). Next, we graph the inequality $2x - y \leq 4$ as the shaded region in Figure 30(b). Now superimpose the two graphs, as shown in Figure 30(c). The points that are in both shaded regions [the overlapping, purple region in Figure 30(c)] are the solutions we seek to the system, because they simultaneously satisfy each linear inequality. ■

Figure 30

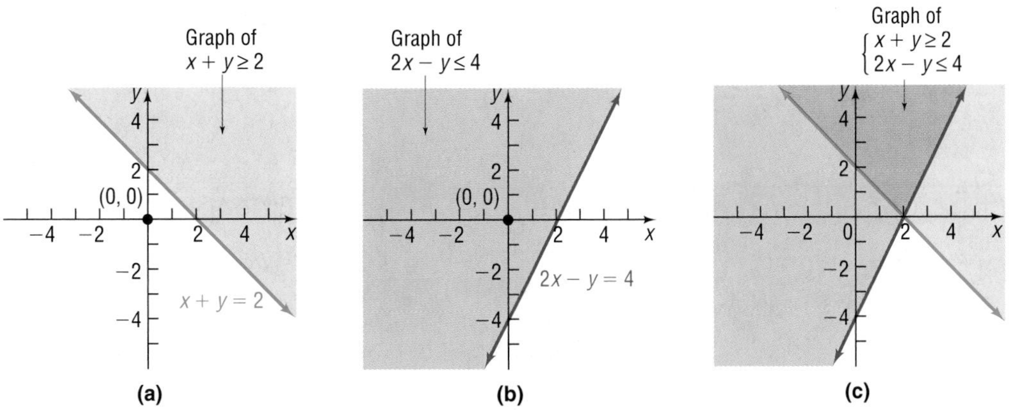

(a) (b) (c)

EXAMPLE 5 Graphing a System of Linear Inequalities Using a Graphing Utility

Graph the system: $\begin{cases} x + y \geq 2 \\ 2x - y \leq 4 \end{cases}$

Solution First, we graph the lines $x + y = 2$ $(Y_1 = -x + 2)$ and $2x - y = 4$ $(Y_2 = 2x - 4)$. See Figure 31.

Figure 31

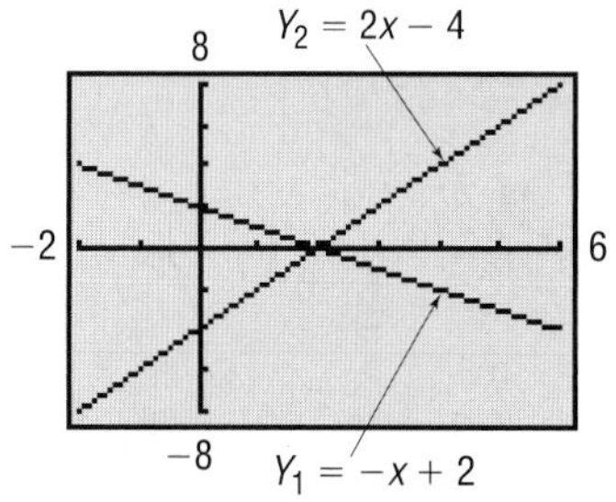

Notice that the graphs divide the viewing window into four regions. We select a test point for each region and determine whether the point makes *both* inequalities true. We choose to test (0, 0), (2, 3), (4, 0), and (2, −2). Figure 32(a) shows that (2, 3) is the only point for which both inequalities are true. Thus, we obtain the graph shown in Figure 32(b).

Figure 32

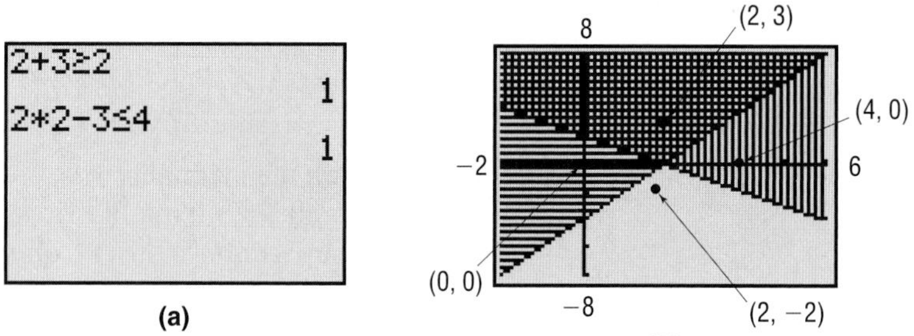

(a) (b)

Notice in Figure 32(b) that the darker region, the graph of the system of inequalities, is the intersection of the graphs of the single inequalities $x + y \geq 2$ and $2x - y \leq 4$. Obtaining Figure 32(b) by this method is sometimes faster than using test points.

NOW WORK PROBLEM 11.

EXAMPLE 6 Graphing a System of Linear Inequalities

Graph the system: $\begin{cases} x + y \leq 2 \\ x + y \geq 0 \end{cases}$

Solution See Figure 33. The overlapping purple-shaded region between the two boundary lines is the graph of the system.

Figure 33

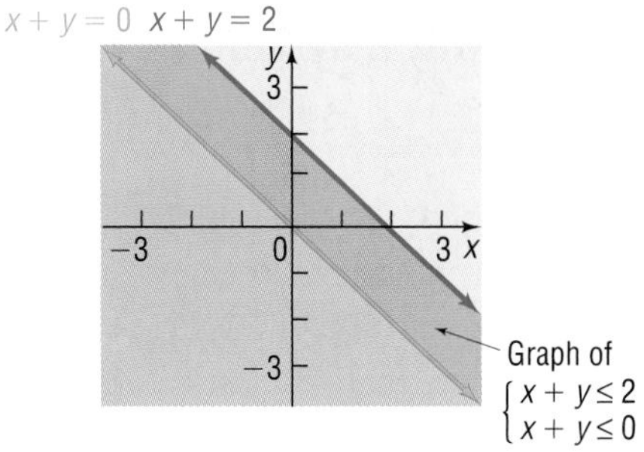

NOW WORK PROBLEM 15.

EXAMPLE 7 Graphing a System of Linear Inequalities

Graph the system: $\begin{cases} 2x - y \geq 2 \\ 2x - y \geq 0 \end{cases}$

Solution See Figure 34. The overlapping purple-shaded region is the graph of the system. Note that the graph of the system is identical to the graph of the single inequality $2x - y \geq 2$.

Figure 34

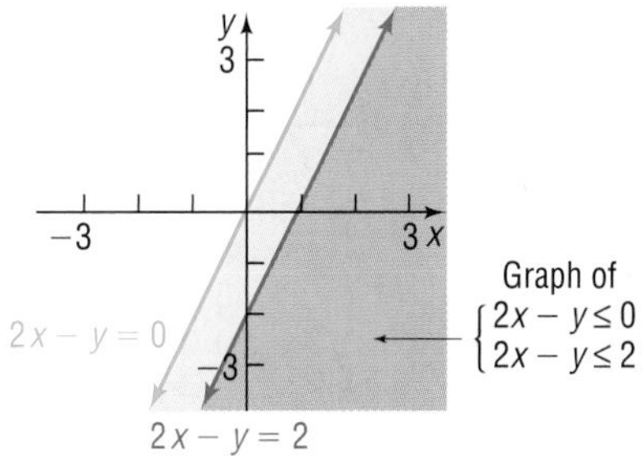

EXAMPLE 8 Graphing a System of Linear Inequalities

Graph the system: $\begin{cases} x + 2y \leq 2 \\ x + 2y \geq 6 \end{cases}$

Solution See Figure 35. Because no overlapping region results, there are no points in the xy-plane that simultaneously satisfy each inequality. Hence, the system has no solution.

Figure 35

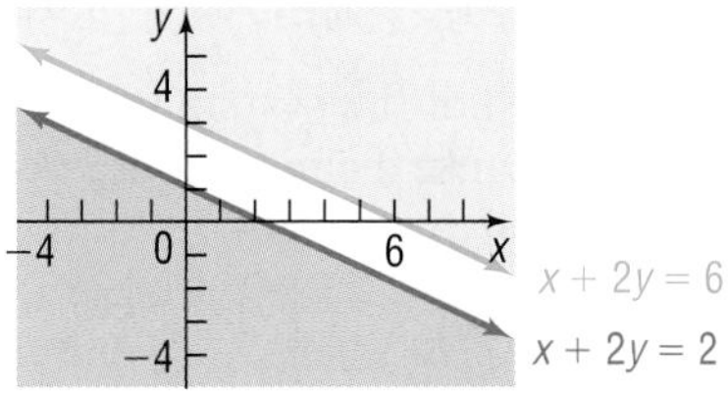

EXAMPLE 9 Graphing a System of Four Linear Inequalities

Graph the system: $\begin{cases} x + y \le 3 \\ 2x + y \le 4 \\ x \ge 0 \\ y \ge 0 \end{cases}$

Solution The two inequalities $x \ge 0$ and $y \ge 0$ require that the graph be in quadrant I. We set our viewing window accordingly. Figure 36 shows the graph of $x + y = 3\ (Y_1 = -x + 3)$ and $2x + y = 4\ (Y_2 = -2x + 4)$. The graphs divide the viewing window into four regions.

Figure 36

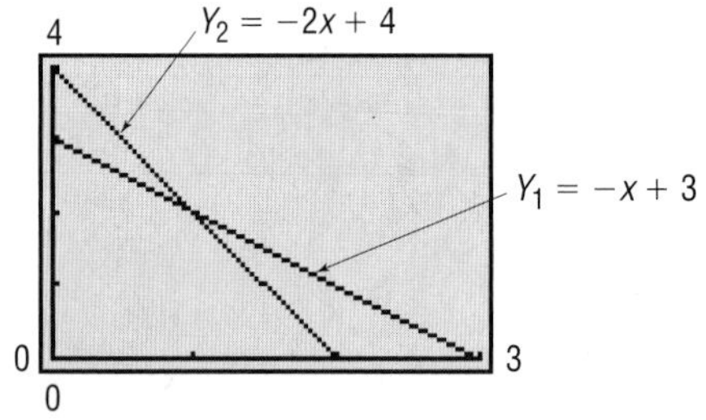

The graph of the inequality $x + y \le 3$ consists of the half-plane below Y_1, and the graph of $2x + y \le 4$ consists of the half-plane below Y_2, so we shade accordingly. See Figures 37(a) and (b).

Figure 37

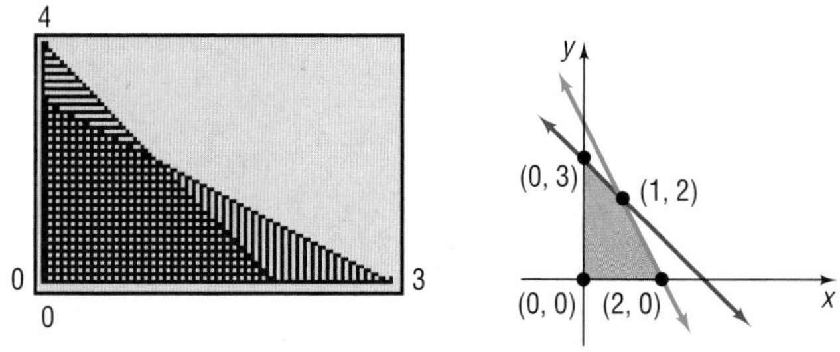

(a) (b)

Notice in Figure 37(b) that those points belonging to the graph that are also points of intersection of boundary lines have been plotted. Such points are referred to as **vertices** or **corner points** of the graph. The corner points $(0, 0)$, $(0, 3)$, $(1, 2)$ and $(2, 0)$ are found by solving the system of equations formed by the two boundary lines. So the corner points are found by solving the following systems:

$\begin{cases} x = 0 \\ y = 0 \end{cases}$ Corner point: (0, 0)

$\begin{cases} x + y = 3 \\ x = 0 \end{cases}$ Corner point: (0, 3)

$\begin{cases} x + y = 3 \\ 2x + y = 4 \end{cases}$ Corner point: (1, 2)

$\begin{cases} 2x + y = 4 \\ y = 0 \end{cases}$ Corner point: (2, 0)

The graph of the system of linear inequalities in Figure 37 is said to be **bounded,** because it can be contained within some circle of sufficiently large radius. A graph that cannot be contained in any circle is said to be **unbounded.** For example, the graph of the system of linear inequalities in Figure 34 is unbounded, since it extends indefinitely in a particular direction.

NOW WORK PROBLEM 23.

LINEAR PROGRAMMING

4 Historically, linear programming evolved as a technique for solving problems involving resource allocation of goods and materials for the U.S. Air Force during World War II. Today, linear programming techniques are used to solve a wide variety of problems, such as optimizing airline scheduling and establishing telephone lines. Although most practical linear programming problems involve systems of several hundred linear inequalities containing several hundred variables, we will limit our discussion to problems containing only two variables, because we can solve such problems using graphing techniques.*

EXAMPLE 10 Financial Planning

A retired couple has up to \$25,000 to invest. As their financial adviser, you recommend that they place at least \$15,000 in Treasury bills yielding 6% and at most \$5000 in corporate bonds yielding 9%. How much money should be placed in each investment so that income is maximized? ■

The problem given here is typical of a *linear programming problem.* The problem requires that a certain linear expression, the income, be maximized. If I represents income, x the amount invested in Treasury bills at 6%, and y the amount invested in corporate bonds at 9%, then

$$I = 0.06x + 0.09y$$

This linear expression is called the **objective function.** Furthermore, the problem requires that the maximum income be achieved under certain conditions or **constraints,** each of which is a linear inequality involving the variables.

The linear programming problem given in Example 10 may be restated as

$$\text{Maximize:} \quad I = 0.06x + 0.09y$$

subject to the conditions that

$$x \geq 0, \qquad y \geq 0$$

$$x + y \leq 25{,}000 \quad \text{Up to \$25,000 to invest}$$

$$x \geq 15{,}000 \quad \text{At least \$15,000 in Treasury bills}$$

$$y \leq 5000 \quad \text{At most \$5000 in Corporate bonds}$$

In general, every linear programming problem has two components:

1. A linear objective function that is to be maximized or minimized
2. A collection of linear inequalities that must be satisfied simultaneously

*The **simplex method** is a way to solve linear programming problems involving many inequalities and variables. This method was developed by George Dantzig in 1946 and is particularly well suited for computerization. In 1984, Narendra Karmarkar of Bell Laboratories discovered a way of solving large linear programming problems that improves on the simplex method.

A **linear programming problem** in two variables x and y consists of maximizing (or minimizing) a linear objective function

$$z = Ax + By$$

where A and B are real numbers, not both 0, subject to certain conditions, or constraints, expressible as linear inequalities in x and y.

To maximize (or minimize) the quantity $z = Ax + By$, we need to identify points (x, y) that make the expression for z the largest (or smallest) possible. But not all points (x, y) are eligible; only those that also satisfy each linear inequality (constraint) can be used. We refer to each point (x, y) that satisfies the system of linear inequalities (the constraints) as a **feasible point.** Thus, in a linear programming problem, we seek the feasible point(s) that maximizes (or minimizes) the objective function.

Let's look again at the linear programming problem in Example 10.

EXAMPLE 11 Analyzing a Linear Programming Problem

Consider the following linear programming problem:

$$\text{Maximize:} \quad I = 0.06x + 0.09y$$

subject to the conditions that

$$x \geq 0, \qquad y \geq 0$$
$$x + y \leq 25{,}000$$
$$x \geq 15{,}000$$
$$y \leq 5000$$

Graph the constraints. Then graph the objective function for $I = 0, 0.9, 1.35, 1.65$, and 1.8.

Solution Figure 38 shows the graph of the constraints. We superimpose on this graph the graph of the objective function for the given values of I.

Figure 38

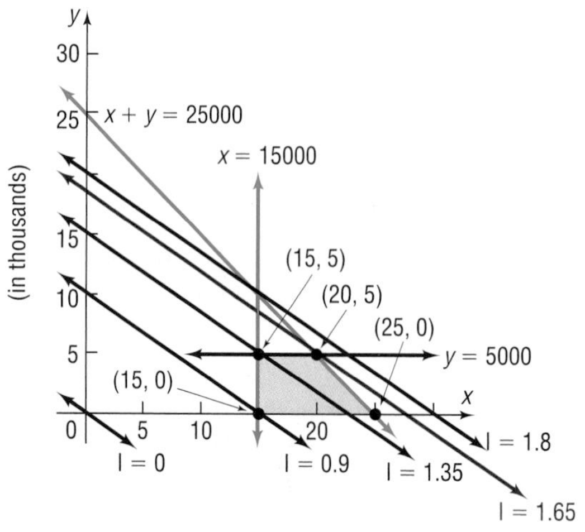

For $I = 0$, the objective function is the line $0 = 0.06x + 0.09y$.
For $I = 0.9$, the objective function is the line $0.9 = 0.06x + 0.09y$.
For $I = 1.35$, the objective function is the line $1.35 = 0.06x + 0.09y$.
For $I = 1.65$, the objective function is the line $1.65 = 0.06x + 0.09y$.
For $I = 1.8$, the objective function is the line $1.8 = 0.06x + 0.09y$. ■

A **solution** to a linear programming problem consists of the feasible point(s) that maximizes (or minimizes) the objective function, together with the corresponding value of the objective function.

5 One condition for a linear programming problem in two variables to have a solution is that the graph of the feasible points be bounded. (Refer to page 794).

If none of the feasible points maximizes (or minimizes) the objective function or if there are no feasible points, then the linear programming problem has no solution.

Consider the linear programming problem stated in Example 11, and look again at Figure 38. The feasible points are the points that lie in the shaded region. For example, $(20, 3)$ is a feasible point, as is $(15, 5)$, $(20, 5)$, $(18, 4)$, and so on. To find the solution of the problem requires that we find a feasible point (x, y) that makes $I = 0.06x + 0.09y$ as large as possible. Notice that, as I increases in value from $I = 0$ to $I = 0.9$ to $I = 1.35$ to $I = 1.65$ to $I = 1.8$, we obtain a collection of parallel lines. Furthermore, notice that the largest value of I that can be obtained while feasible points are present is $I = 1.65$, which corresponds to the line $1.65 = 0.06x + 0.09y$. Any larger value of I results in a line that does not pass through any feasible points. Finally, notice that the feasible point that yields $I = 1.65$ is the point $(20, 5)$, a corner point. These observations form the basis of the following result, which we state without proof.

Theorem **Location of the Solution of a Linear Programming Problem**

If a linear programming problem has a solution, it is located at a corner point of the graph of the feasible points.

If a linear programming problem has multiple solutions, at least one of them is located at a corner point of the graph of the feasible points.

In either case, the corresponding value of the objective function is unique.

■

We shall not consider here linear programming problems that have no solution. As a result, we can outline the procedure for solving a linear programming problem as follows:

PROCEDURE FOR SOLVING A LINEAR PROGRAMMING PROBLEM

STEP 1: Write an expression for the quantity to be maximized (or minimized). This expression is the objective function.

STEP 2: Write all the constraints as a system of linear inequalities and graph the system.

STEP 3: List the corner points of the graph of the feasible points.

STEP 4: List the corresponding values of the objective function at each corner point. The largest (or smallest) of these is the solution.

EXAMPLE 12 Solving a Minimum Linear Programming Problem

Minimize the expression

$$z = 2x + 3y$$

subject to the constraints

$$y \leq 5 \qquad x \leq 6 \qquad x + y \geq 2 \qquad x \geq 0 \qquad y \geq 0$$

Solution The objective function is $z = 2x + 3y$. We seek the smallest value of z that can occur if x and y are solutions of the system of linear inequalities

$$\begin{cases} y \leq 5 \\ x \leq 6 \\ x + y \geq 2 \\ x \geq 0 \\ y \geq 0 \end{cases}$$

The graph of this system (the feasible points) is shown as the shaded region in Figure 39. We have also plotted the corner points. Table 1 lists the corner points and the corresponding values of the objective function. From the table, we can see that the minimum value of z is 4, and it occurs at the point (2, 0).

Figure 39

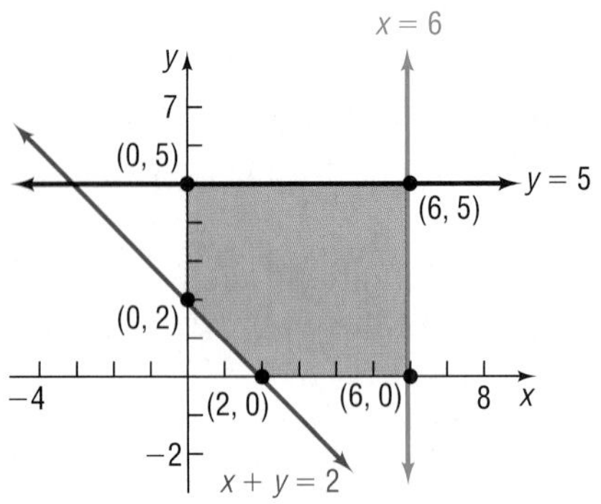

TABLE 1

Corner Point	Value of the Objective Function
(x, y)	$z = 2x + 3y$
(0, 2)	$z = 2(0) + 3(2) = 6$
(0, 5)	$z = 2(0) + 3(5) = 15$
(6, 5)	$z = 2(6) + 3(5) = 27$
(6, 0)	$z = 2(6) + 3(0) = 12$
(2, 0)	$z = 2(2) + 3(0) = 4$

■

NOW WORK PROBLEMS 37 AND 43.

EXAMPLE 13 Maximizing Profit

At the end of every month, after filling orders for its regular customers, a coffee company has some pure Colombian coffee and some special-blend coffee remaining. The practice of the company has been to package a mixture of the two coffees into 1-pound packages as follows: a low-grade mixture containing 4 ounces of Colombian coffee and 12 ounces of special-blend coffee and a high-grade mixture containing 8 ounces of Colombian and 8 ounces of special-blend coffee. A profit of \$0.30 per package is made on the low-grade mixture, whereas a profit of \$0.40 per package is made on the high-grade mixture. This month, 120 pounds of special-blend coffee and 100 pounds of pure Colombian coffee remain. How many packages of each mixture should be prepared to achieve a maximum profit? Assume that all packages prepared can be sold.

Solution We begin by assigning symbols for the two variables.

$$x = \text{Number of packages of the low-grade mixture}$$
$$y = \text{Number of packages of the high-grade mixture}$$

If P denotes the profit, then

$$P = \$0.30x + \$0.40y$$

This expression is the objective function. We seek to maximize P subject to certain constraints on x and y. Because x and y represent numbers of packages, the only meaningful values for x and y are nonnegative integers. Thus, we have the two constraints

$$x \geq 0 \qquad y \geq 0 \quad \text{Nonnegative constraints}$$

We also have only so much of each type of coffee available. For example, the total amount of Colombian coffee used in the two mixtures cannot exceed 100 pounds, or 1600 ounces. Because we use 4 ounces in each low-grade package and 8 ounces in each high-grade package, we are led to the constraint

$$4x + 8y \leq 1600 \quad \text{Colombian coffee constraint}$$

Similarly, the supply of 120 pounds, or 1920 ounces, of special-blend coffee leads to the constraint

$$12x + 8y \leq 1920 \quad \text{Special-blend coffee constraint}$$

The linear programming problem may be stated as

$$\text{Maximize:} \quad P = 0.3x + 0.4y$$

subject to the constraints

$$x \geq 0 \qquad y \geq 0 \qquad 4x + 8y \leq 1600 \qquad 12x + 8y \leq 1920$$

The graph of the constraints (the feasible points) is illustrated in Figure 40. We list the corner points and evaluate the objective function at each one. In Table 2, we can see that the maximum profit, \$84, is achieved with 40 packages of the low-grade mixture and 180 packages of the high-grade mixture.

Figure 40

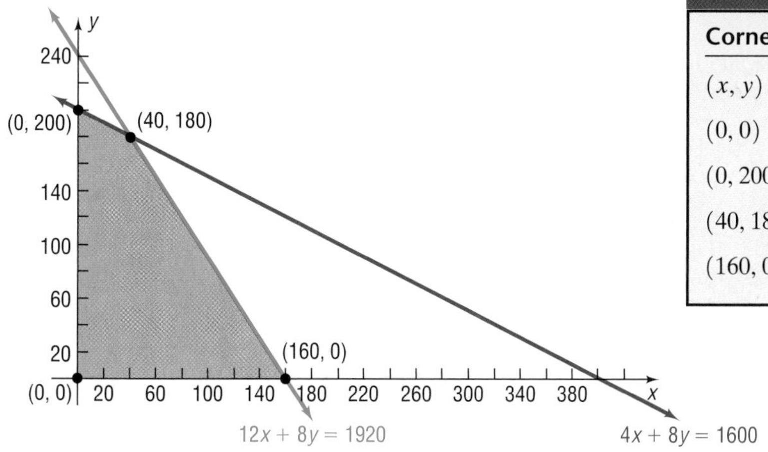

TABLE 2

Corner Point	Value of Profit
(x, y)	$P = 0.3x + 0.4y$
$(0, 0)$	$P = 0$
$(0, 200)$	$P = 0.3(0) + 0.4(200) = \$80$
$(40, 180)$	$P = 0.3(40) + 0.4(180) = \$84$
$(160, 0)$	$P = 0.3(160) + 0.4(0) = \$48$

NOW WORK PROBLEM **51**.

11.8 EXERCISES

In Problems 1–8, graph each inequality by hand.

1. $x \geq 0$ **2.** $y \geq 0$ 3. $x \geq 4$ **4.** $y \leq 2$

5. $x + y > 1$ **6.** $x + y \leq 9$ **7.** $2x + y \geq 6$ **8.** $3x + 2y \leq 6$

In Problems 9–20, graph each system of inequalities by hand.

9. $\begin{cases} x + y \leq 2 \\ 2x + y \geq 4 \end{cases}$ **10.** $\begin{cases} 3x - y \geq 6 \\ x + 2y \leq 2 \end{cases}$ 11. $\begin{cases} 2x - y \leq 4 \\ 3x + 2y \geq -6 \end{cases}$ **12.** $\begin{cases} 4x - 5y \leq 0 \\ 2x - y \geq 2 \end{cases}$

13. $\begin{cases} 2x - 3y \leq 0 \\ 3x + 2y \leq 6 \end{cases}$ **14.** $\begin{cases} 4x - y \geq 2 \\ x + 2y \geq 2 \end{cases}$ 15. $\begin{cases} x - 2y \leq 6 \\ 2x - 4y \geq 0 \end{cases}$ **16.** $\begin{cases} x + 4y \leq 8 \\ x + 4y \geq 4 \end{cases}$

17. $\begin{cases} 2x + y \geq -2 \\ 2x + y \geq 2 \end{cases}$ **18.** $\begin{cases} x - 4y \leq 4 \\ x - 4y \geq 0 \end{cases}$ **19.** $\begin{cases} 2x + 3y \geq 6 \\ 2x + 3y \leq 0 \end{cases}$ **20.** $\begin{cases} 2x + y \geq 0 \\ 2x + y \geq 2 \end{cases}$

In Problems 21–30, graph each system of linear inequalities by hand. Tell whether the graph is bounded or unbounded, and label the corner points.

21. $\begin{cases} x \geq 0 \\ y \geq 0 \\ 2x + y \leq 6 \\ x + 2y \leq 6 \end{cases}$ **22.** $\begin{cases} x \geq 0 \\ y \geq 0 \\ x + y \geq 4 \\ 2x + 3y \geq 6 \end{cases}$ 23. $\begin{cases} x \geq 0 \\ y \geq 0 \\ x + y \geq 2 \\ 2x + y \geq 4 \end{cases}$ **24.** $\begin{cases} x \geq 0 \\ y \geq 0 \\ 3x + y \leq 6 \\ 2x + y \leq 2 \end{cases}$ **25.** $\begin{cases} x \geq 0 \\ y \geq 0 \\ x + y \geq 2 \\ 2x + 3y \leq 12 \\ 3x + y \leq 12 \end{cases}$

26. $\begin{cases} x \geq 0 \\ y \geq 0 \\ x + y \geq 2 \\ x + y \leq 10 \\ 2x + y \leq 3 \end{cases}$ **27.** $\begin{cases} x \geq 0 \\ y \geq 0 \\ x + y \geq 2 \\ x + y \leq 8 \\ 2x + y \leq 10 \end{cases}$ **28.** $\begin{cases} x \geq 0 \\ y \geq 0 \\ x + y \geq 2 \\ x + y \leq 8 \\ x + 2y \geq 1 \end{cases}$ **29.** $\begin{cases} x \geq 0 \\ y \geq 0 \\ x + 2y \geq 1 \\ x + 2y \leq 10 \end{cases}$ **30.** $\begin{cases} x \geq 0 \\ y \geq 0 \\ x + 2y \geq 1 \\ x + 2y \leq 10 \\ x + y \geq 2 \\ x + y \leq 8 \end{cases}$

In Problems 31–34, write a system of linear inequalities that has the given graph.

31.

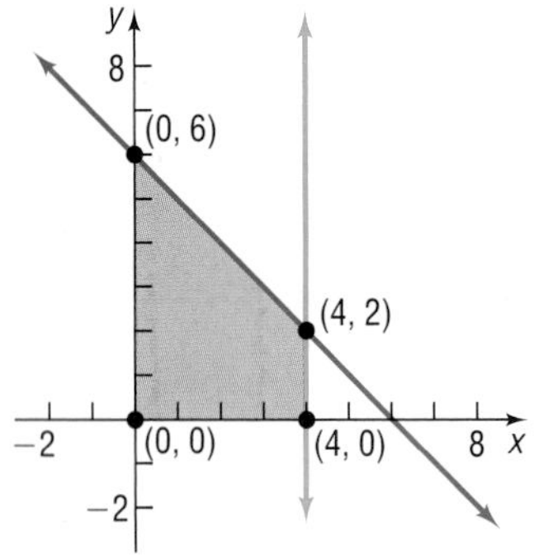

32.

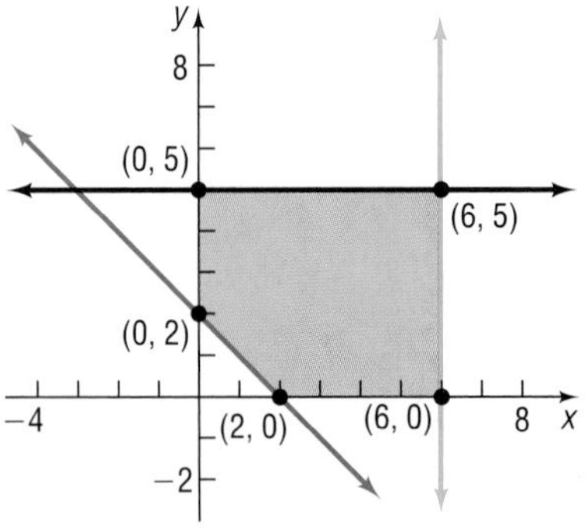

33.

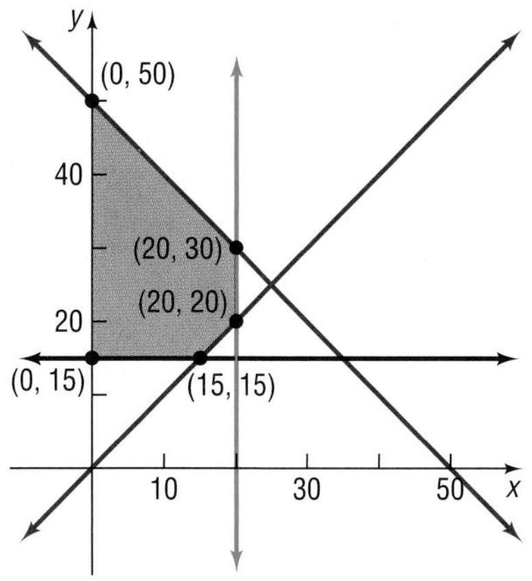

34.

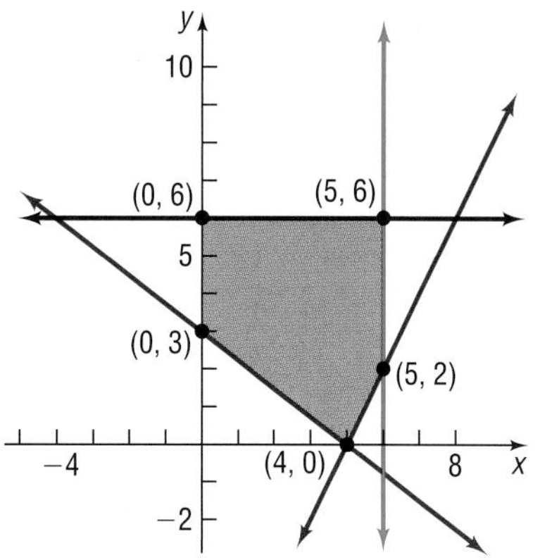

In Problems 35–40, find the maximum and minimum value of the given objective function of a linear programming problem. The figure illustrates the graph of the feasible points.

35. $z = x + y$ **36.** $z = 2x + 3y$ **37.** $z = x + 10y$

38. $z = 10x + y$ **39.** $z = 5x + 7y$ **40.** $z = 7x + 5y$

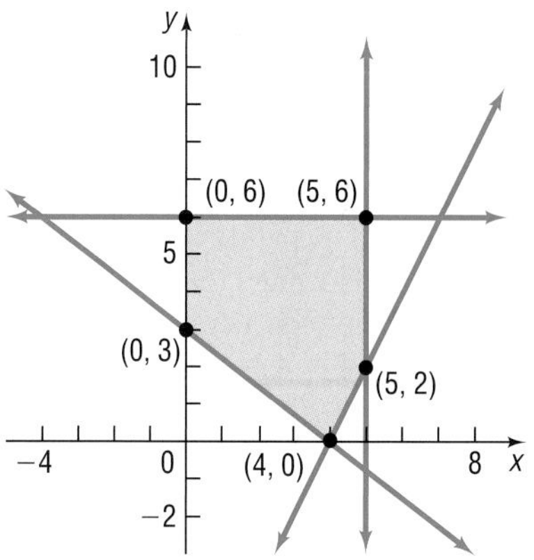

In Problems 41–50, solve each linear programming problem.

41. Maximize $z = 2x + y$ subject to $x \geq 0$, $y \geq 0$, $x + y \leq 6$, $x + y \geq 1$

42. Maximize $z = x + 3y$ subject to $x \geq 0$, $y \geq 0$, $x + y \geq 3$, $x \leq 5$, $y \leq 7$

43. Minimize $z = 2x + 5y$ subject to $x \geq 0$, $y \geq 0$, $x + y \geq 2$, $x \leq 5$, $y \leq 3$

44. Minimize $z = 3x + 4y$ subject to $x \geq 0$, $y \geq 0$, $2x + 3y \geq 6$, $x + y \leq 8$

45. Maximize $z = 3x + 5y$ subject to $x \geq 0$, $y \geq 0$, $x + y \geq 2$, $2x + 3y \leq 12$, $3x + 2y \leq 12$

46. Maximize $z = 5x + 3y$ subject to $x \geq 0$, $y \geq 0$, $x + y \geq 2$, $x + y \leq 8$, $2x + y \leq 10$

47. Minimize $z = 5x + 4y$ subject to $x \geq 0$, $y \geq 0$, $x + y \geq 2$, $2x + 3y \leq 12$, $3x + y \leq 12$

48. Minimize $z = 2x + 3y$ subject to $x \geq 0$, $y \geq 0$, $x + y \geq 3$, $x + y \leq 9$, $x + 3y \geq 6$

49. Maximize $z = 5x + 2y$ subject to $x \geq 0$, $y \geq 0$, $x + y \leq 10$, $2x + y \geq 10$, $x + 2y \geq 10$

50. Maximize $z = 2x + 4y$ subject to $x \geq 0$, $y \geq 0$, $2x + y \geq 4$, $x + y \leq 9$

51. Maximizing Profit A manufacturer of skis produces two types: downhill and cross country. Use the following table to determine how many of each kind of ski should be produced to achieve a maximum profit. What is the maximum profit? What would the maximum profit be if the maximum time available for manufacturing is increased to 48 hours?

	Downhill	Cross Country	Maximum Time Available
Manufacturing time per ski	2 hours	1 hour	40 hours
Finishing time per ski	1 hour	1 hour	32 hours
Profit per ski	$70	$50	

52. Farm Management A farmer has 70 acres of land available for planting either soybeans or wheat. The cost of preparing the soil, the workdays required, and the expected profit per acre planted for each type of crop are given in the following table:

	Soybeans	Wheat
Preparation cost per acre	\$60	\$30
Workdays required per acre	3	4
Profit per acre	\$180	\$100

The farmer cannot spend more than \$1800 in preparation costs nor use more than a total of 120 workdays. How many acres of each crop should be planted in order to maximize the profit? What is the maximum profit? What is the maximum profit if the farmer is willing to spend no more than \$2400 on preparation?

53. Farm Management A small farm in Illinois has 100 acres of land available on which to grow corn and soybeans. The following table shows the cultivation cost per acre, the labor cost per acre, and the expected profit per acre. The column on the right shows the amount of money available for each of these expenses. Find the number of acres of each crop that should be planted in order to maximize profit.

	Soybeans	Corn	Money Available
Cultivation cost per acre	\$40	\$60	\$1800
Labor cost per acre	\$60	\$60	\$2400
Profit per acre	\$200	\$250	

54. Dietary Requirements A certain diet requires at least 60 units of carbohydrates, 45 units of protein, and 30 units of fat each day. Each ounce of Supplement A provides 5 units of carbohydrates, 3 units of protein, and 4 units of fat. Each ounce of Supplement B provides 2 units of carbohydrates, 2 units of protein, and 1 unit of fat. If Supplement A costs \$1.50 per ounce and Supplement B costs \$1.00 per ounce, how many ounces of each supplement should be taken daily to minimize the cost of the diet?

55. Production Scheduling In a factory, machine 1 produces 8-inch pliers at the rate of 60 units per hour and 6-inch pliers at the rate of 70 units per hour. Machine 2 produces 8-inch pliers at the rate of 40 units per hour and 6-inch pliers at the rate of 20 units per hour. It costs \$50 per hour to operate machine 1, while machine 2 costs \$30 per hour to operate. The production schedule requires that at least 240 units of 8-inch pliers and at least 140 units of 6-inch pliers must be produced during each 10-hour day. Which combination of machines will cost the least money to operate?

56. Farm Management An owner of a fruit orchard hires a crew of workers to prune at least 25 of his 50 fruit trees. Each newer tree requires 1 hour to prune, while each older tree needs 1.5 hours. The crew contracts to work for at least 30 hours and charge \$15 for each newer tree and \$20 for each older tree. To minimize his cost, how many of each kind of tree will the orchard owner have pruned? What will be the cost?

57. Managing a Meat Market A meat market combines ground beef and ground pork in a single package for meat loaf. The ground beef is 75% lean (75% beef, 25% fat) and costs the market \$0.75 per pound. The ground pork is 60% lean and costs the market \$0.45 per pound. The meat loaf must be at least 70% lean. If the market wants to use at least 50 pounds of its available pork, but no more than 200 pounds of its available ground beef, how much ground beef should be mixed with ground pork so that the cost is minimized?

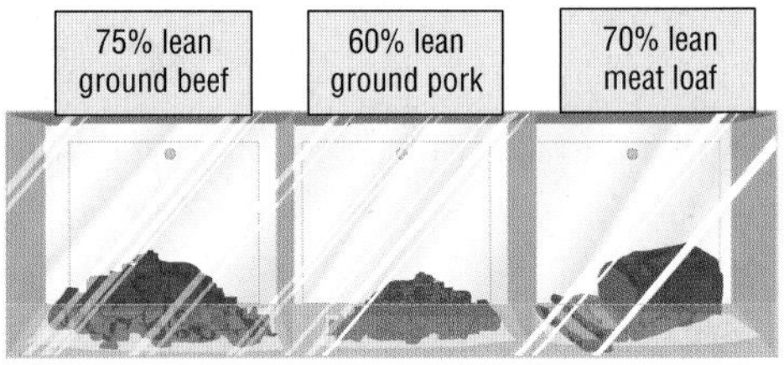

58. Return on Investment An investment broker is instructed by her client to invest up to \$20,000, some in a junk bond yielding 9% per annum and some in Treasury bills yielding 7% per annum. The client wants to invest at least \$8000 in T-bills and no more than \$12,000 in the junk bond.

(a) How much should the broker recommend that the client place in each investment to maximize income if the client insists that the amount invested in T-bills must equal or exceed the amount placed in junk bonds?

(b) How much should the broker recommend that the client place in each investment to maximize income if the client insists that the amount invested in T-bills must not exceed the amount placed in junk bonds?

59. Maximizing Profit on Ice Skates A factory manufactures two kinds of ice skates: racing skates and figure skates. The racing skates require 6 work-hours in the fabrication department, whereas the figure skates require 4 work-hours there. The racing skates require 1 work-hour in the finishing department, whereas the figure skates require 2 work-hours there. The fabricating department has available at most 120 work-hours per day, and the finishing department has no more than 40 work-hours per day available. If the profit on each racing skate is \$10 and the profit on each figure skate is \$12, how many of each should be manufactured each day to maximize profit? (Assume that all skates made are sold.)

60. Financial Planning A retired couple has up to \$50,000 to place in fixed-income securities. Their financial advisor suggests two securities to them: one is a AAA bond that yields 8% per annum; the other is a Certificate of

Deposit (CD) that yields 4%. After careful consideration of the alternatives, the couple decides to place at most $20,000 in the AAA bond and at least $15,000 in the CD. They also instruct the financial adviser to place at least as much in the CD as in the AAA bond. How should the financial adviser proceed to maximize the return on their investment?

61. Product Design An entrepreneur is having a design group produce at least six samples of a new kind of fastener that he wants to market. It costs $9.00 to produce each metal fastener and $4.00 to produce each plastic fastener. He wants to have at least two of each version of the fastener and needs to have all the samples 24 hours from now. It takes 4 hours to produce each metal sample and 2 hours to produce each plastic sample. To minimize the cost of the samples, how many of each kind should the entrepreneur order? What will be the cost of the samples?

62. Animal Nutrition Kevin's dog Amadeus likes two kinds of canned dog food. Gourmet Dog costs 40 cents a can and has 20 units of a vitamin complex; the calorie content is 75 calories. Chow Hound costs 32 cents a can and has 35 units of vitamins and 50 calories. Kevin likes Amadeus to have at least 1175 units of vitamins a month and at least 2375 calories during the same time period. Kevin has space to store only 60 cans of dog food at a time. How much of each kind of dog food should Kevin buy once a month in order to minimize his cost?

63. Airline Revenue An airline has two classes of service: first class and coach. Management's experience has been that each aircraft should have at least 8 but not more than 16 first-class seats and at least 80 but not more than 120 coach seats.

(a) If management decides that the ratio of first-class to coach seats should never exceed 1:12, with how many of each type seat should an aircraft be configured to maximize revenue?

(b) If management decides that the ratio of first-class to coach seats should never exceed 1:8, with how many of each type seat should an aircraft be configured to maximize revenue?

(c) If you were management, what would you do?

[**Hint:** Assume that the airline charges $C for a coach seat and $F for a first-class seat; $C > 0, F > 0$.]

64. Minimizing Cost A firm that specializes in raising frying chickens supplements the regular chicken feed with four vitamins. The owner wants the supplemental food to contain at least 50 units of vitamin I, 90 units of vitamin II, 60 units of vitamin III, and 100 units of vitamin IV per 100 ounces of feed. Two supplements are available: supplement A, which contains 5 units of vitamin I, 25 units of vitamin II, 10 units of vitamin III, and 35 units of vitamin IV per ounce; and supplement B, which contains 25 units of vitamin I, 10 units of vitamin II, 10 units of vitamin III, and 20 units of vitamin IV per ounce. If supplement A costs $0.06 per ounce and supplement B costs $0.08 per ounce, how much of each supplement should the manager of the farm buy to add to each 100 ounces of feed in order to keep the total cost at a minimum, while still meeting the owner's vitamin specifications?

65. Explain in your own words what a linear programming problem is and how it can be solved.

CHAPTER REVIEW

Things To Know

Systems of equations (page 711)

Systems with no solutions are inconsistent. Systems with a solution are consistent.

Consistent systems of linear equations have either a unique solution or an infinite number of solutions.

Matrix (pages 727 and 751)	Rectangular array of numbers, called entries
m by n matrix (page 752)	Matrix with m rows and n columns
Identity matrix I (page 760)	Square matrix whose diagonal entries are 1's, while all other entries are 0's
Inverse of a matrix (page 761)	A^{-1} is the inverse of A if $AA^{-1} = A^{-1}A = I$
Nonsingular matrix (page 761)	A matrix that has an inverse

Linear programming problem (page 796)

Maximize (or minimize) a linear objective function, $z = Ax + By$, subject to certain conditions, or constraints, expressible as linear inequalities in x and y

Feasible point (page 796)

A point (x, y) that satisfies the constraints of a linear programming problem

Location of solution (page 797)

If a linear programming problem has a solution, it is located at a corner point of the graph of the feasible points.

If a linear programming problem has multiple solutions, at least one of them is located at a corner point of the graph of the feasible points.

In either case, the corresponding value of the objective function is unique.

How To

Solve systems of equations by substitution (p. 713)
Solve systems of equations by elimination (p. 715)
Identify inconsistent systems (pp. 716 and 723)
Identify dependent equations (p. 716 and 724)
Solve systems of three equations containing three variables (p. 722)
Write the augmented matrix of a system of linear equations (p. 727)
Write the system from the augmented matrix (p. 728)
Perform row operations on a matrix (p. 729)
Solve systems of linear equations using matrices (p. 730)
Use a graphing utility to solve a system of linear equations (p. 732)

Evaluate 2 by 2 determinants (p. 740)

Use Cramer's Rule to solve a system of equations (pp. 742 and 746)

Evaluate 3 by 3 determinants (p. 744)

Know properties of determinants (p. 748)

Work with equality and addition of matrices (p. 752)

Know properties of matrices (p. 754)

Know how to multiply matrices (p. 756)

Find the inverse of a matrix (p. 761)

Solve systems of equations using inverse matrices (p. 765)

Decompose P/Q, where Q has only nonrepeated linear factors (p. 770)

Decompose P/Q, where Q has repeated linear factors (p. 772)

Decompose P/Q, where Q has only nonrepeated irreducible quadratic factors (p. 774)

Decompose P/Q, where Q has repeated irreducible quadratic factors (p. 775)

Solve a system of nonlinear equations using substitution (p. 776)

Solve a system of nonlinear equations using elimination (p. 778)

Graph a linear inequality by hand and using a graphing utility (pp. 788 and 789)

Graph a system of linear inequalities (p. 791)

Set up and solve a linear programming problem (pp. 795 and 797)

Fill-in-the-Blank Items

1. If a system of equations has no solution, it is said to be ______________.
2. An m by n rectangular array of numbers is called a(n) ______________.
3. Cramer's Rule uses ______________ to solve a system of linear equations.
4. The matrix used to represent a system of linear equations is called a(n) ______________ matrix.
5. A matrix B, for which $AB = I_n$, the identity matrix, is called the ______________ of A.
6. A matrix that has the same number of rows as columns is called a(n) ______________ matrix.
7. In the algebra of matrices, the matrix that has properties similar to the number 1 is called the ______________ matrix.
8. The graph of a linear inequality is called a(n) ______________.
9. A linear programming problem requires that a linear expression, called the ______________ ______________, be maximized or minimized.
10. Each point that satisfies the constraints of a linear programming problem is called a(n) ______________ ______________.
11. A rational function is called ______________ if the degree of its numerator is less than the degree of its denominator.

True/False Items

T F 1. A system of two linear equations containing two unknowns always has at least one solution.
T F 2. The augmented matrix of a system of two equations containing three variables has two rows and four columns.
T F 3. A 3 by 3 determinant can never equal 0.
T F 4. A consistent system of equations will have exactly one solution.
T F 5. Every square matrix has an inverse.
T F 6. Matrix multiplication is commutative.
T F 7. Any pair of matrices can be multiplied.
T F 8. The graph of a linear inequality is a half-plane.
T F 9. The graph of a system of linear inequalities is sometimes unbounded.
T F 10. If a linear programming problem has a solution, it is located at a corner point of the graph of the feasible points.
T F 11. The factors of the denominator of a rational expression are used to arrive at the partial fraction decomposition.

Review Exercises

Blue problem numbers indicate the authors' suggestions for use in a Practice Test.

In Problems 1–20, solve each system of equations algebraically using the method of substitution or the method of elimination. If the system has no solution, say that it is inconsistent. Verify your result using a graphing utility.

1. $\begin{cases} 2x - y = 5 \\ 5x + 2y = 8 \end{cases}$

2. $\begin{cases} 2x + 3y = 2 \\ 7x - y = 3 \end{cases}$

3. $\begin{cases} 3x - 4y = 4 \\ x - 3y = \frac{1}{2} \end{cases}$

4. $\begin{cases} 2x + y = 0 \\ 5x - 4y = -\frac{13}{2} \end{cases}$

5. $\begin{cases} x - 2y - 4 = 0 \\ 3x + 2y - 4 = 0 \end{cases}$

6. $\begin{cases} x - 3y + 5 = 0 \\ 2x + 3y - 5 = 0 \end{cases}$

7. $\begin{cases} y = 2x - 5 \\ x = 3y + 4 \end{cases}$

8. $\begin{cases} x = 5y + 2 \\ y = 5x + 2 \end{cases}$

9. $\begin{cases} x - y + 4 = 0 \\ \frac{1}{2}x + \frac{1}{6}y + \frac{2}{5} = 0 \end{cases}$

10. $\begin{cases} x + \frac{1}{4}y = 2 \\ y + 4x + 2 = 0 \end{cases}$

11. $\begin{cases} x - 2y - 8 = 0 \\ 2x + 2y - 10 = 0 \end{cases}$

12. $\begin{cases} x - 3y + \frac{7}{2} = 0 \\ \frac{1}{2}x + 3y - 5 = 0 \end{cases}$

13. $\begin{cases} y - 2x = 11 \\ 2y - 3x = 18 \end{cases}$

14. $\begin{cases} 3x - 4y - 12 = 0 \\ 5x + 2y + 6 = 0 \end{cases}$

15. $\begin{cases} 2x + 3y - 13 = 0 \\ 3x - 2y = 0 \end{cases}$

16. $\begin{cases} 4x + 5y = 21 \\ 5x + 6y = 42 \end{cases}$

17. $\begin{cases} 3x - 2y = 8 \\ x - \frac{2}{3}y = 12 \end{cases}$

18. $\begin{cases} 2x + 5y = 10 \\ 4x + 10y = 15 \end{cases}$

19. $\begin{cases} x + 2y - z = 6 \\ 2x - y + 3z = -13 \\ 3x - 2y + 3z = -16 \end{cases}$

20. $\begin{cases} x + 5y - z = 2 \\ 2x + y + z = 7 \\ x - y + 2z = 11 \end{cases}$

In Problems 21–28, use the following matrices to compute each expression. Verify your result using a graphing utility.

$$A = \begin{bmatrix} 1 & 0 \\ 2 & 4 \\ -1 & 2 \end{bmatrix} \quad B = \begin{bmatrix} 4 & -3 & 0 \\ 1 & 1 & -2 \end{bmatrix} \quad C = \begin{bmatrix} 3 & -4 \\ 1 & 5 \\ 5 & -2 \end{bmatrix}$$

21. $A + C$ **22.** $A - C$ **23.** $6A$ **24.** $-4B$

25. AB **26.** BA **27.** CB **28.** BC

In Problems 29–34, find the inverse of each matrix algebraically, if there is one. If there is not an inverse, say that the matrix is singular. Verify your result using a graphing utility.

29. $\begin{bmatrix} 4 & 6 \\ 1 & 3 \end{bmatrix}$

30. $\begin{bmatrix} -3 & 2 \\ 1 & -2 \end{bmatrix}$

31. $\begin{bmatrix} 1 & 3 & 3 \\ 1 & 2 & 1 \\ 1 & -1 & 2 \end{bmatrix}$

32. $\begin{bmatrix} 3 & 1 & 2 \\ 3 & 2 & -1 \\ 1 & 1 & 1 \end{bmatrix}$

33. $\begin{bmatrix} 4 & -8 \\ -1 & 2 \end{bmatrix}$

34. $\begin{bmatrix} -3 & 1 \\ -6 & 2 \end{bmatrix}$

In Problems 35–44, solve each system of equations algebraically using matrices. If the system has no solution, say that it is inconsistent. Verify your result using a graphing utility.

35. $\begin{cases} 3x - 2y = 1 \\ 10x + 10y = 5 \end{cases}$

36. $\begin{cases} 3x + 2y = 6 \\ x - y = -\frac{1}{2} \end{cases}$

37. $\begin{cases} 5x + 6y - 3z = 6 \\ 4x - 7y - 2z = -3 \\ 3x + y - 7z = 1 \end{cases}$

38. $\begin{cases} 2x + y + z = 5 \\ 4x - y - 3z = 1 \\ 8x + y - z = 5 \end{cases}$

39. $\begin{cases} x - 2z = 1 \\ 2x + 3y = -3 \\ 4x - 3y - 4z = 3 \end{cases}$

40. $\begin{cases} x + 2y - z = 2 \\ 2x - 2y + z = -1 \\ 6x + 4y + 3z = 5 \end{cases}$

41. $\begin{cases} x - y + z = 0 \\ x - y - 5z - 6 = 0 \\ 2x - 2y + z - 1 = 0 \end{cases}$

42. $\begin{cases} 4x - 3y + 5z = 0 \\ 2x + 4y - 3z = 0 \\ 6x + 2y + z = 0 \end{cases}$

43. $\begin{cases} x - y - z - t = 1 \\ 2x + y - z + 2t = 3 \\ x - 2y - 2z - 3t = 0 \\ 3x - 4y + z + 5t = -3 \end{cases}$

44. $\begin{cases} x - 3y + 3z - t = 4 \\ x + 2y - z = -3 \\ x + 3z + 2t = 3 \\ x + y + 5z = 6 \end{cases}$

In Problems 45–50, find the value of each determinant algebraically. Verify your result using a graphing utility.

45. $\begin{vmatrix} 3 & 4 \\ 1 & 3 \end{vmatrix}$

46. $\begin{vmatrix} -4 & 0 \\ 1 & 3 \end{vmatrix}$

47. $\begin{vmatrix} 1 & 4 & 0 \\ -1 & 2 & 6 \\ 4 & 1 & 3 \end{vmatrix}$

48. $\begin{vmatrix} 2 & 3 & 10 \\ 0 & 1 & 5 \\ -1 & 2 & 3 \end{vmatrix}$

49. $\begin{vmatrix} 2 & 1 & -3 \\ 5 & 0 & 1 \\ 2 & 6 & 0 \end{vmatrix}$

50. $\begin{vmatrix} -2 & 1 & 0 \\ 1 & 2 & 3 \\ -1 & 4 & 2 \end{vmatrix}$

In Problems 51–56, use Cramer's Rule, if applicable, to solve each system.

51. $\begin{cases} x - 2y = 4 \\ 3x + 2y = 4 \end{cases}$

52. $\begin{cases} x - 3y = -5 \\ 2x + 3y = 5 \end{cases}$

53. $\begin{cases} 2x + 3y - 13 = 0 \\ 3x - 2y = 0 \end{cases}$

54. $\begin{cases} 3x - 4y - 12 = 0 \\ 5x + 2y + 6 = 0 \end{cases}$

55. $\begin{cases} x + 2y - z = 6 \\ 2x - y + 3z = -13 \\ 3x - 2y + 3z = -16 \end{cases}$

56. $\begin{cases} x - y + z = 8 \\ 2x + 3y - z = -2 \\ 3x - y - 9z = 9 \end{cases}$

In Problems 57–66, write the partial decomposition of each rational expression.

57. $\dfrac{6}{x(x-4)}$

58. $\dfrac{x}{(x+2)(x-3)}$

59. $\dfrac{x-4}{x^2(x-1)}$

60. $\dfrac{2x-6}{(x-2)^2(x-1)}$

61. $\dfrac{x}{(x^2+9)(x+1)}$

62. $\dfrac{3x}{(x-2)(x^2+1)}$

63. $\dfrac{x^3}{(x^2+4)^2}$

64. $\dfrac{x^3+1}{(x^2+16)^2}$

65. $\dfrac{x^2}{(x^2+1)(x^2-1)}$

66. $\dfrac{4}{(x^2+4)(x^2-1)}$

In Problems 67–76, solve each system of equations algebraically. Verify your result using a graphing utility.

67. $\begin{cases} 2x + y + 3 = 0 \\ x^2 + y^2 = 5 \end{cases}$

68. $\begin{cases} x^2 + y^2 = 16 \\ 2x - y^2 = -8 \end{cases}$

69. $\begin{cases} 2xy + y^2 = 10 \\ 3y^2 - xy = 2 \end{cases}$

70. $\begin{cases} 3x^2 - y^2 = 1 \\ 7x^2 - 2y^2 - 5 = 0 \end{cases}$

71. $\begin{cases} x^2 + y^2 = 6y \\ x^2 = 3y \end{cases}$

72. $\begin{cases} 2x^2 + y^2 = 9 \\ x^2 + y^2 = 9 \end{cases}$

73. $\begin{cases} 3x^2 + 4xy + 5y^2 = 8 \\ x^2 + 3xy + 2y^2 = 0 \end{cases}$

74. $\begin{cases} 3x^2 + 2xy - 2y^2 = 6 \\ xy - 2y^2 + 4 = 0 \end{cases}$

75. $\begin{cases} x^2 - 3x + y^2 + y = -2 \\ \dfrac{x^2 - x}{y} + y + 1 = 0 \end{cases}$

76. $\begin{cases} x^2 + x + y^2 = y + 2 \\ x + 1 = \dfrac{2-y}{x} \end{cases}$

In Problems 77–82, graph each system of inequalities. Tell whether the graph is bounded or unbounded, and label the corner points.

77. $\begin{cases} -2x + y \le 2 \\ x + y \ge 2 \end{cases}$

78. $\begin{cases} x - 2y \le 6 \\ 2x + y \ge 2 \end{cases}$

79. $\begin{cases} x \ge 0 \\ y \ge 0 \\ x + y \le 4 \\ 2x + 3y \le 6 \end{cases}$

80. $\begin{cases} x \ge 0 \\ y \ge 0 \\ 3x + y \ge 6 \\ 2x + y \ge 2 \end{cases}$

81. $\begin{cases} x \ge 0 \\ y \ge 0 \\ 2x + y \le 8 \\ x + 2y \ge 2 \end{cases}$

82. $\begin{cases} x \ge 0 \\ y \ge 0 \\ 3x + y \le 9 \\ 2x + 3y \ge 6 \end{cases}$

In Problems 83–88, solve each linear programming problem.

83. Maximize $z = 3x + 4y$ subject to $x \ge 0$, $y \ge 0$, $3x + 2y \ge 6$, $x + y \le 8$

84. Maximize $z = 2x + 4y$ subject to $x \ge 0$, $y \ge 0$, $x + y \le 6$, $x \ge 2$

85. Minimize $z = 3x + 5y$ subject to $x \ge 0$, $y \ge 0$, $x + y \ge 1$, $3x + 2y \le 12$, $x + 3y \le 12$

86. Minimize $z = 3x + y$ subject to $x \ge 0$, $y \ge 0$, $x \le 8$, $y \le 6$, $2x + y \ge 4$

87. Maximize $z = 5x + 4y$ subject to $x \ge 0$, $y \ge 0$, $x + 2y \ge 2$, $3x + 4y \le 12$, $y \ge x$

88. Maximize $z = 4x + 5y$ subject to $x \ge 0$, $y \ge 0$, $2x + 3y \ge 6$, $x \ge 4$, $2x + y \le 12$

89. Find A such that the system of equations has infinitely many solutions.

$$\begin{cases} 2x + 5y = 5 \\ 4x + 10y = A \end{cases}$$

90. Find A such that the system in Problem 89 is inconsistent.

91. Curve Fitting Find the quadratic function $y = ax^2 + bx + c$ that passes through the three points $(0, 1)$, $(1, 0)$, and $(-2, 1)$.

92. Curve Fitting Find the general equation of the circle that passes through the three points $(0, 1)$, $(1, 0)$, and $(-2, 1)$.

[**Hint:** The general equation of a circle is $x^2 + y^2 + Dx + Ey + F = 0$.]

93. Blending Coffee A coffee distributor is blending a new coffee that will cost \$3.90 per pound. It will consist of a blend of \$3.00 per pound coffee and \$6.00 per pound coffee. What amounts of each type of coffee should be mixed to achieve the desired blend?

[**Hint:** Assume that the weight of the blended coffee is 100 pounds.]

94. Farming A 1000-acre farm in Illinois is used to raise corn and soy beans. The cost per acre for raising corn is \$65 and the cost per acre for soy beans is \$45. If \$54,325 has been budgeted for costs and all the acreage is to be used, how many acres should be allocated for each crop?

95. Cookie Orders A cookie company makes three kinds of cookies, oatmeal raisin, chocolate chip, and shortbread, packaged in small, medium, and large boxes. The small box contains 1 dozen oatmeal raisin and 1 dozen chocolate chip; the medium box has 2 dozen oatmeal raisin, 1 dozen chocolate chip, and 1 dozen shortbread; the large box contains 2 dozen oatmeal raisin, 2 dozen chocolate chip, and 3 dozen shortbread. If you require exactly 15 dozen oatmeal raisin, 10 dozen chocolate chip, and 11 dozen shortbread, how many of each size box should you buy?

96. Mixed Nuts A store that specializes in selling nuts has 72 pounds of cashews and 120 pounds of peanuts available. These are to be mixed in 12-ounce packages as follows: a lower-priced package containing 8 ounces of peanuts and 4 ounces of cashews and a quality package containing 6 ounces of peanuts and 6 ounces of cashews.

(a) Use x to denote the number of lower-priced packages, and use y to denote the number of quality packages. Write a system of linear inequalities that describes the possible number of each kind of package.

(b) Graph the system and label the corner points.

97. A small rectangular lot has a perimeter of 68 feet. If its diagonal is 26 feet, what are the dimensions of the lot?

98. The area of a rectangular window is 4 square feet. If the diagonal measures $2\sqrt{2}$ feet, what are the dimensions of the window?

99. Geometry A certain right triangle has a perimeter of 14 inches. If the hypotenuse is 6 inches long, what are the lengths of the legs?

100. Geometry A certain isosceles triangle has a perimeter of 18 inches. If the altitude is 6 inches, what is the length of the base?

101. Building a Fence How much fence is required to enclose 5000 square feet by two squares whose sides are in the ratio of 1:2?

102. Mixing Acids A chemistry laboratory has three containers of hydrochloric acid, HCl. One container holds a solution with a concentration of 10% HCl, the second holds 25% HCl, and the third holds 40% HCl. How many liters of each should be mixed to obtain 100 liters of a solution with a concentration of 30% HCl? Construct a table showing some of the possible combinations.

103. Calculating Allowances Katy, Mike, Danny, and Colleen agreed to do yard work at home for \$45 to be split among them. After they finished, their father determined that Mike deserves twice what Katy gets, Katy and Colleen deserve the same amount, and Danny deserves half of what Katy gets. How much does each receive?

104. Finding the Speed of the Jet Stream On a flight between Midway Airport in Chicago and Ft. Lauderdale, Florida, a Boeing 737 jet maintains an airspeed of 475 miles per hour. If the trip from Chicago to Ft. Lauderdale takes 2 hours, 30 minutes and the return flight takes 2 hours, 50 minutes, what is the speed of the jet stream? (Assume that the speed of the jet stream remains constant at the various altitudes of the plane and that the plane flies with the jet stream one way and against it the other way.)

105. Constant Rate Jobs If Bruce and Bryce work together for 1 hour and 20 minutes, they will finish a certain job. If Bryce and Marty work together for 1 hour and 36 minutes, the same job can be finished. If Marty and Bruce work together, they can complete this job in 2 hours and 40 minutes. How long will it take each of them working alone to finish the job?

106. Maximizing Profit on Figurines A factory manufactures two kinds of ceramic figurines: a dancing girl and a mermaid. Each requires three processes: molding, painting, and glazing. The daily labor available for molding is no more than 90 work-hours, labor available for painting does not exceed 120 work-hours, and labor available for glazing is no more than 60 work-hours. The dancing girl requires 3 work-hours for molding, 6 work-hours for painting, and 2 work-hours for glazing. The mermaid requires 3 work-hours for molding, 4 work-hours for painting, and 3 work-hours for glazing. If the profit on each figurine is \$25 for dancing girls and \$30 for mermaids, how many of each should be produced each day to maximize profit? If management decides to produce the number of each figurine that maximizes profit, determine which of these processes has excess work-hours assigned to it.

107. Minimizing Production Cost A factory produces gasoline engines and diesel engines. Each week the factory is obligated to deliver at least 20 gasoline engines and at least 15 diesel engines. Due to physical limitations, however, the factory cannot make more than 60 gasoline engines nor more than 40 diesel engines. Finally, to prevent layoffs, a total of at least 50 engines must be produced. If gasoline engines cost \$450 each to produce and diesel engines cost \$550 each to produce, how many of each should be produced per week to minimize the cost? What is the excess capacity of the factory; that is, how many of each kind of engine are being produced in excess of the number that the factory is obligated to deliver?

108. Describe four ways of solving a system of three linear equations containing three variables. Which method do you prefer? Why?

PROJECT AT MOTOROLA

Error Control Coding

Digital cell phones communicate using the language of computers, with voice and data sent as groups of information bits (zeros and ones) over the air. Wireless communication is very difficult—the signals are reflected off buildings, absorbed by trees and rain, and scattered by street signs before being received by the antenna on the phone. Because of these impairments, it is difficult to receive all of the information bits correctly—some will have invariably flipped from 0 to 1 or vice versa. On a voice call, the effect is digital distortion, which may make the call sound like it is coming from underwater, for example. On a data call, the effect changes the data present in the transferred files. One of the primary reasons digital phones are so powerful is that Motorola phone designers can use *error control coding* to ensure that your phone call is clear and error free.

In practice, error control coding adds extra bits (called *parity*) to the information bits so that errors can be corrected. The process of *encoding* the information by adding these extra bits and *decoding* to detect and correct any errors is called **error control coding**. Let's see how using matrix algebra to perform encoding and decoding makes your phones calls clear and error free.

1. A set of information bits plus parity bits forms a *codeword*, and the set of all possible codewords forms a *code*. Let us encode 4 information bits with 3 parity bits to form 7 bit codewords (such as 0000 000, 0001 110, etc). This is known as a (7,4) code. How many codewords are in a (7,4) code? (**Hint**: the number of codewords only depends on the number of information bits.)
2. Matrix algebra provides a succinct representation of a code by describing an encoder for the code. The *generator matrix* G encodes an information vector u into codeword v by performing the matrix multiplication $v = uG$. Use the matrix G below and vector-matrix multiplication to list all the codewords of this (7,4) code. Note that the algebra must be performed mod 2 so that the codewords are composed of zeros and ones (i.e., vector 1001 is encoded to codeword 1001 001, not 1001 221).

$$G = \begin{bmatrix} 1 & 0 & 0 & 0 & 1 & 1 & 1 \\ 0 & 1 & 0 & 0 & 0 & 1 & 1 \\ 0 & 0 & 1 & 0 & 1 & 0 & 1 \\ 0 & 0 & 0 & 1 & 1 & 1 & 0 \end{bmatrix}.$$

3. A code that can be generated with matrix multiplication is a linear code. A linear code always contains the all zero vector (e.g., 0000 000), and has the property that the sum (mod 2) of any two codewords equals another codeword. Pick two non-zero codewords of the code and show that the sum is another codeword.
4. The generator matrix G has a corresponding *parity check* matrix H. The parity check matrix has the property that for any codeword v, $vH = 0$. Explain why the following H satisfies this property. (Hint: note that the codeword $v = uG$ for some information vector u, and perform the matrix multiplication GH).

$$H = \begin{bmatrix} 1 & 1 & 1 \\ 0 & 1 & 1 \\ 1 & 0 & 1 \\ 1 & 1 & 0 \\ 1 & 0 & 0 \\ 0 & 1 & 0 \\ 0 & 0 & 1 \end{bmatrix}.$$

5. The code described by G and H is used to correct single bit errors—for example, if the codeword 0000 000 is sent and the word 1000 000 is received, the decoder will correct the error and produce 0000 000. The decoder multiplies the received matrix r with the parity check matrix H to determine an error pattern, which is added to the received matrix to produce the original codeword. Use matrix multiplication and the following table to produce the codeword associated with the received matrix $r = [0101\ 000]$. (**Hint**: Your answer should appear in the codeword list from question 2.)

rH	error pattern
[000]	0000 000
[001]	0000 001
[010]	0000 010
[011]	0100 000
[100]	0000 100
[101]	0010 000
[110]	0001 000
[111]	1000 000

CHAPTER

12 Sequences; Induction; The Binomial Theorem

Outline

FIELD TRIP TO MOTOROLA

The world around us is filled with examples of vibration and impact. Speakers and microphones in a cellular phone, guitars, undersea acoustic imaging, athletes doing a high jump, airplanes landing, and earthquakes are all examples of vibration behavior. To solve and understand these problems one must be able to work with complex numbers and series. These skills enable you to relate physical phenomena to mathematical models.

INTERVIEW AT MOTOROLA

Andrew Skipor is a Principal Staff Engineer with the Design Reliability Group in Motorola Lab's, Motorola Advanced Technology Center. He has been with Motorola over 16 years. His interests include development of methods to evaluate package reliability, numerical simulation, experimental methods such as Moiré interferometry, nonlinear solder mechanics, fatigue, and PCB/HDI mechanics. Andrew has five patents, an engineering award, and has authored numerous internal and external publications. He is the Motorola Chairman of the Motorola-IEEE/CPMT Graduate Student Fellowship for Research in Electronic Packaging. He received his Ph.D. in engineering mechanics from the University of Illinois at Chicago, supported in part by the Motorola Distinguished Scholar Program. Throughout his career at Motorola, he has used advanced topics in mathematics that directly depend on a sound foundation in algebra, trigonometry, and analytic geometry.

PREPARING FOR THIS SECTION

Before getting started, review the following concepts:

- ✓ Evaluating Functions (p. 102)
- ✓ Compound Interest (Section 5.6)
- ✓ Integer Exponents (Appendix, Section 2)

12.1 SEQUENCES

OBJECTIVES

1. Write the First Several Terms of a Sequence
2. Write the Terms of a Sequence Defined by a Recursion Formula
3. Write a Sequence in Summation Notation
4. Find the Sum of a Sequence by Hand and by Using a Graphing Utility
5. Solve Annuity and Amortization Problems

A **sequence** is a function whose domain is the set of positive integers.

Because a sequence is a function, it will have a graph. In Figure 1(a), you will recognize the graph of the function $f(x) = 1/x$, $x > 0$. If all the points on this graph were removed except those whose x-coordinates are positive integers, that is, if all points were removed except $(1, 1)$, $(2, \frac{1}{2})$, $(3, \frac{1}{3})$, and so on, the remaining points would be the graph of the sequence $f(n) = 1/n$, as shown in Figure 1(b).

Figure 1

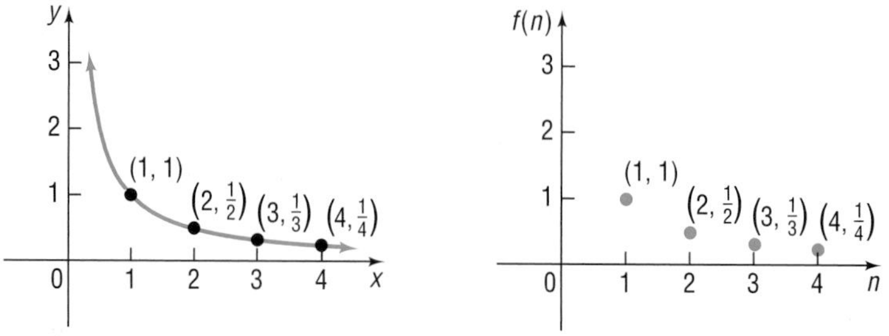

(a) $f(x) = \frac{1}{x}$, $x > 0$ **(b)** $f(n) = \frac{1}{n}$

A sequence is usually represented by listing its values in order. For example, the sequence whose graph is given in Figure 1(b) might be represented as

$$f(1), f(2), f(3), f(4), \ldots \quad \text{or} \quad 1, \frac{1}{2}, \frac{1}{3}, \frac{1}{4}, \ldots$$

The list never ends, as the ellipsis dots indicate. The numbers in this ordered list are called the **terms** of the sequence.

1 In dealing with sequences, we usually use subscripted letters, such as a_1, to represent the first term, a_2 for the second term, a_3 for the third term, and so on. For the sequence $f(n) = 1/n$, we write

$$a_1 = f(1) = 1, \quad a_2 = f(2) = \frac{1}{2}, \quad a_3 = f(3) = \frac{1}{3}, \quad a_4 = f(4) = \frac{1}{4}, \ldots, \quad a_n = f(n) = \frac{1}{n}, \ldots$$

In other words, we usually do not use the traditional function notation $f(n)$ for sequences. For this particular sequence, we have a rule for the nth term, which is $a_n = 1/n$, so it is easy to find any term of the sequence.

When a formula for the nth term of a sequence is known, rather than write out the terms of the sequence, we usually represent the entire sequence by placing braces around the formula for the nth term. For example, the sequence whose nth term is $b_n = \left(\frac{1}{2}\right)^n$ may be represented as

$$\{b_n\} = \left\{\left(\frac{1}{2}\right)^n\right\}$$

or by

$$b_1 = \frac{1}{2}, \quad b_2 = \frac{1}{4}, \quad b_3 = \frac{1}{8}, \quad \ldots, \quad b_n = \left(\frac{1}{2}\right)^n, \quad \ldots$$

EXAMPLE 1 Writing the First Several Terms of a Sequence

Write down the first six terms of the following sequence and graph it.

$$\{a_n\} = \left\{\frac{n-1}{n}\right\}$$

Solution The first six terms of the sequence are

$$a_1 = 0, \quad a_2 = \frac{1}{2}, \quad a_3 = \frac{2}{3}, \quad a_4 = \frac{3}{4}, \quad a_5 = \frac{4}{5}, \quad a_6 = \frac{5}{6}$$

See Figure 2 for the graph.

Figure 2

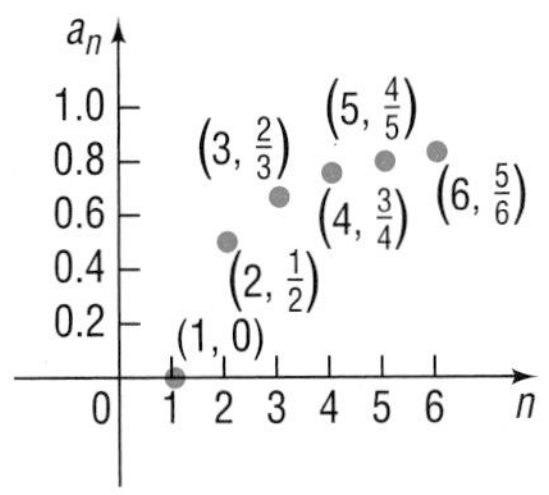

■

Graphing utilities can be used to write the terms of a sequence and graph them, as the following example illustrates.

EXAMPLE 2 Using a Graphing Utility to Write the First Several Terms of a Sequence

Use a graphing utility to write the first six terms of the following sequence and graph it.

$$\{a_n\} = \left\{\frac{n-1}{n}\right\}$$

Solution Figure 3 shows the sequence generated on a TI-83 graphing calculator. We can see the first few terms of the sequence on the screen. You need to press the right arrow key to scroll right in order to see the remaining terms of the sequence.

Figure 3

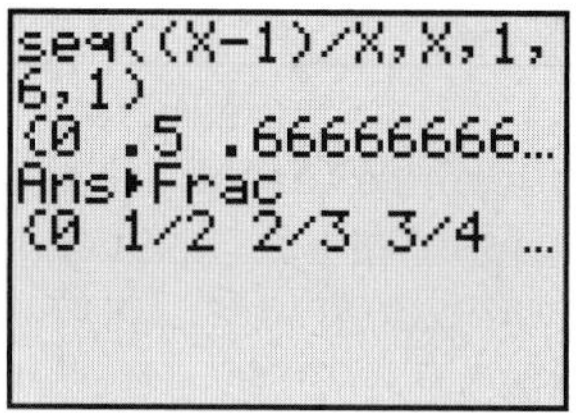

We could also obtain the terms of the sequence using the TABLE feature. First, put the graphing utility in SEQuence mode. Using $Y =$, enter the formula for the sequence into the graphing utility. See Figure 4. Set up the table with TblStart $= 1$ and ΔTbl $= 1$. See Table 1. Finally, we can graph the sequence. See Figure 5. Notice that the first term of the sequence is not visible since it lies on the x-axis. TRACEing the graph will allow you to determine the terms of the sequence.

Figure 4

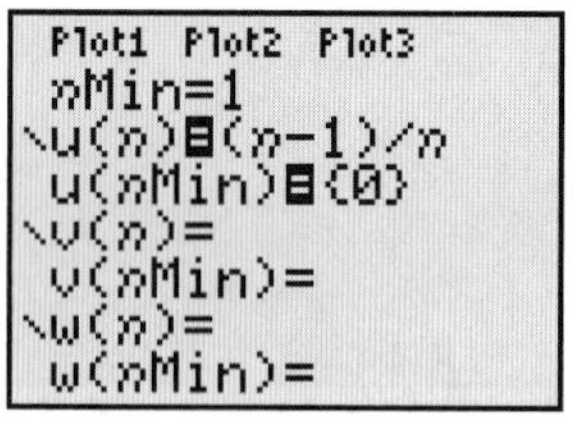

TABLE 1

n	u(n)
1	0
2	.5
3	.66667
4	.75
5	.8
6	.83333
7	.85714

u(n)=(n-1)/n

Figure 5

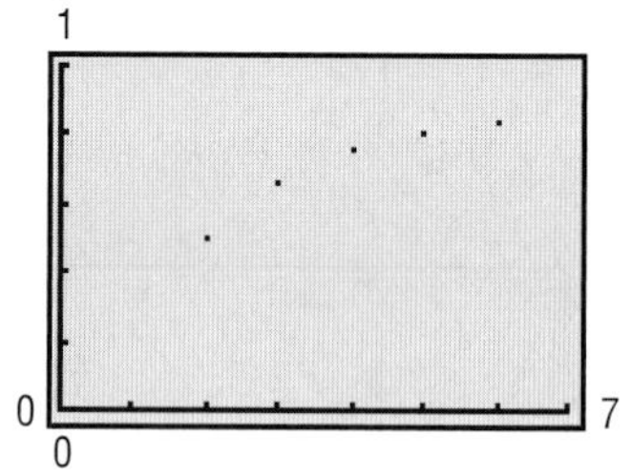

We will usually provide solutions done by hand. The reader is encouraged to verify solutions using a graphing utility.

EXAMPLE 3 Writing the First Several Terms of a Sequence

Write down the first six terms of the following sequence and graph it by hand.

$$\{b_n\} = \left\{(-1)^{n-1}\left(\frac{2}{n}\right)\right\}$$

Solution The first six terms of the sequence are

$$b_1 = 2, \quad b_2 = -1, \quad b_3 = \frac{2}{3}, \quad b_4 = -\frac{1}{2}, \quad b_5 = \frac{2}{5}, \quad b_6 = -\frac{1}{3}$$

See Figure 6 for the graph.

Figure 6

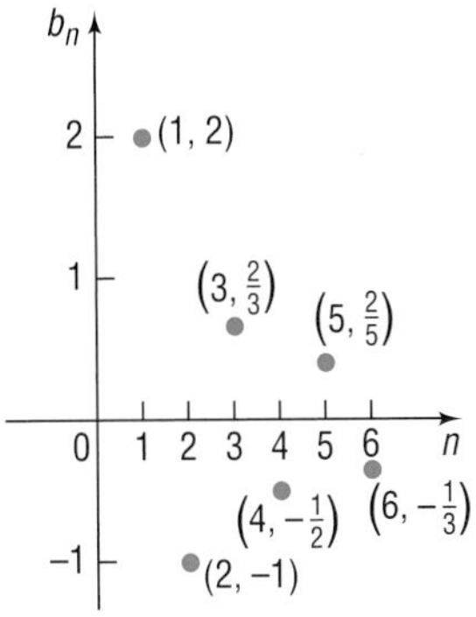

EXAMPLE 4 Writing the First Several Terms of a Sequence

Write down the first six terms of the following sequence and graph it by hand.

$$\{c_n\} = \begin{Bmatrix} n & \text{if } n \text{ is even} \\ 1/n & \text{if } n \text{ is odd} \end{Bmatrix}$$

Solution The first six terms of the sequence are

$$c_1 = 1, \quad c_2 = 2, \quad c_3 = \frac{1}{3}, \quad c_4 = 4, \quad c_5 = \frac{1}{5}, \quad c_6 = 6$$

See Figure 7 for the graph.

Figure 7

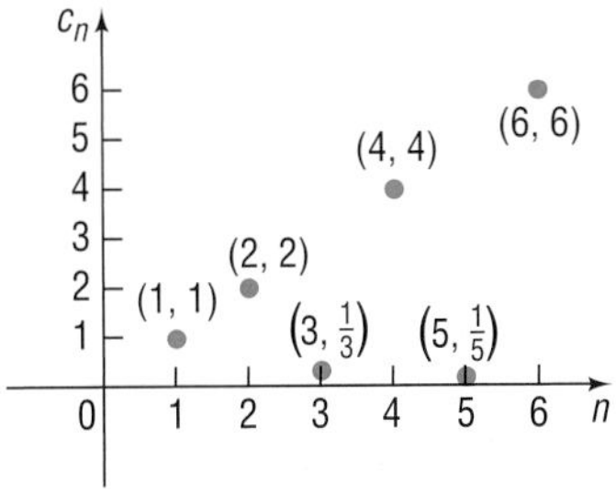

NOW WORK PROBLEMS 3 AND 5.

Sometimes a sequence is indicated by an observed pattern in the first few terms that makes it possible to infer the makeup of the nth term. In the example that follows, a sufficient number of terms of the sequence is given so that a natural choice for the nth term is suggested.

EXAMPLE 5 Determining a Sequence from a Pattern

(a) $e, \frac{e^2}{2}, \frac{e^3}{3}, \frac{e^4}{4}, \ldots \qquad a_n = \frac{e^n}{n}$

(b) $1, \frac{1}{3}, \frac{1}{9}, \frac{1}{27}, \ldots \qquad b_n = \frac{1}{3^{n-1}}$

(c) $1, 3, 5, 7, \dots$ $\quad c_n = 2n - 1$

(d) $1, 4, 9, 16, 25, \dots$ $\quad d_n = n^2$

(e) $1, -\frac{1}{2}, \frac{1}{3}, -\frac{1}{4}, \frac{1}{5}, \dots$ $\quad e_n = (-1)^{n+1}\left(\frac{1}{n}\right)$ ■

Notice in the sequence $\{e_n\}$ in Example 5(e) that the signs of the terms **alternate.** When this occurs, we use factors such as $(-1)^{n+1}$, which equals 1 if n is odd and -1 if n is even, or $(-1)^n$, which equals -1 if n is odd and 1 if n is even.

NOW WORK PROBLEM 13.

The Factorial Symbol

If $n \geq 0$ is an integer, the **factorial symbol** $\boldsymbol{n!}$ is defined as follows:

$$0! = 1 \qquad 1! = 1$$
$$n! = n(n-1) \cdot \ldots \cdot 3 \cdot 2 \cdot 1 \qquad \text{if } n \geq 2$$

For example, $2! = 2 \cdot 1 = 2$, $3! = 3 \cdot 2 \cdot 1 = 6$, $4! = 4 \cdot 3 \cdot 2 \cdot 1 = 24$, and so on. Table 2 lists the values of $n!$ for $0 \leq n \leq 6$.

TABLE 2

n	0	1	2	3	4	5	6
$n!$	1	1	2	6	24	120	720

Because

$$n! = n\underbrace{(n-1)(n-2) \cdot \ldots \cdot 3 \cdot 2 \cdot 1}_{(n-1)!}$$

we can use the formula

$$n! = n(n-1)!$$

to find successive factorials. For example, because $6! = 720$, we have

$$7! = 7 \cdot 6! = 7(720) = 5040$$

and

$$8! = 8 \cdot 7! = 8(5040) = 40{,}320$$

COMMENT: Your calculator has a factorial key. Use it to see how fast factorials increase in value. Find the value of 69!. What happens when you try to find 70!? In fact, 70! is larger than 10^{100} (a *googol*), the largest number most calculators can display. ■

Recursion Formulas

2 A second way of defining a sequence is to assign a value to the first (or the first few) terms and specify the nth term by a formula or equation that involves one or more of the terms preceding it. Sequences defined this way

are said to be defined **recursively,** and the rule or formula is called a **recursive formula.**

EXAMPLE 6 Writing the Terms of a Recursively Defined Sequence

Write down the first five terms of the following recursively defined sequence.

$$s_1 = 1, \qquad s_n = 4s_{n-1}$$

Solution The first term is given as $s_1 = 1$. To get the second term, we use $n = 2$ in the formula to get $s_2 = 4s_1 = 4 \cdot 1 = 4$. To get the third term, we use $n = 3$ in the formula to get $s_3 = 4s_2 = 4 \cdot 4 = 16$. To get a new term requires that we know the value of the preceding term. The first five terms are

$$s_1 = 1$$
$$s_2 = 4 \cdot 1 = 4$$
$$s_3 = 4 \cdot 4 = 16$$
$$s_4 = 4 \cdot 16 = 64$$
$$s_5 = 4 \cdot 64 = 256$$

Graphing utilities can be used to generate recursively defined sequences.

EXAMPLE 7 Using a Graphing Utility to Write the Terms of a Recursively Defined Sequence

Use a graphing utility to write down the first five terms of the following recursively defined sequence.

$$s_1 = 1, \qquad s_n = 4s_{n-1}$$

Solution First, put the graphing utility into SEQuence mode. Using $Y=$, enter the recursive formula into the graphing utility. See Figure 8(a). Next, set up the viewing window to generate the desired sequence. Finally, graph the recursion relation and use TRACE to determine the terms in the sequence. See Figure 8(b). For example, we see that the fourth term of the sequence is 64. Table 3 also shows the terms of the sequence.

Figure 8

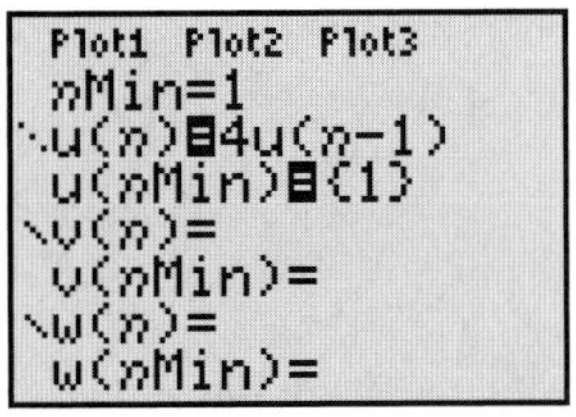

(a)

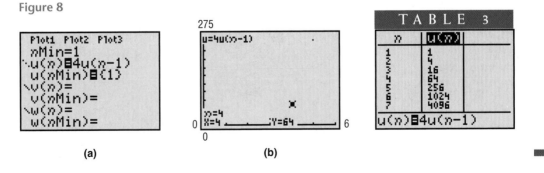

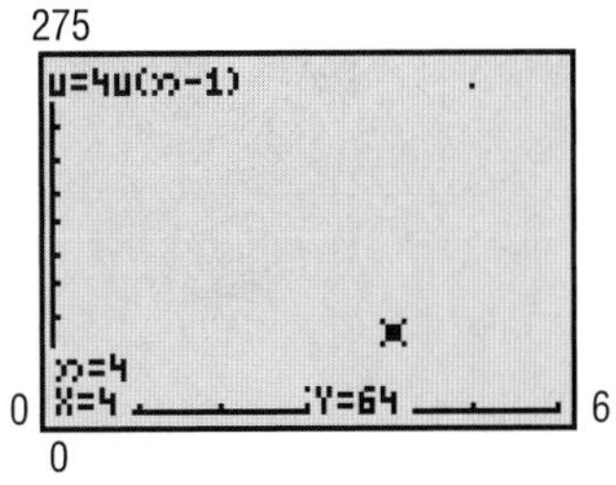

(b)

TABLE 3

n	u(n)
1	1
2	4
3	16
4	64
5	256
6	1024
7	4096

u(n)■4u(n-1)

EXAMPLE 8 Writing the Terms of a Recursively Defined Sequence

Write down the first five terms of the following recursively defined sequence.

$$u_1 = 1, \qquad u_2 = 1, \qquad u_{n+2} = u_n + u_{n+1}$$

Solution We are given the first two terms. To get the third term requires that we know each of the previous two terms. Thus,

$$\begin{aligned} u_1 &= 1 \\ u_2 &= 1 \\ u_3 &= u_1 + u_2 = 1 + 1 = 2 \\ u_4 &= u_2 + u_3 = 1 + 2 = 3 \\ u_5 &= u_3 + u_4 = 2 + 3 = 5 \end{aligned}$$

The sequence defined in Example 8 is called a **Fibonacci sequence,** and the terms of this sequence are called **Fibonacci numbers.** These numbers appear in a wide variety of applications (see Problems 77–80).

NOW WORK PROBLEMS 21 AND 29.

Summation Notation

3 It is often important to be able to find the sum of the first n terms of a sequence $\{a_n\}$, that is,

$$a_1 + a_2 + a_3 + \cdots + a_n \tag{1}$$

Rather than write down all these terms, we introduce a more concise way to express the sum, called **summation notation.** Using summation notation, we would write the sum (1) as

$$a_1 + a_2 + a_3 + \cdots + a_n = \sum_{k=1}^{n} a_k$$

The symbol Σ (a stylized version of the Greek letter sigma, which is an S in our alphabet) is simply an instruction to sum, or add up, the terms. The integer k is called the **index** of the sum; it tells you where to start the sum and where to end it. Therefore, the expression

$$\sum_{k=1}^{n} a_k \tag{2}$$

is an instruction to add the terms a_k of the sequence $\{a_n\}$ from $k = 1$ through $k = n$. We read expression (2) as "the sum of a_k from $k = 1$ to $k = n$."

EXAMPLE 9 Expanding Summation Notation

Write out each sum.

(a) $\sum_{k=1}^{n} \frac{1}{k}$ (b) $\sum_{k=1}^{n} k!$

Solution (a) $\sum_{k=1}^{n} \frac{1}{k} = 1 + \frac{1}{2} + \frac{1}{3} + \cdots + \frac{1}{n}$ (b) $\sum_{k=1}^{n} k! = 1! + 2! + \cdots + n!$

EXAMPLE 10 Writing a Sum in Summation Notation

Express each sum using summation notation.

(a) $1^2 + 2^2 + 3^2 + \cdots + 9^2$ (b) $1 + \frac{1}{2} + \frac{1}{4} + \frac{1}{8} + \cdots + \frac{1}{2^{n-1}}$

Solution (a) The sum $1^2 + 2^2 + 3^2 + \cdots + 9^2$ has 9 terms, each of the form k^2, and starts at $k = 1$ and ends at $k = 9$:

$$1^2 + 2^2 + 3^2 + \cdots + 9^2 = \sum_{k=1}^{9} k^2$$

(b) The sum

$$1 + \frac{1}{2} + \frac{1}{4} + \frac{1}{8} + \cdots + \frac{1}{2^{n-1}}$$

has n terms, each of the form $1/2^{k-1}$, and starts at $k = 1$ and ends at $k = n$:

$$1 + \frac{1}{2} + \frac{1}{4} + \frac{1}{8} + \cdots + \frac{1}{2^{n-1}} = \sum_{k=1}^{n} \frac{1}{2^{k-1}}$$

■

The index of summation need not always begin at 1 or end at n; for example, we could have expressed the sum in Example 10(b) as

$$\sum_{k=0}^{n-1} \frac{1}{2^k} = 1 + \frac{1}{2} + \frac{1}{4} + \cdots + \frac{1}{2^{n-1}}$$

Letters other than k may be used as the index. For example,

$$\sum_{j=1}^{n} j! \quad \text{and} \quad \sum_{i=1}^{n} i!$$

each represent the same sum as the one given in Example 9(b).

NOW WORK PROBLEMS 49 AND 59.

ADDING THE FIRST n TERMS OF A SEQUENCE

4 Next, we list some properties of sequences using summation notation. These properties are useful for adding the terms of a sequence.

Theorem

Properties of Sequences

If $\{a_n\}$ and $\{b_n\}$ are two sequences and c is a real number, then:

1. $\displaystyle\sum_{k=1}^{n} c = c \cdot n$
2. $\displaystyle\sum_{k=1}^{n} ca_k = c\sum_{k=1}^{n} a_k$
3. $\displaystyle\sum_{k=1}^{n} (a_k + b_k) = \sum_{k=1}^{n} a_k + \sum_{k=1}^{n} b_k$
4. $\displaystyle\sum_{k=1}^{n} (a_k - b_k) = \sum_{k=1}^{n} a_k - \sum_{k=1}^{n} b_k$
5. $\displaystyle\sum_{k=1}^{n} a_k = \sum_{k=1}^{j} a_k + \sum_{k=j+1}^{n} a_k$, when $1 < j < n$
6. $\displaystyle\sum_{k=1}^{n} k = 1 + 2 + 3 + \cdots + n = \frac{n(n+1)}{2}$

7. $$\sum_{k=1}^{n} k^2 = 1^2 + 2^2 + 3^2 + \cdots + n^2 = \frac{n(n+1)(2n+1)}{6}$$

8. $$\sum_{k=1}^{n} k^3 = 1^3 + 2^3 + 3^3 + \cdots + n^3 = \left(\frac{n(n+1)}{2}\right)^2$$

We shall not prove these properties. The proofs of 1 through 5 are based on properties of real numbers; the proofs of 6 through 8 require mathematical induction, which is discussed in Section 12.4.

EXAMPLE 11 Finding the Sum of a Sequence

Find the sum of each sequence.

(a) $\sum_{k=1}^{5} 3k$ (b) $\sum_{k=1}^{3} (k^3 + 1)$ (c) $\sum_{k=1}^{4} (k^2 - 7k + 2)$

Solution

(a) $$\sum_{k=1}^{5} 3k = 3\sum_{k=1}^{5} k = 3\left(\frac{5(5+1)}{2}\right) = 3(15) = 45$$

Property 2 Property 6

(b) $$\sum_{k=1}^{3} (k^3 + 1) = \sum_{k=1}^{3} k^3 + \sum_{k=1}^{3} 1$$ Property 3

$$= \left(\frac{3(3+1)}{2}\right)^2 + 1(3)$$ Properties 1, 8

$$= 36 + 3$$

$$= 39$$

(c) $$\sum_{k=1}^{4} (k^2 - 7k + 2) = \sum_{k=1}^{4} k^2 - \sum_{k=1}^{4} 7k + \sum_{k=1}^{4} 2$$ Properties 3, 4

$$= \sum_{k=1}^{4} k^2 - 7\sum_{k=1}^{4} k + \sum_{k=1}^{4} 2$$ Property 2

$$= \frac{4(4+1)(2 \cdot 4 + 1)}{6} - 7\left(\frac{4(4+1)}{2}\right) + 2(4)$$ Properties 1, 6, 7

$$= 30 - 70 + 8$$

$$= -32$$

EXAMPLE 12 Using a Graphing Utility to Find the Sum of a Sequence

Using a graphing utility, find the sum of each sequence.

(a) $\sum_{k=1}^{5} 3k$ (b) $\sum_{k=1}^{3} (k^3 + 1)$ (c) $\sum_{k=1}^{4} (k^2 - 7k + 2)$

Solution (a) Figure 9 shows the solution using a TI-83 graphing calculator.

$$\sum_{k=1}^{5} 3k = 45$$

(b) Figure 10 shows the solution using a TI-83 graphing calculator.

$$\sum_{k=1}^{3} (k^3 + 1) = 39$$

(c) Figure 11 shows the solution using a TI-83 graphing calculator.

$$\sum_{k=1}^{4} (k^2 - 7k + 2) = -32$$

Figure 9

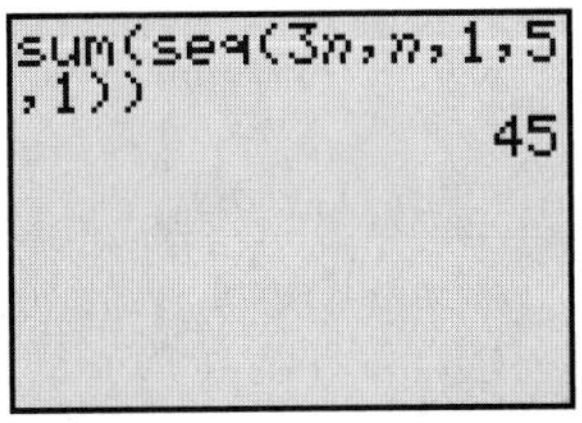

Figure 10

Figure 11

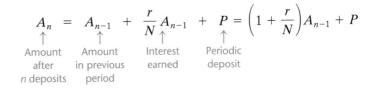

NOW WORK PROBLEM 39.

ANNUITIES; AMORTIZATION

5 In Section 5.6 we developed the compound interest formula, which gives the future value when a fixed amount of money is deposited in an account that pays interest compounded periodically. Often, though, money is invested in small amounts at periodic intervals. An **annuity** is a sequence of equal periodic deposits. The periodic deposits may be made annually, quarterly, monthly, or daily.

When deposits are made at the same time that the interest is credited, the annuity is called **ordinary.** We will only deal with ordinary annuities here. The **amount of an annuity** is the sum of all deposits made plus all interest paid.

Suppose that the initial amount deposited in an annuity is $\$M$, the periodic deposit is $\$P$, and the per annum rate of interest is $r\%$ (expressed as a decimal) compounded N times per year. The periodic deposit is made at the same time that the interest is credited, so N deposits are made per year. The amount A_n of the annuity after n deposits will equal the amount of the annuity after $n - 1$ deposits, A_{n-1}, plus the interest earned on this amount plus the periodic deposit P. That is,

$$\underbrace{A_n}_{\text{Amount after } n \text{ deposits}} = \underbrace{A_{n-1}}_{\text{Amount in previous period}} + \underbrace{\frac{r}{N}A_{n-1}}_{\text{Interest earned}} + \underbrace{P}_{\text{Periodic deposit}} = \left(1 + \frac{r}{N}\right)A_{n-1} + P$$

We have established the following result:

Theorem

Annuity Formula

If $A_0 = M$ represents the initial amount deposited in an annuity that earns $r\%$ per annum compounded N times per year, and if P is the periodic deposit made at each payment period, then the amount A_n of the annuity after n deposits is given by the recursive sequence

$$A_0 = M, \qquad A_n = \left(1 + \frac{r}{N}\right)A_{n-1} + P, \qquad n \geq 1 \qquad \textbf{(3)}$$

Formula (3) may be explained as follows: the money in the account initially, A_0, is $\$M$; the money in the account after $n - 1$ payments $\left(A_{n-1}\right)$ earns interest r/N during the nth period; so when the periodic payment of P dollars is added, the amount after n payments, A_n, is obtained.

EXAMPLE 13 Saving for Spring Break

A trip to Cancun during spring break will cost \$450 and full payment is due March 2. To have the money, a student, on September 1, deposits \$100 in a savings account that pays 4% per annum compounded monthly. On the first of each month, the student deposits \$50 in this account.

(a) Find a recursive sequence that explains how much is in the account after n months.

(b) Use the TABLE feature to list the amounts of the annuity for the first 6 months.

(c) After the deposit on March 1 is made, is there enough in the account to pay for the Cancun trip?

(d) If the student deposits \$60 each month, will there be enough for the trip?

Solution

TABLE 4

n	u(n)
0	100
1	150.33
2	200.83
3	251.5
4	302.34
5	353.35
6	404.53

u(n)■(1+.04/12)...

TABLE 5

n	u(n)
0	100
1	160.33
2	220.87
3	281.6
4	342.54
5	403.68
6	465.03

u(n)■(1+.04/12)...

(a) The initial amount deposited in the account is $A_0 = \$100$. The monthly deposit is $P = \$50$, and the per annum rate of interest is $r = 0.04$ compounded $N = 12$ times per year. The amount A_n in the account after n monthly deposits is given by the recursive sequence

$$A_0 = 100, \qquad A_n = \left(1 + \frac{r}{N}\right)A_{n-1} + P = \left(1 + \frac{0.04}{12}\right)A_{n-1} + 50$$

(b) In SEQuence mode on a TI-83, enter the sequence $\{A_n\}$ and create Table 4. On September 1 ($n = 0$), there is \$100 in the account. After the first payment on October 1, the value of the account is \$150.33. After the second payment on November 1, the value of the account is \$200.83. After the third payment on December 1, the value of the account is \$251.50, and so on.

(c) On March 1 ($n = 6$), there is only 404.53, not enough to pay for the trip to Cancun.

(d) If the periodic deposit, P, is \$60, then on March 1, there is 465.03 in the account, enough for the trip. See Table 5.

AMORTIZATION

Recursive sequences can also be used to compute information about loans. When equal periodic payments are made to pay off a loan, the loan is said to be **amortized**.

Theorem

Amortization Formula

If $\$B$ is borrowed at an interest rate of $r\%$ (expressed as a decimal) per annum compounded monthly, the balance A_n due after n monthly payments of $\$P$ is given by the recursive sequence

$$A_0 = B, \qquad A_n = \left(1 + \frac{r}{12}\right)A_{n-1} - P, \qquad n \geq 1 \tag{4}$$

Formula (4) may be explained as follows: The initial loan balance is $\$B$. The balance due A_n after n payments will equal the balance due previously, A_{n-1}, plus the interest charged on that amount reduced by the periodic payment P.

EXAMPLE 14 Mortgage Payments

John and Wanda borrowed $180,000 at 7% per annum compounded monthly for 30 years to purchase a home. Their monthly payment is determined to be $1197.54.

(a) Find a recursive formula that represents their balance after each payment of $1197.54 has been made.

(b) Determine their balance after the first payment is made.

(c) When will their balance be below $170,000?

Solution (a) We use formula (4) with $A_0 = 180{,}000$, $r = 0.07$, and $P = \$1197.54$. Then

$$A_0 = 180{,}000 \quad A_n = \left(1 + \frac{0.07}{12}\right)A_{n-1} - 1197.54$$

(b) In SEQuence mode on a TI-83, enter the sequence $\{A_n\}$ and create Table 6. After the first payment is made, the balance is $A_1 = \$179{,}852$.

(c) Scroll down until the balance is below $170,000. See Table 7. After the fifty-eighth payment is made ($n = 58$), the balance is below $170,000.

TABLE 6

n	u(n)
0	180000
1	179852
2	179704
3	179555
4	179405
5	179254
6	179102

u(n)■(1+.07/12)...

TABLE 7

n	u(n)
52	171067
53	170868
54	170667
55	170465
56	170262
57	170057
58	169852

u(n)■(1+.07/12)...

12.1 EXERCISES

In Problems 1–12, write down the first five terms of each sequence by hand. Verify your results using a graphing utility.

1. $\{n\}$

2. $\{n^2 + 1\}$

3. $\left\{\dfrac{n}{n+2}\right\}$

4. $\left\{\dfrac{2n+1}{2n}\right\}$

5. $\{(-1)^{n+1}n^2\}$

6. $\left\{(-1)^{n-1}\left(\dfrac{n}{2n-1}\right)\right\}$

7. $\left\{\dfrac{2^n}{3^n+1}\right\}$

8. $\left\{\left(\dfrac{4}{3}\right)^n\right\}$

9. $\left\{\dfrac{(-1)^n}{(n+1)(n+2)}\right\}$ **10.** $\left\{\dfrac{3^n}{n}\right\}$ **11.** $\left\{\dfrac{n}{e^n}\right\}$ **12.** $\left\{\dfrac{n^2}{2^n}\right\}$

In Problems 13–20, the given pattern continues. Write down the nth term of each sequence suggested by the pattern.

13. $\dfrac{1}{2}, \dfrac{2}{3}, \dfrac{3}{4}, \dfrac{4}{5}, \dots$

14. $\dfrac{1}{1 \cdot 2}, \dfrac{1}{2 \cdot 3}, \dfrac{1}{3 \cdot 4}, \dfrac{1}{4 \cdot 5}, \dots$

15. $1, \dfrac{1}{2}, \dfrac{1}{4}, \dfrac{1}{8}, \dots$

16. $\dfrac{2}{3}, \dfrac{4}{9}, \dfrac{8}{27}, \dfrac{16}{81}, \dots$

17. $1, -1, 1, -1, 1, -1, \dots$

18. $1, \dfrac{1}{2}, 3, \dfrac{1}{4}, 5, \dfrac{1}{6}, 7, \dfrac{1}{8}, \dots$

19. $1, -2, 3, -4, 5, -6, \dots$

20. $2, -4, 6, -8, 10, \dots$

In Problems 21–34, a sequence is defined recursively. Write the first five terms by hand. Use a graphing utility to verify your results.

21. $a_1 = 2; \quad a_n = 3 + a_{n-1}$

22. $a_1 = 3; \quad a_n = 4 - a_{n-1}$

23. $a_1 = -2; \quad a_n = n + a_{n-1}$

24. $a_1 = 1; \quad a_n = n - a_{n-1}$

25. $a_1 = 5; \quad a_n = 2a_{n-1}$

26. $a_1 = 2; \quad a_n = -a_{n-1}$

27. $a_1 = 3; \quad a_n = \dfrac{a_{n-1}}{n}$

28. $a_1 = -2; \quad a_n = n + 3a_{n-1}$

29. $a_1 = 1; \quad a_2 = 2; \quad a_n = a_{n-1} \cdot a_{n-2}$

30. $a_1 = -1; \quad a_2 = 1; \quad a_n = a_{n-2} + na_{n-1}$

31. $a_1 = A; \quad a_n = a_{n-1} + d$

32. $a_1 = A; \quad a_n = ra_{n-1}, \quad r \neq 0$

33. $a_1 = \sqrt{2}; \quad a_n = \sqrt{2 + a_{n-1}}$

34. $a_1 = \sqrt{2}; \quad a_n = \sqrt{a_{n-1}/2}$

In Problems 35–46, find the sum of each sequence. Verify your results using a graphing utility.

35. $\sum_{k=1}^{10} 5$ **36.** $\sum_{k=1}^{20} 8$ **37.** $\sum_{k=1}^{6} k$ **38.** $\sum_{k=1}^{4} (-k)$

39. $\sum_{k=1}^{5} (5k + 3)$ **40.** $\sum_{k=1}^{6} (3k - 7)$ **41.** $\sum_{k=1}^{3} (k^2 + 4)$ **42.** $\sum_{k=0}^{4} (k^2 - 4)$

43. $\sum_{k=1}^{6} (-1)^k 2^k$ **44.** $\sum_{k=1}^{4} (-1)^k 3^k$ **45.** $\sum_{k=1}^{4} (k^3 - 1)$ **46.** $\sum_{k=0}^{3} (k^3 + 2)$

In Problems 47–56, write out each sum.

47. $\sum_{k=1}^{n} (k + 2)$ **48.** $\sum_{k=1}^{n} (2k + 1)$ **49.** $\sum_{k=1}^{n} \dfrac{k^2}{2}$ **50.** $\sum_{k=1}^{n} (k + 1)^2$

51. $\sum_{k=0}^{n} \dfrac{1}{3^k}$ **52.** $\sum_{k=0}^{n} \left(\dfrac{3}{2}\right)^k$ **53.** $\sum_{k=0}^{n-1} \dfrac{1}{3^{k+1}}$ **54.** $\sum_{k=0}^{n-1} (2k + 1)$

55. $\sum_{k=2}^{n} (-1)^k \ln k$ **56.** $\sum_{k=3}^{n} (-1)^{k+1} 2^k$

In Problems 57–66, express each sum using summation notation.

57. $1 + 2 + 3 + \cdots + 20$

58. $1^3 + 2^3 + 3^3 + \cdots + 8^3$

59. $\dfrac{1}{2} + \dfrac{2}{3} + \dfrac{3}{4} + \cdots + \dfrac{13}{13 + 1}$

60. $1 + 3 + 5 + 7 + \cdots + [2(12) - 1]$

61. $1 - \dfrac{1}{3} + \dfrac{1}{9} - \dfrac{1}{27} + \cdots + (-1)^6\left(\dfrac{1}{3^6}\right)$

62. $\dfrac{2}{3} - \dfrac{4}{9} + \dfrac{8}{27} - \cdots + (-1)^{11+1}\left(\dfrac{2}{3}\right)^{11}$

63. $3 + \dfrac{3^2}{2} + \dfrac{3^3}{3} + \cdots + \dfrac{3^n}{n}$

64. $\dfrac{1}{e} + \dfrac{2}{e^2} + \dfrac{3}{e^3} + \cdots + \dfrac{n}{e^n}$

65. $a + (a + d) + (a + 2d) + \cdots + (a + nd)$

66. $a + ar + ar^2 + \cdots + ar^{n-1}$

67. **Credit Card Debt** John has a balance of $3000 on his Discover card that charges 1% interest per month on any unpaid balance. John can afford to pay $100 toward the balance each month. His balance each month after making a $100 payment is given by the recursively defined sequence

$$B_0 = \$3000, \qquad B_n = 1.01B_{n-1} - 100$$

(a) Determine John's balance after making the first payment. That is, determine B_1.
(b) Using a graphing utility, determine when John's balance will be below $2000. How many payments of $100 have been made?
(c) Using a graphing utility, determine when John will pay off the balance. What is the total of all the payments?
(d) What was John's interest expense?

68. **Car Loans** Phil bought a car by taking out a loan for $18,500 at 0.5% interest per month. Phil's normal monthly payment is $434.47 per month, but he decides that he can afford to pay $100 extra toward the balance each month. His balance each month is given by the recursively defined sequence

$$B_0 = \$18{,}500, \qquad B_n = 1.005B_{n-1} - 534.47$$

(a) Determine Phil's balance after making the first payment. That is, determine B_1.
(b) Using a graphing utility, determine when Phil's balance will be below $10,000. How many payments of $534.47 have been made?
(c) Using a graphing utility, determine when Phil will pay off the balance. What is the total of all the payments?
(d) What was Phil's interest expense?

69. **Trout Population** A pond currently has 2000 trout in it. A fish hatchery decides to add an additional 20 trout each month. In addition, it is known that the trout population is growing 3% per year. The size of the population after n months is given by the recursively defined sequence

$$p_0 = 2000, \qquad p_n = 1.03p_{n-1} + 20$$

(a) How many trout are in the pond at the beginning of the second month? That is, what is p_2?
(b) Using a graphing utility, determine how long it will be before the trout population reaches 5000.

70. **Environmental Control** The Environmental Protection Agency (EPA) determines that Maple Lake has 250 tons of pollutants as a result of industrial waste and that 10% of the pollutant present is neutralized by solar oxidation every year. The EPA imposes new pollution control laws that result in 15 tons of new pollutant entering the lake each year. The amount of pollutant in the lake at the beginning of each year is given by the recursively defined sequence

$$p_0 = 250, \qquad p_n = 0.9p_{n-1} + 15$$

(a) Determine the amount of pollutant in the lake at the beginning of the second year. That is, determine p_2.
(b) Using a graphing utility, provide pollutant amounts for the next 20 years.
(c) What is the equilibrium level of pollution in Maple Lake? That is, what is $\lim_{n\to\infty} p_n$?

71. **Retirement** Christine currently has $5000 in her 401(k) and plans to contribute $100 each month for the next 30 years into it. What will be the value of Christine's 401(k) in 30 years if the per annum rate of return is assumed to be 12% compounded monthly?

72. **Saving for a Home** Jolene wants to purchase a new home. She currently has $2000 in her mutual fund and plans to invest $400 per month into it. If the per annum rate of return of the mutual fund is assumed to be 10% compounded monthly, how much will Jolene have for a down payment after 3 years?

73. **Roth IRA** On January 1, 1999, Bob decides to place $500 at the end of each quarter into a Roth Individual Retirement Account.
(a) Find a recursive formula that represents Bob's balance at the end of each quarter if the rate of return is assumed to be 8% per annum compounded quarterly.
(b) How long will it be before the value of the account exceeds $100,000?
(c) What will be the value of the account in 25 years when Bob retires?

74. **Education IRA** On January 1, 1999, John's parents decide to place $45 at the end of each month into an Education IRA.
(a) Find a recursive formula that represents the balance at the end of each month if the rate of return is assumed to be 6% per annum compounded monthly.
(b) How long will it be before the value of the account exceeds $4000?
(c) What will be the value of the account in 16 years when John goes to college?

75. **Home Loan** Bill and Laura borrowed $150,000 at 6% per annum compounded monthly for 30 years to purchase a home. Their monthly payment is determined to be $899.33.
(a) Find a recursive formula for their balance after each monthly payment has been made.
(b) Determine Bill and Laura's balance after the first payment.
(c) Using a graphing utility, create a table showing Bill and Laura's balance after each monthly payment.
(d) Using a graphing utility, determine when Bill and Laura's balance will be below $140,000.
(e) Using a graphing utility, determine when Bill and Laura will pay off the balance.
(f) Determine Bill and Laura's interest expense when the loan is paid.
(g) Suppose that Bill and Laura decide to pay an additional $100 each month on their loan. Answer parts (a) to (f) under this scenario.
(h) Is it worthwhile for Bill and Laura to pay the additional $100? Explain.

76. Home Loan Jodi and Jeff borrowed \$120,000 at 6.5% per annum compounded monthly for 30 years to purchase a home. Their monthly payment is determined to be \$758.48.
(a) Find a recursive formula for their balance after each monthly payment has been made.
(b) Determine Jodi and Jeff's balance after the first payment.
(c) Using a graphing utility, create a table showing Jodi and Jeff's balance after each monthly payment.
(d) Using a graphing utility, determine when Jodi and Jeff's balance will be below \$100,000.
(e) Using a graphing utility, determine when Jodi and Jeff will pay off the balance.
(f) Determine Jodi and Jeff's interest expense when the loan is paid.
(g) Suppose that Jodi and Jeff decide to pay an additional \$100 each month on their loan. Answer parts (a) to (f) under this scenario.
(h) Is it worthwhile for Jodi and Jeff to pay the additional \$100? Explain.

77. Growth of a Rabbit Colony A colony of rabbits begins with one pair of mature rabbits, which will produce a pair of offspring (one male, one female) each month. Assume that all rabbits mature in 1 month and produce a pair of offspring (one male, one female) after 2 months. If no rabbits ever die, how many pairs of mature rabbits are there after 7 months?
[**Hint:** A Fibonacci sequence models this colony. Do you see why?]

78. Fibonacci Sequence Let

$$u_n = \frac{(1 + \sqrt{5})^n - (1 - \sqrt{5})^n}{2^n\sqrt{5}}$$

define the nth term of a sequence.
(a) Show that $u_1 = 1$ and $u_2 = 1$.
(b) Show that $u_{n+2} = u_{n+1} + u_n$.
(c) Draw the conclusion that $\{u_n\}$ is a Fibonacci sequence.

79. Pascal's Triangle Divide the triangular array shown (called Pascal's triangle) using diagonal lines as indicated. Find the sum of the numbers in each of these diagonal rows. Do you recognize this sequence?

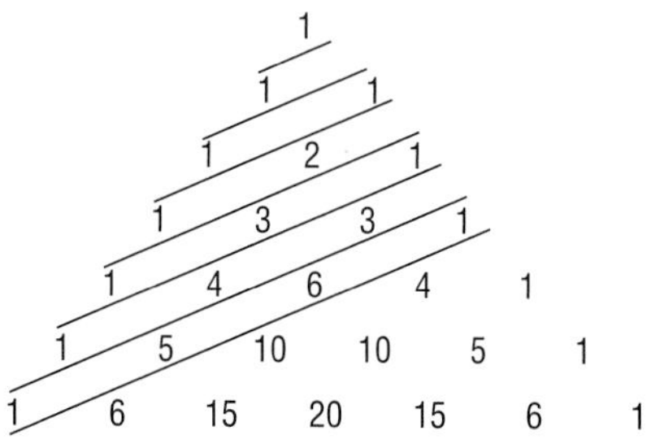

80. Fibonacci Sequence Use the result of Problem 78 to do the following problems:
(a) Write the first 10 terms of the Fibonacci sequence.
(b) Compute the ratio $\frac{u_{n+1}}{u_n}$ for the first 10 terms.
(c) As n gets large, what number does the ratio approach? This number is referred to as the **golden ratio.** Rectangles whose sides are in this ratio were considered pleasing to the eye by the Greeks. For example, the facade of the Parthenon was constructed using the golden ratio.
(d) Compute the ratio $\frac{u_n}{u_{n+1}}$ for the first 10 terms.
(e) As n gets large, what number does the ratio approach? This number is also referred to as the **golden ratio.** This ratio is believed to have been used in the construction of the Great Pyramid in Egypt. The ratio equals the sum of the areas of the four face triangles divided by the total surface area of the Great Pyramid.

81. Investigate various applications that lead to a Fibonacci sequence, such as art, architecture, or financial markets. Write an essay on these applications.

12.2 ARITHMETIC SEQUENCES

OBJECTIVES
1. Determine If a Sequence Is Arithmetic
2. Find a Formula for an Arithmetic Sequence
3. Find the Sum of an Arithmetic Sequence

1 When the difference between successive terms of a sequence is always the same number, the sequence is called **arithmetic.** Thus, an **arithmetic sequence*** may be defined recursively as $a_1 = a, a_n - a_{n-1} = d$, or as

$$a_1 = a, \qquad a_n = a_{n-1} + d \tag{1}$$

where $a = a_1$ and d are real numbers. The number a is the first term, and the number d is called the **common difference.**

The terms of an arithmetic sequence with first term a and common difference d follow the pattern

$$a, \quad a + d, \quad a + 2d, \quad a + 3d, \ldots$$

EXAMPLE 1 Determining If a Sequence Is Arithmetic

The sequence

$$4, 7, 10, 13, \ldots$$

is arithmetic since the difference of successive terms is 3. The first term is 4, and the common difference is 3. ■

EXAMPLE 2 Determining If a Sequence Is Arithmetic

Show that the following sequence is arithmetic. Find the first term and the common difference.

$$\{s_n\} = \{3n + 5\}$$

Solution The first term is $s_1 = 3 \cdot 1 + 5 = 8$. The nth and $(n - 1)$st terms of the sequence $\{s_n\}$ are

$$s_n = 3n + 5 \quad \text{and} \quad s_{n-1} = 3(n - 1) + 5 = 3n + 2$$

Their difference is

$$s_n - s_{n-1} = (3n + 5) - (3n + 2) = 5 - 2 = 3$$

Since the difference of two successive terms does not depend on n, the common difference is 3 and the sequence is arithmetic. ■

EXAMPLE 3 Determining If a Sequence Is Arithmetic

Show that the sequence $\{t_n\} = \{4 - n\}$ is arithmetic. Find the first term and the common difference.

*Sometimes called an **arithmetic progression.**

Solution The first term is $t_1 = 4 - 1 = 3$. The nth and $(n-1)$st terms are

$$t_n = 4 - n \quad \text{and} \quad t_{n-1} = 4 - (n-1) = 5 - n$$

Their difference is

$$t_n - t_{n-1} = (4-n) - (5-n) = 4 - 5 = -1$$

The difference of two successive terms does not depend on n; it always equals the same number, -1. Hence, $\{t_n\}$ is an arithmetic sequence whose common difference is -1. ■

NOW WORK PROBLEM 3.

2 Suppose that a is the first term of an arithmetic sequence whose common difference is d. We seek a formula for the nth term, a_n. To see the pattern, we write down the first few terms:

$$a_1 = a$$

$$a_2 = a_1 + d = a + 1 \cdot d$$

$$a_3 = a_2 + d = (a + d) + d = a + 2 \cdot d$$

$$a_4 = a_3 + d = (a + 2 \cdot d) + d = a + 3 \cdot d$$

$$a_5 = a_4 + d = (a + 3 \cdot d) + d = a + 4 \cdot d$$

$$\vdots$$

$$a_n = a_{n-1} + d = [a + (n-2)d] + d = a + (n-1)d$$

We are led to the following result:

Theorem

nth Term of an Arithmetic Sequence

For an arithmetic sequence $\{a_n\}$ whose first term is a and whose common difference is d, the nth term is determined by the formula

$$a_n = a + (n-1)d \qquad \textbf{(2)}$$

■

EXAMPLE 4 Finding a Particular Term of an Arithmetic Sequence

Find the thirteenth term of the arithmetic sequence: 2, 6, 10, 14, 18, ...

Solution The first term of this arithmetic sequence is $a = 2$, and the common difference is 4. By formula (2), the nth term is

$$a_n = 2 + (n-1)4$$

Hence, the thirteenth term is

$$a_{13} = 2 + 12 \cdot 4 = 50$$ ■

EXPLORATION Use a graphing utility to find the thirteenth term of the sequence given in Example 4. Use it to find the twentieth and fiftieth terms. ■

EXAMPLE 5 Finding a Recursive Formula for an Arithmetic Sequence

The eighth term of an arithmetic sequence is 75, and the twentieth term is 39. Find the first term and the common difference. Give a recursive formula for the sequence.

Solution By formula (2), we know that $a_n = a + (n - 1)d$. As a result,

$$\begin{cases} a_8 = a + 7d = 75 \\ a_{20} = a + 19d = 39 \end{cases}$$

This is a system of two linear equations containing two variables, a and d, which we can solve by elimination. Subtracting the second equation from the first equation, we get

$$-12d = 36$$

$$d = -3$$

With $d = -3$, we find that $a = 75 - 7d = 75 - 7(-3) = 96$. The first term is $a = 96$, and the common difference is $d = -3$. A recursive formula for this sequence is found using formula (1).

$$a_1 = 96, \qquad a_n = a_{n-1} - 3$$ ■

Based on formula (2), a formula for the nth term of the sequence $\{a_n\}$ in Example 5 is

$$a_n = a + (n - 1)d = 96 + (n - 1)(-3) = 99 - 3n$$

NOW WORK PROBLEMS 19 AND 25.

EXPLORATION Graph the recursive formula from Example 5, $a_1 = 96$, $a_n = a_{n-1} - 3$, using a graphing utility. Conclude that the graph of the recursive formula behaves like the graph of a linear function. How is d, the common difference, related to m, the slope of a line? ■

ADDING THE FIRST n TERMS OF AN ARITHMETIC SEQUENCE

3 The next result gives a formula for finding the sum of the first n terms of an arithmetic sequence.

Theorem **Sum of n Terms of an Arithmetic Sequence**

Let $\{a_n\}$ be an arithmetic sequence with first term a and common difference d. The sum S_n of the first n terms of $\{a_n\}$ is

$$S_n = \frac{n}{2}[2a + (n - 1)d] = \frac{n}{2}(a + a_n) \qquad \textbf{(3)}$$

Proof

$$
\begin{aligned}
S_n &= a_1 + a_2 + a_3 + \cdots + a_n && \text{Sum of first n terms}\\
&= a + (a + d) + (a + 2d) + \cdots + [a + (n-1)d] && \text{Formula (2)}\\
&= \underbrace{(a + a + \cdots + a)}_{n \text{ terms}} + [d + 2d + \cdots + (n-1)d] && \text{Rearrange terms}\\
&= na + d[1 + 2 + \cdots + (n-1)]\\
&= na + d\left[\frac{(n-1)n}{2}\right] && \text{Property 6, Section 12.1}\\
&= na + \frac{n}{2}(n-1)d\\
&= \frac{n}{2}[2a + (n-1)d] && \text{Factor out n/2} \quad (4)\\
&= \frac{n}{2}[a + a + (n-1)d]\\
&= \frac{n}{2}(a + a_n) && \text{Formula (2)} \quad (5)
\end{aligned}
$$

■

Formula (3) provides two ways to find the sum of the first n terms of an arithmetic sequence. Notice that (4) involves the first term and common difference, while (5) involves the first term and the nth term. Use whichever form is easier.

EXAMPLE 6 Finding the Sum of n Terms of an Arithmetic Sequence

Find the sum S_n of the first n terms of the sequence $\{3n + 5\}$; that is, find

$$8 + 11 + 14 + \cdots + (3n + 5)$$

Solution The sequence $\{3n + 5\}$ is an arithmetic sequence with first term $a = 8$ and the nth term $(3n + 5)$. To find the sum S_n, we use formula (3), as given in (5).

$$S_n = \frac{n}{2}(a + a_n) = \frac{n}{2}[8 + (3n + 5)] = \frac{n}{2}(3n + 13)$$

■

NOW WORK PROBLEM 33.

EXAMPLE 7 Using a Graphing Utility to Find the Sum of 20 Terms of an Arithmetic Sequence

Use a graphing utility to find the sum S_n of the first 20 terms of the sequence $\{9.5n + 2.6\}$.

Solution Figure 12 shows the results obtained using a TI-83 graphing calculator.

Figure 12

The sum of the first 20 terms of the sequence $\{9.5n + 2.6\}$ is 2047. ■

NOW WORK PROBLEM 41.

EXAMPLE 8 Creating a Floor Design

A ceramic tile floor is designed in the shape of a trapezoid 20 feet wide at the base and 10 feet wide at the top. See Figure 13. The tiles, 12 inches by 12 inches, are to be placed so that each successive row contains one less tile than the preceding row. How many tiles will be required?

Figure 13

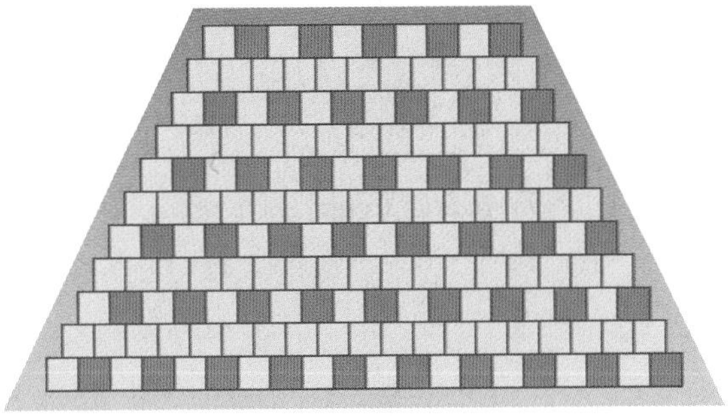

Solution The bottom row requires 20 tiles and the top row, 10 tiles. Since each successive row requires one less tile, the total number of tiles required is

$$S = 20 + 19 + 18 + \cdots + 11 + 10$$

This is the sum of an arithmetic sequence; the common difference is -1. The number of terms to be added is $n = 11$, with the first term $a = 20$ and the last term $a_{11} = 10$. The sum S is

$$S = \frac{n}{2}\left(a + a_{11}\right) = \frac{11}{2}(20 + 10) = 165$$

Thus, 165 tiles will be required. ■

12.2 EXERCISES

In Problems 1–10, an arithmetic sequence is given. Find the common difference and write out the first four terms.

1. $\{n + 4\}$ **2.** $\{n - 5\}$ 3. $\{2n - 5\}$ **4.** $\{3n + 1\}$ **5.** $\{6 - 2n\}$

6. $\{4 - 2n\}$ **7.** $\left\{\frac{1}{2} - \frac{1}{3}n\right\}$ **8.** $\left\{\frac{2}{3} + \frac{n}{4}\right\}$ **9.** $\{\ln 3^n\}$ **10.** $\{e^{\ln n}\}$

In Problems 11–18, find the nth term of the arithmetic sequence whose initial term a and common difference d are given. What is the fifth term?

11. $a = 2;\ d = 3$ **12.** $a = -2;\ d = 4$ **13.** $a = 5;\ d = -3$ **14.** $a = 6;\ d = -2$

15. $a = 0;\ d = \frac{1}{2}$ **16.** $a = 1;\ d = -\frac{1}{3}$ **17.** $a = \sqrt{2};\ d = \sqrt{2}$ **18.** $a = 0;\ d = \pi$

In Problems 19–24, find the indicated term in each arithmetic sequence.

19. 12th term of $2, 4, 6, \ldots$ **20.** 8th term of $-1, 1, 3, \ldots$

21. 10th term of $1, -2, -5, \ldots$ **22.** 9th term of $5, 0, -5, \ldots$

23. 8th term of $a, a + b, a + 2b, \ldots$ **24.** 7th term of $2\sqrt{5}, 4\sqrt{5}, 6\sqrt{5}, \ldots$

In Problems 25–32, find the first term and the common difference of the arithmetic sequence described. Give a recursive formula for the sequence.

25. 8th term is 8; 20th term is 44 **26.** 4th term is 3; 20th term is 35

27. 9th term is -5; 15th term is 31 **28.** 8th term is 4; 18th term is -96

29. 15th term is 0; 40th term is -50

30. 5th term is -2; 13th term is 30

31. 14th term is -1; 18th term is -9

32. 12th term is 4; 18th term is 28

In Problems 33–40, find the sum.

33. $1 + 3 + 5 + \cdots + (2n - 1)$

34. $2 + 4 + 6 + \cdots + 2n$

35. $7 + 12 + 17 + \cdots + (2 + 5n)$

36. $-1 + 3 + 7 + \cdots + (4n - 5)$

37. $2 + 4 + 6 + \cdots + 70$

38. $1 + 3 + 5 + \cdots + 59$

39. $5 + 9 + 13 + \cdots + 49$

40. $2 + 5 + 8 + \cdots + 41$

For Problems 41–46, use a graphing utility to find the sum of each sequence.

41. $\{3.45n + 4.12\}, \quad n = 20$

42. $\{2.67n - 1.23\}, \quad n = 25$

43. $2.8 + 5.2 + 7.6 + \cdots + 36.4$

44. $5.4 + 7.3 + 9.2 + \cdots + 32$

45. $4.9 + 7.48 + 10.06 + \cdots + 66.82$

46. $3.71 + 6.9 + 10.09 + \cdots + 80.27$

47. Find x so that $x + 3$, $2x + 1$, and $5x + 2$ are consecutive terms of an arithmetic sequence.

48. Find x so that $2x$, $3x + 2$, and $5x + 3$ are consecutive terms of an arithmetic sequence.

49. Drury Lane Theater The Drury Lane Theater has 25 seats in the first row and 30 rows in all. Each successive row contains one additional seat. How many seats are in the theater?

50. Football Stadium The corner section of a football stadium has 15 seats in the first row and 40 rows in all. Each successive row contains two additional seats. How many seats are in this section?

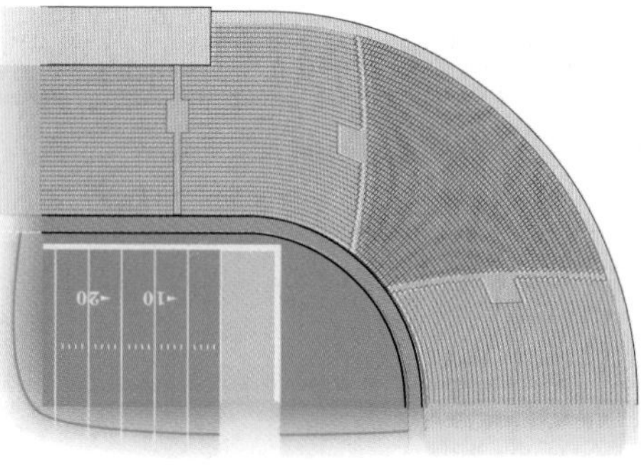

51. Creating a Mosaic A mosaic is designed in the shape of an equilateral triangle, 20 feet on each side. Each tile in the mosaic is in the shape of an equilateral triangle, 12 inches to a side. The tiles are to alternate in color as shown in the illustration. How many tiles of each color will be required?

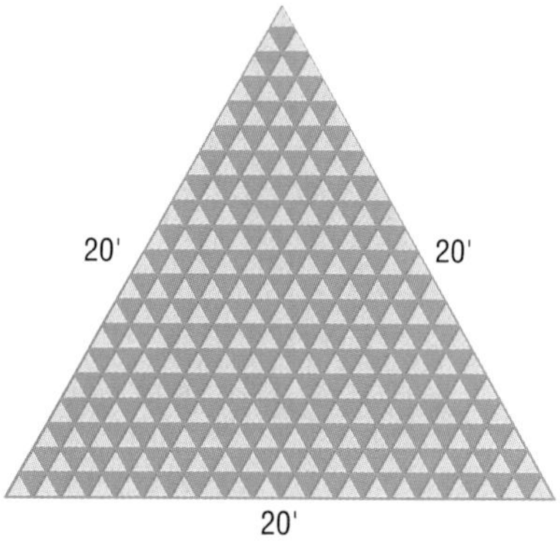

52. Constructing a Brick Staircase A brick staircase has a total of 30 steps. The bottom step requires 100 bricks. Each successive step requires two less bricks than the prior step.
(a) How many bricks are required for the top step?
(b) How many bricks are required to build the staircase?

53. Stadium Construction How many rows are in the corner section of a stadium containing 2040 seats if the first row has 10 seats and each successive row has 4 additional seats?

54. Salary Suppose that you just received a job offer with a starting salary of \$35,000 per year and a guaranteed raise of \$1400 per year. How many years will it take before your aggregate salary is \$280,000?
[**Hint:** Your aggregate salary after two years is \$35,000 + (\$35,000 + \$1400).]

55. Make up an arithmetic sequence. Give it to a friend and ask for its twentieth term.

12.3 GEOMETRIC SEQUENCES; GEOMETRIC SERIES

OBJECTIVES
1. Determine If a Sequence Is Geometric
2. Find a Formula for a Geometric Sequence
3. Find the Sum of a Geometric Sequence
4. Find the Sum of a Geometric Series

1 When the ratio of successive terms of a sequence is always the same nonzero number, the sequence is called **geometric.** A **geometric sequence*** may be defined recursively as $a_1 = a$, $a_n/a_{n-1} = r$, or as

$$a_1 = a, \qquad a_n = ra_{n-1} \tag{1}$$

where $a_1 = a$ and $r \neq 0$ are real numbers. The number a is the first term, and the nonzero number r is called the **common ratio.**

The terms of a geometric sequence with first term a and common ratio r follow the pattern

$$a, \quad ar, \quad ar^2, \quad ar^3, \ldots$$

EXAMPLE 1 Determining If a Sequence Is Geometric

The sequence

$$2, 6, 18, 54, 162, \ldots$$

is geometric since the ratio of successive terms is 3 $\left(\frac{6}{2} = \frac{18}{6} = \cdots = 3\right)$. The first term is 2, and the common ratio is 3. ■

EXAMPLE 2 Determining If a Sequence Is Geometric

Show that the following sequence is geometric. Find the first term and the common ratio.

$$\{s_n\} = 2^{-n}$$

Solution The first term is $s_1 = 2^{-1} = \frac{1}{2}$. The nth and $(n-1)$st terms of the sequence $\{s_n\}$ are

$$s_n = 2^{-n} \quad \text{and} \quad s_{n-1} = 2^{-(n-1)}$$

Their ratio is

$$\frac{s_n}{s_{n-1}} = \frac{2^{-n}}{2^{-(n-1)}} = 2^{-n+(n-1)} = 2^{-1} = \frac{1}{2}$$

Because the ratio of successive terms is a nonzero number independent of n, the sequence $\{s_n\}$ is geometric with common ratio $\frac{1}{2}$. ■

EXAMPLE 3 Determining If a Sequence Is Geometric

Show that the following sequence is geometric. Find the first term and the common ratio.

$$\{t_n\} = \{4^n\}$$

*Sometimes called a **geometric progression.**

Solution The first term is $t_1 = 4^1 = 4$. The nth and $(n - 1)$st terms are

$$t_n = 4^n \quad \text{and} \quad t_{n-1} = 4^{n-1}$$

Their ratio is

$$\frac{t_n}{t_{n-1}} = \frac{4^n}{4^{n-1}} = 4^{n-(n-1)} = 4$$

Thus, $\{t_n\}$ is a geometric sequence with common ratio 4. ■

NOW WORK PROBLEM 3.

2 Suppose that a is the first term of a geometric sequence with common ratio $r \neq 0$. We seek a formula for the nth term a_n. To see the pattern, we write down the first few terms:

$$\begin{aligned} a_1 &= 1a = ar^0 \\ a_2 &= ra_1 = ar^1 \\ a_3 &= ra_2 = r(ar) = ar^2 \\ a_4 &= ra_3 = r(ar^2) = ar^3 \\ a_5 &= ra_4 = r(ar^3) = ar^4 \\ &\vdots \\ a_n &= ra_{n-1} = r(ar^{n-2}) = ar^{n-1} \end{aligned}$$

We are led to the following result:

Theorem

nth Term of a Geometric Sequence

For a geometric sequence $\{a_n\}$ whose first term is a and whose common ratio is r, the nth term is determined by the formula

$$a_n = ar^{n-1}, \qquad r \neq 0 \qquad \textbf{(2)}$$

■

EXAMPLE 4 Finding a Particular Term of a Geometric Sequence

(a) Find the ninth term of the geometric sequence: $10,\ 9,\ \frac{81}{10},\ \frac{729}{100} \ldots$

(b) Find a recursive formula for this sequence.

Solution (a) The first term of this geometric sequence is $a = 10$ and the common ratio is $9/10$. (Use $9/10$, or $\frac{81/10}{9} = \frac{9}{10}$, or any two successive terms.) By formula (2), the nth term is

$$a_n = 10\left(\frac{9}{10}\right)^{n-1}$$

Hence, the ninth term is

$$a_9 = 10\left(\frac{9}{10}\right)^{9-1} = 10\left(\frac{9}{10}\right)^8 = 4.3046721$$

(b) The first term in the sequence is 10 and the common ratio is $r = 9/10$. Using formula (1), the recursive formula is $a_1 = 10$, $a_n = \frac{9}{10}a_{n-1}$. ■

EXPLORATION Use a graphing utility to find the ninth term of the sequence given in Example 4. Use it to find the twentieth and fiftieth terms. Now use a graphing utility to graph the recursive formula found in Example 4(b). Conclude that the graph of the recursive formula behaves like the graph of an exponential function. How is r, the common ratio, related to a, the base of the exponential function $y = a^x$? ■

NOW WORK PROBLEMS 25 AND 33.

ADDING THE FIRST n TERMS OF A GEOMETRIC SEQUENCE

3 The next result gives us a formula for finding the sum of the first n terms of a geometric sequence.

Theorem **Sum of n Terms of a Geometric Sequence**

Let $\{a_n\}$ be a geometric sequence with first term a and common ratio r where $r \neq 0, r \neq 1$. The sum S_n of the first n terms of $\{a_n\}$ is

$$S_n = a\frac{1 - r^n}{1 - r}, \qquad r \neq 0, 1 \tag{3}$$

■

Proof The sum S_n of the first n terms of $\{a_n\} = \{ar^{n-1}\}$ is

$$S_n = a + ar + \cdots + ar^{n-1} \tag{4}$$

Multiply each side by r to obtain

$$rS_n = ar + ar^2 + \cdots + ar^n \tag{5}$$

Now, subtract (5) from (4). The result is

$$S_n - rS_n = a - ar^n$$
$$(1 - r)S_n = a(1 - r^n)$$

Since $r \neq 1$, we can solve for S_n.

$$S_n = a\frac{1 - r^n}{1 - r}$$ ■

EXAMPLE 5 **Finding the Sum of n Terms of a Geometric Sequence**

Find the sum S_n of the first n terms of the sequence $\left\{\left(\frac{1}{2}\right)^n\right\}$; that is, find

$$\frac{1}{2} + \frac{1}{4} + \frac{1}{8} + \cdots + \left(\frac{1}{2}\right)^n$$

Solution The sequence $\left\{\left(\frac{1}{2}\right)^n\right\}$ is a geometric sequence with $a = \frac{1}{2}$ and $r = \frac{1}{2}$. The sum S_n that we seek is the sum of the first n terms of the sequence, so we use formula (3) to get

$$S_n = \sum_{k=1}^{n} \left(\frac{1}{2}\right)^k = \frac{1}{2} + \frac{1}{4} + \frac{1}{8} + \cdots + \left(\frac{1}{2}\right)^n$$

$$= \frac{1}{2}\left[\frac{1 - \left(\frac{1}{2}\right)^n}{1 - \frac{1}{2}}\right] \quad \text{Formula (3)}$$

$$= \frac{1}{2}\left[\frac{1 - \left(\frac{1}{2}\right)^n}{\frac{1}{2}}\right]$$

$$= 1 - \left(\frac{1}{2}\right)^n$$

NOW WORK PROBLEM 39.

EXAMPLE 6 Using a Graphing Utility to Find the Sum of a Geometric Sequence

Use a graphing utility to find the sum S_n of the first 15 terms of the sequence $\left\{\left(\frac{1}{3}\right)^n\right\}$; that is, find

$$\frac{1}{3} + \frac{1}{9} + \frac{1}{27} + \cdots + \left(\frac{1}{3}\right)^{15}$$

Solution Figure 14 shows the result obtained using a TI-83 graphing calculator.

Figure 14

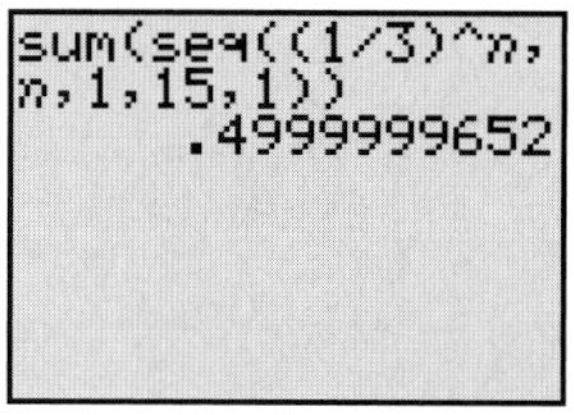

The sum of the first 15 terms of the sequence $\left\{\left(\frac{1}{3}\right)^n\right\}$ is 0.4999999652.

NOW WORK PROBLEM 45.

Geometric Series

An infinite sum of the form

$$a + ar + ar^2 + \cdots + ar^{n-1} + \cdots$$

with first term a and common ratio r, is called an **infinite geometric series** and is denoted by

$$\sum_{k=1}^{\infty} ar^{k-1}$$

4 Based on formula (3), the sum S_n of the first n terms of a geometric series is

$$S_n = a\frac{1 - r^n}{1 - r} = \frac{a}{1 - r} - \frac{ar^n}{1 - r} \tag{6}$$

If this finite sum S_n approaches a number L as $n \to \infty$, then we call L the **sum of the infinite geometric series,** and we write

$$L = \sum_{k=1}^{\infty} ar^{k-1}$$

Theorem

Sum of an Infinite Geometric Series

If $|r| < 1$, the sum of the infinite geometric series $\sum_{k=1}^{\infty} ar^{k-1}$ is

$$\sum_{k=1}^{\infty} ar^{k-1} = \frac{a}{1 - r} \tag{7}$$

■

Intuitive Proof Since $|r| < 1$, it follows that $|r^n|$ approaches 0 as $n \to \infty$. Then, based on formula (6), the sum S_n approaches $a/(1 - r)$ as $n \to \infty$. ■

EXAMPLE 7 Finding the Sum of a Geometric Series

Find the sum of the geometric series: $2 + \frac{4}{3} + \frac{8}{9} + \cdots$

Solution The first term is $a = 2$, and the common ratio is

$$r = \frac{\frac{4}{3}}{2} = \frac{4}{6} = \frac{2}{3}$$

Since $|r| < 1$, we use formula (7) to find that

$$2 + \frac{4}{3} + \frac{8}{9} + \cdots = \frac{2}{1 - \frac{2}{3}} = 6$$

■

NOW WORK PROBLEM 51.

Exploration Use a graphing utility to graph $U_n = 2\left(\frac{2}{3}\right)^{n-1}$ in sequence mode. TRACE the graph for large values of n. What happens to the value of U_n as n increases without bound? What can you conclude about $\sum_{n=1}^{\infty} 2\left(\frac{2}{3}\right)^{(n-1)}$? ■

EXAMPLE 8 Repeating Decimals

Show that the repeating decimal 0.999... equals 1.

Solution

$$0.999\ldots = \frac{9}{10} + \frac{9}{100} + \frac{9}{1000} + \cdots$$

Thus, 0.999... is a geometric series with first term $\frac{9}{10}$ and common ratio $\frac{1}{10}$. Hence,

$$0.999\ldots = \frac{\frac{9}{10}}{1 - \frac{1}{10}} = \frac{\frac{9}{10}}{\frac{9}{10}} = 1$$

■

EXAMPLE 9 Pendulum Swings

Figure 15

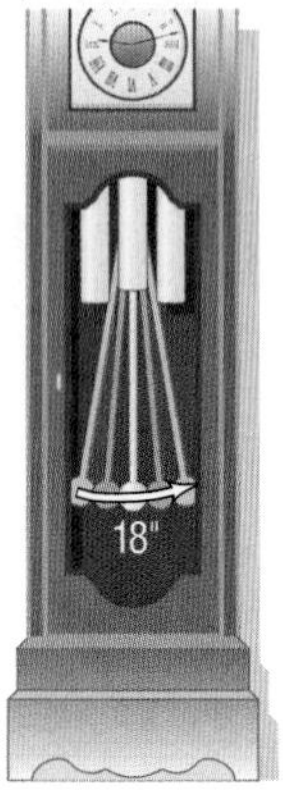

Initially, a pendulum swings through an arc of 18 inches. See Figure 15. On each successive swing, the length of the arc is 0.98 of the previous length.

(a) What is the length of the arc after 10 swings?

(b) On which swing is the length of the arc first less than 12 inches?

(c) After 15 swings, what total distance will the pendulum have swung?

(d) When it stops, what total distance will the pendulum have swung?

Solution (a) The length of the first swing is 18 inches. The length of the second swing is 0.98(18) inches; the length of the third swing is $0.98(0.98)(18) = 0.98^2(18)$ inches. The length of arc of the tenth swing is

$$(0.98)^9(18) = 15.007 \text{ inches}$$

(b) The length of arc of the nth swing is $(0.98)^{n-1}(18)$. For this to be exactly 12 inches requires that

$$(0.98)^{n-1}(18) = 12$$

$$(0.98)^{n-1} = \frac{12}{18} = \frac{2}{3}$$

$$n - 1 = \log_{0.98}\left(\frac{2}{3}\right)$$

$$n = 1 + \frac{\ln\left(\frac{2}{3}\right)}{\ln 0.98} = 1 + 20.07 = 21.07$$

The length of arc of the pendulum exceeds 12 inches on the twenty-first swing and is first less than 12 inches on the twenty-second swing.

(c) After 15 swings, the pendulum will have swung the following total distance L:

$$L = \underset{\text{1st}}{18} + \underset{\text{2nd}}{0.98(18)} + \underset{\text{3rd}}{(0.98)^2(18)} + \underset{\text{4th}}{(0.98)^3(18)} + \cdots + \underset{\text{15th}}{(0.98)^{14}(18)}$$

This is the sum of a geometric sequence. The common ratio is 0.98; the first term is 18. The sum has 15 terms, so

$$L = 18\,\frac{1 - 0.98^{15}}{1 - 0.98} = 18(13.07) = 235.29 \text{ inches}$$

The pendulum will have swung through 235.29 inches after 15 swings.

(d) When the pendulum stops, it will have swung the following total distance T:

$$T = 18 + 0.98(18) + (0.98)^2(18) + (0.98)^3(18) + \cdots$$

This is the sum of a geometric series. The common ratio is $r = 0.98$; the first term is $a = 18$. The sum is

$$T = \frac{a}{1 - r} = \frac{18}{1 - 0.98} = 900$$

The pendulum will have swung a total of 900 inches when it finally stops. ■

HISTORICAL FEATURE

Fibonacci

Sequences are among the oldest objects of mathematical investigation, having been studied for over 3500 years. After the initial steps, however, little progress was made until about 1600.

Arithmetic and geometric sequences appear in the Rhind papyrus, a mathematical text containing 85 problems copied around 1650 BC by the Egyptian scribe Ahmes from an earlier work (see Historical Problem 1). Fibonacci (AD 1220) wrote about problems similar to those found in the Rhind papyrus, leading one to suspect that Fibonacci may have had material available that is now lost. This material would have been in the non-Euclidean Greek tradition of Heron (about AD 75) and Diophantus (about AD 250). One problem, again modified slightly, is still with us in the familiar puzzle rhyme "As I was going to St. Ives..." (see Historical Problem 2).

The Rhind papyrus indicates that the Egyptians knew how to add up the terms of an arithmetic or geometric sequence, as did the Babylonians. The rule for summing up a geometric sequence is found in Euclid's *Elements* (book IX, 35, 36), where, like all Euclid's algebra, it is presented in a geometric form.

Investigations of other kinds of sequences began in the 1500s, when algebra became sufficiently developed to handle the more complicated problems. The development of calculus in the 1600s added a powerful new tool, especially for finding the sum of infinite series, and the subject continues to flourish today.

HISTORICAL PROBLEMS

1. *Arithmetic sequence problem from the Rhind papyrus (statement modified slightly for clarity)* One hundred loaves of bread are to be divided among five people so that the amounts that they receive form an arithmetic sequence. The first two together receive one-seventh of what the last three receive. How many does each receive? [*Partial answer:* First person receives $1\frac{2}{3}$ loaves.]
2. The following old English children's rhyme resembles one of the Rhind papyrus problems.

   As I was going to St. Ives
   I met a man with seven wives
   Each wife had seven sacks
   Each sack had seven cats
   Each cat had seven kits [kittens]
   Kits, cats, sacks, wives
   How many were going to St. Ives?

   (a) Assuming that the speaker and the cat fanciers met by traveling in opposite directions, what is the answer?
   (b) How many kittens are being transported?
   (c) Kits, cats, sacks, wives; how many?

[**Hint:** It is easier to include the man, find the sum with the formula, and then subtract 1 for the man.]

12.3 EXERCISES

In Problems 1–10, a geometric sequence is given. Find the common ratio and write out the first four terms.

1. $\{3^n\}$ **2.** $\{(-5)^n\}$ 3. $\left\{-3\left(\frac{1}{2}\right)^n\right\}$ **4.** $\left\{\left(\frac{5}{2}\right)^n\right\}$ **5.** $\left\{\frac{2^{n-1}}{4}\right\}$

6. $\left\{\frac{3^n}{9}\right\}$ **7.** $\{2^{n/3}\}$ **8.** $\{3^{2n}\}$ **9.** $\left\{\frac{3^{n-1}}{2^n}\right\}$ **10.** $\left\{\frac{2^n}{3^{n-1}}\right\}$

In Problems 11–24, determine whether the given sequence is arithmetic, geometric, or neither. If the sequence is arithmetic, find the common difference; if it is geometric, find the common ratio.

11. $\{n + 2\}$ **12.** $\{2n - 5\}$ **13.** $\{4n^2\}$ **14.** $\{5n^2 + 1\}$ **15.** $\{3 - \frac{2}{3}n\}$

16. $\{8 - \frac{3}{4}n\}$ **17.** $1, 3, 6, 10, \ldots$ **18.** $2, 4, 6, 8, \ldots$ **19.** $\{(\frac{2}{3})^n\}$ **20.** $\{(\frac{5}{4})^n\}$

21. $-1, -2, -4, -8, \ldots$ **22.** $1, 1, 2, 3, 5, 8, \ldots$ **23.** $\{3^{n/2}\}$ **24.** $\{(-1)^n\}$

In Problems 25–32, find the fifth term and the nth term of the geometric sequence whose initial term a and common ratio r are given.

25. $a = 2;\ r = 3$ **26.** $a = -2;\ r = 4$ **27.** $a = 5;\ r = -1$ **28.** $a = 6;\ r = -2$

29. $a = 0;\ r = \frac{1}{2}$ **30.** $a = 1;\ r = -\frac{1}{3}$ **31.** $a = \sqrt{2};\ r = \sqrt{2}$ **32.** $a = 0;\ r = 1/\pi$

In Problems 33–38, find the indicated term of each geometric sequence.

33. 7th term of $1, \frac{1}{2}, \frac{1}{4}, \ldots$ **34.** 8th term of $1, 3, 9, \ldots$ **35.** 9th term of $1, -1, 1, \ldots$

36. 10th term of $-1, 2, -4, \ldots$ **37.** 8th term of $0.4, 0.04, 0.004, \ldots$ **38.** 7th term of $0.1, 1.0, 10.0, \ldots$

In Problems 39–44, find the sum.

39. $\frac{1}{4} + \frac{2}{4} + \frac{2^2}{4} + \frac{2^3}{4} + \cdots + \frac{2^{n-1}}{4}$ **40.** $\frac{3}{9} + \frac{3^2}{9} + \frac{3^3}{9} + \cdots + \frac{3^n}{9}$ **41.** $\sum_{k=1}^{n} \left(\frac{2}{3}\right)^k$

42. $\sum_{k=1}^{n} 4 \cdot 3^{k-1}$ **43.** $-1 - 2 - 4 - 8 - \cdots - (2^{n-1})$ **44.** $2 + \frac{6}{5} + \frac{18}{25} + \cdots + 2\left(\frac{3}{5}\right)^n$

For Problems 45–50, use a graphing utility to find the sum of each geometric sequence.

45. $\frac{1}{4} + \frac{2}{4} + \frac{2^2}{4} + \frac{2^3}{4} + \cdots + \frac{2^{14}}{4}$ **46.** $\frac{3}{9} + \frac{3^2}{9} + \frac{3^3}{9} + \cdots + \frac{3^{15}}{9}$ **47.** $\sum_{n=1}^{15} \left(\frac{2}{3}\right)^n$

48. $\sum_{n=1}^{15} 4 \cdot 3^{n-1}$ **49.** $-1 - 2 - 4 - 8 - \cdots - 2^{14}$ **50.** $2 + \frac{6}{5} + \frac{18}{25} + \cdots + 2\left(\frac{3}{5}\right)^{15}$

In Problems 51–60, find the sum of each infinite geometric series.

51. $1 + \frac{1}{3} + \frac{1}{9} + \cdots$ **52.** $2 + \frac{4}{3} + \frac{8}{9} + \cdots$ **53.** $8 + 4 + 2 + \cdots$ **54.** $6 + 2 + \frac{2}{3} + \cdots$

55. $2 - \frac{1}{2} + \frac{1}{8} - \frac{1}{32} + \cdots$ **56.** $1 - \frac{3}{4} + \frac{9}{16} - \frac{27}{64} + \cdots$ **57.** $\sum_{k=1}^{\infty} 5\left(\frac{1}{4}\right)^{k-1}$ **58.** $\sum_{k=1}^{\infty} 8\left(\frac{1}{3}\right)^{k-1}$

59. $\sum_{k=1}^{\infty} 6\left(-\frac{2}{3}\right)^{k-1}$ **60.** $\sum_{k=1}^{\infty} 4\left(-\frac{1}{2}\right)^{k-1}$

61. Find x so that x, $x + 2$, and $x + 3$ are consecutive terms of a geometric sequence.

62. Find x so that $x - 1$, x, and $x + 2$ are consecutive terms of a geometric sequence.

63. Multiplier Suppose that, throughout the U.S. economy, individuals spend 90% of every additional dollar that they earn. Economists would say that an individual's **marginal propensity to consume** is 0.90. For example, if Jane earns an additional dollar, she will spend $0.9(1) = \$0.90$ of it. The individual that earns \$0.90 (from Jane) will spend 90% of it or \$0.81. This process of spending continues and results in an infinite geometric series as follows:

$$1, 0.90, 0.90^2, 0.90^3, 0.90^4, \ldots$$

The sum of this infinite geometric series is called the **multiplier.** What is the multiplier if individuals spend 90% of every additional dollar that they earn?

64. Multiplier Refer to Problem 63. Suppose that the marginal propensity to consume throughout the U.S. economy is 0.95. What is the multiplier for the U.S. economy?

65. Stock Price One method of pricing a stock is to discount the stream of future dividends of the stock. Suppose that a stock pays $\$P$ per year in dividends and, historically, the dividend has been increased $i\%$ per year. If you desire an annual rate of return of $r\%$, this method of pricing a stock states that the price that you should pay is the present value of an infinite stream of payments.

$$\text{Price} = P + P\frac{1+i}{1+r} + P\left(\frac{1+i}{1+r}\right)^2 + P\left(\frac{1+i}{1+r}\right)^3 + \cdots$$

The price of the stock is the sum of an infinite geometric series. Suppose that a stock pays an annual dividend of \$4.00 and, historically, the dividend has been increased 3% per year. You desire an annual rate of return of 9%. What is the most that you should pay for the stock?

66. Stock Price Refer to Problem 65. Suppose that a stock pays an annual dividend of \$2.50 and, historically, the dividend has increased 4% per year. You desire an annual rate of return of 11%. What is the most that you should pay for the stock?

67. Pendulum Swings Initially, a pendulum swings through an arc of 2 feet. On each successive swing, the length of arc is 0.9 of the previous length.
(a) What is the length of arc after 10 swings?
(b) On which swing is the length of arc first less than 1 foot?
(c) After 15 swings, what total length will the pendulum have swung?
(d) When it stops, what total length will the pendulum have swung?

68. Bouncing Balls A ball is dropped from a height of 30 feet. Each time that it strikes the ground, it bounces up to 0.8 of the previous height.

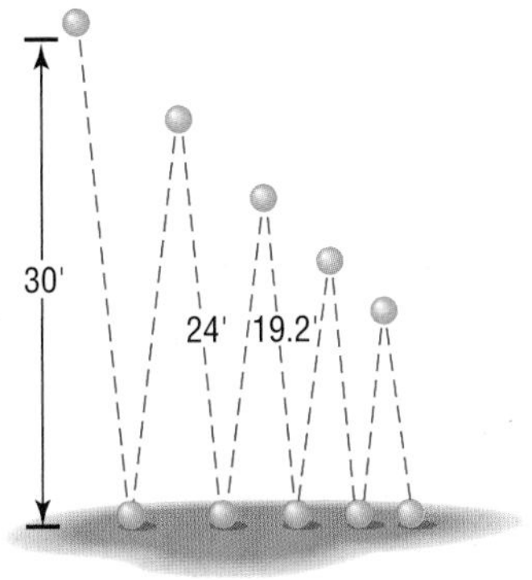

(a) What height will the ball bounce up to after it strikes the ground for the third time?
(b) What is its height after it strikes the ground for the nth time?
(c) How many times does the ball need to strike the ground before its height is less than 6 inches?
(d) What total distance does the ball travel before it stops bouncing?

69. Salary Increases Suppose that you have just been hired at an annual salary of \$18,000 and expect to receive annual increases of 5%. What will your salary be when you begin your fifth year?

70. Equipment Depreciation A new piece of equipment costs a company \$15,000. Each year, for tax purposes, the company depreciates the value by 15%. What value should the company give the equipment after 5 years?

71. Critical Thinking You have just signed a 7-year professional football league contract with a beginning salary of \$2,000,000 per year. Management gives you the following options with regard to your salary over the 7 years.
(1) A bonus of \$100,000 each year
(2) An annual increase of 4.5% per year beginning after 1 year
(3) An annual increase of \$95,000 per year beginning after 1 year
Which option provides the most money over the 7 year period? Which the least? Which would you choose? Why?

72. A Rich Man's Promise A rich man promises to give you \$1000 on September 1, 1998. Each day thereafter he will give you $\frac{9}{10}$ of what he gave you the previous day. What is the first date on which the amount that you receive is less than 1¢? How much have you received when this happens?

73. Grains of Wheat on a Chess Board In an old fable, a commoner who had just saved the king's life was told he could ask the king for any just reward. Being a shrewd man, the commoner said, "A simple wish, sire. Place one grain of wheat on the first square of a chessboard, two grains on the second square, four grains on the third square, continuing until you have filled the board. This is all I seek." Compute the total number of grains needed to do this to see why the request, seemingly simple, could not be granted. (A chessboard consists of $8 \times 8 = 64$ squares.)

74. Can a sequence be both arithmetic and geometric? Give reasons for your answer.

75. Make up a geometric sequence. Give it to a friend and ask for its twentieth term.

76. Make up two infinite geometric series, one that has a sum and one that does not. Give them to a friend and ask for the sum of each series.

77. If $x < 1$, then $1 + x + x^2 + x^3 + \cdots + x^n + \cdots = 1/(1 - x)$. Make up a table of values using $x = 0.1$, $x = 0.25$, $x = 0.5$, $x = 0.75$, and $x = 0.9$ to compute $1/(1 - x)$. Now determine how many terms are needed in the expansion $1 + x + x^2 + x^3 + \cdots + x^n + \cdots$ before it approximates $1/(1 - x)$ correct to two decimal places. For example, if $x = 0.1$, then $1/(1 - x) = 1/(1 - 0.1) = 10/9 = 1.111\ldots$. The expansion requires three terms.

78. Which of the following choices, *A* or *B*, results in more money?

A: To receive \$1000 on day 1, \$999 on day 2, \$998 on day 3, with the process to end after 1000 days

B: To receive \$1 on day 1, \$2 on day 2, \$4 on day 3, for 19 days

79. You are interviewing for a job and receive two offers:

A: \$20,000 to start with guaranteed annual increases of 6% for the first 5 years

B: \$22,000 to start with guaranteed annual increases of 3% for the first 5 years

Which offer is best if your goal is to be making as much as possible after 5 years? Which is best if your goal is to make as much money as possible over the contract (5 years)?

80. Look at the following figure. What fraction of the square is eventually shaded if the indicated shading process continues indefinitely?

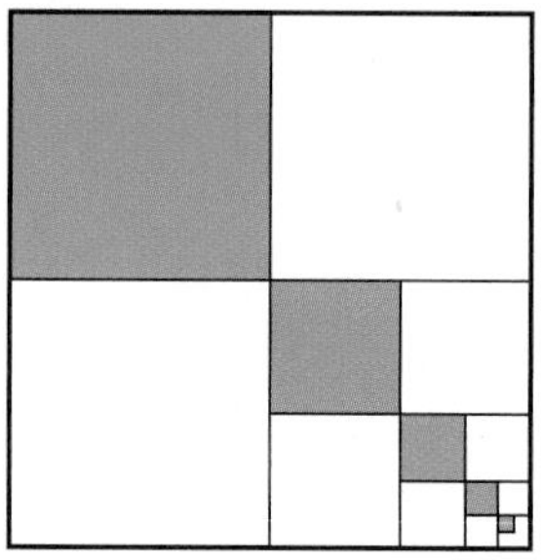

12.4 MATHEMATICAL INDUCTION

OBJECTIVE 1 Prove Statements Using Mathematical Induction

1 *Mathematical induction* is a method for proving that statements involving natural numbers are true for all natural numbers.* For example, the statement "$2n$ is always an even integer" can be proved true for all natural numbers by using mathematical induction. Also, the statement "the sum of the first n positive odd integers equals n^2," that is,

$$1 + 3 + 5 + \cdots + (2n - 1) = n^2 \qquad \textbf{(1)}$$

can be proved for all natural numbers n by using mathematical induction.

Before stating the method of mathematical induction, let's try to gain a sense of the power of the method. We shall use the statement in equation (1) for this purpose by restating it for various values of $n = 1, 2, 3, \ldots$.

$n = 1$ The sum of the first positive odd integer is 1^2; $1 = 1^2$.

$n = 2$ The sum of the first 2 positive odd integers is 2^2; $1 + 3 = 4 = 2^2$.

$n = 3$ The sum of the first 3 positive odd integers is 3^2; $1 + 3 + 5 = 9 = 3^2$.

$n = 4$ The sum of the first 4 positive odd integers is 4^2; $1 + 3 + 5 + 7 = 16 = 4^2$.

Although from this pattern we might conjecture that statement (1) is true for any choice of n, can we really be sure that it does not fail for some choice of n? The method of proof by mathematical induction will, in fact, prove that the statement is true for all n.

*Recall from the Appendix, Section 1, that the natural numbers are the numbers 1, 2, 3, 4, ... In other words, the terms *natural numbers* and *positive integers* are synonymous.

Theorem

The Principle of Mathematical Induction

Suppose that the following two conditions are satisfied with regard to a statement about natural numbers:

CONDITION I: The statement is true for the natural number 1.

CONDITION II: If the statement is true for some natural number k, it is also true for the next natural number $k + 1$.

Then the statement is true for all natural numbers.

■

Figure 16

We shall not prove this principle. However, we can provide a physical interpretation that will help us to see why the principle works. Think of a collection of natural numbers obeying a statement as a collection of infinitely many dominoes (see Figure 16).

Now, suppose that we are told two facts:

1. The first domino is pushed over.
2. If one of the dominoes falls over, say the kth domino, then so will the next one, the $(k + 1)$st domino.

Is it safe to conclude that *all* the dominoes fall over? The answer is yes, because if the first one falls (Condition I), then the second one does also (by Condition II); and if the second one falls, then so does the third (by Condition II); and so on.

Now let's prove some statements about natural numbers using mathematical induction.

EXAMPLE 1 Using Mathematical Induction

Show that the following statement is true for all natural numbers n.

$$1 + 3 + 5 + \cdots + (2n - 1) = n^2 \qquad \textbf{(2)}$$

Solution We need to show first that statement (2) holds for $n = 1$. Because $1 = 1^2$, statement (2) is true for $n = 1$. Thus, Condition I holds.

Next, we need to show that Condition II holds. Suppose that we know for some k that

$$1 + 3 + \cdots + (2k - 1) = k^2 \qquad \textbf{(3)}$$

We wish to show that, based on equation (3), statement (2) holds for $k + 1$. Thus, we look at the sum of the first $k + 1$ positive odd integers to determine whether this sum equals $(k + 1)^2$.

$$\begin{aligned} 1 + 3 + \cdots + (2k - 1) + (2k + 1) &= \underbrace{[1 + 3 + \cdots + (2k - 1)]}_{= k^2 \text{ by equation (3)}} + (2k + 1) \\ &= k^2 + (2k + 1) \\ &= k^2 + 2k + 1 = (k + 1)^2 \end{aligned}$$

Conditions I and II are satisfied; by the Principle of Mathematical Induction, statement (2) is true for all natural numbers. ■

EXAMPLE 2 Using Mathematical Induction

Show that the following statement is true for all natural numbers n.

$$2^n > n$$

Solution First, we show that the statement $2^n > n$ holds when $n = 1$. Because $2^1 = 2 > 1$, the inequality is true for $n = 1$. Thus, Condition I holds.

Next, we assume, for some natural number k, that $2^k > k$. We wish to show that the formula holds for $k + 1$; that is, we wish to show that $2^{k+1} > k + 1$. Now,

$$2^{k+1} = 2 \cdot 2^k \underset{\substack{\uparrow \\ \text{We know that} \\ 2^k > k.}}{>} 2 \cdot k = k + k \underset{\substack{\uparrow \\ k \geq 1.}}{\geq} k + 1$$

Thus, if $2^k > k$, then $2^{k+1} > k + 1$, so Condition II of the Principle of Mathematical Induction is satisfied. Hence, the statement $2^n > n$ is true for all natural numbers n. ■

EXAMPLE 3 Using Mathematical Induction

Show that the following formula is true for all natural numbers n.

$$1 + 2 + 3 + \cdots + n = \frac{n(n + 1)}{2} \tag{4}$$

Solution First, we show that formula (4) is true when $n = 1$. Because

$$\frac{1(1 + 1)}{2} = \frac{1(2)}{2} = 1$$

Condition I of the Principle of Mathematical Induction holds.

Next, we assume that formula (4) holds for some k, and we determine whether the formula then holds for $k + 1$. Thus, we assume that

$$1 + 2 + 3 + \cdots + k = \frac{k(k + 1)}{2} \quad \text{for some } k \tag{5}$$

Now we need to show that

$$1 + 2 + 3 + \cdots + k + (k + 1) = \frac{(k + 1)(k + 1 + 1)}{2} = \frac{(k + 1)(k + 2)}{2}$$

We do this as follows:

$$\begin{aligned} 1 + 2 + 3 + \cdots + k + (k + 1) &= \underbrace{[1 + 2 + 3 + \cdots + k]}_{= \frac{k(k+1)}{2} \text{ by equation (5)}} + (k + 1) \\ &= \frac{k(k + 1)}{2} + (k + 1) \\ &= \frac{k^2 + k + 2k + 2}{2} \\ &= \frac{k^2 + 3k + 2}{2} = \frac{(k + 1)(k + 2)}{2} \end{aligned}$$

Condition II also holds. As a result, formula (4) is true for all natural numbers. ■

NOW WORK PROBLEM 1.

EXAMPLE 4 Using Mathematical Induction

Show that $3^n - 1$ is divisible by 2 for all natural numbers n.

Solution First, we show that the statement is true when $n = 1$. Because $3^1 - 1 = 3 - 1 = 2$ is divisible by 2, the statement is true when $n = 1$. Thus, Condition I is satisfied.

Next, we assume that the statement holds for some k, and we determine whether the statement then holds for $k + 1$. Thus, we assume that $3^k - 1$ is divisible by 2 for some k. We need to show that $3^{k+1} - 1$ is divisible by 2. Now,

$$\begin{aligned} 3^{k+1} - 1 &= 3^{k+1} - 3^k + 3^k - 1 \\ &= 3^k(3 - 1) + (3^k - 1) = 3^k \cdot 2 + (3^k - 1) \end{aligned}$$

Because $3^k \cdot 2$ is divisible by 2 and $3^k - 1$ is divisible by 2, it follows that $3^k \cdot 2 + (3^k - 1) = 3^{k+1} - 1$ is divisible by 2. Thus, Condition II is also satisfied. As a result, the statement "$3^n - 1$ is divisible by 2" is true for all natural numbers n. ■

WARNING: The conclusion that a statement involving natural numbers is true for all natural numbers is made only after *both* Conditions I and II of the Principle of Mathematical Induction have been satisfied. Problem 27 demonstrates a statement for which only Condition I holds, but the statement is not true for all natural numbers. Problem 28 demonstrates a statement for which only Condition II holds, but the statement is *not* true for any natural number. ■

12.4 EXERCISES

In Problems 1–26, use the Principle of Mathematical Induction to show that the given statement is true for all natural numbers.

1. $2 + 4 + 6 + \cdots + 2n = n(n + 1)$

2. $1 + 5 + 9 + \cdots + (4n - 3) = n(2n - 1)$

3. $3 + 4 + 5 + \cdots + (n + 2) = \frac{1}{2}n(n + 5)$

4. $3 + 5 + 7 + \cdots + (2n + 1) = n(n + 2)$

5. $2 + 5 + 8 + \cdots + (3n - 1) = \frac{1}{2}n(3n + 1)$

6. $1 + 4 + 7 + \cdots + (3n - 2) = \frac{1}{2}n(3n - 1)$

7. $1 + 2 + 2^2 + \cdots + 2^{n-1} = 2^n - 1$

8. $1 + 3 + 3^2 + \cdots + 3^{n-1} = \frac{1}{2}(3^n - 1)$

9. $1 + 4 + 4^2 + \cdots + 4^{n-1} = \frac{1}{3}(4^n - 1)$

10. $1 + 5 + 5^2 + \cdots + 5^{n-1} = \frac{1}{4}(5^n - 1)$

11. $\dfrac{1}{1 \cdot 2} + \dfrac{1}{2 \cdot 3} + \dfrac{1}{3 \cdot 4} + \cdots + \dfrac{1}{n(n + 1)} = \dfrac{n}{n + 1}$

12. $\dfrac{1}{1 \cdot 3} + \dfrac{1}{3 \cdot 5} + \dfrac{1}{5 \cdot 7} + \cdots + \dfrac{1}{(2n - 1)(2n + 1)} = \dfrac{n}{2n + 1}$

13. $1^2 + 2^2 + 3^2 + \cdots + n^2 = \frac{1}{6}n(n + 1)(2n + 1)$

14. $1^3 + 2^3 + 3^3 + \cdots + n^3 = \frac{1}{4}n^2(n + 1)^2$

15. $4 + 3 + 2 + \cdots + (5 - n) = \frac{1}{2}n(9 - n)$

16. $-2 - 3 - 4 - \cdots - (n + 1) = -\frac{1}{2}n(n + 3)$

17. $1 \cdot 2 + 2 \cdot 3 + 3 \cdot 4 + \cdots + n(n + 1) = \frac{1}{3}n(n + 1)(n + 2)$

18. $1 \cdot 2 + 3 \cdot 4 + 5 \cdot 6 + \cdots + (2n - 1)(2n) = \frac{1}{3}n(n + 1)(4n - 1)$

19. $n^2 + n$ is divisible by 2.

20. $n^3 + 2n$ is divisible by 3.

21. $n^2 - n + 2$ is divisible by 2.

22. $n(n + 1)(n + 2)$ is divisible by 6.

23. If $x > 1$, then $x^n > 1$.

24. If $0 < x < 1$, then $0 < x^n < 1$.

25. $a - b$ is a factor of $a^n - b^n$.
[**Hint:** $a^{k+1} - b^{k+1} = a(a^k - b^k) + b^k(a - b)$]

26. $a + b$ is a factor of $a^{2n+1} + b^{2n+1}$.

27. Show that the statement "$n^2 - n + 41$ is a prime number" is true for $n = 1$, but is not true for $n = 41$.

28. Show that the formula

$$2 + 4 + 6 + \cdots + 2n = n^2 + n + 2$$

obeys Condition II of the Principle of Mathematical Induction. That is, show that if the formula is true for some k it is also true for $k + 1$. Then show that the formula is false for $n = 1$ (or for any other choice of n).

29. Use mathematical induction to prove that if $r \neq 1$ then

$$a + ar + ar^2 + \cdots + ar^{n-1} = a\frac{1 - r^n}{1 - r}$$

30. Use mathematical induction to prove that

$$a + (a + d) + (a + 2d) + \cdots + [a + (n - 1)d] = na + d\frac{n(n - 1)}{2}$$

31. Geometry Use mathematical induction to show that the sum of the interior angles of a convex polygon of n sides equals $(n - 2) \cdot 180°$.

32. Extended Principle of Mathematical Induction The Extended Principle of Mathematical Induction states that if Conditions I and II hold, that is,

(I) A statement is true for a natural number j.

(II) If the statement is true for some natural number $k > j$, then it is also true for the next natural number $k + 1$.

then the statement is true for *all* natural numbers $\geq j$.

Use the Extended Principle of Mathematical Induction to show that the number of diagonals in a convex polygon of n sides is $\frac{1}{2}n(n - 3)$.

[**Hint:** Begin by showing that the result is true when $n = 4$ (Condition I).]

33. How would you explain to a friend the Principle of Mathematical Induction?

12.5 THE BINOMIAL THEOREM

OBJECTIVES 1 Evaluate a Binomial Coefficient

2 Expand a Binomial

In earlier algebra courses, formulas are given for expanding $(x + a)^n$ for $n = 2$ and $n = 3$. The *Binomial Theorem** is a formula for the expansion of $(x + a)^n$ for n any positive integer. If $n = 1, 2, 3$, and 4, the expansion of $(x + a)^n$ is straightforward:

$(x + a)^1 = x + a$ — Two terms, beginning with x^1 and ending with a^1

$(x + a)^2 = x^2 + 2ax + a^2$ — Three terms, beginning with x^2 and ending with a^2

$(x + a)^3 = x^3 + 3ax^2 + 3a^2x + a^3$ — Four terms, beginning with x^3 and ending with a^3

$(x + a)^4 = x^4 + 4ax^3 + 6a^2x^2 + 4a^3x + a^4$ — Five terms, beginning with x^4 and ending with a^4

Notice that each expansion of $(x + a)^n$ begins with x^n and ends with a^n. As you read from left to right, the powers of x are decreasing, while the powers of a are increasing. Also, the number of terms that appears equals $n + 1$. Notice, too, that the degree of each monomial in the expansion equals n. For example, in the expansion of $(x + a)^3$, each monomial $(x^3, 3ax^2, 3a^2x, a^3)$ is of degree 3. As a result, we might conjecture that the expansion of $(x + a)^n$ would look like this:

$$(x + a)^n = x^n + \_ax^{n-1} + \_a^2x^{n-2} + \cdots + \_a^{n-1}x + a^n$$

where the blanks are numbers to be found. This is, in fact, the case, as we shall see shortly.

First, we need to introduce a symbol.

*The name *binomial* derives from the fact that $x + a$ is a binomial, that is, contains two terms.

THE SYMBOL $\binom{n}{j}$

1 We define the symbol $\binom{n}{j}$, read "n taken j at a time," as follows:

If j and n are integers with $0 \le j \le n$, the **symbol** $\binom{n}{j}$ is defined as

$$\binom{n}{j} = \frac{n!}{j!(n-j)!} \tag{1}$$

COMMENT: On a graphing calculator, the symbol $\binom{n}{j}$ may be denoted by the key nCr.

EXAMPLE 1 Evaluating $\binom{n}{j}$

Find:

(a) $\binom{3}{1}$ (b) $\binom{4}{2}$ (c) $\binom{8}{7}$ (d) $\binom{65}{15}$

Solution

(a) $\binom{3}{1} = \dfrac{3!}{1!(3-1)!} = \dfrac{3!}{1!2!} = \dfrac{3 \cdot 2 \cdot 1}{1(2 \cdot 1)} = \dfrac{6}{2} = 3$

(b) $\binom{4}{2} = \dfrac{4!}{2!(4-2)!} = \dfrac{4!}{2!2!} = \dfrac{4 \cdot 3 \cdot 2 \cdot 1}{(2 \cdot 1)(2 \cdot 1)} = \dfrac{24}{4} = 6$

(c) $\binom{8}{7} = \dfrac{8!}{7!(8-7)!} = \dfrac{8!}{7!1!} = \dfrac{8 \cdot \cancel{7!}}{\cancel{7!} \cdot 1!} = \dfrac{8}{1} = 8$

↑ $8! = 8 \cdot 7!$

Figure 17

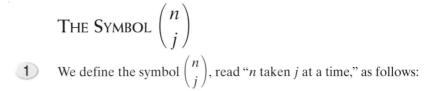

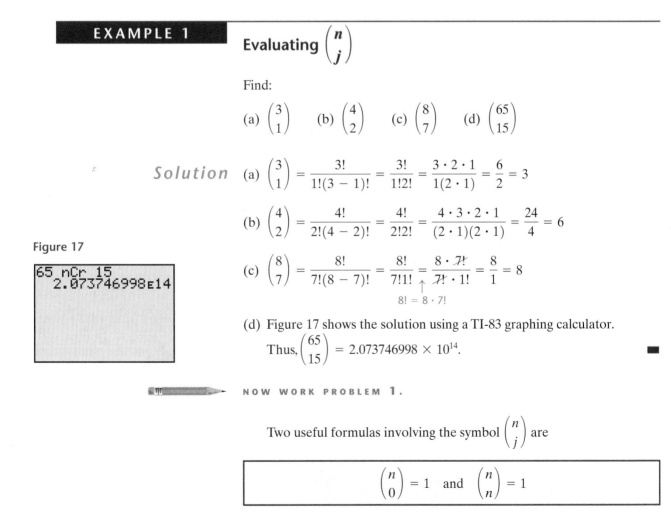

(d) Figure 17 shows the solution using a TI-83 graphing calculator.
Thus, $\binom{65}{15} = 2.073746998 \times 10^{14}$.

NOW WORK PROBLEM 1.

Two useful formulas involving the symbol $\binom{n}{j}$ are

$$\binom{n}{0} = 1 \quad \text{and} \quad \binom{n}{n} = 1$$

Proof

$$\binom{n}{0} = \frac{n!}{0!(n-0)!} = \frac{n!}{0!n!} = \frac{1}{1} = 1$$

You are asked to show that $\binom{n}{n} = 1$ in Problem 41 at the end of this section. ∎

Suppose that we arrange the various values of the symbol $\binom{n}{j}$ in a triangular display, as shown next and in Figure 18.

$$\binom{0}{0}$$
$$\binom{1}{0} \quad \binom{1}{1}$$
$$\binom{2}{0} \quad \binom{2}{1} \quad \binom{2}{2}$$
$$\binom{3}{0} \quad \binom{3}{1} \quad \binom{3}{2} \quad \binom{3}{3}$$
$$\binom{4}{0} \quad \binom{4}{1} \quad \binom{4}{2} \quad \binom{4}{3} \quad \binom{4}{4}$$
$$\binom{5}{0} \quad \binom{5}{1} \quad \binom{5}{2} \quad \binom{5}{3} \quad \binom{5}{4} \quad \binom{5}{5}$$

Figure 18
Pascal triangle

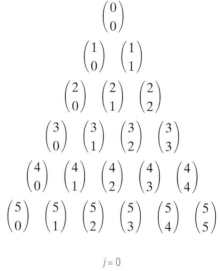

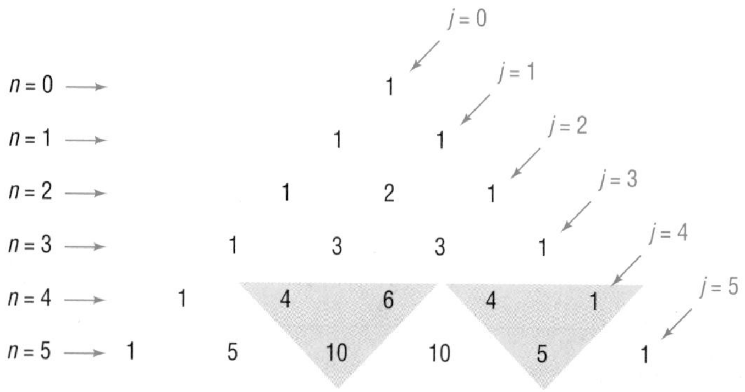

This display is called the **Pascal triangle,** named after Blaise Pascal (1623–1662), a French mathematician.

The Pascal triangle has 1's down the sides. To get any other entry, merely add the two nearest entries in the row above it. The shaded triangles in Figure 18 illustrate this feature of the Pascal triangle. Based on this feature, the row corresponding to $n = 6$ is found as follows:

$n = 5 \rightarrow$ 1 5 10 10 5 1

$n = 6 \rightarrow$ 1 6 15 20 15 6 1

Later we shall prove that this addition always works (see the theorem on page 849).

Although the Pascal triangle provides an interesting and organized display of the symbol $\binom{n}{j}$, in practice it is not all that helpful. For example, if you wanted to know the value of $\binom{12}{5}$, you would need to produce 12 rows of the triangle before seeing the answer. It is much faster instead to use the definition (1).

BINOMIAL THEOREM

2 Now we are ready to state the **Binomial Theorem.** A proof is given at the end of this section.

Theorem **Binomial Theorem**

Let x and a be real numbers. For any positive integer n, we have

$$(x+a)^n = \binom{n}{0}x^n + \binom{n}{1}ax^{n-1} + \cdots + \binom{n}{j}a^j x^{n-j} + \cdots + \binom{n}{n}a^n$$
$$= \sum_{j=0}^{n}\binom{n}{j}x^{n-j}a^j \qquad \textbf{(2)}$$

Now you know why we needed to introduce the symbol $\binom{n}{j}$; these symbols are the numerical coefficients that appear in the expansion of $(x + a)^n$. Because of this, the symbol $\binom{n}{j}$ is called the **binomial coefficient.**

EXAMPLE 2 Expanding a Binomial

Use the Binomial Theorem to expand $(x + 2)^5$.

Solution In the Binomial Theorem, let $a = 2$ and $n = 5$. Then

$$(x+2)^5 = \binom{5}{0}x^5 + \binom{5}{1}2x^4 + \binom{5}{2}2^2x^3 + \binom{5}{3}2^3x^2 + \binom{5}{4}2^4x + \binom{5}{5}2^5$$

↑ Use equation (2).

$$= 1 \cdot x^5 + 5 \cdot 2x^4 + 10 \cdot 4x^3 + 10 \cdot 8x^2 + 5 \cdot 16x + 1 \cdot 32$$

↑ Use row $n = 5$ of the Pascal triangle or formula (1) for $\binom{n}{j}$.

$$= x^5 + 10x^4 + 40x^3 + 80x^2 + 80x + 32$$

EXAMPLE 3 Expanding a Binomial

Expand $(2y - 3)^4$ using the Binomial Theorem.

Solution First, we rewrite the expression $(2y - 3)^4$ as $[2y + (-3)]^4$. Now we use the Binomial Theorem with $n = 4$, $x = 2y$, and $a = -3$.

$$[2y + (-3)]^4 = \binom{4}{0}(2y)^4 + \binom{4}{1}(-3)(2y)^3 + \binom{4}{2}(-3)^2(2y)^2 + \binom{4}{3}(-3)^3(2y) + \binom{4}{4}(-3)^4$$

$$= 1 \cdot 16y^4 + 4(-3)8y^3 + 6 \cdot 9 \cdot 4y^2 + 4(-27)2y + 1 \cdot 81$$

↑ Use row $n = 4$ of the Pascal triangle or formula (1) for $\binom{n}{j}$.

$$= 16y^4 - 96y^3 + 216y^2 - 216y + 81$$

In this expansion, note that the signs alternate due to the fact that $a = -3 < 0$. ■

NOW WORK PROBLEM 17.

EXAMPLE 4 Finding a Particular Coefficient in a Binomial Expansion

Find the coefficient of y^8 in the expansion of $(2y + 3)^{10}$.

Solution We write out the expansion using the Binomial Theorem.

$$(2y+3)^{10} = \binom{10}{0}(2y)^{10} + \binom{10}{1}(2y)^9(3)^1 + \binom{10}{2}(2y)^8(3)^2 + \binom{10}{3}(2y)^7(3)^3$$
$$+ \binom{10}{4}(2y)^6(3)^4 + \cdots + \binom{10}{9}(2y)(3)^9 + \binom{10}{10}(3)^{10}$$

From the third term in the expansion, the coefficient of y^8 is

$$\binom{10}{2}(2)^8(3)^2 = \frac{10!}{2!8!} \cdot 2^8 \cdot 9 = \frac{10 \cdot 9 \cdot \cancel{8!}}{2 \cdot \cancel{8!}} \cdot 2^8 \cdot 9 = 103{,}680$$ ■

As this solution demonstrates, we can use the Binomial Theorem to find a particular term in an expansion without writing the entire expansion. Based on the expansion of $(x + a)^n$, the term containing x^j is

$$\binom{n}{n-j} a^{n-j} x^j \qquad \textbf{(3)}$$

For example, we can solve Example 4 by using formula (3) with $n = 10$, $a = 3$, $x = 2y$, and $j = 8$. Then the term containing y^8 is

$$\binom{10}{10-8} 3^{10-8}(2y)^8 = \binom{10}{2} \cdot 3^2 \cdot 2^8 \cdot y^8 = \frac{10!}{2!8!} \cdot 9 \cdot 2^8 y^8$$
$$= \frac{10 \cdot 9 \cdot \cancel{8!}}{2!\cancel{8!}} \cdot 9 \cdot 2^8 y^8 = 103{,}680 y^8$$

EXAMPLE 5 Finding a Particular Term in a Binomial Expansion

Find the sixth term in the expansion of $(x + 2)^9$.

Solution A We expand using the Binomial Theorem until the sixth term is reached.

$$(x+2)^9 = \binom{9}{0}x^9 + \binom{9}{1}x^8 \cdot 2 + \binom{9}{2}x^7 \cdot 2^2 + \binom{9}{3}x^6 \cdot 2^3 + \binom{9}{4}x^5 \cdot 2^4$$
$$+ \binom{9}{5}x^4 \cdot 2^5 + \cdots$$

The sixth term is

$$\binom{9}{5}x^4 \cdot 2^5 = \frac{9!}{5!4!} \cdot x^4 \cdot 32 = 4032x^4$$

Solution B The sixth term in the expansion of $(x + 2)^9$, which has 10 terms total, contains x^4. (Do you see why?) By formula (3), the sixth term is

$$\binom{9}{9-4}2^{9-4}x^4 = \binom{9}{5}2^5x^4 = \frac{9!}{5!4!} \cdot 32x^4 = 4032x^4$$

NOW WORK PROBLEMS **25** AND **31**.

Next we show that the *triangular addition* feature of the Pascal triangle illustrated in Figure 18 always works.

Theorem If n and j are integers with $1 \leq j \leq n$, then

$$\binom{n}{j-1} + \binom{n}{j} = \binom{n+1}{j} \tag{4}$$

Proof

$$\binom{n}{j-1} + \binom{n}{j} = \frac{n!}{(j-1)![n-(j-1)]!} + \frac{n!}{j!(n-j)!}$$

$$= \frac{n!}{(j-1)!(n-j+1)!} + \frac{n!}{j!(n-j)!}$$

$$= \frac{jn!}{j(j-1)!(n-j+1)!} + \frac{(n-j+1)n!}{j!(n-j+1)(n-j)!}$$

Multiply the first term by j/j and the second terms by $(n-j+1)/(n-j+1)$.

$$= \frac{jn!}{j!(n-j+1)!} + \frac{(n-j+1)n!}{j!(n-j+1)!}$$

Now the denominators are equal.

$$= \frac{jn! + (n-j+1)n!}{j!(n-j+1)!}$$

$$= \frac{n!(j+n-j+1)}{j!(n-j+1)!}$$

$$= \frac{n!(n+1)}{j!(n-j+1)!} = \frac{(n+1)!}{j![(n+1)-j]!} = \binom{n+1}{j}$$

Proof of the Binomial Theorem We use Mathematical Induction to prove the Binomial Theorem. First, we show that formula (2) is true for $n = 1$.

$$(x+a)^1 = x + a = \binom{1}{0}x^1 + \binom{1}{1}a^1$$

Next we suppose that formula (2) is true for some k. That is, we assume that

$$(x+a)^k = \binom{k}{0}x^k + \binom{k}{1}ax^{k-1} + \cdots + \binom{k}{j-1}a^{j-1}x^{k-j+1} + \binom{k}{j}a^jx^{k-j} + \cdots + \binom{k}{k}a^k \tag{5}$$

Now we calculate $(x + a)^{k+1}$.

$$(x + a)^{k+1} = (x + a)(x + a)^k = x(x + a)^k + a(x + a)^k$$

$$= x\left[\binom{k}{0}x^k + \binom{k}{1}ax^{k-1} + \cdots + \binom{k}{j-1}a^{j-1}x^{k-j+1} + \binom{k}{j}a^jx^{k-j} + \cdots + \binom{k}{k}a^k\right]$$

Use ↑ Equation (5).

$$+ a\left[\binom{k}{0}x^k + \binom{k}{1}ax^{k-1} + \cdots + \binom{k}{j-1}a^{j-1}x^{k-j+1} + \binom{k}{j}a^jx^{k-j} + \cdots + \binom{k}{k-1}a^{k-1}x + \binom{k}{k}a^k\right]$$

$$= \binom{k}{0}x^{k+1} + \binom{k}{1}ax^k + \cdots + \binom{k}{j-1}a^{j-1}x^{k-j+2} + \binom{k}{j}a^jx^{k-j+1} + \cdots + \binom{k}{k}a^kx$$

$$+ \binom{k}{0}ax^k + \binom{k}{1}a^2x^{k-1} + \cdots + \binom{k}{j-1}a^jx^{k-j+1} + \binom{k}{j}a^{j+1}x^{k-j} + \cdots + \binom{k}{k-1}a^kx + \binom{k}{k}a^{k+1}$$

$$= \binom{k}{0}x^{k+1} + \left[\binom{k}{1} + \binom{k}{0}\right]ax^k + \cdots + \left[\binom{k}{j} + \binom{k}{j-1}\right]a^jx^{k-j+1}$$

$$+ \cdots + \left[\binom{k}{k} + \binom{k}{k-1}\right]a^kx + \binom{k}{k}a^{k+1}$$

Because

$$\binom{k}{0} = 1 = \binom{k+1}{0}, \quad \binom{k}{1} + \binom{k}{0} \underset{\substack{\uparrow \\ (4)}}{=} \binom{k+1}{1}, \ldots,$$

$$\binom{k}{j} + \binom{k}{j-1} \underset{\substack{\uparrow \\ (4)}}{=} \binom{k+1}{j}, \ldots, \binom{k}{k} = 1 = \binom{k+1}{k+1}$$

we have

$$(x + a)^{k+1} = \binom{k+1}{0}x^{k+1} + \binom{k+1}{1}ax^k + \cdots + \binom{k+1}{j}a^jx^{k-j+1} + \cdots + \binom{k+1}{k+1}a^{k+1}$$

Thus, Conditions I and II of the Principle of Mathematical Induction are satisfied, and formula (2) is therefore true for all n. ■

HISTORICAL FEATURE

Omar Khayyám (1044–1123)

The case $n = 2$ of the Binomial Theorem, $(a + b)^2$, was known to Euclid in 300 BC, but the general law seems to have been discovered by the Persian mathematician and astronomer Omar Khayyám (1044–1123), who is also well known as the author of the *Rubáiyát,* a collection of four-line poems making observations on the human condition. Omar Khayyám did not state the Binomial Theorem explicitly, but he claimed to have a method for extracting third, fourth, fifth roots, and so on. A little study shows that one must know the Binomial Theorem to create such a method.

The heart of the Binomial Theorem is the formula for the numerical coefficients, and, as we saw, they can be written out in a symmetric triangular form. The Pascal triangle appears first in the books of Yang Hui (about 1270) and Chu Shihchie (1303). Pascal's name is attached to the triangle because of the many applications he made of it, especially to counting and probability. In establishing these results, he was one of the earliest users of mathematical induction.

Many people worked on the proof of the Binomial Theorem, which was finally completed for all n (including complex numbers) by Niels Abel (1802–1829).

12.5 EXERCISES

In Problems 1–12, evaluate each expression by hand. Use a graphing utility to verify your answer.

1. $\binom{5}{3}$ **2.** $\binom{7}{3}$ **3.** $\binom{7}{5}$ **4.** $\binom{9}{7}$ **5.** $\binom{50}{49}$ **6.** $\binom{100}{98}$

7. $\binom{1000}{1000}$ **8.** $\binom{1000}{0}$ **9.** $\binom{55}{23}$ **10.** $\binom{60}{20}$ **11.** $\binom{47}{25}$ **12.** $\binom{37}{19}$

In Problems 13–24, expand each expression using the Binomial Theorem.

13. $(x + 1)^5$ **14.** $(x - 1)^5$ **15.** $(x - 2)^6$ **16.** $(x + 3)^5$ **17.** $(3x + 1)^4$ **18.** $(2x + 3)^5$

19. $(x^2 + y^2)^5$ **20.** $(x^2 - y^2)^6$ **21.** $(\sqrt{x} + \sqrt{2})^6$ **22.** $(\sqrt{x} - \sqrt{3})^4$ **23.** $(ax + by)^5$ **24.** $(ax - by)^4$

In Problems 25–38, use the Binomial Theorem to find the indicated coefficient or term.

25. The coefficient of x^6 in the expansion of $(x + 3)^{10}$

26. The coefficient of x^3 in the expansion of $(x - 3)^{10}$

27. The coefficient of x^7 in the expansion of $(2x - 1)^{12}$

28. The coefficient of x^3 in the expansion of $(2x + 1)^{12}$

29. The coefficient of x^7 in the expansion of $(2x + 3)^9$

30. The coefficient of x^2 in the expansion of $(2x - 3)^9$

31. The fifth term in the expansion of $(x + 3)^7$

32. The third term in the expansion of $(x - 3)^7$

33. The third term in the expansion of $(3x - 2)^9$

34. The sixth term in the expansion of $(3x + 2)^8$

35. The coefficient of x^0 in the expansion of $\left(x^2 + \frac{1}{x}\right)^{12}$

36. The coefficient of x^0 in the expansion of $\left(x - \frac{1}{x^2}\right)^9$

37. The coefficient of x^4 in the expansion of $\left(x - \frac{2}{\sqrt{x}}\right)^{10}$

38. The coefficient of x^2 in the expansion of $\left(\sqrt{x} + \frac{3}{\sqrt{x}}\right)^8$

39. Use the Binomial Theorem to find the numerical value of $(1.001)^5$ correct to five decimal places.
[**Hint:** $(1.001)^5 = (1 + 10^{-3})^5$]

40. Use the Binomial Theorem to find the numerical value of $(0.998)^6$ correct to five decimal places.

41. Show that $\binom{n}{n} = 1$.

42. Show that if n and j are integers with $0 \le j \le n$ then

$$\binom{n}{j} = \binom{n}{n-j}$$

Thus, conclude that the Pascal triangle is symmetric with respect to a vertical line drawn from the topmost entry.

43. If n is a positive integer, show that

$$\binom{n}{0} + \binom{n}{1} + \cdots + \binom{n}{n} = 2^n$$

[**Hint:** $2^n = (1 + 1)^n$; now use the Binomial Theorem.]

44. If n is a positive integer, show that

$$\binom{n}{0} - \binom{n}{1} + \binom{n}{2} - \cdots + (-1)^n\binom{n}{n} = 0$$

45. $$\binom{5}{0}\left(\frac{1}{4}\right)^5 + \binom{5}{1}\left(\frac{1}{4}\right)^4\left(\frac{3}{4}\right) + \binom{5}{2}\left(\frac{1}{4}\right)^3\left(\frac{3}{4}\right)^2 + \binom{5}{3}\left(\frac{1}{4}\right)^2\left(\frac{3}{4}\right)^3 + \binom{5}{4}\left(\frac{1}{4}\right)\left(\frac{3}{4}\right)^4 + \binom{5}{5}\left(\frac{3}{4}\right)^5 = ?$$

46. *Stirling's formula* for approximating $n!$ when n is large is given by

$$n! \approx \sqrt{2n\pi}\left(\frac{n}{e}\right)^n\left(1 + \frac{1}{12n - 1}\right)$$

Calculate 12!, 20!, and 25!. Then use Stirling's formula to approximate 12!, 20!, and 25!.

CHAPTER REVIEW

Things To Know

Sequence (p. 810)	A function whose domain is the set of positive integers.
Factorials (p. 814)	$0! = 1, 1! = 1, n! = n(n-1) \cdot \ldots \cdot 3 \cdot 2 \cdot 1$ if $n \geq 2$
Amount of an annuity (p. 820)	$A_0 = M, A_n = \left(1 + \frac{r}{N}\right)A_{n-1} + P$
Arithmetic sequence (p. 825)	$a_1 = a, a_n = a_{n-1} + d$, where a = first term, d = common difference, $a_n = a + (n-1)d$
Sum of the first n terms of an arithmetic sequence (p. 827)	$S_n = \frac{n}{2}[2a + (n-1)d] = \frac{n}{2}(a + a_n)$
Geometric sequence (p. 831)	$a_1 = a, \quad a_n = ra_{n-1}$, where a = first term, r = common ratio, $a_n = ar^{n-1}, \quad r \neq 0$
Sum of the first n terms of a geometric sequence (p. 833)	$S_n = a\frac{1-r^n}{1-r}, \quad r \neq 0, 1$
Infinite geometric series (p. 834)	$a + ar + \cdots + ar^{n-1} + \cdots = \sum_{k=1}^{\infty} ar^{k-1}$
Sum of an infinite geometric series (p. 835)	$\sum_{k=1}^{\infty} ar^{k-1} = \frac{a}{1-r}, \quad \lvert r \rvert < 1$
Principle of Mathematical Induction (p. 841)	Condition I: The statement is true for the natural number 1. Condition II: If the statement is true for some natural number k, it is also true for $k + 1$. Then the statement is true for all natural numbers.
Binomial coefficient (p. 845)	$\binom{n}{j} = \frac{n!}{j!(n-j)!}$
Pascal triangle (p. 846)	See Figure 18.
Binomial Theorem (p. 847)	$(x + a)^n = \binom{n}{0}x^n + \binom{n}{1}ax^{n-1} + \cdots + \binom{n}{j}a^j x^{n-j} + \cdots + \binom{n}{n}a^n$

How To

Write the first several terms of a sequence (p. 811)

Write the terms of a sequence defined by a recursion formula (p. 814)

Write a sequence in summation notation (p. 816)

Find the sum of a sequence by hand and by using a graphing utility (p. 817)

Solve annuity and amortization problems (p. 819)

Determine if a sequence is arithmetic (p. 825)

Find a formula for an arithmetic sequence (p. 826)

Find the sum of an arithmetic sequence (p. 827)

Determine if a sequence is geometric (p. 831)

Find a formula for a geometric sequence (p. 832)

Find the sum of a geometric sequence (p. 833)

Find the sum of a geometric series (p. 834)

Prove statements using mathematical induction (p. 840)

Evaluate a binomial coefficient (p. 845)

Expand a binomial (p. 847)

Fill-in-the-Blank Items

1. A(n) ______________ is a function whose domain is the set of positive integers.
2. In a(n) ______________ sequence, the difference between successive terms is always the same number.
3. In a(n) ______________ sequence, the ratio of successive terms is always the same number.
4. The ______________ ______________ is a triangular display of the binomial coefficients.
5. $\binom{6}{2} =$ ______________.

True/False Items

T F 1. A sequence is a function.

T F 2. For arithmetic sequences, the difference of successive terms is always the same number.

T F 3. For geometric sequences, the ratio of successive terms is always the same number.

T F 4. Mathematical induction can sometimes be used to prove theorems that involve natural numbers.

T F 5. $\binom{n}{j} = \dfrac{j!}{n!(n-j)!}$

T F 6. The expansion of $(x + a)^n$ contains n terms.

T F 7. $\sum_{i=1}^{n+1} i = 1 + 2 + 3 + \cdots + n$

Review Exercises

Blue problem numbers indicate the authors' suggestions for use in a Practice Test.

In Problems 1–8, write down the first five terms of each sequence.

1. $\left\{(-1)^n\left(\dfrac{n+3}{n+2}\right)\right\}$
2. $\{(-1)^{n+1}(2n+3)\}$
3. $\left\{\dfrac{2^n}{n^2}\right\}$
4. $\left\{\dfrac{e^n}{n}\right\}$
5. $a_1 = 3;\quad a_n = \frac{2}{3}a_{n-1}$
6. $a_1 = 4;\quad a_n = -\frac{1}{4}a_{n-1}$
7. $a_1 = 2;\quad a_n = 2 - a_{n-1}$
8. $a_1 = -3;\quad a_n = 4 + a_{n-1}$

In Problems 9–20, determine whether the given sequence is arithmetic, geometric, or neither. If the sequence is arithmetic, find the common difference and the sum of the first n terms. If the sequence is geometric, find the common ratio and the sum of the first n terms.

9. $\{n + 5\}$
10. $\{4n + 3\}$
11. $\{2n^3\}$
12. $\{2n^2 - 1\}$
13. $\{2^{3n}\}$
14. $\{3^{2n}\}$
15. $0, 4, 8, 12, \ldots$
16. $1, -3, -7, -11, \ldots$
17. $3, \frac{3}{2}, \frac{3}{4}, \frac{3}{8}, \frac{3}{16}, \ldots$
18. $5, -\frac{5}{3}, \frac{5}{9}, -\frac{5}{27}, \frac{5}{81}, \ldots$
19. $\frac{2}{3}, \frac{3}{4}, \frac{4}{5}, \frac{5}{6}, \ldots$
20. $\frac{3}{2}, \frac{5}{4}, \frac{7}{6}, \frac{9}{8}, \frac{11}{10}, \ldots$

In Problems 21–26, evaluate each sum. Verify your results using a graphing utility.

21. $\sum_{k=1}^{5}(k^2 + 12)$
22. $\sum_{k=1}^{3}(k + 2)^2$
23. $\sum_{k=1}^{10}(3k - 9)$
24. $\sum_{k=1}^{9}(-2k + 8)$
25. $\sum_{k=1}^{7}\left(\dfrac{1}{3}\right)^k$
26. $\sum_{k=1}^{10}(-2)^k$

In Problems 27–32, find the indicated term in each sequence (a) by hand and (b) using a graphing utility.

27. 9th term of $3, 7, 11, 15, \ldots$

28. 8th term of $1, -1, -3, -5, \ldots$

29. 11th term of $1, \frac{1}{10}, \frac{1}{100}, \ldots$

30. 11th term of $1, 2, 4, 8, \ldots$

31. 9th term of $\sqrt{2}, 2\sqrt{2}, 3\sqrt{2}, \ldots$

32. 9th term of $\sqrt{2}, 2, 2^{3/2}, \ldots$

In Problems 33–36, find a general formula for each arithmetic sequence.

33. 7th term is 31; 20th term is 96

34. 8th term is −20; 17th term is −47

35. 10th term is 0; 18th term is 8

36. 12th term is 30; 22nd term is 50

In Problems 37–42, find the sum of each infinite geometric series.

37. $3 + 1 + \frac{1}{3} + \frac{1}{9} + \cdots$

38. $2 + 1 + \frac{1}{2} + \frac{1}{4} + \cdots$

39. $2 - 1 + \frac{1}{2} - \frac{1}{4} + \cdots$

40. $6 - 4 + \frac{8}{3} - \frac{16}{9} + \cdots$

41. $\sum_{k=1}^{\infty} 4\left(\frac{1}{2}\right)^{k-1}$

42. $\sum_{k=1}^{\infty} 3\left(-\frac{3}{4}\right)^{k-1}$

In Problems 43–48, use the Principle of Mathematical Induction to show that the given statement is true for all natural numbers.

43. $3 + 6 + 9 + \cdots + 3n = \frac{3n}{2}(n + 1)$

44. $2 + 6 + 10 + \cdots + (4n - 2) = 2n^2$

45. $2 + 6 + 18 + \cdots + 2 \cdot 3^{n-1} = 3^n - 1$

46. $3 + 6 + 12 + \cdots + 3 \cdot 2^{n-1} = 3(2^n - 1)$

47. $1^2 + 4^2 + 7^2 + \cdots + (3n - 2)^2 = \frac{1}{2}n(6n^2 - 3n - 1)$

48. $1 \cdot 3 + 2 \cdot 4 + 3 \cdot 5 + \cdots + n(n + 2) = \frac{n}{6}(n + 1)(2n + 7)$

In Problems 49–52, expand each expression using the Binomial Theorem.

49. $(x + 2)^5$

50. $(x - 3)^4$

51. $(2x + 3)^5$

52. $(3x - 4)^4$

53. Find the coefficient of x^7 in the expansion of $(x + 2)^9$.

54. Find the coefficient of x^3 in the expansion of $(x - 3)^8$.

55. Find the coefficient of x^2 in the expansion of $(2x + 1)^7$.

56. Find the coefficient of x^6 in the expansion of $(2x + 1)^8$.

57. Constructing a Brick Staircase A brick staircase has a total of 25 steps. The bottom step requires 80 bricks. Each successive step requires three less bricks than the prior step.
(a) How many bricks are required for the top step?
(b) How many bricks are required to build the staircase?

58. Creating a Floor Design A mosaic tile floor is designed in the shape of a trapezoid 30 feet wide at the base and 15 feet wide at the top. The tiles, 12 inches by 12 inches, are to be placed so that each successive row contains one less tile than the row below. How many tiles will be required? [**Hint:** Refer to Figure 13].

59. Retirement Planning Chris gets paid once a month and contributes $200 each pay period into his 401(k). If Chris plans on retiring in 20 years, what will the value of his 401(k) be if the per annum rate of return of the 401(k) is 10% compounded monthly?

60. Retirement Planning Jacky contributes $500 every quarter to an IRA. If Jacky plans on retiring in 30 years, what will the value of the IRA be if the per annum rate of return of the IRA is 8% compounded quarterly?

61. Bouncing Balls A ball is dropped from a height of 20 feet. Each time it strikes the ground, it bounces up to three-quarters of the previous height.
(a) What height will the ball bounce up to after it strikes the ground for the third time?
(b) What is its height after it strikes the ground for the nth time?
(c) How many times does the ball need to strike the ground before its height is less than 6 inches?
(d) What total distance does the ball travel before it stops bouncing?

62. Salary Increases Your friend has just been hired at an annual salary of $20,000. If she expects to receive annual increases of 4%, what will her salary be as she begins her fifth year?

63. Home Loan Mike and Yola borrowed $190,000 at 6.75% per annum compounded monthly for 30 years to purchase a home. Their monthly payment is determined to be $1232.34.
(a) Find a recursive formula for their balance after each monthly payment has been made.
(b) Determine Mike and Yola's balance after the first payment.
(c) Using a graphing utility, create a table showing Mike and Yola's balance after each monthly payment.
(d) Using a graphing utility, determine when Mike and Yola's balance will be below $100,000.
(e) Using a graphing utility, determine when Mike and Yola will pay off the balance.
(f) Determine Mike and Yola's interest expense when the loan is paid.
(g) Suppose that Mike and Yola decide to pay an additional $100 each month on their loan. Answer parts (a) to (f) under this scenario.

PROJECT AT MOTOROLA

Series Solutions In Engineering Applications

Many engineering problems can be solved using complex numbers and series solutions. Figure 1 illustrates the vibration of a thin plate. Although the analysis of this phenomenon requires advanced topics in mathematics, we can present here some of the background needed.

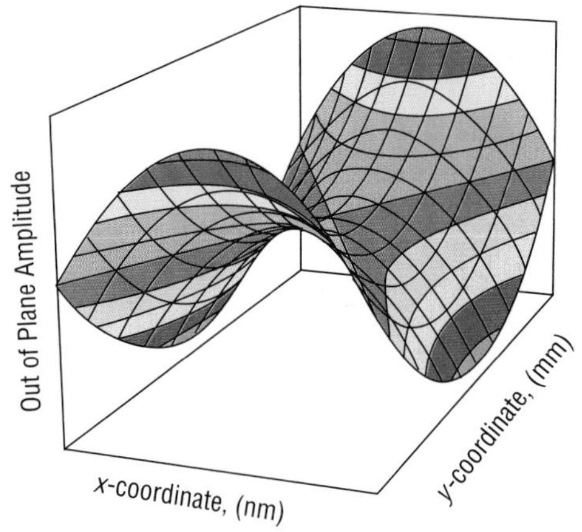

Figure 1

Very often it is difficult to find an analytical solution to an engineering problem. But more often one can find a series solution. Given the current computational capabilities of computers, it requires very little computational time to obtain the solution, that is, adding up enough terms so that the solution converges. To introduce the concept of a series solution it is worthwhile to investigate a simple application that has both an analytical solution and one that can be obtained by a series solution.

Consider the damped vibration of a simple mass, spring, and damper system as shown in Figure 2. A simple example of this is a car, where the mass is the car, supported by a system of springs, and the damper is the system of four shock absorbers.

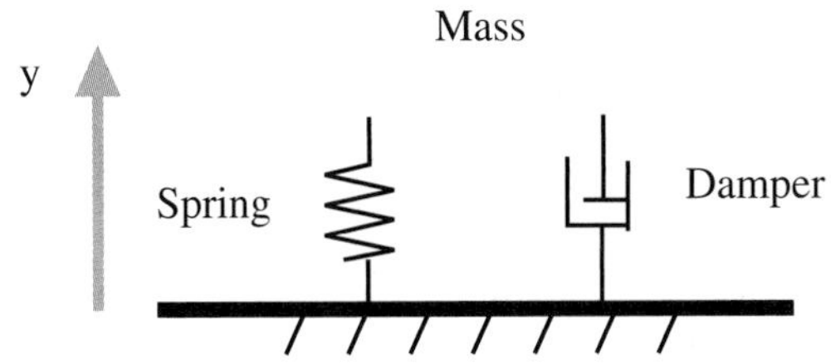

Figure 2 Mass, spring, and damper vibration system

If there was no damping in the system, no energy would be dissipated and the mass would oscillate at a natural frequency in a sinusoidal manner as shown in Figure 3. If we add damping to the system, the response decays, in other words, the energy is dissipated and the amplitude decreases with time, as shown in Figure 4. Equation (1) provides a function that explains a system with damping motion.

$$y(t) = Ae^{-\zeta\omega_n t}\cos(\omega_d t - \phi) \quad \textbf{(1)}$$

where y is the height of the mass at time t, A is the amplitude, $\omega_n = \sqrt{\dfrac{k}{m}}$ is the natural frequency, k is the spring stiffness, m is the mass, ϕ is the phase angle, t is time, c is the damping constant, $\zeta = \dfrac{c}{2m\omega_n}$ is the viscous damping factor, $\omega_d = \omega_n\sqrt{1 - \zeta^2}$ is the frequency of the damped vibration. When there is no damping ($c = 0$), the behavior is sinusoidal free vibration (Figure 3). The effect of the damping constant is to decay the free vibration solution, resulting in Figure 4.

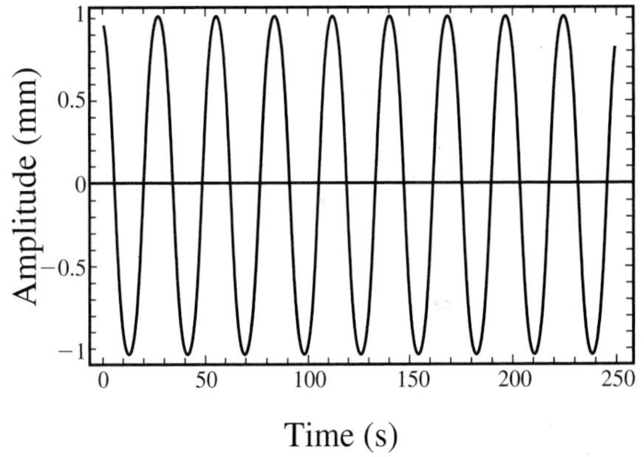

Figure 3 Vibration of system with no damping

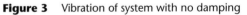

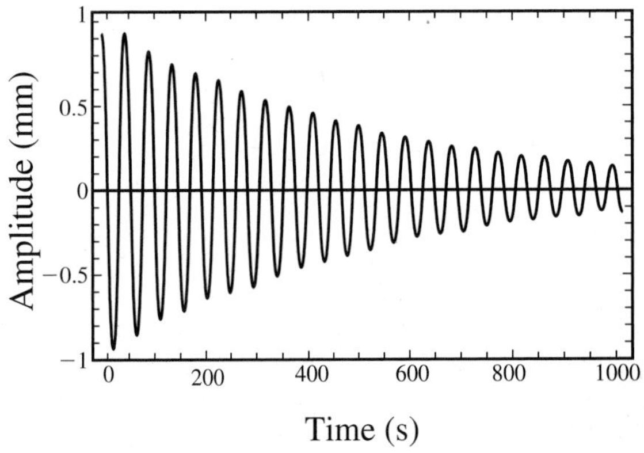

Figure 4 Vibration of same system with damping

To introduce a series solution we can begin with what appears to be a more complicated form of Equation (1) using complex numbers.

$$y(t) = e^{-\zeta\omega_n t}\left(A_1 e^{i\omega_d t} + A_2 e^{-i\omega_d t}\right) \qquad \textbf{(2)}$$

Note that each of the solutions given by (1) and (2) has the same term for the exponential decay (real) part. So

$$\cos(\omega_d t - \phi) = \left(A_1 e^{i\omega_d t} + A_2 e^{-i\omega_d t}\right)$$

From DeMoivre's Theorem and some calculus, it turns out that

$$e^{i\omega_d t} = \cos(\omega_d t) + i\sin(\omega_d t)$$

and

$$e^{-i\omega_d t} = \cos(\omega_d t) - i\sin(\omega_d t)$$

We use recursive formulae for the complex exponential terms in Equation (2) and derive the more simple solution given by Equation (1).

1. Write the first six terms of $e^{i\omega_d t} = \sum_{n=0}^{\infty} \frac{(i\omega_d t)^n}{n!}$ and $e^{-i\omega_d t} = \sum_{n=0}^{\infty} \frac{(-i\omega_d t)^n}{n!}$.

2. Write the first six terms of $\cos(\omega_d t) = \sum_{n=0}^{\infty} \frac{(-1)^{n+1}(i\omega_d t)^{2n-2}}{(2n-2)!}$.

3. Write the first six terms of $\sin(\omega_d t) = \sum_{n=0}^{\infty} \frac{(-1)^{n+1}(i\omega_d t)^{2n-1}}{(2n-1)!}$.

4. Combine the results from Questions 2 and 3 and you should begin to see terms found in Question 1.

5. A few tricks will help you obtain the exact solution given by Equation 1. [**Hint:** $A_1 + A_2 = A\cos\phi$ and $i(A_1 - A_2) = A\sin\phi$]

CHAPTER

13 Counting and Probability

Outline

FIELD TRIP TO MOTOROLA

How would you like to design products that allow you to talk on a cellular phone from anywhere in the world or surf the web while riding in a car? Motorola designs and manufactures products that do these things and more. Probability of error is a performance measure that is important to the design of wireless communication systems and products. Many parts of a wireless system are designed to decrease the probability of error. This is what allows you to hear someone clearly on your cellular phone or surf the web more quickly. In the chapter project, you will learn how speech signals are coded for transmission and error control.

INTERVIEW AT MOTOROLA

Theresa Fry received a Ph.D. in Electrical Engineering from Northwestern University with a concentration in networking and telecommunications. Her graduate studies concentrated on the performance modeling of digital wireless communication systems. She is currently working at Motorola Labs, where her research involves the capacity analysis and design of future wireless systems. She uses mathematics to design models that represent real-life situations, to determine whether a proposed new type of wireless phone will work in any environment. She has also needed math to understand such natural phenomena as the transmission and reception of sound waves, and the results of the mixture of many sound waves in the air. Mathematics form the basis of communications engineering, which is an exciting and dynamic technology.

13.1 SETS AND COUNTING

OBJECTIVES
1. Find All the Subsets of a Set
2. Find the Intersection and Union of Sets
3. Find the Complement of a Set
4. Count the Number of Elements in a Set

SETS

A **set** is a well-defined collection of distinct objects. The objects of a set are called its **elements.** By **well-defined,** we mean that there is a rule that enables us to determine whether a given object is an element of the set. If a set has no elements, it is called the **empty set,** or **null set,** and is denoted by the symbol $\varnothing$.

Because the elements of a set are distinct, we never repeat elements. For example, we would never write $\{1, 2, 3, 2\}$; the correct listing is $\{1, 2, 3\}$. Because a set is a collection, the order in which the elements are listed is immaterial. Thus, $\{1, 2, 3\}$, $\{1, 3, 2\}$, $\{2, 1, 3\}$, and so on, all represent the same set.

EXAMPLE 1 **Writing the Elements of a Set**

Write the set consisting of the possible results (outcomes) from tossing a coin twice. Use H for *heads* and T for *tails.*

Solution In tossing a coin twice, we can get heads each time, HH; or heads the first time and tails the second, HT; or tails the first time and heads the second, TH; or tails each time, TT. Because no other possibilities exist, the set of outcomes is

$$\{\text{HH}, \text{HT}, \text{TH}, \text{TT}\}$$

■

1 If two sets A and B have precisely the same elements, we say that A and B are **equal** and write $A = B$.

If each element of a set A is also an element of a set B, we say that A is a **subset** of B and write $A \subseteq B$.

If $A \subseteq B$ and $A \neq B$, then we say that A is a **proper subset** of B and write $A \subset B$.

Thus, if $A \subseteq B$, every element in set A is also in set B, but B may or may not have additional elements. If $A \subset B$, every element in A is also in B, and B has at least one element not found in A.

Finally, we agree that the empty set is a subset of every set; that is,

$$\varnothing \subseteq A \qquad \text{for any set } A$$

EXAMPLE 2 **Finding All the Subsets of a Set**

Write down all the subsets of the set $\{a, b, c\}$.

Solution To organize our work, we write down all the subsets with no elements, then those with one element, then those with two elements, and finally those with three elements. These will give us all the subsets. Do you see why?

0 Elements	1 Element	2 Elements	3 Elements
$\varnothing$	$\{a\}, \{b\}, \{c\}$	$\{a, b\}, \{b, c\}, \{a, c\}$	$\{a, b, c\}$

NOW WORK PROBLEM 21.

2 If A and B are sets, the **intersection** of A with B, denoted $A \cap B$, is the set consisting of elements that belong to both A and B. The **union** of A with B, denoted $A \cup B$, is the set consisting of elements that belong to either A or B, or both.

EXAMPLE 3 **Finding the Intersection and Union of Sets**

Let $A = \{1, 3, 5, 8\}$, $B = \{3, 5, 7\}$, and $C = \{2, 4, 6, 8\}$. Find:

(a) $A \cap B$ (b) $A \cup B$ (c) $B \cap (A \cup C)$

Solution (a) $A \cap B = \{1, 3, 5, 8\} \cap \{3, 5, 7\} = \{3, 5\}$

(b) $A \cup B = \{1, 3, 5, 8\} \cup \{3, 5, 7\} = \{1, 3, 5, 7, 8\}$

(c) $B \cap (A \cup C) = \{3, 5, 7\} \cap [\{1, 3, 5, 8\} \cup \{2, 4, 6, 8\}]$
$= \{3, 5, 7\} \cap \{1, 2, 3, 4, 5, 6, 8\} = \{3, 5\}$

NOW WORK PROBLEM 5.

3 Usually, in working with sets, we designate a **universal set,** the set consisting of all the elements that we wish to consider. Once a universal set has been designated, we can consider elements of the universal set not found in a given set.

If A is a set, the **complement** of A, denoted $\bar{A}$, is the set consisting of all the elements in the universal set that are not in A.

EXAMPLE 4 **Finding the Complement of a Set**

If the universal set is $U = \{1, 2, 3, 4, 5, 6, 7, 8, 9\}$ and if $A = \{1, 3, 5, 7, 9\}$, then $\bar{A} = \{2,4,6,8\}$.

Notice that $A \cup \bar{A} = U$ and $A \cap \bar{A} = \varnothing$.

NOW WORK PROBLEM 13.

Figure 1

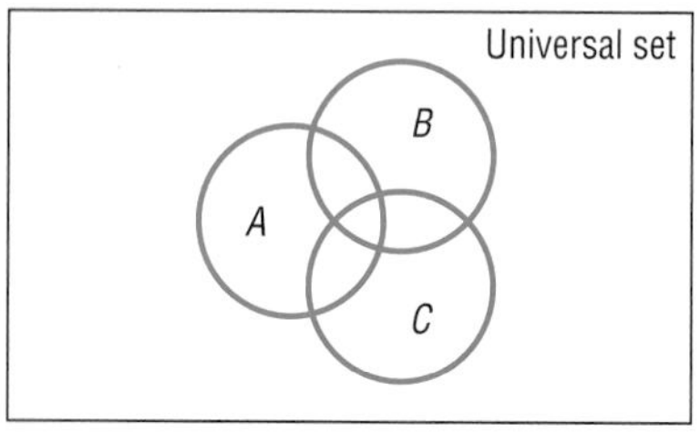

It is often helpful to draw pictures of sets. Such pictures, called **Venn diagrams,** represent sets as circles enclosed in a rectangle, which represents the universal set. Such diagrams often help us to visualize various relationships among sets. See Figure 1.

If we know that $A \subseteq B$, we might use the Venn diagram in Figure 2(a). If we know that A and B have no elements in common, that is, if $A \cap B = \varnothing$, we might use the Venn diagram in Figure 2(b). The sets A and B in Figure 2(b) are said to be **disjoint.**

Figure 2

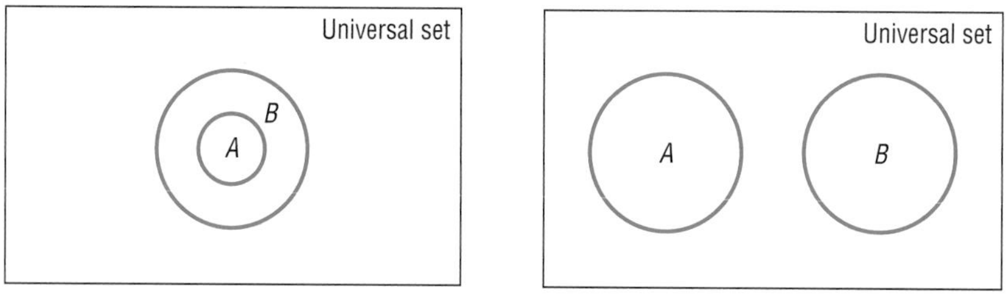

(a) $A \subseteq B$
subset

(b) $A \cap B = \varnothing$
disjoint sets

Figures 3(a), 3(b), and 3(c) use Venn diagrams to illustrate the definitions of intersection, union, and complement, respectively.

Figure 3

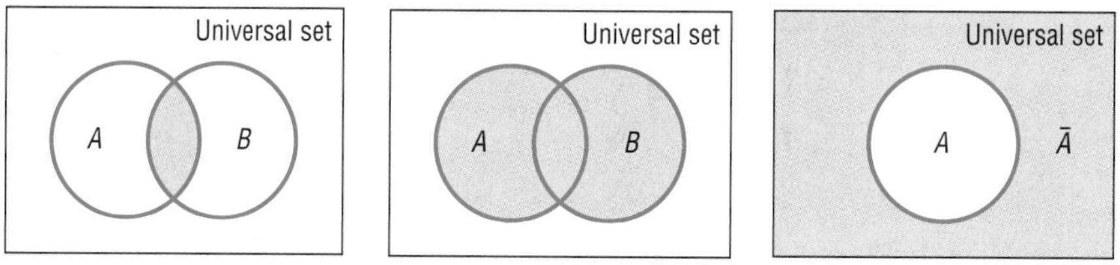

(a) $A \cap B$
intersection

(b) $A \cup B$
union

(c) $\bar{A}$
complement

COUNTING

4 As you count the number of students in a classroom or the number of pennies in your pocket, what you are really doing is matching, on a one-to-one basis, each object to be counted with the counting numbers $1, 2, 3, \ldots, n$, for some number n. If a set A matched up in this fashion with the set $\{1, 2, \ldots, 25\}$, you would conclude that there are 25 elements in the set A. We use the notation $n(A) = 25$ to indicate that there are 25 elements in the set A.

Because the empty set has no elements, we write

$$n(\varnothing) = 0$$

If the number of elements in a set is a nonnegative integer, we say that the set is **finite.** Otherwise, it is **infinite.** We shall concern ourselves only with finite sets.

From Example 2, we can see that a set with 3 elements has $2^3 = 8$ subsets. This result can be generalized.

If A is a set with n elements, then A has 2^n subsets.

For example, the set $\{a, b, c, d, e\}$ has $2^5 = 32$ subsets.

EXAMPLE 5 Analyzing Survey Data

In a survey of 100 college students, 35 were registered in College Algebra, 52 were registered in Computer Science I, and 18 were registered in both courses.

(a) How many students were registered in College Algebra or Computer Science I?

(b) How many were registered in neither course?

Solution (a) First, let A = set of students in College Algebra
B = set of students in Computer Science I

Then the given information tells us that

$$n(A) = 35, \qquad n(B) = 52, \qquad n(A \cap B) = 18$$

Figure 4

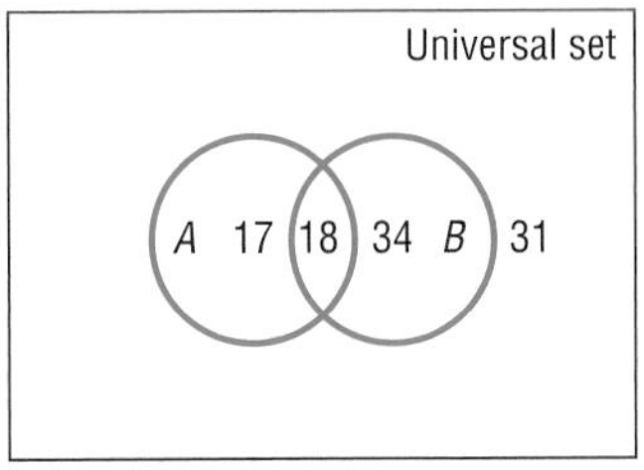

Refer to Figure 4. Since $n(A \cap B) = 18$, we know that the common part of the circles representing set A and set B has 18 elements. In addition, we know that the remaining portion of the circle representing set A will have $35 - 18 = 17$ elements. Similarly, we know that the remaining portion of the circle representing set B has $52 - 18 = 34$ elements. We conclude that $17 + 18 + 34 = 69$ students were registered in College Algebra or Computer Science I.

(b) Since 100 students were surveyed, it follows that $100 - 69 = 31$ were registered in neither course. ■

NOW WORK PROBLEM 35.

The solution to Example 5 contains the basis for a general counting formula. If we count the elements in each of two sets A and B, we necessarily count twice any elements that are in both A and B, that is, those elements in $A \cap B$. Thus, to count correctly the elements that are in A or B, that is, to find $n(A \cup B)$, we need to subtract those in $A \cap B$ from $n(A) + n(B)$.

Theorem **Counting Formula**

If A and B are finite sets, then

$$n(A \cup B) = n(A) + n(B) - n(A \cap B) \qquad (1)$$

■

Refer back to Example 5. Using (1), we have

$$\begin{aligned} n(A \cup B) &= n(A) + n(B) - n(A \cap B) \\ &= 35 + 52 - 18 \\ &= 69 \end{aligned}$$

There are 69 students registered in College Algebra or Computer Science I.

A special case of the counting formula (1) occurs if A and B have no elements in common. In this case, $A \cap B = \varnothing$, so $n(A \cap B) = 0$.

Theorem

Addition Principle of Counting

If two sets A and B have no elements in common, then

$$n(A \cup B) = n(A) + n(B) \qquad (2)$$

■

We can generalize formula (2).

Theorem

General Addition Principle of Counting

If, for n sets $A_1, A_2, \ldots, A_n$, no two have elements in common, then

$$n(A_1 \cup A_2 \cup \cdots \cup A_n) = n(A_1) + n(A_2) + \cdots + n(A_n) \qquad (3)$$

■

EXAMPLE 6 Counting

In 1996 there were 738,028 full-time sworn law-enforcement officers in the United States. Table 1 lists the type of law-enforcement agencies and the corresponding number of full-time sworn officers from each agency.

TABLE 1

Type of Agency	Number of Full-Time Sworn Officers
Local police	410,956
Sheriff	152,922
State police	54,587
Special police	43,082
Texas constable	1,988
Federal	74,493

Source: Bureau of Justice Statistics

(a) How many full-time sworn law-enforcement officers in the United States were local police or sheriffs?

(b) How many full-time sworn law-enforcement officers in the United States were local police, sheriffs, or state police?

Solution Let A represent the set of local police, B represent the set of sheriffs, and C represent the set of state police. No two of the sets A, B, and C have elements in common since a single officer cannot be classified in more than one type of agency.

(a) Using formula (2), we have

$$n(A \cup B) = n(A) + n(B) = 410{,}956 + 152{,}922 = 563{,}878$$

There were 563,878 officers that were local police or sheriffs.

(b) Using formula (3), we have

$$n(A \cup B \cup C) = n(A) + n(B) + n(C) = 410{,}956 + 152{,}922 + 54{,}587 = 618{,}465$$

There were 618,465 officers that were local police, sheriffs, or state police. ■

NOW WORK PROBLEM 39.

13.1 EXERCISES

In Problems 1–10, use $A = \{1, 3, 5, 7, 9\}$, $B = \{1, 5, 6, 7\}$, *and* $C = \{1, 2, 4, 6, 8, 9\}$ *to find each set.*

1. $A \cup B$ **2.** $A \cup C$ **3.** $A \cap B$ **4.** $A \cap C$ 5. $(A \cup B) \cap C$

6. $(A \cap C) \cup (B \cap C)$ **7.** $(A \cap B) \cup C$ **8.** $(A \cup B) \cup C$ **9.** $(A \cup C) \cap (B \cup C)$ **10.** $(A \cap B) \cap C$

In Problems 11–20, use U = *universal set* = $\{0, 1, 2, 3, 4, 5, 6, 7, 8, 9\}$, $A = \{1, 3, 4, 5, 9\}$, $B = \{2, 4, 6, 7, 8\}$, *and* $C = \{1, 3, 4, 6\}$ *to find each set.*

11. $\overline{A}$ **12.** $\overline{C}$ 13. $\overline{A \cap B}$ **14.** $\overline{B \cup C}$ **15.** $\overline{A} \cup \overline{B}$

16. $\overline{B} \cap \overline{C}$ **17.** $\overline{A \cap \overline{C}}$ **18.** $\overline{\overline{B} \cup C}$ **19.** $\overline{A \cup B \cup C}$ **20.** $\overline{A \cap B \cap C}$

21. Write down all the subsets of $\{a, b, c, d\}$.

22. Write down all the subsets of $\{a, b, c, d, e\}$.

23. If $n(A) = 15$, $n(B) = 20$, and $n(A \cap B) = 10$, find $n(A \cup B)$.

24. If $n(A) = 20$, $n(B) = 40$, and $n(A \cup B) = 35$, find $n(A \cap B)$.

25. If $n(A \cup B) = 50$, $n(A \cap B) = 10$, and $n(B) = 20$, find $n(A)$.

26. If $n(A \cup B) = 60$, $n(A \cap B) = 40$, and $n(A) = n(B)$, find $n(A)$.

In Problems 27–34, use the information given in the figure.

27. How many are in set A?

28. How many are in set B?

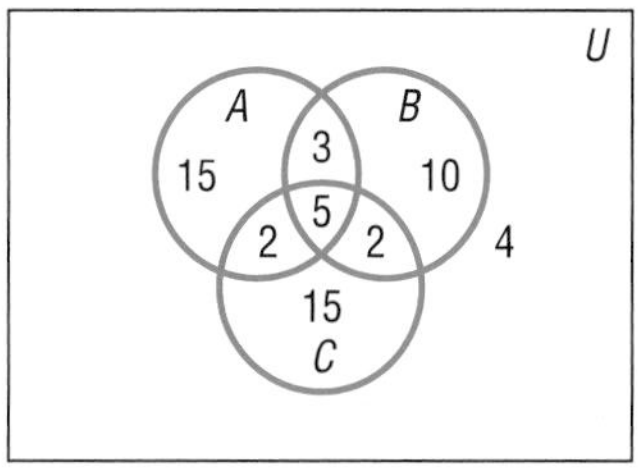

29. How many are in A or B?

30. How many are in A and B?

31. How many are in A but not C?

32. How many are not in A?

33. How many are in A and B and C?

34. How many are in A or B or C?

35. **Analyzing Survey Data** In a consumer survey of 500 people, 200 indicated that they would be buying a major appliance within the next month; 150 indicated that they would buy a car, and 25 said that they would purchase both a major appliance and a car. How many will purchase neither? How many will purchase only a car?

36. Analyzing Survey Data In a student survey, 200 indicated that they would attend Summer Session I and 150 indicated Summer Session II. If 75 students plan to attend both summer sessions and 275 indicated that they would attend neither session, how many students participated in the survey?

37. Analyzing Survey Data In a survey of 100 investors in the stock market,

50 owned shares in IBM
40 owned shares in AT&T
45 owned shares in GE
20 owned shares in both IBM and GE
15 owned shares in both AT&T and GE
20 owned shares in both IBM and AT&T
5 owned shares in all three

(a) How many of the investors surveyed did not have shares in any of the three companies?
(b) How many owned just IBM shares?
(c) How many owned just GE shares?
(d) How many owned neither IBM nor GE?
(e) How many owned either IBM or AT&T but no GE?

38. Classifying Blood Types Human blood is classified as either Rh+ or Rh−. Blood is also classified by type: A, if it contains an A antigen; B, if it contains a B antigen; AB, if it contains both A and B antigens; and O, if it contains neither antigen. Draw a Venn diagram illustrating the various blood types. Based on this classification, how many different kinds of blood are there?

39. The following data represent the marital status of males 18 years old and older in March 1997.

Marital Status	Number (in thousands)
Married, spouse present	54,654
Married, spouse absent	3,232
Widowed	2,686
Divorced	8,208
Never married	25,375

Source: Current Population Survey

(a) Determine the number of males 18 years old and older who are married.
(b) Determine the number of males 18 years old and older who are widowed or divorced.
(c) Determine the number of males 18 years old and older who are married, spouse absent, widowed, or divorced.

40. The following data represent the marital status of females 18 years old and older in March 1997.

Marital Status	Number (in thousands)
Married, spouse present	54,626
Married, spouse absent	4,122
Widowed	11,056
Divorced	11,107
Never married	20,503

Source: Current Population Survey

(a) Determine the number of females 18 years old and older who are married.
(b) Determine the number of females 18 years old and older who are widowed or divorced.
(c) Determine the number of females 18 years old and older who are married, spouse absent, widowed, or divorced.

41. Make up a problem different from any found in the text that requires the addition principle of counting to solve. Give it to a friend to solve and critique.

42. Investigate the notion of counting as it relates to infinite sets. Write an essay on your findings.

PREPARING FOR THIS SECTION

Before getting started, review the following concepts:

✓ Factorial (p. 814)

13.2 PERMUTATIONS AND COMBINATIONS

OBJECTIVES
1 Solve Counting Problems Using the Multiplication Principle
2 Solve Counting Problems Using Permutations
3 Solve Counting Problems Using Combinations

1 Counting plays a major role in many diverse areas, such as probability, statistics, and computer science. In this section we shall look at special types of counting problems and develop general formulas for solving them.

We begin with an example that will demonstrate a general counting principle.

EXAMPLE 1 Counting the Number of Possible Meals

The fixed-price dinner at Mabenka Restaurant provides the following choices:

Appetizer:	soup or salad
Entree:	baked chicken, broiled beef patty, baby beef liver, or roast beef au jus
Dessert:	ice cream or cheese cake

How many different meals can be ordered?

Solution Ordering such a meal requires three separate decisions:

Choose an Appetizer	**Choose an Entree**	**Choose a Dessert**
2 choices	4 choices	2 choices

Look at the **tree diagram** in Figure 5. We see that, for each choice of appetizer, there are 4 choices of entrees. And for each of these $2 \cdot 4 = 8$ choices, there are 2 choices for dessert. Thus, a total of

$$2 \cdot 4 \cdot 2 = 16$$

different meals can be ordered.

Figure 5

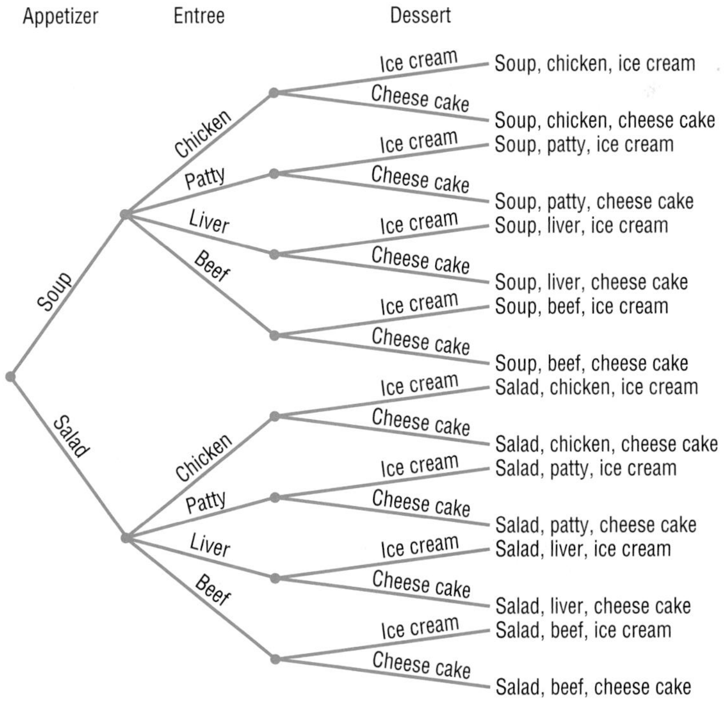

Example 1 illustrates a general counting principle.

Theorem **Multiplication Principle of Counting**

If a task consists of a sequence of choices in which there are p selections for the first choice, q selections for the second choice, r selections for the third choice, and so on, then the task of making these selections can be done in

$$p \cdot q \cdot r \cdot \ldots$$

different ways. ■

EXAMPLE 2 Counting Airport Codes

The International Airline Transportation Association (IATA) assigns three-letter codes to represent airport locations. For example, JFK represents Kennedy International in New York. How many different airport codes are possible?

Solution The task consists of making three selections. Each selection requires choosing a letter of the alphabet (26 choices). By the Multiplication Principle, there could be

$$26 \cdot 26 \cdot 26 = 17{,}576$$

different airport codes. ■

2 In Example 2, we were allowed to repeat a letter. For example, a valid airport code is FLL (Ft. Lauderdale International Airport), in which the letter L appears twice. In the next example, such repetition is not allowed.

EXAMPLE 3 Counting without Repetition

Suppose that we wish to establish a three-letter code using any of the 26 letters of the alphabet, but we require that no letter be used more than once. How many different three-letter codes are there?

Solution The task consists of making three selections. The first selection requires choosing from 26 letters. Because no letter can be used more than once, the second selection requires choosing from 25 letters. The third selection requires choosing from 24 letters. (Do you see why?) By the Multiplication Principle, there are

$$26 \cdot 25 \cdot 24 = 15{,}600$$

different three-letter codes with no letter repeated. ■

EXAMPLE 4 Birthday Problem

How many ways can 4 people have different birthdays? Assume that there are 365 days in a year.

Solution Once a birthday is selected, that birthday will not be repeated. Using the Multiplication Principle, the first person's birthday can be any one of 365 days, the second person's birthday can be any one of 364 days (we exclude the birthday of the first person), the third person's birthday can be any one of 363 days, and finally the fourth person's birthday can be any one of 362 days. Thus, there are $365 \cdot 364 \cdot 363 \cdot 362 = 17{,}458{,}601{,}160$ ways that four people can have different birthdays. ■

NOW WORK PROBLEMS 29 AND 47.

Examples 3 and 4 illustrate a type of counting problem referred to as a *permutation.*

A **permutation** is an ordered arrangement of n distinct objects without repetitions. The symbol $P(n, r)$ represents the number of permutations of n distinct objects, taken r at a time, where $r \leq n$.

For example, the question posed in Example 3 asks for the number of ways that the 26 letters of the alphabet can be arranged using three nonrepeated letters. The answer is

$$P(26, 3) = 26 \cdot 25 \cdot 24 = 15{,}600$$

To arrive at a formula for $P(n, r)$, we note that the task of obtaining an ordered arrangement of n objects in which only $r \leq n$ of them are used, without repeating any of them, requires making r selections. For the first selection, there are n choices; for the second selection, there are $n - 1$ choices; for the third selection, there are $n - 2$ choices; ...; for the rth selection, there are $n - (r - 1)$ choices. By the Multiplication Principle, we have

$$\begin{aligned} &\quad\;\; \text{1st} \quad\;\; \text{2nd} \qquad\quad \text{3rd} \qquad\qquad\qquad\quad r\text{th} \\ P(n, r) &= n \cdot (n - 1) \cdot (n - 2) \cdot \ldots \cdot [n - (r - 1)] \\ &= n \cdot (n - 1) \cdot (n - 2) \cdot \ldots \cdot (n - r + 1) \end{aligned}$$

This formula for $P(n, r)$ can be compactly written using factorial notation.*

$$\begin{aligned} P(n, r) &= n \cdot (n - 1) \cdot (n - 2) \cdot \ldots \cdot (n - r + 1) \\ &= n \cdot (n - 1) \cdot (n - 2) \cdot \ldots \cdot (n - r + 1) \cdot \frac{(n - r) \cdot \ldots \cdot 3 \cdot 2 \cdot 1}{(n - r) \cdot \ldots \cdot 3 \cdot 2 \cdot 1} = \frac{n!}{(n - r)!} \end{aligned}$$

Theorem **Number of Permutations of n Distinct Objects Taken r at a Time**

The number of different arrangements of n objects using $r \leq n$ of them, in which

1. the n objects are distinct,
2. once an object is used it cannot be repeated, and
3. order is important,

is given by the formula

$$P(n, r) = \frac{n!}{(n - r)!} \qquad \textbf{(1)}$$

■

*Recall that $0! = 1, 1! = 1, 2! = 2 \cdot 1, \ldots, n! = n(n - 1) \cdot \ldots \cdot 3 \cdot 2 \cdot 1$.

EXAMPLE 5 Computing Permutations

Evaluate: (a) $P(7, 3)$ (b) $P(6, 1)$ (c) $P(52, 5)$

Solution We shall work parts (a) and (b) in two ways.

(a)
$$P(7, 3) = \underbrace{7 \cdot 6 \cdot 5}_{\text{3 factors}} = 210$$

or

$$P(7, 3) = \frac{7!}{(7 - 3)!} = \frac{7!}{4!} = \frac{7 \cdot 6 \cdot 5 \cdot \cancel{4!}}{\cancel{4!}} = 210$$

(b)
$$P(6, 1) = \underbrace{6}_{\text{1 factor}} = 6$$

or

$$P(6, 1) = \frac{6!}{(6 - 1)!} = \frac{6!}{5!} = \frac{6 \cdot \cancel{5!}}{\cancel{5!}} = 6$$

(c) Figure 6 shows the solution using a TI-83 graphing calculator. Thus, $P(52, 5) = 311{,}875{,}200$.

Figure 6

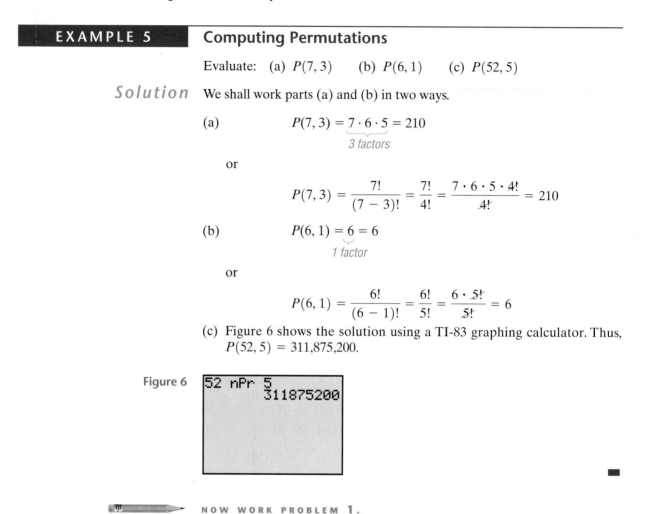

■

NOW WORK PROBLEM 1.

EXAMPLE 6 Lining Up People

In how many ways can 5 people be lined up?

Solution The 5 people are distinct. Once a person is in line, that person will not be repeated elsewhere in the line; and, in lining up people, order is important. Thus, we have a permutation of 5 objects taken 5 at a time. We can line up 5 people in

$$P(5, 5) = \underbrace{5 \cdot 4 \cdot 3 \cdot 2 \cdot 1}_{\text{5 factors}} = 5! = 120 \text{ ways}$$

■

NOW WORK PROBLEM 31.

COMBINATIONS

3 In a permutation, order is important; for example, the arrangements ABC, CAB, BAC, ... are considered different arrangements of the letters A, B, and C. In many situations, though, order is unimportant. For example, in the card game of poker, the order in which the cards are received does not matter; it is the *combination* of the cards that matters.

A **combination** is an arrangement, without regard to order, of n distinct objects without repetitions. The symbol $C(n, r)$ represents the number of combinations of n distinct objects taken r at a time, where $r \leq n$.

EXAMPLE 7 Listing Combinations

List all the combinations of the 4 objects a, b, c, d taken 2 at a time. What is $C(4, 2)$?

Solution One combination of a, b, c, d taken 2 at a time is

$$ab$$

The object ba is excluded, because order is not important in a combination. The list of all such combinations (convince yourself of this) is

$$ab, \quad ac, \quad ad, \quad bc, \quad bd, \quad cd$$

Thus,

$$C(4, 2) = 6$$

■

We can find a formula for $C(n, r)$ by noting that the only difference between a permutation and a combination is that we disregard order in combinations. Thus, to determine $C(n, r)$, we need only eliminate from the formula for $P(n, r)$ the number of permutations that were simply rearrangements of a given set of r objects. This can be determined from the formula for $P(n, r)$ by calculating $P(r, r) = r!$. So, if we divide $P(n, r)$ by $r!$, we will have the desired formula for $C(n, r)$:

$$C(n, r) = \frac{P(n, r)}{r!} \underset{\text{Use formula (1).}}{\overset{\uparrow}{=}} \frac{n!/(n - r)!}{r!} = \frac{n!}{(n - r)!r!}$$

We have proved the following result:

Theorem

Number of Combinations of *n* Distinct Objects Taken *r* at a Time

The number of different arrangements of n objects using $r \leq n$ of them, in which

1. the n objects are distinct,
2. once an object is used, it cannot be repeated, and
3. order is not important,

is given by the formula

$$C(n, r) = \frac{n!}{(n - r)!r!} \tag{2}$$

■

Based on formula (2), we discover that the symbol $C(n, r)$ and the symbol $\binom{n}{r}$ for the binomial coefficients are, in fact, the same. Thus, the Pascal triangle (see Section 12.5) can be used to find the value of $C(n, r)$. However, because it is more practical and convenient, we will use formula (2) instead.

EXAMPLE 8 Using Formula (2)

Use formula (2) to find the value of each expression.

(a) $C(3,1)$ (b) $C(6,3)$ (c) $C(n,n)$ (d) $C(n,0)$ (e) $C(52,5)$

Solution

(a) $C(3,1) = \dfrac{3!}{(3-1)!\,1!} = \dfrac{3!}{2!\,1!} = \dfrac{3 \cdot \cancel{2} \cdot 1}{\cancel{2} \cdot 1 \cdot 1} = 3$

(b) $C(6,3) = \dfrac{6!}{(6-3)!\,3!} = \dfrac{6 \cdot 5 \cdot 4 \cdot \cancel{3!}}{3! \cdot \cancel{3!}} = \dfrac{\cancel{6} \cdot 5 \cdot 4}{\cancel{6}} = 20$

Figure 7

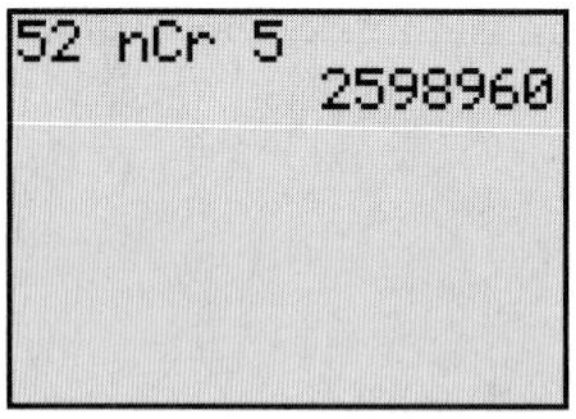

(c) $C(n,n) = \dfrac{n!}{(n-n)!\,n!} = \dfrac{\cancel{n!}}{0!\,\cancel{n!}} = \dfrac{1}{1} = 1$

(d) $C(n,0) = \dfrac{n!}{(n-0)!\,0!} = \dfrac{\cancel{n!}}{\cancel{n!}\,0!} = \dfrac{1}{1} = 1$

(e) Figure 7 shows the solution using a TI-83 graphing calculator. Thus, $C(52, 5) = 2{,}598{,}960$. ▬

NOW WORK PROBLEM 9.

EXAMPLE 9 Forming Committees

How many different committees of 3 people can be formed from a pool of 7 people?

Solution The 7 people are distinct. More important, though, is the observation that the order of being selected for a committee is not significant. Thus, the problem asks for the number of combinations of 7 objects taken 3 at a time.

$$C(7,3) = \frac{7!}{4!\,3!} = \frac{7 \cdot 6 \cdot 5 \cdot \cancel{4!}}{\cancel{4!}3!} = \frac{7 \cdot \cancel{6} \cdot 5}{\cancel{6}} = 35$$ ▬

EXAMPLE 10 Forming Committees

In how many ways can a committee consisting of 2 faculty members and 3 students be formed if 6 faculty members and 10 students are eligible to serve on the committee?

Solution The problem can be separated into two parts: the number of ways that the faculty members can be chosen, $C(6, 2)$, and the number of ways that the student members can be chosen, $C(10, 3)$. By the Multiplication Principle, the committee can be formed in

$$C(6,2) \cdot C(10,3) = \frac{6!}{4!\,2!} \cdot \frac{10!}{7!\,3!} = \frac{6 \cdot 5 \cdot \cancel{4!}}{\cancel{4!}2!} \cdot \frac{10 \cdot 9 \cdot 8 \cdot \cancel{7!}}{\cancel{7!}3!}$$

$$= \frac{30}{2} \cdot \frac{720}{6} = 1800 \text{ ways}$$ ▬

NOW WORK PROBLEM 49.

Permutations with Repetition

Recall that a permutation involves counting *distinct* objects. A permutation in which some of the objects are repeated is called a **permutation with repetition.** Some books refer to this as a **nondistinguishable permutation.**

Let's look at an example.

EXAMPLE 11 Forming Different Words

How many different words can be formed using all the letters in the word REARRANGE?

Solution Each word formed will have 9 letters: 3 R's, 2 A's, 2 E's, 1 N, and 1 G. To construct each word, we need to fill in 9 positions with the 9 letters:

$$\overline{1}\ \overline{2}\ \overline{3}\ \overline{4}\ \overline{5}\ \overline{6}\ \overline{7}\ \overline{8}\ \overline{9}$$

The process of forming a word consists of five tasks:

Task 1: Choose the positions for the 3 R's.

Task 2: Choose the positions for the 2 A's.

Task 3: Choose the positions for the 2 E's.

Task 4: Choose the position for the 1 N.

Task 5: Choose the position for the 1 G.

Task 1 can be done in $C(9, 3)$ ways. There then remain 6 positions to be filled, so Task 2 can be done in $C(6, 2)$ ways. There remain 4 positions to be filled, so Task 3 can be done in $C(4, 2)$ ways. There remain 2 positions to be filled, so Task 4 can be done in $C(2, 1)$ ways. The last position can be filled in $C(1, 1)$ way. Using the Multiplication Principle, the number of possible words that can be formed is

$$C(9,3)\cdot C(6,2)\cdot C(4,2)\cdot C(2,1)\cdot C(1,1) = \frac{9!}{3!\cdot \cancel{6!}}\cdot\frac{\cancel{6!}}{2!\cdot \cancel{4!}}\cdot\frac{\cancel{4!}}{2!\cdot \cancel{2!}}\cdot\frac{\cancel{2!}}{1!\cdot \cancel{1!}}\cdot\frac{\cancel{1!}}{0!\cdot 1!}$$

$$= \frac{9!}{3!\cdot 2!\cdot 2!\cdot 1!\cdot 1!}$$

■

The form of the answer to Example 11 is suggestive of a general result. Had the letters in REARRANGE each been different, there would have been $P(9, 9) = 9!$ possible words formed. This is the numerator of the answer. The presence of 3 R's, 2 A's, and 2 E's reduces the number of different words, as the entries in the denominator illustrate. We are led to the following result:

Theorem **Permutations with Repetition**

The number of permutations of n objects of which n_1 are of one kind, n_2 are of a second kind, ..., and n_k are of a kth kind is given by

$$\frac{n!}{n_1!\cdot n_2!\cdot \ldots \cdot n_k!} \qquad \textbf{(3)}$$

where $n = n_1 + n_2 + \cdots + n_k$.

■

EXAMPLE 12 Arranging Flags

How many different vertical arrangements are there of 8 flags if 4 are white, 3 are blue, and 1 is red?

Solution We seek the number of permutations of 8 objects, of which 4 are of one kind, 3 of a second kind, and 1 of a third kind. Using formula (3), we find that there are

$$\frac{8!}{4! \cdot 3! \cdot 1!} = \frac{8 \cdot 7 \cdot 6 \cdot 5 \cdot \cancel{4!}}{\cancel{4!} \cdot 3! \cdot 1!} = 280 \text{ different arrangements}$$

NOW WORK PROBLEM 55.

13.2 EXERCISES

In Problems 1–8, find the value of each permutation. Verify your results using a graphing calculator.

1. $P(6,2)$
2. $P(7,2)$
3. $P(4,4)$
4. $P(8,8)$
5. $P(7,0)$
6. $P(9,0)$
7. $P(8,4)$
8. $P(8,3)$

In Problems 9–16, use formula (2) to find the value of each combination. Verify your results using a graphing calculator.

9. $C(8,2)$
10. $C(8,6)$
11. $C(7,4)$
12. $C(6,2)$
13. $C(15,15)$
14. $C(18,1)$
15. $C(26,13)$
16. $C(18,9)$

17. List all the permutations of 5 objects a, b, c, d, and e taken 3 at a time. What is $P(5,3)$?
18. List all the permutations of 5 objects a, b, c, d, and e taken 2 at a time. What is $P(5,2)$?
19. List all the permutations of 4 objects 1, 2, 3, and 4 taken 3 at a time. What is $P(4,3)$?
20. List all the permutations of 6 objects 1, 2, 3, 4, 5, and 6 taken 3 at a time. What is $P(6,3)$?
21. List all the combinations of 5 objects a, b, c, d, and e taken 3 at a time. What is $C(5,3)$?
22. List all the combinations of 5 objects a, b, c, d, and e taken 2 at a time. What is $C(5,2)$?
23. List all the combinations of 4 objects 1, 2, 3, and 4 taken 3 at a time. What is $C(4,3)$?
24. List all the combinations of 6 objects 1, 2, 3, 4, 5, and 6 taken 3 at a time. What is $C(6,3)$?
25. A man has 5 shirts and 3 ties. How many different shirt and tie combinations can he wear?
26. A woman has 3 blouses and 5 skirts. How many different outfits can she wear?
27. **Forming Codes** How many two-letter codes can be formed using the letters A, B, C, and D? Repeated letters are allowed.
28. **Forming Codes** How many two-letter codes can be formed using the letters A, B, C, D, and E? Repeated letters are allowed.
29. **Forming Numbers** How many three-digit numbers can be formed using the digits 0 and 1? Repeated digits are allowed.
30. **Forming Numbers** How many three-digit numbers can be formed using the digits 0, 1, 2, 3, 4, 5, 6, 7, 8, and 9? Repeated digits are allowed.
31. In how many ways can 4 people be lined up?
32. In how many ways can 5 different boxes be stacked?
33. **Forming Codes** How many different three-letter codes are there if only the letters A, B, C, D, and E can be used and no letter can be used more than once?
34. **Forming Codes** How many different four-letter codes are there if only the letters A, B, C, D, E, and F can be used and no letter can be used more than once?
35. **Stocks on the NYSE** Companies whose stocks are listed on the New York Stock Exchange (NYSE) have their company name represented by either 1, 2, or 3 letters (repetition of letters is allowed). What is the maximum number of companies that can be listed on the New York Stock Exchange?
36. **Stocks on the NASDAQ** Companies whose stocks are listed on the NASDAQ stock exchange have their company name represented by either 4 or 5 letters (repetition of letters is allowed). What is the maximum number of companies that can be listed on the NASDAQ?
37. **Establishing Committees** In how many ways can a committee of 4 students be formed from a pool of 7 students?
38. **Establishing Committees** In how many ways can a committee of 3 professors be formed from a department having 8 professors?
39. **Possible Answers on a True/False Test** How many arrangements of answers are possible for a true/false test with 10 questions?
40. **Possible Answers on a Multiple-choice Test** How many arrangements of answers are possible in a multiple-choice test with 5 questions, each of which has 4 possible answers?

41. How many four-digit numbers can be formed using the digits 0, 1, 2, 3, 4, 5, 6, 7, 8, and 9 if the first digit cannot be 0? Repeated digits are allowed.

42. How many five-digit numbers can be formed using the digits 0, 1, 2, 3, 4, 5, 6, 7, 8, and 9 if the first digit cannot be 0 or 1? Repeated digits are allowed.

43. **Arranging Books** Five different mathematics books are to be arranged on a student's desk. How many arrangements are possible?

44. **Forming License Plate Numbers** How many different license plate numbers can be made using 2 letters followed by 4 digits selected from the digits 0 through 9, if
(a) Letters and digits may be repeated?
(b) Letters may be repeated, but digits may not be repeated?
(c) Neither letters nor digits may be repeated?

45. **Stock Portfolios** As a financial planner, you are asked to select one stock each from the following groups: 8 DOW stocks, 15 NASDAQ stocks, and 4 global stocks. How many different portfolios are possible?

46. **Combination Locks** A combination lock has 50 numbers on it. To open it, you turn to a number, then rotate clockwise to a second number, and then counterclockwise to the third number. How many different lock combinations are there?

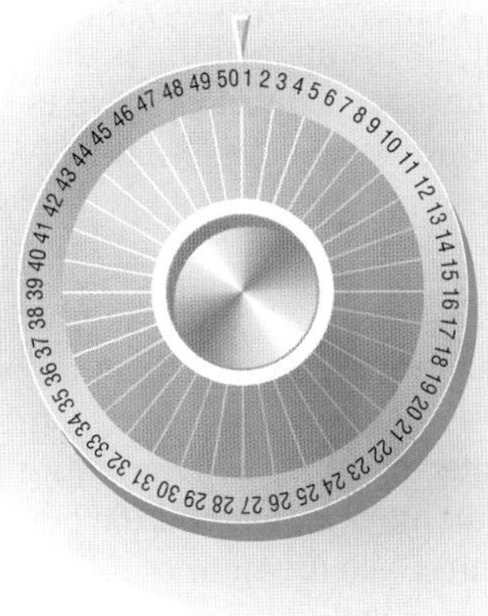

47. **Birthday Problem** In how many ways can 3 people have different birthdays? Assume that there are 365 days in a year.

48. **Birthday Problem** In how many ways can 5 people have different birthdays? Assume that there are 365 days in a year.

49. A student dance committee is to be formed consisting of 2 boys and 3 girls. If the membership is to be chosen from 4 boys and 8 girls, how many different committees are possible?

50. **Baseball Teams** A baseball team has 15 members. Four of the players are pitchers, and the remaining 11 members can play any position. How many different teams of 9 players can be formed?

51. The student relations committee of a college consists of 2 administrators, 3 faculty members, and 5 students. Four administrators, 8 faculty members, and 20 students are eligible to serve. How many different committees are possible?

52. **Football Teams** A defensive football squad consists of 25 players. Of these, 10 are linemen, 10 are linebackers, and 5 are safeties. How many different teams of 5 linemen, 3 linebackers, and 3 safeties can be formed?

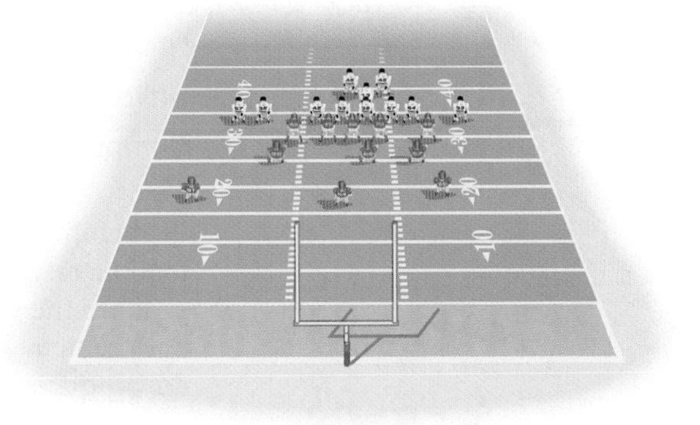

53. **Baseball** In the American Baseball League, a designated hitter may be used. How many batting orders is it possible for a manager to use? (There are 9 regular players on a team.)

54. **Baseball** In the National Baseball League, the pitcher usually bats ninth. If this is the case, how many batting orders is it possible for a manager to use?

55. **Forming Words** How many different 9-letter words (real or imaginary) can be formed from the letters in the word ECONOMICS?

56. **Forming Words** How many different 11-letter words (real or imaginary) can be formed from the letters in the word MATHEMATICS?

57. **Senate Committees** The U.S. Senate has 100 members. Suppose that it is desired to place each senator on exactly 1 of 7 possible committees. The first committee has 22 members, the second has 13, the third has 10, the fourth has 5, the fifth has 16, and the sixth and seventh have 17 apiece. In how many ways can these committees be formed?

58. **World Series** In the World Series the American League team (A) and the National League team (N) play until

one team wins four games. If the sequence of winners is designed by letters (for example, *NAAAA* means that the National League team won the first game and the American League won the next four), how many different sequences are possible?

59. Basketball Teams A basketball team has 6 players who play guard (2 of 5 starting positions). How many different teams are possible, assuming that the remaining 3 positions are filled and it is not possible to distinguish a left guard from a right guard?

60. Basketball Teams On a basketball team of 12 players, 2 only play center, 3 only play guard, and the rest play forward (5 players on a team: 2 forwards, 2 guards, and 1 center). How many different teams are possible, assuming that it is not possible to distinguish left and right guards and left and right forwards?

61. Selecting Objects An urn contains 7 white balls and 3 red balls. Three balls are selected. In how many ways can the 3 balls be drawn from the total of 10 balls:
(a) If 2 balls are white and 1 is red?
(b) If all 3 balls are white?
(c) If all 3 balls are red?

62. Selecting Objects An urn contains 15 red balls and 10 white balls. Five balls are selected. In how many ways can the 5 balls be drawn from the total of 25 balls:
(a) If all 5 balls are red?
(b) If 3 balls are red and 2 are white?
(c) If at least 4 are red balls?

13.3 PROBABILITY OF EQUALLY LIKELY OUTCOMES

OBJECTIVES
1. Construct Probability Models
2. Compute Probabilities of Equally Likely Outcomes
3. Utilize the Addition Rule to Find Probabilities
4. Utilize the Complement Rule to Find Probabilities
5. Compute Probabilities Using Permutations and Combinations

Probability is an area of mathematics that deals with experiments that yield random results, yet admit a certain regularity. Such experiments do not always produce the same result or outcome, so the result of any one observation is not predictable. However, the results of the experiment over a long period do produce regular patterns that enable us to predict with remarkable accuracy.

EXAMPLE 1 Tossing a Fair Coin

In tossing a fair coin, we know that the outcome is either a head or a tail. On any particular throw, we cannot predict what will happen, but, if we toss the coin many times, we observe that the number of times that a head comes up is approximately equal to the number of times that we get a tail. It seems reasonable, therefore, to assign a probability of $\frac{1}{2}$ that a head comes up and a probability of $\frac{1}{2}$ that a tail comes up. ■

PROBABILITY MODELS

1 The discussion in Example 1 constitutes the construction of a **probability model** for the experiment of tossing a fair coin once. A probability model has two components: a sample space and an assignment of probabilities. A **sample space** S is a set whose elements represent all the possibilities that can occur as a result of the experiment. Each element of S is called an **outcome.** To each outcome, we assign a number, called the **probability** of that outcome, which has two properties:

1. Each probability is nonnegative.
2. The sum of all the probabilities equals 1.

If a probability model has the sample space

$$S = \{e_1, e_2, \ldots, e_n\}$$

where $e_1, e_2, \ldots, e_n$ are the possible outcomes, and if $P(e_1), P(e_2), \ldots, P(e_n)$ denote the respective probabilities of these outcomes, then

$$P(e_1) \geq 0, P(e_2) \geq 0, \ldots, P(e_n) \geq 0 \qquad \textbf{(1)}$$

$$\sum_{i=1}^{n} P(e_i) = P(e_1) + P(e_2) + \cdots + P(e_n) = 1 \qquad \textbf{(2)}$$

EXAMPLE 2 Determining Probability Models

In a bag of M&Ms the candies are colored red, green, blue, brown, yellow, and orange. Suppose that a candy is drawn from the bag and the color is recorded. The sample space of this experiment is {red, green, blue, brown, yellow, orange}. Determine which of the following are probability models.

(a)

Outcome	**Probability**
{red}	0.3
{green}	0.15
{blue}	0
{brown}	0.15
{yellow}	0.2
{orange}	0.2

(b)

Outcome	**Probability**
{red}	0.1
{green}	0.1
{blue}	0.1
{brown}	0.4
{yellow}	0.2
{orange}	0.3

(c)

Outcome	**Probability**
{red}	0.3
{green}	−0.3
{blue}	0.2
{brown}	0.4
{yellow}	0.2
{orange}	0.2

(d)

Outcome	**Probability**
{red}	0
{green}	0
{blue}	0
{brown}	0
{yellow}	1
{orange}	0

Solution

(a) This model is a probability model since all the outcomes have probabilities that are nonnegative and the sum of the probabilities is 1.

(b) This model is not a probability model because the sum of the probabilities is not 1.

(c) This model is not a probability model because $P(\text{green})$ is less than 0. Recall, all probabilities must be nonnegative.

(d) This model is a probability model because all the outcomes have probabilities that are nonnegative, and the sum of the probabilities is 1. Notice that $P(\text{yellow}) = 1$, meaning that this outcome will occur with 100% certainty each time that the experiment is repeated. This means that the entire bag of M&Ms has yellow candies. ■

NOW WORK PROBLEM 3.

Let's look at an example of constructing a probability model.

EXAMPLE 3 Constructing a Probability Model

An experiment consists of rolling a fair die once.* Construct a probability model for this experiment.

Figure 8

Solution A sample space S consists of all the possibilities that can occur. Because rolling the die will result in one of six faces showing, the sample space S consists of

$$S = \{1, 2, 3, 4, 5, 6\}$$

Because the die is fair, one face is no more likely to occur than another. As a result, our assignment of probabilities is

$$P(1) = \tfrac{1}{6} \qquad P(2) = \tfrac{1}{6}$$
$$P(3) = \tfrac{1}{6} \qquad P(4) = \tfrac{1}{6}$$
$$P(5) = \tfrac{1}{6} \qquad P(6) = \tfrac{1}{6}$$

■

Now suppose that a die is loaded so that the probability assignments are

$$P(1) = 0, \quad P(2) = 0, \quad P(3) = \frac{1}{3}, \quad P(4) = \frac{2}{3}, \quad P(5) = 0, \quad P(6) = 0$$

This assignment would be made if the die were loaded so that only a 3 or 4 could occur and the 4 is twice as likely as the 3 to occur. This assignment is consistent with the definition, since each assignment is nonnegative and the sum of all the probability assignments equals 1.

NOW WORK PROBLEM **19.**

EXAMPLE 4 Constructing a Probability Model

An experiment consists of tossing a coin. The coin is weighted so that heads (H) is three times as likely to occur as tails (T). Construct a probability model for this experiment.

Solution The sample space S is $S = \{\text{H}, \text{T}\}$. If x denotes the probability that a tail occurs, then

$$P(\text{T}) = x \quad \text{and} \quad P(\text{H}) = 3x$$

Since the sum of the probabilities of the possible outcomes must equal 1, we have

$$P(\text{T}) + P(\text{H}) = x + 3x = 1$$
$$4x = 1$$
$$x = \frac{1}{4}$$

Thus, we assign the probabilities

$$P(\text{T}) = \frac{1}{4} \qquad \text{P(H)} = \frac{3}{4}$$

■

NOW WORK PROBLEM **23.**

* A die is a cube with each face having either 1, 2, 3, 4, 5, or 6 dots on it. See Figure 8.

In working with probability models, the term **event** is used to describe a set of possible outcomes of the experiment. Thus, an event E is some subset of the sample space S. The **probability of an event** E, $E \neq \varnothing$, denoted by $P(E)$, is defined as the sum of the probabilities of the outcomes in E. We can also think of the probability of an event E as the likelihood that the event E occurs. If $E = \varnothing$, then $P(E) = 0$; if $E = S$, then $P(E) = P(S) = 1$.

Equally Likely Outcomes

2 When the same probability is assigned to each outcome of the sample space, the experiment is said to have **equally likely outcomes.**

Theorem **Probability for Equally Likely Outcomes**

If an experiment has n equally likely outcomes and if the number of ways that an event E can occur is m, then the probability of E is

$$P(E) = \frac{\text{Number of ways that } E \text{ can occur}}{\text{Number of all logical possibilities}} = \frac{m}{n} \qquad \textbf{(3)}$$

Thus, if S is the sample space of this experiment, then

$$P(E) = \frac{n(E)}{n(S)} \qquad \textbf{(4)}$$

■

EXAMPLE 5 Calculating Probabilities of Equally Likely Events

Calculate the probability that in a 3-child family there are 2 boys and 1 girl. Assume equally likely outcomes.

Solution We begin by constructing a tree diagram to help in listing the possible outcomes of the experiment. See Figure 9, where B stands for boy and G for girl. The sample space S of this experiment is

Figure 9

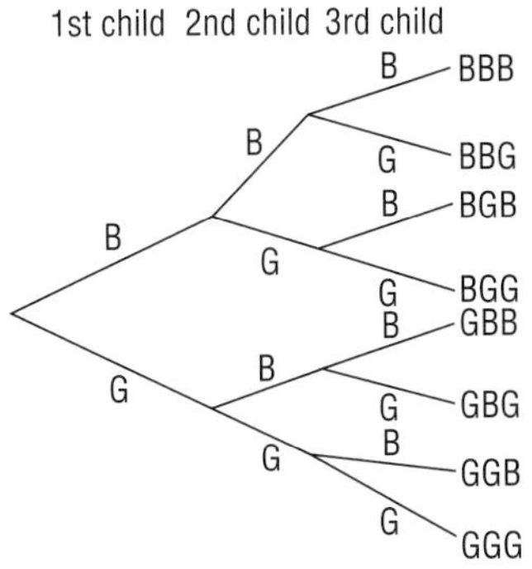

$$S = \{\text{BBB, BBG, BGB, BGG, GBB, GBG, GGB, GGG}\}$$

so $n(S) = 8$.

We wish to know the probability of the event E: "having two boys and one girl." From Figure 9, we conclude that $E = \{\text{BBG, BGB, GBB}\}$, so $n(E) = 3$. Since the outcomes are equally likely, the probability of E is

$$P(E) = \frac{n(E)}{n(S)} = \frac{3}{8}$$

■

NOW WORK PROBLEM 33.

Compound Probabilities

Thus far, we have calculated probabilities of single events. We will now compute probabilities of multiple events, called **compound probabilities.**

EXAMPLE 6 Computing Compound Probabilities

Consider the experiment of rolling a single fair die. Let E represent the event "roll an odd number" and let F represent the event "roll a 1 or 2."

(a) Write the event E and F.
(b) Write the event E or F.
(c) Compute $P(E)$ and $P(F)$.
(d) Compute $P(E \cap F)$.
(e) Compute $P(E \cup F)$.

Solution The sample space S of the experiment is $\{1, 2, 3, 4, 5, 6\}$, so $n(S) = 6$. Since the die is fair, the outcomes are equally likely. The event E: "roll an odd number" is $\{1, 3, 5\}$, and the event F: "roll a 1 or 2" is $\{1, 2\}$, so $n(E) = 3$ and $n(F) = 2$.

(a) The word *and* in probability means the intersection of two events. The event E and F is

$$E \cap F = \{1, 3, 5\} \cap \{1, 2\} = \{1\}, \qquad n(E \cap F) = 1$$

(b) The word *or* in probability means the union of the two events. The event E or F is

$$E \cup F = \{1, 3, 5\} \cup \{1, 2\} = \{1, 2, 3, 5\}, \qquad n(E \cup F) = 4$$

(c) We use formula (4).

$$P(E) = \frac{n(E)}{n(S)} = \frac{3}{6} = \frac{1}{2}, \qquad P(F) = \frac{n(F)}{n(S)} = \frac{2}{6} = \frac{1}{3}$$

(d) $P(E \cap F) = \dfrac{n(E \cap F)}{n(S)} = \dfrac{1}{6}$

(e) $P(E \cup F) = \dfrac{n(E \cup F)}{n(S)} = \dfrac{4}{6} = \dfrac{2}{3}$ ■

3 The **Addition Rule** can be used to find the probability of the union of two events.

Theorem **Addition Rule**

For any two events E and F,

$$P(E \cup F) = P(E) + P(F) - P(E \cap F) \qquad \textbf{(5)}$$

■

For example, we can use the Addition Rule to find $P(E \cup F)$ in Example 6(e). Then

$$P(E \cup F) = P(E) + P(F) - P(E \cap F) = \frac{1}{2} + \frac{1}{3} - \frac{1}{6} = \frac{3}{6} + \frac{2}{6} - \frac{1}{6} = \frac{4}{6} = \frac{2}{3}$$

as before.

EXAMPLE 7

Computing Probabilities of Compound Events Using the Addition Rule

If $P(E) = 0.2$, $P(F) = 0.3$, and $P(E \cap F) = 0.1$, find $P(E \cup F)$.

Solution We use the Addition Rule, formula (5).

$$P(E \cup F) = P(E) + P(F) - P(E \cap F) = 0.2 + 0.3 - 0.1 = 0.4$$

A Venn diagram can sometimes be used to obtain probabilities. To construct a Venn diagram representing the information in Example 7, we draw two sets E and F. We begin with the fact that $P(E \cap F) = 0.1$. See Figure 10(a). Then, since $P(E) = 0.2$ and $P(F) = 0.3$, we fill in E with $0.2 - 0.1 = 0.1$ and F with $0.3 - 0.1 = 0.2$. See Figure 10(b). Since $P(S) = 1$, we complete the diagram by inserting $1 - [0.1 + 0.1 + 0.2] = 0.6$. See Figure 10(c). Now it is easy to see, for example, that the probability of F, but not E, is 0.2. Also, the probability of neither E nor F is 0.6.

Figure 10

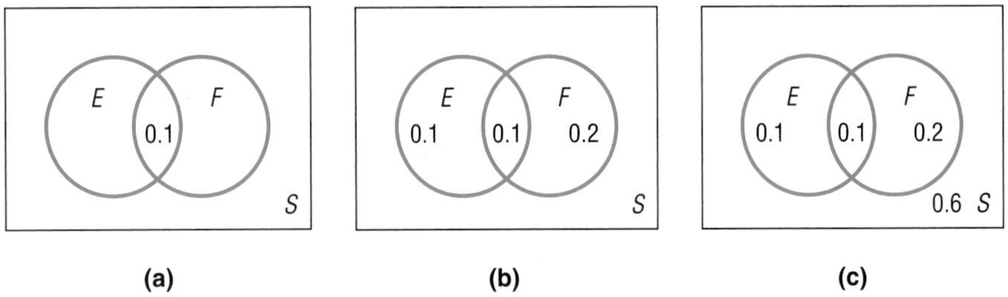

NOW WORK PROBLEM 41.

If events E and F are disjoint so that $E \cap F = \varnothing$, we say they are **mutually exclusive.** In this case, $P(E \cap F) = 0$, and the Addition Rule takes the following form:

Theorem **Mutually Exclusive Events**

If E and F are **mutually exclusive events,** then

$$P(E \cup F) = P(E) + P(F) \qquad \textbf{(6)}$$

EXAMPLE 8

Computing Compound Probabilities of Mutually Exclusive Events

If $P(E) = 0.4$ and $P(F) = 0.25$, and E and F are mutually exclusive, find $P(E \cup F)$.

Solution Since E and F are mutually exclusive, we use formula (6).

$$P(E \cup F) = P(E) + P(F) = 0.4 + 0.25 = 0.65$$

NOW WORK PROBLEM 43.

COMPLEMENTS

4 Recall, if A is a set, the complement of A, denoted $\bar{A}$, is the set of all elements in the universal set U not in A. We similarly define the complement of an event.

Complement of an Event

Let S denote the sample space of an experiment and let E denote an event. The **complement of E,** denoted $\bar{E}$, is the set of all outcomes in the sample space S that are not outcomes in the event E.

The complement $\bar{E}$ of an event E in a sample space S has the following two properties:

$$E \cap \bar{E} = \varnothing \qquad E \cup \bar{E} = S$$

Since E and $\bar{E}$ are mutually exclusive, it follows from (6) that

$$P(E \cup \bar{E}) = P(S) = 1 \qquad P(E) + P(\bar{E}) = 1 \qquad P(\bar{E}) = 1 - P(E)$$

Thus, we have the following result.

Theorem

Computing Probabilities of Complementary Events

If E represents any event and $\bar{E}$ represents the complement of E, then

$$P(\bar{E}) = 1 - P(E) \tag{7}$$

EXAMPLE 9 Computing Probabilities Using Complements

On the local news the weather reporter stated that the probability of rain is 40%. What is the probability that it will not rain?

Solution The complement of the event "rain" is "no rain." Thus,

$$P(\text{no rain}) = 1 - P(\text{rain}) = 1 - 0.4 = 0.6$$

There is a 60% chance of no rain tomorrow.

NOW WORK PROBLEM 47.

EXAMPLE 10 Birthday Problem

What is the probability that in a group of 10 people at least 2 people have the same birthday? Assume that there are 365 days in a year.

Solution We assume that a person is as likely to be born on one day as another, so we have equally likely outcomes.

We first determine the number of outcomes in the sample space S. There are 365 possibilities for each person's birthday. Since there are 10 people in the group, there are 365^{10} possibilities for the birthdays. [For one person in the group, there are 365 days on which his or her birthday can fall; for two people, there are $(365)(365) = 365^2$ pairs of days; and, in general, using the Multiplication Principle, for n people there are 365^n possibilities.] So

$$n(S) = 365^{10}$$

We wish to find the probability of the event E: "at least two people have the same birthday." It is difficult to count the elements in this set; it is much easier to count the elements of the complementary event $\bar{E}$: "no two people have the same birthday."

We find $n(\bar{E})$ as follows: Choose one person at random. There are 365 possibilities for his or her birthday. Choose a second person. There are 364 possibilities for this birthday, if no two people are to have the same birthday. Choose a third person. There are 363 possibilities left for this birthday. Finally, we arrive at the tenth person. There are 356 possibilities left for this birthday. By the Multiplication Principle, the total number of possibilities is

$$n(\bar{E}) = 365 \cdot 364 \cdot 363 \cdot \ldots \cdot 356$$

Hence, the probability of event $\bar{E}$ is

$$P(\bar{E}) = \frac{n(\bar{E})}{n(S)} = \frac{365 \cdot 364 \cdot 363 \cdot \ldots \cdot 356}{365^{10}} \approx 0.883$$

The probability of two or more people in a group of 10 people having the same birthday is then

$$P(E) = 1 - P(\bar{E}) = 1 - 0.883 = 0.117$$

The birthday problem can be solved for any group size. The following table gives the probabilities for two or more people having the same birthday for various group sizes. Notice that the probability is greater than $\frac{1}{2}$ for any group of 23 or more people.

	Number of People															
	5	10	15	20	21	22	23	24	25	30	40	50	60	70	80	90
Probability That Two or More Have the Same Birthday	0.027	0.117	0.253	0.411	0.444	0.476	0.507	0.538	0.569	0.706	0.891	0.970	0.994	0.99916	0.99991	0.99999

NOW WORK PROBLEM 65.

5 PROBABILITIES INVOLVING COMBINATIONS AND PERMUTATIONS

EXAMPLE 11 Computing Probabilities

Because of a mistake in packaging, 5 defective phones were packaged with 15 good ones. All phones look alike and have equal probability of being chosen. Three phones are selected.

(a) What is the probability that all 3 are defective?
(b) What is the probability that exactly 2 are defective?
(c) What is the probability that at least 2 are defective?

Solution The sample space S consists of the number of ways that 3 objects can be selected from 20 objects, that is, the number of combinations of 20 things taken 3 at a time.

$$n(S) = C(20, 3) = \frac{20!}{17! \cdot 3!} = \frac{20 \cdot 19 \cdot 18}{6} = 1140$$

Each of these outcomes is equally likely to occur.

(a) If E is the event "3 are defective," the number of elements in E is the number of ways the 3 defective phones can be chosen from the 5 defective phones: $C(5, 3) = 10$. Thus, the probability of E is

$$P(E) = \frac{n(E)}{n(S)} = \frac{10}{1140} \approx 0.0088$$

(b) If F is the event "exactly 2 are defective" and 3 phones are selected, the number of elements in F is the number of ways to select 2 defective phones from the 5 defective phones and 1 good phone from the 15 good ones. The first of these can be done in $C(5, 2)$ ways and the second in $C(15, 1)$ ways. By the Multiplication Principle, the event F can occur in

$$C(5, 2) \cdot C(15, 1) = \frac{5!}{3! \cdot 2!} \cdot \frac{15!}{14! \cdot 1!} = 10 \cdot 15 = 150 \text{ ways}$$

The probability of F is therefore

$$P(F) = \frac{n(F)}{n(S)} = \frac{150}{1140} \approx 0.1316$$

(c) The event G, "at least two are defective," when 3 are chosen is equivalent to requiring that either exactly 2 defective are chosen or exactly 3 defective are chosen. That is, $G = E \cup F$. Since E and F are mutually exclusive (it is not possible to select 2 defective phones and, at the same time, select 3 defective phones), we find that

$$P(G) = P(E) + P(F) \approx 0.0088 + 0.1316 = 0.1404$$ ■

NOW WORK PROBLEM 73.

EXAMPLE 12 Tossing a Coin

A fair coin is tossed 6 times.

(a) What is the probability of obtaining exactly 5 heads and 1 tail?

(b) What is the probability of obtaining between 4 and 6 heads, inclusive?

Solution The number of elements in the sample space S is found using the Multiplication Principle. Each toss results in a head (H) or a tail (T). Since the coin is tossed 6 times, we have

$$n(S) = \underbrace{2 \cdot 2 \cdot \ldots \cdot 2}_{\text{6 tosses}} = 2^6 = 64$$

The outcomes are equally likely since the coin is fair.

(a) Any sequence that contains 5 heads and 1 tail is determined once the position of the 5 heads (or 1 tail) is known. The number of ways that we can position 5 heads in a sequence of 6 slots is $C(6,5) = 6$. The probability of the event E, exactly 5 heads and 1 tail, is

$$P(E) = \frac{n(E)}{n(S)} = \frac{C(6,5)}{2^6} = \frac{6}{64} \approx 0.0938$$

(b) Let F be the event "between 4 and 6 heads, inclusive." To obtain between 4 and 6 heads is equivalent to the event "either 4 heads or 5 heads or 6 heads." Since each of these is mutually exclusive (it is impossible to obtain both 4 heads and 5 heads when tossing a coin 6 times), we have

$$\begin{aligned} P(F) &= P(4 \text{ heads or } 5 \text{ heads or } 6 \text{ heads}) \\ &= P(4 \text{ heads}) + P(5 \text{ heads}) + P(6 \text{ heads}) \end{aligned}$$

The probabilities on the right are obtained as in part (a). Thus,

$$P(F) = \frac{C(6,4)}{2^6} + \frac{C(6,5)}{2^6} + \frac{C(6,6)}{2^6} = \frac{15}{64} + \frac{6}{64} + \frac{1}{64} = \frac{22}{64} \approx 0.3438$$ ■

HISTORICAL FEATURE

Blaise Pascal (1623–1662)

Set theory, counting, and probability first took form as a systematic theory in an exchange of letters (1654) between Pierre de Fermat (1601–1665) and Blaise Pascal (1623–1662). They discussed the problem of how to divide the stakes in a game that is interrupted before completion, knowing how many points each player needs to win. Fermat solved the problem by listing all possibilities and counting the favorable ones, whereas Pascal made use of the triangle that now bears his name. As mentioned in the text, the entries in Pascal's triangle are equivalent to $C(n, r)$. This recognition of the role of $C(n, r)$ in counting is the foundation of all further developments.

The first book on probability, the work of Christian Huygens (1629–1695), appeared in 1657. In it, the notion of mathematical expectation is explored. This allows the calculation of the profit or loss that a gambler might expect, knowing the probabilities involved in the game (see the Historical Problems that follow).

Although Girolamo Cardano (1501–1576) wrote a treatise on probability, it was not published until 1663 in Cardano's collected works, and this was too late to have any effect on the development of the theory.

In 1713, the posthumously published *Ars Conjectandi* of Jakob Bernoulli (1654–1705) gave the theory the form it would have until 1900. In the current century, both combinatorics (counting) and probability have undergone rapid development due to the use of computers.

A final comment about notation. The notations $C(n, r)$ and $P(n, r)$ are variants of a form of notation developed in England after 1830. The notation $\binom{n}{r}$ for $C(n, r)$ goes back to Leonhard Euler (1707–1783), but is now losing ground because it has no clearly related symbolism of the same type for permutations. The set symbols $\cup$ and $\cap$ were introduced by Giuseppe Peano (1858–1932) in 1888 in a slightly different context. The inclusion symbol $\subset$ was introduced by E. Schroeder (1841–1902) about 1890. The treatment of set theory in the text is due to George Boole (1815–1864), who wrote $A + B$ for $A \cup B$ and AB for $A \cap B$ (statisticians still use AB for $A \cap B$).

HISTORICAL PROBLEMS

1. *The Problem Discussed by Fermat and Pascal* A game between two equally skilled players, A and B, is interrupted when A needs 2 points to win and B needs 3 points. In what proportion would the stakes be divided?

   [**Note:** If each play results in 1 point for either player, at most four more plays will decide the game.]

   (a) *Fermat's solution* List all possible outcomes that will end the game to form the sample space (for example, ABA, $ABBB$, etc.). The probabilities for A to win and B to win then determine how the stakes should be divided.

   (b) *Pascal's solution* Use combinations to determine the number of ways that the 2 points needed for A to win could occur in four plays. Then use combinations to determine the number of ways that the 3 points needed for B to win could occur. This is trickier than it looks, since A can win with 2 points in either two plays, three plays, or four plays. Compute the probabilities and compare with the results in part (a).

2. *Huygen's Mathematical Expectation* In a game with n possible outcomes with probabilities $p_1, p_2, \ldots, p_n$, suppose that the *net* winnings are $w_1, w_2, \ldots, w_n$, respectively. Then the mathematical expectation is

$$E = p_1 w_1 + p_2 w_2 + \cdots + p_n w_n$$

   The number E represents the profit or loss per game in the long run. The following problems are a modification of those of Huygens.

   (a) A fair die is tossed. A gambler wins \$3 if he throws a 6 and \$6 if he throws a 5. What is his expectation?

   [**Note:** $w_1 = w_2 = w_3 = w_4 = 0$]

   (b) A gambler plays the same game as in part (a), but now the gambler must pay \$1 to play. This means that $w_5 = \$5$, $w_6 = \$2$, and $w_1 = w_2 = w_3 = w_4 = -\$1$. What is the expectation?

13.3 EXERCISES

1. In a probability model, which of the following numbers could be the probability of an outcome:
$0, \; 0.01, \; 0.35, \; -0.4, \; 1, \; 1.4$?

2. In a probability model, which of the following numbers could be the probability of an outcome:
$1.5, \; \frac{1}{2}, \; \frac{3}{4}, \; \frac{2}{3}, \; 0, \; -\frac{1}{4}$?

3. Determine whether the following is a probability model.

Outcome	Probability
{1}	0.2
{2}	0.3
{3}	0.1
{4}	0.4

4. Determine whether the following is a probability model.

Outcome	Probability
{Jim}	0.4
{Bob}	0.3
{Faye}	0.1
{Patricia}	0.2

5. Determine whether the following is a probability model.

Outcome	Probability
{Linda}	0.3
{Jean}	0.2
{Grant}	0.1
{Ron}	0.3

6. Determine whether the following is a probability model.

Outcome	Probability
{Lanny}	0.3
{Joanne}	0.2
{Nelson}	0.1
{Rich}	0.5
{Judy}	−0.1

In Problems 7–12, construct a probability model for each experiment.

7. Tossing a fair coin twice

8. Tossing two fair coins once

9. Tossing two fair coins, then a fair die

10. Tossing a fair coin, a fair die, and then a fair coin

11. Tossing three fair coins once

12. Tossing one fair coin three times

In Problems 13–18, use the following spinners to construct a probability model for each experiment.

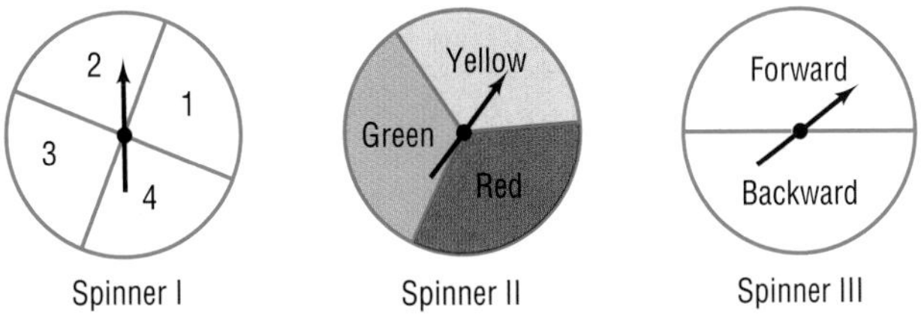

13. Spin spinner I, then spinner II. What is the probability of getting a 2 or a 4, followed by Red?

14. Spin spinner III, then spinner II. What is the probability of getting Forward, followed by Yellow or Green?

15. Spin spinner I, then II, then III. What is the probability of getting a 1, followed by Red or Green, followed by Backward?

16. Spin spinner II, then I, then III. What is the probability of getting Yellow, followed by a 2 or a 4, followed by Forward?

17. Spin spinner I twice, then spinner II. What is the probability of getting a 2, followed by a 2 or a 4, followed by Red or Green?

18. Spin spinner III, then spinner I twice. What is the probability of getting Forward, followed by a 1 or a 3, followed by a 2 or a 4?

In Problems 19–22, consider the experiment of tossing a coin twice. The table lists six possible assignments of probabilities for this experiment. Using this table, answer the following questions.

19. Which of the assignments of probabilities are consistent with the definition of a probability model?

20. Which of the assignments of probabilities should be used if the coin is known to be fair?

21. Which of the assignments of probabilities should be used if the coin is known to always come up tails?

22. Which of the assignments of probabilities should be used if tails is twice as likely as heads to occur?

	Sample Space			
Assignments	**HH**	**HT**	**TH**	**TT**
A	$\frac{1}{4}$	$\frac{1}{4}$	$\frac{1}{4}$	$\frac{1}{4}$
B	0	0	0	1
C	$\frac{3}{16}$	$\frac{5}{16}$	$\frac{5}{16}$	$\frac{3}{16}$
D	$\frac{1}{2}$	$\frac{1}{2}$	$-\frac{1}{2}$	$\frac{1}{2}$
E	$\frac{1}{4}$	$\frac{1}{4}$	$\frac{1}{4}$	$\frac{1}{8}$
F	$\frac{1}{9}$	$\frac{2}{9}$	$\frac{2}{9}$	$\frac{4}{9}$

23. Assigning Probabilities A coin is weighted so that heads is four times as likely as tails to occur. What probability should we assign to heads? to tails?

24. Assigning Probabilities A coin is weighted so that tails is twice as likely as heads to occur. What probability should we assign to heads? to tails?

25. Assigning Probabilities A die is weighted so that an odd-numbered face is twice as likely as an even-numbered face. What probability should we assign to each face?

26. Assigning Probabilities A die is weighted so that a six cannot appear. The other faces occur with the same probability. What probability should we assign to each face?

For Problems 27–30, let the sample space be $S = \{1, 2, 3, 4, 5, 6, 7, 8, 9, 10\}$. Suppose that the outcomes are equally likely.

27. Compute the probability of the event $E = \{1, 2, 3\}$.

28. Compute the probability of the event $F = \{3, 5, 9, 10\}$.

29. Compute the probability of the event E: "an even number."

30. Compute the probability of the event F: "an odd number."

For Problems 31 and 32, an urn contains 5 white marbles, 10 green marbles, 8 yellow marbles, and 7 black marbles.

31. If one marble is selected, determine the probability that it is white.

32. If one marble is selected, determine the probability that it is black.

In Problems 33–36, assume equally likely outcomes.

33. Determine the probability of having 3 boys in a 3-child family.

34. Determine the probability of having 3 girls in a 3-child family.

35. Determine the probability of having 1 girl and 3 boys in a 4-child family.

36. Determine the probability of having 2 girls and 2 boys in a 4-child family.

For Problems 37–40, two fair die are rolled.

37. Determine the probability that the sum of the two die is 7.

38. Determine the probability that the sum of the two die is 11.

39. Determine the probability that the sum of the two die is 3.

40. Determine the probability that the sum of the two die is 12.

In Problems 41–44, find the probability of the indicated event if $P(A) = 0.25$ and $P(B) = 0.45$.

41. $P(A \cup B)$ if $P(A \cap B) = 0.15$

42. $P(A \cap B)$ if $P(A \cup B) = 0.6$

43. $P(A \cup B)$ if A, B are mutually exclusive

44. $P(A \cap B)$ if A, B are mutually exclusive

45. If $P(A) = 0.60$, $P(A \cup B) = 0.85$, and $P(A \cap B) = 0.05$, find $P(B)$.

46. If $P(B) = 0.30$, $P(A \cup B) = 0.65$, and $P(A \cap B) = 0.15$, find $P(A)$.

47. According to the Federal Bureau of Investigation, in 1997 there was a 25.3% probability of theft involving a motor vehicle. If a victim of theft is randomly selected, what is the probability that he or she was not a victim of motor vehicle theft?

48. According to the Federal Bureau of Investigation, in 1997 there was a 5.6% probability of theft involving a bicycle. If a victim of theft is randomly selected, what is the probability that he or she was not a victim of bicycle theft?

49. In Chicago, there is a 30% probability that Memorial Day will have a high temperature in the 70s. What is the probability that next Memorial Day will not have a high temperature in the 70s in Chicago?

50. In Chicago, there is a 4% probability that Memorial Day will have a low temperature in the 30s. What is the probability that next Memorial Day will not have a low temperature in the 30s in Chicago?

For Problems 51–54, a golf ball is selected at random from a container. If the container has 9 white balls, 8 green balls, and 3 orange balls, find the probability of each event.

51. The golf ball is white or green.

52. The golf ball is white or orange.

53. The golf ball is not white.

54. The golf ball is not green.

55. On the "Price is Right" there is a game in which a bag is filled with 3 strike chips and 5 numbers. Let's say that the numbers in the bag are 0, 1, 3, 6, and 9. What is the probability of selecting a strike chip or the number 1?

56. Another game on the "Price is Right" requires the contestant to spin a wheel with numbers $5, 10, 15, 20, \ldots, 100$. What is the probability that the contestant spins 100 or 30?

Problems 57–60 are based on a consumer survey of annual incomes in 100 households. The following table gives the data.

Income	\$0–9999	\$10,000–19,999	\$20,000–29,999	\$30,000–39,999	\$40,000 or more
Number of households	5	35	30	20	10

57. What is the probability that a household has an annual income of \$30,000 or more?

58. What is the probability that a household has an annual income between \$10,000 and \$29,999, inclusive?

59. What is the probability that a household has an annual income of less than \$20,000?

60. What is the probability that a household has an annual income of \$20,000 or more?

61. Surveys In a survey about the number of TV sets in a house, the following probability table was constructed:

Number of TV sets	0	1	2	3	4 or more
Probability	0.05	0.24	0.33	0.21	0.17

Find the probability of a house having:
(a) 1 or 2 TV sets
(b) 1 or more TV sets
(c) 3 or fewer TV sets
(d) 3 or more TV sets
(e) Less than 2 TV sets
(f) Less than 1 TV set
(g) 1, 2, or 3 TV sets
(h) 2 or more TV sets

62. Checkout Lines Through observation it has been determined that the probability for a given number of people waiting in line at the "5 items or less" checkout register of a supermarket is:

Number waiting in line	0	1	2	3	4 or more
Probability	0.10	0.15	0.20	0.24	0.31

Find the probability of:
(a) At most 2 people in line
(b) At least 2 people in line
(c) At least 1 person in line

63. In a certain College Algebra class, there are 18 freshmen and 15 sophomores. Of the 18 freshmen, 10 are male, and of the 15 sophomores, 8 are male. Find the probability that a randomly selected student is:
(a) A freshman or female
(b) A sophomore or male

64. The faculty of the mathematics department at Joliet Junior College is composed of 4 females and 9 males. Of the 4 females, 2 are under the age of 40, and 3 of the males are under age 40. Find the probability that a randomly selected faculty member is:
(a) Female or under age 40
(b) Male or over age 40

65. Birthday Problem What is the probability that at least 2 people have the same birthday in a group of 12 people? Assume that there are 365 days in a year.

66. Birthday Problem What is the probability that at least 2 people have the same birthday in a group of 35 people? Assume that there are 365 days in a year.

67. Winning a Lottery In a certain lottery, there are ten balls, numbered 1, 2, 3, 4, 5, 6, 7, 8, 9, 10. Of these, five are drawn in order. If you pick five numbers that match those drawn in the correct order, you win \$1,000,000. What is the probability of winning such a lottery?

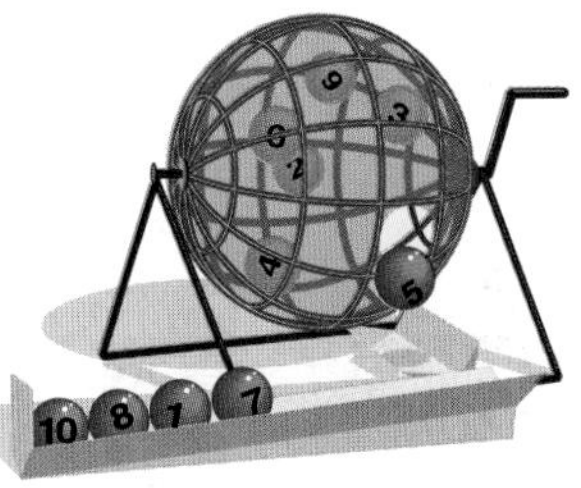

68. A committee of 6 people is to be chosen at random from a group of 14 people consisting of 2 supervisors, 5 skilled laborers, and 7 unskilled laborers. What is the probability that the committee chosen consists of 2 skilled and 4 unskilled laborers?

69. A fair coin is tossed 5 times.
(a) Find the probability that exactly 3 heads appear.
(b) Find the probability that no heads appear.

70. A fair coin is tossed 4 times.
(a) Find the probability that exactly 1 tail appears.
(b) Find the probability that no more than 1 tail appears.

71. A pair of fair dice is tossed 3 times.
(a) Find the probability that the sum of 7 appears 3 times.
(b) Find the probability that a sum of 7 or 11 appears at least twice.

72. A pair of fair dice is tossed 5 times.
(a) Find the probability that the sum is never 2.
(b) Find the probability that the sum is never 7.

73. Through a mix-up on the production line, 5 defective TVs were shipped out with 25 good ones. If 5 are selected at random, what is the probability that all 5 are defective? What is the probability that at least 2 of them are defective?

74. In a shipment of 50 transformers, 10 are known to be defective. If 30 transformers are picked at random, what is the probability that all 30 are nondefective? Assume that all transformers look alike and have an equal probability of being chosen.

75. In a promotion, 50 silver dollars are placed in a bag, one of which is valued at more than \$10,000. The winner of the promotion is given the opportunity to reach into the bag, while blindfolded, and pull out 5 coins. What is the probability that 1 of the 5 coins is the one valued at more than \$10,000?

13.4 OBTAINING PROBABILITIES FROM DATA

OBJECTIVES 1 Compute Probabilities from Data
2 Simulate Probabilities

1 In Section 13.3, probabilities were computed by counting the number of equally likely ways that an event E could occur and dividing this result by the number of possible outcomes of the experiment. Thus, we obtained the probability of an event without actually conducting the experiment. For example, in Example 5, the probability of the event E, "having two boys and one girl," was determined to be $3/8$ without actually observing three-child families. A second method for computing probabilities relies on the performance of the experiment or collection of data. The probability of an event E is then computed by determining the *relative frequency* of the event. The **relative frequency** of an event is found by dividing the number of elements in an event or category by the total number of items or repetitions of the experiment. Probabilities assigned in this way are called **empirical probabilities.**

EXAMPLE 1 Computing Empirical Probabilities from Data

TABLE 2

Temperature (in degrees Fahrenheit)	Frequency
5-9	1
10-14	8
15-19	13
20-24	28
25-29	77
30-34	145
35-39	221
40-44	252
45-49	151
50-54	114
55-59	73
60-64	44
65-69	16

Source: Chicago Tribune

The data in Table 2 represent the daily highs in Chicago from November 22 to November 30 for the years 1872 through 1998.

Construct a probability model for the daily high temperatures for a day between November 22 and 30 in Chicago.

Solution There are a total of $1 + 8 + 13 + \cdots + 16 = 1143$ days of data. To construct a probability model, we compute the relative frequency of each category of data. The relative frequency of the first category, 5° to 9°F, is found by dividing the number of days for which a high temperature between 5° and 9°F is observed by the total number of days. The relative frequency of this category, and therefore the probability, is $1/1143 \approx 0.0009 = 0.09\%$. Following this procedure for the remaining categories of temperature, we obtain Table 3.

TABLE 3

Temperature (in degrees Fahrenheit)	Frequency	Probability
5-9	1	0.09%
10-14	8	0.70%
15-19	13	1.14%
20-24	28	2.45%
25-29	77	6.74%
30-34	145	12.69%
35-39	221	19.34%
40-44	252	22.05%
45-49	151	13.21%
50-54	114	9.97%
55-59	73	6.39%
60-64	44	3.85%
65-69	16	1.40%

From Table 3, we see the probability that a randomly selected day in Chicago from November 22 to November 30 has a high temperature between 40° and 44°F is 22.05%.

NOW WORK PROBLEM 1.

EXAMPLE 2

Determining the Probability of an Event from an Experiment

A mathematics professor at Joliet Junior College randomly selects 40 currently enrolled students and finds that 25 of them pay their own tuition without assistance. Find the probability that a randomly selected student pays his or her own tuition.

Solution Let E represent the event "pays own tuition." The experiment involves asking students, "Do you pay your own tuition?" The experiment is repeated 40 times and the frequency of "yes" responses is 25, so

$$P(E) = \frac{25}{40} = \frac{5}{8} = 0.625$$

There is a 62.5% probability that a randomly selected student pays his or her own tuition.

Seeing the Concept: Perform the preceding experiment yourself by asking 10 students, "Do you pay your own tuition?" Repeat the experiment three times. Are your probabilities the same? Why might they be different?

Simulation

2 Earlier we discussed the experiment of tossing a coin. We said then that if we assumed that the coin was fair then the outcomes were equally likely. Based on this, we assigned probabilities of $1/2$ for heads and $1/2$ for tails.

We could also assign probabilities by actually tossing the coin, say 100 times, and recording the frequencies of heads and tails. Perhaps we obtain 47 heads and 53 tails. Then, based on the experiment, we would use empirical probability and assign $P(\text{H}) = 0.47$ and $P(\text{T}) = 0.53$.

Instead of actually physically tossing the coin, we could use simulation; that is, we could use a graphing utility to replicate the experiment of tossing a coin.

EXPLORATION (a) Simulate tossing a coin 100 times. What percent of the time do you obtain heads? (b) Simulate tossing the coin 100 times again. Are the results the same as the first 100 flips? (c) Repeat the simulation by tossing the coin 250 times. Does the percent of heads get closer to $\frac{1}{2}$?

Figure 11

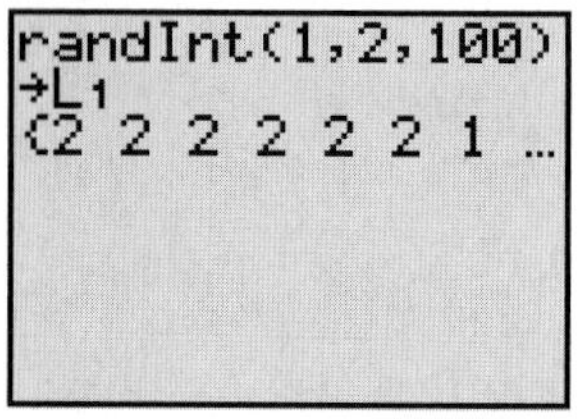

RESULT (a) Using the randInt* feature on a TI-83, we can simulate flipping a coin 100 times by letting 1 represent heads and 2 represent tails and then storing the random integers in L1. See Figure 11. Figure 12 shows the number of heads and the number of tails that result from flipping the coin 100 times. Based on this simulation, we would assign probabilities as

$$P(1) = P(\text{heads}) = \frac{48}{100} = 0.48 \quad \text{and} \quad P(2) = P(\text{tails}) = \frac{52}{100} = 0.52$$

(b) Figure 13 shows the number of heads and the number of tails that result from a second simulation of flipping the coin 100 times. Based on this simulation, we would assign probabilities as

$$P(1) = P(\text{heads}) = \frac{47}{100} = 0.47 \quad \text{and} \quad P(2) = P(\text{tails}) = \frac{53}{100} = 0.53$$

Since repetitions of an experiment (simulation) do not necessarily result in the same outcomes, the empirical probabilities from two experiments will generally be different.

(c) Figure 14 shows the number of heads and the number of tails that result from a third simulation of flipping the coin 250 times. Based on this simulation, we would assign probabilities as

$$P(1) = P(\text{heads}) = \frac{124}{250} = 0.496 \quad \text{and} \quad P(2) = P(\text{tails}) = \frac{126}{250} = 0.504$$

Figure 12

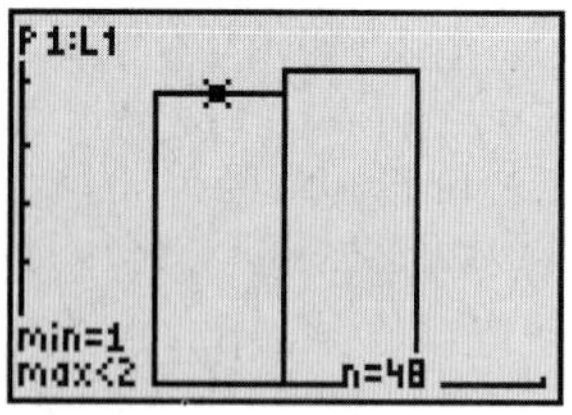

(a)

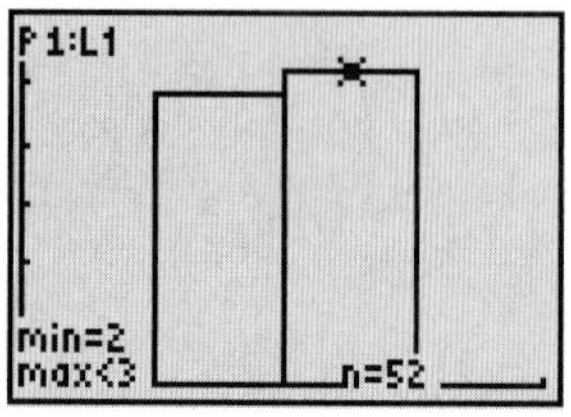

(b)

Figure 13

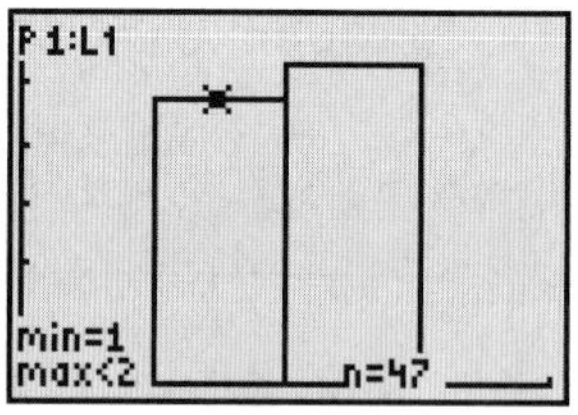

(a)

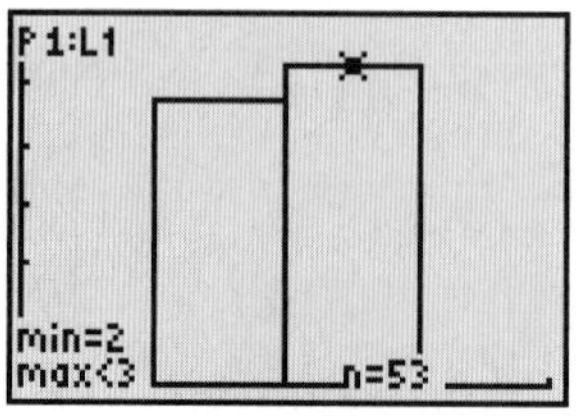

(b)

Figure 14

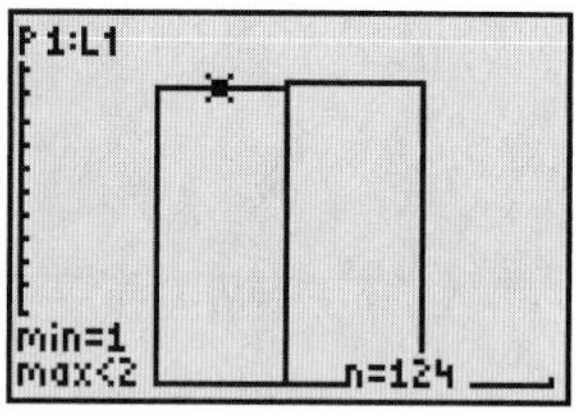

(a)

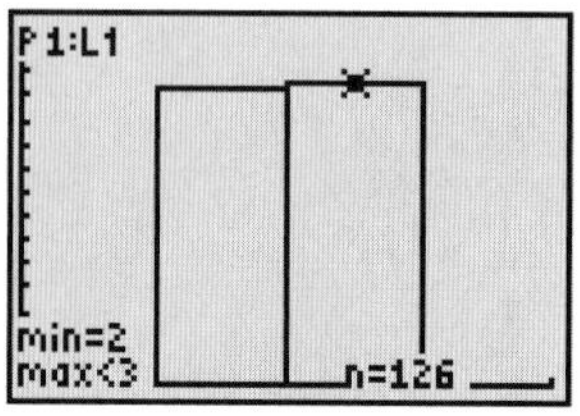

(b)

*The randInt feature involves a mathematical formula that uses a seed number to generate a sequence of random integers. Consult your owner's manual for setting the seed so that the same random numbers are not generated each time that you repeat this exploration.

Notice that the result of flipping the coin more often gives probability assignments that get closer to the ideal of equally likely outcomes. This demonstrates the **Law of Large Numbers,** which states that the more times an experiment involving equally likely outcomes is performed, the closer the empirical probability will come to the probability predicted by equally likely outcomes.

13.4 EXERCISES

1. **Causes of Death** The data below represent the causes of death for 15 to 24 year olds in 1995.
(a) Construct a probability model for these data.
(b) What is the probability that the cause of death for a randomly selected 15 to 24 year old was suicide?
(c) What is the probability that the cause of death for a randomly selected 15 to 24 year old was suicide or malignant neoplasms?
(d) What is the probability that the cause of death for a randomly selected 15 to 24 year old was not suicide nor malignant neoplasms?

Cause of Death	Number
Accidents and adverse effects	13,532
Homicide and legal intervention	6,827
Suicide	4,789
Malignant neoplasms	1,599
Diseases of heart	964
Human immunodeficiency virus infection	643
Congenital anomalies	425
Chronic obstructive pulmonary diseases	220
Pneumonia and influenza	193
Cerebrovascular diseases	166
All other causes	4,211

Source: National Center for Health Statistics, 1996

2. **Licensed Drivers** The following data represent the number of licensed drivers in the Great Lake States in 1994.

State	Number of Licensed Drivers
Illinois	7,502,201
Indiana	3,806,329
Michigan	6,601,924
Minnesota	2,705,701
Ohio	7,142,173
Wisconsin	3,554,003

Source: Each state's authorities

(a) Construct a probability model for these data.
(b) What is the probability that a randomly selected licensed driver from the Great Lakes States is from Michigan?
(c) What is the probability that a randomly selected licensed driver in the Great Lakes States is from Illinois or Indiana?
(d) What is the probability that a randomly selected licensed driver in the Great Lakes States is not from Illinois or Indiana?

3. **Licensed Drivers in Tennessee** The following data provide the number of licensed drivers between the ages of 20 and 84 in the state of Tennessee in 1994.
(a) Construct a probability model for these data.
(b) What is the probability that a randomly selected licensed driver in Tennessee is 45 to 49?
(c) What is the probability that a randomly selected licensed driver in Tennessee is 50 to 54 or 55 to 59?
(d) What is the probability that a randomly selected licensed driver in Tennessee is not 45 to 49?

Age	Number of Licensed Drivers
20–24	345,941
25–29	374,629
30–34	428,748
35–39	439,137
40–44	414,344
45–49	372,814
50–54	292,460
55–59	233,615
60–64	204,235
65–69	181,977
70–74	150,347
75–79	100,068
80–84	50,190

Source: FHWA

4. **Licensed Drivers in Hawaii** The following data provide the number of licensed drivers between the ages of 20 and 84 in the state of Hawaii in 1994.

Age	Number of Licensed Drivers
20–24	65,951
25–29	78,119
30–34	91,976
35–39	92,557
40–44	87,430
45–49	75,978
50–54	55,199
55–59	39,678
60–64	35,650
65–69	33,885
70–74	26,125
75–79	14,990
80–84	6,952

Source: FHWA

(a) Construct a probability model for these data.
(b) What is the probability that a randomly selected licensed driver in Hawaii is 45 to 49?
(c) What is the probability that a randomly selected licensed driver in Hawaii is 50 to 54 or 55 to 59?
(d) What is the probability that a randomly selected licensed driver in Hawaii is not 45 to 49?

5. Undergradute Tuition The following data represent the cost of undergraduate tuition at four-year colleges for 1992–1993 having tuition amounts ranging from $0 through $14,999.

Tuition (Dollars)	Number of 4-Year Colleges
0–999	10
1000–1999	7
2000–2999	45
3000–3999	66
4000–4999	84
5000–5999	84
6000–6999	97
7000–7999	118
8000–8999	138
9000–9999	110
10,000–10,999	104
11,000–11,999	82
12,000–12,999	61
13,000–13,999	34
14,000–14,999	29

Source: The College Board, New York, NY, Annual Survey of Colleges 1992 and 1993.

(a) Construct a probability model for these data.
(b) What is the probability that a randomly selected four-year college has undergraduate tuition between $11,000 and $11,999?
(c) What is the probability that a randomly selected four-year college has undergraduate tuition between $8000 and $10,999?
(d) What is the probability that a randomly selected four-year college has undergraduate tuition that is not $11,000 to $11,999?

6. Undergraduate Tuition The following data represent the cost of undergraduate tuition at four-year colleges for 1993–1994 having tuition amounts ranging from $0 through $14,999.

Tuition (Dollars)	Number of 4-Year Colleges
0–999	8
1000–1999	5
2000–2999	28
3000–3999	48
4000–4999	76
5000–5999	65
6000–6999	81
7000–7999	96
8000–8999	112
9000–9999	118
10,000–10,999	106
11,000–11,999	90
12,000–12,999	70
13,000–13,999	59
14,000–14,999	23

Source: The College Board, New York, NY, Annual Survey of Colleges 1992 and 1993.

(a) Construct a probability model for these data.
(b) What is the probability that a randomly selected four-year college has undergraduate tuition between $11,000 and $11,999?
(c) What is the probability that a randomly selected four-year college has undergraduate tuition between $8000 and $10,999?
(d) What is the probability that a randomly selected four-year college has undergraduate tuition that is not $11,000–$11,999?

7. The following data represent the marital status of males 18 years old and older in March 1997.

Marital Status	Number (in thousands)
Married, spouse present	54,654
Married, spouse absent	3,232
Widowed	2,686
Divorced	8,208
Never married	25,375

Source: Current Population Survey

(a) Construct a probability model for these data.
(b) Determine the probability that a randomly selected male age 18 years old or older is married, spouse present.
(c) Determine the probability that a randomly selected male age 18 years old or older has never been married.
(d) Determine the probability that a randomly selected male age 18 years old or older is married, spouse present, or married, spouse absent.
(e) What is the probability that the marital status of a randomly selected male 18 years old and older is not "married, spouse present."

8. The following data represent the marital status of females 18 years old and older in March 1997.

Marital Status	Number (in thousands)
Married, spouse present	54,626
Married, spouse absent	4,122
Widowed	11,056
Divorced	11,107
Never married	20,503

Source: Current Population Survey

(a) Construct a probability model for these data.
(b) Determine the probability that a randomly selected female age 18 years old or older is married, spouse present.
(c) Determine the probability that a randomly selected female age 18 years old or older has never been married.
(d) Determine the probability that a randomly selected female 18 years old or older is married, spouse present, or married, spouse absent.
(e) What is the probability that the marital status of a randomly selected female 18 years old and older is not "married, spouse present."

9. While golfing, you had the good fortune of finding 10 golf balls, 4 of which were Titleists. Based on this, what is the probability that a golfer plays with Titleist golf balls?

10. In a recent survey, 200 people were asked if they favor a shopping mall in their neighborhood. Of the 200 people questioned, 120 were in favor of the shopping mall. What is the probability that a randomly selected individual in this neighborhood would favor the shopping mall?

11. In a survey of 50 families with 3 children, it was determined that 20 of them had 2 boys and 1 girl. Based on the results of this survey, what is the probability of having 2 boys and 1 girl?

12. Compaq computer just received a shipment of 500 hard disk drives, 12 of which were defective. What is the probability that a randomly selected disk drive is defective?

13. Of the 28,538 tornadoes that occurred between 1916 and 1985 in the United States, 3262 of them occurred between the hours of 5 and 6 PM. What is the probability that a tornado will occur between 5 and 6 PM? What is the probability that tornado will not occur between 5 and 6 PM? (*Source:* U.S. Tornadoes Part 1, Dr. T. Fujita)

14. Of the 28,538 tornadoes that occurred between 1916 and 1985 in the United States, 324 of them occurred between the hours of 5 and 6 AM. What is the probability that a tornado will occur between 5 and 6 AM? What is the probability that tornado will not occur between 5 and 6 AM? (*Source:* U.S. Tornadoes Part 1, Dr. T. Fujita)

15. In Chicago, during the month of June, 11.5 days are partly cloudy, 7.3 days are clear, and 11.2 days are cloudy. What is the probability that a randomly selected day in Chicago in June is cloudy? What is the probability that a randomly selected day is clear? What is the probability that a randomly selected day is not clear?

16. In Chicago, during the month of July, 12.1 days are partly cloudy, 8.0 days are clear, and 10.3 days are cloudy. What is the probability that a randomly selected day in Chicago in July is cloudy? What is the probability that a randomly selected day is clear? What is the probability that a randomly selected day is not clear?

17. In Chicago, during the month of June, 6.4 days are thunderstorm days. What is the probability that a randomly selected day in June will be a thunderstorm day in Chicago? What is the probability that a randomly selected day in June will not be a thunderstorm day in Chicago?

18. In Chicago, during the month of July, 6.0 days are thunderstorm days. What is the probability that a randomly selected day in July will be a thunderstorm day in Chicago? What is the probability that randomly selected day in July will not be a thunderstorm day in Chicago?

19. On September 8, 1998, Mark McGuire hit his sixty-second home run of the season. Of the 62 home runs he hit, 26 went to left field, 21 went to left center, 12 went to center field, 3 went to right center field, and 0 went to right field.
(a) Construct a probability model for Mark McGuire's home runs.
(b) What is the probability that a randomly selected home run was hit to left field?
(c) What is the probability that a randomly selected home run was hit to center field?
(d) What is the probability that a randomly selected home run was hit to right field?
(e) Is it impossible for Mark McGuire to hit a home run to right field? Explain.

20. The following data represent various types of computer abuse based on a survey of companies from 1/97–1/98.

Type of Abuse	Number of Incidents
Virus	380
Insider abuse of Internet access	353
Laptop theft	297
Unauthorized access by insiders	203
System penetration	108
Theft of proprietary information	82
Telecom fraud	75
Financial fraud	68
Sabotage	66
Telecom eavesdropping	45
Active wiretap	5

Source: Computer Security Institute

(a) Construct a probability model of these data.
(b) What is the probability that a randomly selected case of computer abuse is laptop theft?
(c) What is the probability that a randomly selected case of computer abuse is sabotage?

21. Conduct a survey in your school by randomly asking 50 students whether they drive to school. Based on the results of the survey, determine the probability that a randomly selected student drives to school.

22. **Simulation** Use a graphing utility to simulate rolling a six-sided die 100 times.
(a) Use the results to compute the probability of obtaining a 1.
(b) Repeat the simulation. Compute the probability of obtaining a 1.
(c) Simulate rolling a six-sided die 500 times. Compute the probability of obtaining a 1.
(d) Which simulation resulted in the closest estimate to the probability that would be obtained using equally likely outcomes?

CHAPTER REVIEW

Things To Know

Set (p. 858)		Well-defined collection of distinct objects, called elements
Null set (p. 858)	$\varnothing$	Set that has no elements
Equality (p. 858)	$A = B$	A and B have the same elements
Subset (p. 858)	$A \subseteq B$	Each element of A is also an element of B.
Intersection (p. 859)	$A \cap B$	Set consisting of elements that belong to both A and B
Union (p. 859)	$A \cup B$	Set consisting of elements that belong to either A or B, or both
Universal set (p. 859)	U	Set consisting of all the elements that we wish to consider
Complement (p. 859)	$\overline{A}$	Set consisting of elements of the universal set that are not in A
Finite set (p. 860)		The number of elements in the set is a nonnegative integer
Infinite set (p. 860)		A set that is not finite
Counting formula (p. 861)	$n(A \cup B) = n(A) + n(B) - n(A \cap B)$	
Addition Principle (p. 862)		If $A \cap B = \varnothing$, then $n(A \cup B) = n(A) + n(B)$.
Multiplication Principle (p. 866)		If a task consists of a sequence of choices in which there are p selections for the first choice, q selections for the second choice, and so on, then the task of making these selections can be done in $p \cdot q \cdot \ldots$ different ways.
Permutation (p. 867)	$P(n, r) = n(n-1) \cdot \ldots \cdot [n - (r-1)]$ $= \dfrac{n!}{(n-r)!}$	An ordered arrangement of n distinct objects without repetition
Combination (p. 869)	$C(n, r) = \dfrac{P(n, r)}{r!}$ $= \dfrac{n!}{(n-r)!\,r!}$	An arrangement, without regard to order, of n distinct objects without repetition
Permutations with repetition (p. 871)	$\dfrac{n!}{n_1!\,n_2!\cdots n_k!}$	The number of permutations of n objects of which n_1 are of one kind, n_2 are of a second kind, ..., and n_k are of a kth kind, where $n = n_1 + n_2 + \cdots + n_k$

Sample space (p. 874)		Set whose elements represent all the logical possibilities that can occur as a result of an experiment
Probability (p. 874)		A nonnegative number assigned to each outcome of a sample space; the sum of all the probabilities of the outcomes equals 1.
Equally likely outcomes (p. 877)	$P(E) = \dfrac{n(E)}{n(S)}$	The same probability is assigned to each outcome.
Addition Rule (p. 878)	$P(E \cup F) = P(E) + P(F) - P(E \cap F)$	
Complement of an event (p. 880)	$P(\overline{E}) = 1 - P(E)$	

How To

Find all the subsets of a set (p. 858)

Find the intersection and union of sets (p. 859)

Find the complement of a set (p. 859)

Count the number of elements in a set (p. 860)

Solve counting problems using the Multiplication Principle (p. 864)

Solve counting problems using permutations (p. 866)

Solve counting problems using combinations (p. 868)

Construct probability models (p. 874)

Compute probabilities of equally likely outcomes (p. 877)

Utilize the addition rule to find probabilities (p. 878)

Utilize the complement rule to find probabilities (p. 880)

Compute probabilities using permutations and combinations (p. 881)

Compute probabilities from data (p. 888)

Simulate probabilities (p. 889)

Fill-in-the-Blank Items

1. The ______________ of A with B consists of all elements in either A or B or both; the ______________ of A with B consists of all elements in both A and B.
2. $P(5, 2) =$ ______________; $C(5, 2) =$ ______________.
3. A(n) ______________ is an ordered arrangement of n distinct objects.
4. A(n) ______________ is an arrangement of n distinct objects without regard to order.
5. When the same probability is assigned to each outcome of a sample space, the experiment is said to have ______________ ______________ outcomes.
6. The ______________ of an event E is the set of all outcomes in the sample space S that are not outcomes in the event E.

True/False Items

T F 1. The intersection of two sets is always a subset of their union.

T F 2. $P(n, r) = \dfrac{n!}{r!}$

T F 3. In a combination problem, order is not important.

T F 4. In a permutation problem, once an object is used, it cannot be repeated.

T F 5. The probability of an event can never equal 0.

T F 6. In a probability model, the sum of all probabilities is 1.

Review Exercises

Blue problem numbers indicate the authors' suggestions for use in a Practice Test.

In Problems 1–8, use U = *universal set* $= \{1, 2, 3, 4, 5, 6, 7, 8, 9\}$, $A = \{1, 3, 5, 7\}$, $B = \{3, 5, 6, 7, 8\}$, *and* $C = \{2, 3, 7, 8, 9\}$ *to find each set.*

1. $A \cup B$
2. $B \cup C$
3. $A \cap C$
4. $A \cap B$
5. $\overline{A} \cup \overline{B}$
6. $\overline{B} \cap \overline{C}$
7. $\overline{B \cap C}$
8. $\overline{A \cup B}$
9. If $n(A) = 8$, $n(B) = 12$, and $n(A \cap B) = 3$, find $n(A \cup B)$.
10. If $n(A) = 12$, $n(A \cup B) = 30$, and $n(A \cap B) = 6$, find $n(B)$.

In Problems 11–16, use the information supplied in the figure:

11. How many are in A?

12. How many are in A or B?

13. How many are in A and C?

14. How many are not in B?

15. How many are in neither A nor C?

16. How many are in B but not in C?

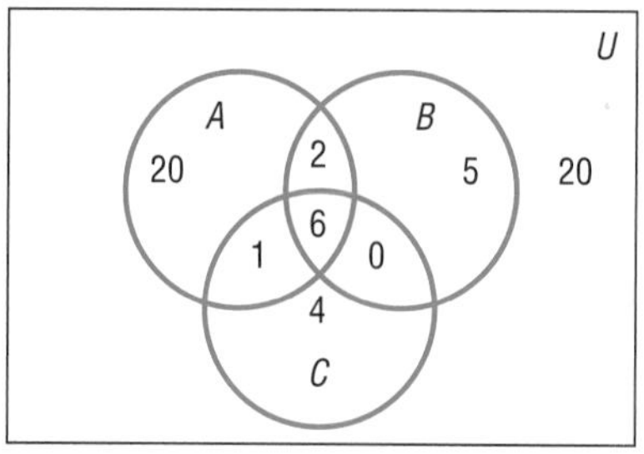

In Problems 17–22, compute the given expression.

17. $5!$ **18.** $6!$ **19.** $P(8, 3)$ **20.** $P(7, 3)$ **21.** $C(8, 3)$ **22.** $C(7, 3)$

23. A clothing store sells pure wool and polyester–wool suits. Each suit comes in 3 colors and 10 sizes. How many suits are required for a complete assortment?

24. In connecting a certain electrical device, 5 wires are to be connected to 5 different terminals. How many different wirings are possible if 1 wire is connected to each terminal?

25. Baseball On a given day, the American Baseball League schedules 7 games. How many different outcomes are possible, assuming that each game is played to completion?

26. Baseball On a given day, the National Baseball League schedules 6 games. How many different outcomes are possible, assuming that each game is played to completion?

27. If 4 people enter a bus having 9 vacant seats, in how many ways can they be seated?

28. How many different arrangements are there of the letters in the word ROSE?

29. In how many ways can a squad of 4 relay runners be chosen from a track team of 8 runners?

30. A professor has 10 similar problems to put on a test with 3 problems. How many different tests can she design?

31. Baseball In how many different ways can 14 baseball teams in the American League be paired without regard to which team is at home?

32. Arranging Books on a Shelf There are 5 different French books and 5 different Spanish books. How many ways are there to arrange them on a shelf if:
(a) Books of the same language must be grouped together, French on the left, Spanish on the right?
(b) French and Spanish books must alternate in the grouping, beginning with a French book?

33. Telephone Numbers Using the digits $0, 1, 2, \ldots, 9$, how many 7-digit numbers can be formed if the first digit cannot be 0 or 9 and if the last digit is greater than or equal to 2 and less than or equal to 3? Repeated digits are allowed.

34. Home Choices A contractor constructs homes with 5 different choices of exterior finish, 3 different roof arrangements, and 4 different window designs. How many different types of homes can be built?

35. License Plate Possibilities A license plate consists of 1 letter, excluding O and I, followed by a 4-digit number that cannot have a 0 in the lead position. How many different plates are possible?

36. Using the digits 0 and 1, how many different numbers consisting of 8 digits can be formed?

37. Forming Different Words How many different words can be formed using all the letters in the word MISSING?

38. Arranging Flags How many different vertical arrangements are there of 10 flags if 4 are white, 3 are blue, 2 are green, and 1 is red?

39. Forming Committees A group of 9 people is going to be formed into committees of 4, 3, and 2 people. How many committees can be formed if:
(a) A person can serve on any number of committees?
(b) No person can serve on more than one committee?

40. Forming Committees A group consists of 5 men and 8 women. A committee of 4 is to be formed from this group, and policy dictates that at least 1 woman be on this committee.
(a) How many committees can be formed that contain exactly 1 man?
(b) How many committees can be formed that contain exactly 2 women?
(c) How many committees can be formed that contain at least 1 man?

41. Male Doctorates in Mathematics The following data represent the ethnicity of male doctoral recipients in mathematics in 1995–1996.

Ethnic Group	Number of Doctoral Recipients
Asian, Pacific Islander	318
Black	15
American Indian, Eskimo, Aleut	1
Mexican American, Puerto Rican, or other Hispanic	36
White (non-Hispanic)	527
Unknown	6

Source: 1996 AMS-IMS-MAA Annual Survey

(a) Construct a probability model for the data.
(b) What is the probability that a randomly selected male doctoral recipient is Asian, Pacific Islander?
(c) What is the probability that a randomly selected male doctoral recipient is Asian, Pacific Islander or Mexican American, Puerto Rican, or other Hispanic?
(d) What is the probability that a randomly selected male doctoral recipient is not Asian, Pacific Islander?

42. Female Doctorates in Mathematics The following data represent the ethnicity of female doctoral recipients in mathematics in 1995–1996.

Ethnic Group	Number of Doctoral Recipients
Asian, Pacific Islander	89
Black	4
American Indian, Eskimo, Aleut	0
Mexican American, Puerto Rican, or other Hispanic	7
White (non-Hispanic)	149
Unknown	1

Source: 1996 AMS-IMS-MAA Annual Survey

(a) Construct a probability model for the data.
(b) What is the probability that a randomly selected female doctoral recipient is Asian, Pacific Islander?
(c) What is the probability that a randomly selected female doctoral recipient is Asian, Pacific Islander or Mexican American, Puerto Rican, or other Hispanic?
(d) What is the probability that a randomly selected female doctoral recipient is not Asian, Pacific Islander?

43. Age and DUI in 1980 The following data represent the estimated number of arrests per 100,000 drivers for driving under the influence for individuals 25 to 64 years of age in 1980.

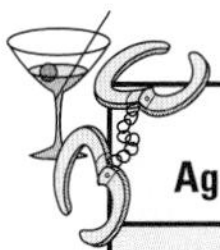

Age	Estimated Number of Arrests
25–29	1347
30–34	1076
35–39	996
40–44	944
45–49	837
50–54	686
55–59	509
60–64	335

Source: U.S. Bureau of Justice Statistics

(a) Construct a probability model for the data.
(b) What is the probability that the age of a randomly selected DUI arrest is 35 to 39?
(c) What is the probability that the age of a randomly selected DUI arrest is 35 to 39 or 55 to 59?
(d) What is the probability that the age of a randomly selected DUI arrest is not 35 to 39?

44. Age and DUI in 1989 The following data represent the estimated number of arrests per 100,000 drivers for driving under the influence for individuals 25 to 64 years of age in 1989.

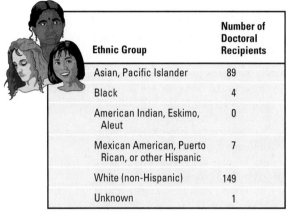

Age	Estimated Number of Arrests
25–29	1869
30–34	1486
35–39	1123
40–44	872
45–49	725
50–54	558
55–59	400
60–64	262

Source: U.S. Bureau of Justice Statistics

(a) Construct a probability model for the data.
(b) What is the probability that the age of a randomly selected DUI arrest is 35 to 39?
(c) What is the probability that the age of a randomly selected DUI arrest is 35 to 39 or 55 to 59?
(d) What is the probability that the age of a randomly selected DUI arrest is not 35 to 39?

45. Birthday Problem For this problem, assume that the year has 365 days.
(a) How many ways can 18 people have different birthdays?
(b) What is the probability that nobody has the same birthday in a group of 18 people?
(c) What is the probability in a group of 18 people that at least 2 people have the same birthday?

46. Death Rates According to the U.S. National Center for Health Statistics, 32.1% of all deaths in 1994 were due to heart disease.
(a) What is the probability that a randomly selected death in 1994 was due to heart disease?
(b) What is the probability that a randomly selected death in 1994 was not due to heart disease?

47. Unemployment According to the U.S. Bureau of Labor Statistics, 5.4% of the U.S. labor force was unemployed in 1996.
(a) What is the probability that a randomly selected member of the U.S. labor force was unemployed in 1996?
(b) What is the probability that a randomly selected member of the U.S. labor force was not unemployed in 1996?

48. From a box containing three 40-watt bulbs, six 60-watt bulbs, and eleven 75-watt bulbs, a bulb is drawn at random. What is the probability that the bulb is 40 watts? What is the probability that it is not a 75-watt bulb?

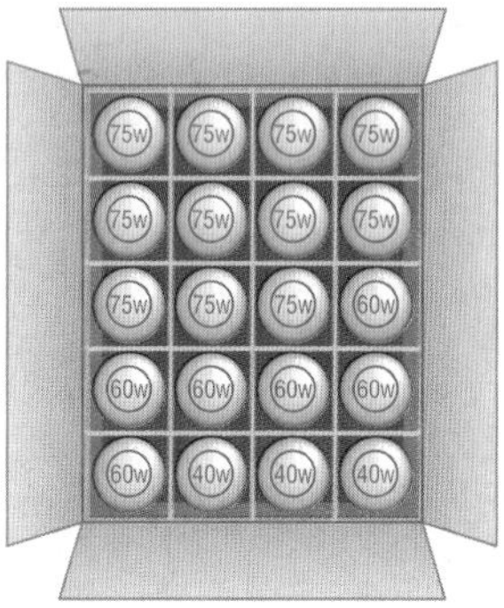

49. You have four \$1 bills, three \$5 bills, and two \$10 bills in your wallet. If you pick a bill at random, what is the probability that it will be a \$1 bill?

50. Each of the letters in the word ROSE is written on an index card and the cards are then shuffled. What is the probability that, when the cards are dealt out, they spell the word ROSE?

51. Each of the numbers, $1, 2, \ldots, 100$ is written on an index card and the cards are then shuffled. If a card is selected at random, what is the probability that the number on the card is divisible by 5? What is the probability that the card selected is either a 1 or names a prime number?

52. Computing Probabilities Because of a mistake in packaging, a case of 12 bottles of red wine contained 5 Merlot and 7 Cabernet, each without labels. All the bottles look alike and have equal probability of being chosen. Three bottles are selected.
(a) What is the probability that all 3 are Merlot?
(b) What is the probability that exactly 2 are Merlot?
(c) What is the probability that none is a Merlot?

53. **Tossing a Coin** A fair coin is tossed 10 times.
(a) What is the probability of obtaining exactly 5 heads?
(b) What is the probability of obtaining all heads?
(c) Use a graphing utility to simulate this experiment. What is the empirical probability of obtaining exactly 5 heads? All heads?

54. At the Milex tune-up and brake repair shop, the manager has found that a car will require a tune-up with a probability of 0.6, a brake job with a probability of 0.1, and both with a probability of 0.02.
(a) What is the probability that a car requires either a tune-up or a brake job?
(b) What is the probability that a car requires a tune-up but not a brake job?
(c) What is the probability that a car requires neither type of repair?

PROJECT AT MOTOROLA

Probability of error in digital wireless communications

Motorola uses digital communication technology in the design of cellular phones. In digital communications, a speech or data signal is converted into bits by a mobile phone and transmitted over the air to a base station. A bit is a unit of information that can be either 0 or 1. Thus, a single bit can represent two levels of information. A speech signal is converted into a digital signal by taking samples of the speech signal at specified time intervals, and then quantizing these samples. A *sample* is simply the value of the speech signal at a single time instant. To *quantize* the sample, a number of acceptable levels is defined, and the level that is closest to the value of the speech signal at a sampling instant is assigned to that sample. A symbol consisting of multiple bits is assigned to each level, and these resulting symbols are transmitted over the air. The number of levels defines the length of the symbols. For instance, a symbol of length one bit represents only two levels (0 and 1). A symbol of length two bits represents 4 levels (00, 01, 10, 11).

1. A code has two quantization levels when the symbol length is one, and four levels when the symbol length is two. Thus, if the symbol lengths are $1, 2, 3, 4, \ldots$, then the number of quantization levels of the corresponding codes form a sequence where the first two terms are 2 and 4. Write the next four terms of this sequence.

2. Write an expression for this sequence, where the number of bits per symbol (i.e. the symbol length) is n. What type of sequence is this? Write a recursive expression for the sequence.

3. How many bits per symbol are needed to have 256 quantization levels?

Each symbol is sent by the cellular phone over the air and received by the base station. Because of many factors, including other cellular phone transmissions and obstacles in the path of the transmitted signal, the symbols may be received incorrectly. If one or more bits of the symbol are changed (i.e. a 0 becomes a 1 or vice versa), then the symbol is received in error. If a symbol is two bits in length and the probability that a single bit is received in error is equal to k, then the probability that both bits are received in error is equal to the probability that the first bit is received in error multiplied by the probability that the second bit is received in error: $k \cdot k = k^2$.

4. Assume the transmitted symbol is 1011. Construct the sample space of possible symbols received, including symbols with 0, 1, 2, 3, or 4 bit errors.

5. Suppose the transmitted symbol is 1011 and the probability that a single bit is changed is equal to 1/3. What is the probability that the symbol is received correctly? [HINT: What is the probability the first bit is correct? The second? The third? The fourth?]

6. Assume that a symbol is eight bits long. How many combinations of received symbols with two bit errors are there? If the probability that a single bit is changed is equal to 1/3, what is the probability that the message is received correctly? Based on this result, what is the probability it is received incorrectly?

In some cases, there is a probability that a transmitted symbol is received in error. Error detection coding is used to determine whether errors have occurred in the received symbol. In some cases, detected errors can also be corrected so that the decoded message is exactly what was sent. A simple type of error detection code adds a single bit to the end of each transmitted symbol, to make the number of ones in the symbol an even number. This is called an **even parity** code. At the receiver, the number of ones in the received symbol is counted. If the number of ones is odd then the receiver knows that there was at least one error in the symbol transmission. This type of code can detect an odd number of bit errors.

7. Assume that a symbol is seven bits long, and a parity code bit is added so that the transmitted symbol is eight bits in length. The probability that a single bit is changed is equal to 1/3. What is the probability that an error is detected in the received symbol? What is the probability that an error occurred, but is not detected? [HINT: Research the Binomial Probability Theorem in a text on Finite Mathematics or Elementary Statistics].

CHAPTER

14 A Preview of Calculus: The Limit, Derivative, and Integral of a Function

Outline

FIELD TRIP TO MOTOROLA

Silicone-based sealants are commonly used in automotive electronics to protect components from the corrosive effects of harsh environments and extend their useful operating life. Silicone sealants that are typically used in industry, or in the home, cure by a chemical reaction involving atmospheric moisture. During the cure process, moisture diffuses into the soft silicone gel and transforms the material into a compliant, heat resistant, seal material. The cure process starts at the outside polymer surface and works its way into the center of the bead.

INTERVIEW AT MOTOROLA

William Olson is a Senior Member of the Technical Staff in Motorola Labs—MATC and is manager of the Design Reliability and Simulation Group, a team which specializes in implementing leading edge packaging and component technologies into consumer electronics. Bill graduated from the University of Wisconsin–Madison with a Ph.D. in Inorganic Chemistry. He joined Motorola's Automotive Industrial Electronics Group in 1991 where he developed high density electronic packaging (DCA/BGA on flex) as well as high volume manufacturing processes for fluxes, sealants, and conformal coating materials. Bill has 13 U.S. patents and over 30 publications to his credit.

PREPARING FOR THIS SECTION

Before getting started, review the following:

✓ Piecewise–Defined Functions (pp. 132–134)

14.1 FINDING LIMITS USING TABLES AND GRAPHS

OBJECTIVES 1 Find a Limit Using a Table
2 Find a Limit Using a Graph

1 The idea of the limit of a function is what connects algebra and geometry to the mathematics of calculus. In working with the limit of a function, we encounter notation of the form

$$\lim_{x \to c} f(x) = N$$

This is read as "the limit of $f(x)$ as x approaches c equals the number N." Here f is a function defined on some open interval containing the number c; f need not be defined at c, however.

We may describe the meaning of $\lim_{x \to c} f(x) = N$ as follows:

> For all x approximately equal to c, with $x \neq c$, the corresponding value $f(x)$ is approximately equal to N.

Another description of $\lim_{x \to c} f(x) = N$ is

> As x gets closer to c, but remains unequal to c, the corresponding values of $f(x)$ get closer to N.

Tables generated with the help of a calculator are useful for finding limits.

EXAMPLE 1 Finding a Limit Using a Table

Find: $\lim_{x \to 3}(5x^2)$

Solution Here $f(x) = 5x^2$ and $c = 3$. We choose values of x close to 3, arbitrarily starting with 2.99. Then we select additional numbers that get closer to 3, but remain less than 3. Next we choose values of x greater than 3, starting with 3.01, that get closer to 3. Finally, we evaluate f at each choice to obtain Table 1.

TABLE 1

X	Y1
2.99	44.701
2.999	44.97
2.9999	44.997
3.0001	45.003
3.001	45.03
3.01	45.301

Y1 = 5X²

From Table 1, we infer that as x gets closer to 3 the value of $f(x) = 5x^2$ gets closer to 45. That is,

$$\lim_{x \to 3}(5x^2) = 45$$

When choosing the values of x in a table, the number to start with and the subsequent entries are arbitrary. However, the entries should be chosen so that the table makes it clear what the corresponding values of f are getting close to.

NOW WORK PROBLEM 1.

EXAMPLE 2 Finding a Limit Using a Table

Find: (a) $\lim_{x \to 2} \dfrac{x^2 - 4}{x - 2}$ (b) $\lim_{x \to 2}(x + 2)$

Solution (a) Here $f(x) = \dfrac{x^2 - 4}{x - 2}$ and $c = 2$. Notice that the domain of f is $\{x \mid x \neq 2\}$, so f is not defined at 2. We proceed to choose values of x close to 2 and evaluate f at each choice, as shown in Table 2.

TABLE 2

X	Y1
1.99	3.99
1.999	3.999
1.9999	3.9999
2.0001	4.0001
2.001	4.001
2.01	4.01

Y1 = (X²−4)/(X−2)

We infer that as x gets closer to 2, the value of $f(x) = \dfrac{x^2 - 4}{x - 2}$ gets closer to 4. That is,

$$\lim_{x \to 2} \frac{x^2 - 4}{x - 2} = 4$$

(b) Here $g(x) = x + 2$ and $c = 2$. The domain of g is all real numbers. See Table 3.

TABLE 3

X	Y1
1.99	3.99
1.999	3.999
1.9999	3.9999
2.0001	4.0001
2.001	4.001
2.01	4.01

Y1 = X+2

We infer that as x gets closer to 2 the value of $g(x)$ gets closer to 4. That is,

$$\lim_{x \to 2}(x + 2) = 4$$

The conclusion that $\lim_{x \to 2}(x + 2) = 4$ could have been obtained without the use of Table 3; as x gets closer to 2, it follows that $x + 2$ will get closer to 4.

Also, for part (a), you are right if you make the observation that, since $x \neq 2$, then

$$f(x) = \frac{x^2 - 4}{x - 2} = \frac{\cancel{(x - 2)}(x + 2)}{\cancel{x - 2}} = x + 2, \qquad x \neq 2$$

Now it is easy to conclude that

$$\lim_{x \to 2} \frac{x^2 - 4}{x - 2} = \lim_{x \to 2} (x + 2) = 4$$

Let's look at an example for which the factoring technique used above does not work.

EXAMPLE 3   **Finding a Limit Using a Table**

Find: $\displaystyle\lim_{x \to 0} \frac{\sin x}{x}$

Solution   First, we observe that the domain of the function $f(x) = \dfrac{\sin x}{x}$ is $\{x \,|\, x \neq 0\}$.

We create Table 4, where x is measured in radians. We infer from Table 4 that $\displaystyle\lim_{x \to 0} \frac{\sin x}{x} = 1$.   ∎

TABLE 4

X	Y1
-.05	.99958
-.02	.99993
-.01	.99998
.01	.99998
.02	.99993
.05	.99958

Y1◻sin(X)/X

② The graph of a function f can also be of help in finding limits. See Figure 1.

Figure 1

(a) $f(c) = N$; $\displaystyle\lim_{x \to c} f(x) = N$

(b) $f(c) \neq N$; $\displaystyle\lim_{x \to c} f(x) = N$

(c) $f(c)$ not defined; $\displaystyle\lim_{x \to c} f(x) = N$

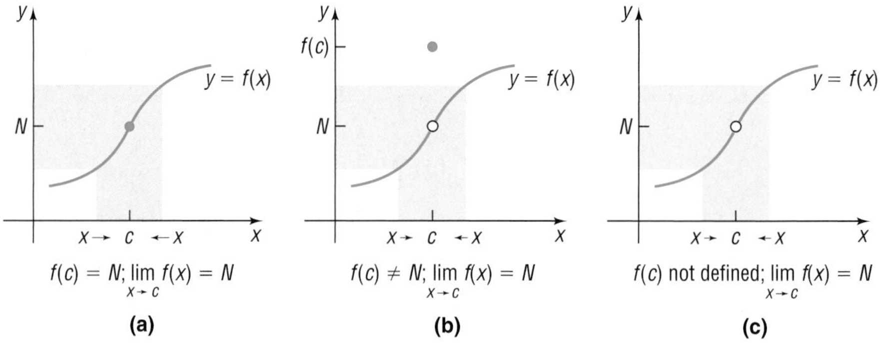

In each graph, notice that, as x gets closer to c, the values of f get closer to the number N. We conclude that

$$\lim_{x \to c} f(x) = N$$

This is the conclusion regardless of the value of f at c. In Figure 1(a), $f(c) = N$, and in Figure 1(b), $f(c) \neq N$. Figure 1(c) illustrates that $\displaystyle\lim_{x \to c} f(x) = N$, even if f is not defined at c.

EXAMPLE 4   **Finding a Limit by Graphing**

Find: $\displaystyle\lim_{x \to 2} f(x)$ if $f(x) = \begin{cases} 3x - 2 & \text{if } x \neq 2 \\ 3 & \text{if } x = 2 \end{cases}$

Solution   The function f is a piecewise-defined function. Its graph is shown in Figure 2(a) drawn by hand. Figure 2(b) shows the graph using a graphing utility.

Figure 2

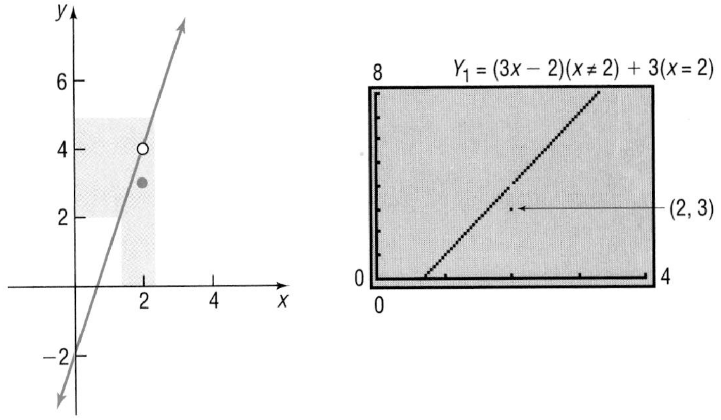

(a) (b)

Upon ZOOMing-IN around $x = 2$ and TRACEing, we conclude that $\lim_{x \to 2} f(x) = 4$. See Figure 3.

Figure 3

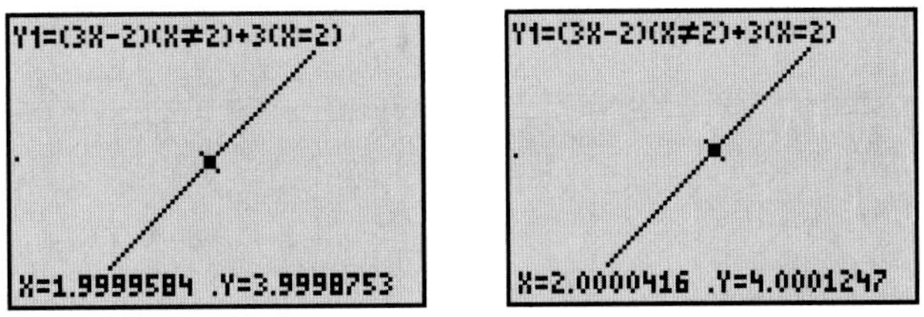

Notice in Example 4 that the value of f at 2, that is, $f(2) = 3$, plays no role in the conclusion that $\lim_{x \to 2} f(x) = 4$. In fact, even if f were undefined at 2, it would still be true that $\lim_{x \to 2} f(x) = 4$.

NOW WORK PROBLEM 17.

Sometimes there is no *single* number that the values of f get closer to as x gets closer to c. In this case, we say that ***f* has no limit as *x* approaches *c*** or that $\lim_{x \to c} f(x)$ **does not exist.**

EXAMPLE 5 A Function That Has No Limit at 0

Find: $\lim_{x \to 0} f(x)$ if $f(x) = \begin{cases} x & \text{if } x \le 0 \\ 1 & \text{if } x > 0 \end{cases}$

Figure 4

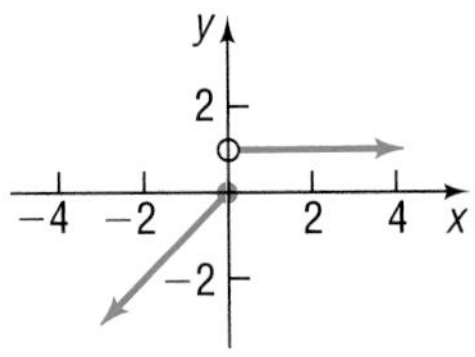

Solution See Figure 4. As x gets closer to 0, but remains negative, the value of f is also getting closer to 0. As x gets closer to 0, but remains positive, the value of f always equals 1. Since there is no single number that the values of f are close to when x is close to 0, we conclude that $\lim_{x \to 0} f(x)$ does not exist.

NOW WORK PROBLEM 31.

EXAMPLE 6 Using a Graphing Utility to Find a Limit

Find: $\lim_{x\to 2} \dfrac{x^3 - 2x^2 + 4x - 8}{x^4 - 2x^3 + x - 2}$

Solution Table 5 shows the solution, from which we conclude that

$$\lim_{x\to 2} \frac{x^3 - 2x^2 + 4x - 8}{x^4 - 2x^3 + x - 2} = 0.889$$

rounded to three decimal places.

TABLE 5

X	Y1
1	2.5
1.5	1.4286
1.8	1.0597
1.9	.96832
1.99	.89635
1.999	.88963
1.9999	.88896

Y1■(X^3−2X²+4X−...

X	Y1
3	.46429
2.5	.61654
2.3	.70555
2.1	.81961
2.01	.88153
2.001	.88815
2.0001	.88881

Y1■(X^3−2X²+4X−...

NOW WORK PROBLEM 37.

In the next section, we will see how algebra can be used to obtain exact solutions to limits like the one in Example 6.

14.1 EXERCISES

In Problems 1–10, use the TABLE feature of a graphing utility to find the indicated limit.

1. $\lim_{x\to 2}(4x^3)$
2. $\lim_{x\to 3}(2x^2 + 1)$
3. $\lim_{x\to 0}\dfrac{x+1}{x^2+1}$
4. $\lim_{x\to 0}\dfrac{2-x}{x^2+4}$
5. $\lim_{x\to 4}\dfrac{x^2-4x}{x-4}$
6. $\lim_{x\to 3}\dfrac{x^2-9}{x^2-3x}$
7. $\lim_{x\to 0}(e^x + 1)$
8. $\lim_{x\to 0}\dfrac{e^x - e^{-x}}{2}$
9. $\lim_{x\to 0}\dfrac{\cos x - 1}{x}$, x in radians
10. $\lim_{x\to 0}\dfrac{\tan x}{x}$, x in radians

In Problems 11–16, use the graph shown to determine if the limit exists. If it does, find its value.

11. 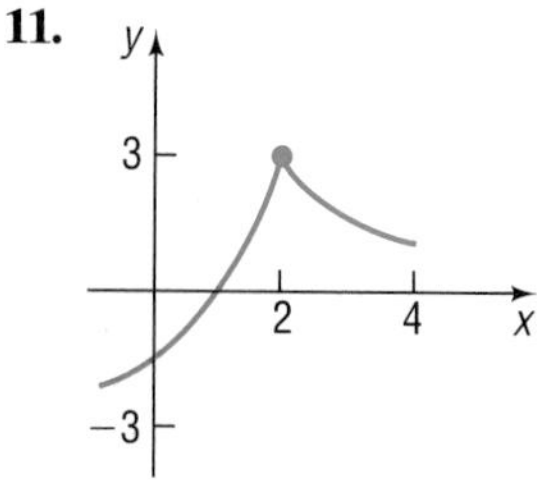

$\lim_{x\to 2} f(x)$

12. 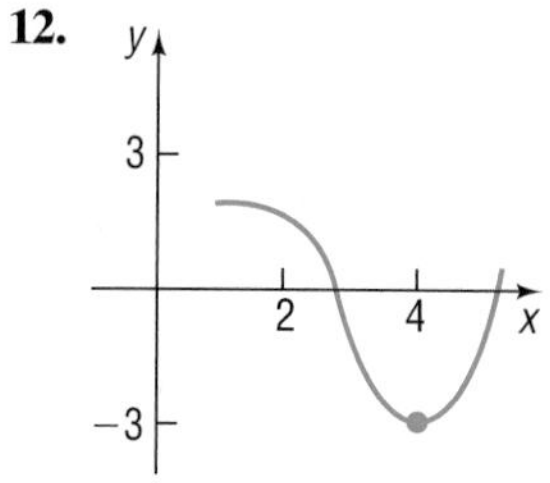

$\lim_{x\to 4} f(x)$

13. 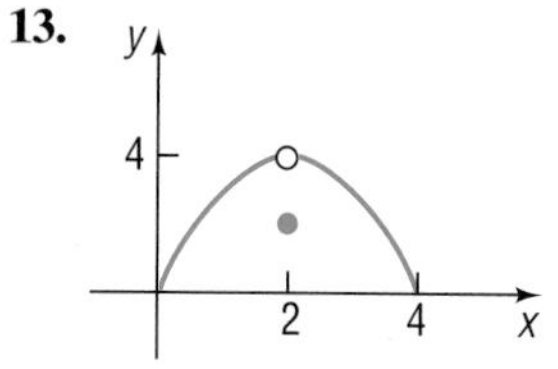

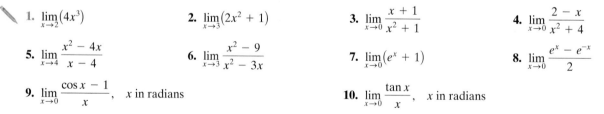

$\lim_{x\to 2} f(x)$

14.

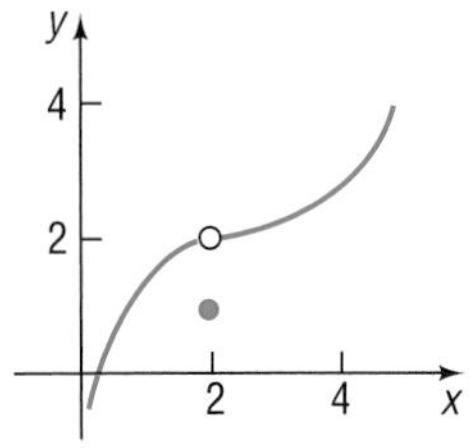

$\lim_{x \to 2} f(x)$

15.

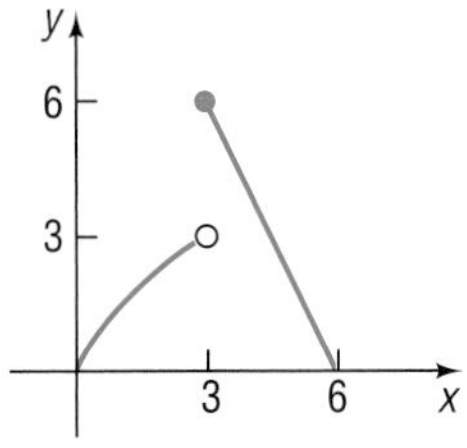

$\lim_{x \to 3} f(x)$

16.

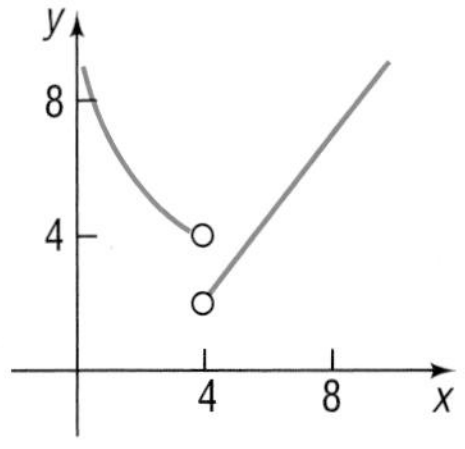

$\lim_{x \to 4} f(x)$

In Problems 17–36, graph each function. Use the graph to find the indicated limit, if it exists.

17. $\lim_{x \to 4} f(x), \quad f(x) = 3x + 1$

18. $\lim_{x \to -1} f(x), \quad f(x) = 2x - 1$

19. $\lim_{x \to 2} f(x), \quad f(x) = 1 - x^2$

20. $\lim_{x \to -1} f(x), \quad f(x) = x^3 - 1$

21. $\lim_{x \to -3} f(x), \quad f(x) = |2x|$

22. $\lim_{x \to 4} f(x), \quad f(x) = 3\sqrt{x}$

23. $\lim_{x \to \pi/2} f(x), \quad f(x) = \sin x$

24. $\lim_{x \to \pi} f(x), \quad f(x) = \cos x$

25. $\lim_{x \to 0} f(x), \quad f(x) = e^x$

26. $\lim_{x \to 1} f(x), \quad f(x) = \ln x$

27. $\lim_{x \to -1} f(x), \quad f(x) = 1/x$

28. $\lim_{x \to 2} f(x), \quad f(x) = 1/x^2$

29. $\lim_{x \to 0} f(x), \quad f(x) = \begin{cases} x^2 & x \geq 0 \\ 2x & x < 0 \end{cases}$

30. $\lim_{x \to 0} f(x), \quad f(x) = \begin{cases} x - 1 & x < 0 \\ 3x - 1 & x \geq 0 \end{cases}$

31. $\lim_{x \to 1} f(x), \quad f(x) = \begin{cases} 3x & x \leq 1 \\ x + 1 & x > 1 \end{cases}$

32. $\lim_{x \to 2} f(x), \quad f(x) = \begin{cases} x^2 & x \leq 2 \\ 2x - 1 & x > 2 \end{cases}$

33. $\lim_{x \to 0} f(x), \quad f(x) = \begin{cases} x & x < 0 \\ 1 & x = 0 \\ 3x & x > 0 \end{cases}$

34. $\lim_{x \to 0} f(x), \quad f(x) = \begin{cases} 1 & x < 0 \\ -1 & x > 0 \end{cases}$

35. $\lim_{x \to 0} f(x), \quad f(x) = \begin{cases} \sin x & x \leq 0 \\ x^2 & x > 0 \end{cases}$

36. $\lim_{x \to 0} f(x), \quad f(x) = \begin{cases} e^x & x > 0 \\ 1 - x & x \leq 0 \end{cases}$

In Problems 37–42, use a graphing utility to find the indicated limit rounded to two decimal places.

37. $\lim_{x \to 1} \frac{x^3 - x^2 + x - 1}{x^4 - x^3 + 2x - 2}$

38. $\lim_{x \to -1} \frac{x^3 + x^2 + 3x + 3}{x^4 + x^3 + 2x + 2}$

39. $\lim_{x \to 2} \frac{x^3 - 2x^2 + 4x - 8}{x^2 + x - 6}$

40. $\lim_{x \to 1} \frac{x^3 - x^2 + 3x - 3}{x^2 + 3x - 4}$

41. $\lim_{x \to -1} \frac{x^3 + 2x^2 + x}{x^4 + x^3 + 2x + 2}$

42. $\lim_{x \to 3} \frac{x^3 - 3x^2 + 4x - 12}{x^4 - 3x^3 + x - 3}$

PREPARING FOR THIS SECTION

Before getting started, review the following:

- ✓ Evaluating Functions (pp. 101–103)
- ✓ Average Rate of Change (pp. 116–118)
- ✓ Rational Expressions (Appendix, Section 7)

14.2 ALGEBRA TECHNIQUES FOR FINDING LIMITS

OBJECTIVES
1. Find the Limit of a Sum, a Difference, a Product, and a Quotient
2. Find the Limit of a Polynomial
3. Find the Limit of a Power or a Root
4. Find the Limit of an Average Rate of Change

We mentioned in the previous section that algebra can sometimes be used to find the exact value of a limit. This is accomplished by developing two formulas involving limits and several properties of limits.

Theorem **Two Formulas: $\lim_{x \to c} A$ and $\lim_{x \to c} x$**

Limit of a Constant

For the constant function $f(x) = A$,

$$\lim_{x \to c} f(x) = \lim_{x \to c} A = A \tag{1}$$

where c is any number.

Limit of x

For the identity function $f(x) = x$,

$$\lim_{x \to c} f(x) = \lim_{x \to c} x = c \tag{2}$$

where c is any number.

See Figure 5. Since the graph of a constant function is a horizontal line, it follows that, no matter how close x is to c, the corresponding value of $f(x)$ equals A. That is, $\lim_{x \to c} A = A$.

See Figure 6. For any choice of c, as x gets closer to c, the corresponding value of $f(x)$ is just as close to c. That is, $\lim_{x \to c} x = c$.

Figure 5

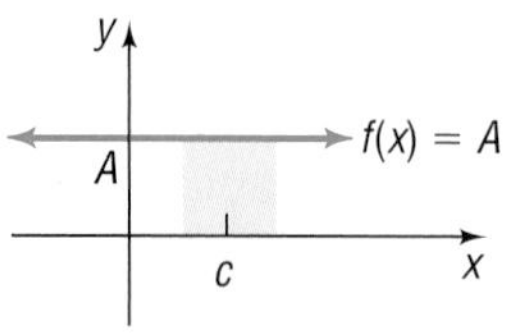

Figure 6

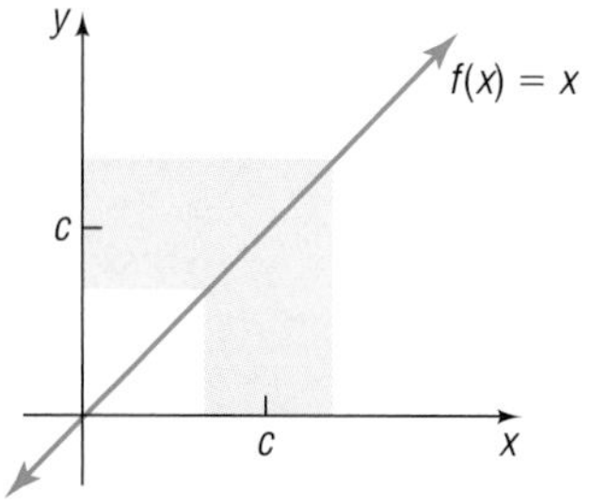

EXAMPLE 1 Using Formulas (1) and (2)

(a) $\lim_{x\to 3} 5 = 5$ (b) $\lim_{x\to 3} x = 3$ (c) $\lim_{x\to 0} -8 = -8$ (d) $\lim_{x\to -1/2} x = -\frac{1}{2}$ ■

NOW WORK PROBLEM 1.

Formulas (1) and (2), when used with the algebra properties that follow, enable us to evaluate limits of more complicated functions.

ALGEBRAIC PROPERTIES OF LIMITS

1 In the following properties, we assume that f and g are two functions for which both $\lim_{x\to c} f(x)$ and $\lim_{x\to c} g(x)$ exist.

Theorem **Limit of a Sum**

$$\lim_{x\to c}[f(x) + g(x)] = \lim_{x\to c} f(x) + \lim_{x\to c} g(x) \quad \textbf{(3)}$$

In words, the limit of the sum of two functions equals the sum of their limits.

■

EXAMPLE 2 Finding the Limit of a Sum

Find: $\lim_{x\to -3}[x + 4]$

Solution The limit we seek is the sum of two functions $f(x) = x$ and $g(x) = 4$. From formulas (1) and (2), we know that

$$\lim_{x\to -3} f(x) = \lim_{x\to -3} x = -3 \quad \text{and} \quad \lim_{x\to -3} g(x) = \lim_{x\to -3} 4 = 4$$

From formula (3), it follows that

$$\lim_{x\to -3}[x + 4] = \lim_{x\to -3} x + \lim_{x\to -3} 4 = -3 + 4 = 1$$ ■

Theorem **Limit of a Difference**

$$\lim_{x\to c}[f(x) - g(x)] = \lim_{x\to c} f(x) - \lim_{x\to c} g(x) \quad \textbf{(4)}$$

In words, the limit of the difference of two functions equals the difference of their limits.

■

EXAMPLE 3 Finding the Limit of a Difference

Find: $\lim_{x\to 4}[6 - x]$

Solution The limit we seek is the difference of two functions $f(x) = 6$ and $g(x) = x$. From formulas (1) and (2), we know that

$$\lim_{x\to 4} f(x) = \lim_{x\to 4} 6 = 6 \quad \text{and} \quad \lim_{x\to 4} g(x) = \lim_{x\to 4} x = 4$$

From formula (4), it follows that

$$\lim_{x\to 4}[6 - x] = \lim_{x\to 4} 6 - \lim_{x\to 4} x = 6 - 4 = 2$$

Theorem **Limit of a Product**

$$\lim_{x\to c}[f(x) \cdot g(x)] = \left[\lim_{x\to c} f(x)\right]\left[\lim_{x\to c} g(x)\right] \qquad \textbf{(5)}$$

In words, the limit of the product of two functions equals the product of their limits.

EXAMPLE 4 Finding the Limit of a Product

Find: $\lim_{x\to -5}(-4x)$

Solution The limit we seek is the product of two functions $f(x) = -4$ and $g(x) = x$. From formulas (1) and (2), we know that

$$\lim_{x\to -5} f(x) = \lim_{x\to -5} -4 = -4 \quad \text{and} \quad \lim_{x\to -5} g(x) = \lim_{x\to -5} x = -5$$

From formula (5), it follows that

$$\lim_{x\to -5}(-4x) = \left[\lim_{x\to -5} -4\right]\left[\lim_{x\to -5} x\right] = (-4)(-5) = 20$$

EXAMPLE 5 Finding Limits Using Algebraic Properties

Find: (a) $\lim_{x\to -2}(3x - 5)$ (b) $\lim_{x\to 2}(5x^2)$

Solution (a) $\lim_{x\to -2}(3x - 5) = \lim_{x\to -2}(3x) - \lim_{x\to -2} 5 = \left[\lim_{x\to -2} 3\right]\left[\lim_{x\to -2} x\right] - \lim_{x\to -2} 5$

$= (3)(-2) - 5 = -6 - 5 = -11$

(b) $\lim_{x\to 2}(5x^2) = \left[\lim_{x\to 2} 5\right]\left[\lim_{x\to 2} x^2\right] = 5 \lim_{x\to 2}(x \cdot x) = 5\left[\lim_{x\to 2} x\right]\left[\lim_{x\to 2} x\right]$

$= 5 \cdot 2 \cdot 2 = 20$

NOW WORK PROBLEM 5.

Notice in the solution to part (b) that $\lim_{x\to 2}(5x^2) = 5 \cdot 2^2$.

Theorem **Limit of a Monomial**

If $n \geq 1$ is a positive integer and a is a constant, then

$$\lim_{x\to c}(ax^n) = ac^n \qquad \textbf{(6)}$$

for any number c.

Proof

$$\lim_{x \to c} (ax^n) = [\lim_{x \to c} a][\lim_{x \to c} x^n] = a[\lim_{x \to c} (\underbrace{x \cdot x \cdot x \cdot \ldots \cdot x}_{n \text{ factors}})]$$

$$= a\underbrace{[\lim_{x \to c} x][\lim_{x \to c} x][\lim_{x \to c} x] \ldots [\lim_{x \to c} x]}_{n \text{ factors}}$$

$$= a \cdot \underbrace{c \cdot c \cdot c \cdot \ldots \cdot c}_{n \text{ factors}} = ac^n$$

■

EXAMPLE 6 Finding the Limit of a Monomial

Find: $\lim_{x \to 2} (-4x^3)$

Solution $\lim_{x \to 2} (-4x^3) = -4 \cdot 2^3 = -4 \cdot 8 = -32$ ■

2 Since a polynomial is a sum of monomials, we can use formula (6) and repeated use of formula (3) to obtain the following result:

Theorem **Limit of a Polynomial**

If P is a polynomial function, then

$$\lim_{x \to c} P(x) = P(c) \quad \textbf{(7)}$$

for any number c.

■

Proof If P is a polynomial function, that is, if

$$P(x) = a_n x^n + a_{n-1} x^{n-1} + \cdots + a_1 x + a_0$$

then

$$\lim_{x \to c} P(x) = \lim_{x \to c} \left[a_n x^n + a_{n-1} x^{n-1} + \cdots + a_1 x + a_0\right]$$

$$= \lim_{x \to c} (a_n x^n) + \lim_{x \to c} (a_{n-1} x^{n-1}) + \cdots + \lim_{x \to c} (a_1 x) + \lim_{x \to c} a_0$$

$$= a_n c^n + a_{n-1} c^{n-1} + \cdots + a_1 c + a_0$$

$$= P(c)$$

■

Formula (7) states that to find the limit of a polynomial as x approaches c all we need to do is to evaluate the polynomial at c.

EXAMPLE 7 Finding the Limit of a Polynomial

Find: $\lim_{x \to 2} \left[5x^4 - 6x^3 + 3x^2 + 4x - 2\right]$

Solution $\lim_{x \to 2} \left[5x^4 - 6x^3 + 3x^2 + 4x - 2\right] = 5 \cdot 2^4 - 6 \cdot 2^3 + 3 \cdot 2^2 + 4 \cdot 2 - 2$

$$= 5 \cdot 16 - 6 \cdot 8 + 3 \cdot 4 + 8 - 2$$

$$= 80 - 48 + 12 + 6 = 50$$

■

NOW WORK PROBLEM 7.

Theorem **Limit of a Power or Root**

If $\lim_{x \to c} f(x)$ is known and if $n \geq 2$ is a positive integer, then

$$\lim_{x \to c}\left[f(x)\right]^n = \left[\lim_{x \to c} f(x)\right]^n \qquad \textbf{(8)}$$

and

$$\lim_{x \to c} \sqrt[n]{f(x)} = \sqrt[n]{\lim_{x \to c} f(x)} \qquad \textbf{(9)}$$

In formula (9), we require that both $\sqrt[n]{f(x)}$ and $\sqrt[n]{\lim_{x \to c} f(x)}$ be defined.

EXAMPLE 8 Finding the Limit of a Power or a Root

3 Find:

(a) $\lim_{x \to 1}(3x - 5)^4$ (b) $\lim_{x \to 0}\sqrt{5x^2 + 8}$ (c) $\lim_{x \to -1}(5x^3 - x + 3)^{4/3}$

Solution (a) $\lim_{x \to 1}(3x - 5)^4 = \left[\lim_{x \to 1}(3x - 5)\right]^4 = (-2)^4 = 16$

(b) $\lim_{x \to 0}\sqrt{5x^2 + 8} = \sqrt{\lim_{x \to 0}(5x^2 + 8)} = \sqrt{8} = 2\sqrt{2}$

(c) $\lim_{x \to -1}(5x^3 - x + 3)^{4/3} = \sqrt[3]{\lim_{x \to -1}(5x^3 - x + 3)^4}$

$= \sqrt[3]{\left[\lim_{x \to -1}(5x^3 - x + 3)\right]^4} = \sqrt[3]{(-1)^4} = \sqrt[3]{1} = 1$

NOW WORK PROBLEM 17.

Theorem **Limit of a Quotient**

$$\lim_{x \to c}\left[\frac{f(x)}{g(x)}\right] = \frac{\left[\lim_{x \to c} f(x)\right]}{\left[\lim_{x \to c} g(x)\right]} \qquad \textbf{(10)}$$

provided $\lim_{x \to c} g(x) \neq 0$. In words, the limit of the quotient of two functions equals the quotient of their limits, provided that the limit of the denominator is not zero.

EXAMPLE 9 Finding the Limit of a Quotient

Find: $\lim_{x \to 1}\dfrac{5x^3 - x + 2}{3x + 4}$

Solution The limit we seek is the quotient of two functions: $f(x) = 5x^3 - x + 2$ and $g(x) = 3x + 4$. First, we find the limit of the denominator $g(x)$.

$$\lim_{x \to 1} g(x) = \lim_{x \to 1}(3x + 4) = 7$$

Since the limit of the denominator is not zero, we can proceed to use formula (10).

$$\lim_{x\to 1}\frac{5x^3 - x + 2}{3x + 4} = \frac{\lim\limits_{x\to 1}(5x^3 - x + 2)}{\lim\limits_{x\to 1}(3x + 4)} = \frac{6}{7}$$

NOW WORK PROBLEM 15.

When the limit of the denominator is zero, formula (10) cannot be used. In such cases, other strategies need to be used. Let's look at two examples.

EXAMPLE 10 Finding the Limit of a Quotient

Find: (a) $\lim\limits_{x\to 3}\dfrac{x^2 - x - 6}{x^2 - 9}$ (b) $\lim\limits_{x\to 0}\dfrac{5x - \sin x}{x}$

Solution (a) The limit of the denominator equals zero, so formula (10) cannot be used. Instead, we notice that the expression can be factored as

$$\frac{x^2 - x - 6}{x^2 - 9} = \frac{(x - 3)(x + 2)}{(x - 3)(x + 3)}$$

When we compute a limit as x approaches 3, we are interested in the values of the function when x is close to 3, but unequal to 3. Since $x \neq 3$, we can cancel the $(x - 3)$'s. Formula (10) can then be used.

$$\lim_{x\to 3}\frac{x^2 - x - 6}{x^2 - 9} = \lim_{x\to 3}\frac{\cancel{(x - 3)}(x + 2)}{\cancel{(x - 3)}(x + 3)} = \frac{\lim\limits_{x\to 3}(x + 2)}{\lim\limits_{x\to 3}(x + 3)} = \frac{5}{6}$$

(b) Again, the limit of the denominator is zero. In this situation, we perform the indicated operation and divide by x.

$$\lim_{x\to 0}\frac{5x - \sin x}{x} = \lim_{x\to 0}\left[\frac{5x}{x} - \frac{\sin x}{x}\right] = \lim_{x\to 0}\frac{5\cancel{x}}{\cancel{x}} - \lim_{x\to 0}\frac{\sin x}{x} = 5 - 1 = 4$$

↑ Refer to Example 3, Section 14.1

Let's work Example 6 of the previous section using algebra.

EXAMPLE 11 Finding Limits Using Algebraic Properties

Find: $\lim\limits_{x\to 2}\dfrac{x^3 - 2x^2 + 4x - 8}{x^4 - 2x^3 + x - 2}$

Solution The limit of the denominator is zero, so formula (10) cannot be used. We factor the expression.

$$\frac{x^3 - 2x^2 + 4x - 8}{x^4 - 2x^3 + x - 2} = \frac{x^2(x - 2) + 4(x - 2)}{x^3(x - 2) + 1(x - 2)} = \frac{(x^2 + 4)(x - 2)}{(x^3 + 1)(x - 2)}$$

Thus,

$$\lim_{x\to 2}\frac{x^3 - 2x^2 + 4x - 8}{x^4 - 2x^3 + x - 2} = \lim_{x\to 2}\frac{(x^2 + 4)\cancel{(x - 2)}}{(x^3 + 1)\cancel{(x - 2)}} = \frac{8}{9}$$

which is exact.

Compare the exact solution above with the approximate solution found in Example 6 of the previous section.

EXAMPLE 12 Finding the Limit of an Average Rate of Change

4 Find the limit as x approaches 2 of the average rate of change of the function

$$f(x) = x^2 + 3x$$

from 2 to x.

Solution The average rate of change of f from 2 to x is

$$\frac{\Delta y}{\Delta x} = \frac{f(x) - f(2)}{x - 2} = \frac{(x^2 + 3x) - 10}{x - 2} = \frac{(x + 5)(x - 2)}{x - 2}$$

The limit of the average rate of change is

$$\lim_{x\to 2} \frac{f(x) - f(2)}{x - 2} = \lim_{x\to 2} \frac{(x^2 + 3x) - 10}{x - 2} = \lim_{x\to 2} \frac{(x + 5)\cancel{(x - 2)}}{\cancel{x - 2}} = 7$$

SUMMARY

To find exact values for $\lim_{x\to c} f(x)$, try the following:

1. If f is a polynomial function, then $\lim_{x\to c} f(x) = f(c)$.
2. If f is a polynomial raised to a power or is the root of a polynomial, use formulas (8) and (9) with formula (7).
3. If f is a quotient and the limit of the denominator is not zero, then use the fact that the limit of a quotient is the quotient of the limits.
4. If f is a quotient and the limit of the denominator is zero, then use other techniques, such as factoring.

14.2 EXERCISES

In Problems 1–32, find the limit algebraically.

1. $\lim_{x\to 1} 5$ **2.** $\lim_{x\to 1} (-3)$ **3.** $\lim_{x\to 4} x$ **4.** $\lim_{x\to -3} x$

5. $\lim_{x\to 2} (3x + 2)$ **6.** $\lim_{x\to 3} (2 - 5x)$ **7.** $\lim_{x\to -1} (3x^2 - 5x)$ **8.** $\lim_{x\to 2} (8x^2 - 4)$

9. $\lim_{x\to 1} (5x^4 - 3x^2 + 6x - 9)$ **10.** $\lim_{x\to -1} (8x^5 - 7x^3 + 8x^2 + x - 4)$

11. $\lim_{x\to 1} (x^2 + 1)^3$ **12.** $\lim_{x\to 2} (3x - 4)^2$ **13.** $\lim_{x\to 1} \sqrt{5x + 4}$ **14.** $\lim_{x\to 0} \sqrt{1 - 2x}$

15. $\lim_{x\to 0} \frac{x^2 - 4}{x^2 + 4}$ **16.** $\lim_{x\to 2} \frac{3x + 4}{x^2 + x}$ **17.** $\lim_{x\to 2} (3x - 2)^{5/2}$ **18.** $\lim_{x\to -1} (2x + 1)^{5/3}$

19. $\lim_{x\to 2} \frac{x^2 - 4}{x^2 - 2x}$ **20.** $\lim_{x\to -1} \frac{x^2 + x}{x^2 - 1}$ **21.** $\lim_{x\to -3} \frac{x^2 - x - 12}{x^2 - 9}$ **22.** $\lim_{x\to -3} \frac{x^2 + x - 6}{x^2 + 2x - 3}$

23. $\lim_{x\to 1} \frac{x^3 - 1}{x - 1}$ **24.** $\lim_{x\to 1} \frac{x^4 - 1}{x - 1}$ **25.** $\lim_{x\to -1} \frac{(x + 1)^2}{x^2 - 1}$ **26.** $\lim_{x\to 2} \frac{x^3 - 8}{x^2 - 4}$

27. $\lim_{x\to 1} \frac{x^3 - x^2 + x - 1}{x^4 - x^3 + 2x - 2}$ **28.** $\lim_{x\to -1} \frac{x^3 + x^2 + 3x + 3}{x^4 + x^3 + 2x + 2}$ **29.** $\lim_{x\to 2} \frac{x^3 - 2x^2 + 4x - 8}{x^2 + x - 6}$

30. $\lim_{x \to 1} \frac{x^3 - x^2 + 3x - 3}{x^2 + 3x - 4}$ **31.** $\lim_{x \to -1} \frac{x^3 + 2x^2 + x}{x^4 + x^3 + 2x + 2}$ **32.** $\lim_{x \to 3} \frac{x^3 - 3x^2 + 4x - 12}{x^4 - 3x^3 + x - 3}$

In Problems 33–42, find the limit as x approaches c of the average rate of change of each function from c to x.

33. $c = 2;\ f(x) = 5x - 3$ **34.** $c = -2;\ f(x) = 4 - 3x$ **35.** $c = 3;\ f(x) = x^2$

36. $c = 3;\ f(x) = x^3$ **37.** $c = -1;\ f(x) = x^2 + 2x$ **38.** $c = -1;\ f(x) = 2x^2 - 3x$

39. $c = 0;\ f(x) = 3x^3 - 2x^2 + 4$ **40.** $c = 0;\ f(x) = 4x^3 - 5x + 8$ **41.** $c = 1;\ f(x) = \frac{1}{x}$

42. $c = 1;\ f(x) = \frac{1}{x^2}$

In Problems 43–46, use the properties of limits and the facts that

$$\lim_{x \to 0} \frac{\sin x}{x} = 1 \qquad \lim_{x \to 0} \frac{\cos x - 1}{x} = 0 \qquad \lim_{x \to 0} \sin x = 0 \qquad \lim_{x \to 0} \cos x = 1$$

where x is in radians, to find each limit.

43. $\lim_{x \to 0} \frac{\tan x}{x}$ **44.** $\lim_{x \to 0} \frac{\sin(2x)}{x}$

45. $\lim_{x \to 0} \frac{3\sin x + \cos x - 1}{4x}$ **46.** $\lim_{x \to 0} \frac{\sin^2 x + \sin x(\cos x - 1)}{x^2}$

PREPARING FOR THIS SECTION

Before getting started, review the following:

- ✓ Piecewise–Defined Functions (pp. 132–134)
- ✓ Domain of Rational Functions (p. 209)
- ✓ Library of Functions (pp. 128–131)
- ✓ Properties of the Exponential Function (pp. 294–296)
- ✓ Properties of the Logarithmic Function (p. 308)
- ✓ Properties of the Trigonometric Functions (Section 6.3)

14.3 ONE-SIDED LIMITS; CONTINUOUS FUNCTIONS

OBJECTIVES 1 Find the One-sided Limits of a Function
2 Determine Whether a Function Is Continuous

1 Earlier we described $\lim_{x \to c} f(x) = N$ by saying that as x gets closer to c, but remains unequal to c, the corresponding values of $f(x)$ get closer to N. Whether we use a numerical argument or the graph of the function f, the variable x can get closer to c in only two ways: either by approaching c from the left, through numbers less than c, or by approaching c from the right, through numbers greater than c.

If we only approach c from one side, we have a one-sided limit. The notation

$$\lim_{x \to c^-} f(x) = L$$

sometimes called the **left limit**, read as "the limit of $f(x)$ as x approaches c from the left equals L," may be described by the following statement:

As x gets closer to c, but remains less than c, the corresponding values of $f(x)$ get closer to L.

The notation $x \to c^-$ is used to remind us that x is less than c.

The notation

$$\lim_{x \to c^+} f(x) = R$$

sometimes called the **right limit**, read as "the limit of $f(x)$ as x approaches c from the right equals R," may be described by the following statement:

As x gets closer to c, but remains greater than c, the corresponding values of $f(x)$ get closer to R.

The notation $x \to c^+$ is used to remind us that x is greater than c.

Figure 7 illustrates left and right limits.

Figure 7

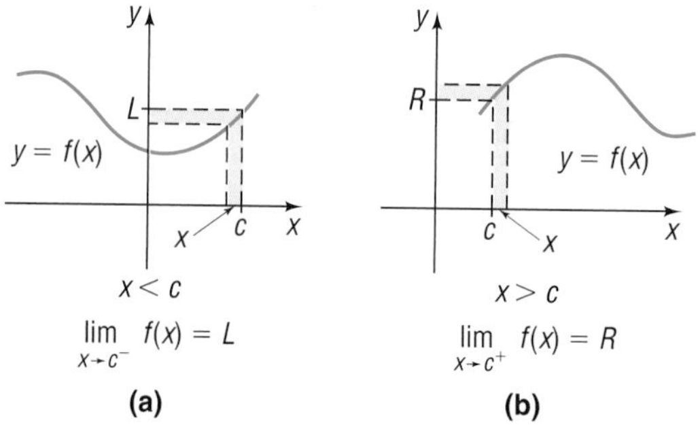

The left and right limits can be used to determine whether $\lim_{x \to c} f(x)$ exists. See Figure 8.

Figure 8

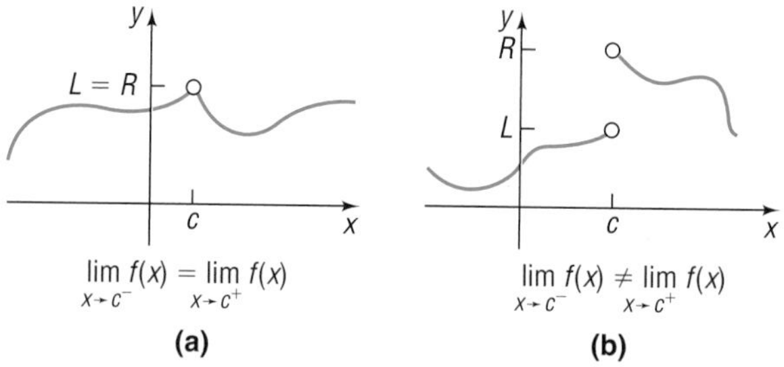

As Figure 8(a) illustrates, $\lim_{x \to c} f(x)$ exists and equals the common value of the left limit and the right limit ($L = R$). In Figure 8(b), we see that $\lim_{x \to c} f(x)$ does not exist and that $L \neq R$. This leads us to the following result:

Theorem Suppose that $\lim_{x \to c^-} f(x) = L$ and $\lim_{x \to c^+} f(x) = R$. Then $\lim_{x \to c} f(x)$ exists if and only if $L = R$. Furthermore, if $L = R$, then $\lim_{x \to c} f(x) = L\ (= R)$.

Collectively, the left and right limits of a function are called **one-sided limits** of the function.

EXAMPLE 1 Finding One-sided Limits of a Function

For the function

$$f(x) = \begin{cases} 2x - 1 & \text{if } x < 2 \\ 1 & \text{if } x = 2 \\ x - 2 & \text{if } x > 2 \end{cases}$$

find: (a) $\lim_{x\to 2^-} f(x)$ (b) $\lim_{x\to 2^+} f(x)$ (c) $\lim_{x\to 2} f(x)$

Figure 9

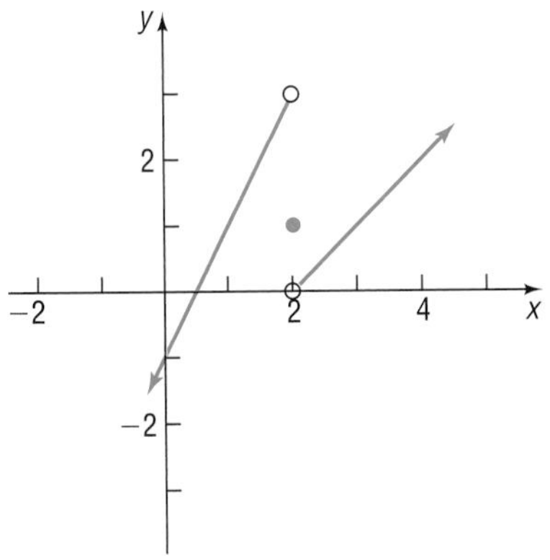

Solution Figure 9 shows the graph of f.

(a) To find $\lim_{x\to 2^-} f(x)$, we look at the values of f when x is close to 2, but less than 2. Since $f(x) = 2x - 1$ for such numbers, we conclude that

$$\lim_{x\to 2^-} f(x) = \lim_{x\to 2^-} (2x - 1) = 3$$

(b) To find $\lim_{x\to 2^+} f(x)$, we look at the values of f when x is close to 2, but greater than 2. Since $f(x) = x - 2$ for such numbers, we conclude that

$$\lim_{x\to 2^+} f(x) = \lim_{x\to 2^+} (x - 2) = 0$$

(c) Since the left and right limits are unequal, $\lim_{x\to 2} f(x)$ does not exist. ■

NOW WORK PROBLEM 49.

CONTINUOUS FUNCTIONS

2 We have observed that the value of a function f at c, $f(c)$, plays no role in determining the one-sided limits of f at c. What is the role of the value of a function at c and its one-sided limits at c? Let's look at some of the possibilities. See Figure 10.

Figure 10

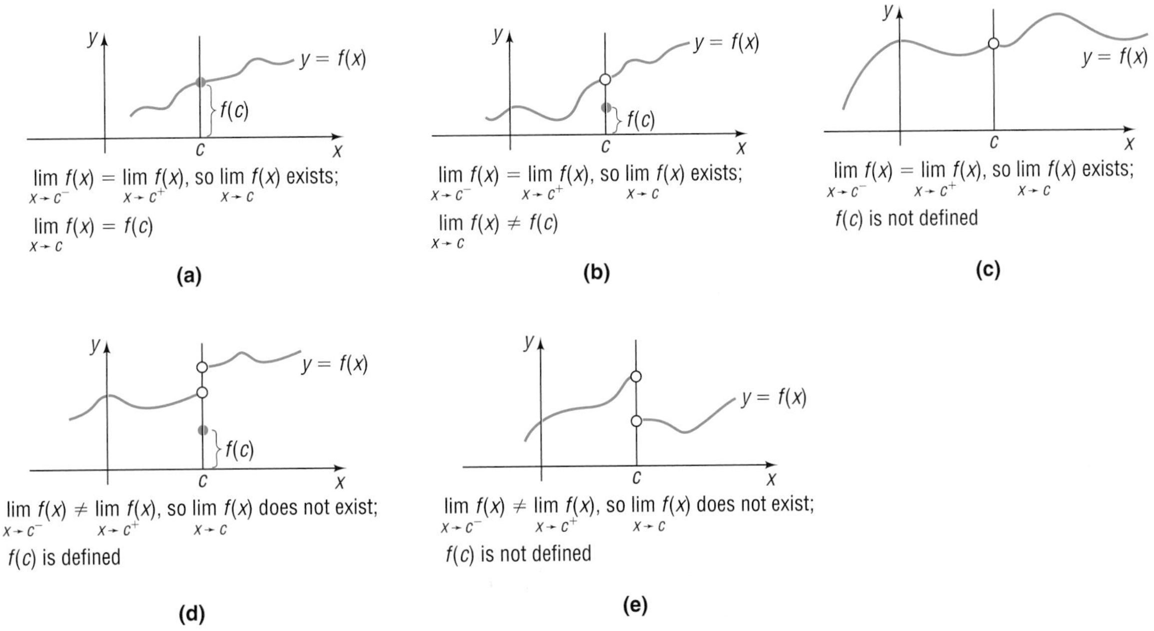

Much earlier in this book, we said that a function f was *continuous* if its graph could be drawn without lifting pencil from paper. In looking at Figure 10, the only graph that has this characteristic is the graph in Figure 10(a), for which the one-sided limits at c each exist and are equal to the value of f at c. This leads us to the following definition:

> A function f is **continuous** at c if:
>
> 1. f is defined at c; that is, c is in the domain of f so that $f(c)$ equals a number.
> 2. $\lim_{x \to c^-} f(x) = f(c)$
> 3. $\lim_{x \to c^+} f(x) = f(c)$
>
> In other words, a function f is continuous at c if
>
> $$\lim_{x \to c} f(x) = f(c)$$

If f is not continuous at c, we say that f is **discontinuous at c.** Each of the functions whose graph appears in Figures 10(b)–10(e) is discontinuous at c.

Look again at formula (7) on page 911 Based on (7), we conclude that a polynomial function is continuous at every number. Look at formula (10). We conclude that a rational function is continuous at every number, except any at which it is not defined. At numbers where a rational function is not defined, either a hole appears in the graph or else an asymptote appears.

NOW WORK PROBLEMS 1 AND 9.

EXAMPLE 2 Determining Whether a Function Is Continuous

Discuss whether the rational function

$$R(x) = \frac{x - 2}{x^2 - 6x + 8}$$

is continuous:

(a) At 2 (b) At 4

Use the one-sided limits of R at 2 and at 4 to analyze the graph of R near 2 and near 4.

Solution Since $R(x) = \dfrac{x - 2}{(x - 2)(x - 4)}$, the domain of R is $\{x \mid x \neq 2, x \neq 4\}$. We conclude that R is discontinuous at both 2 and 4. (Condition 1 of the definition is violated.)

To determine the behavior of the graph near 2 and near 4, we look at $\lim_{x \to 2} R(x)$ and $\lim_{x \to 4} R(x)$.

For $\lim_{x \to 2} R(x)$, we have

$$\lim_{x \to 2} R(x) = \lim_{x \to 2} \frac{\cancel{x - 2}}{\cancel{(x - 2)}(x - 4)} = \lim_{x \to 2} \frac{1}{x - 4} = -\frac{1}{2}$$

As x gets closer to 2, the graph of R gets closer to $-1/2$. Since R is not defined at 2, the graph will have a hole at $(2, -1/2)$.

For $\lim_{x \to 4} R(x)$, we have

$$\lim_{x \to 4} R(x) = \lim_{x \to 4} \frac{\cancel{x-2}}{\cancel{(x-2)}(x-4)} = \lim_{x \to 4} \frac{1}{x-4}$$

TABLE 6

X	Y1
3.9	-10
3.99	-100
3.999	-1000
3.9999	-10000

Y1=(X-2)/(X²-6X...

If $x < 4$ and x is getting closer to 4, the value of $1/(x-4) < 0$ and is becoming unbounded; that is, $\lim_{x \to 4^-} R(x) = -\infty$. See Table 6.

TABLE 7

X	Y1
4.1	10
4.01	100
4.001	1000
4.0001	10000

Y1=(X-2)/(X²-6X...

If $x > 4$ and x is getting closer to 4, the value of $1/(x-4) > 0$ and is becoming unbounded; that is, $\lim_{x \to 4^+} R(x) = \infty$. See Table 7.

Since $|R(x)| \to \infty$ for x close to 4, the graph of R will have a vertical asymptote at $x = 4$.

It is easiest to graph R by observing that

$$\text{if } x \neq 2, \qquad \text{then } R(x) = \frac{\cancel{x-2}}{\cancel{(x-2)}(x-4)} = \frac{1}{x-4}$$

Figure 11

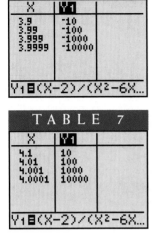

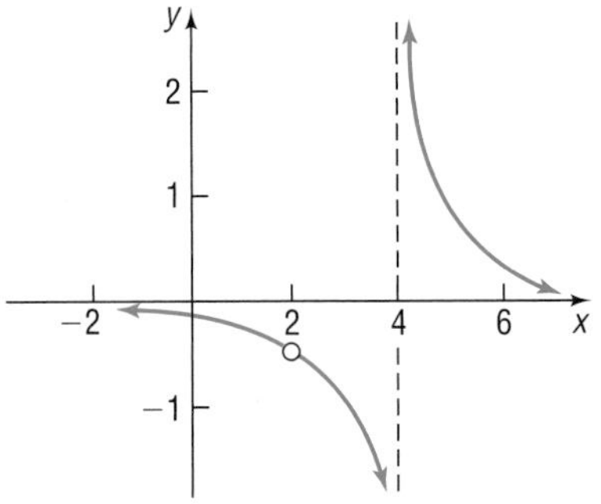

So the graph of R is the graph of $y = 1/x$ shifted to the right 4 units with a hole at $(2, -1/2)$. See Figure 11. ■

NOW WORK PROBLEM 61.

The exponential, logarithmic, sine, and cosine functions are continuous at every number in their domain. The tangent, cotangent, secant, and cosecant functions are continuous except at numbers for which they are not defined, where asymptotes occur. The square root function and absolute value function are continuous at every number in their domain. The function $f(x) = \text{int}(x)$ is continuous except for $x =$ an integer, where a jump occurs in the graph.

Piecewise-defined functions require special attention.

EXAMPLE 3 Determining Where a Piecewise-defined Function Is Continuous

Determine for what numbers x the following function is continuous.

$$f(x) = \begin{cases} x^2 & \text{if } x \leq 0 \\ x + 1 & \text{if } 0 < x < 2 \\ 5 - x & \text{if } 2 \leq x \leq 5 \end{cases}$$

Solution The "pieces" of f, that is, $y = x^2$, $y = x + 1$, and $y = 5 - x$, are each continuous for every number since they are polynomials. In other words, when we graph the pieces, we will not lift our pencil. When we graph the function f, however, we have to be careful, because the pieces change at $x = 0$ and at $x = 2$. So the numbers we need to investigate for f are $x = 0$ and $x = 2$.

For $x = 0$: $\qquad f(0) = 0^2 = 0$

$$\lim_{x \to 0^-} f(x) = \lim_{x \to 0^-} x^2 = 0$$

$$\lim_{x \to 0^+} f(x) = \lim_{x \to 0^+} (x + 1) = 1$$

Since $\lim_{x \to 0^+} f(x) \neq f(0)$, we conclude that f is not continuous at $x = 0$.

Figure 12

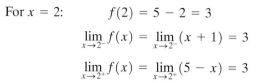

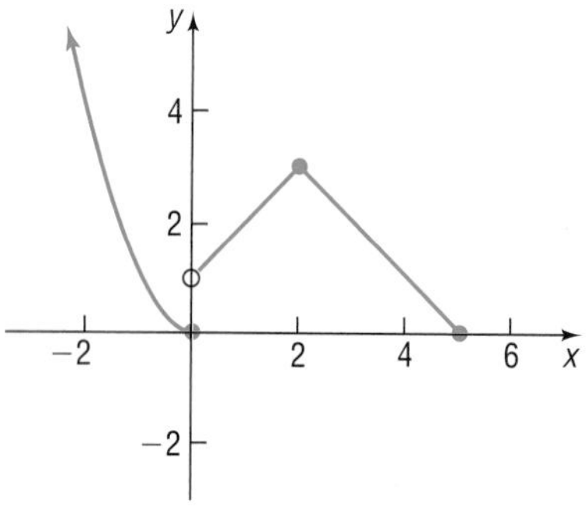

For $x = 2$: $$f(2) = 5 - 2 = 3$$

$$\lim_{x\to 2^-} f(x) = \lim_{x\to 2^-} (x + 1) = 3$$

$$\lim_{x\to 2^+} f(x) = \lim_{x\to 2^+} (5 - x) = 3$$

We conclude that f is continuous at $x = 2$.

The graph of f, given in Figure 12, demonstrates the conclusions drawn above.

NOW WORK PROBLEM 33.

14.3 EXERCISES

In Problems 1–12, find the numbers at which f is continuous. At which numbers is f discontinuous?

1. $f(x) = 2x + 3$
2. $f(x) = 4 - 3x$
3. $f(x) = 3x^2 + x$
4. $f(x) = -3x^3 + 7$
5. $f(x) = 4 \sin x$
6. $f(x) = -2 \cos x$
7. $f(x) = 2 \tan x$
8. $f(x) = 4 \csc x$
9. $f(x) = \dfrac{2x + 5}{x^2 - 4}$
10. $f(x) = \dfrac{x^2 - 4}{x^2 - 9}$
11. $f(x) = \dfrac{x - 3}{\ln x}$
12. $f(x) = \dfrac{\ln x}{x - 3}$

In Problems 13–24, find the one-sided limit.

13. $\lim_{x\to 1^+} (2x + 3)$
14. $\lim_{x\to 2^-} (4 - 2x)$
15. $\lim_{x\to 1^-} (2x^3 + 5x)$
16. $\lim_{x\to -2^+} (3x^2 - 8)$
17. $\lim_{x\to \pi/2^+} \sin x$
18. $\lim_{x\to \pi^-} (3 \cos x)$
19. $\lim_{x\to 2^+} \dfrac{x^2 - 4}{x - 2}$
20. $\lim_{x\to 1^-} \dfrac{x^3 - x}{x - 1}$
21. $\lim_{x\to -1^-} \dfrac{x^2 - 1}{x^3 + 1}$
22. $\lim_{x\to 0^+} \dfrac{x^3 - x^2}{x^4 + x^2}$
23. $\lim_{x\to -2^+} \dfrac{x^2 + x - 2}{x^2 + 2x}$
24. $\lim_{x\to -4^-} \dfrac{x^2 + x - 12}{x^2 + 4x}$

In Problems 25–40, determine whether f is continuous at c.

25. $f(x) = x^3 - 3x^2 + 2x - 6, \quad c = 2$
26. $f(x) = 3x^2 - 6x + 5, \quad c = -3$
27. $f(x) = \dfrac{x^2 + 5}{x - 6}, \quad c = 3$
28. $f(x) = \dfrac{x^3 - 8}{x^2 + 4}, \quad c = 2$
29. $f(x) = \dfrac{x + 3}{x - 3}, \quad c = 3$
30. $f(x) = \dfrac{x - 6}{x + 6}, \quad c = -6$
31. $f(x) = \dfrac{x^3 + 3x}{x^2 - 3x}, \quad c = 0$
32. $f(x) = \dfrac{x^2 - 6x}{x^2 + 6x}, \quad c = 0$
33. $f(x) = \begin{cases} \dfrac{x^3 + 3x}{x^2 - 3x} & \text{if } x \neq 0 \\ 1 & \text{if } x = 0 \end{cases}, \quad c = 0$
34. $f(x) = \begin{cases} \dfrac{x^2 - 6x}{x^2 + 6x} & \text{if } x \neq 0 \\ -2 & \text{if } x = 0 \end{cases}, \quad c = 0$
35. $f(x) = \begin{cases} \dfrac{x^3 + 3x}{x^2 - 3x} & \text{if } x \neq 0 \\ -1 & \text{if } x = 0 \end{cases}, \quad c = 0$
36. $f(x) = \begin{cases} \dfrac{x^2 - 6x}{x^2 + 6x} & \text{if } x \neq 0 \\ -1 & \text{if } x = 0 \end{cases}, \quad c = 0$
37. $f(x) = \begin{cases} \dfrac{x^3 - 1}{x^2 - 1} & \text{if } x < 1 \\ 2 & \text{if } x = 1, \\ \dfrac{3}{x + 1} & \text{if } x > 1 \end{cases} \quad c = 1$
38. $f(x) = \begin{cases} \dfrac{x^2 - 2x}{x - 2} & \text{if } x < 2 \\ 2 & \text{if } x = 2, \\ \dfrac{x - 4}{x - 1} & \text{if } x > 2 \end{cases} \quad c = 2$
39. $f(x) = \begin{cases} 2e^x & \text{if } x < 0 \\ 2 & \text{if } x = 0 \\ \dfrac{x^3 + 2x^2}{x^2} & \text{if } x > 0 \end{cases}, \quad c = 0$
40. $f(x) = \begin{cases} 3 \cos x & \text{if } x < 0 \\ 3 & \text{if } x = 0 \\ \dfrac{x^3 + 3x^2}{x^2} & \text{if } x > 0 \end{cases}, \quad c = 0$

In Problems 41–60, use the accompanying graph of $y = f(x)$.

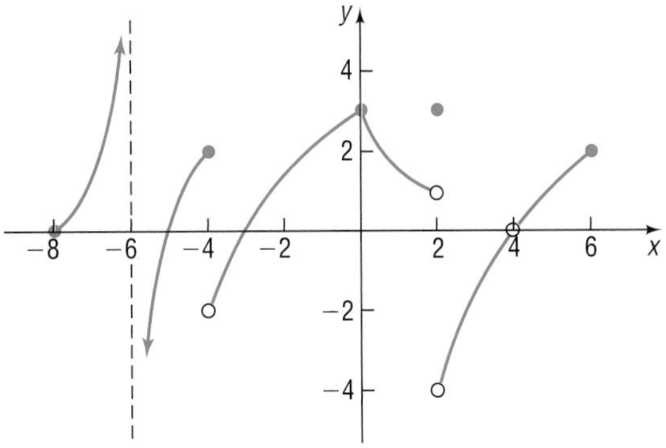

41. What is the domain of f?

42. What is the range of f?

43. Find the x-intercept(s), if any, of f.

44. Find the y-intercept(s), if any, of f.

45. Find $f(-8)$ and $f(-4)$.

46. Find $f(2)$ and $f(6)$.

47. Find $\lim_{x \to -6^-} f(x)$.

48. Find $\lim_{x \to -6^+} f(x)$.

49. Find $\lim_{x \to -4^-} f(x)$.

50. Find $\lim_{x \to -4^+} f(x)$.

51. Find $\lim_{x \to 2^-} f(x)$.

52. Find $\lim_{x \to 2^+} f(x)$.

53. Does $\lim_{x \to 4} f(x)$ exist? If it does, what is it?

54. Does $\lim_{x \to 0} f(x)$ exist? If it does, what is it?

55. Is f continuous at -6?

56. Is f continuous at -4?

57. Is f continuous at 0?

58. Is f continuous at 2?

59. Is f continuous at 4?

60. Is f continuous at 5?

In Problems 61–64, discuss whether R is continuous at c. Use the one-sided limits of R at c to analyze the graph of R. Graph R.

61. $R(x) = \dfrac{x - 1}{x^2 - 1}$ at $c = -1$ and $c = 1$

62. $R(x) = \dfrac{3x + 6}{x^2 - 4}$ at $c = -2$ and $c = 2$

63. $R(x) = \dfrac{x^2 + x}{x^2 - 1}$ at $c = -1$ and $c = 1$

64. $R(x) = \dfrac{x^2 + 4x}{x^2 - 16}$ at $c = -4$ and $c = 4$

In Problems 65–70, determine where each rational function is undefined. Determine whether an asymptote or a hole appears at such numbers. Graph R using a graphing utility to verify your results.

65. $R(x) = \dfrac{x^3 - x^2 + x - 1}{x^4 - x^3 + 2x - 2}$

66. $R(x) = \dfrac{x^3 + x^2 + 3x + 3}{x^4 + x^3 + 2x - 2}$

67. $R(x) = \dfrac{x^3 - 2x^2 + 4x - 8}{x^2 + x - 6}$

68. $R(x) = \dfrac{x^3 - x^2 + 3x - 3}{x^2 + 3x - 4}$

69. $R(x) = \dfrac{x^3 + 2x^2 + x}{x^4 + x^3 + 2x + 2}$

70. $R(x) = \dfrac{x^3 - 3x^2 + 4x - 12}{x^4 - 3x^3 + x - 3}$

PREPARING FOR THIS SECTION

Before getting started, review the following:

✓ Point–slope Form of a Line (p. 66)

✓ Average Rate of Change (pp. 116–118)

14.4 THE TANGENT PROBLEM; THE DERIVATIVE

OBJECTIVES

1. Find an Equation of the Tangent Line to the Graph of a Function
2. Find the Derivative of a Function
3. Find Instantaneous Rates of Change
4. Find the Speed of a Particle

TANGENT PROBLEM

One question that motivated the development of calculus was a geometry problem, the **tangent problem**. This problem asks, "What is the slope of the tangent line to the graph of a function $y = f(x)$ at a point P on its graph?" See Figure 13.

Figure 13

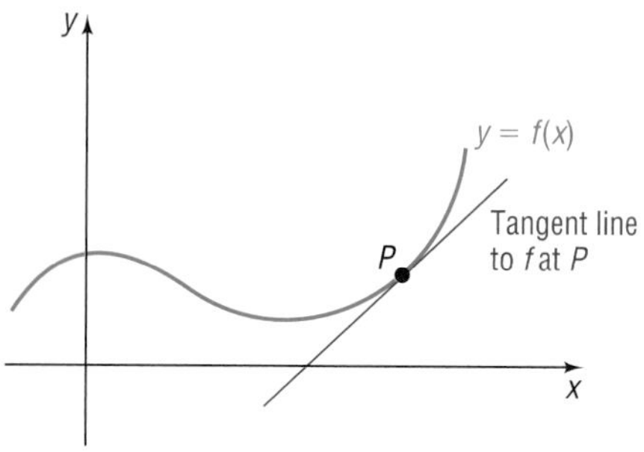

We first need to define what we mean by a *tangent* line. In high school geometry, the tangent line to a circle is defined as the line that intersects the graph in exactly one point. Look at Figure 14. Notice that the tangent line just *touches* the graph of the circle.

This definition, however, does not work in general. Look at Figure 15. The lines L_1 and L_2 only intersect the graph in one point P, but neither touches the graph at P. Additionally, the tangent line L_T shown in Figure 16 touches the graph of f at P, but also intersects the graph elsewhere. So how should we define the tangent line to the graph of f at a point P?

Figure 14

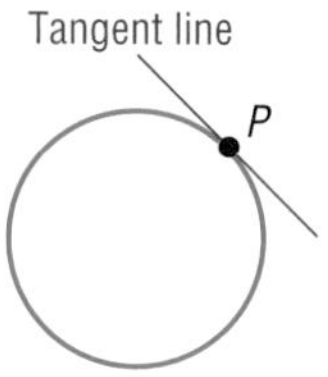

Figure 15

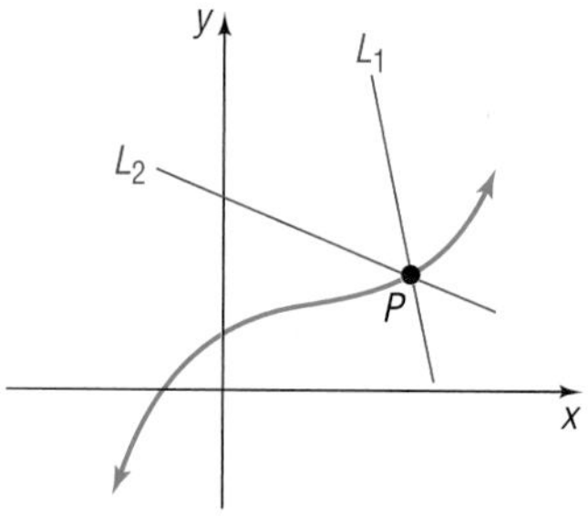

Figure 16

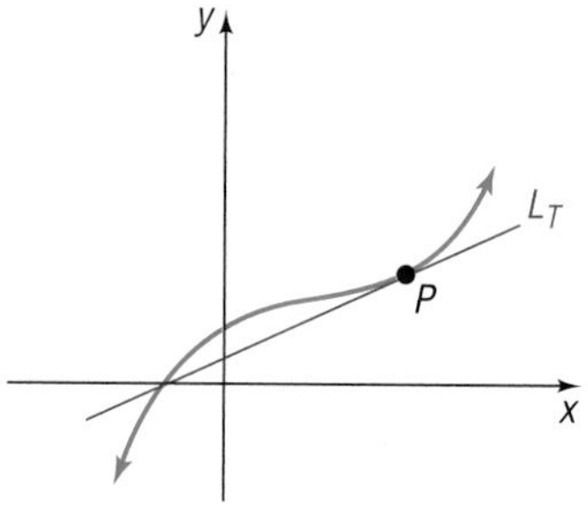

1 The tangent line L_T to the graph of a function $y = f(x)$ at a point P necessarily contains the point P. Because of the point–slope form of the equation of a line, it remains to find the slope $m_{\tan}$ of the tangent line.

Suppose that the coordinates of the point P are $(c, f(c))$. Locate another point $Q = (x, f(x))$ on the graph of f. The line containing P and Q is the secant line. (Refer to Section 2.2.) The slope $m_{\sec}$ of the secant line is

$$m_{\sec} = \frac{f(x) - f(c)}{x - c}$$

Figure 17

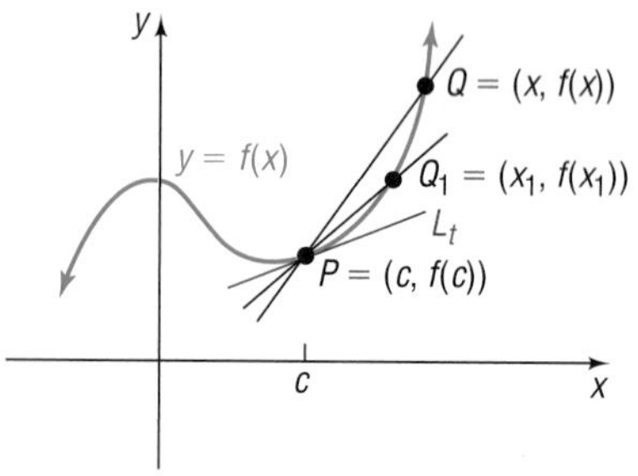

Now look at Figure 17.

As we move along the graph of f from Q toward P, we obtain a succession of secant lines. The closer we get to P, the closer the secant line is to the tangent line. The limiting position of these secant lines is the tangent line. Therefore, the limiting value of the slopes of these secant lines equals the slope of the tangent line. But, as we move from Q toward P, the values of x get closer to c. Therefore, it follows that

$$m_{\tan} = \lim_{x \to c} m_{\sec} = \lim_{x \to c} \frac{f(x) - f(c)}{x - c}$$

The **tangent line** to the graph of a function $y = f(x)$ at a point $P = (c, f(c))$ on its graph is defined as the line containing the point P whose slope is

$$m_{\tan} = \lim_{x \to c} \frac{f(x) - f(c)}{x - c} \quad \textbf{(1)}$$

provided that this limit exists. If $m_{\tan}$ exists, an equation of the tangent line is

$$y - f(c) = m_{\tan}(x - c) \tag{2}$$

EXAMPLE 1 Finding an Equation of the Tangent Line

Find an equation of the tangent line to the graph of $y = x^2/4$ at the point $(1, 1/4)$.

Solution The tangent line contains the point $(1, 1/4)$. The slope of the tangent line to the graph of $y = x^2/4$ at $(1, 1/4)$ is

$$m_{\tan} = \lim_{x\to 1} \frac{f(x) - f(1)}{x - 1} = \lim_{x\to 1} \frac{x^2/4 - 1/4}{x - 1} = \lim_{x\to 1} \frac{\cancel{(x-1)}(x+1)}{4\cancel{(x-1)}}$$

$$= \lim_{x\to 1} \frac{x+1}{4} = \frac{1}{2}$$

An equation of the tangent line is

$$y - \frac{1}{4} = \frac{1}{2}(x - 1) \qquad y - f(c) = m_{\tan}(x - c)$$

$$y = \frac{1}{2}x - \frac{1}{4}$$

Figure 18(a) shows the graph of $y = x^2/4$ and the tangent line at $(1, 1/4)$ drawn by hand. Figure 18(b) shows the graph of $Y_1 = x^2/4$ and the tangent line $Y_2 = (1/2)x - 1/4$ on a graphing utility.

Figure 18

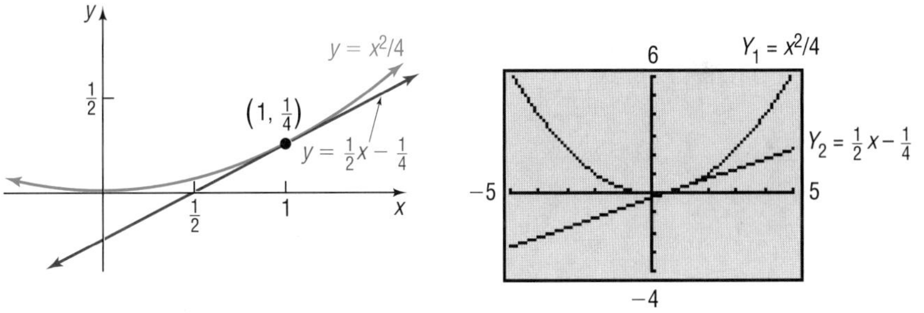

(a) (b)

NOW WORK PROBLEM 3.

2 The limit in formula (1) has an important generalization: it is called the *derivative of f at c*.

Let $y = f(x)$ denote a function f. If c is a number in the domain of f, the **derivative of f at c**, denoted by $f'(c)$, read "f prime of c," is defined as

$$f'(c) = \lim_{x\to c} \frac{f(x) - f(c)}{x - c} \tag{3}$$

provided that this limit exists.

EXAMPLE 2 Finding the Derivative of a Function

Find the derivative of $f(x) = 2x^2 - 5x$ at 2. That is, find $f'(2)$.

Solution Since $f(2) = 2(4) - 5(2) = -2$, we have

$$\frac{f(x) - f(2)}{x - 2} = \frac{(2x^2 - 5x) - (-2)}{x - 2} = \frac{2x^2 - 5x + 2}{x - 2} = \frac{(2x - 1)(x - 2)}{x - 2}$$

The derivative of f at 2 is

$$f'(2) = \lim_{x \to 2} \frac{f(x) - f(2)}{x - 2} = \lim_{x \to 2} \frac{(2x - 1)\cancel{(x - 2)}}{\cancel{x - 2}} = 3$$

NOW WORK PROBLEM 13.

Example 2 provides a way of finding the derivative at 2 analytically. Graphing utilities have built-in procedures to approximate the derivative of a function at any number c. Consult your owner's manual for the appropriate keystrokes.

EXAMPLE 3 Finding the Derivative of a Function Using a Graphing Utility

Use a graphing utility to find the derivative of $f(x) = 2x^2 - 5x$ at 2. That is, find $f'(2)$.

Solution Figure 19 shows the solution using a TI-83 graphing calculator.

Figure 19

So $f'(2) = 3$.

NOW WORK PROBLEM 25.

EXAMPLE 4 Finding the Derivative of a Function

Find the derivative of $f(x) = x^2$ at c. That is, find $f'(c)$.

Solution Since $f(c) = c^2$, we have

$$\frac{f(x) - f(c)}{x - c} = \frac{x^2 - c^2}{x - c} = \frac{(x + c)(x - c)}{x - c}$$

The derivative of f at c is

$$f'(c) = \lim_{x \to c} \frac{f(x) - f(c)}{x - c} = \lim_{x \to c} \frac{(x + c)\cancel{(x - c)}}{\cancel{x - c}} = 2c$$

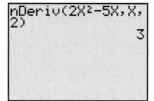

As Example 4 illustrates, the derivative of $f(x) = x^2$ exists and equals $2c$ for any number c. In other words, the derivative is itself a function and, using x for the independent variable, we can write $f'(x) = 2x$. The function f' is called the **derivative function of f** or the **derivative of f**. We also say that f is **differentiable**. The instruction "differentiate f" means "find the derivative of f."

INTERPRETATIONS OF THE DERIVATIVE

3 In Chapter 2 we defined the average rate of change of a function f from c to x as

$$\frac{\Delta y}{\Delta x} = \frac{f(x) - f(c)}{x - c}$$

The limit as x approaches c of the average rate of change of f, based on formula (2), is the derivative of f at c. As a result, we call the derivative of f at c the **instantaneous rate of change of f with respect to x at c**. That is,

$$\left(\begin{array}{c}\text{Instantaneous rate of} \\ \text{change of } f \text{ with respect to } x \text{ at } c\end{array}\right) = f'(c) = \lim_{x \to c} \frac{f(x) - f(c)}{x - c} \qquad \textbf{(4)}$$

EXAMPLE 5 Finding the Instantaneous Rate of Change

The volume V of a right circular cone of height 6 feet and radius r feet is $V = V(r) = 6\pi r^2/3 = 2\pi r^2$. If r is changing, find the instantaneous rate of change of the volume V with respect to the radius r at $r = 3$.

Algebraic Solution The instantaneous rate of change of V with respect to r at $r = 3$ is the derivative $V'(3)$.

$$V'(3) = \lim_{r \to 3} \frac{V(r) - V(3)}{r - 3} = \lim_{r \to 3} \frac{2\pi r^2 - 18\pi}{r - 3} = \lim_{r \to 3} \frac{2\pi(r^2 - 9)}{r - 3}$$

$$= \lim_{r \to 3} [2\pi(r + 3)] = 12\pi$$

Graphing Solution Figure 20 shows the derivative of V at $r = 3$ using a TI-83 graphing calculator.

Figure 20

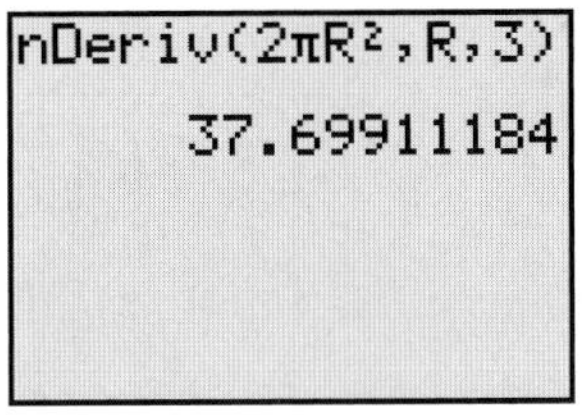

At the instant $r = 3$ feet, the volume of the cone is increasing with respect to r at a rate of $12\pi \approx 37.699$ cubic feet.

NOW WORK PROBLEM 35.

4 If $s = f(t)$ denotes the position of a particle at time t, then the average speed of the particle from c to t is

$$\frac{\text{Change in position}}{\text{Change in time}} = \frac{\Delta s}{\Delta t} = \frac{f(t) - f(c)}{t - c} \qquad \textbf{(5)}$$

The limit as t approaches c of the expression in formula (5) is the **instantaneous speed of the particle at c** or the **velocity of the particle at c.** That is,

$$\left(\begin{array}{c}\text{Instantaneous speed of}\\ \text{a particle at time } c\end{array}\right) = f'(c) = \lim_{t \to c} \frac{f(t) - f(c)}{t - c} \qquad \textbf{(6)}$$

EXAMPLE 6 Finding the Instantaneous Speed of a Particle

In physics it is shown that the height s of a ball thrown straight up with an initial speed of 80 feet per second (ft/sec) from a rooftop 96 feet high is

$$s = s(t) = -16t^2 + 80t + 96$$

where t is the elapsed time that the ball is in the air. The ball misses the rooftop on its way down and eventually strikes the ground. See Figure 21.

Figure 21

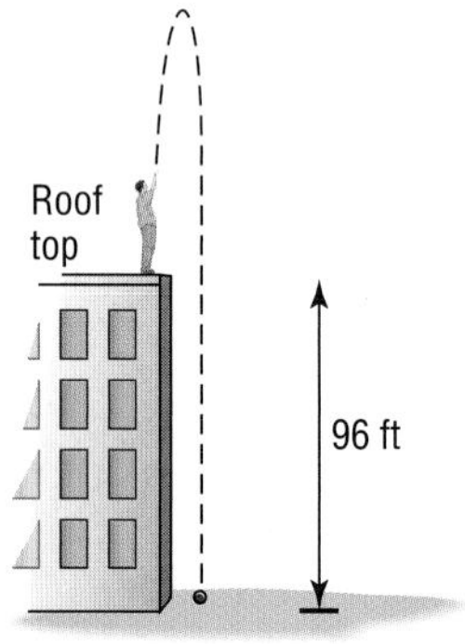

(a) When does the ball strike the ground? That is, how long is the ball in the air?

(b) At what time t will the ball pass the rooftop on its way down?

(c) What is the average speed of the ball from $t = 0$ to $t = 2$?

(d) What is the instantaneous speed of the ball at time t?

(e) What is the instantaneous speed of the ball at $t = 2$?

(f) When is the instantaneous speed of the ball equal to zero?

(g) What is the instantaneous speed of the ball as it passes the rooftop on the way down?

(h) What is the instantaneous speed of the ball when it strikes the ground?

Solution (a) The ball strikes the ground when $s = s(t) = 0$.

$$-16t^2 + 80t + 96 = 0$$
$$t^2 - 5t - 6 = 0$$
$$(t - 6)(t + 1) = 0$$
$$t = 6 \quad \text{or} \quad t = -1$$

We discard the solution $t = -1$. The ball strikes the ground after 6 seconds.

(b) The ball passes the rooftop when $s = s(t) = 96$.

$$-16t^2 + 80t + 96 = 96$$
$$t^2 - 5t = 0$$
$$t(t - 5) = 0$$
$$t = 0 \quad \text{or} \quad t = 5$$

We discard the solution $t = 0$. The ball passes the rooftop on the way down after 5 seconds.

(c) The average speed of the ball from $t = 0$ to $t = 2$ is

$$\frac{\Delta s}{\Delta t} = \frac{s(2) - s(0)}{2 - 0} = \frac{192 - 96}{2} = 48 \text{ ft/sec}$$

(d) The instantaneous speed of the ball at time t_0 is the derivative $s'(t_0)$; that is,

$$\begin{aligned} s'(t_0) &= \lim_{t \to t_0} \frac{s(t) - s(t_0)}{t - t_0} \\ &= \lim_{t \to t_0} \frac{(-16t^2 + 80t + 96) - (-16t_0^2 + 80t_0 + 96)}{t - t_0} \\ &= \lim_{t \to t_0} \frac{-16[t^2 - t_0^2 + 5t + 5t_0]}{t - t_0} \\ &= \lim_{t \to t_0} \frac{-16[(t + t_0)(t - t_0) - 5(t - t_0)]}{t - t_0} \\ &= \lim_{t \to t_0} \frac{-16[(t + t_0 - 5)\cancel{(t - t_0)}]}{\cancel{t - t_0}} = \lim_{t \to t_0} [-16(t + t_0 - 5)] \\ &= -16(2t_0 - 5) \quad \text{ft/sec} \end{aligned}$$

The instantaneous speed of the ball at time t is therefore

$$s'(t) = -16(2t - 5) \quad \text{ft/sec}$$

(e) At $t = 2$ sec, the instantaneous speed of the ball is

$$s'(2) = -16(-1) = 16 \text{ ft/sec}$$

(f) The instantaneous speed of the ball is zero when

$$\begin{aligned} s'(t) &= 0 \\ -16(2t - 5) &= 0 \\ t &= \frac{5}{2} = 2.5 \text{ sec} \end{aligned}$$

(g) The ball passes the rooftop on the way down when $t = 5$. The instantaneous speed at $t = 5$ is

$$s'(5) = -16(10 - 5) = -80 \text{ ft/sec}$$

At $t = 5$ seconds, the ball is traveling -80 ft/sec. When the instantaneous rate of change is negative, it means that the direction of the object is downward. The ball is traveling 80 ft/sec in the downward direction when $t = 5$ seconds.

(h) The ball strikes the ground when $t = 6$. The instantaneous speed when $t = 6$ is

$$s'(6) = -16(12 - 5) = -112 \text{ ft/sec}$$

The speed of the ball at $t = 6$ seconds is -112 ft/sec. Again, the negative value implies that the ball is traveling downward. ■

EXPLORATION: Determine the vertex of the quadratic function given in Example 6. What do you conclude about instantaneous velocity when $s(t)$ is a maximum? ■

SUMMARY

The derivative of a function $y = f(x)$ at c is defined as

$$f'(c) = \lim_{x \to c} \frac{f(x) - f(c)}{x - c}$$

In geometry, $f'(c)$ equals the slope of the tangent line to the graph of f at the point $(c, f(c))$.

In physics, $f'(c)$ equals the instantaneous speed (velocity) of a particle at time c, where $s = f(t)$ is the position of the particle at time t.

In applications, if two variables are related by the function $y = f(x)$, then $f'(c)$ equals the instantaneous rate of change of f with respect to x at c.

14.4 EXERCISES

In Problems 1–12, find the slope of the tangent line to the graph of f at P. Graph f and the tangent line.

1. $f(x) = 3x + 5$ at $(1, 8)$

2. $f(x) = -2x + 1$ at $(-1, 3)$

3. $f(x) = x^2 + 2$ at $(-1, 3)$

4. $f(x) = 3 - x^2$ at $(1, 2)$

5. $f(x) = 3x^2$ at $(2, 12)$

6. $f(x) = -4x^2$ at $(-2, -16)$

7. $f(x) = 2x^2 + x$ at $(1, 3)$

8. $f(x) = 3x^2 - x$ at $(0, 0)$

9. $f(x) = x^2 - 2x + 3$ at $(-1, 6)$

10. $f(x) = -2x^2 + x - 3$ at $(1, -4)$

11. $f(x) = x^3 + x$ at $(2, 10)$

12. $f(x) = x^3 - x^2$ at $(1, 0)$

In Problems 13–24, find the derivative of each function at c.

13. $f(x) = -4x + 5$ at 3

14. $f(x) = -4 + 3x$ at 1

15. $f(x) = x^2 - 3$ at 0

16. $f(x) = 2x^2 + 1$ at -1

17. $f(x) = 2x^2 + 3x$ at 1

18. $f(x) = 3x^2 - 4x$ at 2

19. $f(x) = x^3 + 4x$ at -1

20. $f(x) = 2x^3 - x^2$ at 2

21. $f(x) = x^3 + x^2 - 2x$ at 1

22. $f(x) = x^3 - 2x^2 + x$ at -1

23. $f(x) = \sin x$ at 0

24. $f(x) = \cos x$ at 0

In Problems 25–34, find the derivative of each function at c using a graphing utility.

25. $f(x) = 3x^3 - 6x^2 + 2$ at -2

26. $f(x) = -5x^4 + 6x^2 - 10$ at 5

27. $f(x) = \dfrac{-x^3 + 1}{x^2 + 5x + 7}$ at 8

28. $f(x) = \dfrac{-5x^4 + 9x + 3}{x^3 + 5x^2 - 6}$ at -3

29. $f(x) = x \sin x$ at $\dfrac{\pi}{3}$

30. $f(x) = x \sin x$ at $\dfrac{\pi}{4}$

31. $f(x) = x^2 \sin x$ at $\dfrac{\pi}{3}$

32. $f(x) = x^2 \sin x$ at $\dfrac{\pi}{4}$

33. $f(x) = e^x \sin x$ at 2

34. $f(x) = e^{-x} \sin x$ at 2

35. Instantaneous Rate of Change The volume V of a right circular cylinder of height 9 feet and radius r feet is $V = V(r) = 3\pi r^2$. Find the instantaneous rate of change of the volume with respect to the radius r at $r = 3$.

36. Instantaneous Rate of Change The surface area S of a sphere of radius r feet is $S = S(r) = 4\pi r^2$. Find the instantaneous rate of change of the surface area with respect to the radius r at $r = 2$.

37. Instantaneous Rate of Change The volume V of a sphere of radius r feet is $V = V(r) = \frac{4}{3}\pi r^3$. Find the instantaneous rate of change of the volume with respect to the radius r at $r = 2$.

38. Instantaneous Rate of Change The volume V of a cube of side x meters is $V = V(x) = x^3$. Find the instantaneous rate of change of the volume with respect to the side x at $x = 3$.

39. Instantaneous Speed of a Ball In physics it is shown that the height s of a ball thrown straight up with an initial speed of 96 ft/sec from ground level is

$$s = s(t) = -16t^2 + 96t$$

where t is the elapsed time that the ball is in the air.

(a) When does the ball strike the ground? That is, how long is the ball in the air?
(b) What is the average speed of the ball between $t = 0$ and $t = 2$?
(c) What is the instantaneous speed of the ball at time t?
(d) What is the instantaneous speed of the ball at $t = 2$?
(e) When is the instantaneous speed of the ball equal to zero?
(f) How high is the ball when its instantaneous speed equals zero?
(g) What is the instantaneous speed of the ball when it strikes the ground?

40. Instantaneous Speed of a Ball In physics it is shown that the height s of a ball thrown straight down with an initial speed of 48 ft/sec from a rooftop 160 feet high is

$$s = s(t) = -16t^2 - 48t + 160$$

where t is the elapsed time that the ball is in the air.

(a) When does the ball strike the ground? That is, how long is the ball in the air?
(b) What is the average speed of the ball from $t = 0$ to $t = 1$?
(c) What is the instantaneous speed of the ball at time t?

(d) What is the instantaneous speed of the ball at $t = 1$?
(e) What is the instantaneous speed of the ball when it strikes the ground?

41. Instantaneous Speed on the Moon Neil Armstrong throws a ball down into a crater on the moon. The height s (in feet) of the ball from the bottom of the crater after t seconds is given in the following table:

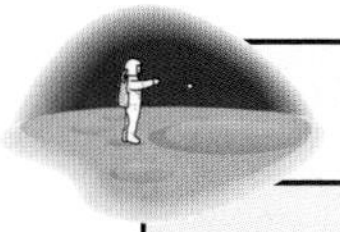

Time, t (in Seconds)	Distance, s (in Feet)
0	1000
1	987
2	969
3	945
4	917
5	883
6	843
7	800
8	749

(a) Find the average rate of change of distance from 1 to 4 seconds.
(b) Find the average rate of change of distance from 1 to 3 seconds.
(c) Find the average rate of change of distance from 1 to 2 seconds.
(d) Using a graphing utility, find the quadratic function of best fit.
(e) Using the function found in part (d), determine the instantaneous speed at $t = 1$ second.

42. Instantaneous Rate of Change The following data represent the total revenue, R (in dollars), received from selling x bicycles at Tunney's Bicycle Shop.

Number of Bicycles, x	Total Revenue, R (in Dollars)
0	0
25	28,000
60	45,000
102	53,400
150	59,160
190	62,360
223	64,835
249	66,525

(a) Find the average rate of change of revenue from 25 to 150 bicycles.
(b) Find the average rate of change in revenue from 25 to 102 bicycles.
(c) Find the average rate of change in revenue from 25 to 60 bicycles.
(d) Using a graphing utility, find the quadratic function of best fit.
(e) Using the function found in part (d), determine the instantaneous rate of change of revenue at $x = 25$ bicycles.

PREPARING FOR THIS SECTION

Before getting started, review the following:

✓ Geometry Formulas (Appendix, Section 9, pp. 996–997)
✓ Summation Notation (pp. 816–819)

14.5 THE AREA PROBLEM; THE INTEGRAL

OBJECTIVES 1 Approximate the Area Under the Graph of a Function
2 Approximate Integrals Using a Graphing Utility

The development of the integral, like that of the derivative, was originally motivated to a large extent by a problem in geometry: the *area problem.*

Area Problem Suppose $y = f(x)$ is a function whose domain is a closed interval $[a, b]$. We assume $f(x) \geq 0$ for all x in $[a, b]$. Find the area enclosed by the graph of f, the x-axis, and the vertical lines $x = a$ and $x = b$.

Figure 22 illustrates the area problem. We refer to the area A shown in Figure 22 as the area under the graph of f from a to b.

For a constant function $f(x) = k$ and for a linear function $f(x) = mx + B$, we can solve the area problem using formulas from geometry. See Figures 23(a) and (b).

Figure 22

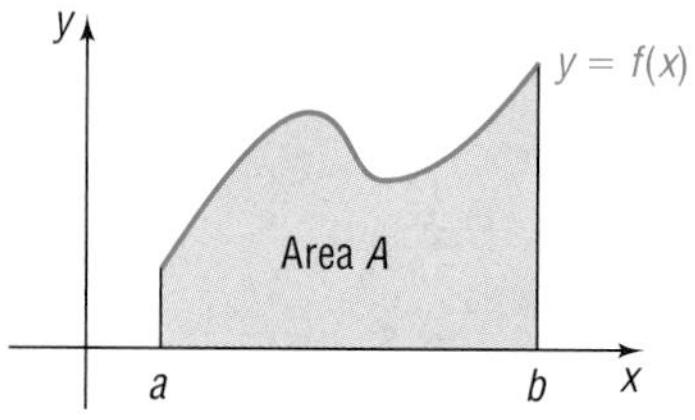

Area A = area under the graph of f from a to b

Figure 23

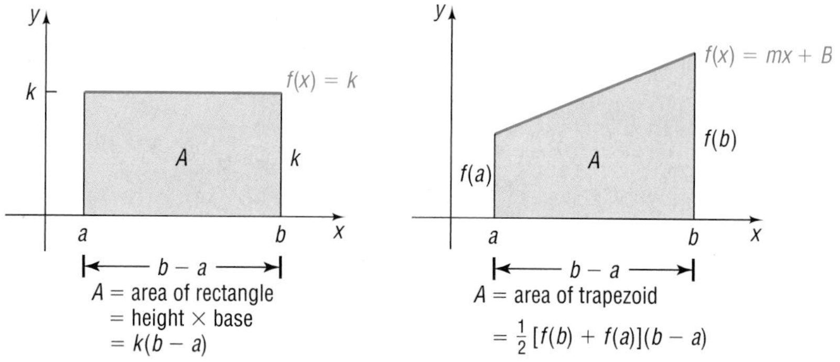

(a) (b)

For most other functions there are no geometry formulas available.

We begin by discussing a way to approximate the area under the graph of a function f from a to b.

APPROXIMATING AREA

1 We use rectangles to approximate the area under the graph of a function f. We do this by *partitioning* or dividing the interval $[a, b]$ into subintervals of equal length. On each subinterval, we form a rectangle whose base is the length of the subinterval and whose height is $f(u)$ for some number u in the subinterval. Look at Figure 24.

Figure 24

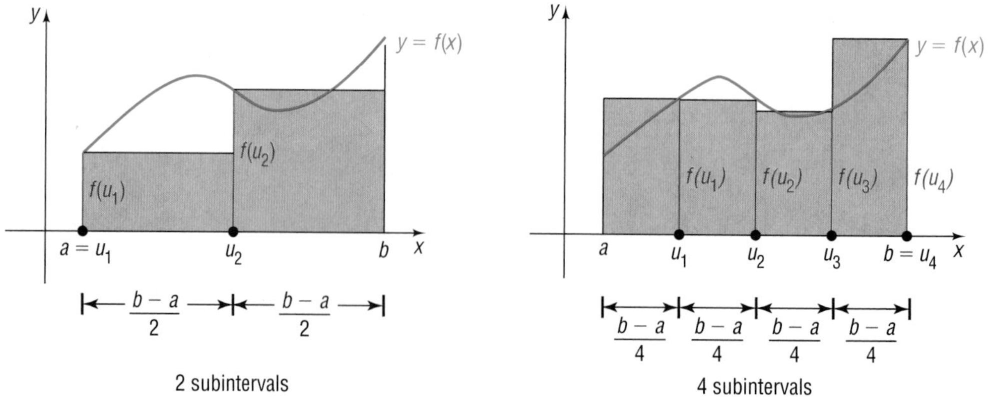

(a) (b)

In Figure 24(a), the interval $[a, b]$ is partitioned into two subintervals, each of length $(b - a)/2$, and the number u is chosen as the left endpoint of each subinterval. In Figure 24(b), the interval $[a, b]$ is partitioned into four subintervals, each of length $(b - a)/4$, and the number u is chosen as the right endpoint of each subintervals. We approximate the area A under f from a to b by adding the areas of the rectangles formed by the partition. Using Figure 24(a):

$$\text{Area } A \approx \text{area of first rectangle} + \text{area of second rectangle}$$
$$= f(u_1)\frac{b-a}{2} + f(u_2)\frac{b-a}{2}$$

Using Figure 24(b):

$$\text{Area } A \approx \text{area of first rectangle} + \text{area of second rectangle}$$
$$+ \text{area of third rectangle} + \text{area of fourth rectangle}$$
$$= f(u_1)\frac{b-a}{4} + f(u_2)\frac{b-a}{4} + f(u_3)\frac{b-a}{4} + f(u_4)\frac{b-a}{4}$$

In approximating the area under the graph of a function f from a to b, the choice of the number u in each subinterval is arbitrary. For convenience, we shall always pick u as either the left endpoint of each subinterval or the right endpoint. The choice of how many subintervals to use is also arbitrary. In general, the more subintervals used, the better the approximation will be. Let's look at a specific example.

EXAMPLE 1 Approximating the Area Under the Graph of $f(x) = 2x$ from 0 to 1

Approximate the area A under the graph of $f(x) = 2x$ from 0 to 1 as follows:

(a) By partitioning $[0, 1]$ into 2 subintervals of equal length and choosing u as left endpoint.

(b) By partitioning $[0, 1]$ into 2 subintervals of equal length and choosing u as right endpoint.

(c) By partitioning $[0, 1]$ into 4 subintervals of equal length and choosing u as left endpoint.

(d) By partitioning $[0, 1]$ into 4 subintervals of equal length and choosing u as right endpoint.

(e) Compare the approximations found in (a)–(d) with the actual area.

Figure 25(a)

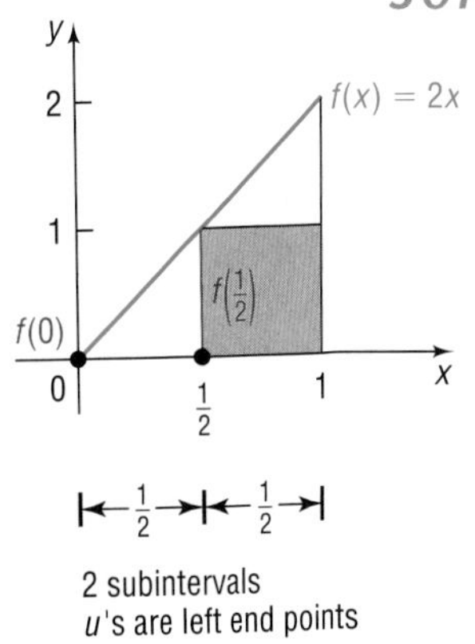

Solution (a) We partition $[0, 1]$ into 2 subintervals, each of length $1/2$, and choose u as left endpoint. See Figure 25(a).
The area A is approximated as

$$A \approx f(0)(1/2) + f(1/2)(1/2)$$
$$= (0)(1/2) + (1)(1/2)$$
$$= 1/2 = 0.5$$

Figure 25(b)

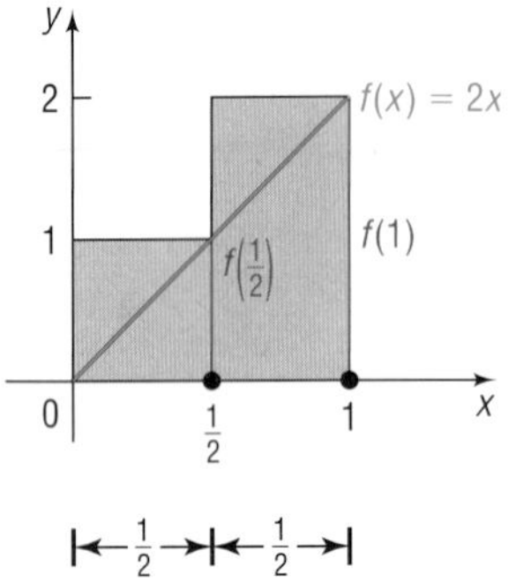

2 subintervals
u's are right end points

(b) We partition $[0, 1]$ into 2 subintervals, each of length $1/2$, and choose u as right endpoint. See Figure 25(b).
The area A is approximated as

$$\begin{aligned} A &\approx f(1/2)(1/2) + f(1)(1/2) \\ &= (1)(1/2) + (2)(1/2) \\ &= 3/2 = 1.5 \end{aligned}$$

Figure 25(c)

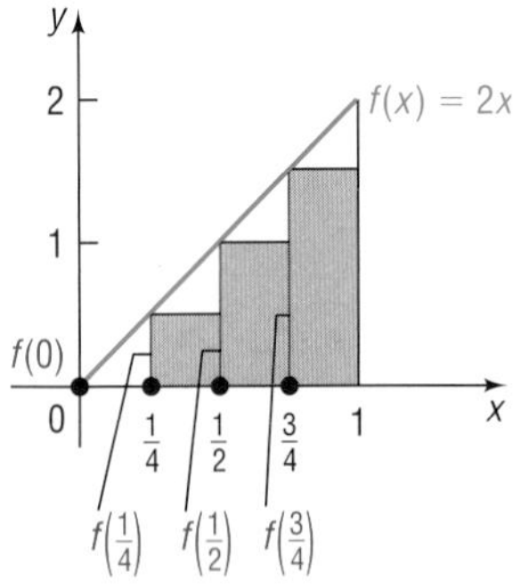

4 subintervals
u's are left end points

(c) We partition $[0, 1]$ into 4 subintervals, each of length $1/4$, and choose u as left endpoint. See Figure 25(c).
The area A is approximated as

$$\begin{aligned} A &\approx f(0)(1/4) + f(1/4)(1/4) + f(1/2)(1/4) + f(3/4)(1/4) \\ &= (0)(1/4) + (1/2)(1/4) + (1)(1/4) + (3/2)(1/4) \\ &= 3/4 = 0.75 \end{aligned}$$

Figure 25(d)

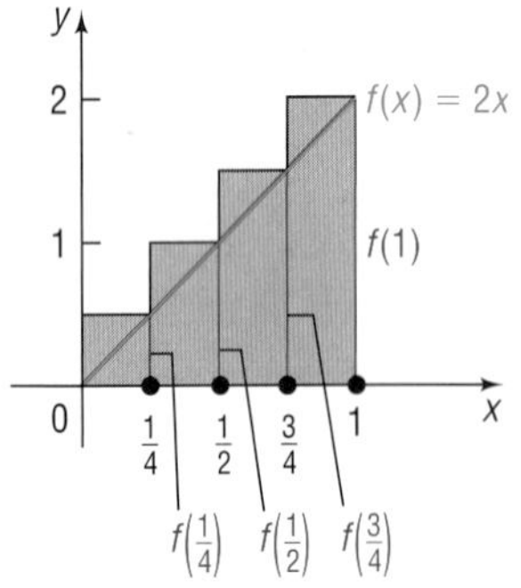

4 subintervals
u's are right end points

(d) We partition $[0, 1]$ into 4 subintervals, each of length $1/4$, and choose u as right endpoint. See Figure 25(d).
The area A is approximated as

$$\begin{aligned} A &\approx f(1/4)(1/4) + f(1/2)(1/4) + f(3/4)(1/4) + f(1)(1/4) \\ &= (1/2)(1/4) + (1)(1/4) + (3/2)(1/4) + (2)(1/4) \\ &= 5/4 = 1.25 \end{aligned}$$

(e) The actual area under the graph of $f(x) = 2x$ from 0 to 1 is the area of a right triangle whose base is of length 1 and whose height is 2. The actual area A is therefore

$$A = 1/2 \text{ base} \times \text{height} = (1/2)(1)(2) = 1$$

Now look at Table 8, which shows the approximations to the area under the graph of $f(x) = 2x$ from 0 to 1 for $n = 2, 4, 10$, and 100 subintervals. Notice that the approximations to the actual area improved as the number of subintervals increased.

TABLE 8

Using left endpoints:	n	2	4	10	100
	area	0.5	0.75	0.9	0.99
Using right endpoints:	n	2	4	10	100
	area	1.5	1.25	1.1	1.01

You are asked to confirm the entries in Table 8 in Problem 27.

There is another useful observation about Example 1. Look again at Figures 25(a)–(d) and at Table 8. Since the graph of $f(x) = 2x$ is increasing on $[0, 1]$, the choice of u as left endpoint gives a lower estimate to the actual area, while choosing u as the right endpoint gives an upper estimate. Do you see why?

NOW WORK PROBLEM 5.

EXAMPLE 2 Approximating the Area Under the Graph of $f(x) = x^2$

Approximate the area under the graph of $f(x) = x^2$ from 1 to 5 as follows:

(a) Using 4 subintervals of equal length.
(b) Using 8 subintervals of equal length.

In each case, choose the number u to be the left endpoint of each subinterval.

Figure 26(a)

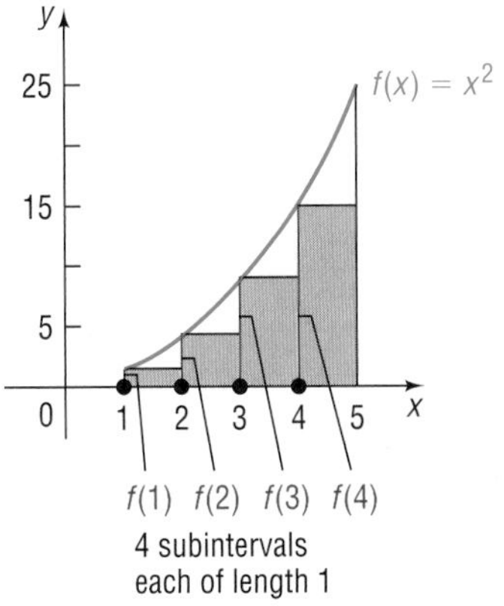

Solution (a) See Figure 26(a). Using 4 subintervals of equal length, the interval $[1, 5]$ is partitioned into subintervals of length $(5 - 1)/4 = 1$ as follows:

$$[1, 2] \quad [2, 3] \quad [3, 4] \quad [4, 5]$$

Each of these subintervals is of length one. Choosing u as the left endpoint of each subinterval, the area A under the graph of $f(x) = x^2$ is approximated by

$$\text{area } A = f(1)(1) + f(2)(1) + f(3)(1) + f(4)(1)$$
$$= 1 + 4 + 9 + 16 = 30$$

Figure 26(b)

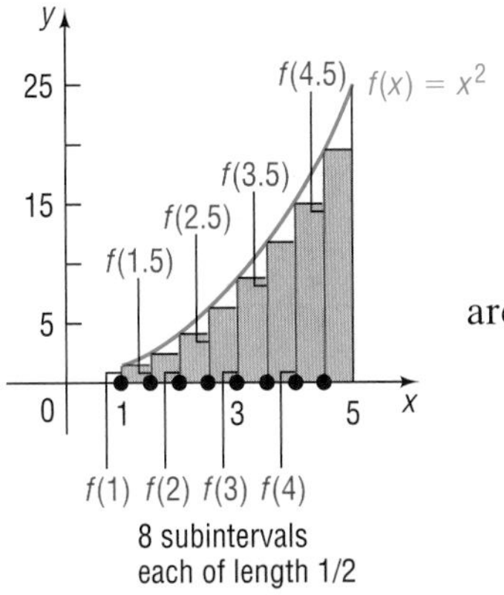

(b) See Figure 26(b). Using 8 subintervals of equal length, the interval $[1, 5]$ is partitioned into subintervals of length $(5 - 1)/8 = 0.5$ as follows:

$$[1, 1.5] \quad [1.5, 2] \quad [2, 2.5] \quad [2.5, 3] \quad [3, 3.5] \quad [3.5, 4] \quad [4, 4.5] \quad [4.5, 5]$$

Each of these subintervals is of length 0.5. Choosing u as the left endpoint of each subinterval, the area A under the graph of $f(x) = x^2$ is approximated by

$$\text{area } A \approx f(1)(0.5) + f(1.5)(0.5) + f(2)(0.5) + f(2.5)(0.5)$$
$$+ f(3)(0.5) + f(3.5)(0.5) + f(4)(0.5) + f(4.5)(0.5)$$
$$= [f(1) + f(1.5) + f(2) + f(2.5) + f(3) + f(3.5) + f(4) + f(4.5)](0.5)$$

Figure 27

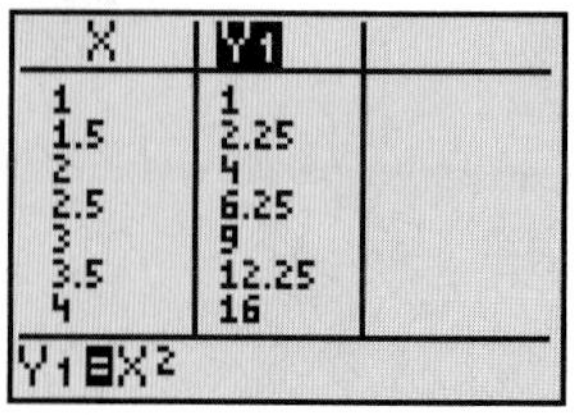

X	Y1
1	1
1.5	2.25
2	4
2.5	6.25
3	9
3.5	12.25
4	16

Y1=X²

We use the TABLE feature to evaluate $f(x) = x^2$. See Figure 27. Then

$$\text{area } A \approx [1 + 2.25 + 4 + 6.25 + 9 + 12.25 + 16 + 20.25](0.5)$$
$$= 35.5$$

In general, we approximate the area under the graph of a function $y = f(x)$ from a to b as follows:

1. Partition the interval $[a, b]$ into n subintervals of equal length. The length Δx of each subinterval is then
$$\Delta x = (b - a)/n$$
2. In each of these subintervals, pick a number u and evaluate the function f at each u. This results in n numbers $u_1, u_2, \ldots, u_n$, and n functional values $f(u_1), f(u_2), \ldots, f(u_n)$.
3. Form n rectangles with base equal to Δx, the length of each subinterval, and with height equal to functional value $f(u_i)$, $i = 1, 2, \ldots, n$. See Figure 28.
4. Add up the areas of the n rectangles
$$A_1 + A_2 + \ldots + A_n = f(u_1)\Delta x + f(u_2)\Delta x + \ldots + f(u_n)\Delta x$$
$$= \sum_{i=1}^{n} f(u_i)\Delta x$$

Figure 28

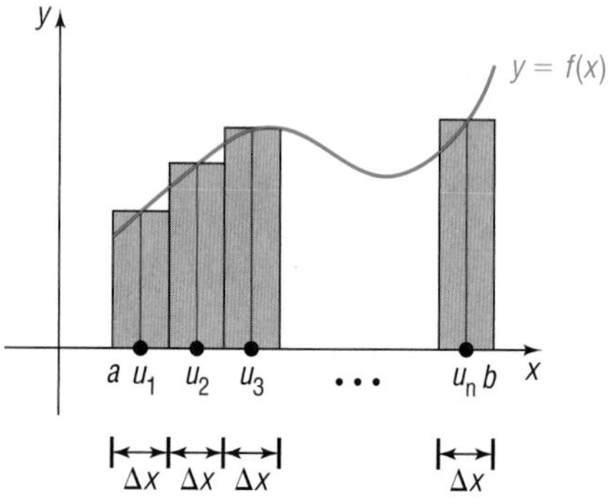

This number is the approximation to the area under the graph of f from a to b.

Definition of Area

We have observed that the larger the number n of subintervals used, the better the approximation to the area. If we let n become unbounded, we obtain the exact area under the graph of f from a to b.

Area Under a Graph

Let f denote a function whose domain is a closed interval $[a, b]$. Partition $[a, b]$ into n subintervals, each of length $\Delta x = (b - a)/n$. In each subinterval, pick a number u_i, $i = 1, 2, \ldots, n$, and evaluate $f(u_i)$. Form the products $f(u_i)\Delta x$ and add them up obtaining the sum

$$\sum_{i=1}^{n} f(u_i)\Delta x$$

If the limit of this sum exists as $n \to \infty$, that is,

$$\text{if } \lim_{n\to\infty} \sum_{i=1}^{n} f(u_i)\Delta x \text{ exists,}$$

it is defined as the area under the graph of f from a to b. Further, if this limit exists, it is denoted by the symbol

$$\int_a^b f(x)\,dx$$

read as "the integral from a to b of $f(x)$".

2 We can use a graphing utility to approximate integrals.

EXAMPLE 3 Using a Graphing Utility to Approximate an Integral

Use a graphing utility to approximate the area under the graph of $f(x) = x^2$ from 1 to 5. That is, evaluate the integral

$$\int_1^5 x^2\,dx$$

Solution Figure 29 shows the result using a TI–83 calculator. Consult your owner's manual for the proper keystrokes.

Figure 29

In calculus, techniques are given for evaluating integrals to obtain exact answers.

14.5 EXERCISES

In Problems 1 and 2, refer to the illustration that shows the area A under the graph of f from 1 to 3. The interval $[1, 3]$ *is partitioned into two subintervals* $[1, 2]$ *and* $[2, 3]$.

1. Approximate the area A choosing u as the left endpoint of each subinterval.
2. Approximate the area A choosing u as the right endpoint of each subinterval.

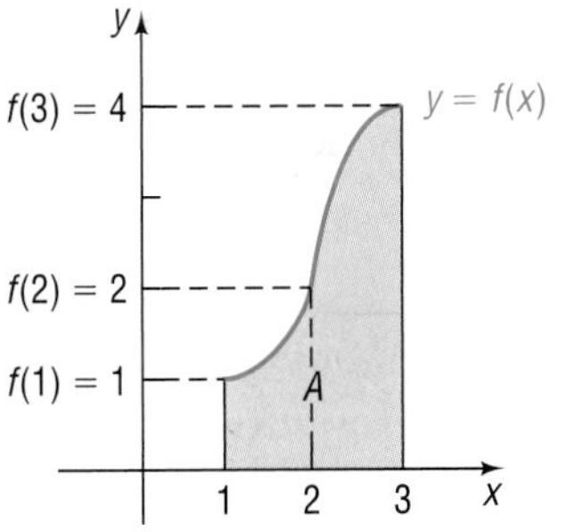

$f(1) = 1,\ f(2) = 2,\ f(3) = 4$

In Problems 3 and 4, refer to the illustration that shows the area A under the graph of f from 0 to 8. The interval $[0, 8]$ *is partitioned into four subintervals* $[0, 2]$, $[2, 4]$, $[4, 6]$, *and* $[6, 8]$.

3. Approximate the area A choosing u as the left endpoint of each subinterval.
4. Approximate the area A choosing u as the right endpoint of each subinterval.

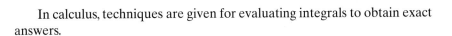

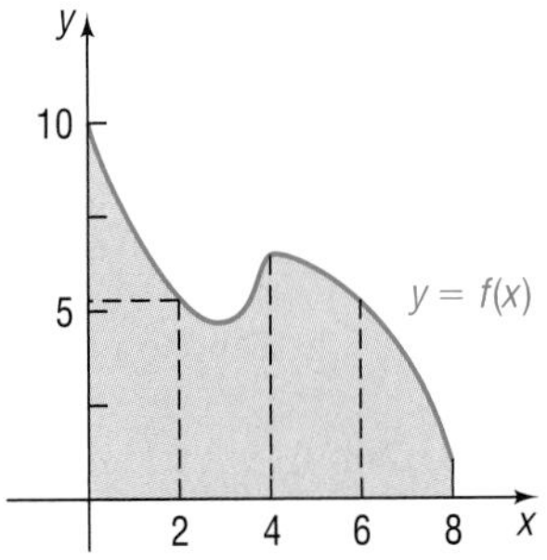

$f(0) = 10,\ f(2) = 6,\ f(4) = 7$
$f(6) = 5,\ f(8) = 1$

5. The function $f(x) = 3x$ is defined on the interval $[0, 6]$.
(a) Graph f
In (b)–(e), approximate the area A under f from 0 to 6 as follows:
(b) By partitioning $[0, 6]$ into three subintervals of equal length and choosing u as the left endpoint of each subinterval.
(c) By partitioning $[0, 6]$ into three subintervals of equal length and choosing u as the right endpoint of each subinterval.
(d) By partitioning $[0, 6]$ into six subintervals of equal length and choosing u as the left endpoint of each subinterval.
(e) By partitioning $[0, 6]$ into six subintervals of equal length and choosing u as the right endpoint of each subinterval.
(f) What is the actual area A?

6. Repeat Problem 5 for $f(x) = 4x$.

7. The function $f(x) = -3x + 9$ is defined on the interval $[0, 3]$.
(a) Graph f
In (b)–(e), approximate the area A under f from 0 to 3 as follows:
(b) By partitioning $[0, 3]$ into three subintervals of equal length and choosing u as the left endpoint of each subinterval.
(c) By partitioning $[0, 3]$ into three subintervals of equal length and choosing u as the right endpoint of each subinterval.
(d) By partitioning $[0, 3]$ into six subintervals of equal length and choosing u as the left endpoint of each subinterval.
(e) By partitioning $[0, 3]$ into six subintervals of equal length and choosing u as the right endpoint of each subinterval.
(f) What is the actual area A?

8. Repeat Problem 7 for $f(x) = -2x + 8$.

In Problems 9–18, a function f is defined over an interval $[a, b]$.
(a) Graph f indicating the area A under f from a to b.
(b) Approximate the area A by partitioning $[a, b]$ into 4 subintervals of equal length and choosing u as the left endpoint of each subinterval.
(c) Approximate the area A by partitioning $[a, b]$ into 8 subintervals of equal length and choosing u as the left endpoint of each subinterval.
(d) Express the area A as an integral.
(e) Use a graphing utility to approximate the integral.

9. $f(x) = x^2 + 2, \quad [0, 4]$
10. $f(x) = x^2 - 4, \quad [2, 6]$
11. $f(x) = x^3, \quad [0, 4]$
12. $f(x) = x^3, \quad [1, 5]$
13. $f(x) = 1/x, \quad [1, 5]$
14. $f(x) = \sqrt{x}, \quad [0, 4]$
15. $f(x) = e^x, [-1, 3]$
16. $f(x) = \ln x, \quad [3, 7]$
17. $f(x) = \sin x, \quad [0, \pi]$
18. $f(x) = \cos x, \quad [0, \pi/2]$

In Problems 19–26, an integral is given.
(a) What area does the integral represent?
(b) Provide a graph that illustrates this area.
(c) Use a graphing utility to approximate this area.

19. $\int_0^4 (3x + 1)\,dx$
20. $\int_1^3 (-2x + 7)\,dx$
21. $\int_2^5 (x^2 - 1)\,dx$
22. $\int_0^4 (16 - x^2)\,dx$
23. $\int_0^{\pi/2} \sin x\,dx$
24. $\int_{-\pi/4}^{\pi/4} \cos x\,dx$
25. $\int_0^2 e^x\,dx$
26. $\int_e^{2e} \ln x\,dx$

27. Confirm the entries in Table 8.
[**Hint:** review the formula for the sum of an arithmetic sequence]

28. Consider the function $f(x) = \sqrt{1 - x^2}$ whose domain is the interval $[-1, 1]$.
(a) Graph f.
(b) Approximate the area under the graph of f from -1 to 1 by dividing $[-1, 1]$ into 5 subintervals each of equal length.
(c) Approximate the area under the graph of f from -1 to 1 by dividing $[-1, 1]$ into 10 subintervals each of equal length.
(d) Express the area as an integral.
(e) Evaluate the integral using a graphing utility.
(f) What is the actual area?

CHAPTER REVIEW

Things To Know

Limit (p. 902)

$\lim_{x \to c} f(x) = N$ — As x gets closer to c, $x \neq c$, the values of f get closer to N.

Limit Formulas (p. 908)

$\lim_{x \to c} A = A$ — The limit of a constant is the constant.

$\lim_{x \to c} x = c$ — The limit of x as x approaches c is c.

Limit Properties

$\lim_{x \to c}[f(x) + g(x)] = \lim_{x \to c} f(x) + \lim_{x \to c} g(x)$ (p. 909) — The limit of a sum equals the sum of the limits.

$\lim_{x \to c}[f(x) - g(x)] = \lim_{x \to c} f(x) - \lim_{x \to c} g(x)$ (p. 909) — The limit of a difference equals the difference of the limits.

$\lim_{x \to c}[f(x) \cdot g(x)] = \lim_{x \to c} f(x) \cdot \lim_{x \to c} g(x)$ (p. 910) — The limit of a product equals the product of the limits.

$\lim_{x \to c}[f(x)/g(x)] = [\lim_{x \to c} f(x)]/[\lim_{x \to c} g(x)]$ (p. 912) provided $\lim_{x \to c} g(x) \neq 0$ — The limit of a quotient equals the quotient of the limits, provided that the limit of the denominator is not zero.

Limit of a Polynomial (p. 911)

$\lim_{x \to c} P(x) = P(c)$, where P is a polynomial

Derivative of a Function (p. 923)

$f'(c) = \lim_{x \to c} \dfrac{f(x) - f(c)}{x - c}$, provided that the limit exists

Continuous Function (p. 918)

$\lim_{x \to c} f(x) = f(c)$

Area Under a Graph (p. 934)

$\int_a^b f(x)\,dx = \lim_{n \to \infty} \sum_{i=1}^{n} f(u_i)\,\Delta x$, provided the limit exists

How To

Find a limit using a table (p. 902)

Find a limit using a graph (p. 904)

Find the limit of a sum, a difference, a product, and a quotient (p. 909)

Find the limit of polynomial (p. 911)

Find the limit of a power or a root (p. 912)

Find the limit of an average rate of change (p. 914)

Find the one-sided limits of a function (p. 915)

Determine whether a function is continuous (p. 917)

Find an equation of the tangent line to the graph of a function (p. 922)

Find the derivative of a function (p. 923)

Find instantaneous rates of change (p. 925)

Find the speed of a particle (p. 925)

Approximate the area under the graph of a function (p. 930)

Approximate integrals using a graphing utility (p. 935)

Fill-in-the-Blank Items

1. The notation ________ may be described by saying, "For x approximately equal to c, but $x \neq c$, the value $f(x)$ is approximately equal to N."

2. If $\lim_{x \to c} f(x) = N$ and f is continuous at c, then $f(c)$ ________ N.

3. If there is no single number that the value of f approaches when x is close to c, then $\lim_{x \to c} f(x)$ does ________ ________.

4. When $\lim_{x \to c} f(x) = f(c)$, we say that f is ________ at c.

5. $\lim_{x \to c} \dfrac{f(x)}{g(x)} = \dfrac{\lim_{x \to c} f(x)}{\lim_{x \to c} g(x)}$, provided that $\lim_{x \to c} f(x)$ and $\lim_{x \to c} g(x)$ each exist and $\lim_{x \to c} g(x)$ ________ 0.

6. If $\lim_{x\to c^-} f(x) = L$ and $\lim_{x\to c^+} f(x) = R$, then $\lim_{x\to c} f(x)$ exists provided that L ________ R.

7. The derivative of f at c equals the slope of the ________ line to the graph of f at $(c, f(c))$.

8. The area under the graph of $f(x) = \sqrt{x^2+1}$ from 0 to 2 may be symbolized by the integral ________.

True/False Items

T F **1.** The limit of the sum of two functions equals the sum of their limits, provided that each limit exists.

T F **2.** The limit of a function f as x approaches c always equals $f(c)$.

T F **3.** $\lim_{x\to 4} \dfrac{x^2-16}{x-4} = 8$

T F **4.** The function $f(x) = \dfrac{5x^2}{x^2+4}$ is continuous at $x = -2$.

T F **5.** The limit of a quotient of two functions equals the quotient of their limits, provided that each limit exists and the limit of the denominator is not zero.

T F **6.** The derivative of a function is the limit of an average rate of change.

T F **7.** The area under the graph of $f(x) = x^4$ from 0 to 2 equals $\int_0^2 x^4\,dx$.

Review Exercises

Blue problem numbers indicate the authors' suggestions for use in a Practice Test.

In Problems 1–22, find the limit.

1. $\lim_{x\to 2}(3x^2 - 2x + 1)$
2. $\lim_{x\to 1}(-2x^3 + x + 4)$
3. $\lim_{x\to -2}(x^2+1)^2$
4. $\lim_{x\to -2}(x^3+1)^2$
5. $\lim_{x\to 3}\sqrt{x^2+7}$
6. $\lim_{x\to -2}\sqrt[3]{x+10}$
7. $\lim_{x\to 1^-}\sqrt{1-x^2}$
8. $\lim_{x\to 2^+}\sqrt{3x-2}$
9. $\lim_{x\to 2}(5x+6)^{3/2}$
10. $\lim_{x\to -3}(15-3x)^{-3/2}$
11. $\lim_{x\to -1}\dfrac{x^2+x+2}{x^2-9}$
12. $\lim_{x\to 3}\dfrac{3x+4}{x^2+1}$
13. $\lim_{x\to 1}\dfrac{x-1}{x^3-1}$
14. $\lim_{x\to -1}\dfrac{x^2-1}{x^2+x}$
15. $\lim_{x\to -3}\dfrac{x^2-9}{x^2-x-12}$
16. $\lim_{x\to -3}\dfrac{x^2+2x-3}{x^2-9}$
17. $\lim_{x\to -1^-}\dfrac{x^2-1}{x^3-1}$
18. $\lim_{x\to 2^+}\dfrac{x^2-4}{x^3-8}$
19. $\lim_{x\to 2}\dfrac{x^3-8}{x^3-2x^2+4x-8}$
20. $\lim_{x\to 1}\dfrac{x^3-1}{x^3-x^2+3x-3}$
21. $\lim_{x\to 3}\dfrac{x^4-3x^3+x-3}{x^3-3x^2+2x-6}$
22. $\lim_{x\to -1}\dfrac{x^4+x^3+2x+2}{x^3+x^2}$

In Problems 23–30, determine whether f is continuous at c.

23. $f(x) = 3x^4 - x^2 + 2, \quad c = 5$
24. $f(x) = \dfrac{x^2-9}{x+10}, \quad c = 2$
25. $f(x) = \dfrac{x^2-4}{x+2}, \quad c = -2$
26. $f(x) = \dfrac{x^2+6x}{x^2-6x}, \quad c = 0$
27. $f(x) = \begin{cases} \dfrac{x^2-4}{x+2} & \text{if } x \neq -2 \\ 4 & \text{if } x = -2 \end{cases}, \quad c = -2$
28. $f(x) = \begin{cases} \dfrac{x^2+6x}{x^2-6x} & \text{if } x \neq 0 \\ 1 & \text{if } x = 0 \end{cases}, \quad c = 0$
29. $f(x) = \begin{cases} \dfrac{x^2-4}{x+2} & \text{if } x \neq -2 \\ -4 & \text{if } x = -2 \end{cases}, \quad c = -2$
30. $f(x) = \begin{cases} \dfrac{x^2+6x}{x^2-6x} & \text{if } x \neq 0 \\ -1 & \text{if } x = 0 \end{cases}, \quad c = 0$

In Problems 31–50, use the accompanying graph of $y = f(x)$.

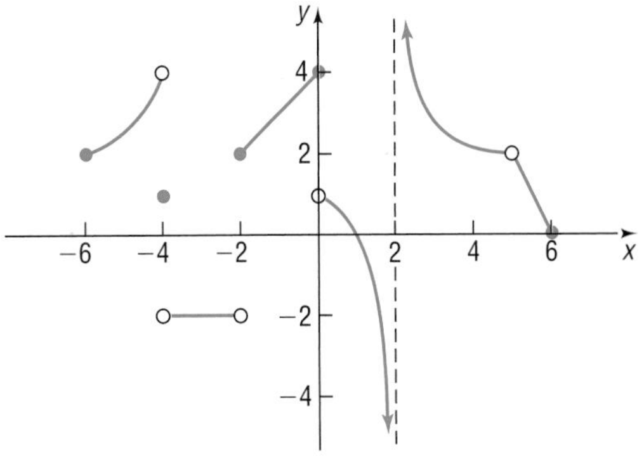

31. What is the domain of f?

32. What is the range of f?

33. Find the x-intercept(s), if any, of f.

34. Find the y-intercept(s), if any, of f.

35. Find $f(-6)$ and $f(-4)$.

36. Find $f(-2)$ and $f(6)$.

37. Find $\lim_{x\to -4^-} f(x)$.

38. Find $\lim_{x\to -4^+} f(x)$.

39. Find $\lim_{x\to -2^-} f(x)$.

40. Find $\lim_{x\to -2^+} f(x)$.

41. Find $\lim_{x\to 2^-} f(x)$.

42. Find $\lim_{x\to 2^+} f(x)$.

43. Does $\lim_{x\to 0} f(x)$ exist? If it does, what is it?

44. Does $\lim_{x\to 2} f(x)$ exist? If it does, what is it?

45. Is f continuous at -2?

46. Is f continuous at -4?

47. Is f continuous at 0?

48. Is f continuous at 2?

49. Is f continuous at 4?

50. Is f continuous at 5?

In Problems 51 and 52, discuss whether R is continuous at c. Use the one-sided limits of R at c to analyze the graph of R. Graph R.

51. $R(x) = \dfrac{x + 4}{x^2 - 16}$ at $c = -4$ and $c = 4$

52. $R(x) = \dfrac{3x^2 + 6x}{x^2 - 4}$ at $c = -2$ and $c = 2$

In Problems 53 and 54, determine where each rational function is undefined. Determine whether an asymptote or a hole appears at such numbers. Graph R using a graphing utility to verify your answers.

53. $R(x) = \dfrac{x^3 - 2x^2 + 4x - 8}{x^2 - 11x + 18}$

54. $R(x) = \dfrac{x^3 + 3x^2 - 2x - 6}{x^2 + x - 6}$

In Problems 55–60, find the slope of the tangent line to the graph of f at P. Graph f and the tangent line.

55. $f(x) = 2x^2 + 8x$ at $(1, 10)$

56. $f(x) = 3x^2 - 6x$ at $(0, 0)$

57. $f(x) = x^2 + 2x - 3$ at $(-1, -4)$

58. $f(x) = 2x^2 + 5x - 3$ at $(1, 4)$

59. $f(x) = x^3 + x^2$ at $(2, 12)$

60. $f(x) = x^3 - x^2$ at $(1, 0)$

In Problems 61–66, find the derivative of each function at the number indicated.

61. $f(x) = -4x^2 + 5$ at 3

62. $f(x) = -4 + 3x^2$ at 1

63. $f(x) = x^2 - 3x$ at 0

64. $f(x) = 2x^2 + 4x$ at -1

65. $f(x) = 2x^2 + 3x + 2$ at 1

66. $f(x) = 3x^2 - 4x + 1$ at 2

In Problems 67–70, find the derivative of each function at c using a graphing utility.

67. $f(x) = 4x^4 - 3x^3 + 6x - 9$ at -2

68. $f(x) = \dfrac{-6x^3 + 9x - 2}{8x^2 + 6x - 1}$ at 5

69. $f(x) = x^3 \tan x$ at $\dfrac{\pi}{6}$

70. $f(x) = x \sec x$ at $\dfrac{\pi}{6}$

71. **Instantaneous Speed of a Ball** In physics it is shown that the height s of a ball thrown straight up with an initial speed of 96 ft/sec from a rooftop 112 feet high is

$$s = s(t) = -16t^2 + 96t + 112$$

where t is the elapsed time that the ball is in the air. The ball misses the rooftop on its way down and eventually strikes the ground.
(a) When does the ball strike the ground? That is, how long is the ball in the air?
(b) At what time t will the ball pass the rooftop on its way down?
(c) What is the average speed of the ball between $t = 0$ and $t = 2$?
(d) What is the instantaneous speed of the ball at time t?
(e) What is the instantaneous speed of the ball at $t = 2$?
(f) When is the instantaneous speed of the ball equal to zero?
(g) What is the instantaneous speed of the ball as it passes the rooftop on the way down?
(h) What is the instantaneous speed of the ball when it strikes the ground?

72. **Finding an Instantaneous Rate of Change** The area A of a circle is πr^2. Find the instantaneous rate of change of area with respect to r at $r = 2$ feet. What is the average rate of change between 2 and 3? What is the average rate of change between 2 and 2.5? Between 2 and 2.1?

73. **Instantaneous Rate of Change** The following data represent the revenue, R (in dollars), received from selling x wristwatches at Wilk's Watch Shop.
(a) Find the average rate of change of revenue from 25 to 130 wristwatches.
(b) Find the average rate of change of revenue from 25 to 90 wristwatches.
(c) Find the average rate of change of revenue from 25 to 50 wristwatches.
(d) Using a graphing utility, find the quadratic function of best fit.
(e) Using the function found in part (d), determine the instantaneous rate of change of revenue at $x = 25$ wristwatches.

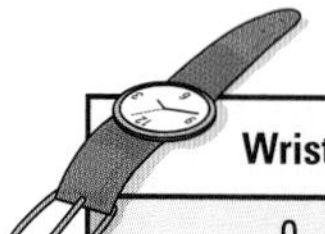

Wristwatches, x	Revenue, R
0	0
25	2340
40	3600
50	4375
90	6975
130	8775
160	9600
200	10,000
220	9900
250	9375

74. **Instantaneous Speed of a Parachutist** The following data represent the distances s (in feet) that a parachutist has fallen over time t (in seconds).

Time, t (in Seconds)	Distance, s (in Feet)
1	16
2	64
3	144
4	256
5	400

(a) Find the average rate of change of distance from 1 to 4 seconds.
(b) Find the average rate of change of distance from 1 to 3 seconds.
(c) Find the average rate of change of distance from 1 to 2 seconds.
(d) Using a graphing utility, find the power function of best fit.
(e) Using the function found in part (d), determine the instantaneous speed at $t = 1$ second.

75. The function $f(x) = 2x + 3$ is defined on the interval $[0, 4]$.
(a) Graph f
In (b)–(e), approximate the area A under f from 0 to 4 as follows:
(b) By partitioning $[0, 4]$ into four subintervals of equal length and choosing u as the left endpoint of each subinterval.
(c) By partitioning $[0, 4]$ into four subintervals of equal length and choosing u as the right endpoint of each subinterval.
(d) By partitioning $[0, 4]$ into eight subintervals of equal length and choosing u as the left endpoint of each subinterval.
(e) By partitioning $[0, 4]$ into eight subintervals of equal length and choosing u as the right endpoint of each subinterval.
(f) What is the actual area A?

76. Repeat Problem 75 for $f(x) = -2x + 8$.

In Problems 77–80, a function f is defined over an interval [a, b].

(a) Graph f indicating the area A under f from a to b.

(b) Approximate the area A by partitioning [a, b] into 3 subintervals of equal length and choosing u as the left endpoint of each subinterval.

(c) Approximate the area A by partitioning [a, b] into 6 subintervals of equal length and choosing u as the left endpoint of each subinterval.

(d) Express the area A as an integral.

(e) Use a graphing utility to approximate the integral.

77. $f(x) = 4 - x^2, \quad [-1, 2]$

78. $f(x) = x^2 + 3, \quad [0, 6]$

79. $f(x) = 1/x^2, \quad [1, 4]$

80. $f(x) = e^x, \quad [0, 6]$

In Problems 81–84, an integral is given.

(a) What area does the integral represent?

(b) Provide a graph that illustrates this area.

(c) Use a graphing utility to approximate this area.

81. $\int_{-1}^{3} (9 - x^2)\,dx$

82. $\int_{1}^{4} \sqrt{x}\,dx$

83. $\int_{-1}^{1} e^x\,dx$

84. $\int_{\pi/3}^{2\pi/3} \sin x\,dx$

PROJECT AT MOTOROLA

A simple experiment was devised to determine how long it would take to cure a sealant used in the manufacture of an electronic module. A bead of silicone was placed in a controlled moisture environment and the depth of cure measured as a function of time at 20°C and three relative humidity levels (RH). Data from the depth of cure experiment follows:

T(°C) RH	20 40%	20 60%	20 80%
Time (hrs)	**Depth of Cure (m)**		
1	0.0011	0.0014	0.0017
2	0.0014	0.0017	0.002
5	0.0017	0.002	0.0023
24	0.003	0.0035	0.004
48	0.004	0.005	0.006
RH	**Relative Humidity**		**2339Pa**

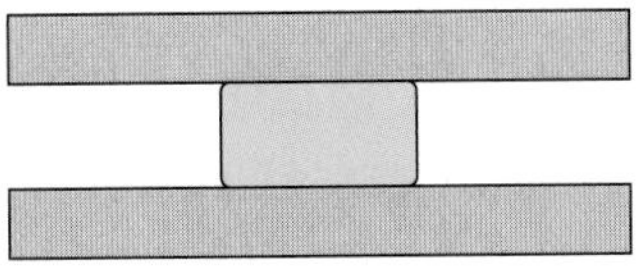

$SiNHR + H_2O(g) \Rightarrow SiOH + H_2NR(g)$

Given the information provided above, determine the following:

1. Draw a scatter diagram treating the depth of cure as the independent variable and time as the dependent variable at 20°C and 40% RH.
2. Calculate the average rate of change in depth of cure from 1 to 2 hours, from 2 to 5 hours, and from 24 to 48 hours.
3. Draw a scatter diagram treating the depth of cure as the independent variable and time as the dependent variable at 20°C and 80% RH.
4. Calculate the average rate of change in depth of cure from 1 to 2 hours, from 2 to 5 hours, and from 24 to 48 hours.
5. Compare the average rate of change in depth of cure at 40% and 80% humidity for each time segment. Which rate is highest? Which is lowest? Why?
6. Using a graphing utility, find the quadratic function of best fit for cure depth as a function of time at 20°C and 40% RH.
7. Using the function found in question 6, determine the instantaneous rate of change of cure depth at 20°C and 40% relative humidity at 2 and at 24 hours.
8. Repeat questions 6 and 7 with the the temperature at 20°C and a relative humidity of 80%. Compare the results with those obtained in question 7.

APPENDIX
Review

Outline

1 ALGEBRA REVIEW

OBJECTIVES

1. Classify Numbers
2. Graph Inequalities
3. Find Distance on a Real Number Line
4. Evaluate Algebraic Expressions

SETS

When we want to treat a collection of similar but distinct objects as a whole, we use the idea of a **set**. For example, the set of *digits* consists of the collection of numbers 0, 1, 2, 3, 4, 5, 6, 7, 8, and 9. If we use the symbol D to denote the set of digits, then we can write

$$D = \{0, 1, 2, 3, 4, 5, 6, 7, 8, 9\}$$

In this notation, the braces { } are used to enclose the objects, or **elements**, in the set. This method of denoting a set is called the **roster method**. A second way to denote a set is to use **set-builder notation**, where the set D of digits is written as

$$D = \{ \quad x \quad | \quad x \text{ is a digit}\}$$

Read as "D is the set of all x such that x is a digit."

EXAMPLE 1 **Using Set-builder Notation and the Roster Method**

(a) $E = \{x | x \text{ is an even digit}\} = \{0, 2, 4, 6, 8\}$

(b) $O = \{x | x \text{ is an odd digit}\} = \{1, 3, 5, 7, 9\}$ ■

In listing the elements of a set, we do not list an element more than once because the elements of a set are distinct. Also, the order in which the elements are listed is not relevant. For example, $\{2, 3\}$ and $\{3, 2\}$ both represent the same set.

If every element of a set A is also an element of a set B, then we say that A is a **subset** of B. If two sets A and B have the same elements, then we say that A **equals** B. For example, $\{1, 2, 3\}$ is a subset of $\{1, 2, 3, 4, 5\}$; and $\{1, 2, 3\}$ equals $\{2, 3, 1\}$.

CLASSIFICATION OF NUMBERS

1 It is helpful to classify the various kinds of numbers that we deal with as sets. The **counting numbers**, or **natural numbers**, are the set of numbers

$\{1, 2, 3, 4, \ldots\}$. (The three dots, called an **ellipsis**, indicate that the pattern continues indefinitely.) As their name implies, these numbers are often used to count things. For example, there are 26 letters in our alphabet; there are 100 cents in a dollar. The **whole numbers** are the set of numbers $\{0, 1, 2, 3, \ldots\}$ that is, the counting numbers together with 0.

> The **integers** are the numbers in the set $\{\ldots, -3, -2, -1, 0, 1, 2, 3, \ldots\}$.

These numbers prove useful in many situations. For example, if your checking account has \$10 in it and you write a check for \$15, you can represent the current balance as −\$5.

Notice that the set of counting numbers is a subset of the set of whole numbers. Each time we expand a number system, such as from the whole numbers to the integers, we do so in order to be able to handle new, and usually more complicated, problems. The integers allow us to solve problems requiring both positive and negative counting numbers, such as profit/loss, height above/below sea level, temperature above/below 0°F, and so on.

But integers alone are not sufficient for *all* problems. For example, they do not answer the question "What part of a dollar is 38 cents?" To answer such a question, we enlarge our number system to include *rational numbers*. For example, $\frac{38}{100}$ answers the question "What part of a dollar is 38 cents?"

> A **rational number** is a number that can be expressed as a quotient a/b of two integers. The integer a is called the **numerator**, and the integer b, which cannot be 0, is called the **denominator**. The rational numbers are the numbers in the set $\left\{x \mid = \dfrac{a}{b}, \text{ where } a, b \neq 0 \text{ are integers}\right\}$.

Examples of rational numbers are $\frac{3}{4}, \frac{5}{2}, \frac{9}{4}, -\frac{2}{3}$, and $\frac{100}{3}$. Since $a/1 = a$ for any integer a, it follows that the set of integers is a subset of the set of rational numbers.

Rational numbers may be represented as **decimals**. For example, the rational numbers $\frac{3}{4}, \frac{5}{2}, -\frac{2}{3}$, and $\frac{7}{66}$ may be represented as decimals by merely carrying out the indicated division:

$$\tfrac{3}{4} = 0.75 \qquad \tfrac{5}{2} = 2.5 \qquad -\tfrac{2}{3} = -0.666\ldots \qquad \tfrac{7}{66} = 0.1060606\ldots$$

Notice that the decimal representations of $\frac{3}{4}$ and $\frac{5}{2}$ terminate, or end. The decimal representations of $-\frac{2}{3}$ and $\frac{7}{66}$ do not terminate, but they do exhibit a pattern of repetition. For $-\frac{2}{3}$, the 6 repeats indefinitely; for $\frac{7}{66}$, the block 06 repeats indefinitely. It can be shown that every rational number may be represented by a decimal that either terminates or is nonterminating with a repeating block of digits, and vice versa.

On the other hand, there are decimals that do not fit into either of these categories. Such decimals represent **irrational numbers**. Every irrational number may be represented by a decimal that neither repeats nor terminates.

Irrational numbers occur naturally. For example, consider the isosceles right triangle whose legs are each of length 1. See Figure 1. The length of the hypotenuse is $\sqrt{2}$, an irrational number.

Figure 1

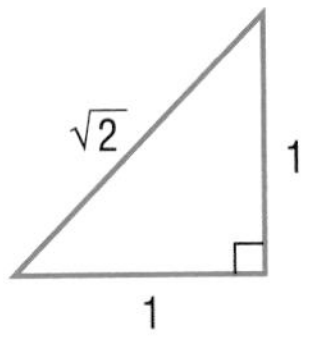

Also, the number that equals the ratio of the circumference C to the diameter d of any circle, denoted by the symbol π (the Greek letter pi), is an irrational number. See Figure 2.

Figure 2

$$\pi = \frac{C}{d}$$

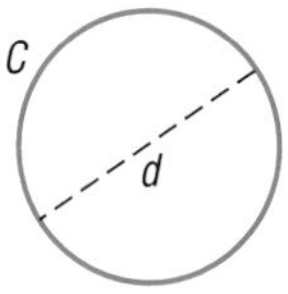

The irrational numbers $\sqrt{2}$ and π have decimal representations that begin as follows:

$$\sqrt{2} = 1.414213\ldots \qquad \pi = 3.14159\ldots$$

In practice, irrational numbers are generally represented by approximations. For example, using the symbol $\approx$ (read as "approximately equal to"), we can write

$$\sqrt{2} \approx 1.4142 \qquad \pi \approx 3.1416$$

Together, the rational numbers and irrational numbers form the set of **real numbers**.

Every decimal may be represented by a real number (either rational or irrational), and every real number may be represented by a decimal. Real numbers in the form of decimals provide a convenient way to measure quantities as they change.

EXAMPLE 2 **Classifying the Numbers in a Set**

List the numbers in the set

$$\left\{-3, \tfrac{4}{3}, 0.12, \sqrt{2}, \pi, 2.151515\ldots \text{ (where the block 15 repeats)}, 10\right\}$$

that are:

(a) Natural numbers (b) Integers (c) Rational numbers
(d) Irrational numbers (e) Real numbers

Solution
(a) 10 is the only natural number.
(b) -3 and 10 are integers.
(c) $-3, \frac{4}{3}, 0.12, 2.151515\ldots$, and 10 are rational numbers.
(d) $\sqrt{2}$ and π are irrational numbers.
(e) All the numbers listed are real numbers. ■

NOW WORK PROBLEM 3.

The Real Number Line

The real numbers can be represented by points on a line called the **real number line**. There is a one-to-one correspondence between real numbers and points on a line. That is, every real number corresponds to a point on the line, and each point on the line has a unique real number associated with it.

Pick a point on the line somewhere near the center, and label it O. This point, called the **origin**, corresponds to the real number 0. See Figure 3. The point 1 unit to the right of O corresponds to the number 1. The distance between 0 and 1 determines the **scale** of the number line. For example, the point associated with the number 2 is twice as far from O as 1 is. Notice that an arrowhead on the right end of the line indicates the direction in which the numbers increase. Figure 3 also shows the points associated with the irrational numbers $\sqrt{2}$ and π. Points to the left of the origin correspond to the real numbers $-1, -2$, and so on.

Figure 3
Real number line

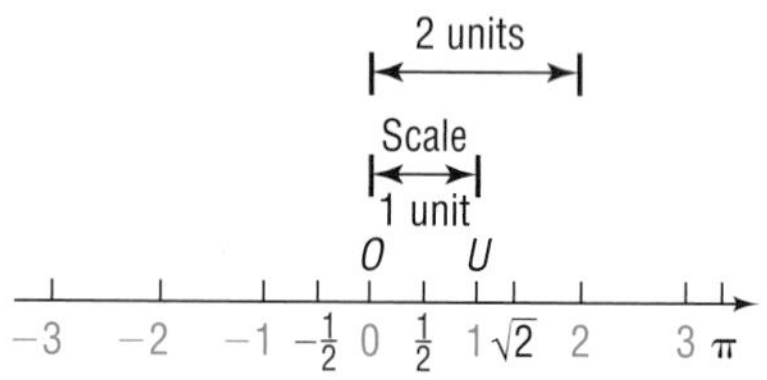

The real number associated with a point P is called the **coordinate** of P, and the line whose points have been assigned coordinates is called the **real number line**.

NOW WORK PROBLEM 7.

The real number line divides the real numbers into three classes, as shown in Figure 4.

Figure 4

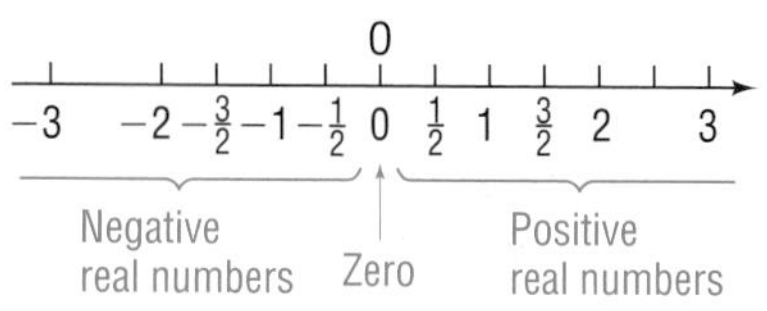

1. The **negative real numbers** are the coordinates of points to the left of the origin O.
2. The real number **zero** is the coordinate of the origin O.
3. The **positive real numbers** are the coordinates of points to the right of the origin O.

Negative and positive numbers have the following multiplication properties:

Multiplication Properties of Positive and Negative Numbers

1. The product of two positive numbers is a positive number.
2. The product of two negative numbers is a positive number.
3. The product of a positive number and a negative number is a negative number.

INEQUALITIES

An important property of the real number line follows from the fact that, given two numbers (points) a and b, either a is to the left of b, a equals b, or a is to the right of b. See Figure 5.

Figure 5

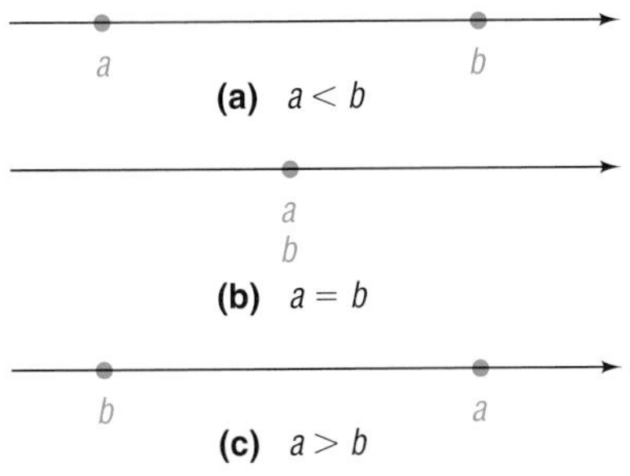

If a is to the left of b, we say that "a is less than b" and write $a < b$. If a is to the right of b, we say that "a is greater than b" and write $a > b$. If a equals b, we write $a = b$. If a is either less than or equal to b, we write $a \leq b$. Similarly, $a \geq b$ means that a is either greater than or equal to b. Collectively, the symbols $<, >, \leq$, and $\geq$ are called **inequality symbols**.

Note that $a < b$ and $b > a$ mean the same thing. It does not matter whether we write $2 < 3$ or $3 > 2$.

Furthermore, if $a < b$ or if $b > a$, then the difference $b - a$ is positive. Do you see why?

EXAMPLE 3 Using Inequality Symbols

(a) $3 < 7$	(b) $-8 > -16$	(c) $-6 < 0$
(d) $-8 < -4$	(e) $4 > -1$	(f) $8 > 0$

In Example 3(a), we conclude that $3 < 7$ either because 3 is to the left of 7 on the real number line or because the difference $7 - 3 = 4$, a positive real number.

Similarly, we conclude in Example 3(b) that $-8 > -16$ either because -8 lies to the right of -16 on the real number line or because the difference $-8 - (-16) = -8 + 16 = 8$, a positive real number.

Look again at Example 3. Note that the inequality symbol always points in the direction of the smaller number.

Statements of the form $a < b$ or $b > a$ are called **strict inequalities**, whereas statements of the form $a \leq b$ or $b \geq a$ are called **nonstrict inequalities**. An **inequality** is a statement in which two expressions are related by an inequality symbol. The expressions are referred to as the **sides** of the inequality.

Based on the discussion thus far, we conclude that

$a > 0$	is equivalent to	a is positive
$a < 0$	is equivalent to	a is negative

We sometimes read $a > 0$ by saying that "a is positive." If $a \geq 0$, then either $a > 0$ or $a = 0$, and we may read this as "a is nonnegative."

NOW WORK PROBLEMS 11 AND 21.

GRAPHING INEQUALITIES

2 We shall find it useful in later work to graph inequalities on the real number line.

EXAMPLE 4 Graphing Inequalities

(a) On the real number line, graph all numbers x for which $x > 4$.
(b) On the real number line, graph all numbers x for which $x \leq 5$.

Solution (a) See Figure 6. Notice that we use a left parenthesis to indicate that the number 4 is *not* part of the graph.
(b) See Figure 7. Notice that we use a right bracket to indicate that the number 5 is part of the graph.

Figure 6

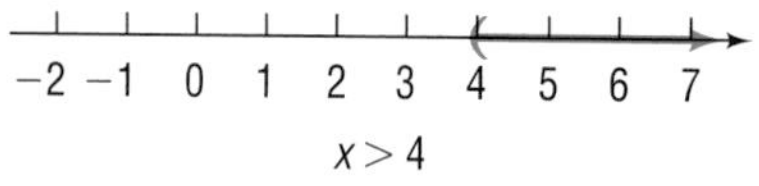

Figure 7

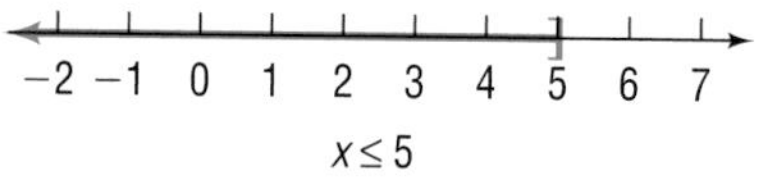

Inequalities are often combined.

EXAMPLE 5 Graphing Combined Inequalities

On the real number line, graph all numbers x for which $x > 4$ and $x < 6$.

Solution We first graph each inequality separately, as illustrated in Figure 8(a). Then it is easy to see that the numbers that belong to *both* the graph of $x > 4$ *and* the graph of $x < 6$ are as shown in Figure 8(b). For example, 5 is part of the graph because $5 > 4$ and $5 < 6$; 7 is not on the graph because 7 is not less than 6.

Figure 8

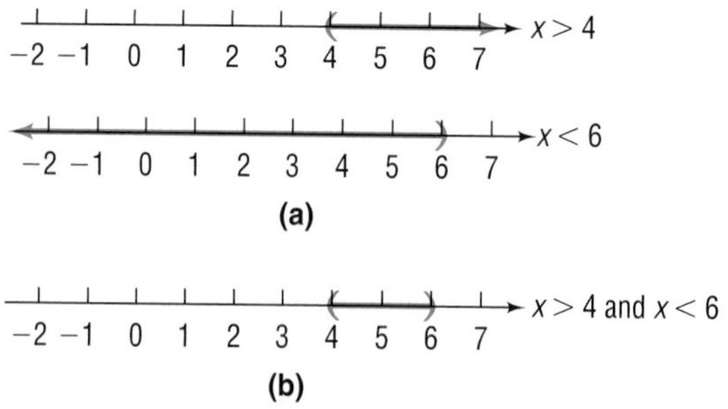

EXAMPLE 6 Graphing Combined Inequalities

On the real number line, graph all numbers x for which $x > 4$ or $x \leq -1$.

Solution See Figure 9(a), where each inequality is graphed separately. The numbers that belong to *either* the graph of $x > 4$ *or* the graph of $x \leq -1$ are graphed in Figure 9(b). For example, 5 is part of the graph because $5 > 4$; 2 is not on the graph because 2 is not greater than 4 nor is 2 less than or equal to -1.

Figure 9

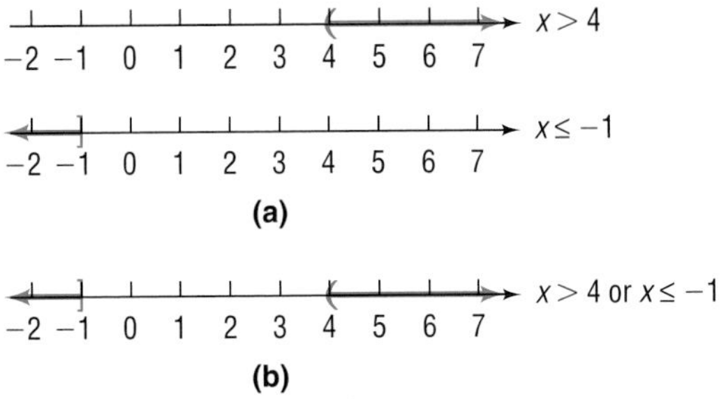

NOW WORK PROBLEM 29.

ABSOLUTE VALUE

Figure 10

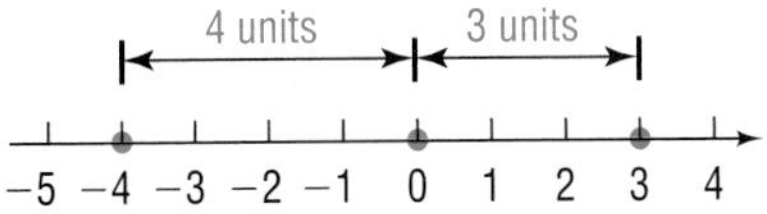

The *absolute value* of a number a is the distance from 0 to a on the number line. For example, −4 is 4 units from 0; and 3 is 3 units from 0. See Figure 10. Thus, the absolute value of −4 is 4, and the absolute value of 3 is 3.

A more formal definition of absolute value is given next.

The **absolute value** of a real number a, denoted by the symbol $|a|$, is defined by the rules

$$|a| = a \quad \text{if } a \geq 0 \qquad \text{and} \qquad |a| = -a \quad \text{if } a < 0$$

For example, since $-4 < 0$, the second rule must be used to get $|-4| = -(-4) = 4$.

EXAMPLE 7 **Computing Absolute Value**

(a) $|8| = 8$ (b) $|0| = 0$ (c) $|-15| = -(-15) = 15$

3 Look again at Figure 10. The distance from −4 to 3 is 7 units. This distance is the difference $3 - (-4)$, obtained by subtracting the smaller coordinate from the larger. However, since $|3 - (-4)| = |7| = 7$ and $|-4 - 3| = |-7| = 7$, we can use absolute value to calculate the distance between two points without being concerned about which is smaller.

If P and Q are two points on a real number line with coordinates a and b, respectively, the **distance between P and Q**, denoted by $d(P, Q)$, is

$$d(P, Q) = |b - a|$$

Since $|b - a| = |a - b|$, it follows that $d(P, Q) = d(Q, P)$.

EXAMPLE 8 **Finding Distance on a Number Line**

Let P, Q, and R be points on a real number line with coordinates −5, 7, and −3, respectively. Find the distance

(a) between P and Q (b) between Q and R

Solution (a) $d(P, Q) = |7 - (-5)| = |12| = 12$

(b) $d(Q, R) = |-3 - 7| = |-10| = 10$ (see Figure 11)

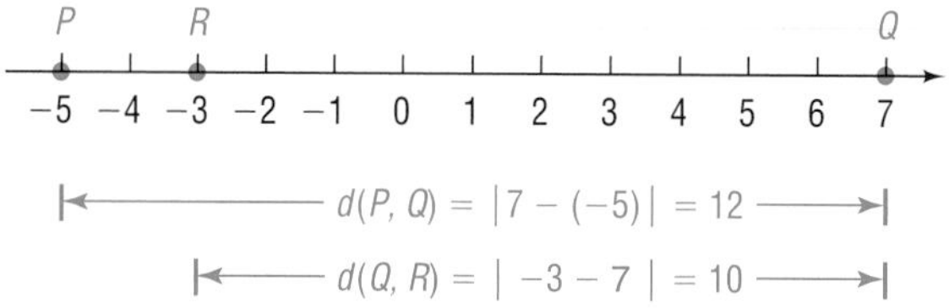

Figure 11

NOW WORK PROBLEM 37.

CONSTANTS AND VARIABLES

In algebra we use letters such as x, y, a, b, and c to represent numbers. If the letter used is to represent *any* number from a given set of numbers, it is called a **variable**. A **constant** is either a fixed number, such as 5 or $\sqrt{3}$, or a letter that represents a fixed (possibly unspecified) number.

Constants and variables are combined using the operations of addition, subtraction, multiplication, and division to form *algebraic expressions*. Examples of algebraic expressions include

$$x + 3 \qquad \frac{3}{1 - t} \qquad 7x - 2y$$

4 To evaluate an algebraic expression, substitute for each variable its numerical value.

EXAMPLE 9 **Evaluating an Algebraic Expression**

Evaluate each expression if $x = 3$ and $y = -1$.

(a) $x + 3y$ (b) $5xy$ (c) $\dfrac{3y}{2 - 2x}$ (d) $|-4x + y|$

Solution (a) Substitute 3 for x and -1 for y in the expression $x + 3y$.

$$x + 3y = 3 + 3(-1) = 3 + (-3) = 0$$

↑ $x = 3$, $y = -1$

(b) If $x = 3$ and $y = -1$, then

$$5xy = 5(3)(-1) = -15$$

(c) If $x = 3$ and $y = -1$, then

$$\frac{3y}{2 - 2x} = \frac{3(-1)}{2 - 2(3)} = \frac{-3}{2 - 6} = \frac{-3}{-4} = \frac{3}{4}$$

(d) If $x = 3$ and $y = -1$, then

$$|-4x + y| = |-4(3) + (-1)| = |-12 + (-1)| = |-13| = 13$$

Graphing calculators can be used to evaluate algebraic expressions. Figure 12 shows the results of Example 9 using a TI-83.

Figure 12

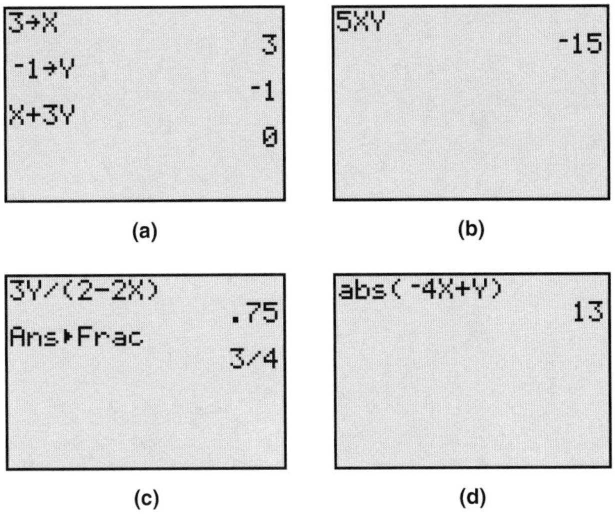

NOW WORK PROBLEMS 39 AND 47.

1 EXERCISES

In Problems 1–6, list the numbers in each set that are (a) natural numbers, (b) integers, (c) rational numbers, (d) irrational numbers, (e) real numbers.

1. $A = \{-6, \frac{1}{2}, -1.333\ldots \text{ (the 3's repeat)}, \pi, 2, 5\}$
2. $B = \{-\frac{5}{3}, 2.060606\ldots \text{ (the block 06 repeats)}, 1.25, 0, 1, \sqrt{5}\}$
3. $C = \{0, 1, \frac{1}{2}, \frac{1}{3}, \frac{1}{4}\}$
4. $D = \{-1, -1.1, -1.2, -1.3\}$
5. $E = \{\sqrt{2}, \pi, \sqrt{2} + 1, \pi + \frac{1}{2}\}$
6. $F = \{-\sqrt{2}, \pi + \sqrt{2}, \frac{1}{2} + 10.3\}$
7. On the real number line, label the points with coordinates $0, 1, -1, \frac{5}{2}, -2.5, \frac{3}{4}$, and 0.25.
8. Repeat Problem 7 for the coordinates $0, -2, 2, -1.5, \frac{3}{2}, \frac{1}{3}$, and $\frac{2}{3}$.

In Problems 9–18 replace the question mark by <, >, or =, whichever is correct.

9. $\frac{1}{2}$? 0
10. 5 ? 6
11. −1 ? −2
12. −3 ? $-\frac{5}{2}$
13. π ? 3.14
14. $\sqrt{2}$? 1.41
15. $\frac{1}{2}$? 0.5
16. $\frac{1}{3}$? 0.33
17. $\frac{2}{3}$? 0.67
18. $\frac{1}{4}$? 0.25

In Problems 19–26, write each statement as an inequality.

19. x is positive
20. z is negative
21. x is less than 2
22. y is greater than −5
23. x is less than or equal to 1
24. x is greater than or equal to 2
25. x is less than 5 and x is greater than 2
26. y is less than or equal to 2 and y is greater than 0

In Problems 27–34 graph the numbers x, if any, on the real number line.

27. $x \geq -2$
28. $x < 4$
29. $x \geq 4$ and $x < 6$
30. $x > 3$ and $x \leq 7$
31. $x \leq 0$ or $x < 6$
32. $x > 0$ or $x \geq 5$
33. $x \leq -2$ or $x > 1$
34. $x \geq 4$ or $x < -2$

In Problems 35–38 use the real number line below to compute each distance.

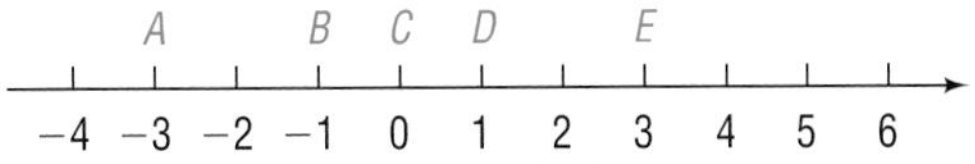

35. $d(D, E)$
36. $d(C, E)$
37. $d(A, E)$
38. $d(D, B)$

In Problems 39–46, evaluate each expression if $x = -2$ and $y = 3$. Check your answer using a graphing utility.

39. $x + 2y$
40. $3x + y$
41. $5xy + 2$
42. $-2x + xy$
43. $\frac{2x}{x - y}$
44. $\frac{x + y}{x - y}$
45. $\frac{3x + 2y}{2 + y}$
46. $\frac{2x - 3}{y}$

In Problems 47–56, find the value of each expression if $x = 3$ and $y = -2$. Check your answers using a graphing utility.

47. $|x + y|$
48. $|x - y|$
49. $|x| + |y|$
50. $|x| - |y|$
51. $\frac{|x|}{x}$
52. $\frac{|y|}{y}$
53. $|4x - 5y|$
54. $|3x + 2y|$
55. $||4x| - |5y||$
56. $3|x| + 2|y|$

In Problems 57–60, use the formula $C = \frac{5}{9}(F - 32)$ for converting degrees Fahrenheit into degrees Celsius to find the Celsius measure of each Fahrenheit temperature.

57. $F = 32°$
58. $F = 212°$
59. $F = 77°$
60. $F = -4°$

61. **Manufacturing Cost** The weekly production cost C of manufacturing x watches is given by the formula $C = 4000 + 2x$, where the variable C is in dollars.
(a) What is the cost of producing 1000 watches?
(b) What is the cost of producing 2000 watches?

62. **Balancing a Checkbook** At the beginning of the month, Mike had a balance of \$210 in his checking account. During the next month, he deposited \$80, wrote a check for \$120, made another deposit of \$25, wrote two checks for \$60 and \$32, and was assessed a monthly service charge of \$5. What was his balance at the end of the month?

63. **U.S. Voltage** In the United States, normal household voltage is 115 volts. It is acceptable for the actual voltage x to differ from normal by at most 5 volts. A formula that describes this is

$$|x - 115| \leq 5$$

(a) Show that a voltage of 113 volts is acceptable.
(b) Show that a voltage of 109 volts is not acceptable.

64. **Foreign Voltage** In other countries, normal household voltage is 220 volts. It is acceptable for the actual voltage x to differ from normal by at most 8 volts. A formula that describes this is

$$|x - 220| \leq 8$$

(a) Show that a voltage of 214 volts is acceptable.
(b) Show that a voltage of 209 volts is not acceptable.

65. **Making Precision Ball Bearings** The FireBall Company manufactures ball bearings for precision equipment. One of their products is a ball bearing with a stated radius of 3 centimeters (cm). Only ball bearings with a radius within 0.01 cm of this stated radius are acceptable. If x is the radius of a ball bearing, a formula describing this situation is

$$|x - 3| \leq 0.01$$

(a) Is a ball bearing of radius $x = 2.999$ acceptable?
(b) Is a ball bearing of radius $x = 2.89$ acceptable?

66. **Body Temperature** Normal human body temperature is 98.6°F. A temperature x that differs from normal by at least 1.5°F is considered unhealthy. A formula that describes this is

$$|x - 98.6| \geq 1.5$$

(a) Show that a temperature of 97°F is unhealthy.
(b) Show that a temperature of 100°F is not unhealthy.

67. Does $\frac{1}{3}$ equal 0.333? If not, which is larger? By how much?

68. Does $\frac{2}{3}$ equal 0.666? If not, which is larger? By how much?

69. Are there any real numbers that are both rational and irrational? Are there any real numbers that are neither? Explain your reasoning.

70. Explain why the sum of a rational number and an irrational number must be irrational.

71. What rational number does the repeating decimal 0.9999... equal?

72. Is there a positive real number "closest" to 0?

73. I'm thinking of a number! It lies between 1 and 10; its square is rational and lies between 1 and 10. The number is larger than π. Correct to two decimal places, name the number. Now think of your own number, describe it, and challenge a fellow student to name it.

74. Write a brief paragraph that illustrates the similarities and differences between "less than" (<) and "less than or equal" ($\leq$).

2 INTEGER EXPONENTS

OBJECTIVES
1. Evaluate Expressions Containing Exponents
2. Work with the Laws of Exponents
3. Use a Calculator to Evaluate Exponents
4. Use Scientific Notation

Integer exponents provide a shorthand device for representing repeated multiplications of a real number. For example,

$$3^4 = 3 \cdot 3 \cdot 3 \cdot 3 = 81$$

Additionally, many formulas have exponents. For example,

- The formula for the horsepower rating H of an engine is

$$H = \frac{D^2N}{2.5}$$

  where D is the diameter of a cylinder and N is the number of cylinders.
- A formula for the resistance R of blood flowing in a blood vessel is

$$R = C\frac{L}{r^4}$$

  where L is the length of the blood vessel, r is the radius, and C is a positive constant.

If a is a real number and n is a positive integer, then the **symbol a^n** represents the product of n factors of a. That is,

$$a^n = \underbrace{a \cdot a \cdot \ldots \cdot a}_{n \text{ factors}} \qquad \textbf{(1)}$$

Here it is understood that $a^1 = a$.

In particular, we have

$$a^1 = a$$
$$a^2 = a \cdot a$$
$$a^3 = a \cdot a \cdot a$$

and so on.

1 In the expression a^n, a is called the **base** and n is called the **exponent**, or **power**. We read a^n as "a raised to the power n" or as "a to the nth power." We usually read a^2 as "a squared" and a^3 as "a cubed."

EXAMPLE 1 Evaluating Expressions Containing Exponents

(a) $2^3 = 2 \cdot 2 \cdot 2 = 8$
(b) $5^2 = 5 \cdot 5 = 25$
(c) $10^1 = 10$
(d) $(-2)^4 = (-2)(-2)(-2)(-2) = 16$
(e) $-2^4 = -(2 \cdot 2 \cdot 2 \cdot 2) = -16$ ■

Notice the difference between Examples 1(d) and 1(e): The exponent applies only to the symbol or parenthetical expression immediately preceding it.

We define a raised to a negative power as follows:

If n is a positive integer and if a is a nonzero real number, then we define

$$a^{-n} = \frac{1}{a^n} \quad \text{if } a \neq 0 \qquad \textbf{(2)}$$

Whenever you encounter a negative exponent, think "reciprocal."

EXAMPLE 2 Evaluating Expressions Containing Negative Exponents

(a) $2^{-3} = \frac{1}{2^3} = \frac{1}{8}$

(b) $x^{-4} = \frac{1}{x^4}$

(c) $\left(\frac{1}{5}\right)^{-2} = \frac{1}{\left(\frac{1}{5}\right)^2} = \frac{1}{\frac{1}{25}} = 25$ ■

NOW WORK PROBLEM 3.

If a is a nonzero number, we define

$$a^0 = 1 \quad \text{if } a \neq 0 \qquad \textbf{(3)}$$

Notice that we do not allow the base a to be 0 in a^{-n} or in a^0.

LAWS OF EXPONENTS

2 Several general rules can be used when dealing with exponents. We refer to the rules in the following as the Laws of Exponents.

Laws of Exponents

If a, b are real numbers and m, n are integers, then

$$a^m a^n = a^{m+n} \qquad (a^m)^n = a^{mn} \qquad (ab)^n = a^n b^n$$

$$\left(\frac{a}{b}\right)^n = \frac{a^n}{b^n} \qquad \frac{a^m}{a^n} = a^{m-n} = \frac{1}{a^{n-m}}$$

where it is assumed that all expressions used are defined and no denominator is 0.

EXAMPLE 3 Using the Laws of Exponents

(a) $x^{-3} \cdot x^5 = x^{-3+5} = x^2$ (b) $\left[(-2)^3\right]^2 = (-2)^{3 \cdot 2} = (-2)^6 = 64$

(c) $(-2x)^0 = 1$ (d) $\dfrac{x^{-2}}{x^{-5}} = x^{-2-(-5)} = x^3$

(e) $\left(\dfrac{x^{-3}}{3y^{-1}}\right)^{-2} = \dfrac{(x^{-3})^{-2}}{(3y^{-1})^{-2}} = \dfrac{x^6}{3^{-2}(y^{-1})^{-2}} = \dfrac{3^2x^6}{y^2} = \dfrac{9x^6}{y^2}$ ■

NOW WORK PROBLEMS 35 AND 49.

CALCULATOR USE

3 Your graphing calculator has the caret ^ key, which is used for computations involving exponents. The next example shows how this key is used.

EXAMPLE 4 Exponents on a Graphing Calculator

Evaluate: $(2.3)^5$

Solution Figure 13 shows the result using a TI-83 graphing calculator.

Figure 13

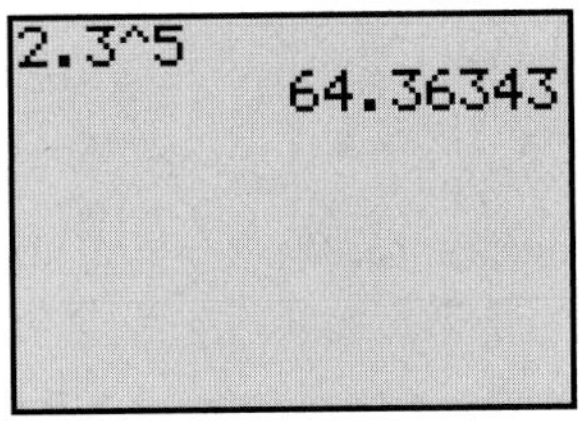

■

NOW WORK PROBLEM 63.

SCIENTIFIC NOTATION

4 Measurements of physical quantities can range from very small to very large. For example, the mass of a proton is approximately 0.00000000000000000000000000167 kilogram and the mass of Earth is about 5,980,000,000,000,000,000,000,000 kilograms. These numbers obviously are tedious to write down and difficult to read, so we use exponents to rewrite each.

When a number has been written as the product of a number x where $1 \le x < 10$, times a power of 10, it is said to be written in **scientific notation**.

In scientific notation,

$$\text{Mass of a proton} = 1.67 \times 10^{-27} \text{ kilograms}$$

$$\text{Mass of Earth} = 5.98 \times 10^{24} \text{ kilograms}$$

CONVERTING A DECIMAL TO SCIENTIFIC NOTATION

To change a positive number into scientific notation:

1. Count the number N of places that the decimal point must be moved in order to arrive at a number x, where $1 \le x < 10$.
2. If the original number is greater than or equal to 1, the scientific notation is $x \times 10^N$. If the original number is between 0 and 1, the scientific notation is $x \times 10^{-N}$.

EXAMPLE 5 Using Scientific Notation

Write each number in scientific notation.

(a) 9582 (b) 1.245 (c) 0.285 (d) 0.000561

Solution (a) The decimal point in 9582 follows the 2. Thus, we count

9 5 8 2 .
3 2 1

stopping after three moves because 9.582 is a number between 1 and 10. Since 9582 is greater than 1, we write

$$9582 = 9.582 \times 10^3$$

(b) The decimal point in 1.245 is between the 1 and 2. Since the number is already between 1 and 10, the scientific notation for it is $1.245 \times 10^0 = 1.245$.

(c) The decimal point in 0.285 is between the 0 and the 2. Thus, we count

0 . 2 8 5
1

stopping after one move because 2.85 is a number between 1 and 10. Since 0.285 is between 0 and 1, we write

$$0.285 = 2.85 \times 10^{-1}$$

(d) The decimal point in 0.000561 is moved as follows:

0 . 0 0 0 5 6 1
1 2 3 4

Thus,

$$0.000561 = 5.61 \times 10^{-4}$$

NOW WORK PROBLEM 69.

EXAMPLE 6 Changing from Scientific Notation to Decimals

Write each number as a decimal.

(a) 2.1×10^4 (b) 3.26×10^{-5} (c) 1×10^{-2}

Solution   (a) $2.1 \times 10^4 = 2 \underset{\substack{\uparrow\;\uparrow\;\uparrow\;\uparrow \\ 1\;\;2\;\;3\;\;4}}{1\;0\;0\;0} \times 10^4 = 21{,}000$

(b) $3.26 \times 10^{-5} = 0 \underset{\substack{\uparrow\;\uparrow\;\uparrow\;\uparrow\;\uparrow \\ 5\;\;4\;\;3\;\;2\;\;1}}{0\;0\;0\;0\;3}.2\;6 \times 10^{-5} = 0.0000326$

(c) $1 \times 10^{-2} = 0 \underset{\substack{\uparrow\;\uparrow\;\leftarrow \\ 2\;\;1}}{0\;1}. \times 10^{-2} = 0.01$ ■

On a graphing calculator, a number such as 3.615×10^{12} is displayed as $\boxed{3.615\text{E}12.}$

 NOW WORK PROBLEM 77.

EXAMPLE 7   **Using Scientific Notation**

(a) The diameter of the smallest living cell is only about 0.00001 centimeter (cm).* Express this number in scientific notation.
(b) The surface area of Earth is about 1.97×10^8 square miles.† Express the surface area as a whole number.

Solution   (a) 0.00001 cm $= 1 \times 10^{-5}$ cm because the decimal point is moved five places and the number is less than 1.
(b) 1.97×10^8 square miles $= 197{,}000{,}000$ square miles. ■

NOW WORK PROBLEM 85.

---

* Powers of Ten, Philip and Phylis Morrison.
† 1998 Information Please Almanac.

HISTORICAL FEATURE

René Descartes (1596–1650)

Our method of writing exponents originated with René Descartes (1596–1650), although the concept goes back in various forms to the ancient Babylonians. Even after its introduction by Descartes in 1637, the method took a remarkable amount of time to become completely standardized, and expressions like $aaaaa + 3aaaa + 2aaa - 4aa + 2a + 1$ remained common until 1750. The concept of a rational exponent (see Section 8 of this Appendix) was known by 1400, although inconvenient notation prevented the development of any extensive theory. John Wallis (1616–1703), in 1655, was the first to give a fairly complete explanation of negative and rational exponents, and Sir Isaac Newton's (1642–1727) use of them made exponents standard in their current form.

---

2 EXERCISES

In Problems 1–24, simplify each expression.

1. 4^2

2. -4^2

3. 4^{-2}

4. $(-4)^2$

5. -4^{-2}

6. $(-4)^{-2}$

7. $4^0 \cdot 2^{-3}$

8. $(-2)^{-3} \cdot 3^0$

9. $2^{-3} + \left(\dfrac{1}{2}\right)^3$

10. $3^{-2} + \left(\dfrac{1}{3}\right)^2$

11. $3^{-6} \cdot 3^4$ **12.** $4^{-2} \cdot 4^3$ **13.** $\dfrac{(3^2)^2}{(2^3)^2}$ **14.** $\dfrac{(2^3)^3}{(2^2)^3}$ **15.** $\left(\dfrac{2}{3}\right)^{-3}$

16. $\left(\dfrac{3}{2}\right)^{-2}$ **17.** $\dfrac{2^3 \cdot 3^2}{2^4 \cdot 3^{-2}}$ **18.** $\dfrac{3^{-2} \cdot 5^3}{3^2 \cdot 5}$ **19.** $\left(\dfrac{9}{2}\right)^{-2}$ **20.** $\left(\dfrac{6}{5}\right)^{-3}$

21. $\dfrac{2^{-2}}{3}$ **22.** $\dfrac{3^{-2}}{2}$ **23.** $\dfrac{-3^{-1}}{2^{-1}}$ **24.** $\dfrac{-2^{-3}}{-1}$

In Problems 25–54, simplify each expression so that all exponents are positive. Whenever an exponent is negative or 0, we assume that the base does not equal 0.

25. x^0y^2 **26.** $x^{-1}y$ **27.** xy^{-2} **28.** x^0y^4 **29.** $(8x^3)^{-2}$

30. $(-8x^3)^{-2}$ **31.** $-4x^{-1}$ **32.** $(-4x)^{-1}$ **33.** $3x^0$ **34.** $(3x)^0$

35. $\dfrac{x^{-2}y^3}{xy^4}$ **36.** $\dfrac{x^{-2}y}{xy^2}$ **37.** $x^{-1}y^{-1}$ **38.** $\dfrac{x^{-2}y^{-3}}{x}$ **39.** $\dfrac{x^{-1}}{y^{-1}}$

40. $\left(\dfrac{2x}{3}\right)^{-1}$ **41.** $\left(\dfrac{4y}{5x}\right)^{-2}$ **42.** $(x^2y)^{-2}$ **43.** $x^{-2}y^{-2}$ **44.** $x^{-1}y^{-1}$

45. $\dfrac{x^{-1}y^{-2}z^3}{x^2yz^3}$ **46.** $\dfrac{3x^{-2}yz^2}{x^4y^{-3}z^2}$ **47.** $\dfrac{(-2)^3x^4(yz)^2}{3^2xy^3z^4}$ **48.** $\dfrac{4x^{-2}(yz)^{-1}}{(-5)^2x^4y^2z^{-2}}$ **49.** $\left(\dfrac{3x^{-1}}{4y^{-1}}\right)^{-2}$

50. $\left(\dfrac{5x^{-2}}{6y^{-2}}\right)^{-3}$ **51.** $\dfrac{(xy^{-1})^{-2}}{xy^3}$ **52.** $\dfrac{(3xy^{-1})^2}{(2x^{-1}y)^3}$ **53.** $\left(\dfrac{x}{y^2}\right)^{-2} \cdot (y^2)^{-1}$ **54.** $\dfrac{(x^2)^{-3}y^3}{(x^3y)^{-2}}$

55. Find the value of the expression $2x^3 - 3x^2 + 5x - 4$ if $x = 2$. If $x = 1$.

56. Find the value of the expression $4x^3 + 3x^2 - x + 2$ if $x = 1$. If $x = 2$.

57. Find the value of $(0.3)^4$.
Hint: $0.3 = 3/10$.

58. Find the value of 30^4.
Hint: $30 = 3 \cdot 10$.

59. What is the value of $\dfrac{(666)^4}{(222)^4}$?

60. What is the value of $(0.1)^3(20)^3$?

In Problems 61–68, use a graphing utility to evaluate each expression. Round your answers to three decimal places.

61. $(8.2)^6$ **62.** $(3.7)^5$ **63.** $(6.1)^{-3}$ **64.** $(2.2)^{-5}$

65. $(-2.8)^6$ **66.** $-(2.8)^6$ **67.** $(-8.11)^{-4}$ **68.** $-(8.11)^{-4}$

In Problems 69–76, write each number in scientific notation.

69. 454.2 **70.** 32.14 **71.** 0.013 **72.** 0.00421

73. 32,155 **74.** 21,210 **75.** 0.000423 **76.** 0.0514

In Problems 77–84, write each number as a decimal.

77. 6.15×10^4 **78.** 9.7×10^3 **79.** 1.214×10^{-3} **80.** 9.88×10^{-4}

81. 1.1×10^8 **82.** 4.112×10^2 **83.** 8.1×10^{-2} **84.** 6.453×10^{-1}

85. Distance from Earth to Its Moon The distance from Earth to the Moon is about 4×10^8 meters.* Express this distance as a whole number.

86. Height of Mt. Everest The height of Mt. Everest is 8872 meters.* Express this height in scientific notation.

87. Wavelength of Visible Light The wavelength of visible light is about 5×10^{-7} meter.* Express this wavelength as a whole number.

88. Diameter of an Atom The diameter of an atom is about 1×10^{-10} meters.* Express this diameter as a whole number.

* *Powers of Ten,* Philip and Phylis Morrison.

89. Diameter of Copper Wire The smallest commercial copper wire is about 0.0005 inch in diameter.† Express this diameter using scientific notation.

90. Smallest Motor The smallest motor ever made is less than 0.05 centimeter wide.† Express this width using scientific notation.

91. World Oil Production In 1996, world oil production averaged 64,000,000 barrels per day.† How many barrels on average were produced in the month of April 1996? Express your answer in scientific notation.

92. U.S. Consumption Oil In 1995, the United States consumed 17,640,000 barrels of petroleum products per day.† If there are 42 gallons in one barrel, how many gallons did the United States consume per day? Express your answer in scientific notation.

93. Astronomy One light-year is defined by astronomers to be the distance a beam of light will travel in 1 year (365 days). If the speed of light is 186,000 miles per second, how many miles are in a light-year? Express your answer in scientific notation.

94. Astronomy How long does it take a beam of light to reach Earth from the Sun, when the Sun is 93,000,000 miles from Earth? Express your answer in seconds, using scientific notation.

95. Look at the summary box where the Laws of Exponents are given. List them in the order of most importance to you. Write a brief position paper defending your ordering.

96. Write a paragraph to justify the definition given in the text that $a^0 = 1, a \neq 0$.

† *1998 Information Please Almanac*

3 POLYNOMIALS

OBJECTIVES
1. Recognize Monomials
2. Recognize Polynomials
3. Add and Subtract Polynomials
4. Multiply Polynomials
5. Know Formulas for Special Products

We have described algebra as a generalization of arithmetic in which letters are used to represent real numbers. From now on, we shall use the letters at the end of the alphabet, such as x, y, and z, to represent variables and the letters at the beginning of the alphabet, such as a, b, and c, to represent constants. Thus, in the expressions $3x + 5$ and $ax + b$, it is understood that x is a variable and that a and b are constants, even though the constants a and b are unspecified. As you will find out, the context usually makes the intended meaning clear.

1 Now we introduce some basic vocabulary.

A **monomial** in one variable is the product of a constant times a variable raised to a nonnegative integer power. Thus, a monomial is of the form

$$ax^k$$

where a is a constant, x is a variable, and $k \geq 0$ is an integer. The constant a is called the **coefficient** of the monomial. If $a \neq 0$, then k is called the **degree** of the monomial.

EXAMPLE 1 Examples of Monomials

Monomial	Coefficient	Degree	
(a) $6x^2$	6	2	
(b) $-\sqrt{2}x^3$	$-\sqrt{2}$	3	
(c) 3	3	0	Since $3 = 3 \cdot 1 = 3x^0, \quad x \neq 0$
(d) $-5x$	-5	1	Since $-5x = -5x^1$
(e) x^4	1	4	Since $x^4 = 1 \cdot x^4$

NOW WORK PROBLEM 1.

Two monomials with the same variable raised to the same power are called **like terms**. For example, $2x^4$ and $-5x^4$ are like terms. In contrast, the monomials $2x^3$ and $2x^5$ are not like terms.

We can add or subtract like terms using the distributive property. For example,

$$2x^2 + 5x^2 = (2 + 5)x^2 = 7x^2 \quad \text{and} \quad 8x^3 - 5x^3 = (8 - 5)x^3 = 3x^3$$

2 The sum or difference of two monomials having different degrees is called a **binomial**. The sum or difference of three monomials with three different degrees is called a **trinomial**. For example,

$x^2 - 2$ is a binomial.

$x^3 - 3x + 5$ is a trinomial.

$2x^2 + 5x^2 + 2 = 7x^2 + 2$ is a binomial.

A **polynomial** in one variable is an algebraic expression of the form

$$a_n x^n + a_{n-1} x^{n-1} + \cdots + a_1 x + a_0 \qquad \textbf{(1)}$$

where $a_n, a_{n-1}, \ldots, a_1, a_0$ are constants* called the **coefficients** of the polynomial, $n \geq 0$ is an integer, and x is a variable. If $a_n \neq 0$, it is called the **leading coefficient**, and n is called the **degree** of the polynomial.

The monomials that make up a polynomial are called its **terms**. If all the coefficients are 0, the polynomial is called the **zero polynomial**, which has no degree.

Polynomials are usually written in **standard form**, beginning with the nonzero term of highest degree and continuing with terms in descending order according to degree.

EXAMPLE 2 Examples of Polynomials

Polynomial	Coefficients	Degree
$3x^2 - 5 = 3x^2 + 0 \cdot x + (-5)$	$3, 0, -5$	2
$8 - 2x + x^2 = 1 \cdot x^2 - 2x + 8$	$1, -2, 8$	2
$5x + \sqrt{2} = 5x^1 + \sqrt{2}$	$5, \sqrt{2}$	1
$3 = 3 \cdot 1 = 3 \cdot x^0$	3	0
0	0	No degree

Although we have been using x to represent the variable, letters such as y or z are also commonly used. Thus,

$3x^4 - x^2 + 2$ is a polynomial (in x) of degree 4.

$9y^3 - 2y^2 + y - 3$ is a polynomial (in y) of degree 3.

$z^5 + \pi$ is a polynomial (in z) of degree 5.

*The notation a_n is read as "a sub n." The number n is called a **subscript** and should not be confused with an exponent. We use subscripts in order to distinguish one constant from another when a large or undetermined number of constants is required.

Algebraic expressions such as

$$\frac{1}{x} \quad \text{and} \quad \frac{x^2 + 1}{x + 5}$$

are not polynomials. The first is not a polynomial because $1/x = x^{-1}$ has an exponent that is not a nonnegative integer. Although the second expression is the quotient of two polynomials, the polynomial in the denominator has degree greater than 0, so the expression cannot be a polynomial.

NOW WORK PROBLEM 11.

ADDING AND SUBTRACTING POLYNOMIALS

3 Polynomials are added and subtracted by combining like terms.

EXAMPLE 3 Adding Polynomials

Find the sum of the polynomials

$$8x^3 - 2x^2 + 6x - 2 \quad \text{and} \quad 3x^4 - 2x^3 + x^2 + x$$

Solution The idea here is to group the like terms and then combine them.

$$\begin{aligned}(8x^3 - 2x^2 + 6x - 2) + (3x^4 - 2x^3 + x^2 + x) \\ &= 3x^4 + (8x^3 - 2x^3) + (-2x^2 + x^2) + (6x + x) - 2 \\ &= 3x^4 + 6x^3 - x^2 + 7x - 2\end{aligned}$$

■

EXAMPLE 4 Subtracting Polynomials

Find the difference: $(3x^4 - 4x^3 + 6x^2 - 1) - (2x^4 - 8x^2 - 6x + 5)$

Solution $(3x^4 - 4x^3 + 6x^2 - 1) - (2x^4 - 8x^2 - 6x + 5)$

$$= 3x^4 - 4x^3 + 6x^2 - 1 + \underbrace{(-2x^4 + 8x^2 + 6x - 5)}$$

Be sure to change the sign of each term in the second polynomial.

$$= (3x^4 - 2x^4) + (-4x^3) + (6x^2 + 8x^2) + 6x + (-1 - 5)$$

↑ Group like terms.

$$= x^4 - 4x^3 + 14x^2 + 6x - 6$$

■

NOW WORK PROBLEM 23.

MULTIPLYING POLYNOMIALS

4 Two monomials may be multiplied using the laws of exponents and commutative and associative properties. For example,

$$(2x^3) \cdot (5x^4) = (2 \cdot 5) \cdot (x^3 \cdot x^4) = 10x^{3+4} = 10x^7$$

Products of polynomials are found by repeated use of the distributive property and the laws of exponents.

EXAMPLE 5 Multiplying Polynomials

Find the product: $(2x+5)(x^2-x+2)$

Solution *Horizontal Multiplication*

$$
\begin{aligned}
(2x+5)(x^2-x+2) &= 2x(x^2-x+2) + 5(x^2-x+2) && \text{Distributive property}\\
&= (2x\cdot x^2 - 2x\cdot x + 2x\cdot 2) + (5\cdot x^2 - 5\cdot x + 5\cdot 2) && \text{Distributive property}\\
&= (2x^3 - 2x^2 + 4x) + (5x^2 - 5x + 10) && \text{Law of Exponents}\\
&= 2x^3 + 3x^2 - x + 10 && \text{Combine like terms}
\end{aligned}
$$

Vertical Multiplication: The idea here is very much like multiplying a two-digit number by a three-digit number.

$$
\begin{array}{rrrrl}
 & x^2 & - x & + 2 & \\
 & & 2x & + 5 & \\
\hline
2x^3 & - 2x^2 & + 4x & & \text{This line is } 2x(x^2-x+2).\\
(+) & 5x^2 & - 5x & + 10 & \text{This line is } 5(x^2-x+2).\\
\hline
2x^3 & + 3x^2 & - x & + 10 & \text{Sum of the above two lines.}
\end{array}
$$

NOW WORK PROBLEM 31.

SPECIAL PRODUCTS

5 Certain products, which we call **special products**, occur frequently in algebra. We can calculate them easily using the **FOIL** (**F**irst, **O**uter, **I**nner, **L**ast) method of multiplying two binomials:

$$
\begin{aligned}
(ax+b)(cx+d) &= ax(cx+d) + b(cx+d)\\
&= \underbrace{ax\cdot cx}_{\text{First}} + \underbrace{ax\cdot d}_{\text{Outer}} + \underbrace{b\cdot cx}_{\text{Inner}} + \underbrace{b\cdot d}_{\text{Last}}\\
&= acx^2 + adx + bcx + bd\\
&= acx^2 + (ad+bc)x + bd
\end{aligned}
$$

EXAMPLE 6 Using FOIL to Find Products of the Form $(x-a)(x+a)$

$$(x-3)(x+3) = \underset{F}{x^2} + \underset{O}{3x} - \underset{I}{3x} - \underset{L}{9} = x^2 - 9$$

EXAMPLE 7 Using FOIL to Find the Square of a Binomial

(a) $(x+2)^2 = (x+2)(x+2) = x^2 + 2x + 2x + 4 = x^2 + 4x + 4$

(b) $(x-3)^2 = (x-3)(x-3) = x^2 - 3x - 3x + 9 = x^2 - 6x + 9$

EXAMPLE 8

Using FOIL to Find the Product of Two Binomials

(a) $(x + 3)(x + 1) = x^2 + x + 3x + 3 = x^2 + 4x + 3$

(b) $(2x + 1)(3x + 4) = 6x^2 + 8x + 3x + 4 = 6x^2 + 11x + 4$ ■

NOW WORK PROBLEMS 33 AND 41.

Some special products occur so frequently that they have been given names. The following special products are based on Examples 6 and 7.

Difference of Two Squares

$$(x - a)(x + a) = x^2 - a^2 \qquad \textbf{(2)}$$

Squares of Binomials, or Perfect Squares

$$(x + a)^2 = x^2 + 2ax + a^2 \qquad \textbf{(3a)}$$

$$(x - a)^2 = x^2 - 2ax + a^2 \qquad \textbf{(3b)}$$

EXAMPLE 9

Using Special Product Formulas

(a) $(x - 5)(x + 5) = x^2 - 5^2 = x^2 - 25$ Difference of two squares

(b) $(x + 7)^2 = x^2 + 2 \cdot 7 \cdot x + 7^2 = x^2 + 14x + 49$ Square of a binomial ■

EXAMPLE 10

Using Special Product Formulas

(a) $(2x + 1)^2 = (2x)^2 + 2 \cdot 1 \cdot 2x + 1^2 = 4x^2 + 4x + 1$ Notice that we used $2x$ in place of x in formula (3a).

(b) $(3x - 4)^2 = (3x)^2 - 2 \cdot 3x \cdot 4 + 4^2 = 9x^2 - 24x + 16$ Replace x by $3x$ in formula (3b). ■

NOW WORK PROBLEM 47.

Let's look at some more examples that lead to general formulas.

EXAMPLE 11

Cubing a Binomial

(a)
$$\begin{aligned}(x + 2)^3 = (x + 2)(x + 2)^2 &= (x + 2)(x^2 + 4x + 4) \quad \text{Formula (3a)}\\ &= (x^3 + 4x^2 + 4x) + (2x^2 + 8x + 8)\\ &= x^3 + 6x^2 + 12x + 8\end{aligned}$$

(b)
$$\begin{aligned}(x - 1)^3 = (x - 1)(x - 1)^2 &= (x - 1)(x^2 - 2x + 1) \quad \text{Formula (3b)}\\ &= (x^3 - 2x^2 + x) - (x^2 - 2x + 1)\\ &= x^3 - 3x^2 + 3x - 1\end{aligned}$$ ■

Cubes of Binomials, or Perfect Cubes

$$(x + a)^3 = x^3 + 3ax^2 + 3a^2x + a^3 \quad \textbf{(4a)}$$
$$(x - a)^3 = x^3 - 3ax^2 + 3a^2x - a^3 \quad \textbf{(4b)}$$

NOW WORK PROBLEM 55.

EXAMPLE 12 Forming the Difference of Two Cubes

$$\begin{aligned}(x - 1)(x^2 + x + 1) &= x(x^2 + x + 1) - 1(x^2 + x + 1)\\ &= x^3 + x^2 + x - x^2 - x - 1\\ &= x^3 - 1\end{aligned}$$

EXAMPLE 13 Forming the Sum of Two Cubes

$$\begin{aligned}(x + 2)(x^2 - 2x + 4) &= x(x^2 - 2x + 4) + 2(x^2 - 2x + 4)\\ &= (x^3 - 2x^2 + 4x) + (2x^2 - 4x + 8)\\ &= x^3 + 8\end{aligned}$$

Examples 12 and 13 lead to two more special products.

Difference of Two Cubes

$$(x - a)(x^2 + ax + a^2) = x^3 - a^3 \quad \textbf{(5)}$$

Sum of Two Cubes

$$(x + a)(x^2 - ax + a^2) = x^3 + a^3 \quad \textbf{(6)}$$

3 EXERCISES

In Problems 1–10, tell whether the expression is a monomial. If it is, name the variable(s) and the coefficient and give the degree of the monomial.

1. $2x^3$ **2.** $-4x^2$ **3.** $8/x$ **4.** $-2x^{-3}$ **5.** $-2x$

6. $5x^2$ **7.** $8x/5$ **8.** $-2x^2/9$ **9.** $x^2 + 9$ **10.** $3x^2 + 4$

In Problems 11–20, tell whether the expression is a polynomial. If it is, give its degree.

11. $3x^2 - 5$ **12.** $1 - 4x$ **13.** 5 **14.** $-\pi$ **15.** $3x^2 - \dfrac{5}{x}$

16. $\dfrac{3}{x} + 2$ **17.** $2y^3 - \sqrt{2}$ **18.** $10z^2 + z$ **19.** $\dfrac{x^2 + 5}{x^3 - 1}$ **20.** $\dfrac{3x^3 + 2x - 1}{x^2 + x + 1}$

In Problems 21–32, add, subtract, or multiply, as indicated. Express your answer as a single polynomial.

21. $(x^2 + 4x + 5) + (3x - 3)$ **22.** $(x^3 + 3x^2 + 2) + (x^2 - 4x + 4)$

23. $(x^3 - 2x^2 + 5x + 10) - (2x^2 - 4x + 3)$
24. $(x^2 - 3x - 4) - (x^3 - 3x^2 + x + 5)$
25. $(x^2 - 3x + 1) + 2(3x^2 + x - 4)$
26. $-2(x^2 + x + 1) + (-5x^2 - x + 2)$
27. $6(x^3 + x^2 - 3) - 4(2x^3 - 3x^2)$
28. $8(4x^3 - 3x^2 - 1) - 6(4x^3 + 8x - 2)$
29. $x(x^2 + x - 4)$
30. $4x^2(x^3 - x + 2)$
31. $(x + 1)(x^2 + 2x - 4)$
32. $(2x - 3)(x^2 + x + 1)$

In Problems 33–44, multiply the polynomials using the FOIL method. Express your answer as a single polynomial.

33. $(x + 2)(x + 4)$
34. $(x + 3)(x + 5)$
35. $(2x + 5)(x + 2)$
36. $(3x + 1)(2x + 1)$
37. $(x - 4)(x + 2)$
38. $(x + 4)(x - 2)$
39. $(x - 3)(x - 2)$
40. $(x - 5)(x - 1)$
41. $(2x + 3)(x - 2)$
42. $(2x - 4)(3x + 1)$
43. $(-2x + 3)(x - 4)$
44. $(-3x - 1)(x + 1)$

In Problems 45–58, multiply the polynomials using the Special Product Formulas.

45. $(x - 7)(x + 7)$
46. $(x - 1)(x + 1)$
47. $(2x + 3)(2x - 3)$
48. $(3x + 2)(3x - 2)$
49. $(x + 4)^2$
50. $(x + 5)^2$
51. $(x - 4)^2$
52. $(x - 5)^2$
53. $(2x - 3)^2$
54. $(3x - 4)^2$
55. $(x - 2)^3$
56. $(x + 1)^3$
57. $(2x + 1)^3$
58. $(3x - 2)^3$

59. Explain why the degree of the product of two polynomials equals the sum of their degrees.

60. Explain why the degree of the sum of two polynomials equals the larger of their degrees.

61. Do you prefer multiplying two polynomials using the horizontal method or the vertical method? Write a brief position paper defending your choice.

62. Do you prefer to memorize the rule for the square of a binomial $(x + a)^2$ or to use FOIL to obtain the product? Write a brief position paper defending your choice.

4 POLYNOMIAL DIVISION; SYNTHETIC DIVISION

OBJECTIVES
1. Divide Polynomials Using Long Division
2. Divide Polynomials Using Synthetic Division

LONG DIVISION

1 The procedure for dividing two polynomials is similar to the procedure for dividing two integers. This process should be familiar to you, but we review it briefly next.

EXAMPLE 1 Dividing Two Integers

Divide 842 by 15.

Solution

	56	← Quotient
Divisor →	$15\overline{)842}$	← Dividend
	75	← 5 · 15 (subtract)
	92	
	90	← 6 · 15 (subtract)
	2	← Remainder

Thus, $\dfrac{842}{15} = 56 + \dfrac{2}{15}$. ■

In the long division process detailed in Example 1, the number 15 is called the **divisor**, the number 842 is the **dividend**, the number 56 is the **quotient**, and the number 2 the **remainder**.

To check the answer obtained in a division problem, multiply the quotient by the divisor and add the remainder. The answer should be the dividend.

(Quotient)(Divisor) + Remainder = Dividend

For example, we can check the results obtained in Example 1 as follows:

$$(56)(15) + 2 = 840 + 2 = 842$$

To divide two polynomials, we first must write each polynomial in standard form, that is, in descending powers of the variable. The process then follows a pattern similar to that of Example 1. The next example illustrates the procedure.

EXAMPLE 2 Dividing Two Polynomials

Find the quotient and the remainder when

$$3x^3 + 4x^2 + x + 7 \quad \text{is divided by} \quad x^2 + 1$$

Solution Each polynomial is in standard form. The dividend is $3x^3 + 4x^2 + x + 7$, and the divisor is $x^2 + 1$.

STEP 1: Divide the leading term of the dividend, $3x^3$, by the leading term of the divisor, x^2. Enter the result, $3x$, over the term $3x^3$, as follows:

$$\begin{array}{r} 3x \\ x^2 + 1 \overline{)\, 3x^3 + 4x^2 + x + 7} \end{array}$$

STEP 2: Multiply $3x$ by $x^2 + 1$ and enter the result below the dividend.

$$\begin{array}{rl} 3x & \\ x^2 + 1 \overline{)\, 3x^3 + 4x^2 + x + 7} & \\ \underline{3x^3 + 3x } & \leftarrow 3x \cdot (x^2 + 1) = 3x^3 + 3x \end{array}$$

↑ Notice that we align the $3x$ term under the x to make the next step easier.

STEP 3: Subtract and bring down the remaining terms.

$$\begin{array}{rl} 3x & \\ x^2 + 1 \overline{)\, 3x^3 + 4x^2 + x + 7} & \\ \underline{3x^3 + 3x } & \leftarrow \text{Subtract.} \\ 4x^2 - 2x + 7 & \leftarrow \text{Bring down the } 4x^2 \text{ and the 7.} \end{array}$$

STEP 4: Repeat Steps 1–3 using $4x^2 - 2x + 7$ as the dividend.

$$\begin{array}{rl} 3x + 4 & \leftarrow \text{Divide } 4x^2 \text{ by } x^2 \text{ to get 4.} \\ x^2 + 1 \overline{)\, 3x^3 + 4x^2 + x + 7} & \\ \underline{3x^3 + 3x } & \\ 4x^2 - 2x + 7 & \\ \underline{4x^2 + 4} & \leftarrow \text{Multiply } x^2 + 1 \text{ by 4; subtract.} \\ -2x + 3 & \end{array}$$

Since x^2 does not divide $-2x$ evenly (that is, the result is not a monomial), the process ends. The quotient is $3x + 4$, and the remainder is $-2x + 3$.

Check: (Quotient)(Divisor) + Remainder

$$\begin{aligned} &= (3x + 4)(x^2 + 1) + (-2x + 3) \\ &= 3x^3 + 4x^2 + 3x + 4 + (-2x + 3) \\ &= 3x^3 + 4x^2 + x + 7 = \text{Dividend} \end{aligned}$$

Thus,

$$\frac{3x^3 + 4x^2 + x + 7}{x^2 + 1} = 3x + 4 + \frac{-2x + 3}{x^2 + 1}$$

The next example combines the steps involved in long division.

EXAMPLE 3 Dividing Two Polynomials

Find the quotient and the remainder when

$$x^4 - 3x^3 + 2x - 5 \quad \text{is divided by} \quad x^2 - x + 1$$

Solution In setting up this division problem, it is necessary to leave a space for the missing x^2 term in the dividend.

$$\begin{array}{rrrrrrl}
 & & x^2 & -2x & -3 & & \leftarrow \text{Quotient} \\
\text{Divisor} \rightarrow x^2 - x + 1 \,\big) & x^4 & -3x^3 & & +2x & -5 & \leftarrow \text{Dividend} \\
\text{Subtract} \rightarrow & x^4 & -x^3 & +x^2 & & & \\
 & & -2x^3 & -x^2 & +2x & -5 & \\
\text{Subtract} \rightarrow & & -2x^3 & +2x^2 & -2x & & \\
 & & & -3x^2 & +4x & -5 & \\
\text{Subtract} \rightarrow & & & -3x^2 & +3x & -3 & \\
 & & & & x & -2 & \leftarrow \text{Remainder}
\end{array}$$

Check: (Quotient)(Divisor) + Remainder

$$\begin{aligned} &= (x^2 - 2x - 3)(x^2 - x + 1) + x - 2 \\ &= x^4 - x^3 + x^2 - 2x^3 + 2x^2 - 2x - 3x^2 + 3x - 3 + x - 2 \\ &= x^4 - 3x^3 + 2x - 5 = \text{Dividend} \end{aligned}$$

Thus,

$$\frac{x^4 - 3x^3 + 2x - 5}{x^2 - x + 1} = x^2 - 2x - 3 + \frac{x - 2}{x^2 - x + 1}$$

The process of dividing two polynomials leads to the following result:

Theorem The remainder after dividing two polynomials is either the zero polynomial or a polynomial of degree less than the degree of the divisor.

NOW WORK PROBLEM 5.

SYNTHETIC DIVISION

2 To find the quotient as well as the remainder when a polynomial function f of degree 1 or higher is divided by $g(x) = x - c$, a shortened version of long division, called **synthetic division**, makes the task simpler.

To see how synthetic division works, we will use long division to divide the polynomial $f(x) = 2x^3 - x^2 + 3$ by $g(x) = x - 3$.

$$
\begin{array}{r l}
 & 2x^2 + 5x + 15 \quad \leftarrow \text{Quotient} \\
x - 3 \,\big) & 2x^3 - x^2 \qquad\quad + 3 \\
 & 2x^3 - 6x^2 \\
 & \qquad 5x^2 \\
 & \qquad 5x^2 - 15x \\
 & \qquad\qquad 15x + 3 \\
 & \qquad\qquad 15x - 45 \\
 & \qquad\qquad\qquad 48 \quad \leftarrow \text{Remainder}
\end{array}
$$

Check: (Divisor) · (Quotient) + Remainder

$$
\begin{aligned}
&= (x - 3)(2x^2 + 5x + 15) + 48 \\
&= 2x^3 + 5x^2 + 15x - 6x^2 - 15x - 45 + 48 \\
&= 2x^3 - x^2 + 3
\end{aligned}
$$

The process of synthetic division arises from rewriting the long division in a more compact form, using simpler notation. For example, in the long division above, the terms in color are not really necessary because they are identical to the terms directly above them. With these terms removed, we have

$$
\begin{array}{r l}
 & 2x^2 + 5x + 15 \\
x - 3 \,\big) & 2x^3 - x^2 \qquad\quad + 3 \\
 & \quad - 6x^2 \\
 & \qquad 5x^2 \\
 & \qquad\quad - 15x \\
 & \qquad\qquad 15x \\
 & \qquad\qquad\quad - 45 \\
 & \qquad\qquad\qquad 48
\end{array}
$$

Most of the x's that appear in this process can also be removed, provided that we are careful about positioning each coefficient. In this regard, we will need to use 0 as the coefficient of x in the dividend, because that power of x is missing. Now we have

$$
\begin{array}{r l}
 & 2x^2 + 5x + 15 \\
x - 3 \,\big) & 2 \quad -1 \quad 0 \quad 3 \\
 & \quad\ \ -6 \\
 & \qquad\ 5 \\
 & \qquad\quad -15 \\
 & \qquad\qquad 15 \\
 & \qquad\qquad\quad -45 \\
 & \qquad\qquad\qquad 48
\end{array}
$$

We can make this display more compact by moving the lines up until the numbers in color align horizontally:

$$\begin{array}{rrrrrl} & & 2x^2 & +\,5x & +\,15 & \text{Row 1} \\ x-3\,) & 2 & -1 & 0 & 3 & \text{Row 2} \\ & & -6 & -15 & -45 & \text{Row 3} \\ \hline & \bigcirc & 5 & 15 & 48 & \text{Row 4} \end{array}$$

Now, if we place the leading coefficient of the quotient (2) in the circled position, the first three numbers in row 4 are precisely the coefficients of the quotient, and the last number in row 4 is the remainder. Thus, row 1 is not really needed, so we can compress the process to three rows, where the bottom row contains the coefficients of both the quotient and the remainder.

$$\begin{array}{rrrrrl} x-3\,) & 2 & -1 & 0 & 3 & \text{Row 1} \\ & & -6 & -15 & -45 & \text{Row 2 (subtract)} \\ \hline & 2 & 5 & 15 & 48 & \text{Row 3} \end{array}$$

Recall that the entries in row 3 are obtained by subtracting the entries in row 2 from those in row 1. Rather than subtracting the entries in row 2, we can change the sign of each entry and add. With this modification, our display will look like this:

$$\begin{array}{rrrrrl} x-3\,) & 2 & -1 & 0 & 3 & \text{Row 1} \\ & & 6 & 15 & 45 & \text{Row 2 (add)} \\ \hline & 2 & 5 & 15 & 48 & \text{Row 3} \end{array}$$

Notice that the entries in row 2 are three times the prior entries in row 3. Our last modification to the display replaces the $x - 3$ by 3. The entries in row 3 give the quotient and the remainder, as shown next.

$$\begin{array}{rrrrrl} 3\,) & 2 & -1 & 0 & 3 & \text{Row 1} \\ & & 6 & 15 & 45 & \text{Row 2 (add)} \\ \hline & 2 & 5 & 15 & 48 & \text{Row 3} \end{array}$$

Quotient: 2 5 15 — Remainder: 48

Let's go through another example step by step.

EXAMPLE 4 Using Synthetic Division to Find the Quotient and Remainder

Use synthetic division to find the quotient and remainder when

$$f(x) = 3x^4 + 8x^2 - 7x + 4 \quad \text{is divided by} \quad g(x) = x - 1$$

Solution **STEP 1:** Write the dividend in descending powers of x. Then copy the coefficients, remembering to insert a 0 for any missing powers of x.

$$\begin{array}{rrrrrl} 3 & 0 & 8 & -7 & 4 & \text{Row 1} \end{array}$$

STEP 2: Insert the usual division symbol. Since the divisor is $x - 1$ we insert 1 to the left of the division symbol

$$\begin{array}{rrrrrrl} 1\,) & 3 & 0 & 8 & -7 & 4 & \text{Row 1} \end{array}$$

STEP 3: Bring the 3 down two rows, and enter it in row 3.

$$\begin{array}{rrrrrrl} 1\,) & 3 & 0 & 8 & -7 & 4 & \text{Row 1} \\ & \downarrow & & & & & \text{Row 2} \\ \hline & 3 & & & & & \text{Row 3} \end{array}$$

STEP 4: Multiply the latest entry in row 3 by 1, and place the result in row 2, but one column over to the right.

1)	3	0	8	−7	4	Row 1
		3				Row 2
	3					Row 3

STEP 5: Add the entry in row 2 to the entry above it in row 1, and enter the sum to row 3.

1)	3	0	8	−7	4	Row 1
		3				Row 2
	3	3				Row 3

STEP 6: Repeat Steps 4 and 5 until no more entries are available in row 1.

1)	3	0	8	−7	4	Row 1
		3	3	11	4	Row 2 (add)
	3	3	11	4	8	Row 3

STEP 7: The final entry in row 3, an 8, is the remainder: the other entries in row 3 (3, 3, 11, and 4) are the coefficients (in descending order) of a polynomial whose degree is 1 less than that of the dividend; this is the quotient. Thus,

$$\text{Quotient} = 3x^3 + 3x^2 + 11x + 4 \qquad \text{Remainder} = 8$$ ■

Check: (Divisor)(Quotient) + Remainder

$$= (x - 1)(3x^3 + 3x^2 + 11x + 4) + 8$$

$$= 3x^4 + 3x^3 + 11x^2 + 4x - 3x^3 - 3x^2 - 11x - 4 + 8$$

$$= 3x^4 + 8x^2 - 7x + 4 = \text{Dividend}$$ ■

Let's do an example in which all seven steps are combined.

EXAMPLE 5 Using Synthetic Division to Verify a Factor

Use synthetic division to show that $g(x) = x + 3$ is a factor of

$$f(x) = 2x^5 + 5x^4 - 2x^3 + 2x^2 - 2x + 3$$

Solution The divisor is $x + 3 = x - (-3)$, so the row 3 entries will be multiplied by −3, entered in row 2, and added to row 1.

−3)	2	5	−2	2	−2	3	Row 1
		−6	3	−3	3	−3	Row 2
	2	−1	1	−1	1	0	Row 3

Because the remainder is 0, we have

(Divisor)(Quotient) + Remainder

$$= (x + 3)(2x^4 - x^3 + x^2 - x + 1) = 2x^5 + 5x^4 - 2x^3 + 2x^2 - 2x + 3$$

As we see, $x + 3$ is a factor of $2x^5 + 5x^4 - 2x^3 + 2x^2 - 2x + 3$. ■

As Example 5 illustrates, the remainder after division gives information about whether the divisor is or is not a factor.

NOW WORK PROBLEM 39.

4 EXERCISES

In Problems 1–26, find the quotient and the remainder. Check your work by verifying that

$$(Quotient)(Divisor) + Remainder = Dividend$$

1. $4x^3 - 3x^2 + x + 1$ divided by $x + 2$
2. $3x^3 - x^2 + x - 2$ divided by $x + 2$
3. $4x^3 - 3x^2 + x + 1$ divided by $x - 4$
4. $3x^3 - x^2 + x - 2$ divided by $x - 4$
5. $4x^3 - 3x^2 + x + 1$ divided by $x^2 + 2$
6. $3x^3 - x^2 + x - 2$ divided by $x^2 + 2$
7. $4x^3 - 3x^2 + x + 1$ divided by $2x^3 - 1$
8. $3x^3 - x^2 + x - 2$ divided by $3x^3 - 1$
9. $4x^3 - 3x^2 + x + 1$ divided by $2x^2 + x + 1$
10. $3x^3 - x^2 + x - 2$ divided by $3x^2 + x + 1$
11. $4x^3 - 3x^2 + x + 1$ divided by $4x^2 + 1$
12. $3x^3 - x^2 + x - 2$ divided by $3x - 1$
13. $x^4 - 1$ divided by $x - 1$
14. $x^4 - 1$ divided by $x + 1$
15. $x^4 - 1$ divided by $x^2 - 1$
16. $x^4 - 1$ divided by $x^2 + 1$
17. $-4x^3 + x^2 - 4$ divided by $x - 1$
18. $-3x^4 - 2x - 1$ divided by $x - 1$
19. $1 - x^2 + x^4$ divided by $x^2 + x + 1$
20. $1 - x^2 + x^4$ divided by $x^2 - x + 1$
21. $1 - x^2 + x^4$ divided by $1 - x^2$
22. $1 - x^2 + x^4$ divided by $1 + x^2$
23. $x^3 - a^3$ divided by $x - a$
24. $x^3 + a^3$ divided by $x + a$
25. $x^4 - a^4$ divided by $x - a$
26. $x^5 - a^5$ divided by $x - a$

In Problems 27–38, use synthetic division to find the quotient $q(x)$ and remainder R when $f(x)$ is divided by $g(x)$.

27. $f(x) = x^3 - x^2 + 2x + 4;\quad g(x) = x - 2$
28. $f(x) = x^3 + 2x^2 - 3x + 1;\quad g(x) = x + 1$
29. $f(x) = 3x^3 + 2x^2 - x + 3;\quad g(x) = x - 3$
30. $f(x) = -4x^3 + 2x^2 - x + 1;\quad g(x) = x + 2$
31. $f(x) = x^5 - 4x^3 + x;\quad g(x) = x + 3$
32. $f(x) = x^4 + x^2 + 2;\quad g(x) = x - 2$
33. $f(x) = 4x^6 - 3x^4 + x^2 + 5;\quad g(x) = x - 1$
34. $f(x) = x^5 + 5x^3 - 10;\quad g(x) = x + 1$
35. $f(x) = 0.1x^3 + 0.2x;\quad g(x) = x + 1.1$
36. $f(x) = 0.1x^2 - 0.2;\quad g(x) = x + 2.1$
37. $f(x) = x^5 - 1;\quad g(x) = x - 1$
38. $f(x) = x^5 + 1;\quad g(x) = x + 1$

In Problems 39–48, use synthetic division to determine whether $x - c$ is a factor of $f(x)$.

39. $f(x) = 4x^3 - 3x^2 - 8x + 4;\quad x - 2$
40. $f(x) = -4x^3 + 5x^2 + 8;\quad x + 3$
41. $f(x) = 3x^4 - 6x^3 - 5x + 10;\quad x - 2$
42. $f(x) = 4x^4 - 15x^2 - 4;\quad x - 2$
43. $f(x) = 3x^6 + 82x^3 + 27;\quad x + 3$
44. $f(x) = 2x^6 - 18x^4 + x^2 - 9;\quad x + 3$
45. $f(x) = 4x^6 - 64x^4 + x^2 - 15;\quad x + 4$
46. $f(x) = x^6 - 16x^4 + x^2 - 16;\quad x + 4$
47. $f(x) = 2x^4 - x^3 + 2x - 1;\quad x - 1/2$
48. $f(x) = 3x^4 + x^3 - 3x + 1;\quad x + 1/3$
49. When dividing a polynomial by $x - c$, do you prefer to use long division or synthetic division? Does the value of c make a difference to you in choosing? Give reasons.
50. Find the sum of a, b, c, and d if

$$\frac{x^3 - 2x^2 + 3x + 5}{x + 2} = ax^2 + bx + c + \frac{d}{x + 2}$$

5 FACTORING POLYNOMIALS

OBJECTIVES

1. Factor the Difference of Two Squares and the Sum and the Difference of Two Cubes
2. Factor Perfect Squares
3. Factor a Second-degree Polynomial: $x^2 + Bx + C$
4. Factor by Grouping
5. Factor a Second-degree Polynomial: $Ax^2 + Bx + C$

Consider the following product:

$$(2x + 3)(x - 4) = 2x^2 - 5x - 12$$

The two polynomials on the left side are called **factors** of the polynomial on the right side. Expressing a given polynomial as a product of other polynomials, that is, finding the factors of a polynomial, is called **factoring**.

We shall restrict our discussion here to factoring polynomials in one variable into products of polynomials in one variable, where all coefficients are integers. We call this **factoring over the integers**.

Any polynomial can be written as the product of 1 times itself or as -1 times its additive inverse. If a polynomial cannot be written as the product of two other polynomials (excluding 1 and -1), then the polynomial is said to be **prime**. When a polynomial has been written as a product consisting only of prime factors, it is said to be **factored completely**. Examples of prime polynomials are

$$2, \quad 3, \quad 5, \quad x, \quad x + 1, \quad x - 1, \quad 3x + 4$$

The first factor to look for in a factoring problem is a common monomial factor present in each term of the polynomial. If one is present, use the distributive property to factor it out.

EXAMPLE 1 Identifying Common Monomial Factors

Polynomial	Common Monomial Factor	Remaining Factor	Factored Form
$2x + 4$	2	$x + 2$	$2x + 4 = 2(x + 2)$
$3x - 6$	3	$x - 2$	$3x - 6 = 3(x - 2)$
$2x^2 - 4x + 8$	2	$x^2 - 2x + 4$	$2x^2 - 4x + 8 = 2(x^2 - 2x + 4)$
$8x - 12$	4	$2x - 3$	$8x - 12 = 4(2x - 3)$
$x^2 + x$	x	$x + 1$	$x^2 + x = x(x + 1)$
$x^3 - 3x^2$	x^2	$x - 3$	$x^3 - 3x^2 = x^2(x - 3)$
$6x^2 + 9x$	$3x$	$2x + 3$	$6x^2 + 9x = 3x(2x + 3)$ ■

Notice that once all common monomial factors have been removed from a polynomial, the remaining factor is either a prime polynomial of degree 1 or a polynomial of degree 2 or higher. (Do you see why?)

NOW WORK PROBLEM 1.

SPECIAL FORMULAS

1 When you factor a polynomial, first check whether you can use one of the special formulas discussed in the previous section.

Difference of Two Squares	$x^2 - a^2 = (x - a)(x + a)$
Perfect Squares	$x^2 + 2ax + a^2 = (x + a)^2$
	$x^2 - 2ax + a^2 = (x - a)^2$
Sum of Two Cubes	$x^3 + a^3 = (x + a)(x^2 - ax + a^2)$
Difference of Two Cubes	$x^3 - a^3 = (x - a)(x^2 + ax + a^2)$

EXAMPLE 2 **Factoring the Difference of Two Squares**

Factor completely: $x^2 - 4$

Solution We notice that $x^2 - 4$ is the difference of two squares, x^2 and 2^2. Thus,

$$x^2 - 4 = (x - 2)(x + 2)$$

EXAMPLE 3 **Factoring the Sum of Two Cubes**

Factor completely: $x^3 + 8$

Solution Because $x^3 + 8$ is the sum of two cubes, x^3 and 2^3, we have

$$x^3 + 8 = (x + 2)(x^2 - 2x + 4)$$

EXAMPLE 4 **Factoring the Difference of Two Squares**

Factor completely: $x^4 - 16$

Solution Because $x^4 - 16$ is the difference of two squares, $x^4 = (x^2)^2$ and $16 = 4^2$, we have

$$x^4 - 16 = (x^2 - 4)(x^2 + 4)$$

But $x^2 - 4$ is also the difference of two squares. Thus,

$$x^4 - 16 = (x^2 - 4)(x^2 + 4) = (x - 2)(x + 2)(x^2 + 4)$$

NOW WORK PROBLEMS 9 AND 25.

2 When the first term and third term of a trinomial are both positive and are perfect squares, such as x^2, $9x^2$, 1, and 4, check to see whether the trinomial is a perfect square.

EXAMPLE 5 **Factoring Perfect Squares**

Factor completely: $9x^2 - 6x + 1$

Solution The first term, $9x^2 = (3x)^2$, and the third term, $1 = 1^2$, are perfect squares. Because the middle term $-6x$ is -2 times the product of $3x$ and 1, we have a perfect square.

$$9x^2 - 6x + 1 = (3x - 1)^2$$

NOW WORK PROBLEMS 21 AND 75.

If a trinomial is not a perfect square, it may be possible to factor it using the technique discussed next.

FACTORING A SECOND-DEGREE POLYNOMIAL: $x^2 + Bx + C$

3 The idea behind factoring a second-degree polynomial like $x^2 + Bx + C$ is to see whether it can be made equal to the product of two, possibly equal, first-degree polynomials.

For example, we know that

$$(x + 3)(x + 4) = x^2 + 7x + 12$$

The factors of $x^2 + 7x + 12$ are $x + 3$ and $x + 4$. Notice the following:

$$x^2 + 7x + 12 = (x + 3)(x + 4)$$

3 and 4 are factors of 12

7 is the sum of 3 and 4

In general, if $x^2 + Bx + C = (x + a)(x + b)$, then $ab = C$ and $a + b = B$.

To factor a second-degree polynomial $x^2 + Bx + C$, find factors of C whose sum is B. That is, if there are numbers a, b, where $ab = C$ and $a + b = B$, then

$$x^2 + Bx + C = (x + a)(x + b)$$

EXAMPLE 6

Factoring Trinomials

Factor completely: $x^2 - 6x + 8$

Solution First, determine all possible factors of the constant term 8 and then compute each sum.

Factors of 8	1, 8	−1, −8	2, 4	−2, −4
Sum	9	−9	6	−6

Since −6 is the coefficient of the middle term,

$$x^2 - 6x + 8 = (x - 2)(x - 4)$$

EXAMPLE 7

Factoring Trinomials

Factor completely: $x^2 + 4x - 12$

Solution The factors −2 and 6 of −12 have the sum 4. Thus,

$$x^2 + 4x - 12 = (x - 2)(x + 6)$$

To avoid errors in factoring, always check your answer by multiplying it out to see if the result equals the original expression.

When none of the possibilities works, the polynomial is prime.

EXAMPLE 8

Identifying Prime Polynomials

Show that $x^2 + 9$ is prime.

Solution First, list the factors of 9 and then compute their sums.

Factors of 9	1, 9	−1, −9	3, 3	−3, −3
Sum	10	−10	6	−6

Since the coefficient of the middle term in $x^2 + 9 = x^2 + 0x + 9$ is 0 and none of the sums equals 0, we conclude that $x^2 + 9$ is prime. ■

Example 8 demonstrates a more general result.

Theorem Any polynomial of the form $x^2 + a^2$, a real, is prime. ■

NOW WORK PROBLEMS 31 AND 59.

FACTORING BY GROUPING

4 Sometimes a common factor does not occur in every term of the polynomial, but in each of several groups of terms that together make up the polynomial. When this happens, the common factor can be factored out of each group by means of the distributive property. This technique is called **factoring by grouping**.

EXAMPLE 9 Factoring by Grouping

Factor completely by grouping: $(x^2 + 2)x + (x^2 + 2) \cdot 3$

Solution Notice the common factor $x^2 + 2$. By applying the distributive property, we have

$$(x^2 + 2)x + (x^2 + 2) \cdot 3 = (x^2 + 2)(x + 3)$$

Since $x^2 + 2$ and $x + 3$ are prime, the factorization is complete. ■

EXAMPLE 10 Factoring by Grouping

Factor completely by grouping: $x^3 - 4x^2 + 2x - 8$

Solution To see if factoring by grouping will work, group the first two terms and the last two terms. Then look for a common factor in each group. In this example, we can factor x^2 from $x^3 - 4x^2$ and 2 from $2x - 8$. The remaining factor in each case is the same, $x - 4$. This means that factoring by grouping will work, as follows:

$$\begin{aligned} x^3 - 4x^2 + 2x - 8 &= (x^3 - 4x^2) + (2x - 8) \\ &= x^2(x - 4) + 2(x - 4) \\ &= (x - 4)(x^2 + 2) \end{aligned}$$

Since $x - 4$ and $x^2 + 2$ are prime, the factorization is complete. ■

NOW WORK PROBLEM 39.

5 FACTORING A SECOND-DEGREE POLYNOMIAL: $Ax^2 + Bx + C$

To factor a second-degree polynomial $Ax^2 + Bx + C$, when $A \neq 1$, follow these steps:

STEPS FOR FACTORING $Ax^2 + Bx + C$, $A \neq 1$

STEP 1: Find the value of AC.

STEP 2: Find the factors of AC that add up to B. That is, find a and b so that $ab = AC$ and $a + b = B$.

STEP 3: Write $Ax^2 + Bx + C = Ax^2 + ax + bx + C$.

STEP 4: Factor this last expression by grouping.

EXAMPLE 11 Factoring Trinomials

Factor completely: $2x^2 - x - 6$

Solution Comparing $2x^2 - x - 6$ to $Ax^2 + Bx + C$, we find that $A = 2$, $B = -1$, and $C = -6$.

STEP 1: The value of AC is $2 \cdot (-6) = -12$.

STEP 2: Determine the factors of $AC = -12$ and compute their sums.

Factors of −12	1, −12	−1, 12	2, −6	−2, 6	3, −4	−3, 4
Sum	−11	11	−4	4	−1	1

STEP 3: The factors of -12 that add up to $B = -1$ are -4 and 3.

$$2x^2 - x - 6 = 2x^2 - 4x + 3x - 6$$

STEP 4: Factor by grouping.

$$\begin{aligned} 2x^2 - 4x + 3x - 6 &= (2x^2 - 4x) + (3x - 6) \\ &= 2x(x - 2) + 3(x - 2) \\ &= (x - 2)(2x + 3) \end{aligned}$$

Thus,

$$2x^2 - x - 6 = (x - 2)(2x + 3)$$

NOW WORK PROBLEM 45.

SUMMARY

We close this section with a capsule summary.

Type of Polynomial	Method	Example
Any polynomial	Look for common monomial factors. (Always do this first!)	$6x^2 + 9x = 3x(2x + 3)$
Binomials of degree 2 or higher	Check for a special product: Difference of two squares, $x^2 - a^2$	$x^2 - 16 = (x - 4)(x + 4)$

	Difference of two cubes, $x^3 - a^3$	$x^3 - 64 = (x - 4)(x^2 + 4x + 16)$
	Sum of two cubes, $x^3 + a^3$	$x^3 + 27 = (x + 3)(x^2 - 3x + 9)$
Trinomials of degree 2	Check for a perfect square	$x^2 + 8x + 16 = (x + 4)^2$
	Follow the STEPS on page 974.	$6x^2 + x - 1 = (2x + 1)(3x - 1)$
Three or more terms	Grouping	$2x^3 - 3x^2 + 4x - 6 = (2x - 3)(x^2 + 2)$

5 EXERCISES

In Problems 1–8, factor each polynomial by removing the common monomial factor.

1. $3x + 6$ **2.** $7x - 14$ **3.** $ax^2 + a$ **4.** $ax - a$
5. $x^3 + x^2 + x$ **6.** $x^3 - x^2 + x$ **7.** $2x^2 - 2x$ **8.** $3x^2 - 3x$

In Problems 9–16, factor the difference of two squares.

9. $x^2 - 1$ **10.** $x^2 - 4$ **11.** $4x^2 - 1$ **12.** $9x^2 - 1$
13. $x^2 - 16$ **14.** $x^2 - 25$ **15.** $25x^2 - 4$ **16.** $36x^2 - 9$

In Problems 17–24, factor the perfect squares.

17. $x^2 + 2x + 1$ **18.** $x^2 - 4x + 4$ **19.** $x^2 - 10x + 25$ **20.** $x^2 + 10x + 25$
21. $4x^2 + 4x + 1$ **22.** $9x^2 + 6x + 1$ **23.** $16x^2 + 8x + 1$ **24.** $25x^2 + 10x + 1$

In Problems 25–30, factor the sum or difference of two cubes.

25. $x^3 - 27$ **26.** $x^3 + 125$ **27.** $x^3 + 27$
28. $27 - 8x^3$ **29.** $8x^3 + 27$ **30.** $64 - 27x^3$

In Problems 31–38, factor each polynomial.

31. $x^2 + 5x + 6$ **32.** $x^2 + 6x + 8$ **33.** $x^2 + 7x + 10$ **34.** $x^2 + 11x + 10$
35. $x^2 - 10x + 16$ **36.** $x^2 - 17x + 16$ **37.** $x^2 - 7x - 8$ **38.** $x^2 - 2x - 8$

In Problems 39–44, factor by grouping.

39. $2x^2 + 4x + 3x + 6$ **40.** $3x^2 - 3x + 2x - 2$ **41.** $2x^2 - 4x + x - 2$
42. $3x^2 + 6x - x - 2$ **43.** $6x^2 + 9x + 4x + 6$ **44.** $9x^2 - 6x + 3x - 2$

In Problems 45–52, factor each polynomial.

45. $3x^2 + 4x + 1$ **46.** $2x^2 + 3x + 1$ **47.** $2z^2 + 5z + 3$
48. $6z^2 + 5z + 1$ **49.** $3x^2 - 2x - 8$ **50.** $3x^2 - 10x + 8$
51. $3x^2 + 10x - 8$ **52.** $3x^2 - 10x - 8$

In Problems 53–92, factor completely each polynomial. If the polynomial cannot be factored, say it is prime.

53. $x^2 - 36$ **54.** $x^2 - 9$ **55.** $1 - 4x^2$ **56.** $1 - 9x^2$
57. $x^2 + 7x + 10$ **58.** $x^2 + 5x + 4$ **59.** $x^2 - 2x + 8$ **60.** $x^2 - 4x + 5$
61. $x^2 + 4x + 16$ **62.** $x^2 + 12x + 36$ **63.** $15 + 2x - x^2$ **64.** $14 + 6x - x^2$
65. $3x^2 - 12x - 36$ **66.** $x^3 + 8x^2 - 20x$ **67.** $y^4 + 11y^3 + 30y^2$ **68.** $3y^3 - 18y^2 - 48y$
69. $4x^2 + 12x + 9$ **70.** $9x^2 - 12x + 4$ **71.** $3x^2 + 4x + 1$ **72.** $4x^2 + 3x - 1$
73. $x^4 - 81$ **74.** $x^4 - 1$ **75.** $x^6 - 2x^3 + 1$ **76.** $x^6 + 2x^3 + 1$
77. $x^7 - x^5$ **78.** $x^8 - x^5$ **79.** $5 + 16x - 16x^2$ **80.** $5 + 11x - 16x^2$
81. $4y^2 - 16y + 15$ **82.** $9y^2 + 9y - 4$ **83.** $1 - 8x^2 - 9x^4$ **84.** $4 - 14x^2 - 8x^4$
85. $x(x + 3) - 6(x + 3)$ **86.** $5(3x - 7) + x(3x - 7)$ **87.** $(x + 2)^2 - 5(x + 2)$ **88.** $(x - 1)^2 - 2(x - 1)$
89. $6x(2 - x)^4 - 9x^2(2 - x)^3$ **90.** $6x(1 - x^2)^4 - 24x^3(1 - x^2)^3$
91. $x^3 + 2x^2 - x - 2$ **92.** $x^3 - 3x^2 - x + 3$ **93.** $x^4 - x^3 + x - 1$ **94.** $x^4 + x^3 + x + 1$

95. Show that $x^2 + 4$ is prime.

96. Show that $x^2 + x + 1$ is prime.

97. Make up a polynomial that factors into a perfect square.

98. Explain to a fellow student what you look for first when presented with a factoring problem. What do you do next?

6 SOLVING EQUATIONS

OBJECTIVE 1 Solve Equations in One Variable

1 An **equation in one variable** is a statement in which two expressions, at least one containing the variable, are equal. The expressions are called the **sides** of the equation. Since an equation is a statement, it may be true or false, depending on the value of the variable. The values of the variable, if any, that result in a true statement are called **solutions**, or **roots**, of the equation. To **solve an equation** means to find all the solutions of the equation.

For example, the following are all equations in one variable, x:

$$x + 5 = 9 \qquad x^2 + 5x = 2x - 2 \qquad \frac{x^2 - 4}{x + 1} = 0 \qquad x^2 + 9 = 5$$

The first of these statements, $x + 5 = 9$, is true when $x = 4$ and false for any other choice of x. Thus, 4 is a solution of the equation $x + 5 = 9$. We also say that 4 **satisfies** the equation $x + 5 = 9$.

Sometimes an equation will have more than one solution. For example, the equation

$$x^2 - 4 = 0$$

has either $x = -2$ or $x = 2$ as a solution.

Sometimes we will write the solution of an equation in set notation. This set is called the **solution set** of the equation. For example, the solution set of the equation $x^2 - 9 = 0$ is $\{-3, 3\}$.

Unless indicated otherwise, we will limit ourselves to real solutions. Some equations have no real solution. For example, $x^2 + 9 = 5$ has no real solution, because there is no real number whose square when added to 9 equals 5.

An equation that is satisfied for every choice of the variable for which both sides are defined is called an **identity**. For example, the equation

$$3x + 5 = x + 3 + 2x + 2$$

is an identity, because this statement is true for any real number x.

Two or more equations that have precisely the same solutions are called **equivalent equations**. For example, all the following equations are equivalent, because each has only the solution $x = 5$.

$$2x + 3 = 13$$
$$2x = 10$$
$$x = 5$$

These three equations illustrate one method for solving many types of equations: Replace the original equation by an equivalent equation, and continue until an equation with an obvious solution, such as $x = 5$, is reached. The question, though, is "How do I obtain an equivalent equation?" In general, there are five ways to do so.

PROCEDURES THAT RESULT IN EQUIVALENT EQUATIONS

1. Interchange the two sides of the equation:

   Replace $3 = x$ by $x = 3$

2. Simplify the sides of the equation by combining like terms, eliminating parentheses, and so on:

   Replace $(x + 2) + 6 = 2x + 5(x + 1)$

   by $x + 8 = 7x + 5$

3. Add or subtract the same expression on both sides of the equation:

   Replace $3x - 5 = 4$

   by $(3x - 5) + 5 = 4 + 5$

4. Multiply or divide both sides of the equation by the same nonzero expression.

   Replace $3x = 6$

   by $\frac{1}{3} \cdot 3x = \frac{1}{3} \cdot 6$

5. If one side of the equation is 0 and the other side can be factored, then we may use the Zero-Product Property* and set each factor equal to 0:

   Replace $x(x - 3) = 0$

   by $x = 0$ or $x - 3 = 0$

Whenever it is possible to solve an equation in your head, do so. For example:

The solution of $2x = 8$ is $x = 4$.
The solution of $3x - 15 = 0$ is $x = 5$.

Often, though, some rearrangement is necessary.

EXAMPLE 1 Solving an Equation

Solve the equation: $3x - 5 = 4$

Solution We replace the original equation by a succession of equivalent equations.

$$3x - 5 = 4$$

$$(3x - 5) + 5 = 4 + 5 \quad \text{Add 5 to both sides.}$$

$$3x = 9 \quad \text{Simplify.}$$

$$\frac{3x}{3} = \frac{9}{3} \quad \text{Divide both sides by 3.}$$

$$x = 3 \quad \text{Simplify.}$$

The last equation, $x = 3$, has the single solution 3. All these equations are equivalent, so 3 is the only solution of the original equation $3x - 5 = 4$. ■

NOW WORK PROBLEM 7.

*The Zero-Product Property says that if $ab = 0$ then $a = 0$ or $b = 0$ or both equal 0. We show how this property is used in Example 2.

One property of real numbers that we often use to solve equations is the Zero-Product Property:

Zero-Product Property

If $ab = 0$, then $a = 0$ or $b = 0$, or both $a = 0$ and $b = 0$.

EXAMPLE 2 **Solving Equations by Factoring**

Solve the equations: (a) $x^2 = 4x$ (b) $2x^2 - x - 3 = 0$

Solution (a) We begin by collecting all terms on one side. This results in 0 on one side and an expression to be factored on the other.

$$x^2 = 4x$$
$$x^2 - 4x = 0$$
$$x(x - 4) = 0 \quad \text{Factor.}$$
$$x = 0 \text{ or } x - 4 = 0 \quad \text{Apply the Zero-Product Property}$$
$$x = 0 \text{ or } x = 4$$

The solution set is $\{0, 4\}$.

(b)
$$2x^2 - x - 3 = 0$$
$$(2x - 3)(x + 1) = 0 \quad \text{Factor.}$$
$$2x - 3 = 0 \text{ or } x + 1 = 0 \quad \text{Zero-Product Property}$$
$$x = 3/2 \text{ or } x = -1 \quad \text{Solve.}$$

The solution set is $\{-1, 3/2\}$.

NOW WORK PROBLEM 19.

6 EXERCISES

In Problems 1–18, solve each equation.

1. $3x + 2 = x$
2. $2x + 7 = 3x$
3. $2t - 6 = 3 - t$
4. $5y + 6 = -18 - y$
5. $6 - x = 2x + 9$
6. $3 - 2x = 2 - x$
7. $3 + 2n = 5n + 7$
8. $3 - 2m = 3m + 1$
9. $2(3 + 2x) = 3(x - 4)$
10. $3(2 - x) = 2x - 1$
11. $8x - (3x + 2) = 3x - 10$
12. $7 - (2x - 1) = 10$
13. $\frac{3}{2}x + 2 = \frac{1}{2} - \frac{1}{2}x$
14. $\frac{1}{3}x = 2 - \frac{2}{3}x$
15. $\frac{1}{2}x - 5 = \frac{3}{4}x$
16. $1 - \frac{1}{2}x = 6$
17. $\frac{2}{3}p = \frac{1}{2}p + \frac{1}{3}$
18. $\frac{1}{2} - \frac{1}{3}p = \frac{4}{3}$

In Problems 19–38, find the real solutions of each equation by factoring.

19. $x^2 - 7x + 12 = 0$
20. $x^2 - x - 6 = 0$
21. $2x^2 + 5x - 3 = 0$
22. $3x^2 + 5x - 2 = 0$
23. $x^2 - 9 = 0$
24. $x^2 + 9 = 0$
25. $x^3 - 27 = 0$
26. $x^3 + 27 = 0$
27. $x^3 + x^2 - 20x = 0$
28. $x^3 + 6x^2 - 7x = 0$
29. $4x^2 = x$
30. $x^2 = 4x$
31. $x^3 = 9x$
32. $4x^3 = x$
33. $x^4 = x^2$
34. $x^4 = 4x^2$
35. $x^3 + x^2 + x + 1 = 0$
36. $x^3 + 4x^2 + x + 4 = 0$
37. $x^3 - 2x^2 - 4x + 8 = 0$
38. $x^3 - 3x^2 - x + 3 = 0$

7 RATIONAL EXPRESSIONS

OBJECTIVES
1. Find the Domain of a Variable
2. Reduce a Rational Expression to Lowest Terms
3. Multiply and Divide Rational Expressions
4. Add and Subtract Rational Expressions
5. Use the Least Common Multiple Method
6. Simplify Mixed Quotients

If we form the quotient of two polynomials, the result is called a **rational expression**. Some examples of rational expressions are

$$\text{(a)}\ \frac{x^3+1}{x} \qquad \text{(b)}\ \frac{3x^2+x-2}{x^2+5} \qquad \text{(c)}\ \frac{x}{x^2-1} \qquad \text{(d)}\ \frac{9x}{(x-3)^2}$$

Rational expressions are described in the same manner as rational numbers. Thus, in expression (a), the polynomial $x^3 + 1$ is called the **numerator**, and x is called the **denominator**. When the numerator and denominator of a rational expression contain no common factors (except 1 and -1), we say that the rational expression is **reduced to lowest terms**, or **simplified**.

1 The polynomial in the denominator of a rational expression cannot be equal to 0 because division by 0 is not defined. For example, for the expression $\frac{x^3+1}{x}$, the variable x cannot take on the value 0.

> The set of values that a variable in an expression may assume is called the **domain** of the variable.

EXAMPLE 1 Finding the Domain of a Variable

The domain of the variable x in the rational expression

$$\frac{5}{x-2}$$

is $\{x \mid x \neq 2\}$, since, if $x = 2$, the denominator becomes 0, which is not defined. ■

In describing the domain of a variable, we may use either set notation or words, whichever is more convenient.

NOW WORK PROBLEM 1.

2 A rational expression is reduced to lowest terms by factoring completely the numerator and the denominator and canceling any common factors by using the cancellation property.

$$\frac{a\cancel{c}}{b\cancel{c}} = \frac{a}{b} \qquad \text{if } b \neq 0,\quad c \neq 0 \tag{1}$$

EXAMPLE 2

Reducing Rational Expressions to Lowest Terms

Reduce each rational expression to lowest terms.

(a) $\dfrac{x^3 - 8}{x^3 - 2x^2}$ (b) $\dfrac{8 - 2x}{x^2 - x - 12}$

Solution (a) $\dfrac{x^3 - 8}{x^3 - 2x^2} = \dfrac{\cancel{(x-2)}(x^2 + 2x + 4)}{x^2\cancel{(x-2)}} = \dfrac{x^2 + 2x + 4}{x^2}, \quad x \neq 0, 2$

(b) $\dfrac{8 - 2x}{x^2 - x - 12} = \dfrac{2(4 - x)}{(x - 4)(x + 3)} = \dfrac{2(-1)\cancel{(x-4)}}{\cancel{(x-4)}(x + 3)} = \dfrac{-2}{x + 3},$

$x \neq -3, 4$ ■

WARNING: Apply the cancellation property only to rational expressions written in factored form. Be sure to cancel only common factors! ■

NOW WORK PROBLEM 9.

Multiplying and Dividing Rational Expressions

3 The rules for multiplying and dividing rational expressions are the same as the rules for multiplying and dividing rational numbers. If $\dfrac{a}{b}$ and $\dfrac{c}{d}$, $b \neq 0$, $d \neq 0$, are two rational expressions, then

$$\frac{a}{b} \cdot \frac{c}{d} = \frac{ac}{bd} \quad \text{if } b \neq 0, d \neq 0 \qquad \textbf{(2)}$$

$$\frac{\frac{a}{b}}{\frac{c}{d}} = \frac{a}{b} \cdot \frac{d}{c} = \frac{ad}{bc} \quad \text{if } b \neq 0, c \neq 0, d \neq 0 \qquad \textbf{(3)}$$

In using equations (2) and (3) with rational expressions, be sure first to factor each polynomial completely so that common factors can be canceled. Leave your answer in factored form.

EXAMPLE 3

Multiplying and Dividing Rational Expressions

Perform the indicated operation and simplify the result. Leave your answer in factored form.

(a) $\dfrac{x^2 - 2x + 1}{x^3 + x} \cdot \dfrac{4x^2 + 4}{x^2 + x - 2}$ (b) $\dfrac{\dfrac{x + 3}{x^2 - 4}}{\dfrac{x^2 - x - 12}{x^3 - 8}}$

Solution (a)

$$\frac{x^2-2x+1}{x^3+x}\cdot\frac{4x^2+4}{x^2+x-2}=\frac{(x-1)^2}{x(x^2+1)}\cdot\frac{4(x^2+1)}{(x+2)(x-1)}$$

$$=\frac{(x-1)^{\cancel{2}}(4)\cancel{(x^2+1)}}{x\cancel{(x^2+1)}(x+2)\cancel{(x-1)}}$$

$$=\frac{4(x-1)}{x(x+2)},\qquad x\neq -2, 0, 1$$

(b)

$$\frac{\dfrac{x+3}{x^2-4}}{\dfrac{x^2-x-12}{x^3-8}}=\frac{x+3}{x^2-4}\cdot\frac{x^3-8}{x^2-x-12}$$

$$=\frac{x+3}{(x-2)(x+2)}\cdot\frac{(x-2)(x^2+2x+4)}{(x-4)(x+3)}$$

$$=\frac{\cancel{(x+3)}\cancel{(x-2)}(x^2+2x+4)}{\cancel{(x-2)}(x+2)(x-4)\cancel{(x+3)}}$$

$$=\frac{x^2+2x+4}{(x+2)(x-4)},\qquad x\neq -3,-2,2,4$$

NOW WORK PROBLEMS 23 AND 29.

ADDING AND SUBTRACTING RATIONAL EXPRESSIONS

4 The rules for adding and subtracting rational expressions are the same as the rules for adding and subtracting rational numbers. If the denominators of two rational expressions to be added (or subtracted) are equal, we add (or subtract) the numerators and keep the common denominator.

If $\frac{a}{b}$ and $\frac{c}{b}$ are two rational expressions, then

$$\frac{a}{b}+\frac{c}{b}=\frac{a+c}{b}\qquad \frac{a}{b}-\frac{c}{b}=\frac{a-c}{b}\qquad \text{if } b\neq 0 \tag{4}$$

EXAMPLE 4 **Adding and Subtracting Rational Expressions with Equal Denominators**

Perform the indicated operation and simplify the result. Leave your answer in factored form.

(a) $\dfrac{2x^2-4}{2x+5}+\dfrac{x+3}{2x+5},\quad x\neq -\dfrac{5}{2}$ (b) $\dfrac{x}{x-3}-\dfrac{3x+2}{x-3},\quad x\neq 3$

Solution (a)

$$\frac{2x^2-4}{2x+5}+\frac{x+3}{2x+5}=\frac{(2x^2-4)+(x+3)}{2x+5}$$

$$=\frac{2x^2+x-1}{2x+5}=\frac{(2x-1)(x+1)}{(2x+5)}$$

$$\text{(b)}\quad \frac{x}{x-3} - \frac{3x+2}{x-3} = \frac{x-(3x+2)}{x-3} = \frac{x-3x-2}{x-3}$$
$$= \frac{-2x-2}{x-3} = \frac{-2(x+1)}{x-3}$$

■

If the denominators of two rational expressions to be added or subtracted are not equal, we can use the general formulas given next.

$$\frac{a}{b} + \frac{c}{d} = \frac{a \cdot d}{b \cdot d} + \frac{b \cdot c}{b \cdot d} = \frac{ad + bc}{bd} \quad \text{if } b \neq 0, d \neq 0 \qquad \textbf{(5)}$$

$$\frac{a}{b} - \frac{c}{d} = \frac{a \cdot d}{b \cdot d} - \frac{b \cdot c}{b \cdot d} = \frac{ad - bc}{bd} \quad \text{if } b \neq 0, d \neq 0 \qquad \textbf{(6)}$$

EXAMPLE 5 Adding and Subtracting Rational Expressions with Unequal Denominators

Perform the indicated operation and simplify the result. Leave your answer in factored form.

(a) $\dfrac{x-3}{x+4} + \dfrac{x}{x-2}, \quad x \neq -4, 2$ (b) $\dfrac{x^2}{x^2-4} - \dfrac{1}{x}, \quad x \neq -2, 0, 2$

Solution (a)

$$\frac{x-3}{x+4} + \frac{x}{x-2} \underset{(5)}{=} \frac{x-3}{x+4} \cdot \frac{x-2}{x-2} + \frac{x+4}{x+4} \cdot \frac{x}{x-2}$$
$$= \frac{(x-3)(x-2) + (x+4)(x)}{(x+4)(x-2)}$$
$$= \frac{x^2 - 5x + 6 + x^2 + 4x}{(x+4)(x-2)} = \frac{2x^2 - x + 6}{(x+4)(x-2)}$$

(b)

$$\frac{x^2}{x^2-4} - \frac{1}{x} = \frac{x^2}{x^2-4} \cdot \frac{x}{x} - \frac{x^2-4}{x^2-4} \cdot \frac{1}{x} = \frac{x^2(x) - (x^2-4)(1)}{(x^2-4)(x)}$$
$$= \frac{x^3 - x^2 + 4}{(x-2)(x+2)(x)}$$

■

NOW WORK PROBLEM 43.

LEAST COMMON MULTIPLE (LCM)

5 If the denominators of two rational expressions to be added (or subtracted) have common factors, we usually do not use the general rules given by equation (5). Just as with fractions, we apply the **least common multiple (LCM) method.** The LCM method uses the polynomial of least degree that contains each denominator polynomial as a factor.

THE LCM METHOD FOR ADDING OR SUBTRACTING RATIONAL EXPRESSIONS

The Least Common Multiple Method (LCM) requires four steps:

STEP 1: Factor completely the polynomial in the denominator of each rational expression.

STEP 2: The LCM of the denominator is the product of each of these factors raised to a power equal to the greatest number of times that the factor occurs in the polynomials.

STEP 3: Write each rational expression using the LCM as the common denominator.

STEP 4: Add or subtract the rational expressions using equation (4).

Let's work an example that only requires Steps 1 and 2.

EXAMPLE 6 Finding the Least Common Multiple

Find the least common multiple of the following pair of polynomials:

$$x(x-1)^2(x+1) \quad \text{and} \quad 4(x-1)(x+1)^3$$

Solution **STEP 1:** The polynomials are already factored completely as

$$x(x-1)^2(x+1) \quad \text{and} \quad 4(x-1)(x+1)^3$$

STEP 2: Start by writing the factors of the left-hand polynomial. (Or you could start with the one on the right.)

$$x(x-1)^2(x+1)$$

Now look at the right-hand polynomial. Its first factor, 4, does not appear in our list, so we insert it.

$$4x(x-1)^2(x+1)$$

The next factor, $x - 1$, is already in our list, so no change is necessary. The final factor is $(x+1)^3$. Since our list has $x + 1$ to the first power only, we replace $x + 1$ in the list by $(x+1)^3$. The LCM is

$$4x(x-1)^2(x+1)^3$$

■

Notice that the LCM is, in fact, the polynomial of least degree that contains $x(x-1)^2(x+1)$ and $4(x-1)(x+1)^3$ as factors.

EXAMPLE 7 Using the Least Common Multiple to Add Rational Expressions

Perform the indicated operation and simplify the result. Leave your answer in factored form.

$$\frac{x}{x^2+3x+2} + \frac{2x-3}{x^2-1}, \qquad x \neq -2, -1, 1$$

Solution **STEP 1:** Factor completely the polynomials in the denominators.

$$x^2 + 3x + 2 = (x+2)(x+1)$$
$$x^2 - 1 = (x-1)(x+1)$$

STEP 2: The LCM is $(x + 2)(x + 1)(x - 1)$. Do you see why?

STEP 3: Write each rational expression using the LCM as the denominator.

$$\frac{x}{x^2 + 3x + 2} = \frac{x}{(x + 2)(x + 1)} = \frac{x}{(x + 2)(x + 1)} \cdot \frac{x - 1}{x - 1} = \frac{x(x - 1)}{(x + 2)(x + 1)(x - 1)}$$

↑ Multiply numerator and denominator by $x - 1$ to get the LCM in the denominator.

$$\frac{2x - 3}{x^2 - 1} = \frac{2x - 3}{(x - 1)(x + 1)} = \frac{2x - 3}{(x - 1)(x + 1)} \cdot \frac{x + 2}{x + 2} = \frac{(2x - 3)(x + 2)}{(x - 1)(x + 1)(x + 2)}$$

↑ Multiply numerator and denominator by $x + 2$ to get the LCM in the denominator.

STEP 4: Now we can add by using equation (4).

$$\frac{x}{x^2 + 3x + 2} + \frac{2x - 3}{x^2 - 1} = \frac{x(x - 1)}{(x + 2)(x + 1)(x - 1)} + \frac{(2x - 3)(x + 2)}{(x + 2)(x + 1)(x - 1)}$$

$$= \frac{(x^2 - x) + (2x^2 + x - 6)}{(x + 2)(x + 1)(x - 1)}$$

$$= \frac{3x^2 - 6}{(x + 2)(x + 1)(x - 1)} = \frac{3(x^2 - 2)}{(x + 2)(x + 1)(x - 1)}$$ ■

NOW WORK PROBLEM 53.

MIXED QUOTIENTS

6 When sums and/or differences of rational expressions appear as the numerator and/or denominator of a quotient, the quotient is called a **mixed quotient.*** For example,

$$\frac{1 + \frac{1}{x}}{1 - \frac{1}{x}} \quad \text{and} \quad \frac{\frac{x^2}{x^2 - 4} - 3}{\frac{x - 3}{x + 2} - 1}$$

are mixed quotients. To **simplify** a mixed quotient means to write it as a rational expression reduced to lowest terms. This can be accomplished in either of two ways.

SIMPLIFYING A MIXED QUOTIENT

METHOD 1: Treat the numerator and denominator of the mixed quotient separately, performing whatever operations are indicated and simplifying the results. Follow this by simplifying the resulting rational expression.

METHOD 2: Find the LCM of the denominators of all rational expressions that appear in the mixed quotient. Multiply the numerator and denominator of the mixed quotient by the LCM and simplify the result.

We will use both methods in the next example. By carefully studying each method, you can discover situations in which one method may be easier to use than the other.

*Some texts use the term **complex fraction.**

EXAMPLE 8 Simplifying a Mixed Quotient

Simplify: $\dfrac{\dfrac{1}{2}+\dfrac{3}{x}}{\dfrac{x+3}{4}}, \qquad x \neq -3, 0$

Solution METHOD 1: First, we perform the indicated operation in the numerator, and then we divide.

$$\frac{\frac{1}{2}+\frac{3}{x}}{\frac{x+3}{4}} = \frac{\frac{1\cdot x+2\cdot 3}{2\cdot x}}{\frac{x+3}{4}} = \frac{\frac{x+6}{2x}}{\frac{x+3}{4}} = \frac{x+6}{2x}\cdot\frac{4}{x+3}$$

↑ Rule for adding quotients ↑ Rule for dividing quotients

$$= \frac{(x+6)\cdot 4}{2\cdot x\cdot(x+3)} = \frac{\cancel{2}\cdot 2\cdot(x+6)}{\cancel{2}\cdot x\cdot(x+3)} = \frac{2(x+6)}{x(x+3)}$$

↑ Rule for multiplying quotients

METHOD 2: The rational expressions that appear in the mixed quotient are

$$\frac{1}{2}, \quad \frac{3}{x}, \quad \frac{x+3}{4}$$

The LCM of their denominators is $4x$. We multiply the numerator and denominator of the mixed quotient by $4x$ and then simplify.

$$\frac{\frac{1}{2}+\frac{3}{x}}{\frac{x+3}{4}} = \frac{4x\cdot\left(\frac{1}{2}+\frac{3}{x}\right)}{4x\cdot\left(\frac{x+3}{4}\right)} = \frac{4x\cdot\frac{1}{2}+4x\cdot\frac{3}{x}}{\frac{4x\cdot(x+3)}{4}}$$

↑ Multiply by $4x$ ↑ Distributive property in numerator

$$= \frac{\cancel{2}\cdot 2x\cdot\frac{1}{\cancel{2}}+4\cancel{x}\cdot\frac{3}{\cancel{x}}}{\frac{\cancel{4}x\cdot(x+3)}{\cancel{4}}} = \frac{2x+12}{x(x+3)} = \frac{2(x+6)}{x(x+3)}$$

↑ Simplify. ↑ Factor. ■

NOW WORK PROBLEM 57.

7 EXERCISES

In Problems 1–8, determine which of the value(s) given, if any, must be excluded from the domain of the variable in each rational expression:

(a) $x = 3$ *(b)* $x = 1$ *(c)* $x = 0$ *(d)* $x = -1$

1. $\dfrac{x^2-1}{x}$ **2.** $\dfrac{x^2+1}{x}$ **3.** $\dfrac{x}{x^2-9}$ **4.** $\dfrac{x}{x^2+9}$

5. $\dfrac{x^2}{x^2+1}$ **6.** $\dfrac{x^3}{x^2-1}$ **7.** $\dfrac{x^2+5x-10}{x^3-x}$ **8.** $\dfrac{-9x^2-x+1}{x^3+x}$

In Problems 9–22, reduce each rational expression to lowest terms.

9. $\dfrac{3x+9}{x^2-9}$

10. $\dfrac{4x^2+8x}{12x+24}$

11. $\dfrac{x^2-2x}{3x-6}$

12. $\dfrac{15x^2+24x}{3x^2}$

13. $\dfrac{24x^2}{12x^2-6x}$

14. $\dfrac{x^2+4x+4}{x^2-16}$

15. $\dfrac{x^2+4x-5}{x^2-2x+1}$

16. $\dfrac{x-x^2}{x^2+x-2}$

17. $\dfrac{x^2-4}{3x^2+5x-2}$

18. $\dfrac{2x^2+7x+3}{9-x^2}$

19. $\dfrac{2x^2-3x-2}{2-x}$

20. $\dfrac{2x^2+5x-3}{1-2x}$

21. $\dfrac{(x+5)^2-2(x+5)(x-2)}{(x+5)^4}$

22. $\dfrac{6x(x^2+5)^2-12x^3(x^2+5)}{(x^2+5)^4}$

In Problems 23–34, perform the indicated operation and simplify the result. Leave your answer in factored form.

23. $\dfrac{3x+6}{5x^2}\cdot\dfrac{x}{x^2-4}$

24. $\dfrac{3}{2x}\cdot\dfrac{x^2}{6x+10}$

25. $\dfrac{4x^2}{x^2-16}\cdot\dfrac{x-4}{2x}$

26. $\dfrac{12}{x^2-x}\cdot\dfrac{x^2-1}{4x-2}$

27. $\dfrac{2x^2+3x-2}{x^2+2x-35}\cdot\dfrac{x^2-49}{2x^2+5x+2}$

28. $\dfrac{6x^2+x-1}{x^2+4x-5}\cdot\dfrac{x^2-25}{2x^2+5x+1}$

29. $\dfrac{\dfrac{6x}{x^2-4}}{\dfrac{3x-9}{2x+4}}$

30. $\dfrac{\dfrac{12x}{5x+20}}{\dfrac{4x^2}{x^2-16}}$

31. $\dfrac{\dfrac{x^2+7x+12}{x^2-7x+12}}{\dfrac{x^2+x-12}{x^2-x-12}}$

32. $\dfrac{\dfrac{x^2+7x+6}{x^2+x-6}}{\dfrac{x^2+5x-6}{x^2+5x+6}}$

33. $\dfrac{\dfrac{2x^2-x-28}{3x^2-x-2}}{\dfrac{4x^2+16x+7}{3x^2+11x+6}}$

34. $\dfrac{\dfrac{9x^2+3x-2}{12x^2+5x-2}}{\dfrac{9x^2-6x+1}{8x^2-10x-3}}$

In Problems 35–48, perform the indicated operations and simplify the result. Leave your answer in factored form.

35. $\dfrac{x+1}{x-3}+\dfrac{2x-3}{x-3}$

36. $\dfrac{2x-5}{3x+2}+\dfrac{x+4}{3x+2}$

37. $\dfrac{3x+5}{2x-1}-\dfrac{2x-4}{2x-1}$

38. $\dfrac{5x-4}{3x+4}-\dfrac{x+1}{3x+4}$

39. $\dfrac{4}{x-2}+\dfrac{x}{2-x}$

40. $\dfrac{6}{x-1}-\dfrac{x}{1-x}$

41. $\dfrac{4}{x-1}-\dfrac{2}{x+2}$

42. $\dfrac{2}{x+5}-\dfrac{5}{x-5}$

43. $\dfrac{x}{x+1}+\dfrac{2x-3}{x-1}$

44. $\dfrac{3x}{x-4}+\dfrac{2x}{x+3}$

45. $\dfrac{x-3}{x+2}-\dfrac{x+4}{x-2}$

46. $\dfrac{2x-3}{x-1}-\dfrac{2x+1}{x+1}$

47. $\dfrac{x^3}{(x-1)^2}-\dfrac{x^2+1}{x}$

48. $\dfrac{3x^2}{4}-\dfrac{x^3}{x^2-1}$

In Problems 49–56, perform the indicated operations and simplify the result. Leave your answer in factored form.

49. $\dfrac{x}{x^2-7x+6}-\dfrac{x}{x^2-2x-24}$

50. $\dfrac{x}{x-3}-\dfrac{x+1}{x^2+5x-24}$

51. $\dfrac{4x}{x^2-4}-\dfrac{2}{x^2+x-6}$

52. $\dfrac{3x}{x-1}-\dfrac{x-4}{x^2-2x+1}$

53. $\dfrac{x+4}{x^2-x-2}-\dfrac{2x+3}{x^2+2x-8}$

54. $\dfrac{2x-3}{x^2+8x+7}-\dfrac{x-2}{(x+1)^2}$

55. $\dfrac{1}{h}\left(\dfrac{1}{x+h}-\dfrac{1}{x}\right)$

56. $\dfrac{1}{h}\left[\dfrac{1}{(x+h)^2}-\dfrac{1}{x^2}\right]$

In Problems 57–64, perform the indicated operations and simplify the result. Leave your answer in factored form.

57. $\dfrac{1 + \dfrac{1}{x}}{1 - \dfrac{1}{x}}$

58. $\dfrac{4 + \dfrac{1}{x^2}}{3 - \dfrac{1}{x^2}}$

59. $\dfrac{x - \dfrac{1}{x}}{x + \dfrac{1}{x}}$

60. $\dfrac{1 - \dfrac{x}{x + 1}}{2 - \dfrac{x - 1}{x}}$

61. $\dfrac{\dfrac{x + 4}{x - 2} - \dfrac{x - 3}{x + 1}}{x + 1}$

62. $\dfrac{\dfrac{x - 2}{x + 1} - \dfrac{x}{x - 2}}{x + 3}$

63. $\dfrac{\dfrac{x - 2}{x + 2} + \dfrac{x - 1}{x + 1}}{\dfrac{x}{x + 1} - \dfrac{2x - 3}{x}}$

64. $\dfrac{\dfrac{2x + 5}{x} - \dfrac{x}{x - 3}}{\dfrac{x^2}{x - 3} - \dfrac{(x + 1)^2}{x + 3}}$

65. **The Lensmaker's Equation** The focal length f of a lens with index of refraction n is

$$\frac{1}{f} = (n - 1)\left[\frac{1}{R_1} + \frac{1}{R_2}\right]$$

where R_1 and R_2 are the radii of curvature of the front and back surfaces of the lens. Express f as a rational expression. Evaluate the rational expression for $n = 1.5$, $R_1 = 0.1$ meter, and $R_2 = 0.2$ meter.

66. **Electrical Circuits** An electrical circuit contains three resistors connected in parallel. If the resistance of each is R_1, R_2, and R_3 ohms, respectively, their combined resistance R is given by the formula

$$\frac{1}{R} = \frac{1}{R_1} + \frac{1}{R_2} + \frac{1}{R_3}$$

Express R as a rational expression. Evaluate R for $R_1 = 5$ ohms, $R_2 = 4$ ohms, and $R_3 = 10$ ohms.

67. The following expressions are called **continued fractions:**

$$1 + \frac{1}{x}, \quad 1 + \frac{1}{1 + \frac{1}{x}},$$

$$1 + \frac{1}{1 + \frac{1}{1 + \frac{1}{x}}}, \quad 1 + \frac{1}{1 + \frac{1}{1 + \frac{1}{1 + \frac{1}{x}}}}, \ldots$$

Each simplifies to an expression of the form

$$\frac{ax + b}{bx + c}$$

Trace the successive values of a, b, and c as you "continue" the fraction. Can you discover the patterns that these values follow? Go to the library and research Fibonacci numbers. Write a report on your findings.

68. Explain to a fellow student when you would use the LCM method to add two rational expressions. Give two examples of adding two rational expressions, one in which you use the LCM and the other in which you do not.

69. Which of the two methods given in the text for simplifying mixed quotients do you prefer? Write a brief paragraph stating the reasons for your choice.

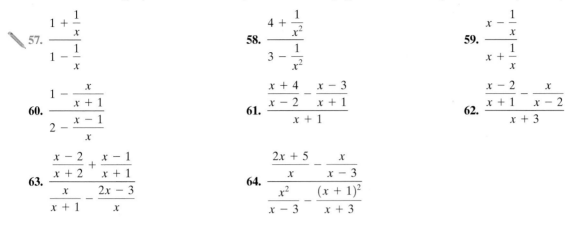

8 RADICALS; RATIONAL EXPONENTS

OBJECTIVES
1. Work with Square Roots and *n*th Roots
2. Simplify Radicals
3. Rationalize Denominators
4. Simplify Expressions with Rational Exponents

SQUARE ROOTS

1 A real number is squared when it is raised to the power 2. The inverse of squaring is finding a **square root**. For example, since $6^2 = 36$ and $(-6)^2 = 36$, the numbers 6 and -6 are square roots of 36.

The symbol $\sqrt{\ }$, called a **radical sign**, is used to denote the **principal**, or nonnegative, square root. For example, $\sqrt{36} = 6$.

> In general, if a is a nonnegative real number, the nonnegative number b such that $b^2 = a$ is the **principal square root** of a and is denoted by $b = \sqrt{a}$.

The following comments are noteworthy:

1. Negative numbers do not have square roots (in the real number system), because the square of any real number is *nonnegative*. For example, $\sqrt{-4}$ is not a real number, because there is no real number whose square is -4.
2. The principal square root of 0 is 0, since $0^2 = 0$. That is, $\sqrt{0} = 0$.
3. The principal square root of a positive number is positive.
4. If $c \geq 0$, then $(\sqrt{c})^2 = c$. For example, $(\sqrt{2})^2 = 2$ and $(\sqrt{3})^2 = 3$.

EXAMPLE 1 **Evaluating Square Roots**

(a) $\sqrt{64} = 8$ (b) $\sqrt{\frac{1}{16}} = \frac{1}{4}$

(c) $(\sqrt{1.4})^2 = 1.4$ (d) $\sqrt{(-3)^2} = |-3| = 3$ ■

Examples 1(a) and (b) are examples of **perfect square roots**. Thus, 64 is a **perfect square**, since $64 = 8^2$; and $\frac{1}{16}$ is a perfect square, since $\frac{1}{16} = \left(\frac{1}{4}\right)^2$.

Notice the need for the absolute value in Example 1(d). Since $a^2 \geq 0$, the principal square root of a^2 is defined whether $a > 0$ or $a < 0$. However, since the principal square root is nonnegative, we need the absolute value to ensure the nonnegative result.

In general, we have

$$\sqrt{a^2} = |a| \qquad \textbf{(1)}$$

EXAMPLE 2 **Using Equation (1)**

(a) $\sqrt{(2.3)^2} = |2.3| = 2.3$ (b) $\sqrt{(-2.3)^2} = |-2.3| = 2.3$

(c) $\sqrt{x^2} = |x|$ ■

*n*th ROOTS

The **principal *n*th root of a real number *a***, symbolized by $\sqrt[n]{a}$, where $n \geq 2$ is an integer, is defined as follows:

$$\sqrt[n]{a} = b \quad \text{means} \quad a = b^n$$

where $a \geq 0$ and $b \geq 0$ if $n \geq 2$ is even and a, b are any real numbers if $n \geq 3$ is odd.

Notice that if a is negative and n is even then $\sqrt[n]{a}$ is not defined. When it is defined, the principal nth root of a number is unique.

The symbol $\sqrt[n]{a}$ for the principal nth root of a is sometimes called a **radical**; the integer n is called the **index**, and a is called the **radicand**. If the index of a radical is 2, we call $\sqrt[2]{a}$ the **square root** of a and omit the index 2 by simply writing $\sqrt{a}$. If the index is 3, we call $\sqrt[3]{a}$ the **cube root** of a.

EXAMPLE 3 **Evaluating Principal *n*th Roots**

(a) $\sqrt[3]{8} = \sqrt[3]{2^3} = 2$

(b) $\sqrt[3]{-64} = \sqrt[3]{(-4)^3} = -4$

(c) $\sqrt[4]{\frac{1}{16}} = \sqrt[4]{\left(\frac{1}{2}\right)^4} = \frac{1}{2}$

(d) $\sqrt{(-2)^2} = |-2| = 2$ ■

Equations (a), (b), and (c) are examples of **perfect roots**. Thus, 8 and -64 are perfect cubes, since $8 = 2^3$ and $-64 = (-4)^3$; $\frac{1}{2}$ is a perfect fourth root of $\frac{1}{16}$, since $\frac{1}{16} = \left(\frac{1}{2}\right)^4$.

Notice the need for the absolute value in Example 3(d). If n is even, then a^n is positive whether $a > 0$ or $a < 0$. But the principal nth root must be nonnegative. Hence, the reason for using the absolute value—it gives a nonnegative result.

In general, if $n \geq 2$ is a positive integer and a is a real number, we have

$$\sqrt[n]{a^n} = a, \quad \text{if } n \geq 3 \text{ is odd} \tag{2a}$$

$$\sqrt[n]{a^n} = |a|, \quad \text{if } n \geq 2 \text{ is even} \tag{2b}$$

PROPERTIES OF RADICALS

Let $n \geq 2$ and $m \geq 2$ denote positive integers, and let a and b represent real numbers. Assuming that all radicals are defined, we have the following properties:

$$\sqrt[n]{ab} = \sqrt[n]{a}\,\sqrt[n]{b} \tag{3a}$$

$$\sqrt[n]{\frac{a}{b}} = \frac{\sqrt[n]{a}}{\sqrt[n]{b}} \tag{3b}$$

$$\sqrt[n]{a^m} = \left(\sqrt[n]{a}\right)^m \tag{3c}$$

$$\sqrt[m]{\sqrt[n]{a}} = \sqrt[mn]{a} \tag{3d}$$

2 When used in reference to radicals, the direction to "simplify" will mean to remove from the radicals any perfect roots that occur as factors. Let's look

at some examples of how the rules listed in the preceding boxes are applied to simplify radicals.

EXAMPLE 4 Simplifying Radicals

Simplify each expression. Assume that all variables are positive when they appear.

(a) $\sqrt{32}$ (b) $\sqrt[3]{8x^4}$ (c) $\sqrt{\sqrt[3]{x^7}}$

Solution (a) $\sqrt{32} = \sqrt{16 \cdot 2} = \sqrt{16}\sqrt{2} = 4\sqrt{2}$

↑ 16 is a perfect square. ↑ (3a)

(b) $\sqrt[3]{8x^4} = \sqrt[3]{8x^3 \cdot x} = \sqrt[3]{(2x)^3 \cdot x} = \sqrt[3]{(2x)^3}\sqrt[3]{x} = 2x\sqrt[3]{x}$

↑ Factor out perfect cube. ↑ (3a) ↑ (2a)

(c) $\sqrt{\sqrt[3]{x^7}} = \sqrt[6]{x^7} = \sqrt[6]{x^6 \cdot x} = \sqrt[6]{x^6} \cdot \sqrt[6]{x} = |x|\sqrt[6]{x}$

↑ (3d) ↑ (2b)

■

NOW WORK PROBLEMS 1 AND 15.

RATIONALIZING

3 When radicals occur in quotients, it has become common practice to rewrite the quotient so that the denominator contains no radicals. This process is referred to as **rationalizing the denominator**.

The idea is to find an appropriate expression so that, when it is multiplied by the radical in the denominator, the new denominator that results contains no radicals. For example,

If Radical Is	Multiply by	To Get Product Free of Radicals
$\sqrt{3}$	$\sqrt{3}$	$\sqrt{9} = 3$
$\sqrt[3]{4}$	$\sqrt[3]{2}$	$\sqrt[3]{8} = 2$
$\sqrt{3} + 1$	$\sqrt{3} - 1$	$(\sqrt{3})^2 - 1^2 = 3 - 1 = 2$
$\sqrt{2} - 3$	$\sqrt{2} + 3$	$(\sqrt{2})^2 - 3^2 = 2 - 9 = -7$
$\sqrt{5} - \sqrt{3}$	$\sqrt{5} + \sqrt{3}$	$(\sqrt{5})^2 - (\sqrt{3})^2 = 5 - 3 = 2$

You are correct if you observed in this list that, after the second type of radical, the special product for differences of squares is the basis for determining by what to multiply.

EXAMPLE 5 Rationalizing Denominators

Rationalize the denominator of each expression.

(a) $\dfrac{4}{\sqrt{2}}$ (b) $\dfrac{\sqrt{3}}{\sqrt[3]{2}}$ (c) $\dfrac{\sqrt{x} - 2}{\sqrt{x} + 2}$, $x \geq 0$

Solution (a)
$$\frac{4}{\sqrt{2}} = \underset{\substack{\uparrow \\ \text{Multiply by } \frac{\sqrt{2}}{\sqrt{2}}}}{\frac{4}{\sqrt{2}} \cdot \frac{\sqrt{2}}{\sqrt{2}}} = \frac{4\sqrt{2}}{(\sqrt{2})^2} = \frac{4\sqrt{2}}{2} = 2\sqrt{2}$$

(b)
$$\frac{\sqrt{3}}{\sqrt[3]{2}} = \underset{\substack{\uparrow \\ \text{Multiply by } \frac{\sqrt[3]{4}}{\sqrt[3]{4}}}}{\frac{\sqrt{3}}{\sqrt[3]{2}} \cdot \frac{\sqrt[3]{4}}{\sqrt[3]{4}}} = \frac{\sqrt{3}\sqrt[3]{4}}{\sqrt[3]{8}} = \frac{\sqrt{3}\sqrt[3]{4}}{2}$$

(c)
$$\frac{\sqrt{x}-2}{\sqrt{x}+2} = \frac{\sqrt{x}-2}{\sqrt{x}+2} \cdot \frac{\sqrt{x}-2}{\sqrt{x}-2} = \frac{(\sqrt{x}-2)^2}{(\sqrt{x})^2 - 2^2}$$
$$= \frac{(\sqrt{x})^2 - 4\sqrt{x} + 4}{x-4} = \frac{x - 4\sqrt{x} + 4}{x-4}$$

In calculus, sometimes the numerator must be rationalized.

EXAMPLE 6 Rationalizing Numerators

Rationalize the numerator: $\dfrac{\sqrt{x}-2}{\sqrt{x}+1}, \quad x \geq 0$

Solution We multiply by $\dfrac{\sqrt{x}+2}{\sqrt{x}+2}$.

$$\frac{\sqrt{x}-2}{\sqrt{x}+1} = \frac{\sqrt{x}-2}{\sqrt{x}+1} \cdot \frac{\sqrt{x}+2}{\sqrt{x}+2} = \frac{(\sqrt{x})^2 - 2^2}{(\sqrt{x}+1)(\sqrt{x}+2)} = \frac{x-4}{x+3\sqrt{x}+2}$$

NOW WORK PROBLEM 29.

RATIONAL EXPONENTS

4 Radicals are used to define rational exponents.

If a is a real number and $n \geq 2$ is an integer, then

$$a^{1/n} = \sqrt[n]{a} \qquad \textbf{(4)}$$

provided that $\sqrt[n]{a}$ exists.

EXAMPLE 7 Using Equation (4)

(a) $4^{1/2} = \sqrt{4} = 2$

(b) $(-27)^{1/3} = \sqrt[3]{-27} = -3$

(c) $8^{1/2} = \sqrt{8} = 2\sqrt{2}$

(d) $16^{1/3} = \sqrt[3]{16} = 2\sqrt[3]{2}$

If a is a real number and m and n are integers containing no common factors with $n \geq 2$, then

$$a^{m/n} = \sqrt[n]{a^m} = \left(\sqrt[n]{a}\right)^m \tag{5}$$

provided that $\sqrt[n]{a}$ exists.

We have two comments about equation (5):

1. The exponent m/n must be in lowest terms and n must be positive.
2. In simplifying $a^{m/n}$, either $\sqrt[n]{a^m}$ or $\left(\sqrt[n]{a}\right)^m$ may be used. Generally, taking the root first is preferred.

It can be shown that the laws of exponents hold for rational exponents.

EXAMPLE 8 Simplifying Expressions with Rational Exponents

Simplify each expression. Express your answer so that only positive exponents occur. Assume that the variables are positive.

(a) $\left(\dfrac{2x^{1/3}}{y^{2/3}}\right)^{-3}$ (b) $\left(x^{2/3}y^{-3/4}\right)\left(x^{-2}y\right)^{1/2}$

Solution (a) $\left(\dfrac{2x^{1/3}}{y^{2/3}}\right)^{-3} = \left(\dfrac{y^{2/3}}{2x^{1/3}}\right)^{3} = \dfrac{\left(y^{2/3}\right)^3}{\left(2x^{1/3}\right)^3} = \dfrac{y^2}{2^3\left(x^{1/3}\right)^3} = \dfrac{y^2}{8x}$

(b)
$$\begin{aligned}\left(x^{2/3}y^{-3/4}\right)\left(x^{-2}y\right)^{1/2} &= \left(x^{2/3}y^{-3/4}\right)\left[\left(x^{-2}\right)^{1/2}y^{1/2}\right]\\ &= x^{2/3}y^{-3/4}x^{-1}y^{1/2} = \left(x^{2/3}x^{-1}\right)\left(y^{-3/4}y^{1/2}\right)\\ &= x^{-1/3}y^{-1/4} = \frac{1}{x^{1/3}y^{1/4}}\end{aligned}$$

■

NOW WORK PROBLEM 51.

The next two examples illustrate some algebra that you will need to know for certain calculus problems.

EXAMPLE 9 Writing an Expression as a Single Quotient

Write the following expression as a single quotient in which only positive exponents appear.

$$\left(x^2+1\right)^{1/2} + x \cdot \frac{1}{2}\left(x^2+1\right)^{-1/2} \cdot 2x$$

Solution

$$\begin{aligned}\left(x^2+1\right)^{1/2} + x \cdot \frac{1}{\cancel{2}}\left(x^2+1\right)^{-1/2} \cdot \cancel{2}x &= \left(x^2+1\right)^{1/2} + \frac{x^2}{\left(x^2+1\right)^{1/2}}\\ &= \frac{\left(x^2+1\right)^{1/2}\left(x^2+1\right)^{1/2} + x^2}{\left(x^2+1\right)^{1/2}}\\ &= \frac{\left(x^2+1\right) + x^2}{\left(x^2+1\right)^{1/2}}\\ &= \frac{2x^2+1}{\left(x^2+1\right)^{1/2}}\end{aligned}$$

■

EXAMPLE 10 Factoring an Expression Containing Rational Exponents

Factor: $4x^{1/3}(2x + 1) + 2x^{4/3}$

Solution We begin by looking for factors that are common to the two terms. Notice that 2 and $x^{1/3}$ are common factors. Thus,

$$\begin{aligned} 4x^{1/3}(2x + 1) + 2x^{4/3} &= 2x^{1/3}[2(2x + 1) + x] \\ &= 2x^{1/3}(5x + 2) \end{aligned}$$

8 EXERCISES

In Problems 1–18, simplify each expression. Assume that all variables are positive when they appear.

1. $\sqrt{8}$
2. $\sqrt[4]{32}$
3. $\sqrt[3]{16x^4}$
4. $\sqrt{27x^3}$
5. $\sqrt[3]{\sqrt{x^6}}$
6. $\sqrt{\sqrt{x^6}}$
7. $\sqrt{\dfrac{32x^3}{9x}}$
8. $\sqrt[3]{\dfrac{x}{8x^4}}$
9. $\sqrt[4]{x^{12}y^8}$
10. $\sqrt[5]{x^{10}y^5}$
11. $\sqrt[4]{\dfrac{x^9y^7}{xy^3}}$
12. $\sqrt[3]{\dfrac{3xy^2}{81x^4y^2}}$
13. $\sqrt{36x}$
14. $\sqrt{9x^5}$
15. $\sqrt{3x^2}\sqrt{12x}$
16. $\sqrt{5x}\sqrt{20x^3}$
17. $(\sqrt{5}\sqrt[3]{9})^2$
18. $(\sqrt[3]{3}\sqrt{10})^4$

In Problems 19–24, perform the indicated operation and simplify the result. Assume that all variables are positive when they appear.

19. $(3\sqrt{6})(2\sqrt{2})$
20. $(5\sqrt{8})(-3\sqrt{3})$
21. $(\sqrt{3} + 3)(\sqrt{3} - 1)$
22. $(\sqrt{5} - 2)(\sqrt{5} + 3)$
23. $(\sqrt{x} - 1)^2$
24. $(\sqrt{x} + \sqrt{5})^2$

In Problems 25–34, rationalize the denominator of each expression. Assume that all variables are positive when they appear.

25. $\dfrac{1}{\sqrt{2}}$
26. $\dfrac{6}{\sqrt[3]{4}}$
27. $\dfrac{-\sqrt{3}}{\sqrt{5}}$
28. $\dfrac{-\sqrt[3]{3}}{\sqrt{8}}$
29. $\dfrac{\sqrt{3}}{5 - \sqrt{2}}$
30. $\dfrac{\sqrt{2}}{\sqrt{7} + 2}$
31. $\dfrac{2 - \sqrt{5}}{2 + 3\sqrt{5}}$
32. $\dfrac{\sqrt{3} - 1}{2\sqrt{3} + 3}$
33. $\dfrac{\sqrt{x + h} - \sqrt{x}}{\sqrt{x + h} + \sqrt{x}}$
34. $\dfrac{\sqrt{x + h} + \sqrt{x - h}}{\sqrt{x + h} - \sqrt{x - h}}$

In Problems 35–46, simplify each expression.

35. $8^{2/3}$
36. $4^{3/2}$
37. $(-27)^{1/3}$
38. $16^{3/4}$
39. $16^{3/2}$
40. $64^{3/2}$
41. $9^{-3/2}$
42. $25^{-5/2}$
43. $\left(\dfrac{9}{8}\right)^{3/2}$
44. $\left(\dfrac{27}{8}\right)^{2/3}$
45. $\left(\dfrac{8}{9}\right)^{-3/2}$
46. $\left(\dfrac{8}{27}\right)^{-2/3}$

In Problems 47–54, simplify each expression. Express your answer so that only positive exponents occur. Assume that the variables are positive.

47. $x^{3/4}x^{1/3}x^{-1/2}$
48. $x^{2/3}x^{1/2}x^{-1/4}$
49. $(x^3y^6)^{1/3}$
50. $(x^4y^8)^{3/4}$
51. $(x^2y)^{1/3}(xy^2)^{2/3}$
52. $(xy)^{1/4}(x^2y^2)^{1/2}$
53. $(16x^2y^{-1/3})^{3/4}$
54. $(4x^{-1}y^{1/3})^{3/2}$

In Problems 55–60, write each expression as a single quotient in which only positive exponents and/or radicals appear.

55. $\dfrac{x}{(1+x)^{1/2}} + 2(1+x)^{1/2}$

56. $\dfrac{1+x}{2x^{1/2}} + x^{1/2}$

57. $\dfrac{\sqrt{1+x} - x \cdot \dfrac{1}{2\sqrt{1+x}}}{1+x}$

58. $\dfrac{\sqrt{x^2+1} - x \cdot \dfrac{2x}{2\sqrt{x^2+1}}}{x^2+1}$

59. $\dfrac{(x+4)^{1/2} - 2x(x+4)^{-1/2}}{x+4}$

60. $\dfrac{(9-x^2)^{1/2} + x^2(9-x^2)^{-1/2}}{9-x^2}$

In Problems 61–66, factor each expression.

61. $(x+1)^{3/2} + x \cdot \frac{3}{2}(x+1)^{1/2}$

62. $(x^2+4)^{4/3} + x \cdot \frac{4}{3}(x^2+4)^{1/3} \cdot 2x$

63. $6x^{1/2}(x^2+x) - 8x^{3/2} - 8x^{1/2}$

64. $6x^{1/2}(2x+3) + x^{3/2} \cdot 8$

65. $x\left(\frac{1}{2}\right)(8-x^2)^{-1/2}(-2x) + (8-x^2)^{1/2}$

66. $2x(1-x^2)^{3/2} + x^2\left(\frac{3}{2}\right)(1-x^2)^{1/2}(-2x)$

9 GEOMETRY REVIEW

OBJECTIVES
1. Use the Pythagorean Theorem and Its Converse
2. Know Geometry Formulas

In this section we review some topics studied in geometry that we shall need for our study of algebra.

Pythagorean Theorem

Figure 14

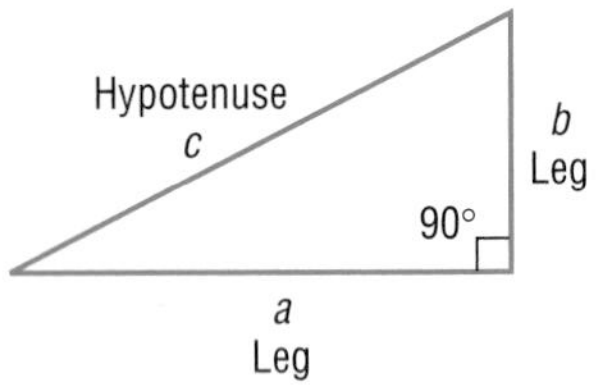

1 The *Pythagorean Theorem* is a statement about *right triangles*. A **right triangle** is one that contains a **right angle**, that is, an angle of 90°. The side of the triangle opposite the 90° angle is called the **hypotenuse**; the remaining two sides are called **legs**. In Figure 14 we have used c to represent the length of the hypotenuse and a and b to represent the lengths of the legs. Notice the use of the symbol ⌐ to show the 90° angle. We now state the Pythagorean Theorem.

Pythagorean Theorem

In a right triangle, the square of the length of the hypotenuse is equal to the sum of the squares of the lengths of the legs. That is, in the right triangle shown in Figure 14,

$$c^2 = a^2 + b^2 \qquad \textbf{(1)}$$

EXAMPLE 1 Finding the Hypotenuse of a Right Triangle

In a right triangle, one leg is of length 3 and the other is of length 9. What is the length of the hypotenuse?

Solution Since the triangle is a right triangle, we use the Pythagorean Theorem with $a = 3$ and $b = 9$ to find the length c of the hypotenuse. From equation (1), we have

$$c^2 = a^2 + b^2$$
$$c^2 = 3^2 + 9^2 = 9 + 81 = 90$$
$$c = \sqrt{90} = 3\sqrt{10}$$

NOW WORK PROBLEM 3.

The converse of the Pythagorean Theorem is also true.

Converse of the Pythagorean Theorem In a triangle, if the square of the length of one side equals the sum of the squares of the lengths of the other two sides, then the triangle is a right triangle. The 90° angle is opposite the longest side.

EXAMPLE 2 Verifying That a Triangle Is a Right Triangle

Show that a triangle whose sides are of lengths 5, 12, and 13 is a right triangle. Identify the hypotenuse.

Figure 15

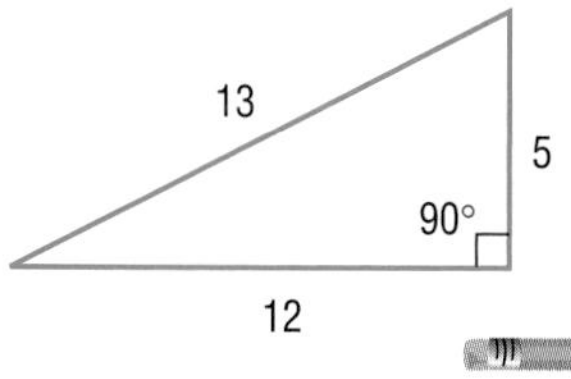

Solution We square the lengths of the sides.

$$5^2 = 25, \qquad 12^2 = 144, \qquad 13^2 = 169$$

Notice that the sum of the first two squares (25 and 144) equals the third square (169). Hence, the triangle is a right triangle. The longest side, 13, is the hypotenuse. See Figure 15.

NOW WORK PROBLEM 11.

EXAMPLE 3 Applying the Pythagorean Theorem

The tallest inhabited building in the world is the Sears Tower in Chicago.* If the observation tower is 1450 feet above ground level, how far can a person standing in the observation tower see (with the aid of a telescope)? Use 3960 miles for the radius of Earth. See Figure 16.

Figure 16

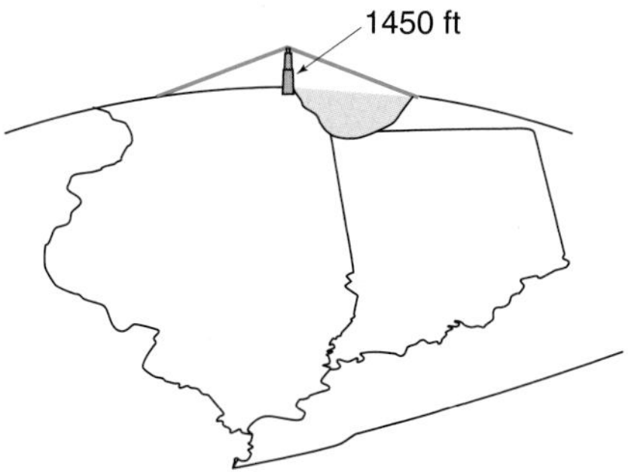

[**Note:** 1 mile = 5280 feet]

* *Source:* Council on Tall Buildings and Urban Habitat (1997): Sears Tower No. 1 for tallest roof (1450 ft) and tallest occupied floor (1431 ft).

Figure 17

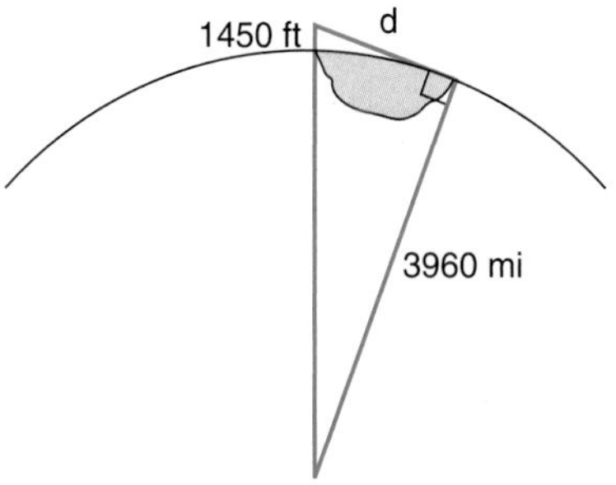

Solution From the center of Earth, draw two radii: one through the Sears Tower and the other to the farthest point a person can see from the tower. See Figure 17. Apply the Pythagorean Theorem to the right triangle. Since 1450 feet = 1450/5280 miles, we have

$$d^2 + (3960)^2 = \left(3960 + \frac{1450}{5280}\right)^2$$

$$d^2 = \left(3960 + \frac{1450}{5280}\right)^2 - (3960)^2 \approx 2175.08$$

$$d \approx 46.64$$

A person can see about 47 miles from the observation tower. ■

Geometry Formulas

2 Certain formulas from geometry are useful in solving algebra problems. We list some of these formulas next.

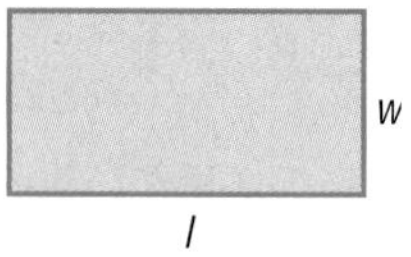

For a rectangle of length l and width w,

Area $= lw$ Perimeter $= 2l + 2w$

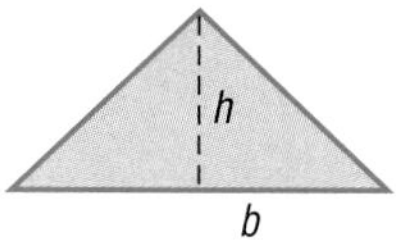

For a triangle with base b and altitude h,

Area $= \frac{1}{2}bh$

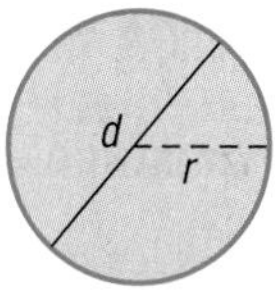

For a circle of radius r (diameter $d = 2r$),

Area $= \pi r^2$ Circumference $= 2\pi r = \pi d$

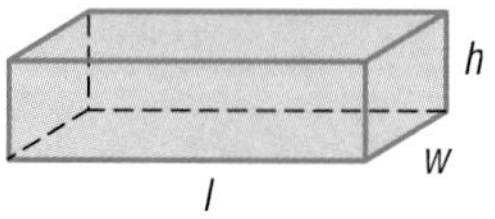

For a rectangular box of length l, width w, and height h,

Volume $= lwh$

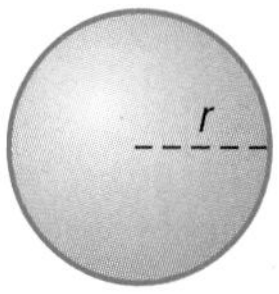

For a sphere of radius r,

Volume $= \frac{4}{3}\pi r^3$ Surface area $= 4\pi r^2$

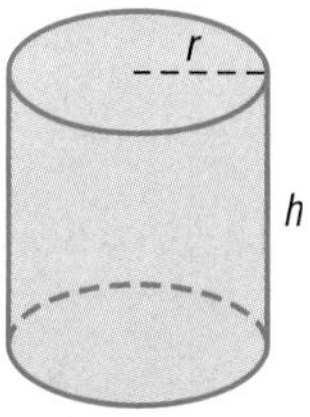

For a right circular cylinder of height h and radius r,

Volume $= \pi r^2 h$

EXAMPLE 4 Using Geometry Formulas

A Christmas tree ornament is in the shape of a semicircle on top of a triangle. How many square centimeters (cm^2) of copper is required to make the ornament if the height of the triangle is 6 cm and the base is 4 cm?

Figure 18

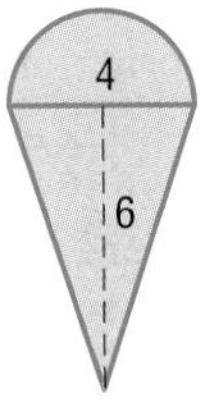

Solution See Figure 18. The amount of copper required equals the shaded area. This area is the sum of the area of the triangle and the semicircle. Thus,

$$\text{Area} = \text{Area of triangle} + \text{Area of semicircle}$$
$$= \tfrac{1}{2}bh + \tfrac{1}{2}\pi r^2 = \tfrac{1}{2}(4)(6) + \tfrac{1}{2}\pi 2^2 \qquad \text{Diameter of semicircle} = 4. \text{ Radius} = 2.$$
$$= 12 + 2\pi \approx 18.28 \text{ cm}^2$$

About 18.28 cm^2 of copper is required. ■

NOW WORK PROBLEM 33.

9 EXERCISES

In Problems 1–6, the lengths of the legs of a right triangle are given. Find the hypotenuse.

1. $a = 5, \ b = 12$
2. $a = 6, \ b = 8$
3. $a = 10, \ b = 24$
4. $a = 4, \ b = 3$
5. $a = 7, \ b = 24$
6. $a = 14, \ b = 48$

In Problems 7–14, the lengths of the sides of a triangle are given. Determine which are right triangles. For those that are, identify the hypotenuse.

7. 3, 4, 5
8. 6, 8, 10
9. 4, 5, 6
10. 2, 2, 3
11. 7, 24, 25
12. 10, 24, 26
13. 6, 4, 3
14. 5, 4, 7

15. Find the area A of a rectangle with length 4 inches and width 2 inches.

16. Find the area A of a rectangle with length 9 centimeters and width 4 centimeters.

17. Find the area A of a triangle with height 4 inches and base 2 inches.

18. Find the area A of a triangle with height 9 centimeters and base 4 centimeters.

19. Find the area A and circumference C of a circle of radius 5 meters.

20. Find the area A and circumference C of a circle of radius 2 feet.

21. Find the volume V of a rectangular box with length 8 feet, width 4 feet, and height 7 feet.

22. Find the volume V of a rectangular box with length 9 inches, width 4 inches, and height 8 inches.

23. Find the volume V and surface area S of a sphere of radius 4 centimeters.

24. Find the volume V and surface area S of a sphere of radius 3 feet.

25. Find the volume V of a right circular cylinder with radius 9 inches and height 8 inches.

26. Find the volume V of a right circular cylinder with radius 8 inches and height 9 inches.

In Problems 27–30, find the area of the shaded region.

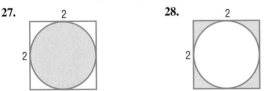

27.

28.

29.

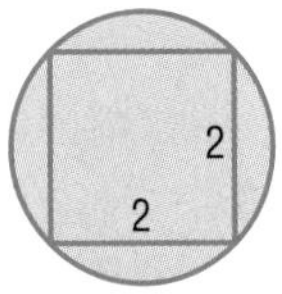

30.

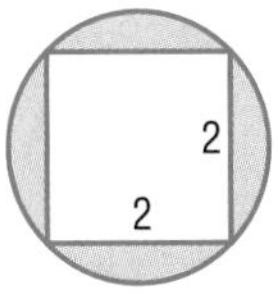

31. How many feet does a wheel with a diameter of 16 inches travel after four revolutions?

32. How many revolutions will a circular disk with a diameter of 4 feet have completed after it has rolled 20 feet?

33. In the figure, $ABCD$ is a square, with each side of length 6 feet. The width of the border (shaded portion) between the outer square $EFGH$ and $ABCD$ is 2 feet. Find the area of the border.

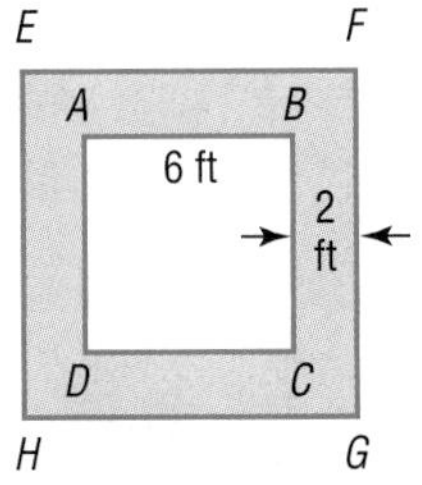

34. Refer to the figure. Square $ABCD$ has an area of 100 square feet; square $BEFG$ has an area of 16 square feet. What is the area of the triangle CGF?

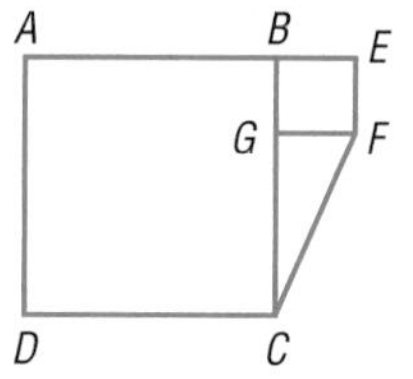

35. Architecture A Norman window consists of a rectangle surmounted by a semicircle. Find the area of the Norman window shown in the illustration. How much wood frame is needed to enclose the window?

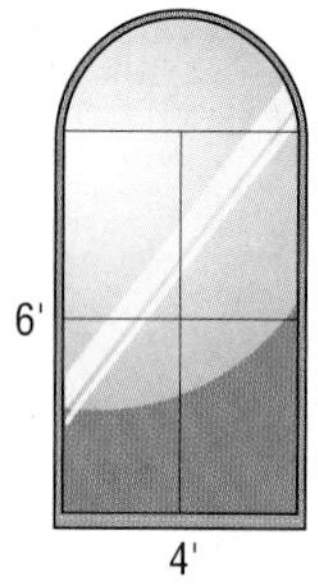

36. Construction A circular swimming pool, 20 feet in diameter, is enclosed by a wooden deck that is 3 feet wide. What is the area of the deck? How much fence is required to enclose the deck?

In Problems 37–39, use the facts that the radius of Earth is 3960 miles and 1 mile = 5280 feet.

37. How Far Can You See? The conning tower of the U.S.S. *Silversides*, a World War II submarine now permanently stationed in Muskegon, Michigan, is approximately 20 feet above sea level. How far can one see from the conning tower?

38. How Far Can You See? A person who is 6 feet tall is standing on the beach in Fort Lauderdale, Florida, and looks out onto the Atlantic Ocean. Suddenly, a ship appears on the horizon. How far is the ship from shore?

39. How Far Can You See? The deck of a destroyer is 100 feet above sea level. How far can a person see from the deck? How far can a person see from the bridge, which is 150 feet above sea level?

40. You have 1000 feet of flexible pool siding and wish to construct a swimming pool. Experiment with rectangular shaped pools with perimeters of 1000 feet. How do their areas vary? What is the shape of the rectangle with the largest area? Now compute the area enclosed by a circular pool with a perimeter (circumference) of 1000 feet. What would be your choice of shape for the pool? If rectangular, what is your preference for dimensions? Justify your choice. If your only consideration is to have a pool that encloses the most area, what shape should you use?

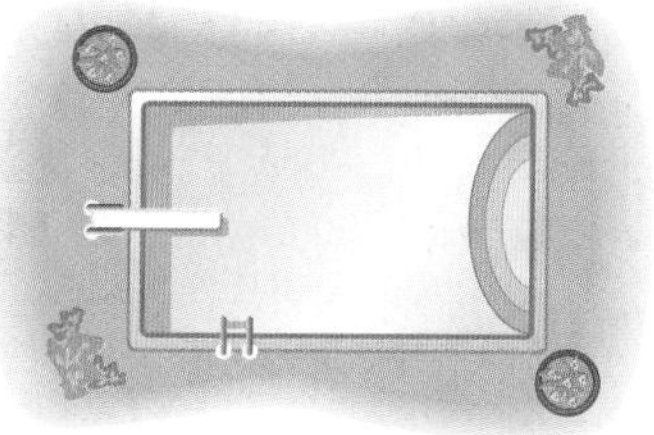

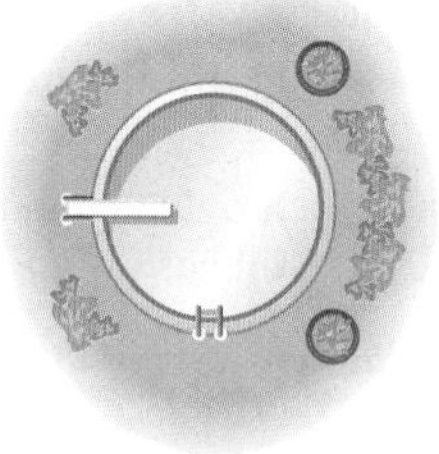

10 COMPLETING THE SQUARE; THE QUADRATIC FORMULA

OBJECTIVES
1. Complete the Square of an Equation
2. Solve a Quadratic Equation by Completing the Square
3. Obtain the Quadratic Formula

COMPLETING THE SQUARE

1 The idea behind the method of **completing the square** is to "adjust" a second-degree polynomial, $ax^2 + bx + c$, so that it becomes a perfect square (the

square of a first-degree polynomial). For example, $x^2 + 6x + 9$ and $x^2 - 4x + 4$ are perfect squares because

$$x^2 + 6x + 9 = (x + 3)^2 \quad \text{and} \quad x^2 - 4x + 4 = (x - 2)^2$$

How do we adjust the second-degree polynomial? We do it by adding the appropriate number to create a perfect square. For example, to make $x^2 + 6x$ a perfect square, we add 9.

Let's look at several examples of completing the square when the coefficient of x^2 is 1.

Start	Add	Result
$x^2 + 4x$	4	$x^2 + 4x + 4 = (x + 2)^2$
$x^2 + 12x$	36	$x^2 + 12x + 36 = (x + 6)^2$
$x^2 - 6x$	9	$x^2 - 6x + 9 = (x - 3)^2$
$x^2 + x$	$\frac{1}{4}$	$x^2 + x + \frac{1}{4} = (x + \frac{1}{2})^2$

Do you see the pattern? Provided that the coefficient of x^2 is 1, we complete the square by adding the square of one-half the coefficient of x.

Start	Add	Result
$x^2 + mx$	$\left(\frac{m}{2}\right)^2$	$x^2 + mx + \left(\frac{m}{2}\right)^2 = \left(x + \frac{m}{2}\right)^2$

NOW WORK PROBLEM 1.

EXAMPLE 1 Completing the Square of an Equation Containing Two Variables

Complete the squares of x and y in the equation

$$x^2 + y^2 - 2x + 4y - 4 = 0$$

Solution We rearrange the equation, grouping the terms involving the variable x and the variable y.

$$(x^2 - 2x) + (y^2 + 4y) = 4$$

Next, we complete the square of each parenthetical expression. Of course, since we want an equivalent equation, whatever we add to the left side, we also add to the right side.

$$(x^2 - 2x + 1) + (y^2 + 4y + 4) = 4 + 1 + 4$$
$$(x - 1)^2 + (y + 2)^2 = 9$$

The terms involving the variables x and y now appear as perfect squares. ■

NOW WORK PROBLEM 7.

Solving a Quadratic Equation by Completing the Square

2 We begin with a preliminary result. Suppose that we wish to solve the quadratic equation

$$x^2 = p \qquad \textbf{(1)}$$

where $p \geq 0$ is a nonnegative number. We proceed as follows:

$$x^2 - p = 0 \qquad \text{Put in standard form.}$$
$$(x - \sqrt{p})(x + \sqrt{p}) = 0 \qquad \text{Factor (over the real numbers).}$$
$$x = \sqrt{p} \quad \text{or} \quad x = -\sqrt{p} \qquad \text{Solve.}$$

We have the following result:

$$\text{If } x^2 = p \text{ and } p \geq 0, \text{ then } x = \sqrt{p} \text{ or } x = -\sqrt{p}. \qquad \textbf{(2)}$$

Note that if $p > 0$ the equation $x^2 = p$ has two solutions: $x = \sqrt{p}$ and $x = -\sqrt{p}$. We usually abbreviate these solutions as $x = \pm\sqrt{p}$, read as "x equals plus or minus the square root of p." For example, the two solutions of the equation

$$x^2 = 4$$

are

$$x = \pm\sqrt{4}$$

and, since $\sqrt{4} = 2$, we have

$$x = \pm 2$$

The solution set is $\{-2, 2\}$.

EXAMPLE 2 **Solving a Quadratic Equation by Completing the Square**

Solve by completing the square: $x^2 + 5x + 4 = 0$

Solution We always begin this procedure by rearranging the equation so that the constant is on the right side.

$$x^2 + 5x + 4 = 0$$
$$x^2 + 5x = -4$$

Since the coefficient of x^2 is 1, we can complete the square on the left side by adding $\left(\frac{1}{2} \cdot 5\right)^2 = \frac{25}{4}$. Of course, in an equation, whatever we add to the left side must also be added to the right side. Thus, we add $\frac{25}{4}$ to *both* sides.

$$x^2 + 5x + \frac{25}{4} = -4 + \frac{25}{4} \qquad \text{Add } \tfrac{25}{4} \text{ to both sides.}$$
$$\left(x + \frac{5}{2}\right)^2 = \frac{9}{4} \qquad \text{Factor; simplify.}$$
$$x + \frac{5}{2} = \pm\sqrt{\frac{9}{4}} \qquad \text{Apply Statement (2).}$$
$$x + \frac{5}{2} = \pm\frac{3}{2}$$
$$x = -\frac{5}{2} \pm \frac{3}{2}$$
$$x = -\frac{5}{2} + \frac{3}{2} = -1 \quad \text{or} \quad x = -\frac{5}{2} - \frac{3}{2} = -4$$

The solution set is $\{-4, -1\}$. ■

NOW WORK PROBLEM 13.

THE QUADRATIC FORMULA

3 We can use the method of completing the square to obtain a general formula for solving the quadratic equation

$$ax^2 + bx + c = 0, \qquad a \neq 0$$

We begin by rearranging the terms as

$$ax^2 + bx = -c$$

Since $a \neq 0$, we can divide both sides by a to get

$$x^2 + \frac{b}{a}x = -\frac{c}{a}$$

Now the coefficient of x^2 is 1. To complete the square on the left side, add the square of $\frac{1}{2}$ the coefficient of x; that is, add

$$\left(\frac{1}{2} \cdot \frac{b}{a}\right)^2 = \frac{b^2}{4a^2}$$

to each side. Then

$$x^2 + \frac{b}{a}x + \frac{b^2}{4a^2} = \frac{b^2}{4a^2} - \frac{c}{a}$$

$$\left(x + \frac{b}{2a}\right)^2 = \frac{b^2 - 4ac}{4a^2} \qquad \frac{b^2}{4a^2} - \frac{c}{a} = \frac{b^2}{4a^2} - \frac{4ac}{4a^2} = \frac{b^2 - 4ac}{4a^2} \qquad (3)$$

Provided that $b^2 - 4ac \geq 0$, we now can apply the result in statement (2) to get

$$x + \frac{b}{2a} = \pm\sqrt{\frac{b^2 - 4ac}{4a^2}}$$

$$x = -\frac{b}{2a} \pm \frac{\sqrt{b^2 - 4ac}}{2a} = \frac{-b \pm \sqrt{b^2 - 4ac}}{2a}$$

What if $b^2 - 4ac$ is negative? Then equation (3) states that the left expression (a real number squared) equals the right expression (a negative number). Since this occurrence is impossible for real numbers, we conclude that if $b^2 - 4ac < 0$ the quadratic equation has no *real* solution. (We discuss quadratic equations for which the quantity $b^2 - 4ac < 0$ in detail in Section 4.2.*)

We now state the *quadratic formula.*

Theorem **Quadratic Formula**

Consider the quadratic equation

$$ax^2 + bx + c = 0 \qquad a \neq 0$$

If $b^2 - 4ac < 0$, this equation has no real solution.
If $b^2 - 4ac \geq 0$, the real solution(s) of this equation is (are) given by the **quadratic formula**:

$$x = \frac{-b \pm \sqrt{b^2 - 4ac}}{2a}$$

* Section 4.2 may be covered anytime after completing Section 1.3 without any loss of continuity.

10 EXERCISES

In Problems 1–6, tell what number should be added to complete the square of each expression.

1. $x^2 - 4x$
2. $x^2 - 2x$
3. $x^2 + \frac{1}{2}x$
4. $x^2 - \frac{1}{3}x$
5. $x^2 - \frac{2}{3}x$
6. $x^2 - \frac{2}{5}x$

In Problems 7–12, complete the square of x and y in each equation.

7. $x^2 + y^2 - 4x + 4y - 1 = 0$
8. $x^2 + y^2 + 4x + 4y - 8 = 0$
9. $x^2 + y^2 + 6x - 2y + 1 = 0$
10. $x^2 + y^2 - 8x + 2y + 1 = 0$
11. $x^2 + y^2 - x - y - \frac{1}{2} = 0$
12. $x^2 + y^2 - x + y - \frac{3}{2} = 0$

In Problems 13–18, solve each equation by completing the square.

13. $x^2 + 4x - 21 = 0$
14. $x^2 - 6x = 13$
15. $x^2 - \frac{1}{2}x = \frac{3}{16}$
16. $x^2 + \frac{2}{3}x = \frac{1}{3}$
17. $3x^2 + x - \frac{1}{2} = 0$
18. $2x^2 - 3x = 1$

ANSWERS

CHAPTER 1 Graphs

1.1 Exercises

1. **(a)** Quadrant II **(b)** Positive x-axis **(c)** Quadrant III **(d)** Quadrant I **(e)** Negative y-axis **(f)** Quadrant IV

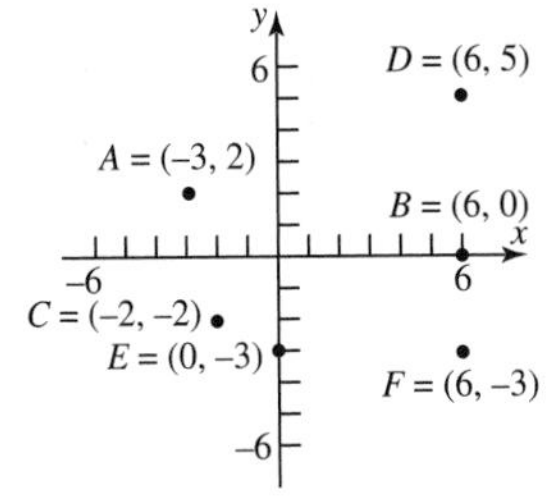

3. The points will be on a vertical line that is 2 units to the right of the y-axis

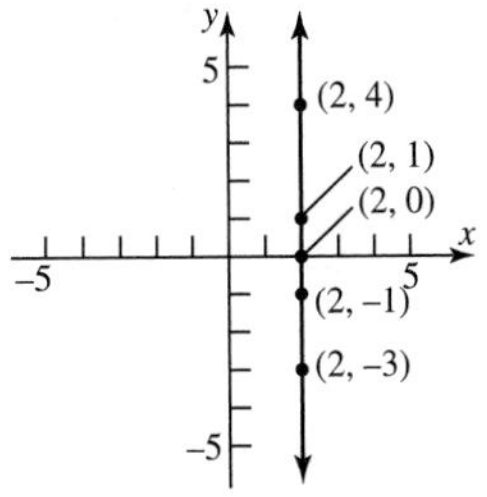

5. $(-1, 4)$ **7.** $(3, 1)$ **9.** $X\min = -11, X\max = 5, X\text{scl} = 1, Y\min = -3, Y\max = 6, Y\text{scl} = 1$
11. $X\min = -30, X\max = 50, X\text{scl} = 10, Y\min = -90, Y\max = 50, Y\text{scl} = 10$
13. $X\min = -10, X\max = 110, X\text{scl} = 10, Y\min = -10, Y\max = 160, Y\text{scl} = 10$
15. $X\min = -6, X\max = 6, X\text{scl} = 2, Y\min = -4, Y\max = 4, Y\text{scl} = 2$
17. $X\min = -6, X\max = 6, X\text{scl} = 2, Y\min = -1, Y\max = 3, Y\text{scl} = 1$
19. $X\min = 3, X\max = 9, X\text{scl} = 1, Y\min = 2, Y\max = 10, Y\text{scl} = 2$
21. $\sqrt{5}$ **23.** $\sqrt{10}$ **25.** $2\sqrt{17}$ **27.** $\sqrt{85}$ **29.** $\sqrt{53}$ **31.** $\sqrt{6.89} \approx 2.625$ **33.** $\sqrt{a^2 + b^2}$ **35.** $4\sqrt{10}$ **37.** $2\sqrt{65}$

39. $d(A, B) = \sqrt{13}$
$d(B, C) = \sqrt{13}$
$d(A, C) = \sqrt{26}$
$(\sqrt{13})^2 + (\sqrt{13})^2 = (\sqrt{26})^2$
Area $= \dfrac{13}{2}$ square units

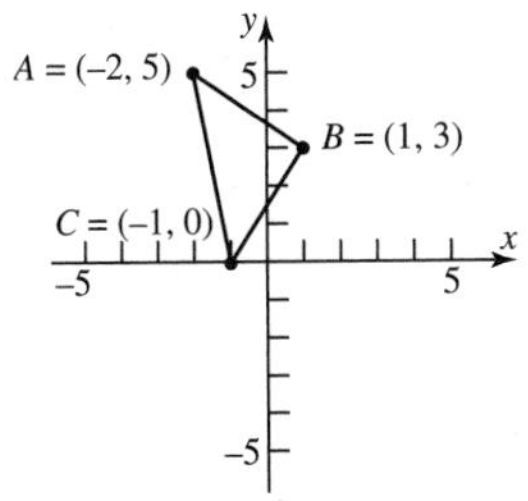

41. $d(A, B) = \sqrt{130}$
$d(B, C) = \sqrt{26}$
$d(A, C) = \sqrt{104}$
$(\sqrt{26})^2 + (\sqrt{104})^2 = (\sqrt{130})^2$
Area $= 26$ square units

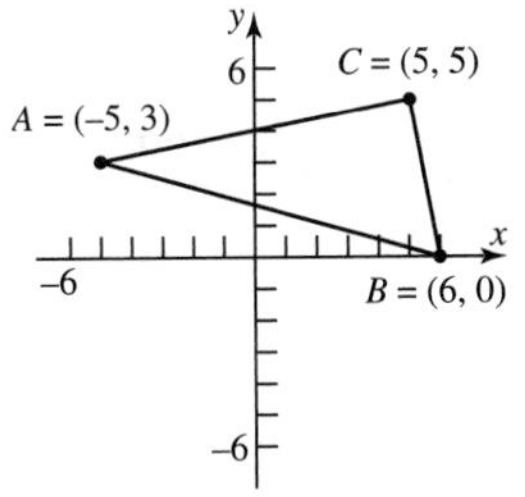

43. $d(A, B) = 4$
$d(A, C) = 5$
$d(B, C) = \sqrt{41}$
$4^2 + 5^2 = 16 + 25 = (\sqrt{41})^2$
Area $= 10$ square units

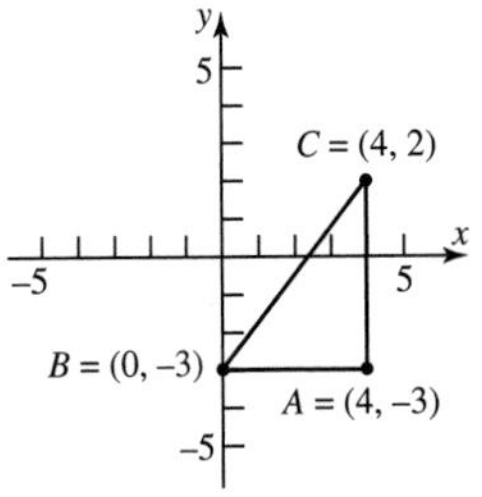

45. $(2, 2); (2, -4)$ **47.** $(0, 0); (8, 0)$ **49.** $(4, -1)$ **51.** $\left(\dfrac{3}{2}, 1\right)$ **53.** $(5, -1)$ **55.** $(1.05, 0.7)$ **57.** $\left(\dfrac{a}{2}, \dfrac{b}{2}\right)$ **59.** $\sqrt{17}; 2\sqrt{5}; \sqrt{29}$
61. $d(P_1, P_2) = 6; d(P_2, P_3) = 4; d(P_1, P_3) = 2\sqrt{13}$; right triangle
63. $d(P_1, P_2) = \sqrt{68}; d(P_2, P_3) = \sqrt{34}; d(P_1, P_3) = \sqrt{34}$; isosceles right triangle **65.** Midpoint of diagonals: $\left(\dfrac{S}{2}, \dfrac{S}{2}\right)$
67. $90\sqrt{2} \approx 127.28$ ft **69.** **(a)** $(90, 0), (90, 90), (0, 90)$ **(b)** approx. 232.4 ft **(c)** approx. 366.2 ft **71.** $d = 50t$ mi

1.2 Exercises

1. (0, 0), (1, 1) **3.** (0, 3) **5.** (0, 2), $(\sqrt{2}, \sqrt{2})$

7. (a)
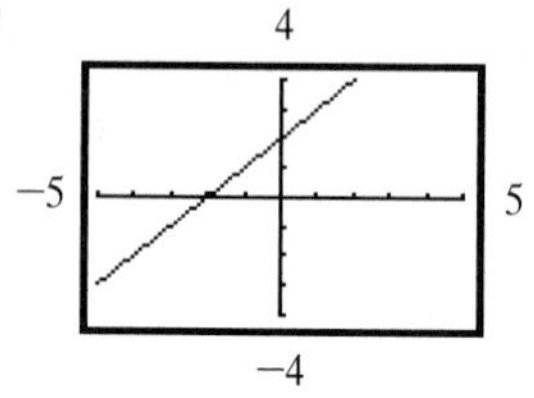

(b)
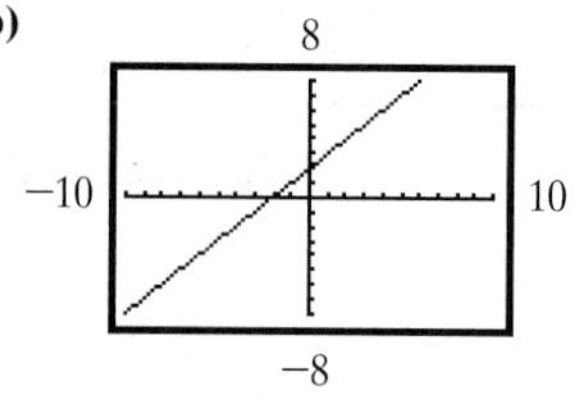

(c)
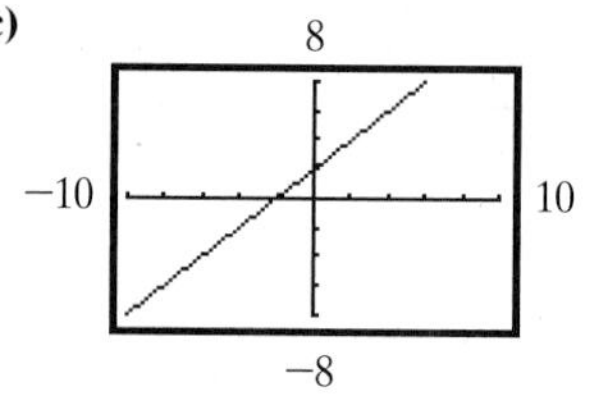

(d)
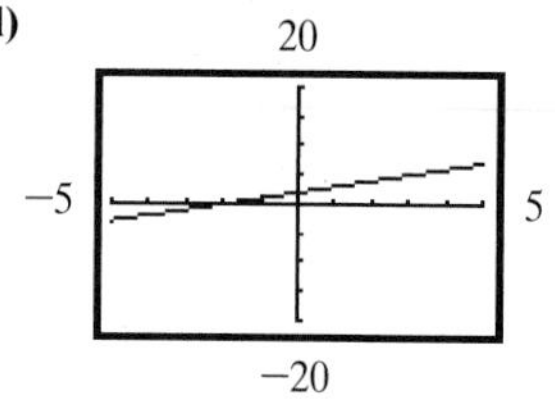

9. (a)
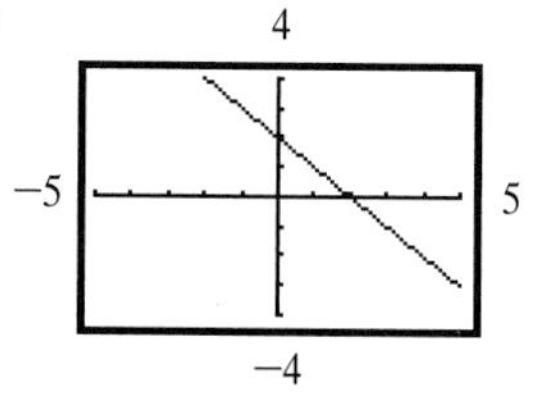

(b)
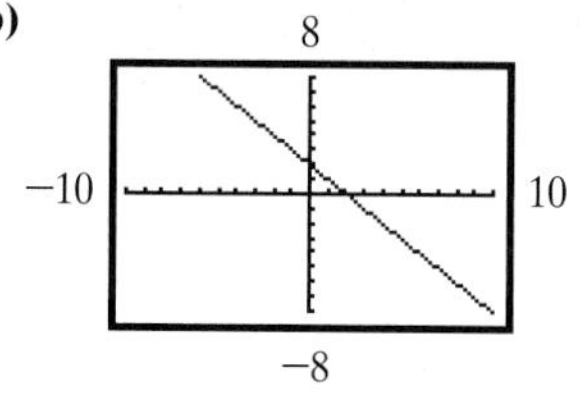

(c)
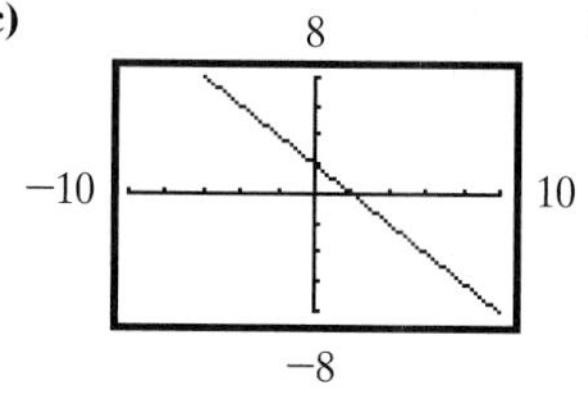

(d)
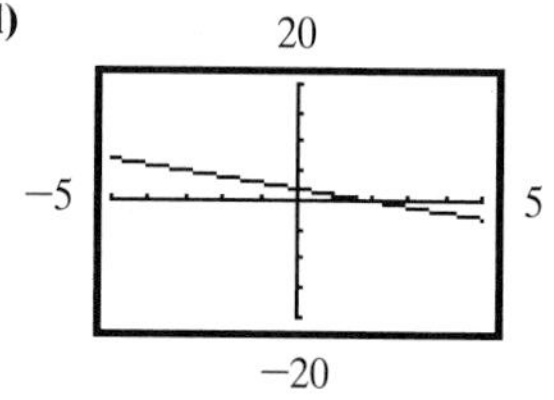

11. (a)
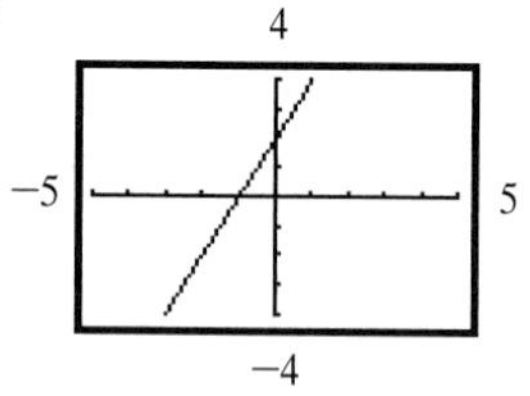

(b)
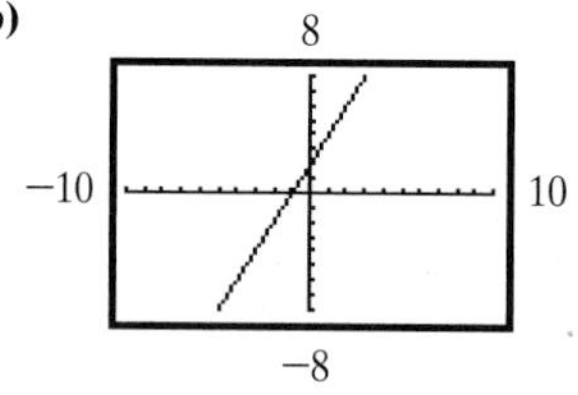

(c)
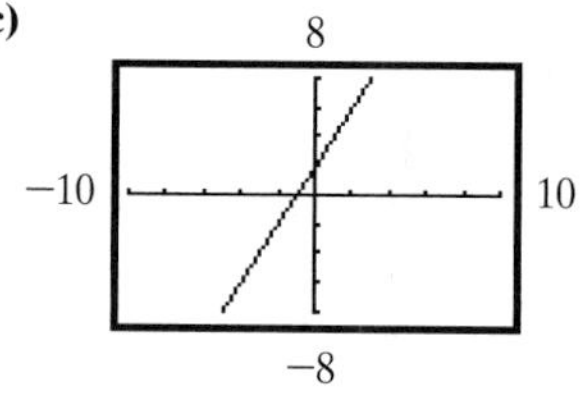

(d)
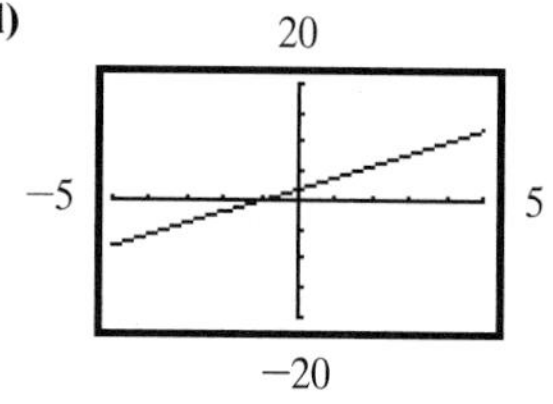

13. (a)
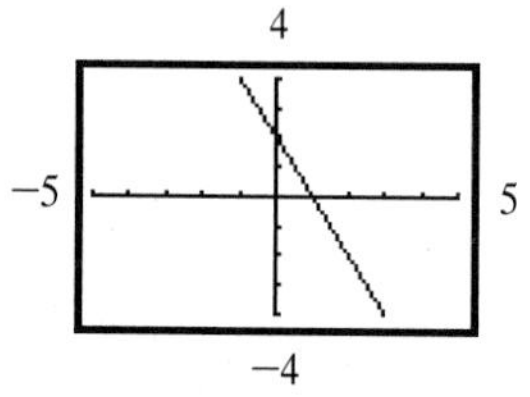

(b)
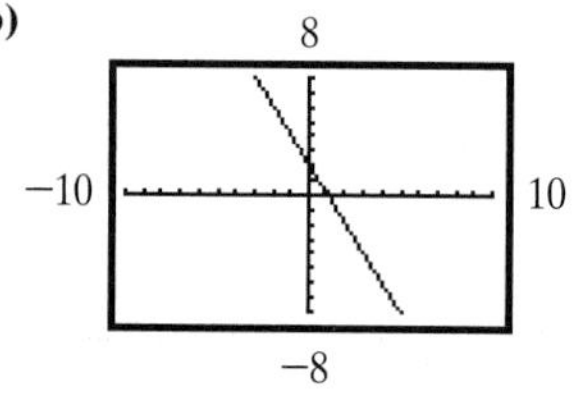

(c)
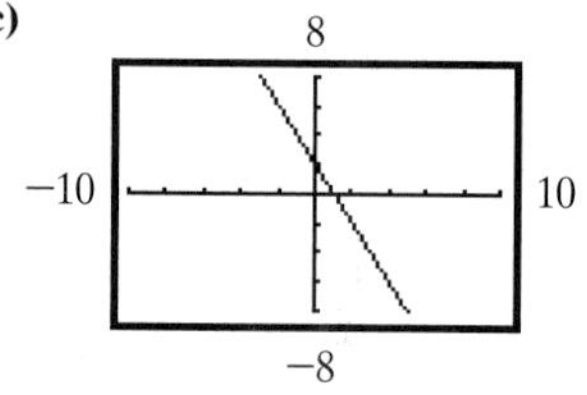

(d)
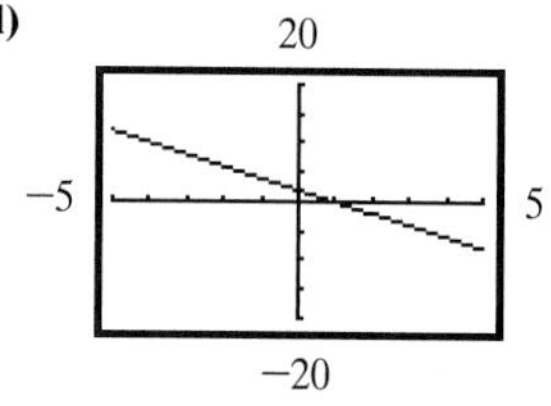

15. (a)
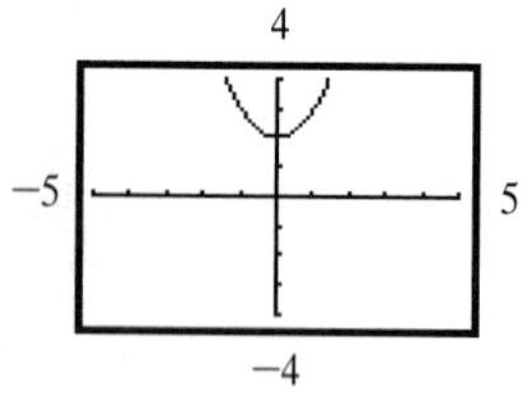

(b)
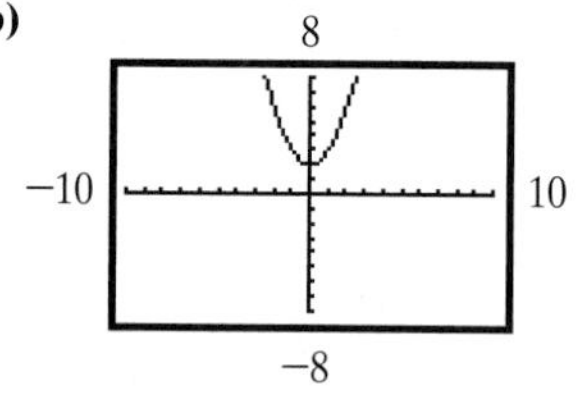

(c)
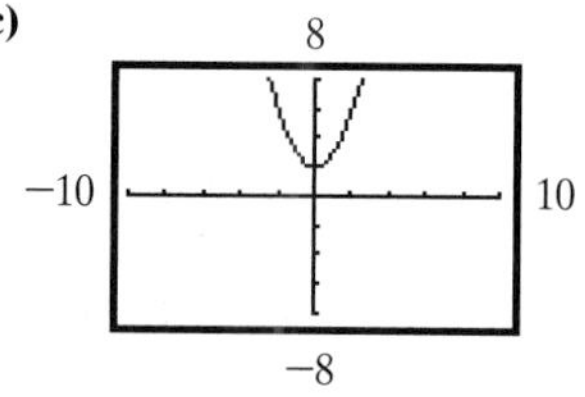

(d)
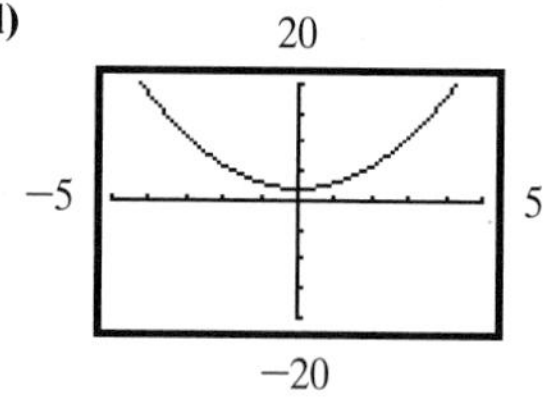

17. (a)
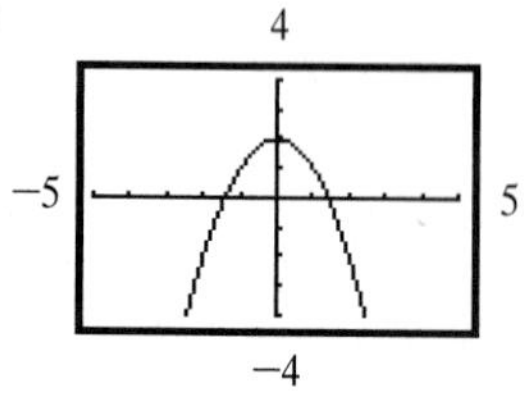

(b)
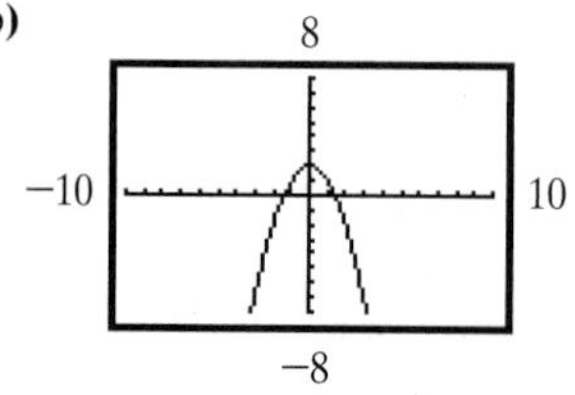

(c)
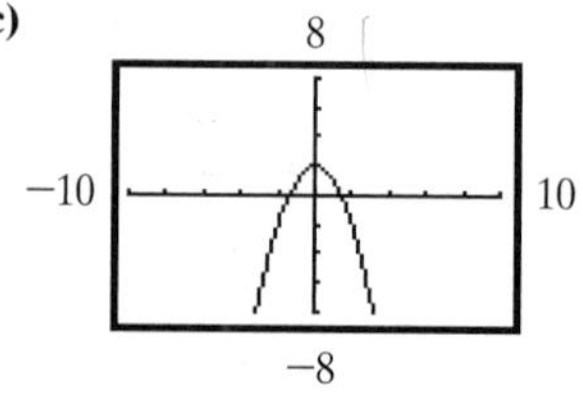

(d)
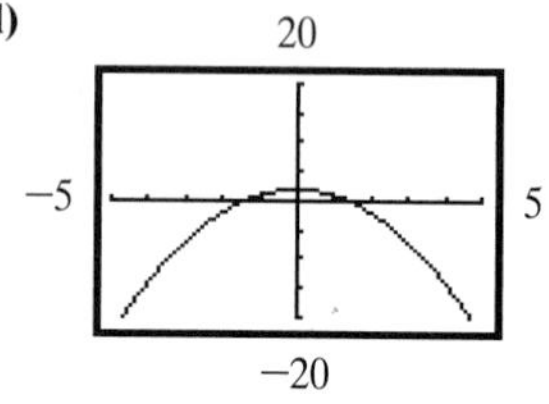

19. (a)
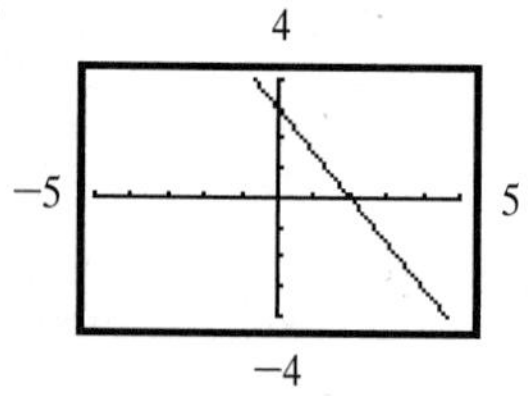

(b)
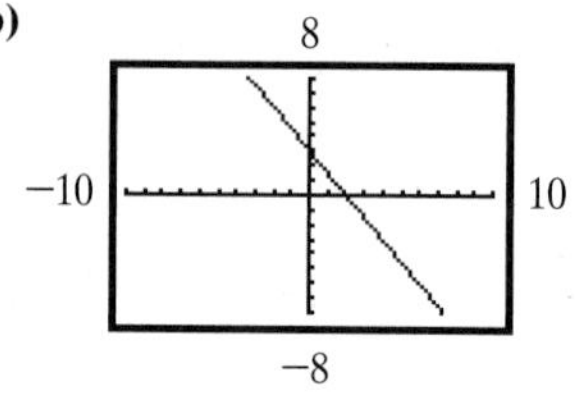

(c)
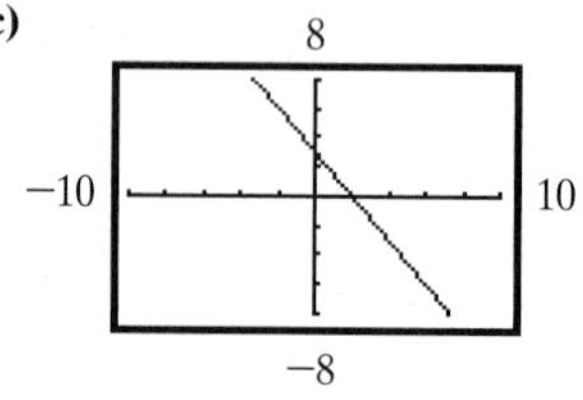

(d)
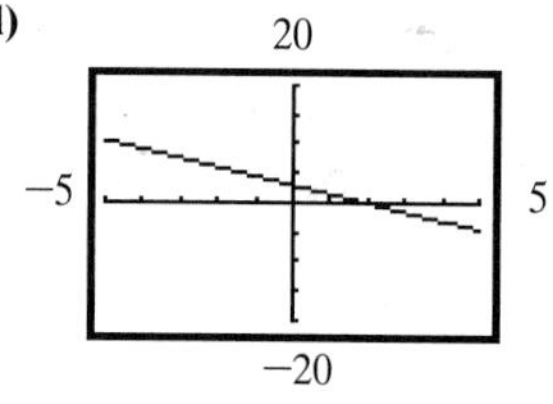

21. **(a)** **(b)** **(c)** **(d)**

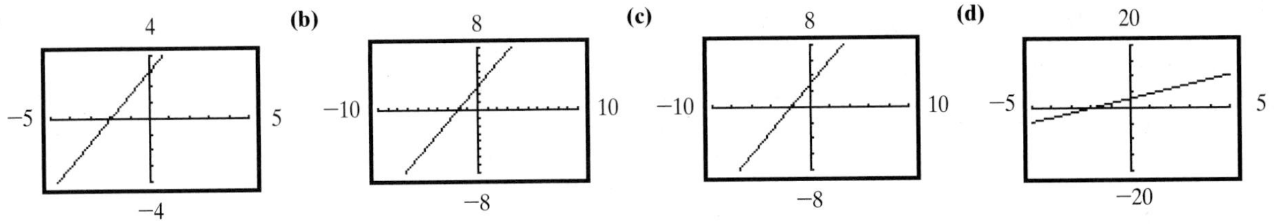

23.

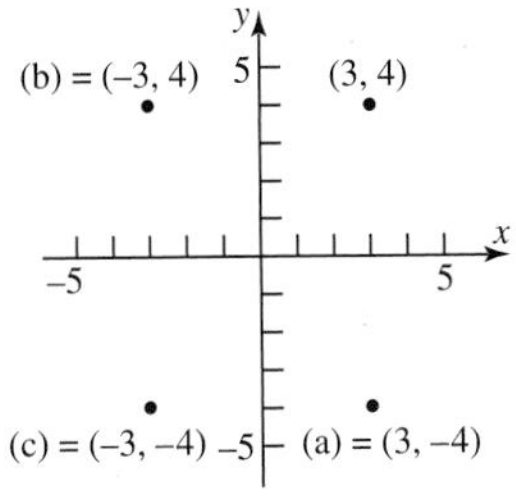

25.

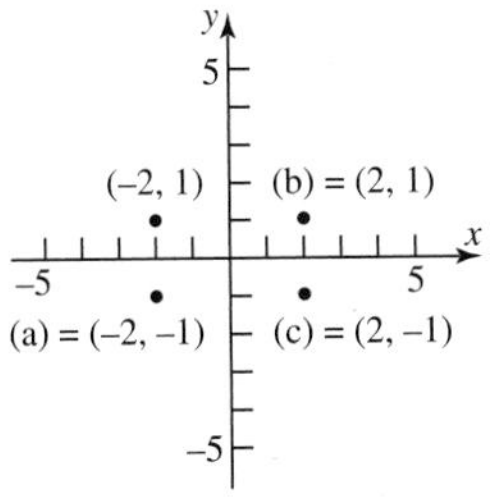

27.

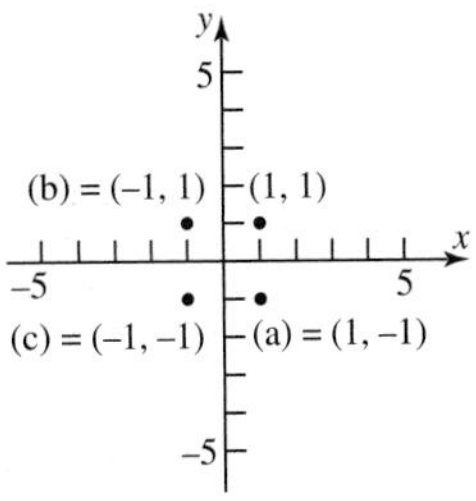

29.

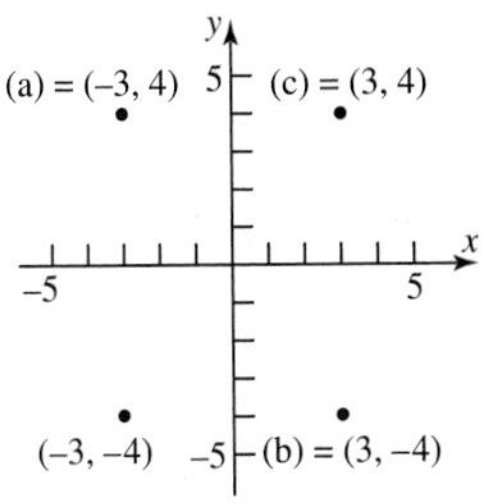

31.

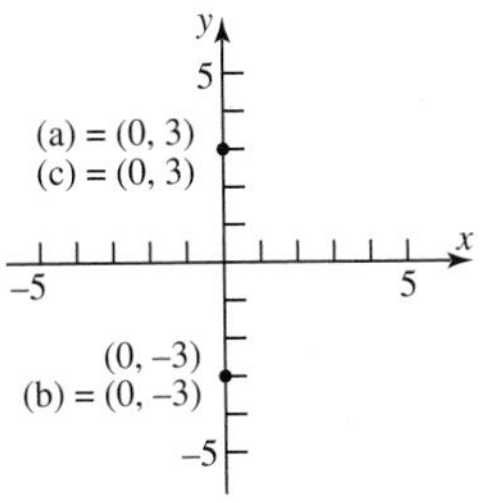

33. **(a)** $(-1, 0), (1, 0)$ **(b)** x-axis, y-axis, origin

35. **(a)** $\left(-\frac{\pi}{2}, 0\right), (0, 1), \left(\frac{\pi}{2}, 0\right)$ **(b)** y-axis

37. **(a)** $(0, 0)$ **(b)** x-axis

39. **(a)** $(1, 0)$ **(b)** none

41. **(a)** $(-1.5, 0), (0, -2), (1.5, 0)$ **(b)** y-axis

43. **(a)** none **(b)** origin

45. $-\frac{2}{5}$ **47.** $2a + 3b = 6$

49.

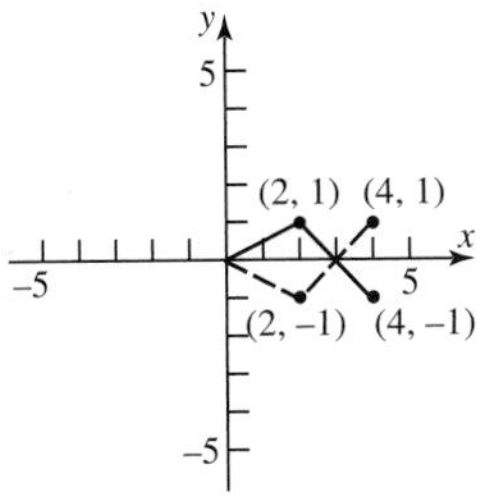

51.

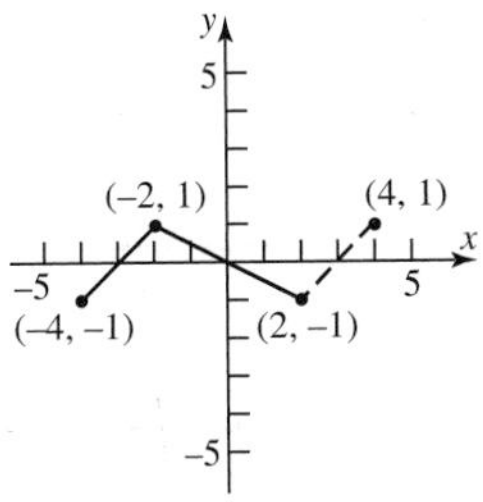

53.

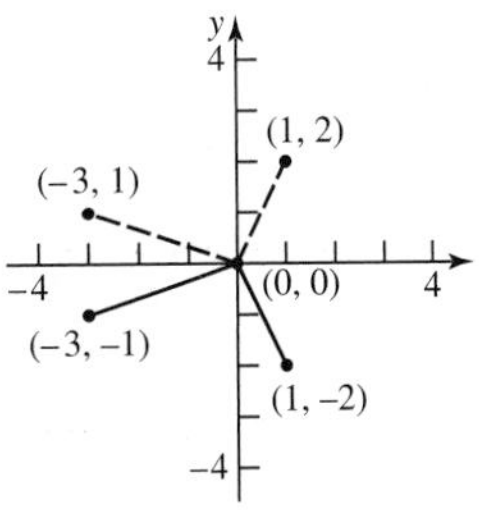

55.

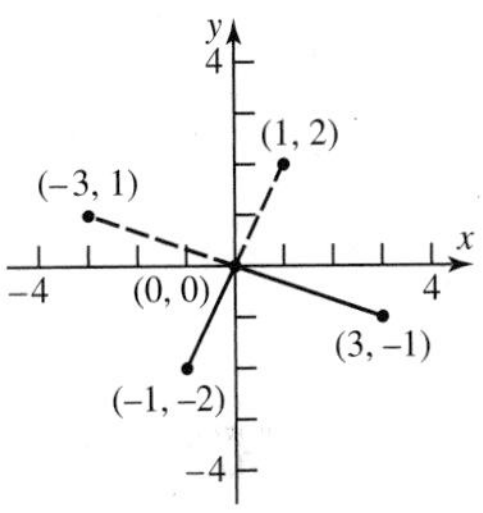

57. $(0, 0)$; symmetric with respect to the y-axis **59.** $(0, 0)$; symmetric with respect to the origin **61.** $(0, 9), (9, 0)$; no symmetry **63.** $(0, 9), (-2, 0), (2, 0)$; symmetric with respect to the y-axis. **65.** $(0, -27), (3, 0)$; no symmetry **67.** $(0, -4), (4, 0), (-1, 0)$; no symmetry **69.** $(0, 0)$; symmetric with respect to the origin

71. **(a)**

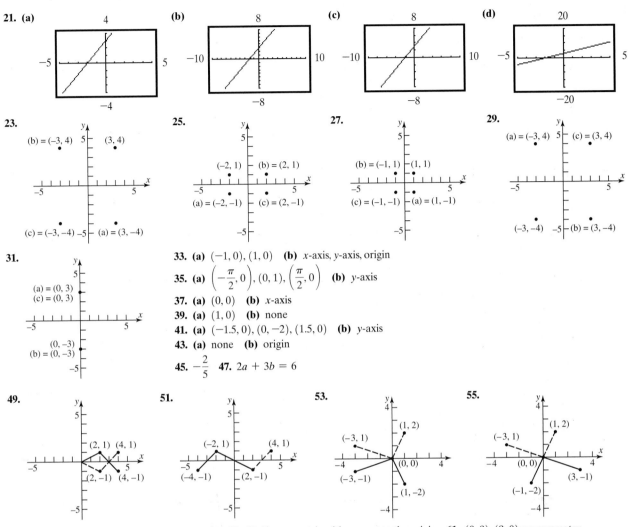

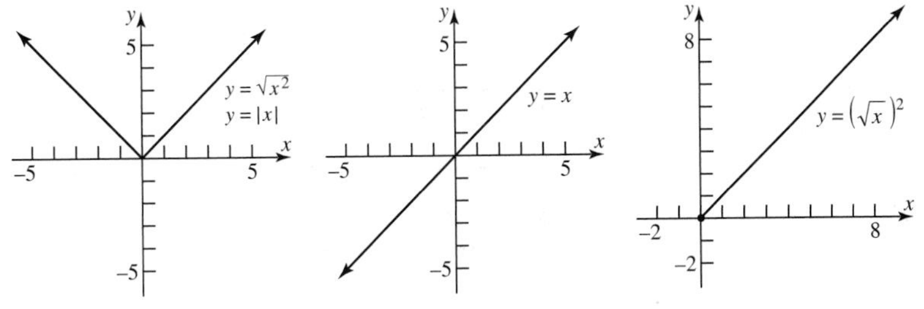

(b) Since $\sqrt{x^2} = |x|$, for all x, the graphs of $y = \sqrt{x^2}$ and $y = |x|$ are the same.

(c) For $y = (\sqrt{x})^2$, the domain of the variable x is $x \geq 0$; for $y = x$, the domain of the variable x is all real numbers. Thus, $(\sqrt{x})^2 = x$ only for $x \geq 0$.

(d) For $y = \sqrt{x^2}$, the range of the variable y is $y \geq 0$; for $y = x$, the range of the variable y is all real numbers. Also, $\sqrt{x^2} = |x|$, which equals x only if $x \geq 0$.

1.3 Exercises

1. 2 **3.** 3 **5.** −1 **7.** $-\frac{4}{3}$ **9.** −18 **11.** −3 **13.** 2 **15.** 0.5 **17.** $\frac{46}{5}$ **19.** 2 **21.** $\{0, 9\}$ **23.** $\{-5, 5\}$ **25.** $\{-4, 3\}$ **27.** $\left\{-\frac{1}{2}, 3\right\}$ **29.** $\{-4, 4\}$ **31.** $\{3, 4\}$ **33.** $\frac{3}{2}$ **35.** $\left\{-\frac{2}{3}, \frac{3}{2}\right\}$ **37.** $\left\{-\frac{2}{3}, \frac{3}{2}\right\}$ **39.** 2 **41.** −1 **43.** 1 **45.** No real solution **47.** −13 **49.** 3 **51.** 2 **53.** 8 **55.** $\{-1, 3\}$ **57.** $\{1, 5\}$ **59.** $\{-4, 4\}$ **61.** $\{-4, 1\}$ **63.** $\left\{-1, \frac{3}{2}\right\}$ **65.** $\{-4, 4\}$ **67.** 2 **69.** $\{-12, 12\}$ **71.** $\left\{-\frac{36}{5}, \frac{24}{5}\right\}$

73. $\dfrac{b+c}{a}$ **75.** $\dfrac{abc}{a+b}$ **77.** $\{2-\sqrt{2}, 2+\sqrt{2}\}$ **79.** $\{2-\sqrt{5}, 2+\sqrt{5}\}$ **81.** $\left\{1, \dfrac{3}{2}\right\}$ **83.** No real solution

85. $\left\{\dfrac{-1-\sqrt{5}}{4}, \dfrac{-1+\sqrt{5}}{4}\right\}$ **87.** $\left\{\dfrac{1}{3}\right\}$ **89.** $\{0.59, 3.41\}$ **91.** $\{-2.80, 1.07\}$ **93.** $\{-0.85, 1.17\}$ **95.** $R = \dfrac{R_1R_2}{R_1+R_2}$ **97.** $R = \dfrac{mv^2}{F}$

99. $r = \dfrac{S-a}{S}$ **101.** 229.94 ft

1.4 Exercises

1. $A = \pi r^2$; r = Radius, A = Area **3.** $A = s^2$; A = Area, s = Length of a side **5.** $F = ma$; F = Force, m = Mass, a = Acceleration
7. $W = Fd$; W = Work, F = Force, d = Distance **9.** $C = 150x$; C = Total cost, x = number of dishwashers
11. \$11,000 will be invested in bonds and \$9000 in CDs. **13.** David will receive \$400,000, Paige \$300,000, and Dan \$200,000.
15. The length is 19 ft; the width is 11 ft **17.** Brooke needs a score of 85.
19. The original price was \$147,058.82; purchasing the model saves \$22,058.82. **21.** The bookstore paid \$44.80.
23. Invest \$31,250 in bonds and \$18,750 in CDs. **25.** \$11,600 was loaned out at 8%.
27. Mix 75 lb of Earl Gray tea with 25 lb of Orange Pekoe tea. **29.** Mix 40 lb of cashews with the peanuts. **31.** Add $\dfrac{20}{3}$ oz of pure water.
33. The Metra commuter averages 30 mph; the Amtrak averages 80 mph. **35.** The speed of the current is 2.286 mph.
37. Working together, it takes 12 min. **39.** The dimensions are 11 ft by 13 ft **41.** The dimensions are 5 m by 8 m
43. The dimensions should be 4 ft by 4 ft. **45.** The speed of the current is 5 mph. **47.** **(a)** The dimensions are 10 ft by 5 ft. **(b)** The area is 50 sq ft. **(c)** The dimensions would be 7.5 ft by 7.5 ft. **(d)** The area would be 56.25 sq ft.
49. The border will be 2.71 ft wide. **51.** The border will be 2.56 ft wide. **53.** Add $\dfrac{2}{3}$ gal of water. **55.** 5 lb must be added.
57. The dimensions should be 11.55 cm by 6.55 cm by 3 cm. **59.** 40 g of 12 karat gold should be mixed with 20 g of pure gold.
61. Mike passes Dan $\dfrac{1}{3}$ mi from the start, 2 min from the time Mike started to race. **63.** Start the auxiliary pump at 9:45 AM.
65. 60 minutes **67.** The most you can invest in the CD is \$66,667. **69.** The average speed is 49.5 mph.
71. Set the original price at \$40. At 50% off, there will be no profit at all.

1.5 Exercises

1. $[0, 4]$

3. $[4, 6)$

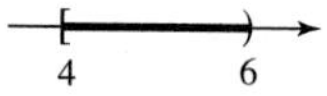

5. $[4, \infty)$

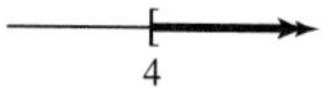

7. $(-\infty, -4)$

9. $2 \le x \le 5$

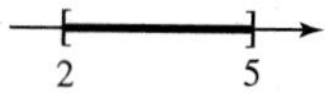

11. $-3 < x < -2$

13. $x \ge 4$

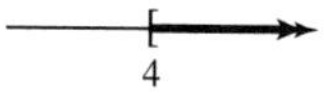

15. $x < -3$

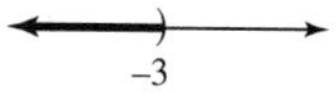

17. $<$ **19.** $>$ **21.** $\ge$ **23.** $>$ **25.** $\le$ **27.** $>$ **29.** $\ge$

31. $\{x \mid x < 4\}$ or $(-\infty, 4)$

33. $\{x \mid x \ge -1\}$ or $[-1, \infty)$

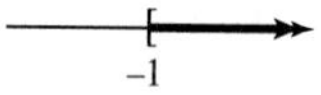

35. $\{x \mid x > 3\}$ or $(3, \infty)$

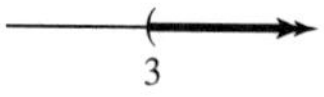

37. $\{x \mid x \ge 2\}$ or $[2, \infty)$

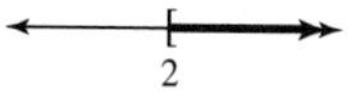

39. $\{x \mid x > -7\}$ or $(-7, \infty)$

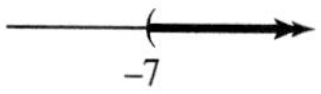

41. $\left\{x \middle| x \le \dfrac{2}{3}\right\}$ or $\left(-\infty, \dfrac{2}{3}\right]$

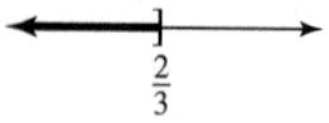

43. $\{x \mid x < -20\}$ or $(-\infty, -20)$

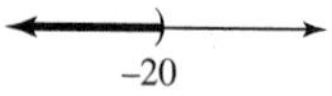

45. $\left\{x \middle| x \ge \dfrac{4}{3}\right\}$ or $\left[\dfrac{4}{3}, \infty\right)$

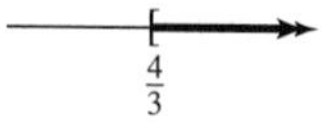

47. $\{x \mid 3 \le x \le 5\}$ or $[3, 5]$

49. $\left\{x \middle| \dfrac{2}{3} \le x \le 3\right\}$ or $\left[\dfrac{2}{3}, 3\right]$

51. $\left\{x \middle| -\dfrac{11}{2} < x < \dfrac{1}{2}\right\}$ or $\left(-\dfrac{11}{2}, \dfrac{1}{2}\right)$

53. $\{x \mid -6 < x < 0\}$ or $(-6, 0)$

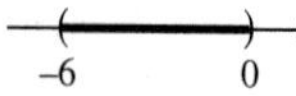

55. $\{x \mid x < -5\}$ or $(-\infty, -5)$

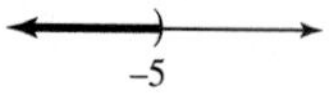

57. $\{x \mid x \ge -1\}$ or $[-1, \infty)$

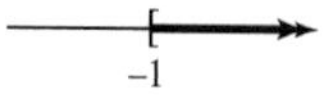

59. $\left\{x \middle| \dfrac{1}{2} \le x < \dfrac{5}{4}\right\}$ or $\left[\dfrac{1}{2}, \dfrac{5}{4}\right)$

61. $(-4, 4)$

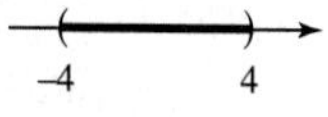

63. $(-\infty, -4)$ or $(4, \infty)$

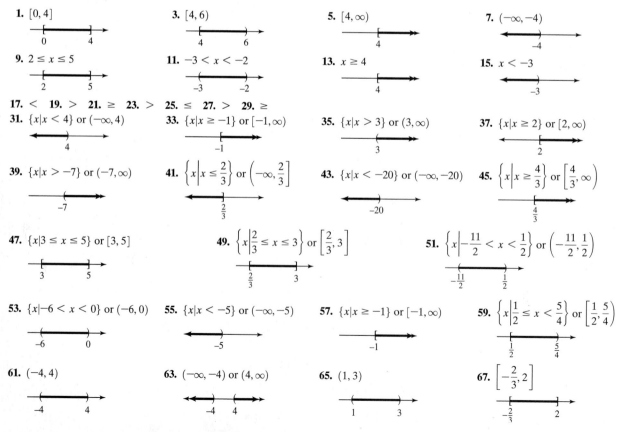

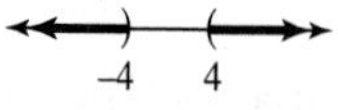

65. $(1, 3)$

67. $\left[-\dfrac{2}{3}, 2\right]$

69. $(-\infty, 1]$ or $[5, \infty)$

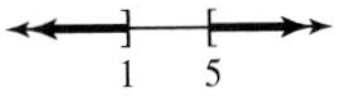

71. $\left(-1, \dfrac{3}{2}\right)$

73. $(-\infty, -1)$ or $(2, \infty)$

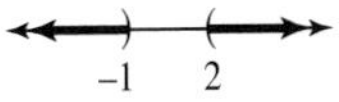

75. $\left|x - 2\right| < \dfrac{1}{2}; \left\{x \,\middle|\, \dfrac{3}{2} < x < \dfrac{5}{2}\right\}$

77. $|x + 3| > 2; \{x|x < -5 \text{ or } x > -1\}$ **79.** $21 < \text{age} < 30$ **81.** $|x - 98.6| \geq 1.5; x \leq 97.1°\text{F}$ or $x \geq 100.1°\text{F}$

83. (a) Male ≥ 73.4 **(b)** Female ≥ 79.7 **(c)** A female can expect to live at least 6.3 years longer.

85. The agent's commission ranges from \$45,000 to \$95,000, inclusive. As a percent of selling price, the commission ranges from 5% to 8.6%, inclusive. **87.** The amount withheld varies from \$72.14 to \$93.14, inclusive.

89. The usage varies from 675.43 kWhr to 2500.86 kWhr, inclusive. **91.** The dealer's cost varies from \$7457.63 to \$7857.14, inclusive.

93. You need at least a 74 on the last test. **95.** The amount of gasoline ranged from 12 to 20 gal, inclusive.

97. $\dfrac{a+b}{2} - a = \dfrac{a+b-2a}{2} = \dfrac{b-a}{2} > 0$; therefore, $a < \dfrac{a+b}{2}$

$b - \dfrac{a+b}{2} = \dfrac{2b-a-b}{2} = \dfrac{b-a}{2} > 0$; therefore, $b > \dfrac{a+b}{2}$

99. $(\sqrt{ab})^2 - a^2 = ab - a^2 = a(b-a) > 0$; thus, $(\sqrt{ab})^2 > a^2$ and $\sqrt{ab} > a$

$b^2 - (\sqrt{ab})^2 = b^2 - ab = b(b-a) > 0$; thus $b^2 > (\sqrt{ab})^2$ and $b > \sqrt{ab}$

101. $h - a = \dfrac{2ab}{a+b} - a = \dfrac{ab - a^2}{a+b} = \dfrac{a(b-a)}{a+b} > 0$; thus, $h > a$

$b - h = b - \dfrac{2ab}{a+b} = \dfrac{b^2 - ab}{a+b} = \dfrac{b(b-a)}{a+b} > 0$; thus $h < b$

1.6 Exercises

1. (a) $\dfrac{1}{2}$ **(b)** If x increases by 2 units, y will increase by 1 unit. **3. (a)** $-\dfrac{1}{3}$ **(b)** If x increases by 3 units, y will decrease by 1 unit.

5. Slope $= -\dfrac{3}{2}$

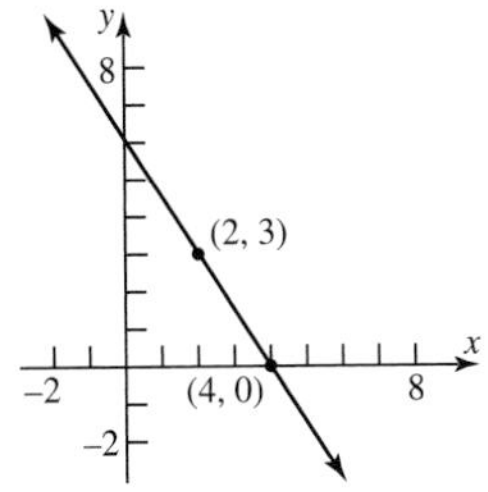

7. Slope $= -\dfrac{1}{2}$

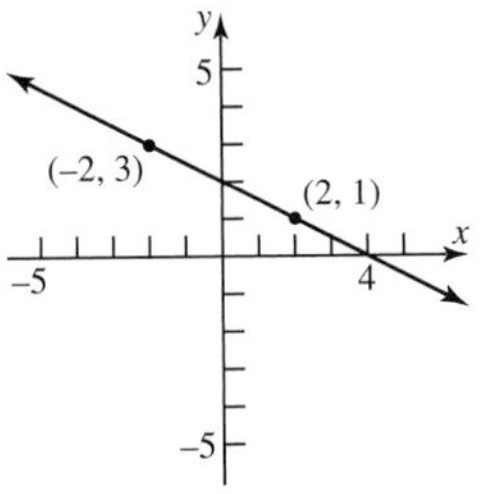

9. Slope $= 0$

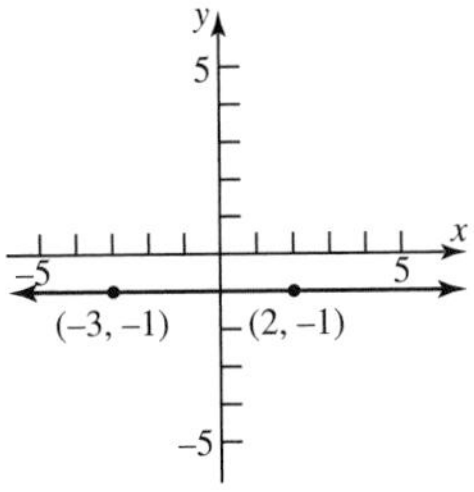

11. Slope undefined

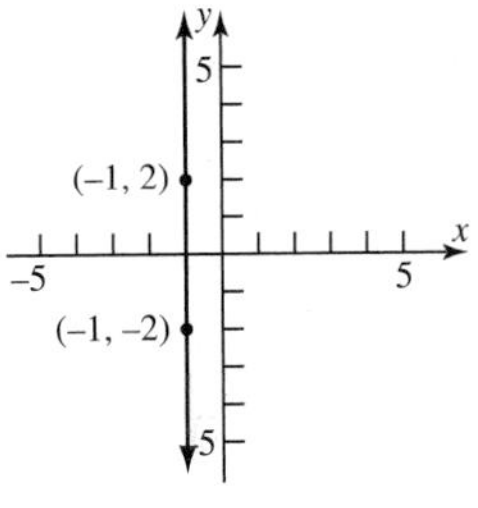

13. Slope $= \dfrac{\sqrt{3} - 3}{1 - \sqrt{2}} \approx 3.06$

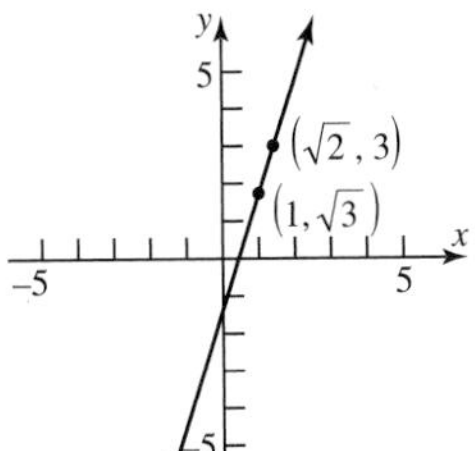

15.

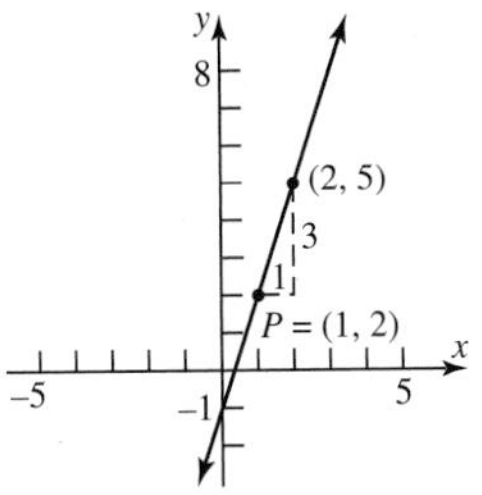

17.

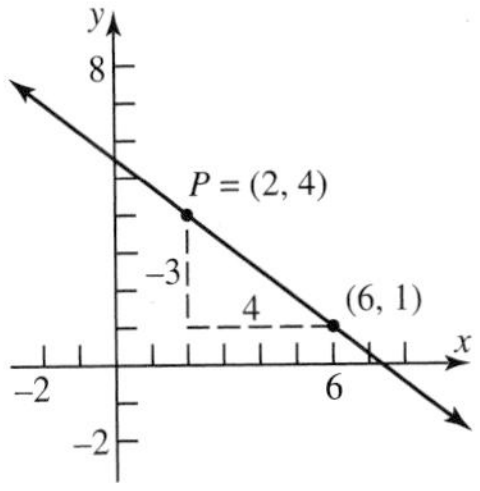

19.

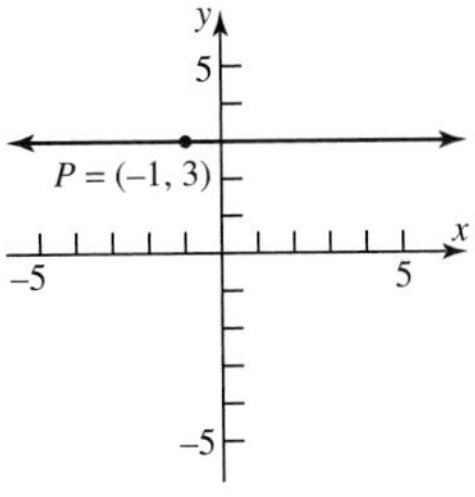

21.

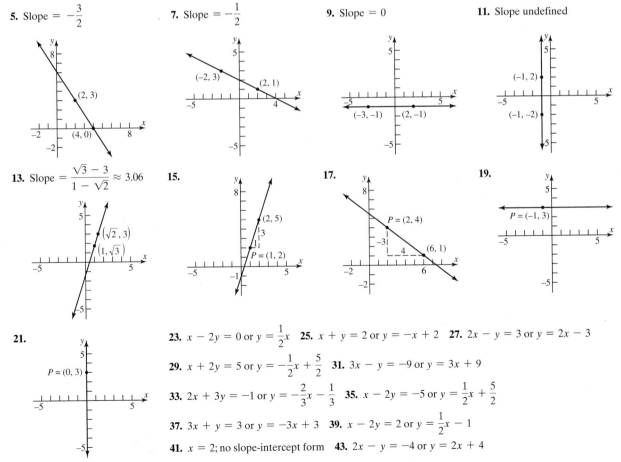

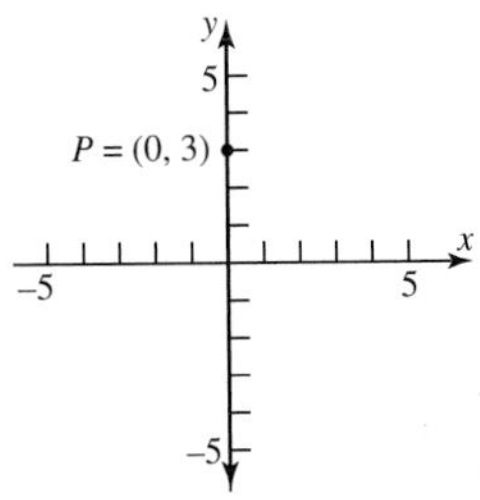

23. $x - 2y = 0$ or $y = \dfrac{1}{2}x$ **25.** $x + y = 2$ or $y = -x + 2$ **27.** $2x - y = 3$ or $y = 2x - 3$

29. $x + 2y = 5$ or $y = -\dfrac{1}{2}x + \dfrac{5}{2}$ **31.** $3x - y = -9$ or $y = 3x + 9$

33. $2x + 3y = -1$ or $y = -\dfrac{2}{3}x - \dfrac{1}{3}$ **35.** $x - 2y = -5$ or $y = \dfrac{1}{2}x + \dfrac{5}{2}$

37. $3x + y = 3$ or $y = -3x + 3$ **39.** $x - 2y = 2$ or $y = \dfrac{1}{2}x - 1$

41. $x = 2$; no slope-intercept form **43.** $2x - y = -4$ or $y = 2x + 4$

45. $2x - y = 0$ or $y = 2x$ **47.** $x = 4$; no slope–intercept form **49.** $2x + y = 0$ or $y = -2x$

51. $x - 2y = -3$ or $y = \frac{1}{2}x + \frac{3}{2}$ **53.** $y = 4$; no slope-intercept form

55. Slope $= 2$; y-intercept $= 3$

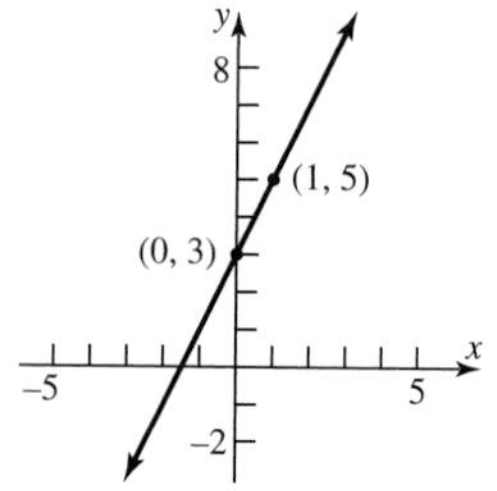

57. $y = 2x - 2$; Slope $= 2$; y-intercept $= -2$

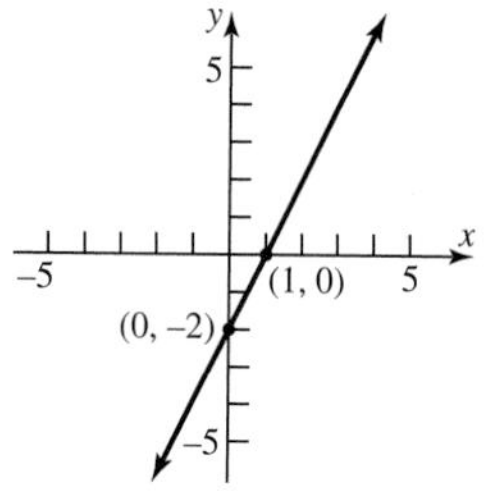

59. Slope $= \frac{1}{2}$; y-intercept $= 2$

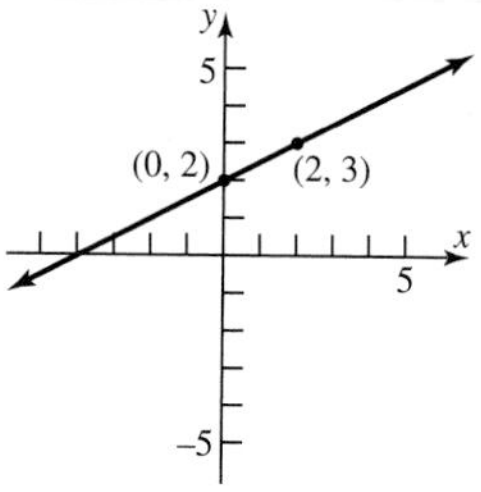

61. $y = -\frac{1}{2}x + 2$; Slope $= -\frac{1}{2}$; y-intercept $= 2$

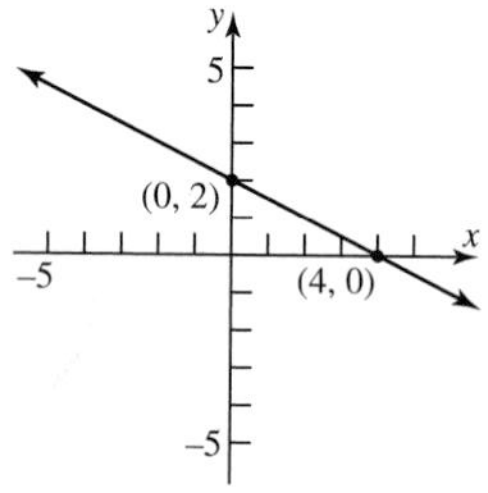

63. $y = \frac{2}{3}x - 2$; Slope $= \frac{2}{3}$; y-intercept $= -2$

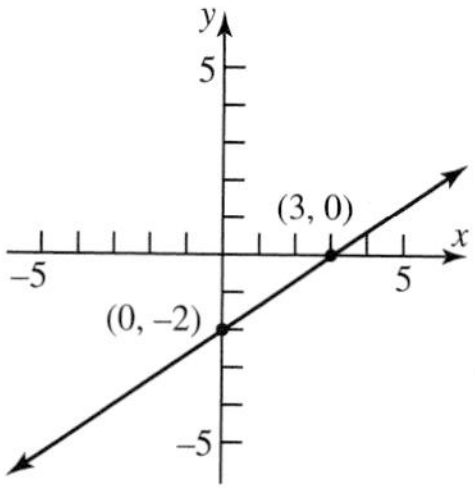

65. $y = -x + 1$; Slope $= -1$; y-intercept $= 1$

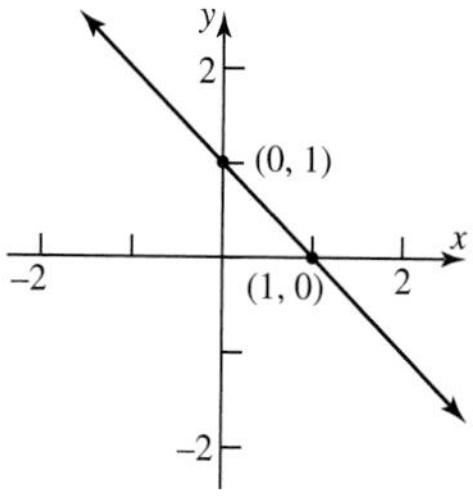

67. Slope undefined; no y-intercept

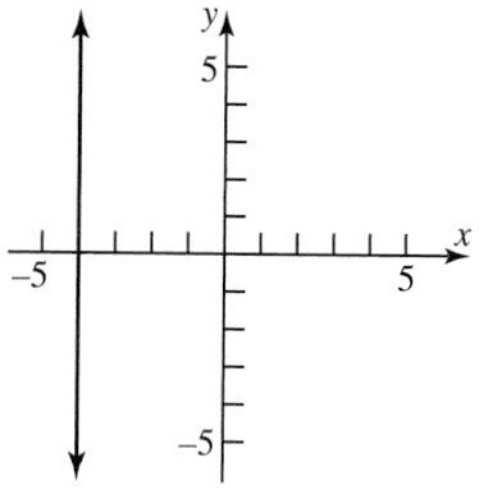

69. Slope $= 0$; y-intercept $= 5$

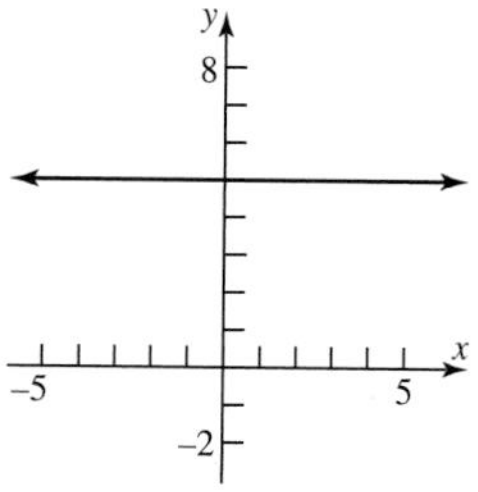

71. $y = x$; Slope $= 1$; y-intercept $= 0$

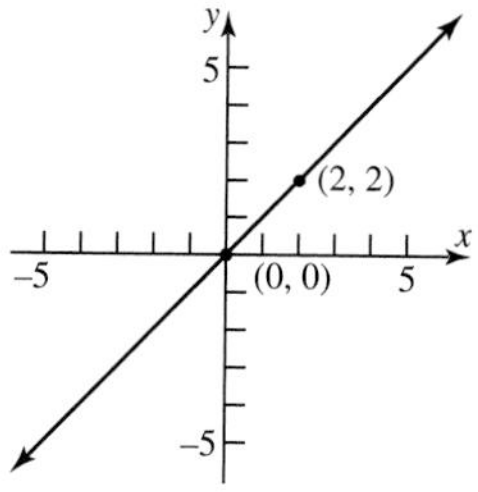

73. $y = \frac{3}{2}x$; Slope $= \frac{3}{2}$; y-intercept $= 0$

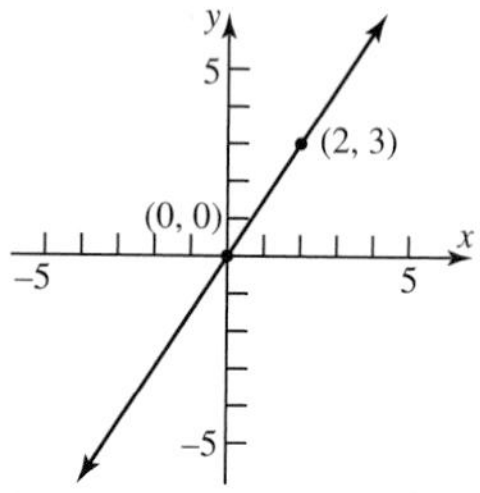

75. $y = 0$
77. (b)
79. (d)
81. $x - y = -2$ or $y = x + 2$
83. $x + 3y = 3$ or $y = -\frac{1}{3}x + 1$
85. $°C = \frac{5}{9}(°F - 32)$; approximately 21°C

87. **(a)** $P = 0.5x - 100$ **(b)** \$400 **(c)** \$2400

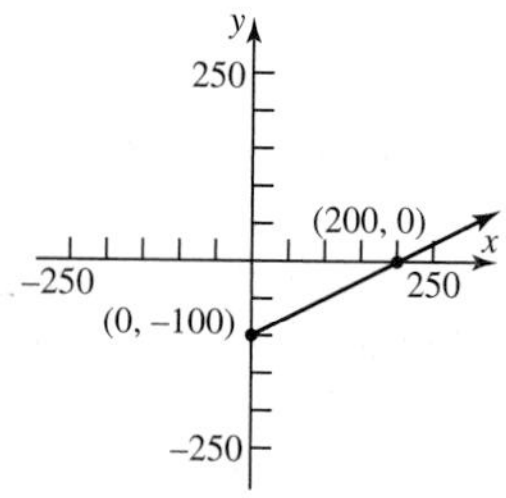

89. $C = 0.06543x + 5.65$; $C = \$25.28$; $C = \$54.72$

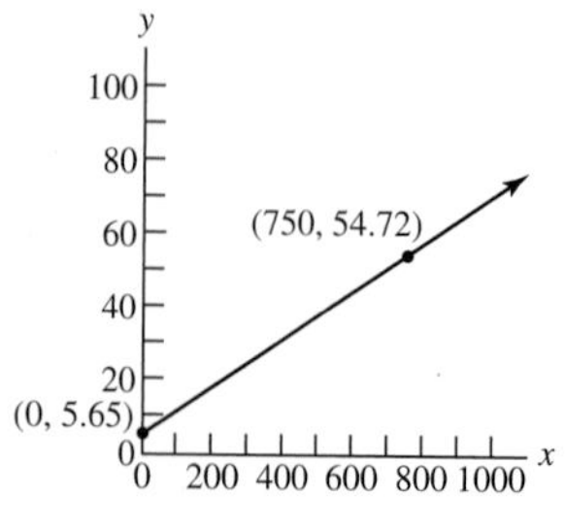

91. All have the same slope, 2; the lines are parallel.

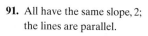

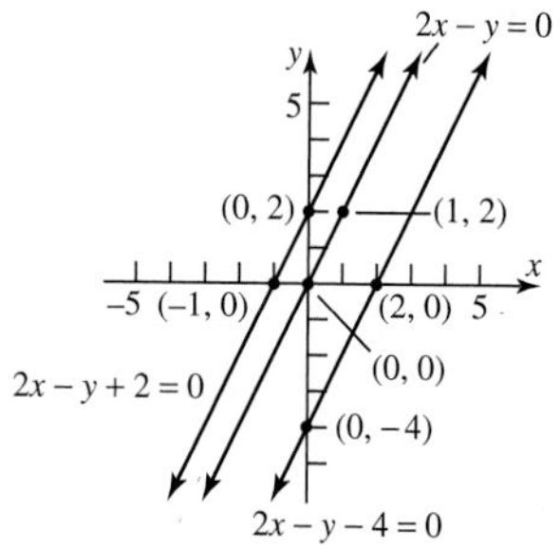

1.7 Exercises

1. Linear relation **3.** Linear relation **5.** Nonlinear relation

7. (a)

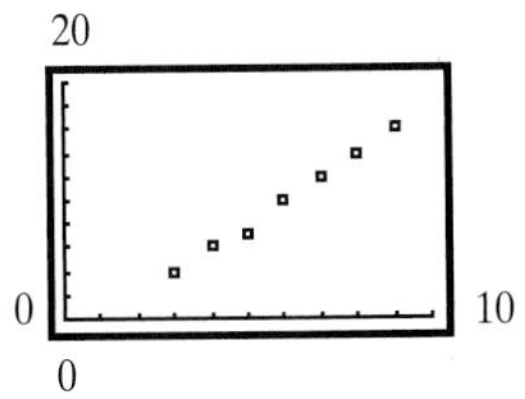

(b) Answers will vary.
Using (4, 6) and (8, 14):
$y = 2x - 2$

(c)

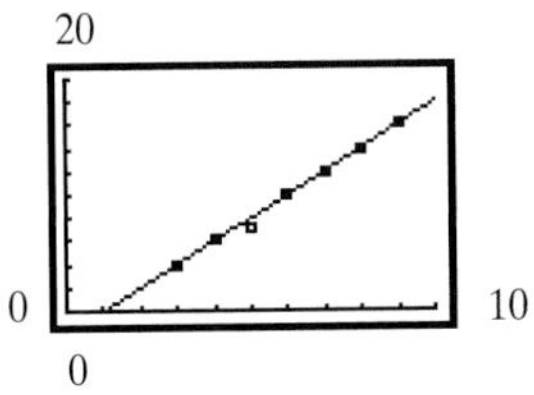

(d) $y = 2.0357x - 2.3571$

(e)

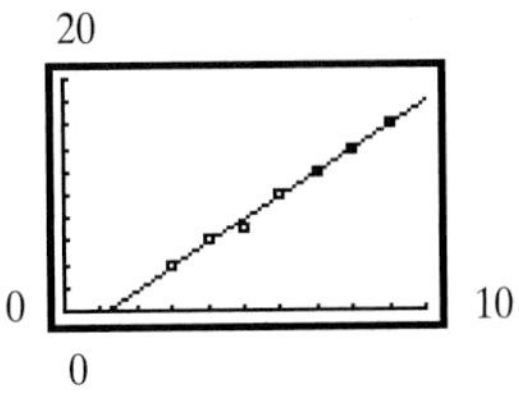

9. (a)

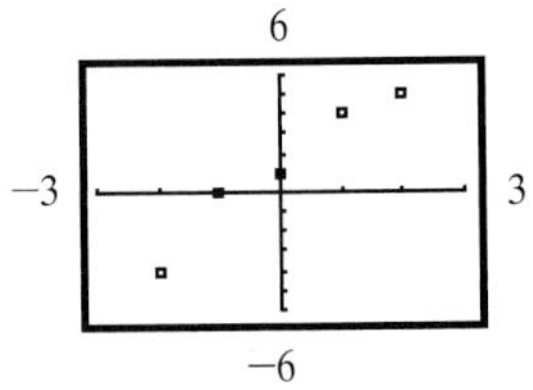

(b) Answers will vary.
Using (−2, −4) and (1, 4):
$y = \frac{8}{3}x + \frac{4}{3}$

(c)

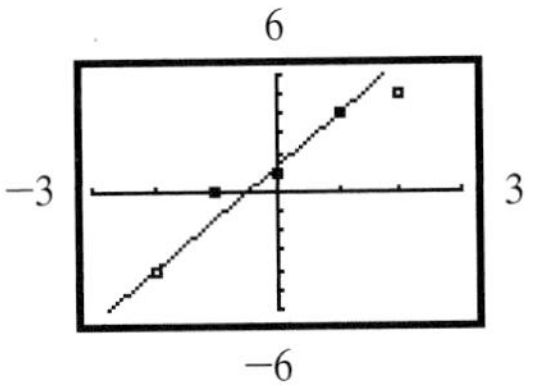

(d) $y = 2.2x + 1.2$

(e)

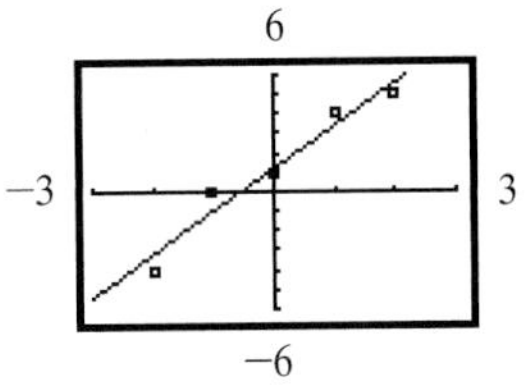

11. (a)

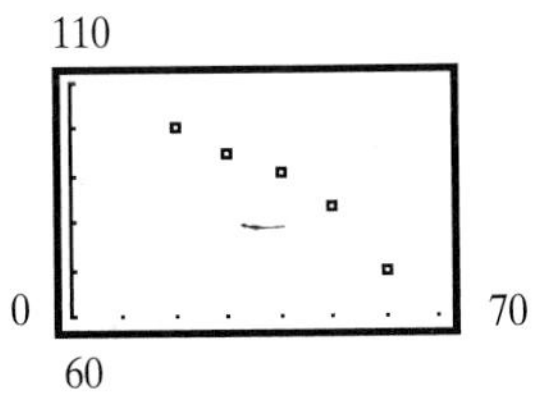

(b) Answers will vary.
Using (30, 95) and (60, 70):
$y = -\frac{5}{6}x + 120$

(c)

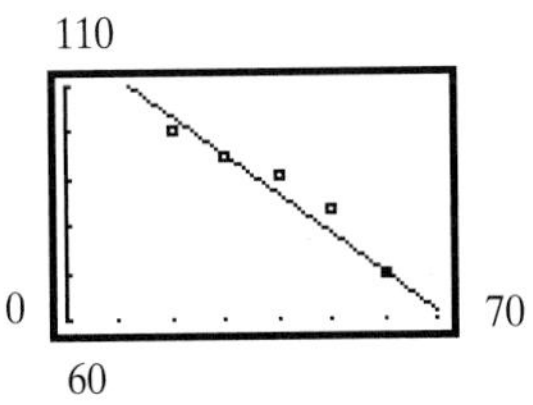

(d) $y = -0.72x + 116.6$

(e)

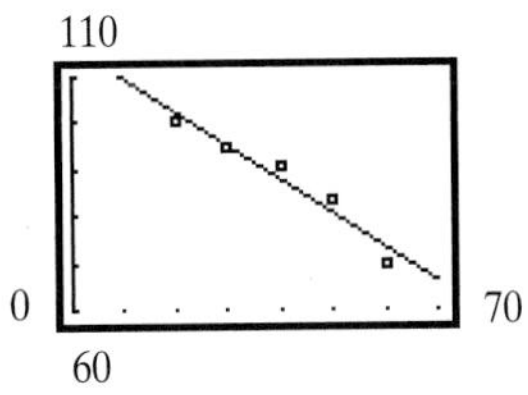

13. (a)

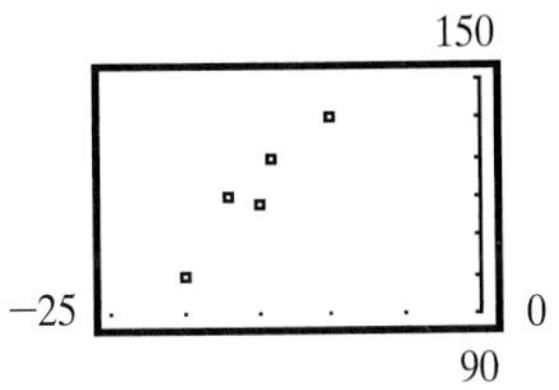

(b) Answers will vary.
Using (−20, 100) and (−15, 118):
$y = \frac{18}{5}x + 172$

(c)

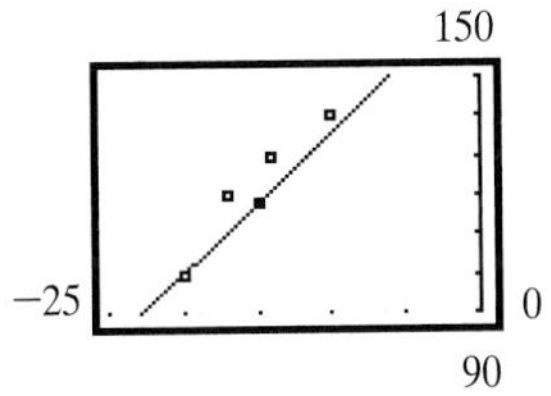

(d) $y = 3.8613x + 180.292$

(e)

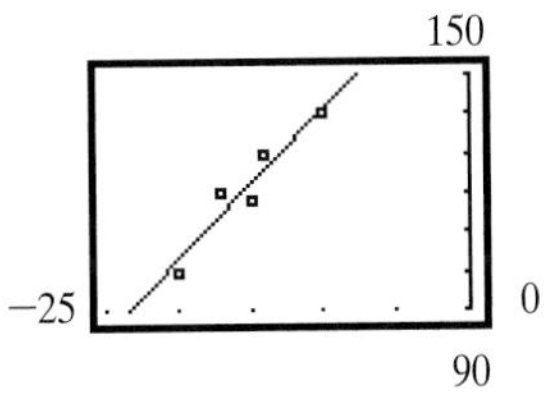

15. (a)

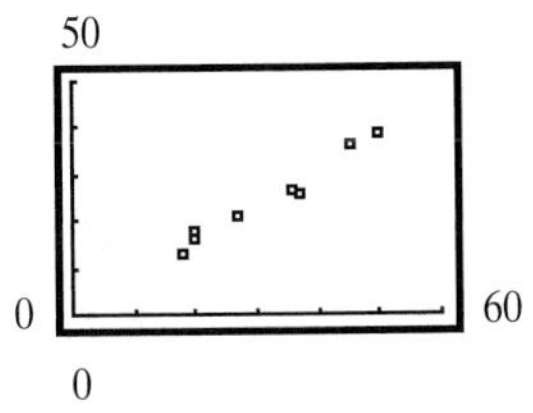

(b) $C = 0.755I + 0.6266$
(c) As disposable income increases by \$1000, consumption increases by about \$755
(d) \$32,337

17. (a)

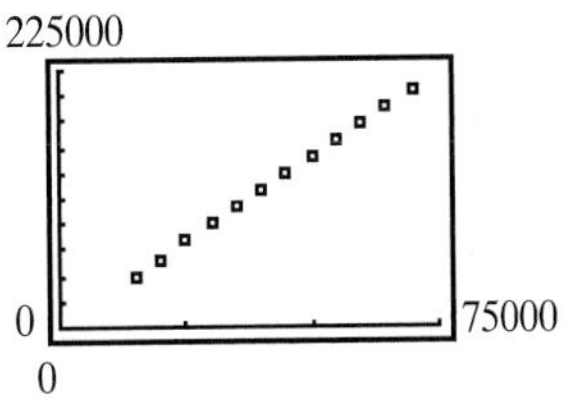

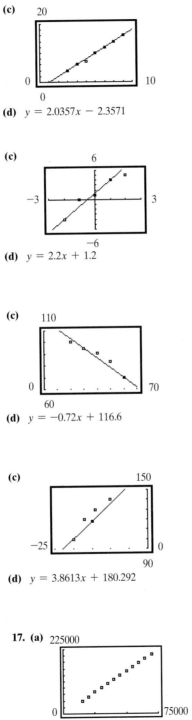

(b) $L = 2.98I - 76.11$

(c)

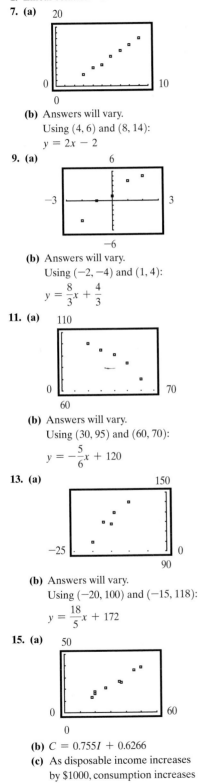

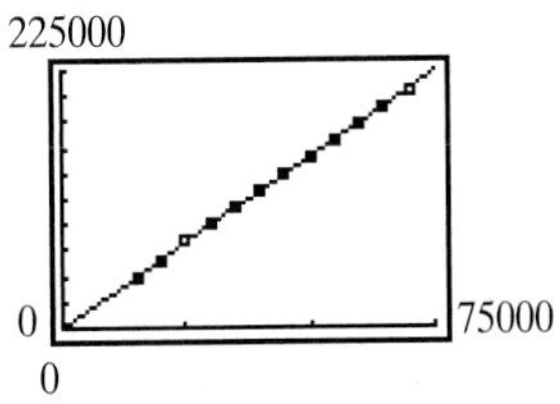

(d) For each additional \$1 in income, the loan amount increases by \$2.98.
(e) \$125,084

19. (a)

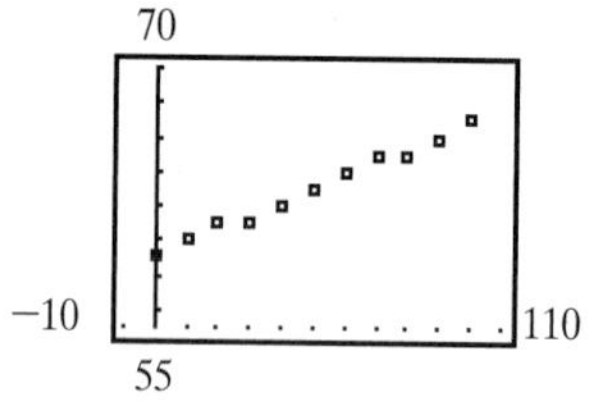

(b) $T = 0.0782h + 59.0909$

(c)

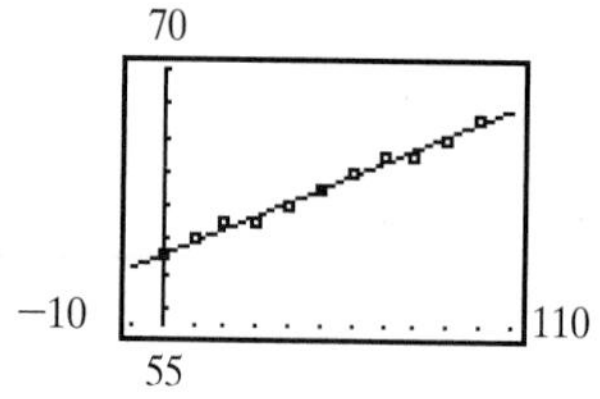

(d) If relative humidity increases by 1%, the apparent temperature increases by 0.0782 degrees Fahrenheit.

(e) 65°F

21. (a)

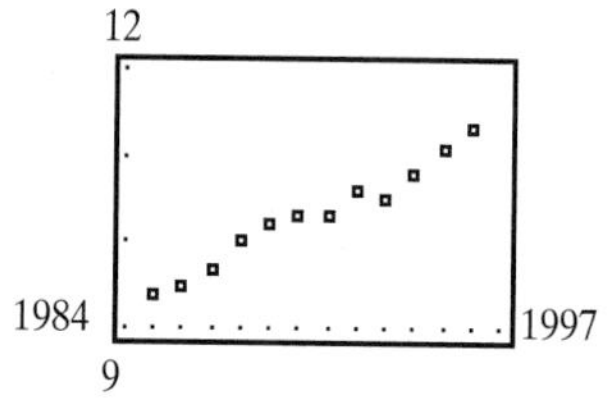

(b) $M = 0.163x - 314.714$

(c)

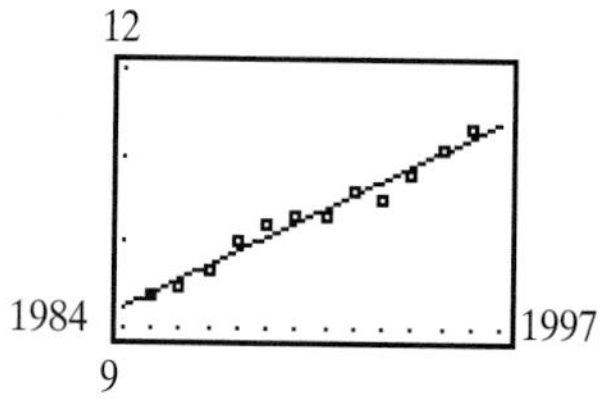

(d) The average miles per car has increased by 0.163 thousand miles per year on average between 1985 and 1996.

(e) 10.8 thousand miles

23. (a)

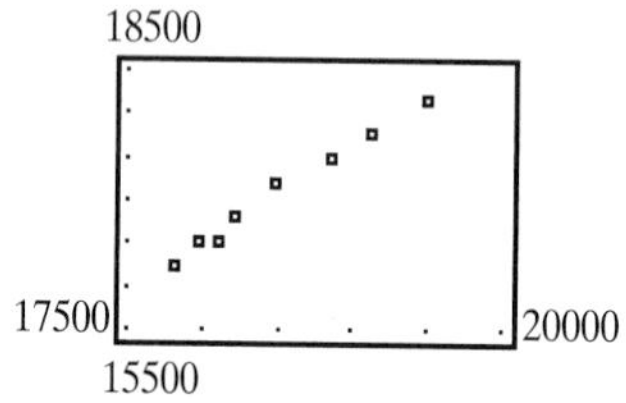

(b) $y = 1.12x - 3687.54$

(c)

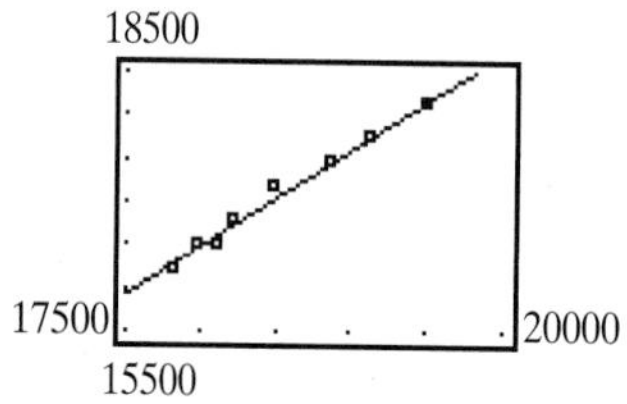

(d) If per capital personal income increase \$1, per capita personal consumption expenditures increase \$1.12.

(e) \$17,876

1.8 Exercises

1. Center (2, 1); Radius 2; $(x - 2)^2 + (y - 1)^2 = 4$ **3.** Center $\left(\frac{5}{2}, 2\right)$; Radius $\frac{3}{2}$; $\left(x - \frac{5}{2}\right)^2 + (y - 2)^2 = \frac{9}{4}$

5. $(x - 1)^2 + (y + 1)^2 = 1$; $x^2 + y^2 - 2x + 2y + 1 = 0$

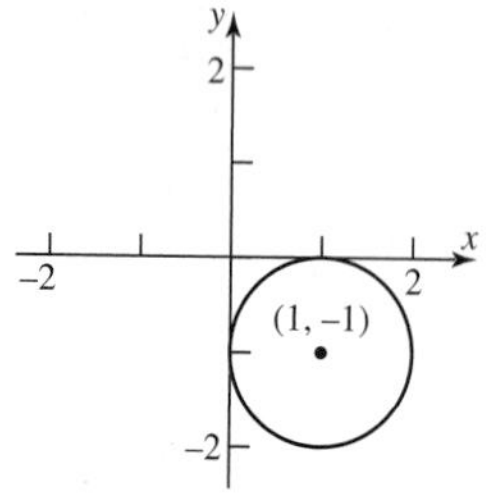

7. $x^2 + (y - 2)^2 = 4$; $x^2 + y^2 - 4y = 0$

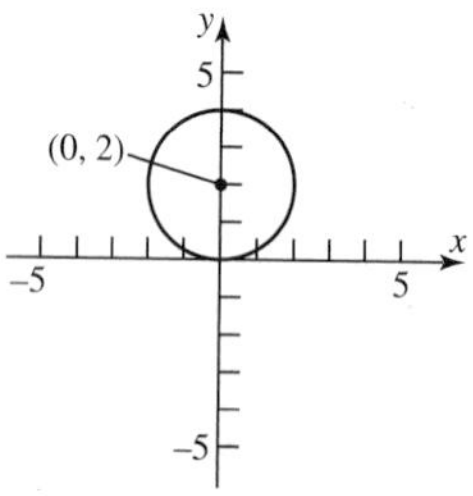

9. $(x - 4)^2 + (y + 3)^2 = 25$; $x^2 + y^2 - 8x + 6y = 0$

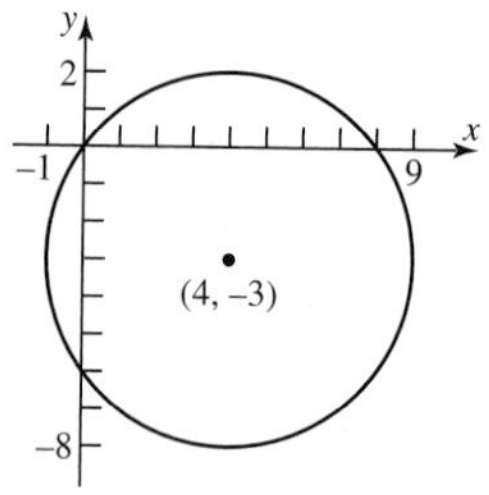

11. $x^2 + y^2 = 4$; $x^2 + y^2 - 4 = 0$

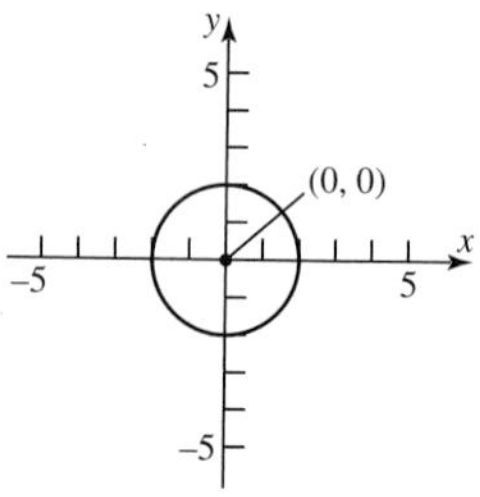

13. $\left(x - \frac{1}{2}\right)^2 + y^2 = \frac{1}{4}$; $x^2 + y^2 - x = 0$

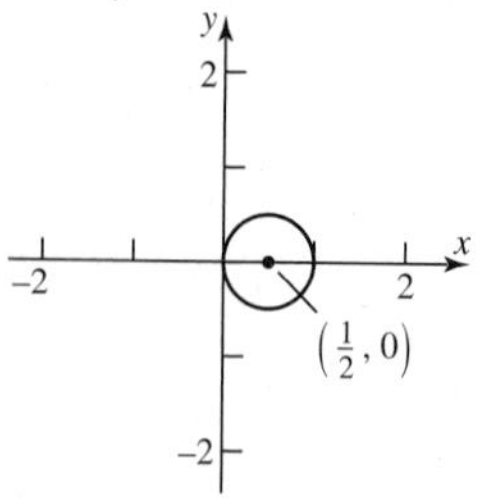

15. $r = 2$; $(h, k) = (0, 0)$

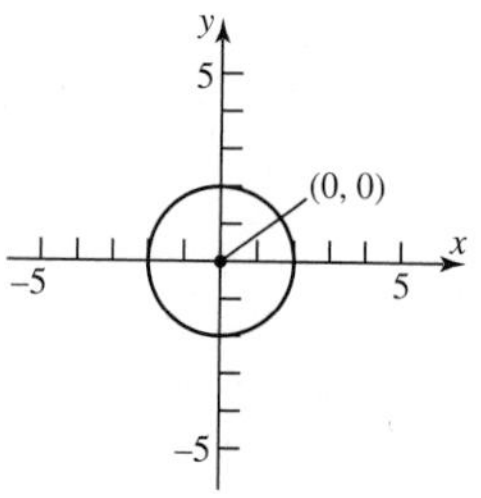

17. $r = 2$; $(h, k) = (3, 0)$

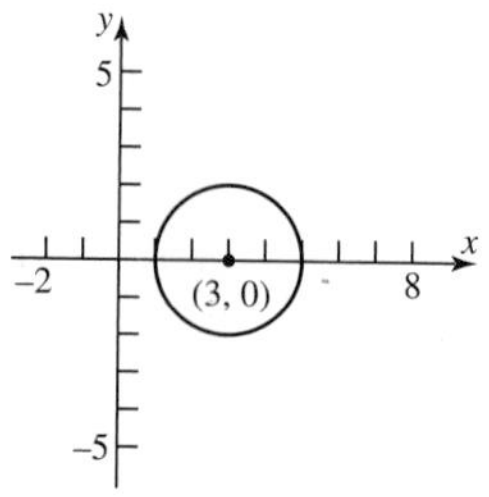

19. $r = 3$; $(h, k) = (-2, 2)$

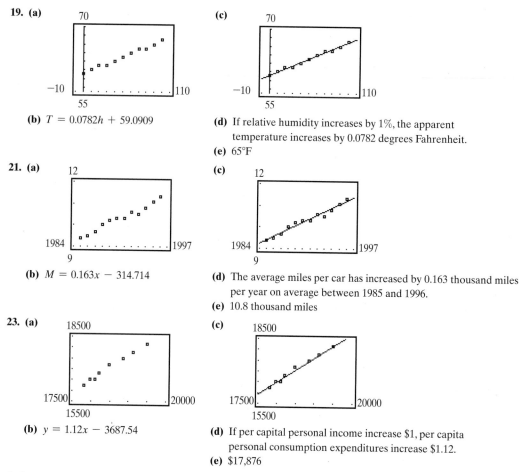

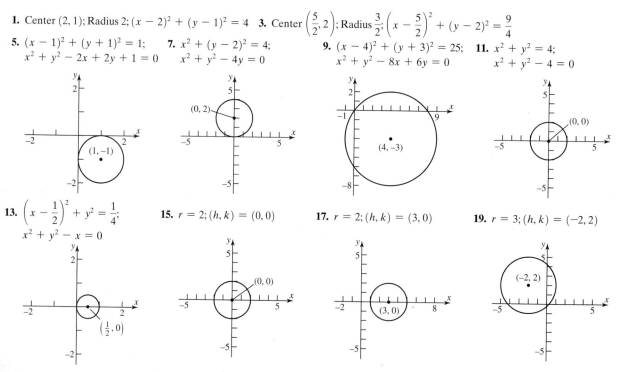

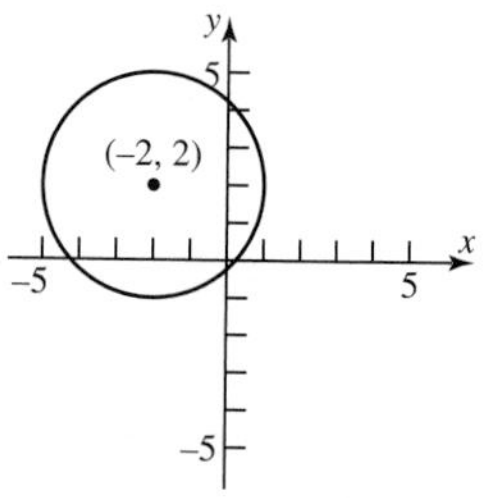

21. $r = \frac{1}{2}; (h, k) = \left(\frac{1}{2}, -1\right)$

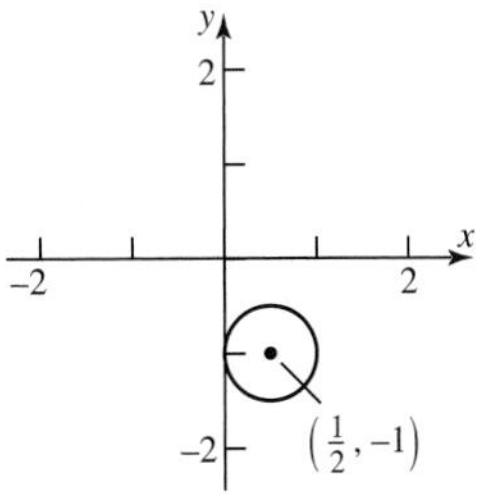

23. $r = 5; (h, k) = (3, -2)$

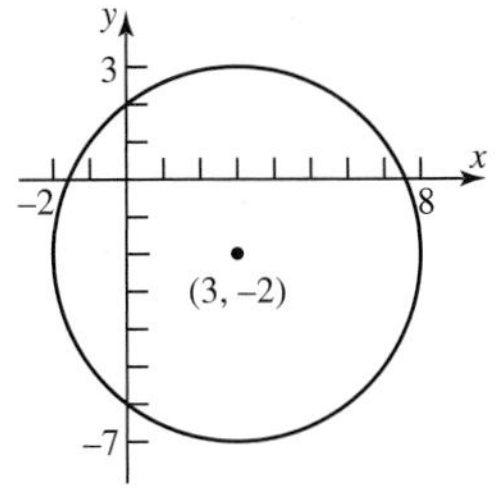

25. $x^2 + y^2 - 13 = 0$

27. $x^2 + y^2 - 4x - 6y + 4 = 0$

29. $x^2 + y^2 + 2x - 6y + 5 = 0$

31. (c)

33. (b)

35. $(x + 3)^2 + (y - 1)^2 = 16$

37. $(x - 2)^2 + (y - 2)^2 = 9$

39. **(a)** $x^2 + (mx + b)^2 = r^2$

$(1 + m^2)x^2 + 2mbx + b^2 - r^2 = 0$

One solution if and only if discriminant $= 0$

$(2mb)^2 - 4(1 + m^2)(b^2 - r^2) = 0$

$-4b^2 + 4r^2 + 4m^2r^2 = 0$

$r^2(1 + m^2) = b^2$

(b) $x = \frac{-2mb}{2(1 + m^2)} = \frac{-2mb}{2b^2/r^2} = -\frac{r^2m}{b}$

$y = m\left(-\frac{r^2m}{b}\right) + b = -\frac{r^2m^2}{b} + b = \frac{-r^2m^2 + b^2}{b} = \frac{r^2}{b}$

(c) Slope of tangent line $= m$

Slope of line joining center to point of tangency $= \frac{r^2/b}{-r^2m/b} = -\frac{1}{m}$

41. $\sqrt{2}x + 4y - 11\sqrt{2} + 12 = 0$

43. $x + 5y + 13 = 0$

Fill-in-the-Blank Items

1. Abscissa; ordinate **3.** y-axis **5.** Extraneous **7.** $-2; 2$ **9.** Undefined; 0

True/False Items

1. False **3.** True **5.** False **7.** True **9.** True **11.** True

Review Exercises

1. -12 **3.** 6 **5.** $\frac{1}{5}$ **7.** -5 **9.** $\{-2, 3\}$ **11.** $\frac{11}{8}$ **13.** $\left\{-2, \frac{3}{2}\right\}$ **15.** $\left\{-\frac{1}{2}, \frac{3}{2}\right\}$ **17.** 9 **19.** $\{-2, 1\}$ **21.** 2 **23.** $\frac{\sqrt{5}}{2}$ **25.** $\{-5, 2\}$

27. $\left\{-\frac{5}{3}, 3\right\}$

29. $\{x \mid 14 \le x < \infty\}$ or $[14, \infty)$

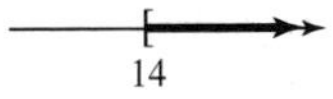

31. $\left\{x \middle| -\frac{31}{2} \le x \le \frac{33}{2}\right\}$ or $\left[-\frac{31}{2}, \frac{33}{2}\right]$

33. $\{x \mid -23 < x < -7\}$ or $(-23, -7)$

35. $\left\{x \middle| -\frac{3}{2} < x < -\frac{7}{6}\right\}$ or $\left(-\frac{3}{2}, -\frac{7}{6}\right)$

37. $\{x \mid -\infty < x \le -2 \text{ or } 7 \le x < \infty\}$ or $(-\infty, -2]$ or $[7, \infty)$

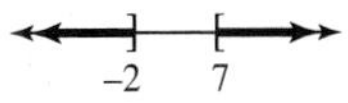

39. $2x + y - 5 = 0$ or $y = -2x + 5$ **41.** $x + 3 = 0$; no slope-intercept form **43.** $x + 5y + 10 = 0$ or $y = -\frac{1}{5}x - 2$

45. $2x - 3y + 19 = 0$ or $y = \frac{2}{3}x + \frac{19}{3}$ **47.** $-x + y + 7 = 0$ or $y = x - 7$

49. $4x - 5y = -20$

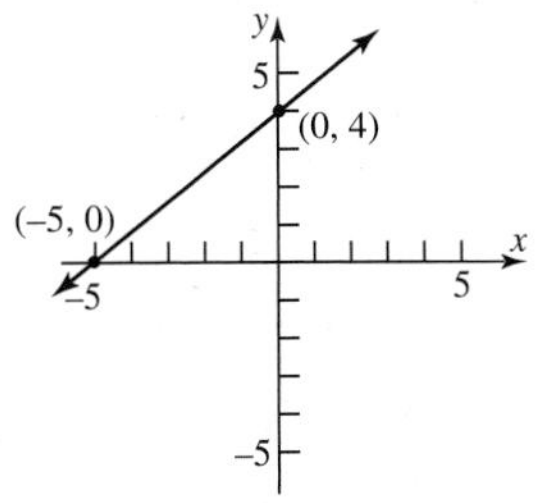

51. $\frac{1}{2}x - \frac{1}{3}y = -\frac{1}{6}$

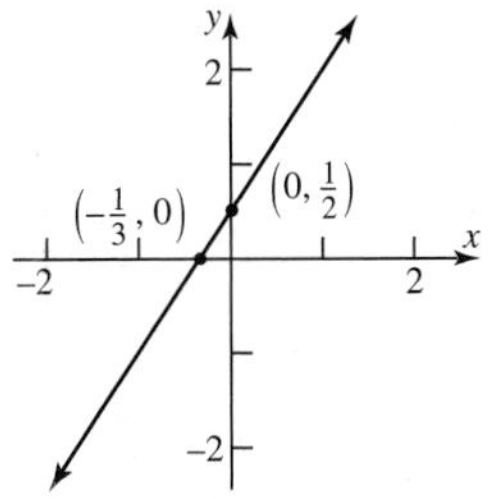

53. $\sqrt{2}x + \sqrt{3}y = \sqrt{6}$

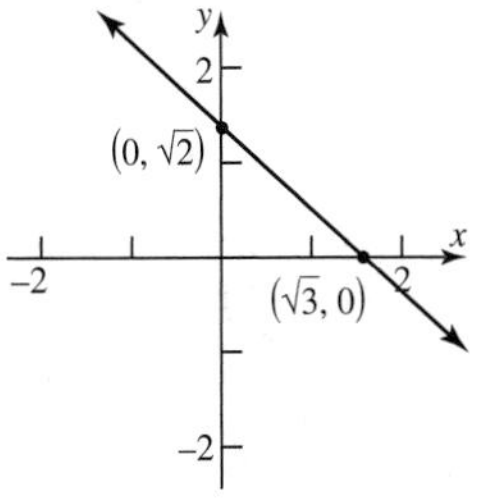

55. Intercept: $(0, 0)$; symmetric with respect to the x-axis

57. Intercepts: $\left(\frac{1}{2}, 0\right), \left(-\frac{1}{2}, 0\right), (0, 1), (0, -1)$; symmetric with respect to the x-axis, y-axis, and origin

59. Intercept: $(0, 1)$; symmetric with respect to the y-axis **61.** Intercepts: $(0, 0), (0, -2), (-1, 0)$; no symmetry

63. Center $(1, -2)$; Radius $= 3$

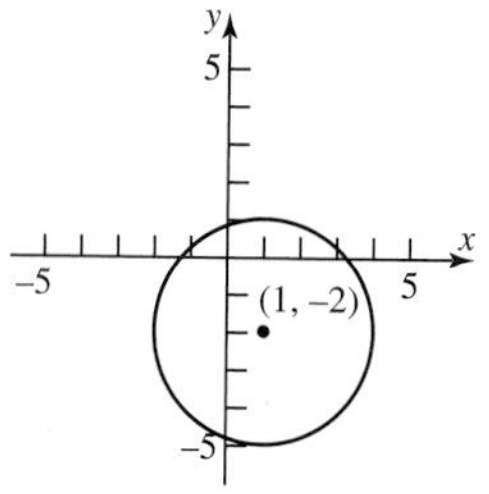

65. Center $(1, -2)$; Radius $= \sqrt{5}$

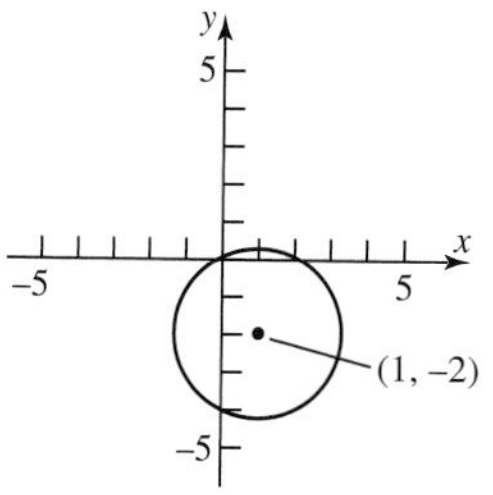

67. Slope $= \frac{1}{5}$; distance $= 2\sqrt{26}$; midpoint $= (2, 3)$

69. $d(A, B) = \sqrt{13}$; $d(B, C) = \sqrt{13}$

71. $M_{AB} = -1$; $M_{BC} = -1$

73. Center $(1, -2)$; radius $= 4\sqrt{2}$; $x^2 + y^2 - 2x + 4y - 27 = 0$

75. (a)

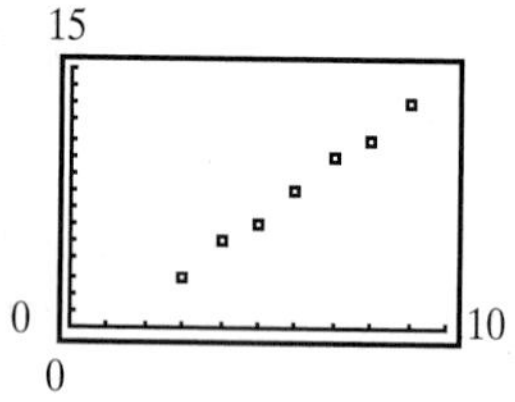

(b) $y = 1.643x - 1.857$

(c) As x increases by 1, y increases by 1.643.

77. (a)

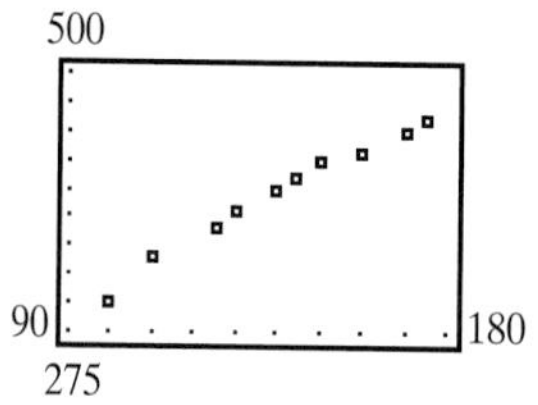

(b) $y = 2.018x + 112.433$

(c) As x increases by 1, y increases by 2.018.

79. The storm is 3300 ft away.

81. The search plane can go as far as 616 mi.

83. The helicopter will reach the life raft in a little less than 1 hr, 35 min.

85. It takes Clarissa 10 days by herself.

87. Mix 30cc of 15% HCI with 70cc of 5% HCI.

89. There were 3260 adults.

91. The freight train is 190.67 ft long.

93. 36 seniors went on the trip; each one paid $13.40.

95. (a)

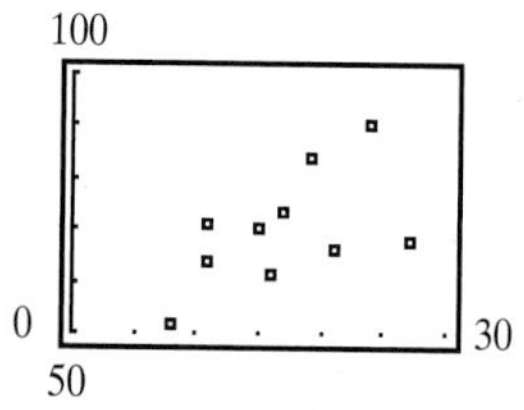

(b) Calculus $= 1.018$algebra $+ 52.69$

(c) As the algebra score increases by 1 point, the calculus score increases

(d) 73

97. (a)

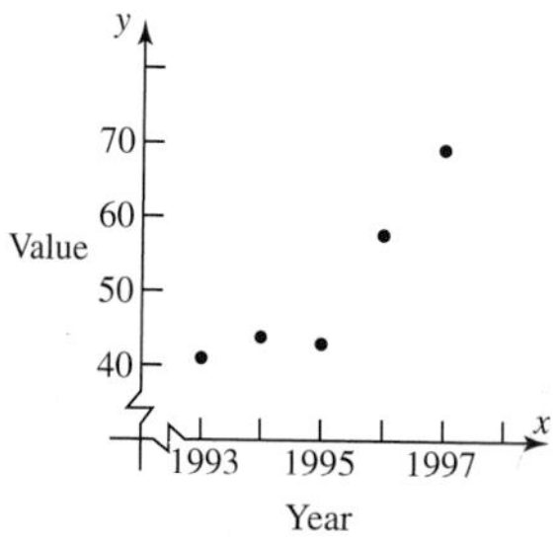

(b) As time passes, the value of the portfolio increases.

(c)

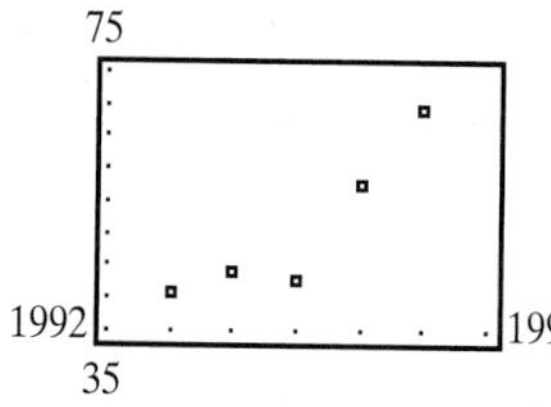

(d) 1

(e) Between the years 1993 and 1995, the value of the portfolio increases $1 per year on average.

(f) 13.1

(g) Between the years 1995 and 1997 the value of the portfolio increased $13.10 per year on average.

CHAPTER 2 Functions and Models

2.1 Exercises

1. Function; Domain: $\{20, 25, 30, 40\}$, Range: $\{\$200, \$300, \$350, \$425\}$ **3.** Not a function
5. Function; Domain: $\{-3, 1, 2, 4\}$, Range: $\{6, 9, 10\}$ **7.** Function; Domain: $\{1, 2, 3, 4\}$; Range: $\{3\}$ **9.** Not a function
11. Function; Domain: $\{-2, -1, 0, 1\}$, Range: $\{0, 1, 4\}$ **13. (a)** -4 **(b)** 1 **(c)** -3 **(d)** $3x^2 - 2x - 4$ **(e)** $-3x^2 - 2x + 4$
(f) $3x^2 + 8x + 1$ **(g)** $f(2x) = 12x^2 + 4x - 4$ **(h)** $f(x + h) = 3x^2 + 6xh + 3h^2 + 2x + 2h - 4$ **15. (a)** 0 **(b)** $\frac{1}{2}$ **(c)** $-\frac{1}{2}$
(d) $\frac{-x}{x^2 + 1}$ **(e)** $\frac{-x}{x^2 + 1}$ **(f)** $\frac{x + 1}{x^2 + 2x + 2}$ **(g)** $f(2x) = \frac{2x}{4x^2 + 1}$ **(h)** $f(x + h) = \frac{x + h}{x^2 + 2xh + h^2 + 1}$
17. (a) 4 **(b)** 5 **(c)** 5 **(d)** $|x| + 4$ **(e)** $-|x| - 4$ **(f)** $|x + 1| + 4$ **(g)** $f(2x) = 2|x| + 4$ **(h)** $f(x + h) = |x + h| + 4$
19. (a) $-\frac{1}{5}$ **(b)** $-\frac{3}{2}$ **(c)** $\frac{1}{8}$ **(d)** $\frac{-2x + 1}{-3x - 5}$ **(e)** $\frac{-2x - 1}{3x - 5}$ **(f)** $\frac{2x + 3}{3x - 2}$ **(g)** $f(2x) = \frac{4x + 1}{6x - 5}$ **(h)** $f(x + h) = \frac{2x + 2h + 1}{3x + 3h - 5}$

21. $f(0) = 3; f(-6) = -3$ **23.** Positive **25.** $-3, 6$ and 10 **27.** $\{x|-6 \le x \le 11\}$ **29.** $(-3, 0), (6, 0), (10, 0)$ **31.** 3 times

33. $0, 4$ **35.** **(a)** No **(b)** $-3; (4, -3)$ **(c)** $14; (14, 2)$ **(d)** $\{x|x \ne 6\}$ **37.** **(a)** Yes **(b)** $\frac{8}{17}; \left(2, \frac{8}{17}\right)$ **(c)** $-1, 1; (-1, 1), (1, 1)$ **(d)** All real numbers **39.** Not a function **41.** Function **(a)** Domain: $\{x|-\pi \le x \le \pi\}$; Range: $\{y|-1 \le y \le 1\}$ **(b)** Intercepts: $\left(-\frac{\pi}{2}, 0\right), \left(\frac{\pi}{2}, 0\right), (0, 1)$ **(c)** y-axis **43.** Not a function **45.** Function **(a)** Domain: $\{x|x > 0\}$; Range: All real numbers **(b)** Intercept: $(1, 0)$ **(c)** None **47.** Function **(a)** Domain: all real numbers; Range: $\{y|y \le 2\}$ **(b)** Intercept: $(-3, 0), (3, 0), (0, 2)$ **(c)** y-axis **49.** Function **(a)** Domain: all real numbers; Range: $\{y|y \ge -3\}$ **(b)** Intercept: $(0, 9), (1, 0), (3, 0)$ **(c)** None **51.** All real numbers **53.** All real numbers **55.** $\{x|x \ne -4, x \ne 4\}$ **57.** $\{x|x \ne 0\}$ **59.** $\{x|x \ge 4\}$ **61.** $\{x|x > 9\}$ **63.** $\{x|x > 1\}$ **65.** Function **67.** Function **69.** Not a function **71.** Function **73.** **(a)** III **(b)** IV **(c)** I **(d)** V **(e)** II

75.

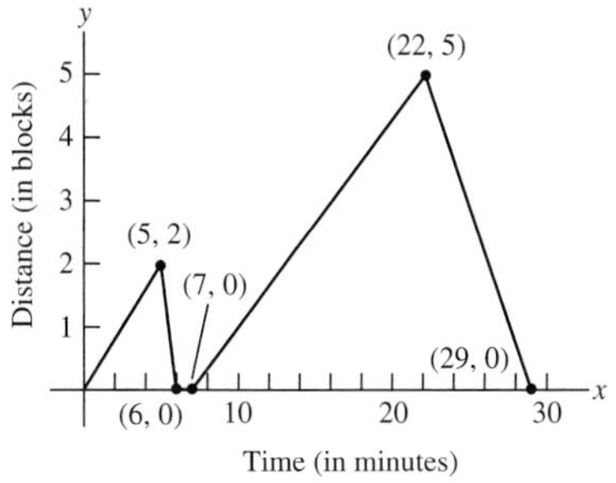

77. **(a)** 4 times
(b) Kevin traveled 3 miles at a constant speed.
(c) Kevin remained in the same location.
(d) Kevin returned home at a constant speed.
(e) Kevin remained at home.
(f) Kevin traveled 2.8 miles with his speed decreasing with time.
(g) Kevin remained in the same location.
(h) Kevin returned home with his speed decreasing with time.
(i) 3 miles

79. $A = -\frac{7}{2}$ **81.** $A = -4$ **83.** $A = 8$; undefined at $x = 3$

85. **(a)** No
(b)

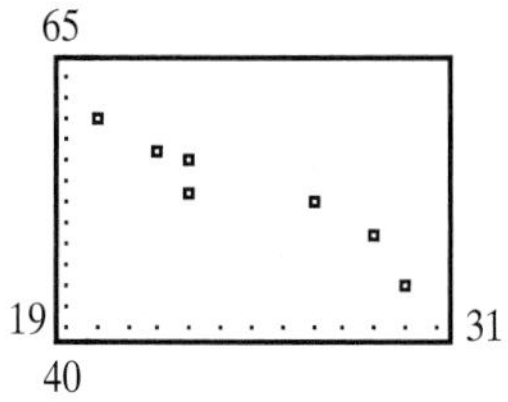

(c) $D = -1.34p + 86.20$
(d) If price increases \$1, quantity demanded will decrease 1.34.
(e) $D(p) = -1.34p + 86.20$
(f) $\{p|p > 0\}$
(g) About 49 pairs

87. **(a)** Yes
(b)

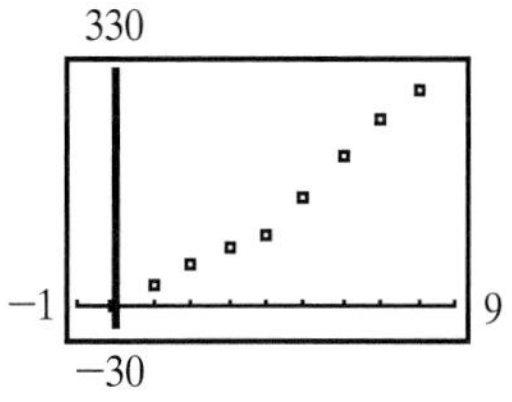

(c) $s = 37.78t - 19.13$
(d) If t increases by 1 hour, distance increases by 37.78 miles.
(e) $s(t) = 37.78t - 19.13$
(f) Since time must be nonnegative, the domain is $\{t|t \ge 0\}$.
(g) About 396 miles.

89. **(a)**

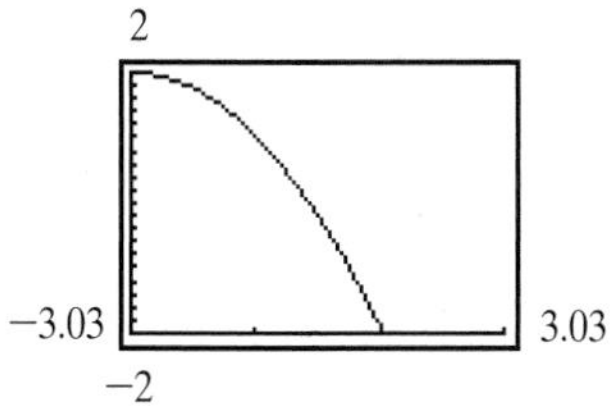

(b) 15.1 m, 14.07 m, 12.94 m, 11.72 m
(c) 1.01 sec, 1.42 sec, 1.74 sec
(d) 2.02 sec

91. $A(x) = \frac{1}{2}x^2$ **93.** $G(x) = 10x$

95. **(a)** 81.1 ft **(b)** 129.6 ft **(c)** 26.6 ft
(d)

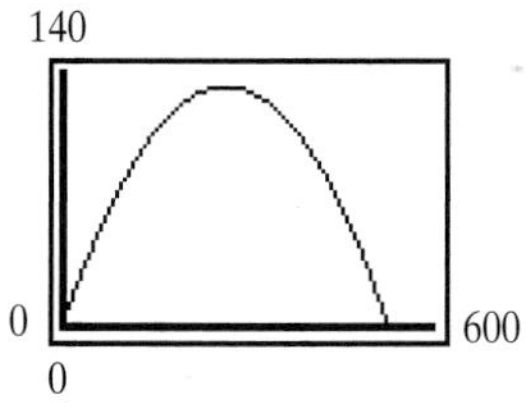

(e) Approximately 115.07 feet and 413.05 feet.
(f) 275 ft **(g)** 264 ft **(h)** Domain: $\{0 \le x \le 528.125\}$

97. **(a)** \$222 **(b)** \$225 **(c)** \$220 **(d)** \$230
(e)

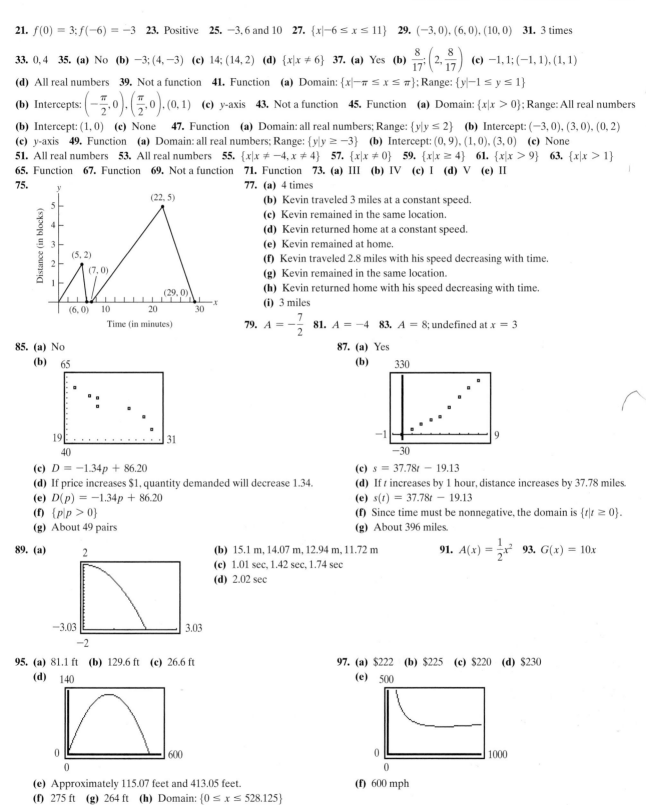

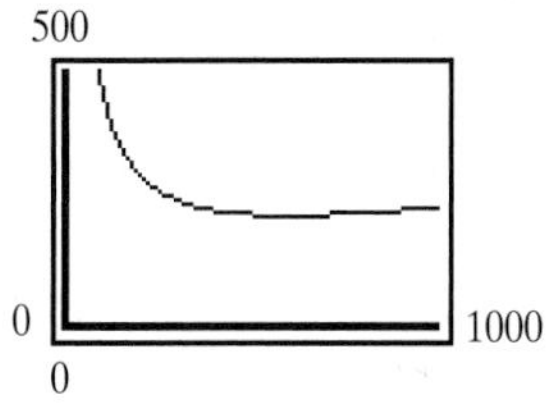

(f) 600 mph

99. Only $h(x) = 2x$ **101.** No $f(x)$ has a domain of all real numbers, while $g(x)$ has a domain of $\{x|x \ne -1\}$.

2.2 Exercises

1. Yes **3.** No **5.** $(-8,-2)$; $(0,2)$; $(5,10)$ **7.** Yes; $f(2)=10$ **9.** -2 and 2; $f(-2)=6$ and $f(2)=10$

11. (a) $(-2,0)$, $(0,3)$, $(2,0)$ **(b)** Domain $\{x|-4\le x\le 4\}$ or $[-4,4]$; Range: $\{y|0\le y\le 3\}$ or $[0,3]$
(c) Increasing on $(-2,0)$ and $(2,4)$; Decreasing on $(-4,-2)$ and $(0,2)$. **(d)** Even

13. (a) $(0,1)$ **(b)** Domain: all real numbers; Range: $\{y|y>0\}$ or $(0,\infty)$. **(c)** Increasing on $(-\infty,\infty)$ **(d)** Neither

15. (a) $(-\pi,0)$, $(0,0)$, $(\pi,0)$ **(b)** Domain: $\{x|-\pi\le x\le \pi\}$ or $[-\pi,\pi]$; Range: $\{y|-1\le y\le 1\}$ or $[-1,1]$.
(c) Increasing on $\left(-\frac{\pi}{2},\frac{\pi}{2}\right)$; Decreasing on $\left(-\pi,-\frac{\pi}{2}\right)$ and on $\left(\frac{\pi}{2},\pi\right)$ **(d)** Odd

17. (a) $\left(0,\frac{1}{2}\right)$, $\left(\frac{1}{2},0\right)$, $(2.5,0)$ **(b)** Domain: $\{x|-3\le x\le 3\}$ or $[-3,3]$; Range: $\{y|-1\le y\le 2\}$ or $[-1,2]$
(c) Increasing on $(2,3)$; Decreasing on $(-1,1)$; Constant on $(-3,-1)$ and on $(1,2)$. **(d)** Neither

19. (a) $(-2,0)$, $(0,2)$, $(2,0)$ **(b)** Domain: $\{x|-4\le x\le 4\}$ or $[-4,4]$; Range: $\{y|0\le y\le 2\}$ or $[0,2]$
(c) Increasing on $(-2,0)$ and on $(2,4)$; Decreasing on $(-4,-2)$ and on $(0,2)$. **(d)** Even

21. (a) 0; $f(0)=3$ **(b)** -2 and 2; $f(-2)=0$ and $f(2)=0$ **23. (a)** $\frac{\pi}{2}$; $f\left(\frac{\pi}{2}\right)=1$ **(b)** $-\frac{\pi}{2}$; $f\left(-\frac{\pi}{2}\right)=-1$

25. (a) 5 **(b)** 5 **(c)** $y=5x$
(d)
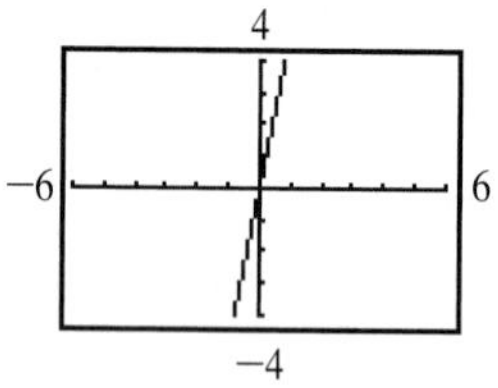

27. (a) -3 **(b)** -3 **(c)** $y=-3x+1$
(d)
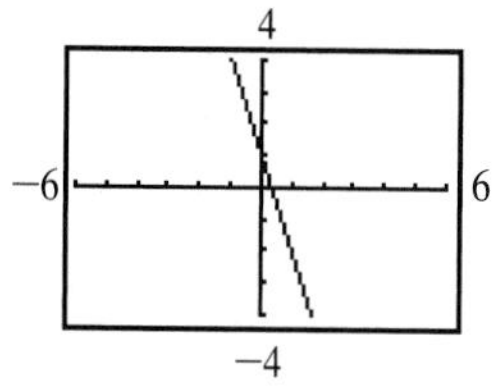

29. (a) $x-1$ **(b)** 1 **(c)** $y=x-2$
(d)
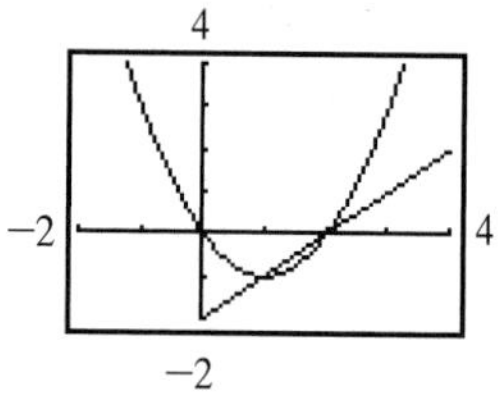

31. (a) $x(x+1)$ **(b)** 6 **(c)** $y=6x-6$
(d)
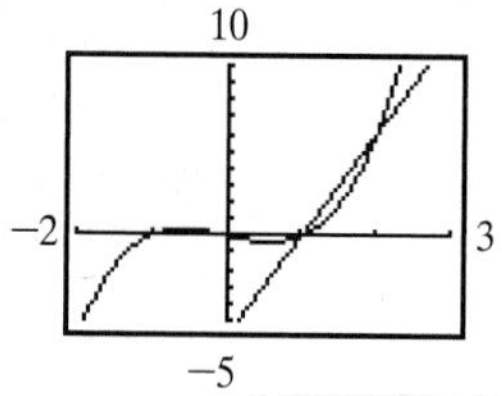

33. (a) $-\frac{1}{x+1}$ **(b)** $-\frac{1}{3}$
(c) $y=-\frac{1}{3}x+\frac{4}{3}$
(d)
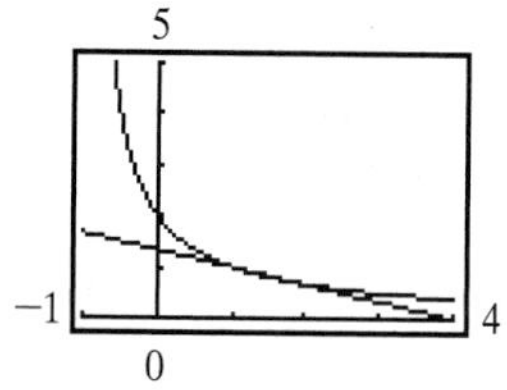

35. (a) $\frac{1}{\sqrt{x}+1}$ **(b)** $\frac{1}{\sqrt{2}+1}$
(c) $y=\frac{1}{\sqrt{2}+1}x-\frac{1}{\sqrt{2}+1}+1$
(d)
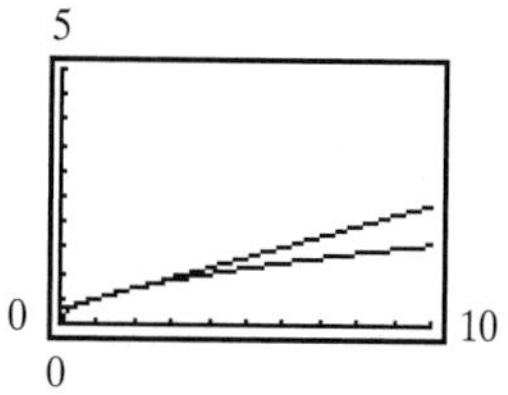

37. Odd **39.** Even **41.** Odd **43.** Neither **45.** Even **47.** Odd **49.** At most one

51.
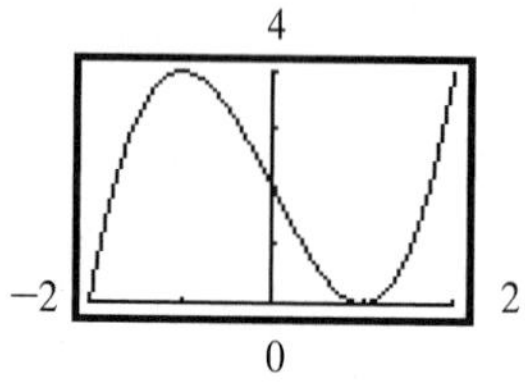

Increasing: $(-2,-1)$, $(1,2)$
Decreasing: $(-1,1)$
Local maximum: $(-1,4)$
Local minimum: $(1,0)$

53.
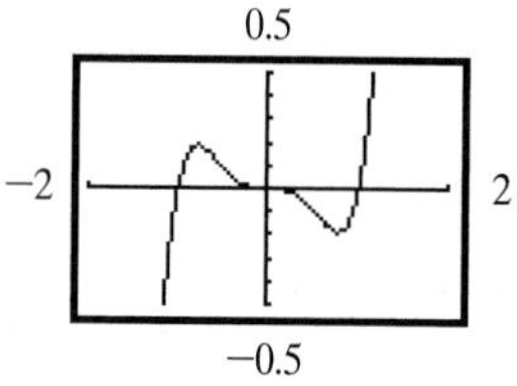

Increasing: $(-2,-0.77)$, $(0.77,2)$
Decreasing: $(-0.77,0.77)$
Local maximum: $(-0.77,0.19)$
Local minimum: $(0.77,-0.19)$

55.
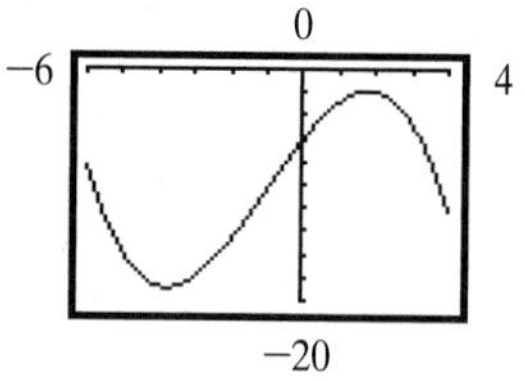

Increasing: $(-3.77,1.77)$
Decreasing: $(-6,-3.77)$, $(1.77,4)$
Local maximum: $(1.77,-1.91)$
Local minimum: $(-3.77,-18.89)$

57.
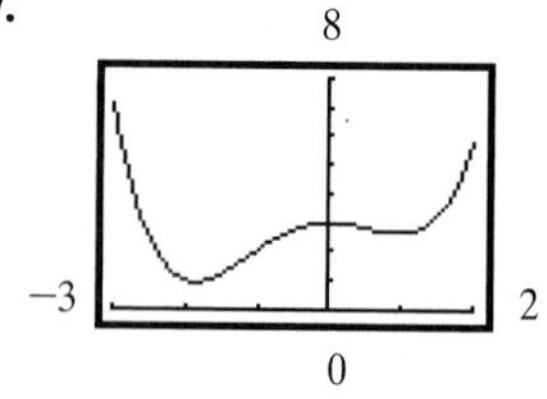

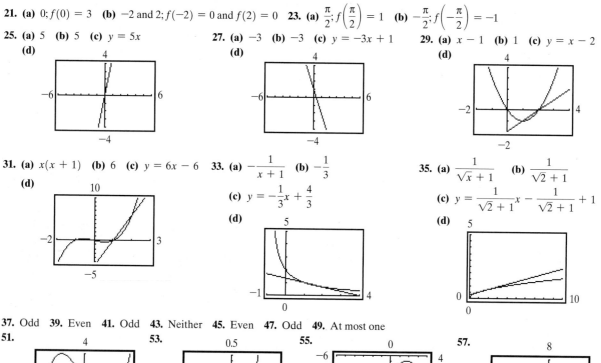

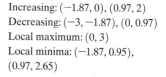

Increasing: $(-1.87,0)$, $(0.97,2)$
Decreasing: $(-3,-1.87)$, $(0,0.97)$
Local maximum: $(0,3)$
Local minima: $(-1.87,0.95)$, $(0.97,2.65)$

59. **(a)** 2 **(b)** 2; 2; 2 $m_{\text{sec}} = 2$
(c) $y = 2x + 5$
(d)

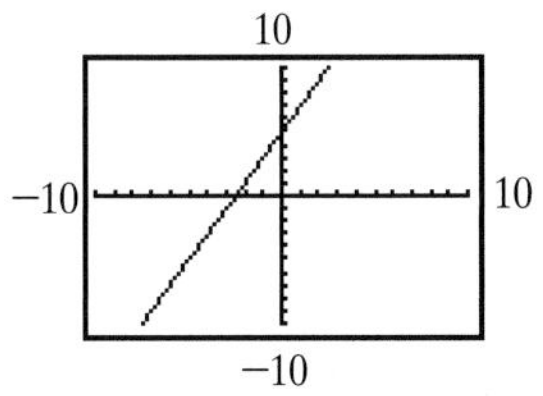

61. **(a)** $2x + h + 2$ **(b)** 4.5, 4.1, 4.01, 4
(c) $y = 4.01x - 1.01$
(d)

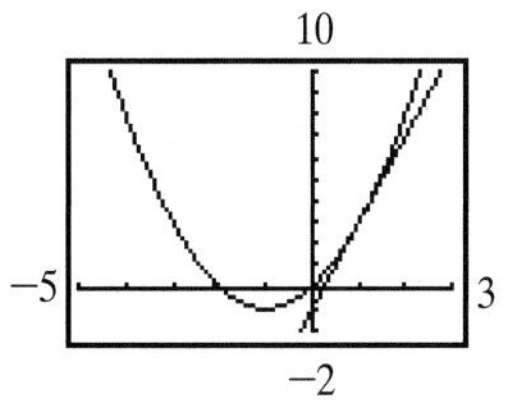

63. **(a)** $-\dfrac{1}{(x+1)(x+h+1)}$
(b) $-0.2, -\dfrac{5}{21}, -\dfrac{50}{201}, -\dfrac{1}{4}$
(c) $y = -\dfrac{50}{201}x + \dfrac{301}{402}$
(d)

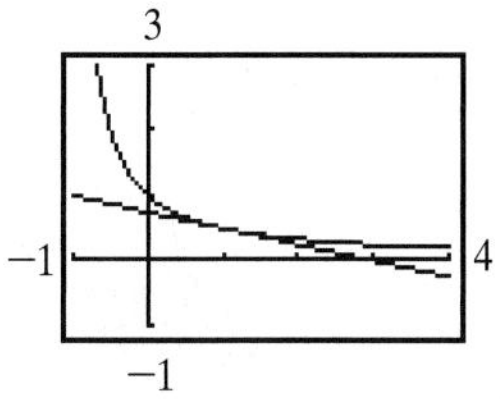

65. **(a)**, **(b)**, **(e)**

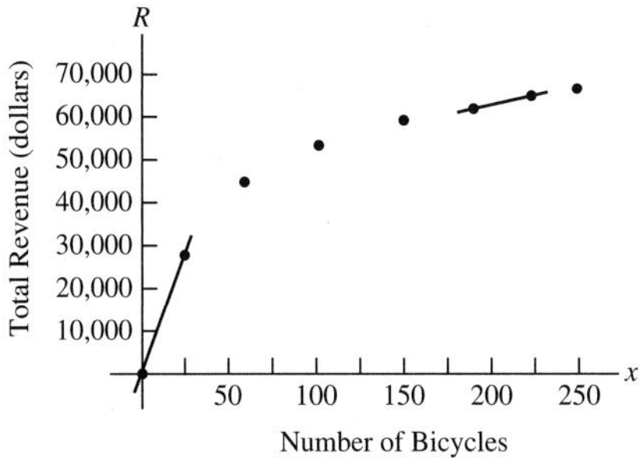

(c) 1120 dollars/bicycle
(d) For each additional bicycle sold between 0 and 25 bicycles, total revenue increases, on average, by $1120.
(f) 75 dollars/bicycle
(g) For each additional bicycle sold between 190 and 223 bicycles, total revenue increases, on average, by $75.

67. **(a)**, **(b)**, **(e)**

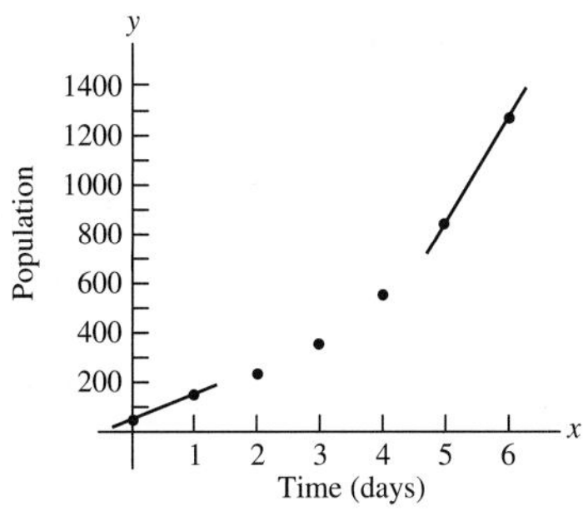

(c) 103 bacteria/day
(d) The population is increasing at an average rate of 103 bacteria per day between day 0 and day 1.
(f) 441 bacteria/day
(g) The population is increasing at an average rate of 441 bacteria per day between day 5 and day 6.
(h) The average rate of change is increasing.

69.

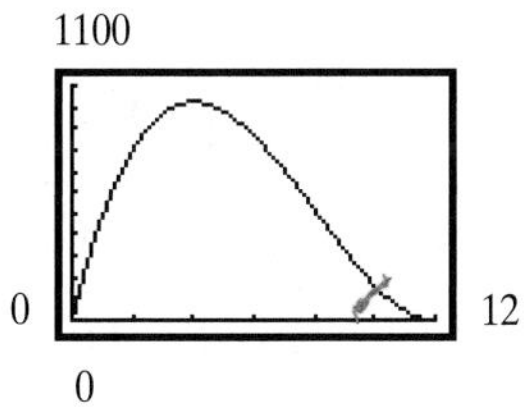

The volume is largest at $x = 4$ inches.

71. **(a)**

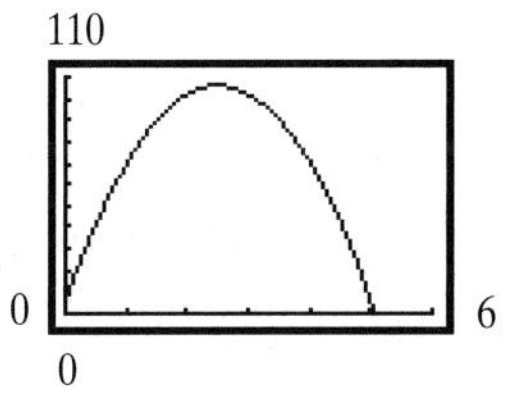

(b) 2.5 seconds **(c)** 106 feet

2.3 Exercises

1. C
3. E
5. B
7. F

9.

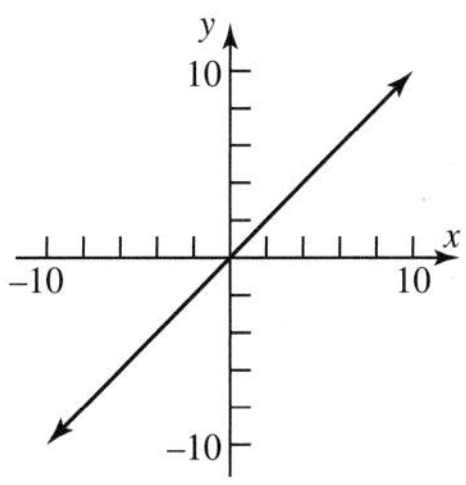

11.

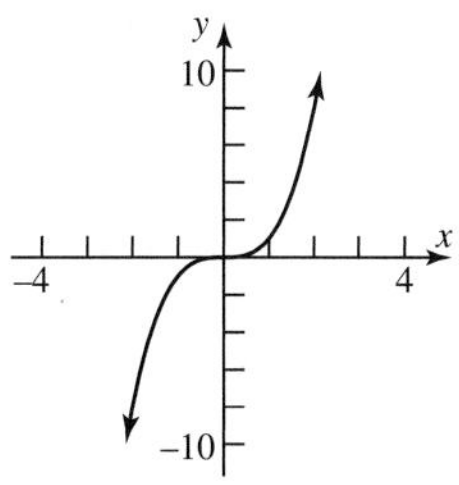

13.

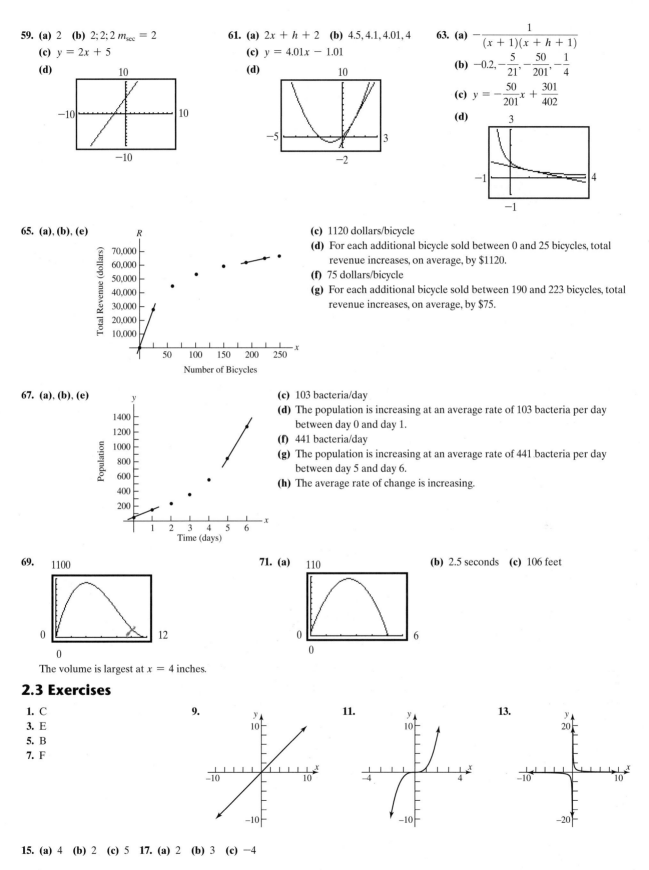

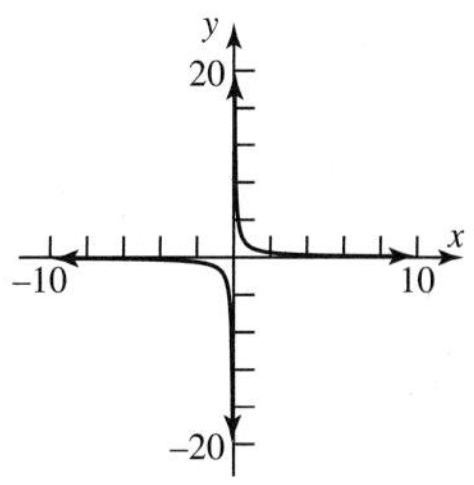

15. **(a)** 4 **(b)** 2 **(c)** 5 **17.** **(a)** 2 **(b)** 3 **(c)** −4

19. **(a)** All real numbers
(b) $(0, 1)$
(c)

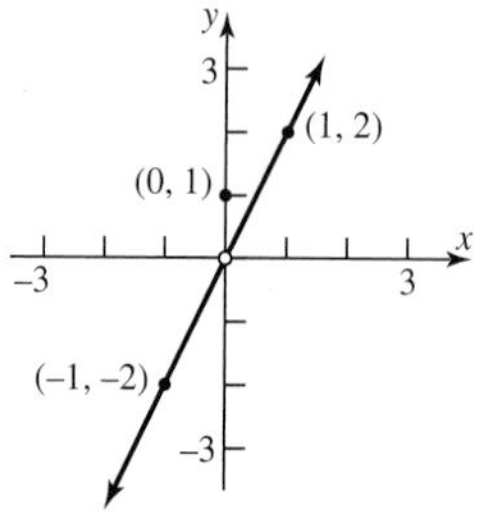

(d) $\{y|y \neq 0\}$ or $(-\infty, 0)$ or $(0, \infty)$

21. **(a)** All real numbers
(b) $(0, 3)$
(c)

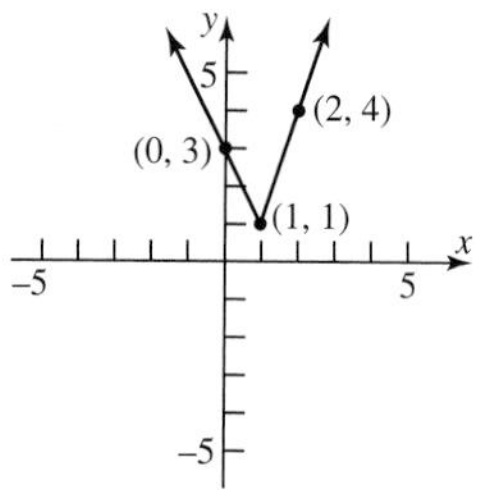

(d) $\{y|y \geq 1\}$ or $[1, \infty]$

23. **(a)** $\{x|x \geq -2\}$ or $[-2, \infty)$
(b) $(0, 3)$, $(2, 0)$
(c)

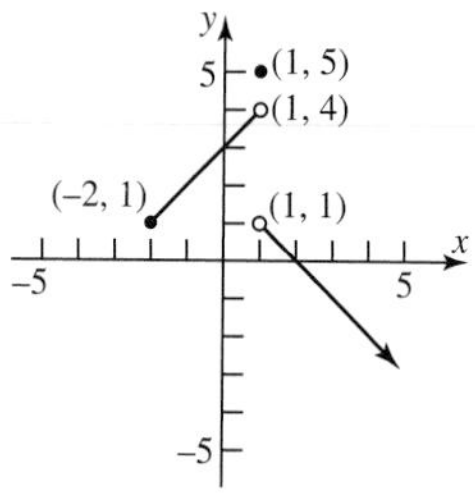

(d) $\{y|y < 4, y = 5\}$

25. **(a)** All real numbers
(b) $(-1, 0)$, $(0, 0)$
(c)

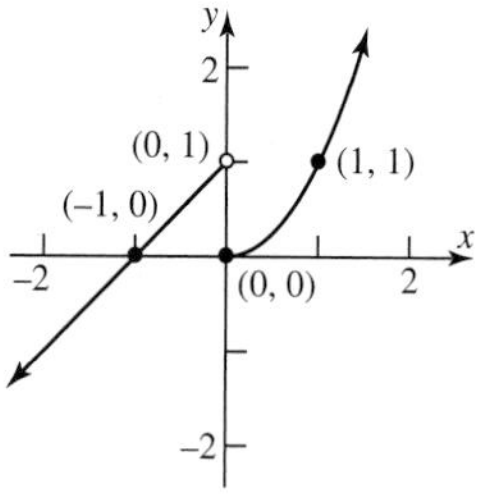

(d) All real numbers

27. **(a)** $\{x|x \geq -2\}$
(b) $(0, 1)$
(c)

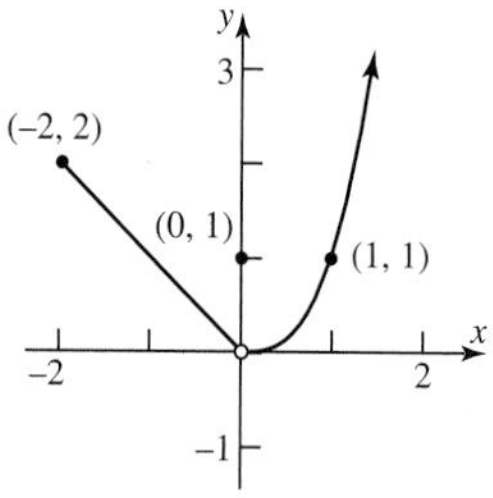

(d) $\{y|y > 0\}$ or $(0, \infty)$

29. **(a)** All real numbers
(b) $(x, 0)$ for $0 \leq x < 1$
(c)

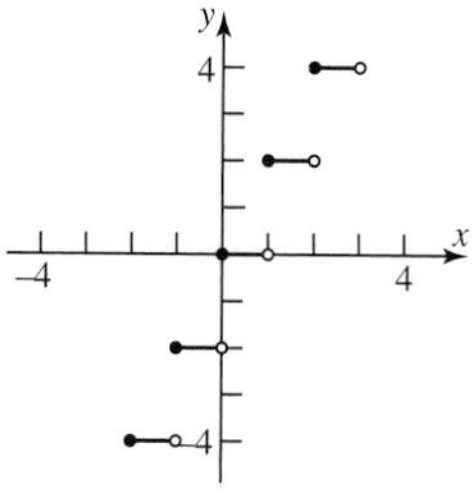

(d) Set of even integers

31. $f(x) = \begin{cases} -x & \text{if } -1 \leq x \leq 0 \\ \frac{1}{2}x & \text{if } 0 < x \leq 2 \end{cases}$ (Other answers are possible.)

33. $f(x) = \begin{cases} -x & \text{if } x \leq 0 \\ -x + 2 & \text{if } 0 < x \leq 2 \end{cases}$ (Other answers are possible.)

35. **(a)** \$43.28 **(b)** \$235.54
(c) $C = \begin{cases} 9.45 + 0.67655x & \text{if } 0 \leq x \leq 50 \\ 43.2775 + .42725(x - 50) & \text{if } x > 50 \end{cases}$
(d)

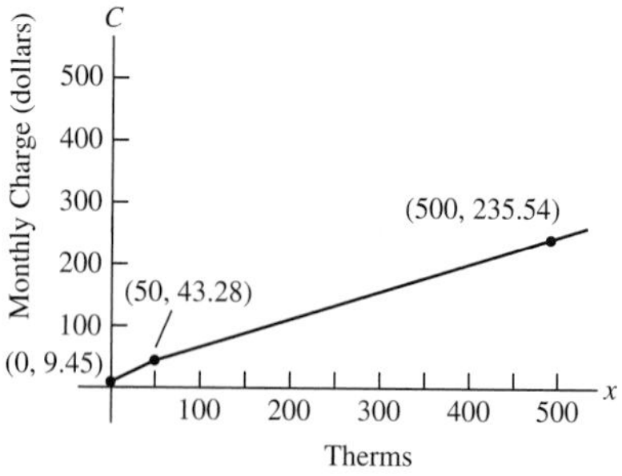

37. **(a)** 10°C **(b)** 4°C **(c)** −3°C **(d)** −4°C
(e) The wind chill is equal to the air temperature.
(f) At wind speed greater than 20 meters per second, the wind chill factor does not change.

39. Each graph is that of $y = x^2$, but shifted vertically. If $y = x^2 + k$, $k > 0$, the shift is up k units; if $y = x^2 + k$, $k < 0$, the shift is down $|k|$ units.

41. Each graph is that of $y = |x|$, but either compressed or stretched. If $y = k|x|$ and $k > 1$, the graph is stretched vertically; if $y = k|x|$, $0 < k < 1$, the graph is compressed vertically.

43. The graph of $y = f(-x)$ is the reflection about the y-axis of the graph of $y = f(x)$.

45. They are all ∪-shaped and open upward. All three go through the points $(-1, 1)$, $(0, 0)$ and $(1, 1)$. As the exponent increases, the steepness of the curve increases (except near $x = 0$).

2.4 Exercises

1. B **3.** H **5.** I **7.** L **9.** F **11.** G **13.** C **15.** B **17.** $y = (x - 4)^3$ **19.** $y = x^3 + 4$ **21.** $y = -x^3$ **23.** $y = 4x^3$
25. (1) $y = \sqrt{x} + 2$; (2) $y = -(\sqrt{x} + 2)$; (3) $y = -(\sqrt{-x} + 2)$ **27.** (1) $y = -\sqrt{x}$; (2) $y = -\sqrt{x} + 2$; (3) $y = -\sqrt{x + 3} + 2$
29. (c) **31.** (c)

33.

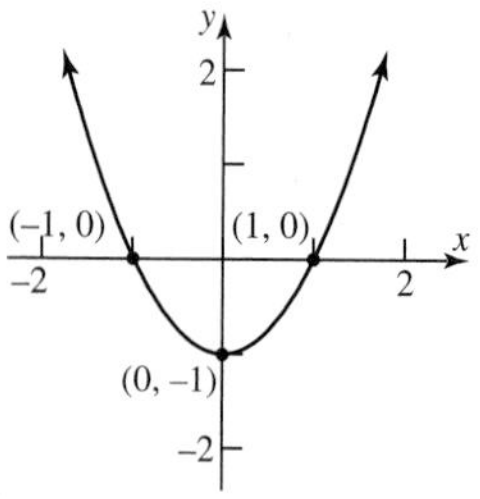

35.

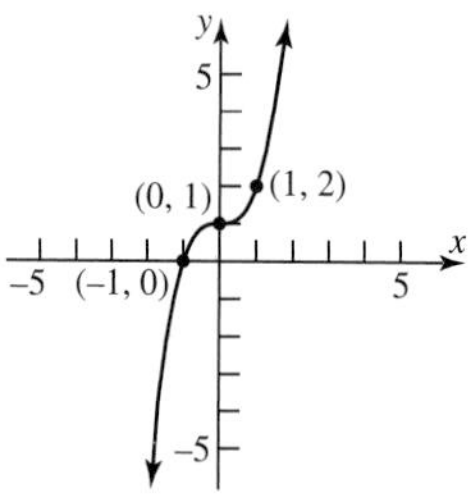

37.

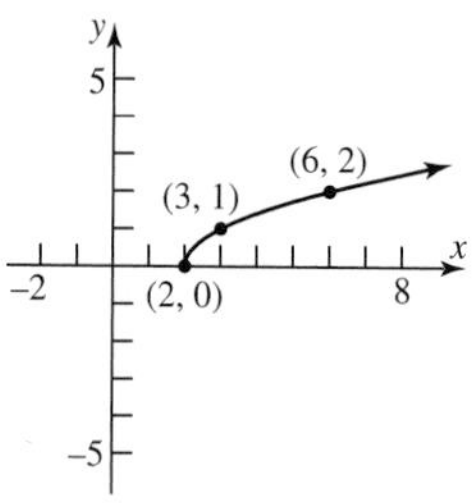

39.

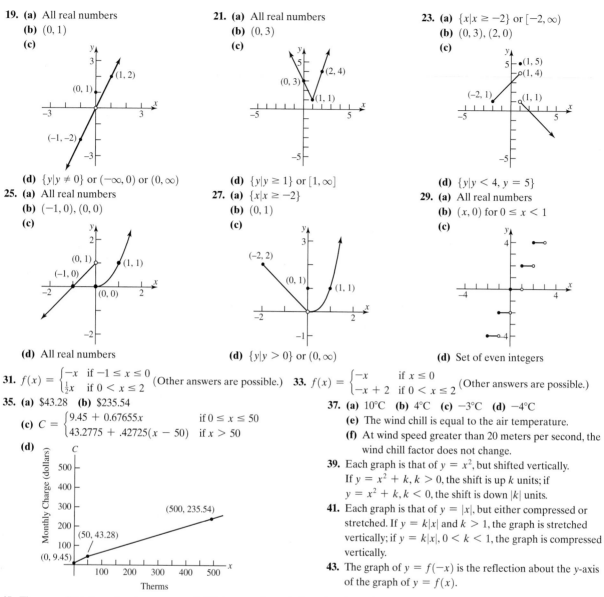

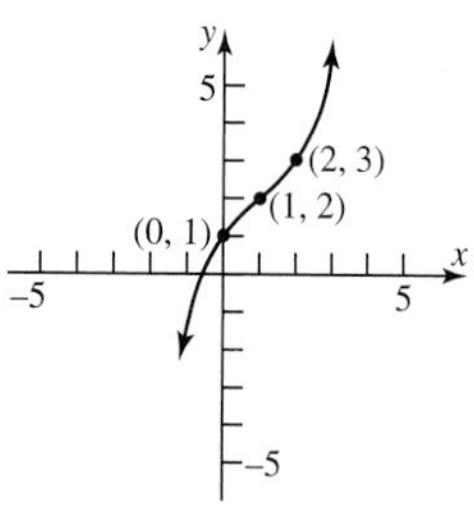

41.

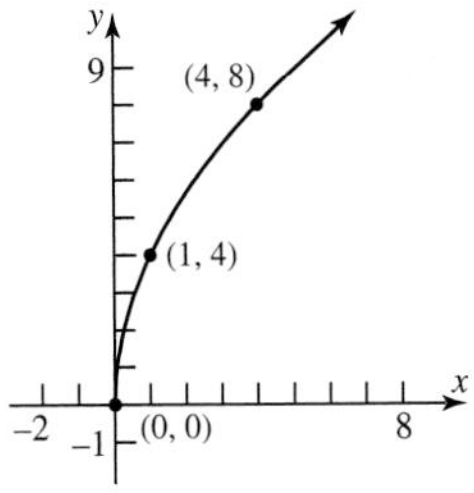

43.

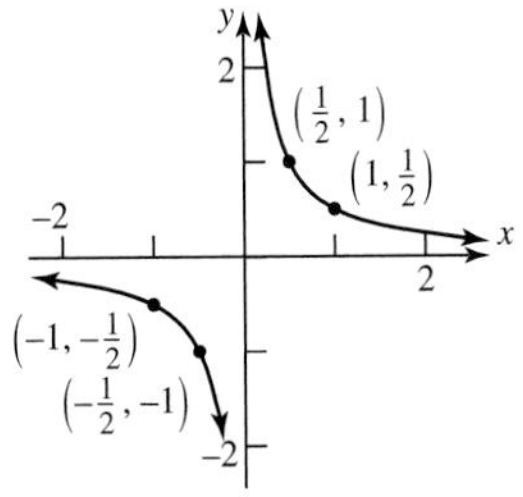

45.

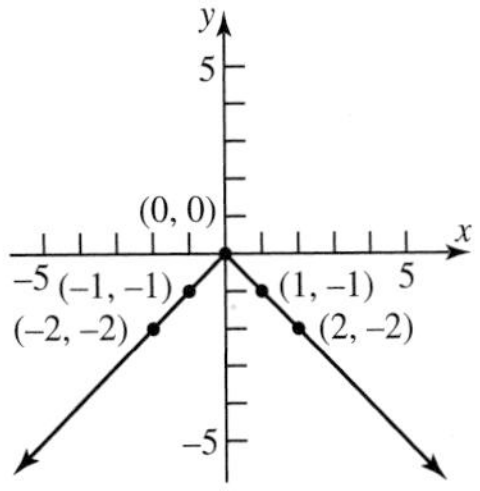

47.

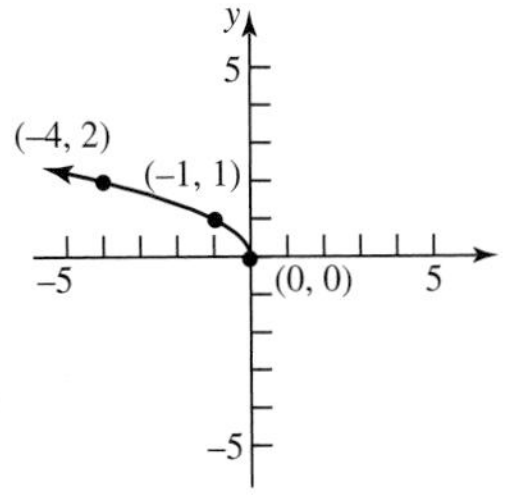

49.

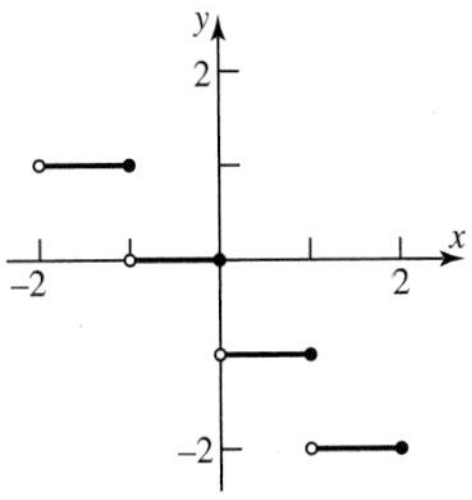

51.

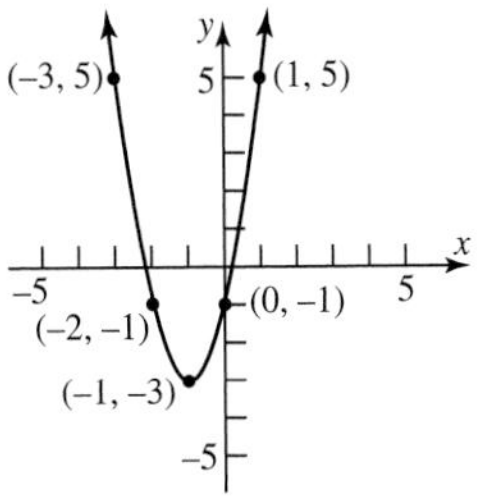

53.

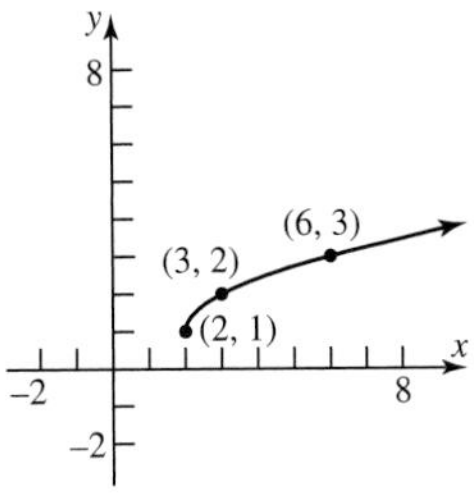

55.

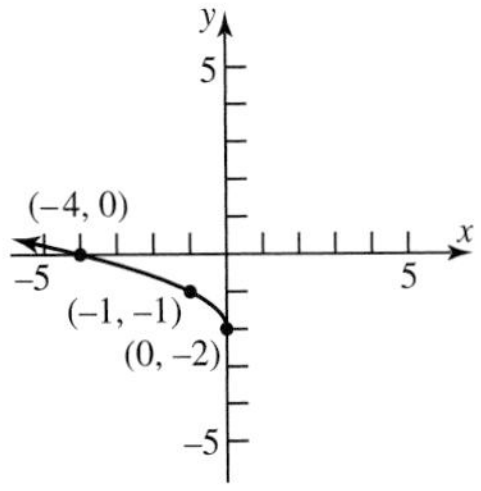

57.

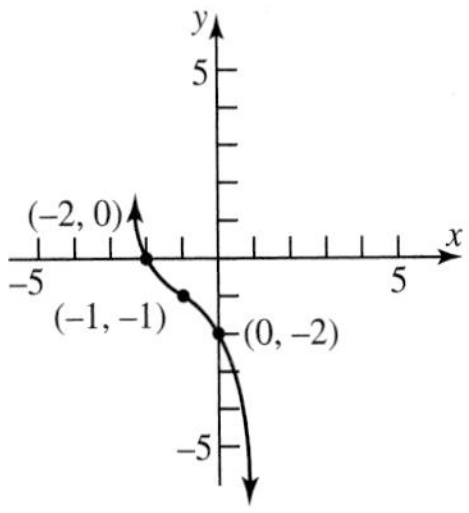

59.

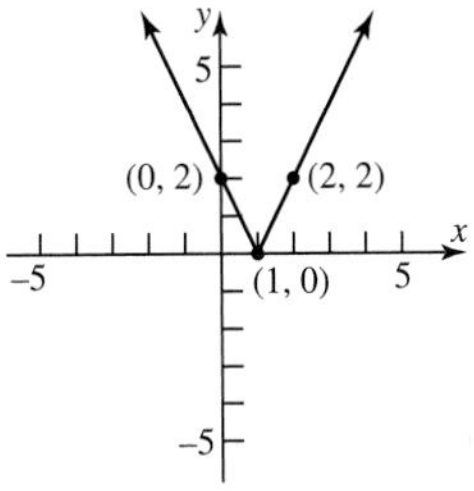

61.

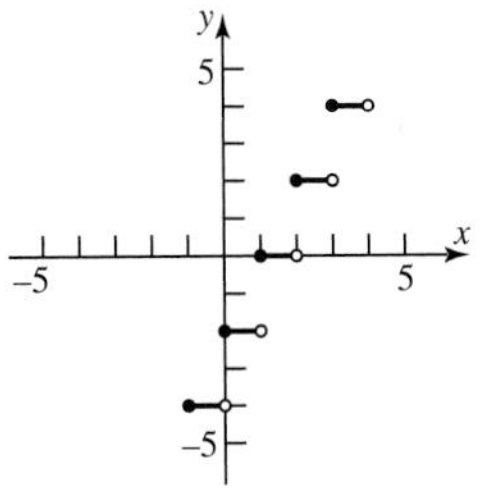

63. (a) $F(x) = f(x) + 3$

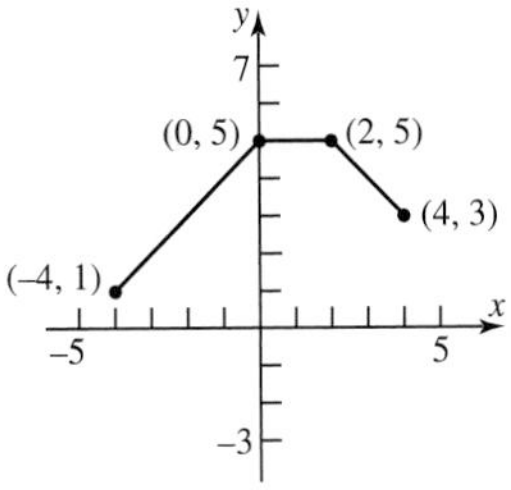

(b) $G(x) = f(x + 2)$

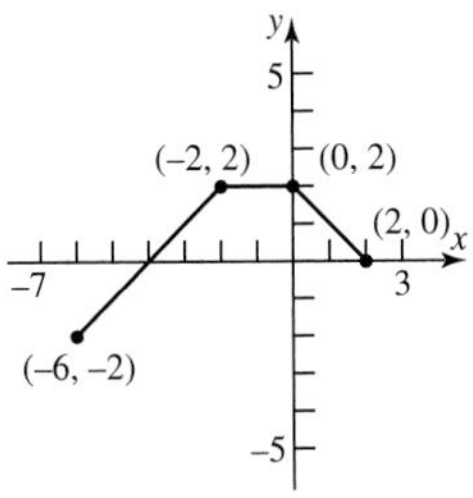

(c) $P(x) = -f(x)$

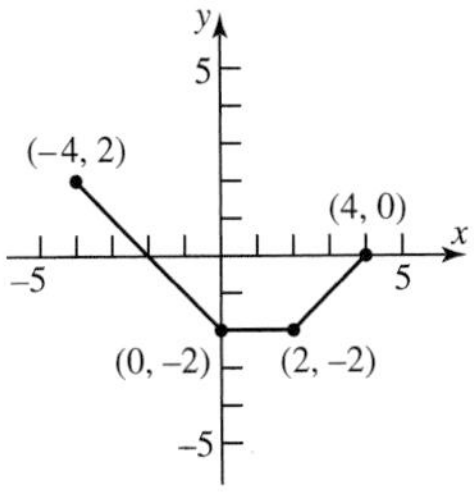

(d) $H(x) = f(x + 1) - 2$

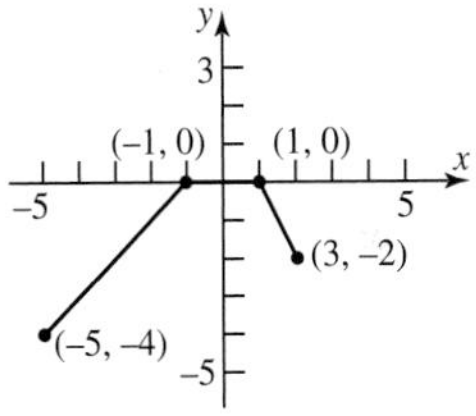

(e) $Q(x) = \frac{1}{2}f(x)$

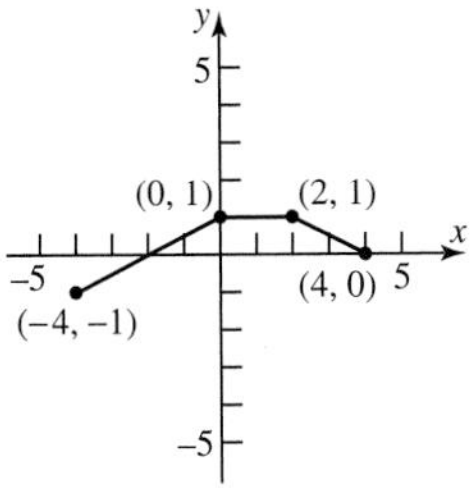

(f) $g(x) = f(-x)$

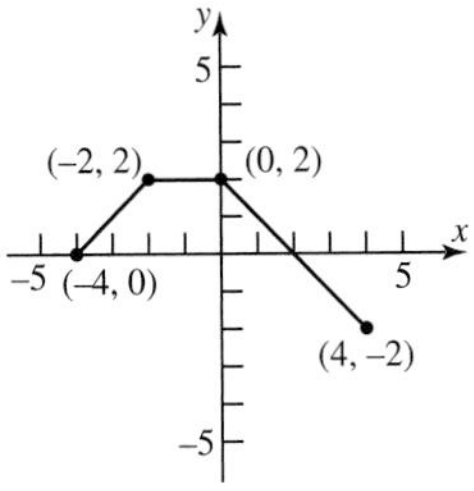

(g) $h(x) = f(2x)$

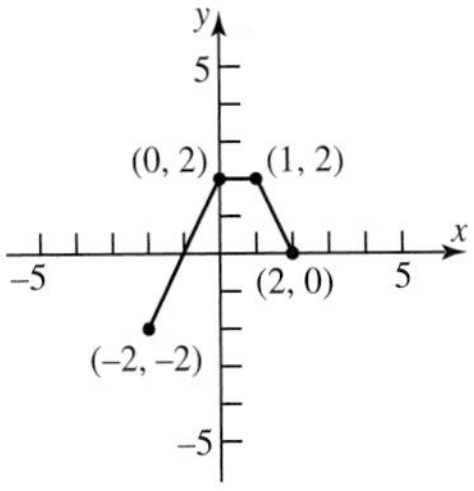

65. (a) $F(x) = f(x) + 3$

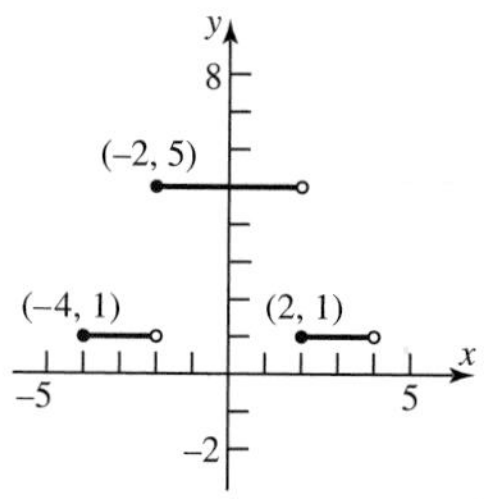

(b) $G(x) = f(x + 2)$

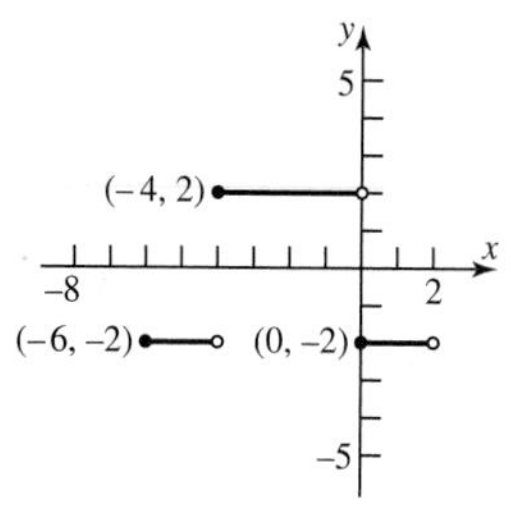

(c) $P(x) = -f(x)$

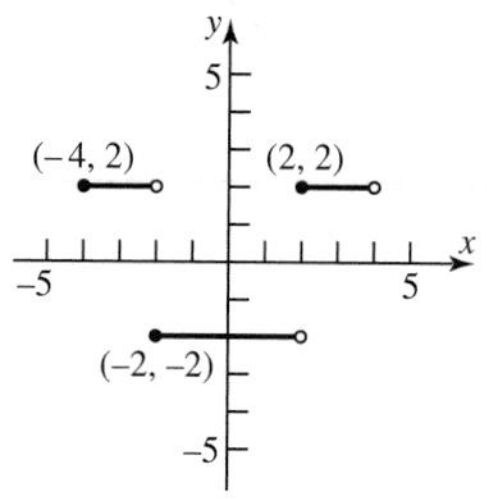

(d) $H(x) = f(x + 1) - 2$

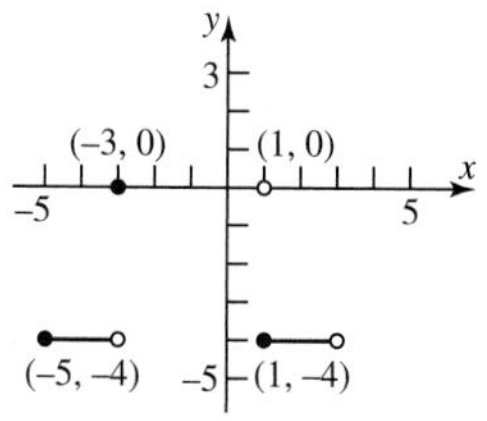

(e) $Q(x) = \frac{1}{2}f(x)$

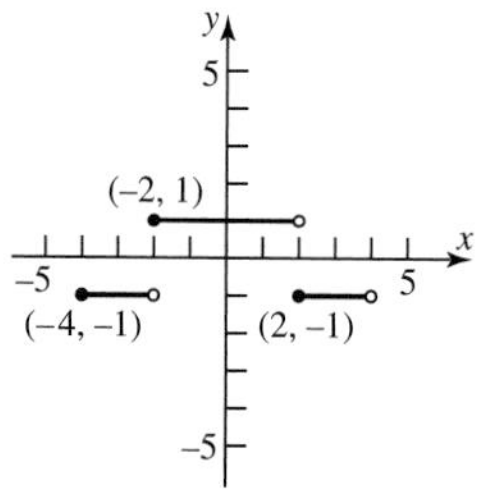

(f) $g(x) = f(-x)$

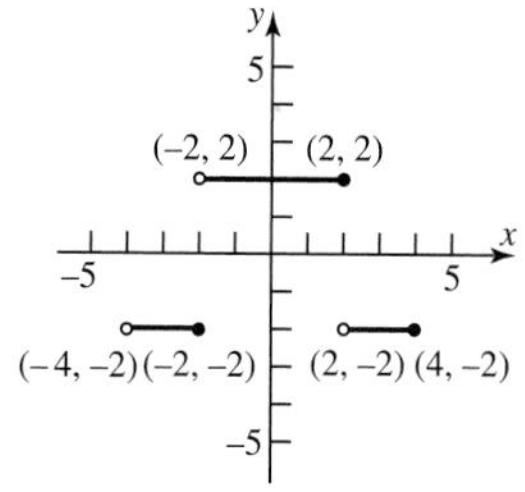

(g) $h(x) = f(2x)$

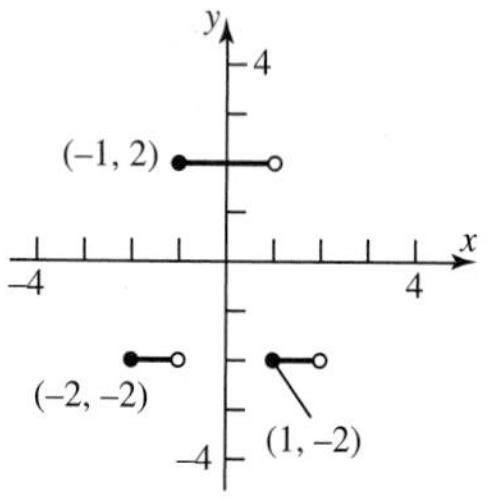

67. (a) $F(x) = f(x) + 3$

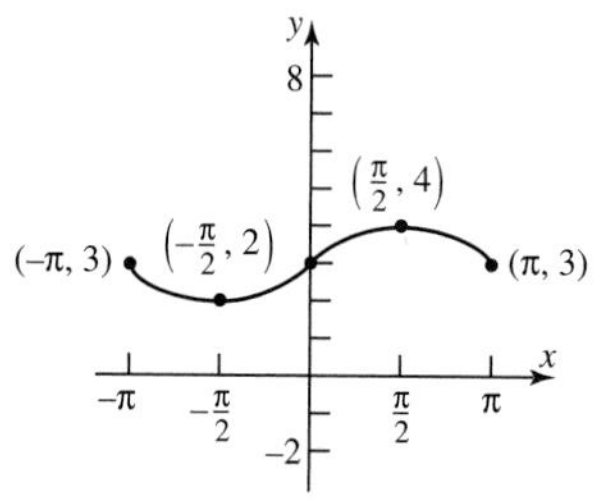

(b) $G(x) = f(x + 2)$

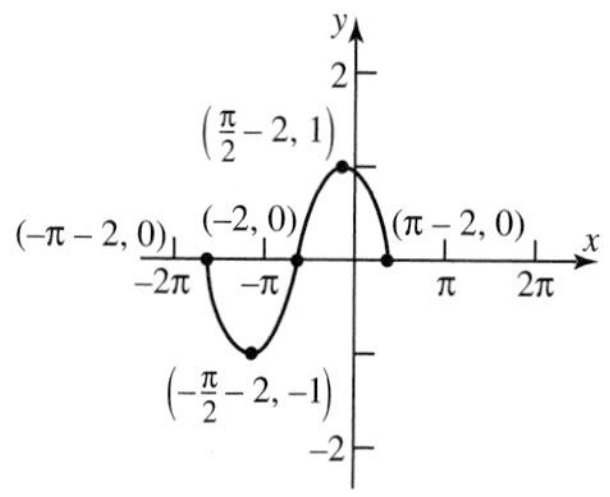

(c) $P(x) = -f(x)$

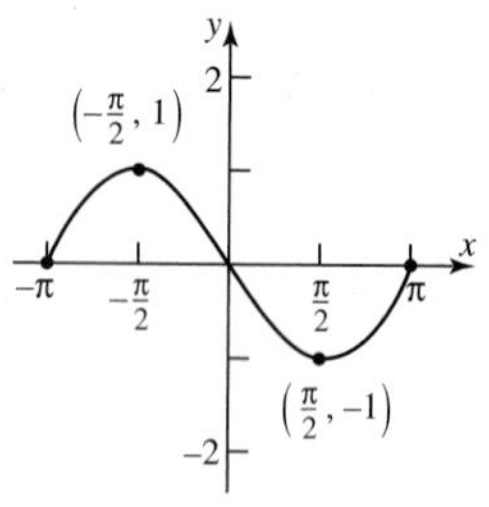

(d) $H(x) = f(x + 1) - 2$

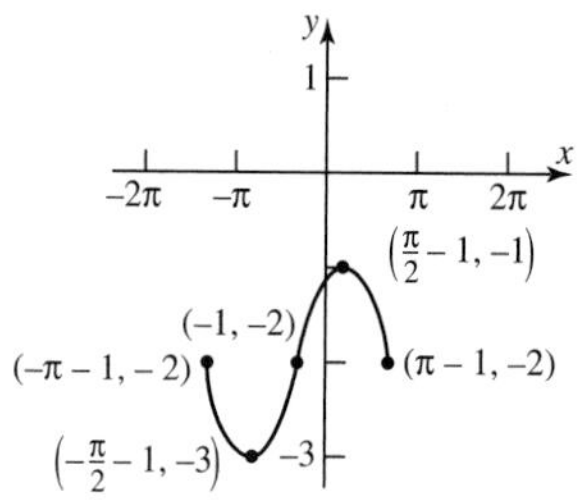

(e) $Q(x) = \frac{1}{2}f(x)$

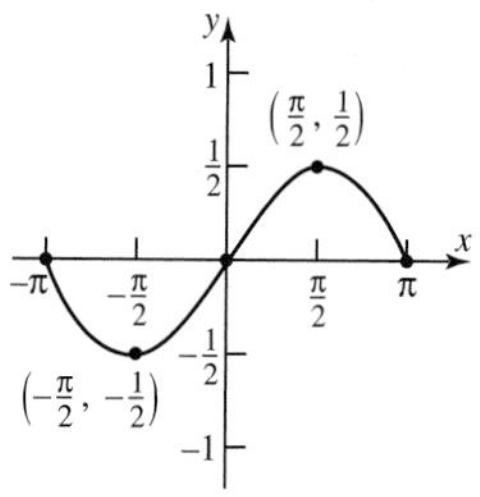

(f) $g(x) = f(-x)$

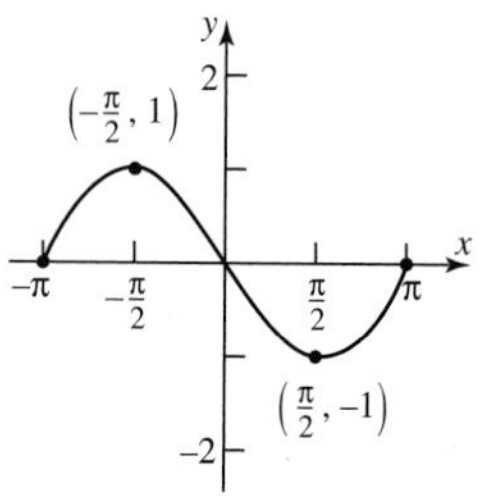

(g) $h(x) = f(2x)$

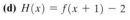

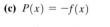

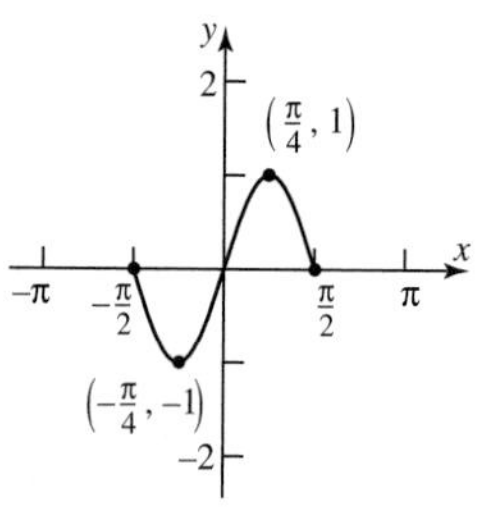

69. (a)

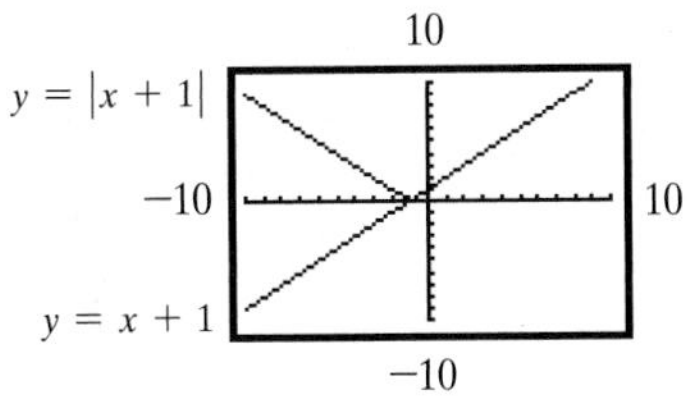

(b)

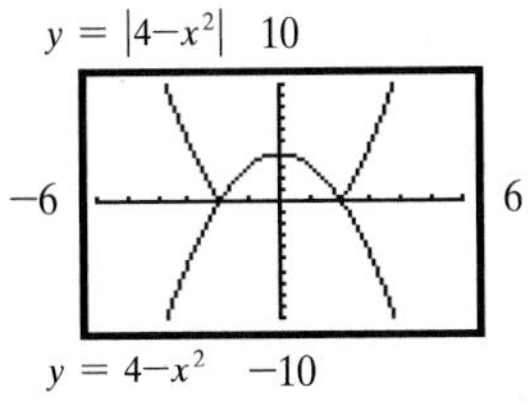

(c)

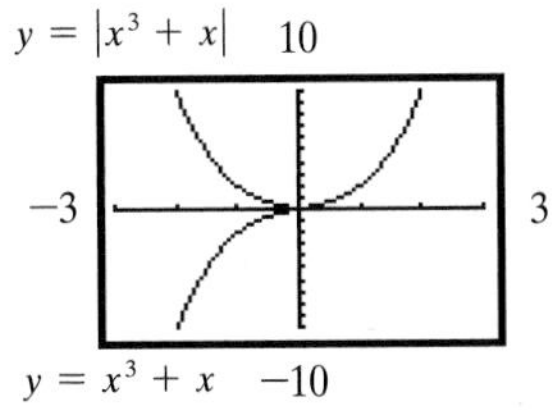

(d) Any part of the graph of $y = f(x)$ that lies below the x-axis is reflected about the x-axis to obtain the graph of $y = |f(x)|$.

71. (a)

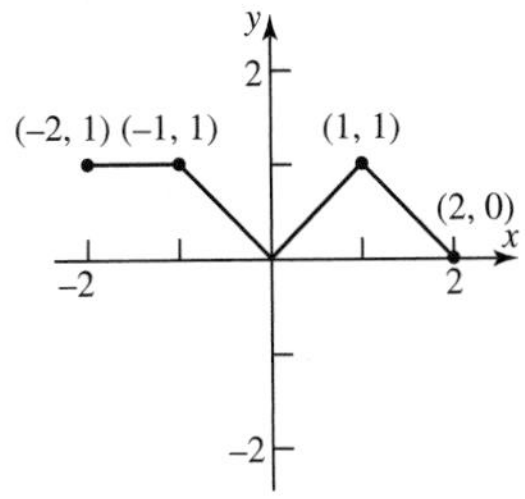

(b)

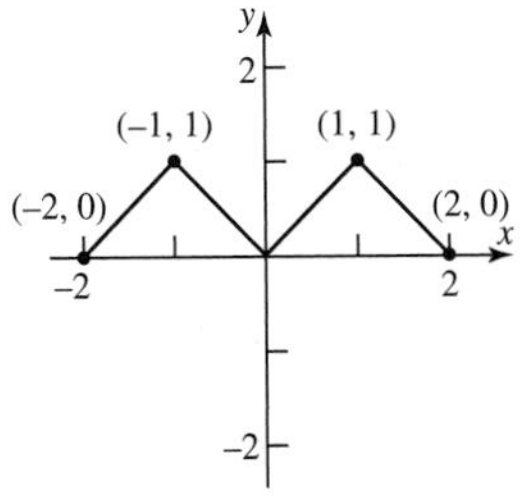

73. $f(x) = (x + 1)^2 - 1$

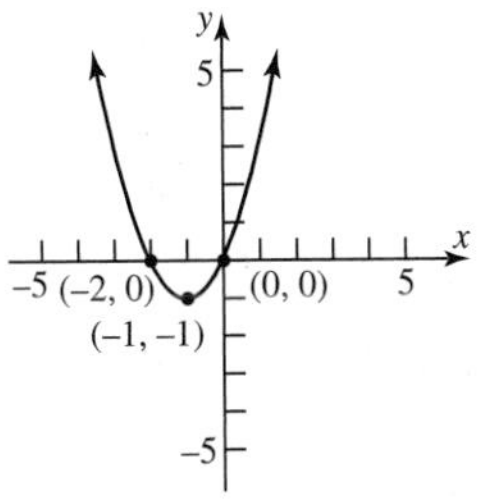

75. $f(x) = (x - 4)^2 - 15$

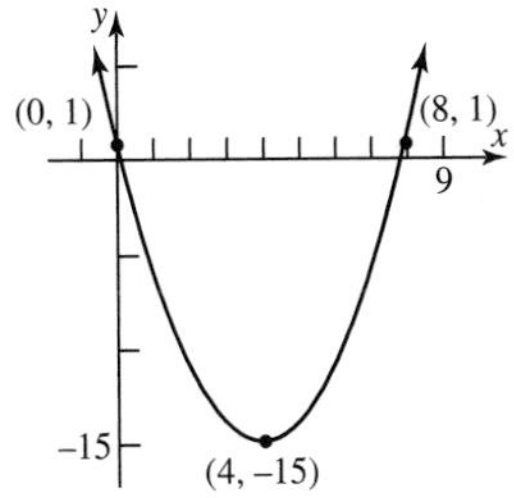

77. $f(x) = \left(x + \frac{1}{2}\right)^2 + \frac{3}{4}$

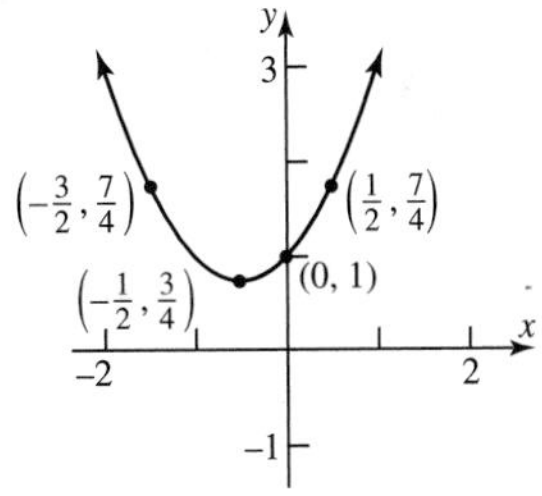

79.

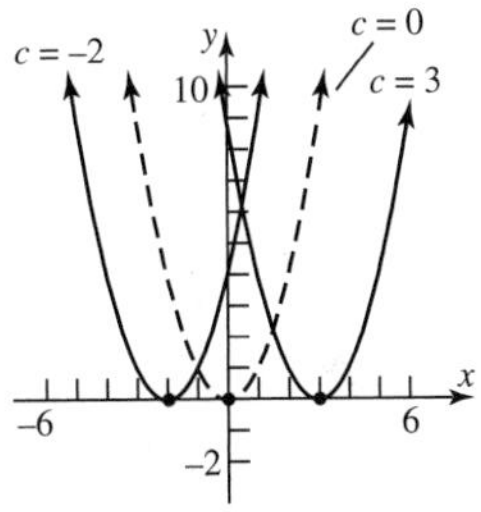

81.

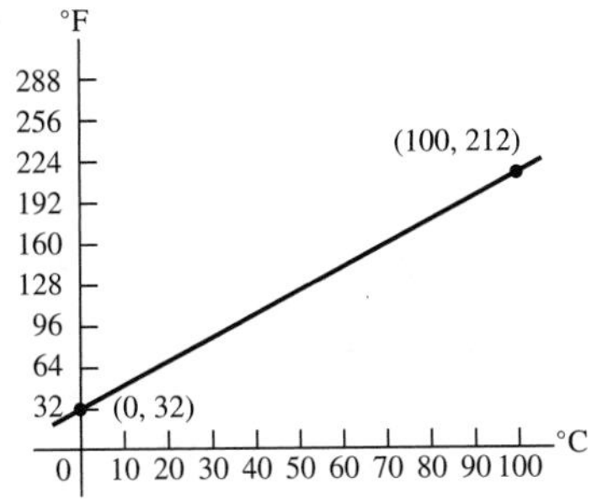

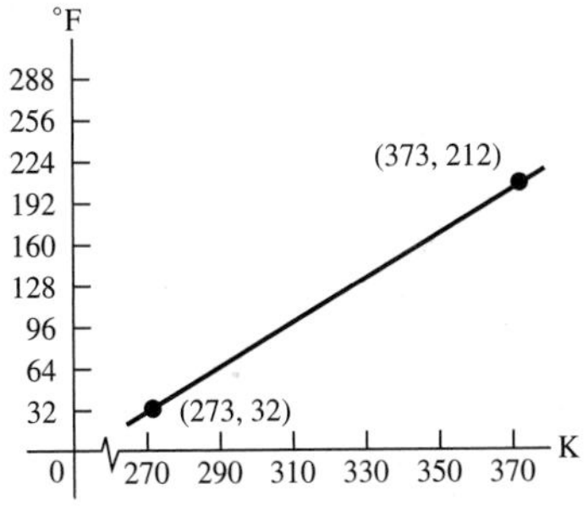

83. (a)

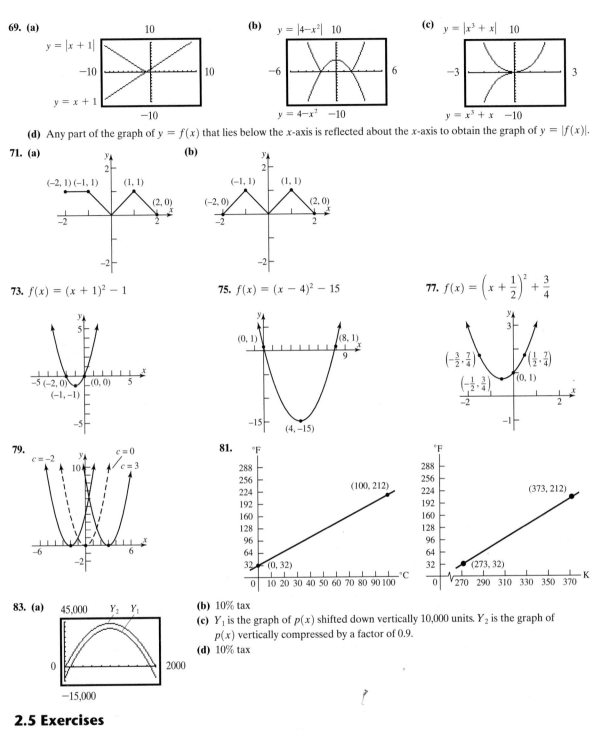

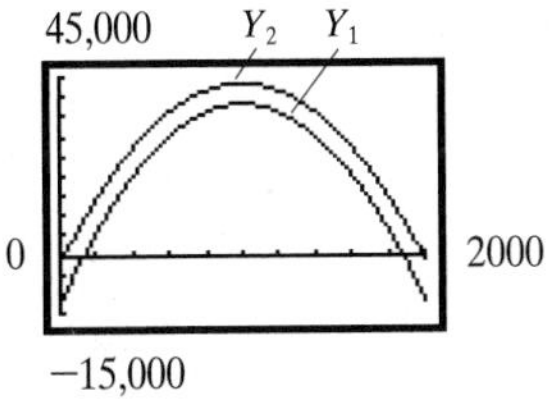

(b) 10% tax
(c) Y_1 is the graph of $p(x)$ shifted down vertically 10,000 units. Y_2 is the graph of $p(x)$ vertically compressed by a factor of 0.9.
(d) 10% tax

2.5 Exercises

1. (a) $(f + g)(x) = 5x + 1$; All real numbers **(b)** $(f - g)(x) = x + 7$; All real numbers
(c) $(f \cdot g)(x) = 6x^2 - x - 12$; All real numbers **(d)** $\left(\frac{f}{g}\right)(x) = \frac{3x + 4}{2x - 3}$; $\left\{x \middle| x \neq \frac{3}{2}\right\}$

3. (a) $(f + g)(x) = 2x^2 + x - 1$; All real numbers **(b)** $(f - g)(x) = -2x^2 + x - 1$; All real numbers
(c) $(f \cdot g)(x) = 2x^3 - 2x^2$; All real numbers **(d)** $\left(\frac{f}{g}\right)(x) = \frac{x - 1}{2x^2}$; $\{x|x \neq 0\}$ **5. (a)** $(f + g)(x) = \sqrt{x} + 3x - 5$; $\{x|x \geq 0\}$
(b) $(f - g)(x) = \sqrt{x} - 3x + 5$; $\{x|x \geq 0\}$ **(c)** $(f \cdot g)(x) = 3x\sqrt{x} - 5\sqrt{x}$; $\{x|x \geq 0\}$ **(d)** $\left(\frac{f}{g}\right)(x) = \frac{\sqrt{x}}{3x - 5}$; $\left\{x \middle| x \geq 0, x \neq \frac{5}{3}\right\}$

7. (a) $(f+g)(x) = 1 + \frac{2}{x}$; $\{x|x \neq 0\}$ **(b)** $(f-g)(x) = 1$; $\{x|x \neq 0\}$ **(c)** $(f \cdot g)(x) = \frac{1}{x} + \frac{1}{x^2}$; $\{x|x \neq 0\}$ **(d)** $\left(\frac{f}{g}\right)(x) = x + 1$; $\{x|x \neq 0\}$ **9. (a)** $(f+g)(x) = \frac{6x+3}{3x-2}$; $\left\{x \middle| x \neq \frac{2}{3}\right\}$ **(b)** $(f-g)(x) = \frac{-2x+3}{3x-2}$; $\left\{x \middle| x \neq \frac{2}{3}\right\}$ **(c)** $(f \cdot g)(x) = \frac{8x^2+12x}{(3x-2)^2}$; $\left\{x \middle| x \neq \frac{2}{3}\right\}$ **(d)** $\left(\frac{f}{g}\right)(x) = \frac{2x+3}{4x}$; $\left\{x \middle| x \neq 0, x \neq \frac{2}{3}\right\}$ **11.** $g(x) = 5 - \frac{7}{2}x$

13. (a) 98 **(b)** 49 **(c)** 4 **(d)** 4 **15. (a)** 97 **(b)** $-\frac{163}{2}$ **(c)** 1 **(d)** $-\frac{3}{2}$ **17. (a)** $2\sqrt{2}$ **(b)** $2\sqrt{2}$ **(c)** 1 **(d)** 0

19. (a) $\frac{1}{17}$ **(b)** $\frac{1}{5}$ **(c)** 1 **(d)** $\frac{1}{2}$ **21. (a)** $\frac{3}{5}$ **(b)** $\frac{\sqrt{15}}{5}$ **(c)** $\frac{12}{13}$ **(d)** 0 **23.** $\{x|x \neq 0, x \neq 2\}$ **25.** $\{x|x \neq -4, x \neq 0\}$

27. $\left\{x \middle| x \geq -\frac{3}{2}\right\}$ **29.** $\{x|x < -1 \text{ or } x > 1\}$ **31. (a)** $(f \circ g)(x) = 6x + 3$; All real numbers **(b)** $(g \circ f)(x) = 6x + 9$; All real numbers **(c)** $(f \circ f)(x) = 4x + 9$; All real numbers **(d)** $(g \circ g)(x) = 9x$; All real numbers **33. (a)** $(f \circ g)(x) = 3x^2 + 1$; All real numbers **(b)** $(g \circ f)(x) = 9x^2 + 6x + 1$; All real numbers **(c)** $(f \circ f)(x) = 9x + 4$; All real numbers **(d)** $(g \circ g)(x) = x^4$; All real numbers **35. (a)** $(f \circ g)(x) = x^4 + 8x^2 + 16$; All real numbers **(b)** $(g \circ f)(x) = x^4 + 4$; All real numbers **(c)** $(f \circ f)(x) = x^4$; All real numbers **(d)** $(g \circ g)(x) = x^4 + 8x^2 + 20$; All real numbers **37. (a)** $(f \circ g)(x) = \frac{3x}{2-x}$; $\{x|x \neq 0, x \neq 2\}$ **(b)** $(g \circ f)(x) = \frac{2(x-1)}{3}$; $\{x|x \neq 1\}$ **(c)** $(f \circ f)(x) = \frac{3(x-1)}{4-x}$; $\{x|x \neq 1, x \neq 4\}$ **(d)** $(g \circ g)(x) = x$; $\{x|x \neq 0\}$

39. (a) $(f \circ g)(x) = \frac{4}{4+x}$; $\{x|x \neq -4, x \neq 0\}$ **(b)** $(g \circ f)(x) = \frac{-4(x-1)}{x}$; $\{x|x \neq 0, x \neq 1\}$ **(c)** $(f \circ f)(x) = x$; $\{x|x \neq 1\}$ **(d)** $(g \circ g)(x) = x$; $\{x|x \neq 0\}$ **41. (a)** $(f \circ g)(x) = \sqrt{2x+3}$; $\left\{x \middle| x \geq -\frac{3}{2}\right\}$ **(b)** $(g \circ f)(x) = 2\sqrt{x} + 3$; $\{x|x \geq 0\}$ **(c)** $(f \circ f)(x) = \sqrt[4]{x}$; $\{x|x \geq 0\}$ **(d)** $(g \circ g)(x) = 4x + 9$; All real numbers **43. (a)** $(f \circ g)(x) = \sqrt{\frac{1+x}{x-1}}$; $\{x|x \leq -1 \text{ or } x > 1\}$ **(b)** $(g \circ f)(x) = \frac{2}{\sqrt{x+1}-1}$; $\{x|x \geq -1, x \neq 0\}$ **(c)** $(f \circ f)(x) = \sqrt{\sqrt{x+1}+1}$; $\{x|x \geq -1\}$ **(d)** $(g \circ g)(x) = \frac{2(x-1)}{3-x}$; $\{x|x \neq 1, x \neq 3\}$ **45. (a)** $(f \circ g)(x) = acx + ad + b$; All real numbers **(b)** $(g \circ f)(x) = acx + bc + d$; All real numbers **(c)** $(f \circ f)(x) = a^2x + ab + b$; All real numbers **(d)** $(g \circ g)(x) = c^2x + cd + d$; All real numbers

47. $(f \circ g)(x) = f(g(x)) = f\left(\frac{1}{2}x\right) = 2\left(\frac{1}{2}x\right) = x$; $(g \circ f)(x) = g(f(x)) = g(2x) = \frac{1}{2}(2x) = x$

49. $(f \circ g)(x) = f(g(x)) = f(\sqrt[3]{x}) = (\sqrt[3]{x})^3 = x$; $(g \circ f)(x) = g(f(x)) = g(x^3) = \sqrt[3]{x^3} = x$

51. $(f \circ g)(x) = f(g(x)) = f\left(\frac{1}{2}(x+6)\right) = 2\left[\frac{1}{2}(x+6)\right] - 6 = x + 6 - 6 = x$;

$(g \circ f)(x) = g(f(x)) = g(2x-6) = \frac{1}{2}(2x - 6 + 6) = x$

53. $(f \circ g)(x) = f(g(x)) = f\left(\frac{1}{a}(x-b)\right) = a\left[\frac{1}{a}(x-b)\right] + b = x$; $(g \circ f)(x) = g(f(x)) = g(ax+b) = \frac{1}{a}(ax+b-b) = x$

55. $f(x) = x^4$; $g(x) = 2x + 3$ (Other answers are possible.) **57.** $f(x) = \sqrt{x}$; $g(x) = x^2 + 1$ (Other answers are possible.)

59. $f(x) = |x|$; $g(x) = 2x + 1$ (Other answers are possible.) **61.** $(f \circ g)(x) = 11$; $(g \circ f)(x) = 2$ **63.** $-3, 3$ **65.** $S(r(t)) = \frac{16}{9}\pi t^6$

67. $C(N(t)) = 15{,}000 + 800{,}000t - 40{,}000t^2$ **69.** $C = \frac{2\sqrt{100-p}}{25} + 600$ **71.** $V(r) = 2\pi r^3$

2.6 Exercises

1. $V(r) = 2\pi r^3$ **3. (a)** $R(x) = -\frac{1}{6}x^2 + 100x$ **(b)** \$13,333 **(c)**

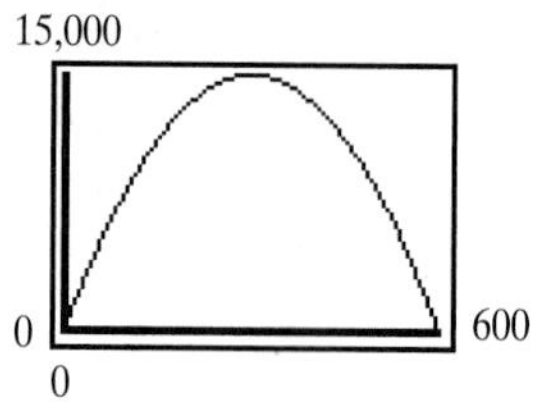

(d) 300; \$15,000 **(e)** \$50

5. (a) $R(x) = -\frac{1}{5}x^2 + 20x$ **(b)** \$255 **(c)**

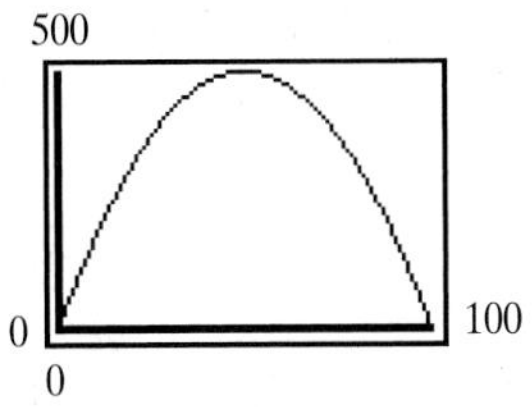

(d) 50; \$500 **(e)** \$10

7. (a) $A(x) = -x^2 + 200x$ **(b)** $0 < x < 200$ **(c)** A is largest when $x = 100$ yards

9. (a) $d(x) = \sqrt{x^4 - 15x^2 + 64}$
(b) $d(0) = 8$ **(c)** $d(1) = \sqrt{50} \approx 7.07$
(d)

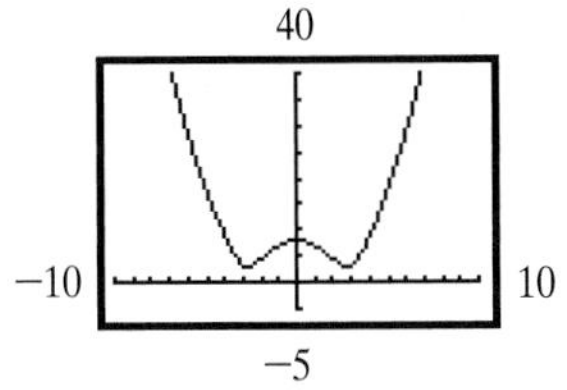

(e) d is smallest when $x \approx -2.74$ or $x \approx 2.74$.

11. (a) $d(x) = \sqrt{x^2 - x + 1}$
(b)

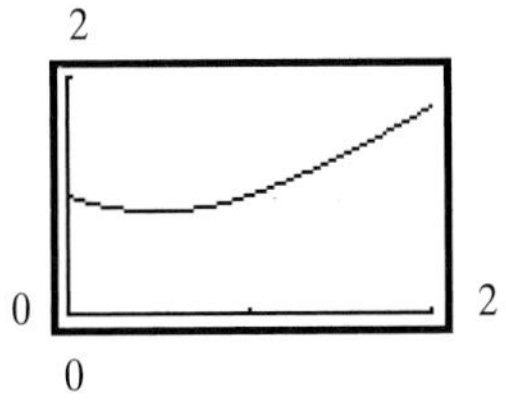

(c) $x = \frac{1}{2}$

13. $A(x) = \frac{1}{2}x^4$

15. (a) $A(x) = x(16 - x^2)$
(b) Domain: $\{x|0 < x < 4\}$
(c) The area is largest when $x \approx 2.31$.

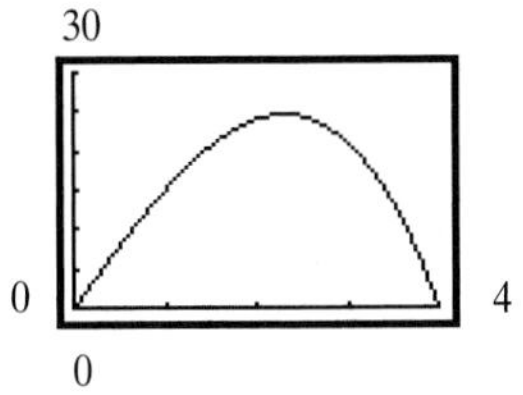

17. (a) $A(x) = 4x\sqrt{4 - x^2}$ **(b)** $p(x) = 4x + 4\sqrt{4 - x^2}$
(c) The area is largest when $x \approx 1.41$.

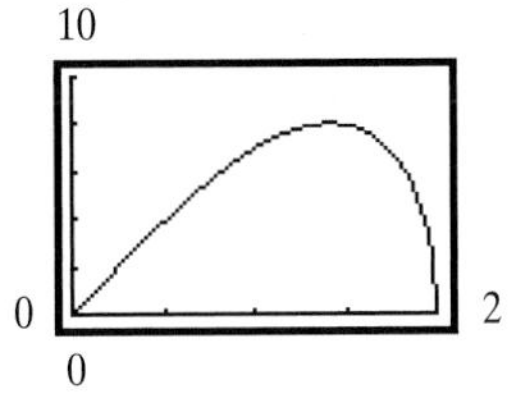

(d) The perimeter is largest when $x \approx 1.41$.

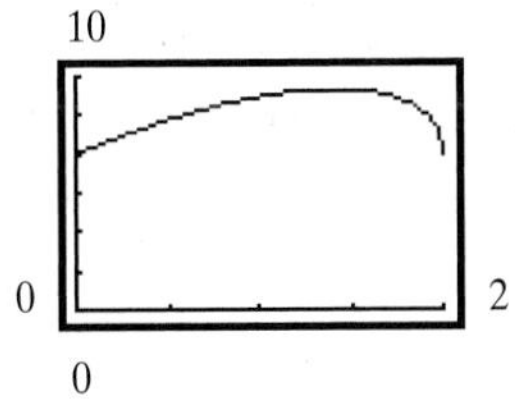

19. (a) $A(x) = x^2 + \dfrac{25 - 20x + 4x^2}{\pi}$
(b) Domain: $\{x|\ 0 < x < 2.5\}$
(c) The area is smallest when $x \approx 1.40$ meters.

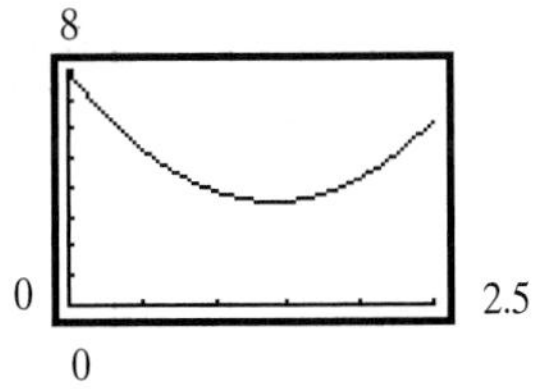

21. (a) $C(x) = x$ **(b)** $A(x) = \dfrac{x^2}{4\pi}$

23. (a) $A(r) = 2r^2$ **(b)** $p(r) = 6r$

25. $A(x) = \left(\dfrac{\pi}{3} - \dfrac{\sqrt{3}}{4}\right)x^2$

27. $C = \begin{cases} 95 & \text{if } x = 7 \\ 119 & \text{if } 7 < x \leq 8 \\ 143 & \text{if } 8 < x \leq 9 \\ 167 & \text{if } 9 < x \leq 10 \\ 190 & \text{if } 10 < x \leq 14 \end{cases}$

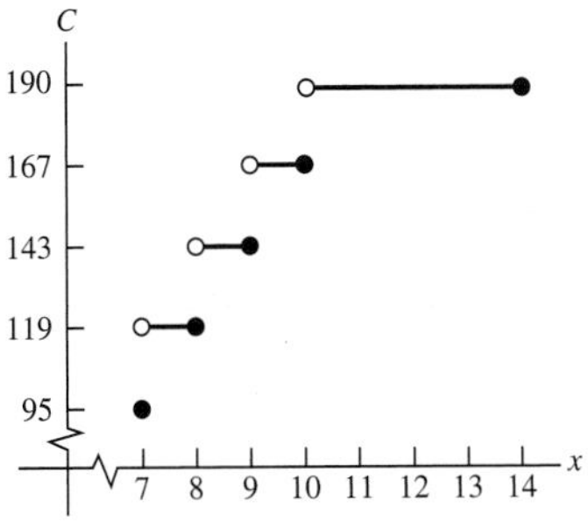

29. $d(t) = 50t$

31. (a) $V(x) = x(24 - 2x)^2$
(b) 972 in^3 **(c)** 160 in^3
(d) The volume is largest when $x = 4$.

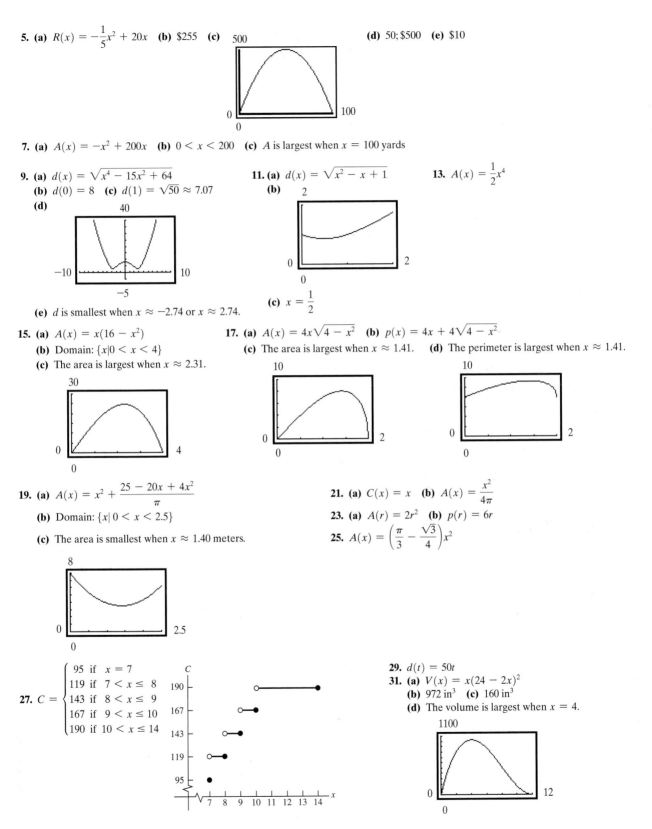

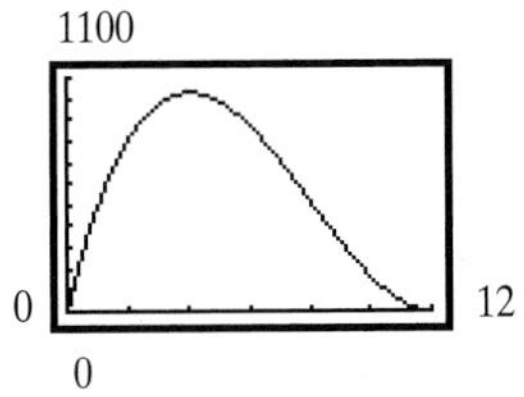

33. $V(h) = \pi h\left(R^2 - \frac{h^2}{4}\right)$

35. **(a)** $C(x) = 10x + 14\sqrt{x^2 - 10x + 29}, 0 \le x \le 5$
(b) $C(1) = \$72.61$
(c) $C(3) = \$69.60$
(d)

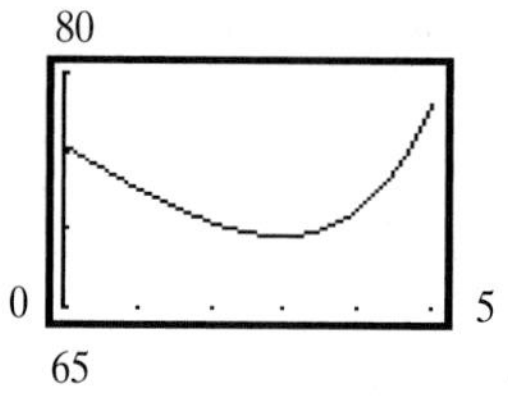

(e) Least cost: $x \approx 2.96$ miles

37. For schedule X: $f(x) = \begin{cases} 0.15x & \text{if } 0 \le x \le 25{,}750 \\ 3862.50 + 0.28(x - 25{,}750) & \text{if } 25{,}750 < x \le 64{,}450 \\ 14{,}138.50 + 0.31(x - 62{,}450) & \text{if } 64{,}450 < x \le 130{,}250 \\ 35{,}156.50 + 0.36(x - 130{,}250) & \text{if } 130{,}250 < x \le 283{,}150 \\ 90{,}200.50 + 0.396(x - 283{,}150) & \text{if } x > 283{,}150 \end{cases}$

For schedule Y-1: $f(x) = \begin{cases} 0.15x & \text{if } 0 \le x \le 43{,}050 \\ 6457.50 + 0.28(x - 43{,}050) & \text{if } 43{,}050 < x \le 104{,}050 \\ 23{,}537.50 + 0.31(x - 104{,}050) & \text{if } 104{,}050 < x \le 158{,}550 \\ 40{,}432.50 + 0.36(x - 158{,}550) & \text{if } 158{,}550 < x \le 283{,}150 \\ 85{,}288.50 + 0.396(x - 283{,}150) & \text{if } x > 283{,}150 \end{cases}$

Fill-in-the-Blank Items

1. independent; dependent **3.** slope **5.** horizontal; right

True/False Items

1. False **3.** True **5.** False **7.** False

Review Exercises

1. $f(x) = -2x + 3$ **3.** $A = 11$ **5.** b, c, d

7. **(a)** $f(-x) = \frac{-3x}{x^2 - 4}$ **(b)** $-f(x) = \frac{-3x}{x^2 - 4}$ **(c)** $f(x + 2) = \frac{3x + 6}{x^2 + 4x}$ **(d)** $f(x - 2) = \frac{3x - 6}{x^2 - 4x}$ **(e)** $f(2x) = \frac{6x}{4x^2 - 4}$

9. **(a)** $f(-x) = \sqrt{x^2 - 4}$ **(b)** $-f(x) = -\sqrt{x^2 - 4}$ **(c)** $f(x + 2) = \sqrt{x^2 + 4x}$ **(d)** $f(x - 2) = \sqrt{x^2 - 4x}$ **(e)** $f(2x) = 2\sqrt{x^2 - 1}$

11. **(a)** $f(-x) = \frac{x^2 - 4}{x^2}$ **(b)** $-f(x) = -\frac{x^2 - 4}{x^2}$ **(c)** $f(x + 2) = \frac{x^2 + 4x}{x^2 + 4x + 4}$ **(d)** $f(x - 2) = \frac{x^2 - 4x}{x^2 - 4x + 4}$ **(e)** $f(2x) = \frac{x^2 - 1}{x^2}$

13. $\{x | x \ne -3, x \ne 3\}$ **15.** $\{x | x \le 2\}$ **17.** $\{x | x > 0\}$ **19.** $\{x | x \ne -3, x \ne 1\}$

21. **(a)** Domain: $\{x | x > -2\}$ or $(-2, \infty)$
(b) $(0, 0)$
(c)

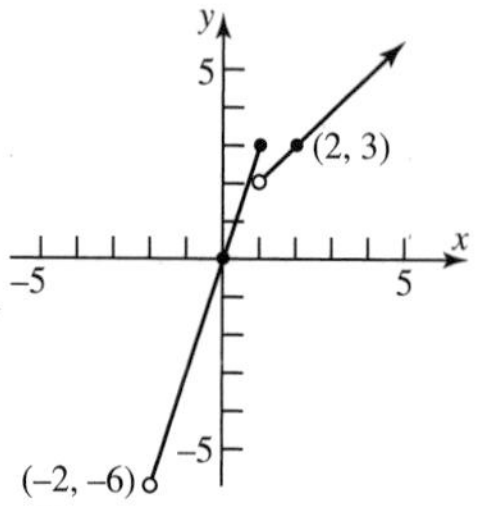

(d) All real numbers

23. **(a)** Domain: $\{x | -4 \le x < \infty\}$ or $[-4, \infty)$
(b) $(0, 1)$
(c)

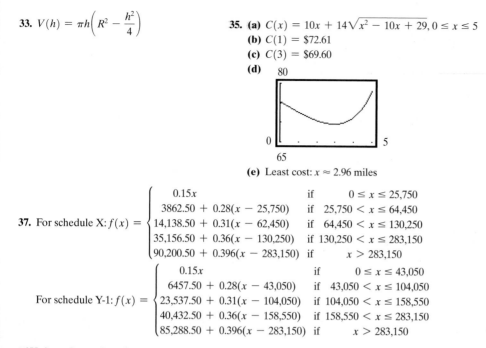

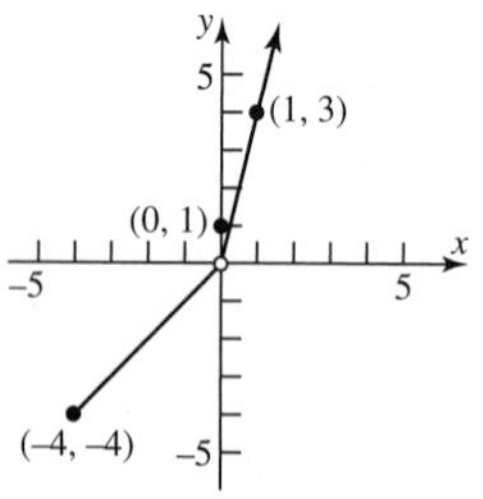

(d) All real numbers, except $y = 0$ or using interval notation $(-\infty, 0)$ or $(0, \infty)$

25. -5
27. $-4x - 5$
29. Odd
31. Even
33. Neither
35. Odd

37.

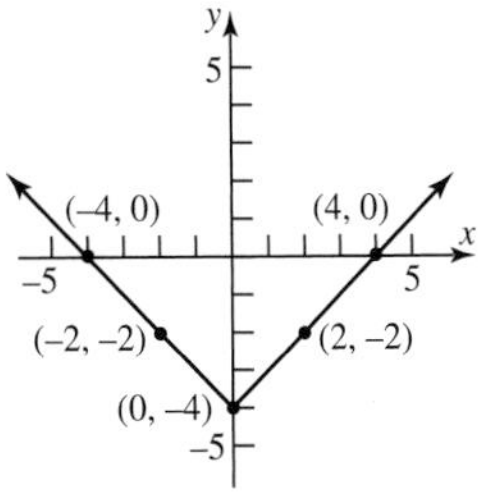

Intercepts: $(-4, 0)$, $(4, 0)$, $(0, -4)$
Domain: all real numbers
Range: $\{y|y \geq -4\}$

39.

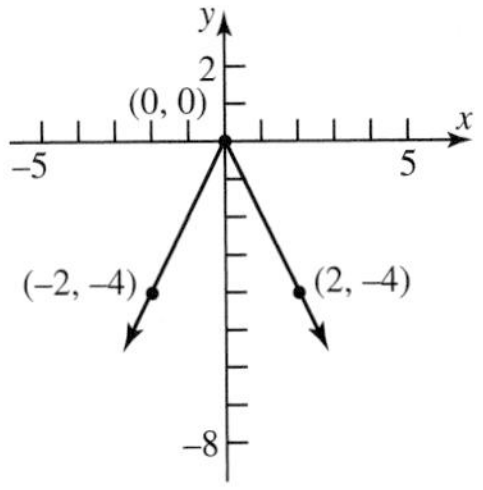

Intercept: $(0, 0)$
Domain: all real numbers
Range: $\{y|y \leq 0\}$

41.

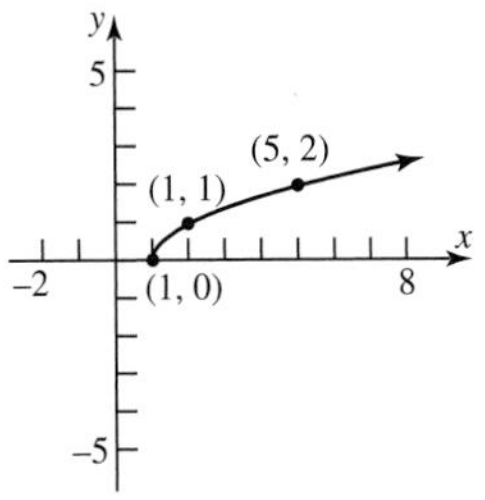

Intercept: $(1, 0)$
Domain: $\{x|x \geq 1\}$
Range: $\{y|y \geq 0\}$

43.

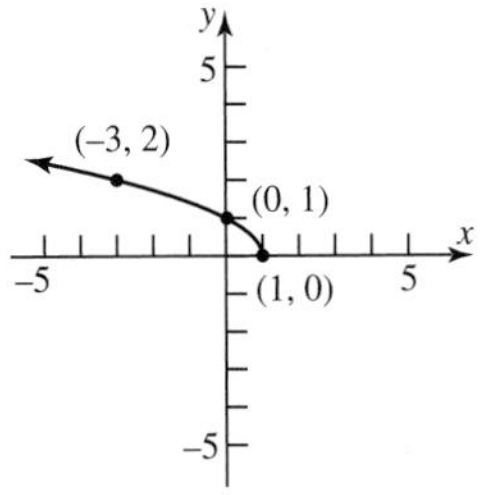

Intercepts: $(0, 1)$, $(1, 0)$
Domain: $\{x|\ x \leq 1\}$
Range: $\{y|y \geq 0\}$

45.

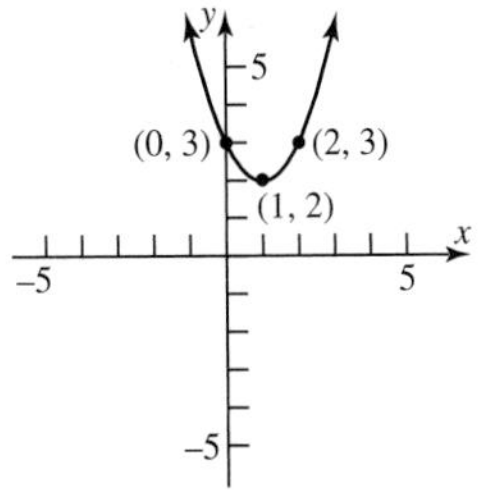

Intercept: $(0, 3)$
Domain: all real numbers
Range: $\{y|y \geq 2\}$

47.

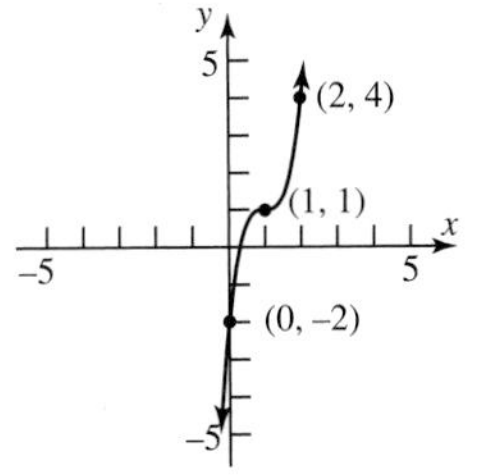

Intercept: $(0, -2)$
Domain: all real numbers
Range: all real numbers

49.

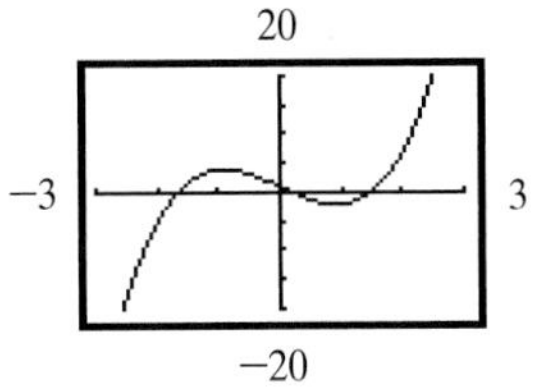

Local maximum: $(-0.91, 4.04)$
Local minimum: $(0.91, -2.04)$
Increasing: $(-3, -0.91)$; $(0.91, 3)$
Decreasing: $(-0.91, 0.91)$

51.

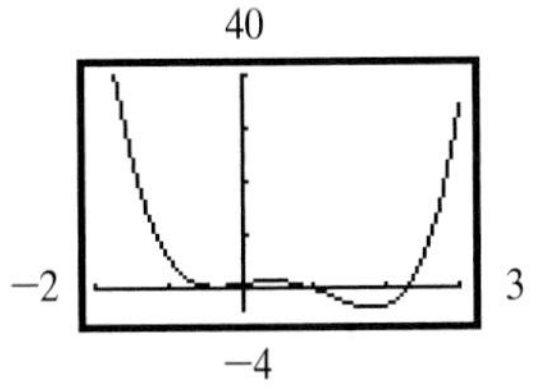

Local maximum: $(0.41, 1.53)$
Local minimum: $(-0.34, 0.54)$; $(1.80, -3.56)$
Increasing: $(-0.34, 0.41)$; $(1.80, 3)$
Decreasing: $(-2, -0.34)$; $(0.41, 1.80)$

53. **(a)** -26 **(b)** -241 **(c)** 16 **(d)** -1

55. **(a)** $\sqrt{11}$ **(b)** 1 **(c)** $\sqrt{\sqrt{6} + 2}$ **(d)** 19

57. **(a)** $\frac{1}{20}$ **(b)** $-\frac{13}{8}$ **(c)** $\frac{400}{1601}$ **(d)** -17

59. $(f \circ g)(x) = 1 - 3x$; All real numbers; $(g \circ f)(x) = 7 - 3x$; All real numbers; $(f \circ f)(x) = x$; All real numbers; $(g \circ g)(x) = 9x + 4$; All real numbers

61. $(f \circ g)(x) = 27x^2 + 3|x| + 1$; All real numbers; $(g \circ f)(x) = 3|3x^2 + x + 1|$; All real numbers; $(f \circ f)(x) = 3(3x^2 + x + 1)^2 + 3x^2 + x + 2$; All real numbers; $(g \circ g)(x) = 9|x|$; All real numbers

63. $(f \circ g)(x) = \dfrac{1 + x}{1 - x}$; $\{x|\ x \neq 0, x \neq 1\}$; $(g \circ f)(x) = \dfrac{x - 1}{x + 1}$; $\{x|\ x \neq -1, x \neq 1\}$; $(f \circ f)(x) = x$; $\{x|x \neq 1\}$; $(g \circ g)(x) = x$; $\{x|x \neq 0\}$

65. **(a)**

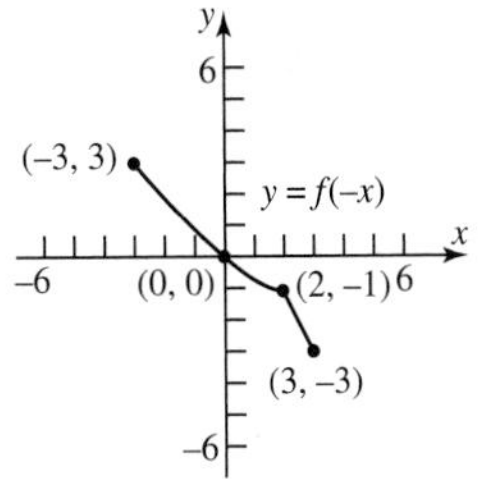

(b)

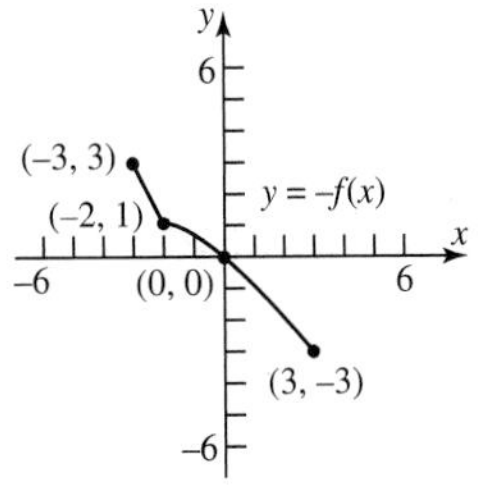

(c)

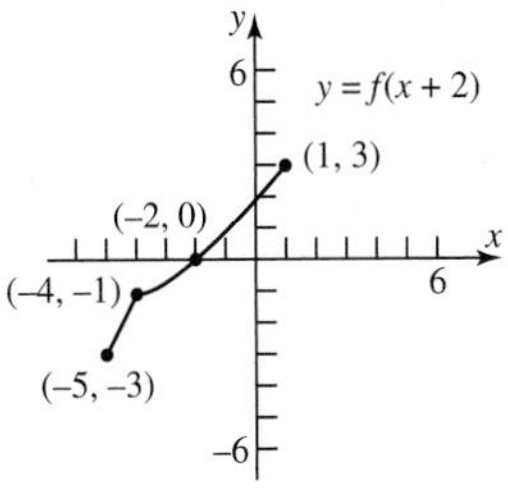

(d)

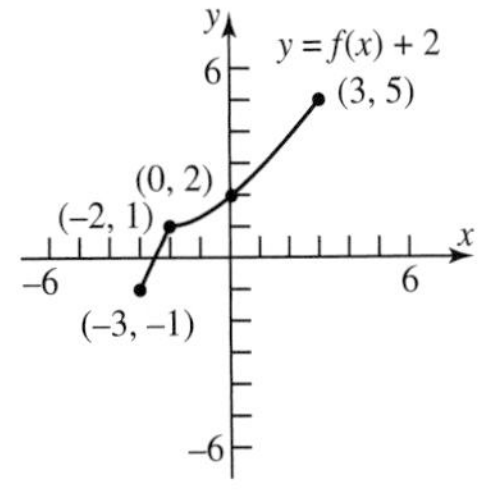

(e)

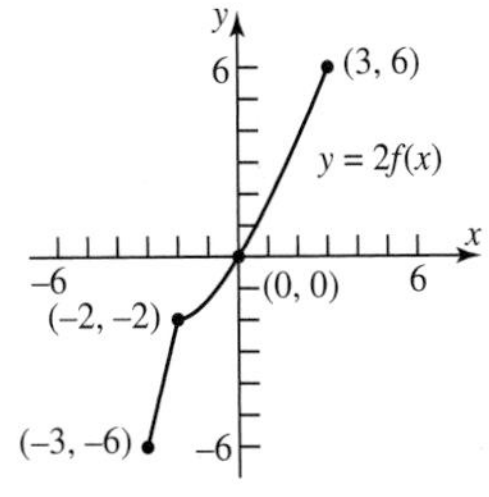

(f)

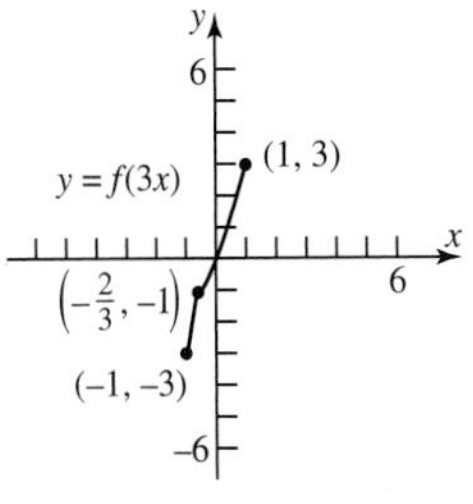

67. (2, 2) **69.** 25 square units **71. (a)** $V(S) = \dfrac{S}{6}\sqrt{\dfrac{S}{\pi}}$ **(b)** $V = \dfrac{1}{3}rS$; if the surface area doubles, the volume doubles.

CHAPTER 3 Polynomial and Rational Functions

3.1 Exercises

1. D **3.** A **5.** B **7.** E **9.** D **11.** B

13.

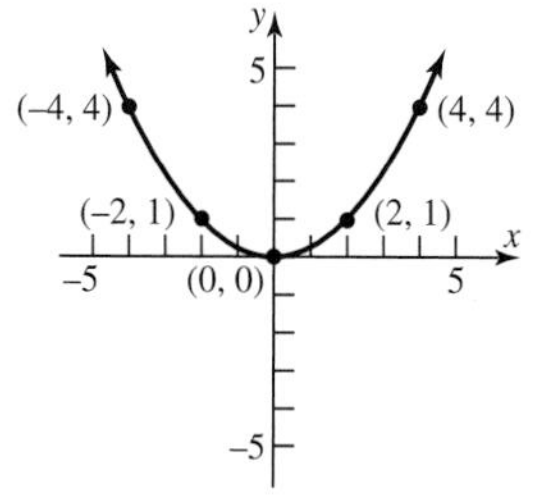

15.

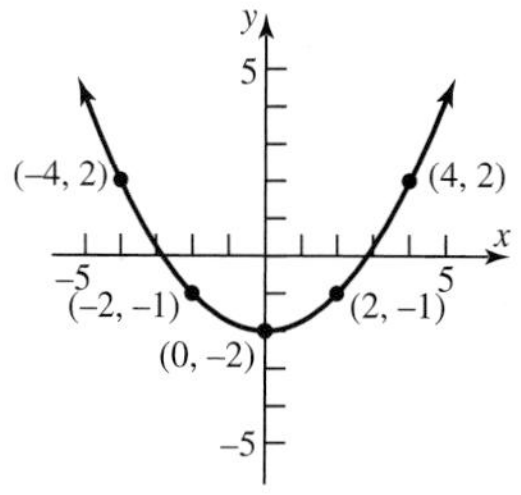

17.

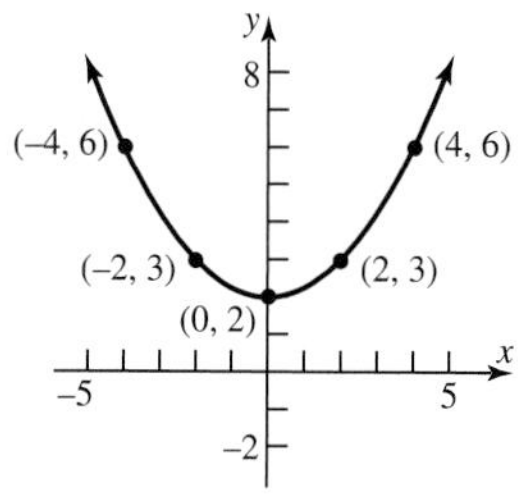

19.

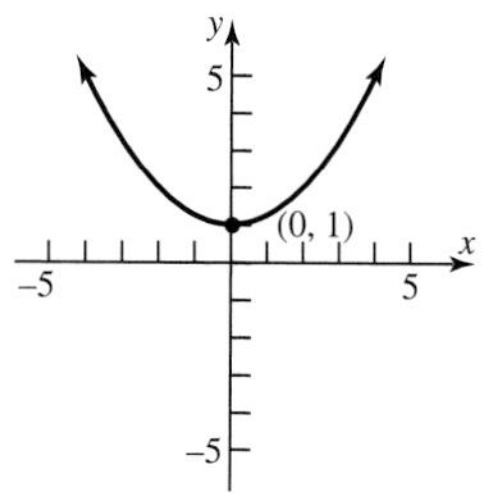

21. $f(x) = (x + 2)^2 - 2$

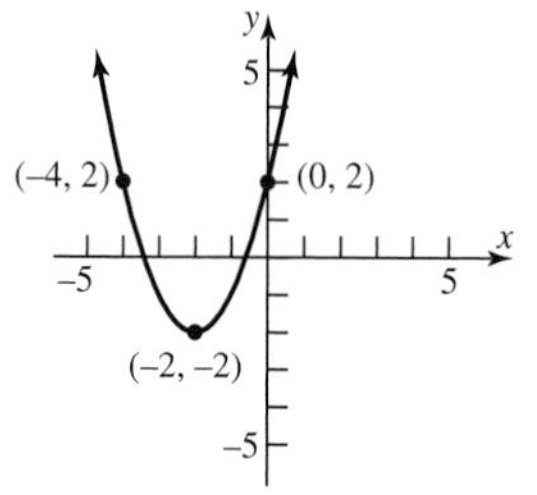

23. $f(x) = 2(x - 1)^2 - 1$

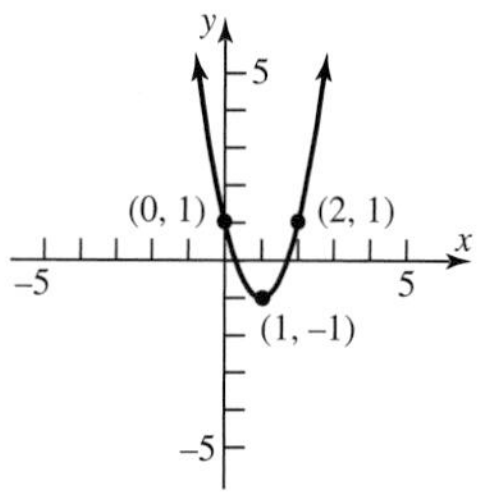

25. $f(x) = -(x + 1)^2 + 1$

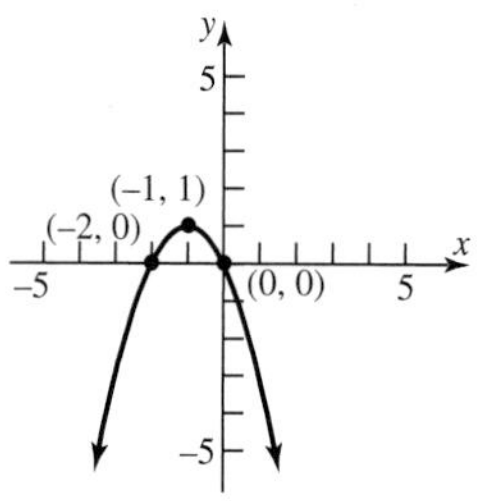

27. $f(x) = \dfrac{1}{2}(x + 1)^2 - \dfrac{3}{2}$

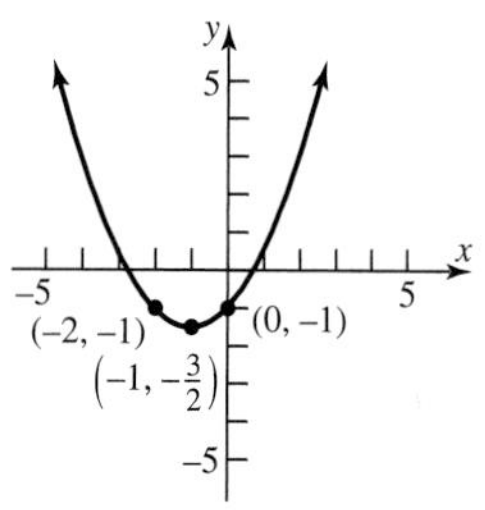

29.

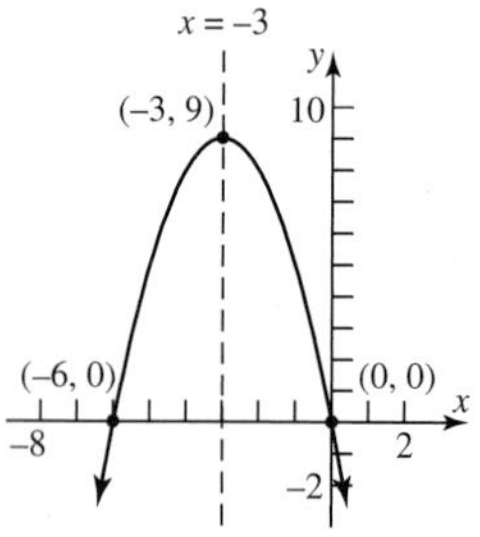

31.

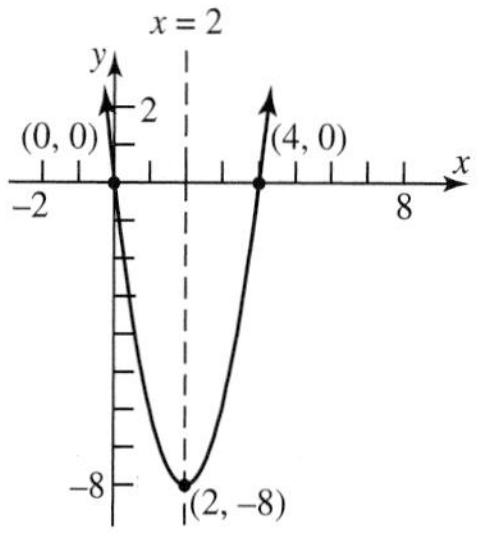

33.

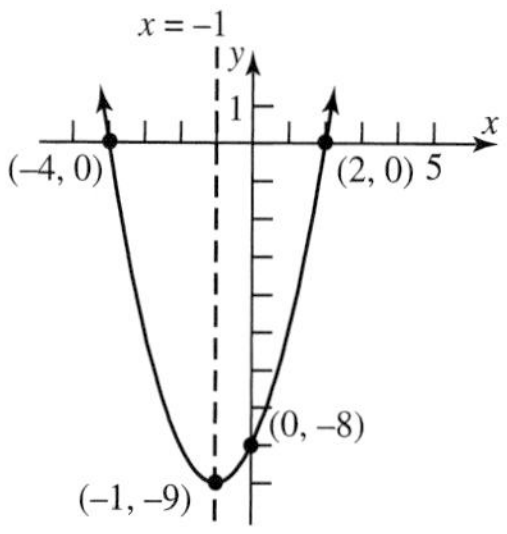

35.

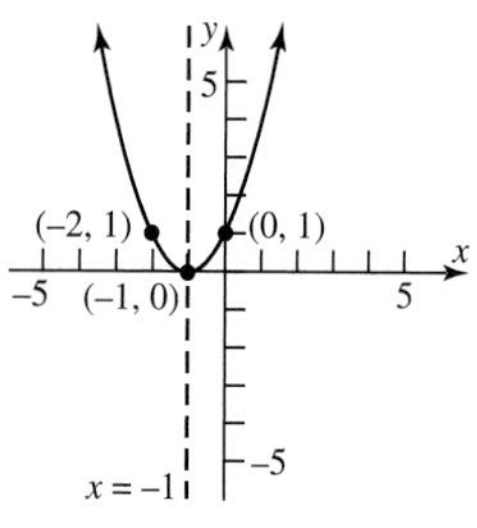

37.

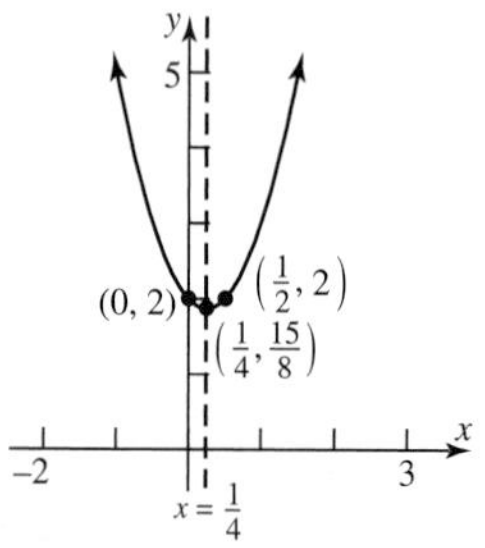

39.

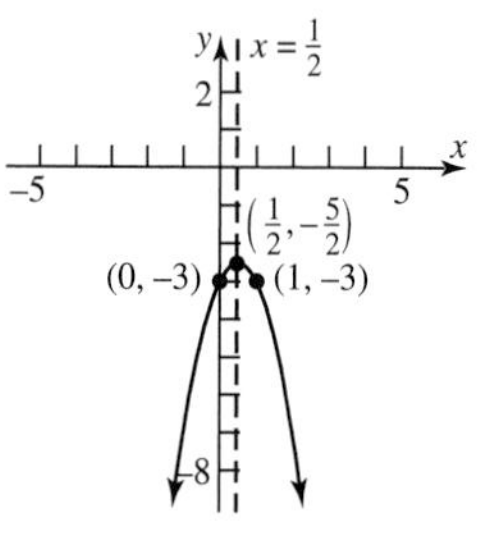

41.

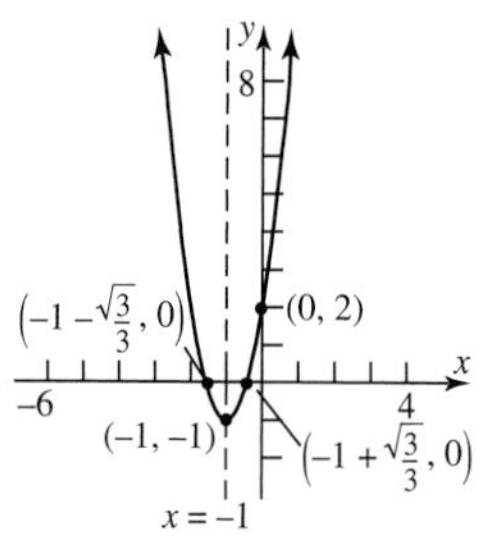

43.

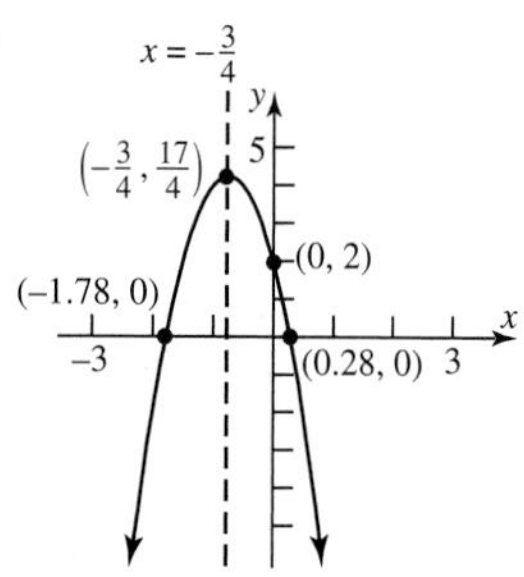

45. Minimum value; −18 **47.** Minimum value; −21 **49.** Maximum value; 21 **51.** Maximum value; 13

53. (a) $a = 1: f(x) = (x + 3)(x - 1) = x^2 + 2x - 3$
$a = 2: f(x) = 2(x + 3)(x - 1) = 2x^2 + 4x - 6$
$a = -2: f(x) = -2(x + 3)(x - 1) = -2x^2 - 4x + 6$
$a = 5: f(x) = 5(x + 3)(x - 1) = 5x^2 + 10x - 15$
(b) The value of a does not affect the intercepts
(c) The value of a does not affect the axis of symmetry. It is $x = -1$ for all values of a.
(d) The value of a does not affect the x-coordinate of the vertex. However, the y-coordinate of the vertex is multiplied by a.
(e) The midpoint of the x-intercepts is the x-coordinate of the vertex.

55. \$500; \$1,000,000

57. **(a)** $R(x) = -\frac{1}{6}x^2 + 100x$
(b) \$13,333
(c) 300; \$15,000
(d) \$50

59. **(a)** $R(x) = -\frac{1}{5}x^2 + 20x$
(b) \$255
(c) 50; \$500
(d) \$10

61. **(a)** $A(x) = -x^2 + 200x$
(b) A is largest when $x = 100$ yd.
(c) 10,000 sq yd

63. 2,000,000 m^2

65. **(a)** $\frac{625}{16} \approx 39$ ft **(b)** 219.5 ft **(c)** 170 ft
(d) When the height is 100 ft, the projectile is 135.7 ft from the cliff.
(e)

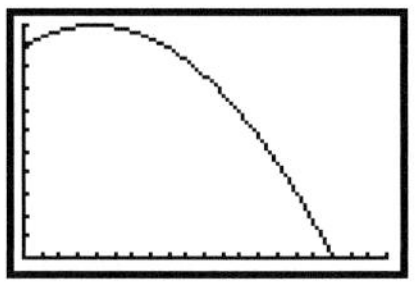

67. 18.75 m

69. 3 in.

71. $\frac{750}{\pi}$ by 375 m

73. **(a)** Quadratic, $a < 0$.

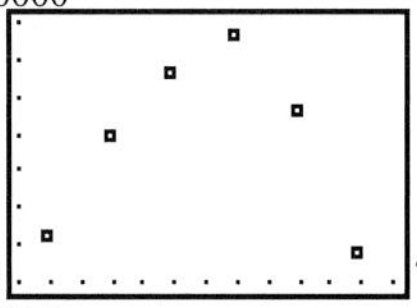

(b) $I(x) = -42.5884x^2 + 3805.5270x - 38{,}526.0037$
(c)

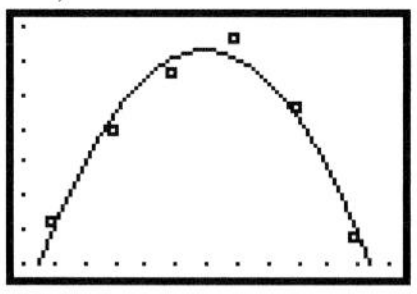

(d) 44.7 years old **(e)** \$46,486

75. **(a)** Quadratic, $a > 0$.

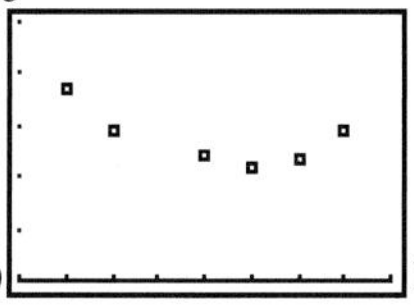

(b) $I(x) = 0.1019x^2 - 406.3402x + 405{,}067.7207$
(c)

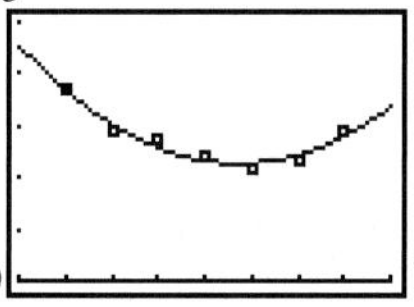

(d) \$3.37 **(e)** 1994

77. **(a)** Quadratic, $a > 0$.

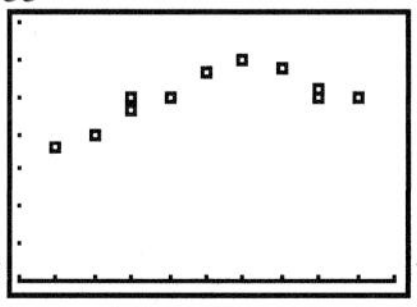

(b) $M(s) = 0.0175s^2 + 1.9346s - 25.341$
(c)

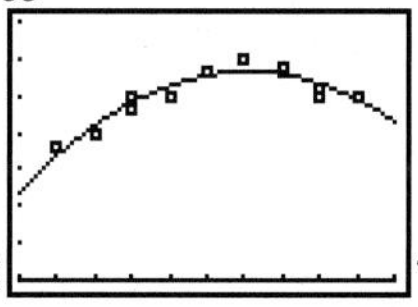

(d) 55 miles per hour **(f)** 27 miles per gallon

79. $x = \frac{a}{2}$ **81.** $\left.\begin{aligned} ah^2 - bh + c &= y_0 \\ c &= y_1 \\ ah^2 + bh + c &= y_2 \end{aligned}\right\} \left.\begin{aligned} y_0 + y_2 &= 2ah^2 + 2c \\ 4y_1 &= 4c \end{aligned}\right\}$ Area $= \frac{h}{3}(2ah^2 + 6c) = \frac{h}{3}(y_0 + 4y_1 + y_2)$ **83.** $\frac{128}{3}$ **85.** $\frac{22}{3}$

3.2 Exercises

1.

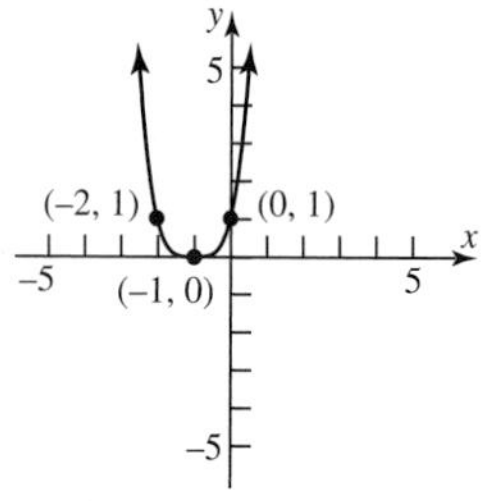

3.

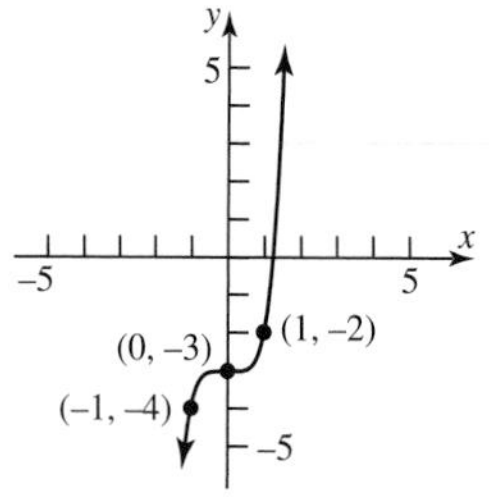

5.

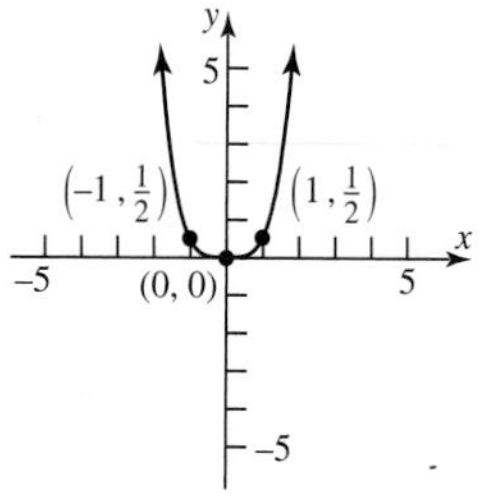

7.

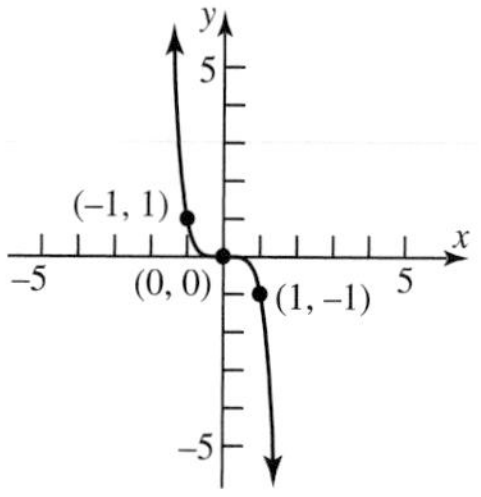

9.

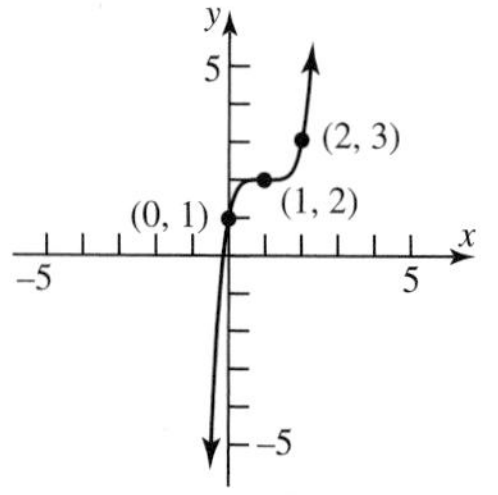

11.

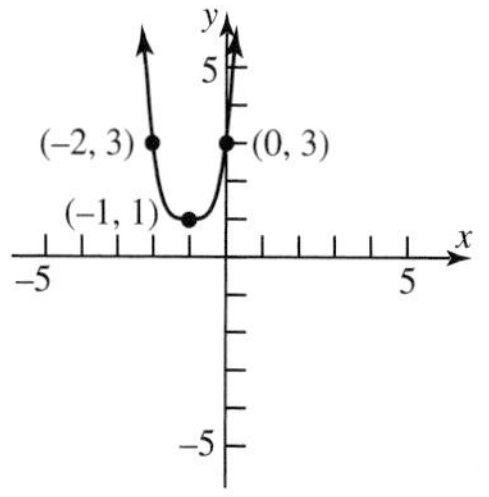

13.

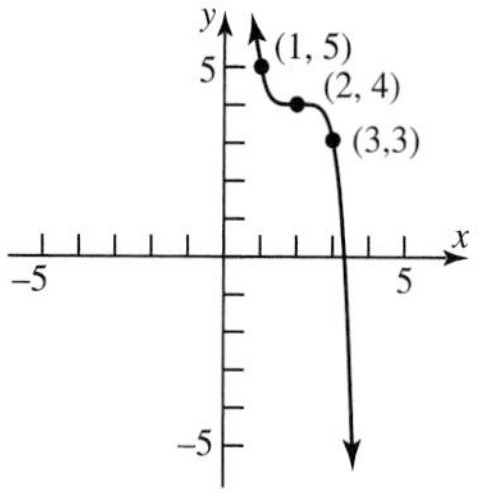

15.

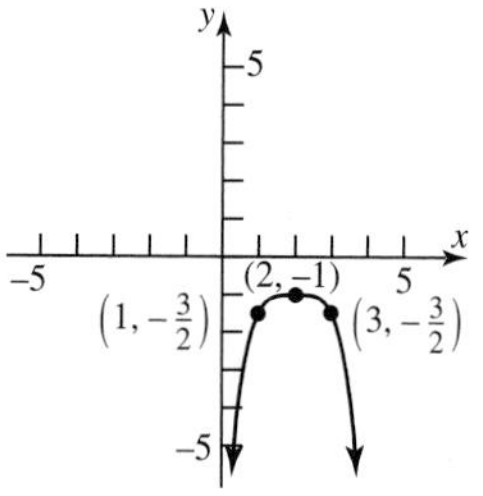

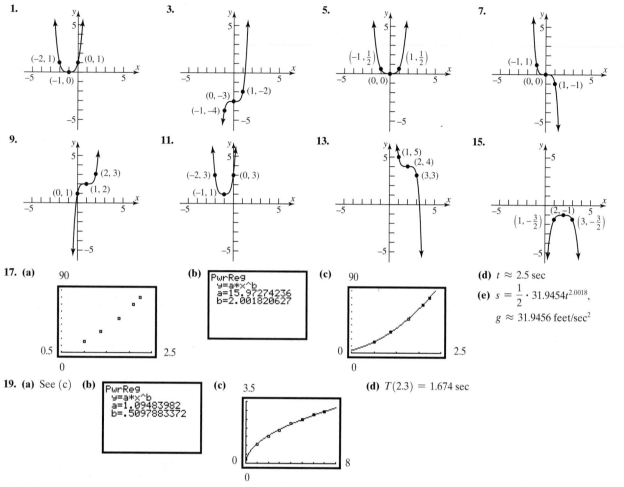

17. (a)

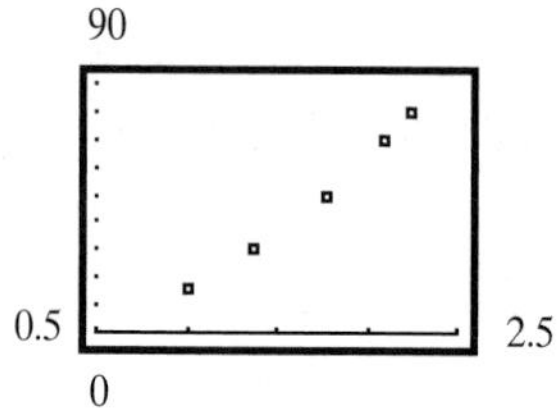

(b)

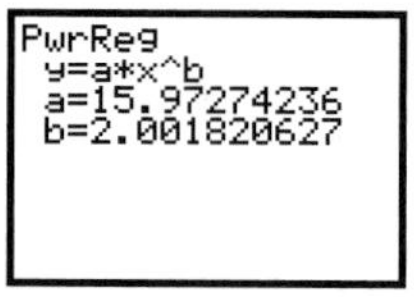
PwrReg
y=a*x^b
a=15.97274236
b=2.001820627

(c)

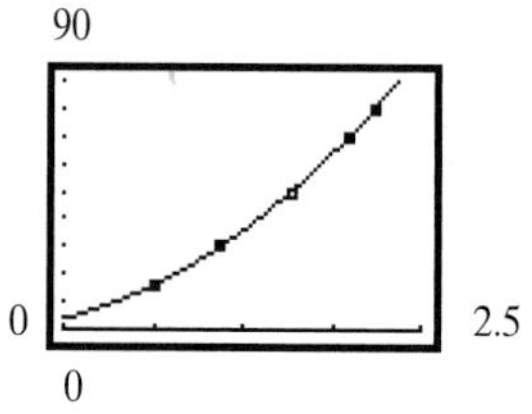

(d) $t \approx 2.5$ sec

(e) $s = \frac{1}{2} \cdot 31.9454t^{2.0018}$, $g \approx 31.9456$ feet/sec^2

19. (a) See (c) **(b)**

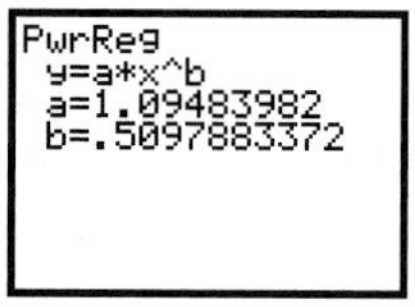
PwrReg
y=a*x^b
a=1.09483982
b=.5097883372

(c)

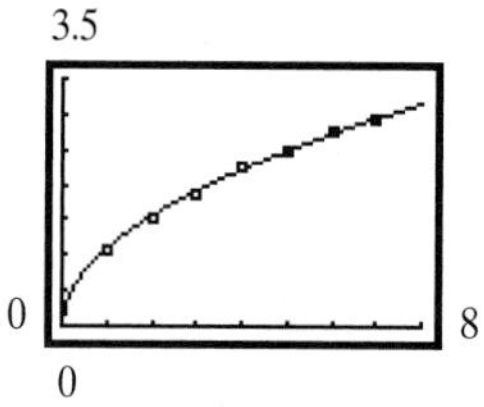

(d) $T(2.3) = 1.674$ sec

3.3 Exercises

1. Yes; degree 3 **3.** Yes; degree 2 **5.** No; x is raised to the -1 power. **7.** No; x is raised to the $\frac{3}{2}$ power. **9.** Yes; degree 4

11. $f(x) = x^3 - 3x^2 - x + 3$ for $a = 1$ **13.** $f(x) = x^3 - x^2 - 12x$ for $a = 1$ **15.** $f(x) = x^4 - 15x^2 + 10x + 24$ for $a = 1$

17. 7, multiplicity 1; -3, multiplicity 2; graph touches the x-axis at -3 and crosses it at 7 **19.** 2, multiplicity 3; graph crosses the x-axis at 2

21. $-\frac{1}{2}$, multiplicity 2; graph touches the x-axis at $-\frac{1}{2}$

23. 5, multiplicity 3; -4, multiplicity 2; graph touches the x-axis at -4 and crosses it at 5

25. No real zeros; graph neither crosses nor touches the x-axis

27. (a)

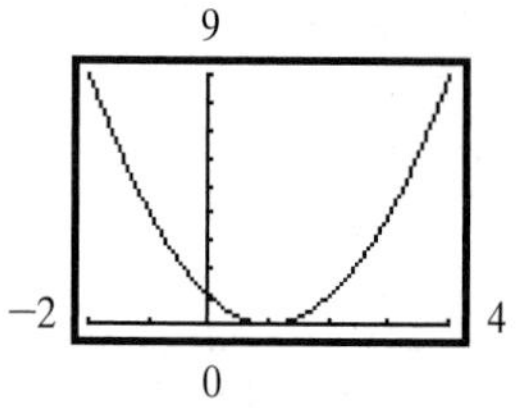

(b) x-intercept: 1; y-intercept: 1
(c) 1: Even
(d) $y = x^2$
(e) 1
(f) Local minimum: (1, 0); Local maxima: None

29. (a)

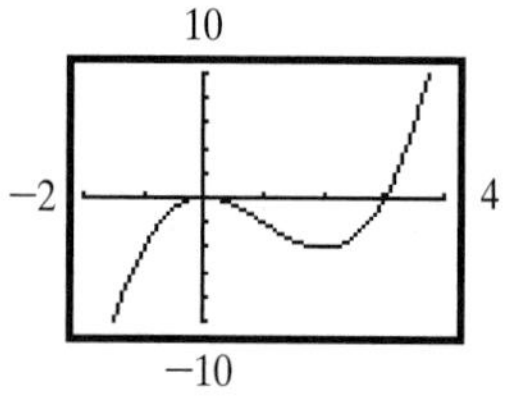

(b) x-intercepts: 0, 3; y-intercept: 0
(c) 0: Even; 3: Odd
(d) $y = x^3$
(e) 2
(f) Local minimum: (2, −4); Local maximum: (0, 0)

31. (a)

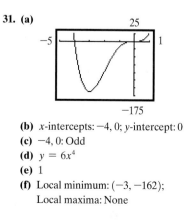

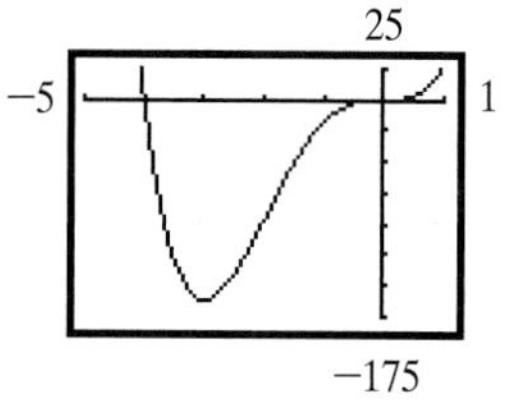

(b) x-intercepts: −4, 0; y-intercept: 0
(c) −4, 0: Odd
(d) $y = 6x^4$
(e) 1
(f) Local minimum: (−3, −162); Local maxima: None

33. (a)

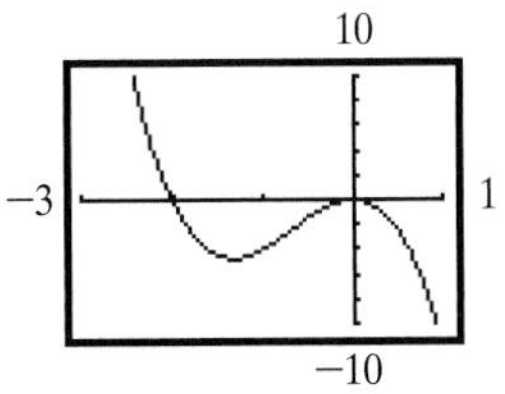

(b) x-intercepts: $-2, 0$; y-intercept: 0
(c) -2: Odd; 0: Even
(d) $y = -4x^3$
(e) 2
(f) Local minimum: $(-1.33, -4.74)$;
Local maximum: $(0, 0)$

35. (a)

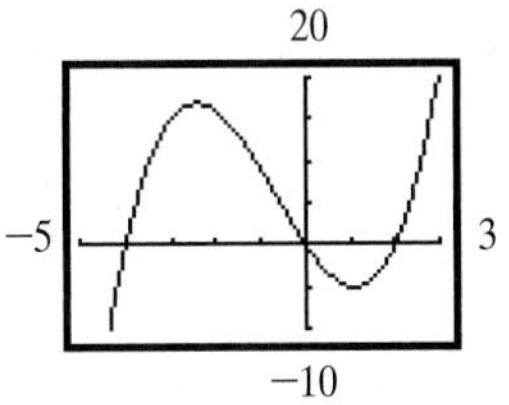

(b) x-intercepts: $-4, 0, 2$; y-intercept: 0
(c) $-4, 0, 2$: Odd
(d) $y = x^3$
(e) 2
(f) Local minimum: $(1.10, -5.05)$;
Local maximum: $(-2.43, 16.90)$

37. (a)

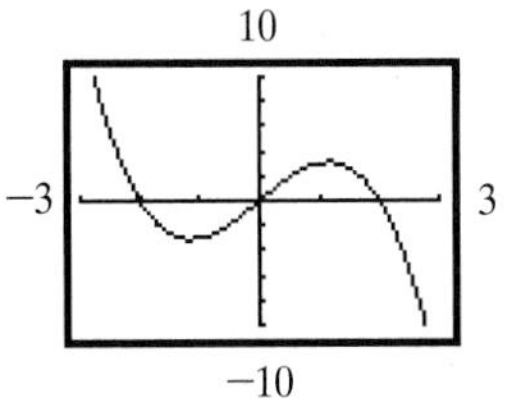

(b) x-intercepts: $-2, 0, 2$; y-intercept: 0
(c) $-2, 0, 2$: Odd
(d) $y = -x^3$
(e) 2
(f) Local minimum: $(-1.15, -3.08)$;
Local maximum: $(1.15, 3.08)$

39. (a)

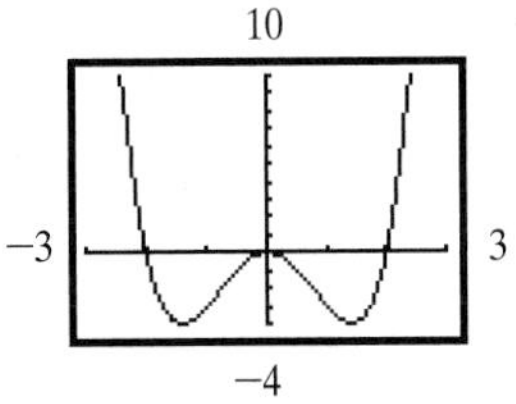

(b) x-intercepts: $-2, 0, 2$; y-intercept: 0
(c) $-2, 2$: Odd; 0: Even
(d) $y = x^4$
(e) 3
(f) Local minima: $(-1.41, -4)$, $(1.41, -4)$;
Local maximum: $(0, 0)$

41. (a)

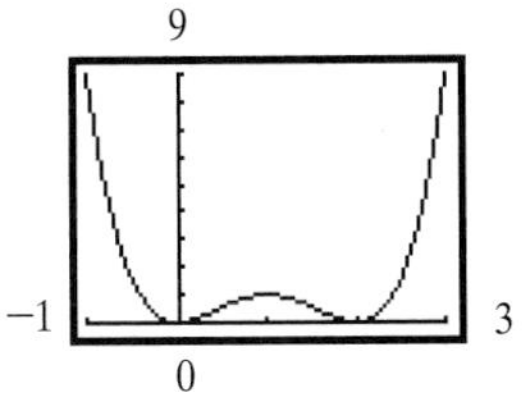

(b) x-intercepts: $0, 2$; y-intercept: 0
(c) $0, 2$: Even
(d) $y = x^4$
(e) 3
(f) Local minima: $(0, 0)$, $(2, 0)$;
Local maximum: $(1, 1)$

43. (a)

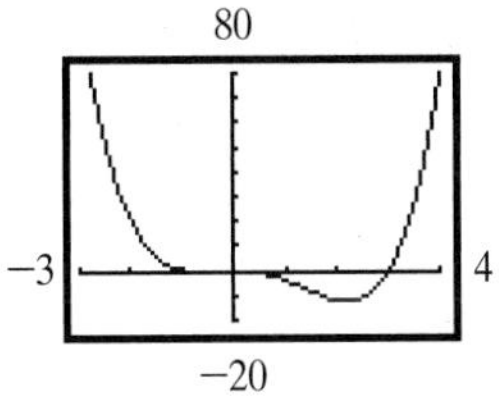

(b) x-intercepts: $-1, 0, 3$; y-intercept: 0
(c) $-1, 3$: Odd; 0: Even
(d) $y = x^4$
(e) 3
(f) Local minima: $(2.19, -12.39)$,
$(-0.69, -0.54)$;
Local maximum: $(0, 0)$

45. (a)

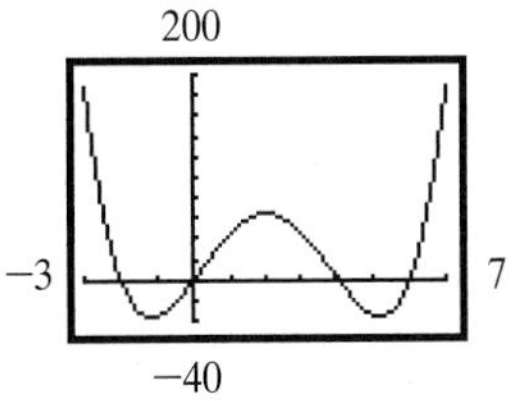

(b) x-intercepts: $-2, 0, 4, 6$; y-intercept: 0
(c) $-2, 0, 4, 6$: Odd
(d) $y = x^4$
(e) 3
(f) Local minima: $(-1.16, -36)$,
$(5.16, -36)$;
Local maximum: $(2, 64)$

47. (a)

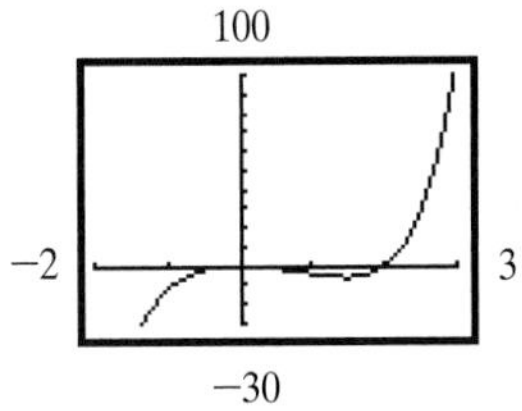

(b) x-intercepts: $0, 2$; y-intercept: 0
(c) 0: Even; 2: Odd
(d) $y = x^5$
(e) 2
(f) Local minimum: $(1.48, -5.91)$;
Local maximum: $(0, 0)$

49. (a)

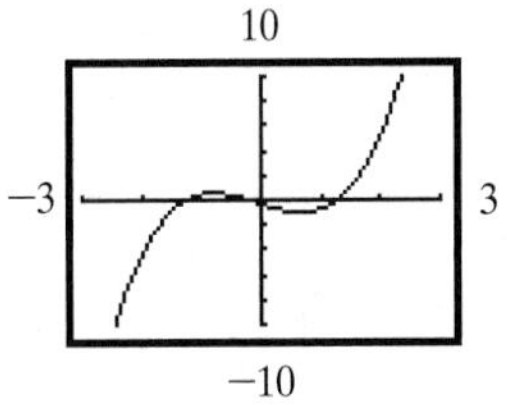

(b) x-intercepts: $-1.26, -0.20, 1.26$;
y-intercept: -0.31752
(c) $-1.26, -0.20, 1.26$: Odd
(d) $y = x^3$
(e) 2
(f) Local minimum: $(0.66, -0.99)$;
Local maximum: $(-0.80, 0.57)$

51. (a)

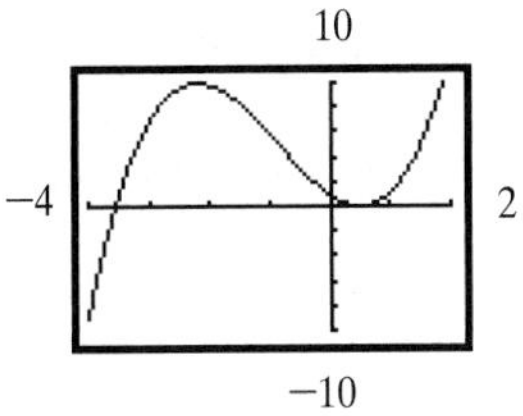

(b) x-intercepts: $-3.56, 0.50$;
y-intercept: 0.89
(c) -3.56: Odd; 0.50: Even
(d) $y = x^3$
(e) 2
(f) Local minimum: $(0.50, 0)$;
Local maximum: $(-2.21, 9.91)$

53. (a)

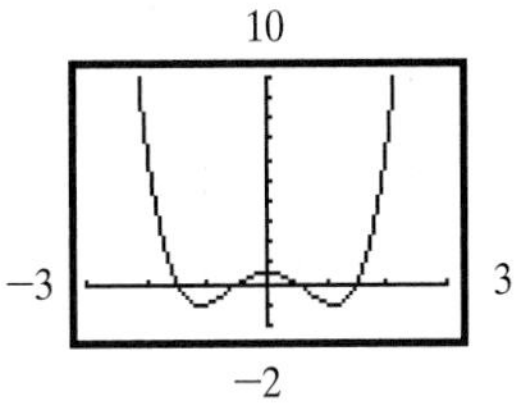

(b) x-intercepts: $-1.50, -0.50, 0.50, 1.50$;
y-intercept: 0.5625
(c) $-1.50, -0.50, 0.50, 1.50$: Odd
(d) $y = x^4$
(e) 3
(f) Local minima: $(-1.12, -1)$,
$(1.12, -1)$;
Local maximum: $(0, 0.5625)$

55. (a)

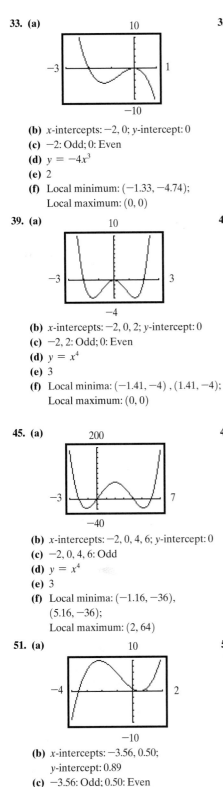

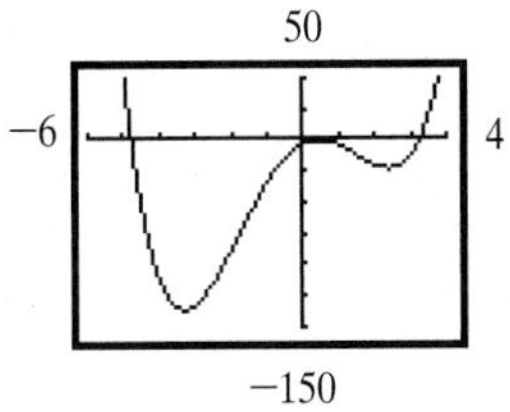

(b) x-intercepts: $-4.78, 0.45, 3.23$;
y-intercept: -3.1264785
(c) $-4.78, 3.23$: Odd; 0.45: Even
(d) $y = x^4$
(e) 3
(f) Local minima: $(-3.32, -135.92)$,
$(2.38, -22.67)$;
Local maximum: $(0.45, 0)$

57. (a)

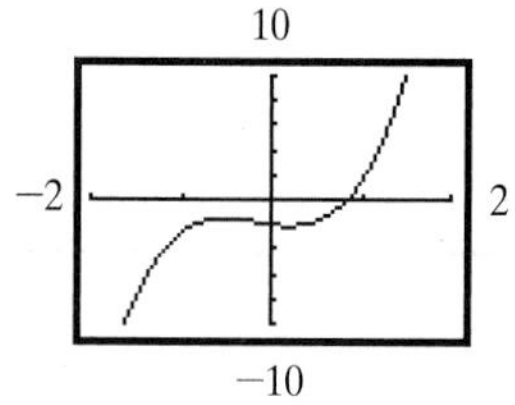

(b) x-intercept: 0.84; y-intercept: −2
(c) 0.84: Odd
(d) $y = \pi x^3$
(e) 2
(f) Local minimum: $(0.21, -2.12)$;
Local maximum: $(-0.51, -1.54)$

59. (a)

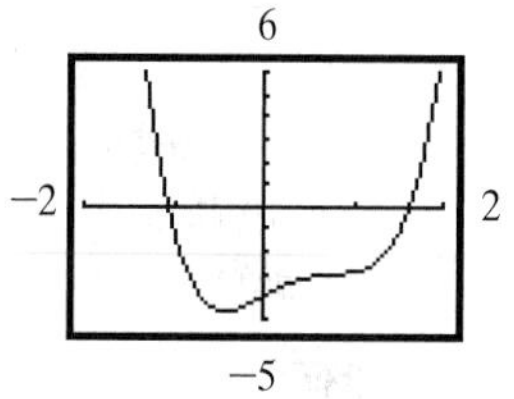

(b) x-intercepts: −1.07, 1.62;
y-intercept: −4
(c) −1.07, 1.62: Odd
(d) $y = 2x^4$
(e) 1
(f) Local minimum: $(-0.42, -4.64)$;
Local maxima: None

61. (a)

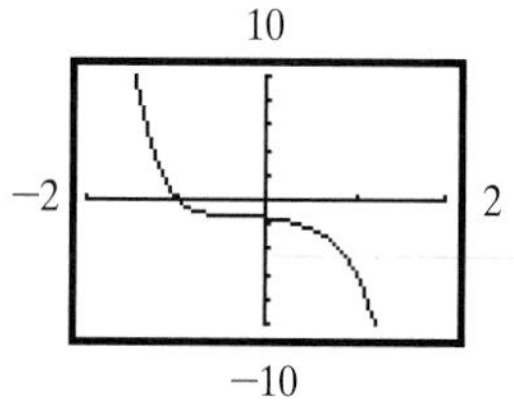

(b) x-intercept: −0.98; y-intercept: $-\sqrt{2}$
(c) −0.98: Odd
(d) $y = -2x^5$
(e) 0
(f) None

63. (a) Answers will vary. Possibilities include $f(x) = x(x-1)(x-2)$; $f(x) = x^3(x-1)(x-2)$
(b) Answers will vary. Possibilities include $f(x) = x(x-1)^2(x-2)$
(c) Answers will vary. Possibilities include $f(x) = -\frac{1}{2}(x+1)(x-1)(x-2)$
(d) Answers will vary. Possibilities include $f(x) = -\frac{1}{2}(x+1)(x-1)^2(x-2)$

65. (a)

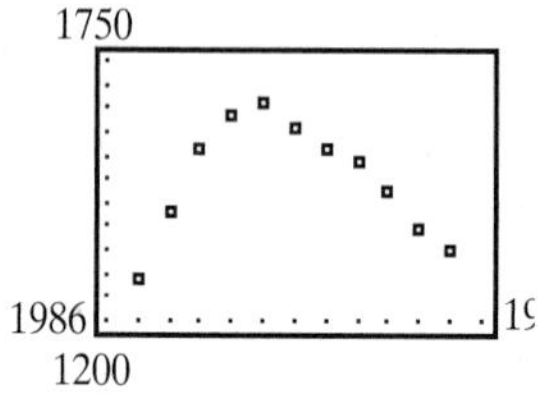

(b) $T(x) = 1.5243x^3 - 9121.4767x^2 + 18{,}194{,}595.386x - 12{,}097{,}532{,}838.668$

(c)

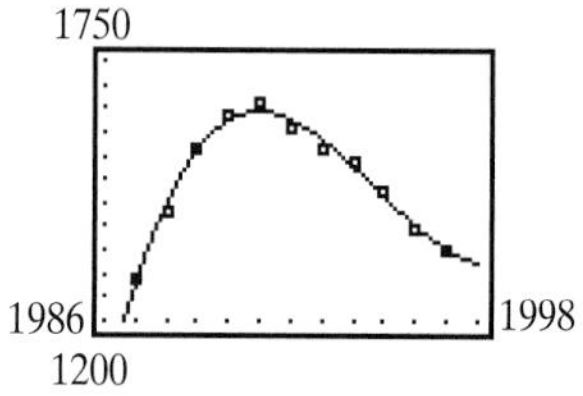

(d) 1324

67. (a) Cubic, $a > 0$

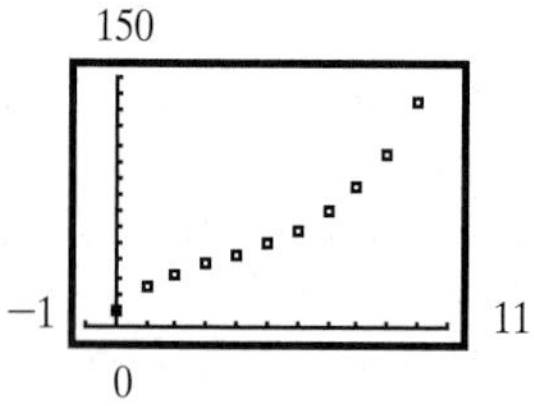

(b) \$7000 per car
(c) \$20,000 per car
(d)

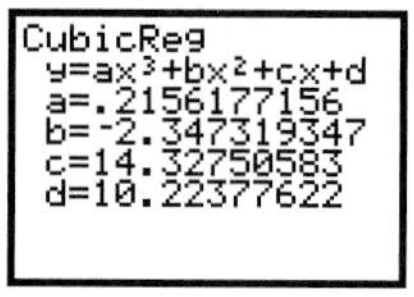

(e)

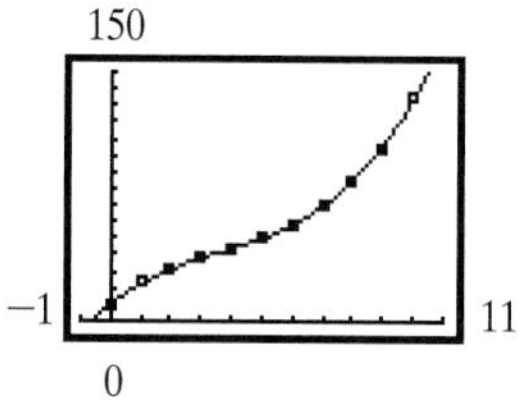

(f) ≈ \$170,800
(g) Fixed costs of ≈ \$10,200

71. $f(x) = (x+2)(x-1)^2$; $g(x) = (x+2)^3(x-1)^2$ **73.** a, b, c, d

3.4 Exercises

1. All real numbers except 3. **3.** All real numbers except 2 and −4. **5.** All real numbers except $-\frac{1}{2}$ and 3. **7.** All real numbers except 2.
9. All real numbers **11.** All real numbers except −3 and 3. **13. (a)** Domain: $\{x \mid x \neq 2\}$; Range: $\{y \mid y \neq 1\}$ **(b)** $(0, 0)$ **(c)** $y = 1$ **(d)** $x = 2$ **(e)** None **15. (a)** Domain: $\{x \mid x \neq 0\}$; Range: all real numbers **(b)** $(-1, 0)$, $(1, 0)$ **(c)** None **(d)** $x = 0$ **(e)** $y = 2x$
17. (a) Domain: $\{x \mid x \neq -2, x \neq 2\}$; Range: $\{y \mid -\infty < y \leq 0 \text{ or } 1 < y < \infty\}$ **(b)** $(0, 0)$ **(c)** $y = 1$ **(d)** $x = -2$, $x = 2$ **(e)** None
19. (a) Domain: $\{x \mid x \neq -1\}$; Range: $\{y \mid y \neq 2\}$ **(b)** $(-1.5, 0)$, $(0, 3)$ **(c)** $y = 2$ **(d)** $x = -1$ **(e)** None
21. (a) Domain: $\{x \mid x \neq -4, x \neq 3\}$; Range: All real numbers **(b)** $(0, 0)$ **(c)** $y = 0$ **(d)** $x = -4$; $x = 3$ **(e)** None

23.

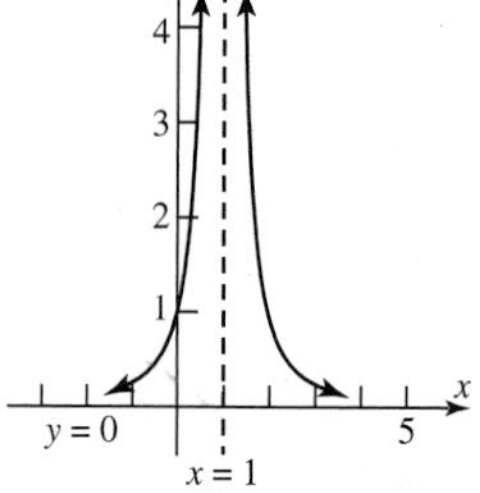

25.

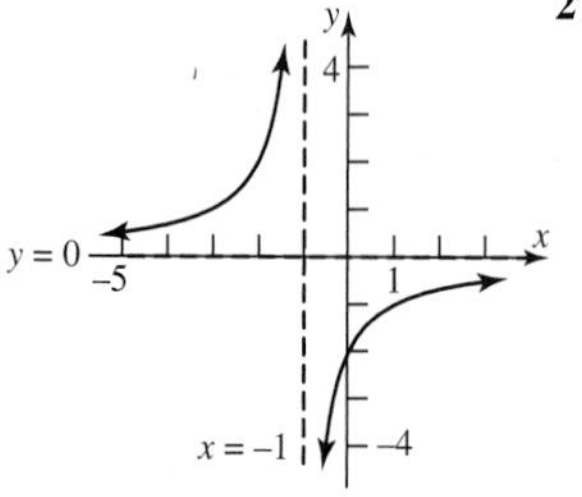

27.

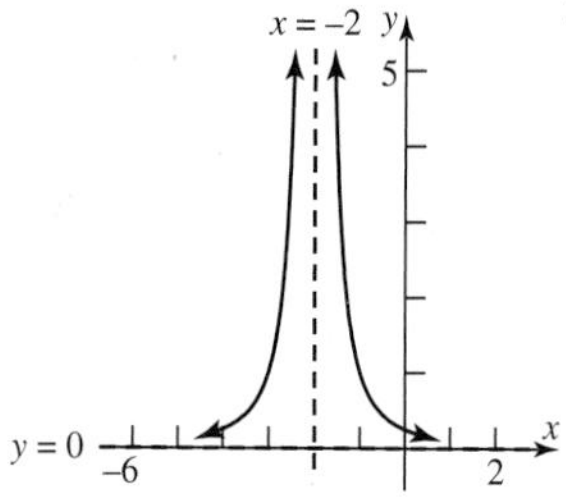

29.

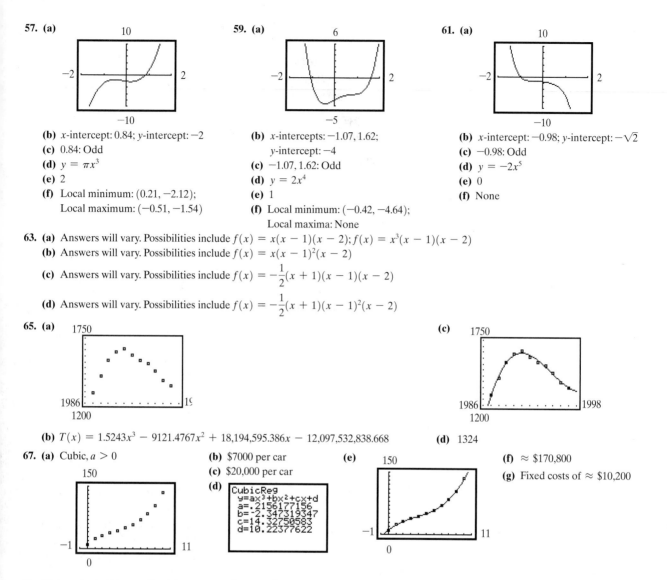

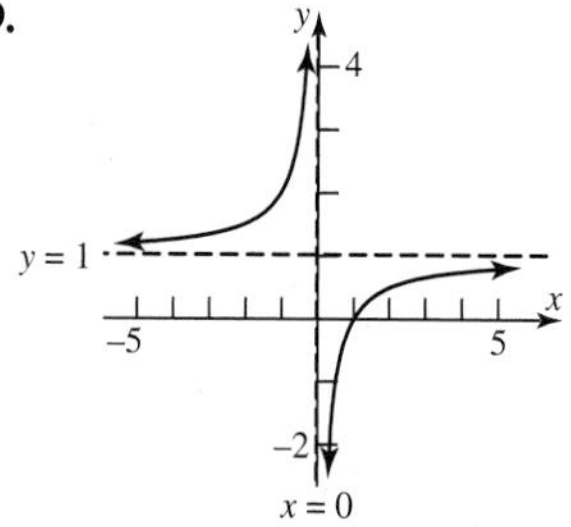

31.

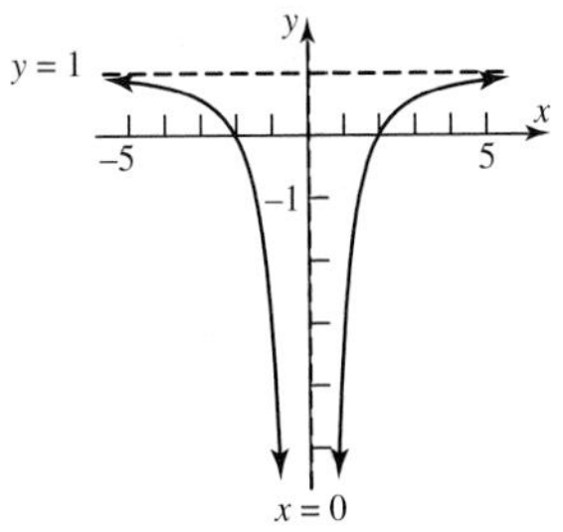

33.

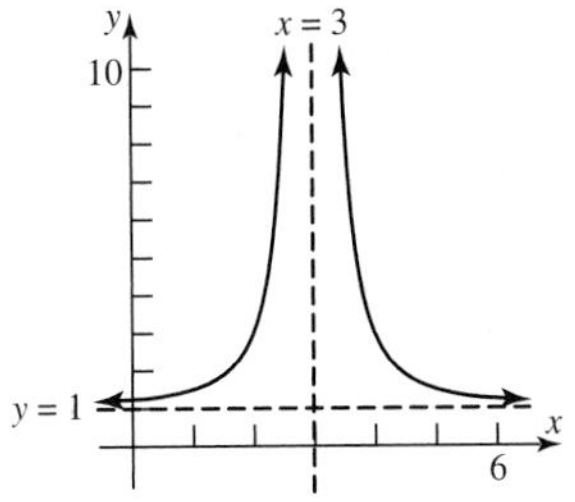

35. Horizontal asymptote: $y = 3$; vertical asymptote: $x = -4$
37. No asymptotes
39. Horizontal asymptote: $y = 0$; vertical asymptotes: $x = 1$, $x = -1$
41. Horizontal asymptote: $y = 0$; vertical asymptote: $x = 0$
43. Oblique asymptote: $y = 3x$; vertical asymptote: $x = 0$
45. Oblique asymptote: $y = -(x + 1)$; vertical asymptote: $x = 0$

3.5 Exercises

1. 1. Domain: $\{x \mid x \neq 0, x \neq -4\}$
2. x-intercept: -1; no y-intercept
3. No symmetry
4. Vertical asymptotes: $x = 0$, $x = -4$
5. Horizontal asymptote: $y = 0$, intersected at $(-1, 0)$
6.

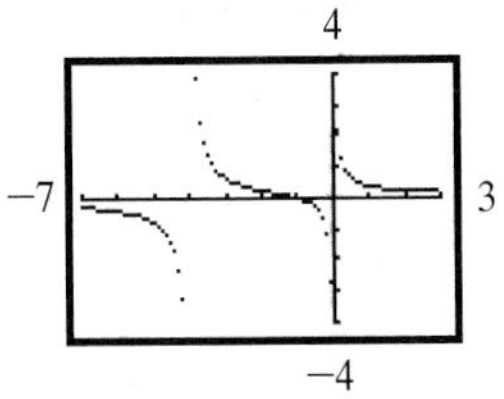

7.

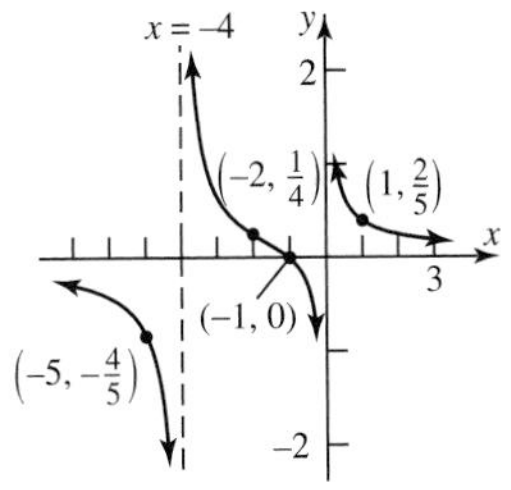

3. 1. Domain: $\{x \mid x \neq -2\}$
2. x-intercept: -1; y-intercept: $\frac{3}{4}$
3. No symmetry
4. Vertical asymptote: $x = -2$
5. Horizontal asymptote: $y = \frac{3}{2}$, not intersected
6.

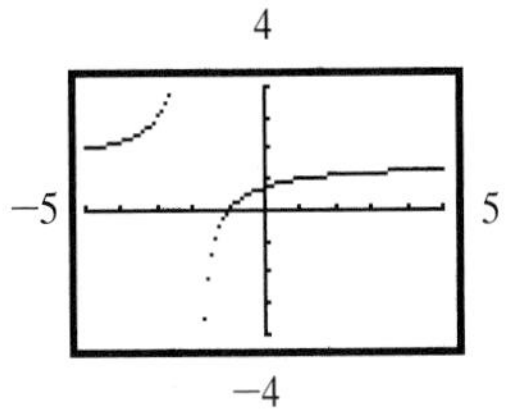

7.

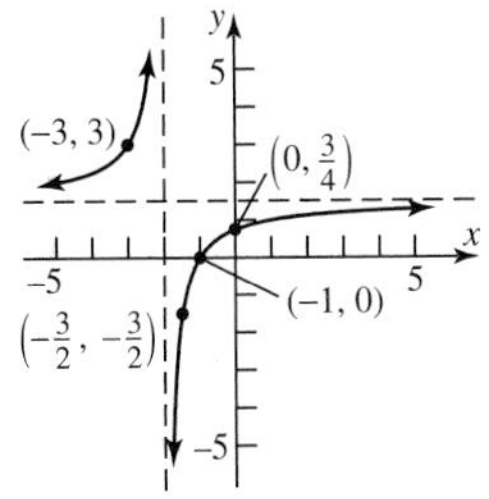

5. 1. Domain: $\{x \mid x \neq -2, x \neq 2\}$
2. No x-intercept; y-intercept: $-\frac{3}{4}$
3. Symmetric with respect to y-axis
4. Vertical asymptotes: $x = 2$, $x = -2$
5. Horizontal asymptote: $y = 0$, not intersected
6.

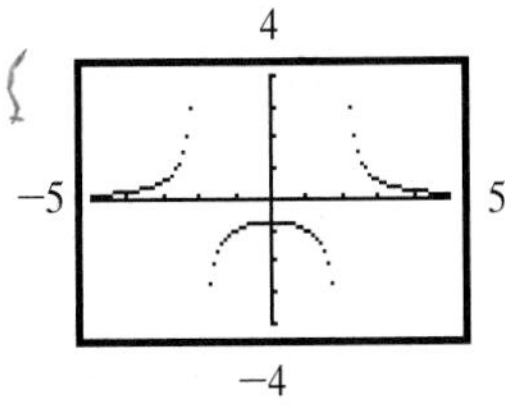

7.

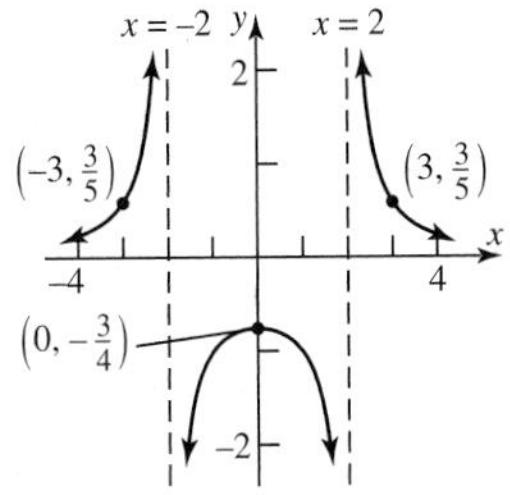

7. 1. Domain: $\{x|x \neq -1, x \neq 1\}$
2. No x-intercept; y-intercept: -1
3. Symmetric with respect to y-axis
4. Vertical asymptotes: $x = -1$, $x = 1$
5. No horizontal or oblique asymptotes
6.

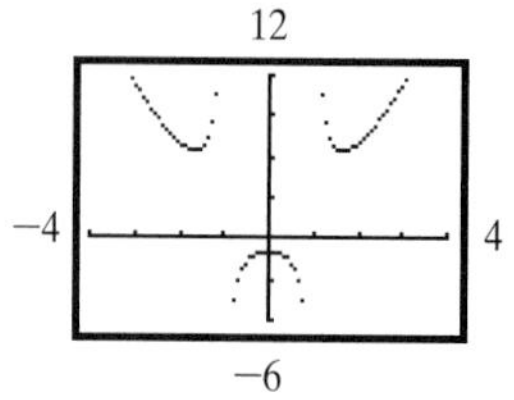

7.

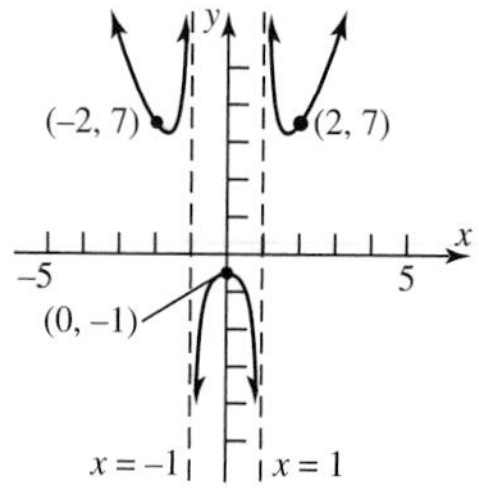

9. 1. Domain: $\{x|x \neq -3, x \neq 3\}$
2. x-intercept: 1; y-intercept: $\frac{1}{9}$
3. No symmetry
4. Vertical asymptotes: $x = 3$, $x = -3$
5. Oblique asymptote: $y = x$, intersected at $\left(\frac{1}{9}, \frac{1}{9}\right)$
6.

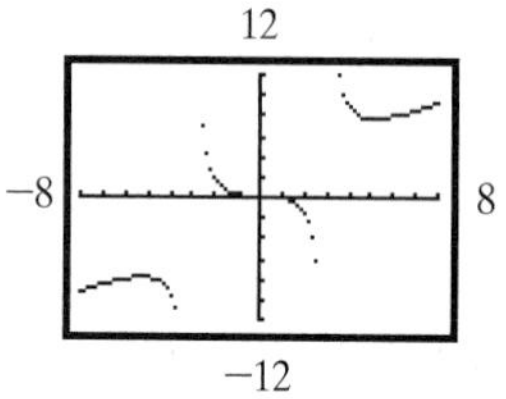

7.

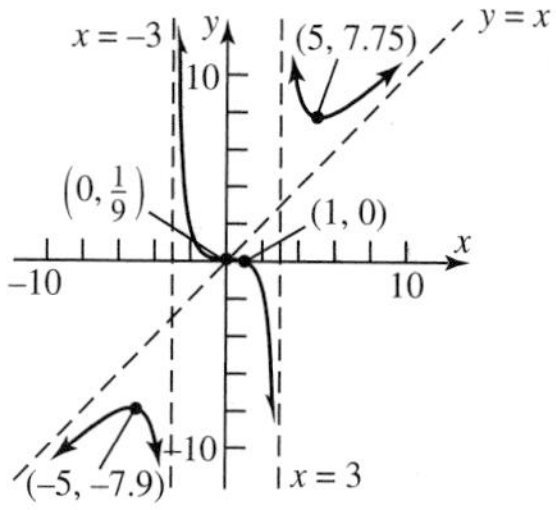

11. 1. Domain: $\{x \neq -3, x \neq 2\}$
2. Intercept: $(0, 0)$
3. No symmetry
4. Vertical asymptotes: $x = 2$, $x = -3$
5. Horizontal asymptote: $y = 1$, intersected at $(6, 1)$
6.

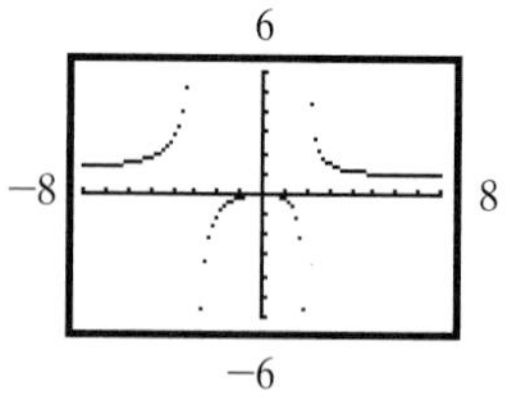

7.

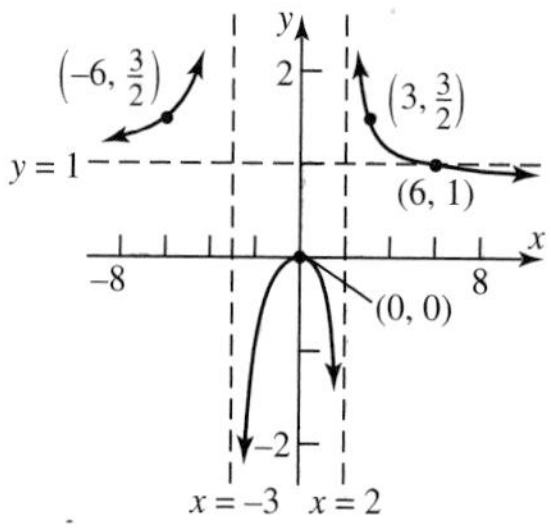

13. 1. Domain: $\{x|x \neq -2, x \neq 2\}$
2. Intercept: $(0, 0)$
3. Symmetry with respect to origin
4. Vertical asymptotes: $x = -2$, $x = 2$
5. Horizontal asymptote: $y = 0$, intersected at $(0, 0)$
6.

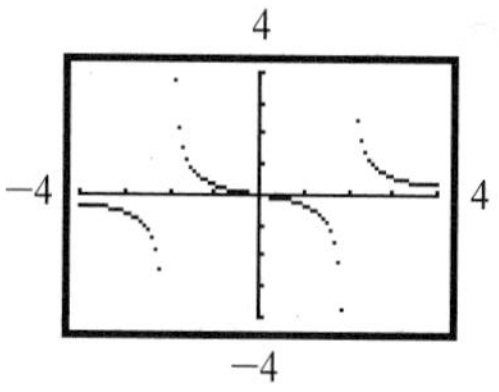

7.

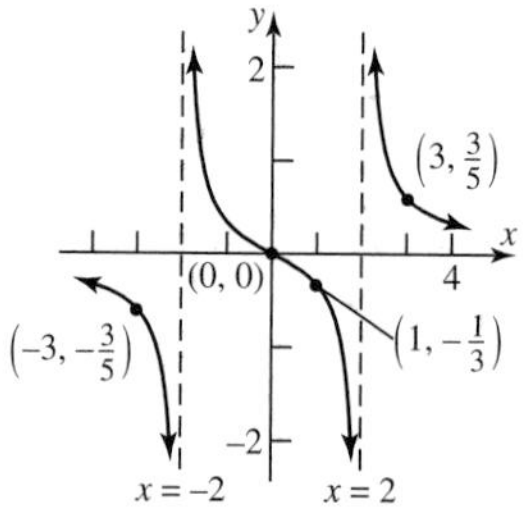

15. 1. Domain: $\{x|x \neq 1, x \neq -2, x \neq 2\}$

2. No x-intercept; y-intercept: $\frac{3}{4}$
3. No symmetry
4. Vertical asymptotes: $x = -2, x = 1, x = 2$
5. Horizontal asymptote: $y = 0$, not intersected
6.

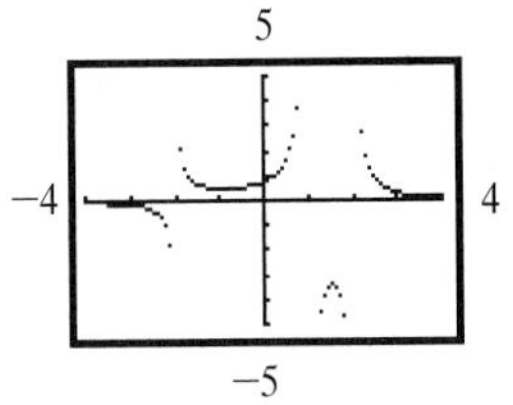

7.

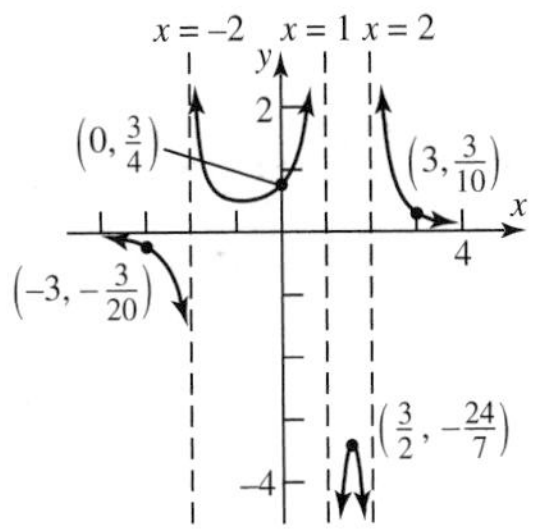

17. 1. Domain: $\{x|x \neq -2, x \neq 2\}$

2. x-intercepts: $-1, 1$; y-intercept: $\frac{1}{4}$
3. Symmetry with respect to y-axis
4. Vertical asymptotes: $x = -2, x = 2$
5. Horizontal asymptote: $y = 0$, intersected at $(-1, 0)$ and $(1, 0)$
6.

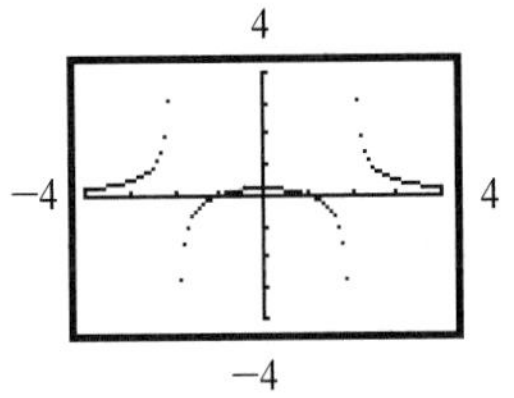

7.

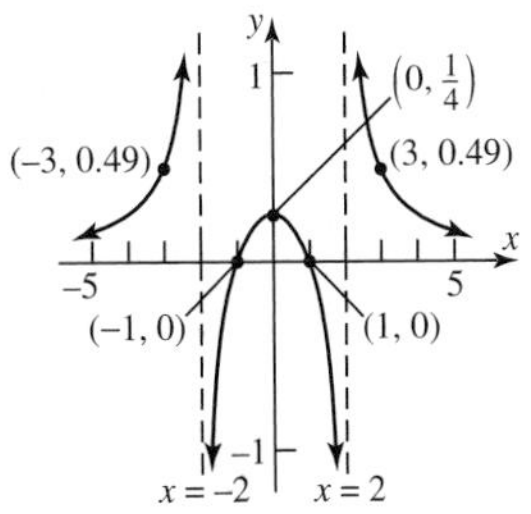

19. 1. Domain: $\{x|x \neq -2\}$

2. x-intercepts: $-1, 4$; y-intercept: -2
3. No symmetry
4. Vertical asymptote: $x = -2$
5. Oblique asymptote: $y = x - 5$, not intersected
6.

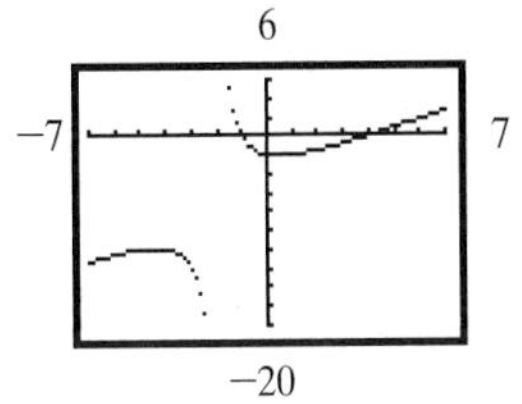

7.

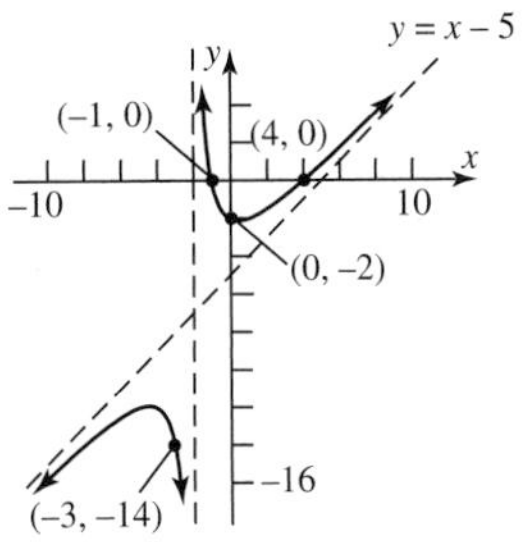

21. 1. Domain: $\{x|x \neq 4\}$

2. x-intercepts: $-4, 3$; y-intercept: 3
3. No symmetry
4. Vertical asymptote: $x = 4$
5. Oblique asymptote: $y = x + 5$, not intersected
6.

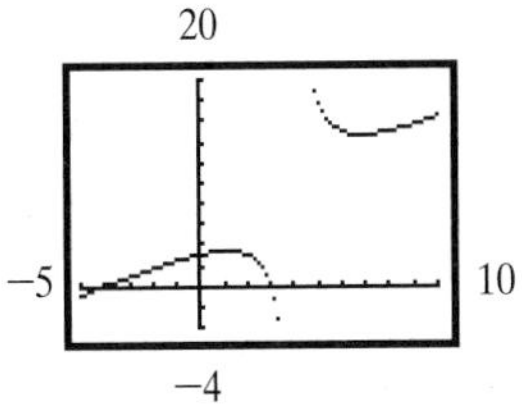

7.

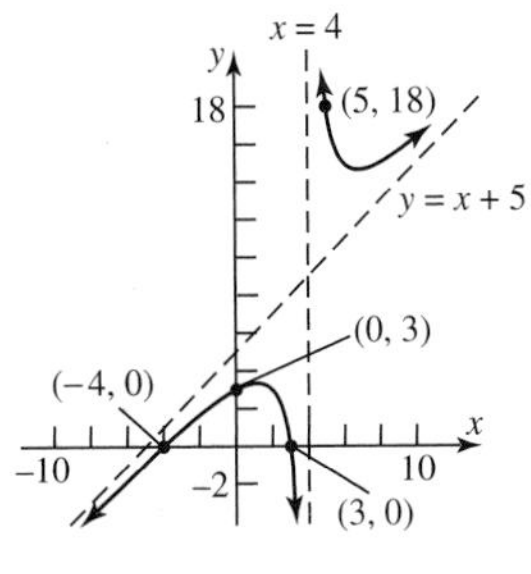

23. 1. Domain: $\{x|x \neq -2\}$

2. x-intercepts: -4, 3; y-intercept: -6
3. No symmetry
4. Vertical asymptote: $x = -2$
5. Oblique asymptote: $y = x - 1$, not intersected
6.

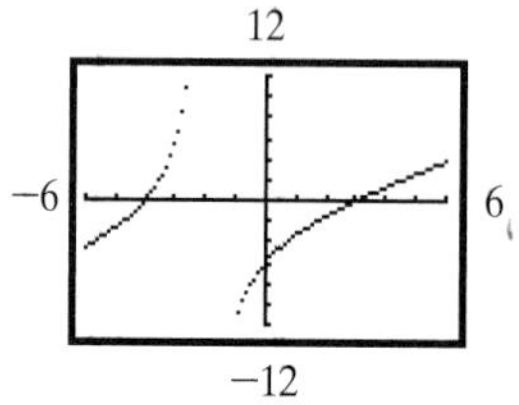

7.

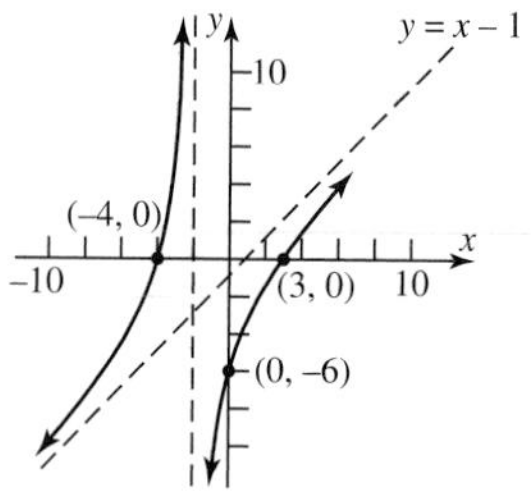

25. 1. Domain: $\{x|x \neq -3\}$

2. x-intercepts: 0, 1; y-intercept: 0
3. No symmetry
4. Vertical asymptote: $x = -3$
5. Horizontal asymptote: $y = 1$, not intersected
6.

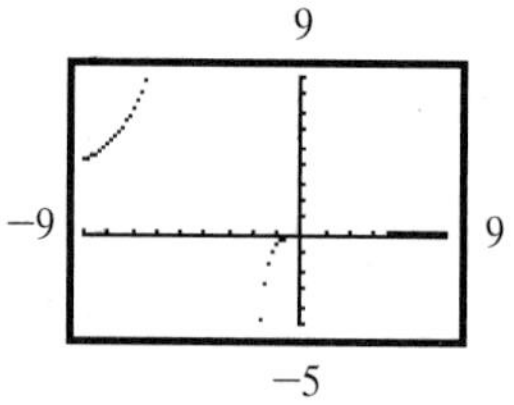

7.

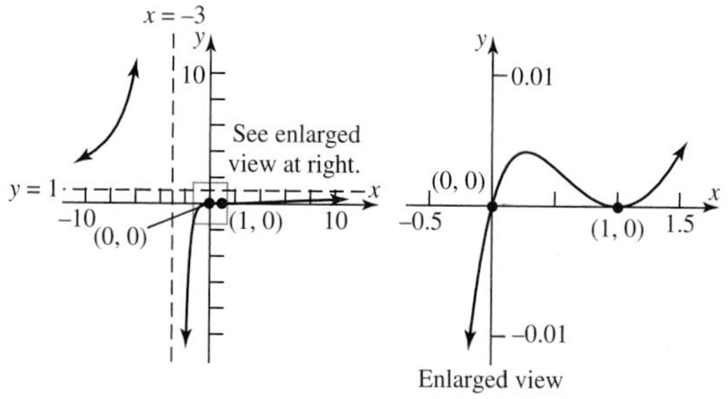

27. 1. Domain: $\{x|x \neq -2, x \neq 3\}$

2. x-intercept: -4; y-intercept: 2
3. No symmetry
4. Vertical asymptote: $x = -2$; hole at $\left(3, \frac{7}{5}\right)$
5. Horizontal asymptote: $y = 1$, not intersected
6.

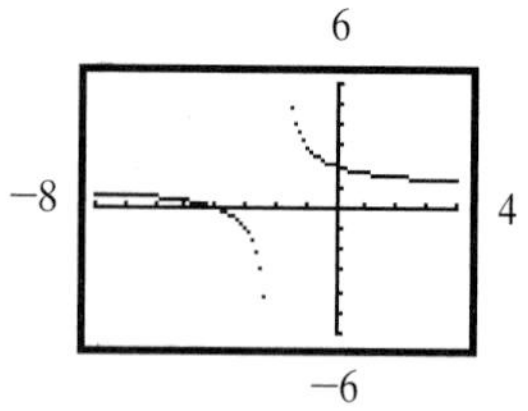

7.

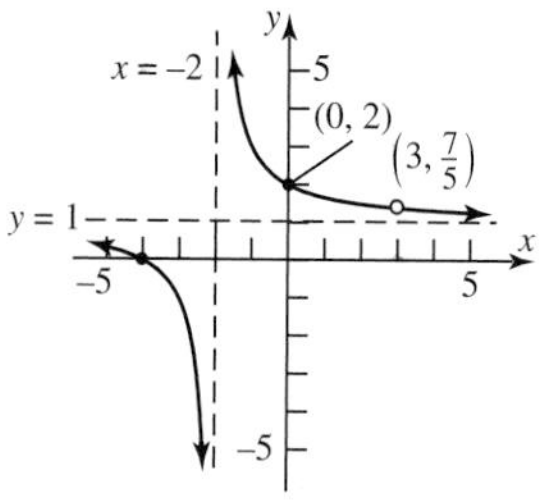

29. 1. Domain: $\left\{x \middle| x \neq \frac{3}{2}, x \neq 2\right\}$

2. x-intercept: $-\frac{1}{3}$; y-intercept: $-\frac{1}{2}$
3. No symmetry
4. Vertical asymptote: $x = 2$; hole at $\left(\frac{3}{2}, -11\right)$
5. Horizontal asymptote: $y = 3$, not intersected
6.

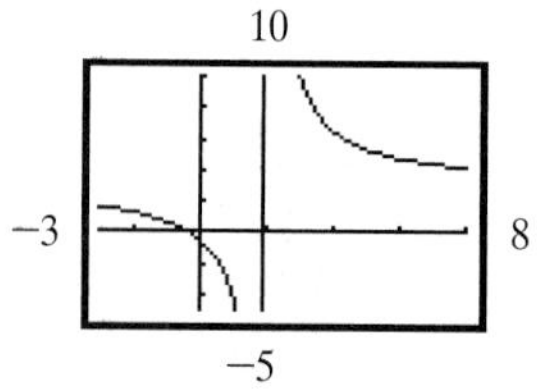

7.

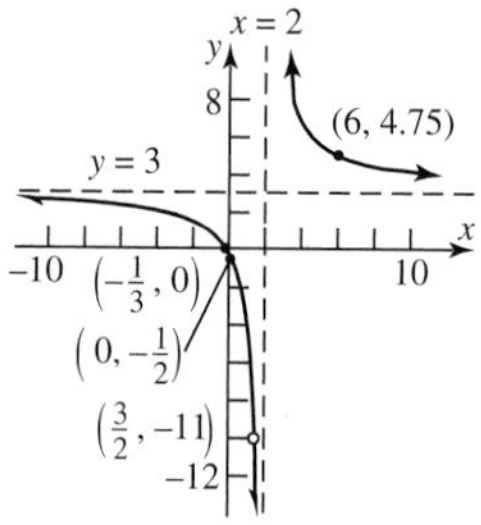

31. 1. Domain: $\{x|x \neq -3\}$
2. x-intercept: -2; y-intercept: 2
3. No symmetry
4. Vertical asymptote: None
5. Oblique asymptote: $y = x + 2$
6.

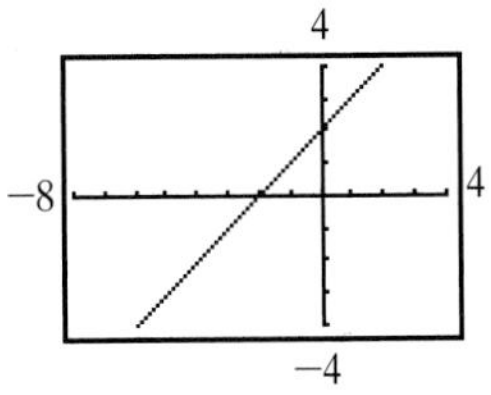

7.

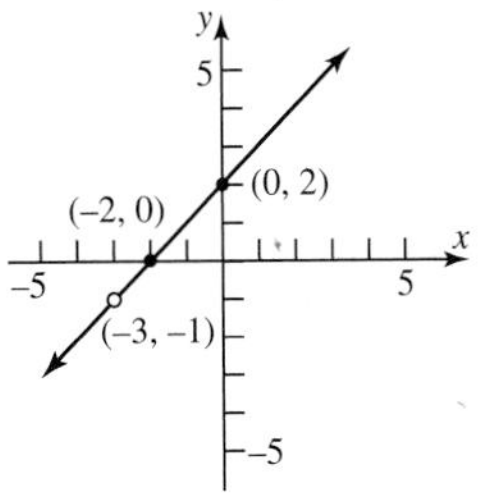

33. 1. Domain: $\{x|x \neq 0\}$
2. No x-intercepts; no y-intercepts
3. Symmetric about the origin
4. Vertical asymptote: $x = 0$
5. Oblique asymptotes: $y = x$, not intersected
6.

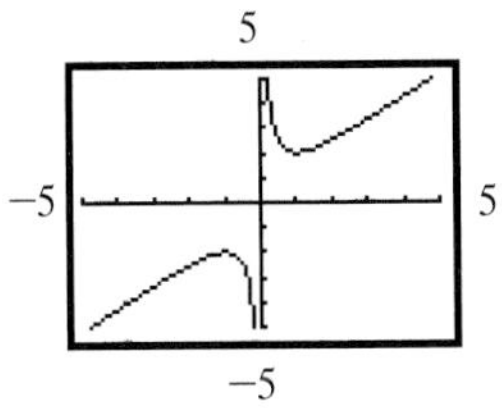

7.

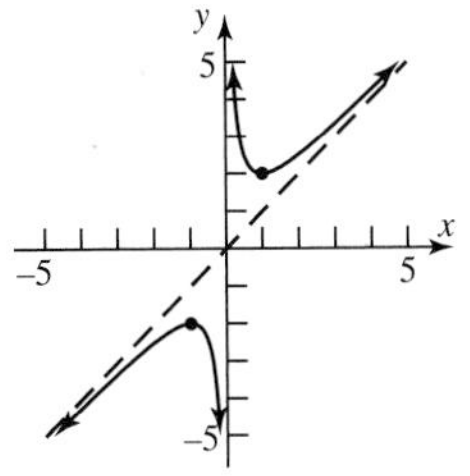

35. 1. Domain: $\{x|x \neq 0\}$
2. x-intercept: -1; no y-intercepts
3. No symmetry
4. Vertical asymptote: $x = 0$
5. No horizontal or oblique asymptotes
6.

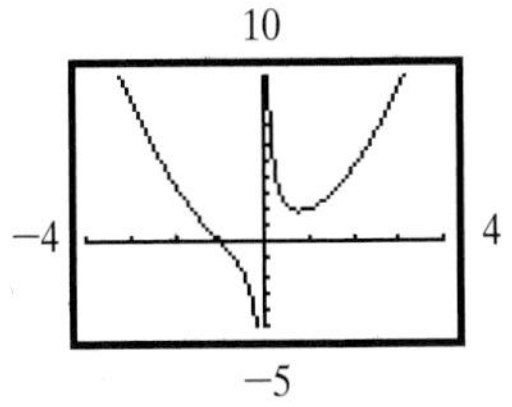

7.

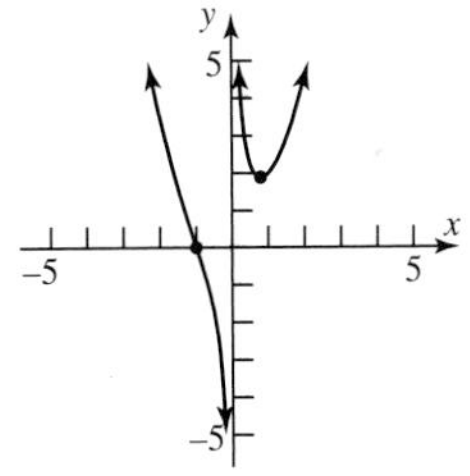

37. 1. Domain: $\{x|x \neq 0\}$
2. No x-intercepts; no y-intercepts
3. Symmetric about the origin
4. Vertical asymptote: $x = 0$
5. Oblique asymptote: $y = x$, not intersected
6.

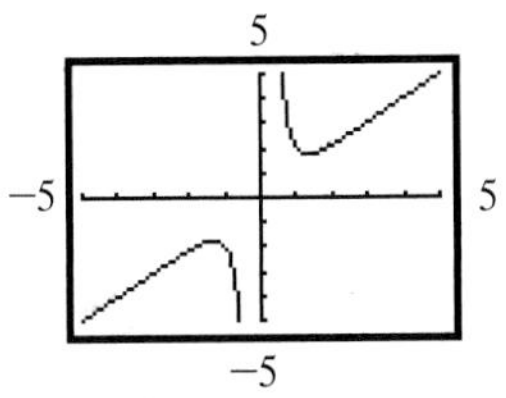

7.

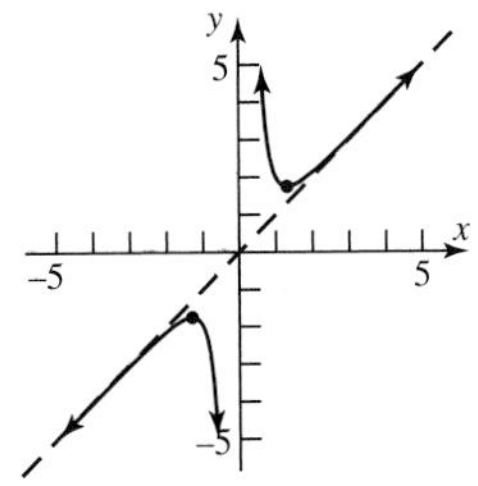

39. **(a)** One possibility: $R(x) = \dfrac{x^2}{x^2 - 4}$ **(b)** One possibility: $R(x) = -\dfrac{3x}{x^2 - 1}$ **(c)** One possibility: $R(x) = \dfrac{(x-1)(x-3)(x^2 + \frac{4}{3})}{(x+1)^2(x-2)^2}$

(d) One possibility: $R(x) = \dfrac{3(x+2)(x-1)^2}{(x+3)(x-4)^2}$ **41.** $R(x) = \dfrac{3(x-2)(x+1)^2}{(x+5)^2(x-6)}$

43. **(a)** 9.82 m/sec^2 **(b)** 9.8195 m/sec^2
(c) 9.7936 m/sec^2 **(d)** h-axis
(e)

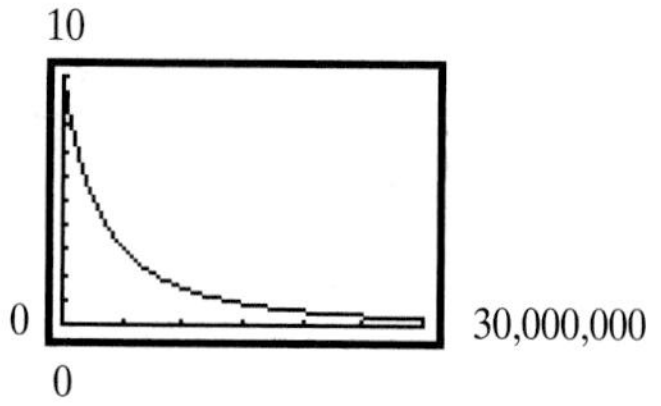

(f) $g(h)$ is never equal to 0, but $g(h) \to 0$ as $h \to \infty$. The further away from sea level you get, the lower the acceleration due to gravity.

45. **(a)**

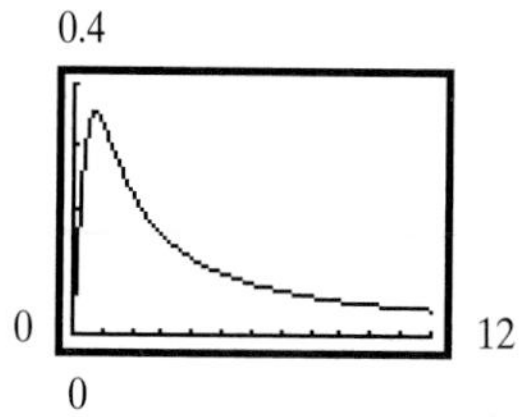

(b) 0.71 hr **(c)** t-axis; $C(t) \to 0$

47. **(a)** $\overline{C}(x) = \dfrac{0.2x^3 - 2.3x^2 + 14.3x + 10.2}{x}$
(b) $\overline{C}(6) = \$9400$ per car
(c) $\overline{C}(9) = \$10{,}933$ per car
(d)

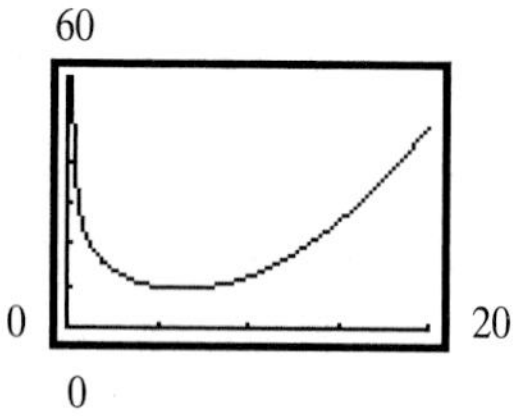

(e) 6 cars
(f) $9400 per car

49. **(a)** $S(x) = 2x^2 + \dfrac{40{,}000}{x}$
(b)

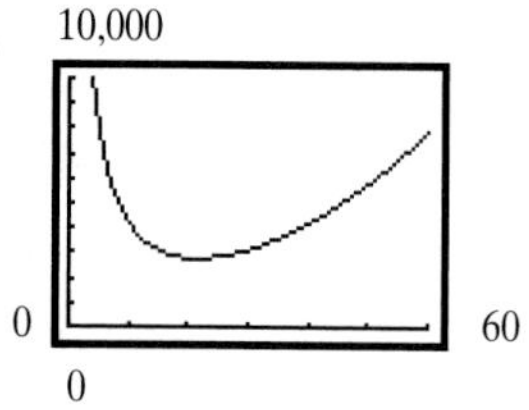

(c) 2784.95 sq in.
(d) 21.54 in. × 21.54 in. × 21.54 in.
(e) Minimizing area of cardboard will minimize their cost.

51. **(a)** $C(r) = 12\pi r^2 + \dfrac{4000}{r}$
(b) $r \approx 3.76$ cm

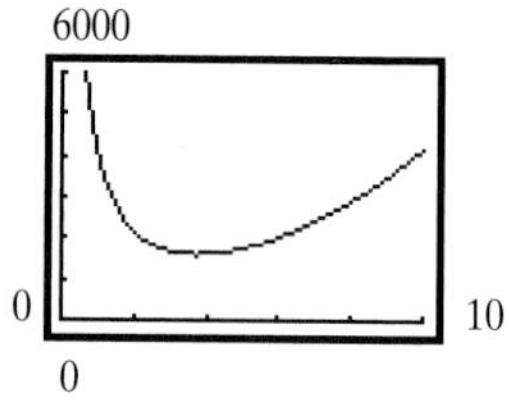

Fill-in-the-Blank Items

1. parabola **3.** zero **5.** $x = -1$

True/False Items

1. True **3.** True **5.** False

Review Exercises

1.

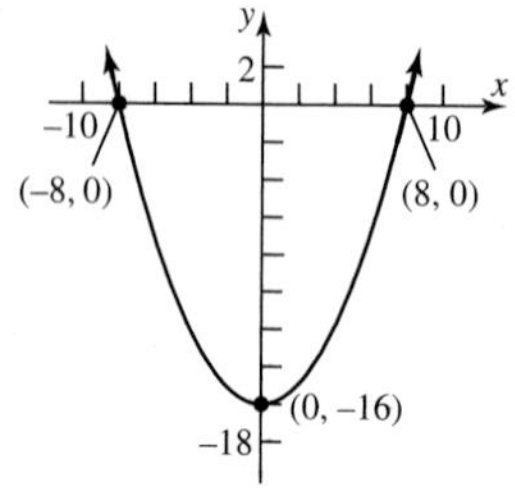

3.

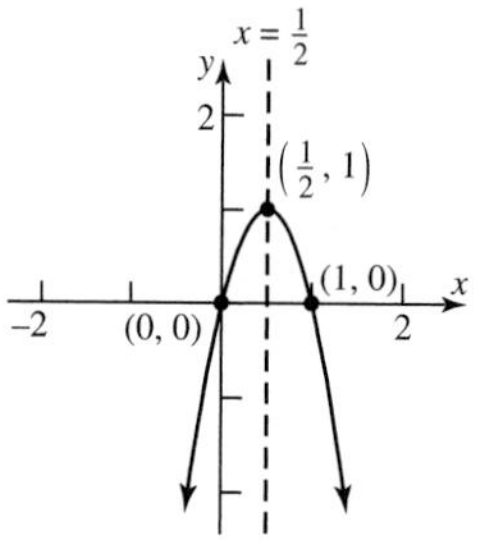

5.

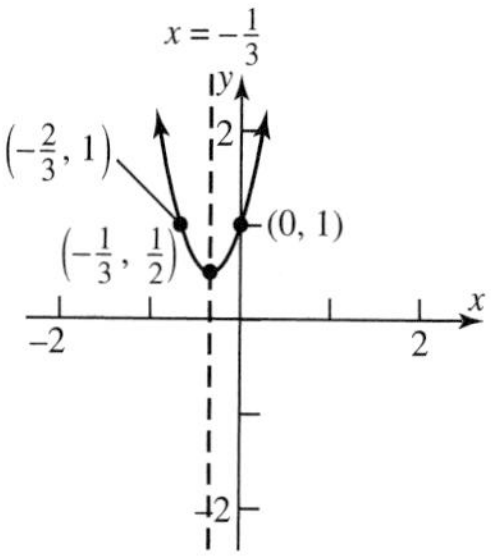

7.

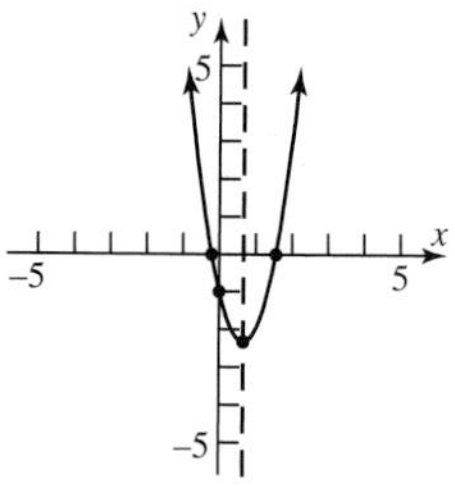

9.

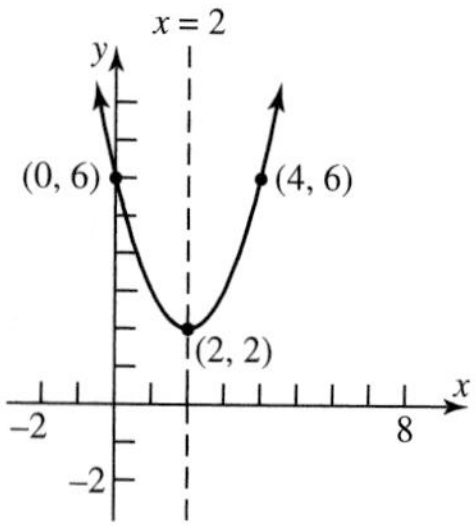

11.

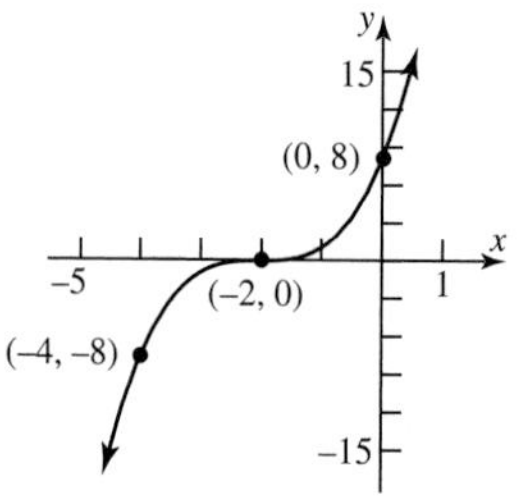

13.

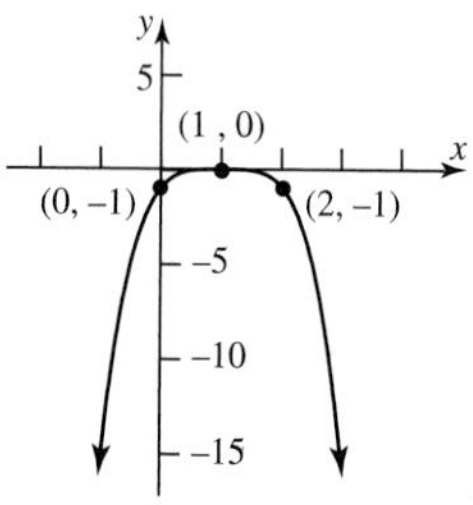

15.

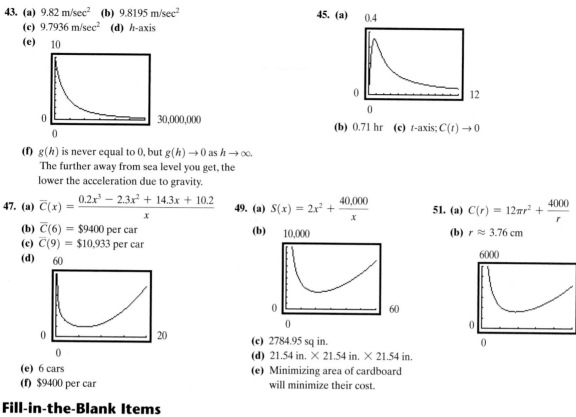

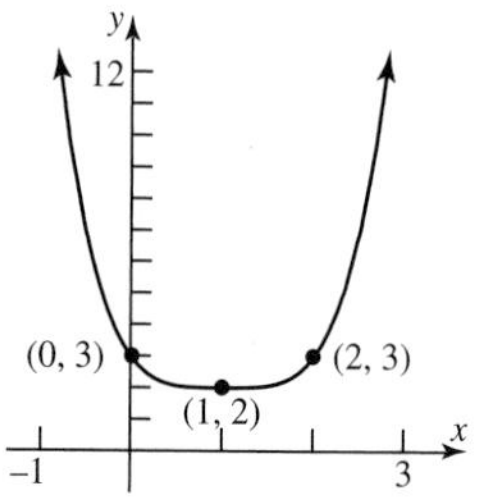

17. Minimum value; 1 **19.** Maximum value; 12 **21.** Maximum value; 16

23. (a)

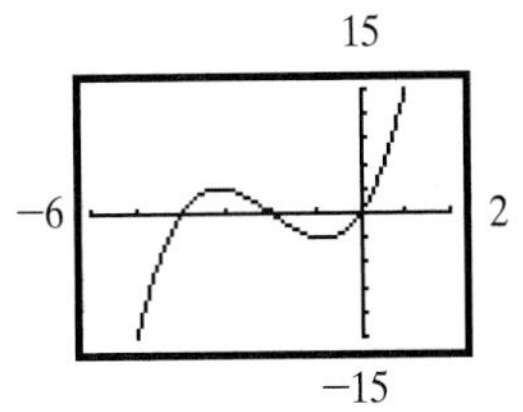

(b) x-intercepts: $-4, -2, 0$; y-intercept: 0
(c) $-4, -2, 0$: Odd
(d) $y = x^3$
(e) 2
(f) Local minimum: $(-0.85, -3.08)$
Local maximum: $(-3.15, 3.08)$

25. (a)

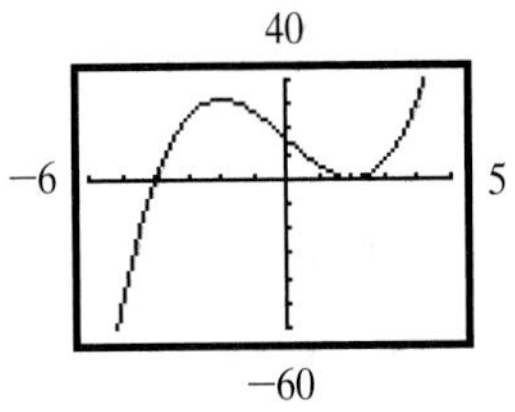

(b) x-intercepts: $-4, 2$; y-intercept: 16
(c) -4: Odd; 2: Even
(d) $y = x^3$
(e) 2
(f) Local minimum: $(2, 0)$
Local maximum: $(-2, 32)$

27. $f(x) = x^3 - 4x^2 = x^2(x - 4)$
(a)

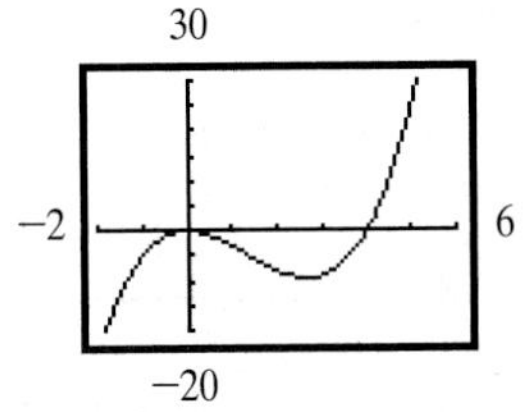

(b) x-intercepts: 0, 4; y-intercept: 0
(c) 0: Even; 4: Odd
(d) $y = x^3$
(e) 2
(f) Local minimum: $(2.67, -9.48)$
Local maximum: $(0.00, 0.00)$

29. (a)

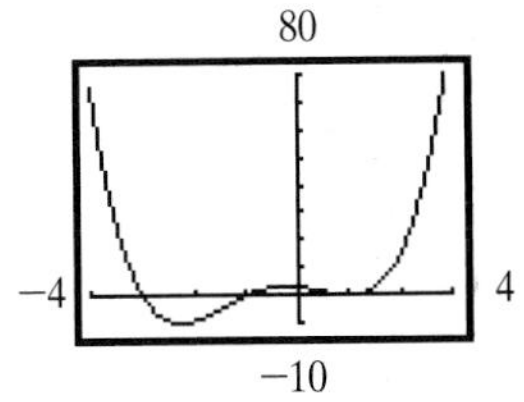

(b) x-intercepts: $-3, -1, 1$; y-intercept: 3
(c) $-3, -1$: Odd; 1: Even
(d) $y = x^4$
(e) 3
(f) Local minima: $(-2.28, -9.91)$, $(1.00, 0.00)$
Local maximum: $(-0.22, 3.23)$

31. 1. Domain: $\{x \mid x \neq 0\}$
2. x-intercept: 3; no y-intercepts
3. No symmetry
4. Vertical asymptote: $x = 0$
5. Horizontal asymptote: $y = 2$, not intersected
6.

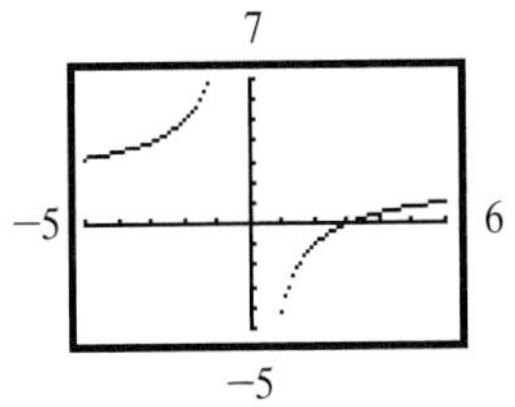

7.

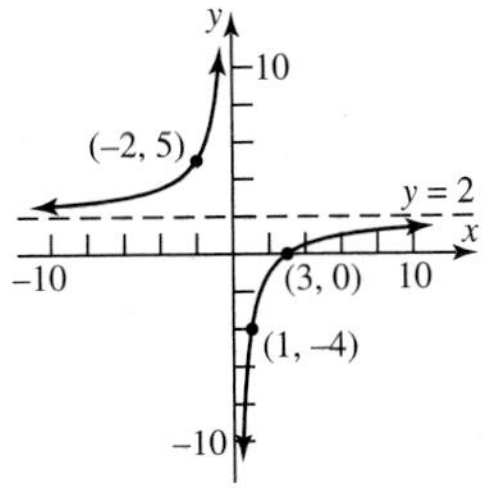

33. 1. Domain: $\{x \mid x \neq 0, x \neq 2\}$
2. x-intercept: -2; no y-intercept
3. No symmetry
4. Vertical asymptotes: $x = 0, x = 2$
5. Horizontal asymptote: $y = 0$, intersected at $(-2, 0)$
6.

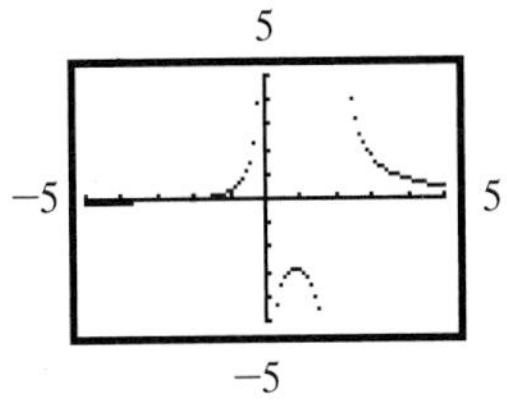

7.

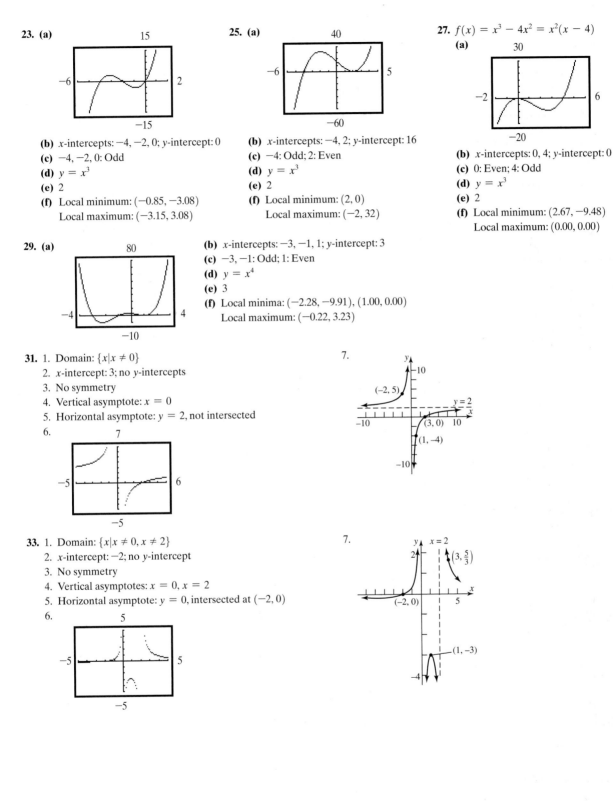

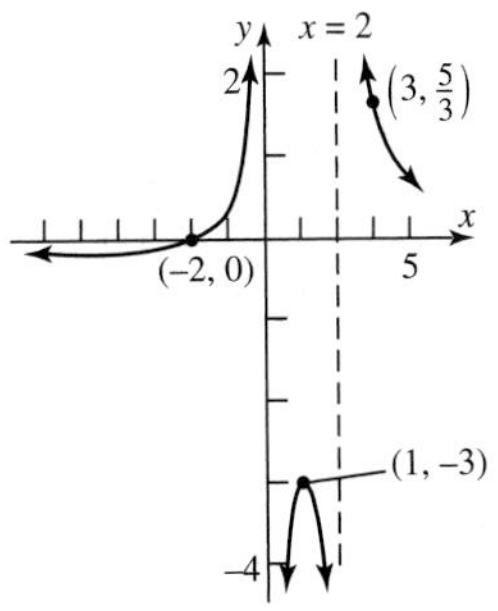

35. 1. Domain: $\{x | x \neq -2, x \neq 3\}$
2. x-intercepts: $-3, 2$; y-intercept: 1
3. No symmetry
4. Vertical asymptote: $x = -2$, $x = 3$
5. Horizontal asymptote: $y = 1$, intersected at $(0, 1)$
6.

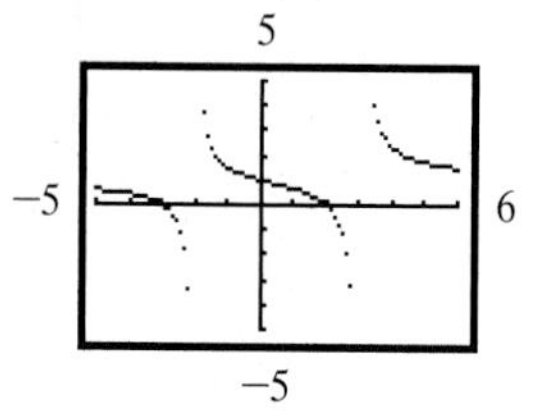

7.

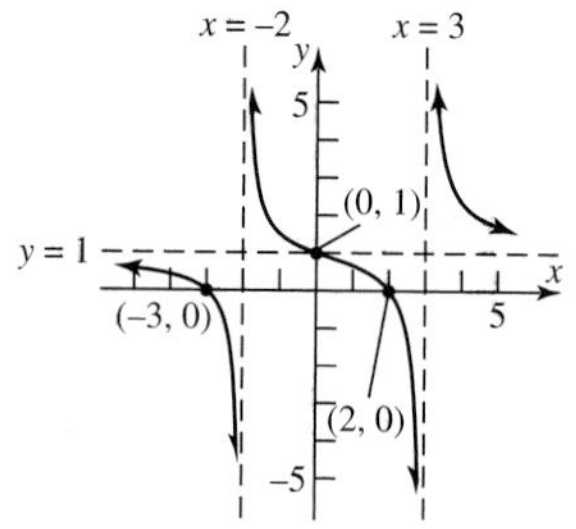

37. 1. Domain: $\{x | x \neq -2, x \neq 2\}$
2. Intercept: $(0, 0)$
3. Symmetric with respect to the origin
4. Vertical asymptotes: $x = -2$, $x = 2$
5. Oblique asymptote: $y = x$, intersected at $(0, 0)$
6.

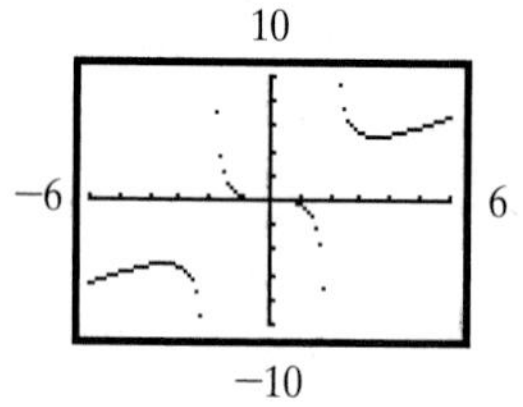

7.

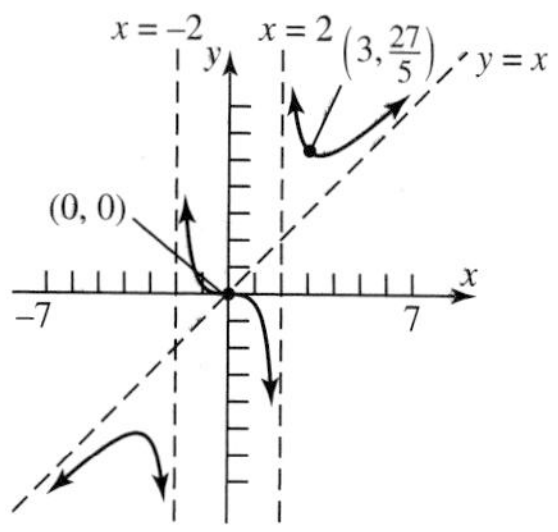

39. 1. Domain: $\{x | x \neq 1\}$
2. Intercept: $(0, 0)$
3. No symmetry
4. Vertical asymptote: $x = 1$
5. No oblique or horizontal asymptote
6.

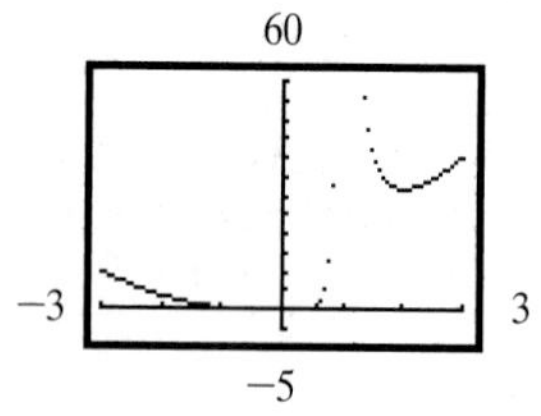

7.

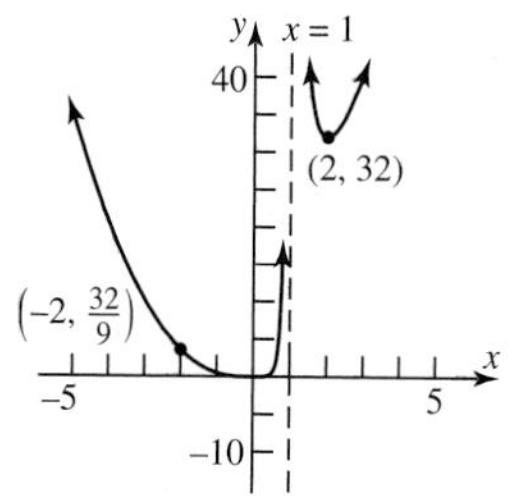

41. 1. Domain: $\{x | x \neq -1, x \neq 2\}$
2. x-intercept: -2; y-intercept: 2
3. No symmetry
4. Vertical asymptote: $x = -1$
5. Horizontal asymptote: $y = 1$, not intersected
6.

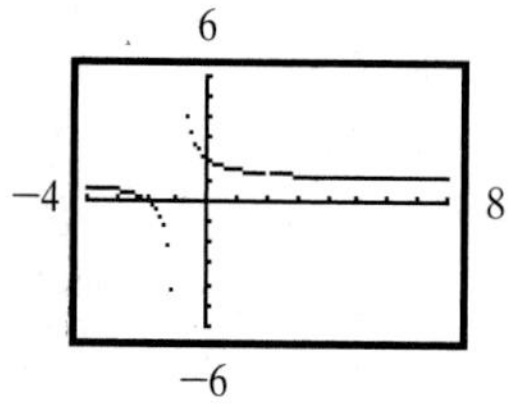

7.

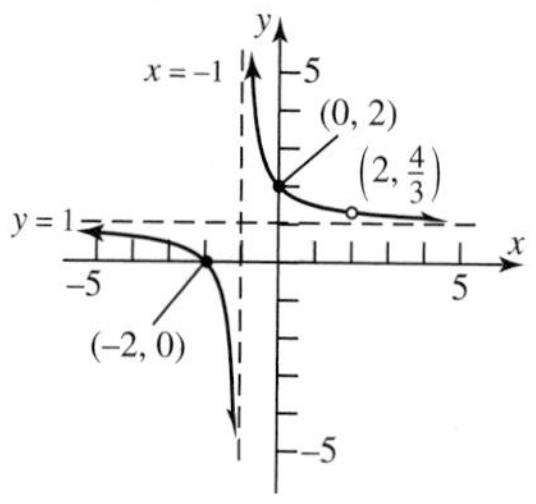

43. 50 ft by 50 ft **45.** 25 sq units **47. (a)** 5000 **(b)** $\frac{C(5000)}{500} = \$7.36$ **(c)** about $111,770

49. (a)

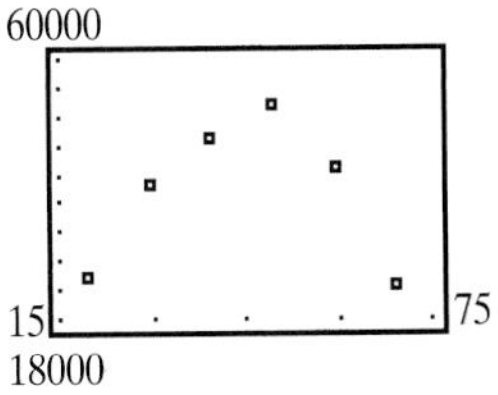

(b) $I(x) = -45.6689x^2 + 4081.5404x - 41{,}021.2126$

(c)

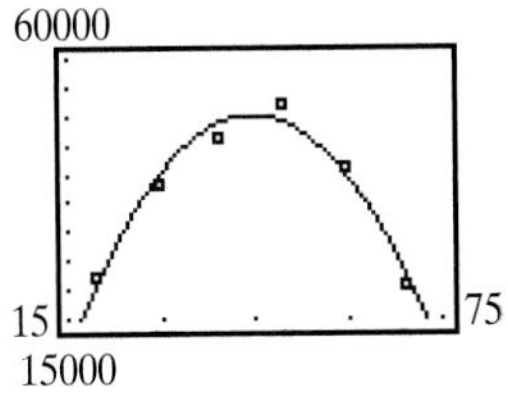

(d) 45 years **(e)** \$50,173

51. (a)

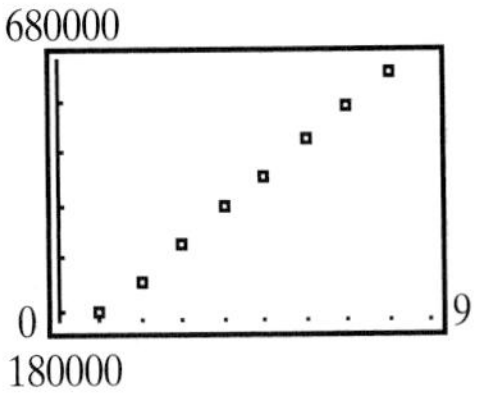

(b) $A(t) = -212.0076t^3 + 2429.132t^2 + 59{,}568.8539t + 130{,}003.1429$

(c)

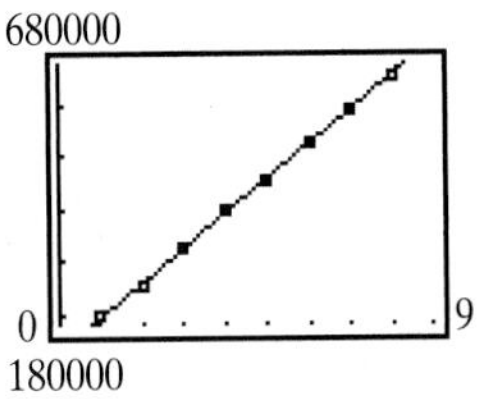

(d) 797,003 **(e)** 2002

CHAPTER 4 The Zeros of a Polynomial Function

4.1 Exercises

1. No; $f(3) = 61$ **3.** No; $f(1) = 2$ **5.** Yes; $f(-2) = 0$ **7.** Yes; $f(4) = 0$ **9.** No; $f\left(-\frac{1}{2}\right) = -\frac{7}{4}$ **11.** 7; 3 or 1 positive, 2 or 0 negative **13.** 6; 2 or 0 positive, 2 or 0 negative **15.** 3; 2 or 0 positive, 1 negative **17.** 4; 2 or 0 positive, 2 or 0 negative **19.** 5; 0 positive, 3 or 1 negative **21.** 6; 1 positive, 1 negative **23.** $\pm 1, \pm\frac{1}{3}$ **25.** $\pm 1, \pm 3$ **27.** $\pm 1, \pm 2, \pm\frac{1}{4}, \pm\frac{1}{2}$ **29.** $\pm 1, \pm 2, \pm\frac{1}{3}, \pm\frac{2}{3}$ **31.** $\pm 1, \pm 2, \pm 4, \pm\frac{1}{2}$ **33.** $\pm 1, \pm 2, \pm\frac{1}{6}, \pm\frac{1}{3}, \pm\frac{1}{2}, \pm\frac{2}{3}$

35. -1 and 1

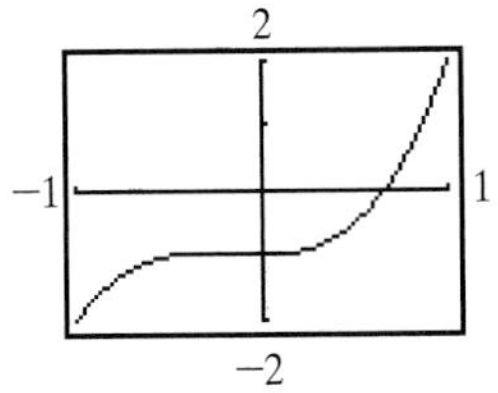

37. -12 and 12

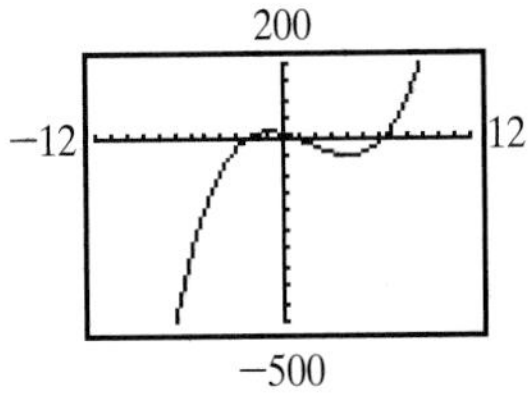

39. -10 and 10

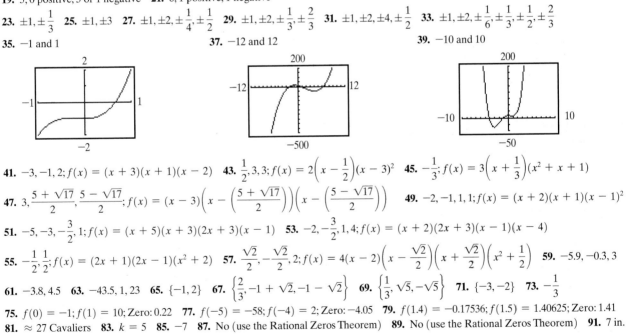

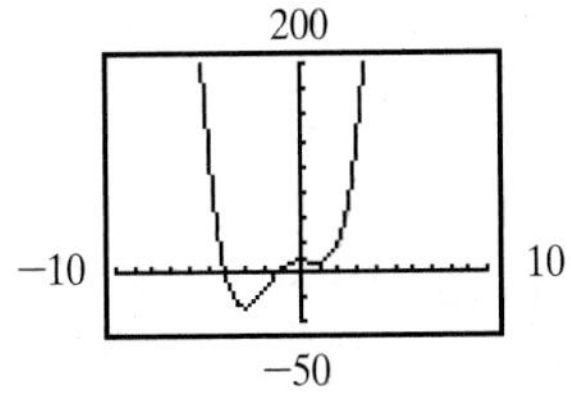

41. $-3, -1, 2$; $f(x) = (x + 3)(x + 1)(x - 2)$ **43.** $\frac{1}{2}, 3, 3$; $f(x) = 2\left(x - \frac{1}{2}\right)(x - 3)^2$ **45.** $-\frac{1}{3}$; $f(x) = 3\left(x + \frac{1}{3}\right)(x^2 + x + 1)$ **47.** $3, \frac{5 + \sqrt{17}}{2}, \frac{5 - \sqrt{17}}{2}$; $f(x) = (x - 3)\left(x - \left(\frac{5 + \sqrt{17}}{2}\right)\right)\left(x - \left(\frac{5 - \sqrt{17}}{2}\right)\right)$ **49.** $-2, -1, 1, 1$; $f(x) = (x + 2)(x + 1)(x - 1)^2$ **51.** $-5, -3, -\frac{3}{2}, 1$; $f(x) = (x + 5)(x + 3)(2x + 3)(x - 1)$ **53.** $-2, -\frac{3}{2}, 1, 4$; $f(x) = (x + 2)(2x + 3)(x - 1)(x - 4)$ **55.** $-\frac{1}{2}, \frac{1}{2}$; $f(x) = (2x + 1)(2x - 1)(x^2 + 2)$ **57.** $\frac{\sqrt{2}}{2}, -\frac{\sqrt{2}}{2}, 2$; $f(x) = 4(x - 2)\left(x - \frac{\sqrt{2}}{2}\right)\left(x + \frac{\sqrt{2}}{2}\right)\left(x^2 + \frac{1}{2}\right)$ **59.** $-5.9, -0.3, 3$ **61.** $-3.8, 4.5$ **63.** $-43.5, 1, 23$ **65.** $\{-1, 2\}$ **67.** $\left\{\frac{2}{3}, -1 + \sqrt{2}, -1 - \sqrt{2}\right\}$ **69.** $\left\{\frac{1}{3}, \sqrt{5}, -\sqrt{5}\right\}$ **71.** $\{-3, -2\}$ **73.** $-\frac{1}{3}$ **75.** $f(0) = -1$; $f(1) = 10$; Zero: 0.22 **77.** $f(-5) = -58$; $f(-4) = 2$; Zero: -4.05 **79.** $f(1.4) = -0.17536$; $f(1.5) = 1.40625$; Zero: 1.41 **81.** $\approx$ 27 Cavaliers **83.** $k = 5$ **85.** -7 **87.** No (use the Rational Zeros Theorem) **89.** No (use the Rational Zeros Theorem) **91.** 7 in.

93.

$$y^3 + by^2 + cy + d = 0$$

$$\left(x - \frac{b}{3}\right)^3 + b\left(x - \frac{b}{3}\right)^2 + c\left(x - \frac{b}{3}\right) + d = 0$$

$$x^3 - \frac{3b}{3}x^2 + 3\left(\frac{b^2}{9}\right)x - \frac{b^3}{27} + b\left(x^2 - \frac{2bx}{3} + \frac{b^2}{9}\right) + cx - \frac{bc}{3} + d = 0$$

$$x^3 - \frac{b^2}{3}x + cx - \frac{b^3}{27} + \frac{b^3}{9} - \frac{bc}{3} + d = 0$$

$$x^3 + \left(c - \frac{b^2}{3}\right)x + \left(\frac{2b^3}{27} - \frac{bc}{3} + d\right) = 0$$

95. $K = \dfrac{-p}{3H}$

$$H^3 + \left(\frac{-p}{3H}\right)^3 = -q$$

$$H^6 + qH^3 - \frac{p^3}{27} = 0$$

$$H^3 = \frac{-q \pm \sqrt{q^2 + \dfrac{4p^3}{27}}}{2} \quad \text{Choose + sign}$$

$$H = \sqrt[3]{\frac{-q}{2} + \sqrt{\frac{q^2}{4} + \frac{p^3}{27}}}$$

97. $x = H + K$; now use results from Problems 95 and 96 **99.** $p = 3, q = -14; x = \sqrt[3]{7 + 5\sqrt{2}} + \sqrt[3]{7 - 5\sqrt{2}}$ **101.** 8

4.2 Exercises

1. $8 + 5i$ **3.** $-7 + 6i$ **5.** $-6 - 11i$ **7.** $6 - 18i$ **9.** $6 + 4i$ **11.** $10 - 5i$ **13.** 37 **15.** $\frac{6}{5} + \frac{8}{5}i$ **17.** $1 - 2i$ **19.** $\frac{5}{2} - \frac{7}{2}i$
21. $-\frac{1}{2} + \frac{\sqrt{3}}{2}i$ **23.** $2i$ **25.** $-i$ **27.** i **29.** -6 **31.** $-10i$ **33.** $-2 + 2i$ **35.** 0 **37.** 0 **39.** $2i$ **41.** $5i$ **43.** $5i$ **45.** $\{-2i, 2i\}$
47. $\{-4, 4\}$ **49.** $\{3 - 2i, 3 + 2i\}$ **51.** $\{3 - i, 3 + i\}$ **53.** $\left\{\frac{1}{4} - \frac{1}{4}i, \frac{1}{4} + \frac{1}{4}i\right\}$ **55.** $\left\{-\frac{1}{5} - \frac{2}{5}i, -\frac{1}{5} + \frac{2}{5}i\right\}$
57. $\left\{-\frac{1}{2} - \frac{\sqrt{3}}{2}i, -\frac{1}{2} + \frac{\sqrt{3}}{2}i\right\}$ **59.** $\{2, -1 - \sqrt{3}i, -1 + \sqrt{3}i\}$ **61.** $\{-2, 2, -2i, 2i\}$ **63.** $\{-3i, -2i, 2i, 3i\}$ **65.** Two complex solutions.
67. Two unequal real solutions. **69.** A repeated real solution. **71.** $2 - 3i$ **73.** 6 **75.** 25
77. $z + \overline{z} = (a + bi) + (a - bi) = 2a; z - \overline{z} = (a + bi) - (a - bi) = 2bi$
79. $\overline{z + w} = \overline{(a + bi) + (c + di)} = \overline{(a + c) + (b + d)i} = (a + c) - (b + d)i = (a - bi) + (c - di) = \overline{z} + \overline{w}$

4.3 Exercises

1. $4 + i$ **3.** $-i, 1 - i$ **5.** $-i, -2i$ **7.** $-i$ **9.** $2 - i, -3 + i$ **11.** $f(x) = x^4 - 14x^3 + 77x^2 - 200x + 208; a = 1$
13. $f(x) = x^5 - 4x^4 + 7x^3 - 8x^2 + 6x - 4; a = 1$ **15.** $f(x) = x^4 - 6x^3 + 10x^2 - 6x + 9; a = 1$ **17.** $-2i, 4$ **19.** $2i, -3, \frac{1}{2}$
21. $3 + 2i, -2, 5$ **23.** $4i, -\sqrt{11}, \sqrt{11}, -\frac{2}{3}$ **25.** $1, -\frac{1}{2} - \frac{\sqrt{3}}{2}i, -\frac{1}{2} + \frac{\sqrt{3}}{2}i$ **27.** $2, 3 - 2i, 3 + 2i$ **29.** $-i, i, -2i, 2i$ **31.** $-3. 1, -5i, 5i$
33. $-4, \frac{1}{3}, 2 - 3i, 2 + 3i$ **35.** Zeros that are complex numbers must occur in conjugate pairs; or a polynomial with real coefficients of odd

degree must have at least one real zero. **37.** If the remaining zero were a complex number, then its conjugate would also be a zero.

4.4 Exercises

1. $\{x|-2 < x < 5\}$ **3.** $\{x|-\infty < x < 0 \text{ or } 4 < x < \infty\}$ **5.** $\{x|-3 < x < 3\}$ **7.** $\{x|-\infty < x < -4 \text{ or } 3 < x < \infty\}$
9. $\left\{x \middle| -\frac{1}{2} < x < 3\right\}$ **11.** $\{x|-\infty < x < -1 \text{ or } 8 < x < \infty\}$ **13.** No real solution **15.** $\left\{x \middle| -\infty < x < -\frac{2}{3} \text{ or } \frac{3}{2} < x < \infty\right\}$
17. $\{x|1 < x < \infty\}$ **19.** $\{x|-\infty < x < 1 \text{ or } 2 < x < 3\}$ **21.** $\{x|-1 < x < 0 \text{ or } 3 < x < \infty\}$ **23.** $\{x|-\infty < x < -1 \text{ or } 1 < x < \infty\}$
25. $\{x|1 < x < \infty\}$ **27.** $\{x|-\infty < x < -1 \text{ or } 1 < x < \infty\}$ **29.** $\{x|-1 < x < 8\}$ **31.** $\{x|2.14 \le x < \infty\}$
33. $\{x|-\infty < x < -2 \text{ or } 2 < x < \infty\}$ **35.** $\{x|-1 \le x \le 2 - \sqrt{5} \text{ or } 2 + \sqrt{5} \le x < \infty\}$ **37.** $\{x|-\infty < x < -1 \text{ or } 1 < x < \infty\}$
39. $\{x|-\infty < x < -1 \text{ or } 0 < x < 1\}$ **41.** $\{x|-\infty < x < -1 \text{ or } 1 < x < \infty\}$ **43.** $\left\{x \middle| -\infty < x < -\frac{2}{3} \text{ or } 0 < x < \frac{3}{2}\right\}$
45. $\{x|-\infty < x < 2\}$ **47.** $\{x|-2 < x \le 9\}$ **49.** $\{x|-\infty < x < 2 \text{ or } 3 < x < 5\}$ **51.** $\{x|-\infty < x < -3 \text{ or } -1 < x < 1 \text{ or } 2 < x < \infty\}$
53. $\{x|-\infty < x < -5 \text{ or } -4 < x < -3 \text{ or } 1 < x < \infty\}$ **55.** $\{x|-\infty < x \le -0.5 \text{ or } 1 \le x < 4\}$
57. $\left\{x \middle| \frac{-3 - \sqrt{13}}{2} < x < -3 \text{ or } \frac{-3 + \sqrt{13}}{2} < x < \infty\right\}$ **59.** $\{x|4 < x < \infty\}$ **61.** $\{x|-\infty < x \le -4 \text{ or } 4 \le x < \infty\}$
63. $\{x|-\infty < x < -4 \text{ or } 2 \le x < \infty\}$

65. **(a)** The ball is more than 96 feet above the ground from time t between 2 and 3 seconds, $2 < t < 3$.
(b)

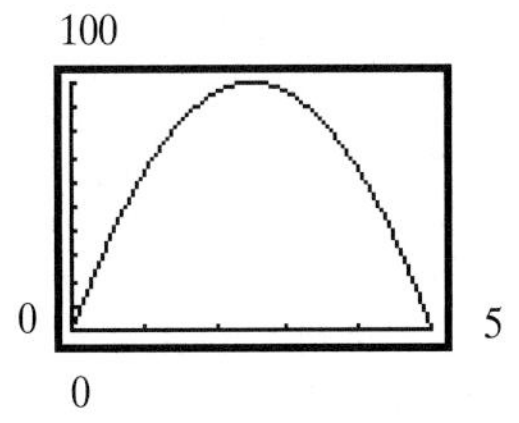

(c) 100 ft **(d)** 2.5 sec

67. **(a)** For a profit of at least \$50, between 8 and 32 watches must be sold, $8 \le x \le 32$.
(b)

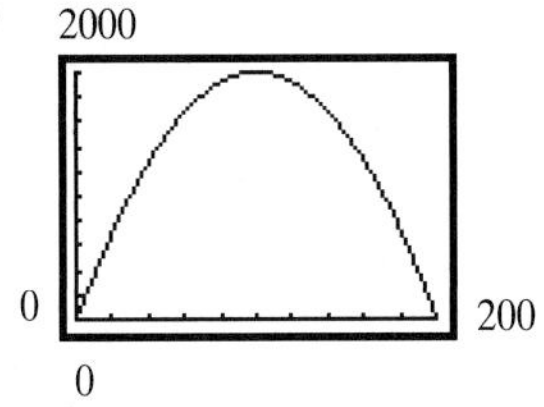

(c) \$2000 **(d)** 100
(e)

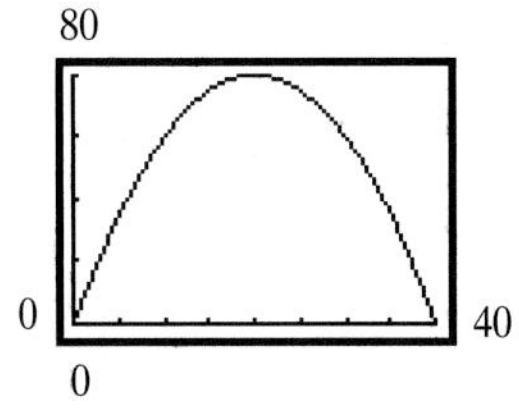

(f) \$80 **(g)** 20

69. Chevy can produce at most 8 Cavaliers in a day, assuming that cars cannot be partially completed in a day.

71. **(a)**

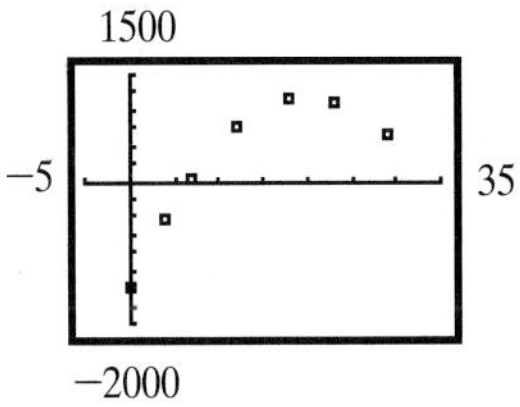

(b)

(c) Between 15 and 25 computers.
(d) 20 computers
(e) \$1202.76

73. $b - a = (\sqrt{b} - \sqrt{a})(\sqrt{b} + \sqrt{a})$; since $a \ge 0$ and $b \ge 0$, then $\sqrt{a} \ge 0$ and $\sqrt{b} \ge 0$ so that $\sqrt{b} + \sqrt{a} \ge 0$; thus, $b - a \ge 0$ is equivalent to $\sqrt{b} - \sqrt{a} \ge 0$ and $a \le b$ is equivalent to $\sqrt{a} \le \sqrt{b}$ **75.** $x^2 + 1$ is always positive, hence can never be less than -5.

Fill-in-the-Blank Items

1. Remainder; dividend **3.** $f(c) = 0$ **5.** $\pm 1, \pm \frac{1}{2}$ **7.** $3 - 4i$ **9.** -4

True/False Items

1. False **3.** True **5.** True

Review Exercises

1. 4, 2 or 0 positive zeros; 2 or 0 negative zeros **3.** $\pm\frac{1}{12}, \pm\frac{1}{6}, \pm\frac{1}{4}, \pm\frac{1}{3}, \pm\frac{1}{2}, \pm\frac{3}{2}, \pm\frac{3}{4}, \pm 1, \pm 3,$ **5.** $-2, 1, 4$ **7.** $\frac{1}{2}, -2$ **9.** 2

11. $-2.5, 3.1, 5.32$ **13.** $-11.3, -0.6, 4, 9.33$ **15.** $-3.67, 1.33$ **17.** $\{-3, 2\}$ **19.** $\left\{-3, -1, -\frac{1}{2}, 1\right\}$

21. -5 and 5

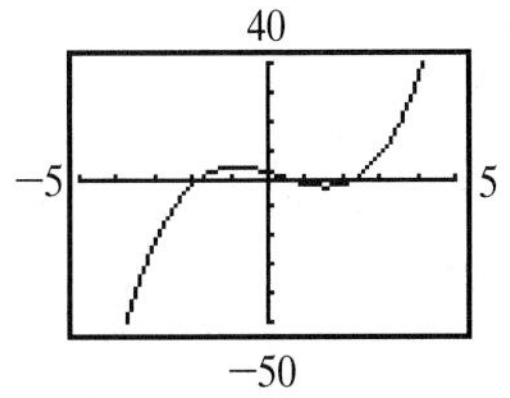

23. $-\frac{37}{2}$ and $\frac{37}{2}$

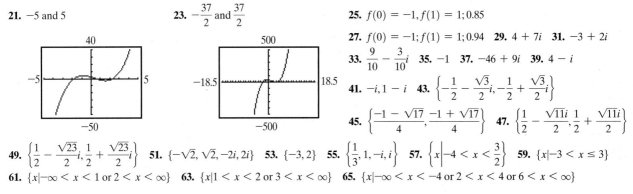

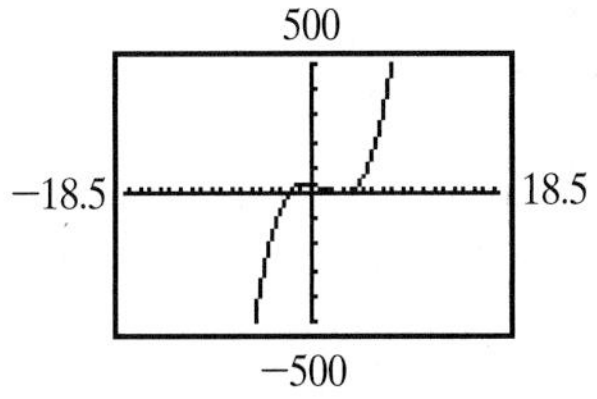

25. $f(0) = -1, f(1) = 1; 0.85$

27. $f(0) = -1; f(1) = 1; 0.94$ **29.** $4 + 7i$ **31.** $-3 + 2i$

33. $\frac{9}{10} - \frac{3}{10}i$ **35.** -1 **37.** $-46 + 9i$ **39.** $4 - i$

41. $-i, 1 - i$ **43.** $\left\{-\frac{1}{2} - \frac{\sqrt{3}}{2}i, -\frac{1}{2} + \frac{\sqrt{3}}{2}i\right\}$

45. $\left\{\frac{-1 - \sqrt{17}}{4}, \frac{-1 + \sqrt{17}}{4}\right\}$ **47.** $\left\{\frac{1}{2} - \frac{\sqrt{11}i}{2}, \frac{1}{2} + \frac{\sqrt{11}i}{2}\right\}$

49. $\left\{\frac{1}{2} - \frac{\sqrt{23}}{2}i, \frac{1}{2} + \frac{\sqrt{23}}{2}i\right\}$ **51.** $\{-\sqrt{2}, \sqrt{2}, -2i, 2i\}$ **53.** $\{-3, 2\}$ **55.** $\left\{\frac{1}{3}, 1, -i, i\right\}$ **57.** $\left\{x \middle| -4 < x < \frac{3}{2}\right\}$ **59.** $\{x | -3 < x \le 3\}$

61. $\{x | -\infty < x < 1 \text{ or } 2 < x < \infty\}$ **63.** $\{x | 1 < x < 2 \text{ or } 3 < x < \infty\}$ **65.** $\{x | -\infty < x < -4 \text{ or } 2 < x < 4 \text{ or } 6 < x < \infty\}$

CHAPTER 5 Exponential and Logarithmic Functions

5.1 Exercises

1. (a)

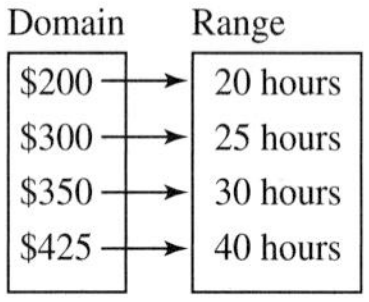

(b) Inverse is a function

3. (a)

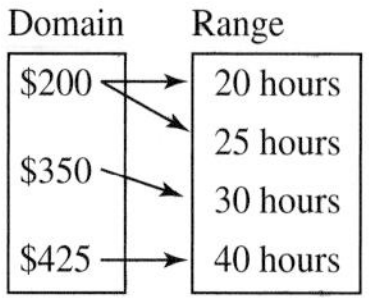

(b) Inverse is not a function

5. (a) $\{(6, 2), (6, -3), (9, 4), (10, 1)\}$
(b) Inverse is not a function

7. (a) $\{(0, 0), (1, 1), (16, 2), (81, 3)\}$
(c) Inverse is a function

9. One-to-one **11.** Not one-to-one

13. One-to-one

15.

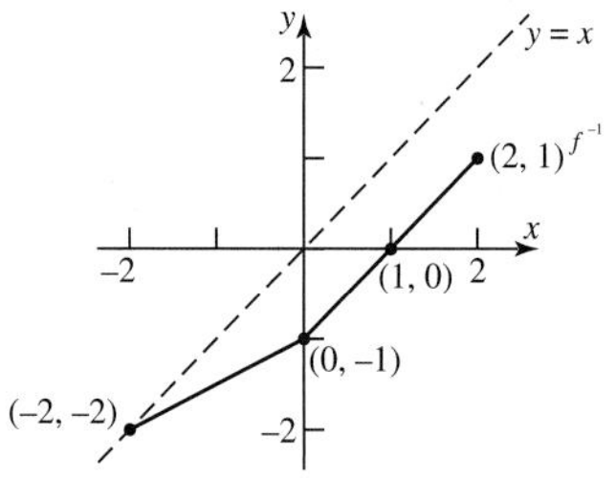

17.

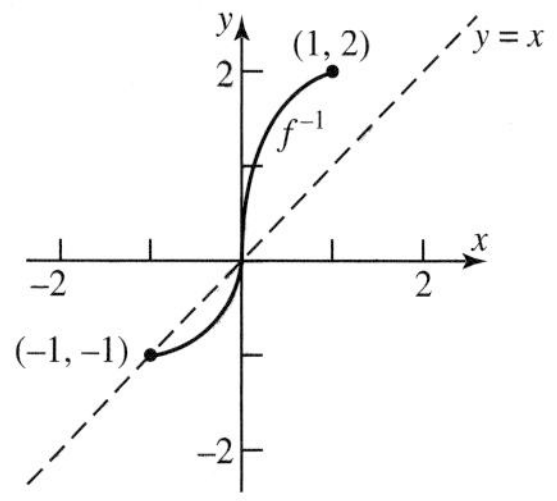

19.

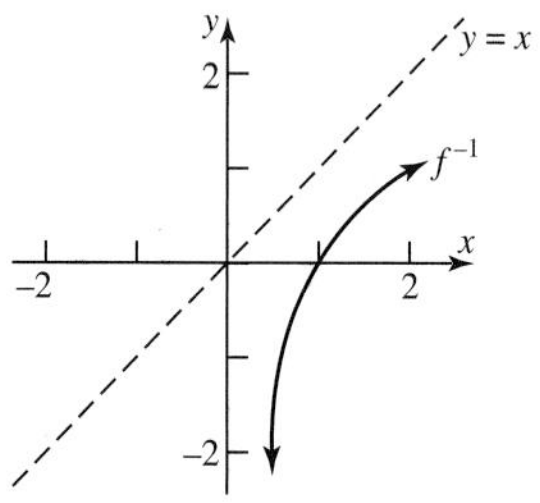

21. $f(g(x)) = f\left(\frac{1}{3}(x-4)\right) = 3\left[\frac{1}{3}(x-4)\right] + 4 = x;$
$g(f(x)) = g(3x+4) = \frac{1}{3}[(3x+4) - 4] = x$

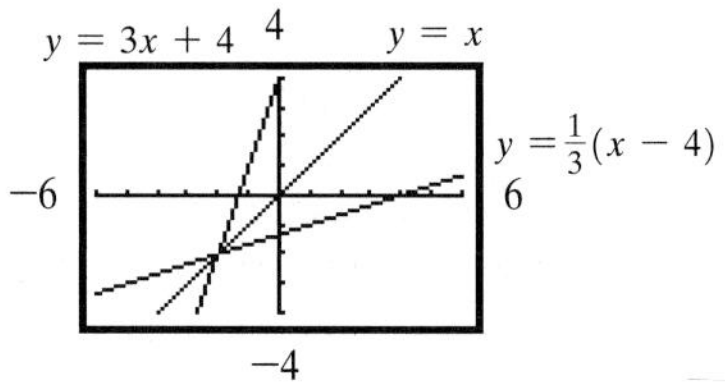

23. $f(g(x)) = 4\left[\frac{x}{4} + 2\right] - 8 = x;$
$g(f(x)) = \frac{4x-8}{4} + 2 = x$

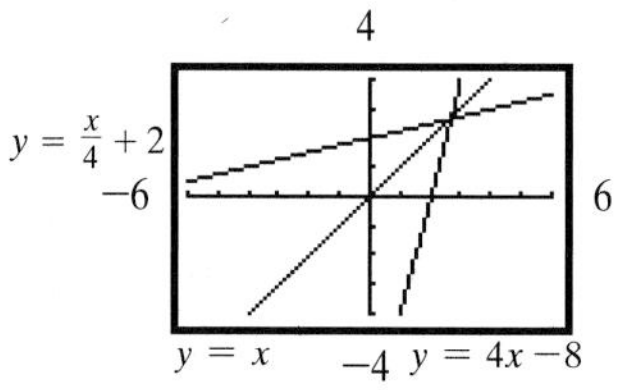

25. $f(g(x)) = (\sqrt[3]{x+8})^3 - 8 = x;$
$g(f(x)) = \sqrt[3]{(x^3-8)+8} = x$

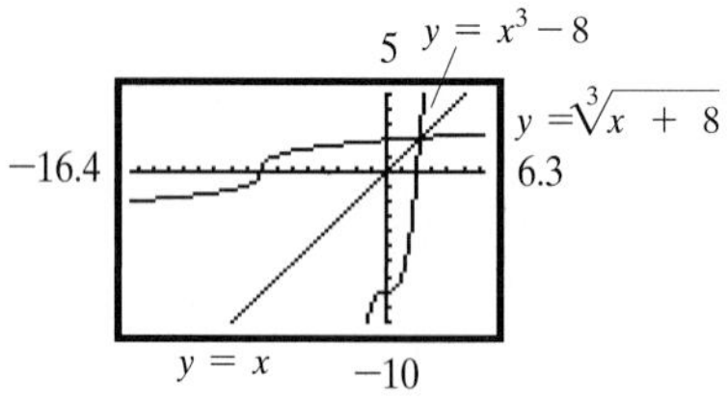

27. $f(g(x)) = \dfrac{1}{\left(\dfrac{1}{x}\right)} = x;$
$g(f(x)) = \dfrac{1}{\left(\dfrac{1}{x}\right)} = x$

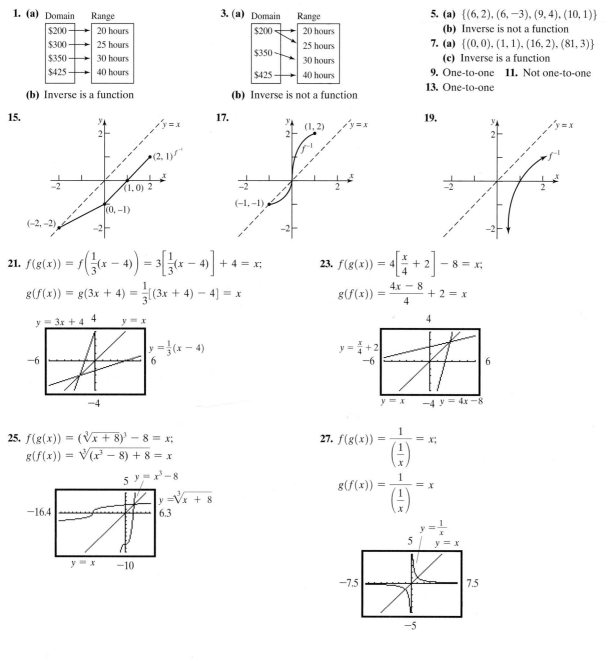

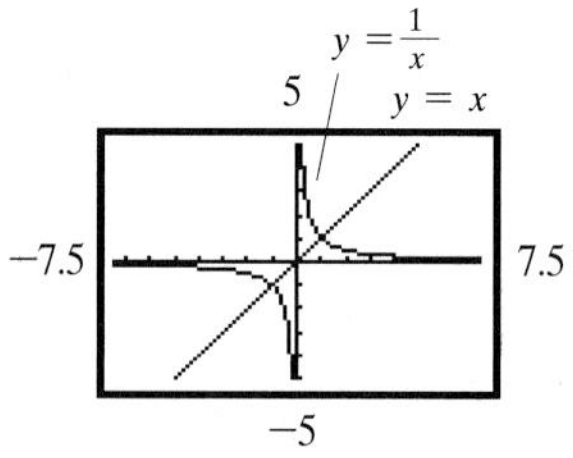

29. $f(g(x)) = \dfrac{2\left(\dfrac{4x-3}{2-x}\right)+3}{\dfrac{4x-3}{2-x}+4} = x;$

$g(f(x)) = \dfrac{4\left(\dfrac{2x+3}{x+4}\right)-3}{2-\dfrac{2x+3}{x+4}} = x$

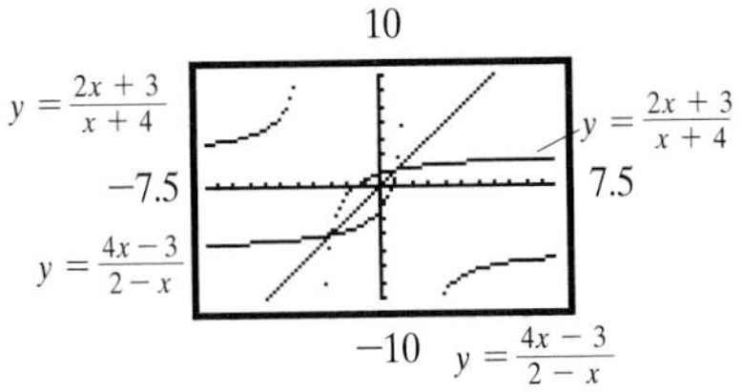

31. $f^{-1}(x) = \dfrac{1}{3}x$

$f(f^{-1}(x)) = 3\left(\dfrac{1}{3}x\right) = x$

$f^{-1}(f(x)) = \dfrac{1}{3}(3x) = x$

Domain f = Range $f^{-1} = (-\infty, \infty)$
Range f = Domain $f^{-1} = (-\infty, \infty)$

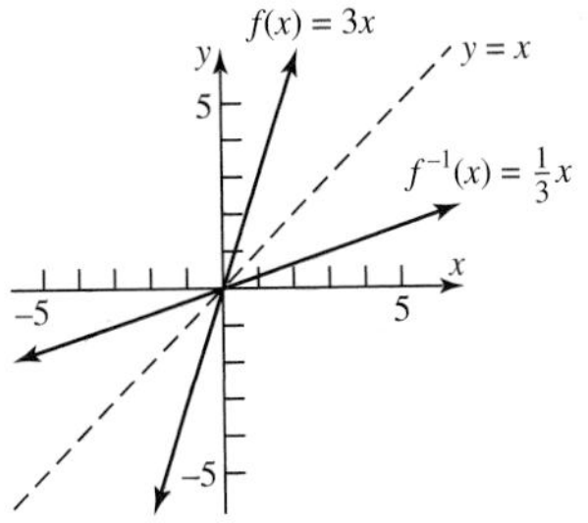

33. $f^{-1}(x) = \dfrac{x}{4} - \dfrac{1}{2}$

$f(f^{-1}(x)) = 4\left(\dfrac{x}{4} - \dfrac{1}{2}\right) + 2 = x$

$f^{-1}(f(x)) = \dfrac{4x+2}{4} - \dfrac{1}{2} = x$

Domain f = Range $f^{-1} = (-\infty, \infty)$
Range f = Domain $f^{-1} = (-\infty, \infty)$

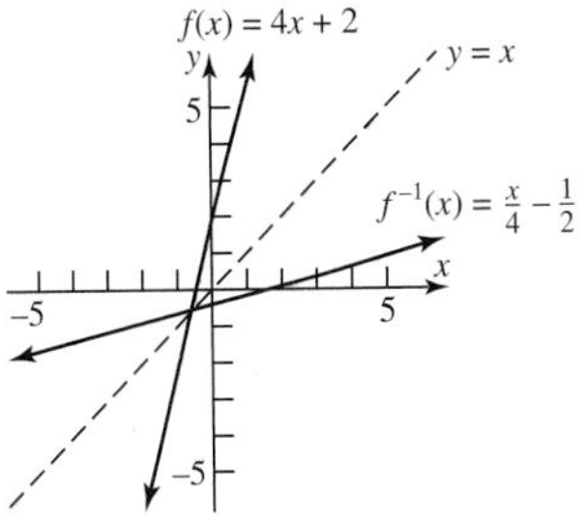

35. $f^{-1}(x) = \sqrt[3]{x+1}$
$f(f^{-1}(x)) = (\sqrt[3]{x+1})^3 - 1 = x$
$f^{-1}(f(x)) = \sqrt[3]{(x^3-1)+1} = x$
Domain f = Range $f^{-1} = (-\infty, \infty)$
Range f = Domain $f^{-1} = (-\infty, \infty)$

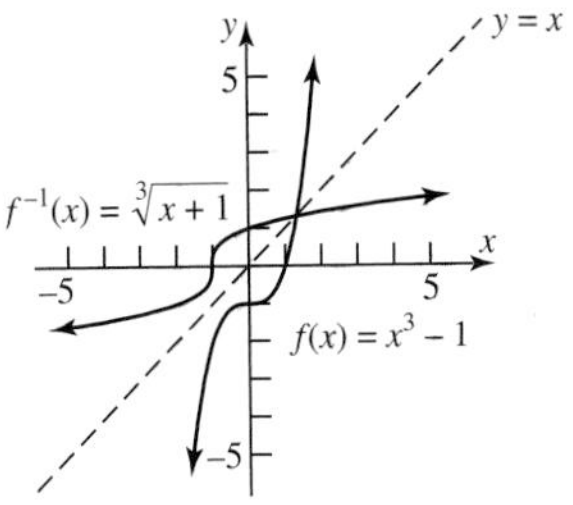

37. $f^{-1}(x) = \sqrt{x-4}$
$f(f^{-1}(x)) = (\sqrt{x-4})^2 + 4 = x$
$f^{-1}(f(x)) = \sqrt{(x^2+4)-4} = \sqrt{x^2} = |x|$
$= x$
Domain f = Range $f^{-1} = [0, \infty)$
Range f = Domain $f^{-1} = [4, \infty)$

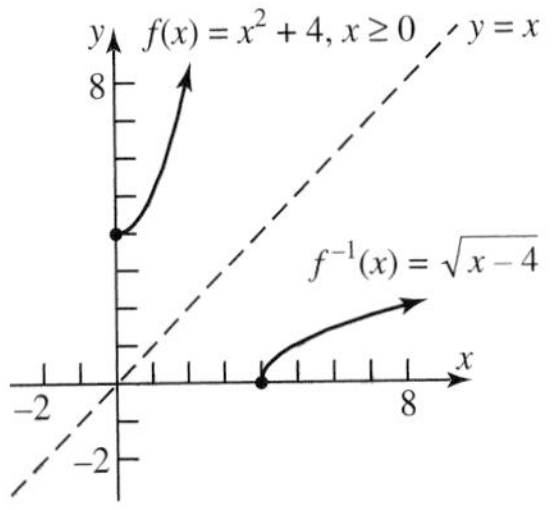

39. $f^{-1}(x) = \dfrac{4}{x}$

$f(f^{-1}(x)) = \dfrac{4}{\dfrac{4}{x}} = x$

$f^{-1}(f(x)) = \dfrac{4}{\dfrac{4}{x}} = x$

Domain f = Range f^{-1} = All real numbers except 0
Range f = Domain f^{-1} = All real numbers except 0

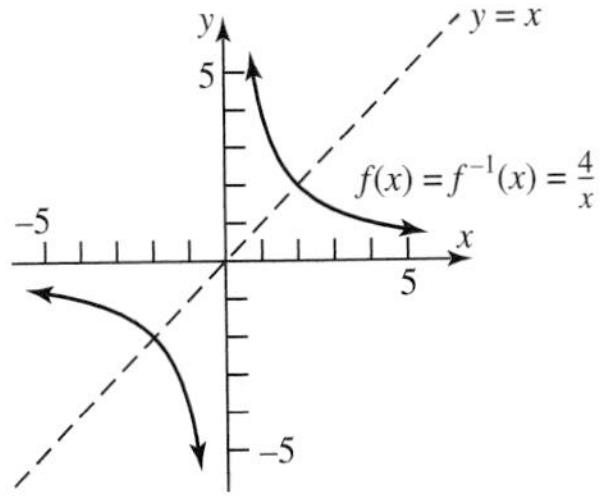

41. $f^{-1}(x) = \dfrac{2x+1}{x}$

$f(f^{-1}(x)) = \dfrac{1}{\dfrac{2x+1}{x} - 2} = x$

$f^{-1}(f(x)) = \dfrac{2\left(\dfrac{1}{x-2}\right)+1}{\dfrac{1}{x-2}} = x$

Domain f = Range f^{-1} = All real numbers except 2
Range f = Domain f^{-1} = All real numbers except 0

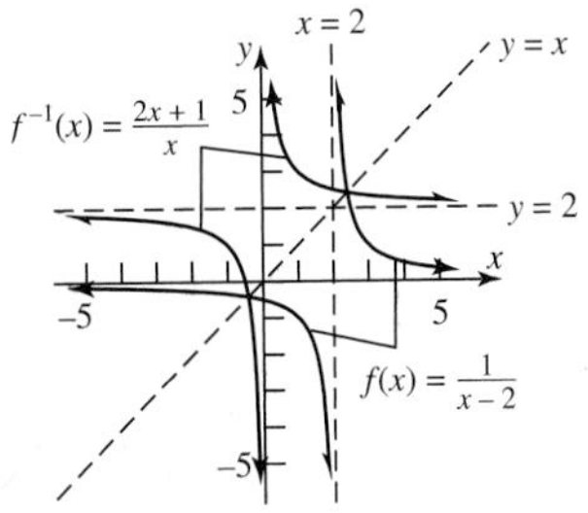

43. $f^{-1}(x) = \dfrac{2 - 3x}{x}$

$$f(f^{-1}(x)) = \frac{2}{3 + \dfrac{2 - 3x}{x}} = x$$

$$f^{-1}(f(x)) = \frac{2 - 3\left(\dfrac{2}{3 + x}\right)}{\dfrac{2}{3 + x}} = x$$

Domain f = Range f^{-1} = All real numbers except -3
Range f = Domain f^{-1} = All real numbers except 0

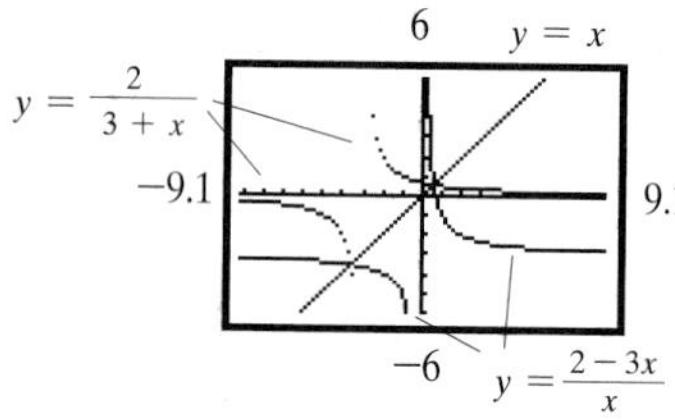

45. $f^{-1}(x) = \sqrt{x} - 2$

$f(f^{-1}(x)) = (\sqrt{x} - 2 + 2)^2 = x$

$f^{-1}(f(x)) = \sqrt{(x + 2)^2} - 2 = |x + 2| - 2 = x, x \geq -2$

Domain f = Range $f^{-1} = [-2, \infty)$
Range f = Domain $f^{-1} = [0, \infty)$

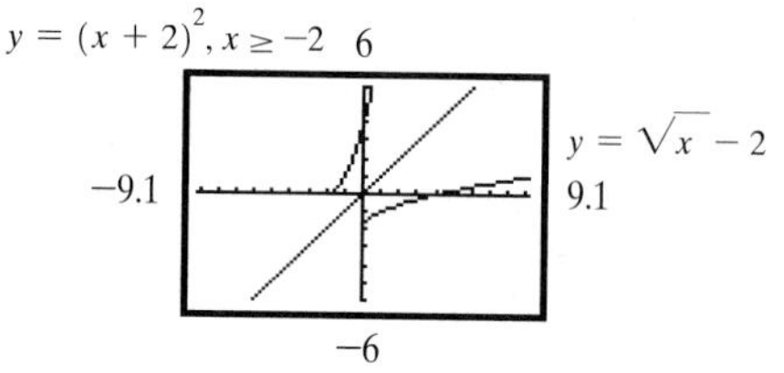

47. $f^{-1}(x) = \dfrac{x}{x - 2}$

$$f(f^{-1}(x)) = \frac{2\left(\dfrac{x}{x - 2}\right)}{\dfrac{x}{x - 2} - 1} = x$$

$$f^{-1}(f(x)) = \frac{\dfrac{2x}{x - 1}}{\dfrac{2x}{x - 1} - 2} = x$$

Domain f = Range f^{-1} = All real numbers except 1
Range f = Domain f^{-1} = All real numbers except 2

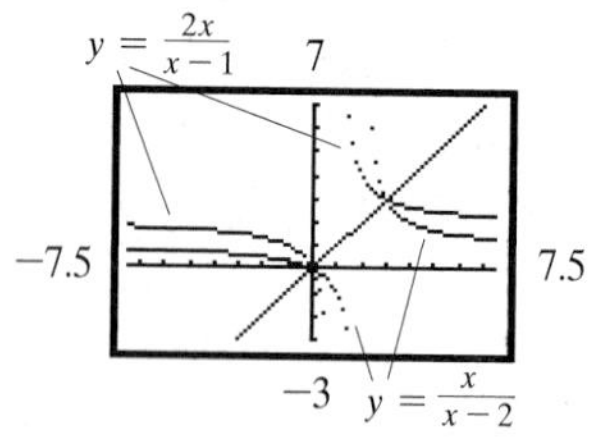

49. $f^{-1}(x) = \dfrac{3x + 4}{2x - 3}$

$$f(f^{-1}(x)) = \frac{3\left(\dfrac{3x + 4}{2x - 3}\right) + 4}{2\left(\dfrac{3x + 4}{2x - 3}\right) - 3} = x$$

$$f^{-1}(f(x)) = \frac{3\left(\dfrac{3x + 4}{2x - 3}\right) + 4}{2\left(\dfrac{3x + 4}{2x - 3}\right) - 3} = x$$

Domain f = Range f^{-1} = All real numbers except $\dfrac{3}{2}$

Range f = Domain f^{-1} = All real numbers except $\dfrac{3}{2}$

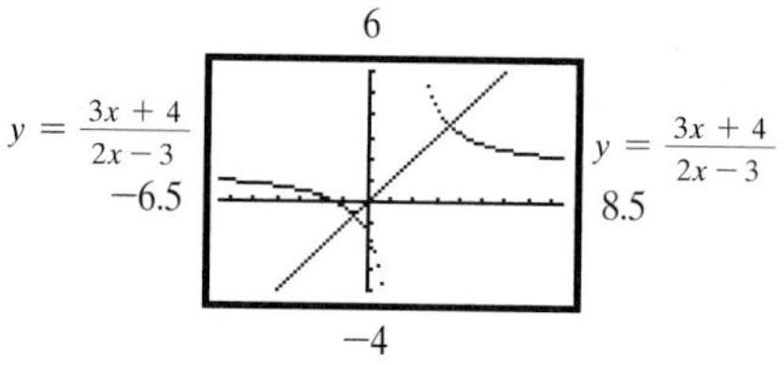

51. $f^{-1}(x) = \dfrac{-2x + 3}{x - 2}$

$$f(f^{-1}(x)) = \frac{2\left(\dfrac{-2x + 3}{x - 2}\right) + 3}{\dfrac{-2x + 3}{x - 2} + 2} = x$$

$$f^{-1}(f(x)) = \frac{-2\left(\dfrac{2x + 3}{x + 2}\right) + 3}{\dfrac{2x + 3}{x + 2} - 2} = x$$

Domain f = Range f^{-1} = All real numbers except -2
Range f = Domain f^{-1} = All real numbers except 2

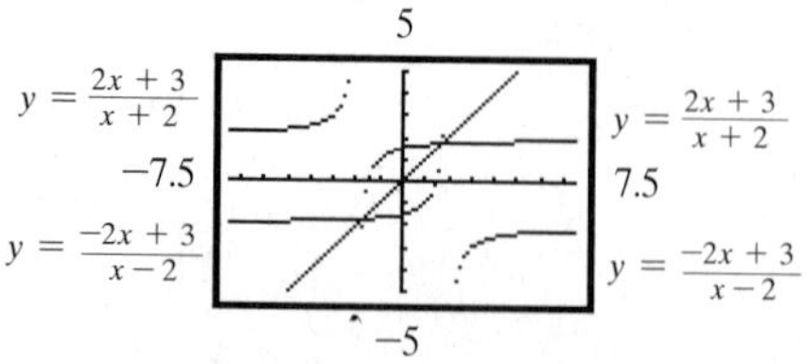

53. $f^{-1}(x) = \dfrac{x^3}{8}$

$f(f^{-1}(x)) = 2\sqrt[3]{\dfrac{x^3}{8}} = x$

$f^{-1}(f(x)) = \dfrac{(2\sqrt[3]{x})^3}{8} = x$

Domain f = Range $f^{-1} = (-\infty, \infty)$
Range f = Domain $f^{-1} = (-\infty, \infty)$

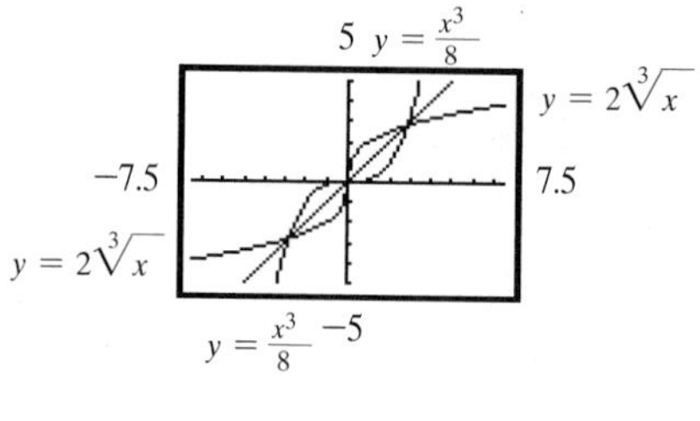

55. $f^{-1}(x) = \frac{1}{m}(x - b), m \neq 0$ **59.** Quadrant I **61.** $f(x) = |x|, x \geq 0$, is one-to-one; $f^{-1}(x) = x, x \geq 0$

63. $f(g(x)) = \frac{9}{5}\left[\frac{5}{9}(x - 32)\right] + 32 = x$; $g(f(x)) = \frac{5}{9}\left[\left(\frac{9}{5}x + 32\right) - 32\right] = x$ **65.** $l(T) = \frac{gT^2}{4\pi^2}, T > 0$

67. $f^{-1}(x) = \frac{-dx + b}{cx - a}$; $f = f^{-1}$ if $a = -d$

5.2 Exercises

1. **(a)** 11.212 **(b)** 11.587 **(c)** 11.664 **(d)** 11.665 **3.** **(a)** 8.815 **(b)** 8.821 **(c)** 8.824 **(d)** 8.825
5. **(a)** 21.217 **(b)** 22.217 **(c)** 22.440 **(d)** 22.459 **7.** 3.320 **9.** 0.427 **11.** B **13.** D **15.** A **17.** E **19.** A **21.** E **23.** B

25.

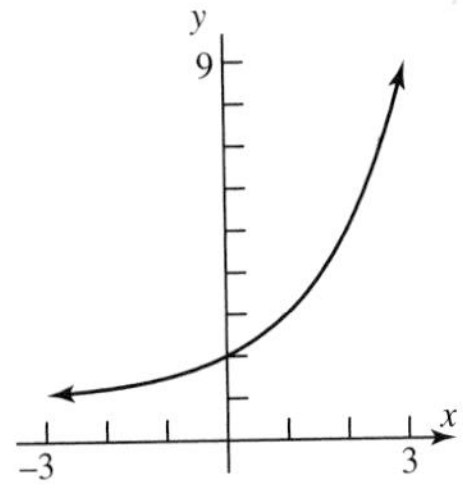

Domain: All real numbers
Range: $\{y|y > 1\}$ or $(1, \infty)$
Horizontal asymptote: $y = 1$

27.

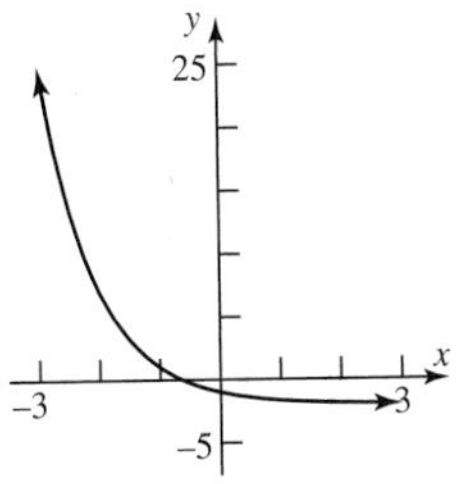

Domain: All real numbers
Range: $\{y|y > -2\}$ or $(-2, \infty)$
Horizontal asymptote: $y = -2$

29.

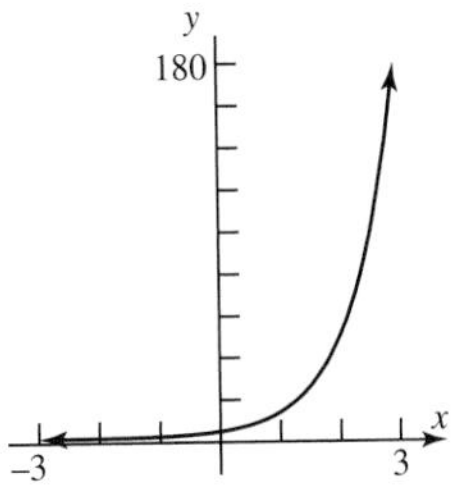

Domain: All real numbers
Range: $\{y|y > 2\}$ or $(2, \infty)$
Horizontal asymptote: $y = 2$

31.

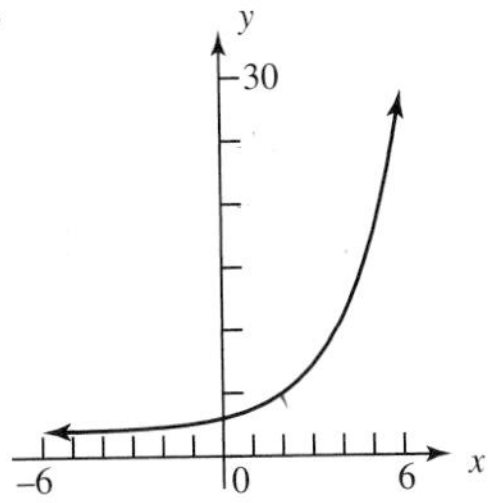

Domain: All real numbers
Range: $\{y|y > 2\}$ or $(2, \infty)$
Horizontal asymptote: $y = 2$

33.

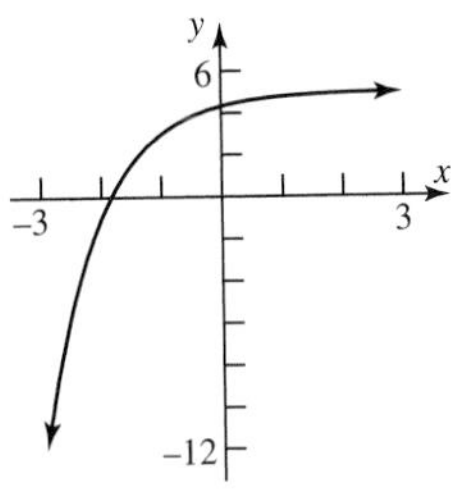

Domain: All real numbers
Range: $\{y|y < 5\}$ or $(-\infty, 5)$
Horizontal asymptote: $y = 5$

35.

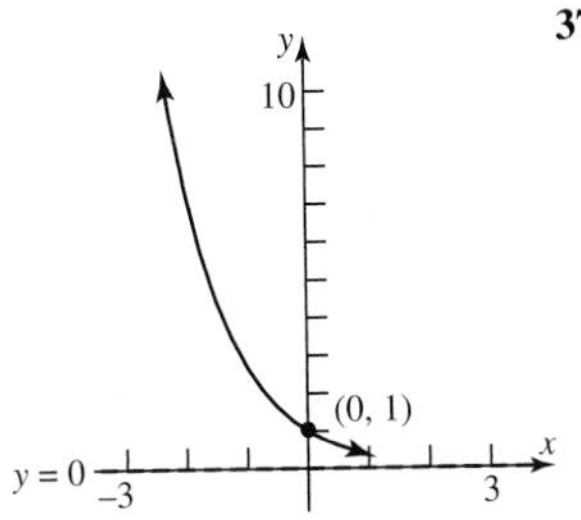

Domain: $(-\infty, \infty)$
Range: $(0, \infty)$
Horizontal asymptote: $y = 0$

37.

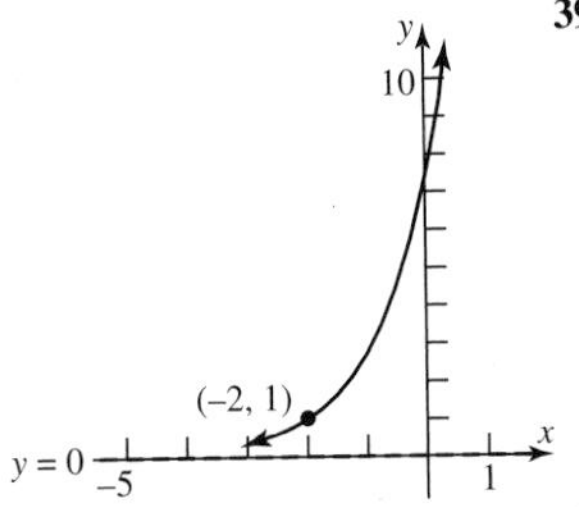

Domain: $(-\infty, \infty)$
Range: $(0, \infty)$
Horizontal asymptote: $y = 0$

39.

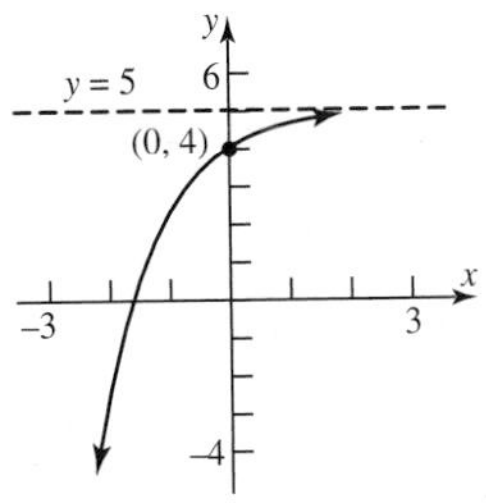

Domain: $(-\infty, \infty)$
Range: $(-\infty, 5)$
Horizontal asymptote: $y = 5$

41.

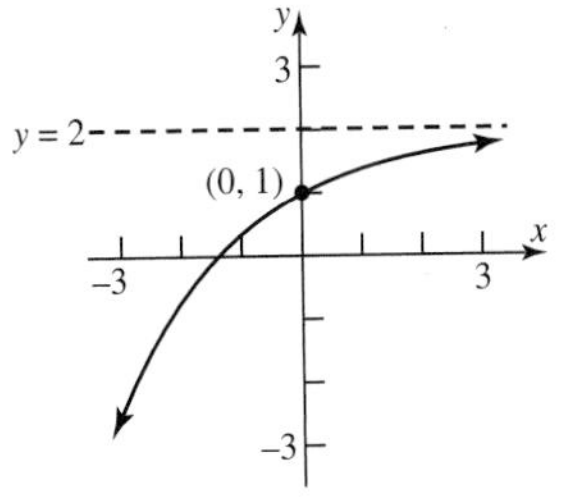

Domain: $(-\infty, \infty)$
Range: $(-\infty, 2)$
Horizontal asymptote: $y = 2$

43. $\frac{1}{2}$ **45.** $\{-\sqrt{2}, 0, \sqrt{2}\}$ **47.** $\left\{1 - \frac{\sqrt{6}}{3}, 1 + \frac{\sqrt{6}}{3}\right\}$ **49.** 0 **51.** 4

53. $\frac{3}{2}$ **55.** $\{1, 2\}$ **57.** $\frac{1}{49}$ **59.** $\frac{1}{4}$ **61.** **(a)** 74% **(b)** 47%

63. **(a)** 44 watts **(b)** 11.6 watts **65.** 3.35 milligrams; 0.45 milligrams

67. **(a)** 0.63 **(b)** 0.98 **(c)**

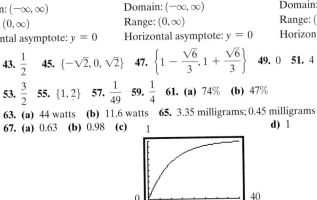

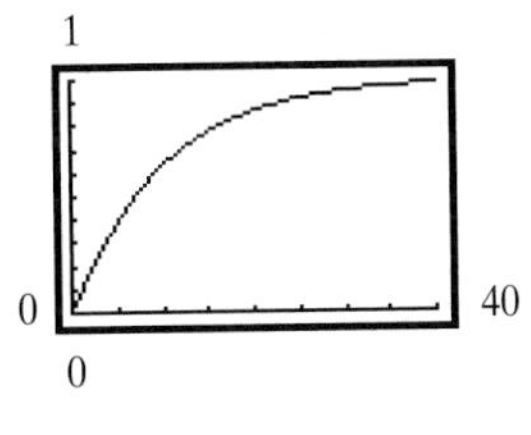

(d) 1

69. **(a)** 5.16% **(b)** 8.88% **71.** **(a)** 71% **(b)** 73% **(c)** 100%

73. **(a)** 5.414 amperes, 7.585 amperes, 10.376 amperes **(b)** 12 amperes
(d) 3.343 amperes, 5.309 amperes, 9.443 amperes **(e)** 24 amperes
(c), (f)

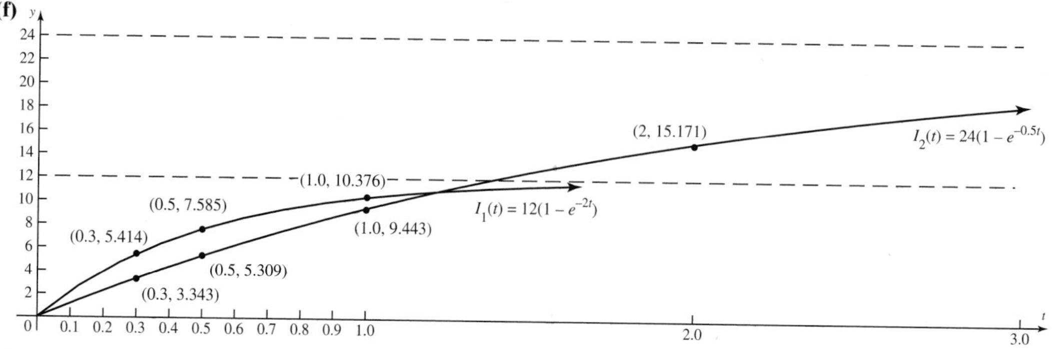

75. $n = 4: 2.7083; n = 6: 2.7181; n = 8: 2.7182788; n = 10: 2.7182818$

77. $\dfrac{f(x+h) = f(x)}{h} = \dfrac{a^{x+h} - a^x}{h} = \dfrac{a^x a^h - a^x}{h} = \dfrac{a^x(a^h - 1)}{h}$ **79.** $f(-x) = a^{-x} = \dfrac{1}{a^x} = \dfrac{1}{f(x)}$

81. $f(-x) = \dfrac{1}{2}(e^{-x} - e^{-(-x)}) = \dfrac{1}{2}(e^{-x} - e^{x}) = -\dfrac{1}{2}(e^{x} - e^{-x})$
$= -f(x)$

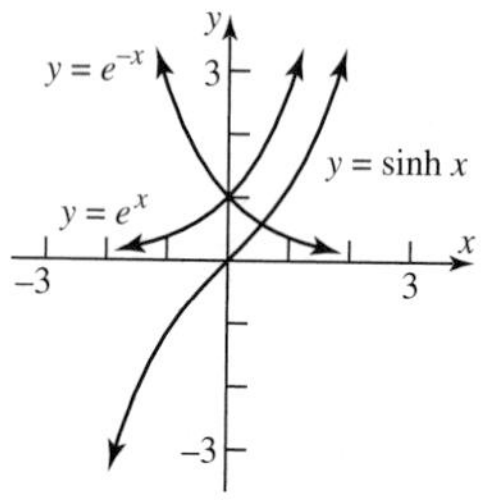

83. $f(1) = 5, f(2) = 17, f(3) = 257, f(4) = 65{,}537,$
$f(5) = 4{,}294{,}967{,}297 = 641 \times 6{,}700{,}417$

5.3 Exercises

1. $2 = \log_3 9$ **3.** $2 = \log_a 1.6$ **5.** $2 = \log_{1.1} M$ **7.** $x = \log_2 7.2$ **9.** $\sqrt{2} = \log_x \pi$ **11.** $x = \ln 8$ **13.** $2^3 = 8$ **15.** $a^6 = 3$ **17.** $3^x = 2$ **19.** $2^{1.3} = M$ **21.** $(\sqrt{2})^x = \pi$ **23.** $e^x = 4$ **25.** 0 **27.** 2 **29.** -4 **31.** $\dfrac{1}{2}$ **33.** 4 **35.** $\dfrac{1}{2}$ **37.** $\{x|x > 3\}$ **39.** All real numbers except 0 **41.** $\{x|x \neq 1\}$ **43.** $\{x|x > -1\}$ **45.** $\{x|x < -1 \text{ or } x > 0\}$ **47.** 0.511 **49.** 30.099 **51.** $\sqrt{2}$ **53.** B **55.** D **57.** A **59.** E

61.

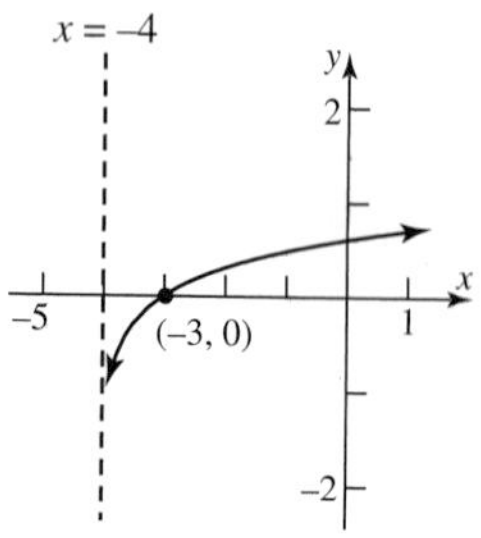

Domain: $(-4, \infty)$
Range: $(-\infty, \infty)$
Vertical asymptote: $x = -4$

63.

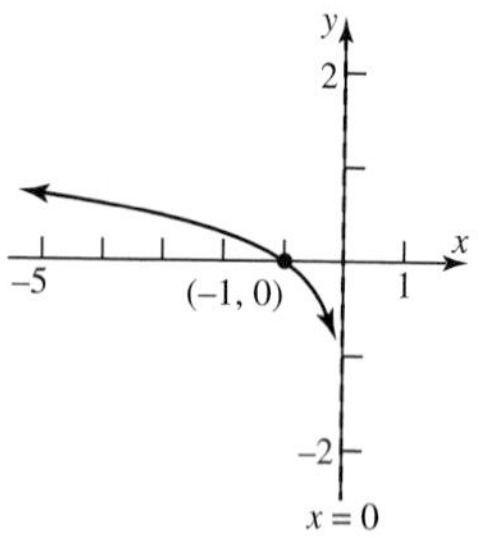

Domain: $(-\infty, 0)$
Range: $(-\infty, \infty)$
Vertical asymptote: $x = 0$

65.

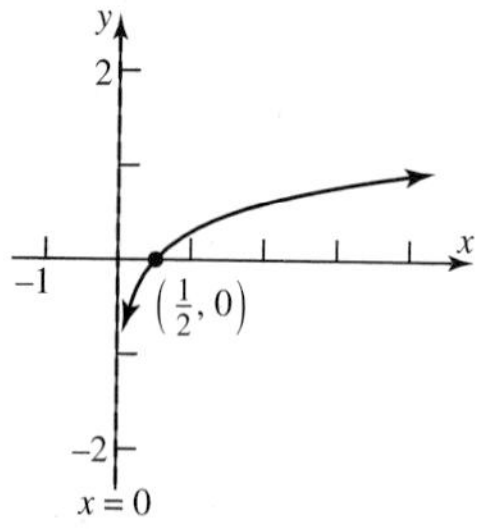

Domain: $(0, \infty)$
Range: $(-\infty, \infty)$
Vertical asymptote: $x = 0$

67

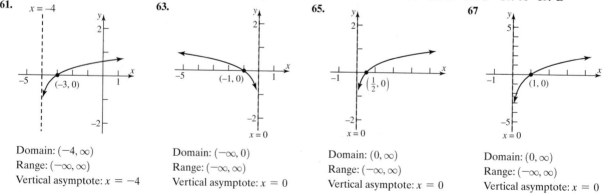

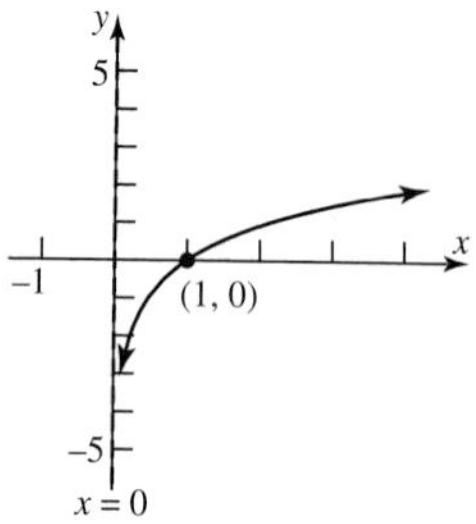

Domain: $(0, \infty)$
Range: $(-\infty, \infty)$
Vertical asymptote: $x = 0$

69.

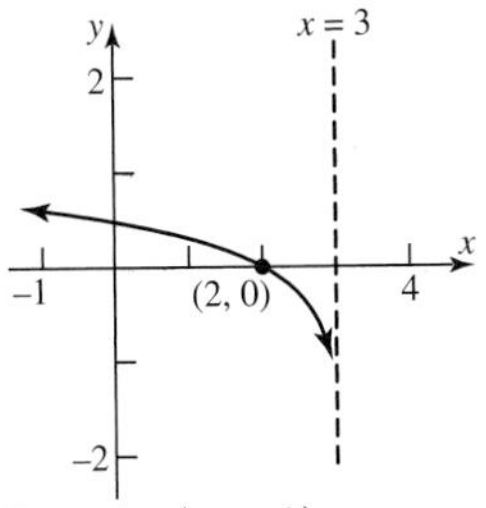

Domain: $(-\infty, 3)$
Range: $(-\infty, \infty)$
Vertical asymptote: $x = 3$

71.

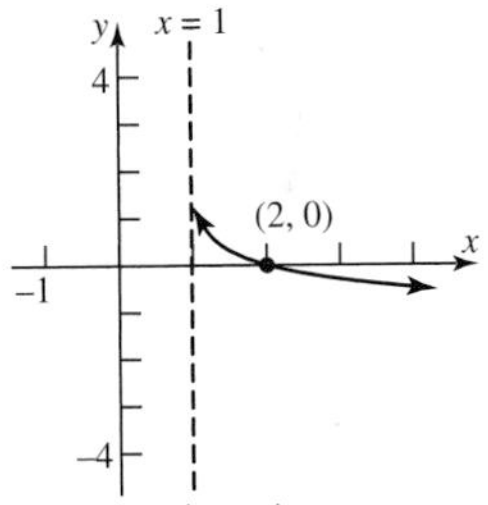

Domain: $(1, \infty)$
Range: $(-\infty, \infty)$
Vertical asymptote: $x = 1$

73. 9 **75.** $\frac{7}{2}$ **77.** 2 **79.** 5

81. 3 **83.** 2 **85.** $\frac{\ln 10}{3}$ **87.** $\frac{\ln 8 - 5}{2}$

89. $\{-2\sqrt{2}, 2\sqrt{2}\}$ **91.** -1

93. (a)

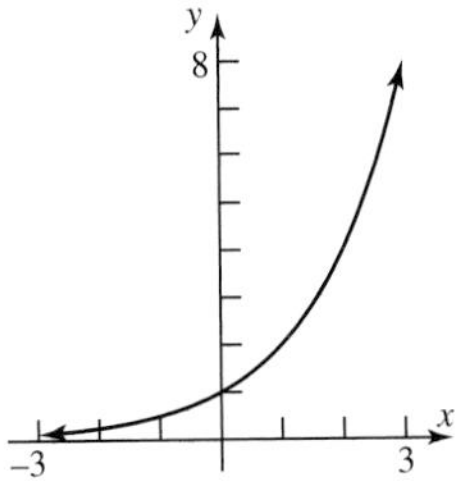

Domain: All real numbers
Range: $\{y|y > 0\}$ or $(0, \infty)$
Horizontal asymptote: $y = 0$

(b) $f^{-1}(x) = \log_2 x$

(c)

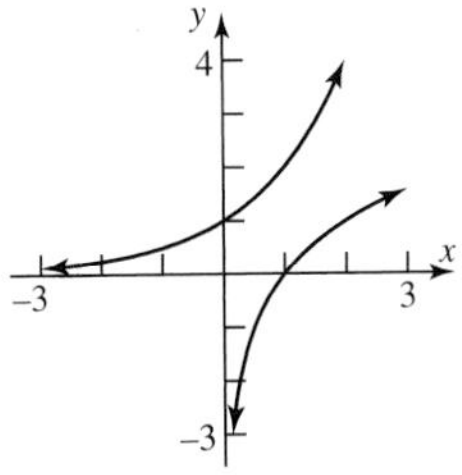

Domain of f^{-1} = Range of f = $(0, \infty)$
Range of f^{-1} = Domain of f = All real numbers
Vertical asymptote of f^{-1}: $x = 0$

95. (a)

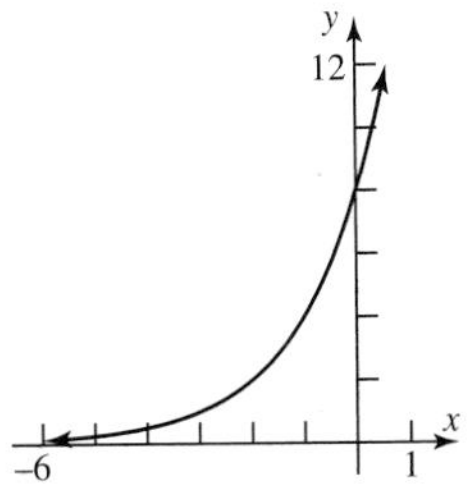

Domain: All real numbers
Range: $\{y|y > 0\}$ or $(0, \infty)$
Horizontal asymptote: $y = 0$

(b) $f^{-1}(x) = \log_2 x - 3$

(c)

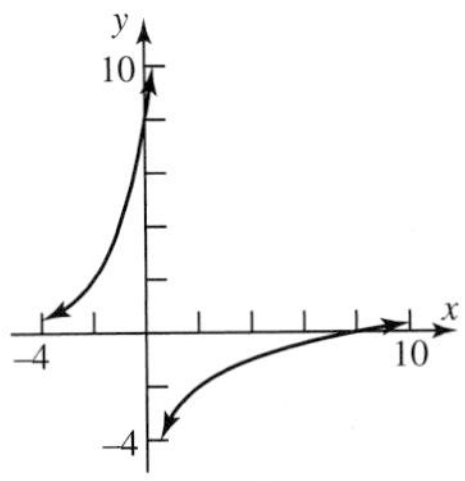

Domain of f^{-1} = Range of f = $(0, \infty)$
Range of f^{-1} = Domain of f = All real numbers
Vertical asymptote of f^{-1}: $x = 0$

97. (a)

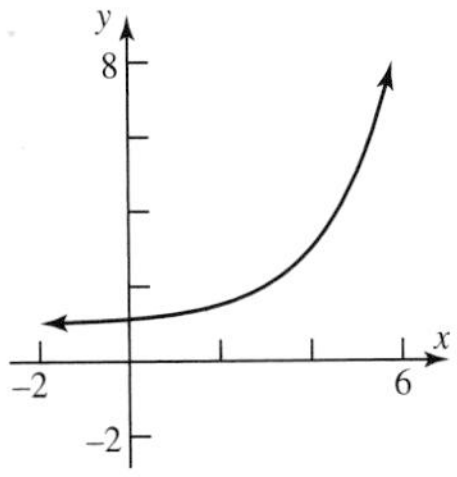

Domain: All real numbers
Range: $\{y|y > 1\}$ or $(1, \infty)$
Horizontal asymptote: $y = 1$

(b) $f^{-1}(x) = \log_2(x - 1) + 3$

(c)

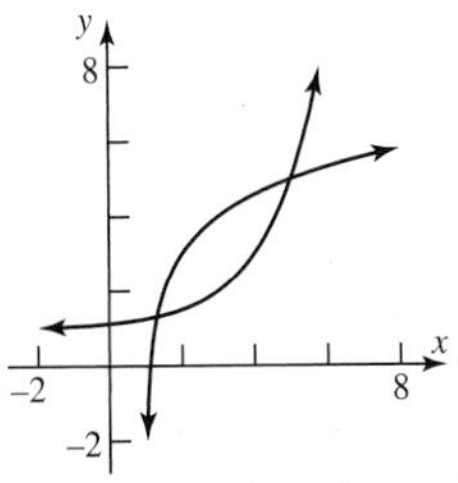

Domain of f^{-1} = Range of f = $(1, \infty)$
Range of f^{-1} = Domain of f = All real numbers
Vertical asymptote of f^{-1}: $x = 1$

99. (a) $n \approx 6.93$ so 7 panes are necessary **(b)** $n \approx 13.86$ so 14 panes are necessary

101. (a) $d \approx 127.7$ so it takes about 128 days **(b)** $d \approx 575.6$ so it takes about 576 days

103. (a) 6.93 min **(b)** 16.09 min **(c)** No, since $F(t)$ can never equal one.

105. $h \approx 2.29$ so the time between injections is about 2 hours, 17 minutes

107. 0.2695 seconds
0.8959 seconds

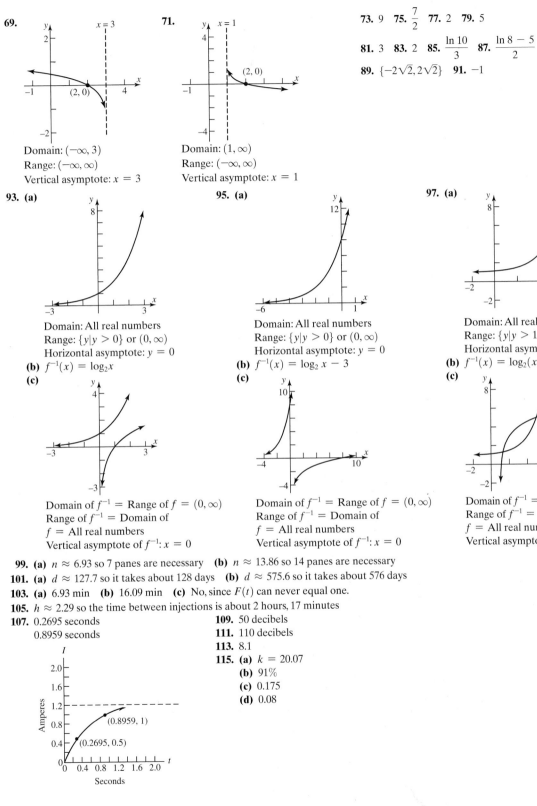

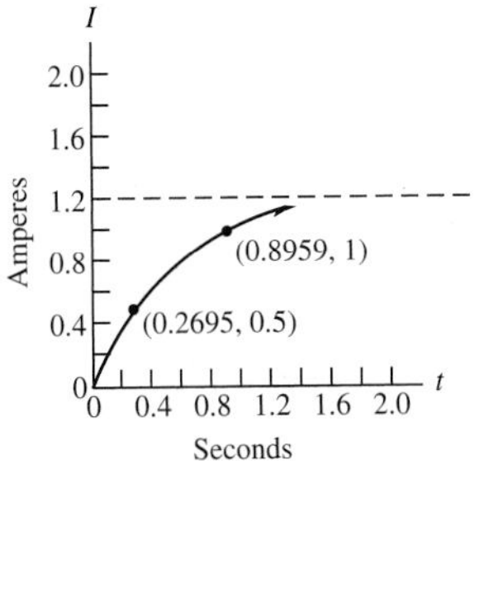

109. 50 decibels

111. 110 decibels

113. 8.1

115. (a) $k = 20.07$
(b) 91%
(c) 0.175
(d) 0.08

5.4 Exercises

1. $a + b$ **3.** $b - a$ **5.** $a + 1$ **7.** $2a + b$ **9.** $\frac{1}{5}(a + 2b)$ **11.** $\frac{b}{a}$ **13.** $2 \log_a u + 3 \log_a v$ **15.** $-3 \log m$ **17.** $\frac{1}{2}[3 \log_5 - \log_5 b]$

19. $2 \ln x + \frac{1}{2} \ln(1 - x)$ **21.** $3 \log_2 x - \log_2(x - 3)$ **23.** $\log x + \log(x + 2) - 2 \log(x + 3)$

25. $\frac{1}{3} \ln(x - 2) + \frac{1}{3} \ln(x + 1) - \frac{2}{3} \ln(x + 4)$ **27.** $\ln 5 + \ln x + \frac{1}{2} \ln(1 - 3x) - 3 \ln(x - 4)$ **29.** $\log_5 u^3v^4$ **31.** $-\frac{5}{2} \log_{1/2} x$

33. $-2 \ln(x - 1)$ **35.** $\log_2[x(3x - 2)^4]$ **37.** $\log_a\left(\frac{25x^6}{\sqrt{2x + 3}}\right)$ **39.** $\ln y = \ln a + (\ln b)x$ **41.** $\frac{5}{4}$ **43.** 4 **45.** 2.771 **47.** -3.880

49. 5.615 **51.** 0.874

53. $y = \left(\frac{\log x}{\log 4}\right)$

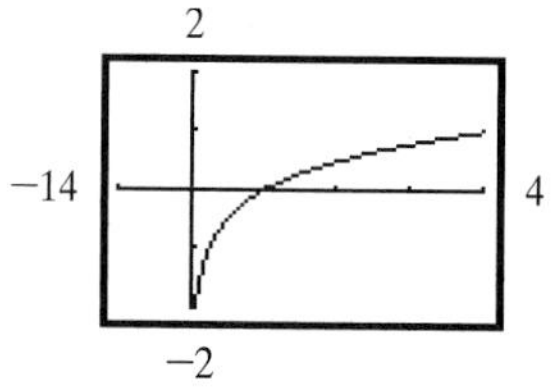

55. $y = \left(\frac{\log(x + 2)}{\log 2}\right)$

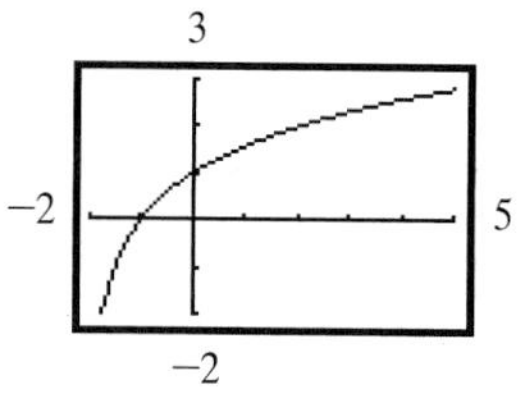

57. $y = \frac{\log(x + 1)}{\log(x - 1)}$

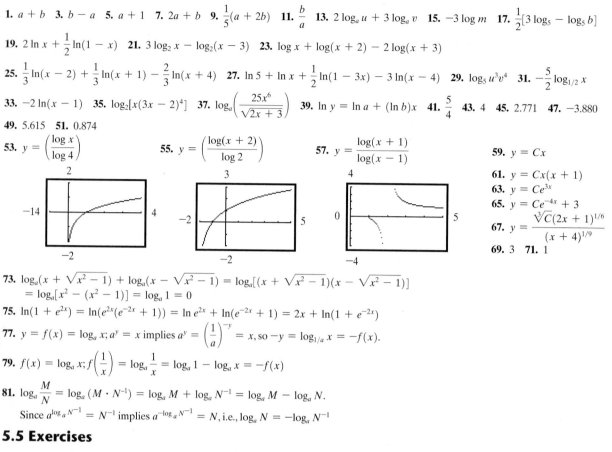

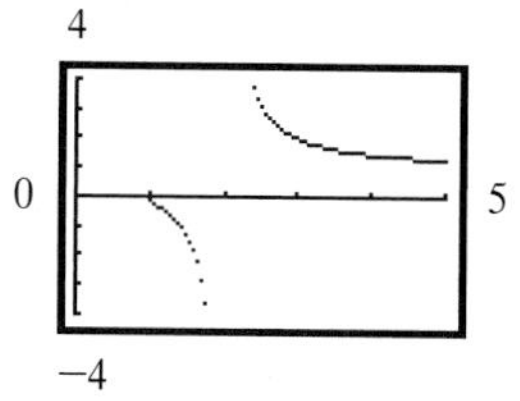

59. $y = Cx$

61. $y = Cx(x + 1)$

63. $y = Ce^{3x}$

65. $y = Ce^{-4x} + 3$

67. $y = \frac{\sqrt[3]{C}(2x + 1)^{1/6}}{(x + 4)^{1/9}}$

69. 3 **71.** 1

73. $\log_a(x + \sqrt{x^2 - 1}) + \log_a(x - \sqrt{x^2 - 1}) = \log_a[(x + \sqrt{x^2 - 1})(x - \sqrt{x^2 - 1})]$
$= \log_a[x^2 - (x^2 - 1)] = \log_a 1 = 0$

75. $\ln(1 + e^{2x}) = \ln(e^{2x}(e^{-2x} + 1)) = \ln e^{2x} + \ln(e^{-2x} + 1) = 2x + \ln(1 + e^{-2x})$

77. $y = f(x) = \log_a x$; $a^y = x$ implies $a^y = \left(\frac{1}{a}\right)^{-y} = x$, so $-y = \log_{1/a} x = -f(x)$.

79. $f(x) = \log_a x$; $f\left(\frac{1}{x}\right) = \log_a \frac{1}{x} = \log_a 1 - \log_a x = -f(x)$

81. $\log_a \frac{M}{N} = \log_a (M \cdot N^{-1}) = \log_a M + \log_a N^{-1} = \log_a M - \log_a N$.
Since $a^{\log_a N^{-1}} = N^{-1}$ implies $a^{-\log_a N^{-1}} = N$, i.e., $\log_a N = -\log_a N^{-1}$

5.5 Exercises

1. 6 **3.** 16 **5.** 8 **7.** 3 **9.** 5 **11.** $\{-1 + \sqrt{1 + e^4}, -1 - \sqrt{1 + e^4}\}$ **13.** $\frac{\ln 3}{\ln 2} \approx 1.585$ **15.** 0 **17.** $\frac{\ln 10}{\ln 2} \approx 3.322$

19. $-\frac{\ln 1.2}{\ln 8} \approx -0.088$ **21.** $\frac{\ln 3}{2 \ln 3 + \ln 4} \approx 0.307$ **23.** $\frac{\ln 7}{\ln 0.6 + \ln 7} \approx 1.356$ **25.** 0 **27.** $\frac{\ln \pi}{1 + \ln \pi} \approx 0.534$ **29.** $\frac{\ln 1.6}{3 \ln 2} \approx 0.226$ **31.** $\frac{9}{2}$

33. 2 **35.** 1 **37.** 16 **39.** $\left\{-1, \frac{2}{3}\right\}$ **41.** 0 **43.** $\ln(2 + \sqrt{5})$ **45.** 1.92 **47.** 2.79 **49.** -0.57 **51.** -0.70 **53.** 0.57 **55.** $\{0.39, 1.00\}$

57. 1.32 **59.** 1.31

5.6 Exercises

1. \$108.29 **3.** \$609.50 **5.** \$697.09 **7.** \$12.46 **9.** \$125.23 **11.** \$88.72 **13.** \$860.72 **15.** \$554.09 **17.** \$59.71 **19.** \$361.93 **21.** 5.35%

23. 26% **25.** $6\frac{1}{4}$% compounded annually **27.** 9% compounded monthly **29.** 104.32 months; 103.97 months **31.** 61.02 months; 60.82 months

33. 15.27 years or 15 years, 4 months **35.** \$104,335 **37.** \$12,910.62 **39.** About \$30.17 per share or \$3017 **41.** 9.35%

43. Not quite. Jim will have \$1057.60. The second bank gives a better deal, since Jim will have \$1060.62 after 1 year.

45. Will has \$11,632.73; Henry has \$10,947.89. **47.** **(a)** Interest is \$30,000 **(b)** Interest is \$38,613.59 **(c)** Interest is \$37,752.73. Simple interest at 12% is best. **49.** **(a)** \$1364.62 **(b)** \$1353.35 **51.** \$4631.93

55. **(a)** 6.1 years **(b)** 18.45 years **(c)**

$$mP = P\left(1 + \frac{r}{n}\right)$$

$$m = \left(1 + \frac{r}{n}\right)^{nt}$$

$$\ln m = \ln\left(1 + \frac{r}{n}\right)^{nt} = nt \ln\left(1 + \frac{r}{n}\right)$$

$$t = \frac{\ln m}{n \ln\left(1 + \frac{r}{n}\right)}$$

5.7 Exercises

1. (a) 34.7 days; 69.3 days
(b)

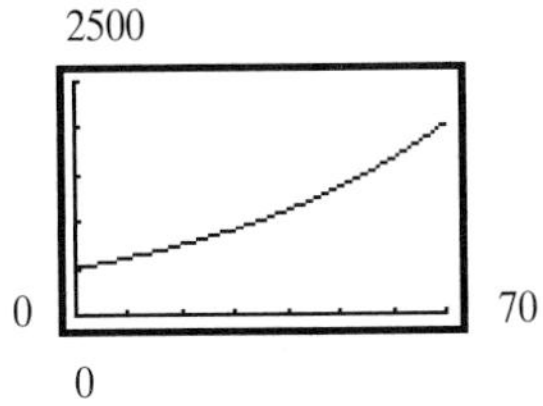

3. 28.4 years

5. (a) 94.4 years
(b)

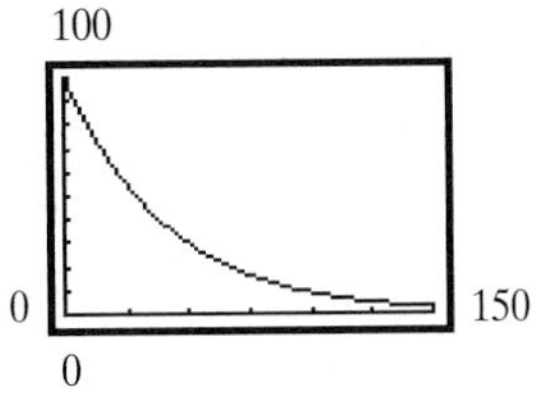

7. 5832; 3.9 days
9. 25,198
11. 9.797 grams

13. (a) 9727 years ago
(b)

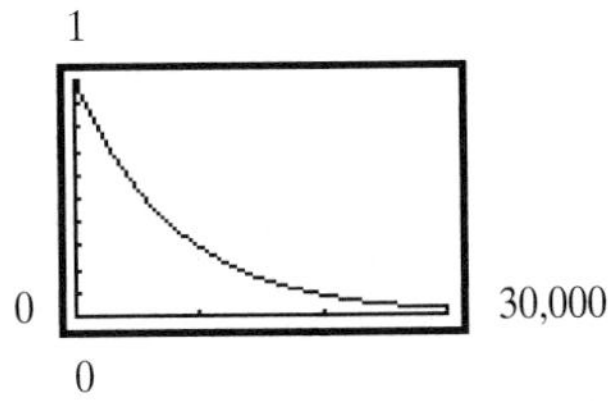

(c) 5600 years

15. (a) 5:18 PM
(b)

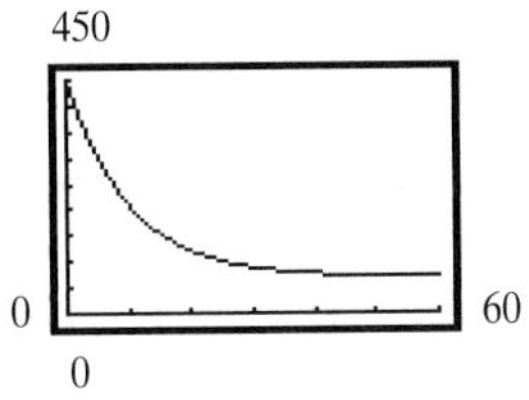

(c) After 14.3 minutes, the pizza will be 160°F.
(d) As time passes, the temperature of the pizza gets closer to 70°F.

17. (a) 18.63°C; 25.1°C
(b)

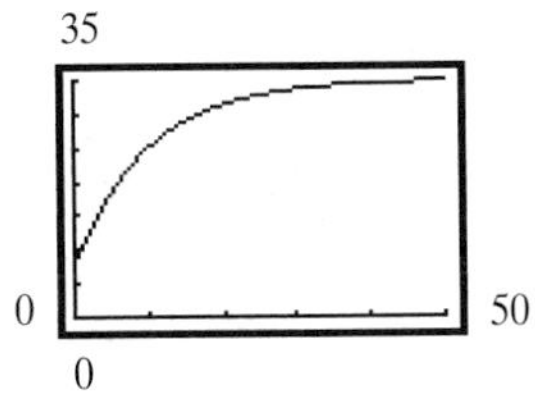

19. 7.34 kg; 76.6 hours
21. 26.6 days

23. (a) 0.1286 **(b)** 0.9
(c)

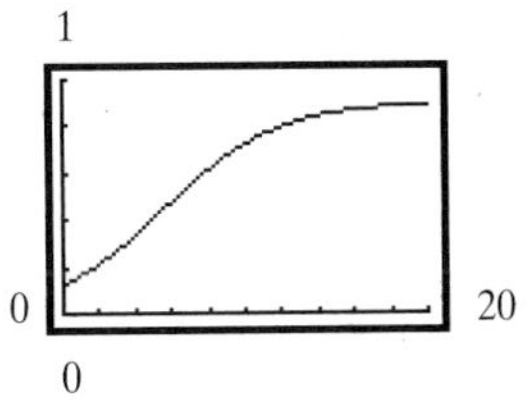

(d) 1996

25. (a)

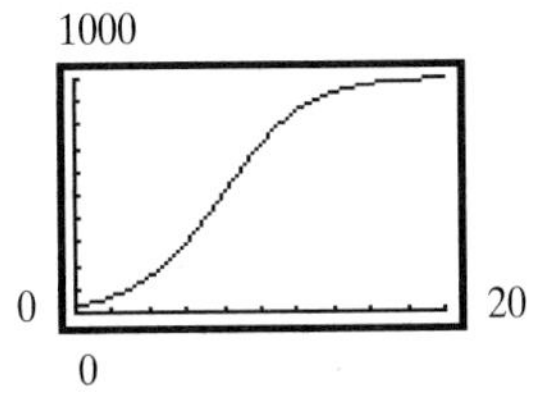

(b) 1000 **(c)** 30 **(d)** 11.076 hours

27. (a) 9.23×10^{-3}, or about 0
(b) 0.81, or about 1 **(c)** 5.01, or about 5
(d) 57.91°, 43.99°, 30.07°

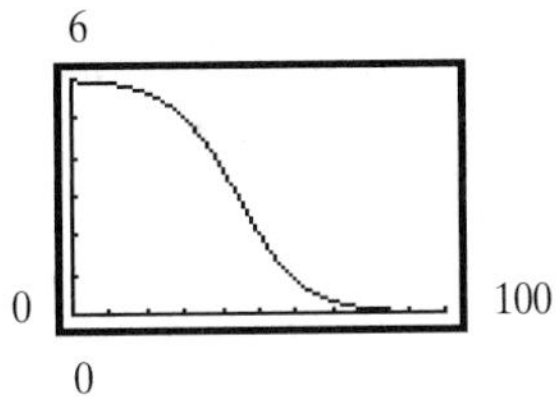

5.8 Exercises

1. (a)

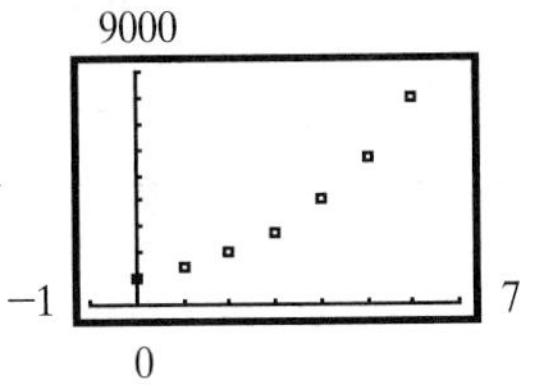

(b) $y = 1000.188(1.414)^x$
(c) $N = 1000.188e^{0.3464t}$

(d)

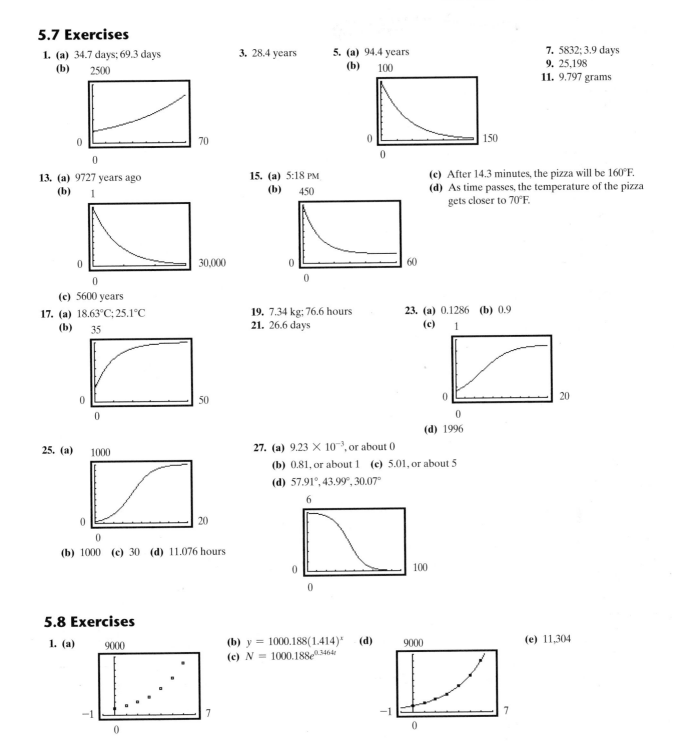

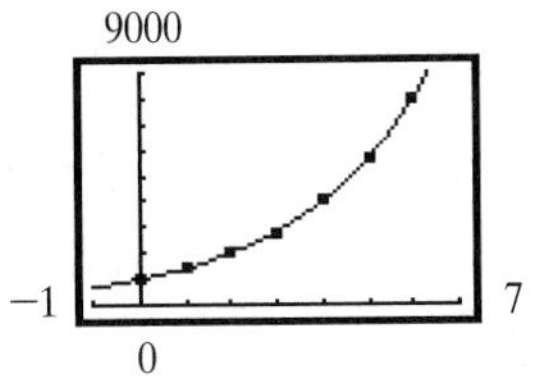

(e) 11,304

3. (a)

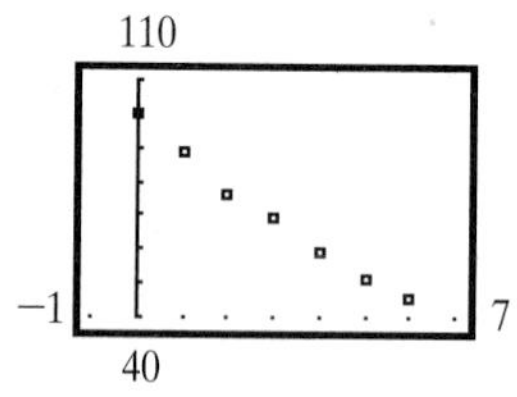

(b) $y = 100.326(0.877)^x$
(c) $A = 100.326e^{-.1312t}$
(d)

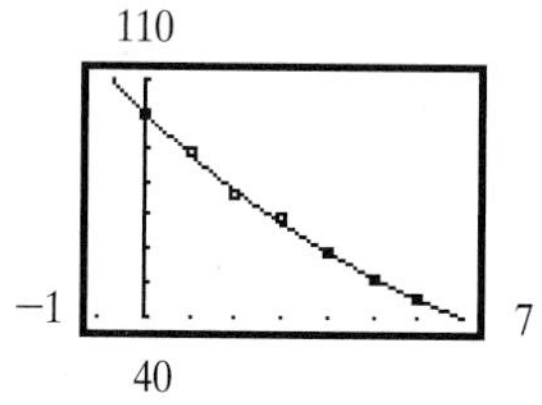

(e) 5.3 weeks **(f)** 0.14 grams

5. (a)

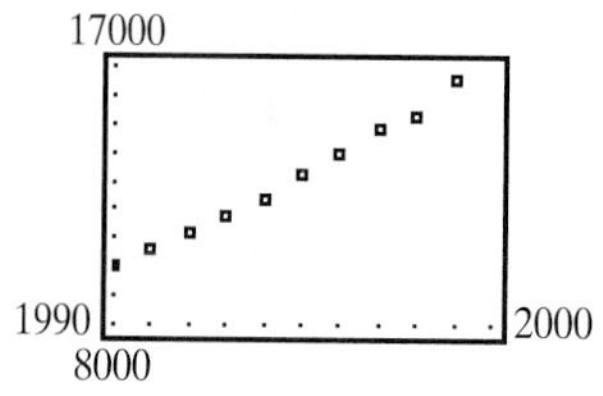

(b) value $= 2.8546x\ 10^{-44}(1.056554737)^{\text{year}}$
(c) 5.640%
(d) $52,166

7. (a)

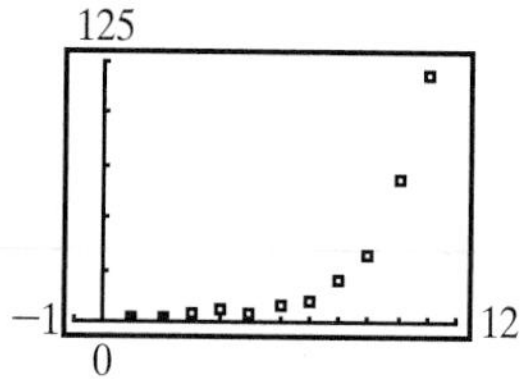

(b) price $= 1.245(1.52556)^t$
(c) price $= 1.245e^{0.42236t}$
(d)

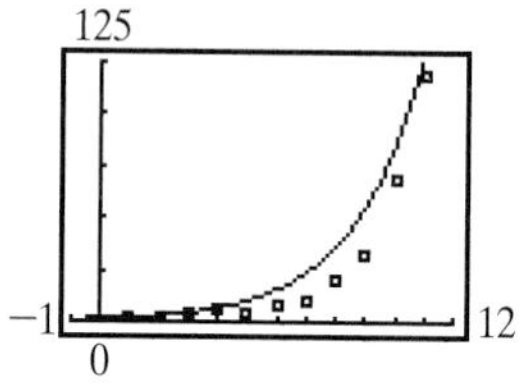

(e) approximately $198

9. (a)

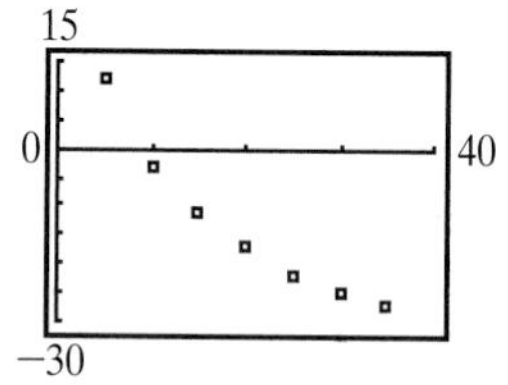

(b) wind chill $= 44.198 - 20.331 \ln$ (wind speed)
(c)

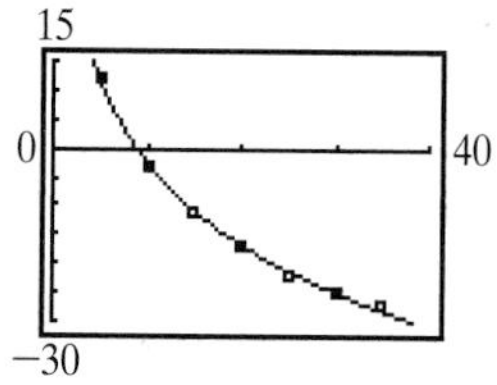

(d) approximately −20 degrees Fahrenheit

11. (a)

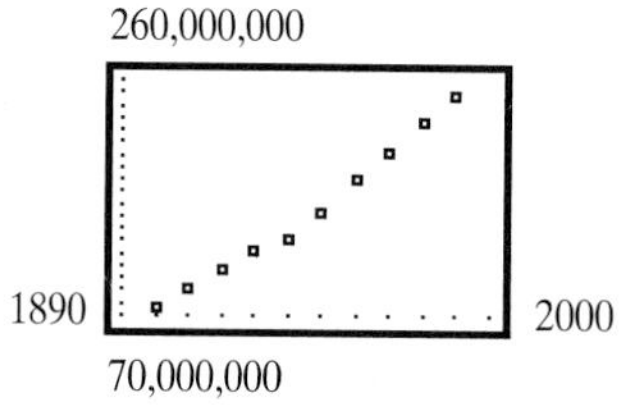

(b) $y = \dfrac{695{,}129{,}657.2}{1 + 4.1136 \times 10^{14} e^{-0.0166x}}$

(c)

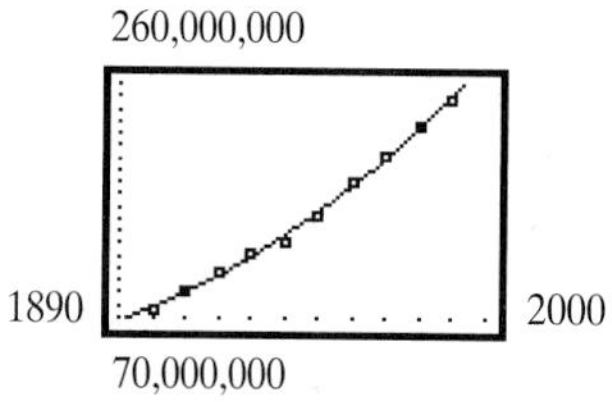

(d) 695,129,657
(e) 270,575,260

13. (a)

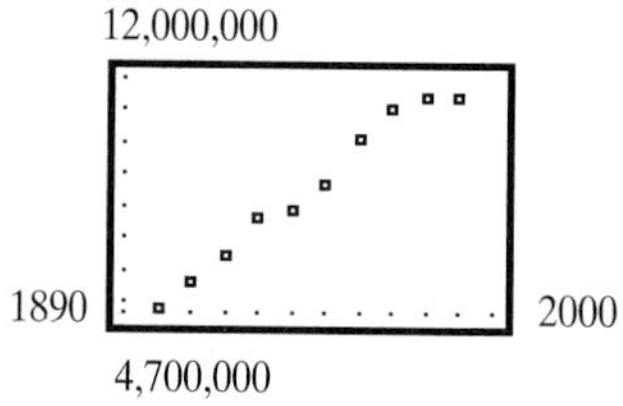

(b) $y = \dfrac{14{,}168{,}687.34}{1 + 1.5075745 \times 10^{21} e^{-0.02531x}}$

(c)

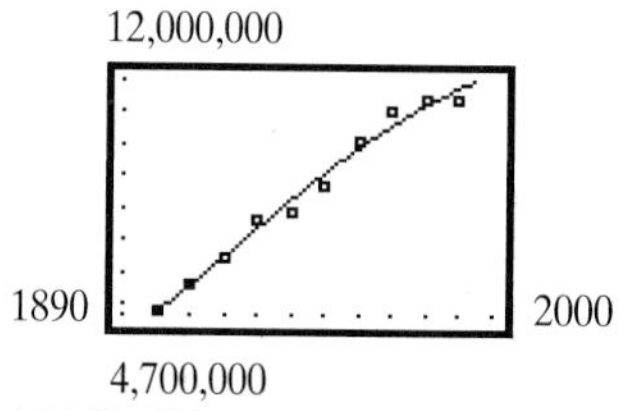

(d) 14,168,687
(e) Approximately 12,252,199

15. (a)

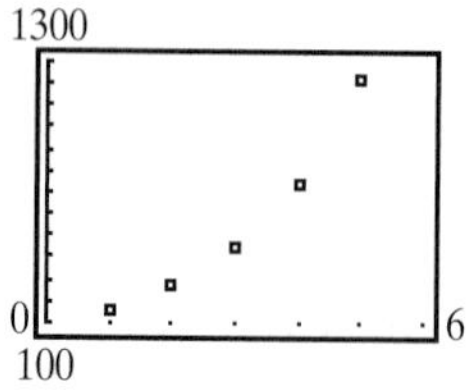

(b) Exponential model:
$y = 99.06645375\ (1.652917358)^x$

17. (a)

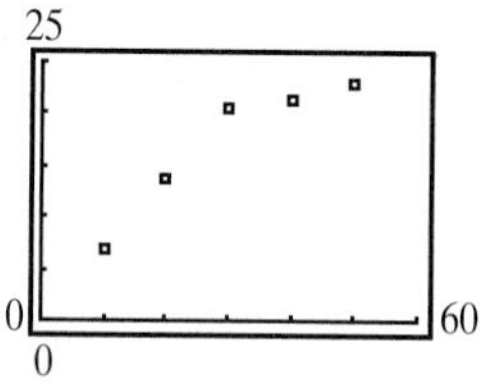

(b) Logarithmic model:
$y = 16.55688464 + 10.31473128 \ln x$

19. (a)

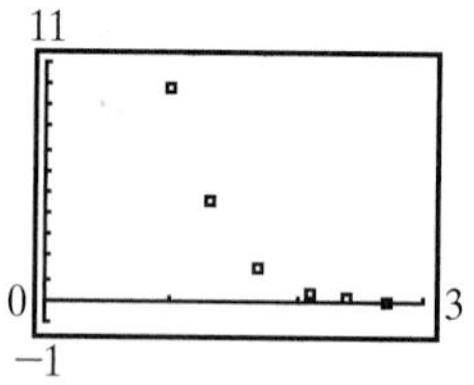

(b) Exponential model:
$y = 346.113717\ (0.0352268374)^x$

21. (a)

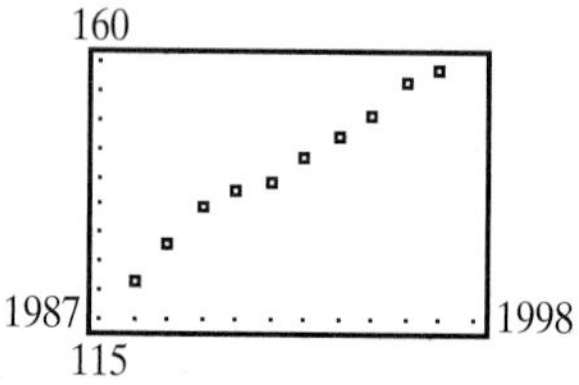

(b) Logarithmic model:
CPI $= -59445.77 + 7843.39 \ln$ (year)
(c) 163.2

23. (a)

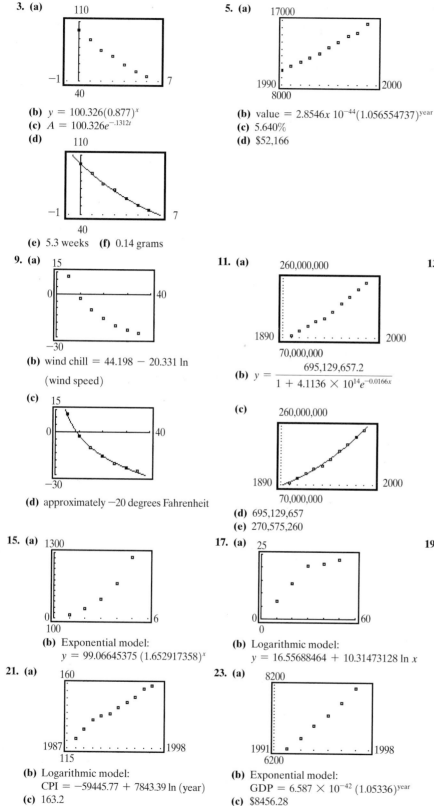

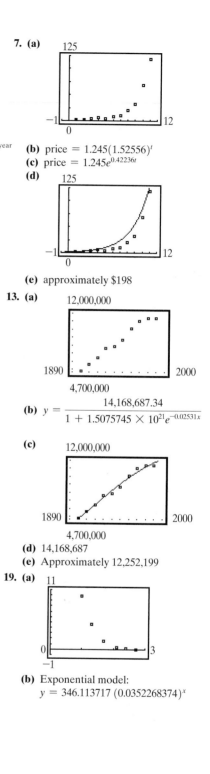

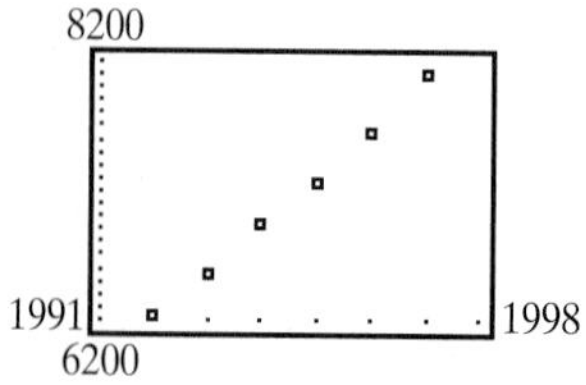

(b) Exponential model:
GDP $= 6.587 \times 10^{-42}\ (1.05336)^{\text{year}}$
(c) $8456.28

Fill-in-the Blank Items

1. One-to-one **3.** $(0, 1)$ and $(1, a)$ **5.** 4 **7.** 1 **9.** All real numbers greater than 0 **11.** 1

True/False Items

1. False **3.** True **5.** False **7.** True **9.** False

Review Exercises

1. $f^{-1}(x) = \dfrac{2x+3}{5x-2}$; $f(f^{-1}(x)) = \dfrac{2\left(\dfrac{2x+3}{5x-2}\right)+3}{5\left(\dfrac{2x+3}{5x-2}\right)-2} = x$;

$f^{-1}(f(x)) = \dfrac{2\left(\dfrac{2x+3}{5x-2}\right)+3}{5\left(\dfrac{2x+3}{5x-2}\right)-2} = x$;

Domain f = Range f^{-1} = All real numbers except $\frac{2}{5}$;

Range f = Domain f^{-1} = All real numbers except $\frac{2}{5}$

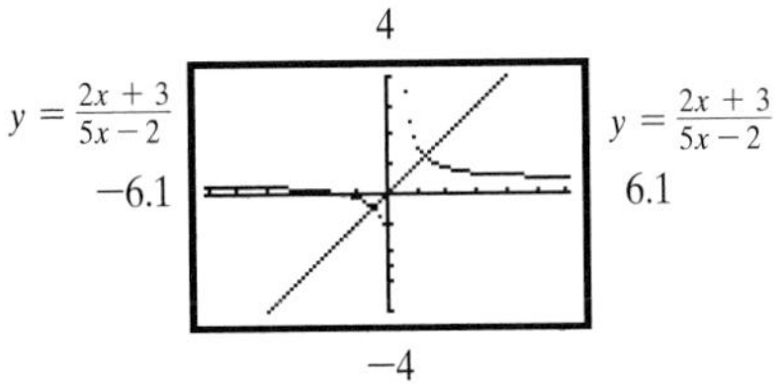

3. $f^{-1}(x) = \dfrac{x+1}{x}$; $f(f^{-1}(x)) = \dfrac{1}{\dfrac{x+1}{x}-1} = x$;

$f^{-1}(f(x)) = \dfrac{\dfrac{1}{x-1}+1}{\dfrac{1}{x-1}} = x$;

Domain f = Range f^{-1} = All real numbers except 1;

Range f = Domain f^{-1} = All real numbers except 0

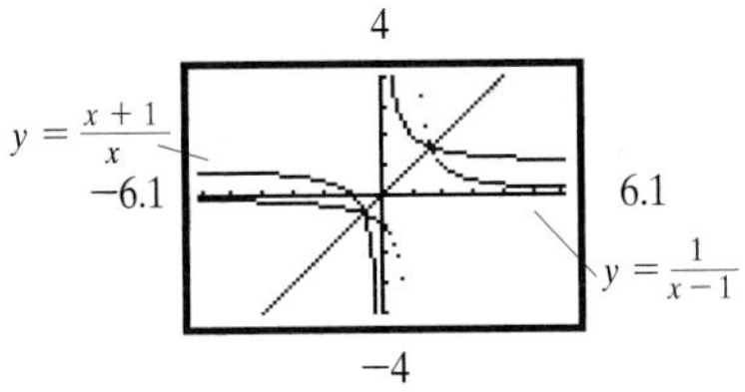

5. $f^{-1}(x) = \dfrac{27}{x^3}$; $f(f^{-1}(x)) = \dfrac{3}{\left(\dfrac{27}{x^3}\right)^{1/3}} = x$;

$f^{-1}(f(x)) = \dfrac{27}{\left(\dfrac{3}{x^{1/3}}\right)^3} = x$;

Domain f = Range f^{-1} = All real numbers except 0;
Range f = Domain f^{-1} = All real numbers except 0

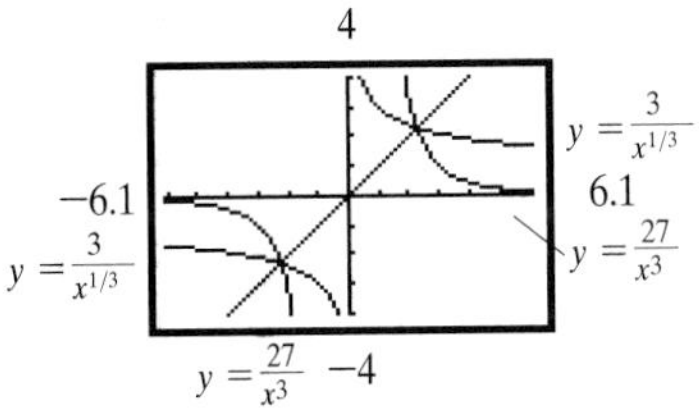

7. -3 **9.** $\sqrt{2}$ **11.** 0.4

13. $\log_3 u + 2\log_3 v - \log_3 w$

15. $2\log x + \frac{1}{2}\log(x^3+1)$

17. $\ln x + \frac{1}{3}\ln(x^2+1) - \ln(x-3)$

19. $\frac{25}{4}\log_4 x$ **21.** $-2\ln(x+1)$

23. $\log\left(\dfrac{4x^3}{[(x+3)(x-2)]^{1/2}}\right)$

25. 2.124 **27.** $y = Ce^{2x^2}$

29. $y = \sqrt{e^{x+C}+9}$

31. $y = \ln(x^2+4) - C$

33.

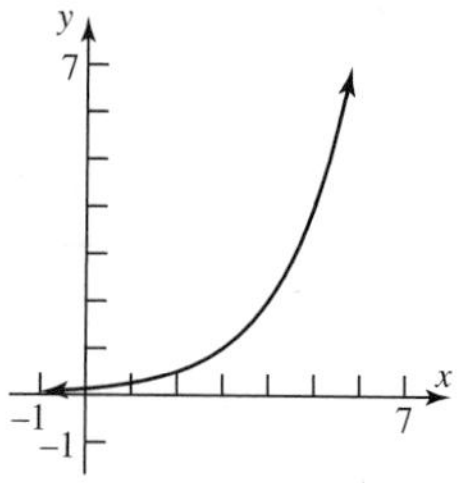

Domain: All real numbers
Range: $\{y|y > 0\}$ or $(0, \infty)$
Horizontal Asymptote: $y = 0$

35.

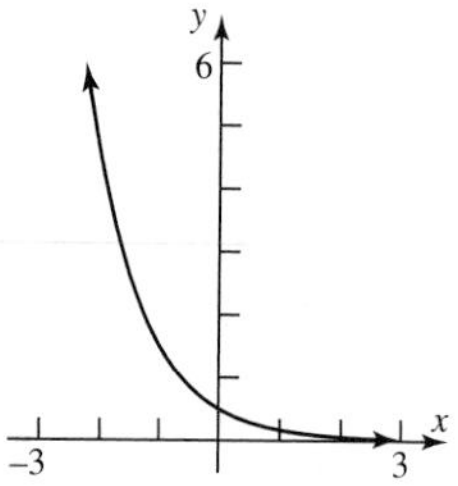

Domain: All real numbers
Range: $\{y|y > 0\}$ or $(0, \infty)$
Horizontal Asymptote: $y = 0$

37.

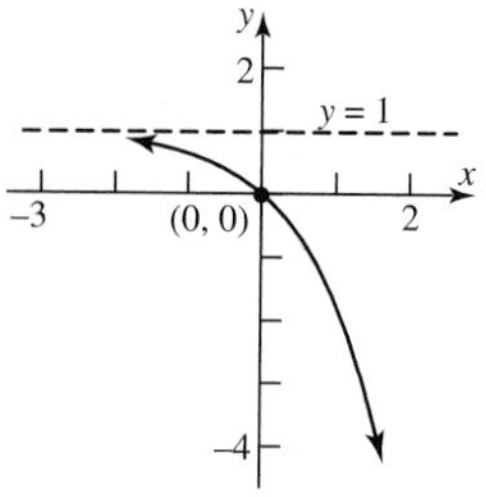

Domain: $(-\infty, \infty)$
Range: $(-\infty, 1)$
Asymptote: $y = 1$

39.

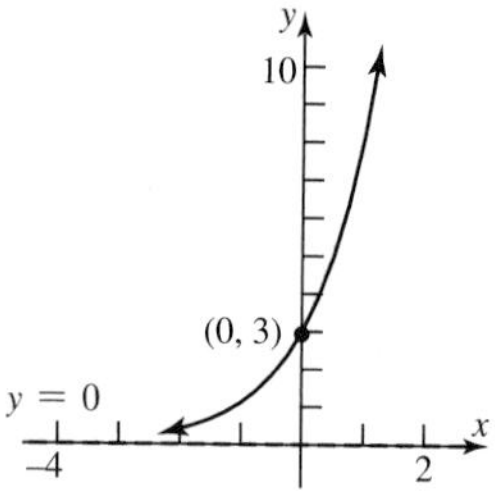

Domain: $(-\infty, \infty)$
Range: $(0, \infty)$
Asymptote: $y = 0$

41.

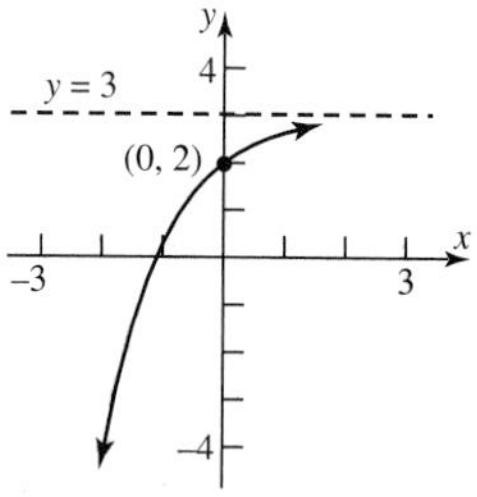

Domain: $(-\infty, \infty)$
Range: $(-\infty, 3)$
Asymptote: $y = 3$

43. $\frac{1}{4}$ **45.** $\left\{\frac{-1-\sqrt{3}}{2}, \frac{-1+\sqrt{3}}{2}\right\}$ **47.** $\frac{1}{4}$ **49.** $\frac{2 \ln 3}{\ln 5 - \ln 3} \approx 4.301$ **51.** $\frac{12}{5}$

53. 83 **55.** $\left\{\frac{1}{2}, -3\right\}$ **57.** -1 **59.** $1 - \ln 5 \approx -0.609$ **61.** $\frac{\ln 3}{3 \ln 2 - 2 \ln 3} \approx -9.327$

63. 3229.5 meters **65.** **(a)** 37.3 watts **(b)** 6.9 decibels

67. **(a)** 9.85 years **(b)** 4.27 years **69.** \$41,669 **71.** 24,203 years ago

73. 6,078,190,457

75. **(a)** 0.3
(b) 0.8
(c)

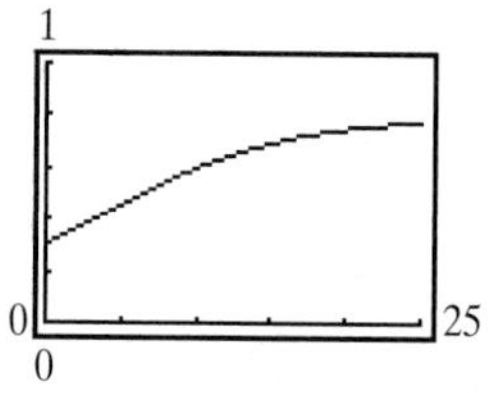

d) About 20.1 years

77. **(a)**

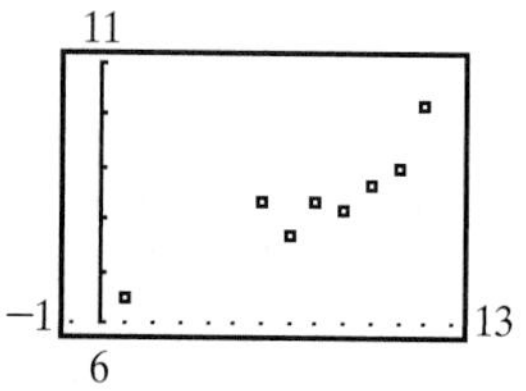

(b) Exponential model: Per capita usage $= 6.28 \times (1.03557)^{\text{year}}$
(c) 9.9 carrots

CHAPTER 6 Trigonometric Functions

6.1 Exercises

1.

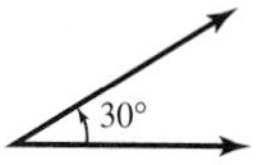

3.

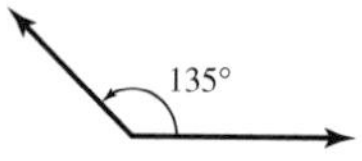

5.

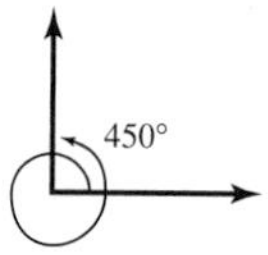

7.

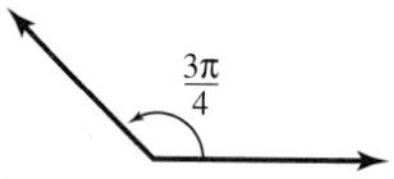

9.

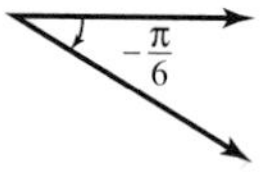

11.

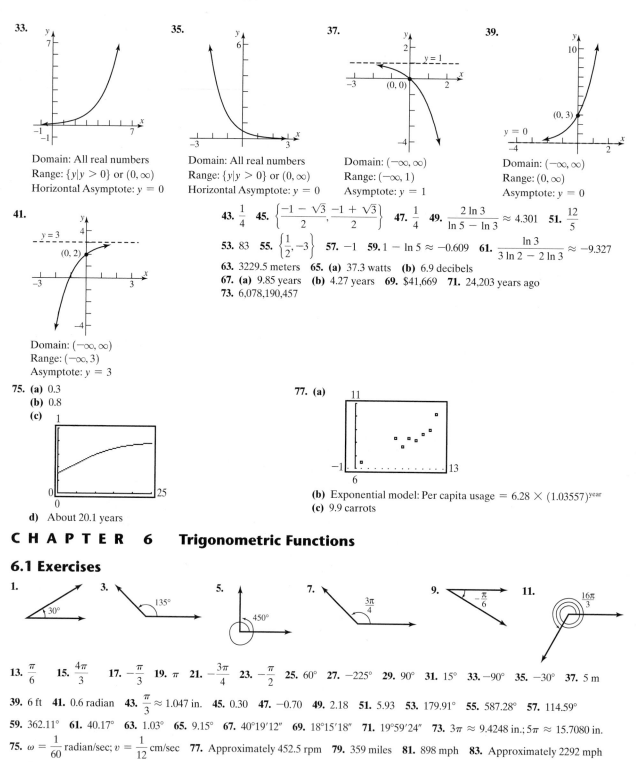

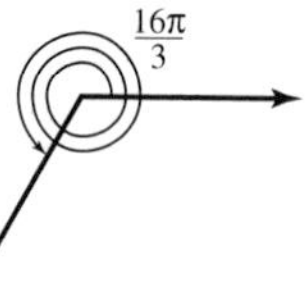

13. $\frac{\pi}{6}$ **15.** $\frac{4\pi}{3}$ **17.** $-\frac{\pi}{3}$ **19.** π **21.** $-\frac{3\pi}{4}$ **23.** $-\frac{\pi}{2}$ **25.** 60° **27.** −225° **29.** 90° **31.** 15° **33.** −90° **35.** −30° **37.** 5 m

39. 6 ft **41.** 0.6 radian **43.** $\frac{\pi}{3} \approx 1.047$ in. **45.** 0.30 **47.** −0.70 **49.** 2.18 **51.** 5.93 **53.** 179.91° **55.** 587.28° **57.** 114.59°

59. 362.11° **61.** 40.17° **63.** 1.03° **65.** 9.15° **67.** 40°19′12″ **69.** 18°15′18″ **71.** 19°59′24″ **73.** $3\pi \approx 9.4248$ in.; $5\pi \approx 15.7080$ in.

75. $\omega = \frac{1}{60}$ radian/sec; $v = \frac{1}{12}$ cm/sec **77.** Approximately 452.5 rpm **79.** 359 miles **81.** 898 mph **83.** Approximately 2292 mph

85. $\frac{3}{4}$ rpm **87.** Approximately 2.86 mph **89.** Approximately 31.47 rpm **91.** Approximately 1037 mph

93. $v_1 = r_1 w_1$, $v_2 = r_2 w_2$, and $v_1 = v_2$ so $r_1 w_1 = r_2 w_2 \Rightarrow \frac{r_1}{r_2} = \frac{w_2}{w_1}$.

6.2 Exercises

1. $\sin t = \frac{\sqrt{15}}{4}$; $\cos t = \frac{1}{4}$; $\tan t = \sqrt{15}$; $\csc t = \frac{4\sqrt{15}}{15}$; $\sec t = 4$; $\cot t = \frac{\sqrt{15}}{15}$

3. $\sin t = \frac{\sqrt{21}}{5}$; $\cos t = -\frac{2}{5}$; $\tan t = \frac{-\sqrt{21}}{2}$; $\csc t = \frac{5\sqrt{21}}{21}$; $\sec t = -\frac{5}{2}$; $\cot t = \frac{-2\sqrt{21}}{21}$

5. $\sin t = -\frac{1}{6}$; $\cos t = \frac{-\sqrt{35}}{6}$; $\tan t = \frac{\sqrt{35}}{35}$; $\csc t = -6$; $\sec t = \frac{-6\sqrt{35}}{35}$; $\cot t = \sqrt{35}$

7. $\sin t = -\frac{1}{3}$; $\cos t = \frac{2\sqrt{2}}{3}$; $\tan t = \frac{-\sqrt{2}}{4}$; $\csc t = -3$; $\sec t = \frac{3\sqrt{2}}{4}$; $\cot t = -2\sqrt{2}$

9. $\sin t = \frac{2}{7}$; $\cos t = -\frac{3\sqrt{5}}{7}$; $\tan t = -\frac{2\sqrt{5}}{15}$; $\csc t = \frac{7}{2}$; $\sec t = -\frac{7\sqrt{5}}{15}$; $\cot t = -\frac{3\sqrt{5}}{2}$ **11.** $\frac{1}{2}(\sqrt{2}+1)$

13. 2 **15.** $\frac{1}{2}$ **17.** $\sqrt{6}$ **19.** 4 **21.** 0 **23.** 0 **25.** $2\sqrt{2} + \frac{4\sqrt{3}}{3}$ **27.** -1 **29.** 1

31. $\sin\left(\frac{2\pi}{3}\right) = \frac{\sqrt{3}}{2}$; $\cos\left(\frac{2\pi}{3}\right) = -\frac{1}{2}$; $\tan\left(\frac{2\pi}{3}\right) = -\sqrt{3}$; $\csc\left(\frac{2\pi}{3}\right) = \frac{2\sqrt{3}}{3}$; $\sec\left(\frac{2\pi}{3}\right) = -2$; $\cot\left(\frac{2\pi}{3}\right) = -\frac{\sqrt{3}}{3}$

33. $\sin 210° = -\frac{1}{2}$; $\cos 210° = -\frac{\sqrt{3}}{2}$; $\tan 210° = \frac{\sqrt{3}}{3}$; $\csc 210° = -2$; $\sec 210° = -\frac{2\sqrt{3}}{3}$; $\cot 210° = \sqrt{3}$

35. $\sin\frac{5\pi}{3} = -\frac{\sqrt{3}}{2}$; $\cos\frac{5\pi}{3} = \frac{1}{2}$; $\tan\frac{5\pi}{3} = -\sqrt{3}$; $\csc\frac{5\pi}{3} = \frac{-2\sqrt{3}}{3}$; $\sec\frac{5\pi}{3} = 2$; $\cot\frac{5\pi}{3} = -\frac{\sqrt{3}}{3}$

37. $\sin\frac{7\pi}{3} = \frac{\sqrt{3}}{2}$; $\cos\frac{7\pi}{3} = \frac{1}{2}$; $\tan\frac{7\pi}{3} = \sqrt{3}$; $\csc\frac{7\pi}{3} = \frac{2\sqrt{3}}{3}$; $\sec\frac{7\pi}{3} = 2$; $\cot\frac{7\pi}{3} = \frac{\sqrt{3}}{3}$

39. $\sin 405° = \frac{\sqrt{2}}{2}$; $\cos 405° = \frac{\sqrt{2}}{2}$; $\tan 405° = 1$; $\csc 405° = \sqrt{2}$; $\sec 405° = \sqrt{2}$; $\cot 405° = 1$

41. $\sin\left(-\frac{\pi}{6}\right) = -\frac{1}{2}$; $\cos\left(-\frac{\pi}{6}\right) = \frac{\sqrt{3}}{2}$; $\tan\left(-\frac{\pi}{6}\right) = -\frac{\sqrt{3}}{3}$; $\csc\left(-\frac{\pi}{6}\right) = -2$; $\sec\left(-\frac{\pi}{6}\right) = \frac{2\sqrt{3}}{3}$; $\cot\left(-\frac{\pi}{6}\right) = -\sqrt{3}$

43. $\sin(-45°) = -\frac{\sqrt{2}}{2}$; $\cos(-45°) = \frac{\sqrt{2}}{2}$; $\tan(-45°) = -1$; $\csc(-45°) = -\sqrt{2}$; $\sec(-45°) = \sqrt{2}$; $\cot(-45°) = -1$

45. $\sin\left(\frac{5\pi}{2}\right) = 1$; $\cos\left(\frac{5\pi}{2}\right) = 0$; $\tan\left(\frac{5\pi}{2}\right)$ is not defined; $\csc\left(\frac{5\pi}{2}\right) = 1$; $\sec\left(\frac{5\pi}{2}\right)$ is not defined; $\cot\left(\frac{5\pi}{2}\right) = 0$

47. $\sin(-180°) = 0$; $\cos(-180°) = -1$; $\tan(-180°) = 0$; $\csc(-180°)$ is not defined; $\sec(-180°) = -1$; $\cot(-180°)$ is not defined

49. $\sin\left(-\frac{\pi}{2}\right) = -1$; $\cos\left(-\frac{\pi}{2}\right) = 0$; $\tan\left(-\frac{\pi}{2}\right)$ is not defined; $\csc\left(-\frac{\pi}{2}\right) = -1$; $\sec\left(-\frac{\pi}{2}\right)$ is not defined; $\cot\left(-\frac{\pi}{2}\right) = 0$

51. $\sin 480° = \frac{\sqrt{3}}{2}$; $\cos 480° = -\frac{1}{2}$; $\tan 480° = -\sqrt{3}$; $\csc 480° = \frac{2\sqrt{3}}{3}$; $\sec 480° = -2$; $\cot 480° = -\frac{\sqrt{3}}{3}$ **53.** 0.47 **55.** 0.38 **57.** 1.33

59. 0.36 **61.** 0.31 **63.** 3.73 **65.** 1.04 **67.** 0.84 **69.** 0.02 **71.** $\frac{\sqrt{3}}{2}$ **73.** $\frac{1}{2}$ **75.** $\frac{3}{4}$ **77.** $\frac{\sqrt{3}}{2}$ **79.** $\sqrt{3}$ **81.** $-\frac{\sqrt{3}}{2}$ **83.** 1

85. **(a)** $\sin 1 \approx 0.8$; $\cos 1 \approx 0.5$; $\tan 1 \approx 1.6$; $\csc 1 \approx 1.25$; $\sec 1 \approx 2$; $\cot 1 \approx 0.625$
(b) $\sin 5.1 \approx -0.9$; $\cos 5.1 \approx 0.4$; $\tan 5.1 \approx -2.4$; $\csc 5.1 \approx -1.1$; $\sec 5.1 \approx 2.5$; $\cot 5.1 \approx 0.4$
(c) $\sin 2.4 \approx 0.7$; $\cos 2.4 \approx -0.7$; $\tan 2.4 \approx -1$; $\csc 2.4 \approx 1.4$; $\sec 2.4 \approx -1.4$; $\cot 2.4 \approx -1$

87. **(a)** $\sin 1.5 \approx 1$; $\cos 1.5 \approx 0.1$; $\tan 1.5 = 10$; $\csc 1.5 = 1$; $\sec 1.5 = 10$; $\cot 1.5 = 0.1$
(b) $\sin 4.3 = -0.9$; $\cos 4.3 = -0.4$; $\tan 4.3 = 2.25$; $\csc 4.3 = -1.1$; $\sec 4.3 = -2.5$; $\cot 4.3 = 0.4$
(c) $\sin 5.3 = -0.8$; $\cos 5.3 = 0.6$; $\tan 5.3 = -1.3$; $\csc 5.3 = -1.25$; $\sec 5.3 = 1.7$; $\cot 5.3 = -0.75$

89. $\sin\theta = \frac{4}{5}$; $\cos\theta = -\frac{3}{5}$; $\tan\theta = -\frac{4}{3}$; $\csc\theta = \frac{5}{4}$; $\sec\theta = -\frac{5}{3}$; $\cot\theta = -\frac{3}{4}$

91. $\sin\theta = -\frac{3\sqrt{13}}{13}$; $\cos\theta = \frac{2\sqrt{13}}{13}$; $\tan\theta = -\frac{3}{2}$; $\csc\theta = -\frac{\sqrt{13}}{3}$; $\sec\theta = \frac{\sqrt{13}}{2}$; $\cot\theta = -\frac{2}{3}$

93. $\sin\theta = -\frac{\sqrt{2}}{2}$; $\cos\theta = -\frac{\sqrt{2}}{2}$; $\tan\theta = 1$; $\csc\theta = -\sqrt{2}$; $\sec\theta = -\sqrt{2}$; $\cot\theta = 1$

95. $\sin\theta = -\frac{2\sqrt{13}}{13}$; $\cos\theta = -\frac{3\sqrt{13}}{13}$; $\tan\theta = \frac{2}{3}$; $\csc\theta = -\frac{\sqrt{13}}{2}$; $\sec\theta = -\frac{\sqrt{13}}{3}$; $\cot\theta = \frac{3}{2}$

97. $\sin\theta = -\frac{3}{5}$; $\cos\theta = \frac{4}{5}$; $\tan\theta = -\frac{3}{4}$; $\csc\theta = -\frac{5}{3}$; $\sec\theta = \frac{5}{4}$; $\cot\theta = -\frac{4}{3}$ **99.** 0 **101.** -0.1 **103.** 3 **105.** 5

107. $R \approx 310.56$ ft; $H \approx 77.64$ ft **109.** $R \approx 19{,}542$ m; $H \approx 2278$ m **111.** **(a)** 1.2 sec **(b)** 1.11 sec **(c)** 1.2 sec
113. **(a)** 1.9 hr; 0.57 hr **(b)** 1.69 hr; 0.75 hr **(c)** 1.63 hr; 0.86 hr **(d)** 1.67 hr

115. (a) 16.6 ft **(b)**

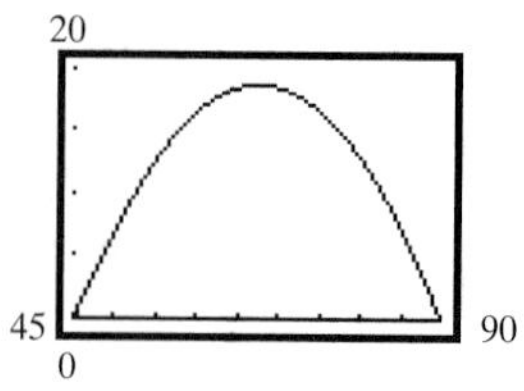

(c) 67.5°

6.3 Exercises

1. $\frac{\sqrt{2}}{2}$ **3.** 1 **5.** 1 **7.** $\sqrt{3}$ **9.** $\frac{\sqrt{2}}{2}$ **11.** 0 **13.** $\sqrt{2}$ **15.** $\frac{\sqrt{3}}{3}$ **17.** II **19.** IV **21.** IV **23.** II

25. $\tan\theta = 2$; $\cot\theta = \frac{1}{2}$; $\sec\theta = \sqrt{5}$; $\csc\theta = \frac{\sqrt{5}}{2}$ **27.** $\tan\theta = \frac{\sqrt{3}}{3}$; $\cot\theta = \sqrt{3}$; $\sec\theta = \frac{2\sqrt{3}}{3}$; $\csc\theta = 2$

29. $\tan\theta = -\frac{\sqrt{2}}{4}$; $\cot\theta = -2\sqrt{2}$; $\sec\theta = \frac{3\sqrt{2}}{4}$; $\csc\theta = -3$ **31.** $\tan\theta \approx 0.2679$; $\cot\theta \approx 3.7322$; $\sec\theta \approx 1.0353$; $\csc\theta \approx 3.8640$

33. $\cos\theta = -\frac{5}{13}$; $\tan\theta = -\frac{12}{5}$; $\csc\theta = \frac{13}{12}$; $\sec\theta = -\frac{13}{5}$; $\cot\theta = -\frac{5}{12}$ **35.** $\sin\theta = -\frac{3}{5}$; $\tan\theta = \frac{3}{4}$; $\csc\theta = -\frac{5}{3}$; $\sec\theta = -\frac{5}{4}$; $\cot\theta = \frac{4}{3}$

37. $\cos\theta = -\frac{12}{13}$; $\tan\theta = -\frac{5}{12}$; $\cot\theta = -\frac{12}{5}$; $\sec\theta = -\frac{13}{12}$; $\csc\theta = \frac{13}{5}$

39. $\sin\theta = \frac{2\sqrt{2}}{3}$; $\tan\theta = -2\sqrt{2}$; $\cot\theta = -\frac{\sqrt{2}}{4}$; $\sec\theta = -3$; $\csc\theta = \frac{3\sqrt{2}}{4}$

41. $\cos\theta = -\frac{\sqrt{5}}{3}$; $\tan\theta = -\frac{2\sqrt{5}}{5}$; $\cot\theta = -\frac{\sqrt{5}}{2}$; $\sec\theta = -\frac{3\sqrt{5}}{5}$; $\csc\theta = \frac{3}{2}$

43. $\sin\theta = -\frac{\sqrt{3}}{2}$; $\cos\theta = \frac{1}{2}$; $\tan\theta = -\sqrt{3}$; $\cot\theta = -\frac{\sqrt{3}}{3}$; $\csc\theta = -\frac{2\sqrt{3}}{3}$

45. $\sin\theta = -\frac{3}{5}$; $\cos\theta = -\frac{4}{5}$; $\cot\theta = \frac{4}{3}$; $\sec\theta = -\frac{5}{4}$; $\csc\theta = -\frac{5}{3}$

47. $\sin\theta = \frac{\sqrt{10}}{10}$; $\cos\theta = -\frac{3\sqrt{10}}{10}$; $\cot\theta = -3$; $\sec\theta = -\frac{\sqrt{10}}{3}$; $\csc\theta = \sqrt{10}$ **49.** $-\frac{\sqrt{3}}{2}$ **51.** $-\frac{\sqrt{3}}{3}$ **53.** 2 **55.** -1 **57.** -1 **59.** $\frac{\sqrt{2}}{2}$

61. 0 **63.** $-\sqrt{2}$ **65.** $\frac{2\sqrt{3}}{3}$ **67.** -1 **69.** -2 **71.** $\frac{2-\sqrt{2}}{2}$ **73.** 1 **75.** 1 **77.** 0 **79.** 0.9 **81.** 9 **83.** 0 **85.** All real numbers

87. Odd multiples of $\frac{\pi}{2}$ **89.** Odd multiples of $\frac{\pi}{2}$ **91.** $-1 \le y \le 1$ **93.** All real numbers **95.** $|y| \ge 1$ **97.** Odd; Yes; Origin

99. Odd; Yes; Origin **101.** Even; Yes; y-axis **103. (a)** $-\frac{1}{3}$ **(b)** 1 **105. (a)** -2 **(b)** 6 **107. (a)** -4 **(b)** -12 **109.** 15.8 min

111. Let $P = (x, y)$ be the point on the unit circle that corresponds to θ. Consider the equation $\tan\theta = \frac{y}{x} = a$. Then $y = ax$. But $x^2 + y^2 = 1$ so that $x^2 + a^2x^2 = 1$. Thus, $x = \frac{1}{\sqrt{1+a^2}}$ and $y = \frac{a}{\sqrt{1+a^2}}$; that is, for any real number a, there is a point $P = (x, y)$ in the unit circle for which $\tan\theta = a$. In other words, $-\infty < \tan\theta < \infty$, and the range of the tangent function is the set of all real numbers.

113. Suppose there is a number p, $0 < p < 2\pi$, for which $\sin(\theta + p) = \sin\theta$ for all θ. If $\theta = 0$, then $\sin(0 + p) = \sin p = \sin 0 = 0$; so that $p = \pi$. If $\theta = \frac{\pi}{2}$, then $\sin\left(\frac{\pi}{2} + p\right) = \sin\left(\frac{\pi}{2}\right)$. But $p = \pi$. Thus, $\sin\left(\frac{3\pi}{2}\right) = -1 = \sin\left(\frac{\pi}{2}\right) = 1$. This is impossible. The smallest positive number p for which $\sin(\theta + p) = \sin\theta$ for all θ is therefore $p = 2\pi$.

115. $\sec\theta = \frac{1}{(\cos\theta)}$; since $\cos\theta$ has period 2π, so does $\sec\theta$

117. If $P = (a, b)$ is the point on the unit circle corresponding to θ, then $Q = (-a, -b)$ is the point on the unit circle corresponding to $\theta + \pi$. Thus $\tan(\theta + \pi) = \frac{(-b)}{(-a)} = \frac{b}{a} = \tan\theta$; that is, the period of the tangent function is π.

119. Let $P = (a, b)$ be the point on the unit circle corresponding to θ. Then $\csc\theta = \frac{1}{b} = \frac{1}{(\sin\theta)}$; $\sec\theta = \frac{1}{a} = \frac{1}{(\cos\theta)}$; $\cot\theta = \frac{a}{b} = \frac{1}{b/a} = \frac{1}{(\tan\theta)}$.

121.
$$(\sin\theta\cos\phi)^2 + (\sin\theta\sin\phi)^2 + \cos^2\theta = \sin^2\theta\cos^2\phi + \sin^2\theta\sin^2\phi + \cos^2\theta$$
$$= \sin^2\theta(\cos^2\phi + \sin^2\phi) + \cos^2\theta = \sin^2\theta + \cos^2\theta = 1$$

6.4 Exercises

1. 0 **3.** $-\frac{\pi}{2} \le x \le \frac{\pi}{2}$ **5.** 1 **7.** $0, \pi, 2\pi$ **9.** $\sin x = 1$ for $x = -\frac{3\pi}{2}, \frac{\pi}{2}$; $\sin x = -1$ for $x = -\frac{\pi}{2}, \frac{3\pi}{2}$ **11.** B, C, F **13.** C **15.** D

17.

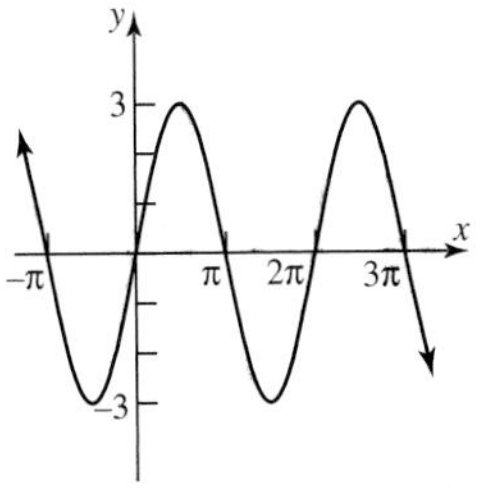

19.

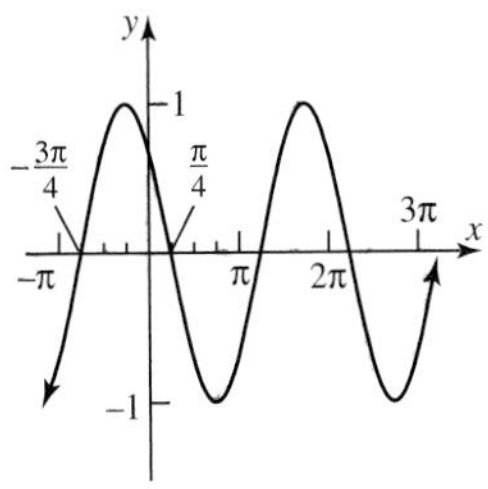

21.

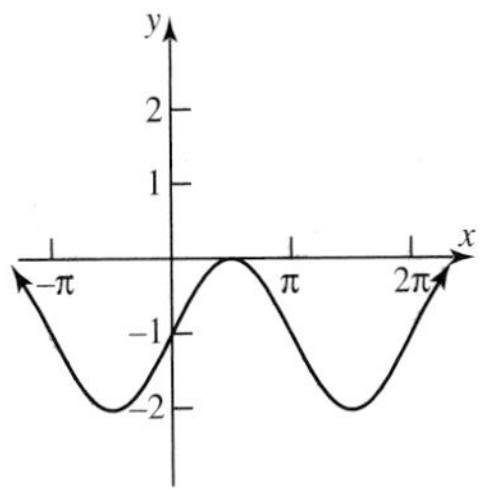

23.

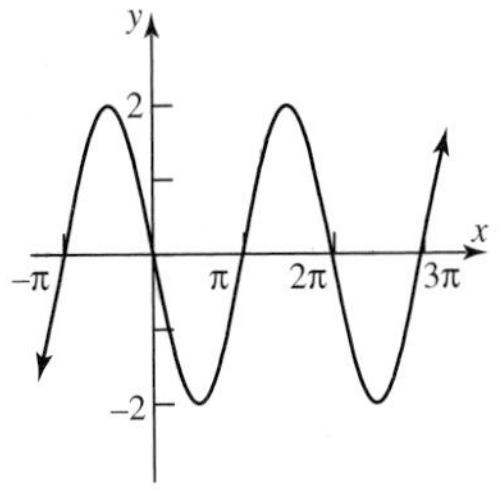

25.

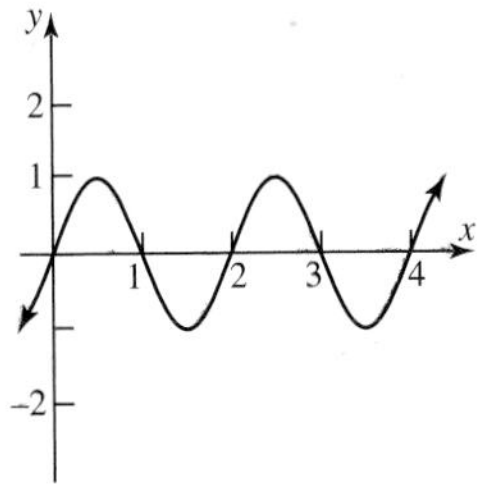

27.

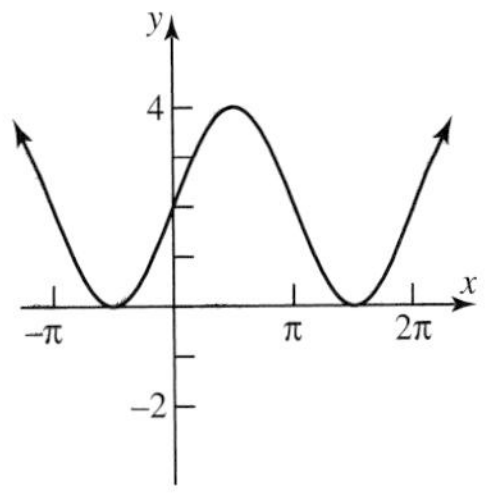

29.

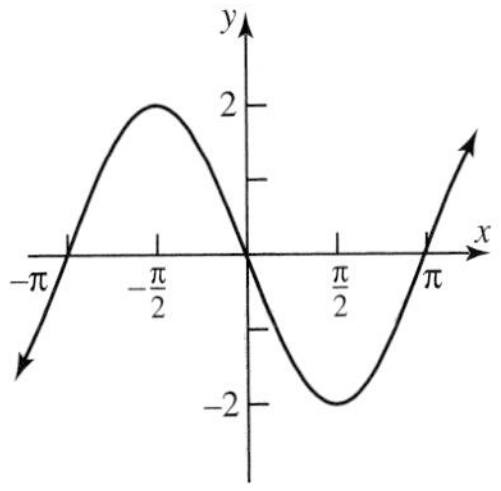

31.

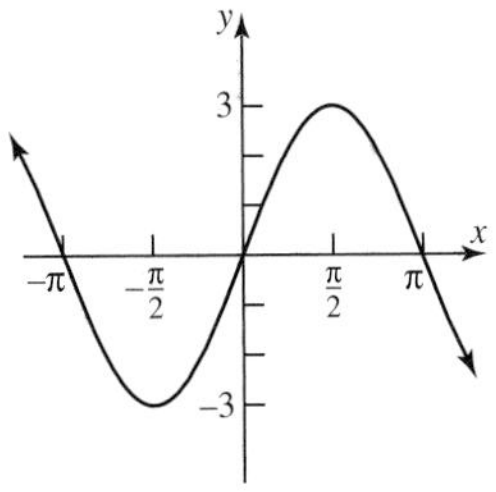

33. The graph of $y = A \sin x$ is a vertical stretch of the graph of $y = \sin x$ by a factor of A.

35. The graph of $y = \sin(x - \phi), \phi > 0$ is the graph of $y = \sin x$ shifted ϕ units to the right.

37.

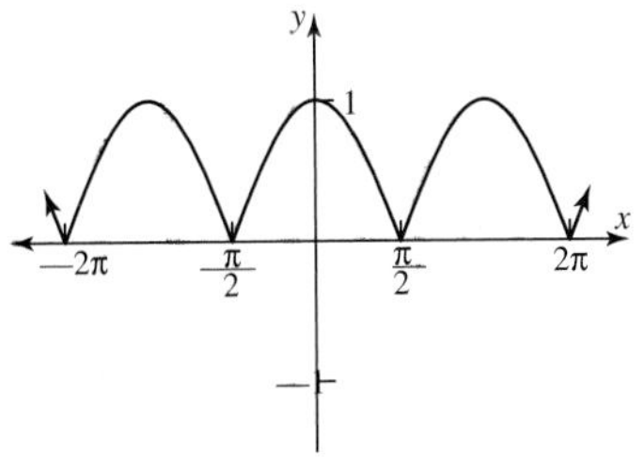

6.5 Exercises

1. 0 **3.** 1 **5.** $\sec x = 1$ for $x = -2\pi, 0, 2\pi$; $\sec x = -1$ for $x = -\pi, \pi$ **7.** $-\frac{3\pi}{2}, -\frac{\pi}{2}, \frac{\pi}{2}, \frac{3\pi}{2}$ **9.** $-\frac{3\pi}{2}, -\frac{\pi}{2}, \frac{\pi}{2}, \frac{3\pi}{2}$ **11.** B **13.** A

15.

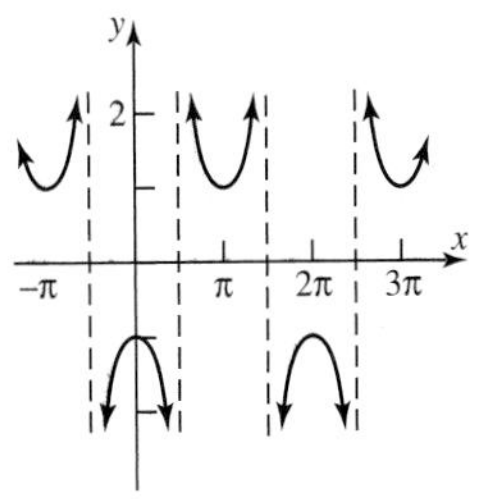

17.

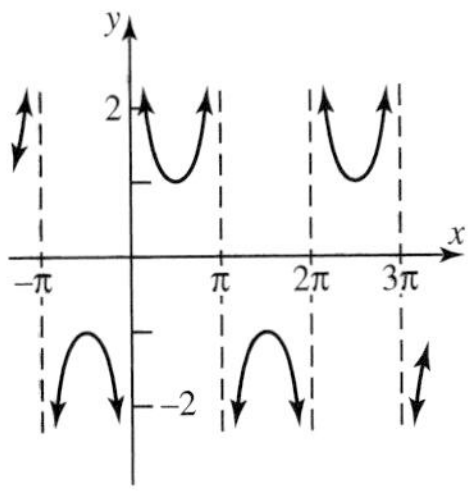

19.

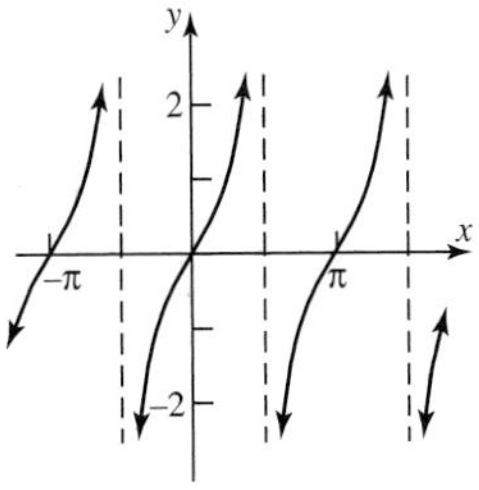

21.

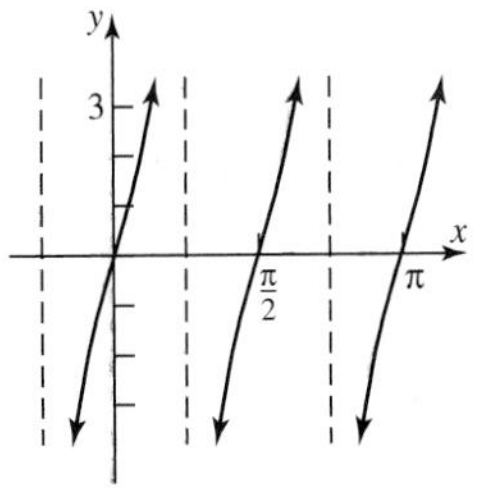

23.

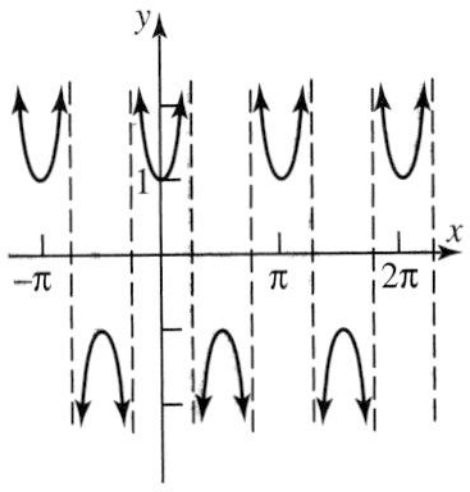

25.

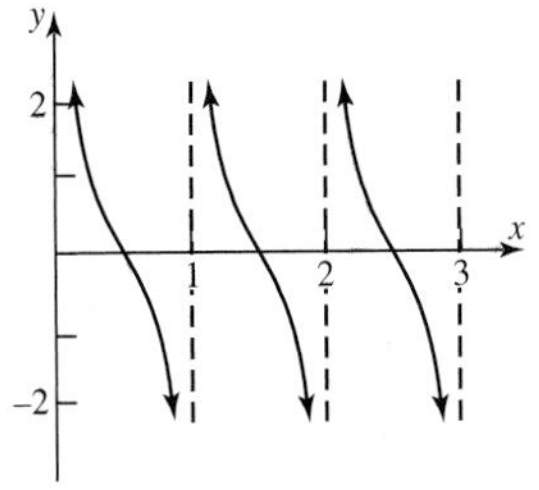

27.

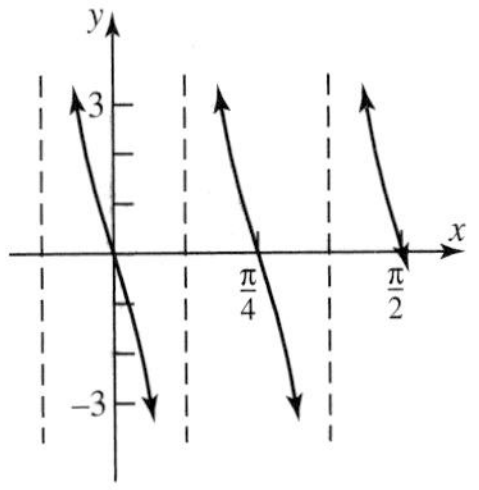

29.

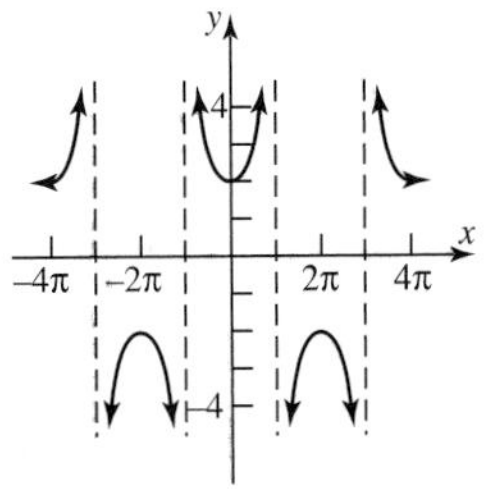

31.

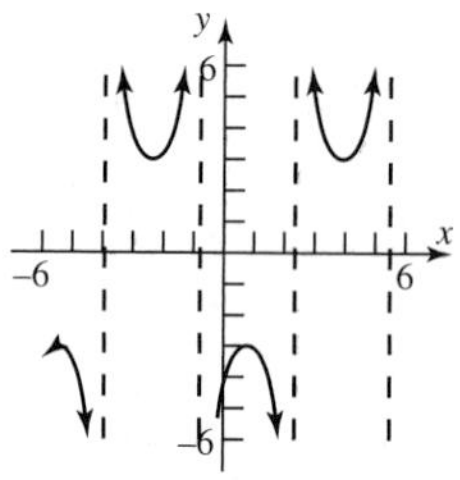

33.

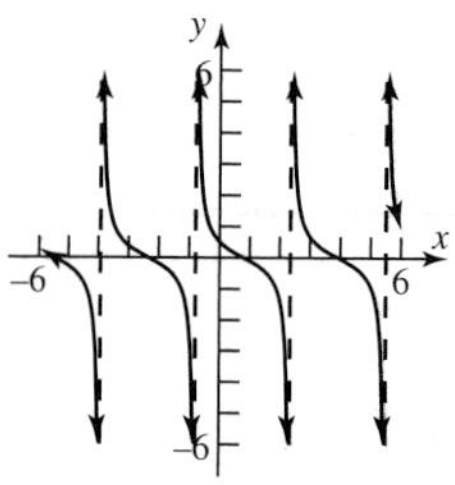

35. **(a)** $L = \dfrac{3}{\cos\theta} + \dfrac{4}{\sin\theta} = 3\sec\theta + 4\csc\theta$

(b)

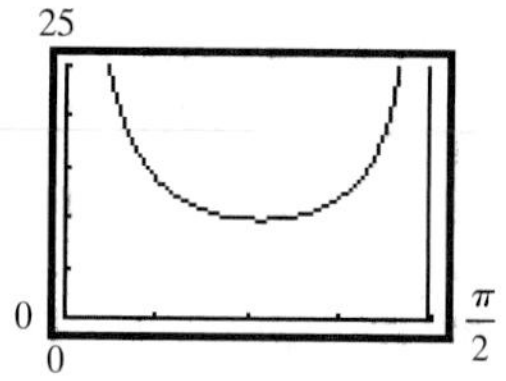

(c) L is the least when $\theta = 0.83$

(d) $L(0.83) = 9.87$ ft

6.6 Exercises

1. Amplitude = 2; Period = 2π **3.** Amplitude = 4; Period = π **5.** Amplitude = 6; Period = 2 **7.** Amplitude = $\frac{1}{2}$; Period = $\frac{4\pi}{3}$

9. Amplitude = $\frac{5}{3}$; Period = 3 **11.** F **13.** A **15.** H **17.** C **19.** J **21.** A **23.** D **25.** B

27.

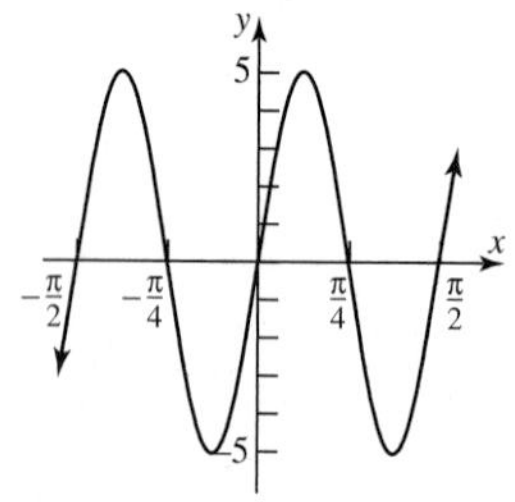

29.

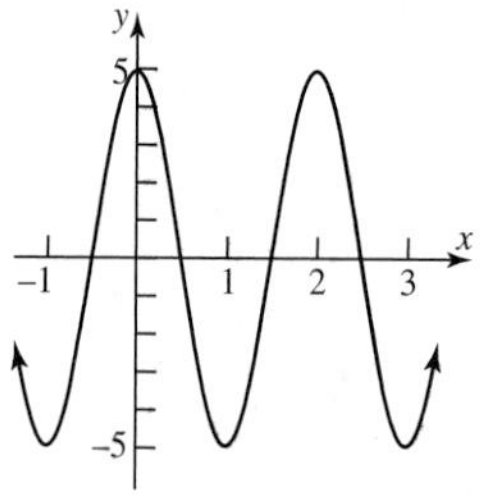

31.

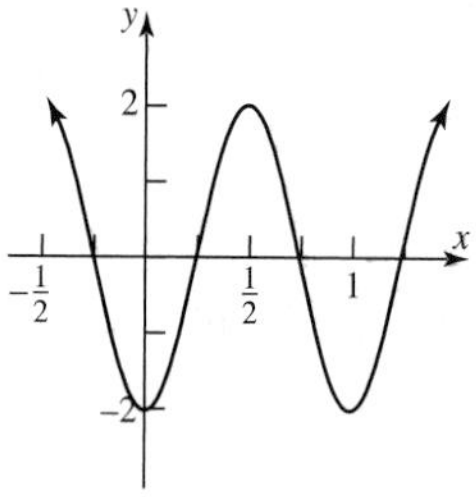

33.

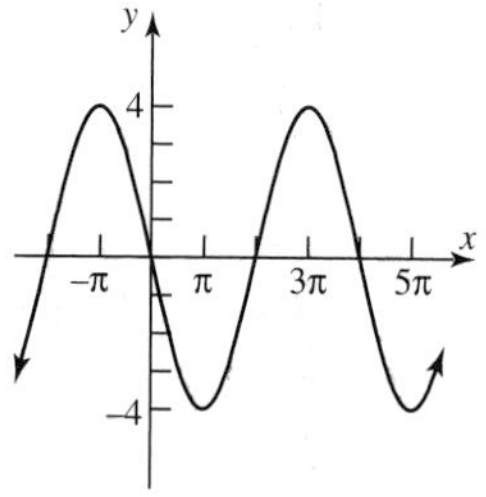

35.

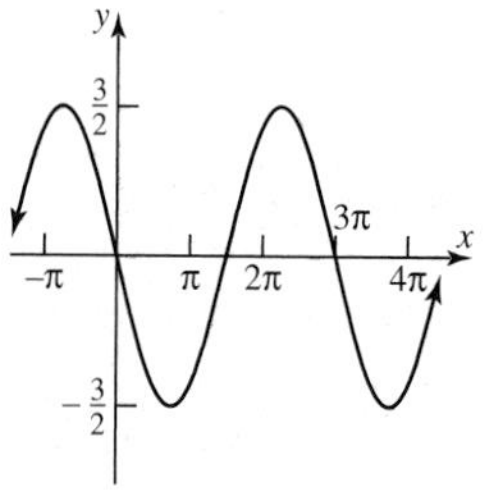

37. $y = 5\cos\frac{\pi}{4}x$ **39.** $y = -3\cos\frac{1}{2}x$ **41.** $y = \frac{3}{4}\sin 2\pi x$ **43.** $y = -\sin\frac{3}{2}x$

45. $y = -2\cos\frac{3\pi}{2}x$ **47.** $y = 3\sin\frac{\pi}{2}x$ **49.** $y = -4\cos 3x$

51. Amplitude = 4

Period = π

Phase shift = $\frac{\pi}{2}$

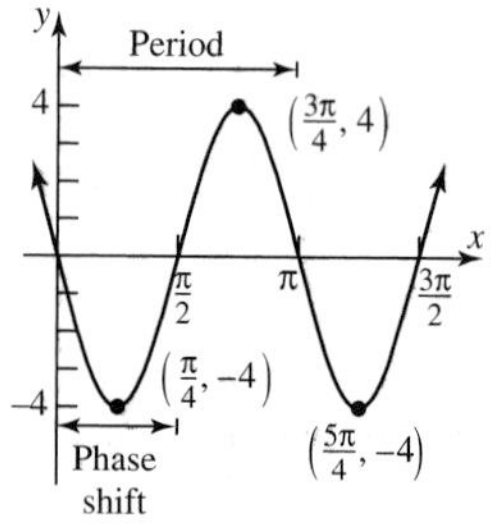

53. Amplitude = 2

Period = $\frac{2\pi}{3}$

Phase shift = $-\frac{\pi}{6}$

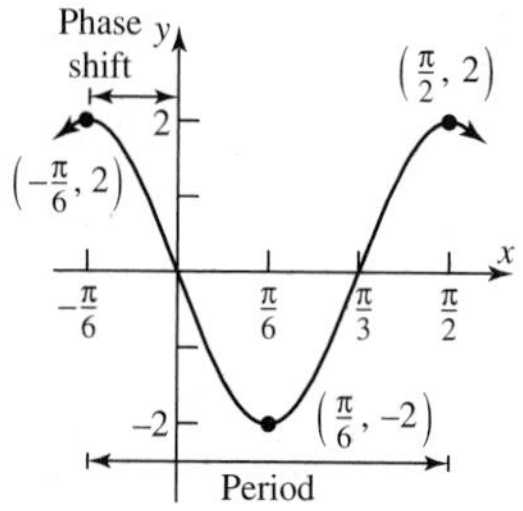

55. Amplitude = 3

Period = π

Phase shift = $-\frac{\pi}{4}$

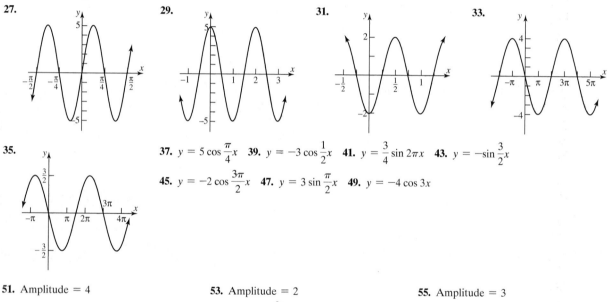

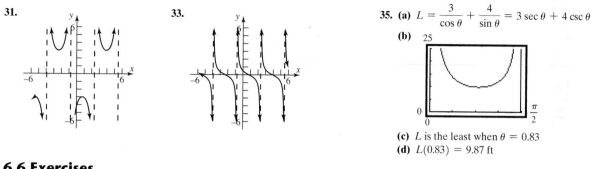

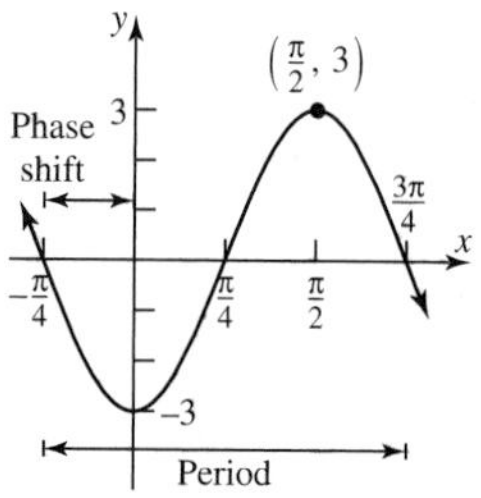

57. Amplitude = 4
Period = 2

Phase shift = $-\dfrac{2}{\pi}$

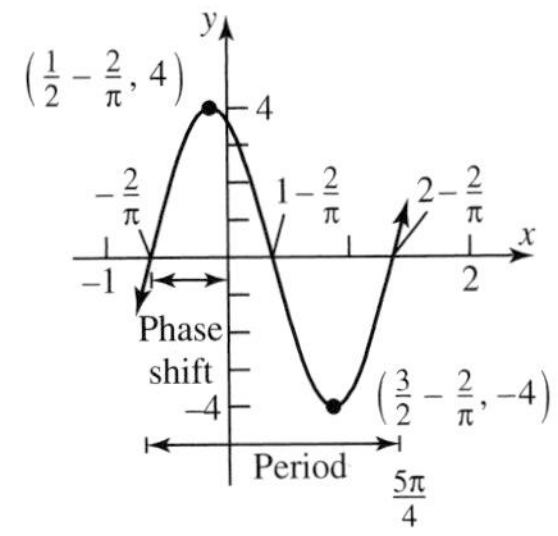

59. Amplitude = 3
Period = 2

Phase shift = $\dfrac{2}{\pi}$

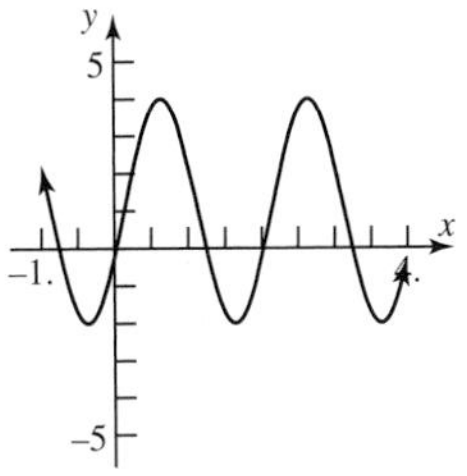

61. Amplitude = 3
Period = π

Phase shift = $\dfrac{\pi}{4}$

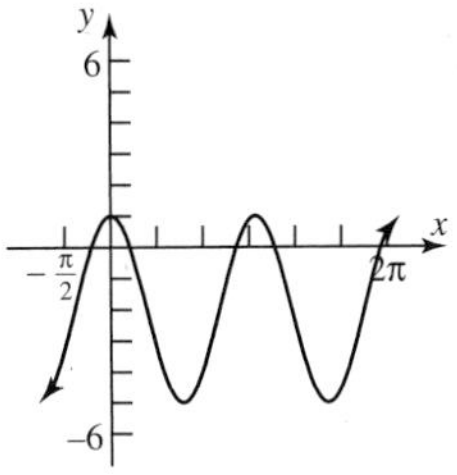

63. $y = 3\sin(2x)$ **65.** $y = 3\sin(\pi x)$ **67.** $y = 2\sin\left[2\left(x - \dfrac{1}{2}\right)\right]$ **69.** $y = 3\sin\left[\dfrac{2}{3}\left(x + \dfrac{1}{3}\right)\right]$

71. Period = $\dfrac{1}{30}$, Amplitude = 220

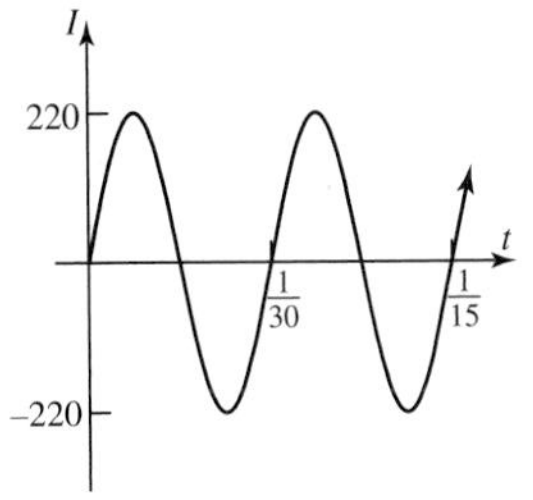

73. Period = $\dfrac{1}{15}$, Amplitude = 120, Phase shift = $\dfrac{1}{90}$

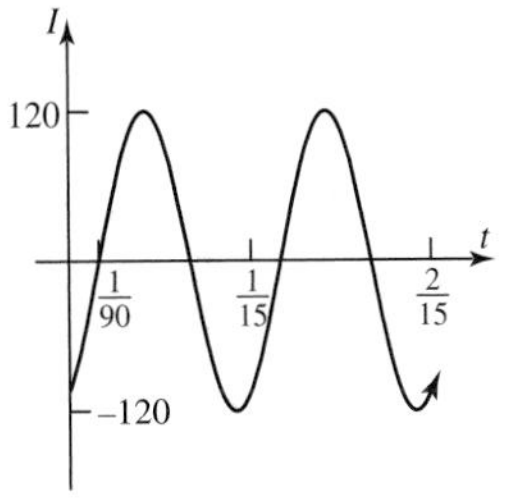

75. (a) Amplitude = 220, period = $\dfrac{1}{60}$

(b), (e)

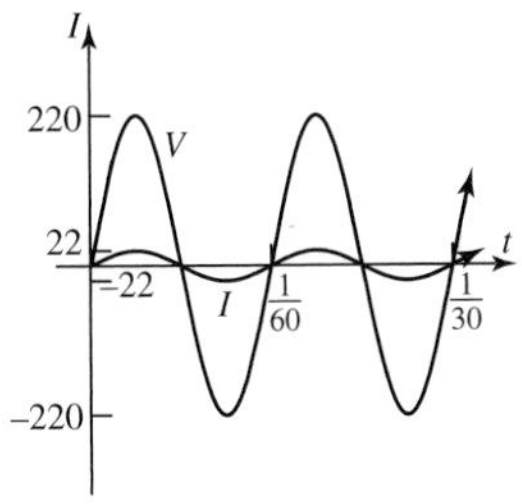

(c) $I = 22\sin 120\pi t$

(d) Amplitude = 22, Period = $\dfrac{1}{60}$

77. (a) $P = \dfrac{(V_0 \sin 2\pi f t)^2}{R} = \dfrac{V_0^2}{R}\sin^2 2\pi f t$

(b) $P = \dfrac{V_0^2}{R}\dfrac{1}{2}(1 - \cos 4\pi f t))$

79. (a)

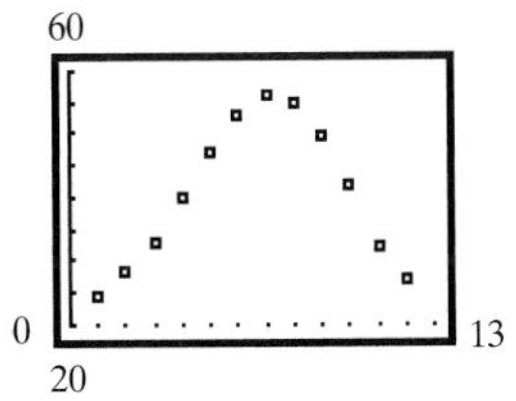

(c)

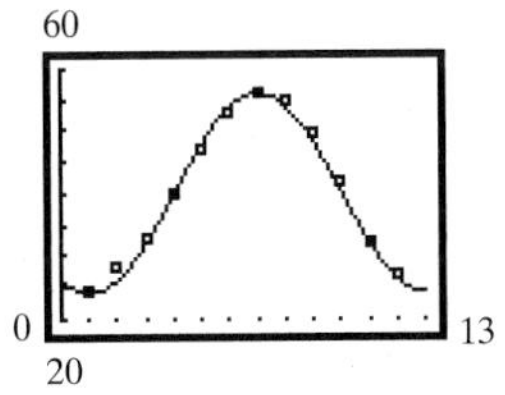

(e)

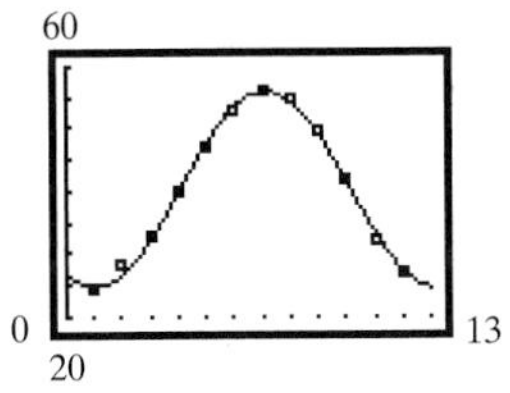

(b) $y = 15.9\sin\left(\dfrac{\pi}{6}x - \dfrac{2\pi}{3}\right) + 40.1$

(d) $y = 15.62\sin(0.517x - 2.096) + 40.377$

81. (a)

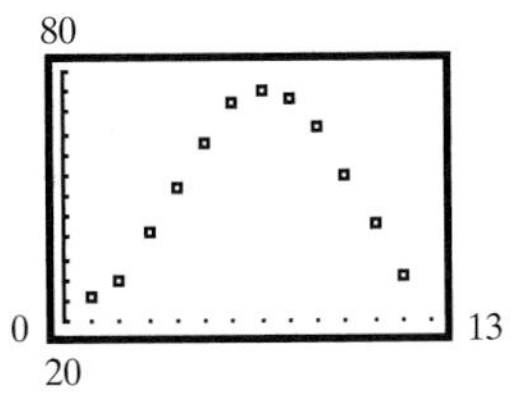

(c)

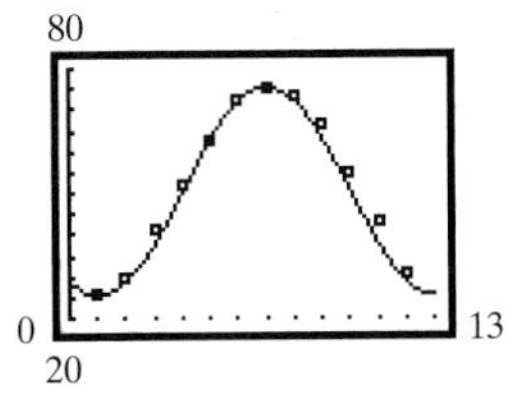

(e)

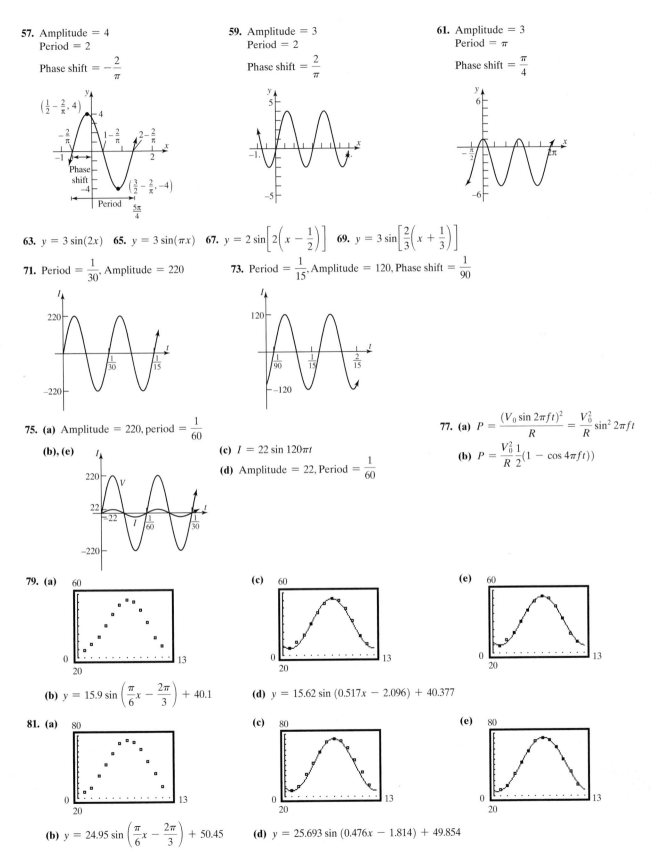

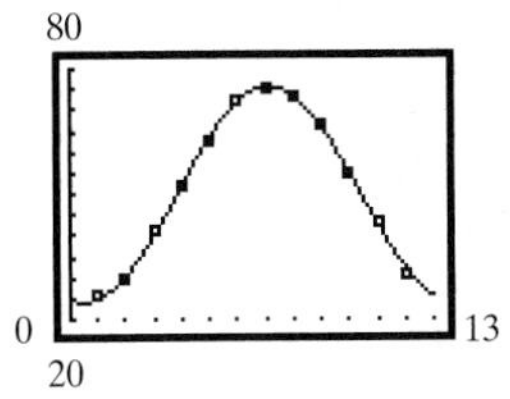

(b) $y = 24.95\sin\left(\dfrac{\pi}{6}x - \dfrac{2\pi}{3}\right) + 50.45$

(d) $y = 25.693\sin(0.476x - 1.814) + 49.854$

83. **(a)** 4:08 P.M.

(b) $y = 4.4 \sin\left(\frac{\pi}{6.25}x - 6.6643\right) + 3.8$

(c)

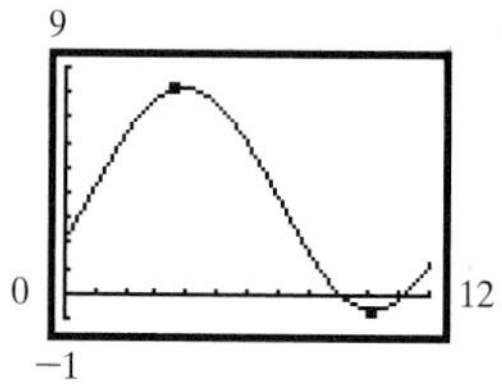

(d) 8.2 ft

85. **(a)** $y = 1.0835 \sin\left(\frac{2\pi}{365}x - 7.6812\right) + 11.6665$

(b)

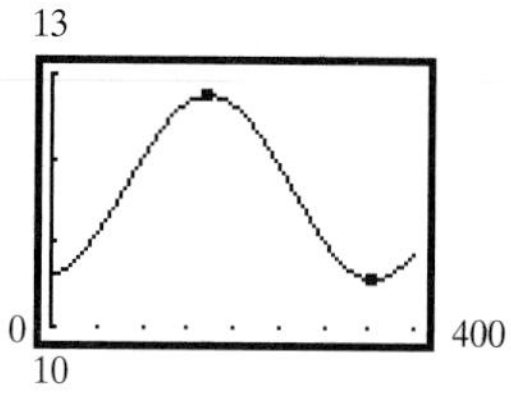

(c) 11.85 hrs

87. **(a)** $y = 5.3915 \sin\left(\frac{2\pi}{365}x - 7.6812\right) + 10.8415$ **(b)**

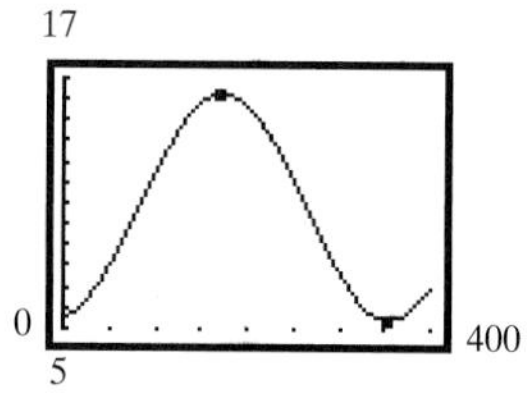

(c) 11.75 hrs

Fill-in-the Blanks

1. Angle; initial side; terminal side **3.** π **5.** 2π; π **7.** Sine, tangent, cosecant, cotangent **9.** $3; \frac{\pi}{3}$

True/False Items

1. False **3.** False **5.** True **7.** False

Review Exercises

1. $\frac{3\pi}{4}$ **3.** $\frac{\pi}{10}$ **5.** 135° **7.** −450° **9.** $\frac{1}{2}$ **11.** $\frac{3\sqrt{2}}{2} - \frac{4\sqrt{3}}{3}$ **13.** $-3\sqrt{2} - 2\sqrt{3}$ **15.** 3 **17.** 0 **19.** 0 **21.** 1 **23.** 1 **25.** 1

27. $\cos\theta = \frac{3}{5}$; $\tan\theta = -\frac{4}{3}$; $\csc\theta = -\frac{5}{4}$; $\sec\theta = \frac{5}{3}$; $\cot\theta = -\frac{3}{4}$ **29.** $\sin\theta = -\frac{12}{13}$; $\cos\theta = -\frac{5}{13}$; $\csc\theta = -\frac{13}{12}$; $\sec\theta = -\frac{13}{5}$; $\cot\theta = \frac{5}{12}$

31. $\sin\theta = \frac{3}{5}$; $\cos\theta = -\frac{4}{5}$; $\tan\theta = -\frac{3}{4}$; $\csc\theta = \frac{5}{3}$; $\cot\theta = -\frac{4}{3}$ **33.** $\cos\theta = -\frac{5}{13}$; $\tan\theta = -\frac{12}{5}$; $\csc\theta = \frac{13}{12}$; $\sec\theta = -\frac{13}{5}$; $\cot\theta = -\frac{5}{12}$

35. $\cos\theta = \frac{12}{13}$; $\tan\theta = -\frac{5}{12}$; $\csc\theta = -\frac{13}{5}$; $\sec\theta = \frac{13}{12}$; $\cot\theta = -\frac{12}{5}$

37. $\sin\theta = -\frac{\sqrt{10}}{10}$; $\cos\theta = -\frac{3\sqrt{10}}{10}$; $\csc\theta = -\sqrt{10}$; $\sec\theta = -\frac{\sqrt{10}}{3}$; $\cot\theta = 3$

39. $\sin\theta = -\frac{2\sqrt{2}}{3}$; $\cos\theta = \frac{1}{3}$; $\tan\theta = -2\sqrt{2}$; $\csc\theta = -\frac{3\sqrt{2}}{4}$; $\cot\theta = -\frac{\sqrt{2}}{4}$

41. $\sin\theta = \frac{\sqrt{5}}{5}$; $\cos\theta = -\frac{2\sqrt{5}}{5}$; $\tan\theta = -\frac{1}{2}$; $\csc\theta = \sqrt{5}$; $\sec\theta = -\frac{\sqrt{5}}{2}$

43.

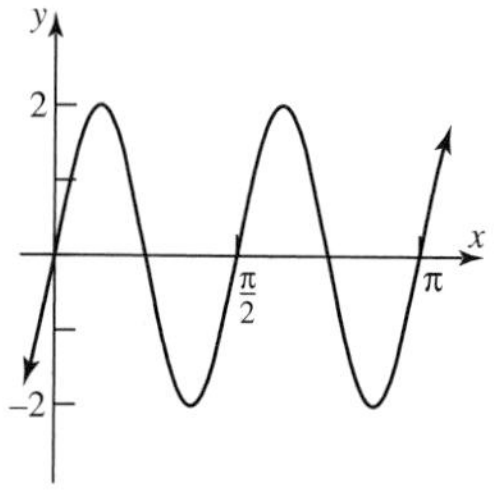

45.

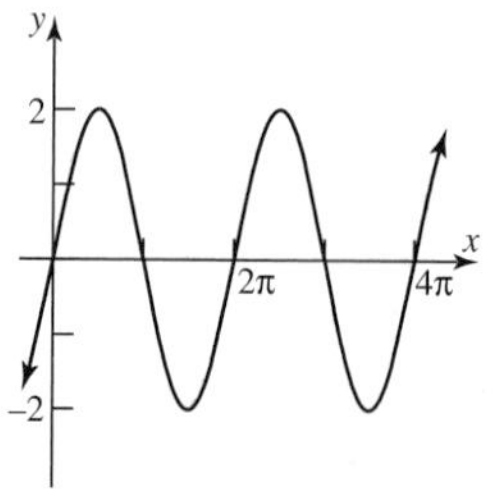

47.

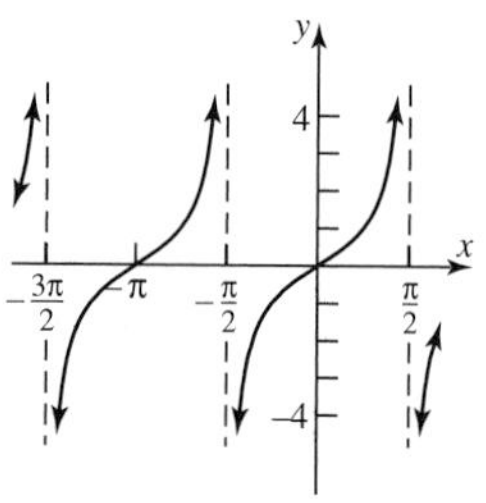

49.

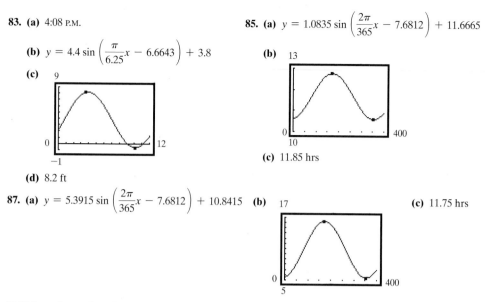

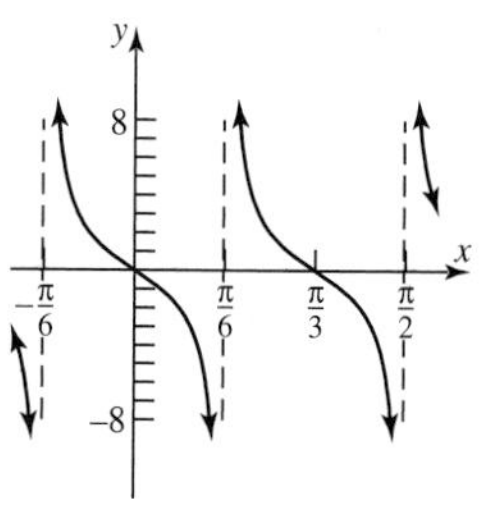

51.

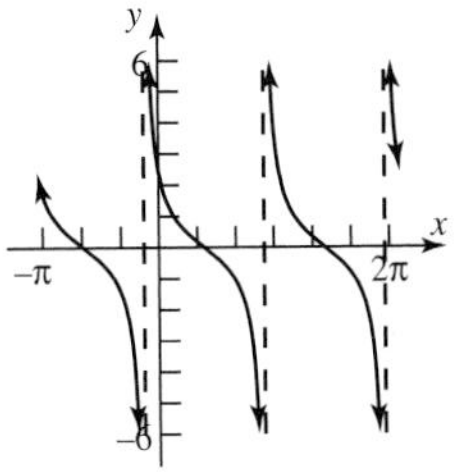

53.

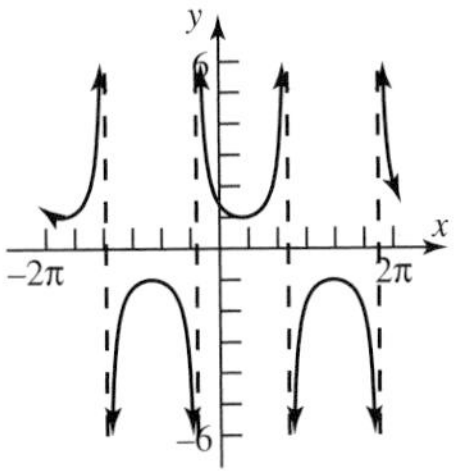

55. Amplitude $= 4$; Period $= 2\pi$
57. Amplitude $= 8$; Period $= 4$

59. Amplitude $= 4$

Period $= \dfrac{2\pi}{3}$

Phase shift $= 0$

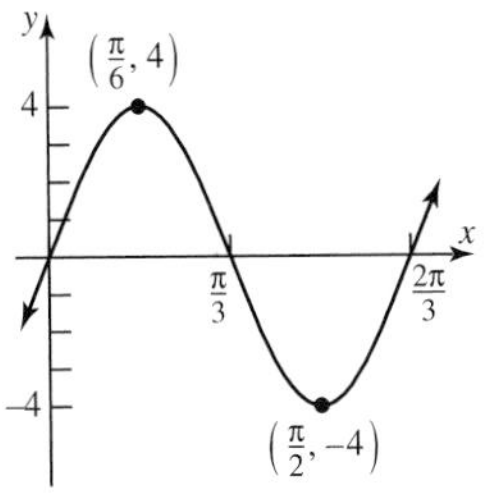

61. Amplitude $= 2$

Period $= 4$

Phase shift $= -\dfrac{1}{\pi}$

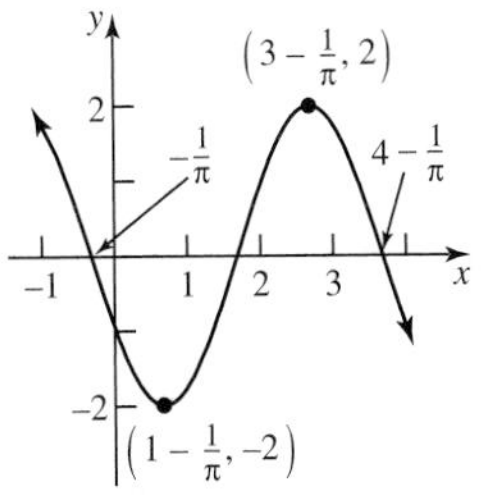

63. Amplitude $= \dfrac{1}{2}$

Period $= \dfrac{4\pi}{3}$

Phase shift $= \dfrac{2\pi}{3}$

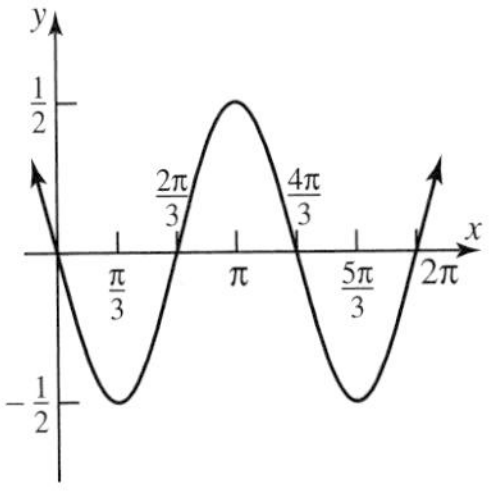

65. Amplitude $= \dfrac{2}{3}$

Period $= 2$

Phase shift $= \dfrac{6}{\pi}$

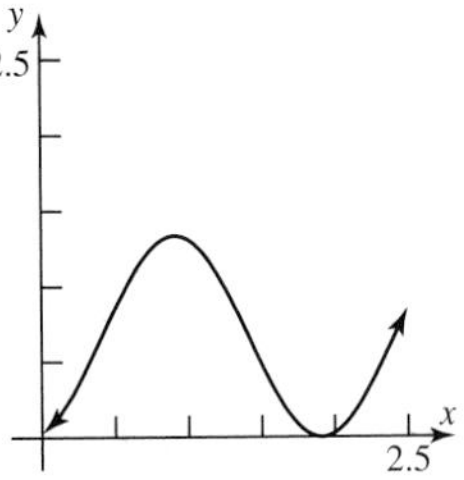

67. $y = 5\cos\dfrac{x}{4}$ **69.** $y = -6\cos\dfrac{\pi}{4}x$ **71.** $\dfrac{\pi}{3}$ ft **73.** Approximately 114.59 revolutions/hr **75.** 0.1 revolutions/sec $= \dfrac{\pi}{5}$ radians/sec

77. **(a)** 120 **(b)** $\dfrac{1}{60}$ **(c)**

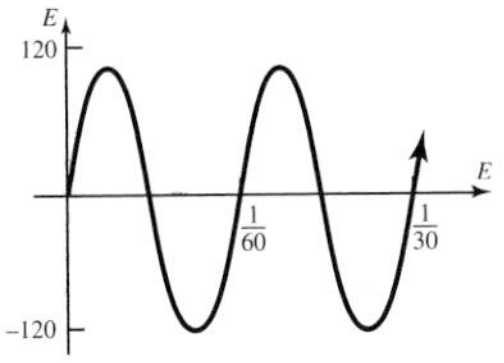

79. **(a)**

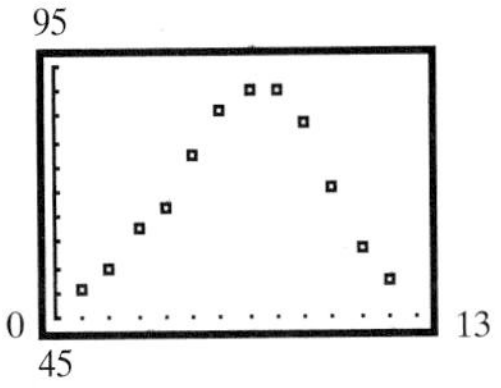

(c)

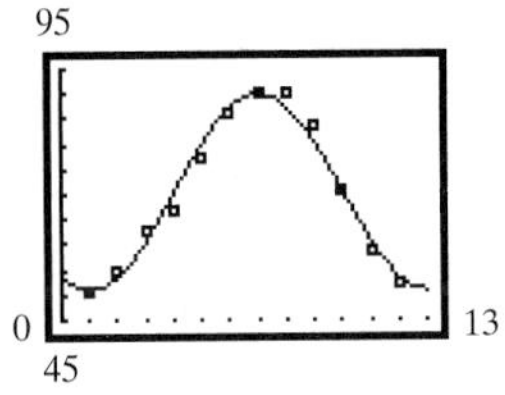

(e)

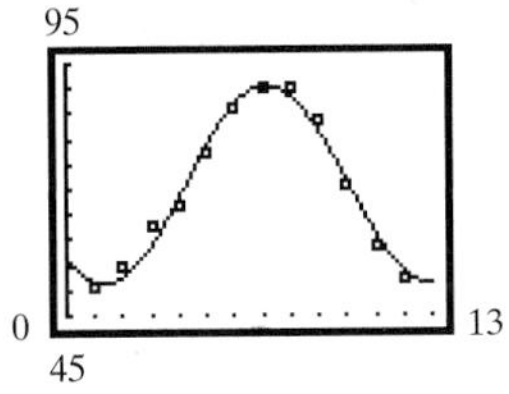

(b) $y = 19.5\sin\left(\dfrac{\pi}{6}x - \dfrac{2\pi}{3}\right) + 70.5$ **(d)** $y = 19.52\sin(0.54x - 2.28) + 71.01$

81. **(a)** $y = 1.85\sin\left(\dfrac{2\pi}{365}x - 7.6812\right) + 11.517$ **(b)**

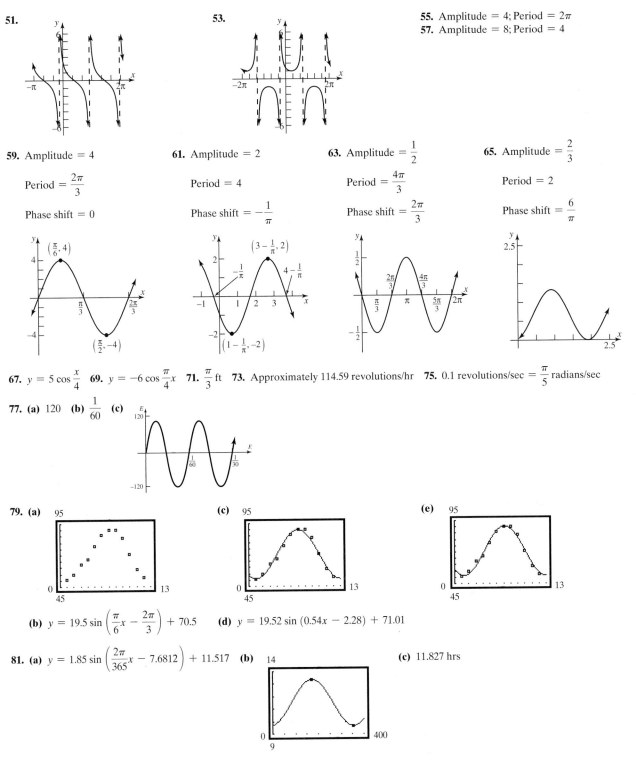

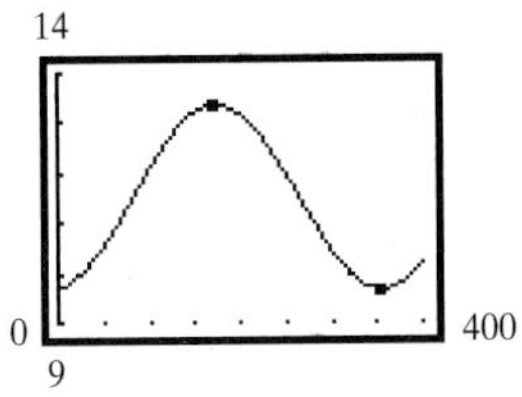

(c) 11.827 hrs

CHAPTER 7 Analytic Trigonometry

7.1 Exercises

1. $\csc\theta \cdot \cos\theta = \frac{1}{\sin\theta}\cdot\cos\theta = \frac{\cos\theta}{\sin\theta} = \cot\theta$ **3.** $1 + \tan^2(-\theta) = 1 + (-\tan\theta)^2 = 1 + \tan^2\theta = \sec^2\theta$

5. $\cos\theta(\tan\theta + \cot\theta) = \cos\theta\left(\frac{\sin\theta}{\cos\theta} + \frac{\cos\theta}{\sin\theta}\right) = \cos\theta\left(\frac{\sin^2\theta + \cos^2\theta}{\cos\theta\sin\theta}\right) = \frac{1}{\sin\theta} = \csc\theta$

7. $\tan\theta\cot\theta - \cos^2\theta = \frac{\sin\theta}{\cos\theta}\cdot\frac{\cos\theta}{\sin\theta} - \cos^2\theta = 1 - \cos^2\theta = \sin^2\theta$ **9.** $(\sec\theta - 1)(\sec\theta + 1) = \sec^2\theta - 1 = \tan^2\theta$

11. $(\sec\theta + \tan\theta)(\sec\theta - \tan\theta) = \sec^2\theta - \tan^2\theta = 1$ **13.** $\cos^2\theta(1 + \tan^2\theta) = \cos^2\theta + \cos^2\theta\cdot\frac{\sin^2\theta}{\cos^2\theta} = \cos^2\theta + \sin^2\theta = 1$

15. $(\sin\theta + \cos\theta)^2 + (\sin\theta - \cos\theta)^2 = \sin^2\theta + 2\sin\theta\cos\theta + \cos^2\theta + \sin^2\theta - 2\sin\theta\cos\theta + \cos^2\theta$
$= \sin^2\theta + \cos^2\theta + \sin^2\theta + \cos^2\theta = 1 + 1 = 2$

17. $\sec^4\theta - \sec^2\theta = \sec^2\theta(\sec^2\theta - 1) = (1 + \tan^2\theta)\tan^2\theta = \tan^4\theta + \tan^2\theta$

19. $\sec\theta - \tan\theta = \frac{1}{\cos\theta} - \frac{\sin\theta}{\cos\theta} = \frac{1 - \sin\theta}{\cos\theta}\cdot\frac{1 + \sin\theta}{1 + \sin\theta} = \frac{1 - \sin^2\theta}{\cos\theta(1 + \sin\theta)} = \frac{\cos^2\theta}{\cos\theta(1 + \sin\theta)} = \frac{\cos\theta}{1 + \sin\theta}$

21. $3\sin^2\theta + 4\cos^2\theta = 3\sin^2\theta + 3\cos^2\theta + \cos^2\theta = 3(\sin^2\theta + \cos^2\theta) + \cos^2\theta = 3 + \cos^2\theta$

23. $1 - \frac{\cos^2\theta}{1 + \sin\theta} = 1 - \frac{1 - \sin^2\theta}{1 + \sin\theta} = 1 - (1 - \sin\theta) = \sin\theta$ **25.** $\frac{1 + \tan\theta}{1 - \tan\theta} = \frac{1 + \frac{1}{\cot\theta}}{1 - \frac{1}{\cot\theta}} = \frac{\frac{\cot\theta + 1}{\cot\theta}}{\frac{\cot\theta - 1}{\cot\theta}} = \frac{\cot\theta + 1}{\cot\theta - 1}$

27. $\frac{\sec\theta}{\csc\theta} + \frac{\sin\theta}{\cos\theta} = \frac{\frac{1}{\cos\theta}}{\frac{1}{\sin\theta}} + \tan\theta = \frac{\sin\theta}{\cos\theta} + \tan\theta = \tan\theta + \tan\theta = 2\tan\theta$ **29.** $\frac{1 + \sin\theta}{1 - \sin\theta} = \frac{1 + \frac{1}{\csc\theta}}{1 - \frac{1}{\csc\theta}} = \frac{\frac{\csc\theta + 1}{\csc\theta}}{\frac{\csc\theta - 1}{\csc\theta}} = \frac{\csc\theta + 1}{\csc\theta - 1}$

31. $\frac{1 - \sin\theta}{\cos\theta} + \frac{\cos\theta}{1 - \sin\theta} = \frac{(1 - \sin\theta)^2 + \cos^2\theta}{\cos\theta(1 - \sin\theta)} = \frac{1 - 2\sin\theta + \sin^2\theta + \cos^2\theta}{\cos\theta(1 - \sin\theta)} = \frac{2 - 2\sin\theta}{\cos\theta(1 - \sin\theta)} = \frac{2(1 - \sin\theta)}{\cos\theta(1 - \sin\theta)} = \frac{2}{\cos\theta}$
$= 2\sec\theta$

33. $\frac{\sin\theta}{\sin\theta - \cos\theta} = \frac{1}{\frac{\sin\theta - \cos\theta}{\sin\theta}} = \frac{1}{1 - \frac{\cos\theta}{\sin\theta}} = \frac{1}{1 - \cot\theta}$

35. $(\sec\theta - \tan\theta)^2 = \sec^2\theta - 2\sec\theta\tan\theta + \tan^2\theta = \frac{1}{\cos^2\theta} - \frac{2\sin\theta}{\cos^2\theta} + \frac{\sin^2\theta}{\cos^2\theta} = \frac{1 - 2\sin\theta + \sin^2\theta}{\cos^2\theta} = \frac{(1 - \sin\theta)^2}{1 - \sin^2\theta}$
$= \frac{(1 - \sin\theta)^2}{(1 - \sin\theta)(1 + \sin\theta)} = \frac{1 - \sin\theta}{1 + \sin\theta}$

37. $\frac{\cos\theta}{1 - \tan\theta} + \frac{\sin\theta}{1 - \cot\theta} = \frac{\cos\theta}{1 - \frac{\sin\theta}{\cos\theta}} + \frac{\sin\theta}{1 - \frac{\cos\theta}{\sin\theta}} = \frac{\cos\theta}{\frac{\cos\theta - \sin\theta}{\cos\theta}} + \frac{\sin\theta}{\frac{\sin\theta - \cos\theta}{\sin\theta}} = \frac{\cos^2\theta}{\cos\theta - \sin\theta} + \frac{\sin^2\theta}{\sin\theta - \cos\theta}$
$= \frac{\cos^2\theta - \sin^2\theta}{\cos\theta - \sin\theta} = \frac{(\cos\theta - \sin\theta)(\cos\theta + \sin\theta)}{\cos\theta - \sin\theta} = \sin\theta + \cos\theta$

39. $\tan\theta + \frac{\cos\theta}{1 + \sin\theta} = \frac{\sin\theta}{\cos\theta} + \frac{\cos\theta}{1 + \sin\theta} = \frac{\sin\theta(1 + \sin\theta) + \cos^2\theta}{\cos\theta(1 + \sin\theta)} = \frac{\sin\theta + \sin^2\theta + \cos^2\theta}{\cos\theta(1 + \sin\theta)} = \frac{\sin\theta + 1}{\cos\theta(1 + \sin\theta)} = \frac{1}{\cos\theta} = \sec\theta$

41. $\frac{\tan\theta + \sec\theta - 1}{\tan\theta - \sec\theta + 1} = \frac{\tan\theta + (\sec\theta - 1)}{\tan\theta - (\sec\theta - 1)}\cdot\frac{\tan\theta + (\sec\theta - 1)}{\tan\theta + (\sec\theta - 1)} = \frac{\tan^2\theta + 2\tan\theta(\sec\theta - 1) + \sec^2\theta - 2\sec\theta + 1}{\tan^2\theta - (\sec^2\theta - 2\sec\theta + 1)}$
$= \frac{\sec^2\theta - 1 + 2\tan\theta(\sec\theta - 1) + \sec^2\theta - 2\sec\theta + 1}{\sec^2\theta - 1 - \sec^2\theta + 2\sec\theta - 1} = \frac{2\sec^2\theta - 2\sec\theta + 2\tan\theta(\sec\theta - 1)}{-2 + 2\sec\theta}$
$= \frac{2\sec\theta(\sec\theta - 1) + 2\tan\theta(\sec\theta - 1)}{2(\sec\theta - 1)} = \frac{2(\sec\theta - 1)(\sec\theta + \tan\theta)}{2(\sec\theta - 1)} = \tan\theta + \sec\theta$

43. $\frac{\tan\theta - \cot\theta}{\tan\theta + \cot\theta} = \frac{\frac{\sin\theta}{\cos\theta} - \frac{\cos\theta}{\sin\theta}}{\frac{\sin\theta}{\cos\theta} + \frac{\cos\theta}{\sin\theta}} = \frac{\frac{\sin^2\theta - \cos^2\theta}{\cos\theta\sin\theta}}{\frac{\sin^2\theta + \cos^2\theta}{\cos\theta\sin\theta}} = \frac{\sin^2\theta - \cos^2\theta}{1} = \sin^2\theta - \cos^2\theta$

45. $\dfrac{\tan\theta - \cot\theta}{\tan\theta + \cot\theta} + 1 = \dfrac{\dfrac{\sin\theta}{\cos\theta} - \dfrac{\cos\theta}{\sin\theta}}{\dfrac{\sin\theta}{\cos\theta} + \dfrac{\cos\theta}{\sin\theta}} + 1 = \dfrac{\dfrac{\sin^2\theta - \cos^2\theta}{\cos\theta\sin\theta}}{\dfrac{\sin^2\theta + \cos^2\theta}{\cos\theta\sin\theta}} + 1 = \sin^2\theta - \cos^2\theta + 1 = \sin^2\theta + (1 - \cos^2\theta) = 2\sin^2\theta$

47. $\dfrac{\sec\theta + \tan\theta}{\cot\theta + \cos\theta} = \dfrac{\dfrac{1}{\cos\theta} + \dfrac{\sin\theta}{\cos\theta}}{\dfrac{\cos\theta}{\sin\theta} + \dfrac{\cos\theta\sin\theta}{\sin\theta}} = \dfrac{\dfrac{1 + \sin\theta}{\cos\theta}}{\dfrac{\cos\theta + \cos\theta\sin\theta}{\sin\theta}} = \dfrac{1 + \sin\theta}{\cos\theta} \cdot \dfrac{\sin\theta}{\cos\theta(1 + \sin\theta)} = \dfrac{\sin\theta}{\cos\theta} \cdot \dfrac{1}{\cos\theta} = \tan\theta\sec\theta$

49. $\dfrac{1 - \tan^2\theta}{1 + \tan^2\theta} + 1 = \dfrac{1 - \tan^2\theta}{\sec^2\theta} + 1 = \dfrac{1}{\sec^2\theta} - \dfrac{\tan^2\theta}{\sec^2\theta} + 1 = \cos^2\theta - \dfrac{\dfrac{\sin^2\theta}{\cos^2\theta}}{\dfrac{1}{\cos^2\theta}} + 1 = \cos^2\theta - \sin^2\theta + 1$

$= \cos^2\theta + (1 - \sin^2\theta) = 2\cos^2\theta$

51. $\dfrac{\sec\theta - \csc\theta}{\sec\theta\csc\theta} = \dfrac{\dfrac{1}{\cos\theta} - \dfrac{1}{\sin\theta}}{\dfrac{1}{\cos\theta} \cdot \dfrac{1}{\sin\theta}} = \dfrac{\dfrac{\sin\theta - \cos\theta}{\cos\theta\sin\theta}}{\dfrac{1}{\cos\theta\sin\theta}} = \sin\theta - \cos\theta$

53. $\sec\theta - \cos\theta - \sin\theta\tan\theta = \left(\dfrac{1}{\cos\theta} - \cos\theta\right) - \sin\theta \cdot \dfrac{\sin\theta}{\cos\theta} = \dfrac{1 - \cos^2\theta}{\cos\theta} - \dfrac{\sin^2\theta}{\cos\theta} = \dfrac{\sin^2\theta}{\cos\theta} - \dfrac{\sin^2\theta}{\cos\theta} = 0$

55. $\dfrac{1}{1 - \sin\theta} + \dfrac{1}{1 + \sin\theta} = \dfrac{1 + \sin\theta + 1 - \sin\theta}{(1 + \sin\theta)(1 - \sin\theta)} = \dfrac{2}{1 - \sin^2\theta} = \dfrac{2}{\cos^2\theta} = 2\sec^2\theta$

57. $\dfrac{\sec\theta}{1 - \sin\theta} = \dfrac{\sec\theta}{1 - \sin\theta} \cdot \dfrac{1 + \sin\theta}{1 + \sin\theta} = \dfrac{\sec\theta(1 + \sin\theta)}{1 - \sin^2\theta} = \dfrac{\sec\theta(1 + \sin\theta)}{\cos^2\theta} = \dfrac{1 + \sin\theta}{\cos^3\theta}$

59. $\dfrac{(\sec\theta - \tan\theta)^2 + 1}{\csc\theta(\sec\theta - \tan\theta)} = \dfrac{\sec^2\theta - 2\sec\theta\tan\theta + \tan^2\theta + 1}{\dfrac{1}{\sin\theta}\left(\dfrac{1}{\cos\theta} - \dfrac{\sin\theta}{\cos\theta}\right)} = \dfrac{2\sec^2\theta - 2\sec\theta\tan\theta}{\dfrac{1}{\sin\theta}\left(\dfrac{1 - \sin\theta}{\cos\theta}\right)} = \dfrac{\dfrac{2}{\cos^2\theta} - \dfrac{2\sin\theta}{\cos^2\theta}}{\dfrac{1 - \sin\theta}{\sin\theta\cos\theta}} = \dfrac{2 - 2\sin\theta}{\cos^2\theta} \cdot \dfrac{\sin\theta\cos\theta}{1 - \sin\theta}$

$= \dfrac{2(1 - \sin\theta)}{\cos\theta} \cdot \dfrac{\sin\theta}{1 - \sin\theta} = \dfrac{2\sin\theta}{\cos\theta} = 2\tan\theta$

61. $\dfrac{\sin\theta + \cos\theta}{\cos\theta} - \dfrac{\sin\theta - \cos\theta}{\sin\theta} = \dfrac{\sin\theta(\sin\theta + \cos\theta) - \cos\theta(\sin\theta - \cos\theta)}{\cos\theta\sin\theta} = \dfrac{\sin^2\theta + \sin\theta\cos\theta - \sin\theta\cos\theta + \cos^2\theta}{\cos\theta\sin\theta}$

$= \dfrac{1}{\cos\theta\sin\theta} = \sec\theta\csc\theta$

63. $\dfrac{\sin^3\theta + \cos^3\theta}{\sin\theta + \cos\theta} = \dfrac{(\sin\theta + \cos\theta)(\sin^2\theta - \sin\theta\cos\theta + \cos^2\theta)}{\sin\theta + \cos\theta} = \sin^2\theta + \cos^2\theta - \sin\theta\cos\theta = 1 - \sin\theta\cos\theta$

65. $\dfrac{\cos^2\theta - \sin^2\theta}{1 - \tan^2\theta} = \dfrac{\cos^2\theta - \sin^2\theta}{1 - \dfrac{\sin^2\theta}{\cos^2\theta}} = \dfrac{\cos^2\theta - \sin^2\theta}{\dfrac{\cos^2\theta - \sin^2\theta}{\cos^2\theta}} = \cos^2\theta$

67. $\dfrac{(2\cos^2\theta - 1)^2}{\cos^4\theta - \sin^4\theta} = \dfrac{[2\cos^2\theta - (\sin^2\theta + \cos^2\theta)]^2}{(\cos^2\theta - \sin^2\theta)(\cos^2\theta + \sin^2\theta)} = \cos^2\theta - \sin^2\theta = (1 - \sin^2\theta) - \sin^2\theta = 1 - 2\sin^2\theta$

69. $\dfrac{1 + \sin\theta + \cos\theta}{1 + \sin\theta - \cos\theta} = \dfrac{(1 + \sin\theta) + \cos\theta}{(1 + \sin\theta) - \cos\theta} \cdot \dfrac{(1 + \sin\theta) + \cos\theta}{(1 + \sin\theta) + \cos\theta} = \dfrac{1 + 2\sin\theta + \sin^2\theta + 2(1 + \sin\theta)(\cos\theta) + \cos^2\theta}{1 + 2\sin\theta + \sin^2\theta - \cos^2\theta}$

$= \dfrac{1 + 2\sin\theta + \sin^2\theta + 2(1 + \sin\theta)(\cos\theta) + (1 - \sin^2\theta)}{1 + 2\sin\theta + \sin^2\theta - (1 - \sin^2\theta)} = \dfrac{2 + 2\sin\theta + 2(1 + \sin\theta)(\cos\theta)}{2\sin\theta + 2\sin^2\theta}$

$= \dfrac{2(1 + \sin\theta) + 2(1 + \sin\theta)(\cos\theta)}{2\sin\theta(1 + \sin\theta)} = \dfrac{2(1 + \sin\theta)(1 + \cos\theta)}{2\sin\theta(1 + \sin\theta)} = \dfrac{1 + \cos\theta}{\sin\theta}$

71. $(a\sin\theta + b\cos\theta)^2 + (a\cos\theta - b\sin\theta)^2 = a^2\sin^2\theta + 2ab\sin\theta\cos\theta + b^2\cos^2\theta + a^2\cos^2\theta - 2ab\sin\theta\cos\theta + b^2\sin^2\theta$

$= a^2(\sin^2\theta + \cos^2\theta) + b^2(\cos^2\theta + \sin^2\theta) = a^2 + b^2$

73. $\dfrac{\tan\alpha + \tan\beta}{\cot\alpha + \cot\beta} = \dfrac{\tan\alpha + \tan\beta}{\dfrac{1}{\tan\alpha} + \dfrac{1}{\tan\beta}} = \dfrac{\tan\alpha + \tan\beta}{\dfrac{\tan\beta + \tan\alpha}{\tan\alpha\tan\beta}} = (\tan\alpha + \tan\beta) \cdot \dfrac{\tan\alpha\tan\beta}{\tan\alpha + \tan\beta} = \tan\alpha\tan\beta$

75. $(\sin\alpha + \cos\beta)^2 + (\cos\beta + \sin\alpha)(\cos\beta - \sin\alpha) = (\sin^2\alpha + 2\sin\alpha\cos\beta + \cos^2\beta) + (\cos^2\beta - \sin^2\alpha)$

$= 2\cos^2\beta + 2\sin\alpha\cos\beta = 2\cos\beta(\cos\beta + \sin\alpha)$

77. $\ln|\sec\theta| = \ln|\cos\theta|^{-1} = -\ln|\cos\theta|$

79. $\ln|1 + \cos\theta| + \ln|1 - \cos\theta| = \ln(|1 + \cos\theta|\,|1 - \cos\theta|) = \ln|1 - \cos^2\theta| = \ln|\sin^2\theta| = 2\ln|\sin\theta|$

7.2 Exercises

1. $\frac{1}{4}(\sqrt{6}+\sqrt{2})$ **3.** $\frac{1}{4}(\sqrt{2}-\sqrt{6})$ **5.** $-\frac{1}{4}(\sqrt{2}+\sqrt{6})$ **7.** $\frac{\sqrt{3}-1}{1+\sqrt{3}}=2-\sqrt{3}$ **9.** $-\frac{1}{4}(\sqrt{6}+\sqrt{2})$ **11.** $\frac{4}{\sqrt{6}+\sqrt{2}}=\sqrt{6}-\sqrt{2}$

13. $\frac{1}{2}$ **15.** 0 **17.** 1 **19.** -1 **21.** $\frac{1}{2}$ **23. (a)** $\frac{2\sqrt{5}}{25}$ **(b)** $\frac{11\sqrt{5}}{25}$ **(c)** $\frac{2\sqrt{5}}{5}$ **(d)** 2 **25. (a)** $\frac{4-3\sqrt{3}}{10}$ **(b)** $\frac{-3-4\sqrt{3}}{10}$ **(c)** $\frac{4+3\sqrt{3}}{10}$

(d) $\frac{4+3\sqrt{3}}{4\sqrt{3}-3}=\frac{25\sqrt{3}+48}{39}$ **27. (a)** $-\frac{1}{26}(5+12\sqrt{3})$ **(b)** $\frac{1}{26}(12-5\sqrt{3})$ **(c)** $-\frac{1}{26}(5-12\sqrt{3})$

(d) $\frac{-5+12\sqrt{3}}{12+5\sqrt{3}}=\frac{-240+169\sqrt{3}}{69}$ **29. (a)** $-\frac{2\sqrt{2}}{3}$ **(b)** $\frac{-2\sqrt{2}+\sqrt{3}}{6}$ **(c)** $\frac{-2\sqrt{2}+\sqrt{3}}{6}$ **(d)** $\frac{2\sqrt{2}-1}{2\sqrt{2}+1}=\frac{9-4\sqrt{2}}{7}$

31. $\sin\left(\frac{\pi}{2}+\theta\right)=\sin\frac{\pi}{2}\cos\theta+\cos\frac{\pi}{2}\sin\theta=1\cdot\cos\theta+0\cdot\sin\theta=\cos\theta$

33. $\sin(\pi-\theta)=\sin\pi\cos\theta-\cos\pi\sin\theta=0\cdot\cos\theta-(-1)\sin\theta=\sin\theta$

35. $\sin(\pi+\theta)=\sin\pi\cos\theta+\cos\pi\sin\theta=0\cdot\cos\theta+(-1)\sin\theta=-\sin\theta$

37. $\tan(\pi-\theta)=\frac{\tan\pi-\tan\theta}{1+\tan\pi\tan\theta}=\frac{0-\tan\theta}{1+0}=-\tan\theta$

39. $\sin\left(\frac{3\pi}{2}+\theta\right)=\sin\frac{3\pi}{2}\cos\theta+\cos\frac{3\pi}{2}\sin\theta=(-1)\cos\theta+0\cdot\sin\theta=-\cos\theta$

41. $\sin(\alpha+\beta)+\sin(\alpha-\beta)=\sin\alpha\cos\beta+\cos\alpha\sin\beta+\sin\alpha\cos\beta-\cos\alpha\sin\beta=2\sin\alpha\cos\beta$

43. $\frac{\sin(\alpha+\beta)}{\sin\alpha\cos\beta}=\frac{\sin\alpha\cos\beta+\cos\alpha\sin\beta}{\sin\alpha\cos\beta}=\frac{\sin\alpha\cos\beta}{\sin\alpha\cos\beta}+\frac{\cos\alpha\sin\beta}{\sin\alpha\cos\beta}=1+\cot\alpha\tan\beta$

45. $\frac{\cos(\alpha+\beta)}{\cos\alpha\cos\beta}=\frac{\cos\alpha\cos\beta-\sin\alpha\sin\beta}{\cos\alpha\cos\beta}=\frac{\cos\alpha\cos\beta}{\cos\alpha\cos\beta}-\frac{\sin\alpha\sin\beta}{\cos\alpha\cos\beta}=1-\tan\alpha\tan\beta$

47. $\frac{\sin(\alpha+\beta)}{\sin(\alpha-\beta)}=\frac{\sin\alpha\cos\beta+\cos\alpha\sin\beta}{\sin\alpha\cos\beta-\cos\alpha\sin\beta}=\frac{\frac{\sin\alpha\cos\beta+\cos\alpha\sin\beta}{\cos\alpha\cos\beta}}{\frac{\sin\alpha\cos\beta-\cos\alpha\sin\beta}{\cos\alpha\cos\beta}}=\frac{\frac{\sin\alpha\cos\beta}{\cos\alpha\cos\beta}+\frac{\cos\alpha\sin\beta}{\cos\alpha\cos\beta}}{\frac{\sin\alpha\cos\beta}{\cos\alpha\cos\beta}-\frac{\cos\alpha\sin\beta}{\cos\alpha\cos\beta}}=\frac{\tan\alpha+\tan\beta}{\tan\alpha-\tan\beta}$

49. $\cot(\alpha+\beta)=\frac{\cos(\alpha+\beta)}{\sin(\alpha+\beta)}=\frac{\cos\alpha\cos\beta-\sin\alpha\sin\beta}{\sin\alpha\cos\beta+\cos\alpha\sin\beta}=\frac{\frac{\cos\alpha\cos\beta-\sin\alpha\sin\beta}{\sin\alpha\sin\beta}}{\frac{\sin\alpha\cos\beta+\cos\alpha\sin\beta}{\sin\alpha\sin\beta}}=\frac{\frac{\cos\alpha\cos\beta}{\sin\alpha\sin\beta}-\frac{\sin\alpha\sin\beta}{\sin\alpha\sin\beta}}{\frac{\sin\alpha\cos\beta}{\sin\alpha\sin\beta}+\frac{\cos\alpha\sin\beta}{\sin\alpha\sin\beta}}=\frac{\cot\alpha\cot\beta-1}{\cot\beta+\cot\alpha}$

51. $\sec(\alpha+\beta)=\frac{1}{\cos(\alpha+\beta)}=\frac{1}{\cos\alpha\cos\beta-\sin\alpha\sin\beta}=\frac{\frac{1}{\sin\alpha\sin\beta}}{\frac{\cos\alpha\cos\beta-\sin\alpha\sin\beta}{\sin\alpha\sin\beta}}=\frac{\frac{1}{\sin\alpha}\cdot\frac{1}{\sin\beta}}{\frac{\cos\alpha\cos\beta}{\sin\alpha\sin\beta}-\frac{\sin\alpha\sin\beta}{\sin\alpha\sin\beta}}=\frac{\csc\alpha\csc\beta}{\cot\alpha\cot\beta-1}$

53. $\sin(\alpha-\beta)\sin(\alpha+\beta)=(\sin\alpha\cos\beta-\cos\alpha\sin\beta)(\sin\alpha\cos\beta+\cos\alpha\sin\beta)=\sin^2\alpha\cos^2\beta-\cos^2\alpha\sin^2\beta$
$=(\sin^2\alpha)(1-\sin^2\beta)-(1-\sin^2\alpha)(\sin^2\beta)=\sin^2\alpha-\sin^2\beta$

55. $\sin(\theta+k\pi)=\sin\theta\cos k\pi+\cos\theta\sin k\pi=(\sin\theta)(-1)^k+(\cos\theta)(0)=(-1)^k\sin\theta$, k any integer

57. $\frac{\sin(x+h)-\sin x}{h}=\frac{\sin x\cos h+\cos x\sin h-\sin x}{h}=\frac{\cos x\sin h-\sin x(1-\cos h)}{h}=\cos x\cdot\frac{\sin h}{h}-\sin x\cdot\frac{1-\cos h}{h}$

59. $\tan\frac{\pi}{2}$ is not defined; $\tan\left(\frac{\pi}{2}-\theta\right)=\frac{\sin\left(\frac{\pi}{2}-\theta\right)}{\cos\left(\frac{\pi}{2}-\theta\right)}=\frac{\cos\theta}{\sin\theta}=\cot\theta$ **61.** $\tan\theta=\tan(\theta_2-\theta_1)=\frac{\tan\theta_2-\tan\theta_1}{1+\tan\theta_1\tan\theta_2}=\frac{m_2-m_1}{1+m_1m_2}$

7.3 Exercises

1. (a) $\frac{24}{25}$ **(b)** $\frac{7}{25}$ **(c)** $\frac{\sqrt{10}}{10}$ **(d)** $\frac{3\sqrt{10}}{10}$ **3. (a)** $\frac{24}{25}$ **(b)** $-\frac{7}{25}$ **(c)** $\frac{2\sqrt{5}}{5}$ **(d)** $-\frac{\sqrt{5}}{5}$

5. (a) $-\frac{2\sqrt{2}}{3}$ **(b)** $\frac{1}{3}$ **(c)** $\sqrt{\frac{3+\sqrt{6}}{6}}$ **(d)** $\sqrt{\frac{3-\sqrt{6}}{6}}$ **7. (a)** $\frac{4\sqrt{2}}{9}$ **(b)** $-\frac{7}{9}$ **(c)** $\frac{\sqrt{3}}{3}$ **(d)** $\frac{\sqrt{6}}{3}$

9. (a) $-\frac{4}{5}$ **(b)** $\frac{3}{5}$ **(c)** $\sqrt{\frac{5+2\sqrt{5}}{10}}$ **(d)** $\sqrt{\frac{5-2\sqrt{5}}{10}}$ **11. (a)** $-\frac{3}{5}$ **(b)** $-\frac{4}{5}$ **(c)** $\frac{1}{2}\sqrt{\frac{10-\sqrt{10}}{5}}$ **(d)** $-\frac{1}{2}\sqrt{\frac{10+\sqrt{10}}{5}}$

13. $\frac{\sqrt{2-\sqrt{2}}}{2}$ **15.** $1-\sqrt{2}$ **17.** $-\frac{\sqrt{2+\sqrt{3}}}{2}$ **19.** $\frac{2}{\sqrt{2+\sqrt{2}}}=(2-\sqrt{2})\sqrt{2+\sqrt{2}}$ **21.** $-\frac{\sqrt{2-\sqrt{2}}}{2}$

23. $\sin^4\theta=(\sin^2\theta)^2=\left(\frac{1-\cos 2\theta}{2}\right)^2=\frac{1}{4}(1-2\cos 2\theta+\cos^2 2\theta)=\frac{1}{4}-\frac{1}{2}\cos 2\theta+\frac{1}{4}\cos^2 2\theta$
$=\frac{1}{4}-\frac{1}{2}\cos 2\theta+\frac{1}{4}\left(\frac{1+\cos 4\theta}{2}\right)=\frac{1}{4}-\frac{1}{2}\cos 2\theta+\frac{1}{8}+\frac{1}{8}\cos 4\theta=\frac{3}{8}-\frac{1}{2}\cos 2\theta+\frac{1}{8}\cos 4\theta$

25. $\sin(4\theta) = \sin[2(2\theta)] = 2\sin(2\theta)\cos(2\theta) = (4\sin\theta\cos\theta)(1 - 2\sin^2\theta) = 4\sin\theta\cos\theta - 8\sin^3\theta\cos\theta = (\cos\theta)(4\sin\theta - 8\sin^3\theta)$
27. $16\sin^5\theta - 20\sin^3\theta + 5\sin\theta$ **29.** $\cos^4\theta - \sin^4\theta = (\cos^2\theta + \sin^2\theta)(\cos^2\theta - \sin^2\theta) = \cos 2\theta$

31. $$\cot(2\theta) = \frac{1}{\tan(2\theta)} = \frac{1}{\dfrac{2\tan\theta}{1-\tan^2\theta}} = \frac{1-\tan^2\theta}{2\tan\theta} = \frac{1 - \dfrac{1}{\cot^2\theta}}{2\left(\dfrac{1}{\cot\theta}\right)} = \frac{\dfrac{\cot^2\theta - 1}{\cot^2\theta}}{\dfrac{2}{\cot\theta}} = \frac{\cot^2\theta - 1}{\cot^2\theta}\cdot\frac{\cot\theta}{2} = \frac{\cot^2\theta - 1}{2\cot\theta}$$

33. $$\sec(2\theta) = \frac{1}{\cos(2\theta)} = \frac{1}{2\cos^2\theta - 1} = \frac{1}{\dfrac{2}{\sec^2\theta} - 1} = \frac{1}{\dfrac{2-\sec^2\theta}{\sec^2\theta}} = \frac{\sec^2\theta}{2-\sec^2\theta}$$

35. $\cos^2(2\theta) - \sin^2(2\theta) = \cos[2(2\theta)] = \cos(4\theta)$

37. $$\frac{\cos(2\theta)}{1+\sin(2\theta)} = \frac{\cos^2\theta - \sin^2\theta}{1 + 2\sin\theta\cos\theta} = \frac{(\cos\theta - \sin\theta)(\cos\theta + \sin\theta)}{\sin^2\theta + \cos^2\theta + 2\sin\theta\cos\theta} = \frac{(\cos\theta - \sin\theta)(\cos\theta + \sin\theta)}{(\sin\theta + \cos\theta)(\sin\theta + \cos\theta)} = \frac{\cos\theta - \sin\theta}{\cos\theta + \sin\theta}$$
$$= \frac{\dfrac{\cos\theta - \sin\theta}{\sin\theta}}{\dfrac{\cos\theta + \sin\theta}{\sin\theta}} = \frac{\dfrac{\cos\theta}{\sin\theta} - \dfrac{\sin\theta}{\sin\theta}}{\dfrac{\cos\theta}{\sin\theta} + \dfrac{\sin\theta}{\sin\theta}} = \frac{\cot\theta - 1}{\cot\theta + 1}$$

39. $$\sec^2\frac{\theta}{2} = \frac{1}{\cos^2\left(\dfrac{\theta}{2}\right)} = \frac{1}{\dfrac{1+\cos\theta}{2}} = \frac{2}{1+\cos\theta}$$

41. $$\cot^2\frac{\theta}{2} = \frac{1}{\tan^2\left(\dfrac{\theta}{2}\right)} = \frac{1}{\dfrac{1-\cos\theta}{1+\cos\theta}} = \frac{1+\cos\theta}{1-\cos\theta} = \frac{1 + \dfrac{1}{\sec\theta}}{1 - \dfrac{1}{\sec\theta}} = \frac{\dfrac{\sec\theta + 1}{\sec\theta}}{\dfrac{\sec\theta - 1}{\sec\theta}} = \frac{\sec\theta + 1}{\sec\theta}\cdot\frac{\sec\theta}{\sec\theta - 1} = \frac{\sec\theta + 1}{\sec\theta - 1}$$

43. $$\frac{1 - \tan^2\left(\dfrac{\theta}{2}\right)}{1 + \tan^2\left(\dfrac{\theta}{2}\right)} = \frac{1 - \dfrac{1-\cos\theta}{1+\cos\theta}}{1 + \dfrac{1-\cos\theta}{1+\cos\theta}} = \frac{\dfrac{1+\cos\theta - (1-\cos\theta)}{1+\cos\theta}}{\dfrac{1+\cos\theta + 1 - \cos\theta}{1+\cos\theta}} = \frac{2\cos\theta}{1+\cos\theta}\cdot\frac{1+\cos\theta}{2} = \cos\theta$$

45. $$\frac{\sin(3\theta)}{\sin\theta} - \frac{\cos(3\theta)}{\cos\theta} = \frac{\sin(3\theta)\cos\theta - \cos(3\theta)\sin\theta}{\sin\theta\cos\theta} = \frac{\sin(3\theta - \theta)}{\dfrac{1}{2}(2\sin\theta\cos\theta)} = \frac{2\sin(2\theta)}{\sin(2\theta)} = 2$$

47. $$\tan(3\theta) = \tan(\theta + 2\theta) = \frac{\tan\theta + \tan(2\theta)}{1 - \tan\theta\tan(2\theta)} = \frac{\tan\theta + \dfrac{2\tan\theta}{1-\tan^2\theta}}{1 - \dfrac{\tan\theta(2\tan\theta)}{1-\tan^2\theta}} = \frac{\tan\theta - \tan^3\theta + 2\tan\theta}{1 - \tan^2\theta - 2\tan^2\theta} = \frac{3\tan\theta - \tan^3\theta}{1 - 3\tan^2\theta}$$

49. $\sin(2\theta) = \dfrac{4x}{4+x^2}$ **51.** $-\dfrac{1}{4}$

53.

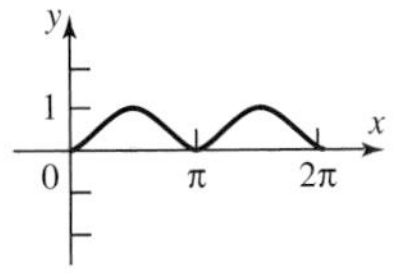

55. $\sin\dfrac{\pi}{24} = \dfrac{\sqrt{2}}{4}\sqrt{4 - \sqrt{6} - \sqrt{2}}$; $\cos\dfrac{\pi}{24} = \dfrac{\sqrt{2}}{4}\sqrt{4 + \sqrt{6} + \sqrt{2}}$

57. $\sin^3\theta + \sin^3(\theta + 120°) + \sin^3(\theta + 240°)$
$= \sin^3\theta + (\sin\theta\cos 120° + \cos\theta\sin 120°)^3 + (\sin\theta\cos 240° + \cos\theta\sin 240°)^3$
$= \sin^3\theta + \left(-\dfrac{1}{2}\sin\theta + \dfrac{\sqrt{3}}{2}\cos\theta\right)^3 + \left(-\dfrac{1}{2}\sin\theta - \dfrac{\sqrt{3}}{2}\cos\theta\right)^3$
$= \sin^3\theta + \dfrac{1}{8}(3\sqrt{3}\cos^3\theta - 9\cos^2\theta\sin\theta + 3\sqrt{3}\cos\theta\sin^2\theta - \sin^3\theta) - \dfrac{1}{8}(\sin^3\theta + 3\sqrt{3}\sin^2\theta\cos\theta + 9\sin\theta\cos^2\theta + 3\sqrt{3}\cos^3\theta)$
$= \dfrac{3}{4}\sin^3\theta - \dfrac{9}{4}\cos^2\theta\sin\theta = \dfrac{3}{4}[\sin^3\theta - 3\sin\theta(1 - \sin^2\theta)] = \dfrac{3}{4}(4\sin^3\theta - 3\sin\theta) = -\dfrac{3}{4}\sin(3\theta)$ (from Example 2)

59. $\dfrac{1}{2}(\ln|1 - \cos(2\theta)| - \ln 2) = \ln\left(\dfrac{|1 - \cos 2\theta|}{2}\right)^{1/2} = \ln|\sin^2\theta|^{1/2} = \ln|\sin\theta|$

61. (a) $R = \dfrac{v_0^2\sqrt{2}}{16}(\sin\theta\cos\theta - \cos^2\theta) = \dfrac{v_0^2\sqrt{2}}{16}\left(\dfrac{1}{2}\sin(2\theta) - \dfrac{1+\cos(2\theta)}{2}\right) = \dfrac{v_0^2\sqrt{2}}{32}(\sin 2\theta - \cos(2\theta) - 1)$

(b)

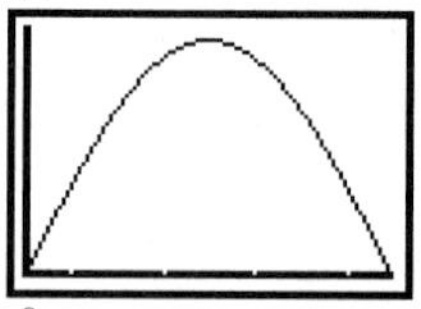

(c) $\theta = 67.50$ makes R largest

63. $\dfrac{2z}{1+z^2} = \dfrac{2\tan\left(\frac{\alpha}{2}\right)}{1+\tan^2\left(\frac{\alpha}{2}\right)} = \dfrac{2\tan\left(\frac{\alpha}{2}\right)}{\sec^2\left(\frac{\alpha}{2}\right)} = \dfrac{\dfrac{2\sin\left(\frac{\alpha}{2}\right)}{\cos\left(\frac{\alpha}{2}\right)}}{\dfrac{1}{\cos^2\left(\frac{\alpha}{2}\right)}} = 2\sin\dfrac{\alpha}{2}\cos\dfrac{\alpha}{2} = \sin\left(2\cdot\dfrac{\alpha}{2}\right) = \sin\alpha$

7.4 Exercises

1. $\frac{1}{2}(\cos(2\theta) - \cos(6\theta))$ **3.** $\frac{1}{2}(\sin(6\theta) + \sin(2\theta))$ **5.** $\frac{1}{2}(\cos(8\theta) + \cos(2\theta))$ **7.** $\frac{1}{2}(\cos\theta - \cos(3\theta))$ **9.** $\frac{1}{2}(\sin(2\theta) + \sin\theta)$

11. $2\sin\theta\cos(3\theta)$ **13.** $2\cos(3\theta)\cos\theta$ **15.** $2\sin(2\theta)\cos\theta$ **17.** $2\sin\theta\sin\left(\frac{\theta}{2}\right)$

19. $\dfrac{\sin\theta + \sin(3\theta)}{2\sin(2\theta)} = \dfrac{2\sin 2\theta\cos(-\theta)}{2\sin(2\theta)} = \cos(-\theta) = \cos\theta$ **21.** $\dfrac{\sin(4\theta)+\sin(2\theta)}{\cos(4\theta)+\cos(2\theta)} = \dfrac{2\sin(3\theta)\cos\theta}{2\cos(3\theta)\cos\theta} = \dfrac{\sin(3\theta)}{\cos(3\theta)} = \tan(3\theta)$

23. $\dfrac{\cos\theta - \cos(3\theta)}{\sin\theta + \sin(3\theta)} = \dfrac{-2\sin(2\theta)\sin(-\theta)}{2\sin(2\theta)\cos(-\theta)} = \dfrac{\sin\theta}{\cos\theta} = \tan\theta$

25. $\sin\theta(\sin\theta + \sin(3\theta)) = \sin\theta[2\sin(2\theta)\cos(-\theta)] = 2\sin(2\theta)\sin\theta\cos\theta = \cos\theta(2\sin(2\theta)\sin\theta) = \cos\theta\left[2\cdot\frac{1}{2}(\cos\theta - \cos(3\theta))\right]$
$= \cos\theta(\cos\theta - \cos 3\theta)$

27. $\dfrac{\sin(4\theta)+\sin(8\theta)}{\cos(4\theta)+\cos(8\theta)} = \dfrac{2\sin(6\theta)\cos(-2\theta)}{2\cos(6\theta)\cos(-2\theta)} = \dfrac{\sin(6\theta)}{\cos(6\theta)} = \tan(6\theta)$

29. $\dfrac{\sin(4\theta)+\sin(8\theta)}{\sin(4\theta)-\sin(8\theta)} = \dfrac{2\sin(6\theta)\cos(-2\theta)}{2\sin(-2\theta)\cos(6\theta)} = \dfrac{\sin(6\theta)}{\cos(6\theta)}\cdot\dfrac{\cos(2\theta)}{-\sin(2\theta)} = \tan(6\theta)(-\cot(2\theta)) = -\dfrac{\tan(6\theta)}{\tan(2\theta)}$

31. $\dfrac{\sin\alpha+\sin\beta}{\sin\alpha-\sin\beta} = \dfrac{2\sin\frac{\alpha+\beta}{2}\cos\frac{\alpha-\beta}{2}}{2\sin\frac{\alpha-\beta}{2}\cos\frac{\alpha+\beta}{2}} = \dfrac{\sin\frac{\alpha+\beta}{2}}{\cos\frac{\alpha+\beta}{2}}\cdot\dfrac{\cos\frac{\alpha-\beta}{2}}{\sin\frac{\alpha-\beta}{2}} = \tan\dfrac{\alpha+\beta}{2}\cot\dfrac{\alpha-\beta}{2}$

33. $\dfrac{\sin\alpha+\sin\beta}{\cos\alpha+\cos\beta} = \dfrac{2\sin\frac{\alpha+\beta}{2}\cos\frac{\alpha-\beta}{2}}{2\cos\frac{\alpha+\beta}{2}\cos\frac{\alpha-\beta}{2}} = \dfrac{\sin\frac{\alpha+\beta}{2}}{\cos\frac{\alpha+\beta}{2}} = \tan\dfrac{\alpha+\beta}{2}$

35. $1 + \cos(2\theta) + \cos(4\theta) + \cos(6\theta) = (1 + \cos(6\theta)) + (\cos(2\theta) + \cos(4\theta))$
$= 2\cos^2(3\theta) + 2\cos(3\theta)\cos(-\theta)$
$= 2\cos(3\theta)(\cos(3\theta) + \cos\theta)$
$= 2\cos(3\theta)(2\cos(2\theta)\cos\theta)$
$= 4\cos\theta\cos(2\theta)\cos(3\theta)$

37. (a) $y = 2\sin(2061\pi t)\cos(357\pi t)$ **(b)** $y_{max} = 2$ **(c)**

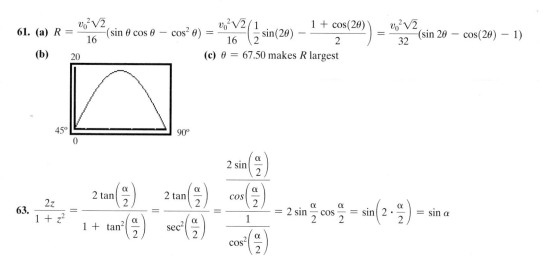

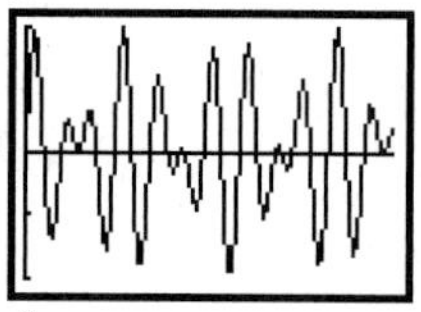

39. $$\begin{aligned}\sin(2\alpha)+\sin(2\beta)+\sin(2\gamma) &= 2\sin(\alpha+\beta)\cos(\alpha-\beta)+\sin 2\gamma = 2\sin(\alpha+\beta)\cos(\alpha-\beta)+2\sin\gamma\cos\gamma\\ &= 2\sin(\pi-\gamma)\cos(\alpha-\beta)+2\sin\gamma\cos\gamma = 2\sin\gamma\cos(\alpha-\beta)+2\sin\gamma\cos\gamma\\ &= 2\sin\gamma[\cos(\alpha-\beta)+\cos\gamma] = 2\sin\gamma\left(2\cos\frac{\alpha-\beta+\gamma}{2}\cos\frac{\alpha-\beta-\gamma}{2}\right)\\ &= 4\sin\gamma\cos\frac{\pi-2\beta}{2}\cos\frac{2\alpha-\pi}{2} = 4\sin\gamma\cos\left(\frac{\pi}{2}-\beta\right)\cos\left(\alpha-\frac{\pi}{2}\right) = 4\sin\gamma\sin\beta\sin\alpha\end{aligned}$$

41. $\sin(\alpha-\beta) = \sin\alpha\cos\beta - \cos\alpha\sin\beta$
$\sin(\alpha+\beta) = \sin\alpha\cos\beta + \cos\alpha\sin\beta$
$\sin(\alpha-\beta)+\sin(\alpha+\beta) = 2\sin\alpha\cos\beta$

43. $2\cos\dfrac{\alpha+\beta}{2}\cos\dfrac{\alpha-\beta}{2} = 2\cdot\dfrac{1}{2}\left[\cos\left(\dfrac{\alpha+\beta}{2}+\dfrac{\alpha-\beta}{2}\right)+\cos\left(\dfrac{\alpha+\beta}{2}-\dfrac{\alpha-\beta}{2}\right)\right] = \cos\dfrac{2\alpha}{2}+\cos\dfrac{2\beta}{2} = \cos\alpha+\cos\beta$

7.5 Exercises

1. 0 **3.** $-\frac{\pi}{2}$ **5.** 0 **7.** $\frac{\pi}{4}$ **9.** $\frac{\pi}{3}$ **11.** $\frac{5\pi}{6}$ **13.** $\frac{\pi}{6}$ **15.** $-\frac{\pi}{2}$ **17.** $\frac{\pi}{6}$ **19.** $\frac{2\pi}{3}$ **21.** 0.10 **23.** 1.37 **25.** 0.51 **27.** −0.38 **29.** −0.12 **31.** 1.08 **33.** 1.32 **35.** 0.46 **37.** −0.34 **39.** 2.72 **41.** −0.73 **43.** 2.55 **45.** 0.54 **47.** $\frac{4\pi}{5}$ **49.** −3.5 **51.** $-\frac{3\pi}{7}$ **53.** 3.35 min **55.** $-1 \le x \le 1$ **57.** $0 \le x \le \pi$ **59.**

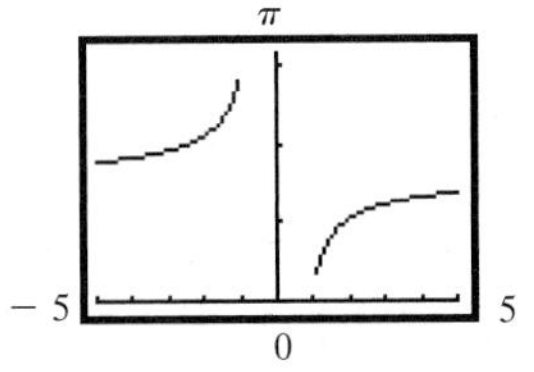

7.6 Exercises

1. $\frac{\sqrt{2}}{2}$ **3.** $-\frac{\sqrt{3}}{3}$ **5.** 2 **7.** $\sqrt{2}$ **9.** $-\frac{\sqrt{2}}{2}$ **11.** $\frac{2\sqrt{3}}{3}$ **13.** $\frac{3\pi}{4}$ **15.** $\frac{\pi}{6}$ **17.** $\frac{\sqrt{2}}{4}$ **19.** $\frac{\sqrt{5}}{2}$ **21.** $-\frac{\sqrt{14}}{2}$ **23.** $-\frac{3\sqrt{10}}{10}$ **25.** $\sqrt{5}$ **27.** $-\frac{\pi}{4}$ **29.** $\frac{\sqrt{3}}{2}$ **31.** $-\frac{24}{25}$ **33.** $-\frac{33}{65}$ **35.** $\frac{65}{63}$ **37.** $\frac{1}{39}(48-25\sqrt{3})$ **39.** $\frac{\sqrt{3}}{2}$ **41.** $\frac{7}{25}$ **43.** $\frac{24}{7}$ **45.** $\frac{24}{25}$ **47.** $\frac{1}{5}$ **49.** $\frac{25}{7}$ **51.** 4

53. $u\sqrt{1-v^2}-v\sqrt{1-u^2}$ **55.** $\dfrac{u\sqrt{1-v^2}-v}{\sqrt{1+u^2}}$ **57.** $\dfrac{uv-\sqrt{1-u^2}\sqrt{1-v^2}}{v\sqrt{1-u^2}+u\sqrt{1-v^2}}$

59. Let $\theta = \tan^{-1} v$. Then $\tan\theta = v, -\frac{\pi}{2} < \theta < \frac{\pi}{2}$. Now, $\sec\theta > 0$ and $\tan^2\theta + 1 = \sec^2\theta$. Thus $\sec(\tan^{-1} v) = \sec\theta = \sqrt{1+v^2}$.

61. Let $\theta = \cos^{-1} v$. Then $\cos\theta = v, 0 \le \theta \le \pi$, and $\tan(\cos^{-1} v) = \tan\theta = \dfrac{\sin\theta}{\cos\theta} = \dfrac{\sqrt{1-\cos^2\theta}}{\cos\theta} = \dfrac{\sqrt{1-v^2}}{v}$.

63. Let $\theta = \sin^{-1} v$. Then $\sin\theta = v, -\frac{\pi}{2} \le \theta \le \frac{\pi}{2}$, and $\cos(\sin^{-1} v) = \cos\theta = \sqrt{1-\sin^2\theta} = \sqrt{1-v^2}$.

65. Let $\alpha = \sin^{-1} v$ and $\beta = \cos^{-1} v$. Then $\sin\alpha = \cos\beta = v$, and since $\sin\alpha = \cos\left(\frac{\pi}{2}-\alpha\right)$, $\cos\left(\frac{\pi}{2}-\alpha\right) = \cos\beta$. If $v \ge 0$, then $0 \le \alpha \le \frac{\pi}{2}$, so that $\left(\frac{\pi}{2}-\alpha\right)$ and β both lie on $\left[0, \frac{\pi}{2}\right]$. If $v < 0$, then $-\frac{\pi}{2} \le \alpha < 0$, so that $\left(\frac{\pi}{2}-\alpha\right)$ and β both lie on $\left(\frac{\pi}{2}, \pi\right]$. Either way, $\cos\left(\frac{\pi}{2}-\alpha\right) = \cos\beta$ implies $\frac{\pi}{2}-\alpha = \beta$, or $\alpha+\beta = \frac{\pi}{2}$.

67. Let $\alpha = \tan^{-1}\frac{1}{v}$, and $\beta = \tan^{-1} v$. Because $\frac{1}{v}$ must be defined, $v \ne 0$ and so $\alpha, \beta \ne 0$. Then $\tan\alpha = \frac{1}{v} = \frac{1}{\tan\beta} = \cot\beta$, and since $\tan\alpha = \cot\left(\frac{\pi}{2}-\alpha\right)$, $\cot\left(\frac{\pi}{2}-\alpha\right) = \cot\beta$. Because $v > 0, 0 < \alpha < \frac{\pi}{2}$ and so $\frac{\pi}{2}-\alpha$ and β both lie on $\left(0, \frac{\pi}{2}\right)$. Then $\cot\left(\frac{\pi}{2}-\alpha\right) = \cot\beta$ implies $\frac{\pi}{2}-\alpha = \beta$, or $\alpha+\beta = \frac{\pi}{2}$.

69. $\sin(\sin^{-1} v + \cos^{-1} v) = \sin(\sin^{-1} v)\cos(\cos^{-1} v) + \cos(\sin^{-1} v)\sin(\cos^{-1} v) = (v)(v) + \sqrt{1-v^2}\sqrt{1-v^2} = v^2 + 1 - v^2 = 1$

71. **(a)** 13.92 hours or 13 hours, 55 minutes **(b)** 12 hours **(c)** 13.85 hours or 13 hours, 51 minutes

73. **(a)** 13.3 hours or 13 hours, 18 minutes **(b)** 12 hours **(c)** 13.25 hours or 13 hours, 15 minutes

75. **(a)** 12 hours **(b)** 12 hours **(c)** 12 hours **(d)** It's 12 hours

7.7 Exercises

1. $\frac{\pi}{6}, \frac{5\pi}{6}$ **3.** $\frac{5\pi}{6}, \frac{11\pi}{6}$ **5.** $\frac{\pi}{2}, \frac{3\pi}{2}$ **7.** $\frac{\pi}{2}, \frac{7\pi}{6}, \frac{11\pi}{6}$ **9.** $\frac{\pi}{3}, \frac{2\pi}{3}, \frac{4\pi}{3}, \frac{5\pi}{3}$ **11.** $\frac{4\pi}{9}, \frac{8\pi}{9}, \frac{16\pi}{9}$ **13.** $\frac{3\pi}{4}, \frac{7\pi}{4}$ **15.** $\frac{11\pi}{6}$ **17.** $\frac{7\pi}{6}, \frac{11\pi}{6}$ **19.** $\frac{3\pi}{4}, \frac{7\pi}{4}$ **21.** $\frac{2\pi}{3}, \frac{4\pi}{3}$ **23.** $\frac{3\pi}{4}, \frac{5\pi}{4}$ **25.** 0.41, 2.73 **27.** 1.37, 4.51 **29.** 2.69, 3.59 **31.** 1.82, 4.46 **33.** 28.9° **35.** Yes; it varies from 1.28 to 1.34 **37.** 1.47

39. If θ is the original angle of incidence and ϕ is the angle of refraction, then $\frac{\sin\theta}{\sin\phi} = n_2$. The angle of incidence of the emerging beam is also ϕ, and the index of refraction is $\frac{1}{n_2}$. Thus, θ is the angle of refraction of the emerging beam.

7.8 Exercises

1. $\frac{\pi}{2}, \frac{2\pi}{3}, \frac{4\pi}{3}, \frac{3\pi}{2}$ **3.** $\frac{\pi}{2}, \frac{7\pi}{6}, \frac{11\pi}{6}$ **5.** $0, \frac{\pi}{4}, \frac{5\pi}{4}$ **7.** $\frac{\pi}{4}, \frac{5\pi}{4}$ **9.** $0, \frac{\pi}{3}, \pi, \frac{5\pi}{3}$ **11.** $\frac{\pi}{2}, \frac{3\pi}{2}$ **13.** $0, \frac{2\pi}{3}, \frac{4\pi}{3}$ **15.** $0, \frac{\pi}{3}, \frac{\pi}{2}, \frac{2\pi}{3}, \pi, \frac{4\pi}{3}, \frac{3\pi}{2}, \frac{5\pi}{3}$ **17.** $0, \frac{\pi}{5}, \frac{2\pi}{5}, \frac{3\pi}{5}, \frac{4\pi}{5}, \pi, \frac{6\pi}{5}, \frac{7\pi}{5}, \frac{8\pi}{5}, \frac{9\pi}{5}$ **19.** $\frac{\pi}{6}, \frac{5\pi}{6}, \frac{3\pi}{2}$ **21.** $\frac{\pi}{3}, \frac{5\pi}{3}$ **23.** No real solutions **25.** No real solutions **27.** $\frac{\pi}{2}, \frac{7\pi}{6}$ **29.** $0, \frac{\pi}{3}, \pi, \frac{5\pi}{3}$ **31.** $\frac{\pi}{4}$

33. −1.29, 0

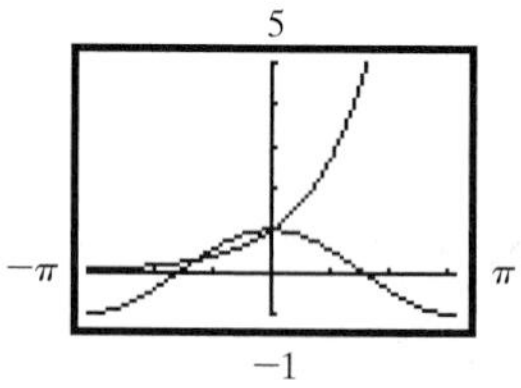

35. −2.24, 0, 2.24

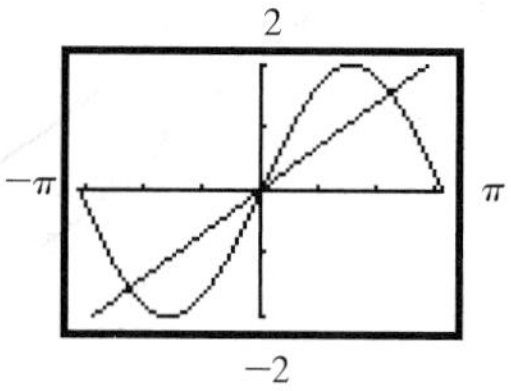

37. −0.82, 0.82

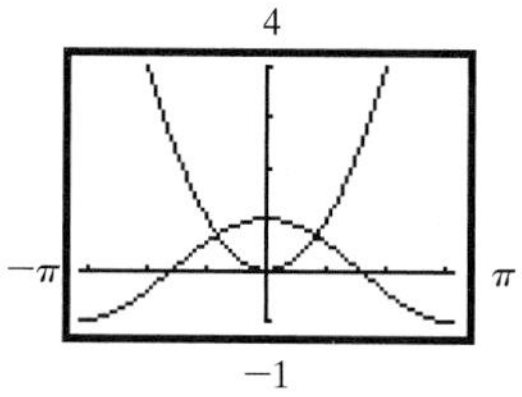

39. −1.31, 1.98, 3.84
41. 0.52
43. 1.26
45. −1.02, 1.02
47. 0, 2.15
49. 0.76, 1.35

51. **(a)** 60° **(b)** 60°
(c) $A(60°) = 12\sqrt{3}$ sq in.
(d)

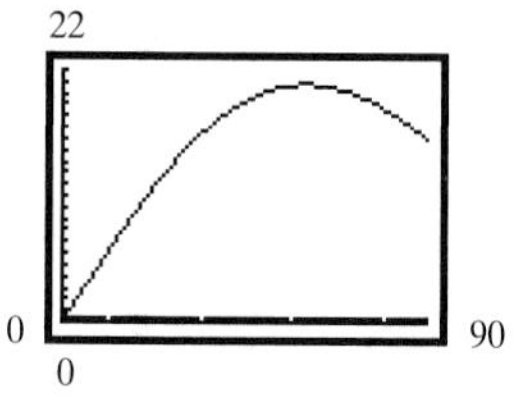

$\theta_{max} = 60°$
Maximum Area = 20.78 sq in.

53. 2.03, 4.91

55. **(a)** 29.99° or 60.01°
(b) 123.6 m
(c)

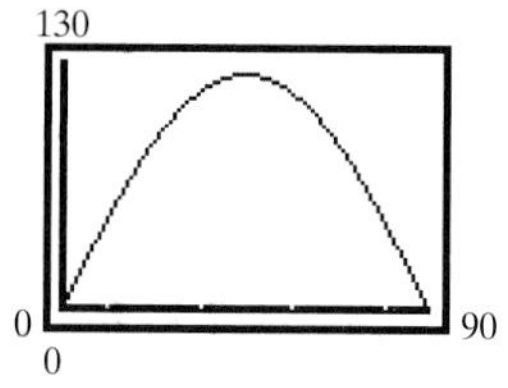

Fill-in-the-Blank Items

1. Identity; conditional **3.** + **5.** $1 - \cos\alpha$ **7.** 0

True/False Items

1. True **3.** True **5.** True **7.** False

Review Exercises

1. $\tan\theta\cot\theta - \sin^2\theta = 1 - \sin^2\theta = \cos^2\theta$ **3.** $\cos^2\theta(1 + \tan^2\theta) = \cos^2\theta\sec^2\theta = 1$

5. $4\cos^2\theta + 3\sin^2\theta = \cos^2\theta + 3(\cos^2\theta + \sin^2\theta) = 3 + \cos^2\theta$

7. $\frac{1-\cos\theta}{\sin\theta} + \frac{\sin\theta}{1-\cos\theta} = \frac{(1-\cos\theta)^2 + \sin^2\theta}{\sin\theta(1-\cos\theta)} = \frac{1 - 2\cos\theta + \cos^2\theta + \sin^2\theta}{\sin\theta(1-\cos\theta)} = \frac{2(1-\cos\theta)}{\sin\theta(1-\cos\theta)} = 2\csc\theta$

9. $\frac{\cos\theta}{\cos\theta - \sin\theta} = \frac{\frac{\cos\theta}{\cos\theta}}{\frac{\cos\theta - \sin\theta}{\cos\theta}} = \frac{1}{1 - \frac{\sin\theta}{\cos\theta}} = \frac{1}{1 - \tan\theta}$

11. $\dfrac{\csc\theta}{1+\csc\theta} = \dfrac{\frac{1}{\sin\theta}}{1+\frac{1}{\sin\theta}} = \dfrac{1}{1+\sin\theta} = \dfrac{1}{1+\sin\theta}\cdot\dfrac{1-\sin\theta}{1-\sin\theta} = \dfrac{1-\sin\theta}{1-\sin^2\theta} = \dfrac{1-\sin\theta}{\cos^2\theta}$

13. $\csc\theta - \sin\theta = \dfrac{1}{\sin\theta} - \sin\theta = \dfrac{1-\sin^2\theta}{\sin\theta} = \dfrac{\cos^2\theta}{\sin\theta} = \cos\theta\cdot\dfrac{\cos\theta}{\sin\theta} = \cos\theta\cot\theta$

15. $\dfrac{1-\sin\theta}{\sec\theta} = \cos\theta(1-\sin\theta)\cdot\dfrac{1+\sin\theta}{1+\sin\theta} = \dfrac{\cos\theta(1-\sin^2\theta)}{1+\sin\theta} = \dfrac{\cos^3\theta}{1+\sin\theta}$

17. $\cot\theta - \tan\theta = \dfrac{\cos\theta}{\sin\theta} - \dfrac{\sin\theta}{\cos\theta} = \dfrac{\cos^2\theta-\sin^2\theta}{\sin\theta\cos\theta} = \dfrac{1-2\sin^2\theta}{\sin\theta\cos\theta}$

19. $\dfrac{\cos(\alpha+\beta)}{\cos\alpha\sin\beta} = \dfrac{\cos\alpha\cos\beta-\sin\alpha\sin\beta}{\cos\alpha\sin\beta} = \dfrac{\cos\alpha\cos\beta}{\cos\alpha\sin\beta} - \dfrac{\sin\alpha\sin\beta}{\cos\alpha\sin\beta} = \cot\beta - \tan\alpha$

21. $\dfrac{\cos(\alpha-\beta)}{\cos\alpha\cos\beta} = \dfrac{\cos\alpha\cos\beta+\sin\alpha\sin\beta}{\cos\alpha\cos\beta} = \dfrac{\cos\alpha\cos\beta}{\cos\alpha\cos\beta} + \dfrac{\sin\alpha\sin\beta}{\cos\alpha\cos\beta} = 1 + \tan\alpha\tan\beta$

23. $(1+\cos\theta)\left(\tan\dfrac{\theta}{2}\right) = \left(2\cos^2\dfrac{\theta}{2}\right)\dfrac{\sin\left(\frac{\theta}{2}\right)}{\cos\left(\frac{\theta}{2}\right)} = 2\sin\dfrac{\theta}{2}\cos\dfrac{\theta}{2} = \sin\theta$

25. $2\cot\theta\cot 2\theta = 2\left(\dfrac{\cos\theta}{\sin\theta}\right)\left(\dfrac{\cos 2\theta}{\sin 2\theta}\right) = \dfrac{2\cos\theta(\cos^2\theta-\sin^2\theta)}{2\sin^2\theta\cos\theta} = \dfrac{\cos^2\theta-\sin^2\theta}{\sin^2\theta} = \cot^2\theta - 1$

27. $1 - 8\sin^2\theta\cos^2\theta = 1 - 2(2\sin\theta\cos\theta)^2 = 1 - 2\sin^2 2\theta = \cos 4\theta$

29. $\dfrac{\sin 2\theta+\sin 4\theta}{\cos 2\theta+\cos 4\theta} = \dfrac{2\sin 3\theta\cos(-\theta)}{2\cos 3\theta\cos(-\theta)} = \tan 3\theta$

31. $\dfrac{\cos 2\theta-\cos 4\theta}{\cos 2\theta+\cos 4\theta} - \tan\theta\tan 3\theta = \dfrac{-2\sin 3\theta\sin(-\theta)}{2\cos 3\theta\cos(-\theta)} - \tan\theta\tan 3\theta = \tan 3\theta\tan\theta - \tan\theta\tan 3\theta = 0$

33. $\frac{1}{4}(\sqrt{6}-\sqrt{2})$ **35.** $\frac{1}{4}(\sqrt{6}-\sqrt{2})$ **37.** $\frac{1}{2}$ **39.** $\sqrt{\dfrac{2-\sqrt{2}}{2+\sqrt{2}}} = \sqrt{2}-1$

41. **(a)** $-\frac{33}{65}$ **(b)** $-\frac{56}{65}$ **(c)** $-\frac{63}{65}$ **(d)** $\frac{33}{56}$ **(e)** $\frac{24}{25}$ **(f)** $\frac{119}{169}$ **(g)** $\frac{5\sqrt{26}}{26}$ **(h)** $\frac{2\sqrt{5}}{5}$

43. **(a)** $-\frac{16}{65}$ **(b)** $-\frac{63}{65}$ **(c)** $-\frac{56}{65}$ **(d)** $\frac{16}{63}$ **(e)** $\frac{24}{25}$ **(f)** $\frac{119}{169}$ **(g)** $\frac{\sqrt{26}}{26}$ **(h)** $-\frac{\sqrt{10}}{10}$

45. **(a)** $-\frac{63}{65}$ **(b)** $\frac{16}{65}$ **(c)** $\frac{33}{65}$ **(d)** $-\frac{63}{16}$ **(e)** $\frac{24}{25}$ **(f)** $-\frac{119}{169}$ **(g)** $\frac{2\sqrt{13}}{13}$ **(h)** $-\frac{\sqrt{10}}{10}$

47. **(a)** $\frac{(-\sqrt{3}-2\sqrt{2})}{6}$ **(b)** $\frac{(1-2\sqrt{6})}{6}$ **(c)** $\frac{(-\sqrt{3}+2\sqrt{2})}{6}$ **(d)** $\frac{(-\sqrt{3}-2\sqrt{2})}{(1-2\sqrt{6})} = \frac{(8\sqrt{2}+9\sqrt{3})}{23}$ **(e)** $-\frac{\sqrt{3}}{2}$ **(f)** $-\frac{7}{9}$ **(g)** $\frac{\sqrt{3}}{3}$ **(h)** $\frac{\sqrt{3}}{2}$ **49.** **(a)** 1 **(b)** 0 **(c)** $-\frac{1}{9}$ **(d)** Not defined **(e)** $\frac{4\sqrt{5}}{9}$ **(f)** $-\frac{1}{9}$ **(g)** $\frac{\sqrt{30}}{6}$ **(h)** $-\frac{\sqrt{6}\sqrt{3-\sqrt{5}}}{6}$ **51.** $\frac{\pi}{2}$ **53.** $\frac{\pi}{4}$ **55.** $\frac{5\pi}{6}$ **57.** $\frac{\sqrt{2}}{2}$ **59.** $-\sqrt{3}$ **61.** $\frac{2\sqrt{3}}{3}$ **63.** $\frac{3}{5}$ **65.** $-\frac{4}{3}$ **67.** $-\frac{\pi}{6}$ **69.** $-\frac{\pi}{4}$ **71.** $\frac{4+3\sqrt{3}}{10}$ **73.** $\frac{3\sqrt{3}+4}{3-4\sqrt{3}} = \frac{48+25\sqrt{3}}{-39}$ **75.** $-\frac{24}{25}$ **77.** $\frac{\pi}{3}, \frac{5\pi}{3}$ **79.** $\frac{3\pi}{4}, \frac{5\pi}{4}$ **81.** $\frac{3\pi}{4}, \frac{7\pi}{4}$ **83.** $0, \frac{\pi}{2}, \pi, \frac{3\pi}{2}$ **85.** $1.12, \pi - 1.12$ **87.** $0, \pi$ **89.** $0, \frac{2\pi}{3}, \pi, \frac{4\pi}{3}$ **91.** $0, \frac{\pi}{6}, \frac{5\pi}{6}$ **93.** $\frac{\pi}{6}, \frac{\pi}{2}, \frac{5\pi}{6}$ **95.** $\frac{\pi}{2}, \pi$ **97.** 1.11 **99.** 0.87 **101.** 2.22

CHAPTER 8 Applications of Trigonometric Functions

8.1 Exercises

1. $\sin\theta = \frac{5}{13}$; $\cos\theta = \frac{12}{13}$; $\tan\theta = \frac{5}{12}$; $\csc\theta = \frac{13}{5}$; $\sec\theta = \frac{13}{12}$; $\cot\theta = \frac{12}{5}$

3. $\sin\theta = \frac{2\sqrt{13}}{13}$; $\cos\theta = \frac{3\sqrt{13}}{13}$; $\tan\theta = \frac{2}{3}$; $\csc\theta = \frac{\sqrt{13}}{2}$; $\sec\theta = \frac{\sqrt{13}}{3}$; $\cot\theta = \frac{3}{2}$

5. $\sin\theta = \frac{\sqrt{3}}{2}$; $\cos\theta = \frac{1}{2}$; $\tan\theta = \sqrt{3}$; $\csc\theta = \frac{2\sqrt{3}}{3}$; $\sec\theta = 2$; $\cot\theta = \frac{\sqrt{3}}{3}$

7. $\sin\theta = \frac{\sqrt{6}}{3}$; $\cos\theta = \frac{\sqrt{3}}{3}$; $\tan\theta = \sqrt{2}$; $\csc\theta = \frac{\sqrt{6}}{2}$; $\sec\theta = \sqrt{3}$; $\cot\theta = \frac{\sqrt{2}}{2}$

9. $\sin\theta = \frac{\sqrt{5}}{5}$; $\cos\theta = \frac{2\sqrt{5}}{5}$; $\tan\theta = \frac{1}{2}$; $\csc\theta = \sqrt{5}$; $\sec\theta = \frac{\sqrt{5}}{2}$; $\cot\theta = 2$ **11.** 0 **13.** 1 **15.** 0 **17.** 0 **19.** 1

21. **(a)** $\frac{1}{3}$ **(b)** $\frac{8}{9}$ **(c)** 3 **(d)** 3 **23.** **(a)** 17 **(b)** $\frac{1}{4}$ **(c)** 4 **(d)** $\frac{17}{16}$ **25.** **(a)** $\frac{1}{4}$ **(b)** 15 **(c)** 4 **(d)** $\frac{16}{15}$ **27.** 0.6
29. $a \approx 13.74, c \approx 14.62, \alpha = 70°$ **31.** $b \approx 5.03, c \approx 7.83, \alpha \approx 50°$ **33.** $a \approx 0.71, c \approx 4.06, \beta = 80°$
35. $c \approx 5.83, \alpha \approx 59.0°, \beta \approx 31.0°$ **37.** $b \approx 4.58, \alpha \approx 23.6°, \beta \approx 66.4°$ **39.** 4.59 in., 6.55 in. **41.** 5.52 in. or 11.83 in. **43.** 23.6° and 66.4°
45. 70.02 ft **47.** 985.91 ft **49.** 137.37 m **51.** 20.67 ft **53.** 15.9° **55.** 60.27 ft **57.** 530.18 ft **59.** 554.52 ft
61. **(a)** 111.96 ft/sec or 76.3 mph **(b)** 82.42 ft/sec or 56.2 mph **(c)** Under 18.8° **63.** S76.6°E **65.** 14.9°

67. **(a)** $T(\theta) = \dfrac{2}{3 \sin \theta} - \dfrac{1}{4 \tan \theta} + 1$ **(b)** 68°; 1.62 hours; 1.12 hours

69. **(a)** 10 min **(b)** 20 min **(c)** $T(\theta) = 5 - \dfrac{5}{3 \tan \theta} + \dfrac{5}{\sin \theta}$ **(d)** 10.4 min **(e)** 70.5°; 9.71 min; 79.2 ft

71. 3.83 mi **73.** No; About 1 foot **75.** Line of sight: 146 miles; 146 nautical miles; 168 statute miles

77. **(a)** $|OA| = |OC| = 1$; angle OAC + angle $OAC + 180° - \theta = 180°$; angle $OAC = \dfrac{\theta}{2}$

(b) $\sin \theta = \dfrac{|CD|}{|OC|} = |CD|$; $\cos \theta = \dfrac{|OD|}{|OC|} = |OD|$ **(c)** $\tan \dfrac{\theta}{2} = \dfrac{|CD|}{|AD|} = \dfrac{\sin \theta}{1 + |OD|} = \dfrac{\sin \theta}{1 + \cos \theta}$

79. $h = x \tan \theta$ and $h = (1 - x) \tan n\theta$; thus, $x \tan \theta = (1 - x) \tan n\theta$

$$x = \frac{\tan n\theta}{\tan \theta + \tan n\theta}$$

81. **(a)** $A(\theta) = 2 \sin \theta \cos \theta$ **(b)** From double-angle formula, since $2 \sin \theta \cos \theta = \sin(2\theta)$ **(c)** $\theta = 45°$ **(d)** $\dfrac{\sqrt{2}}{2}$ by $\sqrt{2}$

8.2 Exercises

1. $a \approx 3.23, b \approx 3.55, \alpha = 40°$ **3.** $a \approx 3.25, c \approx 4.23, \beta = 45°$ **5.** $\gamma = 95°, c \approx 9.86, a \approx 6.36$ **7.** $\alpha = 40°, a = 2, c \approx 3.06$
9. $\gamma = 120°, b \approx 1.06, c \approx 2.69$ **11.** $\alpha = 100°, a \approx 5.24, c \approx 0.92$ **13.** $\beta = 40°, a \approx 5.64, b \approx 3.86$ **15.** $\gamma = 100°, a \approx 1.31, b \approx 1.31$
17. One triangle; $\beta \approx 30.7°, \gamma \approx 99.3°, c \approx 3.86$ **19.** One triangle; $\gamma \approx 36.2°, \alpha \approx 43.8°, a \approx 3.51$ **21.** No triangle
23. Two triangles; $\gamma_1 \approx 30.9°, \alpha_1 \approx 129.1°, a_1 \approx 9.08$ or $\gamma_2 \approx 149.1°, \alpha_2 \approx 10.9°, a_2 \approx 2.21$ **25.** No triangle
27. Two triangles; $\alpha_1 \approx 57.7°, \beta_1 \approx 97.3°, b_1 \approx 2.35$ or $\alpha_2 \approx 122.3°, \beta_2 \approx 32.7°, b_2 \approx 1.28$
29. **(a)** Station Able is about 143.33 mi from the ship; Station Baker is about 135.58 mi from the ship. **(b)** Approx. 41 min **31.** 1490.48 ft
33. 381.69 ft **35.** **(a)** 169.18 mi **(b)** 161.3° **37.** 84.7°; 183.72 ft **39.** 2.64 mi **41.** 1.88 mi **43.** 449.36 ft **45.** 39.39 ft **47.** 29.97 ft

49. $$\frac{a-b}{c} = \frac{a}{c} - \frac{b}{c} = \frac{\sin \alpha}{\sin \gamma} - \frac{\sin \beta}{\sin \gamma} = \frac{\sin \alpha - \sin \beta}{\sin \gamma} = \frac{2 \sin \frac{\alpha - \beta}{2} \cos \frac{\alpha + \beta}{2}}{2 \sin \frac{\gamma}{2} \cos \frac{\gamma}{2}} = \frac{\sin \frac{\alpha - \beta}{2} \cos\left(\frac{\pi}{2} - \frac{\gamma}{2}\right)}{\sin \frac{\gamma}{2} \cos \frac{\gamma}{2}} = \frac{\sin \frac{1}{2}(\alpha - \beta)}{\cos \frac{1}{2}\gamma}$$

51. $$\frac{a-b}{a+b} = \frac{\frac{a-b}{c}}{\frac{a+b}{c}} = \frac{\dfrac{\sin \frac{1}{2}(\alpha - \beta)}{\cos \frac{1}{2}\gamma}}{\dfrac{\cos \frac{1}{2}(\alpha - \beta)}{\sin \frac{1}{2}\gamma}} = \frac{\tan \frac{1}{2}(\alpha - \beta)}{\cot \frac{1}{2}\gamma} = \frac{\tan \frac{1}{2}(\alpha - \beta)}{\tan\left(\frac{\pi}{2} - \frac{\gamma}{2}\right)} = \frac{\tan \frac{1}{2}(\alpha - \beta)}{\tan \frac{1}{2}(\alpha + \beta)}$$

8.3 Exercises

1. $b \approx 2.95, \alpha \approx 28.7°, \gamma \approx 106.3°$ **3.** $c \approx 3.75, \alpha \approx 32.1°, \beta \approx 52.9°$ **5.** $\alpha \approx 48.5°, \beta \approx 38.6°, \gamma \approx 92.9°$
7. $\alpha \approx 127.2°, \beta \approx 32.1°, \gamma \approx 20.7°$ **9.** $c \approx 2.57, \alpha \approx 48.6°, \beta \approx 91.4°$ **11.** $a \approx 2.99, \beta \approx 19.2°, \gamma \approx 80.8°$
13. $b \approx 4.14, \alpha \approx 43.0°, \gamma \approx 27.0°$ **15.** $c \approx 1.69, \alpha \approx 65.0°, \beta \approx 65.0°$ **17.** $\alpha \approx 67.4°, \beta = 90°, \gamma \approx 22.6°$
19. $\alpha = 60°, \beta = 60°, \gamma = 60°$ **21.** $\alpha \approx 33.6°, \beta \approx 62.2°, \gamma \approx 84.3°$ **23.** $\alpha \approx 97.9°, \beta \approx 52.4°, \gamma \approx 29.7°$
25. 70.75 ft **27.** **(a)** 12.0° **(b)** 220.8 mph **29.** **(a)** 63.7 ft **(b)** 66.8 ft **(c)** 92.8° **31.** **(a)** 492.6 ft **(b)** 269.3 ft **33.** 342.3 ft

35. Using the Law of Cosines:
$L^2 = x^2 + r^2 - 2rx \cos \theta$
$x^2 - 2rx \cos \theta + r^2 - L^2 = 0$
Then, using the quadratic formula:
$x = r \cos \theta + \sqrt{r^2 \cos^2 \theta + L^2 - r^2}$

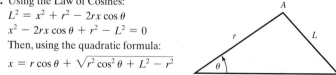

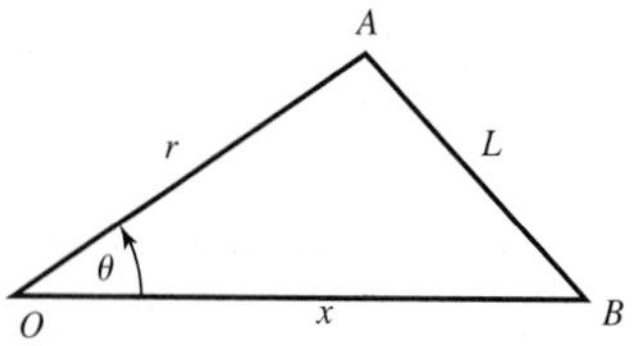

37. $\cos\frac{\gamma}{2} = \sqrt{\frac{1+\cos\gamma}{2}} = \sqrt{\frac{1+\frac{a^2+b^2-c^2}{2ab}}{2}} = \sqrt{\frac{2ab+a^2+b^2-c^2}{4ab}} = \sqrt{\frac{(a+b)^2-c^2}{4ab}} = \sqrt{\frac{(a+b+c)(a+b-c)}{4ab}}$
$= \sqrt{\frac{2s(2s-2c)}{4ab}} = \sqrt{\frac{s(s-c)}{ab}}$

39. $\frac{\cos\alpha}{a} + \frac{\cos\beta}{b} + \frac{\cos\gamma}{c} = \frac{b^2+c^2-a^2}{2abc} + \frac{a^2+c^2-b^2}{2abc} + \frac{a^2+b^2-c^2}{2abc} = \frac{b^2+c^2-a^2+a^2+c^2-b^2+a^2+b^2-c^2}{2abc}$
$= \frac{a^2+b^2+c^2}{2abc}$

8.4 Exercises

1. 2.83 **3.** 3 **5.** 14.98 **7.** 9.56 **9.** 3.86 **11.** 1.48 **13.** 2.82 **15.** 1.53 **17.** 30 **19.** 1.73 **21.** 19.90 **23.** 19.81 **25.** 9.03 sq ft **27.** \$5446.38 **29.** 9.26 sq cm **31.** $A = \frac{1}{2}ab\sin\gamma = \frac{1}{2}a\sin\gamma\left(\frac{a\sin\beta}{\sin\alpha}\right) = \frac{a^2\sin\beta\sin\gamma}{2\sin\alpha}$ **33.** 0.92 **35.** 2.27 **37.** 5.44 **39.** $A = \frac{1}{2}r^2(\theta + \sin\theta)$

41. (a) Area $\triangle OAC = \frac{1}{2}|OC||AC| = \frac{1}{2}\cdot\frac{|OC|}{1}\cdot\frac{|AC|}{1} = \frac{1}{2}\sin\alpha\cos\alpha$

(b) Area $\triangle OCB = \frac{1}{2}|BC||OC| = \frac{1}{2}|OB|^2\frac{|BC|}{|OB|}\cdot\frac{|OC|}{|OB|} = \frac{1}{2}|OB|^2\sin\beta\cos\beta$

(c) Area $\triangle OAB = \frac{1}{2}|BD||OA| = \frac{1}{2}|OB|\frac{|BD|}{|OB|} = \frac{1}{2}|OB|\sin(\alpha+\beta)$

(d) $\frac{\cos\alpha}{\cos\beta} = \frac{\frac{|OC|}{1}}{\frac{|OC|}{|OB|}} = |OB|$ **(e)** Use the hint and above results.

43. 31,145.15 sq ft **45.** $h_1 = 2\frac{K}{a}, h_2 = 2\frac{K}{b}, h_3 = 2\frac{K}{c}$. Then $\frac{1}{h_1} + \frac{1}{h_2} + \frac{1}{h_3} = \frac{a}{2K} + \frac{b}{2K} + \frac{c}{2K} = \frac{a+b+c}{2K} = \frac{2s}{2K} = \frac{s}{K}$

47. Angle AOB measures $180 - \left(\frac{\alpha}{2} + \frac{\beta}{2}\right) = 180 - \frac{1}{2}(180 - \gamma) = 90 + \frac{\gamma}{2}$, so $r = \frac{c\sin(\alpha/2)\sin(\beta/2)}{\sin(90+\gamma/2)} = \frac{c\sin(\alpha/2)\sin(\beta/2)}{\cos(\gamma/2)}$

49. $\cot\frac{\alpha}{2} + \cot\frac{\beta}{2} + \cot\frac{\gamma}{2} = \frac{s-a}{r} + \frac{s-b}{r} + \frac{s-c}{r} = \frac{3s-(a+b+c)}{r} = \frac{3s-2s}{r} = \frac{s}{r}$

8.5 Exercises

1. $d = -5\cos\pi t$ **3.** $d = -6\cos 2t$ **5.** $d = -5\sin\pi t$ **7.** $d = -6\sin 2t$ **9. (a)** Simple harmonic **(b)** 5 m **(c)** $\frac{2\pi}{3}$ sec **(d)** $\frac{3}{2\pi}$ oscillation/sec **11. (a)** Simple harmonic **(b)** 6 m **(c)** 2 sec **(d)** $\frac{1}{2}$ oscillation/sec **13. (a)** Simple harmonic **(b)** 3 m **(c)** 4π sec **(d)** $\frac{1}{4\pi}$ oscillation/sec **15. (a)** Simple harmonic **(b)** 2 m **(c)** 1 sec **(d)** 1 oscillation/sec

17. (a) $d = -10e^{-0.7t/50}\cos\left(\sqrt{\left(\frac{2\pi}{5}\right)^2 - \frac{(0.7)^2}{4(625)}}\,t\right)$

(b)

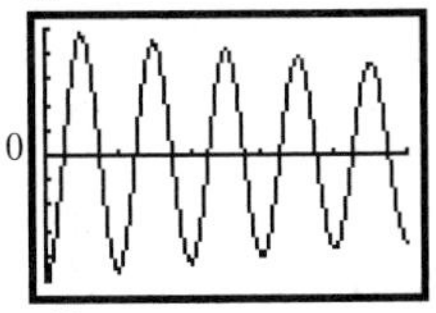

19. (a) $d = -18e^{-0.6t/60}\cos\left(\sqrt{\left(\frac{\pi}{2}\right)^2 - \frac{(0.6)^2}{4(900)}}\,t\right)$

(b)

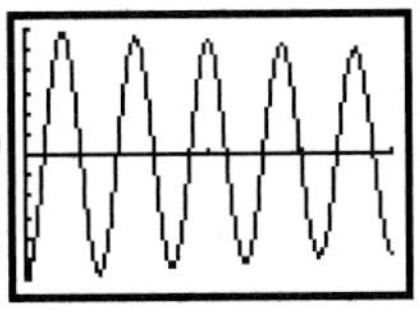

21. (a) $d = -5e^{-0.8t/20}\cos\left(\sqrt{\left(\frac{2\pi}{3}\right)^2 - \frac{(0.8)^2}{4(100)}}\,t\right)$ **(b)**

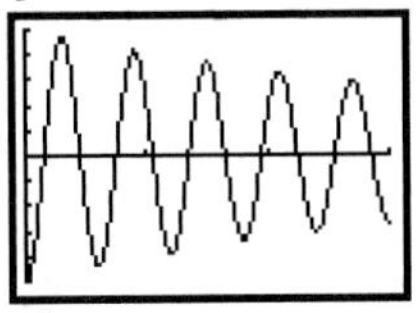

23. (a) The motion is damped. The bob has mass $m = 20$ kg with a damping factor of 0.7 kg/sec.
(b) 20 m downward
(c)

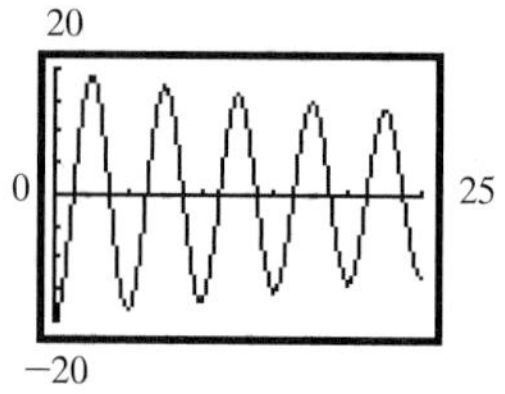

(d) 18.33 m **(e)** $d \to 0$

25. (a) The motion is damped. The bob has mass $m = 40$ kg with a damping factor of 0.6 kg/sec.
(b) 30 m downward
(c)

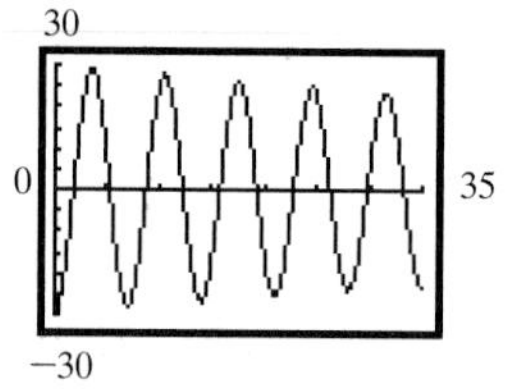

(d) 28.47 m **(e)** $d \to 0$

27. (a) The motion is damped. The bob has mass $m = 15$ kg with a damping factor of 0.9 kg/sec.
(b) 15 m downward
(c)

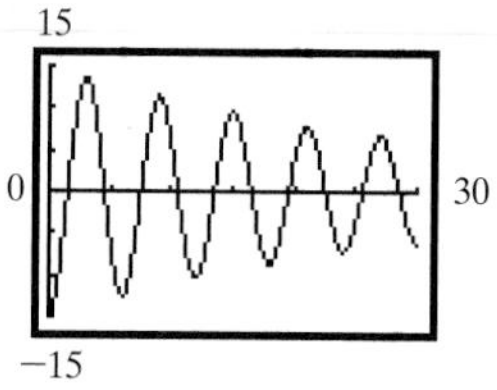

(d) 12.53 m **(e)** $d \to 0$

29. (a)

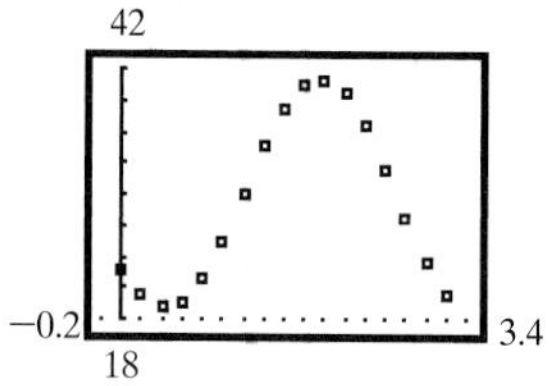

(b) $y = 10.935 \sin\left(\frac{2\pi}{3}x - 2.4086\right) + 29.995$, assuming $T = 3$ seconds

(c)

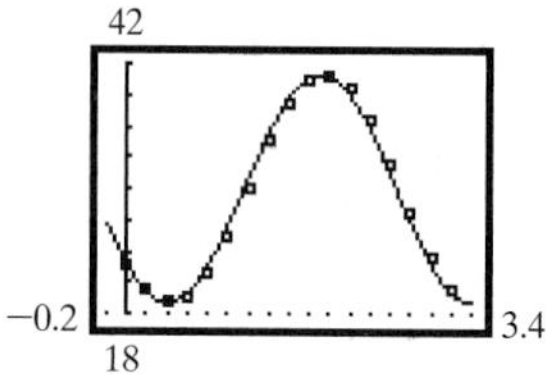

(d) $y = 11.043 \sin(2.076x - 2.472) + 29.877$ **(f)** 3.03 seconds

(e)

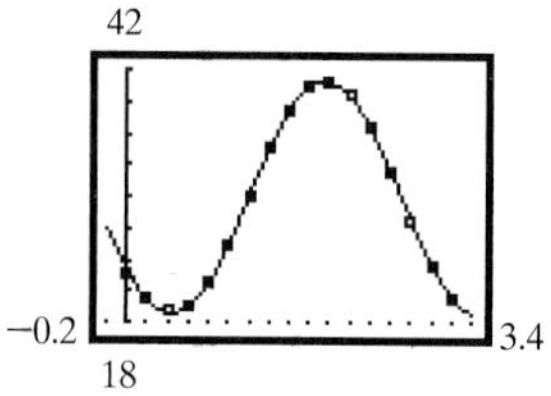

31. (a)

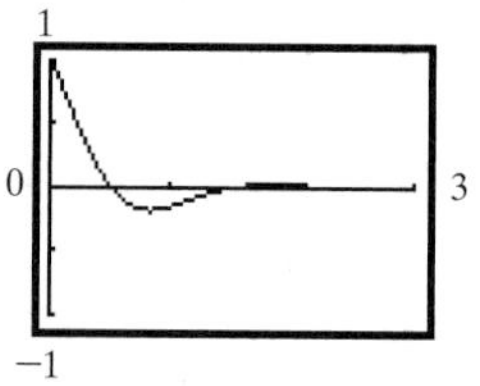

(b) The graph of V touches the graph of $y = e^{-1.9t}$ when $t = 0, 2, 4, \cdots$. The graph of V touches the graph of $y = -e^{-1.9t}$ when $t = 1, 3, 5, \cdots$.
(c) $-0.1 < V < 0.1$ for $t > 1.15$

Fill-in-the Blank Items

1. complementary **3.** Sines **5.** Heron's **7.** 5; 0.5

True/False Items

1. False **3.** True **5.** True

Review Exercises

1. $\alpha = 70°, b \approx 3.42, a \approx 9.4$ **3.** $a \approx 4.58, \alpha = 66.4°, \beta \approx 23.6°$ **5.** $\gamma = 100°, b \approx 0.65, c \approx 1.29$ **7.** $\beta \approx 56.8°, \gamma \approx 23.2°, b \approx 4.25$
9. No triangle **11.** $b \approx 3.32, \alpha \approx 62.8°, \gamma \approx 17.2°$ **13.** No triangle **15.** $c \approx 2.32, \alpha \approx 16.1°, \beta \approx 123.9°$
17. $\beta = 36.2°, \gamma = 63.8°, c = 4.55$ **19.** $\alpha = 39.6°, \beta = 18.5°, \gamma = 121.9°$
21. Two triangles: $\beta_1 \approx 13.4°, \gamma_1 \approx 156.6°, c_1 \approx 6.86$ or $\beta_2 \approx 166.6°, \gamma_2 \approx 3.4°, c_2 \approx 1.02$ **23.** $a = 5.23, \beta = 46.0°, \gamma = 64.0°$
25. 1.93 **27.** 18.79 **29.** 6 **31.** 3.80 **33.** 0.32 **35.** 839.10 ft **37.** 23.32 ft **39.** 2.15 mi **41.** 204.07 mi
43. (a) 2.59 mi **(b)** 2.92 mi **(c)** 2.53 mi **45. (a)** 131.8 mi **(b)** 23.1° **(c)** 0.21 hr **47.** 8798.67 sq ft **49.** 15.71 sq in. **51.** 76.94 in.
53. (a) Simple harmonic **(b)** 6 ft **(c)** π sec **(d)** $\frac{1}{\pi}$ oscillation/sec **55. (a)** Simple harmonic **(b)** 2 ft **(c)** 2 sec **(d)** $\frac{1}{2}$ oscillation/sec

57. (a) $d = -15e^{-0.75t/80} \cos\left(\sqrt{\left(\frac{2\pi}{5}\right)^2 - \frac{(0.75)^2}{4(40)^2}}\, t\right)$ **(b)**

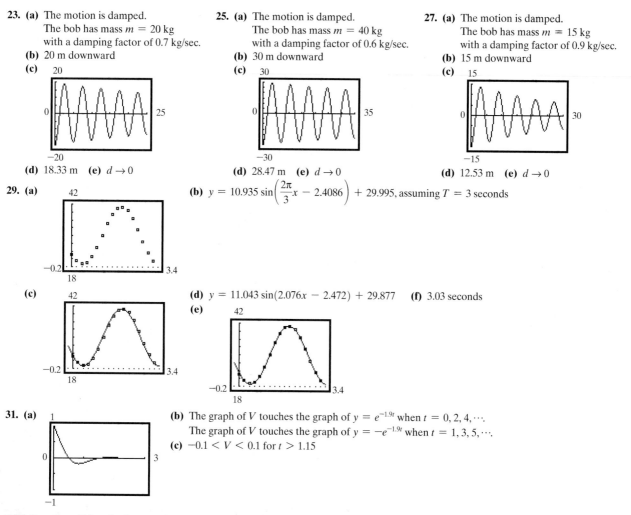

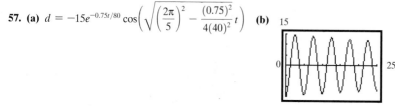

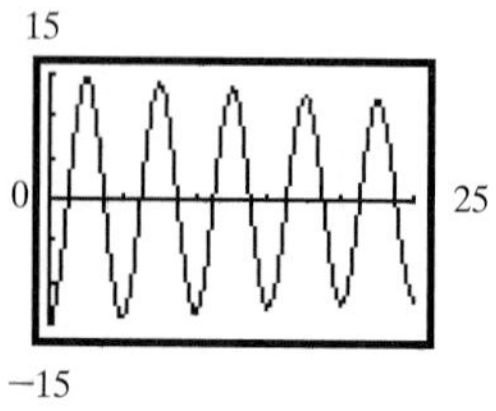

59. (a) The motion is damped. The bob has mass $m = 20$ kg with a damping factor of 0.6 kg/sec.
(b) 15 m downward **(c)** 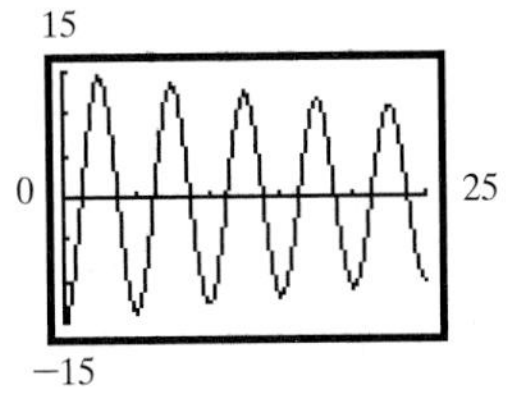

(d) 13.92 m **(e)** $d \to 0$

CHAPTER 9 Polar Coordinates; Vectors

9.1 Exercises

1. A **3.** C **5.** B **7.** A

9. 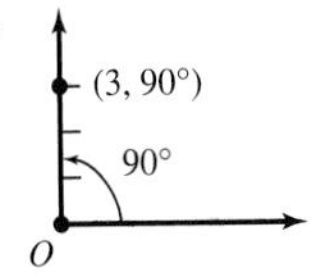

11. 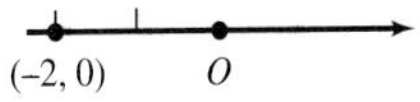

13. 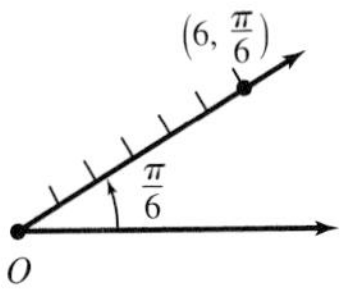

15. 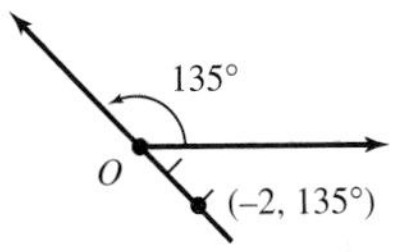

17. 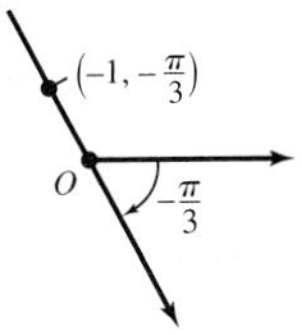

19. 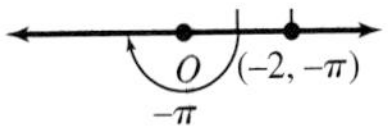

21. 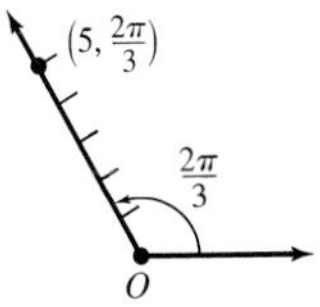

(a) $\left(5, -\frac{4\pi}{3}\right)$ **(b)** $\left(-5, \frac{5\pi}{3}\right)$ **(c)** $\left(5, \frac{8\pi}{3}\right)$

23. 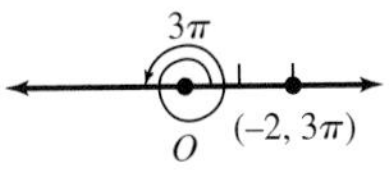

(a) $(2, -2\pi)$ **(b)** $(-2, \pi)$ **(c)** $(2, 2\pi)$

25. 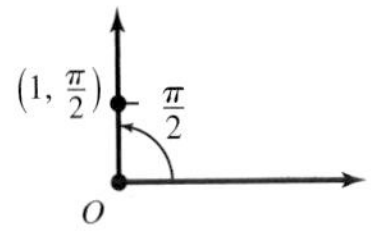

(a) $\left(1, -\frac{3\pi}{2}\right)$ **(b)** $\left(-1, \frac{3\pi}{2}\right)$ **(c)** $\left(1, \frac{5\pi}{2}\right)$

27. 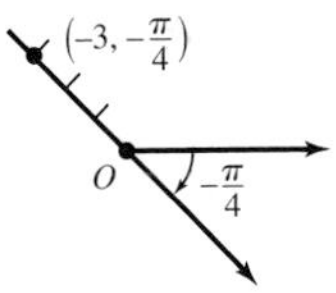

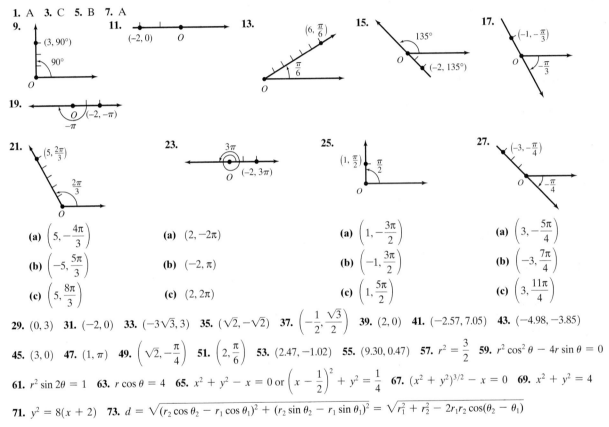

(a) $\left(3, -\frac{5\pi}{4}\right)$ **(b)** $\left(-3, \frac{7\pi}{4}\right)$ **(c)** $\left(3, \frac{11\pi}{4}\right)$

29. $(0, 3)$ **31.** $(-2, 0)$ **33.** $(-3\sqrt{3}, 3)$ **35.** $(\sqrt{2}, -\sqrt{2})$ **37.** $\left(-\frac{1}{2}, \frac{\sqrt{3}}{2}\right)$ **39.** $(2, 0)$ **41.** $(-2.57, 7.05)$ **43.** $(-4.98, -3.85)$

45. $(3, 0)$ **47.** $(1, \pi)$ **49.** $\left(\sqrt{2}, -\frac{\pi}{4}\right)$ **51.** $\left(2, \frac{\pi}{6}\right)$ **53.** $(2.47, -1.02)$ **55.** $(9.30, 0.47)$ **57.** $r^2 = \frac{3}{2}$ **59.** $r^2 \cos^2 \theta - 4r \sin \theta = 0$

61. $r^2 \sin 2\theta = 1$ **63.** $r \cos \theta = 4$ **65.** $x^2 + y^2 - x = 0$ or $\left(x - \frac{1}{2}\right)^2 + y^2 = \frac{1}{4}$ **67.** $(x^2 + y^2)^{3/2} - x = 0$ **69.** $x^2 + y^2 = 4$

71. $y^2 = 8(x + 2)$ **73.** $d = \sqrt{(r_2 \cos \theta_2 - r_1 \cos \theta_1)^2 + (r_2 \sin \theta_2 - r_1 \sin \theta_1)^2} = \sqrt{r_1^2 + r_2^2 - 2r_1 r_2 \cos(\theta_2 - \theta_1)}$

9.2 Exercises

1. Circle, radius 4, center at pole

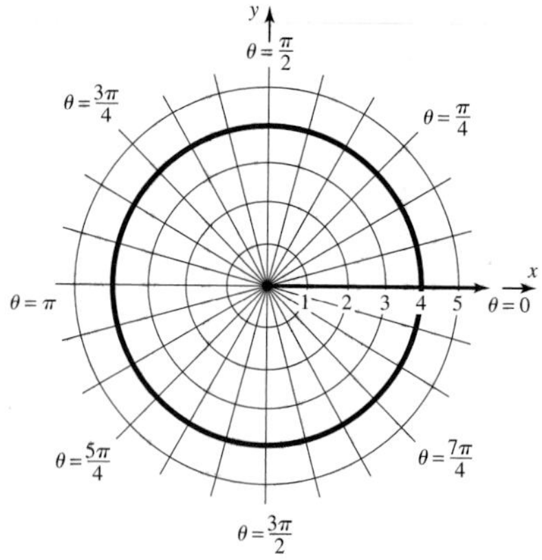

3. Line through pole, making an angle of $\frac{\pi}{3}$ with polar axis

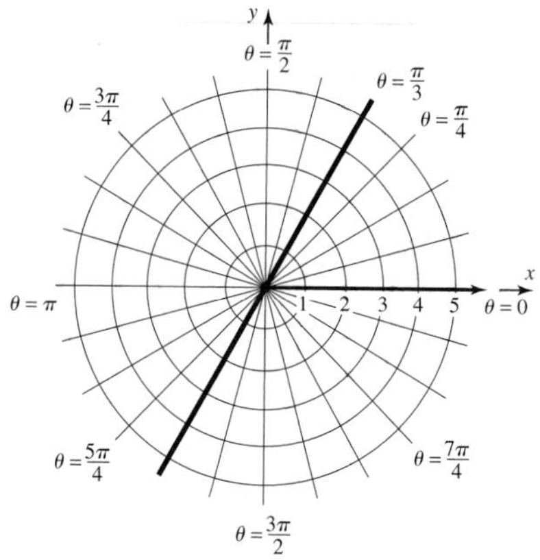

5. Horizontal line 4 units above the pole

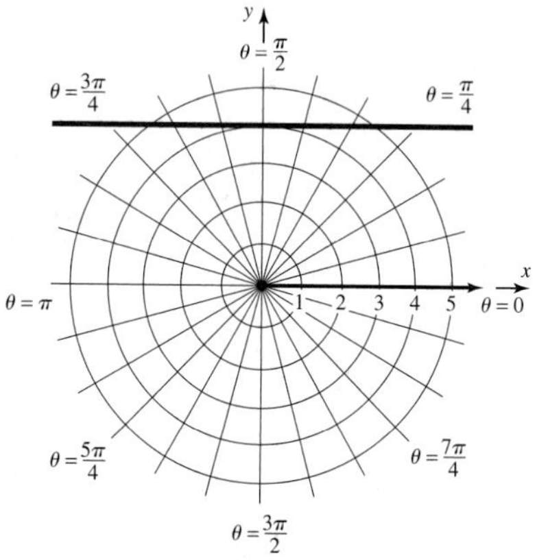

7. Vertical line 2 units to the left of the pole

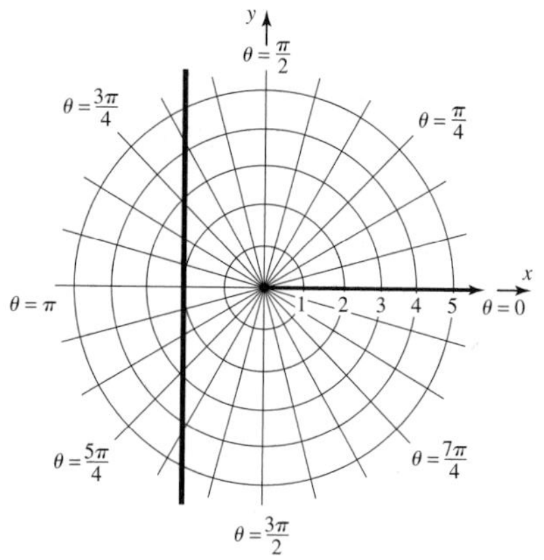

9. Circle, radius 1, center $(1, 0)$ in rectangular coordinates

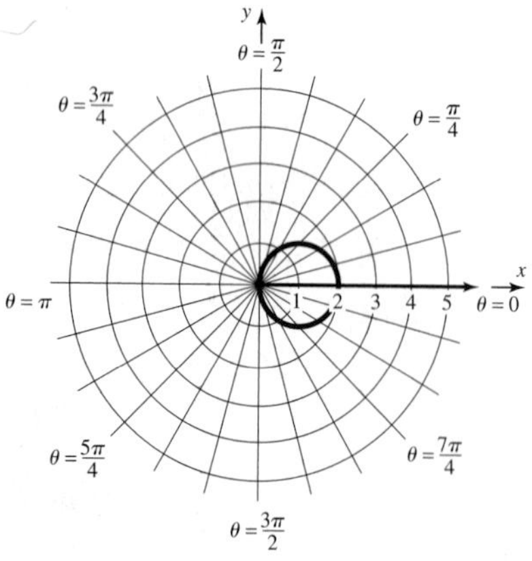

11. Circle, radius 2, center at $(0, -2)$ in rectangular coordinates

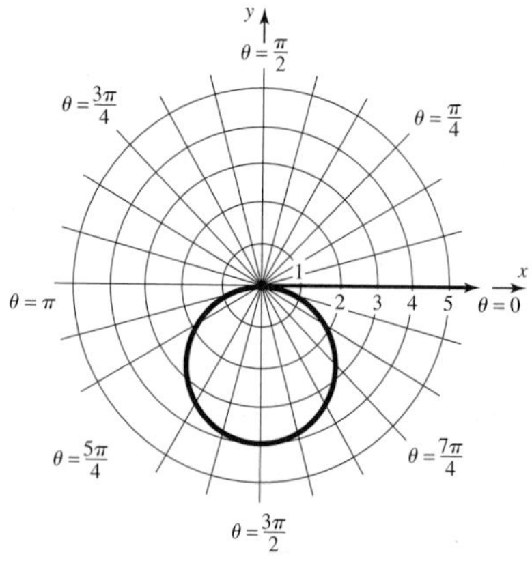

13. Circle, radius 2, center at (2, 0) in rectangular coordinates

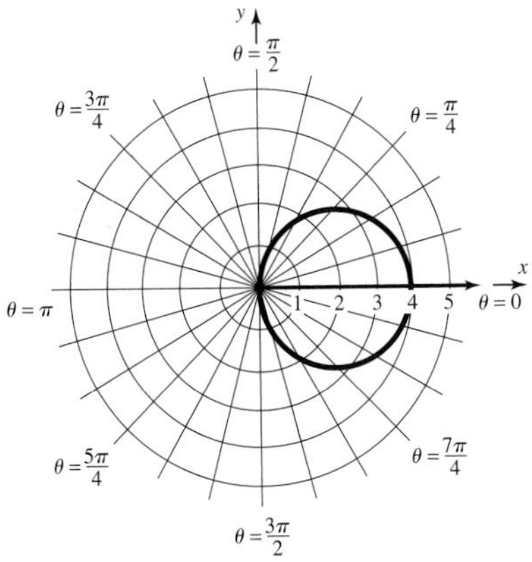

15. Circle, radius 1, center at (0, −1) in rectangular coordinates

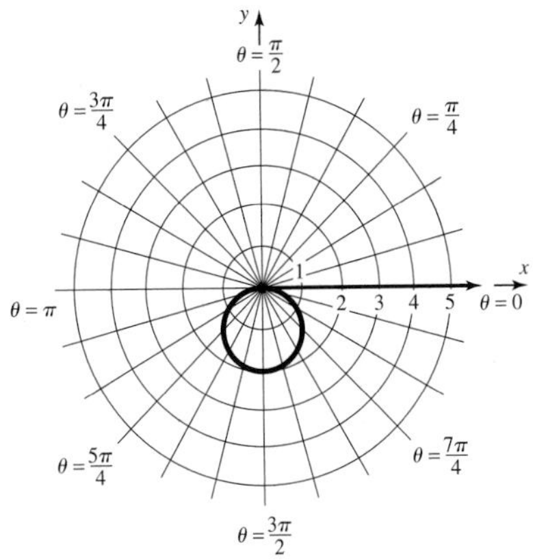

17. **(E)** **19.** **(F)** **21.** **(H)** **23.** **(D)** **25.** **(D)** **27.** **(F)** **29.** **(A)**

31. Cardioid

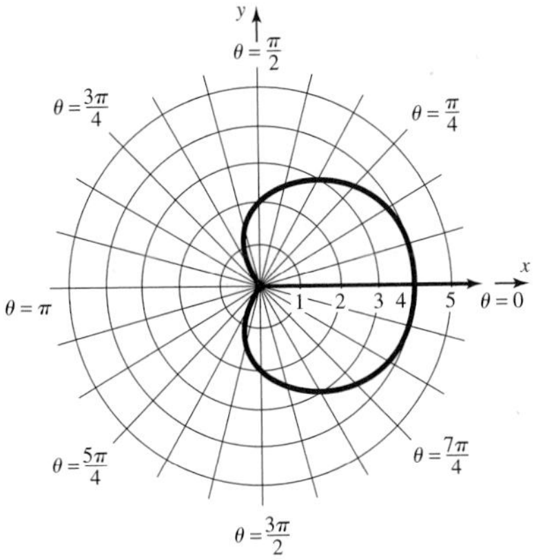

33. Cardioid

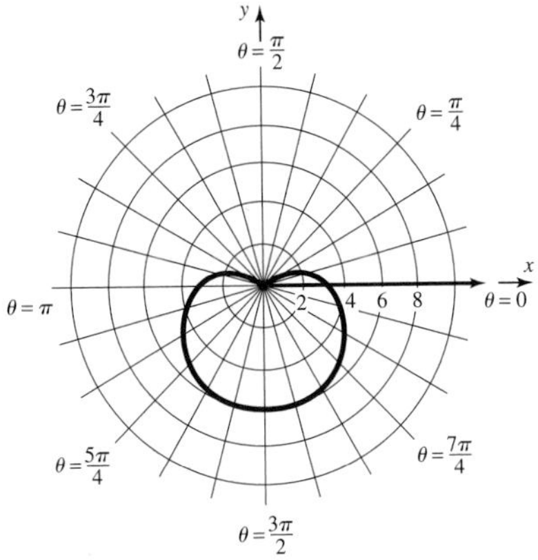

35. Limaçon without inner loop

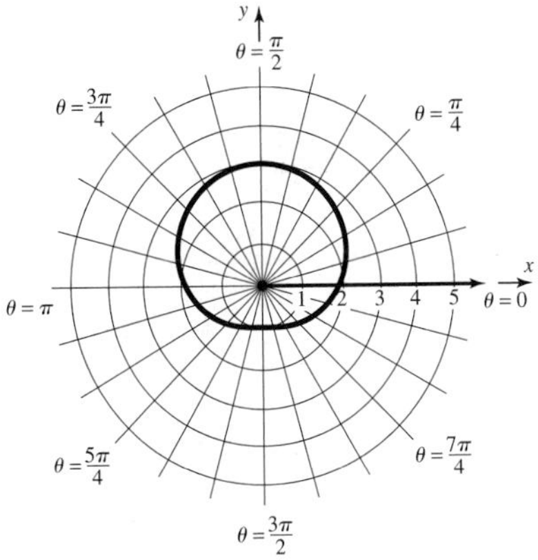

37. Limaçon without inner loop

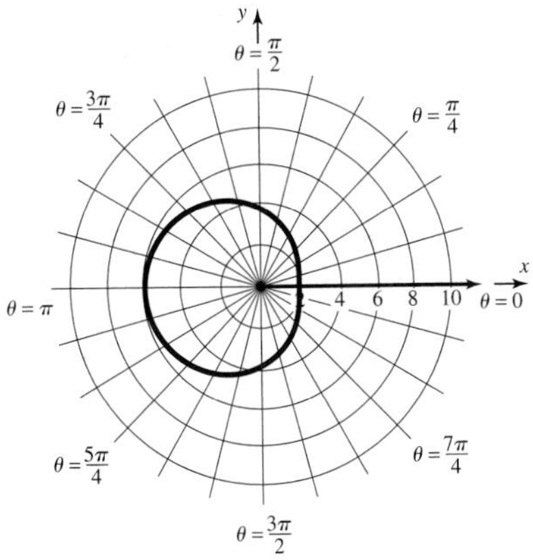

39. Limaçon with inner loop

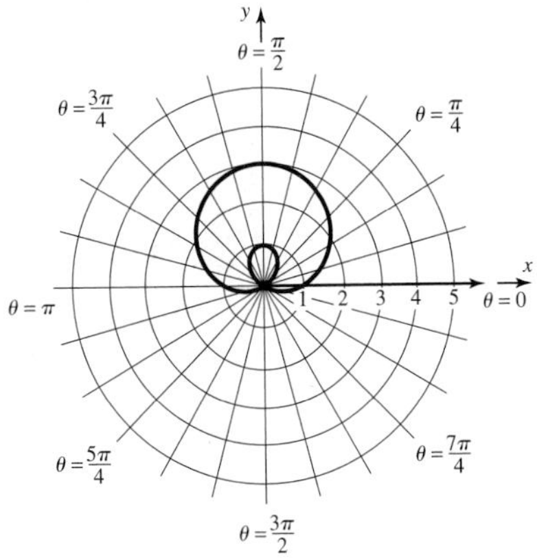

41. Limaçon with inner loop

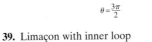

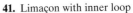

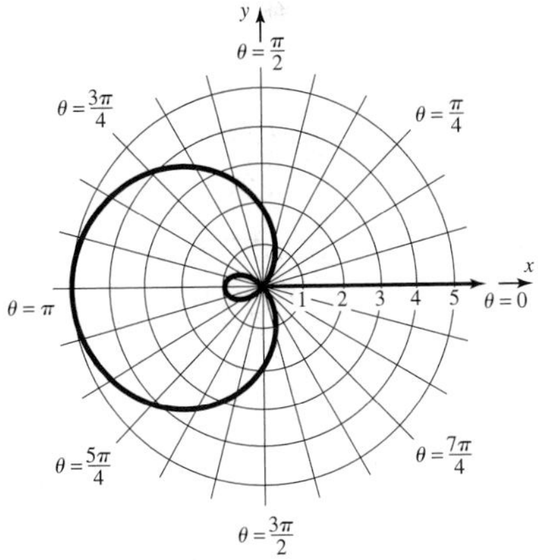

43. Rose

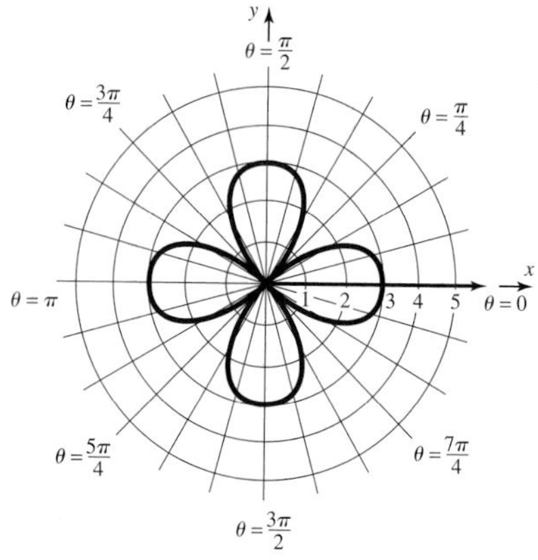

45. Rose

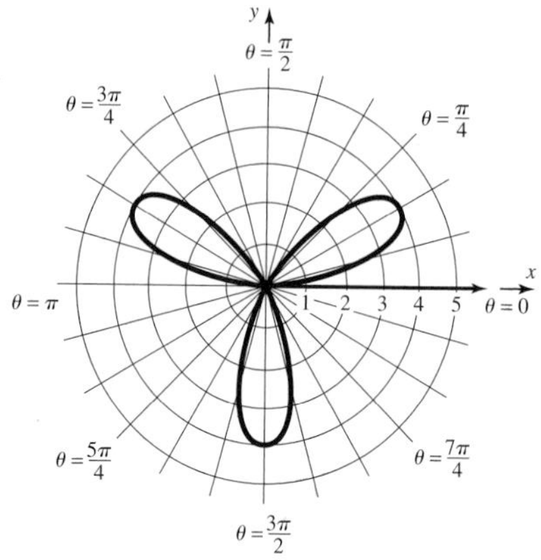

47. Lemniscate

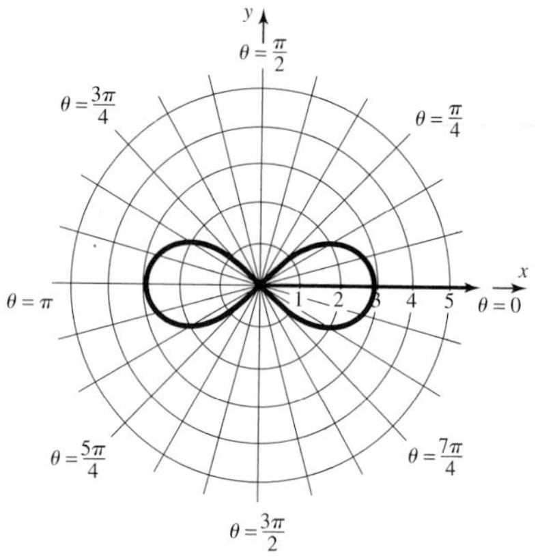

49. Spiral

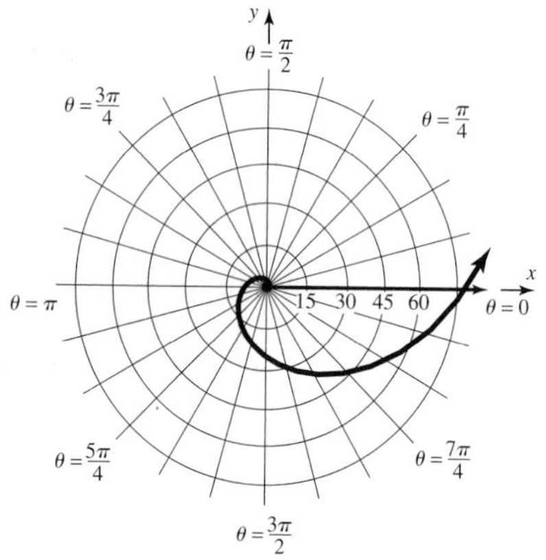

51. Cardioid

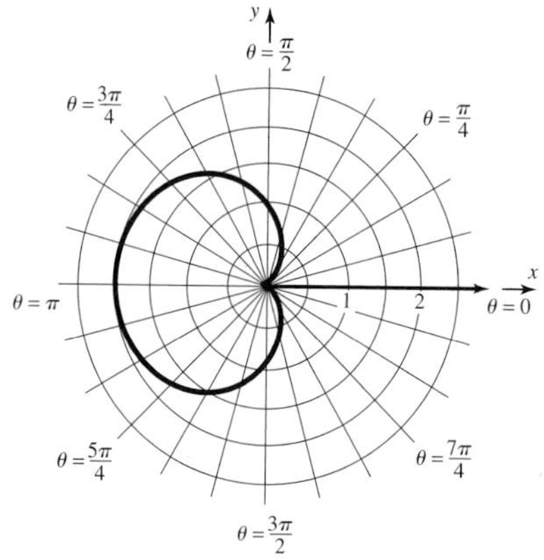

53. Limaçon with inner loop

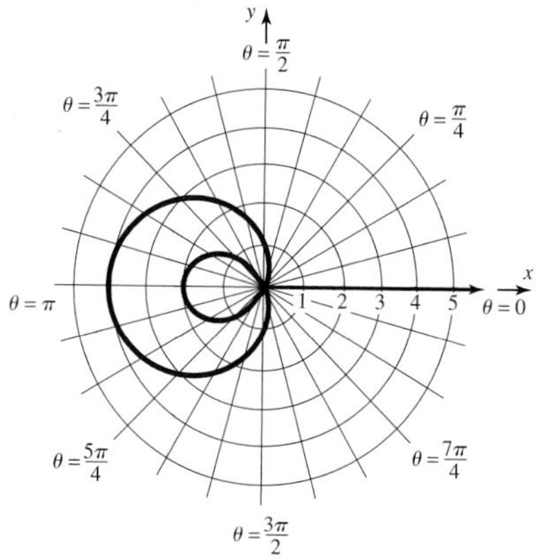

55.

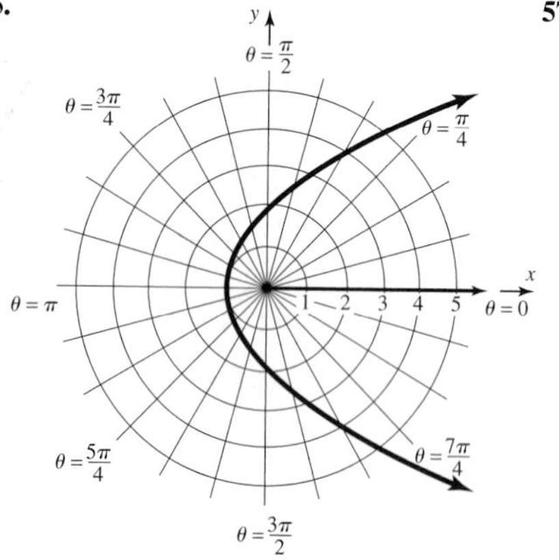

57.

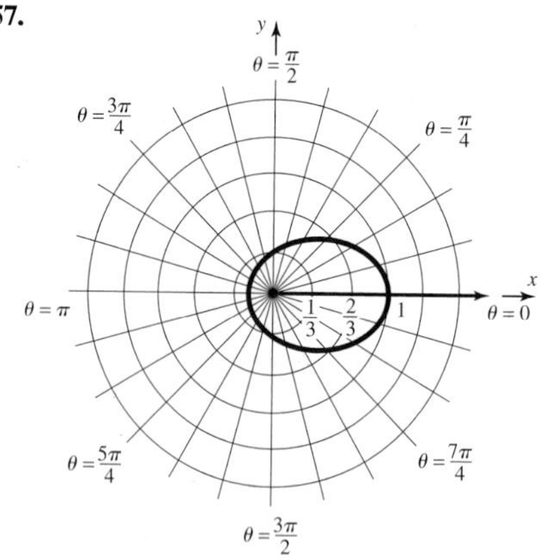

59.

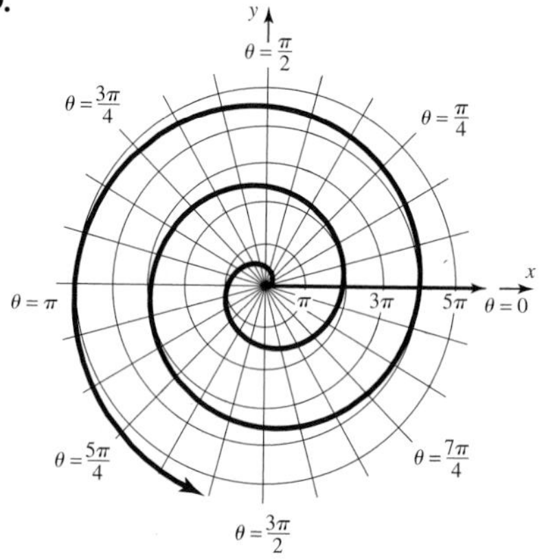

61.

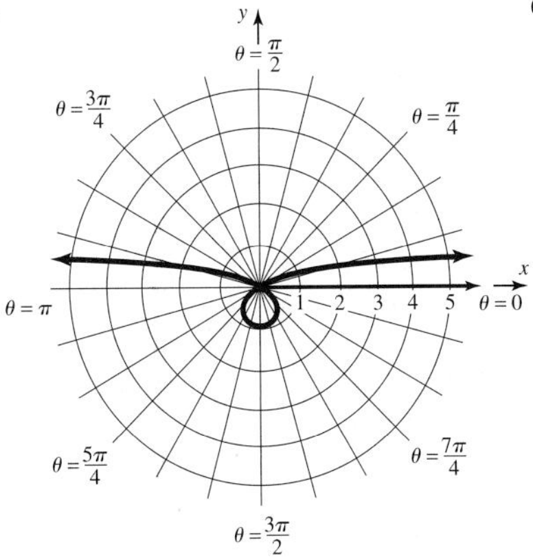

63.

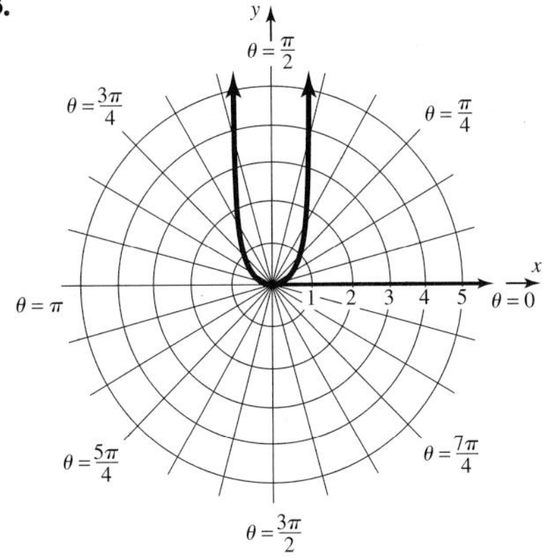

65. $r \sin \theta = a$
$y = a$

67. $r = 2a \sin \theta$
$r^2 = 2ar \sin \theta$
$x^2 + y^2 = 2ay$
$x^2 + y^2 - 2ay = 0$
$x^2 + (y - a)^2 = a^2$
Circle, radius a, center at $(0, a)$ in rectangular coordinates

69. $r = 2a \cos \theta$
$r^2 = 2ar \cos \theta$
$x^2 + y^2 = 2ax$
$x^2 - 2ax + y^2 = 0$
$(x - a)^2 + y^2 = a^2$
Circle, radius a, center at $(a, 0)$ in rectangular coordinates

71. (a) $r^2 = \cos \theta$: $r^2 = \cos(\pi - \theta)$
$r^2 = -\cos \theta$
Not equivalent; test fails
$(-r)^2 = \cos(-\theta)$
$r^2 = \cos \theta$
New test works

(b) $r^2 = \sin \theta$: $r^2 = \sin(\pi - \theta)$
$r^2 = \sin \theta$
Test works
$(-r)^2 = \sin(-\theta)$
$r^2 = -\sin \theta$
Not equivalent; new test fails

Historical Problems

1. (a) $1 + 4i, 1 + i$ **(b)** $-1, 2 + i$

9.3 Exercises

1.

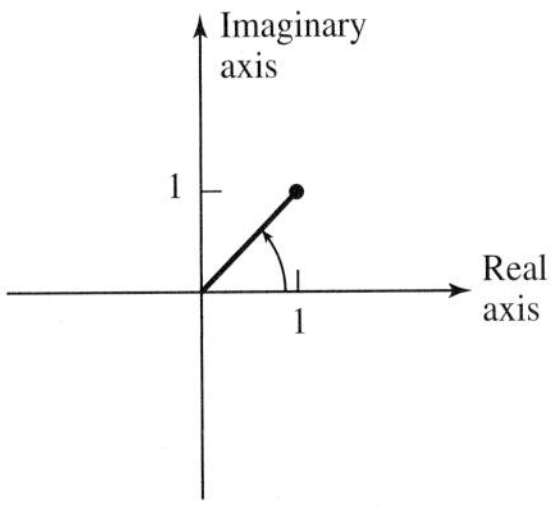

$\sqrt{2}(\cos 45^\circ + i \sin 45^\circ)$

3.

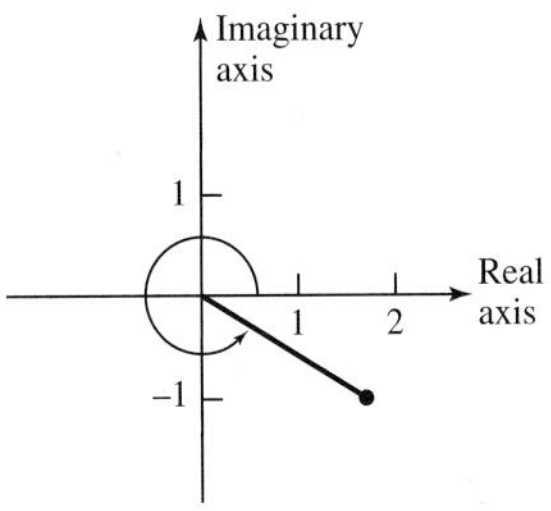

$2(\cos 330^\circ + i \sin 330^\circ)$

5.

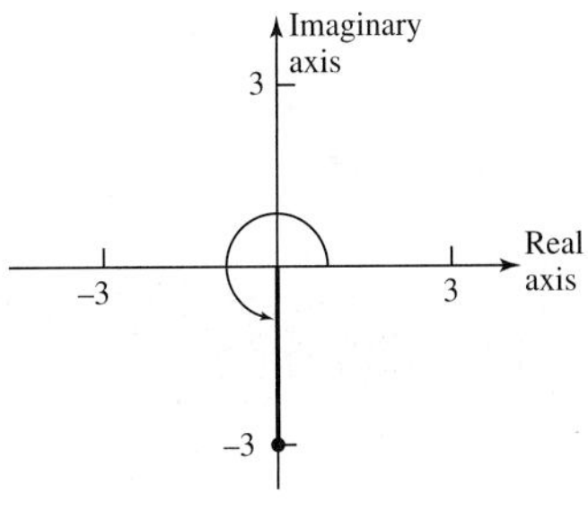

$3(\cos 270^\circ + i \sin 270^\circ)$

7.

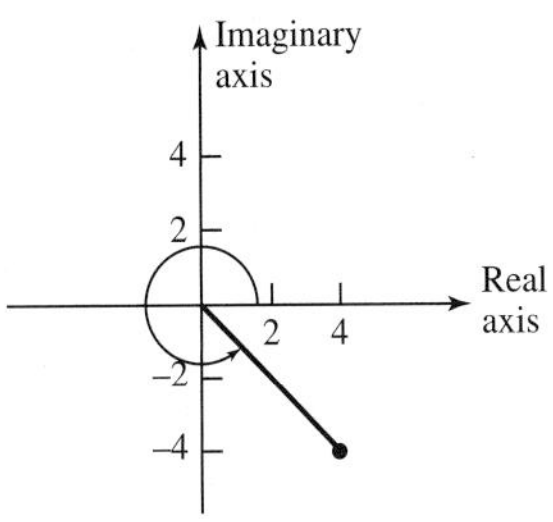

$4\sqrt{2}(\cos 315^\circ + i \sin 315^\circ)$

9.

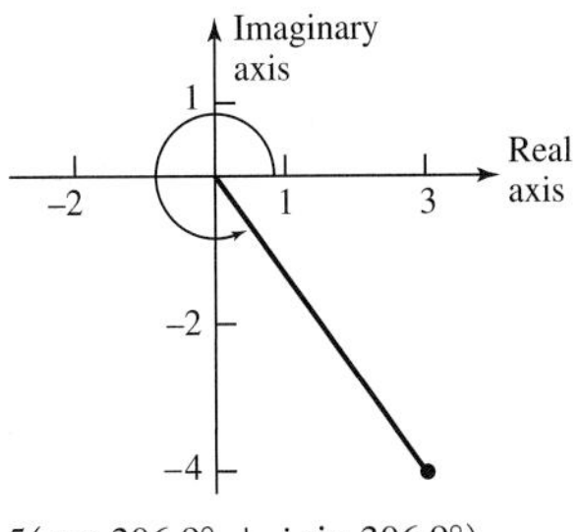

$5(\cos 306.9^\circ + i \sin 306.9^\circ)$

11.

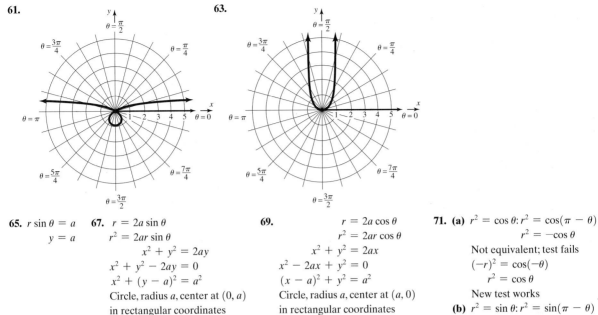

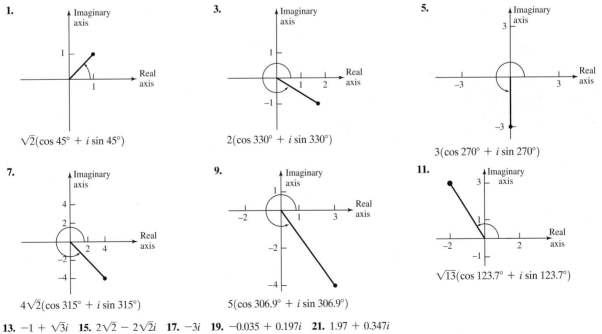

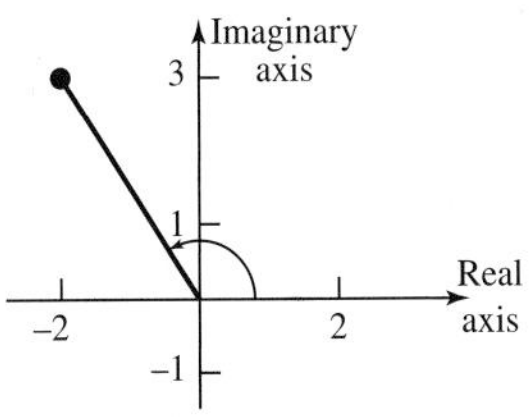

$\sqrt{13}(\cos 123.7^\circ + i \sin 123.7^\circ)$

13. $-1 + \sqrt{3}i$ **15.** $2\sqrt{2} - 2\sqrt{2}i$ **17.** $-3i$ **19.** $-0.035 + 0.197i$ **21.** $1.97 + 0.347i$

23. $zw = 8(\cos 60^\circ + i \sin 60^\circ)$; $\dfrac{z}{w} = \dfrac{1}{2}(\cos 20^\circ + i \sin 20^\circ)$ **25.** $zw = 12(\cos 40^\circ + i \sin 40^\circ)$; $\dfrac{z}{w} = \dfrac{3}{4}(\cos 220^\circ + i \sin 220^\circ)$

27. $zw = 4\left(\cos\frac{9\pi}{40} + i\sin\frac{9\pi}{40}\right); \frac{z}{w} = \cos\frac{\pi}{40} + i\sin\frac{\pi}{40}$ **29.** $zw = 4\sqrt{2}(\cos 15° + i\sin 15°); \frac{z}{w} = \sqrt{2}(\cos 75° + i\sin 75°)$

31. $-32 + 32\sqrt{3}i$ **33.** $32i$ **35.** $\frac{27}{2} + \frac{27\sqrt{3}}{2}i$ **37.** $-\frac{25\sqrt{2}}{2} + \frac{25\sqrt{2}}{2}i$ **39.** $-4 + 4i$ **41.** $-23 + 14.14i$

43. $\sqrt[6]{2}(\cos 15° + i\sin 15°), \sqrt[6]{2}(\cos 135° + i\sin 135°), \sqrt[6]{2}(\cos 255° + i\sin 255°)$

45. $\sqrt[4]{8}(\cos 75° + i\sin 75°), \sqrt[4]{8}(\cos 165° + i\sin 165°), \sqrt[4]{8}(\cos 255° + i\sin 255°), \sqrt[4]{8}(\cos 345° + i\sin 345°)$

47. $2(\cos 67.5° + i\sin 67.5°), 2(\cos 157.5° + i\sin 157.5°), 2(\cos 247.5° + i\sin 247.5°), 2(\cos 337.5° + i\sin 337.5°)$

49. $\cos 18° + i\sin 18°, \cos 90° + i\sin 90°, \cos 162° + i\sin 162°, \cos 234° + i\sin 234°, \cos 306° + i\sin 306°$

51. $1, i, -1, -i$

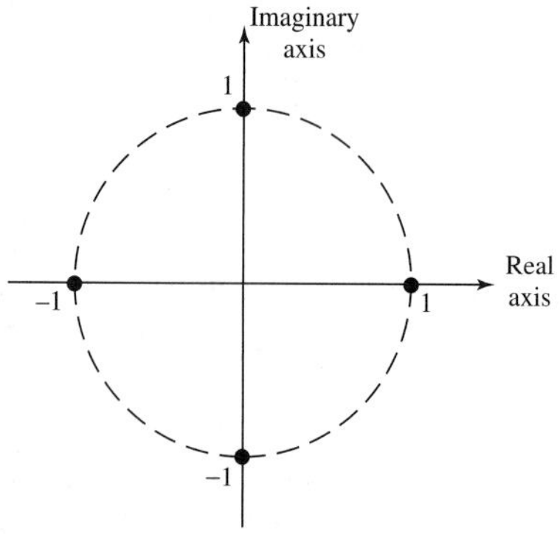

53. Look at formula (8); $|z_k| = \sqrt[n]{r}$ for all k.

55. Look at formula (8). The z_k are spaced apart by an angle of $\frac{2\pi}{n}$.

9.4 Exercises

1.

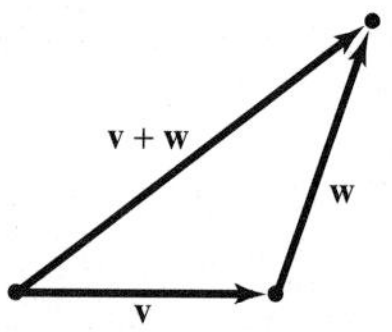

3.

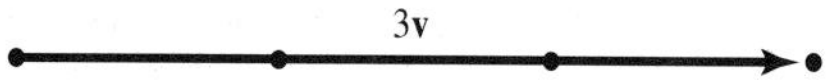

5.

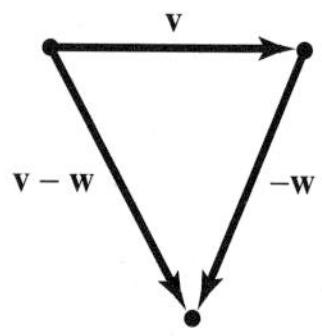

7.

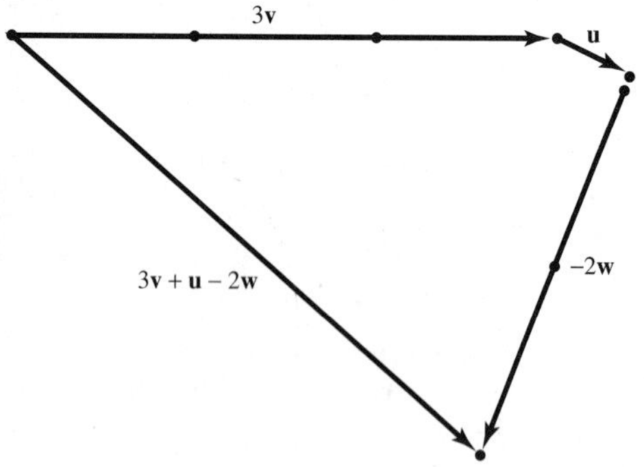

9. T **11.** F **13.** F **15.** T **17.** 12 **19.** $\mathbf{v} = 3\mathbf{i} + 4\mathbf{j}$

21. $\mathbf{v} = 2\mathbf{i} + 4\mathbf{j}$ **23.** $\mathbf{v} = 8\mathbf{i} - \mathbf{j}$ **25.** $\mathbf{v} = -\mathbf{i} + \mathbf{j}$ **27.** 5

29. $\sqrt{2}$ **31.** $\sqrt{13}$ **33.** $-\mathbf{j}$ **35.** $\sqrt{89}$ **37.** $\sqrt{34} - \sqrt{13}$ **39.** $\mathbf{i}$

41. $\frac{3}{5}\mathbf{i} - \frac{4}{5}\mathbf{j}$ **43.** $\frac{\sqrt{2}}{2}\mathbf{i} - \frac{\sqrt{2}}{2}\mathbf{j}$

45. $\mathbf{v} = \frac{8\sqrt{5}}{5}\mathbf{i} + \frac{4\sqrt{5}}{5}\mathbf{j}$ or $\mathbf{v} = -\frac{8\sqrt{5}}{5}\mathbf{i} - \frac{4\sqrt{5}}{5}\mathbf{j}$

47. $\{-2 + \sqrt{21}, -2 - \sqrt{21}\}$ **49.** $\mathbf{v} = \frac{5}{2}(\mathbf{i} + \sqrt{3}\mathbf{j})$

51. $\mathbf{v} = 7(-\mathbf{i} + \sqrt{3}\mathbf{j})$ **53.** $\mathbf{v} = \frac{25}{2}(\sqrt{3}\mathbf{i} - \mathbf{j})$

55. $\mathbf{F} = 20(\sqrt{3}\mathbf{i} + \mathbf{j})$

57. $\mathbf{F} = (20\sqrt{3} + 30\sqrt{2})\mathbf{i} + (20 - 30\sqrt{2})\mathbf{j}$ **59.** Tension in left cable: 1000 lbs; Tension in right cable: 845.2 lbs

61. Tension in left part: 1088.4 lbs; Tension in right part: 1089.1 lbs **63.**

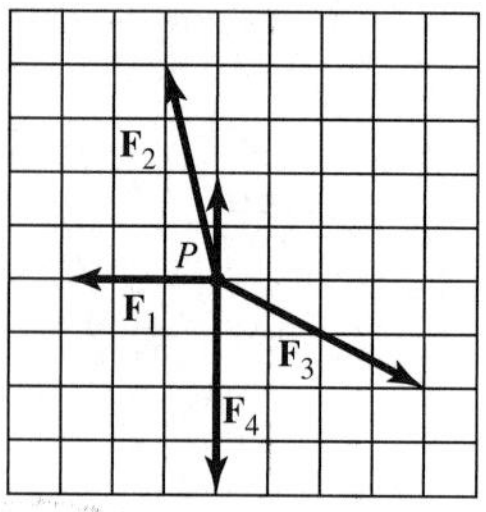

Historical Problems

1. $(a\mathbf{i} + b\mathbf{j}) \cdot (c\mathbf{i} + d\mathbf{j}) = ac + bd$
real part $[(\overline{a + bi})(c + di)]$ = real part$[(a - bi)(c + di)]$ = real part$[ac + adi - bci - bdi^2] = ac + bd$

9.5 Exercises

1. 0; 90°; orthogonal **3.** 4; 36.9°; neither **5.** $\sqrt{3} - 1$; 75°; neither **7.** 24; 16.3°; neither **9.** 0; 90°; orthogonal **11.** $\frac{2}{3}$

13. $\mathbf{v}_1 = \text{proj}_\mathbf{w}\mathbf{v} = \frac{5}{2}\mathbf{i} - \frac{5}{2}\mathbf{j}, \mathbf{v}_2 = -\frac{1}{2}\mathbf{i} - \frac{1}{2}\mathbf{j}$ **15.** $\mathbf{v}_1 = \text{proj}_\mathbf{w}\mathbf{v} = -\frac{1}{5}\mathbf{i} - \frac{2}{5}\mathbf{j}, \mathbf{v}_2 = \frac{6}{5}\mathbf{i} - \frac{3}{5}\mathbf{j}$

17. $\mathbf{v}_1 = \text{proj}_\mathbf{w}\mathbf{v} = \frac{14}{5}\mathbf{i} + \frac{7}{5}\mathbf{j}, \mathbf{v}_2 = \frac{1}{5}\mathbf{i} - \frac{2}{5}\mathbf{j}$ **19.** 496.7 mph; 38.5° west of south
21. 8.6° off direct heading across the current, upstream; 1.52 min
23. Force required to keep Sienne from rolling down the hill: 737.6 lbs. Force perpendicular to the hill: 5352.1 lbs.
25. $\mathbf{v} = (250\sqrt{2} - 30)\mathbf{i} + (250\sqrt{2} + 30\sqrt{3})\mathbf{j}$; 518.8 km/hr; N38.6°E
27. $\mathbf{v} = 3i + 20\mathbf{j}$; 20.22 mph; N8.5°E (Assuming boat traveling north and current traveling east).
29. 3 ft-lb **31.** $1000\sqrt{3}$ ft-lb $\approx$ 1732 ft-lb **33.** Let $\mathbf{u} = a_1\mathbf{i} + b_1\mathbf{j}, \mathbf{v} = a_2\mathbf{i} + b_2\mathbf{j}, \mathbf{w} = a_3\mathbf{i} + b_3\mathbf{j}$. Compute $\mathbf{u} \cdot (\mathbf{v} + \mathbf{w})$ and $\mathbf{u} \cdot \mathbf{v} + \mathbf{u} \cdot \mathbf{w}$.

35. $\cos\alpha = \frac{\mathbf{v} \cdot \mathbf{i}}{\|\mathbf{v}\|\,\|\mathbf{i}\|} = \mathbf{v} \cdot \mathbf{i}$; if $\mathbf{v} = x\mathbf{i} + y\mathbf{j}$, then $\mathbf{v} \cdot \mathbf{i} = x = \cos\alpha$ and $\mathbf{v} \cdot \mathbf{j} = y = \cos\left(\frac{\pi}{2} - \alpha\right) = \sin\alpha$.

37. $\mathbf{v} = a\mathbf{i} + b\mathbf{j}$; $\text{proj}_\mathbf{i}\mathbf{v} = \frac{\mathbf{v} \cdot \mathbf{i}}{\|\mathbf{i}\|^2}\mathbf{i} = (\mathbf{v} \cdot \mathbf{i})\mathbf{i}$; $\mathbf{v} \cdot \mathbf{i} = a, \mathbf{v} \cdot \mathbf{j} = b$, so $\mathbf{v} = (\mathbf{v} \cdot \mathbf{i})\mathbf{i} + (\mathbf{v} \cdot \mathbf{j})\mathbf{j}$.

39. $(\mathbf{v} - \alpha\mathbf{w}) \cdot \mathbf{w} = \mathbf{v} \cdot \mathbf{w} - \alpha\mathbf{w} \cdot \mathbf{w} = \alpha\|\mathbf{w}\|^2 - \alpha\|\mathbf{w}\|^2 = 0$ **41.** $W = \mathbf{F} \cdot \overrightarrow{AB} = 0$ when $\mathbf{F}$ is orthogonal to $\overrightarrow{AB}$

9.6 Exercises

1. All points of the form $(x, 0, z)$. **3.** All points of the form $(x, y, 2)$. **5.** All points of the form $(-4, y, z)$.
7. All points of the form $(1, 2, z)$. **9.** $\sqrt{21}$ **11.** $\sqrt{33}$ **13.** $\sqrt{26}$ **15.** (2, 0, 0); (2, 1, 0); (0, 1, 0); (2, 0, 3); (0, 1, 3); (0, 0, 3)
17. (1, 4, 3); (3, 2, 3); (3, 4, 3); (3, 2, 5); (1, 4, 5); (1, 2, 5) **19.** (−1, 2, 2); (4, 0, 2); (4, 2, 2); (−1, 2, 5); (4, 0, 5); (−1, 0, 5)
21. $\mathbf{v} = 3\mathbf{i} + 4\mathbf{j} - \mathbf{k}$ **23.** $\mathbf{v} = 2\mathbf{i} + 4\mathbf{j} + \mathbf{k}$ **25.** $\mathbf{v} = 8\mathbf{i} - \mathbf{j}$ **27.** 7 **29.** $\sqrt{3}$ **31.** $\sqrt{22}$ **33.** $-\mathbf{j} - 2\mathbf{k}$ **35.** $\sqrt{105}$ **37.** $\sqrt{38} - \sqrt{17}$

39. $\frac{\mathbf{v}}{\|\mathbf{v}\|} = \mathbf{i}$ **41.** $\frac{\mathbf{v}}{\|\mathbf{v}\|} = \frac{3}{7}\mathbf{i} - \frac{6}{7}\mathbf{j} - \frac{2}{7}\mathbf{k}$ **43.** $\frac{\mathbf{v}}{\|\mathbf{v}\|} = \frac{\sqrt{3}}{3}\mathbf{i} + \frac{\sqrt{3}}{3}\mathbf{j} + \frac{\sqrt{3}}{3}\mathbf{k}$ **45.** $\mathbf{v} \cdot \mathbf{w} = 0;\ \theta = 90°$
47. $\mathbf{v} \cdot \mathbf{w} = -2, \theta \approx 100°$ **49.** $\mathbf{v} \cdot \mathbf{w} = 0;\ \theta = 90°$ **51.** $\mathbf{v} \cdot \mathbf{w} = 52;\ \theta = 0°$
53. $\alpha \approx 65°; \beta \approx 149°;\ \gamma \approx 107°; \mathbf{v} = 7(\cos 65°\mathbf{i} + \cos 149°\mathbf{j} + \cos 107°\mathbf{k})$
55. $\alpha \approx 55°; \beta \approx 55°;\ \gamma \approx 55°; \mathbf{v} = \sqrt{3}(\cos 55°\mathbf{i} + \cos 55°\mathbf{j} + \cos 55°\mathbf{k})$
57. $\alpha = 45°; \beta = 45°;\ \gamma = 90°; \mathbf{v} = \sqrt{2}(\cos 45°\mathbf{i} + \cos 45°\mathbf{j} + \cos 90°\mathbf{k})$
59. $\alpha \approx 61°; \beta \approx 144°;\ \gamma \approx 71°; \mathbf{v} = \sqrt{38}(\cos 61°\mathbf{i} + \cos 144°\mathbf{j} + \cos 71°\mathbf{k})$ **63.** $(x - 1)^2 + (y - 2)^2 + (z - 2)^2 = 4$
65. radius = 3, center (−1, 1, 0) **67.** radius = 3, center (2, −2, −1) **69.** radius $= \frac{3\sqrt{2}}{2}$, center (2, 0, −1) **71.** 2 joules **73.** 9

9.7 Exercises

1. 2 **3.** 4 **5.** $-11A - 2B + 5C$ **7.** $-6A - 23B - 15C$ **9. (a)** $5\mathbf{i} + 5\mathbf{j} + 5\mathbf{k}$ **(b)** $-5\mathbf{i} - 5\mathbf{j} - 5\mathbf{k}$ **(c)** $0\mathbf{i} + 0\mathbf{j} + 0\mathbf{k}$
(d) $0\mathbf{i} + 0\mathbf{i} + 0\mathbf{k}$ **11. (a)** $1\mathbf{i} - 1\mathbf{j} - 1\mathbf{k}$ **(b)** $-1\mathbf{i} + 1\mathbf{j} + 1\mathbf{k}$ **(c)** $0\mathbf{i} + 0\mathbf{j} + 0\mathbf{k}$ **(d)** $0\mathbf{i} + 0\mathbf{j} + 0\mathbf{k}$ **13. (a)** $-1\mathbf{i} + 2\mathbf{j} + 2\mathbf{k}$
(b) $1\mathbf{i} - 2\mathbf{j} - 2\mathbf{k}$ **(c)** $0\mathbf{i} + 0\mathbf{j} + 0\mathbf{k}$ **(d)** $0\mathbf{i} + 0\mathbf{j} + 0\mathbf{k}$ **15. (a)** $3\mathbf{i} - 1\mathbf{j} + 4\mathbf{k}$ **(b)** $-3\mathbf{i} + 1\mathbf{j} - 4\mathbf{k}$ **(c)** $0\mathbf{i} + 0\mathbf{j} + 0\mathbf{k}$
(d) $0\mathbf{i} + 0\mathbf{j} + 0\mathbf{k}$ **17.** $-9\mathbf{i} - 7\mathbf{j} - 3\mathbf{k}$ **19.** $9\mathbf{i} + 7\mathbf{j} + 3\mathbf{k}$ **21.** $0\mathbf{i} + 0\mathbf{j} + 0\mathbf{k}$ **23.** $-27\mathbf{i} - 21\mathbf{j} - 9\mathbf{k}$ **25.** $-18\mathbf{i} - 14\mathbf{j} - 6\mathbf{k}$ **27.** 0
29. −25 **31.** 25 **33.** $0\mathbf{i} + 0\mathbf{j} + 0\mathbf{k}$ **35.** $-9\mathbf{i} - 7\mathbf{j} - 3\mathbf{k}$ **37.** $-1\mathbf{i} + 1\mathbf{j} + 5\mathbf{k}$ **39.** $\sqrt{166}$ **41.** $\sqrt{555}$ **43.** $\sqrt{34}$ **45.** $\sqrt{998}$

47. $\frac{11}{\sqrt{171}}\mathbf{i} + \frac{1}{\sqrt{171}}\mathbf{j} + \frac{7}{\sqrt{171}}\mathbf{k}$ **49.**
$$\mathbf{u} \times \mathbf{v} = \begin{vmatrix} \mathbf{i} & \mathbf{i} & \mathbf{k} \\ a_1 & b_1 & c_1 \\ a_2 & b_2 & c_2 \end{vmatrix} = (b_1c_2 - b_2c_1)\mathbf{i} - (a_1c_2 - a_2c_1)\mathbf{j} + (a_1b_2 - a_2b_1)\mathbf{k}$$
$$= -[b_2c_1 - b_1c_2)\mathbf{i} - (a_2c_1 - a_1c_2)\mathbf{j} + (a_2b_1 - a_1b_2)\mathbf{k}]$$
$$= -\begin{vmatrix} \mathbf{i} & \mathbf{j} & \mathbf{k} \\ a_2 & b_2 & c_2 \\ a_1 & b_2 & c_1 \end{vmatrix} = -(\mathbf{v} \times \mathbf{u})$$

51. $\mathbf{u} \times \mathbf{v} = \begin{vmatrix} \mathbf{i} & \mathbf{j} & \mathbf{k} \\ a_1 & b_1 & c_1 \\ a_2 & b_2 & c_2 \end{vmatrix} = (b_1c_2 - b_2c_1)\mathbf{i} - (a_1c_2 - a_2c_1)\mathbf{j} + (a_1b_2 - a_2b_1)\mathbf{k}$

$\|\mathbf{u} \times \mathbf{v}\|^2 = (\sqrt{(b_1c_2 - b_2c_1)^2 + (a_1c_2 - a_2c_1)^2 + (a_1b_2 - a_2b_1)^2})^2$

$= b_1^2c_2^2 - 2b_1b_2c_1c_2 + b_2^2c_1^2 + a_1^2c_2^2 - 2a_1a_2c_1c_2 + a_2^2c_1^2 + a_1^2b_2^2 - 2a_1a_2b_1b_2 + a_2^2b_1^2$

$\|\mathbf{u}\|^2 = a_1^2 + b_1^2 + c_1^2,\ \|\mathbf{v}\|^2 = a_2^2 + b_2^2 + c_2^2$

$\|\mathbf{u}\|^2\|\mathbf{v}\|^2 = (a_1^2 + b_1^2 + c_1^2)(a_2^2 + b_2^2 + c_2^2) = a_1^2a_2^2 + a_1^2b_2^2 + a_1^2c_2^2 + b_1^2a_2^2 + b_1^2b_2^2 + b_1^2c_2^2 + a_2^2c_1^2 + b_2^2c_1^2 + c_1^2c_2^2$

$(\mathbf{u} \cdot \mathbf{v})^2 = (a_1a_2 + b_1b_2 + c_1c_2)^2 = (a_1a_2 + b_1b_2 + c_1c_2)(a_1a_2 + b_1b_2 + c_1c_2)$

$= a_1^2a_2^2 + a_1a_2b_1b_2 + a_1a_2c_1c_2 + b_1b_2c_1c_2 + b_1b_2a_1a_2 + b_1^2b_2^2 + b_1b_2c_1c_2 + a_1a_2c_1c_2 + b_1b_2c_1c_2 + c_1^2c_2^2$

$= a_1^2a_2^2 + b_1^2b_2^2 + c_1^2c_2^2 + 2a_1a_2b_1b_2 + 2b_1b_2c_1c_2 + 2a_1a_2c_1c_2$

$\|\mathbf{u}\|^2\|\mathbf{v}\|^2 - (\mathbf{u} \cdot \mathbf{v})^2 = a_1^2b_2^2 + a_1^2c_2^2 + b_1^2a_2^2 + b_1^2c_2^2 + a_2^2c_1^2 + b_2^2c_1^2 - 2a_1a_2b_1b_2 - 2b_1b_2c_1c_2 - 2a_12a_2c_1c_2$

53. $\mathbf{u} \cdot \mathbf{v} = a_1a_2 + b_1b_2 + c_1c_2 = 0$

$\|\mathbf{u} \times \mathbf{v}\| = \sqrt{(b_1c_2 - b_2c_1)^2 + (a_1c_2 - a_2c_1)^2 + (a_1b_2 - a_2b_1)^2}$

$= \sqrt{b_1^2c_2^2 - 2b_1b_2c_1c_2 + b_2^2c_1^2 + a_1^2c_2^2 - 2a_1a_2c_1c_2 + a_2^2c_1^2 + a_1^2b_2^2 - 2a_1a_2b_1b_2 + a_2^2b_1^2}$

$\|\mathbf{u}\|\|\mathbf{v}\| = \sqrt{a_1^2 + b_1^2 + c_1^2}\sqrt{a_2^2 + b_2^2 + c_2^2}$

$= \sqrt{a_1^2a_2^2 + a_1^2b_2^2 + a_1^2c_2^2 + a_2^2b_1^2 + b_1^2b_2^2 + b_1^2c_2^2 + a_2^2c_1^2 + b_2^2c_1^2 + c_1^2c_2^2}$

$(\mathbf{u} \cdot \mathbf{v})^2 = a_1^2a_2^2 + b_1^2b_2^2 + c_1^2c_2^2 + 2a_1a_2b_1b_2 + 2a_1a_2c_1c_2 + 2b_1b_2c_2c_2 = 0$

Square both sides of $\|\mathbf{u} \times \mathbf{v}\| = \|\mathbf{u}\|\|\mathbf{v}\|$ to obtain result.

Fill-in-the-Blank Items

1. pole; polar axis **3.** $r = 2\cos\theta$ **5.** magnitude or modulus; argument **7.** 0

True/False Items

1. False **3.** False **5.** True **7.** True

Review Exercises

1. $\left(\frac{3\sqrt{3}}{2}, \frac{3}{2}\right)$

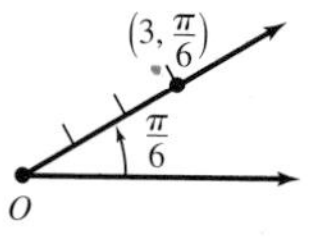

3. $(1, \sqrt{3})$

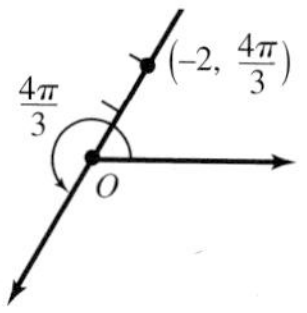

5. $(0, 3)$

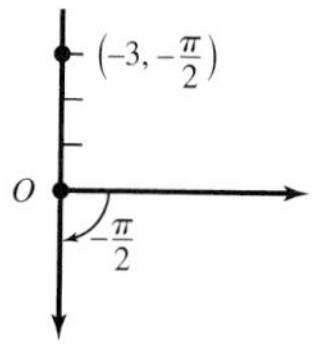

7. $\left(3\sqrt{2}, \frac{3\pi}{4}\right), \left(-3\sqrt{2}, -\frac{\pi}{4}\right)$ **9.** $\left(2, -\frac{\pi}{2}\right), \left(-2, \frac{\pi}{2}\right)$ **11.** $(5, 0.93), (-5, 4.07)$ **13.** $3r^2 - 6r\sin\theta = 0$ **15.** $r^2(2\cos^2\theta - \sin^2\theta) - \tan\theta = 0$ **17.** $r^3\cos\theta = 4$ **19.** $x^2 + y^2 - 2y = 0$ **21.** $x^2 + y^2 = 25$ **23.** $x + 3y = 6$

25. Circle: radius 2, center at (2, 0) in rectangular coordinates

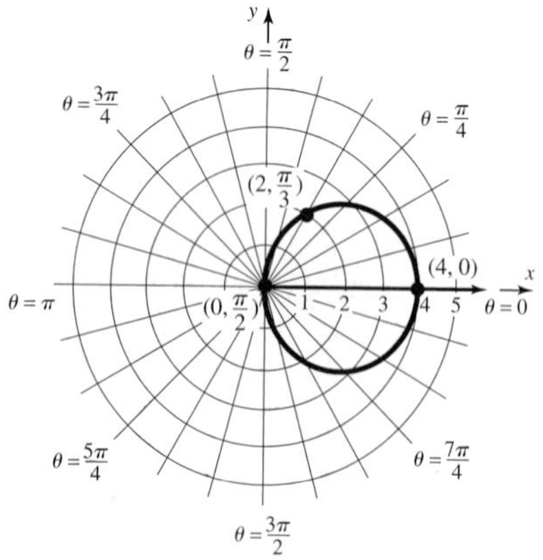

27. Cardioid

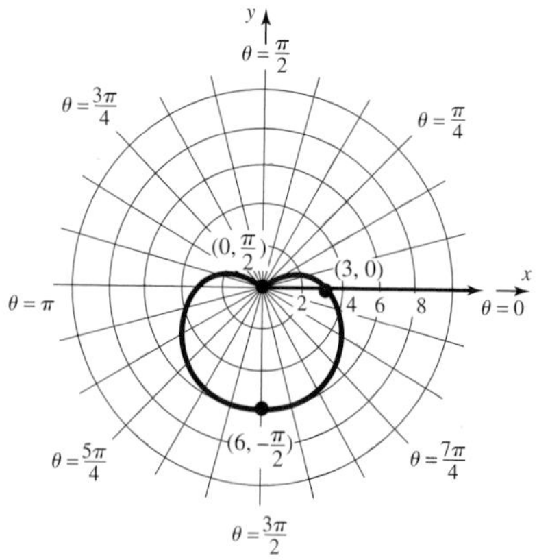

29. Limaçon without inner loop

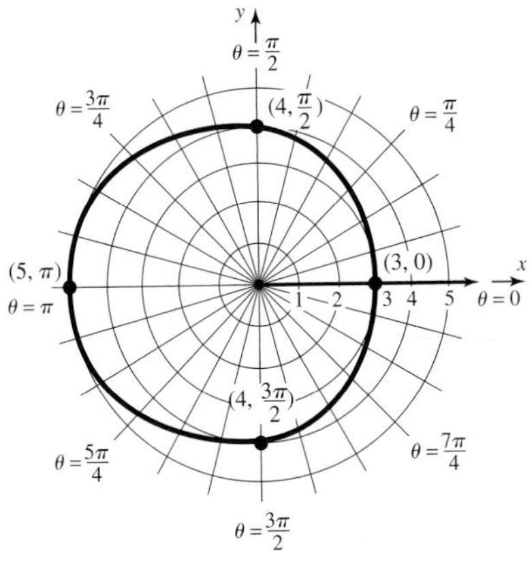

31. $\sqrt{2}(\cos 225° + i \sin 225°)$ **33.** $5(\cos 323.1° + i \sin 323.1°)$ **35.** $-\sqrt{3} + i$

37. $-\frac{3}{2} + \left(\frac{3\sqrt{3}}{2}\right)i$ **39.** $0.098 - 0.017\mathbf{i}$

41. $zw = \cos 130° + i \sin 130°; \frac{z}{w} = \cos 30° + i \sin 30°$

43. $zw = 6(\cos 0 + i \sin 0); \frac{z}{w} = \frac{3}{2}\left(\cos\frac{8\pi}{5} + i \sin\frac{8\pi}{5}\right)$

45. $zw = 5(\cos 5° + i \sin 5°); \frac{z}{w} = 5(\cos 15° + i \sin 15°)$ **47.** $\frac{27}{2} + \frac{27\sqrt{3}}{2}i$

49. $4i$ **51.** 64 **53.** $-527 - 336i$

55. $3, 3(\cos 120° + i \sin 120°), 3(\cos 240° + i \sin 240°)$

57. $\mathbf{v} = 2\mathbf{i} - 4\mathbf{j}; \|\mathbf{v}\| = 2\sqrt{5}$ **59.** $\mathbf{v} = -\mathbf{i} + 3\mathbf{j}; \|\mathbf{v}\| = \sqrt{10}$

61. $\mathbf{v} = -3\mathbf{i} - 2\mathbf{j} + \mathbf{k}; \|\mathbf{v}\| = \sqrt{14}$ **63.** $\mathbf{v} = 3\mathbf{i} - \mathbf{k}; \|\mathbf{v}\| = \sqrt{10}$

65. $-20\mathbf{i} + 13\mathbf{j}$ **67.** $\sqrt{5}$ **69.** $\sqrt{5} + 5 \approx 7.24$ **71.** $\frac{-2\sqrt{5}}{5}\mathbf{i} + \frac{\sqrt{5}}{5}\mathbf{j}$

73. $21\mathbf{i} - 2\mathbf{j} - 5\mathbf{k}$ **75.** $\sqrt{38}$ **77.** 0 **79.** $3\mathbf{i} + 9\mathbf{j} + 9\mathbf{k}$

81. $\frac{3\sqrt{14}}{14}\mathbf{i} + \frac{\sqrt{14}}{14}\mathbf{j} - \frac{\sqrt{14}}{7}\mathbf{k}; -\frac{3\sqrt{14}}{14}\mathbf{i} - \frac{\sqrt{14}}{14}\mathbf{j} + \frac{\sqrt{14}}{7}\mathbf{k}$ **83.** $\mathbf{v} \cdot \mathbf{w} = -11; \theta \approx 169.7°$ **85.** $\mathbf{v} \cdot \mathbf{w} = -4; \theta \approx 153.4°$

87. $\mathbf{v} \cdot \mathbf{w} = 1; \theta \approx 70.5$ **89.** $\mathbf{v} \cdot \mathbf{w} = 0; \theta = 90°$ **91.** $\text{proj}_{\mathbf{w}}\mathbf{v} = \frac{9}{10}(3\mathbf{i} + \mathbf{j})$ **93.** $\alpha \approx 56.1°; \beta \approx 138°; \gamma \approx 68.2°$ **95.** $\sqrt{332}$

97. $\sqrt{29} \approx 5.39$ mph; 0.4 mi **99.** Left cable: 1843.2 lbs; right cable: 1630.4 lbs

CHAPTER 10 Analytic Geometry

10.2 Exercises

1. B **3.** E **5.** H **7.** C **9.** F **11.** G **13.** D **15.** B

17. $y^2 = 16x$

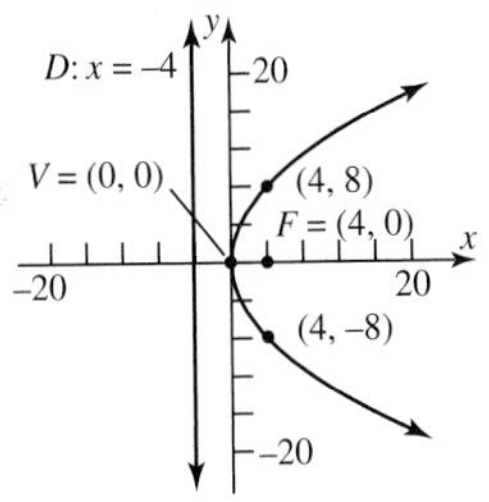

19. $x^2 = -12y$

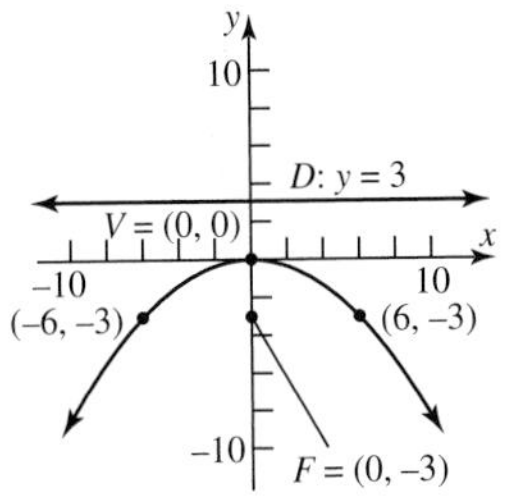

21. $y^2 = -8x$

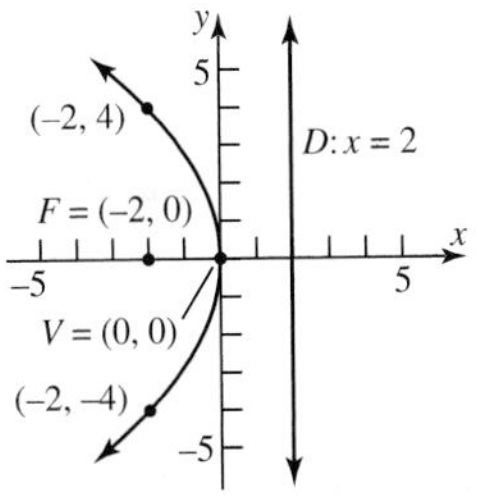

23. $x^2 = 2y$

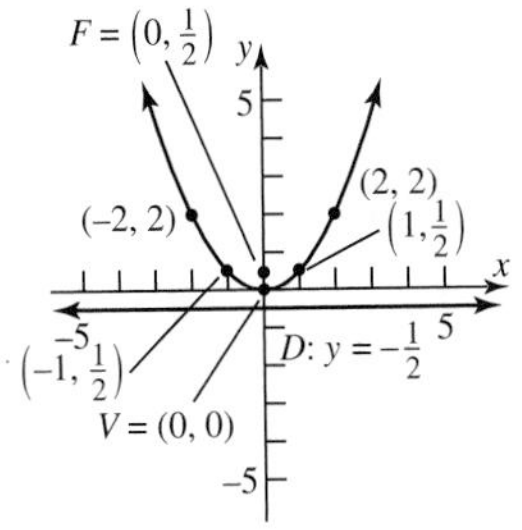

25. $(x - 2)^2 = -8(y + 3)$

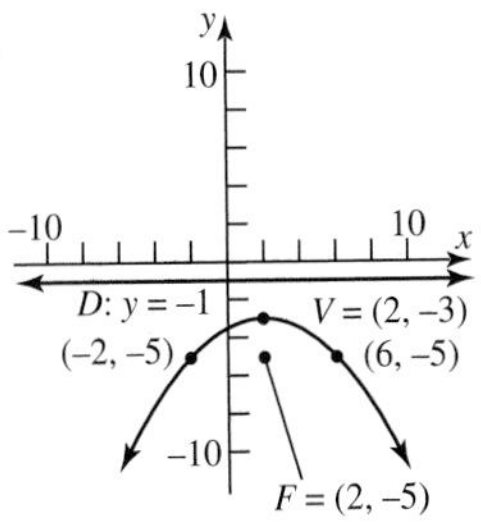

27. $x^2 = \frac{4}{3}y$

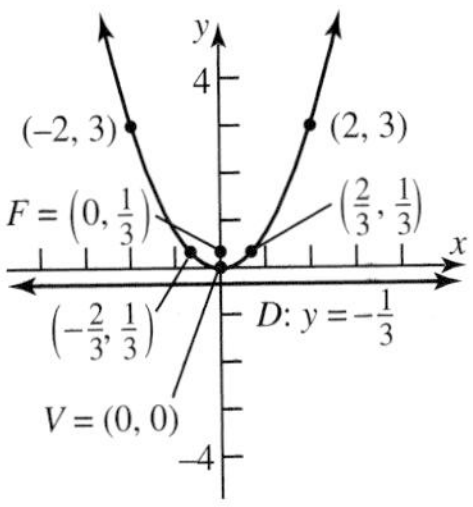

29. $(x+3)^2 = 4(y-3)$

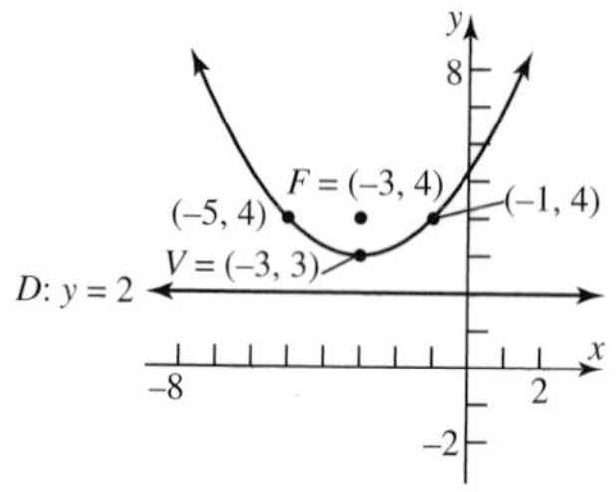

31. $(y+2)^2 = -8(x+1)$

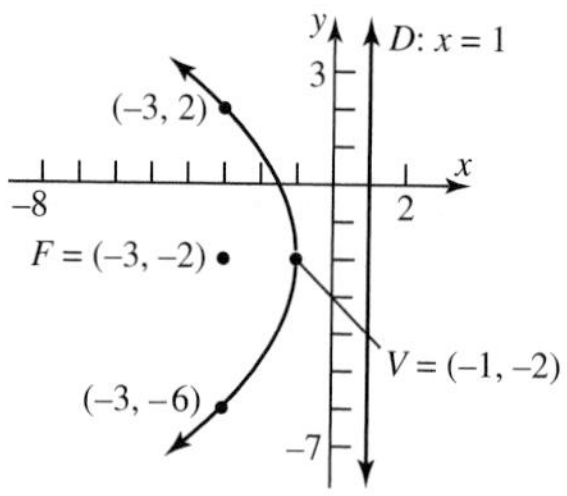

33. Vertex: $(0, 0)$; Focus: $(0, 1)$; Directrix: $y = -1$

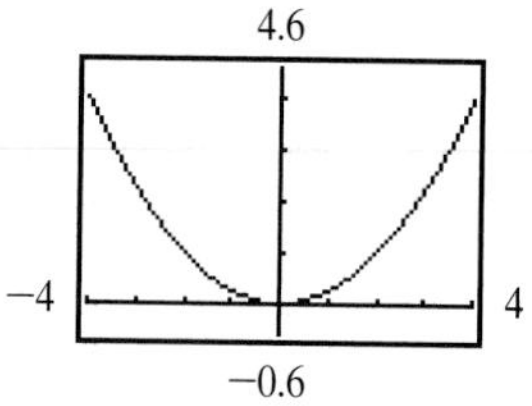

35. Vertex: $(0, 0)$; Focus: $(-4, 0)$;

Directrix: $x = 4$

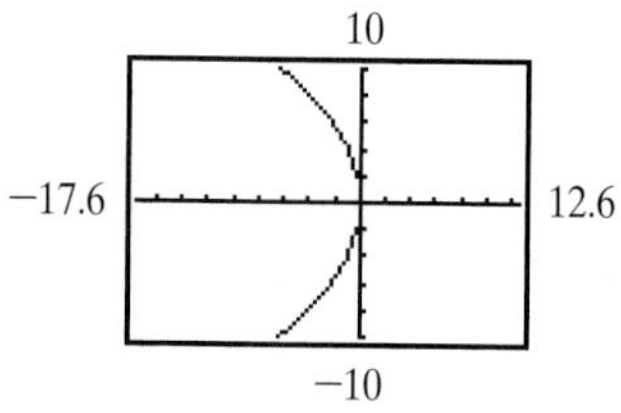

37. Vertex: $(-1, 2)$; Focus: $(1, 2)$;

Directrix: $x = -3$

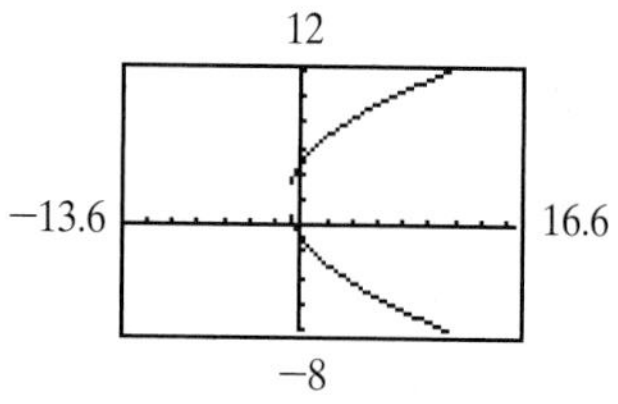

39. Vertex: $(3, -1)$; Focus: $\left(3, -\frac{5}{4}\right)$;

Directrix: $y = -\frac{3}{4}$

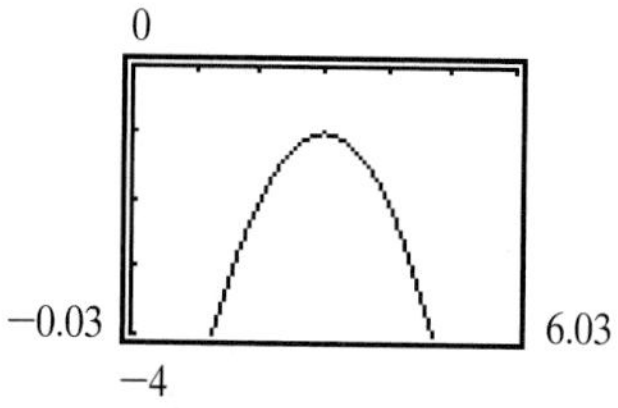

41. Vertex: $(2, -3)$; Focus: $(4, -3)$; Directrix: $x = 0$

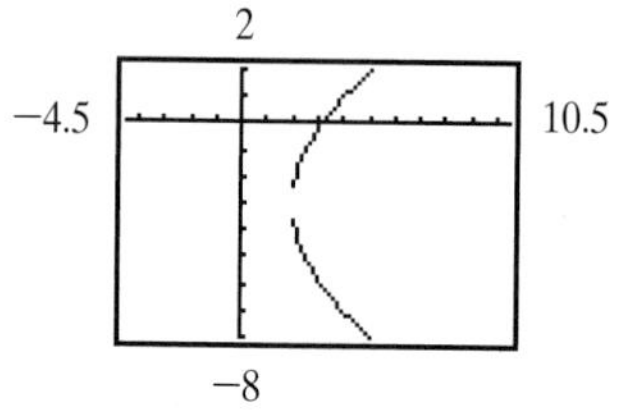

43. Vertex: $(0, 2)$; Focus: $(-1, 2)$; Directrix: $x = 1$

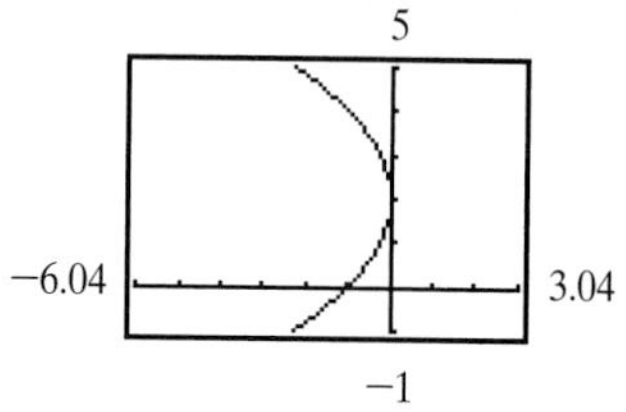

45. Vertex: $(-4, -2)$; Focus: $(-4, -1)$; Directrix: $y = -3$

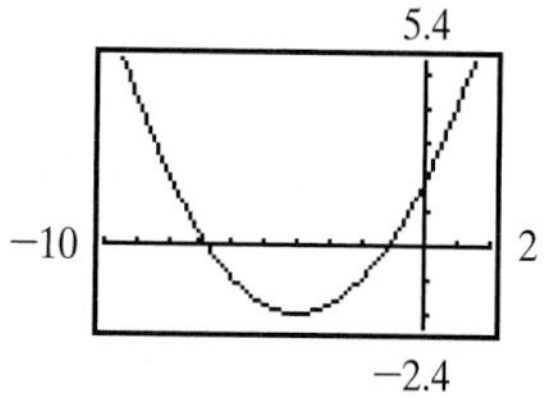

47. Vertex: $(-1, -1)$; Focus: $\left(-\frac{3}{4}, -1\right)$;

Directrix: $x = -\frac{5}{4}$

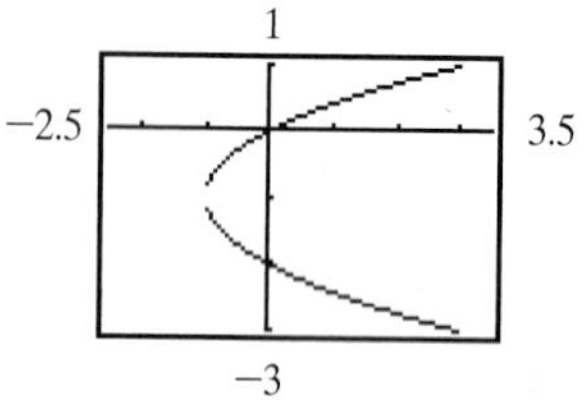

49. Vertex: $(2, -8)$; Focus: $\left(2, -\frac{31}{4}\right)$;

Directrix: $y = -\frac{33}{4}$

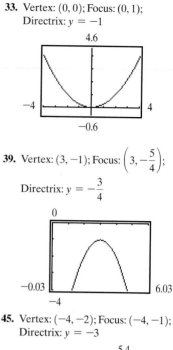

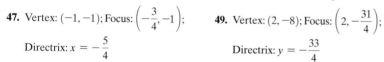

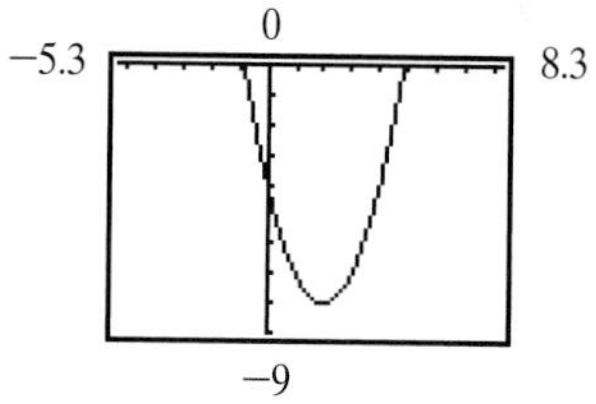

51. $(y-1)^2 = x$ **53.** $(y-1)^2 = -(x-2)$ **55.** $x^2 = 4(y-1)$ **57.** $y^2 = \frac{1}{2}(x+2)$
59. 1.5625 ft from the base of the dish, along the axis of symmetry **61.** 1 in. from the vertex **63.** 20 ft **65.** 0.78125 ft
67. 4.17 ft from the base along the axis of symmetry **69.** 24.31 ft, 18.75 ft, 7.64 ft
71. $Ax^2 + Ey = 0$ This is the equation of a parabola with vertex at $(0, 0)$ and axis of symmetry the y-axis. The focus is
$x^2 = -\frac{E}{A}y$ $\left(0, -\frac{E}{4A}\right)$; the directrix is the line $y = \frac{E}{4A}$.

73. $Ax^2 + Dx + Ey + F = 0, A \neq 0$

$$Ax^2 + Dx = -Ey - F$$

$$x^2 + \frac{D}{A}x = -\frac{E}{A}y - \frac{F}{A}$$

$$\left(x + \frac{D}{2A}\right)^2 = -\frac{E}{A}y - \frac{F}{A} + \frac{D^2}{4A^2}$$

$$\left(x + \frac{D}{2A}\right)^2 = -\frac{E}{A}y + \frac{D^2 - 4AF}{4A^2}$$

(a) If $E \neq 0$, then the equation may be written as

$$\left(x + \frac{D}{2A}\right)^2 = -\frac{E}{A}\left(y - \frac{D^2 - 4AF}{4AE}\right)$$

This is the equation of a parabola with vertex at

$\left(-\frac{D}{2A}, \frac{D^2 - 4AF}{4AE}\right)$ and axis of symmetry parallel to the y-axis.

(b)–(d) If $E = 0$, the graph of the equation contains no points if $D^2 - 4AF < 0$, is a single vertical line if $D^2 - 4AF = 0$, and is two vertical lines if $D^2 - 4AF > 0$.

10.3 Exercises

1. C **3.** B **5.** C **7.** D

9. Vertices: $(-5, 0), (5, 0)$
Foci: $(-\sqrt{21}, 0), (\sqrt{21}, 0)$

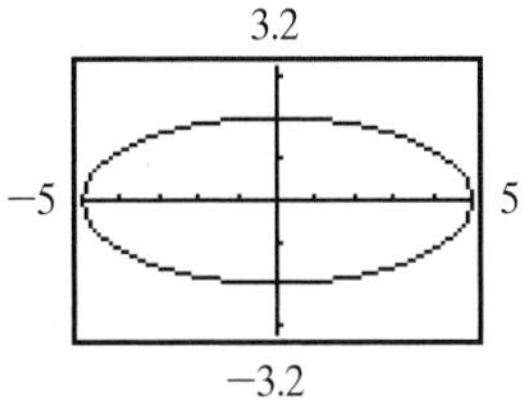

11. Vertices: $(0, -5), (0, 5)$
Foci: $(0, -4), (0, 4)$

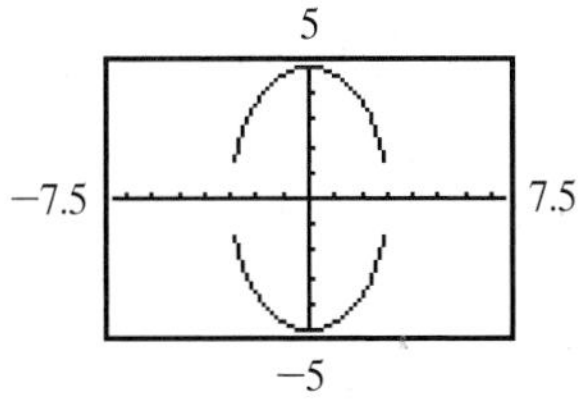

13. Vertices: $(0, -4), (0, 4)$
Foci: $(0, -2\sqrt{3}), (0, 2\sqrt{3})$

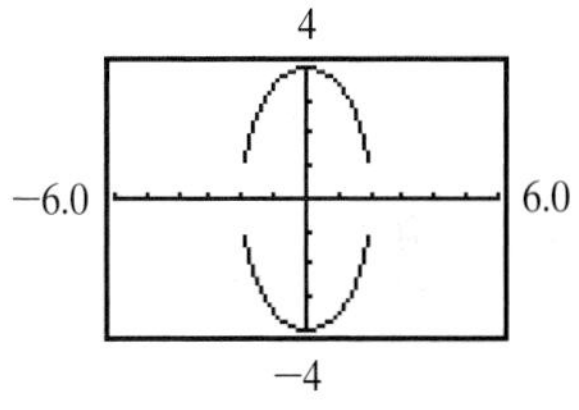

15. Vertices: $(-2\sqrt{2}, 0), (2\sqrt{2}, 0)$
Foci: $(-\sqrt{6}, 0), (\sqrt{6}, 0)$

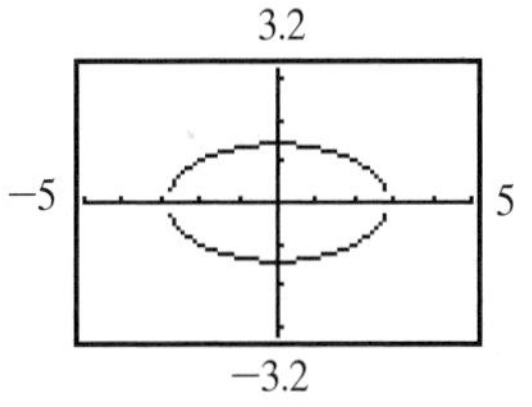

17. Vertices: $(-4, 0), (4, 0), (0, -4), (0, 4)$
Focus: $(0, 0)$

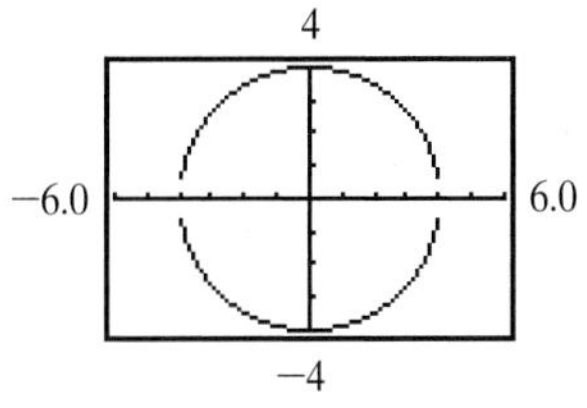

19. $\frac{x^2}{25} + \frac{y^2}{16} = 1$

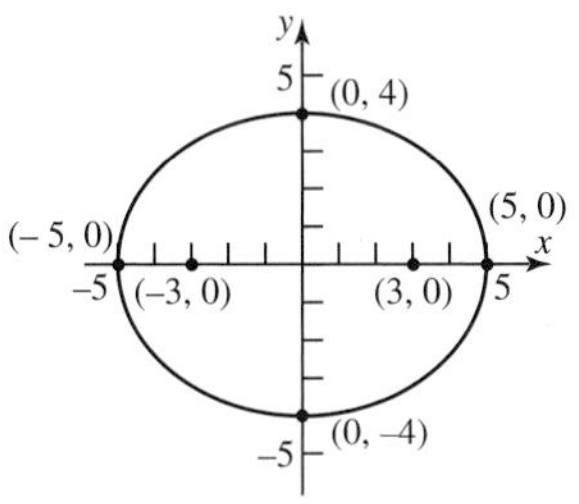

21. $\frac{x^2}{9} + \frac{y^2}{25} = 1$

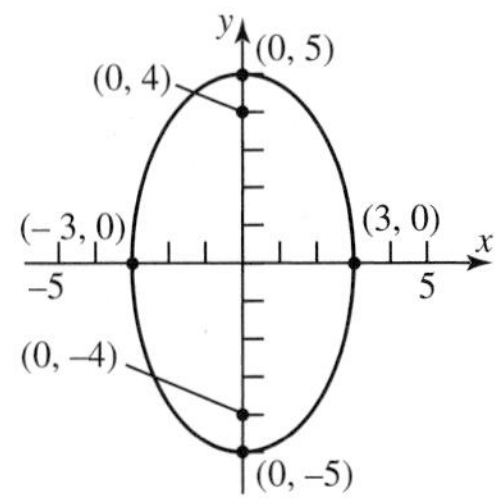

23. $\frac{x^2}{9} + \frac{y^2}{5} = 1$

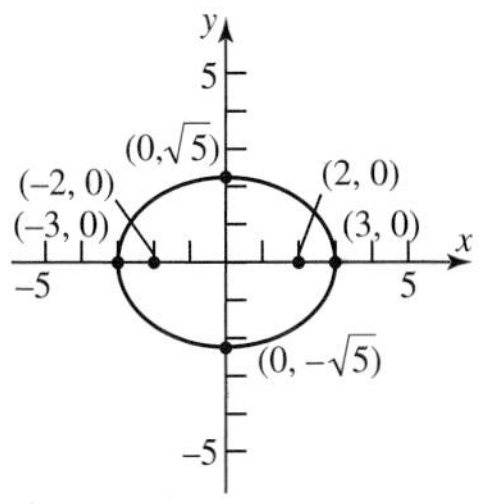

25. $\frac{x^2}{4} + \frac{y^2}{13} = 1$

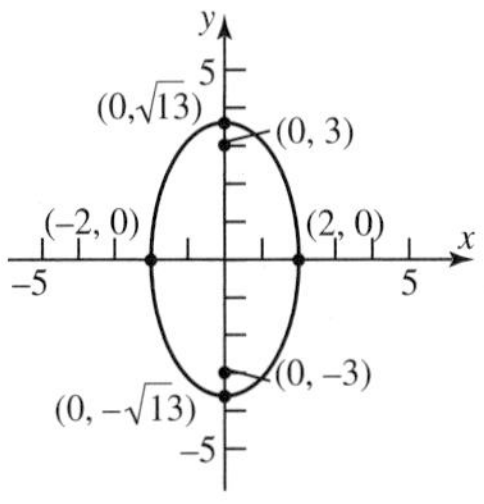

27. $x^2 + \frac{y^2}{16} = 1$

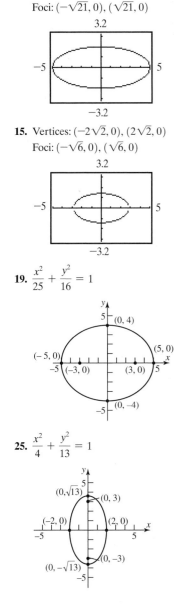

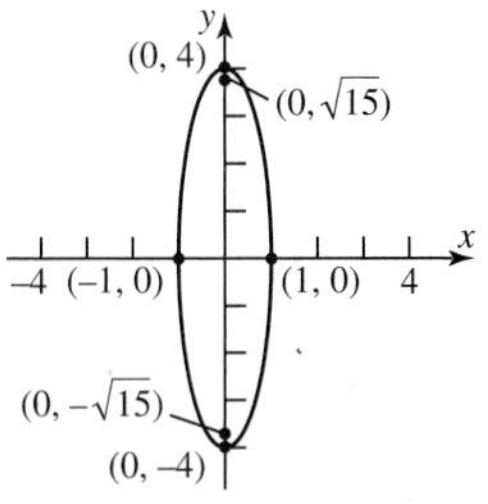

29. $\frac{(x + 1)^2}{4} + (y - 1)^2 = 1$ **31.** $(x - 1)^2 + \frac{y^2}{4} = 1$

33. Center: $(3, -1)$
Vertices: $(3, -4)$, $(3, 2)$
Foci: $(3, -1 - \sqrt{5})$, $(3, -1 + \sqrt{5})$

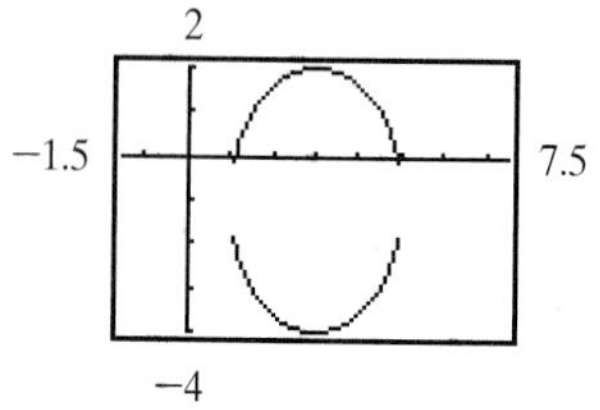

35. Center: $(-5, 4)$
Vertices: $(-9, 4)$, $(-1, 4)$
Foci: $(-5 - 2\sqrt{3}, 4)$, $(-5 + 2\sqrt{3}, 4)$

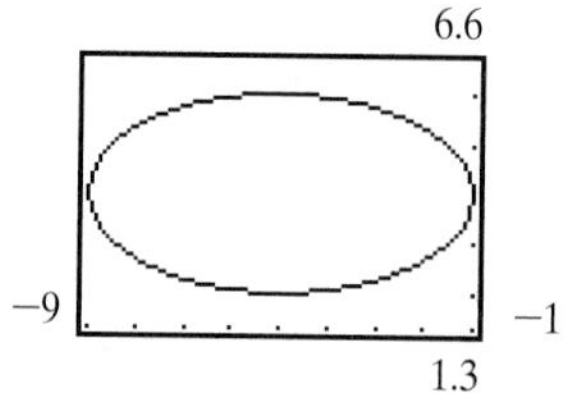

37. Center: $(-2, 1)$
Vertices: $(-4, 1)$, $(0, 1)$
Foci: $(-2 - \sqrt{3}, 1)$, $(-2 + \sqrt{3}, 1)$

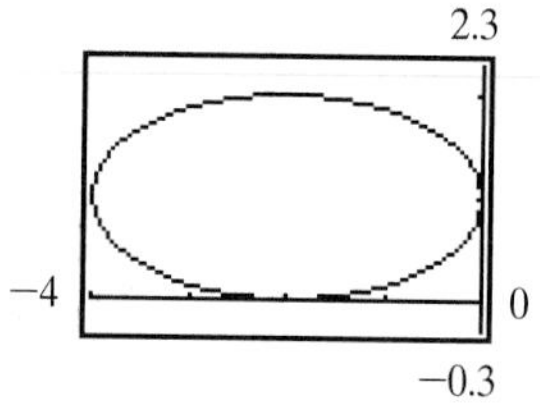

39. Center: $(2, -1)$
Vertices: $(2 - \sqrt{3}, -1)$, $(2 + \sqrt{3}, -1)$
Foci: $(1, -1)$, $(3, -1)$

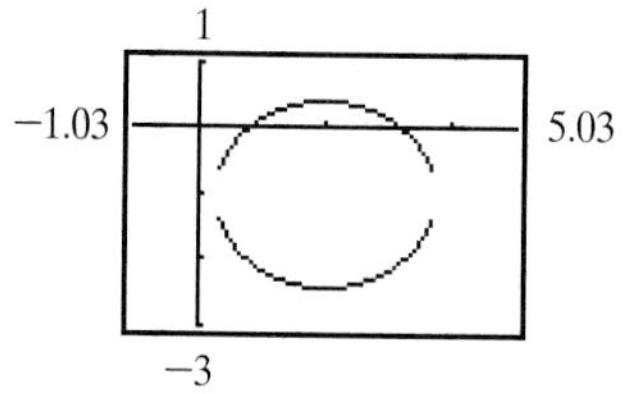

41. Center: $(1, -2)$
Vertices: $(1, -5)$, $(1, 1)$
Foci: $(1, -2 - \sqrt{5})$, $(1, -2 + \sqrt{5})$

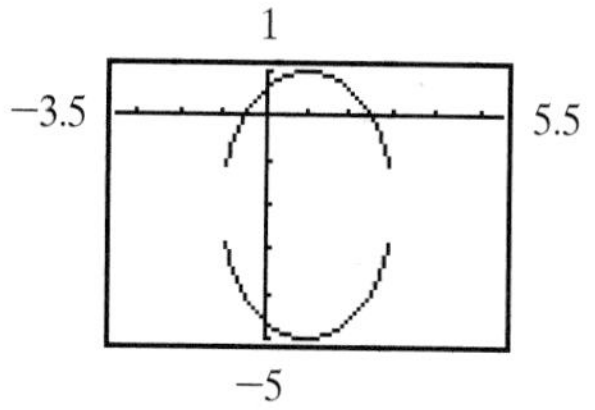

43. Center: $(0, -2)$
Vertices: $(0, -4)$, $(0, 0)$
Foci: $(0, -2 - \sqrt{3})$, $(0, -2 + \sqrt{3})$

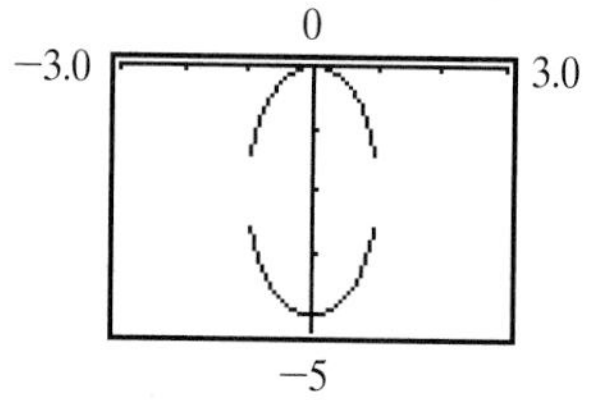

45. $\dfrac{(x - 2)^2}{25} + \dfrac{(y + 2)^2}{21} = 1$

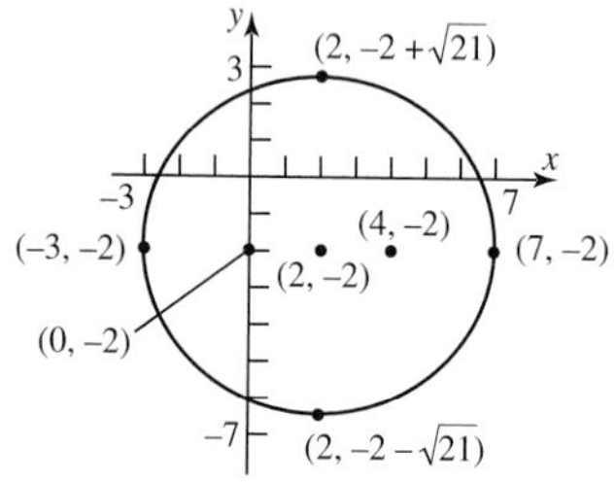

47. $\dfrac{(x - 4)^2}{5} + \dfrac{(y - 6)^2}{9} = 1$

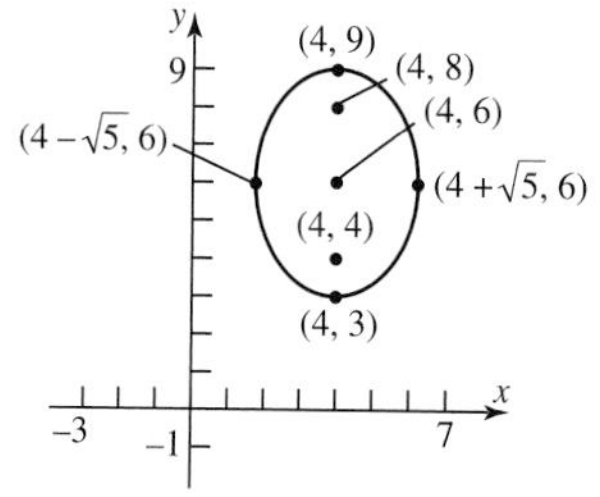

49. $\dfrac{(x - 2)^2}{16} + \dfrac{(y - 1)^2}{7} = 1$

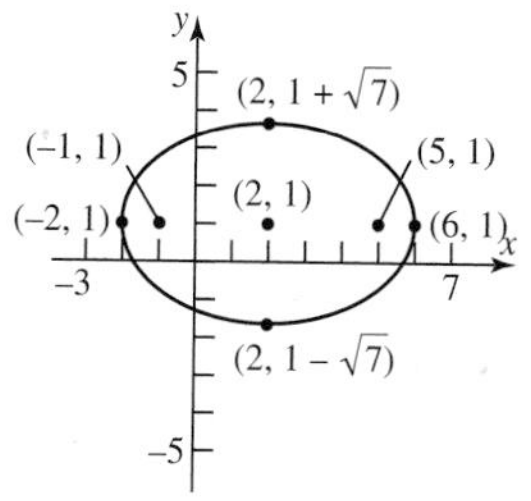

51. $\dfrac{(x - 1)^2}{10} + (y - 2)^2 = 1$

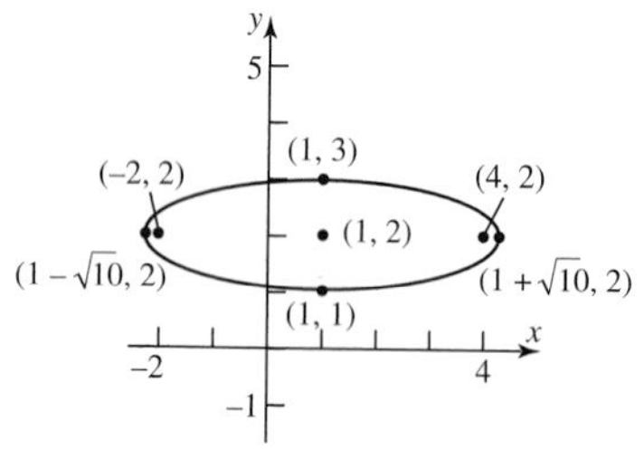

53. $\dfrac{(x - 1)^2}{9} + (y - 2)^2 = 1$

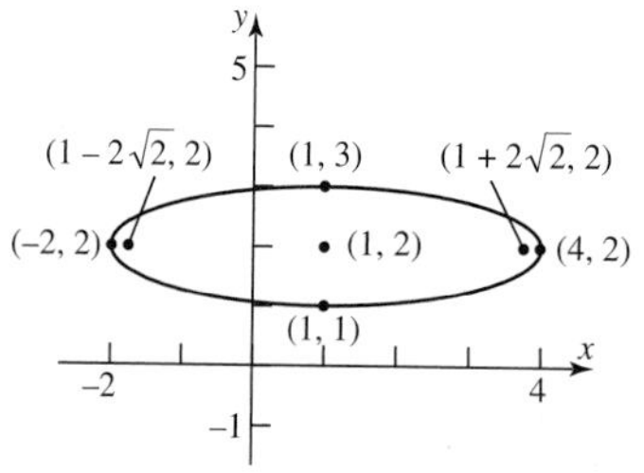

55.

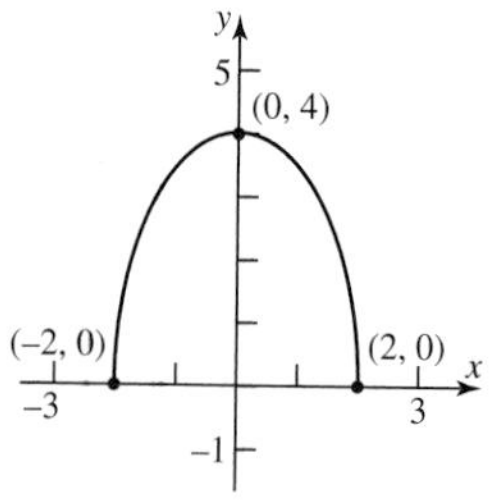

57.

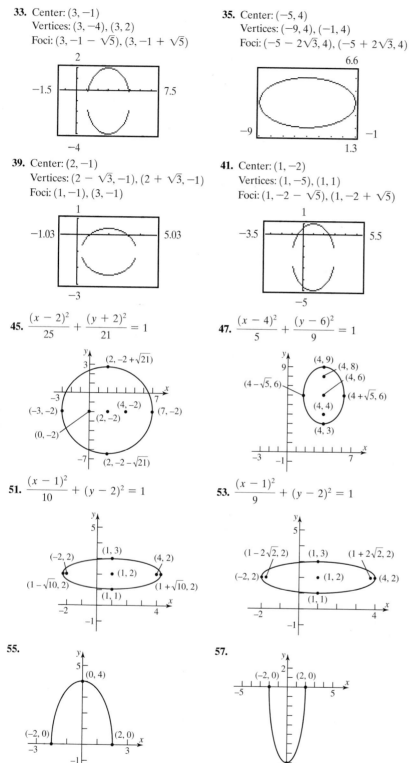

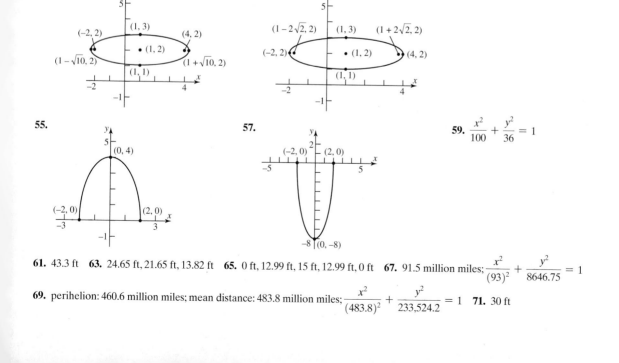

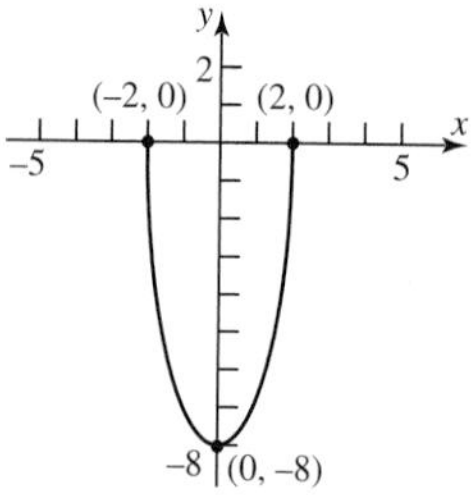

59. $\dfrac{x^2}{100} + \dfrac{y^2}{36} = 1$

61. 43.3 ft **63.** 24.65 ft, 21.65 ft, 13.82 ft **65.** 0 ft, 12.99 ft, 15 ft, 12.99 ft, 0 ft **67.** 91.5 million miles; $\dfrac{x^2}{(93)^2} + \dfrac{y^2}{8646.75} = 1$

69. perihelion: 460.6 million miles; mean distance: 483.8 million miles; $\dfrac{x^2}{(483.8)^2} + \dfrac{y^2}{233{,}524.2} = 1$ **71.** 30 ft

73. (a) $Ax^2 + Cy^2 + F = 0$

$Ax^2 + Cy^2 = -F$

If A and C are of the same and F is of opposite sign, then the equation takes the form $\dfrac{x^2}{\left(-\dfrac{F}{A}\right)} + \dfrac{y^2}{\left(-\dfrac{F}{C}\right)} = 1$, where $-\dfrac{F}{A}$ and $-\dfrac{F}{C}$ are positive. This is the equation of an ellipse with center at $(0, 0)$.

(b) If $A = C$, the equation may be written as $x^2 + y^2 = -\dfrac{F}{A}$. This is the equation of a circle with center at $(0, 0)$ and radius equal to $\sqrt{-\dfrac{F}{A}}$.

10.4 Exercises

1. B **3.** A **5.** B **7.** C

9. $x^2 - \dfrac{y^2}{8} = 1$

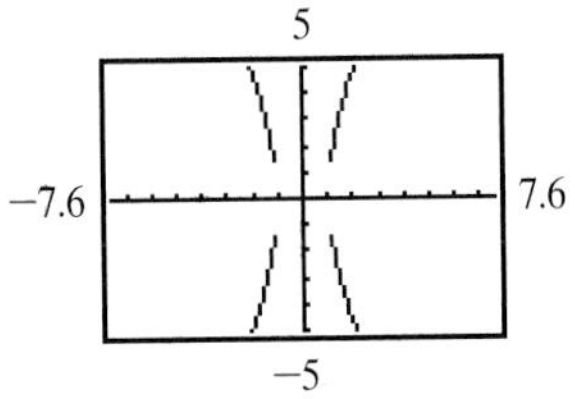

11. $\dfrac{y^2}{16} - \dfrac{x^2}{20} = 1$

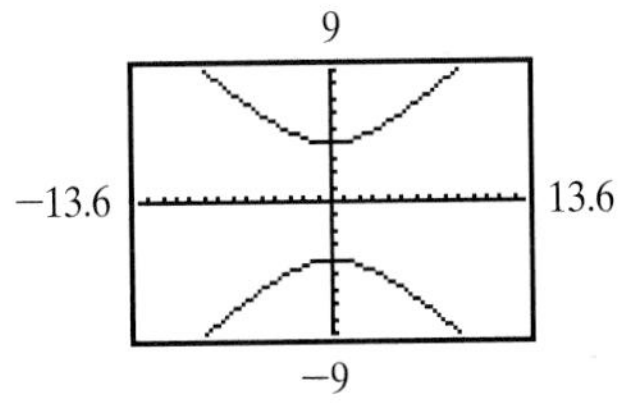

13. $\dfrac{x^2}{9} - \dfrac{y^2}{16} = 1$

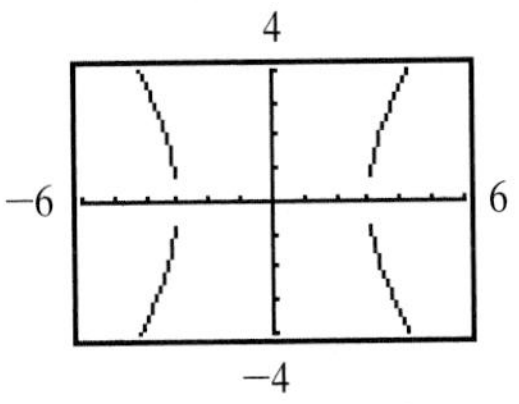

15. $\dfrac{y^2}{36} - \dfrac{x^2}{9} = 1$

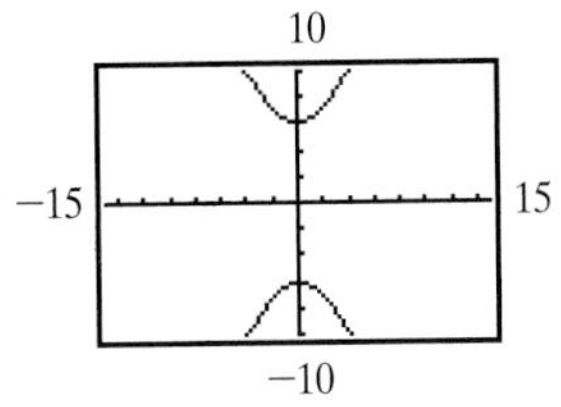

17. $\dfrac{x^2}{8} - \dfrac{y^2}{8} = 1$

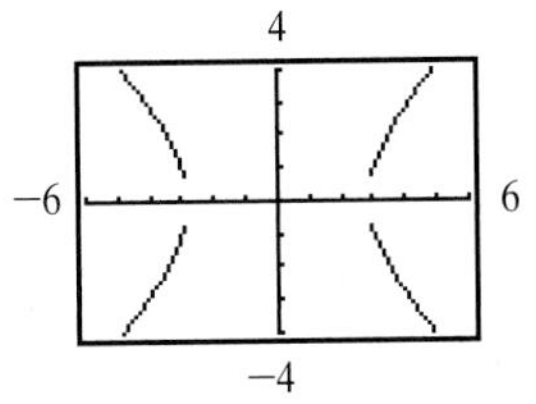

19. Center: $(0, 0)$
Transverse axis: x-axis
Vertices: $(-5, 0)$, $(5, 0)$
Foci: $(-\sqrt{34}, 0)$, $(\sqrt{34}, 0)$
Asymptotes: $y = \pm\dfrac{3}{5}x$

(a)

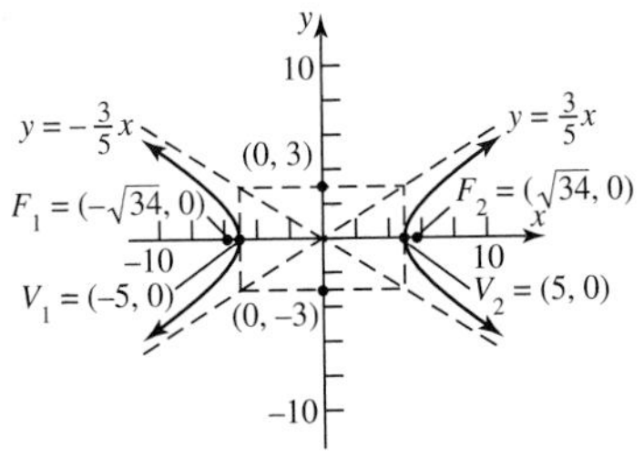

(b)

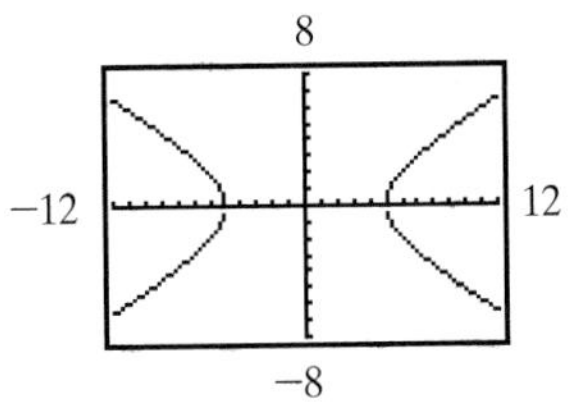

21. Center: $(0, 0)$
Transverse axis: x-axis
Vertices: $(-2, 0)$, $(2, 0)$
Foci: $(-2\sqrt{5}, 0)$, $(2\sqrt{5}, 0)$

Asymptotes: $y = \pm 2x$

(a)

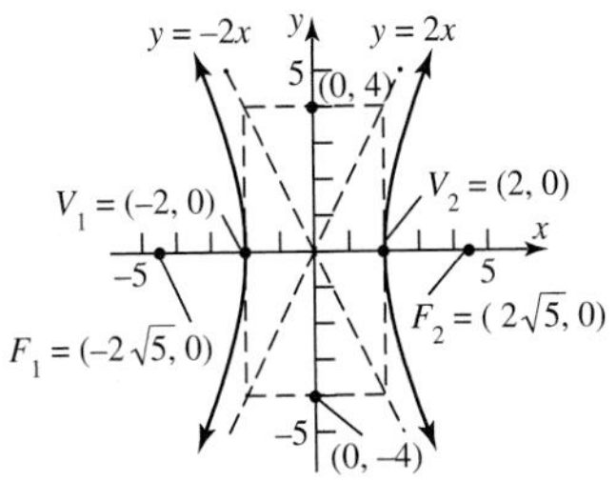

(b)

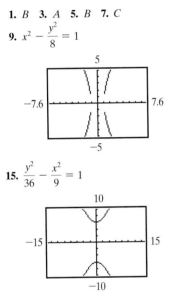

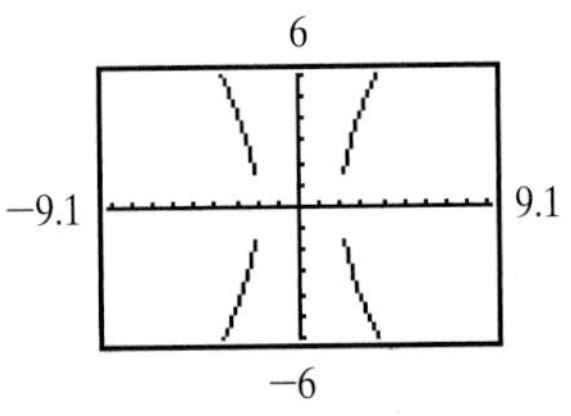

23. Center: $(0, 0)$
Transverse axis: y-axis
Vertices: $(0, -3)$, $(0, 3)$
Foci: $(0, -\sqrt{10})$, $(0, \sqrt{10})$
Asymptotes: $y = \pm 3x$

(a)

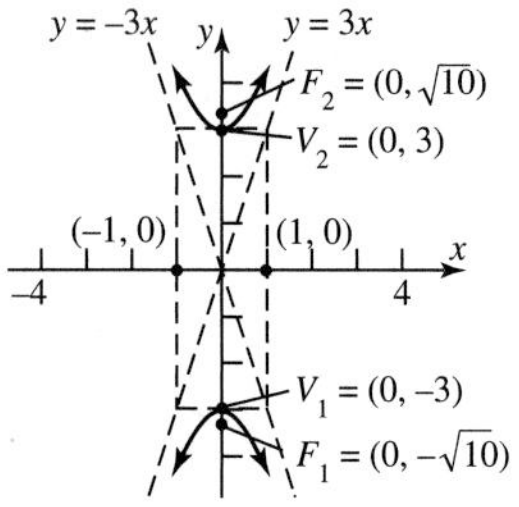

(b)

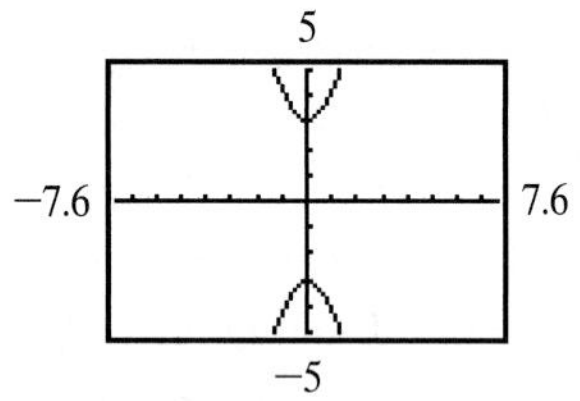

25. Center: $(0, 0)$
Transverse axis: y-axis
Vertices: $(0, -5)$, $(0, 5)$
Foci: $(0, -5\sqrt{2})$, $(0, 5\sqrt{2})$
Asymptotes: $y = \pm x$

(a)

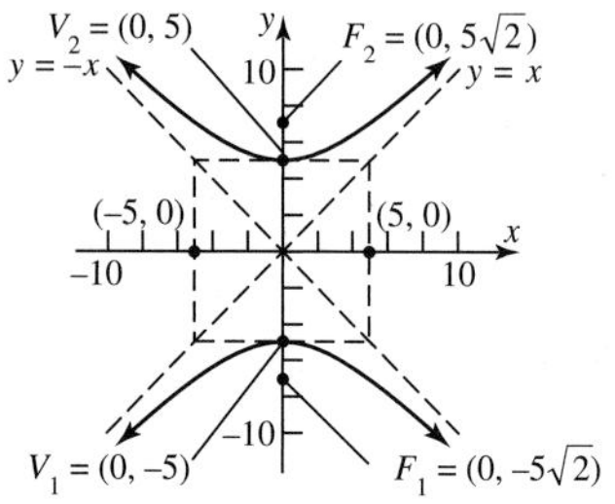

(b)

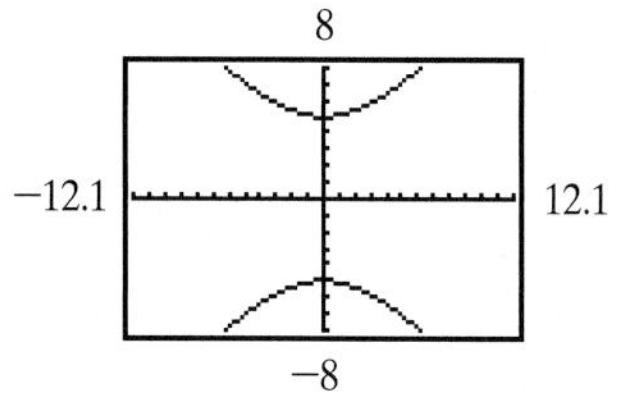

27. $x^2 - y^2 = 1$ **29.** $\dfrac{y^2}{36} - \dfrac{x^2}{9} = 1$

31. $\dfrac{(x - 4)^2}{4} - \dfrac{(y + 1)^2}{5} = 1$

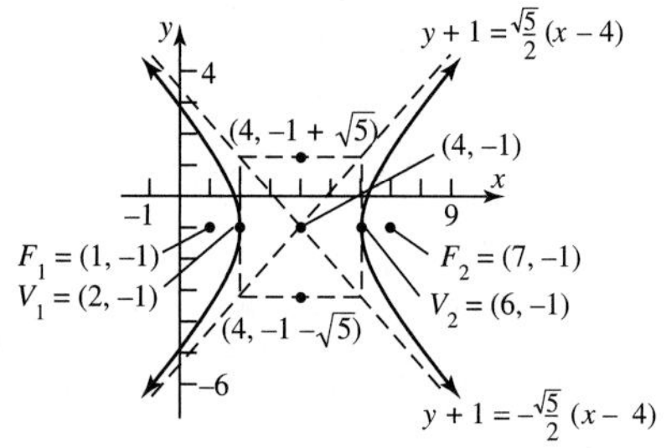

33. $\dfrac{(y + 4)^2}{4} - \dfrac{(x + 3)^2}{12} = 1$

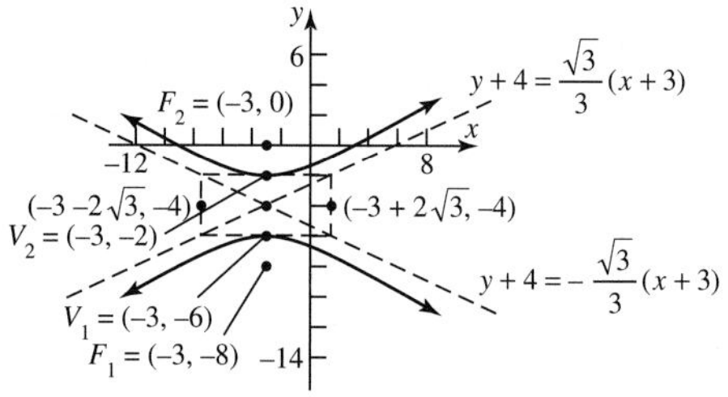

35. $(x - 5)^2 - \dfrac{(y - 7)^2}{3} = 1$

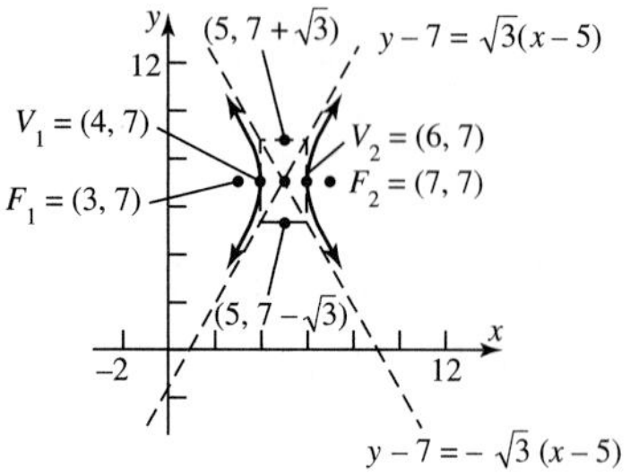

37. $\dfrac{(x - 1)^2}{4} - \dfrac{(y + 1)^2}{9} = 1$

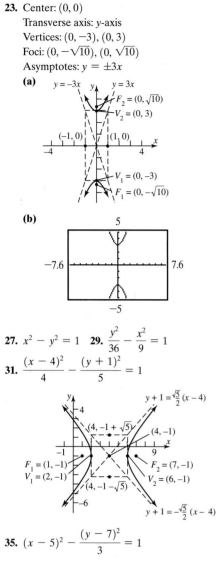

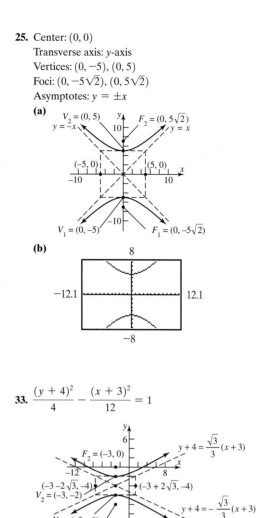

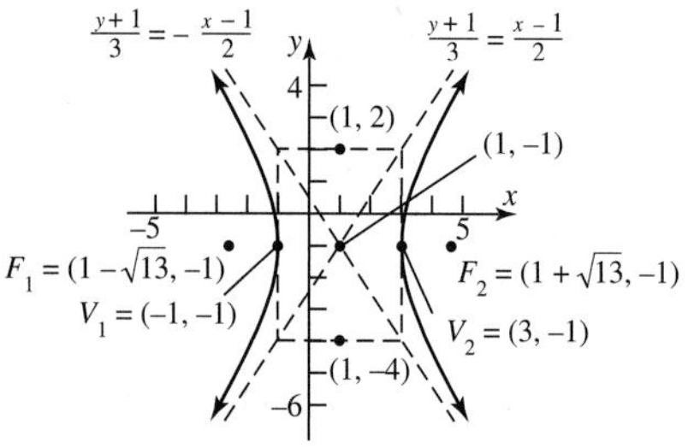

39. Center: $(2, -3)$
Transverse axis: Parallel to x-axis
Vertices: $(0, -3)$, $(4, -3)$
Foci: $(2 - \sqrt{13}, -3)$, $(2 + \sqrt{13}, -3)$
Asymptotes: $y + 3 = \pm\frac{3}{2}(x - 2)$

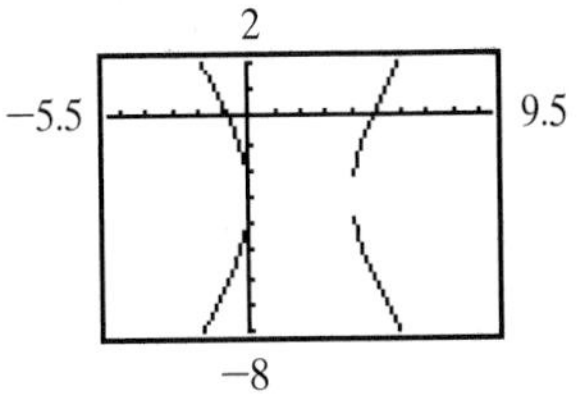

41. Center: $(-2, 2)$
Transverse axis: Parallel to y-axis
Vertices: $(-2, 0)$, $(-2, 4)$
Foci: $(-2, 2 - \sqrt{5})$, $(-2, 2 + \sqrt{5})$
Asymptotes: $y - 2 = \pm 2(x + 2)$

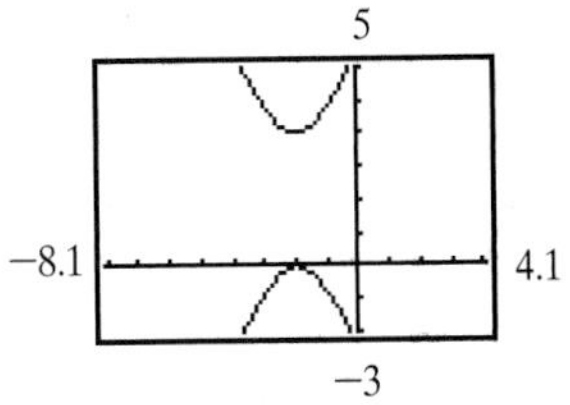

43. Center: $(-1, -2)$
Transverse axis: Parallel to x-axis
Vertices: $(-3, -2)$, $(1, -2)$
Foci: $(-1 - 2\sqrt{2}, -2)$, $(-1 + 2\sqrt{2}, -2)$
Asymptotes: $y + 2 = \pm(x + 1)$

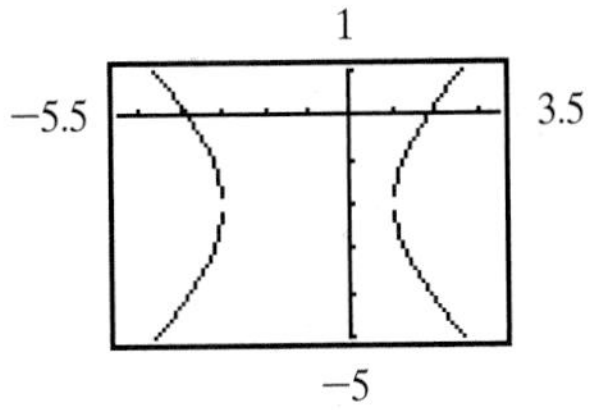

45. Center: $(1, -1)$
Transverse axis: Parallel to x-axis
Vertices: $(0, -1)$, $(2, -1)$
Foci: $(1 - \sqrt{2}, -1)$, $(1 + \sqrt{2}, -1)$
Asymptotes: $y + 1 = \pm(x - 1)$

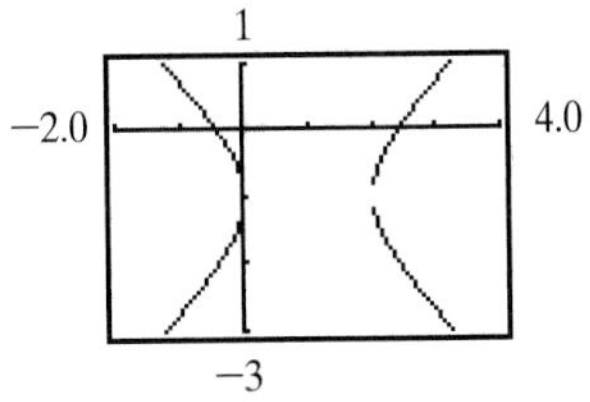

47. Center: $(-1, 2)$
Transverse axis: Parallel to y-axis
Vertices: $(-1, 0)$, $(-1, 4)$
Foci: $(-1, 2 - \sqrt{5})$, $(-1, 2 + \sqrt{5})$
Asymptotes: $y - 2 = \pm 2(x + 1)$

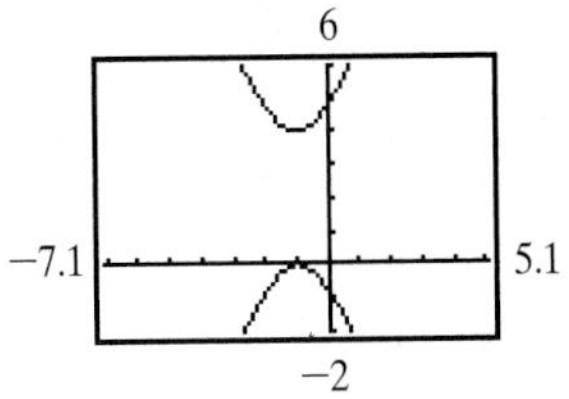

49. Center: $(3, -2)$
Transverse axis: Parallel to x-axis
Vertices: $(1, -2)$, $(5, -2)$
Foci: $(3 - 2\sqrt{5}, -2)$, $(3 + 2\sqrt{5}, -2)$
Asymptotes: $y + 2 = \pm 2(x - 3)$

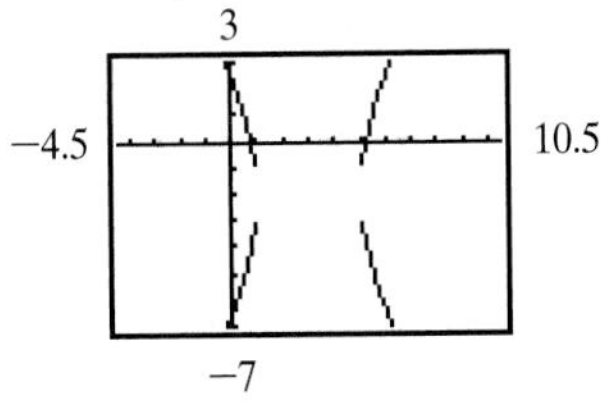

51. Center: $(-2, 1)$
Transverse axis: Parallel to y-axis
Vertices: $(-2, -1)$, $(-2, 3)$
Foci: $(-2, 1 - \sqrt{5})$, $(-2, 1 + \sqrt{5})$
Asymptotes: $y - 1 = \pm 2(x + 2)$

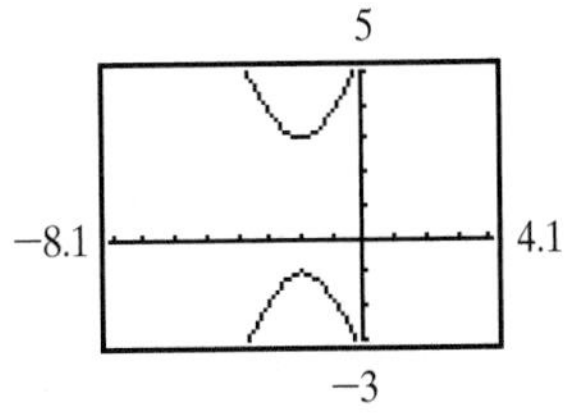

53.

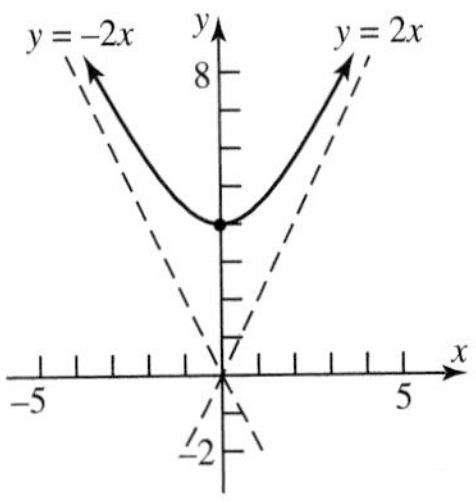

55.

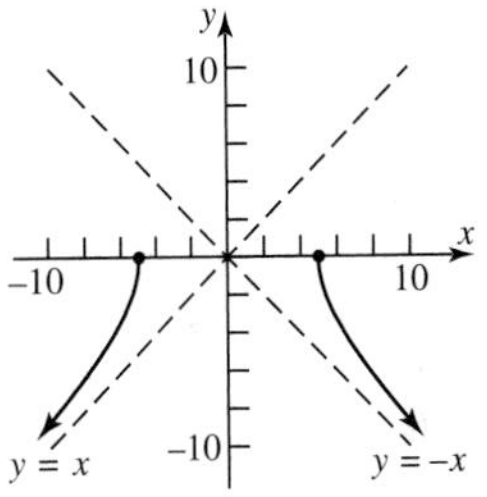

57. **(a)** The ship would reach shore 64.66 miles from the master station. **(b)** 0.00086 sec **(c)** (104, 50)
59. **(a)** 450 ft **61.** If e is close to 1, narrow hyperbola; if e is very large, wide hyperbola
63. $\frac{x^2}{4} - y^2 = 1$; asymptotes $y = \pm\frac{1}{2}x$, $y^2 - \frac{x^2}{4} = 1$; asymptotes $y = \pm\frac{1}{2}x$

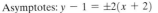

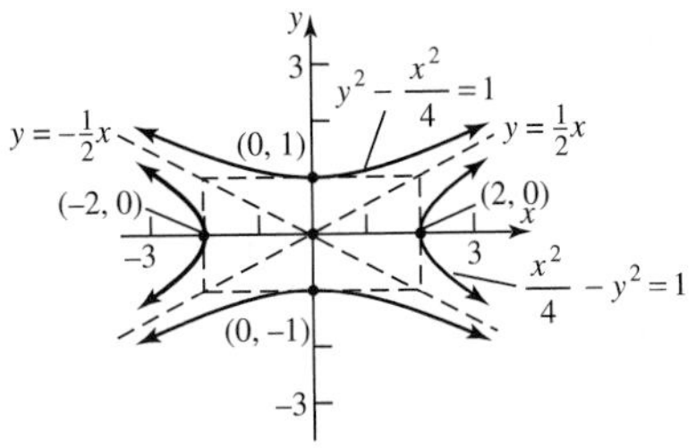

65. $Ax^2 + Cy^2 + F = 0$ If A and C are of opposite sign and $F \neq 0$, this equation may be written as $\dfrac{x^2}{\left(-\dfrac{F}{A}\right)} + \dfrac{y^2}{\left(-\dfrac{F}{C}\right)} = 1,$

$Ax^2 + Cy^2 = -F$ where $-\dfrac{F}{A}$ and $-\dfrac{F}{C}$ are opposite in sign. This is the equation of a hyperbola with center $(0, 0)$.

The transverse axis is the x-axis if $-\dfrac{F}{A} > 0$; the transverse axis is the y-axis if $-\dfrac{F}{A} < 0$.

10.5 Exercises

1. Parabola **3.** Ellipse **5.** Hyperbola **7.** Hyperbola **9.** Circle **11.** $x = \frac{\sqrt{2}}{2}(x' - y'), y = \frac{\sqrt{2}}{2}(x' + y')$

13. $x = \frac{\sqrt{2}}{2}(x' - y'), y = \frac{\sqrt{2}}{2}(x' + y')$ **15.** $x = \frac{1}{2}(x' - \sqrt{3}y'), y = \frac{1}{2}(\sqrt{3}x' + y')$

17. $x = \frac{\sqrt{5}}{5}(x' - 2y'), y = \frac{\sqrt{5}}{5}(2x' + y')$ **19.** $x = \frac{\sqrt{13}}{13}(3x' - 2y'), y = \frac{\sqrt{13}}{13}(2x' + 3y')$

21.

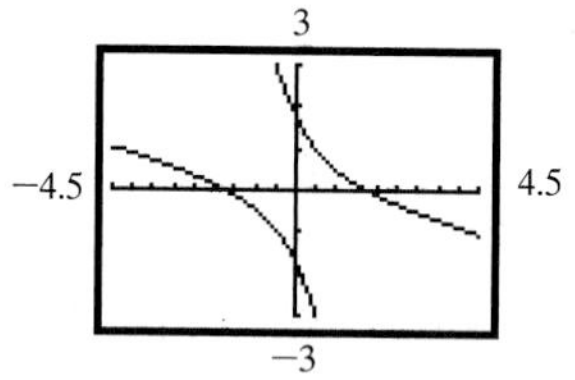

$\theta = 45^\circ$ (see Problem 11)

$x'^2 - \frac{y'^2}{3} = 1$

Hyperbola
Center at $(0, 0)$
Transverse axis is the x'-axis.
Vertices at $(\pm 1, 0)$

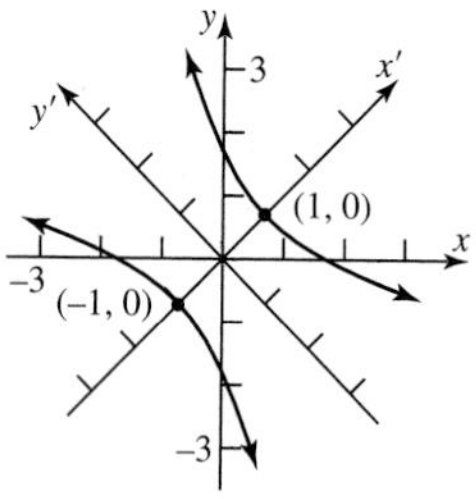

23.

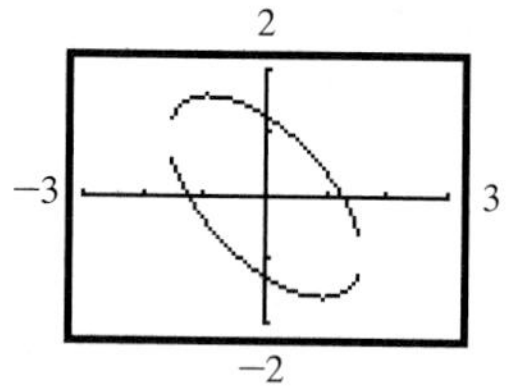

$\theta = 45^\circ$ (see Problem 13)

$x'^2 + \frac{y'^2}{4} = 1$

Ellipse
Center at $(0, 0)$
Major axis is the y'-axis.
Vertices at $(0, \pm 2)$

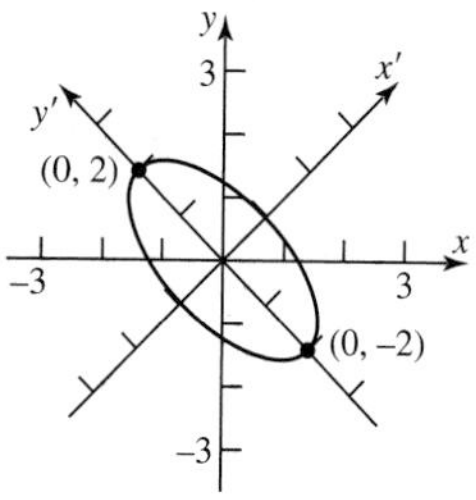

25.

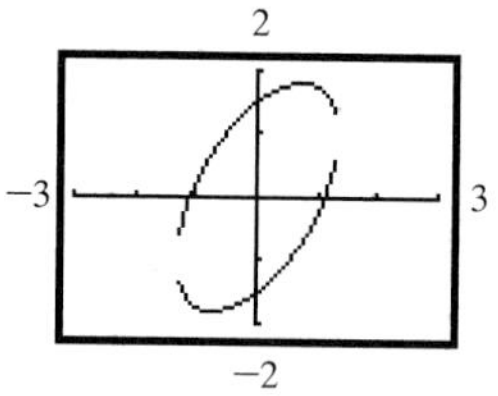

$\theta = 60^\circ$ (see Problem 15)

$\frac{x'^2}{4} + y'^2 = 1$

Ellipse
Center at $(0, 0)$
Major axis is the x'-axis.
Vertices at $(\pm 2, 0)$

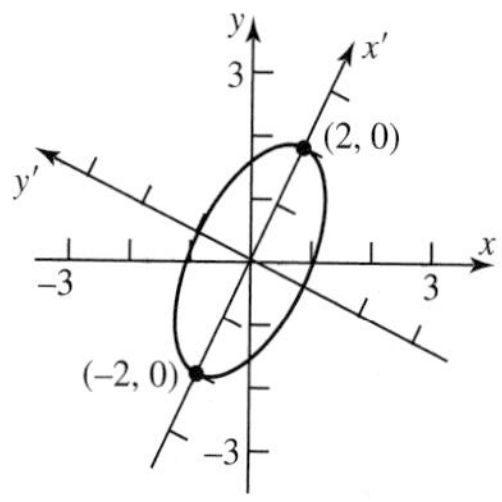

27.

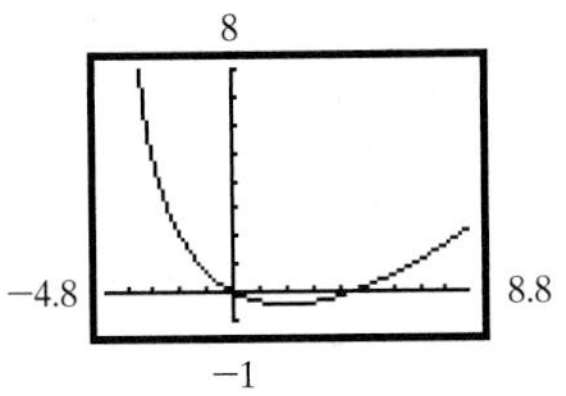

$\theta = 63°$ (see Problem 17)

$y'^2 = 8x'$

Parabola

Vertex at $(0, 0)$

Focus at $(2, 0)$

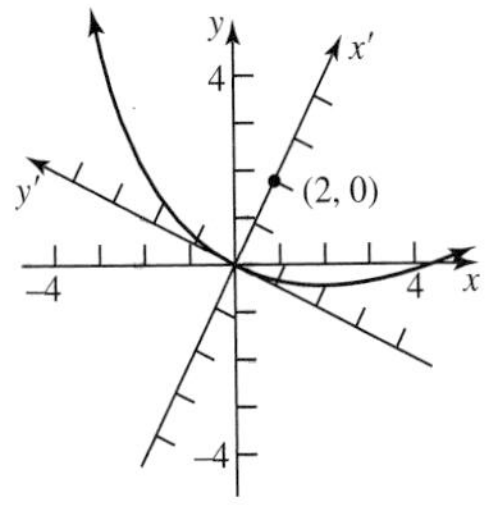

29.

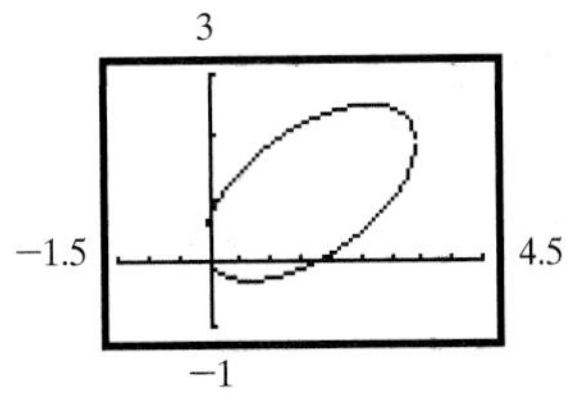

$\theta = 34°$ (see Problem 19)

$$\frac{(x' - 2)^2}{4} + y'^2 = 1$$

Ellipse

Center at $(2, 0)$

Major axis is the x'-axis.

Vertices at $(4, 0)$ and $(0, 0)$

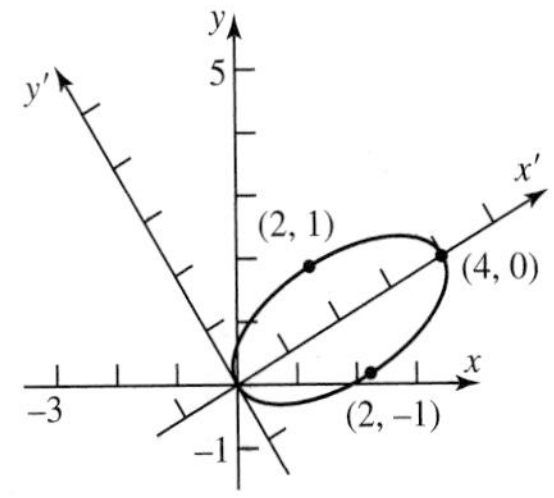

31.

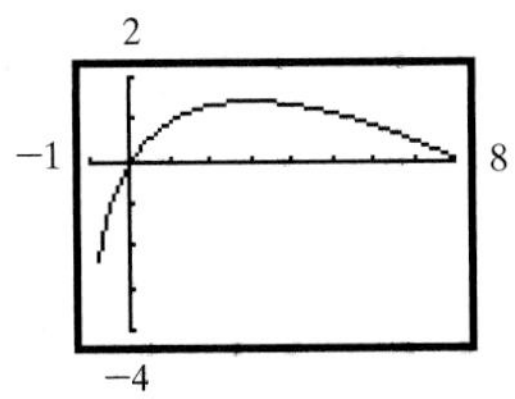

$\cot 2\theta = \frac{7}{24}; \theta \approx 37°$

$$(x' - 1)^2 = -6\left(y' - \frac{1}{6}\right)$$

Parabola

Vertex at $\left(1, \frac{1}{6}\right)$

Focus at $\left(1, -\frac{4}{3}\right)$

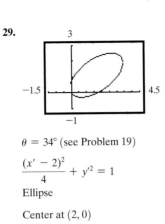

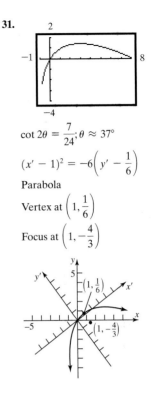

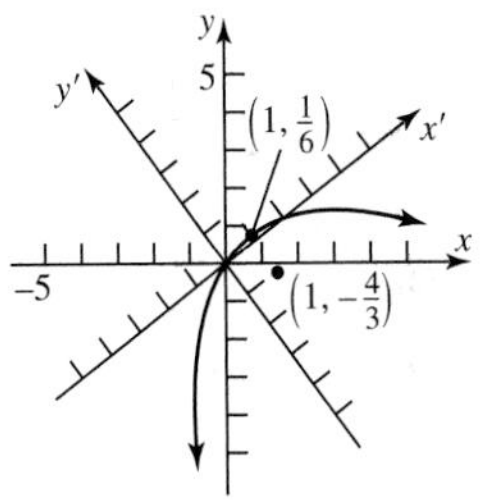

33. Hyperbola **35.** Hyperbola **37.** Parabola **39.** Ellipse **41.** Ellipse

43. Refer to equation (6):

$$\begin{aligned} A' &= A\cos^2\theta + B\sin\theta\cos\theta + C\sin^2\theta \\ B' &= B(\cos^2\theta - \sin^2\theta) + 2(C - A)(\sin\theta\cos\theta) \\ C' &= A\sin^2\theta - B\sin\theta\cos\theta + C\cos^2\theta \\ D' &= D\cos\theta + E\sin\theta \\ E' &= -D\sin\theta + E\cos\theta \\ F' &= F \end{aligned}$$

45. Use Problem 43 to find $B'^2 - 4A'C'$. After much cancellation, $B'^2 - 4A'C' = B^2 - 4AC$.

47. Use formula (5) and find $d^2 = (x_2 - x_1)^2 + (y_2 - y_1)^2$. After simplifying, $(x_2 - x_1)^2 + (y_2 - y_1)^2 = (x'_2 - x'_1)^2 + (y'_2 - y'_1)^2$.

10.6 Exercises

1. Parabola; directrix is perpendicular to the polar axis 1 unit to the right of the pole.

3. Hyperbola; directrix is parallel to the polar axis $\frac{4}{3}$ units below the pole.

5. Ellipse; directrix is perpendicular to the polar axis $\frac{3}{2}$ units to the left of the pole.

7.

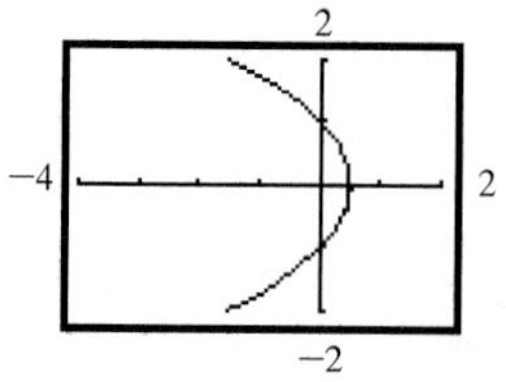

Parabola; directrix is perpendicular to the polar axis 1 unit to the right of the pole.

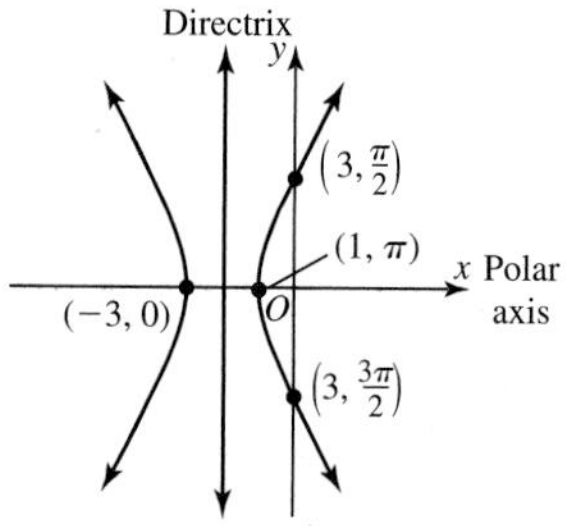

9.

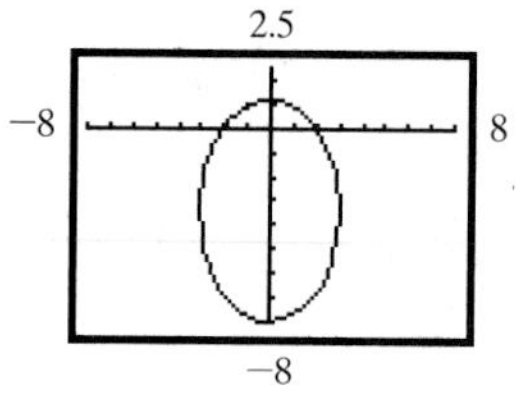

Ellipse; directrix is parallel to the polar axis $\frac{8}{3}$ units above the pole; vertices are at $\left(\frac{8}{7}, \frac{\pi}{2}\right)$ and $\left(8, \frac{3\pi}{2}\right)$.

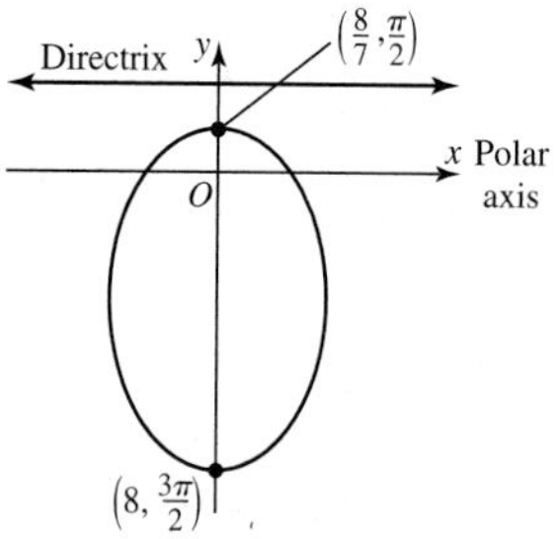

11.

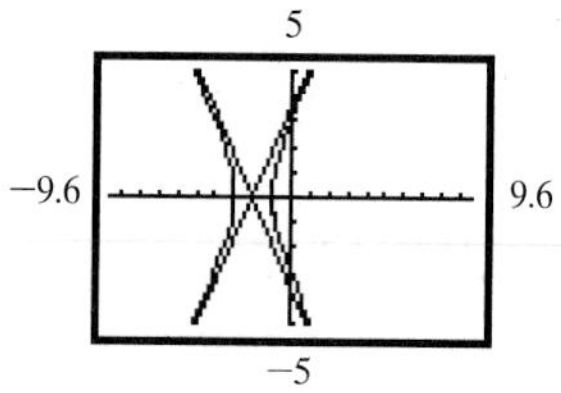

Hyperbola; directrix is perpendicular to the polar axis $\frac{3}{2}$ units to the left of the pole; vertices are at $(1, \pi)$ and $(-3, 0)$.

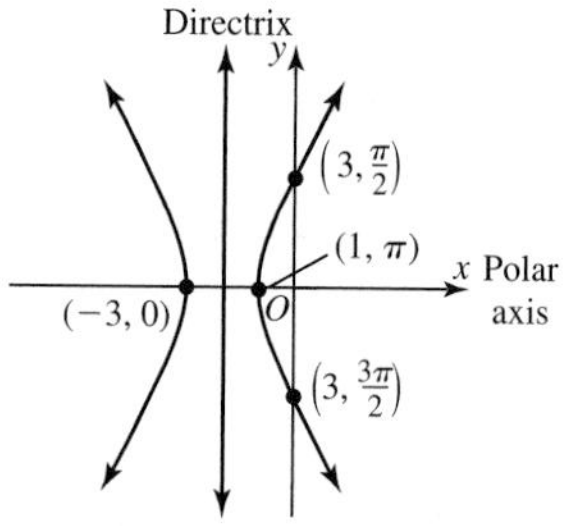

13.

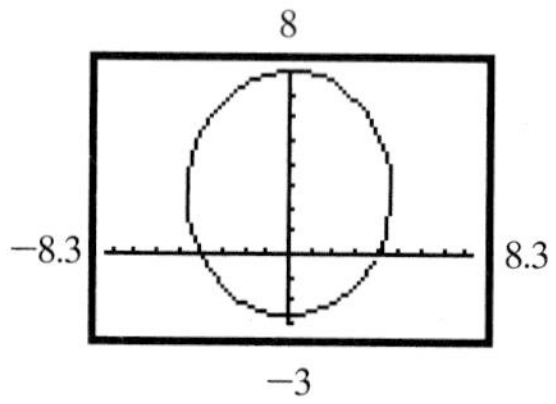

Ellipse; directrix is parallel to the polar axis 8 units below the pole; vertices are at $\left(8, \frac{\pi}{2}\right)$ and $\left(\frac{8}{3}, \frac{3\pi}{2}\right)$.

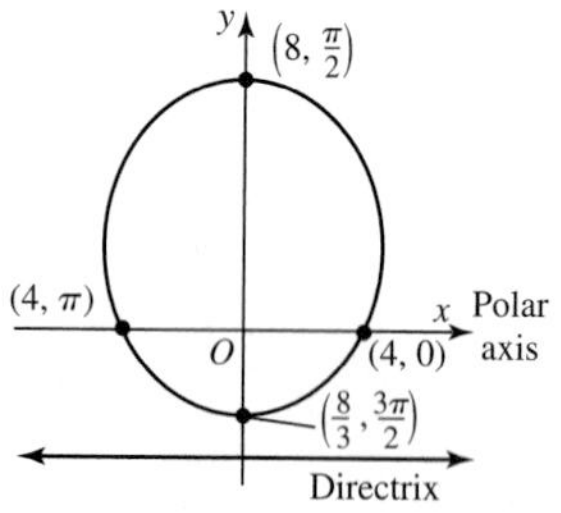

15.

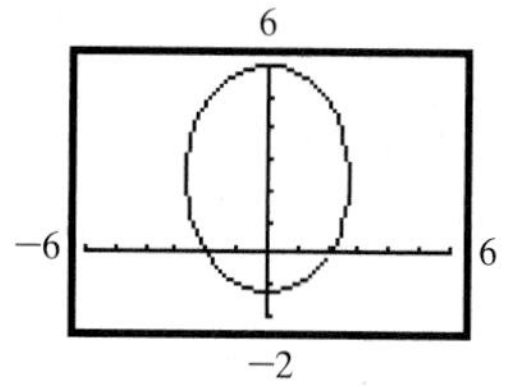

Ellipse; directrix is parallel to the polar axis 3 units below the pole; vertices are at $\left(6, \frac{\pi}{2}\right)$ and $\left(\frac{6}{5}, \frac{3\pi}{2}\right)$.

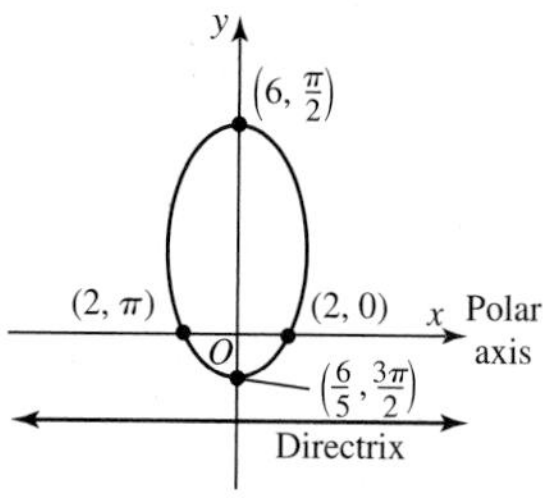

17.

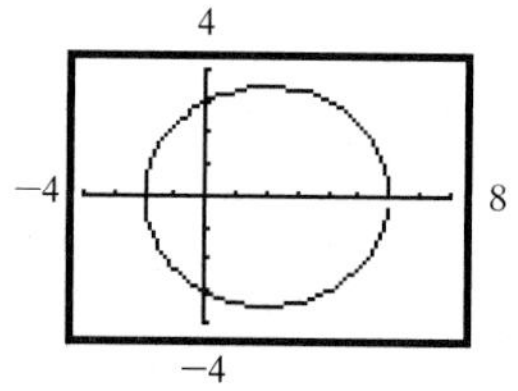

Ellipse; directrix is perpendicular to the polar axis 6 units to the left of the pole; vertices are at $(6, 0)$ and $(2, \pi)$.

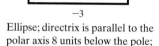

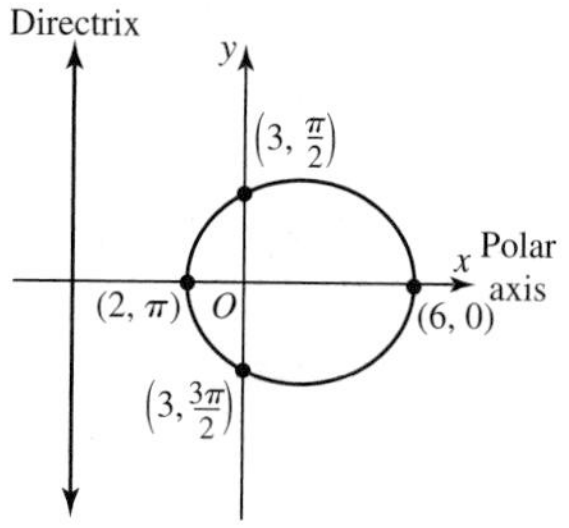

19. $y^2 + 2x - 1 = 0$ **21.** $16x^2 + 7y^2 + 48y - 64 = 0$ **23.** $3x^2 - y^2 + 12x + 9 = 0$ **25.** $4x^2 + 3y^2 - 16y - 64 = 0$

27. $9x^2 + 5y^2 - 24y - 36 = 0$ **29.** $3x^2 + 4y^2 - 12x - 36 = 0$ **31.** $r = \frac{1}{1 + \sin\theta}$ **33.** $r = \frac{12}{5 - 4\cos\theta}$ **35.** $r = \frac{12}{1 - 6\sin\theta}$

37. Use $d(D, P) = p - r\cos\theta$ in the derivation of equation (a) in Table 5.

39. Use $d(D, P) = p + r\sin\theta$ in the derivation of equation (a) in Table 5.

10.7 Exercises

1.

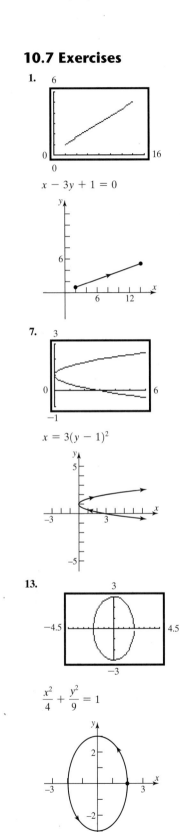

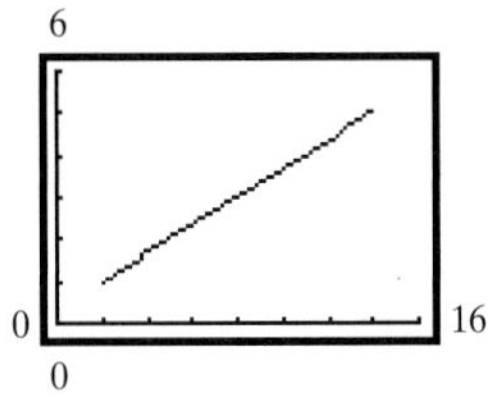

$x - 3y + 1 = 0$

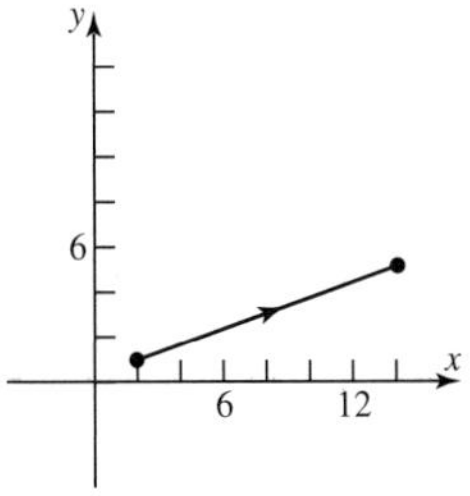

3.

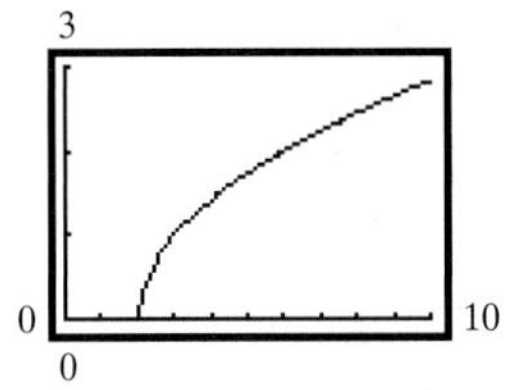

$y = \sqrt{x - 2}$

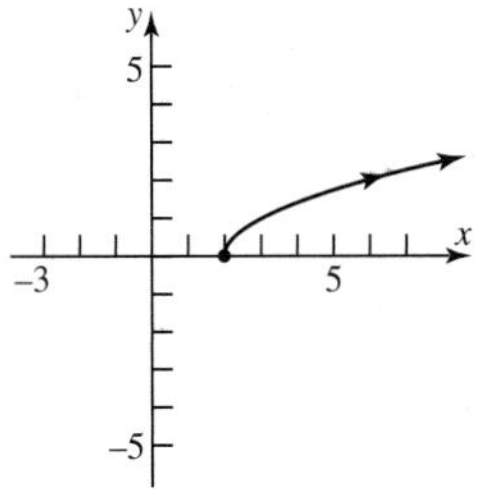

5.

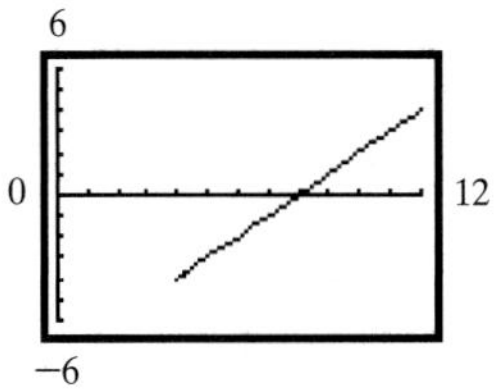

$x = y + 8$

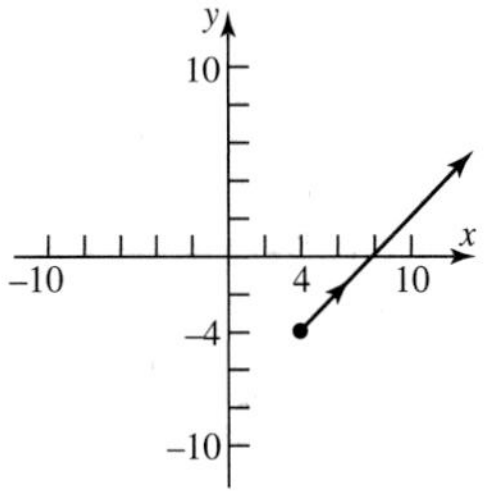

7.

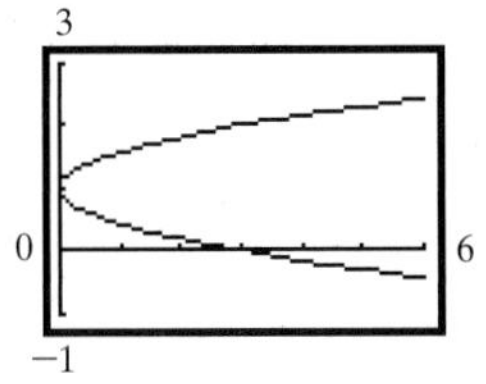

$x = 3(y - 1)^2$

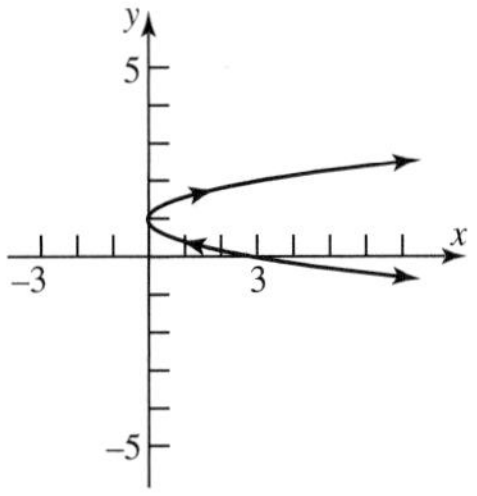

9.

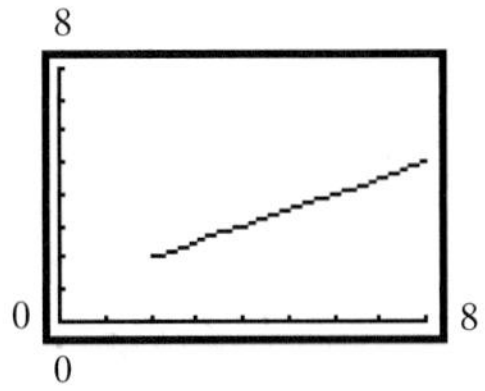

$2y = 2 + x$

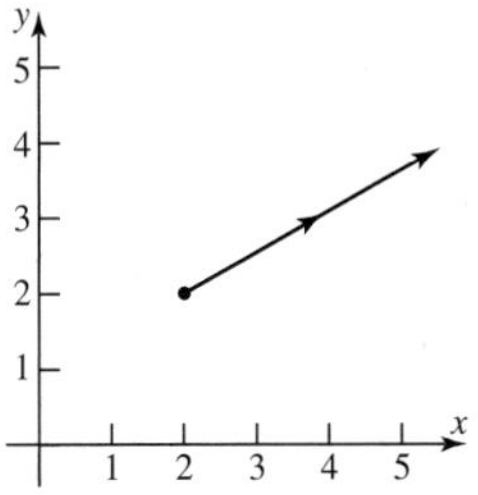

11.

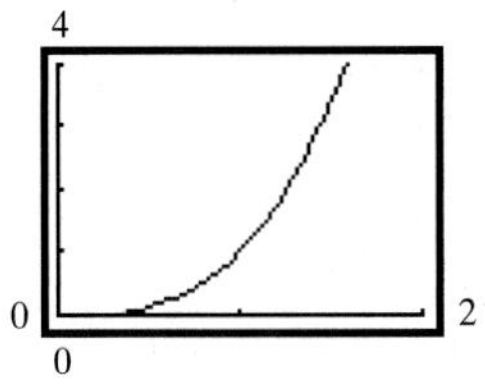

$y = x^3$

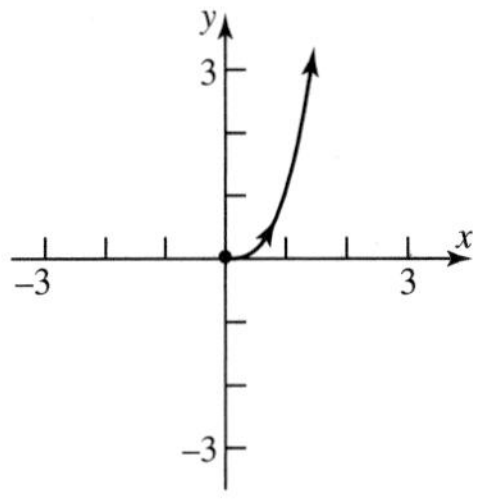

13.

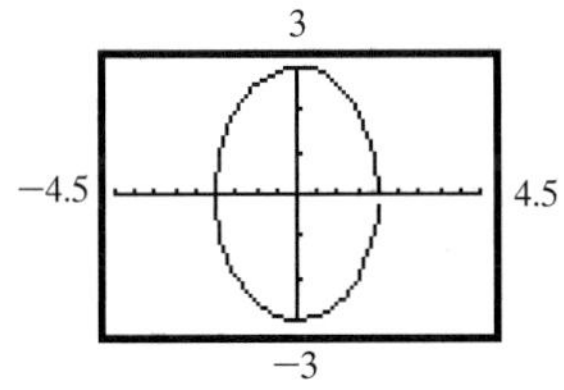

$\frac{x^2}{4} + \frac{y^2}{9} = 1$

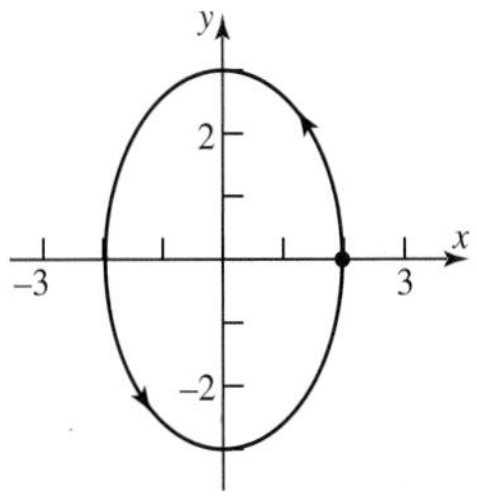

15.

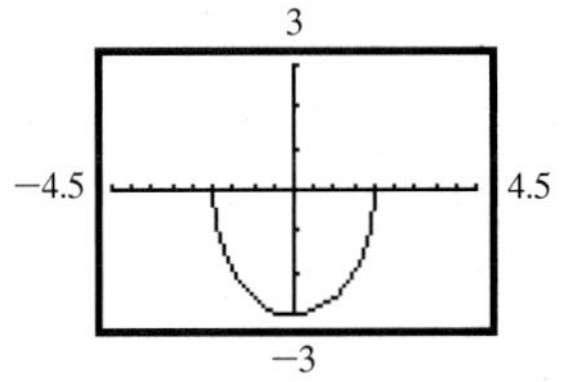

$\frac{x^2}{4} + \frac{y^2}{9} = 1$

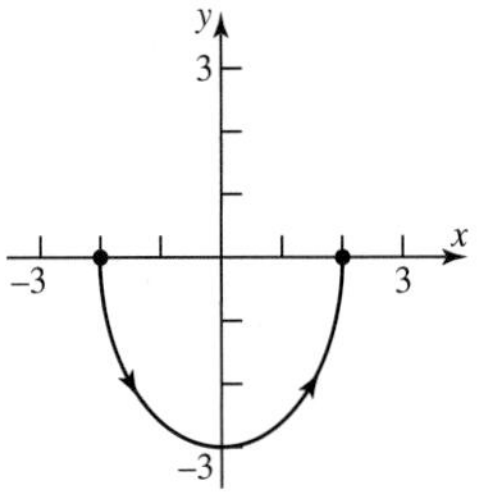

17.

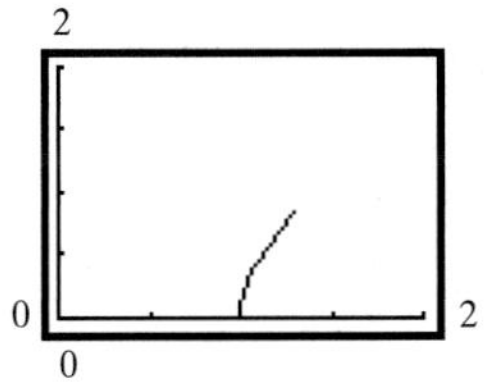

$x^2 - y^2 = 1$

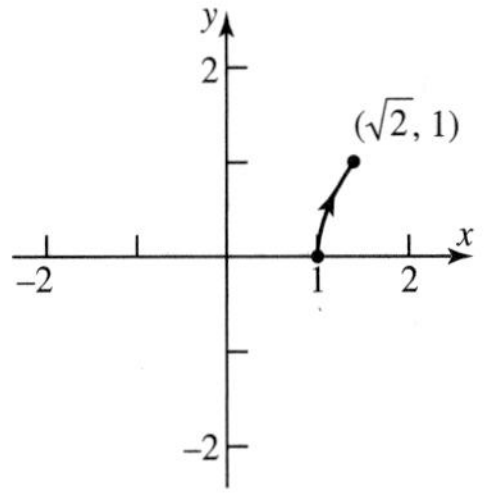

19.

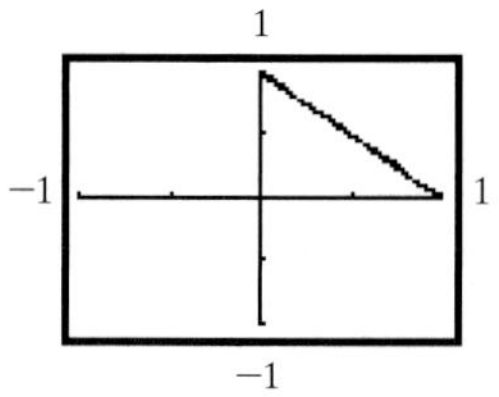

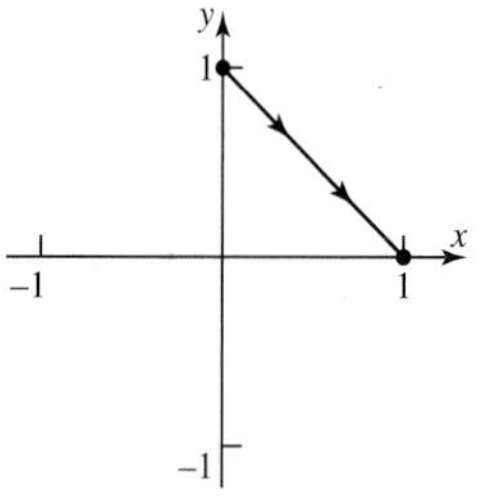

$x + y = 1$

21. (a) $x = 0$
$y = -16t^2 + 50t + 6$
(b) 3.24 seconds
(c) 1.5625 seconds; 45.0625 feet
(d) To see the vertical motion better, use $x = 3$.

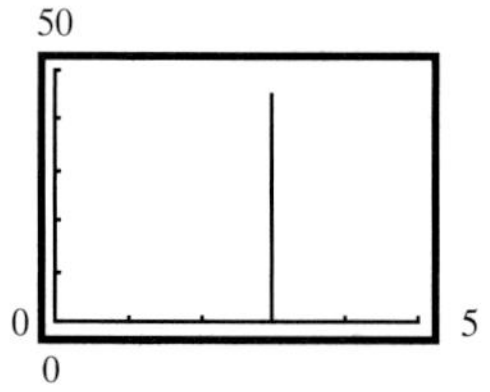

23. (a) Train: $x_1 = t^2$ Bill: $x_2 = 5(t - 5)$
$y_1 = 1$ $y_2 = 3$
(b) Bill won't catch the train.
(c)

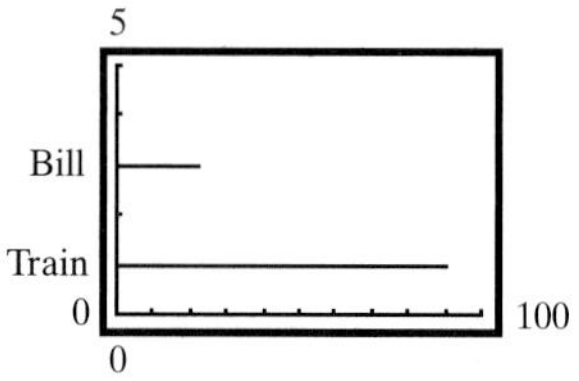

25. (a) $x = (145 \cos 20°)t$
$y = -16t^2 + (145 \sin 20°)t + 5$
(b) 3.197 seconds
(c) 1.55 seconds; 43.43 feet
(d) 435.65 feet
(e)

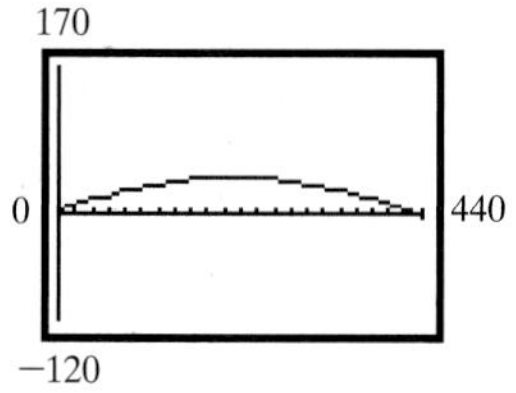

27. (a) $x = 20\sqrt{2}t$
$y = -4.9t^2 + 20\sqrt{2}t + 300$
(b) 11.226 seconds
(c) 2.886 seconds; 340.82 meters
(d) 317.52 meters
(e)

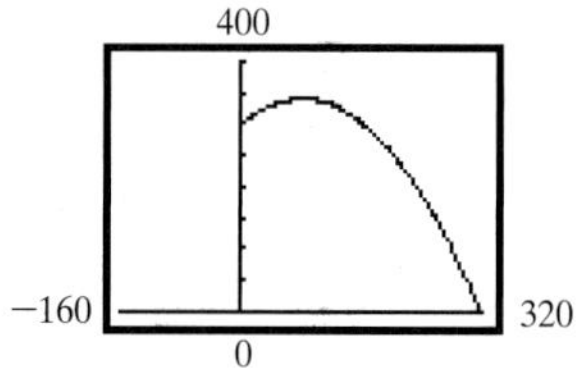

29. (a) $d = \sqrt{(40t - 5)^2 + (30t - 4)^2}$
(b)

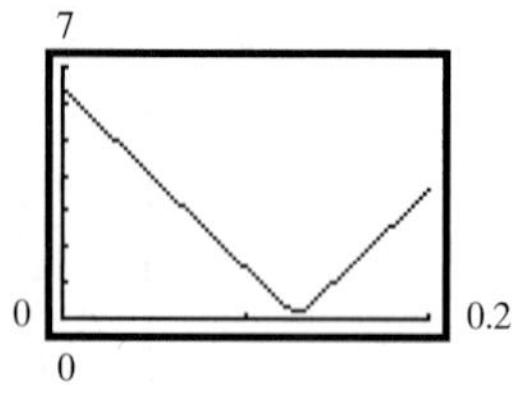

(c) 0.2 miles; 0.128 seconds
(d) Paseo: $x = 40t - 5$ Bonneville: $x = 0$
$y = 0$ $y = 30t - 4$
(e) Turn axes off to see the graph:

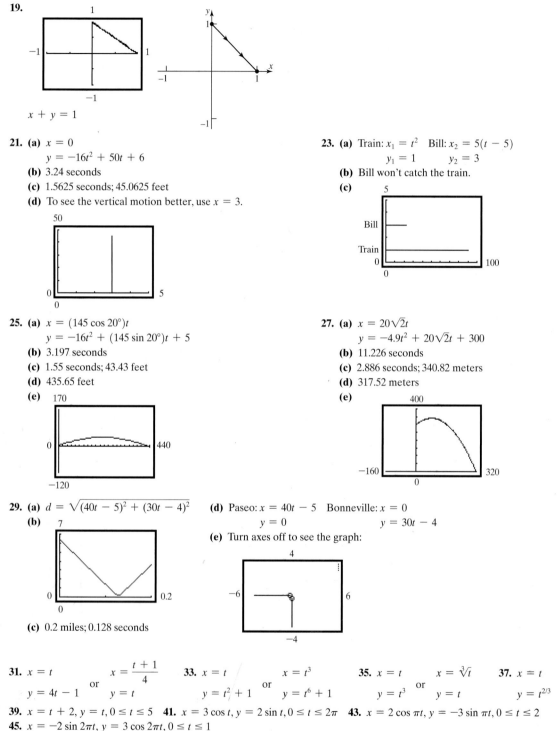

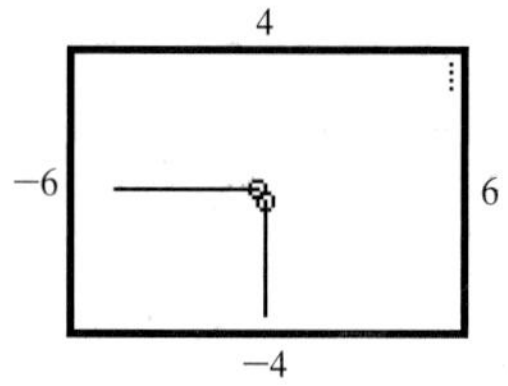

31. $x = t$, $y = 4t - 1$ or $x = \dfrac{t+1}{4}$, $y = t$
33. $x = t$, $y = t^2 + 1$ or $x = t^3$, $y = t^6 + 1$
35. $x = t$, $y = t^3$ or $x = \sqrt[3]{t}$, $y = t$
37. $x = t$, $y = t^{2/3}$ or $x = t^3$, $y = t^2$
39. $x = t + 2, y = t, 0 \le t \le 5$ **41.** $x = 3 \cos t, y = 2 \sin t, 0 \le t \le 2\pi$ **43.** $x = 2 \cos \pi t, y = -3 \sin \pi t, 0 \le t \le 2$
45. $x = -2 \sin 2\pi t, y = 3 \cos 2\pi t, 0 \le t \le 1$

47.

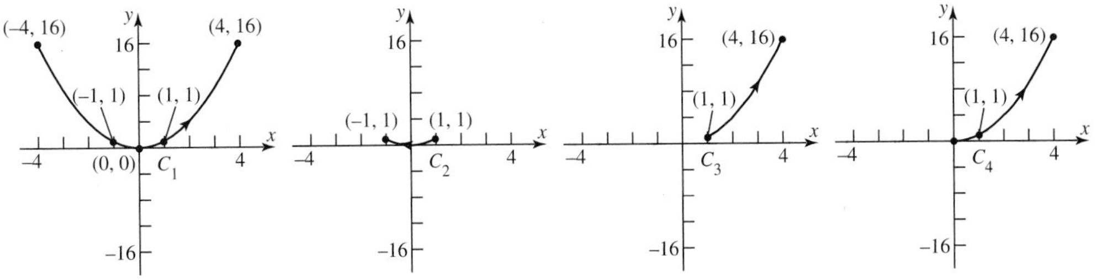

49. The orientation is from (x_1, y_1) to (x_2, y_2).

51.

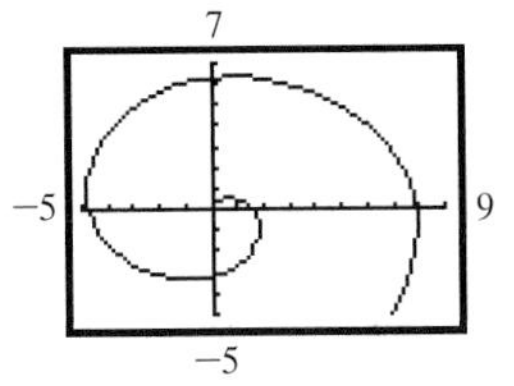

53.

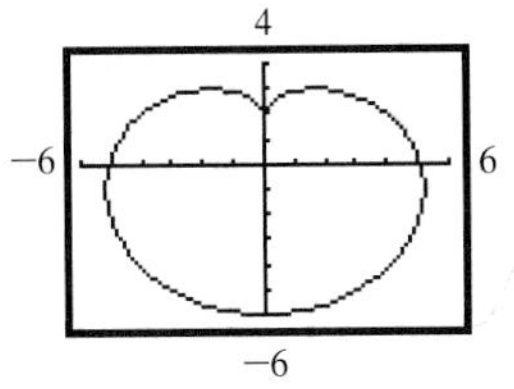

55. (a)

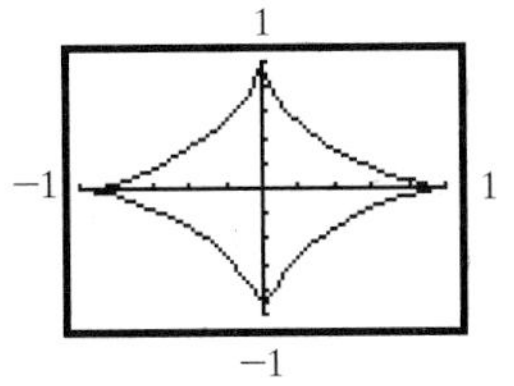

(b) $x^{2/3} + y^{2/3} = 1$

Fill-in-the-Blank Items

1. Parabola **2.** Ellipse **3.** Hyperbola **4.** Major; transverse **5.** y-axis **6.** $\frac{y}{3} = \frac{x}{2}; \frac{y}{3} = -\frac{x}{2}$ **7.** $\cot 2\theta = \frac{A - C}{B}$

8. $\frac{1}{2}$; ellipse; parallel; 4; below **9.** Ellipse

True/False Items

1. True **2.** False **3.** True **4.** True **5.** True **6.** True **7.** True **8.** True **9.** True **10.** False

Review Exercises

1. Parabola; vertex $(0, 0)$, focus $(-4, 0)$, directrix $x = 4$

3. Hyperbola; center $(0, 0)$, vertices $(5, 0)$ and $(-5, 0)$, foci $(\sqrt{26}, 0)$ and $(-\sqrt{26}, 0)$, asymptotes $y = \frac{1}{5}x$ and $y = -\frac{1}{5}x$

5. Ellipse; center $(0, 0)$, vertices $(0, 5)$ and $(0, -5)$, foci $(0, 3)$ and $(0, -3)$

7. $x^2 = -4(y - 1)$: Parabola; vertex $(0, 1)$, focus $(0, 0)$, directrix $y = 2$

9. $\frac{x^2}{2} - \frac{y^2}{8} = 1$: Hyperbola; center $(0, 0)$, vertices $(\sqrt{2}, 0)$ and $(-\sqrt{2}, 0)$, foci $(\sqrt{10}, 0)$ and $(-\sqrt{10}, 0)$, asymptotes $y = 2x$ and $y = -2x$

11. $(x - 2)^2 = 2(y + 2)$: Parabola; vertex $(2, -2)$, focus $\left(2, -\frac{3}{2}\right)$, directrix $y = -\frac{5}{2}$

13. $\frac{(y - 2)^2}{4} - (x - 1)^2 = 1$: Hyperbola; center $(1, 2)$, vertices $(1, 4)$ and $(1, 0)$, foci $(1, 2 + \sqrt{5})$ and $(1, 2 - \sqrt{5})$, asymptotes $y - 2 = \pm 2(x - 1)$

15. $\frac{(x - 2)^2}{9} + \frac{(y - 1)^2}{4} = 1$: Ellipse; center $(2, 1)$, vertices $(5, 1)$ and $(-1, 1)$, foci $(2 + \sqrt{5}, 1)$ and $(2 - \sqrt{5}, 1)$

17. $(x - 2)^2 = -4(y + 1)$: Parabola; vertex $(2, -1)$, focus $(2, -2)$, directrix $y = 0$

19. $\frac{(x - 1)^2}{4} + \frac{(y + 1)^2}{9} = 1$: Ellipse; center $(1, -1)$, vertices $(1, 2)$ and $(1, -4)$, foci $(1, -1 + \sqrt{5})$ and $(1, -1 - \sqrt{5})$

21. $y^2 = -8x$

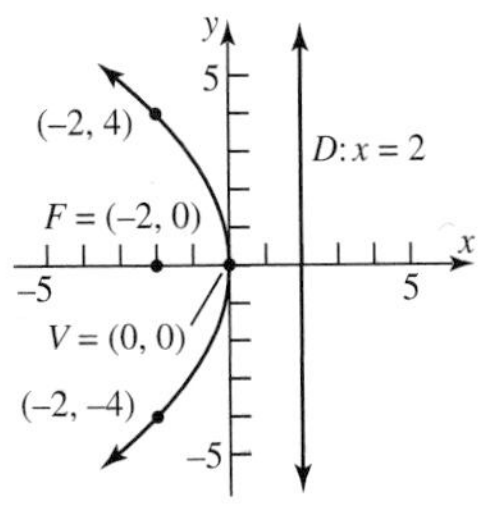

23. $\frac{y^2}{4} - \frac{x^2}{12} = 1$

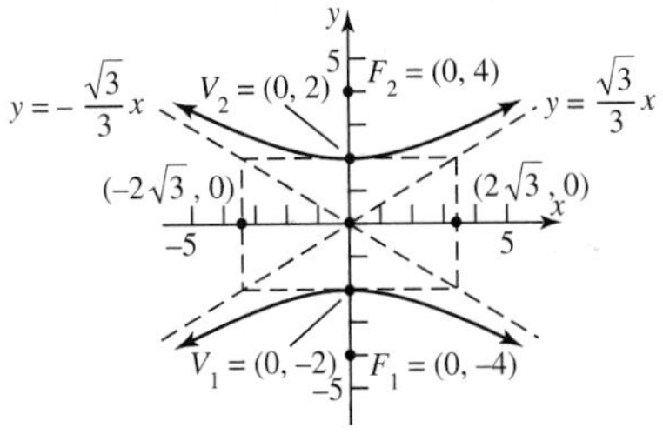

25. $\frac{x^2}{16} + \frac{y^2}{7} = 1$

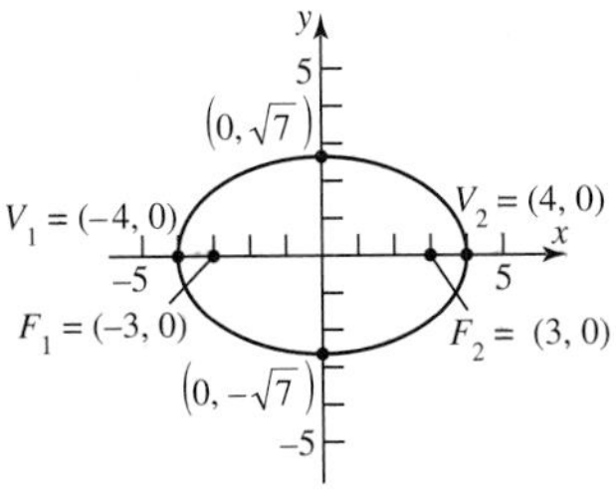

27. $(x-2)^2 = -4(y+3)$

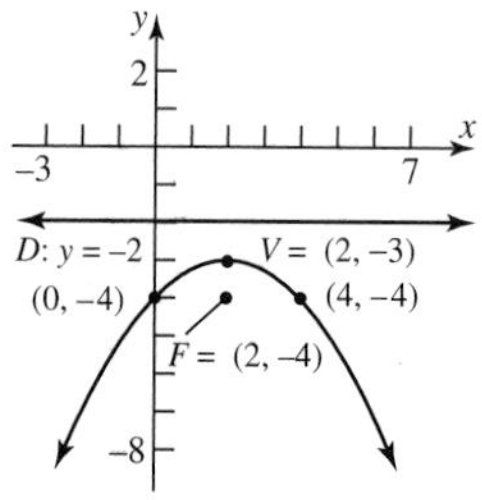

29. $(x+2)^2 - \dfrac{(y+3)^2}{3} = 1$

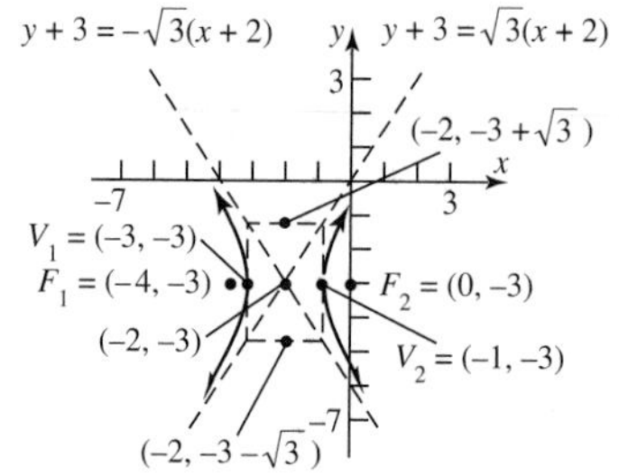

31. $\dfrac{(x+4)^2}{16} + \dfrac{(y-5)^2}{25} = 1$

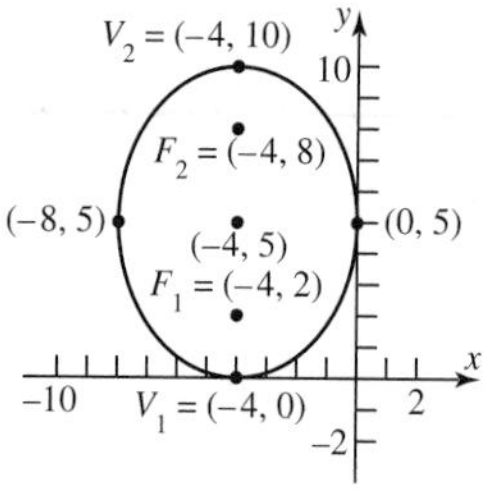

33. $\dfrac{(x+1)^2}{9} - \dfrac{(y-2)^2}{7} = 1$

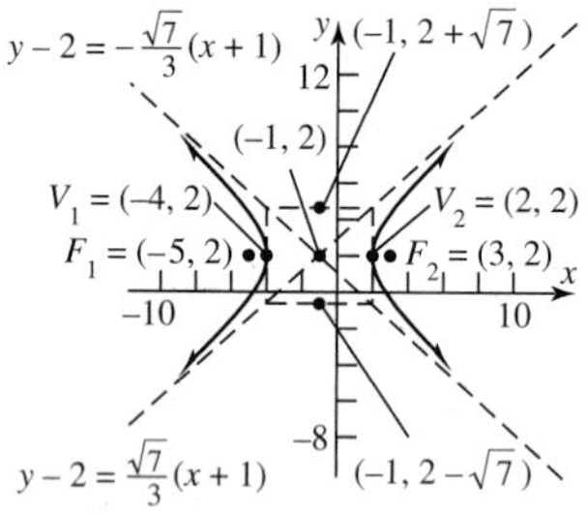

35. $\dfrac{(x-3)^2}{9} - \dfrac{(y-1)^2}{4} = 1$

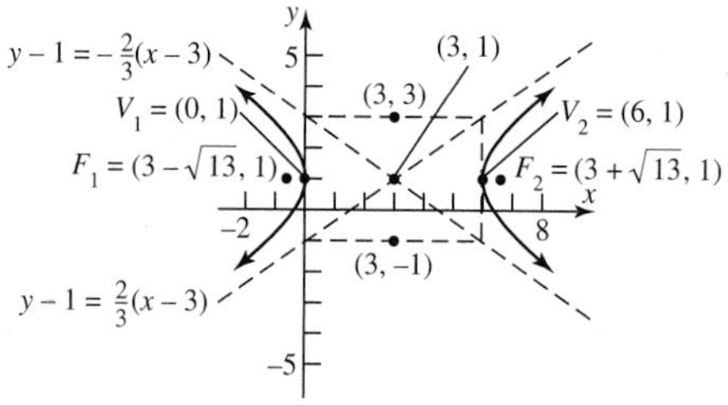

37. Parabola **39.** Ellipse **41.** Parabola **43.** Hyperbola **45.** Ellipse

47. $x'^2 - \dfrac{y'^2}{9} = 1$

Hyperbola
Center at the origin

Transverse axis the x'-axis

Vertices at $(\pm 1, 0)$

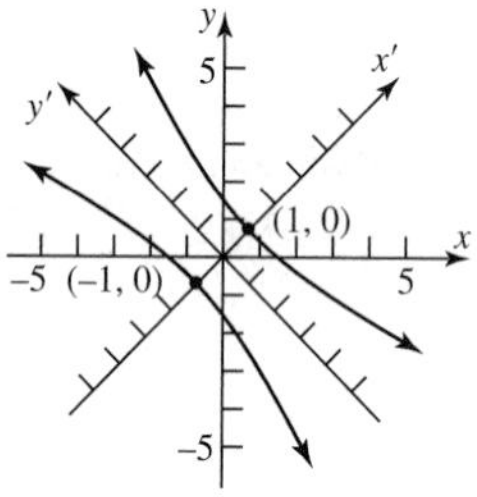

49. $\dfrac{x'^2}{2} + \dfrac{y'^2}{4} = 1$

Ellipse
Center at origin

Major axis the y'-axis

Vertices at $(0, \pm 2)$

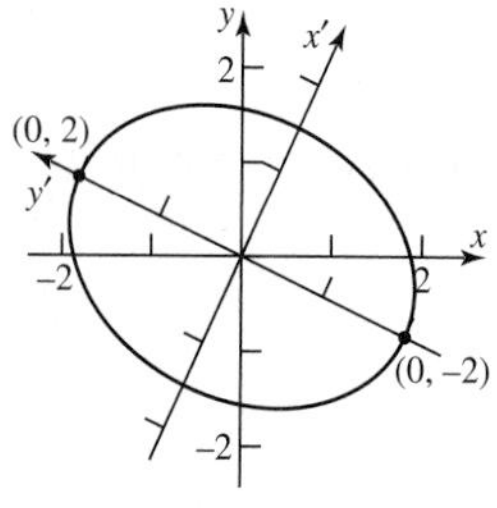

51. $y'^2 = -\dfrac{4\sqrt{13}}{13}x'$

Parabola
Vertex at the origin

Focus on the x'-axis at $\left(-\dfrac{\sqrt{13}}{13}, 0\right)$

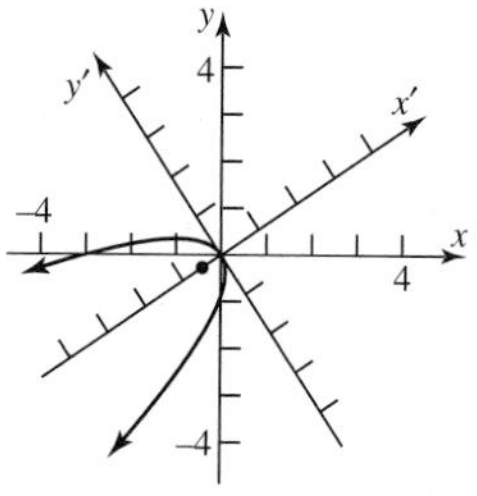

53. Parabola; directrix is perpendicular to the polar axis 4 units to the left of the pole; vertex at $(2, \pi)$.

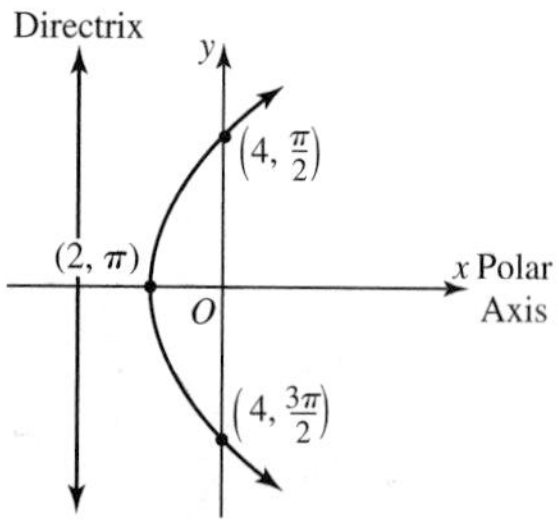

55. Ellipse; directrix is parallel to the polar axis 6 units below the pole; vertices are $\left(6, \dfrac{\pi}{2}\right)$ and $\left(2, \dfrac{3\pi}{2}\right)$.

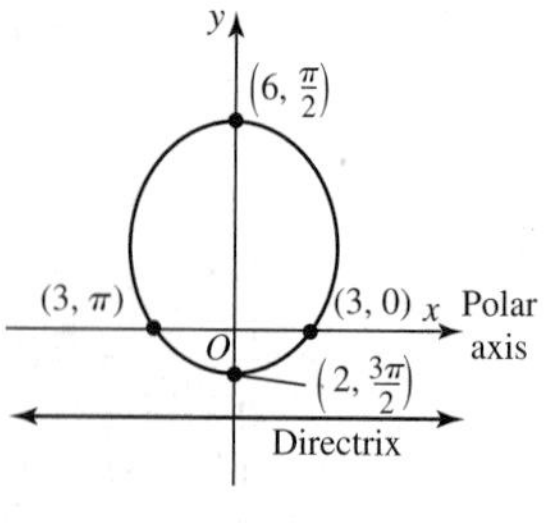

57. Hyperbola; directrix is perpendicular to the polar axis 1 unit to the right of the pole; vertices are at $\left(\dfrac{2}{3}, 0\right)$ and $(-2, \pi)$.

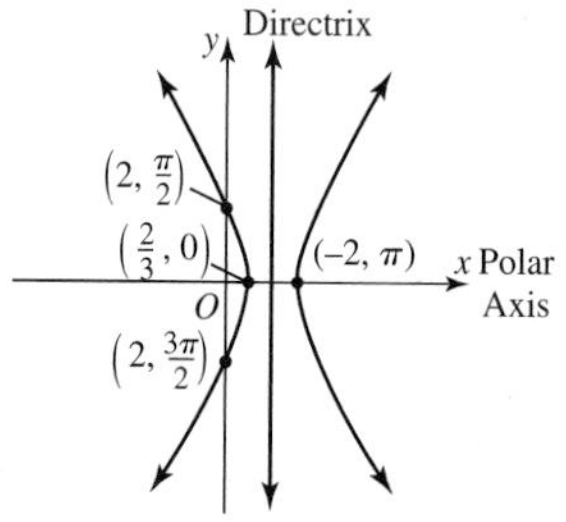

59. $y^2 - 8x - 16 = 0$ **61.** $3x^2 - y^2 - 8x + 4 = 0$

63.

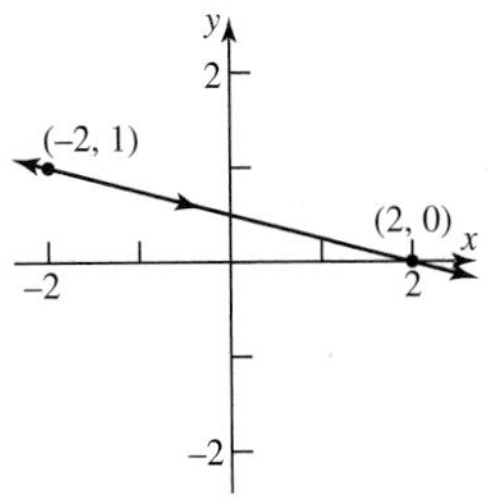

$x + 4y = 2$

65.

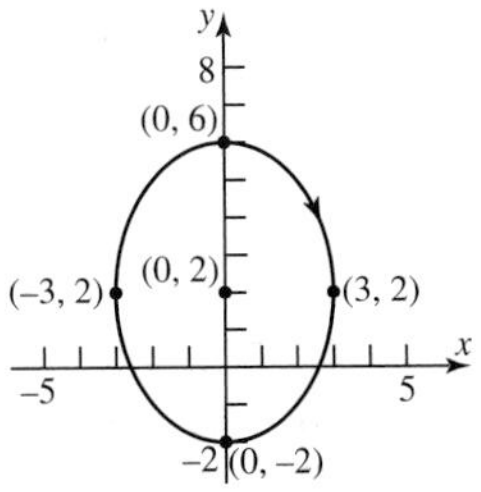

$\frac{x^2}{9} + \frac{(y-2)^2}{16} = 1$

67.

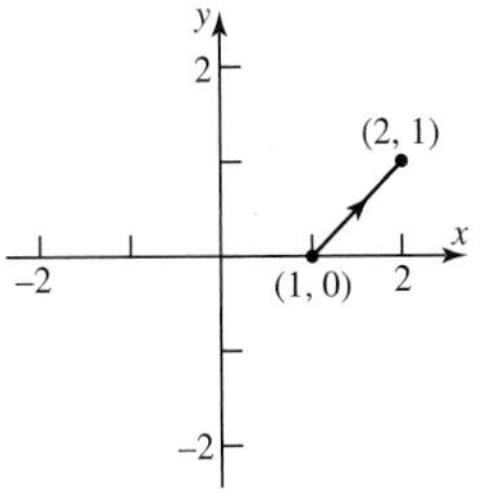

$1 + y = x$

69. $\frac{x^2}{5} - \frac{y^2}{4} = 1$

71. The ellipse $\frac{x^2}{16} + \frac{y^2}{7} = 1$ **73.** $\frac{1}{4}$ ft or 3 in. **75.** 19.72 ft, 18.86 ft, 14.91 ft

77. (a) 45.24 miles from the Master Station **(b)** 0.000645 sec **(c)** (66, 20)

79. (a) $x = (100 \cos 35°)t$
$y = -16t^2 + (100 \sin 35°)t + 6$
(b) 3.6866 sec **(c)** 1.7924 sec; 57.4 ft **(d)** 302 ft **(e)**

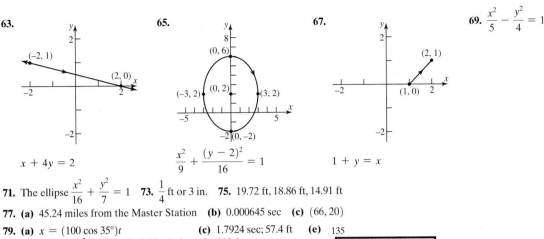

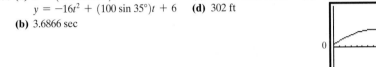

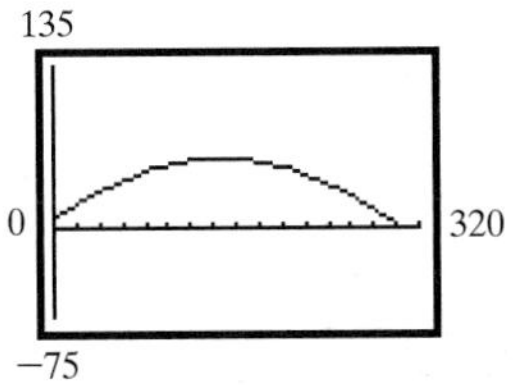

CHAPTER 11 Systems of Equations and Inequalities

11.1 Exercises

1. $2(2) - (-1) = 5$ and $5(2) + 2(-1) = 8$ **3.** $3(2) - 4\left(\frac{1}{2}\right) = 4$ and $\frac{1}{2}(2) - 3\left(\frac{1}{2}\right) = -\frac{1}{2}$ **5.** $4 - 1 = 3$ and $\frac{1}{2}(4) + 1 = 3$

7. $3(1) + 3(-1) + 2(2) = 4$, $1 - (-1) - 2 = 0$, and $2(-1) - 3(2) = -8$ **9.** $x = 6, y = 2$ **11.** $x = 3, y = 2$ **13.** $x = 8, y = -4$

15. $x = \frac{1}{3}, y = -\frac{1}{6}$ **17.** Inconsistent **19.** $x = 1, y = 2$ **21.** $x = 4 - 2y$, where y is any real number **23.** $x = 1, y = 1$

25. $x = \frac{3}{2}, y = 1$ **27.** $x = 4, y = 3$ **29.** $x = \frac{4}{3}, y = \frac{1}{5}$ **31.** $x = \frac{1}{5}, y = \frac{1}{3}$ **33.** $x = 48.15, y = 15.18$ **35.** $x = -21.48, y = 16.12$

37. $x = 0.26, y = 0.07$ **39.** $p = \$16, q = 600$ T-shirts **41.** Length 30 feet; width 15 feet **43.** Cheeseburger \$1.55; shake \$0.85

45. 22.5 pounds **47.** Average wind speed 25 mph; average airspeed 175 mph **49.** 80 \$25 sets and 120 \$45 sets **51.** \$5.56 **53.** 5000

55. $b = -\frac{1}{2}, c = \frac{3}{2}$

11.2 Exercises

1. $x = 8, y = 2, z = 0$ **3.** $x = 2, y = -1, z = 1$ **5.** Inconsistent **7.** $x = 5z - 2, y = 4z - 3$, where z is any real number, or $x = \frac{5}{4}y + \frac{7}{4}, z = \frac{1}{4}y + \frac{3}{4}$, where y is any real number, or $y = \frac{4}{5}x - \frac{7}{5}, z = \frac{1}{5}x + \frac{2}{5}$, where x is any real number. **9.** Inconsistent

11. $x = 1, y = 3, z = -2$ **13.** $x = -3, y = \frac{1}{2}, z = 1$ **15.** $a = \frac{4}{3}, b = -\frac{5}{3}, c = 1$ **17.** $I_1 = \frac{10}{71}, I_2 = \frac{65}{71}, I_3 = \frac{55}{71}$

19. 100 orchestra, 210 main, and 190 balcony seats **21.** 1.5 chicken, 1 corn, 2 milks

11.3 Exercises

1. $\left[\begin{array}{rr|r} 1 & -5 & 5 \\ 4 & 3 & 6 \end{array}\right]$ **3.** $\left[\begin{array}{rr|r} 2 & 3 & 6 \\ 4 & -6 & -2 \end{array}\right]$ **5.** $\left[\begin{array}{rr|r} 0.01 & -0.03 & 0.06 \\ 0.13 & 0.10 & 0.20 \end{array}\right]$ **7.** $\left[\begin{array}{rrr|r} 1 & -1 & 1 & 10 \\ 3 & 3 & 0 & 5 \\ 1 & 1 & 2 & 2 \end{array}\right]$ **9.** $\left[\begin{array}{rrr|r} 1 & 1 & -1 & 2 \\ 3 & -2 & 0 & 2 \\ 5 & 3 & -1 & 1 \end{array}\right]$

11. $\left[\begin{array}{rrr|r} 1 & -3 & -5 & -2 \\ 0 & 1 & 6 & 9 \\ 0 & 0 & 13 & 36 \end{array}\right]$ **13.** $\left[\begin{array}{rrr|r} 1 & -3 & 4 & 3 \\ 0 & 1 & -2 & 0 \\ 0 & 0 & 4 & 15 \end{array}\right]$ **15.** $\left[\begin{array}{rrr|r} 1 & -3 & 2 & -6 \\ 0 & 1 & -1 & 8 \\ 0 & 0 & -5 & 108 \end{array}\right]$ **17.** $\left[\begin{array}{rrr|r} 1 & -3 & 1 & -2 \\ 0 & 1 & 4 & 2 \\ 0 & 0 & 39 & 16 \end{array}\right]$ **19.** $\left[\begin{array}{rrr|r} 1 & -3 & -2 & 3 \\ 0 & 1 & 6 & -7 \\ 0 & 0 & 64 & -62 \end{array}\right]$

21. $\begin{cases} x = 5 \\ y = -1 \end{cases}$ consistent; $x = 5, y = -1$ **23.** $\begin{cases} x = 1 \\ y = 2 \\ 0 = 3 \end{cases}$ inconsistent **25.** $\begin{cases} x + 2z = -1 \\ y - 4z = -2 \\ 0 = 0 \end{cases}$ consistent; $x = -1 - 2z$, $y = -2 + 4z$, z is any real number **27.** $x = 6, y = 2$ **29.** $x = 2, y = 3$ **31.** $x = 4, y = -2$ **33.** Inconsistent **35.** $x = \frac{1}{2}, y = \frac{3}{4}$ **37.** $x = 4 - 2y$, where y is any real number **39.** $x = \frac{3}{2}, y = 1$ **41.** $x = \frac{4}{3}, y = \frac{1}{5}$ **43.** $x = 8, y = 2, z = 0$

45. $x = 2, y = -1, z = 1$ **47.** Inconsistent **49.** $x = 5z - 2, y = 4z - 3$ where z is any real number, or $x = \frac{5}{4}y + \frac{7}{4}, z = \frac{1}{4}y + \frac{3}{4}$, where y is any real number, or $y = \frac{4}{5}x - \frac{7}{5}, z = \frac{1}{5}x + \frac{2}{5}$ where x is any real number **51.** Inconsistent **53.** $x = 1, y = 3, z = -2$

55. $x = -3, y = \frac{1}{2}, z = 1$ **57.** $x = \frac{1}{3}, y = \frac{2}{3}, z = 1$ **59.** $y = 0, z = 1 - x$, x is any real number **61.** $x = 1, y = 2, z = 0, w = 1$

63. $y = -2x^2 + x + 3$ **65.** $f(x) = 3x^3 - 4x^2 + 5$ **67.** 1.5 salmon steak, 2 baked eggs, 1 acorn squash

69. \$4000 in Treasury bills, \$4000 in Treasury bonds, \$2000 in corporate bonds **71.** 8 Deltas, 5 Betas, 10 Sigmas

73. $I_1 = \frac{44}{23}, I_2 = 2, I_3 = \frac{16}{23}, I_4 = \frac{28}{23}$

11.4 Exercises

1. 2 **3.** 22 **5.** −2 **7.** 10 **9.** −26 **11.** $x = 6, y = 2$ **13.** $x = 3, y = 2$ **15.** $x = 8, y = -4$ **17.** $x = 4, y = -2$ **19.** Not applicable **21.** $x = \frac{1}{2}, y = \frac{3}{4}$ **23.** $x = \frac{1}{10}, y = \frac{2}{5}$ **25.** $x = \frac{3}{2}, y = 1$ **27.** $x = \frac{4}{3}, y = \frac{1}{5}$ **29.** $x = 1, y = 3, z = -2$ **31.** $x = -3, y = \frac{1}{2}, z = 1$

33. Not applicable **35.** $x = 0, y = 0, z = 0$ **37.** Not applicable **39.** $x = \frac{1}{5}, y = \frac{1}{3}$ **41.** −5 **43.** $\frac{13}{11}$ **45.** 0 or −9 **47.** −4 **49.** 12

51. 8 **53.** 8

55.
$$\begin{aligned} (y_1 - y_2)x - (x_1 - x_2)y + (x_1y_2 - x_2y_1) &= 0 \\ (y_1 - y_2)x + (x_2 - x_1)y &= x_2y_1 - x_1y_2 \\ (x_2 - x_1)y - (x_2 - x_1)y_1 &= (y_2 - y_1)x + x_2y_1 - x_1y_2 - (x_2 - x_1)y_1 \\ (x_2 - x_1)(y - y_1) &= (y_2 - y_1)x - (y_2 - y_1)x_1 \\ y - y_1 &= \frac{y_2 - y_1}{x_2 - x_1}(x - x_1) \end{aligned}$$

57.
$$\begin{vmatrix} x^2 & x & 1 \\ y^2 & y & 1 \\ z^2 & z & 1 \end{vmatrix} = x^2\begin{vmatrix} y & 1 \\ z & 1 \end{vmatrix} - x\begin{vmatrix} y^2 & 1 \\ z^2 & 1 \end{vmatrix} + \begin{vmatrix} y^2 & y \\ z^2 & z \end{vmatrix} = x^2(y - z) - x(y^2 - z^2) + yz(y - z)$$
$$= (y - z)[x^2 - x(y + z) + yz] = (y - z)[(x^2 - xy) - (xz - yz)] = (y - z)[x(x - y) - z(x - y)]$$
$$= (y - z)(x - y)(x - z)$$

59.
$$\begin{vmatrix} a_{13} & a_{12} & a_{11} \\ a_{23} & a_{22} & a_{21} \\ a_{33} & a_{32} & a_{31} \end{vmatrix} = a_{13}(a_{22}a_{31} - a_{32}a_{21}) - a_{12}(a_{23}a_{31} - a_{33}a_{21}) + a_{11}(a_{23}a_{32} - a_{33}a_{22})$$
$$= (-1)[a_{11}(a_{22}a_{33} - a_{32}a_{23}) - a_{12}(a_{21}a_{33} - a_{31}a_{23}) + a_{13}(a_{21}a_{32} - a_{31}a_{22})] = -\begin{vmatrix} a_{11} & a_{12} & a_{13} \\ a_{21} & a_{22} & a_{23} \\ a_{31} & a_{32} & a_{33} \end{vmatrix}$$

61.
$$\begin{vmatrix} a_{11} & a_{12} & a_{11} \\ a_{21} & a_{22} & a_{21} \\ a_{31} & a_{32} & a_{31} \end{vmatrix} = a_{11}(a_{22}a_{31} - a_{32}a_{21}) - a_{12}(a_{21}a_{31} - a_{31}a_{21}) + a_{11}(a_{21}a_{32} - a_{31}a_{22})$$
$$= a_{11}a_{22}a_{31} - a_{11}a_{32}a_{21} - a_{12}(0) + a_{11}a_{21}a_{32} - a_{11}a_{31}a_{22} = 0$$

Historical Problems

1. (a) $2 - 5i \longleftrightarrow \begin{bmatrix} 2 & -5 \\ 5 & 2 \end{bmatrix}, 1 + 3i \longleftrightarrow \begin{bmatrix} 1 & 3 \\ -3 & 1 \end{bmatrix}$ **(b)** $\begin{bmatrix} 2 & -5 \\ 5 & 2 \end{bmatrix}\begin{bmatrix} 1 & 3 \\ -3 & 1 \end{bmatrix} = \begin{bmatrix} 17 & 1 \\ -1 & 17 \end{bmatrix}$ **(c)** $17 + i$ **(d)** $17 + i$

11.5 Exercises

1. $\begin{bmatrix} 4 & 4 & -5 \\ -1 & 5 & 4 \end{bmatrix}$ **3.** $\begin{bmatrix} 0 & 12 & -20 \\ 4 & 8 & 24 \end{bmatrix}$ **5.** $\begin{bmatrix} -8 & 7 & -15 \\ 7 & 0 & 22 \end{bmatrix}$ **7.** $\begin{bmatrix} 28 & -9 \\ 4 & 23 \end{bmatrix}$ **9.** $\begin{bmatrix} 1 & 14 & -14 \\ 2 & 22 & -18 \\ 3 & 0 & 28 \end{bmatrix}$ **11.** $\begin{bmatrix} 15 & 21 & -16 \\ 22 & 34 & -22 \\ -11 & 7 & 22 \end{bmatrix}$

13. $\begin{bmatrix} 25 & -9 \\ 4 & 20 \end{bmatrix}$ **15.** $\begin{bmatrix} -13 & 7 & -12 \\ -18 & 10 & -14 \\ 17 & -7 & 34 \end{bmatrix}$ **17.** $\begin{bmatrix} -2 & 4 & 2 & 8 \\ 2 & 1 & 4 & 6 \end{bmatrix}$ **19.** $\begin{bmatrix} 9 & 2 \\ 34 & 13 \\ 47 & 20 \end{bmatrix}$ **21.** $\begin{bmatrix} 1 & -1 \\ -1 & 2 \end{bmatrix}$ **23.** $\begin{bmatrix} 1 & -\frac{5}{2} \\ -1 & 3 \end{bmatrix}$ **25.** $\begin{bmatrix} 1 & \frac{-1}{a} \\ -1 & \frac{2}{a} \end{bmatrix}$

27. $\begin{bmatrix} 3 & -3 & 1 \\ -2 & 2 & -1 \\ -4 & 5 & -2 \end{bmatrix}$ **29.** $\begin{bmatrix} -\frac{5}{7} & \frac{1}{7} & \frac{3}{7} \\ \frac{9}{7} & \frac{1}{7} & -\frac{4}{7} \\ \frac{3}{7} & -\frac{2}{7} & \frac{1}{7} \end{bmatrix}$ **31.** $x = 3, y = 2$ **33.** $x = -5, y = 10$ **35.** $x = 2, y = -1$ **37.** $x = \frac{1}{2}, y = 2$

39. $x = -2, y = 1$ **41.** $x = \frac{2}{a}, y = \frac{3}{a}$ **43.** $x = -2, y = 3, z = 5$ **45.** $x = \frac{1}{2}, y = -\frac{1}{2}, z = 1$ **47.** $x = -\frac{34}{7}, y = \frac{85}{7}, z = \frac{12}{7}$

49. $x = \frac{1}{3}, y = 1, z = \frac{2}{3}$

51. $\left[\begin{array}{cc|cc} 4 & 2 & 1 & 0 \\ 2 & 1 & 0 & 1 \end{array}\right] \to \left[\begin{array}{cc|cc} 1 & \frac{1}{2} & \frac{1}{4} & 0 \\ 2 & 1 & 0 & 1 \end{array}\right] \to \left[\begin{array}{cc|cc} 1 & \frac{1}{2} & \frac{1}{4} & 0 \\ 0 & 0 & -\frac{1}{2} & 1 \end{array}\right]$ **53.** $\left[\begin{array}{cc|cc} 15 & 3 & 1 & 0 \\ 10 & 2 & 0 & 1 \end{array}\right] \to \left[\begin{array}{cc|cc} 1 & \frac{1}{5} & \frac{1}{15} & 0 \\ 10 & 2 & 0 & 1 \end{array}\right] \to \left[\begin{array}{cc|cc} 1 & \frac{1}{5} & \frac{1}{15} & 0 \\ 0 & 0 & -\frac{2}{3} & 1 \end{array}\right]$

55. $\left[\begin{array}{ccc|ccc} -3 & 1 & -1 & 1 & 0 & 0 \\ 1 & -4 & -7 & 0 & 1 & 0 \\ 1 & 2 & 5 & 0 & 0 & 1 \end{array}\right] \to \left[\begin{array}{ccc|ccc} 1 & 2 & 5 & 0 & 0 & 1 \\ 1 & -4 & -7 & 0 & 1 & 0 \\ -3 & 1 & -1 & 1 & 0 & 0 \end{array}\right] \to \left[\begin{array}{ccc|ccc} 1 & 2 & 5 & 0 & 0 & 1 \\ 0 & -6 & -12 & 0 & 1 & -1 \\ 0 & 7 & 14 & 1 & 0 & 3 \end{array}\right] \to \left[\begin{array}{ccc|ccc} 1 & 2 & 5 & 0 & 0 & 1 \\ 0 & 1 & 2 & 0 & -\frac{1}{6} & \frac{1}{6} \\ 0 & 1 & 2 & \frac{1}{7} & 0 & \frac{3}{7} \end{array}\right]$

$\to \left[\begin{array}{ccc|ccc} 1 & 2 & 5 & 0 & 0 & 1 \\ 0 & 1 & 2 & 0 & -\frac{1}{6} & \frac{1}{6} \\ 0 & 0 & 0 & \frac{1}{7} & \frac{1}{6} & \frac{11}{42} \end{array}\right]$ **57.** $\begin{bmatrix} 0.01 & 0.05 & -0.01 \\ 0.01 & -0.02 & 0.01 \\ -0.02 & 0.01 & 0.03 \end{bmatrix}$ **59.** $\begin{bmatrix} 0.02 & -0.04 & -0.01 & 0.01 \\ -0.02 & 0.05 & 0.03 & -0.03 \\ 0.02 & 0.01 & -0.04 & 0.00 \\ -0.02 & 0.06 & 0.07 & 0.06 \end{bmatrix}$

61. $x = 4.57, y = -6.44, z = -24.07$ **63.** $x = -1.19, y = 2.46, z = 8.27$

65. **(a)** $\begin{bmatrix} 500 & 350 & 400 \\ 700 & 500 & 850 \end{bmatrix}$; $\begin{bmatrix} 500 & 700 \\ 350 & 500 \\ 400 & 850 \end{bmatrix}$ **(b)** $\begin{bmatrix} 15 \\ 8 \\ 3 \end{bmatrix}$ **(c)** $\begin{bmatrix} 11{,}500 \\ 17{,}050 \end{bmatrix}$ **(d)** $[0.10 \quad 0.05]$ **(e)** \$2002.50

11.6 Exercises

1. Proper **3.** Improper; $1 + \frac{9}{x^2 - 4}$ **5.** Improper; $5x + \frac{22x - 1}{x^2 - 4}$ **7.** Improper; $1 + \frac{-2(x - 6)}{(x + 4)(x - 3)}$ **9.** $\frac{-4}{x} + \frac{4}{x - 1}$

11. $\frac{1}{x} + \frac{-x}{x^2 + 1}$ **13.** $\frac{-1}{x - 1} + \frac{2}{x - 2}$ **15.** $\frac{\frac{1}{4}}{x + 1} + \frac{\frac{3}{4}}{x - 1} + \frac{\frac{1}{2}}{(x - 1)^2}$ **17.** $\frac{\frac{1}{12}}{x - 2} + \frac{-\frac{1}{12}(x + 4)}{x^2 + 2x + 4}$

19. $\frac{\frac{1}{4}}{x - 1} + \frac{\frac{1}{4}}{(x - 1)^2} + \frac{-\frac{1}{4}}{x + 1} + \frac{\frac{1}{4}}{(x + 1)^2}$ **21.** $\frac{-5}{x + 2} + \frac{5}{x + 1} + \frac{-4}{(x + 1)^2}$ **23.** $\frac{\frac{1}{4}}{x} + \frac{1}{x^2} + \frac{-\frac{1}{4}(x + 4)}{x^2 + 4}$

25. $\frac{\frac{2}{3}}{x + 1} + \frac{\frac{1}{3}(x + 1)}{x^2 + 2x + 4}$ **27.** $\frac{\frac{2}{7}}{3x - 2} + \frac{\frac{1}{7}}{2x + 1}$ **29.** $\frac{\frac{3}{4}}{x + 3} + \frac{\frac{1}{4}}{x - 1}$ **31.** $\frac{1}{x^2 + 4} + \frac{2x - 1}{(x^2 + 4)^2}$ **33.** $\frac{-1}{x} + \frac{2}{x - 3} + \frac{-1}{x + 1}$

35. $\frac{4}{x - 2} + \frac{-3}{x - 1} + \frac{-1}{(x - 1)^2}$ **37.** $\frac{x}{(x^2 + 16)^2} + \frac{-16x}{(x^2 + 16)^3}$ **39.** $\frac{-\frac{8}{7}}{2x + 1} + \frac{\frac{4}{7}}{x - 3}$ **41.** $\frac{-\frac{2}{9}}{x} + \frac{-\frac{1}{3}}{x^2} + \frac{\frac{1}{6}}{x - 3} + \frac{\frac{1}{18}}{x + 3}$

Historical Problem

1. 6 units, 8 units

11.7 Exercises

1.

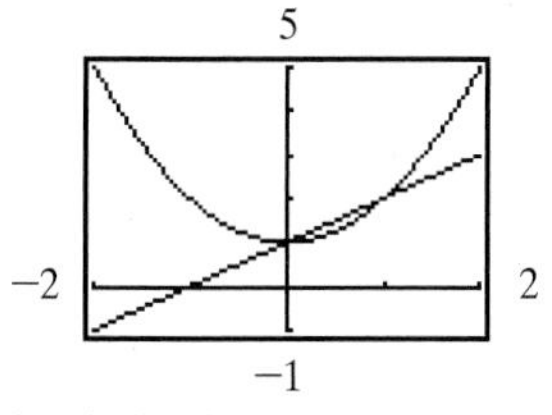

(0, 1); (1, 2)

3.

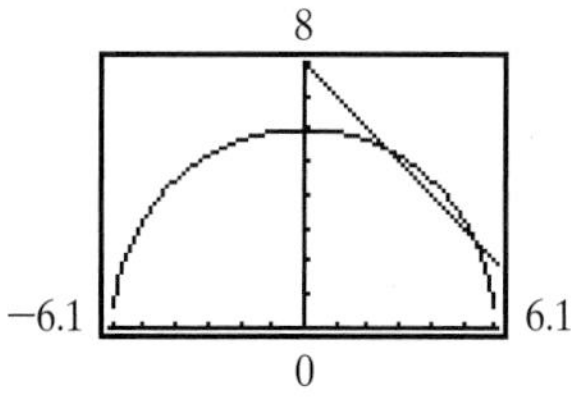

(2.59, 5.41); (5.41, 2.59)
$(4 - \sqrt{2}, 4 + \sqrt{2}); (4 + \sqrt{2}, 4 - \sqrt{2})$

5.

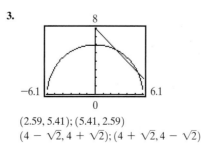

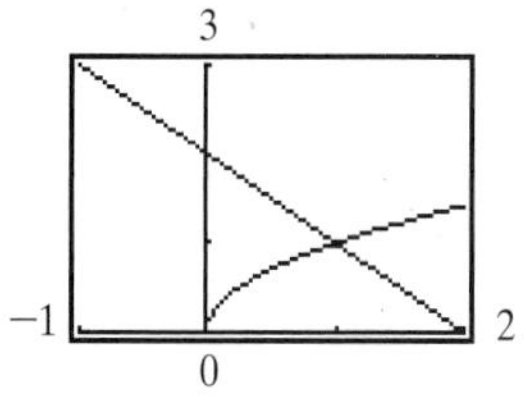

(1, 1)

7.

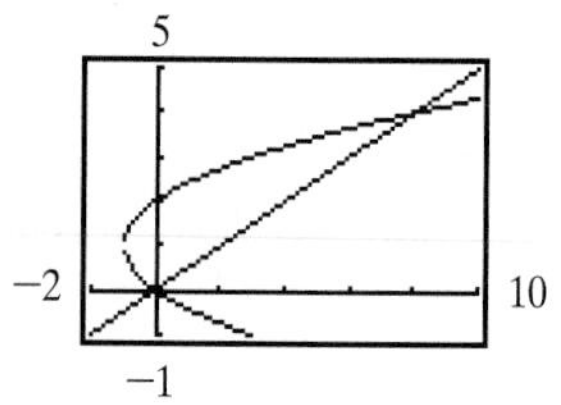

(0, 0); (8, 4)

9.

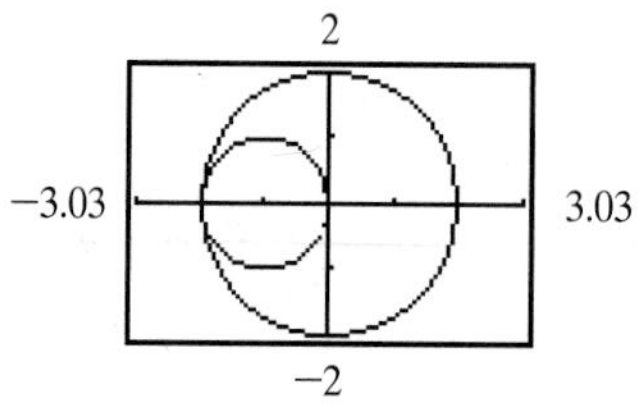

(−2, 0)

11.

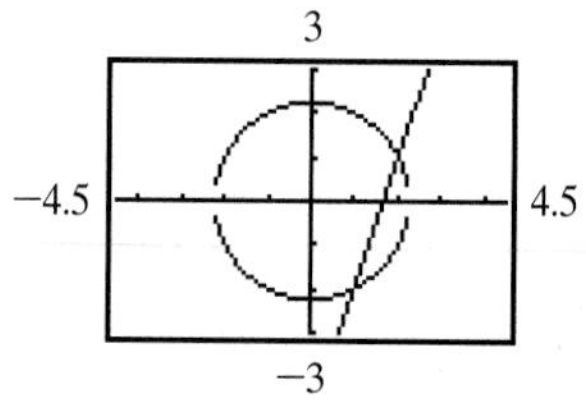

(1, −2); (2, 1)

13.

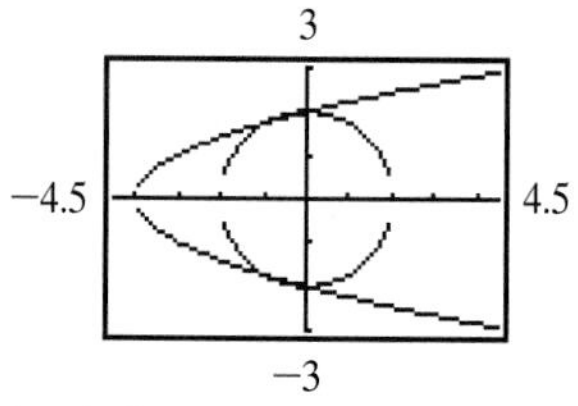

(0, 2); (0, −2); (−1, −1.73); (−1, 1.73)
(0, 2); (0, −2); $(-1, -\sqrt{3})$; $(-1, \sqrt{3})$

15.

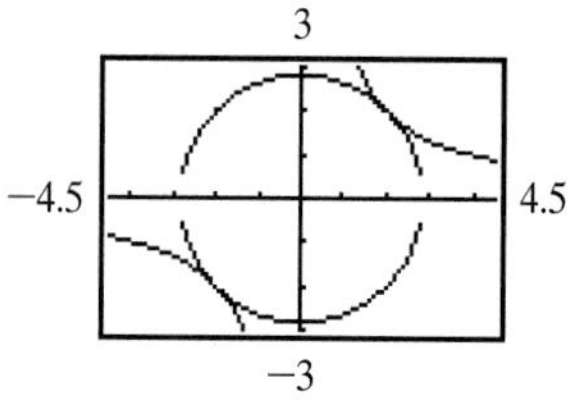

(2, 2); (−2, −2)

17.

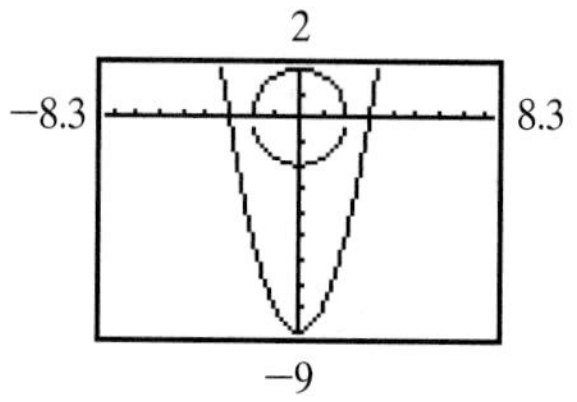

No solution; Inconsistent

19.

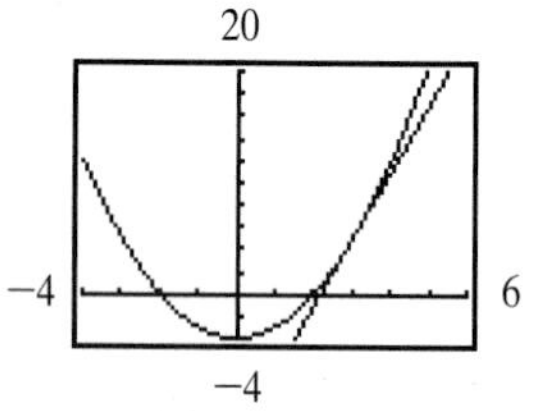

(3, 5)

21. $x = 1, y = 4; x = -1, y = -4; x = 2\sqrt{2}, y = \sqrt{2}; x = -2\sqrt{2}, y = -\sqrt{2}$

23. $x = 0, y = 1; x = -\frac{2}{3}, y = -\frac{1}{3}$ **25.** $x = 0, y = -1; x = \frac{5}{2}, y = -\frac{7}{2}$

27. $x = 2, y = \frac{1}{3}; x = \frac{1}{2}, y = \frac{4}{3}$

29. $x = 3, y = 2; x = 3, y = -2; x = -3, y = 2; x = -3, y = -2$

31. $x = \frac{1}{2}, y = \frac{3}{2}; x = \frac{1}{2}, y = -\frac{3}{2}; x = -\frac{1}{2}, y = \frac{3}{2}; x = -\frac{1}{2}, y = -\frac{3}{2}$

33. $x = \sqrt{2}, y = 2\sqrt{2}; x = -\sqrt{2}, y = -2\sqrt{2}$ **35.** No solution; system is inconsistent

37. $x = \frac{8}{3}, y = \frac{2\sqrt{10}}{3}; x = -\frac{8}{3}, y = \frac{2\sqrt{10}}{3}; x = \frac{8}{3}, y = -\frac{2\sqrt{10}}{3}; x = -\frac{8}{3}, y = -\frac{2\sqrt{10}}{3}$

39. $x = 1, y = \frac{1}{2}; x = -1, y = \frac{1}{2}; x = 1, y = -\frac{1}{2}; x = -1, y = -\frac{1}{2}$ **41.** System is inconsistent

43. $x = \sqrt{3}, y = \sqrt{3}; x = -\sqrt{3}, y = -\sqrt{3}; x = 2, y = 1; x = -2, y = -1$

45. $x = 3, y = 2; x = -3, y = -2; x = 2, y = \frac{1}{2}; x = -2, y = -\frac{1}{2}$

47. $x = 3, y = 1; x = -1, y = -3$ **49.** $x = 0, y = -2; x = 0, y = 1; x = 2, y = -1$ **51.** $x = 2, y = 8$ **53.** $x = 0.48, y = 0.62$

55. $x = -1.65, y = -0.89$ **57.** $x = 0.58, y = 1.86; x = 1.81, y = 1.05; x = 0.58, y = -1.86, x = 1.81, y = -1.05$

59. $x = 2.35, y = 0.86$ **61.** 3 and 1; −3 and −1 **63.** 2 and 2; −2 and −2 **65.** $\frac{1}{2}$ and $\frac{1}{3}$ **67.** 5

69.

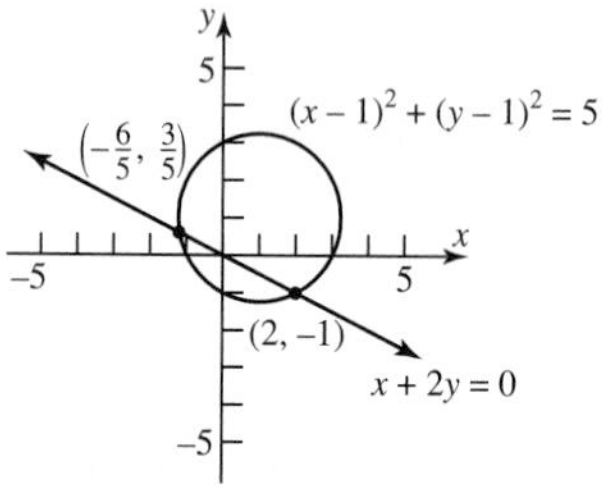

71. Solutions: $(0, -\sqrt{3} - 2)$, $(0, \sqrt{3} - 2)$, (1, 0), (1, −4)

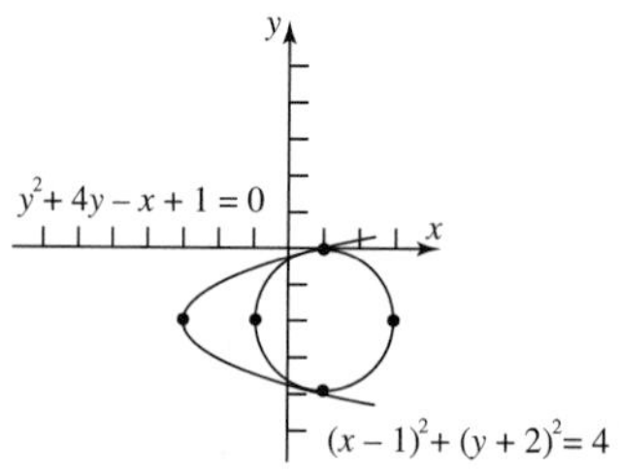

73.

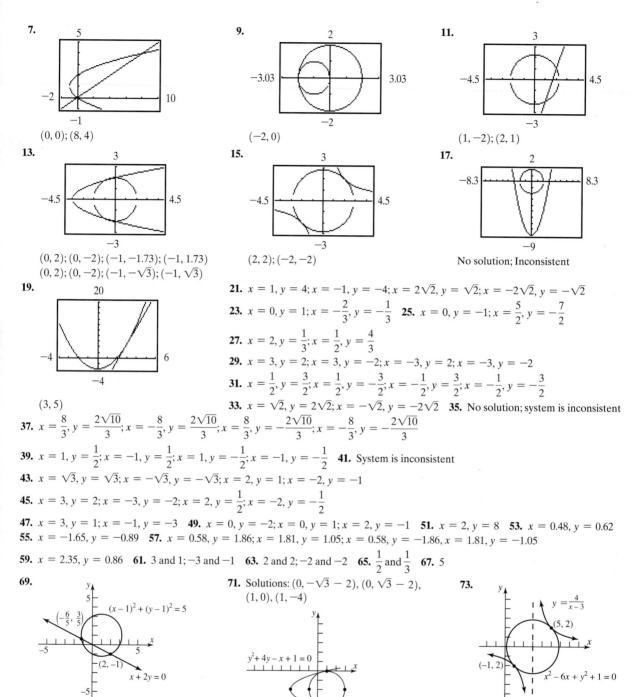

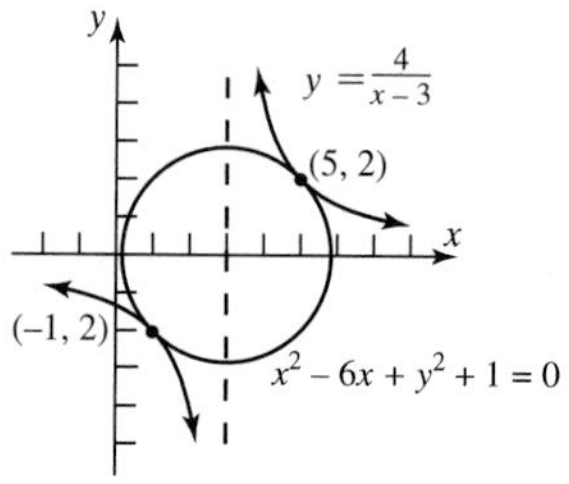

75. 5 in. by 3 in. **77.** 2 cm and 4 cm **79.** Tortoise: 7 mph, hare: $7\frac{1}{2}$ mph **81.** 12 cm by 18 cm **83.** $x = 60$ ft; $y = 30$ ft

85. $l = \frac{P + \sqrt{P^2 - 16A}}{4}; w = \frac{P - \sqrt{P^2 - 16A}}{4}$ **87.** $y = 4x - 4$ **89.** $y = 2x + 1$ **91.** $y = -\frac{1}{3}x + \frac{7}{3}$ **93.** $y = 2x - 3$

95. $r_1 = \frac{-b + \sqrt{b^2 - 4ac}}{2a}; r_2 = \frac{-b - \sqrt{b^2 - 4ac}}{2a}$

11.8 Exercises

1.

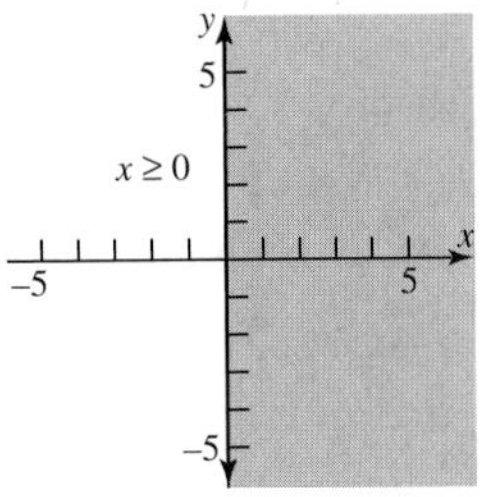

3.

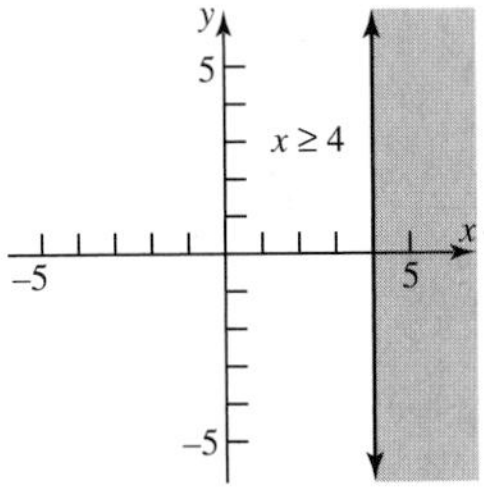

5.

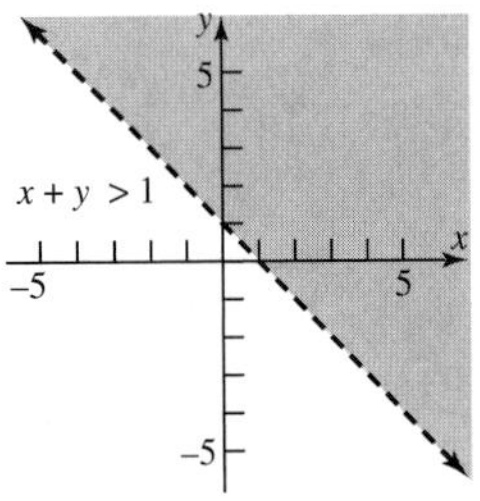

7.

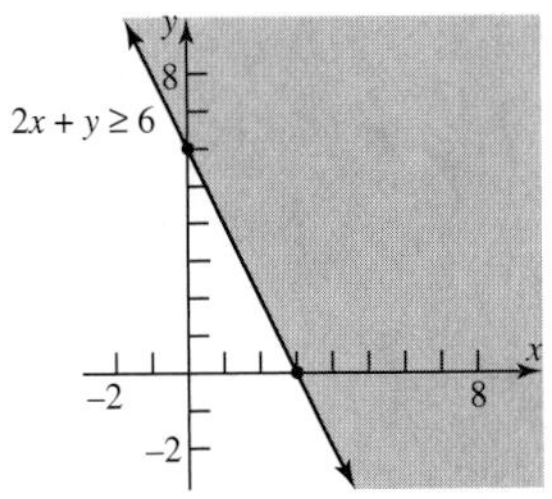

9.

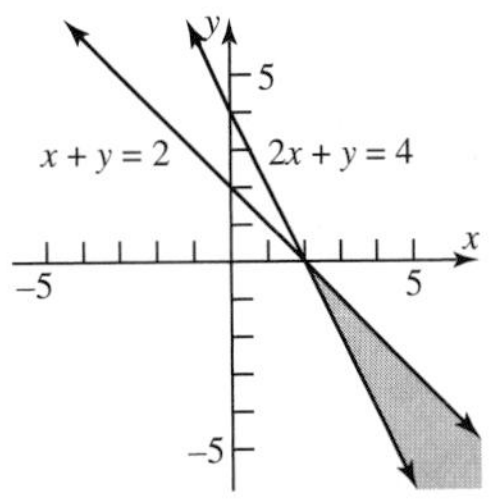

11.

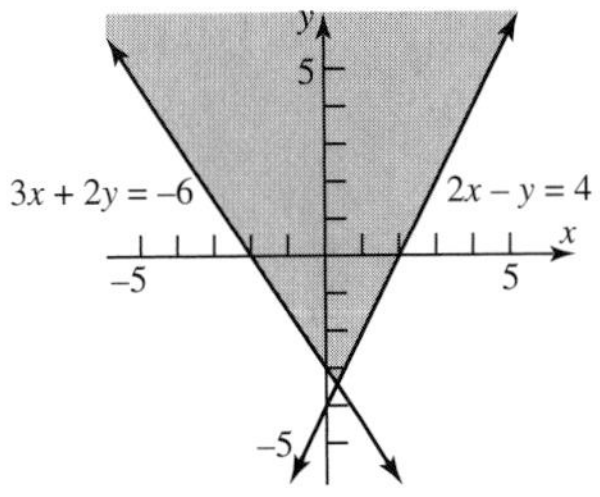

13.

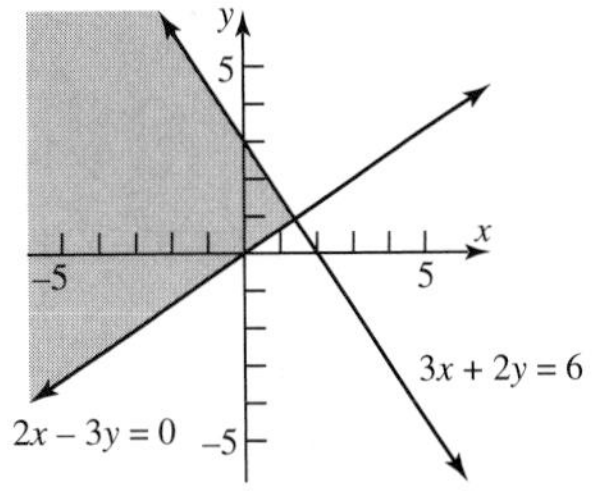

15.

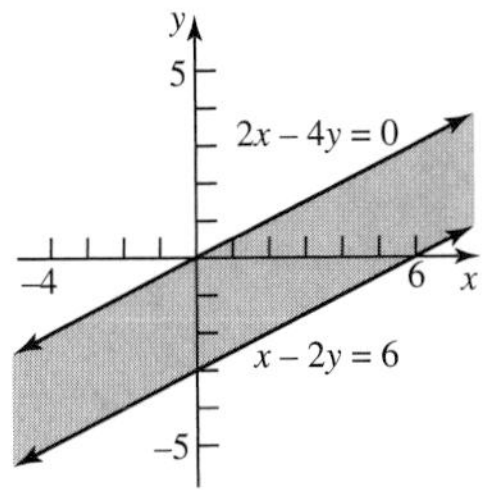

17.

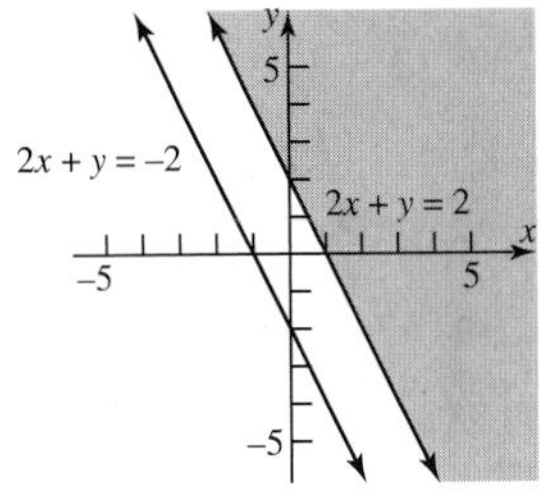

19. No solution

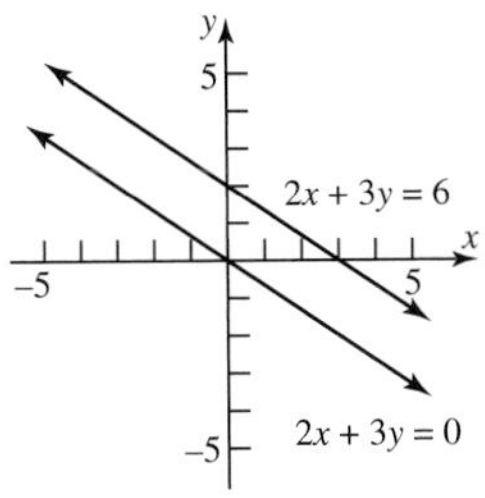

21. Bounded; corner points $(0, 0), (3, 0), (2, 2), (0, 3)$

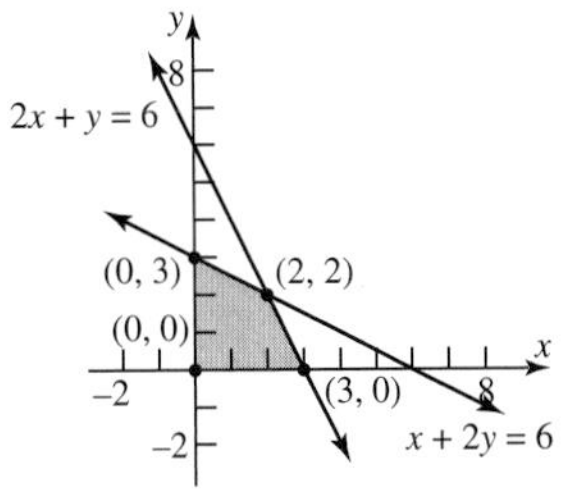

23. Unbounded; corner points $(2, 0), (0, 4)$

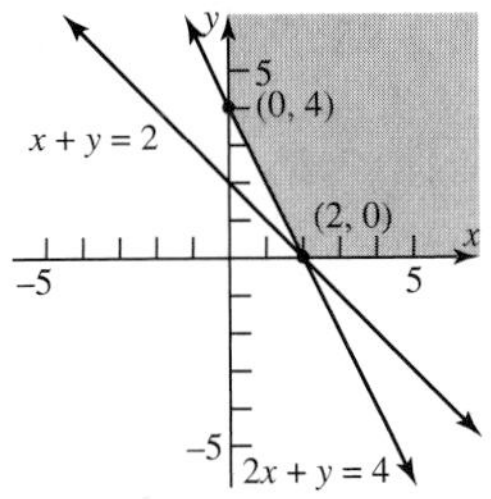

25. Bounded; corner points $(2, 0), (4, 0), \left(\frac{24}{7}, \frac{12}{7}\right), (0, 4), (0, 2)$

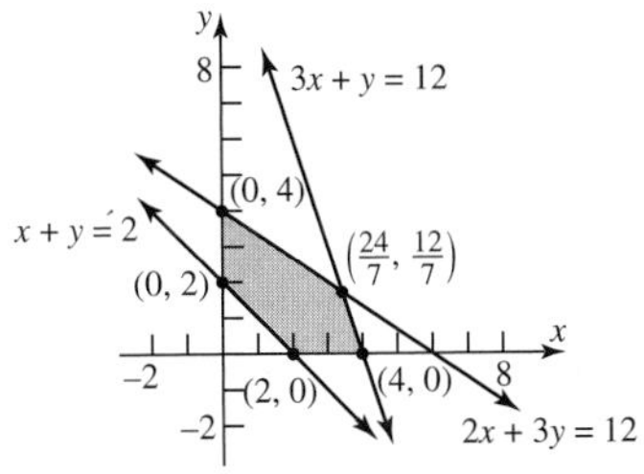

27. Bounded; corner points $(2, 0), (5, 0), (2, 6), (0, 8), (0, 2)$

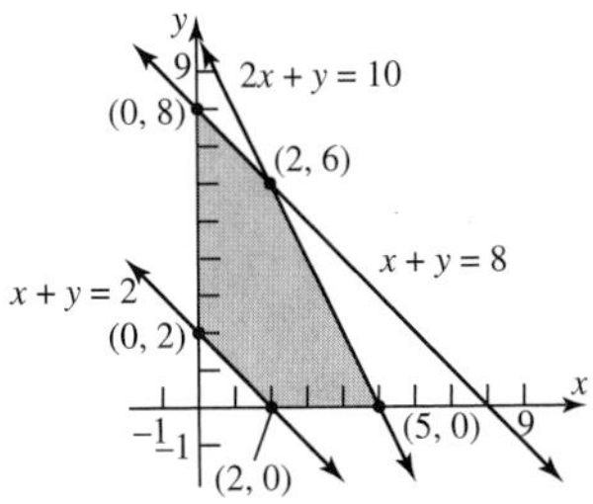

29. Bounded; corner points $(1, 0), (10, 0), (0, 5), \left(0, \frac{1}{2}\right)$

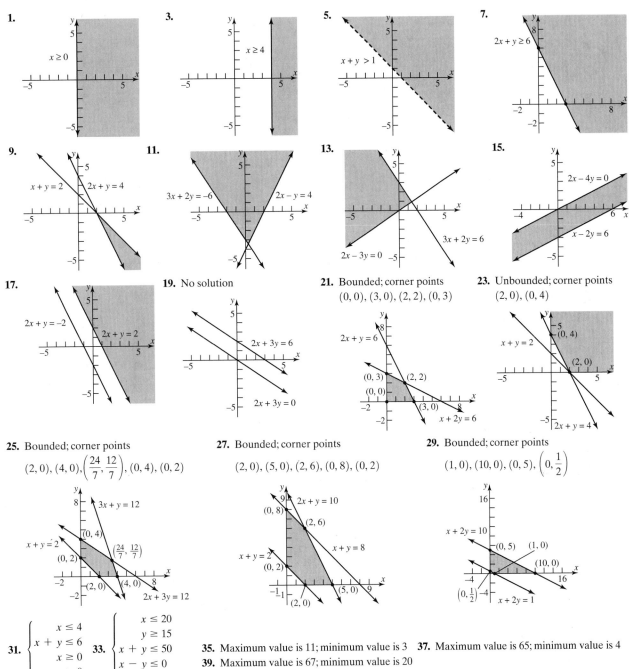

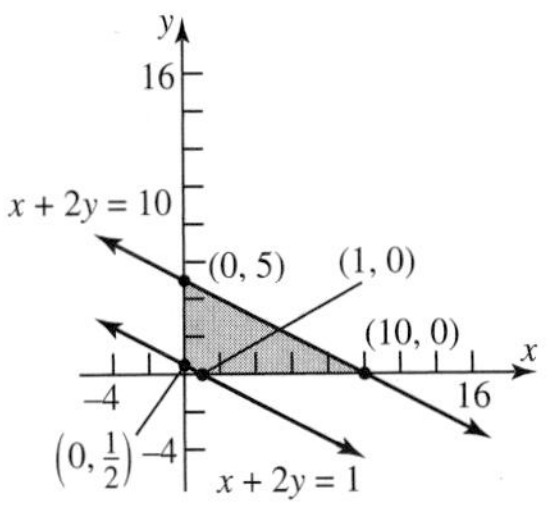

31. $\begin{cases} x \leq 4 \\ x + y \leq 6 \\ x \geq 0 \\ y \geq 0 \end{cases}$ **33.** $\begin{cases} x \leq 20 \\ y \geq 15 \\ x + y \leq 50 \\ x - y \leq 0 \\ x \geq 0 \end{cases}$ **35.** Maximum value is 11; minimum value is 3 **37.** Maximum value is 65; minimum value is 4 **39.** Maximum value is 67; minimum value is 20 **41.** The maximum value of z is 12, and it occurs at the point $(6, 0)$. **43.** The minimum value of z is 4, and it occurs at the point $(2, 0)$. **45.** The maximum value of z is 20, and it occurs at the point $(0, 4)$. **47.** The minimum value of z is 8, and it occurs at the point $(0, 2)$. **49.** The maximum value of z is 50, and it occurs at the point $(10, 0)$. **51.** 8 downhill, 24 cross-country; \$1760; \$1920 **53.** 30 acres of soybeans and 10 acres of corn. **55.** $\frac{1}{2}$ hour on machine 1; $5\frac{1}{4}$ hours on machine 2 **57.** 100 pounds of ground beef and 50 pounds of pork **59.** 10 racing skates, 15 figure skates **61.** 2 metal samples, 2 plastic samples; \$26 **63.** **(a)** 10 first class, 120 coach **(b)** 15 first class, 120 coach

Fill-in-the-Blank Items

1. inconsistent **3.** determinants **5.** inverse **7.** identity **9.** objective function **11.** proper

True/False Items

1. False **3.** False **5.** False **7.** False **9.** True **11.** True

Review Exercises

1. $x = 2, y = -1$ **3.** $x = 2, y = \frac{1}{2}$ **5.** $x = 2, y = -1$ **7.** $x = \frac{11}{5}, y = -\frac{3}{5}$ **9.** $x = -\frac{8}{5}, y = \frac{12}{5}$ **11.** $x = 6, y = -1$

13. $x = -4, y = 3$ **15.** $x = 2, y = 3$ **17.** Inconsistent **19.** $x = -1, y = 2, z = -3$ **21.** $\begin{bmatrix} 4 & -4 \\ 3 & 9 \\ 4 & 0 \end{bmatrix}$ **23.** $\begin{bmatrix} 6 & 0 \\ 12 & 24 \\ -6 & 12 \end{bmatrix}$

25. $\begin{bmatrix} 4 & -3 & 0 \\ 12 & -2 & -8 \\ -2 & 5 & -4 \end{bmatrix}$ **27.** $\begin{bmatrix} 8 & -13 & 8 \\ 9 & 2 & -10 \\ 18 & -17 & 4 \end{bmatrix}$ **29.** $\begin{bmatrix} \frac{1}{2} & -1 \\ -\frac{1}{6} & \frac{2}{3} \end{bmatrix}$ **31.** $\begin{bmatrix} -\frac{5}{7} & \frac{9}{7} & \frac{3}{7} \\ \frac{1}{7} & \frac{1}{7} & -\frac{2}{7} \\ \frac{3}{7} & -\frac{4}{7} & \frac{1}{7} \end{bmatrix}$ **33.** Singular **35.** $x = \frac{2}{5}, y = \frac{1}{10}$

37. $x = \frac{1}{2}, y = \frac{2}{3}, z = \frac{1}{6}$ **39.** $x = -\frac{1}{2}, y = -\frac{2}{3}, z = -\frac{3}{4}$ **41.** $z = -1, x = y + 1$, where y is any real number

43. $x = 4, y = 2, z = 3, t = -1$ **45.** 5 **47.** 108 **49.** -100 **51.** $x = 2, y = -1$ **53.** $x = 2, y = 3$ **55.** $x = -1, y = 2, z = -3$

57. $\dfrac{-\frac{3}{2}}{x} + \dfrac{\frac{3}{2}}{x-4}$ **59.** $\dfrac{-3}{x-1} + \dfrac{3}{x} + \dfrac{4}{x^2}$ **61.** $\dfrac{-\frac{1}{10}}{x+1} + \dfrac{\frac{1}{10}x + \frac{9}{10}}{x^2+9}$ **63.** $\dfrac{x}{x^2+4} + \dfrac{-4x}{(x^2+4)^2}$ **65.** $\dfrac{\frac{1}{2}}{x^2+1} + \dfrac{\frac{1}{4}}{x-1} + \dfrac{-\frac{1}{4}}{x+1}$

67. $x = -\frac{2}{5}, y = -\frac{11}{5}; x = -2, y = 1$ **69.** $x = 2\sqrt{2}, y = \sqrt{2}; x = -2\sqrt{2}, y = -\sqrt{2}$

71. $x = 0, y = 0; x = -3, y = 3; x = 3, y = 3$

73. $x = \sqrt{2}, y = -\sqrt{2}; x = -\sqrt{2}, y = \sqrt{2}; x = \frac{4}{3}\sqrt{2}, y = -\frac{2}{3}\sqrt{2}; x = -\frac{4}{3}\sqrt{2}, y = \frac{2}{3}\sqrt{2}$ **75.** $x = 1, y = -1$

77. Unbounded; corner point (0, 2)

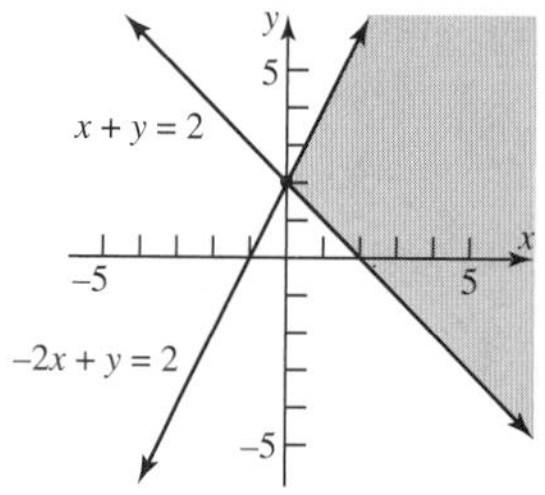

79. Bounded; corner points (0, 0), (0, 2), (3, 0)

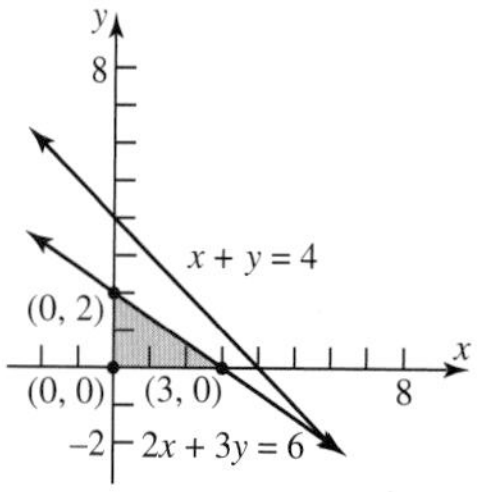

81. Bounded; corner points (0, 1), (0, 8), (4, 0), (2, 0)

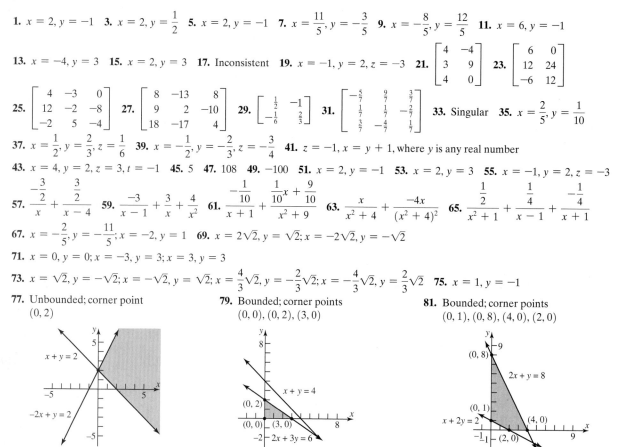

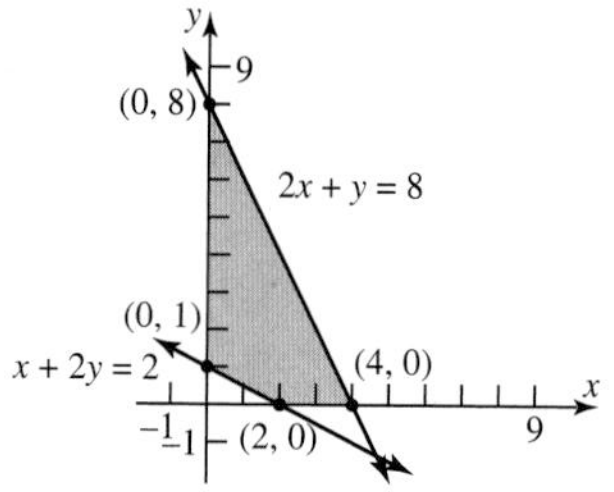

83. The maximum value is 32, and it occurs at the point (0, 8). **85.** The minimum value is 3, and it occurs at the point (1, 0).

87. The minimum value is $\frac{108}{7}$, and it occurs at the point $\left(\frac{12}{7}, \frac{12}{7}\right)$. **89.** 10 **91.** $y = -\frac{1}{3}x^2 - \frac{2}{3}x + 1$

93. 70 pounds of \$3 coffee and 30 pounds of \$6 coffee **95.** 1 small, 5 medium, 2 large **97.** 24 feet by 10 feet

99. $4 + \sqrt{2}$ inches and $4 - \sqrt{2}$ inches **101.** $100\sqrt{10}$ feet **103.** Katy gets \$10, Mike gets \$20, Danny gets \$5, Colleen gets \$10

105. Bruce: 4 hours; Bryce: 2 hours; Marty: 8 hours **107.** 35 gasoline engines, 15 diesel engines; 15 gasoline engines, 0 diesel engines

CHAPTER 12 Sequences; Induction; The Binomial Theorem

12.1 Exercises

1. 1, 2, 3, 4, 5 **3.** $\frac{1}{3}, \frac{2}{4}, \frac{3}{5}, \frac{4}{6}, \frac{5}{7}$ **5.** 1, −4, 9, −16, 25 **7.** $\frac{1}{2}, \frac{2}{5}, \frac{2}{7}, \frac{8}{41}, \frac{8}{61}$ **9.** $-\frac{1}{6}, \frac{1}{12}, -\frac{1}{20}, \frac{1}{30}, -\frac{1}{42}$ **11.** $\frac{1}{e}, \frac{2}{e^2}, \frac{3}{e^3}, \frac{4}{e^4}, \frac{5}{e^5}$

13. $\dfrac{n}{n+1}$ **15.** $\dfrac{1}{2^{n-1}}$ **17.** $(-1)^{n+1}$ **19.** $(-1)^{n+1}n$ **21.** $a_1 = 2, a_2 = 5, a_3 = 8, a_4 = 11, a_5 = 14$ **23.** $a_1 = -2, a_2 = 0, a_3 = 3, a_4 = 7, a_5 = 12$ **25.** $a_1 = 5, a_2 = 10, a_3 = 20, a_4 = 40, a_5 = 80$ **27.** $a_1 = 3, a_2 = \frac{3}{2}, a_3 = \frac{1}{2}, a_4 = \frac{1}{8}, a_5 = \frac{1}{40}$

29. $a_1 = 1, a_2 = 2, a_3 = 2, a_4 = 4, a_5 = 8$ **31.** $a_1 = A, a_2 = A + d, a_3 = A + 2d, a_4 = A + 3d, a_5 = A + 4d$

33. $a_1 = \sqrt{2}, a_2 = \sqrt{2 + \sqrt{2}}, a_3 = \sqrt{2 + \sqrt{2 + \sqrt{2}}}, a_4 = \sqrt{2 + \sqrt{2 + \sqrt{2 + \sqrt{2}}}}, a_5 = \sqrt{2 + \sqrt{2 + \sqrt{2 + \sqrt{2 + \sqrt{2}}}}}$

35. 50 **37.** 21 **39.** 90 **41.** 26 **43.** 42 **45.** 96 **47.** $3 + 4 + \cdots + (n + 2)$ **49.** $\frac{1}{2} + 2 + \frac{9}{2} + \cdots + \frac{n^2}{2}$ **51.** $1 + \frac{1}{3} + \frac{1}{9} + \cdots + \frac{1}{3^n}$

53. $\frac{1}{3} + \frac{1}{9} + \cdots + \frac{1}{3^n}$ **55.** $\ln 2 - \ln 3 + \ln 4 - \cdots + (-1)^n \ln n$ **57.** $\sum_{k=1}^{20} k$ **59.** $\sum_{k=1}^{13} \frac{k}{k+1}$ **61.** $\sum_{k=0}^{6} (-1)^k\left(\frac{1}{3^k}\right)$ **63.** $\sum_{k=1}^{n} \frac{3^k}{k}$

65. $\sum_{k=0}^{n} (a + kd)$ or $\sum_{k=1}^{n+1} [a + (k-1)d]$ **67. (a)** $2930 **(b)** 14 payments have been made. **(c)** 36 payments. $3584.62 **(d)** $584.62

69. (a) 2080 **(b)** After 26 months. **71.** $529,244.62 **73. (a)** $a_0 = 0, a_n = (1.02)a_{n-1} + 500$ **(b)** After 82 quarters **(c)** $156,116.15

75. (a) $a_0 = 150{,}000, a_n = (1.005)a_{n-1} - 899.33$
(b) $149,850.67
(c)

n	u(n)
0	150000
1	149851
2	149701
3	149550
4	149398
5	149246
6	149093

n=0

(d) After 58 payments
(e) 30 years
(f) $173,758.80

(g) (a) $a_0 = 150{,}000, a_n = (1.005)a_{n-1} - 999.33$
(b) $149,750.67
(c)

n	u(n)
0	150000
1	149751
2	149500
3	149248
4	148995
5	148741
6	148485

n=0

(d) After 37 months
(e) After 279 payments, last payment: $353.69(1.005) = $355.46
(f) $128,169.20

77. 21
79. A Fibonacci sequence

12.2 Exercises

1. $d = 1; 5, 6, 7, 8$ **3.** $d = 2; -3, -1, 1, 3$ **5.** $d = -2; 4, 2, 0, -2$ **7.** $d = -\frac{1}{3}; \frac{1}{6}, -\frac{1}{6}, -\frac{1}{2}, -\frac{5}{6}$ **9.** $d = \ln 3; \ln 3, 2\ln 3, 3\ln 3, 4\ln 3$

11. $a_n = 3n - 1; a_5 = 14$ **13.** $a_n = 8 - 3n; a_5 = -7$ **15.** $a_n = \frac{1}{2}(n-1); a_5 = 2$ **17.** $a_n = \sqrt{2}n; a_5 = 5\sqrt{2}$ **19.** $a_{12} = 24$

21. $a_{10} = -26$ **23.** $a_8 = a + 7b$ **25.** $a_1 = -13; d = 3; a_n = a_{n-1} + 3$ **27.** $a_1 = -53, d = 6; a_n = a_{n-1} + 6$

29. $a_1 = 28; d = -2; a_n = a_{n-1} - 2$ **31.** $a_1 = 25; d = -2; a_n = a_{n-1} - 2$ **33.** n^2 **35.** $\frac{n}{2}(9 + 5n)$ **37.** 1260 **39.** 324

41.
```
sum(seq(3.45n+4.
12,n,1,20,1))
           806.9
```
43.
```
sum(seq(2.4n+.4,
n,1,15,1))
             294
```
45.
```
sum(seq(2.58n+2.
32,n,1,25,1))
           896.5
```

47. $-\frac{3}{2}$
49. 1185 seats
51. 210 of beige and 190 blue
53. 30 rows

Historical Problems

1. $1\frac{2}{3}, 10\frac{5}{6}, 20, 29\frac{1}{6}, 38\frac{1}{3}$

12.3 Exercises

1. $r = 3; 3, 9, 27, 81$ **3.** $r = \frac{1}{2}; -\frac{3}{2}, -\frac{3}{4}, -\frac{3}{8}, -\frac{3}{16}$ **5.** $r = 2; \frac{1}{4}, \frac{1}{2}, 1, 2$ **7.** $r = 2^{1/3}; 2^{1/3}, 2^{2/3}, 2, 2^{4/3}$ **9.** $r = \frac{3}{2}; \frac{1}{2}, \frac{3}{4}, \frac{9}{8}, \frac{27}{16}$

11. Arithmetic; $d = 1$ **13.** Neither **15.** Arithmetic; $d = -\frac{2}{3}$ **17.** Neither **19.** Geometric; $r = \frac{2}{3}$ **21.** Geometric; $r = 2$

23. Geometric; $r = 3^{1/2}$ **25.** $a_5 = 162; a_n = 2 \cdot 3^{n-1}$ **27.** $a_5 = 5; a_n = 5 \cdot (-1)^{n-1}$ **29.** $a_5 = 0; a_n = 0$ **31.** $a_5 = 4\sqrt{2}; a_n = (\sqrt{2})^n$

33. $a_7 = \frac{1}{64}$ **35.** $a_9 = 1$ **37.** $a_8 = 0.00000004$ **39.** $-\frac{1}{4}(1 - 2^n)$ **41.** $2\left[1 - \left(\frac{2}{3}\right)^n\right]$ **43.** $1 - 2^n$

45.
```
(1/4)sum(seq(2^n
,n,0,14,1))
         8191.75
```
47.
```
sum(seq((2/3)^n,
n,1,15,1))
     1.995432683
```
49.
```
-1sum(seq(2^n,n,
0,14,1))
          -32767
```

51. $\frac{3}{2}$ **53.** 16 **55.** $\frac{8}{5}$
57. $\frac{20}{3}$ **59.** $\frac{18}{5}$ **61.** -4

63. 10 **65.** $72.67 **67. (a)** 0.775 ft **(b)** 8th **(c)** 15.88 ft **(d)** 20 ft **69.** $21,879

71. Option 2 results in the most: $16,038,304; Option 1 results in the least: $14,700,000 **73.** 1.845×10^{19} **77.** 3, 5, 9, 24, 73

79. *A*: $25,250 per year in 5th year, $112,742 total; *B*: $24,761 per year in 5th year, $116,801 total

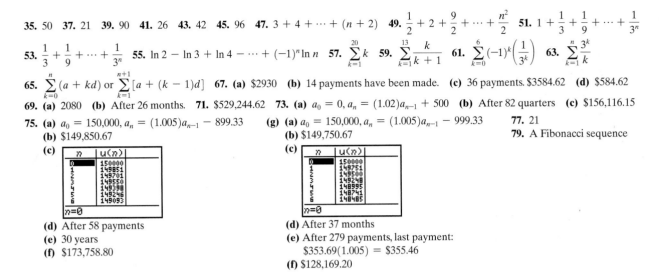

12.4 Exercises

1. (I) $n = 1: 2 \cdot 1 = 2$ and $1(1 + 1) = 2$

(II) If $2 + 4 + 6 + \cdots + 2k = k(k + 1)$, then $2 + 4 + 6 + \cdots + 2k + 2(k + 1) = (2 + 4 + 6 + \cdots + 2k) + 2(k + 1)$
$= k(k + 1) + 2(k + 1)$
$= k^2 + 3k + 2$
$= (k + 1)(k + 2)$.

3. (I) $n = 1: 1 + 2 = 3$ and $\frac{1}{2}(1)(1 + 5) = \frac{1}{2}(6) = 3$

(II) If $3 + 4 + 5 + \cdots + (k + 2) = \frac{1}{2}k(k + 5)$, then $3 + 4 + 5 + \cdots + (k + 2) + [(k + 1) + 2]$

$= [3 + 4 + 5 + \cdots + (k + 2)] + (k + 3) = \frac{1}{2}k(k + 5) + k + 3 = \frac{1}{2}(k^2 + 7k + 6) = \frac{1}{2}(k + 1)(k + 6)$

5. (I) $n = 1: 3 \cdot 1 - 1 = 2$ and $\frac{1}{2}(1)[3(1) + 1] = \frac{1}{2}(4) = 2$

(II) If $2 + 5 + 8 + \cdots + (3k - 1) = \frac{1}{2}k(3k + 1)$, then $2 + 5 + 8 + \cdots + (3k - 1) + [3(k + 1) - 1]$

$= [2 + 5 + 8 + \cdots + (3k - 1)] + 3k + 2 = \frac{1}{2}k(3k + 1) + (3k + 2) = \frac{1}{2}(3k^2 + 7k + 4) = \frac{1}{2}(k + 1)(3k + 4)$

7. (I) $n = 1: 2^{1-1} = 1$ and $2^1 - 1 = 1$

(II) If $1 + 2 + 2^2 + \cdots + 2^{k-1} = 2^k - 1$, then $1 + 2 + 2^2 + \cdots + 2^{k-1} + 2^{(k+1)-1} = (1 + 2 + 2^2 + \cdots + 2^{k-1}) + 2^k$
$= 2^k - 1 + 2^k = 2(2^k) - 1 = 2^{k+1} - 1$.

9. (I) $n = 1: 4^{1-1} = 1$ and $\frac{1}{3}(4^1 - 1) = \frac{1}{3}(3) = 1$

(II) If $1 + 4 + 4^2 + \cdots + 4^{k-1} = \frac{1}{3}(4^k - 1)$, then $1 + 4 + 4^2 + \cdots + 4^{k-1} + 4^{(k+1)-1} = (1 + 4 + 4^2 + \cdots + 4^{k-1}) + 4^k$

$= \frac{1}{3}(4^k - 1) + 4^k = \frac{1}{3}[4^k - 1 + 3(4^k)] = \frac{1}{3}[4(4^k) - 1] = \frac{1}{3}(4^{k+1} - 1)$.

11. (I) $n = 1: \frac{1}{1 \cdot 2} = \frac{1}{2}$ and $\frac{1}{1 + 1} = \frac{1}{2}$

(II) If $\frac{1}{1 \cdot 2} + \frac{1}{2 \cdot 3} + \frac{1}{3 \cdot 4} + \cdots + \frac{1}{k(k + 1)} = \frac{k}{k + 1}$, then $\frac{1}{1 \cdot 2} + \frac{1}{2 \cdot 3} + \frac{1}{3 \cdot 4} + \cdots + \frac{1}{k(k + 1)} + \frac{1}{(k + 1)[(k + 1) + 1]}$
$= \left[\frac{1}{1 \cdot 2} + \frac{1}{2 \cdot 3} + \frac{1}{3 \cdot 4} + \cdots + \frac{1}{k(k + 1)}\right] + \frac{1}{(k + 1)(k + 2)} = \frac{k}{k + 1} + \frac{1}{(k + 1)(k + 2)} = \frac{k^2 + 2k + 1}{(k + 1)(k + 2)} = \frac{k + 1}{k + 2}$.

13. (I) $n = 1: 1^2 = 1$ and $\frac{1}{6} \cdot 1 \cdot 2 \cdot 3 = 1$

(II) If $1^2 + 2^2 + 3^2 + \cdots + k^2 = \frac{1}{6}k(k + 1)(2k + 1)$, then $1^2 + 2^2 + 3^2 + \cdots + k^2 + (k + 1)^2$

$= (1^2 + 2^2 + 3^2 + \cdots + k^2) + (k + 1)^2 = \frac{1}{6}k(k + 1)(2k + 1) + (k + 1)^2 = \frac{1}{6}(2k^3 + 9k^2 + 13k + 6)$

$= \frac{1}{6}(k + 1)(k + 2)(2k + 3)$.

15. (I) $n = 1: 5 - 1 = 4$ and $\frac{1}{2}(1)(9 - 1) = \frac{1}{2} \cdot 8 = 4$

(II) If $4 + 3 + 2 + \cdots + (5 - k) = \frac{1}{2}k(9 - k)$, then $4 + 3 + 2 + \cdots + (5 - k) + [5 - (k + 1)]$

$= [4 + 3 + 2 + \cdots + (5 - k)] + 4 - k = \frac{1}{2}k(9 - k) + 4 - k = \frac{1}{2}(-k^2 + 7k + 8) = \frac{1}{2}(k + 1)(8 - k)$

$= \frac{1}{2}(k + 1)[9 - (k + 1)]$.

17. (I) $n = 1: 1 \cdot (1 + 1) = 2$ and $\frac{1}{3} \cdot 1 \cdot 2 \cdot 3 = 2$

(II) If $1 \cdot 2 + 2 \cdot 3 + 3 \cdot 4 + \cdots + k(k + 1) = \frac{1}{3}k(k + 1)(k + 2)$, then

$1 \cdot 2 + 2 \cdot 3 + 3 \cdot 4 + \cdots + k(k + 1) + (k + 1)(k + 2) = [1 \cdot 2 + 2 \cdot 3 + 3 \cdot 4 + \cdots + k(k + 1)] + (k + 1)(k + 2)$
$= \frac{1}{3}k(k + 1)(k + 2) + (k + 1)(k + 2) = \frac{1}{3}(k + 1)(k + 2)(k + 3)$

19. (I) $n = 1: 1^2 + 1 = 2$ is divisible by 2.
(II) If $k^2 + k$ is divisible by 2, then $(k + 1)^2 + (k + 1) = k^2 + 2k + 1 + k + 1 = (k^2 + k) + 2k + 2$. Since $k^2 + k$ is divisible by 2 and $2k + 2$ is divisible by 2, therefore, $(k + 1)^2 + k + 1$ is divisible by 2.

21. (I) $n = 1: 1^2 - 1 + 2 = 2$ is divisible by 2.
(II) If $k^2 - k + 2$ is divisible by 2, then $(k + 1)^2 - (k + 1) + 2 = k^2 + 2k + 1 - k - 1 + 2 = (k^2 - k + 2) + 2k$. Since $k^2 - k + 2$ is divisible by 2 and $2k$ is divisible by 2, therefore, $(k + 1)^2 - (k + 1) + 2$ is divisible by 2.

23. (I) $n = 1$: If $x > 1$, then $x^1 = x > 1$.
(II) Assume, for any natural number k, that if $x > 1$, then $x^k > 1$. Show that if $x > 1$, then $x^{k+1} > 1$:
$$x^{k+1} = x^k \cdot x^1 > 1 \cdot x = x > 1$$
$$\uparrow$$
$$x^k > 1$$

25. (I) $n = 1$: $a - b$ is a factor of $a^1 - b^1 = a - b$.
(II) If $a - b$ is a factor of $a^k - b^k$, show that $a - b$ is a factor of $a^{k+1} - b^{k+1}$: $a^{k+1} - b^{k+1} = a(a^k - b^k) + b^k(a - b)$. Since $a - b$ is a factor of $a^k - b^k$ and $a - b$ is a factor of $a - b$, therefore, $a - b$ is a factor of $a^{k+1} - b^{k+1}$.

27. $n = 1: 1^2 - 1 + 41 = 41$ is a prime number.
$n = 41: 41^2 - 41 + 41 = 1681 = 41^2$ is not prime.

29. (I) $n = 1$: $ar^{1-1} = a \cdot 1 = a$ and $a \cdot \dfrac{1 - r^1}{1 - r} = a$, because $r \neq 1$.
(II) If $a + ar + ar^2 + \cdots + ar^{k-1} = a\left(\dfrac{1 - r^k}{1 - r}\right)$, then $a + ar + ar^2 + \cdots + ar^{k-1} + ar^{(k+1)-1} = (a + ar + ar^2 + \cdots + ar^{k-1}) + ar^k$
$$= a\left(\frac{1 - r^k}{1 - r}\right) + ar^k = \frac{a(1 - r^k) + ar^k(1 - r)}{1 - r} = \frac{a - ar^k + ar^k - ar^{k+1}}{1 - r} = a\left(\frac{1 - r^{k+1}}{1 - r}\right)$$

31. (I) $n = 3$: The sum of the angles of a triangle is $(3 - 2) \cdot 180° = 180°$.
(II) Assume for any k that the sum of the angles of a convex polygon of k sides is $(k - 2) \cdot 180°$. A convex polygon of $k + 1$ sides consists of a convex polygon of k sides plus a triangle (see the illustration). The sum of the angles is
$(k - 2) \cdot 180° + 180° = (k - 1) \cdot 180°$.

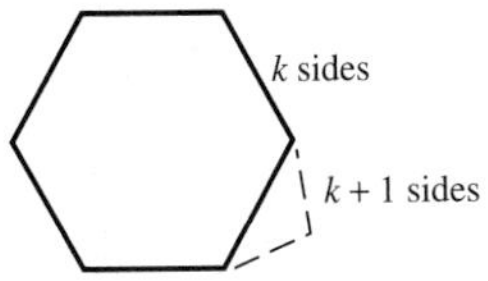

12.5 Exercises

1. 10 **3.** 21 **5.** 50 **7.** 1 **9.** 1.866×10^{15} **11.** 1.483×10^{13} **13.** $x^5 + 5x^4 + 10x^3 + 10x^2 + 5x + 1$
15. $x^6 - 12x^5 + 60x^4 - 160x^3 + 240x^2 - 192x + 64$ **17.** $81x^4 + 108x^3 + 54x^2 + 12x + 1$
19. $x^{10} + 5y^2x^8 + 10y^4x^6 + 10y^6x^4 + 5y^8x^2 + y^{10}$ **21.** $x^3 + 6\sqrt{2}x^{5/2} + 30x^2 + 40\sqrt{2}x^{3/2} + 60x + 24\sqrt{2}x^{1/2} + 8$
23. $(ax)^5 + 5by(ax)^4 + 10(by)^2(ax)^3 + 10(by)^3(ax)^2 + 5(by)^4(ax) + (by)^5$ **25.** 17,010 **27.** −101,376 **29.** 41,472 **31.** $2835x^3$
33. $314{,}928x^7$ **35.** 495 **37.** 3360 **39.** 1.00501 **41.** $\dbinom{n}{n} = \dfrac{n!}{n!(n - n)!} = \dfrac{n!}{n!0!} = \dfrac{n!}{n!} = 1$
43. $2^n = (1 + 1)^n = \dbinom{n}{0}1^n + \dbinom{n}{1}(1)(1)^{n-1} + \cdots + \dbinom{n}{n}1^n = \dbinom{n}{0} + \dbinom{n}{1} + \cdots + \dbinom{n}{n}$ **45.** 1

Fill-in-the-Blank Items

1. sequence **3.** geometric **5.** 15

True/False Items

1. True **3.** True **5.** False **7.** False

Review Exercises

1. $-\dfrac{4}{3}, \dfrac{5}{4}, -\dfrac{6}{5}, \dfrac{7}{6}, -\dfrac{8}{7}$ **3.** $2, 1, \dfrac{8}{9}, 1, \dfrac{32}{25}$ **5.** $3, 2, \dfrac{4}{3}, \dfrac{8}{9}, \dfrac{16}{27}$ **7.** 2, 0, 2, 0, 2 **9.** Arithmetic; $d = 1$; $\dfrac{n}{2}(n + 11)$ **11.** Neither
13. Geometric; $r = 8$; $\dfrac{8}{7}(8^n - 1)$ **15.** Arithmetic; $d = 4$; $2n(n - 1)$ **17.** Geometric; $r = \dfrac{1}{2}$; $6\left[1 - \left(\dfrac{1}{2}\right)^n\right]$ **19.** Neither **21.** 115
23. 75 **25.** 0.49977 **27.** 35 **29.** $\dfrac{1}{10^{10}}$ **31.** $9\sqrt{2}$ **33.** $5n - 4$ **35.** $n - 10$ **37.** $\dfrac{9}{2}$ **39.** $\dfrac{4}{3}$ **41.** 8

43. (I) $n = 1: 3 \cdot 1 = 3$ and $\frac{3 \cdot 1}{2}(2) = 3$

(II) If $3 + 6 + 9 + \cdots + 3k = \frac{3k}{2}(k+1)$, then $3 + 6 + 9 + \cdots + 3k + 3(k+1) = (3 + 6 + 9 + \cdots + 3k) + (3k+3)$

$= \frac{3k}{2}(k+1) + (3k+3) = \frac{3k^2}{2} + \frac{9k}{2} + \frac{6}{2} = \frac{3}{2}(k^2 + 3k + 2) = \frac{3}{2}(k+1)(k+2) = \frac{3(k+1)}{2}[(k+1)+1]$.

45. (I) $n = 1: 2 \cdot 3^{1-1} = 2$ and $3^1 - 1 = 2$

(II) If $2 + 6 + 18 + \cdots + 2 \cdot 3^{k-1} = 3^k - 1$, then $2 + 6 + 18 + \cdots + 2 \cdot 3^{k-1} + 2 \cdot 3^{(k+1)-1}$

$= (2 + 6 + 18 + \cdots + 2 \cdot 3^{k-1}) + 2 \cdot 3^k = 3^k - 1 + 2 \cdot 3^k = 3 \cdot 3^k - 1 = 3^{k+1} - 1$.

47. (I) $n = 1: 1^2 = 1$ and $\frac{1}{2}(6 - 3 - 1) = \frac{1}{2}(2) = 1$

(II) If $1^2 + 4^2 + 7^2 + \cdots + (3k-2)^2 = \frac{1}{2}k(6k^2 - 3k - 1)$, then

$1^2 + 4^2 + 7^2 + \cdots + (3k-2)^2 + [3(k+1) - 2]^2 = [1^2 + 4^2 + 7^2 + \cdots + (3k-2)^2] + (3k+1)^2$

$= \frac{1}{2}k(6k^2 - 3k - 1) + (3k+1)^2$

$= \frac{1}{2}(6k^3 + 15k^2 + 11k + 2) = \frac{1}{2}(k+1)(6k^2 + 9k + 2) = \frac{1}{2}(k+1)[6(k+1)^2 - 3(k+1) - 1]$.

49. $x^5 + 10x^4 + 40x^3 + 80x^2 + 80x + 32$ **51.** $32x^5 + 240x^4 + 720x^3 + 1080x^2 + 810x + 243$ **53.** 144 **55.** 84

57. (a) 8 **(b)** 1100 **59.** $151,873.77 **61. (a)** $\left(\frac{3}{4}\right)^3 \cdot 20 = \frac{135}{16}$ ft **(b)** $20\left(\frac{3}{4}\right)^n$ ft **(c)** after the 13th time **(d)** 140 ft

63. (a) $A_0 = 190{,}000$; $A_n = \left(1 + \frac{0.0675}{12}\right)A_{n-1} - 1232.34$

(b) $189,836.41

(c)

n	u(n)
0	190000
1	189836
2	189672
3	189506
4	189340
5	189173
6	189005

n=0

(d) After 252 months

(e) After 30 years (360 payments)

(f) $253,642.40

(g) (a) $A_0 = 190{,}000$; $A_n = \left(1 + \frac{0.0675}{12}\right)A_{n-1} - 1332.34$

(b) $189,736.41

(c)

n	u(n)
0	190000
1	189736
2	189471
3	189205
4	188937
5	188667
6	188396

n=0

(d) After 192 months

(e) After 289 months

(f) $194,863.14

CHAPTER 13 Counting and Probability

13.1 Exercises

1. {1, 3, 5, 6, 7, 9} **3.** {1, 5, 7} **5.** {1, 6, 9} **7.** {1, 2, 4, 5, 6, 7, 8, 9} **9.** {1, 2, 4, 5, 6, 7, 8, 9} **11.** {0, 2, 6, 7, 8} **13.** {0, 1, 2, 3, 5, 6, 7, 8, 9} **15.** {0, 1, 2, 3, 5, 6, 7, 8, 9} **17.** {0, 1, 2, 3, 4, 6, 7, 8} **19.** {0} **21.** ∅, {*a*}, {*b*}, {*c*}, {*d*}, {*a*, *b*}, {*a*, *c*}, {*a*, *d*}, {*b*, *c*}, {*b*, *d*}, {*c*, *d*}, {*a*, *b*, *c*}, {*b*, *c*, *d*}, {*a*, *c*, *d*}, {*a*, *b*, *d*}, {*a*, *b*, *c*, *d*} **23.** 25 **25.** 40 **27.** 25 **29.** 37 **31.** 18 **33.** 5 **35.** 175; 125 **37. (a)** 15 **(b)** 15 **(c)** 15 **(d)** 25 **(e)** 40 **39. (a)** 57,886 thousand **(b)** 10,894 thousand **(c)** 14,126 thousand

13.2 Exercises

1. 30 **3.** 24 **5.** 1 **7.** 1680 **9.** 28 **11.** 35 **13.** 1 **15.** 10,400,600

17. {*abc, abd, abe, acb, acd, ace, adb, adc, ade, aeb, aec, aed*
bac, bad, bae, bca, bcd, bce, bda, bdc, bde, bea, bec, bed
cab, cad, cae, cba, cbd, cbe, cda, cdb, cde, cea, ceb, ced
dab, dac, dae, dba, dbc, dbe, dca, dcb, dce, dea, deb, dec
eab, eac, ead, eba, ebc, ebd, eca, ecb, ecd, eda, edb, edc}; 60

19. {123, 124, 132, 134, 142, 143, 213, 214, 231, 234, 241, 243, 312, 314, 321, 324, 341, 342, 412, 413, 421, 423, 431, 432}; 24

21. {*abc, abd, abe, acd, ace, ade, bcd, bce, bde, cde*}; 10 **23.** {123, 124, 134, 234}; 4 **25.** 15 **27.** 16 **29.** 8 **31.** 24 **33.** 60 **35.** 18,278 **37.** 35 **39.** 1024 **41.** 9000 **43.** 120 **45.** 480 **47.** 48,228,180 **49.** 336 **51.** 5,209,344 **53.** 362,880 **55.** 90,720 **57.** 1.156×10^{76} **59.** 15 **61. (a)** 63 **(b)** 35 **(c)** 1

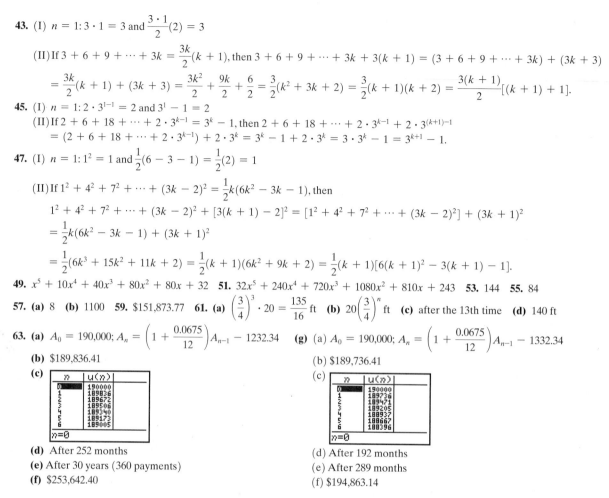

Historical Problems

1. (a) $\{AA, ABA, BAA, ABBA, BBAA, BABA, BBB, ABBB, BABB, BBAB\}$

(b) $P(A \text{ wins}) = \dfrac{C(4,2) + C(4,3) + C(4,4)}{2^4} = \dfrac{6+4+1}{16} = \dfrac{11}{16}$

$P(B \text{ wins}) = \dfrac{C(4,3) + C(4,4)}{2^4} = \dfrac{4+1}{16} = \dfrac{5}{16}$

The outcomes listed in part (a) are not equally likely.

13.3 Exercises

1. 0, 0.01, 0.35, 1 **3.** Probability model **5.** Not a probability model

7. $S = \{\text{HH, HT, TH, TT}\}$; $P(\text{HH}) = \frac{1}{4}$, $P(\text{HT}) = \frac{1}{4}$, $P(\text{TH}) = \frac{1}{4}$, $P(\text{TT}) = \frac{1}{4}$

9. $S = \{$HH1, HH2, HH3, HH4, HH5, HH6, HT1, HT2, HT3, HT4, HT5, HT6, TH1, TH2, TH3, TH4, TH5, TH6, TT1, TT2, TT3, TT4, TT5, TT6$\}$; each outcome has the probability of $\frac{1}{24}$.

11. $S = \{$HHH, HHT, HTH, HTT, THH, THT, TTH, TTT$\}$; each outcome has the probability of $\frac{1}{8}$.

13. $S = \{$1 Yellow, 1 Red, 1 Green, 2 Yellow, 2 Red, 2 Green, 3 Yellow, 3 Red, 3 Green, 4 Yellow, 4 Red, 4 Green$\}$; each outcome has the probability of $\frac{1}{12}$; thus, $P(\text{2 Red}) + P(\text{4 Red}) = \frac{1}{12} + \frac{1}{12} = \frac{1}{6}$.

15. $S = \{$1 Yellow Forward, 1 Yellow Backward, 1 Red Forward, 1 Red Backward, 1 Green Forward, 1 Green Backward, 2 Yellow Forward, 2 Yellow Backward, 2 Red Forward, 2 Red Backward, 2 Green Forward, 2 Green Backward, 3 Yellow Forward, 3 Yellow Backward, 3 Red Forward, 3 Red Backward, 3 Green Forward, 3 Green Backward, 4 Yellow Forward, 4 Yellow Backward, 4 Red Forward, 4 Red Backward, 4 Green Forward, 4 Green Backward$\}$; each outcome has the probability of $\frac{1}{24}$; thus, $P(\text{1 Red Backward}) + P(\text{1 Green Backward}) = \frac{1}{24} + \frac{1}{24} = \frac{1}{12}$.

17. $S = \{$11 Red, 11 Yellow, 11 Green, 12 Red, 12 Yellow, 12 Green, 13 Red, 13 Yellow, 13 Green, 14 Red, 14 Yellow, 14 Green, 21 Red, 21 Yellow, 21 Green, 22 Red, 22 Yellow, 22 Green, 23 Red, 23 Yellow, 23 Green, 24 Red, 24 Yellow, 24 Green, 31 Red, 31 Yellow, 31 Green, 32 Red, 32 Yellow, 32 Green, 33 Red, 33 Yellow, 33 Green, 34 Red, 34 Yellow, 34 Green, 41 Red, 41 Yellow, 41 Green, 42 Red, 42 Yellow, 42 Green, 43 Red, 43 Yellow, 43 Green, 44 Red, 44 Yellow, 44 Green$\}$; each outcome has the probability of $\frac{1}{48}$; thus, $E = \{$22 Red, 22 Green, 24 Red, 24 Green$\}$; $P(E) = \dfrac{n(E)}{n(S)} = \dfrac{4}{48} = \dfrac{1}{12}$.

19. A, B, C, F **21.** B **23.** $\frac{4}{5}; \frac{1}{5}$ **25.** $P(1) = P(3) = P(5) = \frac{2}{9}$; $P(2) = P(4) = P(6) = \frac{1}{9}$ **27.** $\frac{3}{10}$ **29.** $\frac{1}{2}$ **31.** $\frac{1}{6}$ **33.** $\frac{1}{8}$ **35.** $\frac{1}{4}$

37. $\frac{1}{6}$ **39.** $\frac{1}{18}$ **41.** 0.55 **43.** 0.70 **45.** 0.30 **47.** 0.747 **49.** 0.7 **51.** $\frac{17}{20}$ **53.** $\frac{11}{20}$ **55.** $\frac{1}{2}$ **57.** $\frac{3}{10}$ **59.** $\frac{2}{5}$

61. (a) 0.57 **(b)** 0.95 **(c)** 0.83 **(d)** 0.38 **(e)** 0.29 **(f)** 0.05 **(g)** 0.78 **(h)** 0.71 **63. (a)** $\frac{25}{33}$ **(b)** $\frac{25}{33}$ **65.** 0.167 **67.** 0.000033069

69. (a) $\frac{5}{16}$ **(b)** $\frac{1}{32}$ **71. (a)** 0.00463 **(b)** 0.049 **73.** 7.02×10^{-6}; 0.183 **75.** 0.1

13.4 Exercises

1. (a)

Cause of Death	Probability
Accidents and adverse effects	0.403
Homicide and legal intervention	0.203
Suicide	0.143
Malignant neoplasms	0.048
Diseases of heart	0.029
Human immunodeficiency virus infection	0.019
Congenital anomalies	0.013
Chronic obstructive pulmonary diseases	0.007
Pneumonia and influenza	0.006
Cerebrovascular diseases	0.005
All other causes	0.125

(b) 14.3% **(c)** 19.1% **(d)** 80.9%

3. (a)

Age	Probability
20–24	0.0964
25–29	0.1044
30–34	0.1195
35–39	0.1224
40–44	0.1155
45–49	0.1039
50–54	0.0815
55–59	0.0651
60–64	0.0569
65–69	0.0507
70–74	0.0419
75–79	0.0279
80–84	0.0140

(b) 10.39% **(c)** 14.66% **(d)** 89.61%

5. (a)

Tuition	Probability
0–999	0.0094
1000–1999	0.0065
2000–2999	0.0421
3000–3999	0.0617
4000–4999	0.0786
5000–5999	0.0786
6000–6999	0.0907
7000–7999	0.1104
8000–8999	0.1291
9000–9999	0.1029
10,000–10,999	0.0973
11,000–11,999	0.0767
12,000–12,999	0.0571
13,000–13,999	0.0318
14,000–14,999	0.0271

(b) 7.67% **(c)** 32.93% **(d)** 92.33%

7. (a)

Marital Status	Probability
Married, spouse present	0.5805
Married, spouse absent	0.0343
Widowed	0.0285
Divorced	0.0872
Never married	0.2695

(b) 58.05% **(c)** 26.95% **(d)** 61.48% **(e)** 41.95%

9. 40%
11. 40%
13. 11.43%; 88.57%
15. 37.33%; 24.33%; 75.67%
17. 21.33%; 78.67%

19. (a)

Location	Probability
Left	0.4194
Left center	0.3387
Center	0.1935
Right center	0.0484
Right	0

(b) 41.94% **(c)** 19.35% **(d)** 0%

Fill-in-the-Blank Items

1. union; intersection **3.** permutation **5.** equally likely

True/False Items

1. True **3.** True **5.** False

Review Exercises

1. {1, 3, 5, 6, 7, 8} **3.** {3, 7} **5.** {1, 2, 4, 6, 8, 9} **7.** {1, 2, 4, 5, 6, 9} **9.** 17 **11.** 29 **13.** 7 **15.** 25 **17.** 120 **19.** 336 **21.** 56 **23.** 60 **25.** 128 **27.** 3024 **29.** 70 **31.** 91 **33.** 1,600,000 **35.** 216,000 **37.** 1260 **39. (a)** 381,024 **(b)** 1260

41. (a)

Ethnic Group	Probability
Asian, Pacific Islander	0.3522
Black	0.0166
American Indian, Eskimo, Aleut	0.0011
Mexican-American, Puerto Rican, or other Hispanic	0.0399
White (non-Hispanic)	0.5836
Unknown	0.0066

(b) 35.22% **(c)** 39.21% **(d)** 64.78%

43. (a)

Age	Probability
25–29	0.2001
30–34	0.1599
35–39	0.1480
40–44	0.1403
45–49	0.1244
50–54	0.1019
55–59	0.0756
60–64	0.0498

(b) 14.80% **(c)** 22.36% **(d)** 85.20%

45. (a) $8.634628387 \times 10^{45}$ **(b)** 65.31% **(c)** 34.69%
47. (a) 5.4% **(b)** 94.6%
49. $\frac{4}{9}$
51. 0.2; 0.26
53. (a) 0.246 **(b)** 9.8×10^{-4}

CHAPTER 14 A Preview of Calculus: The Limit and the Derivative of a Function

14.1 Exercises

1. 32 **3.** 1 **5.** 4 **7.** 2 **9.** 0 **11.** 3 **13.** 4 **15.** Does not exist

17.

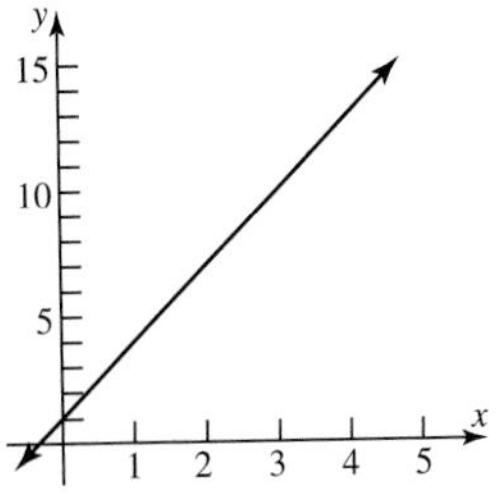

$\lim_{x\to 4} f(x) = 13$

19.

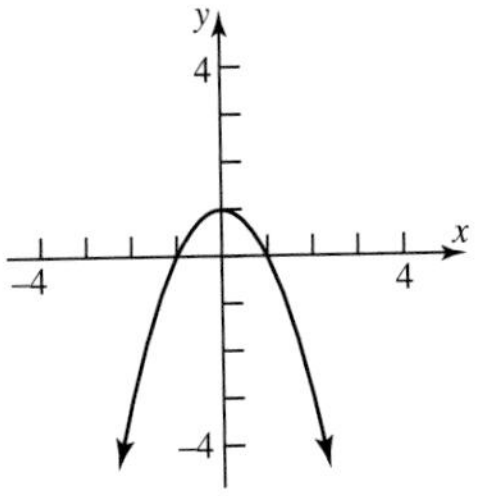

$\lim_{x\to 2} f(x) = -3$

21.

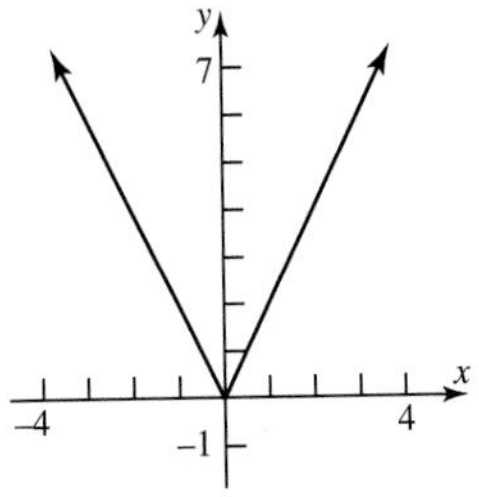

$\lim_{x\to -3} f(x) = 6$

23.

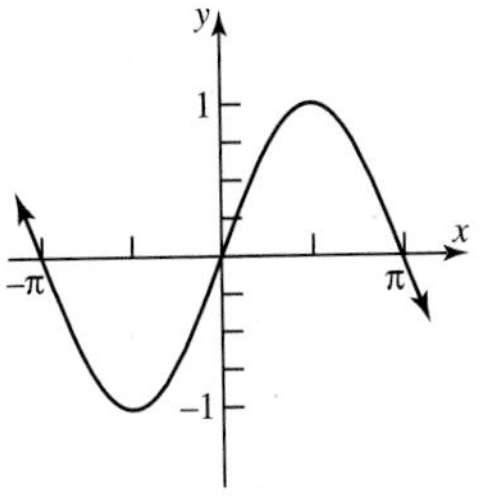

$\lim_{x\to \pi/2} f(x) = 1$

25.

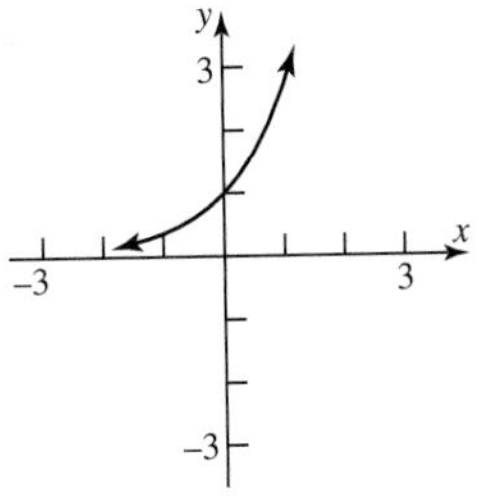

$\lim_{x\to 0} f(x) = 1$

27.

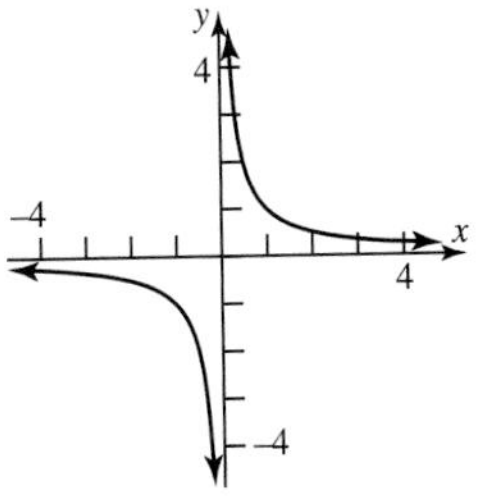

$\lim_{x\to -1} f(x) = -1$

29.

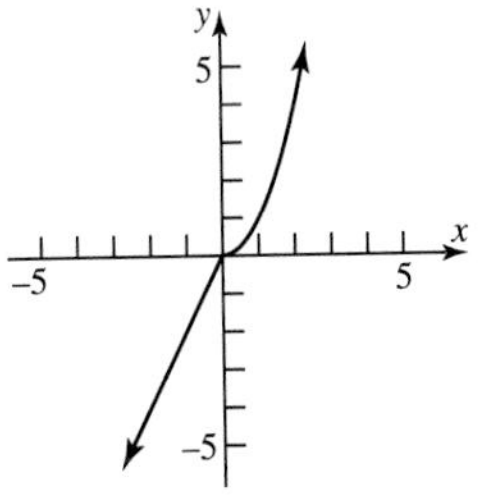

$\lim_{x\to 0} f(x) = 0$

31.

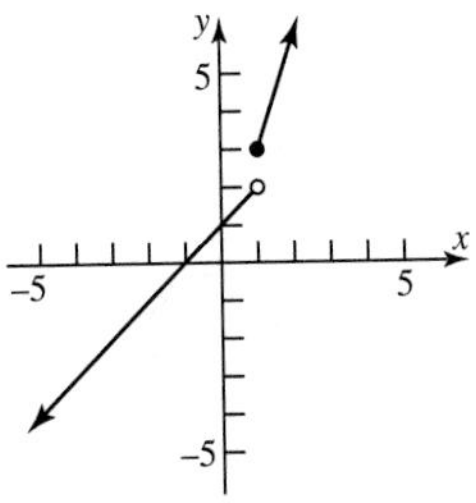

$\lim_{x\to 1} f(x)$ does not exist

33.

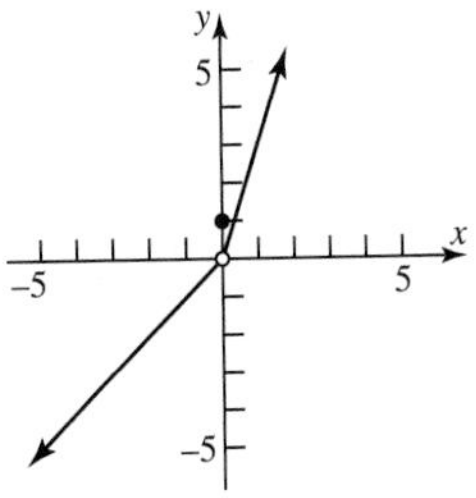

$\lim_{x\to 0} f(x) = 0$

35.

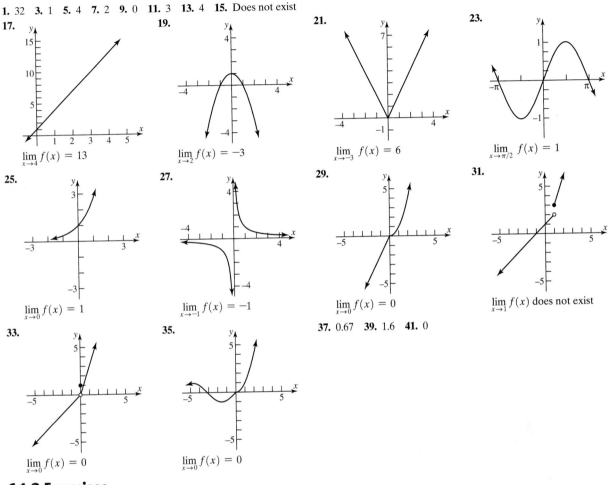

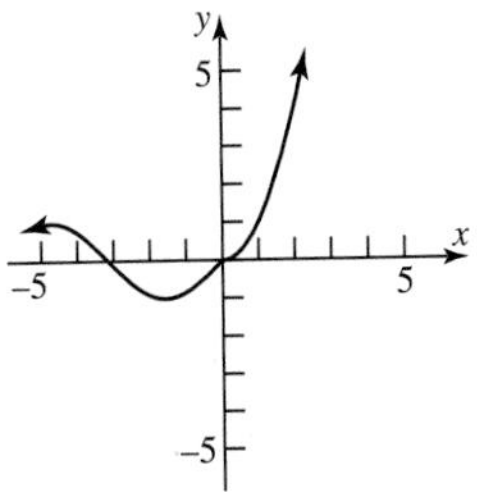

$\lim_{x\to 0} f(x) = 0$

37. 0.67 **39.** 1.6 **41.** 0

14.2 Exercises

1. 5 **3.** 4 **5.** 8 **7.** 8 **9.** −1 **11.** 8 **13.** 3 **15.** −1 **17.** 32 **19.** 2 **21.** $\frac{7}{6}$ **23.** 3 **25.** 0 **27.** $\frac{2}{3}$ **29.** $\frac{8}{5}$ **31.** 0 **33.** 5 **35.** 6 **37.** 0 **39.** 0 **41.** −1 **43.** 1 **45.** $\frac{3}{4}$

14.3 Exercises

1. Continuous for all real numbers. **3.** Continuous for all real numbers. **5.** Continuous for all real numbers. **7.** Continuous for all real numbers except $x = \frac{k\pi}{2}$, k is an odd integer. **9.** Continuous for all real numbers except $x = -2$ and $x = 2$. **11.** Continuous for all positive real numbers except $x = 1$. **13.** 5 **15.** 7 **17.** 1 **19.** 4 **21.** $-\frac{2}{3}$ **23.** $\frac{3}{2}$ **25.** Continuous **27.** Continuous **29.** Not continuous **31.** Not continuous **33.** Not continuous **35.** Continuous **37.** Not continuous **39.** Continuous **41.** $\{x \mid -8 \le x < -6 \text{ or } -6 < x < 4 \text{ or } 4 < x \le 6\}$ **43.** −8, −5, −3 **45.** $f(-8) = 0$; $f(-4) = 2$ **47.** ∞ **49.** 2 **51.** 1 **53.** Limit exists; 0 **55.** No **57.** Yes **59.** No

61. Discontinuous at $x = -1$ and $x = 1$.

$\lim_{x \to 1} R(x) = \frac{1}{2}$: Hole at $\left(1, \frac{1}{2}\right)$

$\lim_{x \to -1^-} R(x) = -\infty$; $\lim_{x \to -1^+} R(x) = \infty$:

Vertical Asymptote at $x = -1$

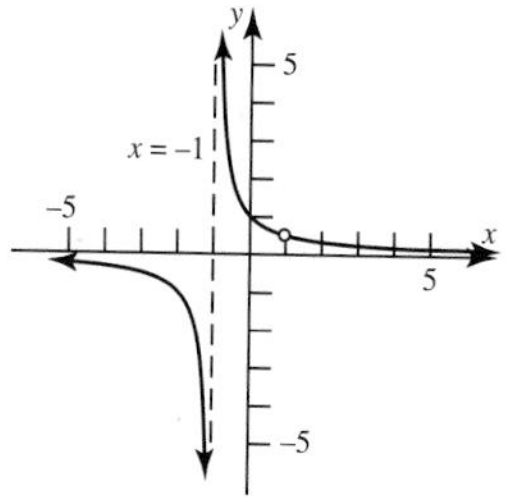

63. Discontinuous at $x = -1$ and $x = 1$.

$\lim_{x \to -1} R(x) = \frac{1}{2}$: Hole at $\left(-1, \frac{1}{2}\right)$

$\lim_{x \to 1^-} R(x) = -\infty$; $\lim_{x \to 1^+} R(x) = \infty$:

Vertical Asymptote at $x = 1$

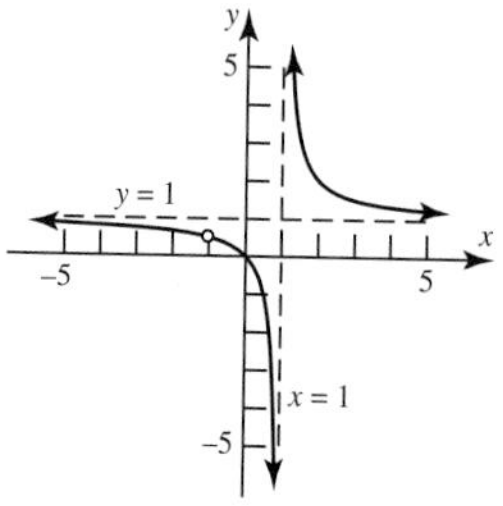

65. $x = -\sqrt[3]{2}$: Asymptote; $x = 1$: Hole

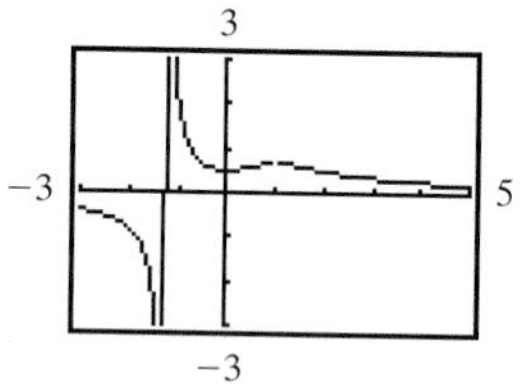

67. $x = -3$: Asymptote; $x = 2$: Hole

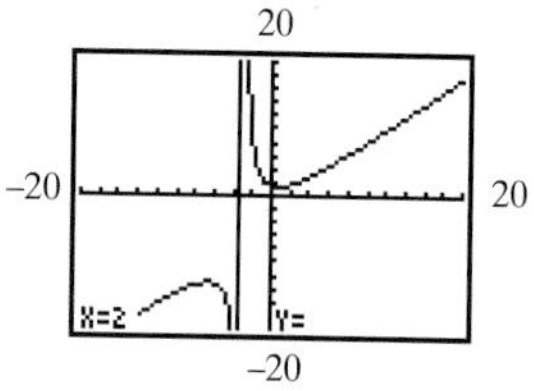

69. $x = -\sqrt[3]{2}$: Asymptote; $x = -1$: Hole

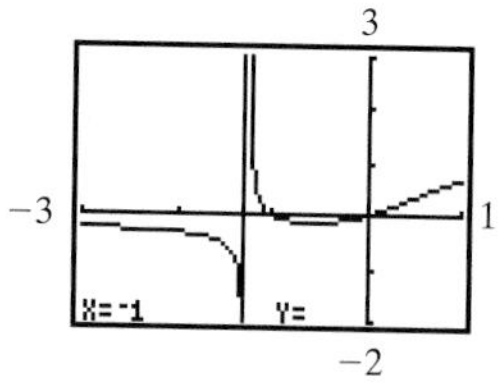

14.4 Exercises

1. $m_{\text{tan}} = 3$

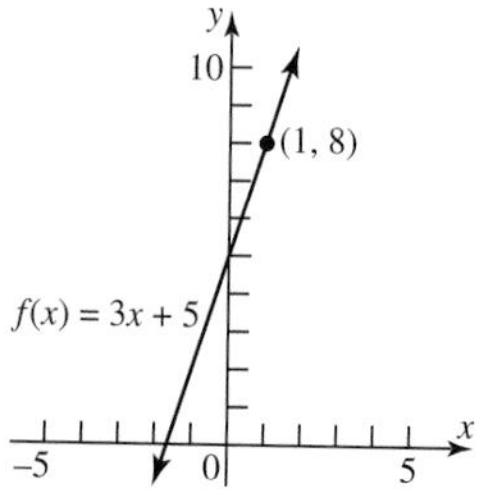

3. $m_{\text{tan}} = -2$

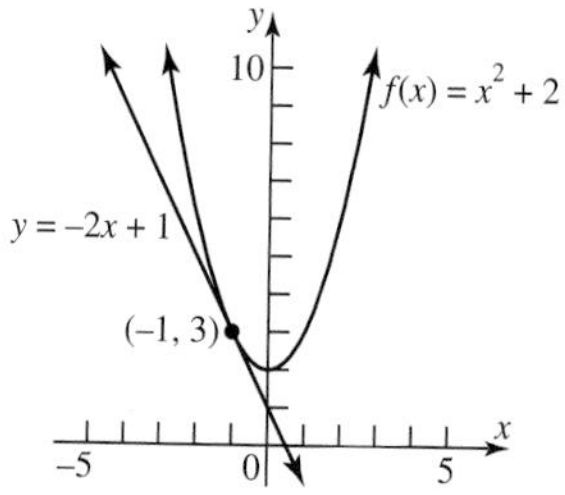

5. $m_{\text{tan}} = 12$

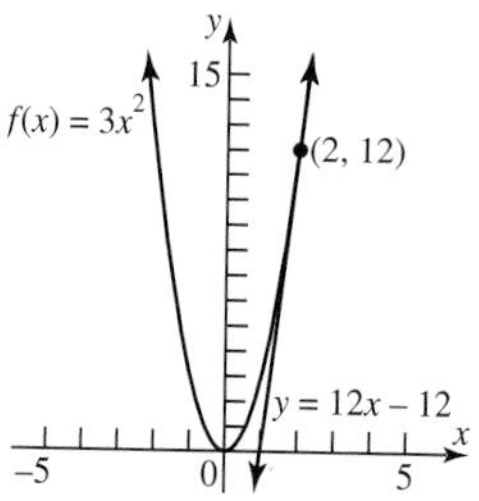

7. $m_{\text{tan}} = 5$

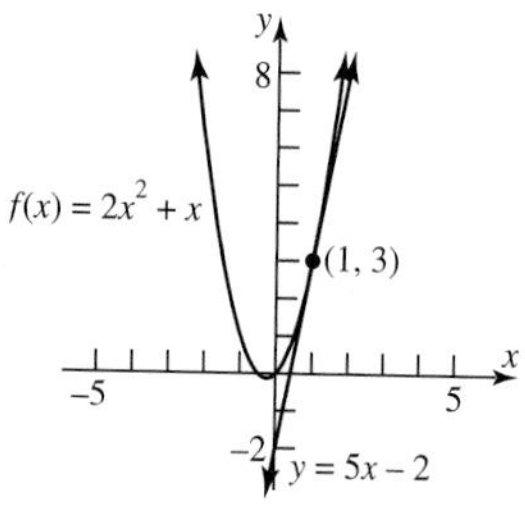

9. $m_{\text{tan}} = -4$

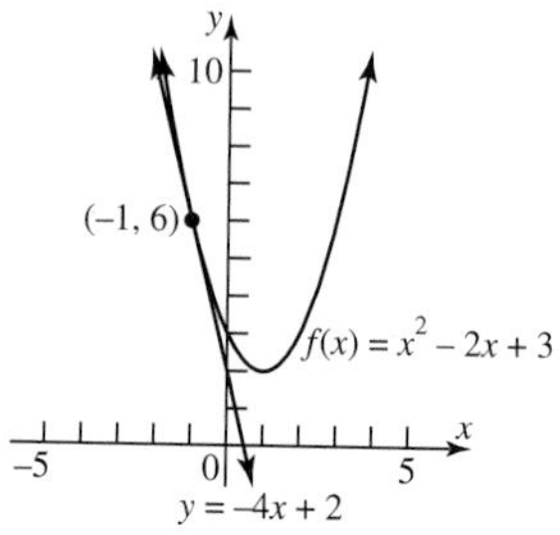

11. $m_{\text{tan}} = 13$

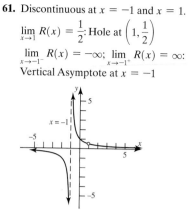

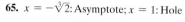

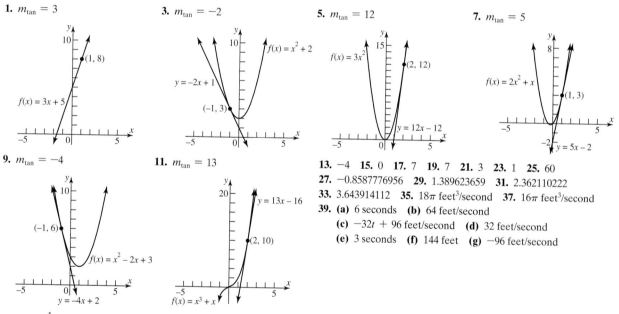

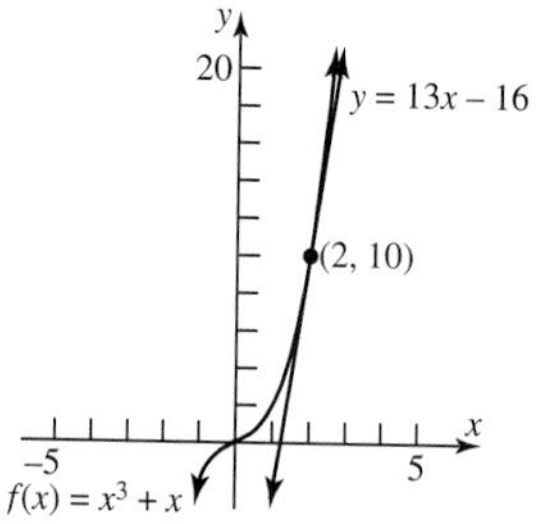

13. -4 **15.** 0 **17.** 7 **19.** 7 **21.** 3 **23.** 1 **25.** 60
27. -0.8587776956 **29.** 1.389623659 **31.** 2.362110222
33. 3.643914112 **35.** 18π feet^3/second **37.** 16π feet^3/second
39. **(a)** 6 seconds **(b)** 64 feet/second
(c) $-32t + 96$ feet/second **(d)** 32 feet/second
(e) 3 seconds **(f)** 144 feet **(g)** -96 feet/second

41. **(a)** $-23\frac{1}{3}$ feet/second **(b)** -21 feet/second **(c)** -18 feet/second **(d)** $-2.631t^2 - 10.269t + 999.933$ **(e)** -15.531 feet/second

14.5 Exercises

1. 3 **3.** 56 **5. (a)**

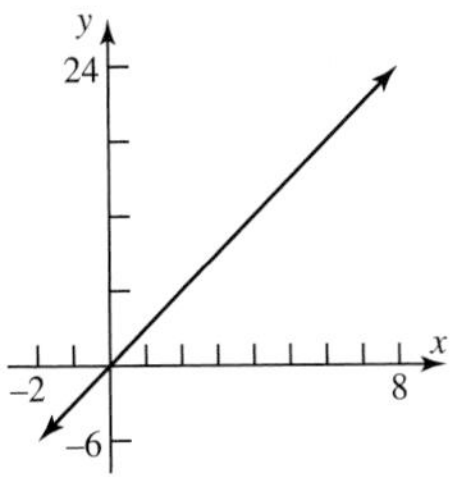

(b) 36
(c) 72
(d) 45
(e) 63
(f) 54

7. (a)

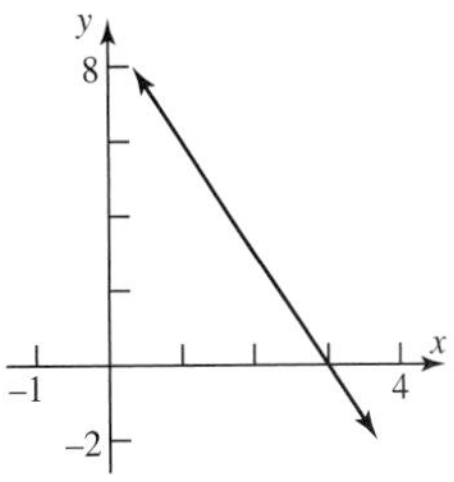

(b) 18
(c) 9
(d) $\frac{63}{4}$
(e) $\frac{45}{4}$
(f) $\frac{27}{2}$

9. (a)

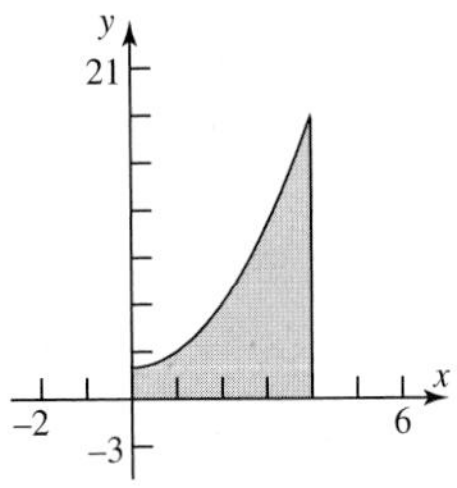

(b) 22
(c) 25.5
(d) $\int_0^4 (x^2 + 2)\,dx$
(e) 29.33

11. (a)

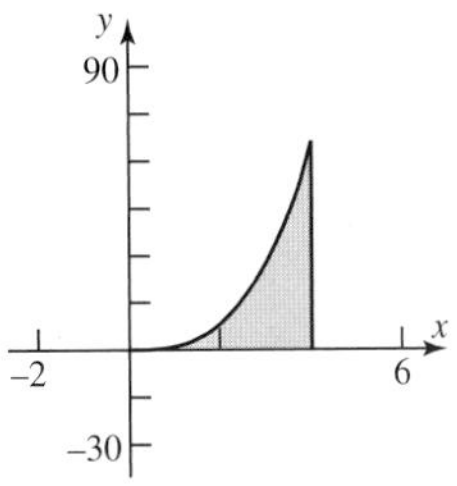

(b) 36
(c) 49
(d) $\int_0^4 x^3\,dx$
(e) 64

13. (a)

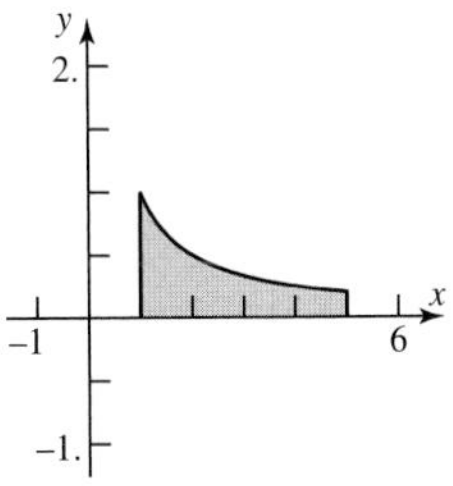

(b) $\frac{25}{12}$
(c) 1.83
(d) $\int_1^5 \frac{1}{x}\,dx$
(e) 1.609

15. (a)

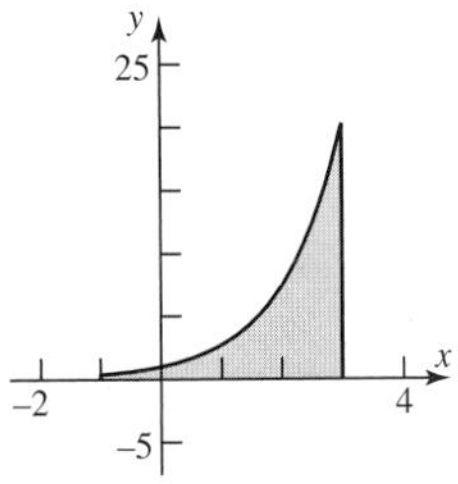

(b) 11.48
(c) 15.2
(d) $\int_{-1}^3 e^x\,dx$
(e) 19.72

17. (a)

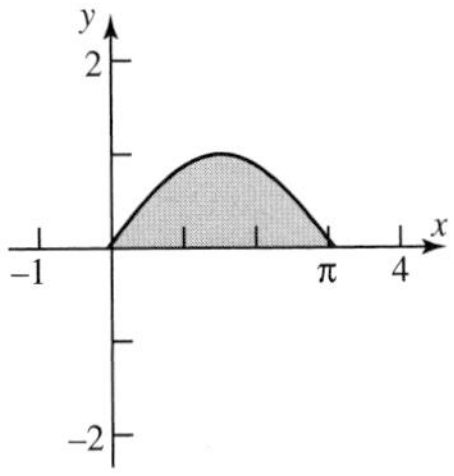

(b) 1.89
(c) 1.97
(d) $\int_0^\pi \sin x\,dx$
(e) 2

19. (a) Under the graph of $f(x) = 3x + 1$ from 0 to 4
(b)

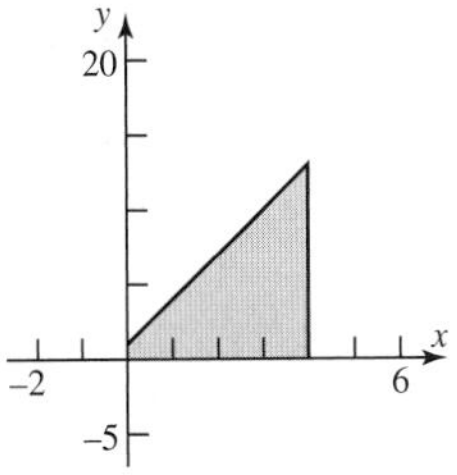

(c) 28

21. (a) Under the graph of $f(x) = x^2 - 1$ from 2 to 5
(b)

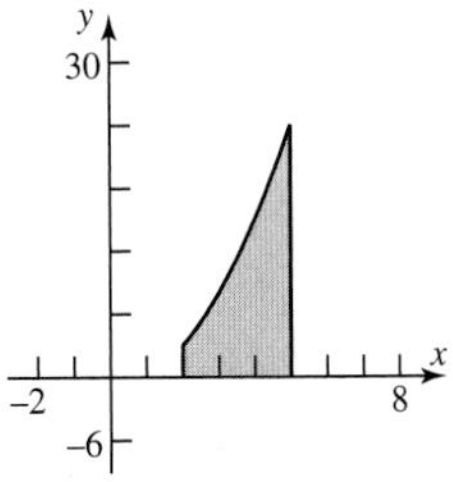

(c) 36

23. (a) Under the graph of $f(x) = \sin x$ from 0 to $\frac{\pi}{2}$
(b)

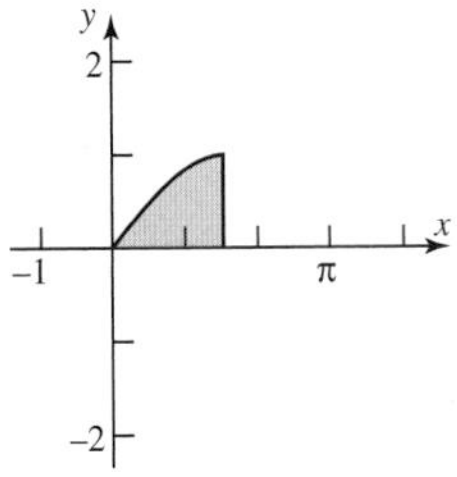

(c) 1

25. (a) Under the graph of $f(x) = e^x$ from 0 to 2 **(b)**

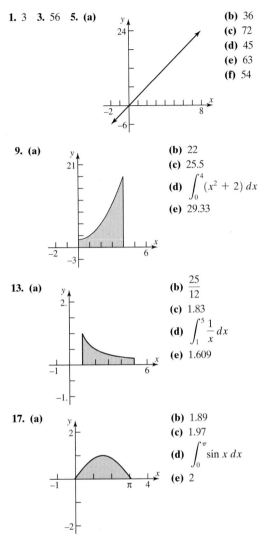

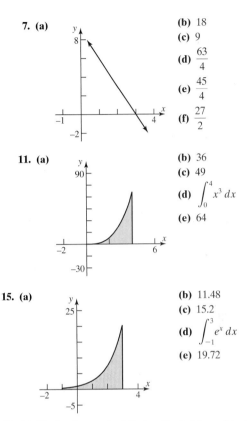

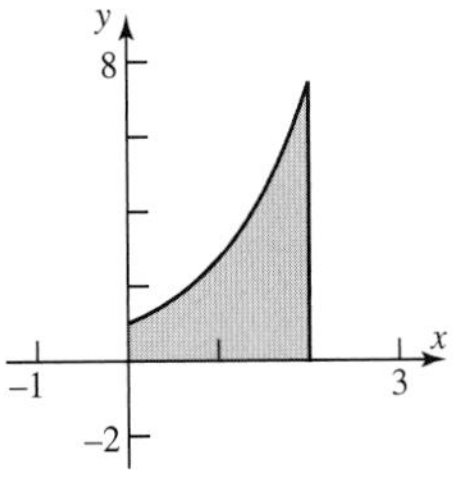

(c) 6.39

Fill-in-the-Blank Items

1. $\lim_{x \to c} f(x) = N$ **3.** not exist **5.** $\neq$ **7.** tangent

True/False Items

1. T **3.** T **5.** T **7.** T

Review Exercises

1. 9 **3.** 25 **5.** 4 **7.** 0 **9.** 64 **11.** $-\frac{1}{4}$ **13.** $\frac{1}{3}$ **15.** $\frac{6}{7}$ **17.** 0 **19.** $\frac{3}{2}$ **21.** $\frac{28}{11}$ **23.** Continuous **25.** Not continuous **27.** Not continuous **29.** Continuous **31.** $\{x \mid -6 \le x < 2 \text{ or } 2 < x < 5 \text{ or } 5 < x \le 6\}$ **33.** 1, 6 **35.** $f(-6) = 2; f(-4) = 1$ **37.** 4 **39.** -2 **41.** $-\infty$ **43.** No **45.** No **47.** No **49.** Yes

51. R is discontinuous at $x = -4$ and $x = 4$.

$\lim_{x \to -4} R(x) = -\frac{1}{8}$: Hole at $\left(-4, -\frac{1}{8}\right)$

$\lim_{x \to 4^-} R(x) = -\infty$; $\lim_{x \to 4^+} R(x) = \infty$:

The graph of R has a vertical asymptote at $x = 4$.

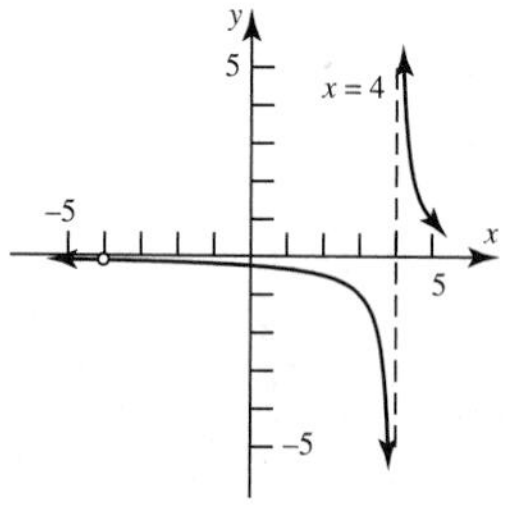

53. Undefined at $x = 2$ and $x = 9$.

R has a hole at $x = 2$ and a vertical asymptote at $x = 9$.

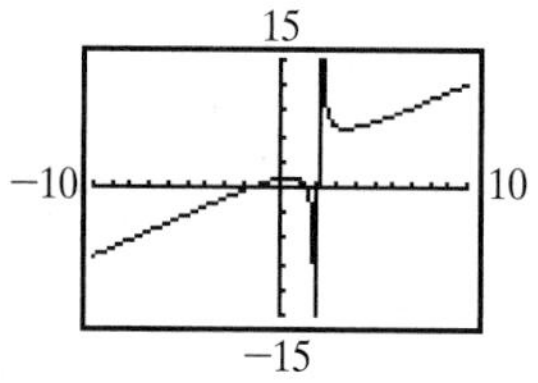

55. $m_{\tan} = 12$

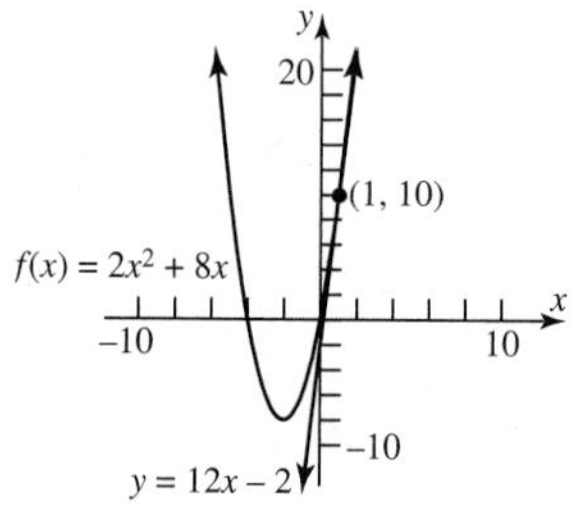

57. $m_{\tan} = 0$

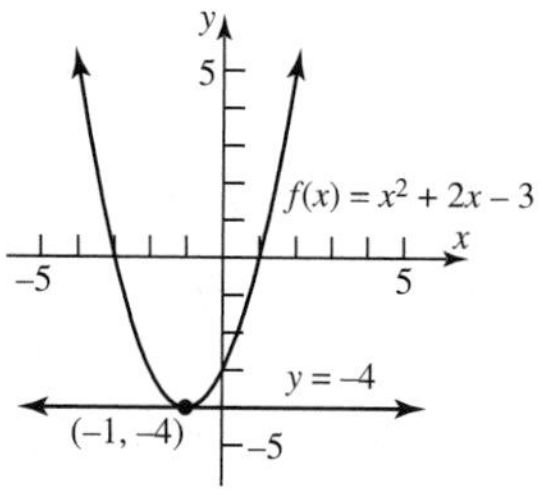

59. $m_{\tan} = 16$

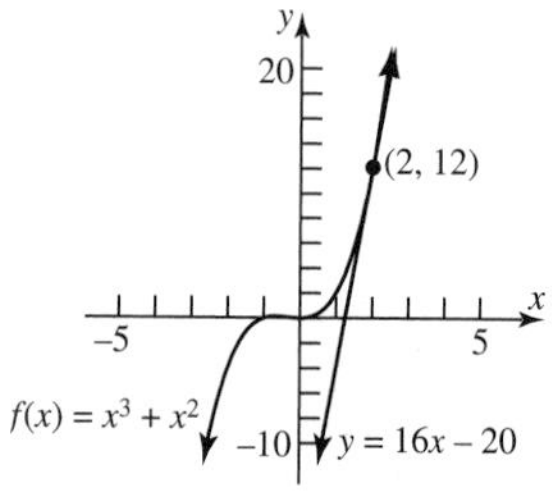

61. -24 **63.** -3 **65.** 7 **67.** -158.000035 **69.** 0.6662517653 **71.** **(a)** 7 seconds **(b)** 6 seconds **(c)** 64 feet/second **(d)** $-32t + 96$ feet/second **(e)** 32 feet/second **(f)** 3 seconds **(g)** -96 feet/second **(h)** -128 feet/second
73. **(a)** \$61.29/watch **(b)** \$71.31/watch **(c)** \$81.40/watch **(d)** $R(x) = -0.25x^2 + 100.01x - 1.24$ **(e)** \$87.51/watch

75. **(a)**

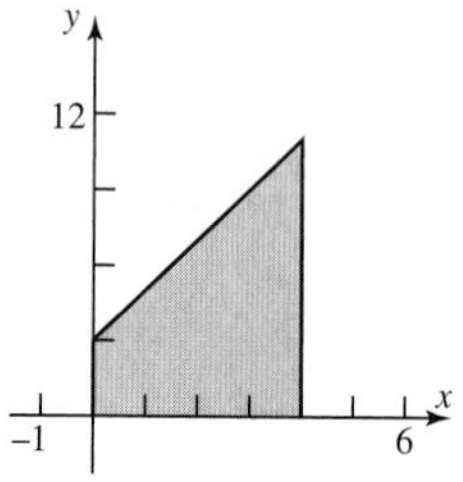

(b) 24
(c) 32
(d) 26
(e) 30
(f) 28

77. **(a)**

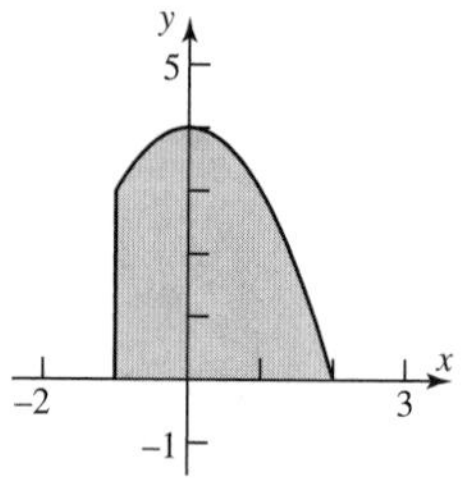

79. **(a)**

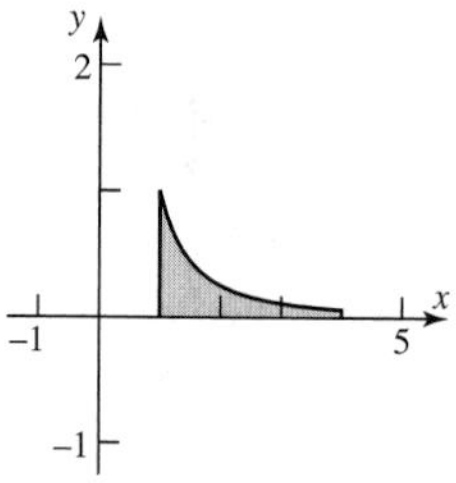

(b) $\frac{49}{36} \approx 1.36$
(c) 1.02
(d) $\int_1^4 \frac{1}{x^2}\,dx$
(e) 0.75

81. **(a)** Under the graph of $f(x) = 9 - x^2$ from -1 to 3

(b)

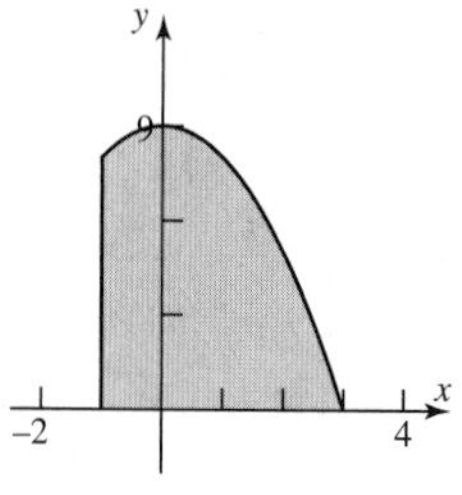

(c) 26.67

83. **(a)** Under the graph of $f(x) = e^x$ from -1 to 1 **(b)**

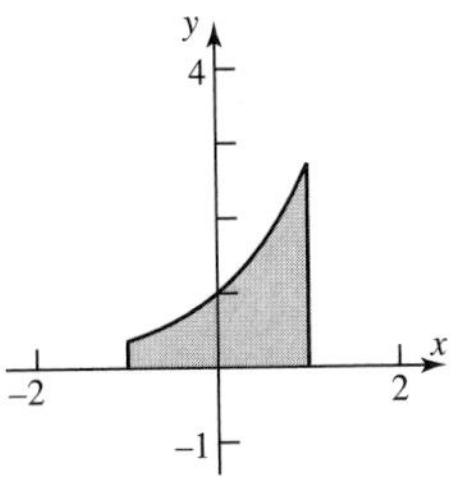

(c) 2.35

APPENDIX Review

1 Exercises

1. **(a)** $\{2, 5\}$ **(b)** $\{-6, 2, 5\}$ **(c)** $\left\{-6, \frac{1}{2}, -1.333..., 2, 5\right\}$ **(d)** $\{\pi\}$ **(e)** $\left\{-6, \frac{1}{2}, -1.333..., \pi, 2, 5\right\}$

3. **(a)** $\{1\}$ **(b)** $\{0, 1\}$ **(c)** $\left\{0, 1, \frac{1}{2}, \frac{1}{3}, \frac{1}{4}\right\}$ **(d)** None **(e)** $\left\{0, 1, \frac{1}{2}, \frac{1}{3}, \frac{1}{4}\right\}$

5. **(a)** None **(b)** None **(c)** None **(d)** $\left\{\sqrt{2}, \pi, \sqrt{2} + 1, \pi + \frac{1}{2}\right\}$ **(e)** $\left\{\sqrt{2}, \pi, \sqrt{2} + 1, \pi + \frac{1}{2}\right\}$

7. 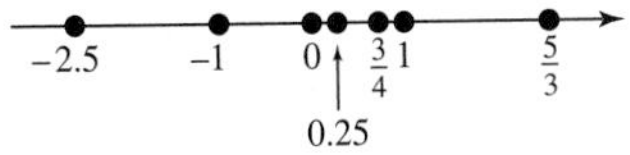

9. $>$ **11.** $>$ **13.** $>$ **15.** $=$ **17.** $<$ **19.** $x > 0$ **21.** $x < 2$ **23.** $x \le 1$ **25.** $2 < x < 5$ **27.** 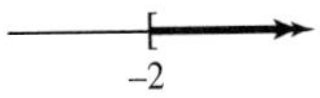

29. 

31. 

33. 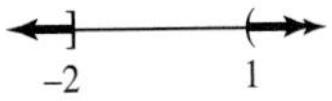

35. 2 **37.** 6 **39.** 4 **41.** −28 **43.** $\frac{4}{5}$ **45.** 0 **47.** 1 **49.** 5 **51.** 1 **53.** 22 **55.** 2 **57.** 0°C **59.** 25°C

61. **(a)** \$6000 **(b)** \$8000 **63.** **(a)** $2 \le 5$ **(b)** $6 > 5$ **65.** **(a)** Acceptable **(b)** Not acceptable **67.** No; 0.000333... **69.** No; no **71.** 1 **73.** 3.15 or 3.16

2 Exercises

1. 16 **3.** $\frac{1}{16}$ **5.** $-\frac{1}{16}$ **7.** $\frac{1}{8}$ **9.** $\frac{1}{4}$ **11.** $\frac{1}{9}$ **13.** $\frac{81}{64}$ **15.** $\frac{27}{8}$ **17.** $\frac{81}{2}$ **19.** $\frac{4}{81}$ **21.** $\frac{1}{12}$ **23.** $-\frac{2}{3}$ **25.** y^2 **27.** $\frac{x}{y^2}$ **29.** $\frac{1}{64x^6}$ **31.** $-\frac{4}{x}$

33. 3 **35.** $\frac{1}{x^3y}$ **37.** $\frac{1}{xy}$ **39.** $\frac{y}{x}$ **41.** $\frac{25x^2}{16y^2}$ **43.** $\frac{1}{x^2y^2}$ **45.** $\frac{1}{x^3y^3}$ **47.** $-\frac{8x^3}{9yz^2}$ **49.** $\frac{16x^2}{9y^2}$ **51.** $\frac{y}{x^3}$ **53.** $\frac{y^2}{x^2}$ **55.** 10; 0 **57.** 0.0081 **59.** 81

61. 304,006.671 **63.** 0.004 **65.** 481.890 **67.** 0.000 **69.** 4.542×10^2 **71.** 1.3×10^{-2} **73.** 3.2155×10^4 **75.** 4.23×10^{-4} **77.** 61,500 **79.** 0.001214 **81.** 110,000,000 **83.** 0.081 **85.** 400,000,000 meters **87.** 0.0000005 meter **89.** 5×10^{-4} inch **91.** 1.92×10^9 barrels **93.** 5.866×10^{12} miles

3 Exercises

1. Monomial; Variable: x; Coefficient: 2; Degree: 3 **3.** Not a monomial **5.** Monomial; Variable: x; Coefficient: −2; Degree: 1

7. Monomial; Variable: x; Coefficient: $\frac{8}{5}$; Degree: 1 **9.** Not a monomial **11.** Yes; 2 **13.** Yes; 0 **15.** No **17.** Yes; 3 **19.** No

21. $x^2 + 7x + 2$ **23.** $x^3 - 4x^2 + 9x + 7$ **25.** $7x^2 - x - 7$ **27.** $-2x^3 + 18x^2 - 18$ **29.** $x^3 + x^2 - 4x$ **31.** $x^3 + 3x^2 - 2x - 4$ **33.** $x^2 + 6x + 8$ **35.** $2x^2 + 9x + 10$ **37.** $x^2 - 2x - 8$ **39.** $x^2 - 5x + 6$ **41.** $2x^2 - x - 6$ **43.** $-2x^2 + 11x - 12$ **45.** $x^2 - 49$ **47.** $4x^2 - 9$ **49.** $x^2 + 8x + 16$ **51.** $x^2 - 8x + 16$ **53.** $4x^2 - 12x + 9$ **55.** $x^3 - 6x^2 + 12x - 8$ **57.** $8x^3 + 12x^2 + 6x + 1$

59. The degree of the product equals the degree of the product of the leading terms:
$(a_nx^n + a_{n-1}x^{n-1} + ... + a_1x + a_0)(b_mx^m + b_{m-1}x^{m-1} + ... + b_1x + b_0) = a_nb_mx^{n+m} + (a_nb_{m-1} + a_{n-1}b_m)x^{n+m-1} + ... + (a_1b_0 + a_0b_1)x + a_0b_0$, which is of degree $m + n$

4 Exercises

1. Quotient: $4x^2 - 11x + 23$; Remainder: -45 **3.** Quotient: $4x^2 + 13x + 53$; Remainder: 213

5. Quotient: $4x - 3$; Remainder: $-7x + 7$ **7.** Quotient: 2; Remainder: $-3x^2 + x + 3$ **9.** Quotient: $2x - \frac{5}{2}$ Remainder: $\frac{3}{2}x + \frac{7}{2}$

11. Quotient: $x - \frac{3}{4}$; Remainder: $\frac{7}{4}$ **13.** Quotient: $x^3 + x^2 + x + 1$; Remainder: 0 **15.** Quotient: $x^2 + 1$; Remainder: 0

17. Quotient: $-4x^2 - 3x - 3$; Remainder: -7 **19.** Quotient: $x^2 - x - 1$; Remainder: $2x + 2$ **21.** Quotient: $-x^2$; Remainder: 1 **23.** Quotient: $x^2 + ax + a^2$; Remainder: 0 **25.** Quotient: $x^3 + ax^2 + a^2x + a^3$; Remainder: 0 **27.** $q(x) = x^2 + x + 4$; $R = 12$ **29.** $q(x) = 3x^2 + 11x + 32$; $R = 99$ **31.** $q(x) = x^4 - 3x^3 + 5x^2 - 15x + 46$; $R = -138$ **33.** $q(x) = 4x^5 + 4x^4 + x^3 + x^2 + 2x + 2$; $R = 7$ **35.** $q(x) = 0.1x^2 - 0.11x + 0.321$; $R = 0.3531$

37. $q(x) = x^4 + x^3 + x^2 + x + 1; R = 0$ **39.** No; $f(2) = 8$ **41.** Yes; $f(2) = 0$ **43.** Yes; $f(-3) = 0$ **45.** No; $f(-4) = 1$
47. Yes; $f\left(\frac{1}{2}\right) = 0$

5 Exercises

1. $3(x + 2)$ **3.** $a(x^2 + 1)$ **5.** $x(x^2 + x + 1)$ **7.** $2x(x - 1)$ **9.** $(x - 1)(x + 1)$ **11.** $(2x + 1)(2x - 1)$ **13.** $(x + 4)(x - 4)$ **15.** $(5x + 2)(5x - 2)$ **17.** $(x + 1)^2$ **19.** $(x - 5)^2$ **21.** $(2x + 1)^2$ **23.** $(4x + 1)^2$ **25.** $(x - 3)(x^2 + 3x + 9)$ **27.** $(x + 3)(x^2 - 3x + 9)$ **29.** $(2x + 3)(4x^2 - 6x + 9)$ **31.** $(x + 2)(x + 3)$ **33.** $(x + 5)(x + 2)$ **35.** $(x - 8)(x - 2)$ **37.** $(x - 8)(x + 1)$ **39.** $(x + 2)(2x + 3)$ **41.** $(x - 2)(2x + 1)$ **43.** $(2x + 3)(3x + 2)$ **45.** $(3x + 1)(x + 1)$ **47.** $(z + 1)(2z + 3)$ **49.** $(x - 2)(3x + 4)$ **51.** $(x + 4)(3x - 2)$ **53.** $(x + 6)(x - 6)$ **55.** $(1 + 2x)(1 - 2x)$ **57.** $(x + 2)(x + 5)$ **59.** Prime **61.** Prime **63.** $-(x - 5)(x + 3)$ **65.** $3(x + 2)(x - 6)$ **67.** $y^2(y + 5)(y + 6)$ **69.** $(2x + 3)^2$ **71.** $(3x + 1)(x + 1)$ **73.** $(x - 3)(x + 3)(x^2 + 9)$ **75.** $(x - 1)^2(x^2 + x + 1)^2$ **77.** $x^5(x - 1)(x + 1)$ **79.** $-(4x - 5)(4x + 1)$ **81.** $(2y - 5)(2y - 3)$ **83.** $-(3x - 1)(3x + 1)(x^2 + 1)$ **85.** $(x + 3)(x - 6)$ **87.** $(x + 2)(x - 3)$ **89.** $3x(4 - 5x)(2 - x)^3$ **91.** $(x - 1)(x + 1)(x + 2)$ **93.** $(x - 1)(x + 1)(x^2 - x + 1)$
95. The possibilities are $(x \pm 1)(x \pm 4) = x^2 \pm 5x + 4$ or $(x \pm 2)(x \pm 2) = x^2 \pm 4x + 4$, none of which equal $x^2 + 4$

6 Exercises

1. -1 **3.** 3 **5.** -1 **7.** $-\frac{4}{3}$ **9.** -18 **11.** -4 **13.** $-\frac{3}{4}$ **15.** -20 **17.** 2 **19.** $\{3, 4\}$ **21.** $\left\{-3, \frac{1}{2}\right\}$ **23.** $\{-3, 3\}$ **25.** 3 **27.** $\{-5, 0, 4\}$
29. $\left\{0, \frac{1}{4}\right\}$ **31.** $\{-3, 0, 3\}$ **33.** $\{-1, 0, 1\}$ **35.** -1 **37.** $\{-2, 2\}$

7 Exercises

1. 0 **3.** 3 **5.** None **7.** $0, 1, -1$ **9.** $\frac{3}{x - 3}$ **11.** $\frac{x}{3}$ **13.** $\frac{4x}{2x - 1}$ **15.** $\frac{x + 5}{x - 1}$ **17.** $\frac{x - 2}{3x - 1}$ **19.** $-(2x + 1)$ **21.** $\frac{-x + 9}{(x + 5)^3}$
23. $\frac{3}{5x(x - 2)}$ **25.** $\frac{2x}{x + 4}$ **27.** $\frac{(2x - 1)(x - 7)}{(x - 5)(2x + 1)}$ **29.** $\frac{4x}{(x - 2)(x - 3)}$ **31.** $\frac{(x + 3)^2}{(x - 3)^2}$ **33.** $\frac{(x - 4)(x + 3)}{(x - 1)(2x + 1)}$ **35.** $\frac{3x - 2}{x - 3}$
37. $\frac{x + 9}{2x - 1}$ **39.** $\frac{4 - x}{x - 2}$ **41.** $\frac{2(x + 5)}{(x - 1)(x + 2)}$ **43.** $\frac{3x^2 - 2x - 3}{(x + 1)(x - 1)}$ **45.** $\frac{-11x - 2}{(x + 2)(x - 2)}$ **47.** $\frac{2x^3 - 2x^2 + 2x - 1}{x(x - 1)^2}$
49. $\frac{5x}{(x - 6)(x - 1)(x + 4)}$ **51.** $\frac{2(2x^2 + 5x - 2)}{(x - 2)(x + 2)(x + 3)}$ **53.** $\frac{-x^2 + 3x + 13}{(x - 2)(x + 1)(x + 4)}$ **55.** $\frac{-1}{x(x + h)}$ **57.** $\frac{x + 1}{x - 1}$
59. $\frac{(x - 1)(x + 1)}{x^2 + 1}$ **61.** $\frac{2(5x - 1)}{(x - 2)(x + 1)^2}$ **63.** $\frac{-2x(x^2 - 2)}{(x + 2)(x^2 - x - 3)}$ **65.** $f = \frac{R_1 \cdot R_2}{(n - 1)(R_1 + R_2)}; \frac{2}{15}$ m

8 Exercises

1. $2\sqrt{2}$ **3.** $2x\sqrt[3]{2x}$ **5.** x^2 **7.** $\frac{4\sqrt{2x}}{3}$ **9.** x^3y^2 **11.** x^2y **13.** $6\sqrt{x}$ **15.** $6x\sqrt{x}$ **17.** $15\sqrt[3]{3}$ **19.** $12\sqrt{3}$ **21.** $2\sqrt{3}$
23. $x - 2\sqrt{x} + 1$ **25.** $\frac{\sqrt{2}}{2}$ **27.** $\frac{-\sqrt{15}}{5}$ **29.** $\frac{\sqrt{3}(5 + \sqrt{2})}{23}$ **31.** $\frac{-19 + 8\sqrt{5}}{41}$ **33.** $\frac{2x + h - 2\sqrt{x(x + h)}}{h}$ **35.** 4 **37.** -3 **39.** 64
41. $\frac{1}{27}$ **43.** $\frac{27\sqrt{2}}{32}$ **45.** $\frac{27\sqrt{2}}{32}$ **47.** $x^{7/12}$ **49.** xy^2 **51.** $x^{4/3}y^{5/3}$ **53.** $\frac{8x^{3/2}}{y^{1/4}}$ **55.** $\frac{3x + 2}{(1 + x)^{1/2}}$ **57.** $\frac{2 + x}{2(1 + x)^{3/2}}$ **59.** $\frac{4 - x}{(x + 4)^{3/2}}$
61. $\frac{1}{2}(5x + 2)(x + 1)^{1/2}$ **63.** $2x^{1/2}(3x - 4)(x + 1)$ **65.** $\frac{2(4 - x^2)}{(8 - x^2)^{1/2}}$

9 Exercises

1. 13 **3.** 26 **5.** 25 **7.** Right triangle; 5 **9.** Not a right triangle **11.** Right triangle; 25 **13.** Not a right triangle **15.** 8 in^2 **17.** 4 in^2
19. $A = 25\pi$ m^2; $C = 10\pi$ m **21.** 224 ft^3 **23.** $V = \frac{256}{3}\pi$ cm^3; $S = 64\pi$ cm^2 **25.** 648π in^3 **27.** π square units **29.** 2π square units
31. About 16.8 ft **33.** 64 square feet **35.** $24 + 2\pi \approx 30.28$ ft^2; $16 + 2\pi \approx 22.28$ ft **37.** About 5.477 mi
39. From 100 ft: 12.247 mi, From 150 ft: 15 mi

10 Exercises

1. 4 **3.** $\frac{1}{16}$ **5.** $\frac{1}{9}$ **7.** $(x - 2)^2 + (y + 2)^2 = 9$ **9.** $(x + 3)^2 + (y - 1)^2 = 9$ **11.** $\left(x + \frac{1}{2}\right)^2 + \left(y - \frac{1}{2}\right)^2 = 1$ **13.** $\{-7, 3\}$
15. $\left\{-\frac{1}{4}, \frac{3}{4}\right\}$ **17.** $\left\{\frac{-1 - \sqrt{7}}{6}, \frac{-1 + \sqrt{7}}{6}\right\}$

INDEX

CONICS

Parabola

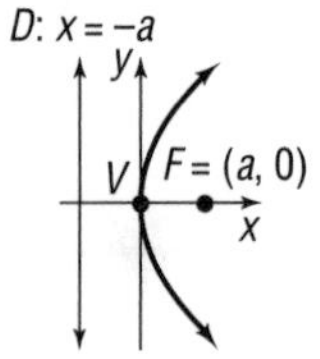

$y^2 = 4ax$

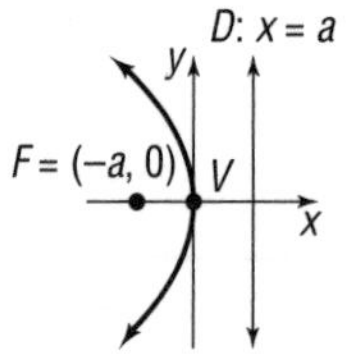

$y^2 = -4ax$

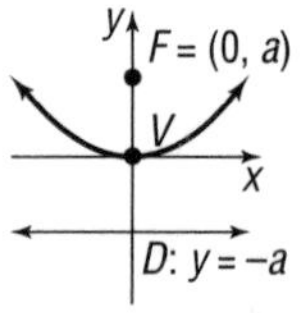

$x^2 = 4ay$

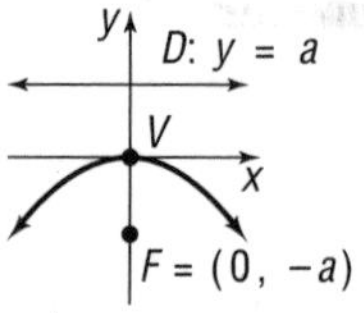

$x^2 = -4ay$

Ellipse

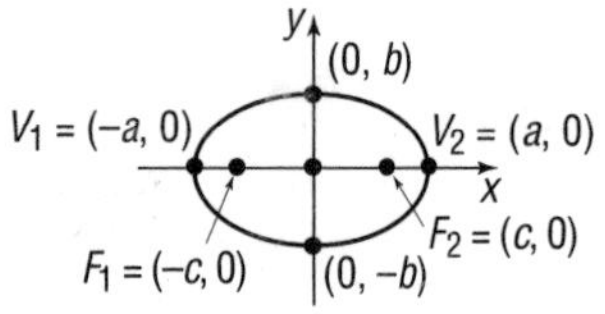

$$\frac{x^2}{a^2} + \frac{y^2}{b^2} = 1, \quad c^2 = a^2 - b^2$$

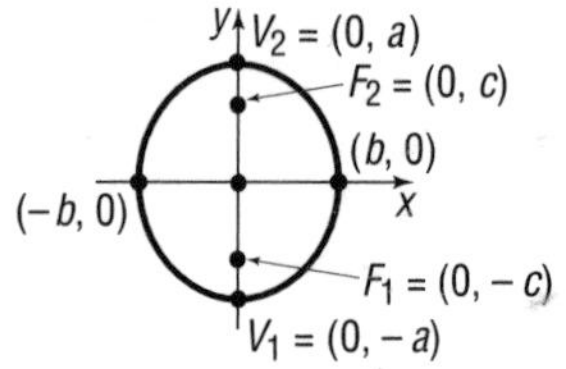

$$\frac{x^2}{b^2} + \frac{y^2}{a^2} = 1, \quad c^2 = a^2 - b^2$$

Hyperbola

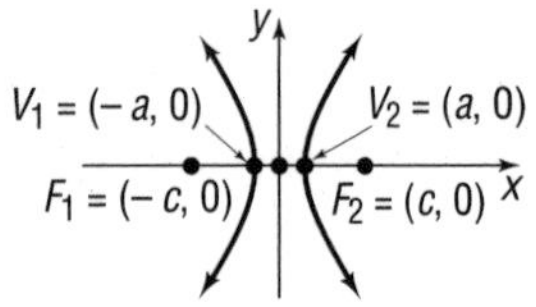

$$\frac{x^2}{a^2} - \frac{y^2}{b^2} = 1, \quad c^2 = a^2 + b^2$$

Asymptotes: $y = \frac{b}{a}x, \quad y = -\frac{b}{a}x$

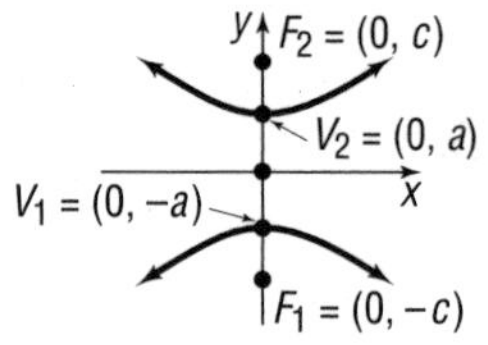

$$\frac{y^2}{a^2} - \frac{x^2}{b^2} = 1, \quad c^2 = a^2 + b^2$$

Asymptotes: $y = \frac{a}{b}x, \quad y = -\frac{a}{b}x$

PROPERTIES OF LOGARITHMS

$$\log_a(MN) = \log_a M + \log_a N$$

$$\log_a\left(\frac{M}{N}\right) = \log_a M - \log_a N$$

$$\log_a M^r = r\log_a M$$

$$\log_a M = \frac{\log M}{\log a} = \frac{\ln M}{\ln a}$$

PERMUTATIONS/COMBINATIONS

$$0! = 1 \qquad 1! = 1$$

$$n! = n(n-1)\cdot\ldots\cdot(3)(2)(1)$$

$$P(n, r) = \frac{n!}{(n-r)!}$$

$$C(n, r) = \binom{n}{r} = \frac{n!}{(n-r)!\,r!}$$

BINOMIAL THEOREM

$$(x+a)^n = \binom{n}{0}x^n + \binom{n}{1}ax^{n-1} + \binom{n}{2}a^2x^{n-2} + \cdots + \binom{n}{n-1}a^{n-1}x + \binom{n}{n}a^n$$

ARITHMETIC SEQUENCE

$$a + (a+d) + (a+2d) + \cdots + [a + (n-1)d] = \frac{n}{2}[2a + (n-1)d]$$

GEOMETRIC SEQUENCE

$$a + ar + ar^2 + \cdots + ar^{n-1} = a\frac{1-r^n}{1-r}$$

GEOMETRIC SERIES

$$\text{If } |r| < 1, \quad a + ar + ar^2 + \cdots = \sum_{k=1}^{\infty} ar^{k-1} = \frac{a}{1-r}$$